HARRIS' SHOCK AND VIBRATION HANDBOOK

ABOUT THE EDITORS

Allan G. Piersol was an engineer in private practice, specializing in analysis of data from and the design of structures for shock, vibration, and acoustical environments. He was a licensed engineer in both mechanical and safety engineering. The author of five books and chapter author of five additional handbooks dealing with these subjects, Mr. Piersol also taught graduate courses in mechanical shock and vibration at Loyola Marymount University in Los Angeles, California.

Thomas L. Paez, recently a distinguished member of the technical staff at Sandia National Laboratories, works as a consultant specializing in probabilistic structural dynamics and validation of mathematical models. He is the author of a text on random vibrations and many chapters and papers dealing with random vibrations, mechanical shock, and model validation. Mr. Paez frequently teaches short courses on random vibrations and mechanical shock.

HARRIS' SHOCK AND VIBRATION HANDBOOK

Allan G. Piersol

Thomas L. Paez

Sixth Edition

McGraw Hill

New York Chicago San Francisco Lisbon London Madrid
Mexico City Milan New Delhi San Juan Seoul
Singapore Sydney Toronto

The McGraw-Hill Companies

Cataloging-in-Publication Data is on file with the Library of Congress.

Harris' Shock and Vibration Handbook, Sixth Edition

1 2 3 4 5 6 7 8 9 0 DOC/DOC 0 14 3 2 1 0 9

ISBN 978-0-07-150819-3
MHID 0-07-150819-8

This book is printed on acid-free paper.

Sponsoring Editor
Larry S. Hager

Acquisitions Coordinator
Alexis Richard

Editorial Supervisor
David E. Fogarty

Project Manager
Jan Bedger

Copy Editor
Virginia Carroll

Proofreaders
Linda Enterline, Virginia Landis, Sue Miller, Stewart Smith

Indexer
Tanya Lawyer

Production Supervisor
Pamela A. Pelton

Composition
North Market Street Graphics

Art Director, Cover
Jeff Weeks

Allan Piersol dedicated his entire professional life to development of shock and vibration and acoustical theory and practice. He made substantial and critical contributions to signal analysis and estimation, design principles of complex, practical systems, test criteria and specification, the understanding of pyroshock, and many other engineering fields. Allan was an author of five books and many textbook chapters and a superior teacher who gave his time and imagination freely to those who sought his guidance. He was the primary editor of this handbook; it has been completed because of his leadership. He will be missed by the educational, scientific, and professional engineering communities.

CONTENTS

CONTRIBUTORS

Kjell Ahlin Professor Emeritus of Mechanical Engineering, Blekinge Institute of Technology, Karlskrona, Sweden (CHAP. 20)

Randall J. Allemang Professor of Mechanical Engineering and Director, Structural Dynamics Research Laboratory, University of Cincinnati, Cincinnati, Ohio (CHAP. 21)

Vesta I. Bateman Mechanical Shock Consultant, Albuquerque, New Mexico (CHAPS. 27, 28)

Ralph E. Blake Late Consultant, Technical Center of the Silicon Valley, San Jose, California (CHAP. 2)

Robert D. Blevins Consultant, San Diego, California (CHAP. 30)

Anthony J. Brammer Professor of Medicine, Ergonomic Technology Center, University of Connecticut Health Center, Farmington, Connecticut, and Guest Worker, Institute for Microstructural Sciences, National Research Council, Ottawa, Ontario, Canada (CHAP. 41)

David L. Brown Professor Emeritus, Structural Dynamics Research Laboratory, University of Cincinnati, Cincinnati, Ohio (CHAP. 21)

Anthony S. Chu Business Unit Director, Vibration Sensors, Measurement Specialties, Inc., Aliso Viejo, California (CHAP. 10)

Robert N. Coppolino Chief Scientist, Measurement Analysis Corporation, Torrance, California (CHAP. 23)

Stephen H. Crandall Ford Professor of Engineering Emeritus, Massachusetts Institute of Technology, Cambridge, Massachusetts (CHAP. 22)

Alan G. Davenport Late Founding Director, Boundary Layer Wind Tunnel, and Professor Emeritus of Civil Engineering, University of Western Ontario, London, Ontario, Canada (CHAP. 31)

Neil T. Davie Principal Member of the Technical Staff, Sandia National Laboratories, Albuquerque, New Mexico (CHAP. 28)

Richard G. DeJong Professor of Engineering, Calvin College, Grand Rapids, Michigan (CHAP. 24)

Jeffrey Dosch Technical Director, PCB Piezotronics, Depew, New York (CHAP. 11)

Fredric Ehrich Senior Lecturer, Massachusetts Institute of Technology, Cambridge, Massachusetts (CHAPS. 4, 5)

Amr S. Elnashai William and Elaine Hall Endowed Professor of Civil Engineering, Director, Mid-America Earthquake Center, and Director, NEES@UIUC Simulation Facility, University of Illinois, Urbana, Illinois (CHAP. 29)

Ronald L. Eshleman Director, Vibration Institute, Willowbrook, Illinois (CHAPS. 16, 37)

David J. Evans Mechanical Engineer, National Institute of Standards and Technology, Gaithersburg, Maryland (CHAP. 17)

William J. Hall Professor Emeritus of Civil Engineering, University of Illinois, Urbana, Illinois (CHAP. 29)

Cyril M. Harris Charles Batchelor Professor Emeritus of Electrical Engineering, Columbia University, New York, New York (CHAPS. 1, 15)

Harry Himelblau Consultant, Los Angeles, California (CHAP. 3)

Elmer L. Hixson Professor Emeritus of Electrical Engineering, University of Texas, Austin, Texas (CHAP. 9)

Charles A. Hull Senior Staff Engineer, Lockheed Martin Corporation, Syracuse, New York (CHAP. 40)

David I. G. Jones Consultant, D/Tech Consulting, Chandler, Arizona (CHAP. 36)

Keith T. Kedward Professor of Mechanical Engineering, University of California, Santa Barbara, California (CHAP. 34)

J. Peter C. King Managing Director, Alan G. Davenport Wind Engineering Group, Boundary Layer Wind Tunnel, University of Western Ontario, London, Ontario, Canada (CHAP. 31)

Herb LeKuch Consultant, SVP, Inc., New York, New York (CHAP. 39)

Robert B. McCalley, Jr. Retired Engineering Manager, General Electric Company, Schenectady, New York (CHAP. 22)

Ronald G. Merritt Mechanical Engineer, Naval Air Warfare Center, China Lake, California (CHAP. 7)

M. R. Mitchell Adjunct Professor of Mechanical Engineering, Northern Arizona University, Flagstaff, Arizona (CHAP. 33)

C. Nataraj Professor and Chair, Department of Mechanical Engineering, and Director, Center for Nonlinear Dynamics and Control, Villanova University, Villanova, Pennsylvania (CHAP. 4)

Thomas L. Paez Consultant and Former Distinguished Member of the Technical Staff, Validation and Uncertainty Quantification Department, Sandia National Laboratories, Albuquerque, New Mexico (CHAP. 8)

Allan G. Piersol Late Consultant, Piersol Engineering Company, Woodland Hills, California (CHAPS. 1, 18, 19, 40)

Henry C. Pusey Manager of Technical Services, Shock and Vibration Information Analysis Center, Winchester, Virginia (CHAP. 17)

Robert B. Randall Associate Professor of Engineering, University of New South Wales, Sydney, New South Wales, Australia (CHAP. 14)

Sheldon Rubin Consultant, Rubin Engineering Company, Sherman Oaks, California (CHAPS. 3, 6, 20)

David O. Smallwood Former Distinguished Member of the Technical Staff, Sandia National Laboratories, Albuquerque, New Mexico (CHAP. 25)

Strether Smith President, Structural/Signal Analysis Consultants, Cupertino, California (CHAP. 13)

Karl A. Sweitzer Senior Staff Systems Engineer, ITT Corporation, Space Systems Division, Rochester, New York (CHAP. 40)

Michael A. Talley Research Engineer, Central Shock Group, Northrop Grumman Shipbuilding, Newport News, Virginia (CHAP. 38)

Peter J. Torvik Professor Emeritus of Aerospace Engineering and Engineering Mechanics, Air Force Institute of Technology, Xenia, Ohio (CHAP. 35)

Marcos A. Underwood President, Tu'tuli Enterprises, Gualala, California (CHAP. 26)

Patrick L. Walter Professor of Professional Practice in Engineering, Texas Christian University, Fort Worth, Texas (CHAP. 12)

John F. Wilby Consultant, Wilby Associates, Calabasas, California (CHAP. 32)

PREFACE

The first edition of this handbook, then titled simply the *Shock and Vibration Handbook,* was published in 1961, with Cyril M. Harris, Charles Batchelor Professor of Engineering at Columbia University, New York, New York, and Charles E. Crede, Professor of Mechanical Engineering, California Institute of Technology, Pasadena, California, as the editors. The handbook brought together a comprehensive summary of basic shock and vibration theory and the applications of that theory to contemporary engineering practice. In so doing, it quickly found a wide international audience that continued to expand with each new edition. Unfortunately, Charles Crede, one of the world's most respected shock and vibration engineers of his day, passed away shortly after the publication of the first edition, but his name was still carried as the coeditor of the second edition, published in 1976. The third and fourth editions of the handbook were published in 1988 and 1996, respectively, with Cyril Harris as the sole editor. For the fifth edition of the handbook published in 2002, Cyril Harris brought in Allan G. Piersol, Consultant, Piersol Engineering Company, Los Angeles, California, to be his coeditor. It was also at that time that the handbook was renamed *Harris' Shock and Vibration Handbook.* Cyril Harris has now fully retired, so for this sixth edition of the handbook, Allan Piersol brought in Thomas L. Paez, Consultant and Former Distinguished Member of the Technical Staff, Sandia National Laboratories, Albuquerque, New Mexico, to be his coeditor.

This sixth edition of the handbook represents a major revision of the material in prior editions in four important ways. First, several chapters in the fifth edition that covered material that is either obsolete or of secondary interest (e.g., "Mechanical Properties of Rubber") have been deleted to make room for the coverage of new technologies that have become important since the publication of the fifth edition, as well as to reduce the number of pages in the handbook. Second, with only one exception, every chapter retained from the fifth edition with a deceased author has been edited or completely rewritten by a new author who is a contemporary authority on the subject matter of the chapter. If the modifications to the chapter are minor, an acknowledgment is given to the original author. The exception is Chap. 2, "Basic Vibration Theory," which was beautifully written by the late Ralph Blake and presents fundamental material that does not become dated or require references. Third, the numerical values in most of the text, tables, and figures in the handbook are now presented in customary (usually English) units followed in parentheses by SI units, as detailed in Chap. 1. Finally, again as detailed in Chap. 1, the chapters have been reordered to group together those chapters covering related subjects; for example, Chaps. 29 through 32 cover the response of structures induced by (a) ground motion, (b) fluid flow, (c) wind, and (d) sound.

As for previous editions, this sixth edition of the handbook is written primarily to provide practical guidance to working engineers and scientists actively involved in solving shock and vibration problems. However, the discussions of all engineering applications are preceded by a presentation of basic theoretical background material. Hence, as for previous editions of the handbook, it is likely that this sixth edition will often find its way into higher education classrooms to support the teaching of various aspects of shock and vibration engineering, particularly at graduate school level. The extensive and fully updated references in all chapters further enhance the handbook's usefulness as a supporting text for teaching purposes. (*Note:* Text citations of the fifth edition of this handbook refer to Cyril M. Harris and Allan J. Piersol, *Harris' Shock and Vibration Handbook,* Fifth Edition, McGraw-Hill, New York 2001.)

Finally, we wish to thank all the contributors, in particular, the thirteen new authors, for their skill and dedication in preparing this sixth edition of the handbook. We are also very grateful to Cyril Harris for his support and, as always, all the involved personnel at McGraw-Hill for their excellent work in preparing this new edition.

Allan G. Piersol
Thomas L. Paez

CHAPTER 1
INTRODUCTION TO THE HANDBOOK

Cyril M. Harris
Allan G. Piersol

CONCEPTS IN SHOCK AND VIBRATION

The terms *shock* and *vibration* are generally used to refer to the dynamic mechanical excitation that may cause a dynamic response of a physical system, usually a mechanical structure that is exposed to that excitation. To be more specific, a shock is a dynamic excitation with a relatively short duration, and a vibration is a dynamic excitation with a relatively long duration as compared to the time required for a physical system exposed to that excitation to fully respond. Both shock and vibration excitations can appear either as an input motion or force at the mounting points or as a pressure field over the exterior surface of the physical system of interest. In either case, the basic description of a shock or vibration is given by the instantaneous magnitude of the excitation as a function of time, which is called a *time history.*

Shock and vibration excitations can be broadly classified as being either *deterministic* or *random* (also called *stochastic*). A deterministic excitation is one where, using analytical calculations based upon fundamental physics or repeated observations of the excitation produced under identical circumstances, the exact time history of the excitation in the future can be predicted with only minor errors. For example, a step input with a fixed magnitude at the mounting points of an equipment item would constitute a deterministic shock, while the excitation produced by an unbalanced shaft rotating at constant speed would produce a deterministic vibration. On the other hand, a random excitation is one where neither analytical calculations nor previous observations of the excitation produced under identical circumstances will allow the prediction of the exact time history of the excitation in the future. For example, a chemical explosion produces a pressure time history with detailed characteristics that are unique to that particular explosion, while the vibration of a pipe produced by the turbulence in the boundary layer between the pipe and the high-velocity flow of a fluid through the pipe will also be random in character.

The simplest model for a physical system that will respond to a shock or vibration excitation is given by a rigid mass supported by a linear spring, commonly referred to as a *single-degree-of-freedom-system.* The vibration of such a model, or system, may be "free" or "forced." In *free vibration,* there is no energy added to the system but rather the vibration is the continuing result of an initial disturbance. An *ideal system* may be considered undamped for mathematical purposes; in such a system the free vibration is assumed to continue indefinitely. In any *real system,* damping (i.e., energy dissipation) causes the amplitude of free vibration to decay continuously to a negligible value. Such free vibration sometimes is referred to as *transient vibration. Forced vibration,* in contrast to free vibration, continues under "steady-state" conditions because energy is supplied to the system continuously to compensate for that dissipated by damping in the system. In general, the frequency at which energy is supplied (i.e., the forcing frequency) appears in the vibration of the system. Forced vibration may be either deterministic or random. In either instance, the vibration of the system depends upon the relation of the excitation or forcing function to the properties of the system. This relationship is a prominent feature of the analytical aspects of vibration.

The technology of shock and vibration embodies both theoretical and experimental facets prominently. Thus, methods of analysis and instruments for the measurement of shock and vibration are of primary significance. The results of analysis and measurement are used to evaluate shock and vibration environments, to devise testing procedures and testing machines, and to design and operate equipment and machinery. Shock and/or vibration may be either wanted or unwanted, depending upon circumstances. For example, vibration is involved in the primary mode of operation of such equipment as conveying and screening machines; the setting of rivets depends upon the application of impact or shock. More frequently, however, shock and vibration are unwanted. Then the objective is to eliminate or reduce their severity or, alternatively, to design equipment to withstand their influences. These procedures are embodied in the control of shock and vibration. Methods of control are emphasized throughout this handbook.

CONTROL OF SHOCK AND VIBRATION

Methods of shock and vibration control may be grouped into three broad categories:

1. **Reduction at the source**
   a. Balancing of moving masses. Where the vibration originates in rotating or reciprocating members, the magnitude of a vibratory force frequently can be reduced or possibly eliminated by balancing or counterbalancing. For example, during the manufacture of fans and blowers, it is common practice to rotate each rotor and to add or subtract material as necessary to achieve balance.
   b. Balancing of magnetic forces. Vibratory forces arising in magnetic effects of electrical machinery sometimes can be reduced by modification of the magnetic path. For example, the vibration originating in an electric motor can be reduced by skewing the slots in the armature laminations.
   c. Control of clearances. Vibration and shock frequently result from impacts involved in operation of machinery. In some instances, the impacts result from inferior design or manufacture, such as excessive clearances in bearings, and can be reduced by closer attention to dimensions. In other instances, such as the movable armature of a relay, the shock can be decreased by employing a rubber bumper to cushion motion of the plunger at the limit of travel.
2. **Isolation**
   a. Isolation of source. Where a machine creates significant shock or vibration during its normal operation, it may be supported upon isolators to protect

other machinery and personnel from shock and vibration. For example, a forging hammer tends to create shock of a magnitude great enough to interfere with the operation of delicate apparatus in the vicinity of the hammer. This condition may be alleviated by mounting the forging hammer upon isolators.

b. Isolation of sensitive equipment. Equipment often is required to operate in an environment characterized by severe shock or vibration. The equipment may be protected from these environmental influences by mounting it upon isolators. For example, equipment mounted in ships of the navy is subjected to shock of great severity during naval warfare and may be protected from damage by mounting it upon isolators.

3. Reduction of the response

a. Alteration of natural frequency. If the natural frequency of the structure of an equipment coincides with the frequency of the applied vibration, the vibration condition may be made much worse as a result of resonance. Under such circumstances, if the frequency of the excitation is substantially constant, it often is possible to alleviate the vibration by changing the natural frequency of such structure. For example, the vibration of a fan blade was reduced substantially by modifying a stiffener on the blade, thereby changing its natural frequency and avoiding resonance with the frequency of rotation of the blade. Similar results are attainable by modifying the mass rather than the stiffness.

b. Energy dissipation. If the vibration frequency is not constant or if the vibration involves a large number of frequencies, the desired reduction of vibration may not be attainable by altering the natural frequency of the responding system. It may be possible to achieve equivalent results by the dissipation of energy to eliminate the severe effects of resonance. For example, the housing of a washing machine may be made less susceptible to vibration by applying a coating of damping material on the inner face of the housing.

c. Auxiliary mass. Another method of reducing the vibration of the responding system is to attach an auxiliary mass to the system by a spring; with proper tuning the mass vibrates and reduces the vibration of the system to which it is attached. For example, the vibration of a textile-mill building subjected to the influence of several hundred looms was reduced by attaching large masses to a wall of the building by means of springs; then the masses vibrated with a relatively large motion, and the vibration of the wall was reduced. The incorporation of damping in this auxiliary mass system may further increase its effectiveness.

CONTENT OF HANDBOOK

Each chapter of this handbook deals with a discrete phase of the subject of shock and vibration. Frequent references are made from one chapter to another, to refer to basic theory in other chapters, to call attention to supplementary information, and to give illustrations and examples. Therefore, each chapter, when read with other referenced chapters, presents one complete facet of the subject of shock and vibration.

Chapters dealing with similar subject matter are grouped together. The first eight chapters following this introductory chapter deal with fundamental concepts of shock and vibration. Chapter 2 discusses the free and forced vibration of linear systems that can be defined by lumped parameters with similar types of coordinates. The properties of rigid bodies are discussed in Chap. 3, together with the vibration of resiliently supported rigid bodies wherein several modes of vibration are coupled.

Nonlinear vibration is discussed in Chap. 4, and self-excited vibration in Chap. 5. Chapter 6 discusses two degree-of-freedom systems in detail—including both the basic theory and the application of such theory to dynamic absorbers and auxiliary mass dampers. The vibration of systems defined by distributed parameters—notably, beams and plates—is discussed in Chap. 7. Chapter 8 discusses the response of lumped parameter systems to step- and pulse-type excitations, while Chap. 9 discusses applications of the use of mechanical impedance and mechanical admittance methods.

The second group of chapters is concerned with instrumentation for the measurement of shock and vibration. Chapter 10 discusses not only piezoelectric and piezo resistive transducers, but also other types such as force transducers, although strain gages are described separately in Chap. 12. The calibration of shock and vibration transducers is detailed in Chap. 11, and the electrical instruments to which such transducers are connected (including various types of amplifiers, signal conditioners, analog-to-digital conversion, and data storage) are considered in detail in Chap. 13. Chapter 14 is devoted to the important topics of spectrum analysis instrumentation and techniques. The use of all such equipment in making vibration measurements in the field is described in Chap. 15. The specific application of vibration measurement equipment for monitoring the mechanical condition of machinery, as an aid in preventive maintenance, is the subject of Chap. 16.

The third group of chapters covers the selection of shock and vibration test criteria and data analysis procedures. Specifically, Chap. 17 summarizes national and international standards and test codes related to shock and vibration, while Chap. 18 details the procedures for deriving shock and vibration test specifications from measured or predicted data. Chapters 19 and 20 then summarize the procedures for computing the important properties of measured vibration and shock data, respectively. This is followed by four chapters that detail procedures for the experimental and analytical methods for determining the dynamic characteristics of structures. Chapter 21 details experimental modal analysis procedures, while Chaps. 22 through 24 cover the most widely used analytical procedures—namely, matrix methods, finite element methods, and statistical energy methods of analysis.

The next four chapters are concerned with shock and vibration testing machines and procedures. Chapter 25 covers vibration testing machines, while Chap. 26 fully elaborates on the digital control systems used for electrodynamic and electrohydraulic testing machines. Chapters 27 and 28 then cover conventional shock and pyroshock testing machines, respectively. This material is followed by four chapters that discuss the response of structures to four important and common sources of shock and vibration—namely, ground motion in Chap. 29, fluid flow in Chap. 30, wind loads in Chap. 31, and acoustic environments in Chap. 32. The next group of chapters covers the mechanical properties and potential shock- and vibration-induced failure mechanisms of metals in Chap. 33 and composites in Chap. 34. Material and slip damping is then covered in Chap. 35, followed by applied damping treatments in Chap. 36.

The last five chapters address specialized issues of importance. Specifically, torsional vibration is discussed in Chap. 37, with particular applications to internal combustion engines and rotating machines. The theory of shock and vibration isolation is discussed in detail in Chap. 38, and various types of isolators for shock and vibration are described in Chap. 39, along with the selection and practical application of such isolators. Chapter 40 describes procedures for the design of equipment to withstand shock and vibration environments, including simple techniques to facilitate preliminary design. Finally, a comprehensive discussion of the human aspects of shock and vibration is considered in Chap. 41, which describes the effects of shock and vibration on people.

SYMBOLS AND ACRONYMS

This section includes a list of symbols and acronyms generally used in the handbook. Symbols of special or limited application are defined in the respective chapters as they are used.

Symbol	Meaning
a	radius
a	acceleration
A/D	analog-to-digital
ANSI	American National Standards Institute
ASTM	American Society for Testing and Materials
B	bandwidth
B	magnetic flux density
c	damping coefficient
c	velocity of sound
c_c	critical damping coefficient
C	capacitance
CSIRO	Commonwealth Scientific and Industrial Research Organisation
D	diameter
D/A	digital-to-analog
DFT	discrete Fourier transform
DSP	discrete signal processor
e	electrical voltage
e	eccentricity
E	energy
E	modulus of elasticity in tension and compression (Young's modulus)
f	frequency
f_n	undamped natural frequency
f_i	undamped natural frequencies in a multiple-degree-of-freedom system, where $i = 1, 2, \ldots$
f_d	damped natural frequency
f_r	resonance frequency
F	force
f_f	coulomb friction force
FEM	finite element method, finite element model
FFT	fast Fourier transform
g	acceleration of gravity
G	modulus of elasticity in shear
h	height, depth
H	magnetic field strength
Hz	hertz
i	electric current
I_i	area or mass moment of inertia (subscript indicates axis)
I_p	polar moment of inertia
I_{ij}	area or mass product of inertia (subscripts indicate axes)
IC	integrated circuit
ISO	International Standards Organization
$\mathcal{I}$	imaginary part of
j	$\sqrt{-1}$
J	inertia constant (weight moment of inertia)
J	impulse
k	spring constant, stiffness, stiffness constant

k_t	rotational (torsional) stiffness
l	length
L	inductance
m	mass
m_u	unbalanced mass
M	torque
M	mutual inductance
$\mathfrak{M}$	mobility
MIMO	multiple input/multiple output
n	number of coils, supports, etc.
NEMA	National Electrical Manufacturers Association
NIST	National Institute of Standards and Technology
p	alternating pressure
p	probability density
P	probability distribution
P	static pressure
q	electric charge
Q	resonance factor (also ratio of reactance to resistance)
r	electrical resistance
R	radius
$\mathfrak{R}$	real part of
s	arc length
S	area of diaphragm, tube, etc.
SEA	statistical energy analysis
SIMO	single input, multiple output
SCC	Standards Council of Canada
t	thickness
t	time
T	transmissibility
T	kinetic energy
v	linear velocity
V	potential energy
w	width
W	weight
W	power
W_e	spectral density of the excitation
W_r	spectral density of the response
x	linear displacement in direction of X axis
$\dot{x}$	first time derivative of x
$\ddot{x}$	second time derivative of x
y	linear displacement in direction of Y axis
z	linear displacement in direction of Z axis
Z	impedance
α	rotational displacement about X axis
β	rotational displacement about Y axis
γ	rotational displacement about Z axis
γ	shear strain
γ	weight density
δ	deflection
δ_{st}	static deflection
Δ	logarithmic decrement
ϵ	tension or compression strain
ζ	fraction of critical damping
η	stiffness ratio, loss factor
θ	phase angle
λ	wavelength

μ	coefficient of friction
μ	mass density
μ	mean value
ν	Poisson's ratio
ρ	mass density
ρ_i	radius of gyration (subscript indicates axis)
σ	Poisson's ratio
σ	normal stress
σ	standard deviation
τ	period
τ	shear stress
ϕ	magnetic flux
Φ	phase angle
ψ	phase angle
Ψ	root-mean-square (rms) value
ω	forcing frequency—angular
ω_n	undamped natural frequency—angular
ω_i	undamped natural frequencies—angular—in a multiple-degree-of-freedom system, where $i = 1, 2, \ldots$
ω_d	damped natural frequency—angular
ω_r	resonance frequency—angular
Ω	rotational speed
$\simeq$	approximately equal to

CHARACTERISTICS OF HARMONIC MOTION

Harmonic functions are employed frequently in the analysis of shock and vibration. A body that experiences simple harmonic motion follows a displacement pattern defined by

$$x = x_0 \sin (2\pi ft) = x_0 \sin \omega t \tag{1.1}$$

where f is the *frequency* of the simple harmonic motion, $\omega = 2\pi f$ is the corresponding *angular frequency,* and x_0 is the *amplitude* of the displacement.

The velocity $\dot{x}$ and acceleration $\ddot{x}$ of the body are found by differentiating the displacement once and twice, respectively:

$$\dot{x} = x_0(2\pi f) \cos 2\pi ft = x_0\omega \cos \omega t \tag{1.2}$$

$$\ddot{x} = -x_0(2\pi f)^2 \sin 2\pi ft = -x_0\omega^2 \sin \omega t \tag{1.3}$$

The maximum absolute values of the displacement, velocity, and acceleration of a body undergoing harmonic motion occur when the trigonometric functions in Eqs. (1.1) to (1.3) are numerically equal to unity. These values are known, respectively, as displacement, velocity, and acceleration amplitudes; they are defined mathematically as follows:

$$x_0 = x_0 \qquad \dot{x}_0 = (2\pi f)x_0 \qquad \ddot{x}_0 = (2\pi f)^2 x_0 \tag{1.4}$$

For certain purposes in analysis, it is convenient to express the amplitude in terms of the average value of the harmonic function, the root-mean-square (rms) value, or 2 times the amplitude (i.e., peak-to-peak value). These terms are defined mathematically in Chap. 19; numerical conversion factors are set forth in Table 1.1 for ready reference.

TABLE 1.1 Conversion Factors for Simple Harmonic Motion

Multiply numerical value in terms of → By ↘ To obtain value in terms of ↓	Amplitude	Average value	Root-mean-square (rms) value	Peak-to-peak value
Amplitude	1	1.571	1.414	0.500
Average value	0.637	1	0.900	0.318
Root-mean-square (rms) value	0.707	1.111	1	0.354
Peak-to-peak value	2.000	3.142	2.828	1

MEASUREMENT UNITS

With only a few exceptions, the measurement units throughout this handbook are presented in *customary* (usually English) units followed in parentheses by *Standard International* (SI) units. The few exceptions occur in complicated figures—in particular, three-dimensional figures—where it would be confusing to present the dupli-

TABLE 1.2 Conversion Factors for English to SI Units

Measurement (symbol)	English units (symbol)	SI units (symbol)	Conversion factor (multiply English units by)
Linear displacement (*x, y, z*)	Inches (in.)	Meters (m)	0.0254
	Feet (ft)	Meters (m)	0.3048
Linear velocity (*v*)	Inches per second (in./sec)	Meters per second (m/s)	0.0254
	Feet per second (ft/sec)	Meters per second (m/s)	0.3048
Linear acceleration (*a*)	Inches per second squared (in./sec^2)	Meters per second squared (m/s^2)	0.0254
	Feet per second squared (ft/sec^2)	Meters per second squared (m/s^2)	0.3048
Force (*F*)	Pounds (lb)	Newtons (N)	4.448
Mass (*m*)	Pounds (lb)	Kilograms (kg)	0.4536
	Slugs—weight/g where g is in ft/sec^2 (lb-sec^2/ft)	Kilograms (kg)	14.59
Pressure (*p*)	Pounds per square inch (lb/in^2)	Pascals (Pa)	6895

cate axes necessary to display the results in both English and SI units. In a few other cases, the customary units are SI units; for example, the reference pressure for sound pressure levels expressed in decibels (dB) is universally 20 μPa in air. A brief list of the conversion factors relating English to SI units for the primary measurements of interest in shock and vibration are summarized in Table 1.2. More detailed unit conversion factors are available from *Marks' Standard Handbook for Mechanical Engineers,* 11th edition, McGraw-Hill, New York, 2007.

APPENDIX 1.1 NATURAL FREQUENCIES OF COMMONLY USED SYSTEMS

The most important aspect of vibration analysis often is the calculation or measurement of the natural frequencies of mechanical systems. Natural frequencies are discussed prominently in many chapters of the handbook. Appendix 1.1 includes in tabular form, convenient for ready reference, a compilation of frequently used expressions for the natural frequencies of common mechanical systems:

1. Mass-spring systems in translation
2. Rotor-shaft systems
3. Massless beams with concentrated mass loads
4. Beams of uniform section and uniformly distributed load
5. Thin, flat plates of uniform thickness
6. Miscellaneous systems

The data for beams and plates are abstracted from Chap. 7.

APPENDIX 1.2 TERMINOLOGY

For convenience, definitions of terms which are used frequently in the field of shock and vibration are assembled here. Many of these are identical with those developed by technical committees of the International Standards Organization (ISO) and the International Electrotechnical Commission (IEC) in cooperation with the American National Standards Institute (ANSI). Copies of standards publications may be obtained from the Standards Secretariat, Acoustical Society of America, 35 Pinelawn Road, Suite 114E, Melville, NY 11747; the e-mail address is asastds@aip.org. In addition to the following definitions, many more terms used in shock and vibration are defined throughout the handbook—far too many to include in this appendix. The reader is referred to the index.

acceleration Acceleration is a vector quantity that specifies the time rate of change of velocity.

acceleration of gravity (See *g*.)

accelerometer An accelerometer is a transducer whose output is proportional to the acceleration input.

ambient vibration Ambient vibration is the all-encompassing vibration associated with a given environment, being usually a composite of vibration from many sources, near and far.

amplitude Amplitude is the maximum value of a sinusoidal quantity.

MASS-SPRING SYSTEMS IN TRANSLATION
(RIGID MASS AND MASSLESS SPRING)

k = SPRING STIFFNESS, LB/IN. (N/M)
m = MASS, LB-SEC^2/IN. (KG)
ω_n = ANGULAR NATURAL FREQUENCY, RAD/SEC

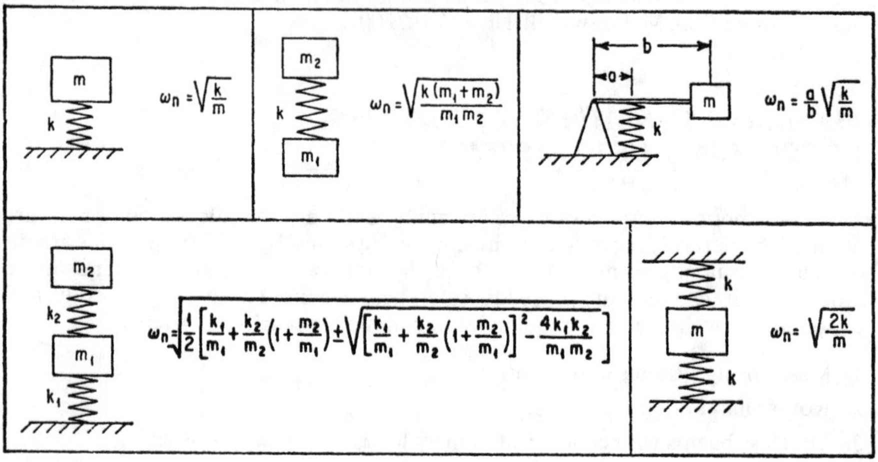

SPRINGS IN COMBINATION

k_r = RESULTANT STIFFNESS OF COMBINATION

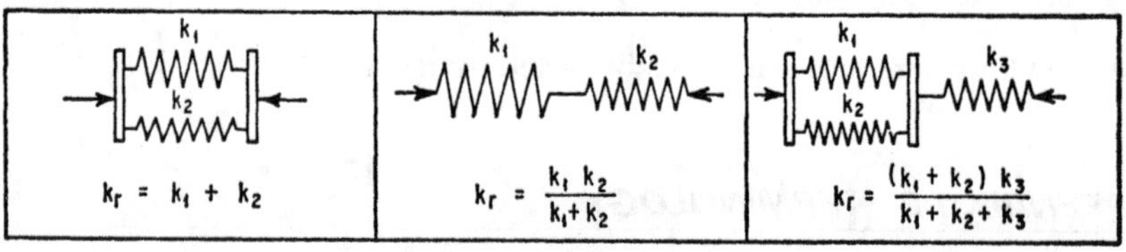

HELICAL SPRINGS

d = WIRE DIAMETER, IN. (M)
D = MEAN COIL DIAMETER, IN. (M)
n = NUMBER OF ACTIVE COILS
E = YOUNG'S MODULUS, LB/IN.^{2} (PA)
G = MODULUS OF ELASTICITY IN SHEAR, LB/IN.^{2} (PA)

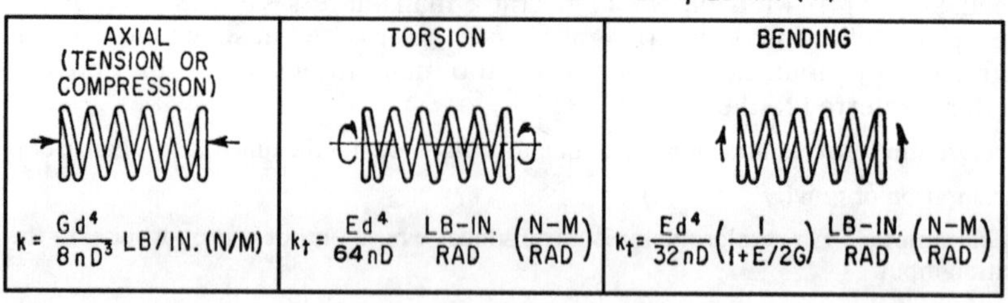

ROTOR-SHAFT SYSTEMS
(RIGID ROTOR AND MASSLESS SHAFT)

k_t = TORSIONAL STIFFNESS OF SHAFT, LB-IN./RAD (N-M/RAD)

I = MASS MOMENT OF INERTIA OF ROTOR, LB-IN.-SEC^2 (N-M-SEC^2)

ω_n = ANGULAR NATURAL FREQUENCY, RAD/SEC

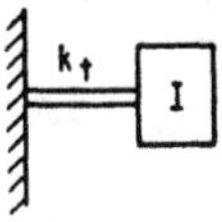

$$\omega_n = \sqrt{\frac{k_t}{I}}$$

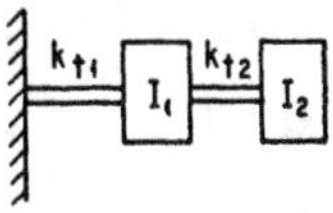

$$\omega_n = \sqrt{\frac{1}{2}\left[\frac{k_{t1}}{I_1} + \frac{k_{t2}}{I_2}\left(1+\frac{I_2}{I_1}\right) \pm \sqrt{\left[\frac{k_{t1}}{I_1} + \frac{k_{t2}}{I_2}\left(1+\frac{I_2}{I_1}\right)\right]^2 - \frac{4k_{t1}k_{t2}}{I_1 I_2}}\,\right]}$$

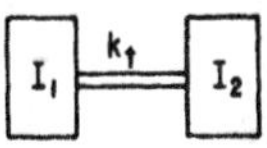

$$\omega_n = \sqrt{\frac{k_t(I_1+I_2)}{I_1 I_2}}$$

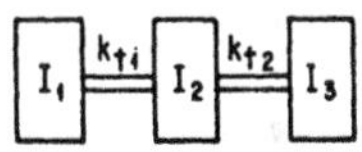

$$\omega_n = \sqrt{\frac{1}{2}\left[B \pm \sqrt{B^2 - \frac{4k_{t1}k_{t2}}{I_1 I_2 I_3}(I_1+I_2+I_3)}\,\right]}$$

WHERE $B = \frac{k_{t1}}{I_1} + \frac{k_{t2}}{I_3} + \frac{k_{t1}+k_{t2}}{I_2}$

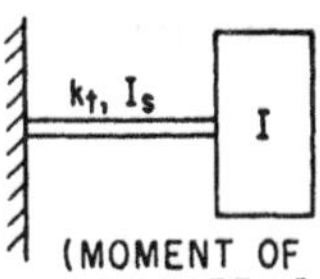

(MOMENT OF INERTIA OF SHAFT = I_s)

$$\omega_n = \sqrt{\frac{k_t}{I + \frac{1}{3}I_s}}$$

GEARED SYSTEM WITH MASSLESS GEARS (SPEED OF ROTOR 2 IS n TIMES SPEED OF ROTOR 1)

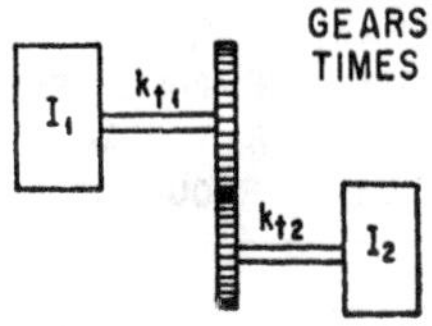

$$\omega_n = \sqrt{\frac{k_1 k_2 (I_1 + n^2 I_2)}{I_1 I_2 (n^2 k_2 + k_1)}}$$

STIFFNESS OF SHAFTS IN TORSION

G = MODULUS OF ELASTICITY IN SHEAR, LB/IN.^{2} (PA)

l = LENGTH OF SHAFT, IN. (M)

I_p = POLAR MOMENT OF INERTIA OF SHAFT CROSS-SECTION, IN.^{4} (M^4)

SOLID CIRCULAR

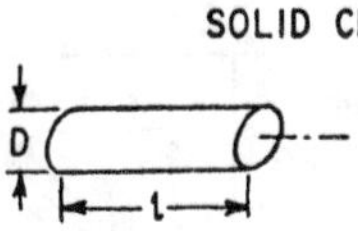

$$k_t = \frac{\pi G D^4}{32 l}$$

HOLLOW CIRCULAR

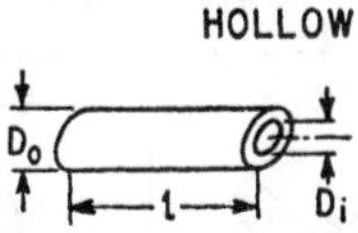

$$k_t = \frac{\pi G(D_o^4 - D_i^4)}{32 l}$$

RECTANGULAR

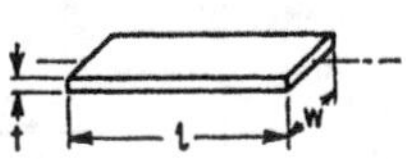

$$k_t = \frac{G w t^3}{3 l}$$

ANY SOLID SECTION

S = AREA

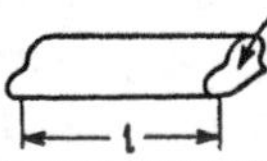

$$k_t = \frac{G S^4}{4\pi^2 l I_p}$$

MASSLESS BEAMS WITH CONCENTRATED MASS LOADS

m = MASS OF LOAD, LB-SEC^2/IN. (KG)
ℓ = LENGTH OF BEAM, IN. (M)
I = AREA MOMENT OF INERTIA OF BEAM CROSS SECTION, IN.^{4} (M^4)
E = YOUNG'S MODULUS, LB/IN.^{2} (PA)
ω_n = ANGULAR NATURAL FREQUENCY, RAD/SEC

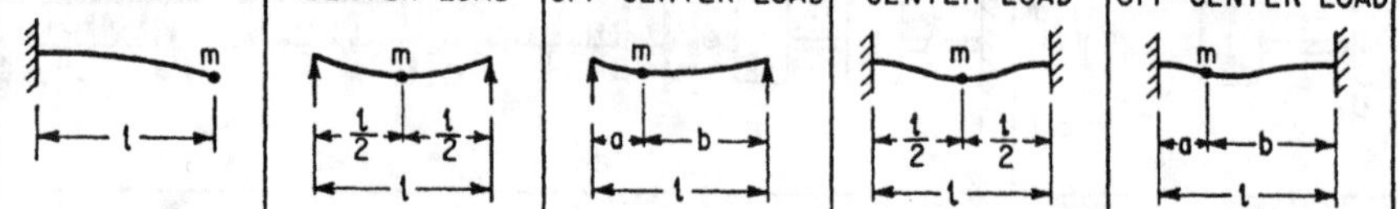

FIXED-FREE END LOAD	HINGED-HINGED CENTER LOAD	HINGED-HINGED OFF-CENTER LOAD	FIXED-FIXED CENTER LOAD	FIXED-FIXED OFF-CENTER LOAD
$\omega_n = \sqrt{\frac{3EI}{m\ell^3}}$	$\omega_n = 4\sqrt{\frac{3EI}{m\ell^3}}$	$\omega_n = \frac{1}{ab}\sqrt{\frac{3EI\ell}{m}}$	$\omega_n = 8\sqrt{\frac{3EI}{m\ell^3}}$	$\omega_n = \frac{1}{ab}\sqrt{\frac{3EI\ell^3}{mab}}$

MASSIVE SPRINGS (BEAMS) WITH CONCENTRATED MASS LOADS

m = MASS OF LOAD, LB-SEC^2/IN. (KG)
$m_s(m_b)$ = MASS OF SPRING (BEAM), LB-SEC^2/IN. (KG)
k = STIFFNESS OF SPRING LB/IN. (N/M)
ℓ = LENGTH OF BEAM, IN. (M)
I = AREA MOMENT OF INERTIA OF BEAM CROSS SECTION, IN.^{4} (M^4)
E = YOUNG'S MODULUS, LB/IN.^{2} (PA)
ω_n = ANGULAR NATURAL FREQUENCY, RAD/SEC

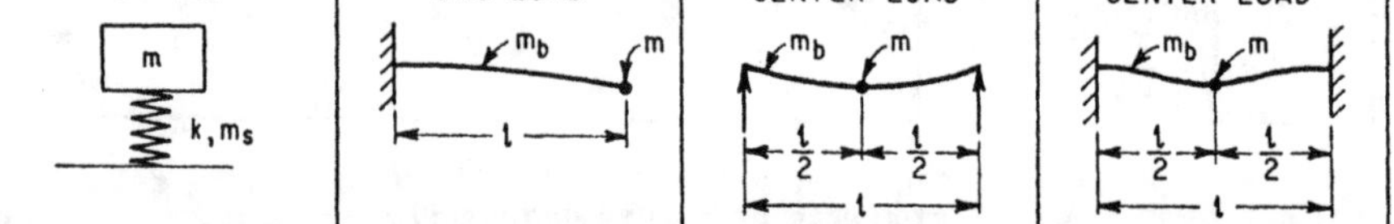

MASS-HELICAL SPRING	FIXED-FREE END LOAD	HINGED-HINGED CENTER LOAD	FIXED-FIXED CENTER LOAD
$\omega_n = \sqrt{\frac{k}{(m + \frac{m_s}{3})}}$	$\omega_n = \sqrt{\frac{3EI}{\ell^3(m+0.23m_b)}}$	$\omega_n = \sqrt{\frac{48EI}{\ell^3(m+0.5m_b)}}$	$\omega_n = 14\sqrt{\frac{EI}{\ell^3(m+0.375m_b)}}$

AREA MOMENT OF INERTIA OF BEAM SECTIONS
(WITH RESPECT TO AXIS a-a)

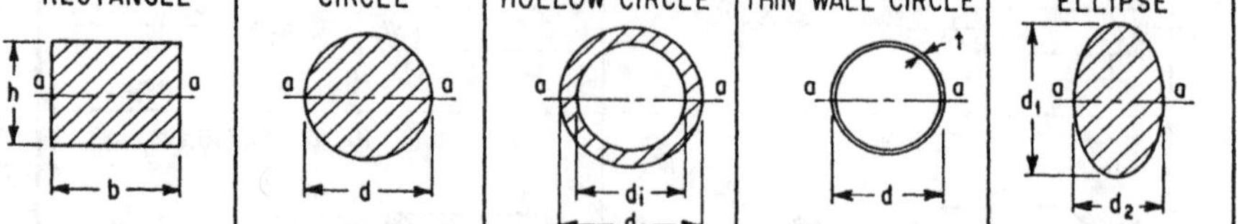

RECTANGLE	CIRCLE	HOLLOW CIRCLE	THIN WALL CIRCLE	ELLIPSE
$I = \frac{bh^3}{12}$	$I = \frac{\pi d^4}{64}$	$I = \frac{\pi}{64}(d_o^4 - d_i^4)$	$I = \frac{\pi d^3 t}{8}$	$I = \frac{\pi d_2 d_1^3}{64}$

BEAMS OF UNIFORM CROSS SECTION

ANGULAR NATURAL FREQUENCY $\omega_n = A\sqrt{\frac{EI}{\mu l^4}}$ RAD/SEC

WHERE E = YOUNG'S MODULUS, LB/IN.^{2} (PA)

I = AREA MOMENT OF INERTIA OF BEAM CROSS SECTION, IN.^{4} (M^4)

l = LENGTH OF BEAM, IN. (M)

μ = MASS PER UNIT LENGTH OF BEAM, LB-SEC^2/IN.^{2} (KG/M)

A = COEFFICIENT FROM TABLE BELOW

NODES ARE INDICATED IN TABLE BELOW AS A PROPORTION OF LENGTH l MEASURED FROM LEFT END

FIXED-FREE (CANTILEVER)	A = 3.52	0.774 A = 22.0	0.500 0.868 A = 61.7	0.356 0.644 0.906 A = 121.0	0.279 0.500 0.723 0.926 A = 200.0
HINGED-HINGED (SIMPLE)	A = 9.87	0.500 A = 39.5	0.333 0.667 A = 88.9	0.25 0.50 0.75 A = 158	0.20 0.40 0.60 0.80 A = 247
FIXED-FIXED (BUILT-IN)	A = 22.4	0.500 A = 61.7	0.359 0.641 A = 121	0.278 0.500 0.722 A = 200	0.227 0.409 0.591 0.773 A = 298
FREE-FREE	0.224 0.776 A = 22.4	0.132 0.500 0.868 A = 61.7	0.094 0.356 0.644 0.906 A = 121	0.073 0.277 0.500 0.723 0.927 A = 200	0.060 0.227 0.409 0.591 0.773 0.940 A = 298
FIXED-HINGED	A = 15.4	0.560 A = 50.0	0.384 0.692 A = 104	0.294 0.529 0.765 A = 178	0.238 0.429 0.619 0.810 A = 272
HINGED-FREE	0.736 A = 15.4	0.446 0.853 A = 50.0	0.308 0.616 0.898 A = 104	0.235 0.471 0.707 0.922 A = 178	0.190 0.381 0.581 0.763 0.937 A = 272

NATURAL FREQUENCIES OF THIN FLAT PLATES OF UNIFORM THICKNESS

$$\omega_n = B\sqrt{\frac{Et^2}{\rho a^4(1-\nu^2)}} \text{ RAD/SEC}$$

E = YOUNG'S MODULUS, LB/IN.^{2} (PA)
t = THICKNESS OF PLATE, IN. (M)
ρ = MASS DENSITY, LB-SEC^2/IN.^{4} (KG/M)
a = DIAMETER OF CIRCULAR PLATE OR SIDE OF SQUARE PLATE, IN. (M)
ν = POISSON'S RATIO

SHAPE OF PLATE	DIAGRAM	EDGE CONDITIONS	VALUE OF B FOR MODE:							
			1	2	3	4	5	6	7	8
CIRCULAR	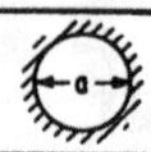	CLAMPED AT EDGE	11.84	24.61	40.41	46.14	103.12			
CIRCULAR		FREE	6.09	10.53	14.19	23.80	40.88	44.68	61.38	69.44
CIRCULAR	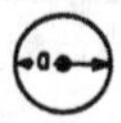	CLAMPED AT CENTER	4.35	24.26	70.39	138.85				
CIRCULAR	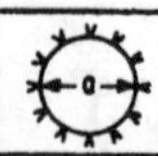	SIMPLY SUPPORTED AT EDGE	5.90							
SQUARE		ONE EDGE CLAMPED–THREE EDGES FREE	1.01	2.47	6.20	7.94	9.01			
SQUARE	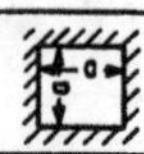	ALL EDGES CLAMPED	10.40	21.21	31.29	38.04	38.22	47.73		
SQUARE		TWO EDGES CLAMPED–TWO EDGES FREE	2.01	6.96	7.74	13.89	18.25			
SQUARE	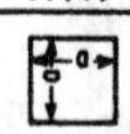	ALL EDGES FREE	4.07	5.94	6.91	10.39	17.80	18.85		
SQUARE	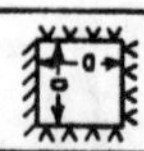	ONE EDGE CLAMPED–THREE EDGES SIMPLY SUPPORTED	6.83	14.94	16.95	24.89	28.99	32.71		
SQUARE	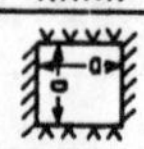	TWO EDGES CLAMPED–TWO EDGES SIMPLY SUPPORTED	8.37	15.82	20.03	27.34	29.54	37.31		
SQUARE	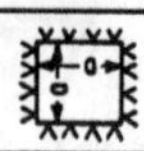	ALL EDGES SIMPLY SUPPORTED	5.70	14.26	22.82	28.52	37.08	48.49		

MASSLESS CIRCULAR PLATE WITH CONCENTRATED CENTER MASS

CLAMPED EDGES

$$\omega_n = 4.09\sqrt{\frac{Eh^3}{ma^2(1-\nu^2)}}$$

SIMPLY SUPPORTED EDGES

$$\omega_n = 4.09\sqrt{\frac{Eh^3}{ma^2(1-\nu)(3+\nu)}}$$

NATURAL FREQUENCIES OF MISCELLANEOUS SYSTEMS

(ω_n = ANGULAR NATURAL FREQUENCY, RAD/SEC)

SIMPLE PENDULUM

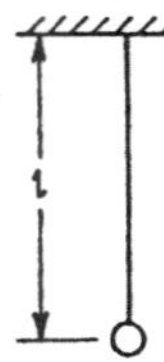

$$\omega_n = \sqrt{\frac{g}{\ell}}$$

g = ACCELERATION OF GRAVITY
(g AND ℓ IN CONSISTENT UNITS)

COMPOUND PENDULUM

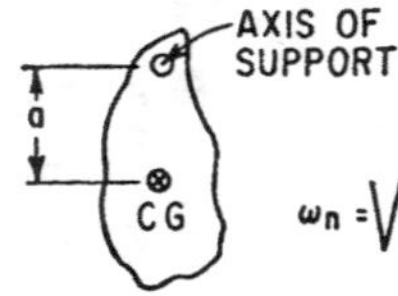

$$\omega_n = \sqrt{\frac{ag}{\rho_0^2}} = \sqrt{\frac{aW}{I_0}}$$

ρ_0 = RADIUS OF GYRATION ABOUT AXIS OF SUPPORT

g = ACCELERATION OF GRAVITY

W = WEIGHT OF PENDULUM

I_0 = MOMENT OF INERTIA ABOUT AXIS OF SUPPORT

SPHERE IN CYLINDRICAL TRACK

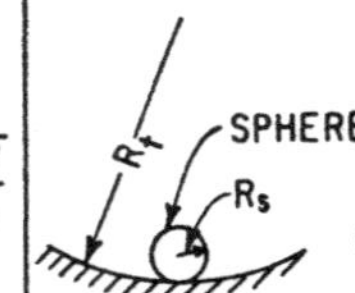

$$\omega_n = \sqrt{\frac{5g}{7(R_t - R_s)}}$$

g = ACCELERATION OF GRAVITY

R_t = RADIUS OF TRACK

R_s = RADIUS OF SPHERE

PNEUMATIC SYSTEM

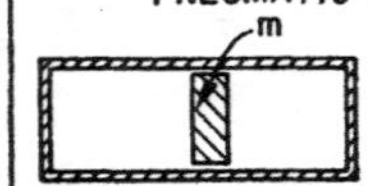

$$\omega_n = \sqrt{\frac{2pS^2}{mV_0}}$$

p = PRESSURE AT EACH END OF CYLINDER, LB/IN.^{2} (PA)

S = AREA OF PISTON, IN.^{2} (M^2)

m = MASS OF PISTON, LB-SEC^2/IN. (KG)

V_0 = VOLUME OF EACH END OF CYLINDER, IN.^{3} (M^3)

U-TUBE WITH LIQUID

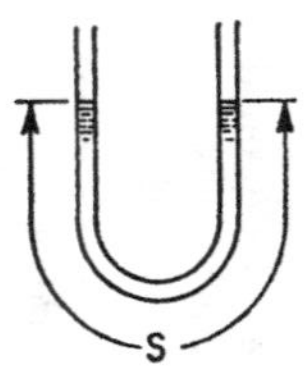

$$\omega_n = \sqrt{\frac{2g}{S}}$$

g = ACCELERATION OF GRAVITY

PLANK ON ROTATING DRUMS

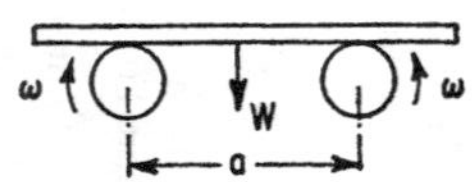

$$\omega_n = \sqrt{\frac{2\mu g}{a}}$$

g = ACCELERATION OF GRAVITY

μ = COEFFICIENT OF FRICTION BETWEEN PLANK AND DRUM

TANKS WITH CONNECTING CONDUIT

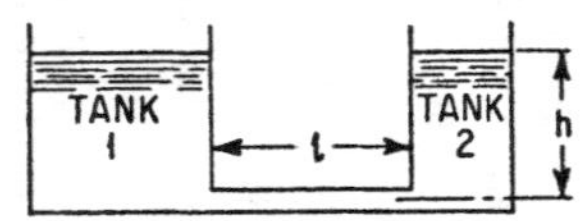

$$\omega_n = \sqrt{\frac{g(1 + S_1/S_2)}{h(1 + S_1/S_2) + \ell(S_1/S_0)}}$$

g = ACCELERATION OF GRAVITY

S_1 = AREA OF TANK 1

S_2 = AREA OF TANK 2

S_0 = AREA OF CONDUIT

CYLINDER IN CYLINDRICAL TRACK

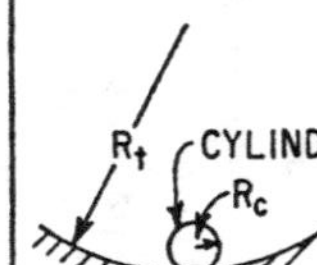

$$\omega_n = \sqrt{\frac{2g}{3(R_t - R_c)}}$$

g = ACCELERATION OF GRAVITY

R_t = RADIUS OF TRACK

R_c = RADIUS OF CYLINDER

analog If a first quantity or structural element is analogous to a second quantity or structural element belonging in another field of knowledge, the second quantity is called the analog of the first, and vice versa.

analogy An analogy is a recognized relationship of consistent mutual similarity between the equations and structures appearing within two or more fields of knowledge, and an identification and association of the quantities and structural elements that play mutually similar roles in these equations and structures, for the purpose of facilitating transfer of knowledge of mathematical procedures of analysis and behavior of the structures between these fields.

angular frequency (circular frequency) The angular frequency of a periodic quantity, in radians per unit time, is the frequency multiplied by 2π.

angular mechanical impedance (rotational mechanical impedance) Angular mechanical impedance is the impedance involving the ratio of torque to angular velocity. (See *impedance.*)

antinode (loop) An antinode is a point, line, or surface in a standing wave where some characteristic of the wave field has maximum amplitude.

antiresonance For a system in forced oscillation, antiresonance exists at a point when any change, however small, in the frequency of excitation causes an increase in the response at this point.

aperiodic motion A vibration that is not periodic.

apparent mass (See *effective mass.*)

audio frequency An audio frequency is any frequency corresponding to a normally audible sound wave.

autocorrelation coefficient The autocorrelation coefficient of a signal is the ratio of the autocorrelation function to the mean-square value of the signal:

$$R(\tau) = \overline{x(t)x(t+\tau)}/\overline{[x(t)]^2}$$

autocorrelation function The autocorrelation function of a signal is the average of the product of the value of the signal at time t with the value at time $t + \tau$:

$$R(\tau) = \overline{x(t)x(t+\tau)}$$

For a stationary random signal of infinite duration, the power spectral density (except for a constant factor) is the cosine Fourier transform of the autocorrelation function.

autospectral density The limiting mean-square value (e.g., of acceleration, velocity, displacement, stress, or other random variable) per unit bandwidth, i.e., the limit of the mean-square value in a given rectangular bandwidth divided by the bandwidth, as the bandwidth approaches zero. Also called *power spectral density.*

auxiliary mass damper (damped vibration absorber) An auxiliary mass damper is a system consisting of a mass, spring, and damper which tends to reduce vibration by the dissipation of energy in the damper as a result of relative motion between the mass and the structure to which the damper is attached.

background noise Background noise is the total of all sources of interference in a system used for the production, detection, measurement, or recording of a signal, independent of the presence of the signal.

balancing Balancing is a procedure for adjusting the mass distribution of a rotor so that vibration of the journals, or the forces on the bearings at once-per-revolution, are reduced or controlled. (See Chap. 39 for a complete list of definitions related to *balancing.*)

bandpass filter A bandpass filter is a wave filter that has a single transmission band extending from a lower cutoff frequency greater than zero to a finite upper cutoff frequency.

bandwidth, effective (See *effective bandwidth.*)

beat frequency The absolute value of the difference in frequency of two oscillators of slightly different frequency.

beats Beats are periodic variations that result from the superposition of two simple harmonic quantities of different frequencies f_1 and f_2. They involve the periodic increase and decrease of amplitude at the beat frequency $(f_1 - f_2)$.

broadband random vibration Broadband random vibration is random vibration having its frequency components distributed over a broad frequency band. (See *random vibration.*)

calibration factor The average sensitivity of a transducer over a specified frequency range.

center of gravity Center of gravity is the point through which passes the resultant of the weights of its component particles for all orientations of the body with respect to a gravitational field; if the gravitational field is uniform, the center of gravity corresponds with the *center of mass.*

circular frequency (See *angular frequency.*)

complex angular frequency As applied to a function $\alpha = Ae^{\sigma t} \sin(\omega t - \phi)$, where σ, ω, and ϕ are constant, the quantity $\omega_c = \sigma + j\omega$ is the complex angular frequency where j is an operator with rules of addition, multiplication, and division as suggested by the symbol $\sqrt{-1}$. If the signal decreases with time, σ must be negative.

complex function A complex function is a function having real and imaginary parts.

complex vibration Complex vibration is vibration whose components are sinusoids not harmonically related to one another. (See *harmonic.*)

compliance Compliance is the reciprocal of stiffness.

compressional wave A compressional wave is one of compressive or tensile stresses propagated in an elastic medium.

continuous system (distributed system) A continuous system is one that is considered to have an infinite number of possible independent displacements. Its configuration is specified by a function of a continuous spatial variable or variables in contrast to a discrete or lumped parameter system which requires only a finite number of coordinates to specify its configuration.

correlation coefficient The correlation coefficient of two variables is the ratio of the correlation function to the product of the averages of the variables:

$$\overline{x_1(t) \cdot x_2(t)}/\overline{x_1(t)} \cdot \overline{x_2(t)}$$

correlation function The correlation function of two variables is the average value of their product:

$$\overline{x_1(t) \cdot x_2(t)}$$

coulomb damping (dry friction damping) Coulomb damping is the dissipation of energy that occurs when a particle in a vibrating system is resisted by a force whose magnitude is a constant independent of displacement and velocity and whose direction is opposite to the direction of the velocity of the particle.

coupled modes Coupled modes are modes of vibration that are not independent but which influence one another because of energy transfer from one mode to the other. (See *mode of vibration.*)

coupling factor, electromechanical The electromechanical coupling factor is a factor used to characterize the extent to which the electrical characteristics of a transducer are modified by a coupled mechanical system, and vice versa.

crest factor The crest factor is the ratio of the peak value to the root-mean-square value.

critical damping Critical damping is the minimum viscous damping that will allow a displaced system to return to its initial position without oscillation.

critical speed Critical speed is the speed of a rotating system that corresponds to a resonance frequency of the system.

cross-talk The signal observed in one channel due to a signal in another channel.

cycle A cycle is the complete sequence of values of a periodic quantity that occur during a period.

damped natural frequency The damped natural frequency is the frequency of free vibration of a damped linear system. The free vibration of a damped system may be considered periodic in the limited sense that the time interval between zero crossings in the same direction is constant, even though successive amplitudes decrease progressively. The frequency of the vibration is the reciprocal of this time interval.

damper A damper is a device used to reduce the magnitude of a shock or vibration by one or more energy dissipation methods.

damping Damping is the dissipation of energy with time or distance.

damping ratio (See *fraction of critical damping.*)

decibel (dB) The decibel is a unit which denotes the magnitude of a quantity with respect to an arbitrarily established reference value of the quantity, in terms of the logarithm (to the base 10) of the ratio of the quantities. For example, in electrical transmission circuits a value of power may be expressed in terms of a power level in decibels; the power level is given by 10 times the logarithm (to the base 10) of the ratio of the actual power to a reference power (which corresponds to 0 dB).

degrees of freedom The number of degrees of freedom of a mechanical system is equal to the minimum number of independent coordinates required to define completely the positions of all parts of the system at any instant of time. In general, it is equal to the number of independent displacements that are possible.

deterministic function A deterministic function is one whose value at any time can be predicted from its value at any other time.

displacement Displacement is a vector quantity that specifies the change of position of a body or particle and is usually measured from the mean position or position of rest. In general, it can be represented as a rotation vector or a translation vector, or both.

displacement pickup Displacement pickup is a transducer that converts an input displacement to an output that is proportional to the input displacement.

distortion Distortion is an undesired change in waveform. Noise and certain desired changes in waveform, such as those resulting from modulation or detection, are not usually classed as distortion.

distributed system (See *continuous system.*)

driving point impedance Driving point impedance is the impedance involving the ratio of force to velocity when both the force and velocity are measured at the same point and in the same direction. (See *impedance.*)

dry friction damping (See *coulomb damping.*)

duration of shock pulse The duration of a shock pulse is the time required for the acceleration of the pulse to rise from some stated fraction of the maximum amplitude and to decay to this value. (See *shock pulse.*)

dynamic stiffness Dynamic stiffness is the ratio of the change of force to the change of displacement under dynamic conditions.

dynamic vibration absorber (tuned damper) A dynamic vibration absorber is an auxiliary mass-spring system which tends to neutralize vibration of a structure to which it is attached. The basic principle of operation is vibration out of phase with the vibration of such structure, thereby applying a counteracting force.

effective bandwidth The effective bandwidth of a specified transmission system is the bandwidth of an ideal system which (1) has uniform transmission in its passband equal to the maximum transmission of the specified system and (2) transmits the same power as the specified system when the two systems are receiving equal input signals having a uniform distribution of energy at all frequencies.

effective mass (apparent mass) The complex ratio of force to acceleration during simple harmonic motion.

electromechanical coupling factor (See *coupling factor, electromechanical.*)

electrostriction Electrostriction is the phenomenon wherein some dielectric materials experience an elastic strain when subjected to an electric field, this strain being independent of the polarity of the field.

ensemble A collection of signals. (See also *process.*)

environment (See *natural environments* and *induced environments.*)

equivalent system An equivalent system is one that may be substituted for another system for the purpose of analysis. Many types of equivalence are common in vibration and shock technology: (1) equivalent stiffness, (2) equivalent damping, (3) torsional system equivalent to a translational system, (4) electrical or acoustical system equivalent to a mechanical system, etc.

equivalent viscous damping Equivalent viscous damping is a value of viscous damping assumed for the purpose of analysis of a vibratory motion, such that the dissipation of energy per cycle at resonance is the same for either the assumed or actual damping force.

ergodic process An ergodic process is a random process that is stationary and of such a nature that all possible time averages performed on one signal are independent of the signal chosen and hence are representative of the time averages of each of the other signals of the entire random process.

excitation (stimulus) Excitation is an external force (or other input) applied to a system that causes the system to respond in some way.

filter A filter is a device for separating waves on the basis of their frequency. It introduces relatively small insertion loss to waves in one or more frequency bands and relatively large insertion loss to waves of other frequencies. (See *insertion loss.*)

force factor The force factor of an electromechanical transducer is (1) the complex quotient of the force required to block the mechanical system divided by the corresponding current in the electric system and (2) the complex quotient of the resulting open-circuit voltage in the electric system divided by the velocity in the mechanical system. Force factors (1) and (2) have the same magnitude when consistent units are used and the transducer satisfies the principle of reciprocity. It is sometimes convenient in an electrostatic or piezoelectric transducer to use the ratios between force and charge or electric displacement, or between voltage and mechanical displacement.

forced vibration (forced oscillation) The oscillation of a system is forced if the response is imposed by the excitation. If the excitation is periodic and continuing, the oscillation is steady-state.

foundation (support) A foundation is a structure that supports the gravity load of a mechanical system. It may be fixed in space, or it may undergo a motion that provides excitation for the supported system.

fraction of critical damping The fraction of critical damping (damping ratio) for a system with viscous damping is the ratio of actual damping coefficient c to the critical damping coefficient c_c.

free vibration Free vibration is that which occurs after the removal of an excitation or restraint.

frequency The frequency of a function periodic in time is the reciprocal of the period. The unit is the cycle per unit time and must be specified; the unit *cycle per second* is called *hertz* (Hz).

frequency, angular (See *angular frequency.*)

fundamental frequency (1) The fundamental frequency of a periodic quantity is the frequency of a sinusoidal quantity which has the same period as the periodic quantity. (2) The fundamental frequency of an oscillating system is the lowest natural frequency. The normal mode of vibration associated with this frequency is known as the fundamental mode.

fundamental mode of vibration The fundamental mode of vibration of a system is the mode having the lowest natural frequency.

g The quantity g is the acceleration produced by the force of gravity, which varies with the latitude and elevation of the point of observation. By international agreement, the value 980.665 cm/sec^2 = 386.087 in./sec^2 = 32.1739 ft/sec^2 has been chosen as the standard acceleration due to gravity.

harmonic A harmonic is a sinusoidal quantity having a frequency that is an integral multiple of the frequency of a periodic quantity to which it is related.

harmonic motion (See *simple harmonic motion.*)

harmonic response Harmonic response is the periodic response of a vibrating system exhibiting the characteristics of resonance at a frequency that is a multiple of the excitation frequency.

high-pass filter A high-pass filter is a wave filter having a single transmission band extending from some critical or cutoff frequency, not zero, up to infinite frequency.

image impedances The image impedances of a structure or device are the impedances that will simultaneously terminate all of its inputs and outputs in such a way that at each of its inputs and outputs the impedances in both directions are equal.

impact An impact is a single collision of one mass in motion with a second mass which may be either in motion or at rest.

impedance Mechanical impedance is the ratio of a force-like quantity to a velocity-like quantity when the arguments of the real (or imaginary) parts of the quantities increase linearly with time. Examples of force-like quantities are force, sound pressure, voltage, temperature. Examples of velocity-like quantities are velocity, volume velocity, current, heat flow. *Impedance* is the reciprocal of *mobility.* (See also *angular mechanical impedance, linear mechanical impedance, driving point impedance,* and *transfer impedance.*)

impulse Impulse is the product of a force and the time during which the force is applied; more specifically, the impulse is $\int_{t_1}^{t_2} F dt$ where the force F is time dependent and equal to zero before time t_1 and after time t_2.

induced environments Induced environments are those conditions generated as a result of the operation of a structure or equipment.

insertion loss The insertion loss, in decibels, resulting from insertion of an element in a transmission system is 10 times the logarithm to the base 10 of the ratio of the power delivered to that part of the system that will follow the element, before the insertion of the element, to the power delivered to that same part of the system after insertion of the element.

isolation Isolation is a reduction in the capacity of a system to respond to an excitation, attained by the use of a resilient support. In steady-state forced vibration, isolation is expressed quantitatively as the complement of transmissibility.

isolator (See *vibration isolator.*)

jerk Jerk is a vector that specifies the time rate of change of acceleration; jerk is the third derivative of displacement with respect to time.

level Level is the logarithm of the ratio of a given quantity to a reference quantity of the same kind; the base of the logarithm, the reference quantity, and the kind of level must be indicated.

(The type of level is indicated by the use of a compound term such as *vibration velocity level.* The level of the reference quantity remains unchanged whether the chosen quantity is peak, rms, or otherwise.) Unit: decibel. Unit symbol: dB.

line spectrum A line spectrum is a spectrum whose components occur at a number of discrete frequencies.

linear mechanical impedance Linear mechanical impedance is the impedance involving the ratio of force to linear velocity. (See *impedance.*)

linear system A system is linear if for every element in the system the response is proportional to the excitation. This definition implies that the dynamic properties of each element in the system can be represented by a set of linear differential equations with constant coefficients, and that for the system as a whole superposition holds.

logarithmic decrement The logarithmic decrement is the natural logarithm of the ratio of any two successive amplitudes of like sign, in the decay of a single-frequency oscillation.

longitudinal wave A longitudinal wave in a medium is a wave in which the direction of displacement at each point of the medium is normal to the wave front.

low-pass filter A low-pass filter is a wave filter having a single transmission band extending from zero frequency up to some critical or cutoff frequency which is not infinite.

magnetostriction Magnetostriction is the phenomenon wherein ferromagnetic materials experience an elastic strain when subjected to an external magnetic field. Also, magnetostriction is the converse phenomenon in which mechanical stresses cause a change in the magnetic induction of a ferromagnetic material.

maximum value The maximum value is the value of a function when any small change in the independent variable causes a decrease in the value of the function.

mechanical admittance (See *mobility.*)

mechanical impedance (See *impedance.*)

mechanical shock Mechanical shock is a nonperiodic excitation (e.g., a motion of the foundation or an applied force) of a mechanical system that is characterized by suddenness and severity and usually causes significant relative displacements in the system.

mechanical system A mechanical system is an aggregate of matter comprising a defined configuration of mass, stiffness, and damping.

mobility (mechanical admittance) Mobility is the ratio of a velocity-like quantity to a force-like quantity when the arguments of the real (or imaginary) parts of the quantities increase linearly with time. *Mobility* is the reciprocal of *impedance.* The terms *angular mobility, linear mobility, driving point mobility,* and *transfer mobility* are used in the same sense as corresponding impedances.

modal numbers When the normal modes of a system are related by a set of ordered integers, these integers are called modal numbers.

mode of vibration In a system undergoing vibration, a mode of vibration is a characteristic pattern assumed by the system in which the motion of every particle is simple harmonic with the same frequency. Two or more modes may exist concurrently in a multiple-degree-of-freedom system.

modulation Modulation is the variation in the value of some parameter which characterizes a periodic oscillation. Thus, amplitude modulation of a sinusoidal oscillation is a variation in the amplitude of the sinusoidal oscillation.

multiple-degree-of-freedom system A multiple-degree-of-freedom system is one for which two or more coordinates are required to define completely the position of the system at any instant.

narrowband random vibration Narrowband random vibration is random vibration having frequency components only within a narrow band. It has the appearance of a sine wave whose amplitude varies in an unpredictable manner. (See *random vibration.*)

natural environments Natural environments are those conditions generated by the forces of nature and whose effects are experienced when the equipment or structure is at rest as well as when it is in operation.

natural frequency Natural frequency is the frequency of free vibration of a system. For a multiple-degree-of-freedom system, the natural frequencies are the frequencies of the normal modes of vibration.

natural mode of vibration The natural mode of vibration is a mode of vibration assumed by a system when vibrating freely.

neutral surface That surface of a beam, in simple flexure, over which there is no longitudinal stress.

node A node is a point, line, or surface in a standing wave where some characteristic of the wave field has essentially zero amplitude.

noise Noise is any undesired signal. By extension, noise is any unwanted disturbance within a useful frequency band, such as undesired electric waves in a transmission channel or device.

nominal bandwidth The nominal bandwidth of a filter is the difference between the nominal upper and lower cutoff frequencies. The difference may be expressed (1) in cycles per second, (2) as a percentage of the passband center frequency, or (3) in octaves.

nominal passband center frequency The nominal passband center frequency is the geometric mean of the nominal cutoff frequencies.

nominal upper and lower cutoff frequencies The nominal upper and lower cutoff frequencies of a filter passband are those frequencies above and below the frequency of maximum response of a filter at which the response to a sinusoidal signal is 3 dB below the maximum response.

nonlinear damping Nonlinear damping is damping due to a damping force that is not proportional to velocity.

normal mode of vibration A normal mode of vibration is a mode of vibration that is uncoupled from (i.e., can exist independently of) other modes of vibration of a system. When vibration of the system is defined as an eigenvalue problem, the normal modes are the eigenvectors and the normal mode frequencies are the eigenvalues. The term *classical normal mode* is sometimes applied to the normal modes of a vibrating system characterized by vibration of each element of the system at the same frequency and phase. In general, classical normal modes exist only in systems having no damping or having particular types of damping.

octave The interval between two frequencies that have a frequency ratio of two.

oscillation Oscillation is the variation, usually with time, of the magnitude of a quantity with respect to a specified reference when the magnitude is alternately greater and smaller than the reference.

partial node A partial node is the point, line, or surface in a standing-wave system where some characteristic of the wave field has a minimum amplitude differing from zero. The appropriate modifier should be used with the words *partial node* to signify the type that is intended; e.g., displacement partial node, velocity partial node, pressure partial node.

peak-to-peak value The peak-to-peak value of a vibrating quantity is the algebraic difference between the extremes of the quantity.

peak value Peak value is the maximum value of a vibration during a given interval, usually considered to be the maximum deviation of that vibration from the mean value.

period The period of a periodic quantity is the smallest increment of the independent variable for which the function repeats itself.

periodic quantity A periodic quantity is an oscillating quantity whose values recur for certain increments of the independent variable.

phase of a periodic quantity The phase of a periodic quantity, for a particular value of the independent variable, is the fractional part of a period through which the independent variable has advanced, measured from an arbitrary reference.

pickup (See *transducer.*)

piezoelectric (crystal) (ceramic) transducer A piezoelectric transducer is a transducer that depends for its operation on the interaction between the electric charge and the deformation of certain asymmetric crystals having piezoelectric properties.

piezoelectricity Piezoelectricity is the property exhibited by some asymmetrical crystalline materials which when subjected to strain in suitable directions develop electric polarization proportional to the strain. Inverse piezoelectricity is the effect in which mechanical strain is produced in certain asymmetrical crystalline materials when subjected to an external electric field; the strain is proportional to the electric field.

power spectral density Power spectral density is the limiting mean-square value (e.g., of acceleration, velocity, displacement, stress, or other random variable) per unit bandwidth, i.e., the limit of the mean-square value in a given rectangular bandwidth divided by the bandwidth, as the bandwidth approaches zero. Also called *autospectral density.*

power spectral density level The spectrum level of a specified signal at a particular frequency is the level in decibels of that part of the signal contained within a band 1 cycle per second wide, centered at the particular frequency. Ordinarily this has significance only for a signal having a continuous distribution of components within the frequency range under consideration.

power spectrum A spectrum of mean-squared spectral density values.

process A process is a collection of signals. The word *process* rather than the word *ensemble* ordinarily is used when it is desired to emphasize the properties the signals have or do not have as a group. Thus, one speaks of a stationary process rather than a stationary ensemble.

pulse rise time The pulse rise time is the interval of time required for the leading edge of a pulse to rise from some specified small fraction to some specified larger fraction of the maximum value.

***Q* (quality factor)** The quantity Q is a measure of the sharpness of resonance or frequency selectivity of a resonant vibratory system having a single degree of freedom, either mechanical or electrical. In a mechanical system, this quantity is equal to one-half the reciprocal of the damping ratio. It is commonly used only with reference to a lightly damped system and is then approximately equal to the following: (1) Transmissibility at resonance, (2) π/logarithmic decrement, (3) $2\pi W/\Delta W$ where W is the stored energy and ΔW the energy dissipation per cycle, and (4) $f_r/\Delta f$ where f_r is the resonance frequency and Δf is the bandwidth between the half-power points.

quasi-ergodic process A quasi-ergodic process is a random process which is not necessarily stationary but of such a nature that some time averages performed on a signal are independent of the signal chosen.

quasi-periodic signal A quasi-periodic signal is one consisting only of quasi-sinusoids.

quasi-sinusoid A quasi-sinusoid is a function of the form $\alpha = A \sin (2\pi ft - \phi)$ where either A or f, or both, is not a constant but may be expressed readily as a function of time. Ordinarily ϕ is considered constant.

random vibration Random vibration is vibration whose instantaneous magnitude is not specified for any given instant of time. The instantaneous magnitudes of a random vibration are specified only by probability distribution functions giving the probable fraction of the total time that the magnitude (or some sequence of magnitudes) lies within a specified range. Random vibration contains no periodic or quasi-periodic constituents. If random vibration has instantaneous magnitudes that occur according to the gaussian distribution, it is called *gaussian random vibration.*

ratio of critical damping (See *fraction of critical damping.*)

Rayleigh wave A Rayleigh wave is a surface wave associated with the free boundary of a solid, such that a surface particle describes an ellipse whose major axis is normal to the surface, and whose center is at the undisturbed surface. At maximum particle displacement away from the solid surface the motion of the particle is opposite to that of the wave.

recording channel The term *recording channel* refers to one of a number of independent recorders in a recording system or to independent recording tracks on a recording medium.

recording system A recording system is a combination of transducing devices and associated equipment suitable for storing signals in a form capable of subsequent reproduction.

rectangular shock pulse An ideal shock pulse for which motion rises instantaneously to a given value, remains constant for the duration of the pulse, then drops to zero instantaneously.

relaxation time Relaxation time is the time taken by an exponentially decaying quantity to decrease in amplitude by a factor of $1/e = 0.3679$.

resonance Resonance of a system in forced vibration exists when any change, however small, in the frequency of excitation causes a decrease in the response of the system.

resonance frequency Resonance frequency is a frequency at which resonance exists.

response The response of a device or system is the motion (or other output) resulting from an excitation (stimulus) under specified conditions.

response spectrum (See *shock response spectrum.*)

rotational mechanical impedance (See *angular mechanical impedance.*)

seismic pickup; seismic transducer A seismic pickup or transducer is a device consisting of a seismic system in which the differential movement between the mass and the base of the system produces a measurable indication of such movement.

seismic system A seismic system is one consisting of a mass attached to a reference base by one or more flexible elements. Damping is usually included.

self-induced (self-excited) vibration The vibration of a mechanical system is self-induced if it results from conversion, within the system, of nonoscillatory excitation to oscillatory excitation.

sensing element That part of a transducer which is activated by the input excitation and supplies the output signal.

sensitivity The sensitivity of a transducer is the ratio of a specified output quantity to a specified input quantity.

shear wave (rotational wave) A shear wave is a wave in an elastic medium which causes an element of the medium to change its shape without a change of volume.

shock (See *mechanical shock.*)

shock absorber A shock absorber is a device which dissipates energy to modify the response of a mechanical system to applied shock.

shock excitation An excitation, applied to a mechanical system, that produces a mechanical shock.

shock isolator (shock mount) A shock isolator is a resilient support that tends to isolate a system from a shock motion.

shock machine A shock machine is a device for subjecting a system to controlled and reproducible mechanical shock.

shock motion Shock motion is an excitation involving motion of a foundation. (See *foundation* and *mechanical shock.*)

shock mount (See *shock isolator.*)

shock pulse A shock pulse is a substantial disturbance characterized by a rise of acceleration from a constant value and decay of acceleration to the constant value in a short period of time. Shock pulses are normally displayed graphically as curves of acceleration as functions of time.

shock-pulse duration (See *duration of shock pulse.*)

shock response spectrum (SRS) A shock spectrum is a plot of the maximum response experienced by a single-degree-of-freedom system, as a function of its own natural frequency, in response to an applied shock. The response may be expressed in terms of acceleration, velocity, or displacement.

shock testing machine; shock machine A shock testing machine is a device for subjecting a mechanical system to controlled and reproducible mechanical shock.

signal A signal is (1) a disturbance used to convey information; (2) the information to be conveyed over a communication system.

simple harmonic motion A simple harmonic motion is a motion such that the displacement is a sinusoidal function of time; sometimes it is designated merely by the term *harmonic motion.*

single-degree-of-freedom system A single-degree-of-freedom system is one for which only one coordinate is required to define completely the configuration of the system at any instant.

sinusoidal motion (See *simple harmonic motion.*)

snubber A snubber is a device used to increase the stiffness of an elastic system (usually by a large factor) whenever the displacement becomes larger than a specified value.

spectrum A spectrum is a definition of the magnitude of the frequency components that constitute a quantity.

spectrum density (See *power spectral density.*)

standard deviation Standard deviation is the square root of the variance; i.e., the square root of the mean of the squares of the deviations from the mean value of a vibrating quantity.

standing wave A standing wave is a periodic wave having a fixed distribution in space which is the result of interference of progressive waves of the same frequency and kind. Such waves are characterized by the existence of nodes or partial nodes and antinodes that are fixed in space.

stationary process A stationary process is an ensemble of signals such that an average of values over the ensemble at any given time is independent of time.

stationary signal A stationary signal is a random signal of such nature that averages over samples of finite time intervals are independent of the time at which the sample occurs.

steady-state vibration Steady-state vibration exists in a system if the velocity of each particle is a continuing periodic quantity.

stiffness Stiffness is the ratio of change of force (or torque) to the corresponding change on translational (or rotational) deflection of an elastic element.

subharmonic A subharmonic is a sinusoidal quantity having a frequency that is an integral submultiple of the fundamental frequency of a periodic quantity to which it is related.

subharmonic response Subharmonic response is the periodic response of a mechanical system exhibiting the characteristic of resonance at a frequency that is a submultiple of the frequency of the periodic excitation.

superharmonic response Superharmonic response is a term sometimes used to denote a particular type of harmonic response which dominates the total response of the system; it frequently occurs when the excitation frequency is a submultiple of the frequency of the fundamental resonance.

time history The magnitude of a quantity expressed as a function of time.

transducer (pickup) A transducer is a device which converts shock or vibratory motion into an optical, a mechanical, or most commonly to an electrical signal that is proportional to a parameter of the experienced motion.

transfer impedance Transfer impedance between two points is the impedance involving the ratio of force to velocity when force is measured at one point and velocity at the other point. The term *transfer impedance* also is used to denote the ratio of force to velocity measured at the same point but in different directions. (See *impedance.*)

transient vibration Transient vibration is temporarily sustained vibration of a mechanical system. It may consist of forced or free vibration or both.

transmissibility Transmissibility is the nondimensional ratio of the response amplitude of a system in steady-state forced vibration to the excitation amplitude. The ratio may be one of forces, displacements, velocities, or accelerations.

transmission loss Transmission loss is the reduction in the magnitude of some characteristic of a signal, between two stated points in a transmission system.

transverse wave A transverse wave is a wave in which the direction of displacement at each point of the medium is parallel to the wavefront.

tuned damper (See *dynamic vibration absorber.*)

uncorrelated Two signals or variables $\alpha_1(t)$ and $\alpha_2(t)$ are said to be uncorrelated if the average value of their product is zero: $\overline{\alpha_1(t) \cdot \alpha_2(t)} = 0$. If the correlation coefficient is equal to unity, the variables are said to be completely correlated. If the coefficient is less than unity but larger than zero, they are said to be partially correlated. (See *correlation coefficient.*)

uncoupled mode An uncoupled mode of vibration is a mode that can exist in a system concurrently with and independently of other modes.

undamped natural frequency The undamped natural frequency of a mechanical system is the frequency of free vibration resulting from only elastic and inertial forces of the system.

variance Variance is the mean of the squares of the deviations from the mean value of a vibrating quantity.

velocity Velocity is a vector quantity that specifies the time rate of change of displacement with respect to a reference frame. If the reference frame is not inertial, the velocity is often designated "relative velocity."

velocity pickup A velocity pickup is a transducer that converts an input velocity to an output (usually electrical) that is proportional to the input velocity.

velocity shock Velocity shock is a particular type of shock motion characterized by a sudden velocity change of the foundation. (See *foundation* and *mechanical shock.*)

vibration Vibration is an oscillation wherein the quantity is a parameter that defines the motion of a mechanical system. (See *oscillation.*)

vibration acceleration Vibration acceleration is the rate of change of speed and direction of a vibration, in a specified direction. The frequency bandwidth must be identified. Unit meter per second squared. Unit symbol: m/s^2.

vibration acceleration level The vibration acceleration level is 10 times the logarithm (to the base 10) of the ratio of the square of a given vibration acceleration to the square of a reference acceleration, commonly 1*g* or 1 m/s^2. Unit: decibel. Unit symbol: dB.

vibration isolator A vibration isolator is a resilient support that tends to isolate a system from steady-state excitation.

vibration machine A vibration machine is a device for subjecting a mechanical system to controlled and reproducible mechanical vibration.

vibration meter A vibration meter is an apparatus for the measurement of displacement, velocity, or acceleration of a vibrating body.

vibration mount (See *vibration isolator.*)

vibration pickup (See *transducer.*)

vibrometer An instrument capable of indicating some measure of the magnitude (such as rms acceleration) on a scale.

viscous damping Viscous damping is the dissipation of energy that occurs when a particle in a vibrating system is resisted by a force that has a magnitude proportional to the magnitude of the velocity of the particle and direction opposite to the direction of the particle.

viscous damping, equivalent (See *equivalent viscous damping.*)

wave A wave is a disturbance which is propagated in a medium in such a manner that at any point in the medium the quantity serving as measure of disturbance is a function of the time, while at any instant the displacement at a point is a function of the position of the point. Any physical quantity that has the same relationship to some independent variable (usually time) that a propagated disturbance has, at a particular instant, with respect to space, may be called a wave.

wave interference Wave interference is the phenomenon which results when waves of the same or nearly the same frequency are superposed; it is characterized by a spatial or temporal distribution of amplitude of some specified characteristic differing from that of the individual superposed waves.

wavelength The wavelength of a periodic wave in an isotropic medium is the perpendicular distance between two wave fronts in which the displacements have a difference in phase of one complete period.

white noise White noise is a noise whose power spectral density is substantially independent of frequency over a specified range.

CHAPTER 2
BASIC VIBRATION THEORY

Ralph E. Blake

INTRODUCTION

This chapter presents the theory of free and forced steady-state vibration of single-degree-of-freedom systems. Undamped systems and systems having viscous damping and structural damping are included. Multiple-degree-of-freedom systems are discussed, including the normal-mode theory of linear elastic structures and Lagrange's equations.

ELEMENTARY PARTS OF VIBRATORY SYSTEMS

Vibratory systems comprise means for storing potential energy (spring), means for storing kinetic energy (mass or inertia), and means by which the energy is gradually lost (damper). The vibration of a system involves the alternating transfer of energy between its potential and kinetic forms. In a damped system, some energy is dissipated at each cycle of vibration and must be replaced from an external source if a steady vibration is to be maintained. Although a single physical structure may store both kinetic and potential energy, and may dissipate energy, this chapter considers only *lumped parameter systems* composed of ideal springs, masses, and dampers wherein each element has only a single function. In translational motion, displacements are defined as linear distances; in rotational motion, displacements are defined as angular motions.

TRANSLATIONAL MOTION

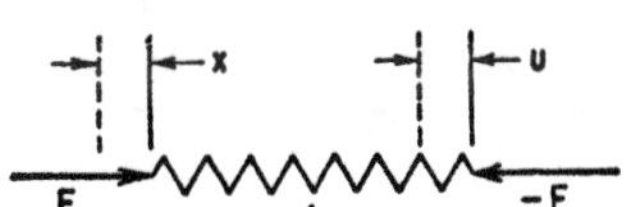

FIGURE 2.1 Linear spring.

Spring. In the linear spring shown in Fig. 2.1, the change in the length of the spring is proportional to the force acting along its length:

$$F = k(x - u) \tag{2.1}$$

The ideal spring is considered to have no mass; thus, the force acting on one end is equal and opposite to the force acting on the other end. The constant of proportionality k is the *spring constant* or *stiffness.*

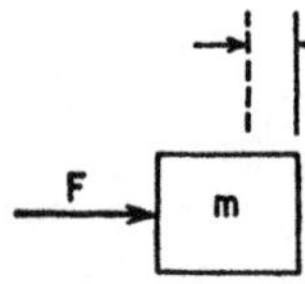

FIGURE 2.2 Rigid mass.

Mass. A mass is a rigid body (Fig. 2.2) whose acceleration $\ddot{x}$ according to Newton's second law is proportional to the resultant F of all forces acting on the mass:*

$$F = m\ddot{x} \tag{2.2}$$

Damper. In the viscous damper shown in Fig. 2.3, the applied force is proportional to the relative velocity of its connection points:

$$F = c(\dot{x} - \dot{u}) \tag{2.3}$$

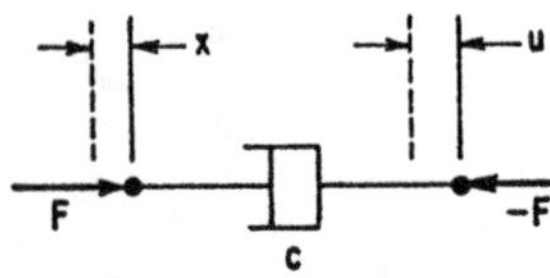

FIGURE 2.3 Viscous damper.

The constant c is the *damping coefficient,* the characteristic parameter of the damper. The ideal damper is considered to have no mass; thus, the force at one end is equal and opposite to the force at the other end. *Structural damping* is considered below.

ROTATIONAL MOTION

The elements of a mechanical system which moves with pure rotation of the parts are wholly analogous to the elements of a system that moves with pure translation. The property of a rotational system which stores kinetic energy is inertia; stiffness and damping coefficients are defined with reference to angular displacement and angular velocity, respectively. The analogous quantities and equations are listed in Table 2.1.

Inasmuch as the mathematical equations for a rotational system can be written by analogy from the equations for a translational system, only the latter are discussed in

TABLE 2.1 Analogous Quantities in Translational and Rotational Vibrating Systems

Translational quantity	Rotational quantity
Linear displacement x	Angular displacement α
Force F	Torque M
Spring constant k	Spring constant k_r
Damping constant c	Damping constant c_r
Mass m	Moment of inertia I
Spring law $F = k(x_1 - x_2)$	Spring law $M = k_r(\alpha_1 - \alpha_2)$
Damping law $F = c(\dot{x}_1 - \dot{x}_2)$	Damping law $M = c_r(\dot{\alpha}_1 - \dot{\alpha}_2)$
Inertia law $F = m\ddot{x}$	Inertia law $M = I\ddot{\alpha}$

* It is common to use the word *mass* in a general sense to designate a rigid body. Mathematically, the mass of the rigid body is defined by m in Eq. (2.2).

detail. Whenever translational systems are discussed, it is understood that corresponding equations apply to the analogous rotational system, as indicated in Table 2.1.

SINGLE-DEGREE-OF-FREEDOM SYSTEM

The simplest possible vibratory system is shown in Fig. 2.4; it consists of a mass m attached by means of a spring k to an immovable support. The mass is constrained to translational motion in the direction of the X axis so that its change of position from an initial reference is described fully by the value of a single quantity x. For this reason it is called a *single-degree-of-freedom system.* If the mass m is displaced from its equilibrium position and then allowed to vibrate free from further external forces, it is said to have *free vibration.* The vibration also may be forced; i.e., a continuing force acts upon the mass or the foundation experiences a continuing motion. Free and forced vibration are discussed below.

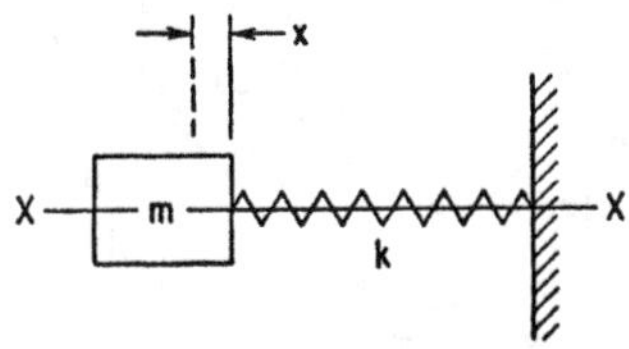

FIGURE 2.4 Undamped single-degree-of-freedom system.

FREE VIBRATION WITHOUT DAMPING

Considering first the free vibration of the undamped system of Fig. 2.4, Newton's equation is written for the mass m. The force $m\ddot{x}$ exerted by the mass on the spring is equal and opposite to the force kx applied by the spring on the mass:

$$m\ddot{x} + kx = 0 \tag{2.4}$$

where $x = 0$ defines the equilibrium position of the mass.

The solution of Eq. (2.4) is

$$x = A \sin \sqrt{\frac{k}{m}}\, t + B \cos \sqrt{\frac{k}{m}}\, t \tag{2.5}$$

where the term $\sqrt{k/m}$ is the *angular natural frequency* defined by

$$\omega_n = \sqrt{\frac{k}{m}} \qquad \text{rad/sec} \tag{2.6}$$

The sinusoidal oscillation of the mass repeats continuously, and the time interval to complete one cycle is the *period:*

$$\tau = \frac{2\pi}{\omega_n} \tag{2.7}$$

The reciprocal of the period is the *natural frequency:*

$$f_n = \frac{1}{\tau} = \frac{\omega_n}{2\pi} = \frac{1}{2\pi}\sqrt{\frac{k}{m}} = \frac{1}{2\pi}\sqrt{\frac{kg}{W}} \tag{2.8}$$

where $W = mg$ is the weight of the rigid body forming the mass of the system shown in Fig. 2.4. The relations of Eq. (2.8) are shown by the solid lines in Fig. 2.5.

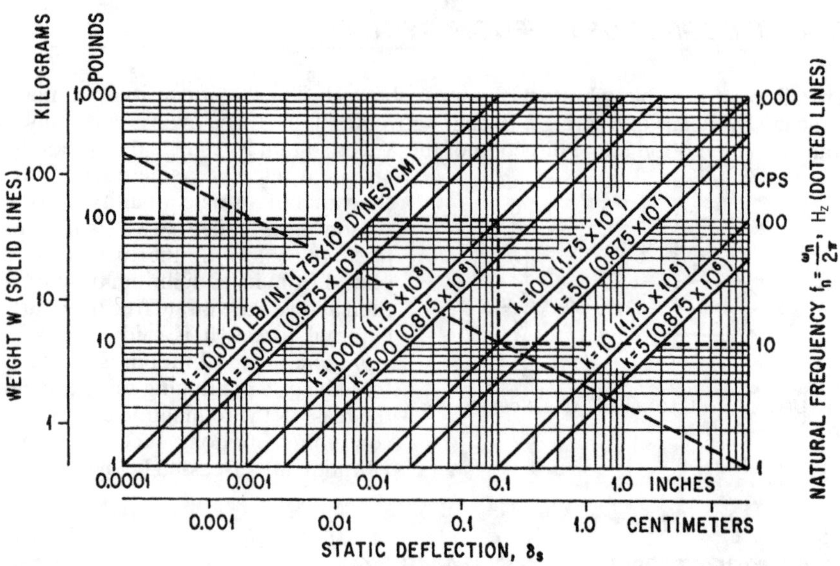

FIGURE 2.5 Natural frequency relations for a single-degree-of-freedom system. Relation of natural frequency to weight of supported body and stiffness of spring [Eq. (2.8)] is shown by solid lines. Relation of natural frequency to static deflection [Eq. (2.10)] is shown by diagonal-dashed line. Example: To find natural frequency of system with $W = 100$ lb and $k = 1000$ lb/in., enter at $W = 100$ on left ordinate scale; follow the dashed line horizontally to solid line $k = 1000$, then vertically down to diagonal-dashed line, and finally horizontally to read $f_n = 10$ Hz from right ordinate scale.

Initial Conditions. In Eq. (2.5), B is the value of x at time $t = 0$, and the value of A is equal to $\dot{x}/\omega_n$ at time $t = 0$. Thus, the conditions of displacement and velocity which exist at zero time determine the subsequent oscillation completely.

Phase Angle. Equation (2.5) for the displacement in oscillatory motion can be written, introducing the frequency relation of Eq. (2.6),

$$x = A \sin \omega_n t + B \cos \omega_n t = C \sin (\omega_n t + \theta) \tag{2.9}$$

where $C = (A^2 + B^2)^{1/2}$ and $\theta = \tan^{-1}(B/A)$. The angle θ is called the *phase angle.*

Static Deflection. The static deflection of a simple mass-spring system is the deflection of spring k as a result of the gravity force of the mass $\delta_{st} = mg/k$. (For example, the system of Fig. 2.4 would be oriented with the mass m vertically above the spring k.) Substituting this relation in Eq. (2.8),

$$f_n = \frac{1}{2\pi}\sqrt{\frac{g}{\delta_{st}}} \tag{2.10}$$

The relation of Eq. (2.10) is shown by the diagonal-dashed line in Fig. 2.5. This relation applies only when the system under consideration is both linear and elastic. For example, rubber springs tend to be nonlinear or exhibit a dynamic stiffness which differs from the static stiffness; hence, Eq. (2.10) is not applicable.

FREE VIBRATION WITH VISCOUS DAMPING

Figure 2.6 shows a single-degree-of-freedom system with a viscous damper. The differential equation of motion of mass *m*, corresponding to Eq. (2.4) for the undamped system, is

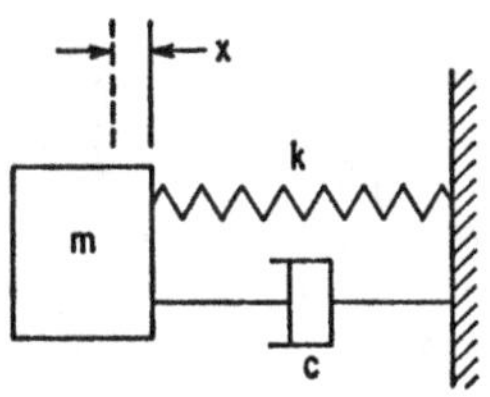

FIGURE 2.6 Single-degree-of-freedom system with viscous damper.

$$m\ddot{x} + c\dot{x} + kx = 0 \qquad (2.11)$$

The form of the solution of this equation depends upon whether the damping coefficient is equal to, greater than, or less than the *critical damping coefficient* c_c:

$$c_c = 2\sqrt{km} = 2m\omega_n \qquad (2.12)$$

The ratio $\zeta = c/c_c$ is defined as the *fraction of critical damping.*

Less-Than-Critical Damping. If the damping of the system is less than critical, $\zeta < 1$; then the solution of Eq. (2.11) is

$$\begin{aligned} x &= e^{-ct/2m}(A \sin \omega_d t + B \cos \omega_d t) \\ &= Ce^{-ct/2m} \sin (\omega_d t + \theta) \end{aligned} \qquad (2.13)$$

where C and θ are defined with reference to Eq. (2.9). The *damped natural frequency* is related to the undamped natural frequency of Eq. (2.6) by the equation

$$\omega_d = \omega_n(1 - \zeta^2)^{1/2} \quad \text{rad/sec} \qquad (2.14)$$

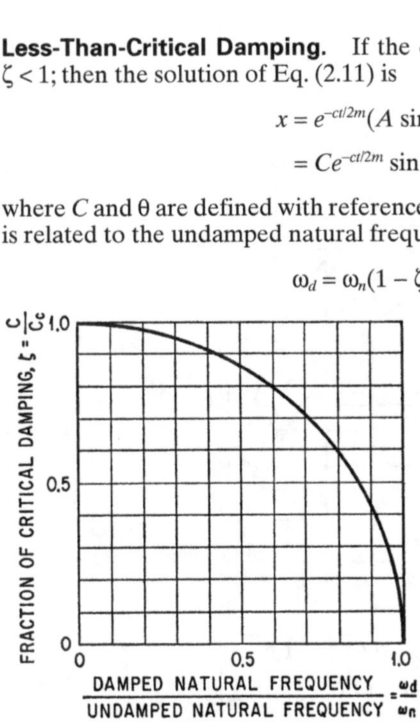

FIGURE 2.7 Damped natural frequency as a function of undamped natural frequency and fraction of critical damping.

Equation (2.14), relating the damped and undamped natural frequencies, is plotted in Fig. 2.7.

Critical Damping. When $c = c_c$, there is no oscillation and the solution of Eq. (2.11) is

$$x = (A + Bt)e^{-ct/2m} \qquad (2.15)$$

Greater-Than-Critical Damping. When $\zeta > 1$, the solution of Eq. (2.11) is

$$x = e^{-ct/2m}\left(Ae^{\omega_n\sqrt{\zeta^2 - 1}\, t} + Be^{-\omega_n\sqrt{\zeta^2 - 1}\, t}\right) \qquad (2.16)$$

This is a nonoscillatory motion; if the system is displaced from its equilibrium position, it tends to return gradually.

Logarithmic Decrement. The degree of damping in a system having $\zeta < 1$ may be defined in terms of successive peak values in a record of a free oscillation. Substituting the expression for critical damping from Eq. (2.12), the expression for free vibration of a damped system, Eq. (2.13), becomes

$$x = Ce^{-\zeta\omega_n t} \sin (\omega_d t + \theta) \tag{2.17}$$

Consider any two maxima (i.e., value of x when $dx/dt = 0$) separated by n cycles of oscillation, as shown in Fig. 2.8. Then the ratio of these maxima is

$$\frac{x_n}{x_0} = e^{-2\pi n\zeta/(1-\zeta^2)^{1/2}} \tag{2.18}$$

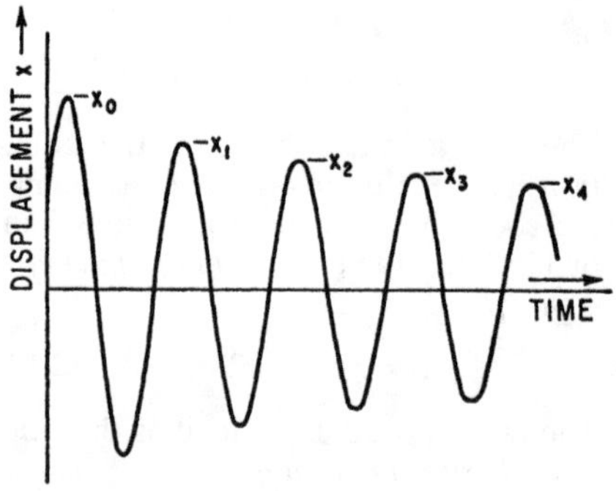

FIGURE 2.8 Trace of damped free vibration showing amplitudes of displacement maxima.

Values of x_n/x_0 are plotted in Fig. 2.9 for several values of n over the range of ζ from 0.001 to 0.10.

The *logarithmic decrement* Δ is the natural logarithm of the ratio of the amplitudes of two successive cycles of the damped free vibration:

$$\Delta = \ln \frac{x_1}{x_2} \quad \text{or} \quad \frac{x_2}{x_1} = e^{-\Delta} \tag{2.19}$$

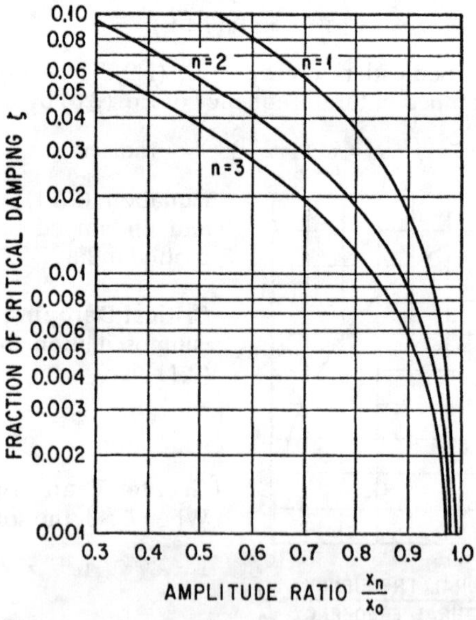

FIGURE 2.9 Effect of damping upon the ratio of displacement maxima of a damped free vibration.

[See also Eq. (37.10).] A comparison of this relation with Eq. (2.18) when $n = 1$ gives the following expression for Δ:

$$\Delta = \frac{2\pi\zeta}{(1 - \zeta^2)^{1/2}} \tag{2.20}$$

The logarithmic decrement can be expressed in terms of the difference of successive amplitudes by writing Eq. (2.19) as follows:

$$\frac{x_1 - x_2}{x_1} = 1 - \frac{x_2}{x_1} = 1 - e^{-\Delta}$$

Writing $e^{-\Delta}$ in terms of its infinite series, the following expression is obtained, which gives a good approximation for $\Delta < 0.2$:

$$\frac{x_1 - x_2}{x_1} = \Delta \tag{2.21}$$

For small values of ζ (less than about 0.10), an approximate relation between the fraction of critical damping and the logarithmic decrement, from Eq. (2.20), is

$$\Delta \simeq 2\pi\zeta \tag{2.22}$$

FORCED VIBRATION

Forced vibration in this chapter refers to the motion of the system which occurs in response to a continuing excitation whose magnitude varies sinusoidally with time. (See Chap. 20 for a treatment of the response of a simple system to step, pulse, and transient vibration excitations.) The excitation may be, alternatively, force applied to the system (generally, the force is applied to the mass of a single-degree-of-freedom system) or motion of the foundation that supports the system. The resulting response of the system can be expressed in different ways, depending upon the nature of the excitation and the use to be made of the result:

1. If the excitation is a force applied to the mass of the system shown in Fig. 2.4, the result may be expressed in terms of (*a*) the amplitude of the resulting motion of the mass or (*b*) the fraction of the applied force amplitude that is transmitted through the system to the support. The former is termed the *motion response* and the latter is termed the *force transmissibility.*
2. If the excitation is a motion of the foundation, the resulting response usually is expressed in terms of the amplitude of the motion of the mass relative to the amplitude of the motion of the foundation. This is termed the *motion transmissibility* for the system.

In general, the response and transmissibility relations are functions of the forcing frequency and vary with different types and degrees of damping. Results are presented in this chapter for undamped systems and for systems with either viscous or structural damping.

FORCED VIBRATION WITHOUT DAMPING

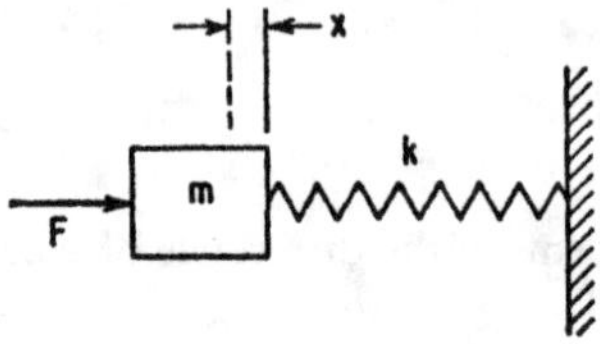

FIGURE 2.10 Undamped single-degree-of-freedom system excited in forced vibration by force acting on mass.

Force Applied to Mass. When the sinusoidal force $F = F_0 \sin \omega t$ is applied to the mass of the undamped single-degree-of-freedom system shown in Fig. 2.10, the differential equation of motion is

$$m\ddot{x} + kx = F_0 \sin \omega t \tag{2.23}$$

The solution of this equation is

$$x = A \sin \omega_n t + B \cos \omega_n t + \frac{F_0/k}{1 - \omega^2/\omega_n^2} \sin \omega t \tag{2.24}$$

where $\omega_n = \sqrt{k/m}$. The first two terms represent an oscillation at the undamped natural frequency ω_n. The coefficient B is the value of x at time $t = 0$, and the coefficient A may be found from the velocity at time $t = 0$. Differentiating Eq. (2.24) and setting $t = 0$,

$$\dot{x}(0) = A\omega_n + \frac{\omega F_0/k}{1 - \omega^2/\omega_n^2} \tag{2.25}$$

The value of A is found from Eq. (2.25).

The oscillation at the natural frequency ω_n gradually decays to zero in physical systems because of damping. The steady-state oscillation at forcing frequency ω is

$$x = \frac{F_0/k}{1 - \omega^2/\omega_n^2} \sin \omega t \tag{2.26}$$

This oscillation exists after a condition of equilibrium has been established by decay of the oscillation at the natural frequency ω_n and persists as long as the force F is applied.

The force transmitted to the foundation is directly proportional to the spring deflection: $F_t = kx$. Substituting x from Eq. (2.26) and defining transmissibility $T = F_t/F$,

$$T = \frac{1}{1 - \omega^2/\omega_n^2} \tag{2.27}$$

If the mass is initially at rest in the equilibrium position of the system (i.e., $x = 0$ and $\dot{x} = 0$) at time $t = 0$, the ensuing motion at time $t > 0$ is

$$x = \frac{F_0/k}{1 - \omega^2/\omega_n^2} \left(\sin \omega t - \frac{\omega}{\omega_n} \sin \omega_n t\right) \tag{2.28}$$

For large values of time, the second term disappears because of the damping inherent in any physical system, and Eq. (2.28) becomes identical to Eq. (2.26).

When the forcing frequency coincides with the natural frequency, $\omega = \omega_n$ and a condition of resonance exists. Then Eq. (2.28) is indeterminate and the expression for x may be written as

$$x = -\frac{F_0\omega}{2k} t \cos \omega t + \frac{F_0}{2k} \sin \omega t \tag{2.29}$$

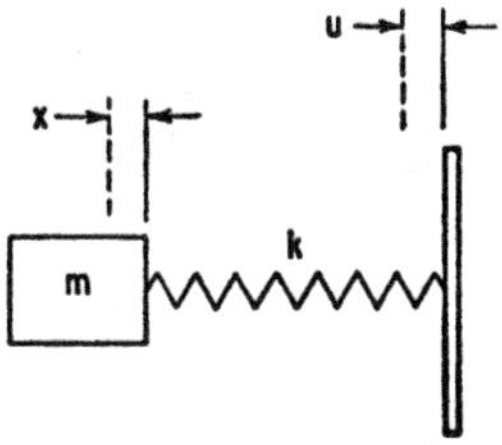

FIGURE 2.11 Undamped single-degree-of-freedom system excited in forced vibration by motion of foundation.

According to Eq. (2.29), the amplitude x increases continuously with time, reaching an infinitely great value only after an infinitely great time.

Motion of Foundation. The differential equation of motion for the system of Fig. 2.11 excited by a continuing motion $u = u_0 \sin \omega t$ of the foundation is

$$m\ddot{x} = -k(x - u_0 \sin \omega t)$$

The solution of this equation is

$$x = A_1 \sin \omega_n t + B_2 \cos \omega_n t + \frac{u_0}{1 - \omega^2/\omega_n^2} \sin \omega t$$

where $\omega_n = k/m$ and the coefficients A_1, B_1 are determined by the velocity and displacement of the mass, respectively, at time $t = 0$. The terms representing oscillation at the natural frequency are damped out ultimately, and the ratio of amplitudes is defined in terms of transmissibility T:

$$\frac{x_0}{u_0} = T = \frac{1}{1 - \omega^2/\omega_n^2} \tag{2.30}$$

where $x = x_0 \sin \omega t$. Thus, in the forced vibration of an undamped single-degree-of-freedom system, the motion response, the force transmissibility, and the motion transmissibility are numerically equal.

FORCED VIBRATION WITH VISCOUS DAMPING

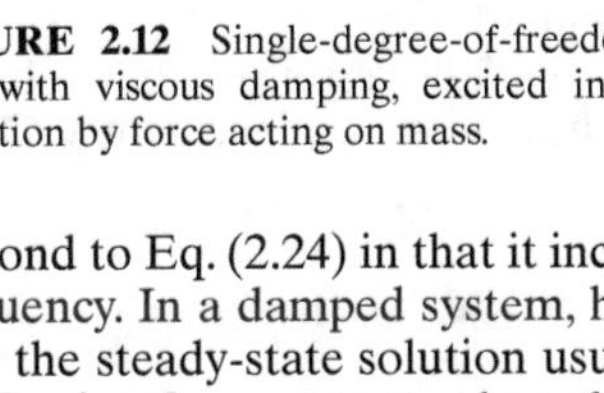

FIGURE 2.12 Single-degree-of-freedom system with viscous damping, excited in forced vibration by force acting on mass.

Force Applied to Mass. The differential equation of motion for the single-degree-of-freedom system with viscous damping shown in Fig. 2.12, when the excitation is a force $F = F_0 \sin \omega t$ applied to the mass, is

$$m\ddot{x} + c\dot{x} + kx = F_0 \sin \omega t \tag{2.31}$$

Equation (2.31) corresponds to Eq. (2.23) for forced vibration of an undamped system; its solution would correspond to Eq. (2.24) in that it includes terms representing oscillation at the natural frequency. In a damped system, however, these terms are damped out rapidly and only the steady-state solution usually is considered. The resulting motion occurs at the forcing frequency ω; when the damping coefficient c is greater than zero, the phase between the force and resulting motion is different than zero. Thus, the response may be written

$$x = R \sin (\omega t - \theta) = A_1 \sin \omega t + B_1 \cos \omega t \tag{2.32}$$

Substituting this relation in Eq. (2.31), the following result is obtained:

$$\frac{x}{F_0/k} = \frac{\sin(\omega t - \theta)}{\sqrt{(1-\omega^2/\omega_n^2)^2 + (2\zeta\omega/\omega_n)^2}} = R_d \sin(\omega t - \theta) \tag{2.33}$$

where $$\theta = \tan^{-1}\left(\frac{2\zeta\omega/\omega_n}{1-\omega^2/\omega_n^2}\right)$$

and R_d is a *dimensionless response factor* giving the ratio of the amplitude of the vibratory displacement to the spring displacement that would occur if the force F were applied statically. At very low frequencies R_d is approximately equal to 1; it rises to a peak near ω_n and approaches zero as ω becomes very large. The displacement response is defined at these frequency conditions as follows:

$$\begin{aligned} x &\simeq \left(\frac{F_0}{k}\right)\sin \omega t && [\omega \ll \omega_n] \\ x &= \frac{F_0}{2k\zeta}\sin\left(\omega_n t + \frac{\pi}{2}\right) = -\frac{F_0 \cos \omega_n t}{c\omega_n} && [\omega = \omega_n] \\ x &\simeq \frac{\omega_n^2 F_0}{\omega^2 k}\sin(\omega t + \pi) = \frac{F_0}{m\omega^2}\sin \omega t && [\omega \gg \omega_n] \end{aligned} \tag{2.34}$$

For the above three frequency conditions, the vibrating system is sometimes described as *spring-controlled, damper-controlled,* and *mass-controlled,* respectively, depending on which element is primarily responsible for the system behavior.

Curves showing the dimensionless response factor R_d as a function of the frequency ratio ω/ω_n are plotted in Fig. 2.13 on the coordinate lines having a positive 45° slope. Curves of the phase angle θ are plotted in Fig. 2.14. A phase angle between 180° and 360° cannot exist in this case since this would mean that the damper is furnishing energy to the system rather than dissipating it.

An alternative form of Eqs. (2.33) and (2.34) is

$$\begin{aligned} \frac{x}{F_0/k} &= \frac{(1-\omega^2/\omega n^2)\sin \omega t - 2\zeta(\omega/\omega_n)\cos \omega t}{(1-\omega^2/\omega_n^2)^2 + (2\zeta\omega/\omega_n)^2} \\ &= (R_d)_x \sin \omega t + (R_d)_R \cos \omega t \end{aligned} \tag{2.35}$$

This shows the components of the response which are in phase $[(R_d)_x \sin \omega t]$ and 90° out of phase $[(R_d)_R \cos \omega t]$ with the force. Curves of $(R_d)_x$ and $(R_d)_R$ are plotted as a function of the frequency ratio ω/ω_n in Figs. 2.15 and 2.16.

Velocity and Acceleration Response. The shape of the response curves changes distinctly if velocity $\dot{x}$ or acceleration $\ddot{x}$ is plotted instead of displacement x. Differentiating Eq. (2.33),

$$\frac{\dot{x}}{F_0/\sqrt{km}} = \frac{\omega}{\omega_n} R_d \cos(\omega t - \theta) = R_v \cos(\omega t - \theta) \tag{2.36}$$

The acceleration response is obtained by differentiating Eq. (2.36):

$$\frac{\ddot{x}}{F_0/m} = -\frac{\omega^2}{\omega_n^2} R_d \sin(\omega t - \theta) = -R_a \sin(\omega t - \theta) \tag{2.37}$$

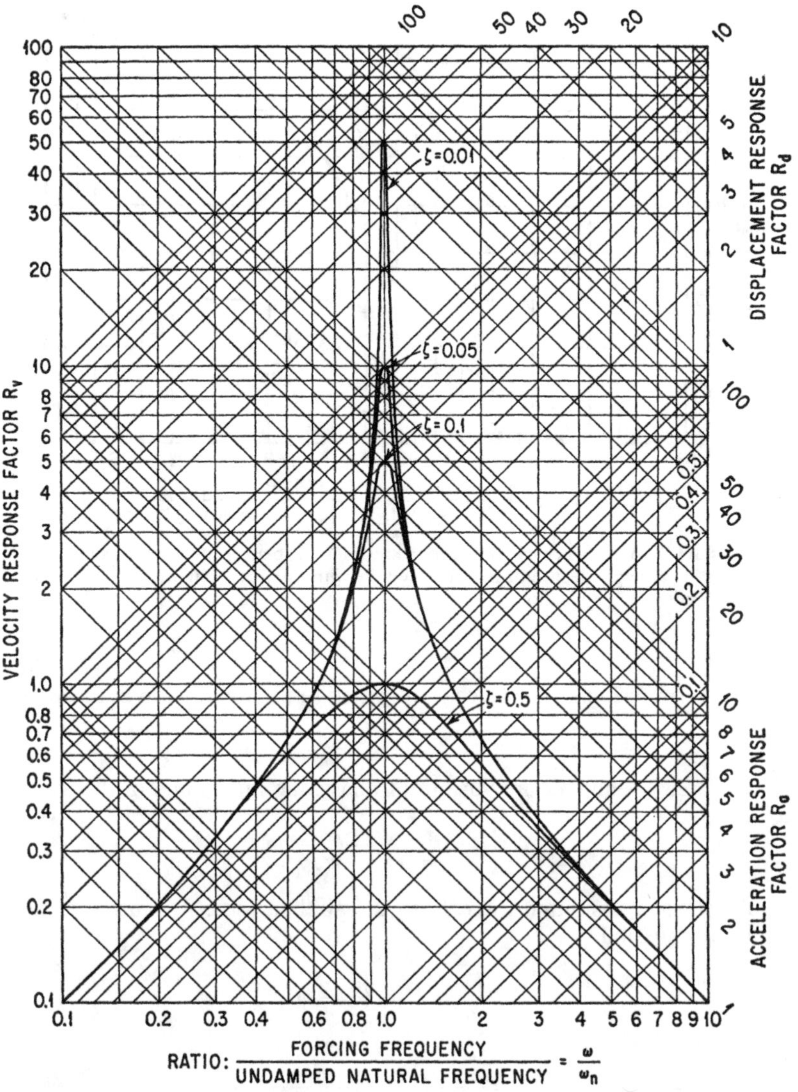

FIGURE 2.13 Response factors for a viscous-damped single-degree-of-freedom system excited in forced vibration by a force acting on the mass. The velocity response factor shown by horizontal lines is defined by Eq. (2.36), the displacement response factor shown by diagonal lines of positive slope is defined by Eq. (2.33), and the acceleration response factor shown by diagonal lines of negative slope is defined by Eq. (2.37).

The velocity and acceleration response factors defined by Eqs. (2.36) and (2.37) are shown graphically in Fig. 2.13, the former to the horizontal coordinates and the latter to the coordinates having a negative 45° slope. Note that the velocity response factor approaches zero as $\omega \to 0$ and $\omega \to \infty$, whereas the acceleration response factor approaches 0 as $\omega \to 0$ and approaches unity as $\omega \to \infty$.

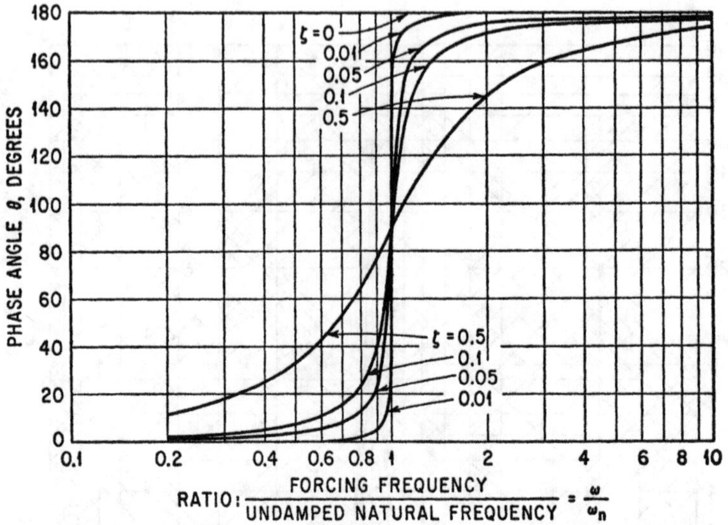

FIGURE 2.14 Phase angle between the response displacement and the excitation force for a single-degree-of-freedom system with viscous damping, excited by a force acting on the mass of the system.

Force Transmission. The force transmitted to the foundation of the system is

$$F_T = c\dot{x} + kx \tag{2.38}$$

Since the forces $c\dot{x}$ and kx are 90° out of phase, the magnitude of the transmitted force is

$$|F_T| = \sqrt{c^2\dot{x}^2 + k^2x^2} \tag{2.39}$$

The ratio of the transmitted force F_T to the applied force F_0 can be expressed in terms of transmissibility T:

$$\frac{F_T}{F_0} = T \sin (\omega t - \psi) \tag{2.40}$$

where

$$T = \sqrt{\frac{1 + (2\zeta\omega/\omega_n)^2}{(1 - \omega^2/\omega_n^2)^2 + (2\zeta\omega/\omega_n)^2}} \tag{2.41}$$

and

$$\psi = \tan^{-1} \frac{2\zeta(\omega/\omega_n)^3}{1 - \omega^2/\omega_n^2 + 4\zeta^2\omega^2/\omega_n^2}$$

The transmissibility T and phase angle ψ are shown in Figs. 2.17 and 2.18, respectively, as a function of the frequency ratio ω/ω_n and for several values of the fraction of critical damping ζ.

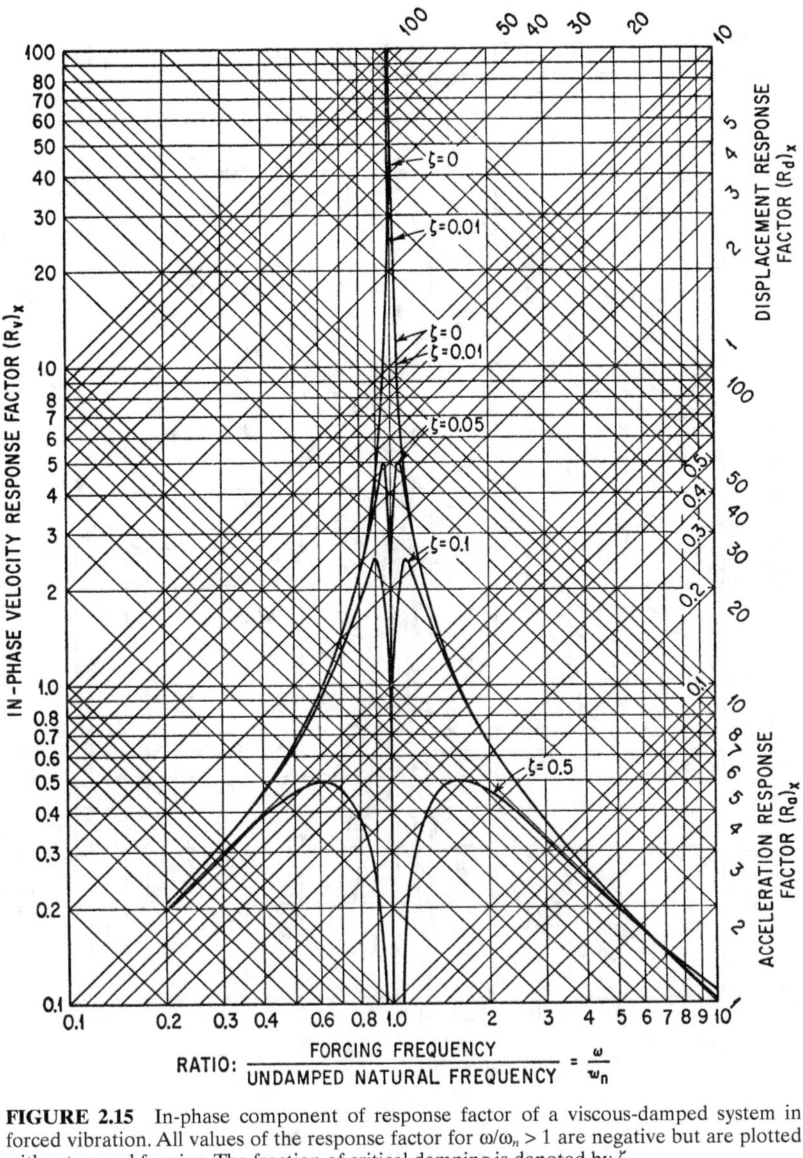

FIGURE 2.15 In-phase component of response factor of a viscous-damped system in forced vibration. All values of the response factor for $\omega/\omega_n > 1$ are negative but are plotted without regard for sign. The fraction of critical damping is denoted by ζ.

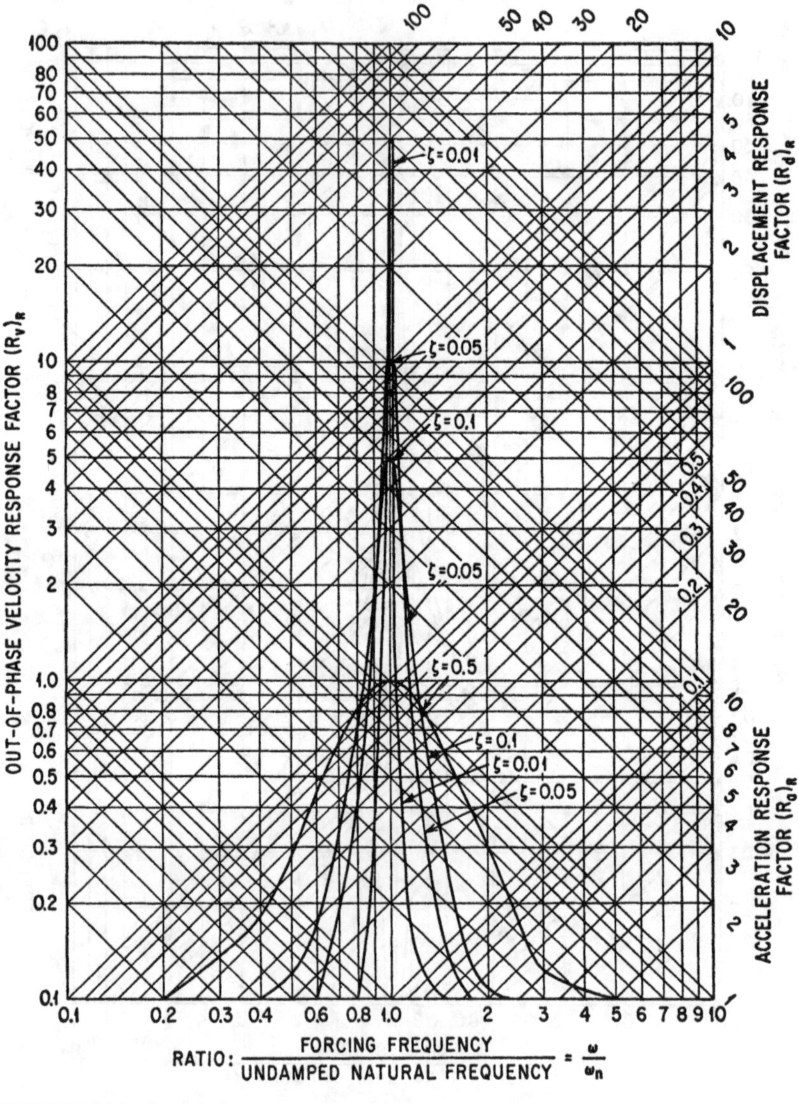

FIGURE 2.16 Out-of-phase component of response factor of a viscous-damped system in forced vibration. The fraction of critical damping is denoted by ζ.

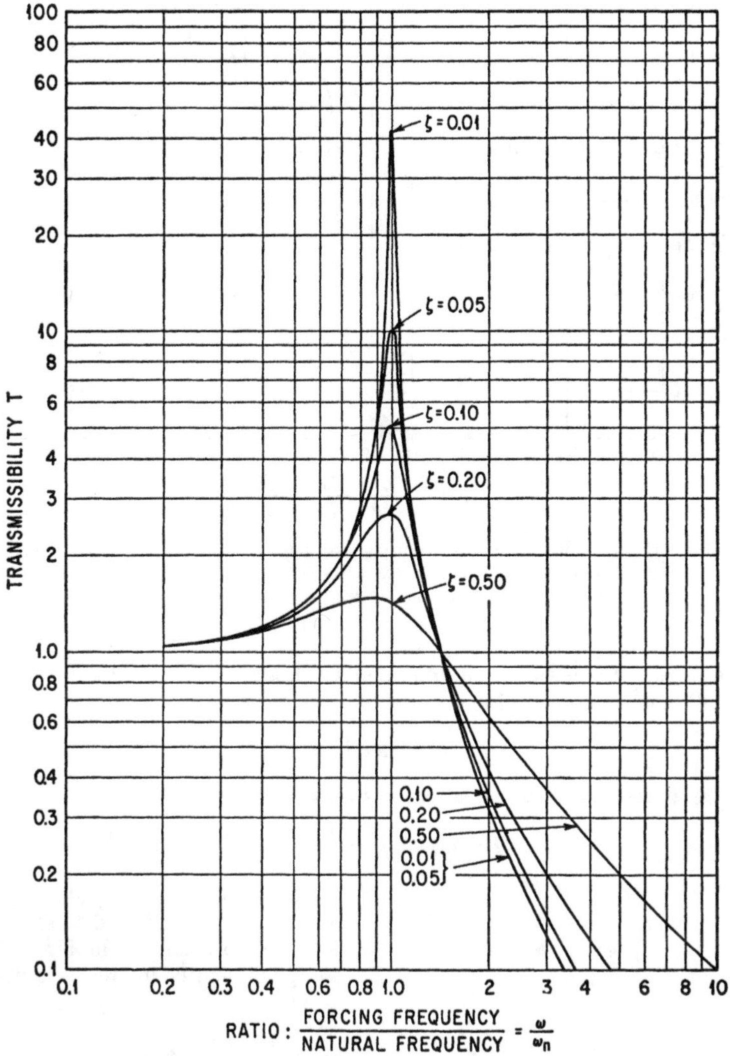

FIGURE 2.17 Transmissibility of a viscous-damped system. Force transmissibility and motion transmissibility are identical numerically. The fraction of critical damping is denoted by ζ.

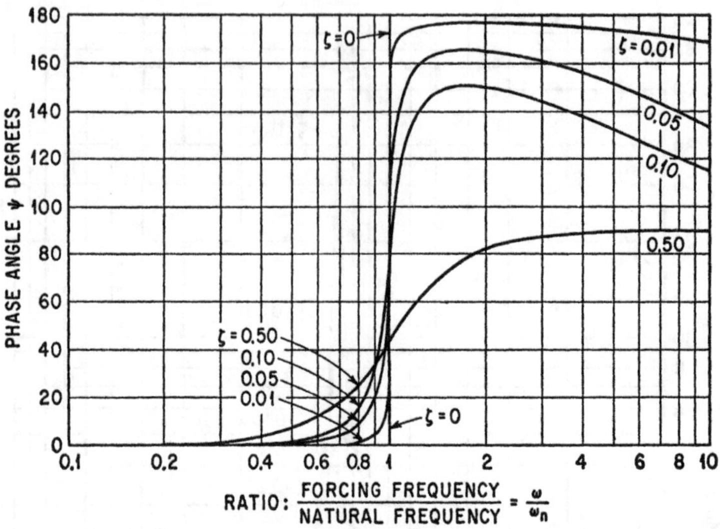

FIGURE 2.18 Phase angle of force transmission (or motion transmission) of a viscous-damped system excited (1) by force acting on mass and (2) by motion of foundation. The fraction of critical damping is denoted by ζ.

Hysteresis. When the viscous damped, single-degree-of-freedom system shown in Fig. 2.12 undergoes vibration defined by

$$x = x_0 \sin \omega t \tag{2.42}$$

the net force exerted on the mass by the spring and damper is

$$F = kx_0 \sin \omega t + c\omega x_0 \cos \omega t \tag{2.43}$$

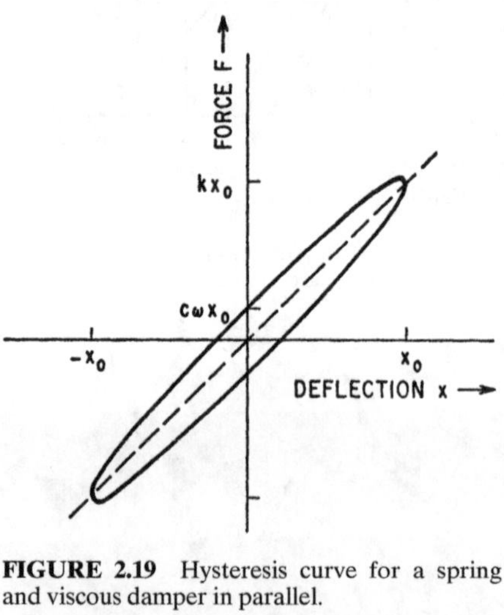

FIGURE 2.19 Hysteresis curve for a spring and viscous damper in parallel.

Equations (2.42) and (2.43) define the relation between F and x; this relation is the ellipse shown in Fig. 2.19. The energy dissipated in one cycle of oscillation is

$$W = \int_{T}^{T + 2\pi/\omega} F \frac{dx}{dt}\, dt = \pi c \omega x_0^2 \tag{2.44}$$

Motion of Foundation. The excitation for the elastic system shown in Fig. 2.20 may be a motion $u(t)$ of the foundation. The differential equation of motion for the system is

$$m\ddot{x} + c(\dot{x} - \dot{u}) + k(x - u) = 0 \tag{2.45}$$

Consider the motion of the foundation to be a displacement that varies sinu-

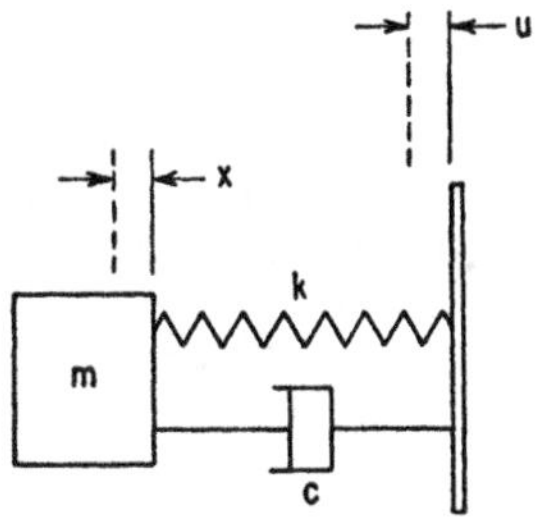

FIGURE 2.20 Single-degree-of-freedom system with viscous damper, excited in forced vibration by foundation motion.

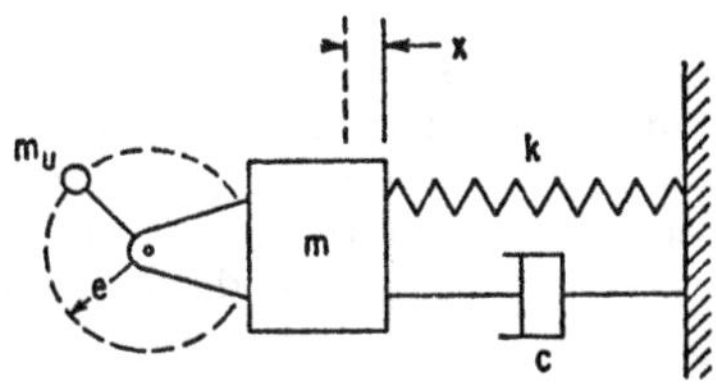

FIGURE 2.21 Single-degree-of-freedom system with viscous damper, excited in forced vibration by rotating eccentric weight.

soidally with time, $u = u_0 \sin \omega t$. A steady-state condition exists after the oscillations at the natural frequency ω_n are damped out, defined by the displacement x of mass m:

$$x = Tu_0 \sin (\omega t - \psi) \qquad (2.46)$$

where T and ψ are defined in connection with Eq. (2.40) and are shown graphically in Figs. 2.17 and 2.18, respectively. Thus, the motion transmissibility T in Eq. (2.46) is identical numerically to the force transmissibility T in Eq. (2.40). The motion of the foundation and of the mass m may be expressed in any consistent units, such as displacement, velocity, or acceleration, and the same expression for T applies in each case.

Vibration Due to a Rotating Eccentric Weight. In the mass-spring-damper system shown in Fig. 2.21, a mass m_u is mounted by a shaft and bearings to the mass m. The mass m_u follows a circular path of radius e with respect to the bearings. The component of displacement in the X direction of m_u relative to m is

$$x_3 - x_1 = e \sin \omega t \qquad (2.47)$$

where x_3 and x_1 are the absolute displacements of m_u and m, respectively, in the X direction; e is the length of the arm supporting the mass m_u; and ω is the angular velocity of the arm in radians per second. The differential equation of motion for the system is

$$m\ddot{x}_1 + m_u \ddot{x}_3 + c\dot{x}_1 + kx_1 = 0 \qquad (2.48)$$

Differentiating Eq. (2.47) with respect to time, solving for $\ddot{x}_3$, and substituting in Eq. (2.48):

$$(m + m_u)\ddot{x}_1 + c\dot{x}_1 + kx_1 = m_u e\omega^2 \sin \omega t \qquad (2.49)$$

Equation (2.49) is of the same form as Eq. (2.31); thus, the response relations of Eqs. (2.33), (2.36), and (2.37) apply by substituting $(m + m_u)$ for m and $m_u e\omega^2$ for F_0. The resulting displacement, velocity, and acceleration responses are

$$\frac{x_1}{m_u e\omega^2} = R_d \sin (\omega t - \theta) \qquad \frac{\dot{x}_1\sqrt{km}}{m_u e\omega^2} = R_v \cos (\omega t - \theta)$$
$$\frac{\ddot{x}_1 m}{m_u e\omega^2} = -R_a \sin (\omega t - \theta) \qquad (2.50)$$

Resonance Frequencies. The peak values of the displacement, velocity, and acceleration response of a system undergoing forced, steady-state vibration occur at slightly different forcing frequencies. Since a *resonance frequency* is defined as the frequency for which the response is a maximum, a simple system has three resonance frequencies if defined only generally. The natural frequency is different from any of the resonance frequencies. The relations among the several resonance frequencies, the damped natural frequency, and the undamped natural frequency ω_n are:

Displacement resonance frequency: $\omega_n(1 - 2\zeta^2)^{1/2}$
Velocity resonance frequency: ω_n
Acceleration resonance frequency: $\omega_n/(1 - 2\zeta^2)^{1/2}$
Damped natural frequency: $\omega_n(1 - \zeta^2)^{1/2}$

For the degree of damping usually embodied in physical systems, the difference among the three resonance frequencies is negligible.

Resonance, Bandwidth, and the Quality Factor *Q*. Damping in a system can be determined by noting the maximum response, i.e., the response at the resonance frequency as indicated by the maximum value of R_v in Eq. (2.36). This is defined by the factor Q sometimes used in electrical engineering terminology and defined with respect to mechanical vibration as

$$Q = (R_v)_{\max} = 1/2\zeta$$

The maximum acceleration and displacement responses are slightly larger, being

$$(R_d)_{\max} = (R_a)_{\max} = \frac{(R_v)_{\max}}{(1 - \zeta^2)^{1/2}}$$

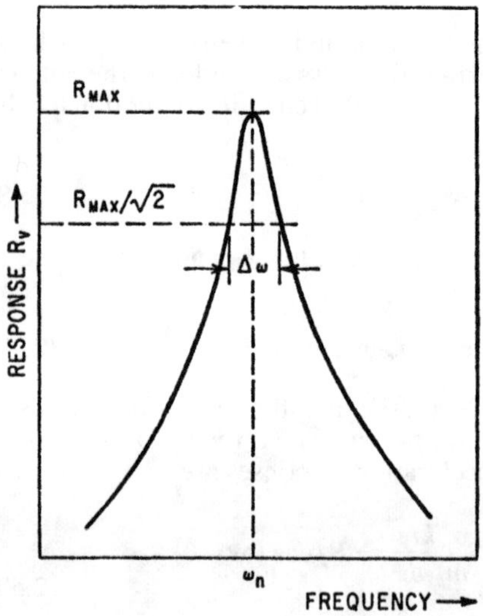

FIGURE 2.22 Response curve showing bandwidth at "half-power point."

The damping in a system is also indicated by the sharpness or width of the response curve in the vicinity of a resonance frequency ω_n. Designating the width as a frequency increment $\Delta\omega$ measured at the "half-power point" (i.e., at a value of R equal to $R_{\max}/2$), as illustrated in Fig. 2.22, the damping of the system is defined to a good approximation by

$$\frac{\Delta\omega}{\omega_n} = \frac{1}{Q} = 2\zeta \tag{2.51}$$

for values of ζ less than 0.1. The quantity $\Delta\omega$, known as the *bandwidth*, is commonly represented by the letter *B*.

Structural Damping. The energy dissipated by the damper is known as *hysteresis loss;* as indicated by Eq. (2.44), it is proportional to the forcing frequency ω. However, the hysteresis loss of many engineering structures has been found to be independent of frequency. To provide a better model for defining the *structural damping* experienced during vibration, an arbitrary damping term $kg = c\omega$ is introduced. In effect, this defines the damping force as being equal to the viscous damping force at some frequency, depending upon the value of g, but being invariant with frequency. The relation of the damping force F to the displacement x is defined by an ellipse similar to Fig. 2.19, and the displacement response of the system is described by an expression corresponding to Eq. (2.33) as follows:

$$\frac{x}{F_0/k} = R_g \sin(\omega t - \theta) = \frac{\sin(\omega t - \theta)}{\sqrt{(1 - \omega^2/\omega_n^2)^2 + g^2}} \tag{2.52}$$

where $g = 2\zeta\omega/\omega_n$. The resonance frequency is ω_n, and the value of R_g at resonance is $1/g = Q$.

The equations for the hysteresis ellipse for structural damping are

$$\begin{aligned} F &= kx_0 (\sin \omega t + g \cos \omega t) \\ x &= x_0 \sin \omega t \end{aligned} \tag{2.53}$$

UNDAMPED MULTIPLE-DEGREE-OF-FREEDOM SYSTEMS

An elastic system sometimes cannot be described adequately by a model having only one mass but rather must be represented by a system of two or more masses considered to be *point masses* or *particles* having no rotational inertia. If a group of particles is bound together by essentially rigid connections, it behaves as a rigid body having both mass (significant for translational motion) and moment of inertia (significant for rotational motion). There is no limit to the number of masses that may be used to represent a system. For example, each mass in a model representing a beam may be an infinitely thin slice representing a cross section of the beam; a differential equation is required to treat this continuous distribution of mass.

DEGREES OF FREEDOM

The number of independent parameters required to define the distance of all the masses from their reference positions is called the number of *degrees of freedom N*. For example, if there are N masses in a system constrained to move only in translation in the X and Y directions, the system has $2N$ degrees of freedom. A continuous system such as a beam has an infinitely large number of degrees of freedom.

For each degree of freedom (each coordinate of motion of each mass) a differential equation can be written in one of the following alternative forms:

$$m_j \ddot{x}_j = F_{xj} \qquad I_k \ddot{\alpha}_k = M_{\alpha k} \tag{2.54}$$

where F_{xj} is the component in the X direction of all external, spring, and damper forces acting on the mass having the jth degree of freedom, and $M_{\alpha k}$ is the component about the α axis of all torques acting on the body having the kth degree of freedom. The moment of inertia of the mass about the α axis is designated by I_k. (This is assumed for the present analysis to be a principal axis of inertia, and product of inertia terms are neglected. See Chap. 3 for a more detailed discussion.) Equations (2.54) are identical in form and can be represented by

$$m_j \ddot{x}_j = F_j \tag{2.55}$$

where F_j is the resultant of all forces (or torques) acting on the system in the jth degree of freedom, $\ddot{x}_j$ is the acceleration (translational or rotational) of the system in the jth degree of freedom, and m_j is the mass (or moment of inertia) in the jth degree of freedom. Thus, the terms defining the motion of the system (displacement, velocity, and acceleration) and the deflections of structures may be either translational or rotational, depending upon the type of coordinate. Similarly, the "force" acting on a system may be either a force or a torque, depending upon the type of coordinate. For example, if a system has n bodies each free to move in three translational modes and three rotational modes, there would be $6n$ equations of the form of Eq. (2.55), one for each degree of freedom.

DEFINING A SYSTEM AND ITS EXCITATION

The first step in analyzing any physical structure is to represent it by a mathematical model which will have essentially the same dynamic behavior. A suitable number and distribution of masses, springs, and dampers must be chosen, and the input forces or foundation motions must be defined. The model should have sufficient degrees of freedom to determine the modes which will have significant response to the exciting force or motion.

The properties of a system that must be known are the natural frequencies ω_n, the normal mode shapes D_{jn}, the damping of the respective modes, and the mass distribution m_j. The detailed distributions of stiffness and damping of a system are not used directly but rather appear indirectly as the properties of the respective modes. The characteristic properties of the modes may be determined experimentally as well as analytically.

STIFFNESS COEFFICIENTS

The spring system of a structure of N degrees of freedom can be defined completely by a set of N^2 *stiffness coefficients.* A stiffness coefficient K_{jk} is the change in spring force acting on the jth degree of freedom when only the kth degree of freedom is slowly displaced a unit amount in the negative direction. This definition is a generalization of the linear, elastic spring defined by Eq. (2.1). Stiffness coefficients have the characteristic of reciprocity, i.e., $K_{jk} = K_{kj}$. The number of independent stiffness coefficients is $(N^2 + N)/2$.

The total elastic force acting on the jth degree of freedom is the sum of the effects of the displacements in all of the degrees of freedom:

$$F_{el} = -\sum_{k=1}^{N} K_{jk} x_k \tag{2.56}$$

Inserting the spring force F_{el} from Eq. (2.56) in Eq. (2.55) together with the external forces F_j results in the n equations:

$$m_j \ddot{x}_j = F_j - \sum_k K_{jk} x_k \tag{2.56a}$$

FREE VIBRATION

When the external forces are zero, the preceding equations become

$$m_j \ddot{x}_j + \sum_k K_{jk} x_k = 0 \tag{2.57}$$

Solutions of Eq. (2.57) have the form

$$x_j = D_j \sin(\omega t + \theta) \tag{2.58}$$

Substituting Eq. (2.58) in Eq. (2.57),

$$m_j \omega^2 D_j = \sum_k K_{jk} D_k \tag{2.59}$$

This is a set of n linear algebraic equations with n unknown values of D. A solution of these equations for values of D other than zero can be obtained only if the determinant of the coefficients of the D's is zero:

$$\begin{vmatrix} (m_1\omega^2 - K_{11}) & -K_{12} & \cdot & \cdot & -K_{in} \\ -K_{21} & (m_2\omega^2 - K_{22}) & \cdot & \cdot & \cdot \\ \cdot & \cdot & \cdot & \cdot & \cdot \\ \cdot & \cdot & \cdot & \cdot & \cdot \\ -K_{ni} & \cdot & \cdot & \cdot & (m_n\omega^2 - K_{nn}) \end{vmatrix} = 0 \tag{2.60}$$

Equation (2.60) is an algebraic equation of the nth degree in ω^2; it is called the *frequency equation* since it defines n values of ω which satisfy Eq. (2.57). The roots are all real; some may be equal, and others may be zero. These values of frequency determined from Eq. (2.60) are the frequencies at which the system can oscillate in the absence of external forces. These frequencies are the *natural frequencies* ω_n of the system. Depending upon the initial conditions under which vibration of the system is initiated, the oscillations may occur at any or all of the natural frequencies and at any amplitude.

Example 2.1. Consider the three-degree-of-freedom system shown in Fig. 2.23; it consists of three equal masses m and a foundation connected in series by three equal springs k. The absolute displacements of the masses are x_1, x_2, and x_3. The stiffness coefficients (see section entitled "Stiffness Coefficients") are thus $K_{11} = 2k$,

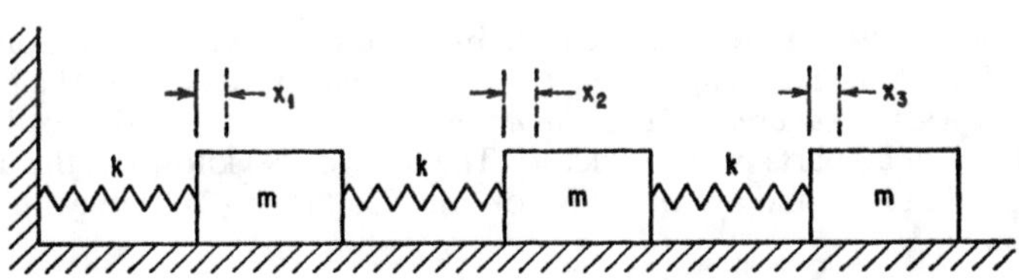

FIGURE 2.23 Undamped three-degree-of-freedom system on foundation.

$K_{22} = 2k$, $K_{33} = k$, $K_{12} = K_{21} = -k$, $K_{23} = K_{32} = -k$, and $K_{13} = K_{31} = 0$. The frequency equation is given by the determinant, Eq. (2.60),

$$\begin{vmatrix} (m\omega^2 - 2k) & k & 0 \\ k & (m\omega^2 - 2k) & k \\ 0 & k & (m\omega^2 - k) \end{vmatrix} = 0$$

The determinant expands to the following polynomial:

$$\left(\frac{m\omega^2}{k}\right)^3 - 5\left(\frac{m\omega^2}{k}\right)^2 + 6\left(\frac{m\omega^2}{k}\right) - 1 = 0$$

Solving for ω,

$$\omega = 0.445\sqrt{\frac{k}{m}}, \qquad 1.25\sqrt{\frac{k}{m}}, \qquad 1.80\sqrt{\frac{k}{m}}$$

Normal Modes of Vibration. A structure vibrating at only one of its natural frequencies ω_n does so with a characteristic pattern of amplitude distribution called a *normal mode of vibration.* A normal mode is defined by a set of values of D_{jn} [see Eq. (2.58)] which satisfy Eq. (2.59) when $\omega = \omega_n$:

$$\omega_n^2 m_j D_{jn} = \sum_k K_{jn} D_{kn} \tag{2.61}$$

A set of values of D_{jn}, which form a normal mode, is independent of the absolute values of D_{jn} but depends only on their relative values. To define a mode shape by a unique set of numbers, any arbitrary *normalizing condition* which is desired can be used. A condition often used is to set $D_{1n} = 1$ but $\sum_j m_j D_{jn}^2 = 1$ and $\sum_j m_j D_{jn}^2 = \sum_j m_j$ also may be found convenient.

Orthogonality of Normal Modes. The usefulness of normal modes in dealing with multiple-degree-of-freedom systems is due largely to the orthogonality of the normal modes. It can be shown that the set of inertia forces $\omega_n^2 m_j D_{jn}$ for one mode does not work on the set of deflections D_{jm} of another mode of the structure:

$$\sum_j m_j D_{jm} D_{jn} = 0 \qquad [m \neq n] \tag{2.62}$$

This is the *orthogonality condition.*

Normal Modes and Generalized Coordinates. Any set of N deflections x_j can be expressed as the sum of normal mode amplitudes:

$$x_j = \sum_{n=1}^{N} q_n D_{jn} \tag{2.63}$$

The numerical values of the D_{jn}'s are fixed by some normalizing condition, and a set of values of the N variables q_n can be found to match any set of x_j's. The N values of q_n constitute a set of *generalized coordinates* which can be used to define the position coordinates x_j of all parts of the structure. The q's are also known as the amplitudes of the normal modes, and are functions of time. Equation (2.63) may be differentiated to obtain

$$\ddot{x}_j = \sum_{n=1}^{N} \ddot{q}_n D_{jn} \tag{2.64}$$

Any quantity which is distributed over the j coordinates can be represented by a linear transformation similar to Eq. (2.63). It is convenient now to introduce the parameter γ_n relating D_{jn} and F_j/m_j as follows:

$$\frac{F_j}{m_j} = \sum_n \gamma_n D_{jn} \tag{2.65}$$

where F_j may be zero for certain values of n.

FORCED MOTION

Substituting the expressions in generalized coordinates, Eqs. (2.63) to (2.65), in the basic equation of motion, Eq. (2.56*a*),

$$m_j \sum_n \ddot{q}_n D_{jn} + \sum_k k_{jk} \sum_n q_n D_{kn} - m_j \sum_n \gamma_n D_{jn} = 0 \tag{2.66}$$

The center term in Eq. (2.66) may be simplified by applying Eq. (2.61) and the equation rewritten as follows:

$$\sum_n (\ddot{q}_n + \omega_n^2 q_n - \gamma_n) m_j D_{jn} = 0 \tag{2.67}$$

Multiplying Eqs. (2.67) by D_{jm} and taking the sum over j (i.e., adding all the equations together),

$$\sum_n (\ddot{q}_n + \omega_n^2 q_n - \gamma_n) \sum_j m_j D_{jn} D_{jm} = 0$$

All terms of the sum over n are zero, except for the term for which $m = n$, according to the orthogonality condition of Eq. (2.62). Then since $\sum_j m_j D_{jn}^2$ is not zero, it follows that

$$\ddot{q}_n + \omega_n^2 q_n - \gamma_n = 0$$

for every value of n from 1 to N.

An expression for γ_n may be found by using the orthogonality condition again. Multiplying Eq. (2.65) by $m_j D_{jm}$ and taking the sum taken over j,

$$\sum_j F_j D_{jm} = \sum_n \gamma_n \sum_j m_j D_{jn} D_{jm} \tag{2.68}$$

All the terms of the sum over n are zero except when $n = m$, according to Eq. (2.62), and Eq. (2.68) reduces to

$$\gamma_n = \frac{\sum_j F_j D_{jn}}{\sum_j m_j D_{jn}^2} \tag{2.69}$$

Then the differential equation for the response of any generalized coordinate to the externally applied forces F_j is

$$\ddot{q}_n + \omega_n^2 q_n = \gamma_n = \frac{\sum_j F_j D_{jn}}{\sum_j m_j D_{jn}^2} \tag{2.70}$$

where $\Sigma F_j D_{jn}$ is the generalized force, i.e., the total work done by all external forces during a small displacement δq_n divided by δq_n, and $\Sigma m_j D_{jn}^2$ is the generalized mass.

Thus the amplitude q_n of each normal mode is governed by its own equation, independent of the other normal modes, and responds as a simple mass-spring system. Equation (2.70) is a generalized form of Eq. (2.23).

The forces F_j may be any functions of time. Any equation for the response of an undamped mass-spring system applies to each mode of a complex structure by substituting:

$$\begin{aligned} &\text{The } \textit{generalized coordinate } q_n \text{ for } x \\ &\text{The } \textit{generalized force } \sum_j F_j D_{jn} \text{ for } F \\ &\text{The } \textit{generalized mass } \sum_j m_j D_{jn} \text{ for } m \\ &\text{The } \textit{mode natural frequency } \omega_n \text{ for } \omega_n \end{aligned} \tag{2.71}$$

Response to Sinusoidal Forces. If a system is subjected to one or more sinusoidal forces $F_j = F_{0j} \sin \omega t$, the response is found from Eq. (2.26) by noting that $k = m\omega_n^2$ [Eq. (2.6)] and then substituting from Eq. (2.71):

$$q_n = \frac{\sum_j F_{0j} D_{jn}}{\omega_n^2 \sum_j m_j D_{jn}^2} \frac{\sin \omega t}{(1 - \omega^2/\omega_n^2)} \tag{2.72}$$

Then the displacement of the kth degree-of-freedom, from Eq. (2.63), is

$$x_k = \sum_{n=1}^{N} \frac{D_{kn} \sum_j F_{0j} D_{jn} \sin \omega t}{\omega_n^2 \sum_j m_j D_{jn}^2 (1 - \omega^2/\omega_n^2)} \tag{2.73}$$

This is the general equation for the response to sinusoidal forces of an undamped system of N degrees of freedom. The application of the equation to systems free in space or attached to immovable foundations is discussed below.

Example 2.2. Consider the system shown in Fig. 2.24; it consists of three equal masses m connected in series by two equal springs k. The system is free in space and a force $F \sin \omega t$ acts on the first mass. Absolute displacements of the masses are x_1, x_2, and x_3. Determine the acceleration $\ddot{x}_3$. The stiffness coefficients (see section enti-

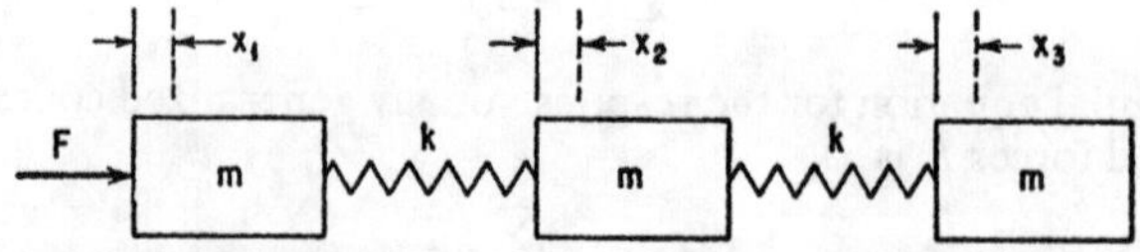

FIGURE 2.24 Undamped three-degree-of-freedom system acted on by sinusoidal force.

tled "Stiffness Coefficients") are $K_{11} = K_{33} = k$, $K_{22} = 2k$, $K_{12} = K_{21} = -k$, $K_{13} = K_{31} = 0$, and $K_{23} = K_{32} = -k$. Substituting in Eq. (2.60), the frequency equation is

$$\begin{vmatrix} (m\omega^2 - k) & k & 0 \\ k & (m\omega^2 - 2k) & k \\ 0 & k & (m\omega^2 - k) \end{vmatrix} = 0$$

The roots are $\omega_1 = 0$, $\omega_2 = \sqrt{k/m}$, and $\omega_3 = \sqrt{3k/m}$. The zero value for one of the natural frequencies indicates that the entire system translates without deflection of the springs. The mode shapes are now determined by substituting from Eq. (2.58) in Eq. (2.57), noting that $\ddot{x} = -D\omega^2$, and writing Eq. (2.59) for each of the three masses in each of the oscillatory modes 2 and 3:

$$mD_{21}\left(\frac{k}{m}\right) = K_{11}D_{21} + K_{21}D_{22} + K_{31}D_{23}$$

$$mD_{22}\left(\frac{k}{m}\right) = K_{12}D_{21} + K_{22}D_{22} + K_{32}D_{23}$$

$$mD_{23}\left(\frac{k}{m}\right) = K_{13}D_{21} + K_{23}D_{22} + K_{33}D_{23}$$

$$mD_{31}\left(\frac{3k}{m}\right) = K_{11}D_{31} + K_{21}D_{32} + K_{31}D_{33}$$

$$mD_{32}\left(\frac{3k}{m}\right) = K_{12}D_{31} + K_{22}D_{32} + K_{32}D_{33}$$

$$mD_{33}\left(\frac{3k}{m}\right) = K_{13}D_{31} + K_{23}D_{32} + K_{33}D_{33}$$

where the first subscript on the D's indicates the mode number (according to ω_1 and ω_2 above) and the second subscript indicates the displacement amplitude of the particular mass. The values of the stiffness coefficients K are calculated above. The mode shapes are defined by the relative displacements of the masses. Thus, assigning values of unit displacement to the first mass (i.e., $D_{21} = D_{31} = 1$), the above equations may be solved simultaneously for the D's:

$$D_{21} = 1 \qquad D_{22} = 0 \qquad D_{23} = -1$$

$$D_{31} = 1 \qquad D_{32} = -2 \qquad D_{33} = 1$$

Substituting these values of D in Eq. (2.71), the generalized masses are determined: $M_2 = 2m$, $M_3 = 6m$.

Equation (2.73) then can be used to write the expression for acceleration $\ddot{x}_3$:

$$\ddot{x}_3 = \left[\frac{1}{3m} + \frac{(\omega^2/\omega_2^2)(-1)(+1)}{2m(1 - \omega^2/\omega_2^2)} + \frac{(\omega^2/\omega_3^2)(+1)(+1)}{6m(1 - \omega^2/\omega_3^2)}\right] F_1 \sin \omega t$$

Free and Fixed Systems. For a structure which is free in space, there are six "normal modes" corresponding to $\omega_n = 0$. These represent motion of the structure without relative motion of its parts; this is rigid-body motion with six degrees of freedom.

The rigid-body modes all may be described by equations of the form

$$D_{jm} = a_{jm} D_m \qquad [m = 1,2,\ldots,6]$$

where D_m is a motion of the rigid body in the m coordinate and a is the displacement of the jth degree of freedom when D_m is moved a unit amount. The geometry of the structure determines the nature of a_{jm}. For example, if D_m is a rotation about the Z axis, $a_{jm} = 0$ for all modes of motion in which j represents rotation about the X or Y axis and $a_{jm} = 0$ if j represents translation parallel to the Z axis. If D_{jm} is a translational mode of motion parallel to X or Y, it is necessary that a_{jm} be proportional to the distance r_j of m_j from the Z axis and to the sine of the angle between r_j and the jth direction. The above relations may be applied to an elastic body. Such a body moves as a rigid body in the gross sense in that all particles of the body move together generally but may experience relative vibratory motion. The orthogonality condition applied to the relation between any rigid-body mode D_{jm} and any oscillatory mode D_{jn} yields

$$\sum_j m_j D_{jn} D_{jm} = \sum_j m_j a_{jm} D_{jn} = 0 \qquad \begin{bmatrix} m \le 6 \\ n > 6 \end{bmatrix} \tag{2.74}$$

These relations are used in computations of oscillatory modes and show that normal modes of vibration involve no net translation or rotation of a body.

A system attached to a fixed foundation may be considered as a system free in space in which one or more "foundation" masses or moments of inertia are infinite. Motion of the system as a rigid body is determined entirely by the motion of the foundation. The amplitude of an oscillatory mode representing motion of the foundation is zero; i.e., $M_j D_{jn}^2 = 0$ for the infinite mass. However, Eq. (2.73) applies equally well regardless of the size of the masses.

Foundation Motion. If a system is small relative to its foundation, it may be assumed to have no effect on the motion of the foundation. Consider a foundation of large but unknown mass m_0 having a motion $x_0 \sin \omega t$, the consequence of some unknown force

$$F_0 \sin \omega t = -m_0 x_0 \omega^2 \sin \omega t \tag{2.75}$$

acting on m_0 in the x_0 direction. Equation (2.73) is applicable to this case upon substituting

$$-m_0 x_0 \omega^2 D_{0n} = \sum_j F_{0j} D_{jn} \tag{2.76}$$

where D_{0n} is the amplitude of the foundation (the 0 degree of freedom) in the nth mode.

The oscillatory modes of the system are subject to Eqs. (2.74):

$$\sum_j = 0\ m_j a_{jm} D_{jn} = 0$$

Separating the 0th degree of freedom from the other degrees of freedom:

$$\sum_{j=0} m_j a_{jm} D_{jn} = m_0 a_{0m} D_{0n} + \sum_{j=1} m_j a_{jm} D_{jn}$$

If m_0 approaches infinity as a limit, D_{0n} approaches zero and motion of the system as a rigid body is identical with the motion of the foundation. Thus, a_{0m} approaches unity for motion in which $m = 0$, and approaches zero for motion in which $m \neq 0$. In the limit:

$$\lim_{m_0 \to \infty} m_0 D_{0n} = -\sum_j m_j a_{j0} D_{jn} \tag{2.77}$$

Substituting this result in Eq. (2.76),

$$\lim_{m_0 \to \infty} \sum_j F_{0j} D_{jn} = x_0 \omega^2 \sum_j m_j a_{j0} D_{jn} \tag{2.78}$$

The generalized mass in Eq. (2.73) includes the term $m_0 D_{0n}^2$, but this becomes zero as m_0 becomes infinite.

The equation for response of a system to motion of its foundation is obtained by substituting Eq. (2.78) in Eq. (2.73):

$$x_k = \sum_{n=1}^{N} \frac{\omega^2}{\omega_n^2} D_{kn} \frac{\sum_j m_j a_{j0} D_{jn} x_0 \sin \omega t}{\sum_j m_j D_{jn}^2 (1 - \omega^2/\omega_n^2)} + x_0 \sin \omega t \tag{2.79}$$

DAMPED MULTIPLE-DEGREE-OF-FREEDOM SYSTEMS

Consider a set of masses interconnected by a network of springs and acted upon by external forces, with a network of dampers acting in parallel with the springs. The viscous dampers produce forces on the masses which are determined in a manner analogous to that used to determine spring forces and summarized by Eq. (2.56). The damping force acting on the jth degree of freedom is

$$(F_d)_j = -\sum_k C_{jk} \dot{x}_k \tag{2.80}$$

where C_{jk} is the resultant force on the jth degree of freedom due to a unit velocity of the kth degree of freedom.

In general, the distribution of damper sizes in a system need not be related to the spring or mass sizes. Thus, the dampers may couple the normal modes together, allowing motion of one mode to affect that of another. Then the equations of response are not easily separable into independent normal mode equations. However, there are two types of damping distribution which do not couple the normal modes. These are known as *uniform viscous damping* and *uniform mass damping.*

UNIFORM VISCOUS DAMPING

Uniform damping is an appropriate model for systems in which the damping effect is an inherent property of the spring material. Each spring is considered to have a damper acting in parallel with it, and the ratio of damping coefficient to stiffness coefficient is the same for each spring of the system. Thus, for all values of j and k,

$$\frac{C_{jk}}{k_{jk}} = 2\mathcal{G} \tag{2.81}$$

where $\mathcal{G}$ is a constant.

Substituting from Eq. (2.81) in Eq. (2.80),

$$-(F_d)_j = \sum_k C_{jk}\dot{x}_k = 2\vartheta \sum_k k_{jk}\dot{x}_k \tag{2.82}$$

Since the damping forces are "external" forces with respect to the mass-spring system, the forces $(F_d)_j$ can be added to the external forces in Eq. (2.70) to form the equation of motion:

$$\ddot{q}_n + \omega_n^2 q_n = \frac{\sum_j (F_d)_j D_{jn} + \sum_j F_j D_{jn}}{\sum_j m_j D_{jn}^2} \tag{2.83}$$

Combining Eqs. (2.61), (2.63), and (2.82), the summation involving $(F_d)_j$ in Eq. (2.83) may be written as follows:

$$\sum_j (F_d)_j D_{jn} = -2\vartheta\omega_n^2 \dot{q}_n \sum_j m_j D_{jn}^2 \tag{2.84}$$

Substituting Eq. (2.84) in Eq. (2.83),

$$\ddot{q}_n + 2\vartheta\omega_n^2 \dot{q}_n + \omega_n^2 q_n = \gamma_n \tag{2.85}$$

Comparison of Eq. (2.85) with Eq. (2.31) shows that each mode of the system responds as a simple damped oscillator.

The damping term $2\vartheta\omega_n^2$ in Eq. (2.85) corresponds to $2\zeta\omega_n$ in Eq. (2.31) for a simple system. Thus, $\vartheta\omega_n$ may be considered the critical damping ratio of each mode. Note that the effective damping for a particular mode varies directly as the natural frequency of the mode.

Free Vibration. If a system with uniform viscous damping is disturbed from its equilibrium position and released at time $t = 0$ to vibrate freely, the applicable equation of motion is obtained from Eq. (2.85) by substituting $2\zeta\omega$ for $2\vartheta\omega_n^2$ and letting $\gamma_n = 0$:

$$\ddot{q}_n + 2\zeta\omega_n\dot{q}_n + \omega_n^2 q_n = 0 \tag{2.86}$$

The solution of Eq. (2.86) for less than critical damping is

$$x_j(t) = \sum_n D_{jn} e^{-\zeta\omega_n t}(A_n \sin \omega_d t + B_n \cos \omega_d t) \tag{2.87}$$

where $\omega_d = \omega_n(1 - \zeta^2)^{1/2}$.

The values of A and B are determined by the displacement $x_j(0)$ and velocity $\dot{x}_j(0)$ at time $t = 0$:

$$x_j(0) = \sum_n B_n D_{jn}$$

$$\dot{x}_j(0) = \sum_n (A_n\omega_{dn} - B_n\zeta\omega_n) D_{jn}$$

Applying the orthogonality relation of Eq. (2.62) in the manner used to derive Eq. (2.69),

$$B_n = \frac{\sum_j x_j(0) m_j D_{jn}}{\sum_j m_j D_{jn}^2}$$
$$A_n \omega_{dn} - B_n \zeta \omega_{dn} = \frac{\sum_j \dot{x}_j(0) m_j D_{jn}}{\sum_j m_j D_{jn}^2} \tag{2.88}$$

Thus, each mode undergoes a decaying oscillation at the damped natural frequency for the particular mode, and the amplitude of each mode decays from its initial value, which is determined by the initial displacements and velocities.

UNIFORM STRUCTURAL DAMPING

To avoid the dependence of viscous damping upon frequency, as indicated by Eq. (2.85), the uniform viscous damping factor $\mathcal{G}$ is replaced by g/ω for uniform structural damping. This corresponds to the structural damping parameter g in Eqs. (2.52) and (2.53) for sinusoidal vibration of a simple system. Thus, Eq. (2.85) for the response of a mode to a sinusoidal force of frequency ω is

$$\ddot{q}_n + \frac{2\mathrm{g}}{\omega}\, \omega_n^2 \dot{q}_n + \omega_n^2 q_n = \gamma_n \tag{2.89}$$

The amplification factor at resonance ($Q = 1/\mathrm{g}$) has the same value in all modes.

UNIFORM MASS DAMPING

If the damping force on each mass is proportional to the magnitude of the mass,

$$(F_d)_j = -B m_j \dot{x}_j \tag{2.90}$$

where B is a constant. For example, Eq. (2.90) would apply to a uniform beam immersed in a viscous fluid.

Substituting as $\dot{x}_j$ in Eq. (2.90) the derivative of Eq. (2.63),

$$\Sigma(F_d)_j D_{jn} = -B \sum_j m_j D_{jn} \sum_m \dot{q}_m D_{jm} \tag{2.91}$$

Because of the orthogonality condition, Eq. (2.62):

$$\Sigma(F_d)_j D_{jn} = -B \dot{q}_n \sum_j m_j D_{jn}^2$$

Substituting from Eq. (2.91) in Eq. (2.83), the differential equation for the system is

$$\ddot{q}_n + B\dot{q}_n + \omega_n^2 q_n = \gamma_n \tag{2.92}$$

where the damping term B corresponds to $2\zeta\omega$ for a simple oscillator, Eq. (2.31). Then $B/2\omega_n$ represents the fraction of critical damping for each mode, a quantity which diminishes with increasing frequency.

GENERAL EQUATION FOR FORCED VIBRATION

All the equations for response of a linear system to a sinusoidal excitation may be regarded as special cases of the following general equation:

$$x_k = \sum_{n=1}^{N} \frac{D_{kn}}{\omega_n^2} \frac{F_n}{m_n} R_n \sin(\omega t - \theta_n) \tag{2.93}$$

where x_k = displacement of structure in kth degree of freedom
N = number of degrees of freedom, including those of the foundation
D_{kn} = amplitude of kth degree of freedom in nth normal mode
F_n = generalized force for nth mode
m_n = generalized mass for nth mode
R_n = response factor, a function of the frequency ratio ω/ω_n (Fig. 2.13)
θ_n = phase angle (Fig. 2.14)

Equation (2.93) is of sufficient generality to cover a wide variety of cases, including excitation by external forces or foundation motion, viscous or structural damping, rotational and translational degrees of freedom, and from one to an infinite number of degrees of freedom.

LAGRANGIAN EQUATIONS

The differential equations of motion for a vibrating system sometimes are derived more conveniently in terms of kinetic and potential energies of the system than by the application of Newton's laws of motion in a form requiring the determination of the forces acting on each mass of the system. The formulation of the equations in terms of the energies, known as lagrangian equations, is expressed as follows:

$$\frac{d}{dt}\frac{\partial T}{\partial \dot{q}_n} - \frac{\partial T}{\partial q_n} + \frac{\partial V}{\partial q_n} = F_n \tag{2.94}$$

where T = total kinetic energy of system
V = total potential energy of system
q_n = generalized coordinate—a displacement
$\dot{q}_n$ = velocity at generalized coordinate q_n
F_n = generalized force, the portion of the total forces not related to the potential energy of the system (gravity and spring forces appear in the potential energy expressions and are not included here)

The method of applying Eq. (2.94) is to select a number of independent coordinates (generalized coordinates) equal to the number of degrees of freedom, and to write expressions for total kinetic energy T and total potential energy V. Differentiation of these expressions successively with respect to each of the chosen coordinates leads to a number of equations similar to Eq. (2.94), one for each coordinate (degree of freedom). These are the applicable differential equations and may be solved by any suitable method.

Example 2.3. Consider free vibration of the three-degree-of-freedom system shown in Fig. 2.23; it consists of three equal masses m connected in tandem by equal springs k. Take as coordinates the three absolute displacements x_1, x_2, and x_3. The kinetic energy of the system is

$$T = \tfrac{1}{2}m(\dot{x}_1^2 + \dot{x}_2^2 + \dot{x}_3^2)$$

The potential energy of the system is

$$V = \frac{k}{2}\,[x_1^2 + (x_1 - x_2)^2 + (x_2 - x_3)^2] = \frac{k}{2}\,(2x_1^2 + 2x_2^2 + x_3^2 - 2x_1x_2 - 2x_2x_3)$$

Differentiating the expression for the kinetic energy successively with respect to the velocities,

$$\frac{\partial T}{\partial \dot{x}_1} = m\dot{x}_1 \qquad \frac{\partial T}{\partial \dot{x}_2} = m\dot{x}_2 \qquad \frac{\partial T}{\partial \dot{x}_3} = m\dot{x}_3$$

The kinetic energy is not a function of displacement; therefore, the second term in Eq. (2.94) is zero. The partial derivatives with respect to the displacement coordinates are

$$\frac{\partial V}{\partial x_1} = 2kx_1 - kx_2 \qquad \frac{\partial V}{\partial x_2} = 2kx_2 - kx_1 - kx_3 \qquad \frac{\partial V}{\partial x_3} = kx_3 - kx_2$$

In free vibration, the generalized force term in Eq. (2.93) is zero. Then, substituting the derivatives of the kinetic and potential energies from above into Eq. (2.94),

$$m\ddot{x}_1 + 2kx_1 - kx_2 = 0$$

$$m\ddot{x}_2 + 2kx_2 - kx_1 - kx_3 = 0$$

$$m\ddot{x}_3 + kx_3 - kx_2 = 0$$

The natural frequencies of the system may be determined by placing the preceding set of simultaneous equations in determinant form, in accordance with Eq. (2.60):

$$\begin{vmatrix} (m\omega^2 - 2k) & k & 0 \\ k & (m\omega^2 - 2k) & k \\ 0 & k & (m\omega^2 - k) \end{vmatrix} = 0$$

The natural frequencies are equal to the values of ω that satisfy the preceding determinant equation.

Example 2.4. Consider the compound pendulum of mass m shown in Fig. 2.25, having its center of gravity located a distance l from the axis of rotation. The moment of inertia is I about an axis through the center of gravity. The position of the mass is defined by three coordinates, x and y to define the location of the center of gravity and θ to define the angle of rotation.

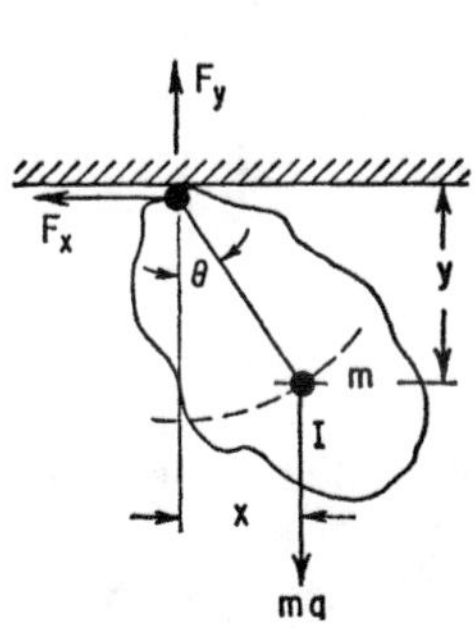

FIGURE 2.25 Forces and motions of a compound pendulum.

The *equations of constraint* are $y = l\cos\theta$; $x = l\sin\theta$. Each equation of constraint reduces the number of degrees of freedom by 1; thus the pendulum is a one-degree-of-freedom system whose position is defined uniquely by θ alone.

The kinetic energy of the pendulum is

$$T = \tfrac{1}{2}(I + ml^2)\dot{\theta}^2$$

The potential energy is

$$V = mgl(1 - \cos\theta)$$

Then

$$\frac{\partial T}{\partial \dot{\theta}} = (I + ml^2)\dot{\theta} \qquad \frac{d}{dt}\left(\frac{\partial T}{\partial \dot{\theta}}\right) = (I + ml^2)\ddot{\theta}$$

$$\frac{\partial T}{\partial \theta} = 0 \qquad \frac{\partial V}{\partial \theta} = mgl\sin\theta$$

Substituting these expressions in Eq. (2.94), the differential equation for the pendulum is

$$(I + ml^2)\ddot{\theta} + mgl\sin\theta = 0$$

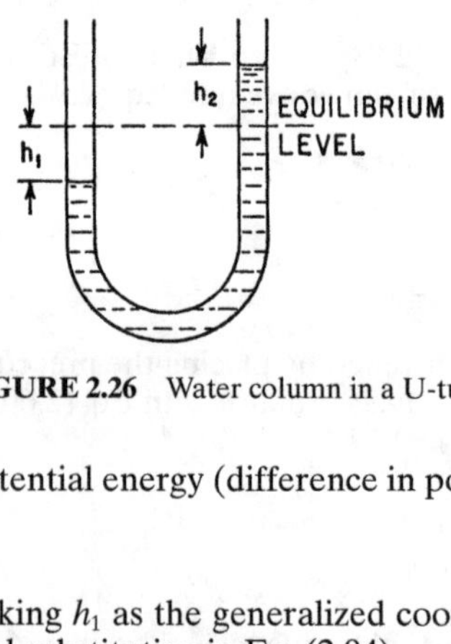

FIGURE 2.26 Water column in a U-tube.

Example 2.5. Consider oscillation of the water in the U-tube shown in Fig. 2.26. If the displacements of the water levels in the arms of a uniform-diameter U-tube are h_1 and h_2, then conservation of matter requires that $h_1 = -h_2$. The kinetic energy of the water flowing in the tube with velocity $\dot{h}_1$ is

$$T = \tfrac{1}{2}\rho S l \dot{h}_1^2$$

where ρ is the water density, S is the cross-section area of the tube, and l is the developed length of the water column. The potential energy (difference in potential energy between arms of tube) is

$$V = S\rho g h_1^2$$

Taking h_1 as the generalized coordinate, differentiating the expressions for energy, and substituting in Eq. (2.94),

$$S\rho l\ddot{h}_1 + 2\rho g S h_1 = 0$$

Dividing through by $\rho S l$,

$$\ddot{h}_1 + \frac{2g}{l}h_1 = 0$$

This is the differential equation for a simple oscillating system of natural frequency ω_n, where

$$\omega_n = \sqrt{\frac{2g}{l}}$$

CHAPTER 3
VIBRATION OF A RESILIENTLY SUPPORTED RIGID BODY

Harry Himelblau
Sheldon Rubin

INTRODUCTION

This chapter discusses the vibration of a rigid body on resilient supporting elements, including (1) methods of determining the inertial properties of a rigid body, (2) discussion of the dynamic properties of resilient elements, and (3) motion of a single rigid body on resilient supporting elements for various dynamic excitations and degrees of symmetry.

The general equations of motion for a rigid body on linear massless resilient supports are given; these equations are general in that they include any configuration of the rigid body and any configuration and location of the supports. They involve six simultaneous equations with numerous terms, for which a general solution is impracticable without the use of high-speed automatic computing equipment. Various degrees of simplification are introduced by assuming certain symmetry, and results useful for engineering purposes are presented. Several topics are considered: (1) determination of undamped natural frequencies and discussion of coupling of modes of vibration, (2) forced vibration where the excitation is a vibratory motion of the foundation, (3) forced vibration where the excitation is a vibratory force or moment generated within the body, and (4) free vibration caused by an instantaneous change in velocity of the system (velocity shock). Results are presented mathematically and, where feasible, graphically.

SYSTEM OF COORDINATES

The motion of the rigid body is referred to a fixed "inertial" frame of reference. The inertial frame is represented by a system of cartesian coordinates $\bar{X},\bar{Y},\bar{Z}$. A similar system of coordinates X, Y, Z fixed in the body has its origin at the center of mass. The two sets of coordinates are coincident when the body is in equilibrium under the

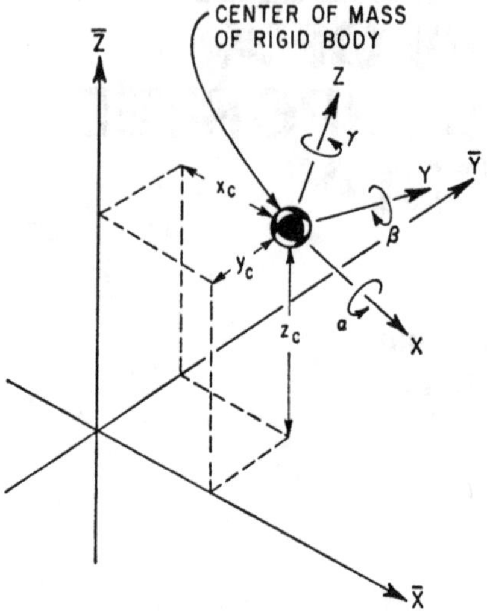

FIGURE 3.1 System of coordinates for the motion of a rigid body consisting of a fixed inertial set of reference axes ($\bar{X}, \bar{Y}, \bar{Z}$) and a set of axes ($X, Y, Z$) fixed in the moving body with its origin at the center of mass. The axes $\bar{X}, \bar{Y}, \bar{Z}$ and X, Y, Z are coincident when the body is in equilibrium under the action of gravity alone. The displacement of the center of mass is given by the translational displacements x_c, y_c, z_c and the rotational displacements α, β, γ as shown. A positive rotation about an axis is one which advances a right-handed screw in the positive direction of the axis.

action of gravity alone. The motions of the body are described by giving the displacement of the body axes relative to the inertial axes. The translational displacements of the center of mass of the body are x_c, y_c, z_c in the $\bar{X}, \bar{Y}, \bar{Z}$ directions, respectively. The rotational displacements of the body are characterized by the angles of rotation α, β, γ of the body axes about the $\bar{X}, \bar{Y}, \bar{Z}$ axes, respectively. These displacements are shown graphically in Fig. 3.1.

Only small translations and rotations are considered. Hence, the rotations are commutative (i.e., the resulting position is independent of the order of the component rotations) and the angles of rotation about the body axes are equal to those about the inertial axes. Therefore, the displacements of a point b in the body (with the coordinates b_x, b_y, b_z in the X, Y, Z directions, respectively) are the sums of the components of the center-of-mass displacement in the directions of the $\bar{X}, \bar{Y}, \bar{Z}$ axes plus the tangential components of the rotational displacement of the body:

$$
\begin{aligned}
x_b &= x_c + b_z\beta - b_y\gamma \\
y_b &= y_c - b_z\alpha + b_x\gamma \\
z_b &= z_c - b_x\beta + b_y\alpha
\end{aligned}
\tag{3.1}
$$

EQUATIONS OF SMALL MOTION OF A RIGID BODY

The equations of motion for the translation of a rigid body are

$$
m\ddot{x}_c = \mathbf{F}_x \qquad m\ddot{y}_c = \mathbf{F}_y \qquad m\ddot{z}_c = \mathbf{F}_z \tag{3.2}
$$

where m is the mass of the body, $\mathbf{F}_x$, $\mathbf{F}_y$, $\mathbf{F}_z$ are the summation of all forces acting on the body, and $\ddot{x}_c$, $\ddot{y}_c$, $\ddot{z}_c$ are the accelerations of the center of mass of the body in the $\bar{X}, \bar{Y}, \bar{Z}$ directions, respectively. The motion of the center of mass of a rigid body is the same as the motion of a particle having a mass equal to the total mass of the body and acted upon by the resultant external force.

The equations of motion for the rotation of a rigid body are

$$
\begin{aligned}
I_{xx}\ddot{\alpha} - I_{xy}\ddot{\beta} - I_{xz}\ddot{\gamma} &= \mathbf{M}_x \\
-I_{xy}\ddot{\alpha} + I_{yy}\ddot{\beta} - I_{yz}\ddot{\gamma} &= \mathbf{M}_y \\
-I_{xz}\ddot{\alpha} - I_{yz}\ddot{\beta} + I_{zz}\ddot{\gamma} &= \mathbf{M}_z
\end{aligned}
\tag{3.3}
$$

where $\ddot{\alpha}, \ddot{\beta}, \ddot{\gamma}$ are the rotational accelerations about the X, Y, Z axes, as shown in Fig. 3.1; $\mathbf{M}_x$, $\mathbf{M}_y$, $\mathbf{M}_z$ are the summation of torques acting on the rigid body about the axes X, Y, Z, respectively; and $I_{xx} \ldots, I_{xy} \ldots$ are the moments and products of inertia of the rigid body as defined below.

INERTIAL PROPERTIES OF A RIGID BODY

The properties of a rigid body that are significant in dynamics and vibration are the mass, the position of the center of mass (or center of gravity), the moments of inertia, the products of inertia, and the directions of the principal inertial axes. This section discusses the properties of a rigid body, together with computational and experimental methods for determining the properties.

MASS

Computation of Mass. The mass of a body is computed by integrating the product of mass density $\rho(V)$ and elemental volume dV over the body:

$$m = \int_V \rho(V)\, dV \tag{3.4}$$

If the body is made up of a number of elements, each having constant or an average density, the mass is

$$m = \rho_1 V_1 + \rho_2 V_2 + \cdots + \rho_n V_n \tag{3.5}$$

where ρ_1 is the density of the element V_1, etc. Densities of various materials may be found in handbooks containing properties of materials.[1]

If a rigid body has a common geometrical shape, or if it is an assembly of subbodies having common geometrical shapes, the volume may be found from compilations of formulas. Typical formulas are included in Tables 3.1 and 3.2. Tables of areas of plane sections as well as volumes of solid bodies are useful.

If the volume of an element of the body is not given in such a table, the integration of Eq. (3.4) may be carried out analytically, graphically, or numerically. A graphical approach may be used if the shape is so complicated that the analytical expression for its boundaries is not available or is not readily integrable. This is accomplished by graphically dividing the body into smaller parts, each of whose boundaries may be altered slightly (without change to the area) in such a manner that the volume is readily calculable or measurable.

The weight W of a body of mass m is a function of the acceleration of gravity g at the particular location of the body in space:

$$W = mg \tag{3.6}$$

Unless otherwise stated, it is understood that the weight of a body is given for an average value of the acceleration of gravity on the surface of the earth. For engineering purposes, $g = 32.2$ ft/sec^2 or 386 in./sec^2 (9.81 m/sec^2) is usually used.

Experimental Determination of Mass. Although Newton's second law of motion, $F = m\ddot{x}$, may be used to measure mass, this usually is not convenient. The mass of a body is most easily measured by performing a static measurement of the weight of the body and converting the result to mass. This is done by use of the value of the acceleration of gravity at the measurement location [Eq. (3.6)].

TABLE 3.1 Properties of Plane Sections (*After G. W. Housner and D. E. Hudson.*[2])

The dimensions X_c, Y_c are the X, Y coordinates of the centroid, A is the area, I_x . . . is the area moment of inertia with respect to the X . . . axis, ρ_x . . . is the radius of gyration with respect to the X . . . axis; uniform solid cylindrical bodies of length l in the Z direction having the various plane sections as their cross sections have mass moment and product of inertia values about the Z axis equal to ρl times the values given in the table, where ρ is the mass density of the body; the radii of gyration are unchanged.

	Plane section	Area and centroid	Area moment of inertia	Square of radius of gyration	Area product of inertia
1	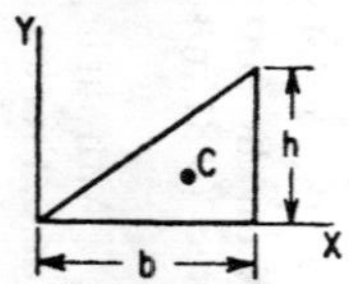 Right triangle	$A = \frac{1}{2}bh$ $X_c = \frac{2}{3}b$ $Y_c = \frac{1}{3}h$	$I_{x_c} = \frac{bh^3}{36}$ $I_{y_c} = \frac{b^3h}{36}$	$\rho_{x_c}{}^2 = \frac{1}{18}h^2$ $\rho_{y_c}{}^2 = \frac{1}{18}b^2$	$I_{x_cy_c} = \frac{A}{36}hb = \frac{h^2b^2}{72}$
2	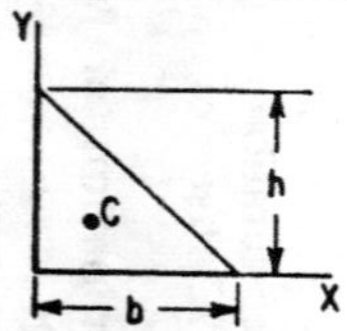	$A = \frac{1}{2}bh$ $X_c = \frac{1}{3}b$ $Y_c = \frac{1}{3}h$	$I_{x_c} = \frac{bh^3}{36}$ $I_{y_c} = \frac{b^3h}{36}$	$\rho_{x_c}{}^2 = \frac{1}{18}h^2$ $\rho_{y_c}{}^2 = \frac{1}{18}b^2$	$I_{x_cy_c} = -\frac{A}{36}hb = -\frac{h^2b^2}{72}$
3	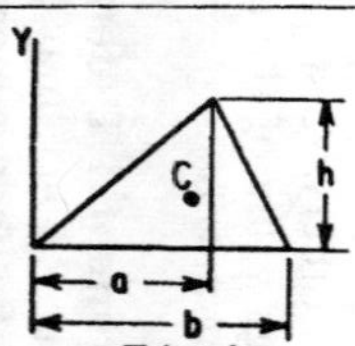 Triangle	$A = \frac{1}{2}bh$ $X_c = \frac{1}{3}(a + b)$ $Y_c = \frac{1}{3}h$	$I_{x_c} = \frac{bh^3}{36}$ $I_{y_c} = \frac{bh}{36}(b^2 - ab + a^2)$	$\rho_{x_c}{}^2 = \frac{1}{18}h^2$ $\rho_{y_c}{}^2 = \frac{1}{18}(b^2 - ab + a^2)$	$I_{x_cy_c} = \frac{Ah}{36}(2a - b) = \frac{bh^2}{72}(2a - b)$

4	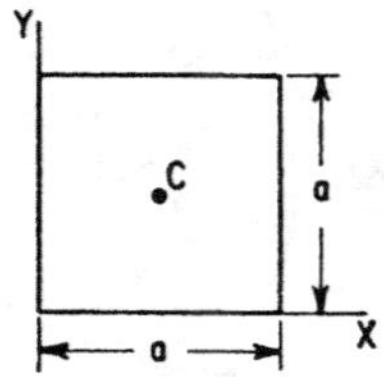 Square	$A = a^2$ $X_c = \frac{1}{2}a$ $Y_c = \frac{1}{2}a$	$I_{x_c} = I_{y_c} = \frac{a^4}{12}$	$\rho_{x_c}{}^2 = \rho_{y_c}{}^2 = \frac{1}{12}a^2$	$I_{x_c y_c} = 0$
5	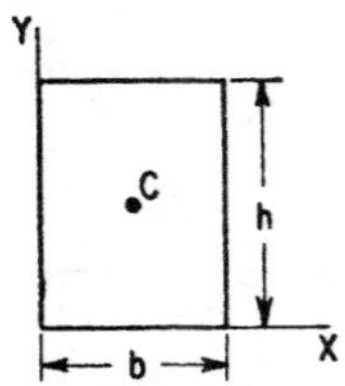 Rectangle	$A = bh$ $X_c = \frac{1}{2}b$ $Y_c = \frac{1}{2}h$	$I_{x_c} = \frac{bh^3}{12}$ $I_{y_c} = \frac{b^3h}{12}$	$\rho_{x_c}{}^2 = \frac{1}{12}h^2$ $\rho_{y_c}{}^2 = \frac{1}{12}h^2$	$I_{x_c y_c} = 0$
6	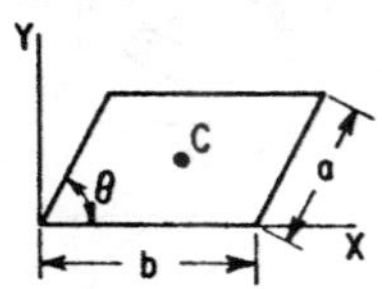 Parallelogram	$A = ab \sin \theta$ $X_c = \frac{1}{2}(b + a \cos \theta)$ $Y_c = \frac{1}{2}(a \sin \theta)$	$I_{x_c} = \frac{a^3b}{12} \sin^3 \theta$ $I_{y_c} = \frac{ab}{12} \sin \theta \, (b^2 + a^2 \cos^2 \theta)$	$\rho_{x_c}{}^2 = \frac{1}{12}(a \sin \theta)^2$ $\rho_{y_c}{}^2 = \frac{1}{12}(b^2 + a^2 \cos^2 \theta)$	$I_{x_c y_c} = \frac{a^3b}{12} \sin^2 \theta \cos \theta$
7	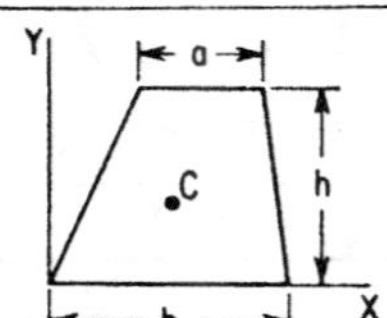 Trapezoid	$A = \frac{1}{2}h(a + b)$ $Y_c = \frac{1}{3}h \left(\frac{2a + b}{a + b}\right)$	$I_{x_c} = \frac{h^3(a^2 + 4ab + b^2)}{36(a + b)}$	$\rho_{x_c}{}^2 = \frac{h^2(a^2 + 4ab + b^2)}{18(a + b)^2}$	

TABLE 3.1 Properties of Plane Sections (*Continued*)

	Plane Section	Area and centroid	Area moment of inertia	Square of radius of gyration	Area product of inertia
8	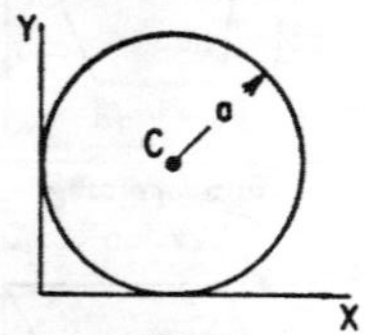 Circle	$A = \pi a^2$ $X_c = a$ $Y_c = a$	$I_{x_c} = I_{y_c} = \frac{1}{4}\pi a^4$	${\rho_{x_c}}^2 = {\rho_{y_c}}^2 = \frac{1}{4}a^2$	$I_{x_c y_c} = 0$
9	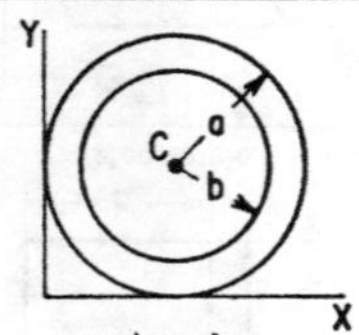 Annulus	$A = \pi(a^2 - b^2)$ $X_c = a$ $Y_c = a$	$I_{x_c} = I_{y_c} = \frac{\pi}{4}(a^4 - b^4)$	${\rho_{x_c}}^2 = {\rho_{y_c}}^2 = \frac{1}{4}(a^2 + b^2)$	$I_{x_c y_c} = 0$
10	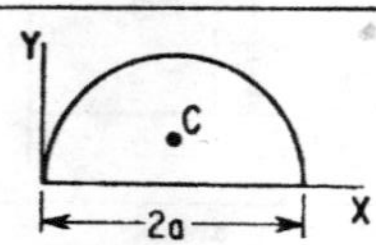 Semicircle	$A = \frac{1}{2}\pi a^2$ $X_c = a$ $Y_c = \frac{4a}{3\pi}$	$I_{x_c} = \frac{a^4(9\pi^2 - 64)}{72\pi}$ $I_{y_c} = \frac{1}{8}\pi a^4$	${\rho_{x_c}}^2 = \frac{a^2(9\pi^2 - 64)}{36\pi^2}$ ${\rho_{y_c}}^2 = \frac{1}{4}a^2$	$I_{x_c y_c} = 0$
11	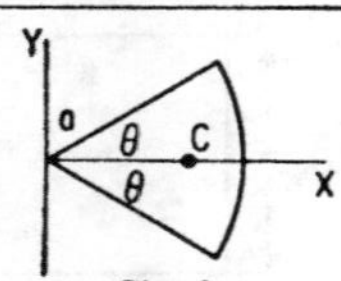 Circular sector	$A = a^2\theta$ $X_c = \frac{2a}{3}\frac{\sin\theta}{\theta}$ $Y_c = 0$	$I_x = \frac{1}{4}a^4(\theta - \sin\theta\cos\theta)$ $I_y = \frac{1}{4}a^4(\theta + \sin\theta\cos\theta)$	${\rho_x}^2 = \frac{1}{4}a^2\left(\frac{\theta - \sin\theta\cos\theta}{\theta}\right)$ ${\rho_y}^2 = \frac{1}{4}a^2\left(\frac{\theta + \sin\theta\cos\theta}{\theta}\right)$	$I_{x_c y_c} = 0$ $I_{xy} = 0$

No.	Shape	Area and centroid	Moments of inertia	Radii of gyration	Products of inertia
12	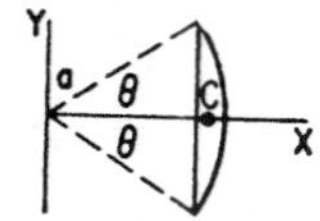 Circular segment	$A = a^2(\theta - \frac{1}{2}\sin 2\theta)$ $X_c = \frac{2a}{3}\left(\frac{\sin^3\theta}{\theta - \sin\theta\cos\theta}\right)$ $Y_c = 0$	$I_x = \frac{Aa^2}{4}\left[1 - \frac{2\sin^3\theta\cos\theta}{3(\theta - \sin\theta\cos\theta)}\right]$ $I_y = \frac{Aa^2}{4}\left[1 + \frac{2\sin^3\theta\cos\theta}{\theta - \sin\theta\cos\theta}\right]$	$\rho_x^2 = \frac{a^2}{4}\left[1 - \frac{2\sin^3\theta\cos\theta}{3(\theta - \sin\theta\cos\theta)}\right]$ $\rho_y^2 = \frac{a^2}{4}\left[1 + \frac{2\sin^3\theta\cos\theta}{\theta - \sin\theta\cos\theta}\right]$	$I_{x_c y_c} = 0$ $I_{xy} = 0$
13	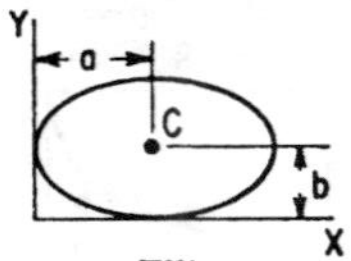 Ellipse	$A = \pi ab$ $X_c = a$ $Y_c = b$	$I_{x_c} = \frac{\pi}{4}ab^3$ $I_{y_c} = \frac{\pi}{4}a^3b$	$\rho_{x_c}^2 = \frac{1}{4}b^2$ $\rho_{y_c}^2 = \frac{1}{4}a^2$	$I_{x_c y_c} = 0$
14	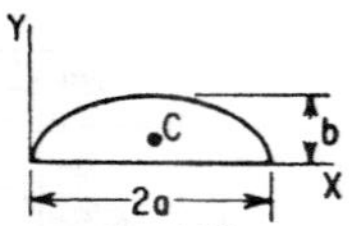 Semiellipse	$A = \frac{1}{2}\pi ab$ $X_c = a$ $Y_c = \frac{4b}{3\pi}$	$I_{x_c} = \frac{ab^3}{72\pi}(9\pi^2 - 64)$ $I_{y_c} = \frac{\pi}{8}a^3b$	$\rho_{x_c}^2 = \frac{b^2}{36\pi^2}(9\pi^2 - 64)$ $\rho_{y_c}^2 = \frac{1}{4}a^2$	$I_{x_c y_c} = 0$
15	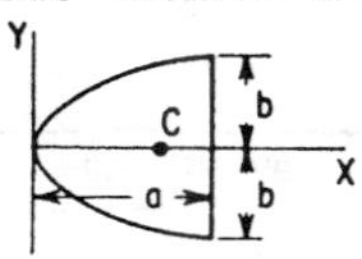 Parabola	$A = \frac{4}{3}ab$ $X_c = \frac{3}{5}a$ $Y_c = 0$	$I_{x_c} = \frac{4}{15}ab^3$ $I_{y_c} = \frac{16}{175}a^3b$	$\rho_{x_c}^2 = \frac{1}{5}b^2$ $\rho_{y_c}^2 = \frac{12}{175}a^2$	$I_{x_c y_c} = 0$
16	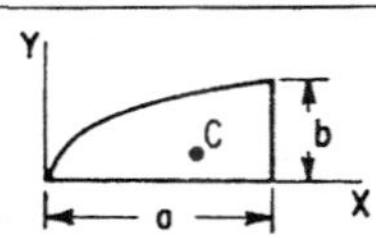 Semiparabola	$A = \frac{2}{3}ab$ $X_c = \frac{3}{5}a$ $Y_c = \frac{3}{8}b$	$I_x = \frac{2}{15}ab^3$ $I_y = \frac{2}{7}a^3b$	$\rho_x^2 = \frac{1}{5}b^2$ $\rho_y^2 = \frac{3}{7}a^2$	$I_{xy} = \frac{A}{4}ab = \frac{1}{6}a^2b^2$

TABLE 3.1 Properties of Plane Sections (*Continued*)

	Plane section	Area and centroid	Area moment of inertia	Square of radius of gyration	Area product of inertia
17	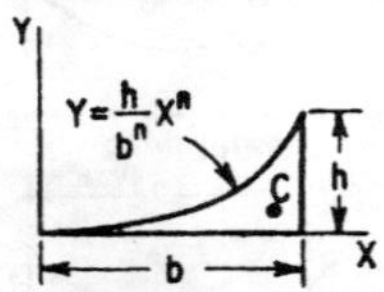 *n*th-degree parabola	$A = \frac{bh}{n+1}$ $X_c = \frac{n+1}{n+2}b$ $Y_c = \frac{h}{2}\left(\frac{n+1}{2n+1}\right)$	$I_x = \frac{bh^3}{3(3n+1)}$ $I_y = \frac{hb^3}{n+3}$	$\rho_x^2 = \frac{h^2(n+1)}{3(3n+1)}$ $\rho_y^2 = \frac{n+1}{n+3}b^2$	
18	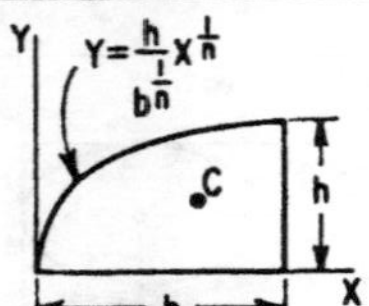 *n*th-degree parabola	$A = \frac{n}{n+1}bh$ $X_c = \frac{n+1}{2n+1}b$ $Y_c = \frac{n+1}{2(n+2)}h$	$I_x = \frac{n}{3(n+3)}bh^3$ $I_y = \frac{n}{3n+1}hb^3$	$\rho_x^2 = \frac{n+1}{3(n+3)}h^2$ $\rho_y^2 = \frac{n+1}{3n+1}b^2$	

TABLE 3.2 Properties of Homogeneous Solid Bodies (*After G. W. Housner and D. E. Hudson.*[2])

The dimensions X_c, Y_c, Z_c are the X, Y, Z coordinates of the centroid, S is the cross-sectional area of the thin rod or hoop in cases 1 to 3, V is the volume, $I_x \ldots$ is the mass moment of inertia with respect to the $X \ldots$ axis, $\rho_x \ldots$ is the radius of gyration with respect to the $X \ldots$ axis, ρ is the mass density of the body.

	Solid body	Volume and centroid	Mass moment of inertia	Radius of gyration squared	Mass product of inertia
1	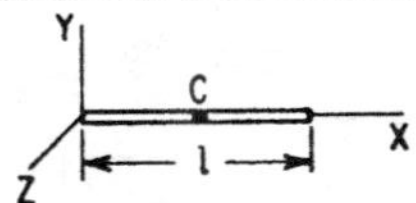 Thin rod	$V = Sl$ $X_c = \frac{1}{2}l$ $Y_c = 0$ $Z_c = 0$	$I_{x_c} = 0$ $I_{y_c} = I_{z_c} = \frac{\rho V}{12} l^2$	${\rho_{x_c}}^2 = 0$ ${\rho_{y_c}}^2 = {\rho_{z_c}}^2 = \frac{1}{12}l^2$	$I_{x_c y_c}$, etc. $= 0$
2	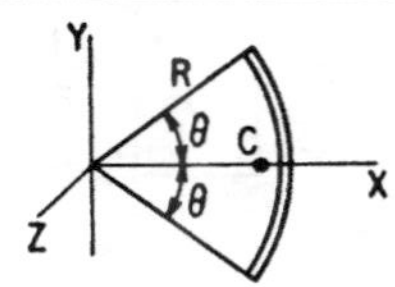 Thin circular rod	$V = 2SR\theta$ $X_c = \frac{R \sin\theta}{\theta}$ $Y_c = 0$ $Z_c = 0$	$I_x = I_{x_c}$ $= \frac{\rho V R^2(\theta - \sin\theta\cos\theta)}{2\theta}$ $I_y = \frac{\rho V R^2(\theta + \sin\theta\cos\theta)}{2\theta}$ $I_z = \rho V R^2$	${\rho_x}^2 = {\rho_{x_c}}^2 = \frac{R^2(\theta - \sin\theta\cos\theta)}{2\theta}$ ${r_y}^2 = \frac{R^2(\theta + \sin\theta\cos\theta)}{2\theta}$ ${\rho_z}^2 = R^2$	$I_{x_c y_c}$, etc. $= 0$ I_{xy}, etc. $= 0$
3	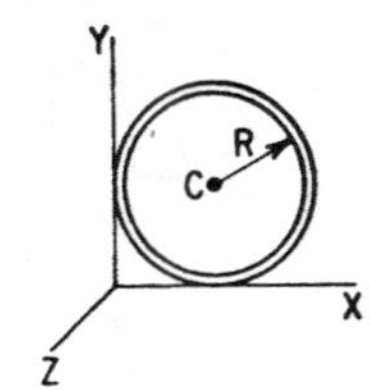	$V = 2\pi SR$ $X_c = R$ $Y_c = R$ $Z_c = 0$	$I_{x_c} = I_{y_c} = \frac{\rho V}{2} R^2$ $I_{z_c} = \rho V R^2$	${\rho_{x_c}}^2 = {\rho_{y_c}}^2 = \frac{1}{2}R^2$ ${\rho_{z_c}}^2 = R^2$	$I_{x_c y_c}$, etc. $= 0$

TABLE 3.2 Properties of Homogeneous Solid Bodies (*Continued*)

	Solid body	Volume and centroid	Mass moment of inertia	Radius of gyration squared	Mass product of inertia
4	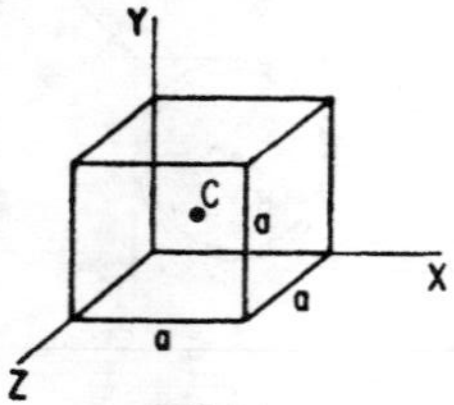 Cube	$V = a^3$ $X_c = \frac{1}{2}a$ $Y_c = \frac{1}{2}a$ $Z_c = \frac{1}{2}a$	$I_{x_c} = I_{y_c} = I_{z_c} = \frac{1}{6}\rho V a^2$	$\rho_{x_c}{}^2 = \rho_{y_c}{}^2 = \rho_{z_c}{}^2 = \frac{1}{6}a^2$	$I_{x_c y_c}$, etc. $= 0$
5	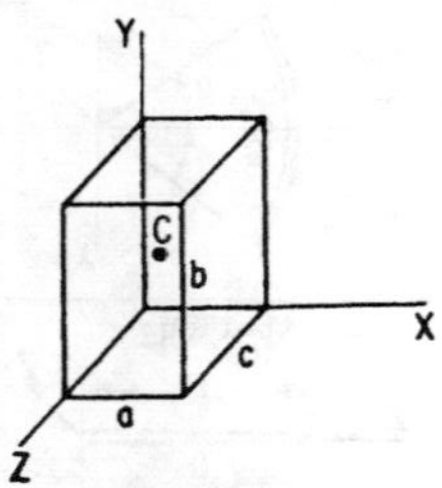 Rectangular prism	$V = abc$ $X_c = \frac{1}{2}a$ $Y_c = \frac{1}{2}b$ $Z_c = \frac{1}{2}c$	$I_{x_c} = \frac{1}{12}\rho V(b^2 + c^2)$	$\rho_{x_c}{}^2 = \frac{1}{12}(b^2 + c^2)$	$I_{x_c y_c}$, etc. $= 0$
6	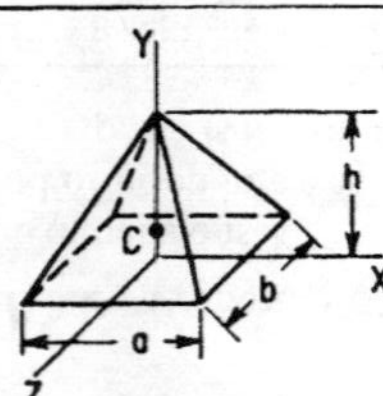 Right rectangular pyramid	$V = \frac{1}{3}abh$ $X_c = 0$ $Y_c = \frac{1}{4}h$ $Z_c = 0$	$I_{x_c} = \frac{1}{80}\rho V(4b^2 + 3h^2)$ $I_{y_c} = \frac{1}{20}\rho V(a^2 + b^2)$	$\rho_{x_c}{}^2 = \frac{1}{80}(4b^2 + 3h^2)$ $\rho_{y_c}{}^2 = \frac{1}{20}(a^2 + b^2)$	$I_{x_c y_c}$, etc. $= 0$

7 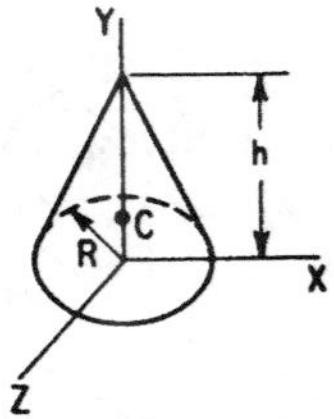 Right circular cone	$V = \frac{1}{3}\pi R^2 h$ $X_c = 0$ $Y_c = \frac{1}{4}h$ $Z_c = 0$	$I_{x_c} = I_{z_c} = \frac{3\rho V}{80}(4R^2 + h^2)$ $I_{y_c} = \frac{3}{10}\rho V R^2$	$\rho_{x_c}^2 = \rho_{z_c}^2 = \frac{3}{80}(4R^2 + h^2)$ $\rho_{y_c}^2 = \frac{3}{10}R^2$	$I_{x_c y_c}$, etc. $= 0$
8 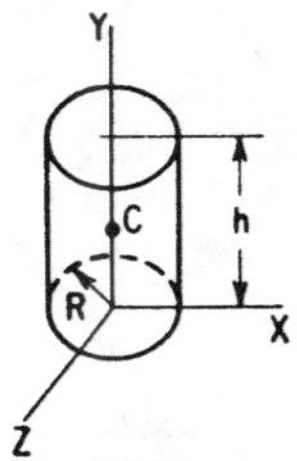 Right circular cylinder	$V = \pi R^2 h$ $X_c = 0$ $Y_c = \frac{1}{2}h$ $Z_c = 0$	$I_{x_c} = I_{z_c} = \frac{1}{12}\rho V(3R^2 + h^2)$ $I_{y_c} = \frac{1}{2}\rho V R^2$	$\rho_{x_c}^2 = \rho_{z_c}^2 = \frac{1}{12}(3R^2 + h^2)$ $\rho_{y_c}^2 = \frac{1}{2}R^2$	$I_{x_c y_c}$, etc. $= 0$
9 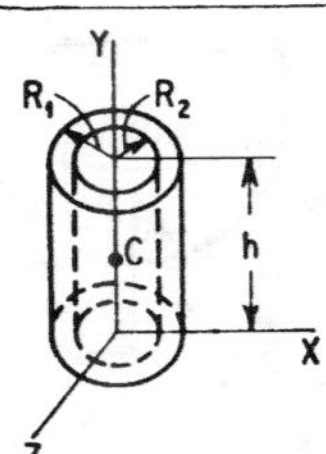 Hollow right circular cylinder	$V = \pi h(R_1^2 - R_2^2)$ $X_c = 0$ $Y_c = \frac{1}{2}h$ $Z_c = 0$	$I_{x_c} = I_{z_c}$ $= \frac{1}{12}\rho V(3R_1^2 + 3R_2^2 + h^2)$ $I_{y_c} = \frac{1}{2}\rho V(R_1^2 + R_2^2)$	$\rho_{x_c}^2 = \rho_{z_c}^2 = \frac{1}{12}(3R_1^2 + 3R_2^2 + h^2)$ $\rho_{y_c}^2 = \frac{1}{2}(R_1^2 + R_2^2)$	$I_{x_c y_c}$, etc. $= 0$

TABLE 3.2 Properties of Homogeneous Solid Bodies (*Continued*)

	Solid body	Volume and centroid	Mass moment of inertia	Radius of gyration squared	Mass product of inertia
10	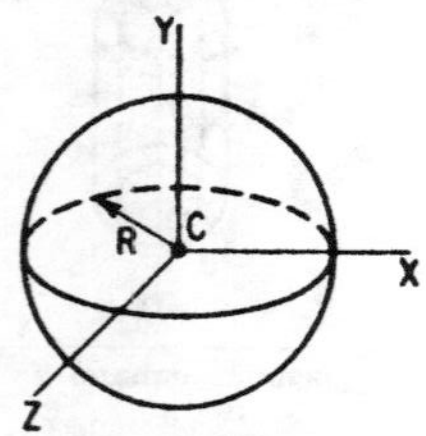 Sphere	$V = \frac{4}{3}\pi R^3$ $X_c = 0$ $Y_c = 0$ $Z_c = 0$	$I_{x_c} = \frac{2}{5}\rho V R^2$ $I_{y_c} = \frac{2}{5}\rho V R^2$ $I_{z_c} = \frac{2}{5}\rho V R^2$	$\rho_{x_c}{}^2 = \frac{2}{5}R^2$ $\rho_{y_c}{}^2 = \frac{2}{5}R^2$ $\rho_{z_c}{}^2 = \frac{2}{5}R^2$	$I_{x_c y_c}$, etc. $= 0$
11	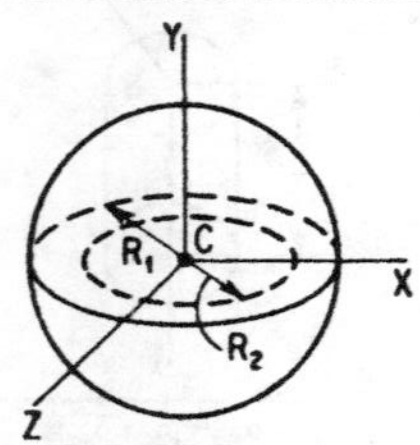 Hollow sphere	$V = \frac{4}{3}\pi(R_1{}^3 - R_2{}^3)$ $X_c = 0$ $Y_c = 0$ $Z_c = 0$	$I_x = I_y = I_z$ $= \frac{2}{5}\rho V \frac{R_1{}^5 - R_2{}^5}{R_1{}^3 - R_2{}^3}$	$\rho_x{}^2 = \rho_y{}^2 = \rho_z{}^2$ $= \frac{2}{5}\frac{R_1{}^5 - R_2{}^5}{R_1{}^3 - R_2{}^3}$	I_{xy}, etc. $= 0$
12	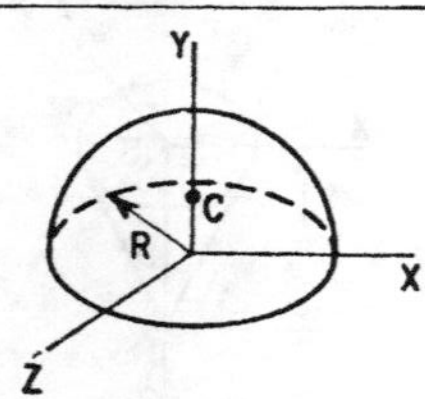 Hemisphere	$V = \frac{2}{3}\pi R^3$ $X_c = 0$ $Y_c = \frac{3}{8}R$ $Z_c = 0$	$I_x = I_y = I_z = \frac{2}{5}\rho V R^2$	$\rho_x{}^2 = \rho_y{}^2 = \rho_z{}^2 = \frac{2}{5}R^2$	$I_{x_c y_c}$, etc. $= 0$ I_{xy}, etc. $= 0$

13	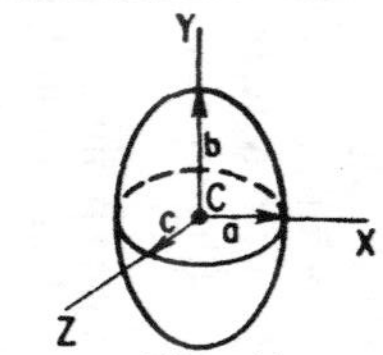 Ellipsoid	$V = \frac{4}{3}\pi abc$ $X_c = 0$ $Y_c = 0$ $Z_c = 0$	$I_x = \frac{1}{5}\rho V(b^2 + c^2)$ $I_y = \frac{1}{5}\rho V(a^2 + c^2)$ $I_z = \frac{1}{5}\rho V(a^2 + b^2)$	$\rho_x^2 = \frac{1}{5}(b^2 + c^2)$ $\rho_y^2 = \frac{1}{5}(a^2 + c^2)$ $\rho_z^2 = \frac{1}{5}(a^2 + b^2)$	I_{xy}, etc. $= 0$
14	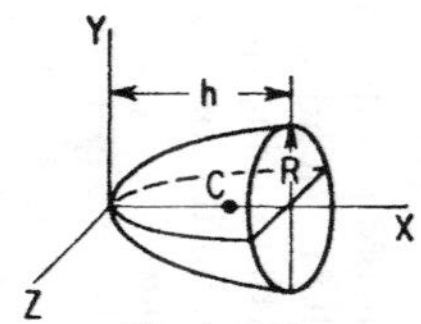 Paraboloid of revolution	$V = \frac{1}{2}\pi R^2 h$ $X_c = \frac{2}{3}h$ $Y_c = 0$ $Z_c = 0$	$I_{x_c} = \frac{1}{3}\rho V R^2$ $I_{y_c} = I_{z_c} = \frac{1}{18}\rho V(3R^2 + h^2)$	${\rho_{x_c}}^2 = \frac{1}{3}R^2$ ${\rho_{y_c}}^2 = {\rho_{z_c}}^2 = \frac{1}{18}(3R^2 + h^2)$	$I_{x_c y_c}$, etc. $= 0$
15	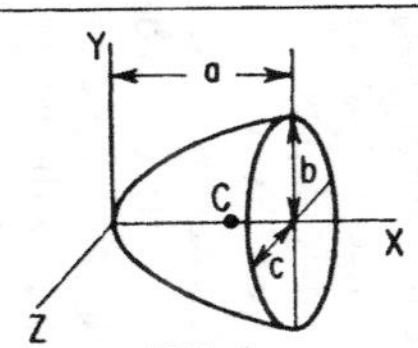 Elliptic paraboloid	$V = \frac{1}{2}\pi abc$ $X_c = \frac{2}{3}a$ $Y_c = 0$ $Z_c = 0$	$I_{x_c} = \frac{1}{6}\rho V(b^2 + c^2)$ $I_{y_c} = \frac{1}{18}\rho V(3c^2 + a^2)$ $I_{z_c} = \frac{1}{18}\rho V(3b^2 + a^2)$	${\rho_{x_c}}^2 = \frac{1}{6}(b^2 + c^2)$ ${\rho_{y_c}}^2 = \frac{1}{18}(3c^2 + a^2)$ ${\rho_{z_c}}^2 = \frac{1}{18}(3b^2 + a^2)$	$I_{x_c y_c}$, etc. $= 0$

CENTER OF MASS

Computation of Center of Mass. The center of mass (or center of gravity) is that point located by the vector

$$\mathbf{r}_c = \frac{1}{m}\int_m \mathbf{r}(m)\,dm \tag{3.7}$$

where $\mathbf{r}(m)$ is the radius vector of the element of mass dm. The center of mass of a body in a cartesian coordinate system X, Y, Z is located at

$$X_c = \frac{1}{m}\int_V X(V)\rho(V)\,dV$$

$$Y_c = \frac{1}{m}\int_V Y(V)\rho(V)\,dV \tag{3.8}$$

$$Z_c = \frac{1}{m}\int_V Z(V)\rho(V)\,dV$$

where $X(V)$, $Y(V)$, $Z(V)$ are the X, Y, Z coordinates of the element of volume dV and m is the mass of the body.

If the body can be divided into elements whose centers of mass are known, the center of mass of the entire body having a mass m is located by equations of the following type:

$$X_c = \frac{1}{m}(X_{c1}m_1 + X_{c2}m_2 + \cdots + X_{cn}m_n), \text{ etc.} \tag{3.9}$$

where X_{c1} is the X coordinate of the center of mass of element m_1. Tables (see Tables 3.1 and 3.2) which specify the location of centers of area and volume (called centroids) for simple sections and solid bodies often are an aid in dividing the body into the submasses indicated in the above equation. The centroid and center of mass of an element are coincident when the density of the material is uniform throughout the element.

Experimental Determination of Center of Mass. The location of the center of mass is normally measured indirectly by locating the center of gravity of the body, and may be found in various ways. Theoretically, if the body is suspended by a flexible wire attached successively at different points on the body, all lines represented by the wire in its various positions when extended inwardly into the body intersect at the center of gravity. Two such lines determine the center of gravity, but more may be used as a check. There are practical limitations to this method in that the point of intersection often is difficult to designate.

Other techniques are based on the balancing of the body on point or line supports. A point support locates the center of gravity along a vertical line through the point; a line support locates it in a vertical plane through the line. The intersection of such lines or planes determined with the body in various positions locates the center of gravity. The greatest difficulty with this technique is the maintenance of the stability of the

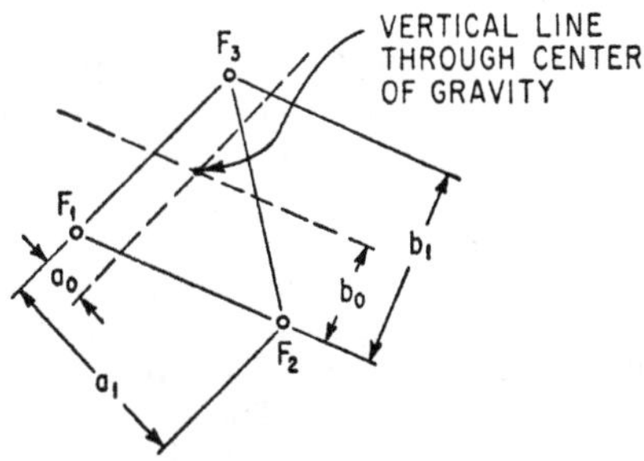

FIGURE 3.2 Three-scale method of locating the center of gravity of a body. The vertical forces F_1, F_2, F_3 at the scales result from the weight of the body. The vertical line located by the distances a_0 and b_0 [see Eqs. (3.10)] passes through the center of gravity of the body.

body while it is balanced, particularly where the height of the body is great relative to a horizontal dimension. If a perfect point or edge support is used, the equilibrium position is inherently unstable. It is only if the support has width that some degree of stability can be achieved, but then a resulting error in the location of the line or plane containing the center of gravity can be expected.

Another method of locating the center of gravity is to place the body in a stable position on three scales. From static moments the vector weight of the body is the resultant of the measured forces at the scales, as shown in Fig. 3.2. The vertical line through the center of gravity is located by the distances a_0 and b_0:

$$a_0 = \frac{F_2}{F_1 + F_2 + F_3} a_1 \qquad b_0 = \frac{F_3}{F_1 + F_2 + F_3} b_1 \tag{3.10}$$

This method cannot be used with more than three scales.

MOMENT AND PRODUCT OF INERTIA

Computation of Moment and Product of Inertia.[2,3] The moments of inertia of a rigid body with respect to the orthogonal axes X, Y, Z fixed in the body are

$$I_{xx} = \int_m (Y^2 + Z^2)\, dm \qquad I_{yy} = \int_m (X^2 + Z^2)\, dm \qquad I_{zz} = \int_m (X^2 + Y^2)\, dm \tag{3.11}$$

where dm is the infinitesimal element of mass located at the coordinate distances X, Y, Z; and the integration is taken over the mass of the body. Similarly, the products of inertia are

$$I_{xy} = \int_m XY\, dm \qquad I_{xz} = \int_m XZ\, dm \qquad I_{yz} = \int_m YZ\, dm \tag{3.12}$$

It is conventional in rigid-body mechanics to take the center of coordinates at the center of mass of the body. Unless otherwise specified, this location is assumed, and the moments of inertia and products of inertia refer to axes through the center of mass of the body. For a unique set of axes, the products of inertia vanish. These axes are called the principal inertial axes of the body. The moments of inertia about these axes are called the principal moments of inertia. The moments of inertia of a rigid body can be defined in terms of radii of gyration as follows:

$$I_{xx} = m\rho_x^2 \qquad I_{yy} = m\rho_y^2 \qquad I_{zz} = m\rho_z^2 \tag{3.13}$$

where I_{xx}, . . . are the moments of inertia of the body as defined by Eqs. (3.11), m is the mass of the body, and ρ_x, . . . are the radii of gyration. The radius of gyration has the dimension of length, and often leads to convenient expressions in dynamics of rigid bodies when distances are normalized to an appropriate radius of gyration. Solid bodies of various shapes have characteristic radii of gyration which sometimes are useful intuitively in evaluating dynamic conditions.

Unless the body has a very simple shape, it is laborious to evaluate the integrals of Eqs. (3.11) and (3.12). The problem is made easier by subdividing the body into parts for which simplified calculations are possible. The moments and products of inertia of the body are found by first determining the moments and products of inertia for the individual parts with respect to appropriate reference axes chosen in the parts, and then summing the contributions of the parts. This is done by selecting axes through the centers of mass of the parts, and then determining the moments and products of inertia of the parts relative to these axes. Then the moments and products of inertia are transferred to the axes chosen through the center of mass of the whole body, and the transferred quantities summed. In general, the transfer involves two sets of nonparallel coordinates whose centers are displaced. Two transformations are required as follows.

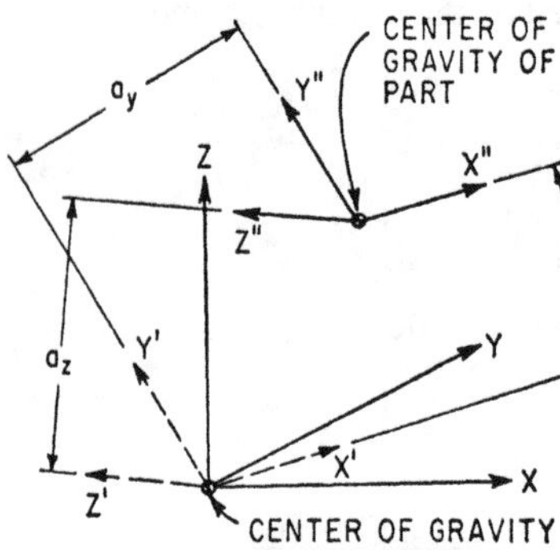

FIGURE 3.3 Axes required for moment and product of inertia transformations. Moments and products of inertia with respect to the axes X'', Y'', Z'' are transferred to the mutually parallel axes X', Y', Z' by Eqs. (3.14) and (3.15), and then to the inclined axes X, Y, Z by Eqs. (3.16) and (3.17).

Transformation to Parallel Axes. Referring to Fig. 3.3, suppose that X, Y, Z is a convenient set of axes for the moment of inertia of the whole body with its origin at the center of mass. The moments and products of inertia for a part of the body are $I_{x''x''}$, $I_{y''y''}$, $I_{z''z''}$, $I_{x''y''}$, $I_{x''z''}$, and $I_{y''z''}$, taken with respect to a set of axes X'', Y'', Z'' fixed in the part and having their center at the center of mass of the part. The axes X', Y', Z' are chosen parallel to X'', Y'', Z'' with their origin at the center of mass of the body. The perpendicular distance between the X'' and X' axes is a_x; that between Y'' and Y' is a_y; that between Z'' and Z' is a_z. The moments and products of inertia of the part of mass m_n with respect to the X', Y', Z' axes are

$$\begin{aligned} I_{x'x'} &= I_{x''x''} + m_n a_x^2 \\ I_{y'y'} &= I_{y''y''} + m_n a_y^2 \\ I_{z'z'} &= I_{z''z''} + m_n a_z^2 \end{aligned} \tag{3.14}$$

The corresponding products of inertia are

$$\begin{aligned} I_{x'y'} &= I_{x''y''} + m_n a_x a_y \\ I_{x'z'} &= I_{x''z''} + m_n a_x a_z \\ I_{y'z'} &= I_{y''z''} + m_n a_y a_z \end{aligned} \tag{3.15}$$

If X'', Y'', Z'' are the principal axes of the part, the product of inertia terms on the right-hand side of Eqs. (3.15) are zero.

Transformation to Inclined Axes. The desired moments and products of inertia with respect to axes X, Y, Z are now obtained by a transformation theorem relating the properties of bodies with respect to inclined sets of axes whose centers coincide. This theorem makes use of the direction cosines λ for the respective sets of axes. For example, $\lambda_{xx'}$ is the cosine of the angle between the X and X' axes. The expressions for the moments of inertia are

$$\begin{aligned}
I_{xx} &= \lambda_{xx'}^2 I_{x'x'} + \lambda_{xy'}^2 I_{y'y'} + \lambda_{xz'}^2 I_{z'z'} - 2\lambda_{xx'}\lambda_{xy'} I_{x'y'} - 2\lambda_{xx'}\lambda_{xz'} I_{x'z'} - 2\lambda_{xy'}\lambda_{xz'} I_{y'z'} \\
I_{yy} &= \lambda_{yx'}^2 I_{x'x'} + \lambda_{yy'}^2 I_{y'y'} + \lambda_{yz'}^2 I_{z'z'} - 2\lambda_{yx'}\lambda_{yy'} I_{x'y'} - 2\lambda_{yx'}\lambda_{yz'} I_{x'z'} - 2\lambda_{yy'}\lambda_{yz'} I_{y'z'} \\
I_{zz} &= \lambda_{zx'}^2 I_{x'x'} + \lambda_{zy'}^2 I_{y'y'} + \lambda_{zz'}^2 I_{z'z'} - 2\lambda_{zx'}\lambda_{zy'} I_{x'y'} - 2\lambda_{zx'}\lambda_{zz'} I_{x'z'} - 2\lambda_{zy'}\lambda_{zz'} I_{y'z'}
\end{aligned} \tag{3.16}$$

The corresponding products of inertia are

$$\begin{aligned}
-I_{xy} &= \lambda_{xx'}\lambda_{yx'} I_{x'x'} + \lambda_{xy'}\lambda_{yy'} I_{y'y'} + \lambda_{xz'}\lambda_{yz'} I_{z'z'} - (\lambda_{xx'}\lambda_{yy'} + \lambda_{xy'}\lambda_{yx'}) I_{x'y'} \\
&\quad - (\lambda_{xy'}\lambda_{yz'} + \lambda_{xz'}\lambda_{yy'}) I_{y'z'} - (\lambda_{xz'}\lambda_{yx'} + \lambda_{xx'}\lambda_{yz'}) I_{x'z'} \\
-I_{xz} &= \lambda_{xx'}\lambda_{zx'} I_{x'x'} + \lambda_{xy'}\lambda_{zy'} I_{y'y'} + \lambda_{xz'}\lambda_{zz'} I_{z'z'} - (\lambda_{xx'}\lambda_{zy'} + \lambda_{xy'}\lambda_{zx'}) I_{x'y'} \\
&\quad - (\lambda_{xy'}\lambda_{zz'} + \lambda_{xz'}\lambda_{zy'}) I_{y'z'} - (\lambda_{xx'}\lambda_{zz'} + \lambda_{xz'}\lambda_{zx'}) I_{x'z'} \\
-I_{yz} &= \lambda_{yx'}\lambda_{zx'} I_{x'x'} + \lambda_{yy'}\lambda_{zy'} I_{y'y'} + \lambda_{yz'}\lambda_{zz'} I_{z'z'} - (\lambda_{yx'}\lambda_{zy'} + \lambda_{yy'}\lambda_{zx'}) I_{x'y'} \\
&\quad - (\lambda_{yy'}\lambda_{zz'} + \lambda_{yz'}\lambda_{zy'}) I_{y'z'} - (\lambda_{yz'}\lambda_{zx'} + \lambda_{yx'}\lambda_{zz'}) I_{x'z'}
\end{aligned} \tag{3.17}$$

Experimental Determination of Moments of Inertia. The moment of inertia of a body about a given axis may be found experimentally by suspending the body as a pendulum so that rotational oscillations about that axis can occur. The period of free oscillation is then measured, and is used with the geometry of the pendulum to calculate the moment of inertia.

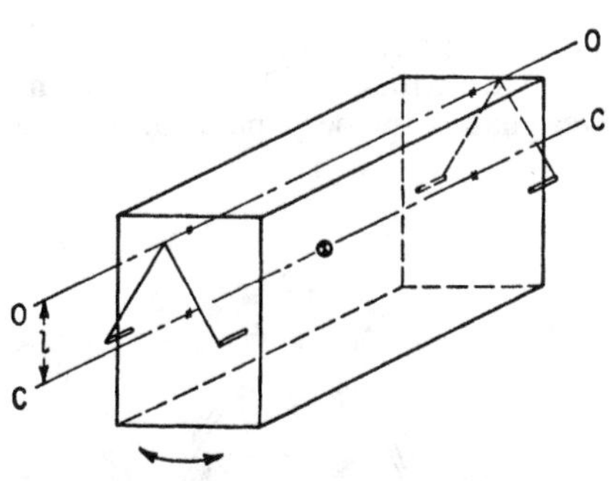

FIGURE 3.4 Compound pendulum method of determining moment of inertia. The period of oscillation of the test body about the horizontal axis O–O and the perpendicular distance l between the axis O–O and the parallel axis C–C through the center of gravity of the test body give I_{cc} by Eq. (3.18).

Two types of pendulums are useful: the compound pendulum and the torsional pendulum. When using the compound pendulum, the body is supported from two overhead points by wires, illustrated in Fig. 3.4. The distance l is measured between the axis of support O–O and a parallel axis C–C through the center of gravity of the body. The moment of inertia about C–C is given by

$$I_{cc} = ml^2\left[\left(\frac{\tau_0}{2\pi}\right)^2\left(\frac{g}{l}\right) - 1\right] \tag{3.18}$$

where τ_0 is the period of oscillation in seconds, l is the pendulum length in inches, g is the gravitational acceleration in in./sec^2, and m is the mass in lb-sec^2/in., yielding a moment of inertia in lb-in./sec^2.

The accuracy of the above method is dependent upon the accuracy with which the distance l is known. Since the center of gravity often is an inaccessible point, a direct measurement of l may not be practicable. However, a change in l can be measured quite readily. If the experiment is repeated with a different support axis O'–O', the length l becomes $l + \Delta l$ and the period of oscillation becomes τ_0'. Then, the distance l can be written in terms of Δl and the two periods τ_0, τ_0':

$$l = \Delta l\left[\frac{(\tau_0'^2/4\pi^2)(g/\Delta l) - 1}{[(\tau_0^2 - \tau_0'^2)/4\pi^2][g/\Delta l] - 1}\right] \tag{3.19}$$

This value of l can be substituted into Eq. (3.18) to compute I_{cc}.

Note that accuracy is not achieved if l is much larger than the radius of gyration ρ_c of the body about the axis C–C ($I_{cc} = m\rho_c^2$). If l is large, then $(\tau_0/2\pi)^2 \simeq l/g$ and the expression in brackets in Eq. (3.18) is very small; thus, it is sensitive to small errors in the measurement of both τ_0 and l. Consequently, it is highly desirable that the distance l be chosen as small as convenient, preferably with the axis O–O passing through the body.

A torsional pendulum may be constructed with the test body suspended by a single torsional spring (in practice, a rod or wire) of known stiffness, or by three flexible wires. A solid body supported by a single torsional spring is shown in Fig. 3.5. From the known torsional stiffness k_t and the measured period of torsional oscillation τ, the moment of inertia of the body about the vertical torsional axis is

$$I_{cc} = \frac{k_t\tau^2}{4\pi^2} \tag{3.20}$$

A platform may be constructed below the torsional spring to carry the bodies to be measured, as shown in Fig. 3.6. By repeating the experiment with two different bodies placed on the platform, it becomes unnecessary to measure the torsional stiffness k_t. If a body with a *known* moment of inertia I_1 is placed on the platform and an oscillation period τ_1 results, the moment of inertia I_2 of a body which produces a period τ_2 is given by

$$I_2 = I_1\left[\frac{(\tau_2/\tau_0)^2 - 1}{(\tau_1/\tau_0)^2 - 1}\right] \tag{3.21}$$

where τ_0 is the period of the pendulum composed of platform alone.

A body suspended by three flexible wires, called a trifilar pendulum, as shown in Fig. 3.7, offers some utilitarian advantages. Designating the perpendicular distances

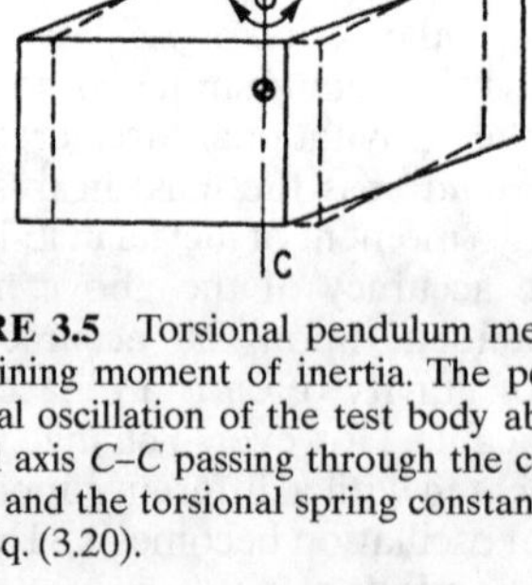

FIGURE 3.5 Torsional pendulum method of determining moment of inertia. The period of torsional oscillation of the test body about the vertical axis C–C passing through the center of gravity and the torsional spring constant k_t give I_{cc} by Eq. (3.20).

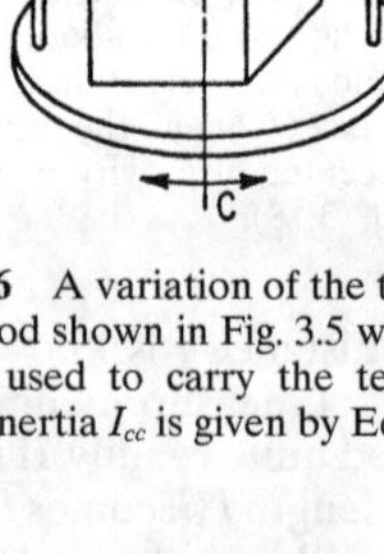

FIGURE 3.6 A variation of the torsional pendulum method shown in Fig. 3.5 wherein a light platform is used to carry the test body. The moment of inertia I_{cc} is given by Eq. (3.20).

of the wires to the vertical axis C–C through the center of gravity of the body by R_1, R_2, R_3, the angles between wires by ϕ_1, ϕ_2, ϕ_3, and the length of each wire by l, the moment of inertia about axis C–C is

$$I_{cc} = \frac{mgR_1R_2R_3\tau^2}{4\pi^2 l} \frac{R_1 \sin \phi_1 + R_2 \sin \phi_2 + R_3 \sin \phi_3}{R_2R_3 \sin \phi_1 + R_1R_3 \sin \phi_2 + R_1R_2 \sin \phi_3} \tag{3.22}$$

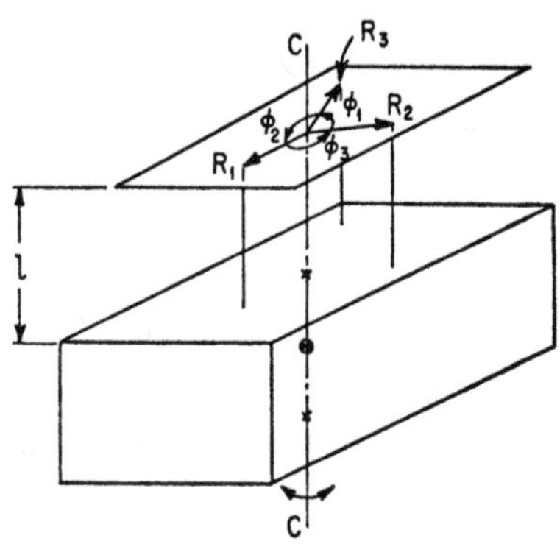

FIGURE 3.7 Trifilar pendulum method of determining moment of inertia. The period of torsional oscillation of the test body about the vertical axis C–C passing through the center of gravity and the geometry of the pendulum give I_{cc} by Eq. (3.22); with a simpler geometry, I_{cc} is given by Eq. (3.23).

Apparatus that is more convenient for repeated use embodies a light platform supported by three equally spaced wires. The body whose moment of inertia is to be measured is placed on the platform with its center of gravity equidistant from the wires. Thus $R_1 = R_2 = R_3 = R$ and $\phi_1 = \phi_2 = \phi_3 = 120°$. Substituting these relations in Eq. (3.22), the moment of inertia about the vertical axis C–C is

$$I_{cc} = \frac{mgR^2\tau^2}{4\pi^2 l} \tag{3.23}$$

where the mass m is the sum of the masses of the test body and the platform. The moment of inertia of the platform is subtracted from the test result to obtain the moment of inertia of the body being measured. It becomes unnecessary to know the distances R and l in Eq. (3.23) if the period of oscillation is measured with the platform empty, with the body being measured on the platform, and with a second body of known mass m_1 and known moment of inertia I_1 on the platform. Then the desired moment of inertia I_2 is

$$I_2 = I_1 \left[\frac{[1 + (m_2/m_0)][\tau_2/\tau_0]^2 - 1}{[1 + (m_1/m_0)][\tau_1/\tau_0]^2 - 1} \right] \tag{3.24}$$

where m_0 is the mass of the unloaded platform, m_2 is the mass of the body being measured, τ_0 is the period of oscillation with the platform unloaded, τ_1 is the period when loaded with known body of mass m_1, and τ_2 is the period when loaded with the unknown body of mass m_2.

Experimental Determination of Product of Inertia. The experimental determination of a product of inertia usually requires the measurement of moments of inertia. (An exception is the balancing machine technique described later.) If possible, symmetry of the body is used to locate directions of principal inertial axes, thereby simplifying the relationship between the moments of inertia as known and the products of inertia to be found. Several alternative procedures are described below, depending on the number of principal inertia axes whose directions are known. Knowledge of two principal axes implies a knowledge of all three since they are mutually perpendicular.

If the directions of all three principal axes (X', Y', Z') are known and it is desirable to use another set of axes (X, Y, Z), Eqs. (3.16) and (3.17) may be simplified because the products of inertia with respect to the principal directions are zero. First, the three principal moments of inertia ($I_{x'x'}$, $I_{y'y'}$, $I_{z'z'}$) are measured by one of the above techniques; then the moments of inertia with respect to the *X, Y, Z* axes are

$$\begin{aligned} I_{xx} &= \lambda_{xx'}^2 I_{x'x'} + \lambda_{xy'}^2 I_{y'y'} + \lambda_{xz'}^2 I_{z'z'} \\ I_{yy} &= \lambda_{yx'}^2 I_{x'x'} + \lambda_{yy'}^2 I_{y'y'} + \lambda_{yz'}^2 I_{z'z'} \\ I_{zz} &= \lambda_{zx'}^2 I_{x'x'} + \lambda_{zy'}^2 I_{y'y'} + \lambda_{zz'}^2 I_{z'z'} \end{aligned} \tag{3.25}$$

The products of inertia with respect to the *X, Y, Z* axes are

$$\begin{aligned} -I_{xy} &= \lambda_{xx'}\lambda_{yx'} I_{x'x'} + \lambda_{xy'}\lambda_{yy'} I_{y'y'} + \lambda_{xz'}\lambda_{yz'} I_{z'z'} \\ -I_{xz} &= \lambda_{xx'}\lambda_{zx'} I_{x'x'} + \lambda_{xy'}\lambda_{zy'} I_{y'y'} + \lambda_{xz'}\lambda_{zz'} I_{z'z'} \\ -I_{yz} &= \lambda_{yx'}\lambda_{zx'} I_{x'x'} + \lambda_{yy'}\lambda_{zy'} I_{y'y'} + \lambda_{yz'}\lambda_{zz'} I_{z'z'} \end{aligned} \tag{3.26}$$

The direction of one principal axis *Z* may be known from symmetry. The axis through the center of gravity perpendicular to the plane of symmetry is a principal axis. The product of inertia with respect to *X* and *Y* axes, located in the plane of symmetry, is determined by first establishing another axis *X′* at a counterclockwise angle θ from *X*, as shown in Fig. 3.8. If the three moments of inertia I_{xx}, $I_{x'x'}$, and I_{yy} are measured by any applicable means, the product of inertia I_{xy} is

$$I_{xy} = \frac{I_{xx}\cos^2\theta + I_{yy}\sin^2\theta - I_{x'x'}}{\sin 2\theta} \tag{3.27}$$

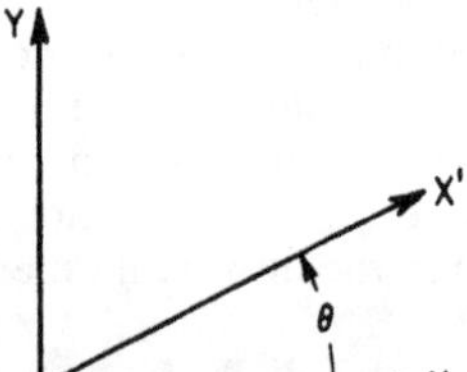

FIGURE 3.8 Axes required for determining the product of inertia with respect to the axes *X* and *Y* when *Z* is a principal axis of inertia. The moments of inertia about the axes *X, Y,* and *X′,* where *X′* is in the plane of *X* and *Y* at a counterclockwise angle θ from *X*, give I_{xy} by Eq. (3.27).

where $0 < \theta < \pi$. For optimum accuracy, θ should be approximately $\pi/4$ or $3\pi/4$. Since the third axis *Z* is a principal axis, I_{xz} and I_{yz} are zero.

Another method is illustrated in Fig. 3.9.[4,5] The plane of the *X* and *Z* axes is a plane of symmetry, or the *Y* axis is otherwise known to be a principal axis of inertia. For determining I_{xz}, the body is suspended by a cable so that the *Y* axis is horizontal and the *Z* axis is vertical. Torsional stiffness about the *Z* axis is provided by four springs acting in the *Y* direction at the points shown. The body is oscillated about the *Z* axis with various positions of the springs so that the angle θ can be varied. The spring stiffnesses and locations must be such that there is no net force in the *Y* direction due to a rotation about the *Z* axis. In general, there is coupling between rotations about the *X* and *Z* axes, with the result that oscillations about both axes occur as a result of an initial rotational displacement about the *Z* axis. At some particular value of $\theta = \theta_0$, the two rotations are uncoupled; i.e., oscillation about the *Z* axis does not cause oscillation about the *X* axis. Then

$$I_{xz} = I_{zz}\tan\theta_0 \tag{3.28}$$

The moment of inertia I_{zz} can be determined by one of the methods described under "Experimental Determination of Moments of Inertia."

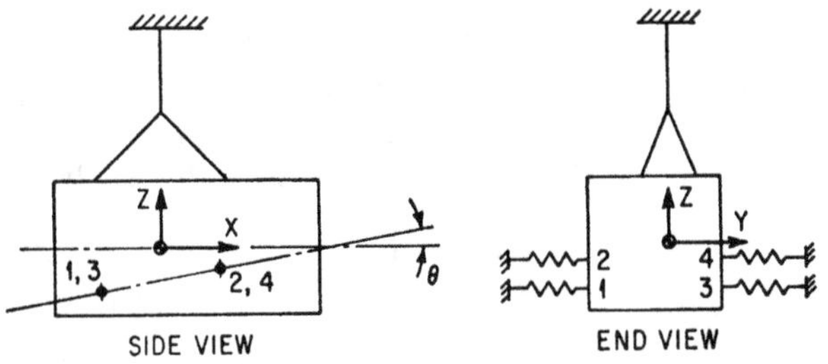

FIGURE 3.9 Method of determining the product of inertia with respect to the axes X and Z when Y is a principal axis of inertia. The test body is oscillated about the vertical Z axis with torsional stiffness provided by the four springs acting in the Y direction at the points shown. There should be no net force on the test body in the Y direction due to a rotation about the Z axis. The angle θ is varied until, at some value of $\theta = \theta_0$, oscillations about X and Z are uncoupled. The angle θ_0 and the moment of inertia about the Z axis give I_{xz} by Eq. (3.28).

When the moments and product of inertia with respect to a pair of axes X and Z in a principal plane of inertia XZ are known, the orientation of a principal axis P is given by

$$\theta_p = \tfrac{1}{2} \tan^{-1}\left(\frac{2I_{xz}}{I_{zz} - I_{xx}}\right) \tag{3.29}$$

where θ_p is the counterclockwise angle from the X axis to the P axis. The second principal axis in this plane is at $\theta_p + 90°$.

Consider the determination of products of inertia when the directions of all principal axes of inertia are unknown. In one method, the moments of inertia about two independent sets of three mutually perpendicular axes are measured, and the direction cosines between these sets of axes are known from the positions of the axes. The values for the six moments of inertia and the nine direction cosines are then substituted into Eqs. (3.16) and (3.17). The result is six linear equations in the six unknown products of inertia, from which the values of the desired products of inertia may be found by simultaneous solution of the equations. This method leads to experimental errors of relatively large magnitude because each product of inertia is, in general, a function of all six moments of inertia, each of which contains an experimental error.

An alternative method is based upon the knowledge that one of the principal moments of inertia of a body is the largest and another is the smallest that can be obtained for any axis through the center of gravity. A trial-and-error procedure can be used to locate the orientation of the axis through the center of gravity having the maximum and/or minimum moment of inertia. After one or both are located, the moments and products of inertia for any set of axes are found by the techniques previously discussed.

The products of inertia of a body also may be determined by rotating the body at a constant angular velocity Ω about an axis passing through the center of gravity, as illustrated in Fig. 3.10. This method is similar to the balancing machine technique used to balance a body dynamically. If the bearings are a distance l apart and the dynamic reactions F_x and F_y are measured, the products of inertia are

$$I_{xz} = -\frac{F_x l}{\Omega^2} \qquad I_{yz} = -\frac{F_y l}{\Omega^2} \tag{3.30}$$

Limitations to this method are (1) the size of the body that can be accommodated by the balancing machine and (2) the angular velocity that the body can withstand without damage from centrifugal forces. If the angle between the Z axis and a principal axis of inertia is small, high rotational speeds may be necessary to measure the reaction forces accurately.

PROPERTIES OF RESILIENT SUPPORTS

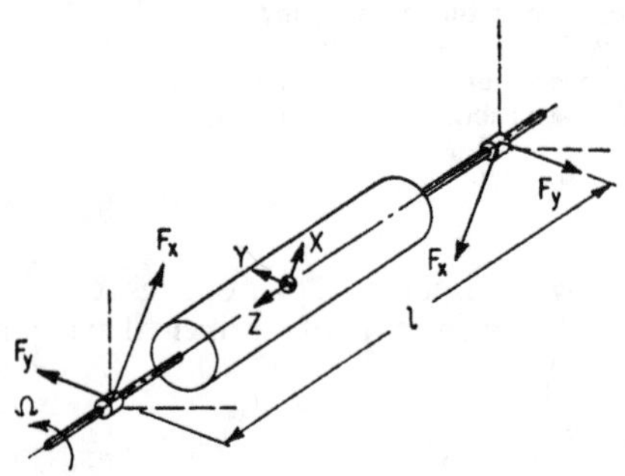

FIGURE 3.10 Balancing machine technique for determining products of inertia. The test body is rotated about the Z axis with angular velocity Ω. The dynamic reactions F_x and F_y measured at the bearings, which are a distance l apart, give I_{xz} and I_{yz} by Eq. (3.30).

A resilient support is considered to be a three-dimensional element having two terminals or end connections. When the end connections are moved one relative to the other in any direction, the element resists such motion. In this chapter, the element is considered to be massless; the force that resists relative motion across the element is considered to consist of a spring force that is directly proportional to the relative displacement (deflection across the element) and a damping force that is directly proportional to the relative velocity (velocity across the element). Such an element is defined as a *linear resilient support.* Nonlinear elements are discussed in Chap. 4, and nonlinear damping is discussed in Chap. 2.

In a single-degree-of-freedom system or in a system having constraints on the paths of motion of elements of the system (Chap. 2), the resilient element is constrained to deflect in a given direction and the properties of the element are defined with respect to the force opposing motion in this direction. In the absence of such constraints, the application of a force to a resilient element generally causes a motion in a different direction. The *principal elastic axes* of a resilient element are those axes for which the element, when unconstrained, experiences a deflection colineal with the direction of the applied force. Any axis of symmetry is a principal elastic axis.

In rigid-body dynamics, the rigid body sometimes vibrates in modes that are coupled by the properties of the resilient elements as well as by their location. For example, if the body experiences a static displacement x in the direction of the X axis only, a resilient element opposes this motion by exerting a force $k_{xx}x$ on the body in the direction of the X axis, where one subscript on the spring constant k indicates the direction of the force exerted by the element and the other subscript indicates the direction of the deflection. If the X direction is not a principal elastic direction of the element and the body experiences a static displacement x in the X direction, the body is acted upon by a force $k_{yx}x$ in the Y direction if no displacement y is permitted. The stiffnesses have reciprocal properties; i.e., $k_{xy} = k_{yx}$. In general, the stiffnesses in the directions of the coordinate axes can be expressed in terms of (1) principal stiffnesses and (2) the angles between the coordinate axes and the principal elastic axes of the element. Therefore, the stiffness of a resilient element can be represented pictorially by the combination of three mutually perpendicular, idealized springs oriented along the principal elastic directions of the resilient element. Each spring has a stiffness equal to the principal stiffness represented.

A resilient element is assumed to have damping properties such that each spring representing a value of principal stiffness is paralleled by an idealized viscous damper, each damper representing a value of principal damping. Hence, coupling through damping exists in a manner similar to coupling through stiffness. Consequently, the viscous damping coefficient c is analogous to the spring coefficient k; i.e., the force exerted by the damping of the resilient element in response to a velocity $\dot{x}$ is $c_{xx}\dot{x}$ in the direction of the X axis and $c_{yx}\dot{x}$ in the direction of the Y axis if $\dot{y}$ is zero. Reciprocity exists; i.e., $c_{xy} = c_{yx}$.

The point of intersection of the principal elastic axes of a resilient element is designated as the *elastic center of the resilient element.* The elastic center is important since it defines the theoretical point location of the resilient element for use in the equations of motion of a resiliently supported rigid body. For example, the torque on the rigid body about the Y axis due to a force $k_{xx}x$ transmitted by a resilient element in the X direction is $k_{xx}a_z x$, where a_z is the Z coordinate of the elastic center of the resilient element.

In general, it is assumed that a resilient element is attached to the rigid body by means of "ball joints"; i.e., the resilient element is incapable of applying a couple to the body. If this assumption is not made, a resilient element would be represented not only by translational springs and dampers along the principal elastic axes but also by torsional springs and dampers resisting rotation about the principal elastic directions.

Figure 3.11 shows that the torsional elements usually can be neglected. The torque which acts on the rigid body due to a rotation β of the body and a rotation $\boldsymbol{\beta}$ of the support is $(k_t + a_z^2 k_x)(\beta - \boldsymbol{\beta})$, where k_t is the torsional spring constant in the β direction. The torsional stiffness k_t usually is much smaller than $a_z^2 k_x$ and can be neglected. Treatment of the general case indicates that if the torsional stiffnesses of the resilient element are small compared with the product of the translational stiffnesses times the square of distances from the elastic center of the resilient element to the center of gravity of the rigid body, the torsional stiffnesses have a negligible effect on the vibrational behavior of the body. The treatment of torsional dampers is completely analogous.

EQUATIONS OF MOTION FOR A RESILIENTLY SUPPORTED RIGID BODY

The differential equations of motion for the rigid body are given by Eqs. (3.2) and (3.3), where the **F**'s and **M**'s represent the forces and moments acting on the body, either directly or through the resilient supporting elements. Figure 3.12 shows a view of a rigid body at rest with an inertial set of axes $\bar{X}, \bar{Y}, \bar{Z}$ and a coincident set of axes X, Y, Z fixed in the rigid body, both sets of axes passing through the center of mass. A typical resilient element (2) is represented by parallel spring and viscous damper combinations arranged respectively parallel with the $\bar{X}, \bar{Y}, \bar{Z}$ axes. Another resilient element (1) is shown with its principal axes not parallel with $\bar{X}, \bar{Y}, \bar{Z}$.

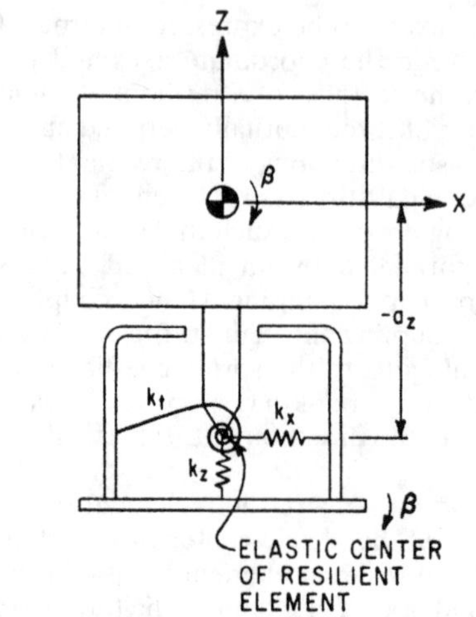

FIGURE 3.11 Pictorial representation of the properties of an undamped resilient element in the *XZ* plane including a torsional spring k_t. An analysis of the motion of the supported body in the *XZ* plane shows that the torsional spring can be neglected if $k_t << a_z^2 k_x$.

The displacement of the center of gravity of the body in the $\bar{X}, \bar{Y}, \bar{Z}$ directions is in Fig. 3.1 indicated by x_c, y_c, z_c, respectively; and rotation of the rigid body about these axes is indicated by α, β, γ, respectively. In Fig. 3.12, each resilient element is represented by three mutually perpendicular spring-damper combinations. One end of each such combination is attached to the rigid body; the other end is considered to be attached to a foundation whose corresponding translational displacement is defined by u, v, w in the $\bar{X}, \bar{Y}, \bar{Z}$ directions, respectively, and whose rotational displacement about these axes is defined by $\boldsymbol{\alpha}, \boldsymbol{\beta}, \boldsymbol{\gamma}$, respectively. The point of attachment of each of the idealized resilient elements is located at the coordinate distances a_x, a_y, a_z of the elastic center of the resilient element.

Consider the rigid body to experience a translational displacement x_c of its center of gravity and no other displacement, and neglect the effects of the viscous dampers. The force developed by a resilient element has the effect of a force $-k_{xx}(x_c - u)$ in the X direction, a moment $k_{xx}(x_c - u)a_y$ in the γ coordinate (about the Z axis), and a moment $-k_{xx}(x_c - u)a_z$ in the β coordinate (about the Y axis). Furthermore, the coupling stiffness causes a force $-k_{xy}(x_c - u)$ in the Y direction and a force $-k_{xz}(x_c - u)$ in the Z direction. These forces have the moments $k_{xy}(x_c - u)a_z$ in the α coordinate; $-k_{xy}(x_c - u)a_x$ in the γ coordinate; $k_{xz}(x_c - u)a_x$ in the β coordinate; and $-k_{xz}(x_c - u)a_y$ in the α coordinate. By considering in a similar manner the forces and moments developed by a resilient element for successive displacements of the rigid body in the three translational and three rotational coordinates, and summing over the number of resilient elements, the equations of motion are written as follows:[6,7]

$$m\ddot{x}_c + \Sigma k_{xx}(x_c - u) + \Sigma k_{xy}(y_c - v) + \Sigma k_{xz}(z_c - w) + \Sigma(k_{xz}a_y - k_{xy}a_z)(\alpha - \boldsymbol{\alpha}) + \Sigma(k_{xx}a_z - k_{xz}a_x)(\beta - \boldsymbol{\beta}) + \Sigma(k_{xy}a_x - k_{xx}a_y)(\gamma - \boldsymbol{\gamma}) = F_x \quad (3.31a)$$

$$I_{xx}\ddot{\alpha} - I_{xy}\ddot{\beta} - I_{xz}\ddot{\gamma} + \Sigma(k_{xz}a_y - k_{xy}a_z)(x_c - u) + \Sigma(k_{yz}a_y - k_{yy}a_z)(y_c - v) + \Sigma(k_{zz}a_y - k_{yz}a_z)(z_c - w) + \Sigma(k_{yy}a_z^2 + k_{zz}a_y^2 - 2k_{yz}a_ya_z)(\alpha - \boldsymbol{\alpha}) + \Sigma(k_{xz}a_ya_z + k_{yz}a_xa_z - k_{zz}a_xa_y - k_{xy}a_z^2)(\beta - \boldsymbol{\beta}) + \Sigma(k_{xy}a_ya_z + k_{yz}a_xa_y - k_{yy}a_xa_z - k_{xz}a_y^2)(\gamma - \boldsymbol{\gamma}) = M_x \quad (3.31b)$$

$$m\ddot{y}_c + \Sigma k_{xy}(x_c - u) + \Sigma k_{yy}(y_c - v) + \Sigma k_{yz}(z_c - w) + \Sigma(k_{yz}a_y - k_{yy}a_z)(\alpha - \boldsymbol{\alpha}) + \Sigma(k_{xy}a_z - k_{yz}a_x)(\beta - \boldsymbol{\beta}) + \Sigma(k_{yy}a_x - k_{xy}a_y)(\gamma - \boldsymbol{\gamma}) = F_y \quad (3.31c)$$

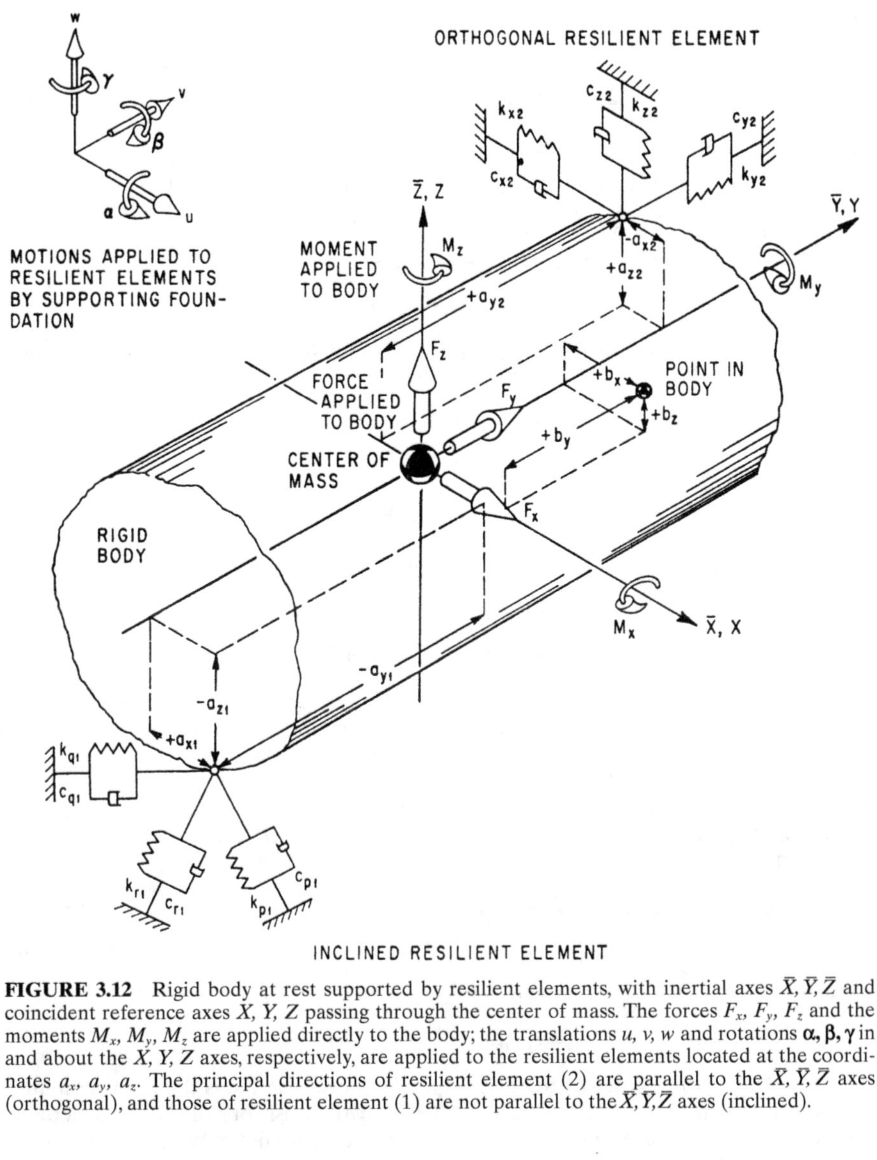

FIGURE 3.12 Rigid body at rest supported by resilient elements, with inertial axes $\bar{X}, \bar{Y}, \bar{Z}$ and coincident reference axes X, Y, Z passing through the center of mass. The forces F_x, F_y, F_z and the moments M_x, M_y, M_z are applied directly to the body; the translations u, v, w and rotations $\boldsymbol{\alpha}, \boldsymbol{\beta}, \boldsymbol{\gamma}$ in and about the X, Y, Z axes, respectively, are applied to the resilient elements located at the coordinates a_x, a_y, a_z. The principal directions of resilient element (2) are parallel to the $\bar{X}, \bar{Y}, \bar{Z}$ axes (orthogonal), and those of resilient element (1) are not parallel to the $\bar{X}, \bar{Y}, \bar{Z}$ axes (inclined).

$$\begin{aligned}
I_{yy}\ddot{\beta} - I_{xy}\ddot{\alpha} - I_{yz}\ddot{\gamma} &+ \Sigma(k_{xx}a_z - k_{xz}a_x)(x_c - u) \\
&+ \Sigma(k_{xy}a_z - k_{yz}a_x)(y_c - v) + \Sigma(k_{xz}a_z - k_{zz}a_x)(z_c - w) \\
&+ \Sigma(k_{xz}a_ya_z + k_{yz}a_xa_z - k_{zz}a_xa_y - k_{xy}a_z^2)(\alpha - \boldsymbol{\alpha}) \\
&+ \Sigma(k_{xx}a_z^2 + k_{zz}a_x^2 - 2k_{xz}a_xa_z)(\beta - \boldsymbol{\beta}) \\
&+ \Sigma(k_{xy}a_xa_z + k_{xz}a_xa_y - k_{xx}a_ya_z - k_{yz}a_x^2)(\gamma - \boldsymbol{\gamma}) = M_y
\end{aligned} \tag{3.31d}$$

$$m\ddot{z}_c + \Sigma k_{xz}(x_c - u) + \Sigma k_{yz}(y_c - v) + \Sigma k_{zz}(z_c - w)$$
$$+ \Sigma(k_{zz}a_y - k_{yz}a_z)(\alpha - \boldsymbol{\alpha}) + \Sigma(k_{xz}a_z - k_{zz}a_x)(\beta - \boldsymbol{\beta})$$
$$+ \Sigma(k_{yz}a_x - k_{xz}a_y)(\gamma - \boldsymbol{\gamma}) = F_z \tag{3.31e}$$

$$I_{zz}\ddot{\gamma} - I_{xz}\ddot{\alpha} - I_{yz}\ddot{\beta} + \Sigma(k_{xy}a_x - k_{xx}a_y)(x_c - u)$$
$$+ \Sigma(k_{yy}a_x - k_{xy}a_y)(y_c - v) + \Sigma(k_{yz}a_x - k_{xz}a_y)(z_c - w)$$
$$+ \Sigma(k_{xy}a_ya_z + k_{yz}a_xa_y - k_{yy}a_xa_z - k_{xz}a_y^2)(\alpha - \boldsymbol{\alpha})$$
$$+ \Sigma(k_{xy}a_xa_z + k_{xz}a_xa_y - k_{xx}a_ya_z - k_{yz}a_x^2)(\beta - \boldsymbol{\beta})$$
$$+ \Sigma(k_{xx}a_y^2 + k_{yy}a_x^2 - 2k_{xy}a_xa_y)(\gamma - \boldsymbol{\gamma}) = M_z \tag{3.31f}$$

where the moments and products of inertia are defined by Eqs. (3.11) and (3.12) and the stiffness coefficients are defined as follows:

$$\begin{aligned} k_{xx} &= k_p\lambda_{xp}^2 + k_q\lambda_{xq}^2 + k_r\lambda_{xr}^2 \\ k_{yy} &= k_p\lambda_{yp}^2 + k_q\lambda_{yq}^2 + k_r\lambda_{yr}^2 \\ k_{zz} &= k_p\lambda_{zp}^2 + k_q\lambda_{zq}^2 + k_r\lambda_{zr}^2 \\ k_{xy} &= k_p\lambda_{xp}\lambda_{yp} + k_q\lambda_{xq}\lambda_{yq} + k_r\lambda_{xr}\lambda_{yr} \\ k_{xz} &= k_p\lambda_{xp}\lambda_{zp} + k_q\lambda_{xq}\lambda_{zq} + k_r\lambda_{xr}\lambda_{zr} \\ k_{yz} &= k_p\lambda_{yp}\lambda_{zp} + k_q\lambda_{yq}\lambda_{zq} + k_r\lambda_{yr}\lambda_{zr} \end{aligned} \tag{3.32}$$

where the λ's are the cosines of the angles between the principal elastic axes of the resilient supporting elements and the coordinate axes. For example, λ_{xp} is the cosine of the angle between the X axis and the P axis of principal stiffness.

The equations of motion, Eqs. (3.31), do not include forces applied to the rigid body by damping forces from the resilient elements. To include damping, appropriate damping terms analogous to the corresponding stiffness terms are added to each equation. For example, Eq. (3.31a) would become

$$m\ddot{x}_c + \Sigma c_{xx}(\dot{x}_c - \dot{u}) + \Sigma k_{xx}(x_c - u) + \cdots$$
$$+ \Sigma(c_{xz}a_y - c_{xy}a_z)(\dot{\alpha} - \dot{\boldsymbol{\alpha}}) + \Sigma(k_{xz}a_y - k_{xy}a_z)(\alpha - \boldsymbol{\alpha}) + \cdots = F_x \tag{3.31a'}$$

where
$$c_{xx} = c_p\lambda_{xp}^2 + c_q\lambda_{xq}^2 + c_r\lambda_{xr}^2$$
$$c_{xy} = c_p\lambda_{xp}\lambda_{yp} + c_q\lambda_{xq}\lambda_{yq} + c_r\lambda_{xr}\lambda_{yr}$$

The number of degrees of freedom of a vibrational system is the minimum number of coordinates necessary to define completely the positions of the mass elements of the system in space. The system of Fig. 3.12 requires a minimum of six coordinates ($x_c, y_c, z_c, \alpha, \beta, \gamma$) to define the position of the rigid body in space; thus, the system is said to vibrate in six degrees of freedom. Equations (3.31) may be solved simultaneously for the three components x_c, y_c, z_c of the center-of-gravity displacement and the three components α, β, γ of the rotational displacement of the rigid body. In most practical instances, the equations are simplified considerably by one or more of the following simplifying conditions:

1. The reference axes X, Y, Z are selected to coincide with the principal inertial axes of the body; then

$$I_{xy} = I_{xz} = I_{yz} = 0 \tag{3.33}$$

2. The resilient supporting elements are so arranged that one or more planes of symmetry exist; i.e., motion parallel to the plane of symmetry has no tendency to excite motion perpendicular to it, or rotation about an axis lying in the plane does not excite motion parallel to the plane. For example, in Eq. (3.31*a*), motion in the *XY* plane does not tend to excite motion in the *XZ* or *YZ* plane if Σk_{xz}, $\Sigma(k_{xz}a_y - k_{xy}a_z)$, and $\Sigma(k_{xx}a_z - k_{xz}a_x)$ are zero.
3. The principal elastic axes *P, Q, R* of all resilient supporting elements are orthogonal with the reference axes *X, Y, Z* of the body, respectively. Then, in Eqs. (3.32),

$$\begin{aligned} k_{xx} = k_p = k_x \qquad k_{yy} = k_q = k_y \qquad k_{zz} = k_r = k_z \\ k_{xy} = k_{xz} = k_{yz} = 0 \end{aligned} \tag{3.34}$$

where k_x, k_y, k_z are defined for use when orthogonality exists. The supports are then called *orthogonal supports.*
4. The forces F_x, F_y, F_z and moments M_x, M_y, M_z are applied directly to the body and there are no motions ($u = v = w = \alpha = \beta = \gamma = 0$) of the foundation; or alternatively, the forces and moments are zero and excitation results from motion of the foundation.

In general, the effect of these simplifications is to reduce the numbers of terms in the equations and, in some instances, to reduce the number of equations that must be solved simultaneously. Simultaneous equations indicate coupled modes; i.e., motion cannot exist in one coupled mode independently of motion in other modes which are coupled to it.

MODAL COUPLING AND NATURAL FREQUENCIES

Several conditions of symmetry resulting from zero values for the product of inertia terms in Eq. (3.33) are discussed in the following sections.

ONE PLANE OF SYMMETRY WITH ORTHOGONAL RESILIENT SUPPORTS

When the *YZ* plane of the rigid body system in Fig. 3.12 is a plane of symmetry, the following terms in the equations of motion are zero:

$$\Sigma k_{yy} a_x = \Sigma k_{zz} a_x = \Sigma k_{yy} a_x a_z = \Sigma k_{zz} a_x a_y = 0 \tag{3.35}$$

Introducing the further simplification that the principal elastic axes of the resilient elements are parallel with the reference axes, Eqs. (3.34) apply. Then the motions in the three coordinates y_c, z_c, α are coupled but are independent of motion in any of the other coordinates; furthermore, the other three coordinates x_c, β, γ also are coupled. For example, Fig. 3.13 illustrates a resiliently supported rigid body, wherein the *YZ* plane is a plane of symmetry that meets the requirements of Eq. (3.35). The three natural frequencies for the y_c, z_c, α coupled directions are found by solving Eqs.

(3.31*b*), (3.31*c*), and (3.31*e*) [or Eqs. (3.31*a*), (3.31*d*), and (3.31*f*) for the x_c, β, γ coupled directions] simultaneously.[6]

$$\left(\frac{f_n}{f_z}\right)^6 - A\left(\frac{f_n}{f_z}\right)^4 + B\left(\frac{f_n}{f_z}\right)^2 - C = 0 \tag{3.36}$$

where

$$f_z = \frac{1}{2\pi}\sqrt{\frac{\Sigma k_z}{m}} \tag{3.37}$$

is a quantity having mathematical rather than physical significance if translational motion in the direction of the Z axis is coupled to other modes of motion. (Such coupling exists for the system of Fig. 3.13.) The roots f_n represent the natural frequencies of the system in the coupled modes. The coefficients A, B, C for the coupled modes in the y_c, z_c, α coordinates are

$$A_{yz\alpha} = 1 + \frac{\Sigma k_y}{\Sigma k_z} + D_{zx}$$

$$B_{yz\alpha} = D_{zx} + \frac{\Sigma k_y}{\Sigma k_z}(1 + D_{zx}) - \frac{(\Sigma k_y a_z)^2 + (\Sigma k_z a_y)^2}{\rho_x^2(\Sigma k_z)^2}$$

$$C_{yz\alpha} = \frac{\Sigma k_y}{\Sigma k_z}\left(D_{zx} - \frac{(\Sigma k_z a_y)^2}{\rho_x^2(\Sigma k_z)^2}\right) - \frac{(\Sigma k_y a_z)^2}{\rho_x^2(\Sigma k_z)^2}$$

where

$$D_{zx} = \frac{\Sigma k_y a_z^2 + \Sigma k_z a_y^2}{\rho_x^2 \Sigma k_z}$$

and ρ_x is the radius of gyration of the rigid body with respect to the X axis.

The corresponding coefficients for the coupled modes in the x_c, β, γ coordinates are

$$A_{x\beta\gamma} = \frac{\Sigma k_x}{\Sigma k_z} + D_{zy} + D_{zz}$$

$$B_{x\beta\gamma} = \frac{\Sigma k_x}{\Sigma k_z}(D_{zy} + D_{zz}) + D_{zy}D_{zz} - \frac{(\Sigma k_x a_z)^2}{\rho_y^2(\Sigma k_z)^2} - \frac{(\Sigma k_x a_y)^2}{\rho_z^2(\Sigma k_z)^2} - \frac{(\Sigma k_x a_y a_z)^2}{\rho_y^2\rho_z^2(\Sigma k_z)^2}$$

$$C_{x\beta\gamma} = \frac{\Sigma k_x}{\Sigma k_z}\left[D_{zy}D_{zz} - \frac{(\Sigma k_x a_y a_z)^2}{\rho_y^2\rho_z^2(\Sigma k_z)^2}\right] - \frac{(\Sigma k_x a_y)^2}{\rho_z^2(\Sigma k_z)^2}D_{zy} - \frac{(\Sigma k_x a_z)^2}{\rho_y^2(\Sigma k_z)^2}D_{zz} + 2\frac{(\Sigma k_x a_y)(\Sigma k_x a_z)(\Sigma k_x a_y a_z)}{\rho_y^2\rho_z^2(\Sigma k_z)^3}$$

where

$$D_{zy} = \frac{\Sigma k_x a_z^2 + \Sigma k_z a_x^2}{\rho_y^2 \Sigma k_z} \qquad D_{zz} = \frac{\Sigma k_x a_y^2 + \Sigma k_y a_x^2}{\rho_z^2 \Sigma k_z}$$

and ρ_y, ρ_z are the radii of gyration of the rigid body with respect to the Y, Z axes.

The roots of the cubic equation Eq. (3.36) may be found graphically from Fig. 3.14.[6] The coefficients A, B, C are first calculated from the above relations for the appropriate set of coupled coordinates. Figure 3.14 is entered on the abscissa scale at the appropriate value for the quotient B/A^2. Small values of B/A^2 are in Fig. 3.14*A*, and large values in Fig. 3.14*B*. The quotient C/A^3 is the parameter for the family of curves. Upon selecting the appropriate curve, three values of $(f_n/f_z)/\sqrt{A}$

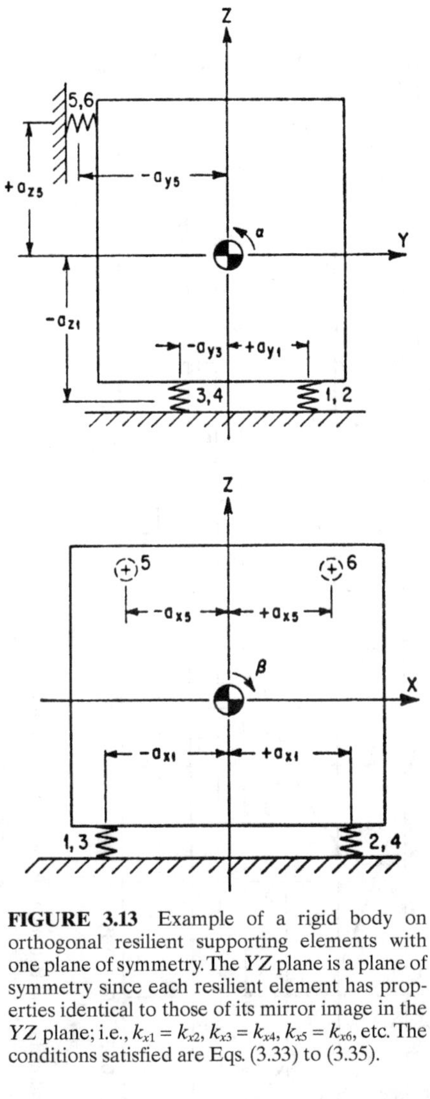

FIGURE 3.13 Example of a rigid body on orthogonal resilient supporting elements with one plane of symmetry. The YZ plane is a plane of symmetry since each resilient element has properties identical to those of its mirror image in the YZ plane; i.e., $k_{x1} = k_{x2}$, $k_{x3} = k_{x4}$, $k_{x5} = k_{x6}$, etc. The conditions satisfied are Eqs. (3.33) to (3.35).

are read from the ordinate and transferred to the left scale of the nomograph in Fig. 3.14*B*. Diagonal lines are drawn for each root to the value of A on the right scale, as indicated by dotted lines, and the roots f_n/f_z of the equation are indicated by the intercept of these dotted lines with the center scale of the nomograph.

The coefficients A, B, C can be simplified if all resilient elements have equal stiffness in the same direction. The stiffness coefficients always appear to be equal powers in numerator and denominator, and lead to dimensionless ratios of stiffness. For n resilient elements, typical terms reduce as follows:

$$\frac{\Sigma k_y}{\Sigma k_z} = \frac{k_y}{k_z} \qquad \frac{\Sigma k_z a_y^2}{\rho_x^2 \Sigma k_z} = \frac{\Sigma a_y^2}{n\rho_x^2}$$

$$\frac{(\Sigma k_x a_y a_z)^2}{\rho_y^2 \rho_z^2 (\Sigma k_z)^2} = \left(\frac{k_x}{nk_z} \frac{\Sigma a_y a_z}{\rho_y \rho_z} \right)^2, \text{ etc.}$$

TWO PLANES OF SYMMETRY WITH ORTHOGONAL RESILIENT SUPPORTS

Two planes of symmetry may be achieved if, in addition to the conditions of Eqs. (3.33) to (3.35), the following terms of Eqs. (3.31) are zero:

$$\Sigma k_{xx} a_y = \Sigma k_{zz} a_y = \Sigma k_{xx} a_y a_z = 0 \tag{3.38}$$

Under these conditions, Eqs. (3.31) separate into two independent equations, Eqs. (3.31*e*) and (3.31*f*), and two sets each consisting of two coupled equations [Eqs. (3.31*a*) and (3.31*d*); Eqs. (3.31*b*) and (3.31*c*)]. The planes of symmetry are the XZ and YZ planes. For example, a common system is illustrated in Fig. 3.15, where four identical resilient supporting elements are located symmetrically about the Z axis in a plane not containing the center of gravity.[6] Coupling exists between translation in the X direction and rotation about the Y axis (x_c, β), as well as between translation in the Y direction and rotation about the X axis (y_c, α). Translation in the Z direction (z_c) and rotation about the Z axis (γ) are each independent of all other modes.

The natural frequency in the Z direction is found by solving Eq. (3.31*e*) to obtain Eq. (3.37), where $\Sigma k_{zz} = 4k_z$. The rotational natural frequency f_γ about the Z axis is found by solving Eq. (3.31*f*); it can be expressed with respect to the natural frequency in the direction of the Z axis:

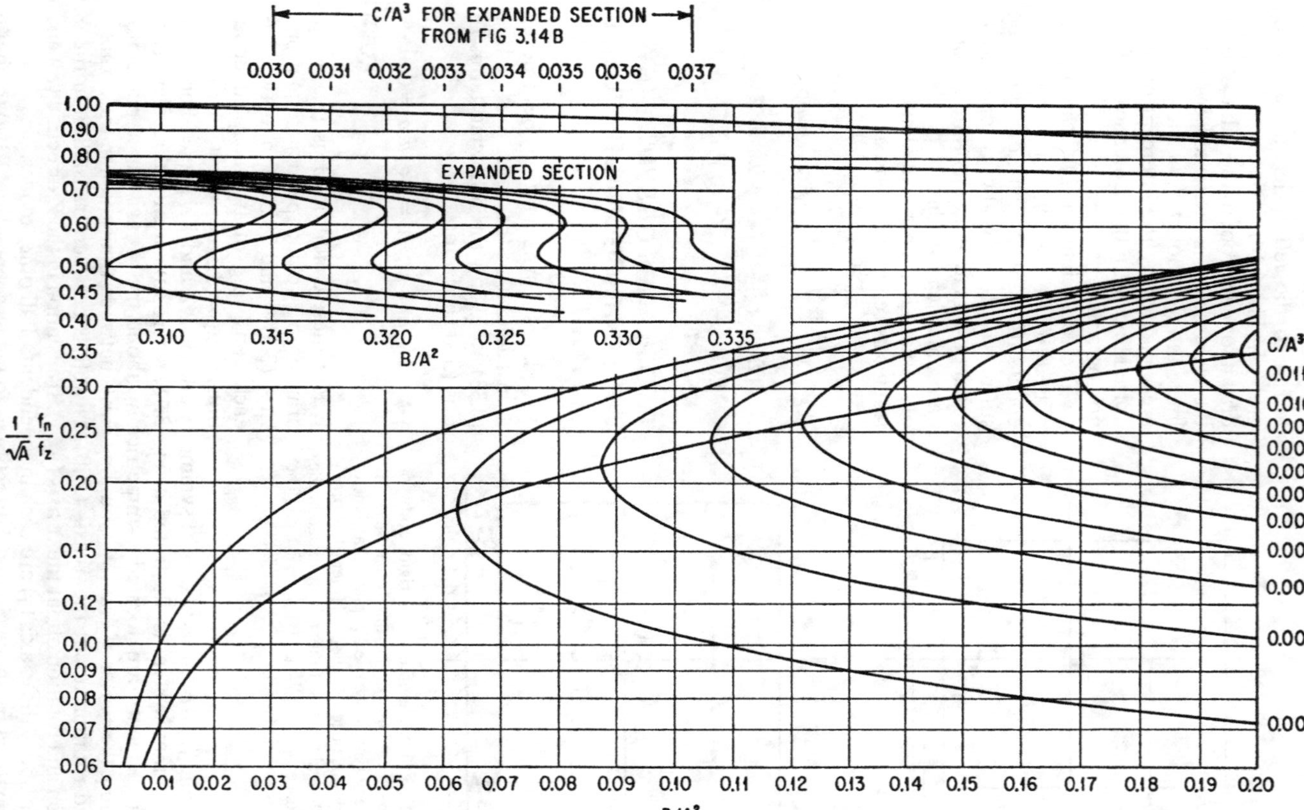

FIGURE 3.14A Graphical method of determining solutions of the cubic Eq. (3.36). Calculate *A, B, C* for the appropriate set of coupled coordinates, enter the abscissa at B/A^2 (values less than 0.2 on Fig. 3.14*A*, values greater than 0.2 on Fig. 3.14*B*), and read three values of $(f_n/f_z)/\sqrt{A}$ from the curve having the appropriate value of C/A^3.

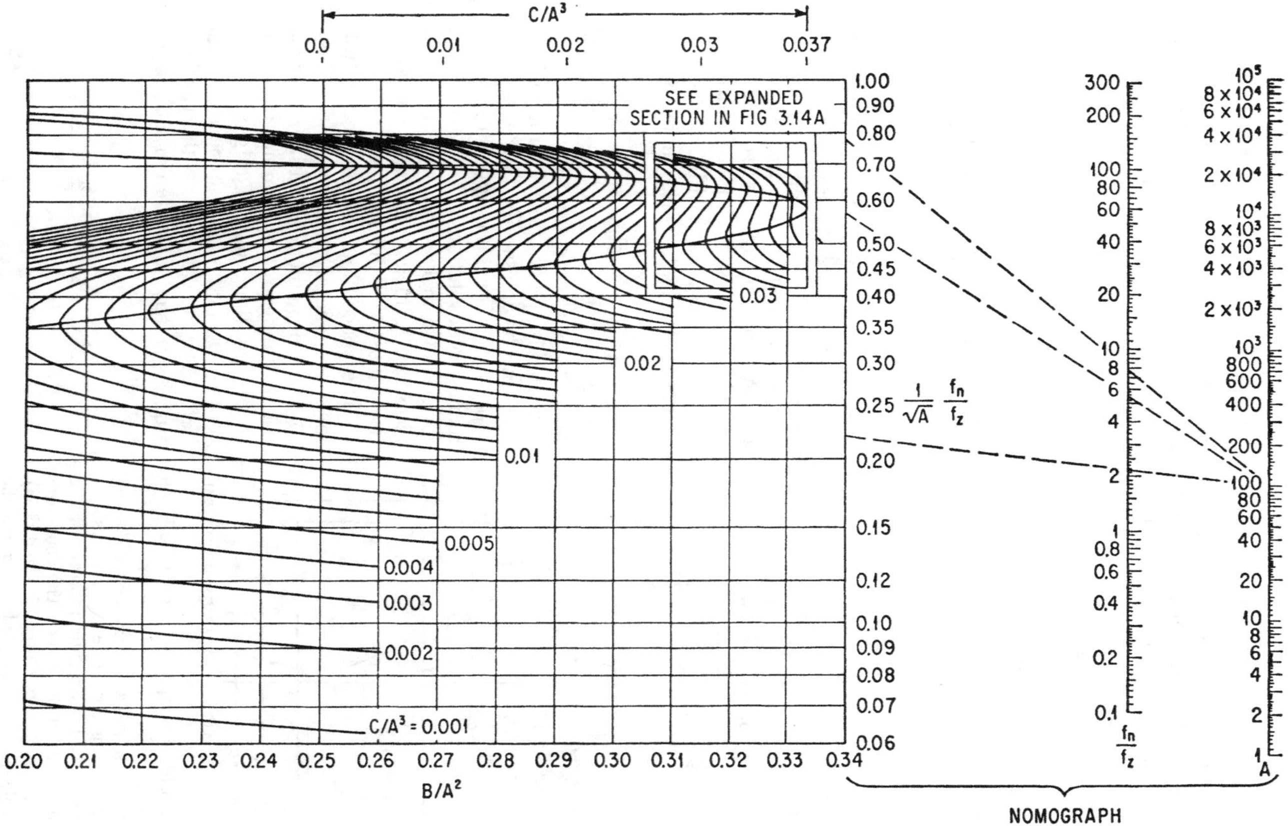

FIGURE 3.14*B* Using the above nomograph with values of $(f_n/f_z)/\sqrt{A}$ (see Fig. 3.14*A*), a diagonal line is drawn from each value of $(f_n/f_z)/\sqrt{A}$ on the left scale of the nomograph to the value of A on the right scale, as indicated by the dotted lines. The three roots f_n/f_z of Eq. (3.36) are given by the intercept of these dotted lines with the center scale of the nomograph. (*After F. F. Vane.*[6])

$$\frac{f_Y}{f_z} = \sqrt{\frac{k_x}{k_z}\left(\frac{a_y}{\rho_z}\right)^2 + \frac{k_y}{k_z}\left(\frac{a_x}{\rho_z}\right)^2} \tag{3.39}$$

where ρ_z is the radius of gyration with respect to the Z axis.

The natural frequencies in the coupled x_c, β modes are found by solving Eqs. (3.31a) and (3.31d) simultaneously; the roots yield the following expression for natural frequency:

$$\frac{f_{x\beta}^2}{f_z^2} = \frac{1}{2}\left\{\frac{k_x}{k_z}\left(1 + \frac{a_z^2}{\rho_y^2}\right) + \frac{a_x^2}{\rho_y^2} \pm \sqrt{\left[\frac{k_x}{k_z}\left(1 + \frac{a_z^2}{\rho_y^2}\right) + \frac{a_x^2}{\rho_y^2}\right]^2 - 4\,\frac{k_x}{k_z}\,\frac{a_x^2}{\rho_y^2}}\right\} \tag{3.40}$$

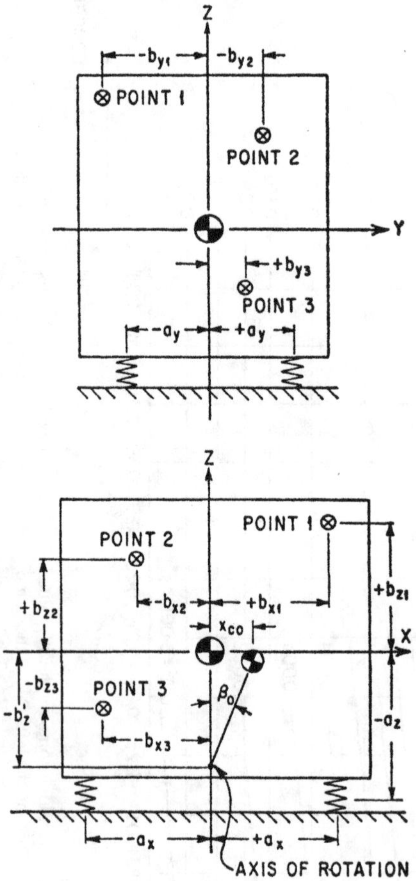

FIGURE 3.15 Example of a rigid body on orthogonal resilient supporting elements with two planes of symmetry. The XZ and YZ planes are planes of symmetry since the four resilient supporting elements are identical and are located symmetrically about the Z axis. The conditions satisfied are Eqs. (3.33), (3.34), (3.35), and (3.38). At any single frequency, coupled vibration in the x_c, β direction due to X vibration of the foundation is equivalent to a pure rotation of the rigid body with respect to an axis of rotation as shown. Points 1, 2, and 3 refer to the example of Fig. 3.26.

Figure 3.16 provides a convenient graphical method for determining the two coupled natural frequencies $f_{x\beta}$. An expression similar to Eq. (3.40) is obtained for $f_{y\alpha}^2/f_z^2$ by solving Eqs. (3.31b) and (3.31d) simultaneously. By replacing ρ_y, a_x, k_x, $f_{x\beta}$ with ρ_x, a_y, k_y, $f_{y\alpha}$, respectively, Fig. 3.16 also can be used to determine the two values of $f_{y\alpha}$.

It may be desirable to select resilient element locations a_x, a_y, a_z which will produce coupled natural frequencies in specified frequency ranges, with resilient elements having specified stiffness ratios k_x/k_z, k_y/k_z. For this purpose it is convenient to plot solutions of Eq. (3.40) in the form shown in Figs. 3.17 to 3.19. These plots are termed *space-plots* and their use is illustrated in Example 3.1.[8]

The space-plots are derived as follows: In general, the two roots of Eq. (3.40) are numerically different, one usually being greater than unity and the other less than unity. Designating the root associated with the positive sign before the radical (higher value) as f_h/f_z, Eq. (3.40) may be written in the following form:

$$\frac{(a_x/\rho_y)^2}{(f_h/f_z)^2} + \frac{(a_z/\rho_y)^2}{(k_z/k_x)(f_h/f_z)^2 - 1} = 1 \tag{3.40a}$$

Equation (3.40a) is shown graphically by the large ellipses about the center of Figs. 3.17 to 3.19, for stiffness ratios k_x/k_z

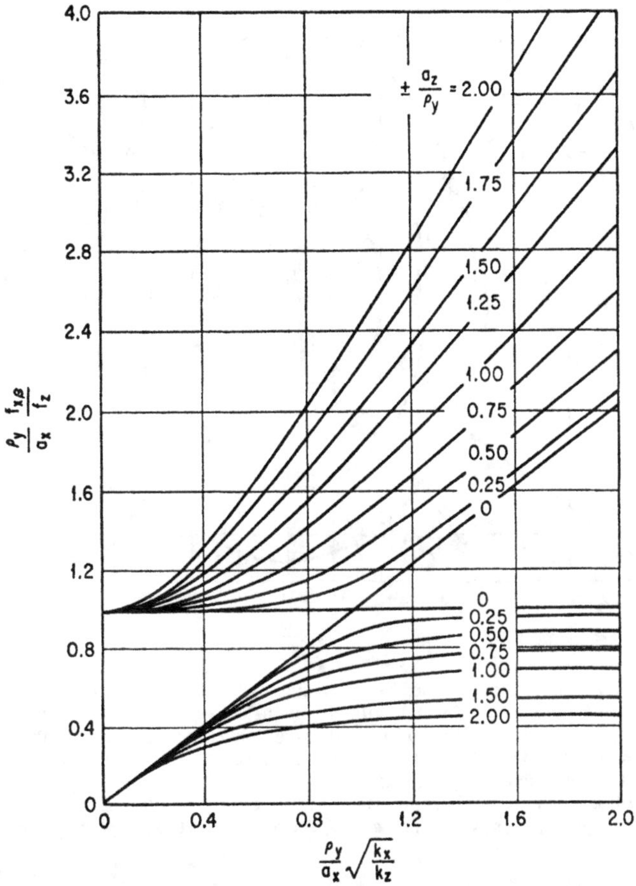

FIGURE 3.16 Curves showing the ratio of each of the two coupled natural frequencies $f_{x\beta}$ to the decoupled natural frequency f_z, for motion in the XZ plane of symmetry for the system in Fig. 3.15 [see Eq. (3.40)]. Calculate the abscissa $(\rho_y/a_x)\sqrt{k_x/k_z}$ and the parameter a_z/ρ_y, where a_x, a_z are indicated in Fig. 3.15; k_x, k_z are the stiffnesses of the resilient supporting elements in the X, Z directions, respectively; and ρ_y is the radius of gyration of the body about the Y axis. The two values read from the ordinate when divided by ρ_y/a_x give the natural frequency ratios $f_{x\beta}/f_z$.

of ½, 1, and 2, respectively. A particular type of resilient element tends to have a constant stiffness ratio k_x/k_z; thus, Figs. 3.17 to 3.19 may be used by cut-and-try methods to find the coordinates a_x, a_z of such elements to attain a desired value of f_h.

Designating the root of Eq. (3.40) associated with the negative sign (lower value) by f_l, Eq. (3.40) may be written as follows:

$$\frac{(a_x/\rho_y)^2}{(f_2/f_x)^2} - \frac{(a_z/\rho_y)^2}{1-(k_z/k_x)(f_l/f_z)^2} = 1 \tag{3.40b}$$

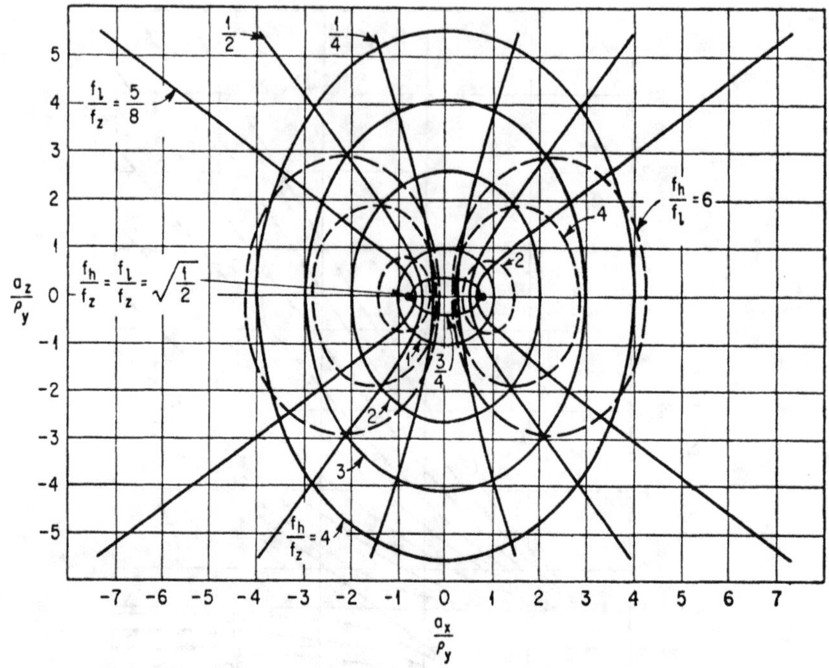

FIGURE 3.17 Space-plot for the system in Fig. 3.15 when the stiffness ratio $k_x/k_z = 0.5$, obtained from Eqs. (3.40*a*) to (3.40*c*). With all dimensions divided by the radius of gyration ρ_y about the *Y* axis, superimpose the outline of the rigid body in the *XZ* plane on the plot; the center of gravity of the body is located at the coordinate center of the plot. The elastic centers of the resilient supporting elements give the natural frequency ratios f_l/f_z, f_h/f_z, and f_h/f_l for x_c, β coupled motion, each ratio being read from one of the three families of curves as indicated on the plot. Replacing k_x, ρ_y, a_x with k_y, ρ_x, a_y, respectively, allows the plot to be applied to motions in the *YZ* plane.

Equation (3.40*b*) is shown graphically by the family of hyperbolas on each side of the center in Figs. 3.17 to 3.19, for values of the stiffness ratio k_x/k_z of ½, 1, and 2.

The two roots f_h/f_z and f_l/f_z of Eq. (3.40) may be expressed as the ratio of one to the other. This relationship is given parametrically as follows:

$$\left[\frac{2\dfrac{a_x}{\rho_y} \pm \sqrt{\dfrac{k_x}{k_z}}\left(\dfrac{f_h}{f_l} + \dfrac{f_l}{f_h}\right)}{\sqrt{\dfrac{k_x}{k_z}}\left(\dfrac{f_h}{f_l} - \dfrac{f_l}{f_h}\right)}\right]^2 + \left[\frac{2\dfrac{a_z}{\rho_y}}{\dfrac{f_h}{f_l} - \dfrac{f_l}{f_h}}\right]^2 = 1 \tag{3.40c}$$

Equation (3.40*c*) is shown graphically by the smaller ellipses (shown dotted) displaced from the vertical center line in Figs. 3.17 to 3.19.

Example 3.1. A rigid body is symmetrical with respect to the *XZ* plane; its width in the *X* direction is 13 in. and its height in the *Z* direction is 12 in. The center of gravity is 5½ in. from the lower side and 6¾ in. from the right side. The radius of gyration about the *Y* axis through the center of gravity is 5.10 in. Use a space-plot to evaluate the effects of the location for attachment of resilient supporting elements having the characteristic stiffness ratio $k_x/k_z = ½$.

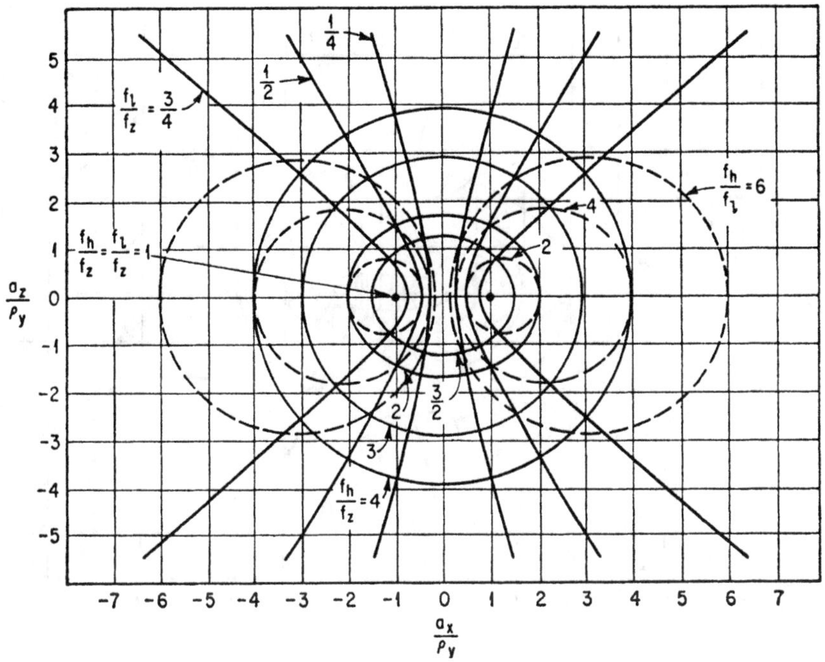

FIGURE 3.18 Space-plot for the system in Fig. 3.15 when the stiffness ratio $k_x/k_z = 1$. See caption for Fig. 3.17.

Superimpose the outline of the body on the space-plot of Fig. 3.20, with its center of gravity at the coordinate center of the plot. (Figure 3.20 is an enlargement of the central portion of Fig. 3.17.) All dimensions are divided by the radius of gyration ρ_y. Thus, the four corners of the body are located at coordinate distances as follows:

Upper right corner:

$$\frac{a_z}{\rho_y} = \frac{+6.50}{5.10} = +1.28 \qquad \frac{a_x}{\rho_y} = \frac{+6.75}{5.10} = +1.32$$

Upper left corner:

$$\frac{a_z}{\rho_y} = \frac{+6.50}{5.10} = +1.28 \qquad \frac{a_x}{\rho_y} = \frac{-6.25}{5.10} = -1.23$$

Lower right corner:

$$\frac{a_z}{\rho_y} = \frac{-5.50}{5.10} = -1.08 \qquad \frac{a_x}{\rho_y} = \frac{+6.75}{5.10} = +1.32$$

Lower left corner:

$$\frac{a_z}{\rho_y} = \frac{-5.50}{5.10} = -1.08 \qquad \frac{a_x}{\rho_y} = \frac{-6.25}{5.10} = -1.23$$

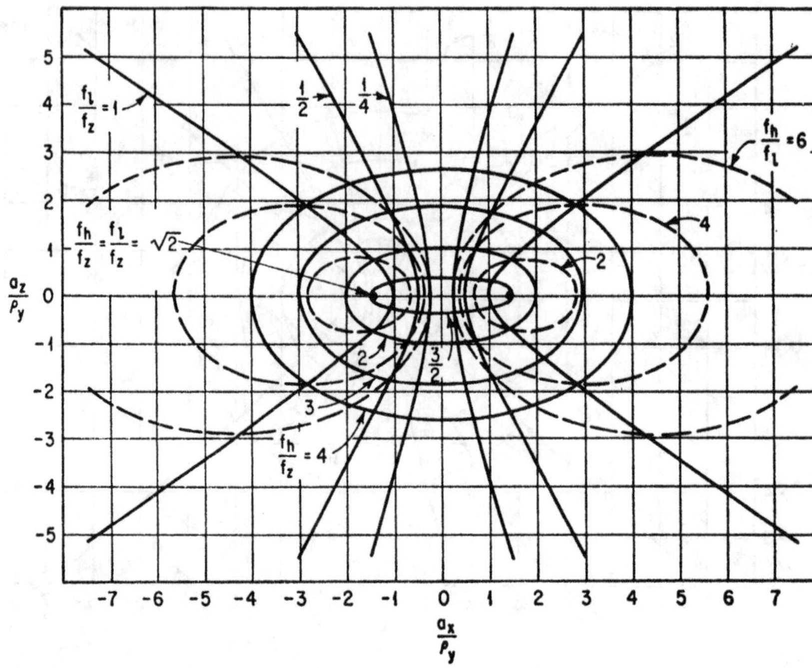

FIGURE 3.19 Space-plot for the system in Fig. 3.15 when the stiffness ratio $k_x/k_z = 2$. See caption for Fig. 3.17.

The resilient supports are shown in heavy outline at *A* in Fig. 3.20, with their elastic centers indicated by the solid dots. The horizontal coordinates of the resilient supports are $a_x/\rho_y = \pm 0.59$, or $a_x = \pm 0.59 \times 5.10 = \pm 3$ in. from the vertical coordinate axis. The corresponding natural frequencies are $f_h/f_z = 1.25$ (from the ellipses) and $f_l/f_z = 0.33$ (from the hyperbolas). An alternative position is indicated by the hollow circles *B*. The natural frequencies for this position are $f_h/f_z = 1.43$ and $f_l/f_z = 0.50$. The natural frequency f_z in vertical translation is found from the mass of the equipment and the summation of stiffnesses in the *Z* direction, using Eq. (3.37). This example shows how space-plots make it possible to determine the locations of the resilient elements required to achieve given values of the coupled natural frequencies with respect to f_z.

THREE PLANES OF SYMMETRY WITH ORTHOGONAL RESILIENT SUPPORTS

A system with three planes of symmetry is defined by six independent equations of motion. A system having this property is sometimes called a *center-of-gravity system.* The equations are derived from Eqs. (3.31) by substituting, in addition to the conditions of Eqs. (3.33), (3.34), (3.35), and (3.38), the following condition:

$$\Sigma k_{xx} a_z = \Sigma k_{yy} a_z = 0 \tag{3.41}$$

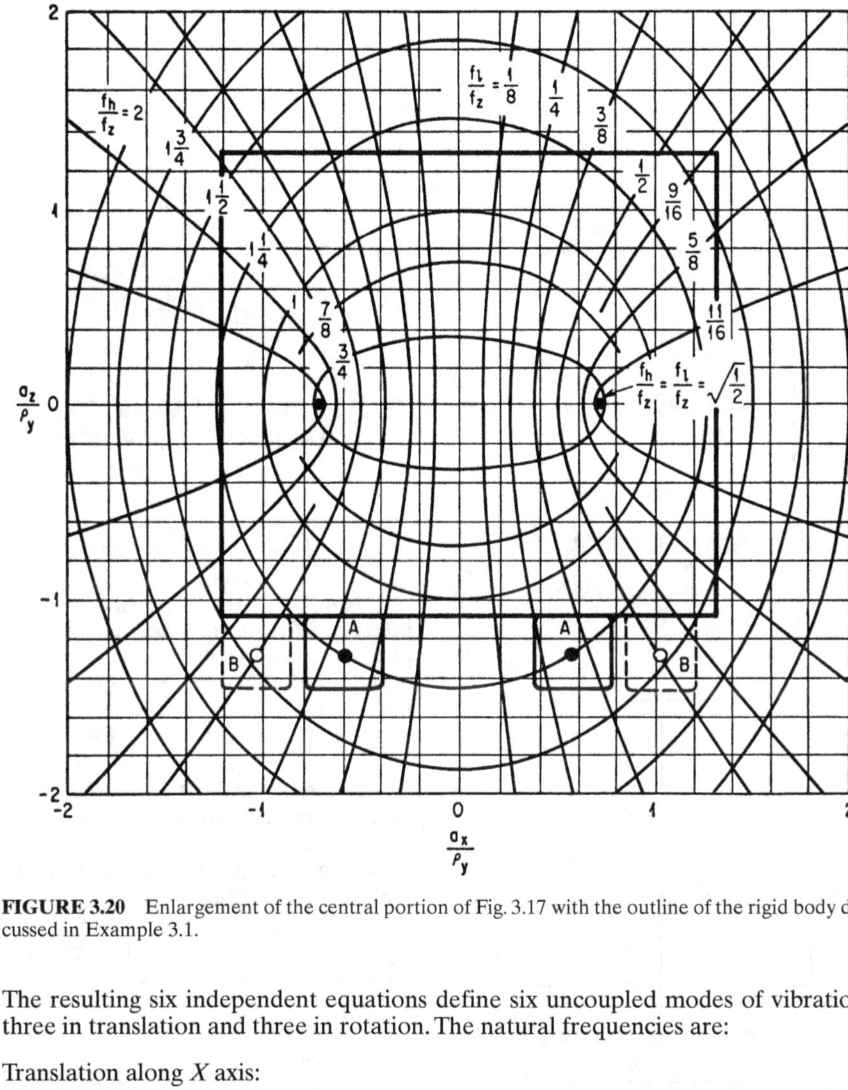

FIGURE 3.20 Enlargement of the central portion of Fig. 3.17 with the outline of the rigid body discussed in Example 3.1.

The resulting six independent equations define six uncoupled modes of vibration, three in translation and three in rotation. The natural frequencies are:

Translation along X axis:

$$f_x = \frac{1}{2\pi} \sqrt{\frac{\Sigma k_x}{m}}$$

Translation along Y axis:

$$f_y = \frac{1}{2\pi} \sqrt{\frac{\Sigma k_y}{m}}$$

Translation along Z axis:

$$f_z = \frac{1}{2\pi} \sqrt{\frac{\Sigma k_z}{m}}$$

Rotation about X axis:

$$f_\alpha = \frac{1}{2\pi}\sqrt{\frac{\Sigma(k_y a_z^2 + k_z a_y^2)}{m\rho_x^2}} \quad (3.42)$$

Rotation about Y axis:

$$f_\beta = \frac{1}{2\pi}\sqrt{\frac{\Sigma(k_x a_z^2 + k_z a_x^2)}{m\rho_y^2}}$$

Rotation about Z axis:

$$f_\gamma = \frac{1}{2\pi}\sqrt{\frac{\Sigma(k_x a_y^2 + k_y a_x^2)}{m\rho_z^2}}$$

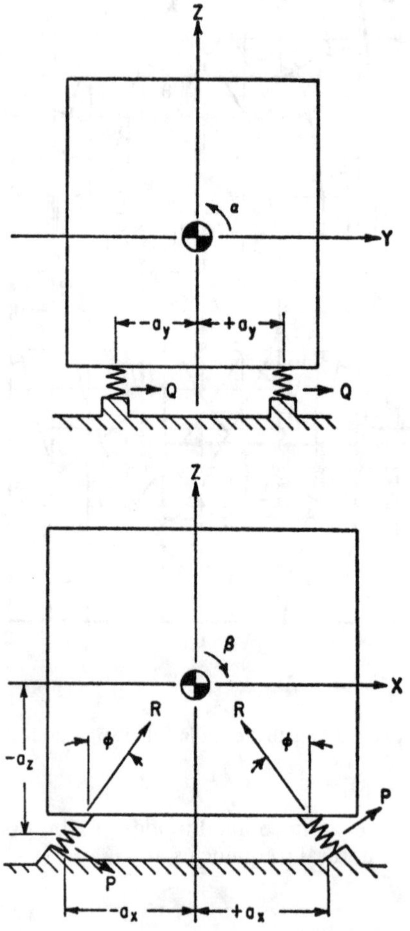

FIGURE 3.21 Example of a rigid body on resilient supporting elements inclined toward the YZ plane. The resilient supporting elements are identical and are located symmetrically about the Z axis, making XZ and YZ planes of symmetry. The principal stiffnesses in the XZ plane are k_p and k_r. The conditions satisfied are Eqs. (3.33), (3.35), and (3.38).

TWO PLANES OF SYMMETRY WITH RESILIENT SUPPORTS INCLINED IN ONE PLANE ONLY

When the principal elastic axes of the resilient supporting elements are inclined with respect to the X, Y, Z axes, the stiffness coefficients k_{xy}, k_{xz}, k_{yz} are nonzero. This introduces elastic coupling, which must be considered in evaluating the equations of motion. Two planes of symmetry may be achieved by meeting the conditions of Eqs. (3.33), (3.35), and (3.38). For example, consider the rigid body supported by four identical resilient supporting elements located symmetrically about the Z axis, as shown in Fig. 3.21. The XZ and the YZ planes are planes of symmetry, and the resilient elements are inclined toward the YZ plane so that one of their principal elastic axes R is inclined at the angle ϕ with the Z direction as shown; hence $k_{yy} = k_q$, and $k_{xy} = k_{yz} = 0$.

Because of symmetry, translational motion z_c in the Z direction and rotation γ about the Z axis are each decoupled from the other modes. The pairs of translational and rotational modes in the x_c, β and y_c, α coordinates are coupled. The natural frequency in the Z direction is

$$\frac{f_z}{f_r} = \sqrt{\frac{k_p}{k_r}\sin^2\phi + \cos^2\phi} \quad (3.43)$$

where f_r is a fictitious natural frequency used for convenience only; it is related to Eq. (3.37) wherein $4k_r$ is substituted for Σk_z:

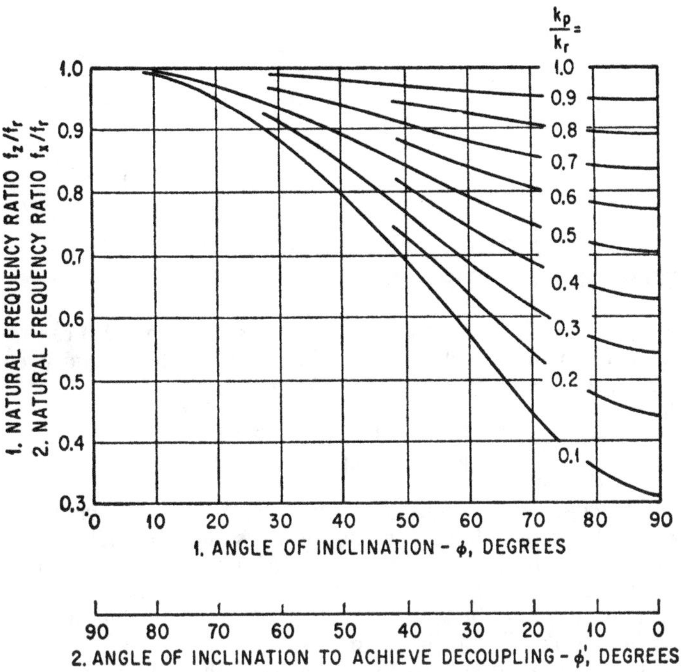

FIGURE 3.22 Curves showing the ratio of the decoupled natural frequency f_z of translation z_c to the fictitious natural frequency f_r for the system shown in Fig. 3.21 [see Eq. (3.43)] when the resilient supporting elements are inclined at the angle ϕ. The curves also indicate the ratio of the decoupled natural frequency f_x of translation x_c to f_r when ϕ has a value ϕ' (use lower abscissa scale) which decouples x_c, β motions [see Eqs. (3.47) and (3.48)].

$$f_r = \frac{1}{2\pi}\sqrt{\frac{4k_r}{m}}$$

Equation (3.43) is plotted in Fig. 3.22, where the angle ϕ is indicated by the upper of the abscissa scales.

The rotational natural frequency about the Z axis is obtained from

$$\frac{f_\gamma}{f_r} = \sqrt{\left(\frac{k_p}{k_r}\cos^2\phi + \sin^2\phi\right)\left(\frac{a_y}{\rho_z}\right)^2 + \frac{k_q}{k_r}\left(\frac{a_x}{\rho_z}\right)^2} \tag{3.44}$$

For the x_c, β coupled mode, the two natural frequencies are

$$\frac{f_{x\beta}}{f_r} = \frac{1}{2}\left[A \pm \sqrt{A^2 - 4\frac{k_p}{k_r}\left(\frac{a_x}{a_y}\right)^2}\right] \tag{3.45}$$

where $$A = \left(\frac{k_p}{k_r}\cos^2\phi + \sin^2\phi\right)\left[1 + \left(\frac{a_z}{\rho_y}\right)^2\right] + \left(\frac{k_p}{k_r}\sin^2\phi + \cos^2\phi\right)\left(\frac{a_x}{\rho_y}\right)^2 + 2\left(1 - \frac{k_p}{k_r}\right)\left|\frac{a_x}{\rho_y}\right|\sin\phi\cos\phi$$

For the y_c, α coupled mode, the natural frequencies are

$$\frac{f_{y\alpha}}{f_r} = \frac{1}{2}\left[B \pm \sqrt{B^2 - 4\,\frac{k_q}{k_r}\left(\frac{k_p}{k_r}\sin^2\phi + \cos^2\phi\right)\left(\frac{a_y}{\rho_x}\right)^2}\right] \tag{3.46}$$

where
$$B = \frac{k_q}{k_r}\left[1 + \left(\frac{a_z}{\rho_x}\right)^2\right] + \left(\frac{k_p}{k_r}\sin^2\phi + \cos^2\phi\right)\left(\frac{a_y}{\rho_x}\right)^2$$

DECOUPLING OF MODES IN A PLANE USING INCLINED RESILIENT SUPPORTS

The angle ϕ of inclination of principal elastic axes (see Fig. 3.21) can be varied to produce changes in the amount of coupling between the x_c and β coordinates. Decoupling of the x_c and β coordinates is effected if

$$\left|\frac{a_z}{a_x}\right| = \frac{[1 - (k_p/k_r)]\cot\phi'}{1 + (k_p/k_r)\cot^2\phi'} \tag{3.47}$$

where ϕ' is the value of the angle of inclination ϕ required to achieve decoupling. When Eq. (3.47) is satisfied, the configuration is sometimes called an "equivalent center-of-gravity system" in the *YZ* plane since all modes of motion in that plane are decoupled. Figure 3.23 is a graphical presentation of Eq. (3.47). There may be two values of ϕ' that decouple the x_c and β modes for any combination of stiffness and location for the resilient supporting elements.

The decoupled natural frequency for translation in the *X* direction is obtained from

$$\frac{f_x}{f_r} = \sqrt{\frac{k_p}{k_r}\cos^2\phi' + \sin^2\phi'} \tag{3.48}$$

The relation of Eq. (3.48) is shown graphically in Fig. 3.22 where the angle ϕ' is indicated by the lower of the abscissa scales. The natural frequency in the β mode is obtained from

$$\frac{f_\beta}{f_r} = \frac{a_x}{\rho_y}\sqrt{\frac{1}{(k_r/k_p)\sin^2\phi' + \cos^2\phi'}} \tag{3.49}$$

COMPLETE DECOUPLING OF MODES USING RADIALLY INCLINED RESILIENT SUPPORTS

In general, the analysis of rigid-body motion with the resilient supporting elements inclined in more than one plane is quite involved. A particular case where sufficient symmetry exists to provide relatively simple yet useful results is the configuration illustrated in Fig. 3.24. From symmetry about the *Z* axis, $I_{xx} = I_{yy}$. Any number n of resilient supporting elements greater than 3 may be used. For clarity of illustration, the rigid body is shown as a right circular cylinder with $n = 3$.

The resilient supporting elements are arranged symmetrically about the *Z* axis; they are attached to one end face of the cylinder at a distance a_r from the *Z* axis and a distance a_z from the *XY* reference plane. The resilient elements are inclined so that their principal elastic axes *R* intersect at a common point on the *Z* axis; thus, the angle between the *Z* axis and the *R* axis for each element is ϕ. The principal elastic axes *P*

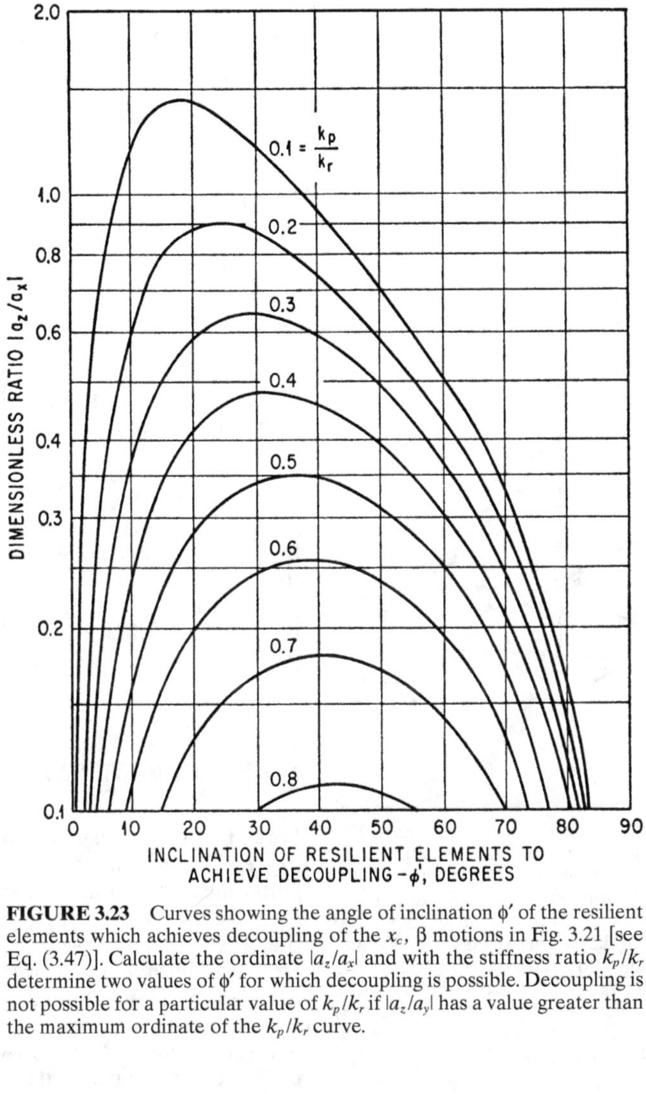

FIGURE 3.23 Curves showing the angle of inclination ϕ' of the resilient elements which achieves decoupling of the x_c, β motions in Fig. 3.21 [see Eq. (3.47)]. Calculate the ordinate $|a_z/a_x|$ and with the stiffness ratio k_p/k_r determine two values of ϕ' for which decoupling is possible. Decoupling is not possible for a particular value of k_p/k_r if $|a_z/a_y|$ has a value greater than the maximum ordinate of the k_p/k_r curve.

also intersect at a common point on the Z axis, the angle between the Z axis and the P axis for each element being $90° - \phi$. Consequently, the Q principal elastic axes are each tangent to the circle of radius a_r which bounds the end face of the cylinder.

The use of such a configuration permits decoupling of all six modes of vibration of the rigid body. This complete decoupling is achieved if the angle of inclination ϕ has the value ϕ' which satisfies the following equation:

$$\left|\frac{a_z}{a_r}\right| = \frac{(\frac{1}{2})[1 - (k_p/k_r)] \sin 2\phi'}{(k_q/k_r) + (k_p/k_r) + [1 - (k_p/k_r)] \sin^2\phi'} \tag{3.50}$$

Since complete decoupling is effected, the system may be termed an "equivalent center-of-gravity system."[9,10] The natural frequencies of the six decoupled modes are

$$\frac{f_x}{f_r} = \frac{f_y}{f_r} = \sqrt{\frac{1}{2}\left(\frac{k_p}{k_r}\cos^2\phi' + \sin^2\phi' + \frac{k_q}{k_r}\right)} \tag{3.51}$$

$$\frac{f_\alpha}{f_r} = \frac{f_\beta}{f_r} = \left\{\frac{a_r}{2\rho_x}\left[\frac{k_p}{k_r}\sin\phi'\left(\frac{a_r}{\rho_x}\sin\phi' + \frac{a_z}{\rho_x}\cos\phi'\right) + \cos\phi'\left(\frac{a_r}{\rho_x}\cos\phi' - \frac{a_z}{\rho_x}\sin\phi'\right)\right]\right\}^{1/2} \tag{3.52}$$

$$\frac{f_\gamma}{f_r} = \sqrt{\frac{k_q}{k_r}\frac{a_r}{\rho_z}} \tag{3.53}$$

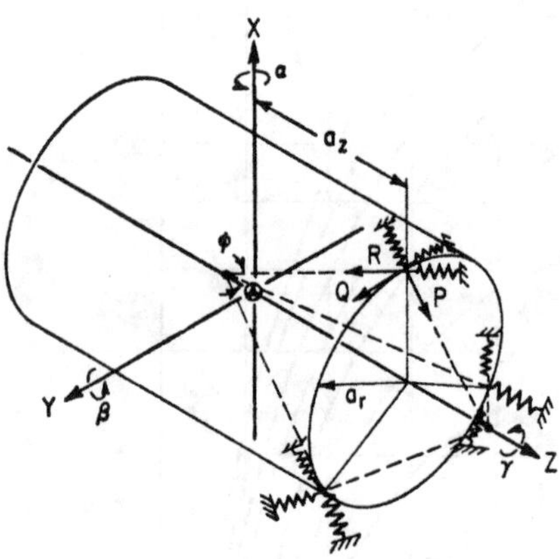

FIGURE 3.24 Example of a rigid cylindrical body on radially inclined resilient supports. The resilient supports are attached symmetrically about the Z axis to one end face of the cylinder at a distance a_r from the Z axis and a distance a_z from the XY plane. The resilient elements are inclined so that their principal elastic axes R and P intersect the Z axis at common points. The angle between the R axes and the Z axis is ϕ; and the angle between the P axis and Z axis is $90° - \phi$. The Q principal elastic axes are each tangent to the circle of radius a_r.

The frequency ratio f_z/f_r is given by Eq. (3.43) or Fig. 3.22. The fictitious natural frequency f_r is given by

$$f_r = (1/2\pi)\sqrt{nk_r/m}$$

Similar solutions are also available for the configuration of four resilient supports located in a rectangular array and inclined to achieve complete decoupling.[11]

FORCED VIBRATION

Forced vibration results from a continuing excitation that varies sinusoidally with time. The excitation may be a vibratory displacement of the foundation for the resiliently supported rigid body (*foundation-induced vibration*), or a force or moment applied to or generated within the rigid body (*body-induced vibration*). These two forms of excitation are considered separately.

FOUNDATION-INDUCED SINUSOIDAL VIBRATION

This section includes an analysis of foundation-induced vibration for two different systems, each having two planes of symmetry. In one system, the principal elastic axes of the resilient elements are parallel to the X, Y, Z axes; in the other system, the principal elastic axes are inclined with respect to two of the axes but in a plane parallel to one of the reference planes. The excitation is translational movement of the foundation in its own plane, without rotation. No forces or moments are applied directly to the rigid body; that is, in the equations of motion [Eqs. (3.31)], the following terms are equal to zero:

$$F_x = F_y = F_z = M_x = M_y = M_z = \alpha = \beta = \gamma = 0 \tag{3.54}$$

Two Planes of Symmetry with Orthogonal Resilient Supports. The system is shown in Fig. 3.15. The excitation is a motion of the foundation in the direction of the X axis defined by $u = u_0 \sin \omega t$. (Alternatively, the excitation may be the displacement $v = v_0 \sin \omega t$ in the direction of the Y axis, and analogous results are obtained.) The resulting motion of the resiliently supported rigid body involves translation x_c and rotation β simultaneously. The conditions of symmetry are defined by Eqs. (3.33), (3.34), (3.35), and (3.38); these conditions decouple Eqs. (3.31) so that only Eqs. (3.31*a*) and (3.31*d*), and Eqs. (3.31*b*) and (3.31*c*), remain coupled. Upon substituting $u = u_0 \sin \omega t$ as the excitation, the response in the coupled modes is of a form $x_c = x_{c0} \sin \omega t$, $\beta = \beta_0 \sin \omega t$ where x_{c0} and β_0 are related to u_0 as follows:

$$\frac{x_{c0}}{u_0} = \frac{\dfrac{k_x}{k_z}\left[\left(\dfrac{a_x}{\rho_y}\right)^2 - \left(\dfrac{f}{f_z}\right)^2\right]}{\left(\dfrac{f}{f_z}\right)^4 - \left[\dfrac{k_x}{k_z} + \dfrac{k_x}{k_z}\left(\dfrac{a_z}{\rho_y}\right)^2 + \left(\dfrac{a_x}{\rho_y}\right)^2\right]\left(\dfrac{f}{f_z}\right)^2 + \dfrac{k_x}{k_z}\left(\dfrac{a_x}{\rho_y}\right)^2} \tag{3.55}$$

$$\frac{\beta_0}{u_0/\rho_y} = \frac{-\dfrac{k_x}{k_z}\dfrac{a_z}{\rho_y}\left(\dfrac{f}{f_z}\right)^2}{\left(\dfrac{f}{f_z}\right)^4 - \left[\dfrac{k_x}{k_z} + \dfrac{k_x}{k_z}\left(\dfrac{a_z}{\rho_y}\right)^2 + \left(\dfrac{a_x}{\rho_y}\right)^2\right]\left(\dfrac{f}{f_z}\right)^2 + \dfrac{k_x}{k_z}\left(\dfrac{a_x}{\rho_y}\right)^2} \tag{3.56}$$

where $f_z = \dfrac{1}{2\pi}\sqrt{4k_z/m}$ in accordance with Eq. (3.37). A similar set of equations applies for vibration in the coupled y_c, α coordinates. There is no response of the system in the z_c or γ modes since there is no net excitation in these directions; that is, $\mathbf{F}_z$ and $\mathbf{M}_z$ are zero.

As indicated by Eqs. (3.1), the displacement at any point in a rigid body is the sum of the displacement at the center of gravity and the displacements resulting from motion of the body in rotation about axes through the center of gravity. Equations (3.55) and (3.56) together with analogous equations for y_{c0}, α_0 provide the basis for calculating these displacements. Care should be taken with phase angles, particularly if two or more excitations u, v, w exist concurrently.

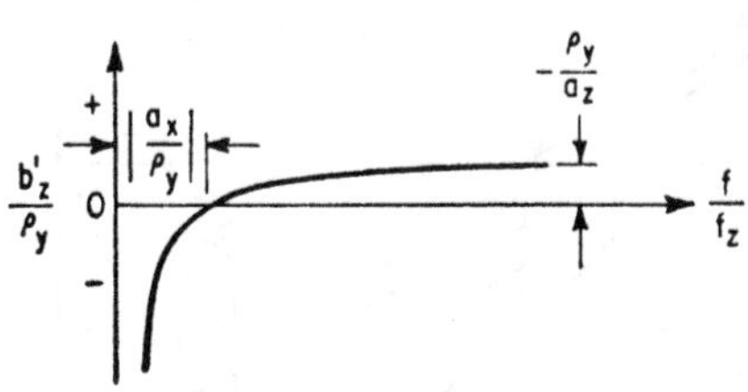

FIGURE 3.25 Curve showing the position of the axis of pure rotation of the rigid body in Fig. 3.15 as a function of the frequency ratio f/f_z when the excitation is sinusoidal motion of the foundation in the X direction [see Eq. (3.57)]. The axis of rotation is parallel to the Y axis and in the XZ plane, and its coordinate along the Z axis is designated by b_z'.

At any single frequency, coupled vibration in the x_c, β modes is equivalent to a pure rotation of the rigid body with respect to an axis parallel to the Y axis, in the YZ plane and displaced from the center of gravity of the body (see Fig. 3.15). As a result, the rigid body has zero displacement x in the horizontal plane containing this axis. Therefore, the Z coordinate of this axis b_z' satisfies $x_{c0} + b_z'\beta_0 = 0$, which is obtained from the first of Eqs. (3.1) by setting $x_b = 0$ (γ_0 motion is not considered). Substituting Eqs. (3.55) and (3.56) for x_{c0} and β_0, respectively, the axis of rotation is located at

$$\frac{b_z'}{\rho_y} = \frac{(a_x/\rho_y)^2 - (f/f_z)^2}{(a_z/\rho_y)(f/f_z)^2} \tag{3.57}$$

Figure 3.25 shows the relation of Eq. (3.57) graphically. At high values of frequency f/f_z, the axis does not change position significantly with frequency; b_z'/ρ_y approaches a positive value as f/f_z becomes large, since a_z is negative (see Fig. 3.15).

When the resilient supporting elements have damping as well as elastic properties, the solution of the equations of motion [see Eq. (3.31a)] becomes too laborious for general use. Responses of systems with damping have been obtained for several typical cases using a digital computer. Figures 3.26 *A*, *B*, and *C* show the response at three points in the body of the system shown in Fig. 3.15, with the excitation $u = u_0 \sin \omega t$. The weight of the body is 45 lb; each of the four resilient supporting elements has a stiffness k_z = 1,050 lb/in. and stiffness ratios $k_x/k_z = k_y/k_z = \frac{1}{2}$. The critical damping coefficients in the *X, Y, Z* directions are taken as $c_{cx} = 2\sqrt{4k_xm}$, $c_{cy} = 2\sqrt{4k_ym}$, $c_{cz} = 2\sqrt{4k_zm}$, respectively, where the expression for c_{cz} follows from the single-degree-of-freedom case defined by Eq. (2.12). The fractions of critical damping are c_x/c_{cx} =

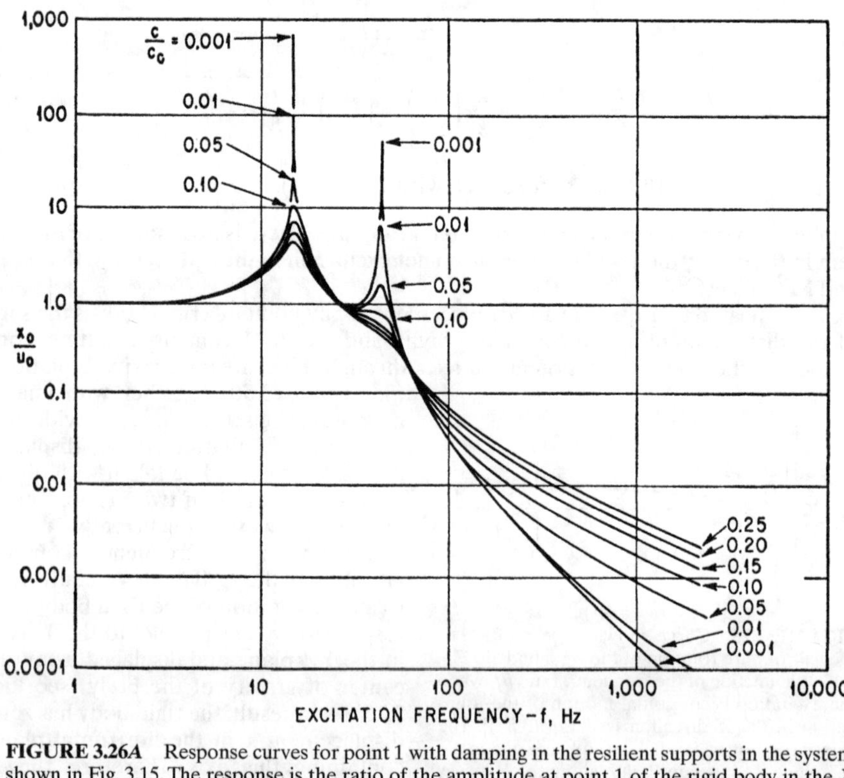

FIGURE 3.26A Response curves for point 1 with damping in the resilient supports in the system shown in Fig. 3.15. The response is the ratio of the amplitude at point 1 of the rigid body in the *X* direction to the amplitude of the foundation in the *X* direction (x_0/u_0). The fraction of critical damping c/c_c is the same in the *X, Y, Z* directions.

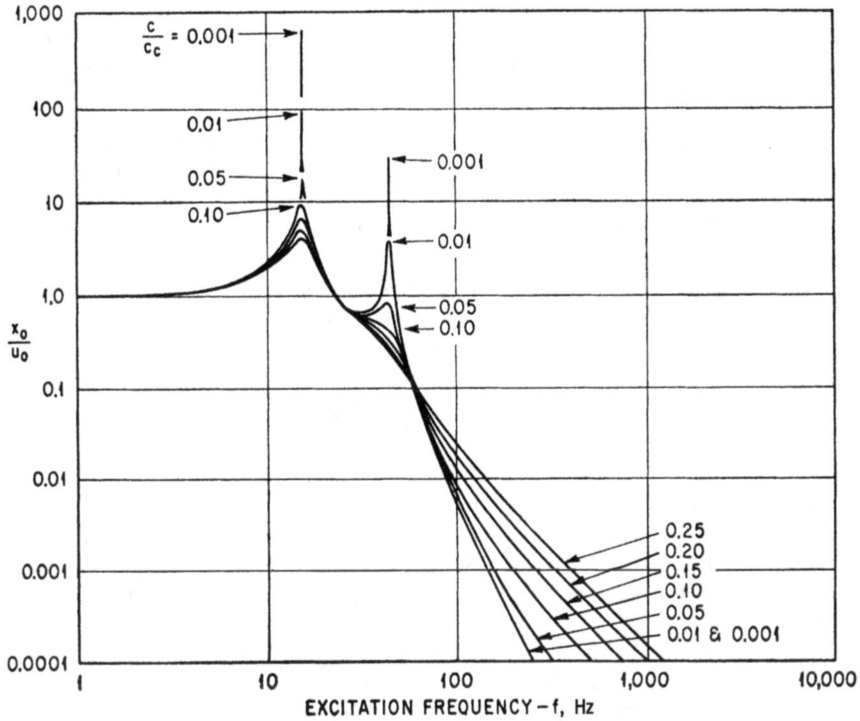

FIGURE 3.26*B* Response curves at point 2 in the system shown in Fig. 3.15. See caption for Fig. 3.26*A*.

$c_y/c_{cy} = c_z/c_{cz} = c/c_c$, the parameter of the curves in Figs. 3.26*A*, *B*, and *C*. Coordinates locating the resilient elements are $a_x = \pm 5.25$ in., $a_y = \pm 3.50$ in., and $a_z = -6.50$ in. The radii of gyration with respect to the *X, Y, Z* axes are $\rho_x = 4.40$ in., $\rho_y = 5.10$ in., and $\rho_z = 4.60$ in.

Natural frequencies calculated from Eqs. (3.37) and (3.40) are $f_z = 30.0$ Hz; $f_{x\beta} = 43.7$ Hz, 15.0 Hz; and $f_{y\alpha} = 43.2$ Hz, 11.7 Hz. The fraction of critical damping c/c_c varies between 0 and 0.25. Certain characteristic features of the response curves in Figs. 3.26*A*, *B*, and *C* are:

1. The relatively small response at the frequency of 24.2 Hz in Fig. 3.26*C* occurs because point 3 lies near the axis of rotation of the rigid body at that frequency. Point 2 lies near the axis of rotation at higher frequencies, and the response becomes correspondingly low, as shown in Fig. 3.26*B*. The position of the axis of rotation changes rapidly for small changes of frequency in the low- and intermediate-frequency range (indicated by the sharp dip in the curves for small damping in Fig. 3.26*C*) and varies asymptotically toward a final position as the forcing frequency increases (see Fig. 3.25).

2. The effect of damping on the magnitude of the response at the higher and lower natural frequencies in coupled modes is illustrated. When the fraction of critical damping is between 0.01 and 0.10, the response at the lower of the coupled natural frequencies is approximately 10 times as great as the response at the higher of the coupled natural frequencies. With greater damping ($c/c_c \geq 0.15$), the effect of resonance in the vicinity of the higher coupled natural frequency becomes so slight as to be hardly discernible.

Two Planes of Symmetry with Resilient Supports Inclined in One Plane Only. The system is shown in Fig. 3.21, and the excitation is $u = u_0 \sin \omega t$. The conditions of symmetry are defined by Eqs. (3.33), (3.35), and (3.38). The response is entirely in the x_c, β coupled mode with the following amplitudes:

$$\frac{x_{c0}}{u_0} = \frac{\dfrac{k_p}{k_r}\left(\dfrac{a_x}{\rho_y}\right)^2 - \left(\dfrac{k_p}{k_r}\cos^2\phi + \sin^2\phi\right)\left(\dfrac{f}{f_r}\right)^2}{\left(\dfrac{f}{f_r}\right)^4 - A\left(\dfrac{f}{f_r}\right)^2 + \dfrac{k_p}{k_r}\left(\dfrac{a_x}{\rho_y}\right)^2}$$

$$\frac{\beta_0}{u_0/\rho_y} = \frac{-\left[\left(\dfrac{k_p}{k_r}\cos^2\phi + \sin^2\phi\right)\left(\dfrac{a_z}{\rho_y}\right) + \left(1 - \dfrac{k_p}{k_r}\right)\left|\dfrac{a_x}{\rho_y}\right|\sin\phi\cos\phi\right]\left(\dfrac{f}{f_r}\right)^2}{\left(\dfrac{f}{f_r}\right)^4 - A\left(\dfrac{f}{f_r}\right)^2 + \dfrac{k_p}{k_r}\left(\dfrac{a_x}{\rho_y}\right)^2} \tag{3.58}$$

where A is defined after Eq. (3.45). A similar set of expressions may be written for the response in the y_c, α coupled mode when the excitation is the motion $v = v_0 \sin \omega t$ of the foundation:

$$\frac{y_{c0}}{v_0} = \frac{\dfrac{k_q}{k_r}\left(\dfrac{k_p}{k_r}\sin^2\phi + \cos^2\phi\right)\left(\dfrac{a_y}{\rho_x}\right)^2 - \dfrac{k_q}{k_r}\left(\dfrac{f}{f_r}\right)^2}{\left(\dfrac{f}{f_r}\right)^4 - B\left(\dfrac{f}{f_r}\right)^2 + \dfrac{k_q}{k_r}\left(\dfrac{k_p}{k_r}\sin^2\phi + \cos^2\phi\right)\left(\dfrac{a_y}{\rho_x}\right)}$$

$$\frac{\alpha_0}{v_0/\rho_x} = \frac{\dfrac{k_q}{k_r}\dfrac{a_z}{\rho_x}\left(\dfrac{f}{f_r}\right)^2}{\left(\dfrac{f}{f_r}\right)^4 - B\left(\dfrac{f}{f_r}\right)^2 + \dfrac{k_q}{k_r}\left(\dfrac{k_p}{k_r}\sin^2\phi + \cos^2\phi\right)\left(\dfrac{a_y}{\rho_x}\right)} \tag{3.59}$$

where B is defined after Eq. (3.46). No motion occurs in the z_c or γ mode since the quantities F_z and M_z are zero in Eqs. (3.31*e*) and (3.31*f*).

Response curves for the system shown in Fig. 3.21 when damping is included are qualitatively similar to those shown in Figs. 3.26. The significant advantage in the use of inclined resilient supports is the additional versatility gained from the ability to vary the angle of inclination ϕ, which directly affects the degree of coupling in the x_c, β coupled mode. For example, a change in the angle ϕ produces a change in the position of the axis of pure rotation of the rigid body. In a manner similar to that used to derive Eq. (3.57), Eqs. (3.58) yield the following expression defining the location of the axis of rotation:

$$\frac{b_z'}{\rho_y} = \frac{\dfrac{k_p}{k_r}\left(\dfrac{a_x}{\rho_y}\right)^2 - \left(\dfrac{k_p}{k_r}\cos^2\phi + \sin^2\phi\right)\left(\dfrac{f}{f_r}\right)^2}{\left[\left(\dfrac{k_p}{k_r}\cos^2\phi + \sin^2\phi\right)\dfrac{a_z}{\rho_y} + \left(1 - \dfrac{k_p}{k_r}\right)\left|\dfrac{a_x}{\rho_y}\right|\sin\phi\cos\phi\right]\left(\dfrac{f}{f_r}\right)^2} \tag{3.60}$$

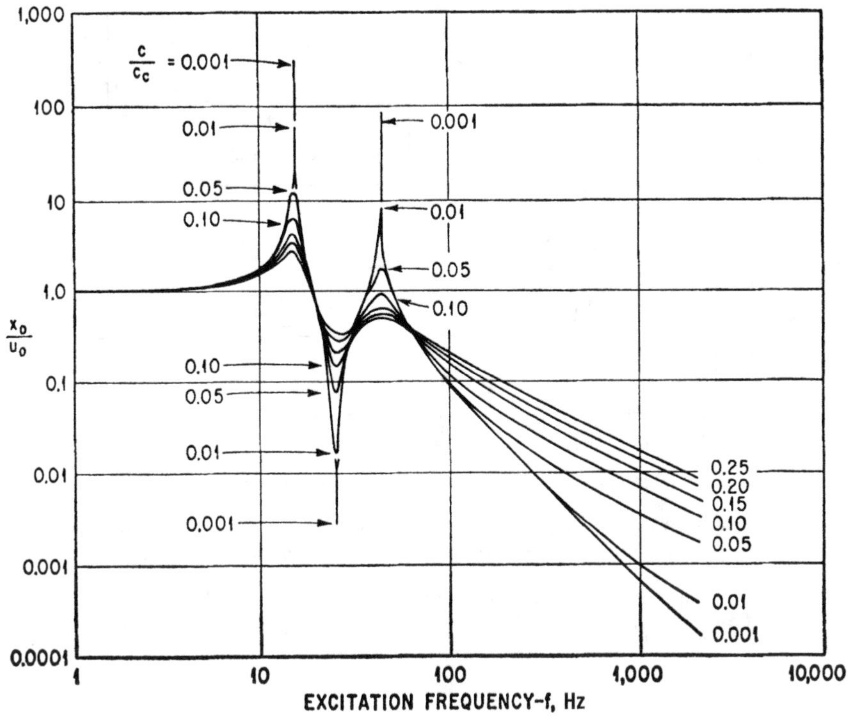

FIGURE 3.26*C* Response curves at point 3 in the system shown in Fig. 3.15. See caption for Fig. 3.26*A*.

BODY-INDUCED SINUSOIDAL VIBRATION

This section includes the analysis of a resiliently supported rigid body wherein the excitation consists of forces and moments applied directly to the rigid body (or originating within the body). The system has two planes of symmetry with orthogonal resilient supports; the modal coupling and natural frequencies for such a system are considered above. Two types of excitation are considered: (1) a force rotating about an axis parallel to one of the principal inertial axes and (2) an oscillatory moment acting about one of the principal inertial axes. There is no motion of the foundation that supports the resilient elements; thus, the following terms in Eqs. (3.31) are equal to zero:

$$u = v = w = \alpha = \beta = \gamma = 0 \tag{3.61}$$

Two Planes of Symmetry with Orthogonal Resilient Elements Excited by a Rotating Force. The system excited by the rotating force is illustrated in Fig. 3.27. The force F_0 rotates at frequency ω about an axis parallel to the Y axis but spaced therefrom by the coordinate distances d_x, d_z; the force is in the XZ plane. The forces and moments applied to the body by the rotating force F_0 are

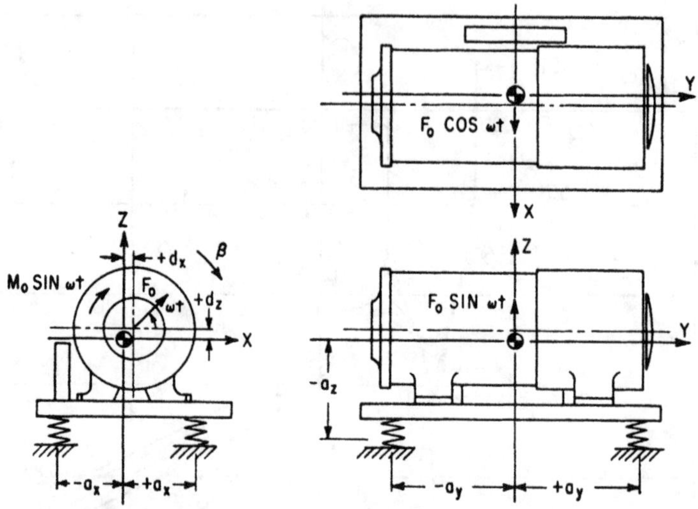

FIGURE 3.27 Example of a rigid body on orthogonal resilient supports with two planes of symmetry, excited by body-induced sinusoidal excitation. Alternative excitations are (1) the force F_0 in the XZ plane rotating with angular velocity ωt about an axis parallel to the Y axis and (2) the oscillatory moment $M_0 \sin \omega t$ acting about the Y axis. There is no motion of the foundation that supports the resilient elements.

$$
\begin{aligned}
F_x &= F_0 \cos \omega t & M_x &= 0 \\
F_y &= 0 & M_y &= F_0(d_z \cos \omega t - d_x \sin \omega t) \\
F_z &= F_0 \sin \omega t & M_z &= 0
\end{aligned}
\tag{3.62}
$$

The conditions of symmetry are defined by Eqs. (3.33), (3.34), (3.35), and (3.38); and the excitation is defined by Eqs. (3.61) and (3.62). Substituting these conditions into the equations of motion, Eqs. (3.31) show that vibration response is not excited in the coupled y_c, α mode or in the γ mode. In the Z direction, the motion z_{c0} of the body and the force F_{tz} transmitted through the resilient elements can be found from Eq. (2.30) and Fig. 2.17 since single-degree-of-freedom behavior is involved. The horizontal displacement amplitude x_{c0} of the center of gravity in the X direction and the rotational displacement amplitude β_0 about the Y axis are given by

$$
\begin{aligned}
\frac{x_{c0}}{F_0/4k_x} &= \frac{k_x}{k_z}\,\frac{\sqrt{\left[\dfrac{k_x}{k_z}\dfrac{a_z}{\rho_y}\left(\dfrac{a_z}{\rho_y}-\dfrac{d_z}{\rho_y}\right)+\left(\dfrac{a_x}{\rho_y}\right)^2-\left(\dfrac{f}{f_z}\right)^2\right]^2+\left[\dfrac{k_x}{k_z}\dfrac{d_x}{\rho_y}\dfrac{a_z}{\rho_y}\right]^2}}{\left(\dfrac{f}{f_z}\right)^4-\left[\dfrac{k_x}{k_z}+\dfrac{k_x}{k_z}\left(\dfrac{a_z}{\rho_y}\right)^2+\left(\dfrac{a_x}{\rho_y}\right)^2\right]\left(\dfrac{f}{f_z}\right)^2+\dfrac{k_x}{k_z}\left(\dfrac{a_x}{\rho_y}\right)^2} \\
\frac{\beta_0}{F_0/4k_x\rho_y} &= \frac{k_x}{k_z}\,\frac{\sqrt{\left[\dfrac{k_x}{k_z}\left(\dfrac{a_z}{\rho_y}-\dfrac{d_z}{\rho_y}\right)+\dfrac{d_z}{\rho_y}\left(\dfrac{f}{f_z}\right)^2\right]^2+\left[\dfrac{d_x}{\rho_y}\left(\dfrac{k_x}{k_z}-\dfrac{f^2}{f_z^2}\right)\right]^2}}{\left(\dfrac{f}{f_z}\right)^4-\left[\dfrac{k_x}{k_z}+\dfrac{k_x}{k_z}\left(\dfrac{a_z}{\rho_y}\right)^2+\left(\dfrac{a_x}{\rho_y}\right)^2\right]\left(\dfrac{f}{f_z}\right)^2+\dfrac{k_x}{k_z}\left(\dfrac{a_x}{\rho_y}\right)^2}
\end{aligned}
\tag{3.63}
$$

where a_x, a_z are location coordinates of the resilient supports, and

$$f_z = \frac{1}{2\pi}\sqrt{\frac{4k_z}{m}} \tag{3.64}$$

The amplitude of the oscillating force F_{tx} in the X direction and the amplitude of the oscillating moment M_{ty} about the Y axis which are transmitted to the foundation by the combination of resilient elements are

$$\begin{aligned} F_{tx} &= 4k_x\sqrt{x_{c0}^2 - 2a_z x_{c0}\beta_0 \cos(\phi_x - \phi_\beta) + a_z^2\beta_0^2} \\ M_{ty} &= 4k_z a_x^2 \beta_0 \end{aligned} \tag{3.65}$$

where F_{tx} is the sum of the forces transmitted by the individual resilient elements and M_{ty} is a moment formed by forces in the Z direction of opposite sign at opposite resilient supports. The angles ϕ_x and ϕ_β are defined by

$$\tan\phi_x = \frac{\dfrac{k_x}{k_z}\dfrac{a_z}{\rho_y}\left(\dfrac{a_z}{\rho_y} - \dfrac{d_z}{\rho_y}\right) + \left(\dfrac{a_x}{\rho_y}\right)^2 - \left(\dfrac{f}{f_z}\right)^2}{\dfrac{k_x}{k_z}\dfrac{a_z}{\rho_y}\dfrac{d_x}{\rho_y}} \qquad [0° \le \phi_x \le 360°]$$

$$\tan\phi_\beta = \frac{\dfrac{k_x}{k_z}\left(\dfrac{a_z}{\rho_y} - \dfrac{d_z}{\rho_y}\right) + \dfrac{d_z}{\rho_y}\left(\dfrac{f}{f_z}\right)^2}{\dfrac{d_x}{\rho_y}\left[\dfrac{k_x}{k_z} - \left(\dfrac{f}{f_z}\right)^2\right]} \qquad [0° \le \phi_\beta \le 360°]$$

To obtain the correct value of $(\phi_x - \phi_\beta)$ in Eq. (3.65), the signs of the numerator and denominator in each tangent term must be inspected to determine the proper quadrant for ϕ_x and ϕ_β.

Example 3.2. Consider an electric motor which has an unbalanced rotor, creating a centrifugal force. The motor weighs 3750 lb and has a radius of gyration ρ_y = 9.10 in. The distances $d_x = d_y = d_z = 0$; that is, the axis of rotation is the Y principal axis and the center of gravity of the rotor is in the XZ plane. The resilient supports each have a stiffness ratio of $k_x/k_z = 1.16$, and are located at $a_z = -14.75$ in., $a_x = \pm 12.00$ in. The resulting displacement amplitudes of the center of gravity, expressed dimensionlessly, are shown in Fig. 3.28; the force and moment amplitudes transmitted to the foundation, expressed dimensionlessly, are shown in Fig. 3.29. The displacements of the center of gravity of the body are dimensionalized with respect to the displacements at zero frequency:

$$\begin{aligned} z_{c0}(0) &= \frac{F_0}{4k_z} \\ x_{c0}(0) &= \frac{F_0}{4k_x}\left[1 + \frac{k_x}{k_z}\left(\frac{a_z}{a_x}\right)^2\right] \\ \beta_0(0) &= \frac{F_0}{4k_z a_z}\left(\frac{a_z}{a_x}\right)^2 \end{aligned} \tag{3.66}$$

At excitation frequencies greater than the higher natural frequency of the x_c, β coupled motion, the displacements, forces, and moment all continuously decrease as the frequency increases.

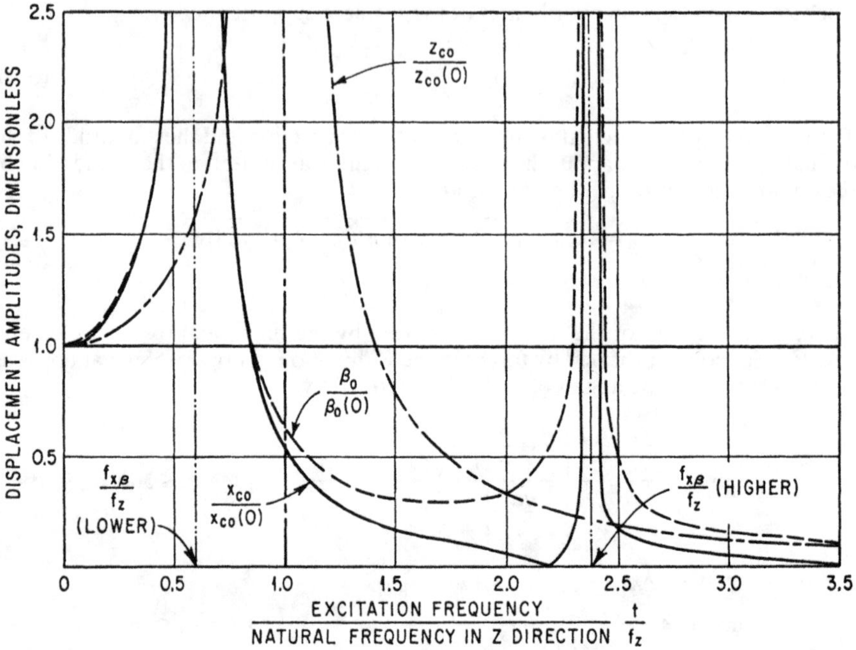

FIGURE 3.28 Response curves for the system shown in Fig. 3.27 when excited by a rotating force F_0 acting about the Y axis. The parameters of the system are $k_x/k_z = 1.16$, $a_x/\rho_y = \pm 1.32$, $a_z/\rho_y = -1.62$. Only x_c, z_c, β displacements of the body are excited [see Eqs. (3.63)]. The displacements are expressed dimensionlessly by employing the displacements at zero frequency [see Eqs. (3.66)].

Two Planes of Symmetry with Orthogonal Resilient Elements Excited by an Oscillating Moment. Consider the oscillatory moment M_0 acting about the Y axis with forcing frequency ω. The resulting applied forces and moments acting on the body are

$$\begin{aligned} M_y &= M_0 \sin \omega t \\ F_x = F_y = F_z &= M_x = M_z = 0 \end{aligned} \tag{3.67}$$

Substituting conditions of symmetry defined by Eqs. (3.33), (3.34), (3.35), and (3.38), and the excitation defined by Eqs. (3.61) and (3.67), the equations of motion [Eqs. (3.31)] show that oscillations are excited only in the x_c, β coupled mode. Solving for the resulting displacements,

$$\begin{aligned} \frac{x_{c0}}{M_0/4k_x\rho_y} &= \frac{\left(\dfrac{k_x}{k_z}\right)^2 \dfrac{a_z}{\rho_y}}{\left(\dfrac{f}{f_z}\right)^4 - \left[\dfrac{k_x}{k_z} + \dfrac{k_x}{k_z}\left(\dfrac{a_z}{\rho_y}\right)^2 + \left(\dfrac{a_x}{\rho_y}\right)^2\right]\left(\dfrac{f}{f_z}\right)^2 + \dfrac{k_x}{k_z}\left(\dfrac{a_x}{\rho_y}\right)^2} \\ \frac{\beta_0}{M_0/4k_x\rho_y^2} &= \frac{\dfrac{k_x}{k_z}\left[\dfrac{k_x}{k_z} - \left(\dfrac{f}{f_z}\right)^2\right]}{\left(\dfrac{f}{f_z}\right)^4 - \left[\dfrac{k_x}{k_z} + \dfrac{k_x}{k_z}\left(\dfrac{a_z}{\rho_y}\right)^2 + \left(\dfrac{a_x}{\rho_y}\right)^2\right]\left(\dfrac{f}{f_z}\right)^2 + \dfrac{k_x}{k_z}\left(\dfrac{a_x}{\rho_y}\right)^2} \end{aligned} \tag{3.68}$$

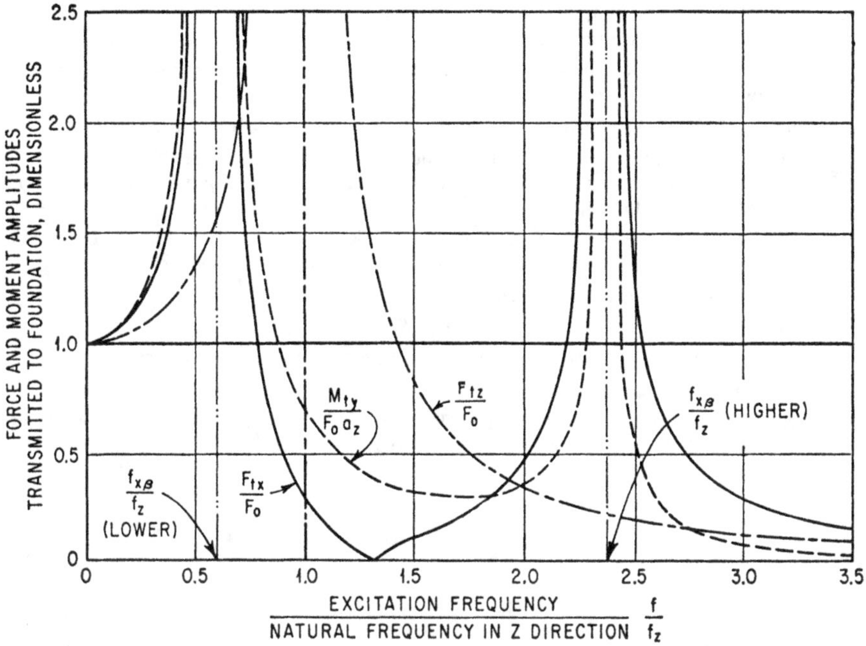

FIGURE 3.29 Force and moment amplitudes transmitted to the foundation for the system shown in Fig. 3.27 when excited by a rotating force F_0 acting about the Y axis. The parameters of the system are $k_x/k_z = 1.16$, $a_x/\rho_y = \pm 1.32$, $a_z/\rho_y = -1.62$. The amplitudes of the oscillating forces in the X and Z directions transmitted to the foundation are F_{tx} and F_{tz}, respectively. The amplitude of the total oscillating moment about the Y axis transmitted to the foundation is M_{ty}.

The amplitude of the oscillating force F_{tx} in the X direction and the amplitude of the oscillating moment M_{ty} about the Y axis transmitted to the foundation by the combination of resilient supports are

$$F_{tx} = 4k_x(x_{c0} - a_z\beta_0)$$
$$M_{ty} = 4k_z a_x^2 \beta_0 \tag{3.69}$$

where F_{tx} and M_{ty} have the same meaning as in Eqs. (3.65). Low vibration transmission of force and moment to the foundation is decreased at the higher frequencies in a manner similar to that shown in Fig. 3.29.

FOUNDATION-INDUCED VELOCITY SHOCK

Discussions of and data for foundation-induced velocity shock may be obtained from Chap. 3 in the first through the fifth editions of this handbook.

REFERENCES

1. Avallone, E. A., and T. Baumeister, III (eds.): "Marks' Standard Handbook for Mechanical Engineers," 10th ed., The McGraw-Hill Companies, New York, 1996.

2. Housner, G. W., and D. E. Hudson: "Applied Mechanics-Dynamics," 2d ed., D. Van Nostrand Company, Princeton, N.J., 1959.
3. Beer, F. P., and E. R. Johnston: "Vector Mechanics for Engineers: Dynamics," 5th ed., McGraw-Hill Book Company, New York, 1988.
4. Boucher, R. W., D. A. Rich, H. L. Crane, and C. E. Matheny: *NACA Tech. Note* 3084, 1954.
5. Woodward, C. R.: "Handbook of Instructions for Experimentally Determining the Moments and Products of Inertia of Aircraft by the Spring Oscillation Method," *WADC Tech. Rept.* 55–415, June 1955.
6. Vane, F. F.: "A Guide for the Selection and Application of Resilient Mountings to Shipboard Equipment—Revised," *David W. Taylor Model Basin Rept.* 880, February 1958.
7. Smollen, L. E.: *J. Acoust. Soc. Amer.,* **40**:195 (1966).
8. Lewis, R. C., and K. Unholtz: *Trans. ASME,* **69**:813 (1947).
9. Taylor, E. S., and K. A. Browne: *J. Aeronaut. Sci.,* **6**:43 (1938).
10. Browne, K. A.: *Trans. SAE,* **44**:185 (1939).
11. Derby, T. F.: "Decoupling the Three Translational Modes from the Three Rotational Modes of a Rigid Body Supported by Four Corner-Located Isolators," *Shock and Vibration Bull.,* 43, pt. 4, June 1973, pp. 91–108.

CHAPTER 4
NONLINEAR VIBRATION

C. Nataraj
Fredric Ehrich

INTRODUCTION

A vast body of scientific knowledge has been developed over a long period of time devoted to a description of natural phenomena. In the field of mechanics, rapid progress in the past two centuries has occurred, due in large measure to the ability of investigators to represent physical laws in terms of rather simple equations. In many cases the governing equations were not so simple; therefore, certain assumptions, more or less consistent with the physical situation, were employed to reduce the equations to types more easily soluble. Thus, the process of linearization has become an intrinsic part of the rational analysis of physical problems. An analysis based on linearized equations, then, may be thought of as an analysis of a corresponding but idealized problem.

In many instances the linear analysis is insufficient to describe the behavior of the physical system adequately. In fact, one of the most fascinating features of a study of nonlinear problems is the occurrence of new and totally unsuspected phenomena; i.e., new in the sense that the phenomena are not predicted, or even hinted at, by the linear theory. On the other hand, certain phenomena observed physically are unexplainable except by giving due consideration to nonlinearities present in the system.

The branch of mechanics that has been subjected to the most intensive attack from the nonlinear viewpoint is the theory of vibration of mechanical and electrical systems. Other branches of mechanics, such as incompressible and compressible fluid flow, elasticity, plasticity, and wave propagation also have been studied as nonlinear problems, but the greatest progress has been made in treating vibration of nonlinear systems. The systems treated in this chapter are systems with a finite number of degrees of freedom which can be defined by a finite number of simultaneous ordinary differential equations; on the other hand, the mechanics of continua involves partial differential equations. Nonlinear ordinary differential equations are easier to handle than nonlinear partial differential equations. An interesting survey of the entire realm of nonlinear mechanics is given in Ref. 1.

This chapter provides information concerning features of nonlinear vibration theory likely to be encountered in practice and methods of nonlinear vibration analysis which find ready application.

EXAMPLES OF SYSTEMS POSSESSING NONLINEAR CHARACTERISTICS

SIMPLE PENDULUM

As a first example of a system possessing nonlinear characteristics, consider a simple pendulum of length l having a bob of mass m, as shown in Fig. 4.1. The well-known differential equation governing free vibration is

$$ml^2\ddot{\theta} + mgl\theta = 0 \tag{4.1}$$

This equation holds only for small oscillations about the position of equilibrium since the actual restoring moment is characterized by the quantity $\sin\theta$. Equation (4.1) thus employs the assumption $\sin\theta \simeq \theta$. The exact, but nonlinear, equation of motion is

$$ml^2\ddot{\theta} + mgl\sin\theta = 0 \tag{4.2}$$

SIMPLE SPRING-MASS SYSTEM

A simple spring-mass system, as shown in Fig. 4.2, is characterized by the equation

$$m\ddot{x} + kx = 0$$

This equation is based on the assumption that the elastic spring obeys Hooke's law; i.e., the characteristic curve of restoring force versus displacement is a straight line. However, many materials do not exhibit such a linear characteristic. Further, in the case of a simple coil spring, a deviation from linearity occurs at large compression as the coils begin to close up, or conversely, when the extension becomes so great that the coils begin to lose their individual identity. In either case, the spring exhibits a characteristic such that the restoring force increases more rapidly than the displacement. Such a characteristic is called *hardening.* In a similar manner, certain systems (e.g., a simple pendulum) exhibit a *softening* characteristic. Both types of characteristic are shown in Fig. 4.3. A simple system with either softening or hardening restoring force may be described approximately by an equation of the form

$$m\ddot{x} + k(x \pm \mu^2 x^3) = 0$$

where the upper sign refers to the hardening characteristic and the lower to the softening characteristic.

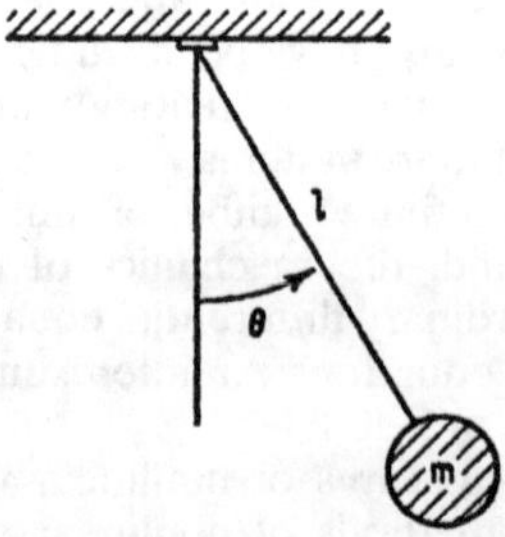

FIGURE 4.1 Simple pendulum.

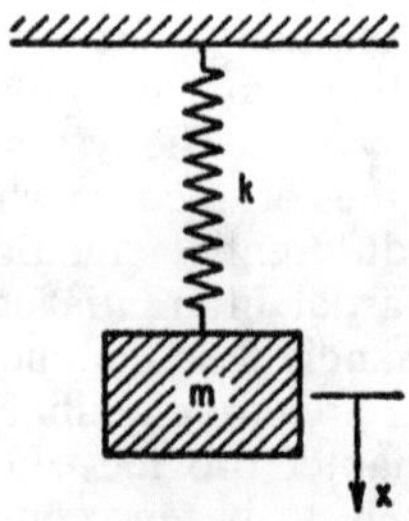

FIGURE 4.2 Simple spring-mass system.

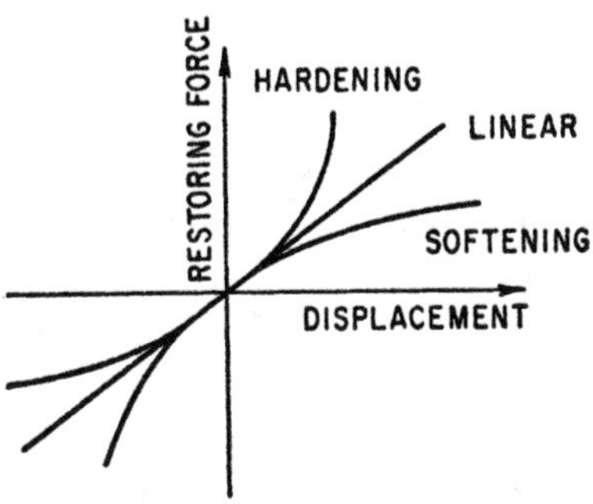

FIGURE 4.3 Restoring force characteristic curves for linear, hardening, and softening vibration systems.

It is possible for a system with only linear components to exhibit nonlinear characteristics, by snubber action for example, as shown in Fig. 4.4. A system undergoing vibration of small amplitude also may exhibit nonlinear characteristics; for example, in the pendulum shown in Fig. 4.5, the length depends on the amplitude.

STRETCHED STRING WITH CONCENTRATED MASS

The large-amplitude vibration of a stretched string with a concentrated mass, as shown in Fig. 4.6, offers another example of a nonlinear system. The governing nonlinear differential equation is, approximately,

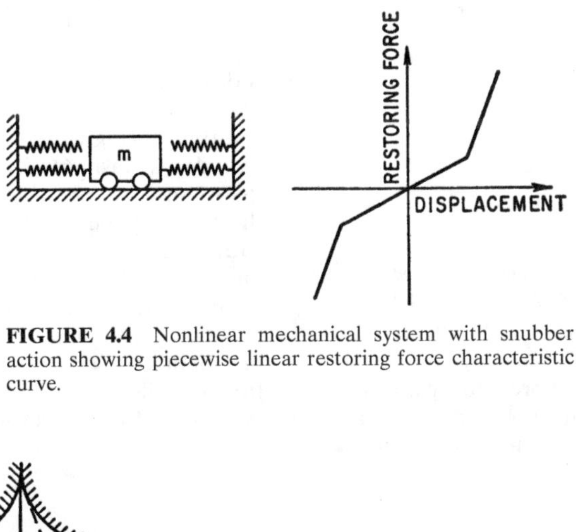

FIGURE 4.4 Nonlinear mechanical system with snubber action showing piecewise linear restoring force characteristic curve.

FIGURE 4.5 Pendulum with nonlinear characteristic resulting from dependence of length on vibration amplitude.

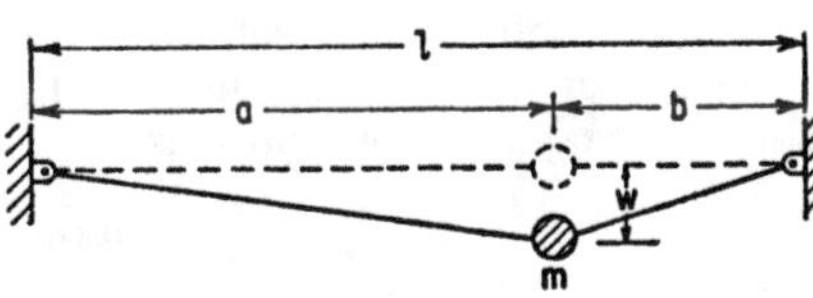

FIGURE 4.6 Vibration of a weighted string as an example of a nonlinear system.

$$m\ddot{w} + F_0\left(\frac{l}{ab}\right)w + (SE - F_0)\left(\frac{a^3 + b^3}{2a^3b^3}\right)w^3 = 0$$

where F_0 is the initial tension, S is the cross-sectional area, and E is the elastic modulus of the string. Consider now the special case of $a = b$ and denote the unstretched length of the half string by l_0. Then the initial tension and the restoring force become

$$F_0 = SE\left(\frac{a - l_0}{l_0}\right)$$

$$F_r \simeq SE\left[2\left(\frac{a}{l_0} - 1\right)\left(\frac{w}{a}\right) + \left(2 - \frac{a}{l_0}\right)\left(\frac{w}{a}\right)^3\right]$$

An interesting feature of this system is that it exhibits a wide variety of either hardening or softening characteristics, depending upon the value of a/l_0, as shown in Fig. 4.7.

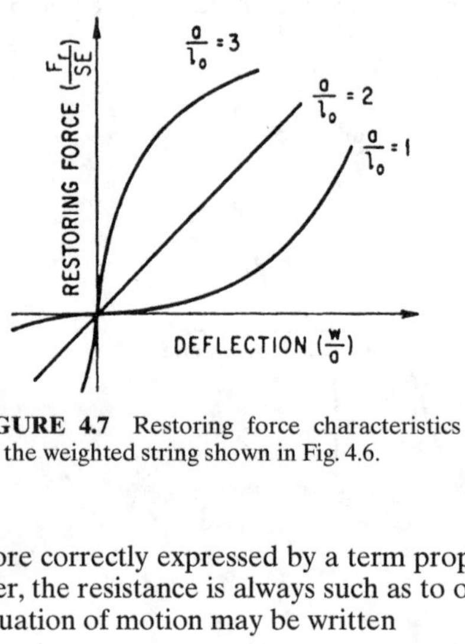

FIGURE 4.7 Restoring force characteristics for the weighted string shown in Fig. 4.6.

SYSTEM WITH VISCOUS DAMPING

The foregoing examples all involve nonlinearities in the elastic components, either as a result of appreciable amplitudes of vibration or as a result of peculiarities of the elastic element. Consider a simple spring-mass system which also includes a dashpot. The usual assumptions pertaining to this system are that the spring is linear and that the motion is sufficiently slow that the viscous resistance provided by the dashpot is proportional to the velocity; therefore, the governing equation of motion is linear. Frequently, the dashpot resistance is more correctly expressed by a term proportional to the square of the velocity. Further, the resistance is always such as to oppose the motion; therefore, the nonlinear equation of motion may be written

$$m\ddot{x} + c|\dot{x}|\dot{x} + kx = 0$$

BELT FRICTION SYSTEM

The system shown in Fig. 4.8*A* involves a nonlinearity depending upon the dry friction between the mass and the moving belt. The belt has a constant speed v_0, and the applicable equation of motion is

$$m\ddot{x} + F(\dot{x}) + kx = 0$$

where the friction force $F(\dot{x})$ is shown in Fig. 4.8*B*. For large values of displacement, the damping term is positive, has positive slope, and removes energy from the sys-

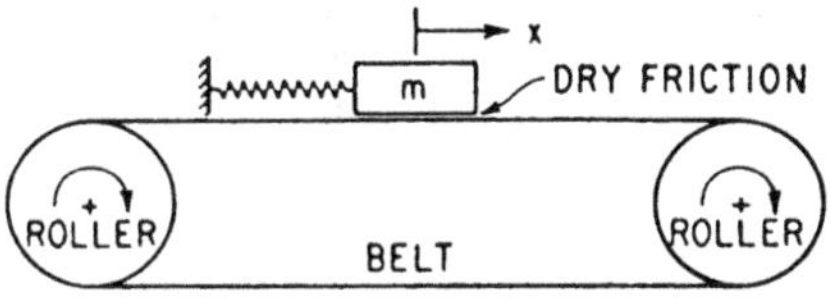

FIGURE 4.8*A* Belt friction system which exhibits self-excited vibration.

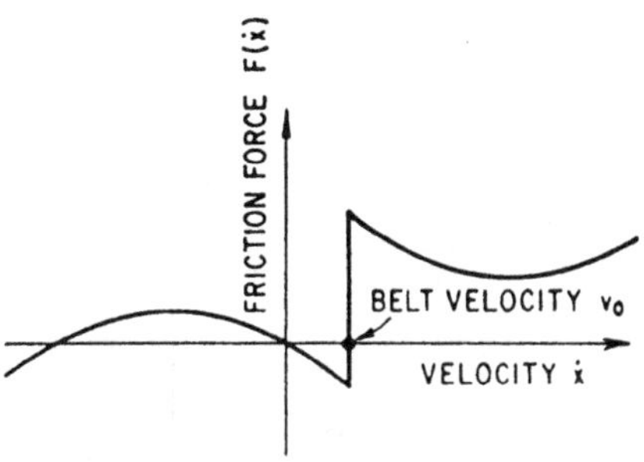

FIGURE 4.8*B* Damping force characteristic curve for the belt friction system shown in Fig. 4.8*A*.

tem; for small values of displacement, the damping term is negative, has negative slope, and actually puts energy into the system. Even though there is no external stimulus, the system can have an oscillatory solution, and thus corresponds to a nonlinear *self-excited* system. (See Chap. 5.)

SYSTEMS WITH ASYMMETRIC STIFFNESS

The aforementioned examples of nonlinear stiffness, typified by the stiffness variations in Figs. 4.3, 4.4, and 4.7, all may be characterized as symmetric. That is, the variation in the absolute value of the restoring force with displacement in the positive direction is identical to the variation in the absolute value of the restoring force with displacement in the negative direction. As will be seen in the following sections, symmetric stiffness distributions result in changes in the shape of the resonant peak of the response curve and slight distortion in the waveform of the dynamic motion without changing the basic synchronism between forcing function and response. But many more diverse phenomena and much more profound changes are encountered when dealing with asymmetric stiffness distributions.

A typical physical situation is encountered in the dynamics of rotating machinery where a softly mounted rotor is located eccentrically within the small clearance of a motion-limiting stiff stator as illustrated in Fig. 4.9*A* and *C*. When rotating with some unbalance in the rotor, the vertical component of the unbalance force will cause intermittent local contact with the stiff stator, resulting in a "bouncing" motion of the rotor. The stiffness characteristic for the vertical motion is asymmetric. In its simplest form, it may be represented as a bilinear relationship—very soft for vertical motion in the upward direction and very stiff for vertical motion in the downward direction, as illustrated in Fig. 4.9*B*. More explicitly,

$$k = K_1 \qquad x > 0$$

$$k = K_2 \qquad x < 0$$

Many other examples of nonlinear systems are given in the references of this chapter.

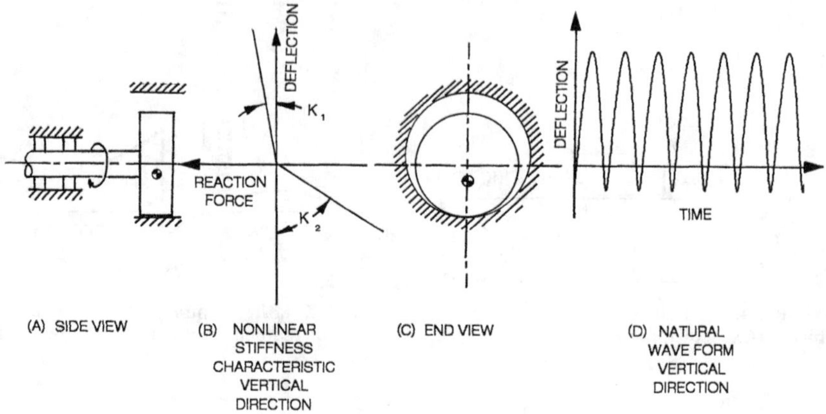

FIGURE 4.9 Nonlinear spring characteristic of a rotor operating with local intermittent contact in a clearance.

DESCRIPTION OF NONLINEAR PHENOMENA

This section describes briefly, largely in nonmathematical terms, certain of the more important features of nonlinear vibration. Further details and methods of analysis are given later.

FREE VIBRATION

Insofar as the free vibration of a system is concerned for systems with symmetric stiffness distributions, one distinguishing feature between linear and nonlinear behavior is the dependence of the period of the motion in nonlinear vibration on the amplitude. For example, the simple pendulum of Fig. 4.1 may be analyzed on the basis of the linearized equation of motion, Eq. (4.1), from which it is found that the period of the vibration is given by the constant value $\tau_o = 2\pi/\omega_n$. An analysis on the basis of the nonlinear equation of motion, Eq. (4.2), leads to an expression for the period of the form

$$\frac{\tau}{\tau_o} = 1 + \tfrac{1}{4}(U)^2 + \tfrac{9}{64}(U)^4 + \tfrac{25}{256}(U)^6 + \cdots \tag{4.3}$$

where U is related to the amplitude of the vibration Θ by the relation $U = \sin(\Theta/2)$. The linear solution thus corresponds to the first term of Eq. (4.3). The dependence of the period of vibration on amplitude is shown in Fig. 4.10. Systems in which the period of vibration is independent of the amplitude are called *isochronous,* while those in which the period τ is dependent on the amplitude are called *nonisochronous.*

The dependence of period on amplitude also may be seen from the vibration trace shown in Fig. 4.11, which corresponds to a solution of the equation

$$m\ddot{x} + c\dot{x} + k(x + \mu^2 x^3) = 0$$

For systems with asymmetric stiffness distributions, free undamped vibration will display significant distortion of the natural waveform. The simple bilinear stiffness distribution of Fig. 4.9*B* will result in the system having a simple harmonic half

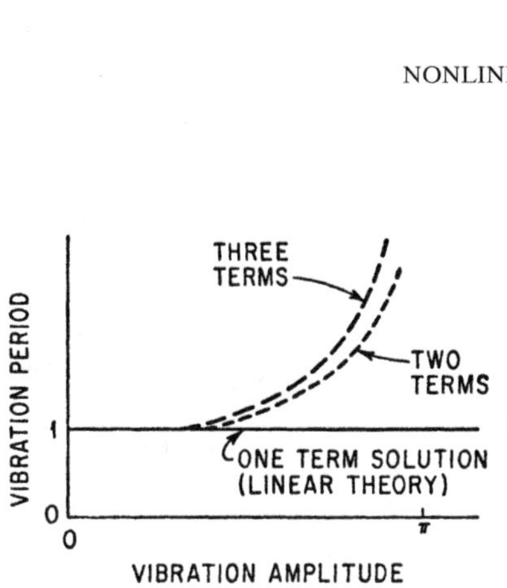

FIGURE 4.10 Period of free vibration of a simple pendulum according to Eq. (4.3) and showing the effect of nonlinear terms.

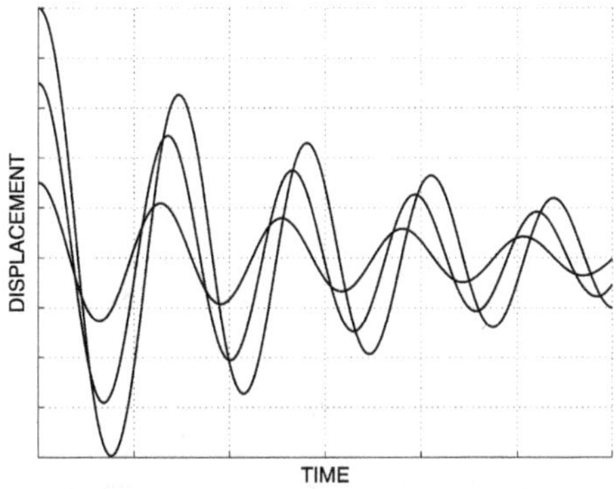

FIGURE 4.11 Deflection time history for free damped vibration of Duffing's equation [Eq. (4.21)].

cycle at relatively low frequency for upward motion and a simple harmonic half cycle at relatively high frequency for downward motion. The overall waveform is then a combination of these two disparate half cycles as represented in Fig. 4.9*D* and suggests a bouncing motion.

RESPONSE CURVES FOR FORCED VIBRATION OF SYSTEMS WITH SYMMETRIC STIFFNESS

Representations of vibration behavior in the form of curves of response amplitude versus exciting frequency are called *response curves.* The response curves for an undamped linear system acted on by a harmonic exciting force of amplitude p and frequency ω may be derived from the equation of motion

$$\ddot{x} + \omega_n^2 x = \frac{p}{m} \cos \omega t \tag{4.4}$$

The solution has the form shown in Fig. 4.12. The vertical line at $\omega = \omega_n$ corresponds not only to resonance but also to free vibration (with $p = 0$); the amplitude in this instance is determined by the initial conditions of the motion. In a nonlinear system the character of the motion is dependent upon the amplitude. This requires that the natural frequency likewise be amplitude-dependent; hence, it follows that the free vibration curve $p = 0$ for nonlinear systems cannot be a straight line. Figure 4.13 shows free vibration curves (i.e., natural frequency as a function of amplitude) for hardening and softening systems.

Figures 4.12 and 4.13 suggest that the forced vibration response curves for systems with nonlinear restoring forces have the general form of those of a linear system but are "swept over" to the right or left, depending on whether the system is hardening or softening. These are shown in Fig. 4.14. The principal effect of damping in forced vibration of a nonlinear system is to limit the amplitude at resonance, as shown in Fig. 4.15.

These rightward- and leftward-leaning resonant response peaks have special meaning to the dynamic response of the system. Consider a hardening system whose

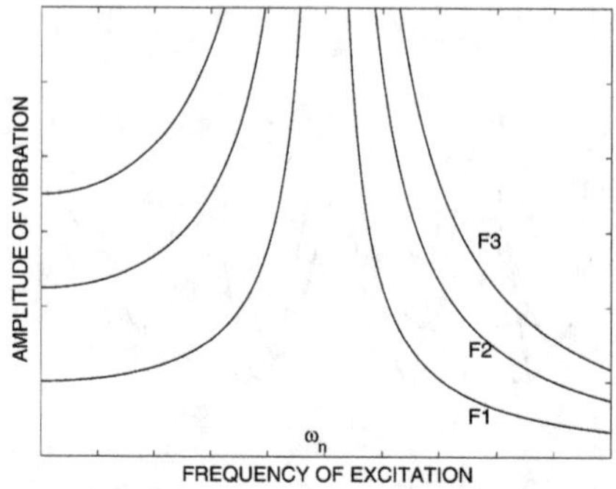

FIGURE 4.12 Family of response curves for the undamped linear system defined by Eq. (4.4).

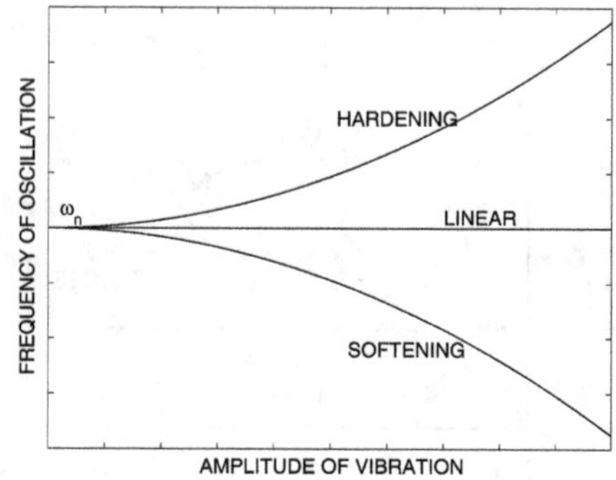

FIGURE 4.13 Free vibration curves (natural frequency as a function of amplitude) in the response diagram for linear, hardening, and softening vibration systems [see Eq. (4.49)].

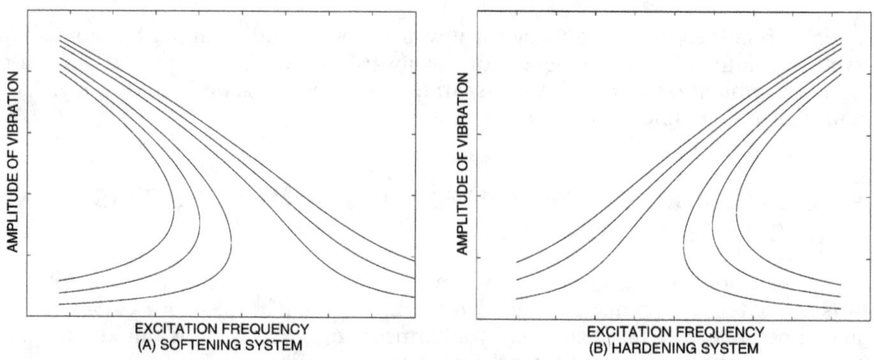

FIGURE 4.14 Response curves for undamped nonlinear systems with hardening and softening restoring force characteristics [see Eq. (4.50)].

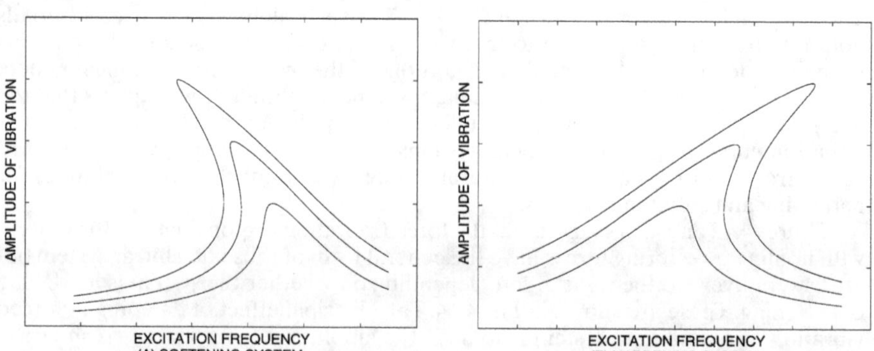

FIGURE 4.15 Response curves for damped nonlinear systems with hardening and softening restoring force characteristics [see Eq. (4.52)].

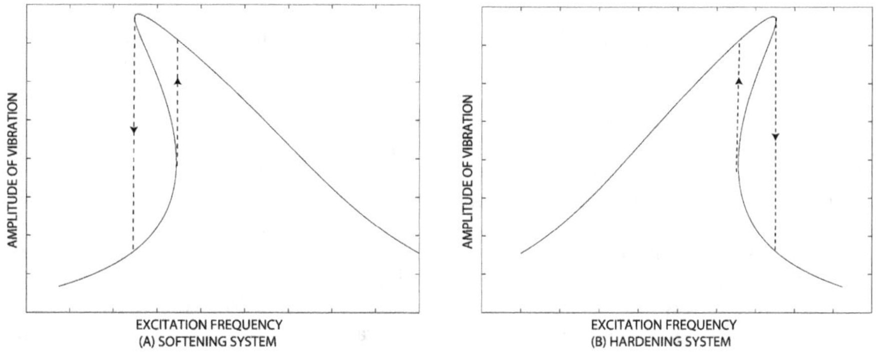

FIGURE 4.16 Jump phenomenon in hardening and softening systems.

response curve is shown in Fig. 4.15. Suppose that the exciting frequency starts at a low value and increases continuously at a slow rate. The amplitude of the vibration also increases, but only up to a point. In particular, at the point of vertical tangency of the response curve, a slight increase in frequency requires that the system perform in an unusual manner; i.e., that it "jump" down in amplitude to the lower branch of the response curve. This experiment may be repeated by starting with a large value of exciting frequency but requiring that the forcing frequency be continuously reduced. A similar situation is encountered; the system must jump up in amplitude in order to meet the conditions of the experiment. This *jump phenomenon* is shown in Fig. 4.16 for both the hardening and softening systems.[2] The jump is not instantaneous in time but requires a few cycles of vibration to establish a steady-state vibration at the new amplitude.

There is a portion of the response curve which is "unattainable"; it is not possible to obtain that particular amplitude by a suitable choice of forcing frequency. Thus, for certain values of ω there appear to be three possible amplitudes of vibration but only the upper and lower can actually exist. If by some means it were possible to initiate a steady-state vibration with just the proper amplitude and frequency to correspond to the middle branch, the condition would be unstable; at the slightest disturbance the motion would jump to either of the other two states of motion. The direction of the jump depends on the direction of the disturbance. Thus, of the three possible states of motion, one in phase and two out of phase with the exciting force, the one having the larger amplitude of the two out-of-phase motions is unstable. This region of instability in the response diagram is defined by the loci of vertical tangents to the response curves.

RESPONSE CURVES FOR FORCED VIBRATION OF SYSTEMS WITH ASYMMETRIC STIFFNESS

The system pictured in Fig. 4.9 is typical of systems with asymmetric stiffness characteristics, and its response[3,4] includes a variety of phenomena, including regions of chaotic response,[1] not observed in systems with symmetric stiffness characteristics.

The equations of motion in the plane normal to the plane of contact, with a stiffness of k_1 when the rotor is deflected from its rest position in the soft direction and a stiffness k_2 when the rotor is deflected from its rest position in the hard direction, may be integrated numerically using a simple trapezoidal integration procedure. The rest position of the rotor is taken at the contact point, so the break point of the bilinear elastic characteristic is at zero deflection. The system is then simply characterized by only two parameters—the ratio of the stiffnesses $\beta = k_1/k_2$ and z_1, the linear damping ratio of the system referred to critical damping of the soft system—when operated at a given rotational speed *s*, which is taken in normalized format as the ratio of rotational frequency to the system natural frequency.

For typical values and z_1 and β at any speed *s*, the numerical model may be used to compute the orbit of the rotor mass point as the orthogonal coordinates of the motion *X* and *Y*, where each of the coordinates is normalized as the ratio of the deflection from the rest position to the unbalance mass eccentricity. In considering the response over a large range of rotational speed, the motion may be simply characterized at any particular speed as Y_p, the local peak value(s) of the normalized amplitude in the direction of the nonlinear stiffness. As shown in Fig. 4.17*A* in comparison with the response of an equivalent system with a linear spring support stiffness, a plot of this parameter over a range of speeds is quite effective in detecting and identifying various different response phenomena.

Superharmonic Response.[5,6] Fig. 4.17*B* characterizes superharmonic response at subcritical speed. Shown here at approximately one-half critical speed, the rotor is bouncing at approximately its natural frequency against the hard surface of the contact point, energized at every other bounce by the component of the unbalance centrifugal force as suggested in Fig. 4.18. The dominant frequency of the response is then precisely 2 times operating speed. Such a pseudo-critical speed is possible for any integer order *M* at approximately 1/*M* times critical speed and with a significant

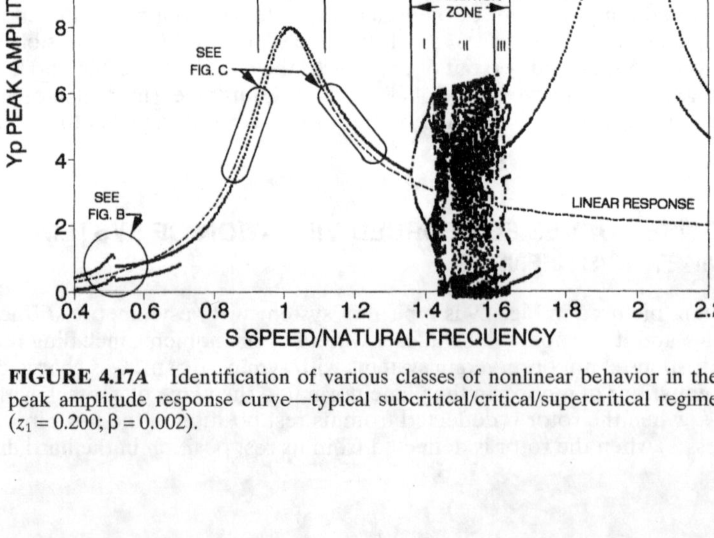

FIGURE 4.17A Identification of various classes of nonlinear behavior in the peak amplitude response curve—typical subcritical/critical/supercritical regime ($z_1 = 0.200$; $\beta = 0.002$).

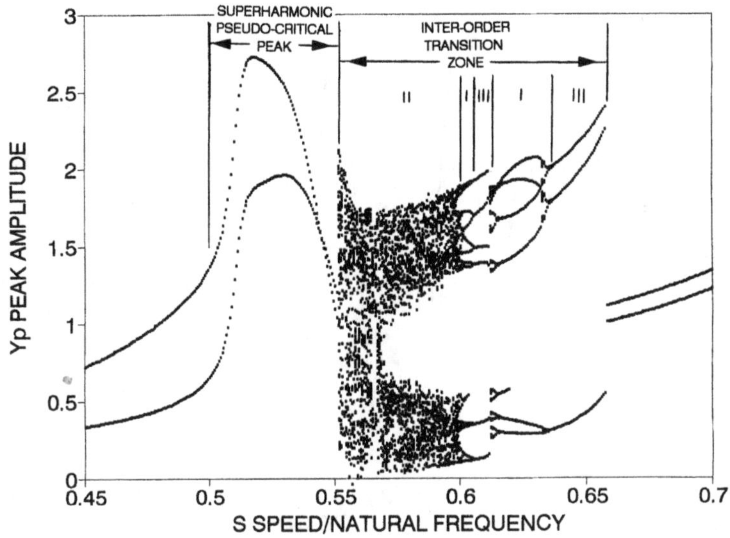

FIGURE 4.17*B* Identification of various classes of nonlinear behavior in the peak amplitude response curve—detail of superharmonic pseudo-critical peak and interorder transition zone ($z_1 = 0.05$; $\beta = 0.005$).

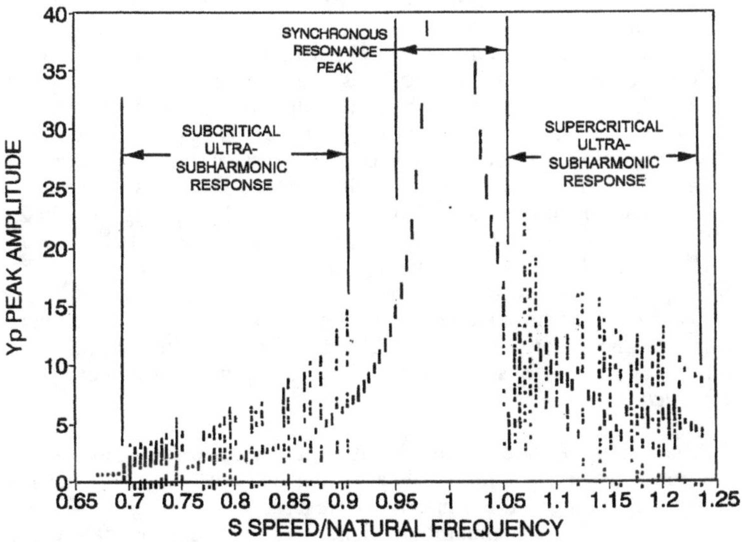

FIGURE 4.17*C* Identification of various classes of nonlinear behavior in the peak amplitude response curve—detail of transcritical ultra-subharmonic response ($z_1 = 0.002$; $\beta = 0.002$).

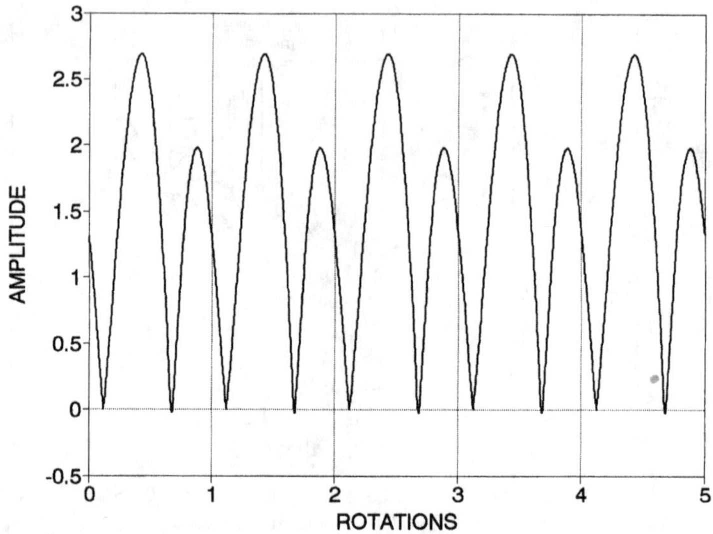

FIGURE 4.18 Subcritical superharmonic response—waveform ($z_1 = 0.050$; $\beta = 0.005$; $s = 0.525$; $M = 2$).

frequency component of precisely M times operating speed or approximately equal to the *natural frequency.*

Transition Between Successive Superharmonic Orders.[6] Between the successive superharmonic response zones [i.e., between the Mth and $(M - 1)$th order superharmonic responses] there may occur a regime of irregular response. Most commonly, the response may be chaotic, as identified as Zone II in Fig. 4.17*B* and shown in Fig. 4.19*A*. For such chaotic motion, the Poincaré section, which is a stroboscopic view of the phase-plane plot of velocity versus displacement at a reference angle of shaft rotation, is effectively a slice of the system's attractor as shown in Fig. 4.19*B*. The chaotic motion may be preceded on one side by a cascade of period-doubling bifurcations in the trace of peak amplitude Y_p, as suggested in Zone I of Fig. 4.17*B*. Another pattern of transition response is periodic in waveform. As shown in Zone III of Fig. 4.17*B*, instead of having an unending series of local peaks with no identifiable periodicity of repetitions as would be the case in truly chaotic motion, the response appears to have clusters of K bounces that actually repeat every L rotations to give a major periodicity of K/L times s. In both the chaotic and periodic transition zones, the response has a significant component at or near the system's *natural frequency.*

Ultra-Subharmonic Response in Transcritical Response (Subcritical).[7,8] A unique response has been identified which appears in very lightly damped, highly nonlinear systems operating in the transcritical range, as shown in Fig. 4.17*C*. It has been observed that one of the dominant sidebands occurs at approximately critical frequency, and the sideband separation is generally a whole-number fraction $|1/(J+1)|$ of the operating speed and, in the Jth order manifestation (that is, $J = -1, -2, -3, \ldots$) of subcritical spontaneous sidebanding when the speed is approximately $(J+1)/J$ times the natural frequency, the dominant frequency is precisely $J/(J + 1)$ times the rotation-

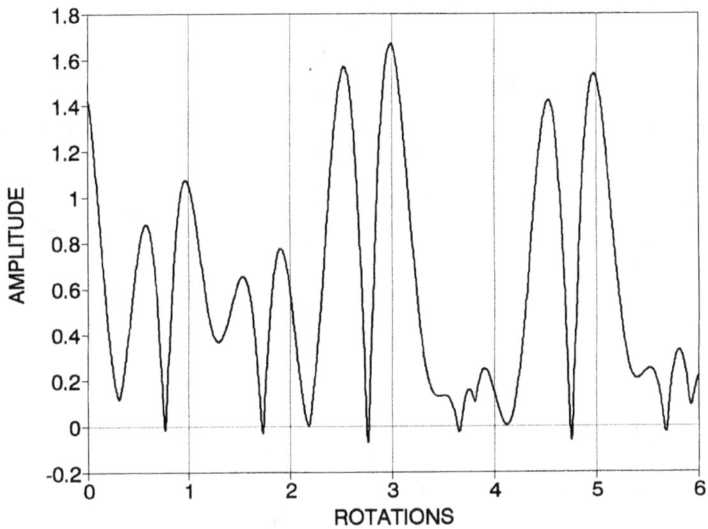

FIGURE 4.19A Subcritical chaotic transition between successive superharmonic orders—waveform ($z_1 = 0.050$; $\beta = 0.005$; $s = 0.560$; $2 < M < 1$).

tive speed or approximately equal to the *natural frequency*. The waveform, shown in Fig. 4.20, is periodic in nature. There appear to be transition zones between successive orders of J when the response has a dominant frequency approximately equal to the system's *natural frequency* and the waveform may be chaotic. The general phenomenon has been referred to as *ultra-subharmonic response*. In a more general

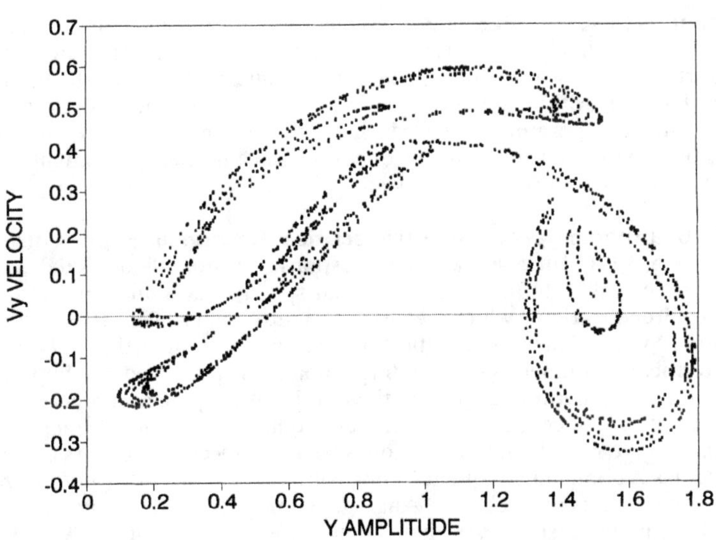

FIGURE 4.19B Subcritical chaotic transition between successive superharmonic orders—Poincaré section ($z_1 = 0.050$; $\beta = 0.005$; $s = 0.560$; $2 < M < 1$).

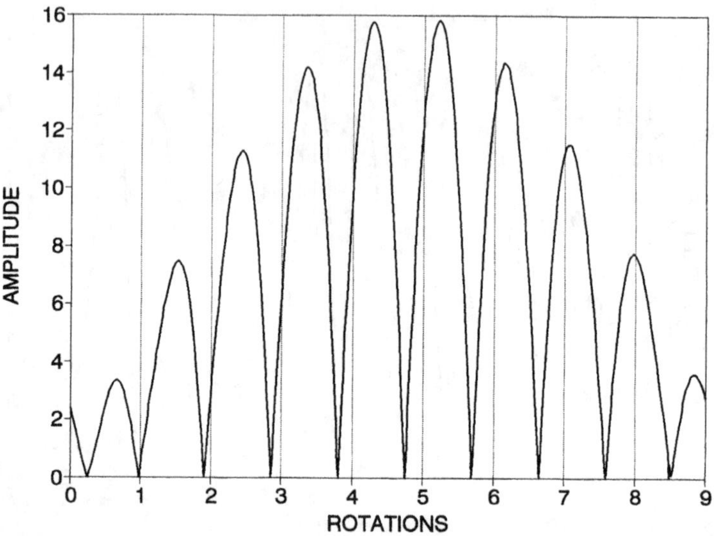

FIGURE 4.20 Transcritical ultra-subharmonic response (z_1 = 0.001; β = 0; s = 0.900; $J = -10$).

formulation,[9] it has been noted that such ultra-subharmonic response can be found in a speed range just below the Mth-order subharmonic peak at a rotational speed which is approximately $(MJ + 1)/J$ times the natural frequency (where $J = -1, -2, -3, \ldots$) with a dominant response frequency precisely equal to $J/(MJ + 1)$ times the rotational speed.

Synchronous Resonant Response. Synchronous critical response in the nonlinear system, shown in Fig. 4.17*A*, is very similar to that of the linear system except for the distortion of the waveform reflecting the bouncing nature of the motion illustrated in Fig. 4.21. Although the dominant frequency component is that of the forcing frequency or operating speed which is close to the *natural frequency* of the system, the bouncing waveform produces significant spectral content at whole-number multiples of the operating speed.

Ultra-Subharmonic Response in Transcritical Response (Supercritical).[8] As shown in Fig. 4.17*C*, ultra-subharmonic response or spontaneous sidebanding can occur at speeds slightly higher than critical speed, very similar in nature to the response already noted, which occurs at slightly subcritical speeds. Again, the waveform is periodic in nature. In the Jth-order manifestation (that is, $J = 1, 2, 3, \ldots$) of supercritical spontaneous sidebanding, when the rotative speed is approximately $(J + 1)/J$ times the natural frequency, the dominant frequency is precisely $J/(J + 1)$ times the rotative speed, or approximately equal to the *natural frequency.* Once again, there appears to be transition zones between successive orders of J when the response has a dominant frequency approximately equal to the *natural frequency* and the waveform may be chaotic. Analogous to the general finding for subcritical ultra-subharmonic response, it has been noted[9] that such ultra-subharmonic response can be found in a speed range just above the Mth-order subharmonic

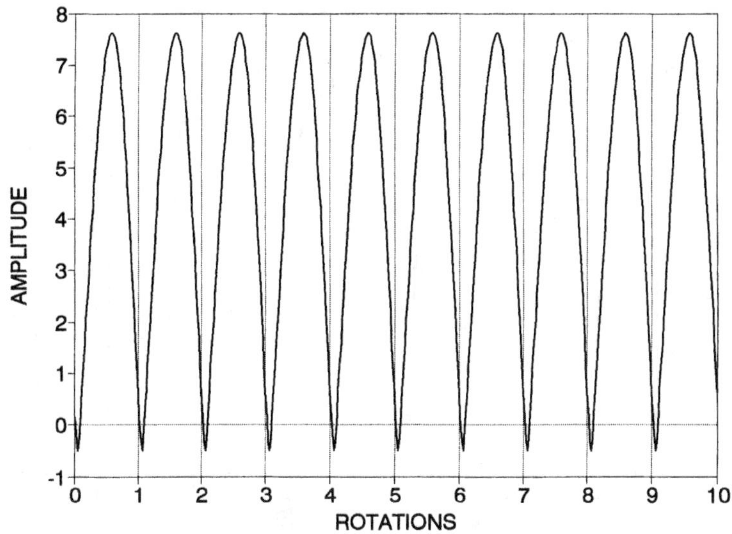

FIGURE 4.21 Critical synchronous resonant response—waveform ($z_1 = 0.200$; $\beta = 0.005$; $s = 1.050$).

peak at a rotational speed which is approximately $(MJ + 1)/J$ times the natural frequency (where $J = 1, 2, 3, \ldots$) with a dominant response frequency precisely equal to $J/(MJ + 1)$ times the rotational speed.

Subharmonic Response.[10–14] The pseudo-critical peak at 2 times critical speeds shown in Fig. 4.17*A* exemplifies subharmonic response at supercritical speed. With a peak amplitude of the same order of magnitude as critical response, the rotor is bouncing at its natural frequency against the hard surface of the contact point, as depicted in Fig. 4.22, and is subjected to the periodic component of the unbalance centrifugal force twice every bounce. Only one of the two pulses of unbalance force is effective in energizing the bouncing motion in the course of each bounce, so the dominant frequency of the response is then precisely one-half the operating speed. Such a pseudo-critical is possible for any integer order N at a rotational speed approximately N times critical speed and with a dominant frequency of precisely $1/N$ times operating speed or approximately the system *natural frequency.*

Transition Between Successive Subharmonic Orders. The transition response between successive subharmonic orders is quite analogous to the transition response between successive superharmonic orders previously noted. Between the successive subharmonic response zones (i.e., between the Nth- and $(N + 1)$th-order subharmonic responses) there may occur a regime of irregular response. The response has been noted by many researchers to be chaotic,[1,15–22] as identified as Zone II in Fig. 4.17*A* and illustrated in Fig. 4.23*A*. The chaotic motion may be preceded on one side by a cascade of period-doubling bifurcations in the trace of peak amplitude Y_p, as suggested in Zone I of Fig. 4.17*A*. Another pattern of transition response is periodic in waveform. As shown in Zone III of Fig. 4.17*A*, instead of having an unending series of local peaks with no identifiable periodicity of repetitions as would be the case in

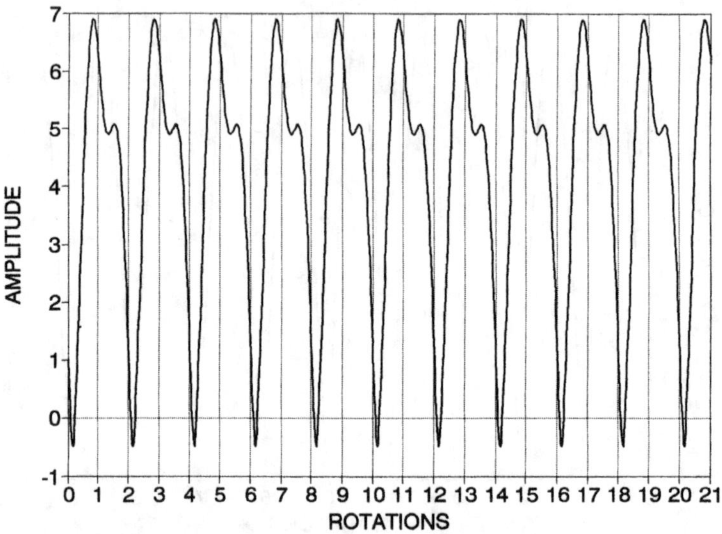

FIGURE 4.22 Supercritical subharmonic response—waveform (z_1 = 0.200; β = 0.005; s = 2.150; N = 2).

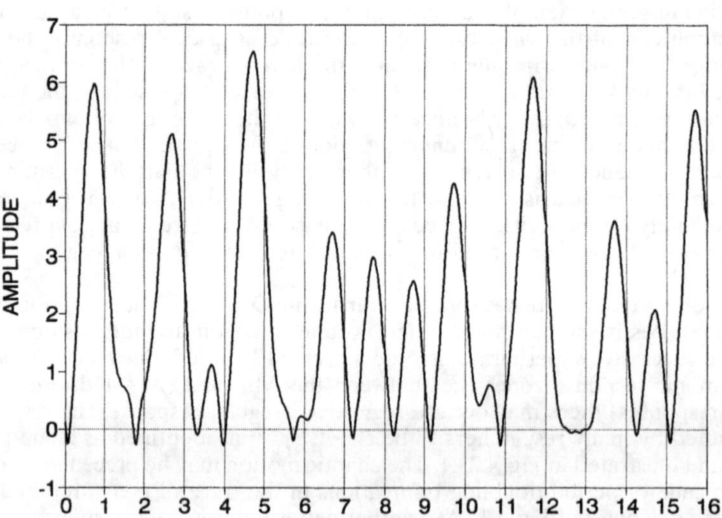

FIGURE 4.23A Supercritical chaotic transition between successive subharmonic orders—waveform (z_1 = 0.200; β = 0.005; s = 1.600; 1 < N < 2).

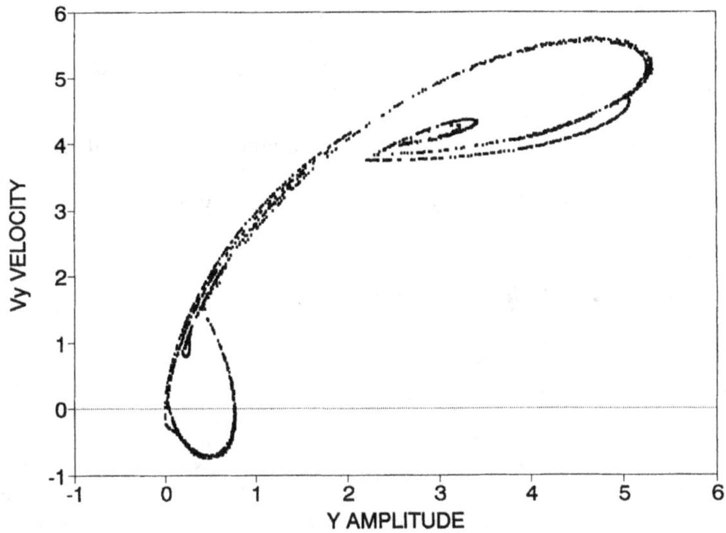

FIGURE 4.23*B* Supercritical chaotic transition between successive subharmonic orders—Poincaré section ($z_1 = 0.200$; $\beta = 0.005$; $s = 1.600$; $1 < N < 2$).

truly chaotic motion, the response appears to have clusters of K bounces that actually repeat every L rotations to give a major periodicity of K/L times s. In both the chaotic and the periodic transition zones, the response has a significant component at or near the system's *natural frequency.* As with subcritical chaotic transition zones, a Poincaré section of chaotic motion in a supercritical chaotic transition zone is effectively a slice of the system's attractor, as shown in Fig. 4.23*B*.

OTHER PHENOMENA

Self-Excited Vibration. A general treatment of self-excited vibration, including reference to bistable vibration (and the Van der Pol and relaxation oscillators) is given in Chap. 5.

Asynchronous Excitation and Quenching. In linear systems, the principle of superposition is valid, and there is no interaction between different oscillations. Moreover, the mathematical existence of a periodic solution always indicates the existence of a periodic phenomenon. In nonlinear systems, there is an interaction between oscillations; the mathematical existence of a periodic solution is only a necessary condition for the existence of corresponding physical phenomena. When supplemented by the condition of stability, the conditions become both necessary and sufficient for the appearance of the physical oscillation. Therefore, it is conceivable that under these conditions the appearance of one oscillation may either create or destroy the stability condition for another oscillation. In the first case, the other oscillation appears (*asynchronous excitation*), and in the second case, disappears (*asynchronous quenching*). The term *asynchronous* is used to indicate that there is no relation between the frequencies of these two oscillations.

Entrainment of Frequency. According to linear theory, if two frequencies ω_1 and ω_2 are caused to beat in a system, the period of beating increases indefinitely as ω_2 approaches ω_1. In nonlinear systems, the beats disappear as ω_2 reaches certain values. Thus, the frequency ω_1 falls in synchronism with, or is entrained by, the frequency ω_2 within a certain range of values. This is called *entrainment of frequency,* and the band of frequencies in which entrainment occurs is called the zone of entrainment or the interval of synchronization. In this region, the frequencies ω_1 and ω_2 combine and only vibration at a single frequency ensues.

OVERVIEW OF NONLINEAR ANALYSIS

As must be clear from the examples presented so far, there are a number of nonlinear relationships that occur in practically every vibrating system. Stiffness, damping, and rigid-body motions can all lead to nonlinear expressions in addition to geometric nonlinearities. For systems where the relationships dominating the dynamic behavior are weakly nonlinear or when the range of realistic operating conditions is narrow, the dynamic behavior outlined in this presentation may not be present. This situation is indeed the case for many operating mechanical systems. However, there are a number of systems where steady-state multiplicity and limit cycle behavior do occur under practical operating conditions.

The techniques outlined in this section allow one to characterize the local and global dynamic behavior associated with the existence of a single, unique, steady-state equilibrium point or a multiplicity of steady-state operating points over some range of process parameters. In this section, we provide a brief introduction to the nonlinear dynamic behavior of a generic second-order system. A simple oscillator with nonlinear stiffness or damping is of this kind. A discussion of the vocabulary and phenomena characteristic of nonlinear systems is also provided. A more rigorous mathematical presentation of nonlinear dynamic analysis is available from many excellent applied mathematical texts.[23–25] It should be noted that multiple-degree-of-freedom systems are significantly more complicated and can reveal more complex phenomena than illustrated here. Quantitative techniques have been developed[26] using substructuring techniques integrated with the method of weighted residuals to render the solutions somewhat tractable for practical vibration problems.

Consider the free vibration of a one-degree-of-freedom system with nonlinearities; it can be written as the following set of coupled, nonlinear, first-order ordinary differential equations:

$$\dot{x}_1 = P(x_1, x_2) \qquad \dot{x}_2 = Q(x_1, x_2) \tag{4.5}$$

where x_1 and x_2 are the two *states* of the system. For example, x_1 could be the displacement and x_2 could be the velocity. Here, time does not appear explicitly in these state equations. Such systems are referred to as *autonomous* systems and are the kind that we consider in this introduction. Nonlinear dynamic analysis of this system normally follows these steps.

- Determine the *fixed points* of the system. Fixed points are what are normally termed "equilibrium points" in vibrations.
- Determine the *local* dynamic character of the system in the immediate neighborhood of the fixed points from a *linear* analysis.
- Determine the *global* topological character of the nonlinear dynamic behavior from a *nonlinear* analysis.

Each of these determinations would be carried out as a function of the physical parameters of interest in the model, such as mass, stiffness, and damping parameters, or suitably defined nondimensional parameters, in order to determine the full spectrum of the predicted behavior of the system under study.

FIXED POINTS

The first step is the determination of the fixed points, which are always obtained from the solution to the nonlinear system of algebraic equations that result from setting the time derivatives to zero in Eq. (4.5).

$$P(x_1, x_2) = 0 \qquad Q(x_1, x_2) = 0 \tag{4.6}$$

The solutions to these equations are also referred to as *equilibrium points* or *singular points* of the system. Multiple solutions to Eq. (4.6) are possible which would then result in multiple fixed points of the nonlinear dynamic system.

LINEAR OR LOCAL ANALYSIS

The second step in the analysis of a nonlinear system is a *linearized* analysis in the neighborhood of each equilibrium point and is often called *local analysis.* Linear analysis is performed by linearizing the original nonlinear system in Eq. (4.5) about the fixed points. A first-order Taylor series is used to obtain a linear dynamic system of the form

$$\begin{aligned} \frac{d}{dt}(x_1 - \bar{x}_1) &= \left.\frac{\partial P}{\partial x_1}\right|_{\bar{x}_1,\bar{x}_2}(x_1 - \bar{x}_1) + \left.\frac{\partial P}{\partial x_2}\right|_{\bar{x}_1,\bar{x}_2}(x_2 - \bar{x}_2) \\ \frac{d}{dt}(x_2 - \bar{x}_2) &= \left.\frac{\partial Q}{\partial x_1}\right|_{\bar{x}_1,\bar{x}_2}(x_1 - \bar{x}_1) + \left.\frac{\partial Q}{\partial x_2}\right|_{\bar{x}_1,\bar{x}_2}(x_2 - \bar{x}_2) \end{aligned} \tag{4.7}$$

or

$$\begin{aligned} \delta\dot{x}_1 &= a\delta x_1 + b\delta x_2 \\ \delta\dot{x}_2 &= c\delta x_1 + d\delta x_2 \end{aligned} \tag{4.8}$$

where $\bar{x}_1$ and $\bar{x}_2$ represent the fixed point, δx_1 and δx_2 are the deviations from the fixed point, and the constant coefficients (a, b, c, d) are the corresponding first partial derivatives evaluated at the fixed point. The origin of the resulting linear system in the transformed variables represents the fixed point of the original system (without any loss of generality). The resulting linear system of equations can also be represented in matrix form, where the matrix A is referred to as the *jacobian matrix* of the nonlinear system model in Eq. (4.5).

$$\delta\dot{x} = A\delta x \qquad A = \begin{bmatrix} a & b \\ c & d \end{bmatrix} \tag{4.9}$$

The solution to the linearized system in Eq. (4.7), or equivalently in Eq. (4.9), is of the form

$$\delta x = r_1 e^{\lambda_1 t} + r_2 e^{\lambda_1 t} \qquad \lambda^2 - (a + d)\lambda + (ad - bc) = 0 \tag{4.10}$$

where the quadratic equation in λ is referred to as the *characteristic equation* of the linear system and λ_i are the roots of this characteristic equation, which are also the eigenvalues of the jacobian matrix. It is not necessary to solve this linear system in order to determine the interesting qualitative aspects of the solution. Insight into the local dynamic behavior of the original *nonlinear* system, including stability, can be determined from the value of the roots of the characteristic equation as determined by the parameters p and q, defined by the following relationships:

$$p = (a + d) \qquad q = (ad - bc) \tag{4.11}$$

The character of each of the possible solutions is discussed in this section, with the results being displayed in the *phase plane*. A phase plane plot is constructed by taking time as the parameter and examining the solution with the two states plotted on the two axes. Note that this procedure is a *local analysis* in that it provides information on the dynamic behavior of the original nonlinear system (and not just the approximate linearized system) in the immediate neighborhood of the fixed point.

Center. A center occurs when $p = 0$ and $q > 0$. In this case, the phase plane plot results in closed orbits about the fixed point (origin), as shown in Fig. 4.24. An undamped mass-spring system is an example of the center in which the response oscillates forever without dying or increasing.

Focus. A focus occurs when $p^2 - 4q < 0$ and $q > 0$. For a *stable focus*, $p < 0$ and all orbits tend to converge to the fixed point (origin) without a limiting direction, as shown in Fig. 4.25. For an *unstable focus*, $p > 0$ and all orbits tend to diverge from the fixed point (origin) without a limiting direction, as shown in Fig. 4.26. An underdamped linear mass-spring oscillator behaves as a stable focus.

Node. A node occurs when $p^2 - 4q > 0$ and $q > 0$. For a *stable node*, $p < 0$ and all orbits converge to the fixed point (origin) with a limiting direction. For an *unstable node*, $p > 0$ and all orbits diverge from the fixed point (origin), as illustrated in Fig. 4.27. The phase plane plot of a stable node is similar to Fig. 4.27 except that the arrows would point into the origin. This is the behavior of an overdamped mass-spring system, where the response dies out without oscillations.

Saddle Point. A saddle point occurs when $q < 0$. Only four orbits connect with the fixed point (origin), as shown in Fig. 4.28. An orbit that converges to the saddle point

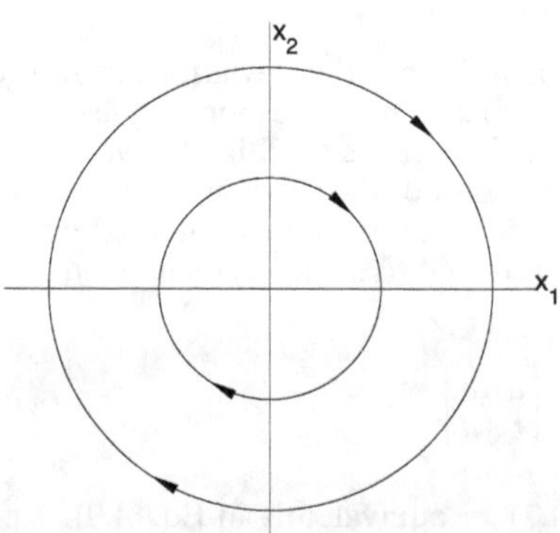

FIGURE 4.24 Phase plane orbits for a center.

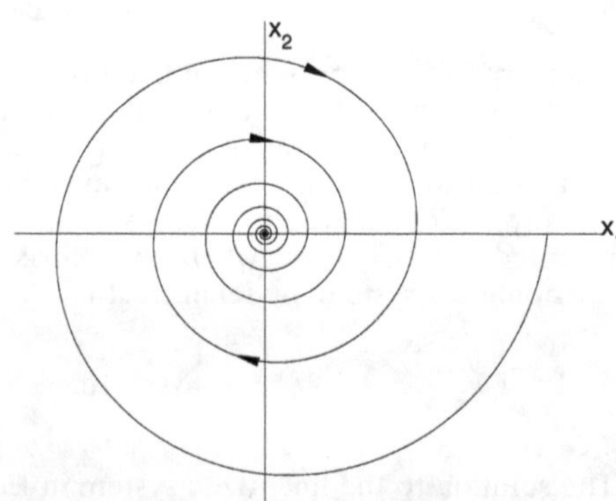

FIGURE 4.25 Phase plane orbits for a stable focus.

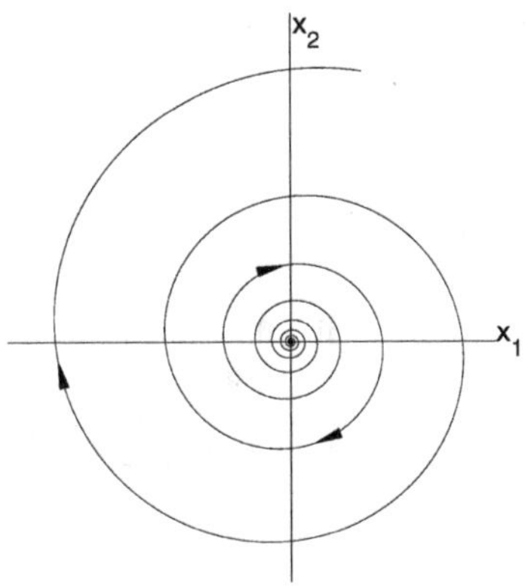

FIGURE 4.26 Phase plane orbits for an unstable focus.

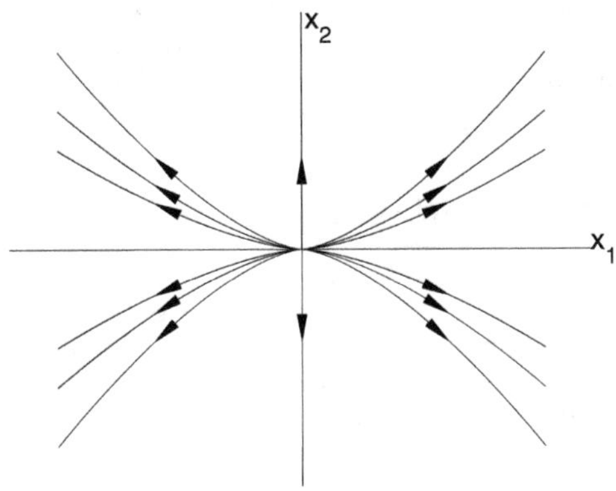

FIGURE 4.27 Phase plane orbits for an unstable node.

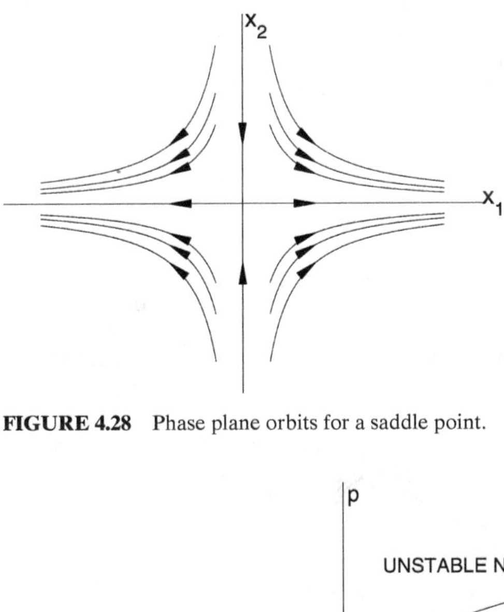

FIGURE 4.28 Phase plane orbits for a saddle point.

is called an *incoming separatrix*, and an orbit that diverges is called an *outgoing separatrix*. An example of a saddle point is the upright equilibrium point of a pendulum; a practical example would be the vertical position of a crane or the torso of a human being, both of which are inherently unstable and need a stabilizing system (such as a hydraulic actuator).

These results for the local dynamic behavior can be summarized on the parameter plane (p, q), as shown in Fig. 4.29. The lines dividing the regions of topo-

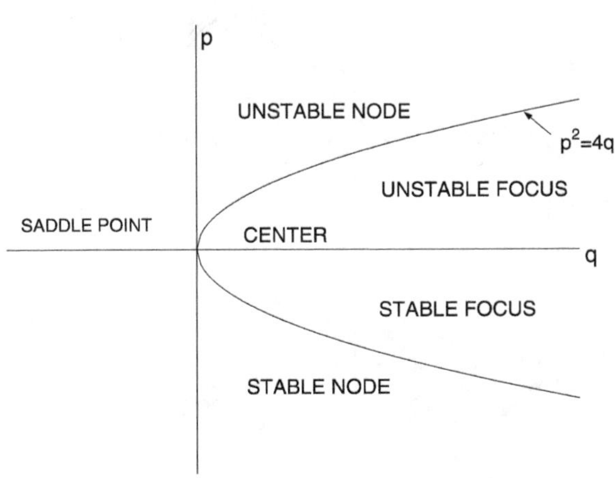

FIGURE 4.29 Summary of solutions in the parameter plane.

logically different orbits are called *bifurcation lines* in the parameter space. When a parameter changes in such a way that a bifurcation line is crossed, the qualitative nature of the dynamic response changes, and the system is said to go through a *dynamic phase transition.*

GLOBAL ANALYSIS

The preceding discussion on local analysis dealt with the local behavior of the system in a neighborhood around each fixed point. In order to obtain a complete understanding of the dynamic behavior of the system, a *global analysis* is necessary. This analysis is not always easy to perform—or even possible. A systematic approach begins with the equation

$$\frac{dx_1}{dx_2} = \frac{P(x_1,x_2)}{Q(x_1,x_2)} \tag{4.12}$$

which determines the tangent of the orbit or state trajectory at any point in the state space. The determination of the local slope of the orbits at a variety of points, combined with the fixed-point information gained from a local linear analysis, can lead to an estimate of the global phase portrait. Numerical integration techniques are typically used to determine these global state trajectories. This investigation is typically carried out only in regions of the state space of interest, resulting in substantial savings in time and computational effort.

LIMIT CYCLE

One important example of nonlinear dynamic behavior determined from a global analysis is the *limit cycle.* A limit cycle is a closed, periodic state trajectory or orbit. If the limit cycle is stable, neighboring orbits tend toward it as $t \to \infty$, resulting in sustained periodic or cyclic oscillations in the dynamic response of the system. If the limit

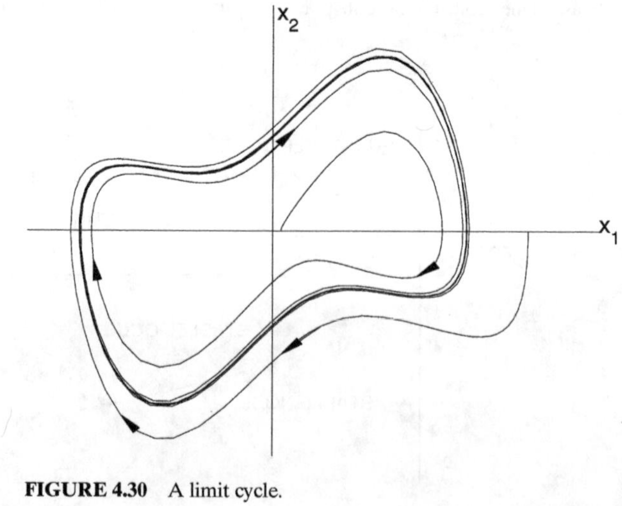

FIGURE 4.30 A limit cycle.

cycle is unstable, neighboring orbits tend away from it. Because neighboring orbits tend away from an unstable limit cycle, it would not be observed as cyclic oscillations in the physical system. However, it is still important to know whether this behavior exists in regions of the parameter space. An example of an unstable fixed point at (0,0), along with a stable limit cycle to which trajectories from both the inside and the outside converge, is shown in Fig. 4.30. Note that a local linear analysis would predict the unstable fixed point but would not be able to predict the limit cycle that would, nevertheless, be observed in practice. Limit cycles are an important phenomenon in the full panoply of nonlinear dynamics. There are a number of practical examples of limit cycles in vibrating systems such as flutter of aircraft wings, hammering of water pipes, chattering in machining operations, and brake disc vibration. In the larger context, stable nodes, foci, and limit cycles represent the simplest forms of *attractors*. When all state trajectories that start within a neighborhood of an equilibrium point or periodic solution converge to that solution, it is referred to as an *attractor*.

BIFURCATIONS

A *bifurcation* refers to a qualitative change in the nature of the system dynamics as one or more model parameters are changed. The term bifurcation is applied to the following two distinct situations as a model parameter is changed: (1) a change in the number or type of fixed points (equilibrium points) and (2) a change in the global phase portrait or global dynamic behavior. Of particular interest in the nonlinear dynamic analysis of a system is the occurrence of a change in the stability of a fixed point and the occurrence of a limit cycle in the phase plane trajectories as a model parameter is varied.

In this brief introduction, a very simple single-degree-of-freedom vibrating system with a parameter c is used to illustrate the concepts.

$$\dot{x} = F(x,c) = (x-1)^2 - c - 2 \tag{4.13}$$

The fixed points, or equilibrium states, of this simple autonomous system are determined from the following equation

$$\dot{x} = 0 \qquad F(x,c) = 0 \qquad \text{for } x = x_s(c) \tag{4.14}$$

where c is the control parameter and the fixed points $x_s(c)$ are a function of the value of the control parameter. If the number of fixed points changes as c is varied, the system goes through a bifurcation as shown in the *control-phase space* in Fig. 4.31. When $c < -2$, there are no fixed points. When $c = c_0 = -2$, two fixed points come into existence and the point c_0 is called a *bifurcation point* of the system model. A bifurcation point occurs when there are two or more distinct steady-state solutions $F(x_1,c) = 0$ and $F(x_2,c) = 0$ in a neighborhood of the fixed point $F(x_s(c_0),c_0) = 0$. When $\frac{\partial F}{\partial x} \neq 0$ for any x_s such that $F(x_s(c),c) = 0$, this value of c does not represent a bifurcation point. This necessary condition for a bifurcation point is a result of the *implicit function theorem.*[24]

Example 4.1. An interesting example of a *pitchfork bifurcation* in a vibration system is a uniform rotating rigid rod hinged at one end (O), as shown in Fig. 4.32. The equation of motion can be shown to be

$$\ddot{\theta} + \frac{3g}{2\ell}\sin\theta - \Omega^2 \cos\theta \sin\theta = 0 \tag{4.15}$$

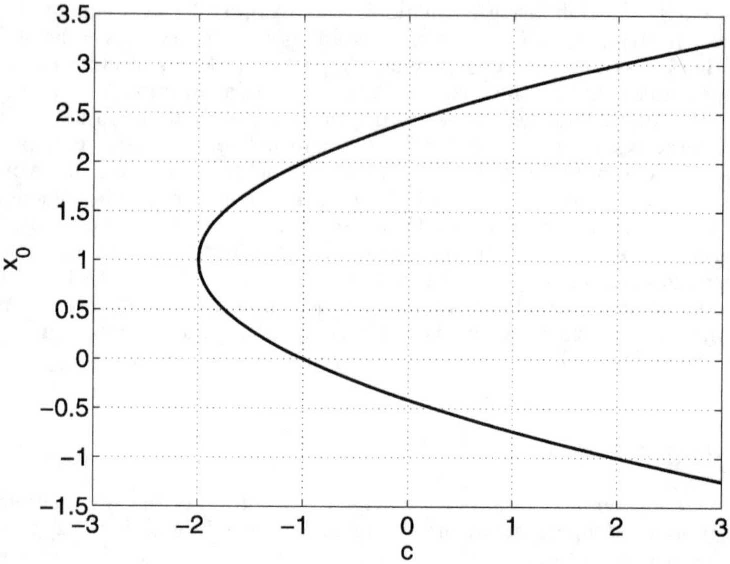

FIGURE 4.31 Bifurcation example.

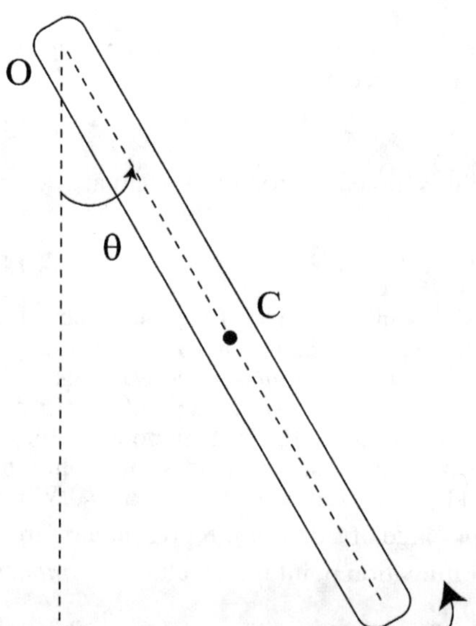

FIGURE 4.32 Rotating rigid rod.

Application of the analysis procedures already discussed leads to the following results, as illustrated in Fig. 4.33. For low speeds, the vertically hanging position ($\theta_0 = 0$) is a stable equilibrium position; hence, small deviations will return the rod to that equilibrium. As the speed is increased, at a critical speed given by $\Omega_c = \sqrt{3g/2\ell}$, a supercritical pitchfork bifurcation occurs, giving rise to two new stable equilibrium positions. At the same time, the zero (vertically hanging) position becomes unstable. The new equilibrium positions are dependent on the speed and are given by $\theta_0 = \pm\cos^{-1}(3g/2\ell\Omega^2)$. This means that for speeds above the critical speed, after being perturbed, the spinning rod moves away from its hanging position and settles in one of these two nonzero equilibrium positions.

Andronov-Hopf Bifurcation. A bifurcation that involves the change in the stability of a focus, along with the birth of a limit cycle as a control param-

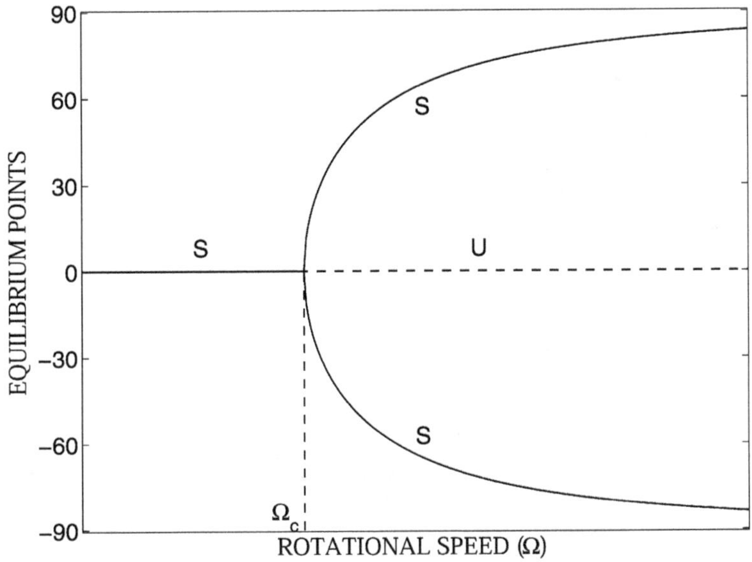

FIGURE 4.33 Bifurcation diagram for the rotating rigid rod; S—stable, U—unstable.

eter is varied, is called an *Andronov-Hopf bifurcation,* or simply a *Hopf bifurcation.* This bifurcation can be visualized using a three-dimensional perspective, with the control parameter axis being added to the phase plane. In this space, there is a surface that contains only periodic solutions of the nonlinear equations, referred to as a *center manifold,* in the control-phase space. These periodic solutions exist a finite distance from the fixed point independent of the initial conditions leading to the limit cycle, unlike a *center* where the distance from the fixed point is determined by the initial condition.

The periodic solutions arising from a Hopf bifurcation are either stable or unstable limit cycles. If the surface consists of stable limit cycles, the bifurcations are said to be *supercritical,* or *soft,* as shown in Fig. 4.34. If the center manifold consists of unstable limit cycles, the bifurcation point is called *subcritical,* or *hard,* as illustrated in Fig. 4.35. In this case, even if the fixed point is stable, a small finite perturbation can take the solution well outside the unstable limit cycle and into a solution space far away from the fixed point, hence, the word *hard.* This type of instability, which cannot be detected by a linear analysis, is especially important because it can produce drastic effects, even for small perturbations, that can have profound practical implications. In fluid mechanics, the transition from laminar to turbulent flow is an example of this phenomenon. The Hopf bifurcation concerns the creation of a limit cycle around a fixed point and is a *localized* phenomenon. A *global* method of determining the existence of periodic solutions is given by a mathematical result called the Poincaré-Bendixson theorem.[25]

Global Bifurcations. The topological character of the phase portraits of a system can change when the control parameter is varied (leading to a bifurcation) without changing the type or stability of any of the fixed points of the system. In such

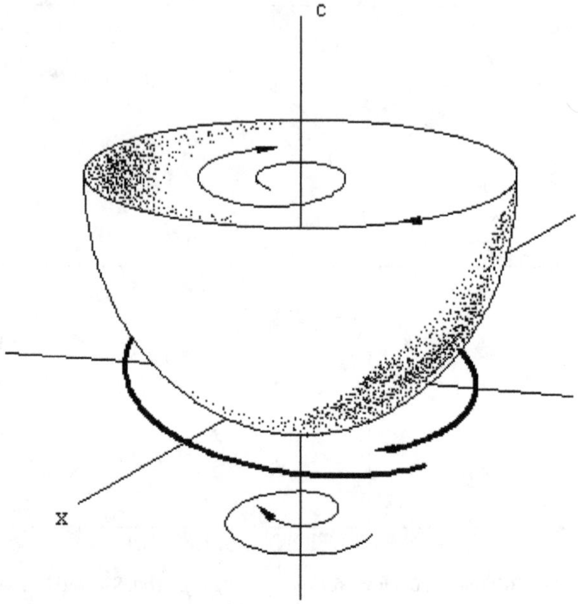

FIGURE 4.34 Supercritical bifurcation.

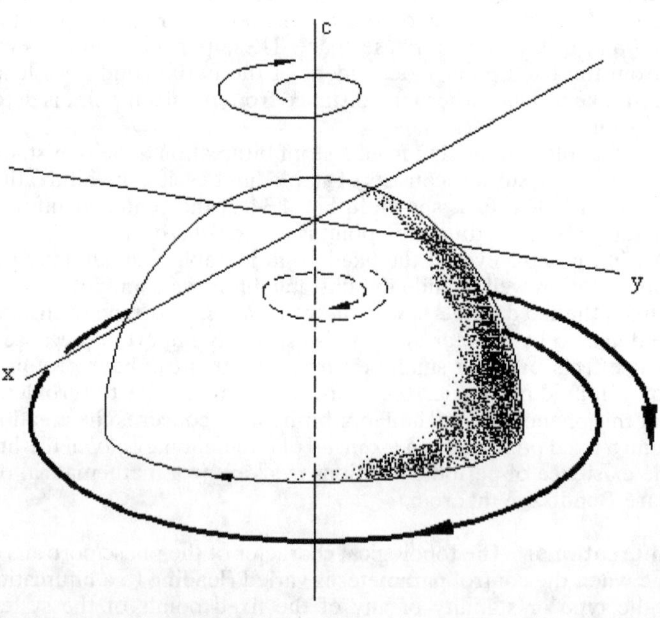

FIGURE 4.35 Subcritical bifurcation.

instances, the change in the phase portraits is not observable in the immediate neighborhood of any fixed point, but can only be discerned on a global scale. Such occurrences are *global bifurcations* and are very difficult to determine because they cannot be detected using the elementary theories outlined here. A judicious use of computer solutions, along with some analytical tools, is then necessary.

As an illustration of the complexities of global bifurcations, consider the following nonlinear vibrating system example.[27]

$$\ddot{x} = c_1 + c_2 x + x^2 + x\dot{x} \tag{4.16}$$

The fixed points of this system are

$$x_0 = -\frac{1}{2}c_2 \pm \frac{1}{2}\sqrt{c_2^2 - 4c_1} \tag{4.17}$$

when $c_2^2 > 4c_1$. Along the curve $c_2^2 = 4c_1$ in the control space, the fixed points are degenerate. If the parameter values cross this curve, there is a localized change in the phase portrait: the two fixed points come together and vanish. It can be shown that these two simple singularities must be a saddle point and a node or a focus. Such a birth of a saddle point and a node is called a *saddle-node bifurcation*.

In addition, there is a curve associated with an Andronov-Hopf bifurcation and a curve associated with a global bifurcation, as shown in Fig. 4.36. The bifurcation lines are labeled SN^-, SN^+, AH, and SC, which divide the control space region into the regions A, B, C, and D. The local saddle-node and Hopf bifurcations occur across the lines $SN^\pm$ and AH, respectively, whereas the global bifurcation called a *saddle connection* occurs across the curve SC. Starting with region A, where $4c_1 > c_2^2$, and moving clockwise around the origin, the first fixed points occur upon crossing SN^+. This crossing gives birth to a saddle point and an unstable node. Moving from B_1 to B_2, the spiral on the node tightens and a Hopf bifurcation occurs on AH. The result is a

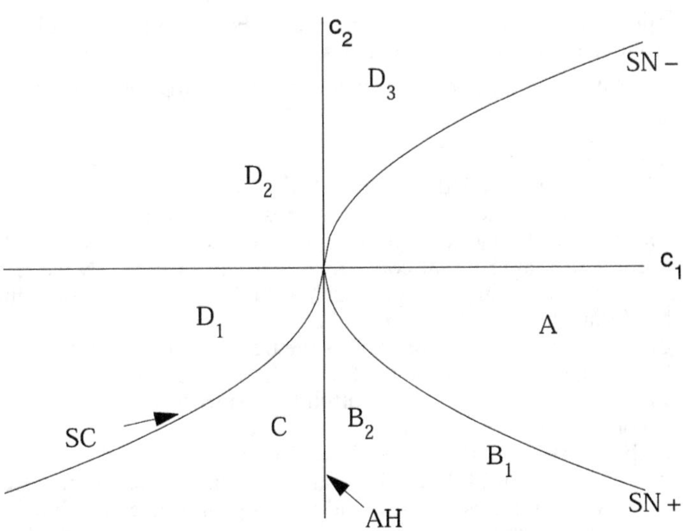

FIGURE 4.36 Global bifurcation example.

limit cycle in region C in addition to a stable node and a saddle point. The saddle connection occurs along SC. Crossing into region D, there is no longer a limit cycle and one branch of the saddle point flows into the second fixed point. Crossing the curve SC does not change the saddle point and the stable node.

CHAOS

When a nonlinear system is driven by a periodic forcing function, such as can occur with harmonic excitation, chaotic dynamic behavior is possible depending on the nature of the nonlinear system and the frequency and amplitude of the driving force. Attractors in the form of points or limit cycles in the phase plane are associated with stable steady-state dynamics and periodic oscillations. *Strange attractors* are associated with chaotic dynamic behavior. This behavior arises from the convergence of trajectories originating from the exterior of a strange attractor to its interior and the divergence of neighboring trajectories within the interior of a strange attractor away from each other. A strange attractor is therefore stable, but the motion within the attractor is unstable. The result of the divergence of neighboring trajectories is an extremely sensitive dependence on the initial conditions of the dynamic system.

Chaotic dynamic systems are often classified into categories such as self-excited and polynomial oscillators. Self-excited systems are systems that are capable of sustaining limit cycles without any external excitation (such as the van der Pol equation); polynomial oscillators are oscillators with polynomial terms added to them (like Duffing's equation and the simple pendulum). There are indeed mathematical theorems that can be used to establish the existence of chaos under certain specific conditions. Although such a systematic understanding is not available at present to apply to all cases, the following general forms describe the resulting chaotic dynamic behavior.

1. Periodic oscillations with harmonics, subharmonics, and ultraharmonics
2. Almost periodic oscillations representable by Fourier series with incommensurable frequencies (when the frequencies are not related by integers)
3. Coexisting (multistable) periodic oscillations and nonperiodic and unstable solutions

Note that individual solutions, or trajectories, are deterministic and smooth, but a family of solutions can be called "chaotic." The transition of solutions from the first or the second category to the third is quite interesting and is the subject of much current research. The most concise definition of a chaotic solution is a family of solutions with nearly the same initial conditions that can produce dynamic behaviors that are very dissimilar; an intuitive practical example is the infinite sequence of coin tosses. The fact that slightly differing initial conditions can evolve into very different states is indeed a dramatic result that upsets the traditional view of dynamic systems. Fig. 4.37 shows the divergent response for the famous Lorenz attractor (modeling convection rolls), where the solid and dotted lines start from nonidentical, but very close, initial conditions.[25]

Chaotic vibrations are characterized by an irregular or ragged waveform, such as illustrated in Figs. 4.19*A* and 4.23*A*. Although there may be recurrent patterns in the waveform, they are not precisely alike, and they repeat at irregular intervals, so the motion is truly nonperiodic, as is implied in Zone II of Figs. 4.17*A* and 4.17*B*. Indeed,

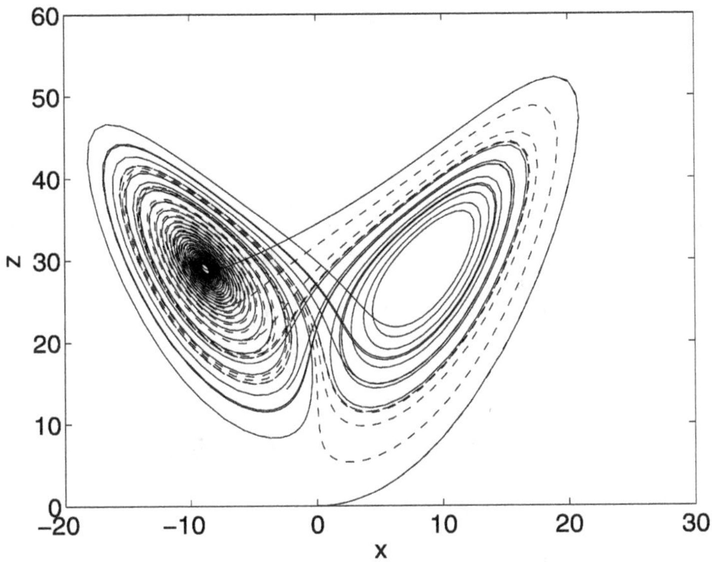

FIGURE 4.37 Lorenz attractor: response for slightly different initial conditions.

care must be taken in characterizing vibrations as chaotic since there are irregular motions which mimic chaotic response but in which there *are* recurrent patterns which repeat at regular intervals, such as are implied in Zone III of Figs. 4.17*A* and 4.17*B*.

For all its irregularity, there is a certain basic structure and patternation implicit in chaotic vibration. As one can infer from the response curves of local peak amplitude for chaotic vibration shown in Zone II of Figs. 4.17*A* and 4.17*B*, the maximum amplitude is bounded.

A remarkable response behavior associated with chaotic vibration is the cascade of period-doubling bifurcations or tree-like structure in the peak amplitude response curve (illustrated in Zone I of Figs. 4.17*A* and *B*) that may take place in transition from simple periodic response to chaotic response.

But the most remarkable property of chaotic vibrations is evident in the Poincaré section of the motion shown typically in Figs. 4.19*B* and 4.23*B*. The Poincaré section contains a large number of discrete points of velocity plotted as a function of the displacement of the chaotic motion, where the points are sampled *stroboscopically* with reference to a particular phase angle of the forcing periodic function. Rather than a random scatter of points, the Poincaré section generally reveals striking patterns. The Poincaré section is sometimes referred to as an *attractor*.

Chaotic vibration also differs from random motion in that the power frequency spectrum (see Chap. 19 of this handbook) generally has distinct peaks rather than consisting of random motion with a broadband spectrum. There will often be not only synchronous response peaks at the forcing function frequency as in the response of linear systems, but also significant asynchronous response peak (or peaks) at the system's natural frequency (or frequencies).

Chaotic vibration has been observed and numerically predicted in many practical machine components such as bearings.[28] More details on this fascinating topic can be found in general texts on chaotic dynamics.[20,24]

EXACT SOLUTIONS

It is possible to obtain exact solutions for only a relatively few second-order nonlinear differential equations. In this section, some of the more important of these exact solutions are listed. They are exact in the sense that the solution is given either in closed form or in an expression that can be evaluated numerically to any desired degree of accuracy. Some general examples follow.

FREE VIBRATION

Consider the free vibration of an undamped system with a general restoring force $f(x)$ as governed by the differential equation

$$\ddot{x} + \kappa^2 f(x) = 0$$

This can be rewritten as

$$\frac{d(\dot{x}^2)}{dx} + 2\kappa^2 f(x) = 0 \tag{4.18}$$

and integrated to yield

$$\dot{x}^2 = 2\kappa^2 \int_x^X f(\xi)\, d\xi$$

where ξ is an integration variable and X is the value of the displacement when $\dot{x} = 0$. Thus

$$|\dot{x}| = \kappa\sqrt{2}\sqrt{\int_x^X f(\xi)\, d\xi}$$

This may be integrated again to yield

$$t - t_0 = \frac{1}{\kappa\sqrt{2}} \int_0^x \frac{d\zeta}{\sqrt{\int_\zeta^X f(\xi)\, d\xi}} \tag{4.19}$$

where ζ is an integration variable and t_0 corresponds to the time when $x = 0$. The displacement-time relation may be obtained by inverting this result. Considering the restoring force term to be an odd function, i.e.,

$$f(-x) = -f(x)$$

and considering Eq. (4.19) to apply to the time from zero displacement to maximum displacement, the period τ of the vibration is

$$\tau = \frac{4}{\kappa\sqrt{2}} \int_0^X \frac{d\zeta}{\sqrt{\int_\zeta^X f(\xi)\, d\xi}} \tag{4.20}$$

Exact solutions can be obtained in all cases where the integrals in Eq. (4.20) can be expressed explicitly in terms of X.

NUMERICAL METHODS

Although exact solutions and asymptotic methods can provide highly satisfying broad-spectrum results that are valid over large parameter ranges, they are usually valid for restrictive conditions (such as small nonlinearities); moreover, they can only reveal certain limited aspects of nonlinear behavior. Certain intrinsically nonlinear phenomena such as bifurcations and chaos can be predicted and verified only by numerical methods.

For obtaining steady-state or equilibrium solutions, one should be cognizant of the fact that the convergence to a particular solution depends on the initial conditions (based on the domain of attraction of the fixed point). The difficult aspect of determining the integration step size (to trade off computation time and accuracy) has been mitigated significantly in recent years by the development of highly efficient adaptive solution algorithms and powerful desktop computers. Still, numerical algorithms should never be used blindly with the faith that they are automatic systems that will always provide correct solutions with no active involvement of the user. On the other hand, knowledge of the nonlinear phenomena such as those presented in this chapter should be used as a guide in the employment of numerical techniques.

Straightforward numerical integration can often fail near singularities and bifurcation points. Normal form theory is then often employed to derive polynomial representations to describe the dynamics near singularities of certain simplified nonlinear models. For these simpler models, direct analytical solution of the nonlinear steady-state equations for bifurcation analysis is often possible. For more complex and higher dimensional system models, alternate numerical solution techniques must be employed.

Most root-finding numerical algorithms for nonlinear systems of equations are not useful for bifurcation analysis of the steady-state equations. These algorithms find only a single solution that is a function of the initial estimate, making it difficult to ensure that all solutions have been found. In order to reliably find all solutions and bifurcation points (such as the illustrations in the preceding section), analytic continuation methods are typically employed to compute bifurcation diagrams. Some specialized public domain software packages (AUTO[29] is an example) have been developed recently.

When numerical simulation is employed to determine and characterize chaotic vibrations, care must be taken to ensure that modern adaptive schemes with adequate numerical precision are used, since the solution obtained can be sensitive to this choice.

APPROXIMATE ANALYTICAL METHODS

A large number of approximate analytical methods of nonlinear vibration analysis exist, each of which may or may not possess advantages for certain classes of problems. Some of these are restricted techniques which may work well with some types of equations but not with others. The methods which are outlined below are among the better known and possess certain advantages as to ranges of applicability.

Approximate analytical methods, while useful for yielding insights into basic mechanisms and relative influence of independent variables, have been largely displaced by numerical methods which are capable of giving very precise results for very much more complex models by exploiting the enormous power of modern computers.

DUFFING'S METHOD

Consider the nonlinear differential equation (known as Duffing's equation)

$$\ddot{x} + \kappa^2(x \pm \mu^2 x^3) = p \cos \omega t \tag{4.21}$$

where the ± sign indicates either a hardening or softening system. As a first approximation to a harmonic solution, assume that

$$x_1 = A \cos \omega t \tag{4.22}$$

and rewrite Eq. (4.21) to obtain an equation for the second approximation:

$$\ddot{x}_2 = -(\kappa^2 A \pm \tfrac{3}{4}\kappa^2\mu^2 A^3 - p) \cos \omega t - \tfrac{1}{4}\kappa^2\mu^2 A^3 \cos \omega t$$

This equation may now be integrated to yield

$$x_2 = \frac{1}{\omega^2}(\kappa^2 A \pm \tfrac{3}{4}\kappa^2\mu^2 A^3 - p) \cos \omega t + \tfrac{1}{36}\kappa^2\mu^2 A^3 \cos 3\omega t \tag{4.23}$$

where the constants of integration have been taken as zero to ensure periodicity of the solution.

This may be regarded as an iteration procedure by reinserting each successive approximation into Eq. (4.21) and obtaining a new approximation. For this iteration procedure to be convergent, the nonlinearity must be small; i.e., κ^2, μ^2, A, and p must be small quantities. This restricts the study to motions in the neighborhood of linear vibration (but not near $\omega = \kappa$, since A would then be large); thus, Eq. (4.22) must represent a reasonable first approximation. It follows that the coefficient of the cos ωt term in Eq. (4.23) must be a good second approximation and should not be far different from the first approximation.[30] Since this procedure furnishes the exact result in the linear case, it might be expected to yield good results for the "slightly nonlinear" case. Thus, a relation between frequency and amplitude is found by equating the coefficients of the first and second approximations:

$$\omega^2 = \kappa^2(1 \pm \tfrac{3}{4}\mu^2 A^2) - \frac{p}{A} \tag{4.24}$$

This relation describes the response curves, as shown in Fig. 4.14.

This method applies equally well when linear velocity damping is included.

THE PERTURBATION METHOD

In one of the most common methods of nonlinear vibration analysis, the desired quantities are developed in powers of some parameter which is considered small; then the coefficients of the resulting power series are determined in a stepwise manner. The method is straightforward, although it becomes cumbersome for actual computations if many terms in the perturbation series are required to achieve a desired degree of accuracy.

Consider Duffing's equation, Eq. (4.21), in the form

$$\omega^2 x'' + \kappa^2(x + \mu^2 x^3) - p \cos \phi = 0 \tag{4.25}$$

where $\phi = \omega t$ and primes denote differentiation with respect to ϕ. The conditions at time $t = 0$ are $x(0) = A$ and $x'(0) = 0$, corresponding to harmonic solutions of period $2\pi/\omega$. Assume that μ^2 and p are small quantities, and define $\kappa^2\mu^2 \equiv \varepsilon, p \equiv \varepsilon p_0$. The displacement $x(\phi)$ and the frequency ω may now be expanded in terms of the small quantity ε:

$$\begin{aligned} x(\phi) &= x_0(\phi) + \varepsilon x_1(\phi) + \varepsilon^2 x_2(\phi) + \cdots \\ \omega &= \omega_0 + \varepsilon\omega_1 + \varepsilon^2\omega_2 + \cdots \end{aligned} \tag{4.26}$$

The initial conditions are taken as $x_i(0) = x_i'(0) = 0$ $[i = 1,2, \ldots]$.

Introducing Eq. (4.26) into Eq. (4.25) and collecting terms of zero order in ε gives the linear differential equation

$$\omega_0^2 x_0'' + \kappa^2 x_0 = 0$$

Introducing the initial conditions into the solution of this linear equation gives $x_0 = A \cos \omega t$ and $\omega_0 = \kappa$. Collecting terms of the first order in ε,

$$\omega_0^2 x_1'' + \kappa^2 x_1 - (2\omega_0\omega_1 A - \tfrac{3}{4}A^3 + p_0) \cos \phi + \tfrac{1}{4}A^3 \cos 3\phi = 0 \tag{4.27}$$

The solution of this differential equation has a nonharmonic term of the form $\phi \cos \phi$, but since only harmonic solutions are desired, the coefficient of this term is made to vanish so that

$$\omega_1 = \frac{1}{2\kappa}\left(\tfrac{3}{4}A^2 - \frac{p_0}{A}\right)$$

Using this result and the appropriate initial conditions, the solution of Eq. (4.27) is

$$x_1 = \frac{A^3}{32\kappa^2}(\cos 3\phi - \cos \phi)$$

To the first order in ε, the solution of Duffing's equation, Eq. (4.25), is

$$x = A \cos \omega t + \varepsilon \frac{A^3}{32\kappa^2}(\cos 3\omega t - \cos \omega t)$$

$$\omega = \kappa + \frac{\varepsilon}{2\kappa}\left(\tfrac{3}{4}A^2 - \frac{p_0}{A}\right)$$

This agrees with the results obtained previously [Eqs. (4.23) and (4.24)]. The analysis may be carried beyond this point, if desired, by application of the same general procedures.

As a further example of the perturbation method, consider the self-excited system described by Van der Pol's equation

$$\ddot{x} - \varepsilon(1 - x^2)\dot{x} + \kappa^2 x = 0 \tag{4.28}$$

where the initial conditions are $x(0) = 0, \dot{x}(0) = A\kappa_0$. Assume that

$$x = x_0 + \varepsilon x_1 + \varepsilon^2 x_2 + \cdots$$

$$\kappa^2 = \kappa_0^2 + \varepsilon\kappa_1^2 + \varepsilon^2\kappa_2^2 + \cdots$$

Inserting these series into Eq. (4.28) and equating coefficients of like terms, the result to the order ε^2 is

$$x = \left(2 - \frac{29\varepsilon^2}{96\kappa_0^2}\right)\sin \kappa_0 t + \frac{\varepsilon}{4\kappa_0}\cos \kappa_0 t + \frac{\varepsilon}{4\kappa_0}\left(\frac{3\varepsilon}{4\kappa_0}\sin 3\kappa_0 t - \cos 3\kappa_0 t\right) - \frac{5\varepsilon^2}{124\kappa_0^2}\sin 5\kappa_0 t \tag{4.29}$$

THE METHOD OF KRYLOFF AND BOGOLIUBOFF[31]

Consider the general autonomous differential equation

$$\ddot{x} + F(x, \dot{x}) = 0$$

which can be rewritten in the form

$$\ddot{x} + \kappa^2 x + \varepsilon f(x, \dot{x}) = 0 \qquad [\varepsilon << 1] \tag{4.30}$$

For the corresponding linear problem ($\varepsilon \equiv 0$), the solution is

$$x = A \sin(\kappa t + \theta) \tag{4.31}$$

where A and θ are constants.

The procedure employed often is used in the theory of ordinary linear differential equations and is known variously as the method of variation of parameters or the method of Lagrange. In the application of this procedure to a nonlinear equation of the form of Eq. (4.30), assume the solution to be of the form of Eq. (4.31) but with A and θ as time-dependent functions rather than constants. This procedure, however, introduces an excessive variability into the solution; consequently, an additional restriction may be introduced. The assumed solution, of the form of Eq. (4.30), is differentiated once considering A and θ as time-dependent functions; this is made equal to the corresponding relation from the linear theory (A and θ constant) so that the additional restriction

$$\dot{A}(t) \sin[\kappa t + \theta(t)] + \dot{\theta}(t)A(t) \cos[\kappa t + \theta(t)] = 0 \tag{4.32}$$

is placed on the solution. The second derivative of the assumed solution is now formed and these relations are introduced into the differential equation, Eq. (4.30). Combining this result with Eq. (4.32),

$$\dot{A}(t) = -\left(\frac{\varepsilon}{\kappa}\right) f[A(t) \sin \Phi, A(t)\kappa \cos \Phi] \cos \Phi$$

$$\dot{\theta}(t) = \frac{\varepsilon}{\kappa A(t)} f[A(t) \sin \Phi, A(t)\kappa \cos \Phi] \sin \Phi$$

where $$\Phi = \kappa t + \theta(t)$$

Thus, the second-order differential equation, Eq. (4.30), has been transformed into two first-order differential equations for $A(t)$ and $\theta(t)$.

The expressions for $\dot{A}(t)$ and $\dot{\theta}(t)$ may now be expanded in Fourier series:

$$\dot{A}(t) = -\left(\frac{\varepsilon}{\kappa}\right)\left\{K_0(A) + \sum_{n=1}^{r} [K_n(A)\cos n\Phi + L_n(A)\sin n\Phi]\right\}$$
$$\dot{\theta}(t) = \frac{\varepsilon}{\kappa A}\left\{P_0(A) + \sum_{n=1}^{r} [P_n(A)\cos n\Phi + Q_n(A)\sin n\Phi]\right\} \tag{4.33}$$

where

$$K_0(A) = \frac{1}{2\pi}\int_0^{2\pi} f[A\sin\Phi, A\kappa\cos\Phi]\cos\Phi\, d\Phi$$

$$P_0(A) = \frac{1}{2\pi}\int_0^{2\pi} f[A\sin\Phi, A\kappa\cos\Phi]\sin\Phi\, d\Phi$$

It is apparent that A and θ are periodic functions of time of period $2\pi/\kappa$; therefore, during one cycle, the variation of $\dot{A}$ and $\dot{\theta}$ is small because of the presence of the small parameter ε in Eqs. (4.33). Hence, the average values of $\dot{A}$ and $\dot{\theta}$ are considered. Since the motion is over a single cycle, and since the terms under the summation signs are of the same period and consequently vanish, then approximately:

$$\dot{A} \simeq -\left(\frac{\varepsilon}{\kappa}\right)K_0(A)$$

$$\dot{\theta} \simeq \frac{\varepsilon}{\kappa A}P_0(A)$$

$$\dot{\Phi} \simeq \kappa + \frac{\varepsilon}{\kappa A}P_0(A)$$

For example, consider Rayleigh's equation

$$\ddot{x} - (\alpha - \beta\dot{x}^2)\dot{x} + \kappa^2 x = 0 \tag{4.34}$$

By application of the above procedures:

$$\dot{A} = -\left(\frac{1}{\kappa}\right)K_0(A) = -\frac{1}{\kappa}\left[\frac{1}{2\pi}\int_0^{2\pi}(-\alpha + \beta A^2\kappa^2\cos^2\Phi)A\kappa\cos^2\Phi\, d\Phi\right]$$
$$= \frac{A}{2}(\alpha - \tfrac{3}{4}\beta A^2\kappa^2) \tag{4.35}$$

Equation (4.35) may be integrated directly:

$$t = 2\int_{A_0}^{A}\frac{dA}{A(\alpha - \gamma A^2)} = \frac{1}{\alpha}\ln\frac{A^2}{\alpha - \gamma A^2}$$

Solving for A,

$$A = \frac{\alpha}{\gamma}\left[\frac{1}{1 + \left(\frac{\alpha}{\gamma A_0^2} - 1\right)e^{-\alpha t}}\right]^{1/2} \tag{4.36}$$

where $$\gamma = \tfrac{3}{4}\beta^2\kappa^2 \tag{4.37}$$

The application of the method to Van der Pol's equation, Eq. (4.28), is easily accomplished and leads to a solution in the first approximation of the form similar to that of the perturbation solution given by Eq. (4.24).

THE RITZ METHOD

In addition to methods of nonlinear vibration analysis stemming from the idea of small nonlinearities and from extensions of methods applicable to linear equations, other methods are based on such ideas as satisfying the equation at certain points of the motion or satisfying the equation in the average. The Ritz method is an example of the latter method and is quite powerful for general studies.

One method of determining such "average" solutions is to multiply the differential equation by some "weight function" $\psi_n(t)$ and then integrate the product over a period of the motion. If the differential equation is denoted by E, this procedure leads to

$$\int_0^{2\pi} E \cdot \psi_n(t)\, dt = 0 \tag{4.38}$$

A second method of obtaining such average solutions can be derived from the calculus of variations by seeking functions that minimize a certain integral:

$$I = \int_{t_0}^{t_1} F(\dot{x},x,t)\, dt = \text{minimum}$$

Consider a function of the form

$$\tilde{x}(t) = a_1\psi_1(t) + a_2\psi_2(t) + \cdots + a_n\psi_n(t)$$

where the $\psi_k(t)$ are prescribed functions. If $\tilde{x}$ is now introduced for x, then

$$I = I(a_1, a_2, \ldots, a_n)$$

and a necessary condition for I to be a minimum is

$$\frac{\partial I}{\partial a_1} = 0, \qquad \frac{\partial I}{\partial a_2} = 0, \ldots, \qquad \frac{\partial I}{\partial a_n} = 0 \tag{4.39}$$

This gives n equations of the form

$$\frac{\partial I}{\partial a_k} = \int_{t_0}^{t_1} \left(\frac{\partial F}{\partial \tilde{x}} \psi_k + \frac{\partial F}{\partial \dot{\tilde{x}}} \dot{\psi}_k \right) dt = 0 \tag{4.40}$$

for determining the n unknown coefficients. Integrating Eq. (4.40),

$$\frac{\partial I}{\partial a_k} = \left[\frac{\partial F}{\partial \dot{\tilde{x}}} \psi_k \right]_{t_0}^{t_1} + \int_{t_0}^{t_1} \left[\frac{\partial F}{\partial \tilde{x}} - \frac{d}{dt}\left(\frac{\partial F}{\partial \dot{\tilde{x}}} \right) \right] \psi_k\, dt = 0$$

The first term is zero because ψ_k must satisfy the boundary conditions; the expression in brackets under the integral in the second term is Euler's equation. The conditions given in Eqs. (4.39) then reduce to

$$\int_{t_0}^{t} E(\tilde{x})\psi_k\, dt = 0 \qquad [k = 1, 2, \ldots, n] \tag{4.41}$$

This is the same as Eq. (4.38); thus, it is not necessary to "know" the variational problem, but only the differential equation. The conditions given in Eqs. (4.41) then yield average solutions based on variational concepts.

Examples. As a first example of the application of the Ritz method, consider the equation

$$\ddot{x} + \kappa^2 x^n = 0$$

Assume a single-term solution of the form

$$\tilde{x} = A \cos \omega t$$

The Ritz procedure, defined by Eq. (4.41), gives

$$\int_0^{2\pi} (-\omega^2 A \cos^2 \omega t + \kappa^2 A^n \cos^{n+1} \omega t)\, d(\omega t) = 0$$

from which

$$\frac{\omega^2}{\kappa^2} = \frac{4}{\pi} A^{n-1} \int_0^{\pi/2} \cos^{n+1} \omega t\, d(\omega t) = A^{n-1} \varphi(n) \tag{4.42}$$

The comparable exact solution is

$$\frac{\omega^2}{\kappa^2} = \left[\frac{\pi^2/4}{\psi^2(n)} \right] X^{n-1} = \Phi(n) X^{n-1} \tag{4.43}$$

Values of $\varphi(n)$ from the approximate analysis and $\Phi(n)$ from the exact analysis are compared directly in Table 4.1, affording an appraisal of the accuracy of the method.

TABLE 4.1 Values of the Functions $\psi(n)$, $\Phi(n)$, $\varphi(n)$*

n	$\psi(n)$	$\Phi(n)$	$\varphi(n)$
0	1.4142	1.2337	1.2732
1	1.5708	1.0000	1.0000
2	1.7157	0.8373	0.8488
3	1.8541	0.7185	0.7500
4	1.9818	0.6282	0.6791
5	2.1035	0.5577	0.6250
6	2.2186	0.5013	0.5820
7	2.3282	0.4552	0.5469

* The mathematical expressions for $\psi(n)$, $\Phi(n)$, and $\varphi(n)$ and the equations to which they refer are:

$$\psi(n) = \sqrt{\frac{n+1}{2}} \int_0^1 \frac{du}{\sqrt{1-u^{n+1}}}$$

$$\Phi(n) = \frac{\pi^2/4}{\psi^2(n)} \qquad \text{[Eq. (4.43)]}$$

$$\varphi(n) = \frac{4}{\pi} \int_0^{\pi/2} \cos^{n+1} \sigma\, d\sigma \qquad \text{[Eq. (4.42)]}$$

Consider now the nonautonomous system described by Duffing's equation

$$E \equiv \ddot{x} + \kappa^2(x + \mu^2 x^3) - p \cos \omega t = 0$$

Assuming

$$\tilde{x} = A \cos \phi, \qquad \phi = \omega t$$

the Ritz condition, Eq. (4.41), leads to

$$\int_0^{2\pi} \{[(1 - \eta^2)A - s] \cos \phi + \mu^2 A^3 \cos^3 \phi\} \cos \phi \, d\phi$$

from which the amplitude-frequency relation is

$$(1 - \eta^2)A + \tfrac{3}{4}\mu^2 A^3 = \pm s \tag{4.44}$$

where $$s = \frac{p}{\kappa^2} \qquad \eta^2 = \frac{\omega^2}{\kappa^2}$$

The upper sign indicates vibration in phase with the exciting force. Equation (4.44) describes the response curves shown in Fig. 4.14 and corresponds to Eq. (4.24) obtained by Duffing's method.

Application of the Ritz method to Van der Pol's equation, Eq. (4.28), leads to the identical result given by Eq. (4.37).

GENERAL EQUATIONS FOR RESPONSE CURVES

The Ritz method has been applied extensively in studies of nonlinear differential equations. Some of the general equations for response curves thereby obtained are given here, both as a further example of the application of the method and as a collection of useful relations.

SYSTEM WITH LINEAR DAMPING AND GENERAL RESTORING FORCES

Consider a system with general elastic restoring force (an odd function) and described by the equation of motion

$$a\ddot{x} + b\dot{x} + cf(x) - P \cos \omega t = 0$$

A solution may be assumed in the form

$$\tilde{x} = A \cos (\omega t - \theta) = B \cos \phi + C \sin \phi \tag{4.45}$$

where $\phi = \omega t$, $B = A \cos \theta$, $C = A \sin \theta$. Introducing Eq. (4.45) according to the Ritz conditions, and recalling that $f(x)$ is to be an odd function,

$$\begin{aligned} -a\omega^2 A \cos \theta + b\omega A \sin \theta + cAF(A) \cos \theta &= P \\ -a\omega^2 A \sin \theta - b\omega A \cos \theta + cAF(A) \sin \theta &= 0 \end{aligned} \tag{4.46}$$

where $$F(A) = \frac{1}{\pi A}\int_0^{2\pi} f(A\cos\sigma)\cos\sigma\, d\sigma$$

and σ is simply an integration variable.

Some algebraic manipulations with Eqs. (4.46) give independent equations for the two unknowns A and θ:

$$[F(A) - \eta^2]^2 + 4D^2\eta^2 = \left(\frac{s}{A}\right)^2 \tag{4.47}$$

$$\tan\theta = \frac{2D\eta}{F(A) - \eta^2} \tag{4.48}$$

where η^2 and s are defined according to Eq. (4.44) and

$$\kappa^2 = \frac{c}{a} \qquad p = \frac{P}{a} \qquad D = \frac{b}{2\sqrt{ac}}$$

Equation (4.47) describes response curves of the form shown in Fig. 4.15, and Eq. (4.48) gives the corresponding phase angle relationships. These two equations also yield other special relations which describe various curves in the response diagram:

Undamped free vibration curve (Fig. 4.13),

$$\eta^2 = F(A) \tag{4.49}$$

Undamped response curves (Fig. 4.14),

$$\eta^2 = F(A) \mp \frac{s}{A} \tag{4.50}$$

Locus of vertical tangents of undamped response curves (Fig. 4.17),

$$\eta^2 = F(A) + A\frac{\partial F(A)}{\partial A} \tag{4.51}$$

Damped response curves (Fig. 4.15),

$$\eta^2 = [F(A) - 2D^2] \mp \sqrt{\left(\frac{s}{A}\right)^2 - 4D^2[F(A) - D^2]} \tag{4.52}$$

Locus of vertical tangents of damped response curves (Fig. 4.17),

$$[F(A) - \eta^2]\left[F(A) + A\frac{\partial F(A)}{\partial A} - \eta^2\right] = -4D^2\eta^2 \tag{4.53}$$

The maximum amplitude of vibration is of interest. The amplitude at the point at which a response curve crosses the free vibration curve is termed the *resonance amplitude,* and is determined in the nonlinear case by solving Eqs. (4.49) and (4.52) simultaneously. This leads to

$$2D\eta = \frac{s}{A} \qquad \theta = \frac{\pi}{2} \tag{4.54}$$

The first of these two equations defines a hyperbola in the response diagram, describing the locus of crossing points, as shown in Fig. 4.38; hence, the intersection

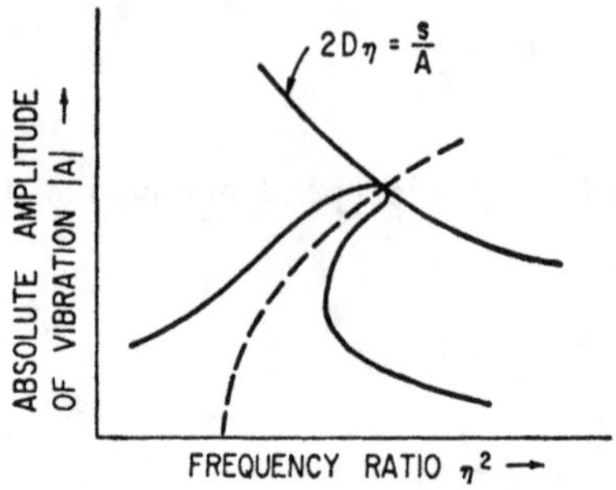

FIGURE 4.38 Determination of the resonant amplitude in accordance with Eq. (4.54).

of this curve with the free vibration curve gives the resonance amplitude. The phase angle at resonance has the value $\pi/2$, as in the linear case. This result is of great help in computing response curves since the effect of damping (except for very large values) is negligible except in the neighborhood of resonance. Therefore, one may compute only the undamped curves (which is not difficult) and the hyperbola (which does not contain the nonlinearity); then, the effect of damping may be sketched in from knowledge of the crossing point.

SYSTEM WITH GENERAL DAMPING AND GENERAL RESTORING FORCES

The preceding analysis may be extended to include the more general differential equation

$$E \equiv \ddot{x} + 2D\kappa g(\dot{x}) + \kappa^2 f(x) - p \cos \omega t = 0$$

By procedures similar to those employed above:

$$[F(A) - \eta^2]^2 + 4D^2S^2(A) = \left(\frac{s}{A}\right)^2 \tag{4.55}$$

$$\tan \theta = \frac{2DS(A)}{F(A) - \eta^2} \tag{4.56}$$

where $$S(A) = \frac{1}{\pi\kappa A} \int_0^{2\pi} g(\omega A \sin \sigma) \sin \sigma \, d\sigma$$

In the case of linear velocity damping, $S(A) = \eta$, and Eqs. (4.55) and (4.56) reduce to Eqs. (4.47) and (4.48). The results for various types of damping forces are:

Coulomb damping:	$g(\dot{x}) = \pm\upsilon_0$	$S(A) = \frac{4}{\pi}\frac{\upsilon_0}{\kappa A}$
Linear velocity damping:	$g(\dot{x}) = \upsilon_1\dot{x}$	$S(A) = \upsilon_1\eta$
Velocity squared damping:	$g(\dot{x}) = \upsilon_2\dot{x}\lvert\dot{x}\rvert$	$S(A) = \frac{8}{3\pi}\upsilon_{2\eta}(A\omega)$
nth-power velocity damping:	$g(\dot{x}) = \upsilon_n\dot{x}\lvert\dot{x}\rvert^{n-1}$	$S(A) = \upsilon_{n\eta}(A\omega)^{n-1}\varphi(n)$

where $\varphi(n)$ is defined in Eq. (4.42) and values are given in Table 4.1.

The locus of resonance amplitudes or crossing points is now given by

$$2DS(A) = \frac{s}{A} \qquad \theta = \frac{\pi}{2}$$

STABILITY OF PERIODIC NONLINEAR VIBRATION

Certain systems having nonlinear restoring forces and undergoing forced vibration exhibit unstable characteristics for certain combinations of amplitude and exciting frequency. The existence of such an instability leads to the "jump phenomenon" shown in Fig. 4.16. To investigate the stability characteristics of the response curves, consider Duffing's equation

$$\ddot{x} + \kappa^2(x + \mu^2 x^3) = p \cos \omega t \tag{4.57}$$

Assume that two solutions of this equation exist and have slightly different initial conditions:

$$x_1 = x_0$$

$$x_2 = x_0 + \delta \qquad [\delta << x_0]$$

Introducing the second of these into Eq. (4.57) and employing the condition that x_0 is also a solution,

$$\ddot{\delta} + \kappa^2(1 + 3\mu^2 x_0^2)\delta = 0 \tag{4.58}$$

Now an expression for x_0 must be obtained; assuming a one-term approximation of the form $x_0 = A \cos \omega t$, Eq. (4.58) becomes

$$\frac{d^2\delta}{d\varphi^2} + (\lambda + \gamma \cos \varphi)\delta = 0 \tag{4.59}$$

where
$$\kappa^2(1 + \tfrac{3}{2}\mu^2 A^2) = 4\omega^2\lambda$$
and
$$\tfrac{3}{2}\kappa^2\mu^2 A^2 = 4\omega^2\gamma \qquad 2\omega t = \varphi \tag{4.60}$$

Equation (4.59) is known as *Mathieu's equation.*

Mathieu's equation has appeared in this analysis as a variational equation characterizing small deviations from the given periodic motion whose stability is to be investigated; thus, the stability of the solutions of Mathieu's equation must be studied. A given periodic motion is stable if *all* solutions of the variational equation associated with it tend toward zero for all positive time and unstable if there is at least one solution which does not tend toward zero. The stability characteristics of Eq. (4.59) often are represented in a chart as shown in Fig. 4.39.

From the response diagram of Duffing's equation, the out-of-phase motion having the larger amplitude appears to be unstable. This portion of the response diagram (Fig. 4.16) corresponds to unstable motion in the Mathieu stability chart (Fig. 4.39), and the locus of vertical tangents of the response curves (considering undamped vibration for simplicity) corresponds exactly to the boundaries between stable and unstable regions in the stability chart. Thus, the region of interest in the response diagram is described by the free vibration

$$\omega^2 = \kappa^2(1 + \tfrac{3}{4}\mu^2 A^2) \tag{4.61}$$

and the locus of vertical tangents

$$\tfrac{3}{2}\kappa^2\mu^2 A^2 + \frac{p}{A} = 0 \tag{4.62}$$

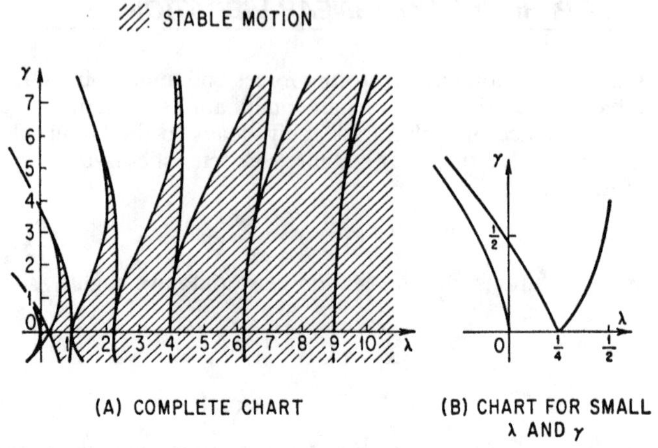

FIGURE 4.39 Stability chart for Mathieu's equation [Eq. (4.59)].

The corresponding curves in the stability chart are taken as those for small positive values of γ and λ which have the approximate equations

$$\gamma = \tfrac{1}{2} - 2\lambda \tag{4.63}$$

$$\gamma = -\tfrac{1}{2} + 2\lambda \tag{4.64}$$

Now, if Eq. (4.61) is introduced into Eqs. (4.60), the resulting equations expanded by the binomial theorem (assuming μ^2 small), and Eq. (4.64) introduced, the result is an identity. Therefore, the free vibration-response curve maps onto the curve of positive slope in the stability chart. The locus of vertical tangents to the response curves maps into the curve of negative slope in the stability chart; this may be seen from the identity obtained by introducing the equations obtained above by binomial expansion into Eq. (4.63) and then employing Eq. (4.62).

In any given case, it can be determined whether a motion is stable or unstable on the basis of the values of γ and λ, according to the location of the point in the stability chart.

The question of stability of response also can be resolved by means of a "stability criterion" developed from the Kryloff-Bogoliuboff procedures. The differential equation of motion is considered in the form

$$\ddot{x} + \kappa^2 x + f(x,\dot{x}) = p \cos \omega t$$

Proceeding in the manner of the Kryloff-Bogoliuboff procedure described earlier,

$$\dot{A} = \frac{1}{\kappa} f(x,\dot{x}) \sin (\kappa t + \theta) - \frac{p}{\kappa} \cos \omega t \sin (\kappa t + \theta)$$

$$\dot{\theta} = \frac{1}{\kappa} f(x,\dot{x}) \cos (\kappa t + \theta) - \frac{p}{A\kappa} \cos \omega t \cos (\kappa t + \theta)$$

Expanding the last terms of these equations, the result contains motions of frequency κ, $\kappa + \omega$, and $\kappa - \omega$. The motion over a long interval of time is of interest, and the motions of frequencies $\kappa + \omega$ and $\kappa - \omega$ may be averaged out; this is accomplished by integrating over the period $2\pi/\omega$:

$$\dot{A} = S(A) - \frac{p}{2\kappa}\sin(\Phi - \omega t)$$

$$\dot{\theta} = \frac{C(A)}{A} - \frac{p}{2\kappa A}\cos(\Phi - \omega t)$$

where $$S(A) = \frac{1}{2\pi\kappa}\int_0^{2\pi} f(A\cos\Phi, -A\kappa\sin\Phi)\sin\Phi\, d\Phi$$

$$C(A) = \frac{1}{2\pi\kappa}\int_0^{2\pi} f(A\cos\Phi, -A\kappa\sin\Phi)\cos\Phi\, d\Phi$$

The steady-state solution may be determined by employing the conditions $A = A_0$, $\psi = \Phi - \omega t = \psi_0$:

$$\frac{p^2}{4\kappa^2} = S^2(A_0) + [C(A_0) + A_0(\kappa - \omega)]^2$$

$$\tan\psi_0 = \frac{S(A_0)}{C(A_0) + A_0(\kappa - \omega)}$$

This steady-state solution will now be perturbed and the stability of the ensuing motion investigated. Let

$$A(t) = A_0 + \xi(t) \qquad [\xi << A_0]$$

$$\psi(t) = \psi_0 + \eta(t) \qquad [\eta << \psi_0]$$

By Taylor's series expansion:

$$\dot{\xi} = \xi S'(A_0) - \frac{p}{2\kappa}\eta\cos\psi_0$$

$$\dot{\eta} = \frac{\xi}{A_0}[(\kappa - \omega) + C'(A_0)] + \frac{p}{2\kappa A_0}\eta\sin\psi_0$$

where primes indicate differentiation with respect to A. These two differential equations are satisfied by the solutions

$$\xi = \mathcal{A}e^{zt} \qquad \eta = \mathcal{B}e^{zt}$$

where $\mathcal{A}$ and $\mathcal{B}$ are arbitrary constants and

$$z = \frac{1}{2A_0}\left\{[S(A_0) + A_0S'(A_0)] \pm \sqrt{[S(A_0) + A_0S'(A_0)]^2 - 4A_0\bar{p}\frac{d\bar{p}}{dA_0}}\right\}$$

and $\bar{p} = p/2\kappa$.

For stability, the real parts of z must be negative; hence, the following criteria can be established:

$$[S(A_0) + A_0S'(A_0)] < 0, \ \frac{d\bar{p}}{dA_0} > 0, \text{ ensures stability}$$

$$[S(A_0) + A_0S'(A_0)] < 0, \frac{d\bar{p}}{dA_0} < 0, \text{ ensures instability}$$

$$[S(A_0) + A_0S'(A_0)] > 0, \frac{d\bar{p}}{dA_0} \gtrless 0, \text{ ensures instability}$$

$$[S(A_0) + A_0S'(A_0)] = 0, \frac{d\bar{p}}{dA_0} > 0, \text{ ensures stability}$$

These criteria can be interpreted in terms of response curves by reference to Fig. 4.14. For systems of this type, $[S(A_0) + A_0S'(A_0)] < 0$; when $d\bar{p}/dA_0 > 0$, $\bar{p}$ increases as A_0 also increases. This does not hold for the middle branch of the response curves, thus confirming the earlier results, namely, that motion along this branch is unstable.

REFERENCES

1. Thompson, J. M. T., and H. B. Stewart: "Nonlinear Dynamics and Chaos," John Wiley & Sons, New York, 1987.
2. Ehrich, F. F.: "Stator Whirl with Rotors in Bearing Clearance," *J. of Engineering for Industry,* **89(B)**(3):381–390 (1967).
3. Ehrich, F. F.: "Rotordynamic Response in Nonlinear Anisotropic Mounting Systems," *Proc. of the 4th Intl. Conf. on Rotor Dynamics,* IFTOMM, 1–6, Chicago, September 7–9, 1994.
4. Ehrich, F. F.: "Nonlinear Phenomena in Dynamic Response of Rotors in Anisotropic Mounting Systems," *J. of Vibration and Acoustics,* **117(B):**117–161 (1995).
5. Choi, Y. S., and S. T. Noah: "Forced Periodic Vibration of Unsymmetric Piecewise-Linear Systems," *J. of Sound and Vibration,* **121**(3):117–126 (1988).
6. Ehrich, F. F.: "Observations of Subcritical Superharmonic and Chaotic Response in Rotordynamics," *J. of Vibration and Acoustics,* **114**(1):93–100 (1992).
7. Nayfeh, A. H., B. Balachandran, M. A. Colbert, and M. A. Nayfeh: "An Experimental Investigation of Complicated Responses of a Two-Degree-of-Freedom Structure," ASME Paper No. 90-WA/APM-24, 1990.
8. Ehrich, F. F.: "Spontaneous Sidebanding in High Speed Rotordynamics," *J. of Vibration and Acoustics,* **114**(4):498–505 (1992).
9. Ehrich, F. F., and M. Berthillier: "Spontaneous Sidebanding at Subharmonic Peaks of Rotordynamic Nonlinear Response," *Proceedings of ASME DETC '97,* Paper No. VIB-4041:1–7 (1997).
10. Ehrich, F. F.: "Subharmonic Vibration of Rotors in Bearing Clearance," ASME Paper No. 66-MD-1, 1966.
11. Bently, D. E.: "Forced Subrotative Speed Dynamic Action of Rotating Machinery," ASME Paper No. 74-Pet-16, 1974.
12. Childs, D. W.: "Fractional Frequency Rotor Motion Due to Nonsymmetric Clearance Effects," *J. of Eng. for Power,* July 1982, pp. 533–541.
13. Muszynska, A.: "Partial Lateral Rotor to Stator Rubs," IMechE Paper No. C281/84, 1984.
14. Ehrich, F. F.: "High Order Subharmonic Response of High Speed Rotors in Bearing Clearance," *J. of Vibration, Acoustics, Stress and Reliability in Design,* **110**(9):9–16 (1988).
15. Masri, S. F.: "Theory of the Dynamic Vibration Neutralizer with Motion Limiting Stops," *J. of Applied Mechanics,* **39:**563–569 (1972).

16. Shaw, S. W., and P. J. Holmes: "A Periodically Forced Piecewise Linear Oscillator," *J. of Sound and Vibration,* **90**(1):129–155 (1983).
17. Shaw, S. W.: "Forced Vibrations of a Beam with One-Sided Amplitude Constraint: Theory and Experiment," *J. of Sound and Vibration,* **99**(2):199–212 (1985).
18. Shaw, S. W.: "The Dynamics of a Harmonically Excited System Having Rigid Amplitude Constraints," *J. of Applied Mechanics,* **52:**459–464 (1985).
19. Choi, Y. S., and S. T. Noah: "Nonlinear Steady-State Response of a Rotor-Support System," *J. of Vibration, Acoustics, Stress and Reliability in Design,* July 1987, pp. 255–261.
20. Moon, F. C.: "Chaotic Vibrations," John Wiley & Sons, New York, 1987.
21. Sharif-Bakhtiar, M., and S. W. Shaw: "The Dynamic Response of a Centrifugal Pendulum Vibration Absorber with Motion Limiting Stops," *J. of Sound and Vibration,* **126**(2):221–235 (1988).
22. Ehrich, F. F.: "Some Observations of Chaotic Vibration Phenomena in High Speed Rotordynamics," *J. of Vibration and Acoustics,* **113**(1):50–57 (1991).
23. Guckenheimer, J., and P. Holmes: "Nonlinear Oscillations, Dynamical Systems, and Bifurcations of Vector Fields," Vol. 42, *Applied Mathematical Sciences,* Springer-Verlag, New York, 1983.
24. Wiggins, S.: "Introduction to Applied Nonlinear Dynamical Systems and Chaos," Vol. 2, *Texts in Applied Mathematics,* Springer-Verlag, New York, 1990.
25. Jackson, E. Atlee: "Perspectives of Nonlinear Dynamics," Vols. 1 and 2, Cambridge University Press, Cambridge, U.K., 1990.
26. Nataraj, C.: "Periodic Oscillations in Nonlinear Mechanical Systems," Ph.D. dissertation, Arizona State University, 1987.
27. Arnold, V. I.: "Lectures on Bifurcations in Versal Families," *Russian Math Survey,* **27:**54–123 (1972).
28. Nataraj, C., and S. P. Harsha: "The Effect of Bearing Cage Run-Out on the Nonlinear Dynamics of a Rotating Shaft," *Communications in Nonlinear Science and Numerical Simulation,* **13:**822–838 (2008).
29. Doedel, E. J., R. C. Paffenroth, A. R. Champneys, T. F. Fairgrieve, Yu. A. Kuznetsov, B. Sandstede, and X. Wang: "AUTO 2000: Continuation and Bifurcation Software for Ordinary Differential Equations (with HomCont)," Technical Report, Caltech, February 2001.
30. Duffing, G.: "Erzwungene Schwingungen bei veränderlicher Eigenfrequenz," F. Vieweg u Sohn, Brunswick, 1918.
31. Kryloff, N., and Bogoliuboff: "Introduction to Nonlinear Mechanics," Princeton University Press, Princeton, N.J., 1943.

CHAPTER 5
SELF-EXCITED VIBRATION

Fredric Ehrich

INTRODUCTION

Self-excited systems begin to vibrate of their own accord spontaneously, the amplitude increasing until some nonlinear effect limits any further increase. The energy supplying these vibrations is obtained from a uniform source of power associated with the system which, due to some mechanism inherent in the system, gives rise to oscillating forces. The nature of self-excited vibration compared to forced vibration is:[1]

In self-excited vibration the alternating force that sustains the motion is created or controlled by the motion itself; when the motion stops, the alternating force disappears.

In a forced vibration the sustaining alternating force exists independent of the motion and persists when the vibratory motion is stopped.

The occurrence of self-excited vibration in a physical system is intimately associated with the stability of equilibrium positions of the system. If the system is disturbed from a position of equilibrium, forces generally appear which cause the system to move either toward the equilibrium position or away from it. In the latter case the equilibrium position is said to be unstable; then the system may either oscillate with increasing amplitude or monotonically recede from the equilibrium position until nonlinear or limiting restraints appear. The equilibrium position is said to be stable if the disturbed system approaches the equilibrium position either in a damped oscillatory fashion or asymptotically.

The forces which appear as the system is displaced from its equilibrium position may depend on the displacement or the velocity, or both. If displacement-dependent forces appear and cause the system to move away from the equilibrium position, the system is said to be statically unstable. For example, an inverted pendulum is statically unstable. Velocity-dependent forces which cause the system to recede from a statically stable equilibrium position lead to dynamic instability.

Self-excited vibrations are characterized by the presence of a mechanism whereby a system will vibrate at its own natural or critical frequency, essentially *independent* of the *frequency* of any external stimulus. In mathematical terms, the motion is described by the unstable *homogeneous* solution to the homogeneous equations of motion. In contradistinction, in the case of "forced," or "resonant," vibrations, the *frequency* of the oscillation is *dependent* on (equal to, or a whole number ratio of) the frequency of a forcing function external to the vibrating system (e.g., shaft rotational speed in the case of rotating shafts). In mathematical terms, the forced vibration is the *particular* solution to the *nonhomogeneous* equations of motion.

Self-excited vibrations pervade all areas of design and operations of physical systems where motion or time-variant parameters are involved—aeromechanical systems (flutter, aircraft flight dynamics), aerodynamics (separation, stall, musical wind instruments, diffuser and inlet chugging), aerothermodynamics (flame instability, combustor screech), mechanical systems (machine-tool chatter), and feedback networks (pneumatic, hydraulic, and electromechanical servomechanisms).

ROTATING MACHINERY

One of the more important manifestations of self-excited vibrations, and the one that is the principal concern in this chapter, is that of rotating machinery, specifically, the self-excitation of lateral, or flexural, vibration of rotating shafts (as distinct from torsional, or longitudinal, vibration).

In addition to the description of a large number of such phenomena in standard vibrations textbooks (most typically and prominently, Ref. 1), the field has been subject to several generalized surveys.[2–4] The mechanisms of self-excitation which have been identified can be categorized as follows:

Whirling or Whipping

Hysteretic whirl

Fluid trapped in the rotor

Dry friction whip

Fluid bearing whip

Seal and blade-tip-clearance effect in turbomachinery

Propeller and turbomachinery whirl

Parametric Instability

Asymmetric shafting

Pulsating torque

Pulsating longitudinal loading

Stick-Slip Rubs and Chatter

Instabilities in Forced Vibrations

Bistable vibration

Unstable imbalance

In each instance, the physical mechanism is described and aspects of its prevention or its diagnosis and correction are given. Some exposition of its mathematical analytic modeling is also included.

WHIRLING OR WHIPPING

ANALYTIC MODELING

In the most important subcategory of instabilities (generally termed whirling or whipping), the unifying generality is the generation of a tangential force, normal to an arbitrary radial deflection of a rotating shaft, whose magnitude is proportional to (or varies monotonically with) that deflection. At some "onset" rotational speed, such a force system will overcome the stabilizing external damping forces which are generally present and induce a whirling motion of ever-increasing amplitude, limited only by nonlinearities which ultimately limit deflections.

A simple mathematical representation of a self-excited vibration in linear systems with constant coefficients, subject to *plane* vibration, may be found in the concept of negative damping. Consider the differential equation for a damped, free vibration:

$$m\ddot{x} + c\dot{x} + kx = 0 \tag{5.1}$$

This is generally solved by assuming a solution of the form

$$x = Ce^{st}$$

Substitution of this solution into Eq. (5.1) yields the characteristic (algebraic) equation

$$s^2 + \frac{c}{m}s + \frac{k}{m} = 0 \tag{5.2}$$

If $c < 2\sqrt{mk}$, the roots are complex:

$$s_{1,2} = -\frac{c}{2m} \pm iq$$

where

$$q = \sqrt{\frac{k}{m} - \left(\frac{c}{2m}\right)^2}$$

The solution takes the form

$$x = e^{-ct/2m}(A\cos qt + B\sin qt) \tag{5.3}$$

This represents a decaying oscillation because the exponential factor is negative, as illustrated in Fig. 5.1*A*. If $c < 0$, the exponential factor has a positive exponent and the vibration appears as shown in Fig. 5.1*B*. The system, initially at rest, begins to oscillate spontaneously with ever-increasing amplitude. Then, in any physical system, some nonlinear effect enters and Eq. (5.1) fails to represent the system realistically. Equation (5.4) defines a nonlinear system with negative damping at small amplitudes but with large positive damping at larger amplitudes, thereby limiting the amplitude to finite values:

$$m\ddot{x} + (-c + ax^2)\dot{x} + kx = 0 \tag{5.4}$$

Thus, the fundamental criterion of stability in linear systems is that the roots of the characteristic equation have negative real parts, thereby producing decaying amplitudes.

In the case of a whirling or whipping shaft, the equations of motion (for an idealized shaft with a single lumped mass m) are more appropriately written in polar coordinates for the radial force balance,

$$-m\omega^2 r + m\ddot{r} + c\dot{r} + kr = 0 \tag{5.5}$$

and for the tangential force balance,

$$2m\omega\dot{r} + c\omega r - F_n = 0 \tag{5.6}$$

where we presume a constant rate of whirl ω.

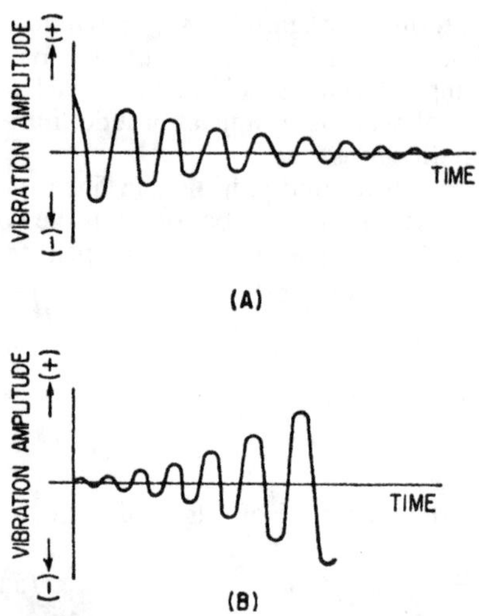

FIGURE 5.1 (*A*) Illustration showing a decaying vibration (stable) corresponding to negative real parts of the complex roots. (*B*) Increasing vibration corresponding to positive real parts of the complex roots (unstable).

In general, the whirling is predicated on the existence of some physical phenomenon which will induce a force F_n that is normal to the radial deflection r and is in the direction of the whirling motion—i.e., in opposition to the damping force, which tends to inhibit the whirling motion. Very often, this normal force can be characterized or approximated as being proportional to the radial deflection:

$$F_n = f_n r \tag{5.7}$$

The solution then takes the form

$$r = r_0 e^{at} \tag{5.8}$$

For the system to be stable, the coefficient of the exponent

$$a = \frac{f_n - c\omega}{2m\omega} \tag{5.9}$$

must be negative, giving the requirement for stable operation as

$$f_n < \omega c \tag{5.10}$$

As a rotating machine increases its rotational speed, the left-hand side of this inequality (which is generally also a function of shaft rotation speed) may exceed the right-hand side, indicative of the onset of instability. At this onset condition,

$$a = 0+ \tag{5.11}$$

so that whirl speed at onset is found to be

$$\omega = \left(\frac{k}{m}\right)^{1/2} \tag{5.12}$$

That is, the whirling speed at onset of instability is the shaft's natural or critical frequency, irrespective of the shaft's rotational speed (rpm). The direction of whirl may be in the same rotational direction as the shaft rotation (*forward* whirl) or opposite to the direction of shaft rotation (*backward* whirl), depending on whether the direction of the destabilizing force F_n is in the direction of rotation or counter to it.

When the system is unstable, the solution for the trajectory of the shaft's mass is, from Eq. (5.8), an exponential spiral as in Fig. 5.2. Any planar component of this two-dimensional trajectory takes the same form as the unstable planar vibration shown in Fig. 5.1*B*.

GENERAL DESCRIPTION

The most important examples of whirling and whipping instabilities are

Hysteretic whirl

Fluid trapped in the rotor

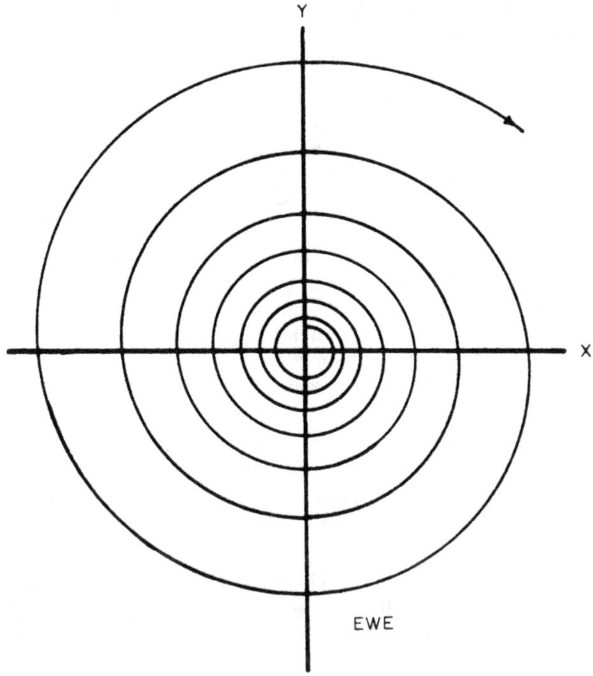

FIGURE 5.2 Trajectory of rotor center of gravity in unstable whirling or whipping.

Dry friction whip
Fluid bearing whip
Seal and blade-tip-clearance effect in turbomachinery
Propeller and turbomachinery whirl

All these self-excitation systems involve friction or fluid energy mechanisms to generate the destabilizing force.

These phenomena are rarer than forced vibration due to unbalance or shaft misalignment, and they are difficult to anticipate before the fact or diagnose after the fact because of their subtlety. Also, self-excited vibrations are potentially more destructive, since the asynchronous whirling of self-excited vibration induces alternating stresses in the rotor and can lead to fatigue failures of rotating components. Synchronous forced vibration typical of unbalance does not involve alternating stresses in the rotor and will rarely involve rotating element failure. The general attributes of these instabilities, insofar as they differ from forced excitations, are summarized in Table 5.1 and Figs. 5.3*A* and 5.3*B*.

HYSTERETIC WHIRL

The mechanism of hysteretic whirl, as observed experimentally,[5] defined analytically,[6] or described in standard texts,[7] may be understood from the schematic representation of Fig. 5.4. With some nominal radial deflection of the shaft, the flexure of the shaft would induce a neutral strain axis normal to the deflection direction. From

TABLE 5.1 Characterization of Two Categories of Vibration of Rotating Shafts

	Forced or resonant vibration	Whirling or whipping
Vibration frequency–rpm relationship	Frequency is equal to (i.e., synchronous with) rpm or a whole number or rational fraction of rpm, as in Fig. 5.3*A*.	Frequency is nearly constant and relatively independent of rotor rotational speed or any external stimulus and is at or near one of the shaft critical or natural frequencies, as in Fig. 5.3*B*.
Vibration amplitude–rpm relationship	Amplitude will peak in a narrow band of rpm wherein the rotor's critical frequency is equal to the rpm or to a whole-number multiple or a rational fraction of the rpm or an external stimulus, as in Fig. 5.3*A*.	Amplitude will suddenly increase at an onset rpm and continue at high or increasing levels as rpm is increased, as in Fig. 5.3*B*.
Influence of damping	Addition of damping may reduce peak amplitude but not materially affect rpm at which peak amplitude occurs, as in Fig. 5.3*A*.	Addition of damping may defer onset to a higher rpm but not materially affect amplitude after onset, as in Fig. 5.3*B*.
System geometry	Excitation level and hence amplitude are dependent on some lack of axial symmetry in the rotor mass distribution or geometry, or external forces applied to the rotor. Amplitudes may be reduced by refining the system to make it more perfectly axisymmetric (i.e., balancing).	Amplitudes are independent of system axial symmetry. Given an infinitesimal deflection to an otherwise symmetric system, the amplitude will self-propagate.
Rotor fiber stress	For synchronous vibration, the rotor vibrates in a frozen, deflected state, without oscillatory fiber stress.	Rotor fibers are subject to oscillatory stress at a frequency equal to the difference between rotor rpm and whirling speed.
Avoidance or elimination	1. Tune the system's critical frequencies to be out of the rpm operating range. 2. Eliminate all deviations from axial symmetry in the system as built or as induced during operation (e.g., balancing). 3. Introduce damping to limit peak amplitudes at critical speeds which must be traversed.	1. Restrict operating rpm to below instability onset rpm. 2. Defeat or eliminate the instability mechanism. 3. Introduce damping to raise the instability onset speed to above the operating speed range. 4. Introduce stiffness anisotropy to the bearing support system.[8]

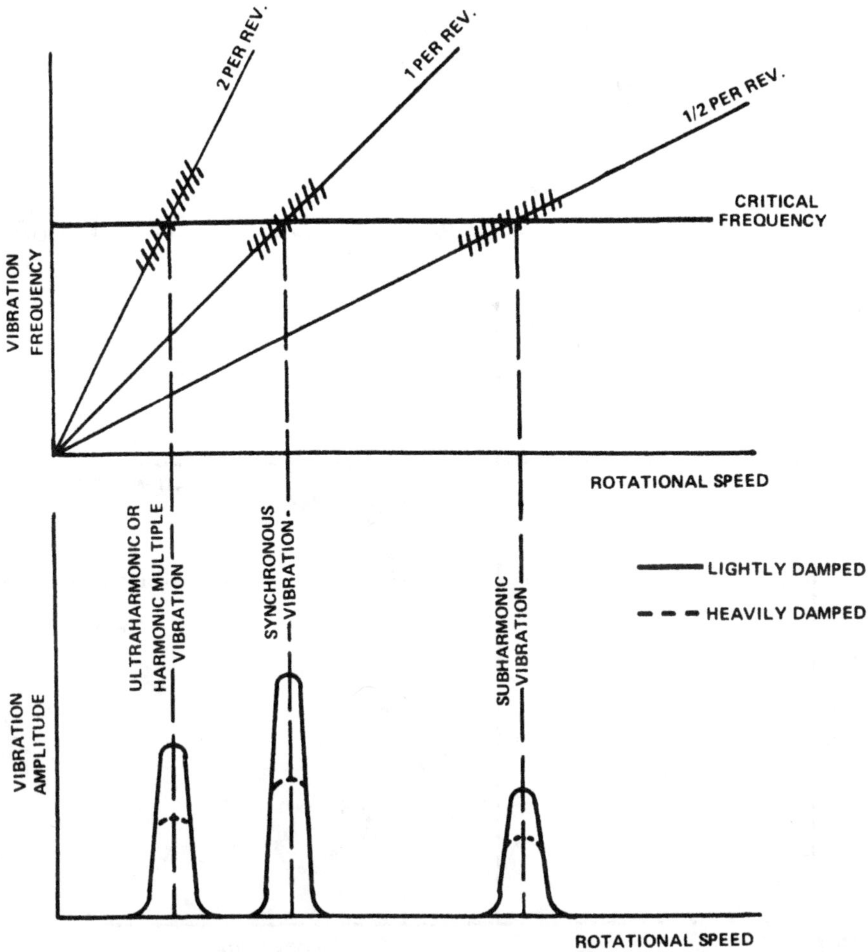

FIGURE 5.3A Attributes of forced vibration or resonance in rotating machinery.

first-order considerations of elastic-beam theory, the neutral axis of stress would be coincident with the neutral axis of strain. The net elastic restoring force would then be perpendicular to the neutral stress axis, i.e., parallel to and opposing the deflection. In actual fact, hysteresis, or internal friction, in the rotating shaft will cause a phase shift in the development of stress as the shaft fibers rotate around through peak strain to the neutral strain axis. The net effect is that the neutral stress axis is displaced in angle orientation from the neutral strain axis, and the resultant force is not parallel to the deflection. In particular, the resultant force has a tangential component *normal* to the deflection, which is the fundamental precondition for whirl. This tangential force component is in the direction of rotation and induces a *forward* whirling motion which increases centrifugal force on the deflected rotor, thereby increasing its deflection. As a consequence, induced stresses are increased, thereby increasing the whirl-inducing force component.

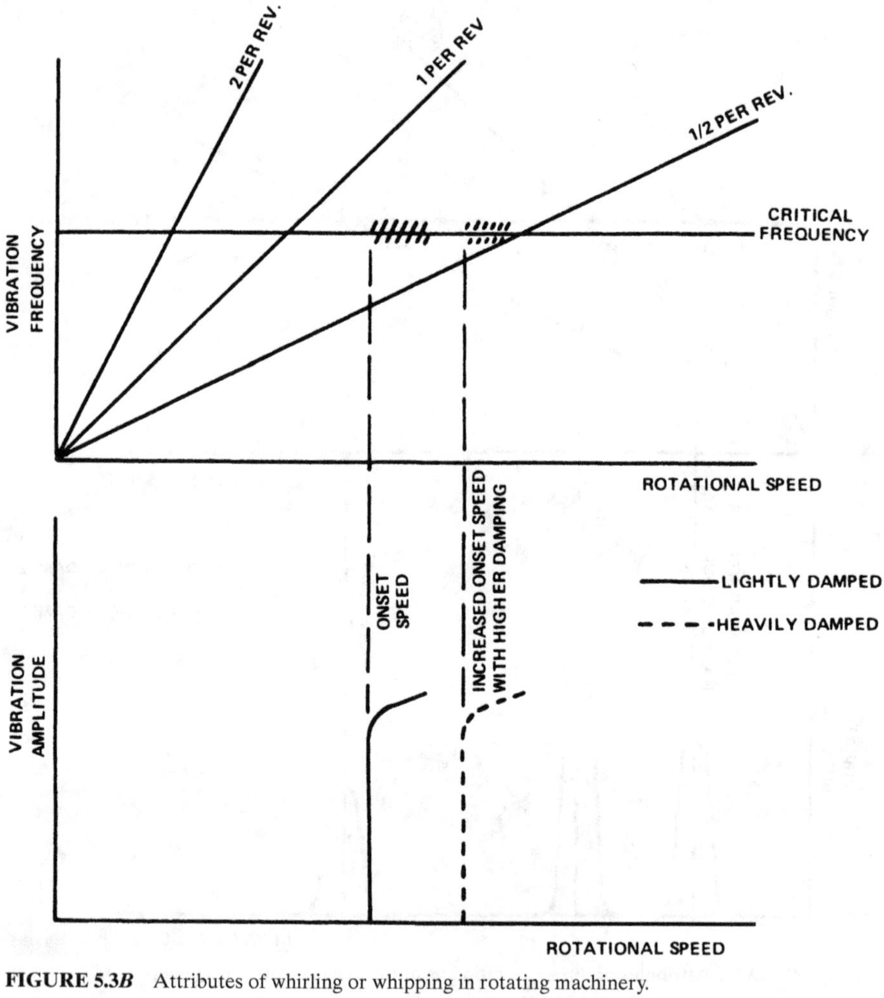

FIGURE 5.3*B* Attributes of whirling or whipping in rotating machinery.

Several surveys and contributions to the understanding of the phenomenon have been published in Refs. 9, 10, 11, and 12. It has generally been recognized that hysteretic whirl can occur only at rotational speeds above the first-shaft critical speed (the lower the hysteretic effect, the higher the attainable whirl-free operating rpm). It has been shown[13] that once whirl has started, the critical whirl speed that will be induced (from among the spectrum of criticals of any given shaft) will have a frequency approximately half the onset rpm.

A straightforward method for hysteretic whirl avoidance is that of limiting shafts to subcritical operation, but this is unnecessarily and undesirably restrictive. A more effective avoidance measure is to limit the hysteretic characteristic of the rotor. Most investigators (e.g., Ref. 5) have suggested that the essential hysteretic effect is

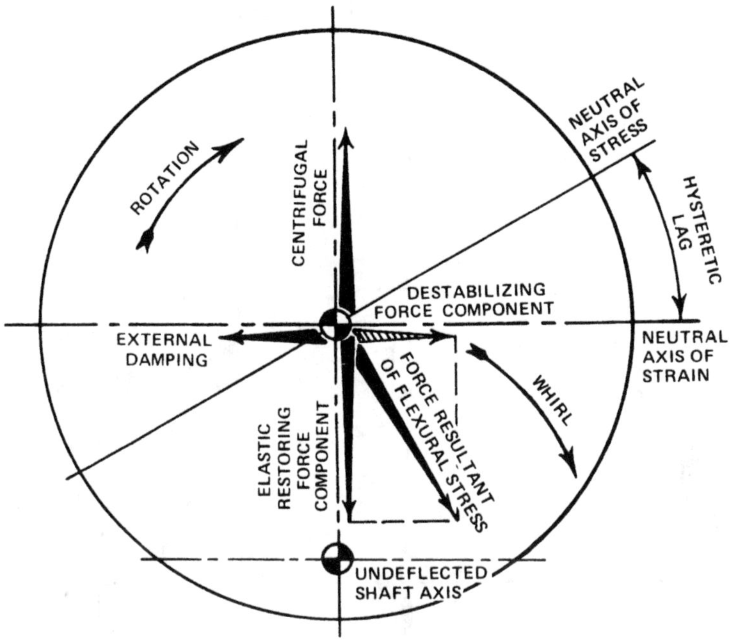

FIGURE 5.4 Hysteretic whirl.

caused by working at the interfaces of joints in a rotor rather than within the material of that rotor's components. Success in avoiding hysteretic whirl has been achieved by minimizing the number of separate elements, restricting the span of concentric rabbets and shrunk fitted parts, and providing secure lockup of assembled elements held together by tie bolts and other compression elements. Bearing-foundation characteristics also play a role in suppression of hysteretic whirl.[9]

WHIRL DUE TO FLUID TRAPPED IN ROTOR

There has always been a general awareness that high-speed centrifuges are subject to a special form of instability. It is now appreciated that the same self-excitation may be experienced more generally in high-speed rotating machinery where liquids (e.g., oil from bearing sumps, steam condensate, etc.) may be inadvertently trapped in the internal cavity of hollow rotors. The mechanism of instability is shown schematically in Fig. 5.5. For some nominal deflection of the rotor, the fluid is flung out radially in the direction of deflection. But the fluid does not remain in simple radial orientation. The spinning surface of the cavity drags the fluid (which has some finite viscosity) in the direction of rotation. This angle of advance results in the centrifugal force on the fluid having a component in the tangential direction in the direction of rotation. This force then is the basis of instability, since it induces forward whirl which increases the centrifugal force on the fluid and thereby increases the whirl-inducing force.

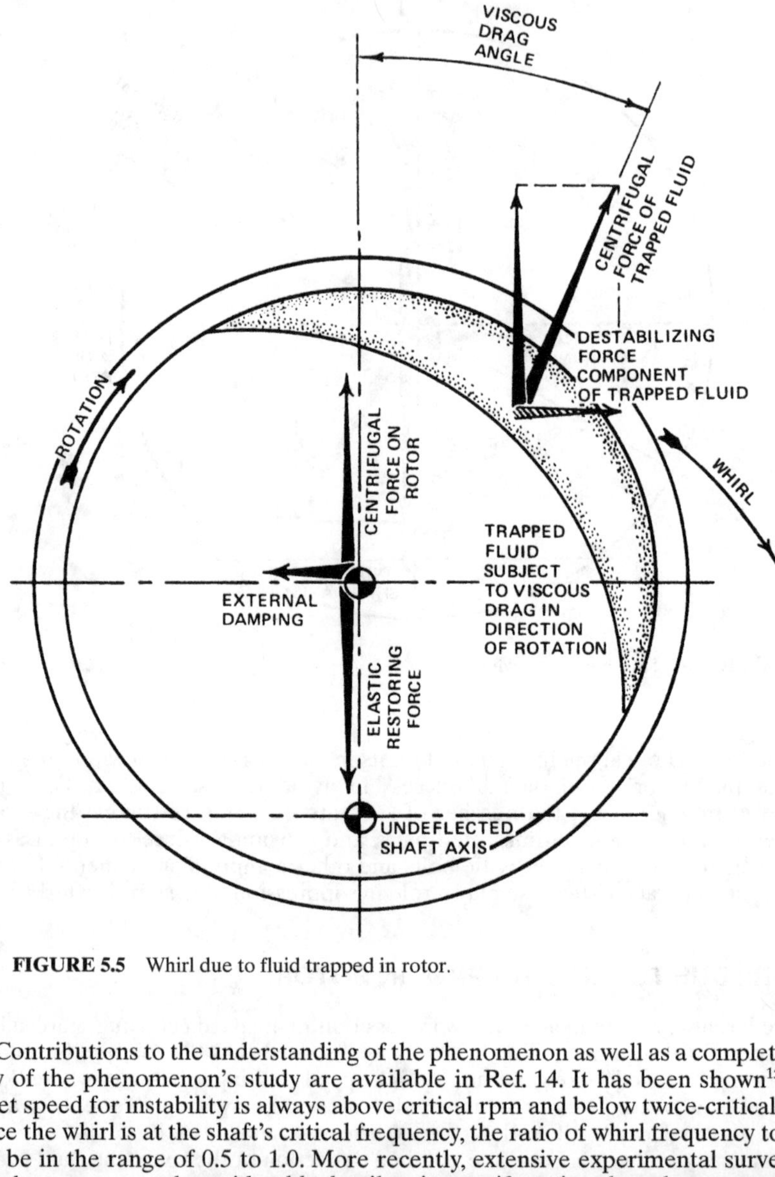

FIGURE 5.5 Whirl due to fluid trapped in rotor.

Contributions to the understanding of the phenomenon as well as a complete history of the phenomenon's study are available in Ref. 14. It has been shown[15] that onset speed for instability is always above critical rpm and below twice-critical rpm. Since the whirl is at the shaft's critical frequency, the ratio of whirl frequency to rpm will be in the range of 0.5 to 1.0. More recently, extensive experimental surveys of the phenomenon and considerable detail on its manifestations have been reported.[16]

Avoidance of this self-excitation can be accomplished by running shafting subcritically, although this is generally undesirable in centrifuge-type applications when further consideration is made of the role of trapped fluids as unbalance in forced vibration of rotating shafts (as described in Ref. 15). Where the trapped fluid is not fundamental to the machine's function, the appropriate avoidance measure, if the particular application permits, is to provide drain holes at the outermost radius of all hollow cavities where fluid might otherwise be trapped.

DRY FRICTION WHIP

As described in standard vibration texts (e.g., Ref. 7), dry friction whip is experienced when the surface of a rotating shaft comes in contact with an unlubricated stationary guide or shroud or stator system. This can occur in an unlubricated journal bearing; or with loss of clearance in a hydrodynamic bearing; or inadvertent closure and contact in the radial clearance of labyrinth seals or turbomachinery blading; or in power screws.[17]

The phenomenon may be understood with reference to Fig. 5.6. When radial contact is made between the surface of the rotating shaft and a static part, coulomb friction will induce a tangential force on the rotor. Since the friction force is approximately proportional to the radial component of the contact force, we have the preconditions for instability. The tangential force induces a whirling motion which induces larger centrifugal force on the rotor, which in turn induces a large radial contact and hence larger whirl-inducing friction force.

It is interesting to note that this whirl system is one of the few phenomena in which the destabilizing force is *counter* to the shaft rotation direction (i.e., *backward* whirl). One may envision the whirling system as the rolling (accompanied by appreciable slipping) of the shaft in the stator system.

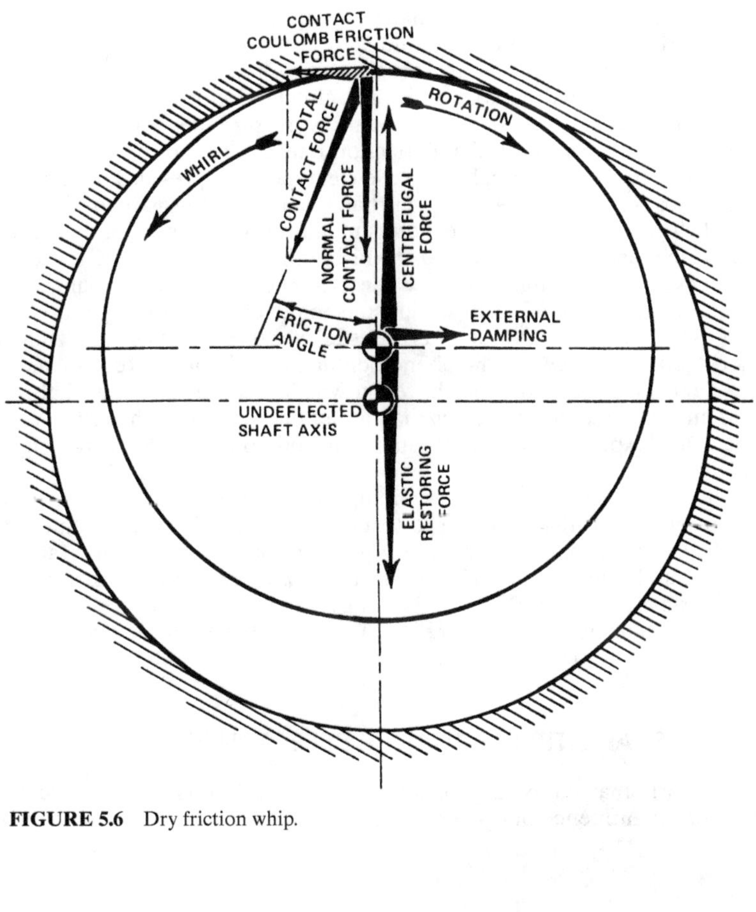

FIGURE 5.6 Dry friction whip.

The same situation can be produced by a thrust bearing where angular deflection is combined with lateral deflection.[18] If contact occurs on the same side of the disc as the virtual pivot point of the deflected disc, then backward whirl will result. Conversely, if contact occurs on the side of the disc opposite to the side where the virtual pivot point of the disc is located, then forward whirl will result.

It has been suggested (but not concluded)[19] that the whirling frequency is generally less than the critical speed.

The vibration is subject to various types of control. If contact between rotor and stator can be avoided or the contact area can be kept well lubricated, no whipping will occur. Where contact must be accommodated, and lubrication is not feasible, whipping may be avoided by providing abradability of the rotor or stator element to allow disengagement before whirl. When dry friction is considered in the context of the dynamics of the stator system in combination with that of the rotor system,[20] it is found that whirl can be inhibited if the independent natural frequencies of the rotor and stator are kept dissimilar, that is, a very stiff rotor should be designed with a very soft mounted stator element that may be subject to rubs. No first-order interdependence of whirl speed with rotational speed has been established.

FLUID BEARING WHIP

As described in experimental and analytic literature,[21] and in standard texts (e.g., Ref. 22), fluid bearing whip can be understood by referring to Fig. 5.7. Consider some nominal radial deflection of a shaft rotating in a fluid (gas- or liquid-) filled clearance. The entrained, viscous fluid will circulate with an average velocity of about half the shaft's surface speed. The bearing pressures developed in the fluid will not be symmetric about the radial deflection line. Because of viscous losses of the bearing fluid circulating through the close clearance, the pressure on the upstream side of the close clearance will be higher than that on the downstream side. Thus, the resultant bearing force will include a tangential force component in the direction of rotation which tends to induce *forward* whirl in the rotor. The tendency to instability is evident when this tangential force exceeds inherent stabilizing damping forces. When this happens, any induced whirl results in increased centrifugal forces; this, in turn, closes the clearance further and results in ever-increasing destabilizing tangential force. Detailed reviews of the phenomenon are available in Refs. 23 and 24.

These and other investigators have shown that to be unstable, shafting must rotate at an rpm equal to or greater than approximately twice the critical speed, so that one would expect the ratio of frequency to rpm to be equal to less than approximately 0.5.

The most obvious measure for avoiding fluid bearing whip is to restrict rotor maximum rpm to less than twice its lowest critical speed. Detailed geometric variations in the bearing runner design, such as grooving and tilt-pad configurations, have also been found effective in inhibiting instability. In extreme cases, use of rolling contact bearings instead of fluid film bearings may be advisable.

Various investigators (e.g., Ref. 25) have noted that fluid seals as well as fluid bearings are subject to this type of instability.

SEAL AND BLADE-TIP-CLEARANCE EFFECT IN TURBOMACHINERY

Axial-flow turbomachinery may be subject to an additional whirl-inducing effect by virtue of the influence of tip clearance on turbopump or compressor or turbine

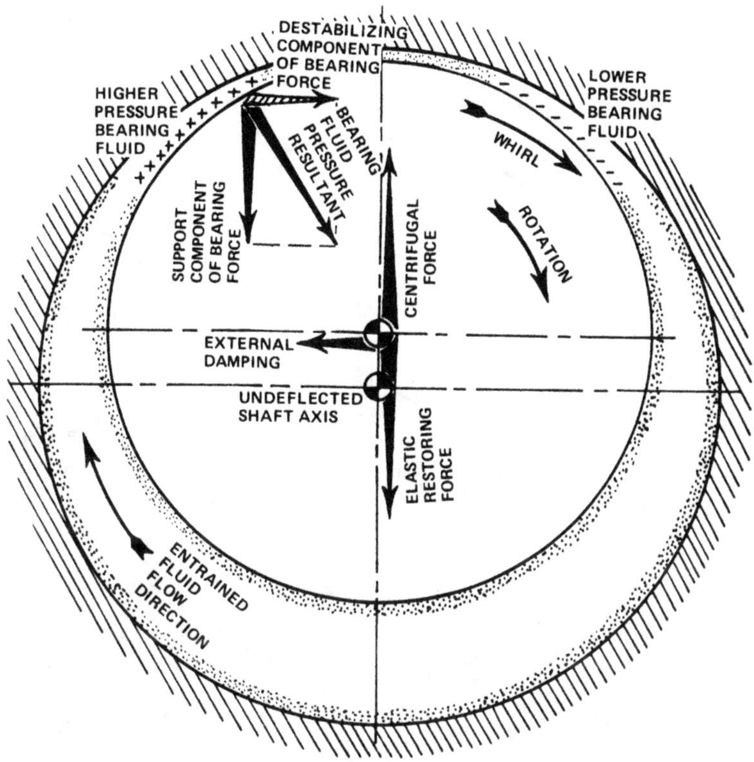

FIGURE 5.7 Fluid bearing whip.

efficiency.[26] As shown schematically in Fig. 5.8, some nominal radial deflection will close the radial clearance on one side of the turbomachinery component and open the clearance 180° away on the opposite side. We would expect the closer clearance zone to operate more efficiently than the open clearance zone. For a turbine, a greater work extraction and blade force level is achieved in the more efficient region for a given average pressure drop so that a resultant net tangential force is generated to induce whirl in the direction of rotor rotation (i.e., forward whirl). For an axial compressor, it has been found[27] that the magnitude and direction of the destabilizing forces are a very strong function of the operating point's proximity to the stall line. For operation close to the stall line, very large negative forces (i.e., inducing backward whirl) are generated. The magnitude of the destabilizing force declines sharply for lower operating lines, and stabilizes at a small positive value (i.e., making a small contribution to inducing forward whirl). In the case of radial-flow turbomachinery, it has been suggested[28] that destabilizing forces are exerted on an eccentric (i.e., dynamically deflected) impeller due to variations of loading of the diffuser vanes.

One text[29] describes several manifestations of this class of instability—in the thrust balance piston of a steam turbine; in the radial labyrinth seal of a radial-flow Ljungstrom counterrotating steam turbine; in the Kingsbury thrust bearing of a

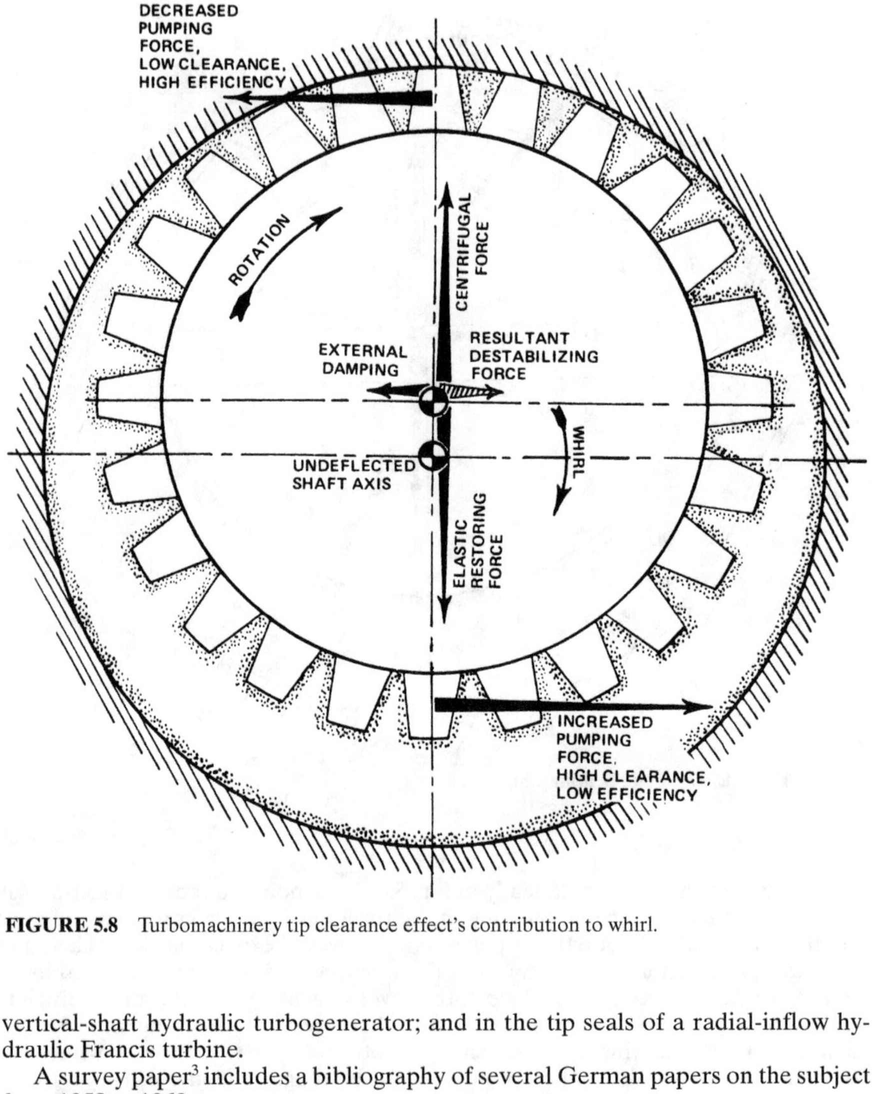

FIGURE 5.8 Turbomachinery tip clearance effect's contribution to whirl.

vertical-shaft hydraulic turbogenerator; and in the tip seals of a radial-inflow hydraulic Francis turbine.

A survey paper[3] includes a bibliography of several German papers on the subject from 1958 to 1969.

An analysis is available[30] dealing with the possibility of stimulating flexural vibrations in the seals themselves, although it is not clear if the solutions pertain to gross deflections of the entire rotor.

It is reasonable to expect that such destabilizing forces may at least contribute to instabilities experienced on high-powered turbomachines. If this mechanism were indeed a key contributor to instability, one would conjecture that very small or very large initial tip clearances would minimize the influence of tip clearance on the unit's performance and, hence, minimize the contribution to destabilizing forces.

PROPELLER AND TURBOMACHINERY WHIRL

Propeller whirl has been identified both analytically[31] and experimentally.[32] In this instance of shaft whirling, a small *angular* deflection of the shaft is hypothesized, as shown schematically in Fig. 5.9. The tilt in the propeller disc plane results at any instant at any blade in a small angle change between the propeller rotation velocity vector and the approach velocity vector associated with the aircraft's speed. The change in local relative velocity angle and magnitude seen by any blade results in an increment in its load magnitude and direction. The cumulative effect of these changes in load on all the blades results in a net moment whose vector has a significant component which is normal to and approximately proportional to the angular deflection vector. By analogy to the destabilizing cross-coupled *deflection* stiffness we noted in previously described instances of whirling and whipping, we have now identified the existence of a cross-

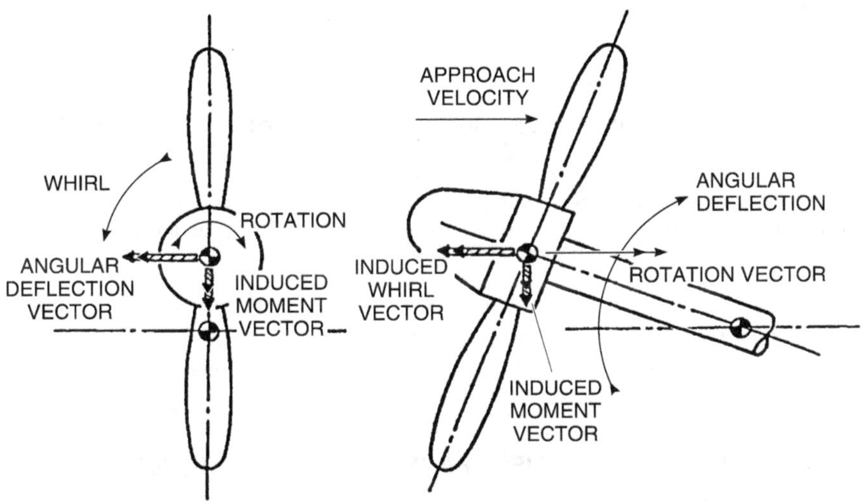

FIGURE 5.9 Propeller whirl.[2]

coupled destabilizing *moment* stiffness. At high airspeeds, the destabilizing moments can grow to the point where they may overcome viscous damping moments to cause destructive whirling of the entire system in a "conical" mode. This *propeller whirl* is generally found to be counter to the shaft rotation direction. It has been suggested[33] that equivalent stimulation is possible in turbomachinery. An attempt has been made[34] to generalize the analysis for axial-flow turbomachinery. Although it has been shown that this analysis is not accurate, the general deduction seems appropriate that forward whirl may also be possible if the virtual pivot point of the deflected rotor is forward of the rotor (i.e., on the side of the approaching fluid).

Instability is found to be load-sensitive in the sense of being a function of the velocity and density of the impinging flow. It is not thought to be sensitive to the torque level of the turbomachine since, for example, experimental work[32] was done on an unloaded windmilling rotor. Corrective action is generally recognized to be stiffening the entire system and manipulating the effective pivot center of the whirling mode to inhibit angular motion of the propeller (or turbomachinery) disc as well as system damping.

PARAMETRIC INSTABILITY

ANALYTIC MODELING

There are systems in engineering and physics which are described by linear differential equations having periodic coefficients

$$\frac{d^2y}{dz^2} + p(z)\frac{dy}{dz} + q(z)y = 0 \tag{5.13}$$

where $p(z)$ and $q(z)$ are periodic in z. These systems also may exhibit self-excited vibrations, but the stability of the system cannot be evaluated by finding the roots of a characteristic equation. A specialized form of this equation, which is representative of a variety of real physical problems in rotating machinery, is Mathieu's equation:

$$\frac{d^2f}{dz^2} + (a - 2q\cos 2z)f = 0 \tag{5.14}$$

Mathematical treatment and applications of Mathieu's equation are given in Refs. 35 and 36.

This general subcategory of self-excited vibrations is termed "parametric instability," since instability is induced by the effective periodic variation of the system's parameters (stiffness, inertia, natural frequency, etc.). Three particular instances of interest in the field of rotating machinery are

Lateral instability due to asymmetric shafting and/or bearing characteristics
Lateral instability due to pulsating torque
Lateral instabilities due to pulsating longitudinal compression

LATERAL INSTABILITY DUE TO ASYMMETRIC SHAFTING

If a rotor or its stator contains sufficient levels of asymmetry in the flexibility associated with its two principal axes of flexure as illustrated in Fig. 5.10, self-excited vibration may take place. This phenomenon is completely independent of any unbalance, and independent of the forced vibrations associated with twice-per-revolution excitation of such shafting mounted horizontally in a gravitational field.

As described in standard vibration texts,[37] we find that presupposing a nominal whirl amplitude of the shaft at some whirl frequency, the rotation of the asymmetric shaft at an rpm different from the whirling speed will appear as periodic change in flexibility in the plane of the whirling shaft's radial deflection. This will result in an instability in certain specific ranges of rpm as a

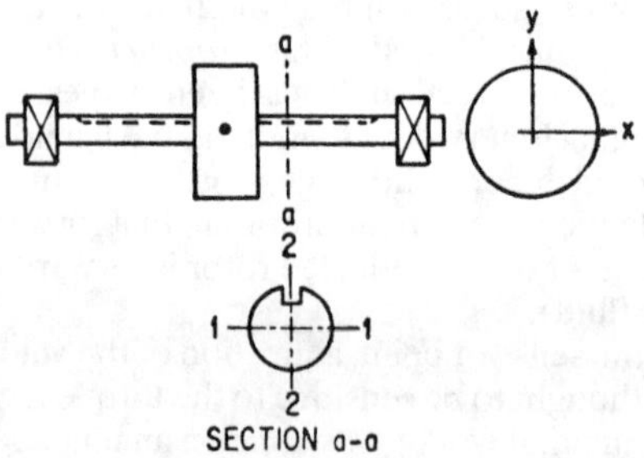

FIGURE 5.10 Shaft system possessing unequal rigidities, leading to a pair of coupled inhomogeneous Mathieu equations.

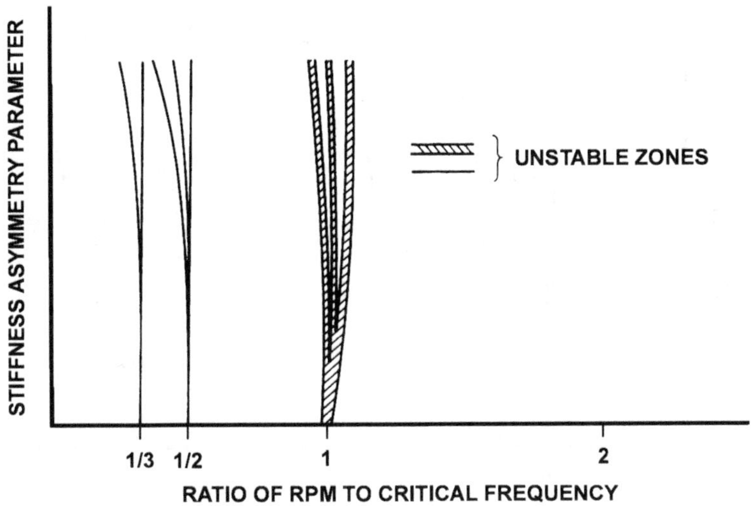

FIGURE 5.11 Instability regimes of rotor system induced by asymmetric stiffness (Ref. 39).

function of the degree of asymmetry. In general, instability is experienced when the rpm is approximately one-third and one-half the critical rpm and approximately equal to the critical rpm (where the critical rpm is defined with the average value of shaft stiffness), as in Fig. 5.11. The ratios of whirl frequency to rotational speed will then be approximately 3.0, 2.0, and 1.0. But with gross asymmetries, and with the additional complication of asymmetrical inertias with principal axes in arbitrary orientation to the shaft's principal axes' flexibility, no simple generalization is possible.

There is a considerable literature dealing with many aspects of the problem and substantial bibliographies.[38–40]

Stability is accomplished by minimizing shaft asymmetries and avoiding rpm ranges of instability.

LATERAL INSTABILITY DUE TO PULSATING TORQUE

Experimental confirmation[41] has been achieved that establishes the possibility of inducing first-order lateral instability in a rotor-disc system by the application of a proper combination of constant and pulsating torque. The application of torque to a shaft in the range of its torsional buckling magnitude affects its natural frequency in lateral vibration so that the instability may also be characterized as "parametric." Analytic formulation and description of the phenomenon are available in Ref. 42 and in the bibliography of Ref. 3. The experimental work (Ref. 41) explored regions of shaft speed where the disc always whirled at the first critical speed of the rotor-disc system, regardless of the torsional forcing frequency or the rotor speed within the unstable region.

It therefore appears that combinations of ranges of steady and pulsating torque, which have been identified[40] as being sufficient to cause instability, should be

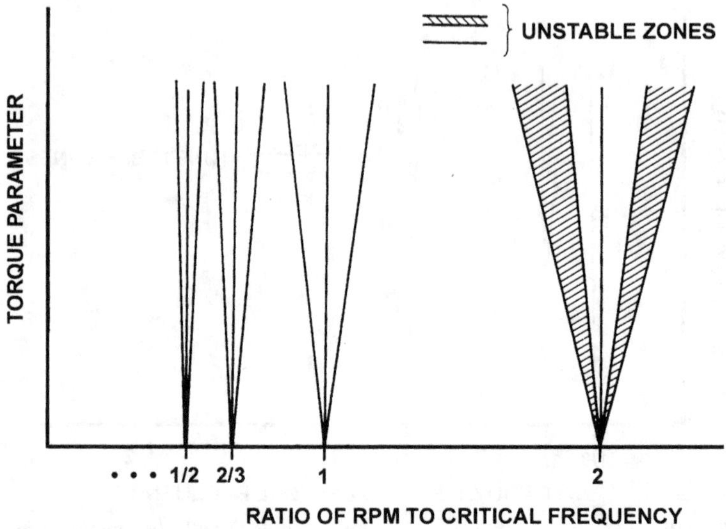

FIGURE 5.12 Instability regimes of rotor system induced by pulsating torque (Ref. 42).

avoided in the narrow-speed bands where instability is possible in the vicinity of twice the critical speed and lesser instabilities at 2/2, 2/3, 2/4, 2/5, . . . times the critical frequency, as in Fig. 5.12, implying frequency/speed ratios of approximately 0.5, 1.0, 1.5, 2.0, 2.5, ...

LATERAL INSTABILITY DUE TO PULSATING LONGITUDINAL LOADS

Longitudinal loads on a shaft which are of an order of magnitude of the buckling will tend to reduce the natural frequency of that lateral, flexural vibration of the shaft. Indeed, when the compressive buckling load is reached, the natural frequency goes to zero. Therefore pulsating longitudinal loads effectively cause a periodic variation in stiffness, and they are capable of inducing "parametric instability" in rotating as well as stationary shafts,[43] as noted in Fig. 5.13.

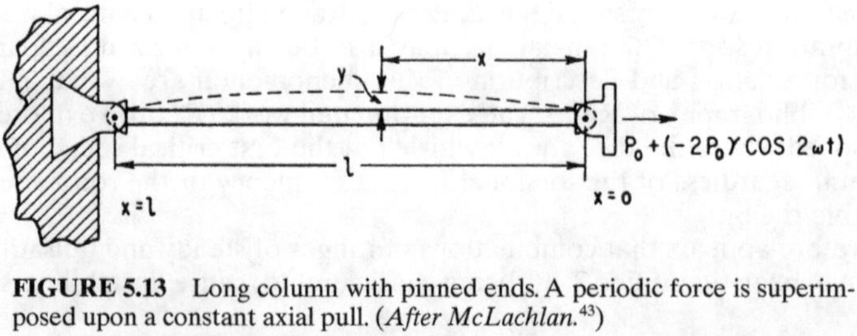

FIGURE 5.13 Long column with pinned ends. A periodic force is superimposed upon a constant axial pull. (*After McLachlan.*[43])

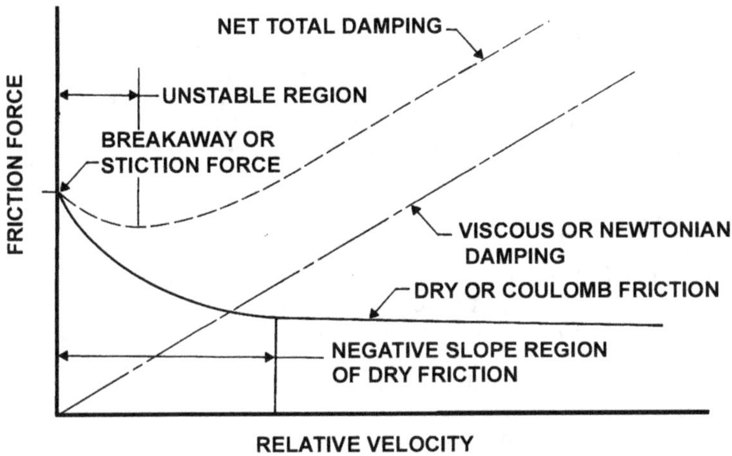

FIGURE 5.14 Dry friction characteristic giving rise to stick-slip rubs or chatter.

STICK-SLIP RUBS AND CHATTER

Mention is appropriate of another family of instability phenomena—stick-slip or chatter. Though the instability mechanism is associated with the dry friction contact force at the point of rubbing between a rotating shaft and a stationary element, it must not be confused with dry friction whip, previously discussed. In the case of stick-slip, as is described in standard texts (e.g., Ref. 44), the instability is caused by the irregular nature of the friction force developed at very low rubbing speeds.

At high velocities, the friction force is essentially independent of contact speed. But at very low contact speeds we encounter the phenomenon of "stiction," or breakaway friction, where higher levels of friction force are encountered, as in Fig. 5.14. Any periodic motion of the rotor's point of contact, superimposed on the basic relative contact velocity, will be self-excited. In effect, there is negative damping (as illustrated in Fig. 5.1*B*) since motion of the rotor's contact point in the direction of rotation will increase relative contact velocity and reduce stiction and the net force resisting motion. Rotor motion counter to the contact velocity will reduce relative velocity and increase friction force, again reinforcing the periodic motion. The ratio of vibration frequency to rotation speed will be much larger than unity.

While the vibration associated with stick-slip or chatter is often reported to be torsional, planar lateral vibrations can also occur. Surveys of the phenomenon are included in Refs. 45 and 46. The phenomenon is closely related to *chatter* in machine tools.

Measures for avoidance are similar to those prescribed for dry friction whip: avoid contact where feasible and lubricate the contact point where contact is essential to the function of the apparatus.

INSTABILITIES IN FORCED VIBRATIONS

In a middle ground between the generic categories of force vibrations and self-excited vibrations is the category of *instabilities in force vibrations.* These instabilities are characterized by forced vibration at a frequency equal to rotor rotation (generally induced by unbalance), but with the amplitude of that vibration being unsteady or unstable. Such unsteadiness or instability is induced by the interaction of the forced vibration on the mechanics of the system's response, or on the unbalance itself. Two manifestations of such instabilities and unsteadiness have been identified in the literature—bistable vibration and unstable imbalance.

BISTABLE VIBRATION

A classical model of one type of unstable motion is the relaxation oscillator, or *multivibrator.* A system subject to relaxation oscillation has *two* fairly stable states, separated by a zone where stable operation is impossible.[47] Furthermore, in each of the stable states, a mechanism exists which will induce the system to drift toward the unstable state. The system will develop a periodic motion of the general form shown in Fig. 5.15.

An idealized formulation of this class of vibration with nonlinear damping is[48]

$$m\ddot{x} + c(x^2 - 1)\dot{x} + kx = 0 \tag{5.15}$$

When the deflection amplitude x is greater than +1 or less than −1, as in A-B and C-D, the damping coefficient is positive, and the system is stable, although presence of a spring system k will always tend to drag the mass to a smaller absolute deflection amplitude. When the deflection amplitude lies between −1 and +1, as in B-C or D-A, the damping coefficient is negative and the system will move violently until it stabilizes in one of the damped stable zones.

While such systems are common in electronic circuitry, where they may be referred to as *flip-flop* circuits, they are rather rare in the field of rotating machinery. One instance has been observed[49] in a rotor system supported by rolling element bearings with finite internal clearance. In this situation, the effective stiffness of the rotor is small for small deflections (within the clearance) but large for large deflections (when full contact is made between the rollers and the rotor and stator). Such a

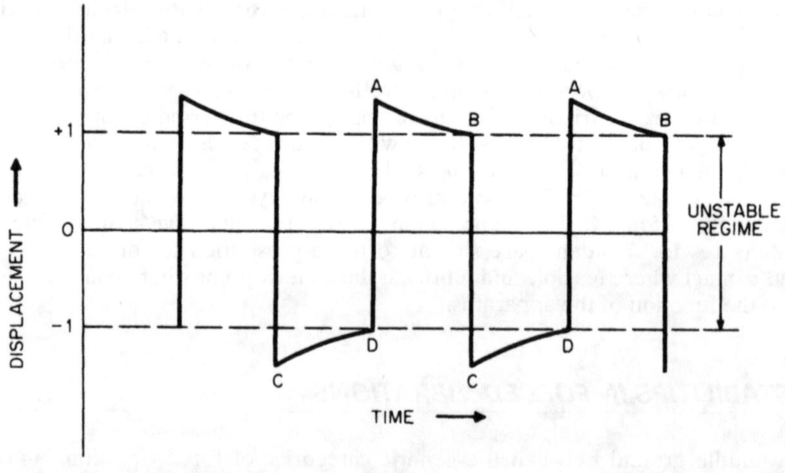

FIGURE 5.15 General form of relaxation oscillations.

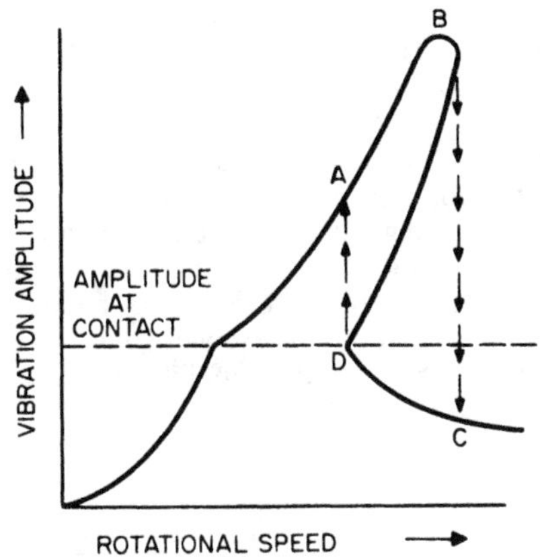

FIGURE 5.16 Response of a rotor, in bearings with (constant) internal clearance, to unbalance excitation in the vicinity of its critical speed.

nonlinearity in stiffness causes a "rightward leaning" peak in the response curve when the rotor is operating in the vicinity of its critical speed and being stimulated by unbalance. In this region, two stable modes of operation are possible, as in Fig. 5.16. In region A-B, the rotor and stator are in solid contact through the rollers. In region C-D, the rotor is whirling within the clearance, out of contact. A jump in amplitude is experienced when operating from B to C or D to A. When operating at constant speed, either of the nominally stable states can drift toward instability by virtue of thermal effects on the rollers. When the rollers are unloaded, they will skid and heat up, thereby reducing the clearance. When the rollers are loaded, they will be cooled by lubrication and will tend to contract and increase clearance. In combination, these mechanisms are sufficient to cause a relaxation oscillation in the amplitude of the forced vibration.

The remedy for this type of self-excited vibration is to eliminate the precondition of skidding rollers by reducing bearing geometric clearance, by preloading the bearing, or by increasing the temperature of any recirculating lubricant.

UNSTABLE IMBALANCE

A standard text[50] describes the occurrence of unstable vibration of steam turbines where the rotor "would vibrate with the frequency of its rotation, obviously caused by unbalance, but the intensity of the vibration would vary periodically and extremely slowly." The instability in the vibration amplitude is attributable to thermal bowing of the shaft, which is caused by the heat input associated with rubbing at the rotor's deflected "high spot," or by the mass of accumulated steam condensate in the inside of a hollow rotor at the rotor's deflected high spot. In either case, there is basis for continuous variation of amplitude, since unbalance gives rise to deflection and the deflection is, in turn, a function of that imbalance.

The phenomenon is sometimes referred to as the Newkirk effect in reference to its early recorded experimental observation.[51] A manifestation of the phenomenon in a steam turbine has been diagnosed and reported in Ref. 52 and a bibliography is available in Ref. 53. An analytic study[54] shows the possibility of both spiraling, oscillating, and constant modes of amplitude variability.

IDENTIFICATION OF SELF-EXCITED VIBRATION

Even with the best of design practice and application of the most effective methods of avoidance, the conditions and mechanisms of self-excited vibrations in rotating machinery are so subtle and pervasive that incidents continue to occur, and the major task for the vibrations engineer is diagnosis and correction.

Figure 5.3*B* suggests the forms for display of experimental data to perceive the patterns characteristic of whirling or whipping, so as to distinguish it from forced vibration, Fig. 5.3*A*. Table 5.2 summarizes particular quantitative measurements that can be made to distinguish between the various types of whirling and whipping, and other types of self-excited vibrations. The table includes the characteristic ratio of whirl speed to rotation speed at onset of vibration, and the direction of whirl with respect to the rotor rotation. The latter parameter can generally be sensed by noting the phase relation between two stationary vibration pickups mounted at 90° to one another at similar radial locations in a plane normal to the rotor's axis of rotation. Table 5.1 and specific prescriptions in the foregoing text and references suggest corrective action based on these diagnoses. Reference 55 gives additional description of corrective actions.

TABLE 5.2 Diagnostic Table of Rotating Machinery Self-excited Vibrations

	R, characteristic ratio: whirl frequency/rpm	Whirl direction
Whirling or whipping:		
Hysteretic whirl	$R \approx 0.5$	Forward
Fluid trapped in rotor	$0.5 < R < 1.0$	Forward
Dry friction whip	No functional relationship; whirl frequency a function of coupled rotor-stator system; onset rpm is a function of rpm at contact	Backward—axial contact on disc side nearest virtual pivot; Forward—axial contact on disc side opposite to virtual pivot; Backward—radial contact
Fluid bearing whip	$R < 0.5$	Forward
Seal and blade-tip-clearance effect in turbomachinery	Load-dependent	Forward—turbine blade tip; Backward—compressor blade tip; Unspecified—for seal clearance
Propeller and turbomachinery whirl	Load-dependent	Backward—virtual pivot aft of rotor; Forward—virtual pivot front of rotor (where front is source of impinging flow)
Parametric instability:		
Asymmetric shafting	$R \approx 1.0, 2.0, 3.0, \ldots$	Unspecified
Pulsating torque	$R \approx 0.5, 1.0, 1.5, 2.0, \ldots$	Unspecified
Pulsating longitudinal load	A function of pulsating load frequency rather than rpm	Unspecified
Stick-slip rubs and chatter	$R \ll 1$	Essentially planar rather than whirl motion
Instabilities in forced vibrations:		
Bistable vibration	$R = 1$ with periodic square wave fluctuations in amplitude of frequency much lower than rotation rate	Forward
Unstable imbalance	$R = 1$ with slow variation in amplitude	Forward

REFERENCES

1. Den Hartog, J. P.: "Mechanical Vibrations," 4th ed., McGraw-Hill Book Company, New York, 1956, p. 346.
2. Ehrich, F. F.: "Handbook of Rotordynamics," 2d ed., Krieger Publishing Co., Malabar, Fla., 1999, pp. 1.72–1.106.
3. Kramer, E.: "Instabilities of Rotating Shafts," *Proc. Conf. Vib. Rotating Systems, Inst. Mech. Eng.,* London, February 1972.
4. Vance, J. M.: "High Speed Rotor Dynamics—Assessment of Current Technology for Small Turboshaft Engines," USAAMRDL-TR-74-66, Ft. Eustis, Va., July 1974.
5. Newkirk, B. L.: *Gen. Elec. Rev.,* **27**:169 (1924).
6. Kimball, A. L.: *Gen. Elec. Rev.,* **17**:244 (1924).
7. Ref. 1, pp. 292–296.
8. Ehrich, F.: *ASME DE,* **18**:1 (September 1989).
9. Gunter, E. J.: "Dynamic Stability of Rotor-Bearing Systems," NASA SP-113, Chap. 4, 1966.
10. Bolatin, V. V.: "Non-Conservative Problems of the Theory of Elastic Stability," Pergamon Press, New York, 1964.
11. Bentley, D. E.: "The Re-Excitation of Balance Resonance Regions by Internal Friction," *ASME Paper* 72-PET-49, September 1972.
12. Vance, J. M., and J. Lee: "Stability of High Speed Rotors with Internal Friction," *ASME Paper* 73-DET-127, September 1973.
13. Ehrich, F. F.: *J. Appl. Mech.,* (E) **31**(2):279 (1964).
14. Wolf, J. A.: "Whirl Dynamics of a Rotor Partially Filled with Liquids," *ASME Paper* 68-WA/APM-25, December 1968.
15. Ehrich, F. F.: *J. Eng. Ind.,* (B) **89**(4):806 (1967).
16. Changsheng, Z.: *J. Vibration and Sound,* **124**(4):483–491 (2002).
17. Sapetta, L. P., and R. J. Harker: "Whirl of Power Screws Excited by Boundary Lubrication at the Interface," *ASME Paper* 67-Vibr-37, March 1967.
18. Ref. 1, pp. 293–295.
19. Begg, I. C.: "Friction Induced Rotor Whirl—A Study in Stability," *ASME Paper* 73-DET-10, September 1973.
20. Ehrich, F. F.: "The Dynamic Stability of Rotor/Stator Radial Rubs in Rotating Machinery," *ASME Paper* 69-Vibr-56, April 1969.
21. Newkirk, B. L., and H. D. Taylor: *Gen. Elec. Rev.,* **28**:559–568 (1925).
22. Ref. 1, pp. 297–298.
23. Ref. 9, Chap. 5.
24. Pinkus, O., and B. Sternlight: "Theory of Hydrodynamic Lubrication," Chap. 8, McGraw-Hill Book Company, New York, 1961.
25. Black, H. F., and D. N. Jenssen: "Effects of High Pressure Seal Rings on Pump Rotor Vibrations," *ASME Paper* 71-WA/FE-38, December 1971.
26. Alford, J. S.: *J. Eng. Power,* **87**(4):333 (October 1965).
27. Ehrich, F.F., et al.: *J. of Turbomachinery,* **123**(3):446–452 (2001).
28. Black, H. F.: "Calculation of Forced Whirling and Stability of Centrifugal Pump Rotor Systems," *ASME Paper* 73-DET-131, September 1973.
29. Ref. 1, pp. 317–321.
30. Ehrich, F. F.: *Trans. ASME,* (A) **90**(4):369 (1968).
31. Taylor, E. S., and K. A. Browne: *J. Aeronaut. Sci.,* **6**(2):43–49 (1938).

32. Houbolt, J. C., and W. H. Reed: "Propeller Nacelle Whirl Flutter," *I.A.S. Paper* 61–34, January 1961.

33. Trent, R., and W. R. Lull: "Design for Control of Dynamic Behavior of Rotating Machinery," *ASME Paper* 72-DE-39, May 1972.

34. Ehrich, F. F.: "An Aeroelastic Whirl Phenomenon in Turbomachinery Rotors," *ASME Paper* 73-DET-97, September 1973.

35. Floquet, G.: *Ann. l'école normale supérieure,* **12**:47 (1883).

36. McLachlan, N. W.: "Theory and Applications of Mathieu Functions," Oxford University Press, New York, 1947, p. 40.

37. Ref. 1, pp. 336–339.

38. Brosens, P. J., and H. S. Crandall: *J. Appl. Mech.,* **83**(4):567 (1961).

39. Messal, E. E., and R. J. Bronthon: "Subharmonic Rotor Instability Due to Elastic Asymmetry," *ASME Paper* 71-Vibr-57, September 1971.

40. Arnold, R. C., and E. E. Haft: "Stability of an Unsymmetrical Rotating Cantilever Shaft Carrying an Unsymmetrical Rotor," *ASME Paper* 71-Vibr-57, September 1971.

41. Eshleman, R. L., and R. A. Eubanks: "Effects of Axial Torque on Rotor Response: An Experimental Investigation," *ASME Paper* 70-WA/DE-14, December 1970.

42. Wehrli, V. C.: "Uber Kritische Drehzahlen unter Pulsierender Torsion," *Ing. Arch.,* **33**:73–84 (1963).

43. Ref. 36, p. 292.

44. Ref. 1, p. 290.

45. Conn, H.: *Tool Eng.,* **45**:61–65 (1960).

46. Sadowy, M.: *Tool Eng.,* **43**:99–103 (1959).

47. Ref. 1, pp. 365–368.

48. Van der Pol, B.: *Phil. Mag.,* **2**:978 (1926).

49. Ehrich, F. F.: "Bi-Stable Vibration of Rotors in Bearing Clearance," *ASME Paper* 65-WA/MD-1, November 1965.

50. Ref. 1, pp. 245–246.

51. Newkirk, B. L.: "Shaft Rubbing," *Mech. Eng.,* **48**:830 (1926).

52. Kroon, R. P., and W. A. Williams: "Spiral Vibration of Rotating Machinery," *5th Int. Congr. Appl. Mech.*, John Wiley & Sons, New York, 1939, p. 712.

53. Dimarogonas, A. D., and G. N. Sander: *Wear,* **14**(3):153 (1969).

54. Dimarogonas, A. D.: "Newkirk Effect: Thermally Induced Dynamic Instability of High-Speed Rotors," *ASME Paper* 73-GT-26, April 1973.

55. Ehrich, F. and D. Childs: *Mech. Eng.,* **106**(5):66 (1984).

CHAPTER 6
DYNAMIC VIBRATION ABSORBERS AND AUXILIARY MASS DAMPERS

Sheldon Rubin

INTRODUCTION

Auxiliary masses can be attached to vibrating systems by elastic and damping devices to assist in reducing the amplitude of vibration of the system. Depending upon the application, these auxiliary systems fall into two distinct classes:

1. If the vibration reduction of the primary system is to be in a narrow band of frequency, then it is possible to use one or more dynamic auxiliary systems, each tuned to the center frequency of the target band so as to magnify the resistance to motion of the primary system in that band. The band may be the centered on an exciting frequency or on the frequency of a vibration mode of the primary system. Such an auxiliary system is traditionally called a *tuned vibration absorber* (TVA). The effectiveness of such an auxiliary system is due to narrowband increase in damping, so it is sometimes called a *tuned mass damper* (TMD).
2. If damping increase of the primary system over a wide range of frequency is the objective, then one or more auxiliary systems involving mass addition may be a practical alternative to modifying the structure of the primary system by other types of damping treatment (Chap. 36). Such an auxiliary system can be called an *auxiliary mass damper.* Tuning may or may not be involved.

FORMS OF AUXILIARY MASS SYSTEMS

In its simplest form, as applied to a single-degree-of-freedom (SDOF) system, the character of a dynamic auxiliary system is the same as that of the primary system.

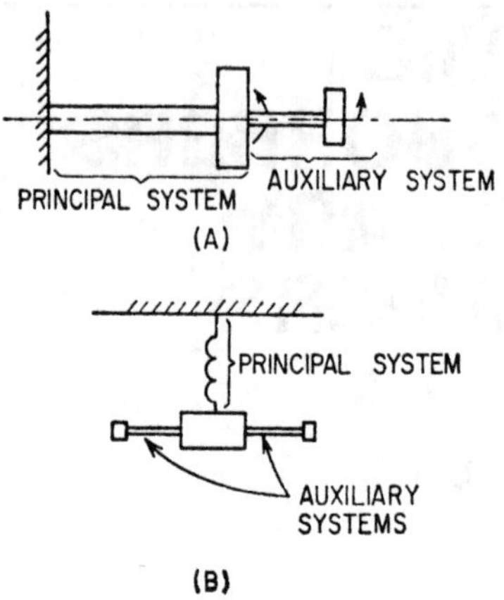

FIGURE 6.1 Typical dynamic vibration absorbers. The principal and auxiliary systems vibrate in torsion in the arrangement at (*A*); the auxiliary system is in the form of masses and beams at (*B*).

Thus, a torsional system has a torsionally connected auxiliary system and a linear system has a linearly connected auxiliary system. Schematics are shown in Fig. 6.1.

Multiple-degree-of-freedom auxiliary systems are often practical to apply and can more effectively address vibration reduction of the primary system. One example is a resiliently connected axisymmetric dynamic absorber placed on a round shaft. The absorber tuning can be in any radial direction to reduce bending vibration, or in the axial direction to reduce axial vibration, or in the torsional direction to address torsional vibration, or in some combination of these directions. Another example is an axially symmetric cantilever beam containing a mass near its end, as depicted in Fig. 6.2. In this case, the auxiliary system is effective for motion in any axis in the plane normal to the TVA axis at its base and for rotation about such an axis. The only directions of base motion not reacted by the tuning influence are the TVA beam axial and torsional directions. Thus, the auxiliary system depicted in Fig. 6.2 is effective in four of the six degrees

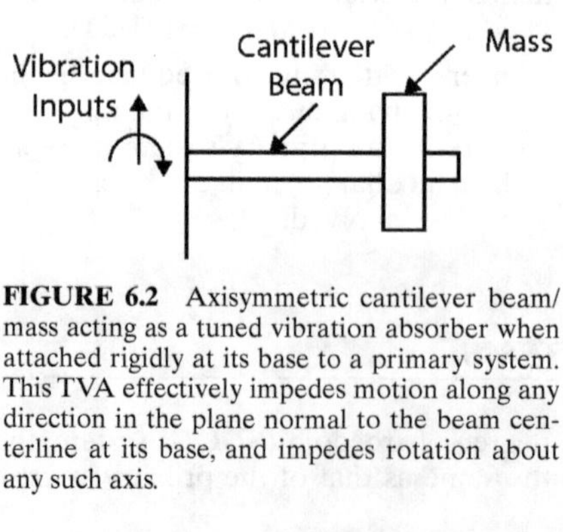

FIGURE 6.2 Axisymmetric cantilever beam/mass acting as a tuned vibration absorber when attached rigidly at its base to a primary system. This TVA effectively impedes motion along any direction in the plane normal to the beam centerline at its base, and impedes rotation about any such axis.

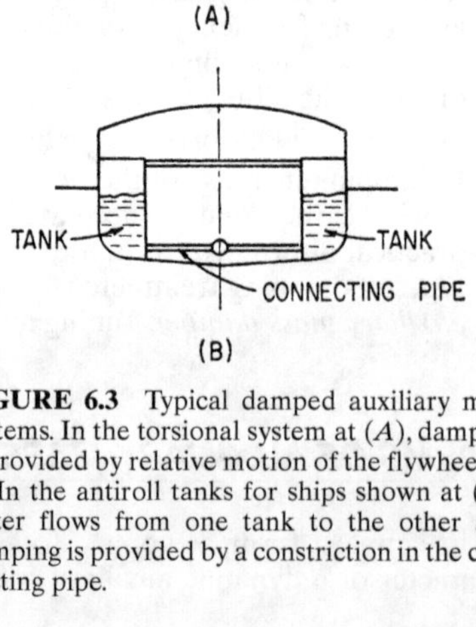

FIGURE 6.3 Typical damped auxiliary mass systems. In the torsional system at (*A*), damping is provided by relative motion of the flywheels J, J_a. In the antiroll tanks for ships shown at (*B*), water flows from one tank to the other and damping is provided by a constriction in the connecting pipe.

of freedom at the base of the beam. The tuning of an auxiliary system on a beam can be adjusted by adding or subtracting mass or by moving the position of the mass along the beam.

An auxiliary mass damper is intended to provide vibration reduction in a range of frequencies by contributing energy dissipation to the primary system. The energy dissipation in the torsional example in Fig. 6.3*A* can be in the form of viscous, elastomeric, or hysteretic damping reacting to relative motion of the primary and auxiliary flywheels. In the ship roll example in Fig. 6.3*B*, the dissipation is by laminar or turbulent flow of fluid through the restriction in the pipe transferring fluid from one tank to the other caused by ship roll motion.

INFLUENCE OF AN AUXILIARY SYSTEM

Single-degree-of-freedom analysis is often a practical approach to preliminary sizing of an auxiliary system. Influence of an auxiliary system is a function of the relative mechanical impedance of the primary and auxiliary systems in the affected degree of freedom. A block diagram of an auxiliary system attached to a primary system that is a source of vibration at its output is shown in Fig. 6.4*A*. The impedance of the primary system Z_0 is the complex ratio of force and resulting velocity applicable to the auxiliary system attachment; the inverse of Z_0 is the mobility $\mathfrak{M}_0$ (Chap. 9). (Another name for mobility is admittance.) The impedance of the auxiliary system Z_a is the corresponding force-to-velocity ratio looking into its attachment point; the inverse of Z_a is the mobility $\mathfrak{M}_a$.

Consider a primary system of any complexity vibrating in response to a source of vibration. Assume that the strength of the vibration source is unaffected by the presence of an attachment to an output point of the primary system. In a single degree of freedom at the attachment point, the equation governing the relation of the complex amplitude of a harmonic force F applied by the primary system and resulting velocity v is

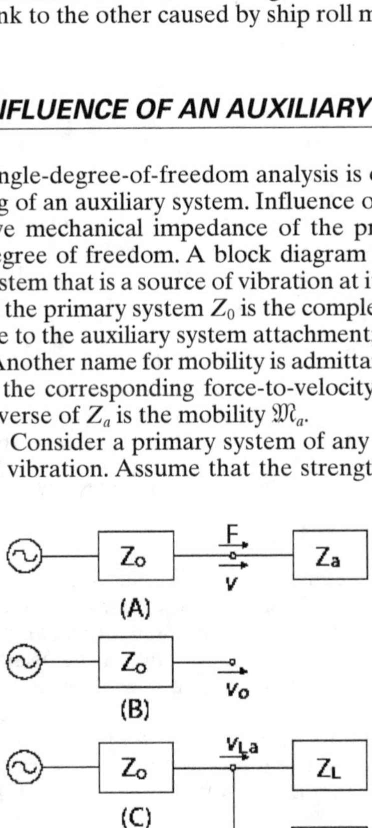

FIGURE 6.4 (*A*) Block diagram of an auxiliary system with impedance Z_a attached to the output of an excited primary system with impedance Z_0; the force F is delivered to the auxiliary system and a velocity v results. (*B*) The primary system without an attachment ($F = 0$) has an output *free velocity* of v_0. (*C*) A general load with impedance Z_L is attached to the primary system and an auxiliary system is also attached to the primary system; the velocity at the output is v_{La}.

$$F = Z_0\,(v_0 - v) \qquad \text{or} \qquad \mathfrak{M}_0\,F = v_0 - v \tag{6.1}$$

The "free velocity" v_0 is the complex amplitude of velocity at the output of the primary system in the absence of the attached system; thus, $v_0 = v$ when $F = 0$ (see Fig. 6.4*B*). Equation (6.1) stems from the linear superposition of the free vibration and the additional vibration caused by the action of the force reacted from an attached auxiliary system.

The equation governing the attached system is

$$F = Z_a\,v \qquad \text{or} \qquad \mathfrak{M}_a\,F = v \tag{6.2}$$

Combining these two equations, the influence of the attached system can be expressed as the complex ratio v/v_0:

$$v/v_0 = 1/(1 + Z_a/Z_0) = 1/(1 + \mathfrak{M}_0/\mathfrak{M}_a) \tag{6.3}$$

Thus, attenuation of the magnitude of the vibration at the attachment point is inversely proportional to the magnitude of $1 + Z_a/Z_0$ or $1 + \mathfrak{M}_0/\mathfrak{M}_a$. A high magnitude of impedance of the attached system relative to the source will lead to large attenuation. Alternatively, a high magnitude of mobility of the source relative to the attached system will lead to large attenuation. The force delivered to the attached system is

$$F = Z_0 v_0 /(1 + Z_a/Z_0) = Z_0 v_0 /(1 + \mathfrak{M}_0/\mathfrak{M}_a) \tag{6.4}$$

If the attached system impedance is very large relative to the source impedance, the delivered force becomes Z_0v_0, referred to as the *blocked force* of the source, since this is the force delivered when the motion of the output is zero; see Eq. (6.1).

A block diagram of an auxiliary system placed at the junction of a source system and a "load" system is shown in Fig. 6.4*C*. The aim of the auxiliary system is to reduce the transmission of vibration into the load whose impedance is Z_L. The impedance of the combination of the load and auxiliary system is simply the sum of their impedances, $Z_L + Z_a$. Corresponding to Eq. (6.3), the ratio of velocity into the load and absorber, v_{La}, to the free velocity is

$$v_{La}/v_0 = 1/[1 + (Z_L + Z_a)/Z_0] \tag{6.5}$$

The result without the absorber is

$$v_L/v_0 = F/F_0 = 1/(1 + Z_L/Z_0) \tag{6.6}$$

So the effect of the absorber in reducing the input to the load is expressed by

$$v_{La}/v_L = 1/[1 + Z_a/(Z_L + Z_0)] \tag{6.7}$$

Thus, high absorber impedance relative to the sum of the source and load impedances will provide large attenuation of the magnitude of the vibration transmitted to the load. An application of Eq. (6.7) to engine vibration transmission into an aircraft fuselage involved experimental mobility for the engine (source) and for the fuselage (load) and theoretical impedance of an SDOF auxiliary system (development follows) tuned to a disturbance frequency.[1]

Equations (6.1) to (6.7) have counterparts for multiple-DOF systems wherein the impedances or mobilities are square matrices and the forces and velocities are column vectors. The generalization of Eq. (6.3) is the relation of the column of modified velocities $\boldsymbol{v}$ to the column of free velocities $\boldsymbol{v}_0$

$$\boldsymbol{v} = (\mathbf{Z}_a + \mathbf{Z}_0)^{-1}\mathbf{Z}_0\, \boldsymbol{v}_0 = (\boldsymbol{\mathfrak{M}}_a + \boldsymbol{\mathfrak{M}}_0)^{-1}\boldsymbol{\mathfrak{M}}_a\, \boldsymbol{v}_0 \tag{6.8}$$

The bold symbols denote matrix or column quantities. $\mathbf{Z}_a$ and $\mathbf{Z}_s$ are the square matrices of auxiliary system and source impedance, respectively; $\boldsymbol{\mathfrak{M}}_a$ and $\boldsymbol{\mathfrak{M}}_0$ are the corresponding mobility matrices that are inverses of the impedance matrices (Chap. 9). Experimental measurements for a multiple-degree-of-freedom system invariably produce elements of the mobility matrix. The generalization of Eq. (6.7) is

$$\boldsymbol{v}_{La} = (\mathbf{Z}_a + \mathbf{Z}_L + \mathbf{Z}_0)^{-1}(\mathbf{Z}_L + \mathbf{Z}_0)\boldsymbol{v}_L \tag{6.9}$$

Effectiveness of the auxiliary system can be expressed in terms of any ratio of modified and source velocities.

IMPEDANCE OF A SINGLE-DEGREE-OF-FREEDOM TUNED ABSORBER

Using the nomenclature in Fig. 6.5, the equation of motion of an idealized SDOF absorber (massless stiffness and viscous damper) is

$$m_a\ddot{x}_a + c_a(\dot{x}_a - \dot{x}_0) + k_a(x_a - x_0) = 0 \tag{6.10}$$

For steady-state harmonic motion at the radian frequency ω,

$$-\omega^2 m_a x_a + (j\omega c_a + k_a)(x_a - x_0) = 0 \tag{6.11}$$

Let F_0 be the force required to be applied to the base of the absorber to produce the motion x_0. Then,

$$F_0 = -(j\omega c_a + k_a)(x_a - x_0) = -\omega^2 m_a x_a \tag{6.12}$$

From Eqs. (6.11) and (6.12), the impedance (Chap. 9) at the base looking toward the absorber mass is

$$Z_a = F_0/j\omega x_0 = j\omega m_a(j\omega c_a + k_a)/(-\omega^2 m_a + j\omega c_a + k_a) \tag{6.13}$$

In nondimensional terms, Eq. (6.13) becomes

$$Z_a = j\omega m_a(1 + j2\xi_a\beta_a)/(1 - \beta_a^2 + j2\xi_a\beta_a) \tag{6.14}$$

where $\omega_a = (k_a/m_a)^{1/2}$, the natural frequency of the absorber (tuning frequency)
$\beta_a = \omega/\omega_a$, the nondimensional frequency
$c_{ca} = 2(k_a m_a)^{1/2} = 2m_a\omega_a$, the critical damping coefficient of the absorber
$\xi_a = c_a/c_{ca}$, the fraction of critical damping

Expressing the normalized impedance of the tuned absorber in real (Re) and imaginary (*Im*) parts,

$$Re(Z_a)/\omega_a m_a = 2\xi_a\beta_a^4/[(1 - \beta_a^2)^2 + (2\xi_a\beta_a)^2] \tag{6.15}$$

$$Im(Z_a)/\omega_a m_a = \beta_a[1 - \beta_a^2 + (2\xi_a\beta_a)^2]/[(1 - \beta_a^2)^2 + (2\xi_a\beta_a)^2] \tag{6.16}$$

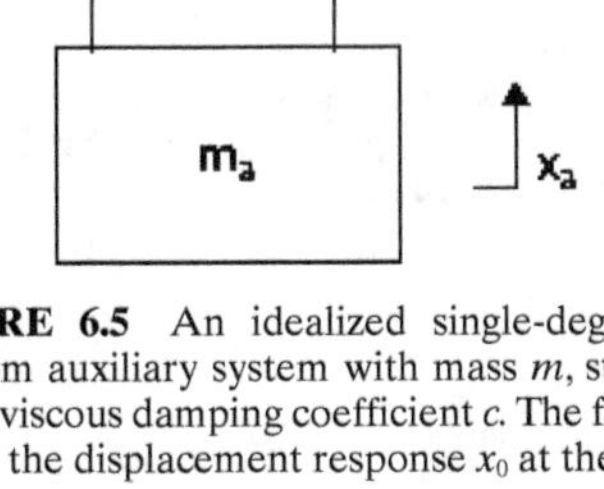

FIGURE 6.5 An idealized single-degree-of-freedom auxiliary system with mass m, stiffness k, and viscous damping coefficient c. The force F_0 causes the displacement response x_0 at the base.

The real part expresses a damping contribution (positive at all frequencies as required for a stable system). The real and imaginary part of the absorber impedance, as a factor on the mass impedance magnitude ($\omega_a m_a$) at its tuning frequency, is plotted in Fig. 6.6*A* and *B*, respectively. When the frequency equals the tuning frequency $\beta_a = 1$, the normalized real part (damping impedance) goes to the resonant magnification ($1/2\xi_a$, also known as the quality factor Q_a):

$$Re(Z_a)/\omega_a m_a = 1/2\xi_a \qquad \text{for } \omega = \omega_a \tag{6.17}$$

In the limit of zero absorber damping, the real part of the impedance goes to

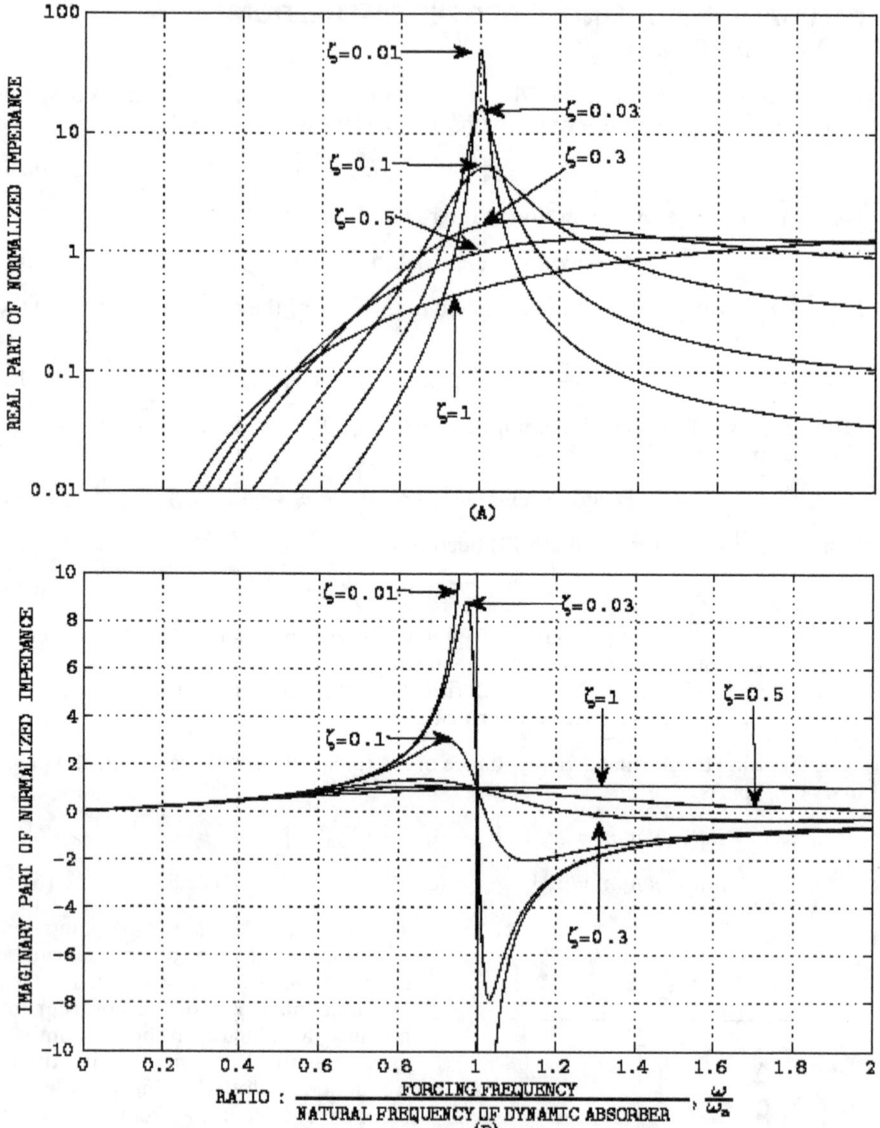

FIGURE 6.6 (*A*) Real part of normalized input impedance (effective damping) of a dynamic absorber versus the frequency relative to the natural frequency of the absorber ω_a. Curves for a sequence of values of fraction of critical damping ξ appear. (*B*) Corresponding curves for the imaginary part (mass-like contribution when positive and spring-like when negative).

infinity; thus, a finite force produces no base motion (i.e., a vibration node is produced at the point and direction of absorber attachment). In the limit of an infinite frequency ($\beta_a = \infty$), the real and imaginary impedances go to the impedance of the spring and dashpot in parallel since the mass becomes fixed in inertial space. Thus,

$$Re(Z_a) = 2\xi_a\omega_a m_a = c_a \qquad \text{and} \qquad Im(Z_a) = -\omega_a^2 m_a/\omega = -k_a/\omega \qquad \text{for } \omega = \infty \quad (6.18)$$

As seen in Fig. 6.6*A*, when ξ_a is less than 0.3 the real impedance (damping) is a maximum at the tuning frequency and continually decreases with increasing frequency. For all ξ, an asymptotic value is reached essentially by three times the natural frequency at a value equal to the viscous damping coefficient per Eq. (6.14). Devices with ξ_a up to 0.1 are typically used for narrowband damping effectiveness and are commonly denoted as tuned vibration absorbers (TVAs). Tuned devices intended for broad damping effectiveness may be targeted to have ξ_a in the range of 0.3 to 1 and may be more appropriately denoted as tuned mass dampers (TMDs).

As seen in Fig. 6.6*B*, the imaginary part of $Z_a/\omega_a m_a$, is positive (mass-like) below the natural frequency and negative (spring-like) above. For ξ_a up to 0.3, a mass-like peak occurs just below the natural frequency and a stiffness-like peak occurs just above the natural frequency. Thus, the dynamic absorber will cause a lowering of system natural frequencies below the tuning frequency due to mass-like addition by the absorber and an increase of system natural frequencies above its tuning frequency due to stiffness-like addition.

IMPEDANCE OF AN ABSORBER WITH MASS AND VISCOUS DAMPER ONLY

Without stiffness, this becomes the limiting case for a zero tuning frequency of a dynamic absorber. Setting $k_a = 0$ in Eq. (6.13), the impedance is

$$Z_a = F_0/j\omega x_0 = j\omega m_a(j\omega c_a)/(-\omega^2 m_a + j\omega c_a) = c_a/(1 - jc_a/m_a\omega) \qquad (6.19)$$

When the product of mass and frequency is high enough to make $c_a/m_a\omega$ small, the impedance is close to that of the viscous damper alone. Effectively, the mass motion becomes small enough so that the input motion to the absorber is close to the relative motion across the damper. An optimum damping for such an auxiliary system attached to an SDOF primary system is described subsequently.

EFFECTIVENESS OF AN SDOF ABSORBER ATTACHED TO AN SDOF PRIMARY SYSTEM

It is instructive to address the interaction between an SDOF dynamic absorber and an SDOF primary system.[2–4] The equation of motion of the SDOF primary system alone, shown in Fig. 6.7*A* with force excitation of the primary mass, is

$$m\ddot{x}_0 + c\dot{x}_0 + kx_0 = F \qquad (6.20)$$

The impedance of this primary system is

$$Z_0 = F/j\omega x_0 = (-\omega^2 m + j\omega c + k)/j\omega \qquad (6.21)$$

In nondimensional terms, Eq. (6.21) becomes

$$Z_0 = m\omega_0^2(1 - \beta^2 + j2\xi\beta)/j\omega \tag{6.22}$$

where $\omega_0 = (k/m)^{1/2}$, the natural frequency of the primary system
$\beta = \omega/\omega_0$, a nondimensional frequency
$c_c = 2(km)^{1/2} = 2m\omega_0$, the critical damping of the primary system
$\xi = c/c_c$, the fraction of critical damping of the primary system

At the source natural frequency, $\omega = \omega_0$ and $\beta = 1$, the magnitude of the source impedance is a minimum equal to the damping coefficient of the primary system:

$$Z_0 = m\omega_0^2\, j2\xi\beta/j\omega = c_c \qquad \text{for } \omega = \omega_0 \tag{6.23}$$

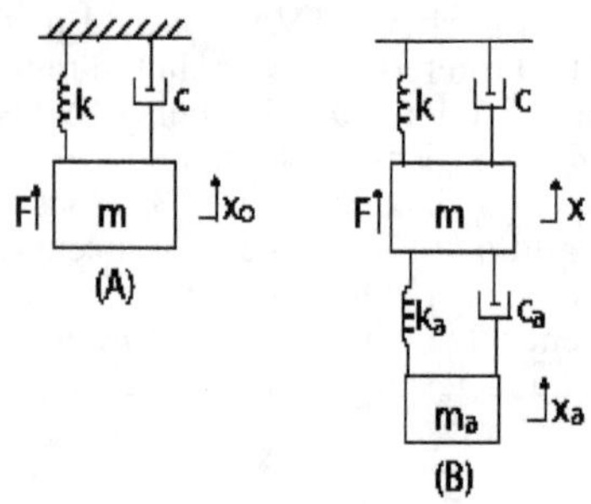

FIGURE 6.7 (*A*) The primary system is excited by a force F on its mass, resulting in a displacement x_0. (*B*) An auxiliary system is attached and the displacement of the primary mass becomes x.

Thus, the primary system impedance vanishes at its natural frequency if it is undamped; that is, an infinitely large velocity will result from a finite force applied to the primary system mass.

A dynamic auxiliary system attached to a primary system is shown in Fig. 6.7*B*. The ratio of the velocity of the primary mass with the absorber attached to the free velocity of the primary mass, v/v_0, is given in terms of impedances in Eq. (6.3). Inserting Eq. (6.23) for the primary system impedance and Eq. (6.14) for the absorber impedance, yields

$$v/v_0 = [1 - \mu\beta^2(1 + j2\xi_a\beta_a)/(1 - \beta_a^2 + j2\xi_a\beta_a)(1 - \beta^2 + j2\xi\beta)]^{-1} \tag{6.24}$$

The ratio of absorber mass to primary mass is $\mu = m_a/m$.

If the absorber were tuned to the primary system natural frequency, $\omega_a = \omega_0$, the velocity ratio at this common frequency becomes

$$v/v_0 = 4\xi_a\xi/[4\xi_a\xi + \mu(1 + j2\xi_a)] \tag{6.25}$$

If $4\xi_a\xi << \mu$ and $4\xi_a^2 << 1$, the absorber effectiveness is closely given by

$$v/v_0 = 4\xi_a\xi/\mu \tag{6.26}$$

For example, with Eq. (6.26), if $\mu = 0.05$ and $\xi = \xi_a = 0.02$, $v/v_0 = 0.032$ versus 0.031 by Eq. (6.25), so that a reduction of the free velocity by a factor of 31 will result.

Free vibration of the composite system obeys $Z_0 + Z_a = 0$ or

$$j\omega m_a(1 + j2\xi_a\beta_a)/(1 - \beta_a^2 = j2\xi_a\beta_a) + m\omega_0^2(1 - \beta^2 + j2\xi\beta)/j\omega = 0 \tag{6.27}$$

In the absence of damping, free vibration occurs at the natural frequencies of this two-DOF system (designated ω_n) which are solutions of

$$(\omega_n^2 - \omega_a^2)(\omega_n^2 - \omega_0^2) - \omega_n^2\omega_a^2 = 0 \tag{6.28}$$

The natural frequencies are found from the roots ω_n^2 of Eq. (6.28):

$$\omega_n^2 = [\omega_a^2(1 + \mu) + \omega_0^2]/2 \pm \{[\omega_a^2(1 + \mu) - \omega_0^2]^2/4 + \omega_a^2\omega_0^2\mu\}^{0.5} \tag{6.29}$$

Consider the case of an absorber intended to attenuate response to an excitation with a moderately variable frequency. A strategy for preliminary design of the absorber mass might be to select a trial series of absorber mass values and tune each mass case to place the square of the target excitation frequency ω_T midway between the square of the higher and lower system resonance frequencies. Namely,

$$\omega_a^2 = (2\omega_T^2 - \omega_0^2)/(1 + \mu) \tag{6.30}$$

Then, to observe whether the absorber mass is sufficient to cause the two system resonances to lie outside the range of the excitation frequency, calculate the system frequencies from Eq. (6.29). Finally, include system damping and absorber damping to achieve the desired attenuation versus excitation frequency.

Minimization of Response Near a Primary System Mode. When a relatively light auxiliary system is tuned to near the natural frequency of the SDOF primary system, the coupled system has a pair of closely spaced modes. The amplification of primary mass response to force input will have two closely spaced peaks as long as the damping in the auxiliary system is not too large; for example, see Fig. 6.8. Note that all curves pass through the two fixed points A and B, independent of damping. Peak amplification can be minimized with proper tuning and damping of the auxiliary system; for example, see Fig. 6.9.

This optimization problem has been investigated extensively for an undamped primary system.[2–6] Let α be the ratio of absorber and primary natural frequencies:

$$\alpha = \omega_a/\omega_0 \tag{6.31}$$

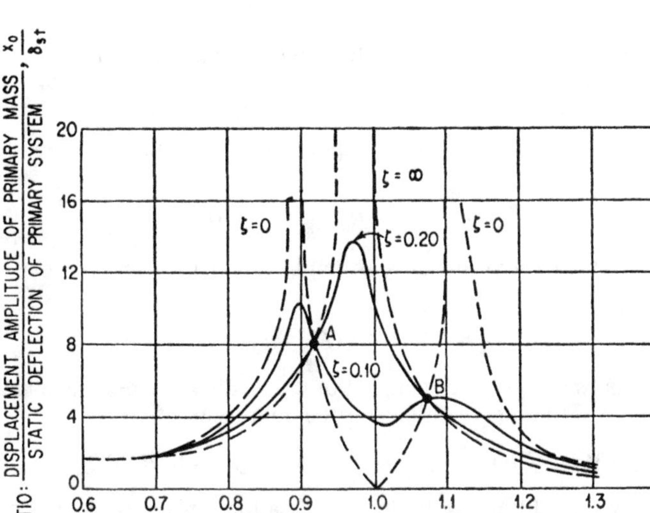

FIGURE 6.8 Curves for auxiliary mass damper showing amplitude of vibration of mass of primary system as a function of the ratio of forcing frequency ω to natural frequency of primary system $\omega = \sqrt{k/m}$. The mass ratio $m_a/m = 0.05$, and the natural frequency ω_a of the auxiliary mass system is equal to the natural frequency ω_0 of the primary system. Curves are included for several values of damping in the auxiliary system.

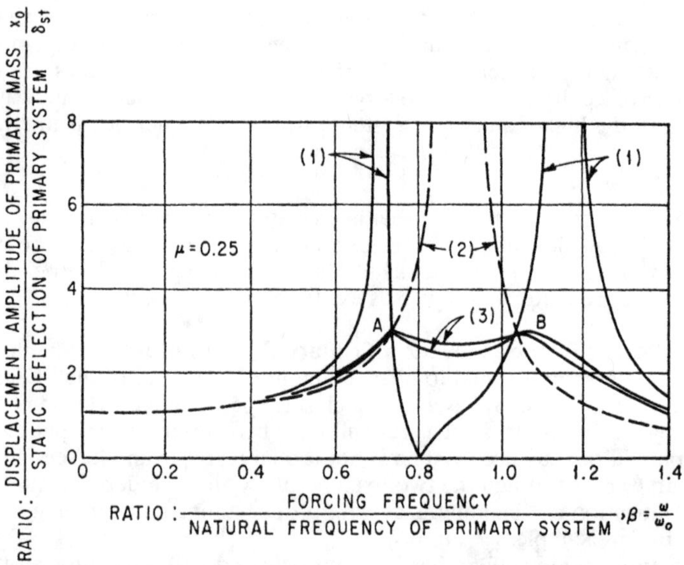

FIGURE 6.9 Curves similar to Fig. 6.8 but with optimum tuning. Curves 1 apply to an undamped absorber, curve 2 represents infinite damping in the auxiliary system, and curves 3 have horizontal tangents at the fixed points A and B.

The condition that the points A and B have the same amplification is

$$\alpha^* = 1/(1 + \mu) \tag{6.32}$$

This optimum tuning leads to equal amplification P^* at the fixed points A and B given by

$$P^* = P_A = P_B = [(2 + \mu)/\mu]^{1/2} \tag{6.33}$$

Given that P^* is the target value of peak amplification, the necessary mass ratio is μ^* from Eq. (6.33):

$$\mu^* = 2/(P^{*2} - 1) \tag{6.34}$$

Finally, two values of optimum damping of the auxiliary system have been derived based on slightly different conditions for the optimization:[2,6]

$$\xi_{a,\text{opt1}} = [(3/8)\mu^*/(1 + \mu^*)]^{0.5} \quad \text{and} \quad \xi_{a,\text{opt2}} = [(1/2)\mu^*/(1 + \mu^*)]^{0.5} \tag{6.35}$$

Figure 6.9 shows the difference in the amplification for the two values of optimum damping for the mass ratio of 0.05. The higher value of damping in Eq. (6.35) has the benefit of a lower relative displacement within the absorber.[6] Studies have also been applied to multimass systems.[7,8]

Optimum Damping for an Auxiliary Mass Absorber Connected to the Primary System with Damping Only. In general, the most effective damping is obtained where the auxiliary mass damping system includes a spring in its connec-

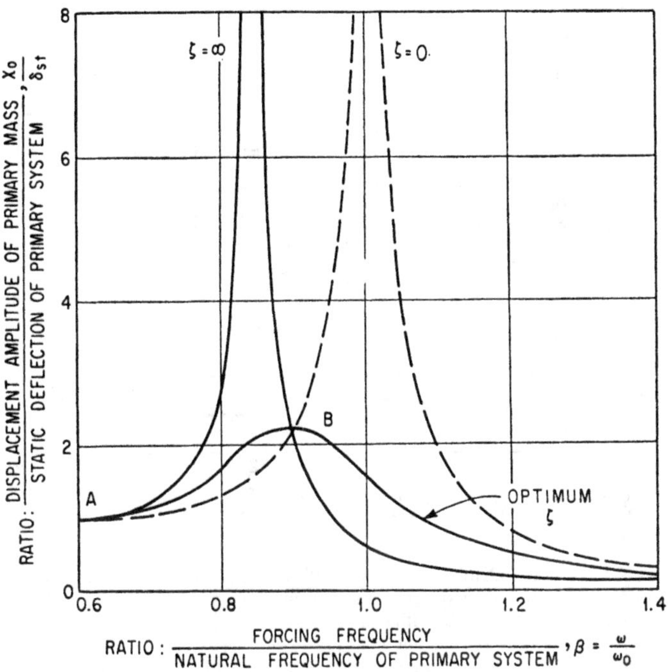

FIGURE 6.10 Curves similar to Fig. 6.9 for system having auxiliary mass coupled by damping only. Several values of damping are included.

tion to the primary system. However, such a design requires a calculation of the optimum stiffness of the spring. Sometimes it is more expedient to add an oversize mass, coupled only by damping to the primary system, than it is to compute the optimum system. However, if use is made of such a simplified damper by taking it from a list of standard dampers and applying it with a minimum of calculations, the stock dampers should be as efficient as the application will permit.

Curves showing the amplification of motion of the primary mass for a mass ratio $\mu = 0.4$ appear in Fig. 6.10. The fixed point A is at zero frequency and B is at the intersection of the curves for $\xi = 0$ (absorber mass is disconnected) and $\xi = \infty$ (absorber mass is rigidly connected). The optimum damping is the one that makes the peak amplification appear at fixed point B. The optimum viscous damping coefficient of the absorber is

$$c_a{}^* = 2m_a\omega_0/[2(2+\mu)(1+\mu)]^{0.5} \tag{6.36}$$

As a function of the ratio of primary to absorber mass (m/m_a), the values for the amplitude of vibration of the primary mass, the relative amplitude between the primary and absorber masses, and optimum damping constants are given in Figs. 6.11, 6.12, and 6.13, respectively.

Auxiliary Mass Damper Using Coulomb Friction Damping.[9] Dampers relying on coulomb friction (i.e., friction whose force is constant) have been widely used. A damper relying on dry friction and connected to its primary system with a

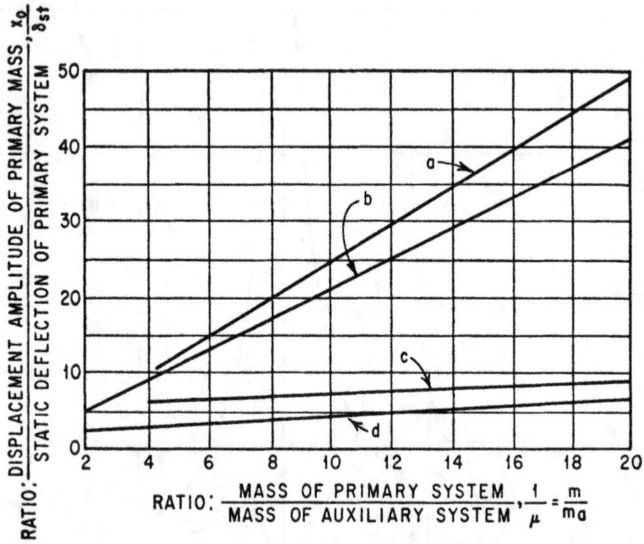

FIGURE 6.11 Displacement amplitude of the primary mass as a function of the size of the auxiliary mass: (*a*) auxiliary system coupled only by coulomb friction ($\alpha = 0$) with optimum damping; (*b*) auxiliary system coupled only by viscous damping ($\alpha = 0$) of optimum value; (*c*) auxiliary system coupled by spring and damper tuned to frequency of primary system ($\alpha = 1$) with optimum damping; (*d*) auxiliary system coupled by spring and damper with optimum tuning [$\alpha = 1/(1 + \mu)$] and optimum damping.

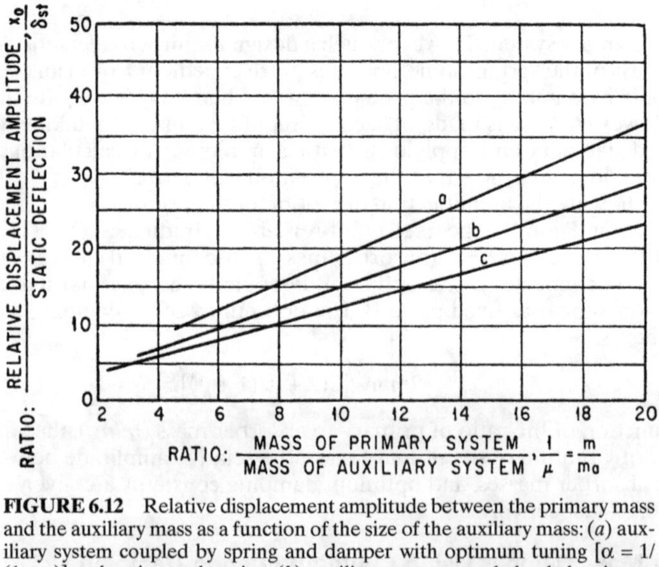

FIGURE 6.12 Relative displacement amplitude between the primary mass and the auxiliary mass as a function of the size of the auxiliary mass: (*a*) auxiliary system coupled by spring and damper with optimum tuning [$\alpha = 1/(1 + \mu)$] and optimum damping; (*b*) auxiliary system coupled only by viscous damping ($\alpha = 0$) of optimum value; (*c*) auxiliary system coupled by spring and damper tuned to frequency of primary system ($\alpha = 1$) with optimum damping.

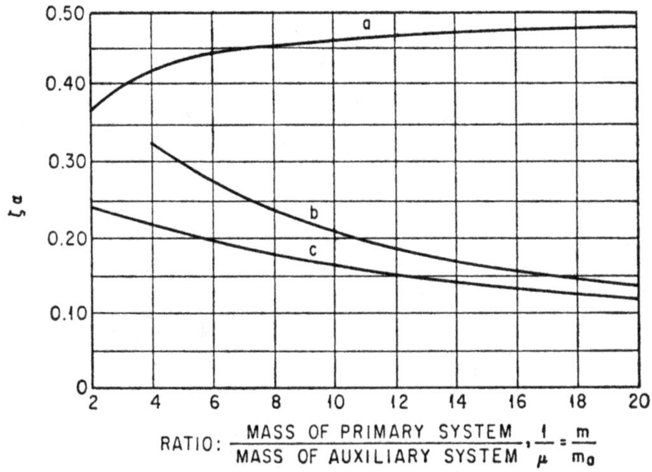

FIGURE 6.13 Curves showing damping required in auxiliary mass systems to minimize vibration amplitude of primary system: (*a*) auxiliary mass coupled by viscous damping only ($\alpha = 0$); (*b*) auxiliary system coupled by spring and damper tuned to frequency of primary system ($\alpha = 1$); (*c*) auxiliary system coupled by spring and damper with optimum tuning [$\alpha = 1/(1 + \mu)$]. The ordinate of the curves is $\zeta\alpha$, where ζ is the fraction of critical damping in the auxiliary system [Eq. (6.4)] and α is the tuning parameter [Eq. (6.31)].

spring is too complicated to be analyzed or to be adjusted by experiment. For this reason, a damper with coulomb friction has been used with only friction damping connecting the seismic mass (usually in a torsional application) to the primary system.[2,4] Because the motion is irregular, it is necessary to use energy methods of analysis. The analysis given here applies to the case of linear vibration. By analogy, the application to torsional or other vibration can be made easily (see Table 2.1 for analogous parameters).

Consider the system shown in Fig. 6.14. It consists of a mass resting on wheels that provide no resistance to motion and are connected through a friction damper to a wall that is moving sinusoidally. The friction damper consists of two friction facings that are held on opposite sides of a plate by a spring that can be adjusted to give a desired clamping force. The maximum force that can be transmitted through each interface of the damper is the product of the normal force and the coefficient of friction; the maximum total force for the damper is the summation over the number of interfaces.

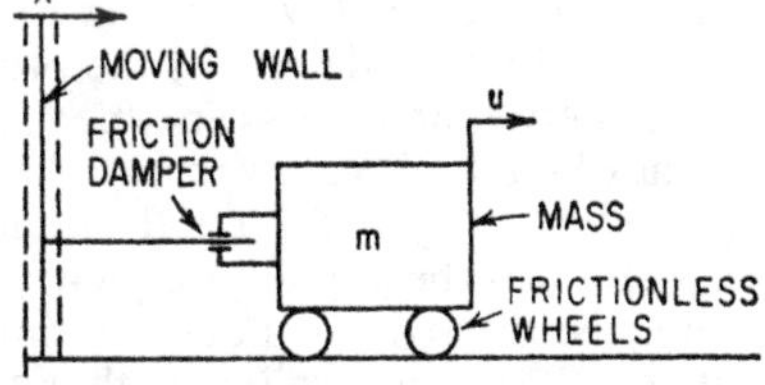

FIGURE 6.14 Schematic diagram of auxiliary mass absorber with coulomb friction damping.

Consider the velocity diagrams shown in Fig. 6.15*A*, *B*, and *C*. In these diagrams the velocity of the moving wall, $\dot{x} = x_0\omega \sin \omega t$, is shown by curve 1; the velocity $\dot{u}$ of the mass is shown by curve 2. The force exerted by the damper when slipping occurs is F_s. When $F_s \geq m\ddot{u}$, the mass moves sinusoidally with the wall. When $F_s < m\ddot{u}$, slipping occurs in the damper and the mass is accelerated at a constant rate. Since a constant acceleration produces a uniform change in velocity, the

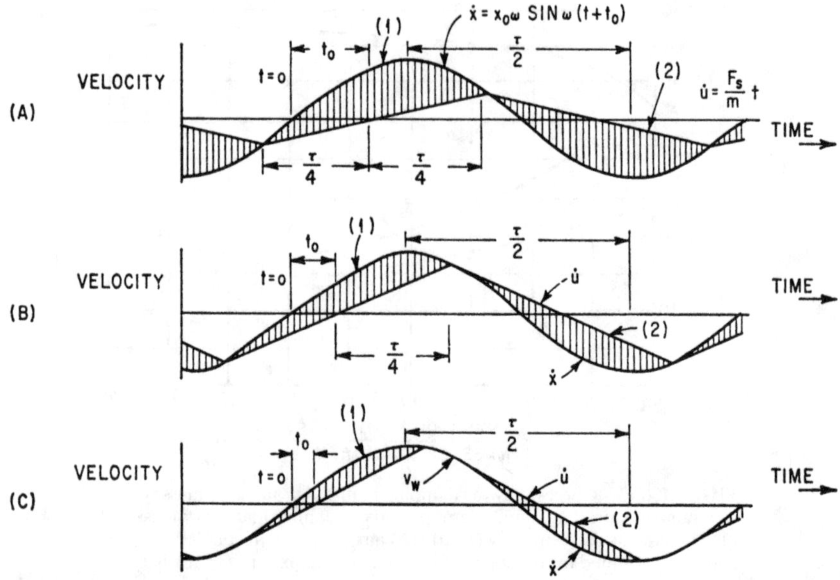

FIGURE 6.15 Velocity-time diagrams for motion of wall (curve 1) and mass (curve 2) of Fig. 6.14. The conditions for a small damping force are shown at (*A*), for an intermediate damping force at (*B*), and for a large damping force at (*C*). The relative velocity between the wall and the mass is indicated by vertical shading.

velocity of the mass when the damper is slipping is shown by straight lines. The relative velocity between the wall and the mass is shown by the vertical shading.

Figure 6.15*A* applies to a damper with a low friction force. The damper slips continuously. In Fig. 6.15*B* the velocities resulting from a larger friction force are shown. Slipping disappears for certain portions of the cycle. Where the wall and the mass have the same velocity, their accelerations also are equal. Slipping occurs when the force transmitted by the damper is not large enough to keep the mass accelerating with the wall. Since at the breakaway point the accelerations of the wall and mass are equal, their velocity-time curves have the same slope; i.e., the curves are tangent at this point. In Fig. 6.15*C*, the damping force is so large that the mass follows the wall for a considerable portion of the cycle and slips only where its acceleration becomes greater than the value of F_s/m. A slight increase in the clamping force or in the coefficient of friction locks the mass to the wall; then there is no relative motion and no damping.

Because of the nature of the damping force, the damping provided by the friction damper can be computed most practically in terms of energy. If the friction force exerted through the damper is F_s, the energy dissipated by the damper is the product of the friction force and the total relative motion between the mass and the moving wall. The time reference is taken at the moment when the auxiliary mass m has a zero velocity and is being accelerated to a positive velocity, Fig. 6.15*A*. Let the period of the vibratory motion of the wall be $\tau = 2\pi/\omega$, where ω is the angular frequency of the wall motion. By symmetry, the points of no slippage in the damper occur at times $-\tau/4$, $\tau/4$, and $3\tau/4$. Let the time when the velocity of the wall is zero be $-t_0$; then the velocity of the wall $\dot{x}$ is

$$\dot{x} = +x_0\omega \sin \omega(t + t_0)$$

The velocity $\dot{u}$ of the mass for $-\tau/4 < t < \tau/4$ is

$$\dot{u} = \ddot{u}t = \frac{F_s}{m}t$$

The velocities of the wall and the mass are equal at time $t = \tau/4$:

$$x_0\omega \sin \omega\left(\frac{\tau}{4} + t_0\right) = \frac{F_s}{m}\frac{\tau}{4}$$

Since $\omega\tau/4 = \pi/2$, $\sin \omega(\tau/4 + t_0) = \cos \omega t_0$. Therefore,

$$\cos \omega t_0 = \frac{F_s}{m}\frac{\pi}{2x_0\omega^2}$$

The relative velocity between the moving wall and the mass is $\dot{x} - \dot{u}$, and the total relative motion is the integral of the relative velocity over a cycle. Note that the area between the two curves for the second half of the cycle is the same as for the first. Hence, the work V per cycle is

$$V = 2\int_{-\tau/4}^{\tau/4} F_s(\dot{x} - \dot{u})\, dt = 4F_s x_0 \sqrt{1 - \left(\frac{F_s\pi}{2mx_0\omega^2}\right)^2} \tag{6.37}$$

Optimum damping occurs when the work per cycle is a maximum. It can be determined by setting the derivative of V with respect to F_s in Eq. (6.37) equal to zero and solving for F_s:

$$(F_s)_{\text{opt}} = \frac{\sqrt{2}}{\pi} m\omega^2 x_0 \tag{6.38}$$

Energy absorption per cycle with optimum damping is, from Eq. (6.37),

$$V_{\text{opt}} = \frac{4}{\pi} m\omega^2 x_0^2 \tag{6.39}$$

A comparison of the effectiveness of the Coulomb friction damper with other types is given in Fig. 6.11.

PRACTICAL APPLICATIONS OF AUXILIARY MASS DAMPERS AND ABSORBERS TO SINGLE-DEGREE-OF-FREEDOM SYSTEMS

THE DYNAMIC ABSORBER

The dynamic absorber, because of its tuning, can be used to eliminate vibration only where the frequency of the vibration is constant. Many pieces of equipment to which it is applied are operated by alternating current. So that it can be used for time keeping, the frequency of this alternating current is held remarkably constant. For this reason, most applications of dynamic absorbers are made to mechanisms that operate in synchronism from an ac power supply.

An application of a dynamic absorber to the pedestal of an ac generator having considerable vibration is shown in Fig. 6.16, where the relative sizes of absorber and pedestal are shown approximately to scale. In this case, the application is made to a complicated structure and the mass of the absorber is much less than that of the pri-

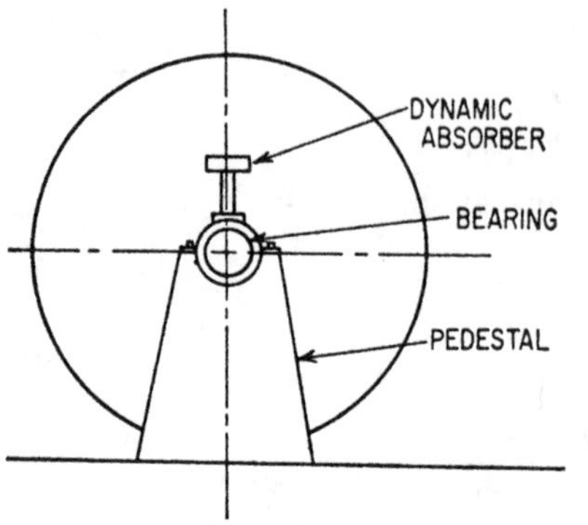

FIGURE 6.16 Application of a dynamic absorber to the bearing pedestal of an ac generator.

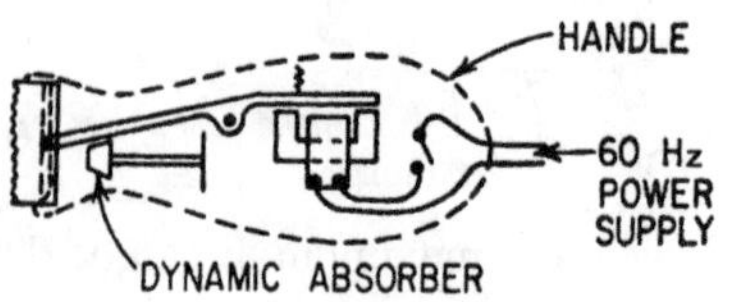

FIGURE 6.17 Application of a dynamic absorber to a hair clipper.

mary system; however, since the frequency of the excitation is constant, the dynamic absorber reduces the vibration. When the mass ratio is small, it is important that the absorber be accurately tuned and that the damping be small. In this case, the excitation was the unbalance in the turbine rotor which was elastically connected to the pedestal through the flexibility of the shaft. If the absorber were ideally effective, there would be no forces at the frequency of the shaft speed; therefore, there would be no displacements from the pedestal where the force is neutralized through the remainder of the structure.

The dynamic absorber has been applied to the electric clipper shown in Fig. 6.17. The structure consisting of the cutter blade and its driving mechanism is actuated by the magnetic field at a frequency of 120 Hz, as a result of the 60-Hz ac power supply. The forces and torques required to move the blade are balanced by reactions on the housing, causing it to vibrate. The dynamic absorber tuned to a frequency of 120 Hz enforces a node at the location of its mass. Since this is approximately the center of gravity of the assembly of the cutter and its driving mechanism, the absorber effectively neutralizes the unbalanced force. The moment caused by the rotation of the moving parts is still unbalanced. A second very small dynamic absorber placed in the handle of the clipper could enforce a node at the handle and substantially eliminate all vibration. The design of these absorbers is simple after the unbalanced forces and torques generated by the cutter mechanism are computed. The sum of the inertia forces generated by the two absorbers, $m_1x_1\omega^2 + m_2x_2\omega^2$ (where m_1 and x_1 are the mass and amplitude of motion of the first absorber, m_2 and x_2 are the corresponding values for the second absorber, and $\omega = 240\pi$), must equal the unbalanced force generated by the clipper mechanism. The torque generated by the two absorbers must balance the torque of the mechanism. Since the value of ω^2 is known, the values of m_1x_1 and m_2x_2 can be determined. Weights that fit into the available space with adequate room to move are chosen, and a spring is designed of such stiffness that the natural frequency is 120 Hz.

Because of the desirable balancing properties of the simple dynamic absorber and the constancy of frequency of ac power, it might be expected that devices operating at a frequency of 120 Hz would be used more widely. However, their application is limited because the frequency of vibration is too high to allow large amplitudes of motion.

REDUCTION OF ROLL OF SHIPS BY AUXILIARY TANKS

An interesting application of auxiliary mass absorbers is found in the auxiliary tanks used to reduce the rolling of ships, as shown in Fig. 6.18. When a ship is heeled, the

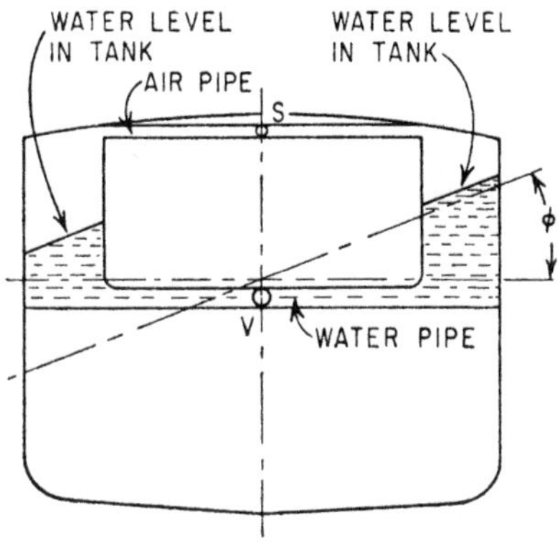

FIGURE 6.18 Cross section of ship equipped with antiroll tanks. The flow of water from one tank to the other tends to counteract rolling of the ship.

restoring moment $k_r\phi$ acting on it is proportional to the angle of heel (or roll). This restoring moment acts to return the ship (and the water that moves with it) to its equilibrium position. If I_s represents the polar moment of inertia of the ship and its entrained water, the differential equation for the rolling motion of the ship is

$$I_s\ddot{\phi} + k_r\phi = M_s \tag{6.40}$$

where M_s represents the rolling moments exerted on the ship, usually by waves.

To reduce rolling of the ship, auxiliary wing tanks connected by pipes are used. The water flowing from one tank to another has a natural frequency that is determined by the length and cross-sectional area of the tube connecting the tanks. The damping is controlled by restricting the flow of water, either with a valve *S* in the line that allows air to flow between the tanks (Fig. 6.18) or with a valve *V* in the water line. Since the tanks occupy valuable space, the mass ratio of the water in the tanks to the ship is small. Fortunately, the excitation from waves generally is not large relative to the restoring moments, and roll becomes objectionable only because the normal damping of a ship in rolling motion is not very large. The use of antirolling tanks in the German luxury liners *Bremen* and *Europa* reduced the maximum roll from 15 to 5°.

REDUCTION OF ROLL OF SHIPS BY GYROSCOPES

A large gyroscope may be used to reduce roll in ships, as shown in Fig. 6.19. In response to the velocity of roll of a ship, the gyroscope precesses in the plane of symmetry of the ship. By braking this precession, energy can be dissipated and the roll reduced. The torque exerted by the gyroscope is proportional to the rate of change of the angular momentum about an axis perpendicular to the torque. Letting I represent the polar moment of inertia of the gyroscope about its spin axis and $\dot{\theta}$ the angular velocity of precession of the gyroscope, then the equation of motion of the ship is

$$I_s\ddot{\phi} + k_r\phi + I\Omega\dot{\theta} = M_s \tag{6.41}$$

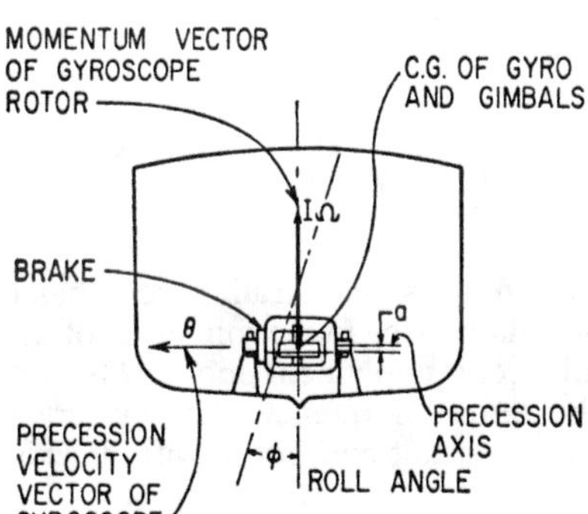

FIGURE 6.19 Application of a gyroscope to a ship to reduce roll.

Assume that the gyroscope has (1) a moment of inertia about the precession axis of I_g, (2) a weight of W, and (3) that its center of gravity is below the gimbal axis (as it must be for the gyro to come to equilibrium in a working position) a distance *a,* as shown in Fig. 6.19. Then the equation of motion of the gyroscope is

$$I_g\ddot{\theta} + W_a\theta + c\dot{\theta} - I\Omega\dot{\phi} = 0 \qquad (6.42)$$

where Ω is the spin velocity of the gyroscope. From Eq. (6.42), for a roll frequency of ω, the angle of precession of the gyroscope is

$$\theta = \frac{jI\Omega\omega\phi}{-I_g\omega^2 + Wa - jc\omega} \qquad (6.43)$$

The torque exerted on the ship is

$$I\Omega\dot{\theta} = \frac{-(I\Omega)^2\omega^2\phi}{-I_g\omega^2 + Wa + jc\omega} \qquad (6.44)$$

The equivalent moment of inertia of the gyroscope system in its reaction on the ship is

$$\frac{I\Omega^2}{-I_g\omega^2 + Wa + cj\omega} \qquad (6.45)$$

It follows that

$$\frac{\phi}{\phi_{st}} = \sqrt{\frac{(1-\beta_g{}^2)^2 + (2\zeta\beta_g)^2}{[(1-\beta_g{}^2)(1-\beta^2) - \beta^2\mu]^2 + (2\zeta\beta_g)^2(1-\beta^2)^2}} \qquad (6.46)$$

where the parameters are defined in terms of ship and gyro constants as follows:

$$\beta_g = \frac{\omega}{\sqrt{Wa/I_g}} \qquad \beta = \frac{\omega}{\sqrt{k_r/I_s}} \qquad \zeta = \frac{c}{2\sqrt{WaI_g}} \qquad \mu = \frac{(I\Omega)^2}{WaI_s} \qquad \phi_{st} = \frac{M_s}{k_r}$$

Because $I\Omega$ can be made large by using a large gyro rotor and spinning it at a high speed, and Wa can be made small by choice of a design, the value of μ can be made quite large even though I_s is large. In one experimental ship, $\mu = 20$ was obtained. Even with this large value of μ, the precession angle of the gyroscope would become very large for optimum damping. Therefore it is necessary to use much more damping than optimum. Gyro stabilizers were used on the Italian ship *Conte di Savoia;* they are sometimes installed on yachts.

Both antirolling tanks and gyro stabilizers are more effective if they are active rather than passive. Activated dampers are considered below.

AUXILIARY MASS DAMPERS APPLIED TO ROTATING MACHINERY

An important industrial use of auxiliary mass systems is to neutralize the unbalance of centrifugal machinery. An application is the balance ring in the spin dryer of home washing machines. The operation of such a balancer is dependent upon the basket of the washer rotating at a speed greater than the natural frequency of its support. The balance ring is attached to the washing machine basket concentric with its axis of rotation, as shown in Fig. 6.20.

Consider the washing machine basket shown in Fig. 6.20. When its center of gravity does not coincide with its axis of rotation and it is rotating at a speed lower than its critical speed (corresponding to the natural frequency in rocking motion about the spherical seat), the centrifugal force tends to pull the rotational axis in the direction of

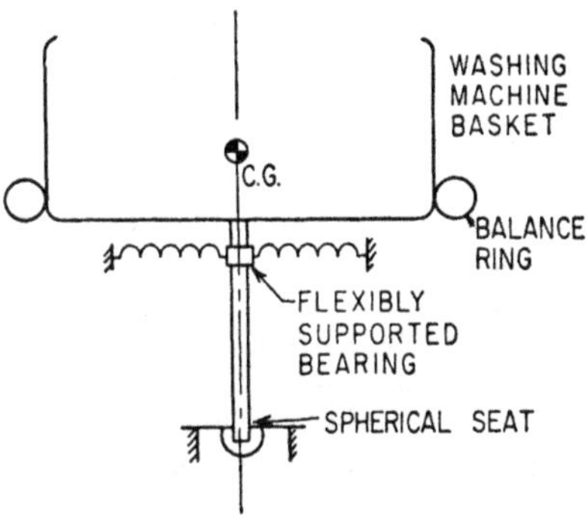

FIGURE 6.20 Schematic diagram showing location of balance ring on basket of a spin dryer.

the unbalance. This effect increases with an increase in rotational speed until the critical speed is reached. At this speed, the amplitude would become infinite if it were not for the damping in the system. Above the critical speed, the phase position of the axis of rotation relative to the center of gravity shifts so that the basket tends to rotate about its center of gravity with the flexibly supported bearing moving in a circle about an axis through the center of gravity. The relative positions of the bearing center and the center of gravity are shown in Fig. 6.21*A* and *B*.

Since the balance ring is circular with a smooth inner surface, any weights or fluids contained in the ring can be acted upon only by forces directed radially. When the ring is rotated about a vertical axis, the weights or fluids will move within the ring in such a manner as to be concentrated on the side farthest from the axis of rotation. If this concentration occurs below the natural frequency (Fig. 6.21*A*), the weights tend to move farther from the axis and the resultant center of gravity is displaced so as to give a greater eccentricity. The points A and G rotate about the axis O at the frequency ω. The initial eccentricity of the center of gravity of the washer basket and its load from the axis of rotation is represented by e, and ρ is the elastic displacement of this center of rotation due to the centrifugal force. Where the off-center rotating weight is W, the unbalanced force is $(W/g)(\rho + e)\omega^2$ [where $\rho = e/(1 - \beta^2)$ and $\beta^2 = \omega^2/\omega_n^2 < 1$] and acts in the direction from A to G.

If the displacement of the weights or fluids in the balance ring occurs above the natural frequency, the center of gravity tends to move closer to the dynamic location of the axis. The action in this case is shown in Fig. 6.21*B*. Then the points A and G rotate about O at the frequency ω. The unbalanced force is $(W/g)(\rho + e)\omega^2$ [where $\rho = e/(1 - \beta^2)$ and $\beta^2 = \omega^2/\omega_n^2 > 1$]. This gives a negative force that acts in a direction from G to A. Thus, the eccentricity is brought toward zero and the rotor is automatically balanced. Because it is necessary to pass through the critical speed in bringing the rotor up to speed and in stopping it, it is desirable to heavily damp the balancing elements, either fluid or weights.

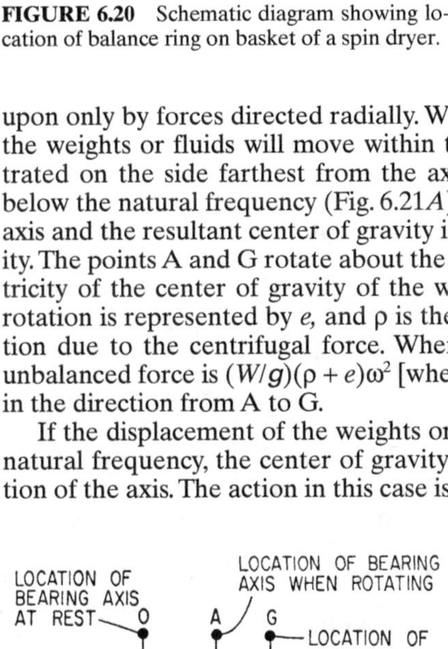

FIGURE 6.21 Diagram in plane normal to axis of rotation of spin dryer in Fig. 6.20. Relative positions of axes when rotating speed is less than natural frequency are shown at (*A*); corresponding diagram for rotation speed greater than natural frequency is shown at (*B*).

In practical applications, the balancing elements can take several forms. The earliest form consisted of two or more spheres or cylinders free to move in a race concentric with the axis of the rotor, as shown in Fig. 6.22*A*. A later modification consists of three annular discs that rotate about an enlarged shaft concentric with the axis, as indicated in Fig. 6.22*B*. These are contained in a sealed compart-

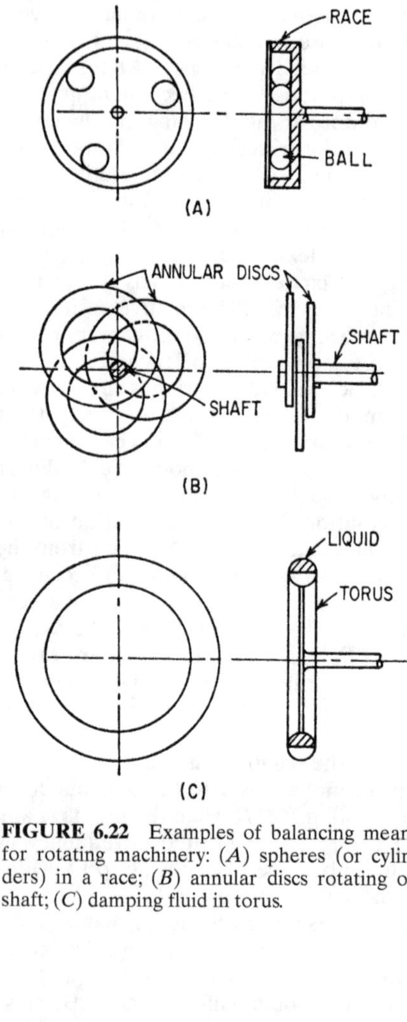

FIGURE 6.22 Examples of balancing means for rotating machinery: (*A*) spheres (or cylinders) in a race; (*B*) annular discs rotating on shaft; (*C*) damping fluid in torus.

ment with oil for lubrication and damping. A fluid type of damper is shown in Fig. 6.22*C*, the fluid usually being a high-density viscous material. With proper damping, mercury would be excellent, but it is too expensive. Therefore a more viscous, high-density halogenated fluid is used.

The balancers must be of sufficient weight and operate at such a radius that the product of their weight and the maximum eccentricity they can attain is equivalent to the unbalanced moment of the load. This requirement makes the use of the spheres or cylinders difficult because they cannot be made large; it makes the annular plates large because they are limited in the amount of eccentricity that can be obtained.

In a cylindrical volume 24 in. (61 cm) in diameter and 2 in. (5 cm) thick, seven spheres 2 in. (5 cm) in diameter can neutralize 98.6 lb-in. (114 kg-cm) of unbalance; three cylinders 4 in. (10 cm) in diameter by 2 in. (5 cm) thick can neutralize 255 lb-in. (295 kg-cm); three annular discs, each ⅝ in. (1.6 cm) thick with an outside diameter of 19.55 in. (50 cm) and an inside diameter of 10.45 in. (26.5 cm) [the optimum for a center post 6 in. (15.2 cm) in diameter], can neutralize 250 lb-in. (290 kg-cm); and half of a 2-in. (5-cm) diameter torus filled with fluid of density 0.2 lb/in^3 (5.5 gram/cm^3) can neutralize 609 lb-in. (700 kg-cm). Only the fluid-filled torus would be initially balanced.

AUXILIARY MASS DAMPERS APPLIED TO TORSIONAL VIBRATION

Dampers and absorbers are used widely for the control of torsional vibration of internal-combustion engines. The most common absorber is the viscous-damped, untuned auxiliary mass unit shown in Fig. 6.23. The device is comprised of a cylindrical housing carrying an inertia mass that is free to rotate. There is a preset clearance between the housing and the inertia mass that is filled with a silicone oil of proper viscosity. Silicone oil is used because of its high viscosity index; i.e., its viscosity changes relatively little with temperature. With the inertia mass and the damping medium contained, the housing is seal-welded to provide a leakproof and simple absorber. However, the silicone oil has poor boundary lubricating prop-

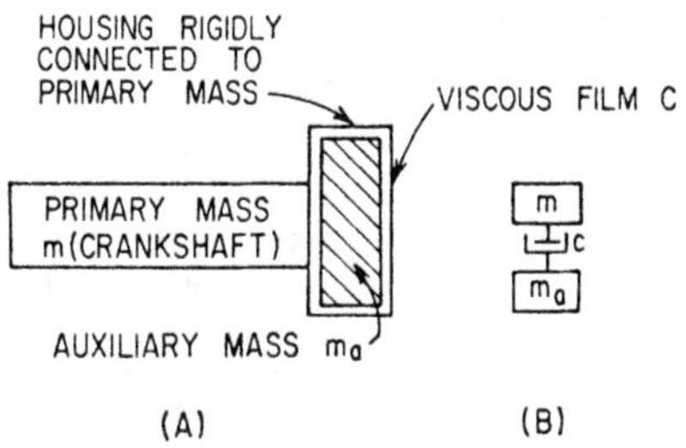

FIGURE 6.23 Untuned auxiliary mass damper with viscous damping. The application to a torsional system is shown at (*A*), and the linear analog at (*B*).

erties and if decomposed by a local hot spot (such as might be caused by a reduced clearance at some particular spot), the decomposed damping fluid is abrasive.

Because of the simplicity of this untuned damper, it is commonly used in preference to the more effective tuned absorber. However, it is possible to use the same construction methods for a tuned damper, as shown in Fig. 6.24. It is also possible to mount the standard damper with the housing for the unsprung inertia mass attached to the main mass by a spring, as shown in Fig. 6.25. If the viscosity of the oil and the dimensions of the masses and the clearance spaces are known, the damping effects of the dampers shown in Figs. 6.23 and 6.25 can be computed directly in terms of the equations previously developed. The damper in Fig. 6.25 can be analyzed by treating the spring and housing as additional elements in the main system and the untuned mass as a viscous damped auxiliary mass. If the inertia of the housing is negligible, the inertia mass is effectively connected to the main mass through a spring and a dashpot in series. The two elements in series can be represented by a complex spring constant equal to

$$\frac{1}{(1/jc\omega) + (1/k)} = \frac{kcj\omega}{k + cj\omega}$$

Where there is no damping in parallel with the spring, the effective mass becomes

$$m_{\text{eq}} = km/(k - m\omega^2)$$

Substituting the complex value of the spring constant, the effective mass is

$$m_{\text{eq}} = \frac{ckj\omega}{k + cj\omega}\left[\frac{m}{-m\omega^2 + cjk\omega/(k + cj\omega)}\right] \tag{6.47}$$

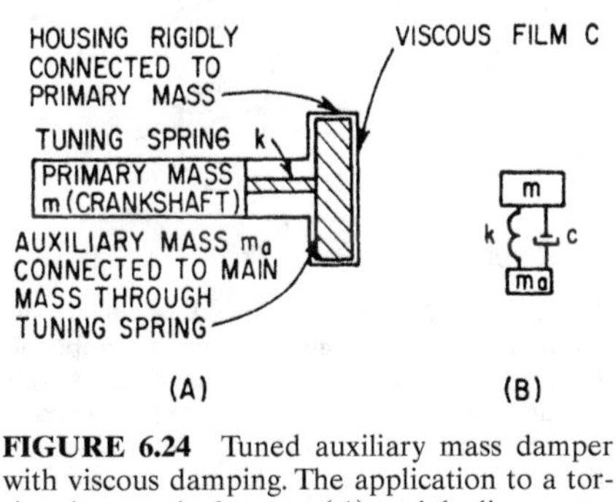

FIGURE 6.24 Tuned auxiliary mass damper with viscous damping. The application to a torsional system is shown at (*A*), and the linear analog at (*B*).

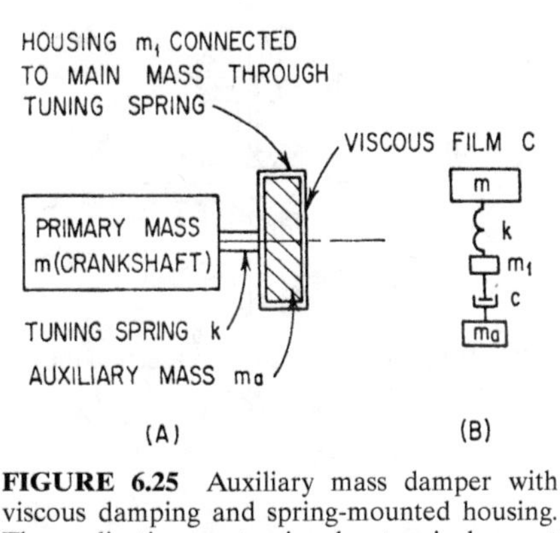

FIGURE 6.25 Auxiliary mass damper with viscous damping and spring-mounted housing. The application to a torsional system is shown at (*A*), and the linear analog at (*B*).

In terms of the nondimensional parameters defined in Eq. (6.14):

$$m_{eq} = \frac{(2\zeta\beta_a)^2(1-\beta_a^2)}{\beta_a^4-(2\zeta\beta_a)^2(1-\beta_a^2)}\,m + \frac{-2\zeta\beta_a^3 m}{\beta_a^4-(2\zeta\beta_a)^2(1-\beta_a^2)}\,j \tag{6.48}$$

Before the advent of silicone oil with its chemical stability and relatively constant viscosity over service temperature conditions, the damper most commonly used for absorbing torsional vibration energy was the dry friction or Lanchester damper shown in Fig. 6.26. The damping is determined by the spring tension and the coefficient of friction at the sliding interfaces. Its optimum value is determined by the equation for a torsional system analogous to Eq. (6.38) for a linear system:

$$(T_s)_{opt} = \frac{\sqrt{2}}{\pi} I\omega^2\theta_0 \tag{6.49}$$

where T_s is the slipping torque, I is the moment of inertia of the flywheels, and θ_0 is the amplitude of angular motion of the primary system. The dry-friction-based Lanchester damper requires frequent adjustment, as the braking material wears, to maintain a constant braking force.

It is possible to use torque-transmitting couplings that can absorb vibration energy, as the spring elements for tuned dampers. The Bibby coupling (Fig. 6.27) is used in this manner. Since the stiffness of this coupling is nonlinear, the optimum tuning of such an absorber is secured for only one amplitude of motion.

A discussion of dampers and of their application to engine systems is given in Chap. 37.

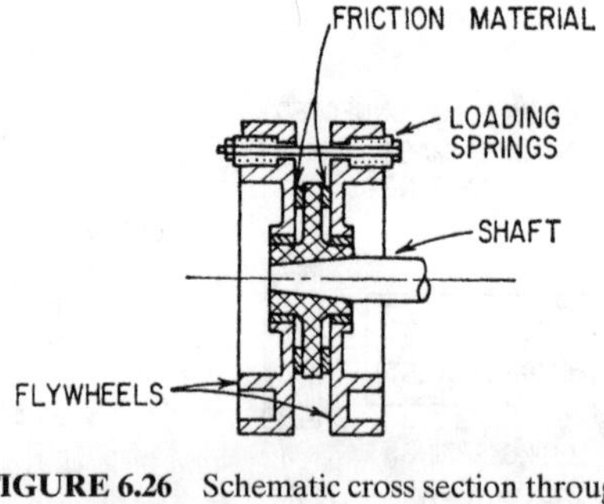

FIGURE 6.26 Schematic cross section through Lanchester damper.

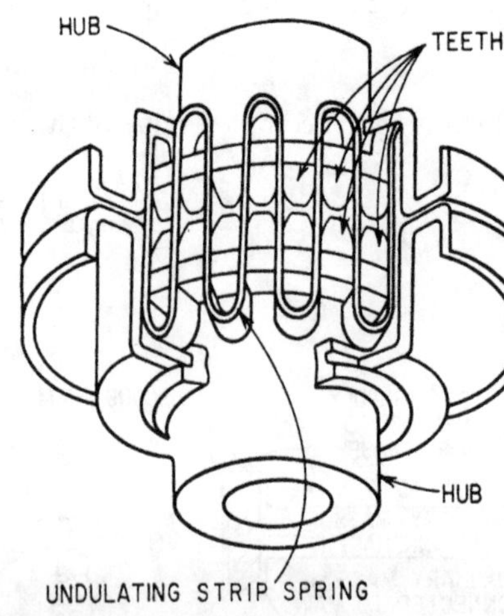

FIGURE 6.27 Coupling used as elastic and damping element in auxiliary mass damper for torsional vibration. The torque is transmitted by an undulating strip of thin steel interposed between the teeth on opposite hubs. The stiffness of the strip is nonlinear, increasing as torque increases. Oil pumped between the strip and teeth dissipates energy.

DYNAMIC ABSORBERS TUNED TO ORDERS OF VIBRATION RATHER THAN CONSTANT FREQUENCIES

In the torsional vibration of rotating machinery, it is generally found that exciting torques and forces occur at the same frequency as the rotational speed or at multiples of this frequency. The ratio of the frequency of vibration to the rotational speed is called the *order of the vibration q*. Thus, a power plant driving a four-bladed propeller may have a torsional vibration whose frequency is 4 times the rotational speed of the drive shaft; sometimes it may have a second torsional vibration whose frequency is 8 times the rotational speed. These are called the fourth-order and eighth-order torsional vibrations.

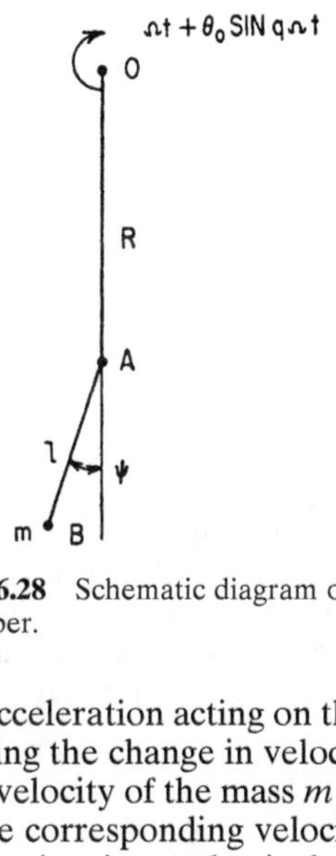

FIGURE 6.28 Schematic diagram of a pendulum absorber.

If a dynamic absorber in the form of a pendulum acting in a centrifugal field is used, then its natural frequency increases linearly with speed. Therefore, it can be used to neutralize an order of vibration.[10–12]

Consider a pendulum of length l and of mass m attached at a distance R from the center of a rotating shaft, as shown in Fig. 6.28. Since the pendulum is excited by torsional vibration in the shaft, let the radius R be rotating at a constant speed Ω with a superposed vibration $\theta = \theta_0 \cos q\Omega t$, where q represents the order of the vibration. Then the angle of R with respect to any desired reference is $\Omega t + \theta_0 \cos q\Omega t$. The angle of the pendulum with respect to the radius R is defined as $\psi = \psi_0 \cos q\Omega t$, as shown by Fig. 6.28.

The acceleration acting on the mass m at position B is most easily ascertained by considering the change in velocity during a short increment of time Δt. The components of velocity of the mass m at time t are shown graphically in Fig. 6.29*A;* at time $t + \Delta t$, the corresponding velocities are shown in Fig. 6.29*B*. The change in velocity during the time interval Δt is shown in Fig. 6.29*C*. Since the acceleration is the change in velocity per unit of time, the accelerations along and perpendicular to l are:

Acceleration along l:

$$\frac{-l(\Omega + \dot\theta + \dot\psi^2)\,\Delta t - R(\Omega + \dot\theta)^2\,\Delta t \cos\psi + R\ddot\theta\,\Delta t \sin\psi}{\Delta t} \tag{6.50}$$

Acceleration perpendicular to l:

$$\frac{l(\ddot\theta + \ddot\psi)\,\Delta t + R(\Omega + \dot\theta)^2\,\Delta t \sin\psi + R\ddot\theta\,\Delta t \cos\psi}{\Delta t} \tag{6.51}$$

Only the force $-F$, directed along the pendulum, acts on the mass m. Therefore the equations of motion are

$$\begin{aligned} -F &= -ml(\Omega + \dot\theta + \dot\psi)^2 - mR(\Omega + \dot\theta)^2 \cos\psi + R\ddot\theta \sin\psi \\ 0 &= ml(\ddot\theta + \ddot\psi) + mR(\Omega + \dot\theta)^2 \sin\psi + mR\ddot\theta \cos\psi \end{aligned} \tag{6.52}$$

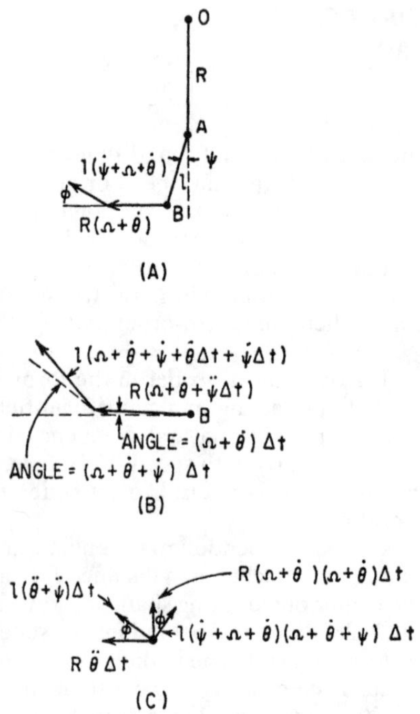

FIGURE 6.29 Velocity vectors for the pendulum absorber: (*A*) velocities at time *t;* (*B*) velocities at time $t + \Delta t$; (*C*) change in velocities during time increment Δt.

Assuming that ψ and θ are small, Eqs. (6.52) simplify to

$$\begin{aligned} Ft &= m(R + l)\Omega^2 \\ l(\ddot{\theta} + \ddot{\psi}) + R\Omega^2\psi + R\ddot{\theta} &= 0 \end{aligned} \tag{6.53}$$

The second of Eqs. (6.53) upon substitution of $\theta = \theta_0 \cos q\Omega t$ and $\psi = \psi_0 \cos q\Omega t$ yields

$$\frac{\psi_0}{\theta_0} = \frac{(q\Omega)^2(l + R)}{-(q\Omega)^2 l + \Omega^2 R} = \frac{q^2(l + R)}{R - q^2 l} \tag{6.54}$$

The torque M exerted at point 0 by the force F is

$$M = RF \sin \psi = RF\psi \qquad \text{when } \psi \text{ is small}$$

From Eqs. (6.53) and (6.54), when ψ is small,

$$M = \frac{mq^2R(R + l)^2\Omega^2}{R - q^2 l} \tag{6.55}$$

If a flywheel having a moment of inertia I is accelerated by a shaft having an amplitude of angular vibratory motion θ_0 and a frequency $q\Omega$, the torque amplitude exerted on the shaft is $I(q\Omega)^2\theta_0$. Therefore, the equivalent moment of inertia I_{eq} of the pendulum is

$$I_{eq} = \frac{mR(R + l)^2}{R - q^2 l} = \frac{m(R + l)^2}{1 - q^2 l/R} \tag{6.56}$$

When

$$\frac{R}{l} = q^2 \tag{6.57}$$

the equivalent inertia is infinite and the pendulum acts as a dynamic absorber by enforcing a node at its point of attachment.

Where the pendulum is damped, the equivalent moment of inertia is given by:

$$\begin{aligned} I_{eq} &= \frac{1 + 2\zeta\upsilon j}{(1 - \upsilon^2) + 2\zeta\upsilon j}\, m(R + l)^2 \\ &= m(R + l)^2 \left[\frac{1 - \upsilon^2 + (2\zeta\upsilon)^2}{(1 - \upsilon^2)^2 + (2\zeta\upsilon)^2} - \frac{2\zeta\upsilon^3 j}{(1 - \upsilon^2)^2 + (2\zeta\upsilon)^2} \right] \end{aligned} \tag{6.58}$$

where $\upsilon^2 = q^2 l/R$ and $\zeta = (c/2m\Omega)\sqrt{l/R}$.

When the pendulum is attached to a single-degree-of-freedom system, as is shown in Fig. 6.30, the amplitude of motion θ_a of the flywheel of inertia I is given by

$$\frac{\theta_a}{\theta_{st}} = \sqrt{\frac{(1-v^2)^2 + (2\zeta v)^2}{[(1-v^2)(1-\beta_p^2) - \beta_p^2\mu]^2 + (2\zeta v)^2[1-\beta_p^2 - \beta_p^2\mu_p]^2}} \tag{6.59}$$

where
$$2\zeta v = \frac{cql}{mR}$$

$$\mu_p = \frac{m(R+l)^2}{I}$$

$$\beta_p = \frac{q}{k_r I}$$

$$\theta_{st} = \frac{m_0}{k_r}$$

The pendulum tends to detune when the amplitude of motion of the pendulum is large, thereby introducing harmonics of the torque that it neutralizes.[10] Suppose the shaft rotates at a constant speed Ω, i.e., $\theta_0 = 0$, and consider the torque exerted on the shaft as m moves through a large amplitude ψ_0 about its equilibrium position. Equations (6.52) become

$$\begin{aligned} F &= ml(\Omega + \dot{\psi})^2 + mR\Omega^2 \cos\psi \\ l\ddot{\psi} &+ R\Omega^2 \sin\psi = 0 \end{aligned} \tag{6.60}$$

A solution for the second of Eqs. (6.60) is

$$\dot{\psi} = \sqrt{\frac{2\Omega^2 R}{l}}\sqrt{\cos\psi - \cos\psi_0} \tag{6.61}$$

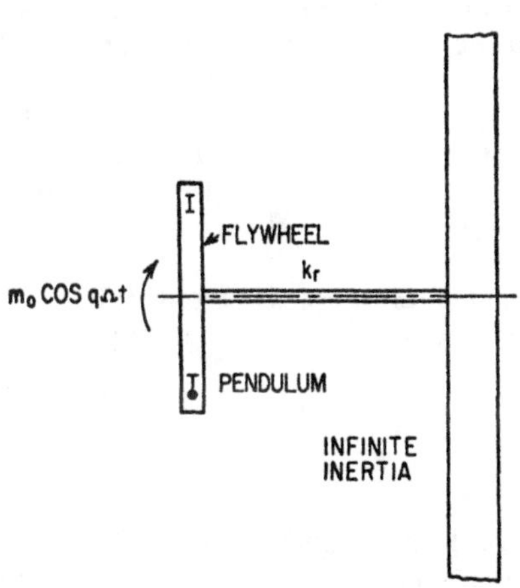

FIGURE 6.30 Application of a pendulum absorber to a rotational single-degree-of-freedom system.

The solution of Eq. (6.61) involves elliptic integrals and is given approximately by

$$\psi = \psi_0 \sin \omega t$$

where $\omega = \sqrt{\dfrac{R}{l}}\, \dfrac{\pi/2}{F(\psi_0/2, \pi/2)}\Omega$

and $F(\psi_0/2, \pi/2)$ is an elliptic function of the first kind whose value may be obtained from tables.

Since $\omega/\Omega = q$ (the order of the disturbance), the tuning of the damper will be changed for large angles and becomes

$$q^2 = \frac{R}{l}\left(\frac{\pi/2}{F(\psi_0/2, \pi/2)}\right)^2 \tag{6.62}$$

The value of $q^2 l/R = v^2$ used in Eqs. (6.58) and (6.59) is given in Fig. 6.31 as a function of the amplitude of the pendulum.

Since the force exerted by the mass m is directed along the rod connecting it to the pivot A (Fig. 6.28), the reactive torque on the shaft is

$$
\begin{aligned}
M &= FR \sin \psi \\
&= mR^2\Omega^2 \left[\frac{l}{R}\left(1 + \frac{\dot{\psi}}{\Omega}\right)^2 \sin \psi + \sin \psi \cos \psi \right] \\
&= mR^2\Omega^2(A_1 \sin q\Omega t + A_2 \sin 2q\Omega t + A_3 \sin 3q\Omega t + \cdots) \qquad (6.63)
\end{aligned}
$$

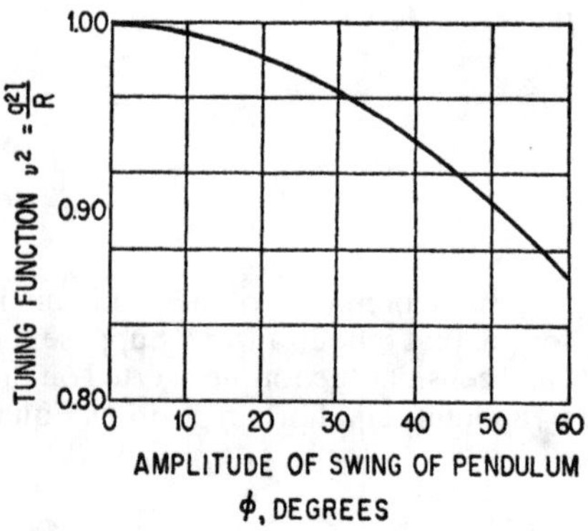

FIGURE 6.31 Tuning function for a pendulum absorber used in Eqs. (6.58) and (6.59).

The values of the fundamental torque corresponding to the tuned frequency and to the second and third harmonics of this tuned frequency are given in Fig. 6.32 as a function of the angle of swing of the pendulum, for a typical installation. In this case, the pendulum is tuned to the 4½ order of vibration. (The 4½ order of vibration is one whose frequency is 4½ times the rotational frequency and 9 times the fundamental frequency. The latter is called the half order and occurs at half of the rotational frequency. This is common in four-cycle engines.)

Two types of pendulum absorber are used. The one most commonly used is shown in Fig. 6.33. The counterweight, which also is used to balance rotating forces in the engine, is suspended from a hub carried by the crankshaft by pins that act through holes with clearance, Fig. 6.33*A*. By suspending the pendulum from two pins, the pendulum when oscillating does not rotate but rather moves as shown in Fig. 6.33*B*. Since it is not subjected to angular acceleration, it may be treated as a particle located at its center of gravity. Referring to Fig. 6.33*A* and *B*, the expression for acceleration [Eqs. (6.50) and (6.51)] and the equations of motion [Eqs. (6.52)] apply if

$$
\begin{aligned}
R &= H_1 + H_2 \\
l &= \frac{D_c + D_p}{2} - D_b
\end{aligned}
\qquad (6.64)
$$

where H_1 = distance from center of rotation to center of holes in crank hub
H_2 = distance from center of holes in pendulum to center of gravity of pendulum
D_c = diameter of hole in crank hub
D_p = diameter of hole in pendulum
D_b = diameter of pin

In practice, difficulty arises from the wear of the holes and the pin. Moreover, the motion on the pins generally is small and the loads due to centrifugal forces are large so that fretting is a problem. Because the radius of motion of the pendulum is short, only a small amount of wear can be tolerated. Hardened pins and bushings are used to reduce the wear.

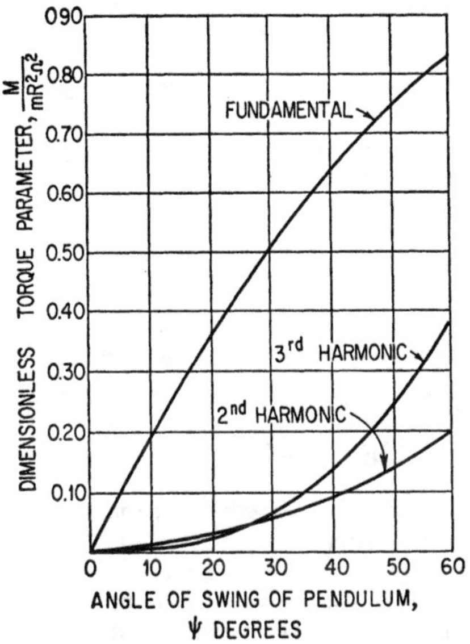

FIGURE 6.32 Harmonic components of torque generated by a pendulum absorber as a function of its angle of swing. The torque is expressed by the parameters used in Eq. (6.63).

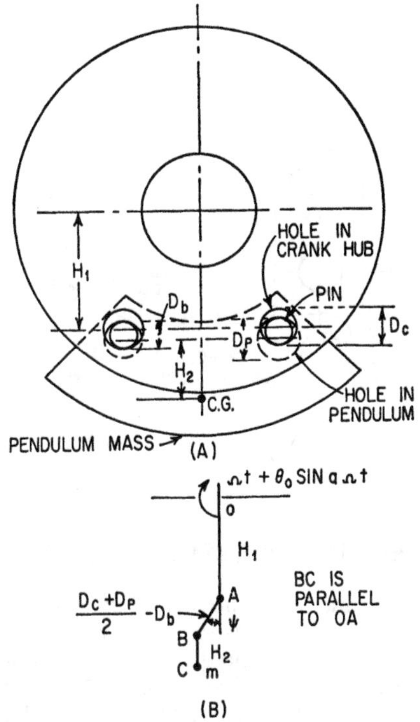

FIGURE 6.33 Bifilar type of pendulum absorber. The mechanical arrangement is shown at (*A*), and a schematic diagram at (*B*).

The pendulum is most easily designed if it is recognized that the inertia torques generated by the pendulum must neutralize the forcing torques. Thus,

$$m\omega^2 l\psi_0 R = M \tag{6.65}$$

The radii l and R are set by the design of the crank and the order of vibration to be neutralized. The original motion ψ_0 is generally limited to a small angle, approximately 20°. It is probable that the most stringent condition is at the lowest operating speed, although the absorber may be required only to avoid difficulty at some particular critical speed. Knowing the excitation M, it is possible to compute the required mass of the pendulum weight.

A second type of pendulum absorber is a cylinder that rolls in a hole in a counterweight, as shown in Fig. 6.34. In this type, the radius of the pendulum corresponds to the difference in the radii of the hole and of the cylinder. It is found, by observing tests and checking the tuning of actual systems using cylindrical pendulums, that the weight rotates with a uniform angular velocity. Therefore, the tuning is independent of the moments of inertia of the cylinder. It is common to allow a larger amplitude of motion with the absorber of Fig. 6.34 than with the absorber of Fig. 6.30.

Applications of pendulum absorbers to torsional-vibration problems are given in Chap. 37.

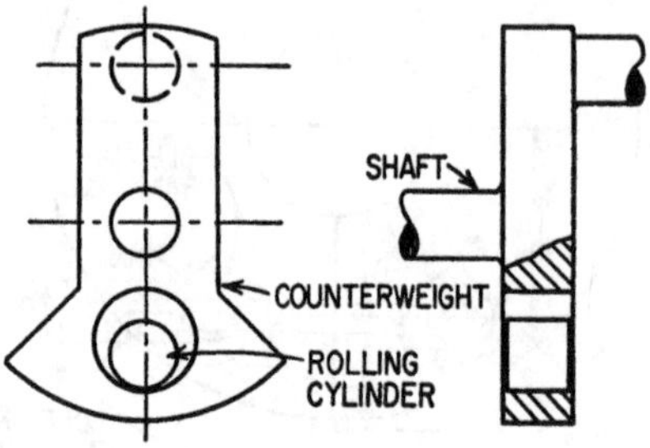

FIGURE 6.34 Roller type of pendulum absorber.

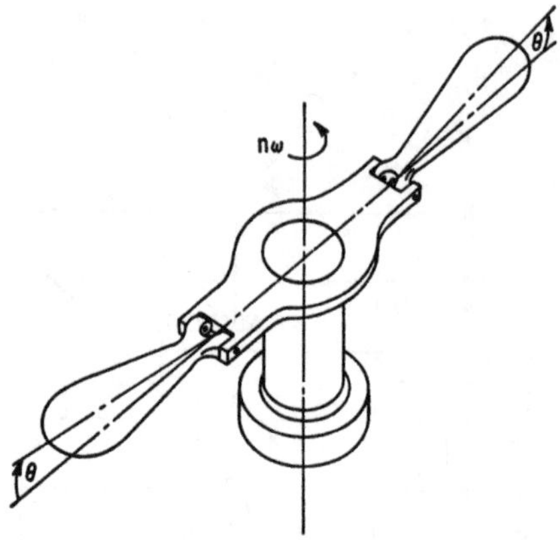

FIGURE 6.35 Application of pendulum absorbers to counteract linear vibration.

PENDULUM ABSORBER FOR LINEAR VIBRATION

The principle of the pendulum absorber can be applied to linear vibration as well as to torsional vibration. To neutralize linear vibration, pendulums are rotated about an axis parallel to the direction of vibration, as shown in Fig. 6.35. This can be accomplished with an absorber mounted on the moving body. Two or more pendulums are used so that centrifugal forces are balanced. Free rotational movement of each pendulum in the plane of the axis allows the axial forces to be neutralized. The pendulum assembly must rotate about the axis at some submultiple of the frequency of vibration. The size of the absorber is determined by the condition that the components of the inertia forces of the weights in the axial direction $[\Sigma m\omega^2 r\theta]$ must balance the exciting forces. This device can be applied where the vibration is generated by the action of rotating members but the magnitude of the vibratory forces is uncertain. A discussion of this absorber, including the influence of moments of inertia and damping of the pendulum, together with some applications to the elimination of vibration in special locations on a ship, is given in Ref. 13.

APPLICATIONS OF DAMPERS TO MULTIPLE-DEGREE-OF-FREEDOM SYSTEMS

Auxiliary mass dampers as applied to systems of several degrees of freedom can be represented most effectively by equivalent masses or moments of inertia. The choice of proper damping constants is more difficult. For the case of torsional vibration, the practical problems of designing dampers and selecting the proper damping are considered in Chap. 37.

There are many applications of dampers to vibrating structures that illustrate the use of different types of auxiliary mass damper. One such application has been to ships.[14] These absorbers had low damping and were designed to be filled with water so that they could be tuned to the objectionable frequencies. In one case, the absorber was located near the propeller (the source of excitation) and when properly tuned was found to be effective in reducing the resonant vibration of the ship. In another case, the absorber was located on an upper deck but was not as effective. It enforced a node at its point of attachment, but, because of the flexibility between the upper deck and the bottom of the ship, there was appreciable motion in the vicinity of the propeller and vibratory energy was fed to the ship's structure. To operate properly, the absorbers must be closely tuned and the propeller speed closely maintained. Because the natural frequencies of the ship vary with the types of loading, it is not sufficient to install a fixed-frequency absorber that is effective at only one natural frequency of the hull, corresponding to a particular loading condition.

An auxiliary mass absorber has been applied to the reduction of vibration in a heavy building that vibrated at a low frequency under the excitation of a number of looms.[15] The frequency of the looms was substantially constant. However, the magnitude of the excitation was variable as the looms came into and out of phase. The dynamic absorber, consisting of a heavy weight hung as a pendulum, was tuned to the frequency of excitation. Because the frequency was low and the forces large, the absorber was quite large. However, it was effective in reducing the amplitude of vibration in the building and was relatively simple to construct.

THE USE OF AUXILIARY MASS DEVICES TO REDUCE TRANSIENT AND SELF-EXCITED VIBRATIONS

Where the vibration is self-excited or caused by repeated impact, it is necessary to have sufficient damping to prevent a serious buildup of vibration amplitude. This damping, which need not always be large, may be provided by a loosely coupled auxiliary mass. A simple application of this type is the ring fitted to the interior of a gear, as shown in Fig. 6.36. By fitting this ring with the proper small clearance so that relative motion occurs between it and the gear, it is possible to obtain enough energy dissipation to damp the high-frequency, low-energy vibration that causes the gear to ring. The rubbery coatings applied to large, thin-metal panels such as automobile doors to give them a solid rather than a "tinny" sound depend for their effectiveness on a proper balance of mass, elasticity, and damping (see Chap. 36).

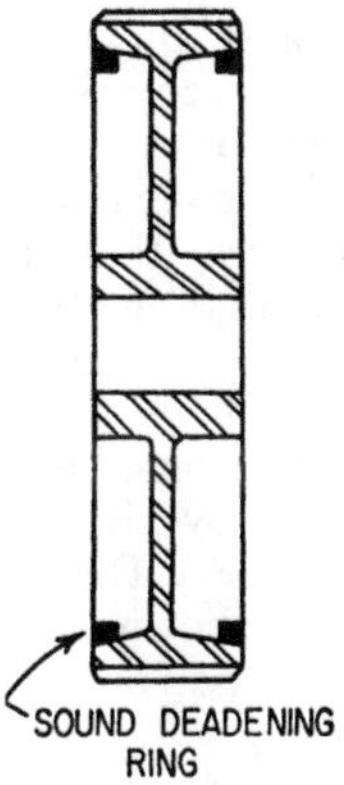

FIGURE 6.36 Application of auxiliary mass damper to deaden noise in gear.

Another application where auxiliary mass dampers are useful is in the prevention of fatigue failures in turbines. At the high-pressure end of an impulse turbine, steam or hot gas is admitted through only a few nozzles. Consequently, as the blade passes the nozzle it is given an impulse by the steam and set into vibration at its natural frequency. It is a characteristic of alloy steels that they have very little internal damping at high operating temperature. For this reason the free vibration persists with only a slightly diminished amplitude until the

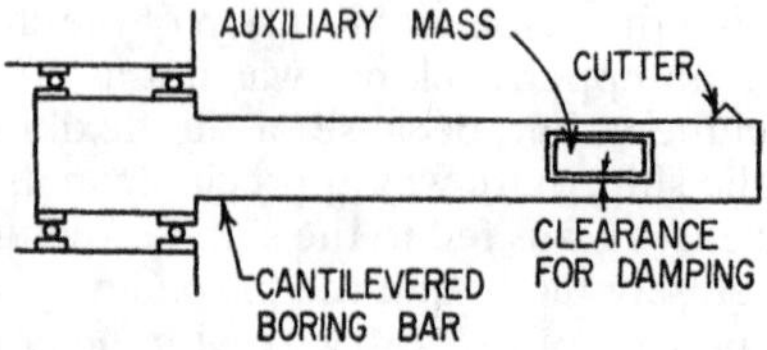

FIGURE 6.37 Application of auxiliary mass damper to reduce chatter in boring bar.

blade again is subjected to the steam impulse. Some of these second impulses will be out of phase with the motion of the blade and will reduce its amplitude; however, successive impulses may increase the amplitude on subsequent passes until failure occurs. Damping can be increased by placing a number of loose wires in a cylindrical hole cut in the blade in a radial direction. The damping of a number of these wires has been computed in terms of the geometry of the application (number of wires, density of wires, size of the hole, radius of the blade, rotational speed, etc.) and the amplitude of vibration.[16] These computations show reasonable agreement with experimental results.

An auxiliary mass has been used to damp the cutting tool chatter set up in a boring bar.[17] Because of the characteristics of the metal-cutting process or of some coupling between motions of the tool parallel and perpendicular to the work face, it is sometimes found that a self-excited vibration is initiated at the natural frequency of the cutter system. Since the self-excitation energy is low, the vibration usually is initiated only if the damping is small. Chatter of the tool is most common in long, poorly supported tools, such as boring bars. To eliminate this chatter, a loose auxiliary mass is incorporated in the boring bar, as shown in Fig. 6.37. This may be air-damped or fluid-damped. Since the excitation is at the natural frequency of the tool, the damping should be such that the tool vibrates with a minimum amplitude at this frequency. The damping requirement can be estimated by

$$\frac{x_0}{\delta_{st}} = \sqrt{\frac{1+4(\zeta\alpha)^2}{4(\zeta\alpha)^2\mu^2}} \tag{6.66}$$

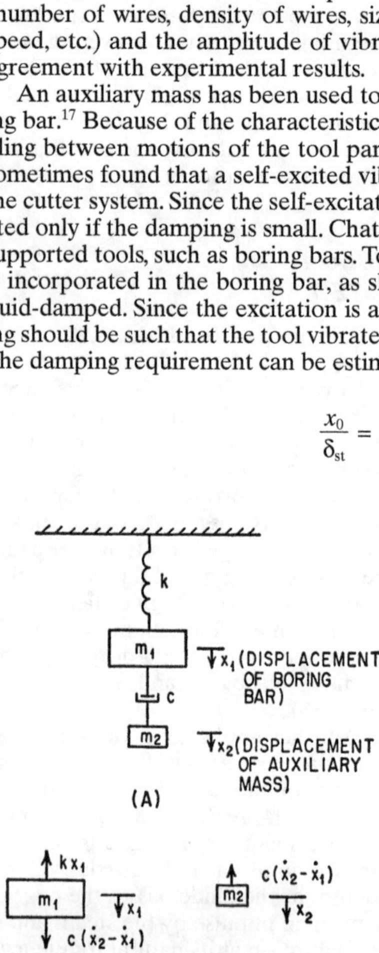

FIGURE 6.38 Schematic diagram of damper shown in Fig. 6.37. The arrangement is shown at (*A*), and the forces acting on the boring bar and auxiliary mass are shown at (*B*).

The optimum value of the parameter $(\zeta\alpha)$ is infinity. Thus, when the frequency of excitation is constant, a greater reduction in amplitude can be obtained by a shift in natural frequency than by damping. However, such a shift cannot be attained because the frequency of the excitation always coincides with the natural frequency of the complete system. Instead, a better technique is to determine the damping that gives the maximum decrement of the free vibration.

Let the boring bar and damper be represented by a single-degree-of-freedom system with a damper mass coupled to the main mass by viscous damping, as shown in Fig. 6.38*A*. The forces acting on the masses are shown in Fig. 6.38*B*. The equations of motion are

$$\begin{aligned} -kx_1 - c\dot{x}_1 + c\dot{x}_2 &= m_1\ddot{x}_1 \\ c\dot{x}_1 - c\dot{x}_2 &= m_2\ddot{x}_2 \end{aligned} \tag{6.67}$$

Substituting $x = Ae^{st}$, the resulting frequency equation is

$$s^3 + \frac{c(m_1 + m_2)}{m_1 m_2} s^2 + \frac{k}{m_1} s + \frac{kc}{m_1 m_2} = 0 \tag{6.68}$$

Where chatter occurs, this equation has three roots, one real and two complex. The complex roots correspond to decaying free vibrations. Let the roots be as follows:

$$\alpha_1, \alpha_2 + j\beta, \alpha_2 - j\beta$$

The value of β determines the frequency of the free vibration, and the value of α_2 determines the decrement (rate of decrease of amplitude) of the free vibration. The decrement α_2 is of primary interest; it is most easily found from the conditions that when the coefficient of s^3 is unity, (1) the sum of the roots is equal to the negative of the coefficient of s^2, (2) the sum of the products of the roots taken two at a time is the negative of the coefficient of s, and (3) the product of the roots is the negative of the constant term. The equations thus obtained are

$$\alpha_1 + 2\alpha_2 = -\frac{c(1 + \mu)}{\mu m_1} \tag{6.69}$$

$$2\alpha_1\alpha_2 + \alpha_2^2 + \beta^2 = -\omega_n^2 \tag{6.70}$$

$$\alpha_1(\alpha_2^2 + \beta^2) = -\omega_n^2 \frac{c}{m_1\mu} \tag{6.71}$$

where $\omega_n^2 = k/m_1$ and $\mu = m_2/m_1$. It is not practical to find the optimum damping by solving these equations for α_2 and then setting the derivative of α_2 with respect to c equal to zero. However, it is possible to find the optimum damping by the following process. Eliminate $(\alpha_2^2 + \beta^2)$ between Eqs. (6.70) and (6.71) to obtain

$$2\alpha_1^2\alpha_2 = \omega_n^2 \left(\frac{c}{\mu m_1} - \alpha_1\right) \tag{6.72}$$

Substituting the value of α_1 from Eq. (6.69) in Eq. (6.72),

$$2\alpha_2 \left[2\alpha_2 + \frac{c(1 + \mu)}{\mu m_1}\right]^2 = \frac{c\omega_n^2}{\mu m_1} + \omega_n^2 \left[2\alpha_2 + \frac{c(1 + \mu)}{\mu m_1}\right] \tag{6.73}$$

To find the damping that gives the maximum decrement, differentiate with respect to c and set $d\alpha_2/dc = 0$:

$$2\alpha_2 \left[2\alpha_2 + \frac{c(1 + \mu)}{\mu m_1}\right] = \tfrac{1}{2}\omega_n^2 \frac{2 + \mu}{1 + \mu} \tag{6.74}$$

Solving Eqs. (6.73) and (6.74) simultaneously,

$$c_{\text{opt}} = \frac{\mu^2 m_1 \omega_n}{2(1 + \mu)^{3/2}} \tag{6.75}$$

$$(\alpha_2)_{\text{opt}} = -\frac{(2 + \mu)\omega_n}{4(1 + \mu)^{1/2}} \tag{6.76}$$

These values may be obtained by proper choice of clearance between the auxiliary mass and the hole in which it is located. Air damping is preferable to oil because it requires less clearance. Therefore, the plug is not immobilized by the centrifugal forces that, with the rotating boring bar, become larger as the clearance is increased.

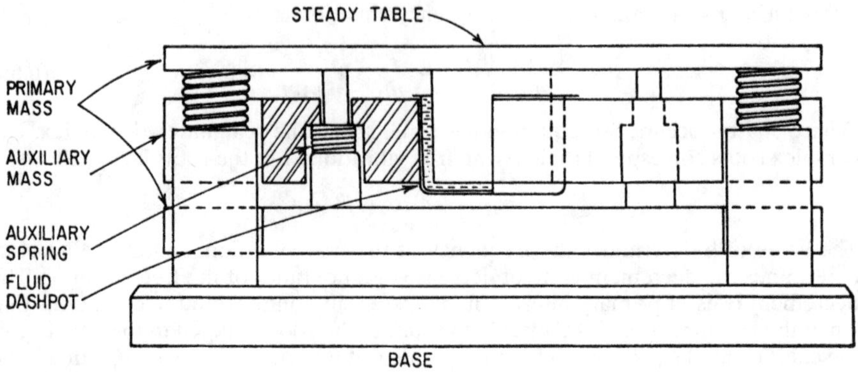

FIGURE 6.39 Application of auxiliary mass to spring-mounted table to reduce vibration of table.[18]

In precision measurements, it is necessary to isolate the instruments from effects of shock and vibration in the earth and to damp any oscillations that might be generated in the measuring instruments. A heavy spring-mounted table fitted with a heavy auxiliary mass that is attached to the table by a spring and submerged in an oil bath (Fig. 6.39) has proved to be effective.[18] In this example, the table has a top surface of 13½ in. (34 cm) by 13½ in. (34 cm) and a height of 6 in. (15 cm). Each auxiliary mass weighs about 70 lb (32 kg). The springs for both the primary table and the auxiliary system are designed to give a natural frequency between 2 and 4 Hz in both the horizontal and the vertical directions. By trying different fluids in the bath, suitable damping may be obtained experimentally.

ACTIVE VIBRATION ABSORBERS

In many cases, the auxiliary mass system must be relatively lightweight and perhaps compact so as not to unduly burden the functionality of the primary system. A semi-active TVA implements small energy expenditure to alter the TVA properties to optimize its effectiveness. A theory for variable damping and variable stiffness has been applied to machine tools.[19] An active TVA implements a force-generating mechanism in parallel with the spring and damper to allow the flexibility to incorporate a control algorithm to provide additional vibration cancellation forcing.[20] Application of an active TVA to large-scale civil engineering structures has been made.[21]

ACKNOWLEDGMENT

Some of the material in this chapter was originally prepared by F. Everett Reed, former president, Research and Engineering Corporation, Littleton, Massachusetts.

REFERENCES

1. Biehl, F. A., and S. Rubin: "Mechanical Impedance Approach to Engine Vibration Transmission into an Aircraft Fuselage," *SAE Trans.*, 1968, pp. 2711–2719.

2. Den Hartog, J. P.: "Mechanical Vibrations," 4th ed., Chap. 3, reprinted by Dover Publications, New York, 1985.
3. Thomson, W. T., and M. D. Dahleh: "Theory of Vibrations with Applications," 5th ed., Prentice-Hall, Englewood Cliffs, New Jersey, 1998.
4. Sun, J. Q., M. R. Jolly, and M. A. Norris: "Passive, Adaptive and Active Tuned Vibration Absorbers—A Survey," *Trans. ASME,* 3d ed., June 1995, pp. 234–242.
5. Brock, J. E.: *J. Appl. Mechanics,* **16**(1):86 (1949).
6. Krenk, S.: "Frequency Analysis of the Tuned Mass Damper," *Trans. ASME,* **72**:936 (2005).
7. Saver, F. M., and C. F. Garland: *J. Appl. Mechanics,* **16**(2):109 (1949).
8. Lewis, F. M.: *J. Appl. Mechanics,* **22**(3):377 (1955).
9. Den Hartog, J. P., and J. Ormondroyd: *Trans. ASME,* **52:**133 (1930).
10. Den Hartog, J. P.: "Stephen Timoshenko 60th Anniversary Volume," Macmillan Company, New York, 1939.
11. Porter, F. P.: "Evaluation of Effects of Torsional Vibration," SAE War Engineering Board, SAE, New York, 1945, p. 269.
12. Crossley, F. R. E.: *J. Appl. Mechanics,* **20**(1):41 (1953).
13. Reed, F. E.: *J. Appl. Mechanics,* **16**:190 (1949).
14. Constanti, M.: *Trans. Inst. of Naval Arch.,* **80**:181 (1938).
15. Crede, C. E.: *Trans. ASME,* **69**:937 (1947).
16. DiTaranto, R. A.: *J. Appl. Mechanics,* **25**(1):21 (1958).
17. Hahn, R. S.: *Trans. ASME,* **73**:331 (1951).
18. Macinante, J. A.: *J. Sci. Instr.,* **35**:224 (1958).
19. Benesho, J. A., and J. G. Bollinger: "Dual Variable Self-Optimizing Vibration Damper Theory," Advances in Machine Tool Design and Research: *Proc. Ninth International Machine Tool Design and Research Conference,* 1968, pp. 635–646.
20. von Flotow, A. H., A. Beard, and D. Bailey: "Adaptive Tuned Vibration Absorbers—Tuning Laws, Tracking Agility, Sizing, and Physical Implementations," *Proc. Noise-Con 94,* pp. 437–454.
21. Chang, J. C. H., and T. T. Soong: "Structural Control Using Active Tuned Mass Dampers," *J. Engrg. Mechanics Div.*, December 1980, pp. 1091–1098.

CHAPTER 7
VIBRATION OF SYSTEMS HAVING DISTRIBUTED MASS AND ELASTICITY

Ronald G. Merritt

INTRODUCTION

This chapter considers the vibration of systems that have mass and elasticity continuously distributed in space. The objectives of this chapter are to (1) provide a basis for continuous system vibration in terms of energy functionals and Hamilton's variational principle (HVP); (2) mention topics related to continuous system vibration such as damping, stochastic loading, wave propagation, and dynamic stability; (3) demonstrate analytical solutions and reference numerical solutions; and (4) present detailed dynamic expressions of four basic continuous system configurations (bars, shafts, beams, and plates) and reference four other configurations (strings, membranes, arches, and shells). The examples, tables, and references challenge readers to go beyond the use of tables and limited assumptions for problem solution. This chapter contains few tables extracted from readily available reference material and encourages engineers to build a library containing practical structure dynamic information. Motivation for study of continuous systems comes from two sources. First, popular numerical methods available for solution of vibration of continuous systems—for example, the finite element method—must be checked against classical solutions. Second, micromachines, nanostructures, and "intrinsically discrete objects only a few atoms in diameter" can be formulated as continuum solutions.[1]

This chapter does not consider certain important solution techniques such as Green's function (integral equation solution) and transforms (Fourier or Laplace) for either finite or infinite extent structures with transient loading. Dynamic loading that can result in dynamic stability problems is not considered. Some potentially useful approximate solution techniques such as the dynamic transfer matrix method are only referenced.

In place of the general term *system,* the term *structure* is used throughout this chapter. *Structure* refers to a finite-extent physical system of simple form having a continuous distribution of mass and elasticity. A structure has a well-defined geometric boundary S and occupies a spatial domain V such that the governing equation(s) of motion satisfy certain conditions on S and in V. Mathematical formulation is in terms of a *boundary-value problem* on S and an *initial-value problem* where time constraints are involved. Mass and elasticity are expressed in terms of quantities representing averages over microproperties of the continuum. This chapter does not consider material behavior that goes beyond Hooke's generalized constitutive law, for example, viscoelastic or plastic behavior.

Basic and extended theories for vibration of structures lie in the area of continuum mechanics,[2] where details of the material kinematic and constitutive relationships are stated and vibration is in the context of wave propagation through a material continuum. Transition to structure macro vibration relationships is demonstrated by vibration of a string whereby the equations of motion and solution can be viewed as either standing waves in the string or vibration of the string.[3,4]

A brief review of the foundations of analytical mechanics provides context for this chapter.[5] Newton's laws provide the origin of analytical mechanics in terms of vector force, mass, and acceleration. Leibniz reformulated analytical mechanics in terms of scalar energy and variational principles. Hamilton's variational principle (HVP), considered the fundamental law of dynamics, encompasses Newton's equations of motion, Lagrange's equations for structure dynamics, D'Alembert's principle, and the principle of virtual work (PVW). Vibration of structures can be conveniently formulated in terms of HVP. At times it may be expeditious to derive the equations of motion for a structure in forms other than HVP. Because HVP is *energy functional* based, the diverse areas of structure dynamics, numerical solutions of partial differential equations, finite element methods, and functional analysis can all be linked in a single development.[6]

HAMILTON'S PRINCIPLE AND THE LAGRANGIAN

Hamilton's variational principle states[7] that for a structure over domain V with boundary S_2, *among all dynamic paths that satisfy the boundary conditions over* S_2 *at all times and that start and end with the actual values at two arbitrary instants of time* t_1 *and* t_2, *at every point of the structure, the "actual" dynamic path of the structure is distinguished by making the Lagrangian functional an extremum.* Structure energy is expressed in terms of the Lagrangian energy functional, and HVP requires that this energy functional have a stationary value. The term *functional* is used to denote a general expression for a continuous function of the domain V of the structure in space and time. Table 7.1 defines the Lagrangian energy functional for a continuous structure.

By convention, L is defined in component parts from Table 7.1 as

$$L = \pi - T + W \tag{7.1}$$

The "action" or "principle function of dynamics"[9] A can be expressed as the time integral of L between two times t_1 and t_2:

$$A = \int_{t_1}^{t_2} L\,dt = \int_{t_1}^{t_2} (\pi - T + W)\,dt \tag{7.2}$$

HVP states that A has a stationary value expressed as $\delta A = 0$, where δ is termed the *variation operator.* Table 7.2 summarizes the basic properties of the variation operator.

The δ operator over a structure domain V implies that complete description of the structure *shape* requires an infinite number of *degrees of freedom*—one degree

TABLE 7.1 Lagrangian Energy Functional L for an Elastic Structure[8]

$$L = \frac{1}{2}\int_V \left\{ \underbrace{\left(\rho \dot{\vec{d}}^T \dot{\vec{d}}\right)}_{T} - \underbrace{\left(\vec{\varepsilon}^T [D] \vec{\varepsilon}\right)}_{\pi} + \underbrace{\left(2\vec{d}^T \vec{\vec{\phi}}\right)}_{-W_{S_1}} \right\} dV + \int_{S_2} \underbrace{\left(\vec{d}^T \vec{\vec{\Phi}}\right)}_{-W_{S_2}} dS_2$$

V structure volume	$\vec{\vec{\phi}}$ average body force vector per unit volume
$S = S_1 + S_2$ total structure surface	$\vec{\vec{\Phi}}$ average surface force vector on S_2
S_1 $\vec{d}$ prescribed for average body forces $\vec{\vec{\phi}}$	$T = \rho \dot{\vec{d}}^T \dot{\vec{d}}$ kinetic energy per unit volume
S_2 average surface forces $\vec{\vec{\Phi}}$ prescribed for $\vec{d}$	$\pi = \vec{\varepsilon}^T [D] \vec{\varepsilon}$ elastic energy per unit volume
ρ mass density	$W_{S_1} = -2\vec{d}^T \vec{\vec{\phi}}$ work energy from average body forces $\vec{\vec{\phi}}$ with $\vec{d}$ prescribed on S_1
$\vec{d}^T = [u\ v\ w]$ displacement vector	
$\dot{\vec{d}}^T = [\dot{u}\ \dot{v}\ \dot{w}]$ velocity vector	
D material constitutive matrix	$W_{S_2} = -\vec{d}^T \vec{\vec{\Phi}}$ work energy from average surface forces $\vec{\vec{\Phi}}$ prescribed on S_2
$\vec{\varepsilon}$ material strain vector	
$\vec{\sigma} - (=[D]\vec{\varepsilon})$ material stress vector	
	$W = W_{S_1} + W_{S_2}$ total work

of freedom for each shape. HVP requires that the action defined in Eq. (7.2) have a minimum or stationary value over all possible structure variations, that is,

$$\delta A = \delta \int_{t_1}^{t_2} L dt = \int_{t_1}^{t_2} \delta L dt = \int_{t_1}^{t_2} \delta\,(\pi - T + W)\, dt = \int_{t_1}^{t_2} [\delta\pi - \delta T + \delta W]\, dt = 0 \quad (7.3)$$

This expression of HVP in terms of δA, utilizing the δ operator properties and integration by parts, provides the (1) differential equation(s) of motion termed the Lagrange's equation(s), (2) boundary displacements (kinematic boundary conditions), (3) boundary forces (natural boundary conditions), and (4) eigenvalue solution form.[10]

Table 7.3 illustrates the general pattern for expression of the structure's stationary value of energy functional to obtain Lagrange's equation(s) and boundary conditions. This derivation generalizes in a computationally intensive way to all structures considered in this chapter. Application of integration by parts imposes certain continuity/differentiability conditions on equations and solutions in addition to implying that the operator form of an equation is self-adjoint.[10,11] In this chapter, for time/space variable a, $\dot{a}$ signifies differentiation with respect to time and a' signifies differentiation with respect to a spatial variable. Table 7.4 illustrates the procedure outlined in Table 7.3 for derivation of the one-dimensional wave equation.

TABLE 7.2 Definition and Properties of δ the Variation Operator

$\delta A = \overline{A}(s) - A(s) = \varepsilon\eta(s)$

$A(s)$ arbitrary scalar functional for $s_1 \le s \le s_2$
$\overline{A}(s)$ infinitesimal change in $A(s)$ at time t
$\eta(s)$ infinitesimal functional change
$\varepsilon > 0$ small number

where δ obeys the following rules for differentiation and integration

$\delta(dA/ds) \equiv d(\delta A)/ds$ and $\delta \int_{t_1}^{t_2} A(s)\, ds \equiv \int_{t_1}^{t_2} \delta A(s)\, ds$

TABLE 7.3 Expression of the Stationary Value of the Action A (HVP)

Form of energy functional:

$L = L(t, u, \dot{u}, u') \qquad t_1 \le t \le t_2$ and $s_1 \le s \le s_2$
$u = u(s,t)$ displacement
$\dot{u} = \partial u/\partial t$ velocity
$u' = \partial u/\partial s$ first spatial derivative

Displacement variation:

$\delta u(s,t) = \tilde{u}(s,t) - u(s,t) = \varepsilon\eta(s,t)$

Variation of action:

$$\delta A = \delta\int_{t_1}^{t_2} L dt = \delta\int_{t_1}^{t_2}\int_{s_1}^{s_2} L(t, u, \dot{u}, u')\, dsdt = \int_{t_1}^{t_2}\int_{s_1}^{s_2} \delta L(t, u, \dot{u}, u')\, dsdt$$

$$= \int_{t_1}^{t_2}\int_{s_1}^{s_2}\left(\frac{\partial L}{\partial u}\delta u + \frac{\partial L}{\partial \dot{u}}\delta \dot{u} + \frac{\partial L}{\partial u'}\delta u'\right) dsdt = 0$$

Integration by parts for *t*:

$$t: \int_{t_1}^{t_2}\int_{s_1}^{s_2}\left(\frac{\partial L}{\partial \dot{u}}\delta \dot{u}\right) dsdt = \int_{s_1}^{s_2}\frac{\partial L}{\partial \dot{u}}\delta u\Big|_{t_1}^{t_2} ds - \int_{t_1}^{t_2}\int_{s_1}^{s_2}\frac{d}{dt}\left(\frac{\partial L}{\partial \dot{u}}\delta u\right) dsdt$$

Integration by parts for *s*:

$$s: \int_{t_1}^{t_2}\int_{s_1}^{s_2}\left(\frac{\partial L}{\partial u'}\delta u'\right) dsdt = \int_{t_1}^{t_2}\frac{\partial L}{\partial u'}\delta u\Big|_{s_1}^{s_2} dt - \int_{t_1}^{t_2}\int_{s_1}^{s_2}\frac{d}{ds}\left(\frac{\partial L}{\partial u'}\delta u\right) dsdt$$

Substitution for δu arbitrary:

$$\delta A = \int_{t_1}^{t_2}\int_{s_1}^{s_2}\left[\frac{\partial L}{\partial u} - \frac{d}{dt}\left(\frac{\partial L}{\partial \dot{u}}\right) - \frac{d}{ds}\left(\frac{\partial L}{\partial u'}\right)\right]\delta u\, dsdt + \int_{s_1}^{s_2}\frac{\partial L}{\partial \dot{u}}\delta u\Big|_{t_1}^{t_2} ds + \int_{t_1}^{t_2}\frac{\partial L}{\partial u'}\delta \hat{u}\Big|_{s_1}^{s_2} dt = 0$$

TABLE 7.4 Derivation of the One-Dimensional Wave Equation Using HVP

Energy functional:

$$L = -\frac{1}{2}\rho\dot{u}^2 + \frac{1}{2}Pu'^2 + fu$$

$u = u(s, t)$ displacement — ρ volume density
$\dot{u} = \partial u/\partial t$ velocity — P tension force
$u' = \partial u/\partial s$ first spatial derivative — f external applied force

For δu arbitrary

$$\left[\frac{\partial L}{\partial u}\delta u - \frac{d}{dt}\left(\frac{\partial L}{\partial \dot{u}}\delta u\right) - \frac{d}{ds}\left(\frac{\partial L}{\partial u'}\delta u\right)\right] = (f + \rho\ddot{u} - Pu'')\,\delta u = 0$$

One-dimensional wave equation ($f = 0$):

$$\frac{\partial^2 u}{\partial t^2} - c^2\frac{\partial^2 u}{\partial s^2} = 0 \text{ for } c = \sqrt{\frac{P}{\rho}}\quad \rho\frac{\partial u}{\partial t}\delta u\Big|_{t_1}^{t_2} = 0,\ P\frac{\partial u}{\partial s}\delta u\Big|_{s_1}^{s_2} = 0$$

Solution:

$u(s, t) = G_1(s - ct) + G_2(s + ct)$ for G_1 and G_2 arbitrary functions dependent upon initial conditions

For each of the structures considered in detail, the energy components are provided for immediate application of Rayleigh's method, Lagrange equation(s), and the boundary conditions (kinematic and natural). This information is adequate for determination of modes of structure vibration. Inclusion of initial conditions implies a specific dynamic solution.

MECHANICAL PRINCIPLES

The static and dynamic behavior of structures is built upon mechanical principles. The most important of these principles—Hamilton's variational principle, D'Alembert's principle, Newton's second law of motion, and the principle of virtual work—involve subtle details in relationship to one another and in application.[5,12] For example, expression of the PVW (controlling the static equilibrium of a structure) illustrates a form of practical subtlety in application. The PVW may be expressed with variation in displacement δu alone (principle of minimum potential energy), variation in stress $\delta\sigma_{ss}$ alone (principle of minimum complementary energy), or variation in combination of displacement and stress (principle of stationary Reissner energy).[8] The overriding importance of the PVW, regardless of expression, is that for a continuum, all the internal forces can be neglected and only the external forces consistent with the structure kinematic constraints need be considered. For kinetic energy $T = 0$ and no moving boundary constraints, HVP reduces to the PVW.

DISSIPATION OF ENERGY (DAMPING) IN STRUCTURES

Energy dissipation is inherent in both structure materials and structure mechanisms such as structure joints. Energy dissipation mechanisms may be difficult to identify and even more difficult to specify mathematically and, if specified mathematically, result in equations of motion for continuous structures that can be solved only by numerical integration.[4,7,13] Energy dissipation considerations play little role in this chapter. Reference 14 on viscoelastic damping and Ref. 15 on material damping, along with Chaps. 36 and 37, provide additional information on damping.

Since the damping effect on structure motion is most prominent at structure resonant frequencies, then for structures with proportional damping[8] approximate representation of damping effects can be obtained by specification of individual single-degree-of-freedom modal viscous damping factors. Such damping factors can then be applied to independent nonoverlapping resonant modes of a multimode structure. Under the discussion of the assumed modes method, a convolution integral is provided, along with a modal damping ratio, to illustrate this approach to structure response determination.

FORCING IN DYNAMIC EQUATIONS OF MOTION

Solution of the *free vibration problem* for a structure is a function of the structure equation of motion and the boundary conditions on the structure. The solution provides (1) characteristic vibration frequencies (*eigenfrequencies*) and (2) a corresponding orthogonal set of shape, or basis, functions (*eigenvectors*) in spatial coordinates. Each characteristic frequency and associated basis function defines a *normal mode* of vibration for the structure. Obtaining the normal mode information for a structure is a first step in assessing the dynamic behavior of a structure under applied external body, surface, or boundary forces. In the assumed modes method discussed in this chapter, the normal modes of vibration for the structure can be used to estimate structure response when the structure is loaded by either deterministic or stochastic force fields. The power of the assumed modes method has been used for understanding the dynamic stability of structures—an important nonlinear vibration topic that is beyond the scope of this chapter.[16]

Forces on a structure are most often assumed to be deterministic—that is, they are functionally described in time. The response of structures under stochastic forces is an evolving field of inquiry and beyond the scope of this chapter. Information in this area can be found in Refs. 4, 17, and 18.

The simplest forced vibration is a time-variant deterministic force field defined by either (1) a single force at a point on the structure or its boundary, (2) a force distributed over a portion of the structure, or (3) a force moving along a structure surface. Two useful deterministic force fields are characterized by being (1) harmonically varying with discrete Fourier series representation or (2) nonperiodic with Fourier integral representation.

SOLUTION OF DYNAMIC EQUATIONS OF MOTION

Knowledge of the form of structure displacement in time and space constitutes solution of the dynamic equations of motion. Solution may refer to (1) determining the *homogeneous equation solution part* represented by knowledge of the first few normal modes of free structure vibration under kinematic boundary conditions with initial conditions or (2) determining the *particular equation solution part* under the applied forces. This chapter focuses primarily on the homogeneous equation solution part, with structure displacement the variable for solution. Estimates of structure modal stresses and strains can be obtained from the modal displacements and constitutive relationships. In the simplest cases of deterministic forced vibration, structure free vibration modes and integration over the structure spatial domain lead to an ordinary differential equation(s) in time and to a series solution for total response displacement. Even though the number of analytical or approximated eigenvalues and eigenvectors may be unlimited, seldom do the assumptions in deriving the equations of motion support use of more than a few of the lowest modes of vibration for accurate structure stress/strain determination.

For general dynamic equation solution (free or forced), there are at least five major categories of tools available. The first category is analytical solution of the Lagrange equations of motion utilizing infinite series. References on both linear and nonlinear partial differential equations summarize most all the problems that can be solved in this manner.[19,20,21] Analytic methods of solution are very limited in cases of spatially coupled partial differential equations and dissipative forces within the governing equations. Occasionally, exact methods for simple structure configurations lend themselves to creating tables that specify natural frequencies and mode shapes for free vibration. The second category is an approximate solution for free vibration for the fundamental natural frequency of a nonenergy dissipating structure using Rayleigh's method.[8,10,22] The third category extending Rayleigh's method for free vibration beyond the fundamental natural frequency to a few higher frequencies is referred to as the Rayleigh-Ritz method[8,10] and entails solution of an algebraic eigenvalue problem. A fourth category, the assumed modes method,[8] is designed to solve problems associated with forced vibration and results in solution of a system of ordinary differential equations whose coefficients are determined by integrating over the structure spatial coordinates using suitable admissible functions. The fifth category represents an assortment of approximation techniques. Researchers generally use a method that has been well developed or resort to some modification of an existing method for solution. Table 7.5 provides a list of several approximate solution techniques and one or more references providing additional information. These

TABLE 7.5 Approximate Solution Methodologies with Selected References

Methodology	Reference
Rayleigh form methods:	
Rayleigh	8, 10, 11, 22, 23, 25, 29
Ritz	6, 7, 8, 10, 11, 22, 25
Rayleigh-Ritz	28, 29
Rayleigh-Schmidt	27
Rayleigh-Ritz-Meirovitch	33
Weinstein	10
Assumed modes method:	8, 11, 23
Weighted residual form methods:	
Galerkin	7, 8, 11, 29
Bubnov-Galerkin	6, 24
Petrov-Galerkin	6
Collocation	6, 7, 8, 11
Subdomain	6, 7, 8, 11
Method of least squares	6, 7, 8, 11
Method of moments	11, 23
Lanczo's orthogonalization	23
Pade's	23
Lumped parameter	11
Courant (Ritz-least squares)	6
Dunkerley's method:	26, 29
Transfer matrix method:	25, 29, 34
Iterative methods:	
Stodola	25
Holzer	25
Holzer-Van Den Dungen	29
Holzer-Myklestad	25
Perturbation method (Taylor series):	11, 31
Finite difference method:	7, 31
WKB and wave methods:	30
Differential quadrature method:	32
Kantorovich method:	6, 7
Trefftz method:	6, 7
Numerical integration methods:	
Linear acceleration	25
Wilson θ	25
Central difference scheme	6
Houbolt (method)	6
Newmark scheme	6

techniques fall under broad categories and may overlap one another in their assumptions and the details of their implementation.

This chapter illustrates Rayleigh's method, the Rayleigh-Ritz method, and the assumed modes method in case of forced vibration of simple structures. Complications arise for structures that have modes of a different nature that couple in more than one dynamic equation of motion. A brief summary of each of these three methods follows.

RAYLEIGH'S PRINCIPLE (METHOD)

Reference 22 expresses Rayleigh's principle as follows: "In the fundamental mode of vibration of an elastic system, the distribution of kinetic and potential energies is such as to make the frequency a minimum." For the free vibration of a structure with no dissipative forces and potential/kinetic energies that vary in time, Rayleigh's method assumes that *Rayleigh's quotient* has a stationary value. Table 7.6 provides the basic mechanics of this method. As Ref. 9 indicates, a rigorous and general proof of the principle behind Rayleigh's method is difficult. It is remarkable that the ratio of the maximum potential energy to the maximum kinetic energy must have a stationary value that represents an upper-bound approximation to the lowest natural frequency of the structure.

The true first normal mode can be approximated with simple functions, and the Rayleigh's quotient is reasonably insensitive to the shape of $X(s)$. The maximum potential energy generally is a function of derivatives of $X(s)$, and this may affect the accuracy of the stationary value.[10,22,33] For higher-order theories, Rayleigh's method is applicable, since evaluation is in the area of energy functionals; however, kinematic and constitutive complexities may not be reflected in selected admissible functions $X(s)$. Reference 8 provides a table of energies for a wide variety of structures that may be used for forming Rayleigh's quotient. In addition, Ref. 8 illustrates Rayleigh's method when nondissipative passive boundary elements are present.

TABLE 7.6 Rayleigh's Method for Approximating an Upper Bound on the Lowest Natural Frequency for a Structure by Rayleigh's Principle

Rayleigh's quotient*:

$$R[X(s)] = \max_t \{\pi_p[X(s)]\} / \max_t \{T[X(s)]\} = \omega^2$$

where generally π_p and T can be computed independent of time considerations

$X(s)$ an admissible function satisfying strucure kinematic boundary conditions and approximating structure mode shape at frequency ω

$\pi_p[X(s)]$ structure potential (elastic) energy including any passive boundary elements

$T[X(s)]$ structure kinetic energy including any passive boundary elements

Assumption:

$\max_t \{T[X(s)]\} = \omega^2 T_{\max} [X(s)]$ by separation of s and t variables

Stationary value assumption for quotient:

$\omega_1^2 = \min \{R\,[X_1\,(s)]\} \geq \omega^2$

*NOTE: Since π_p and T are positive and continuously time varying maximum values exist when evaluated for $X\,(s)$

TABLE 7.7 Rayleigh-Ritz Method for Approximating the Normal Modes for a Structure[8]

Assumption:

$$X(s) = \sum_{i=1}^{n} c_i \phi_i(s)$$

c_i unknown constants (Ritz coefficients) $i = 1,2, \ldots, n$
$\phi_i(s)$ admissible functions $i = 1,2, \ldots, n$

Rayleigh's quotient:

$$R[X(s)] = R(c_1, c_2, \ldots, c_n) \qquad (= \max_t \{\pi_p[X(s)]\} / \max_t \{T[X(s)]\})$$

Solution:

Minimizing Rayleigh's quotient with respect to coefficients c_i

$\dfrac{\partial R}{\partial c_i} = 0, i = 1,2, \ldots, n$ leads to an algebraic eigenvalue problem of order n in structure parameters where the ith eigenvector $\vec{c}^{(i)}$ provides a set of n constants and ith approximate mode shape $X^{(i)}(s) = \sum_{j=1}^{n} c_j^{(i)} \phi_j(s)$. In matrix notation for the ith eigenterm

$$\left[[k] - \lambda_i^{(n)}[m]\right] \vec{c}^{(i)} = \vec{0} \Rightarrow \left|[k] - \lambda^{(n)}[m]\right| = 0 \Rightarrow \omega_i = \sqrt{\lambda_i^{(n)}}, i = 1,2, \ldots, n$$

$[k]$ n by n structure stiffness matrix, $\lambda_i^{(n)}$ eigenvalues $i = 1,2, \ldots, n$
$[m]$ n by n structure mass matrix, $\vec{c}^{(i)}$ eigenvector with eigenvalue $\lambda_i^{(n)}$

RAYLEIGH-RITZ METHOD

The Rayleigh-Ritz method is an extension of Rayleigh's method in that it assumes several independent modes and depends upon a family of admissible functions. Table 7.7 presents the Rayleigh-Ritz method for determining an approximation to the true modal description of a structure.

ASSUMED MODES METHOD

The assumed modes method solves deterministic forced structure vibration problems. In place of unknown constants c_i in the Rayleigh-Ritz method, Table 7.7, unknown functions of time termed generalized coordinates $\eta_i(t)$, are used and a solution to the partial differential equation(s) of the form

$$X(s,t) = \sum_{i=1}^{n} \eta_i(t)\phi_i(s) \tag{7.4}$$

is assumed. It is also assumed that the deterministic structure forcing function can be expressed in terms of the admissible functions and the generalized coordinates. Substitution of Eq. (7.4) into the Lagrange's equation(s) and subsequent integration over the spatial coordinate(s) in the partial differential equation(s) leads to a set of ordinary differential equations (ODEs) in the generalized coordinates $\eta_i(t)$ for $i = 1,2, \ldots, n$. Table 7.8 provides a general expression for the solution in terms of generalized coordinates, a modal damping ratio, and structure initial conditions.

Substitution of $\eta_i(t)$ into Eq. (7.4) provides the overall solution $X(s,t)$. The initial conditions on the structure become a superimposed time invariant *loading*. For deterministic nonharmonic forced structure vibration problems or moving loads on the structure,[8,35] Fourier transform methods must be used.

TABLE 7.8 Solution for Generalized Coordinates with Viscous Damping (Duhamel's Integral)[8]

Assumption:

Uncoupled set of n second-order ODE in $\eta_i(t), \dot{\eta}_i(t)$, and $\ddot{\eta}_i(t)$, $i = 1,2, \ldots, n$ and where

$\eta_i(t)$ ith generalized coordinate ($\eta_i \neq \eta_j$ for $i \neq j$)
$\eta_{i,0} = \eta_i(0)$ ith generalized initial "displacement"
$\dot{\eta}_{i,0} = \dot{\eta}_i(0)$ ith generalized initial "velocity"
ζ_i damping ratio for the ith normal mode
$\omega_{di} = \sqrt{1-\zeta_i^2}\,\omega_i$ damped natural frequency for the ith normal mode with natural frequency ω_i
$Q_i(t)$ ith generalized force
$\phi_i(s)$ ith admissible function

Solution (convolution or Duhamel's integral):

$$\eta_i(t) = \eta_{i,0} e^{-\zeta_i \omega_i t}\left[\cos(\omega_{di} t) + \frac{\zeta_i \omega_i}{\omega_{di}} \sin(\omega_{di} t)\right] + \frac{\dot{\eta}_{i,0}}{\omega_{di}} e^{-\zeta_i \omega_i t} \sin(\omega_{di} t) + \frac{1}{\omega_{di}} \int_0^t Q_i(\tau) e^{-\zeta_i \omega_i (t-\tau)} \sin[\omega_i (t-\tau)]\, d\tau$$

APPLIED DYNAMIC BEHAVIOR OF STRUCTURES

The final portion of this chapter provides practicing engineers insight into solving vibration problems for basic structures. Continuous structure vibration (1) favors a HVP formulation, (2) relies upon references that provide solutions in table form, and (3) uses personal computer (PC) computational power for unique solutions.

Personal computer technology has led to two major advances in structure vibration analysis. First, PC-based Mathematica,[31] MATLAB,[36,37] and FORTRAN[38] provide computing tools for solution of partial differential equations. Second, PC Internet access provides a resource for substantial literature on vibration of structures.

A practicing engineer is able to formulate the equation(s) of motion for a structure, decide on a method of solution, and have access to a library of structure dynamic tables. Basic composition of either (1) lumped mass and stiffness components for simple structures or (2) buildup of simple structures into more complex structures represent options. The power of the finite element method has assisted in vibration solutions for complex structures and made possible solutions for structures with nonuniform properties.[39]

This last chapter section provides detailed information on the vibration of four common structure forms: bars (rods), shafts, beams, and plates. Four remaining structure forms—strings, membranes, arches, and shells—are referenced at the end of the section. Structure forms considered in detail contain (1) expressions for both the structure potential and kinetic energies for direct application of Rayleigh's method, structure energy input from external load, and the equation(s) of motion with both kinematic and natural boundary conditions; (2) tables for various structure configurations; (3) an occasional computational example; and (4) an extended reference guide. Selection of peer-reviewed references reflects no particular pattern but rather demonstrates the diversity of applications. References related to useful handbooks concerning structure dynamic information are highlighted by a dagger (†) placed at the end of the reference in the References section, and engineers are encouraged to begin building a practical library that includes these references. The Rayleigh, Raleigh-Ritz, and assumed modes methods are illustrated by examples.

TABLE 7.9 Definitions for Material/Structure Parameters for Equations

E modulus of elasticity
ν Poisson's ratio
G modulus of rigidity $\left(=\frac{E}{2(1+\nu)}\right)$
$D=\frac{Eh^3}{12(1-\nu^2)}$ modulus of flexural rigidity (plates)
k shear correction factor
l beam length
h plate thickness

A cross-sectional area (bar or beam)
I_b beam section bending area moment of inertia
$I_p=\iint_A (y^2+z^2)\,dA$ polar moment of inertia
ρ_l mass/length
ρ_A mass/area
ρ_m mass/volume
ρ general density (when context is clear)
ψ plane section warping function
C torsional rigidity coefficient
$u(x,t)$ rod or beam longitudinal displacement

$w(x,t)$ beam transverse displacement
$w(x,y,t)$ plate transverse displacement
$\theta(x,t)$ rod rotational coordinate
$f(x,t)$ applied force component in coordinate direction u or w
$m_t(x,t)$ applied torque $\theta(x,t)$ direction
k_a beam foundation deflection stiffness coefficient
k_1 beam foundation rotational stiffness coefficient
$I_g=\frac{1}{2}\int_0^l GI_p\left(\frac{\partial\theta}{\partial x}\right)^2 dx$ shaft strain energy
$I_\psi=\iint_A \psi^2\,dA$ shaft warping function
$I_\theta=\int_0^l \rho\left(\frac{\partial^2\theta}{\partial t\partial x}\right)^2 dx$ shaft rotary inertia
Q plate transverse shearing forces
M plate bending and twisting moments
V Kelvin-Kirchhoff plate edge reactions
l_x and l_y direction cosines in x, y directions, respectively
T_r applied torque

All of the major structure forms have more than one dynamic formulation dependent upon the assumptions made in formulating Hamilton's variational principle. Assumptions underlying HVP are found in most cases in expression of the displacement/strain relationships contained in references. Each dynamic formulation is generally termed a *theory.* Practicing engineers need to identify the theory and understand the scope of the assumptions behind it. No parameters are explicitly stated as functions of spatial variables. If a material parameter [e.g., $E = E(x)$ – (elastic modulus)] or a structure parameter [e.g., $A = A(x)$ – (cross-sectional area)] varies according to spatial dimension, then the chain rule for differentiation for variable products (or quotients) applies and single terms expand into multiple terms, complicating the solution of the resulting differential equation.

Table 7.9 provides reference definitions for all the parameters in the equations in this section. Tables and examples may explicitly define parameters for use with the structure configurations pertinent to the table or example, respectively. Parameter correspondences from table or example parameters to equation parameters should be obvious.

BARS (LONGITUDINAL/TORSIONAL VIBRATION BASIC THEORY)

Equations of Motion. Basic longitudinal (bar) and torsional (shaft) vibration have the same form of governing equation: a one-dimensional wave equation. Bars are characterized by propagation of elastic energy in a single linear or rotational

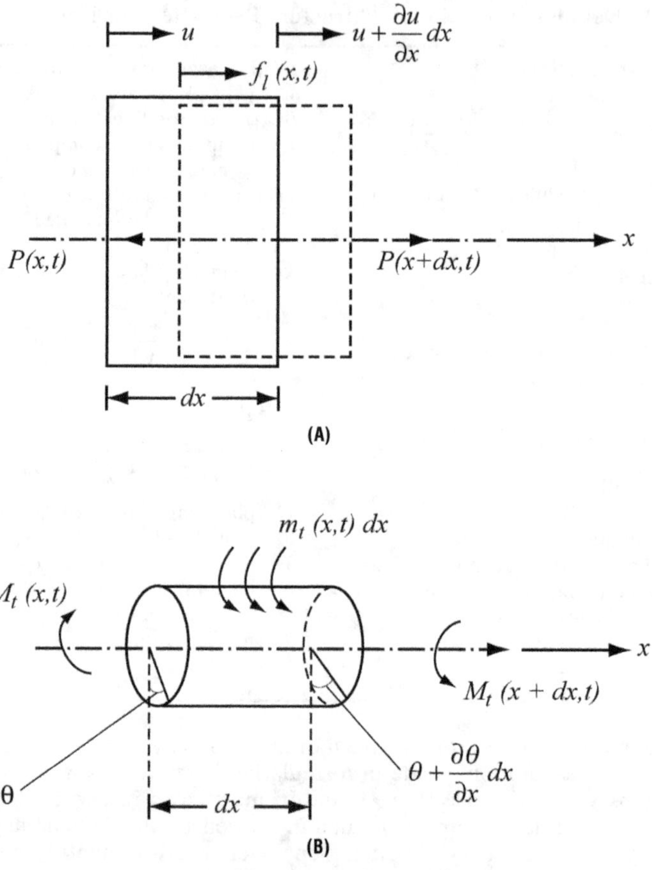

FIGURE 7.1 Loaded bar/shaft configuration: (*A*) longitudinal applied load $f_l(x,t)$; (*B*) rotational applied torque $m_t(x,t)$.

dimension. Figure 7.1 displays a bar/shaft configuration at coordinate x for both longitudinal and rotational response modes.

Equations (7.5) through (7.10) are based upon the HVP development for the bar longitudinal dynamics in terms of displacement along the bar length $u(x,t)$ with applied force $f_l(x,t)$ (force / unit length) and torsional dynamics in terms of angle of rotation about the center of twist of the bar $\theta(x,t)$, with applied moment $m_t(x,t)$ (force·unit length / unit length·radian). The expression on the left-hand side of each equation represents the bar longitudinal dynamics, while the expression on the right-hand side represents bar torsional dynamics. Structure/material properties, forces, and torques may be nonuniform along the bar axis.

Structure potential (elastic) energy for constants in Table 7.9 is given by

$$\underbrace{\pi = \frac{1}{2}\int_0^l \left[EA\left(\frac{\partial u}{\partial x}\right)^2\right] dx}_{\text{longitudinal}} \qquad \underbrace{\pi = \frac{1}{2}\int_0^l \left[GI_p\left(\frac{\partial \theta}{\partial x}\right)^2\right] dx}_{\text{torsional}} \tag{7.5}$$

and structure kinetic energy for mass density ρ_m by the following expressions:

$$\underbrace{T = \frac{1}{2}\int_0^l \left[\rho_m A\left(\frac{\partial u}{\partial t}\right)^2\right] dx}_{\text{longitudinal}} \qquad \underbrace{T = \frac{1}{2}\int_0^l \left[\rho_m I_p\left(\frac{\partial \theta}{\partial t}\right)^2\right] dx}_{\text{torsional}} \tag{7.6}$$

The work performed on the structure by external forces is given by

$$\underbrace{W = \int_0^l [f_l u]\, dx}_{\text{longitudinal}} \qquad \underbrace{W = \int_0^l [m_t \theta]\, dx}_{\text{torsional}} \tag{7.7}$$

Application of integration by parts in HVP provides Lagrange's equations of motion:

$$\underbrace{\frac{\partial}{\partial x}\left(EA\,\frac{\partial u}{\partial x}\right) + f_l = \rho_m A\,\frac{\partial^2 u}{\partial t^2}}_{\text{longitudinal}} \qquad \underbrace{\frac{\partial}{\partial x}\left(GI_p\,\frac{\partial \theta}{\partial x}\right) + m_t = \rho_m I_p\,\frac{\partial^2 \theta}{\partial t^2}}_{\text{torsional}} \tag{7.8}$$

Kinematic and natural boundary conditions are provided in terms of the variation operator δ

$$\underbrace{EA\,\frac{\partial u}{\partial x}\,\delta u\Big|_0^l = 0}_{\text{longitudinal}} \qquad \underbrace{GI_p\,\frac{\partial \theta}{\partial x}\,\delta \theta\Big|_0^l = 0}_{\text{torsional}} \tag{7.9}$$

Finally, initial displacement and velocity can be stated as

$$\underbrace{u(x,\,0) = u_0(x), \dot{u}(x,\,0) = \dot{u}_0(x)}_{\text{longitudinal}} \qquad \underbrace{\theta(x,\,0) = \theta_0(x), \dot{\theta}(x,\,0) = \dot{\theta}_0(x)}_{\text{torsional}} \tag{7.10}$$

for Figure 7.2*A*.

Basic bar boundary conditions are fixed-end (X) or free (R). Figure 7.2*A* generalizes this by attaching spring (linear or rotational) and mass (lumped or disk) elements at the ends of the uniform bar to form a simple combination continuous/discrete system. Table 7.10 summarizes the dynamics of nine unique cases in terms of the characteristic frequency equation and the form of the mode shape. Example 7.1 for the configuration in Fig. 7.2*B* illustrates the Rayleigh-Ritz normal mode formulation along with estimation of the lowest natural frequency using Rayleigh's quotient. This example serves to illustrate use of Table 7.10.

***Example 7.1:* Longitudinal Vibration of a Fixed-End Bar with a Spring/Mass Attached to the Free End**

SOLUTION. The bar in Fig. 7.2*B* has displacement/force boundary conditions in Eq. (7.11) and, for completeness, initial conditions in Eq. (7.12):

$$u(0,t) = 0 \text{ and } EA\,\frac{\partial u(l,t)}{\partial x} = -ku(l,t) - M\,\frac{\partial^2 u(l,t)}{\partial t^2} \tag{7.11}$$

$$u(x,0) = u_0(x) \text{ and } \frac{\partial u(x,0)}{\partial t} = \dot{u}_0(x) \tag{7.12}$$

where $u_0(x)$ and $\dot{u}_0(x)$ are specified functions.

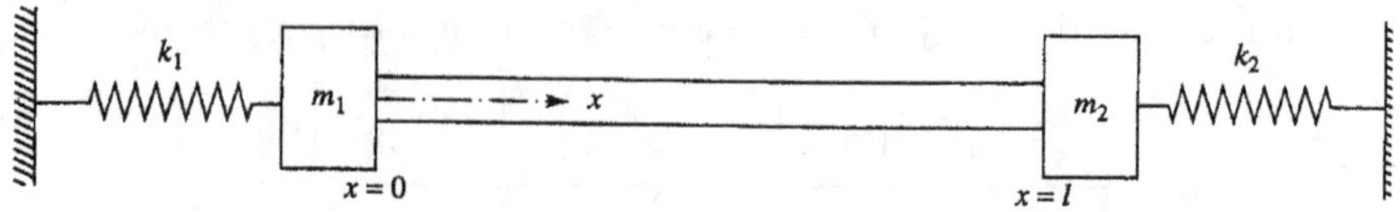

$$EA\frac{\partial u}{\partial x}\bigg|_{(0,t)} = k_1\, u\big|_{(0,t)} + m_1\frac{\partial^2 u}{\partial t^2}\bigg|_{(0,t)} \quad \text{and} \quad EA\frac{\partial u}{\partial x}\bigg|_{(l,t)} = -k_2\, u\big|_{(l,t)} - m_2\frac{\partial^2 u}{\partial t^2}\bigg|_{(l,t)}$$

$$GI_p\frac{\partial \theta}{\partial x}\bigg|_{(0,t)} = k_{t1}\,\theta\big|_{(0,t)} + I_1\frac{\partial^2 \theta}{\partial t^2}\bigg|_{(0,t)} \quad \text{and} \quad GI_p\frac{\partial \theta}{\partial x}\bigg|_{(l,t)} = -k_{t2}\,\theta\big|_{(l,t)} - I_2\frac{\partial^2 \theta}{\partial t^2}\bigg|_{(l,t)}$$

(A)

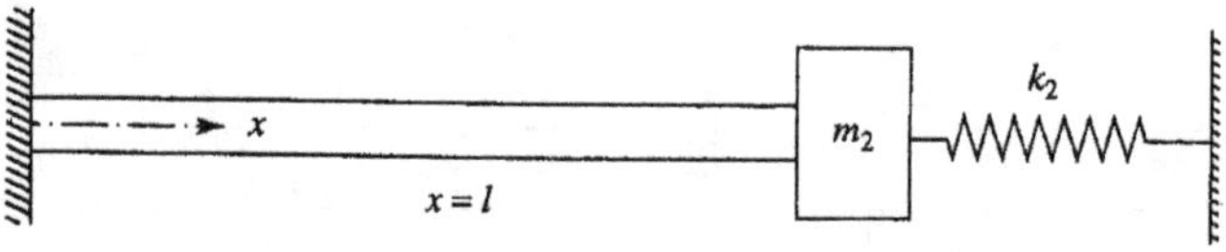

(B)

FIGURE 7.2 Bar with mass and spring boundary conditions: (*A*) longitudinal vibration for two-mass/two-spring continuous bar system; (*B*) longitudinal vibration for continuous bar with mass/spring boundary condition.

Applying the boundary conditions and allowing $\alpha = \omega l/c$, $k = EA/l$, $m = Al\rho$, and $\beta = m/m_2$ provides the following characteristic equation to be solved for $\alpha = \omega l/c$:

$$\alpha \cot\alpha = \frac{\alpha^2}{\beta} - \frac{k_2}{k} \tag{7.13}$$

and solution

$$u(x,t) = \sum_{n=1}^{\infty}\left[\sin\frac{\omega_n x}{c}\,(C_n \cos\omega_n t + D_n \sin\omega_n t)\right] \text{ for } \omega_n = \frac{\alpha_n c}{l} \tag{7.14}$$

where C_n and D_n must be determined from the bar initial conditions:

$$C_n = \frac{2}{ncl}\int_0^l u_0(x)\sin\left(\frac{\omega_n x}{c}\right)dx \qquad D_n = \frac{2}{ncl}\int_0^l \dot{u}_0(x)\sin\left(\frac{\omega_n x}{c}\right)dx \text{ for } n = 1,2,\ldots \tag{7.15}$$

Since the characteristic equation varies with the end configuration at $x = l$, four additional configurations with left end fixed with corresponding characteristic equations and normal modes are:

$$\begin{aligned}
&k_2 \text{ and } m_2 \text{ removed (X/R): } \cos(\alpha) = 0,\ \tilde{B}_n \sin\left[\frac{(2n+1)\pi x}{2l}\right] \text{ for } n = 0,1,2,3,\ldots \\
&k_2 \text{ and } m_2 \text{ removed (X/X): } \sin(\alpha) = 0,\ \tilde{B}_n \sin\left(\frac{n\pi x}{l}\right) \text{ for } n = 1,2,3,\ldots \\
&m_2 \text{ removed, } k_2 > 0\text{(X/-): } \alpha\tan(\alpha) = -\gamma,\ \tilde{B}_n \sin\left(\frac{\alpha_n x}{l}\right) \text{ for } n = 1,2,3,\ldots \\
&k_2 \text{ removed, } m_2 > 0 \text{ (X/-): } \alpha\tan(\alpha) = \beta,\ \tilde{B}_n \sin\left(\frac{\alpha_n x}{l}\right) \text{ for } n = 1,2,3,\ldots
\end{aligned} \tag{7.16}$$

TABLE 7.10 Characteristic Frequencies and Natural Vibration Modes for a Bar Structure System

Configuration	Characteristic equation	Normal mode shape
- k_1 m_1 B m_2 k_2 -	$\tan(\alpha) = \dfrac{k\alpha[(k_1+k_2)-(m_1+m_2)\omega^2]}{(k\alpha)^2-(k_1-m_1\omega^2)(k_2-m_2\omega^2)}$	$\tilde{A}_n\left\{\cos\left(\dfrac{\alpha_n x}{l}\right)+\dfrac{k_1-m_1\omega_n^2}{k\alpha}\sin\left(\dfrac{\alpha_n x}{l}\right)\right\}$ n = 1,2,..
- k_1 - B - k_2 -	$\tan(\alpha) = \dfrac{k\alpha(k_1+k_2)}{(k\alpha)^2-k_1k_2}$	$\tilde{A}_n\left\{\cos\left(\dfrac{\alpha_n x}{l}\right)+\dfrac{k_1}{k\alpha_n}\sin\left(\dfrac{\alpha_n x}{l}\right)\right\}$ n = 1,2,..
- - m_1 B m_2 - -	$\tan(\alpha) = \dfrac{k\alpha(m_1+m_2)\omega^2}{m_1m_2\omega^4-(k\alpha)^2}$	$\tilde{A}_n\left\{\cos\left(\dfrac{\alpha_n x}{l}\right)+\dfrac{m_1k\alpha_n^2}{m}\sin\left(\dfrac{\alpha_n x}{l}\right)\right\}$ n = 1,2,...
- k_1 - B m_2 - -	$\tan(\alpha) = \dfrac{k\alpha(k_1-m_2\omega^2)}{(k\alpha)^2+m_2k_1\omega^2}$	$\tilde{A}_n\left\{\cos\left(\dfrac{\alpha_n x}{l}\right)+\dfrac{k_1}{k\alpha}\sin\left(\dfrac{\alpha_n x}{l}\right)\right\}$ n = 1,2,.
- k_1 - B m_2 k_2 -	$\tan(\alpha) = \dfrac{k\alpha((k_1-k_2)-m_2\omega^2)}{(k\alpha)^2-k_1(k_2-m_2\omega^2)}$	$\tilde{A}_n\left\{\cos\left(\dfrac{\alpha_n x}{l}\right)+\dfrac{k_1}{k\alpha}\sin\left(\dfrac{\alpha_n x}{l}\right)\right\}$ n = 1,2,.
- - - B m_2 k_2 -	$\alpha\tan(\alpha) = \dfrac{k_2-m_2\omega^2}{k}$	$\tilde{A}_n\cos\left(\dfrac{\alpha_n x}{l}\right)$ n = 1,2,..
R - - B - - R	$\sin(\alpha) = 0$	$\tilde{A}_n\cos\left(\dfrac{n\pi x}{l}\right)$ n = 0,1,2,..
R - - B m_2 - -	$\tan(\alpha) = -\alpha/\beta$	$\tilde{A}_n\cos\left(\dfrac{\alpha_n x}{l}\right)$ n = 1,2,..
R - - B - k_2 -	$\cot(\alpha) = \alpha\delta$	$\tilde{A}_n\cos\left(\dfrac{\alpha_n x}{l}\right)$ n = 1,2,..

Parameters:

$$\alpha = \frac{\omega l}{c},\ \omega = \frac{\alpha c}{l},\ m = \rho A l,\ c^2 = \frac{E}{\rho},\ k = \frac{EA}{l},\ \delta = \frac{k}{k_2},\ \beta = \frac{m}{m_2},\ \gamma = \frac{m\omega^2}{k_2}$$

from governing equations longitudinal EA ~ torsional GI_p and longitudinal ρA ~ torsional I_o

Rayleigh's method, assuming $u(x) \approx u_0\, x/l\,; 0 \le x \le 1$ and $u(0) = 0, u(1) = u_0$:

$$T = \frac{1}{2}\int_0^l \rho A\left[\frac{\partial u}{\partial t}\right]^2 dx + \frac{1}{2}\, m_2\left[\frac{\partial u(l,t)}{\partial t}\right]^2 \rightarrow \omega_n^2 T_{\max} = \frac{\rho A}{2}\,\omega_n^2 u_0^2 \frac{l}{3} + \frac{1}{2}\, m_2\omega_n^2 u_0^2 \qquad (7.17)$$

$$\pi = \frac{1}{2}\int_0^l EA\left(\frac{\partial u}{\partial x}\right)^2 dx + \frac{1}{2}\, k_2 u^2(l,t) \rightarrow \pi_{\max} = \frac{EA}{2}\,\frac{u_0^2}{l} + \frac{1}{2}\, k_2 u_0^2 \qquad (7.18)$$

$$\text{Rayleigh's quotient: } R(u_0\, x/l) = \frac{\pi_{\max}}{T_{\max}} = \omega_n^2 \rightarrow \omega_n$$

$$= \sqrt{\frac{EA/l + k_2}{\rho A l/3 + m_2}} = \sqrt{\frac{k + k_2}{m/3 + m_2}} \qquad (7.19)$$

In Table 7.10, the leftmost column describes the remaining unique nine structure system configurations, as illustrated in Fig. 7.2*A*. For Example 7.1 and Fig. 7.2*B*, the structure configuration is denoted [X - - B m_2 k_2 -] implying a bar (B) with a fixed end and a mass and spring attached to the other end. The other four configurations included in this example are not repeated in Table 7.10. The middle column provides the characteristic equation to be solved for the natural frequencies according to index n, where it is generally assumed that the denominator in these equations is bounded away from zero. The rightmost column provides the form of the mode shape in terms of bar reference $u(x,t)$ or $\theta(x,t)$ as a function of coordinate x. Even though information in Table 7.10 is expressed in terms of the bar's longitudinal configuration, the table also contains all information necessary for the torsional configuration, with torsional springs replacing linear springs and discrete mass moments of inertia replacing lumped masses.

Stress Versus Particle Velocity. Hamilton's variational principle leads to the one-dimensional wave equation and the conclusion that the longitudinal stress in a bar is proportional to particle velocity.[7] Table 7.11 derives the relationship.

Hopkinson Bar.[7] A uniform bar, with a longitudinal impact area on one end and a test item on the opposite end, has dynamics that are governed by the one-dimensional wave equation. For such a bar, termed a *Hopkinson bar,* knowing the form of the input

TABLE 7.11 Material Particle Velocity Proportional to Longitudinal Stress in a Bar

1. Displacement field: $u = u\,(x,t) = G_1\,(x - c_0 t) + G_2\,(x + c_0 t)$ for G_1 and G_2 arbitrary

2. Strain and stress/displacement: $\varepsilon_{xx} = \dfrac{\partial u}{\partial x}$, $\sigma_{xx} = E\,\dfrac{\partial u}{\partial x}$

3. One-dimensional wave equation: $E\,\dfrac{\partial^2 u}{\partial x^2} = \rho_m\,\dfrac{\partial^2 u}{\partial t^2}$ and $\dfrac{\partial u}{\partial x} = \dfrac{1}{c_0}\,\dfrac{\partial u}{\partial t}$ [for $u\,(x,t) = G_2\,(x + c_0 t)$]

4. Derivation: $E\,\dfrac{\partial^2 u}{\partial x^2} = \dfrac{\partial \sigma_{xx}}{\partial x} = \rho_m\,\dfrac{\partial^2 u}{\partial t^2} = \rho_m\,\dfrac{\partial}{\partial t}\left(c_0\,\dfrac{\partial u}{\partial x}\right) = \rho_m c_0\,\dfrac{\partial}{\partial x}\left(\dfrac{\partial u}{\partial t}\right)$ then integration with respect to x provides $\sigma_{xx} = \rho_m c_0\,\dfrac{\partial u}{\partial t}$ for $c_0 = \sqrt{E/\rho_m}$

pulse, the base input to the test item can be predicted at the opposite end from a Fourier integral solution of the equation of motion. The wave propagation phenomenon is discussed in Ref. 7.

Structure degradation from large stress can occur under high material velocity.

SELECTED REFERENCE INFORMATION

Bar propagation of elastic energy in a single linear dimension may be accompanied by local rotary and shear effects as a result of the material's Poisson ratio. Displacements in the longitudinal and the cross-sectional axes, that is, $u(x,t)$, $v(x,t)$, and $w(x,t)$, contribute energy in higher-order theories. The shorter a bar in relation to its lateral dimensions, the larger the rotary inertia and shear effects. Two extended theories, Rayleigh and Bishop, for bar longitudinal vibration are discussed in Ref. 8. The Rayleigh theory considers the effect of rotary inertia, while Bishop's theory includes effects of both rotary inertia and shear. Longitudinal vibrations of nonuniform bars of certain form can have analytical solutions as a result of functional transformation.[40,41] Reference 42 examines the impulsive response of variable cross-section bars using Green's function, arriving at an integral equation formulation. Two or more longitudinal bars coupled in longitudinal vibration to compose a structural system with complex modal structure is solved via Green's function in Ref. 43.

BARS (SHAFTS) WITH ROTATIONAL VIBRATION (EXTENDED THEORIES)

Higher-order extended theories of shafts are of greater importance than those for bars because of the common use of shaft structures for transferring rotary motion. These theories become complex when noncircular cross sections are considered because of the difficulty in defining *torsional rigidity* C (defined by the shaft torque divided by the angle of twist). In this section it is assumed that external moments act through the center of twist. The first model comes from St. Venant's theory, designated Σ and includes out-of-plane displacement of plane sections normal to the axis of rotation but neglects inertia due to axial motion. The second model is based upon Love's theory, designated Λ, that is, Σ plus rotary inertia attributed to plane sections. Finally, there is the Timoshenko-Gere theory, designated T, that is, Λ plus the effects of torsional shear for short shafts. Theories Σ, Λ, and T are telescoping in terms of equations and boundary conditions, as is noted in Eqs. (7.22) through (7.26). These equations identify the effect for each term in the equation—for example, the rotary inertia designation represents the effect of rotary inertia. Higher-order theories result in two coupled partial differential equations, with the second equation in terms of a *warping function.* Practical application requires Prandtl's membrane analogy, found in the references.[6,8]

Figure 7.3 provides a schematic of a square shaft with x-axis of rotation and yz-out of plane deformation. It will be assumed that all material and configuration parameters are a function of the rotational coordinate θ. With rotation $\theta(x,t)$ and an arbitrary point in the shaft cross section (x,y,z), then cross-section displacement is defined by a *warping function* $\psi(y,z)$, where for time implicit in Eq. (7.20)

$$u(x,y,z) = \psi(y,z)\,\frac{\partial\theta}{\partial x} \tag{7.20}$$

and other displacements for explicit time are given by

$$v = -z\theta(x,t) \qquad \text{and} \qquad w = y\theta(x,t) \tag{7.21}$$

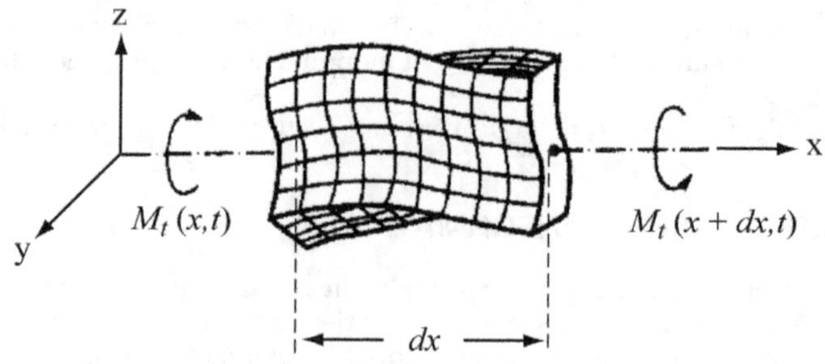

FIGURE 7.3 Schematic for a square shaft under torsion rotating through center of twist. *(Ref. 8.)*

Shaft torsional elastic energy in terms of axial stress or "stretching" and shear is given by

$$\pi = \frac{1}{2}\int_0^l \iint_A \Bigg(\underbrace{\left\{ G\left[\left(\frac{\partial \psi}{\partial y} - z\right)\frac{\partial \theta}{\partial x}\right]^2 + G\left[\left(\frac{\partial \psi}{\partial z} + y\right)\frac{\partial \theta}{\partial x}\right]^2 \right\}}_{\Sigma,\Lambda,\text{T—torsional shear}} + \underbrace{\left\{ E\left(\psi\frac{\partial^2 \theta}{\partial x^2}\right)^2 \right\}}_{\text{T—axial stress}} \Bigg)\, dA\,dx \qquad (7.22)$$

while the corresponding shaft kinetic energy is given by

$$T = \frac{1}{2}\int_0^l \iint_A \rho_m \Bigg[\underbrace{\left(z\frac{\partial \theta}{\partial t}\right)^2 + \left(y\frac{\partial \theta}{\partial t}\right)^2}_{\Sigma,\Lambda,\text{T—rotary inertia}} + \underbrace{\left(\psi\frac{\partial^2 \theta}{\partial x \partial t}\right)^2}_{\Lambda,\text{T—axial inertia}} \Bigg]\, dA\,dx \qquad (7.23)$$

For an applied external torque m_t, the shaft work energy is given as

$$W = \int_0^l \underbrace{[m_t\theta]}_{\Sigma,\Lambda,\text{T—external torque}}\, dx \qquad (7.24)$$

In both Λ and T theories, variables [$\theta(x,t)$ and $\psi(x,t)$] are coupled.

$$\underbrace{-\frac{\partial}{\partial x}\left(C\left(\frac{\partial \theta}{\partial x}\right)\right)}_{\Sigma,\Lambda,\text{T—torsional shear}} + \underbrace{\rho_m I_p \frac{\partial^2 \theta}{\partial t^2}}_{\Sigma,\Lambda,\text{T—rotary inertia}} - \underbrace{\frac{\partial^2}{\partial t \partial x}\left(\rho_m I_\psi \frac{\partial^2 \theta}{\partial t \partial x}\right)}_{\Lambda,\text{T—axial inertia}} + \underbrace{\frac{\partial^2}{\partial x^2}\left(EI_\psi \frac{\partial^2 \theta}{\partial x^2}\right)}_{\text{T—axial stress}} = \underbrace{m(x,t)}_{\Sigma,\Lambda,\text{T—external torque}} \qquad (7.25a)$$

$$\underbrace{\int_0^l G\left(\frac{\partial \theta}{\partial x}\right)^2 dx\left(\frac{\partial^2 \psi}{\partial y^2} + \frac{\partial^2 \psi}{\partial z^2}\right)}_{\Sigma,\Lambda,\text{T—torsional shear}} + \underbrace{\left[\int_0^l \rho_m\left(\frac{\partial^2 \theta}{\partial x \partial t}\right)^2 dx\right]\psi}_{\Lambda,\text{T—rotary inertia}} - \underbrace{\left[\int_0^l E\left(\frac{\partial^2 \theta}{\partial x^2}\right)^2 dx\right]\psi}_{\text{T—axial stress}} = 0 \qquad (7.25b)$$

while the kinematic and natural boundary conditions are likewise coupled

$$\left[\underbrace{C\frac{\partial \theta}{\partial x}}_{\Sigma,\Lambda,\text{T—torsional shear}} + \underbrace{\rho_m I_\psi \frac{\partial^3 \theta}{\partial x \partial t^2}}_{\Lambda,\text{T—rotary inertia}} - \underbrace{\frac{\partial}{\partial x}\left(EI_\psi \frac{\partial^2 \theta}{\partial x^2}\right)}_{\text{T—axial stress}}\right]\delta\theta\Bigg|_0^l + \underbrace{\left(EI_\psi \frac{\partial^2 \theta}{\partial x^2}\right)}_{\text{T—axial stress}}\delta\left(\frac{\partial \theta}{\partial x}\right)\Bigg|_0^l = 0 \qquad (7.26a)$$

$$\underbrace{\left(\frac{\partial\psi}{\partial y} - z\right)l_y + \left(\frac{\partial\psi}{\partial z} + y\right)l_z = 0}_{\Sigma,\Lambda,\mathrm{T}\text{—warping function boundary condition}} \quad \text{for } l_y \text{ and } l_z \text{ boundary direction cosines} \tag{7.26b}$$

Solution in terms of $\theta(x,t)$ and $\psi(y,z)$ is difficult with regard to the boundary condition on ψ. St. Venant's theory for torsion of noncircular shafts [Eqs. (7.25*a*) and (7.25*b*)] replaces C by GI_p. The differential equation for ψ is the second-order Laplace partial differential equation. For practical solution, ψ can be expressed in terms of the Prandtl stress function and the torsional rigidity computed for noncircular shafts by solving Poisson's equation (a nonhomogeneous form of Laplace's equation). Prandtl's membrane analogy based upon the Prandtl stress function relates properties of an *inflated membrane* of the same plan form as the noncircular shaft to torsion characteristics of noncircular shafts. For example, in this analogy the slope at the boundary of the membrane in a given axis is proportional to the shear stress in the specified axis.[8]

Table 7.12 provides torsional rigidity estimates along with the maximum shear stress for several forms of closed cross section. To use this table for determining torsional rigidity defined as $C = T_r/\theta$, the value provided in the second column divided by the torque T_r and inverted is the torsional rigidity constant C. Reference 8 demonstrates derivation of the torsional rigidity constant C for an elliptic and rectangular cross-section shaft.

SELECTED REFERENCE INFORMATION

An early paper examines the effect of warping restraint on torsion of a thin-walled cantilever tube.[44] A simplified time domain model is used in quantifying torsional vibrations and studying shaft breakage in motor drives during start-up.[45] Damping of shaft vibration is considered in Ref. 46. Additional references are provided under the beam dynamics section, where coupling of bending and torsional vibration is of concern.

BEAMS (TRANSVERSE VIBRATION)

Beam structures carry load transverse to the long axis through material bending. As in the case of a bar, rotary inertia and shear of plane sections can be considerations for beam bending if the beam transverse dimension is sizable in comparison with the long axis. Seldom are combined axial and bending vibrations of beam structures a concern; however, thin-walled beam structures may have coupled vibration modes in bending and torsion. Beams considered in this section are described in a rectilinear coordinate system (beams in a curvilinear coordinate system are termed *arches* and are referenced in a separate section) and include (1) beams with diverse boundary conditions, (2) continuous beams, and (3) beams on elastic foundations. Reference 8 discusses in detail other beam configurations such as beams with axial forces (e.g., rotating beams) and beams with combined bending and torsion. Beam dynamics may be expressed in at least 10 theories or variants of major theories.[47] The most common theories of Euler-Bernoulli (designated E), Rayleigh (P), and Timoshenko (T) will be presented here. These theories do not permit distortion of the beam plane cross sections such as in the case of torsional bar dynamics with a cross-section warping function.

TABLE 7.12 Torsional Rigidity and Maximum Shear Stress for Shafts (Adapted from Ref. 8)

<table>
<tr><th>Cross section</th><th>Angle of twist per unit length θ</th><th>Maximum shear stress τ_{max}</th></tr>
<tr><td>1. Thick-walled tube
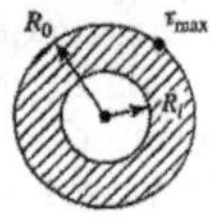</td><td>T_r = torque
G = shear modulus
$\dfrac{2T_r}{\pi G(R_0^4 - R_i^4)}$</td><td>$\dfrac{2T_r R_o}{\pi(R_0^4 - R_i^4)}$</td></tr>
<tr><td>2. Solid elliptic shaft
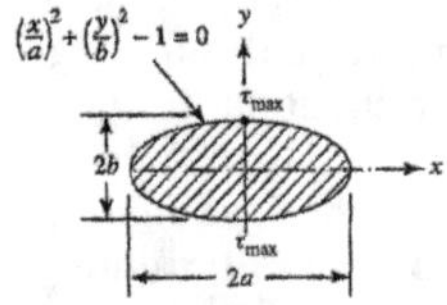</td><td>$\dfrac{(a^2 + b^2)T_r}{\pi G a^3 b^3}$</td><td>$\dfrac{2T_r}{\pi a b^2}$</td></tr>
<tr><td>3. Hollow elliptic tube
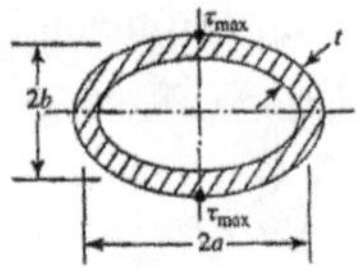</td><td>$\dfrac{\sqrt{2(a^2 + b^2)}T_r}{4\pi G a^2 b^2 t}$</td><td>$\dfrac{T_r}{2\pi abt}$</td></tr>
<tr><td>4. Solid rectangular shaft
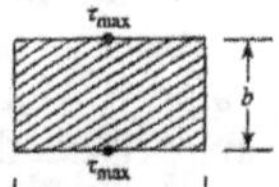</td><td>$\dfrac{T_r}{\alpha G a b^3}$
<table>
<tr><th>$\dfrac{a}{b}$</th><th>α</th><th>β</th></tr>
<tr><td>1.0</td><td>0.141</td><td>0.208</td></tr>
<tr><td>2.0</td><td>0.229</td><td>0.246</td></tr>
<tr><td>3.0</td><td>0.263</td><td>0.267</td></tr>
<tr><td>5.0</td><td>0.291</td><td>0.292</td></tr>
<tr><td>10.0</td><td>0.312</td><td>0.312</td></tr>
<tr><td>∞</td><td>0.333</td><td>0.333</td></tr>
</table></td><td>$\dfrac{T_r}{\beta a b^2}$</td></tr>
<tr><td>5. Thin-walled tube
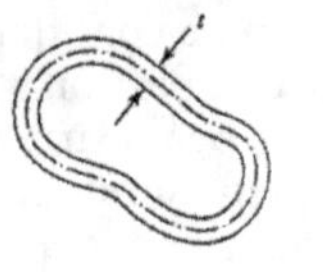</td><td>$\dfrac{T_r S}{4G\tilde{A}^2 t}$
S = circumference of the centerline of the tube (midwall perimeters)
$\tilde{A}$ = area enclosed by the midwall perimeters</td><td>$\dfrac{T_r}{2\tilde{A}t}$</td></tr>
</table>

Significant kinematic/dynamic relationships for higher-order theories P and T are given by rotary inertia dynamics (P) and plane section shear deformation (T) as depicted in Fig. 7.4. If $v = 0$ and $w = w(x,t)$, then

$$\underbrace{u = -z\left(\frac{\partial w}{\partial x}\right)}_{\text{E,P,T—pure bending}} \qquad \underbrace{\frac{\partial u}{\partial t} = -z\left(\frac{\partial^2 w}{\partial t \partial x}\right)}_{\text{P—rotary inertia}} \qquad \underbrace{u = -z\left(\frac{\partial w}{\partial x} - \beta\right) \equiv -z\phi(x,t)}_{\text{T—plane section shear}} \tag{7.27}$$

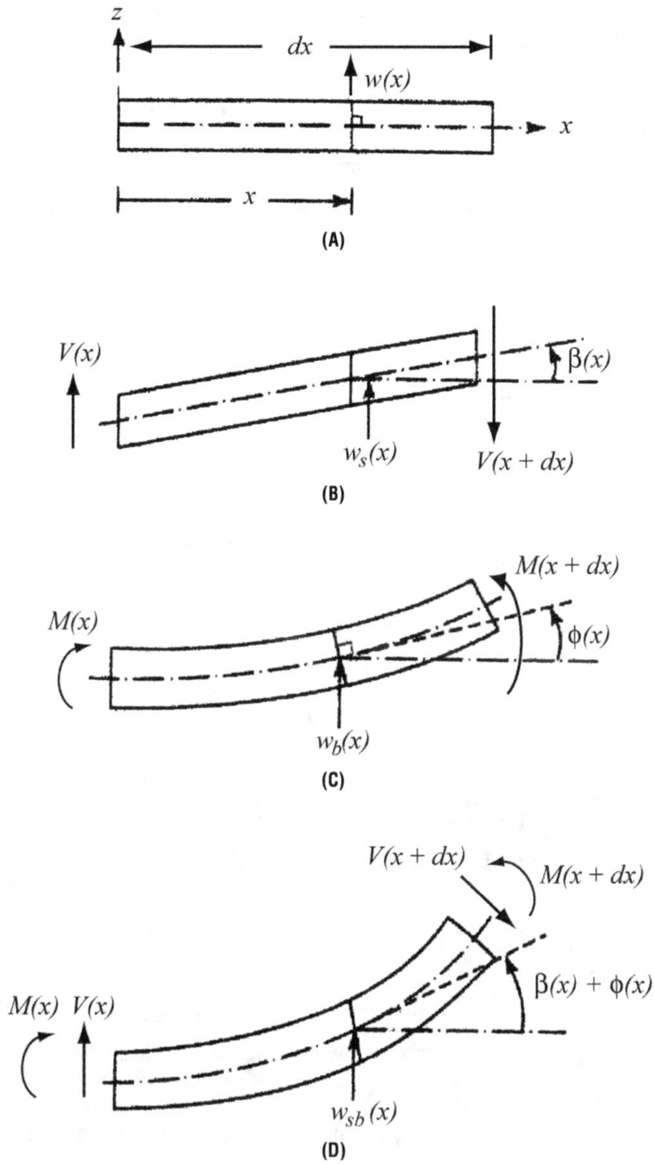

FIGURE 7.4 Bending moment and shear force deformation relationships (*A*) nondeformed; (*B*) plane section translation (shear); (*C*) plane section rotation (bending); (*D*) combined plane section translation and rotation. *(Adapted from Ref. 8.)*

For a beam, the elastic energy including plane section shear with shear correction factor k (a function of cross-section shape), accounting for a nonuniform distribution of shear stress σ_{zx} over the plane section such that $\sigma_{zx} = G\varepsilon_{zx} = kG(\partial w/\partial x)$, is given by

$$\pi = \frac{1}{2}\int_0^l \Bigg[\underbrace{EI_b\left(\frac{\partial \phi}{\partial x}\right)^2}_{\text{K,P,T—bending}} + \underbrace{kAG\left(\frac{\partial w}{\partial x} - \phi\right)^2}_{\text{T—shear}}\Bigg] dx \tag{7.28}$$

The expression for beam kinetic energy including plane section rotational inertia is given as

$$T = \frac{1}{2}\int_0^l \rho_m\Bigg[\underbrace{A\left(\frac{\partial w}{\partial t}\right)^2}_{\substack{\text{E,P,T—bending}\\ \text{inertia}}} + \underbrace{I_b\left(\frac{\partial \phi}{\partial t}\right)^2}_{\substack{\text{P,T—rotary}\\ \text{inertia}}}\Bigg] dx \tag{7.29}$$

Work from forces applied transverse to the beam axis is

$$W = \int_0^l \underbrace{(fw)}_{\substack{\text{E,P,T—}\\ \text{external force}}} dx \tag{7.30}$$

From HVP and integration by parts, Lagrange's differential equations couple displacement w and rotation ϕ.

$$\underbrace{-\frac{\partial}{\partial x}\left[kAG\left(\frac{\partial w}{\partial x} - \phi\right)\right]}_{\text{T—shear stiffness}} + \underbrace{\rho_m A\frac{\partial^2 w}{\partial t^2}}_{\substack{\text{E,P,T—transverse}\\ \text{mass inertia}}} = \underbrace{f}_{\substack{\text{E,P,T—external}\\ \text{transverse force}}} \tag{7.31a}$$

$$\underbrace{-\frac{\partial}{\partial x}\left(EI_b\frac{\partial \phi}{\partial x}\right)}_{\text{E,P,T—bending stiffness}} - \underbrace{kAG\left(\frac{\partial w}{\partial x} - \phi\right)}_{\text{T—shear stiffness}} + \underbrace{\rho_m I_b\frac{\partial^2 \phi}{\partial t^2}}_{\text{P—rotary inertia}} = 0 \tag{7.31b}$$

The w,ϕ coupled boundary conditions for this particular form of beam dynamics are given by

$$\underbrace{\left[kAG\left(\frac{\partial w}{\partial x} - \phi\right)\right]}_{\text{T—shear}}\delta w\Big|_0^l = 0 \quad \text{and} \quad \underbrace{\left(EI\frac{\partial \phi}{\partial x}\right)}_{\text{E,P,T—bending}}\delta\phi\Big|_0^l = 0 \tag{7.32}$$

For governing equation transparency, relative to the three dynamic theories, a beam with uniform structure and mass properties must be assumed and ϕ eliminated from Eqs. (7.31) and (7.32). The resulting equation of motion provides a single fourth-order partial differential equation in $w(x,t)$, requiring spatial derivatives of the applied transverse load $f(x,t)$.

$$\underbrace{EI_b\frac{\partial^4 w}{\partial x^4}}_{\text{E,P,T—bending stiffness}} + \underbrace{\rho_m A\frac{\partial^2 w}{\partial t^2}}_{\substack{\text{E,P,T—transverse}\\ \text{mass inertia}}} - \underbrace{\rho_m I_b\frac{\partial^4 w}{\partial x^2\partial t^2}}_{\substack{\text{P,T—rotary}\\ \text{inertia strain}}} - \underbrace{\frac{\rho_m I_b E}{kG}\frac{\partial^4 w}{\partial x^2\partial t^2}}_{\substack{\text{T—rotary inertia}\\ \text{shear strain}}} + \underbrace{\frac{\rho^2 I}{kG}\frac{\partial^4 w}{\partial t^4}}_{\substack{\text{T—rotary inertia}\\ \text{shear strain}}}$$

$$= -\underbrace{\frac{EI}{kAG}\frac{\partial^2 f}{\partial x^2}}_{\text{T—bending/shear}} + \underbrace{\frac{\rho I}{kAG}\frac{\partial^2 f}{\partial t^2}}_{\text{T—inertia/shear}} + \underbrace{f}_{\text{E,P,T—force}} \tag{7.33}$$

The moment/slope boundary condition is given by Eq. (7.34):

$$\underbrace{EI_b \frac{\partial^2 w}{\partial x^2}}_{\text{E,P,T—bending moment}} \delta\left(\frac{\partial w}{\partial x}\right)\Bigg|_0^l = 0 \tag{7.34}$$

and the shear/displacement boundary condition is given as follows:

$$\left[\underbrace{\frac{\partial}{\partial x}\left(EI_b \frac{\partial^2 w}{\partial x^2}\right)}_{\text{E,P,T—shear}} - \underbrace{\rho_m I_b \frac{\partial^3 w}{\partial x \partial t^2}}_{\text{P,T—rotary shear}}\right] \delta w\Big|_0^l = 0 \tag{7.35}$$

Illustrations that follow are almost exclusively in terms of Euler-Bernoulli beam theory because of its simplicity. Rayleigh theory, with mixed second-order derivatives in the term $\partial^4 w/\partial x^2 \partial t^2$, is often separable in x and t, resulting in an added term to the natural frequency. For Timoshenko theory, bending and shear are coupled, resulting in simultaneously occurring natural frequencies in bending and higher natural frequencies in shear.

EULER-BERNOULLI BEAM THEORY

Normal Modes. For practical application, Euler-Bernoulli beam theory allows separation of variables and a first approximation for beam dynamic behavior under (1) a broad set of kinematic (displacement/slope) and natural (shear/moment) boundary conditions including mass, damper, and spring end loading elements; (2) initial conditions; and (3) transverse dynamic loading.[8] The form of solution to E theory included in Eqs. (7.33) through (7.35) with no external forces is

$$w(x,t) = W(x)T(t) \tag{7.36}$$

where $W(x)$ satisfies

$$\frac{d^4 W(x)}{dx^4} - \kappa^4 W(x) = 0 \text{ for } \kappa^4 = \frac{\omega_n^2 \rho_A A}{EI} \text{ and } \omega_n = \kappa^2 \sqrt{\frac{EI}{\rho_A A}} \tag{7.37}$$

with solution

$$\begin{aligned} W(x) = A(\cos \kappa x + \cosh \kappa x) + B(\cos \kappa x - \cosh \kappa x) + \\ C(\sin \kappa x + \sinh \kappa x) + D(\sin \kappa x - \sinh \kappa x) \end{aligned} \tag{7.38}$$

and $T(t)$ satisfies

$$\frac{d^2 T(t)}{dt^2} + \omega^2 T(t) = 0 \tag{7.39}$$

with solution

$$T(t) = \tilde{A} \cos \omega t + \tilde{B} \sin \omega t \tag{7.40}$$

Table 7.13 provides typical modal information for a uniform beam under five kinematic boundary condition configurations. This table illustrates the fact that as the displacement/slope constraints increase so does the modal frequency.

TABLE 7.13 Natural Frequencies and Normal Modes of Uniform Beams

SUPPORTS	MODE n	(A) SHAPE AND NODES (NUMBERS GIVE LOCATION OF NODES IN FRACTION OF LENGTH FROM LEFT END)	(B) BOUNDARY CONDITIONS EQ (7.38)	(C) FREQUENCY EQUATION	(D) CONSTANTS EQ (7.38)	(E) $\kappa\ell$ EQ (7.37)	(F) R RATIO OF NON-ZERO CONSTANTS COLUMN (D)
HINGED-HINGED	1		$x=0\begin{cases}X=0\\X''=0\end{cases}$	SIN $kl=0$	A = 0	3.1416	1.0000
	2	0.50			B = 0	6.283	1.0000
	3	0.333 0.667			$\frac{C}{D}=1$	9.425	1.0000
	4	0.25 0.50 0.75	$x=l\begin{cases}X=0\\X''=0\end{cases}$			12.566	1.0000
	n>4					$\approx n\pi$	1.0000
CLAMPED-CLAMPED	1		$x=0\begin{cases}X=0\\X'=0\end{cases}$	(COS kl) (COSH kl) = 1	A = 0	4.730	-0.9825
	2	0.50			C = 0	7.853	-1.0008
	3	0.359 0.641			$\frac{D}{B}=R$	10.996	-1.0000-
	4	0.278 0.50 0.722	$x=l\begin{cases}X=0\\X'=0\end{cases}$			14.137	-1.0000+
	n>4					$\approx\frac{(2n+1)\pi}{2}$	-1.0000-
CLAMPED-HINGED	1		$x=0\begin{cases}X=0\\X'=0\end{cases}$	TAN kl = TANH kl	A = 0	3.927	-1.0008
	2	0.558			C = 0	7.069	-1.0000+
	3	0.386 0.692			$\frac{D}{B}=R$	10.210	-1.0000
	4	0.294 0.529 0.765	$x=l\begin{cases}X=0\\X''=0\end{cases}$			13.352	-1.0000
	n>4					$\approx\frac{(4n+1)\pi}{4}$	-1.0000
CLAMPED-FREE	1		$x=0\begin{cases}X=0\\X'=0\end{cases}$	(COS kl) (COSH kl) = -1	A = 0	1.875	-0.7341
	2	0.783			C = 0	4.694	-1.0185
	3	0.504 0.868			$\frac{D}{B}=R$	7.855	-0.9992
	4	0.358 0.644 0.906	$x=l\begin{cases}X''=0\\X'''=0\end{cases}$			10.996	-1.0000+
	n>4					$\approx\frac{(2n-1)\pi}{2}$	-1.0000-
FREE-FREE	1		$x=0\begin{cases}X''=0\\X'''=0\end{cases}$	(COS kl) (COSH kl) = 1	B = 0	0 (REPRESENTS TRANSLATION)	
	2	0.224 0.776			D = 0	4.730	-0.9825
	3	0.132 0.50 0.868			$\frac{C}{A}=R$	7.853	-1.0008
	4	0.094 0.356 0.644 0.906	$x=l\begin{cases}X''=0\\X'''=0\end{cases}$			10.996	-1.0000-
	5	0.0734 0.277 0.50 0.723 0.927				14.137	-1.0000+
	n>5					$\approx\frac{(2n-1)\pi}{2}$	-1.0000-

Figure 7.5 displays a uniform, undamped, simply supported beam with an idealized harmonic point load moving at a constant velocity v_0. This example serves to illustrate some subtleties in the assumed mode solution and necessary assumptions for a well-defined problem.

***Example 7.2:* Assumed Mode Solution for a Uniform Simply Supported Beam with a Harmonic Moving Point Load.** For the Euler-Bernoulli beam theory, the

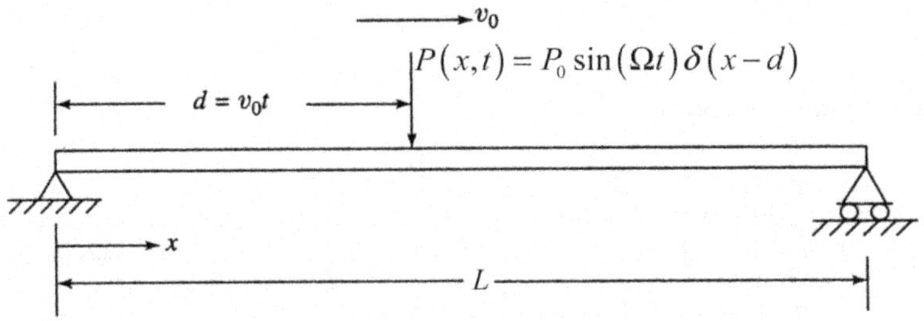

FIGURE 7.5 Simply supported beam with a moving harmonic point load.

equation of motion with kinematic/natural boundary conditions and zero initial conditions is

$$EI_b \frac{\partial^4 w}{\partial x^4} + \rho_A A \frac{\partial^2 w}{\partial t^2} = f(x,t) \text{ for } 0 \le x \le L \text{ and } 0 \le t \le T \text{ and } t_1 = L/v_0 < T \quad (7.41)$$

$$w(0,t) = w(L,t) = 0 \text{ and } EI \frac{\partial^2 w(0,t)}{\partial x^2} = EI \frac{\partial^2 w(L,t)}{\partial x^2} = 0 \quad (7.42)$$

$$w(x,0) = 0 \text{ and } \frac{\partial w}{\partial t}(x,0) = 0 \text{ for } 0 \le x \le L \quad (7.43)$$

Proceeding formally, the assumed modes method solution assumes that

$$w(x,t) = \sum_{i=1}^{\infty} W_i(x)\eta_i(t) \text{ for } 0 \le x \le L \text{ and } 0 \le t \le T \quad (7.44)$$

where simply supported boundary conditions allow the eigenvector form

$$W_i(x) = C_i \sin\left(\frac{i\pi x}{L}\right) \text{ for } 0 \le x \le L \text{ and } C_i \text{ (a normalizing constant)} \quad (7.45)$$

determined from the orthogonality condition

$$\int_0^L \rho_A A W_i(x) W_j(x)\, dx = \begin{cases} 0 \text{ for } i \ne j \\ 1 \text{ for } i = j \end{cases} \text{ and } C_i = \sqrt{2/(\rho_A A L)} \quad (7.46)$$

Separating variables, the ith mode natural frequency of vibration is given by

$$\omega_i = (i\pi)^2 \sqrt{\frac{EI_b}{\rho_A A L^4}} \quad (7.47)$$

Integration of Eq. (7.41) over x for the ith mode leads to the following second-order ordinary differential equation in modal coordinate $\eta_i(t)$, modal force $Q_i(t)$, and corresponding natural frequency ω_i:

$$\frac{d^2\eta_i(t)}{dt^2} + \omega_i^2 \eta_i(t) = Q_i(t) = \int_0^L W_i(x) f(x,t)\, dx \text{ for } i = 1,2,\ldots \quad (7.48)$$

Standard solution for this differential equation is superposition of a homogeneous solution $\eta_h(t)$ and a particular solution $\eta_p(t)$ or expressed in terms of initial conditions and modal force for the ith mode:

$$\eta_i(t) = \cos(\omega_i t)\eta_i(0) + \frac{1}{\omega_i}\sin(\omega_i t)\dot{\eta}_i(0) + \frac{1}{\omega_i}\int_0^t Q_i(\tau)\sin[\omega_i(t-\tau)]\,d\tau \quad (7.49)$$

where from initial conditions

$$\eta_i(0) = \dot{\eta}_i(0) = 0 \quad (7.50)$$

In consideration of the loading imposed in Fig. 7.5, Ref. 48 discusses additional assumptions that need to be made for the problem to be well defined. For the case of $P = P_0$, a constant magnitude point load, after the load has traversed the beam, the beam is in a state of free vibration, with no work having been performed on the beam during the transversal. To resolve this *Timoshenko paradox,* a rolling circular disk of negligible mass must be assumed, with input torque energy equivalent to the energy of the vibrating beam once the load has left the rightmost support (this is also a requirement for the point load to move at constant speed v_0). Structural flexibility may be important but is not considered here. If a finitely distributed load is assumed, complications arise at the supports and any finite distribution must be accounted for in higher vibration modes whose wavelengths may be comparable to the length over which the finite load distribution is defined. The assumption of a time-varying point load is easily accommodated but leads to complex solutions when any of the natural frequencies of the beam ω_i coincide with either the load *transversal frequency* $\omega_v = v_0/L$ or the load harmonic frequency Ω. To simplify the resulting equation, it is assumed that the loading frequencies do not coincide with any of the beam natural frequencies, even though the loading is transient over time $t_1 = L/v_0$. The harmonic point load is initially zero, moves at a constant velocity of v_0, and is only nonzero over a finite time interval. That is,

$$f(x,t) = \begin{cases} P_0 \sin(\Omega t)\delta(x - tv_0) & \text{for } 0 \le t \le L/v_0,\ 0 \le x \le L \\ 0 & \text{for } L/v_0 < t \le T,\ 0 \le x \le \text{L} \end{cases} \quad (7.51)$$

Once the moving load has traversed the beam $0 \le t \le t_1$, the beam enters free vibration for $t_1 < t \le T$, so the general solution must include a transient load time interval along with a free vibration time interval.

Determination of the modal force $Q_i(t)$ for $0 \le t \le t_1$ for the moving harmonic load could employ either a delta function approach or, as will be illustrated here, a Fourier series approach (Ref. 8). First, the point load at $d = tv_0$ for $0 \le t \le t_1$ is expressed as follows:

$$\tilde{f}(x) = \begin{cases} 0 & 0 \le x < d - \Delta x \\ \tilde{f} & d - \Delta x \le x \le d + x \\ 0 & d + \Delta x < x \le L \end{cases} \quad (7.52)$$

with Fourier expansion

$$\tilde{f}(x) = \sum_{j=1}^{\infty} a_j \sin\left(\frac{j\pi x}{L}\right) \text{ where } a_j = \frac{2}{L}\int_0^L f(x)\sin\left(\frac{j\pi x}{L}\right)dx \quad (7.53)$$

that reduces to

$$a_j \approx \int_{d-\Delta x}^{d+\Delta x} \tilde{f}(x)\sin\left(\frac{j\pi x}{L}\right)dx = \frac{2P_0\sin(\Omega t)}{L}\sin\left(\frac{j\pi d}{L}\right)\frac{\sin(j\pi\Delta x/L)}{j\pi\Delta x/L} \quad (7.54)$$

As Δx approaches zero, a_j becomes

$$a_j = \frac{2P_0\sin(\Omega t)}{L}\sin\left(\frac{j\pi d}{L}\right) \quad (7.55)$$

and

$$f(x,t) = \frac{2P_0 \sin(\Omega t)}{L} \sum_{j=1}^{\infty} \sin\left(\frac{j\pi v_0 t}{L}\right) \sin\left(\frac{j\pi x}{L}\right) \tag{7.56}$$

From Eq. (7.48), the ith modal force becomes

$$Q_i(t) = \begin{cases} \int_0^L W_i(x) f(x,t)\, dx = C_i P_0 \sin(\Omega t) \sin\left(\frac{i\pi v_0 t}{L}\right) & \text{for } 0 \le t \le t_1 \\ 0 & \text{for } t_1 < t \le T \end{cases} \tag{7.57}$$

and substituting this force into Eq. (7.49) defines the ith modal coordinate:

$$\eta_i(t) = \frac{1}{\omega_i} \int_0^t C_i P_0 \sin(\Omega \tau) \sin(i\pi\omega_v \tau) \sin[\omega_i(t-\tau)]\, d\tau \text{ for } 0 \le t \le t_1 \tag{7.58}$$

An explicit analytical expression for $\eta_i(t)$ can be obtained by tedious application of trigonometric identities and integration. Evaluating $\eta_i(t_1)$ and $\dot{\eta}_i(t_1)$ provides the beam free vibration initial conditions that apply once the load passes the rightmost support. The final solution is given as

$$w(x,t) = \begin{cases} \sum_{i=1}^{\infty} \sqrt{\frac{2}{\rho_A AL}} \sin\left(\frac{i\pi x}{L}\right) \eta_i(t) \text{ for } 0 \le t \le t_1 \text{ and } 0 \le x \le L \\ \sum_{i=1}^{\infty} \sin\left(\frac{i\pi x}{L}\right) \left[\cos(\omega_i t)\eta_i(t_1) + \frac{1}{\omega_i} \sin(\omega_i t)\dot{\eta}_i(t_1)\right] \text{ for } t_1 \le t \le T \text{ and } 0 \le x \le L \end{cases} \tag{7.59}$$

Varying v_0 and Ω, and examining the resulting beam displacement/velocity/bending moment in time leads to complex results for evaluation.

Continuous Beams. Continuous beams—beams that have multiple supports along the axis of the beam—present no technical problems for solution using the Euler-Bernoulli theory and can be extended to Timoshenko beams.[49,50] Figure 7.6 displays a continuous beam with $n - 1$ segments and n supports.

Solution for the natural frequencies and mode shapes proceeds by considering simultaneous solution of $n - 1$ beam equations for $2n$ deflection/slope boundary conditions. A nontrivial solution for the algebraic set of equations leads to the frequency equation having a determinant of order $4(n - 1)$. Beams with varying properties between supports can be easily incorporated in the formulation. Table 7.14 provides algorithmic equations for establishing the frequency equation and determining the mode shapes. The continuous beam solution procedure is related to the transfer matrix method for complex structures.[34]

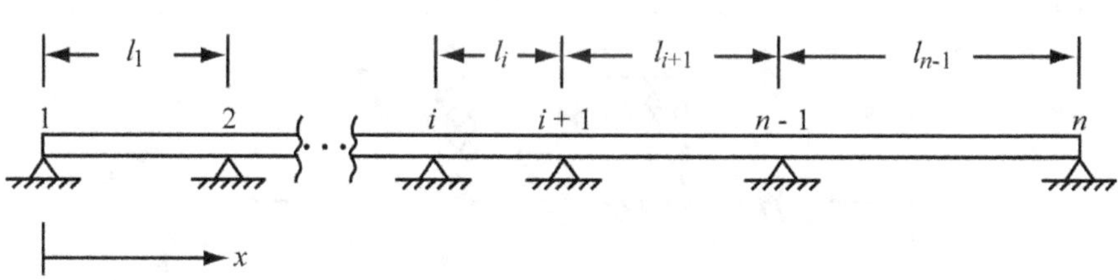

FIGURE 7.6 Continuous beam with $n - 1$ segments.

TABLE 7.14 Continuous Beam Relationships

Theory: Euler-Bernoulli

Solution: $w_i(x) = A_i \cos \kappa_i x + B_i \sin \kappa_i x + C_i \cosh \kappa_i x + D_i \sinh \kappa_i x$

$$\text{for } \kappa_i = \left(\frac{\rho_i A_i \omega^2}{E_i I_i}\right)^{1/4} \qquad i = 1,2,\ldots,n-1$$

Boundary conditions (at ends $i = 1$ and $i = n$):

$$\text{Moment/slope: } EI\frac{\partial^2 w}{\partial x^2}\left(\frac{\partial w}{\partial x}\right) \text{ and Shear/displacement: } \frac{\partial}{\partial x}\left(EI\frac{\partial^2 w}{\partial x^2}\right)w$$

Continuity conditions (at supports):

$$\text{Moment: } E_{i-1}I_{i-1}\left.\frac{\partial^2 w}{\partial x^2}\right|_{x=l_{i-1}} = E_i I_i \left.\frac{\partial^2 w}{\partial x^2}\right|_{x=l_i} \qquad \text{Slope: } \left.\frac{\partial w}{\partial x}\right|_{x=l_{i-1}} = \left.\frac{\partial w}{\partial x}\right|_{x=li} \qquad i = 2,3,\ldots,n-1$$

Auxiliary condition:

Support displacement: $w\big|_{x=l_i} = 0$ for $i = 2,3,\ldots,n-1$

Conditions provide $4(n-1)$ homogeneous algebraic equations in A_i, B_i, C_i, and D_i from which the determinant of the equations must be zero leading to a transcendental equation for an infinite number of modal frequencies ω.

Parameters (for the ith section):

ρ_i mass density	E_i modulus of elasticity
A_i beam cross-section area	I_i bending moment of inertia
ω frequency (rad/sec)	

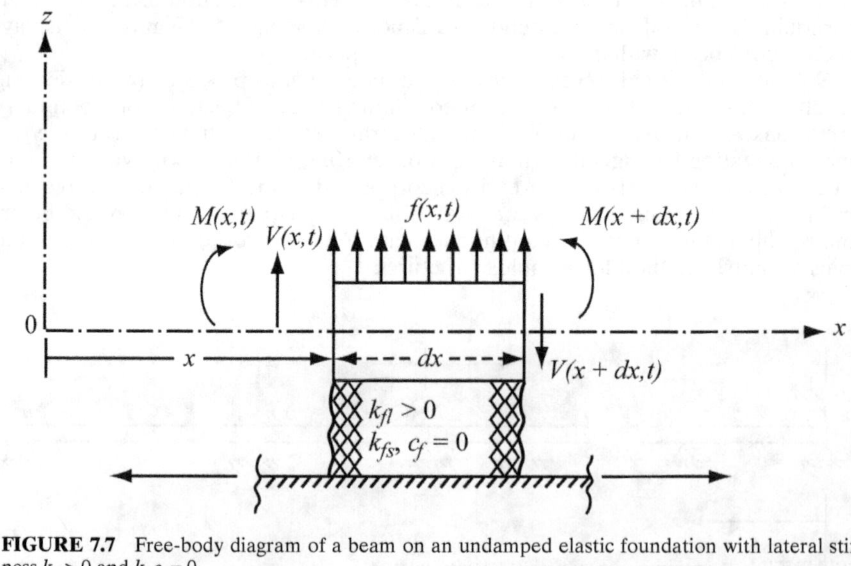

FIGURE 7.7 Free-body diagram of a beam on an undamped elastic foundation with lateral stiffness $k_{fl} > 0$ and $k_{fs}c_f = 0$

Beams on Foundations. Figure 7.7 displays an infinitesimal section of a beam on an elastic foundation that is one case of a *generalized* foundation. A generalized foundation is characterized by a foundation *pressure* and *moment* at each point along the beam. This can be expressed simply as

$$\begin{aligned} &\text{Pressure: } p(x) = k_{fl}w(x) \\ &\text{Moment: } m(x) = k_n \frac{dw(x)}{dn} \\ &(n \text{ normal to the beam axis}) \end{aligned} \tag{7.60}$$

There are at least seven major foundation configurations that modify the dynamics of an elastic beam that rests on any one of them.[47] For free vibration of a uniform Euler-Bernoulli beam on an elastic foundation with foundation modulus k_f, the equation of motion is written as

$$\underbrace{EI_b \frac{\partial^4 w}{\partial x^4}}_{\text{bending stiffness}} + \underbrace{\rho_A A \frac{\partial^2 w}{\partial t^2}}_{\text{transverse mass inertia force}} + \underbrace{k_{fl} w}_{\text{foundation force}} = 0 \tag{7.61}$$

Relying upon separation of variables, the following solution is provided for a simply supported set of end conditions:

$$w(x,t) = \sum_{i=1}^{\infty} C_i \sin \kappa_i x \tag{7.62}$$

and the natural frequencies are given by

$$\omega_i = \frac{i^2\pi^2}{l^2} \sqrt{\frac{EI_b}{\rho_A A}} \sqrt{1 + \frac{k_{fl} l^4}{EI_b i^4 \pi^4}} \quad \text{for } i = 1,2,3,\ldots \tag{7.63}$$

Addition of the elastic foundation for $k_{fl} > 0$ generally increases the natural frequencies of a beam, causing the beam to be stiffer in bending. For the Timoshenko beam, both the shear and the rotary inertia effects should result in higher natural frequencies. If *foundation mass* is included in the problem formulation, then the problem resembles one of a *composite* beam. This problem can be generalized to consideration of a beam on an elastic foundation subject to a moving load.[8]

TIMOSHENKO BEAM THEORY

For an unloaded uniform simply supported Timoshenko beam, the following equation for natural frequencies can be easily derived.[8]

$$\omega_n^4 \frac{\rho_v r^2}{kG} - \omega_n^2 \left(1 + \frac{n^2\pi^2 r^2}{l^2} + \frac{n^2\pi^2 r^2}{l^2} \frac{E}{kG}\right) + \frac{\alpha^2 n^4 \pi^4}{l^4} = 0 \tag{7.64}$$

For Rayleigh's theory, Eq. (7.64) reduces to

$$-\omega_n^2 \left(1 + \frac{n^2\pi^2 r^2}{l^2}\right) + \frac{\alpha^2 n^4 \pi^4}{l^4} = 0 \tag{7.65}$$

and for the Euler-Bernoulli theory

$$-\omega_n^2 + \frac{\alpha^2 n^4 \pi^4}{l^4} = 0 \tag{7.66}$$

TABLE 7.15 Normalized Natural Frequencies of Vibration of a Simply Supported Rectangular Steel Beam Under Three Theories (Frequency Normalization Factor of 703.0149, rad/s) (Table adapted from Ref. 8)

n	Normalized natural frequency (rad/s)			
	Euler-Bernoulli	Rayleigh	Timoshenko	
			Bending	Shear
1	1.0000	0.9909	0.9643	31.6623
2	4.0000	3.8597	3.5182	34.7119
3	9.0000	8.3328	7.0383	39.0408

$l = 39$ in (1 m), $w = 1.95$ in (0.05 m), $t = 5.85$ in (0.15 m)

$E = 30 \times 10^6$ lb/in² (207 × 10⁹ Pa), $G = 12 \times 10^6$ lb/in² (79.3 × 10⁹ Pa),

$\rho_V = 489$ lbf/ft³ (76.5 × 10³ N/m³), $k = 5/6$ (shear correction factor)

Table 7.15 provides natural frequency estimates normalized to the fundamental mode for a particular configuration of steel beam. Since Eq. (7.64) is fourth order, two extra roots in the characteristic equation provide an estimate of the shear vibration frequency that is substantially higher than the bending vibration. Plane section rotary inertia and shear representing higher-order modeling of internal material constraints *lower* the natural bending modal frequency of the beam.

SELECTED REFERENCE INFORMATION

Literature on beam vibration is extensive. Eight modifications to the three beam theories presented here are described in Ref. 47. Thin-walled beams under combination loading are examined in Ref. 51. Reference 52 examines beam plane section warping flexibility and its effects on beam stiffness. A combination of flexural and torsional vibrations is provided for a uniform spinning beam in Ref. 53. Fundamental frequency estimates can be made for beams with polynomial form pressure on an elastic foundation by Rayleigh's method.[54] Coupling in two perpendicular beam bending axes, along with torsion, is considered in Ref. 55. The power of HVP is illustrated for a very general pretwisted Timoshenko beam configuration with time-dependent boundary conditions in Ref. 56. Stationary stochastic loading is applied to a Timoshenko thin-walled beam with flexure and torsion coupling in Ref. 57. Reference 58 considers a general-configuration Euler-Bernoulli beam traversed by a time-varying concentrated force. Knowing the vibration modes of a beam is of use in determining both the modulus of elasticity and the shear modulus experimentally.[59] Reference 60 illustrates use of continuous structure mechanics and Eringen's nonlocal constitutive relationship in nanotechnology as in Ref. 1. A cantilever Timoshenko beam with a rigid mass tip displaying flexural-torsional coupled vibration is analyzed in Ref. 61. The differential quadrature method is used to solve nonlinear equations in Ref. 62, and Ref. 63 provides an example of extended application in consideration of a combination of parametric excitation, a viscoelastic foundation with random parameters, and a moving load. In Ref. 64, a Timoshenko column is considered with a compressive follower load at the ends, and this paper discusses a configuration that has no variational formulation. Reference 65 provides information on a variety of beam vibration and buckling configurations. Reference 66 provides MATLAB code for solving a number of thin beam vibration problems based upon the Euler-Bernoulli formulation.

PLATES (TRANSVERSE VIBRATION)

Thin plates are characterized by two-dimensional in-plane stretching and out-of-plane bending. In addition to these two sources of energy, thick plates may demonstrate the effects of transverse shear and rotary inertia. For a thin plate and rectangular coordinate system, Fig. 7.8*A* displays in-plane and shear forces resulting from external load $f(x,y,t)$, while Fig. 7.8*B* displays the moments.

Classical plate theory considers only plate-bending energy, modification to classical plate theory includes in-plane forces, and the Mindlin theory accounts for shear and rotary inertia for thick plates. Generally, the iterated second-order Poisson operator (i.e., ∇^4) governs plate behavior. This operator is easily expressed in a num-

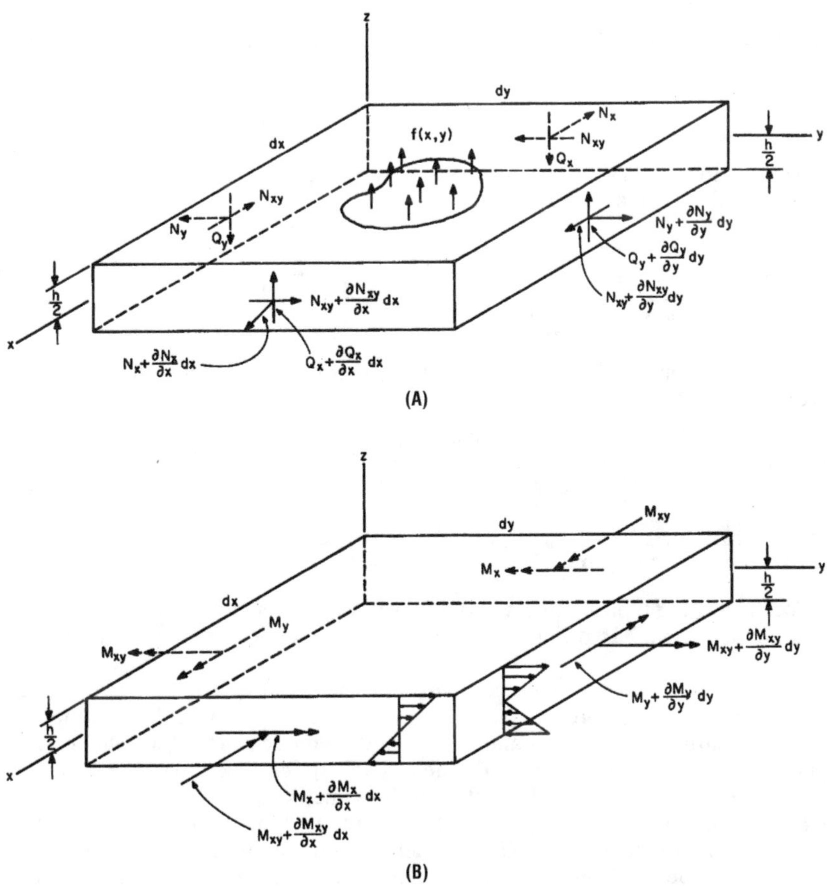

FIGURE 7.8 Forces and moments (intensities) acting on a plate element: (*A*) normal and shear forces with distributed load; (*B*) bending moments.

ber of coordinate systems for determination of governing equations for other plate shapes. For example, $\nabla^4_{r,\theta,z}$ governs behavior of circular plates whose geometry is in terms of a radial coordinate r, an orthogonal angular coordinate θ, and the z-coordinate normal to the plate surface. Transformations in terms of skew and elliptical coordinate systems are available.[67]

Equations (7.67) through (7.71) provide HVP formulation for classical plate theory in rectangular coordinates. Elastic energy in bending is given as

$$\pi = \frac{E}{2(1-\nu^2)} \iint_A \left[\left(\frac{\partial^2 w}{\partial x^2}\right)^2 + \left(\frac{\partial^2 w}{\partial y^2}\right)^2 + 2\nu \frac{\partial^2 w}{\partial x^2}\frac{\partial^2 w}{\partial y^2} + 2(1-\nu)\left(\frac{\partial^2 w}{\partial x \partial y}\right)^2 \right] dA \int_{z=-h/2}^{z=h/2} z^2 dz \tag{7.67}$$

Corresponding plate kinetic energy is simply

$$T = \frac{\rho h}{2} \iint_A \left(\frac{\partial w}{\partial t}\right)^2 dA \tag{7.68}$$

while work performed by force f, perpendicular to the plate surface, is given as

$$W = \iint_A f w dA \tag{7.69}$$

Application of HVP and integration by parts yields Lagrange's equation of motion for the loaded plate:

$$\underbrace{D\nabla^4 w}_{\text{bending effect}} + \underbrace{\rho h \frac{\partial^2 w}{\partial t^2}}_{\text{transverse mass inertia}} = \underbrace{f}_{\text{external transverse force}} \quad \text{for } \nabla^4 = \frac{\partial^4}{\partial x^4} + 2\frac{\partial^4}{\partial x^2 \partial y^2} + \frac{\partial^4}{\partial y^4} = \nabla^2(\nabla^2) \tag{7.70}$$

The corresponding boundary conditions in edge moment and shear are due to Kirchoff.

$$M_x = -D\left(\frac{\partial^2 w}{\partial x^2} + \nu\frac{\partial^2 w}{\partial y^2}\right), M_y = -D\left(\frac{\partial^2 w}{\partial y^2} + \nu\frac{\partial^2 w}{\partial x^2}\right), M_{xy} = M_{yx} = -(1-\nu)\frac{\partial^2 w}{\partial x \partial y} \tag{7.71}$$

$$V = Q_x + \frac{\partial M}{\partial y} = -D\frac{\partial}{\partial x}\left(\frac{\partial^2 w}{\partial x^2} + (2-\nu)\frac{\partial^2 w}{\partial y^2}\right), V = Q_y + \frac{\partial M}{\partial x} = -D\frac{\partial}{\partial y}\left(\frac{\partial^2 w}{\partial y^2} + (2-\nu)\frac{\partial^2 w}{\partial x^2}\right)$$

For illustrating typical effects of boundary conditions on natural frequencies and mode shapes, Table 7.16 provides natural frequency and visual nodal lines for square plates.

Textbooks exhaust the plate configurations that can be solved by separation of variables; however, researchers using advanced methods do provide useful tables for estimating natural frequencies and corresponding mode shapes for many plate configurations. Once a mode shape is defined as an explicit function of the spatial coordinates, then internal plate physical quantities such as stress, strain, moments, and shears can be approximated for the selected mode. Unfortunately, two limitations affect accuracy. First, most practical problems require an infinite-series form of solution (modal superposition) for which it may be difficult to decide where to truncate for practical results. Second, internal plate physical quantities require derivatives of the mode shape functions that are quite sensitive to the form of the mode shape.

TABLE 7.16 Natural Frequencies and Nodal Lines of Square Plates with Various Edge Conditions (Adapted from Ref. 67)

	1ST MODE	2ND MODE	3RD MODE	4TH MODE	5TH MODE	6TH MODE
$\omega_n a^2 \sqrt{\rho/D}$	3.494	8.547	21.44	27.46	31.17	
NODAL LINES	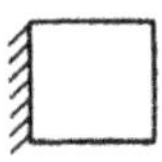	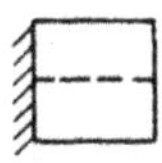	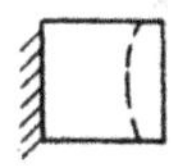	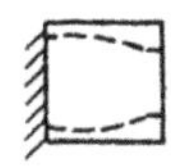	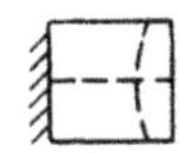	
$\omega_n a^2 \sqrt{\rho/D}$	35.99	73.41	108.27	131.64	132.25	165.15
NODAL LINES	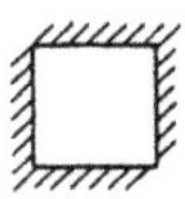	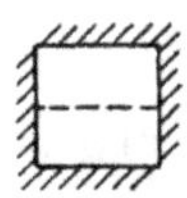	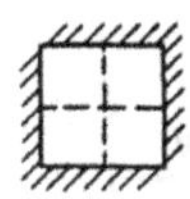	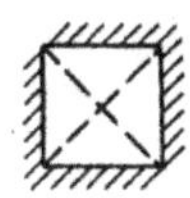	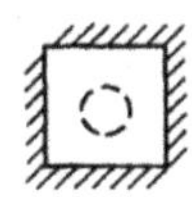	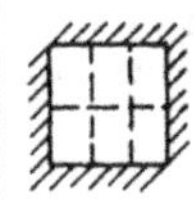
$\omega_n a^2 \sqrt{\rho/D}$	6.958	24.08	26.80	48.05	63.14	
NODAL LINES	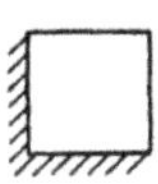	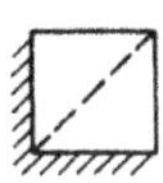	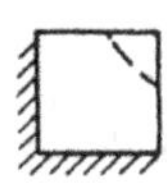	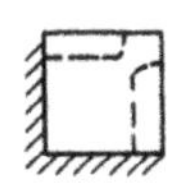		

$\omega_n = 2\pi f_n$ $\qquad$ h = Plate thickness
$D = Eh^3/12(1 - \nu^2)$ $\qquad$ α = Plate length
ρ = Mass density

Table 7.17 summarizes useful plate vibration configuration information from Ref. 67, and this reference provides the basis of solution for the modal frequencies (e.g., two-term Galerkin series) and the mode shape explicitly in terms of spatial coordinates. Generally, the extensive list of references contained in Ref. 67 should be consulted to fully understand the nature of the approximation and solution. It is well to note that Ref. 67 is nearly 40 years old, and a number of vibration configurations for structures have been added to the literature during the intervening years.

Solution of plate vibration under general anisotropic conditions is very difficult, but results are possible for both rectangular and polar orthotropy using the iterated Poisson operator. For rectangular orthotropy such as plate stiffeners in an orthogonal grid, the fourth-order bending operator is defined as follows:

$$\nabla_D^4 w = \underbrace{D_x \frac{\partial^4 w}{\partial x^4} + 2D_{xy} \frac{\partial^4 w}{\partial x^2 \partial y^2} + D_y \frac{\partial^4 w}{\partial y^4}}_{\text{orthotropic bending}} \tag{7.72}$$

for D_x, D_{xy}, and D_y constant coefficients

Equation (7.72) can be substituted into any governing differential equation that contains the iterated fourth-order Poisson operator alone, such as a plate on a uniform elastic foundation.

TABLE 7.17 Plate Vibration Configurations Providing Modal Frequencies and Corresponding Mode Shapes (from Reference 67)

Plate configuration/properties	C	S	F	V	E	D	P	M	I*
Isotropic plates									
Circular plate	x	x	x	x	x			x	
Annular plate	x	x	x	x				x	
Elliptical plate	x		x						
Rectangular	x	x	x	x	x	x	x	x	x
Parallelogram	x	x	x	x				x	
Trapezoidal		x		x					
Triangular	x	x	x	x					
Polygonal			x						
Sectorial	x			x					
Irregular				x					
Anisotropic (orthotropic) plates									
Circular (polar orthotropy)	x	x							
Annular (polar orthotropy)			x						
Rectangular (rectangular orthotropy)	x	x		x					
Circular (rectangular orthotropy)	x								
Elliptical (rectangular orthotropy)	x								
Inplane forces									
Circular	x	x	x	x					
Rectangular	x	x		x					
Polygonal	x								
Triangular	x		x						
Plates with variable thickness									
Circular	x				x				
Annular	x			x					
Rectangular	x	x		x					
Miscellaneous considerations									
Sparse results are provided for plates interacting with surrounding media, plates undergoing large deflections, thick plates with shear deformation and rotary inertia, plates with nonhomogeneous properties									

* Boundary conditions: C(CCCC)—clamped, S(SSSS)—simply supported, F(FFFF)—free, V—varied conditions on boundary, E—elastic boundary support, D—discontinuous support, P—point support; M—added mass; I—internal cutouts.

Dynamic behavior of a plate on a uniform elastic foundation (constant stiffness coefficient k_p) and a plate with external in-plane forces (N_x, N_y, and N_{xy}) contributing to plate bending energy is described by Eqs. (7.73) and (7.74), respectively.

$$\underbrace{D\nabla^4 w}_{\text{bending}} + \underbrace{\rho h \frac{\partial^2 w}{\partial t^2}}_{\text{transverse mass inertia}} = \underbrace{-\,k_p w}_{\text{elastic foundation}} \tag{7.73}$$

$$\underbrace{D\nabla^4 w}_{\text{bending}} + \underbrace{\rho h \frac{\partial^2 w}{\partial t^2}}_{\text{transverse mass inertia}} = \underbrace{N_x \frac{\partial^2 w}{\partial x^2} + 2N_{xy}\frac{\partial^2 w}{\partial x \partial y} + N_y \frac{\partial^2 w}{\partial y^2}}_{\text{in-plane forces}} \tag{7.74}$$

Equation (7.74) for the Mindlin plate with rotary inertia and shear requires spatial derivatives of the applied force f, and substantially complicates the fourth-order operator form of governing equation of motion.

$$\underbrace{\left(\nabla^2 - \frac{\rho}{k^2G}\frac{\partial^2}{\partial t^2}\right)}_{\text{shear}}\underbrace{\left(D\nabla^2 - \frac{\rho h^3}{12}\frac{\partial^2}{\partial t^2}\right)}_{\text{rotary inertia}}w + \underbrace{\rho h\frac{\partial^2 w}{\partial t^2}}_{\text{transverse mass inertia}} = \underbrace{\left(1 - \frac{D}{k^2Gh}\nabla^2 + \frac{\rho h^2}{12k^2G}\frac{\partial^2}{\partial t^2}\right)f}_{\text{externally applied forces}} \quad (7.75)$$

A simple isotropic plate with variable thickness, Eq. (7.68), for $h = h(x,y)$, must take account of the flexible rigidity that also becomes a function of x and y, that is, $D = D(x,y)$ and derivatives of D.

$$\nabla^2(D\nabla^2 w) - (1-\nu)\left(\frac{\partial^2 D}{\partial y^2}\frac{\partial^2 w}{\partial x^2} - 2\frac{\partial^2 w}{\partial x\partial y}\frac{\partial^2 w}{\partial x\partial y} + \frac{\partial^2 D}{\partial x^2}\frac{\partial^2 w}{\partial y^2}\right) + \rho h\frac{\partial^2 w}{\partial t^2} = 0 \quad (7.76)$$

SELECTED REFERENCE INFORMATION

Plate vibration literature is prolific. Reference 68 is useful for a general introduction with discussion on plate loading. Mindlin plate theory related to both rotary inertia and transverse shear is considered in Refs. 69 and 70. Reference 71 applies a polynomial approximation method for solution of a broad variety of Mindlin plates.

OTHER STRUCTURES

There exist four important structures whose development in the area of structural vibration can only be referenced.

Strings (Cables). Strings represent one-, two-, and three-dimensional structures with intractability of equation solution increasing with dimension. Reference 3 provides two-dimensional equations for a string from equilibrium considerations. Substantial application in strings research is in the area of cables and transmission line conductors, reflected in Refs. 72, 73, and 74.

Curved Beams. When the cross-section centerline of a rectilinear beam becomes either a two- or three-dimensional curve, solution is difficult. For the most general case, the curvature expressions for a line in space from elementary differential geometry must be used to establish basic kinematic relationships. Moreover, if external loading of the curved beam is not through the shear center, the kinematic relationships become very complex, requiring simultaneous solution of more than one high-order partial differential equation.[75] Reference 47 provides a practical guide to vibration of a number of arch configurations. The use of a *director approach* has led to insight into dynamics of complex configurations.[76] Elastically coupled concentric rings have been investigated both analytically and experimentally.[77] The effect of simple geometry change for sinusoidal, parabolic, and elliptic-shaped arches is examined in Ref. 78.

Membranes. Membrane structures are characterized by an in-plane force field resisting applied external forces perpendicular to the in-plane force field and are analogous to the one-dimensional string structure with tension force. Membrane structures are important for (1) providing modal information on submembranes delineated by the modal lines of a larger membrane and (2) the Prandtl analogy that provides information on torsion of noncircular cross-section bars. Reference 79 considers vibration of membranes that are nonhomogeneous in density and thickness. Rectangular membranes subject to shear stress and nonuniform tensile stresses are considered in Ref. 80. An integral equation formulation demonstrated in Ref. 81 tends to be more efficient than a variational approach when discontinuous coefficients arise in differential equations as a result of a stepped radial density. One useful area for membrane analysis is the modeling of cable nets.[82]

Shells. Shells—particularly thin shells—represent three-dimensional structures with a well-defined two-dimensional surface having kinematic structure governed by the second fundamental form for surfaces in differential geometry.[83] Because shells carry external loads by virtue of their geometry, which allows for dispersion of stress, vibrations of shells have not been studied as extensively as some other aspects of shell behavior such as stability (buckling) under static or even dynamic load. However, since shells can be effective radiators of acoustic energy, vibration of shells relative to acoustic emission has been important in many industries, including the automotive and aircraft industries. Early shell considerations[84,85] and more recent advanced shell considerations[86,87] have laid the foundation for in-depth understanding of shell vibration.

Shell vibration theory is well documented for simple configurations where the radius of curvature of the shell is constant in space such as cylindrical or spherical shells or, at most, constant varying in one dimension such as the conical shell. Since shell local kinematic information is a function of the second fundamental form in the differential geometry of surfaces, any variation of this fundamental form as a function of coordinates of a global coordinate system complicates solution of the equations of motion immensely because the coefficients of the equation operators become functions of the local coordinates. This is analogous to the curved beam in one dimension, where the local form is the curvature at the point along the axis of the beam. References on the behavior of shells under static loads abound. Reference 88 provides information on the vibration of circular and noncircular cylindrical shells in addition to conical and spherical shells and shells of revolution. The appendix to this reference provides solution of the three-dimensional equations of motion for cylinders. References 89 and 90 provide examples of solution techniques for vibration of a variety of shell configurations. Mechanical and thermal response of thick spherical and cylindrical shells by a generalized Fourier transform method is given in Ref. 91. The effects of axial stress for a thick cylindrical shell are considered in Ref. 92, while regular polygonal prismatic shells are considered from the point of view of beam and plate vibration in Ref. 93. In Ref. 94, not only the vibration dynamics but also the control of shell structures is discussed. Reference 95 considers both free and forced vibration on shallow shells.

REFERENCES

1. Wang, C. M., V.B.C. Tan, and Y.Y. Zhang: "Timoshenko Beam Model for Vibration Analysis of Multi-walled Carbon Nanotubes," *Journal of Sound and Vibration,* **294** (2006).
2. Lai, W. Michael, David Rubin, and Erhard Krempl: "Introduction to Continuum Mechanics," 3d ed., Butterworth-Heinemann Ltd., Oxford, 1993.
3. Weinberger, H. F.: "A First Course in Partial Differential Equations with Complex Variables and Transform Methods," Blaisdell Publishing Company, New York, 1965.

4. Skudrzyk, Eugen: "Simple and Complex Vibratory Systems," The Pennsylvania State University Press, University Park, Pa., 1968.
5. Lanczos, Cornelius: "The Variational Principles of Mechanics," 3d ed., University of Toronto Press, Toronto, 1966.
6. Reddy, J. N.: "Applied Functional Analysis and Variational Methods in Engineering," McGraw-Hill, New York, 1986.
7. Fung, Y. C.: "Foundations of Solid Mechanics," Prentice-Hall, Englewood Cliffs, N.J., 1965.
8. Rao, Singiresu S.: "Vibration of Continuous Systems," John Wiley & Sons, Hoboken, N.J., 2007.
9. Langhaar, Henry L.: "Energy Methods in Applied Mechanics," John Wiley and Sons, Inc., New York, 1962.
10. Gould, S. H.: "Variational Methods for Eigenvalue Problems," 2d ed., Oxford University Press, London, 1966.
11. Meirovitch, Leonard: "Computational Methods in Structural Dynamics," Sijthoff & Noordhoff International Publishers B.V., Alphen aan den Rijn, The Netherlands, 1980.
12. Gladwell, G. M. L., and G. Zimmermann: "On Energy and Complementary Energy Formulations of Acoustic and Structural Vibration Problems," *Journal of Sound and Vibration,* **3**(3) (1966).
13. de Silva, Clarence W.: "Vibration Fundamentals and Practice," CRC Press, Boca Raton, Fla., 2000.
14. Jones, David I. G.: "Handbook of Viscoelastic Vibration Damping," John Wiley & Sons, Baffins Lane, England, 2001.
15. Lazan, Benjamin J.: "Damping of Materials and Members in Structural Mechanics," 1st ed., Pergamon Press, New York, 1968.
16. Bolotin, V. V.: "The Dynamic Stability of Elastic Systems," Holden-Day, San Francisco, 1964.
17. Elishakoff, Isaac: "Probabilistic Methods in the Theory of Structures," John Wiley & Sons, New York, 1983.
18. Bendat, Julius S., and Allan G. Piersol: "Engineering Applications of Correlation and Spectral Analysis," 2d ed., John Wiley & Sons, New York, 1993.
19. Polyanin, Andrei D., and Alexander V. Manzhirov: "Handbook of Mathematics for Engineers and Scientists," Chapman & Hall/CRC, Boca Raton, Fla., 2007.
20. Polyanin, Andrei D.: "Handbook of Linear Partial Differential Equations for Engineers and Scientists," Chapman & Hall/CRC, Boca Raton, 2002.
21. Polyanin, Andrei D., and Valentin F. Zaitsev: "Handbook of Nonlinear Partial Differential Equations," Chapman & Hall/CRC, Boca Raton, 2004.
22. Temple, G., and W. G. Bickley: "Rayleigh's Principle and Its Applications to Engineering," Dover Publications, New York, 1956.
23. Roseau, Maurice: "Vibrations in Mechanical Systems: Analytical Methods and Applications," Springer-Verlag, Berlin, 1987.
24. Bogdanovich, Alexander: "Non-linear Dynamic Problems for Composite Cylindrical Shells," Elsevier Applied Science Publishers, Essex, England, 1993.
25. Clough, Ray W., and Joseph Penzien: "Dynamics of Structures," McGraw-Hill, New York, 1975.
26. Levy, C.: "An Iterative Technique Based on the Dunkerley Method for Determining the Natural Frequencies of Vibrating Systems," *Journal of Sound and Vibration,* **150**(1):111–118 (1991).
27. Laura, P. A. A., R. H. Gutierrez, R. Carnicer, and H. C. Sanzi: "Free Vibrations of a Solid Circular Plate of Linearly Varying Thickness and Attached to a Winkler Foundation," *Journal of Sound and Vibration,* **144**(1):149–161 (1991).

28. Fertis, Demeter G.: "Mechanical and Structural Vibrations," John Wiley & Sons, New York, 1995.

29. Dimarogonas, Andrew D., and Sam Haddad: "Vibration for Engineers," Prentice-Hall, Englewood Cliffs, N.J., 1992.

30. Chen Goong and Jianxin Zhou: "Vibration and Damping in Distributed Systems, Volume II: WKB and Wave Methods, Visualization and Experimentation," CRC Press, Boca Raton, Fla., 1993.

31. Kythe, Prem K., Pratap Puri, and Michael R. Schaferkotter: "Partial Differential Equations and Boundary Value Problems with MATHEMATICA," 2d ed., Chapman & Hall/CRC, Boca Raton, Fla., 2003.

32. Zhong, Hongzhi, and Minmao Liao: "Higher-order Nonlinear Vibration Analysis of Timoshemko Beams by the Spline-based Differential Quadrature Method," *Shock and Vibration,* **14** (2007).

33. Morales, C. A., and R. Goncalves: "Eigenfunction Convergence of the Rayleigh-Ritz-Meirovitch Method and the FEM," *Shock and Vibration,* **14** (2007).

34. Pestel, E. C., and F. A. Leckie: "Matrix Methods in Elastomechanics," McGraw-Hill, New York, 1963.

35. Fryba, Ladislav: "Vibration of Solids and Structures Under Moving Loads," Noordhoff International Publishing, Groningen, The Netherlands, 1972.†

36. Wilson, Howard B., Louis H. Turcotte, and David Halpern: "Advanced Mathematics and Mechanics Applications Using MATLAB," 3d ed., Chapman & Hall/CRC, Boca Raton, Fla., 2003.

37. Cooper, Jeffery: "Introduction to Partial Differential Equations with MATLAB," Birkhauser, Boston, 1998.

38. Schiesser, William E.: "Computational Mathematics in Engineering and Applied Science: ODEs, DAEs and PDE's," CRC Press, Boca Raton, Fla., 1994.

39. Chopra, Anil K.: "Dynamics of Structures: Theory and Applications to Earthquake Engineering," Prentice Hall, Upper Saddle River, N.J., 2001.

40. Li, Q. S.: "Exact Solutions for Free Longitudinal Vibrations of Non-Uniform Rods," *Journal of Sound and Vibration,* **234**(1) (2000).

41. Raj, Anil, and R. I. Sujith: "Closed-form Solutions for the Free Longitudinal Vibration of Inhomogeneous Rods," *Journal of Sound and Vibration,* **283** (2005).

42. Matsuda, H., T. Sakiyama, C. Morita, and M. Kawakami: "Longitudinal Impulsive Response Analysis of Variable Cross-Section Bars," *Journal of Sound and Vibration,* **181**(3) (1995).

43. Kukla, S., J. Przybylski, and L. Tomski: "Longitudinal Vibration of Rods Coupled by Translational Springs," *Journal of Sound and Vibration,* **185**(4) (1995).

44. Lo, Hsu, and Madeline Goulard: "Torsion with Warping Restraint from Hamilton's Principle," *Proc. Second Midwestern Conference on Solid Mechanics,* Research Series No. 129, Engineering Experiment Station, Purdue University, Lafayette, Ind., 1955.

45. Ran, L., R. Yacamini, and K. S. Smith: "Torsional Vibrations in Electrical Induction Motor Drives During Start-up," *Journal of Vibration and Acoustics,* **118** (1996).

46. Shen, I. Y., Weili Guo, and Y. C. Pao: "Torsional Vibration Control of a Shaft Through Active Constrained Layer Damping Treatments," *Journal of Vibration and Acoustics,* **119** (1997).

47. Karnovsky, Igor A., and Olga I. Lebed: "Non-Classical Vibrations of Arches and Beams: Eigenvalues and Eigenfunctions," McGraw-Hill, New York, 2004.

48. Olsson, M.: "On the Fundamental Moving Load Problem," *Journal of Sound and Vibration,* **145**(2) (1991).

49. Wang, T. M.: "Natural Frequencies of Continuous Timoshenko Beams," *Journal of Sound and Vibration,* **13**(4) (1970).

50. Wang, R.-T.: "Vibration of Multi-Span Timoshenko Beams to a Moving Force," *Journal of Sound and Vibration,* **207**(5) (1997).

51. Li, Jun, Rongying Shen, Hongxing Hua, and Xianding Jin: "Coupled Bending and Torsional Vibration of Axially Loaded Thin-Walled Timoshenko Beams," *International Journal of Mechanical Sciences,* **46** (2004).

52. Ewing, M. S.: "Another Second Order Beam Vibration Theory: Explicit Bending and Warping Flexibility and Restraint," *Journal of Sound and Vibration,* **137**(1) (1990).

53. Filipich, C. P., and M. B. Rosales: "Free Flexural-Torsional Vibrations of a Uniform Spinning Beam," *Journal of Sound and Vibration,* **141**(3) (1990).

54. Qaisi, M. I.: "Normal Modes of a Continuous System with Quadratic and Cubic Non-Linearities," *Journal of Sound and Vibration,* **265** (2003).

55. Yaman, Y.: "Vibrations of Open-Section Channels: A Coupled Flexural and Torsional Wave Analysis," *Journal of Sound and Vibration,* **204**(1) (1997).

56. Lin, S. M., and S. Y. Lee: "The Forced Vibration and Boundary Control of Pretwisted Timoshenko Beams with General Time Dependent Elastic Boundary Conditions," *Journal of Sound and Vibration,* **254**(1) (2002).

57. Li, Jun, Hongxing Hua, Rongying Shen, and Xianding Jin: "Stochastic Vibration of Axially Loaded Monosymmetric Timoshenko Thin-Walled Beam," *Journal of Sound and Vibration,* **274** (2004).

58. Gutierrez, R. H., and P. A. A. Laura: "Vibrations of a Beam of Non-Uniform Cross-Section Traversed by a Time Varying Concentrated Force," *Journal of Sound and Vibration,* **207**(3) (1997).

59. Larsson, P.-O.: "Determination of Young's and Shear Moduli from Flexural Vibrations of Beams," *Journal of Sound and Vibration,* **146**(1) (1991).

60. Wang, C. M., Y. Y. Zhang, and X. Q. He: "Vibration of Nonlocal Timoshenko Beams," *Nanotechnology,* **18** (2007).

61. Salarieh, Hassan, and Mehrdaad Ghorashi: "Free Vibration of Timoshenko Beam with Finite Mass Rigid Tip Load and Flexural-Torsional Coupling," *International Journal of Mechanical Sciences,* **48** (2006).

62. Zhong, Hongzhi, and Qiang Guo: "Nonlinear Vibration Analysis of Timoshenko Beams Using the Differential Quadrature Method," *Nonlinear Dynamics,* **32** (2003).

63. Younesian, D., M. H. Kargarnovin, D. J. Thompson, and C. J. C. Jones: "Parametrically Excited Vibration of a Timoshenko Beam on Random Viscoelastic Foundation Subjected to a Harmonic Moving Load," *Nonlinear Dynamics,* **45** (2005).

64. Kounadis, A. N.: "On the Derivation of Equations of Motion for a Vibrating Timoshenko Column," *Journal of Sound and Vibration,* **73**(2) (1980).

65. Blevins, Robert D.: "Formulas for Natural Frequency and Mode Shape," Van Nostrand Reinhold Company, New York, 1979.

66. Magrab, Edward B., et al.: "An Engineer's Guide to MATLAB@ with Applications from Mechanical, Aerospace, Electrical, and Civil Engineering," 2d ed., Pearson Prentice Hall, Pearson Education Inc., Upper Saddle River, N.J.

67. Leissa, Arthur W.: "Vibration of Plates," NASA SP-160, 1969.

68. Timoshenko, S., and S. Woinowsky-Krieger: "Theory of Plates and Shells," 2d ed., McGraw-Hill, New York, 1959.

69. Wittrick, W. H.: "Analytical, Three-Dimensional Elasticity Solutions to Some Plate Problems, and Some Observations on Mindlin's Plate Theory," *International Journal of Solids Structures,* **23**(4) (1987).

70. Xiang, Y., K. M. Liew, and S. Kitipornchai: "Vibration Analysis of Rectangular Mindlin Plates Resting on Elastic Edge Supports," *Journal of Sound and Vibration,* **204** (1997).

71. Liew, K. M., C. M. Wang, Y. Xiang, and S. Kitipornchai: "Vibration of Mindlin Plates Programming the p-Version Ritz Method," Elsevier, Amsterdam, 1998.

72. Irvine, H. M., and T. K. Caughey: "The Linear Theory of Free Vibrations of a Suspended Cable," *Proc. Royal Society of London, Series A, Mathematical and Physical Sciences,* **341**(1626) (1974).

73. Yu, P.: "Explicit Vibration Solutions of a Cable Under Complicated Loads," *ASME Journal of Applied Mechanics,* **64** (1997).

74. Wang, H. Q., J. C. Miao, J. H. Luo, F. Huang, and L. G. Wang: "The Free Vibration of Long-Span Transmission Line Conductors with Dampers," *Journal of Sound and Vibration,* **208**(4) (1997).

75. Rao, S. S.: "Effects of Transverse Shear and Rotatory Inertia on the Coupled Twist-Bending Vibrations of Circular Rings," *Journal of Sound and Vibration,* **16**(4) (1971).

76. Villaggio, Piero: "Mathematical Models for Elastic Structures," Cambridge University Press, Cambridge, England, 1997.

77. Rao, S. S.: "On the Natural Vibrations of Systems of Elastically Connected Concentric Thick Rings," *Journal of Sound and Vibration,* **32**(4) (1974).

78. Oh, S. J., B. K. Lee, and I. W. Lee: "Natural Frequencies of Non-Circular Arches with Rotatory Inertia and Shear Deformation," *Journal of Sound and Vibration,* **219**(1) (1999).

79. Wang, C. Y.: "Some Exact Solutions of the Vibration of Non-Homogeneous Membranes," *Journal of Sound and Vibration,* **210**(4) (1998).

80. Leissa, Arthur W., and Amir Ghamat-Rezaei: "Vibrations of Rectangular Membranes Subjected to Shear and Nonuniform Tensile Stresses," *Journal of the Acoustical Society of America* **88**(1) (1990).

81. Spence, J. P., and C. O. Horgan: "Bounds on Natural Frequencies of Composite Circular Membranes: Integral Equation Methods," *Journal of Sound and Vibration,* **87**(1) (1983).

82. Yamada, G., Y. Kobayashi, and H. Hamaya: "Transient Response of a Hanging Curtain," *Journal of Sound and Vibration,* **130**(2) (1989).

83. Soedel, W.: "Shells," Encyclopedia of Vibration, Volume 3, S. Braun, D. Ewins, and S. S. Rao, eds., Academic Press, San Diego, Calif., 2002.

84. Vlasov, V. Z.: "General Theory of Shells and Its Application in Engineering," NASA TT F-99, Washington D.C., 1964.

85. Kil'chevskiy, N. A.: "Fundamentals of the Analytical Mechanics of Shells," NASA TT F-292, Washington D.C., 1965.

86. Valid, R.: "The Nonlinear Theory of Shells Through Variational Principles: From Elementary Algebra to Differential Geometry," John Wiley & Sons, New York, 1995.

87. Vorovich, I. I.: "Nonlinear Theory of Shallow Shells," Springer-Verlag New York, Inc., New York, 1999.

88. Leissa, Arthur W.: "Vibration of Shells," NASA SP-288, 1973.

89. Kraus, Harry: "Thin Elastic Shells," John Wiley & Sons, New York, 1967.

90. Donnell, Lloyd Hamilton: "Beams, Plates and Shells," McGraw-Hill, New York, 1976.

91. Pilkey, W. D.: "Mechanically and/or Thermally Generated Dynamic Response of Thick Spherical and Cylindrical Shells with Variable Material Properties," *Journal of Sound and Vibration,* **6**(1) (1967).

92. Matsunaga, H.: "Free Vibration of Thick Circular Cylindrical Shells Subjected to Axial Stresses," *Journal of Sound and Vibration,* **211**(1) (1998).

93. Liang, Sen., H. L. Chen, and T. X. Liang: "An Analytical Investigation of Free Vibration for a Thin-Walled Regular Polygonal Prismatic Shell with Simply Supported Odd/Even Number of Sides," *Journal of Sound and Vibration,* **284** (2005).

94. Tzou, H. S., and L. A. Bergman, eds.: "Dynamics and Control of Distributed Systems," Cambridge University Press, Cambridge, 1998.

95. Ventsel, Eduard, and Theodor Krauthammer: "Thin Plates and Shells Theory, Analysis, and Applications," Marcel Dekker, New York, 2001.

CHAPTER 8
TRANSIENT RESPONSE TO STEP AND PULSE FUNCTIONS

Thomas L. Paez

INTRODUCTION

The design of structures to withstand dynamic environments as well as the need to characterize structural behavior requires a theory of structural dynamics. Sometimes the best means for characterizing the behavior of a structure, either an experimental system or a structure modeled with mathematical equations, is to subject it to a simple excitation and characterize the system through its response.

This chapter deals with the computation of structural response to step and pulse functions. The main emphasis of the chapter is to establish the responses of linear single-degree-of-freedom (SDOF) structures to step and pulse excitations. As well, response characteristics such as peak responses and shock response spectra (SRS) are established for various step and pulse inputs. Responses of continuous and multiple-degree-of-freedom (MDOF) linear structures are considered in Chaps. 1, 2, 7, 9, and 21 through 24. Responses of nonlinear structures are considered in Chap. 4. Responses of structures to random excitation are considered in Chaps. 1, 21 through 24, and 29 through 32.

LINEAR SINGLE-DEGREE-OF-FREEDOM STRUCTURES

EQUATION OF MOTION

The system to be considered is a linear single-degree-of-freedom structure.[1,2] (Behavior of the linear SDOF structure and its responses to general excitations are developed, in detail, in Chap. 2 of this handbook.) The SDOF structure may be an idealization of a very simple real structure or a simplified representation of one component (mode) in the response of a more complex real structure. The schematic of an SDOF structure with the forcing excitation applied to the mass is shown in Fig. 8.1*A*. The mass is attached to a rigidly fixed boundary through a

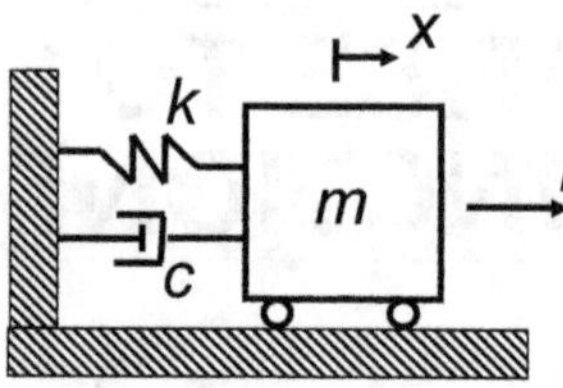

FIGURE 8.1A Force-excited SDOF structure.

spring and a damper. It is assumed that (1) the mass in the structure is rigid and is a known positive constant m; (2) the force in the spring is a linear function of displacement across the spring, and the spring is elastic with a known positive constant k; (3) the force in the damper is a linear function of velocity across the damper and is a known positive constant c (we are interested in SDOF structures for which the damping c is in the interval $[0,2\sqrt{km})$; such structures are called *underdamped,* and they execute oscillatory responses; and (4) the spring and damper act in one dimension, and the force $F(t)$ excites the mass in that dimension. With these assumptions, Newton's second law can be written for equilibrium of the mass as

$$m\ddot{x} = -kx - c\dot{x} + F(t) \tag{8.1a}$$

The force $F(t)$ is assumed known, along with a time interval over which the equation is valid. The quantities $x, \dot{x}, \ddot{x}$, are the scalar absolute displacement measured from the equilibrium position, absolute velocity, and absolute acceleration of the SDOF structure mass in the dimension of its motion, and they are the quantities sought during analysis of the response. Equation (8.1a) is the equation of motion of the SDOF structure; it is a second-order ordinary differential equation (ODE). It can be rewritten in many forms; two traditional forms are

$$m\ddot{x} + c\dot{x} + kx = F(t) \tag{8.1b}$$

$$\ddot{x} + 2\zeta\omega_n\dot{x} + \omega_n^2 x = \frac{F(t)}{m} \tag{8.1c}$$

In Eq. (8.1b) the stiffness and damping restoring force terms have simply been moved to the left side of the equation. Equation (8.1c) divides every term in Eq. (8.1b) by the mass m and defines

$$\omega_n = \sqrt{\frac{k}{m}} \qquad \zeta = \frac{c}{2m\omega_n} \tag{8.1d}$$

The quantity ω_n is known as the undamped natural frequency of the SDOF structure in radians per second, and ζ is known as the damping factor of the SDOF structure. (We are interested in SDOF structures for which the damping factor ζ is in the interval $[0,1)$.) The rationales behind both names will become clear in the following section. The natural period of response of the SDOF structure is $T_n = (2\pi)/\omega_n$.

As stated, the time period for which Eqs. (8.1a) through (8.1c) are valid must be specified, and this is typically determined by the time period over which the response is sought. Before the equation of motion can be solved, the condition of the SDOF structure must be specified at some time during the response. Normally, the displacement and velocity are specified at the start of the response and are known as the *initial conditions.* Therefore, for example, when the time period over which the equation of motion is valid is $[0, \infty)$, the initial conditions might be specified as $x(0) = x_0, \dot{x}(0) = \dot{x}_0$.

An alternative form of the equation of motion governs the response of the base-excited SDOF structure of Fig. 8.1B. The mass is attached with a spring and a damper to a boundary that can move. The first three assumptions made previ-

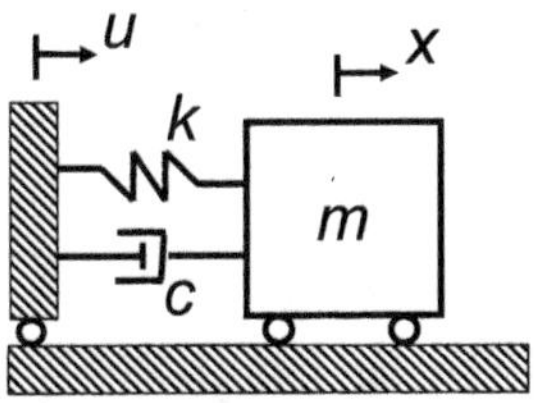

FIGURE 8.1*B* Base-excited SDOF structure.

ously also apply here, and the force applied to the mass comes through the spring and the damper. Newton's second law can be written for equilibrium of the mass as

$$m\ddot{x} = -k\delta - c\dot{\delta} \qquad (8.2a)$$

where

$$\delta = x - u \qquad \dot{\delta} = \dot{x} - \dot{u} \qquad (8.2b)$$

are the displacement and velocity of the mass relative to the base motions $u,\dot{u}$. The base motions are assumed known, and the relative displacement, velocity, and acceleration, or absolute displacement, velocity, and acceleration, are the quantities sought during analysis of the response. This equation of motion can also be modified in many ways. For example, it can be rewritten in either of two ways:

$$m\ddot{\delta} + c\dot{\delta} + k\delta = -m\ddot{u} \qquad (8.2c)$$

$$\ddot{\delta} + 2\zeta\omega_n\dot{\delta} + \omega_n^2\delta = -\ddot{u} \qquad (8.2d)$$

The parameters have the same meanings as previously, and the acceleration of the mass relative to the base motion is defined as

$$\ddot{\delta} = \ddot{x} - \ddot{u} \qquad (8.2e)$$

It is important to note that Eqs. (8.2*c*) and (8.2*d*), as written, govern motion of the mass relative to the motion of the base.

As with Eqs. (8.1*b*) and (8.1*c*), the time period for which Eqs. (8.2*a*), (8.2*c*), and (8.2*d*) are valid must be specified, and the condition of the SDOF structure must be specified at some time during the response. For example, when the time period over which the equation of motion is valid is $[0, \infty)$, the initial conditions might be specified as $\delta(0) = \delta_0, \dot{\delta}(0) = \dot{\delta}_0$.

Although Eqs. (8.2*a*), (8.2*c*), and (8.2*d*) are written for the relative motion of the SDOF structure, the absolute measures of response of the mass can be obtained once the relative motions are established. Because the measures of base motion $u,\dot{u},\ddot{u}$, are assumed known, Eqs. (8.2*b*) and (8.2*e*) can be used to establish $x,\dot{x},\ddot{x}$.

Regardless of the particular SDOF structure whose response is to be evaluated and the appropriate form of the governing equation, the structure shown in Fig. 8.1*A* and governed by Eq. (8.1*b*) or (8.1*c*), or the structure shown in Fig. 8.1*B* and governed by Eq. (8.2*c*) or (8.2*d*), the general form of the equation is

$$a_1\ddot{v} + a_2\dot{v} + a_3v = s(t) \qquad (8.3)$$

where the parameters a_1, a_2, and a_3 are related to the system mass, damping, and stiffness and are known, and the term $s(t)$ represents the scaled externally applied force or the imposed base motion. In view of this fact, the method developed to solve one form of the governing equation can be used to solve any of the other forms.

Some methods for solving the governing equation of motion for the SDOF structure response are presented in the following sections. In addition, some examples of SDOF structure responses to step and pulse excitations are provided, as well as the definition and some examples of SRS. Fourier transform and Laplace transform methods[3] can also be used for solving the equations of motion, as well as direct numerical integration methods.[4]

SOLUTION OF THE EQUATION OF MOTION—CLASSICAL APPROACH

It is useful to think of the single-degree-of-freedom structure as a specific, simple physical structure. Consider an SDOF structure represented by the schematic shown in Fig. 8.1*A*, and governed by Eq. (8.1*c*). In general, there are two things that can excite a response in the system: (1) nonzero initial conditions of displacement and velocity and (2) externally applied force. The solution to the governing differential equation of motion has two parts that correspond to these two factors.[5,6] The part of the solution that corresponds to the response driven by initial conditions is the complementary solution. When a structure responds to initial conditions only, it is said to be in *free vibration.* The part of the solution that corresponds to the response driven by external force or base motion is the particular solution. When a structure responds to applied force or base motions it is said to execute *forced vibrations.*

In order to establish the complementary solution, we solve the governing differential equation with the forcing term (externally applied force or imposed base motion) set to zero and subject the solution to the SDOF structure initial conditions. There are several methods for establishing the particular solution, and we will consider two of them in the following paragraphs and sections. The complete solution of the governing differential equation consists of the sum of the complementary and the particular solutions if the system under consideration is linear.

It can be shown[5,6] that the solutions of all linear ordinary differential equations with constant coefficients can be expressed in terms of exponential functions, and some special forms of the exponential functions are the harmonic functions, that is, sine and cosine functions. Because it can also be shown that the solution to a linear ODE is unique, any and all solutions to the homogeneous equation governing motion of an SDOF structure [Eq. (8.1*c*) with $F(t) = 0$]

$$\ddot{x} + 2\zeta\omega_n\dot{x} + \omega_n^2 x = 0 \tag{8.1e}$$

can be expressed as

$$x(t) = e^{-\zeta\omega_n t}[A\cos(\omega_d t) + B\sin(\omega_d t)] \tag{8.4a}$$

where $\omega_d = \omega_n\sqrt{1-\zeta^2}$ is known as the damped natural frequency of the response, and the damping factor ζ is in the interval [0,1). The terms A and B are known as arbitrary constants and can be evaluated based on the initial conditions of the SDOF structure. The initial conditions can be written $x(0) = x_0, \dot{x}(0) = \dot{x}_0$. Taking the derivative of the displacement in Eq. (8.4*a*) yields the velocity as

$$\dot{x}(t) = e^{-\zeta\omega_n t}[(-\zeta\omega_n A + \omega_d B)\cos(\omega_d t) + (-\omega_d A - \zeta\omega_n B)\sin(\omega_d t)] \tag{8.4b}$$

When the absolute displacement, Eq. (8.4*a*), is evaluated at $t = 0$ and the result equated to x_0, and the absolute velocity, Eq. (8.4*b*), is evaluated at $t = 0$ and the result equated to $\dot{x}_0$, two simultaneous linear equations for the constants A and B are obtained. The equations can be solved to obtain

$$A = x_0 \qquad B = \frac{\zeta}{\sqrt{1-\zeta^2}}\,x_0 + \frac{1}{\omega_d}\,\dot{x}_0 \tag{8.4c}$$

The complementary solution for displacement and velocity of the SDOF structure driven by its initial conditions is obtained by substituting the expressions for A and B into Eqs. (8.4a) and (8.4b).

$$x(t) = e^{-\zeta\omega_n t}\left(x_0 \cos(\omega_d t) + \frac{(\zeta\omega_n x_0 + \dot{x}_0)}{\omega_d} \sin(\omega_d t)\right) \qquad t \geq 0 \qquad (8.4d)$$

$$\dot{x}(t) = e^{-\zeta\omega_n t}\left(\dot{x}_0 \cos(\omega_d t) - \frac{(\omega_n x_0 + \zeta\dot{x}_0)}{\sqrt{1-\zeta^2}} \sin(\omega_d t)\right) \qquad t \geq 0 \qquad (8.4e)$$

Of course, the SDOF structure response velocity can be differentiated to establish the structural acceleration. Because the equation governing unforced, relative motion of the SDOF structure, Eq. (8.2d) with $-\ddot{u} = 0$, is identical in form and parameters to the equation governing unforced absolute motion, Eq. (8.1c) with $F(t)/m = 0$, the complementary solution of the former equation is the same as that provided by Eqs. (8.4d) and (8.4e).

The appearance of the damped natural frequency of motion ω_d in the arguments of the cosine and sine functions in the solution, Eqs. (8.4d) and (8.4e), makes clear the rationale for the terminology *natural frequency,* first introduced in the previous section. When $\zeta = 0$, the damping is zero and $\omega_d = \omega_n$; the quantity ω_n is the undamped natural frequency. Further, the explicit appearance of the damping factor ζ in the exponent of the leading term in Eqs. (8.4d) and (8.4e) shows why it is referred to as the damping factor; it governs how quickly the response amplitude diminishes (or is damped out) with time.

Equations (8.4d) and (8.4e) are applicable when the initial displacement is nonzero and the initial velocity is zero, or when the initial displacement is zero and the initial velocity is nonzero, or when both the initial displacement and the initial velocity are nonzero. The fact that the response expressions in Eqs. (8.4d) and (8.4e) equal sums of the responses for the two cases where the initial conditions are $(x_0 \neq 0, \dot{x}_0 = 0)$ plus $(x_0 = 0, \dot{x}_0 \neq 0)$ indicates that the principle of superposition holds for linear SDOF structure response to initial conditions.

***Example 8.1:* Unit Initial Displacement, Unit Initial Velocity, Nonzero Initial Displacement, and Initial Velocity.** First, consider the case where the initial displacement is one and the initial velocity is zero. The structure's natural frequency is $\omega_n = 2\pi$ rad/sec, and the damping factor ζ takes the values 0, 0.01, 0.05, 0.20, for purposes of comparison. The single-degree-of-freedom structure displacement response is

$$x(t) = e^{-\zeta(2\pi)t}\left\{\cos\left[(2\pi)\sqrt{1-\zeta^2}\, t\right] + \left(\frac{\zeta}{\sqrt{1-\zeta^2}}\right) \sin\left[(2\pi)\sqrt{1-\zeta^2}\, t\right]\right\} \qquad t \geq 0 \quad (8.5a)$$

The responses are graphed for times in the interval [0,10] sec in Fig. 8.2A.

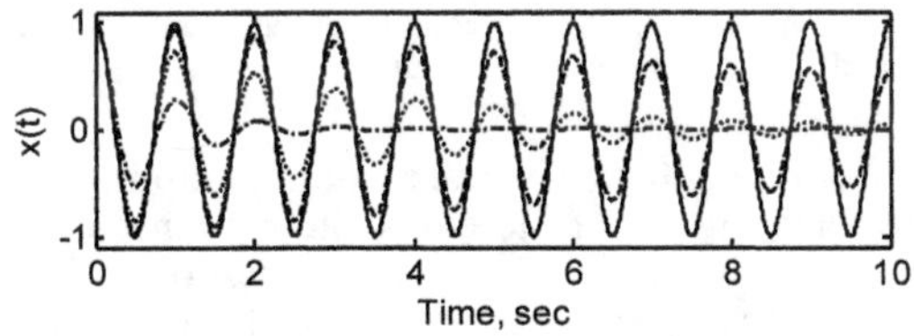

FIGURE 8.2A Responses of SDOF structures to unit initial displacement. (——$\zeta = 0$, – –$\zeta = 0.01$, ···$\zeta = 0.05$, –·–·$\zeta = 0.20$.)

Next, the SDOF structure displacement responses to unit initial velocity and zero initial displacement are computed. The same natural frequency and damping factors are used. The SDOF structure response is

$$x(t) = e^{-\zeta(2\pi)t}\left[\frac{1}{2\pi\sqrt{1-\zeta^2}} \sin\left(2\pi\sqrt{1-\zeta^2}\, t\right)\right] \qquad t \geq 0 \tag{8.5b}$$

The responses are graphed for times in the interval [0,10] sec in Fig. 8.2*B*.

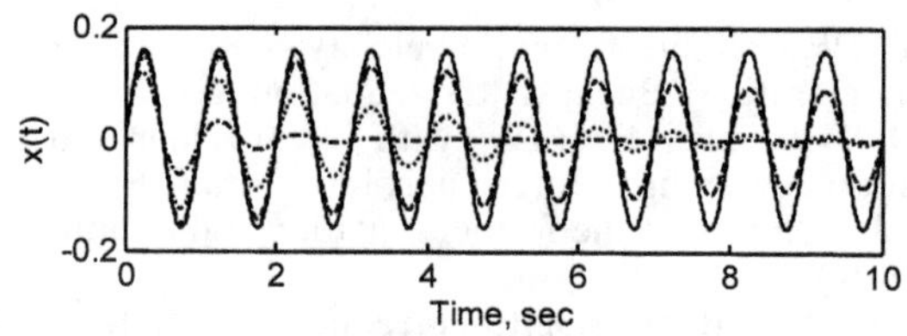

FIGURE 8.2*B* Responses of SDOF structures to unit initial velocity. (——$\zeta = 0$, – –$\zeta = 0.01$, ···$\zeta = 0.05$, –·–·$\zeta = 0.20$.)

Finally, the SDOF structure displacement response to a combined unit initial displacement and initial velocity of 10 units is computed. The same natural frequency and a damping factor of $\zeta = 0.05$ are used. The SDOF structure response is

$$x(t) = e^{-(0.05)(2\pi)t}\left\{\cos\left[(2\pi)\sqrt{1-(0.05)^2}\, t\right] + \left(\frac{(0.05)(2\pi)+10}{(2\pi)\sqrt{1-(0.05)^2}}\right)\sin\left[(2\pi)\sqrt{1-(0.05)^2}\, t\right]\right\} \qquad t \geq 0 \tag{8.5c}$$

The response is graphed for times in the interval [0,10] sec in Fig. 8.2*C*.

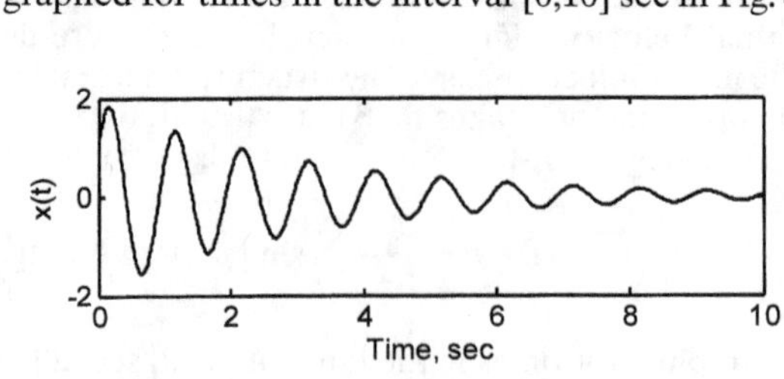

FIGURE 8.2*C* Response of SDOF structure with damping factor $\zeta = 0.05$ to combined unit initial displacement and initial velocity of 10 units.

The effect of the damping factor on the response amplitude is apparent from all the figures. As the damping factor increases, the response decays with increasing rapidity. When the damping factor is $\zeta = 0$, the response does not decay at all. When the damping factor is $\zeta = 0.01$, the response decays about 6.1 percent during each cycle, for a total of 47 percent $(1 - (0.939)^{10})$ during 10 cycles of response. When the damping factor is $\zeta = 0.05$, the response decays about 27 percent during each cycle of response. When the damping factor is $\zeta = 0.20$, the response decays about 71.5 percent during each cycle of response. The damping factor must be in the interval [0,1) in order for the formulas in Eqs. (8.4) and (8.5) to be easily interpreted. Damping factors of 1 percent or less are considered low, though some monolithic structures have damping factors much lower than 1 percent. Damping factors higher than about 15 percent are considered high for passively damped structures. When the damping factor is small, it has relatively little effect on the response frequency.

The second component of any solution to a linear ordinary differential equation is the particular solution, the response to an applied forcing function or to imposed base motions. There are several general methods for establishing the particular solution to an ODE, and one will be developed later. For now, we consider a simple special case that is best demonstrated through an example.

***Example 8.2:* Imposed Step Displacement/Velocity Pulse.** The single-degree-of-freedom structure considered here is shown schematically as the base-excited system of Fig. 8.1*B*. The imposed acceleration excitation is the full sine wave shown in Fig. 8.3*A* and given by

$$\ddot{u}(t) = \begin{cases} a \sin(\omega_0 t) & 0 < t < 2\pi/\omega_0 \\ 0 & t \le 0, t \ge 2\pi/\omega_0 \end{cases} \tag{8.6a}$$

where a is an amplitude coefficient and ω_0 is the frequency of the full-sine pulse. For the pulse in Fig. 8.3*A*, $a = 1$ and $\omega_0 = 2\pi$. The imposed acceleration is a two-sided pulse. There are imposed velocity and displacement conditions corresponding to the imposed acceleration. They can be obtained by integrating the imposed acceleration once, then a second time. They are

$$\dot{u}(t) = \begin{cases} \dfrac{a}{\omega_0}[1 - \cos(\omega_0 t)] & 0 < t < 2\pi/\omega_0 \\ 0 & t \le 0, t \ge 2\pi/\omega_0 \end{cases} \tag{8.6b}$$

$$u(t) = \begin{cases} \dfrac{a}{\omega_0^2}[\omega_0 t - \sin(\omega_0 t)] & 0 < t < 2\pi/\omega_0 \\ 0 & t \le 0 \\ 2\pi a/\omega_0^2 & t \ge 2\pi/\omega_0 \end{cases} \tag{8.6c}$$

and they are graphed in Figs. 8.3*B* and 8.3*C*. The functional form of the velocity is known as a *haversine;* it is a one-sided pulse. The imposed displacement is a step function. It is assumed that the SDOF structure is at rest at the initial time $t = 0$. [Because $y(0) = 0$ and $\dot{y}(0) = 0$, this statement implies that both $x(0) = 0$ and $\dot{x}(0) = 0$, and $\delta(0) = 0$ and $\dot{\delta}(0) = 0$.]

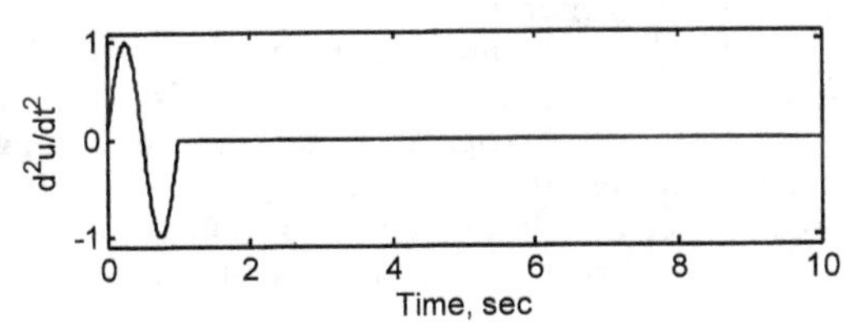

FIGURE 8.3*A* Input to a base-excited SDOF structure for the case where $\omega_0 = 2\pi$: acceleration.

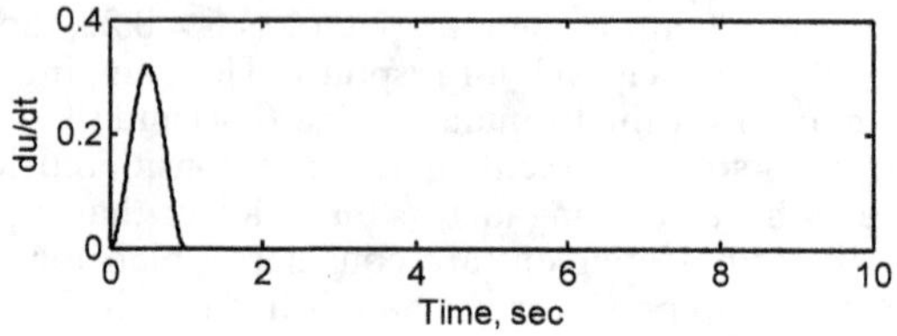

FIGURE 8.3*B* Input to a base-excited SDOF structure for the case where $\omega_0 = 2\pi$: velocity.

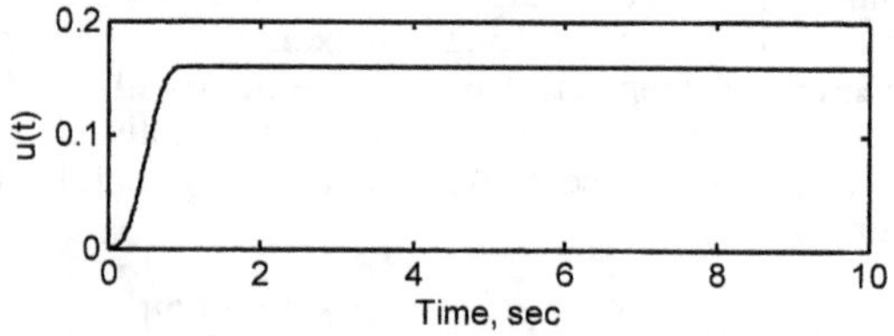

FIGURE 8.3*C* Input to a base-excited SDOF structure for the case where $\omega_0 = 2\pi$: displacement.

The response to the excitation defined by Eq. (8.6) is obtained by assuming a form and using it in the governing equation (8.2*d*). Based on Eq. (8.6*a*), we choose

$$\delta(t) = C \cos(\omega_0 t) + D \sin(\omega_0 t) \qquad 0 \le t \le 2\pi/\omega_0 \tag{8.7a}$$

where C and D are arbitrary constants to be determined. When the expression of Eq. (8.7*a*) is used in Eq. (8.2*d*), the result is a coefficient times cos $(\omega_0 t)$ plus a coefficient times $\sin(\omega_0 t)$. The two coefficients must be equated to the corresponding coefficients on the right-hand side. On the right-hand side, the coefficient of $\cos(\omega_0 t)$ is zero, and the coefficient of $\sin(\omega_0 t)$ is a. The equation operation described here yields two linear equations in C and D, and they can be solved to obtain

$$C = \frac{a 2\zeta\omega_n\omega_0}{(\omega_n^2 - \omega_0^2)^2 + (2\zeta\omega_n\omega_0)^2} \qquad D = \frac{-a(\omega_n^2 - \omega_0^2)}{(\omega_n^2 - \omega_0^2)^2 + (2\zeta\omega_n\omega_0)^2} \tag{8.7b}$$

The particular solution to this problem is obtained by using C and D from Eq. (8.7*b*) in Eq. (8.7*a*).

Note, however, that evaluation of Eq. (8.7*a*) at $t = 0$ indicates that the initial displacement is nonzero, in general. Specifically, from Eq. (8.7*a*), $\delta(0) = C$ and $\dot{\delta}(0) = \omega_0 D$, but we specified that the initial displacement and velocity must be zero. To force the solution to have zero initial conditions, it is necessary to superpose a homogeneous solution on the particular solution. The homogeneous solution needs to have initial conditions $\delta(0) = -C$ and $\dot{\delta}(0) = -\omega_0 D$; these cancel the conditions imposed by the particular solution. Use of $\delta(0)$ in place of x_0 and $\dot{\delta}(0)$ in place of $\dot{x}_0$ in Eq. (8.4*c*) yields the arbitrary constants for the homogeneous solution. They are

$$A = -C \qquad B = \frac{-1}{\omega_d}(\zeta\omega_n C + \omega_0 D) \tag{8.7c}$$

When the constants A and B are used in Eq. (8.4a) and the constants C and D from Eq. (8.7b) are used in Eq. (8.7a), and the two solutions, homogeneous and particular, are superposed, the result is the solution to the ordinary differential equation for times $0 \le t \le 2\pi/\omega_0$. At the end of the time period $t = 2\pi/\omega_0$, the excitation ceases, and the response of the SDOF structure starts a free decay. The initial conditions of the free decay are the relative displacement and relative velocity obtained from the solution described previously at $t = 2\pi/\omega_0$. These initial conditions can be used in Eq. (8.4d) and then time-shifted by replacing every occurrence of t in the equation with $t - 2\pi/\omega_0$.

The relative displacement response is graphed in Fig. 8.4A for structures with natural frequency $\omega_n = 2\pi$ rad/sec, and damping factors ζ of 0.01 and 0.05. The amplitude of the acceleration pulse is $a = 1$, and the frequency of the excitation is $\omega_0 = 2\pi$ rad/sec. The corresponding absolute displacement responses are graphed in Fig. 8.4B; these responses superimpose the base displacement of Fig. 8.3C onto the relative displacement responses of Fig. 8.4A. Because the frequency and duration of the excitation equal the natural frequency and period of the SDOF structure, the dynamics of the excitation matches the dynamics of the SDOF structures in some sense, and this leads to dynamic amplification of the response. To be specific, the value of the base displacement at the top of the ramp is $y = 1/(2\pi) \cong 0.159$, and the value of the absolute displacement of the more heavily damped structure at its peak is $x \cong 0.228$; therefore, the dynamic amplification of the response over the excitation is 1.43.

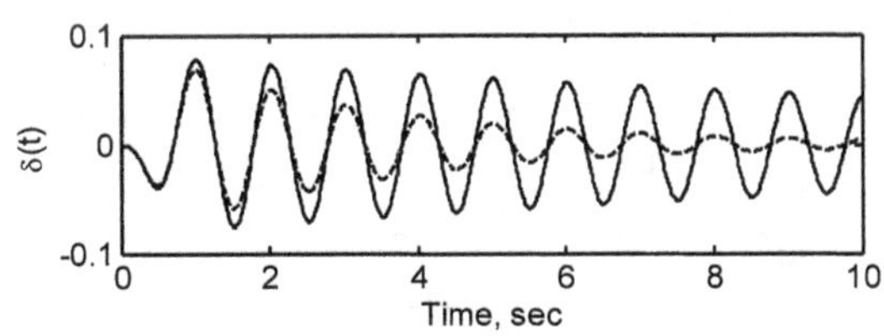

FIGURE 8.4A Relative displacement responses for base-excited linear SDOF structures with $\omega_0 = 2\pi$ and ——$\zeta = 0.01$, – –$\zeta = 0.05$.

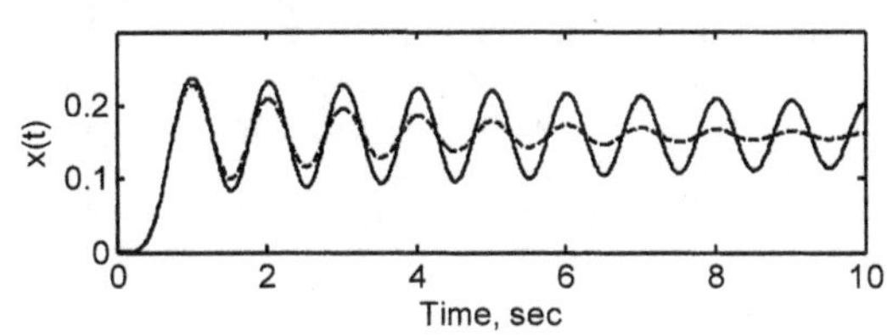

FIGURE 8.4B Corresponding absolute displacement responses.

The relative displacement responses are graphed in Fig. 8.4C for the same structures, but an excitation with acceleration amplitude $a = 1$ and frequency of excitation $\omega_0 = 20\pi$ rad/sec. The corresponding absolute displacement responses are graphed in Fig. 8.4D. The frequency of the excitation is 10 times the natural frequency of the SDOF structure, and its duration is $\frac{1}{10}$ the natural period of the SDOF structure. The largest value of the imposed displacement is $1/(200\pi) = 1.58 \times 10^{-3}$, and the largest value in the absolute displacement response is 4.25×10^{-3}; therefore, the amplification of the response over the excitation is 2.69. The response has a dynamic amplification. Because the input that excites the responses in Figs. 8.4*C* and 8.4*D* has short duration relative to the natural period of the structure, the direct effect caused by the excitation ceases when the SDOF structure has just started to move, as seen by the notch at the start of Fig. 8.4*C*. The effect is something like the application of a pair of impulses, to be described in the following section.

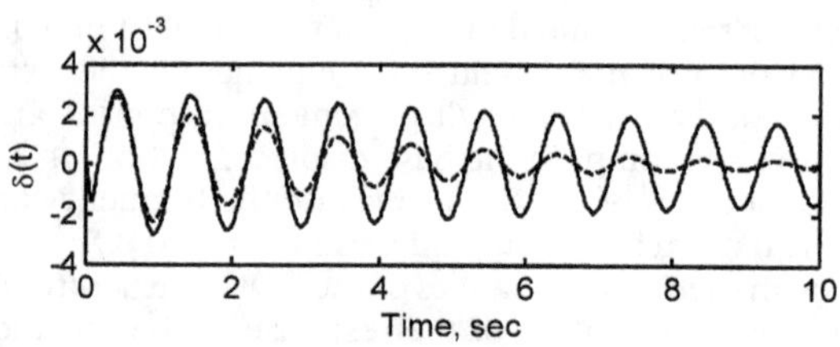

FIGURE 8.4*C* Relative displacement responses for base-excited linear SDOF structures with $\omega_0 = 20\pi$ and the same damping factors used in Fig. 8.4*A*.

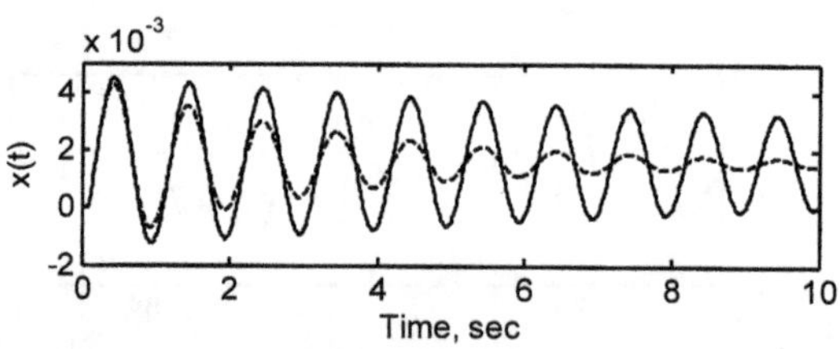

FIGURE 8.4*D* Corresponding absolute displacement responses.

The relative displacement responses are graphed in Fig. 8.4*E* for the same structures, but an excitation with acceleration amplitude $a = 1$ and frequency of excitation $\omega_0 = 0.2\pi$ rad/sec. The corresponding absolute displacement responses are graphed in Fig. 8.4*F*. The structural responses with different damping factors are so near one another that they are almost indistinguishable. The frequency of the excitation is $\frac{1}{10}$ the natural frequency of the SDOF structure, and its duration is 10 times the natural

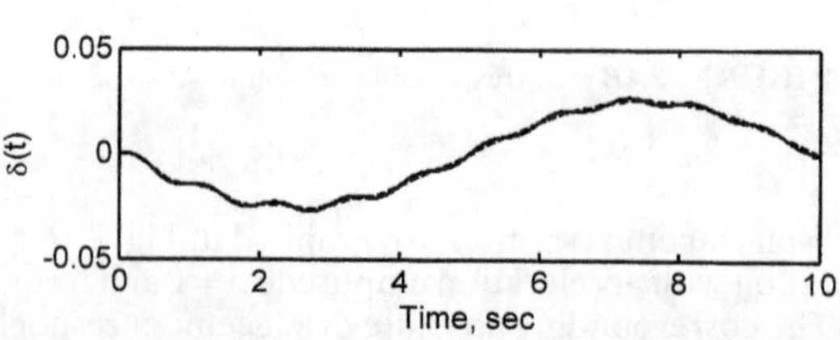

FIGURE 8.4*E* Relative displacement responses for base-excited linear SDOF structures with $\omega_0 = 0.2\pi$ and the same damping factors used in Fig. 8.4*A*.

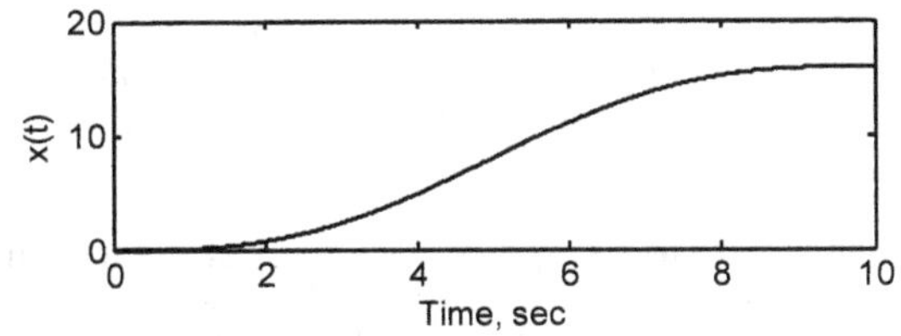

FIGURE 8.4*F* Corresponding absolute displacement responses.

period of the SDOF structure. The largest value of the imposed displacement is $50/\pi = 15.92$, and the largest value in the absolute displacement response is 15.92; therefore, the amplification of the response over the excitation is 1.00. The response is quasi-static. When the frequency content of motion applied at the base of a base-excited structure lies at frequencies that are low relative to the natural frequency of the structure, the response displays little dynamic amplification and is quasi-static. This means that design of such a structure or component can be performed using static analysis.

SOLUTION OF THE EQUATION OF MOTION—CONVOLUTION INTEGRAL

A general expression for the form of the particular solution of an ordinary differential equation with constant coefficients is the *convolution integral*[1,2] (also known as Duhamel's integral or the superposition integral). The convolution integral represents the response of a linear structure as the superposition of the structure responses to a sequence of impulses or short-duration pulses. To see how the convolution integral is constructed, consider an example that develops one of its building blocks, the *impulse response function* (IRF).

***Example 8.3:* Response of Structure to a Short-Duration Pulse.** Consider the force-excited linear single-degree-of-freedom structure of Fig. 8.1*A*. It is excited by the force pulse $F(t) = F_0 w(t - t_0, \Delta T)$ shown in Fig. 8.5, and we seek the absolute displacement response. The pulse is a one-sided square pulse, with magnitude F_0 and duration ΔT, and it starts at time t_0. We define the pulse function

$$w(t,\Delta T) = \begin{cases} 0 & t < 0 \\ 1 & 0 \le t < \Delta T \\ 0 & t \ge \Delta T \end{cases} \tag{8.8}$$

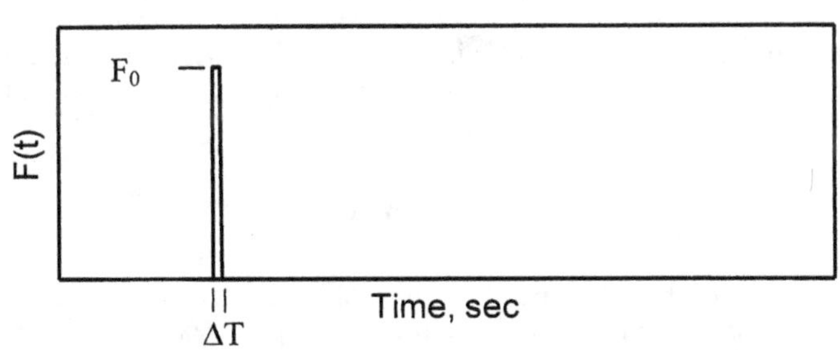

FIGURE 8.5 Force pulse.

The use of $t - t_0$ in the definition of $F(t)$ in place of t shifts the function an amount t_0 to the right. The pulse duration is short compared to the natural period $T_n = 2\pi/\omega_n$ of the structure it will excite, and the pulse is defined to have constant impulse. That is, the quantity $F_0\Delta T$ is a constant. Assume that the SDOF structure response starts from rest.

First, because there is no force prior to time t_0, the structure remains at rest until time t_0. During the very brief time period $[t_0, t_0 + \Delta T)$, when the force is applied to the structure mass, the acceleration of the structure mass is

$$\ddot{x}(t) = \begin{cases} 0 & t < t_0 \\ \dfrac{F_0}{m} & t_0 \le t < t_0 + \Delta T \end{cases} \tag{8.9a}$$

(It is assumed that because the duration of application of the force is very short, the change in displacement and velocity of the mass are very small; therefore, the forces applied to the mass during this time by the spring and damper are negligible.) The velocity of the mass is the integral of its acceleration, that is,

$$\dot{x}(t) = \begin{cases} 0 & t < t_0 \\ \dfrac{F_0}{m}(t - t_0) & t_0 \le t < t_0 + \Delta T \end{cases} \tag{8.9b}$$

and the displacement of the mass is the integral of its velocity, that is,

$$x(t) = \begin{cases} 0 & t < t_0 \\ \dfrac{F_0}{2m}(t - t_0)^2 & t_0 \le t < t_0 + \Delta T \end{cases} \tag{8.9c}$$

Equations (8.9*b*) and (8.9*c*) indicate that the mass motion is zero prior to application of the load and that the mass velocity and displacement following application of the load are $F_0\Delta T/m$ and $F_0\Delta T^2/(2m)$, respectively. Because it was assumed that the force impulse is constant (i.e., $F_0\Delta T = C_0$, where C_0 is the constant) as the load duration approaches zero, $\Delta T \to 0$, the mass velocity at the end of the application of the force pulse approaches C_0/m, and its displacement approaches zero because $F_0\Delta T^2/(2m) = (F_0\Delta T)\Delta T/(2m) = C_0\Delta T/(2m)$, and $\Delta T \to 0$ in the final term on the right-hand side. As $\Delta T \to 0$ the motion of the structure is denoted $(F_0\Delta T)h(t - t_0), t \ge t_0$, where $h(t), t \ge 0$ is called the unit impulse response function of the linear SDOF structure, or simply the impulse response function (IRF).

The function $h(t), t \ge 0$ is the response to a unit impulse, a short-duration high-amplitude pulse, or equivalently, as shown in the preceding, it is the response to the initial conditions of zero initial displacement and unit initial velocity. Therefore, the expression for $h(t), t \ge 0$ can be obtained from Eq. (8.4*d*), giving the response of an SDOF structure to initial conditions. The IRF for absolute displacement response to a unit impulsive force excitation is

$$h_x(t) = \frac{e^{-\zeta\omega_n t}}{m\omega_d} \sin(\omega_d t) \qquad t \ge 0 \tag{8.10}$$

(The subscript x on the IRF indicates that this is the IRF associated with absolute displacement response.) The IRF of the linear force-excited SDOF structure is graphed in Fig. 8.6 for the parameters $m = 1, \omega_n = 2\pi, \zeta = 0.05$.

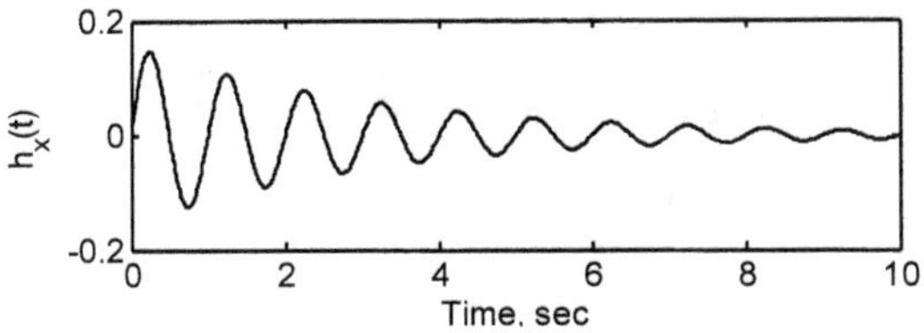

FIGURE 8.6 Absolute displacement impulse response function of a force-excited linear SDOF structure.

Finally, to complete the example, the response of the linear SDOF structure to the short-duration pulse $F(t) = F_0w(t - t_0, \Delta T)$ of Fig. 8.5 is

$$x(t) = \begin{cases} 0 & t < t_0 \\ (F_0\Delta T)h_x(t - t_0) & t \geq t_0 \end{cases} \tag{8.11}$$

This is the response to an impulse with magnitude $F_0\Delta T$.

Consider the response of the linear SDOF structure to a sequence of very short duration impulses $F_jw(t - t_j,\Delta T), j = 0,1, \ldots$, applied to the structure at times $t_j = j\Delta T$, $j = 0,1, \ldots$, where ΔT is a very short duration time increment. Because the pulses in the sequence do not overlap, the forcing function can be expressed as

$$F(t) = \sum_{j=0,1,\ldots} F_jw(t - t_j,\Delta T) \qquad t \geq 0 \tag{8.12}$$

That is, the forcing function is the superposition of a sequence of pulses, each with impulse $F_j\Delta T$. Because the response of a linear structure can be constructed using the principle of superposition, the displacement response of the linear SDOF structure to the sequence of pulses expressed in Eq. (8.12) is

$$x(t) = \sum_{j=0,1,\ldots} (F_j\Delta T)h_x(t - t_j) \qquad t \geq 0 \tag{8.13}$$

where $h_x(t), t \geq 0$ is the absolute displacement response IRF of Eq. (8.10).

In the sum of Eq. (8.13), we can take the limit as $\Delta T \to 0$ and write the sum as an integral. The absolute displacement response to the arbitrary excitation $F(t), t \geq 0$ is

$$x(t) = \int_0^t F(\tau)h_x(t - \tau)d\tau \qquad t \geq 0 \tag{8.14}$$

This is the convolution integral representation of the absolute displacement response of a linear SDOF structure to the specified force when the system starts from rest.

The response velocity can be obtained by differentiating the response displacement. (See Ref. 7 for details regarding differentiation of integrals.) The response velocity is the derivative of the convolution integral of Eq. (8.14), that is,

$$\begin{aligned} \dot{x}(t) &= \frac{d}{dt}\, x(t) = \frac{d}{dt}\left[\int_0^t F(\tau)h_x(t - \tau)d\tau\right] \\ &= F(t)h_x(0) + \int_0^t F(\tau)\, \frac{d}{dt}\, [h_x(t - \tau)]d\tau \qquad t \geq 0 \end{aligned} \tag{8.15}$$

The rule used to compute the derivative of an integral is known as Leibniz's rule. The derivative expressed in Eq. (8.15) can always be computed for linear structures, and for the case of the force-excited linear SDOF structure it is

$$\begin{aligned}\frac{d}{dt}h_x(t) &= \frac{d}{dt}\frac{e^{-\zeta\omega_n t}}{m\omega_d}\sin(\omega_d t)\\ &= \frac{e^{-\zeta\omega_n t}}{m}\left[\cos(\omega_d t) - \frac{\zeta}{\sqrt{1-\zeta^2}}\sin(\omega_d t)\right] \\ &= h_{\dot{x}}(t) \qquad t \geq 0\end{aligned} \tag{8.16}$$

The function $h_{\dot{x}}(t), t \geq 0$ is the IRF associated with the absolute velocity response of a linear force-excited SDOF structure. Further, for the force-excited linear SDOF structure, $h_x(0) = 0$, so the velocity response of the structure is

$$\dot{x}(t) = \int_0^t F(\tau)h_{\dot{x}}(t-\tau)d\tau \qquad t \geq 0 \tag{8.17}$$

When the initial conditions of the system are nonzero, then the convolution integral must be augmented with the complementary solution of the equation of motion. For example, the absolute displacement response would consist of the sum of Eqs. (8.14) and (8.4*d*). The arbitrary constants in the complementary solution would need to be established in terms of the initial displacement and velocity.

Because the form of the equation governing relative motion of the linear base-excited SDOF structure is essentially the same as the equation governing the force-excited SDOF structure, the response of the former structure to a base excitation can be expressed using the convolution integral. The relative displacement response can be written as

$$\delta(t) = \int_0^t [-\ddot{u}(\tau)]h_\delta(t-\tau)d\tau \qquad t \geq 0 \tag{8.18}$$

where $h_\delta(t), t \geq 0$ is the base-excited relative displacement response IRF of a linear SDOF structure. It is given by

$$h_\delta(t) = \frac{e^{-\zeta\omega_n t}}{\omega_d}\sin(\omega_d t) \qquad t \geq 0 \tag{8.19}$$

The convolution integrals for base-excited relative velocity and relative acceleration responses can be developed by differentiating Eq. (8.18) with respect to time. The essential operation is the differentiation of $h_\delta(t), t \geq 0$ to obtain the IRFs of the relative velocity and relative acceleration responses.

Further, the absolute displacement, velocity, and acceleration responses of the base-excited structures can be developed by adding the appropriate base motion to the relative motion. For example, the absolute acceleration response of the base-excited linear SDOF structure can be expressed as

$$\ddot{x}(t) = \int_0^t [-\ddot{u}(\tau)]h_{\ddot{x}}(t-\tau)d\tau \qquad t \geq 0 \tag{8.20}$$

where $h_{\ddot{x}}(t), t \geq 0$ is the IRF of the absolute acceleration response of a linear base-excited SDOF structure. The IRF is given by

$$h_{\ddot{x}}(t) = e^{-\zeta\omega_n t}\left[(-2\zeta\omega_n)\cos(\omega_d t) - \frac{\omega_n(1-2\zeta^2)}{\sqrt{1-\zeta^2}}\sin(\omega_d t)\right] \qquad t \geq 0 \tag{8.21}$$

***Example 8.4:* Absolute Displacement Response of a Linear SDF Structure to a Complex Pulse.** Consider the response of a force-excited linear single-degree-of-freedom structure to the complex input shown in Fig. 8.7. The actual structural excitation is the continuous curve shown in the figure. If we had available a closed-form expression for the excitation (i.e., an expression that could be written in terms of standard mathematical functions), then we could use Eq. (8.14) to solve for the response. In many practical situations, though, only a version of the excitation sampled at discrete times is available. For example, the values of the excitation measured at the times $t_j = (0.1)j$ sec, $j = 0,1, \ldots$ are shown by the dots in Fig. 8.7. In those cases, the pre-integral sum of Eq. (8.13) is used to approximate the response.

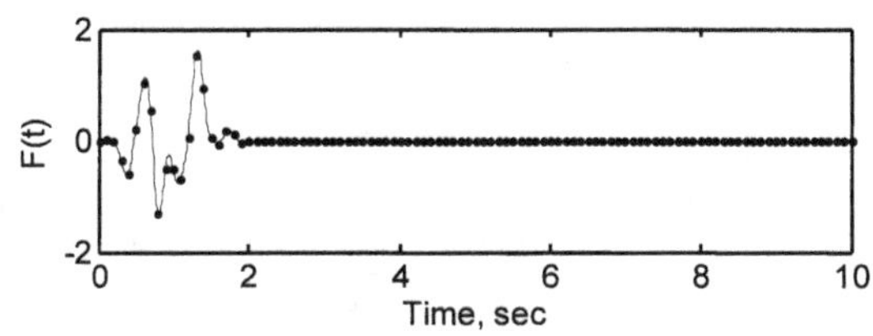

FIGURE 8.7 Complex excitation to a force-excited linear SDOF structure.

Figure 8.8 shows the absolute displacement response of a linear SDOF structure computed using Eq. (8.13) with the forcing function that is the sampled version of the force in Fig. 8.7. The mass of the SDOF structure is $m = 1$, its natural frequency is $\omega_n = 2\pi$ rad/sec, and its damping factor is $\zeta = 0.05$. Because a high-resolution version of the forcing function actually is available in this case, the response computed with a temporal resolution of $\Delta T = 0.1$ sec can be compared to the response computed using a temporal resolution of $\Delta T = 0.001$ sec. The latter response is also shown in Fig. 8.8, but the responses are so close that they are visually indistinguishable. In fact, the maximum difference between the two curves is about 1.4×10^{-3} at the time $t_{14} = 1.3$ sec. The difference is about 4 percent and highlights the fact that care must be taken when discrete approximations to continuous solutions are written.

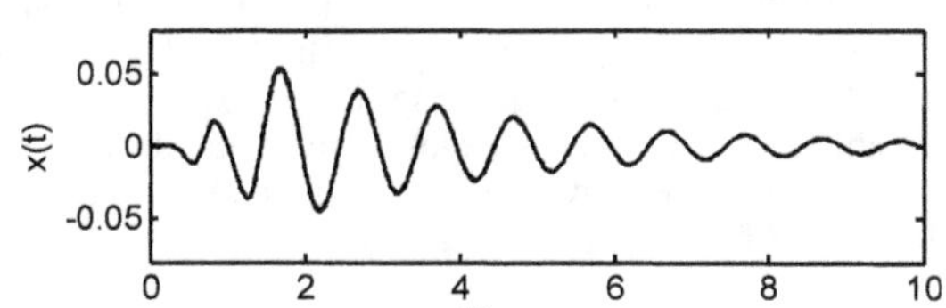

FIGURE 8.8 Linear SDOF structure response to force excitation.

It is interesting to note that the SDOF structure response to the complex excitation of Fig. 8.7 is quite smooth and, to a degree, resembles the structure's impulse response function. This is true, in general, because linear SDOF structures tend to amplify input force or motion near their own natural frequencies but not away from those frequencies.

The convolution integral can also be used to express the response of a structure excited by a relatively simple pulse.

***Example 8.5:* Absolute Acceleration Response of a Base-Excited SDF Structure to a Square Wave.** Consider the response of a base motion–excited linear single-degree-of-freedom structure to the square-wave input shown in Fig. 8.9. In this case, we have available a closed-form expression for the excitation, so we can use Eq. (8.14) to solve for the response. The excitation is

$$\ddot{u}(t) = \begin{cases} A & 0 \le t < T/2 \\ -A & T/2 \le t \le T \\ 0 & t > T \end{cases} \tag{8.22}$$

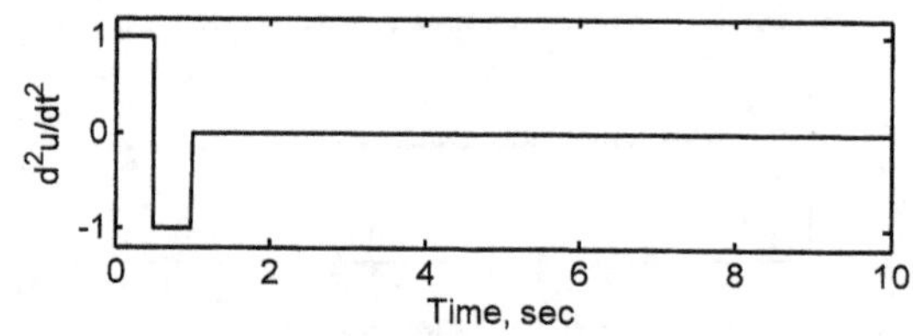

FIGURE 8.9 Square-wave excitation.

The function is graphed in Fig. 8.9 for $A = 1$ and $T = 1$ sec. When this expression is used in Eq. (8.20) along with the impulse response function of Eq. (8.21), the absolute acceleration response of the linear SDOF structure is found to be

$$\ddot{x}(t) = \begin{cases} A\left\{[1 - e^{-\zeta\omega_n t}\cos(\omega_d t)] - \dfrac{\zeta}{\sqrt{1-\zeta^2}}\sin(\omega_d t)\right\} & 0 \le t < T/2 \\ A\left\{[1 - e^{-\zeta\omega_n t}\cos(\omega_d t)] - \dfrac{\zeta}{\sqrt{1-\zeta^2}}\sin(\omega_d t)\right\} & \\ -2A\left\{[1 - e^{-\zeta\omega_n (t-T/2)}\cos(\omega_d (t-T/2))] - \dfrac{\zeta}{\sqrt{1-\zeta^2}}\sin(\omega_d (t-T/2))\right\} & T/2 \le t \le T \end{cases} \tag{8.23}$$

Because the excitation following time T is zero, the response is governed by the complementary solution of Eq. (8.4*a*). The arbitrary constants in the complementary solution are established based on the values of $\dot{x}(T)$ and $\ddot{x}(T)$. The former quantity is found simply by evaluating Eq. (8.23) at $t = T$; the latter quantity is found by taking the derivative of Eq. (8.23) and evaluating it at $t = T$.

The response excited by the base acceleration input described by Eq. (8.22) is graphed in Fig. 8.10 for the case in which $\omega_n = 2\pi$ rad/sec and $\zeta = 0.05$. The response beyond time $t = 1$ sec is in free decay.

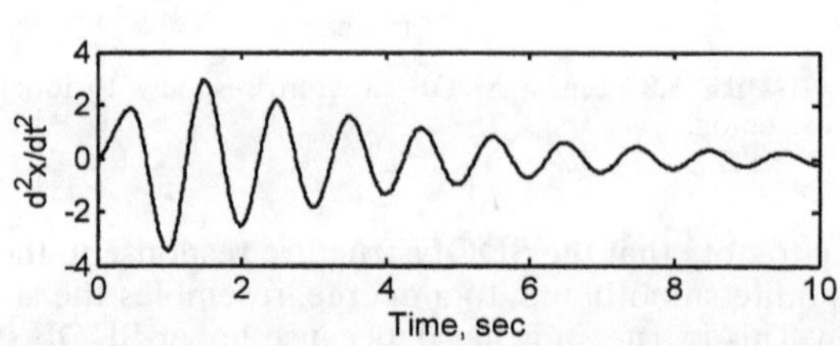

FIGURE 8.10 Response of a base-excited linear SDOF structure to the input of Fig. 8.9.

THE SHOCK RESPONSE SPECTRUM

The subject of this section is an introduction to the shock response spectrum (SRS), which is discussed in detail in Chap. 20 of this handbook. The SRS is a measure of the severity of a shock. It is the sequence of peak responses in a collection of linear single-degree-of-freedom structures excited by the shock. To define the SRS formally but concisely, let $\ddot{u}(t), t \geq 0$ denote a base excitation shock signal whose SRS is sought. Let $S_{\ddot{x}}(f_n,\zeta)$ denote the peak in the absolute value of a response (say, absolute acceleration) over all time excited by the input $\ddot{u}(t), t \geq 0$ in an SDOF structure with natural frequency $f_n = \omega_n/2\pi$, where ω_n is the undamped natural frequency of the structure, in radians per second, and damping factor ζ. Then the absolute acceleration maximax SRS of the shock $\ddot{u}(t), t \geq 0$ is defined as $S_{\ddot{x}}(f,\zeta), 0 < f_{\min} \leq f \leq f_{\max} < \infty$. The SRS is defined as an absolute acceleration SRS because it characterizes peak absolute acceleration responses. The SRS is defined as a maximax SRS because it refers to the peak acceleration over all time. SRS values other than the maximax and based on other measures of response and peak responses that occur within specific time frames can also be defined as shown in Chap. 20. The frequencies f in the interval $[f_{\min}, f_{\max}]$ where the SRS is defined do not usually constitute a continuous set because the peak response cannot be computed as a continuous function of the SDOF structure natural frequency, in general.

The definition makes it clear that a critical element in the computation of an SRS is the efficient computation of the peak responses of many SDOF structures. The convolution integral, whose general form is given in Eq. (8.14), can be used to compute SDOF structure responses, but much more efficient approximate methods are available.[8,9] The number of SDOF structure peak responses used to define the SRS establishes the resolution of the approximation. The SRS is often defined at frequencies that are spaced logarithmically—for example, 6, 8, or 12 samples per octave.

The absolute acceleration maximax SRS of the one-cycle sine-wave acceleration shock defined in Eq. (8.6) is shown in Fig. 8.11. The damping factor of the SDOF structures used to compute the SRS is $\zeta = 0.05$. The specific shock considered here is the one with amplitude $a = 1g$ and frequency $\omega_0 = 2\pi$ rad/sec. The SRS defines the absolute peak responses of a sequence of SDOF structures. Its highest value is 2.81g, and that occurs at a frequency of 1.15 Hz.

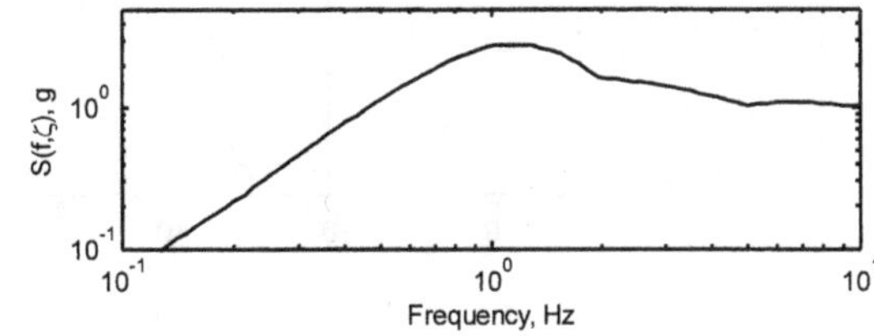

FIGURE 8.11 Absolute acceleration maximax SRS of a one-cycle sine shock. Maximum SRS value is 2.81g and occurs at 1.15 Hz. SRS $\zeta = 0.05$.

The SRS shown in Fig. 8.11 approaches an asymptote of 1g at high frequencies. The reasons are that relatively stiff SDOF structures (structures with a high natural frequency) simply transmit base excitations to the mass of the SDOF structure quasi-statically, and the amplitude of the shock input is $a = 1g$. For this reason, the high-frequency range of an SRS is sometimes called the quasi-static range. The roll-off rate of the SRS at low frequencies is approximately 12 dB/octave; the reason for this is explained, in detail, in Ref. 10. The reason that the roll-off at low frequencies takes the value it does is that the entire excitation appears to the low-frequency SDOF structures to be a single impulse or a sequence of impulses with alternating signs. The one-cycle sine-wave pulse whose SRS is plotted in Fig. 8.11 appears to very low frequency structures to be a sequence of two impulse functions with alternating signs.

Though each shock has a unique SRS, it must be emphasized that the SRS of a shock is not an invertible function. That is, the specific shock that led to an SRS is not obtainable from the SRS only. The reason is that the sequence of peak SDOF structural responses excited by a shock is not unique to a particular shock. In the case of Fig. 8.11, there are other signals that would lead to the same SRS.

There are amplitude and frequency scaling rules for the SRS. When we obtain the SRS for a particular shock signal, we can use it with a simple amplitude scaling to obtain the SRS of any other shock signal that is a constant multiple of the original shock. For example, if the SRS of the shock $\ddot{u}(t), t \geq 0$ is $S_{\ddot{x}}(f,\zeta), 0 < f_{\min} \leq f \leq f_{\max} < \infty$, then the SRS of the shock $C\ddot{u}(t), t \geq 0$ (where C is a constant) is $CS_{\ddot{x}}(f,\zeta), 0 < f_{\min} \leq f \leq f_{\max} < \infty$. Figures 8.12*A*, *B*, and *D* demonstrate this principle for the full-sine shock pulse. Figure 8.12*B* is simply Fig. 8.12*A* multiplied by a factor of 2. Figure 8.12*D* shows the SRS of the two shocks. The upper curve is the SRS of the shock in Fig. 8.12*B*, and the lower curve is the SRS of the shock in Fig. 8.12*A*. The two SRS differ by a factor of 2. All the SRS values in the figures are computed for SDOF structures with damping factors $\zeta = 0.05$.

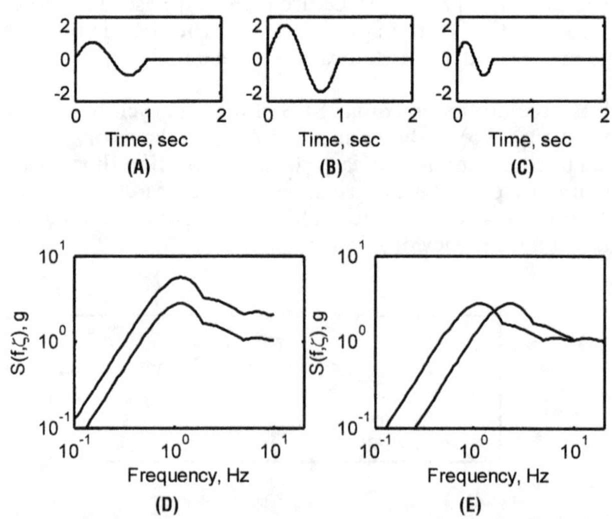

FIGURE 8.12 (*A*) One-cycle sine shock with unit amplitude and unit duration. (*B*) One-cycle sine shock with two-unit amplitude and unit duration. (*C*) One-cycle sine shock with unit amplitude and half-unit duration. (*D*) SRS of shocks *A* and *B*. (*E*) SRS of shocks *A* and *C*. SRS $\zeta = 0.05$.

The SRS of a particular shock can be used to obtain the SRS of another shock that differs from the first by a time/frequency scaling simply by shifting the SRS of the original shock on a log-log graph. For example, if the SRS of the shock $\ddot{u}(t), t \ge 0$ is $S_{\ddot{x}}(f,\zeta), 0 < f_{\min} \le f \le f_{\max} < \infty$, then the SRS of the shock $\ddot{u}(st), t \ge 0$ (where s is a constant) is $S_{\ddot{x}}(f/s,\zeta), 0 < sf_{\min} \le f \le sf_{\max} < \infty$. Figures 8.12*A*, *C*, and *E* demonstrate this principle for the full-sine shock pulse. Figure 8.12*C* is simply Fig. 8.12*A* compressed in time by a factor of 2; that is, $\ddot{u}(2t), t \ge 0$. Figure 8.12*E* shows the SRS of the two shocks. The curve on the left is the SRS of the shock in Fig. 8.12*A*, and the curve on the right is the SRS of the shock in Fig. 8.12*C*. The curves are identical in shape, with the curve on the right shifted up in frequency by a factor of 2; that is, the curve on the right is $S_{\ddot{x}}(f/2,\zeta), 0 < 2f_{\min} \le f \le 2f_{\max} < \infty$, where $S_{\ddot{x}}(f,\zeta), 0 < f_{\min} \le f \le f_{\max} < \infty$ is the curve on the left. The equivalence of SRS shapes shown, for example, in Fig. 8.12*E* occurs only when the curves are plotted above a logarithmically scaled abscissa.

The amplitude and scaling rules for SRS are important because they permit the specification of the SRS of an arbitrarily amplitude- and/or time/frequency-scaled shock from the SRS of the basic shock. The ability to scale SRS is important when a test shock whose SRS is approximately equal to the SRS of shock measured in the field is sought.

The SRS shown in Figs. 8.11 and 8.12 were all computed for SDOF structures with a damping factor of $\zeta = 0.05$. All responses of linear structures are dependent upon structural damping. Therefore, the SRS changes when the damping changes. Figure 8.13 shows five SRS computed for the one-cycle sine-wave shock pulse with damping in the SDOF structures set to (0.001, 0.01, 0.05, 0.10, 0.50). The top curve (at frequency $f = 1$ Hz) is for the SRS computed with $\zeta = 0.001$, and damping increases as the peak value of the curve decreases. The SRS for the first two curves are practically indistinguishable in the graph.

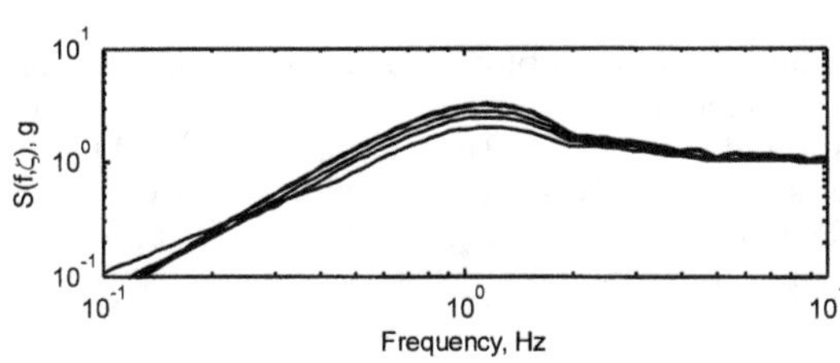

FIGURE 8.13 Five SRS computed for the one-cycle sine-wave shock pulse with damping in the SDOF structures set to (0.001, 0.01, 0.05, 0.10, 0.50), top to bottom.

The SRS computed in the preceding examples are absolute acceleration maximax SRS, and these are used widely for shock signal characterization in the aerospace community. However, other SRS are also used. One of these is the SRS of relative displacement response, denoted $\delta(t)$ in the previous sections. Its maximax relative displacement SRS is denoted $S_\delta(f,\zeta), 0 < f_{\min} \le f \le f_{\max} < \infty$ and is defined in terms of peak responses of the absolute value of relative displacements of SDOF structures with frequencies in the interval $[f_{\min}, f_{\max}]$ and damping factors ζ to the input $\ddot{u}(t), t \ge 0$. Rearrangement of Eq. (8.2*d*) and use of Eq. (8.2*e*) yield the relation

$$\ddot{x} = -2\zeta\omega_n\dot{\delta} - \omega_n^2\delta \tag{8.24}$$

When the damping factor is relatively small or is zero, the relation becomes approximately

$$\ddot{x} \cong -\omega_n^2\delta \tag{8.24a}$$

Based on this relation, relative displacement responses of linear SDOF structures approximately equal $(-1/\omega_n^2)$ times their absolute acceleration responses, at all times. Therefore, the maximum in the absolute value of the relative displacement response approximately equals $(1/\omega_n^2)$ times the maximum in the absolute value of the absolute acceleration response, and the following relation holds for the maximax relative displacement SRS.

$$S_\delta(f,\zeta) \cong \frac{1}{(2\pi f)^2} S_{\ddot{x}}(f,\zeta) \qquad 0 < f_{\min} \le f \le f_{\max} < \infty \tag{8.25}$$

Another response measure, called the pseudo-velocity,[11,12] is defined as $\omega_n\delta$, and its SRS is frequently sought. Based on the amplitude scaling principle previously described, its SRS is defined as

$$S_{\omega_n\delta}(f,\zeta) = (2\pi f)S_\delta(f,\zeta) \cong \frac{1}{(2\pi f)} S_{\ddot{x}}(f,\zeta) \qquad 0 < f_{\min} \le f \le f_{\max} < \infty \tag{8.26}$$

For light damping ($\zeta < 0.1$), the pseudo-velocity SRS provides a close approximation to the relative velocity SRS, which in turn is directly proportional to the modal stress imparted to a structure at its natural frequencies.[13] Hence, it is widely used as a measure of the damaging potential of a transient environment for preliminary design purposes, as detailed in Chap. 40.

We now show the SRS of some acceleration pulse functions. The first pulse is the half-sine pulse, defined as

$$\ddot{u}(t) = \begin{cases} \sin(\pi t) & 0 \le t \le 1 \\ 0 & t > 1 \end{cases} \tag{8.27}$$

The half-sine pulse with an amplitude of 1*g* and its corresponding velocity and displacement are graphed in Fig. 8.14*A–C*. The absolute acceleration maximax SRS of the pulse is graphed in Fig. 8.17. The damping in the SDOF structures used to compute the SRS is $\zeta = 0.05$. (This is the damping factor used in computation of all the remaining SRS.) The peak value of the SRS is 1.65*g*, and it occurs at 0.82 Hz.

The haversine pulse is defined by

$$\ddot{u}(t) = \begin{cases} \frac{1}{2}[1 - \cos(2\pi t)] & 0 \le t \le 1 \\ 0 & t > 1 \end{cases} \tag{8.28}$$

The haversine pulse with an amplitude of 1*g* and its corresponding velocity and displacement are graphed in Fig. 8.15*A–C*. The absolute acceleration maximax SRS of the pulse is graphed in Fig. 8.17. The peak value of the SRS is 1.60*g*, and it occurs at 1.02 Hz.

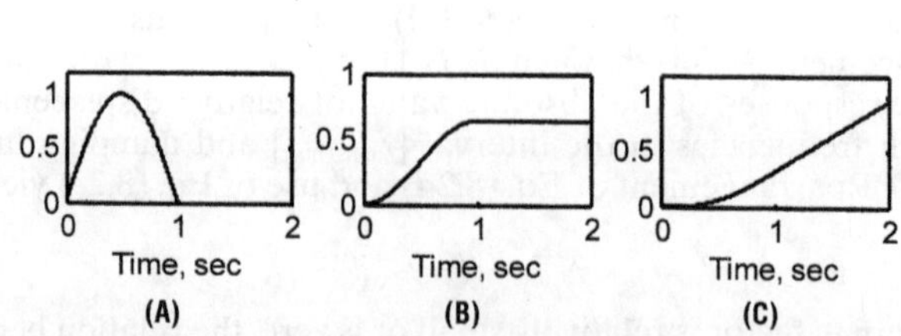

FIGURE 8.14 (*A*) Half-sine acceleration shock pulse and (*B*) its corresponding velocity and (*C*) displacement, in compatible units.

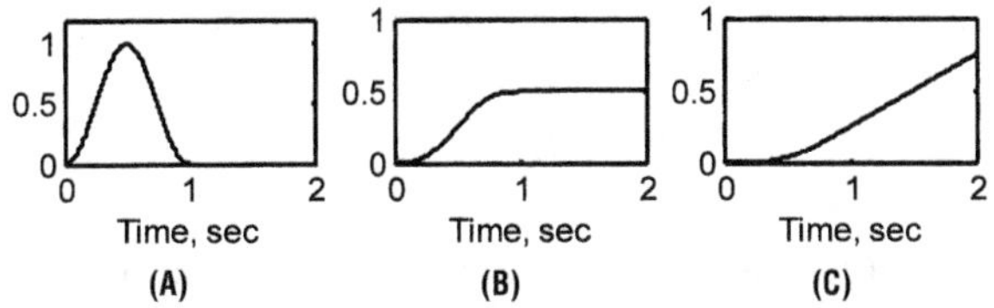

FIGURE 8.15 (*A*) Haversine acceleration shock pulse and (*B*) its corresponding velocity and (*C*) displacement, in compatible units.

The triangle pulse is defined as

$$\ddot{u}(t) = \begin{cases} 2t & 0 \le t < 1/2 \\ 2 - 2t & 1/2 \le t \le 1 \\ 0 & t > 1 \end{cases} \tag{8.29}$$

The triangle pulse with an amplitude of 1*g* and its corresponding velocity and displacement are graphed in Fig. 8.16*A–C*. The absolute acceleration maximax SRS of the pulse is graphed in Fig. 8.17. The peak value of the SRS is 1.42*g*, and it occurs at 0.92 Hz.

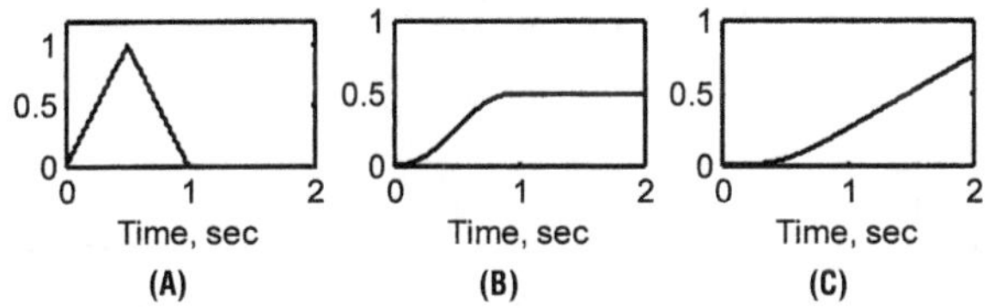

FIGURE 8.16 (*A*) Triangle acceleration shock pulse and (*B*) its corresponding velocity and (*C*) displacement, in compatible units.

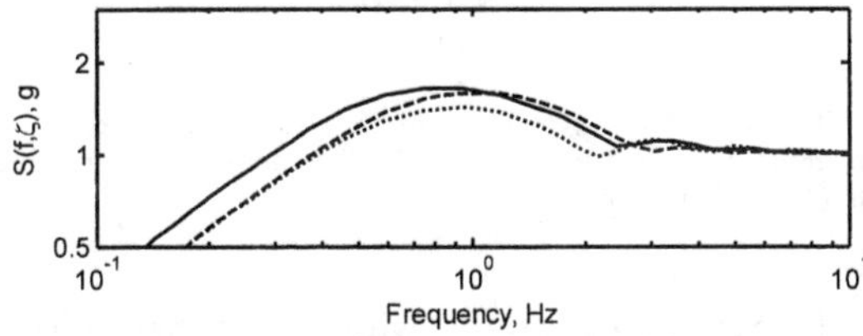

FIGURE 8.17 SRS of half-sine (solid), haversine (dashed), and triangle (dotted) acceleration shock pulses. Maximum SRS of sine pulse is 1.65*g* and occurs at 0.82 Hz. Maximum SRS of haversine pulse is 1.60*g* and occurs at 1.02 Hz. Maximum SRS of triangle pulse is 1.42*g* and occurs at 0.93 Hz. SRS $\zeta = 0.05$.

The three acceleration pulses defined in Eqs. (8.27), (8.28), and (8.29) have several features in common. First, all the pulses have a single lobe; they rise from zero to a finite value, then decrease to zero. The slopes from the start to the peak are all finite. The first integrals of all the pulses are nonzero. For this reason, the pulses represent shocks with an associated velocity change. The rates of change of the displacements past the ends of the acceleration pulses of all the shocks equal the velocities at the ends of the acceleration pulses. The three SRS roll off at low frequencies at a rate of 6 dB/octave.

Some shocks have both first and second integrals that equal zero. An example is the wavelet (or wavsyn) pulse, defined as

$$\ddot{u}(t) = \begin{cases} A \sin(\pi t) \sin(b\pi t) & 0 \le t \le 1 \\ 0 & t > 1 \end{cases} \tag{8.30}$$

where A is an amplitude coefficient and b is an odd integer that establishes the frequency of the shock, in a sense. The wavelet pulse with $A = 1g$ and $b = 3$, and its corresponding velocity and displacement, are graphed in Fig. 8.18*A–C*. The absolute acceleration maximax SRS of the pulse is graphed in Fig. 8.18*D*. The peak value of the SRS is 2.70*g*, and it occurs at 1.79 Hz. Both the ending velocity and the displacement are zero. The roll-off of the SRS at very low frequencies is 18 dB/octave.

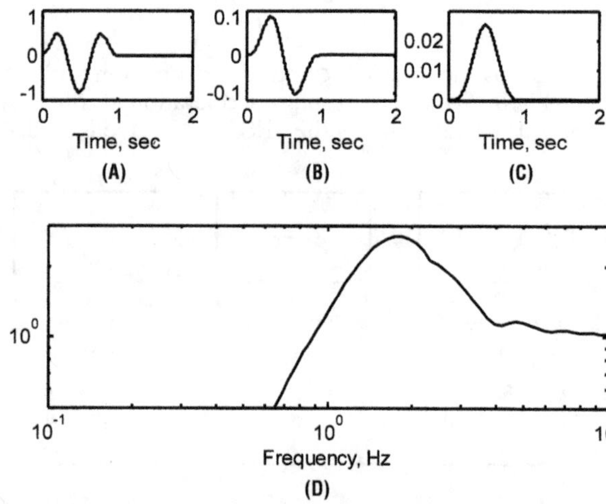

FIGURE 8.18 (*A*) Wavelet acceleration shock pulse and (*B*) its corresponding velocity and (*C*) displacement, in compatible units. (*D*) SRS of acceleration pulse. SRS $\zeta = 0.05$.

Finally, we consider a shock idealization with special characteristics. It is the one-half-cycle square wave, defined by

$$\ddot{u}(t) = \begin{cases} 1 & 0 \le t \le 1 \\ 0 & t > 1 \end{cases} \tag{8.31}$$

The pulse with an amplitude of 1*g* and its corresponding velocity and displacement are graphed in Fig. 8.19*A–C*. The absolute acceleration maximax SRS of the pulse is graphed in Fig. 8.19*D*. The peak value of the SRS is 1.86*g*, and it occurs at 0.5 Hz. The SRS ordinates at all frequencies greater than 0.5 Hz equal 1.86*g*. The reason this SRS does not asymptotically approach the peak value of the input acceleration pulse is that the SRS has no quasi-static region. The perfectly vertical rise at the start of the shock pulse excites dynamic response in all SDOF structures with finite natural frequency. It is for this reason that the one-half-cycle square wave is almost always an unrealistic representation of physical reality. Great care should be taken in using the square wave in analysis. The reason the greatest magnitude of the SRS equals 1.86 is that the peak response of a damped SDOF structure to a unit step acceleration is approximately $1 + \sqrt{1-\zeta^2}e^{-\pi\zeta}$, and this quantity equals 1.86 when the damping factor is $\zeta = 0.05$. If the damping factor were zero, the SRS peak value would be 2.

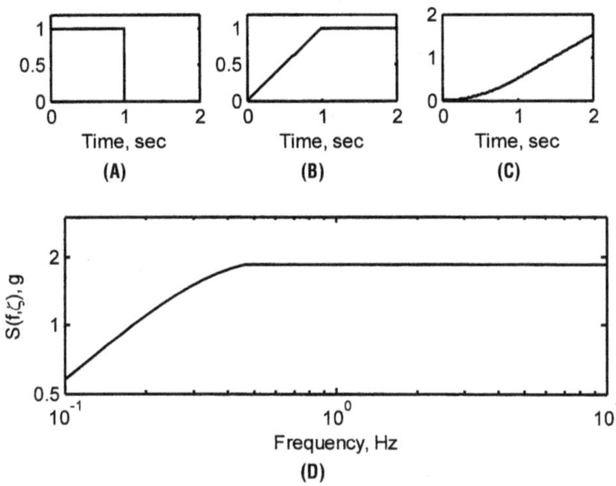

FIGURE 8.19 (*A*) One-half-cycle square-wave acceleration shock pulse and (*B*) its corresponding velocity and (*C*) displacement, in compatible units. (*D*) SRS of acceleration pulse. SRS $\zeta = 0.05$.

REFERENCES

1. Thomson, W., and M. D. Dahleh: "Theory of Vibrations with Applications," 5th ed., Prentice-Hall, Englewood Cliffs, N.J., 1997.
2. Inman, D.: "Engineering Vibration," 3d ed., Prentice-Hall, Englewood Cliffs, N.J., 2008.
3. Meirovitch, L.: "Analytical Methods in Vibrations," The MacMillan Company, London, 1967.
4. Bathe, K. J., and E. L. Wilson: "Numerical Methods in Finite Element Analysis," Prentice-Hall, Englewood Cliffs, N.J., 1976.
5. Zill, D. G.: "A First Course in Differential Equations with Modeling Applications," Brooks/Cole, Pacific Grove, Calif., 2004.
6. Edwards, C. H., and D. E. Penney: "Differential Equations and Linear Algebra," 2d ed., Prentice-Hall, Englewood Cliffs, N.J., 2004.
7. Korn, G. A., and T. M. Korn: "Mathematical Handbook for Scientists and Engineers," Dover Publications, Mineola, N.Y., 2000.
8. Smallwood, D. O.: "An Improved Recursive Formula for Calculating Shock Response Spectra," *Shock and Vibration Bulletin,* **51**(2):211 (1981).
9. Ahlin, K.: "On the Use of Digital Filters for Mechanical System Simulation," *Proc. 74th Shock and Vibration Symposium,* 2003.
10. Smallwood, D. O.: "The Shock Response Spectrum at Low Frequencies," *Shock and Vibration Bulletin,* **56**(1) (1986).
11. Manfredo, G.: "Evaluation of Seismic Energy Demand," *Earthquake Engineering and Structural Dynamics,* **30:**485–499 (2001).
12. Khashaee, P., B. Mahraz, F. Sadek, H. S. Lew, and J. L. Gross: "Distribution of Earthquake Input Energy in Structures," *NISTIR 6903,* NIST, U.S. Department of Commerce, 2003.
13. Gaberson, H. A., and R. H. Chalmers: *Shock and Vibration Bulletin,* **41**(2):31 (1969).

CHAPTER 9
MECHANICAL IMPEDANCE/MOBILITY

Elmer L. Hixson

INTRODUCTION

Impedance methods allow the modeling, analysis, and measurement of linear mechanical dynamical systems. Systems are represented in the frequency domain to predict input characteristics and input/output relationships. The Fourier and Laplace transforms allow the results to be expressed in the time domain. Excitation is usually a pure sinusoid; however, sine sweeps, impulse functions, and random noise can be expressed in the frequency domain with the fast Fourier transform. In the following sections of this chapter, the impedance and its inverse mobility of the basic elements that make up vibratory systems are presented. This is followed by a discussion of combinations of these elements and methods of analysis. Then, various mechanical circuit theorems are described. Such theorems can be used as an aid in the modeling of mechanical circuits and in determining the response of vibratory systems; they are the mechanical equivalents of well-known theorems employed in the analysis of electrical circuits. The analysis of two-port/multiport systems is discussed and some analysis examples are given.

MECHANICAL IMPEDANCE OF VIBRATORY SYSTEMS

The *mechanical impedance* Z of a system is the ratio of a sinusoidal driving force F acting on the system to the resulting velocity v of the system. Its *mechanical mobility* $\mathfrak{M}$ is the reciprocal of the mechanical impedance.

Consider a sinusoidal driving F that has a magnitude F_0 and an angular frequency ω:

$$F = F_0\, e^{j\omega t} \tag{9.1}$$

The application of this force to a linear mechanical system results in a velocity v:

$$v = v_0 e^{j(\omega t + \phi)} \tag{9.2}$$

where v_0 is the magnitude of the velocity and ϕ is the phase angle between F and v.

Then by definition, the mechanical impedance of the system Z (at the point of application of the force) is given by

$$Z = F/v \tag{9.3}$$

BASIC MECHANICAL ELEMENTS

The idealized mechanical systems considered in this chapter are considered to be represented by combinations of basic mechanical elements assembled to form linear mechanical systems. These basic elements are *mechanical resistances* (*dampers*), *springs*, and *masses*. In general, the characteristics of real masses, springs, and mechanical resistance elements differ from those of ideal elements in two respects:

1. A spring may have a nonlinear force-deflection characteristic; a mass may suffer plastic deformation with motion; and the force presented by a resistance may not be exactly proportional to velocity.
2. All materials have some mass; thus, a perfect spring or resistance cannot be made. Some compliance or spring effect is inherent in all elements. Energy can be dissipated in a system in several ways: friction, acoustic radiation, hysteresis, etc. Such a loss can be represented as a resistive component of the element impedance.

Mechanical Resistance (Damper). A mechanical resistance is a device in which the relative velocity between the endpoints is proportional to the force applied to the endpoints. Such a device can be represented by the dashpot of Fig. 9.1*A*, in which the force resisting the extension (or compression) of the dashpot is the result of viscous friction. An ideal resistance is assumed to be made of massless, infinitely rigid elements. The velocity of point *A*, v_1, with respect to the velocity at point *B*, v_2, is

$$v = (v_1 - v_2) = \frac{F_a}{c} \tag{9.4}$$

where c is a constant of proportionality called the *mechanical resistance* or *damping constant.* For there to be a relative velocity v as a result of force at A, there must be an equal reaction force at B. Thus, the transmitted force F_b is equal to F_a. The velocities v_1 and v_2 are measured with respect to the stationary reference G; their difference is the relative velocity v between the end points of the resistance.

With the sinusoidal force of Eq. (9.1) applied to point A with point B attached to a fixed (immovable) point, the velocity v_1 is obtained from Eq. (9.4):

$$v_1 = \frac{F_0 e^{j\omega t}}{c} = v_0 e^{j\omega t} \tag{9.5}$$

Because c is a real number, the force and velocity are said to be "in phase."

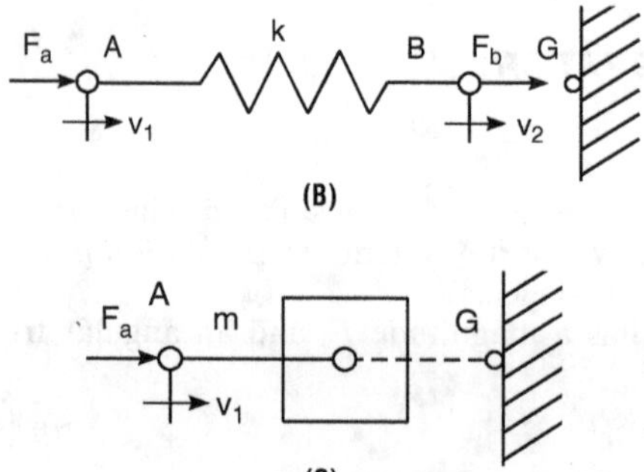

FIGURE 9.1 Schematic representations of basic mechanical elements. (*A*) An ideal mechanical resistance. (*B*) An ideal spring. (*C*) An ideal mass.

The mechanical impedance of the resistance is obtained by substituting from Eqs. (9.1) and (9.5) in Eq. (9.3):

$$Z_c = \frac{F}{v} = c \tag{9.6}$$

The mechanical impedance of a resistance is the value of its damping constant c.

Spring. A linear spring is a device for which the relative displacement between its endpoints is proportional to the force applied. It is illustrated in Fig. 9.1*B* and can be represented mathematically as follows:

$$x_1 - x_2 = \frac{F_a}{k} \tag{9.7}$$

where x_1, x_2 are displacements relative to the reference point G and k is the *spring stiffness.* The stiffness k can be expressed alternately in terms of a *compliance* $C = 1/k$. The spring transmits the applied force, so that $F_b = F_a$.

With the force of Eq. (9.1) applied to point A and with point B fixed, the displacement of point A is given by Eq. (9.7):

$$x_1 = \frac{F_0 e^{j\omega t}}{k} = x_0 e^{j\omega t}$$

The displacement is thus sinusoidal and in phase with the force. The relative velocity of the end connections is required for impedance calculations and is given by the differentiation of x with respect to time:

$$\dot{x} = v = \frac{j\omega F_0 e^{j\omega t}}{k} = \frac{\omega}{k} F_0 e^{j(\omega t + 90°)} \tag{9.8}$$

Substituting Eqs. (9.1) and (9.8) in Eq. (9.3), the impedance of the spring is

$$Z_k = -\frac{jk}{\omega} = \frac{1}{j\omega C} \tag{9.9}$$

Mass. In the ideal mass illustrated in Fig 9.1*C*, the acceleration $\ddot{x}$ of the rigid body is proportional to the applied force F:

$$\ddot{x}_1 = \frac{F_a}{m} \tag{9.10}$$

where m is the mass of the body. By Eq. (9.10), the force F_a is required to give the mass the acceleration $\ddot{x}_1$, and the force F_b is transmitted to the reference G. When a sinusoidal force is applied, Eq. (9.10) becomes

$$\ddot{x}_1 = \frac{F_0 e^{j\omega t}}{m} \tag{9.11}$$

The acceleration is sinusoidal and in phase with the applied force.

Integrating Eq. (9.11) to find velocity,

$$\dot{x} = v = \frac{F}{j\omega m}$$

The mechanical impedance of the mass is the ratio of F to v, so that

$$Z_m = \frac{F_0 e^{j\omega t}}{F_0 e^{j\omega t}/j\omega m} = j\omega m \tag{9.12}$$

Thus, the impedance of a mass is an imaginary quantity that depends on the magnitude of the mass and on the frequency.

COMBINATIONS OF MECHANICAL ELEMENTS

In analyzing the properties of mechanical systems, it is often advantageous to combine groups of basic mechanical elements into single impedances. Methods for calculating the impedances of such combined elements are described in this section. An extensive coverage of mechanical impedance theory and a table of combined elements is given in Ref. 1.

Parallel Elements. Consider the combination of elements shown in Fig. 9.2, a spring and a mechanical resistance. They are said to be in parallel since the same force is applied to both, and both are constrained to have the same relative velocities between their connections. The force F_c required to give the resistance the velocity v is found from Eqs. (9.6) and (9.9).

$$F_c = vZ_c = vc$$

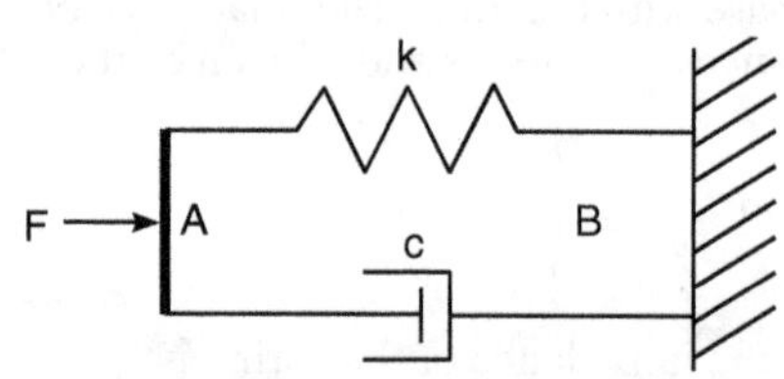

FIGURE 9.2 Schematic representation of a parallel spring-resistance combination.

The force required to give the spring this same velocity is, from Eqs. (9.9)

$$F_k = vZ_k = \frac{vk}{j\omega}$$

The total force F is

$$F = F_c + F_k$$

Since $Z = F/v$,

$$Z = c - j\frac{k}{\omega}$$

Thus, the total mechanical impedance is the sum of the impedances of the two elements.

By extending this concept to any number of parallel elements, the driving force F equals the sum of the resisting forces:

$$F = \sum_{i=1}^{n} vZ_i = v\sum_{i=1}^{n} Z_i \quad \text{and} \quad Z_p = \sum_{i=1}^{n} Z_i \tag{9.13}$$

where Z_p is the total mechanical impedance of the parallel combination of the individual elements Z_i.

Since mobility is the reciprocal of impedance, when the properties of the parallel elements are expressed as mobilities, the total mobility of the combination follows from Eq. (9.13):

$$\frac{1}{\mathfrak{M}_p} = \sum_{i=1}^{n} \frac{1}{\mathfrak{M}_i} \tag{9.14}$$

Series Elements. In Fig. 9.3 a spring and damper are connected so that the applied force passes through both elements to the inertial reference. Then the velocity v is the sum of v_k and v_c. This is a series combination of elements. The method for determining the mechanical impedance of the combination follows.

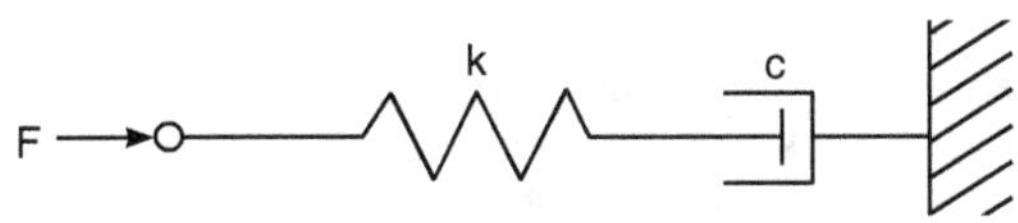

FIGURE 9.3 Schematic representation of a series combination of a spring and a damper.

Consider the more general case of three arbitrary impedances shown in Fig. 9.4. Determine the impedance presented by the end of a number of series-connected elements. Elements Z_1 and Z_2 must have no mass, since a mass always has one end connected to a stationary inertial reference. However, the impedance Z_3 may be a mass. The relative velocities between the end connections of each element are indicated by v_a, v_b, and v_c. The velocities of the connections with respect to the stationary reference point G are indicated by v_1, v_2, and v_3:

$$v_3 = v_c \qquad v_2 = v_3 + (v_2 - v_3) = v_c + v_b$$

$$v_1 = v_2 + (v_1 - v_2) = v_a + v_b + v_c$$

The impedance at point 1 is F/v_1, and the force F is transmitted to all three elements. The relative velocities are

$$v_a = \frac{F}{Z_1} \qquad v_b = \frac{F}{Z_2} \qquad v_c = \frac{F}{Z_3}$$

Thus, the total impedance is defined by

$$\frac{1}{Z} = \frac{F/Z_1 + F/Z_2 + F/Z_3}{F} = \frac{1}{Z_1} + \frac{1}{Z_2} + \frac{1}{Z_3}$$

Extending this principle to any number of massless series elements,

$$\frac{1}{Z_s} = \sum_{i=1}^{n} \frac{1}{Z_i} \tag{9.15}$$

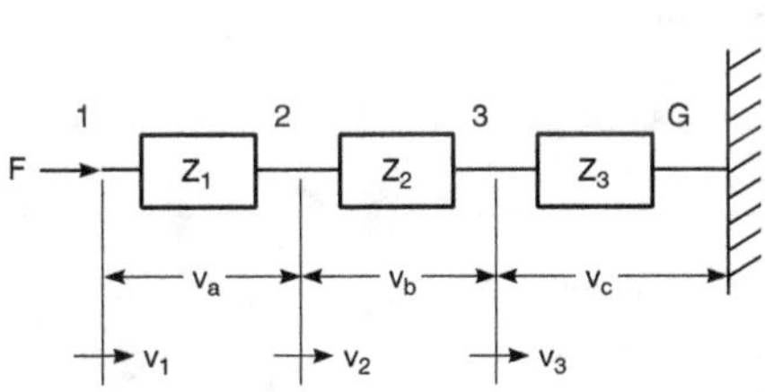

FIGURE 9.4 Generalized three-element system of series-connected mechanical impedances.

where Z_s is the total mechanical impedance of the elements Z_i connected in series.

Since mobility is the reciprocal of impedance, the total mobility of series connected elements (expressed as mobilities) is

$$\mathfrak{M}_s = \sum_{i=1}^{n} \mathfrak{M}_i \tag{9.16}$$

Using Eqs. (9.15) and (9.16), the mobility and impedance for Fig. 9.3 become:

$$\mathfrak{M} = 1/c + j\omega/k \qquad \text{and} \qquad Z = (ck/j\omega)/(c + k/j\omega)$$

MECHANICAL CIRCUIT THEOREMS

The following theorems are the mechanical analogs of theorems widely used in analyzing electric circuits. They are statements of basic principles (or combinations of them) that apply to elements of mechanical systems. In all but Kirchhoff's laws, these theorems apply only to systems composed of linear, bilateral elements. A *linear element* is one in which the magnitudes of the basic elements (c, k, and m) are constant, regardless of the amplitude of motion of the system; a *bilateral element* is one in which forces are transmitted equally well in either direction through its connections.

KIRCHHOFF'S LAWS

1. *The sum of all the forces acting at a point (common connection of several elements) is zero:*

$$\sum_{i}^{n} F_i = 0 \qquad \text{(at a point)} \tag{9.17}$$

This follows directly from the considerations leading to Eq. (9.13).

2. *The sum of the relative velocities across the connections of series mechanical elements taken around a closed loop is zero:*

$$\sum_{i}^{n} v_i = 0 \qquad \text{(around a closed loop)} \tag{9.18}$$

This follows from the considerations leading to Eq. (9.14).

Kirchhoff's laws apply to any system, even when the elements are not linear or bilateral.

Example 9.1. Find the velocity of all the connection points and the forces acting on the elements of the system shown in Fig. 9.5. The system contains two velocity generators v_1 and v_6. Their magnitudes are known, their frequencies are the same, and they are 180° out-of-phase.

A. Using Eq. (9.17), write a force equation for each connection point except *a* and *e*.

At point *b*: $F_1 - F_2 - F_3 = 0$. In terms of velocities and impedances:

$$(v_1 - v_2)Z_1 - (v_2 - v_3)Z_2 - (v_2 - v_4)Z_4 = 0 \tag{a}$$

At point *c*, the two series elements have the same force acting: $F_2 - F_2 = 0$. In terms of velocities and impedances:

$$(v_2 - v_3)Z_2 - (v_3 - v_4)Z_3 = 0 \tag{b}$$

At point *d*: $F_2 + F_3 - F_4 - F_5 = 0$. In terms of velocities and impedances:

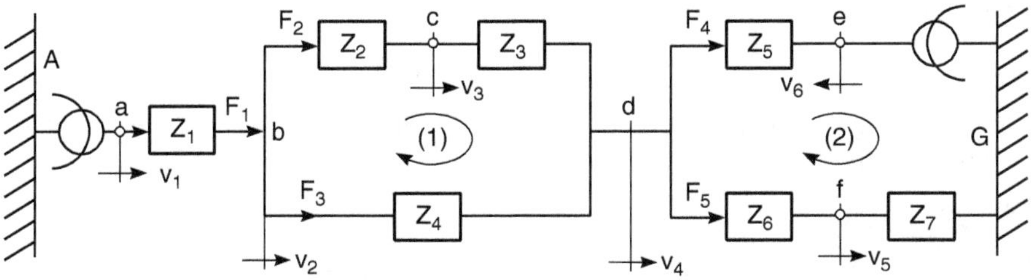

FIGURE 9.5 System of mechanical elements and vibration sources analyzed in Example 9.1 to find the velocity of each connection and the force acting on each element.

$$(v_3 - v_4)Z_3 + (v_2 - v_4)Z_4 - (v_4 + v_6)Z_5 - (v_4 - v_5)Z_6 = 0 \tag{c}$$

Note that v_6 is (+) because of the 180° phase relation to v_1.
At point *f*: $F_5 - F_5 = 0$. In terms of velocities and impedances:

$$(v_4 - v_5)Z_6 - v_5 Z_7 = 0 \tag{d}$$

Since v_1 and v_6 are known, the four unknown velocities v_2, v_3, v_4, and v_5 may be determined by solving the four simultaneous equations above. After the velocities are obtained, the forces may be determined from the following:

$$F_1 = (v_1 - v_2)Z_1 \qquad F_2 = (v_2 - v_3)Z_2 = (v_3 - v_4)Z_3$$

$$F_3 = (v_2 - v_4)Z_4 \qquad F_4 = (v_4 + v_6)Z_5$$

$$F_5 = (v_4 - v_5)Z_6 = v_5 Z_7$$

B. The method of *node forces.* Equations (*a*) through (*d*) above can be rewritten as follows:

$$v_1 Z_1 = (Z_1 + Z_2 + Z_3)v_2 - Z_2 v_3 - Z_4 v_4 \tag{a'}$$

$$0 = -Z_2 v_2 + (Z_2 + Z_3)v_3 - Z_3 v_4 \tag{b'}$$

$$0 = -Z_4 v_2 - Z_3 v_3 + (Z_3 + Z_4 + Z_5 + Z_6)v_4 - Z_6 v_5 \tag{c'}$$

$$-v_6 Z_5 = -Z_6 v_4 + (Z_6 + Z_7)v_5 \tag{d'}$$

These equations can be written by inspection of the schematic diagram by the following rule: *At each point with a common velocity (force node), equate the force generators to the sum of the impedances attached to the node multiplied by the velocity of the node, minus the impedances multiplied by the velocities of their other connection points.*

When the equations are written so that the unknown velocities form columns, the equations are in the proper form for a determinant solution for any of the unknowns. Note that the determinant of the *Z*'s is symmetrical about the main diagonal. This condition always exists and provides a check for the correctness of the equations.

C. Using Eq. (9.18), write a velocity equation in terms of force and mobility around enough closed loops to include each element at least once. In Fig. 9.5, note that

$$F_3 = F_1 - F_2 \qquad \text{and} \qquad F_5 = F_1 - F_4$$

Around loop (1):

$$F_2(\mathfrak{M}_2 + \mathfrak{M}_3) - (F_1 - F_2)\mathfrak{M}_4 = 0 \qquad (e)$$

The minus sign preceding the second term results from going across the element 4 in a direction opposite to the assumed force acting on it.

Around loop (2):

$$F_4\mathfrak{M}_5 - v_6 - (F_1 - F_4)(\mathfrak{M}_6 + \mathfrak{M}_7) = 0 \qquad (f)$$

A summation of velocities from A to G along the upper path forms the following closed loop:

$$v_1 + F_1\mathfrak{M}_1 + F_2(\mathfrak{M}_2 + \mathfrak{M}_3) + F_4\mathfrak{M}_5 - v_6 = 0 \qquad (g)$$

Equations (*e*), (*f*), and (*g*) then may be solved for the unknown forces F_1, F_2, and F_4. The other forces are $F_3 = F_1 - F_2$ and $F_5 = F_1 - F_4$. The velocities are:

$$v_2 = v_1 - F_1\mathfrak{M}_1 \qquad v_3 = v_2 - F_2\mathfrak{M}_2 \qquad v_4 = v_2 - F_3\mathfrak{M}_4 \qquad v_5 = F_5\mathfrak{M}_7$$

When a system includes more than one source of vibration energy, a Kirchhoff's law analysis with impedance methods can be made only if all the sources are operating at the same frequency. This is the case because sinusoidal forces and velocities can add as phasors only when their frequencies are identical. However, they may differ in magnitude and phase. Kirchhoff's laws still hold for instantaneous values and can be used to write the differential equations of motion for any system.

RECIPROCITY THEOREM

If a force generator operating at a particular frequency at some point (1) *in a system of linear bilateral elements produces a velocity at another point* (2), *the generator can be removed from* (1) *and placed at* (2); *then the former velocity at* (2) *will exist at* (1), *provided the impedances at all points in the system are unchanged.* This theorem also can be stated in terms of a vibration generator that produces a certain velocity at its point of attachment (1), regardless of force required, and the force resulting on some element at (2).

Reciprocity is an important characteristic of linear bilateral elements. It indicates that a system of such elements can transmit energy equally well in both directions. It further simplifies the calculation on two-way energy transmission systems since the characteristics need be calculated for only one direction.

SUPERPOSITION THEOREM

If a mechanical system of linear bilateral elements includes more than one vibration source, the force or velocity response at a point in the system can be determined by adding the response to each source, taken one at a time (the other sources supplying no energy but replaced by their internal impedances).

The internal impedance of a vibrational generator is that impedance presented at its connection point when the generator is supplying no energy. This theorem finds useful application in systems having several sources. A very important application arises when the applied force is nonsinusoidal but can be represented by a Fourier series. Each term in the series can be considered a separate sinusoidal generator. The response at any point in the system can be calculated for each generator by using the impedance values at that frequency. Each response term becomes a term in the Fourier series representation of the total response function. The overall response as a function of time then can be synthesized from the series.

Figure 9.6 illustrates an application of superposition. The velocities v_c' and v_c'' can be determined by the methods of Example 9.1. Then the velocity v_c is the sum of v_c' and v_c''.

THÉVENIN'S EQUIVALENT SYSTEM

If a mechanical system of linear bilateral elements contains vibration sources and produces an output to a load at some point at any particular frequency, the whole system can be represented at that frequency by a single constant-force generator F_c in parallel with a single impedance Z_i connected to the load. Thévenin's equivalent-system representation for a physical system may be determined by the following experimental procedure: Denote by F_c the force which is transmitted by the attachment point of the system to an infinitely rigid fixed point; this is called the *blocked force.* When the load connection is disconnected and perfectly free to move, a free velocity v_f is measured. Then the parallel impedance Z_i is F_c/v_f. The impedance Z_i also can be determined by measuring the internal impedance of the system when no source is supplying motional energy.

If the values of all the system elements in terms of ideal elements are known, F_c and Z_i may be determined analytically. A great advantage is derived from this representation in that attention is focused on the characteristics of a system at its output point and not on the details of the elements of the system. This allows an easy prediction of the response when different loads are attached to the output connection. After a final load condition has been determined, the system may be analyzed in detail for strength considerations.

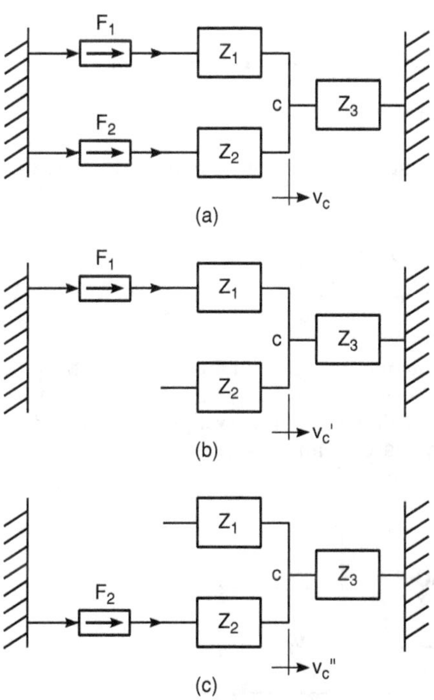

FIGURE 9.6 System of mechanical elements including two force generators used to illustrate the principle of superposition.

NORTON'S EQUIVALENT SYSTEM

A mechanical system of linear bilateral elements having vibration sources and an output connection may be represented at any particular frequency by a single constant-velocity generator v_f in series with an internal impedance Z_i.

This is the series system counterpart of Thévenin's equivalent system where v_f is the free velocity and Z_i is the impedance as defined above. The same advantages in analysis exist as with Thévenin's parallel representation. The most advantageous one depends upon the type of structure to be analyzed. In the experimental determination of an equivalent system, it is usually easier to measure the free velocity than the blocked force on large heavy structures, while the converse is true for light structures. In any case, one representation is easily derived from the other. When v_f and Z_i are determined, $F_c = v_f Z_i$.

MECHANICAL 2-PORTS

Consider the "black box" shown in Fig. 9.7. It may have many elements between terminals (ports) (1) and (2). The forces and velocities at the ports can be determined by the use of 2-port equations in terms of impedances and mobilities. The impedance parameter equations are

$$F_1 = Z_{11}v_1 + Z_{12}v_2 \qquad \text{and} \qquad F_2 = Z_{21}v_1 + Z_{22}v_2$$

The Z parameters can be determined by measurements or from a known circuit model. These parameters are defined as follows:

1. For $v_2 = 0$ (port 2 blocked), $Z_{11} = F_1/v_1$ and $Z_{21} = F_2/v_1$.
2. For $v_1 = 0$ (port 1 blocked), $Z_{12} = F_1/v_2$ and $Z_{22} = F_2/v_2$

These can be called the "blocked impedance parameters."

The mobility parameter equations for this situation are as follows:

$$v_1 = \mathfrak{M}_{11}F_1 + \mathfrak{M}_{12}F_2 \qquad \text{and} \qquad v_2 = \mathfrak{M}_{12}F_1 + \mathfrak{M}_{22}F_2$$

These $\mathfrak{M}$ parameters can be determined by measurement or from a model. The definitions are as follows:

1. For $F_2 = 0$ (port 2 free), $\mathfrak{M}_{11} = v_1/F_1$ and $\mathfrak{M}_{12} = v_2/F_1$.
2. For $F_1 = 0$ (port 1 free), $\mathfrak{M}_{21} = v_1/F_2$ and $\mathfrak{M}_{22} = v_2/F_2$.

These can be called "free mobility parameters."

Note that for large, massive structures, it may be difficult to clamp the ports to measure the impedance parameters. In this case, the mobility parameters requiring free conditions may be more appropriate. Likewise, for very light structures, the impedance parameters may be more appropriate. In any case, one set of parameters can be determined from the other by matrix inversion.

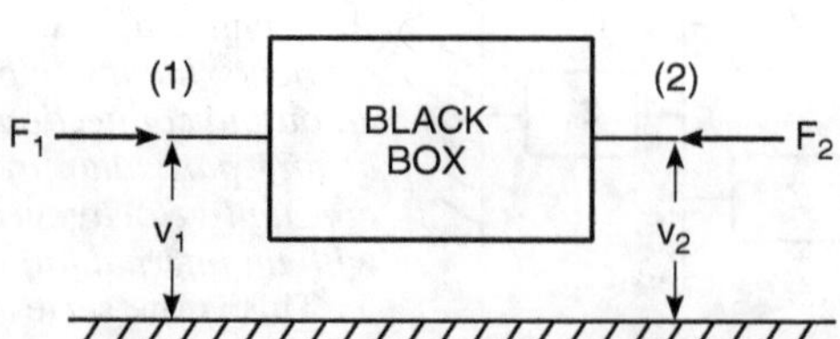

FIGURE 9.7 "Black box" representation of a mechanical system.

Every attachment point on a mechanical system can have six degrees of freedom, transnational and rotational; Reference 2 is a guide to the measurement of the complete mobility matrix of such a system.

MECHANICAL IMPEDANCE MEASUREMENTS AND APPLICATIONS

MEASUREMENTS

Transducers (Chap. 10), instrumentation (Chap. 13), and spectrum analyzers (Chap. 14) are essential subjects related to impedance measurements. Some special considerations are given here. The measurement of mechanical impedance involves the application of a sinusoidal force and the measurement of the complex ratio of force to the resulting velocity. Many combinations of transducers are capable of performing these measurements. However, the most effective method is to use an impedance transducer such as that shown in Fig. 9.8. These devices are available from suppliers of vibration-measuring sensors. As shown in Fig. 9.8, the force supplied by the vibration exciter passes through a force sensor to the unknown Z_x, and the motion is measured by an accelerometer whose output is integrated to obtain velocity. The accelerometer measures the true motion, but the force sensor measures the force required to move the accelerometer and its mounting structure, as well as the force to Z_x. This extra mass is usually called the "mass below the force gage." The impedance is then as follows:

$$Z_x = j\omega(K_f/K_a)(E_f/E_a) - j\omega m_o$$

where E_f and E_a are the force and acceleration phasor potentials, K_f in N/volt is the force gage sensitivity, K_a in m/s^2/volt is the accelerometer sensitivity, and m_0 is mass below the force gage. The ratio K_f/K_a and m_0 can be determined by a calibration as follows:

1. With no attachment, $Z_x = 0$.
Then $m_0 = (K_f/K_a)\,(E_f/E_a)_0$.

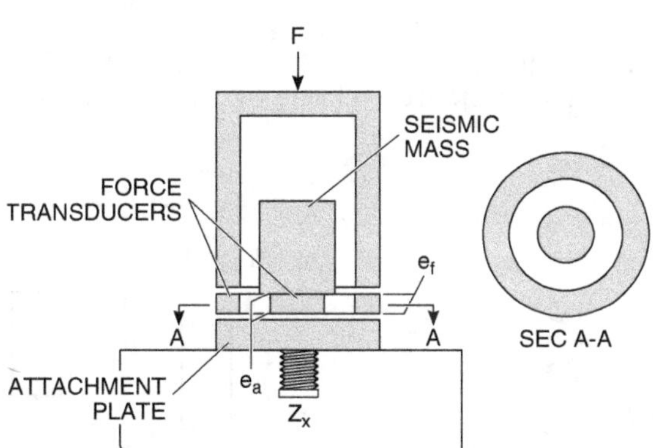

FIGURE 9.8 Device for the measurement of mechanical impedance in which force and acceleration are measured.

2. Attach a known mass M.
   Then $M + m_0 = (K_f/K_a)\,(E_f/E_a)_1$,
   which yields $m_0 = M/[(E_f/E_a)_0/(E_f/E_f/E_a)_1] - 1$.
3. Thus $[K_f/K_a] = m_0/(E_f/E_a)_0$.

With the aid of a two-channel analyzer (see Chap. 14) or appropriate signal processing software (see Chap. 19), forces such as sine sweeps, broad bandwidth random noise, or impacts can be used for these measurements. The Fourier transform of the force and acceleration potentials will provide correct sinusoidal terms. The impact method can be implemented with a hammer equipped with a force gage and accelerometer, as detailed in Chap. 21.

References 3 and 4 are standards for impact transducers and for measurement methods for translational excitation.

APPLICATIONS

The impedance concept is widely used in the study of mechanical systems. Three practical applications are presented here. See Ref. 5.

Application 1. Assume one wishes to determine the free motion at a point on a structure that would be altered by the attachment of a sensor such as an accelerometer. The procedure is illustrated in Fig. 9.9, and involves the following steps.

1. Turn off the source causing the vibration v_f.
2. Measure the internal impedance Z_0 at a point A over the expected frequency range.
3. Attach the measuring device whose known impedance is Z_m and measure v_m.
4. Draw the Norton equivalent circuit at point A with Z_m attached. Note that Z_0 is attached to the reference since it may be masslike.
5. Calculate the free velocity from

$$v_f = v_m Z_m/(Z_0 + Z_m)$$

Application 2. Assume one wishes to choose a vibration isolator between a vibrating machine and a flexible structure. The criteria are to reduce the ratio of the velocity of the structure to the free velocity of the machine below some desired

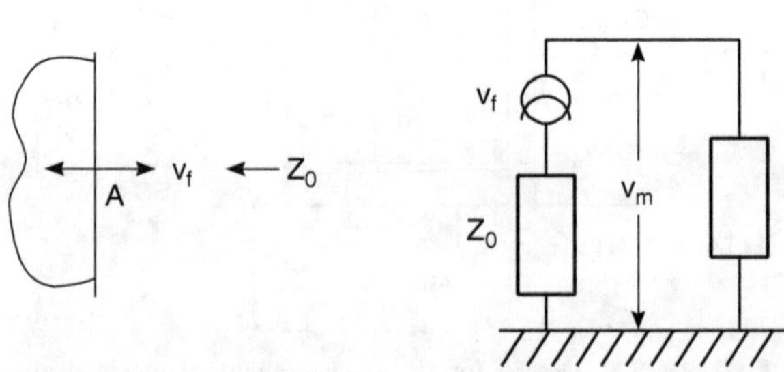

FIGURE 9.9 Measurement of free motion.

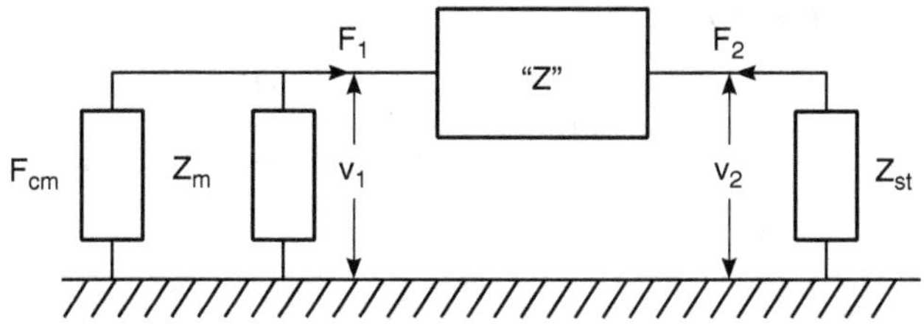

FIGURE 9.10 Vibration isolation application.

value, or to reduce the ratio of the force transmitted to the structure to the blocked force of the machine below some desired value. The procedure is as follows:

1. Model the system as shown in Fig. 9.10, where F_{cm} is the blocked force and Z_m is the impedance at the attachment point. The structural impedance at the attachment point is Z_{st} and "Z" is a set of the blocked Z parameters of the isolator that satisfy

$$F_1 = Z_{11}v_1 + Z_{12}v_2 \qquad \text{and} \qquad F_2 = Z_{21}v_1 + Z_{22}v_2$$

2. Add the source and structure to obtain

$$F_1 = F_{cm} - Z_m v_1 \qquad \text{and} \qquad F_2 = -Z_{st}v_2$$

The system equations then become

$$F_{cm} = (Z_{11} + Z_m)v_1 + Z_{12}v_2 \qquad \text{and} \qquad 0 = Z_{21}v_1 + (Z_{22} + Z_{st})v_1$$

3. Solve for the force to the structure $F_{st} = F_2$ from

$$F_{st}/F_{cm} = Z_{12}Z_{st}/[(Z_{11} + Z_m)(Z_{22} + Z_{st}) - Z_{12}Z_{21}]$$

This result follows from $v_{st} = F_{st}/Z_{st}$ and $v_{fm} = F_{cm}/Z_m$.

4. The ratio of the velocity of the structure to the free velocity of the machine is then given by

$$v_{st}/v_{fm} = Z_{21}Z_m/[(Z_{11} + Z_m)(Z_{22} + Z_{st}) - Z_{12}Z_{21}]$$

Typical vibration isolators can be modeled as shown in Fig. 9.11, where the Z parameters are given by

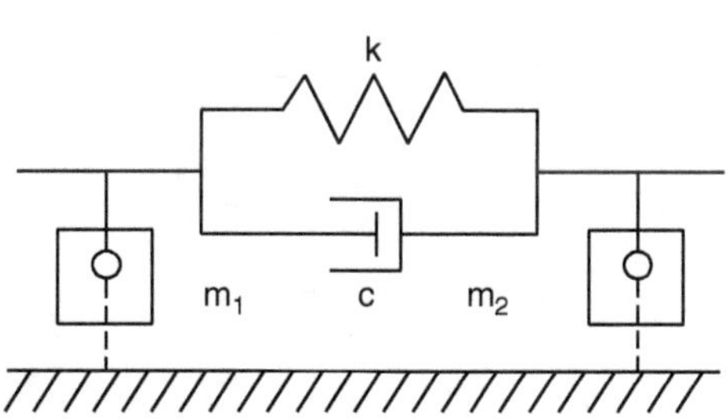

FIGURE 9.11 Vibration isolator model.

$$Z_{11} = c + j\omega m_1 + k/j\omega; \quad Z_{22} = c + j\omega m_2 + k/j\omega; \quad Z_{12} = Z_{21} = c + k/j\omega$$

The values of c, k, m_1, and m_2 should be available from the manufacturer, or they can be measured. Using the measured values of Z_m and Z_{st}, the transmissibilities of the force and velocity can be computed from the expression above, and plots of these functions versus frequency can be compared to the desired criteria.

Application 3. Assume one wishes to isolate a piece of equipment from a vibrating structure. The procedure is essentially the same as detailed in Application 2. Specifically, measure the blocked force F_{st}, or the free velocity v_{st}, of the structure. Then in Fig. 9.10, replace the F_{cm} and Z_m with F_{st} and Z_{st}, and replace Z_{st} with Z_m. Proceed to write the system 2-port equations and solve for the force or velocity transmissibility.

REFERENCES

1. Hixson, E. L.: "Mechanical Impedance and Mobility," Chap. 10, in C. M. Harris and C. E. Crede (eds.), "Shock and Vibration Handbook," 1st ed., McGraw-Hill Book Company, New York, 1961.
2. "A Guide to the Experimental Determination of Rotational Mobility Properties and the Complete Mobility Matrix," American National Standard, ANSI S2.34-1984.
3. "Methods for the Experimental Determination of Mechanical Mobility," Part I, American National Standard, ANSI S2.31-1979.
4. "Methods for the Experimental Determination of Mechanical Mobility," Part II, American National Standard, ANSI S.32-1982.
5. Ewins, D. J., and M. G. Sainsbury, "Mobility Measurements for Vibration Analysis of Connected Structures," *Sound and Vibration Bulletin,* Part I: 105–122 (1972).
6. Carlson, U.: *J. Acous. Soc. Am.,* **97**(2):1345 (1995).

CHAPTER 10
SHOCK AND VIBRATION TRANSDUCERS

Anthony S. Chu

INTRODUCTION

This chapter on vibration transducers is the first in a group of seven chapters on the measurement of shock and vibration. Chapter 13 describes typical instrumentation used in making measurements with such devices; Chap. 15 covers the mounting of vibration transducers and how they may be calibrated under field conditions; more precise calibration under laboratory conditions is described in detail in Chap. 11. The selection of vibration transducers is treated in Chap. 15 and this chapter. This chapter defines the terms and describes the general principles of the most common transducers; it also sets forth the mathematical basis for the use of shock and vibration transducers and includes a brief description of piezoelectric accelerometers, piezoresistive accelerometers, piezoelectric force and impedance gages, and piezoelectric drivers, along with a review of their performance and characteristics. Finally, the following various special types of transducers are considered: laser Doppler vibrometers, fiber-optic reflective displacement sensors, electrodynamic (velocity coil) pickups, differential-transformer (LVDT) pickups, and capacitance-type transducers.

Certain solid-state materials are electrically responsive to mechanical force; they often are used as the mechanical-to-electrical transduction elements in shock and vibration transducers. Generally exhibiting high elastic stiffness, these materials can be divided into two categories: the *self-generating type,* in which electric charge is displaced as a direct result of applied force, and the *passive-circuit type,* in which applied force causes a change in the electrical characteristics of the material.

A *piezoelectric* material is one which displaces an electric charge proportional to the stress applied to it, within its linear elastic range. Piezoelectric materials are of the self-generating type. A *piezoresistive* material is one whose electrical resistance depends upon applied force. Piezoresistive materials are of the passive-circuit type.

A *transducer* (sometimes called a *pickup* or *sensor*) is a device which converts shock or vibratory motion into an optical, a mechanical, or, most commonly, an electrical signal that is proportional to a parameter of the experienced motion.

A *transducing element* is the part of the transducer that accomplishes the conversion of motion into the signal.

A *measuring instrument* or *measuring system* converts shock and vibratory motion into an observable form that is directly proportional to a parameter of the experienced motion. It may consist of a transducer with transducing element, signal-conditioning equipment, and device for displaying the signal. An instrument contains all of these elements in one package, while a system utilizes separate packages.

An *accelerometer* is a transducer whose output is proportional to the acceleration input. The output of a force gage is proportional to the force input; an impedance gage contains both an accelerometer and a force gage.

CLASSIFICATION OF MOTION TRANSDUCERS

In principle, shock and vibration motions are measured with reference to a point fixed in space by either of two fundamentally different types of transducers:

1. *Fixed-reference transducer.* One terminal of the transducer is attached to a point that is fixed in space; the other terminal is attached (e.g., mechanically, electrically, optically) to the point whose motion is to be measured.
2. *Mass-spring transducer (seismic transducer).* The only terminal is the base of a mass-spring system; this base is attached at the point where the shock or vibration is to be measured. The motion at the point is inferred from the motion of the mass relative to the base.

MASS-SPRING TRANSDUCERS (SEISMIC TRANSDUCERS)

In many applications, such as moving vehicles or missiles, it is impossible to establish a fixed reference for shock and vibration measurements. Therefore, many transducers use the response of a mass-spring system to measure shock and vibration. A mass-spring transducer is shown schematically in Fig. 10.1; it consists of a mass m suspended from the transducer case a by a spring of stiffness k. The motion of the mass within the case may be damped by a viscous fluid or electric current, symbolized by a dashpot with damping coefficient c. It is desired to measure the motion of the moving part whose displacement with respect to fixed space is indicated by u. When the transducer case is attached to the moving part, the transducer may be used to measure displacement, velocity, or acceleration, depending on the portion of the frequency range which is utilized and whether the relative displacement or relative velocity $d\delta/dt$ is sensed by the transducing element. The typical response of the mass-spring system is analyzed in the following paragraphs and applied to the interpretation of transducer output.

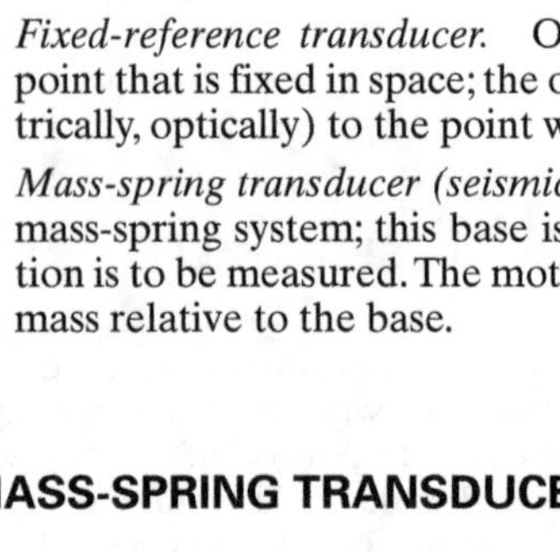

FIGURE 10.1 Mass-spring type of vibration-measuring instrument consisting of a mass m supported by spring k and viscous damper c. The case a of the instrument is attached to the moving part whose vibratory motion u is to be measured. The motion u is inferred from the relative motion δ between the mass m and the case a.[1]

Consider a transducer whose case experiences a displacement motion u, and let the relative displacement between the mass and the case be δ. Then the motion of the mass with respect to a reference fixed in space is $\delta + u$, and the force causing its acceleration is $m[d^2(\delta + u)/dt^2]$. Thus, the force applied by the mass to the spring and dashpot assembly is $-m[d^2(\delta + u)/dt^2]$. The force applied by the spring is $-k\delta$, and the force applied by the damper is $-c(d\delta/dt)$, where c is the damping coefficient. Adding all force terms and equating the sum to zero,

$$-m\frac{d^2(\delta + u)}{dt^2} - c\frac{d\delta}{dt} - k\delta = 0 \tag{10.1}$$

Equation (10.1) may be rearranged:

$$m\frac{d^2\delta}{dt^2} + c\frac{d\delta}{dt} + k\delta = -m\frac{d^2u}{dt^2} \tag{10.2}$$

Assume that the motion u is sinusoidal, $u = u_0 \cos \omega t$, where $\omega = 2\pi f$ is the angular frequency in radians per second and f is expressed in cycles per second. Neglecting transient terms, the response of the instrument is defined by $\delta = \delta_0 \cos(\omega t - \theta)$; then the solution of Eq. (10.2) is

$$\frac{\delta_0}{u_0} = \frac{\omega^2}{\sqrt{\left(\frac{k}{m} - \omega^2\right)^2 + \left(\omega\frac{c}{m}\right)^2}} \tag{10.3}$$

$$\theta = \tan^{-1}\frac{\omega\frac{c}{m}}{\frac{k}{m} - \omega^2} \tag{10.4}$$

The undamped natural frequency f_n of the instrument is the frequency at which

$$\frac{\delta_0}{u_0} = \infty$$

when the damping is zero ($c = 0$), or the frequency at which $\theta = 90°$. From Eqs. (10.3) and (10.4), this occurs when the denominators are zero:

$$\omega_n = 2\pi f_n = \sqrt{\frac{k}{m}} \qquad \text{rad/sec} \tag{10.5}$$

Thus, a stiff spring and/or light mass produces an instrument with a high natural frequency. A heavy mass and/or compliant spring produces an instrument with a low natural frequency.

The damping in a transducer is specified as a *fraction of critical damping.* Critical damping c_c is the minimum level of damping that prevents a mass-spring transducer from oscillating when excited by a step function or other transient. It is defined by

$$c_c = 2\sqrt{km} \tag{10.6}$$

Thus, the fraction of critical damping ζ is

$$\zeta = \frac{c}{c_c} = \frac{c}{2\sqrt{km}} \tag{10.7}$$

It is convenient to define the excitation frequency ω for a transducer in terms of the undamped natural frequency ω_n by using the dimensionless frequency ratio

ω/ω_n. Substituting this ratio and the relation defined by Eq. (10.7), Eqs. (10.3) and (10.4) may be written

$$\frac{\delta_0}{u_0} = \frac{\left(\frac{\omega}{\omega_n}\right)^2}{\sqrt{\left[1-\left(\frac{\omega}{\omega_n}\right)^2\right]^2 + \left(2\zeta\frac{\omega}{\omega_n}\right)^2}} \tag{10.8}$$

$$\theta = \tan^{-1}\frac{2\zeta\frac{\omega}{\omega_n}}{1-\left(\frac{\omega}{\omega_n}\right)^2} \tag{10.9}$$

The response of the mass-spring transducer given by Eq. (10.8) may be expressed in terms of the acceleration $\ddot{u}$ of the moving part by substituting $\ddot{u}_0 = -u_0\omega^2$. Then the ratio of the relative displacement amplitude δ_0 between the mass m and transducer case a to the impressed acceleration amplitude $\ddot{u}_0$ is

$$\frac{\delta_0}{\ddot{u}_0} = -\frac{1}{\omega_n^2}\left[\frac{1}{\sqrt{\left[1-\left(\frac{\omega}{\omega_n}\right)^2\right]^2 + \left(2\zeta\frac{\omega}{\omega_n}\right)^2}}\right] \tag{10.10}$$

The relation between δ_0/u_0 and the frequency ratio ω/ω_n is shown graphically in Fig. 10.2 for several values of the fraction of critical damping ζ. Corresponding curves for $\delta_0/\ddot{u}_0$ are shown in Fig. 10.3. The phase angle θ defined by Eq. (10.9) is shown graphically in Fig. 10.4, using the scale at the left side of the figure. Corresponding phase angles between the relative displacement δ and the velocity $\dot{u}$ and acceleration $\ddot{u}$ are indicated by the scales at the right side of the figure.

ACCELERATION-MEASURING TRANSDUCERS

As indicated in Fig. 10.3, the relative displacement amplitude δ_0 is directly proportional to the acceleration amplitude $\ddot{u}_0 = -u_0\omega^2$ of the sinusoidal vibration being measured, at small values of the frequency ratio ω/ω_n. Thus, when the natural frequency ω_n of the transducer is high, the transducer is an accelerometer. If the transducer is undamped, the response curve of Fig. 10.3 is substantially flat when $\omega/\omega_n < 0.2$, approximately. Consequently, an undamped accelerometer can be used for the measurement of acceleration when the vibration frequency does not exceed approximately 20 percent of the natural frequency of the accelerometer. The range of measurable frequency increases as the damping of the accelerometer is increased, up to an optimum value of damping. When the fraction of critical damping is approximately 0.65, an accelerometer gives accurate results in the measurement of vibration at frequencies as great as approximately 60 percent of the natural frequency of the accelerometer.

As indicated in Fig. 10.3, the useful frequency range of an accelerometer increases as its natural frequency ω_n increases. However, the deflection of the spring in an accelerometer is inversely proportional to the square of the natural frequency; i.e., for a given value of $\ddot{u}_0$, the relative displacement is directly proportional to $1/\omega_n^2$ [see Eq. (10.10)]. As a consequence, the electrical signal from the transducing element may be very small, thereby requiring a large amplification to increase the signal to a level at which recording is feasible. For this reason, a compromise usually is

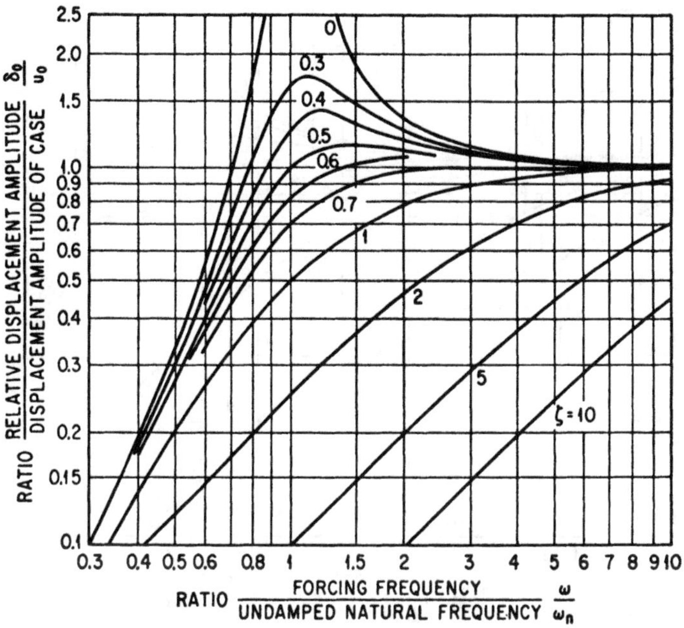

FIGURE 10.2 Displacement response δ_0/u_0 of a mass-spring system subjected to a sinusoidal displacement $\ddot{u} = u_0 \sin \omega t$. The fraction of critical damping ζ is indicated for each curve.

made between high sensitivity and the highest attainable natural frequency, depending upon the desired application.

ACCELEROMETER REQUIREMENTS FOR SHOCK

High-Frequency Response. The capability of an accelerometer to measure shock may be evaluated by observing the response of the accelerometer to acceleration pulses. Ideally, the response of the accelerometer (i.e., the output of the transducing element) should correspond identically with the pulse. In general, this result may be approached but not attained exactly. Three typical pulses and the corresponding responses of accelerometers are shown in Figs. 10.5 to 10.7. The pulses are shown in dashed lines. A sinusoidal pulse is shown in Fig. 10.5, a triangular pulse in Fig. 10.6, and a rectangular pulse in Fig. 10.7. Curves of the response of the accelerometer are shown in solid lines. For each of the three pulse shapes, the response is given for ratios τ_n/τ of 1.014 and 0.203, where τ is the pulse duration and $\tau_n = 1/f_n$ is the natural period of the accelerometer. These response curves, computed for the fraction of critical damping $\zeta = 0, 0.4, 0.7$, and 1.0, indicate the following general relationships:

1. The response of the accelerometer follows the pulse most faithfully when the natural period of the accelerometer is smallest relative to the period of the pulse. For example, the responses at A in Figs. 10.5 to 10.7 show considerable deviation

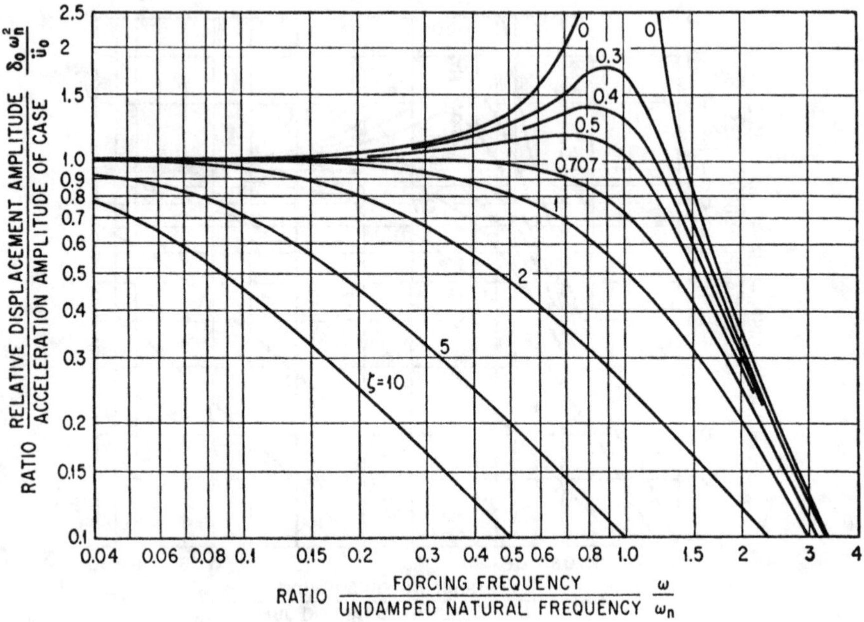

FIGURE 10.3 Relationship between the relative displacement amplitude δ_0 of a mass-spring system and the acceleration amplitude $\ddot{u}_0$ of the case. The fraction of critical damping ζ is indicated for each response curve.

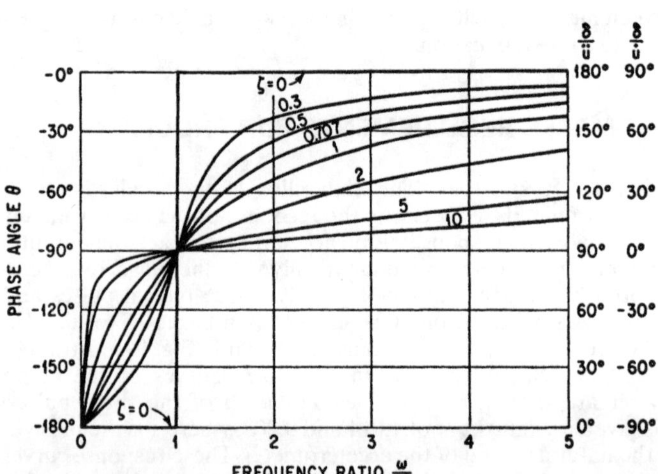

FIGURE 10.4 Phase angle of a mass-spring transducer when used to measure sinusoidal vibration. The phase angle θ on the left-hand scale relates the relative displacement δ to the impressed displacement, as defined by Eq. (10.9). The right-hand scales relate the relative displacement δ to the impressed velocity and acceleration.

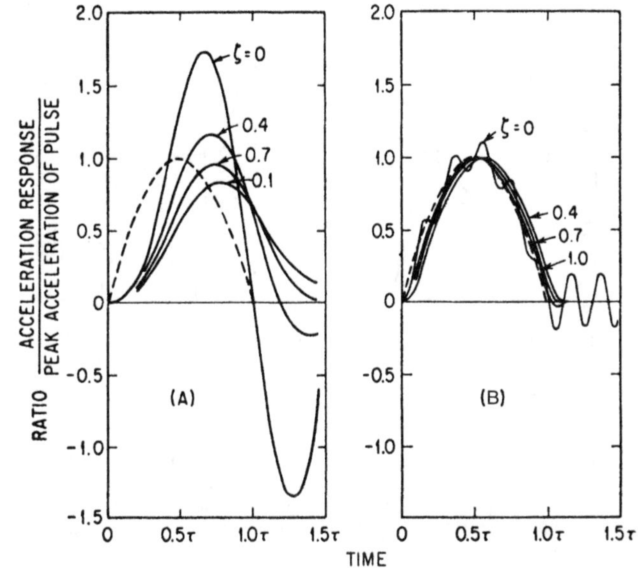

FIGURE 10.5 Acceleration response to a half-sine pulse of acceleration of duration τ (dashed curve) of a mass-spring transducer whose natural period τ_n is equal to: (*A*) 1.014 times the duration of the pulse and (*B*) 0.203 times the duration of the pulse. The fraction of critical damping ζ is indicated for each response curve. (*Levy and Kroll.*[1])

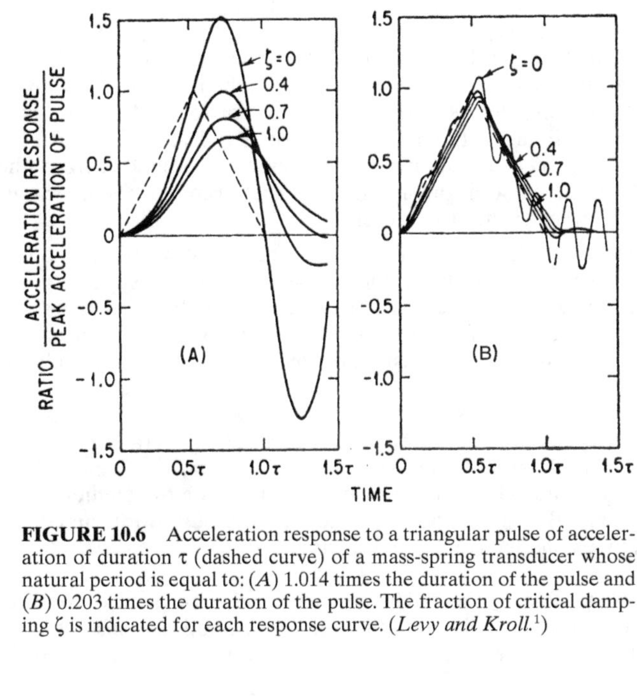

FIGURE 10.6 Acceleration response to a triangular pulse of acceleration of duration τ (dashed curve) of a mass-spring transducer whose natural period is equal to: (*A*) 1.014 times the duration of the pulse and (*B*) 0.203 times the duration of the pulse. The fraction of critical damping ζ is indicated for each response curve. (*Levy and Kroll.*[1])

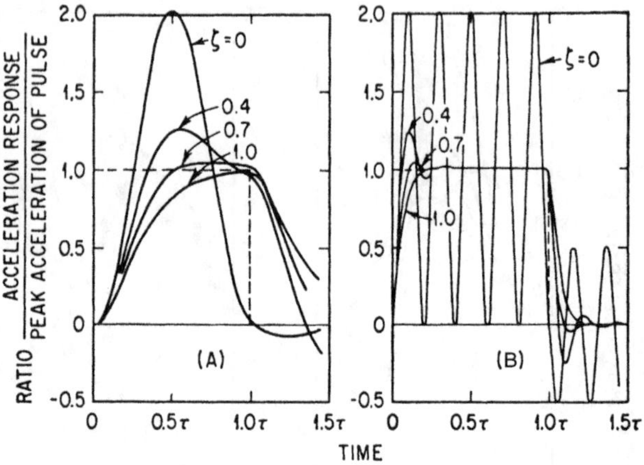

FIGURE 10.7 Acceleration response to a rectangular pulse of acceleration of duration τ (dashed curve) of a mass-spring transducer whose natural period τ_n is equal to: (*A*) 1.014 times the duration of the pulse and (*B*) 0.203 times the duration of the pulse. The fraction of critical damping ζ is indicated for each response curve. (*Levy and Kroll.*[1])

between the pulse and the response; this occurs when τ_n is approximately equal to τ. However, when τ_n is small relative to τ (Figs. 10.5*B* to 10.7*B*), the deviation between the pulse and the response is much smaller. If a shock is generated by metal-to-metal impact or by a pyrotechnic device such as that described in Chap. 28, *and* the response accelerometer is located in close proximity to the excitation source(s), the initial pulses of acceleration may have an extremely fast rise time and high amplitude. In such cases, any type of mass-spring accelerometer may not accurately follow the leading wavefront and characterize the shock inputs faithfully. For example, measurements made in the near field of a high-*g* shock show that undamped piezoresistive accelerometers having resonance above 1 MHz were excited at resonance, thereby invalidating the measured responses. To avoid this effect, accelerometers should be placed as far away as possible, or practical, from the source of excitation. Other considerations related to accelerometer resonance are discussed below in the sections entitled "Zero Shift" and "Survivability."

2. Damping in the transducer reduces the response of the transducer at its own natural frequency; i.e., it reduces the transient vibration superimposed upon the pulse, which is sometimes referred to as *ringing.* For this reason, an accelerometer with internal fluid or gas damping may be ideal for shock measurements when the area of interest lies not in the transient behavior but in the rigid-body motion of the test object. The ringing produced by an undamped accelerometer may induce nonlinear output characteristics internal to the sensing element or drive the signal-conditioning electronics to saturation unknowingly, generating distortion as a by-product.[2] Internal damping effectively isolates the sensing element from unwanted high-frequency inputs which are responsible for setting the element into resonance. It should be noted that the physical protection provided by internal damping cannot be achieved by using electronic postfiltering.

Low-Frequency Response. The measurement of shock requires that the accelerometer and its associated equipment have good response at low frequencies because pulses and other types of shock motions characteristically include low-frequency components. Such pulses can be measured accurately only with an instrumentation system whose response is flat down to the lowest frequency of the spectrum; in general, this lowest frequency is zero for pulses.

The response of an instrumentation system is defined by a plot of output voltage versys excitation frequency. For purposes of shock measurement, the decrease in response at low frequencies is significant. The decrease is defined quantitatively by the frequency f_c, at which the response is down 3 dB or approximately 30 percent below the flat response which exists at the higher frequencies. The distortion which occurs in the measurement of a pulse is related to the frequency f_c, as illustrated in Fig. 10.8.

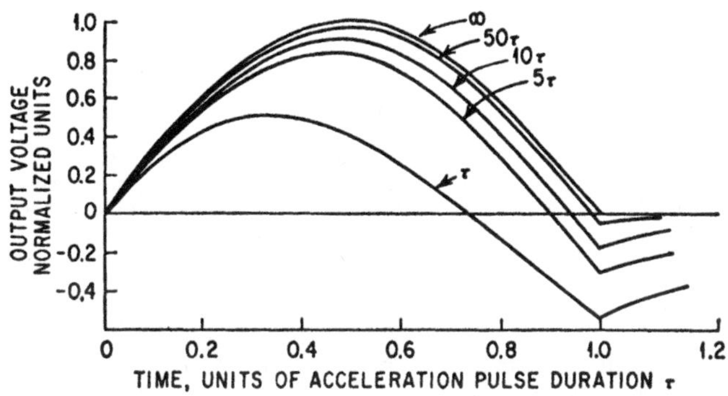

FIGURE 10.8 Response of an accelerometer to a half-sine acceleration pulse for *RC* time constants equal to τ, 5τ, 10τ, 50τ, and ∞, where τ is equal to the duration of the half-sine pulse.[1]

This is particularly important when acceleration data are integrated to obtain velocity, or integrated twice to obtain displacement. A small amount of undershoot shown in Fig. 10.8 may cause a large error after integration. A dc-coupled accelerometer (such as a piezoresistive accelerometer, described later in this chapter) is recommended for this type of application.

Zero Shift. *Zero shift* is the displacement of the zero-reference line of an accelerometer after it has been exposed to a very intense shock. This is illustrated in Fig. 10.9. The loss of zero reference and the apparent dc components in the time history cause a problem in peak-value determination and induce errors in shock response spectrum calculations. Although the accelerometer is not the sole source of zero shift, it is the main contributor.

All piezoelectric shock accelerometers, under extreme stress load (e.g., a sensing element at resonance), will exhibit zero-shift phenomena due either to crystal domain switching or to a sudden change in crystal preload condition.[3] A mechanical filter may be used to protect the crystal element(s) at the expense of a limitation in bandwidth or possible nonlinearity.[4] Piezoresistive shock accelerometers typically produce negligible zero shift.

Survivability. Survivability is the ability of an accelerometer to withstand intense shocks without affecting its performance. An accelerometer is usually rated in terms of the maximum value of acceleration it can withstand. Accelerometers used for shock measurements may have a range of well over many thousands of *g*s. In piezoresistive accelerometers which are excited at resonance, the stress buildup due to high magnitudes of acceleration may lead to fracture of the internal compo-

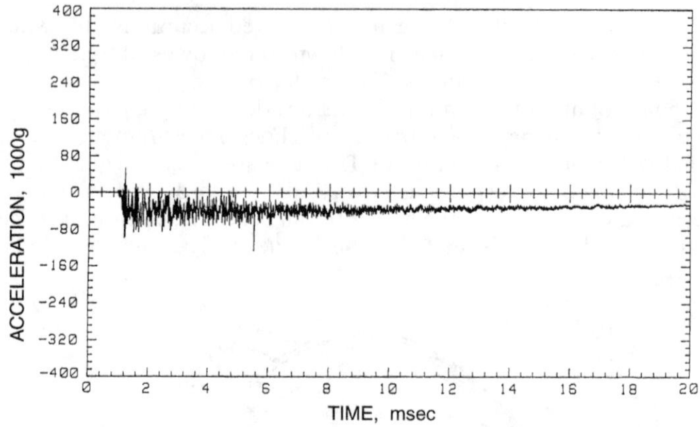

FIGURE 10.9 A time history of an accelerometer that has been exposed to a pyrotechnic shock. Note that there is a shift in the baseline (i.e., the zero reference) of the accelerometer as a result of this shock; the shift may either be positive or negative.

nents. In contrast, piezoelectric accelerometers are more robust than their piezoresistive counterparts due to lower internal stress.

IMPORTANT CHARACTERISTICS OF ACCELEROMETERS

SENSITIVITY

The *sensitivity* of a shock- and vibration-measuring instrument is the ratio of its electrical output to its mechanical input. The output usually is expressed in terms of voltage per unit of displacement, velocity, or acceleration. This specification of sensitivity is sufficient for instruments which generate their own voltage independent of an external voltage power source. However, the sensitivity of an instrument requiring an external voltage usually is specified in terms of output voltage per unit of voltage supplied to the instrument per unit of displacement, velocity, or acceleration, e.g., millivolts per volt per *g* of acceleration. It is important to note the terms in which the respective parameters are expressed, e.g., average, rms, or peak. The relation between these terms is shown in Fig. 10.10.

RESOLUTION

The *resolution* of a transducer is the smallest change in mechanical input (e.g., acceleration) for which a change in the electrical output is discernible. The resolution of an accelerometer is a function of the transducing element and the mechanical design.

Recording equipment, indicating equipment, and other auxiliary equipment used with accelerometers often establish the resolution of the overall measurement sys-

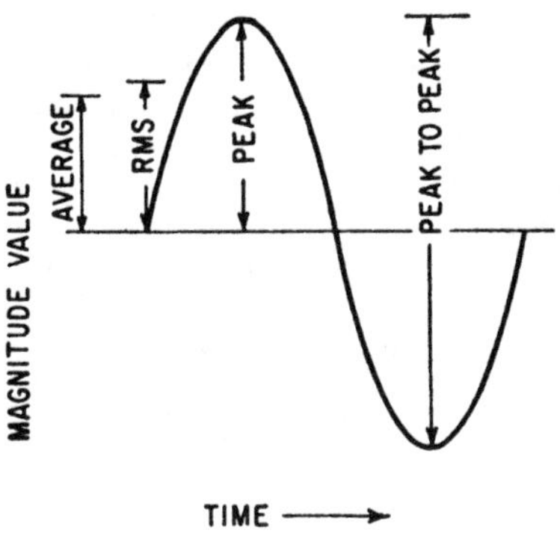

FIGURE 10.10 Relationships between average, rms, peak, and peak-to-peak values for a simple sine wave. These values are used in specifying sensitivities of shock and vibration transducers (e.g., peak millivolts per peak *g*, or rms millivolts per peak-to-peak displacement). These relationships do not hold true for other than simple sine waves.

tem. If the electrical output of an instrument is indicated by a meter, the resolution may be established by the smallest increment that can be read from the meter. Resolution can be limited by noise levels in the instrument or in the system. In general, any signal change smaller than the noise level will be obscured by the noise, thus determining the resolution of the system.

TRANSVERSE SENSITIVITY

If a transducer is subjected to vibration of unit amplitude along its axis of maximum sensitivity, the amplitude of the voltage output e_{max} is the sensitivity. The sensitivity e_θ along the X axis, inclined at an angle θ to the axis of e_{max}, is $e_\theta = e_{max} \cos \theta$, as illustrated in Fig. 10.11. Similarly, the sensitivity along the Y axis is $e_t = e_{max} \sin \theta$. In general, the sensitive axis of a transducer is designated. Ideally, the X axis would be designated the sensitive axis, and the angle θ would be zero. Practically, θ can be made only to approach zero because of manufacturing tolerances and/or unpredictable variations in the characteristics of the transducing element. Then the transverse sensitivity (cross-axis sensitivity) is expressed as the tangent of the angle, i.e., the ratio of e_t to e_θ:

$$\frac{e_t}{e_\theta} = \tan \theta \qquad (10.11)$$

In practice, tan θ is between 0.01 and 0.05 and is expressed as a percentage. For example, if tan θ = 0.05, the transducer is said to have a transverse sensitivity of 5 percent. Figure 10.12 is a typical polar plot of transverse sensitivity.

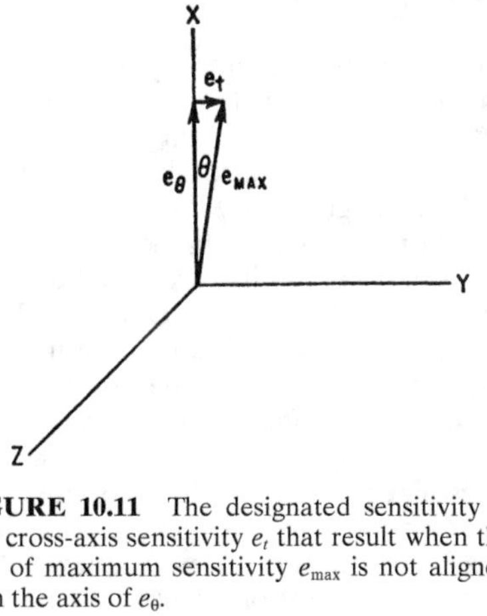

FIGURE 10.11 The designated sensitivity e_θ and cross-axis sensitivity e_t that result when the axis of maximum sensitivity e_{max} is not aligned with the axis of e_θ.

ZERO ACCELERATION OUTPUT (ZAO)

The electrical output indicated by an accelerometer at zero acceleration is commonly referred to as *zero acceleration output* (ZAO), *zero-offset,* or *zero output bias.* With an accelerometer whose output is electrically ac-coupled, such as the piezoelectric type, the zero acceleration reference is at ground potential or some reference dc level called *zero output bias.* With an accelerometer that is capable of responding to static acceleration, such as the piezoresistive type, the

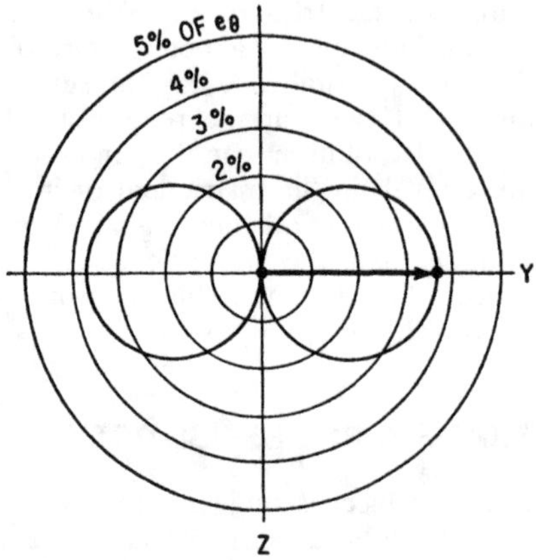

FIGURE 10.12 Plot of transducer sensitivity in all axes normal to the designated axis e_θ plotted according to axes shown in Fig. 10.11. Cross-axis sensitivity reaches a maximum e_t along the Y axis and a minimum value along the Z axis.

zero acceleration reference should ideally be at zero output unit or some specified dc level. But this is technically impractical due to component tolerances. Sensor manufacturers typically specify the ZAO to be within a range, i.e., ±50 mV, and the measured ZAO figure is supplied with the accelerometer as calibration data. ZAO changes with temperature. This will be described later in the chapter in "Environmental Effects."

AMPLITUDE LINEARITY AND LIMITS

When the ratio of the electrical output of a transducer to the mechanical input (i.e., the sensitivity) remains constant within specified limits, the transducer is said to be "linear" within those limits, as illustrated in Fig. 10.13. A transducer is linear only over a certain range of amplitude values. The lower end of this range is determined by the electrical noise of the measurement system.

The upper limit of linearity may be imposed by the electrical characteristics of the transducing element and by the size or the fragility of the instrument. Generally, the greater the sensitivity of a transducer, the more nonlinear it will be. Similarly, for very large acceleration values, the large forces produced by the spring of the mass-spring system may exceed the yield strength of a part of the instrument, causing nonlinear behavior or complete failure.[3]

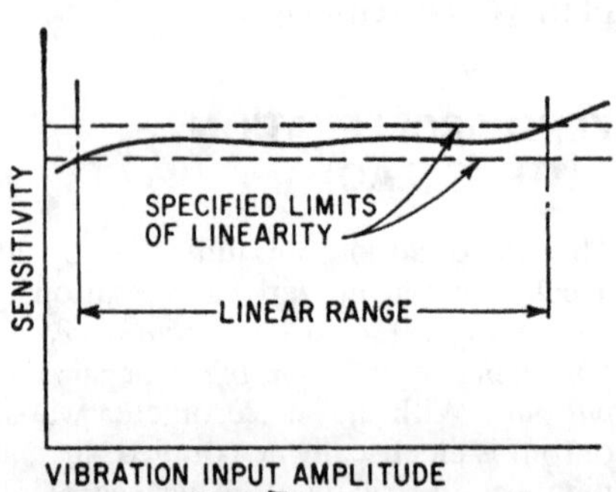

FIGURE 10.13 Typical plot of sensitivity as a function of amplitude for a shock and vibration transducer. The *linear range* is established by the intersection of the sensitivity curve and the specified limits (dashed lines).

FREQUENCY RANGE

The operating frequency range is the range over which the sensitivity of the transducer does not vary more than a stated percentage from the rated sensitivity. This range may be limited by the electrical or mechanical characteristics of the transducer or by its associated auxiliary equipment. These limits can be added to amplitude linearity limits to define completely the operating ranges of the instrument, as illustrated in Fig. 10.14.

Low-Frequency Limit. The mechanical response of a mass-spring transducer does not impose a low-frequency limit for an acceleration transducer because

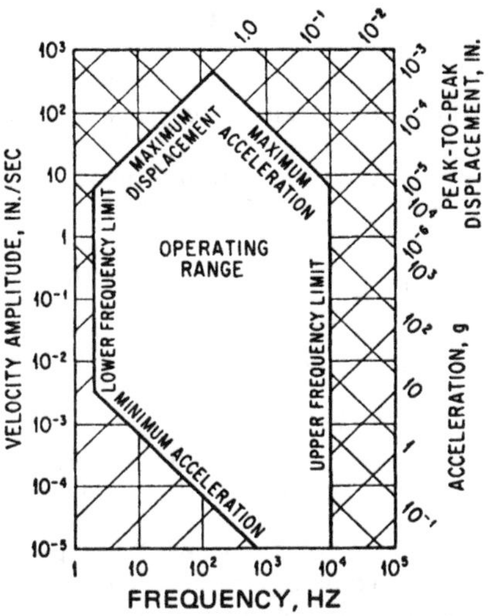

FIGURE 10.14 Linear operating range of a transducer. Amplitude linearity limits are shown as a combination of displacement and acceleration values. The lower amplitude limits usually are expressed in acceleration values as shown.

the transducer responds to vibration with frequencies less than the natural frequency of the transducer.

However, it is necessary to consider the electrical characteristics of both the transducer and the associated electronic equipment in evaluating the low-frequency limit. An accelerometer that is capable of sensing static acceleration is commonly referred to as a *dc accelerometer.* In general, an accelerometer that utilizes external power or a carrier voltage, such as the piezoresistive or variable capacitive designs, is a *dc accelerometer,* which has no low-frequency limit, whereas a self-generating transducer type, such as the piezoelectric design, is not operative at zero frequency. The low-frequency response of a piezoelectric accelerometer is determined solely by the connecting charge amplifier.

High-Frequency Limit. An acceleration transducer (accelerometer) has an upper usable frequency limit because it responds to vibration whose frequency is less than the natural frequency of the transducer. The limit is a function of (1) the natural frequency and (2) the damping of the transducer, as discussed with reference to Fig. 10.3. An attempt to use such a transducer beyond this frequency limit may result in distortion of the signal, as illustrated in Fig. 10.15.

The upper frequency limit for slightly damped vibration-measuring instruments is important because these instruments exaggerate the small amounts of harmonic content that may be contained in the motion, even when the operating frequency is well within the operating range of the instrument. The result of exciting an undamped instrument at its natural frequency may be to either damage the instrument or obscure the desired measurement. Figure 10.15 shows how a small amount of harmonic distortion in the vibratory motion may be exaggerated by an undamped transducer.

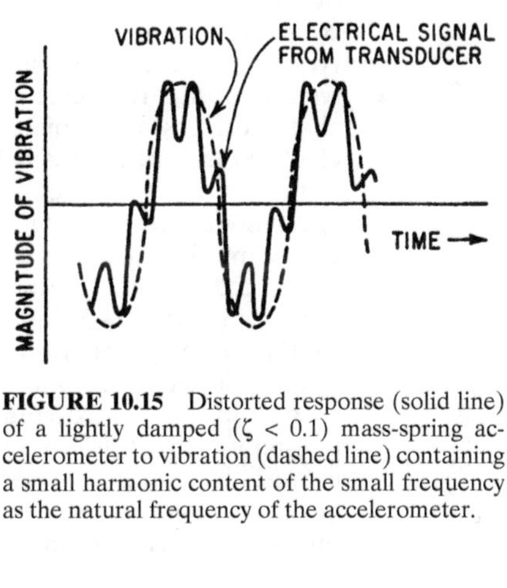

FIGURE 10.15 Distorted response (solid line) of a lightly damped ($\zeta < 0.1$) mass-spring accelerometer to vibration (dashed line) containing a small harmonic content of the small frequency as the natural frequency of the accelerometer.

Phase Shift. Phase shift is the time delay between the mechanical input and the electrical output signal of the instrumentation system. Unless the phase-shift characteristics of an instrumentation system meet certain requirements, a distortion may be introduced that consists of the superposition of vibration at several different frequencies. Consider first an accelerometer, for which the phase angle θ_1 is given by Fig. 10.4. If the accelerometer is undamped, $\theta_1 = 0$ for values of ω/ω_n less than 1.0; thus, the phase of the relative displacement δ is equal to that of the acceleration being measured, for all values of frequency within the useful range of the accelerometer. Therefore, an undamped accelerometer measures acceleration without distortion of phase. If the fraction of critical damping ζ for the accelerometer is 0.65, the phase angle θ_1 increases approximately linearly with the frequency ratio ω/ω_n within the useful frequency range of the accelerometer. Then the expression for the relative displacement may be written

$$\delta = \delta_0 \cos(\omega t - \theta) = \delta_0 \cos(\omega t - a\omega) = \delta_0 \cos \omega(t - a) \tag{10.12}$$

where a is a constant. Thus, the relative motion δ of the instrument is displaced in phase relative to the acceleration $\ddot{u}$ being measured; however, the increment along the time axis is a constant independent of frequency. Consequently, the waveform of the accelerometer output is undistorted but is delayed with respect to the waveform of the vibration being measured. As indicated by Fig. 10.4, any value of damping in an accelerometer other than $\zeta = 0$ or $\zeta = 0.65$ (approximately) results in a nonlinear shift of phase with frequency and a consequent distortion of the waveform.

ENVIRONMENTAL EFFECTS

Temperature. The sensitivity, natural frequency, and damping of a transducer may be affected by temperature. The specific effects produced depend on the type of transducer and the details of its design. The sensitivity may increase or decrease with temperature, or remain relatively constant. The temperature characteristics of an accelerometer may be measured as a function of temperature, if necessary, and appropriate compensations can then be applied to the measured data in real time or after the fact. The compensations in real time can be accomplished passively or actively. To passively compensate a piezoelectric type, a parallel capacitor with opposite temperature characteristics of the piezoelectric element is inserted in the circuit at the factory. For the piezoresistive type, one or multiple resistors are connected to the bridge circuit in various fashions to lessen the temperature effects. Figure 10.16 shows an example of a zero compensation circuit for a piezoresistive accelerometer. Several modern accelerometer designs have incor-

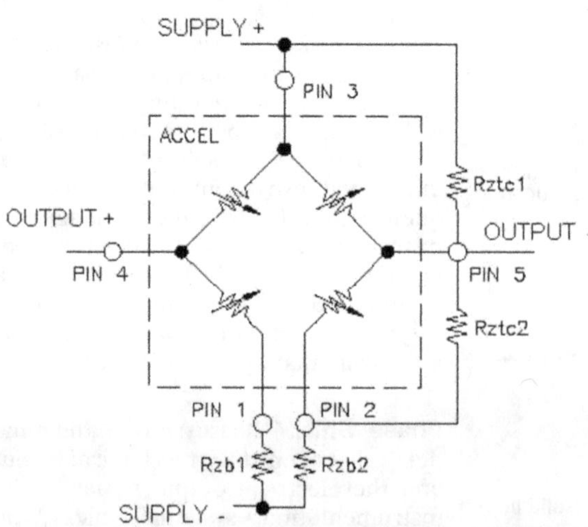

FIGURE 10.16 Typical temperature compensation circuit for zero offset in a piezoresistive full-bridge sensor.[5] Rzb1, Rzb2, Rztc1, and Rztc2 adjust the zero offset and compensate the offset error due to temperature.

porated built-in temperature sensors and microprocessor integrated circuits for active temperature compensation.

Humidity. Humidity may affect the characteristics of certain types of vibration instruments. In general, a transducer which operates at a high electrical impedance is affected by humidity more than a transducer which operates at a low electrical impedance. It usually is impractical to correct the measured data for humidity effects. However, instruments that might otherwise be adversely affected by humidity often are sealed hermetically to protect them from the effects of moisture.

Acoustic Noise. High-intensity sound waves often accompany high-amplitude vibration. If the case of an accelerometer can be set into vibration by acoustic excitation, error signals may result. In general, a well-designed accelerometer will not produce a significant electrical response except at extremely high sound pressure levels. Under such circumstances, it is likely that vibration levels also will be very high, so that the error produced by the accelerometer's exposure to acoustic noise usually is not important.

Strain Sensitivity. An accelerometer may generate a spurious output when its case is strained or distorted. Typically this occurs when the transducer mounting is not flat against the surface to which it is attached, and so this effect is often called *base-bend sensitivity* or *strain sensitivity.* It is usually reported in equivalent g per micro-strain, where 1 *microstrain* is 1×10^{-6} inch per inch. The Instrument Society of America recommends a test procedure that determines strain sensitivity at 250 microstrain.[6]

An accelerometer with a sensing element which is tightly coupled to its base tends to exhibit large strain sensitivity. An error due to strain sensitivity is most likely to occur when the accelerometer is attached to a structure which is subject to large amounts of flexure. In such cases, it is advisable to select an accelerometer with low strain sensitivity.

PHYSICAL PROPERTIES

Size and weight of the transducer are very important considerations in many vibration and shock measurements. A large instrument may require a mounting structure that will change the local vibration characteristics of the structure whose vibration is being measured. Similarly, the added mass of the transducer may also produce substantial changes in the vibratory response of such a structure. Generally, the natural frequency of a structure is lowered by the addition of mass; specifically, for a simple spring-mass structure:

$$\frac{f_n - \Delta f_n}{f_n} = \sqrt{\frac{m}{m + \Delta m}} \tag{10.13}$$

where f_n = natural frequency of structure
Δf_n = change in natural frequency
m = mass of structure
Δm = increase in mass resulting from addition of transducer

In general, for a given type of transducing element, the sensitivity increases approximately in proportion to the mass of the transducer. In most applications, it is more important that the transducer be small in size than that it have high sensitivity because amplification of the signal increases the output to a usable level.

Mass-spring-type transducers for the measurement of displacement usually are larger and heavier than similar transducers for the measurement of acceleration. In the former, the mass must remain substantially stationary in space while the instrument case moves about it; this requirement does not exist with the latter.

For the measurement of shock and vibration in aircraft or missiles, the size and weight of not only the transducer but also the auxiliary equipment are important. In these applications, self-generating instruments that require no external power may have a significant advantage.

PIEZOELECTRIC ACCELEROMETERS

PRINCIPLE OF OPERATION

An accelerometer of the type shown in Fig. 10.17*A* is a linear seismic transducer utilizing a piezoelectric element in such a way that an electric charge is produced which is proportional to the applied acceleration. This "ideal" seismic piezoelectric transducer can be represented (over most of its frequency range) by the elements shown in Fig. 10.17*B*. A mass is supported on a linear spring which is fastened to the frame of the instrument. The piezoelectric crystal which produces the charge acts as the spring. Viscous damping between the mass and the frame is represented by the dashpot c. In Fig. 10.17*C* the frame is given an acceleration upward to a displacement of u, thereby producing a compression in the spring equal to δ. The displacement of the

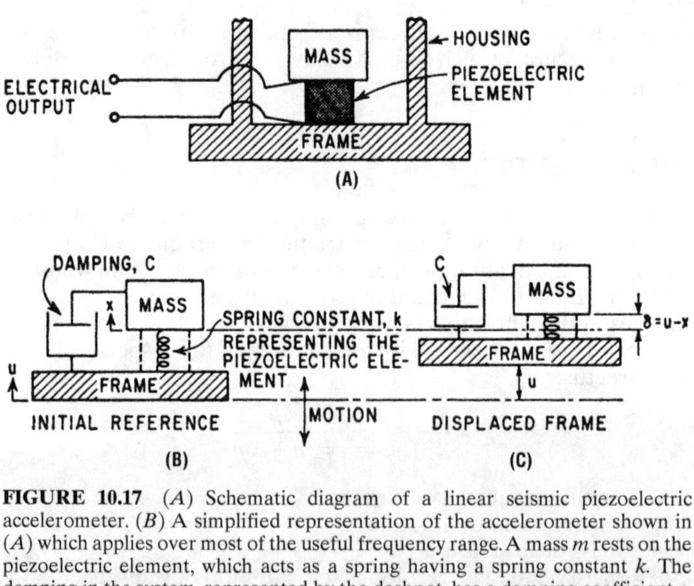

FIGURE 10.17 (*A*) Schematic diagram of a linear seismic piezoelectric accelerometer. (*B*) A simplified representation of the accelerometer shown in (*A*) which applies over most of the useful frequency range. A mass m rests on the piezoelectric element, which acts as a spring having a spring constant k. The damping in the system, represented by the dashpot, has a damping coefficient c. (*C*) The frame is accelerated upward, producing a displacement u of the frame, moving the mass from its initial position by an amount x, and compressing the spring by an amount δ.

mass relative to the frame is dependent upon the applied acceleration of the frame, the spring stiffness, the mass, and the viscous damping between the mass and the frame, as indicated in Eq. (10.10) and illustrated in Fig. 10.3.

For frequencies far below the resonance frequency of the mass and spring, this displacement is directly proportional to the acceleration of the frame and is independent of frequency. At low frequencies, the phase angle of the relative displacement δ, with respect to the applied acceleration, is proportional to frequency. As indicated in Fig. 10.4, for low fractions of critical damping which are characteristic of many piezoelectric accelerometers, the phase angle is proportional to frequency at frequencies below 30 percent of the resonance frequency.

In Fig. 10.17, inertial force of the mass causes a mechanical strain in the piezoelectric element, which produces an electric charge proportional to the stress and, hence, proportional to the strain and acceleration. If the dielectric constant of the piezoelectric material does not change with electric charge, the voltage generated is also proportional to acceleration. Metallic electrodes are applied to the piezoelectric element, and electrical leads are connected to the electrodes for measurement of the electrical output of the piezoelectric element.

In the ideal seismic system shown in Fig. 10.17, the mass and the frame have infinite stiffness, the spring has zero mass, and viscous damping exists only between the mass and the frame. In practical piezoelectric accelerometers, these assumptions cannot be fulfilled. For example, the mass may have as much compliance as the piezoelectric element. In some seismic elements, the mass and spring are inherently a single structure. Furthermore, in many practical designs where the frame is used to hold the mass and piezoelectric element, distortion of the frame may produce mechanical forces upon the seismic element. All these factors may change the performance of the seismic system from those calculated using equations based on an ideal system. In particular, the resonance frequency of the piezoelectric combination may be substantially lower than that indicated by theory. Nevertheless, the equations for an ideal system are useful in both design and application of piezoelectric accelerometers.

Figure 10.18 shows a typical frequency response curve for a piezoelectric accelerometer. In this illustration, the electrical output in millivolts per g acceleration is plotted as a function of frequency. The resonance frequency is denoted by f_n. If the accelerometer is properly mounted on the device being tested, then the upper frequency limit of the useful frequency range usually is taken to be $f_n/3$ for a deviation of 12 percent (1 dB) from the mean value of the response. For a deviation of 6 percent (0.5 dB) from the mean value, the upper frequency limit usually is taken to be $f_n/5$. As indicated in Fig. 10.1, the type of mounting can have a significant effect on the value of f_n.

The decrease in response at low frequencies (i.e., the "roll-off") depends primarily on the characteristics of the preamplifier that follows the accelerometer. The low-frequency limit also is usually expressed in terms of the deviation from the mean value of the response over the flat portion of the response curve, being the frequency at which the response is either 12 percent (1 dB) or 6 percent (0.5 dB) below the mean value.

PIEZOELECTRIC MATERIALS

A polarized ceramic called lead zirconate titanate (PZT) is most commonly used in piezoelectric accelerometers. It is low in cost, high in sensitivity, and useful in the temperature range from −180° to +550°F (−100° to +288°C). Polarized ceramics in the bismuth titanate family have substantially lower sensitivities than PZT, but they also have more stable characteristics and are useful at temperatures as high as 1000°F (538°C).

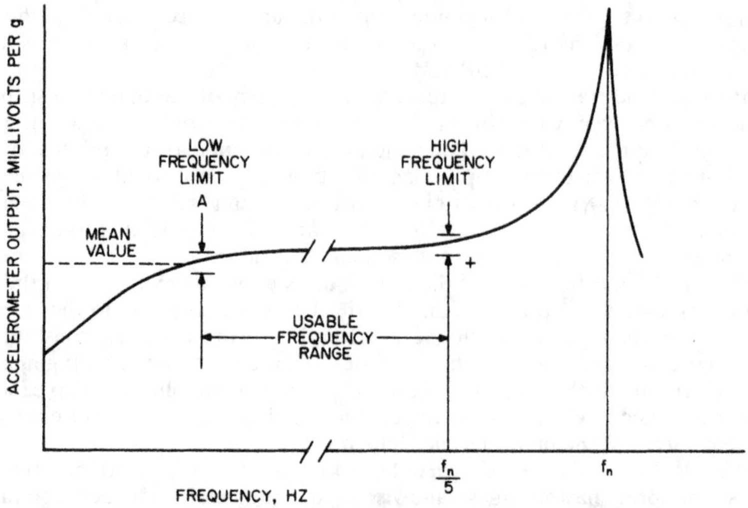

FIGURE 10.18 Typical response curve for a piezoelectric accelerometer. The resonance frequency is denoted by f_n. The useful range depends on the acceptable deviation from the mean value of the response over the "flat" portion of the response curve.

Quartz, the single-crystal material most widely used in accelerometers, has a substantially lower sensitivity than polarized ceramics, but its characteristics are very stable with time and temperature; it has high resistivity. Tourmaline is a single-crystal material that can be used in accelerometers at high temperatures up to 1400°F (760°C). The upper limit of the useful range is usually set by the thermal characteristics of the structural materials rather than by the characteristics of these two crystalline materials.

Polarized polyvinylidene fluoride (PVDF), an engineering plastic similar to Teflon and known as Piezofilm, is used as the sensing element in accelerometers as well as for direct measurement of dynamic strain. It is inexpensive, but it is generally less stable with temperature (and limited in the upper temperature range, normally to around 85 to 125°C) than ceramics or single-crystal materials. It has low mechanical Q and is highly resistant to shock, and thin Piezofilm in compression mode allows very high frequency measurements.

TYPICAL PIEZOELECTRIC ACCELEROMETER CONSTRUCTIONS

Piezoelectric accelerometers utilize a variety of seismic element configurations. Their methods of mounting are described in Chap. 15. See also Ref. 8. Most are constructed of polycrystalline ceramic piezoelectric materials because of their ease of manufacture, high piezoelectric sensitivity, and excellent time and temperature stability. These seismic devices may be classified in two modes of operation: compression- or shear-type accelerometers.

Compression-type Accelerometer. The compression-type seismic accelerometer, in its simplest form, consists of a piezoelectric disc and a mass placed on a frame as shown in Fig. 10.17. Motion in the direction indicated causes compressive (or tensile) forces to act on the piezoelectric element, producing an electrical output pro-

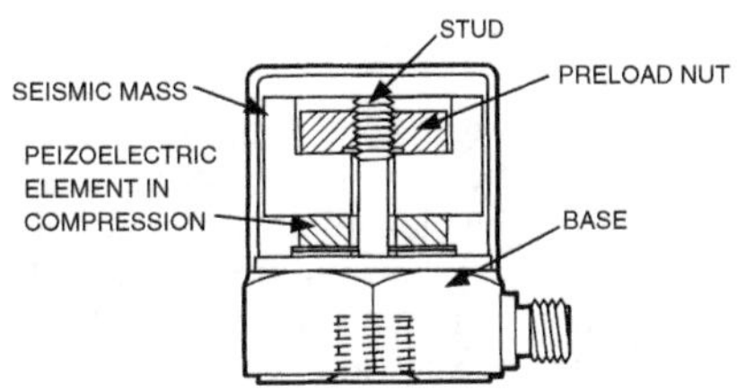

FIGURE 10.19 A typical compression-type piezoelectric accelerometer. The piezoelectric element(s) must be preloaded (biased) to produce an electrical output under both tension forces and compression forces. (*Courtesy of Endevco Corp.*)

portional to acceleration. In this example, the mass is cemented with a conductive material to the piezoelectric element which, in turn, is cemented to the frame. The components must be cemented firmly so as to avoid being separated from each other by the applied acceleration.

In the typical commercial accelerometer shown in Fig. 10.19, the mass is held in place by means of a stud extending from the frame through the ceramic. Accelerometers of this design often use quartz, tourmaline, or ferroelectric ceramics as the sensing material.

This type of accelerometer must be attached to the structure with care in order to minimize distortion of the housing and base, which can cause an electrical output. See the section entitled "Strain Sensitivity."

The temperature characteristics of compression-type accelerometers have been improved greatly in recent years; it is now possible to measure acceleration over a temperature range of –425 to +1400°F (–254 to +760°C). This wider range has been primarily a result of the use of two piezoelectric materials: tourmaline and lithium niobate.

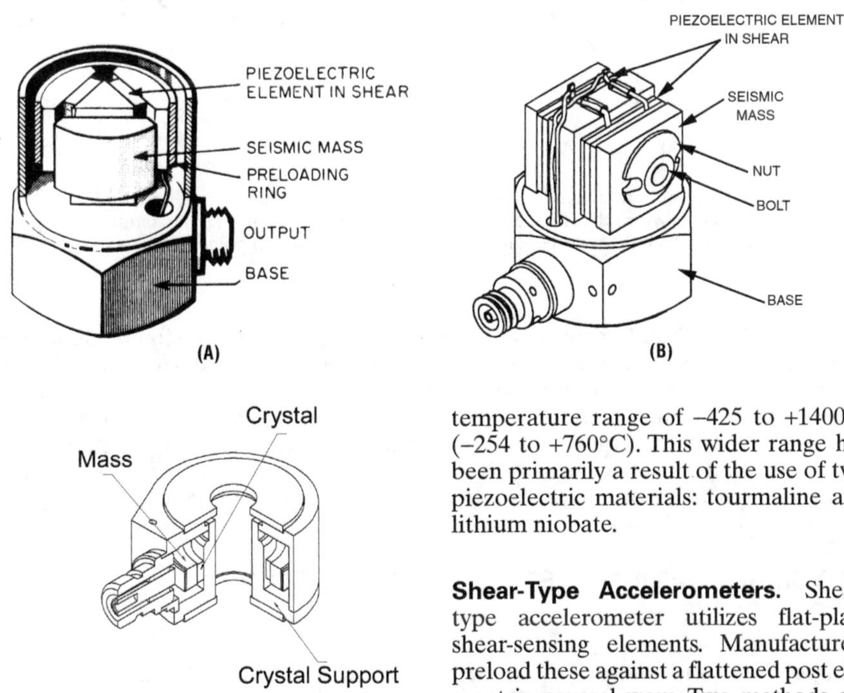

FIGURE 10.20 Piezoelectric accelerometers: (*A*) delta-shear type (*courtesy of Brüel & Kjear*), (*B*) isoshear type (*courtesy of Endevco*), (*C*) annular shear type (*courtesy of Measurement Specialties, Inc.*).

Shear-Type Accelerometers. Shear-type accelerometer utilizes flat-plate shear-sensing elements. Manufacturers preload these against a flattened post element in several ways. Two methods are shown in Fig. 10.20 (*A*, *B*). Accelerometers of this style have low cross-axis response, excellent temperature characteristics, and negligible output from strain sensitivity or base bending. The temperature range of the bolted shear design can be from –425 to +1400°F (–254 to +760°C).

The annular shear type of accelerometer, illustrated in Fig. 10.20*C*, employs a hollow cylindrically shaped piezoelectric element fitted around a middle mounting post; a loading ring (or mass) is affixed to the outer diameter of the piezoelectric element. The ceramic piezoelectric element is polarized along its length; the output voltage of the accelerometer is taken from its inner and outer walls. This type of design allows a mounting screw to be inserted through the center of the accelerometer, which offers a 360-degree connector orientation.

Beam-Type Accelerometers. The beam-type accelerometer is a variation of the compression-type accelerometer. It is usually made from two piezoelectric plates which are rigidly bonded together to form a beam supported at one end, as illustrated in Fig. 10.21. As the beam flexes, the bottom element compresses, so that it increases in thickness. In contrast, the upper element expands, so that it decreases in thickness. Accelerometers of this type generate high electrical output for their size, but are more fragile and have a lower resonance frequency than most other designs.

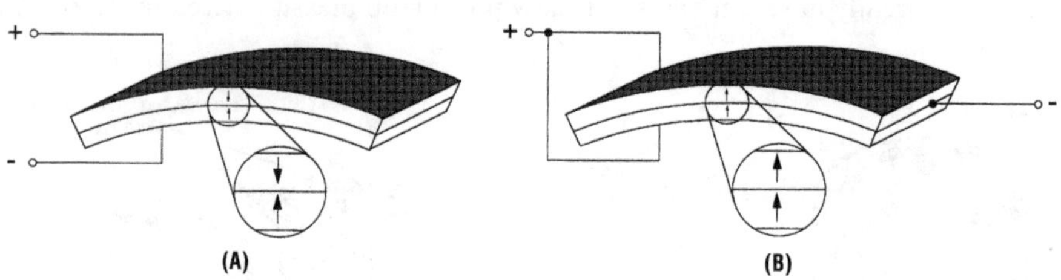

FIGURE 10.21 Configurations of piezoelectric elements in a beam-type accelerometer. (*A*) A series arrangement, in which the two elements have opposing directions of polarization. (*B*) A parallel arrangement, in which the two elements have the same direction of polarization.

Piezofilm-Type Accelerometers. Piezofilm is used in compression mode to produce very sensitive and wide-bandwidth accelerometers. For a device with 10 mV/g open-circuit sensitivity, the resonance frequency may exceed 75 kHz. Because the PVDF sensor element tends to have lower capacitance and, therefore, higher electrical impedance than equivalent piezoceramic designs, an impedance buffer is usually integrated into the device. Piezofilm accelerometers are generally used in low-cost applications where calibration accuracy is not critical. (See Fig. 10.22.)

PHYSICAL CHARACTERISTICS OF PIEZOELECTRIC ACCELEROMETERS

Shape, Size, and Weight. Commercially available piezoelectric accelerometers usually are cylindrical in shape. They are available with both attached and detachable mounting studs at the bottom of the cylinder. A coaxial cable connector is provided at either the top or side of the housing.

Most commercially available piezoelectric accelerometers are relatively light in weight, ranging from approximately 0.005 to 4.2 oz (0.14 to 120 g). Usually, the

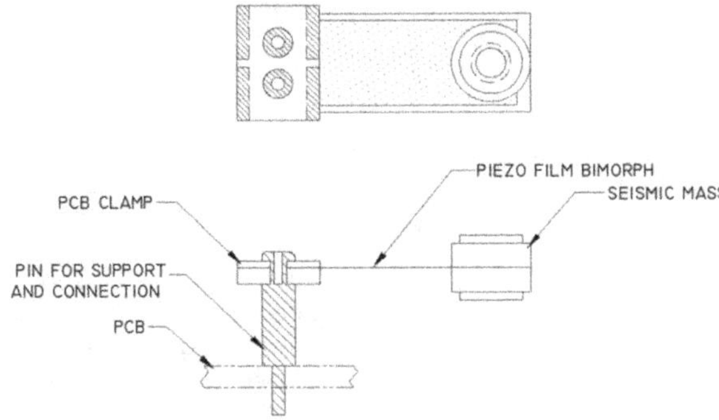

FIGURE 10.22 Piezofilm can be used in length-extension mode as a bimorph cantilever structure, detecting acceleration by inertial response of the free beam. The cantilever length, and seismic mass if deployed, can be varied to achieve a wide continuum of sensitivity and resonant frequency results.

larger the accelerometer, the higher its sensitivity and the lower its resonance frequency. The smallest units have a diameter of less than about 0.2 in. (5 mm); the larger units have a diameter of about 1 in. (25.4 mm) and a height of about 1 in. (25.4 mm).

Resonance Frequency. Typical resonance frequency of an accelerometer may be above 40,000 Hz. The higher the resonance frequency, the lower will be the sensitivity. A typical piezoelectric accelerometer offers flat response (±1 dB) up to 10 kHz.

Damping. Most piezoelectric accelerometers are essentially undamped, having amplification ratios between 20 and 100, or a fraction of critical damping less than 0.1.

ELECTRICAL CHARACTERISTICS OF PIEZOELECTRIC ACCELEROMETERS

Dependence of Voltage Sensitivity on Shunt Capacitance. The *sensitivity* of an accelerometer is defined as the electrical output per unit of applied acceleration. The sensitivity of a piezoelectric accelerometer can be expressed as either a *charge sensitivity* $q/\ddot{x}$ or a *voltage sensitivity* $e/\ddot{x}$. Charge sensitivity usually is expressed in units of coulombs generated per g of applied acceleration; voltage sensitivity usually is expressed in volts per g (where g is the acceleration of gravity). Voltage sensitivity often is expressed as open-circuit voltage sensitivity, i.e., in terms of the voltage produced across the electrical terminals per unit acceleration when the electrical load impedance is infinitely high. Open-circuit voltage sensitivity may be given either with or without the connecting cable.

An electrical capacitance often is placed across the output terminals of a piezoelectric transducer. This added capacitance (called *shunt capacitance*) may result from the connection of an electrical cable between the pickup and other electrical equipment (all electrical cables exhibit interlead capacitance). The effect of shunt capacitance in reducing the sensitivity of a pickup is shown in Fig. 10.23.

The charge equivalent circuits, with shunt capacitance C_S, are shown in Fig. 10.23*A*. The charge sensitivity is not changed by addition of shunt capacitance. The

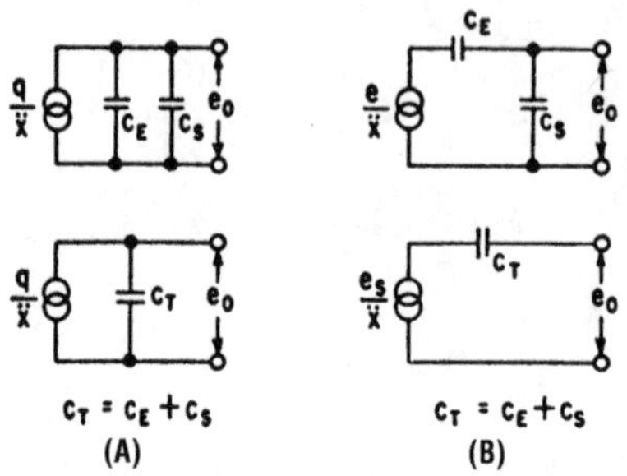

FIGURE 10.23 Equivalent circuits which include shunt capacitance across a piezoelectric pickup. (*A*) Charge equivalent circuit. (*B*) Voltage equivalent circuit.

total capacitance C_T of the pickup including shunt is given by

$$C_T = C_E + C_S \tag{10.14}$$

where C_E is the capacitance of the transducer without shunt capacitance.

The voltage equivalent circuits are shown in Fig. 10.23*B*. With the shunt capacitance C_S, the total capacitance is given by Eq. (10.14) and the open-circuit voltage sensitivity is given by

$$\frac{e_s}{\ddot{x}} = \frac{q_s}{\ddot{x}} \frac{1}{C_E + C_S} \tag{10.15}$$

where $q_s/\ddot{x}$ is the charge sensitivity. The voltage sensitivity without shunt capacitance is given by

$$\frac{e}{\ddot{x}} = \frac{q_s}{\ddot{x}} \frac{1}{C_E} \tag{10.16}$$

Therefore, the effect of the shunt capacitance is to reduce the voltage sensitivity by a factor

$$\frac{e_s/\ddot{x}}{e/\ddot{x}} = \frac{C_E}{C_E + C_S} \tag{10.17}$$

Piezoelectric accelerometers are used with both voltage-sensing and charge-sensing signal conditioners, although charge sensing is by far the most common because the sensitivity does not change with external capacitance (up to a limit). These factors are discussed in Chap. 13. In addition, electronic circuitry can be placed within the case of the accelerometer, as discussed below.

LOW-IMPEDANCE PIEZOELECTRIC ACCELEROMETERS CONTAINING INTERNAL ELECTRONICS

Piezoelectric accelerometers are available with simple electronic circuits internal to their cases to provide signal amplification and low-impedance output. For example, see the charge preamplifier circuit shown in Fig. 13.2. Some designs operate from low-current dc voltage supplies and are designed to be intrinsically safe when coupled by appropriate barrier circuits. Other designs have common power and signal lines and use coaxial cables.

The principal advantages of piezoelectric accelerometers with integral electronics are that they are relatively immune to cable-induced noise and spurious response, they can be used with lower-cost cable, and they have a lower signal conditioning cost. In the simplest case the power supply might consist of a battery, a resistor, and a capacitor. Some such accelerometers provide a velocity or displacement output. These advantages do not come without compromise.[9] Because the impedance-matching circuitry is built into the transducer, gain cannot be adjusted to utilize the wide dynamic range of the basic transducer. Ambient temperature is limited to that which the circuit will withstand, and this is considerably lower than that of the piezoelectric sensor itself. In order to retain the advantages of small size, the integral electronics must be kept relatively simple. This precludes the use of multiple filtering and dynamic overload protection and thus limits their application.

All other things being equal, the *reliability factor* (i.e., the mean time between failures) of any accelerometer with internal electronics is lower than that of an accelerometer with remote electronics, especially if the accelerometer is subject to abnormal environmental conditions. However, if the environmental conditions are fairly normal, accelerometers with internal electronics can provide excellent signal fidelity and immunity from noise. Internal electronics provides a reduction in overall system noise level because it minimizes the cable capacitance between the sensor and the signal-conditioning electronics.

Velocity-Output Piezoelectric Devices. Piezoelectric accelerometers are available with internal electronic circuitry which integrates the output signal provided by the accelerometer, thereby yielding a velocity or displacement output. These transducers have several advantages not possessed by ordinary velocity pickups. They are smaller, have a wider frequency response, have no moving parts, and are relatively unaffected by magnetic fields where measurements are made.

CHARACTERISTICS OF PIEZOELECTRIC ACCELEROMETER

Measurement Range. Piezoelectric accelerometers are generally useful for the measurement of acceleration of magnitudes of from $10^{-6}g$ to more than 10^5g. The lowest value of acceleration which can be measured is approximately that which will produce an output voltage equivalent to the electrical input noise of the coupling amplifier connected to the accelerometer when the pickup is at rest. Over its useful operating range, the output of a piezoelectric accelerometer is directly and continuously proportional to the input acceleration. A single accelerometer often can be used to provide measurements over a dynamic amplitude range of 90 dB or more, which is substantially greater than the dynamic range of some of the associated transmission, recording, and analysis equipment. Commercial accelerometers generally exhibit excellent linearity of electrical output versus input acceleration under normal usage. In fact, the upper dynamic ranges of many piezoelectric accelerometers are actually determined by their output (charge) sensitivities and not by their nonlinearity characteristics. When such an accelerometer is used with a charge amplifier with high-input charge capacity (i.e., over 50,000 pC), the usable dynamic range of the system can easily exceed 110 dB. This is, however, not true with piezoelectric accelerometers with built-in electronics in which the maximum output swing has been predetermined by the internal amplifier at the factory.

Temperature Range. Piezoelectric accelerometers are available which may be used in the temperature range from –425°F (–254°C) to above +1400°F (+760°C) without the aid of external cooling. The voltage sensitivity, charge sensitivity, capacitance, and frequency response depend upon the ambient temperature of the transducer. This temperature dependence is due primarily to variations in the characteristics of the piezoelectric material, but it also may be due to variations in the insulation resistance of cables and connectors—especially at high temperatures.

Effects of Temperature on Charge Sensitivity. The charge sensitivity of a piezoelectric accelerometer is directly proportional to the d piezoelectric constant of the material used in the piezoelectric element. The d constants of most piezoelectric materials vary with temperature.

Effects of Temperature on Voltage Sensitivity. The open-circuit voltage sensitivity of an accelerometer is the ratio of its charge sensitivity to its total capacitance ($C_s + C_E$). Hence, the temperature variation in voltage sensitivity depends on the temperature dependence of both charge sensitivity and capacitance. The voltage sensitivity of most piezoelectric accelerometers decreases with temperature.

Effects of Transient Temperature Changes. A piezoelectric accelerometer that is exposed to transient temperature changes may produce outputs as large as several volts, even if the sensitivity of the accelerometer remains constant. These spurious output voltages arise from

1. Differential thermal expansion of the piezoelectric elements and the structural parts of the accelerometer, which may produce varying mechanical forces on the piezoelectric elements, thereby producing an electrical output.
2. Generation of a charge in response to a change in temperature because the piezoelectric material is inherently pyroelectric. In general, the charge generated is proportional to the temperature change.

Such thermally generated transients tend to generate signals at low frequencies because the accelerometer case acts as a thermal low-pass filter. Therefore, such spurious signals often may be reduced significantly by adding thermal insulation around the accelerometer to minimize the thermal changes and by electrical filtering of low-frequency output signals from the accelerometer.

PIEZORESISTIVE ACCELEROMETERS

PRINCIPLE OF OPERATION

A piezoresistive accelerometer differs from the piezoelectric type in that it is not self-generating. In this type of transducer a semiconductor material, usually silicon, is used as the strain-sensing element. Such a material changes its resistivity in proportion to an applied stress or strain. The equivalent electric circuit of a piezoresistive transducing element is a variable resistor. Piezoresistive elements are almost always arranged in pairs; a given acceleration places one element in tension and the other in compression. This causes the resistance of one element to increase while the resistance of the other decreases. Often two pairs are used and the four elements are connected electrically in a Wheatstone-bridge circuit, as shown in Fig. 10.24*B*. This is called a *full-bridge* configuration. When only one pair is used, it forms half of a Wheatstone bridge (a *half-bridge* configuration), the other half being made up of fixed-value resistors, either in the transducer or in the signal-conditioning equipment. The use of transducing elements by pairs not only increases the sensitivity, but also cancels zero-output errors due to temperature changes, which occur in each resistive element.

Silicon elements are often used as the transducing elements because of their high sensitivity. (Metallic gages made of foil or wire change their resistance with strain because the dimensions change. The resistance of a piezoresistive material changes because the material's electrical nature changes.) Sensitivity is a function of the gage factor; the *gage factor* is the ratio of the fractional change in resistance to the fractional change in length that produced it. The gage factor of a typical wire or foil strain gage is approximately 2.5; the gage factor of silicon is approximately 100.

A major advantage of piezoresistive accelerometers is that they are capable of responding down to dc (0 Hz) along with a relatively good high-frequency response. Today, most piezoresistive accelerometers are constructed using micromachining technology.

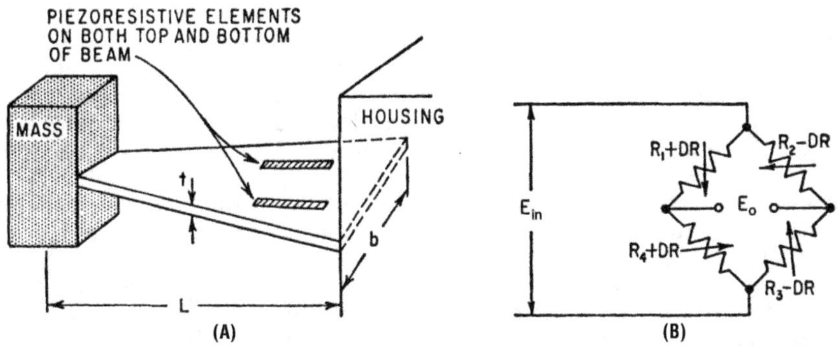

FIGURE 10.24 (*A*) Schematic drawing of a piezoresistive accelerometer of the cantilever-beam type. Four piezoresistive elements are used—two are either cemented to each side of the stressed beam or are diffused or ion implanted into a silicon beam. (*B*) The four piezoresistive elements are connected in a bridge circuit as illustrated.

DESIGN PARAMETERS

Many different configurations are possible for an accelerometer of this type. For purposes of illustration, the design parameters are considered for a piezoresistive accelerometer which has a cantilever arrangement as shown in Fig. 10.24*A*. This uniformly stressed cantilever beam is loaded at its end with mass m. In this arrangement, four identical piezoresistive elements are used—two on each side of the beam, whose length is L in. These elements, whose resistance is R, form the active arms of the balanced bridge shown in Fig. 10.24*B*. A change of length L of the beam produces a change in resistance R in each element. The gage factor K for each of the elements [defined by Eq. (12.1)] is

$$K = \frac{\Delta R/R}{\Delta L/L} = \frac{\Delta R/R}{\epsilon} \tag{10.18}$$

where ε is the strain induced in the beam, expressed in inches/inch, at the surface where the elements are cemented. If the resistances in the four arms of the bridge are equal, then the ratio of the output voltage E_o of the bridge circuit to the input voltage E_i is

$$\frac{E_o}{E_i} = \frac{\Delta R}{R} = \epsilon K \tag{10.19}$$

TYPICAL PIEZORESISTIVE ACCELEROMETER CONSTRUCTIONS

Figure 10.25 shows two basic piezoresistive accelerometer designs.

Bonded Strain Gage, Fluid Damped Type. To provide high output sensitivities and resonance frequencies, discrete semiconductor piezoresistors are bonded firmly to the seismic mass where the strain is most concentrated. This is described by Fig. 10.25*A*. This approach is used to provide sensitivities more suitable for the measurement of acceleration below 1000g. To provide environmental shock resistance, overtravel stops are added. To extend the usable frequency range and enhance shock survivability, damping is added by surrounding the mechanism with silicone oil. The advantages of these designs are high sensitivity, broad frequency response for the sensitivity, and overrange protection. The disadvantages are complexity and limited temperature range. Overrange protection is almost mandatory in sensitive piezoresistive accelerometers; without it they would not survive ordinary shipping and handling. The viscosity of the damping fluid does change with temperature; as a result, the damping coefficient changes significantly with temperature.

Microelectro-Mechanical Systems (MEMS), Gas Damped Type. Also known as a micromachined accelerometer, the entire working mechanism (mass, spring, and support) of a MEMS-type accelerometer is etched from a single crystal of silicon, a process known as *micromachining.* This produces a very tiny and rugged device, shown in Fig. 10.25*C*. The advantages of the MEMS type are very small size, very high resonance frequency, ruggedness, and high range. Accelerometers of such design are used to measure a wide range of accelerations, from below 10*g* to over 200,000*g*. No adhesive is required to bond a strain gage of this type to the structure, which helps to make it a very stable device from a thermal and hysteresis point of

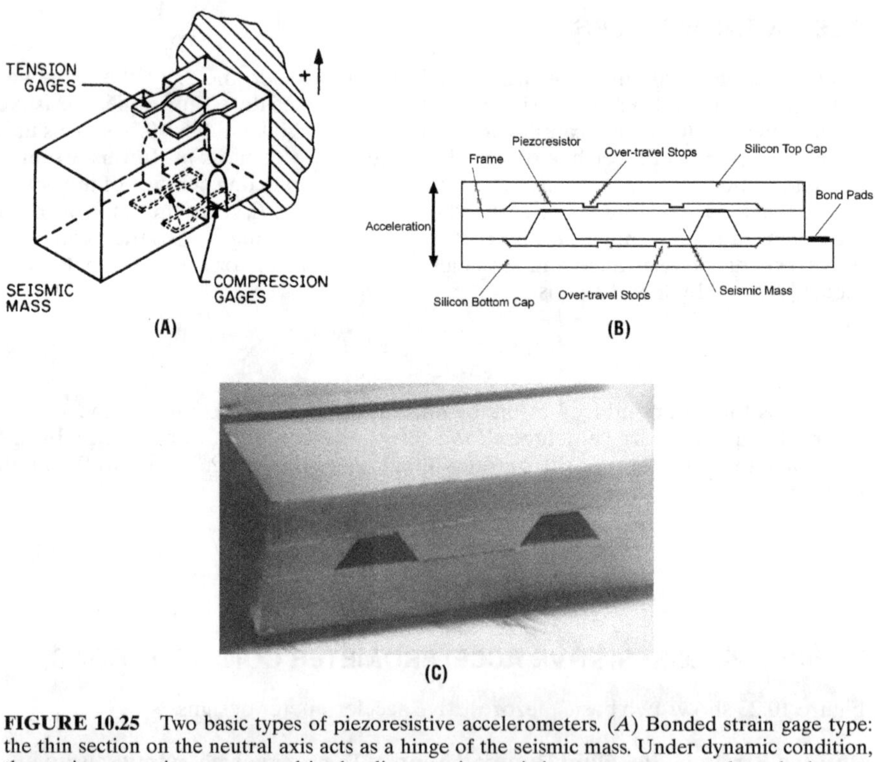

FIGURE 10.25 Two basic types of piezoresistive accelerometers. (*A*) Bonded strain gage type: the thin section on the neutral axis acts as a hinge of the seismic mass. Under dynamic condition, the strain energy is concentrated in the discrete piezoresistive gages on top and on the bottom of the mass. Viscous fluid typically encapsulates the seismic subassembly to provide the necessary damping. (*B*) Microelectro-mechanical systems (MEMS) type: the entire mechanism (seismic mass, hinges/piezoresistors) is etched from a single piece of silicon. The thin sections on the neutral axis near the top of the mass act as hinges; the microns-thick gaps between the mass and the top and bottom caps provide the squeezed-film damping. (*C*) A SEM cross-section view of the accelerometer shown in (*B*), where the mass has a thickness of ~300 μm. (*Courtesy of Measurement Specialties Inc.*.)

view. For shock applications, see the section entitled "Survivability." A few modern MEMS accelerometer designs offer squeezed-film gas damping as an alternative to silicone oil damping. Squeezed-film damping can be observed when a plate moves in close proximity to another solid surface, in effect alternately stretching and squeezing any fluid that may be present in the space between the moving plate and the solid surface. Depending on the range of frequency, this fluid motion can be a significant effect on the damping behaviors of the moving plate.

ELECTRICAL CHARACTERISTICS OF PIEZORESISTIVE ACCELEROMETERS

Excitation. Piezoresistive transducers require an external power supply to provide the necessary current or voltage excitation in order to operate. These energy sources must be well regulated and stable since they may introduce sensitivity errors and secondary effects at the transducer which will result in error signals at the output.

Traditionally, the excitation has been provided by a battery or a constant voltage supply. Other sources of excitation, such as constant current supplies or ac excitation generators, may be used. The sensitivity and temperature response of a piezoresistive transducer may depend on the kind of excitation applied. Therefore, it should be operated in a system which provides the same source of excitation as used during temperature compensation and calibration of the transducer. A common excitation source ranges from 2 to 10 V dc.

Sensitivity. The *sensitivity* of an accelerometer is defined as the ratio of its electrical output to its mechanical input. Specifically, in the case of piezoresistive accelerometers, it is expressed as voltage per unit of acceleration at the rated excitation (i.e., mV/g or peak mV/peak g at 10 volts dc excitation). Most piezoresistive accelerometers are designed in full- or half-bridge configuration, as shown in Fig. 10.24*B*. Their sensitivity is therefore ratiometric, which refers to the output voltage as a ratio of the supply voltage. For example, if the input voltage is doubled, the output voltage is doubled. This relationship is not perfectly linear in practice, but it is a close approximation.

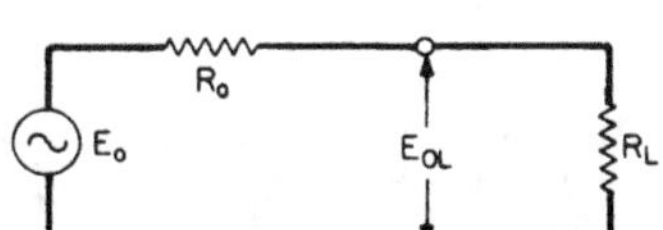

FIGURE 10.26 Loading effects on piezoresistive accelerometers.

Loading Effects. An equivalent circuit of a piezoresistive accelerometer, for use when considering loading effects, is shown in Fig. 10.26. Using the equivalent circuit and the measured output resistance of the transducer, the effect of loading may be directly calculated:

$$E_{oL} = E_o \frac{R_L}{R_o + R_L} \tag{10.20}$$

where R_o = output resistance of accelerometer, including cable resistance
E_o = sensitivity into an infinite load
E_{oL} = loaded output sensitivity
R_L = load resistance

Because the resistance of the strain-gage elements varies with temperature, output resistance should be measured at the operating temperature.

Effect of Cable on Sensitivity. Long cables may result in the following effects:

1. A reduction in sensitivity because of resistance in the input wires. The fractional reduction in sensitivity is equal to

$$\frac{R_i}{R_i + 2R_{ci}} \tag{10.21}$$

where R_i is the input resistance of the transducer and R_{ci} is the resistance of one input (excitation) wire. This effect may be overcome by using remote sensing leads.

2. Signal attenuation resulting from resistance in the output wires. This fractional reduction in signal is given by

$$\frac{R_L}{R_o + R_L + 2R_{co}} \tag{10.22}$$

where R_{co} is the resistance of one output wire between transducer and load.

3. Attenuation of the high-frequency components in the data signal as a result of *R-C* filtering in the shielded instrument leads. The stray and distributed capacitance present in the transducer and a short cable are such that any filtering effect is negligible to frequencies well beyond the usable range of the accelerometer. However, when long leads are connected between transducer and readout equipment, the frequency response at higher frequencies may be affected significantly.

Warmup Time. The excitation voltage across the piezoresistive elements causes a current to flow through each element. The I^2R heating results in an increase in temperature of the elements above ambient, which slightly increases the resistance of the elements. Differentials in this effect may cause the zero acceleration output voltage to vary slightly with time until the temperature is stabilized. Therefore, resistance measurements and shock and vibration data should not be taken until stabilization is reached. In a half-bridge configuration, due to the differences in thermal characteristics between the piezoresistors and the fixed completion resistors, the I^2R heating differentials may cause long warmup time before stabilization can be reached.

Input and Output Resistance. For an equal-arm Wheatstone bridge, the input and output resistances are equal. However, temperature-compensating and zero-balance resistors may be internally connected in series with the input leads or in series with the sensing elements. These additional resistors will usually result in unequal input and output resistance. The resistance of piezoresistive transducers varies with temperature much more than the resistance of metallic strain gages, usually having resistivity temperature coefficients between about 0.17 and 0.95 percent per degree Celsius.

Zero Balance. Although the resistance elements in the bridge of a piezoresistive accelerometer may be closely matched during manufacture, slight differences in resistance will exist. These differences result in a small offset or residual dc voltage at the output of the bridge at zero acceleration. Circuitry within associated signal-conditioning instruments may provide compensation or adjustment of the electrical zero.

Insulation. The case of the accelerometer acts as a mechanical and electrical shield for the sensing elements. Sometimes it is electrically insulated from the elements but connected to the shield of the cable. If the case is grounded at the structure, the shield of the connecting cable may be left floating and should be connected to ground at the end farthest from the accelerometer. When connecting the cable shield at the end away from the accelerometer, care must be taken to prevent ground loops.

Thermal Sensitivity Shift. The sensitivity of a piezoresistive accelerometer varies as a function of temperature. This change in the sensitivity is caused by changes in the gage factor and resistance and is determined by the temperature characteristics of the modulus of elasticity and piezoresistive coefficient of the sensing elements. The sensitivity deviations are minimized by installing compensating resistors in the bridge circuit within the accelerometer.

Thermal Zero Shift. Because of small differences in resistance change of the sensing elements as a function of temperature, the bridge may become slightly unbalanced when subjected to temperature changes. This unbalance produces small changes in the dc voltage output of the bridge. Transducers are usually compensated during manufacture to minimize the change in dc voltage output (zero balance) of the accelerometer with temperature. Adjustment of external balancing circuitry should not be necessary in most applications.

Damping. The frequency response characteristics of piezoresistive accelerometers having damping near zero are similar to those obtained with piezoelectric accelerometers. Viscous damping is provided in accelerometers having relatively low resonance frequencies to increase the useful high-frequency range of the accelerometer and to reduce the output at resonance. At room temperature this damping is usually 0.7 of critical damping or less. With damping, the sensitivity of the accelerometer is "flat" to greater than one-fifth of its resonance frequency.

The piezoresistive accelerometer using fluid damping is intended for use in a limited temperature range, usually +20 to +200°F (−7 to +94°C). At high temperatures the viscosity of the oil decreases, resulting in low damping; and at low temperatures the viscosity increases, which causes high damping. Accordingly, the frequency response characteristics change as a function of temperature. Piezoresistive accelerometers using gas damping have a wider operating temperature range due to significantly less viscosity variation over temperature.

FORCE GAGES AND IMPEDANCE HEADS

MECHANICAL IMPEDANCE MEASUREMENT

Mechanical impedance measurements are made to relate the force applied to a structure to the motion of a point on the structure. If the motion and force are measured at the same point, the relationship is called the *driving-point impedance;* otherwise it is called the *transfer impedance.* Any given point on a structure has six degrees of freedom: translations along three orthogonal axes and rotations around the axes, as explained in Chap. 2. A complete impedance measurement requires measurement of all six excitation forces and response motions. In practice, rotational forces and motions are rarely measured, and translational forces and motions are measured in a single direction, usually normal to the surface of the structure under test.

Mechanical impedance is the ratio of input force to resulting output velocity. *Mobility* is the ratio of output velocity to input force, the reciprocal of mechanical impedance. *Dynamic stiffness* is the ratio of input force to output displacement. *Receptance,* or *admittance,* is the ratio of output displacement to input force, the reciprocal of dynamic stiffness. *Dynamic mass,* or *apparent mass,* is the ratio of input force to output acceleration. All of these quantities are complex and functions of frequency. All are often loosely referred to as impedance measurements. They all require the measurement of input force obtained with a force gage (an instrument which produces an output proportional to the force applied through it). They also require the measurement of output motion. This is usually accomplished with an accelerometer; if velocity or displacement is the desired measure of motion, either can be determined from the acceleration.

Impedance measurements usually are made for one of these reasons:

1. To determine the natural frequencies and mode shapes of a structure (see Chap. 21)
2. To measure a specific property, such as stiffness or damping, of a material or structure
3. To measure the dynamic properties of a structure in order to develop an analytical model of it

The input force (excitation) applied to a structure under test should be capable of exciting the structure over the frequency range of interest. This excitation may be either a vibratory force or a transient impulse force (shock). If vibration excitation is used, the frequency is swept over the range of interest while the output motion (response) is measured. If shock excitation is used, the transient input excitation and resulting transient output response are measured. The frequency spectra of the input and output are then calculated by Fourier analysis.

FORCE GAGES

A force gage measures the force which is being applied to a structural point. Force gages used for impedance measurements use mostly piezoelectric transducing elements, although piezoresistive gages can also be used. A force gage is, in principle, very similar to an accelerometer in operation. The transducing element generates an output charge or voltage proportional to the applied force. Piezoelectric and piezoresistive transducing elements are discussed in detail earlier in this chapter.

TYPICAL FORCE-GAGE AND IMPEDANCE-HEAD CONSTRUCTIONS

Force Gages for Use with Vibration Excitation. Force gages for use with vibration excitation are designed with provision for attaching one end to the structure and the other end to a force driver (vibration exciter). A thin film of oil or grease is often used between the gage and the structure to improve the coupling at high frequencies.

Force Gages for Use with Shock Excitation. Force gages for use with shock excitation are usually built into the head of a hammer. Excitation is provided by striking the structure with the hammer. The hammer is often available with interchangeable faces of various materials to control the waveform of the shock pulse generated. Hard materials produce a short-duration, high-amplitude shock with fast rise and fall times; soft materials produce longer, lower-amplitude shocks with slower rise and fall times. Short-duration shocks have a broad frequency spectrum extending to high frequencies. Long-duration shocks have a narrower spectrum with energy concentrated at lower frequencies.

Shock excitation by a hammer with a built-in force gage requires less equipment than sinusoidal excitation and requires no special preparation of the structure.

Impedance Heads. Impedance heads combine a force gage and an accelerometer in a single instrument. They are convenient for measuring driving-point impedance because only a single instrument is required and the force gage and accelerometer are mounted as nearly as possible at a single point.

FORCE-GAGE CHARACTERISTICS

Amplitude Response, Signal Conditioning, and Environmental Effects. The amplitude response, signal conditioning requirements, and environmental effects associated with force gages are the same as those associated with piezoelectric accelerometers. They are described in detail earlier in this chapter. The sensitivity is expressed as charge or voltage per unit of force, e.g., picocoulomb/newton or millivolt/lb.

Near a resonance, usually a point of particular interest, the input force may be quite low; it is important that the force-gage sensitivity be high enough to provide accurate readings, unobscured by noise.

Frequency Response. A force gage, unlike an accelerometer, does not have an inertial mass attached to the transducing element. Nevertheless, the transducing element is loaded by the mass of the output end of the force gage. This is called the *end dynamic mass.* Therefore, it has a frequency response that is very similar to that of an accelerometer, as described earlier in this chapter.

Effect of Mass Loading. The dynamic mass of a transducer (force gage, accelerometer, or impedance head) affects the motion of the structure to which the transducer is attached. Neglecting the effects of rotary inertia, the motion of the structure with the transducer attached is given by

$$A = A_o \frac{m_s}{m_s + m_t} \tag{10.23}$$

where
a = amplitude of motion with transducer attached
A_o = amplitude of motion without transducer attached
m_s = dynamic mass of structure at point of transducer attachment in direction of sensitive axis of transducer
m_t = dynamic mass of the transducer in its sensitive direction

These are all complex quantities and functions of frequency. Near a resonance the dynamic mass of the structure becomes very small; therefore, the mass of the transducer should be as small as possible. The American National Standards Institute recommends that the dynamic mass of the transducer be less than 10 times the dynamic mass of the structure at resonance.

OPTICAL-ELECTRONIC TRANSDUCER SYSTEMS

LASER DOPPLER VIBROMETERS

The laser Doppler vibrometer (LDV) uses the Doppler shift of laser light which has been backscattered from a vibrating test object to produce a real-time analog signal output that is proportional to instantaneous velocity. The velocity measurement range, typically between a minimum peak value of 0.5 μm/sec and a maximum peak value of 10 m/sec, is illustrated in Fig. 10.27.

An LDV is typically employed in an application where other accelerometers or other types of conventional sensors cannot be used. LDVs' main features are

- There are no transducer mounting or mass loading effects.
- There is no built-in transverse sensitivity or other environmental effects.
- They measure remotely from nearly any standoff distance.
- There is ultra-high spatial resolution with small measurement spot (5 to 100 μm typically).
- They can be easily fitted with fringe-counter electronics for producing absolute calibration of dynamic displacement.
- The laser beam can be automatically scanned to produce full-field vibration pattern images.

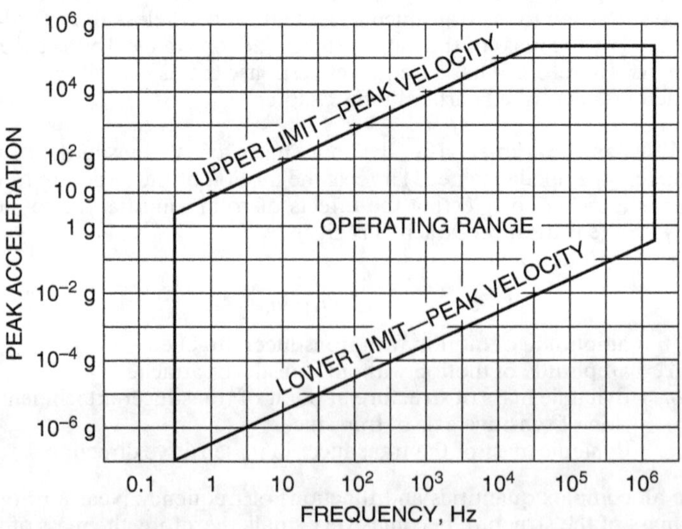

FIGURE 10.27 Typical operating range for a laser Doppler vibrometer. (*Courtesy of Polytec Pi, Inc.*)

Caution must be exercised in the installation and calibration of laser Doppler vibrometers (LDVs). In installing such an optical-electronic transducer system, care must be given to the location unit relative to the location of the target; in many applications, optical alignment can be difficult. Although absolute calibration of the associated electronic system can be carried out, an absolute calibration of the optical system usually cannot be. Thus, the calibration is usually restricted to the range of the secondary standard accelerometer used, which is only a small portion of the dynamic range of the LDV; the secondary standard accelerometer should be calibrated against a National Institute of Standards and Technology (NIST) traceable reference, at least once a year, in compliance with MIL-STD-45662A. Since the application of LDV technology is based on the reflection of coherent light scattered by the target surface, ideally this surface should be flat relative to the wavelength of the light used in the laser. If it is not, the nonuniform surface can result in spurious reflectivity (resulting in noise) or complete loss of reflectivity (signal dropout).

Types of Laser Doppler Vibrometers Four types of laser Doppler vibrometers are illustrated in Fig. 10.28.

Standard (Out of Plane). The standard LDV measures the vibrational component $v_z(t)$ which lies along the laser beam. Triaxial measurements can be obtained by

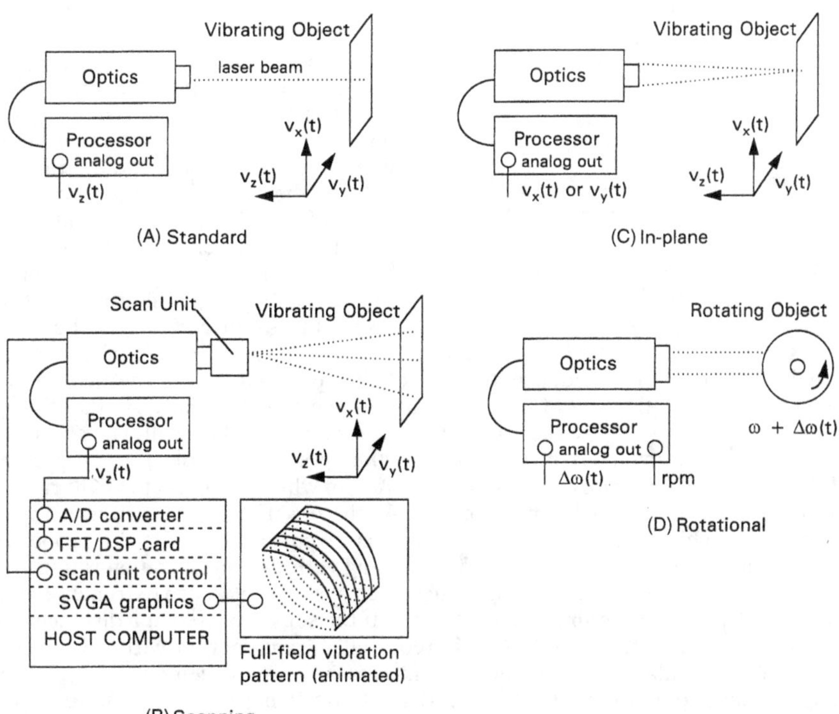

FIGURE 10.28 The four basic types of laser Doppler vibrometer systems. (*Courtesy of Polytec Pi, Inc.*)

approaching the same measurement point from three different directions. This is the most common type of LDV system.

Scanning. An extension of the standard out-of-plane system, the scanning LDV uses computer-controlled deflection mirrors to direct the laser to a user-selected array of measurement points. The system automatically collects and processes vibration data at each point; scales the data in standard displacement, velocity, or acceleration engineering units; performs fast Fourier transform (FFT) or other operations; and displays full-field vibration pattern images and animated operational deflection shapes.

In-plane. A special optics probe emitting two crossed laser beams is directed at normal incidence to the test surface and measures in-plane velocity. By rotating the probe by 90°, $v_x(t)$ or $v_y(t)$ can be measured.

Rotational. Two parallel laser beams from an optics probe measure angular vibration in units of degrees per second. Rotational systems are commonly used for torsional vibration analysis.

FIBER-OPTIC REFLECTIVE DISPLACEMENT SENSOR

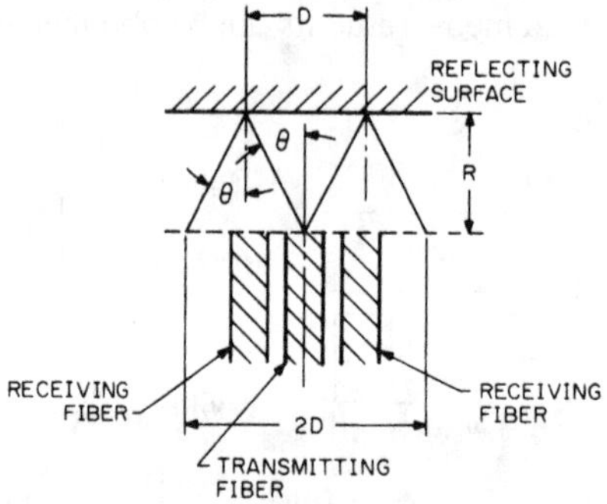

FIGURE 10.29 Fiber-optic displacement sensor. (*Courtesy of EOTEC Corp.*)

A fiber-optic reflective displacement sensor measures the amount of light normal to, and vibrating along, the optical axis of the device. The amount of reflected light is related to the distance between the surface and the fiber-optic transmitting/receiving element, as illustrated in Fig. 10.29. The sensor is composed of two bundles of single optical fibers. One of these bundles transmits light to the reflecting target; the other traps reflected light and transmits it to a detector. The intensity of the detected light depends on how far the reflecting surface is from the fiber-optic probe. Light is transmitted from the bundle of fibers in a solid cone defined by a numerical aperture. Since the angle of reflection is equal to the angle of incidence, the size of the spot that strikes the bundle after reflection is twice the size of the spot that hits the target initially. As the distance from the reflecting surface increases, the spot size increases as well. The amount of reflected light is inversely proportional to the spot size. As the probe tip comes closer to the reflecting target, there is a position in which the reflected light rays are not coupled to the receiving fiber bundle. At the onset of this occurrence, a maximum forms which drops to zero as the reflecting surface contacts the probe. The output-current sensitivity can be varied by using various optical configurations.

While sensitivities approaching 1 microinch are possible, such extreme sensitivities limit the corresponding dynamic range. If the sensor is used at a distance from the reflecting target, a lens system is required in conjunction with a fiber-optic probe. With available lenses, the instruments have displacement measurement ranges from 0 to 0.015 in. (0 to 0.38 mm) and 0 to 5.0 in. (0 to 12.7 cm). Resolution typically is better than 1⁄100 of the full-scale range. The sensor is sensitive to rotation of the reflecting target. For rotations of ±3° or less, the error is less than ±3 percent.

ELECTRODYNAMIC TRANSDUCERS

ELECTRODYNAMIC (VELOCITY COIL) PICKUPS

The output voltage of the electrodynamic pickup is proportional to the relative velocity between the coil and the magnetic flux lines being cut by the coil. For this reason it is commonly called a velocity coil. The principle of operation of the device is illustrated in Fig. 10.30. A magnet has an annular gap in which a coil wound on a hollow cylinder of nonmagnetic material moves. Usually a permanent magnet is used, although an electromagnet may be used. The pickup also can be designed with the coil stationary and the magnet movable. The open-circuit voltage *e* generated in the coil is[2,3]

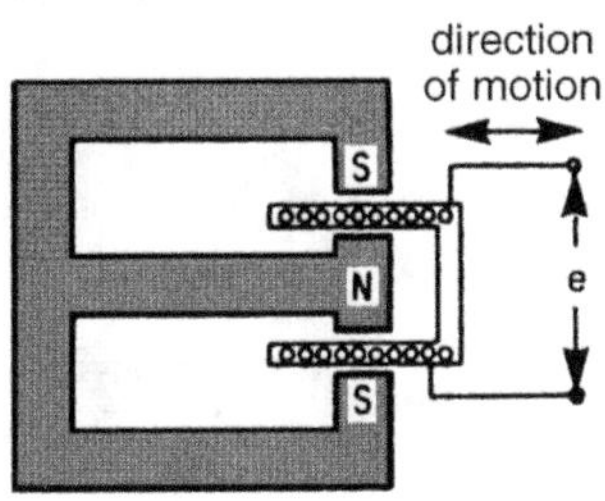

FIGURE 10.30 Principle of operation of an electrodynamic pickup. The voltage *e* generated in the coil is proportional to the velocity of the coil relative to the magnet.

$$e = -Blv(10^{-8}) \qquad \text{volts}$$

where B is the flux density in gausses; l is the total length in centimeters of the conductor in the magnetic field; and v is the relative velocity in centimeters per second between the coil and magnetic field. The magnetic field decreases sharply outside the space between the pole pieces; therefore, the length of coil wire outside the gap generates only a very small portion of the total voltage.

One application of the electrodynamic principle is the velocity-type seismic pickup. Usually the pickup is used only at frequencies above its natural frequency, and it is not very useful at frequencies above several thousand hertz. The sensitivity of most pickups of this type is quite high, particularly at low frequencies where their output voltage is greater than that of many other types of pickups. The coil impedance is low even at relatively high frequencies, so that the output voltage can be measured directly with a high-impedance voltmeter. This type of pickup is designed to measure quite large displacement amplitudes.

LINEAR VARIABLE DIFFERENTIAL TRANSFORMER (LVDT) PICKUPS

The output of a linear variable differential transformer (LVDT) depends on the mutual inductance between a primary and a secondary coil. It is an electromechanical device that produces an electrical output proportional to the displacement of a separate movable core. The device consists of a primary coil and two secondary coils symmetrically spaced on a cylindrical form. A free-moving, rod-shaped magnetic core inside the coil assembly provides a path for the magnetic flux linking the coils. See Fig. 10.31*A*. When the primary coil is energized by an external ac source, voltages are induced in the two secondary coils. These are connected series opposing so the two voltages are of opposite polarity. Therefore, the net output of the transducer is the difference between these voltages, which is zero when the core is at the center or null position. When the core is moved from the null position, the induced voltage in the coil toward which the core is moved increases, while the induced voltage in the opposite coil decreases. This action produces a differential voltage output that varies linearly with changes in core position. See Fig. 10.31*B*.

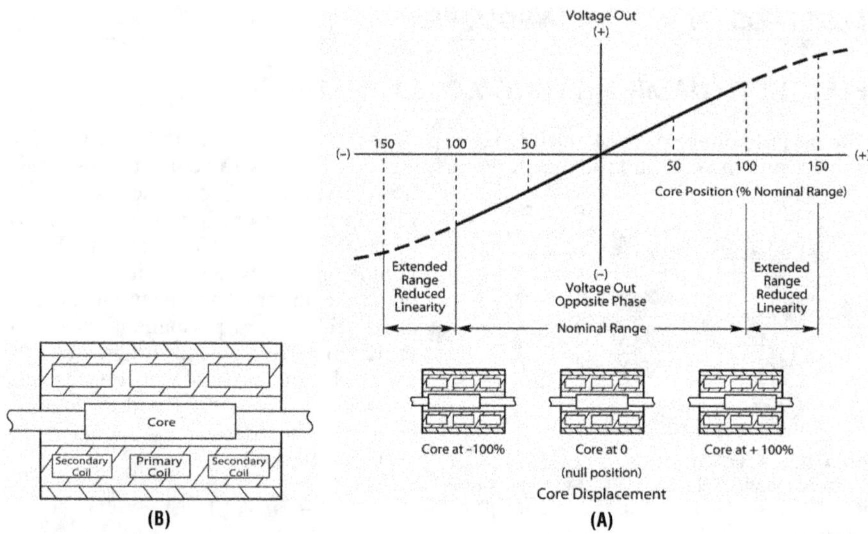

FIGURE 10.31 Operation of a linear variable differential transformer (LVDT). (*A*) Cross section of an LVDT showing the primary and secondary coils and the moving core. (*B*) The phase of this output voltage changes abruptly by 180° as the core in moved from one side of null to the other. The core must always be fully within the coil assembly during operation of the LVDT; otherwise, gross nonlinearity will occur. (*Courtesy of Measurement Specialties Inc.*)

LVDT is used for low-frequency measurements. The sensitivity varies with the carrier frequency of the current in the primary coil. The carrier frequency should be at least 10 times the highest frequency of the motion to be measured. Modern LVDT has a carrier frequency at 10 kHz and a usable bandwidth from 0 to 1 kHz.[10]

CAPACITANCE-TYPE TRANSDUCERS

DISPLACEMENT TRANSDUCER (PROXIMITY PROBE)

The capacitance-type transducer is basically a displacement-sensitive device. Its output is proportional to the change in capacitance between two plates caused by the change of relative displacement between them as a result of the motion to be measured. Appropriate electronic equipment is used to generate a voltage corresponding to the change in capacitance.

The capacitance-type displacement transducer's main advantages are (1) its simplicity in installation, (2) its negligible effect on the operation of the vibrating system since it is a proximity-type pickup which adds no mass or restraints, (3) its extreme sensitivity, (4) its wide displacement range, due to its low background noise, and (5) its wide frequency range, which is limited only by the electric circuit used.

The capacitance-type transducer often is applied to a conducting surface of a vibrating system by using this surface as the ground plate of the capacitor. In this arrangement, the insulated plate of the capacitor should be supported on a rigid structure close to the vibrating system. Figure 10.32*A* shows the construction of a typical capacitance pickup; Fig. 10.32*B, C, D,* and *E* show a number of possible methods of applying this type of transducer. In each of these, the metallic vibrating

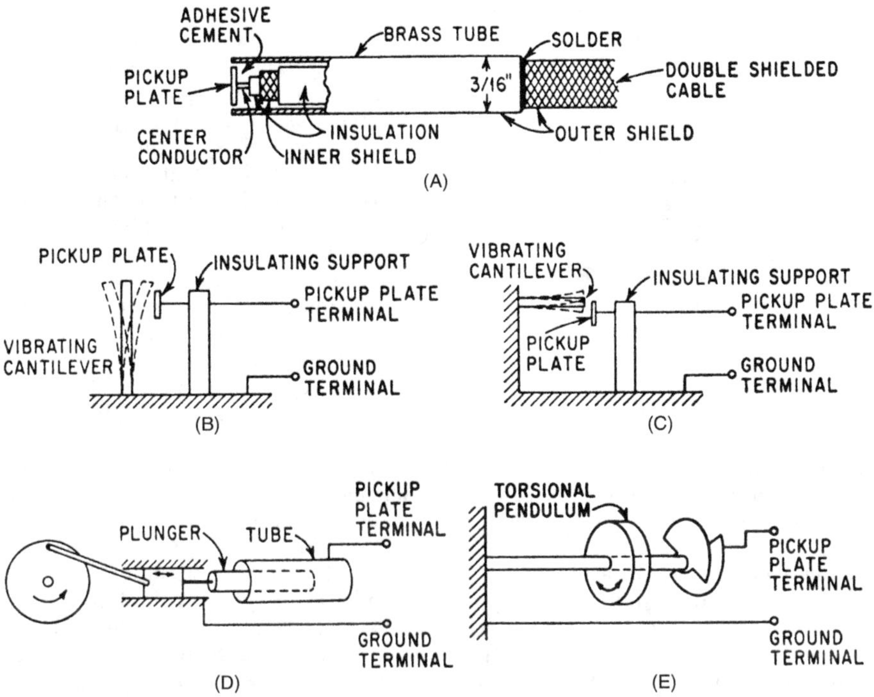

FIGURE 10.32 Capacitance-type transducers and their application: (*A*) construction of typical assembly, (*B*) gap length or spacing sensitive pickup for transverse vibration, (*C*) area sensitive pickup for transverse vibration, (*D*) area sensitive pickup for axial vibration, and (*E*) area sensitive pickup for torsional vibration.

system is the ground plate of the capacitor. Where the vibrating system at the point of instrumentation is an electrical insulator, the surface can be made slightly conducting and grounded by using a metallic paint or by rubbing the surface with graphite.

The maximum operating temperature of the transducer is limited by the insulation breakdown of the plate supports and leads. Bushings made of alumina are commercially available and provide adequate insulation at temperatures as high as 2000°F (1093°C).

VARIABLE-CAPACITANCE-TYPE ACCELEROMETER

Silicon micromachined variable-capacitance technology is utilized to produce miniaturized accelerometers suitable for measuring low-level accelerations (2*g* to 100*g*) and capable of withstanding high-level shocks (5000*g* to 20,000*g*).

Acceleration sensing is accomplished by using a half-bridge variable-capacitance microsensor. The capacitance of one circuit element increases with applied acceleration, while that of the other decreases. With the use of signal conditioning, the accelerometer provides a linearized high-level output.

In the following example, the microsensor is fabricated in an array of three micromachined single-crystal silicon wafers bonded together using an anodic bonding process (see exploded view in Fig. 10.33). The top and bottom wafers contain the fixed capacitor plates (the lid and base, respectively), which are electrically isolated from the middle wafer. The middle wafer contains the inertial mass, the suspension, and the supporting ringframe. The stiffness of the flexure system is controlled by varying the shape, cross-sectional dimensions, and number of suspension beams. Damping is controlled by varying the dimensions of grooves and orifices on the parallel plates. Overrange protection is extended by adding overtravel stops.

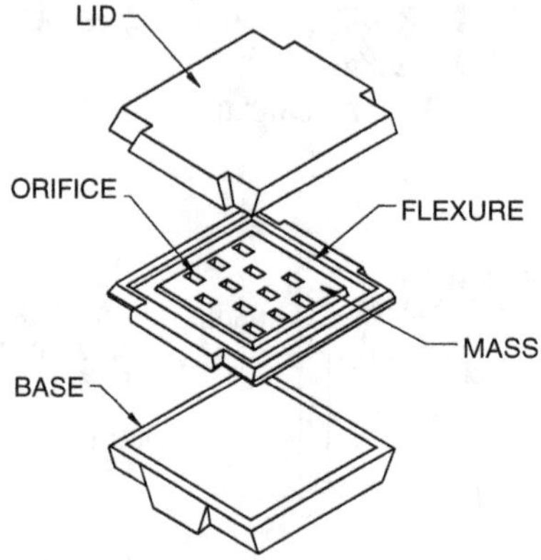

FIGURE 10.33 Exploded view of a micromachined capacitive accelerometer with a trampoline-like seismic mass mechanism.

The full-scale displacement of the seismic mass of the microsensor element is slightly more than 10 microinches. To detect minor capacitance changes in the microsensor due to acceleration, high-precision supporting electronic circuits are required. One approach applies a triangle wave to both capacitive elements of the microsensor. This produces currents through the elements which are proportional to their capacitances. A current detector and subtractor full-wave rectifies the currents and outputs their difference. An operational amplifier then converts this current difference to an output voltage signal. A high-level output is provided that is proportional to input acceleration.

REFERENCES

1. Levy, S., and W. D. Kroll: Research Paper 2138, *J. Research Natl. Bur. Standards,* **45**:4 (1950).
2. Chu, A. S.: "A Shock Amplifier Evaluation," *Proceedings, Institute of Environmental Sciences,* April 1990.
3. Ref. 7, TP290 by A. S. Chu.
4. Ref. 7, TP308 by A. S. Chu.
5. Measurement Specialties, Inc.: "Temperature Compensation Techniques for Piezoresistive Accelerometers," TN-009, Hampton, Va., 1988.
6. *ISA Recommended Practice,* RP37.2, ¶6.6, "Strain Sensitivity," Instrument Society of America, 1964.
7. Technical Papers, Endevco Corp., San Juan Capistrano, CA 92675: TP290, "Zero Shift of Piezoelectric Accelerometers" (1990); TP308, "Problems in High-Shock Measurements" (1993); TP319, "A Guide to Accelerometer Installation" (1999); TP320, "Isotron and Charge Mode Piezoelectric Accelerometers" (2000).
8. Ref. 7, TP319 by A. Coghill.
9. Ref. 7, TP320 by B. Arkell.
10. Schaevitz/Measurement Specialties, Inc.: "LVDT Overview," Position Sensor Product Catalog, Hampton Va., 2008.

CHAPTER 11
CALIBRATION OF SHOCK AND VIBRATION TRANSDUCERS

Jeffrey Dosch

INTRODUCTION

This chapter describes various methods of calibrating shock and vibration transducers, commonly called *vibration pickups.* The objective of calibrating a transducer is to determine its sensitivity or calibration factor, as defined below. The chapter is divided into three major parts, which discuss comparison methods of calibration, absolute methods of calibration, and calibration methods that employ high acceleration and shock. Field calibration techniques are described in Chap. 15.

PICKUP SENSITIVITY, CALIBRATION FACTOR, AND FREQUENCY RESPONSE

As defined in Chap. 10, the *sensitivity* of a vibration pickup is the ratio of electrical output to mechanical input applied along a specified axis.[1,2] The sensitivity of all pickups is a function of frequency, containing both amplitude and phase information, as illustrated in Fig. 11.1, and therefore is usually a complex quantity. If the sensitivity is practically independent of frequency over a range of frequencies, the value of its magnitude is referred to as the *calibration factor* for that range, but it is specified at a discrete frequency. The phase component of the sensitivity function likewise has a constant value in that range of frequencies, usually equal to zero or 180°, but it may also be proportional to frequency, as explained in Chap. 10.

The *frequency response* of a pickup is shown by plotting the magnitude and phase components of its sensitivity versus frequency. This information is usually presented relative to the value of sensitivity at a reference frequency within the flat range. A preferred frequency, internationally accepted, is 160 Hz.

Displacements are usually expressed as single-amplitude (peak) or double-amplitude (peak-to-peak) values, while velocities are usually expressed as peak, root-mean-square (rms), or average values. Acceleration and force generally are expressed as peak or rms values. The electrical output of the vibration pickup may be expressed as peak, rms, or average value. The sensitivity magnitude or calibration

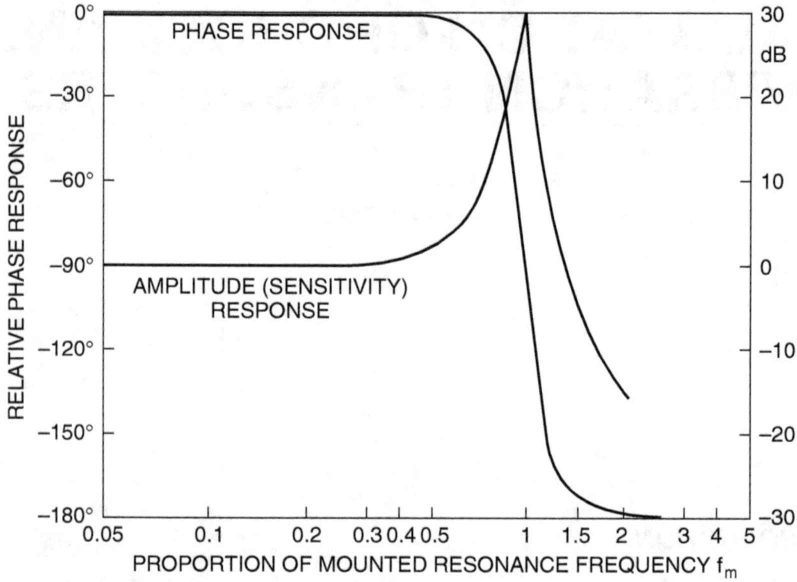

FIGURE 11.1 Pickup amplitude and phase response as functions of frequency. (*After M. Serridge and T.R. Licht.*[3])

factor is commonly stated in similarly expressed values, i.e., the numerator and denominator are both peak or both rms values. Examples of typical sensitivity specifications for an accelerometer: 2 pC/m/s^2, 10 mV/m/s^2, 5 mV/g (where C is the symbol for coulomb, V is the symbol for volt, and g is the standard acceleration due to gravity equal to 9.80665 m/s^2). For some special applications it may be desirable to express the sensitivity in mixed values, such as rms voltage per peak acceleration.

CALIBRATION TRACEABILITY

In calibrating an instrument, one measures the instrument's error relative to a reference which is traceable to the national standard of a country. A calibration is said to be *traceable*[4] to a national or international standard if it can be related to the standard through an unbroken chain of comparisons—all having stated uncertainties. In the U.S.A., for example, national vibration standards are maintained at the National Institute of Standards and Technology in Gaithersburg, Maryland. A number of other national metrology laboratories having known capabilities for maintaining national vibration standards are listed in Table 11.1. Countries whose national laboratories do not provide a national vibration standard may belong to a regional international association, such as NORAMET (North American Metrology Cooperation), EUROMET (European Metrology Cooperation), or OIML (Organization for Legal Metrology) that can assist transducer manufacturers in setting up steps necessary for establishing traceability to a national standard.

Vendors of transducers must be able to show that calibrations of their instruments are traceable to a national standard by means of calibration reports stating

TABLE 11.1 National Standards Laboratories Responsible for the Calibration of Vibration Pickups

Institution	Laboratory	Location	Country
CSIRO	Natl. Measurement Laboratory	Lindfield	Australia
INMETRO	Laboratório de Vibrações	Rio de Janeiro	Brazil
NRC-CNRC	Inst. Natl. Meas. Stds.	Ottawa	Canada
NIM	Vibrations Laboratory	Beijing	China
CMU	Primary Stds. of Kinematics	Prague	Czech Repub.
B&K	Danish Prim. Lab. for Acoustics	Naerum	Denmark
BNM	CEA/CESTA	Belin-Beliet	France
PTB	Fachlabor. Beschleunigung	Braunschweig	Germany
IMGC	Sezione Meccanica	Torino	Italy
NRLM	Mechanical Metrology Dept.	Tsukuba	Japan
KSRI	Division of Appl. Metrology	Daedeog Danji	Rep. of Korea
NMC	SIRIM	Berhad	Malaysia
CENAM	Div. Acustica y Vibraciones	Queretaro	Mexico
DSIR	Measurement Stds. Laboratory	Lower Hutt	New Zealand
VNIIM	Mendeleyev Inst. for Metrology	St. Petersburg	Russia
ITRI	Center for Measurement Stds.	Hsinchu	Taiwan
NIST	Manufacturing Metrology Div.	Gaithersburg	U.S.A.

the value(s) of sensitivity, measurement uncertainty, environmental conditions, and identification of the standard(s) used in the calibration procedure. Depending on the application, there may be one or more links to the national standard.

Primary and Secondary Standards. *Primary standards,*[5,6] maintained at national metrology institutes, are derived from *absolute measurements* of the transducer's sensitivity, measured in terms of seven basic units. For example, the absolute measurement of "speed" must be made in terms of measurements of distance and time, not by a speedometer. Thus, the word *absolute* implies nothing about precision or accuracy. An example of a laboratory setup for the calibration of primary standard accelerometers, derived from absolute measurements, is shown in Fig. 11.2.[7] A vibration exciter

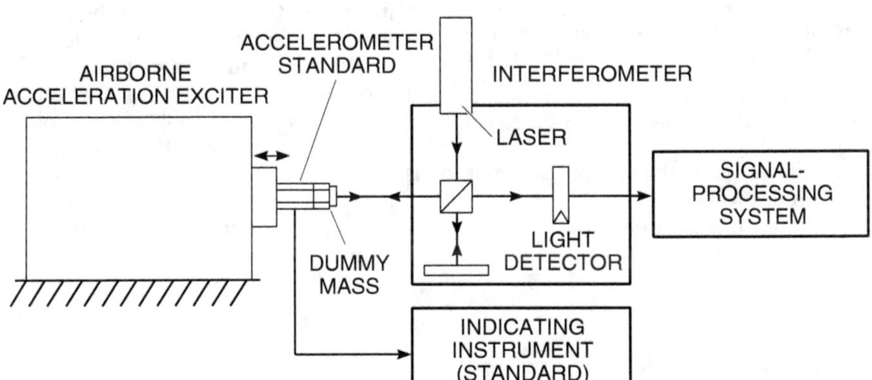

FIGURE 11.2 Primary (absolute) calibration of an accelerometer standard using laser interferometry. (*After von Martens.*[7])

generates sinusoidal motion which is measured by a Michelson interferometer (described later in this chapter). The vibration is applied to the base of the standard accelerometer whose output is measured. A dummy mass, mounted on its top surface, simulates the conditions when this standard accelerometer is used to calibrate a *secondary standard*[8] accelerometer by the comparison method described in the next section. *Secondary standards* (also referred to as *transfer standards* or *working standards*) are maintained at various government laboratories and industrial laboratories. A secondary standard accelerometer may be calibrated either from absolute measurements or from a comparison with a primary standard accelerometer. Such secondary standards are usually used for purposes of comparisons of calibrations between laboratories or for checking production and field units.

COMPARISON METHODS OF CALIBRATION

A rapid and convenient method of measuring the sensitivity of a vibration pickup to be tested is by direct comparison of the pickup's electrical output with that of a second pickup (used as a "reference" standard) that has been calibrated by one of the methods described in this chapter. A comparison method is used in most shock and vibration laboratories, which periodically send their standards to a primary standards laboratory for recalibration. This procedure should be followed on a yearly basis in order to establish a history of the accuracy and quality of its reference standard pickup.

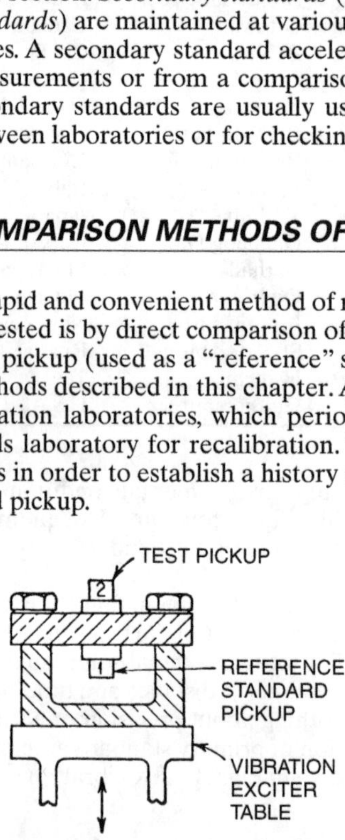

FIGURE 11.3 Comparison method of calibration: Pickup 2 is calibrated against Pickup 1 (the reference standard). The two pickups may be excited by any of the means described in this chapter. (*After ANSI Standard S2.2-1959, R 2006.*[1])

In this method of calibration the two pickups usually are mounted back-to-back on a vibration exciter as shown in Fig. 11.3. It is essential to ensure that each pickup experiences the same motion. Any angular rotation of the table should be small to avoid any difference in excitation between the two pickup locations. The error due to rotation may be reduced by carefully locating the pickups firmly on opposite faces with the center of gravity of the pickups located at the center of the table. Relative differences in pickup excitation may be observed by reversing the pickup locations and observing if the voltage ratio is the same for both positions.

Calibration by the comparison method is limited to the range of frequencies and amplitudes for which the reference standard pickup has been previously calibrated. If both pickups are linear, the sensitivity of the test pickup can be calculated in both magnitude and phase from

$$S_t = \frac{e_t}{e_r} S_r \tag{11.1}$$

where S_t = sensitivity of test pickup
S_r = sensitivity of reference standard pickup
e_t = output voltage from test pickup
e_r = output voltage from reference standard pickup

Several calibration methods described below are variations on the implementation of Eq. (11.1); they differ mainly in the manner of vibration excitation.

USING THE COMPARISON METHOD

A simple and convenient way of performing a comparison calibration is to fix the test pickup and reference standard pickup so they experience identical motion, as in Fig. 11.3. Then, set the frequency of the vibration exciter at a desired value, adjust the amplitude of vibration of the vibration exciter to a desired value, and then compare the electrical outputs of the pickups. Often, instead of making a comparison at a fixed frequency, a graphical plot of the sensitivity versus frequency is obtained by incorporating a swept-frequency signal generator in the calibration system.

RANDOM-EXCITATION-TRANSFER-FUNCTION METHOD

The use of random-vibration-excitation and transfer-function analysis techniques can provide quick and accurate comparison calibrations.[9] The reference standard pickup and the test pickup are mounted back-to-back on a suitable vibration exciter. Their outputs are usually fed into a spectrum analyzer through a pair of low-pass (antialiasing) filters. The bandwidth of the random signal which drives the exciter is determined by settings of the analyzer.

This method provides a nearly continuous calibration over a desired frequency spectrum, with the resulting sensitivity function having both amplitude and phase information. Since purely sinusoidal motion is not a requirement as in the other calibration methods, this lessens the requirements for the power amplifier and exciter to maintain low values of harmonic distortion. A very useful measure of process quality is obtained by computing the input/output coherence function, which requires knowledge of the input and output power spectra, the cross-power spectrum, and the transfer function.

CALIBRATION BY ABSOLUTE METHODS

RECIPROCITY METHOD

The reciprocity calibration method is an absolute means for calibrating vibration exciters that have a velocity coil or reference accelerometer. This method relates the pickup sensitivity to measurements of voltage ratio, resistance, frequency, and mass. For this method to be applicable, it is necessary that the vibration exciter system be linear (e.g., that the displacement, velocity, acceleration, and current in the driver coil each increase linearly with force and driver-coil voltage). The reciprocity method is used chiefly with electrodynamic exciters[11] but also with piezoelectric vibration exciters.[11]

The reciprocity method is applied only under controlled laboratory conditions. Many precautions must be taken, and the process is time-consuming. Several variations of the basic approach have been developed at national standards laboratories.[12,13] The method described here has been used at the National Institute of Standards and Technology.[14–17] The method consists of two laboratory experiments:

1. The measurement of the transfer admittance between the exciter's driver coil and the attached velocity coil or accelerometer.
2. The measurement of the voltage ratio of the open-circuit velocity coil or accelerometer and the driving coil while the exciter is driven by a second external exciter. The use of a piezoelectric accelerometer is assumed here. The electrical connections for the transfer admittance and voltage ratio measurements are shown in Fig. 11.4.

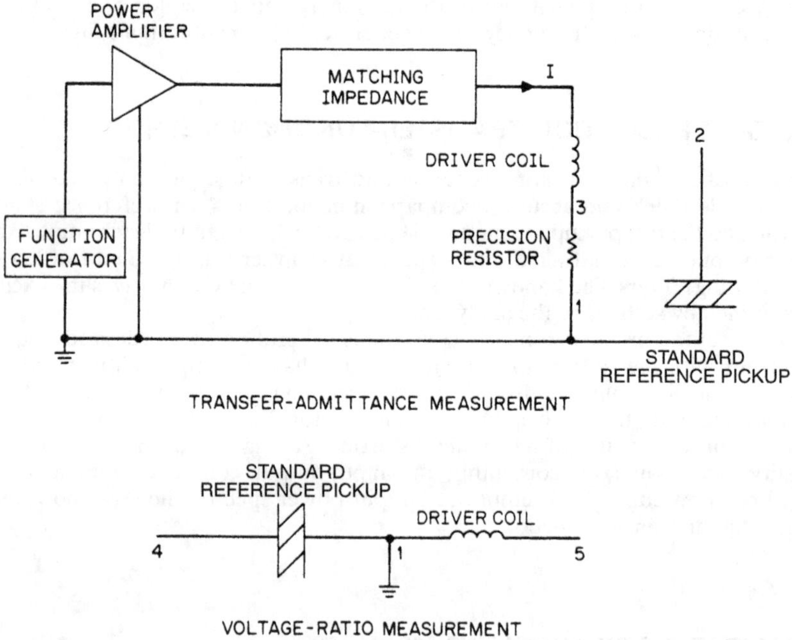

FIGURE 11.4 Transfer-admittance and voltage-ratio-measurement circuit connections for the reciprocity calibration method in the Levy-Bouche realization.[12]

The relationship defining the transfer admittance is

$$\mathbf{Y} = \frac{\mathbf{I}}{\mathbf{e}_{12}} \tag{11.2}$$

where $\mathbf{Y}$ = transfer admittance
$\mathbf{e}_{12}$ = voltage generated in standard accelerometer and amplifier
$\mathbf{I}$ = current in driver coil

and the bold letters denote phasor (complex) quantities. The current is determined by measuring the voltage drop across a standard resistor. The phase, ψ_Y, of $\mathbf{Y}$ is measured with a phase meter having an uncertainty of $\pm 0.1°$ or better. Transfer admittance measurements are made with a series of masses attached, one at a time, to the table of the exciter. Also, a zero-load transfer admittance measurement is made before and after attaching each mass. This zero-load measurement is denoted by $\mathbf{Y}_0$. Using the measured values of $\mathbf{Y}$ and $\mathbf{Y}_0$, graphs of the real and imaginary values of the ratio

$$\mathbf{T}_n = \frac{M_n}{\mathbf{Y} - \mathbf{Y}_0} \tag{11.3}$$

are plotted versus M_n for each frequency, where M_n is the value of the mass attached to the table. The zero intercepts, J_i and J_r, of the resulting nominally straight lines and their slopes, Q_i and Q_r, are computed by a weighted least-squares method.[10] The values of $\mathbf{Y}_0$ used in the calculations are obtained by averaging the values of the $\mathbf{Y}_0$ measurements before and after each measurement of $\mathbf{Y}$ using different masses. These computed values are used in determining the sensitivity of the standard.

The ratio of two voltages, measured while the exciter is driven with an external exciter, is given by

$$\mathbf{R} = \frac{\mathbf{e}_{14}}{\mathbf{e}_{15}} \tag{11.4}$$

where $\mathbf{e}_{14}$ = voltage generated in standard accelerometer and amplifier, and $\mathbf{e}_{15}$ = open-circuit voltage in driving coil.

After $\mathbf{R}$, J_r, J_i, Q_r, and Q_i have been determined for a number of frequencies, f, the sensitivity of the exciter is calculated from the following relationship:[10]

$$\mathbf{S} = \left[\frac{\mathbf{RJ}}{j2\pi f}\right]^{1/2}\left[1 + \frac{\mathbf{MQ}}{\mathbf{J}}\right] \tag{11.5}$$

where j = unit imaginary vector

$\mathbf{J} = J_r + jJ_i$

$\mathbf{Q} = Q_r + jQ_i$

$\mathbf{M} = \dfrac{\mathbf{J}(\mathbf{Y} - \mathbf{Y}_0)}{1 - \mathbf{Q}(\mathbf{Y} - \mathbf{Y}_0)}$

The sensitivity of the exciter is, therefore, determined from the measured quantities **Q, J, T,** and f and from the masses M_n which are attached to the exciter table. The sensitivity as computed from Eq. (11.5) has the units of volts per meter per second squared if the values of the measurements are in the SI system. If the masses M_n are not in kilograms, appropriate conversion factors must be applied to the quantities **J, Q,** and **M.** A commonly used engineering formula,[10] with the mass expressed in pounds and the sensitivity in millivolts per g, is

$$\mathbf{S} = 2635\left[\frac{\mathbf{RJ}}{jf}\right]^{1/2} \tag{11.6}$$

which also assumes that $MQ/J \ll 1$, a condition usually satisfied in practice but which should be verified experimentally. The use of a computer greatly facilitates the application of the reciprocity calibration process.

Assuming the errors to be uncorrelated, a typical estimate of uncertainty expected from a reciprocity calibration method is ±0.5 percent in the frequency range 100 to 1000 Hz. This is a twofold improvement over the earlier systems.[16] The critical component in a reciprocity-based calibration system is the vibration exciter. Electrodynamic exciters utilizing an air bearing are generally superior to other types, for this application.

CALIBRATION USING THE EARTH'S GRAVITATIONAL FIELD

The earth's gravitational field provides a convenient means of applying a small constant acceleration equal to the local value for gravitational deceleration g_l. It is particularly useful in calibrating accelerometers whose frequency range extends down to 0 Hz. A $2g_l$ change in acceleration may be obtained by first aligning the sensitivity axis of the transducer in one direction of the earth's gravitational field, as shown in Fig. 15.5*A*, and then inverting it so the sensitive axis is aligned in the opposite direction. This method of calibration is particularly useful in field work.

Accelerations in the 1–10g range can be generated by several methods, which have been largely replaced by the *structural gravimetric calibrator* described in the next section. In the *tilting-support calibrator*[1] the pickup is fastened to one end of an arm attached to a platform. The arm may be set at any angle between 0° and 180° relative to the vertical, thus yielding different values of acceleration. The *pendulum calibrator*[1] generates transient accelerations as great as 10g for a duration of about one second. In the *rotating-table calibrator*[18,19] the disk on which the test pickup is mounted rotates at a uniform angular rate about a horizontal axis in such a way that the pickup's axis of sensitivity rotates in a vertical plane. This method makes it possible to obtain both the static and dynamic responses of the pickup in the same test setup.

Structural-Gravimetric Calibration. This technique provides a simple, robust, and low-cost method of calibrating pickups.[20,21] The structural-gravimetric-calibration (SGC) method is applicable over a broad frequency range because it relies on a quartz force transducer as the reference pickup and the behavior of the simplest of structures (i.e., a mass behaving as a rigid body). It references the acceleration of gravity and allows the measurement of sensitivity magnitude and phase. The results of calibration using this method agree within a fraction of 1 percent with those obtained by laser interferometry and reciprocity methods. The following steps are the procedure of SGC method:

Step 1. *Determine the acceleration sensitivity* S_r *of the reference force transducer.* Mount the reference force transducer, reference mass (can be built-in or external), and the test pickup to be calibrated on a drop-test fixture, as shown in Fig. 11.5. (For use at higher frequencies it is important to make the reference mass small in size in order to satisfy the rigid-body assumption.) Then subject the mass and the two pickups to a free fall of $1g_l$ by striking the junction of line, which causes the line to relax momentarily and impart a step-function gravitational acceleration to the assembly by allowing it to fall freely. Measure the output of the reference force transducer, e_g;

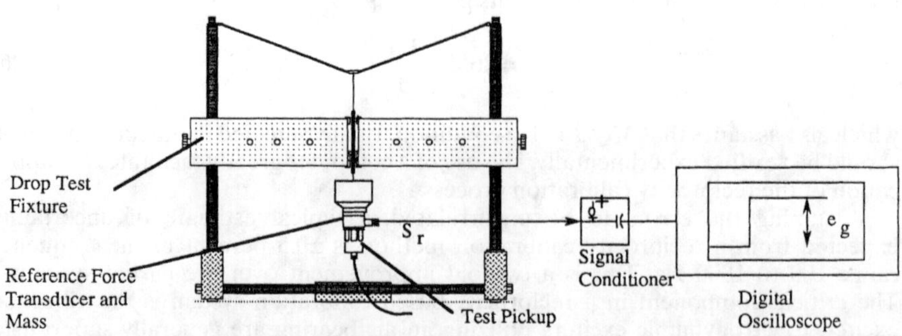

FIGURE 11.5 Gravimetric free-fall calibrator for scaling reference force gage. (*After D. Corelli and R. W. Lally.*[20])

in order to reduce the effect of measurement noise, curve fitting may be used to estimate the step value. Equation (11.7) shows how the sensitivity of the reference force transducer is related to the other parameters of the system.

$$S_r = S_{rf}M = \frac{e_g}{g_l M}M = \frac{e_g}{g_l} \tag{11.7}$$

where S_r = acceleration sensitivity of the reference force transducer, in mV/ms^{-2}
S_{rf} = force sensitivity of the reference force transducer, in mV/N
M = total mass on the force transducer, in kg
e_g = output of the force transducer, in mV
g_l = local gravitational acceleration in ms^{-2}

Step 2. *Measure the voltage ratio* e_t/e_r. Remove the reference force transducer, reference mass, and the pickup being calibrated from the drop-test fixture; then mount them on the vibration exciter, as shown in Fig. 11.6. By measuring the transfer function e_t/e_r (i.e., the ratio of the voltage output of the signal conditioner from the test pickup to the voltage output of the signal conditioner from the reference force transducer, shown in Fig. 11.6) the frequency response of the test pickup can be

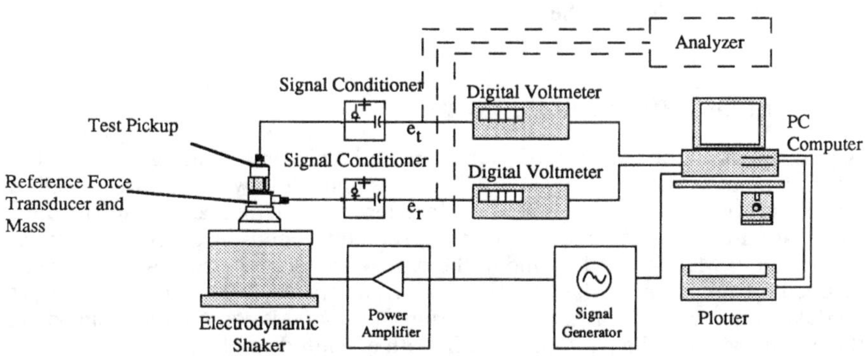

FIGURE 11.6 System configuration for frequency response calibration by measuring acceleration-to-force ratio.

measured over 0.1 to 100,000 Hz, depending upon frequency range of the vibration exciter and signal-to-noise ratio of the system. For use at low frequencies, the discharge time constant of the reference force transducer should be ten times greater than that of the test pickup.

Step 3. *Calculate the sensitivity* S_t *of the test pickup.* If the reference force transducer and the test pickup are linear, the acceleration sensitivity of the test pickup S_t, expressed in the same units as S_r, can be calculated from Eq. 11.1. If either velocity or displacement sensitivity of the test pickup is required, it can be obtained by dividing the acceleration sensitivity by $2\pi f$ or $(2\pi f)^2$, respectively.

CENTRIFUGE CALIBRATOR

A centrifuge provides a convenient means of applying constant acceleration to a pickup. Simple centrifuges can be obtained readily for acceleration levels up to 100g

981 m/s^{2}and can be custom-made for use at much higher values because of the light load requirement by this application. They are particularly useful in calibrating rectilinear accelerometers whose frequency range extends down to 0 Hz and whose sensitivity to rotation is negligible. Centrifuges are mounted so as to rotate about a vertical axis. Cable leads from the pickup, as well as power leads, usually are brought to the table of the centrifuge through specially selected low-noise slip rings and brushes.

To perform a calibration, the accelerometer is mounted on the centrifuge with its axis of sensitivity carefully aligned along a radius of the circle of rotation. If the centrifuge rotates with an angular velocity of ω rad/sec, the acceleration a acting on the pickup is

$$a = \omega^2 r \tag{11.8}$$

where r is the distance from the center of gravity of the mass element of the pickup to the axis of rotation. If the exact location of the center of gravity of the mass in the pickup is not known, the pickup is mounted with its positive sensing axis first outward and then inward; then the average response is compared with the average acceleration acting on the pickup as computed from Eq. (11.8), where r is taken as the mean of the radii to a given point on the pickup case. The calibration factor is determined by plotting the output e of the pickup as a function of the acceleration a given by Eq. (11.8) for successive values of ω and then determining the slope of the straight line fitted through the data.

INTERFEROMETER CALIBRATORS

A primary (absolute) method of calibrating an accelerometer using standard laser interferometry is shown in Fig. 11.2. All systems in the following category of calibrators consist of three stages: modulation, interference, and demodulation. The differences are in the specific type of interferometer that is used (for example, a Michelson or Mach-Zehnder) and in the type of signal processing, which is usually dictated by the nature of the vibration. The vibratory displacement to be measured modulates one of the beams of the interferometer and is consequently encoded in the output signal of the photodetector in both magnitude and phase.

Figure 11.7 shows the principle of operation of the Michelson interferometer. One of the mirrors, D in Fig. 11.7*A*, is attached to the plate on which the device to be calibrated is mounted. Before exciting vibrations, it is necessary to obtain an interference pattern similar to that shown in Fig. 11.7*B*. The relationship underlying the illustrations to be presented is the classical interference formula for the time average intensity I of the light impinging on the photodetector surface.[22,23]

$$I = A + B \cos 4\pi\delta/\lambda \tag{11.9}$$

where A and B are system constants depending on the transfer function of the detector, the intensities of the interfering beams, and alignment of the interferometer. The vibration information is contained in the quantity δ, 2δ being the optical-path difference of the interfering beams. The absoluteness of the measurement comes from λ, the wavelength of the illumination, in terms of which the magnitude of vibratory displacement is expressed. Velocity and acceleration values are obtained from displacement measurements by differentiation with respect to time.

Fringe-Counting Interferometer. An optical interferometer is a natural instrument for measuring vibration displacement. The Michelson and Fizeau interferome-

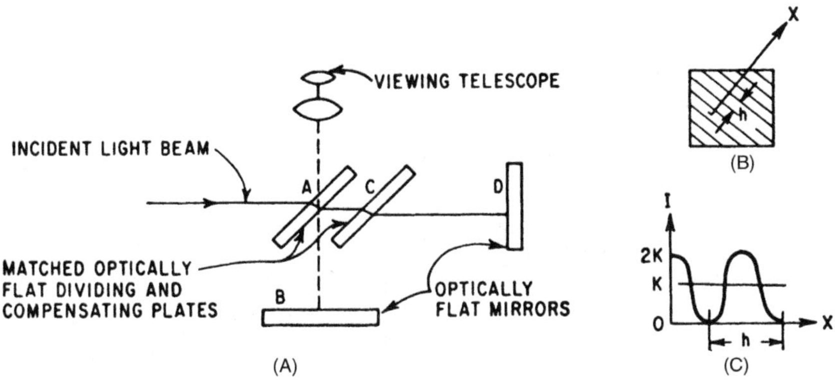

FIGURE 11.7 The principle of operation of a Michelson interferometer: (*A*) Optical system. (*B*) Observed interference pattern. (*C*) Variation of the light intensity along the *X* axis.

ters are the most popular configurations. A modified Michelson interferometer is shown in Fig. 11.8.[24] A corner cube reflector is mounted on the vibration-exciter table. A helium-neon laser is used as a source of illumination. The photodiode and its amplifier must have sufficient bandwidth (as high as 10 MHz) to accommodate the Doppler frequency shift associated with high velocities. An electrical pulse is generated by the photodiode for each optical fringe passing it. The vibratory displacement amplitude is directly proportional to the number of fringes per vibration cycle. For sinusoidal motion, the peak acceleration can be calculated from

$$a = \frac{\lambda \nu \pi^2 f^2}{2} \tag{11.10}$$

where λ = wavelength of light
ν = number of fringes per vibration cycle
f = vibration frequency

Interferometric fringe counting is useful for vibration-displacement measurement in the lower frequency ranges, perhaps to several hundred hertz depending on the characteristics of the vibration exciter.[25,26] At the low end of the frequency spectrum, conventional procedures and commercially available equipment are not able to meet all the present requirements. Low signal-to-noise ratios, cross-axis components of motion, and zero-drifts are some of the problems usually encountered. In response to those restrictions an electrodynamic exciter for the frequency range 0.01 to 20 Hz has been developed. It features a maximum displacement amplitude of 0.5 meter, a transverse sensitivity less than 0.01 percent, and a maximum uncorrected distortion of 2 percent. These characteristics have been achieved by means of a specially designed air bearing, an electro-optic control, and a suitable foundation.

Figure 11.9 shows the main components of a computer-controlled low-frequency calibration system which employs this exciter. Its functions are (1) generation of sinusoidal vibrations, (2) measurement of rms and peak values of voltage and charge, (3) measurement of displacement magnitude and phase response, and (4) control of nonlinear distortion and zero correction for the moving element inside a tubelike magnet. Position of the moving element is measured by a fringe-counting interferometer. Uncertainties in accelerometer calibrations using this system have been reduced to about 0.25 to 0.5 percent, depending on frequency and vibration amplitude.

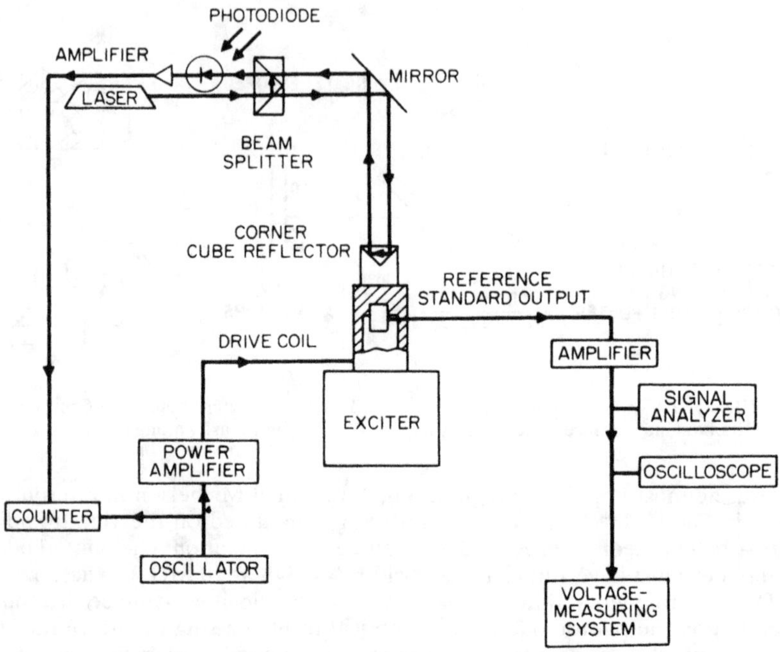

FIGURE 11.8 Typical laboratory setup for interferometric measurement of vibratory displacement by fringe counting. (*After R. S. Koyanagi.*[24])

Fringe-Disappearance Interferometer. The phenomenon of the interference band disappearance in an optical interferometer can be used to establish a precisely known amplitude of motion. Figure 11.7 shows the principle of operation of the Michelson interferometer employed in this technique. One of the mirrors D, in Fig. 11.7*A*, is attached to the mounting plate of the calibrator. Before exciting vibrations it is necessary to obtain an interference pattern similar to that shown in Fig. 11.7*B*.

When the mirror D vibrates sinusoidally[28] with a frequency f and a peak displacement amplitude d, the time average of the light intensity I at position x, measured from a point midway between two dark bands, is given by

$$I = A + BJ_0\left(\frac{4\pi d}{\lambda}\right)\cos\left(\frac{2\pi x}{h}\right) \tag{11.11}$$

where J_0 = zero-order Bessel function of the first kind
A and B = constants of measuring system
h = distance between fringes, as shown in Fig. 11.11*B* and *C*

For certain values of the argument, the Bessel function of zero order is zero; then the fringe pattern disappears and a constant illumination intensity A is present. Electronic methods for more precisely establishing the fringe disappearance value of the vibratory displacement have been successfully used at the National Institute of Standards and Technology[15,29] and elsewhere. The latter method has been fully automated using a desktop computer.

The use of piezoelectric exciters is common for high-frequency calibration of accelerometers.[30] They provide pistonlike motion of relatively high amplitude and

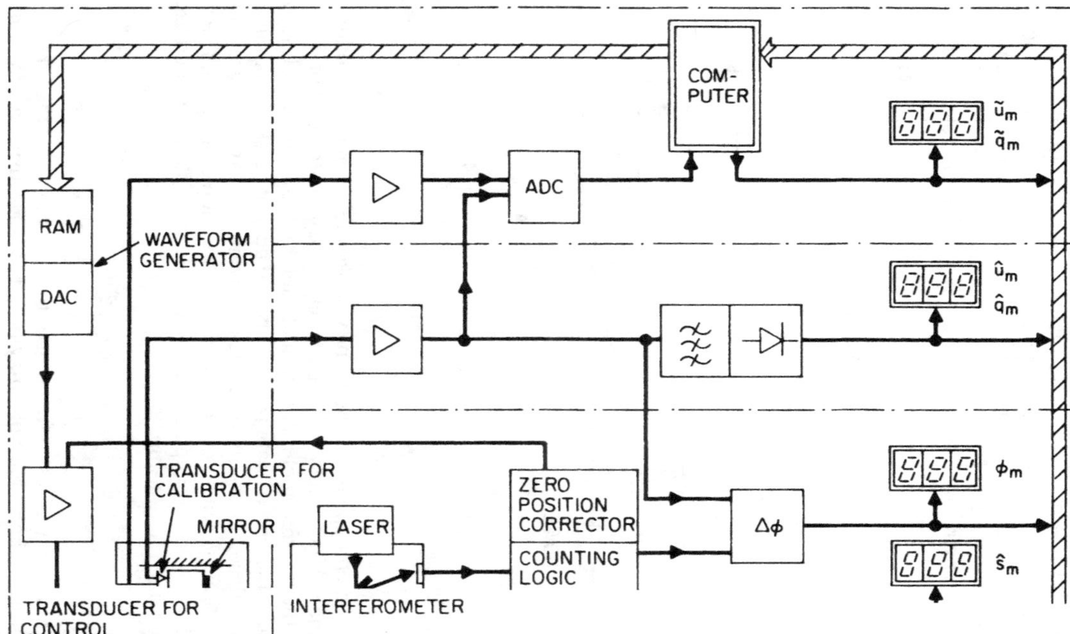

FIGURE 11.9 Simplified block diagram of a low-frequency vibration standard. (*After H. J. von Martens.*[27])

are structurally stiff at the lower frequencies, where displacement noise is bothersome. When electrodynamic exciters are used with fringe disappearance methods, it is generally necessary to stiffen the armature suspensions to reduce the background displacement noise.

Signal-Nulling Interferometer. This method, although mathematically similar to fringe disappearance, relies on finding the nulls in the fundamental frequency component of the signal from a photodetector.[9,23,31] The instrumentation is, therefore, quite different, except for the interferometer. One successful arrangement is shown in Fig. 11.10. Laboratory environmental restrictions are much more severe for this method.

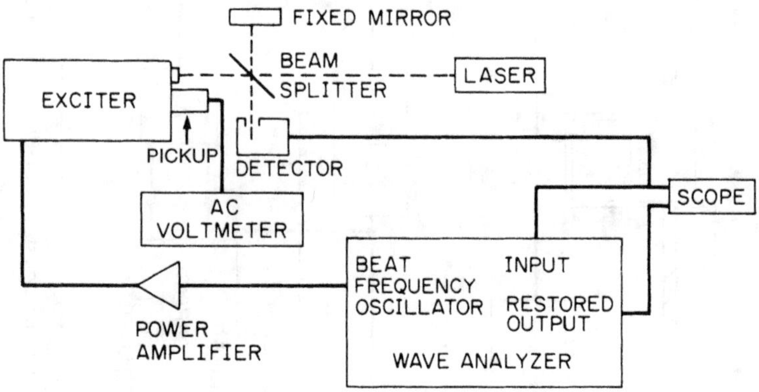

FIGURE 11.10 Interferometric measurement of displacement d as given by $J_1(4\pi d/\lambda) = 0$.

The interferometer apparatus should be well-isolated to ensure stability of the photodetector signals. Air currents in the room may contribute to noise problems by physically moving the interferometer components and by changing the refractive index of the air. An active method of stabilization has also been successfully employed.[32]

To make displacement amplitude measurements, a wave analyzer tuned to the frequency of vibration can be used to filter the photodetector signal. The filtered signal amplitude will pass through nulls as the vibration amplitude is increased, according to the following relationship:

$$I = 2BJ_1\left(\frac{4\pi d}{\lambda}\right) \tag{11.12}$$

where J_1 is the first-order Bessel function of the first kind, and the other terms are as previously defined. The signal nulls may be established using a wave analyzer. The null amplitude will generally be 60 dB below the maximum signal level of the photodetector output.

The accelerometer output may be measured by an accurate voltmeter at the same time that the nulls are obtained. The sensitivity is then calculated by dividing the output voltage by the displacement. Because the filtered output of the photodetector is a replica of the vibrational displacement, a phase calibration of the pickup can also be obtained with this arrangement.

Heterodyne Interferometer. A *homodyne interferometer* is an interferometer in which interfering light beams are created from the same beam by a process of beam splitting. All illumination is at the same optical frequency. In contrast, in the *heterodyne interferometer,*[33] light from a laser-beam source containing two components, each with a unique polarization, is separated into (1) a measurement beam and (2) a reference beam by a polarized beam splitter. When the mounting surface of the device under test is stationary, the interference pattern impinging on the photodetector produces a signal of varying intensity at the beat frequency of the two beams. When surface moves, the frequency of the measurement beam is shifted because of the Doppler effect, but that of the reference beam remains undisturbed. Thus, the photodetector output can be regarded as a carrier that is frequency modulated by the velocity waveform of the motion.

The main advantages of the heterodyne interferometer are greater measurement stability and lower noise susceptibility. Both advantages occur because displacement information is carried on ac waveforms; hence, a change in the average value of beam intensity cannot be interpreted as motion. Digitization and subsequent phase demodulation of the interferometer output reduce measurement uncertainties.[34] This can yield significant improvements in calibration results at high frequencies, where the magnitude of displacement typically is only a few nanometers. As in the case of homodyning, variations of the heterodyning technique have been developed to meet specific needs of calibration laboratories. Reference 35 describes an accelerometer calibration system, applicable in the frequency range from 1 mHz to 25 kHz and at vibration amplitudes from 1 nanometer to 10 meters. The method requires the acquisition of instantaneous position data as a function of the phase angle of the vibration signal and the use of Fourier analysis.

HIGH-ACCELERATION METHODS OF CALIBRATION

Some applications in shock or vibration measurement require that high amplitudes be determined accurately. To ensure that the pickups used in such applications meet certain performance criteria, calibrations must be made at these high amplitudes. The following methods are available for calibrating pickups subject to accelerations in excess of several hundred g.

SINUSOIDAL-EXCITATION METHODS

The use of a metal bar, excited at its fundamental resonance frequency, to apply sinusoidal accelerations for calibration purposes has several advantages: (1) an inherently constant frequency, (2) very large amplitudes of acceleration (as much as 4000g, and (3) low waveform distortion. A disadvantage of this type of calibrator is that calibration is limited to the resonance frequencies of the metal bar.

The bar can be supported at its nodal points, and the pickup to be calibrated can be mounted at its mid-length location. The bar can be energized by a small electromagnet or can be self-excited. Acceleration amplitudes of several thousand g can thus be obtained at frequencies ranging from several hundred to several thousand hertz. The bar also may be calibrated by clamping it at its midpoint and mounting the pickup at one end.[36] The displacement at the point of attachment of the pickup can be measured optically since displacements encountered are adequately large.

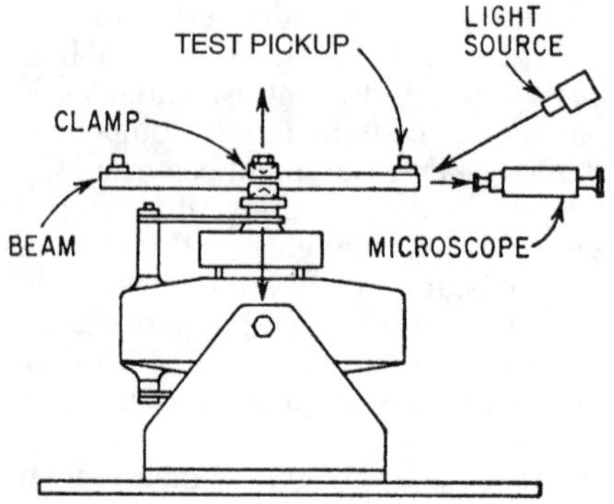

FIGURE 11.11 Resonant-bar calibrator with the pickup mounted at end and a counterbalancing weight at the other. (*After E. I. Feder and A. M. Gillen.*[36])

The resonant-bar calibrator shown in Fig. 11.11 is limited in amplitude primarily by the fatigue resistance of the bar.[36] Accelerations as much as 500*g* have been attained using aluminum bars without special designs. Peak accelerations as large as 4000*g* have been attained using tempered vanadium steel bar. The bar is mounted at its mid-length on a conventional electrodynamic exciter. The accelerometer being calibrated is mounted at one end of the bar, and an equivalent balance weight is mounted at the opposite end in the same relative position.

Axial resonances of long rods have been used to generate motion for accurate calibration of vibration pickups over a frequency range from about 1 to 20 kHz and at accelerations up to 12,000*g*.[37,38] The use of axially driven rods has an advantage over the beams discussed above in that no bending or lateral motion is present. This minimizes errors from the pickup response to such unwanted modes and also from the direct measurement of the displacement having nonrectilinear motion.

SHOCK-EXCITATION METHODS

There are several methods by which a sudden velocity change may be applied to pickups designed for high-frequency acceleration measurement, for example, the ballistic pendulum, drop-test, and drop-ball calibrators, described below. Any method which generates a reproducible velocity change as function of time can be used to obtain the calibration factor.[1] Impact techniques can be employed to obtain calibrations over an amplitude range from a few *g* to over 100,000*g*. An example of the latter is the Hopkinson bar, in which the test pickup is mounted at one end and stress pulses are generated by an air gun firing projectiles impacting at the other end, described below.

An accurate determination of shock performance of an accelerometer depends not only upon the mechanical and electrical characteristics of the test pickup but also upon the characteristics of the instrumentation and recording equipment. It is often best to perform system calibrations to determine the linearity of the test pickup as well as the linearity of the recording instrumentation in the range of intended use. Several of the following methods make use of the fact that the velocity change during a transient pulse is equal to the time integral of acceleration:

$$v = \int_{t_1}^{t_2} a\, dt \tag{11.13}$$

where the initial or final velocity is taken as reference zero, and the integration is performed to or from the time at which the velocity is constant. If the output closely resembles a half-sine pulse, the area is equal to approximately $2h(t_2 - t_1)/\pi$, where h is the height of the pulse, and $(t_2 - t_1)$ is its width.

In this section, several methods for applying known velocity changes v to a pickup are presented. The voltage output e and the acceleration a of the test pickup are related by the following linear relationship:

$$S = \frac{e}{a} \tag{11.14}$$

where S is the pickup calibration factor.

After Eq. (11.14) is substituted into Eq. (11.13), the calibration factor for the test pickup can be expressed as

$$S = \frac{A}{v} \tag{11.15}$$

where

$$A = \int_{t_1}^{t_2} e\, dt \tag{11.16}$$

the area under the acceleration-versus-time curve.

The calibration factor assumes that no significant spectral energy exists beyond the frequency region in which the test pickup has nominally constant complex sensitivity (uniform magnitude and phase response as functions of frequency). In general, this assumption becomes less valid with decreasing pulse duration resulting in increasing bandwidth in the excitation signal.

Sometimes it is convenient to express acceleration as a multiple of g. The corresponding calibration factor S_1 is in volts per g:

$$S_1 = \frac{e}{(a/g)} = \frac{Ag}{v} \tag{11.17}$$

In either case, the integrals representing A and v must first be evaluated. The linear range of a pickup is determined by noting the magnitude of the velocity change v at which the calibration factor S or S_1 begins to deviate from a constant value. The minimum pulse duration is similarly found by shortening the pulse duration and noting when S changes appreciably from previous values.

Hopkinson Bar Calibrator. An apparatus called a *Hopkinson bar*[39–41] provides very high levels of acceleration for use in the calibration and acceptance testing of shock accelerometers. As shown in Fig. 11.12, a controlled-velocity projectile strikes one end of the bar, at $x = 0$; a strain gage is placed at the middle of the bar, at $x = L/2$; and the accelerometer under test is mounted at the other end of the bar, at $x = L$. When the projectile strikes the bar, a strain wave is initiated at $x = 0$. This wave travels along the bar, producing a large acceleration at the accelerometer. The duration and shape of the strain wave can be controlled by varying the geometry and mate-

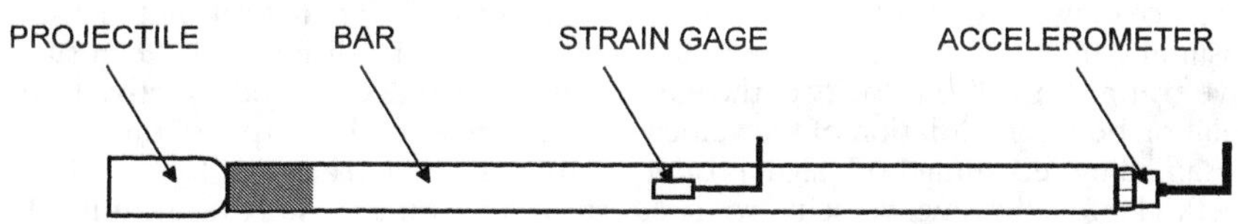

FIGURE 11.12 A Hopkinson bar, showing a projectile striking the bar at $x = 0$; a strain gage mounted on the bar at $x = L/2$; and the accelerometer under test is attached to the bar at $x = L$. Impact of the projectile on the bar generates a strain wave which travels down the bar.

rial of the projectile. And, to a limited extent, the duration of the pulse can be controlled by placing a piece of soft metal or rubber on the bar at the position where the projectile strikes the bar, $x = 0$. The acceleration at the accelerometer may be determined from equations given in Ref. 41, using measured values of strain.

Ballistic Pendulum Calibrator. A ballistic pendulum calibrator provides a means for applying a sudden velocity change to a test pickup. The calibrator consists of two masses which are suspended by wires or metal ribbons. These ribbons restrict the motion of the masses to a common vertical plane.[42] This arrangement, shown in Fig. 11.13, maintains horizontal alignment of the principal axes of the masses in the direction parallel to the direction of motion at impact. The velocity attained by the anvil mass as the result of the sudden impact is determined.

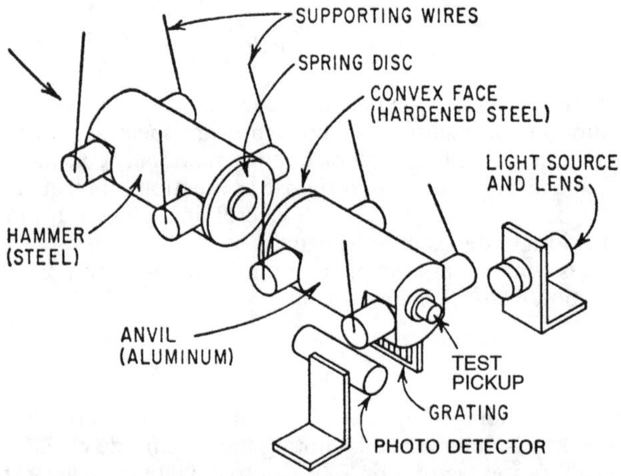

FIGURE 11.13 Components arrangement of the ballistic pendulum with photodetector and light grating to determine the anvil-velocity change during impact. (*After R. W. Conrad and I. Vigness.*[42])

The accelerometer to be calibrated is mounted to an adapter which attaches to the forward face of the anvil. The hammer is raised to a predetermined height and held in the release position by a solenoid-actuated clamp. Since the anvil is at rest prior to impact, it is necessary to record the measurement of the change in velocity of anvil and transient waveform on a calibrated time base. One method of measurement of velocity change is performed by focusing a light beam through a grating attached to the anvil, as shown in Fig. 11.13. The slots modulate the light beam intensity, thus varying the photodetector output, which is recorded with the pickup output. Since the distance between grating lines is known, the velocity of the anvil is calculated directly, assuming that the velocity is essentially constant over the distance between successive grating lines. The velocity of the anvil in each case is determined directly; the time relation between initiation of the velocity and the pulse at the output of the pickup is obtained by recording both signals on the same time base. The most frequently used method infers the anvil velocity from its vertical rise by measuring the maximum horizontal displacement and making use of the geometry of the pendulum system.

The duration of the pulse, which is the time during which the hammer and anvil are in contact, can be varied within close limits.[42] In Fig. 11.13 the hammer nosepiece is a disc with a raised spherical surface. It develops a contact time of 0.55 millisecond. For larger periods, ranging up to 1 millisecond, the stiffness of the nosepiece is decreased by bolting a hollow ring between it and the hammer. A pulse longer than 1 millisecond may be obtained by placing various compliant materials, such as lead, between the contacting surfaces.

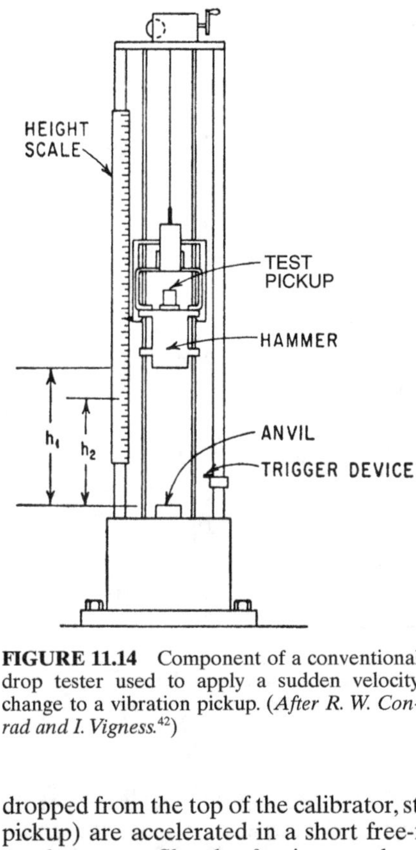

FIGURE 11.14 Component of a conventional drop tester used to apply a sudden velocity change to a vibration pickup. (*After R. W. Conrad and I. Vigness.*[42])

Drop-Test Calibrator. In the drop-test calibrator, shown in Fig. 11.14, the test pickup is attached to the hammer using a suitable adapter plate. An impact is produced as the guided hammer falls under the influence of gravity and strikes the fixed anvil. To determine the velocity change, measurement is made of the time required for a contactor to pass over a known region just prior to and after impact. The pickup output and the contactor indicator are recorded simultaneously in conjunction with a calibrated time base. The velocity change also may be determined by measuring the height h_1 of hammer drop before rebound and the height h_2 of hammer rise after rebound. The total velocity is calculated from the following relationship:

$$v = (2gh_1)^{1/2} + (2gh_2)^{1/2} \quad (11.18)$$

A total velocity change of 40 ft/sec (12.2 m/sec) is typical.

Drop-Ball Shock Calibrator. Figure 11.15 shows a drop-ball shock calibrator.[10,43] The accelerometer is mounted on an anvil which is held in position by a magnet assembly. A large steel ball is dropped from the top of the calibrator, striking the anvil. The anvil (and mounted test pickup) are accelerated in a short free-flight path. A cushion catches the anvil and accelerometer. Shortly after impact, the anvil passes through an optical timing gate of a known distance. From this, the velocity after impact can be calculated. Acceleration amplitudes and pulse durations can be varied by selecting the mass of the anvil, mass of the impacting ball, and resilient pads on top of the anvil where the ball strikes. Common accelerations and durations are 100g at 33 milliseconds, 500g at 1 millisecond, 1000g at 1 millisecond, 5000g at 2 milliseconds, and 10,000g at 0.1 millisecond.[43] With experience and care, shock calibrations can be performed with an uncertainty of about ±5 percent.

INTEGRATION OF ACCELEROMETER OUTPUT

Change-of-velocity methods for calibrating an accelerometer at higher accelerations than obtainable by the methods discussed above have been developed using specially modified ballistic pendulums, air guns, inclined troughs, and other devices.

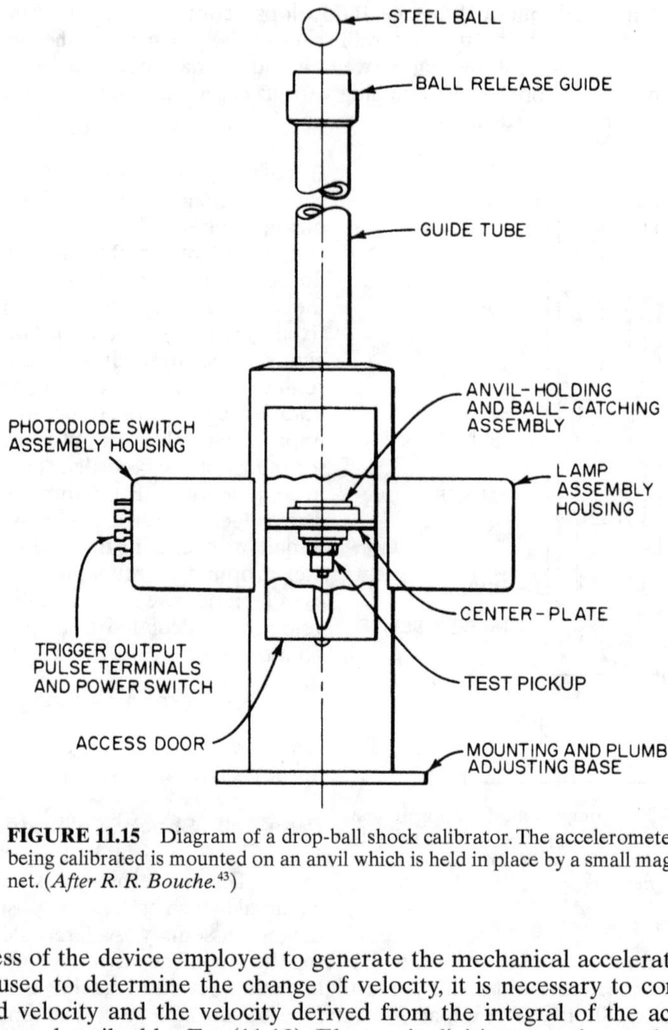

FIGURE 11.15 Diagram of a drop-ball shock calibrator. The accelerometer being calibrated is mounted on an anvil which is held in place by a small magnet. (*After R. R. Bouche.*[43])

Regardless of the device employed to generate the mechanical acceleration or the method used to determine the change of velocity, it is necessary to compare the measured velocity and the velocity derived from the integral of the acceleration waveform as described by Eq. (11.13). Electronic digitizers can be used to capture the waveform and produce a recording. Care must be exercised in selecting the time at which the acceleration waveform is considered complete, and its integral should be compared with the velocity. The calibration factor for the test pickup is computed from Eq. (11.15) or (11.17).

IMPACT-FORCE SHOCK CALIBRATOR

The impact-force shock calibrator has a free-fall carriage and a quartz load cell. The accelerometer to be calibrated is mounted onto the top of the carriage, as shown in Fig. 11.16. The carriage is suspended about ½ to 1 meter above the load cell and allowed to fall freely onto the cell.[44] The carriage's path is guided by a plastic tube. Cushion pads are attached at the top of the load cell to lengthen the impulse duration

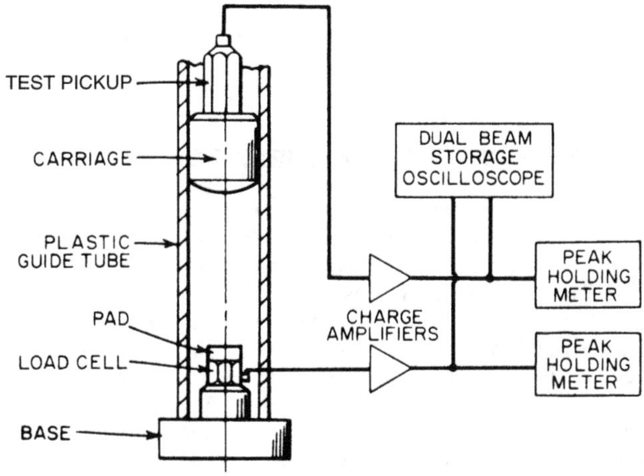

FIGURE 11.16 Impact-force calibrator with auxiliary instruments. (*After W. P. Kistler.*[44])

and to shape the pulse. Approximate haversines are generated by this calibrator. The outputs of the accelerometer and load cell are fed to two nominally identical charge amplifiers or power units. The outputs from load cell and test accelerometer are recorded or measured on a storage-type oscilloscope or peak-holding meters.

During impact, the voltage produced at the output of the accelerometer, $e_a(t)$, is

$$e_a(t) = a(t)S_aH_a \tag{11.19}$$

where $a(t)$ = acceleration, m/s²
S_a = calibration factor for accelerometer, mV/m/s²
H_a = gain of charge amplifier or power unit

The output of load cell $e_f(t)$ is

$$e_f(t) = F(t)S_fH_f \tag{11.20}$$

where $F(t)$ = force, N
S_f = calibration factor for load cell, mV/N
H_f = gain of charge amplifier or power unit

By using the relationship $F(t) = ma(t)$, where m is the falling mass, and combining Eqs. (11.19) and (11.20),

$$\frac{e_a(t)}{e_f(t)} = \frac{a(t)S_aH_a}{ma(t)S_fH_f} \tag{11.21}$$

and hence

$$S_a = \frac{e_a(t)\ H_fm}{e_f(t)\ H_a} S_f \tag{11.22}$$

When calculating the mass, it is necessary to know the mass of the carriage, accelerometer, mounting stud, cable connector, and a short portion of the accelerometer cable. Experience has shown that for small coaxial cables, a length of about 2 to 4 cm is correct. Calibrations by this method can be accomplished with uncertainties generally between ±2 to ±5 percent.

FOURIER-TRANSFORM SHOCK CALIBRATION

The above calibration methods yield the approximate magnitude of the sensitivity function for the accelerometer being tested. For shock standards and other critical applications, more information may be required, for example, the accelerometer's sensitivity, both in magnitude and phase, as a function of frequency.[45–48] The equipment required for obtaining this information usually consists of a mechanical-shock-generating machine and a two-channel signal analyzer, in addition to the accelerometer being tested and a reference accelerometer. For a typical application, a signal analyzer with 12-bit resolution and 5 MHz sampling rate is adequate. The calibration results are obtained from the complex ratios of the output of the test accelerometer to that of the standard accelerometer (see Chap. 14). The magnitude and phase of these ratios represent the sensitivity of the test accelerometer relative to the standard.

The range of usable frequencies is limited by the pulse shape and duration, sampling rate, and analyzer capability. Figure 11.17 shows a typical half-sine shock pulse whose spectral content is predominantly below about 2 kHz, but pulses of shorter duration contain sufficient energy up to 10 kHz, and even 30 kHz.[48] An important advantage of the spectral methods over the time-domain methods is that they do not require the waveform or pulse to be smooth and clean. Modern signal processing equipment has made it possible to calibrate shock accelerometers at amplitudes approaching 1 megameter per sec^2 by using the fast Fourier transform (FFT) method with a Hopkinson bar,[48] shown in Fig. 11.12. The uncertainties in this type of calibration can be as low as 1 percent.[49,50]

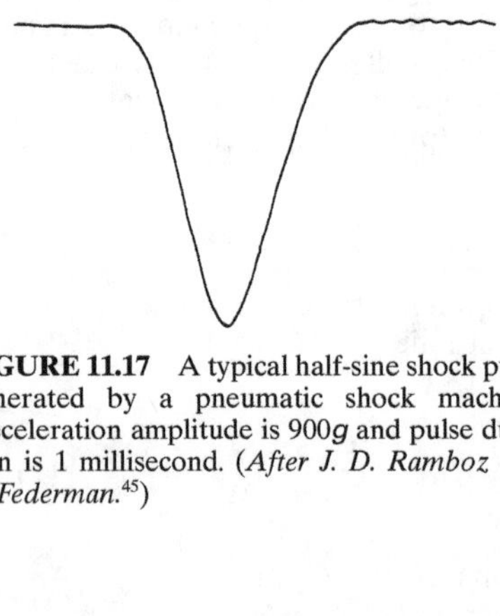

FIGURE 11.17 A typical half-sine shock pulse generated by a pneumatic shock machine. Deceleration amplitude is 900*g* and pulse duration is 1 millisecond. (*After J. D. Ramboz and C. Federman.*[45])

VIBRATION EXCITERS USED FOR CALIBRATION

A vibration exciter that is suitable for calibration of vibration pickups should provide:

- Distortion-free sinusoidal motion
- True rectilinear motion in a direction normal to the vibration-table surface without the presence of any other motion
- A table that is rigid for all design loads at all operating frequencies
- A table that remains at ambient temperature and does not provide either a source or sink for heat regardless of the ambient temperature
- A table whose mounting area is free from electromagnetic disturbances

- Stepless variation of frequency and amplitude of motion within specified limits, which is easily adjustable

ELECTRODYNAMIC EXCITERS

Electrodynamic exciters, described in Chap. 25, satisfactorily meet the requirements of the ideal calibrator, providing a constant-force (acceleration) output with little distortion over a rather wide frequency range from 1 to 10,000 Hz.[51] Ordinarily, to cover this frequency range, more than one exciter is required. Specially designed machines featuring long strokes for very low frequencies or ultralight moving elements for very high frequencies are commercially available. One national standards laboratory has a custom-built vibration exciter that has a low-frequency limit of 20 mHz.[27] This machine employs a special air bearing, real-time electro-optic control, and a suitable foundation.

A shaker system for the calibration of accelerometer sensitivity has been developed at the National Institute of Standards and Technology[52,53] with the goal of reducing the inherent uncertainties in the absolute measurements of accelerometer sensitivity. The shaker has dual retractable magnets equipped with optical ports to allow laser-beam access to the surface upon which the accelerometer is mounted and the one opposite to it. The purpose of the optical ports is to enable interferometric measurement of the surface displacement. The moving element of the shaker is physically compact for directional stability and good high-frequency response. At each end it is equipped with nominally identical coils and axially oriented mounting tables. The driving and sensing coils are located on the same moving element so that a separate shaker external to the calibration shaker is not needed when a reciprocity calibration is performed. The dual-coil feature eliminates complications resulting from mutual mechanical coupling between two separate shakers. Minimal distortion and cross-action motion were two of the most important design requirements of this vibration generator. These parameters are essential for the validity of the assumptions underlying the theory of electromechanical reciprocity.

PIEZOELECTRIC EXCITERS

The piezoelectric exciter (see Fig. 25.9 and Chap. 10) offers a number of advantages in the calibration of vibration pickups, particularly at high frequencies. Calibration is impracticable at low frequencies because of inherently small displacements in this frequency range. A design which has been used at the National Institute of Standards and Technology for many years is described in Ref. 30.

MECHANICAL EXCITERS

Rectilinear motion can be produced by mechanical exciter systems of the type described in Chap. 25 under "Direct-Drive Mechanical Vibration Machine." Their usable frequency range is from few hertz to less than 100 Hz. Despite their relatively low cost, mechanical exciters are no longer used for high-quality calibrations of transducers because of their appreciable waveform distortion and background noise.

For generating vibratory motion at discrete frequencies (below 5 Hz), a linear oscillator can be employed. Reference 54 describes a calibrator consisting of a spring-supported table which is guided vertically by air bearings. Its advantages are a clean waveform, resulting from free vibration, and large rectilinear displacement with little damping, made possible by use of air bearings.

CALIBRATION OF TRANSVERSE SENSITIVITY

The characteristics of a vibration pickup may be such that an extraneous output voltage is generated as a result of vibration which is in a direction at right angles to the axis of designated sensitivity of the pickup. This effect, illustrated in Fig. 10.11, results

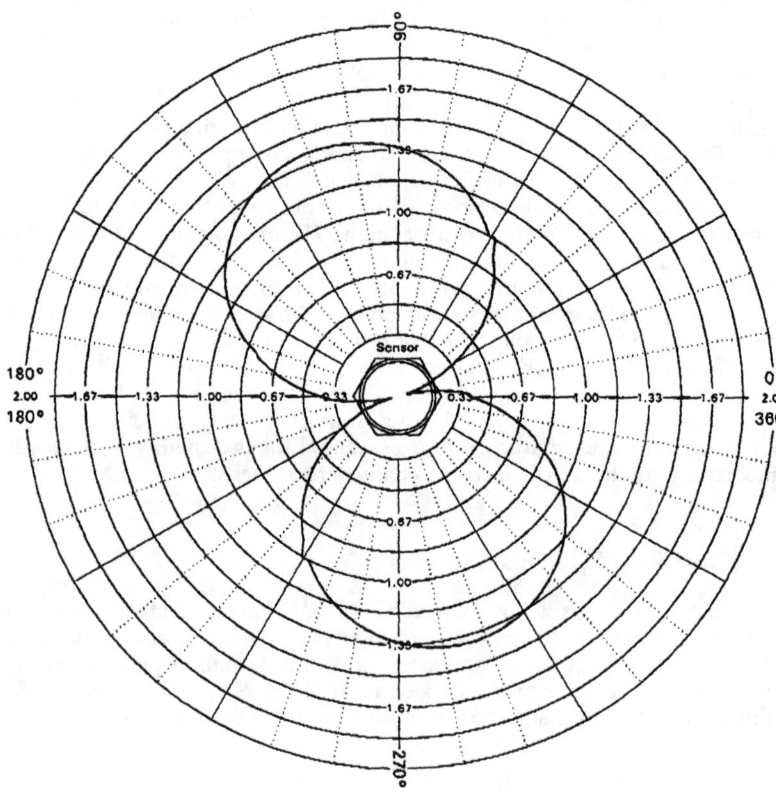

Model Number	353		Test Frequency	700	Hz
Serial Number	1		Test g level	10.93	g's
Sensitivity	20.90	mV/g	Maximum Transverse	1.40% @ 107°	
			Minimum Transverse	0.035% @ 195°	

FIGURE 11.18 Transverse sensitivity of a piezoelectric accelerometer to vibration in the plane normal to the sensitive axis.[55]

in the axis of maximum sensitivity not being aligned with the axis of designated sensitivity. As indicated in Eq. (10.11), the cross-axis or *transverse sensitivity* of a pickup is expressed as the tangent of an angle, i.e., the ratio of the output resulting from the transverse motion divided by the output resulting from motion in the direction of designated sensitivity. This ratio varies with the azimuth angle in the transverse plane, as shown in Fig. 10.12, and also with frequency. In practice, tan θ has a value between 0.01 and 0.05 and is expressed as a percentage. Figure 11.18 presents a typical result of a transverse-sensitivity calibration.[55]

Knowledge of the transverse sensitivity is vitally important in making accurate vibration measurements, particularly at higher frequencies (i.e., at frequencies approaching the mounted resonance frequency of the pickup). Figure 11.19 shows

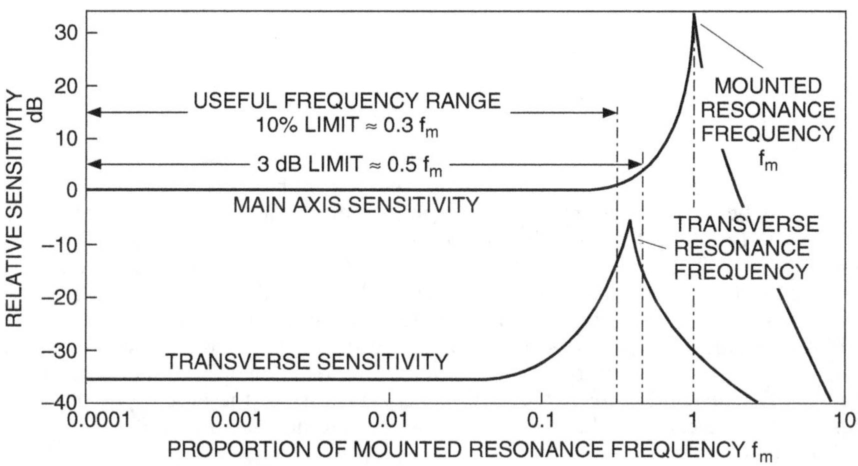

FIGURE 11.19 The relative response of an accelerometer to main-axis and transverse-axis vibrations.[5]

the relative responses of an accelerometer to main-axis and transverse-axis vibration. It is noteworthy that the transverse resonance frequency is lower than the usually specified mounted resonance frequency.

A direct measurement of the transverse sensitivity of a pickup requires a vibration exciter capable of pure unidirectional motion at the frequencies of interest. This usually means that any cross-axis motion of the mounting table should be less than 2 percent of the main-axis motion.[10] Resonance beam exciters[1] and air-bearing shakers[52] have been used for this purpose.

The *resonant-beam* method,[56] used by many testing laboratories to provide the sensitivity of a transducer automatically (in both magnitude and direction) yields a plot of its sensitivity versus angle (similar to the one shown in Fig. 11.18). The accelerometer under test is mounted at the free end of a circular-section steel beam which is cantilevered from a massive base. Motion of the accelerometer is generated by exciting the beam near resonance in its first bending mode, providing a large-amplitude vibration at the free end of the beam, typically at a frequency between 300 and 800 Hz. A pair of vibration exciters, and associated electronic equipment,

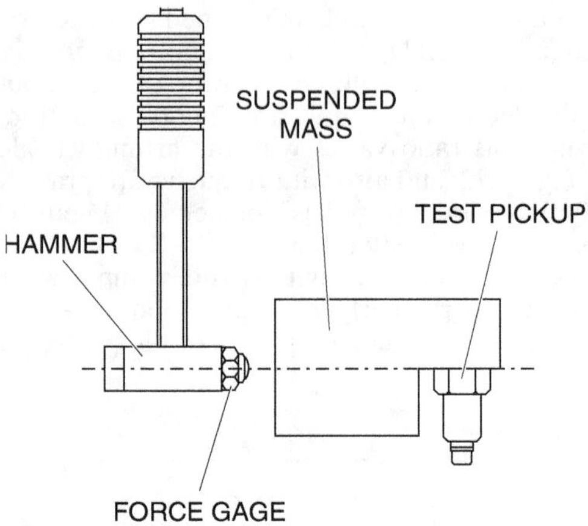

FIGURE 11.20 Schematic diagram for impact hammer method of measuring transverse sensitivity.[55]

permits the beam to be excited in any desired direction. Thus the transverse sensitivity may be obtained at any angle without reorientation of the accelerometer.

Another method for obtaining the transverse sensitivity of a pickup is by use of the impulse technique similar to that used in modal analysis (Chap. 21). An impulse is generated by the impact of a hammer against a suspended mass on which the test pickup is mounted. A force gage is mounted on the hammer, as illustrated in Fig. 11.20. From the characteristics of the force gage and its output when it strikes against the suspended mass, from the output signal of the test pickup, and from the magnitude of the suspended mass, the transverse sensitivity of the accelerometer under test S_{ta} may be calculated according to a procedure described in Ref. 55, using the following formula:

$$S_{ta} = mS_f\left(\frac{e_a}{e_f}\right) \tag{11.23}$$

where m = the mass of the suspended rigid block
S_f = the sensitivity of the force gage
e_a = the output of the accelerometer under test
e_f = the output of the force gage

ACKNOWLEDGMENT

M. Roman Serbyn, Associate Professor Emeritus, Morgan State University, Baltimore, MD, originally prepared some of the material in this chapter.

REFERENCES

1. American National Standards Institute: "Methods for the Calibration of Shock and Vibration Pickups," ANSI S2.2-1959 (R2006), New York.

2. International Organization for Standardization: "Methods for the Calibration of Vibration and Shock Transducers—Basic Concepts," ISO/IS 16063-1, Geneva 1998. (Available from the ANSI, New York.)
3. Serridge, M., and T. R. Licht: "Piezoelectric Accelerometer and Vibration Preamplifier Handbook," Bruel & Kjaer, Naerum, Denmark, 1987.
4. BIPM, IEC, IFCC, ISO, IUPAC, IUPAP, OIML: *International Vocabulary of Basic and General Terms in Metrology* (*VIM*), Geneva, 1993.
5. Clark, N. H.: "First-level Calibrations of Accelerometers," *Metrologia,* **36**:385–389 (1999).
6. International Organization for Standardization: "Methods for the Calibration of Vibration and Shock Transducers—Primary Vibration Calibration by Laser Interferometry," ISO/IS 16063-11, Geneva, 1999. (Available from American National Standards Institute, New York.)
7. von Martens, H. J.: "Current State and Trends of Ensuring Traceability for Vibration and Shock Measurements," *Metrologia,* **36**:357–373 (1999).
8. International Organization for Standardization: "Methods of Calibration of Vibration and Shock Pickups—Secondary Vibration Calibration," ISO/IS 5347-3, Geneva, 1993. (Available from ANSI, New York.)
9. Hartz, K.: *Proc. Natl. Conf. of Stds. Labs. Symposium,* 1984.
10. Bouche, R. R. "Calibration of Shock and Vibration Measuring Transducers," Shock and Vibration Monograph SVM-11, The Shock and Vibration Information Center, Washington, D.C., 1979.
11. Ge, L.-F.: "The Reciprocity Method with Complex Non-Linear Fitting for Primary Vibration Standards," *J. Acoust. Soc. Amer.,* **97**:324–330 (1995).
12. Levy, S., and R. R. Bouche: *J. Res. Natl. Std.,* **57**:227 (1956).
13. Fromentin, J., and M. Fourcade: *Bull. d'Informations Sci. et Tech.,* (201):21, 1975.
14. Payne, B. F.: *Shock and Vibration Bull.,* **36**, pt. 6 (1967).
15. Robinson, D. C., M. R. Serbyn, and B. F. Payne: *Natl. Bur. Std.* (*U.S.*) *Tech. Note* 1232, 1987.
16. Payne, B. F.: "Vibration Laboratory Automation at NIST with Personal Computers," *Proc. 1990 Natl. Conf. Stds. Labs. Workshop & Symposium,* Session 1C-1.
17. Payne, B., and D. J. Evans: "Comparison of Results of Calibrating the Magnitude of the Sensitivity of Accelerometers by Laser Interferometry and Reciprocity," *Metrologia,* **36**: 391–394 (1999).
18. Wildhack, W. A., and R. O. Smith: *Proc. 9th Annu. Meet. Instr. Soc. Am.,* Paper 54-40-3 (1954).
19. Hillten, J. S.: *Natl. Bur. Std.* (*U.S.*) *Tech. Note 517,* March 1970.
20. Corelli, D., and R. W. Lally: "Gravimetric Calibration," *Third Int. Modal Analysis Conf.,* 1985.
21. Lally, R.: "Structural Gravimetric Calibration Technique," master's thesis, University of Cincinnati, 1991.
22. Born, M., and E. Wolf: "Principles of Optics," 5th ed., Pergamon Press, 1975.
23. Hohmann, P., *Akustica,* **26**:122 (1972).
24. Koyanagi, R. S.: *Exp. Mech.,* **15**:443 (1975).
25. Logue, S. H.: "A Laser Interferometer and its Applications to Length, Displacement, and Angle Measurement," *Proc. 14th Ann. Meet. Inst. Environ. Sci.,* 1968, p. 465.
26. Payne, B. F.: "An Automated Fringe Counting Laser Interferometer for Low Frequency Vibration Measurements," *Proc. Instr. Soc. Am. Intern. Instr. Symp.,* May 1986.
27. von Martens, H. J.: "Representation of Low-Frequency Rectilinear Vibrations for High-Accuracy Calibration of Measuring Instruments for Vibration," *Proc. 2nd Symp. IMEKO Tech. Comm. on Metrology-TC8,* Budapest, 1983.

28. C. Candler: "Modern Interferometers," p. 105, Hilger & Watts Ltd., Glasgow, Scotland, 1950.

29. Payne, B. F., and M. R. Serbyn: "An Automated System for the Absolute Measurement of Pickup Sensitivity," *Proc. 1983 Natl. Conf. Stds. Labs. Workshop and Symposium,* Part II, 11.1–11.2, Boulder, Colo., July 18–21, 1983.

30. Jones, E., W. B. Yelon, and S. Edelman: *J. Acoust. Soc. Amer.,* **45**:1556 (1969).

31. Deferrari, H. A., and F. A. Andrews: "Vibration Displacement and Mode-Shape Measurement by a Laser Interferometer," *J. Acoust. Soc. Am.,* **42**:982–990 (1967).

32. Serbyn, M. R., and W. B. Penzes: *Instr. Soc. Amer. Trans,.* **21**:55 (1982).

33. Luxon, J. T., and D. E. Parker: "Industrial Lasers and Their Applications," Chap. 10, Prentice-Hall, Englewood Cliffs, N.J., 1985.

34. Lauer, G.: "Interferometrische Verfahren zum Messen von Schwing- und Stossbewegungen," *Fachkolloq. Experim. Mechanik,* Stuttgart University, October 9–10, 1986.

35. Sutton, C. M.: *Metrologia,* **27**:133–138 (1990).

36. Feder, E. I., and A. M. Gillen: *IRE Trans Instr.,* **6**:1 (1957).

37. Nisbet, J. S., J. N. Brennan, and H. I. Tarpley: *J. Acoust. Soc. Amer.,* **32**:71 (1960).

38. Jones, E., S. Edelman, and K. S. Sizmore: *J. Acoust. Soc. Amer.,* **33**:1462 (1961).

39. Davies, R. M.: *Phil. Trans. A,* **240**:375 (1948).

40. Bateman, V. I., F. A. Brown, and N. T. Davie: "Isolation of a Piezoresistive Accelerometer Used in High-Acceleration Tests," 17th Transducer Workshop, pp. 46–65 (1994).

41. Dosch, J. J., and J. Lin: "Hopkinson Bar Acceptance Testing for Shock Accelerometers," *Sound and Vibration,* **33**:16–21 (1999).

42. Conrad, R. W., and I. Vigness: *Proc. 8th Annu. Meet. Instr. Soc. Amer.,* Paper 11-3 (1953).

43. Bouche, R. R.: *Endevco Corp. Tech. Paper* TP 206, April 1961.

44. Kistler, W. P.: *Shock and Vibration Bull.,* **35**, pt. 4 (1966).

45. Ramboz, J. D., and C. Federman: *Natl. Bur. Std* (*U.S.*) *Rept.* NBSIR74-480, March 1974.

46. Bateman, V., et al.: "Calibration of a Hopkinson Bar with a Transfer Standard," *J. Shock and Vibration,* **1**:145–152 (1993).

47. Dosch, J., and J. Lin: "Application of the Hopkinson Bar Calibrator to the Evaluation of Accelerometers," *Proc. 44 Ann. Mtg. Inst. Env. Sci. And Techn.,* Phoenix, Ariz., 1998, pp. 185–191.

48. Bateman, V., et al.: "Use of a Beryllium Hopkinson Bar to Characterize a Piezoresistive Accelerometer in Shock Environments," *J. Inst. Env. Sci.,* **Nov./Dec.**:33–39 (1996).

49. Link, A., H.-J. von Martens, and W. Wabinski: "New Method for Absolute Shock Calibration of Accelerometers," *Proc. SPIE,* **3411**:224–235 (1998).

50. Ueda, K., A. Umeda, and H. Imai: "Uncertainty Evaluation of a Primary Shock Calibration Method for Accelerometers," *Metrologia,* **37**:187–197 (2000).

51. Dimoff, T.: *J. Acoust. Soc. Amer.,* **40**:671 (1966).

52. Payne, B. F., and G. B. Booth: "NIST Supershaker Project," *Proc. Metrologie,* **95**:296–301 (1995).

53. Payne, B.: *Proc. SPIE 1998,* **3411**:187–195 (1998).

54. O'Toole, K. M., and B. H. Meldrum: *J. Sci. Instr.* (*J. Phys. E*), **1**(2):672 (1968).

55. Lin, J.: "Transverse Response of Piezoelectric Accelerometers," 18th Transducer Workshop, San Diego, Calif., 1995.

56. Dosch, J. J.: "Automated Testing of Accelerometer Transverse Sensitivity," PCB Piezotronics Technical Note A-R 69, November 2000.

CHAPTER 12
STRAIN GAGE INSTRUMENTATION

Patrick L. Walter

INTRODUCTION

This chapter provides an overview of bonded metal strain gage technology with a lesser focus on both piezoresistive (MEMS, defined later) and piezoelectric strain gages. After a brief discussion concerning the evolution of bonded gages, their manufacture is described, along with the necessary mechanical considerations that should occur to ensure their successful application. A description of the electrical circuits in which the bonded gages must function to satisfy the requirements of experimental stress analysis and electromechanical transducer design is then presented. Finally, because entire textbooks have been written to comprehensively cover all the topics associated with strain gage selection and application, a comprehensive literature resource is identified.

Bonded metal foil strain gages are a mature technology. Their importance is routinely encountered in our daily lives every time a weighing process occurs. In addition, in experimental mechanics applications, such as flight qualification of a new aircraft design, approximately one-third of the 1000 or more instrumented data channels are dedicated to strain gage measurements. These gages are used to monitor for material fatigue and structural design margin as well as identify structural frequencies.

The detailed history of the evolution of the strain gage is presented in Ref. 1. On September 10, 1936, an electrical engineering graduate student at the California Institute of Technology by the name of Ed Simmons suggested using bonded wire to measure the dynamic forces generated by an impact testing machine. The professor with whom he was working, Dr. Gottfried Datwyler, bonded 40-gage, cotton-wrapped, insulated, constantan wire, supplied by Mr. Simmons, to a piece of clock spring with Glyptal cement. The spring was mounted as a cantilever beam, and the wire change in resistance was proven to be linear, repeatable, and hysteresis free with applied strain. The bonded wire strain gage was born. Mr. Simmons developed a pulsed current excitation supply to use with these gages, and the experimental work was completed and presented at a meeting of the American Society for Testing and Materials (ASTM) in June 1938. Scant attention was given at this meeting to the strain gage development that supported the experimental work.

In 1938, at MIT, Professor Arthur Ruge was working on a research contract with his graduate assistant, Hans Meir, to measure the stresses induced in water towers under earthquake conditions. On April 3, 1938, while assisting Mr. Meir in his experimental work, Professor Ruge unwound wire from a precision resistor, bonded it to a test beam, and created a strain gage. The importance of this discovery was immediately apparent to Professor Ruge. He encouraged Mr. Meir to divert the emphasis of his graduate work from the water tower to focus on the further development of the strain gage. Professor Ruge and Mr. Meir spent the rest of their lives developing and commercializing the bonded strain gage and transducers based on its operating principle. During a patent search following the 1938 MIT discovery, Mr. Simmons' earlier work was uncovered. As a result, he ultimately received patent number 2,292,549, on August 11, 1942, as the recognized inventor of the bonded resistance strain gage. Today the bonded wire strain gage has been replaced by foil etched gages formed by printed circuit techniques. These manufacturing techniques will be described later. Currently the vast majority of strain gage applications associated with experimental stress analysis are performed using bonded metal foil strain gages.

In the 1940s through the 1950s it was recognized that, when geometrically distorted, the resistance change of semiconductor materials could also be correlated to strain.[2] The semiconductor strain gage became of interest because its sensitivity to strain was about 50 to 200 times that of metal gages. However, whether using p- or n-type silicon, gage sensitivity was discovered to be strongly influenced by both temperature and strain level. For this reason, semiconductor strain gages find principal application only in experimental stress analysis involving small strain differences at controlled temperatures. Colloquially, the term *piezoresistive* strain gage is used synonymously with silicon or semiconductor strain gages.

While not extensively used in experimental stress analysis work, due to their higher strain sensitivity piezoresistive strain gages find significant application in the construction of transducers (e.g., pressure, force, and acceleration) whose output can be thermally compensated. Piezoresistive transducers manufactured in the 1960s first used silicon strain gages fabricated from lightly doped ingots. These ingots were sliced with respect to the crystal axes of the silicon to form small bars or patterns, which became gages. These gages were usually bonded directly to the transducer flexure.

Since the late 1970s there has been a continual evolution of microsensors into the marketplace. Piezoresistive transducers manufactured in this manner use silicon both as their flexural element and as their transduction element (see Chap. 10). The strain gages are diffused directly into the flexure. The most typical fabrication process has the following sequence of events: the single crystal silicon is grown; the ingot is trimmed, sliced, polished, and cleaned; diffusion of a dopant into a surface region of the parent silicon wafer is controlled by a deposited film; a photolithography process includes etching of the film at places defined in the developing process, followed by removal of the photoresist; and isotropic and anisotropic wet chemicals are used for shaping the mechanical microstructure. Both the resultant stress distribution in the microstructure and the dopant control the piezoresistive coefficients of the silicon.

Electrical interconnection of various controlled surfaces formed in the silicon crystal as well as bonding pads are provided by thin-film metallization. The silicon wafer is then separated into individual dies. The dies are bonded by various techniques into the transducer housing, and wire bonding connects the metalized pads to metal terminals in the transducer housing. Sensors fabricated in this manner are known as microelectromechanical systems (MEMS) transducers. Metal strain gage–based transducers typically provide 20 to 30 millivolts (mV) of unamplified full-scale signal; by comparison, MEMS resistance–based transducers produce 100 to 200 mV of unamplified signal. MEMS transducer technology is rapidly expanding in commercial and military applications. The piezoresistive transducer discussions in Chap. 10 specifically include MEMS transducers when discussing the applicability of this technology to mechanical shock measurements. Reference 3 provides an extensive chapter on strain gage–based transducers.

The last strain gage technology to be mentioned is piezoelectric. Strain within a piezoelectric material displaces electrical charges within the strained elements, and the charges accumulate on opposing electrode surfaces. Piezoelectric strain gages do not have response to 0 Hz. Therefore, their application in experimental stress analysis is limited. However, modern gages have integral signal-conditioning electronics (ICP® or IEPE) that greatly enhance the measurement system's signal-to-noise ratio. Five volts of signal can be provided for 100×10^{-6} in./in. (cm/cm) of strain, making this type of gage very desirable for low-level, dynamic strain measurements.

Figures 12.1, 12.2, and 12.3 collectively illustrate all of the just-described technologies. Figure 12.1 shows a traditional metal strain gage, Fig. 12.2 shows a MEMS transducer flexure (in this case, an accelerometer), and Fig. 12.3 shows a piezoelectric strain gage with integral electronics (ICP®).

FIGURE 12.1 Single-element, bonded metal film strain gage.

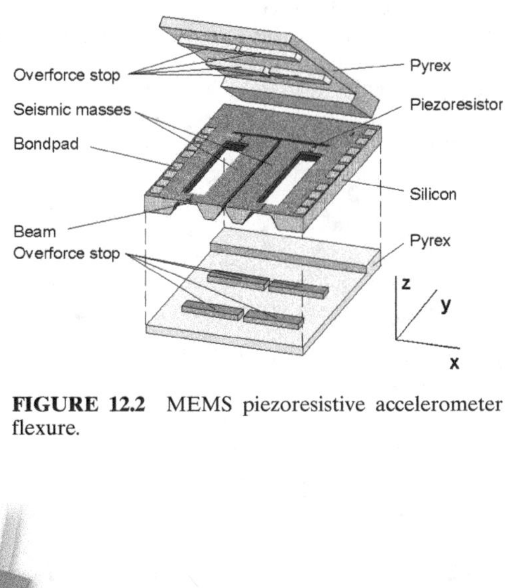

FIGURE 12.2 MEMS piezoresistive accelerometer flexure.

FIGURE 12.3 Piezoelectric strain gage with ICP® signal conditioning.

BONDED METAL STRAIN GAGE MATERIALS AND MANUFACTURE

As noted previously, for experimental stress analysis, the bonded metal foil strain gage is used almost exclusively. The two most common materials are constantan (55 percent copper, 45 percent nickel) and Karma (20 percent chromium, 2.5 percent aluminum, 2.5 percent copper, balance nickel). Both these materials offer (1) atypical resistance versus temperature behavior, (2) malleability sufficient to allow processing into foil less than 0.001 in. (0.025 mm) thick, (3) ease of photochemical machining into accurate configurations, and (4) reasonable cost. Considering gage resistance-temperature behavior, it is desired that a strain gage eliminate false signals due to thermal expansion of the material on which it is mounted. If the increase in gage resistance due to thermal expansion of this material can be offset by a corresponding decrease in gage alloy resistivity, the result will be zero change and no false signal. In reality, a finished gage assembly includes its backing, sealant, and adhesive. All of these materials expand at their own rate and contribute to this false signal. Thermal coefficient of resistance (TCR) values required to achieve thermal compensation tend to range from –25 to +5 ppm/°C. These values are well within the capability of cold-rolled constantan and Karma. Optimum compensation is obtained by heat treatment of the two foils, dependent on the material on which they are mounted. Figure 12.4 shows the typical thermal or false strain compensation that can be achieved by a strain gage in a temperature range around room temperature.

Strain gage manufacturing involves, first, putting the stringently manufactured alloys through closely controlled melting processes, resulting in ingots 14 in. (35.56 cm) in diameter and weighing approximately 1 ton (454 kg mass). After an extended

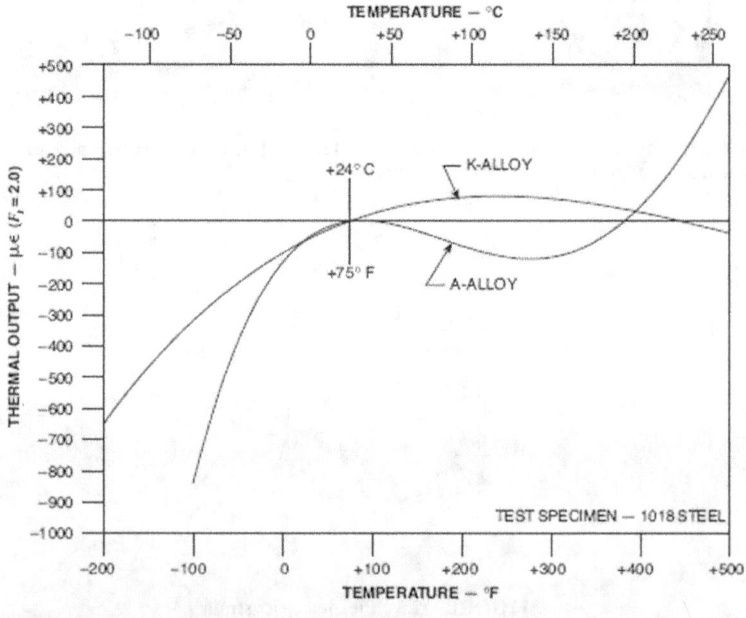

FIGURE 12.4 Typical temperature compensation curves achievable with Constantan (A) and Karma (K) strain gage materials.

high-temperature soak, hot-forging transitions the cast ingot into a slab. The slabs are then cooled and ground to remove surface defects. Next, the slabs are reheated, rolled to about 0.2 in. (5.1 mm), descaled, acid-cleaned, cold-rolled again, surface-ground a second time, cold-rolled again, annealed, leaving the thickness at about 0.06 in. (1.5 mm), and placed in stock. Once removed from stock, additional rolling and cleaning, followed by heat treatment, enable the foil to be bonded to a backing film that acts as a carrier. The foil with the backing is then etched with photochemicals to form the desired gage geometry. Sheets of gages are then cut apart and packaged for sale. Reference 4 provides a detailed description of this process.

MECHANICAL ASPECTS OF GAGE OPERATION

To build effective strain-sensing circuits, one must be aware of the interaction between the gage and the surface of the flexure to which it is mounted. Mechanical aspects of this interaction include the influence of backing material, size, orientation, transverse sensitivity, distance from the surface, bonding, and installation.

BACKING MATERIAL

The purpose of the backing material used in constructing strain gages is to provide support, dimensional stability, and mechanical protection for the grid element. The backing material of the gage element(s) acts as a spring in parallel with the parent material to which it is attached and can potentially modify mechanical behavior. In addition, the temperature operating range of the gage can be constrained by its backing material. Most backings are polyimide or glass fiber–reinforced epoxies. Some gages are encapsulated for chemical and mechanical protection as well as extended fatigue life. For high-temperature applications, some gages have strippable backings for mounting with ceramic adhesives. Still other metal gages can be welded. The frequency response of welded gages, due to uncertainties in dynamic response, is a subject area that still requires investigation.

SIZE

The major factors to be considered in determining the size of strain gage to use are available space for gage mounting, strain gradient at the test location, and character of the material under test. The strain gage must be small enough to be compatible with its mounting location and the concentrated strain field. It must be large enough so that, on metals with large grain size, it measures average strain as opposed to local effects. Grid elements greater than 0.125 in. (3mm) generally have greater fatigue resistance.

TRANSVERSE SENSITIVITY AND ORIENTATION

Strain gage transverse sensitivity and mounting orientation are concurrent considerations. Transverse sensitivity in strain gages is important due to the fact that part of the geometry of the gage grid is oriented in directions other than parallel to the principal gage sensing direction. Values of transverse sensitivities are provided with individual gages but typically vary between fractional and several percent. The position of the strain gage axis relative to the numerically larger principal strain on the surface to which it is mounted will have an influence on indicated strain.

Distance from the Surface. The grid element of a strain gage is separated from the structure under test by its backing material and cement. The grid then responds to strain at a location removed from its mounting surface. The strain on structures such as thin plates in bending can vary considerably from that measured by the strain gage.

Bonding Adhesives. Resistance strain gage performance is entirely dependent on the bond attaching it to the parent material. The grid element must have the strain transmitted to it undiminished by the bonding adhesive. Typical adhesives are as follows.

Epoxy Adhesives. Epoxy adhesives are useful over a temperature range of –270 to +320°C. The two classes are either room-temperature curing or thermal setting type; both are available with various organic fillers to optimize performance for individual test requirements.

Phenolic Adhesives. Bakelite, or phenolic adhesive, requires high bonding pressure and long curing cycles. It is used in some transducer applications because of long-term stability under load. The maximum operating temperature for static loads is 180°C.

Polyimide Adhesives. Polyimide adhesives are used to install gages backed by polyimide carriers or high-temperature epoxies. They are a one-part thermal setting resin and are used from –200 to +400°C.

Ceramic cements (applicable from –270 to +550°C) and welding are other mounting techniques.

ELECTRICAL ASPECTS OF GAGE OPERATION

The resistance strain gage, which manifests a change in resistance proportional to strain, must form part of an electrical circuit such that a current passed through the gage transforms this change in resistance into a current, voltage, or power change to be measured. The electrical aspects of gage operation to be considered include current in the gage, resistance to ground, and shielding.

Strain gages are seldom damaged by excitation voltages in excess of proper values, but their performance degrades. The voltage applied to a strain gage bridge creates a power loss in each arm, which must be dissipated in the form of heat. By its basic design, all of the power input to the bridge is dissipated in the bridge, with none available to the output circuit. The sensing grid of every strain gage then operates at a higher temperature than the structure on which it is mounted. Heat flow into the structure causes a temperature rise, which is a function of its heat sink capacity and gage power level. The optimum excitation level for strain gage applications is a function of the strain gage grid area, gage resistance, heat sink installation, required operational specifications, and installation and wiring techniques. Zero shift versus load and stability under load at the maximum operating temperature are the performance tests most sensitive to excessive excitation voltage.

Resistance to ground is an important parameter in strain gage mounting, since insulation leakage paths produce shunting of the gage resistance between the gage and the metal structure to which it is bonded, producing false compressive strain readings. The ingress of fluids typically leads to this breakdown in resistance-to-ground value and can also change the mechanical properties of the adhesive. A minimum gage–to–mounting surface resistance-to-ground value of 50 MΩ is recommended.

Since signals of interest from strain gage bridges are typically on the order of a few millivolts, shielding of the bridge from stray pickup is important. Gage leads should also be shielded and proper grounding procedures followed. Stray pickup may be introduced by 60-Hz line voltage associated with other electronic equipment, electrical noise from motors, radio frequency interference, and so on. Note that shielding materials for electrical fields are different from those for magnetic fields. Nickel alloy strain gages are particularly susceptible to magnetic fields.

THE WHEATSTONE BRIDGE

Small strains result in small impedance changes in resistive strain gage elements. A Wheatstone bridge circuit can detect a small change in impedance to a high degree of accuracy.

BRIDGE EQUATIONS

The circuit most often used with metal strain gages is a four-arm bridge with a constant-voltage power supply. Figure 12.5 shows a basic bridge configuration. The supply voltage E_{ex} can be either ac or dc, but for now it is assumed to be dc, so equations can be written in terms of resistance R rather than a complex impedance. The condition for a balanced bridge with e_0 equal to zero is:

$$\frac{R_1}{R_2} = \frac{R_4}{R_3} \tag{12.1}$$

FIGURE 12.5 Four-arm Wheatstone bridge with constant voltage (E_{ex}) power supply.

Next, an expression is presented for the bridge output voltage e_0 due to *small* changes in R_1, R_2, R_3, and R_4:

$$e_0 = \left[-\frac{R_3\,dR_4}{(R_3 + R_4)^2} + \frac{R_4\,dR_3}{(R_3 + R_4)^2} - \frac{R_1\,dR_2}{(R_1 + R_2)^2} + \frac{R_2\,dR_1}{(R_1 + R_2)^2} \right] E_{ex} \tag{12.2}$$

In many cases, the bridge circuit is made up of equal resistances. Substituting for individual resistances, with a strain gage resistance R, and using the definition of the gage factor supplied with every gage ($F = (\Delta R/R)/\varepsilon$), Eq. 12.2 becomes:

$$e_0 = \frac{FE_{ex}}{4}(-\varepsilon_4 + \varepsilon_3 - \varepsilon_2 + \varepsilon_1) \tag{12.3}$$

The unbalance of the bridge is seen to be proportional to the sum of the strain (or resistance changes) in opposite arms and to the difference of strain (or resistance changes) in adjacent arms.

Equations (12.2) and (12.3) indicate another technique to compensate strain gage circuits to minimize the influence of false temperature-induced strain. This is referred to as the *dummy gage method.*

Assume that we have a bridge circuit with one active arm, and arbitrarily let this arm be number 4. Equation (12.3) becomes:

$$e_0 = \frac{FE_{\text{ex}}}{4}(-\varepsilon_4) \tag{12.4}$$

Arm 4 responds to the total strain induced in it, which is composed of both thermal (t) and mechanical (m) strain:

$$\varepsilon_4 = \varepsilon_m + \varepsilon_t \tag{12.5}$$

A problem arises if it is desired to isolate the mechanical strain component. One solution is to take another strain gage (the dummy gage) and mount it on a strain-isolated piece of the same material as that on which gage 4 is mounted. If placed in the same thermal environment as gage 4, the output from the dummy gage becomes simply e_t. If the dummy gage is wired in an adjacent bridge arm to 4 (1 or 3), Eq. (12.3) becomes:

$$e_0 = \frac{FE_{\text{ex}}}{4}(-\varepsilon_m - \varepsilon_t + \varepsilon_t) \tag{12.6}$$

Equation (12.6) indicates that thermal strain effects are canceled. In reality, perfect temperature compensation is not achieved, since no two strain gages from a lot track one another identically. However, compensation adequate for many applications can be accomplished.

Equation (12.2) presented the generalized form of the bridge equation for four active arms. If only one arm (e.g., arm 4) is active, this equation becomes:

$$e_0 = \left[\frac{-R_3 dR_4}{(R_3 + R_4)^2}\right] E_{\text{ex}} \tag{12.7}$$

This equation was specifically presented for *small* changes in resistance, such as those associated with metallic strain gages. If the change in resistance in arm 4 is large, Eq. (12.7) is better expressed as:

$$e_0 = \frac{(R_4 + \Delta R_4)E_{\text{ex}}}{(R_4 + \Delta R_4) + R_3} - \frac{R_4 E_{\text{ex}}}{R_4 + R_3} \tag{12.8}$$

For an equal-arm bridge, this becomes:

$$e_0 = \frac{\Delta R E_{\text{ex}}}{4R + 2\Delta R} = \frac{FE_{\text{ex}}\varepsilon}{4 + 2F\varepsilon} \tag{12.9}$$

For an equal-arm bridge, Eq. (12.7) becomes:

$$e_0 = \frac{dR\, E_{\text{ex}}}{4R} = \frac{FE_{\text{ex}}\varepsilon}{4} \tag{12.10}$$

The difference between Eqs. (12.10) and (12.9) is that Eq. (12.10) describes a linear process while Eq. (12.9) describes a nonlinear one. Semiconductor gages, because of their large gage factor, require analysis using Eq. (12.9).

Semiconductor gages may be used in constant-voltage, four-arm bridge circuits when two or four gages are used in adjacent arms and strained so that their outputs are additive. Analysis of the bridge equations for this situation will show that if gages in adjacent arms are subjected to equal but opposite values of ΔR, the output signal is doubled and the nonlinearity in the bridge output is eliminated. Another approach to eliminating this nonlinearity is to design a circuit such that the current through the strain gage remains constant.

The alternating signs in Eq. (12.3) are useful in isolating various strain components when using bridge circuits containing strain gages. Table 12.1 provides generalized bridge equations for one, two, and four equal-active-arm bridges that show

TABLE 12.1 Strain Gage Bridge Configurations That Isolate Various Strain Components

Bridge/Strain Arrangement (Note 1)	Description	Bridge Output, E_o/E mV/V (Notes 2, 3)	Nonlinearity, η Where $E_o/E = K\epsilon \times 10^{-3}(1-\eta)$ (Notes 2, 3)	Corrections (Note 3)
1 	Single active gage in uniaxial tension or compression.	$\frac{E_o}{E} = \frac{F\epsilon \times 10^{-3}}{4 + 2F\epsilon \times 10^{-6}}$	$K = \frac{F}{4}$; $\eta = \frac{F\epsilon \times 10^{-6}}{2 + F\epsilon \times 10^{-6}}$	$\epsilon = \frac{2\epsilon_i}{2 - F\epsilon_i \times 10^{-6}}$
2 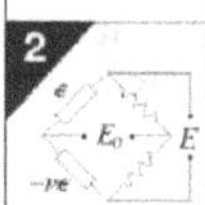	Two active gages in uniaxial stress field — one aligned with maximum principal strain, one "Poisson" gage.	$\frac{E_o}{E} = \frac{F\epsilon(1+\nu) \times 10^{-3}}{4 + 2F\epsilon(1-\nu) \times 10^{-6}}$	$K = \frac{F(1+\nu)}{4}$; $\eta = \frac{F\epsilon(1-\nu) \times 10^{-6}}{2 + F\epsilon(1-\nu) \times 10^{-6}}$	$\epsilon = \frac{2\epsilon_i}{2(1+\nu) - F\epsilon_i(1-\nu) \times 10^{-6}}$
3 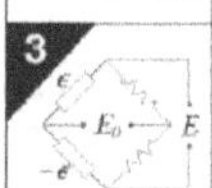	Two active gages with equal and opposite strains — typical of bending-beam arrangement.	$\frac{E_o}{E} = \frac{F\epsilon}{2} \times 10^{-3}$	$K = \frac{F}{2}$; $\eta = 0$	$\epsilon = \frac{\epsilon_i}{2}$
4 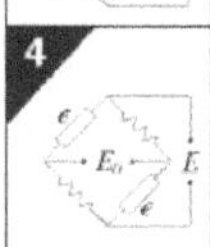	Two active gages with equal strains of same sign — used on opposite sides of column with low temperature gradient (bending cancellation, for instance).	$\frac{E_o}{E} = \frac{F\epsilon \times 10^{-3}}{2 + F\epsilon \times 10^{-6}}$	$K = \frac{F}{2}$; $\eta = \frac{F\epsilon \times 10^{-6}}{2 + F\epsilon \times 10^{-6}}$	$\epsilon = \frac{2\epsilon_i}{4 - F\epsilon_i \times 10^{-6}}$
5 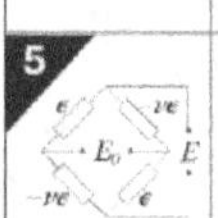	Four active gages in unaxial stress field two aligned with maximum principal strain, two "Poisson" gages (column).	$\frac{E_o}{E} = \frac{F\epsilon(1+\nu) \times 10^{-3}}{2 + F\epsilon(1-\nu) \times 10^{-6}}$	$K = \frac{F(1+\nu)}{2}$; $\eta = \frac{F\epsilon(1-\nu) \times 10^{-6}}{2 + F\epsilon(1-\nu) \times 10^{-6}}$	$\epsilon = \frac{2\epsilon_i}{4(1+\nu) - F\epsilon_i(1-\nu) \times 10^{-6}}$
6 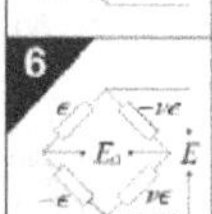	Four active gages in uniaxial stress field — two aligned with maximum principal strain, two "Poisson" gages (beam).	$\frac{E_o}{E} = \frac{F\epsilon(1+\nu) \times 10^{-3}}{2}$	$K = \frac{F(1+\nu)}{2}$; $\eta = 0$	$\epsilon = \frac{\epsilon_i}{2(1+\nu)}$
7 	Four active gages with pairs subjected to equal and opposite strains (beam in bending or shaft in torsion).	$\frac{E_o}{E} = F\epsilon \times 10^{-3}$	$K = F$; $\eta = 0$	$\epsilon = \frac{\epsilon_i}{4}$

Notes: 1. $(R_1/R_2)_{\text{bridge}} = 1$; $(R_3/R_4)_{\text{bridge}} = 1$ when two or less active arms are used.
2. Constant voltage power supply is assumed.
3. ϵ and ϵ_i (strains) are expressed in microstrain units (in/in x 10^{6}).

(Note: These equations are appropriate for both large- and small-strain circuits. For small strains, some contributors to the calculations may be very small as circuit nonlinearities become negligible.)

Courtesy: Vishay Measurements Group.

how these strain components can be isolated. The dimensionless bridge output is presented in millivolts per volts for a constant-voltage power supply. Strain is presented in microstrain. No small-strain assumption is built into these equations. For large strains with semiconductor gages, F may not be a constant and this correction also has to be built into the equations. In this table, the Poisson gage is one which measures the lateral compressive strain accompanying an axial tension strain. As noted earlier, only for two adjacent active gages with equal and opposite strains or for four active gages with pairs subjected to equal and opposite strains is the bridge output a linear function of strain.

OTHER CIRCUIT DESIGN CONSIDERATIONS

While bonded metal foil strain gages are a mature technology and the large numbers of nuances associated with their application have been well studied, this chapter has provided but a brief introduction to considerations necessary for their successful application. During the last decade, the Vishay Micro-Measurements Division has concentrated essentially the entire strain gage manufacturing capability in the United States under one roof, in Raleigh, North Carolina. Their website can be found at http://www.vishay.com/company/brands/micromeasurements/. This site remains a stable, readily available source of technical support including both technical and application notes. Although much more information is contained in this site, key technical notes (TNs) are highlighted as follows:

- Strain Gage Selection: Criteria, Procedures, Recommendations, Tech Note TN-505-4: *Details gage selection considering material type, backing type, size, grid pattern, and more.*
- Strain Gage Rosettes: Selection, Application and Data Reduction, Tech Note 515: *Describes strain gage construction, selection, and data reduction to determine principal stress magnitudes and directions on the surface of a material.*
- Optimizing Strain Gage Excitation Levels, Tech Note TN-502: *Provides guidance for selection of strain gage bridge voltage excitation levels to minimize thermal heating errors due to internal gage power dissipation.*
- Errors Due to Transverse Sensitivity in Strain Gages, Tech Note TN-509: *Identifies problems in gage readings due to transverse sensitivity and provides methods for data correction.*
- Strain Gage Thermal Output and Gage Factor Variation with Temperature, Tech Note-504-1: *Provides a methodology for compensating for false signals due to thermal expansion of the gage as mounted on its parent material.*
- Fatigue Characteristics of Vishay Micro-Measurements Strain Gages, Tech Note TN-508-1: *Provides fatigue exposure limits of gages and their associated backing and mounting.*
- Errors Due to Misalignment of Strain Gages, Tech Note TN-511: *Details errors due to misalignment of gages when mounted, of particular importance in characterizing a biaxial stress field.*
- Errors Due to Wheatstone Bridge Nonlinearity, Tech Note TN-507-1: *Quantifies nonlinearity errors in bridge circuits due to large changes in resistance (e.g., post yield, piezoresistive gages, etc.).*

- Errors Due to Shared Lead Wires in Parallel Strain Gage Circuits, Tech Note TN-516: *Deals with errors that can occur in attempting to share common leads in strain gage circuits.*
- Shunt Calibration of Strain Gage Instrumentation, Tech Note TN-514: *Provides methods to verify accuracy of strain gage read out circuitry.*
- Noise Control in Strain Gage Measurements, Tech Note TN-501-2: *Describes the susceptibility of strain gage circuits to low-level noise and provides preventative measures.*

These TNs provide a total of more than 100 pages of guidance in strain gage technology and should be consulted before initiating important experimental stress analysis assessments for structural systems.

REFERENCES

1. Stein, Peter K.: "1936—A Banner Year for Strain Gages and Experimental Stress Analysis—An Historical Perspective," *Experimental Techniques,* **30**(1): 23–41 (January/February 2006).
2. *Semiconductor Strain Gage Handbook,* Baldwin, Lima, Hamilton, Inc., 1973.
3. Walter, Patrick L.: "Bridge Transducers," Chap. 3 in *Mechanical Engineer's Handbook,* 3d ed., *Part 1, Instrumentation, Systems, Controls, and MEMS,* Myer Kutz, ed., John Wiley and Sons, Upper Saddle River, N.J., 2006 (originally published as *Instrumentation and Control,* Wiley, 1990).
4. Robinson, M.: "Strain Gage Materials Processing, Metallurgy, and Manufacture," *Experimental Techniques,* **30**(1): 42–46 (January/February 2006).

CHAPTER 13
SHOCK AND VIBRATION DATA ACQUISITION

Strether Smith

INTRODUCTION

This chapter discusses the basic functions of systems that are used to acquire and store data from shock and vibration experiments. The discussion concentrates on the measurement and storage of signals in the audiofrequency range defined here as 1 Hz to 100 kHz. Particular attention is paid to the primary features and problems that are relevant to the discrete recording of these signals, specifically, dynamic range, headroom, alias protection, data sparsity, and out-of-band energy.

The signals to be acquired and stored may be generated by a variety of devices that sense acceleration, velocity, or displacement and might employ one of a multitude of transduction mechanisms. This discussion will be restricted to the most common systems used for vibration and shock, specifically, a piezoelectric transduction measuring acceleration or force via appropriate signal conditioning. A description of other devices and their conditioning can be found in Chap. 10 and Ref. 1.

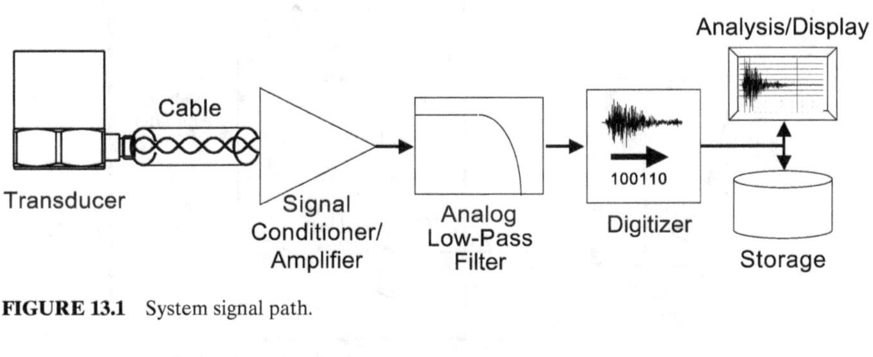

FIGURE 13.1 System signal path.

Figure 13.1 shows a typical instrumentation-signal path, which is made up of the following:

- Transducer
- Cable
- Signal conditioner/amplifier
- Analog low-pass filter
- Analog-to-digital converter/digitizer
- Run-time analysis and display
- Data storage

PIEZOELECTRIC TRANSDUCERS, CABLING, AND SIGNAL CONDITIONING

The basic principles of piezoelectric transducers are described in Chap. 10. These devices produce a charge signal that is proportional to the mechanical input (usually acceleration or force). The charge signal is converted to voltage by either a *charge amplifier* or an *internal electronic piezoelectric* (IEPE) system.[2] Of the two options, charge-based systems offer the most flexibility. Figure 13.2 shows a diagram of a charge-amplifier measurement system. Charge, generated by the piezoelectric sensor in response to a mechanical input, is transmitted to the amplifier with a special *low-noise* or *microdot* cable. The charge amplifier converts the charge to a voltage that is used by the downstream devices. The gain can be easily changed over a wide range to adjust system sensitivity. Because there are no electronic components in the transducer, measurements can be taken at higher temperatures and the transducers have better reliability than IEPE devices. However, applications are normally restricted to short cable runs because the high impedance of the signal circuit makes it very sensitive to induced noise. Also, the requirement of expensive, mechanically fragile microdot cables often limits their practical use to research and other environments where conditions are very well controlled.

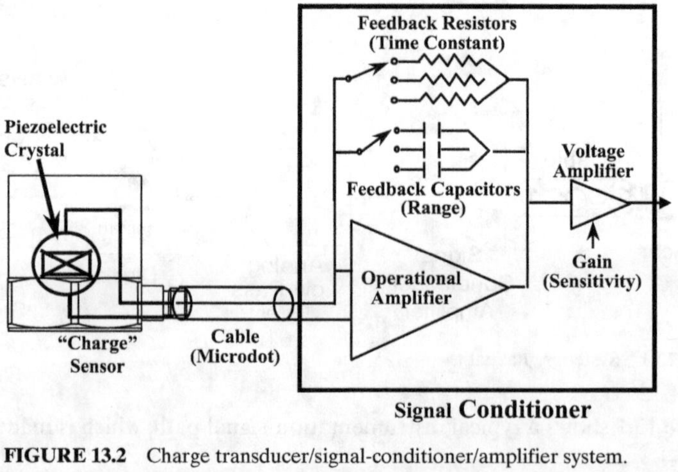

FIGURE 13.2 Charge transducer/signal-conditioner/amplifier system.

Many of the practical limitations of charge-based instrumentation are addressed by IEPE systems. As shown in Fig. 13.3, an IEPE system converts the charge (high-impedance) signal to a modulated current (low-impedance) signal in (or near) the transducer. The modulated current is then converted to voltage in the signal conditioner. This strategy has significant practical advantages over charge-based systems for most testing applications. The primary advantage of an IEPE system is that conversion to a low-impedance signal at, or near, the transducer allows the use of less expensive, more robust cabling systems. Although nearly any cable with two or more conductors will do (including microdot and other coaxial types), the use of twisted-shielded pair produces significantly improved signal-to-noise ratios (SNRs). This is particularly important in facilities that require long (>30-m/100-ft) cables. One fault of the approach is that long lines (with excessive capacitance) and/or inadequate excitation can produce reduced measurement bandwidth.[3]

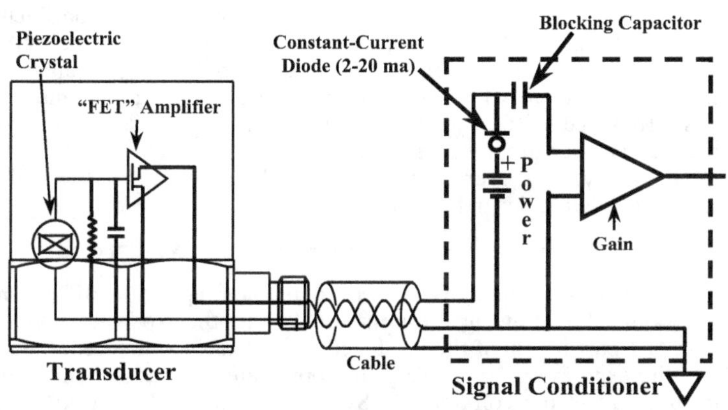

FIGURE 13.3 IEPE transducer/signal-conditioner/amplifier system.

An additional feature that is allowed by the IEPE technology is the *Transducer Electronic Data Sheet* (TEDS).[4] In a TEDS system, the transducers (or in-line conditioners) are equipped with a digital memory that contains basic device information such as manufacturer, model number, and calibration. On command, a matching signal conditioner retrieves the information and passes it back to the data acquisition system for channel identification and characterization. For reasons similar to the IEPE bandwidth limitations already described, TEDS systems also have problems with long analog lines.[5]

THE DIGITAL DATA ACQUISITION SYSTEM

During the past 40 years, strategies for the recording of shock, acoustic, and vibration data have shifted almost entirely from continuous (analog) to discrete (digital) technology. The primary advantages of discrete conversion include better accuracy, higher dynamic range, easier processing of data into meaningful engineering terms, and more straightforward data storage and retrieval. The primary restriction, bandwidth, was resolved about 1995, with the development of high-resolution, continuous-recording, multichannel systems with bandwidths of 100 kHz or greater—enough to cover most signals in the regimens of concern here.[6]

The most significant feature that digital data acquisition offers to signal measurement is greater *dynamic range,* defined as follows:

$$\text{Dynamic range} = \text{Full-scale value/Smallest detectable value} \quad (13.1)$$

Analog measurement devices, such as strip chart recorders and analog tape systems, are inherently limited to a dynamic range of 100/1 (40 dB). Digital systems are available with better accuracy, resolution, and signal-to-noise ratios than the best of the instrumentation and signal conditioning subsystems that feed them.

When a sinusoidal input signal is near full scale, the theoretical dynamic range, or SNR, of a digitizing process is[7]

$$\text{Dynamic range} = \text{Full-scale SNR} = 6.02N + 1.76 \text{ (dB)} \quad (13.2)$$

where N is the number of bits in the digitizer. Of course, systems that include real transducers, cabling, and analog electronics do not perform as well as theoretical ones, but the earliest digital systems (10-bit) offered close to 55 dB (~560/1) and the best 16-bit systems provide over 90 dB (~31,000/1) when characterized in the time domain.[8]

Systems with high dynamic range can use very conservative scaling to provide an input range that is much greater than the anticipated signal. This in turn reduces the chance of saturation in the event of unexpectedly large responses. The scaling margin is called *headroom* and is defined as follows:

$$\text{Headroom} = \text{(System full scale)/(Expected maximum data)} \quad (13.3)$$

For example, if a measurement system with dynamic range of 80 dB is configured with a headroom of 20 (allowing a response of 20 times the expected), it will provide a resolution of 1 part in 500 of the predicted peak. This is well within the expected accuracy of the experiment. It is critical to appreciate that the dynamic range of a system includes all of the components. Systems based on IEPE transducers with cable runs as long as 100 m have been shown to provide 85 dB (18,000/1) of dynamic range.[6] The need for even more conservative scaling is addressed in the discussion of out-of-band energy later in this chapter.

Accuracy and repeatability have been comparably improved. The result is that, when the digital acquisition is done correctly, overall system accuracy is driven almost entirely by the characteristics of the transducer, cabling, and signal conditioning that precede it.

This improvement in performance when compared to analog recording concepts does not come without a price. Figure 13.4 shows the basic process of digitizing an analog signal. The continuous signal is sampled instantaneously at equally spaced

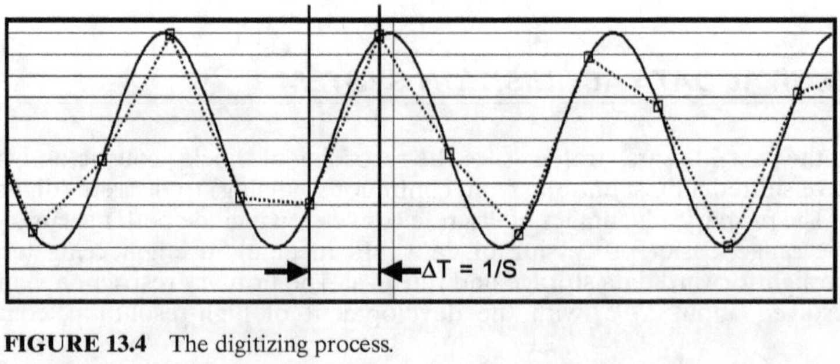

FIGURE 13.4 The digitizing process.

intervals (= ΔT). The fundamental question is: How often do we have to sample the data? In other words, what is the required *sample rate* (R_S)?

The answer was developed about the same time by several investigators,[9] but a paper by Claude Shannon[10] is normally given primary credit. Reference 9 contains a concise discussion of the Nyquist-Shannon sampling theory, where it is stated: "Exact reconstruction of a continuous-time baseband signal from its samples is possible if the signal is band limited and the sampling frequency is greater than twice the signal bandwidth." An alternative statement is: "If we acquire more than two points per cycle of the highest frequency component in the signal, the entire signal can be reproduced exactly."[11] Inspection of either of these statements shows the critical problem with discrete data acquisition. If there is any energy above one-half of the sampling rate, often called the *Nyquist frequency* = $f_A = R_S/2$, Shannon's theorem is violated. In the real world, this will always be the case because there is always energy at frequencies higher than f_A. The violation results in errors called *aliasing*.

The effect of aliasing errors can be viewed in both the time and the frequency domains. Figure 13.5 shows the effect of sampling a 900-Hz sine wave at 1000 samples/sec (violating Shannon's theorem). The discretely acquired signal appears to be at 100 Hz. Figure 13.6 shows the same data viewed in the spectral domain. It

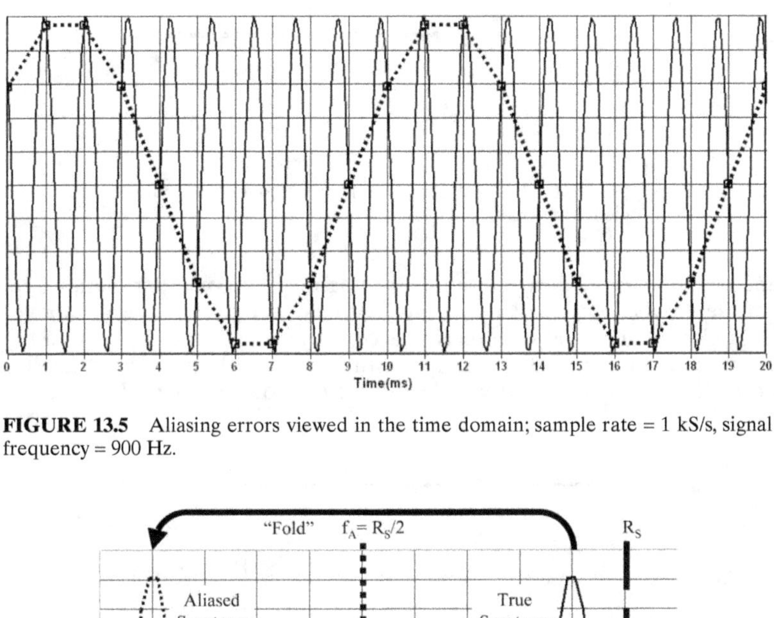

FIGURE 13.5 Aliasing errors viewed in the time domain; sample rate = 1 kS/s, signal frequency = 900 Hz.

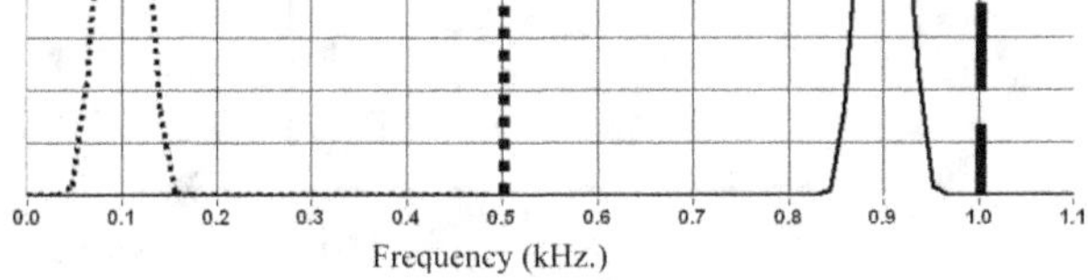

FIGURE 13.6 Aliasing errors viewed in the spectral domain; sample rate = 1 kS/S, signal frequency = 900 Hz.

shows that energy that is in violation of Shannon's theorem appears to "fold" around the Nyquist frequency (f_A) and superimposes itself on top of the energy that is at the folded frequency.

The effect of energy above the sample rate is shown in Fig. 13.7. Folding around the Nyquist frequency (f_A) produces error energy that appears at negative frequency. This, in turn, folds around zero frequency into the desired frequency range. Energy at higher frequencies will alternately fold around the Nyquist frequency and zero until it superimposes itself on the data in the frequency range of interest.

An obvious requirement for a digital data acquisition system is to reduce these errors to an acceptable level. To accomplish this, we must reduce the energy above the Nyquist frequency with a low-pass filter applied before digitization. The filtering strategy required depends on the accuracy and robustness required of the data and the type and amount of distortion that is deemed acceptable.

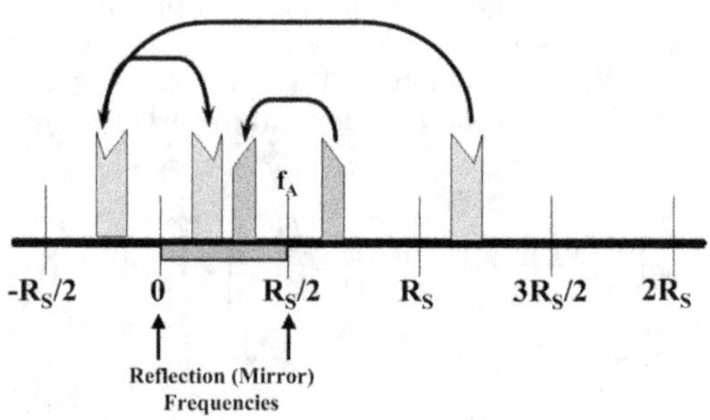

FIGURE 13.7 Aliasing from higher frequencies (multiple folding).

The first step in the design of a measurement system or experiment is the selection of the *desired bandwidth* (B_D). This frequency represents a "wall" above which the data either cannot be seen or is unacceptably aliased (corrupted). In the ideal world, we could apply a low-pass filter with frequency-domain characteristics such as those shown in Fig. 13.8. If we place the cutoff of this "barn-door" filter at the

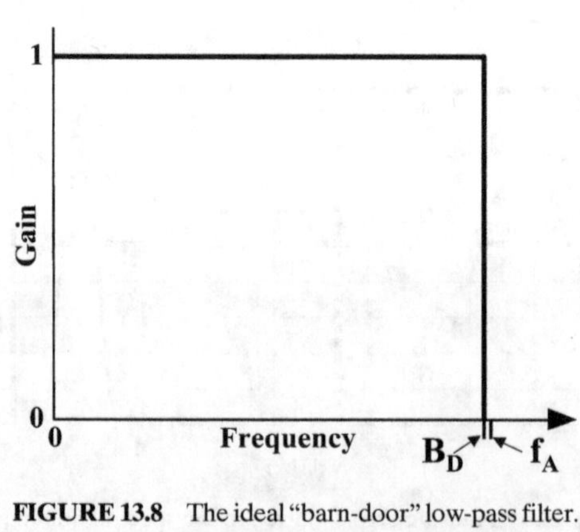

FIGURE 13.8 The ideal "barn-door" low-pass filter.

desired bandwidth (B_D), then Shannon's theorem states that we could get a complete representation of the data by sampling the filtered signal at a rate that is slightly more than $2 \times B_D$. Unfortunately, we cannot build a real filter with these characteristics, and compromises must be made. For example, Fig. 13.9 shows the attenuation (as a function of frequency) of several commercially available analog filters that might be used in systems for vibration and shock testing.

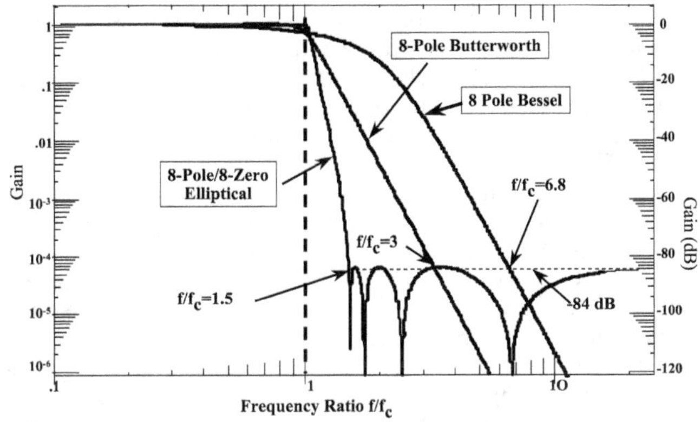

FIGURE 13.9 "Strong" analog filter characteristics.

System designers use *aliasing diagrams* to evaluate the aliasing-error rejection of a candidate filtering/sampling strategy. To create an aliasing diagram, the aliasing attenuation is calculated by folding the filter shape around the Nyquist frequency ($f_A = R_S/2$). Figure 13.10 shows the diagram of a strategy that might be used to provide good performance for a desired bandwidth (B_D) of 10 kHz: an eight-pole But-

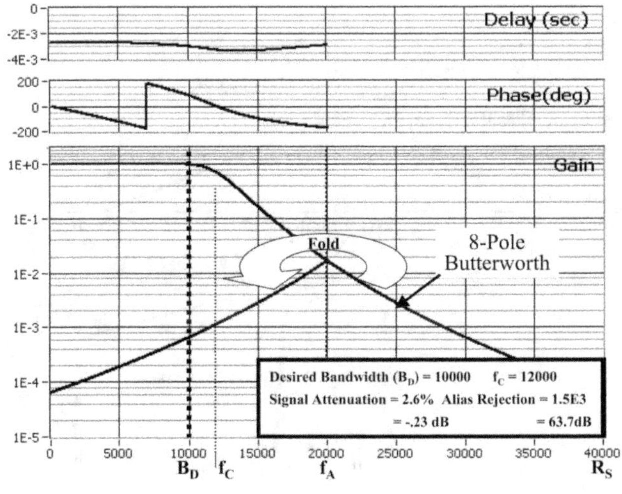

FIGURE 13.10 Aliasing diagram for an eight-pole Butterworth filter.

terworth filter with a *cutoff frequency* (F_C) of 12 kHz and a 40 kS/s sample rate. This combination provides

- Signal attenuation (amplitude distortion) of less than 3 percent in the passband (the frequency range from zero frequency to B_D)
- Aliasing rejection of 1000/1 (60 dB) or more for all frequencies below the desired bandwidth (B_D) of 10 kHz, a result that is normally considered adequate

This *strong analog filter* aliasing-protection strategy was used during the 1970s and 1980s for most shock and vibration systems.[12] These filters provided good aliasing protection (and, hence, relatively low sample rates for a given sample rate and acceptable error) and were followed by multiple-pass or successive-approximation analog-to-digital (A/D) converters. The disadvantage of this approach is that these complex analog filters are physically large, expensive, and inherently limited in cutoff capability and phase matching.

The 1990s brought improvements in sampling technology that enabled the development of a hybrid A/D aliasing-protection/digital-conversion strategy called *oversampling.* The concept, shown in Fig. 13.11, is that the sampling is performed at a multiple (called the *oversample ratio*) of what would be used with conventional systems. This in turn raises the Nyquist frequency and reduces the sharpness (hence, the complexity and cost) of the analog filtering system required to provide adequate alias protection. Once sampled, a digital filter and decimation process provides the final filter shape and sample rate. The objective of the oversampling strategy is to minimize the analog part of the operation and do most of the operation with digital calculations. Figure 13.12 shows the aliasing diagram of an oversampling system that has a 10-kHz bandwidth and uses an oversample ratio of 10. The combination of a three-pole Butterworth filter at 20 kHz and a base sample rate of 25 kS/s provides the same, or better, aliasing rejection (1000/1) below the desired bandwidth (B_D) than the strong analog-filter strategy previously discussed. The oversampled time history is low-pass-filtered with a very sharp digital filter with a cutoff frequency of

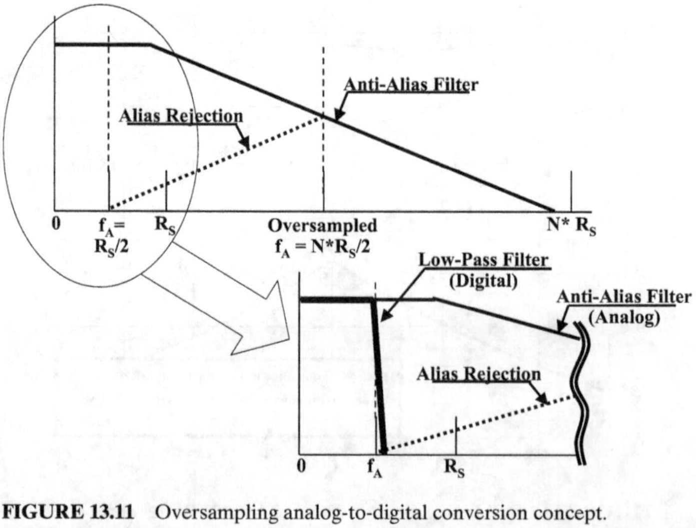

FIGURE 13.11 Oversampling analog-to-digital conversion concept.

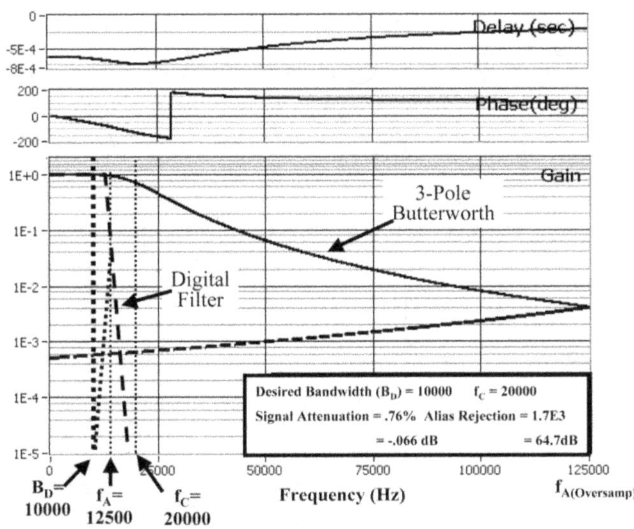

FIGURE 13.12 Oversampled ($N = 10$) aliasing diagram.

slightly more than 10 kHz. It must be sharp enough to attenuate the signal at 15 kHz so that, when the time history is decimated by 10 to produce the output sample rate (25 kS/s), the resulting aliasing error below 10 kHz (B_D) is acceptably small. The basic concept is that a less complex analog filter is used for alias protection and the sharper, more repeatable digital filter does most of the work.

The advantages of the oversampling strategy are fully realized with a high oversample ratio. For example, if an oversample ratio of 128 is used, then a simple, two-pole analog filter at 20 kHz and a base sample rate of 25kS/s will provide more than 88 dB of alias protection in the 10-kHz frequency range of interest. However, this requires a digitization rate of over 3.2 million samples/sec (128 $*$ 25 kS/s). To provide 100-kHz bandwidth, over 32 million samples/sec are required.

The *sigma-delta* ($\Sigma\Delta$) A/D converter was developed to provide these high sampling rates. The concept[13–16] employs a high-speed, low-resolution converter loop followed by a digital low-pass filter/decimator to perform the digitization process. In most systems, a variation on the theme of the 1-bit converter, called a *delta-sigma modulator,* is used (Fig. 13.13). The filtering and decimation process is normally performed by a *finite impulse response* (FIR) filter.[17] This digital filter has several char-

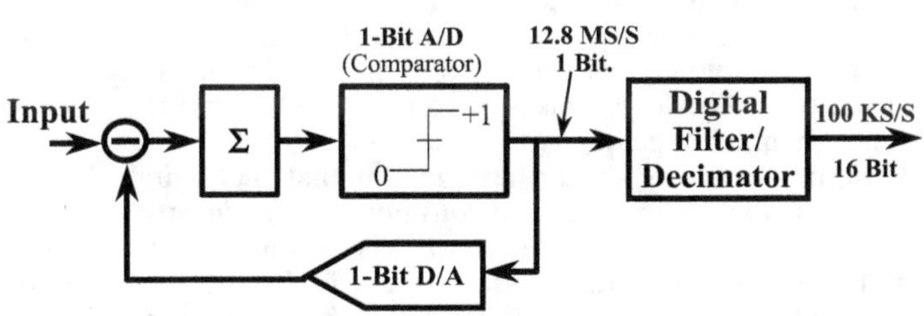

FIGURE 13.13 Sigma-delta oversampling converter.

acteristics that make it ideal for shock and vibration applications. The FIR filter's critical contribution to the shock and vibration measurement problem is shown by the near-perfect gain characteristic shown in Fig. 13.14. Features include

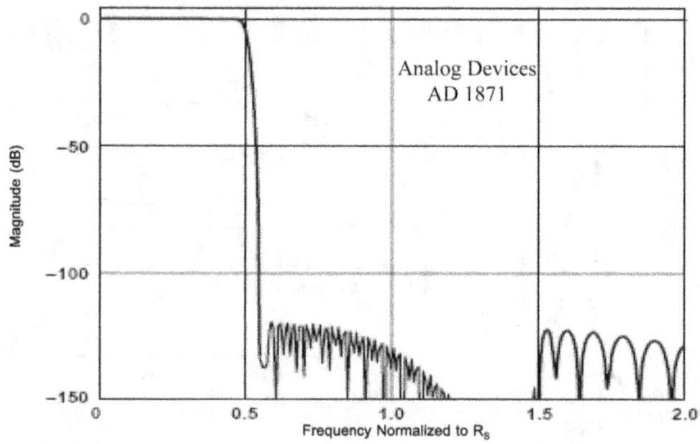

FIGURE 13.14 Typical sigma-delta (ΣΔ) filter shape. *(Copyright Analog Devices, Inc. Used with permission.)*

- Essentially flat (±0.01 db, ~0.1) percent response from zero frequency to 0.453 R_S.
- More than 120 dB of rejection for all frequencies above 0.546 R_S, well below the digitizing noise floor for a 19-bit converter. Thus, frequencies below 0.453 R_S have aliasing errors attenuated by 120 dB or more.

This nearly perfect performance allows alias-free digital data acquisition of signals with frequencies of up to about 45 percent of the sample rate. In other words, from the standpoint of alias protection, the sample rate need be only 2.2 times the maximum desired bandwidth (B_D). This is adequate when spectral analysis is the only requirement. However, for time-domain analysis where resampling is to be performed, a minimum sample rate of three times the desired bandwidth range (three points/cycle) is recommended to provide a conservative margin.

The term *alias free* has been coined to describe systems where aliasing errors are reduced to levels below the noise floor of the system. Strictly interpreted, this means that alias rejection should be greater than the dynamic range of an ideal digitization process indicated by Eq. (13.2) for all frequencies between zero and the Nyquist frequency. However, even for the best of systems, the alias-free frequency range is ~0.45 R_S. Signals in the frequency range between ~0.45 R_S and the Nyquist frequency (0.5 R_S) are unacceptably aliased, so the measurement cannot be truly alias free. Also, available systems with 20 or more bits provide only 120 dB of rejection in the stopband. However, this is not a problem because this rejection level is far below the dynamic range of any real experiment.

An additional feature of the ΣΔ approach is that the digital filter is normally implemented in a way that produces a *constant-delay* or *linear-phase* response. An emulation of the ΣΔ response to a square wave is shown overlaid with the responses from Bessel, Butterworth, and elliptical filters in Fig. 13.15. Note that the ΣΔ response rings at both ends of the transition—a basic characteristic of the digital FIR filter used. A feature of the concept is that, since the ringing energy is split

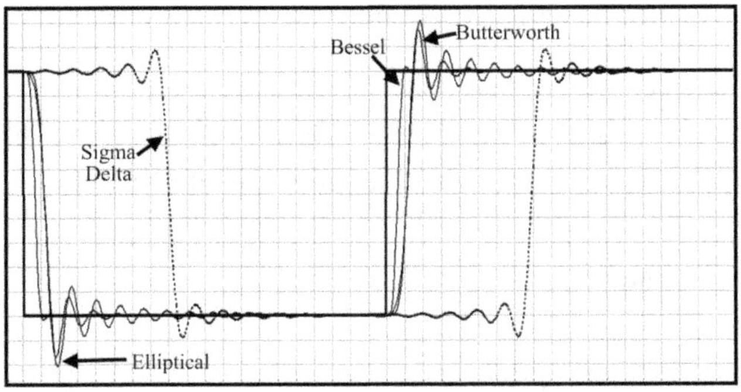

FIGURE 13.15 Square-wave responses of different alias protection strategies.

between the entrance and the exit of the transition, the overshoot is less than with the Butterworth and elliptical filters that do not have frequency-domain characteristics that are as good.

Since most modern systems used for vibration and shock measurements use an oversampling/FIR filter approach with similar strategies, the results will be consistent from machine to machine. If other strategies (e.g., strong analog filter) are used, processes are available that compensate for the differences in distortion to produce consistent results between systems.[18]

Sigma-delta converters must be used in an A/D-converter-per-channel configuration (Fig. 13.16) because their internal filtering disallows multiplexing. This, com-

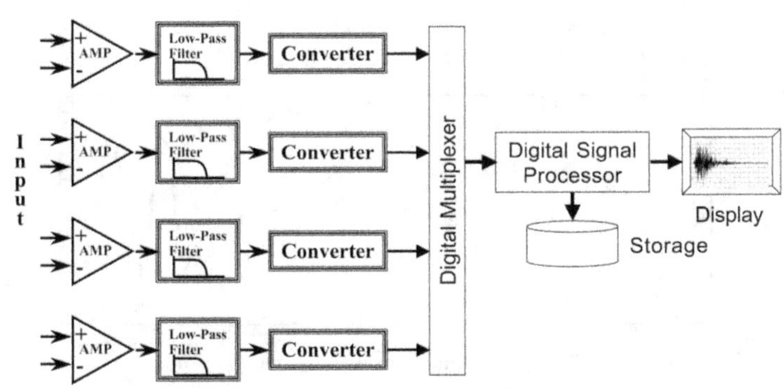

FIGURE 13.16 A/D-per-channel system architecture.

bined with the use of simple, repeatable analog filters, produces multiple-channel data that is almost perfectly simultaneous. The end result is that a properly implemented oversampling strategy provides an almost ideal data acquisition system where

- Aliasing errors are effectively eliminated within the bandwidth of the acquisition.
- The bandwidth is ~90 percent of the Nyquist frequency.

- Amplitude reproduction from zero frequency to the system bandwidth (B_D) is effectively constant.
- Multiple-channel acquisition is essentially simultaneous.
- The systems provide constant delay/linear phase over the system bandwidth.

Sigma-delta systems have one significant limitation for some applications. Because of the delay caused by the FIR filter (approximately ½ of the oversample ratio in samples), the approach is usually not useful in rapid-response applications where timing is critical.

THE DATA SPARSITY PROBLEM

A critical feature of Shannon's theorem is that when proper alias-protection methods are used, a relatively small number of points can be acquired to completely define the signal. If a ΣΔ system is used, we have seen that only 2.2 samples/cycle of the highest desired frequency component (B_D) are required. Figure 13.17 shows a time history where the sample rate is 2.5 times the maximum significant signal frequency (easily satisfying Shannon's theorem), but the signal is obviously not well represented by the raw digital data points. In particular, if the peaks of the data are of interest, the raw data provides a very poor representation.

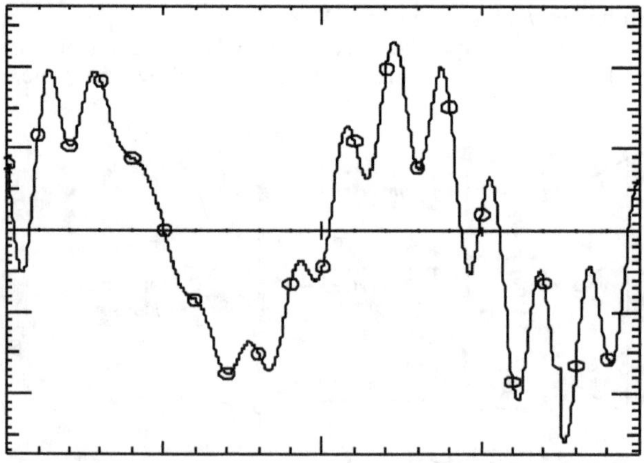

FIGURE 13.17 Sparse data that satisfies Shannon's theorem.

The reconstruction process is called *upsampling,* and the only question is what trick to use. The Whittaker-Shannon interpolation formula[19] is the classical approach, but most applications use one of two strategies that take advantage of modern signal-analysis capabilities:

1. The *time-domain method*[20] is to add zeroes between the acquired points and then apply a very sharp finite impulse response filter just below the Nyquist frequency. This approach is best for continuous signals and is the technique used in commercial audio equipment to provide a higher sample rate (hence, smoother waveform) before sending the signal to the speakers.

2. A *frequency-domain method*[21] that converts a windowed time history to the spectral domain by Fourier transform, adds zeroes to the end of the spectrum, inverse-transforms the extended spectrum to the time domain, and then corrects for the applied window. This approach is best for upsampling of short slices of data. An example of this method of upsampling is shown in Fig. 13.18.

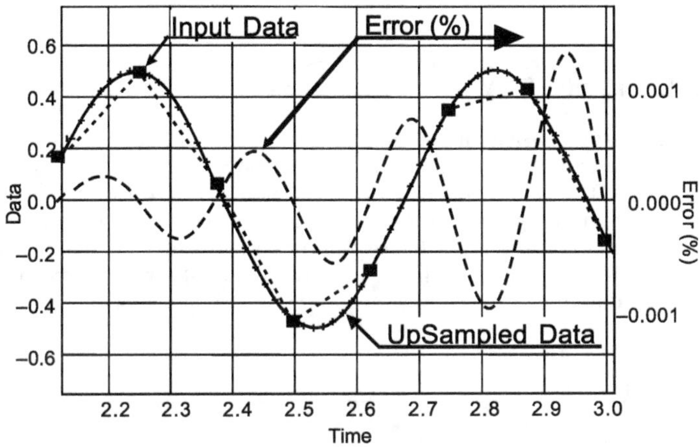

FIGURE 13.18 Upsampled data (frequency-domain method).

The relationship between the number of points per cycle (N) and the peak-determination error when the points are equidistant from the peak (Fig. 13.19) is given by

$$\text{Max error }(\%) = 100[1 - \cos(\Pi/N)] \tag{13.4}$$

$$N = \Pi/(\arccos\{1 - [\text{Max. error }(\%)/100]\}) \tag{13.5}$$

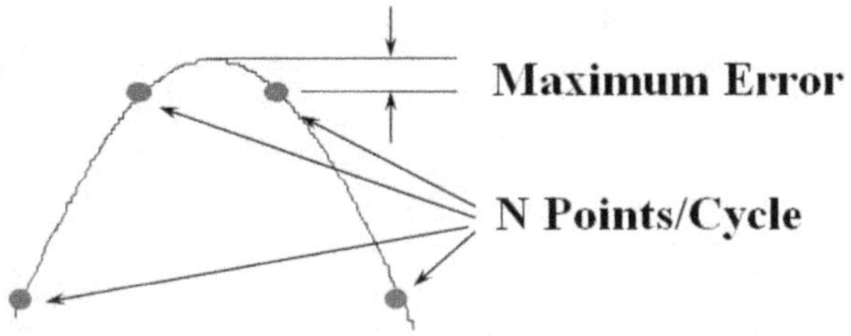

FIGURE 13.19 Peak-detection error.

To evaluate signal peaks to an accuracy of 1 percent, approximately 22 points per cycle of the highest frequency of interest (B_D) are required. For many applications this exceeds the sample-rate or storage capabilities of the available data acquisition system.

A strategy that minimizes the requirements of the data acquisition system is to

- Acquire the data with a sample rate of $3 \times B_D$.
- Upsample by a factor of 10 to produce 30 points/cycle, resulting in approximately 0.5 percent accuracy in peak determination for the highest frequency of interest.

Higher upsample factors will produce smaller errors at the expense of more computation.

OUT-OF-BAND ENERGY

Out-of-band energy consists of signals that are above the desired bandwidth (B_D) and/or above the bandwidth of the data acquisition system. In shock and vibration testing there is often significant energy at frequencies that are in this range. This requires experiment designers to plan for the unknown. To demonstrate the concept of out-of-band energy, the time history and Fourier spectrum from a near-field pyroshock test, acquired at 1 million samples/sec, are shown in gray in Fig. 13.20. Features of the data include

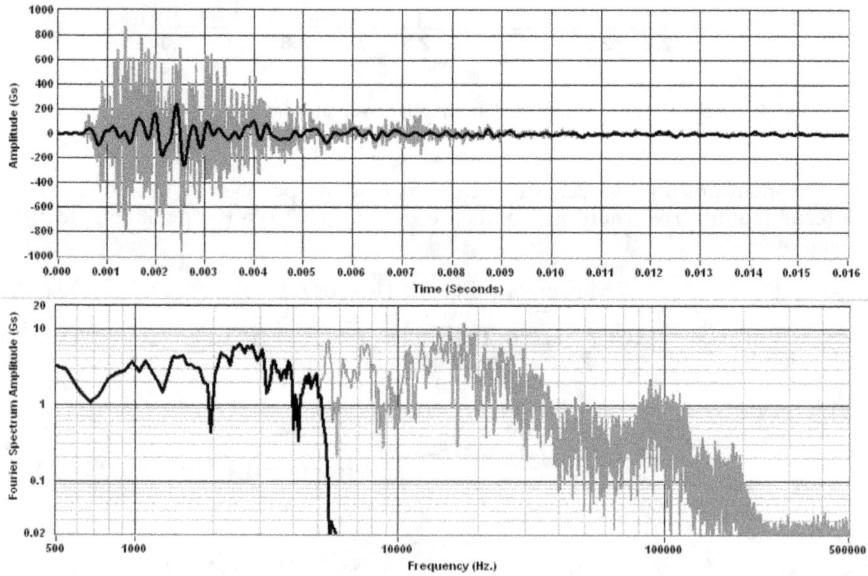

FIGURE 13.20 The effect of out-of-band energy.

- Significant energy between 0 and 200 kHz (including a transducer resonance at about 90 kHz)
- A time-history data range of +877/–954*g* (1831*g* p-p).

This environment is what the transducer and input amplifier (all components upstream of the low-pass filter) experience.

The result of selecting a data acquisition strategy that has a bandwidth of 5 kHz is shown in black. The data range for the band-limited data (low-pass-filtered) is +238/–254*g* (492*g* p-p). This is the data that users see. All evidence of the higher, pre-filtered raw data is lost in the data acquisition process.

Although users see only about 250*g* in the filtered signal, if the transducer and input amplifier did not have a range of 1000*g* or more, they would have been overloaded. This would cause the signal to be clipped before the filter and would result in corrupt data that would be very hard to detect because the clipping would be smoothed by the filtering operation. Therefore, for tests that have significant out-of-band energy, even more headroom is required than was considered in the example discussed earlier. In cases where the expected response is not well known and there is an expectation of significant out-of-band excitation, headroom of up to 50 might be appropriate. In a system with high dynamic range, this can be accomplished through conservative scaling. Refer to the previous discussion of dynamic range. There is also evidence that if the rate of change in the voltage signal (dv/dt) is too large, amplifiers in the system may be saturated in slew, and offsets in the data will result.[22] If this is the case, even greater headroom might be required to reduce the voltage range and change rate.

CONCLUSION

Oversampling digital data acquisition systems provide almost perfect recording capabilities for most applications in the shock and vibration world. The sigma delta ($\Sigma\Delta$) form of this basic strategy is used in most of the commercially available systems that serve the shock and vibration area. These systems (when properly implemented) allow the use of relatively low data acquisition rates, reducing both the data transfer and data storage requirements of the system.

Fundamental steps required to ensure that the acquired data sets are adequate include

- Determining the desired bandwidth (B_D) for the experiment. Data with frequency components above this frequency will be unacceptably aliased and/or hidden.
- Selecting an appropriate sample rate (R_S). For $\Sigma\Delta$-based machines this can be as little as 2.2 B_D.
- Ensuring that, for time-domain data, the sample rate is sufficient to adequately define the signal. This can be done by sampling very fast or sampling relatively slowly (but above 2.2 B_D) and then using upsampling techniques to produce the required data density.

REFERENCES

1. Wilson, Jon: "Sensor Technology Handbook," Elsevier, 2005.
2. Kistler Instrument Corporation: "The Piezoelectric Effect, Theory, Design and Usage," http://www.designinfo.com/kistler/ref/tech_theory_text.htm (accessed July 21, 2008).
3. PCB Piezotronics: "Introduction to Signal Conditioning for ICP® & Charge Piezoelectric Sensors," http://www.pcb.com/techsupport/tech_signal.php (accessed July 21, 2008).

4. National Instruments/IEEE: "An Overview of IEEE 1451.4 Transducer Electronic Data Sheets (TEDS)," http://standards.ieee.org/regauth/1451/IEEE_1451d4_Templates_Tutorial_061804.doc (accessed July 21, 2008).

5. Finney, Stephen H., and Douglas R. Firth: "LDTEDS: A Method for Long Distance Communication to Smart Transducers with TEDS," http://www.pfinc.com/briefs/ldteds.pdf (accessed July 21, 2008).

6. Smith, Strether, et al.: "Developments in Large-Scale Dynamic Data Acquisition Systems for Large-Scale, Structural-Dynamic Testing Facilities," *Sound and Vibration,* April 1998, pp. 18–22. (Also published in *Proc. 68th Shock and Vibration Symposium,* October 1997.)

7. Kester, Walt: "Taking the Mystery out of the Infamous Formula, 'SNR = 6.02N + 1.76dB,' and Why You Should Care," http://www.analog.com/en/analog-to-digital-converters/ad-converters/products/tutorials/CU_tutorials_MT-001/resources/fca.html (accessed July 21, 2008).

8. Smith, Strether: "Digital Signal Processing and Data Analysis," Chap. 12, Course #197 notes, Technical Training Incorporated (TTi), Las Vegas, 2008.

9. Nyquist-Shannon sampling theorem, Wikipedia, http://en.wikipedia.org/wiki/Nyquist%E2%80%93Shannon_sampling_theorem (accessed July 21, 2008).

10. Shannon, Claude E: "Communication in the Presence of Noise," *Proc. Institute of Radio Engineers* **37**(1):10–21. Reprint at http://www.stanford.edu/class/ee104/shannonpaper.pdf. (accessed July 21, 2008).

11. Smith, Strether: "Digital Data Acquisition," Chap. 7, Course #196 notes, Technical Training Incorporated (TTi), Las Vegas, 2008.

12. Smith, Strether, Eric Olson, James Elder, and Jerry Bosco: "A 5,000,000 Sample Per Second Data Acquisition/Analysis/Storage System for a High Intensity Acoustic Environment Testing Facility," *Proc. 30th International Instrumentation Symposium,* Instrument Society of America, Denver, Colorado, May 1984.

13. Hauser, Max W.: "Principles of Oversampling A/D Conversion," *Journal of the Audio Engineering Society,* **39**(1/2):3–26 (January/February 1991).

14. Candy, J. C., and Gabor C. Temes: "Oversampling Delta-Sigma Data Converters," IEEE Press, 1992.

15. Kester, Walt: "ADC Architectures III: Sigma-Delta ADC Basics," http://www.analog.com/en/analog-to-digital-converters/ad-converters/products/tutorials/CU_tutorials_MT-022/resources/fca.html (accessed July 21, 2008).

16. Kester, Walt: "ADC Architectures IV: Sigma-Delta ADC Advanced Concepts and Applications," http://www.analog.com/en/analog-to-digital-converters/ad-converters/products/tutorials/CU_tutorials_MT-023/resources/fca.html (accessed July 21, 2008).

17. Rabiner, Lawrence R., and Bernard Gold: "Theory and Application of Digital Signal Processing," Prentice-Hall, Englewood Cliffs, N.J., 1975.

18. Smith, Strether: "Why Shock Measurements Performed at Different Facilities Don't Agree," *Proc. 66th Shock and Vibration Symposium,* Biloxi, Miss., October 1995.

19. Wikipedia: "Whittaker–Shannon Interpolation Formula," http://en.wikipedia.org/wiki/Whittaker%E2%80%93Shannon_interpolation_formula.

20. Wikipedia: "Upsampling," http://en.wikipedia.org/wiki/Upsampling.

21. Smith, Strether: "Interpolation of Sparse Time History Data," *Proc. 65th Shock and Vibration Symposium,* San Diego, Calif., October 1994.

22. Smith, Strether: "Test Data Anomalies—When Tweaking's OK," *Sensors,* December 2003.

CHAPTER 14
VIBRATION ANALYZERS AND THEIR USE

Robert B. Randall

INTRODUCTION

This chapter deals primarily with frequency analysis, but also some related analysis techniques—namely, synchronous averaging, cepstrum analysis, and demodulation techniques—are considered.

With the increase in availability of signal processing packages, virtually all of the techniques discussed, and a large number of others, can now be directly programmed by the user on a general-purpose computer (see Chap. 20), but dedicated analyzers still have a number of advantages, as follows:

- Dedicated hardware for preprocessing signals before they are actually stored in the analyzer's memory. This includes real-time zoom with decimation to a lower sampling frequency (vastly reducing the amount of data to be stored), real-time digital resampling for order analysis, and even something as trivial as real-time triggering. If the data only has to be processed after the occurrence of some event that can be used as a trigger, the latter can avoid the storage and postprocessing of vast amounts of useless data.
- Fractional octave digital filter analyzers decimate the sampling frequency of low-frequency signal components as part of their operation. If the equivalent analysis over three frequency decades were to be carried out by postprocessing of an already digitized signal, approximately one million samples would be required to obtain a single one-twelfth-octave spectrum with sufficient averaging for a random signal.
- Dedicated analyzers are more likely to provide error-free results in terms of correct scaling as rms spectra, power spectra, power spectral density, or energy spectral density, while compensating for the data windows used. They also often indicate if insufficient averaging has been used for random signals, etc.

Virtually all frequency analysis is now done digitally, using the fast Fourier transform (FFT) for constant bandwidth analysis on a linear frequency scale, and recursive digital filters for constant-percentage bandwidth (fractional octave) analysis on a logarithmic frequency scale; since the latter behave essentially in the same way as analog filters, the chapter starts with a general discussion of filters and their use for frequency analysis, and later covers FFT analysis. Although spectrum analysis can be done in other ways, such as autoregressive (AR) analysis, moving average (MA) analysis, and their combination (ARMA analysis), these methods are not yet incorporated in spectrum analyzers, and so have not been treated in this chapter.

FILTERS

An ideal bandpass filter transmits that part of the input signal within its passband and completely attenuates components at all other frequencies. Practical filters differ slightly from the ideal, as discussed below. Analog filters have now been almost entirely superseded by digital filters.

Digital Filters. Digital filters (in particular, recursive digital filters) are devices which process a continuous digitized signal and provide another signal as an output which is filtered in some way with respect to the original. The relationship between the output and input samples can be expressed as a difference equation (in general, involving previous output and input values) with properties similar to those of a differential equation which might describe an analog filter. Figure 14.1 shows a typical two-pole section used in a one-third-octave digital filter analyzer (three of these are cascaded to give six-pole filtration).

Two ways of changing the properties of a given digital filter circuit such as that shown in Fig. 14.1 are:

1. For a given sampling frequency, the characteristics can be changed by changing the coefficients of the difference equation. (In the circuit of Fig. 14.1 there are three, effectively defining the resonance frequency, damping, and scaling.)
2. For given coefficients, the filter characteristic is defined only with respect to the sampling frequency. Thus, halving the sampling frequency will halve the cutoff

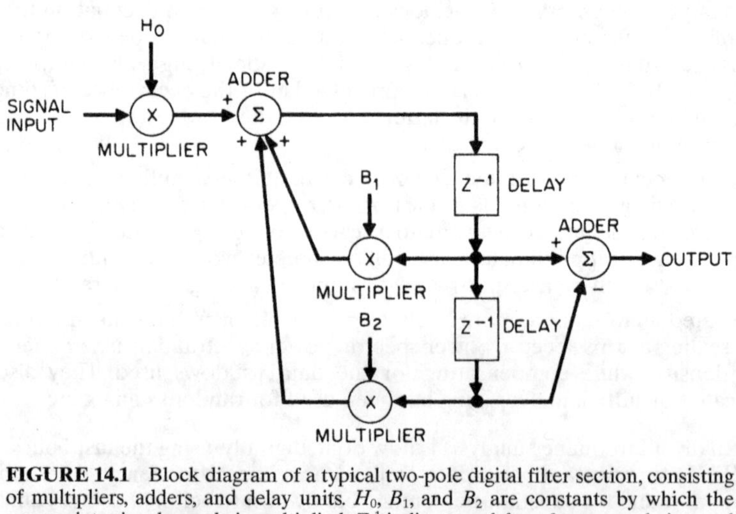

FIGURE 14.1 Block diagram of a typical two-pole digital filter section, consisting of multipliers, adders, and delay units. H_0, B_1, and B_2 are constants by which the appropriate signal sample is multiplied. Z^{-1} indicates a delay of one sample interval before the following operation.

frequencies, center frequencies, and bandwidths; consequently, the constant-percentage characteristics are maintained one octave lower in frequency. For this reason, digital filters are well adapted to constant-percentage bandwidth analysis on a logarithmic (i.e., octave-based) frequency scale.

Thus, the 3 one-third-octave characteristics within each octave are generated by changing coefficients, while the various octaves are covered by repetitively halving the sampling frequency. Every time the sampling frequency is halved, it means that only half the number of samples must be processed in a given time; the total number of samples for all octaves lower than the highest is ($\frac{1}{2} + \frac{1}{4} + \frac{1}{8} + \cdots$), which in the limit is the same as the number in the highest octave. By being able to calculate twice as fast as is necessary for the upper octave alone, it is possible to cover any number of lower octaves in real time. This is the other reason why digital filters are so well adapted to real-time constant-percentage bandwidth analysis over a wide frequency range.

Filter Properties. Figure 14.2 illustrates what is meant by the 3-dB bandwidth and the effective noise bandwidth, the first being most relevant when separating discrete frequencies, and the second when dealing with random signals. For filters having good selectivity (i.e., having steep filter flanks), there is not a great difference between the two values, and so in the following discussion no distinction is made between them.

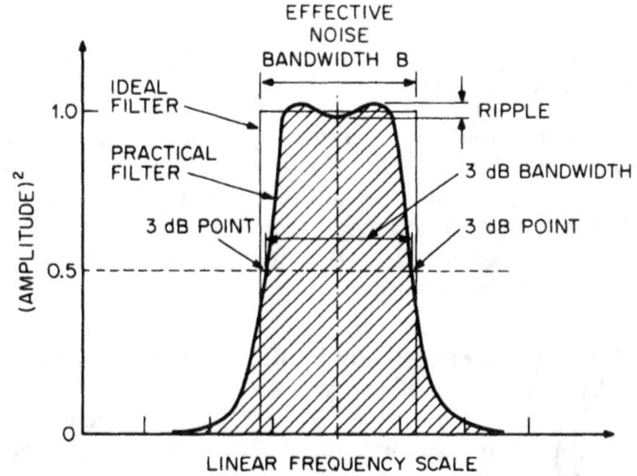

FIGURE 14.2 Bandwidth definitions for a practical filter characteristic. The *3-dB bandwidth* is the width at the 3-dB (half-power) points. The *effective noise bandwidth* is the width of an ideal filter with the same area as the (hatched) area under the practical filter characteristic on an amplitude squared (power) scale.

The response time T_R of a filter of bandwidth B is on the order of $1/B$, as illustrated in Fig. 14.3, and thus the delay introduced by the filter is also on this order. This relationship can be expressed in the form

$$BT_R \approx 1 \tag{14.1}$$

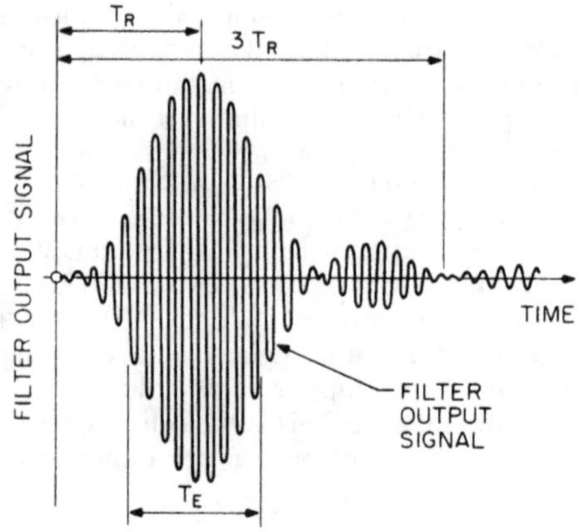

FIGURE 14.3 Typical filter impulse response. T_R = filter-response time ($\approx 1/B$); T_E = effective duration of the impulse ($\approx 1/B$); B = bandwidth.

which is most applicable to constant-bandwidth filters, or in the form

$$bn_r \approx 1 \tag{14.2}$$

where $b = B/f_0$ = relative bandwidth
$n_r = f_0 T_R$ = number of periods of frequency f_0 in time T_R
f_0 = center frequency of filter

This form is more applicable to constant-percentage bandwidth filters. Thus, the response time of a 10-Hz bandwidth filter is approximately 100 milliseconds, while the response time of a 1 percent bandwidth filter is approximately 100 periods. Figure 14.3 also illustrates that the effective length of the impulse T_E is also approximately $1/B$, while to integrate all of the energy contained in the filter impulse response it is necessary to integrate over at least $3T_R$.

Choice of Bandwidth and Frequency Scale. In general it is found that analysis time is governed by expressions of the type $BT \geq K$, where K is a constant [see, for example, Eq. (14.1)] and T is the time required for each measurement with bandwidth B. Thus, it is important to choose the maximum bandwidth which is consistent with obtaining an adequate resolution, because not only is the analysis time per bandwidth proportional to $1/B$ but so is the number of bandwidths required to cover a given frequency range—a squared effect.

It is difficult to give precise rules for the selection of filter bandwidth, but the following discussion provides some general guidelines: For stationary deterministic and, in particular, periodic signals containing equally spaced discrete frequency components, the aim is to separate adjacent components; this can best be done using a constant bandwidth on a linear frequency scale. The bandwidth should, for example, be chosen as one-fifth to one-third of the minimum expected spacing (e.g., the lowest shaft speed, or its half-order if this is to be expected) (see Fig. 14.4*A*). For stationary random or transient signals, the shape of the spectrum will most likely be determined by resonances in the transmission path between the source and the pickup, and the bandwidth B should be chosen so that it is about one-third of the bandwidth B_r of the narrowest resonance peak (Fig. 14.4*B*). For constant damping these tend to have a constant Q or constant-percentage bandwidth character, and thus constant-percentage bandwidth on a logarithmic frequency scale often is most appropriate. See Chap. 19 for further discussions of the desired resolution bandwidth for random data analysis.

A linear frequency scale is normally used together with a constant bandwidth, while a logarithmic frequency scale is normally used together with a constant-percentage bandwidth, as each combination gives uniform resolution along the scale. A logarithmic scale may be selected in order to cover a wide frequency range, and then a constant-percentage bandwidth is virtually obligatory. A logarithmic frequency scale may, however, occasionally be chosen in conjunction with a constant bandwidth (though over a limited frequency range) in order to demonstrate a relationship which is linear on log-log scales (e.g., conversions between acceleration, velocity, and displacement).

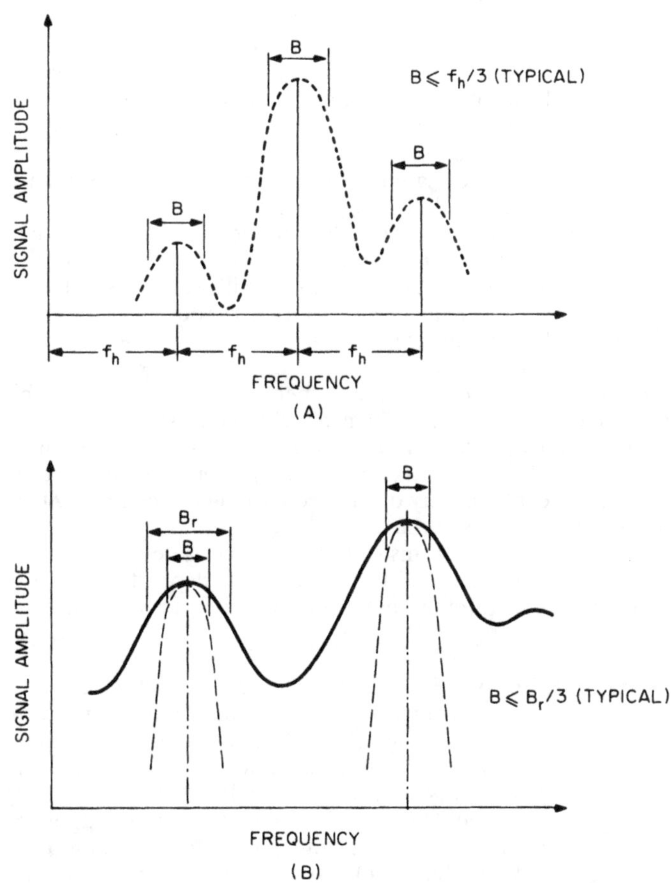

FIGURE 14.4 Choice of filter bandwidth B for different types of signals. (*A*) Discrete frequency signals—harmonic spacing f_h. (*B*) Stationary random and impulsive signals.

Choice of Amplitude Scale. Externally measured vibrations, on a machine casing for example, are almost always the result of internal forces acting on a structure whose frequency response function modifies the result. Because the structural response functions vary over a very wide dynamic range, it is almost always an advantage to depict the vibration spectra on a logarithmic amplitude axis. This applies particularly when the vibration measurements are used as an indicator of machine condition (and thus, internal forces and stresses) since the largest vibration components by no means necessarily represent the largest stresses. Even where the vibration is of direct interest itself, in vibration measurements on humans, the amplitude axis should be logarithmic because this is the way the body perceives the vibration level.

It is a matter of personal choice (though sometimes dictated by standards) whether the logarithmic axes are scaled directly in linear units or in logarithmic units expressed in decibels (dB) relative to a reference value. Another aspect to be considered is dynamic range. The signal from an accelerometer (plus preamplifier) can very easily have a valid dynamic range of 120 dB (and more than 60 dB over three frequency decades when integrated to velocity). The only way to utilize this wide range of information is on a logarithmic amplitude axis. Figure 14.5 illustrates both these considerations; it shows spectra measured at two different points on the same gearbox (and representing the same internal condition) on both logarithmic and linear amplitude axes. The logarithmic representations of the two spectra are quite similar, while the linear representations are not only different but hide a number of components which could be important.

An exception where a linear amplitude scale usually is preferable to a logarithmic scale is in the analysis of relative displacement signals, measured using proximity probes, for the following reasons: (1) The parameter being measured is directly of interest for comparison with the results of rotor dynamic and bearing hydrodynamic calculations. (2) The dynamic range achievable with relative shaft vibration measurements (as limited by mechanical and electrical runout) does not justify or necessitate depiction on a logarithmic axis.

Analysis Speed. There are two basic elements in a filter analyzer which can give rise to significant delays and thus influence the speed of analysis.

The *filter* introduces a delay on the same order as its response time T_R (see Fig. 14.3). This is most likely to dominate in the analysis of stationary deterministic signals, where the filter contains only one discrete frequency component at a time and only a short averaging time is required.

The *detector* which measures rms values introduces a delay on the same order as the averaging time T_A. The choice of averaging time depends on the type of signal being analyzed, namely, stationary deterministic (discrete frequency) or stationary random.

Choice of Averaging Time. For deterministic signals, made up entirely of discrete frequency components, the minimum averaging time required when there is only one component in the filter passband (e.g., for a one-third-octave filter containing the first, second, or third harmonic of shaft speed) comprises three periods of this frequency. However, since a result is obtained only after the filter response time ($1/B$) the averaging time should be set at least equal to this for exponential averaging, or double that value for linear averaging. When a filter contains two to five discrete frequencies (e.g., a one-third-octave filter in the range from the fourth to the twentieth harmonic of shaft speed) there will possibly be a beat frequency equal to the difference between adjacent components (i.e., the shaft speed), and an averaging time five times the beat period (reciprocal of the beat frequency) will be required to

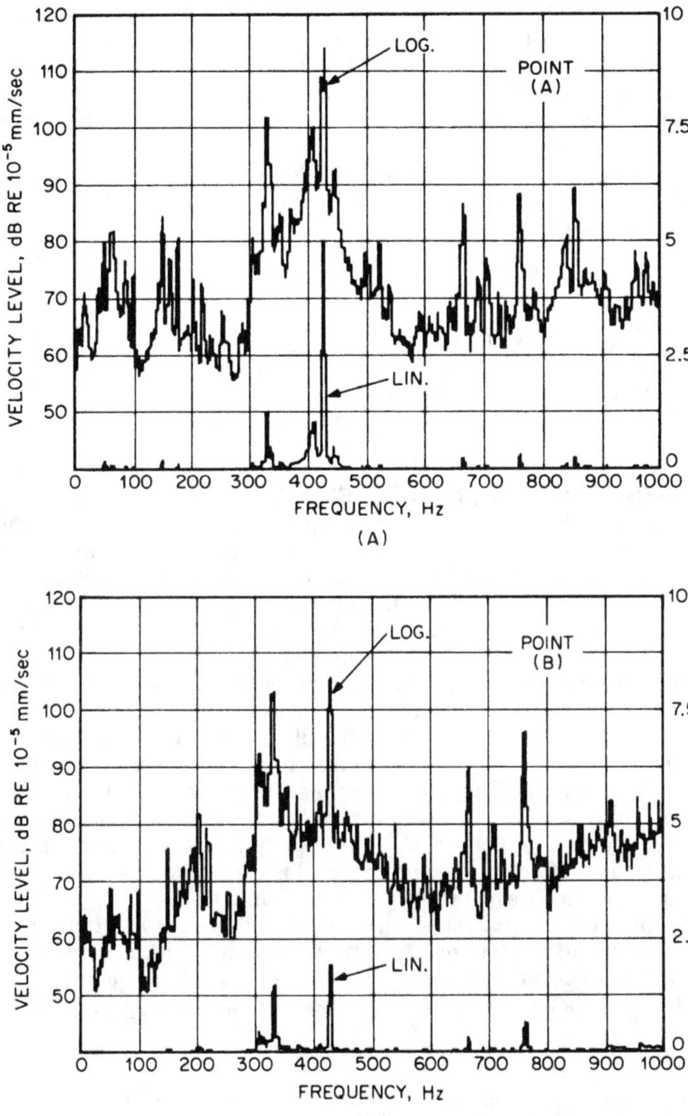

FIGURE 14.5 Comparison of rms logarithmic and rms linear amplitude scales for the depiction of vibration velocity spectra from two measurement points [(*A*) and (*B*)] on the same gearbox (thus representing the same internal condition). The logarithmic representations in terms of velocity level are similar and show all components of interest. The linear spectra in terms of velocity amplitude are quite different, and both hide many components which could be important.

smooth the result. In theory, the same applies with more components in the passband (e.g., a one-third-octave filter at higher frequencies), but the bandpassed signal will then resemble a pseudo-random signal, and can be treated as a truly random signal for analysis. If a single frequency component dominates a higher frequency band (e.g., a gearmesh frequency without sidebands), it is possible to revert to the requirement given above for a single component.

For random signals, it is necessary to limit the standard deviation of the error to an acceptable value. The *normalized random error* as a proportion of the rms value is given by the formula:

$$\varepsilon = \frac{1}{2\sqrt{BT_A}} \tag{14.3}$$

where B is the filter bandwidth, and T_A the averaging time. This error corresponds to approximately 1 dB when the BT_A product is 16. To halve the error, the averaging time must be increased by a factor of 4, etc.

Table 14.1 summarizes the above information; further detailed information is given in Ref. 1.

Scaling and Calibration for Stationary Signals. *Scaling* is the process of determining the correct units for the Y axis of a frequency analysis, while *calibration* is the process of setting and confirming the numerical values along the axis. In the most general case, spectra can be scaled in terms of mean-square or rms values at each frequency (or, strictly speaking, for each filter band). For signals dominated by discrete frequency components, with no more than one component per filter band, this yields the mean-square or rms value of each component.

For random signals, the mean square value within each frequency resolution bandwidth is the most desirable scaling because the mean square value passed by a narrow bandpass filter in proportional to the filter bandwidth. This allows the spectrum of mean square values to be normalized to a *power spectral density* $W(f)$ by dividing by the bandwidth. The results then are independent of the analysis bandwidth, provided the latter is narrower than the width of peaks in the spectrum being analyzed (e.g., following Fig. 14.4*B*). As examples, power spectral density is expressed in g^2 per hertz when the input signal is expressed in *g*s acceleration, and in volts squared per hertz when the input signal is in volts.

The concept of power spectral density is meaningless in connection with discrete frequency components (with infinitely narrow bandwidth); it can be applied only to the random parts of signals containing mixtures of discrete frequency and random components. Nevertheless, it is possible to calibrate a power spectral density scale

TABLE 14.1 Choice of Averaging Time for Filter Analysis of Stationary Signals

	Signal type			
	Deterministic—1 component in band	Deterministic—2–5 components in band	Deterministic—>5 components in band	Random*
Averaging time T_A	$T_A > 3/f_1$ +	$T_A > 5/f_{\text{beat}}$ +		$T_A > 16/B$
Exponential	$T_A > 1/B$	$T_A > 1/B$	Treat as random	Ditto
Linear	$T_A > 2/B$	$T_A > 2/B$		Ditto

Legend: f_1 = single frequency in band, f_{beat} = minimum beat frequency in band, B = filter bandwidth.
* For normalized error ≈ 1 dB.

using a discrete frequency calibration signal. For example, when analyzing a $1g$ sinusoidal signal with a 10-Hz analyzer bandwidth, the height of the discrete frequency peak may be labeled $1^2g^2/10$ Hz $= 0.1g^2$/Hz.

For constant-bandwidth analysis, the scaling thus achieved is valid for all frequencies; for constant-percentage bandwidth analysis, the bandwidth and power spectral density scaling vary with frequency. On log-log axes, it is possible to draw straight lines representing constant power spectral density, which slope upward at 10 dB per frequency decade from the calibration point.

FFT ANALYZERS

Fast Fourier transform analyzers make use of the FFT algorithm[2] to calculate the spectra of blocks of data. The FFT algorithm is an efficient way of calculating the discrete Fourier transform (DFT). As described in Chap. 19, this is a finite, discrete approximation of the Fourier integral transform. The equations given there for the DFT assume real-valued time signals [see Eq. (19.30)]. The FFT algorithm makes use of the following versions, which apply equally to real or complex time series:

$$X(m) = \Delta t \sum_{n=0}^{N-1} x(n\,\Delta t) \exp(-j2\pi m\,\Delta f\,n\,\Delta t) \tag{14.4}$$

$$x(n) = \Delta f \sum_{m=0}^{N-1} X(m\,\Delta f) \exp(j2\pi m\,\Delta f\,n\,\Delta t) \tag{14.5}$$

These equations give the spectrum values $X(m)$ at the N discrete frequencies $m\,\Delta f$ and give the time series $x(n)$ at the N discrete time points $n\,\Delta t$.

Whereas the Fourier transform equations are infinite integrals of continuous functions, the DFT equations are finite sums but otherwise have similar properties. The function being transformed is multiplied by a rotating unit vector $\exp(\pm j2\pi m\,\Delta f\,n\,\Delta t)$, which rotates (in discrete jumps for each increment of the time parameter n) at a speed proportional to the frequency parameter m. The direct calculation of each frequency component from Eq. (14.4) requires N complex multiplications and additions, and so to calculate the whole spectrum requires N^2 complex multiplications and additions.

The FFT algorithm factors the equation in such a way that the same result is achieved in roughly $N \log_2 N$ operations.[1] This represents a speedup by a factor of more than 100 for the typical case where $N = 1024 = 2^{10}$. However, the properties of the FFT result are the same as those of the DFT.

Inherent Properties of the DFT. Figure 14.6 graphically illustrates the differences between the discrete Fourier transform and the Fourier integral transform.

Because the spectrum is available only at discrete frequencies $m\,\Delta f$ (where m is an integer), the time function is implicitly periodic (as for the Fourier series). The periodic time

$$T = N\,\Delta t = 1/\Delta f \tag{14.6}$$

where $N =$ number of samples in time function and frequency spectrum
$T =$ corresponding record length of time function
$\Delta t =$ time sample spacing
$\Delta f =$ frequency line spacing $= 1/T$

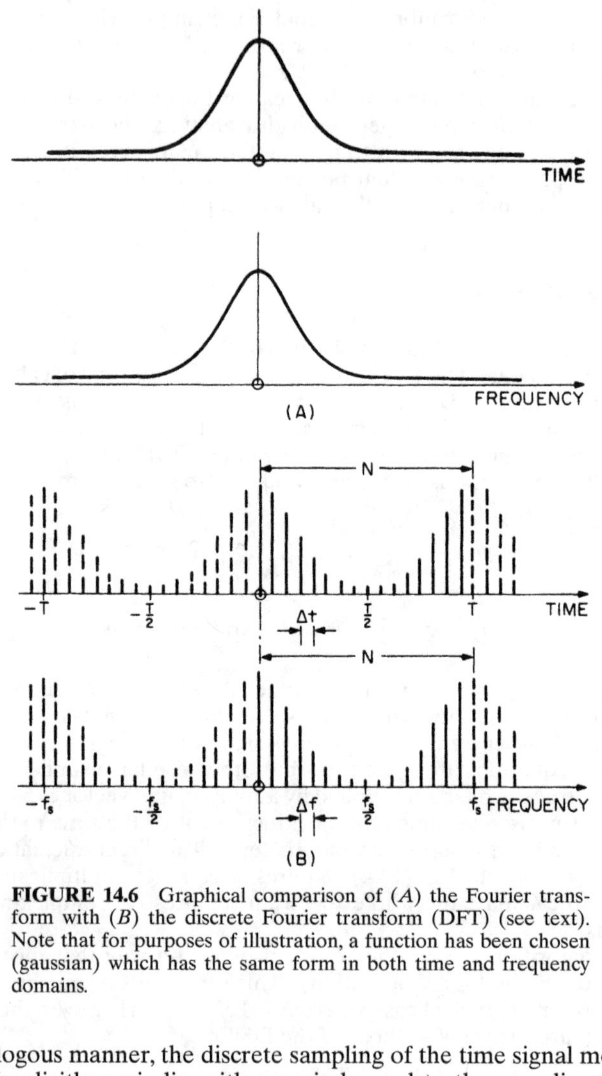

FIGURE 14.6 Graphical comparison of (*A*) the Fourier transform with (*B*) the discrete Fourier transform (DFT) (see text). Note that for purposes of illustration, a function has been chosen (gaussian) which has the same form in both time and frequency domains.

In an analogous manner, the discrete sampling of the time signal means that the spectrum is implicitly periodic, with a period equal to the sampling frequency f_s, where

$$f_s = N\,\Delta f = 1/\Delta t \tag{14.7}$$

Note from Fig. 14.6 that because of the periodicity of the spectrum, the latter half ($m = N/2$ to N) actually represents the negative frequency components ($m = -N/2$ to 0). For real-valued time samples (the usual case), the negative frequency components are determined in relation to the positive frequency components by the equation

$$X(-m) = X^*(m) \tag{14.8}$$

and the spectrum is said to be *conjugate even.*

In the usual case where the $x(n)$ are real, it is only necessary to calculate the spectrum from $m = 0$ to $N/2$, and the transform size may be halved by one of the following two procedures:

1. The N real samples are transformed as though representing $N/2$ complex values, and that result is then manipulated to give the correct result.[3]
2. A zoom analysis (discussed in a later section) is performed which is centered on the middle of the baseband range to achieve the same result.

Thus, most FFT analyzers produce a (complex) spectrum with a number of spectral lines equal to half the number of (real) time samples transformed. To avoid the effects of aliasing (see next section), not all the spectrum values calculated are valid, and it is usual to display, say, 400 lines for a 1024-point transform or 800 lines for a 2048-point transform.

Aliasing. *Aliasing* is an effect introduced by the sampling of the time signal, whereby high frequencies after sampling appear as lower ones (as with a stroboscope). The DFT algorithm of Eq. (14.4) cannot distinguish between a component which rotates, say, seven-eighths of a revolution between samples and one which rotates a negative one-eighth of a revolution. Aliasing is normally prevented by low-pass filtering the time signal before sampling to exclude all frequencies above half the sampling frequency (i.e., $-N/2 < m < N/2$). From Fig. 14.6 it will be seen that this removes the ambiguity. In order to utilize up to 80 percent of the calculated spectrum components (e.g., 400 lines from 512 calculated), it is necessary to use very steep antialiasing filters with a slope of about 120 dB/octave.

Normally, the user does not have to be concerned with aliasing because suitable antialiasing filters automatically are applied by the analyzer. One situation where it does have to be allowed for, however, is in tracking analysis (discussed in a following section) where, for example, the sampling frequency varies in synchronism with machine speed.

Leakage. *Leakage* is an effect whereby the power in a single frequency component appears to leak into adjacent bands. It is caused by the finite length of the record transformed (N samples) whenever the original signal is longer than this; the DFT implicitly assumes that the data record transformed is one period of a periodic signal, and the leakage depends on what is actually captured within the time window, or data window.

Figure 14.7 illustrates this for three different sinusoidal signals. In (A) the data window corresponds to an exact integer number of periods, and a periodic repetition of this produces an infinitely long sinusoid with only one frequency. For (B) and (C) (which have a slightly higher frequency) there is an extra half-period in the data record, which gives a discontinuity where the ends are effectively joined into a loop, and considerable leakage is apparent. The leakage would be somewhat less for intermediate frequencies. The difference between the cases of Fig. 14.7B and C lies in the phase of the signal, and other phases give an intermediate result.

When analyzing a long signal using the DFT, it can be considered to be multiplied by a (rectangular) time window of length T, and its spectrum consequently is convolved with the Fourier spectrum of the rectangular time window,[4] which thus acts like a filter characteristic. The actual filter characteristic depends on how the resulting spectrum is sampled in the frequency domain, as illustrated in Fig. 14.8.

In practice, leakage may be counteracted:

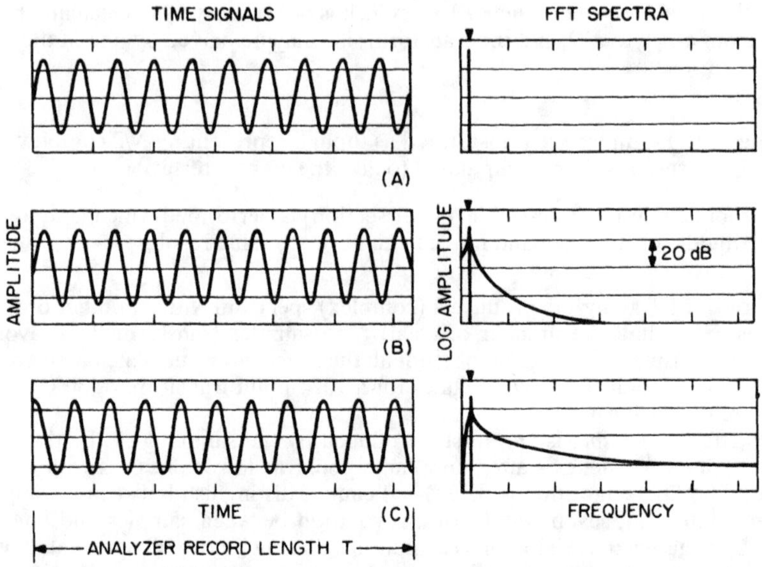

FIGURE 14.7 Time-window effects when analyzing a sinusoidal signal in an FFT analyzer using rectangular weighting. (*A*) Integer number of periods, no discontinuity. (*B*) and (*C*) Half integer number of periods but with different phase relationships, giving a different discontinuity when the ends are joined into a loop.

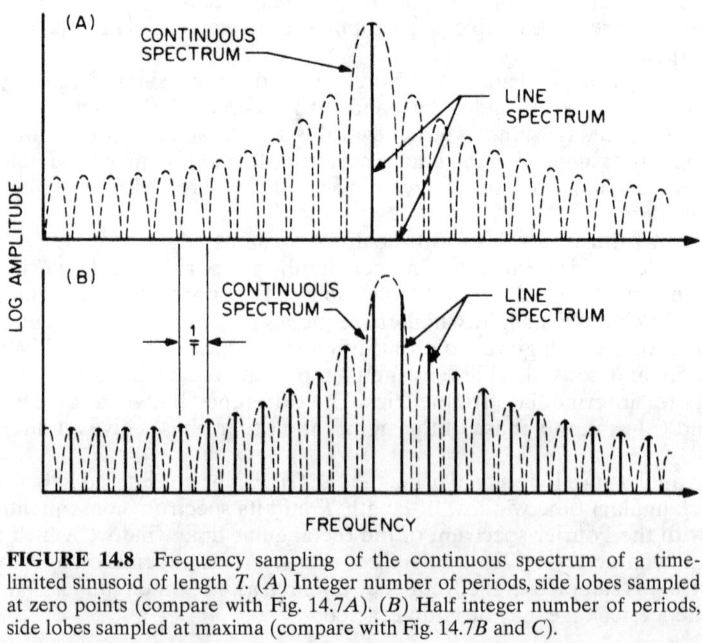

FIGURE 14.8 Frequency sampling of the continuous spectrum of a time-limited sinusoid of length *T*. (*A*) Integer number of periods, side lobes sampled at zero points (compare with Fig. 14.7*A*). (*B*) Half integer number of periods, side lobes sampled at maxima (compare with Fig. 14.7*B* and *C*).

1. By forcing the signal in the data window to correspond to an integer number of periods of all important frequency components. This can be done in tracking analysis (discussed in a later section) and in modal analysis measurements (Chap. 21), for example, where periodic excitation signals can be synchronized with the analyzer cycle.
2. (For long transient signals) By increasing the length of the time window (for example, by zooming) until the entire transient is contained within the data record.
3. By applying a special time window which has better leakage characteristics than the rectangular window already discussed.

Later sections deal with the choice of data windows for both stationary and transient signals.

Picket Fence Effect. The *picket fence effect* is a term used to describe the effects of discrete sampling of the spectrum in the frequency domain. It has two connotations:

1. It results in a nonuniform frequency weighting corresponding to a set of overlapping filter characteristics, the tops of which have the appearance of a picket fence (Fig. 14.9).
2. It is as though the spectrum is viewed through the slits in a picket fence, and thus peak values are not necessarily observed.

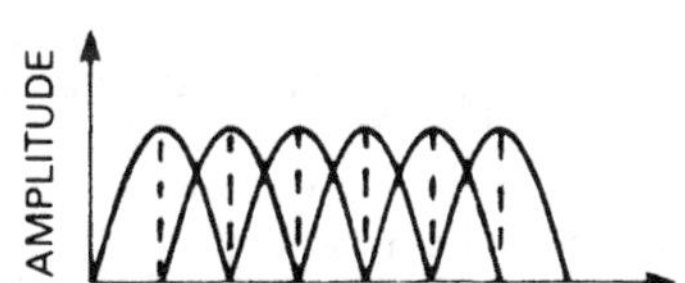

FIGURE 14.9 Illustration of the picket fence effect. Each analysis line has a filter characteristic associated with it which depends on the weighting function used. If a frequency coincides exactly with a line, it is indicated at its full level. If it falls midway between two lines, it is represented in each at a lower level corresponding to the point where the characteristics cross.

One extreme example is in fact shown in Fig. 14.8, where in (*A*) the side lobes are completely missed, while in (*B*) the side lobes are sampled at their maxima and the peak value is missed.

The picket fence effect is not a unique feature of FFT analysis; it occurs whenever discrete fixed filters are used, such as in normal one-third-octave analysis. The maximum amplitude error which can occur depends on the overlap of the adjacent filter characteristics, and this is one of the factors taken into account in the following discussion on the choice of data window.

Data Windows for Analysis of Stationary Signals. A *data window* is a weighting function by which the data record is effectively multiplied before transformation. (It is sometimes more efficient to apply it by convolution in the frequency domain.) The purpose of a data window is to minimize the effects of the discontinuity which occurs when a section of continuous signal is joined into a loop.

For stationary signals, a good choice is the *Hanning window* (one period of a sine squared function), which has a zero value and slope at each end and thus gives a gradual transition over the discontinuity. In Fig. 14.10 it is compared with a rectangular window, in both the time and frequency domains. Even though the main lobe (and thus the bandwidth) of the frequency function is wider, the side lobes fall off much more rapidly and the highest is at –32 dB, compared with –13.4 dB for the rectangular.

Other time-window functions may be chosen, usually with a trade-off between the steepness of filter characteristic on the one hand and effective bandwidth on the other. Table 14.2 compares the time windows most commonly used for stationary

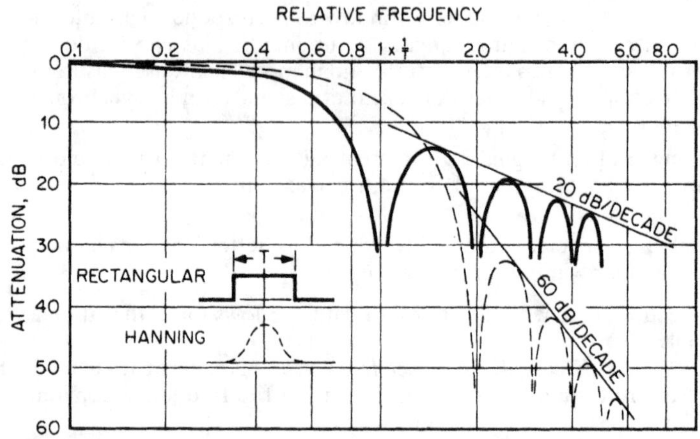

FIGURE 14.10 Comparison of rectangular and Hanning window functions of length T seconds. Full line—rectangular weighting; dotted line—Hanning weighting. The inset shows the weighting functions in the time domain.

signals, and Fig. 14.11 compares the effective filter characteristics of the most important. The most highly selective window, giving the best separation of closely spaced components of widely differing levels, is the *Kaiser-Bessel window.* On the other hand, it is usually possible to separate closely spaced components by zooming, at the expense of a slightly increased analysis time.

Another window, the *flattop window,* is designed specifically to minimize the picket fence effect so that the correct level of sinusoidal components will be indicated, independent of where their frequency falls with respect to the analysis lines. This is particularly useful with calibration signals. Nonetheless, by taking account of the distribution of samples around a spectrum peak, it is possible to compensate for picket fence effects with other windows as well. Figure 14.12, which is specifically for the Hanning window, is a nomogram giving both amplitude and frequency corrections, based on the decibel difference (ΔdB) between the two highest samples around a peak. For stable single-frequency components this allows determination of the frequency to an accuracy of an order of magnitude better than the line spacing.

TABLE 14.2 Properties of Various Data Windows

Window type	Highest side lobe, dB	Side lobe fall-off, dB/decade	Noise bandwidth*	Maximum amplitude error, dB
Rectangular	–13.4	–20	1.00	3.9
Hanning	–32	–60	1.50	1.4
Hamming	–43	–20	1.36	1.8
Kaiser-Bessel	–69	–20	1.80	1.0
Truncated Gaussian	–69	–20	1.90	0.9
Flattop	–93	0	3.70	<0.1

* Relative to line spacing.

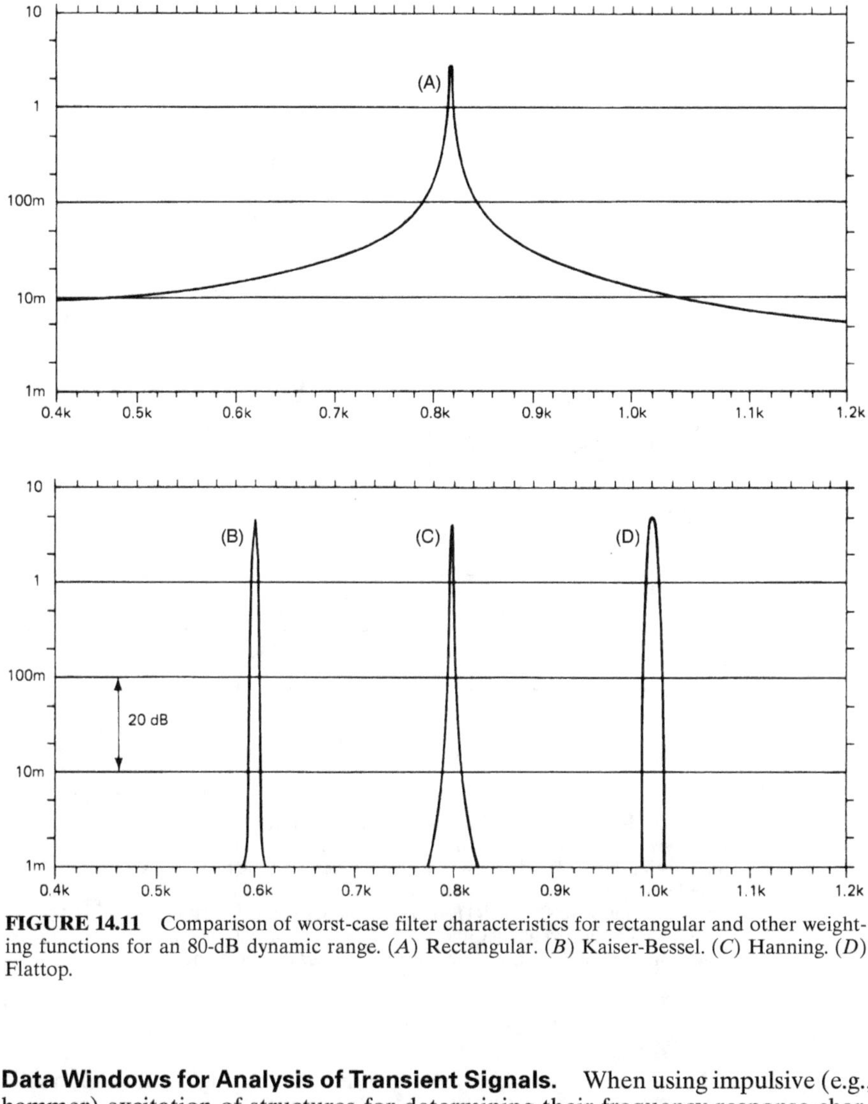

FIGURE 14.11 Comparison of worst-case filter characteristics for rectangular and other weighting functions for an 80-dB dynamic range. (*A*) Rectangular. (*B*) Kaiser-Bessel. (*C*) Hanning. (*D*) Flattop.

Data Windows for Analysis of Transient Signals. When using impulsive (e.g., hammer) excitation of structures for determining their frequency response characteristics (e.g., see Chap. 21), it is common to use the following special data windows:

1. A *short rectangular* window may be applied over the very short excitation impulse in order to exclude noise from the remaining portion of the record.
2. An *exponential* window can be applied where the response is very long (i.e., lightly damped structures) to reduce the signal to practically zero at the end of the record, and thus avoid discontinuities. The effect is the same as adding extra damping which is very precisely known and can be subtracted from the measurement results. A half-Hanning taper is often added to both the leading and trailing

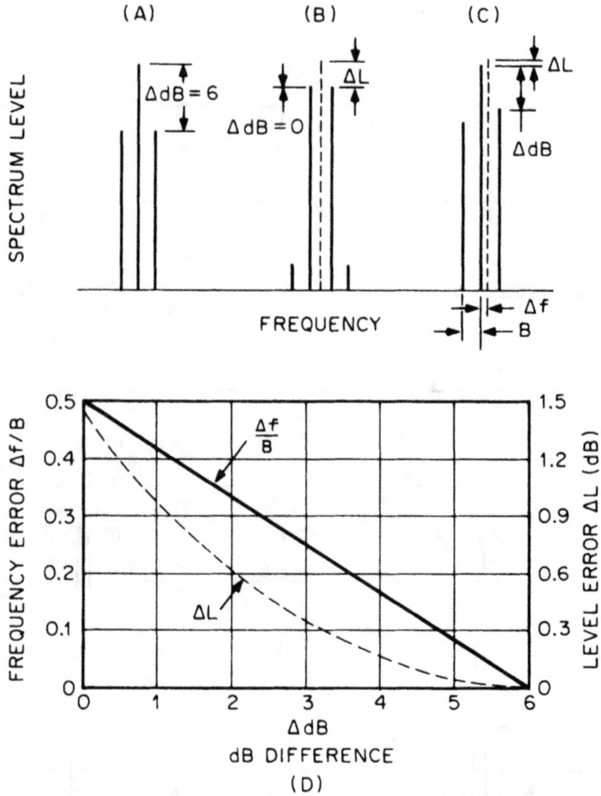

FIGURE 14.12 Picket fence corrections for Hanning weighting, where ΔL = level correction, dB; Δf = frequency correction; Hz; B = line spacing, Hz; ΔdB = difference in decibels between the two highest samples around a peak representing a discrete frequency component. Three examples are shown: (*A*) Actual frequency coincides with center line. (*B*) Actual frequency midway between two lines. (*C*) General situation. Note that the frequency correction $\Delta f/B$ is almost linear.

edges of a short rectangular window, and to the leading edge of an exponential window, to mitigate the effects of the discontinuities.

Zoom Analysis.[5] Zoom analysis is the term given to a spectrum analysis having increased resolution over a restricted part of the frequency range. The following technique is commonly incorporated in FFT analyzers.

Real-time zoom (illustrated in the block diagram of Fig. 14.13) is a zoom process in which the entire signal is first modified to shift its frequency origin to the center of the zoom range. Then it is passed through a low-pass filter (usually a digital filter in real time) which has a passband corresponding to the original zoom-band (Fig. 14.14). Because of the low-pass filtration, the signal then can be resampled at a lower sampling rate without aliasing, and the resampled signal processed by a fast Fourier

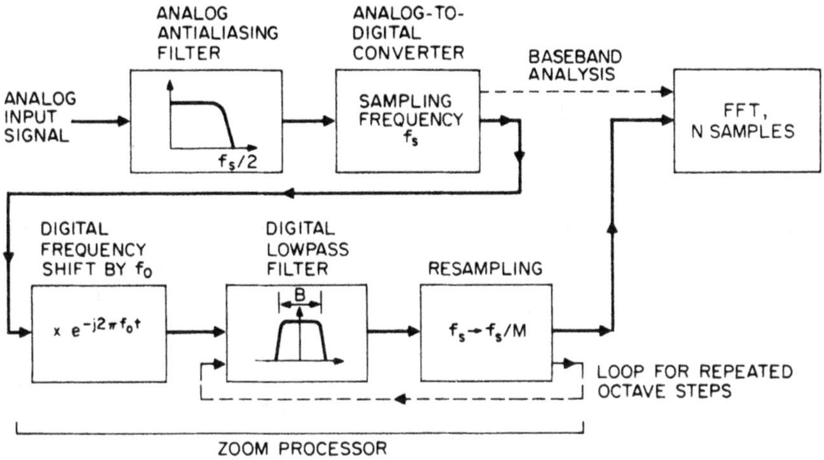

FIGURE 14.13 Block diagram for real-time zoom with bandwidth B centered on frequency f_0. M is the zoom factor and also the factor by which the sampling frequency is reduced.

transform. The original frequency shift is accomplished by multiplying the incoming signal by a unit vector (phasor) rotating at $-f_0$ (thereby subtracting f_0 from all frequencies in it), and the modified time signal is thus complex. This is one situation where the fast Fourier transform of complex data is used. Figure 14.15 gives an example of the use of zoom analysis to show that what appears in a baseband analysis to be the second harmonic of shaft speed actually is dominated by twice the line frequency at 100 Hz and reveals that what appears to be a single-frequency component in a base-band spectrum actually comprises a family of uniformly spaced components, the second highest of which is the second harmonic of the shaft speed.

Most analyzers now provide flexibility in selecting the transform size, so zoom factors of up to 64 or so can be achieved by performing a correspondingly longer transform and viewing only the desired part of the result.

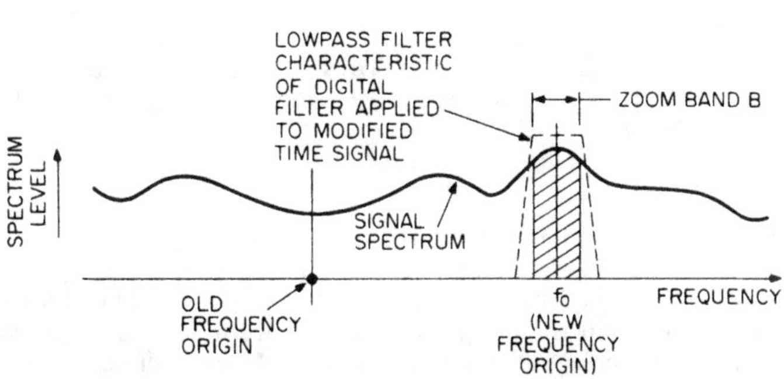

FIGURE 14.14 Principle of real-time zoom, using a low-pass filter to filter out the portion of the original signal in the zoom-band of width B. Prior to this, the frequency origin is shifted to frequency f_0 (the desired center frequency of the zoom-band) by multiplying the (digitized) time signal by $e^{-j2\pi f_0 t}$.

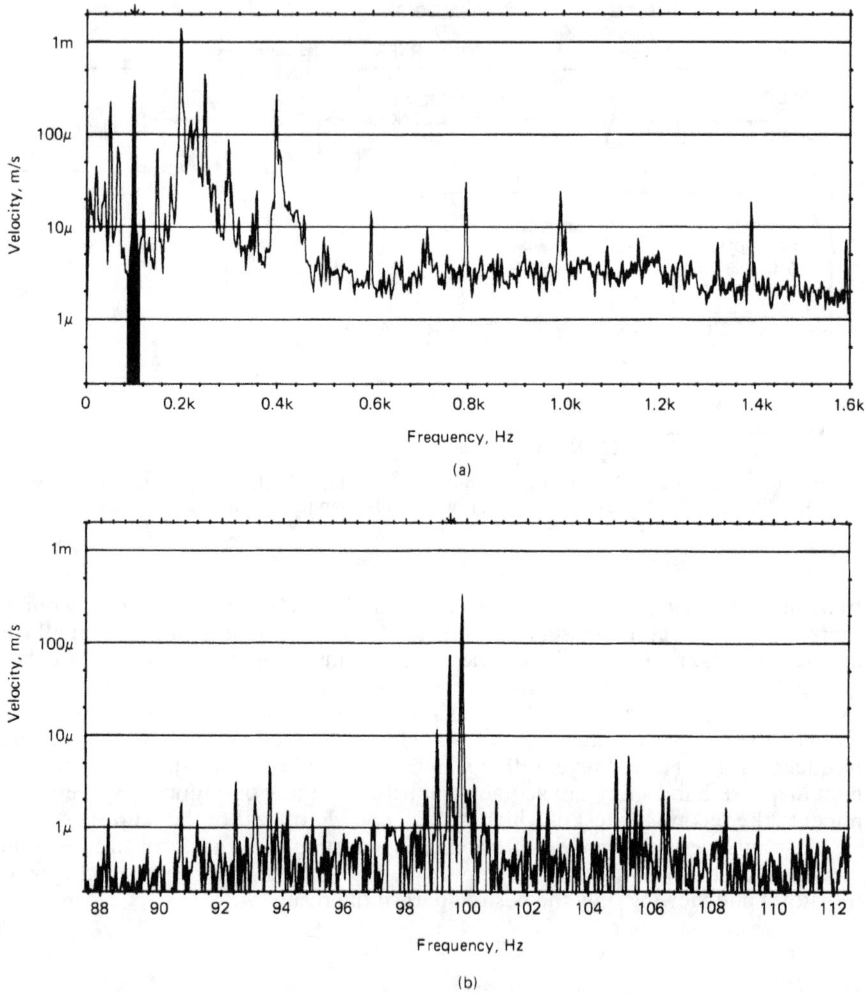

FIGURE 14.15 (*A*) Original baseband spectrum. (*B*) Shaded section of (*A*) zoomed by a factor of 64:1. Highest component at 100 Hz is twice the line frequency. Next highest component on the left is twice the shaft rotational speed.

Real-time zoom has the advantage that the zoom factor obtainable is virtually unlimited. A procedure is often employed (as illustrated in Fig. 14.13) whereby the signal samples are repeatedly circulated around a loop containing a low-pass filter which cuts off at one-half the previous maximum frequency, after which the sampling frequency is halved by dropping every second sample. Each circulation doubles the zoom factor and at the same time doubles the length of original signal required to fill the transform buffer. It is this time requirement which places a limit on the zoom factor, as well as on the stability of the signal itself. A zoom factor of 16 in a 400-line spectrum, for example, gives the equivalent of a 6400-line spectrum; a finer resolution is not required to analyze the vibration spectrum of a machine whose speed fluctuates by, say, 0.1 percent.

Real-time zoom suffers the disadvantage that the entire signal must be reprocessed to zoom in another band. This has two detrimental consequences:

1. For very narrow bandwidths (long record lengths), the analysis time is very long for each zoom analysis.
2. There is no certainty that exactly the same signal is processed each time unless it is stored digitally and reapplied to the analyzer.

On the other hand, using a single, large transform has the advantage that for zoom analysis in different bands, exactly the same data record is used. Thus, it is known, for example, that there will be an exact integer relationship between the various harmonics of a periodic signal. This can be useful, as a typical example, in separating the various harmonics of shaft speed from those of line frequency, in induction motor vibrations. Furthermore, the long analysis time is required only once (to fill the data buffer); further zoom analyses on the same record are purely a matter of viewing a different part of the result.

Thus, both types of zoom are advantageous for different purposes. A large transform is probably best for diagnostic analysis of machine vibration signals, whereas real-time zoom gives more flexibility in frequency response measurements (system frequency response should not change even where the excitation signals change). Real-time zoom also gives the possibility of very large zoom factors when they are required.

In real-time zoom, it is only the preprocessing of the signal which has to be in real time; the actual FFT analysis of the signal, once it is stored in the transform buffer, does not have to be in real time.

ANALYSIS OF STATIONARY SIGNALS USING FFT

Equation (14.6) shows that for a single fast Fourier transform, the product (*bandwidth* times *averaging time*) $BT_A = 1$, at least for rectangular weighting where B is equal to the line spacing Δf (Table 14.2). The same applies for any weighting function, the increased bandwidth being exactly compensated by a corresponding decrease in effective record length.[6]

For *stationary deterministic signals,* a single transform having a BT_A product equal to unity is theoretically adequate, although a small number of averages is sometimes performed if the signal is not completely stable. Figure 14.16 illustrates the effect of averaging for a deterministic signal and demonstrates that the sinusoidal components are unaffected; the only effect is to smooth out the (nondeterministic) noise at the base of the spectrum (Fig. 14.16*B*).

For *stationary random signals,* the normalized random error of the result of rms averaging n independent spectra is given by the equivalent of Eq. (14.3), or

$$\varepsilon = \frac{1}{2\sqrt{n}} \tag{14.9}$$

As is always the case, rms averaging means averaging the squared magnitudes (the mean-square values) of the various different spectra and taking the square root of the result.

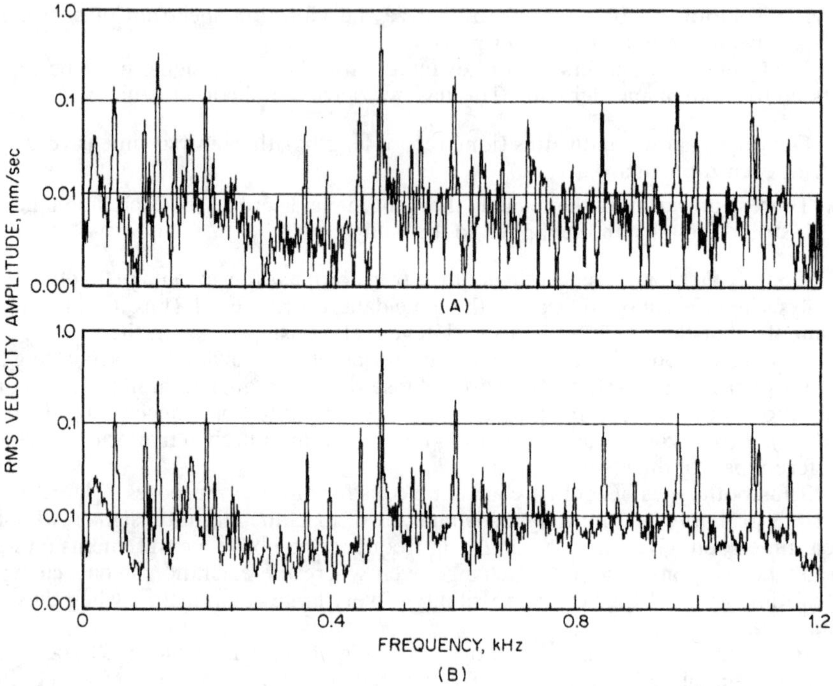

FIGURE 14.16 Effect of averaging with a stationary deterministic signal. (*A*) Instantaneous spectrum (average of 1). (*B*) The linear average of eight spectra.

Figure 14.17 illustrates (*A*) an instantaneous spectrum, (*B*) the average of eight spectra, and (*C*) the average of 128 spectra. The meaning of the standard error ε [Eq. (14.9)] is illustrated in (*B*) and (*C*). Statistically, there is a 68 percent probability that the actual error will be less than ε, a 95.5 percent probability that it will be less than 2ε, and a 99.7 percent probability that it will be less than 3ε.

For rectangular weighting, independent spectra are those from nonoverlapping time records; when other weighting functions are used, the situation is different. For example, Fig. 14.18*A* illustrates the overall (power) weighting obtained when Hanning windows are applied to contiguous records. Note that virtually half of the incoming signal is excluded from the analysis, whereas a 50 percent overlapping of consecutive records regains most of the lost information. Thus, when using window functions similar to Hanning (as recommended for stationary signals), it is almost always advantageous to average the results from 50 percent overlapping records. A method for calculating the effective number of averages obtained in this way is given in Ref. 7; for 50 percent overlapping Hanning windows the error is very small in treating them as independent records.

Real-Time Analysis. A fast Fourier transform analyzer is said to operate in real time when it is able to process all the incoming data, even though presentation of the results is delayed by an amount corresponding to the calculation time. This

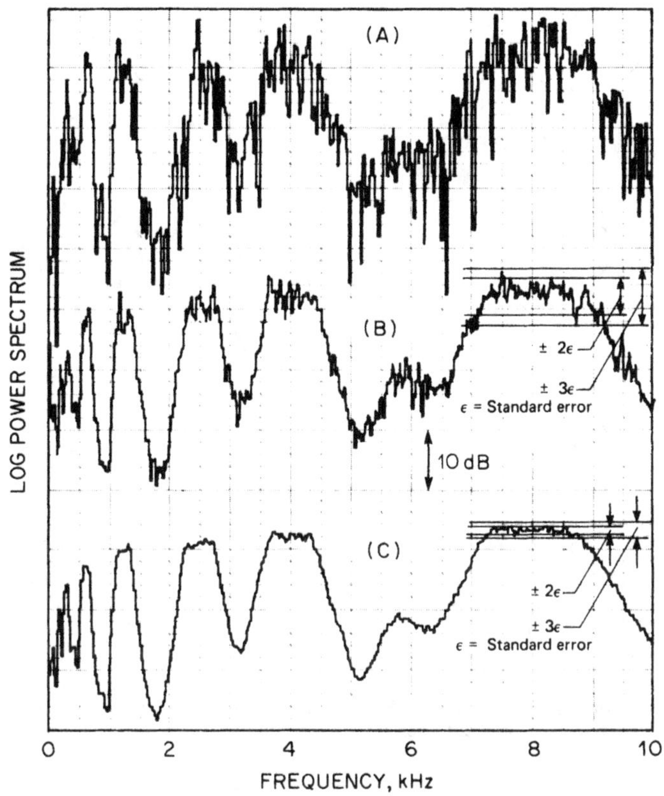

FIGURE 14.17 Effect of averaging with a stationary random signal. (*A*) Instantaneous spectrum. (*B*) Average of eight spectra. (*C*) Average of 128 spectra.

implies that the time taken to analyze a data record, T_a, is less than the time taken to collect the data transformed, T. It also implies that the analysis process should not interrupt the continuous recording of data, so that recording can continue in one part of the memory at the same time as analysis is being performed in another. T is inversely proportional to the selected frequency range, and the highest frequency range for which T_a is less than T is called the *real-time frequency.* This condition will ensure that all the incoming data are analyzed only when rectangular weighting is used. With Hanning weighting, for example, where 50 percent overlap analysis must be employed to analyze all the data, the true real-time frequency will be halved, since twice as many transforms must be performed for the same length of data record. In yet another sense, the analysis is not truly real-time unless the overall weighting function is uniform. As illustrated in Fig. 14.18, the minimum overlap of Hanning windows to achieve this is two-thirds, which reduces the true real-time frequency to one-third of the commonly understood definition given above.

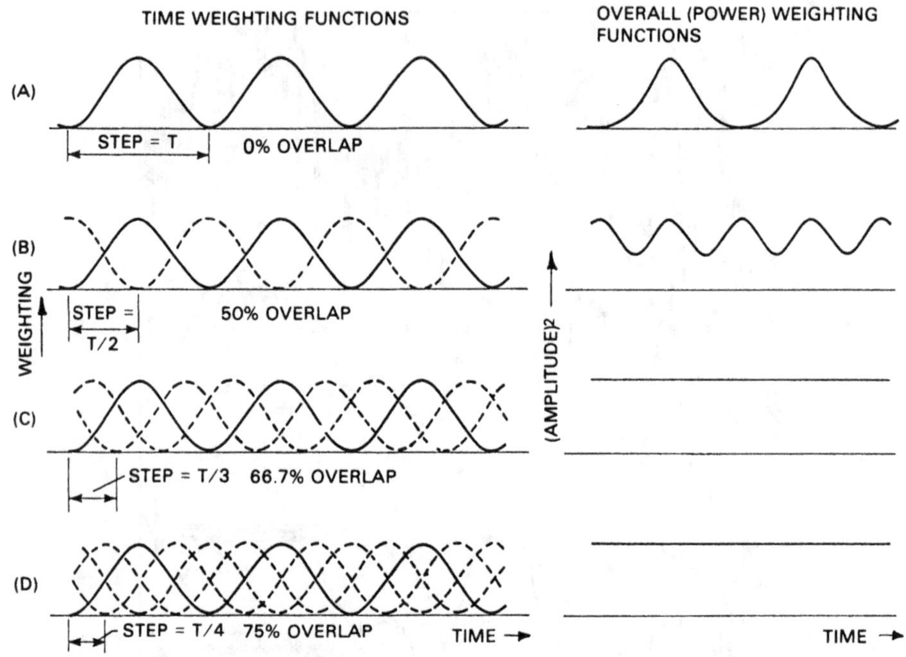

FIGURE 14.18 Overall weighting functions for spectrum averaging with overlapping Hanning windows. (*A*) Zero overlap (step length *T*). (*B*) 50 percent overlap (step length *T*/2). (*C*) 66.7 percent overlap (step length *T*/3). (*D*) 75 percent overlap (step length *T*/4). *T* is the record length for the FFT transform.

In practice, with stationary signals, there is no advantage to more than a 50 percent overlap, since (1) statistical reliability is not significantly improved and (2) all sections of the record are statistically equivalent, so that the overall weighting function is not important. It can be important for nonstationary signals, such as transients, as discussed below. For stationary signals, where any data missed are statistically no different from the data analyzed, the only advantages of real-time analysis are that (1) results with a given accuracy are obtained in the minimum possible time and (2) maximum information is extracted from a record of limited length.

FFT Analysis of Transients. Consider the use of fast Fourier transform analysis when the entire transient fits into the transform size *T* without loss of high-frequency information. Figure 14.19 shows such an example where the duration of the transient is less than the analyzer record length of 2048 samples (2K) in a frequency range which does not exclude high-frequency information in the signal. Rectangular weighting should be used in such a case, where the signal value is zero at each end, so that no discontinuity arises from making the record into a loop (an inherent property of the FFT process). Exponential weighting sometimes may be used to force the signal down to zero at the end of the record, but the frequency spectrum will then include the effects of the extra damping which this represents.

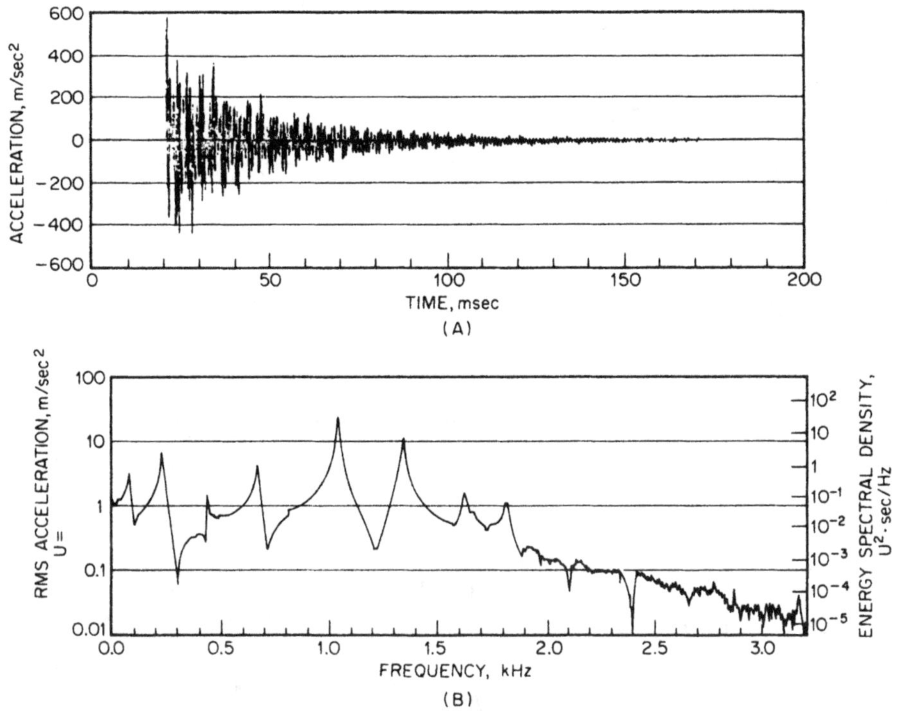

FIGURE 14.19 Example of an FFT analysis of a short transient signal. (*A*) Time signal of length 2048 samples (2K) corresponding to 250 milliseconds ($T = 250$ milliseconds). (*B*) 800-line FFT spectrum with bandwidth $B = \Delta f = 4$ Hz (rectangular weighting). Scaling on left is in rms units. Scaling on right is converted to energy spectral density (ESD) by multiplying mean-square values by T^2.

With rectangular weighting, the analysis bandwidth is equal to the line spacing $1/T$, which is always less than the effective signal bandwidth. Conversion of the results to energy spectral density, therefore, is valid in most practical situations. Some analyzers provide the results in terms of energy spectral density, but if the results are available only in terms of power or mean square value (U^2), they must be multiplied by the time T corresponding to the record length to convert them to energy and divided by the bandwidth $1/T$ to convert them to energy spectral density, expressed in engineering units squared times seconds per hertz. Altogether, this represents a multiplication by T^2.

Where a transient is longer than the normal transform size T, it can be analyzed in one of the following ways:

1. *Zoom FFT* (see *Zoom Analysis,* above, for background information). A suitable zoom factor is chosen such that the transform length ($1/\Delta f$) is greater than the duration of the transient. In the case of real-time zoom, analysis in more than one zoom-band requires that the transient be recorded in an external medium and played back for each zoom analysis. Rectangular weighting should be used (thus $B = \Delta f$) and energy spectral density (as above) is always valid using a value of T corre-

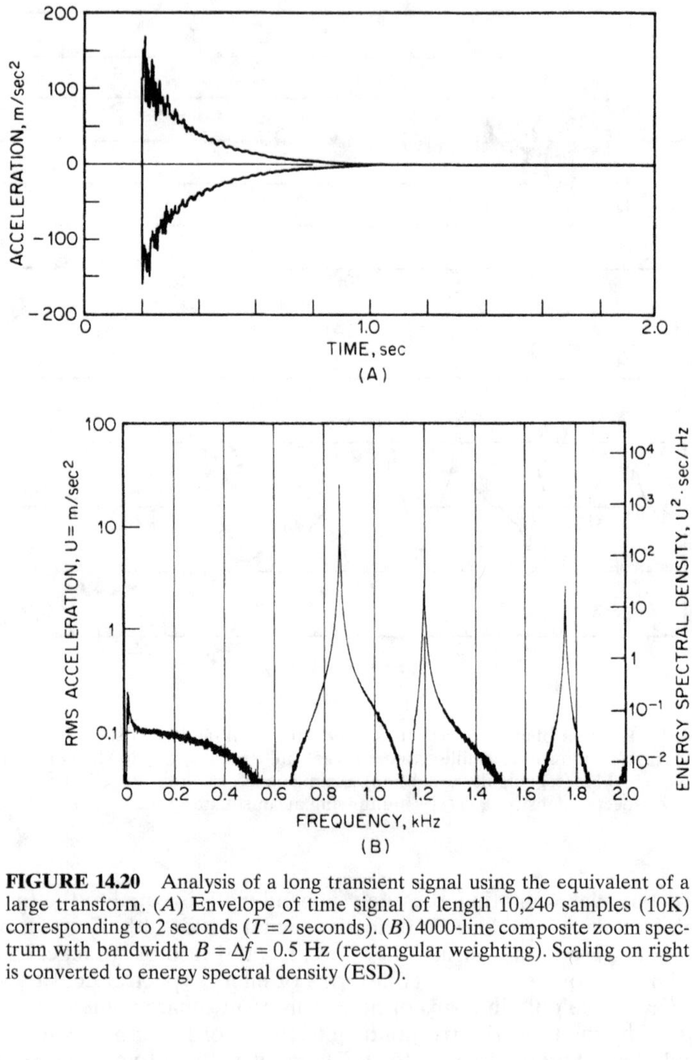

FIGURE 14.20 Analysis of a long transient signal using the equivalent of a large transform. (*A*) Envelope of time signal of length 10,240 samples (10K) corresponding to 2 seconds ($T = 2$ seconds). (*B*) 4000-line composite zoom spectrum with bandwidth $B = \Delta f = 0.5$ Hz (rectangular weighting). Scaling on right is converted to energy spectral density (ESD).

sponding to the zoom record length ($1/B$). The narrow bandwidth may give a restriction of dynamic range of the result. Figure 14.20 shows a typical energy spectrum, obtained by a process giving the same result as a single large transform.

2. *Scan averaging.* When the entire transient is stored in digital form in a long memory, it is possible to obtain its spectrum by scanning a short time window (e.g., a Hanning window) of length T over the entire record; this is done in overlapping steps, and the results are averaged. As already demonstrated for stationary signals (Fig. 14.18), this procedure yields a result with uniform weighting for step lengths $T/3$ and $T/4$. The same applies to step lengths $T/5$, $T/6$, etc., but there is a slight difference with respect to the overall weighting function for the different step lengths. Figure 14.21 illustrates the overall time weighting function for different step lengths T/n (where n is an integer greater than 2) and shows the length of the uniform section (within which

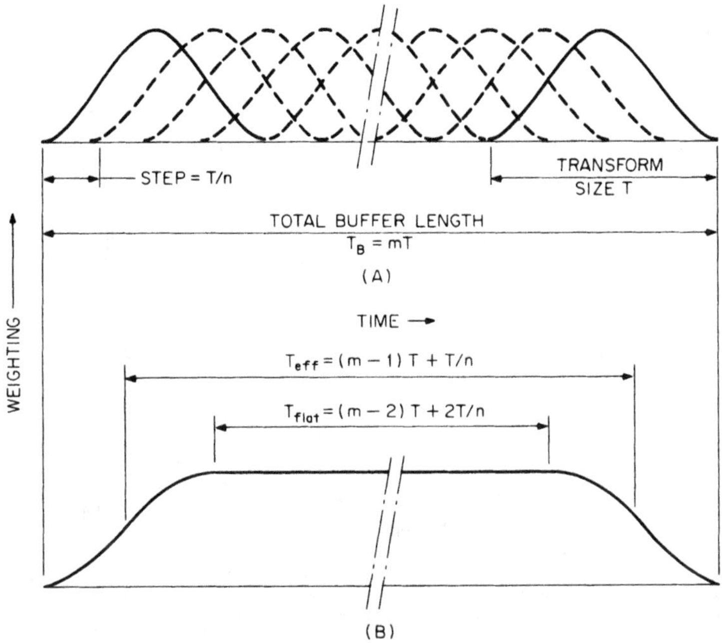

FIGURE 14.21 Overall weighting function for scan averaging of a transient. (*A*) Overlapping Hanning windows of length T with definition of parameters m and n. (*B*) Overall weighting function with indication of T_{eff} and T_{flat} in terms of T, m, and n. T_{eff} is the effective length of the time window for conversion of power to energy units. T_{flat} is the length of the section with uniform weighting within which the transient ideally should be located.

the entire transient should ideally be located) and the effective length T_{eff} by which power units should be multiplied to convert them to energy. For a conversion to energy spectral density to be valid, the width of spectrum peaks must be somewhat greater than the analysis bandwidth; this can be seen by inspection of the analysis results. For example, for the Hanning window, the bandwidth B is 1.5 times the line spacing Δf (see Table 14.2), and so spectrum peaks should have a 3-dB bandwidth of more than five lines.

Even though the broader bandwidth obtained by scan averaging may result in a loss of spectrum detail, it provides considerable improvement in the dynamic range of the result. Figure 14.22 (using scan averaging) illustrates these points for the same signal as Fig. 14.20 (using zoom). The spectrum obtained by scan averaging generally has 12 dB more dynamic range than that obtained by zoom (with factor 10), but the level of peaks does not differ by this amount; this confirms that their resolution is not sufficient to allow scaling in terms of energy spectral density.

To obtain Fig. 14.22, scan averaging with a step length of $T/4$ was used (an overlap of successive records of 75 percent). Even though a step length of $T/3$ (overlap of 66.7 percent) is theoretically more efficient, $T/4$ is usually more convenient because the number of samples in T generally is a power of 2.

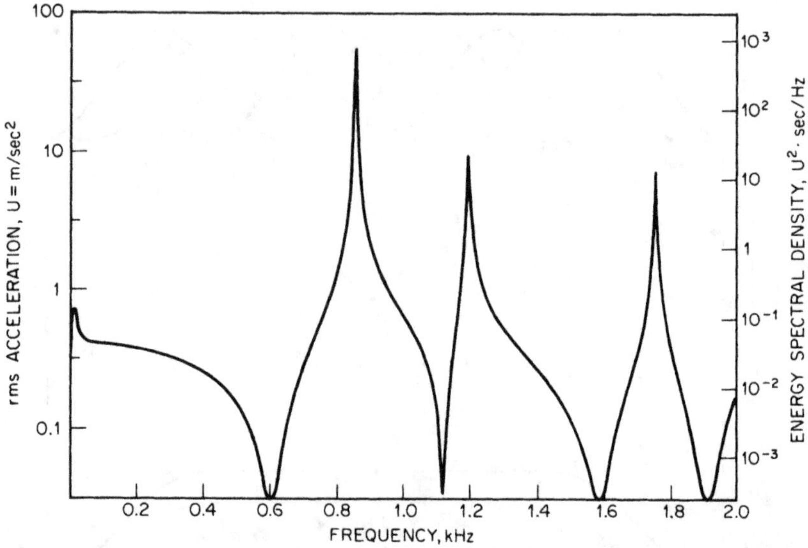

FIGURE 14.22 Analysis of a long transient by scan averaging (same signal as Fig. 14.20). The energy spectral density (ESD) scaling on the right can be compared with that in Fig. 14.20, although the peaks are not valid because of insufficient resolution.

ANALYSIS OF NONSTATIONARY SIGNALS

A typical nonstationary signal results from measurements made during a machine run-up or coast-down (here, the primary cause of the nonstationary signal is a change in shaft speed). The signal can be analyzed by dividing it up into a series of short quasi-stationary time periods (often overlapping), in each of which the speed is roughly constant. The length of the time window used to select a portion of the continuous signal may have to be chosen so as to ensure this. The simplest way to analyze a nonstationary signal of this type is to use a tracking filter tuned to a specific harmonic of shaft speed and to record the results vs. rpm of the machine. If a phase meter is inserted between the filtered signal and the tracking signal, it is possible to record phase as well as amplitude against rpm to give what is called a *Bode plot.*[8]

Using a fast Fourier transform analyzer, the behavior of several harmonics may be studied simultaneously. One way to do this, using an FFT analyzer having a long memory, is with a simple scan analysis; a short Hanning window is scanned through the record (as for a *scan average*), and successive instantaneous spectra (from each window position) are viewed on the display screen. The speed of the scan may be changed by varying the step length; this is one situation (in contrast to scan averaging) where very short step lengths may be of advantage, for example, in slowing down the passage through a resonance.

A highly effective method of representing such a scan analysis is by a "waterfall," or "cascade," plot as shown in Fig. 14.23 (which represents a typical machine run-up). As indicated, the third dimension of such a three-dimensional plot can be either time or rpm; for a simple scan analysis it usually is time, but if the spectra are spaced at equal intervals of rpm, a number of advantages result. Harmonically related components (whose bases follow radial lines) then can be separated easily

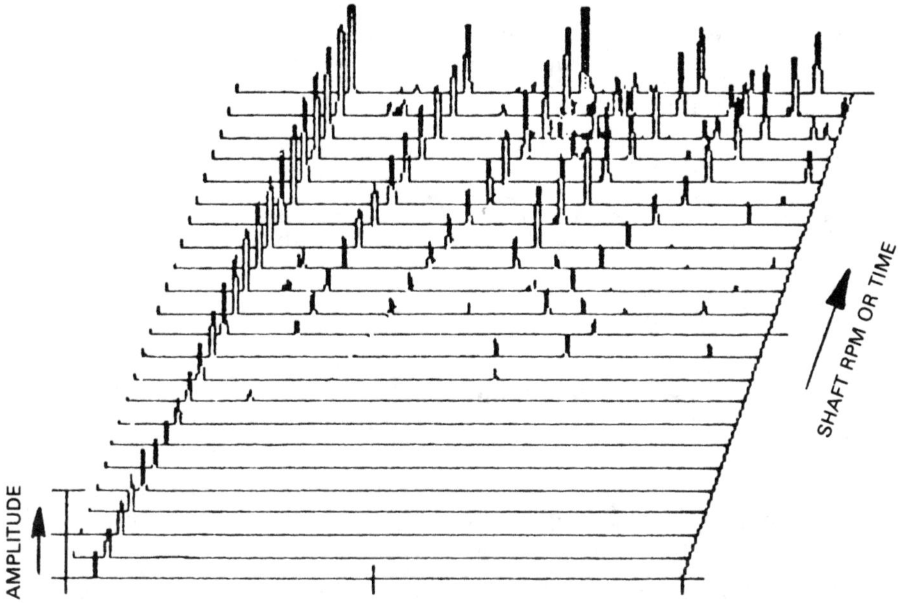

FIGURE 14.23 Three-dimensional spectral map or waterfall plot, showing how spectra change with shaft rpm or time.

from constant-frequency components (e.g., related to line frequency or resonances) whose bases follow lines parallel with the rpm axis. Such a cascade plot, with rpm as the third axis, is sometimes referred to as a *Campbell diagram,* although strictly speaking a Campbell diagram has a vertical frequency axis, a horizontal rpm axis, and a signal amplitude represented as the diameter of a circle (or square) centered on the appropriate point in the diagram.

Ideally, each of the spectra in a cascade plot such as Fig. 14.23 should be obtained with constant shaft speed at the respective rpm. This is sometimes possible, for example, during the very slow start-up of a large steam turbine, but usually each spectrum is a windowed section of a continuously varying signal with a small speed change within the window length. Consequently, the peak corresponding to each harmonic is not always localized in one analysis line; in particular, the higher harmonics are likely to be spread over progressively more lines. Thus, the height of each peak cannot be used directly as a measure of the strength of each component; it would be necessary to integrate over the whole of a distributed peak to measure the total power contained in it.

A way of overcoming this problem is to use *tracking analysis,* where the sampling rate of the FFT analyzer is related directly to shaft speed. A frequency multiplier may be used to produce a sampling frequency signal (controlling the A/D converter of the analyzer) which is a specified multiple of the shaft speed.

Figure 14.24 illustrates the basic principles. Figure 14.24*B* shows a hypothetical signal produced by a rotating shaft during a run-up (in practice, the amplitude normally also would vary with shaft speed). Figure 14.24*A* shows the samples obtained

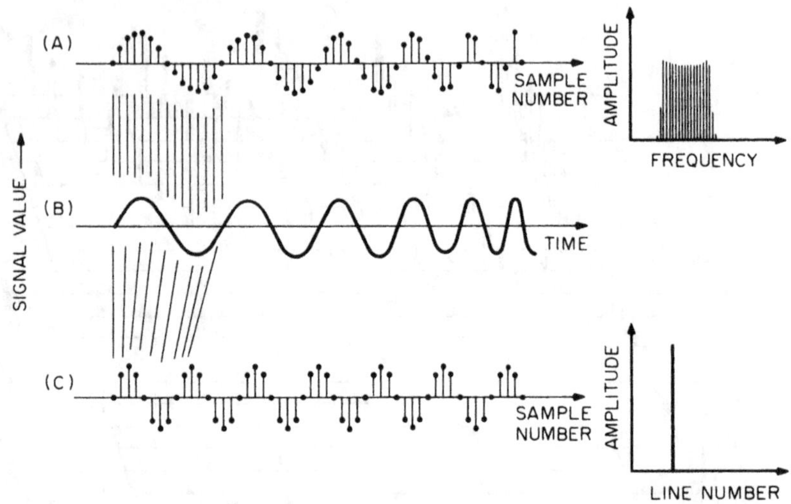

FIGURE 14.24 Analysis of a fundamental component which is increasing in frequency. (*A*) Data record resulting from a uniform sampling rate, and its spectrum, which spreads over a frequency band corresponding to the speed change. (*B*) The original time signal. (*C*) Data record resulting from sampling eight times per fundamental cycle, and its spectrum, which is concentrated in one analysis line.

by sampling the signal value at a constant sampling frequency (as for normal frequency analysis) and the spectrum resulting from FFT analysis of these samples. The spectral peak is seen to spread over a number of lines corresponding to the speed change along the time record. Figure 14.24*C* shows the samples obtained by sampling the signal a fixed number of times per shaft revolution (in this case, eight). The samples are indistinguishable from those obtained from normal analysis of a constant-frequency component, and thus the frequency spectrum is concentrated in one line.

A frequency multiplier, based on a phase-locked loop, suffers from the disadvantage of a finite response time, so that it cannot keep up if the speed is changing rapidly. A better alternative, offered by some analyzers, is based on digital resampling (interpolation) of each record in line with the simultaneously measured tachometer signal.

When the sampling frequency varies with shaft speed, however, special precautions must be taken to avoid problems with aliasing. One possibility is to use a tracking low-pass filter with a cutoff frequency suitably less than half the sampling frequency. Because of the difficulty of obtaining a tracking filter having a very steep roll-off (e.g., 120 dB/octave), it is often simpler to choose one of a series of filters with a fixed cutoff frequency, depending on the current shaft speed. Such a series of filters (in, for example, a 2, 5, 10 sequence) often is available in the analyzer to determine the normal frequency ranges. Taking the case of a 400-line analyzer, for example, all 400 lines in the measured spectrum are valid when the sampling frequency is appropriate to the selected filter (Fig. 14.25*A*). If the sampling frequency is higher than the ideal for a given filter, the upper part of the spectrum is affected by the filter (Fig. 14.25*B*). If it is lower, the upper part of the spectrum may be contaminated by aliasing components (Fig. 14.25*C*). Nevertheless, by arranging for the selection of the optimum filter at all times (either manually or automatically), at

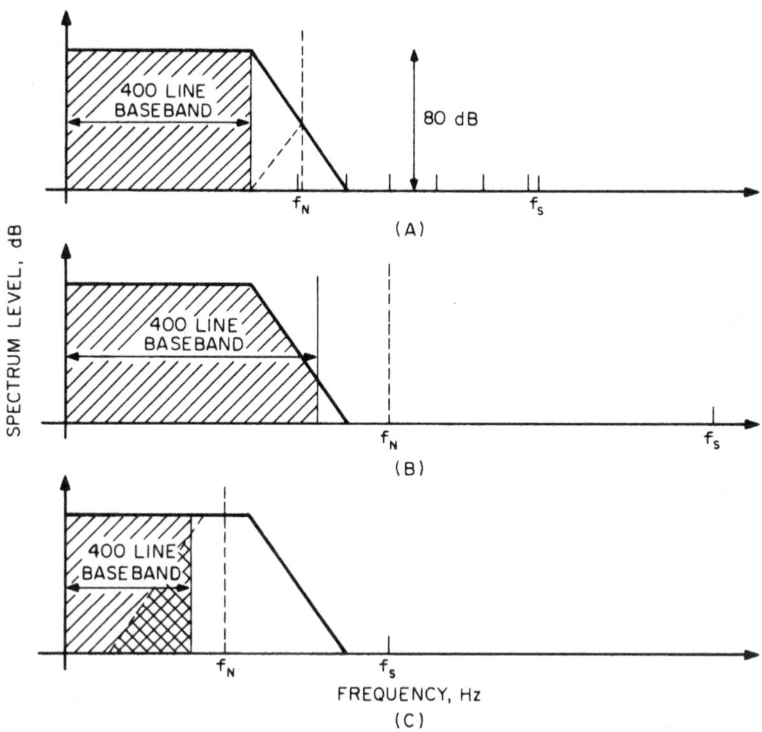

FIGURE 14.25 Effect of sampling frequency on the validity of spectral components, assuming an FFT analyzer with 400 lines and 80-dB dynamic range. f_s = sampling frequency. f_N = Nyquist folding frequency $= f_s/2$. (*A*) Normal situation with optimum choice of sampling frequency for the low-pass filter. (*B*) Situation with increased sampling frequency. The upper lines in the spectrum are influenced by the low-pass filter. (*C*) Situation with decreased sampling frequency. The upper lines in the spectrum are influenced by aliasing components folded around f_N (double cross-hatched area).

least 60 percent of the measured spectrum (i.e., in this case 240 lines) is always valid. The analysis parameters can be selected so that the desired number of harmonics is contained within this range, based on the fact that the line number in the spectrum of a given component is equal to the number of periods it represents in the data record of length N samples. If, for example, the 30th harmonic is to be located in line no. 240, the fundamental must be in line no. 8; there must be eight periods of the fundamental component along the data record. Where the data record contains 1024 samples (i.e., $N = 1024$), the sampling frequency must then be 128 times the shaft speed; thus, a frequency multiplier with a multiplication factor of 128 should be used in this specific case.

For FFT analyzers with zoom, a simpler approach can be used, as illustrated in Fig. 14.26. An analog low-pass filter is applied to the signal with a cutoff frequency corresponding to the highest required harmonic at maximum shaft speed. However, a frequency multiplying factor is chosen so as to make the sampling frequency, say, 10 or 20 times this cutoff frequency (instead of the normal 2.56). The spectrum then is obtained by zooming in a range corresponding to the highest required harmonic.

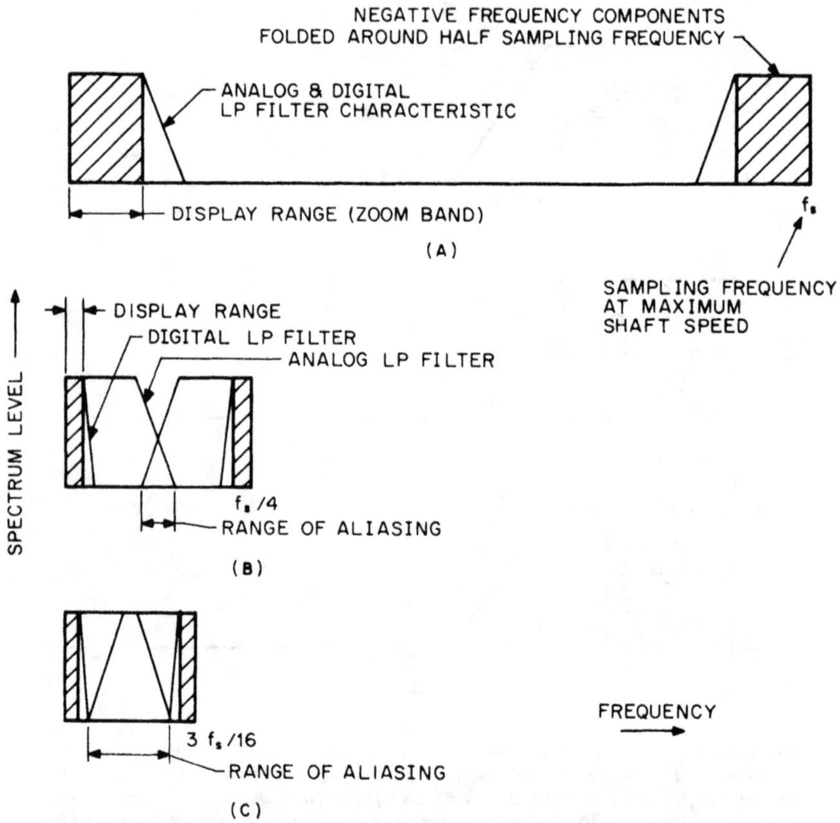

FIGURE 14.26 Use of a fixed low-pass filter to prevent aliasing when tracking with an FFT analyzer employing zoom to analyze in a lower-frequency band. For illustration purposes, the sampling frequency at maximum shaft speed has been made four times greater than that appropriate to the analog LP filter. The shaft speed range could be made proportionally greater by increasing this factor. (*A*) Situation at maximum shaft speed. All harmonics of interest must be contained in the display range. (*B*) Situation at one-fourth maximum shaft speed. The analog filter characteristics overlap, but are well separated from the display range. (*C*) Situation at three-sixteenths maximum shaft speed. The aliasing range almost intrudes on the display range.

As shown in Fig. 14.26, the shaft speed (and thus the sampling frequency) can then be varied over a wide range, without aliasing components affecting the measurement results. A somewhat similar procedure is used in conjunction with the digital resampling technique mentioned above. By using four times oversampling, a maximum speed range of 5.92:1 can be accommodated without changing the decimation rate (i.e., the proportion of samples retained after digital filtration), but an even wider range can be covered, at the expense of small "glitches" at the junctions, if the decimation rate is allowed to change.

Figure 14.27 shows the results of tracking FFT analysis on a large turbogenerator. It was made using nondestructive zoom with zoom factor 10 (equivalent to a large transform size giving 4000 lines). A frequency multiplying factor of 256 was used,

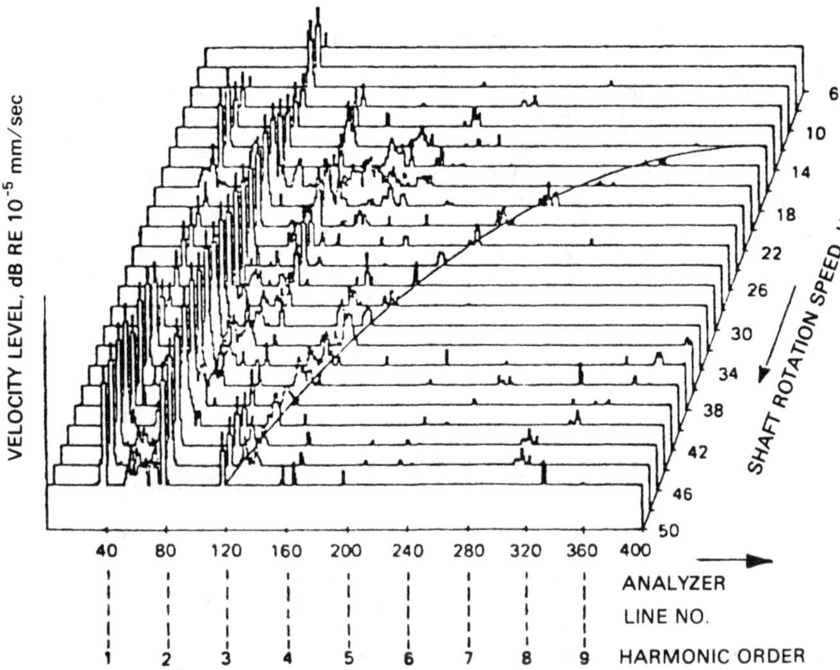

FIGURE 14.27 Tracking FFT analysis of the rundown of a large turbogenerator. The superimposed hyperbolic curve represents a fixed-frequency component at 150 Hz.

giving 40 periods of the fundamental component in the 10K (10,240-point) memory of the FFT analyzer. The fundamental is thus located in line no. 40 of the 400-line zoom spectrum. Because the harmonics coincide exactly with analysis lines, rectangular weighting could have been used in place of the Hanning weighting actually used (all harmonics have exact integer numbers of periods along the record length); Hanning weighting can, however, be advantageous for nonsynchronous components such as constant-frequency components. Such a component at 150 Hz (initially coinciding with the third harmonic of shaft speed) is shown in Fig. 14.27. Constant-frequency components follow a hyperbolic locus in cascade plots employing order tracking. See Chap. 19 for further discussions of nonstationary data analysis.

RELATED ANALYSIS TECHNIQUES

Signal analysis techniques other than those described above, which are useful as an adjunct to frequency analysis, include synchronous averaging, cepstrum analysis, and demodulation.

Synchronous Averaging (Signal Enhancement). *Synchronous averaging* is an averaging of digitized time records, the start of which is defined by a repetitive trigger signal. One example of such a trigger signal is a once-per-revolution synchronizing pulse from a rotating shaft. This process serves to enhance the repetitive part of the signal (whose period coincides with that of the trigger signal) with respect to nonsynchronous effects. That part of the signal which repeats each time adds directly, in proportion to the number of averages, n. The nonsynchronous components, both random noise and periodic signals with a different period, add like noise, with random phase; the amplitude increase is in proportion to $\sqrt{n}$. The overall improvement in the signal-to-noise rms ratio is thus $\sqrt{n}$, resulting in an improvement of $10 \log_{10} n$ dB, i.e., 10 dB for 10 averages, 20 dB for 100, 30 dB for 1000.

Figure 14.28 shows the application of synchronous averaging to vibration signals from similar gearboxes in good and faulty condition. Figure 14.28*A* shows the enhanced time signal (120 averages) for the gear on the output shaft. The signal is fairly uniform and gives evidence of periodicity corresponding to the tooth-meshing.

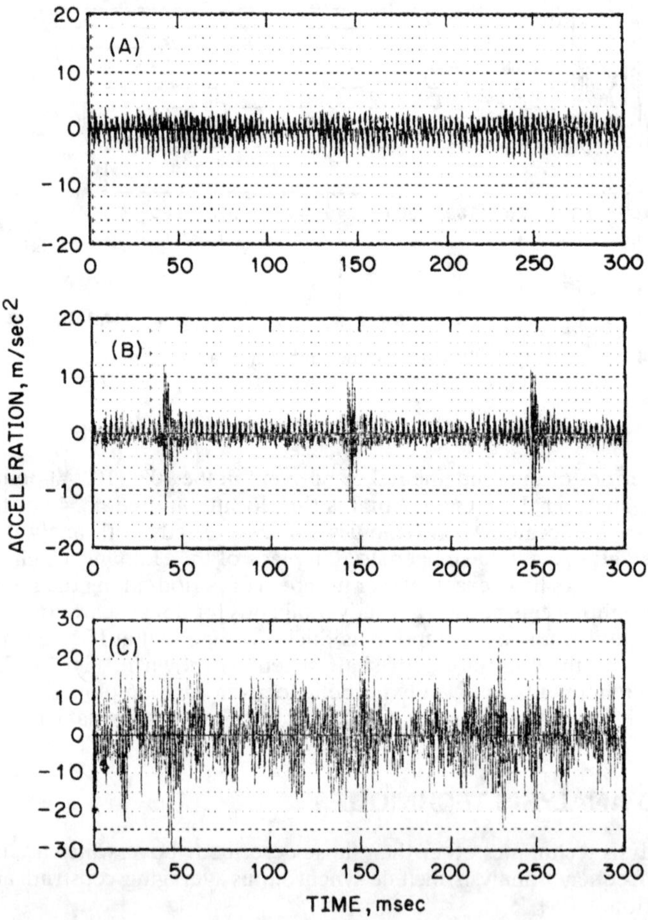

FIGURE 14.28 Use of signal enhancement in gear fault diagnosis. (*A*) Enhanced signal (120 averages) for a gear in normal condition. (*B*) Enhanced signal (120 averages) for a similar gear with a local fault. (*C*) Section of raw signal corresponding to (*B*).

Figure 14.28*B* is a similarly enhanced time signal for a faulty gear; a localized defect on the gear is revealed. By way of comparison, Fig. 14.28*C* shows a single time record, without enhancement, for the same signal as in Fig. 14.28*B;* neither the tooth-meshing effect nor the fault is readily seen.

For best results, synchronous averaging should be combined with tracking. Where there is no synchronization between the digital sampling and the (analog) trigger signal, an uncertainty of up to one sample spacing can occur between successive digitized records. This represents a phase change of 360° at the sampling frequency, and approximately 140° at the highest valid frequency component in the signal, even with perfectly stable speed. Where speed varies, an additional phase shift occurs; for example, a speed fluctuation of 0.1 percent would cause a shift of one sample spacing at the end of a typical 1024-sample record. The use of tracking analysis (generating the sampling frequency from the synchronizing signal) reduces both effects to a minimum.

Cepstrum Analysis. Originally the *cepstrum* was defined as the power spectrum of the logarithmic power spectrum.[9] A number of other terms commonly found in the cepstrum literature (and with an equivalent meaning in the cepstrum domain) are derived in an analogous way, e.g., *ceps*trum from *spec*trum, *quef*rency from *frequ*ency, *rah*monic from *har*monic. The distinguishing feature of the cepstrum is not just that it is a spectrum of a spectrum, but rather that it is the spectrum of a spectrum on a logarithmic amplitude axis; by comparison, the autocorrelation function [see Eq. (19.11)] is the inverse Fourier transform of the power spectrum without logarithmic conversion.

Most commonly, the *power cepstrum* is defined as the inverse Fourier transform of the logarithmic power spectrum,[10] which differs primarily from the original definition in that the result of the second Fourier transformation is not modified by obtaining the amplitude squared at each quefrency; it is thus reversible back to the logarithmic spectrum. Another type of cepstrum, the *complex cepstrum,* discussed in Refs. 10 and 11, is reversible to a time signal.

Figure 14.29, the analysis of a vibration signal from a faulty bearing, shows the advantage of the power cepstrum over the autocorrelation function. In Fig. 14.29*A*, the same power spectrum is depicted on both linear and logarithmic amplitude axes; in (*B*) and (*C*) the autocorrelation and cepstrum, respectively, are shown. In (*C*), the use of the logarithmic power spectrum reveals the existence of a family of harmonics which are concealed in the linear depiction. The presence of the family of harmonics is made evident by a corresponding series of rahmonics in the cepstrum (denoted ①, ②, etc.), but is not detected in the autocorrelation function. The quefrency axis of the cepstrum is a time axis, most closely related to the *X* axis of the autocorrelation function (i.e., time delay or periodic time rather than absolute time). The reciprocal of the quefrency of any component gives the equivalent *frequency spacing* in the spectrum, not the absolute frequency.

Most of the applications of the power cepstrum derive from its ability to detect a periodic structure in the spectrum, for example, families of uniformly spaced harmonics and/or sidebands. The application of the cepstrum to the diagnosis of faults in gears and rolling element bearings is discussed in Ref. 11.

To obtain a distinct peak in the cepstrum, a reasonable number of the members of the corresponding harmonic or sideband family must be present (although the fundamental may be absent). These uniformly spaced components must be adequately resolved in the spectrum. As a guide, the spacing of components to be detected should be a minimum of eight lines in the original spectrum. For this reason, it is often advantageous to perform a cepstrum analysis on a spectrum obtained by *zoom FFT.*[11]

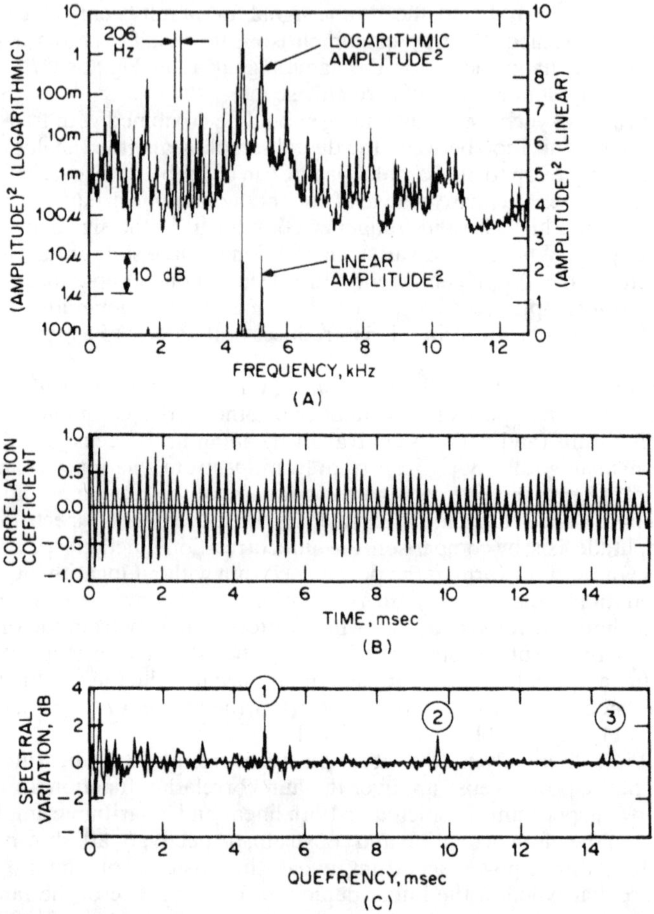

FIGURE 14.29 Effect of linear vs. logarithmic amplitude scale in power spectrum. (*A*) Power spectrum on linear scale (lower curve) and logarithmic scale (upper curve). (*B*) Autocorrelation function (obtained from linear representation). (*C*) Cepstrum (obtained from logarithmic representation)—①, ②, etc., are rahmonics corresponding to harmonic series in spectrum (4.85 milliseconds equivalent to 1/206 Hz). The harmonics result from a fault in a bearing.

Demodulation. Many fast Fourier transform analyzers perform demodulation, as it is a natural function in connection with a real-time zoom processor, and there are many applications in vibration analysis. An example of amplitude demodulation is given by "envelope analysis" in rolling element bearing diagnostics (see Chap. 16) where information about bearing faults is contained in the spectrum of the envelope of the series of high-frequency bursts produced by a fault, but not in the spectrum of the raw signal. The envelope of the signal can be obtained by amplitude demodulation of a frequency band containing resonances excited by the faults.

Phase and frequency demodulation is, for example, important in torsional vibrations, as torsional vibration is a phase modulation (of the nominal shaft speed as carrier frequency) when expressed as variations in angular displacement and is frequency modulation when expressed in terms of angular velocity fluctuations around the fixed carrier. Frequency modulation is the time derivative of phase modulation. A further differentiation of the angular velocity gives the angular acceleration of a shaft, but there is no corresponding modulation term.

A zoom band such as illustrated in Fig. 14.14 would generally represent a "phase coherent" signal such as described by one component of Eq. (19.23). In fact, Eq. (19.23) has a two-sided frequency spectrum, but the zoom band of Fig. 14.14 contains only positive frequencies and can be represented by the complex exponential function

$$x(t) = A(t)\exp[2\pi f_i(t) + \theta_0] \tag{14.10}$$

The actual signal is the real part of this, that is, the projection on the real axis of a rotating vector of time-varying amplitude $A(t)$ and rotational speed $f_i(t)$ with phase angle θ_0 at time zero. Because the spectrum is one-sided, the imaginary part of the complex exponential function would be the Hilbert transform of the real part.[1]

By definition, phase modulation is the variation of phase around the linearly increasing phase of the carrier frequency, which can be expressed as

$$\phi_m(t) = 2\pi[f_i(t) - f_0 t] + \theta_0 \tag{14.11}$$

where f_0 is the carrier frequency and θ_0 has been incorporated into $\phi_m(t)$. Thus, if a zoom analysis is performed around the carrier frequency, this frequency is subtracted from every component in the band, so that the remainder (the zoom processor output) can be expressed as

$$x_m(t) = A(t)\exp[\phi_m(t)] \tag{14.12}$$

where the amplitude modulation information is contained in

$$A(t) = A_0 + A_m(t) \tag{14.13}$$

$A_m(t)$ is the actual amplitude modulation signal and A_0 is a suitable dc offset to ensure that $A(t)$ is always positive. Note that the amplitude modulation information is independent of any frequency shift. The phase of the zoom processor output is directly the phase modulation signal. To obtain the corresponding frequency modulation signal, it should be differentiated and scaled in Hz, as

$$f_m(t) = \frac{1}{2\pi}\frac{d}{dt}[\phi_m(t)] \tag{14.14}$$

This differentiation can be achieved by multiplication by jw in the frequency domain, and some FFT analyzers provide this option.

Note that f_0 can only be chosen as one of the frequency lines in the zoom analysis, and if the actual physical carrier frequency is slightly different, this just means that the phase modulation signal will have a small added slope. Since one line spacing of a fast Fourier transform spectrum corresponds to one rotation (period) in the record length, the maximum slope which can be encountered by choosing the nearest frequency line as carrier will lie in the range $\pm\pi$ radians over the record length. It can be removed by subtracting the linear regression line (trend removal). Where order tracking is used—for example, on a gear signal—a typical carrier frequency (e.g., a gearmesh frequency) will normally correspond to an analysis line. Note also that the scaling of the phase is in radians of the carrier frequency. If this represents the *n*th harmonic of the shaft speed (e.g., a gearmesh frequency), the phase modulation signal should be divided by *n* to scale it in terms of radians of shaft rotation.

REFERENCES

1. Randall, R. B.: "Frequency Analysis," Brüel & Kjaer, Naerum, Denmark, 1987.
2. Cooley, J. W., and J. W. Tukey: *Math. Computing,* **19**(90):297 (1965).
3. Cooley, J. W., P. A. W. Lewis, and P. D. Welch: *J. Sound Vibration,* **12**(3):315 (1970).
4. Brigham, E. O.: "The Fast Fourier Transform," Prentice-Hall, Englewood Cliffs, N.J., 1974.
5. Thrane, N.: "Zoom-FFT," *Brüel & Kjaer Tech. Rev.,* (2) (1980).
6. Sloane, E. A.: *IEEE Trans. Audio Electroacoust.,* **AU-17**(2):133 (1969).
7. Welch, P. D.: *IEEE Trans. Audio Electroacoust.,* **AU-15**(2):70 (1967).
8. Mitchell, J. S.: "An Introduction to Machinery Analysis and Monitoring," Penwell Publishing Company, Tulsa, Okla., 1981.
9. Bogert, B. P., M. J. R. Healy, and J. W. Tukey: In M. Rosenblatt (ed.), "Proceedings of the Symposium on Time Series Analysis," John Wiley & Sons, New York, 1963, pp. 209–243.
10. Childers, D. G., D. P. Skinner, and R. C. Kemerait: *Proc. IEEE,* **65**(10):1428 (1977).
11. Randall, R. B.: "Cepstrum Analysis," *Encyclopedia of Vibration,* S. G. Braun, D. J. Ewins, and S. S. Rao, eds., Academic Press, New York, 2001, pp. 216–227.

CHAPTER 15
MEASUREMENT TECHNIQUES

Cyril M. Harris

INTRODUCTION

Earlier chapters describe equipment used in vibration measurements. For example, detailed information concerning transducers, their characteristics, and how these characteristics are influenced by environmental factors is given in Chap. 10. The various measurement system components and the characteristics which determine their selection are described in Chaps. 13 and 14. The use of such measurement systems in vibration problems may involve only one or two engineers as in monitoring the condition of machinery in a factory (Chap. 16), in some problems in modal testing (Chap. 21), in measurements in building structures (Chap. 29), and in measuring torsional vibration in reciprocating and rotating engines (Chap. 37). In contrast, in the aerospace industry, some measurement problems are so complex that teams of engineers and several divisions of the company may be involved. Yet all these examples share certain basic measurement procedures. It is these basic procedures (rather than measurement details, which vary from problem to problem) that are considered here. Thus, this chapter includes a general discussion of (1) planning measurements to achieve stated objectives, (2) selecting the type of measurements which should be made to achieve these objectives, (3) selecting transducers, (4) mounting transducers, (5) mounting cable and wiring (including shielding and grounding), (6) selecting techniques for the field calibration of the overall measurement system, (7) collecting and logging the data obtained, and (8) conducting a measurement error analysis.

The best method of analyzing the vibration measurement data, once they have been acquired, depends on a number of factors, including the quantity of data to be processed, the objectives of the measurements, test criteria, specifications, and the accuracy required. These factors are discussed in Chaps. 14, 18, 19, 20, and 26.

MEASUREMENT PLANNING

Careful pretest planning (and, in the case of a complex measurement program, detailed documentation) can save much time in making measurements and in ensuring that the most useful information is obtained from the test data. In many cases, as in environmental testing, measurement procedures are contained in test specifications to ensure that a specification or legal requirement has been met. In other cases (as in balancing rotating machinery), measurement procedures are outlined in detail in national or international standards. In general, the first step in planning is to define the purpose of the test and to define what is to be measured. Planning should start with a clear definition of the test objectives, including the required accuracy and reliability. The second step is to define those non-equipment-related factors which influence the selection of measurement equipment and measurement techniques. These include availability of trained personnel; cost considerations; length of time available for measurements; scheduling considerations; and available techniques for data analysis, validation, and presentation.

Next, the various factors listed in Table 15.1 should be considered. For example, it is important to have some estimate of the characteristics of the motion to be measured—e.g., its frequency range, amplitude, dynamic range, duration, and principal direction of motion. Such information is needed to provide the basis for the optimum selection of measurement equipment. Yet often very little is known about the characteristics of the motion to be measured. Previous experience may provide a guide in estimating signal characteristics. Where this is not available, preliminary measurements may be carried out to obtain information which serves as a guide for further measurements. For example, suppose preliminary measurements show a frequency spectrum having considerable content in the region of the lowest frequency measured. This would indicate that the instrumentation capability should be extended to a somewhat lower frequency in subsequent measurements. Thus an iterative process often takes place in a shock and vibration measurement program. To speed this process, it is helpful to employ equipment whose characteristics cover a wide range and which has considerable flexibility. Failure to take this feedback process into account can sometimes result in the acquisition of meaningless test results. For example, a measurement program was carried out by one organization over a period of many weeks. The objective was to correlate building vibration data, measured in the organization's own laboratories, with the acceptability of these laboratories as sites for ultrasensitive galvanometers and other motion-sensitive equipment. No correlation was found, and the entire measurement program was a waste of time, for two reasons: (a) The measurements were made with equipment with a frequency limit which was not sufficiently low, so that important spectral components of building vibration could not be measured. (b) Measurements were made only in the vertical direction, whereas it was the horizontal component which was dominant and which made certain laboratory areas unacceptable for the location of vibration-sensitive equipment.

Many of the various factors, listed in Table 15.1, which should be considered in planning instrumentation for shock and vibration measurements are discussed in earlier chapters and are cross-referenced, rather than repeated, here. For example, Chap. 10 discusses the effects of environmental conditions on transducer characteristics; Chap. 13 describes various components which follow the transducer in a measurement system (such as preamplifiers, signal conditioners, filters, analyzers, and recorders). Chapter 14 describes the selection of the appropriate analyzer bandwidth, frequency scale, amplitude scale, selection of data windows, etc.

Before making measurements, it is usually important to establish a measurement protocol—the more complex the measurements to be made, the more formal and detailed the measurement protocol should be. It is also important to make an *error analysis,* i.e., (a) to estimate the error introduced into the data acquisition and analysis by each individual item of equipment, and (b) to determine the total error by calculating the square root of the sum of the squares of the individual errors. For

TABLE 15.1 Factors Which Are Important Considerations in the Selection of Measurement Equipment and Measurement Techniques for Mechanical Shock and Vibration Measurements

Parameter to be measured	
Acceleration	Strain
Velocity	Force
Displacement	Mechanical impedance
Characteristics of motion to be measured	
Frequency range	Direction of motion
Amplitude range	Transient characteristics
Phase	Duration
Environmental conditions	
Temperature (ambient and transient)	Magnetic and radio-frequency fields
Humidity	Corrosive and abrasive media
Ambient pressure	Nuclear radiation
Acoustic noise	Sustained acceleration
Transducer characteristics	
Electrical characteristics (sensitivity, resolution, transverse sensitivity, amplitude linearity, dynamic range, frequency response, phase response, effects of environment on the transducer)	
Physical characteristics (e.g., size and mass)	
Self-generating or auxiliary power required	
Electrically grounded to case, or isolated	
Self-contained amplifier	
Transducer mountings and locations of mountings	
Effect of mounting on transducer characteristics	
Effect of mounting on vibratory characteristics of item under test	
Number of measurement locations	
Space availability for measurement locations	
Availability of well-regulated power, free of voltage spikes	
Ease of installation	
Possibility of mounting misalignment with respect to intended direction of measurement	
System components (preamplifiers, signal conditioners, filters, analyzers)	
Electrical characteristics (e.g., input and output impedances)	
Power availability	
Noise interference (shielding, avoidance of ground loops)	
Number of channels required for measurement and recording: maximum duration of measurements, tape storage requirements	
Possible requirement for real-time information	
Method of data transmission	
Coaxial cable	
Twisted pair of wires	
Telemetry (channels assigned)	
Optical fiber	
Recording equipment	
Recording-time capability	
Electrical characteristics (e.g., signal-to-noise ratio)	
Portability; power requirements	
Correlation between recorded information and physical phenomena	
Redundancy to minimize the risk of loss of vital information	

TABLE 15.1 Factors Which Are Important Considerations in the Selection of Measurement Equipment and Measurement Techniques for Mechanical Shock and Vibration Measurements (*Continued*)

Field calibration
Transducers
Over-all measurement system
Data analysis, presentation, and validation
Manual or automatic; computer
Type of presentation required

example, such an analysis may discover that an individual item of equipment is primarily responsible for introducing a significant total error, suggesting that perhaps it should be replaced. Furthermore, such a determination will indicate whether the total error is within the bounds of acceptability, thereby avoiding useless measurements.

SELECTION OF THE PARAMETER TO BE MEASURED

Often, the selection of the parameter to be measured (displacement, velocity, acceleration, or strain) is predetermined by specifications or by standards. When this is not the case, it is often helpful to apply the considerations given in Table 15.2 or to apply the *flattest spectrum rule* described in Chap. 16. According to this rule, the best motion parameter to use is the one whose spectrum is closest to being uniform (i.e., the one having the flattest spectrum). This is important for two reasons: if the spectrum is relatively flat, then (1) an increase at any frequency has a roughly even chance of influencing overall vibration levels, and (2) minimum demands are placed on the required dynamic range of the equipment which follows the transducer. For example, Fig. 16.2 shows two spectra obtained under identical conditions—one a velocity spectrum, the other a displacement spectrum. The spectrum obtained using a velocity transducer is the more uniform of the two; therefore, velocity would be the appropriate motion parameter to select.

SELECTING THE TRANSDUCER

In selecting the transducer best suited for a given measurement, the various factors listed in Table 15.1 must be taken into consideration, particularly those under "Parameter to Be Measured," "Characteristics of Motion to Be Measured," "Environmental Conditions," and "Transducer Characteristics." Each of these factors (as well as cost and availability) influences the selection process. If consideration of different factors leads to recommendations which are in opposition, then the relative importance of each factor must be determined and a decision made on this basis. For example, consider two factors which enter into the selection of a piezoelectric accelerometer, *sensitivity* and *mass*. Sensitivity considerations would suggest that a transducer of large size be selected since transducer sensitivity generally increases with size (and therefore with mass) for an accelerometer of this type. In contrast, mass considerations would suggest that a transducer of small size be selected in order to minimize the mass loading on the test item; a small size is advantageous since, as Eq. (10.11) indicates, the nat-

TABLE 15.2 A Guide for the Selection of the Parameter to Be Measured

Acceleration measurements
Used at high frequencies where acceleration measurements provide the highest signal outputs
Used where forces, loads, and stresses must be analyzed—where force is proportional to acceleration (which is not always the case)
Used where a transducer of small size and small mass is required, since accelerometers usually are somewhat smaller than velocity or displacement pickups
Velocity measurements
Used where vibration measurements are to be correlated with acoustic measurements since sound pressure is proportional to the velocity of the vibrating surface
Used at intermediate frequencies where displacement measurements yield transducer outputs which may be too small to measure conveniently
Used extensively in measurements on machinery where the velocity spectrum usually is more uniform than either the displacement or acceleration spectra
Used where vibration measurements on resonant structures are to be correlated with modal stress, since modal stress is proportional to modal velocity at resonance frequencies
Displacement measurements
Used where amplitude of displacement is particularly important—e.g., where vibrating parts must not touch or where displacement beyond a given value results in equipment damage
Used where the magnitude of the displacement may be an indication of stresses to be analyzed
Used at low frequencies, where the output of accelerometers or velocity pickups may be too small for useful measurement
Used to measure relative motion between rotating bodies and structure of a machine
Strain measurements
Used where a portion of the specimen being tested undergoes an appreciable variation in strain caused by vibration—usually limited to low frequencies

ural frequency of a structure is lowered by the addition of mass. Therefore in this case one should choose the most sensitive transducer (and therefore the largest size) which produces no significant mass loading. In special cases, even the smallest transducer may result in an unacceptable load. Then one of the devices described in Chap. 10 which make no contact with the test surface may be selected.

Consider another example. Suppose a specification requires that vibration displacement be measured. It is reasonable to assume that a displacement transducer (such as the one described in Chap. 10) should be chosen since (depending on the frequency spectrum) such a selection could yield the highest signal-to-noise ratio. On the other hand, in many measurement problems it is more convenient and equally satisfactory to select an accelerometer having a wide dynamic range and to employ an electric circuit which obtains displacement by double integration of the signal from the transducer's output.

TRANSDUCER MOUNTINGS

Various methods of mounting a transducer on a test surface include (1) screwing the transducer to the test surface by means of a threaded stud, (2) cementing the transducer to the test surface, (3) mounting the transducer on the test surface by means of a layer of wax, (4) attaching the transducer to a ferromagnetic surface by means of a permanent magnet, (5) mounting the transducer on a bracket which, in turn, is mounted on the test surface, and (6) holding the transducer against the test surface by hand. Several of these mounting techniques are illustrated in Fig. 15.1, and their frequency response characteristics are shown in Fig. 15.2. Two types of mechanical brackets are illustrated in Fig. 15.3.

The method of mounting affects the resonance frequency and, hence, the useful frequency range of the transducer. Therefore it is important to ensure that the frequency response is adequate before measurements are taken. Each of the above methods of mounting has its advantages and disadvantages. The appropriate choice for a given measurement problem depends on a number of factors, including the following:

Effect of the mounting on the useful frequency range of the transducer

Effect of mass loading of the transducer mounting on the test surface

Maximum level of vibration the mounting can withstand

Maximum operating temperature

Measurement accuracy

Repeatability of measurements (Can the transducer be remounted at exactly the same position with the same orientation?)

Stability of the mounting with time

Requirement that the test surface not be damaged by screw holes

Requirement for electrical insulation of the transducer

Time required for preparation of test surface

Time required to prepare mounting

Time required to remove mounting

Difficulty in cleaning the transducer after removal from test surface

Difficulty in cleaning test surface after transducer removed

Skill required to prepare mounting

Cost of mounting

Environmental problems (dirt, dust, oil, moisture)

For example, the above "requirement for electrical insulation of the transducer" would be a major consideration in the selection of a method of mounting if the insulation so obtained would result in the breaking of a ground loop, as explained in a following section.

Stud Mounting. Figure 15.1*A* illustrates a typical stud-mounted transducer; the transducer is fixed to the test surface by means of a threaded metal screw. One method of insulating the stud-mounted transducer from the test surface is shown in Fig. 15.1*B*. The metal stud is replaced with one which is fabricated of insulating material, and a mica washer is inserted between the transducer and the test surface. Other manufacturers employ a threaded, insulated stud with a flange made of the same material; the flange, midway along the length of the stud, serves as the base for the accelerometer. The entire base of the transducer should be in intimate contact with the test surface. The mounting stud must be of the correct length, incorporating a flange to prevent "bottoming" of the stud which may result in strain-induced errors.

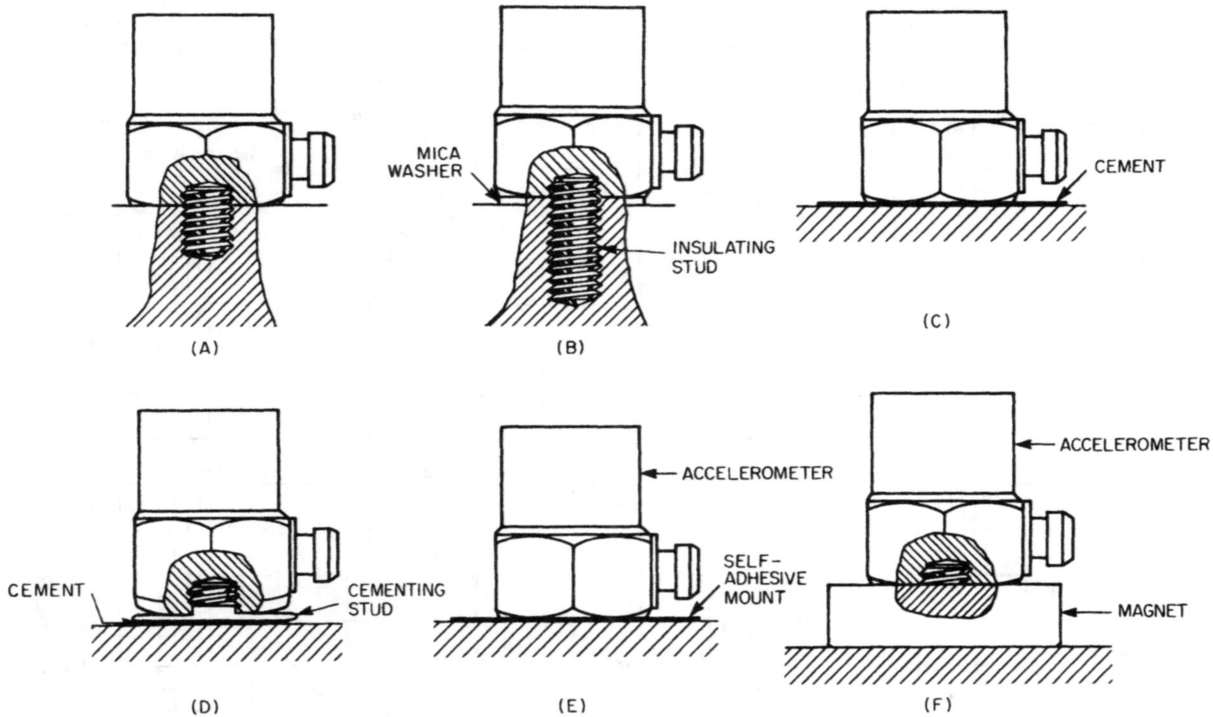

FIGURE 15.1 Various methods of mounting a transducer on a test surface: (*A*) Stud mounting; transducer screws directly to the surface by a threaded stud. (*B*) Same as (*A*) but with a transducer insulated from test surface by use of stud fabricated of insulating material and by a mica washer between the surface and transducer. (*C*) Cement mounting of a transducer; the cement bonds the transducer directly to the surface. (*D*) Similar to (*C*), but here cement bonds the surface to a cementing stud screwed into the transducer. (*E*) Transducer mounted to surface by means of double-sided adhesive tape or disc. (*F*) Transducer mounted to surface by means of a magnet. (*Courtesy of Brüel & Kjaer.*)

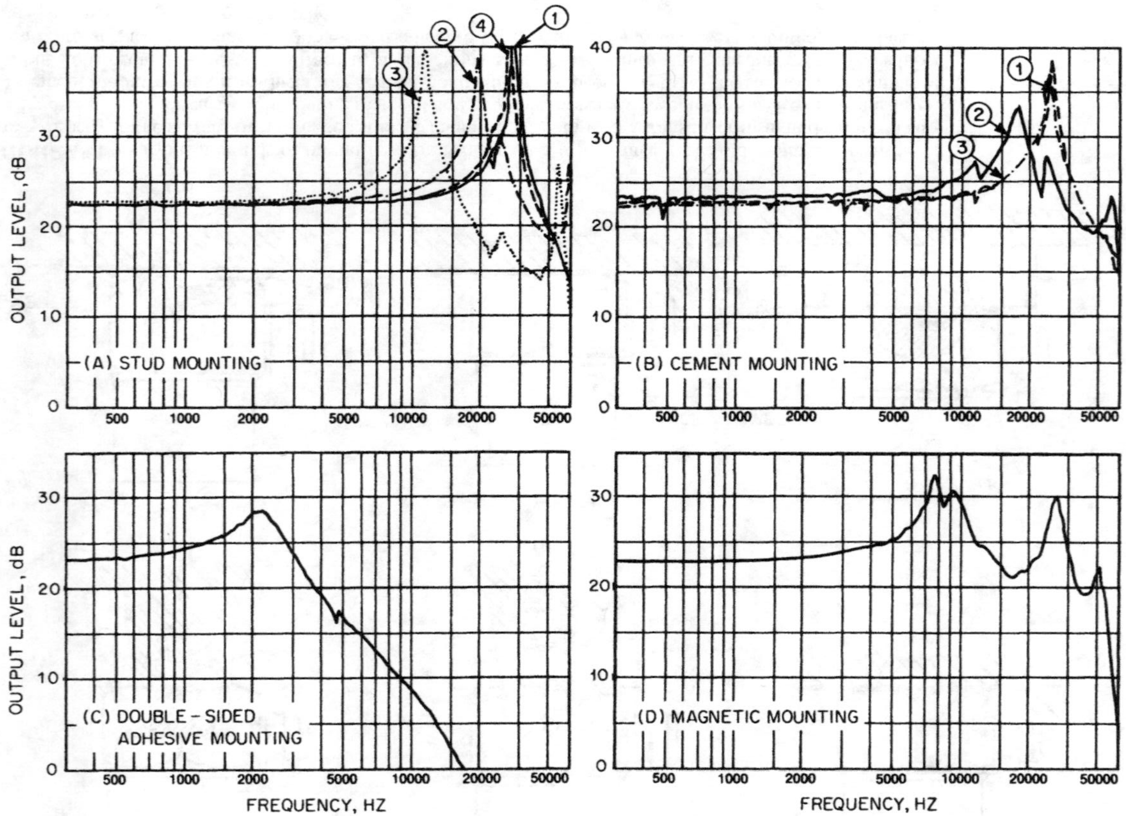

FIGURE 15.2 Frequency-response curves for the same piezoelectric accelerometer mounted by the different methods illustrated in Fig. 15.1: (*A*) stud mounting; (*B*) cement mounting; (*C*) double-sided adhesive mounting; (*D*) magnetic mounting. (*Courtesy of Brüel & Kjaer.*)

Where stud mounting is practical, it is the best type to use for the following reasons:

1. It provides the highest resonance frequency (up to 100 kHz) of any of the mounting techniques and, therefore, the widest possible measurement frequency range (up to 50 kHz).
2. It permits measurements at very high vibration levels without the loosening of the transducer from the test surface.
3. It does not reduce the maximum permissible operating temperature at which measurements can be made.
4. It permits accurate and reproducible results since the measurement position can always be duplicated.

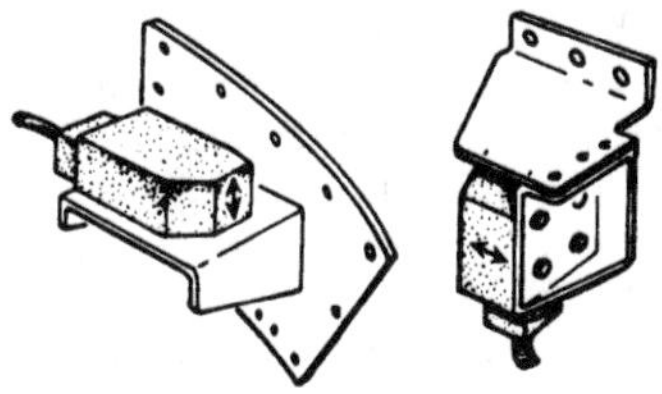

FIGURE 15.3 Two types of mounting brackets. In this example, a velocity-type transducer is shown; the arrows indicate the direction of sensed motion.

In preparing a stud mounting, the test surface must be drilled and tapped. A standard 10-32 thread is widely used. (Also see International Standards Organisation Standard ISO 1101.) Distortion of the transducer as mounted may produce strains that affect the transducer's response. Therefore, it is important (1) to ensure that the test surface is very flat (which can be done by grinding or lapping), (2) to prevent the mounting stud from bottoming in the transducer case—this can lead to strain, and (3) to screw the stud into the hole in the test surface, and then the accelerometer onto the stud using a torque wrench to ensure repeatability in installation of the transducers and to prevent thread damage; use the torque recommended by the transducer's manufacturer. The application of a silicone grease (such as Dow-Corning DC-4) or a light machine oil between the transducer and the test surface usually provides better response at high frequencies—say, above 2000 Hz. The upper temperature limit for the stud mounting of Fig. 15.1*A* is limited only by the accelerometer, but with the mica washer insert shown in Fig. 15.1*B*, the upper limit may be as low as 480°F (250°C).

Figure 15.2*A* shows response curves for a stud-mounted accelerometer for the following conditions: ① spanner tight, which has the highest resonance frequency, ② finger tight, ③ mounted with a mica washer to provide electrical insulation between the transducer and the vibrating surface, and ④ mounted on a somewhat thinner mica washer—which results in a higher resonance frequency than for ③.

Cement Mountings. A *cement* is a substance that bonds two surfaces together when the cement hardens; it acts as an adhesive. Where it is not possible to use a stud mounting, a transducer can be bonded to a clean test surface by means of a thin layer of cement (for example, a cyanoacrylate, dental cement, or epoxy cement), as shown in Fig. 15.1*C*. If the test surface is not flat and a miniature accelerometer is used, it is not difficult to build up a layer of dental cement around the accelerometer so as to provide firm attachment for the accelerometer. In mounting the transducer, it should be pressed firmly against the flat, smooth surface to ensure that the adhesive layer is thin; excess adhesive around the perimeter should then be removed immediately.

The cement method of mounting a transducer provides excellent frequency response, as shown in Fig. 15.2*B* for three conditions: ① accelerometer cemented directly to test surface, ② accelerometer cemented with a "soft" adhesive (not recommended), and ③ accelerometer with a cementing stud which is cemented to the surface with a hard cement.

This type of mounting may be used at high levels of vibration if the cementing surfaces are carefully prepared, *following the manufacturer's instructions.* Cement mounting may or may not provide electrical insulation; if insulation is required, the electrical resistance between the transducer and the test surface should be checked with an ohmmeter. The maximum temperature at which measurements can be made is limited by the physical characteristics of the cement employed—generally about 176°F (80°C). Some cements such as 3M Cyanolite 303 have an upper limit as high as 390°F (200°C). At room temperature, it has the best coupling characteristics over a wide frequency range. This type of mounting has good stability with time. Where a transducer has been attached to a surface by the use of a cement, exercise considerable caution in removing the transducer from the surface to avoid damaging it; application of a solvent to soften the cement is strongly recommended.

Cements 3M Cyanolite 101 and Permabond 747 dry much more rapidly than epoxy cements and therefore require less time to mount a transducer. They may be removed easily and the surface cleaned with a solvent such as acetone. Removal of epoxy from the test surface and from the transducer may be time-consuming. In fact, the epoxy bond may be so good that the transducer can be damaged in removing it from the test surface. When encased in epoxy, an accelerometer may be subject to considerable strain, which will significantly alter its characteristics. On the other hand, unless the cemented surfaces are very smooth, an epoxy can provide a superior bond since it will fill in a rough surface far better than a cyanoacrylate cement. With either bonding agent, the surfaces must be very clean before application of the cement. This mounting technique is not recommended for conditions of prolonged high humidity or for pyroshock measurements.

Commercial adhesives are obtainable for use in very hot or in very cold environments. For cryogenic applications, a two-component epoxy resin, room-temperature-cured, is available that is effective down to −200°C and is able to withstand cryogenic thermal shock without cracking. For use at very high temperatures (up to 700°C) ceramic-based adhesives are available that are effective, but require so high a curing temperature that their use is usually restricted to high-temperature applications. Several epoxy resins are commercially available that are cured at room temperature and can operate at temperatures as high as 260°C.[1]

Wax Mounting. Beeswax or a petroleum-based petrowax may be used to attach a transducer to a flat test surface. If the bonding layer is thin (say, no greater than 0.2 mm), it is possible to obtain a resonance frequency almost as high as that for the stud mounting, but if the test surface is not smooth, a thicker wax layer is required and the resonance frequency will be reduced. If the mating surfaces are very clean and free from moisture, the transducer can be mounted fairly easily, although some practice may be required. The transducer can be removed rapidly with a naphtha-type solvent. Disadvantages include the possibility of disattachment of the transducer at high vibration levels, a temperature limitation because of the relatively low melting point of wax, and poor long-time stability of the mounting. The maximum temperature at which measurements can be made with this mounting technique is usually about 100°F (40°C).

Adhesive-Tape Mounting. An *adhesive* is a substance used to bond two surfaces together. The adhesive is usually applied to a tape or disc. In such application, this term is often used as a synonym for the word "cement." An adhesive film may be used to mount a small transducer on a flat, clean test surface—usually by means of a double-sided adhesive tape. Double-sided adhesive discs are supplied by some transducer manufacturers. This mounting technique, illustrated in Fig. 15.1*E*, is rapid and easy to apply. Furthermore, such a mounting has the advantage of providing electrical insulation between the transducer and the test surface, and it does not require the drilling of a hole in the test surface; it is particularly applicable for use with a transducer having no tapped hole in its base. Such adhesives can provide secure attachment over a limited temperature range, usually below 200°F (95°C). In preparing an adhesive mounting, it is important to clean both the accelerometer and the test surface so that the adhesive will adhere firmly. When this is done, the frequency response can be fairly good, as illustrated in Fig. 15.2*C*, but not as good as with a wax mounting.

Another method of mounting is to use a cementing stud which is threaded into the transducer; the flat side of the stud is then cemented to the test surface as shown in Fig. 15.1*D*. This is a useful technique where repeated measurements at the same point are required. The transducer may be removed for measurements elsewhere, but the cementing stud is left in place. This provides assurance that future measurements will be made at precisely the same point.

Magnetic Mounting. With magnetic mounting, illustrated in Fig. 15.1*F*, a permanent magnet attaches the transducer to the test surface, which must be ferromagnetic, flat, free from dirt particles, and reasonably smooth. Magnetic mounting is useful in measuring low acceleration levels. The transducer can be attached to the test surface easily and moved quickly from one measurement point to another. For example, in a condition-monitoring system (described in Chap. 16) it can be used to determine a suitable measurement location for a transducer to be mounted permanently on a large rotating machine. In a heavy machine of this type, the added mass of the magnet is not important, but in other problems, the additional mass loading on the test surface may make the use of magnetic mounting unacceptable. Furthermore, if the acceleration levels are sufficiently high, as in impact testing, the magnet may become loosened momentarily. This can result in an inaccurate reading and possibly a slight change in the position of the transducer, which would also change the reading. The frequency response for this type of mounting is fair, as shown in Fig. 15.2*D*, but not as good as with the wax mounting. The magnet, often available from the transducer's manufacturer, usually is attached to the transducer by means of (1) a projecting screw on the magnet, which is threaded into the base of the transducer, or (2) a machine screw, one end of which is threaded into the transducer and the other end into the magnet. Application of a light machine oil or silicone grease usually improves the frequency response above about 2,000 Hz. The maximum temperature at which measurements can be made with this mounting technique is usually about 300°F (150°C). In attaching a magnetically mounted transducer to a test surface, the magnetic force that pulls the assembly toward the surface may sometimes be sufficiently high to result in a high level of mechanical shock at the time of contact, causing damage to the sensing elements or its internal electronics.

Mounting Blocks or Brackets. Physical conditions may make it impractical to mount a transducer by any of the above methods. In such cases, a mounting bracket or block that has been especially prepared for use on the test surface may be employed. For example, if the structural surface is rounded, a solid mounting block can be fabricated which is rounded to this same contour on one side and flat on the other side for mounting the transducer. A mounting block also may be useful where the surface is subject to structural bending; in this case, two accelerometers selected to have the same characteristics may be attached to the mounting block to measure bending-induced rotation. The effect of the mass of the mounting block is considered in Eq. (15.1). Two types of mounting brackets are illustrated in Fig. 15.3. Instead of using a triaxial accelerometer, sometimes it is more convenient to mount three transducers on a single block having sensitivities in three orthogonal directions. Any such mounting must couple the transducer to the test surface so that the transducer accurately follows the motion of the surface to which it is attached. This requires that the effective stiffness of the transducer mounting be high so that the mounting does not deflect under the inertial load of the transducer mass. This is not a problem in many transducer installations.

Mounting brackets may have resonance frequencies which are below 2,000 Hz and have little damping. Under such conditions, their use may result in significant measurement error as a result of resonant amplification or because of attenuation of vibration in the mounting. This is illustrated in Fig. 15.4, which shows the frequency response of a transducer mounted on brackets which are identical in geometry but which are fabricated from different materials. Note that a change in material from (*A*) steel to (*B*) a phenolic plastic halves the resonance frequency of the mounting. A change in the method of attachment, from (*B*) screw mounting to (*C*) an epoxy resin adhesive bond, significantly increases the frequency of the mounting resonance. Although these results are not of a general nature, they show that such minor variations in the transducer mounting may produce significant changes in the output characteristics of the transducer. It is good practice to calibrate an accelerometer in combination with its mounting block.

Hand-held Transducer. A transducer which is held against the test surface by hand provides the poorest performance of any of the techniques described here, but it sometimes can be useful in making a rapid survey of a test surface because the measurement location can be changed more rapidly than with any other method of mounting. Usually, a rod (called a *probe*), which is threaded at one end, is screwed into the transducer; the other end has a tip that is pressed against the test surface.

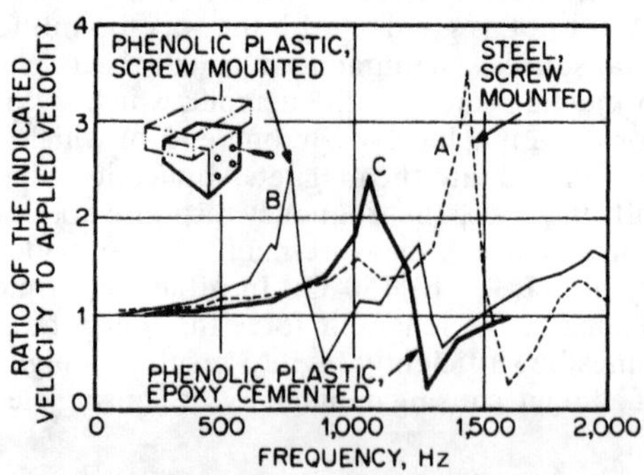

FIGURE 15.4 Relative frequency response of a velocity transducer mounted on three brackets which have identical geometry but are fabricated of different materials: (*A*) steel bracket, screw mounted, (*B*) cloth-reinforced phenolic plastic bracket, screw mounted, and (*C*) same as (*B*) but attached with epoxy resin adhesive.

The frequency response is highly restricted—about 20 to 1,000 Hz; furthermore this technique should not be employed for accelerations greater than 1*g*. Thus, this technique is used when measurement accuracy is not essential, e.g., in finding the nodal points on a vibrating surface.

Mass-Loading. The effect of the mounting on the accuracy of measurement can be estimated roughly if it is assumed that the combination of the transducer (having a mass m) and the mounting (having a stiffness k) behaves as a simple spring-mass system driven at the spring end of the system. Then the acceleration of the transducer $\ddot{x}$ is given by

$$\ddot{x} = \ddot{u}\frac{k}{k + m(2\pi f)^2} \tag{15.1}$$

where $\ddot{u}$ is the acceleration of the test item, and f is its frequency of vibration. If the acceleration of the transducer is to be within 10 percent of the acceleration of the test item, then from Eq. (15.1), k must have a value at least 10 times greater than the term $m(2\pi f)^2$. Since the undamped natural frequency f_n of the transducer-mounting system is given by $f_n = \frac{1}{2}\pi(k/m)^{1/2}$, the value of the natural frequency of the system must be at least 10 times the frequency of vibration of the test item—especially for the measurement of transients.

Alternatively, the unloaded dynamic environment at the mounting point can be calculated from the measured dynamic environment using the mechanical impedance ratio given by Eq. (3.4) of Ref. 1.

FIELD CALIBRATION TECHNIQUES

TRANSDUCERS

Various methods of calibrating transducers are described in Chap. 18. If a transducer is to be used under unusual temperature conditions, it is important to perform the calibration in the temperature range in which it will operate. Of these, the following are particularly convenient for use in the field.

Comparison Method. This is a rapid and convenient method of obtaining the sensitivity of a transducer. It is one of the most commonly used calibration techniques. Calibration is obtained by a direct comparison of the output generated when the transducer is attached to a vibration exciter with the output generated by a secondary standard transducer which is attached to the same vibration exciter and which is subject to precisely the same motion. The two transducers are mounted back to back, as illustrated in Fig. 11.3. Calibration by this method is limited to the frequency and amplitude ranges for which the secondary standard has been calibrated and for which the vibration exciter has adequate rectilinear motion. The secondary standard accelerometer should be calibrated against a National Institute of Standards and Technology (NIST) traceable reference, at least once a year, in compliance with MIL-STD-45662A.

Free-fall Calibration Method. The gravimetric free-fall calibration method (sometimes called a *drop test*) is a simple and rapid method of calibrating motion and force sensors. The transducer under test is allowed to fall freely for an instant of time under the influence of gravity; the peak signal then is measured for an acceleration of gravity having a value of 1*g*. This technique is illustrated in Fig. 11.5.

Earth's Gravitational Field Method. In the following technique (sometimes called the "*inversion method*" of calibration), the sensitive axis of the transducer is first aligned vertically in one direction of the earth's gravitational field, as shown in Fig. 15.5*A*. Then it is inverted so that its sensitive axis is aligned in the opposite direction, as shown in Fig. 15.5*B*. The transducer output is observed for a 2*g* change in acceleration, as shown in Fig. 15.5*C*. This method is limited in application to accelerometers having sensitivity down to 0 Hz; it is not recommended for calibration of accelerometers having significant transverse sensitivity.

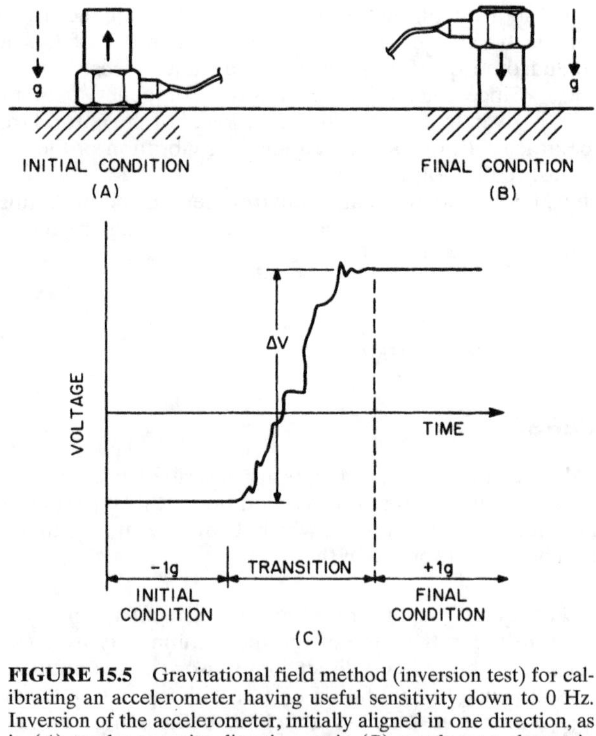

FIGURE 15.5 Gravitational field method (inversion test) for calibrating an accelerometer having useful sensitivity down to 0 Hz. Inversion of the accelerometer, initially aligned in one direction, as in (*A*), to the opposite direction, as in (*B*), produces a change in acceleration of 2*g*. The transducer output for this change is measured in (*C*). (*Courtesy of Quixote Measurement Dynamics, Inc.*)

OVERALL SYSTEM

Calibration of a complete vibration measurement system usually is referred to as *overall calibration* or *end-to-end calibration*. It is good practice to perform such a calibration at periodic intervals—particularly both before and after an extensive series of measurements. In such a calibration, the amplitude characteristics, phase characteristics, and linearity of the overall system are determined when the transducer is subject to a known acceleration, velocity, or displacement, for example, by means of a *field calibrator.*

Field Calibrator. This is a portable device on which a transducer can be mounted and subjected to a known acceleration, velocity, or displacement at a fixed frequency. Such an instrument (essentially a small, portable, battery-powered shaker) provides a convenient means for calibrating a transducer in the field and/or calibrating the overall vibration measurement system. For example, the hand-held device shown in Fig. 15.6 can be used to calibrate a transducer weighing up to 85 grams at a frequency of 79.6 Hz. This device is furnished with an internal oscillator and a stable, built-in reference accelerometer in a feedback loop controlling the electrodynamic exciter; the exciter subjects the transducer under test to a constant rms acceleration amplitude of 1*g*.

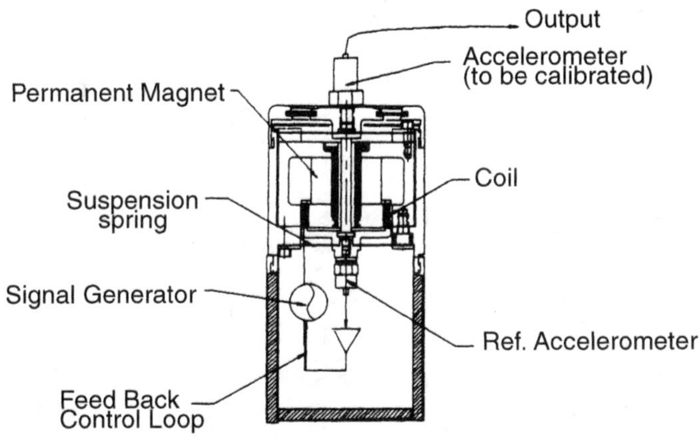

FIGURE 15.6 A handheld vibration calibrator especially designed for field application. (*Courtesy PCB Piezotronics, Inc.*)

Combining Calibration Characteristics of a Measurement System's Components. An overall system can be calibrated by combining the measured electrical characteristics of all components in the measurement system from one end to the other. Obtaining a system calibration in this way circumvents the difficulties of precise field calibration, but it requires that each element in the system be calibrated in the laboratory with extreme care and that the effects of the source and load impedances be completely accounted for. Thus, a system calibration is subject to the sum of the experimental errors introduced by the calibration of each element, in addition to any errors resulting from improper simulation of, or accounting for, loading effects. In general, the calibration of each element is performed before the system is assembled, and so this method is subject to error resulting from (1) undetected damage to components between calibration and use and/or (2) improper connections, misidentifications, or confusion in polarity.

Voltage Substitution Method of Calibration. A suitable simulated transducer for use in field checkout must duplicate the electrical outputs of the actual transducer for the various vibration conditions to be simulated. The simulated transducer must either (1) reproduce the electrical voltage- or current-generating characteris-

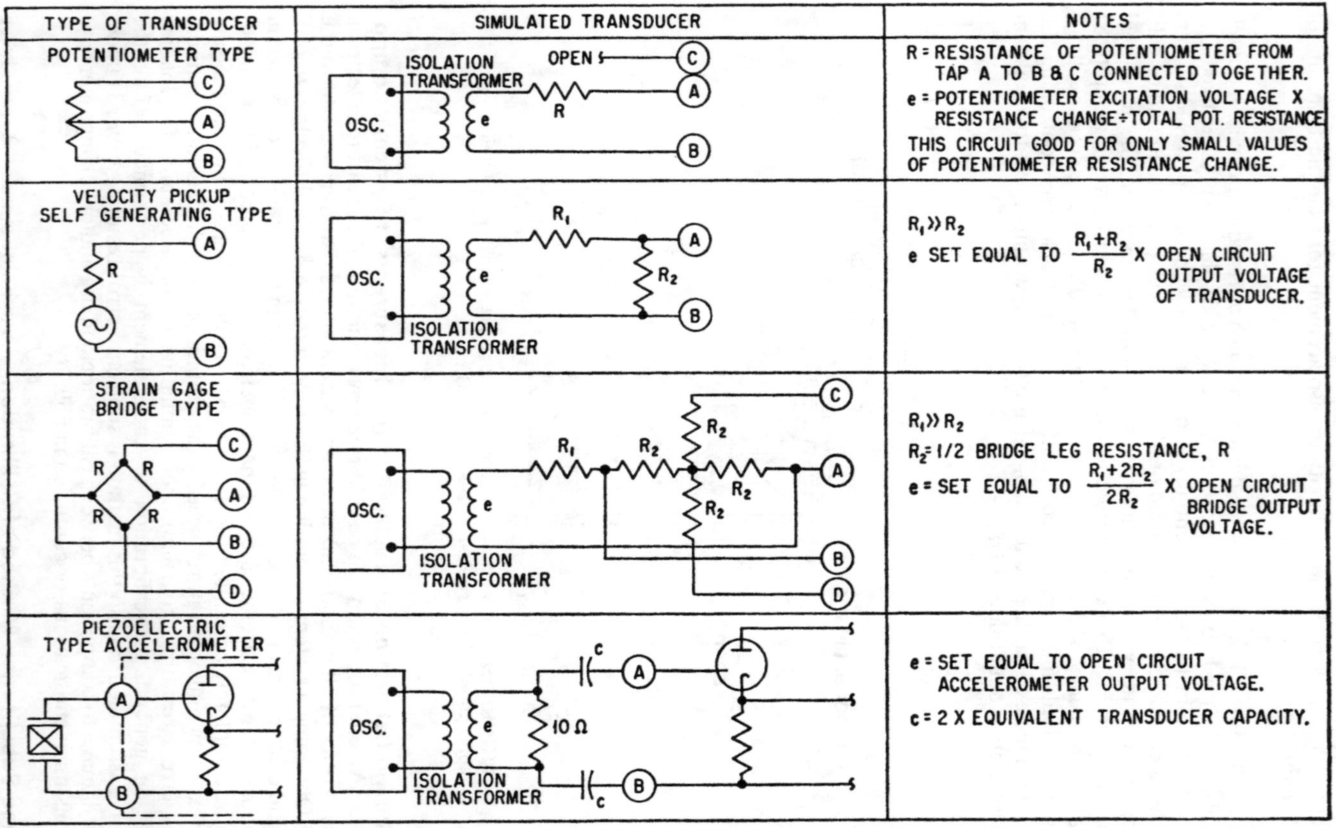

FIGURE 15.7 Electrical schematic diagrams of some common types of transducers and typical circuits used to simulate them during field calibration. Terminals labeled A and B are the signal lead connections to which either the transducer or the simulated transducer is connected.

tics of the actual transducer and have the same output impedance or (2) duplicate the electrical quantity generated by the actual transducer when connected to its load. Failure to meet these conditions will result in a different loading of the actual and simulated transducers and will probably cause calibration errors. It is important that the simulated transducer have the same electrical grounding configuration as the actual transducer; otherwise, electric-circuit noise and cross talk* will not be represented accurately when the simulated transducer is in use.

Typical examples of circuits which simulate transducers are shown in Fig. 15.7. The simulated transducer introduces an electrical signal into the measurement system, thereby simulating the response of the actual transducer.

CABLE AND WIRING CONSIDERATIONS

The method of data transmission between a transducer and the electronic instrumentation which follows it depends on the complexity of the problem. In general, cable is used for most problems, but the aerospace industry often relies on telemetry for data transmission. Many types of cable are available. The choice of a suitable cable depends primarily on the particular application, the transducer, the cable length, whether the transducer is followed by a voltage amplifier or charge amplifier, and environmental conditions. For example, cable jackets may be made of silicone rubber having a useful temperature range from −100 to 500°F (−73 to 260°C), of polyvinylchloride having a useful range from −65 to 175°F (−54 to 79°C), or of fused Teflon having a useful range from −450 to 500°F (−268 to 260°C). Special-purpose cables are available that can be used at much higher temperatures. In general, cable should be as light and flexible as possible—consistent with other requirements. The effect of the shunt capacitance of the cable following the transducer on the sensitivity of the transducer depends on the type of amplifier connected to the cable. If a voltage amplifier is used, there is a reduction in sensitivity of the transducer, given by Eq. (10.17). In contrast, when a charge amplifier is used, the effect of the shunt capacitance of the cable in reducing the sensitivity of the transducer is negligible (although the noise pickup in the high-impedance circuit increases with cable length). See Chapter 10 for details.

In the audio-frequency range, the series inductance L and the shunt leakage G of short, good-quality cables are negligibly small in comparison with other parameters and may be neglected. Figure 15.8*A* shows the equivalent low-frequency representation of a cable with distributed constants. For most purposes the simpler lumped-constant configuration of Fig. 15.8*B* is a sufficiently accurate representation. The quantities R_c and C_c are the total resistance of the conductors and the total capacitance between them, respectively. Values for a typical coaxial cable having a Teflon dielectric are R_c = 0.01 ohm/ft (0.03 ohm/m) and C_c = 29 pF/ft (88 pF/m).

The normal characteristic impedance of about 50 ohms for such cable has no significance in most measurement problems, where cables usually are relatively short. The open-circuit input impedance of the cable is almost exclusively capacitative. When terminated, it takes on the impedance of the load, modified by the series and shunt parameters.

**Cross talk* is the output of one measurement channel when a signal is applied to another measurement channel. Cross talk can be distinguished from other electrical disturbances because it is a function of the applied signal in the other measurement channel and disappears when this applied signal is removed.

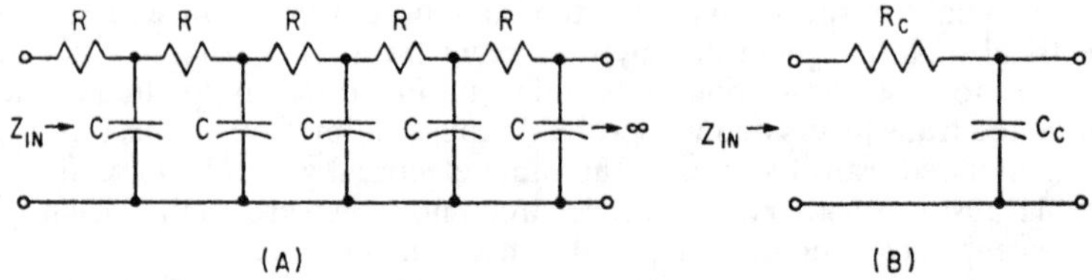

FIGURE 15.8 Successive approximations in the representation of a short, high-quality transmission line at audio frequencies. (*A*) Distributed constant configuration neglecting series inductance and shunt leakage. (*B*) Lumped-constant configuration.

In general, cables should be treated with the same care given transducers in shock and vibration measurement systems. The following are based on recommendations given in Ref. 2; they represent good engineering practice.

1. Attach a coaxial cable to a transducer by turning the connector nut onto the threads of the transducer (not vice versa) to avoid damage to the pins.
2. Avoid cable whip by tying down the cable at a point near the transducer and at regular intervals to avoid induced cable noise.
3. Screw the cable connection to the tightness specified by the manufacturer.
4. Loop the cable near the connector in a high-humidity environment, to allow condensation to drip off before reaching the connector.
5. Clean the cable connector before use (e.g., acetone or chlorothene) to remove contamination as a result of handling; the contamination can create a low impedance between the signal path and ground.
6. Check electrical continuity of cable conductors and shield if intermittent signals are observed. Then, flex the cable—especially near the connector—and observe if the signal is affected by flexing.
7. Select cables that are light and flexible enough to avoid loading the transducer and/or the structure under test, or exerting a force on the transducer.
8. Avoid twisting the cable when it is connected to the transducer.
9. Move the cable back and forth to determine if such movement generates unacceptable electrical noise; if so, tie the cable more securely or replace the cable.

CABLE NOISE GENERATION

When two dissimilar substances are rubbed together, they become oppositely charged—a phenomenon known as *triboelectricity,* illustrated in Fig. 15.9. Thus a charge may be generated when a cable is flexed, bent, struck, squeezed, or otherwise distorted, for then such friction takes place between the dielectric and the outer shield or between the dielectric and the center conductor.[3] A charge is generated across the cable capacitance so that a voltage appears across the termination of the cable.

Another mechanism by which noise may be induced in the cable results from the change in capacitance of the cable when it is flexed. If the transducer produces a charge across the cable, the change in capacitance results in a voltage change across the output of the cable, appearing as noise at the input of a voltage amplifier; it will not produce a similar change if a charge amplifier is used.

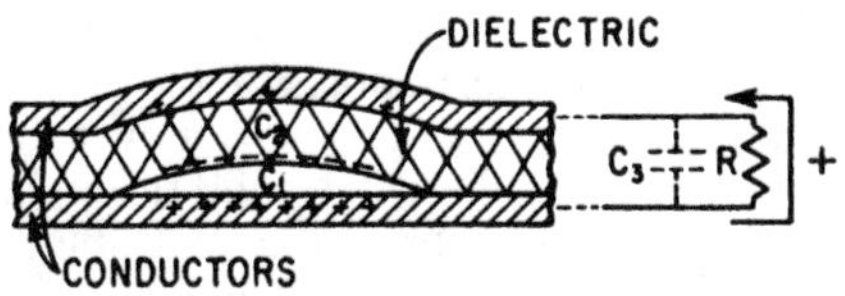

FIGURE 15.9 A section of cable during distortion, showing how separation of triboelectric charge leads to the production of cable noise across the termination resistance. (*After T. T. Perls.*[3])

Suppose the dielectric surfaces within the cable are coated so that an electrical leakage path is provided along the dielectric surface. Then if the cable shield is separated from the outer surface of the dielectric, the charges flow along the surface to the nearest point of contact of the dielectric and shield; without this leakage path, the charges would flow to the terminating impedance, where they would give rise to a noise signal. Such coatings are provided in low-noise cables which are available commercially. Cables of this type are capable of withstanding considerable abuse before becoming noisy. Usually they are tested by the manufacturer continuously along their lengths to assure meeting the low-noise characteristics. It is important in fitting such a cable with a connector, or in splicing such a cable, that no conducting material be allowed to form a leakage path between the conductors. Carbon tetrachloride and xylene are satisfactory solvents and cleaning agents.

NOISE-SUPPRESSION TECHNIQUES

Under certain conditions of use and environment, spurious signals (noise) may be induced in wiring and cables in a measurement system. Then there will be signals at the termination of the system that were not present in the transducer output.

Electrical noise may be generated by motion of some parts of the wiring because of variation in contact resistance in connectors, because of changes in geometry of the wiring, or because of voltages induced by motion through, or changes in, the electrostatic fields or magnetic fields which may be present. No cable should carry wiring *both* for data transmission and for electrical power; all electrical power wiring should be twisted pair. In general, such electrical noise will be reduced if the cable is securely fastened to the structure at frequent intervals and if connectors are provided with mechanical locks and strain-relief loops in their cables. Precautions taken to avoid interference usually include the use of shielding, cables which are only as long as necessary, and proper grounding. Cable jackets must be selected that will not deteriorate under the measurement environment. In addition, the use of a transducer containing an internal amplifier can provide advantages in noise suppression.

Shielding. A change in the electric field or a change in the magnetic field around a circuit or cable may induce a voltage within it and thus be a source of electrical noise. Such electrical interference can be avoided by completely surrounding the circuit or cable with a conductive surface which keeps the space within it free of external electrostatic or magnetic fields. This is called *shielding.* Protection against changes in each type of field is different.

Electrostatic Shields. Electrostatic shields provide a conducting surface for the termination of electrostatic lines of flux. Stranded braid, mesh, and screens of good electrical conductors such as copper or aluminum are good electrostatic shields. Most shielded cables use copper braid as the outer conductor and electrostatic shield. A good magnetic shield is also a good electrostatic shield, but the converse is not true. For installations where cable lengths are especially long, where impedances are high, or where noise interference is highly objectionable, double-shielded cable is sometimes used. In this type of cable, a second shielding braid is woven over the cable jacket, electrically insulating it from the inner shield; the inner braid furnishes additional shielding against electrostatic fields which penetrate the first shield. The shields should be connected to ground at one point only, as explained under "Grounding; Avoiding Ground Loops."

Magnetic Shields. Magnetic shields are effective partly because of the short circuiting of magnetic lines of flux by low-reluctance paths and partly because of the cancellation resulting from opposing fields set up by eddy currents. Accordingly, they are made from high-permeability materials such as Permalloy, are as thick as possible, and contain a minimum of joints, holes, etc.

Magnetic fields associated with current-carrying power lines, electronic equipment, and power transformers are among the most troublesome sources of magnetic interference in instrumentation setups—chiefly at the frequency of the power line and its harmonics. Since these fields attenuate rapidly with distance from the source, the most practical solution for this type of interference usually is to keep the signal cables as far from the power source as possible.

Grounding; Avoiding Ground Loops. A circuit is said to be grounded when one terminal of the circuit is connected to the "earth." Grounding removes the potential difference between that side of the circuit and earth, and the variable stray capacitances which tend to induce voltages in "floating" (i.e., ungrounded) systems. Water pipes make good ground connections because of their intimate contact with the earth.

Ground loops are formed when a common connection in a system is grounded at more than one point, as illustrated in Fig. 15.10, where the cable shield is grounded at both ends. Since it is unlikely that the two grounds will be at a common potential, their potential difference, e_{gnd}, will be the source of circulating currents in the ground loop. Then a signal produced by the transducer will be modulated by the potential e_{gnd}, thereby introducing noise in the measurement system. Such a condition may occur when one end of a cable is connected to one side of the electrical output of a transducer that has been grounded to the transducer's housing and the other end of the cable is connected to a voltage amplifier or signal conditioner which is also grounded (usually to the case of the instrument). Then, a ground loop

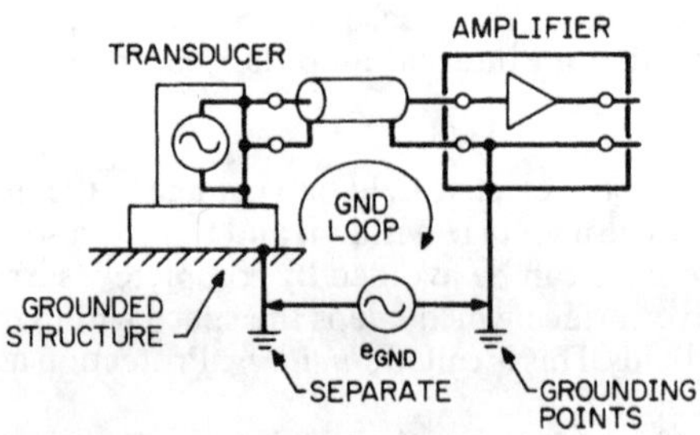

FIGURE 15.10 Ground loop in a system as a result of grounding the cable shield at two points. Then, the input signal e_1 is modulated by the potential difference e_{gnd} which develops between these two points.

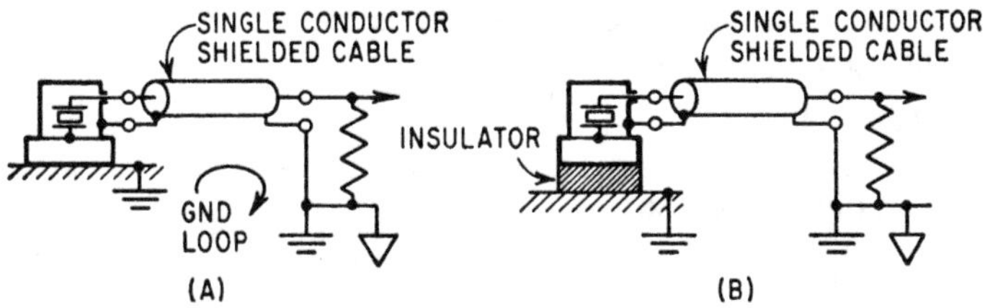

FIGURE 15.11 (*A*) A ground loop formed when the "low" sides of both the transducer and the amplifier are connected to their respective cases, which are grounded. (*B*) The ground loop shown in (*A*) is broken by isolating the case of the transducer from ground.

will be formed. *Such a condition must be avoided by grounding the circuit at only one point.* Thus the circuit shown in Fig. 15.11*A* will result in noise because of the ground loop, but by insulating the transducer as shown in Fig. 15.11*B* the ground loop has been broken.

DATA SHEETS FOR LOGGING TEST INFORMATION

When data are acquired in the field, measurement conditions may be far from ideal; environmental conditions may be unfavorable, and the time available for measurements may be extremely limited. Therefore it is good practice to prepare data sheets that are relatively simple and that require a minimum amount of writing; for example, use multiple-choice entries. The data sheets should include sufficient information so that someone else, at a later time, could duplicate the measurement setup on the basis of information supplied by the data sheets. If there are any anomalies that occur during the test, they should be duly noted. In general, the following information should be included:

Basic data concerning the test measurements:

- Date, times, and duration of test.
- Identification of test by test number.
- Identification of equipment, machine, or device under test.
- Conditions of operation during the measurement.
- Any anomalies in operation and their times of occurrence.
- Location of test, using diagram where appropriate.
- Environmental conditions during test; note anomalies where appropriate.
- Persons participating in the test.

Equipment, including transducers, cables, signal conditions, data recorders, telemeter:

- Type.
- Manufacturer, model number, and serial number.
- Transducer sensitivity, exact location, orientation, and type of mounting.

- Signal conditioner and amplifier gain and attenuator settings; note any changes in these settings during the test.
- Filter settings, if any.
- Recorder speed, number of tracks, tape speed, gain settings; note any changes in these settings during the test.

Calibration information:

- Transducer calibration.
- Overall system (end-to-end) calibration of system.
- Phase of output signal relative to input signal.
- Any changes in calibration between pretest and posttest conditions.

REFERENCES

1. Himelblau, H., and A. G. Piersol, "Handbook for Dynamic Data Acquisition and Analysis," 2d ed., IEST Reference Document DTE012.2, Institute of Environmental Sciences and Technology, Arlington Heights, Ill., 2006.
2. Endevco Corp.: "A Guide for Accelerometer Installation," TP 319, San Juan Capistrano, Calif., 1999.
3. Perls, T. A.: *J. Appl. Phys.,* **23**(6):674 (1952).

CHAPTER 16
CONDITION MONITORING OF MACHINERY

Ronald L. Eshleman

INTRODUCTION

Condition monitoring of machinery is the measurement of various parameters related to the mechanical condition of the machinery (such as vibration, bearing temperature, oil pressure, oil debris, and performance), which makes it possible to determine whether the machinery is in good or bad mechanical condition. If the mechanical condition is bad, then condition monitoring makes it possible to determine the cause of the problem.

Condition monitoring is used in conjunction with *predictive maintenance,* i.e., maintenance of machinery based on an indication that a problem is about to occur. In many plants predictive maintenance is replacing *run-to-breakdown maintenance* and *preventive maintenance* (in which mechanical parts are replaced periodically at fixed time intervals regardless of the machinery's mechanical condition). Predictive maintenance of machinery:

- Avoids unexpected catastrophic breakdowns with expensive or dangerous consequences.
- Reduces the number of overhauls on machines to a minimum, thereby reducing maintenance costs.
- Eliminates unnecessary interventions with the consequent risk of introducing faults on smoothly operating machines.
- Allows spare parts to be ordered in time and thus eliminates costly inventories.
- Reduces the intervention time, thereby minimizing production loss. Because the fault to be repaired is known in advance, overhauls can be scheduled when most convenient.

This chapter describes the use of vibration measurements for monitoring the condition of machinery.[1,2] Vibration is the parameter which can be used to predict the broadest range of faults in machinery most successfully. This description includes:

- Selection of an appropriate type of monitoring system (permanent or periodic)
- Establishment of a condition monitoring program
- Fault detection
- Spectrum interpretation and fault diagnosis
- Special analysis techniques
- Trend analysis
- Correction methods

TYPES OF CONDITION MONITORING SYSTEMS

Condition monitoring systems are of two types: periodic and permanent. In a *periodic monitoring system* (also called an *off-line condition monitoring system*), machinery vibration is measured (or recorded and later analyzed) at selected time intervals in the field; then an analysis is made either in the field or in the laboratory. Advanced analysis techniques usually are required for fault diagnosis and trend analysis. Intermittent monitoring provides information at a very early stage about incipient failure and usually is used where (1) very early warning of faults is required, (2) advanced diagnostics are required, (3) measurements must be made at many locations on a machine, and (4) machines are complex.

In a *permanent monitoring system* (also called an *on-line condition monitoring system*), machinery vibration is measured continuously at selected points of the machine and is constantly compared with acceptable levels of vibration. The principal function of a permanent condition monitoring system is to protect one or more machines by providing a warning that the machine is operating improperly and/or to shut the machine down when a preset safety limit is exceeded, thereby avoiding catastrophic failure and destruction. The measurement system may be permanent (as in parallel acquisition systems where one transducer and one measurement chain are used for each measurement point), or it may be quasi-permanent (as in multiplexed systems where one transducer is used for each measurement point but the rest of the measurement chain is shared between a few points with a multiplexing interval of a few seconds).

In a permanent monitoring system, transducers are mounted permanently at each selected measurement point. These sensors may be hard-wired to a central location or connected through a transmitter-receiver wireless system. Such an approach is normally used in critical applications where: (1) no personnel are available to perform measurements (offshore, remote pumping stations, etc.), (2) it is necessary to stop the machine before a breakdown occurs in order to avoid a catastrophic accident, (3) an instantaneous fault may occur that requires machine shutdown, and (4) the environment (explosive, toxic, or high-temperature) does not permit the human involvement required by intermittent measurements.

Before a permanent monitoring system is selected, preliminary measurements should be made periodically over a period of time to become acquainted with the vibration characteristics of the machine. This procedure will make it possible to select the most appropriate vibration measurement parameter, frequency range, and normal alarm and trip levels.

ESTABLISHING A CONDITION MONITORING PROGRAM

A condition monitoring program may be established to check the satisfactory operation of a single machine or, more usually, it is established to check the operation of a number of machines, perhaps all the machines in an entire plant. The following steps are usually considered in the establishment of such a program, depending on the type of machine and impact of failure on the operation of machines.

Step 1. *Determine the type of condition monitoring system,* described in the preceding section, that best meets the needs of the plant.

Step 2. *Make a list of all of the machines to be monitored* (see, for example, Table 16.1), based on the importance of these machines in the production line.

Step 3. *Tabulate the characteristics of the machines* that are important in conducting vibration analyses of the machines of step 2. These characteristics are associated with machine construction such as the natural frequencies of shafts, casings, and pedestals, and operational and defect responses. A tabulation of machine frequencies is important because fault analysis is conducted by matching machine frequencies to measured frequencies appearing in a spectrum. The following machine characteristics provide the necessary information for fault analysis.

- Shaft rotational speeds, bearing defect frequencies, number of teeth in gears, number of vanes and blades in pumps and fans, and number of motor poles, stator slots and rotor bars.
- Vibration responses due to process changes, such as temperature and pressure.
- Fault responses associated with specific machine types, such as motors, pumps, and fans.
- Sensitivity to instability in components, such as fluid film bearings and seals due to wear and clearance.
- Effects of loads or changes in operating conditions.
- Effects of mass unbalance, misalignment, distortion, and other malfunction/defect excitations on vibration response.

Step 4. *Select the most appropriate vibration measurement parameter.* When an accelerometer is employed as the sensing device in a condition monitoring system,

TABLE 16.1 Machinery Classification for Monitoring

Machinery classification	Result of failure
Critical	Unexpected shutdown or failure causes significant production loss.
Interrupts production	Unexpected shutdown or failure causes minor interruptions in production.
Causes inconvenience	Inconvenience in operation, but no interruption in production.
Noncritical	Production is not affected by failure.

the resulting *acceleration* signal can be electronically integrated to obtain *velocity* or *displacement.* Any one of these three parameters may be used in an analysis. However, the integration process may compromise the data. Measurement sensitivity and dynamic range should be considered. Displacement provides the best sensitivity in the low-frequency range (0 to 10 Hz) and acceleration in the high-frequency range (above 1000 Hz). Velocity usually provides the "flattest" spectrum and, hence, a relatively uniform sensitivity over the medium-frequency range (10 to 1000 Hz).

Step 5. *Select one of the following vibration transducers that will best meet the requirements of step 4.*

Displacement Transducer. A displacement transducer is a transducer that converts an input mechanical displacement into an electrical output that is proportional to the input displacement. Displacement transducer of the eddy-current type (described in Chap. 10), which have noncontacting probes, are commonly used to measure the relative motion between a shaft and its journal bearings. This information can be related directly to physical values such as mechanical clearance or oil-film thickness; that is, it can give an indication of incipient rubbing or excessive pressure on the bearing babbit. Relative shaft vibration provides information on the current machine condition, but is principally used in permanent monitoring systems for machine protection, in other words, immediate shutdown in the event of serious trouble. Noncontacting displacement transducers are sensitive to shaft runout and surface irregularities, either mechanical or electrical.

Accelerometer Transducer. Transducers of this type (see Chap. 10) are used to detect faults through seismic measurements. In the medium-frequency range (10 to 1000 Hz), the acceleration signal is integrated to velocity to detect faults such as mass unbalance, misalignment, and looseness. In the high-frequency range (above 1000 hz), the acceleration signal is used directly to detect rolling element bearing and gear defects.

Step 6. *Select the measurement locations.* When a periodic (off-line) monitoring system is employed, the number of points at which measurements are made is limited only by the requirement for keeping measurement time to a minimum. As a general rule, bearing vibration measurements are made in the radial direction on each accessible bearing, and in the axial direction on thrust bearings. It is not usually necessary to measure bearing vibration in both the horizontal *and* the vertical direction, since both measurements give the same information regarding the forces within the machine; this information is merely transmitted through two different transmission paths. This applies for *detecting* developing faults. It will later be seen, however, that in order subsequently to *diagnose* the origin of the impending fault, measurements in both the horizontal and the vertical direction may give valuable information. When measuring shaft vibrations with permanently mounted proximity transducers, it is convenient to use two probes on each bearing, located at 90° from each other, thereby providing an indication of the orbit of the shaft within the bearing. Axial displacement transducers, programmed to shut the machine down on preset levels, are mounted where a thrust measurement will protect the machine rotating parts, such as blades, from rubbing the stationary casing due to fault-induced axial forces.

When a permanent (on-line) monitoring system is employed using a seismic pickup, the number of measurement points usually is minimized for reasons of economy. Selection must be made following a study of the vibration spectra of different bearings in order to locate those points where all significant components related to the different expected faults are transmitted at measurable vibration levels if full spectrum comparison is performed. If only broadband measurements are monitored, then a further requirement is that all frequency components related to the expected faults must be of approximately the same level within the selected frequency range. Otherwise, measurements must be made in selected frequency bands.

Step 7. *Select the time interval between measurements.* The selection of the time interval between measurements requires knowledge of the specific machine. Some machines develop faults quickly, and others run trouble-free for years. A compromise must be found between the safety of the system and the time taken for measurements and analysis. Measurements should be made frequently in the initial stages of a condition monitoring program to ensure that the vibration levels measured are stable and that no fault is already developing. When a significant change is detected, the time interval between measurements should be reduced sufficiently so as not to risk a breakdown before the next measurement. The trend curve will help in determining when the next measurement should be performed.

Step 8. *Establish an optimum sequence of data acquisition.* The sequence in which data acquired in a condition monitoring program must be planned so that the data are acquired efficiently. For example, the data collection may be planned on the basis of plant layout, on the type of data required, or on the sequence of components in the machine train, from driver to driven components.

Step 9. *Select the data acquisition, storage, and analysis system.* The data acquisition is normally performed sequentially on a preprogrammed route by a small, lightweight special computer called a *data collector.* This device acquires the data from a transducer and can store overall values as well as digitized time waveforms and spectra. It can also be used as an analyzer when used "off route." The results of the route are then downloaded to a host computer that analyzes, trends, and stores the data, providing a report if the machine condition, including trend charts, overall levels, and alarms, is violated. In a permanent monitoring system, the complete data acquisition, analysis, and alarms (including warning and shutdown) are contained in a package tied to the overall plant process system.

FAULT DETECTION IN ROTATING MACHINERY

It is highly desirable to be able to detect all types of faults likely to occur during the operation of rotating machinery. Such faults range from vibrations at very low frequencies (subsynchronous components indicating looseness, oil whirl, faulty belt drive, etc.) to vibrations at very high frequencies (tooth-meshing frequencies, blade-passing frequencies, frequencies of structural resonances excited by faulty rolling-element bearings, etc.). Such detection should be applicable to the complete range of machines in a plant, which operate from very low to very high speed. This requires the selection of equipment and analysis techniques, which cover a very broad frequency range.

Measurements of *absolute* vibration levels of bearings provide no indication of the machine's condition, since they are influenced by the transmission path between the force and the measurement point, which may amplify some frequencies and attenuate others. Bearing vibration levels change from one measurement point to another on a given machine, since the transmission paths are different; they also change for the same reason from machine to machine for measurements made at the same measurement point. Therefore, in estimating the condition of a machine, it is essential to monitor *changes* in vibration from a reference value established when the machine was known to be in good condition. Changes are expressed as a ratio or as a *change of level,* i.e., the logarithm of a ratio, in decibels.

The objective of condition monitoring of a machine is to predict a fault well in advance of its occurrence. Therefore, a measurement of the overall vibration level sometimes will not provide successful prediction because the highest vibration component within the overall frequency range may dominate the measurement and not be the result of the defect (particularly with rolling element bearings). This is illustrated in Fig. 16.1, which shows an example where overall measurements of the vibration velocity resulted in an incorrect prediction with an overestimate of the lead time. The early successful detection of faults in machinery can be ensured only by comparison with a *reference spectrum.* This section compares types of spectrum analysis for this purpose.

FALSE ALARMS

Changes in machinery vibration may result from a number of causes which are not necessarily related to the deterioration of the machine. For example, a change in speed of the machine or a change in the load on the machine may greatly modify the relative amplitudes of the different components of vibration at a fixed transducer location or may modify the relative pattern of vibration at different locations. Depending on the criteria used for fault detection, such changes may result in a false indication of deterioration of the machine. Appropriate selection of the technique employed and/or alarm levels can avoid such false alarms.

HOW SPECTRUM CHANGES ARE RELATED TO THE CONDITION OF A MACHINE

To obtain information about changes in condition of a machine, vibration spectra should be compared only for similar operating conditions. The influence of operating condition of the machine (such as machine speed, load, and temperature) on the vibration parameter being measured varies greatly for different types of machines. Speed changes of up to 10 percent usually can be compensated for, and spectra can be compared. If the speed changes are greater than this value, the operating condition of the machine should be considered to be different and a new reference spectrum used as a basis of comparison. The reference spectrum need not be measured when the machine is new (after allowing for a run-in period). The reference spectrum can be determined at any time during the life of a machine provided the vibrations are stable, since a stable spectrum is a sign of stable operation of the machine. The principal difficulty is to establish when changes in the spectrum are sufficiently large to warrant stopping the machine.

Most national and international standards for the measurement of seismic vibration do not consider frequency spectra; instead, they give values for vibration changes of the rms value of the overall velocity amplitude from 10 to 1000 Hz (or

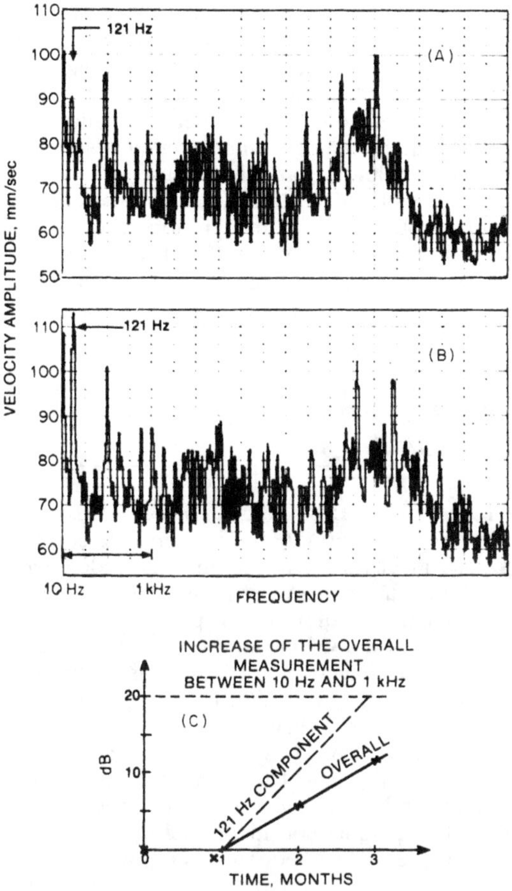

FIGURE 16.1 Trend analysis performed on an overall measurement and on an individual component. (*A*) The velocity spectrum of vibration measured on a gearbox after installation. Note the high amplitude of the 480-Hz component, dominating the reference spectrum. (*B*) The velocity spectrum three months later. Note the dramatic increase in the 121-Hz component, which corresponds to the output shaft speed of the gearbox. (*C*) Curves comparing the increase in the 121-Hz component in the velocity spectrum; the increase in overall velocity in the band from 10 to 1000 Hz indicates a developing fault.

10,000 Hz) for machines in good and bad condition. These ratios have successfully been transposed to characteristic components in the vibration spectrum such as unbalance or misalignment. Usually, a change in the seismic vibration amplitude (measured in terms of acceleration, velocity, or displacement) on any characteristic component from the spectrum by a factor of 2 to 2.5 (6 to 8 dB in vibration level) is considered significant; a change by a factor of 8 to 10 (18 to 20 dB in vibration level) is considered critical, unless specified otherwise by the manufacturer. Limits for shaft vibration measurements, giving the relative motion of the shaft inside the bearing, directly relate to physical bearing clearance in the machine. The required time interval between measurements varies greatly from one machine to another and depends directly on the expected mean time between failure and the deterioration rate of the expected failures; therefore, measurements should be made more frequently as soon as incipient deterioration is noticed.

Successful fault detection in machinery is the first step toward a successful condition monitoring program. Early recognition of deterioration is the key to valuable fault diagnosis and efficient trend analysis. Consequently, this phase of condition monitoring should not be neglected, although sometimes it may seem tedious.

SPECTRUM INTERPRETATION AND FAULT DIAGNOSIS

Commercially available computer-based fast Fourier transform analyzers provide a suitable tool for spectrum interpretation. They provide constant bandwidth (on a linear frequency scale), and, by means of zoom or extended lines of resolution, they also provide very high resolution in any frequency range of interest. This permits (1) early recognition and separation of harmonic patterns or sideband patterns and (2) separation of closely spaced individual components. Fast Fourier transform (FFT) analyzers also may provide diagnostic tools such as synchronous time averaging, cepstrum analysis, peakness analysis, and/or use of the Hilbert transform for amplitude and phase demodulation (see Chap. 14).

Table 16.2 classifies different types of faults and indicates at which frequency the faults are displayed in a vibration spectrum. Although such a table is of considerable help in spectrum interpretation, any such simplified presentation must be used with care, as illustrated by the examples considered below. The various faults can be classified according to their spectral components, as follows.

SUBSYNCHRONOUS COMPONENTS

Subsynchronous components of vibration (at frequencies below the rotational speed of the machine) usually occur where sleeve bearings are used. The most common are the vibrations due to oil whirl, hysteresis whirl, resonant whirl, or mechanical looseness. Figure 16.2 shows a spectrum measured on the journal bearing of a centrifugal compressor with mechanical looseness. A characteristic pattern of half-order harmonics of rotation speed can be clearly seen. Figure 16.3 shows a spectrum of the journal bearing of a pump in which a developing oil whirl shows up clearly at 21 Hz (42 percent of the rotation speed) and its second harmonic.

Both examples clearly indicate how the use of a linear frequency scale facilitates the diagnosis of the fault by providing a clear indication of the different types of harmonic patterns. High resolution is required to separate a half-order harmonic component due to looseness or subharmonic resonance (exactly 50 percent of rotation speed) from a component due to oil whirl (42 to 48 percent of rotation speed).

LOW HARMONICS OF ROTATIONAL SPEED

Low harmonics of the rotational speed are generated by shaft unbalance, misalignment, and eccentricity, as well as cracks in shafts and bent shafts. These various faults may be difficult to distinguish, since they are mechanically related.[4] A bad coupling may result in misalignment. A bent shaft results in unbalance. Even a well-known and well-defined fault such as unbalance may give misleading vibration components. The exciting fault due to eccentric masses is a centrifugal force (thus radial) rotating at the shaft speed and is therefore expected to result in a component in the vibration spectrum at the machine speed and in the radial direction. However, dynamic unbalance may also result in a rocking motion and consequently in vibration in both radial and axial directions. In the same way, if there is a nonlinear transmission path from the point where the force is applied to the point of measurement, a rise in the harmonics of the rotation speed can be observed in the vibration spectrum, due to distortion of the signal.

The phase relationship between bearings provides essential information for differentiating these various types of faults. As an example, unbalance will generate a rotating force, and therefore the phase relationship between bearings can be expected to be identical in both horizontal and vertical directions (in the absence of resonances). For mass unbalance, the phase difference between a vertical and horizontal transducer is 90° on the same bearing. Misalignment, however, does not create a rotating force, and thus the phase relationship between bearings in both vertical and horizontal directions can be vastly different.

HARMONICS OF THE POWER LINE FREQUENCY

Vibrational components, which are related to the frequency of the power line or variable frequency drive, or to the difference between the synchronous frequency and the rotational speed, occur in electric machines such as induction motors or generators. These vibrations are due to electromagnetically induced forces. These forces, which occur in the case of a malfunction in the electric machine, are related to the air gap between the rotor and the stator and to the current. The faults on the electric machine are due either to the stator (called *stationary faults*) or to the rotor (called *rotating faults*). They may originate from either a variation in the air gap or a variation in the current.

Figure 16.4 shows a vibration signal measured on the rolling-element bearing of an asynchronous electric motor. By zooming in the region of the high-level 100-Hz component (i.e., twice the line frequency in Europe), this component can be diagnosed as the pole-passing frequency of 100 Hz and not the 2 × rotation speed at 99.6 Hz which could have been an indication of a faulty alignment. This demonstrates the value of being able to zoom to the frequency region containing the component of interest. The zoom or extended lines of resolution provides sufficient resolution to separate closely spaced components. It is of no help in analyzing synchronous machines or generators, since the rotation speed and the line (mains) frequency are identical. In such a case, the machine should be permitted to coast to a stop. When the power is cut, electrically induced components of vibration disappear, and the harmonics of the rotation speed gradually decrease in frequency and amplitude.

Vibration forces resulting from an effective variation of the reluctance in the magnetic circuit as a function of the rate of the stator and rotor slot passing will be

TABLE 16.2 A Vibration Troubleshooting Chart

Nature of fault	Frequency of dominant vibration, Hz = rpm/60	Direction	Remarks
Rotating members out of balance	1 × rpm	Radial	A common cause of excess vibration in machinery
Misalignment and bent shaft	Usually 1 × rpm Often 2 × rpm Sometimes 3 and 4 × rpm	Radial and axial	A common fault
Damaged rolling element bearings (ball, roller, etc.)	Impact rates for the individual bearing component Also vibrations at high frequencies (2 to 60 kHz) often related to radial resonances in bearings	Radial and axial	Uneven vibration levels, often with shocks 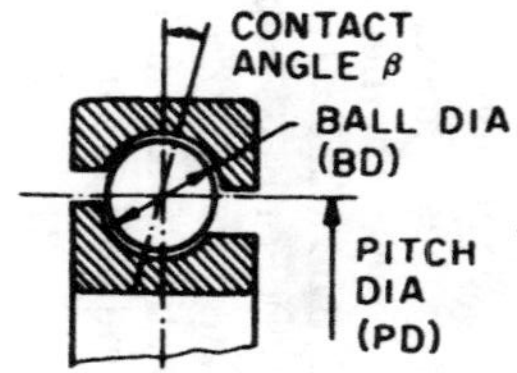 Impact Rates f (Hz): For Outer Race Defect $f(\text{Hz}) = \frac{n}{2} f_r\left(1 - \frac{\text{BD}}{\text{PD}} \cos\beta\right)$ For Inner Race Defect $f(\text{Hz}) = \frac{n}{2} f_r\left(1 + \frac{\text{BD}}{\text{PD}} \cos\beta\right)$ For Ball Defect $f(\text{Hz}) = \frac{\text{PD}}{\text{BD}} f_r\left[1 - \left(\frac{\text{BD}}{\text{PD}} \cos\beta\right)^2\right]$ n = number of balls or rollers f_r = relative rps between inner and outer races
Journal bearings loose in housing	Subharmonics of shaft rpm, exactly ½ or ⅓ × rpm	Primarily radial	Looseness may only develop at operating speed and temperature (e.g., turbomachines)

Oil-film whirl or whip in journal bearings	Slightly less than half shaft speed (42 to 48 percent)	Primarily radial	Applicable to high-speed (e.g., turbo) machines
Hysteresis whirl	Shaft critical speed	Primarily radial	Vibrations excited when passing through critical shaft speed are maintained at higher shaft speeds. Can sometimes be cured by tightening the rotor components.
Damaged or worn gears	Tooth-meshing frequencies (shaft rpm × number of teeth) and harmonics	Radial and axial	Sidebands around tooth-meshing frequencies indicate modulation (e.g., eccentricity) at frequency corresponding to sideband spacings. Normally only detectable with very narrow-band analysis and cepstrum analysis.
Mechanical looseness	2 × rpm		Also sub- and interharmonics, as for loose journal bearings
Faulty belt drive	1, 2, 3, and 4 × rpm of belt	Radial	The precise problem can usually be identified visually with the help of a stroboscope
Unbalanced reciprocating forces and couples	1 × rpm and/or multiples for higher-order unbalance	Primarily radial	
Increased turbulence	Blade & vane passing frequencies and harmonics	Radial and axial	An increased level indicates increased turbulence
Electrically induced vibrations	1 × rpm or 2 times line frequency	Radial and axial	Should disappear when power is turned off

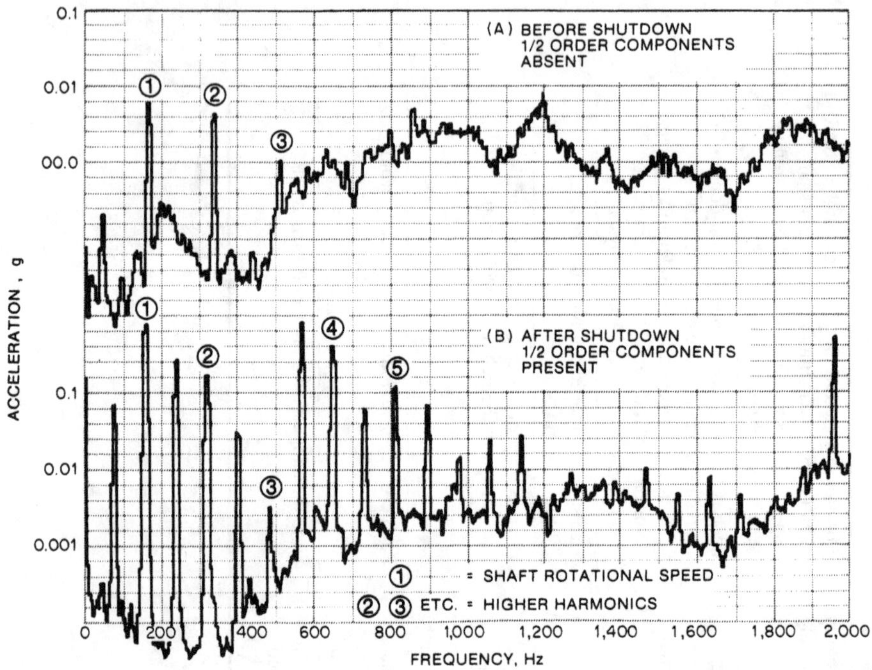

FIGURE 16.2 Acceleration spectra of a journal bearing on a centrifugal compressor. (*A*) Compressor in *good condition.* Before shutdown, the vibration pattern is normal with few harmonics of the compressor's rotation speed and broadband noise at higher frequencies due to inherent turbulences. (*B*) Compressor *with looseness* in the journal bearing. After shutdown, the higher-order harmonics have an increased amplitude, and the presence of half-order harmonics can be observed.

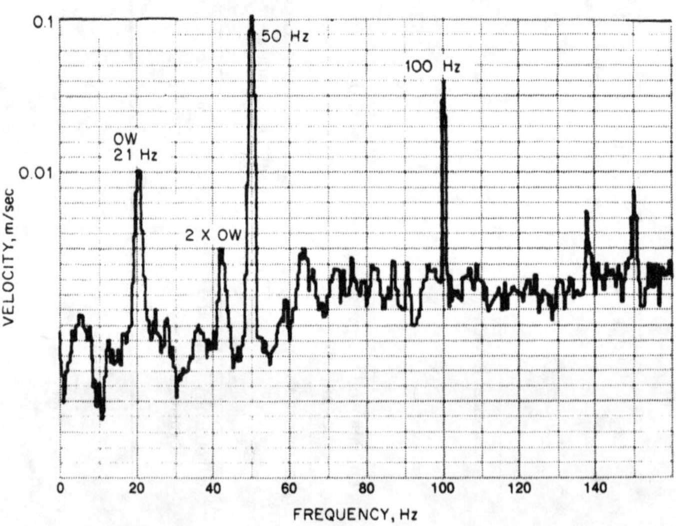

FIGURE 16.3 Spectrum analysis showing component due to oil whirl at 42 percent of the rotation speed measured on the journal bearing of a pump.

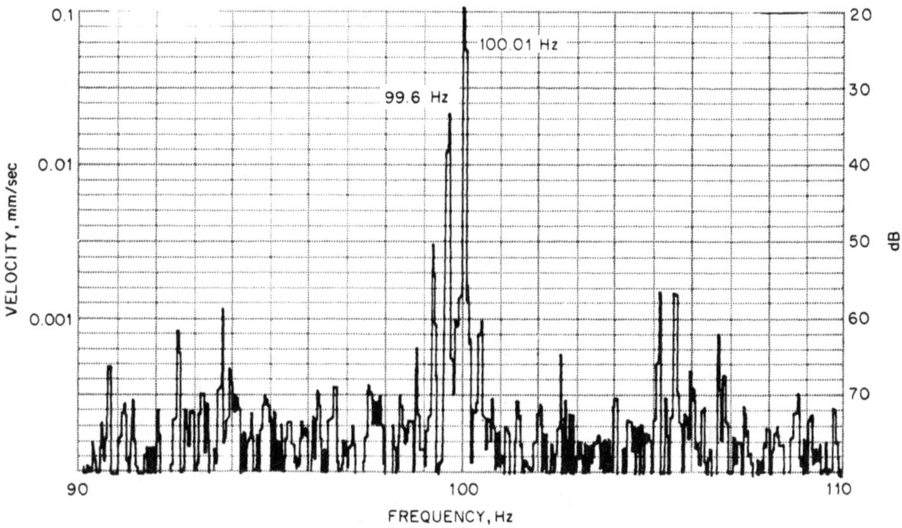

FIGURE 16.4 Spectrum analysis employing zoom frequency analysis around the 100-Hz component, measured on the rolling-element bearing of an asynchronous electric motor.

present even in a motor which is in good condition. These vibrations occur at the slot harmonics given by the following equation:

$$f_{\text{slot}} = R_s f_{\text{rot}} \pm k f_{\text{line}}$$

where f_{slot} = slot passing frequency
R_s = number of rotor slots
f_{rot} = rotating speed
k = zero or even number
f_{line} = power line frequency

The vibration components at low frequency differentiate between stator problems and rotor problems. They do not, however, indicate whether the faults originate from variations in air gaps or current. The components at the slot harmonics, on the other hand, will behave differently depending on whether the fault originates from an air gap or current variation as indicated in Table 16.3.

Figure 16.5 shows that by using a zoom around slot harmonics, sidebands can be observed at twice the slip frequency, thereby permitting the diagnosis of broken rotor bars. For a four-pole motor, sidebands occur at four times the slip frequency.

As an alternative to using signal analysis of vibration, signal analysis of the motor current may be used to monitor certain types of problems. It is a more direct measurement for all electrical problems and, with the help of algorithms, makes it possible, for example, to determine with a certain amount of accuracy the number of broken rotor bars. Reference 4 mentions that mechanical phenomena such as worn gears, tooth wear, and steam packing degradation (in motor-operated valves) can be detected as well. It also mentions the applicability of this technique to dc motors.

TABLE 16.3 Troubleshooting Guide of Induction Motor Vibrations

Static eccentricity	2 × line frequency and components at $\omega \times [nR_s(1-s)/p \pm k_1]$	Radial	Can result from poor internal alignment, bearing wear, or from local stator heating (vibration worsens as motor heats up).
Weakness/looseness of stator support, unbalanced phase resistance or coil sides Shorted stator laminations/turns	2 × line frequency	Radial	Referred to as "loose iron." Difficult to differentiate between this group using only vibration analysis, but they will also be apparent at no load as well as on load.
Loose stator laminations	2 × line frequency and components spaced by 2 × line frequency at around 1 kHz	Radial	Can have high amplitude but not usually destructive. The high-frequency components may be similar to static eccentricity.
Dynamic eccentricity	1 × rpm with 2 × slip-frequency sidebands and components at $\omega \times [((nR_s \pm k_e) \times (1-s)/p) \pm k_1]$	Radial	Can result from rotor bow, rotor runout, or from local rotor heating (vibration worsens as motor heats up).
Broken or cracked rotor bar Loose rotor bar Shorted rotor laminations Poor end-ring joints	1 × rpm with 2 × slip-frequency sidebands and components similar to those given above for dynamic eccentricity with addition of 2 × slip-frequency sidebands around slot harmonics	Radial	The slip sidebands may be low level, requiring a large dynamic range as well as frequency selectivity in measuring instrumentation. Typical spectra show that these components in the region of the principal vibration slot harmonics also have slip-frequency sidebands.

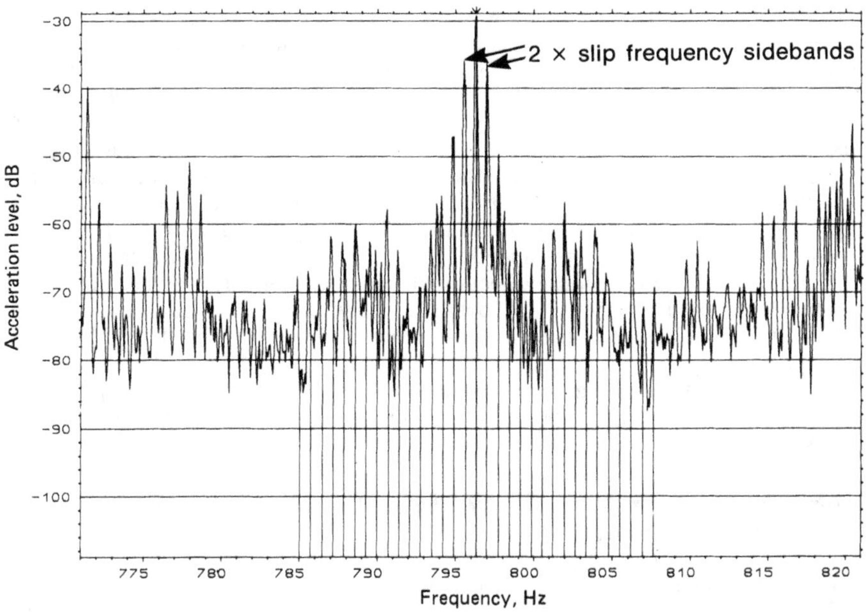

FIGURE 16.5 Zoom spectrum centered around the second principal vibration slot harmonic, showing 2 × slip-frequency sidebands on the component at this frequency.

HIGHER HARMONICS OF THE ROTATIONAL SPEED

Higher harmonics of the rotational speed typically occur where characteristic frequencies are an integral multiple of the rotational speed of the machine, for example, in the case of gearboxes, compressors, and turbines, where vibration occurs in multiples of the number of teeth, blades, lobes, etc. An increase in components, such as tooth-meshing frequencies or blade-passing frequencies, indicates deterioration acting on all teeth or blades, e.g., as uniform wear or increased turbulences, respectively.

"Ghost components" sometimes are observed in vibration spectra obtained from measurements on gearboxes; these components appear as tooth-meshing frequencies, but at frequencies where no gear in the gearbox has the corresponding number of teeth. Such components arise from faults on the gear-cutting equipment which have been transmitted to the new gear. Being geometrical faults, they are not load-sensitive, nor do they increase with wear; rather, as the gear's surface wears, they tend to decrease with time. The frequencies of the components are an integral multiple of the number of teeth on the index wheel and therefore appear as harmonics of the speed of rotation of the faulty gear.

SIDEBAND PATTERNS DUE TO MODULATION

Modulations, frequently seen in vibration measurements on gearboxes, are caused by such faults as eccentricities, varying gear-tooth spacing, pitch errors, varying load, tooth-to-tooth pittting, and uneven wear. Such modulations manifest themselves as families of sidebands around the gear-tooth-meshing frequency with a frequency spacing equal to the modulating frequency (e.g., the rotation speed of the faulty gear in the case of an eccentric gear). Figure 16.6*A* shows the distribution of the sidebands for such a condition. Any gear in a gearbox can be a source of modulation. In order to distinguish all possible sidebands, the analysis must be carried out with sufficient resolution to detect sidebands with a spacing equal to even the lowest rotational speed inside the gearbox, and therefore the zoom feature is indispensable.

Local faults, such as cracked or broken gear teeth, cause bursts of energy where the fault passes through the gearmesh (see Fig. 16.6*B*). The spectrum appears as a family of sidebands with a spacing equal to the rotation speed of the faulty gear, as this induces a change in tooth deflection, during meshing, once per revolution. The sidebands shown in Fig. 16.6*B* are low in level and cover a broad frequency range. Very often the influence of the transmission path will modify the shape of the sideband pattern and does not permit a precise diagnosis. Local faults are best detected in the time waveform of Fig. 16.6*B*.[5] Similarly, sidebands at the rotational speed and slip frequency are quite common in patterns for asynchronous machines.

HARMONIC PATTERNS NOT HARMONICALLY RELATED TO THE ROTATIONAL SPEED

Harmonic patterns which are not harmonically related to the speed of rotation typically appear where there are local faults in rolling-element bearings.[5] A local fault produces an impulse having a repetition rate equal to the characteristic frequencies of the bearing: ball-passing frequency for the outer raceway, ball-passing frequency for the inner raceway, and twice the ball-spin frequency (see Table 16.2). Such faults appear as a series of harmonics separated by the impact frequency with an amplitude proportional to the spectrum of the single impulse. An impact tends to excite bearing defect frequencies or excite structural resonances in the frequency range covered, and the harmonic patterns around these resonances thus are emphasized. This provides two methods of detecting rolling-element bearing faults: (1) by finding

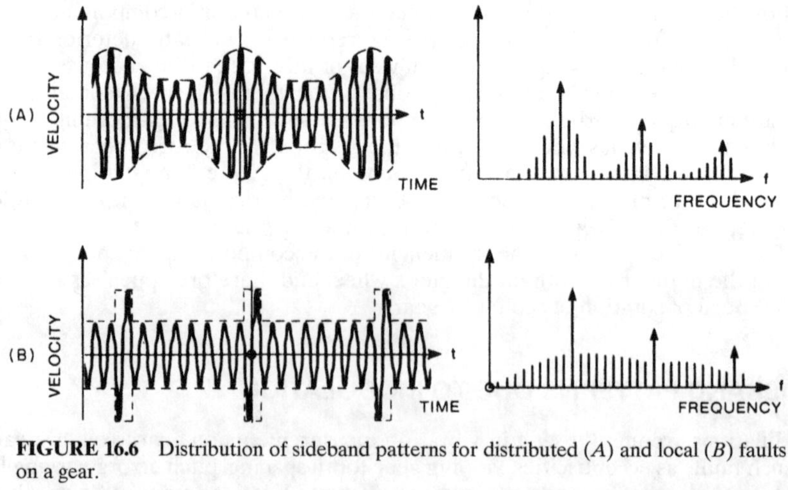

FIGURE 16.6 Distribution of sideband patterns for distributed (*A*) and local (*B*) faults on a gear.

the fundamental of the impact rate in the low-frequency range; and (2) by finding the harmonic pattern at the impact frequency in the high-frequency range, where resonances are excited; this may be difficult because speed fluctuations tend to smear these components.

SPECIAL ANALYSIS TECHNIQUES

Table 16.4 summarizes the applications of the various analysis techniques described below.

TABLE 16.4 Typical Applications of the Various Analysis Techniques

Technique	Application	Fault/machine
Zoom	Separation of closely spaced components Improvement of signal-to-noise ratio, separation of resonances from pure tones	Electrical machines, gearboxes, turbines
Phase	Operational deflection shapes Detection of developing cracks in shafts Balancing	
Time signal	Waveform visualization for identification of distortion	Rubbing, impacts, clipping, cracked teeth
Cepstrum	Identification and separation of families of harmonics Identification and separation of families of sidebands	Rolling elements bearing, bladed machines, gearboxes
Envelope analysis	Amplitude demodulation Observation of a low-frequency amplitude modulation happening at high frequency	Rolling element bearing, electrical machines, gearboxes
Peakness methods	Calculation of high-pass filtered signals	Faults in low-speed machines
Synchronous time averaging	Improving signal-to-noise ratio Waveform analysis Separating effects of adjacent machines Separating effects of different shafts Separating electrically and mechanically induced vibrations	Electrical machines, reciprocating machines, gearboxes, etc.
Impact testing	Resonance testing	Foundations, bearings, couplings, gears
Scan analysis	Analysis of nonstationary signals	Fast run-up/coast down

ENVELOPE DETECTION

Envelope detection (envelope detectors are discussed in Chap. 13) is particularly useful for fault diagnosis in machinery, since it permits elimination of the signal resulting from background vibration and concentrates the analysis in the frequency range placing the greatest emphasis on the harmonic pattern of the impact frequency—a resonance of the structure excited by the impulse. This can be done by either analog or digital means.[6] Figure 16.7 illustrates the analog process. The signal (see Fig. 16.7*A*) is first bandpass-filtered around the frequency range where a significant broadband increase has been detected, as illustrated in Fig. 16.7*B* and *D* (usually one or more resonances between 2 and 20,000 Hz have been excited). The filtered signal (which now contains only the ringing of the selected resonance excited by the repetitive impacts, Fig. 16.7*C*) is rectified and analyzed once again in a low-frequency range in order to determine the repetition frequency of the impacts, as shown in Fig. 16.7*E* and *F*.

The advantages of envelope detection are as follows:

1. The use of bandpass filters eliminates background noise resulting from other vibration sources (for example, from unbalance or gear vibration). All that remains is the repetition rate of the impacts exciting the structural resonance, possibly amplitude-modulated.
2. High-frequency analysis is not required, since only the envelope of the signal is of importance, not the signal itself, which can extend upward to hundreds of kilohertz.
3. Diagnosis is possible, since the impact frequencies are determined and can be related to a specific source (ball-passing frequency for the outer raceway, ball-passing frequency for the inner raceway, ball-spin frequency, fundamental train frequency, or some other source of repetitive impacts, for example, a cracked gear tooth).

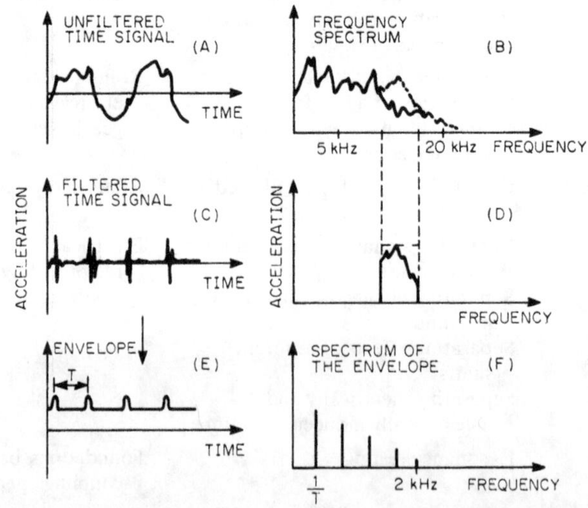

FIGURE 16.7 Principle of analog envelope detection applied to the analysis of impacts due to rolling-element bearing faults.

Figure 16.8*A* and *B* shows the acceleration spectra from 0 to 25 kHz of a good bearing and a faulty bearing. Note that the spectrum is noticeably higher on the good bearing than on the faulty one, which confirms that comparative measurements should not be made between different measurement points or different machines. Absolute vibration levels often do not provide a satisfactory indication of the condition of a machine; only changes in level are relevant. Figure 16.8*C* and *D* shows the analysis of the envelopes on the good and the faulty bearings obtained after zooming around 5400 Hz with an 800-Hz frequency span. The only noticeable pattern on the good bearing comes from the forced lubrication system. In contrast, the result of the envelope analysis on the faulty bearings shows a complex pattern, and frequency information is absolutely necessary to confirm whether or not there is a ball-bearing fault. The following frequencies appear: 5.4 Hz (the repetition rate of the forced lubrication system on the actual bearing, and its harmonics), 6.4 Hz (the repetition range of the forced lubrication system on adjacent bearings, and its harmonics), and 15.43 Hz (the ball-passing frequency for an outer raceway defect, and its harmonics).

CEPSTRUM ANALYSIS

The use of cepstrum analysis (explained in more detail in Chap. 14) is particularly advantageous for detecting periodicities in the power spectrum (e.g., harmonics and sideband patterns), since it provides a precise measure of the frequency spacing between components.[6] Figure 16.9 shows the spectrum and the corresponding cepstrum analysis of a measurement made on an auxiliary gearbox driving a generator on a gas-turbine-driven oil pump. As a fault on one of the bearings develops, the first rahmonic appears and then increases at a quefrency equal to the reciprocal of the spacing in the frequency spectrum which corresponds to an outer raceway defect in one of the bearings. Another advantage of cepstrum analysis is that one component in the cepstrum represents the global "power" content of a whole family of harmon-

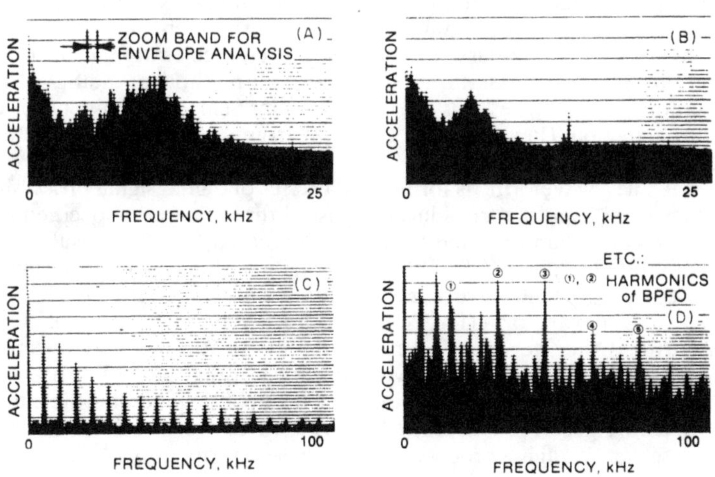

FIGURE 16.8 Envelope analysis of a good and bad bearing in the frequency range from 0 to 25 kHz.

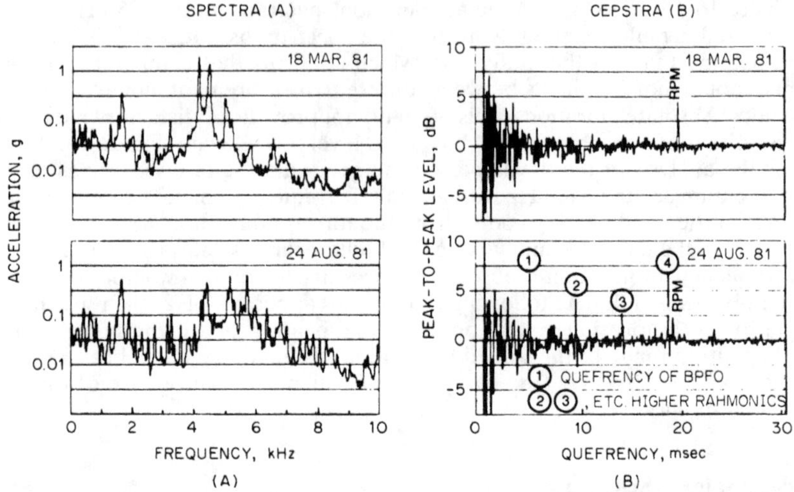

FIGURE 16.9 Analyses of vibration of an auxiliary gearbox before and after the development of a fault on one of the bearings. (*A*) Spectrum analysis; (*B*) the corresponding cepstrum analysis.

ics or sidebands, and this value is practically independent of extraneous factors such as machine-load condition, selection of measurement location, and phasing between amplitude and phase modulation.

PEAKNESS METHODS

It has been found that the analysis of stress waves (short pulses compared to operating speed) generated by metal-to-metal contact indicate such defects as bearing flaws, gear tooth surface deformities, rubs, and insufficient lubrication. Peakness methods are based on high-pass or bandpass-filtered acceleration measurement and sampling; however, high-pass filtering varies from method to method. Instead of demodulation, peakness methods directly use either peak, peak-to-peak, or dynamic crest factor detection. They eliminate all energy from the spectrum except the impulsive activity.

The PeakVue® waveform* is formed by a series of digital signal processing steps. The analog signal from the transducer is passed through a user-specified high-pass filter set at greater than or equal to the user-specified F_{max}. The resultant signal is sampled at a rate of 2.56 × 40,000 kHz. This digital string of data is decimated by a factor of 40,000/F_{max} while saving the absolute peak value out of each decimation step. The PeakVue waveform (Fig. 16.10*A*) is formed by saving a block of peak values of 2.56 times the user-specified number of lines. The PeakVue spectrum (Fig. 16.10*B*) is calculated from the PeakVue waveform using the FFT process with the selected F_{max} and number of lines. In addition, the analyst generates the autocorrelation coefficient (Fig. 16.10*C*) from the PeakVue waveform. The autocorrelation procedure is a scalar averaging process with no phase information.[7]

*PeakVue® is a trademark of Emerson Process Management.

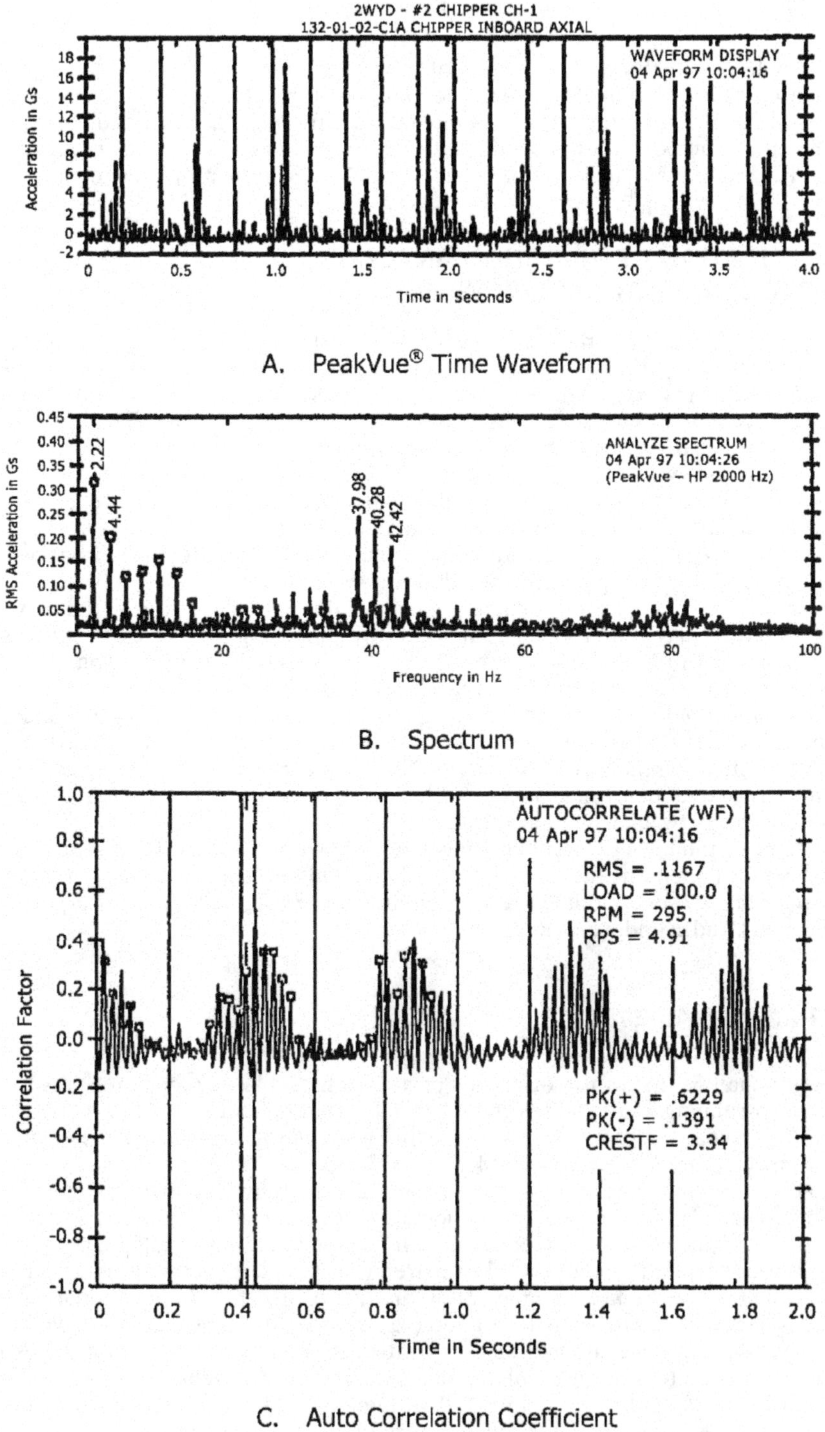

FIGURE 16.10 PeakVue® analysis of a chipper bearing.[7]

Figure 16.10 shows the analysis of a chipper with a defective rolling-element bearing. The spectrum of the PeakVue time waveform (Fig. 16.10*B*) shows frequencies at 2.2 Hz and multiples (fundamental train frequency), which indicates a cage defect. The autocorrelation coefficient (Fig. 16.10*C*) explicitly shows frequencies of 2.2 Hz and 40.2 Hz (ball spin frequency), which indicates a rolling-element defect.

APPLICATION OF VIBRATION ANALYSIS TO RECIPROCATING MACHINES

Vibration signals from reciprocating machines (such as diesel engines, reciprocating compressors, hydraulic pumps, and gas engines) differ from those of rotating machines in that they are not stationary. Instead, they consist of short impulses which occur at different points in time for different events (valves opening and closing, piston slap, combustion, etc.) and are repeated with the same timing for each new machine cycle. If these signals are averaged over a longer period of time, as is common practice in the analysis of rotating machines, these individual events would be averaged out so that changes would go undetected.

In reciprocating machines, different events will excite different resonances of a structure; the resulting frequencies that are generated provide valuable diagnostic information. *Timing* provides equally valuable information because the time when an event occurs may be related to what is actually happening in the cycle of the engine.

In gated vibration analysis, the vibration signal is analyzed at various angles of the crankshaft in order to cover a complete cycle of the machine in a three-dimensional plot.[8] The analyzer is triggered by a once-per-cycle trigger signal; then the delay after triggering is shifted to provide adequate overlap; this procedure continues until a complete cycle is covered. Note that each spectrum represents actually an average over many machine cycles for one time delay. This process averages any differences between machine cycles.

Shock monitoring technology[9] has been shown to be sensitive to reciprocating compressor faults in the early stages. These faults, which have the symptoms of mechanical looseness, relate to loose rod nuts, loose bolts, broken parts, liquid in the cylinders, and cylinder scoring.

TREND ANALYSIS

Trend analysis makes use of graphs of a condition-related parameter versus time (date or running hours) to determine when the parameter is likely to exceed a given limit. The goal of a successful condition monitoring program is to predict the time of an expected breakdown well in advance of its occurrence in order to shut down the machine in ample time, to order spare parts, and thereby to minimize the shutdown time. Since all vibration criteria indicate that equal changes on a log scale correspond to equal changes in severity, data for a trend analysis should be plotted on a logarithmic scale in decibels. A linear trend on a logarithmic scale is found occasionally, but the actual trend may follow another course; for example, when the fault feeds back on the rate of deterioration (e.g., gear wear), the trend, when plotted on a logarithmic scale, may then be exponential. In some cases the fault changes suddenly in finite steps (for example, a spall caused by gradual subsurface fatigue), making it very difficult to extrapolate to determine the date of the shutdown. To ensure accurate trend analysis, the following precautions should be taken:

1. Determine a trend based on measurements of a parameter directly related to a specific type of fault—not on measurements of overall levels.
2. Diagnose faults *before* attempting to interpret a trend curve in order to (*a*) select the appropriate parameter for the type of fault which is being monitored (for example, the parameter may be the level of an individual component, of a cepstrum component, or of a selected frequency range) and (*b*) observe critically the results of the trend analysis so as to determine if the linear or exponential interpolation is adequate.
3. Keep in mind that the best estimate of the lead time will be obtained by employing a trend of the most recent measurements.

CORRECTION METHODS

Condition monitoring of machinery always leads to correction of faults that caused excessive and/or abnormal vibration. Abnormal vibration is the symptom of normal wear, design deficiencies, installation errors, manufacturing tolerances, or process excitation. It is important to correct because excessive vibration destroys bearings, foundations, casings, and shafts and may affect the process and the health of personnel working near the machine. Some vibration is normal because of the process and function of the machine. The common correction methods employed to reduce excessive vibration include:

1. Replacement of defective parts
2. Alignment[10]
3. Rotor mass balancing[11,12]
4. Resonance removal (Chaps. 21 and 23)
5. System redesign

ACKNOWLEDGMENT

Much of the material in this chapter was originally prepared by Joelle Courrech, previously Area Sales Manager, Bruel & Kjaer, Denmark.

REFERENCES

1. International Standard, ISO 13373-1, Condition Monitoring and Diagnostics of Machines—Vibration Condition Monitoring, Part 1: General Principles: 2006.
2. International Standard, ISO 13373-2, Condition Monitoring and Diagnostics of Machines—Vibration Condition Monitoring, Part 2: Processing, Presentation, and Analysis of Vibration Data.
3. Eshleman, R. L., "Machinery Vibration Analysis," Vibration Institute, Willowbrook, Ill., 2006.
4. Kryter R. C. and H. D. Haynes: "Condition Monitoring and Machinery Using Current Signature Analysis," *Sound and Vibration,* September 1989.

5. Taylor, J. I.: "The Vibration Analysis Handbook," Vibration Consultants, Inc., Tampa, Fla., 1994.

6. Courrech, J.: "New Techniques for Fault Diagnostics in Rotating Element Bearings," *Proc. 40th Meeting of the Mechanical Failure Preventive Group,* National Bureau of Standards, Gaithersburg, Md., April 16–18, 1985.

7. Robinson, James C., and R. Cook: "Rolling Element Bearing Defect Analysis Using PeakVue®," *Proc. 24th Annual Meeting of the Vibration Institute,* Willowbrook, Ill., 2000, pp. 163–170.

8. Courrech, J.: "Examples of the Application of Gated Vibration Analysis for the Detection of Faults in Reciprocating Machines," *Noise and Vibration '89 Conference,* Singapore, August 16–18, 1989.

9. Zusman, G., and J. Palm: "Impact Measurement for Reciprocating Compressors," *Proc. 26th Annual Meeting of the Vibration Institute,* 2002, p. 65–71.

10. Piotrowski, J.: "Shaft Alignment Handbook," 3d ed., CRC Press, Taylor and Francis Group, Boca Raton, Fla., 2007

11. International Standard, ISO 1940: "Mechanical Vibration—Balance Quality Requirements of Rigid Rotors," 2003.

12. Eshleman, R. L.: Balancing of Rotating Machinery, Vibration Institute, Willowbrook, Ill., 2006.

CHAPTER 17
SHOCK AND VIBRATION STANDARDS

David J. Evans
Henry C. Pusey

INTRODUCTION

This chapter is concerned with shock and vibration standards covering (1) terminology; (2) use and calibration of transducers and instrumentation; (3) shock and vibration generators; (4) structures and structural systems; (5) vehicles including land-based, airborne, and ocean-going; (6) machines and machinery including testing, condition monitoring, diagnostics, prognostics, and balancing; (7) human exposure to shock and vibration; and (8) testing. These topics may be covered by international, regional, or national documents that are issued as either standards or recommended practices. The dominant international consensus standards bodies concerned with shock and vibration are the International Organization for Standardization (ISO) and the International Electrotechnical Commission (IEC). The U.S. members of ISO and IEC are the American National Standards Institute (ANSI) and the United States National Committee of the International Electrotechnical Commission (USNC/IEC), respectively. The USNC/IEC is a committee of ANSI. Examples of regional standards bodies are the European Committee for Standardization (CEN) and the European Committee for Electrotechnical Standardization (CENELEC). Within the U.S.A., ANSI standards are developed by standards committees following the accredited standards procedures of ANSI. These national committees also often furnish the expert members from the U.S.A. to working groups within ISO and IEC. The national standards committees are typically sponsored by professional societies that have an interest in particular areas of standardization work. Within the U.S.A., additional national consensus standards bodies exist, such as the American Society for Testing and Materials (ASTM), that develop standards by consensus of the members of their society.

STANDARDS ORGANIZATIONS AND COMMITTEES

ISO technical committee (TC) 108 (Mechanical Vibration, Shock and Condition Monitoring) and its five subcommittees (SCs) are predominantly responsible for any international standards activity related to shock and vibration. TC 108 and its subcommittees maintain numerous liaisons with other technical committees and subcommittees within ISO and IEC, including ISO TC 20 (Aircraft and Space Vehicles), ISO TC 43 (Acoustics), ISO TC 45 (Rubber and Rubber Products), ISO TC 159 (Ergonomics), IEC TC 2 (Rotating Machinery), IEC TC 5 (Steam Turbines), and IEC TC 87 (Ultrasonics).

The subcommittees of TC 108 also maintain liaisons with other organizations outside of ISO and IEC that are interested in their work. IEC TC 104 is responsible for standards activities related to environmental testing, including testing using shock and vibration. The primary counterpart to ISO TC 108 within the U.S.A. is ANSI-accredited standards committee S2 (Mechanical Vibration and Shock), which holds the U.S. Technical Advisory Group (TAG) for ISO TC 108 and all of its subcommittees. The ANSI-accredited standards committee S2 and its U.S. TAGs are administered by the Acoustical Society of America Committee on Standards (ASACOS) and the Acoustical Society of America (ASA) Standards Secretariat. The U.S. TAG for IEC TC 104 is administered and managed by the Electronic Industries Alliance (EIA) Corporate Engineering Department. The activities of CEN TC 231 on shock and vibration are reported to ISO TC 108. Much of the standardization work of CEN TC 231 is related to the EU (European Union) Machinery Directive(s).

STANDARDS ACTIVITIES

The various international standards activities related to shock and vibration are summarized in Table 17.1 and discussed in the following sections.

Terminology. Documents on standardized terminology of all aspects of TC 108 and its six subcommittees are coordinated under TC 108. This vocabulary is contained in ISO document *ISO 2041.*

Use and Calibration of Transducers and Instrumentation. The use and calibration of shock and vibration transducers and instrumentation, including standardized calibration methods, measuring instrumentation for human response to vibration, and vibration condition monitoring transducers and instrumentation, is assigned to ISO TC 108/SC 3 (Use and Calibration of Vibration and Shock Measuring Instrumentation). TC 108/SC 3 maintains a liaison with the International Organization of Legal Metrology (OIML). Numerous standards on calibration are contained in the *ISO 5347* series of standards, as well as in the *ISO 16063* series of standards. The ANSI standard on methods of calibration of shock and vibration transducers is *ANSI S2.2.* The ISO standard on measuring instrumentation for human response to vibration is *ISO 8041.* The Instrumentation, Systems, and Automation Society (ISA) administers a number of standards committees, one of which is SP37 on specifications and tests for sensors and transducers used in measurement and control. SP37 has a number of subcommittees that involve transducers

TABLE 17.1 Summary of International Standards Activities

Category	Document series	Responsible ISO TC/SC	Related documents
Vocabulary	*ISO 2041*	TC 108	*ANSI S2.1*
Mobility	*ISO 7626*	TC 108	*ANSI S2.31–34*
Isolators	*ISO 2017*	TC 108	*ANSI S2.8*
Balancing	*ISO 1940*	TC 108	*ANSI S2.19, S2.42,* and *S2.43*
Balancing machines	*ISO 2953*	TC 108	*ANSI S2.38*
Machines/machinery	*ISO 7919* and *10816*	TC 108/SC 2	*ANSI S2.13, S2.40,* and *S2.41*
Vehicles	*ISO 8002*	TC 108/SC 2	
Ships	*ISO 4867, 4868, 6954,* and *10055*	TC 108/SC 2	*ANSI S2.16* and *S2.25; MIL-STD-167*
Buildings	*ISO 4866* and *8569*	TC 108/SC 2	*ANSI S2.47*
Calibration	*ISO 5347* and *16063*	TC 108/SC 3	*ANSI S2.2*
Human response	*ISO 8041*	TC 108/SC 3	
Human exposure	*ISO 2631, 5349, 6897, 8727,* and *13090*	TC 108/SC 4	*ANSI S3.18, S3.29,* and *S3.34*
Generating systems	*ISO 5344, 6070,* and *8626*	TC 108/SC 6	*ANSI S2.5, S2.45, S2.48,* and *S2.58*
Shock machines	*ISO 8568*	TC 108	*ANSI S2.3, S2.14,* and *S2.15*

used in shock and vibration measurements, e.g., strain gages, accelerometers, servo-accelerometers, and force transducers. SP37.20 is a separate subcommittee of SP37 devoted specifically to vibration transducers.

Shock and Vibration Generators. ISO TC 108/SC 6 (Vibration and Shock Generating Systems) has been assigned standards activities related to systems for the generation of shock and vibration and their terminology. TC 108/SC 6 maintains a liaison with IEC TC 104. IEC TC 104 (Environmental Conditions, Classification, and Methods of Test) is concerned with standardized environmental testing, of which shock and vibration are only two of several variables defining a test environment. ANSI has a number of standards related to the specification of the performance of shock- and vibration-testing machines, as well as standards covering the performance characteristics of these machines.

Structures and Structural Systems. ISO TC 108 (Mechanical Vibration and Shock) and TC 108/SC 2 (Measurement and Evaluation of Mechanical Vibration and Shock as Applied to Machines, Vehicles, and Structures) both have items in their program of work related to stationary structures or structural systems. Guidelines on building vibration are contained in *ISO 4866* and *ANSI S2.47.* Work on condition monitoring and assessment of structures and structural systems is ongoing in TC 108.

Vehicles. This comprises a very broad area of standardization with a small, but important, portion of it directly related to shock and vibration. ISO TC 108/SC 2 (Measurement and Evaluation of Mechanical Vibration and Shock as Applied to Machines, Vehicles, and Structures) is involved with the vibration of ships, and *ISO 4867, 4868,* and *6954* specifically address the measurement and reporting of vibration onboard ships. Much of the U.S. participation in this work is contributed by members of the Society of Naval Architects and Marine Engineers (SNAME). *ANSI S2.16* covers the measurement and acceptance criteria for the vibratory noise of shipboard equipment, and *ANSI S2.25* covers the evaluation and reporting of hull and superstructure vibration in ships. ISO TC 108/SC 2 is also involved with vibration of land-based vehicles, and *ISO 8002, 8608,* and *10326* are specifically related to the evaluation and reporting of the vibration associated with either land-based vehicles or road surface profiles. ISO TC 20 (Aircraft and Space Vehicles) is involved with standards related to aerospace vehicles in general, and a number of ISO technical committees exist that generally cover specific types of land-based vehicles, e.g., construction, agricultural, and off-road vehicles. The U.S. TAG for ISO TC 20 and the U.S. TAGs for many of the ISO technical committees on land-based vehicles in general are administered by the Society of Automotive Engineers (SAE). The CEN document *CEN EN 1032* on testing mobile machinery has been published, and work is ongoing within CEN TC 231 with respect to testing mobile machinery to determine whole-body vibration and vibration emission values. CEN TC 231 maintains liaisons with CEN TC 144 and CEN TC 151 on tractors and agricultural machines, and construction equipment, respectively.

Machines and Machinery. Standardization related to the shock and vibration of machines and machinery including balancing, condition monitoring, diagnostics, prognostics, and testing is within the program of work of ISO TC 108 ISO TC 108/SC 2 (Measurement and Evaluation of Mechanical Vibration and Shock as Applied to Machines, Vehicles, and Structures), and ISO TC 108/SC 5 (Condition Monitoring and Diagnostics of Machines). Numerous ISO and ANSI standards exist on balancing, balancing machines, balancing terminology, balance quality, and the measurement and evaluation of mechanical vibration related to various classes of rotating and reciprocating machinery. The National Electrical Manufacturers Association (NEMA), American Petroleum Institute (API), Compressed Air and Gas Institute, and Hydraulic Institute publish standards on motors, generators, turbines, pumps, and compressors that may contain parts that are related to shock and vibration of these machines. TC 108/SC 2 maintains liaisons with more than a dozen different ISO and IEC technical committees and subcommittees including IEC TC 104. TC 108/SC 5 maintains a liaison with IEC TC 2 (Rotating Machinery). ISO TC 118/SC 3 (Pneumatic Tools and Machines) maintains liaisons with ISO TC 108/SC 2 and TC 108/SC 4. CEN TC 231 has a number of published standards related to the vibration of hand-held power tools, as well as guidance on safety standards related to vibration. An additional program of work within CEN TC 231 pertains to the vibration of a variety of hand-held power tools, e.g., grinders, drills and rotary hammers, chipping and riveting hammers, and hammers for construction.

Human Exposure to Shock and Vibration. The program of work on human exposure to shock and vibration is assigned to ISO TC 108/SC 4 (Human Exposure to Mechanical Vibration and Shock). ISO TC 108/SC 4 maintains liaisons with about a dozen ISO technical committees and subcommittees including ISO TC 43 (Acoustics), as well as with other organizations such as the European Committee of Associations of Manufacturers of Agricultural Machinery (CEMA), the International Maritime Organization (IMO), and the International Union of Railways (UIC). There are a number of ISO and ANSI standards on exposure to whole-body and hand-arm vibration including standards covering occupants of fixed-structures, single shocks, guidance on safety aspects of tests and experiments, transmissibility of gloves and resilient materials, and terminology. (See Chap. 41.)

Testing. Numerous standards and handbooks that cover shock and vibration testing have been issued by ISO and IEC, as well as agencies of the U.S. government, in particular the National Aeronautics and Space Administration (NASA) and the Department of Defense (DoD). Although NASA and DoD standards and handbooks are concerned primarily with aerospace vehicles and military hardware, many are sufficiently general to have broad applications to commercial structures, vehicles, and equipment.

International Standards. While IEC TC 104 (Environmental Conditions, Classification, and Methods of Test) has work programs devoted to a number of environmental variables such as temperature and relative humidity, a portion of the work is directed toward testing using shock and vibration. Specifically, a number of documents in the *IEC 60068-2* series of documents cover sinusoidal vibration, broadband random vibration, shock, drop and topple, free fall, and bump testing. ASTM publishes standards that address using shock and vibration to test unpackaged manufactured products, packaging systems, shipping containers, and materials. *ISO 8568* addresses shock testing machines. ISO TC 108 has a work item on the analysis of the mechanical properties of visco-elastic materials using vibration, and there are a number of ANSI-approved standards published on measuring the mechanical properties of visco-elastic materials using vibration.

NASA Standards and Handbooks. NASA has issued three standards (STD) and two handbooks (HDBK) related to shock and vibration testing that are approved for NASA-wide application to launch vehicles and payloads. Descriptions of the scopes of these publications follow. All of these publications are available via the World Wide Web (www) at *standards.nasa.gov.*

The term *vibroacoustics* is defined as an environment induced by high-intensity acoustic noise associated with various segments of the flight profile (see Chap. 32, of this handbook). It manifests itself throughout the launch vehicle and payload structure in the form of transmitted acoustic excitation and as structure-borne random vibration. The NASA standard *NASA-STD-7001,* "Payload Vibroacoustic Test Criteria," specifically addresses the acoustic and random vibration environments and test levels associated with vibroacoustics.

Selected environmental exposure tests are contained in *NASA-STD-7002,* "Payload Test Requirements." This standard includes tests that are generally regarded as the most critical and the ones having the highest cost and schedule impact. The standard also includes functional demonstration tests necessary to verify the capability of the hardware to perform its intended function, with and without environmental exposure. Test levels, factors, margins, durations, and other parameters are specified where appropriate. In some cases, these specifications are expressed statistically or are described by reference to other NASA standards.

NASA-STD-7003, "Pyroshock Test Criteria," provides a consistent methodology for developing pyroshock test criteria for NASA spacecraft, payload, and launch vehicle hardware during all test phases of the verification process. Various aspects of pyroshock testing are discussed, including test environments, methods and facilities, test margins and number of exposures, control tolerances (when applicable), data acquisition and analysis, test tailoring, dynamic analysis, and prediction techniques for pyroshock environments.

The NASA handbook *NASA-HDBK-7004,* "Force Limited Vibration Testing," establishes a methodology for conducting force-limited vibration tests for all NASA flight projects. The methodology in the handbook may be followed by those desiring to use force limiting without having to conduct an extensive literature search or research and development effort before conducting the test. A monograph on force-limited vibration testing is available for reference and is recommended for those needing more detailed technical information (*NASA-RP-1403*).

NASA-HDBK-7005, "Dynamic Environmental Criteria," summarizes procedures for deriving design and test criteria for space vehicles exposed to a wide range of shock and vibration environments. Included in this handbook are detailed discussions of the machines and procedures approved by NASA for the shock and vibration testing of spacecraft and their components. Many of these machines and procedures are equally applicable to the testing of commercial hardware.

DoD Standards. Despite a significant effort to modify or eliminate military (MIL) standards and specifications in favor of commercial standards, a considerable group of MIL standards still remain. In many cases, MIL standards are unique in application and scope and, in some cases, more useful than similar commercial standards. A specific case in point is *MIL-STD-810,* "Environmental Engineering Considerations and Laboratory Tests," now in its "F" revision. This document covers most environments, including shock and vibration. Through its many revisions, the scope of the document has expanded to include new environments and most ground and air platforms. Its principal contribution to product design engineering is its emphasis on *test tailoring,* introduced in the "D" revision and expanded with later revisions. This test concept is not emphasized in any commercial specification and allows *MIL-STD-810* to be used for both defense and commercial applications, and for both U.S. and non-U.S. test programs.

Several useful MIL standards that include shock and vibration requirements are maintained and available. The most widely used are the latest revisions of *MIL-STD-1540* and *MIL-HDBK-340* on space vehicle shock and vibration, *MIL-S-901D* on Navy shock, *MIL-STD-781* on reliability, and *MIL-STD-167* on ship vibration (parts of this standard have been, or are in the process of being, converted to ANSI or ISO standards). Nearly all of these standards can be located at the Document Automation and Production Service DoD Single Stock Point (DoDSSP) web site. A complete collection of DoD specifications and standards is indexed in the Acquisition Streamlining and Standardization Information System (ASSIST), which is managed by the DoDSSP. The ASSIST Shopping Wizard website provides the capability to request DoD standardization documents over the Internet. Users may place orders for documents in paper and CD-ROM formats by establishing a customer account with the DoDSSP. The U.S. Government Printing Office allows the purchase of a variety of DoD and other U.S. Government Agency publications. A catalog of government periodicals and subscription services is available from the Superintendent of Documents, U.S. Government Printing Office. Most DoD standardization documents can also be obtained by contacting the controlling military service. In the case of *MIL-STD-810,* for example, the controlling military service is the U.S. Army.

STANDARDS-DEVELOPING ORGANIZATIONS AND SOURCES

Some societies and organizations involved in the production of standards are given below. Sources for catalogs of standards and for purchasing standards are also given. A significant amount of information concerning standards development, meetings, organizations, catalogs, and procurement is readily available via the World Wide Web (www) at the uniform resource locators (URLs) listed below. This list, while extensive, is not intended to be all inclusive.

Acoustical Society of America (ASA)
Standards Secretariat
35 Pinelawn Road, Suite 114E
Melville, NY 11747 USA
Telephone: +1 631 390 0215
URL: *asa.aip.org*

American National Standards Institute (ANSI)
1819 L Street NW, 6th Floor
Washington, DC 20036 USA
Telephone: +1 202 293 8020
URL: *www.ansi.org*

American Society for Testing and Materials (ASTM)
100 Barr Harbor Drive
West Conshohocken, PA 19428-2959 USA
Telephone: +1 610 832 9500
URL: *www.astm.org*

Electronic Industries Alliance (EIA)
Corporate Engineering Department
2500 Wilson Boulevard
Arlington, VA 22201 USA
Telephone: +1 703 907 7500
URL: *www.eia.org*

European Committee for Standardization (CEN)
Avenue Marnix 17
B 1050 Brussels, Belgium
Telephone: +32 2 550 08 11
URL: *www.cen.eu/cenorm*

Global Engineering Documents
15 Inverness Way East
Englewood, CO 80112 USA
Telephone: +1 800 854 7179
URL: *global.ihs.com*

International Electrotechnical Commission (IEC)
3, rue de Varembé
P.O. Box 131
CH-1211 Geneva 20, Switzerland
Telephone: +41 22 919 02 11
URL: *www.iec.ch*

International Organization for Standardization (ISO)
1, ch. de la Voie-Creuse
Case postale 56
CH-1211 Geneva 20, Switzerland
Telephone: +41 22 749 01 11
URL: *www.iso.ch*

Instrumentation, Systems, and Automation Society (ISA)
67 Alexander Drive
Research Triangle Park, NC 27709 USA
Telephone: +1 919 549 8411
URL: *www.isa.org*

NASA/Marshall Space Flight Center
Mail Code: ED41
Marshall Space Flight Center, AL 35812 USA
Attention: Paul Gill
Telephone: +1 256 544 2557
URL: *standards.nasa.gov*

Society of Automotive Engineers (SAE)
World Headquarters
400 Commonwealth Drive
Warrendale, PA 15096-0001 USA
Telephone: +1 724 776 4841
URL: *www.sae.org*

Society of Naval Architects and Marine Engineers (SNAME)
601 Pavonia Avenue
Jersey City, NJ 07306 USA
Telephone: +1 800 798 2188
URL: *www.sname.org*

U.S. Government Printing Office
Washington, DC 20402 USA
Attention: Superintendent of Documents
Telephone: +1 202 512 1704
URL: *bookstore.gpo.gov/collections*

U.S. National Committee of the IEC (USNC/IEC)
c/o American National Standards Institute
25 West 43d Street, Fourth Floor
New York, NY 10036 USA
Telephone: +1 212 642 4900
URL: *www.ansi.org*

CHAPTER 18
TEST CRITERIA AND SPECIFICATIONS

Allan G. Piersol

INTRODUCTION

This chapter covers the development of shock and vibration test criteria for mechanical, electrical, electronic, or hydraulically powered equipment, for example, an alternator for an automobile or an electronic instrument for an airplane. The emphasis throughout is on the selection of test criteria rather than the formulation of design criteria, but specified shock and vibration test levels and durations are commonly used as design criteria as well. Following a brief overview of environmental specifications, this chapter presents (1) a summary of the descriptions of shock and vibration environments used to establish test criteria, (2) a discussion of the different types of tests used to achieve various objectives, (3) procedures to select shock and vibration test levels, (4) procedures to select vibration test durations, and (5) general testing considerations.

ENVIRONMENTAL SPECIFICATIONS

An *environmental specification* is a written document that details the environmental conditions under which an item of equipment to be purchased must operate during its service life. Several contracting agencies of the U.S. government and various professional societies issue general environmental specifications for particular classes of equipment (see Chap. 17), but deviations from the specified environmental conditions in such documents are permitted when more appropriate conditions can be established by direct measurements or predictions of the environments of concern. An *environmental test specification* is a written document that details the specific criteria for an environmental test, as well as other matters such as the preparation of the test item, identification of all test equipment and instrumentation, description of any test fixtures, instructions for mounting sensors, step-by-step procedures for operating the test item (if operation is required), procedures for taking data on the test item function and the applied environment, and performance acceptability criteria. The test criteria (the magnitude and duration of the test excitation) in environmental test specifications often serve as design criteria as well (see Chap. 40).

GENERAL TYPES OF ENVIRONMENTS

The environments that must be considered in equipment design and testing are listed in Table 18.1. Those printed in boldface, namely, shock and vibration, are the ones of special concern in this handbook. Shock and vibration environments may result from the equipment operation (for example, the vibration caused by shaft unbalance in equipment with a rotating element), but it is the external shock and vibration motions transmitted into the equipment through its mounting points to the structure of the system incorporating the equipment that are of primary interest here. The acoustical, blast, fluid flow, and wind environments noted in Table 18.1 are often the original source of the shock and vibration motions of the system structure that transmit into the equipment, but the original source may also be a direct motion input to the system, for example, earthquake inputs to a building or road roughness inputs to an automobile. Such environments have complicated transmission patterns that are modified or intensified by mechanical resonances of the system structure and, therefore, are appropriately described by frequency-dependent functions, i.e., spectra.

TABLE 18.1 Various Types of Environments to Which Equipment May Be Exposed

Acceleration (sustained)	Fungus	Salt spray
Acoustical noise	Humidity	Temperature (sustained)
Blast	**Mechanical shock**	Temperature cycling
Dust and sand	Pressure (sustained)	**Vibration**
Fluid flow	Rain, hail, and snow	Wind

In practice, for economy of effort, equipment is often designed and tested for exposure to each of the environments listed in Table 18.1 as if they occur separately. However, some of the environments in Table 18.1 may occur simultaneously and have an additive effect; for example, a shock may occur during a period of high static acceleration where the stress in the equipment due to the combination of the two environments is greater than the stress due to either applied separately. Worse yet, two environments may have a synergistic effect; for example, equipment may be subject to high vibration during a period when the temperature exposure is also high, and high temperatures cause a degradation of the equipment strength, making it more vulnerable to vibration-induced failures. These matters must be carefully evaluated during the definition of a test program to determine if simultaneous testing for two or more environments is required.

SHOCK AND VIBRATION ENVIRONMENTS

From a testing viewpoint, it is important to carefully distinguish between a shock environment and a vibration environment. In general, equipment is said to be exposed to *shock* if it is subject to a relatively short-duration (transient) mechanical excitation; equipment is said to be exposed to *vibration* if it is subject to a longer-duration mechanical excitation. If the basic properties of the vibration are time-invariant, it is called *stationary* (or steady-state for periodic vibrations). However, vibration environments are often *nonstationary,* i.e., one or more of their basic properties vary with time. If the properties change slowly relative to the lowest frequency of the vibration, then the vibration can be analyzed to arrive at criteria for a stationary vibration test, as detailed later. Otherwise, the environment must be viewed as a shock. Practical distinctions between shock and vibration environments cannot be made on an absolute basis, independent of the equipment exposed to the environment. To be more specific, any mechanical device that is more or less linear can be characterized by one or more resonance frequencies and damping coefficients (see Chap. 2) or by a corresponding set of decaying transient responses after a momentary excitation. In more analytical terms, the response characteristics of a mechanical device are given by the *unit impulse response function* defined in Chap. 21. From a testing viewpoint, an excitation whose duration is comparable to, or less than, the response (or decay) time of the equipment is considered a shock, while an excitation whose duration is long compared to the response time of the equipment is considered a vibration.

DESCRIPTIONS OF SHOCK AND VIBRATION ENVIRONMENTS

The response of equipment to shock and vibration at its mounting points is dependent on frequency. Hence, shock and vibration environments are usually described by some type of spectrum; a *spectrum* is a description of the magnitude of the frequency components that constitute the shock or vibration. The most common spectral descriptions of both deterministic and random shock and vibration environments are summarized in Table 18.2 (see Chaps. 14, 19, and 20 for details). It is common to present data for test specification purposes in terms of acceleration, primarily because it is convenient to measure acceleration with accelerometers described in Chap. 10. However, for shock data presented in the form of a shock response spectrum, a response in terms of velocity or pseudo-velocity (see Chap. 20) is often preferred to acceleration. This is because the shock response spectrum represents the peak response of a single-degree-of-freedom system, and modal (relative) velocity for such a response has a direct linear relationship to stress[2,3] [see Eq. (27.1)]. Nevertheless, the use of an acceleration parameter for shock response spectra is not a problem in specifying test criteria as long as the criteria simulate the spectrum of the environment, and acceleration is used for both the environmental description and the test criteria.

TABLE 18.2 Common Spectral Descriptions of Shock and Vibration Environments

Environment	Characteristic	Spectral description
Shock	Deterministic	Fourier (integral) spectrum (see Chap. 20) Shock response spectrum (see Chaps. 8 and 20)
	Random	Energy spectral density (see Chap. 20 and Ref. 1) Shock response spectrum (see Chaps. 8 and 20)
Vibration	Deterministic	Line spectrum (see Chaps. 14 and 19)
	Random	Power spectral density (see Chap. 19)

The vibration environment for an item of equipment usually varies in magnitude and spectral content during its service life. Similarly, a shock environment may involve repetitive shocks with different magnitudes and spectral content. For reliability tests discussed later in this chapter, it may be necessary to measure or predict the spectra of the shock and/or vibration environment for all conditions (or a representative sample thereof) throughout the service life and to formulate test criteria that require a series of tests with several different magnitudes and spectral content. For most testing applications, however, a test involving a single spectrum is desired for convenience. To assure that the test produces a conservative result, a maximax spectrum is used; a *maximax spectrum* is the envelope of the spectra for all conditions throughout the service environment. Thus, the maximax spectrum may not equal any of the individual spectra measured or predicted during the service environment, since the maximum value at two different frequencies may occur at different times.

TYPES OF SHOCK AND VIBRATION TESTS

An *environmental test* is any test of a device under specified environmental conditions (or sometimes under the environment generated by a specified testing machine) to determine whether the environment produces any deterioration of performance or any damage or malfunction of the device; an environmental test may also be distinguished by the objectives of the test. In assessing the effects of shock and vibration on equipment, the types of tests most commonly performed fall into the following categories:

1. Development
2. Qualification
3. Acceptance
4. Screening
5. Statistical reliability
6. Reliability growth

DEVELOPMENT TESTS

A *development test* (sometimes called an *analytical test*) is a test performed early in a program to facilitate the design of a device or piece of equipment to withstand its anticipated service environments. It may involve determining the resonance frequency of a constituent component mounted inside the equipment by applying a sinusoidal excitation with a slowing-varying frequency (often called a *swept-sine-wave* test). Sinusoidal vibration is widely used as the excitation for development tests because of its simplicity and well-defined deterministic properties. In contrast, it may involve a more elaborate test to determine the normal modes and damping ratio of the equipment structure as described in Chap. 21. A stationary random vibration or a controlled shock excitation with appropriate data reduction software can greatly reduce the time required to perform a more extensive modal analysis of the equipment. In either type of test, the characteristics and magnitude of the excitation used for the test are not related to the actual shock and/or vibration environment to which the equipment is exposed during its service use.

QUALIFICATION TESTS

A *qualification test* is a test intended to verify that an equipment design is satisfactory for its intended purpose in the anticipated service environments. Such a test is commonly a contractual requirement, and hence, a specific test specification is usually involved. Preliminary qualification tests are sometimes performed on prototype hardware to identify and correct design problems before the formal qualification test is performed. Also, qualification test requirements might be based upon a general environmental specification (see Chap. 17). In some cases, the specification may require a test on a specific type of testing machine that produces a desired qualification environment (see Chap. 27). However, contracts usually allow deviations from the specified test levels and/or test durations in general environmental specifications, if it can be established that different test conditions would be more suitable for the given equipment. In any case, the basic purpose of a qualification test requires that the test conditions conservatively simulate the basic characteristics of the anticipated service environments.

Some years ago, when test facilities were more limited, it was argued that shock and vibration environments for equipment could be simulated for qualification test purposes in terms of the damaging potential of the environment, without the need for an accurate simulation of the detailed characteristics of the environment.[4] For example, it was assumed that random vibration could be simulated with sinusoidal vibration designed to produce the same damage. The validity of such "equivalent damage concepts" requires the assumption of a specific damage model to arrive at an appropriate test level and duration. Since the assumed damage model might be incorrect for the equipment of interest, there is a substantial increase in the risk that the resulting test criteria will severely under- or overtest the equipment. With the increasing size and flexibility of modern test facilities, the use of equivalent damage concepts to arrive at test criteria is rarely required and should be avoided, although equivalent damage concepts are still useful in arriving at criteria for "accelerated tests," as discussed later in this chapter. Whenever feasible, *qualification tests should be performed using an excitation that has the same basic characteristics as the environment of concern; for example, random vibration environments should be simulated with random vibration excitations, shock environments should be simulated with shock excitations of similar duration, etc.*

ACCEPTANCE TESTS

An *acceptance test* (sometimes called a *production test* or a *quality control test*) is a test applied to production items to help ensure that a satisfactory quality of workmanship and materials is maintained. For equipment whose failure in service might result in a major financial loss or personal injury, all production items are subjected to an acceptance test. Otherwise, a statistical sample of production items is selected, and each item is tested in accordance with an acceptance sampling plan that assures an acceptable average outgoing quality.[5] In either case, there are two basic approaches to acceptance testing for shock and vibration environments. The first approach is to design a test that will quickly reveal common workmanship errors and/or material defects as determined from prior experience and studies of failure data for the equipment, independent of the characteristics of the service environment. For example, suppose a specific type of electrical equipment has a history of malfunctions induced by scrap-wire or poorly soldered wire junctions. Then, the application of sinusoidal vibration at the resonance frequencies of wire bundles will quickly reveal such problems and, hence, constitute a good test excitation even though there may be no sinusoidal vibrations in the service environment. The second and more common approach is to apply an excitation that simulates the shock and/or vibration environments anticipated in service, similar to the qualification test but usually at a less conservative (lower) level.

SCREENING TESTS

A *screening test* is a test designed to quickly induce failures due to latent defects that would otherwise occur later during service use so that they can be corrected before delivery of the equipment, i.e., to detect workmanship errors and/or material defects that will not cause an immediate failure, but will cause a failure before the equipment has reached its design service life. Screening tests are similar to acceptance tests, but usually are more severe in level and/or longer in duration. If performed at all, screening tests are usually applied to all production items. Vibration screening tests are commonly performed with the simultaneous application of temperature cycling, a process referred to as *environmental stress screening* (ESS). The vibration environment is sometimes applied using relatively inexpensive, mechanically or pneumatically driven vibration testing machines (often referred to as impact or repetitive shock machines) that allow little or no control over the spectrum of the excitation (see Chap. 25). Hence, except perhaps for the overall level, the screening test environment generally does not represent an accurate simulation of the service environment for the equipment. Although there have been some efforts to standardize screening procedures,[6] most details of a screening program are determined by individual test specification writers and/or laboratory test managers. It should be mentioned that ESS has led to derivative testing procedures commonly referred to as *highly accelerated life testing* (HALT) and *highly accelerated stress screening* (HASS).[7] It should also be mentioned that there has been some controversy over the merits of environmental stress screening as it has evolved over the years.[8]

STATISTICAL RELIABILITY TESTS

A *statistical reliability test* is a test performed on a large sample of production items for a long duration to establish or verify an assigned reliability objective for the equipment operating in its anticipated service environment, where the reliability objective is usually stated in terms of a mean-time-to-failure (MTTF), or if all failures are assumed to be statistically independent, a mean-time-between-failures (MTBF) or failure rate (the reciprocal of MTBF). To provide an accurate indication of reliability, such tests must simulate the equipment shock and vibration environments with great accuracy. In some cases, rather than applying stationary vibration at the measured or predicted maximax levels of the environment, even the nonstationary characteristics of the vibration are reproduced, often in combination with shocks and other environments anticipated during the service life. The determination of reliability is accomplished by evaluating the times to individual failures, if any, by conventional statistical techniques.[9]

RELIABILITY GROWTH TESTS

A *reliability growth test* is a test performed on one or a few prototype items at extreme test levels to quickly cause failures and thus identify weaknesses in the equipment design. In many cases, the test level is increased in a stepwise manner to clearly identify the magnitude of the load needed to cause a specific type of failure. Design changes are then made and the failure rate of the equipment is monitored by either statistical reliability tests in the laboratory or evaluations of failure data from service experience to verify that the design changes produced an improvement in reliability. Unlike statistical reliability tests, reliability growth tests do not simulate the magnitudes of the service environments, although some effort is often made to simulate the general characteristics of the environments; for example, random vibration would be used to test equipment exposed to a random vibration service environment.

SELECTION OF SHOCK AND VIBRATION TEST LEVELS

The *test level* for a shock or vibration test is the spectrum of the excitation applied to the equipment at its mounting points by the test machine. For tests that require a simulation of the actual service shock and vibration environments (qualification, reliability, and some acceptance tests), the selection of test levels involves four steps, as follows:

1. Measurement or prediction of spectra for shock and vibration environments
2. Grouping of measured or predicted spectra into appropriate zones
3. Determination of zone limits
4. Selection of specified test levels

MEASUREMENT OR PREDICTION OF SPECTRA

Where equipment is to be installed in an existing system (for example, a new alternator for an existing automobile), the shock and/or vibration response of the system structure at the mounting points of the equipment can be determined by direct measurements (see Chap. 15). However, where equipment is to be installed in a system that has not yet been built and/or operated, the shock and/or vibration environment at the equipment mounting points must be predicted. Procedures for the prediction of shock and vibration environments vary widely depending upon the characteristics of environment and the system producing it. In general, however, prediction procedures can be divided into the following broad categories:

Lumped-Parameter Modeling Procedures. At least crude predictions for the shock and vibration response of a structural system at the mounting points of equipment can be achieved using the lumped-parameter analysis procedures detailed in Chap. 22. The accuracy of the resulting shock and vibration predictions depends heavily upon the complexity of the system structure being modeled and the exact analytical modeling procedure used.

Finite Element Method (FEM) Procedures. A popular modeling procedure for the prediction of shock and vibration environments is the finite element method (FEM) detailed in Chap. 23. Properly characterized shock and vibration excitations can be applied to an FEM model to predict the structural response at any point of interest. The FEM model can also be used to compute the frequency response functions between excitation and response points needed to make predictions by the frequency response procedures discussed later. Depending on the complexity of the structure being modeled, FEM procedures can generally produce reasonably accurate shock and vibration predictions up to a frequency equivalent to about the 50th normal mode of the structure.

Statistical Energy Analysis (SEA) Procedures. At frequencies above the range where finite element method procedures are accurate, statistical energy analysis (SEA) procedures described in Chap. 24 are commonly used to predict vibration environments. Specifically, as frequency increases, the response of the system structure can be predicted in terms of the space-averaged response for each of a set of individual structural elements that are coupled to collectively describe the system, where each element has near-homogeneous properties and light damping; an example is a constant thickness panel. Such prediction procedures can be applied to a wide range of structural systems if the assumptions detailed in Chap. 24 are satisfied.

Extrapolation Procedures. The spectra of the responses measured on one system during its operation can often be used to predict the spectra in a newer model of the system, assuming the old and new systems have a similar purpose and are of broadly similar design, for example, a new airplane that flies faster but otherwise is similar in structural design to an earlier model of the airplane. In such cases, the shock and/or vibration responses of the new system at the structural locations of equipment can be predicted, at least coarsely, by scaling the measurements made on the previous system based upon the differences in at least two parameters, namely, (1) the magnitude of the original excitation to the system structure and (2) the weight of the system structure at the points where the equipment is mounted. Specifically, as a first order of approximation, the shock and/or vibration magnitude on the new system can be assumed to vary directly with the magnitude of the excitation and inversely with the weight of the system structure. Such extrapolation techniques have been widely used to predict spectra for the vibration response of new aerospace vehicles[3] and can often be applied to other types of systems as well.

GROUPING OF MEASURED OR PREDICTED SPECTRA INTO ZONES

The shock and vibration responses of system structures that support equipment are typically nonhomogeneous in space, sometimes to the extent that the spectra of the responses vary substantially from one mounting point to another for a single item of equipment. At relatively low frequencies, corresponding to frequencies below about the fiftieth normal mode of the system structure (see Chap. 21), finite element method models for the system structure and the mounted equipment can be used to predict the motions at the specific equipment attachment points. It is more common, however, to define shock and vibration environments by making measurements or predictions at selected points on the system structure that do not correspond to the exact mounting points for equipment, or if they do, the equipment is not present during the measurements or accurately modeled for the predictions. Hence, it is necessary to separate the measured or predicted responses at various points on the system structure into groups, where the responses in each group have broadly similar spectra that can be represented for test purposes by a single spectrum. A *zone* is defined as a region on the system structure that includes those points where the measured or predicted shock and/or vibration responses have broadly similar spectra. It is clear that a zone should correspond to a region of interest in the formulation of shock and vibration test criteria for equipment, i.e., a single zone should include all the attachment points for at least one item of equipment, and preferably, for several items of equipment. However, a zone need not be a single contiguous structural region. For example, all frames of a given size in an airplane, no matter where they are located, might constitute a single zone if the responses of those frames are similar.

The determination of zones is usually based upon engineering judgment and experience. For example, given a system with frame-panel construction, engineering judgment dictates that frames and panels should represent different zones, since the responses of light panels will generally be greater than the much heavier frames. Also, the responses perpendicular to the surface of the panels are generally greater than the responses in the plane of the panels, so the responses along these two axes might be divided into separate zones. A visual inspection of the spectra for the measured or predicted responses also can be used to group locations with spectra of similar magnitudes to arrive at appropriate zones. In any case, it is desirable to minimize the number of zones used to describe the shock and vibration responses over those areas of the system structure where equipment will be mounted so as to minimize the number of individual spectra required to test all the equipment for that system.

DETERMINATION OF ZONE LIMITS

A *zone limit* (also called the *maximum expected environment*) is a single spectrum that will conservatively bound the measured or predicted spectra at most or all points within the zone, without severely exceeding the spectrum at any one point. A zone limit may be determined using any one of several procedures.[3] The most common procedure is to envelop the measured or predicted spectra in the zone, but a more rigorous approach is to compute a tolerance limit for the spectra. Specifically, given n measurements of a random variable x, an upper *tolerance limit* is defined as that value of x (denoted by L_x) that will exceed at least β fraction of all values of x with a *confidence coefficient* of γ. The fraction β represents the minimum probability that a randomly selected value of x will be less than L_x; the confidence coefficient γ can be interpreted as the probability that the L_x computed for a future set of data will indeed exceed at least β fraction of all values of x. Tolerance limits are commonly expressed in terms of the ratio $(100\beta)/(100\gamma)$. For example, a tolerance limit determined for $\beta = 0.95$ and $\gamma = 0.50$ is called the 95/50 normal tolerance limit. In the context of shock and/or vibration measurements or predictions, x represents the spectral value at a specific frequency (see Table 18.2) for the response of the system structure at a randomly selected point within a given zone, where x differs from point-to-point within the zone due to the spatial variability of the response. However, x may also differ due to other factors, such as variations in the response from one system to another of the same design or from one environmental exposure to another of the same system. In selecting a sample of measured or predicted spectra to compute a tolerance limit, beyond the spectra at different locations within a zone, it is wise to include spectra from different systems of the same design and different environmental exposures of the same system, if feasible, so that all sources of variability are represented in the measured or predicted spectra.

Tolerance limits are most easily computed when the random variable is *normally distributed* (see Chap. 19). The point-to-point (spatial) variation of the shock and vibration responses of system structures is generally not normally distributed, but there is empirical evidence that the logarithm of the responses does have an approximately normal distribution. Hence, by simply making the logarithmic transformation

$$y = \log_{10} x \tag{18.1}$$

where x is the spectral value at a specific frequency of the response within a zone, the transformed variable y can be assumed to have a normal distribution. For n sample values of y, a normal tolerance limit is given by[5]

$$L_y(n,\beta,\gamma) = \bar{y} + k s_y \tag{18.2}$$

where $\bar{y}$ is the sample average and s_y is the sample standard deviation of the n transformed spectral values computed as follows:

$$\bar{y} = \frac{1}{n}\sum_{i=1}^{n} y_i \qquad s_y = \sqrt{\frac{1}{n-1}\sum_{i=1}^{n}(y_i - \bar{y})^2} \tag{18.3}$$

The term k in Eq. (18.2) is called the *normal tolerance factor* and is a tabulated value; a short tabulation of k for selected values of n, β, and γ, is presented in Table 18.3. The normal tolerance limit for the transformed variable y is converted to the original engineering units of x by

$$L_x(n,\beta,\gamma) = 10^{L_y(n,\beta,\gamma)} \tag{18.4}$$

To simplify test criteria, normal tolerance limits are often smoothed using a series of straight lines, usually no more than seven with slopes of 0, ±3, or ±6 dB.

As an illustration, Fig. 18.1 shows the range of the maximax power spectra for $n = 12$ vibration measurements made at different locations in a selected zone of the structure of a large space vehicle during lift-off. Also shown in this figure are the unsmoothed and smoothed normal tolerance limit versus frequency computed with $\beta = 0.95$ and $\gamma = 0.50$ (the 95/50 limit). Note that the normal tolerance limit at most frequencies is higher than the largest of the 12 spectral values from which the limit is computed. However, a normal tolerance limit could be either higher or lower than the largest spectral values from which the limit is computed, depending on the values of n, β, and γ.

TABLE 18.3 Normal Tolerance Factors for Upper Tolerance Limit

	$\gamma = 0.50$			$\gamma = 0.75$			$\gamma = 0.90$		
n	$\beta = 0.90$	$\beta = 0.95$	$\beta = 0.99$	$\beta = 0.90$	$\beta = 0.95$	$\beta = 0.99$	$\beta = 0.90$	$\beta = 0.95$	$\beta = 0.99$
3	1.50	1.94	2.76	2.50	3.15	4.40	4.26	5.31	7.34
4	1.42	1.83	2.60	2.13	2.68	3.73	3.19	3.96	5.44
5	1.38	1.78	2.53	1.96	2.46	3.42	2.74	3.40	4.67
7	1.35	1.73	2.46	1.79	2.25	3.13	2.33	2.89	3.97
10	1.33	1.71	2.42	1.67	2.10	2.93	2.06	2.57	3.53
15	1.31	1.68	2.39	1.58	1.99	2.78	1.87	2.33	3.21
20	1.30	1.67	2.37	1.53	1.93	2.70	1.76	2.21	3.05
30	1.29	1.66	2.35	1.48	1.87	2.61	1.66	2.08	2.88
50	1.29	1.65	2.34	1.43	1.81	2.54	1.56	1.96	2.74
∞	1.28	1.64	2.33	1.28	1.64	2.33	1.28	1.64	2.33

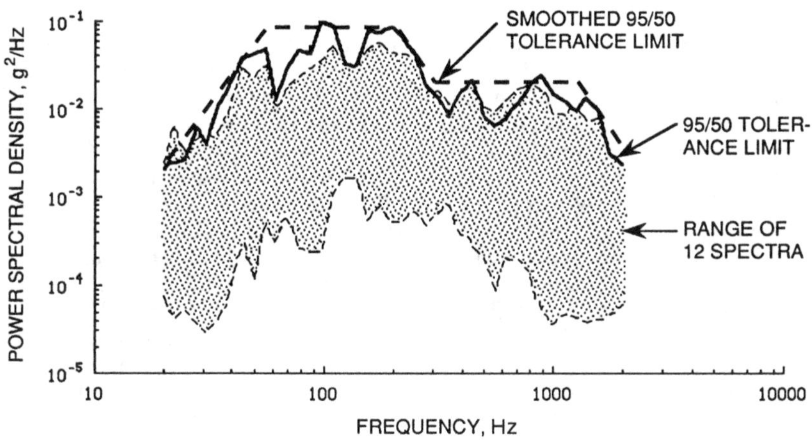

FIGURE 18.1 95/50 normal tolerance limit for spectra of 12 vibration measurements.

SELECTION OF FINAL TEST LEVELS

A *test level* is the spectrum of the shock or vibration environment that is specified for testing purposes, i.e., the spectrum given in a final test specification. The determination of a test level based upon a computed zone limit requires the selection of a value for β, the fraction of the locations within a zone where the spectra of the shock and/or vibration responses of the system structure will be exceeded by the zone (tolerance) limit. This selection is often made somewhat arbitrarily, with values in the range $0.90 \leq \beta \leq 0.99$ being the most common for acceptance and qualification tests. However, the value of β used to arrive at a test level can be optimized based upon an assessment of the adverse consequences (the potential cost) of an undertest versus an overtest. Also, even with an optimum selection, modifications to the test level may be required to account for the interactions of the equipment and the system structure and other considerations.

Optimum Test Level Selection. A number of procedures have been developed[3] that yield an optimum test level for equipment in terms of a percentile of the environmental distribution (which is essentially the value of β for a tolerance limit) as a function of a "cost" ratio C_T/C_F, where C_T is the cost of a test failure and C_F is the cost of a service failure. Some of the procedures assume the equipment being tested has already been manufactured in quantity, raising the possibility that a test failure will lead to refurbishing costs, while others account for a safety factor in the equipment design or a *test factor* based upon the assumed strength of the item being tested. The simplest test level selection rule, which applies to the acceptance testing of a single item of equipment, is given by[3]

$$\beta = \frac{1}{1 + (C_T/C_F)} \tag{18.5}$$

As an illustration, consider an item of equipment where a failure during test could be corrected by a relatively simple replacement of an inexpensive component, but a failure during service would be catastrophic, perhaps resulting in personal injury. According to Eq. (18.5), the item should be tested to a very severe level relative to the measured or predicted shock and/or vibration environment so as to sharply minimize the risk of an undertest; for example, if a service failure is assessed to be 1000 times as costly as a test failure, $\beta = 0.999$. On the other hand, consider an item where a failure in test would lead to a difficult and expensive redesign, but a failure during service would not be catastrophic. According to Eq. (18.5), the test level now should be moderate relative to the measured or predicted shock and/or vibration environment so as to minimize the risk of an overtest; for example, if a service failure is assessed to be only 9 times as costly as a test failure, then $\beta = 0.90$. Note that the selection procedure does not require the determination of quantitative costs in dollars, but only relative costs, which can be interpreted in qualitative terms. This allows such factors as the consequences of a possible delivery delay caused by a test failure or customer dissatisfaction caused by a service failure to be considered. Also, the conservatism of the test level can be further increased or decreased by selecting a larger or smaller value of γ for the tolerance limit computation.

It should be mentioned that there has been a movement in some industries, particularly the aerospace industry,[3] to standardize the computation of maximum expected environments using the 95/50 normal tolerance limit on the logarithms of the measured spectral values in a zone, independent of the cost considerations discussed above. This has been done to make the maximum expected environments computed by different organizations directly comparable in statistical terms. It is commonly argued that the *test margin,* to be discussed shortly, adds sufficient conservatism to the test levels to obscure differences in the selected tolerance factor.

Equipment-Structure Interactions. Test levels are commonly specified in terms of a motion parameter, for example, g^2/Hz versus frequency for a random vibration test. However, at the resonance frequencies of relatively heavy items of equipment, the apparent mass of the equipment dramatically increases, causing the equipment to behave like a dynamic vibration absorber on the system structure to which the equipment is mounted (see Chap. 6). If the test machine is made to deliver the specified motion to the equipment at its resonance frequencies, a severe overtest may occur. This problem is sometimes addressed by placing limits on the response of the equipment or by allowing "notches" in the specified test spectrum to be introduced at the frequencies of strong resonances of the equipment. The best approach, however, is to derive a second spectrum for the limiting force at the mounting points of the equipment and establish criteria for a dual control test that limits both the input force and the input motion to the equipment.[10]

Added Test Level Factors. For qualification tests where the item of equipment being tested will not be used in service, it is common to add a factor (often referred to as a *test margin*) to the derived maximum expected environment to arrive at a final specified test level. Such factors are usually justified to account for uncertainties not considered in the determination of the maximum expected environment, such as unknown variabilities in the equipment strength or its possible service use. These factors are sometimes selected rather arbitrarily, with typical values ranging from 3 to 6 dB above the maximum expected environment.

SELECTION OF VIBRATION TEST DURATIONS

The *test duration* for a vibration test is the total time the excitation is applied to the equipment at its mounting points by the test machine. In some cases, the test duration is not relevant to the purpose of the test, for example, a development test. In many cases, however, an appropriate simulation of the total duration of the vibration environment anticipated in service is an important part of the test criteria. This is particularly true of qualification and statistical reliability tests, where the purpose is to detect design inadequacies that may lead to failures of any type during exposure to the service vibration environment, including "wearout" failures. For shock environments, this means exposing the equipment to repeated simulations of all the shocks anticipated during its service life, which can usually be accomplished in a reasonable period of time. For vibration environments, however, this means exposing the equipment to a simulation of the anticipated service vibration environment for a duration equivalent to the service life of the equipment, which may be thousands of hours. Vibration environments usually vary widely in overall level and perhaps spectral content during the equipment service life, for example, equipment on an automobile or truck in normal service use. As noted earlier in this chapter, statistical reliability tests are sometimes performed with a duration similar to the anticipated service life of the equipment. For qualification tests, however, it is usual to compress a long, time-varying service environment into a stationary test level of much shorter duration.[9] To do this, the following steps are required:

1. Assume a time-dependent failure model for the equipment
2. Compress the time-varying magnitudes of the environment into a single test level corresponding to a conservative estimate of the maximum magnitude of the environment
3. In some cases, increase the test level beyond the maximum magnitude of the environment to further accelerate the test

FAILURE MODELS

A *failure* of an item of equipment is defined as any deterioration of performance or any damage or malfunction that prevents the equipment from accomplishing its intended purpose. There are two basic types of failures that may be caused by vibration:

1. *Hard failure.* A failure involving permanent physical damage that makes the equipment unable to perform its intended purpose, even after the vibration is terminated. Hard failures generally result in observable damage, such as the fracture of a structural element or the permanent disability of an electronic element.
2. *Soft failure.* A failure involving a malfunction or deterioration of performance during the vibration exposure that makes the equipment unable to accomplish its intended purpose, but after the vibration is terminated, the equipment does not reveal any damage and functions properly. Soft failures most commonly occur in electrical, electronic, and/or optical elements, although soft failures may occasionally occur in complex mechanical elements, such as gyroscopic devices.

A *failure mechanism* is the specific means by which an item of equipment is damaged by exposure to an environment. All failure mechanisms are a function of the magnitude of the vibration exposure. A *time-dependent failure mechanism* is a function of both the magnitude and the duration of the vibration exposure. Soft failures during exposure to a vibration environment are rarely time-dependent, i.e., they usually occur immediately at the start of the vibration exposure. On the other hand, hard failures usually are time-dependent, although there are some exceptions. For example, if a vibration environment produces stresses that exceed the ultimate strength of a critical element in the equipment, a fracture will occur immediately at the start of the vibration exposure. See Chaps. 33, 34, and 40 for further discussions of equipment failures.

To establish appropriate test durations for qualification vibration tests, only time-dependent failure mechanisms (usually producing hard failures) are of interest. Common examples of time-dependent failure mechanisms for equipment exposed to vibration environments are fatigue damage, force contact wear, relative velocity wear, and the loosening of bolts or rivets. A *failure model* is an analytical relationship between the time-to-failure of the equipment during exposure to a vibration environment and the magnitude of the vibration environment. For a wide class of time-dependent failure mechanisms, the time-to-failure τ for a stationary vibration excitation can be approximated by the *inverse power law* given by

$$\tau = c\,\sigma^{-b} \tag{18.6}$$

where σ is the stress in the equipment caused by the vibration (or any measure of the vibration magnitude that is linearly related to stress), and b and c are constants related to the specific failure mechanism. From Chap. 33, if the endurance limit is ignored, the fatigue endurance curves for common metals fit the form of Eq. (18.6).

Using Eq. (18.6) and assuming a vibration test is performed that accurately simulates the basic characteristics (for example, random versus periodic) and the spectrum of a service vibration environment, the time required to produce a similar amount of damage in the test environment T_t and the time in the service environment T_e are related by

$$T_t = \left(\frac{\sigma_e}{\sigma_t}\right)^b T_e \tag{18.7}$$

where σ is the rms value of the vibration, and the subscripts t and e denote the test and service environments, respectively. For random vibrations defined in terms of power spectra [i.e., $W(f)$ defined in Chap. 19], Eq. (18.7) becomes

$$T_t = \left(\frac{W_e(f)}{W_t(f)}\right)^{b/2} T_e \tag{18.8}$$

The value of the power b in Eqs. (18.7) and (18.8) varies widely for different failure mechanisms. For metal fatigue damage, a value of $b = 8$ is reasonable for many common materials (see Chap. 33) and is recommended in Ref. 3. However, a value of $b = 4$ is usually more appropriate for the typical failure mechanisms in electrical and electronic equipment.[11]

COMPRESSING TIME-VARYING SERVICE ENVIRONMENTS

For those vibration environments that vary substantially in severity during the equipment service life, the duration of the environment can often be reduced for testing purposes by using Eq. (18.7) to scale the less severe vibration levels to the most severe levels that occur during the service life. Such scaling procedures are most applicable to environments that vary in overall level but not substantially in spectral content. For

TABLE 18.4 Determination of Equivalent Duration for Automobile Equipment Vibration Environment

Type of road segment	Duration on road segment, hours	rms vibration on road segment, *g*	Equivalent duration on road segment A, hours
A. Unpaved secondary roads	40	3	40
B. Improved secondary roads	460	1.4	22
C. Primary roads	1500	0.9	12
D. Major highways	2000	0.7	6
Total equivalent duration on road segment A (hours)			80

example, consider an item of electrical equipment designed for a motor vehicle with a service life of 4000 hours. Assume the anticipated service vibration environment for the vehicle at the equipment mounting points has the rms values summarized in Table 18.4. Further assume $b = 4$ in Eq. (18.7), and the vibrations during the various service conditions have a similar spectral content. Table 18.4 indicates the damage potential of the 4000-hour service vibration environment can be simulated by a vibration test with a duration of 80 hours at the maximum service vibration level.

For those vibration environments where the spectral content and the overall levels change during service operations, the test duration computations illustrated in Table 18.4 must be made on a frequency-by-frequency basis using Eq. (18.8) or a similar expression for the appropriate spectral description in Table 18.2. This will result in a different test duration at each frequency, leading to two possible testing options: (1) a series of tests, each covering a different frequency range with a different test duration or (2) a single test with a test duration equal to the longest test duration computed at any frequency. The second option is usually the more practical and ensures a conservative test.

ACCELERATED TESTS

An *accelerated test* is a test where the test duration is reduced by increasing the test level in a manner that will maintain the same environment-induced damage to the equipment. The determination of a test duration for a stationary vibration test that produces the same damage as a nonstationary vibration environment, as detailed in the preceding section, constitutes the most desirable form of accelerated testing because the test level never exceeds the maximum vibration level that the equipment will experience during its service environment. Furthermore, most of the damage experienced by equipment in service usually occurs during exposure to the maximum vibration level in the service environment, which typically covers a small fraction of the total service duration (see Table 18.4). In such cases, reducing the relatively long durations of the less severe vibrations by scaling to the maximum level according to Eq. (18.7) does not introduce a major error, even if the exponent in Eq. (18.7) is inaccurate.

Highly Accelerated Tests. Situations sometimes arise where scaling the less severe segments of a nonstationary vibration environment to a stationary vibration level corresponding to the maximum level of the environment may yield a test duration that is still too long to be practical; for example, the test duration of 80 hours computed for the 4000-hour service environment in Table 18.4 may still be too long for testing purposes. In such a case, it is common to further reduce the test duration by increasing the test level beyond the maximum level the equipment will experience during its anticipated service environment. Indeed, if no limit is placed on the rms test level in Eq. (18.7), the test duration theoretically can be made as short as desired, provided the ultimate strength of the equipment structure is not exceeded. However, increasing the test level beyond the maximum level during the anticipated service environment introduces major uncertainties in the test results, particularly if the equipment is fabricated using different materials and/or incorporates electrical, electronic, and/or optical elements. The problem is that the failure mechanisms of some elements may not comply with the inverse power law in Eq. (18.6). Furthermore, even if all failure mechanisms do comply with Eq. (18.6), the exponent b may vary from one element to another within the equipment. Hence, increasing the test level to accelerate the test rapidly in compliance with Eq. (18.7) may cause some elements of the equipment to be undertested and others to be overtested. The result could be the occurrence of unrepresentative failures during the accelerated test.[11] It is for these reasons that highly accelerated testing should be pursued only with great caution.

Durability and Functional Tests. A common procedure to suppress unrepresentative failures that may be caused by rapidly accelerating a vibration test of equipment with a long service life is to perform two separate tests, namely, a durability test and a functional test. A *durability test* is intended to reveal only time-dependent failures and is rapidly accelerated to produce the same damage as the entire duration of the service vibration environment based upon a specific damage model, for example, Eq. (18.7). The equipment is not required to function during the durability test, and any failures that are not time-dependent are ignored. A *functional test* is intended to reveal failures that are not time-dependent (i.e., failures related only to the vibration level) and is not accelerated with test levels that exceed the maximum expected vibration level during the service environment. The equipment is required to function during the test, but since the failures of interest are not time-dependent, the test duration is not critical; for example, the test duration is often fixed by the time required to fully operate the equipment and verify that it properly performs its intended purpose.

SHOCK AND VIBRATION TESTING

The laboratory machinery used to perform vibration tests and shock tests are detailed in Chaps. 25 through 28. In all cases, there are several issues that must be carefully considered in performing such tests, the most important being:

1. Identification of test failures
2. Type of excitation to be used
3. Single- versus multiple-axis excitation
4. Test fixtures

IDENTIFICATION OF TEST FAILURES

In all shock and vibration tests of equipment, it is important to carefully establish what types of equipment malfunctions or anomalies will be considered failures. This determination depends heavily on the purpose of the test and sometimes on the judgment of the purchaser of the equipment. Here are a few examples:

1. Since a qualification test is intended to identify design problems, failures during the test that are clearly due to workmanship errors or material defects are usually ignored; i.e., the equipment is repaired and the test is continued.
2. Since the test level for a highly accelerated qualification test is based upon a specific failure model, failures during the test that are not consistent with the failure model should be carefully evaluated and ignored if they are determined to involve a failure mechanism that is not time-dependent.
3. During durability tests of equipment, if a fatigue crack forms in the equipment structure that does not propagate to a fracture, whether the fatigue crack constitutes a failure or the length of the fatigue crack that constitutes a failure must be specified.
4. During functional tests of electrical, electronic, and/or optical equipment, if there is measurable deterioration in the performance of the equipment during the test, the exact degree of deterioration that prevents the equipment from performing its intended purpose must be specified.

TYPES OF EXCITATION

Shock tests are sometimes performed using specified test machines, but more often are performed using more general test machines that can produce transients with a desired shock response spectrum (see Chaps. 27 and 28). Although vibration environments may be simulated by mounting the equipment in a prototype system and reproducing the actual environment for the system, it is more common to apply the vibration directly to the equipment mounting points using vibration testing machines described in Chap. 25.

Random Tests. Random excitations are used to simulate random vibration in those tests where an accurate representation of the environment is desired, specifically, qualification, reliability, and some acceptance tests. The most commonly used random test machines produce a near-gaussian vibration. If the actual environment is random but not gaussian, a gaussian simulation is still usually acceptable since the response of the equipment exposed to the environment will be near-gaussian at its resonance frequencies, assuming the equipment response is linear; this is because equipment resonances constitute narrowband filtering operations that suppress deviations from the gaussian form in the vibration response of the equipment.[12]

Sine-Wave Tests. Sine-wave excitations are used to simulate the fixed-frequency periodic vibrations produced by constant-speed rotating machines and reciprocating engines. Sine-wave excitations are sometimes superimposed on random excitations for those situations where the service vibration environment involves both. Sine-wave excitations fixed sequentially at the resonance frequencies of an equipment item (often referred to as a *dwell sine test*) are sometimes used in development tests, as well as in durability tests, to evaluate the fatigue resistance of the equipment.

Swept-Sine-Wave Tests. Sweep-sine-wave excitations are produced by continuously varying the frequency of a sine wave in a linear or logarithmic manner. Such excitations are used to simulate the vibration environments produced by variable-speed rotating machines and reciprocating engines. The usual approach is to make the sweep rate sufficiently slow to allow the equipment being tested to reach a near-full (steady-state) response as the swept-sine-wave excitation passes through each resonance frequency. Swept-sine-wave excitations are also used for development tests to identify resonance frequencies and sometimes to estimate frequency response functions (see Chap. 21).

MULTIPLE-AXIS EXCITATIONS

Shock and vibration environments are typically multiple-axial; i.e., the excitations occur simultaneously along all three orthogonal axes of the equipment. Multiple-axis shock and vibration test facilities are often used to simulate low-frequency shock and vibration environments, generally below 50 Hz, such as earthquake motions (see Chap. 29). Also, multiple-axis vibration test facilities have been developed for higher-frequency vibration excitations (up to 2000 Hz), but it is more common to perform shock and vibration tests using machines that apply the excitation sequentially along one axis at a time, i.e., machines that deliver rectilinear motion only (see Chaps. 25 through 28). Single-axis testing introduces an additional uncertainty of unknown magnitude in the accuracy of the test simulation, but there is debate as to whether the removal of this uncertainty justifies the high cost and complexity of multiple-axis test facilities.

TEST FIXTURES

A *test fixture* is a special structure that allows the test item to be attached to the table of a shock or vibration test machine. Test fixtures are required for almost all shock and vibration tests of equipment because the mounting hole locations on the equipment and the test machine table do not correspond. For the usual case where the test machine generates rectilinear motion normal to the table surface, a test fixture is also necessary to reorient the equipment relative to the table so that vibratory motion can be delivered along the lateral axes of the equipment, i.e., the axes parallel to the plane of the equipment mounting points. This requires a versatile test fixture between the table and the equipment, or perhaps three different test fixtures. If the direction of gravity is important to the equipment, the test machine must be rotated from vertical to horizontal, or vice versa, to meet the test conditions.

For equipment that is small relative to the test machine table, L-shaped test fixtures with side gussets are commonly used to deliver excitation along the lateral axes of the equipment as illustrated Fig. 18.2. Unless designed with great care, such fixtures are likely to have resonances in the test frequency range. In principle, the consequent spectral peaks and valleys due to fixture resonances can be flattened out by electronic equalization of the test machine table motion (see Chap. 26), but this is difficult if the damping of the fixture is low. The best approach is to design the fixture to have, if possible, no resonances in the test frequency range.

For equipment that is large relative to the test machine table, excitation along the lateral axes of the equipment is commonly achieved by mounting the equipment on a horizontal plate driven by the test machine rotated into the horizontal plane, where the plate is separated from the flat opposing surface of a massive block by an oil film or hydrostatic oil bearings as shown in Fig. 18.3. The oil film or hydrostatic bearings provide little shearing restraint but give great stiffness normal to the surface, the stiffness being distributed uniformly over the complete horizontal area.

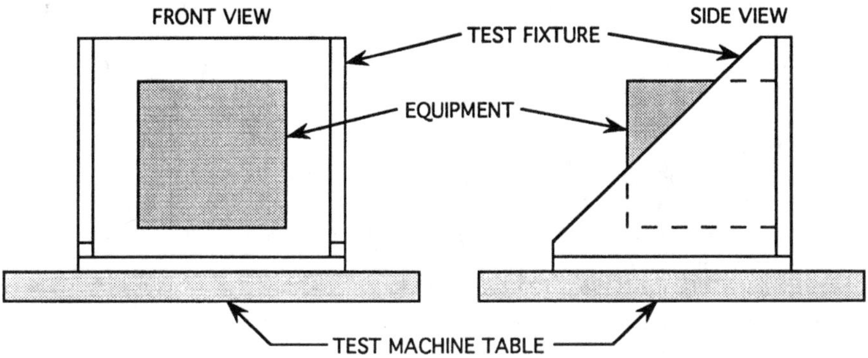

FIGURE 18.2 Test fixture to deliver excitation in the plane of the equipment mounting points.

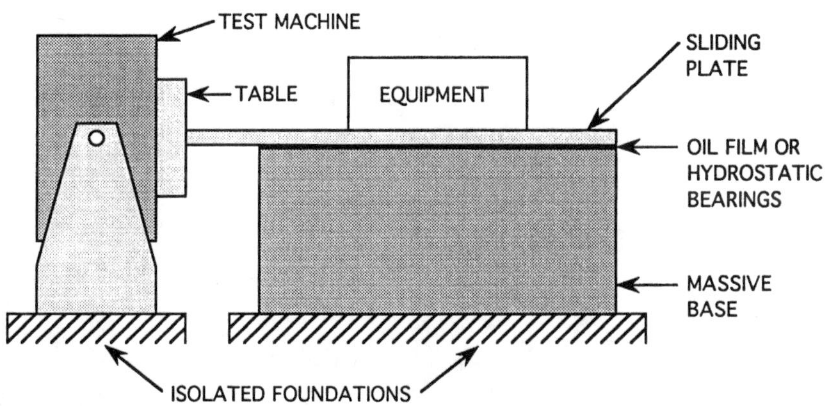

FIGURE 18.3 Horizontal plate to deliver excitation in the plane of the equipment mounting points.

Accordingly, a relatively light moving plate can be vibrated that has the properties of the massive rigid block in the direction normal to its plane. See Refs. 7 and 13 for further discussions of vibration and shock test fixturing.

REFERENCES

1. Bendat, J. S., and A. G. Piersol: "Random Data: Analysis and Measurement Procedures," 3d ed., John Wiley & Sons, New York, 2000.
2. Gaberson, H. A., and R. H. Chalmers: *Shock and Vibration Bull.,* **40**(2):31 (1969).
3. Kern, D. L., et al.: "Dynamic Environmental Criteria," *NASA-HDBK-7005,* 2001.
4. Harris, C. M., and C. E. Crede: "Shock and Vibration Handbook," 1st ed., Chap. 24, McGraw-Hill Book Company, New York, 1961.

5. Hines, W. W., et al., "Probability and Statistics in Engineering," John Wiley & Sons, Hoboken, N.J., 2003.
6. Anon., "Management and Technical Guidelines for ESS Process," *IEST-RP-PR001.1,* Institute of Environmental Sciences and Technology, Rolling Meadows, Ill., 1999.
7. Tustin, W. "Random Vibration and Shock Testing," Equipment Reliability Institute, Santa Barbara, Calif., 2005.
8. Pusey, H. C., "50 Years of Shock and Vibration Technology," *SVM-15,* pp. 342–349, Shock and Vibration Information and Analysis Center, Hi-Test Labs, Richmond, Va., 1996.
9. Lawless, F. E., "Statistical Models and Methods for Lifetime Data," 2d ed., John Wiley & Sons, New York, 2002.
10. Scharton, T. D.: "Force Limited Vibration Testing Monograph," *NASA-RP-1403,* 1997.
11. Kana, D. D., and T. G. Butler: "Reliability Design for Vibroacoustic Environments," *ASME-AMD-9,* p. 139, 1974.
12. Papoulis, A.: "Narrow-Band Systems and Gaussianity," *USAF-RADC-TR-71-225,* 1971.
13. Anon.: "Vibration and Shock Test Fixturing," *IEST-RP-DTE013.1,* Institute of Environmental Sciences and Technology, Mount Prospect, Ill., 1998.

CHAPTER 19
VIBRATION DATA ANALYSIS

Allan G. Piersol

INTRODUCTION

Vibration data are usually acquired in the form of continuous electrical (analog) signals generated by transducers (see Chap. 10), where each analog signal represents the instantaneous value of a strain, pressure, force, or motion parameter (displacement, velocity, or acceleration) as a function of time. Such a signal is commonly referred to as a *time history.* A *sample record* is defined as the time history representing a single vibration measurement $x(t)$ over a finite duration T. Although sample records are usually acquired in the form of time histories, any other variable of interest can replace time t as the independent variable for analysis purposes. For example, road roughness data are commonly acquired as sample records of road elevation x versus distance d, that is, $x(d)$; $0 \leq d < D$, where D is the length of the record. However, for clarity, all discussions and equations in this chapter are presented in terms of sample time-history records, where it is understood that any other variable can be substituted for time.

It should be mentioned that vibration data are sometimes analyzed online by direct operations on the acquired transducer output signals using specialized data analysis instruments. The most common vibration data analysis in such cases usually involves a frequency decomposition of the time history data (the computation of a spectrum), as detailed in Chap. 14. This chapter is concerned with the more detailed data analysis operations that are often performed on stored time histories of the acquired vibration data.

CLASSIFICATIONS OF VIBRATION DATA

The appropriate analysis procedures for vibration environments depend heavily upon certain basic characteristics of the vibration. The most important distinctions

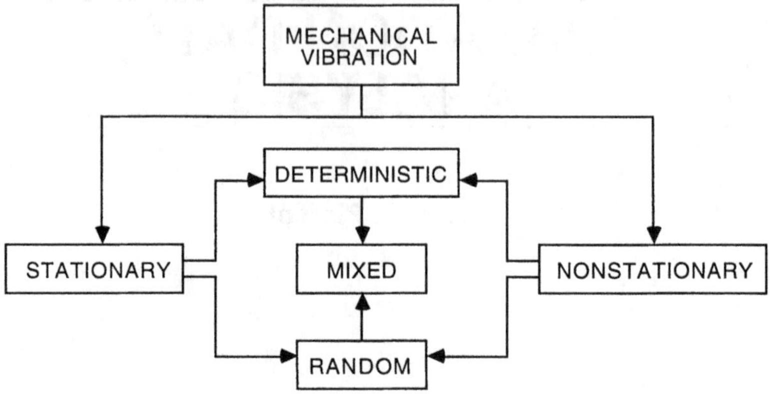

FIGURE 19.1 Classifications of vibration environments.

are defined in Chap. 1 and illustrated in Fig. 19.1. These definitions may be summarized as follows:

1. A *stationary vibration* is one whose basic properties do not vary with time. Stationary vibrations typically occur when the operating and/or environmental conditions producing the vibration are time invariant. For example, the vibration environment for a motor vehicle driving over a homogeneous road at constant speed and with a constant engine rpm will be stationary.

2. A *nonstationary vibration* is one whose basic properties vary with time, but slowly relative to the lowest frequency of the vibration. For example, the vibration environment for a motor vehicle during acceleration from zero to highway speed will be nonstationary. Those dynamic environments that change rapidly relative to the lowest frequency in the environment are considered transients or shocks, which are addressed in Chap. 20.

3. A *deterministic vibration* is one whose value at any time can be predicted from its value at any other time. It follows that sample records of a deterministic vibration collected repeatedly under similar conditions will have similar time histories. For example, the vibration environments of rotating machines and reciprocating engines (see Chap. 37) are generally deterministic.

4. A *random vibration* is one whose instantaneous magnitude is not specified at any given time. The instantaneous magnitudes of a random vibration are specified only by probability functions giving the probable fraction of the total time that the magnitude (or some sequence of magnitudes) lies within a specified range. From another viewpoint, a random vibration can be thought of as a single physical realization, $x(t)$, of a random process, which theoretically is described by an ensemble of all possible physical realizations denoted by $\{x(t)\}$.[1] Virtually all stationary random vibrations can be represented by an *ergodic* random process (see Chap. 1), meaning the properties of the random process $\{x(t)\}$ can be described by time averages over a signal sample record $x(t)$. It follows that the sample records of a stationary random vibration collected repeatedly under similar conditions will have time histories that differ in detail but have the same average properties. For example, the vibrations induced by earthquakes, turbulent flow, wind, and jet noise (see Chaps. 29 through 32) are generally random.

5. A *mixed vibration* is one that includes a combination of deterministic and random components. To some degree, most vibration environments are mixed, although either a deterministic or random component will often dominate.

The next section in this chapter summarizes the quantitative descriptions of vibration environments. This is followed by a discussion of the important preliminary steps in preparing measured vibration data for analysis, and the specific analysis procedures for measured vibration data.

QUANTITATIVE DESCRIPTIONS OF STATIONARY VIBRATIONS

The properties of stationary vibration environments, both deterministic and random, that are of primary interest to engineering applications are now summarized by functional relationships that lead directly to the applied computational algorithms used to compute the desired properties from sample records of measured vibration data.

OVERALL VALUES

The most fundamental descriptions of a stationary vibration with a time history $x(t)$ are given by overall values. In general, various different overall values might be determined, but often the *mean value* μ_x, the *mean-square value* ψ_x^2, and/or the *variance* σ_x^2 are the only overall values of interest. These values for a sample record $x(t)$ with duration T are theoretically given by[1,2]

$$\text{Mean value:} \quad \mu_x = \lim_{T\to\infty} \frac{1}{T}\int_0^T x(t)\,dt$$

$$\text{Mean-square value:} \quad \psi_x^2 = \lim_{T\to\infty} \frac{1}{T}\int_0^T x^2(t)\,dt \qquad (19.1)$$

$$\text{Variance:} \quad \sigma_x^2 = \lim_{T\to\infty} \frac{1}{T}\int_0^T [x(t) - \mu_x]^2\,dt$$

It can be shown[1] that the three quantities defined in Eq. (19.1) are interrelated by

$$\psi_x^2 = \mu_x^2 + \sigma_x^2 \qquad (19.2)$$

Hence, a knowledge of any two quantities determines the third. The positive square root of the mean-square value and the variance, ψ_x and σ_x, are called the *root-mean-square (rms) value* and the *standard deviation,* respectively.

The mean value defines the central tendency (static value) of the vibration, while the standard deviation defines the dispersion of the vibration, each with the same units as the vibration. The rms value is a measure of both the central tendency and dispersion. In many cases, one or more of the following will be true: (*a*) the mean value of the vibration is zero, (*b*) the vibration transducer cannot produce a static (dc) output corresponding to a mean value (e.g., piezoelectric accelerometers), and/or (*c*) a mean value cannot be measured because the data acquisition system is ac coupled, that is, it will not transmit dc. In these cases, the rms value of the vibration is the same as its standard deviation, that is, $\psi_x = \sigma_x$.

FOURIER TRANSFORMS

Since frequency domain descriptions of vibrations are generally of the greatest engineering value, the Fourier transform plays a major role in both the theoretical definitions of properties and the analysis algorithms for vibration data. The Fourier transform of a sample record $x(t)$ of duration T is defined as

$$X(f,T) = \int_0^T x(t)e^{-j2\pi ft}dt = \int_0^T x(t)\cos(2\pi ft)dt - j\int_0^T x(t)\sin(2\pi ft)dt \qquad (19.3)$$

where $j = \sqrt{-1}$. Three properties of the definition in Eq. (19.3) should be noted, as follows:

1. The Fourier transform is generally a complex number that is defined for both positive and negative frequencies, that is, $X(f,T)$; $-\infty < f < \infty$. However, $X(-f,T) = X^*(f,T)$, where the asterisk denotes the complex conjugate, meaning that values at mathematically negative frequencies are redundant and provide no information beyond that provided by the values at positive frequencies. Since engineers typically think of frequency as a positive value, it is common to present finite Fourier transforms as $2X(f,T); 0 < f < \infty$.
2. Fourier transforms are often defined as a function of radial frequency ω in radians/sec, as opposed to cyclical frequency f in Hz, particularly for analytical applications. However, data analysis is usually accomplished in terms of cyclical frequency f, as defined in Eq. (19.3). The two definitions are interrelated by $X(f,T) = 2\pi\, X(\omega,T)$.
3. The Fourier transform $X(f,T)$ is equivalent to the Fourier series of $x(t)$ assumed to have a period T.

See Chap. 14 for details on the computation of Fourier transforms from a sample time-history record, $x(t)$, with a duration of T seconds.

STATIONARY DETERMINISTIC VIBRATIONS

Stationary deterministic vibration environments generally fall into one of two categories, namely, periodic vibrations or almost-periodic vibrations.

Periodic Vibrations. Periodic vibrations are those with time histories that exactly repeat themselves after a time interval T_P, that is, $x(t) = x(t + iT_P)$; $i = 1, 2, 3, \ldots$, where T_P is called the *period* of the vibration. All periodic vibrations can be decomposed into a Fourier series, which consists of a collection of commensurately related sine waves,[1,2] that is,

$$x(t) = a_0 + \sum_k a_k \sin(2\pi k f_1 t + \theta_k) \qquad k = 1, 2, 3, \ldots \qquad (19.4)$$

where a_0 is the mean value, kf_1 is the kth frequency component (harmonic), and a_k and θ_k are the amplitude and phase angle associated with the kth frequency component of the periodic vibration. The $k = 1$ component is called the *fundamental fre-*

quency of the periodic vibration, and is given by $f_1 = 1/T_P$. The magnitudes of the frequency components in Eq. (19.4) are given by

$$L_x(f) = \frac{2|X(f,T_P)|}{T_P} \qquad 0 > f \tag{19.5}$$

where $X(f,T_P)$ is as defined in Eq. (19.3) with $T = T_P$, the period of the vibration. A plot of $L_x(f)$ versus frequency is called a *line spectrum* or a *linear spectrum.* The phase angles, θ_k; $k = 1, 2, 3, \ldots$, are usually ignored, but these phase values should be retained if the time history is not retained, since both the magnitude and phase values in Eq. (19.4) are required to reconstruct the time history.

Periodic vibrations are usually produced by the mechanical excitations of rotating machines and reciprocating engines operating with a constant rotational speed. They are also produced by the aerodynamic excitations from large fans and propellers, again operating at a constant rotational speed. An illustration of the time history and line spectrum for a periodic vibration composed of three harmonic components ($k = 1, 2,$ and 3) is shown in Fig. 19.2.

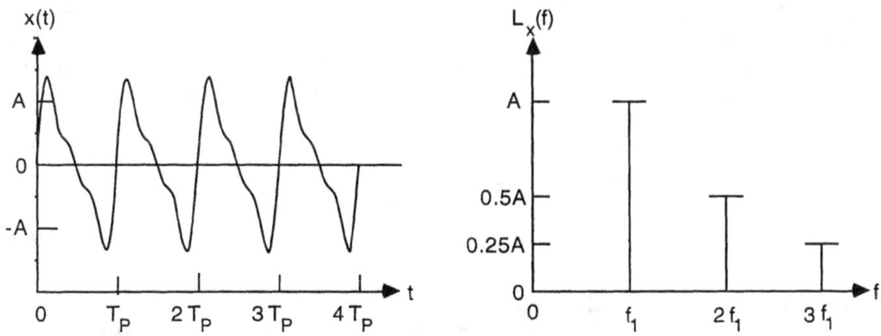

FIGURE 19.2 Time history and line spectrum for periodic vibration.

Almost-Periodic Vibrations. Although periodic vibrations can be decomposed into a collection of commensurately related sine waves, as given by Eq. (19.4), it does not follow that the sum of two or more independent sinusoidal excitations will produce a periodic vibration. In fact, the sum of such independent sine waves will be periodic only if the ratios of all pairs of frequencies create rational numbers. Those deterministic vibrations that do not have commensurately related frequency components are called *almost-periodic*[1] (also called *quasi-periodic* or *complex*) vibrations. Nevertheless, such vibrations can be described by a line spectrum based upon a relationship similar to Eq. (19.4), except the commensurately related frequencies kf_1 are replaced by independent frequencies f_k; $k = 1, 2, 3, \ldots$. As for periodic vibrations, the magnitude of the frequency components for almost-periodic vibrations can be described by a line spectrum defined in Eq. (19.5), except $T_P \to \infty$.

Almost-periodic vibrations often occur when two or more independent periodic excitations are summed. For example, the vibration produced by two independent rotating machines that are not synchronized or geared together will usually be almost-periodic rather than periodic. An illustration of the time history and line spectrum for an almost-periodic vibration composed of the sum of two sine waves that are not commensurately related is shown in Fig. 19.3.

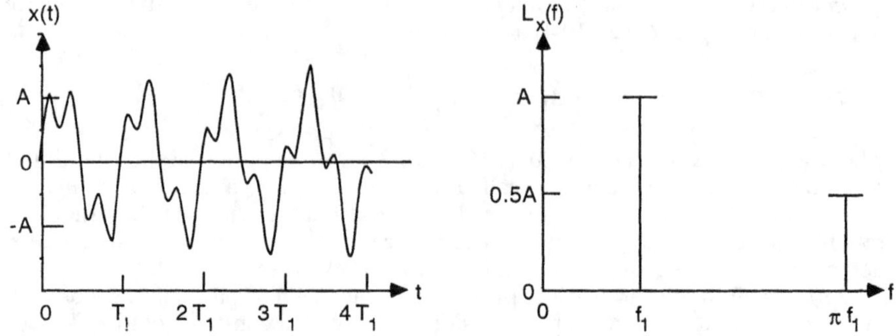

FIGURE 19.3 Time history and line spectrum for almost-periodic vibration.

STATIONARY RANDOM VIBRATIONS

By definition, random vibrations cannot be described by an explicit mathematical function and, hence, must be described in statistical terms. This can be done (*a*) in the amplitude domain by probability functions, (*b*) in the time domain by correlation functions, and/or (*c*) in the frequency domain by spectral density functions.

Probability Density Functions. The *probability density function* of a stationary random vibration $x(t)$ may be defined as

$$p(x) = \lim_{\substack{T \to \infty \\ \Delta x \to 0}} \frac{1}{\Delta x} \frac{T(x,\Delta x)}{T} \tag{19.6}$$

where $T(x,\Delta x)$ is the time that $x(t)$ is within the magnitude interval Δx centered at x during the sample record duration T. The integral of the probability density function between any two magnitudes x_1 and x_2 defines the probability at any future instant that the value of $x(t)$ will fall between x_1 and x_2, that is,

$$\text{Prob}[x_1 < x(t) \le x_2] = \int_{x_1}^{x_2} p(x)dx \tag{19.7}$$

For the special case where the lower limit of integration in Eq. (19.7) is $x_1 = -\infty$, the resulting function is called the *cumulative probability distribution function, P(x)* (often referred to as simply the *probability distribution function*), that is,

$$P(x) = \int_{-\infty}^{x} p(x)dx \tag{19.8}$$

In terms of the probability distribution function, the probability at any future instant that the value of $x(t)$ will fall between x_1 and x_2 is now given by

$$\text{Prob}[x_1 < x(t) \le x_2] = P(x_2) - P(x_1) \tag{19.9}$$

Illustrations of probability density and distribution functions for a typical stationary random vibration are shown in Fig. 19.4. Note that since the limiting operations in Eq. (19.6) can never be achieved in practice, probability density functions and all derivative functions thereof can only be estimated with potential bias and random errors, as discussed later.

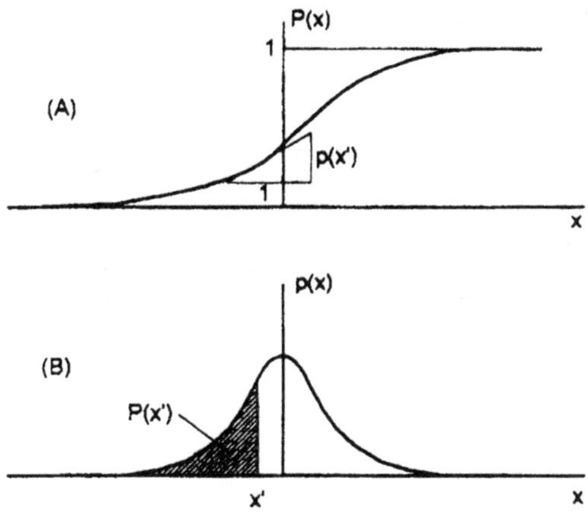

FIGURE 19.4 Examples of the probability distributions of a random variable x. (A) Cumulative (probability) distribution function, $P(x)$. (B) Probability density function $p(x)$.

Due to the practical implications of the central limit theorem in statistics,[1] there is a strong tendency for most stationary random vibration data to have a specific type of probability density function called the *normal* or *gaussian probability density function,* given in normalized form by

$$p(z) = \frac{1}{\sqrt{2\pi}} e^{-z^2/2} \qquad z = (x - \mu_x)/\sigma_x \tag{19.10}$$

where μ_x and σ_x are the mean value and standard deviation, respectively, of the data, as defined in Eqs. (19.1). It can be shown[3] that all linear operations on a gaussian random variable produce another gaussian random variable. Furthermore, all linear operations that limit the frequency range of the input random variable tend to suppress all deviations from the gaussian form in the output random variable.[4] Since the response of most physical systems, like mechanical structures, is dominated by the response of the system at its normal mode frequencies (see Chap. 1), the vibration response of the system is commonly more gaussian in character than its excitation, assuming the system is linear. It is for this reason that the computation of a probability density function is often omitted in the analysis of vibration data representing the response of a physical system of interest; it is simply assumed the response has a gaussian probability density function. However, a computed probability density function can provide a valuable tool for the detection of anomalies in the measured data introduced by data acquisition system errors,[5] as well as the detection of nonlinear characteristics in the system response.[3]

Correlation Functions. Given a stationary random vibration $x(t)$, the *autocorrelation function* $R_{xx}(\tau)$ of $x(t)$ is given by

$$R_{xx}(\tau) = \lim_{T \to \infty} \frac{1}{T} \int_0^T x(t)\, x(t+\tau)\, d\tau \tag{19.11}$$

where τ is a time delay (plus or minus). The autocorrelation function is essentially a measure of the linear relationship (correlation) between the values of the random vibration at any two instances t and $t + \tau$. Note that for $\tau = 0$, the value of the autocorrelation function is simply its mean square value as defined in Eq. (19.1), i.e., $R_{xx}(0) = \psi_x^2$. The autocorrelation function, as defined in Eq. (19.11), is rarely of direct interest in the analysis of stationary random vibration data. However, the Fourier transform of the autocorrelation function yields one of the most important descriptive properties of a stationary random vibration, namely, the *power spectral density function,* as defined in Chap. 14 and discussed next in this chapter.

Given two stationary random vibrations, $x(t)$ and $y(t)$, the *cross-correlation function* $R_{xy}(\tau)$ between $x(t)$ and $y(t)$ is given by

$$R_{xy}(\tau) = \lim_{T \to \infty} \frac{1}{T} \int_0^T x(t)\, y(t+\tau) d\tau \tag{19.12}$$

where, again, τ is a time delay (plus or minus). The cross-correlation function is a measure of the relationship (correlation) between two random vibrations at any instance t with a time delay τ between the two vibration time histories. The cross-correlation function is sometimes of direct interest in the analysis of stationary random vibration data, particularly for defining propagation paths in noise and vibration control problems.[6,7] However, as for the autocorrelation function, the Fourier transform of the cross-correlation function yields what is generally a more important descriptive property of two stationary random vibrations, namely, the *cross-spectral density function,* to be discussed shortly. Note that since the limiting operation in Eqs. (19.11) and (19.12) can never be achieved in practice, correlation functions can only be estimated with a potential random error, to be discussed later.

Power Spectral Density Function. The *power spectral density function* (also called the *autospectral density function*, or more simply the *power spectrum* or *autospectrum*) of a stationary random vibration $x(t)$ may be defined simply as the Fourier transform of the autocorrelation function of $x(t)$, as discussed in the preceding section. From Chap. 14, however, the power spectrum of $x(t)$ may be defined in a manner more relevant to data analysis algorithms by

$$W_{xx}(f) = \lim_{T \to \infty} \frac{2}{T} E[|X(f,T)|^2] \qquad f > 0 \tag{19.13}$$

where $E[\ ]$ denotes the expected value of $[\ ]$, which implies an ensemble average, and $X(f,T)$ is defined in Eq. (19.3). Note that the power spectrum $W_{xx}(f)$ in Eq. (19.13) is defined for positive frequencies only, and is often referred to as a *one-sided* spectrum.

The power spectrum describes the frequency content of the vibration and, hence, is generally the most important and widely used function for engineering applications,[6,8] which are facilitated by three important properties of power spectra, as follows:

1. Given two or more statistically independent vibrations, the power spectrum for the sum of the vibrations is equal to the sum of the power spectra for the individual vibrations, that is,

$$W_{xx}(f) = \sum_i W_{ii}(f) \qquad i = 1, 2, 3, \ldots \tag{19.14}$$

2. The area under the power spectrum between any two frequencies, f_a and f_b, equals the mean square value of the vibration in the frequency range from f_a to f_b, that is,

$$\psi_x^2(f_a, f_b) = \int_{f_a}^{f_b} W_{xx}(f)df \tag{19.15}$$

3. Given an excitation $x(t)$ to a structural system with a frequency response function $H(f)$ (see Chap. 21), the power spectrum of the response $y(t)$ is given by the product of the power spectrum of the excitation and the squared magnitude of the frequency response function, that is,

$$W_{yy}(f) = |H(f)|^2 \, W_{xx}(f) \tag{19.16}$$

Illustrations of the time histories and autospectra for both wide-bandwidth and narrow-bandwidth random vibrations are shown in Fig. 19.5.

Cross-Spectral Density Functions. Given two stationary random vibrations $x(t)$ and $y(t)$, the *cross-spectral density function* (also called the *cross spectrum*) is defined as

$$W_{xy}(f) = \lim_{T \to \infty} \frac{2}{T} \, E[X^*(f,T)Y(f,T)] \qquad f > 0 \tag{19.17}$$

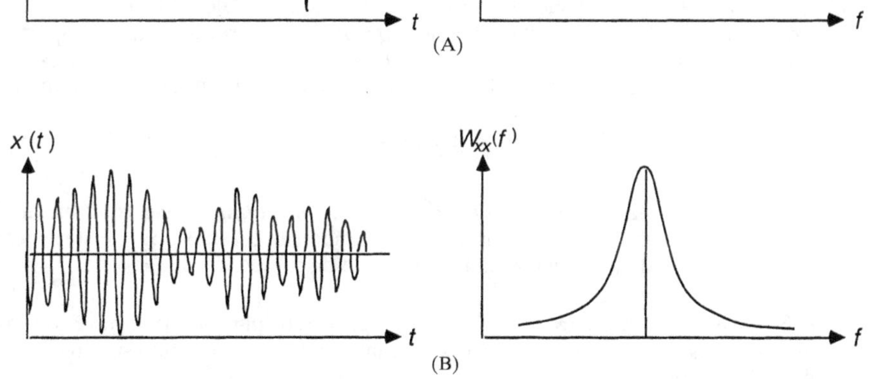

FIGURE 19.5 Time histories and autospectra for wide-bandwidth (*A*) and narrow-bandwidth (*B*) random vibrations.

where $E[\ ]$ is the expected value of [ ], which implies an ensemble average, $X^*(f,T)$ is the complex conjugate of the Fourier transform of $x(t)$, as defined in Eq. (19.3), and $Y(f)$ is the finite Fourier transform of $y(t)$, as defined in Eq. (19.3) with $y(t)$ replacing $x(t)$.

The cross spectrum is generally a complex number that measures the linear relationship between two random vibrations as a function of frequency with a possible phase shift between the vibrations. Specifically, the cross spectrum can be written as

$$W_{xy}(f) = |W_{xy}(f)|e^{-j\theta_{xy}(f)} \qquad \theta_{xy}(f) = 2\pi f\tau(f) \tag{19.18}$$

where $\tau(f)$ is the time delay between $x(t)$ and $y(t)$ at frequency f. An important application of the cross spectrum is as follows. Given a random excitation $x(t)$ to a structure with a frequency response function $H(f)$ (see Chap. 21), the cross spectrum between the excitation $x(t)$ and the response $y(t)$ is given by the product of the power spectrum of the excitation and the frequency response function, $H(f)$, that is,

$$W_{xy}(f) = H(f)W_{xx}(f) \tag{19.19}$$

Note that since the expected value and limiting operations in Eqs. (19.13) and (19.17) can never be achieved in practice, power and cross-spectral density functions and all derivative functions thereof can only be estimated with potential bias and random errors, as discussed later.

Coherence Functions. From Chap. 21, the *coherence function* between two random vibrations $x(t)$ and $y(t)$ is given by

$$\gamma^2_{xy}(f) = \frac{|W_{xy}(f)|^2}{W_{xx}(f)W_{yy}(f)} \qquad f > 0 \tag{19.20}$$

where all terms are as defined in Eqs. (19.13) and (19.17). The coherence function is bounded at all frequencies by zero and unity, where $\gamma^2_{xy}(f) = 0$ means there is no linear relationship between $x(t)$ and $y(t)$ at the frequency f (the two vibrations are uncorrelated) and $\gamma^2_{xy}(f) = 1$ means there is a perfect linear relationship between $x(t)$ and $y(t)$ at the frequency f (one vibration can be exactly predicted from the other). This property leads to an important application of the coherence function. Specifically, given a stationary random vibration $y(t) = x(t) + n(t)$, where $n(t)$ represents extraneous noise, including other vibrations that are not correlated with $x(t)$, then

$$W_{xx}(f) = \gamma^2_{xy}(f)\, W_{yy}(f) \tag{19.21}$$

The result in Eq. (19.22) is referred to as the *coherent output power* relationship.[1] The coherence function is also an important parameter in establishing the statistical sampling errors in various spectral estimates to be discussed later.

Other Functions. There are various other specialized functions that have important applications for certain advanced stationary random data analysis problems, including the following:

1. *Cepstrum functions,* which have important applications to machinery condition monitoring[9]

2. *Hilbert transforms,* which can be used to determine the causality between two measurements[1] and certain properties of modulation processes[1]
3. *Conditioned spectral density* and *coherence functions,* which have important applications to the analysis of structural vibration responses to multiple excitations that are partially correlated,[1,6] as well as to the analysis of the vibration responses of nonlinear systems.[3,6]
4. *Higher-order spectral density functions,* such as bi-spectra and tri-spectra, which have applications to the analysis of the vibration responses of nonlinear systems.[3]
5. *Cyclostationary functions,* which have important applications to machinery fault diagnosis procedures.[10]
6. *Wavelet analysis*, which provides a decomposition of a vibration time-history record into a set of orthogonal time-domain functions that can be used for various advanced analysis operations[11]
7. *Parametric spectral analysis*, which involves fitting a multipole filter describing a power spectrum to the time-history record of the vibration using one of several optimum curve-fitting procedures[12]

QUANTITATIVE DESCRIPTIONS OF NONSTATIONARY VIBRATIONS

Unlike stationary vibrations, the properties of nonstationary vibrations must be described as a function of time, which theoretically requires instantaneous averages computed over an ensemble of sample records, $\{x(t)\}$, acquired under statistically similar conditions. In this context, the overall values for stationary vibrations in Eq. (19.1) are given for nonstationary vibrations by

$$\begin{aligned} \text{Mean value:} \quad & \mu_x(t) = E[x(t)] \\ \text{Mean-square value:} \quad & \psi_x^2(t) = E[x^2(t)] \\ \text{Variance:} \quad & \sigma_x^2(t) = E[\{x(t) - \mu_x(t)\}^2] \end{aligned} \tag{19.22}$$

where $E[\ ]$ denotes the expected value of [ ], which implies an ensemble average. Equation (19.2) applies to the values in Eq. (19.22) at each time t, and the interpretations of these values following Eq. (19.2) apply.

NONSTATIONARY DETERMINISTIC VIBRATIONS

Nonstationary deterministic vibrations are defined here as those vibrations that would be periodic under constant conditions, but where the conditions are time-varying such that the instantaneous magnitude and/or the fundamental frequency of the vibration versus time vary slowly compared to the fundamental frequency of the vibration (often called *phase coherent vibrations*). In other words, the vibration can be described by Eq. (19.4) where the magnitude and phase terms, a_k and θ_k, are replaced by time-varying magnitude and phase terms $a_k(t)$ and $\theta_k(t)$ and/or the fundamental frequency f_1 is replaced by a time-varying fundamental frequency $f_1(t)$, that is,

$$x(t) = a_0(t) + \sum_k a_k(t) \cos [2\pi k f_1(t) + \theta_k(t)] \tag{19.23}$$

A similar nonstationary deterministic vibration is given by Eq. (19.23) with $kf_1(t)$ replaced by $f_k(t)$. Nonstationary deterministic vibrations described by Eq. (19.23) are commonly displayed as a three-dimensional plot of the magnitude of the time-varying coefficients versus time and frequency. Such a plot is often referred to as an *instantaneous line spectrum.* An illustration of the time history and instantaneous line spectrum for a single instantaneous frequency component with linearly increasing magnitude and frequency is shown in Fig. 19.6.

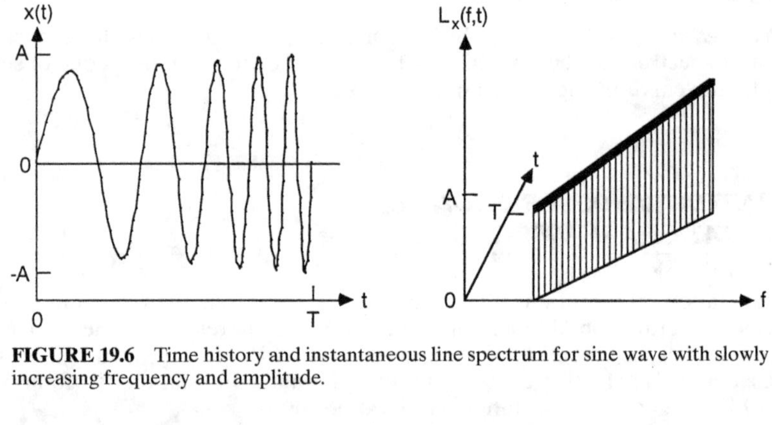

FIGURE 19.6 Time history and instantaneous line spectrum for sine wave with slowly increasing frequency and amplitude.

Another way to describe the frequency-time characteristics of a nonstationary deterministic vibration is by the *Wigner distribution,* defined as[1,13]

$$WD_{xx}(f,t) = \int_{-\infty}^{\infty} x\left(t - \frac{\tau}{2}\right) x\left(t + \frac{\tau}{2}\right) e^{-j2\pi f\tau}\, d\tau \tag{19.24}$$

The Wigner distribution is similar to the instantaneous power spectrum discussed later in this chapter, and has interesting theoretical properties.[13] However, it often produces negative spectral values, which are difficult to interpret for most engineering applications, and offers few advantages over the instantaneous line spectrum given by Eq. (19.23).

NONSTATIONARY RANDOM VIBRATIONS

There are several theoretical ways to describe nonstationary random data,[1] including generalized spectra defined for two frequency variables that provide rigorous excitation-response relationships, even for time-varying linear systems. From a data analysis viewpoint, however, the most useful theoretical description for nonstationary random vibrations is provided by the *instantaneous power spectral density function* (also called the *instantaneous power spectrum* or *instantaneous autospectrum*). The instantaneous power spectrum is defined by[1]

$$W_{xx}(f,t) = \int E\left[x\left(t - \frac{\tau}{2}\right) x\left(t + \frac{\tau}{2}\right)\right] e^{-j2\pi f\tau}\, d\tau \tag{19.25}$$

where $E[\ ]$ denotes the expected value of [ ], which implies an ensemble average. Note that the instantaneous power spectrum is essentially the Wigner distribution defined in Eq. (19.24), except the product of the values of $x(t)$ at two different times is averaged.

Like the Wigner distribution, the instantaneous power spectrum can have negative values at some frequencies and times.[1] For example, let a nonstationary random process be defined as

$$\{x(t)\} = [\cos 2\pi f_0 t]\{u(t)\} \tag{19.26}$$

where $\{u(t)\}$ is a narrow-bandwidth stationary random process with a mean value of zero and a standard deviation of unity, and the cosine term is a modulating function. Substituting Eq. (19.26) for Eq. (19.25) yields

$$W_{xx}(f,t) = \frac{1}{4}\,[W_{uu}(f - f_0) + W_{uu}(f + f_0)] + \frac{1}{2}\cos(4\pi f_0 t) W_{uu}(f) \tag{19.27}$$

where $W_{uu}(f)$ is the power spectrum of the stationary component $\{u(t)\}$. The instantaneous power spectrum given by Eq. (19.27) is plotted in Fig. 19.7. Note that the instantaneous power spectrum consists of two stationary components (often called *sidebands*) that are offset in frequency from the center frequency f_1 of $\{u(t)\}$ by plus and minus the modulating frequency f_0, and a time-varying component at the center

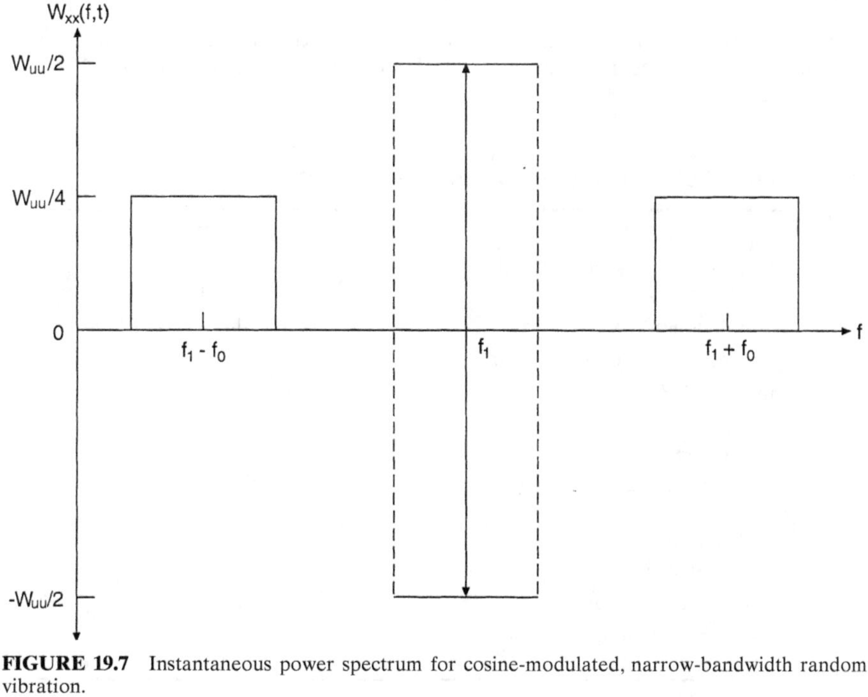

FIGURE 19.7 Instantaneous power spectrum for cosine-modulated, narrow-bandwidth random vibration.

frequency f_1 of $\{u(t)\}$ that oscillates between positive and negative values. Further note that for nonstationary vibration environments, as defined in this chapter, a modulating frequency is small compared to the lowest frequency of the stationary component, that is, $f_0 << f_1 - B/2$, where B is the bandwidth of the stationary component. It follows that the stationary and time-varying spectral components of the instantaneous power spectrum will heavily overlap and, hence, eliminate negative spectral values at most times and frequencies.

PRELIMINARY DATA ANALYSIS CONSIDERATIONS

Before the detailed analysis of vibration data is initiated, careful consideration should be given to the following:

1. Final engineering applications of the analyzed data
2. Stationary sample record durations for the data analysis
3. Validation and editing of the data
4. Data storage
5. Analog-to-digital conversion

The first two matters should actually be considered prior to the acquisition of the data, but in any case should be carefully reviewed prior to the initiation of the data analysis.

ENGINEERING APPLICATIONS OF DATA ANALYSIS

Numerous possible applications might motivate the acquisition and analysis of vibration data, including the applications in this handbook summarized in Table 19.1.

TABLE 19.1 Applications of Analyzed Vibration Data

Application	Chapter
Formulation of test criteria and verification of test results	19, 20
Formulation of design criteria	40
Condition monitoring of machinery	16
Modal analysis and testing	21
Assessing the vibration response of structures	23, 24, 25
Assessing the effects of vibration on humans	41
Prediction of structural failures and fatigue damage	33, 34
Calibration of transducers	11
Evaluation of vibration responses of nonlinear systems	4

The final application for the data is important in determining which properties of the data should be computed. In most cases, the primary property of interest will be some form of a frequency spectrum. However, there may be applications that require other types of analysis. For example, fatigue damage predictions for random vibration environments generally require some form of amplitude distribution analysis, as detailed in Chap. 33. These matters should be thoroughly reviewed prior to initiating data analysis, not only to ensure the needed data properties are computed, but also to avoid computing large amounts of unneeded information.

STATIONARY SAMPLE RECORD DURATIONS

It is clear from the descriptions of vibrations in preceding sections that stationary vibrations are much easier to analyze than nonstationary vibrations. It follows that an effort should be made to collect stationary sample records of vibration data for analysis. This is easily accomplished for the vibration data produced by laboratory experiments, since most such experiments are performed under constant conditions that naturally produce stationary results. On the other hand, the vibration data collected from measurements of actual vibration environments are commonly nonstationary. Even in this case, measurement programs can often be designed to produce stationary data for analysis purposes. For example, the vibration environment for a motor vehicle during normal service operations is generally nonstationary. However, if the vehicle is operated over a homogeneous road at constant speed and engine rpm, the resulting vibration levels will be approximately stationary. It follows that the vibration environment of the vehicle under all conditions can be measured and analyzed from a collection of stationary sample records, each representing a specific road condition, vehicle speed, and/or engine rpm, that together cover all the operating conditions for the vehicle. Whether a laboratory experiment or a field experiment, the vibration data acquired for analysis should be forced to be stationary when possible.

Some vibrations are produced by excitations that cannot be forced to be stationary. Examples include the response of structures to wind loads (see Chap. 31) and ocean waves (see Chap. 30). Even in these cases, however, it is often possible to identify and select piecewise stationary segments from a long sample record for data analysis purposes. On the other hand, there are some types of vibration environments that are inherently nonstationary, for example, a laboratory vibration test involving a sweep-sine excitation (see Chap. 18) or the vibration environment of a space vehicle during launch. In these situations, some type of nonstationary data analysis procedure must be employed.

DATA VALIDATION AND EDITING

Every effort should be made to acquire accurate vibration data, as outlined in Chap. 15. However, all vibration data collected and stored for later analysis should be validated and, if necessary, edited to remove anomalies prior to analysis. The four most common and serious anomalies in acquired vibration data are as follows:[5]

1. *Signal clipping,* which is a limiting on one or both sides of the time-history record, is caused by too high a gain setting on one or more data acquisition instruments.

Severe clipping will reduce the rms value of the data and introduce spurious high-frequency components.

2. *Excessive instrumentation noise,* which appears in the data as broad bandwidth random noise, is caused by too low a gain setting on one or more of the data acquisition instruments. Severe instrumentation noise will sum with random vibration data, increasing the rms value of the data and obscuring the spectral characteristics of the data.
3. *Intermittent noise spikes,* which appear as one or more sharp spikes in the time-history record, are usually caused by a faulty connector in the data acquisition system, but may also occur due to a faulty transmission in telemetry data. Intermittent noise spikes will often severely distort the computed spectral characteristics of the data.
4. *Power-line pickup,* which appears as a sine wave with a frequency of 60 Hz in North America and 50 Hz in many other regions of the world, is caused by faulty shielding and/or grounding of the data acquisition system. Power-line pickup will cause a spectral component in the data at the power-line frequency and, if severe, may saturate one or more of the data acquisition instruments.

These and other anomalies can often be detected by a visual inspection of the time-history record of the measured vibration[5] or, for data at frequencies above 50 Hz, by simply listening to the vibration signal with a headset during the data acquisition or the playback of stored sample records. The hearing system of an experienced vibration data analyst can be a powerful detector of data anomalies.

In many cases, the anomalies in acquired vibration data cannot be corrected, but there are important exceptions. For example, power-line pickup can easily be removed from data by interpolation procedures in the frequency domain, assuming the power-line pickup did not saturate a data acquisition instrument and the data do not include an actual periodic component at the power-line frequency. Similarly, intermittent noise spikes can often be removed from the data by interpolation procedures in the time domain. For stationary random vibration data with even the most severe clipping, accurate spectral information can often be recovered by specialized analysis procedures.[1] See Ref. 5 for details and illustrations.

DATA STORAGE

In some cases, the analysis of sample records of vibration data is accomplished online using real-time data analysis equipment or appropriate online computer programs (see Chap. 14), but it is more common to input the sample records into some storage medium for later analysis.[2] In either case, since virtually all modern vibration data analysis is accomplished using digital techniques, each analog sample record, $x(t)$; $0 \leq t \leq T$, is usually converted immediately to a digital sample record, $x(n\Delta t)$; $n = 0, 1, 2, \ldots, (N-1)$, where Δt is the sampling interval in seconds and $N\Delta t = T$. This translation into a digital format is accomplished using an analog-to-digital converter (see Chap. 13). The storage of digital sample records can then be accomplished by directly inputting the data into the random access memory (RAM) or hard disk (HD) on a digital computer or, for long-term storage, a removable storage medium such as a digital tape recorder, digital video disk (DVD), or compact disk/read-only memory (CD/ROM).

ANALOG-TO-DIGITAL CONVERSION

The analog-to-digital (A/D) conversion operation discussed in Chap. 13 introduces two potential errors that must be carefully suppressed, namely, *aliasing errors* and *quantization errors.*

Aliasing Error. The first potential error arises because at least two sample values are needed to define one cycle of a vibration signal. This imposes an upper frequency limit on the digital data given by[1,2]

$$f_A = 1/(2\Delta t) \tag{19.28}$$

where f_A is called the *Nyquist frequency* in Hz. Any signal content in the sample record above the Nyquist frequency f_A will fold back around f_A and sum with the signal content below f_A, often causing a severe distortion of the data referred to as an *aliasing error.* Aliasing can be suppressed by low-pass filtering the analog signals from the transducers before the A/D conversion, where the low-pass filter cut-off frequency is set at $f_c = 0.5\, f_A$ to $0.8\, f_A$, depending on the roll-off rate of the low-pass filter. See Chap. 13 for details.

Quantization Error. The second potential error arises because a continuous analog signal is being converted into a finite set of numbers. This introduces a round-off error commonly referred to as the *quantization error* or *digital noise.* The round-off error is established by the A/D conversion word size, which is the number of binary digits (bits) used to describe each data value. Specifically, a word size of w provides 2^w discrete values (see Chap. 13). Assuming the full range of the A/D converter is used and allowing one bit for sign designation, the peak signal-to-rms noise ratio of the digitized data in dB is given by[1,2]

$$\text{PS/N(dB)} = 6(w - 1) + 10.8 \tag{19.29}$$

The rms signal-to-noise ratio (S/N) for the converter is then given by Eq. (19.29) minus the peak-to-rms value in dB for the signal being converted. For example, if the vibration signal were a sine wave, 3 dB would be subtracted from Eq. (19.29) to obtain the S/N, since the peak-to-rms ratio for a sine wave is $1/\sqrt{2} = -3$ dB. Modern A/D converters typically employ word sizes of $w \geq 16$ bits, corresponding to a PS/N(dB) ≥ 100 dB. The actual PS/N may be somewhat less than indicated by Eq. (19.29) because of miscellaneous errors in the converter that reduce the effective word size.[1] Nevertheless, if the full range of the converter is used, the digital noise level will usually be sufficiently low for a proper analysis of the data, and often lower than the noise level of the transducer and analog instrumentation preceding the A/D converter. On the other hand, if the full range of the converter is not used, the digital noise could restrict the dynamic range of the analyzed data.

VIBRATION DATA ANALYSIS PROCEDURES

The algorithms for analyzing vibration data evolve directly from the equations for the quantitative descriptions presented earlier, but without the limiting operations. Although usually computed from sample records in the form of a digital time series,

$x(n\Delta t); n = 0, 1, 2 \ldots$, all analysis procedures are presented in terms of both analog equations and digital algorithms for clarity.

THE DISCRETE FOURIER TRANSFORM

Many of the analysis products for both deterministic and random vibration data require the computation of the Fourier transform defined in Eq. (19.3). In digital terms where the sample record $x(t) = x(n\Delta t)$, this Fourier transform, often called a *discrete Fourier transform* (DFT), is given by

$$X(m\Delta f) = \Delta t \sum_{n=0}^{N-1} x(n\Delta t) \exp[-j2\pi m\Delta f\, n\Delta t]; m = 0, 1, 2, \ldots, (N-1) \quad (19.30)$$

As discussed in Chap. 14, the DFT can be computed with remarkable efficiency using a *fast Fourier transform* (FFT) algorithm. Note that the DFT defines N discrete frequency values for N discrete time values with an inherent frequency resolution of

$$\Delta f = \frac{1}{N\Delta t} \quad (19.31)$$

However, the Nyquist frequency defined in Eq. (19.28) occurs at $m = (N/2)$. Hence, only the first $[(N/2) + 1]$ frequency components represent unique values; the last $[(N/2) - 1]$ frequency components constitute the redundant values representing the negative frequency components in Eq. (19.3).

PROCEDURES FOR STATIONARY DETERMINISTIC DATA ANALYSIS

The analog equations and digital algorithms for the analysis of stationary deterministic vibration data are summarized in Table 19.2. The hat (^) over the symbol for each computed parameter in Table 19.2 denotes an estimate as opposed to an exact value.

TABLE 19.2 Summary of Algorithms for Stationary Deterministic Vibration Data Analysis

Function	Analog equation	Digital algorithm
Mean value	$\hat{\mu}_x = \frac{1}{T}\int_0^T x(t)dt$	$\hat{\mu}_x = \frac{1}{N}\sum_{n=0}^{N-1} x(n\Delta t)$
Mean-square value	$\hat{\psi}_x^2 = \frac{1}{T}\int_0^T x^2(t)dt$	$\hat{\psi}_x^2 = \frac{1}{N}\sum_{n=0}^{N-1} x^2(n\Delta t)$
Variance	$\hat{\sigma}_x^2 = \frac{1}{T}\int_0^T [x(t) - \hat{\mu}_x]^2 dt$	$\hat{\sigma}_x^2 = \frac{1}{N-1}\sum_{n=0}^{N-1} [x(n\Delta t) - \hat{\mu}_x]^2$
Line spectrum*	$\hat{L}_x(f) = \frac{2}{T}\lvert X(f,T)\rvert; f > 0$	$\hat{L}_x(m\Delta f) = \frac{2}{N\Delta t}\lvert X(m\Delta f)\rvert;$ $m = 1, 2, \ldots, \left(\frac{N}{2} - 1\right)$

*$X(f,T)$ defined in Eq. (19.3), $X(m\Delta f)$ defined in Eq. (19.30).

Overall Values. The mean, mean-square, and variance values for stationary deterministic vibrations are estimated from a sample record using Eq. (19.1) with a finite value for the averaging time T, as shown in Table 19.2. For periodic data, as defined by Eq. (19.4), the averaging time should ideally cover an integer multiple of periods, that is,

$$T = iT_P \qquad i = 1, 2, 3, \ldots \tag{19.32}$$

where T_P is the period of the data. However, since the period of a measured periodic vibration is probably not known prior to estimating its overall values, it is unlikely in practice that the averaging time will comply with Eq. (19.32). This leads to a truncation error that diminishes as the averaging time T increases, and is generally negligible (less than 3 percent) if $T > 10T_P$. For almost-periodic vibration data, there will always be a truncation error, but again it will be negligible if $T > 10T_1$ where T_1 is the period of the lowest frequency in the data.

Line Spectra. The line spectrum for a periodic signal, as defined in Eq. (19.5), will be exact as long as the averaging time complies with Eq. (19.32). Again, compliance with Eq. (19.32) is unlikely in practice for periodic data and is not possible for almost-periodic data, so a line spectrum estimate will generally involve a truncation error. Specifically, rather than a single spectral line at the frequency of each harmonic component of the periodic vibration, as illustrated in Fig. 19.2, spectral lines will occur at all frequencies given by

$$f_k = k/T \qquad k = 1, 2, 3, \ldots \tag{19.33}$$

where $T \neq iT_P$; $i = 1, 2, 3, \ldots$. The largest spectral lines will fall at those frequencies nearest the frequency of the harmonic components of the vibration, but they will underestimate the magnitudes of the harmonic components. Furthermore, the computed spectral lines will fall off about each harmonic frequency as shown in Fig. 14.8. This allows a second type of error, referred to as the *leakage error*, where the magnitude of any one harmonic component can influence the computed values of neighboring harmonic components. Of course, these errors diminish rapidly as $T >> T_P$ for periodic data, or $T >> T_1$ for almost-periodic data where T_1 is the period of the lowest frequency in the data. In addition, sample record-tapering operations (see Chap. 14) or interpolation algorithms[2] can be used to suppress these errors.

PROCEDURES FOR STATIONARY RANDOM DATA ANALYSIS

The analog equations and digital algorithms for the analysis of stationary random vibration data are summarized in Table 19.3. As before, the hat (^) over the symbol for each computed function in Table 19.3 denotes an estimate as opposed to an exact value. Unlike deterministic data, the estimation of parameters for random vibration data will involve statistical sampling errors of two types, namely, (*a*) a random error and (*b*) a bias (systematic) error. It is convenient to present these errors in normalized terms. Specifically, for an estimate $\hat{\phi}$ of a parameter $\phi \neq 0$,

$$\text{Random error: } \varepsilon_r[\hat{\phi}] = \sigma[\hat{\phi}]/\phi \tag{19.34a}$$

$$\text{Bias error: } \varepsilon_b[\hat{\phi}] = (E[\hat{\phi}] - \phi)/\phi \tag{19.34b}$$

TABLE 19.3 Summary of Algorithms for Stationary Random Vibration Data Analysis

Function	Analog equation*	Digital algorithm*
Mean, mean-square, and variance values	Same as in Table 19.2	Same as in Table 19.2
Probability density function	$\hat{p}(x) = \dfrac{T(x,\Delta x)}{\Delta x\, T}$	$\hat{p}(x) = \dfrac{N(x,\Delta x)}{\Delta x\, N}$
Power spectrum	$\hat{W}_{xx}(f) = \dfrac{2}{n_d T} \sum_{i=1}^{n_d} \lvert X_i(f,T)\rvert^2; f > 0$	$\hat{W}_{xx}(m\Delta f) = \dfrac{2}{n_d N \Delta t} \sum_{i=1}^{n_d} \lvert X_i(m\Delta f)\rvert^2;$ $m = 1,2,\ldots,\left(\dfrac{N}{2} - 1\right)$
Cross-spectrum	$\hat{W}_{xy}(f) = \dfrac{2}{n_d T} \sum_{i=1}^{n_d} X^*(f,T)Y(f,T);$ $f > 0$	$\hat{W}_{xy}(m\Delta f) = \dfrac{2}{n_d N \Delta t} \sum_{i=1}^{n_d} X_i^*(m\Delta f)Y_i(m\Delta f);$ $m = 1,2,\ldots,\left(\dfrac{N}{2} - 1\right)$
Coherence function	$\hat{\gamma}^2_{xy}(f) = \dfrac{\lvert \hat{W}_{xy}(f)\rvert^2}{\hat{W}_{xx}(f)\hat{W}_{yy}(f)}; f > 0$	$\hat{\gamma}^2_{xy}(m\Delta f) = \dfrac{\lvert \hat{W}_{xy}(m\Delta f)\rvert^2}{\hat{W}_{xx}(m\Delta f)\hat{W}_{yy}(m\Delta f)}$ $m = 1,2,\ldots,\left(\dfrac{N}{2} - 1\right)$
Frequency response function	$\hat{H}_{xy}(f) = \dfrac{\hat{W}_{xy}(f)}{\hat{W}_{xx}(f)}; f > 0$	$\hat{H}_{xy}(m\Delta f) = \dfrac{\hat{W}_{xy}(m\Delta f)}{\hat{W}_{xx}(m\Delta f)};$ $m = 1,2,\ldots,\left(\dfrac{N}{2} - 1\right)$
Coherent output power function	$\hat{W}_{xx}(f) = \hat{\gamma}_{xy}(f)\hat{W}_{yy}(f); f > 0$	$\hat{W}_{xx}(m\Delta f) = \hat{\gamma}^2_{xy}(m\Delta f)\hat{W}_{yy}(m\Delta f);$ $m = 1,2,\ldots,\left(\dfrac{N}{2} - 1\right)$

*$X(f,T)$ defined in Eq. (19.3), $X(m\Delta f)$ defined in Eq. (19.30).

where $\sigma[\hat{\phi}]$ is the standard deviation of the estimate $\hat{\phi}$ and $E[\ ]$ denotes the expected value. For example, if the random error for an estimate $\hat{\phi}$ is $\varepsilon_r[\hat{\phi}] = 0.1$, this means that the estimate $\hat{\phi}$ is a random variable with a standard deviation that is 10 percent of the value of the parameter ϕ being estimated. If the bias error is $\varepsilon_b[\hat{\phi}] = -0.1$, this means the estimate $\hat{\phi}$ is systematically 10 percent less than the value of the parameter ϕ being estimated; note that the bias error can be either positive or negative. The random and bias errors for the various estimates in Table 19.3 are summarized in Table 19.4.

TABLE 19.4 Statistical Sampling Errors for Stationary Random Vibration Data Analysis

Function	Random error	Bias error
Mean value	$\varepsilon_r[\hat{\mu}_x] = \frac{1}{\sqrt{2BT}}\left(\frac{\sigma_x}{\mu_x}\right)$	None
Mean-square value	$\varepsilon_r[\hat{\psi}_x] = \frac{1}{\sqrt{BT}}\left(\frac{\sigma_x^2}{\psi_x^2}\right) + \frac{\sqrt{2}}{\sqrt{BT}}\left(\frac{\mu_x\sigma_x}{\psi_x^2}\right)$	None
Variance	$\varepsilon_r[\hat{\sigma}_x^2] = \frac{1}{\sqrt{BT}}$	None
Probability density function	$\varepsilon_r[\hat{p}(x)] \leq \frac{1}{\sqrt{2BT\,\Delta x\,p(x)}}$	$\varepsilon_b[\hat{p}(x)] = \frac{(\Delta x)^2 d^2[p(x)]/dx^2}{24\,p(x)}$
Power spectrum*	$\varepsilon_r[\hat{W}_{xx}(f)] = \frac{1}{\sqrt{n_d}}$	$\varepsilon_b[\hat{W}_{xx}(f)] = -\frac{1}{3}\left(\frac{B_e}{2\zeta f_r}\right)^2$
Cross-spectrum magnitude*	$\varepsilon_r[\lvert\hat{W}_{xy}(f)\rvert] = \frac{1}{\lvert\gamma_{xy}(f)\rvert\sqrt{n_d}}$	$\varepsilon_b[\hat{W}_{xy}(f)] = \frac{B_e d^2\lvert W_{xy}(f)\rvert/df^2}{24\,W_{xy}(f)}$
Cross-spectrum phase*	$\sigma_r[\lvert\hat{\theta}_{xy}(f)\rvert] = \frac{[1-\gamma_{xy}^2(f)]^{1/2}}{\lvert\gamma_{xy}(f)\rvert\sqrt{2n_d}}$	**
Coherence function*	$\varepsilon_r[\lvert\hat{\gamma}_{xy}^2(f)\rvert] = \frac{\sqrt{2}[1-\gamma_{xy}^2(f)]}{\lvert\gamma_{xy}(f)\rvert\sqrt{n_d}}$	$\varepsilon_b[\hat{\gamma}_{xy}^2(f)] = \frac{[1-\gamma_{xy}^2(f)]^2}{\gamma_{xy}^2(f)n_d}$
Frequency response function magnitude*	$\varepsilon_r[\lvert\hat{H}_{xy}(f)\rvert] = \frac{[1-\gamma_{xy}^2(f)]^{1/2}}{\lvert\gamma_{xy}(f)\rvert\sqrt{2n_d}}$	**
Frequency response function phase*	$\sigma_r[\lvert\hat{\phi}_{xy}(f)\rvert] = \frac{[1-\gamma_{xy}^2(f)]^{1/2}}{\lvert\gamma_{xy}(f)\rvert\sqrt{2n_d}}$	**
Coherent output power spectrum*	$\varepsilon_r[\hat{\gamma}_{xy}(f)\hat{W}_{yy}(f)] = \frac{[2-\gamma_{xy}^2(f)]^{1/2}}{\lvert\gamma_{xy}(f)\rvert\sqrt{n_d}}$	**

* n_d can be replaced by B_eT_r when overlapped processing is employed.

** There are several sources of bias errors,[1,14] but they usually will be small if the bias error for the power spectral density estimate is small.

Overall Values. The mean, mean-square, and variance values for a stationary random vibration are estimated from a sample record using Eq. (19.1) with a finite value for the averaging time T in the same way as for stationary deterministic vibration data, as shown in Table 19.2. For random data, however, truncation errors are replaced by the random errors given in Table 19.4, where it is assumed that the data have a uniform power spectrum over a frequency range with a bandwidth B. Since vibration data rarely have uniform power spectra, the error formulas for the overall values provide only coarse approximations for the random errors to be expected. However, for sample records of adequate duration to provide reasonably accurate power spectra estimates, to be detailed shortly, the random error in overall value estimates will generally be negligible.

Probability Density Functions. The probability density function for a stationary random vibration is estimated from a sample record using Eq. (19.6) with finite values for the averaging time T and an amplitude window width Δx, as shown in Table 19.3. In this table, $T(x,\Delta x)$ is the total time the analog record $x(t)$ falls within the amplitude window Δx centered at x, and $N(x,\Delta x)$ is the total number of values of the digital record $x(n\Delta t)$, $n = 0, 1, 2, \ldots$, that fall within the amplitude window Δx centered at x. Probability density estimates for random vibration data will involve both a bias error and a random error, as summarized in Table 19.4. The bias error is a function of the second derivative of the probability density versus amplitude, which generally is not known prior to the analysis. However, if the probability density function is relatively smooth and the analysis is performed with an amplitude window width of $\Delta x \leq 0.1\ \sigma_x$, experience suggests the bias error will typically be less than 5 percent for all values of x. The random error shown in Table 19.4 is only a bound; the actual random error depends on the power spectrum of the data,[1] but in most cases will be small if the sample record duration is adequate to provide accurate power spectra estimates.

Power Spectra. Referring to Table 19.3, there are two basic ways to estimate the power spectrum from a sample record of a stationary random vibration, as follows:

Ensemble-Averaging Procedure. The first approach to the estimation of a power spectrum is based upon the definition in Eq. (19.3), and involves the following primary steps:[1]

1. Given a sample record of total duration $T_r = n_d N\Delta t$, divide the record into an ensemble of n_d contiguous segments, each of duration $T = N\Delta t$.
2. Apply an appropriate tapering operation to each segment of duration $T = N\Delta t$ to suppress side-lobe leakage (see Chap. 14).
3. Compute a "raw" power spectrum from each segment of duration $T = N\Delta t$, which will produce $N/2$ spectral values at positive frequencies with a resolution of $\Delta f = 1/T = 1/(N\Delta t)$.
4. Average the "raw" power spectra values from the n_d segments to obtain a power spectrum estimate with n_d averages and a frequency resolution of $B_e = \Delta f$.

The averaging operation over the ensemble of n_d estimates simulates the expected value operation in Eq. (19.13), and determines the random error in the estimate given in Table 19.4. The resolution bandwidth $B_e = 1/(N\Delta t)$ determines the maximum bias error in the estimate given in Table 19.4, which for structural vibration data typically occurs at peaks and notches in the power spectrum caused by the resonant response of the structure at a frequency f_r with a damping ratio ζ. See Chap. 14 for details on the computation of power spectra for random data, including overlapped processing and "zoom" transform procedures.

Frequency-Averaging Procedure. The ensemble-averaging procedure can be replaced by a frequency-averaging procedure, as follows:[1]

1. Given a sample record of total duration $T_r = n_d N\Delta t$, compute a raw power spectrum over the entire duration of the sample record, which will produce $n_d N/2$

spectral estimates at positive frequencies with a resolution of $B_e = 1/T_r = 1/(n_d N\Delta t)$.

2. Divide the frequency range of the spectral components into a collection of contiguous frequency segments, each containing n_d spectral components.
3. Average the spectral components in each of the frequency segments to obtain the power spectrum estimate.

The averaging over n_d spectral components in a frequency segment produces the same random error in Table 19.4 as averaging over n_d raw power spectra estimates in the ensemble-averaging procedure. In addition, for the same values of N and n_d, the frequency resolution is the same as for the ensemble-averaging procedure, meaning the bias error in Table 19.4 is essentially the same. However, the bandwidth for the various frequency segments need not be a constant. Any desired variation in the bandwidth can be introduced, including a bandwidth that increases linearly with its center frequency (commonly referred to as a *constant percentage* frequency resolution).

Optimum Resolution Bandwidth Selections. A common problem in the estimation of power spectra from sample records of stationary random vibration data is the selection of an appropriate resolution bandwidth, $B_e = 1/T = 1/(N\Delta t)$. One approach to this problem is to select that resolution bandwidth that will minimize the total *mean square error* in the estimate given by

$$\varepsilon^2 = \varepsilon_r^2 + \varepsilon_b^2 \tag{19.35}$$

where ε_r and ε_b are defined in Eq. (19.34). From Table 19.4, the maximum mean-square error for power spectral density estimates of structural vibration data is approximated by

$$\varepsilon^2[\hat{W}_{xx}(f)] = \frac{1}{B_e T_r} + \frac{1}{9}\left(\frac{B_e}{2\zeta f_r}\right)^4 \tag{19.36}$$

where ζ is the damping ratio of the structure at the resonance frequency f_r. Taking the derivative of Eq. (19.36) with respect to B_e and equating to zero yields the resolution bandwidth that will minimize the mean-square error as

$$B_0(f) \cong 2\frac{(\zeta f_r)^{4/5}}{T_r^{1/5}} \tag{19.37}$$

Note in Eq. (19.37) that the optimum resolution bandwidth $B_0(f)$ is a function of the $-\frac{1}{5}$ power of the sample record duration, T_r, meaning the optimum resolution bandwidth is relatively insensitive to the sample record duration. Further, the optimum resolution bandwidth $B_0(f)$ is proportional to the $\frac{4}{5}$ power of the product ζf. Assuming all structural resonances have approximately the same damping, this means a constant percentage resolution bandwidth will provide near-optimum results in terms of a minimum mean square error in the power spectrum estimate. For example, assume the vibration response of a structure exposed to a random excitation is measured with a total sample record duration of $T_r = 10$ sec. Further assume all resonant modes of the structure have a damping ratio of $\zeta = 0.05$. From Eq. (19.37), the optimum resolution bandwidth for the computation of a power spectrum of the structural vibration is $B_0(f) = 0.115 f^{4/5}$. Hence, if the frequency range of the analysis is, say, 10 Hz to 1000 Hz, the optimum resolution bandwidth for the analysis increases from $B_0 = 0.726$ Hz at $f = 10$ Hz $[B_0(f) = 0.0726f]$ to $B_0 = 28.9$ Hz at $f = 1000$

Hz [$B_0(f) = 0.0280\,f$]. It follows that a 1/12 octave bandwidth resolution, which is equivalent to $B_e(f) = 0.058f$, will provide relatively good spectral estimates over the frequency range of interest.

Cross Spectra. Referring to Table 19.3 and Eq. (19.17), the computational approach for estimating the cross-spectrum between two sample records $x(t)$ and $y(t)$ is the same as described for power spectra, except $|X(f)|^2$ is replaced by $X^*(f)Y(f)$. Referring to Table 19.4, the random errors in the magnitude and phase of a cross-spectrum estimate are heavily dependent on the coherence function, as defined in Eq. (19.20). Specifically, if the coherence at any frequency is unity, this means the two sample records, $x(t)$ and $y(t)$, are linearly related and the normalized random error in the estimate is the same as for a power spectrum estimate. On the other hand, if the coherence is zero, then $x(t)$ and $y(t)$ are unrelated and the normalized random error in any estimate that may be computed is infinite. In practice, the true value of the coherence is not known, so sample estimates of the coherence, to be discussed shortly, would be used in the error formula shown in Table 19.4. There are several sources of bias errors for cross-spectra estimates,[1,10] but these bias errors will generally be minor if the bias errors in the power spectra estimates for the two sample records are small and there is no major time delay between the two sample records.

Other Spectral Functions. Referring to Table 19.3, the frequency response, coherence, and coherent output power functions defined in Eqs. (19.19) through (19.21) are estimated from sample records using the appropriate estimates for the power spectra, cross spectra, and coherence functions of the data. From Table 19.4, as for the cross spectrum, the random errors for estimates of these functions are heavily dependent on the coherence function. There are several sources of bias errors in the estimates of these functions,[1,10] but the bias errors will generally be minor if the bias errors in the power spectra estimates used to compute the functions is small and there is no major time delay between the two sample records.

PROCEDURES FOR NONSTATIONARY DATA ANALYSIS

As noted earlier, nonstationary vibration data are defined here as those whose basic properties vary slowly relative to the period of the lowest frequency in the vibration time history. Under this definition, the analog equations and digital algorithms for the analysis of nonstationary vibration data from a single sample record $x(t)$ are essentially the same as summarized in Tables 19.2 and 19.3, except the computations are performed over each of a sequence of short, contiguous segments of the data where each segment is sufficiently short not to smooth out the nonstationary characteristics of the data. In other words, given a nonstationary sample record $x(t)$ of total duration T_r, the record is assumed to be a sequence of piecewise stationary segments, each covering the interval

$$iT \text{ to } (i+1)T = iN\Delta t \text{ to } (i+1)N\Delta t \qquad i = 0, 1, 2, \ldots \tag{19.38}$$

In many cases, rather than computing the estimates over the contiguous segments defined in Eq. (19.38), a new segment is initiated every digital increment Δt such that each covers the interval

$$i\Delta t \text{ to } (i+N)\Delta t \qquad i = 0, 1, 2, \ldots \tag{19.39}$$

The computation of estimates over the intervals defined in either Eq. (19.38) or (19.39) is commonly referred to as a *running average* (also called a *moving average*). Whether the averaging is performed over segments given by Eq. (19.38) or (19.39), the primary problem is to select an appropriate averaging time, $T = N\Delta t$, for the estimates.

Overall Average Values for Deterministic Data. Referring to Table 19.2, the optimum averaging time for the computation of time-varying mean, mean square, and variance values for nonstationary deterministic vibration data is bounded as follows. At the lower end, the averaging time must be at least as long as the period for periodic data or the period of the lowest frequency component for almost-periodic data. At the upper end, the averaging time must be sufficiently short to not smooth out the time-varying properties in the data. This selection is usually accomplished by trial-and-error procedures, as illustrated shortly.

Overall Average Values for Random Data. The optimum averaging time for the computation of time-varying mean, mean square, and variance values for nonstationary random vibration data is bounded as for nonstationary deterministic data with one difference, namely, the computations for random data will involve a statistical sampling (random) error, as summarized in Table 19.4. To minimize these random errors, an averaging time that is as close as feasible to the upper bound noted for deterministic data is desirable. Analytical procedures to select an optimum averaging time that will minimize the mean-square error of the resulting time-varying average value have been formulated,[1] but they require a knowledge of the power spectrum of the data, which is normally not available when overall average values are being estimated. Hence, it is more common to select an averaging time by trial-and-error procedures, as follows:

1. Compute a running average for the overall value of interest using either Eq. (19.38) or (19.39) with an averaging time, $T = N\Delta t$, that is too short to smooth out the variations with time in the overall value being estimated.
2. Continuously recompute the running average with an increasing averaging time until it is clear that the averaging time is smoothing out variations with time in the overall value being estimated.
3. Choose that averaging time for the analysis that is just short of the averaging time that clearly smoothes out variations with time in the overall value being estimated.

This procedure is illustrated in Fig. 19.8, which shows running average estimates for the time-varying mean-square value of a nonstationary random vibration record computed with averaging times of $T = 0.1, 1.0$, and 3.0 sec. Note that the running average estimates with $T = 0.1$ sec reveal substantial random variations from one estimate to the next, indicative of excessive random estimation errors, while the estimates with $T = 3$ sec reveal a clear smoothing of the nonstationary trend in the data, indicative of an excessive time interval bias error. The averaging time of $T = 1$ sec provides a good compromise between the suppression of random and bias errors in the data analysis.

Time-Varying Line Spectra for Deterministic Data. The most common way to analyze the spectral characteristics of time-varying deterministic vibration data is to approximate the instantaneous line spectrum illustrated in Fig. 19.6 by the computation of a sequence of line spectra over the time intervals defined in Eq. (19.38) or (19.39). The resulting collection of line spectra is commonly referred to as a *waterfall plot* or a *cascade plot.* An illustration of a waterfall plot computed from a sample record of nonstationary deterministic vibration data is shown in Fig. 14.23.

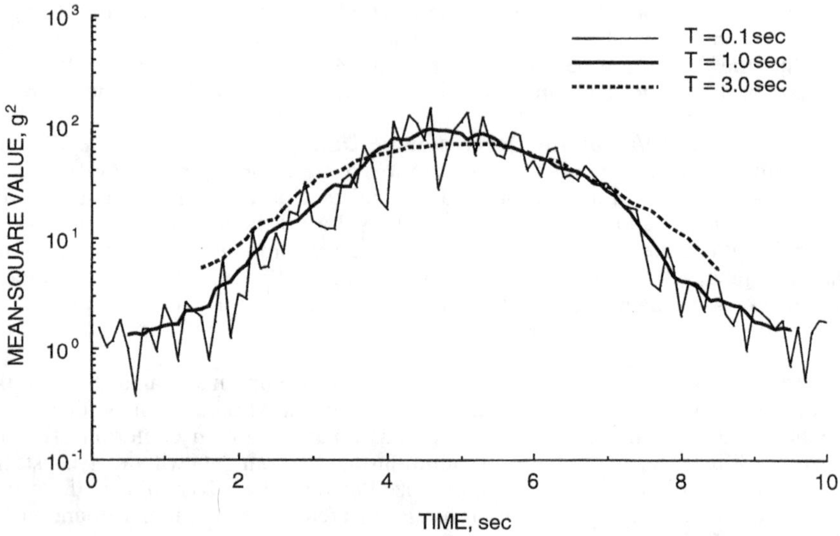

FIGURE 19.8 Running mean-square value estimates for nonstationary vibration data.

For a spectral analysis using Fourier transforms, the averaging time $T = N\Delta t$ and the frequency resolution $\Delta f = 1/T = 1/(N\Delta t)$ are obviously interrelated. It follows that there must always be a compromise between these two analysis parameters. On the one hand, the averaging time must be longer than the period of the lowest instantaneous frequency component in the data at any time covered by the sample record. On the other hand, the frequency resolution must be narrower than the minimum frequency separation of any two instantaneous frequency components in the data at any time covered by the sample record. This compromise will generally be achievable for nonstationary deterministic vibration data that would be periodic if they were stationary. In this case, assuming the maximum period at any time covered by the sample record is T_P, it follows that $\Delta f < 1/T_P$ if $T > T_P$. However, for almost-periodic deterministic vibration data, there may be two spectral components that, at some instant, might be separated by less than $\Delta f = 1/T$ where $T > T_1$. See Chap. 14 for further details on the computation of waterfall plots and other procedures for the analysis of nonstationary deterministic vibration data.

Time-Varying Power Spectra for Random Data. The computation of a time-varying power spectrum for nonstationary random vibration data is essentially the same as for the computation of a time-varying line spectrum for nonstationary deterministic data discussed in the previous section, with one important exception. Referring to the computational algorithm for the power spectrum in Table 19.3 and the estimation errors in Table 19.4, there will be substantial statistical sampling errors in the power spectrum estimate for each of the piecewise stationary segments of duration T, as defined in Eq. (19.38) or (19.39), unless the duration T is relatively long compared to the period of the lowest frequency of interest in the data. Hence, it is critical that the duration T of the piecewise stationary segments be as long as feasible without unduly smoothing the nonstationary trends in the data. A common approach to selecting the segment duration T is to use the maximum value of T for the computation of the time-varying variance of the nonstationary random data by the trial-and-error procedure illustrated in Fig. 19.8.

Concerning the resolution bandwidth B_e for the computation of the power spectrum of each piecewise stationary segment, Eq. (19.37) applies. Hence, it follows that a frequency resolution bandwidth B_e that is approximately proportional to the center frequency of the bandwidth would be a near optimum selection from the viewpoint of minimizing the total mean-square error for bias and random errors in the resulting estimates. This means that the most logical computational procedure for estimating the power spectrum of each segment would be to compute the Fourier transform over the entire segment duration and then use the frequency-averaging procedure described earlier for the analysis of stationary random data.

Finally, it should be noted that there is a more rigorous procedure for the optimum selection of not only the resolution bandwidth, but also the segment duration, that will minimize the total mean square error including both frequency resolution and time resolution bias errors, as well as random errors in nonstationary random vibration data analysis, as detailed in Ref. 1. However, for most nonstationary vibration data acquired in practice, it is rare for one to have a sufficient knowledge of the time-varying characteristics of the data to allow an accurate application of the more rigorous procedure.

REFERENCES

1. Bendat, J. S., and A. G. Piersol: "Random Data: Analysis and Measurement Procedures," 4th ed., John Wiley and Sons, New York, 2010.
2. Himelblau, H., and A. G. Piersol: "Handbook for Dynamic Data Acquisition and Analysis," 2d ed., *IEST Reference Document DTE012.2*, Institute of Environmental Sciences and Technology, Arlington Heights, Ill., 2006.
3. Bendat, J. S.: "Nonlinear Systems Techniques and Applications," John Wiley and Sons, New York, 1998.
4. Papoulis, A.: "Narrow-Band Systems and Gaussianity," *RADC-TR-71–225*, Rome Air Development Center, Griffiss AFB, New York, November 1971.
5. Kern, D. L., et al.: "Dynamic Environmental Criteria," *NASA-TR-7005*, National Aeronautics and Space Administration, Washington, D.C., 2001 (on the Internet).
6. Bendat, J. S., and A. G. Piersol: "Engineering Applications of Correlation and Spectral Analysis," 2d ed., John Wiley and Sons, New York, 1993.
7. Ver, I. L., and L. L. Beranek: "Noise and Vibration Control Engineering," 2d ed., John Wiley and Sons, New York, 2006, pp. 45–70.
8, Wirsching, P. H., T. L. Paez, and K. Ortiz: "Random Vibrations, Theory and Practice," John Wiley and Sons, New York, 1995.
9. Ewins, D. J., S. S. Rao, and S. S. Braun: "Encyclopedia of Vibration," Academic Press, New York, 2001.
10. Gardner, W. A.: "Cyclostationarity in Communications and Signal Processing," IEEE Press, New York, 1994.
11. Newland, D. E.: "Random Vibrations, Spectral & Wavelet Analysis, 3d ed., Longman, Essex, England, 1993.
12. Kay, S. M.: "Modern Spectral Estimation: Theory and Applications," Prentice-Hall, Englewood-Cliffs, N.J., 1988.
13. Cohen, L.: "Time-Frequency Analysis," Prentice-Hall, Upper Saddle River, N.J., 1995.
14. Schmidt, H.: *J. Sound and Vibration*, **101**(3):347 (1985).

CHAPTER 20
SHOCK DATA ANALYSIS

Sheldon Rubin
Kjell Ahlin

INTRODUCTION

This chapter discusses the interpretation of shock measurements and the reduction of data to a form adapted to further engineering use. Methods of data reduction also are discussed. A shock measurement is a trace giving the value of a shock parameter versus time over the duration of the shock, referred to hereafter as a *time history.* The shock parameter may define a motion (such as displacement, velocity, or acceleration) or a load (such as force, pressure, stress, or torque). It is assumed that any corrections that should be applied to eliminate distortions resulting from the instrumentation have been made. The trace may be a pulse or transient. Concepts in vibration data analysis are discussed in Chap. 19.

Examples of sources of shock to which this discussion applies are earthquakes (see Chap. 29), free-fall impacts, collisions, explosions, gunfire, projectile impacts, high-speed fluid entry, aircraft landing and braking loads, and spacecraft launch and staging loads.

BASIC CONSIDERATIONS

Often, a shock measurement in the form of a time history of a motion or loading parameter is not useful directly for engineering purposes. Reduction to a different form is then necessary, the type of data reduction employed depending upon the ultimate use of the data.

Comparison of Measured Results with Theoretical Prediction. The correlation of experimentally determined and theoretically predicted results by comparison of records of time histories is difficult. Generally, it is impractical in theoretical analyses to give consideration to all the effects which may influence the experimentally obtained results. For example, the measured shock often includes the vibrational response of the structure to which the shock-measuring device is attached. Such vibration obscures the determination of the shock input for which an applicable theory is being tested; thus, data reduction is useful in minimizing or eliminating the irrelevancies of the measured data to permit ready comparison of theory with corresponding aspects of the experiment. It often is impossible to make such comparisons on the basis of original time histories.

Calculation of Structural Response. In the design of equipment to withstand shock, the required strength of the equipment is indicated by its response to the shock. The response may be measured in terms of the deflection of a member of the equipment relative to another member or by the magnitude of the dynamic loads imposed upon the equipment. The structural response can be calculated from the time history by known means. If the structure is modeled with the use of finite element methods (see Chap. 23), the calculation time often is considerable. For lumped-parameter models of simple structures see Chaps. 1, 2, and 3. To calculate the structural response in a time-efficient way, a digital filter method combined with modal superposition may be used (see Ref. 1). This is the same method as used for the single-degree-of-freedom (SDOF) system calculation for shock response spectrum.

Laboratory Simulation of Measured Shock. Because of the difficulty of using analytical methods in the design of equipment to withstand shock, it is common practice to prove the design of equipments by laboratory tests that simulate the anticipated actual shock conditions. Unless the shock can be defined by one of a few simple functions, it is not feasible to reproduce in the laboratory the complete time history of the actual shock experienced in service. Instead, the objective is to synthesize a shock having the characteristics and severity considered significant in causing damage to equipment. Then the data reduction method is selected so that it extracts from the original time history the parameters that are useful in specifying an appropriate laboratory shock test. Shock testing machines are discussed in Chaps. 27 and 28.

EXAMPLES OF SHOCK MOTIONS

Five examples of shock motions are illustrated in Fig. 20.1 to show typical characteristics and to aid in the comparison of the various techniques of data reduction. The acceleration impulse and the acceleration step are the classical limiting cases of shock motions. The half-sine pulse of acceleration, the decaying sinusoidal acceleration, and the complex oscillatory-type motion typify shock motions encountered frequently in practice.

In selecting data reduction methods to be used in a particular circumstance, the applicable physical conditions must be considered. The original record, usually a time history, may indicate any of several physical parameters; e.g., acceleration, force, velocity, or pressure. Data reduction methods discussed in subsequent sections of this chapter are applicable to a time history of any parameter. For purposes of illustration in the following examples, the primary time history is that of acceleration; time histories of velocity and displacement are derived therefrom by integration. These examples are included to show characteristic features of typical shock motions and to demonstrate data reduction methods.

ACCELERATION IMPULSE OR STEP VELOCITY

The *delta function* $\delta(t)$ is defined mathematically as a function consisting of an infinite ordinate (acceleration) occurring in a vanishingly small interval of abscissa (time) at time $t = 0$ such that the area under the curve is unity. An acceleration time history of this form is shown diagrammatically in Fig. 20.1*A*. If the velocity and displacement

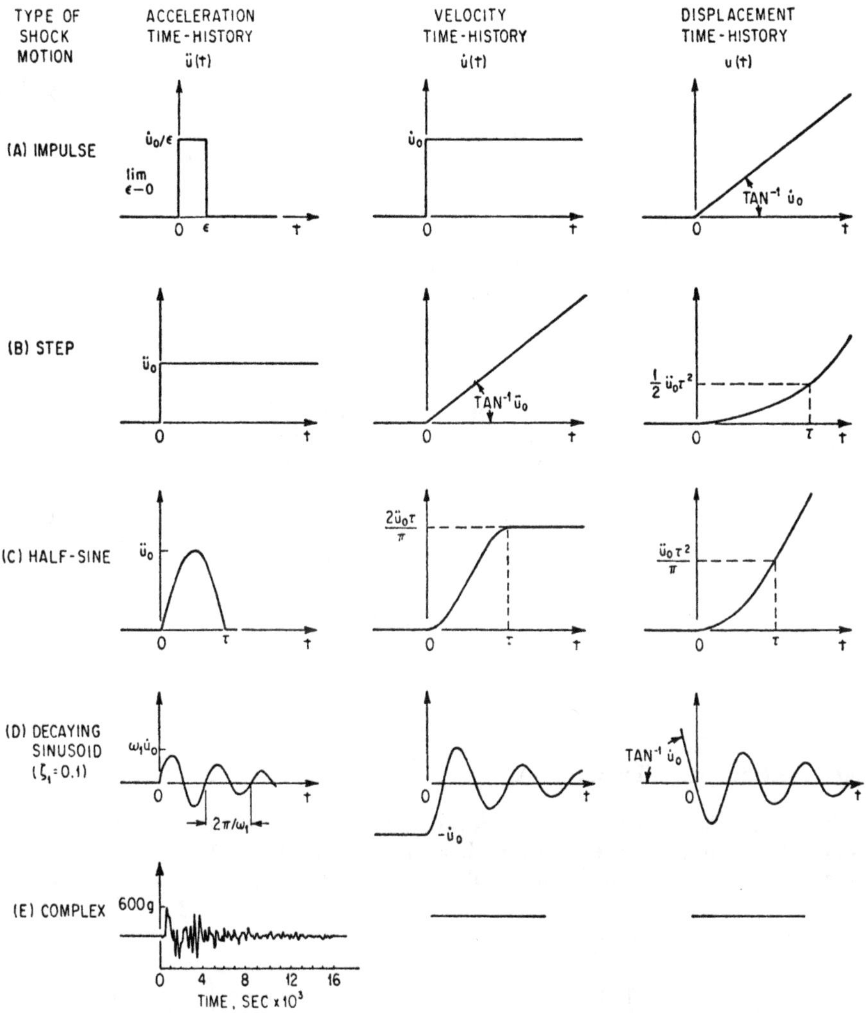

FIGURE 20.1 Five examples of shock motions.

are zero at time $t = 0$, the corresponding velocity time history is the velocity step and the corresponding displacement time history is a line of constant slope, as shown in the figure. The mathematical expressions describing these time histories are

$$\ddot{u}(t) = \dot{u}_0 \boldsymbol{\delta}(t) \tag{20.1}$$

where $\boldsymbol{\delta}(t) = 0$ when $t \neq 0$, $\boldsymbol{\delta}(t) = \infty$ when $t = 0$, and $\int_{-\infty}^{\infty} \boldsymbol{\delta}(t)\, dt = 1$. The acceleration can be expressed alternatively as

$$\ddot{u}(t) = \lim_{\epsilon \to 0} \dot{u}_0/\epsilon \qquad [0 < t < \epsilon] \tag{20.2}$$

where $\ddot{u}(t) = 0$ when $t < 0$ and $t > \epsilon$. The corresponding expressions for velocity and displacement for the initial conditions $u = \dot{u} = 0$ when $t < 0$ are

$$\dot{u}(t) = \dot{u}_0 \qquad [t > 0] \tag{20.3}$$

$$u(t) = \dot{u}_0 t \qquad [t > 0] \tag{20.4}$$

ACCELERATION STEP

The *unit step function* $\mathbf{1}(t)$ is defined mathematically as a function which has a value of zero at time less than zero ($t < 0$) and a value of unity at time greater than zero ($t > 0$). The mathematical expression describing the acceleration step is

$$\ddot{u}(t) = \ddot{u}_0 \mathbf{1}(t) \tag{20.5}$$

where $\mathbf{1}(t) = 1$ for $t > 0$ and $\mathbf{1}(t) = 0$ for $t < 0$. An acceleration time history of the unit step function is shown in Fig. 20.1*B;* the corresponding velocity and displacement time histories are also shown for the initial conditions $u = \dot{u} = 0$ when $t = 0$.

$$\dot{u}(t) = \ddot{u}_0 t \qquad [t > 0] \tag{20.6}$$

$$u(t) = \tfrac{1}{2}\ddot{u}_0 t^2 \qquad [t > 0] \tag{20.7}$$

The unit step function is the time integral of the delta function:

$$\mathbf{1}(t) = \int_{-\infty}^{t} \delta(t)\, dt \qquad [t > 0] \tag{20.8}$$

HALF-SINE ACCELERATION

A half-sine pulse of acceleration of duration τ is shown in Fig. 20.1*C;* the corresponding velocity and displacement time histories also are shown, for the initial conditions $u = \dot{u} = 0$ when $t = 0$. The applicable mathematical expressions are

$$\begin{aligned} \ddot{u}(t) &= \ddot{u}_0 \sin\left(\frac{\pi t}{\tau}\right) && [0 < t < \tau] \\ \ddot{u}(t) &= 0 && \text{when } t < 0 \quad \text{and } t > \tau \end{aligned} \tag{20.9}$$

$$\begin{aligned} \dot{u}(t) &= \frac{\ddot{u}_0 \tau}{\pi}\left(1 - \cos\frac{\pi t}{\tau}\right) && [0 < t < \tau] \\ \dot{u}(t) &= \frac{2\ddot{u}_0 \tau}{\pi} && [t > \tau] \end{aligned} \tag{20.10}$$

$$\begin{aligned} u(t) &= \frac{\ddot{u}_0 \tau^2}{\pi^2}\left(\frac{\pi t}{\tau} - \sin\frac{\pi t}{\tau}\right) && [0 < t < \tau] \\ u(t) &= \frac{\ddot{u}_0 \tau^2}{\pi}\left(\frac{2t}{\tau} - 1\right) && [t > \tau] \end{aligned} \tag{20.11}$$

This example is typical of a class of shock motions in the form of acceleration pulses not having infinite slopes.

DECAYING SINUSOIDAL ACCELERATION

A decaying sinusoidal trace of acceleration is shown in Fig. 20.1*D;* the corresponding time histories of velocity and displacement also are shown for the initial conditions $\dot{u} = -\dot{u}_0$ and $u = 0$ when $t = 0$. The applicable mathematical expression is

$$\ddot{u}(t) = \frac{\dot{u}_0\omega_1}{\sqrt{1-\zeta_1^2}} e^{-\zeta_1\omega_1 t} \sin\left(\sqrt{1-\zeta_1^2}\,\omega_1 t + \sin^{-1}\left(2\zeta_1\sqrt{1-\zeta_1^2}\right)\right) \qquad [t > 0] \qquad (20.12)$$

where ω_1 is the frequency of the vibration and ζ_1 is the fraction of critical damping corresponding to the decrement of the decay. Corresponding expressions for velocity and displacement are

$$\dot{u}(t) = \frac{\dot{u}_0}{\sqrt{1-\zeta_1^2}} e^{-\zeta_1\omega_1 t} \cos\left(\sqrt{1-\zeta_1^2}\,\omega_1 t + \sin^{-1}\zeta_1\right) \qquad [t > 0] \qquad (20.13)$$

where $\dot{u}(t) = -\dot{u}_0$ when $t < 0$.

$$u(t) = -\frac{\dot{u}_0}{\omega_1\sqrt{1-\zeta_1^2}} e^{-\zeta_1\omega_1 t} \sin\left(\sqrt{1-\zeta_1^2}\,\omega_1 t\right) \qquad [t > 0] \qquad (20.14)$$

where $u(t) = -\dot{u}_0 t$ when $t < 0$.

COMPLEX SHOCK MOTION

The trace shown in Fig. 20.1*E* is an acceleration time history representing typical field data. It cannot be defined by an analytic function. Consequently, the corresponding velocity and displacement time histories can be obtained only by integration of the acceleration time history.

CONCEPTS OF DATA REDUCTION

Consideration of the engineering uses of shock measurements indicates two basically different methods for describing a shock: (1) a description of the shock in terms of its inherent properties, in the time domain or in the frequency domain; and (2) a description of the shock in terms of the effect on structures when the shock acts as the excitation. The latter is designated reduction to the response domain. The following sections discuss concepts of data reduction to the frequency and response domains.

Whenever practical, the original time history should be retained even though the information included therein is reduced to another form. The purpose of data reduction is to make the data more useful for some particular application. The reduced data usually have a more limited range of applicability than the original time history. These limitations must be borne in mind if the data are to be applied intelligently.

DATA REDUCTION TO THE FREQUENCY DOMAIN

Any nonperiodic function can be represented as the superposition of sinusoidal components, each with its characteristic amplitude and phase.[2] This superposition is the Fourier spectrum, as defined in Eq. (20.15). It is analogous to the Fourier components of a periodic function (Chap. 19). The Fourier components of a periodic function occur at discrete frequencies, and the composite function is obtained by superposition of components. By contrast, the classical Fourier spectrum for a nonperiodic function is a continuous function of frequency, and the composite function is achieved by integration. Fourier spectra can be defined and computed as a function of either radial frequency ω in radians/sec or cyclical frequency *f* in Hz, that is,

$$\mathbf{F}_1(f) = \int_{-\infty}^{\infty} x(t)e^{-j2\pi ft}\,dt \quad \text{or} \quad \mathbf{F}_2(\omega) = \int_{-\infty}^{\infty} x(t)e^{-j\omega t}\,dt \tag{20.15}$$

Where the two functions are related by $\mathbf{F}_2(\omega) = 2\pi\mathbf{F}_1(f)$. The Fourier spectrum is a complex function denoted by a bold F. The following sections discuss the application of the continuous Fourier spectrum to describe the shock motions illustrated in Fig. 20.1. A discrete realization of the Fourier spectrum is given by Eq. (19.30).

Acceleration Impulse. Using the definition of the acceleration pulse given by Eq. (20.5) and substituting this for $f(t)$ in Eq. (20.15),

$$\mathbf{F}(\omega) = \lim_{\epsilon \to 0} \int_0^{\epsilon} \frac{\dot{u}_0}{\epsilon} e^{-j\omega t}\,dt \tag{20.16}$$

Carrying out the integration,

$$\mathbf{F}(\omega) = \lim_{\epsilon \to 0} \frac{\dot{u}_0(1 - e^{-j\omega\epsilon})}{j\omega\epsilon} = \dot{u}_0 \tag{20.17A}$$

The corresponding amplitude and phase spectra are

$$F(\omega) = \dot{u}_0; \qquad \theta(\omega) = 0 \tag{20.17B}$$

These spectra are shown in Fig. 20.2*A*. The magnitude of the Fourier amplitude spectrum is a constant, independent of frequency, equal to the area under the acceleration-time curve.

The physical significance of the spectra in Fig. 20.2*A* is shown in Fig. 20.3, where the rectangular acceleration pulse of magnitude $\dot{u}_0/\epsilon$ and duration $t = \epsilon$ is shown as approximated by superposed sinusoidal components for several different upper limits of frequency for the components. With the frequency limit $\omega_l = 4/\epsilon$, the pulse has a noticeably rounded contour formed by the superposition of all components whose frequencies are less than ω_l. These components tend to add in the time interval $0 < t < \epsilon$ and, though existing for all time from $-\infty$ to $+\infty$, cancel each other outside this interval, so that $\ddot{u}$ approaches zero. When $\omega_l = 16/\epsilon$, the pulse is more nearly rectangular and $\ddot{u}$ approaches zero more rapidly for time $t < 0$ and $t > \epsilon$. When $\omega_l = \infty$, the superposition of sinusoidal components gives $\ddot{u} = \dot{u}_0/\epsilon$ for the time interval of the pulse, and $\ddot{u} = \dot{u}_0/2\epsilon$ at $t = 0$ and $t = \epsilon$. The components cancel completely for all other times. As $\epsilon \to 0$ and $\omega_l \to \infty$, the infinitely large number of superimposed frequency components gives $\ddot{u} = \infty$ at $t = 0$. The same general result is obtained when the Fourier components of other forms of $\ddot{u}(t)$ are superimposed.

Acceleration Step. The Fourier spectrum of the acceleration step does not exist in the strict sense since the integrand of Eq. (20.15) does not tend to zero as $\omega \to \infty$. Using a convergence factor, the Fourier transform is found by substituting $\ddot{u}(t)$ for $x(t)$ in Eq. (20.15):

$$\mathbf{F}(\omega - ja) = \int_0^{\infty} \ddot{u}_0 e^{-j(\omega - ja)t}\,dt = \frac{\ddot{u}_0}{j(\omega - ja)} \tag{20.18}$$

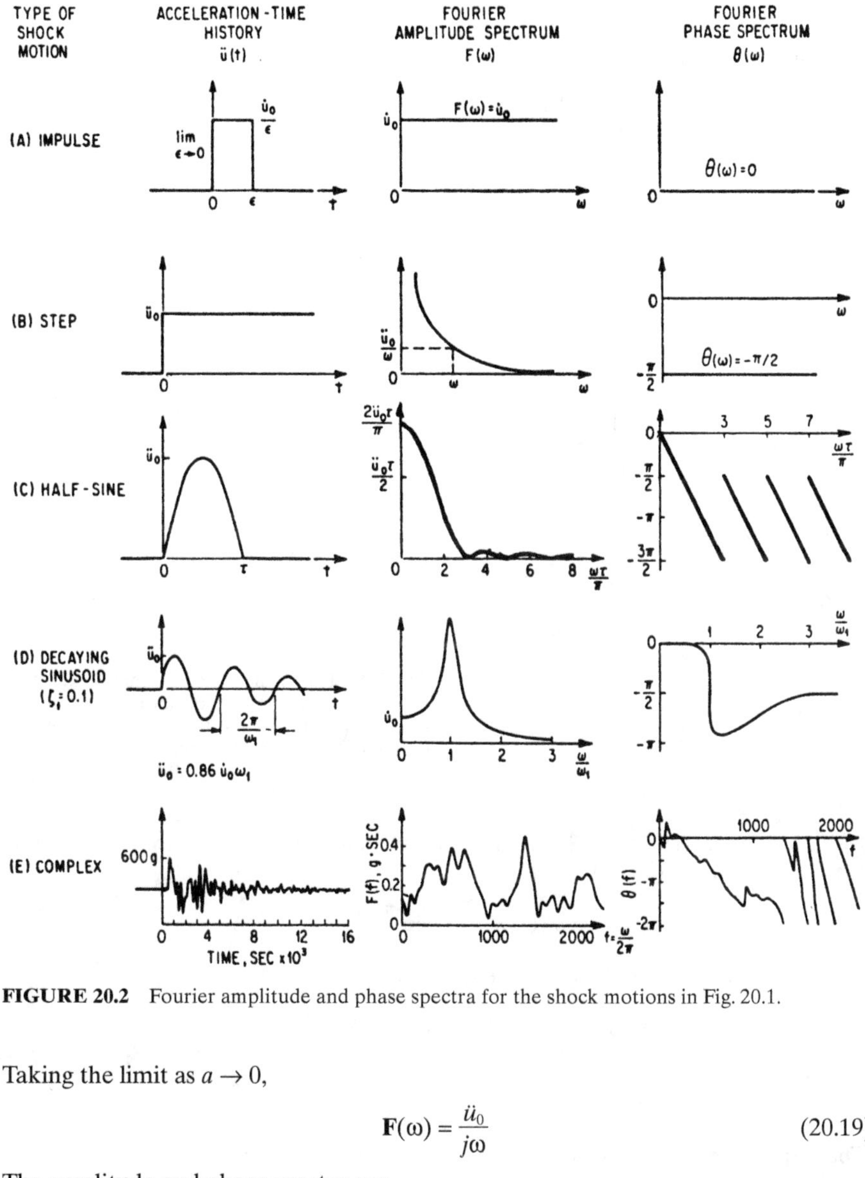

FIGURE 20.2 Fourier amplitude and phase spectra for the shock motions in Fig. 20.1.

Taking the limit as $a \to 0$,

$$\mathbf{F}(\omega) = \frac{\ddot{u}_0}{j\omega} \tag{20.19}$$

The amplitude and phase spectra are

$$F(\omega) = \frac{\ddot{u}_0}{\omega}; \qquad \theta(\omega) = -\frac{\pi}{2} \tag{20.20}$$

These spectra are shown in Fig. 20.2*B;* the amplitude spectrum decreases as frequency increases, whereas the phase is a constant independent of frequency. Note that the spectrum of Eq. (20.19) is $1/j\omega$ times the spectrum for the impulse given by Eq. (20.17*A*).

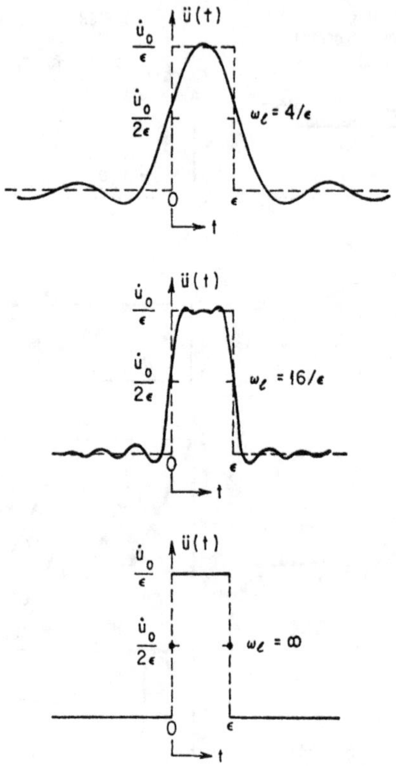

FIGURE 20.3 Time histories which result from the superposition of the Fourier components of a rectangular pulse for several different upper limits of frequency ω_l of the components.

Half-sine Acceleration. Substitution of the half-sine acceleration time history, Eq. (20.9), into Eq. (20.15) gives

$$\mathbf{F}(\omega) = \int_0^{\tau} \ddot{u}_0 \sin \frac{\pi t}{\tau} \, e^{-j\omega t} \, dt \tag{20.21}$$

Performing the indicated integration gives

$$\begin{aligned} \mathbf{F}(\omega) &= \frac{\ddot{u}_0\tau/\pi}{1-(\omega\tau/\pi)^2}\,(1+e^{-j\omega\tau}) && [\omega \neq \pi/\tau] \\ \mathbf{F}(\omega) &= -\frac{j\ddot{u}_0\tau}{2} && [\omega = \pi/\tau] \end{aligned} \tag{20.22}$$

The expressions for the spectra of amplitude and phase are

$$\begin{aligned} F(\omega) &= \frac{2\ddot{u}_0\tau}{\pi}\left|\frac{\cos(\omega\tau/2)}{1-(\omega\tau/\pi)^2}\right| && [\omega \neq \pi/\tau] \\ F(\omega) &= \frac{\ddot{u}_0\tau}{2} && [\omega = \pi/\tau] \end{aligned} \tag{20.23}$$

$$\theta(\omega) = -\frac{\omega\tau}{2} + n\pi \tag{20.24}$$

where n is the smallest integer that prevents $|\theta(\omega)|$ from exceeding $3\pi/2$. The Fourier spectra of the half-sine pulse of acceleration are plotted in Fig. 20.2*C*.

Decaying Sinusoidal Acceleration. The application of Eq. (20.15) to the decaying sinusoidal acceleration defined by Eq. (20.12) gives the following expression for the Fourier spectrum:

$$\mathbf{F}(\omega) = \dot{u}_0 \frac{1 + j2\zeta_1\omega/\omega_1}{(1 - \omega^2/\omega_1^2) + j2\zeta_1\omega/\omega_1} \tag{20.25}$$

This can be converted to a spectrum of absolute values, specifically

$$F(\omega) = \dot{u}_0 \sqrt{\frac{1 + (2\zeta_1\omega/\omega_1)^2}{(1 - \omega^2/\omega_1^2)^2 + (2\zeta_1\omega/\omega_1)^2}} \tag{20.26}$$

Also, a spectrum of phase angle:

$$\theta(\omega) = -\tan^{-1} \frac{2\zeta_1(\omega/\omega_1)^3}{(1 - \omega^2/\omega_1^2) + (2\zeta_1\omega/\omega_1)^2} \tag{20.27}$$

These spectra are shown in Fig. 20.2*D* for a value of $\zeta = 0.1$. The peak in the amplitude spectrum near the frequency ω_1 indicates a strong concentration of Fourier components near the frequency of occurrence of the oscillations in the shock motion.

Complex Shock. The complex shock motion shown in Fig. 20.2*E* is the result of actual measurements; hence, its functional form is unknown. Its Fourier spectrum must be computed numerically. The Fourier spectrum shown in Fig. 20.2*E* was evaluated digitally using 100 time increments of 0.00015-sec duration. The peaks in the amplitude spectrum indicate concentrations of sinusoidal components near the frequencies of various oscillations in the shock motion. The portion of the phase spectrum at the high frequencies creates an appearance of discontinuity. If the phase angle were not returned to zero each time it passed through −360°, as a convenience in plotting, the curve would be continuous.

Application of the Fourier Spectrum. The Fourier spectrum description of a shock is useful in linear analysis when the properties of a structure on which the shock acts are defined as a function of frequency. Such properties are designated by the general term *frequency response function;* in shock and vibration technology, commonly used frequency response functions are mechanical impedance, mobility, and transmissibility. Such functions are often inappropriately called "transfer functions." This terminology should be reserved for functions of the Laplace variable (see Chaps. 8 and 21).

When a shock acts on a structure, the structure responds in a manner that is essentially oscillatory. The frequencies that appear predominantly in the response are (1) the preponderant frequencies of the shock and (2) the natural frequencies of the structure. The Fourier spectrum of the response $\mathbf{R}(\omega)$ is the product of the Fourier spectrum of the shock $\mathbf{F}(\omega)$ and an appropriate frequency response function for the structure. For example, if $\mathbf{F}(\omega)$ and $\mathbf{R}(\omega)$ are Fourier spectra of acceleration, the frequency response function is the transmissibility of the structure, i.e., the ratio of acceleration at the responding station to the acceleration at the driving station, as a function of frequency. However, if $\mathbf{R}(\omega)$ is a Fourier spectrum of velocity and $\mathbf{F}(\omega)$ is a Fourier spectrum of force, the frequency response function is mobility as a function of frequency.

The Fourier spectrum also finds application in evaluating the effect of a load upon a shock source. A source of shock generally consists of a means of shock excitation and a resilient structure through which the excitation is transmitted to a load. Consequently, the character of the shock delivered by the resilient structure of the shock source is influenced by the nature of the load being driven. The characteristics of the source and load may be defined in terms of mechanical impedance or mobility (see Chap. 9). If the shock motion at the source output is measured with no load and expressed in terms of its Fourier spectrum, the effect of the load upon this shock motion can be determined as detailed in Chap. 21. The resultant motion with the load attached is described by its Fourier spectrum.

The frequency response function of a structure may be determined by applying a force to the structure and noting the response. The applied force may be transient, sinusoidal, or random. In the case of a transient force, it is usually applied with the use of a hammer, while the other types of forces are applied using a shaker (see Chap. 21).

DATA REDUCTION TO THE RESPONSE DOMAIN

A structure or physical system has a characteristic response to a particular shock applied as an excitation to the structure. The magnitudes of the response peaks can be used to define certain effects of the shock by considering systematically the properties of the system and relating the peak responses to such properties. This is in contrast to the Fourier spectrum description of a shock in the following respects:

1. Whereas the Fourier spectrum defines the shock in terms of the amplitudes and phase relations of its frequency components, the response spectrum describes only the effect of the shock upon a structure in terms of peak responses. This effect is of considerable significance in the design of equipments and in the specification of laboratory tests.
2. The time history of a shock cannot be determined from the knowledge of the peak responses of a system excited by the shock; i.e., the calculation of peak responses is an irreversible operation. This contrasts with the Fourier spectrum, where the Fourier spectrum can be determined from the time history, and vice versa.

By limiting consideration to the response of a linear, viscously damped single-degree-of-freedom structure with lumped parameters (hereafter referred to as a simple structure and illustrated in Fig. 20.4), there are only two structural parameters upon which the response depends: (1) the undamped natural frequency and (2) the fraction of critical damping, or equivalently, the resonant gain Q. With only two parameters involved, it is feasible to obtain from the shock measurement a systematic presentation of the peak responses of many simple structures. This process is termed *data reduction to the response domain.* This type of reduced data applies directly to a system that responds in an SDOF; it is useful to some extent by normal-mode superposition to evaluate the response of a linear system that responds in more than one DOF. The conditions of a particular application determine the magnitude of errors resulting from superposition.[2–5]

Shock Response Spectrum. The response of a system to a shock can be expressed as the time history of a parameter that describes the motion of the system. For a simple system, the magnitudes of the response peaks can be summarized as a function of the natural frequency or natural period of the responding system, at vari-

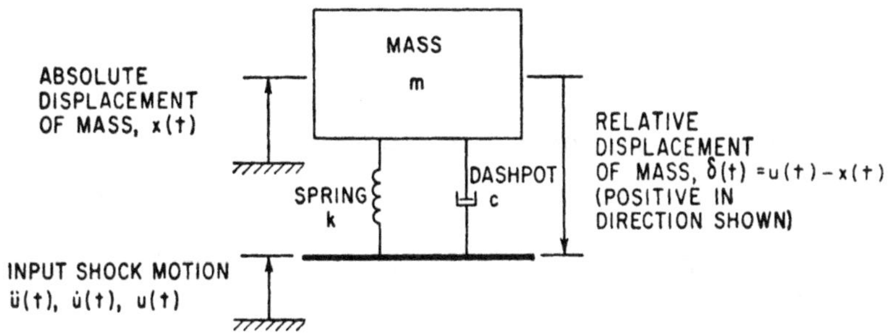

FIGURE 20.4 Representation of a simple structure used to accomplish the data reduction of a shock motion to the response domain.

ous values of the fraction of critical damping. This type of presentation is termed a *shock response spectrum*, or simply a *response spectrum* or a *shock spectrum*.

Parameters for the Shock Response Spectrum. The peak response of the simple structure may be defined, as a function of natural frequency, in terms of any one of several parameters that describe its motion. The parameters often are related to each other by the characteristics of the structure. However, inasmuch as one of the advantages of the shock response spectrum method of data reduction and presentation is convenience of application to physical situations, it is advantageous to give careful consideration in advance to the particular parameter that is best adapted to the attainment of particular objectives. Referring to the simple structure shown in Fig. 20.4, the following significant parameters may be determined directly from measurements on the structure:

1. Absolute displacement $x(t)$ of mass m. This indicates the displacement of the responding structure with reference to an inertial reference plane, i.e., coordinate axes fixed in space.
2. Relative displacement $\delta(t)$ of mass m. This indicates the displacement of the responding structure relative to its support, a quantity useful for evaluating the distortions and strains within the responding structure.
3. Absolute velocity $\dot{x}(t)$ of mass m. This quantity is useful for determining the kinetic energy of the structure.
4. Relative velocity $\dot{\delta}(t)$ of mass m. This quantity is useful for determining the stresses generated within the responding structure due to viscous damping and the maximum energy dissipated by the responding structure.
5. Absolute acceleration $\ddot{x}(t)$ of mass m. This quantity is useful for determining the stresses generated within the responding structure due to the combined elastic and damping reactions of the structure.
6. The relative displacement response may be multiplied by the angular natural frequency ω_n of the simple structure to create a *pseudo-velocity* response.

The *equivalent static acceleration* is that steadily applied acceleration, expressed as a multiple of the acceleration of gravity, which distorts the structure to the maximum distortion resulting from the action of the shock.[6] For the simple structure of Fig. 20.4, the relative displacement response δ indicates the distortion under the shock condition. The corresponding distortion under static conditions, in a $1g$ gravitational field, is

$$\delta_{st} = \frac{mg}{k} = \frac{g}{\omega_n^2} \tag{20.28}$$

By analogy, the maximum distortion under the shock condition is

$$\delta_{\max} = \frac{A_{eq}g}{\omega_n^2} \tag{20.29}$$

where A_{eq} is the equivalent static acceleration in units of gravitational acceleration. From Eq. (20.29),

$$A_{eq} = \frac{\delta_{\max}\omega_n^2}{g} \tag{20.30}$$

The maximum relative displacement $\delta_{\max}$ and the equivalent static acceleration A_{eq} are directly proportional.

If the shock is a loading parameter, such as force, pressure, or torque, as a function of time, the corresponding equivalent static parameter is an equivalent static force, pressure, or torque, respectively. Since the supporting structure is assumed to be motionless when a shock loading acts, the relative response motions and absolute response motions become identical.

The differential equation of motion for the system shown in Fig. 20.4 is

$$-\ddot{x}(t) + 2\zeta\omega_n\dot{\delta}(t) + \omega_n^2\delta(t) = 0 \tag{20.31}$$

where ω_n is the undamped natural frequency and ζ is the fraction of critical damping. When $\zeta = 0$, $\ddot{x}_{\max} = A_{eq}g$; this follows directly from the relation of Eq. (20.29). When $\zeta \neq 0$, the acceleration $\ddot{x}$ experienced by the mass m results from forces transmitted by the spring k and the damper c. Thus, in a damped system, the maximum acceleration of mass m is not exactly equal to the equivalent static acceleration. However, in most mechanical structures, the damping is relatively small; therefore, the equivalent static acceleration and the maximum absolute acceleration often are interchangeable with negligible error.

Calculation of Shock Response Spectrum. The relative displacement response of a simple structure (Fig. 20.4) resulting from a shock defined by the acceleration $\ddot{u}(t)$ of the support is given by the *Duhamel integral*[7]

$$\delta(t) = \frac{1}{\omega_d}\int_0^t \ddot{u}(t_v)e^{-\zeta\omega_n(t - t_v)} \sin \omega_d(t - t_v)\, dt_v \tag{20.32}$$

where $\omega_n = (k/m)^{1/2}$ is the undamped natural frequency, $\zeta = c/2m\omega_n$ is the fraction of critical damping, and $\omega_d = \omega_n(1 - \zeta^2)^{1/2}$ is the damped natural frequency. The excitation $\ddot{u}(t_v)$ is defined as a function of the variable of integration t_v, and the response $\delta(t)$ is a function of time t. The relative displacement δ and relative velocity $\dot{\delta}$ are considered to be zero when $t = 0$. The equivalent static acceleration, defined by Eq. (20.30), as a function of ω_n and ζ is

$$A_{eq}(\omega_n,\zeta) = \frac{\omega_n^2}{g}\, \delta_{\max}(\omega_n,\zeta) \tag{20.33}$$

If a shock loading such as the input force $F(t)$ rather than an input motion acts on the simple structure, the response is

$$\delta(t) = \frac{1}{m\omega_d}\int_0^t F(t_v)e^{-\zeta\omega_n(t - t_v)} \sin \omega_d(t - t_v)\, dt_v \tag{20.34}$$

and an equivalent static force is given by

$$F_{eq}(\omega_n,\zeta) = k\delta_{max}(\omega_n,\zeta) = m\omega_n^2\delta_{max}(\omega_n,\zeta) \tag{20.35}$$

The equivalent static force is related to equivalent static acceleration by

$$F_{eq}(\omega_n,\zeta) = mA_{eq}(\omega_n,\zeta) \tag{20.36}$$

It is often of interest to determine the maximum relative displacement of the simple structure in Fig. 20.4 in both a positive and a negative direction. If $\ddot{u}(t)$ is positive as shown, positive values of $\ddot{x}(t)$ represent upward acceleration of the mass m. Initially, the spring is compressed and the positive direction of $\delta(t)$ is taken to be positive as shown. Conversely, negative values of $\delta(t)$ represent extension of spring k from its original position. It is possible that the ultimate use of the reduced data would require that both extension and compression of spring k be determined. Correspondingly, a positive and a negative sign may be associated with an equivalent static acceleration A_{eq} of the support, so that A_{eq}^+ is an upward acceleration producing a positive deflection δ and A_{eq}^- is a downward acceleration producing a negative deflection δ.

For some purposes it is desirable to distinguish between the maximum response which occurs during the time in which the measured shock acts and the maximum response which occurs during the free vibration existing after the shock has terminated. The shock spectrum based on the former is called a *primary shock response spectrum* and that based on the latter is called a *residual shock response spectrum.* For instance, the response $\delta(t)$ to the half-sine pulse in Fig. 20.1*C* occurring during the period ($t < \tau$) is the primary response and the response $\delta(t)$ occurring during the period ($t > \tau$) is the residual response. Reference is made to primary and residual shock response spectra in the next section, "Examples of Shock Response Spectra" and in the section entitled "Relationship Between Shock Response Spectrum and Fourier Spectrum."

Standard for Calculation of Response Spectra. For calculation of the shock response spectrum there is an ISO standard.[8] In the standard, a shock response spectrum is the response to a given acceleration of a set of single-degree-of-freedom mass-damper-spring oscillators. The given acceleration is applied to the base of all oscillators, and the maximum responses of each oscillator versus the natural frequency make up the spectrum.

Each SDOF system has a unique set of defining parameters: mass m, damping constant c, and spring constant k.

A given acceleration a_1 is applied to the base. If the response is measured as the acceleration of the SDOF mass a_2, then the transfer function $G(s)$ is given by:

$$G(s) = \frac{a_2(s)}{a_1(s)} = \frac{cs + k}{ms^2 + cs + k} \tag{20.37}$$

where s is the Laplace variable (complex frequency s) in radians per second. The SDOF system is normally characterized by its (undamped) natural frequency f_n, in hertz, and the resonance gain Q (Q-factor):

$$f_n = \frac{1}{2\pi}\sqrt{\frac{k}{m}} \tag{20.38}$$

$$Q = \frac{\sqrt{km}}{c} \tag{20.39}$$

The transfer function may then be rewritten as

$$G(s) = \frac{a_2(s)}{a_1(s)} = \frac{\dfrac{\omega_n s}{Q} + \omega_n^2}{s^2 + \dfrac{\omega_n s}{Q} + \omega_n^2} \tag{20.40}$$

with ω_n being the angular natural frequency in radians per second.

Equation (20.40) defines the transfer function used. The maximum is approximately Q, and the maximum occurs approximately at f_n Hz. The approximation is more accurate the larger the Q-value is. Instead of the resonant gain Q, the damping ratio, fraction of critical damping ζ, may be used. ζ is often expressed in percent of critical damping.

$$\zeta = \frac{1}{2Q} = \frac{c}{2\sqrt{km}} \tag{20.41}$$

To calculate the response, a digital filter method is used. In the standard, the filter coefficients are given for many different variations of shock spectra, such as relative displacement and pseudo-velocity. Here only the basic algorithm for acceleration response is given.

The standard deals with the processing of the signal when it exists as a digital record, sampled with a sampling frequency of f_s Hz, corresponding to a time interval between samples of T seconds, $T = 1/f_s$.

The digital filters corresponding to different SDOF system responses are second-order filters, with the general z-transform expression[8]

$$H(z) = \frac{\beta_0 + \beta_1 \cdot z^{-1} + \beta_2 \cdot z^{-2}}{1 + \alpha_1 \cdot z^{-1} + \alpha_2 \cdot z^{-2}} \tag{20.42}$$

The filter expression corresponds to a difference equation describing how to calculate the response time series y_n when the input acceleration time series x_n is given:

$$y_n = \beta_0 \cdot x_n + \beta_1 \cdot x_{n-1} + \beta_2 \cdot x_{n-2} - \alpha_1 \cdot y_{n-1} - \alpha_2 \cdot y_{n-2} \tag{20.43}$$

Filter coefficients for the absolute acceleration response:

$$\begin{aligned}
\beta_0 &= 1 - \exp(-A) \cdot \sin(B)/B \\
\beta_1 &= 2\exp(-A) \cdot \{\sin(B)/B - \cos(B)\} \\
\beta_2 &= \exp(-2A) - \exp(-A) \cdot \sin(B)/B \\
\\
\alpha_1 &= -2\exp(-A) \cdot \cos(B) \\
\alpha_2 &= \exp(-2A)
\end{aligned} \tag{20.44}$$

where

$$A = \frac{\omega_n \cdot T}{2Q}$$

$$B = \omega_n \cdot T \cdot \sqrt{1 - \frac{1}{4Q^2}}$$

Sampling Frequency Consideration. The ramp invariant algorithm contains a bias error, which is dependent on the sampling frequency. There is also an error to consider when the maximum value is to be found in the sampled output. Considering these two error sources, there is a recommendation in the standard. The sampling frequency should be at least 10 times the highest significant frequency content of the input waveform. Formulas for the errors are given in the standard.

Examples of Shock Response Spectra. In this section the shock response spectra are presented for the five acceleration time histories in Fig. 20.1 These spectra, shown in Fig. 20.5, are expressed in terms of equivalent static acceleration for the undamped responding structure, for $\zeta = 0.1, 0.5$, and other selected fractions of critical damping. Both the maximum positive and the maximum negative responses are indicated. In addition, a number of relative displacement response time histories $\delta(t)$ are plotted to show the nature of the responses. A large number of shock response spectra, based on various response parameters, are given in Chap. 8.

ACCELERATION IMPULSE: The application of Eq. (20.32) to the acceleration impulse shown in Fig. 20.1*A* and defined by Eq. (20.1) yields

$$\delta(t) = \frac{\dot{u}_0}{\omega_d} e^{-\zeta\omega_n t} \sin \omega_d t \qquad [\zeta < 1] \tag{20.45}$$

This response is plotted in Fig. 20.5*A* for $\zeta = 0, 0.1$, and 0.5. The response peaks are reached at the times $t = (\cos^{-1} \zeta)/\omega_d$, $\cos^{-1} \zeta$ increasing by π for each succeeding peak. The values of the response at the peaks are

$$\delta_{\max}^{(i)}(\omega_n,\zeta) = \frac{\dot{u}_0}{\omega_n} \exp\left(-\frac{\zeta}{\sqrt{1-\zeta^2}}\,[\cos^{-1}\zeta + (i-1)\pi]\right) \qquad [0 < \cos^{-1}\zeta \leq \pi/2] \tag{20.46}$$

where i is the number of the peak ($i = 1$ for the first positive peak, $i = 2$ for the first negative peak, etc.).

The largest positive response occurs at the first peak, i.e., when $i = 1$, and is shown by the solid dots in Fig. 20.5. Hence, the equivalent static acceleration in the positive direction is obtained by substitution of Eq. (20.45) into Eq. (20.33) with $i = 1$:

$$A_{\mathrm{eq}}^{+}(\omega_n,\zeta) = \frac{\omega_n \dot{u}_0}{g} \exp\left(-\frac{\zeta}{\sqrt{1-\zeta^2}} \cos^{-1}\zeta\right) \tag{20.47}$$

The equivalent static acceleration in the negative direction is calculated from the maximum relative deflection at the second peak, i.e., when $i = 2$, and is shown by the hollow dots in Fig. 20.5*A:*

$$A_{\mathrm{eq}}^{-}(\omega_n,\zeta) = \frac{\omega_n \dot{u}_0}{g} \exp\left(-\frac{\zeta}{\sqrt{1-\zeta^2}} (\cos^{-1}\zeta + \pi)\right) \tag{20.48}$$

The resulting shock spectrum is shown in Fig. 20.5 with curves for $\zeta = 0, 0.1, 0.5$, and 1.0. At any value of damping, a shock response spectrum is a straight line passing through the origin. The peak distortion of the structure $\delta_{\max}$ is inversely proportional to frequency. Thus, the relative displacement of the mass increases as the natural frequency decreases, whereas the equivalent static acceleration has an opposite trend.

ACCELERATION STEP: The response of a simple structure to the acceleration step in Fig. 20.1*B* is found by substituting from Eq. (20.5) in Eq. (20.32) and integrating:

$$\delta(t) = \frac{\ddot{u}_0}{\omega_n^2}\left[1 - \frac{e^{-\zeta\omega_n t}}{\sqrt{1-\zeta^2}} \cos(\omega_d t - \sin^{-1}\zeta)\right] \qquad [\zeta < 1] \tag{20.49}$$

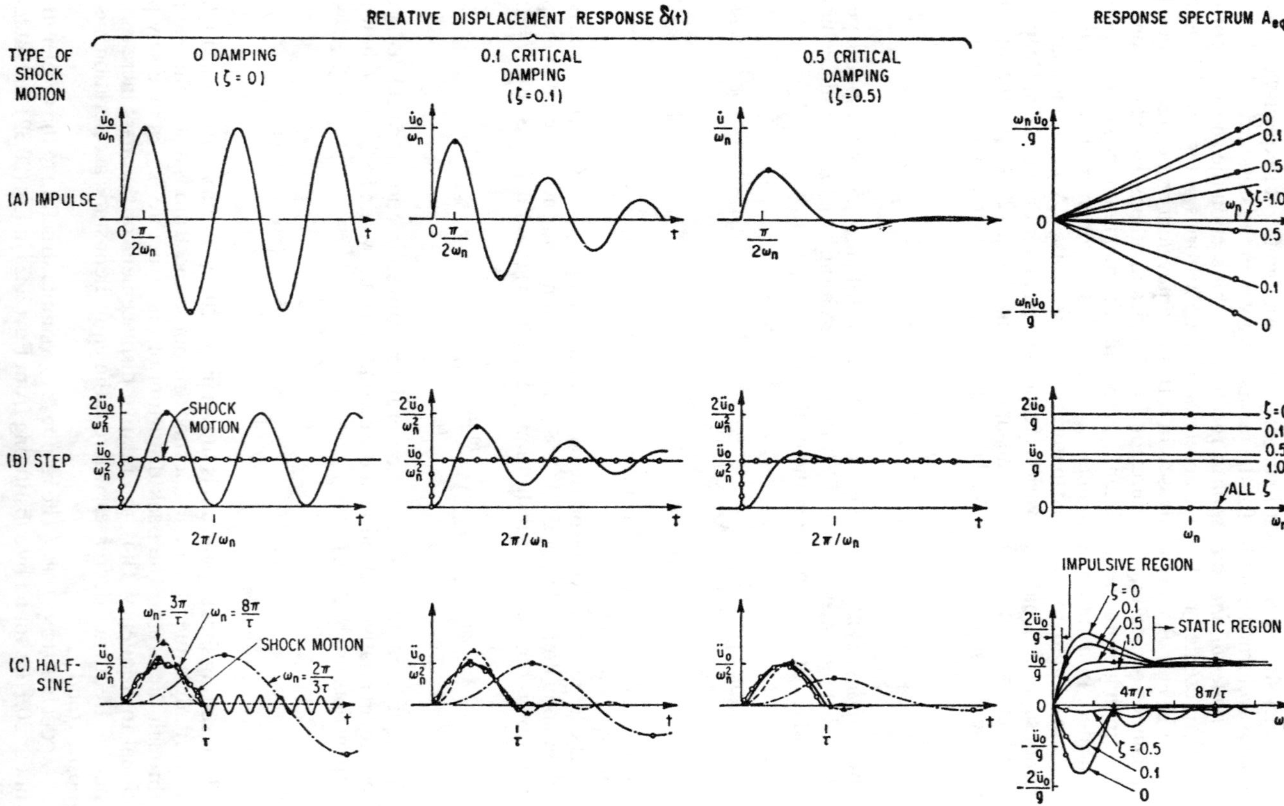
RELATIVE DISPLACEMENT RESPONSE δ(t)
RESPONSE SPECTRUM A_eq
TYPE OF SHOCK MOTION
0 DAMPING (ζ = 0)
0.1 CRITICAL DAMPING (ζ = 0.1)
0.5 CRITICAL DAMPING (ζ = 0.5)
(A) IMPULSE
(B) STEP
(C) HALF-SINE
SHOCK MOTION
ALL ζ
IMPULSIVE REGION
STATIC REGION
ωn = 3π/τ
ωn = 8π/τ
ωn = 2π/3τ
4π/τ
8π/τ
2π/ωn

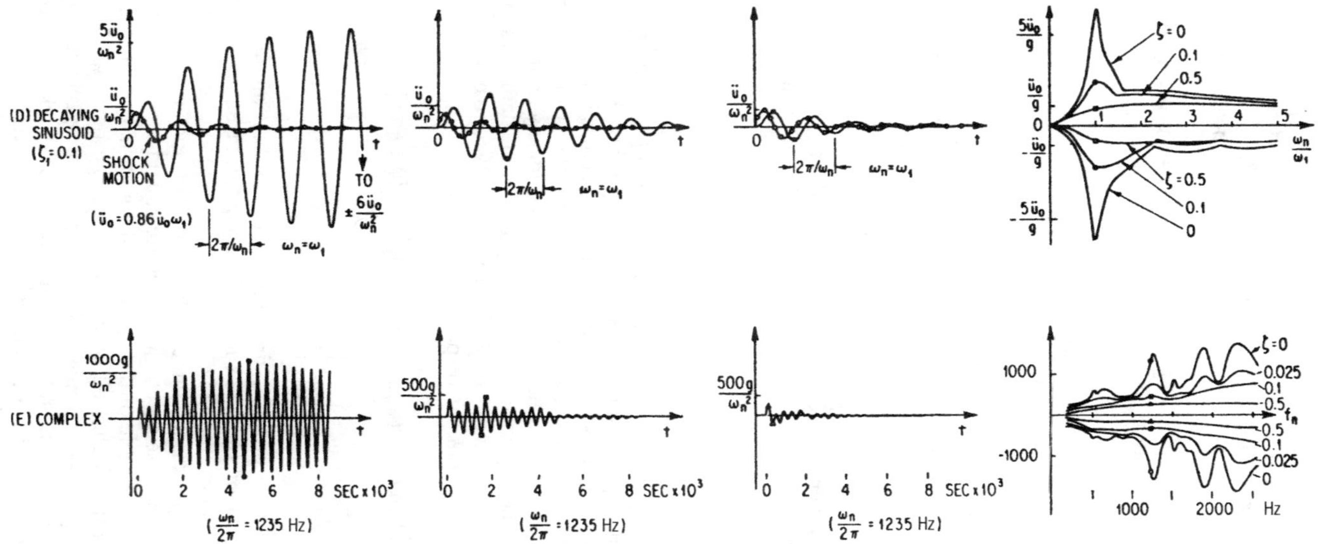

FIGURE 20.5 Time-histories of response to shock motions defined in Fig. 20.1 and corresponding shock response spectra.

The responses $\delta(t)$ are shown in Fig. 20.5*B* for $\zeta = 0, 0.1$, and 0.5. The response overshoots the value $\ddot{u}_0/\omega_n^2$ and then oscillates about this value as a mean with diminishing amplitude as energy is dissipated by damping. An overshoot to $2\ddot{u}_0/\omega_n^2$ occurs for zero damping. A response $\delta = \ddot{u}_0/\omega_n^2$ would result from a steady application of the acceleration $\ddot{u}_0$.

The response maxima and minima occur at the times $t = i\pi/\omega_d$, $i = 0$ providing the first minimum and $i = 1$ the first maximum. The maximum values of the relative displacement response are

$$\delta_{\max}(\omega_n,\zeta) = \frac{\ddot{u}_0}{\omega_n^2}\left[1 + \exp\left(-\frac{\zeta i\pi}{\sqrt{1-\zeta^2}}\right)\right] \qquad [i \text{ odd}] \tag{20.50}$$

The largest positive response occurs at the first maximum, i.e., where $i = 1$, and is shown by the solid symbols in Fig. 20.5*B*. The equivalent static acceleration in the positive direction is obtained by substitution of Eq. (20.50) into Eq. (20.33) with $i = 1$:

$$A^{+}_{\text{eq}}(\omega_n,\zeta) = \frac{\ddot{u}_0}{g}\left[1 + \exp\left(-\frac{\zeta\pi}{\sqrt{1-\zeta^2}}\right)\right] \tag{20.51a}$$

The greatest negative response is zero; it occurs at $t = 0$, independent of the value of damping, as shown by open symbols in Fig. 23.7*B*. Thus, the equivalent static acceleration in the negative direction is

$$A^{-}_{\text{eq}}(\omega_n,\zeta) = 0 \tag{20.51b}$$

Since the equivalent static acceleration is independent of natural frequency, the shock response spectrum curves shown in Fig. 20.5*B* are horizontal lines. The symbols shown on the shock response spectra correspond to the responses shown.

The equivalent static acceleration for an undamped simple structure is twice the value of the acceleration step $\ddot{u}_0/g$. As the damping increases, the overshoot in response decreases; there is no overshoot when the structure is critically damped.

HALF-SINE ACCELERATION: The expressions for the response of the damped simple structure to the half-sine acceleration of Eq. (20.9) are too involved to have general usefulness. For an undamped system, the response $\delta(t)$ is

$$\begin{aligned}
\delta(t) &= \frac{\ddot{u}_0}{\omega_n^2}\left(\frac{(\omega_n\tau/\pi)}{1-(\omega_n\tau/\pi)^2}\right)[\sin \omega_n t - (\omega_n\tau/\pi)\sin(\pi t/\tau)] && [0 < t \le \tau] \\
\delta(t) &= \frac{\ddot{u}_0}{\omega_n^2}\left(\frac{(\omega_n\tau/\pi)}{1-(\omega_n\tau/\pi)^2}\right) 2\cos\left(\frac{\omega_n\tau}{2}\right)\sin\left[\omega_n\left(t-\frac{\tau}{2}\right)\right] && [t > \tau]
\end{aligned} \tag{20.52}$$

For zero damping the residual response is sinusoidal with constant amplitude. The first maximum in the response of a simple structure with natural frequency less than π/τ occurs during the residual response; i.e., after $t = \tau$. As a result, the magnitude of each succeeding response peak is the same as that of the first maximum. Thus the positive and negative shock response spectrum curves are equal for $\omega_n \le \pi/\tau$. The dot-dash curve in Fig. 20.5*C* is an example of the response at a natural frequency of $2\pi/3\tau$. The peak positive response is indicated by a solid circle, the peak negative response by an open circle. The positive and negative shock response spectrum values derived from this response are shown on the undamped ($\zeta = 0$) shock response spectrum curves at the right-hand side of Fig. 20.5*C*, using the same symbols.

At natural frequencies below $\pi/2\tau$, the shock response spectra for an undamped system are very nearly linear with a slope $\pm 2\ddot{u}_0\tau/\pi g$. In this low-frequency region the response is essentially impulsive; i.e., the maximum response is approximately the same as that due to an ideal acceleration impulse (Fig. 20.5*A*) having a velocity change $\dot{u}_0$ equal to the area under the half-sine acceleration time history.

The response at the natural frequency $3\pi/\tau$ is the dotted curve in Fig. 20.5*C*. The displacement and velocity response are both zero at the end of the pulse, and hence no residual response occurs. The solid and open triangles indicate the peak positive and negative response, the latter being zero. The corresponding points appear on the undamped shock response spectrum curves. As shown by the negative undamped shock response spectrum curve, the residual spectrum goes to zero for all odd multiples of π/τ above $3\pi/\tau$.

As the natural frequency increases above $3\pi/\tau$, the response attains the character of relatively low amplitude oscillations occurring with the half-sine pulse shape as a mean. An example of this type of response is shown by the solid curve for $\omega_n = 8\pi/\tau$. The largest positive response is slightly higher than $\ddot{u}_0/\omega_n^2$, and the residual response occurs at a relatively low level. The solid and open square symbols indicate the largest positive and negative responses.

As the natural frequency becomes extremely high, the response follows the half-sine shape very closely. In the limit, the natural frequency becomes infinite and the response approaches the half-sine wave shown in Fig. 20.5*C*. For natural frequencies greater than $5\pi/\tau$, the response tends to follow the input and the largest response is within 20 percent of the response due to a static application of the peak input acceleration. This portion of the shock response spectrum is sometimes referred to as the "static region" (see "Limiting Values of Shock Response Spectrum," below).

The equivalent static acceleration without damping for the positive direction is

$$
\begin{aligned}
A^+_{\text{eq}}(\omega_n,0) &= \frac{\ddot{u}_0}{g}\left(\frac{2(\omega_n\tau/\pi)}{1-(\omega_n\tau/\pi)^2}\right)\cos\left(\frac{\omega_n\tau}{2}\right) && \left[\omega_n \le \frac{\pi}{\tau}\right] \\
A^+_{\text{eq}}(\omega_n,0) &= \frac{\ddot{u}_0}{g}\left(\frac{(\omega_n\tau/\pi)}{(\omega_n\tau/\pi)-1}\right)\sin\left(\frac{2i\pi}{(\omega_n\tau/\pi)+1}\right) && \left[\omega_n > \frac{\pi}{\tau}\right]
\end{aligned}
\tag{20.53}
$$

where i is the positive integer which maximizes the value of the sine term while the argument remains less than π. In the negative direction the peak response always occurs during the residual response; thus, it is given by the absolute value of the first of the expressions in Eq. (20.53):

$$
A^-_{\text{eq}}(\omega_n,0) = \frac{\ddot{u}_0}{g}\left(\frac{2(\omega_n\tau/\pi)}{1-(\omega_n\tau/\pi)^2}\right)\cos\left(\frac{\omega_n\tau}{2}\right) \tag{20.54}
$$

Shock response spectra for damped systems are commonly found by use of a digital computer. Spectra for $\zeta = 0.1$ and 0.5 are shown in Fig. 20.5*C*.

The response of a damped structure whose natural frequency is less than $\pi/2\tau$ is essentially impulsive; i.e., the shock response spectra in this frequency region are substantially identical to the spectra for the acceleration impulse in Fig. 20.5*A*. Except near the zeros in the negative spectrum for an undamped system, damping reduces the peak response. For the positive spectra, the effect is small in the static region since the response tends to follow the input for all values of damping. The greatest effect of damping is seen in the negative spectra because it affects the decay of response oscillations at the natural frequency of the structure.

DECAYING SINUSOIDAL ACCELERATION: Although analytical expressions for the response of a simple structure to the decaying sinusoidal acceleration shown in Fig. 20.1*D* are available, calculation of spectra is impractical without use of a computer. Figure 20.5*D* shows spectra for several values of damping in the decaying sinusoidal acceleration. In the low-frequency region ($\omega_n < 0.2\omega_1$), the response is essentially impulsive. The area under the acceleration time history of the decaying sinusoid is $\dot{u}_0$; hence, the response of a very low frequency structure is similar to the response to an acceleration impulse of magnitude $\dot{u}_0$.

When the natural frequency of the responding system approximates the frequency ω_1 of the oscillations in the decaying sinusoid, a resonant type of buildup tends to occur in the response oscillations. The region in the neighborhood of $\omega_1 = \omega_n$ may be termed a quasi-resonant region of the shock response spectrum. Responses for $\zeta = 0, 0.1$, and 0.5 and $\omega_n = \omega_1$ are shown in Fig. 20.5*D*. In the absence of damping in the responding system, the rate of buildup diminishes with time and the amplitude of the response oscillations levels off as the input acceleration decays to very small values. Small damping in the responding system, e.g., $\zeta = 0.1$, reduces the initial rate of buildup and causes the response to decay to zero after a maximum is reached. When damping is as large as $\zeta = 0.5$, no buildup occurs.

COMPLEX SHOCK: The shock spectra for the complex shock of Fig. 20.2*E* are shown in Fig. 20.5*E*. Time histories of the response of a system with a natural frequency of 1250 Hz also are shown. The ordinate of the spectrum plot is equivalent static acceleration, and the abscissa is the natural frequency in hertz. Three pronounced peaks appear in the spectra for zero damping, at approximately 1250 Hz, 1900 Hz, and 2350 Hz. Such peaks indicate a concentration of frequency content in the shock, similar to the spectra for the decaying sinusoid in Fig. 20.5*D*. Other peaks in the shock spectra for an undamped system indicate less significant oscillatory behavior in the shock. The two lower frequencies at which the pronounced peaks occur correlate with the peaks in the Fourier spectrum of the same shock, as shown in Fig. 20.2*E*. The highest frequency at which a pronounced peak occurs is above the range for which the Fourier spectrum was calculated.

Because of response limitations of the analysis, the shock spectra do not extend below 200 Hz. Since the duration of the complex shock of Fig. 20.1*E* is about 0.016 sec, an impulsive-type response occurs only for natural frequencies well below 200 Hz. As a result, no impulsive region appears in the shock response spectra. There is no static region of the spectra shown because calculations were not extended to a sufficiently high frequency.

In general, the equivalent static acceleration A_{eq} is reduced by additional damping in the responding structure system except in the region of valleys in the shock spectra, where damping may increase the magnitude of the spectrum. Positive and negative spectra tend to be approximately equal in magnitude at any value of damping; thus, the spectra for a complex oscillatory type of shock may be based on peak response independent of sign to a good approximation.

Limiting Values of Shock Response Spectrum. The response data provided by the shock response spectrum sometimes can be abstracted to simplified parameters that are useful for certain purposes. In general, this cannot be done without definite information on the ultimate use of the reduced data, particularly the natural frequencies of the structures upon which the shock acts. Two important cases are discussed in the following sections.

IMPULSE OR VELOCITY CHANGE: The duration of a shock sometimes is much smaller than the natural period of a structure upon which it acts. Then the entire response of the structure is essentially a function of the area under the time history of the shock, described in terms of acceleration or a loading parameter such as force, pressure, or torque. Consequently, the shock has an effect which is equivalent to that produced by an impulse of infinitesimally short duration, i.e., an ideal impulse.

The shock response spectrum of an ideal impulse is shown in Fig. 20.5*A*. All equivalent static acceleration curves are straight lines passing through the origin. The portion of the spectrum exhibiting such straight-line characteristics is termed the

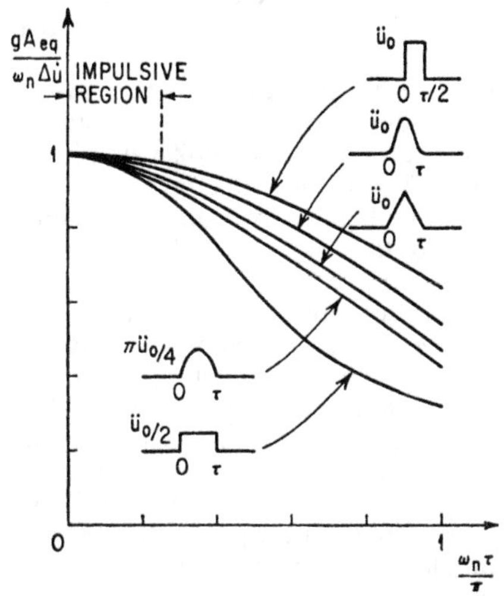

FIGURE 20.6 Portions adjacent to the origin of the positive spectra of an undamped system for several single pulses of acceleration.

impulsive region. The shock response spectrum of the half-sine acceleration pulse has an impulsive region when ω_n is less than approximately $\pi/2\tau$, as shown in Fig. 20.5*C*. If the area under a time history of acceleration or shock loading is not zero or infinite, an impulsive region exists in the shock response spectrum. The extent of the region on the natural frequency axis depends on the shape and duration of the shock.

The portions adjacent to the origin of the positive shock response spectra of an undamped system for several single pulses of acceleration are shown in Fig. 20.6. To illustrate the impulsive nature, each spectrum is normalized with respect to the peak impulsive response $\omega_n\, \Delta\dot{u}/g$, where $\Delta\dot{u}$ is the area under the corresponding acceleration time history. Hence, the spectra indicate an impulsive response where the ordinate is approximately 1. The response to a single pulse of acceleration is impulsive within a tolerance of 10 percent if $\omega_n < \pi/4\tau$; i.e., $f_n < 1/8\tau$, where f_n is the natural frequency of the responding structure in hertz and τ is the pulse duration in seconds. This result also applies when the responding system is damped. Thus, it is possible to reduce the description of a shock pulse to a designated velocity change when the natural frequency of the responding structure is less than a specified value. The magnitude of the velocity change is the area under the acceleration pulse:

$$\Delta\dot{u} = \int_0^{\tau} \ddot{u}(t)\, dt \tag{20.55}$$

PEAK ACCELERATION OR LOADING: The natural frequency of a structure responding to a shock sometimes is sufficiently high that the response oscillations of the structure at its natural frequency have a relatively small amplitude. Examples of such responses are shown in Fig. 20.5*C* for $\omega_n = 8\pi/\tau$ and $\zeta = 0, 0.1, 0.5$. As a result, the maximum response of the structure is approximately equal to the maximum acceleration of the shock and is termed *equivalent static response.* The magnitude of the spectra in such a static region is determined principally by the peak value of the shock acceleration or loading. Portions of the positive spectra of an undamped system in the region of high natural frequencies are shown in Fig. 20.7 for a number of acceleration pulses. Each spectrum is normalized with respect to the maximum acceleration of the pulse. If the ordinate is approximately 1, the shock response spectrum curves behave approximately in a static manner.

The limit of the static region in terms of the natural frequency of the structure is more a function of the slope of the acceleration time history than of the duration of the pulse. Hence, the horizontal axis of the shock response spectra in Fig. 20.7 is given in terms of the ratio of the rise time τ_r to the maximum value of the pulse. As shown in Fig. 20.7, the peak response to a single pulse of acceleration is approxi-

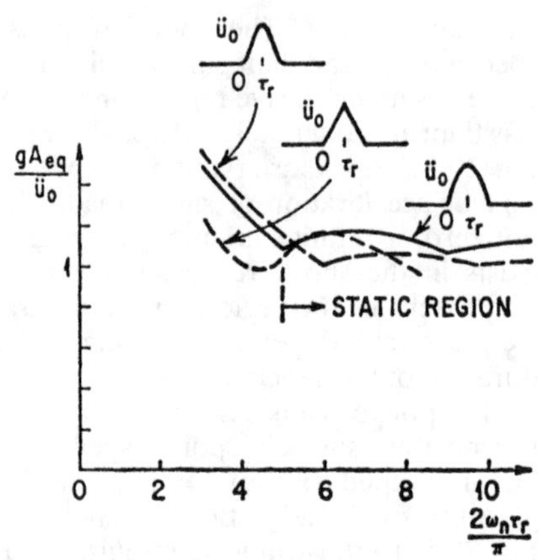

FIGURE 20.7 Portions of the positive shock response spectra of an undamped system with high natural frequencies for several single pulses of acceleration.

mately equal to the maximum acceleration of the pulse, within a tolerance of 20 percent, if $\omega_n > 2.5\pi/\tau_r$; i.e., $f_n > 1.25/\tau_r$, where f_n is the natural frequency of the responding structure in hertz and τ_r is the rise time to the peak value in seconds. The tolerance of 20 percent applies to an undamped system; for a damped system, the tolerance is lower, as indicated in Fig. 20.5*C*.

The concept of the static region also can be applied to complex shocks. Suppose the shock is oscillatory, as shown in Fig. 20.1*E*. If the response to such a shock is to be nearly static, the response to each of the succession of pulses that make up the shock must be nearly static. This is most significant for pulses of large magnitude because they determine the ordinate of the spectrum in the static region. Therefore, the shock response spectrum for a complex shock in the static region is based upon the pulses of greatest magnitude and shortest rise time.

Relationship Between Shock Response Spectrum and Fourier Spectrum. Although the shock response spectrum and the Fourier spectrum are fundamentally different, there is a partial correlation between them. A direct relationship exists between a running Fourier spectrum, to be defined subsequently, and the response of an undamped simple structure. A consequence is a simple relationship between the Fourier spectrum of absolute values and the peak residual response of an undamped simple structure.

For the case of zero damping, Eq. (20.32) provides the relative displacement response

$$\delta(\omega_n,t) = \frac{1}{\omega_n} \int_0^t \ddot{u}(t_v) \sin \omega_n(t - t_v)\, dt_v \tag{20.56}$$

A form better suited to our needs here is

$$\delta(\omega_n,t) = \frac{1}{\omega_n}\, \mathcal{I} \left[e^{j\omega_n t} \int_0^t \ddot{u}(t_v)\, e^{-j\omega_n t_v}\, dt_v \right] \tag{20.57}$$

The integral above is seen to be the Fourier spectrum of the portion of $\ddot{u}(t)$ which lies in the time interval from zero to t, evaluated at the natural frequency ω_n. Such a time-dependent spectrum can be termed a "running Fourier spectrum" and denoted by $\mathbf{F}(\omega,t)$:

$$\mathbf{F}(\omega,t) = \int_0^t \ddot{u}(t_v) e^{-j\omega t_v}\, dt_v \tag{20.58}$$

It is assumed that the excitation vanishes for $t < 0$. The integral in Eq. (20.57) can be replaced by $\mathbf{F}(\omega_n,t)$; and after taking the imaginary part

$$\delta(\omega_n,t) = \frac{1}{\omega_n} F(\omega_n,t) \sin [\omega_n t + \theta(\omega_n,t)] \tag{20.59}$$

where $F(\omega_n,t)$ and $\theta(\omega_n,t)$ are the magnitude and phase of the running Fourier spectrum. Equation (20.59) provides the previously mentioned direct relationship between undamped structural response and the running Fourier spectrum.

When the running time t exceeds τ, the duration of $\ddot{u}(t)$, the running Fourier spectrum becomes the usual spectrum, with τ used in place of the infinite upper limit of the integral. Consequently, Eq. (20.59) yields the sinusoidal residual relative displacement for $t > \tau$:

$$\delta_r(\omega_n,t) = \frac{1}{\omega_n} F(\omega_n) \sin [\omega_n t + \theta(\omega_n)] \tag{20.60}$$

The amplitude of this residual deflection and the corresponding equivalent static acceleration are

$$\begin{aligned} (\delta_r)_{\text{max}} &= \frac{1}{\omega_n} F(\omega_n) \\ (A_{\text{eq}})_r &= \frac{\omega_n^2(\delta_r)_{\text{max}}}{g} = \frac{\omega_n}{g} F(\omega_n) \end{aligned} \tag{20.61}$$

This result is clearly evident for the Fourier spectrum and undamped shock response spectrum of the acceleration impulse. The Fourier spectrum is the horizontal line (independent of frequency) shown in Fig. 20.2*A* and the shock response spectrum is the inclined straight line (increasing linearly with frequency) shown in Fig. 20.5*A*. Since the impulse exists only at $t = 0$, the entire response is residual. The undamped shock spectra in the impulsive region of the half-sine pulse and the decaying sinusoidal acceleration, Fig. 20.5*C* and *D*, respectively, also are related to the Fourier spectra of these shocks, Fig. 20.5*C* and *D*, in a similar manner. This results from the fact that the maximum response occurs in the residual motion for systems with small natural frequencies. Another example is the entire negative shock response spectrum with no damping for the half-sine pulse in Fig. 20.5*C*, whose values are ω_n/g times the values of the Fourier spectrum in Fig. 20.2*C*.

REFERENCES

1. Ahlin, K.: "On the Use of Digital Filters for Mechanical System Simulation," *Proc. 74th Shock and Vibration Symposium*, San Diego, Calif., 2003.
2. Scavuzzo, R. J., and H. C. Pusey: "Principles and Techniques of Shock Data Analysis," *SVM-16*, 2d ed., Shock and Vibration Information Analysis Center, Arlington, Va., 1996.
3. Rubin, S.: *J. Appl. Mechanics*, **25**:501 (1958).
4. Fung, Y. C., and M. V. Barton: *J. Appl. Mechanics*, **25**:365 (1958).
5. Kern, D. L., et al.: "Dynamic Environmental Criteria," *NASA-HDBK-7005*, 2001.
6. Walsh, J. P., and R. E. Blake: *Proc. Soc. Exptl. Stress Anal.*, **6**(2):150 (1948).
7. Weaver, W., Jr., S. P. Timoshenko, and D. H. Young: "Vibration Problems in Engineering," 5th ed., John Wiley & Sons, New York, 1990.
8. ISO Standard 18431–4, 2007: "Mechanical Vibration and Shock—Signal Processing," Part 4: Shock Response Spectrum Analysis.

CHAPTER 21
EXPERIMENTAL MODAL ANALYSIS

Randall J. Allemang
David L. Brown

INTRODUCTION

Experimental modal analysis is the process of determining the modal parameters (natural frequencies, damping factors, modal vectors, and modal scaling) of a linear, time-invariant system. The modal parameters are often determined by analytical means, such as finite element analysis. One common reason for experimental modal analysis is the verification/correction of the results of the analytical approach. Often, an analytical model does not exist and the modal parameters determined experimentally serve as the model for future evaluations such as structural modifications. Predominately, experimental modal analysis is used to explain a dynamics problem (vibration or acoustic) whose solution is not obvious from intuition, analytical models, or previous experience.

The process of determining modal parameters from experimental data involves several phases. The success of the experimental modal analysis process depends upon having very specific goals for the test situation. Every phase of the process is affected by the goals which are established, particularly with respect to the errors associated with that phase. One possible delineation of these phases is as follows:

Modal analysis theory refers to that portion of classical vibrations that explains, theoretically, the existence of natural frequencies, damping factors, mode shapes, and modal scaling for linear systems. This theory includes both lumped-parameter, or discrete, models as well as continuous models that represent the distribution of mass, damping, and stiffness. Since most current modal parameter estimation methods are based upon frequency response functions (FRFs) or impulse response functions (IRFs), modal analysis theory also includes the theoretical definition of these functions with respect to mass, damping, and stiffness as well. Modal analysis theory also includes the concepts of real normal modes as well as complex modes of vibration as possible solutions for the modal parameters.[1–3]

Experimental modal analysis methods involve the theoretical relationship between measured quantities and the classic vibration theory often represented as matrix differential equations. All commonly used methods trace from the matrix differential equations but yield a final mathematical form in terms of measured raw input and output data in the time or frequency domains or some form of processed data such as FRFs or IRFs. Since most current modal parameter estimation methods are based upon FRFs or IRFs, experimental methods that are based upon these functions are of primary concern.

Modal data acquisition involves the practical aspects of acquiring the data that is required to serve as input to the modal parameter estimation phase. This data can be the raw time-domain input and output data or the processed data in terms of FRFs and IRFs. Much care must be taken to ensure that the data match the requirements of the theory as well as the requirements of the numerical algorithm involved in the modal parameter estimation. The theoretical requirements involve concerns such as system linearity as well as time invariance of system parameters. The numerical algorithms are particularly concerned with the bias errors in the data as well as with any overall dynamic range considerations.[4–7]

Modal parameter estimation is concerned with the practical problem of estimating the modal parameters, based upon a choice of mathematical model as justified by the experimental modal analysis method, from the measured data.[8–12]

Modal data presentation/validation is that process of providing a physical view or interpretation of the modal parameters. For example, this may simply

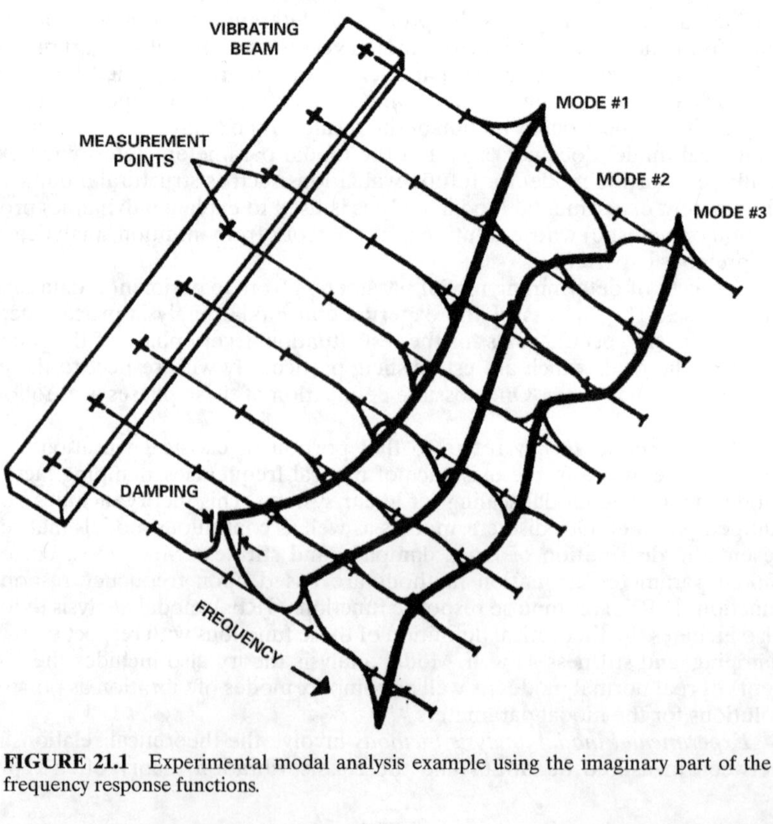

FIGURE 21.1 Experimental modal analysis example using the imaginary part of the frequency response functions.

be the numerical tabulation of the frequency, damping, and modal vectors, along with the associated geometry of the measured degrees of freedom (DOF). More often, modal data presentation involves the plotting and animation of such information.

Figure 21.1 is a representation of all phases of the process. In this example, a continuous beam is being evaluated for the first few modes of vibration. Modal analysis theory explains that this is a linear system and that the modal vectors of this system should be real normal modes. The experimental modal analysis method that has been used is based upon the FRF relationships to the matrix differential equations of motion. At each measured DOF, the imaginary part of the FRF for that measured response DOF and a common input DOF is superimposed perpendicular to the beam. Naturally, the modal data acquisition in this example involves the estimation of FRFs for each DOF shown. The FRFs are complex-valued functions, and only the imaginary portion of each function is shown. One method of modal parameter estimation suggests that for systems with light damping and widely spaced modes, the imaginary part of the FRF, at the damped natural frequency, may be used as an estimate of the modal coefficient for that response DOF. The damped natural frequency can be identified as the frequency of the positive and negative peaks in the imaginary part of the FRFs. The damping can be estimated from the sharpness of the peaks. In this abbreviated way, the modal parameters have been estimated. Modal data presentation for this case is shown as the lines connecting the peaks. While animation is possible, a reasonable interpretation of the modal vector can be gained in this case from plotting alone.

MEASUREMENT DEGREES OF FREEDOM

The development of any theoretical concept in the area of vibrations, including modal analysis, depends upon an understanding of the concept of the number of degrees of freedom (n) of a system. This concept is extremely important to the area of modal analysis, since the number of modes of vibration of a mechanical system is equal to the number of DOF. From a practical point of view, the relationship between this theoretical definition of the number of DOF and the number of *measurement DOF* (N_o, N_i) is often confusing. For this reason, the concept of degree of freedom is reviewed as a preliminary to the following experimental modal analysis material.

To begin with, the basic definition that is normally associated with the concept of the number of DOF involves the following statement: *The number of degrees of freedom for a mechanical system is equal to the number of independent coordinates (or minimum number of coordinates) that is required to locate and orient each mass in the mechanical system at any instant in time.* As this definition is applied to a point mass, three DOF are required, since the location of the point mass involves knowing the x, y, and z translations of the center of gravity of the point mass. As this definition is applied to a rigid-body mass, six DOF are required, since θ_x, θ_y, and θ_z rotations are required in addition to the x, y, and z translations in order to define both the orientation and the location of the rigid-body mass at any instant in time. As this definition is extended to any general deformable body, the number of DOF is essentially infinite. While this is theoretically true, it is quite common, particularly with respect to finite element methods, to view the general deformable body in terms of a large number of physical points of interest, with six DOF for each of the physical points. In this way, the infinite number of DOF can be reduced to a large but finite number.

When measurement limitations are imposed upon this theoretical concept of the number of DOF of a mechanical system, the difference between the theoretical number of DOF (n) and the number of measurement DOF (N_o, N_i) begins to evolve.

Initially, for a general deformable body, the number of DOF (n) can be considered to be infinite or equal to some large finite number if a limited set of physical points of interest is considered, as discussed in the previous paragraph. The first measurement limitation that needs to be considered is that there is normally a limited frequency range that is of interest to the analysis. As this limitation is considered, the number of DOF of this system that are of interest is now reduced from infinity to a reasonable finite number. The next measurement limitation that needs to be considered involves the physical limitation of the measurement system in terms of amplitude. A common limitation of transducers, signal conditioning, and data acquisition systems results in a dynamic range of 80 to 100 dB (10^4 to 10^5) in the measurement. This means that the number of DOF is reduced further due to the dynamic range limitations of the measurement instrumentation. Finally, since few rotational transducers exist at this time, the normal measurements that are made involve only translational quantities (displacement, velocity, acceleration, force) and thus do not include rotational effects, or rotational degrees of freedom (RDOF). In summary, even for the general deformable body, the theoretical number of DOF that are of interest is limited to a very reasonable finite value ($n = 1 - 50$). Therefore, this number of DOF (n) is the number of modes of vibration that are of interest.

Finally, then, the number of *measurement degrees of freedom* (N_o, N_i) can be defined as the number of physical locations at which measurements are made multiplied by the number of measurements made at each physical location. Since the physical locations are chosen somewhat arbitrarily, and certainly without exact knowledge of the modes of vibration that are of interest, there is no specific relationship between the number of DOF (n) and the number of measurement DOF (N_o, N_i). In general, in order to define n modes of vibration of a mechanical system, N_o or N_i must be equal to or larger than n. Note also that even though N_o or N_i is larger than n, this is not a guarantee that n modes of vibration can be found from the measurement DOF. The measurement DOF must include physical locations that allow a unique determination of the n modes of vibration. For example, if none of the measurement DOF are located on a portion of the mechanical system that is active in one of the n modes of vibration, portions of the modal parameters for this mode of vibration cannot be found.

In the development of material in the following text, the assumption is made that a set of measurement DOF exists and allows for n modes of vibration to be determined. In reality, either N_o or N_i is always chosen to be much larger than n, since a prior knowledge of the modes of vibration is not available. If the set of N_o or N_i measurement degrees of freedom is large enough and if the measurement DOF are distributed uniformly over the general deformable body, the n modes of vibration are normally found.

Throughout this experimental modal analysis reference, the frequency response function H_{pq} notation is used to describe the measurement of the response at measurement DOF p resulting from an input applied at measurement DOF q. The single subscript p or q refers to a single sensor aligned in a specific direction ($\pm X$, Y, or Z) at a physical location on or within the structure.

BASIC ASSUMPTIONS

There are four basic assumptions concerning any structure that are made in order to perform an experimental modal analysis.

The first basic assumption is that the structure is linear; that is, the response of the structure to any combination of forces, simultaneously applied, is the sum of the individual responses to each of the forces acting alone. For a wide variety of structures, this is a very good assumption. When a structure is linear, its behavior can be characterized by a controlled excitation experiment in which the forces applied to the structure have a form convenient for measurement and parameter estimation rather than being similar to the forces that are actually applied to the structure in its normal environment. For many important kinds of structures, however, the assumption of linearity is not valid. Where experimental modal analysis is applied in these cases, it is hoped that the linear model that is identified provides a reasonable approximation of the structure's behavior.

The second basic assumption is that the structure is time invariant; that is, the parameters that are to be determined are constants. In general, a system which is not time invariant has components whose mass, stiffness, or damping depends on factors that are not measured or are not included in the model. For example, some components may be temperature dependent. In this case, since temperature effects are not measured, the temperature of the component is an unknown time-varying signal. Hence, the component has time-varying characteristics. Therefore, the modal parameters determined by any measurement and estimation process for this case depend on the time (and the associated temperature dependence) when the measurements are made. If the structure that is tested changes with time, then measurements made at the end of the test period determine a different set of modal parameters than measurements made at the beginning of the test period. Thus, the measurements made at the two different times are inconsistent, violating the assumption of time invariance.

The third basic assumption is that the structure obeys Maxwell's reciprocity; that is, a force applied at degree of freedom p causes a response at DOF q that is the same as the response at DOF p caused by the same force applied at DOF q. With respect to frequency response function measurements, the FRF between points p and q determined by exciting at p and measuring the response at q is the same FRF found by exciting at q and measuring the response at p ($H_{pq} = H_{qp}$).

The fourth basic assumption is that the structure is observable; that is, the input/output measurements that are made contain enough information to generate an adequate behavioral model of the structure. Structures and machines which have loose components, or, more generally, which have DOF of motion that are not measured, are not completely observable. For example, consider the motion of a partially filled tank of liquid when complicated sloshing of the fluid occurs. Sometimes enough data can be collected so that the system is observable under the form chosen for the model, while at other times an impractical amount of data is required. This assumption is particularly relevant to the fact that the data normally describes an incomplete model of the structure. This occurs in at least two different ways. First, the data is normally limited to a minimum and maximum frequency as well as a limited frequency resolution. Second, no information is available relative to local rotations due to a lack of transducers available in this area.

MODAL ANALYSIS THEORY

While modal analysis theory has not changed over the last century, the application of the theory to experimentally measured data has changed significantly. The advances of recent years, with respect to measurement and analysis capabilities, have caused a reevaluation of the aspects of the theory that relate to the practical world of testing. With this in mind, the aspect of transform relationships has taken on renewed importance, since digital forms of the integral transforms are in constant use. The theory from the vibrations point of view involves a more thorough understanding of how the structural parameters of mass, damping, and stiffness relate to the impulse response function (time domain), the frequency response function (Fourier or frequency domain), and the transfer function (Laplace domain) for single- and multiple-degree-of-freedom systems.

SINGLE-DEGREE-OF-FREEDOM SYSTEMS

In order to understand modal analysis, the complete comprehension of single-degree-of-freedom systems is necessary. In particular, the complete familiarity with SDOF systems as presented and evaluated in the time, frequency (Fourier), and Laplace domains serves as the basis for many of the models that are used in modal parameter estimation. This SDOF approach is trivial from a modal analysis perspective, since no modal vectors exist. The true importance of this approach results from the fact that the MDOF case can be viewed as simply a linear superposition of SDOF systems.

The general mathematical representation of an SDOF system is expressed in Eq. (21.1):

$$m\,\ddot{x}(t) + c\,\dot{x}(t) + k\,x(t) = f(t) \tag{21.1}$$

where m = mass constant
c = damping constant
k = stiffness constant

This differential equation yields a characteristic equation of the following form:

$$m\,s^2 + c\,s + k = 0 \tag{21.2}$$

where s = complex-valued frequency variable (Laplace variable)

This characteristic equation of an SDOF system has two roots, λ_1 and λ_2, which are:

$$\lambda_1 = \sigma_1 + j\,\omega_1 \; \lambda_2 = \sigma_2 + j\,\omega_2 \tag{21.3}$$

where σ_1 = damping factor for mode 1
ω_1 = damped natural frequency for mode 1

Thus, the complementary solution of Eq. (21.1) is:

$$x(t) = Ae^{\lambda_1 t} + Be^{\lambda_2 t} \tag{21.4}$$

A and B are complex-valued constants determined from the initial conditions imposed on the system at time $t = 0$.

For most real structures, unless active damping systems are present, the damping factor is negative and the damping ratio is rarely greater than 10 percent. For this reason, all further discussion is restricted to underdamped systems ($\zeta < 1$). With reference to Eq. (21.2), this means that the two roots $\lambda_{1,2}$ are always complex conjugates. Also, the two coefficients A and B are complex conjugates of one another. For an underdamped system, the roots of the characteristic equation can be written as:

$$\lambda_1 = \sigma_1 + j\,\omega_1 \qquad \lambda_1^* = \sigma_1 - j\,\omega_1 \tag{21.5}$$

where σ_1 = damping factor
ω_1 = damped natural frequency

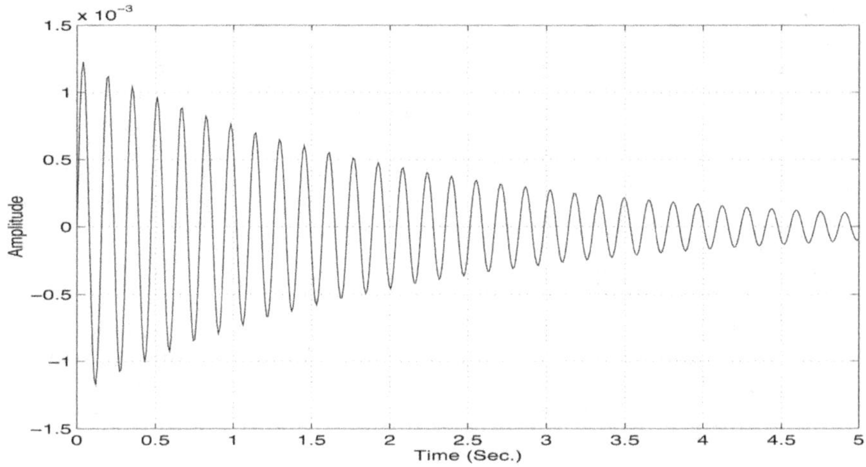

FIGURE 21.2 Single-degree-of-freedom impulse response function.

The roots of characteristic Eq. (21.2) can also be written as:

$$\lambda_1 = -\zeta_1\,\Omega_1 \pm j\,\Omega_1\,\sqrt{1-\zeta_1^2} \tag{21.6}$$

The *damping factor* is defined as the real part of a root of the characteristic equation. The damping factor describes the exponential decay or growth of the harmonic. This parameter has the same units as the imaginary part of the root of the characteristic equation, typically radians per second.

Time Domain: Impulse Response Function. The *impulse response function* of the single-degree-of-freedom system is defined as the time response [$x(t)$] of the system, assuming that the initial conditions are zero and that the system excitation $f(t)$ is a unit impulse. The response of the system $x(t)$ to such a unit impulse is known as the IRF $h(t)$ of the system. Therefore:

$$h(t) = A\,e^{\lambda_1 t} + A^*\,e^{\lambda_1^* t} = e^{\sigma_1 t}[A\,e^{(+j\omega_1 t)} + A^*\,e^{(-j\omega_1 t)}] \tag{21.7}$$

Thus, the residue A controls the amplitude of the impulse response, the real part of the pole is the decay rate, and the imaginary part of the pole is the frequency of oscillation. Figure 21.2 illustrates the IRF for an SDOF system.

Frequency Domain: Frequency Response Function. An equivalent equation of motion for Eq. (21.1) is determined for the Fourier or frequency (ω) domain. This representation has the advantage of converting a differential equation to an algebraic equation. This is accomplished by taking the Fourier transform of Eq. (21.1). Thus, Eq. (21.1) becomes:

$$[-m\,\omega^2 + j\,c\,\omega + k]X(\omega) = F(\omega) \tag{21.8}$$

Restating Eq. (21.8):

$$X(\omega) = H(\omega)\,F(\omega) \tag{21.9}$$

where $H(\omega) = \dfrac{1}{-m\,\omega^2 + j\,c\,\omega + k}$

Equation (21.9) states that the system response $X(\omega)$ is directly related to the system forcing function $F(\omega)$ through the quantity $H(\omega)$. If the system forcing function $F(\omega)$ and its response $X(\omega)$ are known, $H(\omega)$ can be calculated. That is:

$$H(\omega) = \frac{X(\omega)}{F(\omega)} \tag{21.10}$$

The quantity $H(\omega)$ is known as the *frequency response function* of the system. The FRF relates the Fourier transform of the system input to the Fourier transform of the system response.

The denominator of the FRF in Eq. (21.9) contains the *characteristic equation* of the system and is of the same form as Eq. (21.2). Note that the characteristic values of this complex equation are in general complex even though the equation is a function of a real-valued independent variable (ω). The characteristic values of this equation are known as the *complex roots* of the characteristic equation or the *complex poles* of the system. In terms of modal parameters, these characteristic values are also called the *modal frequencies.*

The FRF $H(\omega)$ can now be rewritten as a function of the complex poles as follows:

$$H(\omega) = \frac{1/m}{(j\,\omega - \lambda_1)(j\,\omega - \lambda_1^*)} \tag{21.11}$$

where λ_1 = complex pole
$\lambda_1 = \sigma + j\,\omega_1$
$\lambda_1^* = \sigma - j\,\omega_1$

Since the FRF is a complex-valued function of a real-valued independent variable (ω), the FRF, as shown in Fig. 21.3, is represented by a pair of curves.

Laplace Domain: Transfer Function. Just as in the previous case for the frequency domain, the equivalent information can be presented in the Laplace domain by way of the Laplace transform. The only significant difference in the development concerns the fact that the Fourier transform is defined from negative infinity to positive infinity, while the Laplace transform is defined from zero to positive infinity with initial conditions. The Laplace representation also has the advantage of converting a differential equation to an algebraic equation.

The transfer function is defined in the same way that the frequency response function is defined (assuming zero initial conditions).

$$X(s) = H(s)\,F(s) \tag{21.12}$$

where $H(s) = \dfrac{1}{m\,s^2 + c\,s + k}$

The quantity $H(s)$ is defined as the *transfer function* of the system. The transfer function relates the Laplace transform of the system input to the Laplace transform of the system response. From Eq. (21.12), the transfer function is defined as:

$$H(s) = \frac{X(s)}{F(s)} \tag{21.13}$$

The denominator of the transfer function is once again referred to as the *characteristic equation* of the system. As noted in the previous two cases, the roots of the characteristic equation are given in Eq. (21.5).

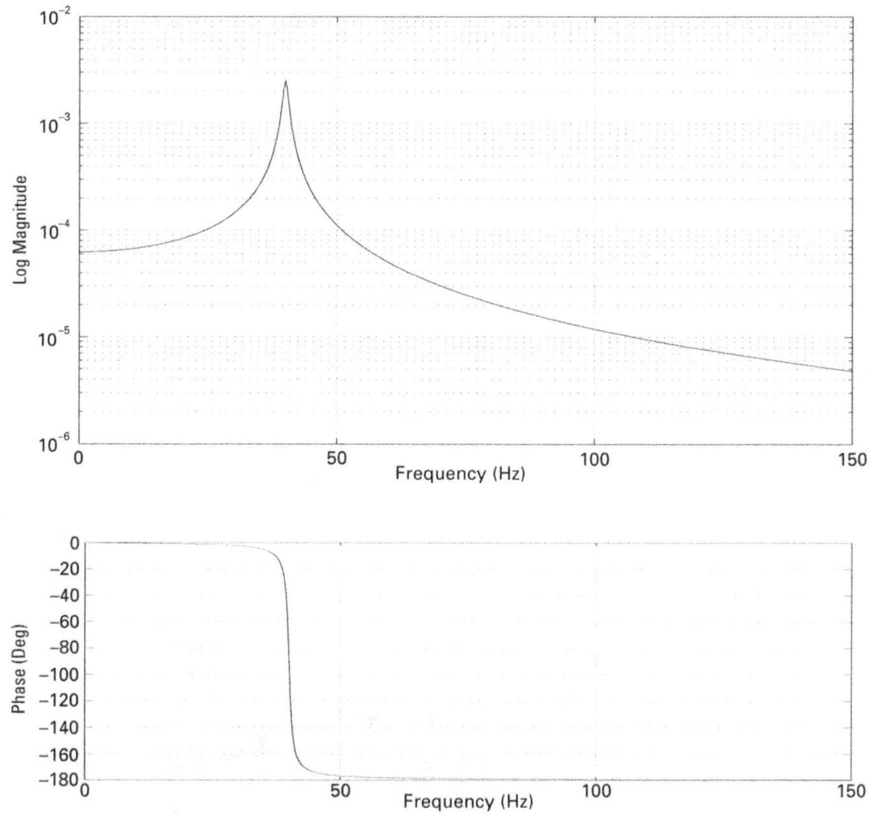

FIGURE 21.3 Single-degree-of-freedom frequency response function (log magnitude/phase format).

The transfer function $H(s)$ is now rewritten, just as in the FRF case, as:

$$H(s) = \frac{1/m}{(s - \lambda_1)(s - \lambda_1^*)} \tag{21.14}$$

Since the transfer function is a complex-valued function of a complex independent variable (s), the transfer function is represented, as shown in Fig. 21.4, as a pair of surfaces.

The definitions of undamped natural frequency, damped natural frequency, damping factor, percent of critical damping, and residue are all relative to the information represented by Fig. 21.4. The projection of this information onto the plane of zero amplitude yields the information as shown in Fig. 21.5.

The concept of residues is now defined in terms of the partial fraction expansion of the transfer function or frequency response function equation. Equation (21.14) is expressed in terms of partial fractions as follows:

$$H(s) = \frac{1/m}{(s - \lambda_1)(s - \mathrm{m}\,\lambda_1^*)} = \frac{A}{(s - \lambda_1)} + \frac{A^*}{(s - \lambda_1^*)} \tag{21.15}$$

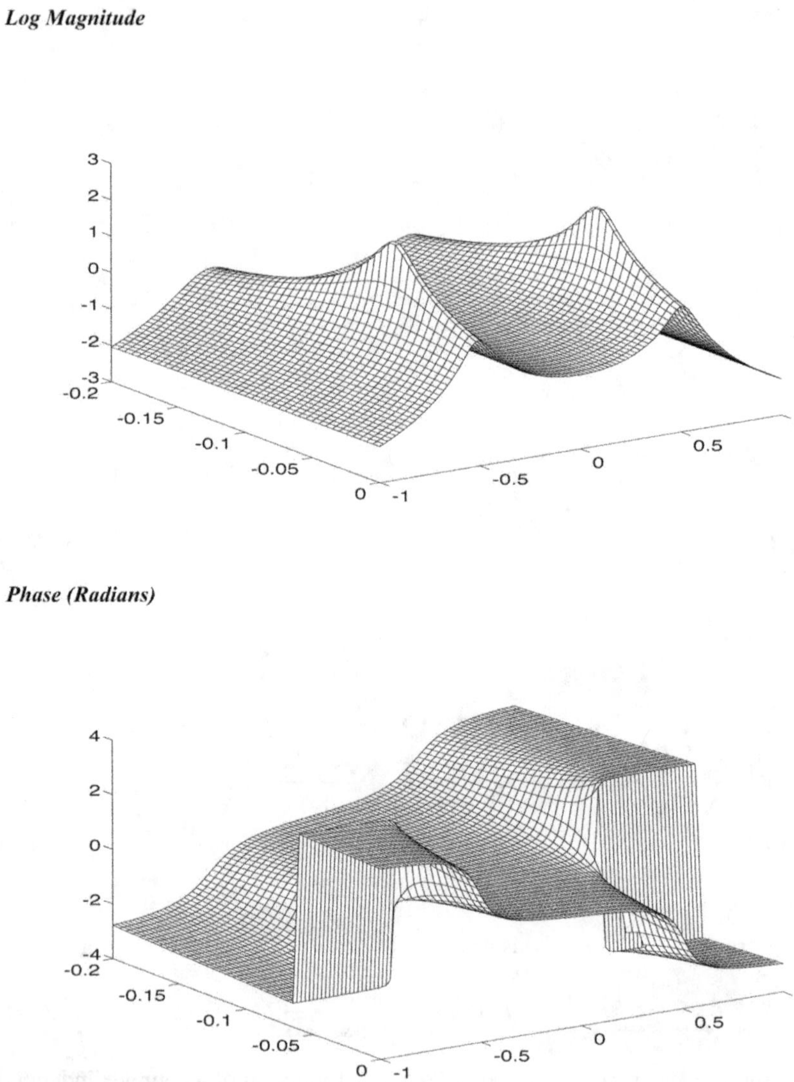

FIGURE 21.4 Single-degree-of-freedom transfer function (log magnitude/phase format).

The *residues* of the transfer function are defined as the constants A and A^*. The terminology and development of residues come from the evaluation of analytic functions in complex analysis. The residues of the transfer function are directly related to the amplitude of the impulse response function. In general, the residue A is a complex quantity. As shown for a single-degree-of-freedom system, A is purely imaginary.

From an experimental point of view, the transfer function is not estimated from measured input/output data. Instead, the FRF is actually estimated via the discrete Fourier transform.

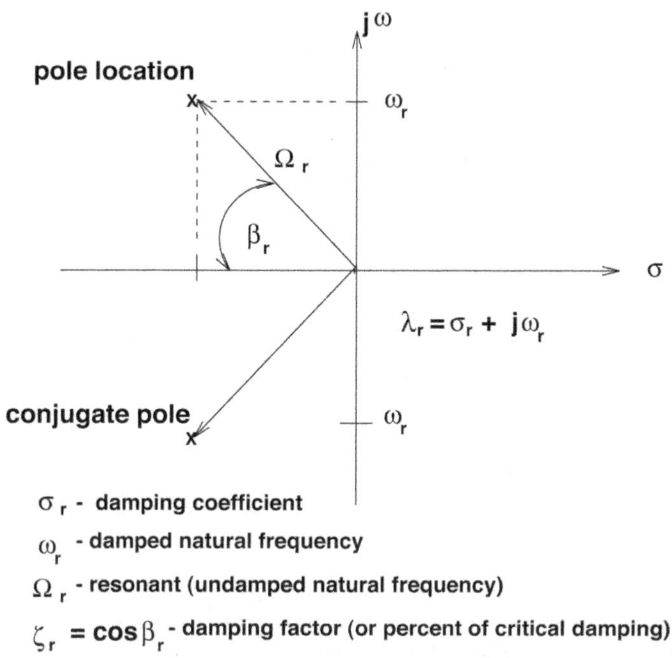

FIGURE 21.5 Transfer function (Laplace domain projection).

MULTIPLE-DEGREE-OF-FREEDOM SYSTEMS

Modal analysis concepts are applied when a continuous, nonhomogeneous structure is described as a lumped-mass, multiple-degree-of-freedom system with more than a single degree of freedom. The modal (natural) frequencies, the modal damping, the modal vectors or relative patterns of motion, and the modal scaling can be found from an estimate of the mass, damping, and stiffness matrices or from the measurement of the associated frequency response functions. From the experimental viewpoint, the relationship of modal parameters with respect to measured FRFs is most important.

The development of the FRF solution for the MDOF case parallels the SDOF case. This development relates the mass, damping, and stiffness matrices to a matrix transfer function model, or matrix frequency response function model, involving MDOF. Just as in the analytical case where the ultimate solution can be described in terms of 1-DOF systems, the FRFs between any input and response DOF can be represented as a linear superposition of the SDOF models derived previously.

As a result of the linear superposition concept, the equations for the impulse response function, the frequency response function, and the transfer function for the MDOF system are defined as follows:

Impulse response function:

$$h_{pq}(t) = \sum_{r=1}^{n} A_{pqr}\, e^{\lambda_r t} + A^*_{pqr}\, e^{\lambda^*_r t} \tag{21.16}$$

Frequency response function:

$$H_{pq}(\omega) = \sum_{r=1}^{n} \frac{A_{pqr}}{j\,\omega - \lambda_r} + \frac{A^*_{pqr}}{j\,\omega - \lambda^*_r} \tag{21.17}$$

Transfer function:

$$H_{pq}(s) = \sum_{r=1}^{n} \frac{A_{pqr}}{s - \lambda_r} + \frac{A^*_{pqr}}{s - \lambda^*_r} \tag{21.18}$$

where
t = time variable
ω = frequency variable
s = Laplace variable
p = measured degree of freedom (response)
q = measured degree of freedom (input)
r = modal vector number
A_{pqr} = residue
$A_{pqr} = Q_r \psi_{pr} \psi_{qr}$
Q_r = modal scaling factor
ψ_{pr} = modal coefficient
λ_r = system pole
n = number of modal frequencies

It is important to note that the residue A_{pqr} in Eqs. (21.16) through (21.18) is the product of the modal deformations at the input q and response p DOF and a modal scaling factor for mode r. Therefore, the product of these three terms is unique, but each of the three terms individually is not unique.

Damping Mechanisms. In order to evaluate multiple-degree-of-freedom systems that are present in the real world, the effect of damping on the complex frequencies and modal vectors must be considered. Many physical mechanisms are needed to describe all of the possible forms of damping that may be present in a particular structure or system. Some of the classical types are (1) structural damping, (2) viscous damping, and (3) coulomb damping. It is generally difficult to ascertain which type of damping is present in any particular structure. Indeed, most structures exhibit damping characteristics that result from a combination of all of these, plus others that have not been described here. (Damping is described in detail in Chap. 36.)

Rather than consider the many different physical mechanisms, the probable location of each mechanism, and the particular mathematical representation of the mechanism of damping that is needed to describe the dissipative energy of the system, a model is used that is only concerned with the resultant mathematical form. This model represents a hypothetical form of damping, which is proportional to the system mass or stiffness matrix. Therefore:

$$[C] = \alpha[M] + \beta[K] \tag{21.19}$$

Under this assumption, *proportional damping,* or what is historically referred to as Rayleigh damping, is the case where the equivalent damping matrix is equal to a linear combination of the mass and stiffness matrices. For this mathematical form of damping, the coordinate transformation that diagonalizes the system mass and stiffness matrices also diagonalizes the system damping matrix. *Nonproportional damping* is the case where this linear combination does not exist.

Therefore, when a system with proportional damping exists, that system of coupled equations of motion can be transformed to a system of equations that represent an uncoupled system of single-degree-of-freedom systems that are easily solved.

With respect to modal parameters, a system with proportional damping has real-valued modal vectors (*real or normal modes*), while a system with nonproportional damping has complex-valued modal vectors (*complex modes*).

Modal Scaling. *Modal scaling* refers to the relationship between the normalized modal vectors and the absolute scaling of the mass matrix (analytical case) and/or the absolute scaling of the residue information (experimental case). Modal scaling is normally presented as *modal mass* or *modal A*. The driving point residue A_{qqr} is particularly important in deriving the modal scaling.

$$A_{qqr} = Q_r \, \psi_{qr} \, \psi_{qr} = Q_r \, \psi_{qr}^2 \tag{21.20}$$

For undamped and proportionally damped systems, the rth modal mass of a multiple-degree-of-freedom system can be defined as:

$$M_r = \frac{1}{j\, 2\, Q_r\, \omega_r} = \frac{\psi_{pr}\, \psi_{qr}}{j\, 2\, A_{pqr}\, \omega_r} \tag{21.21}$$

where M_r = modal mass
Q_r = modal scaling constant
ω_r = damped natural frequency

If the largest scaled modal coefficient is equal to unity, Eq. (21.21) computes a quantity of modal mass that has physical significance. The physical significance is that the quantity of modal mass computed under these conditions is between zero and the total mass of the system. Therefore, under this scaling condition, the modal mass can be viewed as the amount of mass that is participating in each mode of vibration. For a translational rigid-body mode of vibration, the modal mass should be equal to the total mass of the system. The *modal mass* defined in Eq. (21.21) is developed in terms of displacement over force units. If measurements and, therefore, residues are developed in terms of any other units (velocity over force or acceleration over force), Eq. (21.21) has to be altered accordingly.

Once the modal mass is known, the *modal damping* (C_r) and *stiffness* (K_r) can be obtained through the following single-degree-of-freedom equations:

$$C_r = 2\, \sigma_r\, M_r \tag{21.22}$$

$$K_r = (\sigma_r^2 + \omega_r^2)\, M_r = \Omega_r^2\, M_r \tag{21.23}$$

For systems with nonproportional damping, modal mass cannot be used for modal scaling. For this case, and increasingly for undamped and proportionally damped cases as well, the *modal A* scaling factor is used as the basis for the relationship between the scaled modal vectors and the residues determined from the measured frequency response functions. This relationship is as follows:

$$M_{A_r} = \frac{\psi_{pr}\psi_{qr}}{A_{pqr}} = \frac{1}{Q_r} \tag{21.24}$$

Note that this definition of modal A is also developed in terms of displacement over force units. Once the modal A is known, *modal B* (M_{B_r}) can be obtained through the following SDOF equation:

$$M_{B_r} = -\lambda_r\, M_{A_r} \tag{21.25}$$

For undamped and proportionally damped systems, the relationship between the modal mass and the modal A scaling factors can be uniquely determined as

$$M_{A_r} = \pm j 2 M_r \omega_r \tag{21.26}$$

In general, the modal vectors are considered to be dimensionless, since they represent relative patterns of motion. Therefore, the modal mass or modal A scaling terms carry the units of the respective measurement. For example, the development of the frequency response is based upon displacement over force units. The residue must have units of length over force-seconds. Since the modal A scaling coefficient is inversely related to the residue, modal A has units of force-seconds over length. This unit combination is the same as mass over seconds. Likewise, since modal mass is related to modal A, for proportionally damped systems through a direct relationship involving the damped natural frequency, the units on modal mass are mass units as expected.

EXPERIMENTAL MODAL ANALYSIS METHODS

In order to understand the various experimental approaches used to determine the modal parameters of a structure, some sort of outline of the various techniques is helpful in categorizing the different methods that have been developed over the last 50 years. One of several overlapping approaches can be used. One approach is to group the methods according to whether one mode or multiple modes are excited at one time. The terminology that is used for this is:

- Phase resonance (single mode)
- Phase separation (multiple mode)

A slightly more detailed approach is to group the methods according to the type of measured data that is acquired. When this approach is utilized, the relevant terminology is:

- Sinusoidal input/output model (forced normal mode)
- Frequency response function model
- Damped complex exponential response model
- General input/output model

A very common concept in comparing and contrasting experimental modal analysis methodologies that is often used in the literature is based upon the type of model that is used in the modal parameter estimation stage. The relevant nomenclature for this approach is:

- Parametric model
  - Modal model
  - [M], [K], [C] model
- Nonparametric model

The different experimental modal analysis approaches may be grouped according to the domain in which the modal parameter estimation model is formulated. The relevant nomenclature for this approach is:

- Time domain
- Frequency domain
- Spatial domain

Finally, many specialized methods have been developed in order to experimentally estimate modal parameters under difficult measurement conditions. Modal analysis of rotating systems is of great interest to the rotor machinery industries, particularly when high rotation speeds are involved. In this case, the natural frequencies are modulated by the rotational speed and are changed by speed-related stiffening effects. Modal analysis of systems under natural excitation, or what is called *response-only modal analysis,* is of particular interest to the civil engineering area, where structures like buildings, bridges, and off-shore platforms are too large to easily excite with traditional methods. However, these structures are excited by natural excitation such as wind or waves or by operational excitations that are random, such as traffic excitation on a bridge. Under specific assumptions concerning the nature of these excitations, experimental modal analysis can be performed using cross-correlation or cross-spectrum functions. These two methods have received great attention over the past several years, and many references may be found in the published literature on the specialized methods that have been developed for these situations. Further information about these and other specialized experimental modal analysis methods can be found in the *Proceedings of the International Modal Analysis Conference* (http://www.sem.org) and the *Proceedings of the International Conference on Noise and Vibration* (http://www.isma-isaac.be/).

Regardless of the approach used to organize or classify the different approaches to generating modal parameters from experimental data, the fundamental underlying theory is always the same. The differences largely are a matter of logistics, user experience requirements, and numerical or compute limitations rather than a fundamentally superior or inferior method. Most current methodology is based upon measured frequency response or impulse response functions. Further discussion of experimental modal analysis is limited to techniques related to the measurement and use of these functions for determining modal parameters. The most widely utilized methods are discussed in detail in a later section entitled "Modal Parameter Estimation."

MODAL DATA ACQUISITION

Acquisition of data that is used in the formulation of a modal model involves many important technical concerns. The primary concern is the digital signal processing or the converting of analog signals into a corresponding sequence of digital values that accurately describe the time-varying characteristics of the inputs to and responses from a system. Once the data is available in digital form, the most common approach is to transform the data from the time domain to the frequency domain by use of a discrete Fourier transform algorithm. Since this algorithm involves discrete data over a limited time period, there are large potential problems with this approach that must be well understood. Data acquisition and analysis is discussed in detail in Chaps. 10 through 15 and Chap. 19.

MEASUREMENT FORMULATION

For current approaches to experimental modal analysis, the frequency response function is the most important, and most common, measurement to be made. When estimating FRFs, a measurement model is needed that allows the FRF to be estimated from measured input and output data in the presence of noise (errors). These basic error concepts have been discussed in other chapters in great detail.

There are at least four different testing configurations that can be considered with respect to modal parameter estimation. These different testing conditions are largely a function of the number of acquisition channels or excitation sources that are available to the test engineer.

- Single input/single output (SISO)
- Single input/multiple output (SIMO)
- Multiple input/single output (MISO)
- Multiple input/multiple output (MIMO)

In general, the best testing situation is the MIMO configuration, since the data is collected in the shortest possible time with the fewest changes in the test conditions.

FREQUENCY RESPONSE FUNCTION ESTIMATION

The estimation of the frequency response function depends upon the transformation of data from the time to the frequency domain. The Fourier transform is used for this computation. The computation is performed digitally using a fast Fourier transform algorithm. The FRF(s) satisfies the following single- and multiple-input relationships:

Single-input relationship:

$$X_p = H_{pq} F_q \tag{21.27}$$

Multiple-input relationship:

$$\begin{bmatrix} X_1 \\ X_2 \\ \cdot \\ \cdot \\ X_p \end{bmatrix}_{N_o \times 1} = \begin{bmatrix} H_{11} & \cdots\cdots & H_{1q} \\ H_{21} & & \cdot \\ \cdot & & \cdot \\ \cdot & & \cdot \\ H_{p1} & \cdots\cdots & H_{pq} \end{bmatrix}_{N_o \times N_i} \begin{bmatrix} F_1 \\ F_2 \\ \cdot \\ \cdot \\ F_q \end{bmatrix}_{N_i \times 1} \tag{21.28}$$

The most reasonable, and most common, approach to the estimation of FRFs is by use of *least-squares* (LS) or *total-least-squares* (TLS) techniques.[4,8,9] This is a standard technique for estimating parameters in the presence of noise. Least-squares methods minimize the square of the magnitude error and, thus, compute the *best* estimate of the magnitude of the FRF but have little effect on the phase of the FRF. The estimation of FRFs and the practical details concerning the use of these measurements for modal parameter estimation are discussed in detail in Chaps. 13 through 15 and Chap. 19, and also in Refs. 13–18.

MODAL PARAMETER ESTIMATION

Modal parameter estimation is a special case of system identification where the a priori model of the system is known to be in the form of modal parameters. Over the past 20 years, a number of algorithms have been developed to estimate modal param-

TABLE 21.1 Modal Parameter Estimation Algorithm Acronyms

Modal Parameter Estimation Algorithms	
CEA	Complex exponential algorithm[19]
LSCE	Least-squares complex exponential[19]
PTD	Polyreference time domain[20,21]
ITD	Ibrahim time domain[22,23]
MRITD	Multiple-reference Ibrahim time domain[24]
ERA	Eigensystem realization algorithm[25–27]
PFD	Polyreference frequency domain[28–30]
SFD	Simultaneous frequency domain[31]
MRFD	Multireference frequency domain[32]
RFP	Rational fraction polynomial[33]
OP	Orthogonal polynomial[34–39]
PLSCF	Polyreference least-squares complex frequency[40–42]
CMIF	Complex mode indication function[43]

eters from measured frequency or impulse response function data. While most of these individual algorithms, summarized in Table 21.1, are well understood, the comparison of one algorithm to another has become one of the thrusts of current research in this area. Comparison of the different algorithms is possible when the algorithms are reformulated using a common mathematical structure.

This reformulation attempts to characterize different classes of modal parameter estimation techniques in terms of the structure of the underlying matrix polynomials rather than the physically based models used historically. Since the modal parameter estimation process involves a greatly overdetermined problem (more data than independent equations), this reformulation is helpful in understanding the different numerical characteristics of each algorithm and, therefore, the slightly different estimates of modal parameters that each algorithm yields. As a part of this reformulation of the algorithms, the development of a conceptual understanding of modal parameter estimation technology has emerged. This understanding involves the ability to conceptualize the measured data in terms of the concept of *characteristic space,* the data domain (time, frequency, spatial), the evaluation of the order of the problem, the condensation of the data, and a common parameter estimation theory that can serve as the basis for developing any of the algorithms in use today. The following sections review these concepts as applied to the current modal parameter estimation methodology.

DEFINITION OF MODAL PARAMETERS

Modal identification involves estimating the modal parameters of a structural system from measured input/output data. Most current modal parameter estimation is based upon the measured data being the frequency response function or the equivalent impulse response function, typically found by inverse Fourier transforming the FRF. Modal parameters include the complex-valued modal frequencies (λ_r), modal vectors ($\{\psi_r\}$), and modal scaling (modal mass or modal A). Additionally, most current algorithms estimate modal participation vectors ($\{L_r\}$) and residue vectors ($\{A_r\}$) as part of the overall process. Modal participation vectors are a result of multiple reference modal parameter estimation algorithms and relate how well each modal vector is excited from each of the reference locations included in the measured data.

The combination of the modal participation vector ($\{L_r\}$) and the modal vector ($\{\psi_r\}$) for a given mode yields the residue matrix ($[A]_r$) for that mode.

In general, modal parameters are considered to be global properties of the system. The concept of global modal parameters simply means that there is only one answer for each modal parameter and that the modal parameter estimation solution procedure enforces this constraint. Most of the current modal parameter estimation algorithms estimate the modal frequencies and damping in a global sense, but very few estimate the modal vectors in a global sense.

SIMILARITIES IN MODAL PARAMETER ESTIMATION ALGORITHMS

Modal parameter estimation algorithms are similar in more ways than they are different. Fundamentally, all algorithms can be developed beginning with a linear, constant-coefficient, symmetric matrix model involving mass, damping, and stiffness. The common goal in all algorithms, therefore, is the development of a characteristic matrix coefficient equation that describes a linear, time-invariant, reciprocal mechanical system consistent with this theoretical background. This is the rationale behind using the *unified matrix polynomial approach* (UMPA) as the educational basis for demonstrating this common kernel for all modal parameter estimation algorithms.[44–46] The following sections discuss the similar concepts common to all widely used modal parameter estimation algorithms.

Linear Superposition. The current approach in modal identification involves using numerical techniques to separate the contributions of individual modes of vibration in measurements such as frequency response functions. The concept involves estimating the individual single-degree-of-freedom contributions to the multiple-degree-of-freedom measurement.

$$[H(\omega_i)]_{N_o \times N_i} = \sum_{r=1}^{n} \frac{[A_r]_{N_o \times N_i}}{j\omega_i - \lambda_r} + \frac{[A_r^*]_{N_o \times N_i}}{j\omega_i - \lambda_r^*} = \sum_{r=1}^{2n} \frac{[A_r]_{N_o \times N_i}}{j\omega_i - \lambda_r} \tag{21.29}$$

Equation (21.29) represents a mathematical problem that, at first observation, is nonlinear in terms of the unknown modal parameters. Once the modal frequencies (λ_r) are known, the mathematical problem is linear with respect to the remaining unknown modal parameters ($[A_r]$). For this reason, the numerical solution in many algorithms frequently involves two or more linear stages. Typically, the modal frequencies and modal participation vectors are found in a first stage and residues; modal vectors and modal scaling are determined in a second stage. This linear superposition concept is represented mathematically in Eq. (21.29) and graphically in Figs. 21.6 and 21.7.

While the model stated in Eq. (21.29) is fundamental to the linear superposition of individual SDOF contributions, this model is normally limited to being used as the basis for estimating the residues A_{pqr} once the modal frequencies (λ_r) are known.

Data Domain. Modal parameters can be estimated from a variety of different measurements that exist as discrete data in different data domains (time and/or frequency). These measurements can include free decays, forced responses, frequency response functions, or impulse response functions. These measurements can be processed one at a time or in partial or complete sets simultaneously. The measurements can be generated with no measured inputs, a single measured input, or multiple measured inputs. The data can be measured individually or simultaneously. There

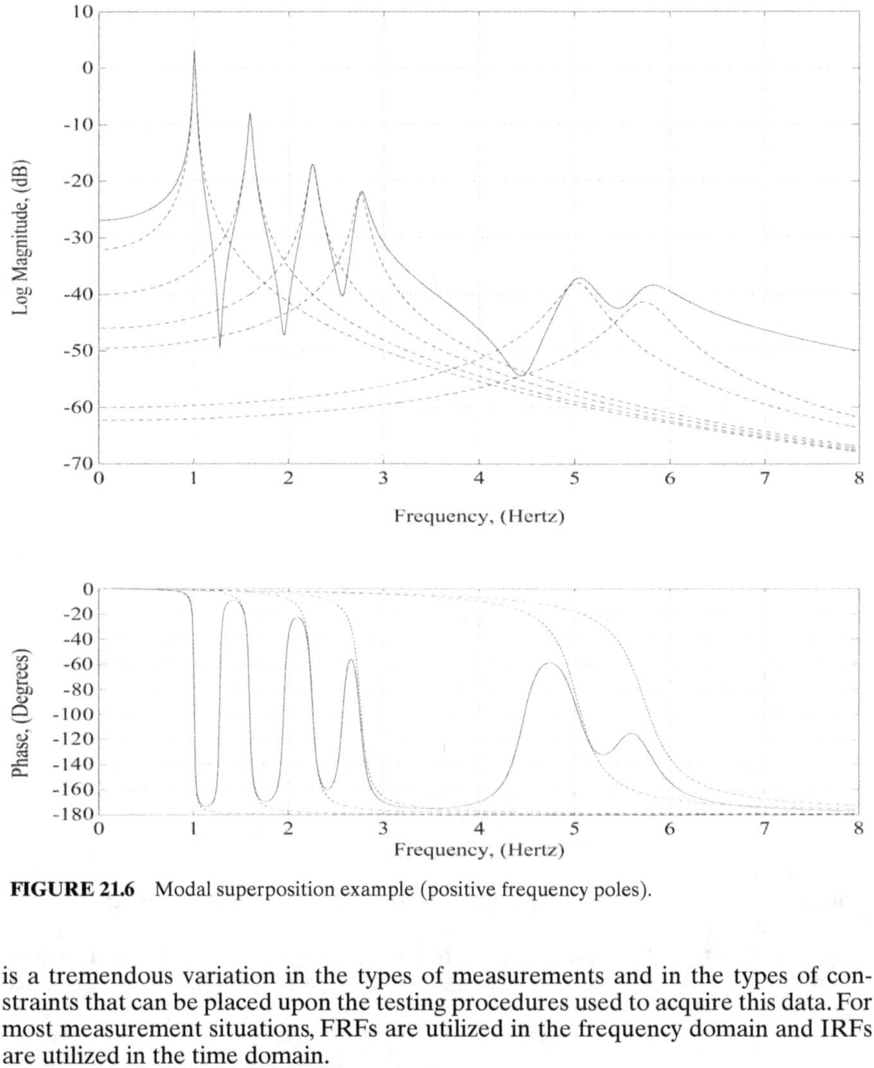

FIGURE 21.6 Modal superposition example (positive frequency poles).

is a tremendous variation in the types of measurements and in the types of constraints that can be placed upon the testing procedures used to acquire this data. For most measurement situations, FRFs are utilized in the frequency domain and IRFs are utilized in the time domain.

In terms of sampled data, the time-domain matrix polynomial results from a set of linear equations where each equation is formulated by choosing various distinct initial times. (Note, however, that the sampled nature of the time data requires that the evaluated coefficients, for each expressed linear equation, be uniformly spaced temporally, i.e., constant Δt.) Analogously, the frequency-domain matrix polynomial results from a set of linear equations where each equation is formulated at one of the frequencies of the measured data. This distinction is important to note, since the roots of the matrix characteristic equation formulated in the time domain are in a mapped complex domain (z_r), which is similar but not identical to the Z domain familiar to control theory. These mapped complex values (z_r) must be converted back to the frequency domain (λ_r), while the roots of the matrix characteristic equation formulated in the frequency domain (λ_r) are already in the desired domain.[46]

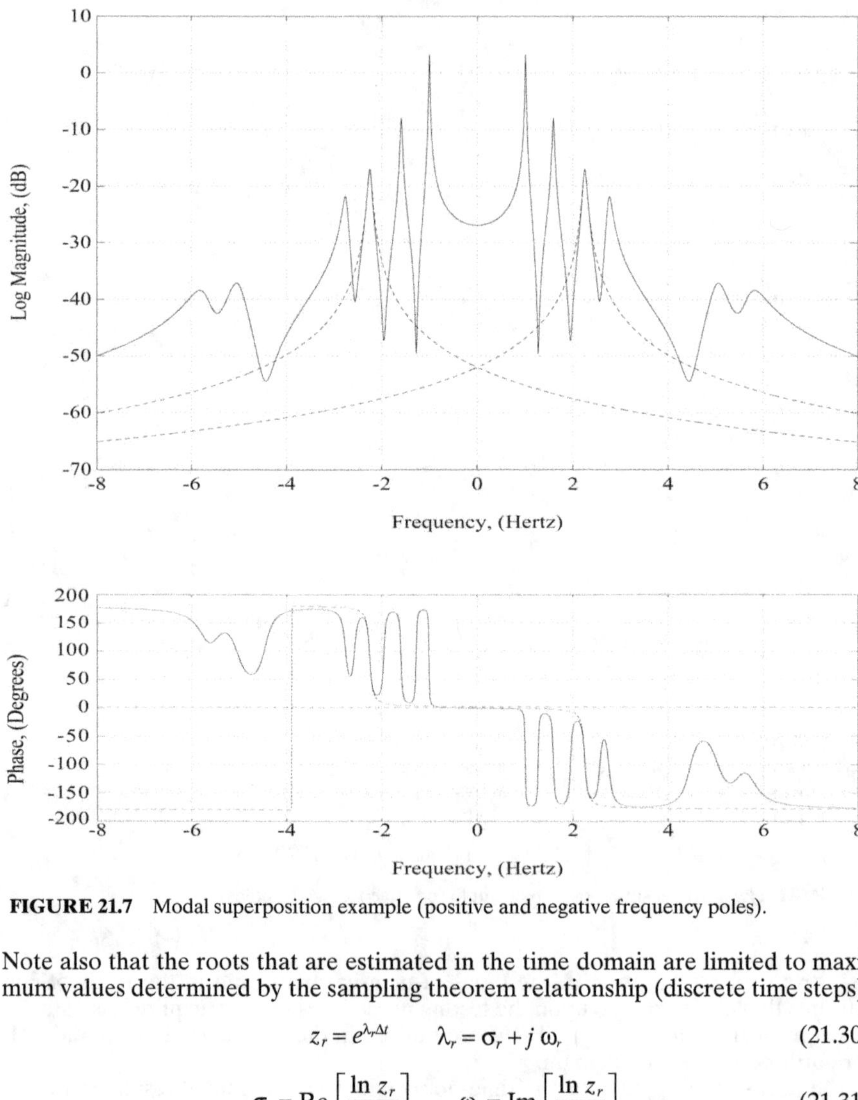

FIGURE 21.7 Modal superposition example (positive and negative frequency poles).

Note also that the roots that are estimated in the time domain are limited to maximum values determined by the sampling theorem relationship (discrete time steps).

$$z_r = e^{\lambda_r \Delta t} \qquad \lambda_r = \sigma_r + j\,\omega_r \tag{21.30}$$

$$\sigma_r = \text{Re}\left[\frac{\ln z_r}{\Delta t}\right] \qquad \omega_r = \text{Im}\left[\frac{\ln z_r}{\Delta t}\right] \tag{21.31}$$

Characteristic Space. From a conceptual viewpoint, the measurement space of a modal identification problem can be visualized as occupying a volume with the coordinate axes defined in terms of the three sets of characteristics. Two axes of the conceptual volume correspond to spatial information, and the third axis corresponds to temporal information. The spatial axes are in terms of the input and output degrees of freedom of the system. The temporal axis is either time or frequency, depending upon the domain of the measurements. These three axes define a 3-D volume, which is referred to as the *characteristic space.* This concept is represented in Fig. 21.8, where

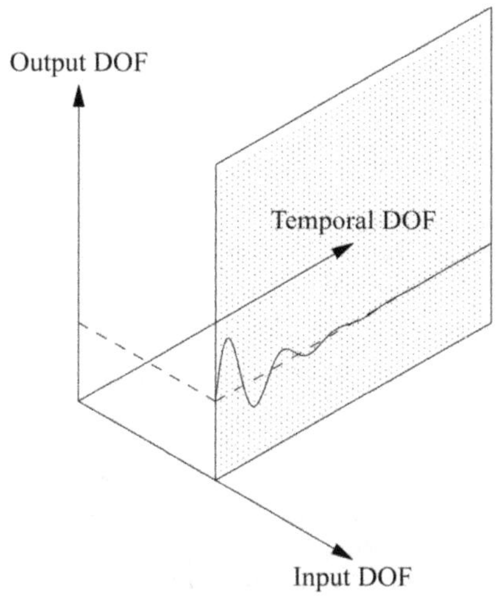

FIGURE 21.8 Conceptualization of modal characteristic space (input DOF axis, output DOF axis, time axis).

the shaded plane, or cut, through the 3-D characteristic space represents the measured (temporal) data from one input location and all output locations. Similar, and multiple, cuts through the 3-D characteristic space at right angles to the axes represent other measurement concepts.

This space or volume represents all possible measurement data. This conceptual representation is very useful in understanding the data subspace that has been measured. Also, this conceptual representation is very useful in recognizing how the data is organized and utilized with respect to different modal parameter estimation algorithms (3D to 2D). Information parallel to one axis consists of a superposition of the characteristics defined by that axis. The other two characteristics determine the scaling of each term in the superposition.

Any structural testing procedure measures a subspace of the total possible data available. Modal parameter estimation algorithms may then use all of this subspace or choose to further limit the data to a more restrictive subspace. It is theoretically possible to estimate the characteristics of the total space by measuring any subspace which samples all three characteristics. Measurement data spaces involving many planes of measured data are the best possible modal identification situations, since the data subspace includes contributions from temporal and spatial characteristics. The particular subspace which is measured and the weighting of the data within the subspace in an algorithm are the main differences between the various modal identification procedures which have been developed.

It should be obvious that the data which defines the subspace needs to be acquired in a consistent measurement process in order for the algorithms to estimate accurate modal parameters. This fact has triggered the need to measure all of the data simultaneously and has led to recent advancements in data acquisition, digital signal processing, and instrumentation designed to facilitate this measurement problem.

Fundamental (Historical) Measurement Models. Most current modal parameter estimation algorithms utilize frequency or impulse response functions as the data, or known information, to solve for modal parameters. The general equation that can be used to represent the relationship between the measured FRF matrix and the modal parameters is shown in Eq. (21.29) or, in the more common matrix product form, in Eqs. (21.32) and (21.33).

$$[H(\omega)]_{N_o \times N_i} = [\psi]_{N_o \times 2n} \left[\frac{1}{j\omega - \lambda_r} \right]_{2n \times 2n} [L]^T_{2n \times N_i} \tag{21.32}$$

$$[H(\omega)]^T_{N_i \times N_o} = [L]_{N_i \times 2n} \left[\frac{1}{j\omega - \lambda_r} \right]_{2n \times 2n} [\psi]^T_{2n \times N_o} \tag{21.33}$$

IRFs are rarely directly measured but are calculated from associated FRFs via the inverse fast Fourter transform (FFT) algorithm. The general equation that can be used to represent the relationship between the IRF matrix and the modal parameters is shown in Eqs. (21.34) and (21.35).

$$[h(t)]_{N_o \times N_i} = [\psi]_{N_o \times 2n} \, [e^{\lambda_r t}]_{2n \times 2n} \, [L]^T{}_{2n \times N_i} \tag{21.34}$$

$$[h(t)]^T{}_{N_i \times N_o} = [L]_{N_i \times 2n} \, [e^{\lambda_r t}]_{2n \times 2n} \, [\psi]^T{}_{2n \times N_o} \tag{21.35}$$

Many modal parameter estimation algorithms have been originally formulated from Eqs. (21.32) through (21.35). However, a more general development for all algorithms is based upon relating these equations to a general matrix coefficient polynomial model.

Fundamental (Current) Modal Identification Models. Rather than using a physically based mathematical model, the common characteristics of different modal parameter estimation algorithms can be more readily identified by using a matrix coefficient polynomial model. One way of understanding the basis of this model can be developed from the polynomial model used historically for the frequency response function. Note the nomenclature in the following equations regarding measured frequency ω_i and generalized frequency s_i. Measured input and response data are always functions of measured frequency, but the generalized frequency variable used in the model may be altered to improve the numerical conditioning. This will become important in a later discussion of generalized frequency involving normalized frequency, orthogonal polynomials, and complex Z mapping.

$$H_{pq}(\omega_i) = \frac{X_p(\omega_i)}{F_q(\omega_i)} = \frac{\beta_n\,(s_i)^n + \beta_{n-1}\,(s_i)^{n-1} + \cdots + \beta_1\,(s_i)^1 + \beta_0\,(s_i)^0}{\alpha_m\,(s_i)^m + \alpha_{m-1}\,(s_i)^{m-1} + \cdots + \alpha_1\,(s_i)^1 + \alpha_0\,(s_i)^0} \tag{21.36}$$

This can be rewritten:

$$H_{pq}(\omega_i) = \frac{X_p(\omega_i)}{F_q(\omega_i)} = \frac{\sum_{k=0}^{n} \beta_k\,(s_i)^k}{\sum_{k=0}^{m} \alpha_k\,(s_i)^k} \tag{21.37}$$

Further rearranging yields the following equation that is linear in the unknown α and β terms:

$$\sum_{k=0}^{m} \alpha_k\,(s_i)^k X_p(\omega_i) = \sum_{k=0}^{n} \beta_k\,(s_i)^k F_q(\omega_i) \tag{21.38}$$

This model can be generalized to represent the general multiple-input/multiple-output case as follows:

$$\sum_{k=0}^{m} \left[[\alpha_k](s_i)^k \right] \{X(\omega_i)\} = \sum_{k=0}^{n} \left[[\beta_k](s_i)^k \right] \{F(\omega_i)\} \tag{21.39}$$

Note that the size of the coefficient matrices $[\alpha_k]$ will normally be $N_i \times N_i$ or $N_o \times N_o$, and the size of the coefficient matrices $[\beta_k]$ will normally be $N_i \times N_o$ or $N_o \times N_i$ when the equations are developed from experimental data.

Rather than developing the basic model in terms of force and response information, the models can be stated in terms of power spectra or frequency response information. First, postmultiply both sides of the equation by $\{F\}^H$:

$$\sum_{k=0}^{m} \left[[\alpha_k](s_i)^k \right] \{X(\omega_i)\}\,\{F(\omega_i)\}^H = \sum_{k=0}^{n} \left[[\beta_k](s_i)^k \right] \{F(\omega_i)\}\,\{F(\omega_i)\}^H \tag{21.40}$$

Now recognize that the product of $\{X(\omega_i)\}\,\{F(\omega_i)\}^H$ is the output/input cross-spectra matrix ($[G_{XF}(\omega_i)]$) for one ensemble and $\{F(\omega_i)\}\,\{F(\omega_i)\}^H$ is the input/input cross-spectra matrix ($[G_{FF}(\omega_i)]$) for one ensemble. With a number of ensembles (averages), these matrices are the common matrices used to estimate the FRFs in a MIMO case. This yields the following cross-spectra model:

$$\sum_{k=0}^{m}\Big[[\alpha_k](s_i)^k\Big]\Big[G_{XF}(\omega_i)\Big] = \sum_{k=0}^{n}\Big[[\beta_k](s_i)^k\Big]\Big[G_{FF}(\omega_i)\Big] \tag{21.41}$$

The previous cross-spectra model can be reformulated to utilize FRF data by post-multiplying both sides of the equation by $[G_{FF}(\omega_i)]^{-1}$:

$$\sum_{k=0}^{m}\Big[[\alpha_k](s_i)^k\Big]\Big[G_{XF}(\omega_i)\Big]\Big[G_{FF}(\omega_i)\Big]^{-1} = \sum_{k=0}^{n}\Big[[\beta_k](s_i)^k\Big]\Big[G_{FF}(\omega_i)\Big]\Big[G_{FF}(\omega_i)\Big]^{-1} \tag{21.42}$$

Therefore, the MIMO FRF model is:

$$m\backslash k = 0\Big[[\alpha_k](s_i)^k\Big]\Big[H(\omega_i)\Big] = \sum_{k=0}^{n}\Big[[\beta_k](s_i)^k\Big][I] \tag{21.43}$$

This model, in the frequency domain, corresponds to an *autoregressive moving average,* or ARMA(m,n), model that is developed from a set of discrete time equations in the time domain. More properly, this model is known as the *autoregressive with exogenous inputs,* or ARX(m,n), model. The general matrix polynomial model concept recognizes that both the time- and the frequency-domain models generate functionally similar matrix polynomial models. For that reason, the *unified matrix polynomial approach* terminology is used to describe both domains, since the ARMA terminology has been connected primarily with the time domain. Additional equations can be developed by repeating Eq. (21.43), Eq. (21.39), or Eq. (21.41) at many frequencies (ω_i) until all data or a sufficient overdetermination factor is achieved. Note that both positive and negative frequencies are required in order to accurately estimate conjugate modal frequencies. Further details concerning specific frequency-domain algorithms can be found in later sections.

Paralleling the development of Eqs. (21.36) through (21.43), a time-domain model representing the relationship between a single-response degree of freedom and a single-input degree of freedom can be stated as follows:

$$\sum_{k=0}^{m}\alpha_k\,x(t_{i+k}) = \sum_{k=0}^{n}\beta_k\,f(t_{i+k}) \tag{21.44}$$

For the general MIMO case:

$$\sum_{k=0}^{m}[\alpha_k]\{x(t_{i+k})\} = \sum_{k=0}^{n}[\beta_k]\,\{f(t_{i+k})\} \tag{21.45}$$

If the discussion is limited to the use of free decay or impulse response function data, the previous time-domain equations can be simplified by noting that the forcing function can be assumed to be zero for all time greater than zero. If this is the case, the $[\beta_k]$ coefficients can be eliminated from the equations.

$$\sum_{k=0}^{m}[\alpha_k]\Big[h(t_{i+k})\Big] = 0 \tag{21.46}$$

Additional equations can be developed by repeating Eq. (21.46) at different time shifts into the data (t_i) until all data or a sufficient overdetermination factor is achieved. Note that at least one time shift is required in order to accurately estimate conjugate modal frequencies. Further details concerning specific time-domain algorithms can be found in later sections.

In light of the preceding discussion, it is now apparent that most of the modal parameter estimation processes available could have been developed by starting from a general matrix polynomial formulation that is justifiable based upon the underlying matrix differential equation. The general matrix polynomial formulation yields essentially the same characteristic matrix polynomial equation for both time- and frequency-domain data. For the frequency-domain data case, this yields:

$$\left| [\alpha_m]\, s^m + [\alpha_{m-1}]\, s^{m-1} + [\alpha_{m-2}]\, s^{m-2} + \cdots + [\alpha_0] \right| = 0 \tag{21.47}$$

For the time-domain data case, this yields:

$$\left| [\alpha_m]\, z^m + [\alpha_{m-1}]\, z^{m-1} + [\alpha_{m-2}]\, z^{m-2} + \cdots + [\alpha_0] \right| = 0 \tag{21.48}$$

Once the matrix coefficients ($[\alpha]$) have been found, the modal frequencies (λ_r or z_r) can be found using a number of numerical techniques. While in certain numerical situations other numerical approaches may be more robust, a companion matrix approach yields a consistent concept for understanding the process. Therefore, the roots of the matrix characteristic equation can be found as the eigenvalues of the associated companion matrix. The companion matrix can be formulated in one of several ways. The most common formulation is as follows:

$$[C] = \begin{bmatrix} -[\alpha]_{m-1} & -[\alpha]_{m-2} & \cdots & \cdots & \cdots & -[\alpha]_1 & -[\alpha]_0 \\ [I] & [0] & \cdots & \cdots & \cdots & [0] & [0] \\ [0] & [I] & \cdots & \cdots & \cdots & [0] & [0] \\ [0] & [0] & \cdots & \cdots & \cdots & [0] & [0] \\ \cdots & \cdots & \cdots & \cdots & \cdots & \cdots & \cdots \\ \cdots & \cdots & \cdots & \cdots & \cdots & \cdots & \cdots \\ \cdots & \cdots & \cdots & \cdots & \cdots & \cdots & \cdots \\ [0] & [0] & \cdots & \cdots & \cdots & [0] & [0] \\ [0] & [0] & \cdots & \cdots & \cdots & [0] & [0] \\ [0] & [0] & \cdots & \cdots & \cdots & [I] & [0] \end{bmatrix} \tag{21.49}$$

Note again that the numerical characteristics of the eigenvalue solution of the companion matrix will be different for low-order cases compared to high-order cases for a given data set. The companion matrix can be used in the following eigenvalue formulation to determine the modal frequencies for the original matrix coefficient equation:

$$[C]\{X\} = \lambda[I]\{X\} \tag{21.50}$$

The eigenvectors that can be found from the eigenvalue-eigenvector solution utilizing the companion matrix may or may not be useful in terms of modal parameters. The eigenvector that is found, associated with each eigenvalue, is of length model order m times matrix coefficient size N_i or N_o. In fact, the unique (meaningful) portion of the eigenvector is of length equal to the size of the coefficient matrices N_i or N_o, and is repeated in the eigenvector $m + 1$ times. For each repetition, the unique portion of the eigenvector is repeated, multiplied by a different complex scalar which is a successively larger integer power of the associated modal frequency. Therefore, the eigenvectors of the companion matrix have the following form:

$$\{\phi\}_r = \begin{Bmatrix} \lambda_r^m\{\psi\}_r \\ \cdot \\ \cdot \\ \cdot \\ \lambda_r^2\{\psi\}_r \\ \lambda_r^1\{\psi\}_r \\ \lambda_r^0\{\psi\}_r \end{Bmatrix}_r \tag{21.51}$$

Note that unless the size of the coefficient matrices is at least as large as the number of measurement degrees of freedom, only a partial set of modal coefficients, the modal participation coefficients (L_{qr}), will be found. For the case involving scalar polynomial coefficients, no meaningful modal coefficients will be found.

If the size of the coefficient matrices, and therefore the modal participation vector, is less than the largest spatial dimension of the problem, then the modal vectors are typically found in a second-stage solution process using one of Eq. (21.29) or Eq. (21.32) through Eq. (21.35). Even if the complete modal vector ($\{\psi\}$) of the system is found from the eigenvectors of the companion matrix approach, the modal scaling and modal participation vectors for each modal frequency are normally found in this second-stage formulation.

Model Order Relationships. From a theoretical consideration, the number of characteristic values (number of modal frequencies, number of roots, number of poles, etc.) that can be determined depends upon the size of the matrix coefficients involved in the model and the order of the polynomial terms in the model. The characteristic matrix polynomial equation, Eq. (21.47) or Eq. (21.48), has a model order of m, and the number of modal frequencies or roots that will be found from this characteristic matrix polynomial equation will be m times the size of the coefficient matrices $[\alpha]$.

For a given algorithm, the size of the matrix coefficients is normally fixed; therefore, determining the model order is directly linked to estimating n, the number of modal frequencies that are of interest in the measured data. As has always been the case, an estimate for the minimum number of modal frequencies can be easily found by counting the number of peaks in the frequency response function in the frequency band of analysis. This is a minimum estimate of n, since the FRF measurement may be at a node of one or more modes of the system, repeated roots may exist, and/or the frequency resolution of the measurement may be too coarse to observe modes that are closely spaced in frequency. Several measurements can be observed and a tabulation of peaks existing in any or all measurements can be used as a more accurate minimum estimate of n. A more automated procedure for including the peaks that are present in several FRFs is to observe the summation of FRF power. This function represents the autopower or automoment of the FRFs summed over a number of response measurements and is normally formulated as follows:

$$H_{\text{power}}(\omega) = \sum_{p=1}^{N_O} \sum_{q=1}^{N_i} H_{pq}(\omega)\, H_{pq}{}^*(\omega) \tag{21.52}$$

These simple techniques are extremely useful but do not provide an accurate estimate of model order when repeated roots exist or when modes are closely spaced in frequency. For these reasons, an appropriate estimate of the order of the model is of prime concern and is the single most important problem in modal parameter estimation.

In order to determine a reasonable estimate of the model order for a set of representative data, a number of techniques have been developed as guides or aids to users. Much of the user interaction involved in modal parameter estimation involves the use of these tools. Most of the techniques that have been developed allow users to establish a maximum model order to be evaluated (in many cases, this is set by the memory limits of the computer algorithm). Data is acquired based upon an assumption that the model order is equal to this maximum. In a sequential fashion, this data is evaluated to determine whether a model order less than the maximum will describe the data sufficiently. This is the point that the user's judgment and the use of various evaluation aids become important. Some of the commonly used techniques are mode indicator functions, consistency (stability) diagrams, and pole surface density plots.

Mode indication functions (MIFs) are normally real-valued, frequency-domain functions that exhibit local minima or maxima at the natural frequencies of the modes. One MIF can be plotted for each reference available in the measured data. The primary MIF will exhibit a local minimum or maximum at each of the natural frequencies of the system under test. The secondary MIF will exhibit a local minimum or maximum at repeated or pseudo-repeated roots of order two or more. Further MIFs yield local minima or maxima for successively higher orders of repeated or pseudo-repeated roots of the system under test.

The *multivariate mode indication function* (MvMIF) is based upon finding a force vector $\{F\}$ that will excite a normal mode at each frequency in the frequency range of interest.[47] If a normal mode can be excited at a particular frequency, the response to such a force vector will exhibit the 90° phase-lag characteristic. Therefore, the real part of the response will be as small as possible, particularly when compared to the imaginary part or the total response. In order to evaluate this possibility, a minimization problem can be formulated as follows:

$$\min_{\|F\|=1} \frac{\{F\}^T [H_{\text{real}}]^T [H_{\text{real}}] \{F\}}{\{F\}^T \left([H_{\text{real}}]^T [H_{\text{real}}] + [H_{\text{imag}}]^T [H_{\text{imag}}] \right) \{F\}} = \varepsilon \tag{21.53}$$

This minimization problem is similar to a Rayleigh quotient, and it can be shown that the solution to the problem is executed by finding the smallest eigenvalue $\varepsilon_{\min}$ and the corresponding eigenvector $\{F\}_{\min}$ of the following problem:

$$[H_{\text{real}}]^T [H_{\text{real}}] \{F\} = \lambda \left([H_{\text{real}}]^T [\text{H}_{\text{real}}] + [H_{\text{imag}}]^T [H_{\text{imag}}] \right) \{F\} \tag{21.54}$$

This eigenvalue problem is formulated at each frequency in the frequency range of interest. Note that the result of the matrix product $[H_{\text{real}}]^T [H_{\text{real}}]$ and $[H_{\text{imag}}]^T [H_{\text{imag}}]$ in each case is a square, real-valued matrix of size equal to the number of references in the measured data $N_i \times N_i$. The resulting plot of the MvMIF for a seven-reference case can be seen in Fig. 21.9.

An algorithm based on singular value decomposition (SVD) methods applied to multiple reference FRF measurements, identified as the *complex mode indication function* (CMIF), was first developed for traditional FRF data in order to identify the proper number of modal frequencies, particularly when there are closely spaced or repeated modal frequencies.[43] Unlike the MvMIF, which indicates the existence of real normal modes, the CMIF indicates the existence of real normal or complex modes and the relative magnitude of each mode. Furthermore, MvMIF yields a set

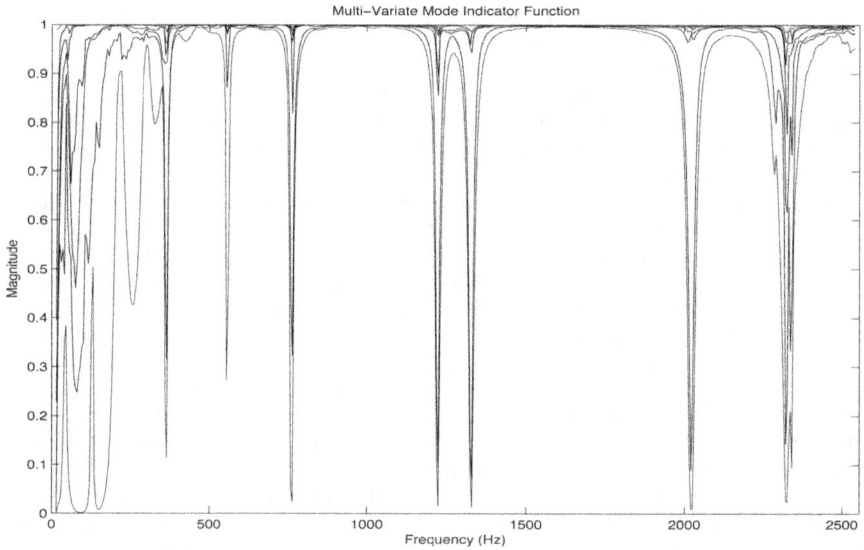

FIGURE 21.9 Multivariate mode indication function: seven-input example.

of force patterns that can best excite the real normal mode, while CMIF yields the corresponding mode shape and modal participation vector.

The CMIF is defined as the economical singular values, computed from the FRF matrix at each spectral line. The CMIF is the plot of these singular values on a log magnitude scale as a function of frequency. The peaks detected in the CMIF plot indicate the existence of modes, and the corresponding frequencies of these peaks give the damped natural frequencies for each mode. In the application of CMIF to traditional modal parameter estimation algorithms, the number of modes detected in CMIF determines the minimum number of degrees of freedom of the system equation for the algorithm. A number of additional DOF may be needed to take care of residual effects and noise contamination.

$$[H(\omega)] = [U(\omega)]\,[\Sigma(\omega)]\,[V(\omega)]^H \tag{21.55}$$

Most often, the number of input points (reference points) N_i is less than the number of response points N_o. In Eq. (21.55), if the number of effective modes is less than or equal to the smaller dimension of the FRF matrix (i.e., $N_{e \le N}i$), the SVD leads to approximate mode shapes (left singular vectors) and approximate modal participation factors (right singular vectors). The singular value is then equivalent to the scaling factor Q_r divided by the difference between the discrete frequency and the modal frequency $(j\omega - \lambda_r)$. For a given mode, since the scaling factor is a constant, the closer the modal frequency is to the discrete frequency, the larger the singular value will be. Therefore, the damped natural frequency is the frequency at which the maximum magnitude of the singular value occurs. If different modes are compared, the stronger the modal contribution (larger residue value), the larger the singular value will be. The peak in the CMIF indicates the location on the frequency axis that is nearest to the pole. The frequency is the estimated damped natural frequency, to within the accuracy of the frequency resolution.

Since the mode shapes that contribute to each peak do not change much around each peak, several adjacent spectral lines from the FRF matrix can be used simultaneously for a better estimation of mode shapes. By including several spectral lines of data in the SVD calculation, the effect of the leakage error can be minimized. If only the quadrature (imaginary) part of the FRF matrix is used in the CMIF, the singular values will be much more distinct. The resulting plot of the CMIF for a seven-reference case can be seen in Fig. 21.10.

Consistency Diagrams. One of the most common methods for determining the number of modes present in the measurement data is the use of consistency, formerly referred to as stability, diagrams. The *consistency diagram* is developed by successively computing different model solutions (utilizing different model order for the characteristic polynomial, different normalization methods for the characteristic matrix coefficient polynomial, different equation condensation methods, and/or different algorithms) and involves tracking the estimates of frequency, damping, and, possibly, modal participation factors as a function of model solution iteration. If only model order is evaluated, recent research[41,42] has shown that the correct choice of normalization of the characteristic matrix coefficient polynomial has a distinct effect on producing clear consistency diagrams. As the number of model solutions is increased, more and more modal frequencies are estimated, but, hopefully, the estimates of the physical modal parameters will be consistently determined as the correct model parameters are found. Nevertheless, the nonphysical (computational) modes will not be estimated in a consistent way during this process and can be sorted out of the modal parameter data set more easily. Note that inconsistencies (frequency shifts, leakage errors, etc.) in the measured data set will obscure this consistency and render the diagram difficult to use. Normally, a tolerance, expressed as a percentage, is given for the consistency of each of the modal parameters that are being evaluated.

Figures 21.11 and 21.12 demonstrate consistency diagrams based upon different normalizations of the characteristic matrix coefficient polynomial. Figure 21.12 shows

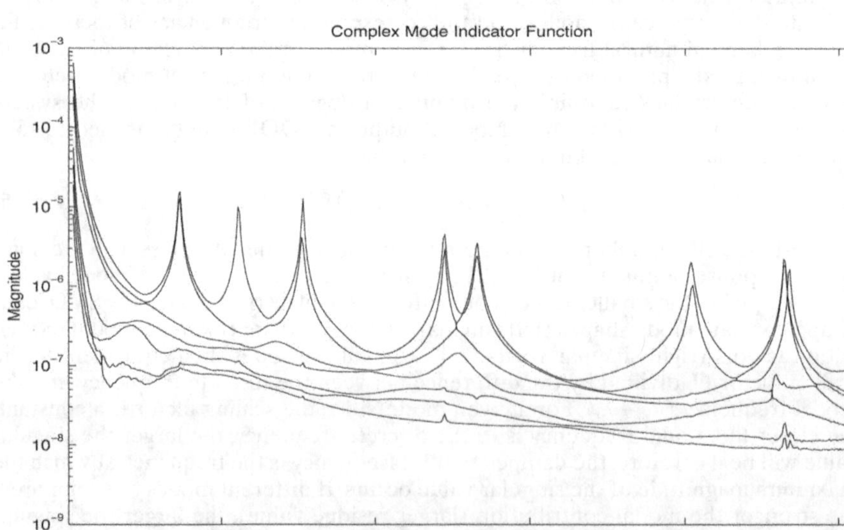

FIGURE 21.10 Complex mode indication function: seven-input example.

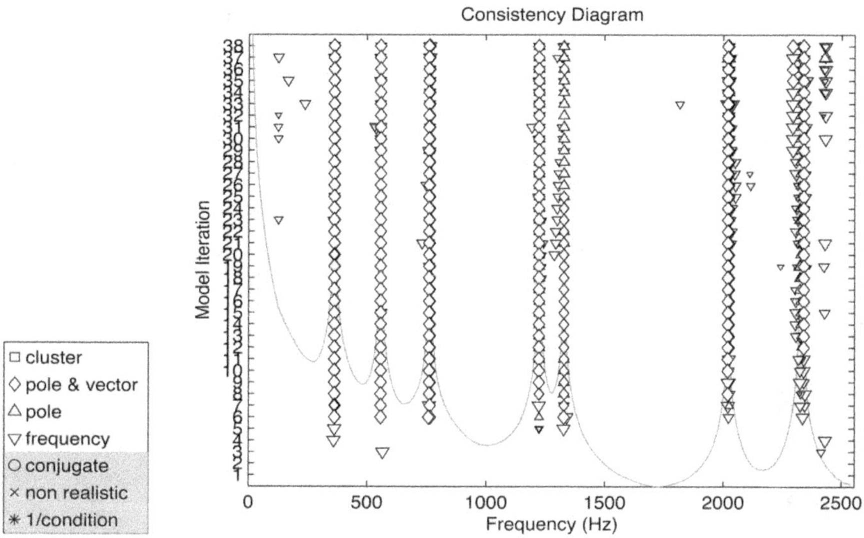

FIGURE 21.11 Consistency diagram—both equation normalizations.

an expanded view of the consistency diagram of Fig. 21.11 in the region of 350 Hz. Note that the multiple symbols indicate nearly repeated natural frequencies with distinctly different mode vectors at the peak in the data, not a single natural frequency that might be thought present from Fig. 21.11. In this consistency diagram, low- and high-order coefficient normalization is alternated at each prospective model order.

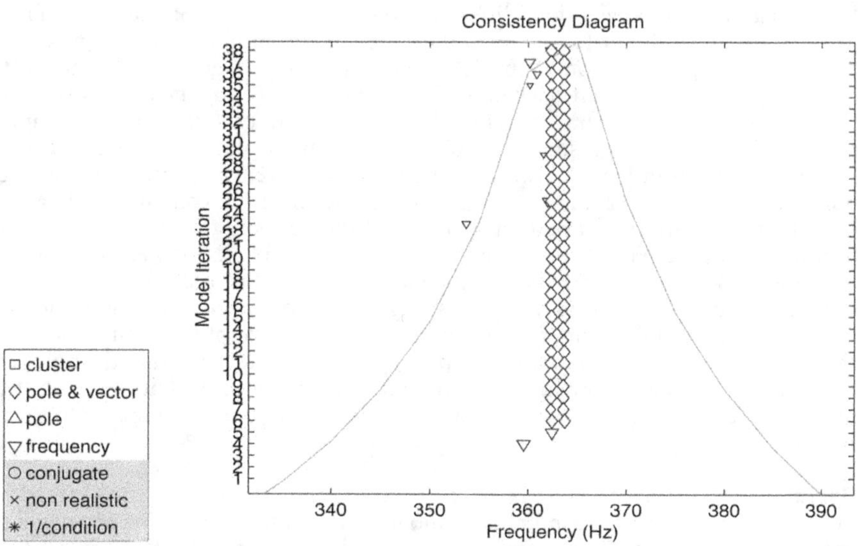

FIGURE 21.12 Consistency diagram—both equation normalizations—expanded.

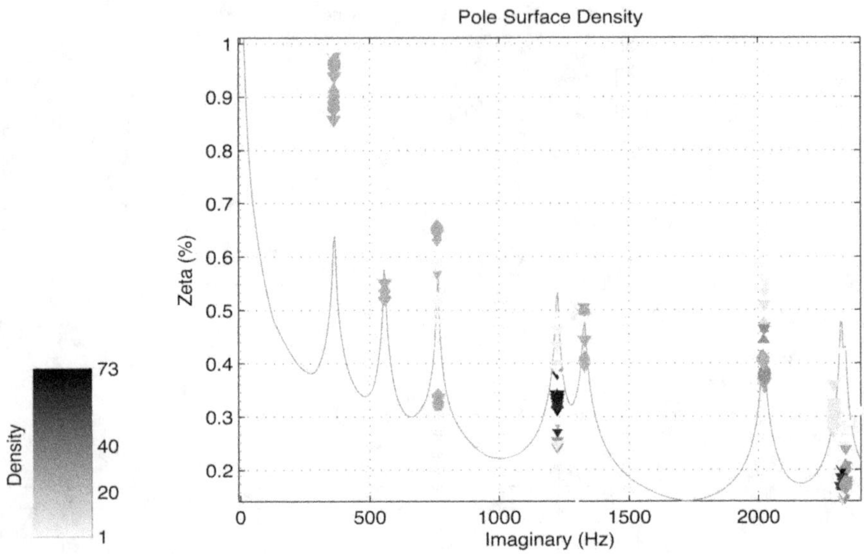

FIGURE 21.13 Pole surface density—both equation normalizations.

The plot in the background is used for visual reference. In this case, the plot is an automoment of the FRFs, but a set of complex mode indicator functions or multivariate mode indicator functions is also frequently used. The consistency diagram also is normally plotted utilizing color for better visual discrimination.

Pole Surface Density Plots. One of the disadvantages of the consistency diagram occurs when different model solution iterations are combined into one consistency diagram. In this case, since different model characteristics are being evaluated at the same time, the order in which the model solution iterations are presented may affect the presentation of consistency (or stability). In general, a clearer estimate of the modal frequencies will be determined by plotting the density of poles found from all model solution iterations in the second quadrant of the complex plane.[48] Figure 21.13 represents the pole surface density plot for the previous consistency diagrams in Fig. 21.11. In this case, the dark areas indicate where a large number of solutions for the natural frequency occur in the consistency diagram. Note that there are two dark areas at many of the peaks in the automoment plot, indicating that there are two natural frequencies at the peaks (350 Hz and 750 Hz, for example) where the frequency is nearly the same but the damping value is quite different. The centroid of the cluster gives a good representation of the natural frequency and damping for each mode, while the size of the cluster indicates the variance in the frequency and damping estimates. Note again that the pole surface density plots are normally in color rather than grayscale, and the ability to move into (expand) and out of (contract) the plot makes the use of these plots very powerful.

Residuals. Continuous systems have an infinite number of degrees of freedom but, in general, only a finite number of modes can be used to describe the dynamic behavior of a system. The theoretical number of DOF can be reduced by using a finite frequency range. Therefore, for example, the frequency response can be broken up into three partial sums, each covering the modal contribution corresponding to modes located in the frequency ranges.

In the frequency range of interest, the modal parameters can be estimated to be consistent with Eq. (21.29). In the lower- and higher-frequency ranges, residual terms can be included to account for modes in these ranges. In this case, Eq. (21.29) can be rewritten for a single frequency response function as:

$$H_{pq}(\omega) = R_{F_{pq}} + \sum_{r=1}^{n} \frac{A_{pqr}}{j\omega - \lambda_r} + \frac{A^*_{pqr}}{j\omega - \lambda^*_r} + R_{I_{pq}}(\omega) \tag{21.56}$$

where $R_{F_{pq}}$ = residual flexibility
$R_{I_{pq}}(\omega)$ = residual inertia

The residual term that compensates for modes below the minimum frequency of interest is called the *inertia restraint* or *residual inertia.* The residual term that compensates for modes above the maximum frequency of interest is called the *residual flexibility.* These residuals are a function of each FRF measurement and are not global properties of the FRF matrix. Therefore, residuals cannot be estimated unless the FRF is measured. In this common formulation of residuals, both terms are real-valued quantities. In general, this is a simplification: the residual effects of modes below and/or above the frequency range of interest cannot be completely represented by such simple mathematical relationships. As the system poles below and above the range of interest are located in the proximity of the boundaries, these effects are not the real-valued quantities noted in Eq. (21.56). In these cases, residual modes may be included in the model to partially account for these effects. When this is done, the modal parameters that are associated with these residual poles have no physical significance but may be required in order to compensate for strong dynamic influences from outside the frequency range of interest. Using the same argument, the lower and upper residuals can take on any mathematical form that is convenient as long as the lack of physical significance is understood. Mathematically, power functions of frequency (zero, first order, and second order) are commonly used within such a limitation. In general, the use of residuals is confined to FRF models. This is due primarily to the difficulty of formulating a reasonable mathematical model and solution procedure in the time domain for the general case that includes residuals.

Generalized Frequency. The fundamental problem with using a high-order frequency-domain (rational fraction polynomial) method can be highlighted by looking at the characteristics of the data matrix involved in estimating the matrix coefficients. These matrices involve power polynomials that are functions of increasing powers of the frequency, typically $s_i = j\omega_i$. These matrices are of the Van der Monde form and are known to be ill conditioned for cases involving wide frequency ranges and high-ordered models.

Van der Monde matrix form:

$$\begin{bmatrix} (s_1)^0 & (s_1)^1 & (s_1)^2 & \cdots & (s_1)^{m-1} \\ (s_2)^0 & (s_2)^1 & (s_2)^2 & \cdots & (s_2)^{m-1} \\ (s_3)^0 & (s_3)^1 & (s_3)^2 & \cdots & (s_3)^{m-1} \\ \cdots & \cdots & \cdots & \cdots & \cdots \\ (s_i)^0 & (s_i)^1 & (s_i)^2 & \cdots & (s_i)^{m-1} \end{bmatrix} \tag{21.57}$$

The ill-conditioning problem can be best understood by evaluating the condition number of the Van der Monde matrix. The *condition number* measures the sensitivity of the solution of linear equations to errors, or small amounts of noise, in the data. The condition number gives an indication of the accuracy of the results from matrix inversion and/or linear equation solution. The condition number for a matrix is computed by taking the ratio of the largest singular value to the smallest singular value. A good condition number is a small number close to unity; a bad condition number is a large number. For the theoretical case of a singular matrix, the condition number is infinite.

The ill-conditioned characteristic of matrices that are of the Van der Monde form can be reduced, but not eliminated, by the following:

- Minimizing the frequency range of the data
- Minimizing the order of the model
- Normalizing the frequency range of the data (−1,1) or (−2,2)
- Using orthogonal polynomials
- Using complex Z mapping

The last three methods involve the concept of *generalized frequency,* whereby the actual frequency response function complex data is not altered in any way but is remapped to a new generalized or virtual frequency, which eliminates or reduces the ill conditioning caused by the weighting of the FRF data by the physical frequency ($s_i = j\omega_i$) in the linear equations. These concepts are briefly explained in the following sections.

Normalized Frequency. The simplest method of using the generalized frequency concept is to normalize the power polynomials by utilizing the following equation:

$$s_i = j^*(\omega_i/\omega_{max}) \tag{21.58}$$

This gives a generalized frequency variable that is bounded by (−1,1) with much better numerical conditioning than utilizing the raw frequency range ($-\omega_{max}, \omega_{max}$). When the modal frequencies are estimated, the corrected modal frequencies must be determined by multiplying by the normalizing frequency (ω_{max}). All frequency-domain algorithms, at a minimum, will use some form of this frequency normalization. The graphical plot of this Van der Monde matrix for orders 0 through 8 is shown in Fig. 21.14.

Orthogonal Polynomials. In the past, the only way to avoid the numerical problems inherent in the frequency-domain methods (Van der Monde matrix), even when normalized frequencies are implemented, is to use a transformation from power polynomials to orthogonal polynomials.[33–39] Any power polynomial series can be represented by an equivalent number of terms in an orthogonal polynomial series. Several orthogonal polynomials have been applied to the ill-conditioning problem, such as Forsythe polynomials[36] and Chebyshev polynomials.[37,38] The orthogonal polynomial concept is represented by the following relationship:

$$\sum_{k=0}^{m} (s_i)^k \alpha_k = \sum_{k=0}^{m} P_k(s_i)\gamma_k \tag{21.59}$$

The orthogonal polynomial series can be formed by the following relationships:

$$P_0(s_i) = 1.0 \tag{21.60}$$

$$P_j(s_i^*) = P_j^*(s_i) \tag{21.61}$$

$$P_{n+1}(s_i) = a_n s_i P_n(s_i) - \sum_{k=0}^{n} b_{n,k}\, P_k(s_i) \tag{21.62}$$

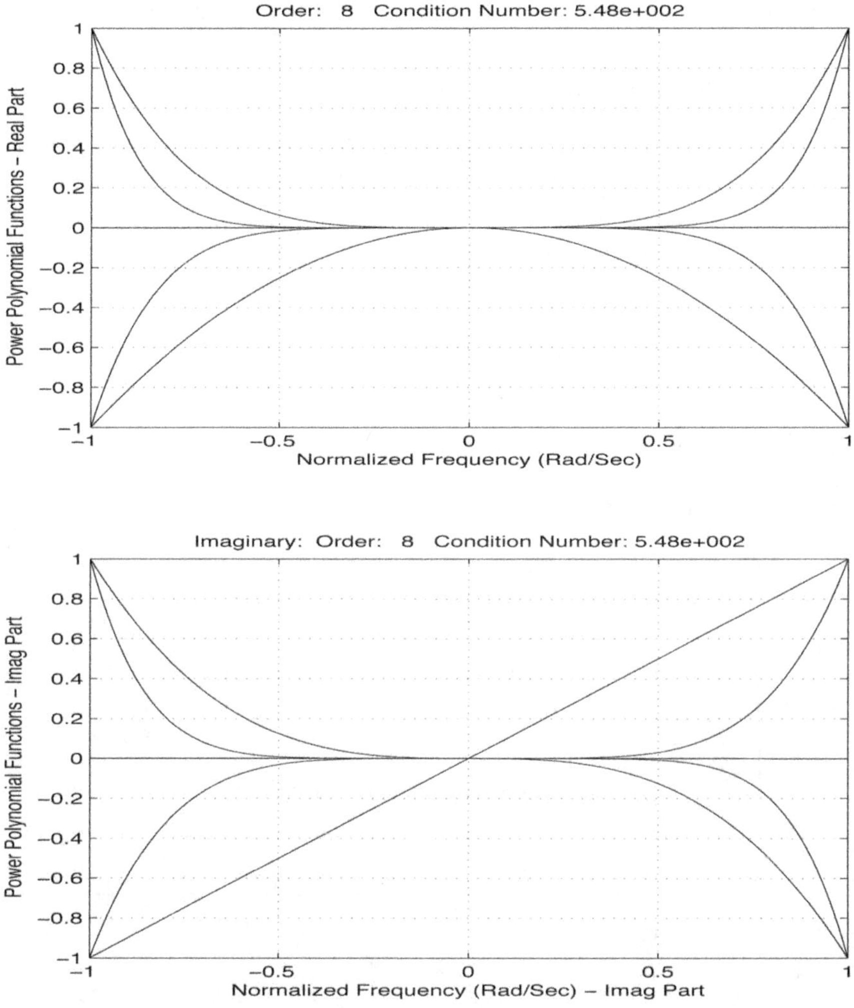

FIGURE 21.14 Van der Monde matrix—normalized frequency—orders 0–8.

This orthogonal polynomial series can be formulated in matrix form by utilizing two weighting matrices involving the coefficients a_n and $b_{n,k}$ as follows:

$$[W_a]=\begin{bmatrix} a_n & 0 & 0 & \cdots & 0 \\ 0 & a_{n-1} & 0 & \cdots & 0 \\ 0 & 0 & a_{n-2} & \cdots & 0 \\ \cdots & \cdots & \cdots & \cdots & \cdots \\ 0 & 0 & 0 & \cdots & a_0 \end{bmatrix} \qquad [W_b]=\begin{bmatrix} b_{n,n} & b_{n,n-1} & b_{n,n-2} & \cdots & b_{n,0} \\ 0 & b_{n-1,n-1} & b_{n-1,n-2} & \cdots & b_{n-1,0} \\ 0 & 0 & b_{n-2,n-2} & \cdots & b_{n-2,0} \\ \cdots & \cdots & \cdots & \cdots & \cdots \\ 0 & 0 & 0 & \cdots & b_{0,0} \end{bmatrix} \tag{21.63}$$

Different orthogonal polynomials are generated using different weighting coefficients and are orthogonal over different ranges. For example, Forsythe orthogonal polynomials are orthogonal over the (–2,2) range, while Chebyshev orthogonal polynomials are orthogonal over the (−1,1) range. In the orthogonal polynomial approach, the original complex-valued FRF data is used together with the orthogonal polynomial coefficients $P_k(s_i)$ in place of the generalized frequency $(s_i)^k$, where s_i is the properly normalized generalized frequency—Eq. (21.58), for example. The unknown matrix coefficients of the matrix orthogonal polynomial (γ_k) are estimated in place of the original matrix coefficients (α_k). These matrix orthogonal polynomial coefficients are then loaded into the companion matrix $[C]$ as before.

When this orthogonal polynomial transformation is used to generate a new generalized frequency, the corrected modal frequencies are determined from a generalized form of the companion matrix solution. The companion matrix $[C]$ is determined

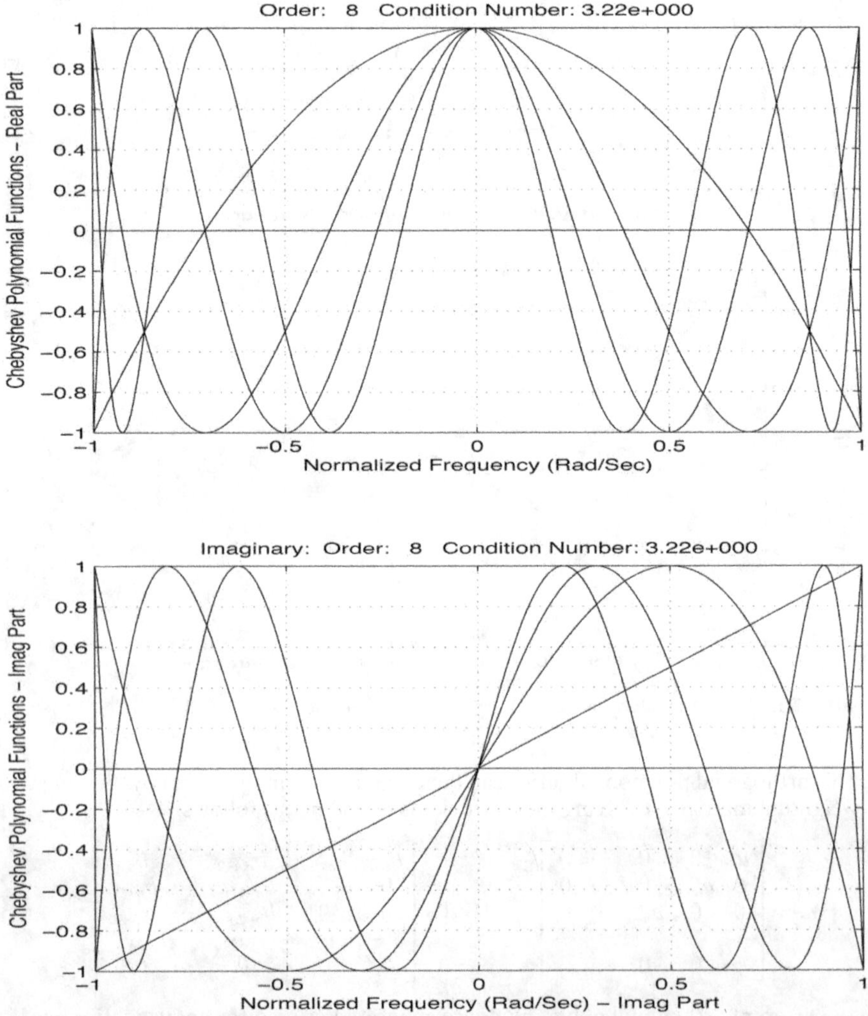

FIGURE 21.15 Van der Monde matrix—Chebyshev orthogonal polynomials—orders 0–8.

in the same way as always, but the solution for the modal frequencies is found from the following equation:

$$\Big[[C] + [W_b]\Big]\{X\} = \lambda\Big[W_a\Big]\{X\} \tag{21.64}$$

The graphical plot of the Van der Monde matrix for orders 0 through 8 for a set of Chebyshev orthogonal polynomials is shown in Fig. 21.15. Note that the use of orthogonal polynomials is the basis for the recently developed Alias Free Polynomial (AF-Poly®) method of modal parameter estimation.[39]

Complex Z Mapping. The important contribution behind the development of the Polyreference Least Squares Complex Frequency (PLSCF)[40–42] method is the recognition of a new method of frequency mapping. The generalized frequency in this approach is a trigonometric mapping function (complex Z) that has numerical conditioning superior to orthogonal polynomials without the added complication of solving a generalized companion matrix eigenvalue problem. This approach can be applied to any frequency-domain method—low-order frequency-domain methods as well as high-order frequency-domain methods—although the numerical advantage is not as profound for the low-order methods.

The basic complex Z mapping function, in the nomenclature of this presentation, is as follows:

$$s_i = z_i = e^{j*\pi*(\omega_i/\omega_{\max})} = e^{j*\omega*\Delta t} \tag{21.65}$$

$$s_i^m = z_i^m = e^{j*\pi*m*(\omega_i/\omega_{\max})} \tag{21.66}$$

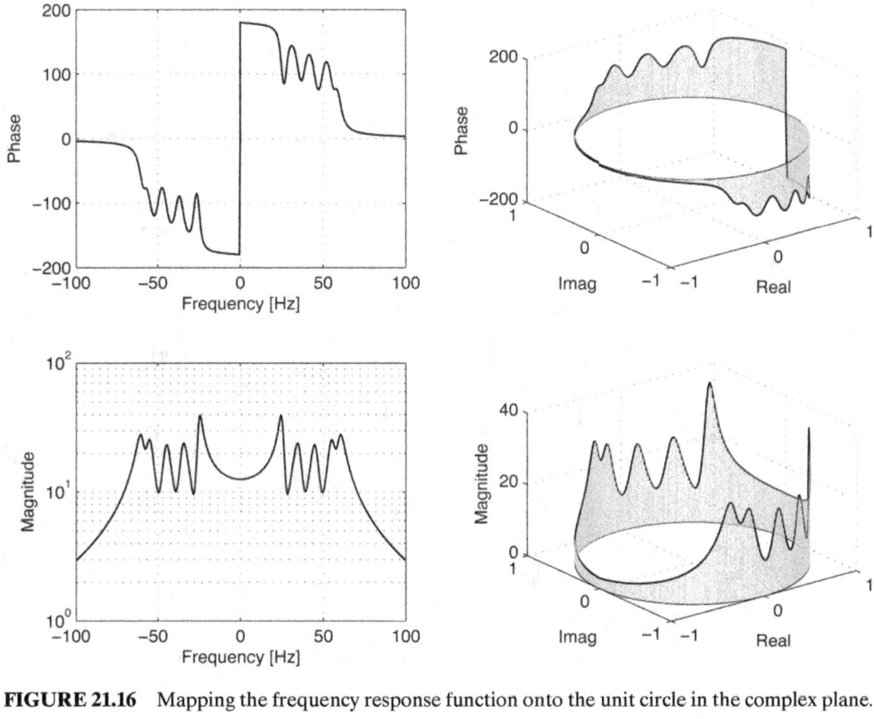

FIGURE 21.16 Mapping the frequency response function onto the unit circle in the complex plane.

This mapping function maps the positive frequency range to the positive unit circle in the complex plane and the negative frequency range to the negative unit circle in the complex plane. This is graphically represented in Fig. 21.16. Note that the use of complex Z mapping is the basis for the recently developed PLSCF, or, commercially, PolyMAX®, method of modal parameter estimation.[40–42]

This effectively yields a real part of the mapping functions, which is cosine terms, and an imaginary part, which is sine functions. Since sine and cosine functions at different frequencies are mathematically orthogonal, the numerical conditioning of this mapping function is quite good. Since these functions are also the basis of the Fourier transform, this method is essentially a rational fraction polynomial (RFP) method with an embedded inverse Fourier transform, yielding a method

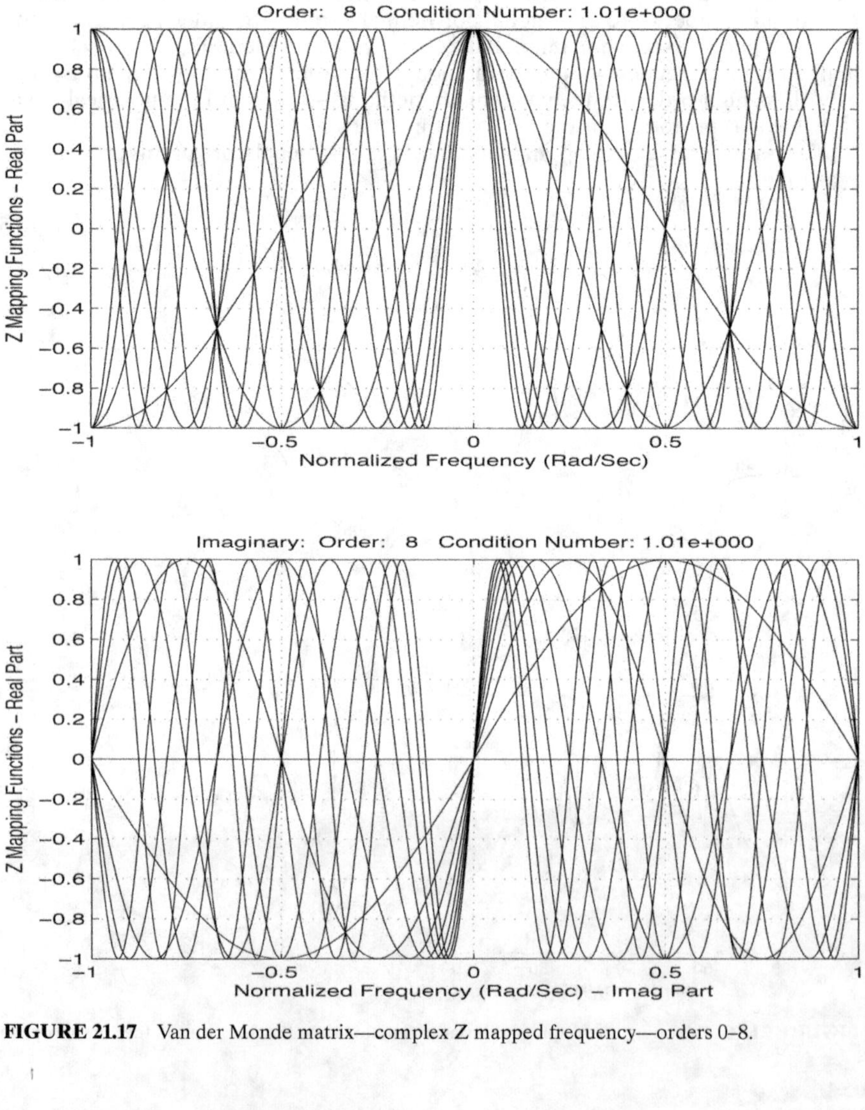

FIGURE 21.17 Van der Monde matrix—complex Z mapped frequency—orders 0–8.

that is very similar to the polyreference time-domain (PTD) method of modal parameter estimation.

The graphical plot of this Van der Monde matrix for orders 0 through 8 is shown in Fig. 21.17. The condition number for this example matrix is 1.01 (Fig. 21.17) compared to a condition number of 548 for the normalized frequency example (Fig. 21.14).

General (Two-Stage) Solution Procedure. Based upon Eq. (21.43) or (21.46), most modern modal identification algorithms are very similar, and all can be outlined as follows:

1. Load measured data into linear equation form.
   a. Frequency domain—utilize generalized frequency concept.
2. Find scalar or matrix coefficients ($[\alpha_k]$).
3. Solve matrix coefficient polynomial for modal frequencies.
   a. Formulate companion matrix.
   b. Obtain eigenvalues of companion matrix (λ_r or z_r).
      (1) Time domain—convert eigenvalues from z_r to λ_r.
      (2) Frequency domain—compensate for generalized frequency.
   c. Obtain modal participation vectors $\{L\}_r$ or modal vectors $\{\psi\}_r$ from eigenvectors of the companion matrix.
4. Find modal vectors and modal scaling from one of Eqs. (21.32) through (21.66).

DIFFERENCES IN MODAL PARAMETER ESTIMATION ALGORITHMS

Modal parameter estimation algorithms typically give slightly different estimates of modal parameters due to the way the frequency response function data is weighted and processed in the computation of the matrix coefficients. Some of the most common variations are discussed in the following sections.

Data Sieving/Filtering/Decimation. For almost all cases of modal identification, a large amount of redundancy or overdetermination exists. This means that the number of equations available compared to the number required to form an exactly determined solution, defined as the *overdetermination factor,* will be quite large. Beyond some value of overdetermination factor, the additional equations contribute little to the result but may add significantly to the noise and, thus, to the solution time. For this reason, the data space is often *filtered* (limited within minimum and maximum temporal axis values), *sieved* (limited to prescribed input degrees of freedom and/or output DOFs), and/or *decimated* (limited number of equations from the allowable temporal data) in order to obtain a reasonable result in the minimum time.

Coefficient Condensation (Virtual DOFs). For the low-order modal identification algorithms, the number of physical coordinates (typically N_o), which dictates the size of the coefficient matrices ($[\alpha_k]$), is often much larger than the number of desired modal frequencies (N). For this situation, the numerical solution procedure is constrained to solve for N_o or $2N_o$ modal frequencies. This can be very time consuming and is unnecessary. One simple approach to reducing the size of the coefficient matrices is to *sieve* the physical degrees of freedom to temporarily reduce the dimension of N_o. Beyond excluding all physical DOFs in a direction, this is difficult to do in an effective manner that will retain the correct information from the frequency response function data matrix.

The number of physical coordinates (N_o) can be reduced to a more reasonable size ($N_e \approx N$ or $N_e \approx 2N$) by using a decomposition transformation from physical coordinates (N_o) to the approximate number of effective modal frequencies (N_e). These resulting N_e transformed coordinates are sometimes referred to as *virtual DOFs.* Currently, singular value decompositions (SVDs) or eigenvalue decompositions (EDs) are used to condense the spatial information while preserving the principal modal information prior to formulating the linear equation solution for unknown matrix coefficients.[8–11,49,50]

It is important to understand that the ED and SVD transformations yield a mathematical transformation that, in general, contains complex-valued vectors as part of the transformation matrix $[T]$. Conceptually, the transformation will work well when these vectors are estimates of the modal vectors of the system—normally, a situation where the vectors can be scaled to real-valued vectors. Essentially, this means that the target goal of the transformation is a transformation from physical space to modal space. As the modal density increases and/or as the level of damping increases, the ED and SVD methods give erroneous results, if the complete $[H]$ matrix is used. Generally, superior results will be obtained when the imaginary part of the $[H]$ matrix is used in the ED or SVD transformation, thus forcing a real-valued transformation matrix $[T]$. Another option is to load both the real and the imaginary portions of the complex data into a real matrix, which will also force a real-valued transformation matrix.[49,50]

In most cases, even when the spatial information must be condensed, it is necessary to use a model order greater than two to compensate for distortion errors or noise in the data and to compensate for the case where the location of the transducers is not sufficient to totally define the structure.

$$[H'] = [T][H] \tag{21.67}$$

where $[H']$ = the transformed (condensed) FRF matrix
$[T]$ = the transformation matrix
$[H]$ = the original FRF matrix

The difference between the two techniques lies in the method of finding the transformation matrix $[T]$. Once $[H]$ has been condensed, however, the parameter estimation procedure is the same as for the full data set. Because the data eliminated from the parameter estimation process ideally corresponds to the noise in the data, the poles of the condensed data are the same as the poles of the full data set. However, the participation factors calculated from the condensed data may need to be expanded back into the full space.

$$[\Psi] = [T]^T[\Psi'] \tag{21.68}$$

where $[\Psi]$ = the full-space participation matrix
$[\Psi']$ = the condensed-space participation matrix

This technique may also be adapted for condensing the input space, as long as the FRF matrix $[H]$ is reshaped (transposed) to an $N_i \times N_o$ matrix at each spectral line.

Equation Condensation. Equation condensation methods are used to reduce the number of equations generated from measured data to more closely match the number of unknowns in the modal parameter estimation algorithms. A large number of condensation algorithms are available. Based upon the modal parameter estimation algorithms in use today, the three types of algorithms most often used are:

- *Least squares:* Least squares (LS), weighted least squares (WLS), total least squares (TLS), or double least squares (DLS) are used to minimize the squared error between the measured data and the estimation model.
- *Transformations:* The measured data is reduced by approximating the data by the superposition of a limited (reduced) set of independent vectors. The number of significant, independent vectors is chosen equal to the maximum number of modes that are expected in the measured data. This set of vectors is used to approximate the measured data and is used as input to the parameter estimation procedures. Singular value decomposition is an example of one of the more popular transformation methods.
- *Coherent averaging:* Coherent averaging is another popular method for reducing the data. In the coherent averaging method, the data is weighted by performing a dot product between the data and a weighting vector (spatial filter). Information in the data which is not coherent with the weighting vectors is averaged out of the data. The most common coherent averaging method utilizes an estimate of one of the modal vectors as the weighting vector to generate data that has one mode dominant.

The LS and the transformation procedures tend to weight those modes of vibration which are well excited. This can be a problem when trying to extract modes which are not well excited. The solution is to use a weighting function for condensation which tends to enhance the mode of interest. This can be accomplished in a number of ways:

- In the time domain, a spatial filter or a coherent averaging process can be used to filter the response to enhance a particular mode or set of modes.
- In the frequency domain, the data can be enhanced in the same manner as in the time domain, plus the data can be additionally enhanced by weighting it in a frequency band near the natural frequency of the mode of interest.

Obviously, the type of equation condensation method that is utilized in a modal identification algorithm has a significant influence on the results.

CURRENT MODAL IDENTIFICATION METHODS

Using the concepts developed in the previous section, the most commonly used modal identification methods can be summarized as shown in Table 21.2. The high-order model is typically used for those cases where the system is undersampled in the spatial domain. For example, the limiting case is when only one measurement is made on the structure. For this case, the left-hand side of the general linear equation corresponds to a scalar polynomial equation with the order equal to or greater than the number of desired modal frequencies.

The low-order model is used for those cases where the spatial information is complete. In other words, the number of physical coordinates is greater than the number of desired modal frequencies. For this case, the order of the left-hand side of the general linear equation, Eq. (21.43) or Eq. (21.46), is equal to 2 and the matrix dimension of the $[\alpha]$ coefficients is much greater than $n \times n$.

TABLE 21.2 Characteristics of Modal Parameter Estimation Algorithms

	Domain		Matrix polynomial order			Coefficients	
Algorithm	Time	Frequency	Zero	Low	High	Scalar	Matrix
CEA	•				•	•	
LSCE	•				•	•	
PTD	•				•		$N_i \times N_i$
ITD	•			•			$N_o \times N_o$
MRITD	•			•			$N_o \times N_o$
ERA	•			•			$N_o \times N_o$
PFD		•		•			$N_o \times N_o$
SFD		•		•			$N_o \times N_o$
MRFD		•		•			$N_o \times N_o$
RFP		•			•	•	$N_i \times N_i$
OP		•			•	•	$N_i \times N_i$
PLSCF		•			•	•	$N_i \times N_i$
CMIF		•	•				$N_o \times N_i$

The zero-order model corresponds to cases where the temporal information is neglected and only the spatial information is used. These methods directly estimate the eigenvectors as a first step. In general, these methods are programmed to process data at a single temporal condition or variable. In this case, the method is essentially equivalent to the single-degree-of-freedom methods which have been used with frequency response functions. In other words, the zero-th-order matrix polynomial model compared to the higher-order matrix polynomial models is similar to the comparison between the SDOF and MDOF methods used historically in modal parameter estimation.

Table 21.3 groups the different methods according to common use characteristics. The methods in each quadrant of this table are used for similar data situations. In general, low-order methods require a large number of measurement locations (input DOFs and/or output DOFs). Time-domain methods work well for low to moderate damping. Conversely, if only a few measurement locations are available, high-order methods will be required. Likewise if the damping is moderate to heavy, frequency-domain methods will be required. When the measured data originates from a system that has light to moderate damping and a large number of input DOFS and/or output DOFs are available, all of the methods should give satisfactory and similar results.

TABLE 21.3 Common Characteristics of Modal Parameter Estimation Algorithms

	Time domain	Frequency domain	
Low-order models	ITD MRITD ERA	PFD-1 PFD-2 PFD-Z	SFD MRFD
High-order models	CEA LSCE PTD	RFP OP PLSCF	RFP-Z PolyMAX AF-Poly

MODAL IDENTIFICATION ALGORITHMS (SDOF)

For any real system, the use of single-degree-of-freedom algorithms to estimate modal parameters is always an approximation, since any realistic structural system has many degrees of freedom. Nevertheless, in cases where the modes are not close in frequency and do not affect one another significantly, SDOF algorithms are very effective. Specifically, SDOF algorithms are quick, rarely involving much mathematical manipulation of the data, and give sufficiently accurate results for most modal parameter requirements. Naturally, most multiple-degree-of-freedom algorithms can be constrained to estimate only a single degree of freedom at a time if further mathematical accuracy is desired. The most commonly used SDOF algorithms involve using the information at a single frequency as an estimate of the modal vector.

Operating Vector Estimation. Technically, when many single-degree-of-freedom approaches are used to estimate modal parameters, sufficient simplifying assumptions are made such that the results are not actually modal parameters. In these cases, the results are often referred to as *operating vectors* rather than modal vectors. This term refers to the fact that if the structural system is excited at this frequency, the resulting motion is a linear combination of the modal vectors rather than a single modal vector. If one mode is dominant, then the operating vector is approximately equal to the mode vector.

The approximate relationships that are used in these cases are represented in the following two equations.

$$H_{pq}(\omega_r) \approx \frac{A_{pqr}}{j\omega_r - \lambda_r} + \frac{A^*_{pqr}}{j\omega_r - \lambda^*_r} \tag{21.69}$$

$$H_{pq}(\omega_r) \approx \frac{A_{pqr}}{-\sigma_r} \tag{21.70}$$

For these less complicated methods, the damped natural frequencies (ω_r) are estimated by observing the maxima in the frequency response functions. The damping factors (σ_r) are estimated using half-power methods.[1] The residues (A_{pqr}) are then estimated from Eq. (21.69) or Eq. (21.70) using the FRF data at the damped natural frequency.

Least-Squares SDOF Methods. The least-squares, local single-degree-of-freedom formulations are simple methods that are based upon using an SDOF model in the vicinity of a resonance frequency. A reasonable estimation of the modal frequency and residue for each mode can be determined under the assumption that modes are not too close together. This method can give erroneous answers when the modal coefficient is near zero. This problem can be detected by comparing the predicted modal frequency to the frequency range of the data used in the algorithm. As long as the predicted modal frequency lies within the frequency band, the estimate of the residue (modal coefficient) should be valid.

The approximate relationship that is used in this case is represented in the following equation. The frequency ω_1 is a frequency near the damped natural frequency ω_r.

$$H_{pq}(\omega_1) \approx \frac{A_{pqr}}{j\omega_1 - \lambda_r} \tag{21.71}$$

$$H_{pq}(\omega_1)(j\omega_1 - \lambda_r) = A_{pqr} \tag{21.72}$$

$$H_{pq}(\omega_1)\,\lambda_r + A_{pqr} = (j\omega_1)H_{pq}(\omega_1) \tag{21.73}$$

Repeating the preceding equation for several frequencies in the vicinity of the peak frequency:

$$\begin{bmatrix} H_{pq}(\omega_1) & 1 \\ H_{pq}(\omega_2) & 1 \\ H_{pq}(\omega_3) & 1 \\ H_{pq}(\omega_p) & 1 \\ \cdots & \cdots \\ H_{pq}(\omega_s) & 1 \end{bmatrix}_{N_s \times 2} \begin{Bmatrix} \lambda_r \\ A_{pqr} \end{Bmatrix}_{2\times 1} = \begin{Bmatrix} (j\omega_1)\, H_{pq}(\omega_1) \\ (j\omega_2)\, H_{pq}(\omega_2) \\ (j\omega_3)\, H_{pq}(\omega_3) \\ (j\omega_p)\, H_{pq}(\omega_p) \\ \cdots \\ (j\omega_s)\, H_{pq}(\omega_s) \end{Bmatrix}_{N_s \times 1} \tag{21.74}$$

The preceding equation again represents an overdetermined set of linear equations that can be solved using any pseudo-inverse or normal equations approach.

Two-Point Finite Difference Formulation. The difference method formulations are methods that are based upon comparing adjacent frequency information in the vicinity of a resonance frequency. When a ratio of this information, together with information from the derivative of the frequency response function at the same frequencies, is formed, a reasonable estimation of the modal frequency and residue for each mode can be determined under the assumption that modes are not too close together. This method can give erroneous answers when the modal coefficient is near zero. This problem can be detected by comparing the predicted modal frequency to the frequency range of the data used in the finite difference algorithm. As long as the predicted modal frequency lies within the frequency band, the estimate of the residue (modal coefficient) should be valid.

The approximate relationships that are used in this case are represented in the following equations. The frequencies noted in these relationships are as follows: ω_1 is a frequency near the damped natural frequency ω_r, and ω_p is the peak frequency close to the damped natural frequency ω_r.

Modal frequency (λ_r):

$$\lambda_r \approx \frac{j\omega_p H_{pq}(\omega_p) - j\omega_1 H_{pq}(\omega_1)}{H_{pq}(\omega_p) - H_{pq}(\omega_1)} \tag{21.75}$$

Residue (A_{pqr}):

$$A_{pqr} \approx \frac{j(\omega_1 - \omega_p) H_{pq}(\omega_1) H_{pq}(\omega_p)}{H_{pq}(\omega_p) - H_{pq}(\omega_1)} \tag{21.76}$$

Since both of the equations that are used to estimate modal frequency λ_r and residue A_{pqr} are linear equations, a least-squares solution can be formed by using other FRF data in the vicinity of the resonance. For this case, additional equations can be developed using $H_{pq}(\omega_2)$ or $H_{pq}(\omega_3)$ in the preceding equations instead of $H_{pq}(\omega_1)$.

MODAL IDENTIFICATION ALGORITHMS (MDOF)

All multiple-degree-of-freedom equations can be represented in a unified matrix polynomial approach. The methods that are summarized in the following sections are listed in Tables 21.1 and 21.2.

High-Order Time-Domain Algorithms. The algorithms that fall into the category of high-order time-domain algorithms include the most commonly used algorithms for determining modal parameters. The least-squares complex exponential

(LSCE) algorithm is the first algorithm to utilize more than one frequency response function, in the form of impulse response functions, in the solution for a global estimate of the modal frequency. The polyreference time-domain (PTD) algorithm is an extension to the LSCE algorithm that allows multiple references to be included in a meaningful way so that the ability to resolve close modal frequencies is enhanced. Since both the LSCE and the PTD algorithms have good numerical characteristics, these algorithms are still the most commonly used algorithms today. The only limitations of these algorithms are the cases involving high damping. As a high-order algorithm, more time-domain information is required compared to the low-order algorithms.

Basic equation—high-order coefficient normalization:

$$\begin{bmatrix}[\alpha_0][\alpha_1] \cdots [\alpha_{m-1}]\end{bmatrix}_{N_i \times mN_i} \begin{bmatrix}[h(t_{i+0})] \\ [h(t_{i+1})] \\ \cdots \\ [h(t_{i+m-1})]\end{bmatrix}_{mN_i \times N_o} = -[h(t_{i+m})]_{N_i \times N_o} \tag{21.77}$$

Basic equation—zero-order coefficient normalization:

$$\begin{bmatrix}[\alpha_1][\alpha_2] \cdots [\alpha_m]\end{bmatrix}_{N_i \times mN_i} \begin{bmatrix}[h(t_{i+1})] \\ [h(t_{i+2})] \\ \cdots \\ [h(t_{i+m})]\end{bmatrix}_{mN_i \times N_o} = -[h(t_{i+0})]_{N_i \times N_o} \tag{21.78}$$

First-Order Time-Domain Algorithms. The first-order time-domain algorithms include several well-known algorithms such as the Ibrahim time-domain (ITD) algorithm and the eigensystem realization algorithm (ERA). These algorithms are essentially a state-space formulation with respect to the second-order time-domain algorithms. The original development of these algorithms was quite different than that presented here, but the resulting solution of linear equations is the same regardless of development. There is a great body of published work on both the ITD and the ERA algorithms, much of which discusses the various approaches for condensing the overdetermined set of equations that results from the data (least squares, double least squares, singular value decomposition). The low-order time-domain algorithms require very few time points in order to generate a solution, due to the increased use of spatial information.

Basic equation—high-order coefficient normalization:

$$\begin{bmatrix}[\alpha_0]\end{bmatrix}_{2N_o \times 2N_o} \begin{bmatrix}[h(t_{i+0})] \\ [h(t_{i+1})]\end{bmatrix}_{2N_o \times N_i} = -\begin{bmatrix}[h(t_{i+1})] \\ [h(t_{i+2})]\end{bmatrix}_{2N_o \times N_i} \tag{21.79}$$

Basic equation—zero-order coefficient normalization:

$$\begin{bmatrix}[\alpha_1]\end{bmatrix}_{2N_o \times 2N_o} \begin{bmatrix}[h(t_{i+1})] \\ [h(t_{i+2})]\end{bmatrix}_{2N_o \times N_i} = -\begin{bmatrix}[h(t_{i+0})] \\ [h(t_{i+1})]\end{bmatrix}_{2N_o \times N_i} \tag{21.80}$$

Second-Order Time-Domain Algorithms. The second-order time-domain algorithm has not been reported in the literature previously but is simply modeled after the second-order matrix differential equation with matrix dimension N_o. Since an impulse response function can be thought to be a linear summation of a number of complementary solutions to such a matrix differential equation, the general second-order matrix form is a natural model that can be used to determine the modal parameters. It is interesting that this method is developed by noting that it is the time-domain equivalent to a frequency-domain algorithm known as the polyreference frequency-domain (PFD) algorithm. Note that the low-order time-domain algorithms require very few time points in order to generate a solution, due to the increased use of spatial information.

Basic equation—high-order coefficient normalization:

$$\left[[\alpha_0][\alpha_1]\right]_{N_o \times 2N_o} \begin{bmatrix} [h(t_{i+i})] \\ [h(t_{i+i})] \end{bmatrix}_{2N_o \times N_i} = -\left[[h(t_{i+i})]\right]_{N_o \times N_i} \tag{21.81}$$

Basic equation—zero-order coefficient normalization:

$$\left[[\alpha_1][\alpha_2]\right]_{N_o \times 2N_o} \begin{bmatrix} [h(t_{i+i})] \\ [h(t_{i+i})] \end{bmatrix}_{2N_o \times N_i} = -\left[[h(t_{i+i})]\right]_{N_o \times N_i} \tag{21.82}$$

High-Order Frequency-Domain Algorithms. The high-order frequency-domain algorithms, in the form of scalar coefficients, are the oldest multiple-degree-of-freedom algorithms utilized to estimate modal parameters from discrete data. These are algorithms like the rational fraction polynomial (RFP), the power polynomial (PP), and the orthogonal polynomial (OP) algorithms. These algorithms work well for narrow-frequency bands and limited numbers of modes but have poor numerical characteristics otherwise. While the use of multiple references reduces the numerical conditioning problem, the problem is still significant and not easily handled. In order to circumvent the poor numerical characteristics, many approaches have been used (frequency normalization, orthogonal polynomials), but the use of low-order frequency-domain models has proven more effective.

Basic equation—high-order coefficient normalization:

$$\left[[\alpha_0][\alpha_1] \cdots [\alpha_{m-1}][\beta_0][\beta_1] \cdots [\beta_n]\right]_{N_i \times mN_i + (n+1)N_o} \begin{bmatrix} (s_i)^0[H(\omega_i)] \\ (s_i)^1[H(\omega_i)] \\ \cdots \\ (s_i)^{m-1}[H(\omega_i)] \\ -(s_i)^0[I] \\ -(s_i)^1[I] \\ \cdots \\ -(s_i)^n[I] \end{bmatrix}_{mN_i + (n+1)N_o \times N_o} = -(s_i)^m[H(\omega_i)]_{N_i \times N_o} \tag{21.83}$$

Basic equation—zero-order coefficient normalization:

$$\left[[\alpha_1][\alpha_2] \cdots [\alpha_m][\beta_0][\beta_1] \cdots [\beta_n]\right]_{N_i \times mN_i + (n+1)N_o} \begin{bmatrix} (s_i)^1[H(\omega_i)] \\ (s_i)^2[H(\omega_i)] \\ \cdots \\ (s_i)^m[H(\omega_i)] \\ -(s_i)^0[I] \\ -(s_i)^1[I] \\ \cdots \\ -(s_i)^n[I] \end{bmatrix}_{mN_i + (n+1)N_o \times N_o} = -(s_i)^0[H(\omega_i)]_{N_i \times N_o} \tag{21.84}$$

First-Order Frequency-Domain Algorithms. Several algorithms have been developed that fall into the category of first-order frequency-domain algorithms, including the simultaneous frequency-domain (SFD) algorithm and the multiple-reference simultaneous frequency-domain algorithm. These algorithms are essentially frequency-domain equivalents to the Ibrahim time-domain and eigensystem realization algorithms and effectively involve a state-space formulation when compared to the second-order frequency-domain algorithms. The state-space formulation utilizes the derivatives of the frequency response functions as well as the FRF in the solution. These algorithms have superior numerical characteristics compared to the high-order frequency-domain algorithms. Unlike the low-order time-domain algorithms, though, sufficient data from across the complete frequency range of interest must be included in order to obtain a satisfactory solution.

Basic equation—high-order coefficient normalization:

$$\begin{bmatrix}[\alpha_0][\beta_0]\end{bmatrix}_{2N_o \times 4N_o} \begin{bmatrix}(s_i)^0[H(\omega_i)] \\ (s_i)^1[H(\omega_i)] \\ -(s_i)^0[I] \\ -(s_i)^1[I]\end{bmatrix}_{4N_o \times N_i} = -\begin{bmatrix}(s_i)^1[H(\omega_i)] \\ (s_i)^2[H(\omega_i)]\end{bmatrix}_{2N_o \times N_i} \tag{21.85}$$

Basic equation—zero-order coefficient normalization:

$$\begin{bmatrix}[\alpha_1][\beta_0]\end{bmatrix}_{2N_o \times 4N_o} \begin{bmatrix}(s_i)^1[H(\omega_i)] \\ (s_i)^2[H(\omega_i)] \\ -(s_i)^0[I] \\ -(s_i)^1[I]\end{bmatrix}_{4N_o \times N_i} = -\begin{bmatrix}(s_i)^0[H(\omega_i)] \\ (s_i)^1[H(\omega_i)]\end{bmatrix}_{2N_o \times N_i} \tag{21.86}$$

Second-Order Frequency-Domain Algorithms. The second-order frequency-domain algorithms include the polyreference frequency-domain (PFD) algorithms. These algorithms have superior numerical characteristics compared to the high-order frequency-domain algorithms. Unlike the low-order time-domain algorithms, though, sufficient data from across the complete frequency range of interest must be included in order to obtain a satisfactory solution.

Basic equation—high-order coefficient normalization:

$$\begin{bmatrix}[\alpha_0][\alpha_1][\beta_0][\beta_1]\end{bmatrix}_{N_o \times 4N_o} \begin{bmatrix}(s_i)^0[H(\omega_i)] \\ (s_i)^1[H(\omega_i)] \\ -(s_i)^0[I] \\ -(s_i)^1[I]\end{bmatrix}_{4N_o \times N_i} = -(s_i)^2[H(\omega_i)]_{N_o \times N_i} \tag{21.87}$$

Basic equation—zero-order coefficient normalization:

$$\begin{bmatrix}[\alpha_1][\alpha_2][\beta_0][\beta_1]\end{bmatrix}_{N_o \times 4N_o} \begin{bmatrix}(s_i)^1[H(\omega_i)] \\ (s_i)^2[H(\omega_i)] \\ -(s_i)^0[I] \\ -(s_i)^1[I]\end{bmatrix}_{4N_o \times N_i} = -(s_i)^0[H(\omega_i)]_{N_o \times N_i} \tag{21.88}$$

Residue Estimation. Once the modal frequencies and modal participation vectors have been estimated, the associated modal vectors and modal scaling (residues) can be found with standard least-squares methods in either the time or the frequency domain. The most common approach is to estimate residues in the frequency domain utilizing residuals, if appropriate.

$$\{H_{pq}(\omega)\}_{N_s \times 1} = \left[\frac{1}{j\omega - \lambda_r}\right]_{N_s \times (2n+2)} \{A_{pqr}\}_{(2n+2)\times 1} \tag{21.89}$$

where N_s = number of spectral lines
$N_s \geq 2n + 2$

$$\left[\frac{1}{j\omega - \lambda_r}\right] = \begin{bmatrix} \frac{1}{j\omega_1 - \lambda_1} & \frac{1}{j\omega_1 - \lambda_2} & \frac{1}{j\omega_1 - \lambda_3} & \cdots & \frac{1}{j\omega_1 - \lambda_{2n}} & \frac{-1}{\omega_1^2} & 1 \\ \frac{1}{j\omega_2 - \lambda_1} & \frac{1}{j\omega_2 - \lambda_2} & \frac{1}{j\omega_2 - \lambda_3} & \cdots & \frac{1}{j\omega_2 - \lambda_{2n}} & \frac{-1}{\omega_2^2} & 1 \\ \frac{1}{j\omega_3 - \lambda_1} & \frac{1}{j\omega_3 - \lambda_2} & \frac{1}{j\omega_3 - \lambda_3} & \cdots & \frac{1}{j\omega_3 - \lambda_{2n}} & \frac{-1}{\omega_3^2} & 1 \\ \cdots & \cdots & \cdots & \cdots & \cdots & \cdots & \cdots \\ \frac{1}{j\omega_{Ns} - \lambda_1} & \frac{1}{j\omega_{Ns} - \lambda_2} & \frac{1}{j\omega_{Ns} - \lambda_3} & \cdots & \frac{1}{j\omega_{Ns} - \lambda_{2n}} & \frac{-1}{\omega_{Ns}^2} & 1 \end{bmatrix}$$

$$\{A_{pqr}\} = \begin{Bmatrix} A_{pq1} \\ A_{pq2} \\ A_{pq3} \\ \cdots \\ A_{pq2n} \\ R_{I_{pq}} \\ R_{F_{pq}} \end{Bmatrix} \qquad \{H_{pq}(\omega)\} = \begin{Bmatrix} H_{pq}(\omega_1) \\ H_{pq}(\omega_2) \\ H_{pq}(\omega_3) \\ \cdots \\ H_{pq}(\omega_{N_s}) \end{Bmatrix}$$

The preceding equation is a linear equation, in terms of the unknown residues, once the modal frequencies are known. Since more frequency information N_s is available from the measured frequency response function than the number of unknowns $2n + 2$, this system of equations is normally solved using the same least-squares methods discussed previously. If multiple-input frequency response function data is available, the preceding equation is modified to find a single set of $2n$ residues representing all of the FRFs for the multiple inputs and a single output.

MODAL DATA PRESENTATION/VALIDATION

Once the modal parameters are determined, several procedures exist that allow for the modal model to be validated. Some of the procedures that are used are:

- Measurement synthesis
- Visual verification (animation)
- Finite element analysis
- Modal vector orthogonality
- Modal vector consistency (modal assurance criterion)

- Modal modification prediction
- Modal complexity
- Modal phase colinearity and mean phase deviation

All of these methods depend upon the evaluation of an assumption concerning the modal model. Unfortunately, the success of the validation method only defines the validity of the assumption; the failure of the modal validation does not generally define what the cause of the problem is.

MEASUREMENT SYNTHESIS

The most common validation procedure is to compare the data synthesized from the modal model with the measured data. This is particularly effective if the measured data is not part of the data used to estimate the modal parameters. This serves as an independent check of the modal parameter estimation process.

The visual match can be given a numerical value if a correlation coefficient, similar to coherence, is estimated. The basic assumption is that the measured frequency response function and the synthesized FRF should be linearly related (unity) at all frequencies.

Synthesis correlation coefficient (SCC):

$$\mathrm{SCC}_{pq} = \Gamma^2_{pq} = \frac{\left| \sum_{\omega=\omega_1}^{\omega_2} H_{pq}(\omega)\, \hat{H}_{pq}^*(\omega) \right|^2}{\sum_{\omega=\omega_1}^{\omega_2} H_{pq}(\omega)\, H_{pq}^*(\omega) \sum_{\omega=\omega_1}^{\omega_2} \hat{H}_{pq}(\omega)\, \hat{H}_{pq}^*(\omega)} \tag{21.90}$$

where $H_{pq}(\omega)$ = measured FRF
$\hat{H}_{pq}(\omega)$ = synthesized FRF

VISUAL VERIFICATION

Another common method of modal model validation is to evaluate the modal vectors visually. While this can be accomplished from plotted modal vectors superimposed upon the undeformed geometry, the modal vectors are normally animated (superimposed upon the undeformed geometry) in order to quickly assess the modal vector. Particularly, modal vectors are evaluated for physically realizable characteristics such as discontinuous motion or out-of-phase problems. Often, rigid-body modes of vibration are evaluated to determine scaling (calibration) errors or invalid measurement degree-of-freedom assignment or orientation. Naturally, if the system under test is believed to be proportionally damped, the modal vectors should be normal modes, and this characteristic can be quickly observed by viewing an animation of the modal vector.

FINITE ELEMENT ANALYSIS

The results of a finite element analysis of the system under test can provide another method of validating the modal model. While the problem of matching the number of analytical degrees of freedom N_a to the number of experimental DOF N_e causes some difficulty, the modal frequencies and modal vectors can be compared visually or through orthogonality or consistency checks. Unfortunately, when the comparison is not sufficiently acceptable, the question of error in the experimental model versus error in the analytical model cannot be easily resolved. Generally, assuming minimal errors and sufficient analysis and test experience, reasonable agreement can be found in the first 10 deformable modal vectors, but agreement for higher modal vectors is more difficult. Finite element analysis is discussed in detail in Chap. 28.

MODAL VECTOR ORTHOGONALITY

Another method that is used to validate an experimental modal model is the weighted orthogonality check. In this case, the experimental modal vectors are used together with a mass matrix normally derived from a finite element model to evaluate orthogonality. The experimental modal vectors are scaled so that the diagonal terms of the modal mass matrix are unity. With this form of scaling, the off-diagonal values in the modal mass matrix are expected to be less than 0.1 (10 percent of the diagonal terms).

Theoretically, for the case of proportional damping, each modal vector of a system is orthogonal to all other modal vectors of that system when weighted by the mass, stiffness, or damping matrix. In practice, these matrices are made available by way of a finite element analysis, and normally the mass matrix is considered to be the most accurate. For this reason, any further discussion of orthogonality is made with respect to mass matrix weighting. As a result, the orthogonality relations can be stated as follows:

Orthogonality of modal vectors:

$$\{\psi_r\}[M]\{\psi_s\} = 0 \qquad r \neq s \tag{21.91}$$

$$\{\psi_r\}[M]\{\psi_s\} = M_r \qquad r = s \tag{21.92}$$

Experimentally, the result of zero for the cross orthogonality [Eq. (21.91)] can rarely be achieved, but values up to one-tenth of the magnitude of the generalized mass of each mode are considered to be acceptable. It is a common procedure to form the modal vectors into a normalized set of mode shape vectors with respect to the mass matrix weighting. The accepted criterion in the aerospace industry, where this confidence check is made most often, is for all of the generalized mass terms to be unity and all cross-orthogonality terms to be less than 0.1. Often, even under this criterion, an attempt is made to adjust the modal vectors so that the cross-orthogonality conditions are satisfied.[51–55]

In Eqs. (21.91) and (21.92) the mass matrix must be an $N_o \times N_o$ matrix corresponding to the measurement locations on the structure. This means that the finite element mass matrix must be modified from whatever size and distribution of grid locations are required in the finite element analysis to the $N_o \times N_o$ square matrix corresponding to the measurement locations. This normally involves some sort of reduction algorithm as well as interpolation of grid locations to match the measurement situation.[54,55]

When Eq. (21.91) is not sufficiently satisfied, one (or more) of three situations may exist. First, the modal vectors can be invalid. This can be due to measurement error or problems with the modal parameter estimation algorithms. This is a very common assumption and many times contributes to the problem. Second, the mass matrix can be invalid. Since the mass matrix is not easily related to the physical properties of the system, this probably contributes significantly to the problem. Third, the reduction of the mass matrix can be invalid. This can certainly be a realistic problem and can cause severe errors. One example of this situation occurs when a relatively large amount of mass is reduced to a measurement location that is highly flexible, such as the center of an unsupported panel. In such a situation, the measurement location is weighted very heavily in the orthogonality calculation of Eq. (21.91) but may represent only incidental motion of the overall modal vector.

In all probability, all three situations contribute to the failure of cross-orthogonality criteria on occasion. When the orthogonality conditions are not satisfied, this result does not indicate where the problem originates. From an experimental point of view, it is important to try to develop methods that indicate confidence that the modal vector is or is not part of the problem.

MODAL VECTOR CONSISTENCY

Since the residue matrix contains redundant information with respect to a modal vector, the consistency of the estimate of the modal vector under varying conditions such as excitation location or modal parameter estimation algorithms can be a valuable confidence factor to be utilized in the process of evaluation of the experimental modal vectors.

The common approach to estimation of modal vectors from the frequency response function method is to measure a complete row or column of the FRF matrix. This gives reasonable definition to those modal vectors that have a nonzero modal coefficient at the excitation location and can be completely uncoupled with the forced normal mode excitation method. When the modal coefficient at the excitation location of a modal vector is zero (very small with respect to the dynamic range of the modal vector) or when the modal vectors cannot be uncoupled, the estimation of the modal vector contains potential bias and variance errors. In such cases, additional rows and/or columns of the FRF matrix are measured to detect such potential problems.

In these cases, information in the residue matrix corresponding to each pole of the system is evaluated to determine separate estimates of the same modal vector. This evaluation consists of the calculation of a complex modal scale factor (relating two modal vectors) and a scalar modal assurance criterion (measuring the consistency between two modal vectors). The function of the *modal scale factor* (MSF) is to provide a means of normalizing all estimates of the same modal vector. When two modal vectors are scaled similarly, elements of each vector can be averaged (with or without weighting), differenced, or sorted to provide a best estimate of the modal vector or to provide an indication of the type of error vector superimposed on the modal vector. In terms of multiple-reference modal parameter estimation algorithms, the MSF is a normalized estimate of the modal participation factor between two references for a specific mode of vibration. The function of the *modal assurance criterion* (MAC) is to provide a measure of consistency between estimates of a modal vector. This provides an additional confidence factor in the evaluation of a modal vector from different excitation locations. The MAC also provides a method of determining the degree of causality between estimates of different modal vectors from the same system.[56,57]

The *modal scale factor* is defined, according to this approach, as follows:

$$\text{MSF}_{cdr} = \frac{\{\psi_{cr}\}^H \{\psi_{dr}\}}{\{\psi_{dr}\}^H \{\psi_{dr}\}} \tag{21.93}$$

Equation (21.93) implies that the modal vector d is the reference to which the modal vector c is compared. In the general case, modal vector c can be considered to be made of two parts. The first part is the part correlated with modal vector d. The second part is the part that is not correlated with modal vector d and is made up of contamination from other modal vectors and of any random contribution. This error vector is considered to be noise. The *modal assurance criterion* is defined as a scalar constant relating the portion of the automoment of the modal vector that is linearly related to the reference modal vector as follows:

$$\mathrm{MAC}_{cdr} = \frac{\left|\{\psi_{cr}\}^H \{\psi_{dr}\}\right|^2}{\{\psi_{cr}\}^H \{\psi_{cr}\}\{\psi_{dr}\}^H \{\psi_{dr}\}} = \frac{\left(\{\psi_{cr}\}^H \{\psi_{dr}\}\right)\left(\{\psi_{dr}\}^H \{\psi_{cr}\}\right)}{\{\psi_{cr}\}^H \{\psi_{cr}\}\{\psi_{dr}\}^H \{\psi_{dr}\}} \tag{21.94}$$

The MAC is a scalar constant relating the causal relationship between two modal vectors. The constant takes on values from 0, representing no consistent correspondence, to 1, representing a consistent correspondence. In this manner, if the modal vectors under consideration truly exhibit a consistent relationship, the MAC should approach unity and the value of the MSF can be considered to be reasonable.

The MAC can indicate only consistency, not validity. If the same errors, random or bias, exist in all modal vector estimates, this is not delineated by the MAC. Invalid assumptions are normally the cause of this sort of potential error. Even though the MAC is unity, the assumptions involving the system or the modal parameter estimation techniques are not necessarily correct. The assumptions may cause consistent errors in all modal vectors under all test conditions verified by the MAC.

A number of other evaluation criteria have been developed based upon the same concept as the MAC.[57–58] The linear regression concept involved in the MAC is very useful whenever a linear relationship between two structural or measurement concepts is anticipated. Another similar concept, the coordinate modal assurance criterion (COMAC) is presented in the next section.

Coordinate Modal Assurance Criterion (COMAC). An extension of the modal assurance criterion is the coordinate modal assurance criterion (COMAC).[58] The COMAC attempts to identify which measurement degrees of freedom contribute negatively to a low value of the MAC. The COMAC is calculated over a set of mode pairs, analytical versus analytical, experimental versus experimental, or experimental versus analytical. The two modal vectors in each mode pair represent the same modal vector, but the set of mode pairs represents all modes of interest in a given frequency range. For two sets of modes that are to be compared, there is a value of the COMAC computed for each (measurement) DOF.

The *coordinate modal assurance criterion* is defined as follows:

$$\mathrm{COMAC}_p = \frac{\left|\sum_{r=1}^{N} \psi_{pr}\,\phi_{pr}\right|^2}{\sum_{r=1}^{N} \psi_{pr}\,\psi_{pr}^* \sum_{r=1}^{N} \phi_{pr}\,\phi_{pr}^*} \tag{21.95}$$

where ψ_{pr} = modal coefficient from (measured) DOF p and modal vector r from one set of modal vectors

ϕ_{pr} = modal coefficient from (measured) DOF p and modal vector r from a second set of modal vectors

Note that the preceding formulation assumes that there is a match for every mode in the two sets. Only those modes that match between the two sets are included in the computation.

MODAL MODIFICATION PREDICTION

The use of a modal model to predict changes in modal parameters caused by a perturbation (modification) of the system is becoming more of a reality as more measured data is acquired simultaneously. In this validation procedure, a modal model is estimated based upon a complete modal test. This modal model is used as the basis to predict a perturbation to the system that is tested, such as the addition of a mass at a particular point on the structure. Then the mass is added to the structure and the perturbed system is retested. The predicted and measured data or modal model can be compared and contrasted as a measure of the validity of the underlying modal model.

MODAL COMPLEXITY

Modal complexity is a variation of the use of sensitivity analysis in the validation of a modal model. When a mass is added to a structure, the modal frequencies should either be unaffected or should shift to a slightly lower frequency. Modal overcomplexity is a summation of this effect over all measured degrees of freedom for each mode. Modal complexity is particularly useful for the case of complex modes in an attempt to quantify whether the mode is genuinely a complex mode, a linear combination of several modes, or a computational artifact. The mode complexity is normally indicated by the *mode overcomplexity value* (MOV), which is the percentage of response points that actually causes the damped natural frequency to decrease when a mass is added compared to the total number of response points. A separate MOV is estimated for each mode of vibration, and the ideal result should be 1.0 (100 percent) for each mode.

MODAL PHASE COLINEARITY AND MEAN PHASE DEVIATION

For proportionally damped systems, each modal coefficient for a specific mode of vibration should differ by 0° or 180°. The *modal phase colinearity* (MPC) is an index expressing the consistency of the linear relationship between the real and imaginary parts of each modal coefficient. This concept is essentially the same as the ordinary coherence function with respect to the linear relationship of the frequency response function for different averages or the modal assurance criterion (MAC) with respect to the modal scale factor between modal vectors. The MPC should be 1.0 (100 percent) for a mode that is essentially a normal mode. A low value of MPC indicates a mode that is complex (after normalization) and is an indication of a nonproportionally damped system or errors in the measured data and/or modal parameter estimation.

Another indicator that defines whether a modal vector is essentially a normal mode is the *mean phase deviation* (MPD). This index is the statistical variance of the phase angles for each mode shape coefficient for a specific modal vector from the mean value of the phase angle. The MPD is an indication of the phase scatter of a modal vector and should be near 0° for a real, normal mode.

REFERENCES

1. Tse, F. S., I. E. Morse, Jr., R. T. Hinkle: "Mechanical Vibrations: Theory and Applications, Second Edition," Prentice-Hall, Englewood Cliffs, N.J., 1978.

2. Craig, R. R., Jr.: "Structural Dynamics: An Introduction to Computer Methods," John Wiley and Sons, New York, 1981.

3. Ewins, D.: "Modal Testing: Theory and Practice," John Wiley and Sons, New York, 1984.

4. Bendat, J. S., and A. G. Piersol: "Random Data: Analysis and Measurement Procedures," John Wiley and Sons, New York, 1971.

5. Bendat, J. S., and A. G. Piersol: "Engineering Applications of Correlation and Spectral Analysis," John Wiley and Sons, New York, 1980.

6. Otnes, R. K., and L. Enochson: "Digital Time Series Analysis," John Wiley and Sons, New York, 1972.

7. Dally, J. W., W. F. Riley, and K. G. McConnell: "Instrumentation for Engineering Measurements," John Wiley & Sons, New York, 1984.

8. Strang, G.: "Linear Algebra and Its Applications, Third Edition," Harcourt Brace Jovanovich, San Diego, 1988.

9. Lawson, C. L., and R. J. Hanson: "Solving Least Squares Problems," Prentice-Hall, Englewood Cliffs, N.J., 1974.

10. Jolliffe, I. T.: "Principal Component Analysis," Springer-Verlag, New York, 1986.

11. Ljung, Lennart: "System Identification: Theory for the User," Prentice-Hall, Englewood Cliffs, N.J., 1987.

12. Wilkinson, J. H.: "The Algebraic Eigenvalue Problem," Clarendon Press, Oxford, U.K., 1965.

13. Allemang, R. J., D. L. Brown, and R. W. Rost: "Dual Input Estimation of Frequency Response Functions for Experimental Modal Analysis of Automotive Structures," *SAE Paper No. 820193,* 1982.

14. Allemang, R. J., R. W. Rost, and D. L. Brown: "Dual Input Estimation of Frequency Response Functions for Experimental Modal Analysis of Aircraft Structures," *Proc. International Modal Analysis Conference,* 1982, pp. 333–340.

15. Potter, R. W.: "Matrix Formulation of Multiple and Partial Coherence," *Journal of the Acoustic Society of America,* **66**(3):776–781 (1977).

16. Brown, D. L., G. Carbon, and R. D. Zimmerman: "Survey of Excitation Techniques Applicable to the Testing of Automotive Structures," *SAE Paper No. 770029,* 1977.

17. Halvorsen, W. G., and D. L. Brown: "Impulse Technique for Structural Frequency Response Testing," *Sound and Vibration,* November 1977, pp. 8–21.

18. Phillips, A. W., and R. J. Allemang: "An Overview of MIMO-FRF Excitation/Averaging/Processing Techniques," *Journal of Sound and Vibration,* **262**(3):651–675 (2003).

19. Brown, D. L., R. J. Allemang, R. D. Zimmerman, and M. Mergeay: "Parameter Estimation Techniques for Modal Analysis," *SAE Transactions,* **88:**828–846 (1979).

20. Vold, H., J. Kundrat, T. Rocklin, and R. Russell: "A Multi-Input Modal Estimation Algorithm for Mini-Computers," *SAE Transactions,* **91**(1):815–821 (1982).

21. Vold, H., and T. Rocklin: "The Numerical Implementation of a Multi-Input Modal Estimation Algorithm for Mini-Computers," *Proc. International Modal Analysis Conference,* 1982, pp. 542–548.

22. Ibrahim, S. R., and E. C. Mikulcik: "A Method for the Direct Identification of Vibration Parameters from the Free Response," *Shock and Vibration Bulletin,* **47**(Part 4):183–198 (1977).

23. Pappa, R. S.: "Some Statistical Performance Characteristics of the ITD Modal Identification Algorithm," *AIAA Paper Number 82-0768,* 1982.

24. Fukuzono, K.: "Investigation of Multiple-Reference Ibrahim Time Domain Modal Parameter Estimation Technique," Master's Thesis, Dept. of Mechanical and Industrial Engineering, University of Cincinnati, Cincinnati, Ohio, 1986.

25. Juang, Jer-Nan, and Richard S. Pappa: "An Eigensystem Realization Algorithm for Modal Parameter Identification and Model Reduction," *AIAA Journal of Guidance, Control, and Dynamics,* **8**(4):620–627 (1985).

26. Juang, J. N.: "Mathematical Correlation of Modal Parameter Identification Methods Via System Realization Theory," *Journal of Analytical and Experimental Modal Analysis,* **2**(1):1–18 (1987).

27. Longman, Richard W., and Jer-Nan Juang: "Recursive Form of the Eigensystem Realization Algorithm for System Identification," *AIAA, Journal of Guidance, Control, and Dynamics,* **12**(5):647–652 (1989).

28. Zhang, L., H. Kanda, D. L. Brown, and R. J. Allemang: "A Polyreference Frequency Domain Method for Modal Parameter Identification," *ASME Paper No. 85-DET-106,* 1985.

29. Lembregts, F., J. Leuridan, L. Zhang, and H. Kanda: "Multiple Input Modal Analysis of Frequency Response Functions based on Direct Parameter Identification," *Proc. International Modal Analysis Conference,* 1986, pp. 589–598.

30. Lembregts, F., J. L. Leuridan, and H. Van Brussel: "Frequency Domain Direct Parameter Identification for Modal Analysis: State Space Formulation," *Mechanical Systems and Signal Processing,* **4**(1):65–76 (1989).

31. Coppolino, R. N.: "A Simultaneous Frequency Domain Technique for Estimation of Modal Parameters from Measured Data," *SAE Paper No. 811046,* 1981.

32. Craig, R. R., A. J. Kurdila, and H. M. Kim: "State-Space Formulation of Multi-Shaker Modal Analysis," *Journal of Analytical and Experimental Modal Analysis,* **5**(3):169–183 (1990).

33. Richardson, M., and D. L. Formenti: "Parameter Estimation from Frequency Response Measurements Using Rational Fraction Polynomials," *Proc. International Modal Analysis Conference,* 1982, pp. 167–182.

34. Vold, H.: "Orthogonal Polynomials in the Polyreference Method," *Proc. International Seminar on Modal Analysis,* Katholieke Universiteit Leuven, Belgium, 1986.

35. Vold, H., and C. Y. Shih: "Numerical Sensitivity of the Characteristic Polynomial," *Proc. International Seminar on Modal Analysis,* Katholieke Universiteit Leuven, Belgium, 1988.

36. Shih, C. Y., Y. G. Tsuei, R. J. Allemang, and D. L. Brown: "A Frequency Domain Global Parameter Estimation Method for Multiple Reference Frequency Response Measurements," *Mechanical System and Signal Processing,* **2**(4):349–365 (1988).

37. Van der Auweraer, H., and J. Leuridan: "Multiple Input Orthogonal Polynomial Parameter Estimation," *Mechanical Systems and Signal Processing,* **1**(3):259–272 (1987).

38. Vold, H.: "Numerically Robust Frequency Domain Modal Parameter Estimation," *Sound and Vibration* **24**(1):38–40 (January 1990).

39. Vold, H., K. Napolitano, D. Hensley, and M. Richardson: "Aliasing in Modal Parameter Estimation, An Historical Look and New Innovations," *Proc. International Modal Analysis Conference,* 2007; *Sound and Vibration* **42**(1):12–17 (January 2008).

40. Van der Auweraer, H., P. Guillaume, P. Verboven, and S. Vanlanduit: "Application of a Fast-Stabilizing Frequency Domain Parameter Estimation Method," *ASME Journal of Dynamic Systems, Measurement and Control,* **123**(4):651–658 (December 2001).

41. Guillaume, P., P. Verboven, S. Vanlanduit, H. Van der Auweraer, and B. Peeters: "A Polyreference Implementation of the Least-Squares Complex Frequency Domain Estimator," *Proc. International Modal Analysis Conference,* 2003.

42. Verboven, P., B. Cauberghe, S. Vanlanduit, E. Parloo, and P. Guillaume: "The Secret Behind Clear Stabilization Diagrams: The Influence of the Parameter Constraint on the Stability of the Poles," *Proc. Society of Experimental Mechanics (SEM) Annual Conference,* 2004.

43. Shih, C. Y., Y. G. Tsuei, R. J. Allemang, and D. L. Brown: "Complex Mode Indication Function and Its Application to Spatial Domain Parameter Estimation," *Mechanical System and Signal Processing,* **2**(4):367–377 (1988).

44. Allemang, R. J., D. L. Brown, and W. Fladung: "Modal Parameter Estimation: A Unified Matrix Polynomial Approach," *Proc. International Modal Analysis Conference,* 1994, pp. 501–514.

45. Allemang, R. J., and D. L. Brown: "A Unified Matrix Polynomial Approach to Modal Identification," *Journal of Sound and Vibration,* **211**(3):301–322 (April 1998).

46. Allemang, R. J., and A. W. Phillips: "The Unified Matrix Polynomial Approach to Understanding Modal Parameter Estimation: An Update," *Proc. International Conference on Noise and Vibration Engineering,* Katholieke Universiteit Leuven, Belgium, 2004.

47. Williams, R., J. Crowley, and H. Vold: "The Multivariable Mode Indicator Function in Modal Analysis," *Proc. International Modal Analysis Conference,* 1985, pp. 66–70.

48. Phillips, A. W., R. J. Allemang, and C. R. Pickrel: "Clustering of Modal Frequency Estimates from Different Solution Sets," *Proc. International Modal Analysis Conference,* 1997, pp. 1053–1063.

49. Dippery, K. D., A. W. Phillips, and R. J. Allemang: "An SVD Condensation of the Spatial Domain in Modal Parameter Estimation," *Proc. International Modal Analysis Conference,* 1994.

50. Dippery, K. D., A. W. Phillips, and R. J. Allemang: "Spectral Decimation in Low Order Frequency Domain Modal Parameter Estimation," *Proc. International Modal Analysis Conference,* 1994.

51. Gravitz, S. I.: "An Analytical Procedure for Orthogonalization of Experimentally Measured Modes," *Journal of the Aero/Space Sciences,* **25:**721–722 (1958).

52. McGrew, J.: "Orthogonalization of Measured Modes and Calculation of Influence Coefficients," *AIAA Journal,* **7**(4):774–776 (1969).

53. Targoff, W. P.: "Orthogonality Check and Correction of Measured Modes," *AIAA Journal,* **14**(2):164–167 (1976).

54. Guyan, R. J.: "Reduction of Stiffness and Mass Matrices," *AIAA Journal,* **3**(2):380 (1965).

55. Irons, B.: "Structural Eigenvalue Problems: Elimination of Unwanted Variables," *AIAA Journal,* **3**(5):961–962 (1965).

56. Allemang, R. J., and D. L. Brown: "A Correlation Coefficient for Modal Vector Analysis," *Proc. International Modal Analysis Conference,* 1982, pp. 110–116.

57. Allemang, R. J.: "The Modal Assurance Criterion (MAC): Twenty Years of Use and Abuse," *Sound and Vibration,* **37**(8):14–23 (August 2003).

58. Lieven, N. A. J., and D. J. Ewins: "Spatial Correlation of Mode Shapes, The Coordinate Modal Assurance Criterion (COMAC)," *Proc. International Modal Analysis Conference,* 1988, pp. 690–695.

CHAPTER 22
MATRIX METHODS OF ANALYSIS

Stephen H. Crandall
Robert B. McCalley, Jr.

INTRODUCTION

The mathematical language which is most convenient for analyzing multiple-degree-of-freedom vibratory systems is that of *matrices.* Matrix notation simplifies the preliminary analytical study, and in situations where particular numerical answers are required, matrices provide a standardized format for organizing the data and the computations. Computations with matrices can be carried out by hand or by digital computers. The availability of programs such as MATLAB makes the solution of many complex problems in vibration analysis a matter of routine.

This chapter describes how matrices are used in vibration analysis. It begins with definitions and rules for operating with matrices. The formulation of vibration problems in matrix notation then is treated. This is followed by general matrix solutions of several important types of vibration problems, including free and forced vibrations of both undamped and damped linear multiple-degree-of-freedom (MDOF) systems.

MATRICES

Matrices are mathematical entities which facilitate the handling of simultaneous equations. They are applied to the differential equations of a vibratory system as follows:

A single-degree-of-freedom (SDOF) system of the type in Fig. 22.1 has the differential equation

$$m\ddot{x} + c\dot{x} + kx = F$$

where m is the mass, c is the damping coefficient, k is the stiffness, F is the applied force, x is the displacement coordinate, and dots denote time derivatives. In Fig. 22.2 a similar three degree-of-freedom system is shown. The equations of motion may be obtained by applying Newton's second law to each mass in turn:

$$\begin{aligned} m\ddot{x}_1 \qquad\qquad\qquad + c\dot{x}_1 \qquad\qquad + 5kx_1 - 2kx_2 \qquad &= F_1 \\ 2m\ddot{x}_2 \qquad\qquad + 2c\dot{x}_2 - 2c\dot{x}_3 - 2kx_1 + 3kx_2 - kx_3 &= F_2 \\ 3m\ddot{x}_3 \qquad - 2c\dot{x}_2 + 2c\dot{x}_3 \qquad\qquad - kx_2 + kx_3 &= F_3 \end{aligned} \tag{22.1}$$

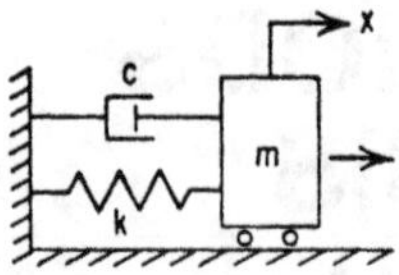

FIGURE 22.1 Single-degree-of-freedom system.

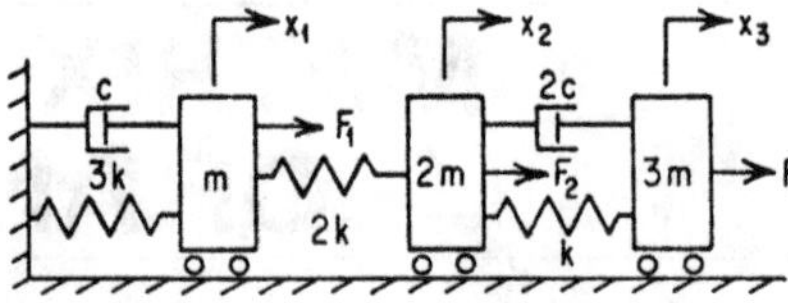

FIGURE 22.2 Three-degree-of-freedom system.

The accelerations, velocities, displacements, and forces may be organized into columns, denoted by single boldface symbols:

$$\ddot{\mathbf{x}} = \begin{bmatrix} \ddot{x}_1 \\ \ddot{x}_2 \\ \ddot{x}_3 \end{bmatrix} \quad \dot{\mathbf{x}} = \begin{bmatrix} \dot{x}_1 \\ \dot{x}_2 \\ \dot{x}_3 \end{bmatrix} \quad \mathbf{x} = \begin{bmatrix} x_1 \\ x_2 \\ x_3 \end{bmatrix} \quad \mathbf{f} = \begin{bmatrix} F_1 \\ F_2 \\ F_3 \end{bmatrix} \tag{22.2}$$

The inertia, damping, and stiffness coefficients may be organized into square arrays:

$$\mathbf{M} = \begin{bmatrix} m & 0 & 0 \\ 0 & 2m & 0 \\ 0 & 0 & 3m \end{bmatrix} \quad \mathbf{C} = \begin{bmatrix} c & 0 & 0 \\ 0 & 2c & -2c \\ 0 & -2c & 2c \end{bmatrix} \quad \mathbf{K} = \begin{bmatrix} 5k & -2k & 0 \\ -2k & 3k & -k \\ 0 & -k & k \end{bmatrix} \tag{22.3}$$

By using these symbols, it is shown below that it is possible to represent the three equations of Eq. (22.1) by the following single equation:

$$\mathbf{M}\ddot{\mathbf{x}} + \mathbf{C}\dot{\mathbf{x}} + \mathbf{K}\mathbf{x} = \mathbf{f} \tag{22.4}$$

Note that this has the same form as the differential equation for the SDOF system of Fig. 22.1. The notation of Eq. (22.4) has the advantage that in systems of many degrees of freedom (DOF) it clearly states the physical principle that at every coordinate the external force is the sum of the inertia, damping, and stiffness forces. Equation (22.4) is an abbreviation for Eq. (22.1). It is necessary to develop the rules of operation with symbols such as those in Eqs. (22.2) and (22.3) to ensure that no ambiguity is involved. The algebra of *matrices* is devised to facilitate manipulations of simultaneous equations such as Eq. (22.1). Matrix algebra does not in any way simplify individual operations such as multiplication or addition of numbers, but it is an organizational tool which permits one to keep track of a complicated sequence of operations in an optimum manner. Matrices are essential elements of linear algebra,[1] and are widely employed in structural analysis[2] and vibration analysis.[3]

DEFINITIONS

A *matrix* is an array of elements arranged systematically in rows and columns. For example, a rectangular matrix **A**, of elements a_{jk}, which has m rows and n columns is

$$\mathbf{A} = [a_{jk}] = \begin{bmatrix} a_{11} & a_{12} & \dots & a_{1n} \\ a_{21} & a_{22} & \dots & a_{2n} \\ \dots & \dots & \dots & \dots \\ a_{m1} & a_{m2} & \dots & a_{mn} \end{bmatrix}$$

The elements a_{jk} are usually numbers or functions, but, in principle, they may be any well-defined quantities. The first subscript j on the element refers to the row number, while the second subscript k refers to the column number. The array is denoted by the single symbol **A**, which can be used as such during operational manipulations in which it is not necessary to specify continually all the elements a_{jk}. When a numerical calculation is finally required, it is necessary to refer back to the explicit specifications of the elements a_{jk}.

A rectangular matrix with m rows and n columns is said to be of order (m,n). A matrix of order (n,n) is a *square matrix* and is said to be simply a square matrix of order n. A matrix of order $(n,1)$ is a *column matrix* and is said to be simply a column matrix of order n. A column matrix is sometimes referred to as a *column vector*. Similarly, a matrix of order $(1,n)$ is a *row matrix* or a *row vector*. Boldface *capital* letters are used here to represent square matrices and *lowercase* boldface letters to represent column matrices or vectors. For example, the matrices in Eq. (22.2) are column matrices of order three and the matrices in Eq. (22.3) are square matrices of order three.

Some special types of matrices are:

1. A *diagonal matrix* is a square matrix **A** whose elements a_{jk} are zero when $j \neq k$. The only nonzero elements are those on the *main diagonal*, where $j = k$. In order to emphasize that a matrix is diagonal, it is often written with small ticks in the direction of the main diagonal:

$$\mathbf{A} = \left\lceil a_{jj} \right\rfloor$$

2. A *unit matrix* or *identity matrix* is a diagonal matrix whose main diagonal elements are each equal to unity. The symbol **I** is used to denote a unit matrix. Examples are

$$\begin{bmatrix} 1 & 0 \\ 0 & 1 \end{bmatrix} \qquad \begin{bmatrix} 1 & 0 & 0 \\ 0 & 1 & 0 \\ 0 & 0 & 1 \end{bmatrix}$$

3. A *null matrix* or *zero matrix* has all its elements equal to zero and is simply written as zero.

4. The *transpose* $\mathbf{A}^T$ of a matrix **A** is a matrix having the same elements but with rows and columns interchanged. Thus, if the original matrix is

$$\mathbf{A} = [a_{jk}]$$

the transpose matrix is

$$\mathbf{A}^T = [a_{jk}]^T = [a_{kj}]$$

For example:

$$\mathbf{A} = \begin{bmatrix} 3 & 2 \\ -1 & 4 \end{bmatrix} \qquad \mathbf{A}^T = \begin{bmatrix} 3 & -1 \\ 2 & 4 \end{bmatrix}$$

The transpose of a square matrix may be visualized as the matrix obtained by rotating the given matrix about its main diagonal as an axis.

The transpose of a column matrix is a row matrix. For example,

$$\mathbf{x} = \begin{bmatrix} 3 \\ 4 \\ -2 \end{bmatrix} \qquad \mathbf{x}^T = [3 \quad 4 \quad -2]$$

Throughout this chapter a row matrix is referred to as the transpose of the corresponding column matrix.

5. A *symmetric matrix* is a square matrix whose off-diagonal elements are symmetric with respect to the main diagonal. A square matrix **A** is symmetric if, for all j and k,

$$a_{jk} = a_{kj}$$

A symmetric matrix is equal to its transpose. For example, all three of the matrices in Eq. (22.3) are symmetric. In addition, the matrix **M** is a diagonal matrix.

MATRIX OPERATIONS

Equality of Matrices. Two matrices of the same order are equal if their corresponding elements are equal. Thus, two matrices **A** and **B** are equal if, for every j and k,

$$a_{jk} = b_{jk}$$

Matrix Addition and Subtraction. Addition or subtraction of matrices of the same order is performed by adding or subtracting corresponding elements. Thus, $\mathbf{A} + \mathbf{B} = \mathbf{C}$ if for every j and k,

$$a_{jk} + b_{jk} = c_{jk}$$

For example, if

$$\mathbf{A} = \begin{bmatrix} 3 & 2 \\ -1 & 4 \end{bmatrix} \qquad \mathbf{B} = \begin{bmatrix} -1 & 2 \\ 5 & 6 \end{bmatrix}$$

then

$$\mathbf{A} + \mathbf{B} = \begin{bmatrix} 2 & 4 \\ 4 & 10 \end{bmatrix} \qquad \mathbf{A} - \mathbf{B} = \begin{bmatrix} 4 & 0 \\ -6 & -2 \end{bmatrix}$$

Multiplication of a Matrix by a Scalar. Multiplication of a matrix by a scalar c multiplies each element of the matrix by c. Thus,

$$c\mathbf{A} = c[a_{jk}] = [ca_{jk}]$$

In particular, the negative of a matrix has the sign of every element changed.

Matrix Multiplication. If **A** is a matrix of order (m,n) and **B** is a matrix of order (n,p), then their *matrix product* $\mathbf{AB} = \mathbf{C}$ is defined to be a matrix **C** of order (m,p) where, for every j and k,

$$c_{jk} = \sum_{r=1}^{n} a_{jr} b_{rk} \tag{22.5}$$

The product of two matrices can be obtained only if they are *conformable*, i.e., if the number of columns in **A** is equal to the number of rows in **B**. The symbolic equation

$$(m,n) \times (n,p) = (m,p)$$

indicates the orders of the matrices involved in a matrix product. Matrix products are not commutative, i.e., in general,

$$\mathbf{AB} \neq \mathbf{BA}$$

The matrix products which appear in this chapter are of the following types:

Square matrix × square matrix = square matrix
Square matrix × column vector = column vector
Row vector × square matrix = row vector
Row vector × column vector = scalar
Column vector × row vector = square matrix

In all cases, the matrices must be conformable. Numerical examples are given below.

$$\mathbf{AB} = \begin{bmatrix} 3 & 2 \\ -1 & 4 \end{bmatrix} \begin{bmatrix} -1 & 2 \\ 5 & 6 \end{bmatrix} = \begin{bmatrix} -(3\times 1) + (2\times 5) & (3\times 2) + (2\times 6) \\ (1\times 1) + (4\times 5) & -(1\times 2) + (4\times 6) \end{bmatrix} = \begin{bmatrix} 7 & 18 \\ 21 & 22 \end{bmatrix}$$

$$\mathbf{Ax} = \begin{bmatrix} 3 & 2 \\ -1 & 4 \end{bmatrix} \begin{bmatrix} 5 \\ 3 \end{bmatrix} = \begin{bmatrix} (3\times 5) + (2\times 3) \\ -(1\times 5) + (4\times 3) \end{bmatrix} = \begin{bmatrix} 21 \\ 7 \end{bmatrix}$$

$$\mathbf{y}^T\mathbf{A} = [-2 \quad 1] \begin{bmatrix} 3 & 2 \\ -1 & 4 \end{bmatrix} = [-(2\times 3) - (1\times 1) - (2\times 2) + (1\times 4)] = [-7 \quad 0]$$

$$\mathbf{y}^T\mathbf{x} = [-2 \quad 1] \begin{bmatrix} 5 \\ 3 \end{bmatrix} = (-10 + 3) = -7$$

$$\mathbf{xy}^T = \begin{bmatrix} 5 \\ 3 \end{bmatrix} [-2 \quad 1] = \begin{bmatrix} -(5\times 2) & (5\times 1) \\ -(3\times 2) & (3\times 1) \end{bmatrix} = \begin{bmatrix} -10 & 5 \\ -6 & 3 \end{bmatrix}$$

The last product always results in a matrix with proportional rows and columns.

The operation of matrix multiplication is particularly suited for representing systems of simultaneous linear equations in a compact form in which the coefficients are gathered into square matrices and the unknowns are placed in column matrices. For example, it is the operation of matrix multiplication which gives unambiguous meaning to the matrix abbreviation in Eq. (22.4) for the three simultaneous differential equations of Eq. (22.1). The two sides of Eq. (22.4) are column matrices of order three whose corresponding elements must be equal. On the right, these elements are simply the external forces at the three masses. On the left, Eq. (22.4) states that the resulting column is the sum of three column matrices, each of which results from the matrix multiplication of a square matrix of coefficients defined in Eq. (22.3) into a column matrix defined in Eq. (22.2). The rules of matrix operation just given ensure that Eq. (22.4) is exactly equivalent to Eq. (22.1).

Premultiplication or postmultiplication of a square matrix by the identity matrix leaves the original matrix unchanged; i.e.,

$$\mathbf{IA} = \mathbf{AI} = \mathbf{A}$$

Two symmetrical matrices multiplied together are generally not symmetric. The product of a matrix and its transpose is symmetric.

Continued matrix products such as **ABC** are defined, provided the number of columns in each matrix is the same as the number of rows in the matrix immediately following it. From the definition of matrix products, it follows that the *associative law* holds for continued products:

$$(\mathbf{AB})\mathbf{C} = \mathbf{A}(\mathbf{BC})$$

A square matrix **A** multiplied by itself yields a square matrix which is called the *square of the matrix* **A** and is denoted by $\mathbf{A}^2$. If $\mathbf{A}^2$ is in turn multiplied by **A**, the resulting matrix is $\mathbf{A}^3 = \mathbf{A}(\mathbf{A}^2) = \mathbf{A}^2(\mathbf{A})$. Extension of this process gives meaning to $\mathbf{A}^m$ for any positive integer *power m.* Powers of symmetric matrices are themselves symmetric.

The rule for *transposition* of matrix products is

$$(\mathbf{AB})^T = \mathbf{B}^T\mathbf{A}^T$$

Inverse or Reciprocal Matrix. If, for a given square matrix **A**, a square matrix $\mathbf{A}^{-1}$ can be found such that

$$\mathbf{A}^{-1}\mathbf{A} = \mathbf{A}\mathbf{A}^{-1} = \mathbf{I} \tag{22.6}$$

then $\mathbf{A}^{-1}$ is called the *inverse* or *reciprocal* of **A**. Not every square matrix **A** possesses an inverse. If the determinant constructed from the elements of a square matrix is zero, the matrix is said to be *singular* and there is no inverse. Every nonsingular matrix possesses a unique inverse. The inverse of a symmetric matrix is symmetric. The rule for the *inverse of a matrix product* is

$$(\mathbf{AB})^{-1} = (\mathbf{B}^{-1})(\mathbf{A}^{-1})$$

The solution to the set of simultaneous equations

$$\mathbf{Ax} = \mathbf{c}$$

where **x** is the unknown vector and **c** is a known input vector can be indicated with the aid of the inverse of **A**. The formal solution for **x** proceeds as follows:

$$\mathbf{A}^{-1}\mathbf{Ax} = \mathbf{A}^{-1}\mathbf{c}$$
$$\mathbf{Ix} = \mathbf{x} = \mathbf{A}^{-1}\mathbf{c}$$

When the inverse $\mathbf{A}^{-1}$ is known, the solution vector **x** is obtained by a simple matrix multiplication of $\mathbf{A}^{-1}$ into the input vector **c**.

Calculation of inverses and the solutions of simultaneous linear equations are readily performed for surprisingly large values of n by programs such as MATLAB. When $n = 2$ and

$$\mathbf{A} = \begin{bmatrix} a_{11} & a_{12} \\ a_{21} & a_{22} \end{bmatrix} \qquad \mathbf{x} = \begin{bmatrix} x_1 \\ x_2 \end{bmatrix} \qquad \mathbf{c} = \begin{bmatrix} c_1 \\ c_2 \end{bmatrix}$$

hand-computation is possible using the following formulas:

$$\mathbf{A}^{-1} = \frac{1}{\Delta}\begin{bmatrix} a_{22} & -a_{12} \\ -a_{21} & a_{11} \end{bmatrix} \qquad x_1 = \frac{\Delta_1}{\Delta} \qquad x_2 = \frac{\Delta_2}{\Delta}$$

where the determinants have the values

$$\Delta = a_{11}a_{22} - a_{12}a_{21} \qquad \Delta_1 = c_1a_{22} - c_2a_{12} \qquad \Delta_2 = c_2a_{11} - c_1a_{21}$$

QUADRATIC FORMS

A general quadratic form Q of order n may be written as

$$Q = \sum_{j=1}^{n} \sum_{k=1}^{n} a_{jk}x_jx_k$$

where the a_{jk} are constants and the x_j are the n variables. The form is quadratic since it is of the second degree in the variables. The laws of matrix multiplication permit Q to be written as

$$Q = [x_1x_2 \ldots x_n] \begin{bmatrix} a_{11} & a_{12} & \ldots & a_{1n} \\ a_{21} & a_{22} & \ldots & a_{2n} \\ \ldots & \ldots & \ldots & \ldots \\ a_{n1} & a_{n2} & \ldots & a_{nn} \end{bmatrix} \begin{bmatrix} x_1 \\ x_2 \\ \ldots \\ x_n \end{bmatrix}$$

which is

$$Q = \mathbf{x}^T\mathbf{A}\mathbf{x}$$

Any quadratic form can be expressed in terms of a symmetric matrix. If the given matrix $\mathbf{A}$ is not symmetric, it can be replaced by the symmetric matrix

$$\mathbf{B} = \tfrac{1}{2}(\mathbf{A} + \mathbf{A}^T)$$

without changing the value of the form.

As an example of a quadratic form, the *potential energy* V for the system of Fig. 22.2 is given by

$$\begin{aligned} 2V &= 3kx_1^2 + 2k(x_2 - x_1)^2 + k(x_3 - x_2)^2 \\ &= 5kx_1x_1 - 2kx_1x_2 \\ &\quad - 2kx_2x_1 + 3kx_2x_2 - kx_2x_3 \\ &\quad - kx_3x_2 + kx_3x_3 \end{aligned}$$

Using the displacement vector $\mathbf{x}$ defined in Eq. (22.2) and the stiffness matrix $\mathbf{K}$ in Eq. (22.3), the potential energy may be written as

$$V = \tfrac{1}{2}\mathbf{x}^T\mathbf{K}\mathbf{x}$$

Similarly, the *kinetic energy* T is given by

$$2T = m\dot{x}_1^2 + 2m\dot{x}_2^2 + 3m\dot{x}_3^2$$

In terms of the inertia matrix $\mathbf{M}$ and the velocity vector $\dot{\mathbf{x}}$ defined in Eqs. (22.3) and (22.2), the kinetic energy may be written as

$$T = \tfrac{1}{2}\dot{\mathbf{x}}^T\mathbf{M}\dot{\mathbf{x}}$$

The *dissipation function* D for the system is given by

$$
\begin{aligned}
2D &= c\dot{x}_1^2 + 2c(\dot{x}_3 - \dot{x}_2)^2 \\
&= c\dot{x}_1\dot{x}_1 \\
&\qquad + 2c\dot{x}_2\dot{x}_2 - 2c\dot{x}_2\dot{x}_3 \\
&\qquad - 2c\dot{x}_3\dot{x}_2 + 2c\dot{x}_3\dot{x}_3
\end{aligned}
$$

In terms of the velocity vector $\dot{\mathbf{x}}$ and the damping matrix $\mathbf{C}$ defined in Eqs. (22.2) and (22.3), the dissipation function may be written as

$$D = \tfrac{1}{2}\dot{\mathbf{x}}^T\mathbf{C}\dot{\mathbf{x}}$$

The dissipation function gives half the rate at which energy is being dissipated in the system.

While quadratic forms assume positive and negative values in general, the three physical forms just defined are intrinsically *positive* for a vibrating system with linear springs, constant masses, and viscous damping; i.e., they can never be negative for a real motion of the system. Kinetic energy is zero only when the system is at rest. The same thing is not necessarily true for potential energy or the dissipation function.

Depending upon the arrangement of springs and dashpots in the system, there may exist motions which do not involve any potential energy or dissipation. For example, in vibratory systems where rigid-body motions are possible (crankshaft torsional systems, free-free beams, etc.), no elastic energy is involved in the rigid-body motions. Also, in Fig. 22.2, if x_1 is zero while x_2 and x_3 have the same motion, there is no energy dissipated and the dissipation function is zero. To distinguish between these two possibilities, a quadratic form is called *positive definite* if it is never negative and if the only time it vanishes is when all the variables are zero. Kinetic energy is always positive definite, while potential energy and the dissipation function are positive but not necessarily positive definite. It depends upon the particular configuration of a given system whether the potential energy and the dissipation function are positive definite or only positive. The terms positive and positive definite are applied also to the matrices from which the quadratic forms are derived. For example, of the three matrices defined in Eq. (22.3), the matrices $\mathbf{M}$ and $\mathbf{K}$ are positive definite, but $\mathbf{C}$ is only positive. It can be shown that a matrix which is positive but not positive definite is *singular.*

Differentiation of Quadratic Forms. In forming Lagrange's equations of motion for a vibrating system,* it is necessary to take derivatives of the potential energy V, the kinetic energy T, and the dissipation function D. When these quadratic forms are represented in matrix notation, it is convenient to have matrix formulas for differentiation. In this paragraph rules are given for differentiating the slightly more general *bilinear form*

$$F = \mathbf{x}^T\mathbf{A}\mathbf{y} = \mathbf{y}^T\mathbf{A}\mathbf{x}$$

where $\mathbf{x}^T$ is a row vector of n variables x_j, $\mathbf{A}$ is a square matrix of constant coefficients, and $\mathbf{y}$ is a column matrix of n variables y_j. In a quadratic form the x_j are identical with the y_j.

For generality it is assumed that the x_j and the y_j are functions of n other variables u_j. In the formulas below, the notation $\mathbf{X}_u$ is used to represent the following square matrix:

* See Chap. 2 for a detailed discussion of Lagrange's equations.

$$\mathbf{X}_u = \begin{bmatrix} \frac{\partial x_1}{\partial u_1} & \frac{\partial x_2}{\partial u_1} & \cdots & \frac{\partial x_n}{\partial u_1} \\ \frac{\partial x_1}{\partial u_2} & \frac{\partial x_2}{\partial u_2} & \cdots & \frac{\partial x_n}{\partial u_2} \\ \cdots & \cdots & \cdots & \cdots \\ \frac{\partial x_1}{\partial u_n} & \frac{\partial x_2}{\partial u_n} & \cdots & \frac{\partial x_n}{\partial u_n} \end{bmatrix}$$

Now letting $\partial/\partial\mathbf{u}$ stand for the column vector whose elements are the partial differential operators with respect to the u_j, the general differentiation formula is

$$\frac{\partial F}{\partial \mathbf{u}} = \begin{bmatrix} \frac{\partial F}{\partial u_1} \\ \frac{\partial F}{\partial u_2} \\ \vdots \\ \frac{\partial F}{\partial u_n} \end{bmatrix} = \mathbf{X}_u\mathbf{A}\mathbf{y} + \mathbf{Y}_u\mathbf{A}^T\mathbf{x}$$

For a quadratic form $Q = \mathbf{x}^T\mathbf{A}\mathbf{x}$ the above formula reduces to

$$\frac{\partial Q}{\partial \mathbf{u}} = \mathbf{X}_u(\mathbf{A} + \mathbf{A}^T)\mathbf{x}$$

Thus, whether $\mathbf{A}$ is symmetric or not, this kind of differentiation produces a *symmetrical* matrix of coefficients $(\mathbf{A} + \mathbf{A}^T)$. It is this fact which ensures that vibration equations in the form obtained from Lagrange's equations always have symmetrical matrices of coefficients. If $\mathbf{A}$ is symmetrical to begin with, the previous formula becomes

$$\frac{\partial Q}{\partial \mathbf{u}} = 2\mathbf{X}_u\mathbf{A}\mathbf{x}$$

Finally, in the important special case where the x_j are identical with the u_j, the matrix $\mathbf{X}_x$ reduces to the identity matrix, yielding

$$\frac{\partial Q}{\partial \mathbf{x}} = 2\mathbf{A}\mathbf{x} \tag{22.7}$$

which is employed in the following section in developing Lagrange's equations.

FORMULATION OF VIBRATION PROBLEMS IN MATRIX FORM

Consider a holonomic linear mechanical system with n degrees of freedom which vibrates about a stable equilibrium configuration. Let the motion of the system be described by *n generalized displacements* $x_j(t)$ which vanish in the equilibrium position. The potential energy V can then be expressed in terms of these displacements as a quadratic form. The kinetic energy T and the dissipation function D can be expressed as quadratic forms in the generalized velocities $\dot{x}_j(t)$.

The equations of motion are obtained by applying Lagrange's equations

$$\frac{d}{dt}\left(\frac{\partial T}{\partial \dot{x}_j}\right) + \frac{\partial D}{\partial \dot{x}_j} + \frac{\partial V}{\partial x_j} = f_j(t) \qquad [j = 1, 2, \ldots, n]$$

The *generalized external force* $f_j(t)$ for each coordinate may be an active force in the usual sense or a force generated by prescribed motion of the coordinates.

If each term in the foregoing equation is taken as the jth element of a column matrix, all n equations can be considered simultaneously and written in matrix form as follows:

$$\frac{d}{dt}\left(\frac{\partial T}{\partial \dot{\mathbf{x}}}\right) + \frac{\partial D}{\partial \dot{\mathbf{x}}} + \frac{\partial V}{\partial \mathbf{x}} = \mathbf{f}$$

The quadratic forms can be expressed in matrix notation as

$$T = \tfrac{1}{2}(\dot{\mathbf{x}}^T\mathbf{M}\dot{\mathbf{x}})$$

$$D = \tfrac{1}{2}(\dot{\mathbf{x}}^T\mathbf{C}\dot{\mathbf{x}})$$

$$V = \tfrac{1}{2}(\mathbf{x}^T\mathbf{K}\mathbf{x})$$

where the *inertia matrix* **M**, the *damping matrix* **C**, and the *stiffness matrix* **K** may be taken as symmetric square matrices of order n. Then the differentiation rule (22.7) yields

$$\frac{d}{dt}(\mathbf{M}\dot{\mathbf{x}}) + \mathbf{C}\dot{\mathbf{x}} + \mathbf{K}\mathbf{x} = \mathbf{f}$$

or simply

$$\mathbf{M}\ddot{\mathbf{x}} + \mathbf{C}\dot{\mathbf{x}} + \mathbf{K}\mathbf{x} = \mathbf{f} \tag{22.8}$$

as the equations of motion in matrix form for a general linear vibratory system with n degrees of freedom. This is a generalization of Eq. (22.4) for the three-DOF system of Fig. 22.2. Equation (22.8) applies to all linear constant-parameter vibratory systems. The specifications of any particular system are contained in the *coefficient matrices* **M**, **C**, and **K**. The type of excitation is described by the column matrix **f**. The individual terms in the coefficient matrices have the following significance:

m_{jk} is the momentum component at j due to a unit velocity at k.

c_{jk} is the damping force at j due to a unit velocity at k.

k_{jk} is the elastic force at j due to a unit displacement at k.

The general solution to Eq. (22.8) contains $2n$ constants of integration which are usually fixed by the n displacements $x_j(t_0)$ and the n velocities $\dot{x}_j(t_0)$ at some initial time t_0. When the excitation matrix **f** is zero, Eq. (22.8) is said to describe the *free vibration* of the system. When **f** is nonzero, Eq. (22.8) describes a *forced vibration.* When the time behavior of **f** is periodic and steady, it is sometimes convenient to divide the solution into a *steady-state response* plus a *transient response* which decays with time. The steady-state response is independent of the initial conditions.

COUPLING OF THE EQUATIONS

The off-diagonal terms in the coefficient matrices are known as *coupling terms.* In general, the equations have inertia, damping, and stiffness coupling; however, it is often possible to obtain equations that have no coupling terms in one or more of the three matrices. If the coupling terms vanish in all three matrices (i.e., if all three square matrices are diagonal matrices), the system of Eq. (22.8) becomes a set of independent uncoupled differential equations for the n generalized displacements $x_j(t)$. Each displacement motion is a single-degree-of-freedom vibration independent of the motion of the other displacements.

The coupling in a system depends on the choice of coordinates used to describe the motion. For example, Figs. 22.3 and 22.4 show the same physical system with two different choices for the displacement coordinates.

The coefficient matrices corresponding to the coordinates shown in Fig. 22.3 are

$$\mathbf{M}=\begin{bmatrix} m_1 & 0 \\ 0 & m_2 \end{bmatrix} \qquad \mathbf{K}=\begin{bmatrix} k_1+k_2 & -k_2 \\ -k_2 & k_2 \end{bmatrix}$$

Here the inertia matrix is uncoupled because the coordinates chosen are the absolute displacements of the masses. The elastic force in the spring k_2 is generated by the relative displacement of the two coordinates, which accounts for the coupling terms in the stiffness matrix.

The coefficient matrices corresponding to the alternative coordinates shown in Fig. 22.4 are

$$\mathbf{M}=\begin{bmatrix} m_1+m_2 & m_2 \\ m_2 & m_2 \end{bmatrix} \qquad \mathbf{K}=\begin{bmatrix} k_1 & 0 \\ 0 & k_2 \end{bmatrix}$$

Here the coordinates chosen relate directly to the extensions of the springs so that the stiffness matrix is uncoupled. The absolute displacement of m_2 is, however, the sum of the coordinates, which accounts for the coupling terms in the inertia matrix.

A fundamental procedure for solving vibration problems in undamped systems may be viewed as the search for a set of coordinates which simultaneously uncouples both the stiffness and inertia matrices. This is always possible. In systems with damping (i.e., with all three coefficient matrices) there exist coordinates which uncouple two of these, but it is not possible to uncouple all three matrices simultaneously, except in the special case, called *proportional damping,* where **C** is a linear combination of **K** and **M**.

The system of Fig. 22.2 provides an example of a three-DOF system with damping. The coefficient matrices are given in Eq. (22.3). The inertia matrix is uncoupled, but the damping and stiffness matrices are coupled.

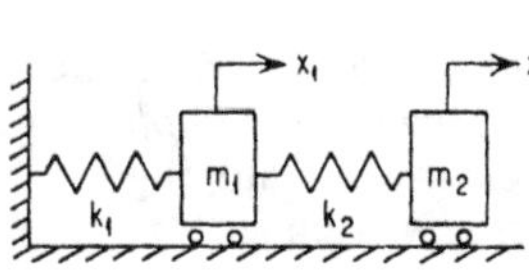

FIGURE 22.3 Coordinates (x_1,x_2) with uncoupled inertia matrix.

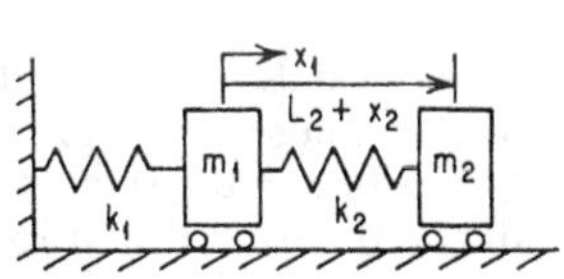

FIGURE 22.4 Coordinates (x_1,x_2) with uncoupled stiffness matrix. The equilibrium length of the spring k_2 is L_2.

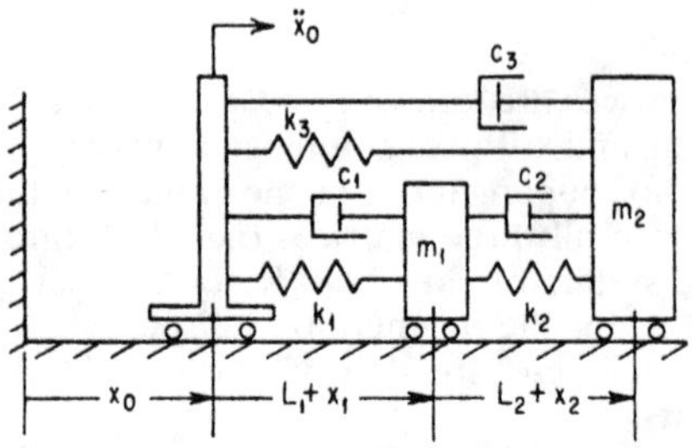

FIGURE 22.5 Two-degree-of-freedom vibratory system. The equilibrium length of the spring k_1 is L_1 and the equilibrium length of the spring k_2 is L_2.

Another example of a system with damping is furnished by the two-DOF system shown in Fig. 22.5. The excitation here is furnished by acceleration $\ddot{x}_0(t)$ of the base. This system is used as the basis for the numerical example at the end of the chapter. With the coordinates chosen as indicated in the figure, all three coefficient matrices have coupling terms. The equations of motion can be placed in the standard form of Eq. (22.8), where the coefficient matrices and the excitation column are as follows:

$$\mathbf{M} = \begin{bmatrix} m_1 + m_2 & m_2 \\ m_2 & m_2 \end{bmatrix} \qquad \mathbf{C} = \begin{bmatrix} c_1 + c_3 & c_3 \\ c_3 & c_2 + c_3 \end{bmatrix}$$
$$\mathbf{K} = \begin{bmatrix} k_1 + k_3 & k_3 \\ k_3 & k_2 + k_3 \end{bmatrix} \qquad \mathbf{f} = -\ddot{x}_0 \begin{bmatrix} m_1 + m_2 \\ m_2 \end{bmatrix} \tag{22.9}$$

THE MATRIX EIGENVALUE PROBLEM

In the following sections the solutions to both free and forced vibration problems are given in terms of solutions to a specialized algebraic problem known as the matrix eigenvalue problem. In the present section a general theoretical discussion of the matrix eigenvalue problem is given.

The free vibration equation for an undamped system,

$$\mathbf{M\ddot{x}} + \mathbf{Kx} = 0 \tag{22.10}$$

follows from Eq. (22.8) when the excitation **f** and the damping **C** vanish. If a solution for **x** is assumed in the form

$$\mathbf{x} = \Re\{\mathbf{v}e^{j\omega t}\}$$

where **v** is a column vector of unknown amplitudes, ω is an unknown frequency, j is the square root of −1, and $\Re\{\ \}$ signifies "the real part of," it is found on substituting in Eq. (22.10) that it is necessary for **v** and ω to satisfy the following algebraic equation:

$$\mathbf{Kv} = \omega^2\mathbf{Mv} \tag{22.11}$$

This algebraic problem is called the *matrix eigenvalue problem.* Where necessary it is called the *real* eigenvalue problem to distinguish it from the *complex* eigenvalue problem described in the section "Vibration of Systems with Damping."

To indicate the formal solution to Eq. (22.11), it is rewritten as

$$(\mathbf{K} - \omega^2\mathbf{M})\mathbf{v} = 0 \tag{22.12}$$

which can be interpreted as a set of n homogeneous algebraic equations for the n elements v_j. This set always has the trivial solution

$$\mathbf{v} = 0$$

It also has nontrivial solutions if the determinant of the matrix multiplying the vector v is zero, i.e., if

$$\det(\mathbf{K} - \omega^2\mathbf{M}) = 0 \tag{22.13}$$

When the determinant is expanded, a polynomial of order n in ω^2 is obtained. Equation (22.13) is known as the *characteristic equation* or *frequency equation.* The restrictions that **M** and **K** be symmetric and that **M** be positive definite are sufficient to ensure that there are n real roots for ω^2. If **K** is singular, at least one root is zero. If **K** is positive definite, all roots are positive. The n roots determine the n *natural frequencies* ω_r ($r = 1, \ldots, n$) of free vibration. These roots of the characteristic equation are also known as *normal values, characteristic values, proper values, latent roots,* or *eigenvalues.* When a natural frequency ω_r is known, it is possible to return to Eq. (22.12) and solve for the corresponding vector $\mathbf{v}_r$ to within a multiplicative constant. The eigenvalue problem does not fix the absolute amplitude of the vectors $\mathbf{v}$, only the relative amplitudes of the n coordinates. There are n independent vectors $\mathbf{v}_r$ corresponding to the n natural frequencies which are known as *natural modes.* These vectors are also known as *normal modes, characteristic vectors, proper vectors, latent vectors,* or *eigenvectors.*

MODAL AND SPECTRAL MATRICES

The complete solution to the eigenvalue problem of Eq. (22.11) consists of n eigenvalues and n corresponding eigenvectors. These can be assembled compactly into matrices. Let the eigenvector $\mathbf{v}_r$ corresponding to the eigenvalue ω_r^2 have elements v_{jr} (the first subscript indicates which row, the second subscript indicates which eigenvector). The n eigenvectors then can be displayed in a single square matrix **V**, each column of which is an eigenvector:

$$\mathbf{V} = [v_{jk}] = \begin{bmatrix} v_{11} & v_{12} & \cdots & v_{1n} \\ v_{21} & v_{22} & \cdots & v_{2n} \\ \cdots & \cdots & \cdots & \cdots \\ v_{n1} & v_{n2} & \cdots & v_{nn} \end{bmatrix}$$

The matrix **V** is called the *modal matrix* for the eigenvalue problem, Eq. (22.11).

The n eigenvalues ω_r^2 can be assembled into a diagonal matrix Ω^2 which is known as the *spectral matrix* of the eigenvalue problem, Eq. (22.11)

$$\mathbf{\Omega}^2 = \lceil \omega_r^2 \rfloor = \begin{bmatrix} \omega_1^2 & 0 & \cdots & 0 \\ 0 & \omega_2^2 & \cdots & 0 \\ \cdots & \cdots & \cdots & \cdots \\ 0 & 0 & \cdots & \omega_n^2 \end{bmatrix}$$

Each eigenvector and corresponding eigenvalue satisfy a relation of the following form:

$$\mathbf{K}\mathbf{v}_r = \mathbf{M}\mathbf{v}_r\omega_r^2$$

By using the modal and spectral matrices it is possible to assemble all of these relations into a single matrix equation

$$\mathbf{KV} = \mathbf{MV}\Omega^2 \tag{22.14}$$

Equation (22.14) provides a compact display of the complete solution to the eigenvalue problem Eq. (22.11).

PROPERTIES OF THE SOLUTION

The eigenvectors corresponding to different eigenvalues can be shown to satisfy the following *orthogonality relations.* When $\omega_r^2 \neq \omega_s^2$,

$$\mathbf{v}_r^T\mathbf{K}\mathbf{v}_s = 0 \qquad \mathbf{v}_r^T\mathbf{M}\mathbf{v}_s = 0 \tag{22.15}$$

In case the characteristic equation has a p-fold multiple root for ω^2, then there is a p-fold infinity of corresponding eigenvectors. In this case, however, it is always possible to choose p of these vectors which mutually satisfy Eq. (22.15) and to express any other eigenvector corresponding to the multiple root as a linear combination of the p vectors selected. If these p vectors are included with the eigenvectors corresponding to the other eigenvalues, a set of n vectors is obtained which satisfies the orthogonality relations of Eq. (22.15) for any $r \neq s$.

The orthogonality of the eigenvectors with respect to **K** and **M** implies that the following square matrices are *diagonal.*

$$\begin{aligned} \mathbf{V}^T\mathbf{K}\mathbf{V} &= \left\lceil \mathbf{v}_r^T\mathbf{K}\mathbf{v}_r \right\rfloor \\ \mathbf{V}^T\mathbf{M}\mathbf{V} &= \left\lceil \mathbf{v}_r^T\mathbf{M}\mathbf{v}_r \right\rfloor \end{aligned} \tag{22.16}$$

The elements $\mathbf{v}_r^T\mathbf{K}\mathbf{v}_r$ along the main diagonal of $\mathbf{V}^T\mathbf{K}\mathbf{V}$ are called the *modal stiffnesses* k_r, and the elements $\mathbf{v}_r^T\mathbf{M}\mathbf{v}_r$ along the main diagonal of $\mathbf{V}^T\mathbf{M}\mathbf{V}$ are called the *modal masses* m_r. Since **M** is positive definite, all modal masses are guaranteed to be positive. When **K** is singular, at least one of the modal stiffnesses will be zero. Each eigenvalue ω_r^2 is the quotient of the corresponding modal stiffness divided by the corresponding modal mass; i.e.,

$$\omega_r^2 = \frac{k_r}{m_r}$$

In numerical work it is sometimes convenient to normalize each eigenvector so that its largest element is *unity.* In other applications it is common to normalize the eigenvectors so that the modal masses m_r all have the *same* value m, where m is some convenient value such as the total mass of the system. In this case,

$$\mathbf{V}^T\mathbf{M}\mathbf{V} = m\mathbf{I} \tag{22.17}$$

and it is possible to express the inverse of the modal matrix **V** simply as

$$\mathbf{V}^{-1} = \frac{1}{m}\,\mathbf{V}^T\mathbf{M}$$

An interpretation of the modal matrix **V** can be given by showing that it defines a set of generalized coordinates for which both the inertia and stiffness matrices are uncoupled. Let $\mathbf{y}(t)$ be a column of displacements related to the original displacements $\mathbf{x}(t)$ by the following simultaneous equations:

$$\mathbf{y} = \mathbf{V}^{-1}\mathbf{x} \qquad \text{or} \qquad \mathbf{x} = \mathbf{V}\mathbf{y}$$

The potential and kinetic energies then take the forms

$$V = \tfrac{1}{2}\mathbf{x}^T\mathbf{K}\mathbf{x} = \tfrac{1}{2}\mathbf{y}^T(\mathbf{V}^T\mathbf{K}\mathbf{V})\mathbf{y}$$

$$T = \tfrac{1}{2}\dot{\mathbf{x}}^T\mathbf{M}\dot{\mathbf{x}} = \tfrac{1}{2}\dot{\mathbf{y}}^T(\mathbf{V}^T\mathbf{M}\mathbf{V})\dot{\mathbf{y}}$$

where, according to Eq. (22.16), the square matrices in parentheses on the right are *diagonal;* i.e., in the y_j coordinate system there is neither stiffness nor inertia coupling.

An alternative method for obtaining the same interpretation is to start from the eigenvalue problem of Eq. (22.11). Consider the structure of the related eigenvalue problem for **w** where again **w** is obtained from **v** by the transformation involving the modal matrix **V**.

$$\mathbf{w} = \mathbf{V}^{-1}\mathbf{v} \qquad \text{or} \qquad \mathbf{v} = \mathbf{V}\mathbf{w}$$

Substituting in Eq. (22.11), premultiplying by $\mathbf{V}^T$, and using Eq. (22.14),

$$\mathbf{K}\mathbf{v} = \omega^2\mathbf{M}\mathbf{v}$$

$$\mathbf{K}\mathbf{V}\mathbf{w} = \omega^2\mathbf{M}\mathbf{V}\mathbf{w}$$

$$\mathbf{V}^T\mathbf{K}\mathbf{V}\mathbf{w} = \omega^2\mathbf{V}^T\mathbf{M}\mathbf{V}\mathbf{w}$$

$$(\mathbf{V}^T\mathbf{M}\mathbf{V})\mathbf{\Omega}^2\mathbf{w} = \omega^2(\mathbf{V}^T\mathbf{M}\mathbf{V})\mathbf{w}$$

Now, since $\mathbf{V}^T\mathbf{M}\mathbf{V}$ is a diagonal matrix of positive elements, it is permissible to cancel it from both sides, which leaves a simple diagonalized eigenvalue problem for **w**:

$$\mathbf{\Omega}^2\mathbf{w} = \omega^2\mathbf{w}$$

A modal matrix for **w** is the identity matrix **I**, and the eigenvalues for **w** are the same as those for **v**.

EIGENVECTOR EXPANSIONS

Any set of n independent vectors can be used as a basis for representing any other vector of order n. In the following sections, the eigenvectors of the eigenvalue problem of Eq. (22.11) are used as such a basis. An eigenvector expansion of an arbitrary vector **y** has the form

$$\mathbf{y} = \sum_{r=1}^{n} \mathbf{v}_r a_r \tag{22.18}$$

where the a_r are scalar *mode multipliers.* When **y** and the $\mathbf{v}_r$ are known, it is possible to evaluate the a_r by premultiplying both sides by $\mathbf{v}_s^T\mathbf{M}$. Because of the orthogonality relations of Eq. (22.15), all the terms on the right vanish except the one for which $r = s$. Inserting the value of the mode multiplier so obtained, the expansion can be rewritten as

$$\mathbf{y} = \sum_{r=1}^{n} \mathbf{v}_r \frac{\mathbf{v}_r^T\mathbf{M}\mathbf{y}}{\mathbf{v}_r^T\mathbf{M}\mathbf{v}_r} \tag{22.19}$$

or alternatively as

$$\mathbf{y} = \sum_{r=1}^{n} \frac{\mathbf{v}_r \mathbf{v}_r^T \mathbf{M}}{\mathbf{v}_r^T \mathbf{M} \mathbf{v}_r} \mathbf{y} \tag{22.20}$$

The form of Eq. (22.19) emphasizes the decomposition into eigenvectors since the fraction on the right is just a scalar. The form of Eq. (22.20) is convenient when a large number of vectors **y** are to be decomposed, since the fractions on the right, which are now square matrices, must be computed only once. The form of Eq. (22.20) becomes more economical of computation time when more than n vectors **y** have to be expanded. A useful check on the calculation of the matrices on the right of Eq. (22.20) is provided by the identity

$$\sum_{r=1}^{n} \frac{\mathbf{v}_r \mathbf{v}_r^T \mathbf{M}}{\mathbf{v}_r^T \mathbf{M} \mathbf{v}_r} = \mathbf{I} \tag{22.21}$$

which follows from Eq. (22.20) because **y** is completely arbitrary.

An alternative expansion which is useful for expanding the excitation vector **f** is

$$\mathbf{f} = \sum_{r=1}^{n} \omega_r^2 \mathbf{M} \mathbf{v}_r a_r = \sum_{r=1}^{n} \mathbf{M} \mathbf{v}_r \frac{\mathbf{v}_r^T \mathbf{f}}{\mathbf{v}_r^T \mathbf{M} \mathbf{v}_r} \tag{22.22}$$

This may be viewed as an expansion of the excitation in terms of the *inertia force* amplitudes of the natural modes. The mode multiplier a_r has been evaluated by premultiplying by $\mathbf{v}_r^T$. A form analogous to Eq. (22.20) and an identity corresponding to Eq. (22.21) can easily be written.

RAYLEIGH'S QUOTIENT

If Eq. (22.11) is premultiplied by $\mathbf{v}^T$, the following scalar equation is obtained:

$$\mathbf{v}^T \mathbf{K} \mathbf{v} = \omega^2 \mathbf{v}^T \mathbf{M} \mathbf{v}$$

The positive definiteness of **M** guarantees that $\mathbf{v}^T \mathbf{M} \mathbf{v}$ is nonzero, so that it is permissible to solve for ω^2.

$$\omega^2 = \frac{\mathbf{v}^T \mathbf{K} \mathbf{v}}{\mathbf{v}^T \mathbf{M} \mathbf{v}} \tag{22.23}$$

This quotient is called "Rayleigh's quotient." It also may be derived by equating time averages of potential and kinetic energy under the assumption that the vibratory system is executing simple harmonic motion at frequency ω with amplitude ratios given by **v** or by equating the maximum value of kinetic energy to the maximum value of potential energy under the same assumption. Rayleigh's quotient has the following interesting properties.

1. When **v** is an eigenvector $\mathbf{v}_r$ of Eq. (22.11), then Rayleigh's quotient is equal to the corresponding eigenvalue ω_r^2.
2. If **v** is an approximation to $\mathbf{v}_r$ with an error which is a *first-order* infinitesimal, then Rayleigh's quotient is an approximation to ω_r^2 with an error which is a *second-order* infinitesimal; i.e., Rayleigh's quotient is *stationary* in the neighborhoods of the true eigenvectors.
3. As **v** varies through all of n-dimensional vector space, Rayleigh's quotient remains bounded between the smallest and largest eigenvalues.

A common engineering application of Rayleigh's quotient involves simply evaluating Eq. (22.23) for a trial vector **v** which is selected on the basis of physical insight. When eigenvectors are obtained by approximate methods, Rayleigh's quotient provides a means of improving the accuracy in the corresponding eigenvalue. If the elements of an approximate eigenvector whose largest element is unity are correct to k decimal places, then Rayleigh's quotient can be expected to be correct to about $2k$ significant decimal places.

Perturbation Formulas. The perturbation formulas which follow provide the basis for estimating the changes in the eigenvalues and the eigenvectors which result from *small* changes in the stiffness and inertia parameters of a system. The formulas are strictly accurate only for infinitesimal changes but are useful approximations for *small* changes. They may be used by the designer to estimate the effects of a proposed change in a vibratory system and may also be used to analyze the effects of minor errors in the measurement of the system properties. Iterative procedures for the solution of eigenvalue problems can be based on these formulas. They are employed here to obtain approximations to the complex eigenvalues and eigenvectors of a lightly damped vibratory system in terms of the corresponding solutions for the same system without damping.

Suppose that the modal matrix **V** and the spectral matrix $\mathbf{\Omega}^2$ for the eigenvalue problem

$$\mathbf{KV} = \mathbf{MV\Omega}^2 \tag{22.14}$$

are known. Consider the perturbed eigenvalue problem

$$\mathbf{K}_*\mathbf{V}_* = \mathbf{M}_*\mathbf{V}_*\mathbf{\Omega}_*^2$$

where

$$\mathbf{K}_* = \mathbf{K} + d\mathbf{K} \qquad \mathbf{M}_* = \mathbf{M} + d\mathbf{M}$$

$$\mathbf{V}_* = \mathbf{V} + d\mathbf{V} \qquad \mathbf{\Omega}_*^2 = \mathbf{\Omega}^2 + d\mathbf{\Omega}^2$$

The perturbation formula for the elements $d\omega_r^2$ of the diagonal matrix $d\Omega^2$ is

$$d\omega_r^2 = \frac{\mathbf{v}_r^T\, d\mathbf{K}\, \mathbf{v}_r - \omega_r^2 \mathbf{v}_r^T\, d\mathbf{M}\, \mathbf{v}_r}{\mathbf{v}_r^T \mathbf{M} \mathbf{v}_r} \tag{22.24}$$

Thus, in order to determine the change in a single eigenvalue due to changes in **M** and **K**, it is necessary to know only the corresponding unperturbed eigenvalue and eigenvector. To determine the change in a single eigenvector, however, it is necessary to know *all* the unperturbed eigenvalues and eigenvectors. The following algorithm may be used to evaluate the perturbations of both the modal matrix and the spectral matrix. Calculate

$$\mathbf{F} = \mathbf{V}^T\, d\mathbf{K}\, \mathbf{V} - \mathbf{V}^T\, d\mathbf{M}\, \mathbf{V\Omega}^2$$

and

$$\mathbf{L} = \mathbf{V}^T\mathbf{MV}$$

The matrix **L** is a diagonal matrix of positive elements and hence is easily inverted. Continue calculating

$$\mathbf{G} = \mathbf{L}^{-1}\mathbf{F} = [g_{jk}] \qquad \text{and} \qquad \mathbf{H} = [h_{jk}]$$

where

$$h_{jk} = \begin{cases} 0 & \text{if } \omega_j^2 = \omega_k^2 \\ \dfrac{g_{jk}}{\omega_k^2 - \omega_j^2} & \text{if } \omega_j^2 \neq \omega_k^2 \end{cases}$$

Then, finally, the perturbations of the modal matrix and the spectral matrix are given by

$$d\mathbf{V} = \mathbf{VH} \qquad d\Omega^2 = \left\lceil g_{jj} \right\rfloor \tag{22.25}$$

These formulas are derived by taking the total differential of Eq. (22.14), premultiplying each term by $\mathbf{V}^T$, and using a relation derived by taking the transpose of Eq. (22.14). An interesting property of the perturbation approximation is that the change in each eigenvector is orthogonal with respect to $\mathbf{M}$ to the corresponding unperturbed eigenvector; i.e.,

$$\mathbf{v}_j{}^T\mathbf{M}\, d\mathbf{v}_j = 0$$

VIBRATIONS OF SYSTEMS WITHOUT DAMPING

In this section the damping matrix $\mathbf{C}$ is neglected in Eq. (22.8), leaving the general formulation in the form

$$\mathbf{M\ddot{x}} + \mathbf{Kx} = \mathbf{f} \tag{22.26}$$

Solutions are outlined for the following three cases: free vibration ($\mathbf{f} = 0$), steady-state forced sinusoidal vibration ($\mathbf{f} = \mathcal{R}\{\mathbf{d}e^{j\omega t}\}$, where $\mathbf{d}$ is a column vector of driving-force amplitudes), and the response to general excitation ($\mathbf{f}$ an arbitrary function of time). The first two cases are contained in the third, but for the sake of clarity each is described separately.

FREE VIBRATION WITH SPECIFIED INITIAL CONDITIONS

It is desired to find the solution $\mathbf{x}(t)$ of Eq. (22.26) when $\mathbf{f} = 0$ which satisfies the initial conditions

$$\mathbf{x} = \mathbf{x}(0) \qquad \mathbf{\dot{x}} = \mathbf{\dot{x}}(0) \tag{22.27}$$

at $t = 0$ where $\mathbf{x}(0)$ and $\mathbf{\dot{x}}(0)$ are columns of prescribed initial displacements and velocities. The differential equation to be solved is identical with Eq. (22.10), which led to the matrix eigenvalue problem in the preceding section. Assuming that the solution of the eigenvalue problem is available, the general solution of the differential equation is given by an arbitrary superposition of the natural modes

$$\mathbf{x} = \sum_{r=1}^{n} \mathbf{v}_r(a_r \cos \omega_r t + b_r \sin \omega_r t)$$

where the $\mathbf{v}_r$ are the eigenvectors or natural modes, the ω_r are the natural frequencies, and the a_r and b_r are $2n$ constants of integration. The corresponding velocity is

$$\dot{\mathbf{x}} = \sum_{r=1}^{n} \mathbf{v}_r\omega_r(-a_r \sin \omega_r t + b_r \cos \omega_r t)$$

Setting $t = 0$ in these expressions and substituting in the initial conditions of Eq. (22.27) provides $2n$ simultaneous equations for determination of the constants of integration.

$$\sum_{r=1}^{n} \mathbf{v}_r a_r = \mathbf{x}(0) \qquad \sum_{r=1}^{n} \mathbf{v}_r\omega_r b_r = \dot{\mathbf{x}}(0)$$

These equations may be interpreted as eigenvector expansions of the initial displacement and velocity. The constants of integration can be evaluated by the same technique used to obtain the mode multipliers in Eq. (22.19). Using the form of Eq. (22.20), the solution of the free vibration problem then becomes

$$\mathbf{x}(t) = \sum_{r=1}^{n} \frac{\mathbf{v}_r\mathbf{v}_r^T\mathbf{M}}{\mathbf{v}_r^T\mathbf{M}\mathbf{v}_r} \left\{\mathbf{x}(0) \cos \omega_r t + \frac{1}{\omega_r} \dot{\mathbf{x}}(0) \sin \omega_r t\right\} \tag{22.28}$$

STEADY-STATE FORCED SINUSOIDAL VIBRATION

It is desired to find the steady-state solution to Eq. (22.26) for single-frequency sinusoidal excitation **f** of the form

$$\mathbf{f} = \Re\{\mathbf{d}e^{j\omega t}\}$$

where **d** is a column vector of driving force amplitudes (these may be complex to permit differences in phase for the various components). The solution obtained is a useful approximation for lightly damped systems provided that the forcing frequency ω is not too close to a natural frequency ω_r. For resonance and near-resonance conditions it is necessary to include the damping as indicated in the section which follows the present discussion.

The steady-state solution desired is assumed to have the form

$$\mathbf{x} = \Re\{\mathbf{a}e^{j\omega t}\}$$

where **a** is an unknown column vector of response amplitudes. When **f** and **x** are inserted in Eq. (22.26), the following set of simultaneous equations for the elements of **a** is obtained:

$$(\mathbf{K} - \omega^2\mathbf{M})\mathbf{a} = \mathbf{d} \tag{22.29}$$

If ω is not a natural frequency, the square matrix $\mathbf{K} - \omega^2\mathbf{M}$ is nonsingular and may be inverted to yield

$$\mathbf{a} = (\mathbf{K} - \omega^2\mathbf{M})^{-1}\mathbf{d}$$

as a complete solution for the response amplitudes in terms of the driving force amplitudes. This solution is useful if several force amplitude distributions are to be studied while the excitation frequency ω is held constant. The process requires repeated inversions if a range of frequencies is to be studied.

An alternative procedure which permits a more thorough study of the effect of frequency variation is available if the natural modes and frequencies are known. The driving force vector **d** is represented by the eigenvector expansion of Eq. (22.22), and the response vector **a** is represented by the eigenvector expansion of Eq. (22.18):

$$\mathbf{d} = \sum_{r=1}^{n} \frac{\mathbf{M}\mathbf{v}_r\mathbf{v}_r^T}{\mathbf{v}_r^T\mathbf{M}\mathbf{v}_r}\,\mathbf{d} \qquad \mathbf{a} = \sum_{r=1}^{n} \mathbf{v}_r c_r$$

where the c_r are unknown coefficients. Substituting these into Eq. (22.29), and making use of the fundamental eigenvalue relation of Eq. (22.11), leads to

$$\sum_{r=1}^{n} (\omega_r^2 - \omega^2)\mathbf{M}\mathbf{v}_r c_r = \sum_{r=1}^{n} \frac{\mathbf{M}\mathbf{v}_r\mathbf{v}_r^T}{\mathbf{v}_r^T\mathbf{M}\mathbf{v}_r}\,\mathbf{d}$$

This equation can be uncoupled by premultiplying both sides by $\mathbf{v}_r^T$ and using the orthogonality condition of Eq. (22.15) to obtain

$$(\omega_r^2 - \omega^2)\mathbf{v}_r^T\mathbf{M}\mathbf{v}_r c_r = \mathbf{v}_r^T\mathbf{d}$$

$$c_r = \frac{1}{\omega_r^2 - \omega^2}\,\frac{\mathbf{v}_r^T\mathbf{d}}{\mathbf{v}_r^T\mathbf{M}\mathbf{v}_r}$$

The final solution is then assembled by inserting the c_r back into **a** and **a** back into **x**.

$$\mathbf{x} = \Re\left\{\sum_{r=1}^{n} \frac{e^{j\omega t}}{\omega_r^2 - \omega^2}\,\frac{\mathbf{v}_r\mathbf{v}_r^T}{\mathbf{v}_r^T\mathbf{M}\mathbf{v}_r}\,\mathbf{d}\right\} \tag{22.30}$$

This form clearly indicates the effect of frequency on the response.

RESPONSE TO GENERAL EXCITATION

It is now desired to obtain the solution to Eq. (22.26) for the general case in which the excitation $\mathbf{f}(t)$ is an arbitrary vector function of time and for which initial displacements $\mathbf{x}(0)$ and velocities $\dot{\mathbf{x}}(0)$ are prescribed. If the natural modes and frequencies of the system are available, it is again possible to split the problem up into n single-degree-of-freedom response problems and to indicate a formal solution.

Following a procedure similar to that just used for steady-state forced sinusoidal vibrations, an eigenvector expansion of the solution is assumed:

$$\mathbf{x}(t) = \sum_{r=1}^{n} \mathbf{v}_r c_r(t)$$

where the c_r are unknown functions of time and the known excitation $\mathbf{f}(t)$ is expanded according to Eq. (22.22). Inserting these into Eq. (22.26) yields

$$\sum_{r=1}^{n} (\mathbf{M}\mathbf{v}_r \ddot{c}_r + \mathbf{K}\mathbf{v}_r c_r) = \sum_{r=1}^{n} \frac{\mathbf{M}\mathbf{v}_r\mathbf{v}_r^T}{\mathbf{v}_r^T\mathbf{M}\mathbf{v}_r}\,\mathbf{f}(t)$$

Using Eq. (22.11) to eliminate **K** and premultiplying by $\mathbf{v}_r^T$ to uncouple the equation,

$$\ddot{c}_r + \omega_r^2 c_r^2 = \frac{\mathbf{v}_r^T\mathbf{f}(t)}{\mathbf{v}_r^T\mathbf{M}\mathbf{v}_r} \tag{22.31}$$

is obtained as a single second-order differential equation for the time behavior of the rth mode multiplier. The initial conditions for c_r can be obtained by making eigenvector expansions of $\mathbf{x}(0)$ and $\dot{\mathbf{x}}(0)$ as was done previously for the free vibration case. Formal solutions to Eq. (22.29) can be obtained by a number of methods, including Laplace transforms and variation of parameters. When these mode multipliers are substituted back to obtain **x**, the general solution has the following appearance:

$$\mathbf{x}(t) = \sum_{r=1}^{n} \frac{\mathbf{v}_r\mathbf{v}_r{}^T\mathbf{M}}{\mathbf{v}_r{}^T\mathbf{M}\mathbf{v}_r} \left\{\mathbf{x}(0) \cos \omega_r t + \frac{1}{\omega_r} \dot{\mathbf{x}}(0) \sin \omega_r t\right\}$$

$$+ \sum_{r=1}^{n} \frac{\mathbf{v}_r\mathbf{v}_r{}^T}{\omega_r\mathbf{v}_r{}^T\mathbf{M}\mathbf{v}_r} \int_0^t \mathbf{f}(t') \sin \{\omega_r(t - t')\}\, dt' \qquad (22.32)$$

The integrals involving the excitation can be evaluated in closed form if the elements $f_j(t)$ of $\mathbf{f}(t)$ are simple (e.g., step functions, ramps, single sine pulses, etc.). When the $f_j(t)$ are more complicated, numerical results can be obtained by using integration software.

VIBRATION OF SYSTEMS WITH DAMPING

In this section solutions to the complete governing equation, Eq. (22.8), are discussed. The results of the preceding section for systems without damping are adequate for many purposes. There are, however, important problems in which it is necessary to include the effect of damping, e.g., problems concerned with resonance, random vibration, etc.

COMPLEX EIGENVALUE PROBLEM

When there is no excitation, Eq. (22.8) becomes

$$\mathbf{M}\ddot{\mathbf{x}} + \mathbf{C}\dot{\mathbf{x}} + \mathbf{K}\mathbf{x} = 0$$

which describes the free vibration of the system. As in the undamped case, there are $2n$ independent solutions which can be superposed to meet $2n$ initial conditions. Assuming a solution in the form

$$\mathbf{x} = \mathbf{u}e^{pt}$$

leads to the following algebraic problem:

$$(p^2\mathbf{M} + p\mathbf{C} + \mathbf{K})\mathbf{u} = 0 \qquad (22.33)$$

for the determination of the vector $\mathbf{u}$ and the scalar p. This is a *complex eigenvalue problem* because the *eigenvalue* p and the elements of the *eigenvector* $\mathbf{u}$ are, in general, complex numbers. The most common technique for solving the nth-order eigenvalue problem, Eq. (22.33), is to transform it to a $2n$th-order problem having the same form as Eq. (22.11). This may be done by introducing the column vector $\tilde{\mathbf{v}}$ of order $2n$ given by

$$\tilde{\mathbf{v}} = \{\mathbf{u} \quad p\mathbf{u}\}^T$$

and the two square matrices of order $2n$ given by

$$\tilde{\mathbf{K}} = \begin{bmatrix} -K & 0 \\ 0 & M \end{bmatrix} \qquad \tilde{\mathbf{M}} = \begin{bmatrix} C & M \\ M & 0 \end{bmatrix}$$

In terms of these, an eigenvalue problem equivalent to Eq. (22.33) is

$$\tilde{\mathbf{K}}\tilde{\mathbf{v}} = p\tilde{\mathbf{M}}\tilde{\mathbf{v}} \tag{22.34}$$

which is similar to Eq. (22.11) except that $\tilde{\mathbf{M}}$ does not have the positive definite property that $\mathbf{M}$ has. As a result, the eigenvalue p and the eigenvector $\mathbf{v}$ are generally complex. Since the computational time for most eigenvalue problems is proportional to n^3, the computational time for the $2n$th-order system of Eq. (22.34) will be about eight times that for the nth-order system of Eq. (22.11).

If the complex eigenvalue $p = -\alpha + j\beta$ together with the complex eigenvector $\mathbf{u} = \mathbf{v} + j\mathbf{w}$ satisfy the eigenvalue problem of Eq. (22.33), then so also does the complex conjugate eigenvalue $p^C = -\alpha - j\beta$ together with the complex conjugate eigenvector $\mathbf{u}^C = \mathbf{v} - j\mathbf{w}$. There are $2n$ eigenvalues which occur in pairs of complex conjugates or as real negative numbers. When the damping is absent all roots lie on the imaginary axis of the complex p-plane; for small damping the roots lie near the imaginary axis. The corresponding $2n$ eigenvectors $\mathbf{u}_r$ satisfy the following *orthogonality* relations:

$$(p_r + p_s)\mathbf{u}_r^T\mathbf{M}\mathbf{u}_s + \mathbf{u}_r^T\mathbf{C}\mathbf{u}_s = 0$$

$$\mathbf{u}_r^T\mathbf{K}\mathbf{u}_s - p_r p_s \mathbf{u}_r^T\mathbf{M}\mathbf{u}_s = 0$$

whenever $p_r \neq p_s$; they can be made to hold for repeated roots by suitable choice of the eigenvectors associated with a multiple root. When p_s is put equal to p_r^C, the orthogonality relations provide convenient formulas for the real and imaginary parts of the eigenvalues in terms of the eigenvectors

$$2\alpha_r = \frac{\mathbf{u}_r^T\mathbf{C}\mathbf{u}_r^C}{\mathbf{u}_r^T\mathbf{M}\mathbf{u}_r^C} = \frac{\mathbf{v}_r^T\mathbf{C}\mathbf{v}_r + \mathbf{w}_r^T\mathbf{C}\mathbf{w}_r}{\mathbf{v}_r^T\mathbf{M}\mathbf{v}_r + \mathbf{w}_r^T\mathbf{M}\mathbf{w}_r}$$

$$\alpha_r^2 + \beta_r^2 = \frac{\mathbf{u}_r^T\mathbf{K}\mathbf{u}_r^C}{\mathbf{u}_r^T\mathbf{M}\mathbf{u}_r^C} = \frac{\mathbf{v}_r^T\mathbf{K}\mathbf{v}_r + \mathbf{w}_r^T\mathbf{K}\mathbf{w}_r}{\mathbf{v}_r^T\mathbf{M}\mathbf{v}_r + \mathbf{w}_r^T\mathbf{M}\mathbf{w}_r}$$

The complex eigenvalue is often represented in the form

$$p_r = \omega_r(-\zeta_r + j\sqrt{1 - \zeta_r^2}) \tag{22.35}$$

where $\omega_r = \sqrt{\alpha_r^2 + \beta_r^2}$ is called the *undamped natural frequency* of the rth mode, and $\zeta_r = \alpha_r/\omega_r$ is called the *critical damping ratio* of the rth mode.

PERTURBATION APPROXIMATION TO COMPLEX EIGENVALUE PROBLEM

The complex eigenvalue problem of Eq. (22.33) can be solved approximately, when the damping is light, by using the perturbation equations of Eqs. (22.24) and (22.25). When $\mathbf{C} = 0$ in Eq. (22.33) the complex eigenvalue problem reduces to the real eigenvalue problem of Eq. (22.11) with $p^2 = -\omega^2$. Suppose that the real eigenvalue ω_r^2 and the real eigenvector $\mathbf{v}_r$ are known. The perturbation of the rth mode due to the addition of small damping $\mathbf{C}$ can be estimated by considering the damping to be a perturbation of the stiffness matrix of the form

$$d\mathbf{K} = j\omega_r\mathbf{C}$$

In this way it is found that the perturbed solution corresponding to the rth mode consists of a pair of complex conjugate eigenvalues

$$p_r = -\alpha_r + j\omega_r \qquad p_r^C = -\alpha_r - j\omega_r$$

and a pair of complex conjugate eigenvectors

$$\mathbf{u}_r = \mathbf{v}_r + j\mathbf{w}_r \qquad \mathbf{u}_r^C = \mathbf{v}_r - j\mathbf{w}_r$$

where ω_r and $\mathbf{v}_r$ are taken directly from the undamped system, and α_r and $\mathbf{w}_r$ are small perturbations which are given below. The superscript C is used to denote the complex conjugate. The real part of the eigenvalue, which describes the rate of decay of the corresponding free motion, is given by the following quotient:

$$2\alpha_r = 2\zeta_r\omega_r = \frac{\mathbf{v}_r^T\mathbf{C}\mathbf{v}_r}{\mathbf{v}_r^T\mathbf{M}\mathbf{v}_r} \tag{22.36}$$

The decay rate α_r for a particular r depends only on the rth mode undamped solution. The imaginary part of the eigenvector $j\mathbf{w}_r$, which describes the perturbations in phase, is more difficult to obtain. All the undamped eigenvalues and eigenvectors must be known. Let $\mathbf{W}$ be a square matrix whose columns are the $\mathbf{w}_r$. The following algorithm may be used to evaluate $\mathbf{W}$ when the undamped modal matrix $\mathbf{V}$ is known. Calculate

$$\mathbf{F} = \mathbf{V}^T\mathbf{C}\mathbf{V}$$

and

$$\mathbf{L} = \mathbf{V}^T\mathbf{M}\mathbf{V}$$

The matrix $\mathbf{L}$ is a diagonal matrix of positive elements and hence is easily inverted. Continue calculating

$$\mathbf{G} = \mathbf{L}^{-1}\mathbf{F} = [g_{jk}] \qquad \text{and} \qquad \mathbf{H} = [h_{jk}]$$

where

$$h_{jk} = \begin{cases} 0 & \text{if } \omega_j^2 = \omega_k^2 \\ \dfrac{g_{jk}\omega_k}{\omega_k^2 - \omega_j^2} & \text{if } \omega_j^2 \neq \omega_k^2 \end{cases}$$

Then, finally, the eigenvector perturbations are given by

$$\mathbf{W} = \mathbf{V}\mathbf{H} \tag{22.37}$$

The individual eigenvector perturbations $\mathbf{w}_r$ obtained in this manner are orthogonal with respect to $\mathbf{M}$ to their corresponding unperturbed eigenvectors $\mathbf{v}_r$; i.e., $\mathbf{w}_r^T\mathbf{M}\mathbf{v}_r = 0$.

FORMAL SOLUTIONS

If the solution to the eigenvalue problem of Eq. (22.33) is available, it is possible to exhibit a general solution to the governing equation

$$\mathbf{M}\mathbf{x} + \mathbf{C}\dot{\mathbf{x}} + \mathbf{K}\mathbf{x} = \mathbf{f} \tag{22.8}$$

for arbitrary excitation $\mathbf{f}(t)$ which meets prescribed initial conditions for $\mathbf{x}(0)$ and $\dot{\mathbf{x}}(0)$ at $t = 0$. The solutions given below apply to the case where the $2n$ eigenvalues occur as n pairs of complex conjugates (which is usually the case when the damping is light). This does, however, restrict the treatment to systems with nonsingular stiffness matrices $\mathbf{K}$ because if $\omega_r^2 = 0$ is an undamped eigenvalue, the corresponding eigenvalues in the presence of damping are real. All quantities in the solutions below are *real.* These forms have been obtained by breaking down complex solutions into real and imaginary parts and recombining. With the notation

$$p_r = -\alpha_r + j\beta_r \qquad \mathbf{u}_r = \mathbf{v}_r + j\mathbf{w}_r$$

for the real and imaginary parts of eigenvalues and eigenvectors, it follows from Eq. (22.35) that

$$\alpha_r = \zeta_r\omega_r \qquad \beta_r = \omega_r\sqrt{1-\zeta_r^2}$$

The general solution to Eq. (22.8) is then

$$\begin{aligned}
\mathbf{x}(t) = &\sum_{r=1}^{n} \frac{2}{a_r^2 + b_r^2}\{\mathbf{G}_r\mathbf{M}\dot{\mathbf{x}}(0) + (-\alpha_r\mathbf{G}_r\mathbf{M} + \beta_r\mathbf{H}_r\mathbf{M} + \mathbf{G}_r\mathbf{C})\mathbf{x}(0)\}e^{-\alpha_r t}\cos\beta_r t \\
&+ \sum_{r=1}^{n} \frac{2}{a_r^2 + b_r^2}\{\mathbf{H}_r\mathbf{M}\dot{\mathbf{x}}(0) + (-\beta_r\mathbf{G}_r\mathbf{M} - \alpha_r\mathbf{H}_r\mathbf{M} + \mathbf{H}_r\mathbf{C})\mathbf{x}(0)\}e^{-\alpha_r t}\sin\beta_r t \\
&+ \sum_{r=1}^{n} \frac{2}{a_r^2 + b_r^2}\mathbf{G}_r\int_0^t \mathbf{f}(t')e^{-\alpha_r(t-t')}\cos\beta_r(t-t')\,dt' \\
&+ \sum_{r=1}^{n} \frac{2}{a_r^2 + b_r^2}\mathbf{H}_r\int_0^t \mathbf{f}(t')e^{-\alpha_r(t-t')}\sin\beta_r(t-t')\,dt'
\end{aligned} \tag{22.38}$$

where

$$a_r = -2\alpha_r(\mathbf{v}_r^T\mathbf{M}\mathbf{v}_r - \mathbf{w}_r^T\mathbf{M}\mathbf{w}_r) - 4\beta_r\mathbf{v}_r^T\mathbf{M}\mathbf{w}_r + \mathbf{v}_r^T\mathbf{C}\mathbf{v}_r - \mathbf{w}_r^T\mathbf{C}\mathbf{w}_r$$

$$b_r = 2\beta_r(\mathbf{v}_r^T\mathbf{M}\mathbf{v}_r - \mathbf{w}_r^T\mathbf{M}\mathbf{w}_r) - 4\alpha_r\mathbf{v}_r^T\mathbf{M}\mathbf{w}_r + 2\mathbf{v}_r^T\mathbf{C}\mathbf{w}_r$$

$$\mathbf{A}_r = \mathbf{v}_r\mathbf{v}_r^T - \mathbf{w}_r\mathbf{w}_r^T \qquad \mathbf{B}_r = \mathbf{v}_r\mathbf{w}_r^T + \mathbf{w}_r\mathbf{v}_r^T$$

$$\mathbf{G}_r = a_r\mathbf{A}_r + b_r\mathbf{B}_r \qquad \mathbf{H}_r = b_r\mathbf{A}_r - a_r\mathbf{B}_r$$

The solution of Eq. (22.38) should be compared with the corresponding solution of Eq. (22.32) for systems without damping. When the damping matrix $\mathbf{C} = 0$, Eq. (22.38) reduces to Eq. (22.32).

For the important special case of steady-state forced sinusoidal excitation of the form

$$\mathbf{f} = \mathcal{R}\{\mathbf{d}e^{j\omega t}\}$$

where $\mathbf{d}$ is a column of driving force amplitudes, the steady-state portion of the response can be written as follows, using the above notation:

$$\mathbf{x}(t) = \mathcal{R}\left\{\sum_{r=1}^{n} \frac{2e^{j\omega t}}{a_r^2 + b_r^2}\,\frac{\alpha_r\mathbf{G}_r + \beta_r\mathbf{H}_r + j\omega\mathbf{G}_r}{\omega_r^2 - \omega^2 + j2\zeta_r\omega_r\omega}\,\mathbf{d}\right\} \tag{22.39}$$

This result reduces to Eq. (22.30) when the damping matrix $\mathbf{C}$ is set equal to zero.

APPROXIMATE SOLUTIONS

For a lightly damped system the exact solutions of Eq. (22.38) and Eq. (22.39) can be abbreviated considerably by making approximations based on the smallness of the damping. A systematic method of doing this is to consider the system without damping as a base upon which an infinitesimal amount of damping is superposed as a perturbation. An approximate solution to the complex eigenvalue problem by this method is provided by Eqs. (22.36) and (22.37). This perturbation approximation can be continued into Eqs. (22.38) and (22.39) by simply neglecting all squares and products of the small quantities α_r, ζ_r, $\mathbf{w}_r$, and $\mathbf{C}$. When this is done it is found that the formulas of Eqs. (22.38) and (22.39) may still be used if the parameters therein are obtained from the simplified expressions below.

$$\begin{aligned} &\alpha_r = \zeta_r\omega_r \qquad \beta_r = \omega_r \\ &a_r = -4\omega_r\mathbf{v}_r^T\mathbf{M}\mathbf{w}_r \qquad b_r = 2\omega_r\mathbf{v}_r^T\mathbf{M}\mathbf{v}_r \\ &a_r^2 + b_r^2 = 4\omega_r^2(\mathbf{v}_r^T\mathbf{M}\mathbf{v}_r)^2 \\ &\mathbf{A}_r = \mathbf{v}_r\mathbf{v}_r^T \qquad \mathbf{B}_r = \mathbf{v}_r\mathbf{w}_r^T + \mathbf{w}_r\mathbf{v}_r^T \\ &\mathbf{G}_r = 2\omega_r(\mathbf{v}_r^T\mathbf{M}\mathbf{v}_r)(\mathbf{v}_r\mathbf{w}_r^T + \mathbf{w}_r\mathbf{v}_r^T) \\ &\mathbf{H}_r = 2\omega_r(\mathbf{v}_r^T\mathbf{M}\mathbf{v}_r)\mathbf{v}_r\mathbf{v}_r^T \end{aligned} \tag{22.40}$$

For example, the steady-state forced sinusoidal solution of Eq. (22.39) takes the following explicit form in the perturbation approximation:

$$\mathbf{x}(t) = \Re\left\{\sum_{r=1}^{n} \frac{e^{j\omega t}}{\mathbf{v}_r^T\mathbf{M}\mathbf{v}_r} \frac{\mathbf{v}_r\mathbf{v}_r^T + \dfrac{j\omega}{\omega_r}\left[\mathbf{v}_r\mathbf{w}_r^T + \mathbf{w}_r\mathbf{v}_r^T\right]}{\omega_r^2 - \omega^2 + j2\zeta_r\omega_r\omega}\mathbf{d}\right\} \tag{22.41}$$

A cruder approximation, which is often used, is based on accepting the complex eigenvalue $p_r = -\alpha_r + j\omega_r$ but completely neglecting the imaginary part $j\mathbf{w}_r$ of the eigenvector $\mathbf{u}_r = \mathbf{v}_r + j\mathbf{w}_r$. It is thus assumed that the undamped mode $\mathbf{v}_r$ still applies for the system with damping. The approximate parameter values of Eq. (22.40) are further simplified by this assumption; e.g., $a_r = 0$, $\mathbf{B}_r = \mathbf{G}_r = 0$. The steady forced sinusoidal response of Eq. (22.41) reduces to

$$\mathbf{x}(t) = \Re\left\{\sum_{r=1}^{n} \frac{e^{j\omega t}}{\omega_r^2 - \omega^2 + j2\zeta_r\omega_r\omega} \frac{\mathbf{v}_r\mathbf{v}_r^T}{\mathbf{v}_r^T\mathbf{M}\mathbf{v}_r}\mathbf{d}\right\} \tag{22.42}$$

This approximation should be compared with the undamped solution of Eq. (22.30), as well as with the exact solution of Eq. (22.39) and the perturbation approximation of Eq. (22.41).

In the special case of proportional damping, the exact eigenvectors are real and Eq. (22.36) produces the exact decay rate $\alpha_r = \zeta_r\omega_r$, so that the response of Eq. (22.42) is an exact result.

Example 22.1. Consider the system of Fig. 22.5 with the following mass, damping, and stiffness coefficients:

$m_1 = 1$ lb-sec²/in. $\qquad m_2 = 2$ lb-sec²/in.

$c_1 = 0.10$ lb-sec/in. $\qquad c_2 = 0.02$ lb-sec/in. $\qquad c_3 = 0.04$ lb-sec/in.

$k_1 = 3$ lb/in. $\qquad k_2 = 0.5$ lb/in. $\qquad k_3 = 1$ lb/in.

The coefficient matrices of Eq. (22.9) then have the following numerical values:

$$\mathbf{M} = \begin{bmatrix} 3 & 2 \\ 2 & 2 \end{bmatrix} \qquad \mathbf{C} = \begin{bmatrix} 0.14 & 0.04 \\ 0.04 & 0.06 \end{bmatrix} \qquad \mathbf{K} = \begin{bmatrix} 4 & 1 \\ 1 & 1.5 \end{bmatrix}$$

Assuming that the numerical values above are exact, the exact solutions to the complex eigenvalue problem of Eq. (22.33) for these values of **M**, **C**, and **K** are, correct to four decimal places,

$$p_r = -\alpha_r + j\beta_r \qquad \mathbf{u}_r = \mathbf{v}_r + j\mathbf{w}_r$$

$$\begin{aligned} 2\alpha_1 &= 0.0279 \qquad & \alpha_1 = \zeta_1\omega_1 &= 0.0139 \qquad & \zeta_1 &= 0.0166 \\ \beta_1 &= 0.8397 & \omega_1 &= 0.8398 & \omega_1^2 &= 0.7053 \\ 2\alpha_2 &= 0.1221 & \alpha_2 = \zeta_2\omega_2 &= 0.0611 & \zeta_2 &= 0.0324 \\ \beta_2 &= 1.8818 & \omega_2 &= 1.8828 & \omega_2^2 &= 3.5449 \end{aligned} \tag{22.43}$$

$$\mathbf{V} = \begin{bmatrix} 0.2179 & -0.9179 \\ 1.0000 & 1.0000 \end{bmatrix} \qquad \mathbf{W} = \begin{bmatrix} 0.0016 & 0.0010 \\ 0 & 0 \end{bmatrix}$$

Note that this is a lightly damped system. The damping ratios in the two modes are 1.66 percent and 3.24 percent, respectively.

For comparison, the solution of the real eigenvalue problem Eq. (22.12) for the corresponding undamped system (i.e., **M** and **K** as above, but $\mathbf{C} = 0$) is, correct to four decimal places,

$$\begin{aligned} \omega_1^2 &= 0.7053 \\ \omega_2^2 &= 3.5447 \end{aligned} \qquad \mathbf{V} = \begin{bmatrix} 0.2179 & -0.9179 \\ 1.0000 & 1.0000 \end{bmatrix}$$

Note that, to this accuracy, there is no discrepancy in the real parts of the eigenvectors. There are, however, small discrepancies in the imaginary parts of the eigenvalues. The difference between β_1 for the damped system and ω_1 for the undamped system is 0.0001, and the corresponding difference between β_2 and ω_2 is 0.0009. The imaginary parts of the eigenvectors and the real parts of the eigenvalues for the damped system are completely absent in the undamped system. They may be approximated by applying the perturbation equations of Eqs. (22.36) and (22.37) to the solution of the eigenvalue problem for the undamped system.

The real parts α_r of the eigenvalues obtained from Eq. (22.36) agree, to four decimal places, with the exact values in Eq. (22.43). The imaginary parts $\mathbf{w}_r$ of the eigenvectors obtained from Eq. (22.37) are

$$\mathbf{w}_1 = \begin{Bmatrix} 0.0013 \\ -0.0014 \end{Bmatrix} \qquad \mathbf{w}_2 = \begin{Bmatrix} 0.0002 \\ 0.0009 \end{Bmatrix}$$

These vectors satisfy the orthogonality conditions $\mathbf{v}_r^T\mathbf{M}\mathbf{w}_r = 0$.

In order to compare these values with Eq. (22.43), it is first necessary to normalize the complete eigenvector $\mathbf{v}_r + j\mathbf{w}_r$, so that its second element is unity. For example, this is done in the case of $r = 1$ by dividing both $\mathbf{v}_1$ and $\mathbf{w}_1$ by $1.0000 - j0.0014$. When this is done, it is found that the perturbation approximation to the eigenvectors agrees, to four decimal places, with the exact solution of Eq. (22.43).

To illustrate the application of the formal solutions given above, consider the steady-state forced oscillation of the system shown in Fig. 22.5 at a frequency ω due to driving force amplitudes d_1 and d_2. Using the exact solution values of Eq. (22.43), the expressions a_r, b_r, $\mathbf{A}_r$, $\mathbf{B}_r$, $\mathbf{G}_r$, and $\mathbf{H}_r$ following Eq. (22.38) are evaluated for $r = 1$ and $r = 2$. With these values, the steady-state response, Eq. (22.39), becomes

$$\begin{bmatrix} x_1 \\ x_2 \end{bmatrix} = \mathcal{R}\left\{ \frac{e^{j\omega t}\left\{\begin{bmatrix} 0.0158 & 0.0723 \\ 0.0723 & 0.3318 \end{bmatrix} + j\omega \begin{bmatrix} 0.0002 & 0.0004 \\ 0.0004 & -0.0011 \end{bmatrix}\right\}}{0.7053 - \omega^2 + 0.0279j\omega} \begin{bmatrix} d_1 \\ d_2 \end{bmatrix} \right.$$

$$\left. + \frac{e^{j\omega t}\left\{\begin{bmatrix} 0.9842 & -1.0724 \\ -1.0724 & 1.1683 \end{bmatrix} + j\omega \begin{bmatrix} -0.0002 & -0.0004 \\ -0.0004 & 0.0011 \end{bmatrix}\right\}}{3.5449 - \omega^2 + 0.1221j\omega} \begin{bmatrix} d_1 \\ d_2 \end{bmatrix} \right\}$$

When the approximation in Eq. (22.41) based on the perturbation solution is evaluated, the result is almost identical to this. A few entries differ by one or two units in the fourth decimal place. The crude approximation, Eq. (22.42), is the same as the perturbation approximation except that the terms in the numerators which are multiplied by $j\omega$ are absent. This means that the relative error between the crude approximation and the exact solution can be large at high frequencies. At low frequencies, however, even the crude approximation provides useful results for lightly damped systems. In the present case, the discrepancy between the crude approximation and the exact solution remains under 1 percent as long as ω is less than ω_2 (the highest natural frequency). At higher frequencies the absolute response level decreases steadily, which tends to undercut the significance of the increasing relative discrepancy between approximations.

REFERENCES

1. Strang, G.: "Introduction to Linear Algebra," 3d ed., Wellesley-Cambridge Press, Wellesley, Mass., 2003.
2. Przemieniecki, J. S.: "Theory of Matrix Structural Analysis," McGraw-Hill Book Company, New York, 1968. See App. A, "Matrix Algebra."
3. Meirovitch, L.: "Fundamentals of Vibration," McGraw-Hill Book Company, Boston, 2001. See App. C, "Linear Algebra."

CHAPTER 23
FINITE ELEMENT METHODS OF ANALYSIS

Robert N. Coppolino

INTRODUCTION

The *finite element method* (FEM), formally introduced by Clough[1] in 1960, has become a mature engineering discipline during the past fifty years. In actual practice, finite element analysis is a systematic applied science, which incorporates (1) the definition of a physical model of a complex system as a collection of building blocks (finite elements), (2) the solution of matrix equations describing the physical model, and (3) the analysis and interpretation of numerical results. The foundations of finite element analysis are (*a*) the design of consistent, robust finite elements[2]; and (*b*) matrix methods of numerical analysis[3–5] (see Chap. 22). Originally developed to address modeling and analysis of complex structures, the finite element approach is now applied to a wide variety of engineering applications including heat transfer, fluid dynamics, and electromagnetics, as well as multiphysics (coupled interaction) applications.

Modern finite element programs include powerful graphical user interface (GUI) driven preprocessors and postprocessors, which automate routine operations required for the definition of models and the interpretation of numerical results, respectively. Moreover, finite element analysis, computer-assisted design and optimization, and laboratory/field testing are viewed as an integrated "concurrent engineering" process. Commercially available products, widely used in industry, include MSC/NASTRAN (a product of MSC.Software), NX/NASTRAN (a product of Siemens), ANSYS (a product family of ANSYS Incorporated), and ABAQUS (a product of Simulia), just to mention a few.

This chapter describes finite element modeling and analysis with an emphasis on its application to the shock and vibration of structures and structures interacting with fluid media. Included are discussions on the theoretical foundations of finite element models, effective modeling guidelines, dynamic system models and analysis strategies, and common industry practice.

THEORETICAL FOUNDATIONS OF FINITE ELEMENT MODELS

APPLICATION OF MINIMAL PRINCIPLES

The matrix equations describing both individual finite elements and complete finite element system models are defined on the basis of minimal principles. In particular, for structural dynamic systems, Hamilton's principle or Lagrange's equations[6] constitute the underlying physical principle. The fundamental statement of *Hamilton's principle* is

$$\delta \int_{t_0}^{t_1} (T + W)dt = 0 \tag{23.1}$$

where T is the system kinetic energy, W is the work performed by internal and external forces, t represents time, and δ is the variational operator. In the case of statics, Hamilton's principle reduces to the *principle of virtual work,* stated mathematically as

$$\delta W = 0 \qquad (\text{if } T = 0) \tag{23.2}$$

For most mechanical systems of interest, W may be expressed in terms of a conservative interior elastic potential energy (U), dissipative interior work (W_D), and the work associated with externally applied forces (W_E). Thus, Hamilton's principle is stated as

$$\int_{t_0}^{t_1} (\delta T - \delta U + \delta W_D + \delta W_E)dt = 0 \tag{23.3}$$

The kinematics of a mechanical system of volume V are described in terms of the displacement field

$$\{u\} = [N_u \; N_q]\begin{Bmatrix} u_i \\ q \end{Bmatrix} \tag{23.4}$$

where $\{u\}$ is the displacement array at any point in V, $\{u_i\}$ is an array of discrete displacements (typically) on the element surface, and $\{q\}$ is an array of generalized displacement coefficients. The transformation matrix partitions N_u and N_q describe assumed shape functions for the particular finite element. The most commonly used elements, namely *H-type* elements, do not include generalized displacement coefficients, $\{q\}$. The more general case element is called a *P-type* element. For simplicity, the subsequent discussion will be limited to H-type elements.

In matrix notation (see Chap. 22), the strain field within the element volume is related to the assumed displacements by the differential operator matrix $[N_{\varepsilon u}]$ as

$$\{\varepsilon(x,y,z,t)\} = \{\varepsilon\} = [N_{\varepsilon u}]\{u\} \tag{23.5}$$

The stress field within the element volume is expressed as

$$\{\sigma(x,y,z,t)\} = \{\sigma\} = [D]\{\varepsilon\} = [D][N_{\varepsilon u}]\{u\} \tag{23.6}$$

In the case of hybrid finite element formulations, for which there is an assumed element stress field other than simply $[D][N_{\varepsilon u}]$, the situation is more involved.

Using the above general expressions, the kinetic and strain energies associated with a finite element are

$$2T = \int_v \{\dot{u}\}^T[N_u]^T[\rho][N_u]\{\dot{u}\}dV = \{\dot{u}\}^T[M_e]\{\dot{u}\} \tag{23.7}$$

$$2U = \int_v \{u\}^T[N_{\varepsilon u}]^T[D][N_{\varepsilon u}]\{u\}dV = \{u\}^T[K_e]\{u\} \tag{23.8}$$

where $[\rho]$ is the material density matrix, $[D]$ is the material elastic matrix, and $[M_e]$ and $[K_e]$ are the individual element mass and stiffness matrices, respectively. The superscript shown as $\{\ \}^T$ and $[\ ]^T$ denotes the transpose of an array and matrix, respectively. In the case of viscous damping (which is a common yet not necessarily realistic assumption), the *element virtual dissipative work* is

$$\delta W_D = \{\delta u\}^T[B_e]\{\dot{u}\} \tag{23.9}$$

where $[B_e]$ is the symmetric element damping matrix.

In order to assemble the mass, stiffness, and damping matrices associated with a complete finite element system model, the displacement array for the entire system, $\{u_g\}$, must first be defined. The individual element contributions to the system are then allocated (and accumulated) to the appropriate rows and columns of the system matrices. This results in the formation of generally sparse, symmetric matrices. The complete system kinetic and strain energies are, respectively,

$$2T_g = \{\dot{u}_g\}^T[M_{gg}]\{\dot{u}_g\} \tag{23.10}$$

$$2U_g = \{u_g\}^T[K_{gg}]\{u_g\} \tag{23.11}$$

where $[M_{gg}]$ and $[K_{gg}]$ are the system mass and stiffness matrices.

For the case of viscous damping, the complete system virtual dissipative work is

$$\delta W_{Dg} = \{\delta u_g\}^T[B_{gg}]\{\dot{u}_g\} \tag{23.12}$$

Finally, the virtual work associated with externally applied forces on the complete system is defined as

$$\delta W_{Eg} = \{\delta u_g\}^T[\Gamma_{ge}]\{F_e\} \tag{23.13}$$

where $[\Gamma_{ge}]$ represents the allocation matrix for externally applied forces $\{F_e\}$, including moments, stresses, and pressures if applicable. Substitution of the above expressions for the complete system energies and virtual work into Hamilton's principle, followed by key manipulations, results in the finite element system differential equations

$$[M_{gg}]\{\ddot{u}_g\} + [B_{gg}]\{\dot{u}_g\} + [K_{gg}]\{u_g\} = [\Gamma_{ge}]\{F_e\} \tag{23.14}$$

The task of defining a finite element model is not yet complete at this point. Constraints and boundary conditions, as required, must now be imposed. The logical sequence of imposed constraint types is (1) *multipoint constraints* (e.g., geometric constraints expressed as algebraic relationships) and (2) *single-point constraints* (e.g., fixed supports). These constraints are described, in summary, by the linear transformation

$$\{u_g\} = [G_{gf}]\{u_f\} \tag{23.15}$$

where $\{u_f\}$ is the array of "free" displacements. By imposing the constraint transformation, $[G_{gf}]$, in a symmetric manner to the system equations [see Eq. (23.14)], the following constrained system equations are formed:

$$[M_{ff}]\{\ddot{u}_f\} + [B_{ff}]\{\dot{u}_f\} + [K_{ff}]\{u_f\} = [\Gamma_{fe}]\{F_e\} \tag{23.16}$$

where

$$\begin{aligned} [M_{ff}] &= [G_{gf}]^T[M_{gg}][G_{gf}], \qquad [B_{ff}] = [G_{gf}]^T[B_{gg}][G_{gf}] \\ [K_{ff}] &= [G_{gf}]^T[K_{gg}][G_{gf}], \qquad [\Gamma_{fe}] = [G_{gf}]^T[\Gamma_{ge}] \end{aligned} \tag{23.17}$$

TYPICAL FINITE ELEMENTS

Commonly used finite elements in commercial codes may be divided into two primary classes, namely, (1) elements based on technical theories, and (2) elements based on three-dimensional continuum theory. The first class of elements includes one-dimensional beam elements. Truss and bar elements are special cases of the general beam element. A modern beam element permits modeling of the shear deformation and warping associated with general cross-section geometry. Beam elements, which may describe a straight or curved segment, are typically described in terms of nodal displacements (three linear and three angular displacements) at the two extremities, as illustrated in Fig. 23.1.

Also within the family of elements based on technical theories are *shell elements.* Membrane and flat plate elements are special cases of the general shell element. Shell elements are typically of triangular or quadrilateral form with straight or

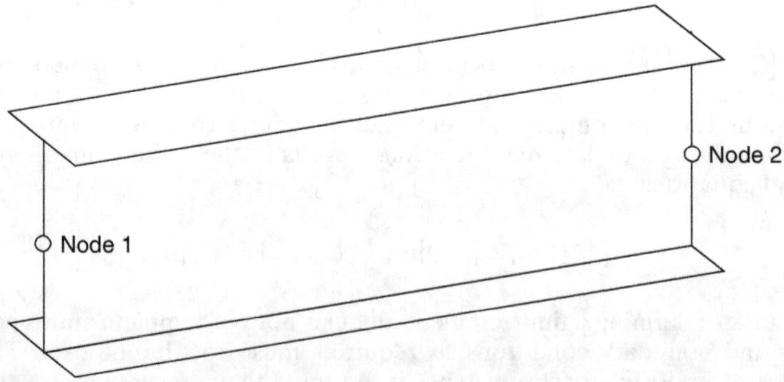

FIGURE 23.1 Typical beam element.

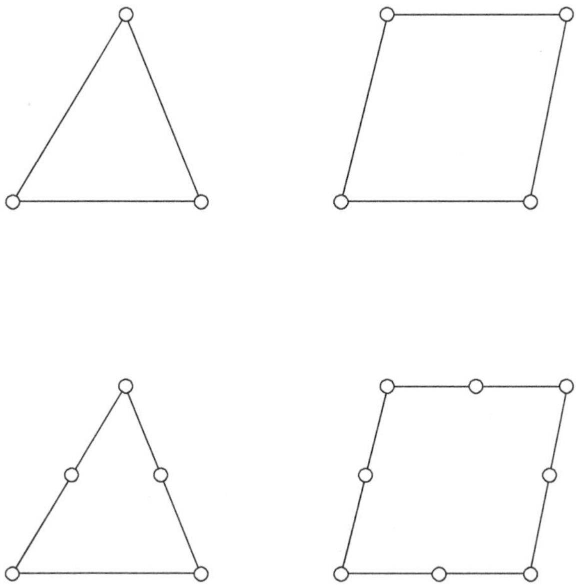

FIGURE 23.2 Typical triangular and quadrilateral shell elements.

curved edges, as illustrated in Fig. 23.2. Common H-type shell elements are defined by nodal displacements (three linear and three angular displacements) at the element corners. Shell elements may also be defined in terms of midside nodal displacements. Modern shell elements may include such features as shear deformation, anisotropic elastic materials, and composite layering.

The family of three-dimensional elastic elements includes tetrahedral, pentahedral, wedge, and hexahedral configurations with straight or curved edges, as illustrated in Fig. 23.3. H-type continuum elements are defined by nodal displacements (three linear) at the element corners. Three-dimensional H-type elements may also be defined in terms of midside nodal displacements. As in the case of shell elements, anisotropic elastic materials may be employed in element formulations.

Effect of Static Loading—Differential Stiffness. The effective stiffness of structures subjected to static loads may be increased or decreased. For example, the lateral stiffness of a column subjected to axial compression decreases, becoming singular if the fundamental buckling load is imposed. In the case of an inflated balloon, the shell-bending stiffness is almost entirely due to significant membrane tension. In each of these situations, the static load–associated differential stiffness derives from a finite geometric change. Modern commercial finite element codes contain the option to include differential stiffness effects in the model definition.

Fluid-Structure Interaction. Linear dynamic models of oscillating (but otherwise assumed stationary) fluids interacting with elastic structures are employed in vibroacoustics, liquid-filled tank vibratory dynamics, and other applications. One popular approach used to describe the fluid medium employs pressure degrees of freedom (DOF). On the basis of complementary energy principles,[7] three-dimensional fluid ele-

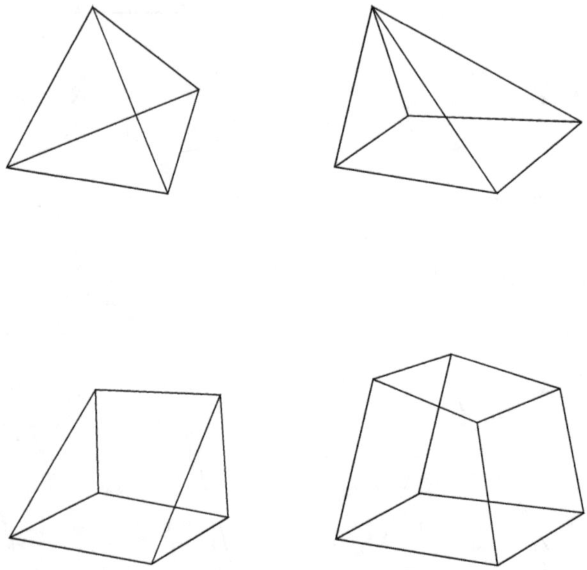

FIGURE 23.3 Typical three-dimensional solid elements.

ments (with the geometric configurations illustrated in Fig. 23.3) are defined. The matrix equations describing dynamics of such a fluid interacting with an elastic structure are of the form

$$\begin{bmatrix} C & A^T \\ 0 & M \end{bmatrix} \begin{Bmatrix} \ddot{P} \\ \ddot{u} \end{Bmatrix} + \begin{bmatrix} S & 0 \\ -A & K \end{bmatrix} \begin{Bmatrix} P \\ u \end{Bmatrix} = \begin{bmatrix} \Gamma_Q & 0 \\ 0 & \Gamma_F \end{bmatrix} \begin{Bmatrix} \ddot{Q}_e \\ F_e \end{Bmatrix} \tag{23.18}$$

where $[C]$ is the fluid compliance matrix, $[S]$ is the fluid susceptance matrix (analogous to the inverse of a mass matrix), and $[A]$ is the fluid-structure interface area matrix. The matrix partitions $[\Gamma_Q]$ and $[\Gamma_F]$ are the fluid volumetric source flow $\{\ddot{Q}_e\}$ and the structural applied load $\{F_e\}$ allocation matrices, respectively. The system of equations is unsymmetric due to the fact that it is based on a blend of standard structural displacement and complementary fluid pressure variational principles.

A variety of algebraic manipulations are used to cast the hydroelastic equations in a conventional symmetric form. In many applications involving approximately incompressible (liquid) fluids, the fluid compliance is ignored. The incompressible hydroelastic equations (without source flow excitation) may then be cast in the symmetric form[7]

$$[M + M_f]\{\ddot{u}\} + [K]\{u\} = [\Gamma_F]\{F_e\} \tag{23.19}$$

where the (generally full) fluid mass matrix is

$$[M_f] = [A][S]^{-1}[A]^T \tag{23.20}$$

Specialized constraints are often required to permit the decomposition of the generally singular fluid susceptance matrix.[7] Moreover, specialized eigenvalue analysis procedures are recommended to efficiently deal with the full fluid mass matrix.

For the most general case of a compressible fluid, introduction of the fluid volumetric strain variable

$$\{v\} = [C]\{P\} \tag{23.21}$$

results in the symmetric equation set

$$\begin{bmatrix} S^{-1} & S^{-1}A^T \\ AS^{-1} & M + AS^{-1}A^T \end{bmatrix}\begin{Bmatrix} \ddot{v} \\ \ddot{u} \end{Bmatrix} + \begin{bmatrix} C^{-1} & 0 \\ 0 & K \end{bmatrix}\begin{Bmatrix} v \\ u \end{Bmatrix} = \begin{bmatrix} S^{-1}\Gamma_Q & 0 \\ -AS^{-1}\Gamma_Q & \Gamma_F \end{bmatrix}\begin{Bmatrix} \ddot{Q}_e \\ F_e \end{Bmatrix} \tag{23.22}$$

As for the incompressible, symmetric formulation, a specialized efficient eigenvalue analysis procedure (based on the subspace iteration algorithm[8]) is recommended to efficiently deal with the full hydroelastic mass matrix.

In situations for which the fluid is a lightweight acoustic gas, a decoupling approximation may provide reasonable, approximate dynamic solutions. The approximation assumes that the acoustic medium is driven by a much heavier structure, which is unaffected by fluid interaction. The decoupled approximate dynamic equations are

$$[M]\{\ddot{u}\} + [K]\{u\} = [\Gamma_F]\{F_e\} \tag{23.23}$$

$$[C]\{\ddot{P}\} + [S]\{P\} = -[A^T]\{\ddot{u}\} + [\Gamma_Q]\{\ddot{Q}_e\} \tag{23.24}$$

Uncoupled modal analyses of the structural and acoustic media are used in the computation of the system dynamic response for this approximate formulation.

General Linear System Dynamic Interaction Considerations. In the previous discussion on fluid-structure interaction, a variety of algebraic manipulations, which transform coupled unsymmetric dynamic equations to a conventional symmetric linear formulation, were described. Transformations resulting in symmetric matrix equations, however, are not possible in more general situations involving dynamic interaction.

Linear systems which include complicating effects due to the interaction with general linear subsystems (e.g., control systems, propulsion systems, and perturbed steady fluid flow) are generally appended with nonsymmetric matrix dynamic relationships. The nonconventional linear dynamic formulation incorporates state equations for the interacting subsystem

$$[A_i]\{q_i\} - \{\dot{q}_i\} = [B_i]\{\dot{u}\} + [K_i]\{u\} \tag{23.25}$$

and the forces of interaction with the structural dynamic system

$$[\Gamma_i]\{F_i\} = [\Gamma_i][C_i]\{q_i\} \tag{23.26}$$

where $\{q_i\}$ are subsystem state variables, $[A_i]$ is the subsystem plant matrix, and $[B_i]$, $[K_i]$, and $[C_i]$ are coupling matrices. The complete dynamic system is described by the state equations

$$\begin{bmatrix} -M^{-1}B & -M^{-1}K & M^{-1}\Gamma_i C_i \\ I & 0 & 0 \\ -B_i & -K_i & A_i \end{bmatrix}\begin{Bmatrix} \dot{u} \\ u \\ q_i \end{Bmatrix} - \begin{Bmatrix} \ddot{u} \\ \dot{u} \\ \dot{q}_i \end{Bmatrix} = \begin{bmatrix} -M^{-1}\Gamma_e \\ 0 \\ 0 \end{bmatrix}\{F_e\} \tag{23.27}$$

The above state equations are of the class

$$[A_{\text{sys}}]\{q_{\text{sysi}}\} - \{\dot{q}_{\text{sys}}\} = [\Gamma_{\text{sysi}}]\{F_{\text{sys}}\} \tag{23.28}$$

Nonlinear Dynamic Systems. The most general type of dynamic system includes nonlinear effects, which may be due to large geometric deformations, nonlinear material behavior, stick-slip friction, gapping, and other complicating effects (see Chap. 4). Fortunately, many dynamic systems are approximately linear. A thorough discussion of nonlinear finite element modeling and analysis techniques is beyond the scope of the present discussion. However, two particularly useful classes of models are pointed out herein, namely, (1) linear systems with physically localized nonlinear features, and (2) general nonlinear systems.

A structural dynamic system with physically localized nonlinear features is described by slightly modified linear matrix equations as

$$[M]\{\ddot{u}\} + [B]\{\dot{u}\} + [K]\{u\} = [\Gamma_N]\{F_N(u_N,\dot{u}_N)\} + [\Gamma_F]\{F_e\} \tag{23.29}$$

where $[\Gamma_N]$ is the allocation matrix for nonlinear features and $\{F_N\}$ are the nonlinear forces related to local displacements and velocities. The local displacements and velocities are related to global displacements and velocities as

$$\{u_N\} = [\Gamma_N]^T\{u\},\ \{\dot{u}_N\} = [\Gamma_N]^T\{\dot{u}\} \tag{23.30}$$

This type of nonlinear dynamic formulation is useful in that the linear portion of the system may be efficiently treated with modal analysis procedures, to be discussed later.

General situations involving extensively distributed nonlinear behavior are described by equations of the type

$$\{\ddot{u}\} = [M]^{-1}\{F(u,\dot{u},t)\} \tag{23.31}$$

or

$$\begin{Bmatrix} \ddot{u} \\ \dot{u} \end{Bmatrix} = \begin{bmatrix} M^{-1} & 0 \\ 0 & I \end{bmatrix} \begin{Bmatrix} F(u,\dot{u},t) \\ \dot{u} \end{Bmatrix} \tag{23.32}$$

Advanced numerical integration procedures are employed to treat general nonlinear dynamic systems. The procedures fall into two distinct classes, namely, (*a*) implicit methods,[9] and (*b*) explicit methods.[4]

EFFECTIVE MODELING GUIDELINES

CUTOFF FREQUENCY AND GRID SPACING

In order to develop a relevant dynamic model, general requirements should be addressed based on

TABLE 23.1 Summary of Typical Dynamic Environments

Environment	Chapter or reference
Seismic excitation	Chap. 29
Fluid flow	Chap. 30
Wind loads	Chap. 31
Sound	Chap. 32
Transportation and handling impact	MIL-STD-810G
Transportation and handling vibration	MIL-STD-810G
Shipboard vibration	MIL-STD-167-1

1. Frequency bandwidth $0 < f < f^*$, and intensity (F^*) of anticipated dynamic environments.
2. General characteristics of structural or mechanical components.

Typical dynamic environments are summarized in Table 23.1. Dynamic environments are generally (*a*) harmonic, (*b*) transient, (*c*) impulsive, or (*d*) random. For all categories, the cutoff frequency (f^*) is reliably determined by shock response spectrum analysis (see Chap. 20). The overall intensity level of a dynamic environment is described by a peak amplitude for harmonic, transient, and impulsive events, or by a statistical amplitude (e.g., mean plus a multiple of the standard deviation) for a long-duration random environment (see Chaps. 19 and 24). With the cutoff frequency (f^*) established, the shortest relevant wavelength of a forced vibration for components in a structural assembly may be calculated. For finite element modeling, the quarter wavelength ($L/4$) is of particular interest, since it defines the grid spacing requirement needed to accurately model the dynamics (note that the quarter wavelength rule is a general guideline, which may be modified based on the performance of specific finite elements). The guidelines for typical structural components are summarized in Table 23.2.

In addition to the above grid spacing guidelines, the engineer must also consider the limitations associated with beam and plate theories. In particular, if the wavelength-to-thickness ratio (L/h) is less than about 10, a higher-order theory or 3D elasticity modeling should be considered. Moreover, modeling requirements for the capture of stress concentration details may call for a finer grid meshing than suggested by the cutoff frequency. Finally, if the dynamic environment is sufficiently high in amplitude, nonlinear modeling may be required, e.g., if plate deflections are greater than the thickness h.

MODAL DENSITY AND EFFECTIVENESS OF FINITE ELEMENT MODELS

Finite element modeling is an effective approach for the study of structural and mechanical system dynamics as long as individual vibration modes have sufficient frequency spacing or low modal density. Modal density is typically described as the number of modes within a ⅓ octave frequency band ($f_0 < f < 1.26 f_0$). When the modal density of a structural component or structural assembly is greater than 10 modes per ⅓ octave band, details of individual vibration modes are not of significance and statistical vibration response characteristics are of primary importance. In such a situation, the statistical energy analysis (SEA) method[10] applies (see Chap. 24). Formulas for modal density[10] as a mathematical derivative, $dn/d\omega$ (n = number of modes, ω = frequency in radians/sec), for typical structural components are summarized in Table 23.3.

TABLE 23.2 Guidelines for Dynamic Finite Element Model Meshing

Component	Mode type	$L/4$	Additional data
String	Lateral	$(\sqrt{T/\rho A})/4f^*$	T = tension, A = area, ρ = mass density
Rod	Axial	$(\sqrt{E/\rho})/4f^*$	E = elastic modulus
Rod	Torsion	$(\sqrt{G/\rho})/4f^*$	G = shear modulus
Beam	Bending	$(\pi/2)(EI/\rho A)^{1/4}/\sqrt{2\pi f^*}$	EI = flexural stiffness
Membrane	Lateral	$(\sqrt{N/\rho h})/4f^*$	N = stress resultant
Plate	Bending	$(\pi/2)(D/\rho h)^{1/4}/\sqrt{2\pi f^*}$	D = plate flexural stiffness, h = plate thickness
3D elastic	Dilatational	$(\sqrt{E/\rho})/4f^*$	
3D elastic	Shear	$(\sqrt{G/\rho})/4f^*$	
Acoustic	Dilatational	$(\sqrt{B/\rho})/4f^*$	B = bulk modulus

DYNAMIC SYSTEM MODELS AND ANALYSIS STRATEGIES

FUNDAMENTAL DYNAMIC FORMULATIONS

Finite element dynamic models fall into a variety of classes, which are expressed by the following general equation sets:

1. Linear structural dynamic systems [see Eq. (23.16)]
2. Linear structural dynamic systems interacting with other media [see Eq. (23.27)]
3. Dynamic systems with localized nonlinear features [see Eqs. (23.29) and (23.30)]
4. Dynamic systems with distributed nonlinear features [see Eqs. (23.31) and (23.32)]

TABLE 23.3 Modal Density for Typical Structural Components

Component	Motion	Modal density, $dn/d\omega$	Additional data
String	Lateral	$L/(\pi\sqrt{T/\rho A})$	T = tension, A = area, ρ = mass density, L = length
Rod	Axial	$L/(\pi\sqrt{E/\rho})$	E = elastic modulus
Rod	Torsion	$L/(\pi\sqrt{G/\rho})$	G = shear modulus
Beam	Bending	$L/(2\pi)(\omega\sqrt{EI/\rho A})^{-1/2}$	EI = flexural stiffness
Membrane	Lateral	$A_s\omega/(2\pi)(N/\rho h)$	N = stress resultant, A_s = surface area
Plate	Bending	$A_s/(4\pi)\sqrt{D/\rho h}$	D = plate flexural stiffness, h = plate thickness
Acoustic	Dilatational	$V_0\omega^2/(2\pi^2)(\sqrt{B/\rho})^3$	B = bulk modulus, V_0 = enclosed volume

The first category represents the type of systems most often dealt with in structural dynamics and mechanical vibration. In the majority of engineering analyses, damping is assumed to be well-distributed in a manner justifying the use of normal mode analysis techniques (see Chaps. 21 and 22). Systems in the first and second categories having more general damping features may be treated by complex modal analysis procedures (see Chap. 22). When localized nonlinear features are present, normal or complex mode analysis procedures may also be applied. The final class, namely dynamic systems with distributed nonlinear features, must be treated using numerical integration procedures. When a nonlinear system is subjected to a slowly applied or moderately low frequency environment, implicit numerical integration is often the preferred numerical integration strategy. Alternatively, when the dynamic environment is suddenly applied, high-frequency and/or short-lived explicit numerical integration is often advantageous.

APPLICATION OF NORMAL MODES IN TRANSIENT DYNAMIC ANALYSIS

The homogeneous form for the conventional linear structural dynamic formulation [see Eq. (23.16)], with damping ignored, defines the real eigenvalue problem, that is,

$$[K]\{\Phi_n\} - [M]\{\Phi_n\}\omega_n^2 = \{0\} \tag{23.33}$$

where

$$\{u\} = \{\Phi_n\} \sin(\omega_n t) \tag{23.34}$$

There are as many distinct eigenvectors or modes $\{\Phi_n\}$ as set degrees of freedom for a well-defined undamped dynamic system. The eigenvalues ω_n^2 (ω_n = natural frequency of mode n), however, are not necessarily all distinct. Individual modes or mode shapes represent displacement patterns of arbitrary amplitude. It is convenient to normalize the mode shapes (to unit modal mass) as follows:

$$\{\Phi_n\}^T[M]\{\Phi_n\} = 1 \tag{23.35}$$

The assembly of all or a truncated set of normalized modes into a modal matrix $[\Phi]$ defines the (orthonormal) modal transformation

$$\{U\} = [\Phi]\{q\} \tag{23.36}$$

where

$$\begin{aligned} [\Phi]^T[M][\Phi] &= [OR] = [I] = \text{diagonal identity matrix} \\ [\Phi]^T[K][\Phi] &= [\Lambda] = [\omega_n^2] = \text{diagonal eigenvalue matrix} \end{aligned} \tag{23.37}$$

The modal transformation produces the mathematically diagonal matrix

$$[\Phi]^T[B][\Phi] = [2\zeta_n\omega_n] = \text{diagonal modal damping matrix} \tag{23.38}$$

only for special forms of the damping matrix. One such form, known as proportional damping, is

$$[B] = \alpha[M] + \beta[K] \tag{23.39}$$

In reality, proportional damping is a mathematical construction that bears little resemblance to physical reality. It is experimentally observed in many situations, however, that the diagonal modal damping matrix is a valid approximation.

Application of the modal transformation to the dynamic equations [see Eq. (23.16)] results in the uncoupled single-DOF dynamic equations

$$\ddot{q}_n + 2\zeta_n\omega_n\dot{q}_n + \omega_n^2 q_n = [\Phi_n^T\Gamma]\{F(t)\} = [\Gamma_{qn}]\{F(t)\} = Q_n(t) \tag{23.40}$$

The symbol ζ_n is the critical damping ratio and $[\Gamma_{qn}] = [\Phi_n^T\Gamma]$ is the modal excitation gain array.

The character and content of an individual normal mode $[\Phi_n]$ is described fundamentally by the geometric distribution of the displacement DOF. Utilizing the mass matrix $[M]$, the modal momentum distribution is

$$\{P_n\} = [M]\{\Phi_n\} \tag{23.41}$$

and the modal kinetic energy distribution is

$$\{E_n\} = \{P_n\} \otimes \{\Phi_n\} = ([M]\{\Phi_n\}) \otimes \{\Phi_n\} \tag{23.42}$$

where $\otimes$ denotes term-by-term multiplication. The sum of the terms in the modal kinetic energy vector $\{E_n\}$ is 1.0 when the mode is normalized to unit modal mass.

Internal structural loads and stresses, relative displacements, strains, and other user-defined terms are calculated as recovery variables. In many cases the recovery variables $\{S\}$ are related to the physical displacements $\{u\}$ through a load transformation matrix $[K_S]$, specifically,

$$\{S\} = [K_S]\{u\} \tag{23.43}$$

A modal (displacement-based) load transformation matrix, defined by substitution of the modal transformation, is

$$\{S\} = [\Phi_{KS}]\{q\} \tag{23.44}$$

where

$$[\Phi_{KS}] = [K_S][\Phi] \tag{23.45}$$

The dynamic response of a structural dynamic system, described in terms of normal modes, is computed as follows:

Step 1. Calculate the modal responses numerically with, for example, the Duhamel integral (see Chap. 8) given by

$$q_n(t) = \int_0^t h_n\,(t-\tau)Q_n(\tau)d\tau \tag{23.46}$$

where

$$h_n(t-\tau) = \frac{\omega_n}{\sqrt{1-\zeta_n^2}}\; e^{-\zeta_n\omega_n(t-\tau)} \sin\left(\left(\omega_n\sqrt{1-\zeta_n^2}\right)(t-\tau)\right) \tag{23.47}$$

Similar relationships exist for modal velocity and acceleration.

Step 2. Calculate the physical displacement, velocity, and acceleration responses by modal superposition using Eq. (23.36) and calculate loads using Eq. (23.44).

It should be noted that the calculation of modal responses to harmonic and random excitation environments follows strategies paralleling steps 1 and 2. These matters will be discussed at the end of this chapter.

MODAL TRUNCATION

A common practice in structural dynamics analysis is to describe a system response in terms of a truncated set of lowest-frequency modes. The selection of an appropriate truncated mode set is accomplished by a normalized displacement, shock response spectrum analysis (see Chap. 20) of each force component in the excitation environment $\{F(t)\}$ and establishment of the cutoff frequency ω^*. All modal responses for systems with a natural frequency $\omega_n > \omega^*$ will respond quasi-statically. Therefore, the dynamic response will be governed by the truncated set of modes $[\Phi_L]$ with natural frequencies below ω^*. The remaining set of high-frequency modes is denoted as $[\Phi_H]$. Therefore, the partitioned modal relationships are

$$\{u\} = [\Phi_L]\{q_L\} + [\Phi_H]\{q_H\}$$

$$\{\ddot{q}_L\} + [2\zeta_L\omega_L]\{\dot{q}_L\} + [\omega_L^2]\{q_L\} = [\Phi_L^T\Gamma]\{F(t)\} \tag{23.48}$$

$$[\omega_H^2]\{q_H\} \approx [\Phi_H^T\Gamma]\{F(t)\}$$

Since the high-frequency modal equations are algebraic, the modal transformation becomes

$$\{u\} = [\Phi_L]\{q_L\} + [\Psi_\rho]\{F(t)\} \tag{23.49}$$

where $[\Psi_\rho]$ is the residual flexibility matrix defined as

$$[\Psi_\rho] = [\Phi_H][\omega_H^2]^{-1}[\Phi_H]^T[\Gamma] \tag{23.50}$$

The computation of structural dynamic response employing a truncated set of modes often is inaccurate if the quasi-static response associated with the high-frequency modes is not accounted for. This being the case, it appears that all modes must be computed as indicated in Eq. (23.50). Such a requirement results in an excessive computational burden for large-order finite element models.

Residual Mode Vectors and Mode Acceleration. The significance of residual flexibility (quasi-static response of high-frequency modes) is well established,[11] as are methods for the efficient definition of residual vectors.[12] The basic definition for residual flexibility, using all of the high-frequency modal vectors, is computationally inefficient for large-order models. Therefore, procedures that do not explicitly require knowledge of the high-frequency modes have been developed.

The most fundamental procedure for deriving residual vectors forms residual shape vectors as the difference between a complete static solution and a static solution based on the low-frequency mode subset. The complete static solution for unit-applied loads, using a shifted stiffness (allowing treatment of an unconstrained structure), is

$$[\Psi_S] = [K + \lambda_S M]^{-1}[\Gamma] \tag{23.51}$$

where λ_S is a small "shift" used for singular stiffness matrices. For nonsingular stiffness, the shift is not required. The corresponding truncated, low-frequency mode static solution is

$$[\Psi_L] = [\Phi_L][\omega_L^2 + \lambda_S]^{-1}[\Phi_L]^T[\Gamma] \tag{23.52}$$

Therefore, the residual vectors are

$$[\Psi_\rho] = [\Psi_S] - [\Psi_L] = [K + \lambda_S M]^{-1}[\Gamma] - [\Phi_L][\omega_L^2 + \lambda_S]^{-1}[\Phi_L]^T[\Gamma] \tag{23.53}$$

Note that the high-frequency modes are not explicitly required in this formulation. Therefore, the excessive computational burden for large-order finite element models is mitigated.

An alternative strategy, which automatically compensates for modal truncation, is the mode acceleration method.[13] The basis for this strategy is the substitution of truncated expressions for acceleration and velocity in the system dynamic equations, which results in

$$[K]\{u\} = [\Gamma]\{F\} - [M][\Phi_L]\{\ddot{q}_L\} - [B][\Phi_L]\{\dot{q}_L\} \tag{23.54}$$

In most applications, the term with modal velocity is ignored. The static solution of the above equation, at each time point, produces physical displacements, which include the quasi-static effects of all high-frequency modes.

Load Transformation Matrices. Recovery of structural loads is often organized by a definition of the *load transformation matrices* (LTMs).[14] When residual mode vectors are employed, Eqs. (23.49) and (23.43) are combined to define the displacement LTM relationship

$$\{S\} = [\mathrm{LTM}_q]\{q\} + [\mathrm{LTM}_F]\{F\} \tag{23.55}$$

where

$$[\text{LTM}_q] = [K_S][\Phi_L], [\text{LTM}_F] = [K_S][\Psi_\rho] \tag{23.56}$$

When the mode acceleration method is employed, Eqs. (23.54) and (23.43) are combined to define the mode acceleration LTM relationship

$$\{S\} = [\text{LTM}_A]\{\ddot{q}\} + [\text{LTM}_V]\{\dot{q}\} + [\text{LTM}_{AF}]\{F\} \tag{23.57}$$

where

$$\begin{aligned}[\text{LTM}_A] &= -[K_S][K^{-1}M\Phi_L]\\ [\text{LTM}_V] &= -[K_S][K^{-1}B\Phi_L]\\ [\text{LTM}_F] &= [K_S][K^{-1}\Gamma]\end{aligned} \tag{23.58}$$

In practice, $[\text{LTM}_V]$ is generally ignored. Mode acceleration LTMs are used extensively in the aeronautical and space vehicle industries, while their mode displacement (and residual vector)–based counterpart is rarely applied.

APPLIED LOADS AND ENFORCED MOTIONS

Dynamic excitation environments sometimes are described in terms of specified foundation or boundary motions, for example, in the study of structural dynamic response to seismic excitations (see Chap. 29). In such situations, the physical displacement array is partitioned into two subsets as follows:

$$\{u\} = \begin{Bmatrix} u_i \\ u_b \end{Bmatrix} = \begin{Bmatrix} \text{interior motions} \\ \text{boundary motions} \end{Bmatrix} \tag{23.59}$$

The conventional linear structural dynamic formulation is expressed in partitioned form as

$$\begin{bmatrix} M_{ii} & M_{ib} \\ M_{bi} & M_{bb} \end{bmatrix}\begin{Bmatrix} \ddot{u}_i \\ \ddot{u}_b \end{Bmatrix} + \begin{bmatrix} B_{ii} & B_{ib} \\ B_{bi} & B_{bb} \end{bmatrix}\begin{Bmatrix} \dot{u}_i \\ \dot{u}_b \end{Bmatrix} + \begin{bmatrix} K_{ii} & K_{ib} \\ K_{bi} & K_{bb} \end{bmatrix}\begin{Bmatrix} u_i \\ u_b \end{Bmatrix} = \begin{Bmatrix} F_i \\ F_b \end{Bmatrix} \tag{23.60}$$

Using the partitioned stiffness matrix, the transformation from absolute to relative response displacements is

$$\begin{Bmatrix} u_i \\ u_b \end{Bmatrix} = \begin{bmatrix} I_{ii} & -K_{ii}^{-1}K_{ib} \\ 0_{bi} & I_{bb} \end{bmatrix}\begin{Bmatrix} u_{ir} \\ u_b \end{Bmatrix} = \begin{bmatrix} I_{ii} & \Psi_{ib} \\ 0_{bi} & I_{bb} \end{bmatrix}\begin{Bmatrix} u_{ir} \\ u_b \end{Bmatrix} \tag{23.61}$$

Moreover, this transformation may be expressed in modal form by substituting the lowest-frequency modes associated with the interior eigenvalue problem, which follows the relationships already discussed in Eqs. (23.33) through (23.38), that is,

$$[K_{ii}]\{\Phi_{in}\} = [M_{ii}]\{\Phi_{in}\}\omega_{in}^2, \{u_i\} = [\Phi_i]\{q_i\} \tag{23.62}$$

By combining Eqs. (23.61) and (23.62), the modal reduction transformation is

$$\begin{Bmatrix} u_i \\ u_b \end{Bmatrix} = \begin{bmatrix} \Phi_i & \Psi_{ib} \\ 0_{bi} & I_{bb} \end{bmatrix} \begin{Bmatrix} q_i \\ u_b \end{Bmatrix} \tag{23.63}$$

Substitution of this transformation into the partitioned dynamic equation set, Eq. (23.60), results in

$$\begin{bmatrix} I_{ii} & P_{ib} \\ P_{bi} & M'_{bb} \end{bmatrix} \begin{Bmatrix} \ddot{q}_i \\ \ddot{u}_b \end{Bmatrix} + \begin{bmatrix} 2\zeta_i\omega_i & 0_{ib} \\ 0_{bi} & B'_{bb} \end{bmatrix} \begin{Bmatrix} \dot{q}_i \\ \dot{u}_b \end{Bmatrix} + \begin{bmatrix} \omega_i^2 & 0_{ib} \\ 0_{bi} & K'_{bb} \end{bmatrix} \begin{Bmatrix} q_i \\ u_b \end{Bmatrix} = \begin{Bmatrix} \Phi_i^T F_i \\ \Psi_{ib}^T F_i + F_b \end{Bmatrix} \tag{23.64}$$

The terms in the above equation set have the following significance:

1. $[P_{ib}]$ is the *modal participation factor matrix.* Its terms express the degree of excitation delivered by individual foundation accelerations. Moreover, its transpose describes the degree of foundation reaction loads associated with individual modal accelerations. The term-by-term product $[P_{ib}] \otimes [P_{ib}]$, called the *modal effective mass matrix,* is often used to evaluate the completeness of a truncated set of modes.
2. $[M'_{bb}]$ is the boundary mass matrix. When the boundary motions are sufficient to impose all six rigid-body motions (in a statically determinate or redundant manner), this matrix expresses the complete rigid-body mass properties of the modeled system.
3. $[K'_{bb}]$ is the boundary stiffness matrix. When the boundary motions are sufficient to impose all six rigid-body motions in a statically determinate manner, this matrix is null. If the boundary is statically indeterminate, the boundary stiffness matrix will have six singularities associated with the six rigid-body motions. In rare situations, additional singularities will (correctly) be present if the structural system includes mechanisms.
4. Critical evaluation of the properties of $[M'_{bb}]$ and $[K'_{bb}]$ is an effective means for model verification.
5. In most situations, damping is not explicitly modeled. Therefore the boundary damping matrix $[B'_{bb}]$ will not be computed.

When the dynamic excitation environment consists entirely of prescribed boundary motions ($\{F_i\} = \{0\}$), Eq. (23.64) may be expressed in the following convenient form:

$$\begin{aligned} \{\ddot{q}_i\} + [2\zeta_i\omega_i]\{\dot{q}_i\} + [\omega_i^2]\{q_i\} &= -[P_{ib}]\{\ddot{u}_b\} && \text{(modal response)} \\ \{F_b\} = [M'_{bb}]\{\ddot{u}_b\} + [K'_{bb}]\{u_b\} &+ [P_{bi}]\{\ddot{q}_i\} && \text{(boundary reactions)} \end{aligned} \tag{23.65}$$

The accurate recovery of structural loads is preferably accomplished with the mode acceleration method. The load transformation matrix relationship for this situation takes the following form (ignoring damping):

$$\{S\} = [\text{LTM}_{\ddot{q}}]\{\ddot{q}\} + [\text{LTM}_{\ddot{u}_b}]\{\ddot{u}_b\} + [\text{LTM}_{u_b}]\{u_b\} + [\text{LTM}_{F_i}]\{F_i\} \tag{23.66}$$

The above relationships are commonly used in seismic structural analysis and equipment shock response analysis.

STRATEGIES FOR DEALING WITH LARGE-ORDER MODELS

The capabilities of computer resources and commercial finite element software have continually increased, making very large-order (~10^6 degrees of freedom or more) finite element models a practical reality. A variety of numerical analysis strategies have been introduced to efficiently deal with these large-order models.

In 1965, what is popularly known as the *Guyan reduction method*[15] was introduced. This method employs a static reduction transformation based on the model stiffness matrix to consistently reduce the mass matrix. By subdividing the model displacements into analysis (a) and omitted (o) subsets, the static reduction transformation is

$$\begin{Bmatrix} u_a \\ u_o \end{Bmatrix} = \begin{bmatrix} I_{aa} \\ -K_{oo}^{-1}K_{oa} \end{bmatrix} \{u_a\} \tag{23.67}$$

By applying this transformation to the dynamic system, an approximate reduced dynamic system for modal analysis is defined as

$$[M_{aa}]\{\ddot{u}_a\} + [K_{aa}]\{u_a\} = \{0\} \tag{23.68}$$

where

$$\begin{aligned} [M_{aa}] &= \begin{bmatrix} I_{aa} \\ -K_{oo}^{-1}K_{oa} \end{bmatrix}^T \begin{bmatrix} M_{aa,o} & M_{ao} \\ M_{oa} & M_{oo} \end{bmatrix} \begin{bmatrix} I_{aa} \\ -K_{oo}^{-1}K_{oa} \end{bmatrix} \\ [K_{aa}] &= \begin{bmatrix} I_{aa} \\ -K_{oo}^{-1}K_{oa} \end{bmatrix}^T \begin{bmatrix} K_{aa,o} & K_{ao} \\ K_{oa} & K_{oo} \end{bmatrix} \begin{bmatrix} I_{aa} \\ -K_{oo}^{-1}K_{oa} \end{bmatrix} \end{aligned} \tag{23.69}$$

The reduced approximate mass and stiffness matrices are generally fully populated, in spite of the fact that the original system matrices are typically quite sparse. The effective selection of an appropriate analysis set $\{u_a\}$ is a process requiring good physical intuition. A recently introduced two-step procedure[16] automatically identifies an appropriate analysis set. The Guyan reduction method is no longer a favored strategy for dealing with large-order dynamic systems due to the development of powerful numerical procedures for very large-order sparse dynamic systems. It continues to be employed, however, for the definition of *test-analysis models* (TAMs) which are used for modal test planning and test-analysis correlation analyses (see Chap. 40). Numerical procedures, which are currently favored for dealing with modern large-order dynamic system modal (eigenvalue) analyses, are (1) the *Lanczos method*[17] (refined and implemented by many other developers) and (2) *subspace iteration.*[8]

Segmentation of Large-Order Dynamic Systems. Many dynamic systems, such as aircraft, launch vehicle–payload assemblies, spacecraft, and automobiles, naturally lend themselves to substructure segmentation (see Fig. 23.4). Numerical analysis strategies, which exploit substructure segmentation, were originally introduced to improve the computational efficiency of large-order dynamic system analysis. However, advances in numerical analysis of very large order dynamic systems have reduced the need for substructure segmentation. The enduring utilization of substructure segmentation, especially in the aerospace industry, is a result of the fact that substructure models provide cooperating organizations with a standard means for sharing and integrating subsystem data. It should also be noted that some research efforts in the area of parallel processing are utilizing mature substructure

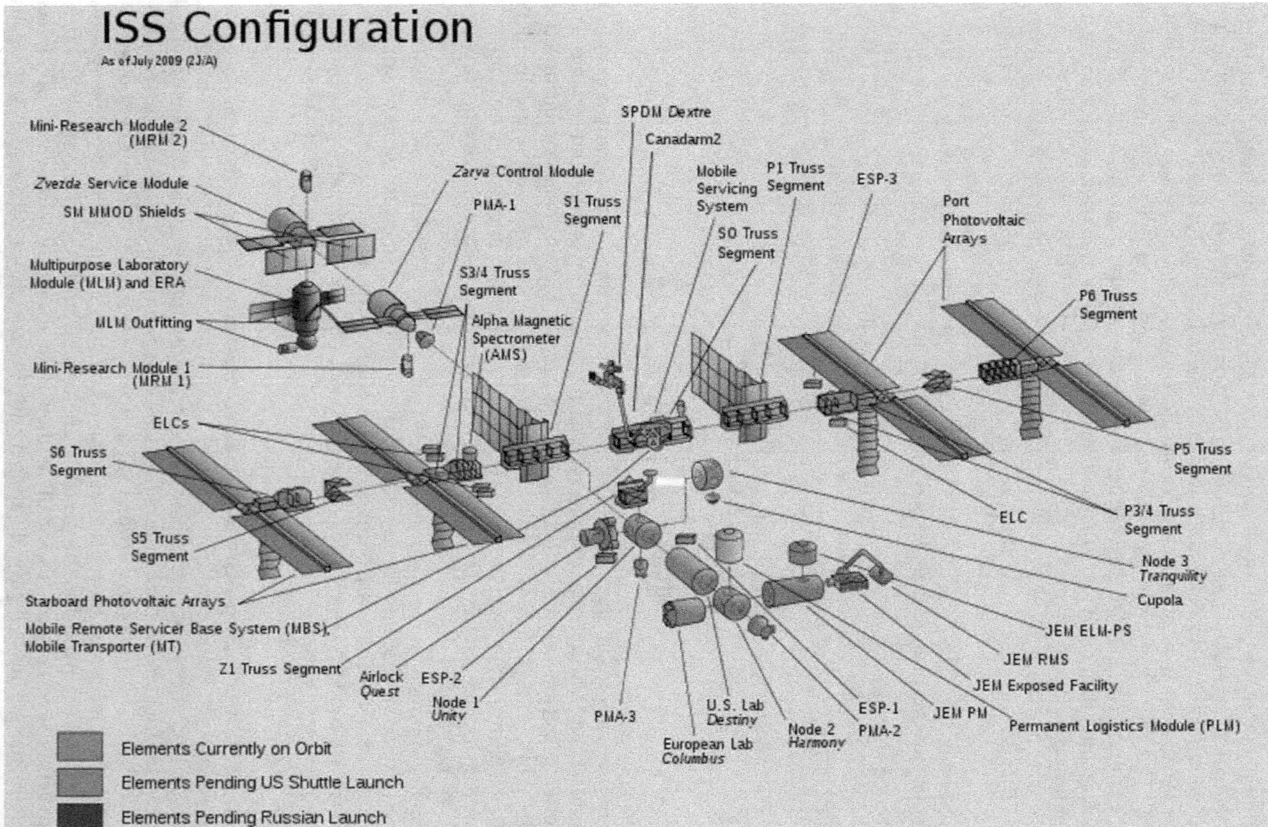

FIGURE 23.4 International space station substructure segmentation.

analysis concepts. Each designated substructure (which also may be termed a superelement) is defined in terms of interior $\{u_i\}$ and boundary $\{u_b\}$ displacement subsets. Specific types of modal analysis strategies are employed to reduce or condense the individual substructures to produce *modal components.*

The Craig-Bampton Modal Component. The most popularly employed modal component type, the *Craig-Bampton*[18] (or *Hurty*[19]) component, is defined by Eqs. (23.59) through (23.64) and (23.66). The undamped key dynamic equations describing this component are as follows:

1. The Craig-Bampton reduction transformation (boundary-fixed interior modes and boundary deflection shapes) is identical to Eq. (23.63), that is,

$$\begin{Bmatrix} u_i \\ u_b \end{Bmatrix} = \begin{bmatrix} \Phi_i & \Psi_{ib} \\ 0_{bi} & I_{bb} \end{bmatrix} \begin{Bmatrix} q_i \\ u_b \end{Bmatrix} \tag{23.70}$$

2. The Craig-Bampton mass and stiffness matrices, from Eq. (23.64), are

$$\begin{bmatrix} I_{ii} & P_{ib} \\ P_{bi} & M'_{bb} \end{bmatrix} \begin{Bmatrix} \ddot{q}_i \\ \ddot{u}_b \end{Bmatrix} + \begin{bmatrix} \omega_i^2 & 0_{ib} \\ 0_{bi} & K'_{bb} \end{bmatrix} \begin{Bmatrix} q_i \\ u_b \end{Bmatrix} = \begin{Bmatrix} 0 \\ 0 \end{Bmatrix} \tag{23.71}$$

The MacNeal-Rubin Modal Component. The *MacNeal-Rubin*[12,20] component reduction transformation consists of a truncated set of free boundary modes and quasi-static residual vectors associated with unit loads applied at the boundary degrees of freedom. The key dynamic equations describing this component are as follows:

1. The MacNeal-Rubin reduction transformation (boundary-free component modes and residual vectors) is

$$\begin{Bmatrix} u_i \\ u_b \end{Bmatrix} = \begin{bmatrix} \Phi_{ii} & \Psi_{i\rho} \\ \Phi_{bi} & \Psi_{b\rho} \end{bmatrix} \begin{Bmatrix} q_i \\ u_\rho \end{Bmatrix} \tag{23.72}$$

Noting that there are as many residual vectors as boundary degrees of freedom, the above transformation may be expressed in terms of the modal and boundary DOF, that is,

$$\begin{Bmatrix} u_i \\ u_b \end{Bmatrix} = \begin{bmatrix} \Phi_{ii} - \Psi_{i\rho}\Psi_{b\rho}^{-1}\Phi_{bi} & \Psi_{i\rho}\Psi_{b\rho}^{-1} \\ 0_{bi} & I_{bb} \end{bmatrix} \begin{Bmatrix} q_i \\ u_b \end{Bmatrix} \tag{23.73}$$

2. The MacNeal-Rubin mass and stiffness matrices: Using the first reduction transformation form [see Eq. (23.72)], the undamped component mode equations are of the form

$$\begin{bmatrix} I_{ii} & 0_{i\rho} \\ 0_{\rho i} & M_{\rho\rho} \end{bmatrix} \begin{Bmatrix} \ddot{q}_i \\ \ddot{u}_\rho \end{Bmatrix} + \begin{bmatrix} \omega_i^2 & 0_{i\rho} \\ 0_{\rho i} & K_{\rho\rho} \end{bmatrix} \begin{Bmatrix} q_i \\ u_\rho \end{Bmatrix} = \begin{Bmatrix} 0 \\ 0 \end{Bmatrix} \tag{23.74}$$

When the second reduction transformation form [see Eq. (23.73)] is employed, the component mode equations are of the fully coupled form

$$\begin{bmatrix} M'_{ii} & M'_{ib} \\ M'_{bi} & M'_{bb} \end{bmatrix} \begin{Bmatrix} \ddot{q}_i \\ \ddot{u}_b \end{Bmatrix} + \begin{bmatrix} K'_{ii} & K'_{ib} \\ K'_{bi} & K'_{bb} \end{bmatrix} \begin{Bmatrix} q_i \\ u_b \end{Bmatrix} = \begin{Bmatrix} 0 \\ 0 \end{Bmatrix} \tag{23.75}$$

The second form of the MacNeal-Rubin mass and stiffness matrices is preferred for automated assembly of modal components.

The Mixed Boundary Modal Component. A more general type of modal component may be defined employing fixed- and free-boundary degree-of-freedom subsets.[21] The reduced component mass and stiffness matrices associated with this component are fully coupled, having a form similar to Eq. (23.75).

Each of the above three modal component types employs a truncated set of subsystem modes. The frequency band, which determines an adequate set of subsystem modes, is related to the base frequency band of the expected dynamic environment. In particular, a generally accepted standard for the modal frequency band defines the cutoff frequency as $1.4f^*$ (see the section titled "Cutoff Frequency and Grid Spacing ").

COMPONENT MODE SYNTHESIS STRATEGIES

Two alternative strategies for component mode synthesis are generally accepted in industry. The first strategy views all substructures as appendages. The second alternative views substructures as appendages, which attach to a common main body.

General Method 1: Assembly of Appendage Substructures. The boundary degrees of freedom for each component of a complete structural assembly map onto an assembled structure boundary (collector, c) array, that is,

$$\{u_b\} = [T_{bc}]\{u_c\} \tag{23.76}$$

Therefore, each component's reduction transformation is expressed in the assembled (collector) DOF as

$$\begin{Bmatrix} u_i \\ u_b \end{Bmatrix} = \begin{bmatrix} \Psi_{ii} & \Psi_{ib}T_{bc} \\ 0_{bi} & T_{bc} \end{bmatrix} \begin{Bmatrix} q_i \\ u_c \end{Bmatrix} \tag{23.77}$$

where Ψ_{ii} represents the upper left modal transformation partition for the particular modal component type. Application of this transformation to Eq. (23.71) or (23.75) results in

$$\begin{bmatrix} M'_{ii} & M'_{ic} \\ M'_{ci} & M'_{cc} \end{bmatrix} \begin{Bmatrix} \ddot{q}_i \\ \ddot{u}_c \end{Bmatrix} + \begin{bmatrix} K'_{ii} & K'_{ic} \\ K'_{ci} & K'_{cc} \end{bmatrix} \begin{Bmatrix} q_i \\ u_c \end{Bmatrix} = \begin{Bmatrix} 0 \\ 0 \end{Bmatrix} \tag{23.78}$$

The format of the assembled system dynamic equations, shown here for an assembly of three components denoted as 1, 2, and 3, is

$$\begin{bmatrix} M'_{11} & & & M'_{1C} \\ & M'_{22} & & M'_{2C} \\ & & M'_{33} & M'_{3C} \\ M'_{C1} & M'_{C2} & M'_{C3} & M'_{CC} \end{bmatrix} \begin{Bmatrix} \ddot{q}_1 \\ \ddot{q}_2 \\ \ddot{q}_3 \\ \ddot{u}_C \end{Bmatrix} + \begin{bmatrix} K'_{11} & & & K'_{1C} \\ & K'_{22} & & K'_{2C} \\ & & K'_{33} & K'_{3C} \\ K'_{C1} & K'_{C2} & K'_{C3} & K'_{CC} \end{bmatrix} \begin{Bmatrix} q_1 \\ q_2 \\ q_3 \\ u_C \end{Bmatrix} = \begin{Bmatrix} 0 \\ 0 \\ 0 \\ 0 \end{Bmatrix} \tag{23.79}$$

The system normal modes are calculated from the above equation where the final system mode transformation (which decouples the system mass and stiffness matrices) is

$$\begin{Bmatrix} q_1 \\ q_2 \\ q_3 \\ u_C \end{Bmatrix} = [\Phi_{\text{sys}}]\{q_{\text{sys}}\} \tag{23.80}$$

General Method 2: Attachment of "Appendage" Substructures to a Main Body. This method of component mode synthesis differs from General Method 1 in that all components are not considered appendages. A simple way to view this approach is to first follow General Method 1 for all appendage substructures up to Eq. (23.79). The boundary collector degrees of freedom, in this case, correspond to those associated with a main body, which is described in terms of main body mass and stiffness matrices $[M_m]$ and $[K_m]$, respectively. The assembled system of appendages and main body are described as

$$\begin{bmatrix} M'_{11} & & & M'_{1C} \\ & M'_{22} & & M'_{2C} \\ & & M'_{33} & M'_{3C} \\ M'_{C1} & M'_{C2} & M'_{C3} & M'_m \end{bmatrix} \begin{Bmatrix} \ddot{q}_1 \\ \ddot{q}_2 \\ \ddot{q}_3 \\ \ddot{u}_C \end{Bmatrix} + \begin{bmatrix} K'_{11} & & & K'_{1C} \\ & K'_{22} & & K'_{2C} \\ & & K'_{33} & K'_{3C} \\ K'_{C1} & K'_{C2} & K'_{C3} & K'_m \end{bmatrix} \begin{Bmatrix} q_1 \\ q_2 \\ q_3 \\ u_C \end{Bmatrix} = \begin{Bmatrix} 0 \\ 0 \\ 0 \\ 0 \end{Bmatrix} \tag{23.81}$$

where the boundary-loaded main body mass and stiffness matrices are

$$[M'_m] = [M'_{cc}] + [M_m], [K'_m] = [K'_{cc}] + [K_m] \tag{23.82}$$

The truncated set of modes associated with the boundary-loaded main body define the intermediate transformation

$$\begin{Bmatrix} q_1 \\ q_2 \\ q_3 \\ u_c \end{Bmatrix} = \begin{bmatrix} I_1 & 0 & 0 & 0 \\ 0 & I_2 & 0 & 0 \\ 0 & 0 & I_3 & 0 \\ 0 & 0 & 0 & \Phi_m \end{bmatrix} \begin{Bmatrix} q_1 \\ q_2 \\ q_3 \\ q_m \end{Bmatrix} \tag{23.83}$$

Application of the above transformation to Eq. (23.82) results in the following modal equations for the system

$$\begin{bmatrix} M'_{11} & & & M''_{1C} \\ & M'_{22} & & M''_{2C} \\ & & M'_{33} & M''_{3C} \\ M''_{C1} & M''_{C2} & M''_{C3} & I_m \end{bmatrix} \begin{Bmatrix} \ddot{q}_1 \\ \ddot{q}_2 \\ \ddot{q}_3 \\ \ddot{q}_m \end{Bmatrix} + \begin{bmatrix} K'_{11} & & & K''_{1C} \\ & K'_{22} & & K''_{2C} \\ & & K'_{33} & K''_{3C} \\ K''_{C1} & K''_{C2} & K''_{C3} & \omega_m^2 \end{bmatrix} \begin{Bmatrix} q_1 \\ q_2 \\ q_3 \\ q_m \end{Bmatrix} = \begin{Bmatrix} 0 \\ 0 \\ 0 \\ 0 \end{Bmatrix} \tag{23.84}$$

If the appendages are all of the Craig-Bampton type, the above equation set reduces to the following *Benfield-Hruda*[22] form

$$\begin{bmatrix} I_1 & & & P_{1C} \\ & I_2 & & P_{2C} \\ & & I_3 & P_{3C} \\ P_{C1} & P_{C2} & P_{C3} & I_m \end{bmatrix} \begin{Bmatrix} \ddot{q}_1 \\ \ddot{q}_2 \\ \ddot{q}_3 \\ \ddot{q}_m \end{Bmatrix} + \begin{bmatrix} \omega_1^2 & & & \\ & \omega_2^2 & & \\ & & \omega_3^2 & \\ & & & \omega_m^2 \end{bmatrix} \begin{Bmatrix} q_1 \\ q_2 \\ q_3 \\ q_m \end{Bmatrix} = \begin{Bmatrix} 0 \\ 0 \\ 0 \\ 0 \end{Bmatrix} \tag{23.85}$$

The mass coupling terms (P_{1C}, etc.) are modal participation factor matrices, which indicate the relative level of excitation delivered to the appendages by main body modal accelerations. This feature of the Benfield-Hruda form is the primary reason for the enduring popularity of the method. Uncoupled system modes are finally computed from the eigenvalue solution of Eq. (23.85). Component mode synthesis procedures are also applied in multilevel cascades when such a strategy is warranted.

DYNAMIC RESPONSE RESULTING FROM VARIOUS ENVIRONMENTS

The response of linear structural dynamic systems to dynamic environments may be computed by either modal or direct methods. Modal methods tend to be computationally efficient when the required number of system modes addressing the dynamic environment frequency band are significantly smaller than the order of the system finite element model. When this is not the case, direct methods may be more efficient. In addition, when transient environments are brief or impulsive, direct integration may be more efficient than modal strategies. The following discussion provides an overview of strategies for the computation of dynamic response to various environments.

Transient Environments. General relationships detailing the modal method of transient dynamic analysis are presented in the section entitled "Application of Normal Modes in Transient Dynamic Analysis." Enhancement of the modal solution accuracy with residual vectors and the mode acceleration method was discussed in the sections entitled "Residual Mode Vectors and Mode Acceleration" and "Load Transformation Matrices," respectively. Direct integration methods employing implicit[9] or explicit[4] numerical strategies may be advantageous when environments are of wide bandwidth and short-lived.

Brief or Impulsive Environments. Brief or impulsive dynamic environments are often described in terms of shock response spectra (see Chap. 20). Peak dynamic responses and structural loads are estimated by employing approximate modal superposition methods utilizing shock response spectra as modal weighting functions.[23] A systematic approach to this process, which incorporates positive and negative spectra and quasi-static residual vectors, is presented in Ref. 11. Approximate shock response spectra–based modal superposition methods are employed in earthquake engineering, equipment (e.g., naval shipboard subsystems) shock survivability prediction, and related applications. This approach is especially appropriate when standard dynamic environments are specified as shock response spectra.

Simple Harmonic Excitation. Computation of the structural dynamic response due to simple harmonic excitation is either an end in itself or a key intermediate step in the computation of the response to random or transient environments. In the case of transient environments, the time-history response may be calculated through application of Fourier transform techniques (see Chap. 20). The applied force and displacement response, respectively, are conveniently expressed in terms of complex exponential functions by

$$\{F\} = F_o(\omega)e^{i\omega t}, \quad \{u\} = \{U(\omega)\}e^{i\omega t}, \quad \{\dot{u}\} = i\omega\{U(\omega)\}e^{i\omega t}, \quad \{\ddot{u}\} = -\omega^2\{U(\omega)\}e^{i\omega t} \tag{23.86}$$

where ω is the forcing frequency in radians per second. Upon substitution of the above relationships into the linear structural dynamic equations [see Eq. (23.16)], the following algebraic matrix equation is defined.

$$[K + i\omega B - \omega^2 M]\{U(\omega)\} = \{\Gamma_F\}F_o(\omega) \tag{23.87}$$

When $F_o(\omega) = 1$, the response quantities are called frequency response functions (see Chap. 21). If the normal mode substitution is employed, the above equation set is diagonalized (assuming modal viscous damping) as follows:

$$\{U(\omega)\} = [\Phi]\{q(\omega)\} \qquad \{\dot{U}(\omega)\} = i\omega[\Phi]\{q(\omega)\} \qquad \{\ddot{U}(\omega)\} = -\omega^2[\Phi]\{q(\omega)\}$$
$$(\omega_n^2 + 2i\zeta_n\omega_n\omega - \omega^2)q_n(\omega) = \{\Phi_n\}^T[\Gamma_F]\{F(\omega)\} \qquad 1 \le n \le n_{\max} \tag{23.88}$$

When the modal method is used, it is recommended that a quasi-static residual vector be employed to mitigate modal truncation errors. This is not required if the direct method, namely, the solution of Eq. (23.87), is employed.

The modal approach to simple harmonic or frequency response analysis is computationally more efficient than the direct method if the number of modes required in a frequency band of interest ($0 \le \omega \le \omega_{\max}$) is much less than the number of finite element model degrees of freedom. When this is not the case, the direct method becomes more efficient since the direct solution for $\{U(\omega)\}$ involves decomposition of a sparse coefficient matrix at each forcing frequency.

When the direct solution procedure is employed, it is most convenient to describe modal damping as complex structural damping (see Chap. 2). In this situation the linear, frequency domain, structural dynamic equations are

$$[(1 + i\eta)K + i\omega B_L - \omega^2 M]\{U(\omega)\} = \{\Gamma_F\}F_o(\omega) \tag{23.89}$$

where the well-known approximate equivalence of structural damping loss factor η and (viscous) modal damping ratio ζ is $\eta \approx 2\zeta$. The advantage associated with structural damping is that the modes need not be explicitly determined in order to account for modal damping effects. The matrix $[B_L]$ is included in the above equation to account for any known discrete viscous damping features.

An important aspect of effective frequency response analysis, regardless of whether the modal or direct method is used, is the selection of a frequency grid for the clear definition of harmonic response peaks. It is generally recommended that solutions be calculated at frequency points capturing at least four points within a modal half-power bandwidth, that is,

$$\Delta\omega = \zeta_n\omega_n/2 = \eta\omega_n/4 \tag{23.90}$$

This guideline suggests a logarithmic frequency grid ($\Delta\omega$ increases with increasing frequency) is desirable.

Random Excitation. In the most common situations, random environments are assumed to be associated with ergodic (see Chap. 1) processes.[24] The computation of structural dynamic response to random excitation, in such a situation, utilizes numerical results from the response to a simple harmonic excitation. If a random environment is imposed at several discrete structural degrees of freedom or as several geometric load patterns, the frequency responses associated with the individual loads are denoted as

$$H_{ij}(\omega) = U_i(\omega)/F_{o,j}(\omega) \tag{23.91}$$

where these functions are computed either by the modal or direct method. Therefore, the frequency-domain response associated with several excitations is

$$U_i(\omega) = \sum_j H_{ij}(\omega) \cdot F_{o,j}(\omega) \tag{23.92}$$

or in matrix form

$$U(\omega) = [H(\omega)]\{F_o(\omega)\} \tag{23.93}$$

Describing the correlated random excitations in terms of the input cross-spectral density matrix, $[G_{FF}(\omega)]$, the response autospectral density is

$$W_{uu}(\omega) = [H(\omega)] \cdot [G_{FF}(\omega)] \cdot [H(\omega)]^{T*} \tag{23.94}$$

where the asterisk $[\ ]^{T*}$ denotes the complex conjugate transpose of a matrix. Finally, the mean square of response is calculated as the integral

$$\Psi_u^2 = \overline{u_i(t)^2} = \int_{\omega_1}^{\omega_2} W_{uu}(\omega)d\omega \tag{23.95}$$

In order to ensure the accurate computation of a mean square response, this integral must be evaluated with a frequency grid with refinement consistent with Eq. (23.90). If too coarse a frequency grid is used, the mean square response may be severely underestimated.

REFERENCES

1. Clough, R. W.: *Proc. 2d ASCE Conf. on Elec. Comp.,* Pittsburgh, 1960, p. 345.
2. MacNeal, R. H.: "Finite Elements: Their Design and Performance," Marcel Dekker, New York, 1994.
3. Strang, G.: "Linear Algebra and Its Applications," Harcourt Brace Jovanovich, San Diego, Calif., 1988.
4. Isaacson, E., and H. B. Keller: "Analysis of Numerical Methods," John Wiley & Sons, New York, 1966.
5. Przemieniecki, J. S.: "Theory of Matrix Structural Analysis," Dover Publications, New York, 1968.
6. Lanczos, C.: "The Variational Principles of Mechanics," 4th ed., Dover Publications, New York, 1986.
7. Coppolino, R. N.: *NASA CR-2662,* 1975.
8. Bathe, K. J., and E. L. Wilson: *Proc. ASCE,* **6**(98):1471 (1972).
9. Bathe, K. J., and E. L. Wilson: "Numerical Methods in Finite Element Analysis," Prentice-Hall, Englewood Cliffs, N.J., 1976.
10. Lyon, R. H., and R. G. DeJong: "Theory and Application of Statistical Energy Analysis," 2d ed., Butterworth-Heinemann, Boston, Mass., 1995.
11. Coppolino, R. N.: *SAE Paper No. 841581,* 1984.
12. MacNeal, R. H.: *Computers in Structures,* **1:**581 (1971).
13. Williams, D.: Great Britain Royal Aircraft Establishment Reports, *SME 3309 and 3316,* 1945.
14. Coppolino, R. N.: "Combined Experimental/Analytical Modeling of Dynamic Structural Systems," *ASME AMD-167,* 79 (1985).
15. Guyan, R. J.: *AIAA Journal,* **3**(2):380 (1965).
16. Coppolino, R. N.: *Proceedings of the 16th IMAC,* **1:**70 (1998).
17. Lanczos, C.: *J. Res. Natl. Bureau of Standards,* **45:**255 (1950).

18. Craig, R. R., and M. D. D. Bampton: *AIAA Journal,* **6**(7):1313 (1968).
19. Hurty, W. C.: *AIAA Journal,* **3**(4):678 (1965).
20. Rubin, S.: *AIAA Journal,* **13**(8):995 (1975).
21. Herting, D. N., and M. J. Morgan: *AIAA/ASME/ASCE/AHS 20th SDM* (1979).
22. Benfield, W. A., and R. F. Hruda: *AIAA Journal,* **9**(7):1255 (1971).
23. Hadjian, A. H.: *Nuclear Engineering and Design,* **66**(2):179 (1981).
24. Bendat, J. S., and A. G. Piersol: "Random Data Analysis and Measurement Procedures," 3d ed., John Wiley & Sons, New York, 2000.

CHAPTER 24
STATISTICAL ENERGY ANALYSIS

Richard G. DeJong

INTRODUCTION

Two situations often occur in which a statistical analysis of vibrating systems is useful. The first occurs when the excitation of a system appears to be random in time, in which case it is convenient to describe the temporal response of the system statistically rather than deterministically. This form of data analysis is presented in Chap. 19. The second situation occurs when a system is complicated enough that its resonant modes appear to be distributed randomly in frequency, in which case it is convenient to describe the frequency response of the system statistically rather than deterministically. This form of analysis is called *statistical energy analysis* (*SEA*)[1] and is presented in this chapter.

In either situation, the randomness need only appear to be so. For example, in random vibration, it may be that the excitation could be calculated exactly if enough information were known. However, if the excitation is adequately described by statistical parameters (such as the mean value and variance), then a statistical analysis of the system response is valid. Similarly, in a complicated system, the modes can presumably be analyzed deterministically. However, if the modal distribution is adequately described by statistical parameters, then an SEA of the system response is valid whether or not the excitation is random.

STATISTICAL RESPONSE OF A SINGLE-DEGREE-OF-FREEDOM SYSTEM

In this section, the single-degree-of-freedom (SDOF) resonator shown in Fig. 24.1 is analyzed to obtain an expression for the mean square response of the mass when the base is subjected to a random vibration. The equation of motion for this system is derived in Chap. 2 as

$$\ddot{z} + \frac{c}{m}\dot{z} + \frac{k}{m}z = -\ddot{y} \tag{24.1}$$

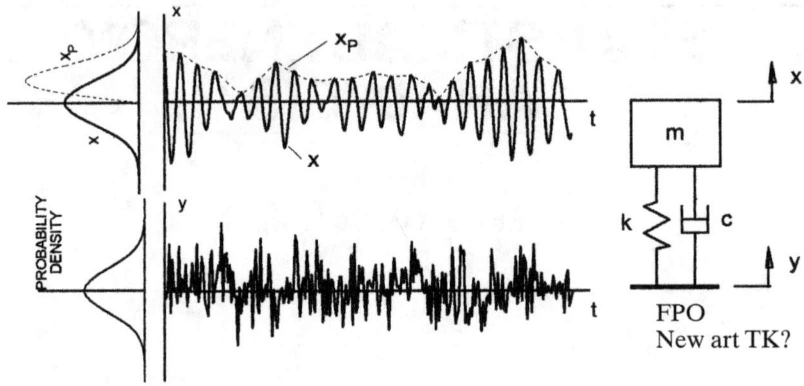

FIGURE 24.1 Example of a resiliently mounted mass m with stiffness k and viscous damper c. When the base is exposed to a broadband random vibration, the mass will have a narrowband random response.

where $z = x - y$ is the motion of the mass relative to the base. This equation is similar in form to the equation for a force excitation $F(t)$ on the mass and a rigid base:

$$\ddot{x} + \frac{c}{m}\dot{x} + \frac{k}{m}x = \frac{F(t)}{m} \tag{24.2}$$

In this chapter, the form of Eq. (24.1) is used, but the results can be transformed to the case of a force excitation by the appropriate substitution of variables. Defining

$$f_n = \frac{1}{2\pi}\sqrt{\frac{k}{m}} = \text{the natural frequency}$$
$$\zeta = \frac{c}{2\sqrt{km}} = \text{the critical damping ratio} \tag{24.3}$$

gives a standard parametric form:

$$\ddot{z} + 4\pi\zeta f_n \dot{z} + (2\pi f_n)^2 z = \ddot{y} \tag{24.4}$$

The response of the system is given in terms of a frequency dependent *transfer function* $H(f)$ (or *frequency response function*) with a magnitude given as

$$|H(f)|^2 = \frac{W_{\ddot{z}}(f)}{W_{\ddot{y}}(f)} = \frac{1}{\left[1 - \left(\frac{f}{f_n}\right)^2\right]^2 + \left(2\zeta\frac{f}{f_n}\right)^2} \tag{24.5}$$

For a broadband random source, if $\zeta << 1$ so that $|H(f)|^2$ is sharply peaked at $f = f_n$ and the source is stationary with a relatively smooth spectrum, as illustrated in Fig. 24.2, then the mean square response of the system is determined by the source spectrum at $f = f_n$ times the area under the $|H(f)|^2$ curve:

$$\overline{\ddot{z}^2} = W_{\ddot{y}}(f_n)\int_0^\infty |H(f)|^2\, df = \frac{\pi f_n}{4\zeta}\, W\ddot{y}(f_n) \tag{24.6}$$

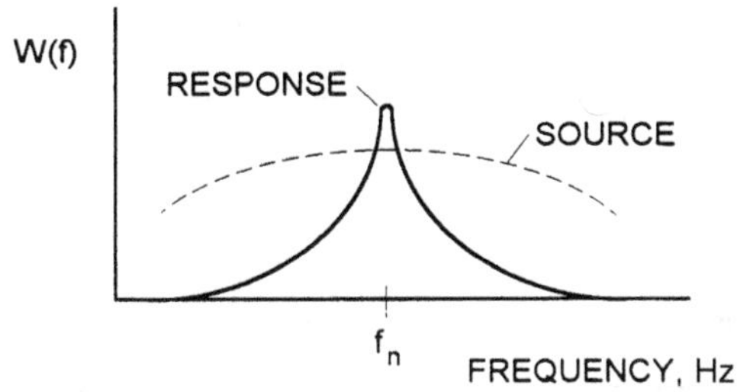

FIGURE 24.2 Power spectral density $W(f)$ of the response of a resonator with $\zeta << 1$ excited by a broadband random source having the spectrum shown by the dashed curve.

The resonance of the system acts as a narrowband filter on the source spectrum, as illustrated in Fig. 24.1. The vibration is essentially at frequency f_n with a gaussian amplitude distribution. The mean square displacement and velocity levels are related to the acceleration response by $\overline{\ddot{z}^2} = (2\pi f_n)^2 \overline{\dot{z}^2} = (2\pi f_n)^4 \overline{z^2}$.

The autocorrelation of the stationary response is found to be

$$R_{\ddot{z}}(\tau) = \overline{\ddot{z}^2}\, e^{-2\pi\zeta f_n \tau}\left[\cos(2\pi f_d t) + \frac{\zeta}{\sqrt{1-\zeta^2}} \sin(2\pi f_d t)\right] \tag{24.7}$$

where $f_d = f_n \sqrt{1-\zeta^2}$

Of interest is the transient response of the resonator when the random source is suddenly turned on and remains stationary thereafter. The transient mean square response starting from rest (for $\zeta << 1$) is then found to be:[2]

$$\overline{\ddot{z}^2}(\tau) = \frac{\pi f_n}{4\zeta}\, W_{\ddot{y}}(f_n)\, (1 - e^{-4\pi\zeta f_n t}) \tag{24.8}$$

The mean square response grows to the steady-state value in the same way that a first-order dynamic system responds to a step input. This is an important result, illustrating that the dynamical power in a vibrating system is transmitted according to the simple first-order diffusion equation with a time constant $\tau = 1/(4\pi\zeta f_n)$.

This result can be used to estimate the shock response spectrum (SRS) of a transient random excitation with a known time-dependent mean square acceleration level $\overline{\ddot{y}^2}(t, \Delta f)$ in the frequency band Δf. The mean square acceleration response $\overline{\ddot{z}^2}$ of a resonator to this excitation can be found by solving the following first-order differential equation either numerically or using the Laplace transform method (see Chap. 8):

$$\frac{d}{dt}\overline{\ddot{z}^2} + (4\pi\zeta f_n)\overline{\ddot{z}^2} = \frac{(2\pi f_n)^2}{4\Delta f}\, \overline{\ddot{y}^2} \tag{24.9}$$

assuming f_n is within the bandwidth Δf.

For example, Fig. 24.3 shows the measured transient acceleration of a concrete floor slab in a building with an operating punch press. As with many transient vibration time histories, the smoothed mean square level can be approximated by

$$\overline{\ddot{z}^2}(\tau) = A t\, e^{-\beta t} \tag{24.10}$$

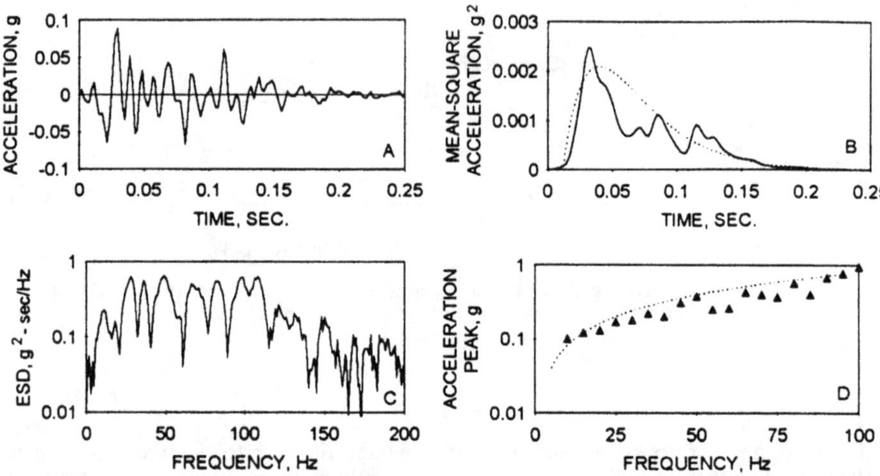

FIGURE 24.3 Transient response of a concrete floor slab with an operating punch press. (*A*) Measured acceleration signal. (*B*) Mean square smoothed signal (solid curve) and curve fit (dashed curve) using Eq. (24.10). (*C*) Measured energy spectral density. (*D*) Computed acceleration shock response spectrum (symbols) and statistical estimate (dashed curve) using Eq. (24.14).

where, for this case, $A \simeq 0.2g^2/s$ and $\beta \simeq 35/s$ with $\Delta f \simeq 80$ Hz. The solution of Eq. (24.9) with this form of excitation is given by

$$\overline{\ddot{z}^2}(\tau) = \frac{(2\pi f_n)^2 A}{4\Delta f}\left[\frac{t(\alpha - \beta)e^{-\beta t} + e^{-\alpha t} - e^{-\beta t}}{(\alpha - \beta)^2}\right] \tag{24.11}$$

where $\alpha = 4\pi\zeta f_n$. The undamped SRS is the maximum response level as $\alpha \to 0$, which is

$$\overline{\ddot{z}^2}_{\max} \to \frac{(2\pi f_n)^2 A}{4\Delta f \beta^2} \tag{24.12}$$

The undamped SRS is the peak root-mean-square (rms) response as a function of f_n (see Chap. 20), which can be estimated with 95 percent certainty as the 2σ level, assuming a gaussian distribution:

$$\ddot{z}_{\text{peak}} \simeq 2\sqrt{\overline{\ddot{z}^2}_{\max}} = \sqrt{\frac{A}{\Delta f}}\,\frac{2\pi f_n}{\beta} \tag{24.13}$$

This result is plotted in Fig. 24.3*D* along with the exact calculation of the SRS at 5-Hz intervals using a particular sample of the acceleration time history.

STATISTICAL RESPONSE OF MULTIPLE-DEGREE-OF-FREEDOM SYSTEMS

Real elastic systems have many degrees of freedom (DOF) and, therefore, many modes of resonance, as discussed in Chaps. 2 and 7. However, each of these *normal modes* ψ_n responds as a simple resonator, and the total response of a system can be obtained by summing the response of all of the modes (*modal superposition*):

$$z(\mathbf{X},t) = \sum_n q_n(t)\psi_n(\mathbf{X}) \tag{24.14}$$

where **X** represents the spatial dimension(s) of the system.

If the damping in the system is distributed proportionately to the mass and stiffness, the normal modes are uncoupled and each has an equation of motion in the form of Eq. (24.4) with a source term given by

$$\ddot{y}_n(t) = \int \ddot{y}(\mathbf{X},t)\psi_n(\mathbf{X})\, d\mathbf{X} \equiv \varphi_n \sqrt{\overline{s^2(\mathbf{X},t)}} \tag{24.15}$$

where φ_n is the modal participation factor of the source. The transfer function for each mode will be of the form of Eq. (24.5) so that the resulting sum of the modal responses gives

$$\overline{\ddot{z}^2} = \sum_n \frac{\pi f_n}{4\zeta n} \varphi_n^2 \psi_n^2 W_s(f_n) \tag{24.16}$$

If the damping is not distributed proportionately but is small ($\zeta \ll 1$), the superposition of normal modes gives approximately correct results. This is illustrated by the two-DOF system shown in Fig. 24.4*A*. An instrument housing (m_1) is resiliently mounted on a vibrating base. A vibration absorber (see Chap. 6) is attached to suppress the vibration of the housing at frequency f_2. Of interest here is the broadband

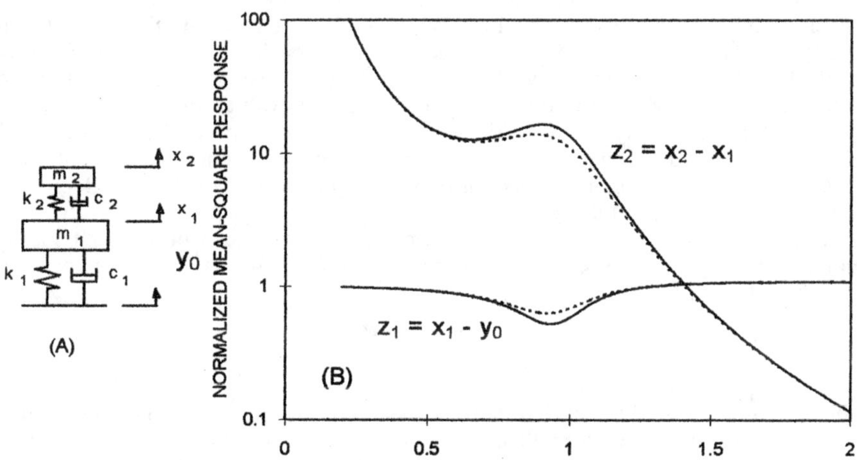

FIGURE 24.4 (*A*) Response of a two-degree-of-freedom system to a base excitation x_0. (*B*) Mean square relative displacement responses z_1 and z_2, normalized to the response of m_1 alone. Solid curves are calculated using the modal summation of Eq. (24.16). Dashed curves are the exact calculations.

response of the system when the base vibration has a uniform acceleration spectral density $W_{\ddot{y}_0}$. The equations for the relative responses, $z_1 = x_1 - y_0$ and $z_2 = x_2 - x_1$ in symmetric, dimensionless form are

$$\begin{bmatrix} 1 & 0 \\ 0 & \dfrac{\mu f_2^2}{f_1^2} \end{bmatrix} \begin{Bmatrix} \ddot{z}_1 \\ \ddot{z}_2 \end{Bmatrix} + \begin{bmatrix} 4\pi\zeta_1 f_1 & -4\pi\mu\zeta_2 f_2 \\ -4\pi\dfrac{\mu f_2^2}{f_1^2}\zeta_1 f_1 & (1+\mu)\dfrac{\mu f_2^2}{f_1^2}(2\pi f_2)^2 \end{bmatrix} \begin{Bmatrix} \dot{z}_1 \\ \dot{z}_2 \end{Bmatrix}$$

$$+ \begin{bmatrix} (2\pi f_1)^2 & -\mu(2\pi f_2)^2 \\ -\mu(2\pi f_2)^2 & (1+\mu)\dfrac{\mu f_2^2}{f_1^2}(2\pi f_2)^2 \end{bmatrix} \begin{Bmatrix} z_1 \\ z_2 \end{Bmatrix} = \begin{Bmatrix} -\ddot{y}_0 \\ 0 \end{Bmatrix} \tag{24.17}$$

where $\mu = m_2/m_1$, $2\pi f_i = \sqrt{k_i/m_i}$, and $4\pi\zeta_i f_i = c_i/m_i$. The damping is symmetric only if $\zeta_1 f_2 = \zeta_2 f_1$.

Consider a specific example where $\mu = 0.04$ and $\zeta_1 = \zeta_2 = 0.05$, so the damping is not symmetric. Figure 24.4 shows the calculated values of the mean square responses $\overline{z_1^2}$ and $\overline{z_2^2}$ as a function of f_2/f_1. The amplitudes are plotted relative to the mean square response that m_1 would have without the attached vibration absorber $\overline{z_{1_o}^2}$, as calculated using Eq. (24.6). The modal superposition calculation ignores the small cross-coupling between the normal modes due to the nonsymmetric damping. These results are compared to the exact solution for the two-DOF system.[3] The mean square response of m_1 is suppressed only when $f_2 \simeq f_1$ and only by about 4 dB.

STATISTICAL ENERGY ANALYSIS OF COMPLEX SYSTEMS

Statistical energy analysis models the vibration response of a complex system as a statistical interaction between groups of modes associated with subsections of the system. While the theoretical development of SEA has its roots in the field of random vibration, it does not require a random excitation for the statistical analysis. Instead, SEA uses the random variation of modal responses in complex systems to obtain statistical response predictions in terms of mean values and variances of the responses. Theoretically, the statistical averaging is over ensembles of nominally identical systems. However, in practice many systems have enough inherent complexity that the variation in the response over frequency and location is adequately represented by the ensemble statistics.

This is seen even in the relatively simple case of the distribution of bending modes in a simply supported rectangular flat plate (Fig. 24.5*A*). The resonance frequencies of the modes are given by

$$f_{m,n} = \left(\frac{\pi}{4\sqrt{3}}\right) h c_L \left[\left(\frac{m}{L_1}\right)^2 + \left(\frac{n}{L_2}\right)^2\right] \tag{24.18}$$

where L_1 and L_2 are the length dimensions, h is the thickness, c_L is the longitudinal wave speed of the plate material, and m and n are integers. The resonance frequencies are seen to follow approximately along a straight line. This slope of this line is the *average frequency spacing* $\overline{\delta f}$ (inverse of modal density per Hz) given by

$$\overline{\delta f} = \frac{h c_L}{\sqrt{3} L_1 L_2} \tag{24.19}$$

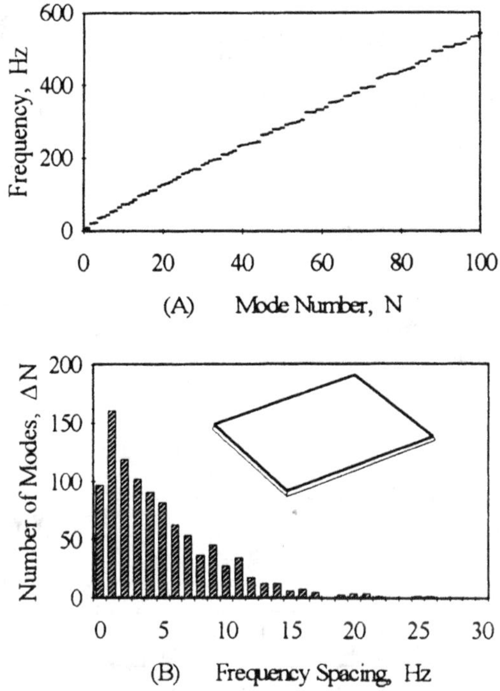

FIGURE 24.5 Mode count of a 2.6- × 2.4- × 0.01-meter simply supported, steel plate. (*A*) Resonance frequencies. (*B*) Distribution of resonance frequency spacings.

One way to represent the variation in the actual resonant frequencies is to plot the distribution in the frequency difference between two successive resonances, which can be plotted as shown in Fig. 24.5*B*. This distribution appears to be Poisson. Repeating this analysis for other plates with the same surface area, thickness, and material (thus having the same δf), but with different values of L_1 and L_2, gives essentially the same results. This indicates that one way of looking at the modes of any one particular plate is to consider it as one realization from an ensemble of plates having the same statistical distribution of resonances. SEA uses this model to develop estimates of the vibration response of systems based on averages over the ensemble of similar systems. However, since the modes are usually a function of the parameter (fL/c), variations in the frequency f in a complex system often have the same statistics as variations in L (dimensions) and c (material properties) in an ensemble of similar systems.

The statistical model of a system is useful in a variety of applications. In the preliminary design phase of a system SEA can be used to obtain quantitative estimates of the vibration response even when all of the details of the design are not completely specified. This is because preliminary SEA estimates can be made using the general characteristics of the system components (overall size, thickness, material properties, etc.) without requiring the details of component shapes and attachments.

SEA is also useful in diagnosing vibration problems. The SEA model can be used to identify the sources and transfer paths of the vibrational energy. When measured data is available, SEA can help to interpret the data, and the measured data can be used to improve the accuracy of a preliminary SEA model. Since the SEA model gives quantitative predictions based on the physical properties of the system, it can be used to evaluate the effectiveness of design modifications. It can also be used with an optimization routine to search for improved design configurations.

SEA MODELING OF SYSTEMS

The statistical energy analysis (SEA) model of a complex system is based on the statistical analysis of the coupling between groups of resonant modes in subsections of the system. The modal coupling is based on the analysis of two coupled resonators, as shown in Fig. 24.6. This is a more general case of the two-degree-of-freedom system analyzed for a random vibration (see Fig. 24.4). Here there are two distinct resonators coupled by stiffness, inertial, and gyroscopic interactions (represented by k_c, m_c, and g_c, respectively). If the two resonators are excited by different broadband force excitations, then the net power flow between them through the coupling is given by

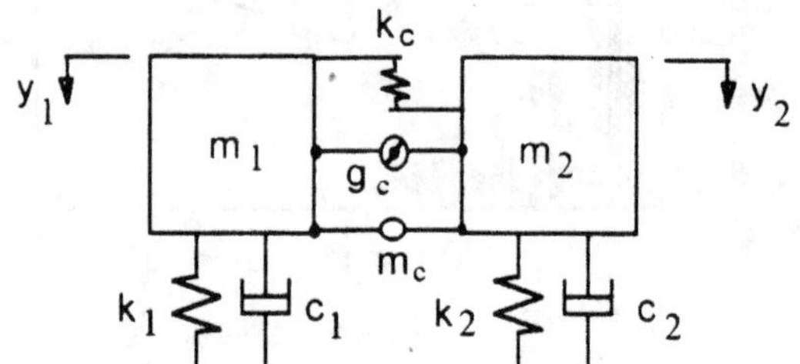

FIGURE 24.6 Two linear, coupled resonators, with displacement *y*, mass *m*, stiffness *k*, damper *c*, and gyroscopic parameter *g*.

$$\Pi_{12} = -k_c\,\overline{z_2\dot{z}_1} - g_c\,\overline{\dot{z}_2\dot{z}_1} + \frac{1}{4}m_c\overline{\ddot{z}_2\dot{z}_1}$$
$$= B\,(E_1 - E_2) \qquad (24.20)$$

where

$$B = \frac{(2\pi\mu)^2}{d}[\Delta_1 f_2^4 + \Delta_2 f_1^4 + f_1 f_2(\Delta_1 f_2^2 + \Delta_2 f_1^2)]$$
$$+\frac{1}{d}[(\gamma^2 + 2\mu\kappa)(\Delta_1 f_2^2 + \Delta_2 f_1^2) + \kappa^2(\Delta_1 + \Delta_2)]$$

$$E_i = (m_i + m_c/4)\overline{\dot{z}_i^2}$$

and

$$d = (1 - \mu^2)[(2\pi)^2(f_1^2 - f_2^2)^2 + (\Delta_1 + \Delta_2)(\Delta_1 f_2^2 + \Delta_2 f_1^2)]$$

$$\Delta_i = c_i/(m_i + m_c/4)$$

$$f_i^2 = (1/2\pi)^2(k_i + k_c)/(m_i + m_c/4)$$

$$\mu = \left(m_c/4\right)\left(m_1 + m_c/4\right)^{-1/2}\left(m_2 + m_c/4\right)^{-1/2}$$

$$\gamma = g_c\left(m_1 + m_c/4\right)^{-1/2}\left(m_2 + m_c/4\right)^{-1/2}$$

$$\kappa = k_c\left(m_1 + m_c/4\right)^{-1/2}\left(m_2 + m_c/4\right)^{-1/2}$$

This result can be interpreted by defining the two individual uncoupled resonators as the subsystems that exist when one of the degrees of freedom is constrained to zero. For either uncoupled resonator, the kinetic energy averaged over a cycle $(m + m_c/4)\overline{\dot{z}_i^2}/2$ is equal to the average potential energy $(k + k_c)\overline{z_i^2}/2$. Equation (24.20) can then be seen to state two important results: (1) the power flow is proportional to the difference in the vibrational energies of the two resonators, and (2) the coupling parameter B is positive definite and symmetrical so the system is reciprocal and power always flows from the more energetic resonator to the less energetic one. As a corollary, when only one resonator is directly excited, the maximum energy level of the second resonator is that of the first resonator.

It should be noted that this analysis is exact for a coupling of arbitrary strength as long as there is no dissipation in the coupling. Even when there is dissipation in the coupling, this analysis is approximately correct as long as the coupling forces due to the dissipation are small compared to the other coupling forces. In practice, when systems have interface damping at the connections between subsystems (such as in bolted or spot welded joints), the associated damping can be split between subsystems and the interface considered damping free.

As an example of how this analysis is extended to a distributed system, consider the two coupled beams in Fig. 24.7*A*. The modes of the system can be obtained from an eigenvalue solution of the complete system, or they can be obtained from a coupled pair of equations for the individual (or uncoupled) straight beam subsystems. The latter case leads to coupled mode equations similar to the ones used for the two coupled resonators. However, in this case each mode in one beam subsystem is coupled to all of the relevant modes in the other beam subsystem. The total power flow between the two beam subsystems is then the sum of the individual mode-to-mode power flows.

If the significant coupling is assumed to occur in a limited frequency range Δf (a good assumption for $\zeta << 1$ and $\Delta f >> \zeta f$), then the average net power flow can be found by averaging the value of B over Δf and using average beam subsystem modal energies in Eq. (24.20). This gives

$$\Pi_{12} = \overline{B} N_1 N_2 \left(\frac{E_1}{N_1} - \frac{E_2}{N_2}\right) \tag{24.21}$$

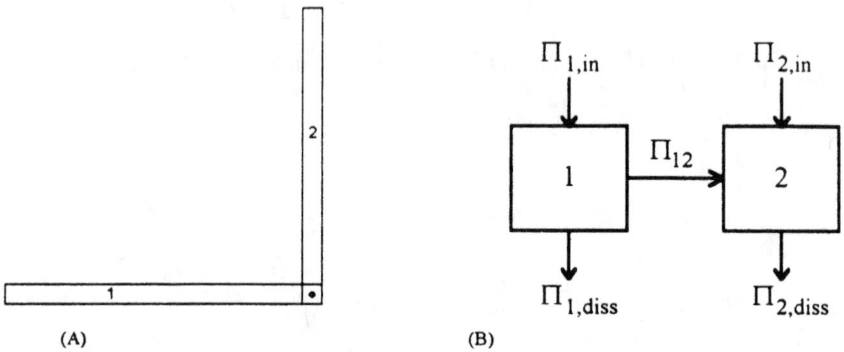

FIGURE 24.7 Modeling of distributed systems. (*A*) Two coupled beams. (*B*) SEA model of two coupled subsystems with power flow Π.

with

$$\overline{B} = \frac{1}{4\Delta f}\left[\mu^2(2\pi f)^2 + (\gamma^2 + 2\mu\kappa) + \frac{\kappa^2}{(2\pi f)^2}\right]$$

where N_1 and N_2 are the number of modes in the two beam subsystems with resonance frequencies in Δf.

For either beam, the total vibrational energy is $E_i = m_i\overline{\dot{z}_i^2}$, where m_i is the total mass of the beam and $\overline{\dot{z}_i^2}$ is the mean square velocity averaged over space and time. Equation (24.21) shows that the power flow between two distributed subsystems is proportional to the difference in the average *modal* energies E_i/N_i, not the difference in the total energies (which are proportional to the vibration level). This means it is possible for a thick beam with fewer resonant modes in a frequency band and a lower vibration level to be the source of power for a connected thinner beam with more resonant modes and a higher vibration level.

A more useful form of Eq. (24.21) is obtained by defining a *coupling loss factor* $\eta_{12} \equiv \overline{B}N_2/(2\pi f)$ (and by reciprocity $\eta_{21} \equiv N_1\eta_{12}/N_2$). The coupling loss factor is analogous to the *damping loss factor* for a subsystem defined by $\eta_i = 2\zeta_i$. The coupling loss factor is a measure of the rate of energy lost by a subsystem through coupling to another subsystem, whereas the damping loss factor is a measure of the rate of energy lost through dissipation. The average power flow is then given by

$$\Pi_{12} = 2\pi f(\eta_{12}E_1 - \eta_{21}E_2) \tag{24.22}$$

Using the equivalent expression for the power dissipated in each subsystem, $\Pi_{i,\text{diss}} = 2\pi f\eta_i E_i$, the following set of equations can be written for the conservation of energy between two coupled subsystems ($\Pi_{\text{in}} = \Pi_{\text{out}} + dE/dt$):

$$\begin{aligned} \Pi_{1,\text{in}} &= 2\pi f(\eta_1 + \eta_{12})E_1 - 2\pi f\eta_{21}\,E_2 + \frac{dE_1}{dt} \\ \Pi_{2,\text{in}} &= -2\pi f\eta_{12}E_1 + 2\pi f(\eta_2 + \eta_{21})E_2 + \frac{dE_2}{dt} \end{aligned} \tag{24.23}$$

where $\Pi_{i,\mathrm{in}}$ is used to denote power supplied by external sources. The SEA block diagram for this power flow model of two coupled subsystems is shown in Fig. 24.7*B*. These equations are first-order differential equations for the diffusion of energy between subsystems. They are in a form analogous to heat flow or fluid potential flow problems. For steady-state problems the dE/dt terms are zero.

For narrowband analysis, the SEA equations can be used to obtain averages in the response of the system over frequency. In this case it is more convenient to use the average frequency spacing between modes $\overline{\delta f} = \Delta f/N$ as the mode count in Eq. (24.21). This gives

$$\overline{\Pi_{12}} = \frac{2\pi f}{\overline{\delta f_1}} \eta_{12}(E_1\overline{\delta f_1} - E_2\overline{\delta f_2}) \tag{24.24}$$

The terms $2\pi E_i\overline{\delta f_i}$ have units of power and are called the *modal power potential.*

The value of η_{12} is difficult to evaluate directly from B in practice. Instead, indirect methods are often used as described in the section "Coupling Loss Factors." The normalized variance in the value of η_{12} averaged over Δf for edge-connected subsystems is given by

$$\frac{\sigma_{\eta 12}^2}{\eta_{12}^2} = \frac{1}{\pi f\left(\dfrac{\eta_1}{\overline{\delta f_1}} + \dfrac{\eta_2}{\overline{\delta f_2}}\right) + \Delta f\left(\dfrac{1}{\overline{\delta f_1}} + \dfrac{1}{\overline{\delta f_2}}\right)} \tag{24.25}$$

The variance in the coupling depends primarily on the system *modal overlap factor* defined by $M_S = \pi f(\eta_1/\overline{\delta f_1} + \eta_2/\overline{\delta f_2})/2$, which is the ratio of the effective modal bandwidth to the average modal frequency spacing. When the system modal overlap factor is less than 1, the variance is larger than the square of the mean value, which may be unacceptably large. This indicates why SEA models tend to converge better with measured results at frequencies above where $M_S = 1$.

Note that the modal overlap in each uncoupled subsystem does not have to be large in order for the variance in the coupling to be small. In fact the SEA model can be used to evaluate the response of a single resonator mode attached to a vibrating flat plate as illustrated in Fig. 24.8. The power flow equations in the form of Eq. (24.23) are used. The uncoupled resonator has one mode at $f_2 = \sqrt{k_2/m_2}$, so $N_2 = 1$. The mean-square vibration velocity level of the plate in a frequency band Δf encompassing f_2 is $\overline{\dot{z}_1^2} = W_{\dot{y}1}(f)\Delta f$. The average number of plate modes resonating in this frequency band is $N_1 = \Delta f/\overline{\delta f_1}$. The coupling loss factor is evaluated to be

$$\eta_{21} = \frac{\pi}{2}\frac{f_2}{\overline{\delta f_1}}\frac{m_2}{m_1} \tag{24.26}$$

Since $\Pi_{12} = \Pi_{2,\mathrm{diss}}$, the mean square response of the resonator mass is given by

$$\overline{\dot{z}_2^2} = \frac{\pi}{2}\frac{f_2 W_{\dot{y}1}(f)}{\eta_{21} + \eta_2} \tag{24.27}$$

Even if the resonator damping goes to zero, its maximum energy level is limited to the average modal energy in the plate:

$$m_2\,\overline{\dot{z}_2^2}_{,\max} = m_1 W_{\dot{y}1}(f)\Delta f \tag{24.28}$$

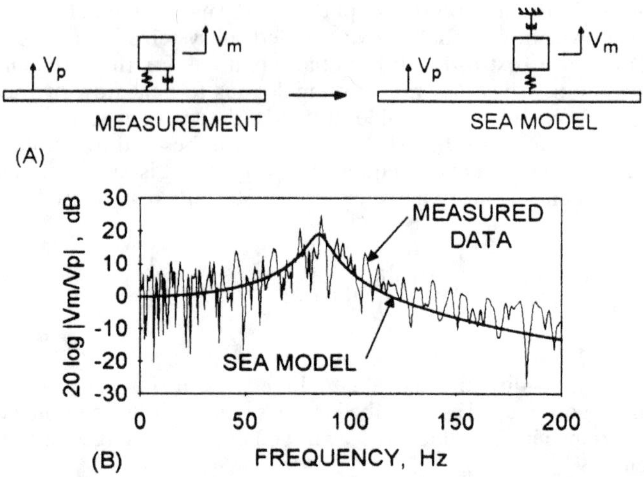

FIGURE 24.8 Response of a resonator with vibration V_m, mounted on a plate with vibration V_p. (*A*) Comparison of the measurement configuration and the SEA model. (*B*) Comparison of the measured response and the SEA predictions.

If the resonator energy momentarily gets higher, it transmits the energy back into the plate. Therefore, the plate acts both as a source of excitation and as a dissipator of energy for the resonator. The effective loss factor for the resonator is $\eta_{21} + \eta_2$.

The frequency response function for the resonator can then be evaluated using Eq. (24.5). Figure 24.8*B* compares this result with the measured narrowband frequency spectrum of a 0.1-kg mass attached to a 2.5-mm steel plate with a resilient mounting having negligible damping and $f_2 = 85$ Hz. The measured response of the mass is multimodal since the resonator responds as a part of all of the modes of the coupled system. However, the statistical average response curve accurately represents the multimodal response. The normalized variance of the narrowband SEA response calculation is estimated from Eq. (24.25) to be 0.5.

For larger systems the following procedure can be used to develop a complete SEA model of the system response to an excitation:

1. Divide the system into a number of coupled subsystems.
2. Determine the mode counts and damping loss factors for the subsystems.
3. Determine the coupling factors between connected subsystems.
4. Determine the subsystem input powers from external sources.
5. Solve the energy equations to determine the subsystem response levels.

The steps in this procedure are described in the following sections of this chapter. When used properly, the SEA model will calculate the distribution of vibration response throughout a system as a result of an excitation. The response distribution is calculated in terms of a mean value and a variance in the vibration response of each subsystem averaged over time and the spatial extent of the subsystem.

MODE COUNTS

In this section the mode counts for a number of idealized subsystem types are given in terms of the average frequency spacing $\overline{\delta f}$ between modal resonances. Experimental and numerical methods for determining the mode counts of more complicated subsystems are also described.

The mode count is sometimes represented by the average number of modes, N or ΔN, resonating in a frequency band, and sometimes by the modal density, represented in cyclical frequency as $n(f) = dN/df$. These are related to the average frequency spacing by

$$n(f) = \frac{1}{\overline{\delta f}} \simeq \frac{\Delta N}{\Delta f} \tag{24.29}$$

For a one-dimensional subsystem, such as a straight beam or bar, with uniform material and cross-sectional properties and with length L, the average frequency spacing between the modal resonances is given by

$$\overline{\delta f}^{1D} = \frac{c_g}{2L} \tag{24.30}$$

where c_g is the energy group speed for the particular wave type being modeled.

For longitudinal waves, c_g is equal to the phase speed $c_L = \sqrt{E/\rho}$, where E is the elastic (Young's) modulus and ρ is the density of the material. For torsional waves c_g is equal to the phase speed $c_T = \sqrt{GJ/\rho I_p}$, where G is the shear modulus of the material, and J and I_p are the torsional moment of rigidity and polar area moment of inertia, respectively, of the cross section. For beam bending waves (with wavelengths long compared to the beam thickness) the group speed is twice the bending phase speed c_B, or $c_g = 2c_B = 2\sqrt{2\pi f \kappa c_L}$, where κ is the radius of gyration of the beam cross section. For a beam of uniform thickness h, $\kappa = h/\sqrt{12}$.

For a two-dimensional subsystem, such as a flat plate, with uniform thickness and material properties and with surface area A, the average frequency spacing between the modal resonances is given by

$$\overline{\delta f}^{2D} = \frac{c_p c_g}{2\pi f A} \tag{24.31}$$

where c_p is the phase speed for the particular wave type being modeled. For plate bending waves (with wavelengths long compared to the plate thickness) $c_g = 2c_p = 2c_{B'} = 2\sqrt{2\pi f \kappa c_{L'}}$, where κ is the radius of gyration, $c_{L'} = \sqrt{E/\rho(1-\mu^2)}$, and μ is Poisson's ratio. For in-plane compression waves $c_g = c_p = c_{L'}$. For in-plane shear waves $c_g = c_p = c_S = \sqrt{G/\rho}$.

For a three-dimensional subsystem, such as an elastic solid, with uniform material properties and with volume V, the average frequency spacing between the modal resonances is given by

$$\overline{\delta f}^{3D} = \frac{c_o^3}{4\pi f^2 V} \tag{24.32}$$

where c_o is the ambient shear or compressional wave speed in the medium.

For more complicated subsystems, the mode counts can be obtained in a number of other ways. Generally, the mode counts only need to be determined within an accuracy of 10 percent in order for any resulting error to be less than 1 dB in the statistical energy analysis model. For more complicated wave types, such as bending in thick beams or plates, the formulas given above for $\overline{\delta f}$ can be used with the correct values of c_g and c_p obtained from the dispersion relation for the medium.

For more complicated geometries, a numerical solution, such as a finite element model, can be used to determine the eigenvalues of the subsystem. Then the values of $\overline{\delta f}$ can be obtained using Eq. (24.29). In this case it is often necessary to average the mode count over a number of particular geometric configurations or boundary conditions in order to obtain an accurate estimate of the average modal spacing.

When a physical sample of the subsystem exists, experimental data can be used to estimate or validate the mode count. For large modal spacing (small modal overlap) the individual modes can sometimes be counted from a frequency response measurement. However, this method usually undercounts the modes because some of them may occur paired too closely together to be distinguished. An alternate experimental procedure is to use the relation between the mode count and the average mobility of a structure:

$$\overline{\delta f} = \frac{1}{4m\overline{G}} \tag{24.33}$$

where m is the mass of the subsystem and $\overline{G}$ is the average real part of the mechanical mobility (ratio of velocity to force at a point excitation; see Chap. 9). As with the numerical method, the experimental measurement should be averaged over a variation in the boundary condition used to support the subsystem since no one static support accurately represents the dynamic boundary condition the subsystem sees when it is part of the full system. Also the measurement of $\overline{G}$ should be averaged over several excitation points.

DAMPING LOSS FACTORS

In this section typical methods for determining the damping loss factor of subsystems are given along with some typical values used in statistical energy analysis models of complex structures. The damping in SEA models is usually specified by the loss factor, which is related to the critical damping ratio ζ and the quality factor Q by

$$\eta = 2\zeta = \frac{1}{Q} \tag{24.34}$$

Chapters 35 and 36 describe the damping mechanisms in structural materials and typical damping treatments. In complex structures the structural material damping is usually small compared to the damping due to slippage at interfaces and added damping treatments. Because the level of added damping is so strongly dependent on the details of the application of a damping treatment, measurements are usually needed to verify analytical calculations of damping levels.

One method to measure the damping of a subsystem is the decay rate method, where the free decay in the vibration level is measured after all excitations are turned off. The initial decay rate DR (in dB/sec) is proportional to the total loss factor for the subsystem:

$$\eta = \frac{DR}{27.3f} \tag{24.35}$$

If the subsystem is attached to other structures, the coupling loss factors will be included in the total loss factor value. Therefore, the subsystem must be tested in a decoupled state. On the other hand, if the connection interfaces provide significant damping due to slippage, then these interfaces must be simulated in the damping test.

Another method of measuring the damping is the half-power bandwidth method illustrated in Fig. 2.22. The width of a resonance Δf in a frequency response measurement is measured 3 dB down from the peak and the damping is determined by

$$\eta = \frac{\Delta f}{f_n} \tag{24.36}$$

As with other measurements of subsystem parameters, the damping measurements must be averaged over multiple excitation points with a variety of boundary conditions.

For preliminary SEA models an empirical database of damping values is useful for initial estimates of the subsystem damping loss factors. Figure 24.9 is an illustration of the typical damping values measured in steel and aluminum machinery structures for different construction methods and different applied damping treatments.

The initial estimates of damping levels in a preliminary SEA model can be improved if measurements of the spatial decay of the vibration levels in the system are available. The spatial decay calculated in the SEA model is quite strongly dependent on the damping values used. Therefore, an accurate estimate of the actual damping can be obtained by comparing the SEA calculations to the measured spatial decay (assuming the other model parameters are correct).

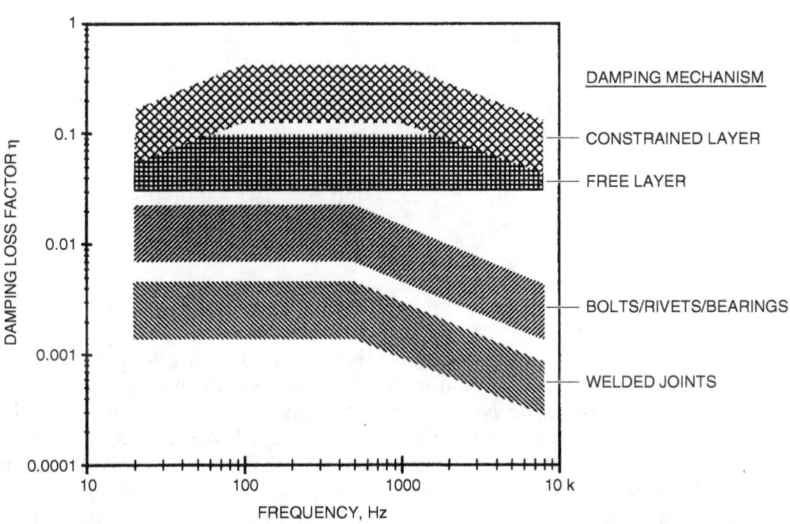

FIGURE 24.9 Empirical values for the damping loss factor η in steel and aluminum machinery structures with different damping mechanisms (assumed to be efficiently applied, but in less than ideal laboratory conditions).

COUPLING LOSS FACTORS

The coupling loss factor is a parameter unique to statistical energy analysis. It is a measure of the rate of energy transfer between coupled modes. However, it is related to the transmission coefficient τ in wave propagation. This can be illustrated with the system shown in Fig. 24.10. For a wave incident on a junction in subsystem 1 with incident power Π_{inc}, the power transmitted to subsystem 2, Π_{tra}, is by definition of the transmission coefficient τ_{12} given by

$$\Pi_{tra} = \tau_{12}\Pi_{inc} \tag{24.37}$$

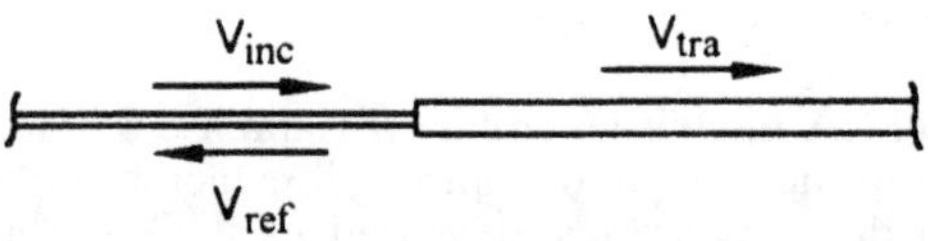

FIGURE 24.10 Evaluation of the coupling loss factor using a wave transmission model for an incident wave V_{inc} at a junction, resulting in a reflected wave V_{ref} and a transmitted wave V_{tra}.

In addition, the junction reflects some power, Π_{ref}, back into subsystem 1 given by

$$\Pi_{ref} = (1 - \tau_{12})\Pi_{inc} \tag{24.38}$$

assuming there is no power dissipated at the junction. The energy density in subsystem 1 is given by $E_1' = c_{g_1}(\Pi_{inc} + \Pi_{ref})$. The corresponding SEA representation of the system is

$$\Pi_{tra} = \Pi_{1\rightarrow 2} = 2\pi f\eta_{12}E_1 \tag{24.39}$$

For a subsystem of length L_1, $\overline{\delta f_1} = c_{g_1}/(2L_1)$ and $E_1 = L_1E_1'$. Solving for the coupling loss factor gives

$$\eta_{12} = \frac{\overline{\delta f_1}}{\pi f}\frac{\tau_{12}}{2 - \tau_{12}} \tag{24.40}$$

A more detailed analysis indicates that this result is valid for point connections in a system with a modal overlap greater than 1. If the system has a constant modal frequency spacing $\overline{\delta f}$, then the Nth mode will occur at $f = N\overline{\delta f}$. If the damping loss factor is η, the system modal overlap is given by $M_S = \pi\eta f/(2\overline{\delta f})$. Then the modal overlap is greater than 1 for frequencies $f > 2\overline{\delta f}/(\pi\eta)$ or for mode numbers $N > 2/(\pi\eta)$. SEA is still valid below this frequency and mode number, but the variance of the model calculations (and in the measured frequency response functions) becomes large.

For point-connected subsystems the transmission coefficient can be evaluated from the junction impedances:[4]

$$\tau_{12} = \frac{4R_1R_2}{|Z_1 + Z_2|^2} \tag{24.41}$$

where R_i is the real part of the impedance Z_i (ratio of force to velocity at a point excitation) at the junction attachment point of subsystem i. When more than two subsystems are connected at a common junction, the denominator of Eq. (24.41) must include the sum of all impedances at the junction.

For subsystems with line and area junctions the analysis of the coupling loss factor is complicated by the distribution of angles of the waves incident on the junction. However, approximate results have been worked out for many important cases. Eq. (24.40) can be generalized for all cases as

$$\eta_{12} = \frac{\overline{\delta f_1}}{\pi f}\,\frac{I_{12}\tau_{12}(0)}{2 - \tau_{12}(0)} \tag{24.42}$$

where $\tau_{12}(0)$ is the normal incidence transmission coefficient for waves traveling perpendicular to the junction, and I_{12} contains the result of an average over all angles of incidence.

For line-connected plates the coupling loss factor between bending modes is found using

$$I_{12} = \frac{L_j}{4}\left(\frac{k_1^{\,4}k_2^{\,4}}{k_1^{\,4} + k_2^{\,4}}\right)^{1/4} \tag{24.43}$$

where L_j is the length of the junction and $k_i = 2\pi f/c_{Bi}$ is the wave number of the modes in subsystem i.

When experimental verification of the evaluation of the coupling loss factor is desired, measurements similar to those used for damping can be used. A decay rate measurement of a subsystem connected to another (heavily damped) subsystem will give a loss factor equal to the sum of the damping and coupling loss factor for the first subsystem. Alternately, subsystem 1 can be excited alone and the spatially averaged response levels of the two connected subsystems can be measured. Using $\Pi_{12} = \Pi_{2,\mathrm{diss}}$, the coupling loss factor is found from

$$\eta_{12} = \frac{\eta_2 E_2}{E_1 - \overline{\delta f_2}E_2/\overline{\delta f_1}} \tag{24.44}$$

This result indicates a potential problem in determining the coupling loss factor from measured results. If $E_2\overline{\delta f_2} \simeq E_1\overline{\delta f_1}$, then taking the difference between their values in Eq. (24.44) will greatly magnify the experimental errors in determining the parameters used in this formula. This indicates why it is mathematically unstable to use measured levels in a multiple subsystem model to back-calculate the coupling loss factors. However, good results can be obtained for a single junction between two subsystems if one is excited and the other is artificially damped in order to increase difference between $E_1\overline{\delta f_1}$ and $E_2\overline{\delta f_2}$. Figure 24.11 shows the results of an experimental validation of Eqs. (24.42) and (24.43) for the coupling loss factor between two plates connected at a point. The experimental error is also included, which even in this idealized laboratory environment is more than 50 percent. While the *back calculation* of the coupling loss factors tends to be unstable, the forward calculation in the SEA model is relatively insensitive to errors in the coupling loss factor values, making the model fairly robust.

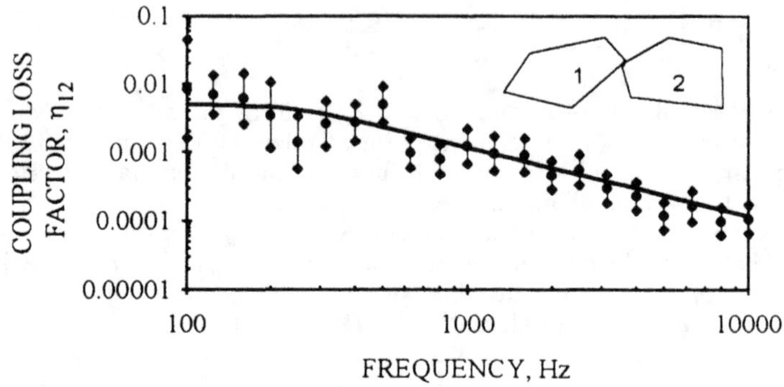

FIGURE 24.11 Coupling loss factor η_{12} for point connected plates; ◆—◆—◆ measured data with 95 percent confidence intervals; —— calculated values using Eqs. (11.80) and (11.81).

MODAL EXCITATIONS

The power put into subsystem modes by the system excitations is needed in order to use the statistical energy analysis model for calculations of absolute response levels. The mode counts, damping, and coupling loss factors can be used to evaluate relative transfer functions in the system for a unit input power. However, for actual response-level calculations the modal input power from the actual excitation sources must be calculated.

For a point force excitation $F(t)$ the average power put into a system is

$$\Pi_{in} = \overline{F^2}\,\overline{G} \tag{24.45}$$

where $\overline{G}$ is the average real part of the mobility at the excitation point. For a prescribed point velocity source $\dot{y}(t)$ the average power put into a system is

$$\Pi_{in} = \overline{\dot{y}^2}\,\overline{R} \tag{24.46}$$

where $\overline{R}$ is the average real part of the impedance at the excitation point.

The normalized variance in the input power due to variations in the mode shapes and frequency response function of the system is approximated by

$$\frac{\sigma_{\Pi_{in}}^{\;2}}{\Pi_{in}^{\;2}} = \frac{3\overline{\delta f}}{\pi f \eta + \Delta f} \tag{24.47}$$

where Δf is the bandwidth of the excitation. For more complicated excitations the input power can be estimated by measuring the response of a system to the excitation and using the SEA model to back-calculate the input power. Alternatively, the measured response levels of the excited subsystem can be used as "source" levels, and the power flow into the rest of the system can be evaluated using the SEA model.

SYSTEM RESPONSE DISTRIBUTION

To solve for the distribution of vibrational energy in a system it is convenient to rewrite Eq. (24.23) in symmetric form:

$$[B]\{\Phi\} + [I]\left\{\frac{dE}{dt}\right\} = \{\Pi_{\text{in}}\} \tag{24.48}$$

where $[I]$ is the identity matrix, $\{\Phi\} = 2\pi\{E/\overline{\delta f}\}$ is the vector of modal power potential, and $[B]$ is the symmetric matrix of coupling and damping terms with off-diagonal terms $B_{ij} = -f\,\eta_{ij}/\overline{\delta f_i}$ and diagonal terms $B_{ii} = (f/\overline{\delta f_i})(\eta_i + \Sigma_j\,\eta_{ij})$. This system of equations can be solved using standard numerical methods. Solving for the values of E gives a mean-value estimate of the energy distribution.

The variance in E is more difficult to evaluate because it depends on the evaluation of the inverse matrix $[B]^{-1}$. If the variance of each term in $[B]$ is small compared to its mean-square value, then the variances in $[B]^{-1}$ can be approximated by

$$[\sigma_{B^{-1}}{}^2] \simeq [(B_{ij}{}^{-1})^2][\sigma_B{}^2][(B_{ij}{}^{-1})^2] \tag{24.49}$$

where the notation $[(B_{ij}{}^{-1})^2]$ refers to a matrix with the squares of the elements in $[B]^{-1}$, term for term.

The subsystem energy values can be converted to dynamic response quantities using the relation $E = m\overline{\dot{z}^2}$. For a narrowband vibration at frequency f_c (which could be each 1/3-octave band response in a broadband analysis) the displacement response is $\overline{z^2} \simeq \overline{\dot{z}^2}/(2\pi f)^2$ and the acceleration response is $\overline{\ddot{z}^2} \simeq (2\pi f)^2\overline{\dot{z}^2}$. The relation between the vibration velocity response and the maximum dynamic strain depends on the type of motion involved. For longitudinal motion, the mean-square strain is $\overline{\epsilon^2} = \overline{\dot{z}^2}/c_L{}^2$. For bending motion of a uniform beam or plate, the maximum strain is $\overline{\epsilon_{\max}{}^2} = 3\overline{\dot{z}^2}/c_L{}^2$.

When the response values in a complex system are plotted on a logarithmic scale, a surprising result occurs. The log-values are distributed with an approximately gaussian distribution over frequency. This is illustrated in Fig. 24.12 for a beam network. The system frequency response function is computed numerically using a transfer impedance model including bending and longitudinal and torsional motions in each of the four beam segments. A histogram of the computed response values on the decibel scale compares very well with a gaussian distribution. This result can be explained by noting that the response value at any particular frequency results from the product of a large number of quantities. Then the logarithm of the response value will be the sum of a large number of terms. If the complexity in the system causes the responses at different frequencies to be independent, then by the central limit theorem the log-values will tend to have a gaussian distribution. This means that the mean square response values will have a log-normal distribution.

The calculated mean values and variances in the SEA model can be converted to the decibel scale as follows. If the mean square velocity $\overline{\dot{z}^2}$ has a log-normal distribution with variance $\sigma_{\dot{z}^2}{}^2$, then the velocity level $L_{\dot{z}} \equiv 10\log_{10}(\dot{z}^2/\dot{z}_{\text{ref}}{}^2)$ has a normal distribution with a mean value and variance given by

$$\overline{L_{\dot{z}}} = 10\log_{10}\left(\frac{\overline{\dot{z}^2}}{\dot{z}_{\text{ref}}{}^2}\right) - 5\log_{10}\left[1 + \frac{\sigma_{\dot{z}^2}{}^2}{(\overline{\dot{z}^2})^2}\right]$$

$$\sigma_{L_{\dot{z}}}{}^2 = 43\log_{10}\left[1 + \frac{\sigma_{\dot{z}^2}{}^2}{(\overline{\dot{z}^2})^2}\right] \tag{24.50}$$

Note that the mean of the decibel levels is not equal to the decibel level of the mean square value.

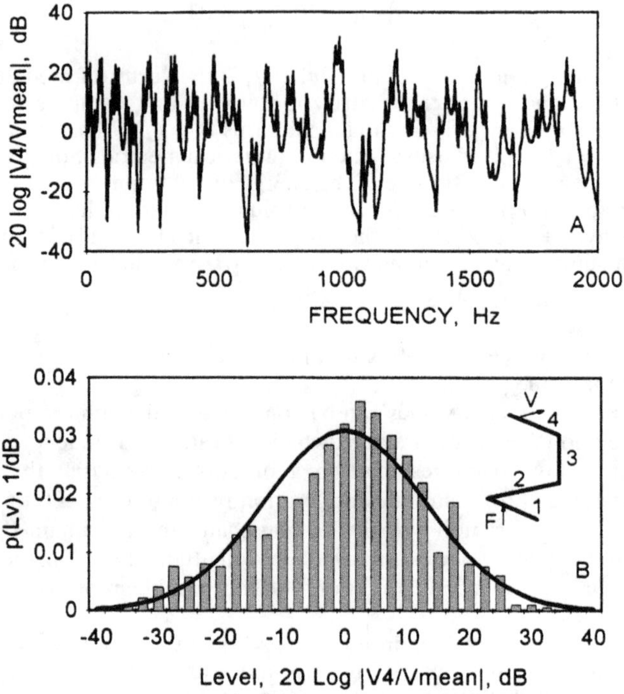

FIGURE 24.12 Numerical calculation of the vibration response of a four-beam network. (*A*) Normalized frequency response function for a point on beam 4. (*B*) Probability density function of log-levels, ▯ Numerical data histogram, — Normal distribution.

TRANSIENT (SHOCK) RESPONSE USING SEA

The statistical energy analysis model can solve for the transient response of a system using Eq. (24.23). The numerical solution methods for equations of this form can be illustrated using the finite difference method. Given an initial energy state $E(0)$, the energy state at a short time later is approximated by

$$E(\Delta t) \simeq E(0) + \frac{dE}{dt}\Delta t \tag{24.51}$$

where $dE/dt = \Pi_{in} - \Pi_{out}$. This new energy distribution is then used to project forward to the next time step, etc. The accuracy of the solution depends on the size of Δt relative to the energy flow time constants in the system, $(2\pi f\eta)^{-1}$. For the finite difference solution, using $\Delta t \leq (6\pi f\eta)^{-1}$ usually provides accurate results.

An example of a transient analysis using SEA is shown in Fig. 24.13. The measured acceleration response of a shelf on an equipment rack for an impact at the leg is shown along with the corresponding transient SEA solution of Eq. (24.23). The energy level of the shelf builds up for the first 0.01 sec before beginning to decay.

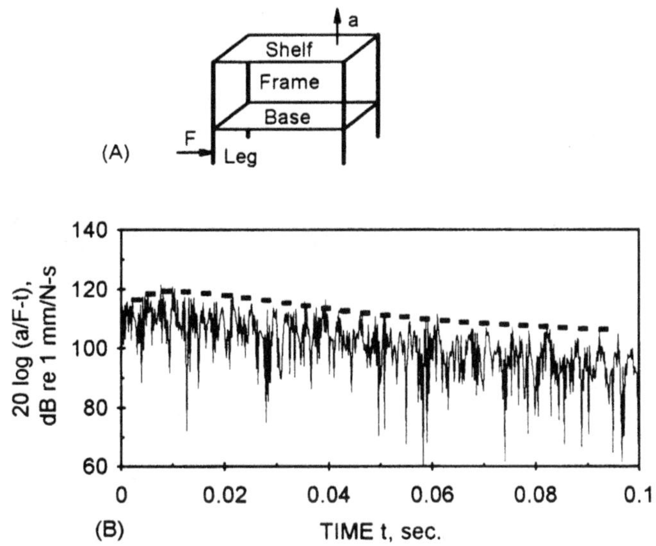

FIGURE 24.13 Transient response of an equipment shelf. (*A*) Experimental structure showing the locations of the impact *F* and the acceleration response *a*. (*B*) Comparison of the transient response of the structure; — Measured data, - - - SEA model.

Modeling the transient mean square response with Eq. (24.10), the undamped shock spectrum for this response signal can be estimated using Eq. (24.13). Alternately, if a shock excitation is modeled as a time-dependent power input to the SEA model, then the peak response spectrum of the system components can be estimated directly from the maximum mean square values in the transient SEA solution.

REFERENCES

1. Lyon, R. H., and R. G. DeJong: "Theory and Application of Statistical Energy Analysis," 2d ed., Butterworth-Heinemann, Boston, Mass., 1995.
2. Caughey, T. K.: "Nonstationary Random Inputs and Responses," Chap. 3, in S. H. Crandall (ed.), *Random Vibration,* Vol. 2, M.I.T. Press, Cambridge, Mass., 1963.
3. Crandall, S. H., and W. D. Mark: "Random Vibration in Mechanical Systems," Academic Press, New York, 1963.
4. Cremer, L., M. Heckl, and E. E. Ungar: "Structure-Borne Sound," 2d ed., Springer-Verlag, Berlin, 1988.

CHAPTER 25

VIBRATION TESTING MACHINES

David O. Smallwood

INTRODUCTION

This chapter describes some of the more common types of vibration testing machines which are used for developmental, simulation, production, or exploratory vibration tests for the purpose of studying the effects of vibration or of evaluating physical properties of materials or structures. A summary of the prominent features of each machine is given. These features should be kept in mind when selecting a vibration testing machine for a specific application. Digital control systems for vibration testing are described in Chap. 26. Applications of vibration testing machines are described in other chapters.

A vibration testing machine (sometimes called a *shake table* or *shaker* and referred to here as a *vibration machine*) is distinguished from a vibration exciter in that it is complete with a mounting table which includes provisions for bolting the test article directly to it. A *vibration exciter,* also called a *vibration generator,* may be part of a vibration machine or it may be a device suitable for transmitting a vibratory force to a structure. A *constant-displacement* vibration machine attempts to maintain constant-displacement amplitude while the frequency is varied. Similarly, a *constant-acceleration* vibration machine attempts to maintain a constant-acceleration amplitude as the frequency is changed.

The *load* of a vibration machine includes the item under test and the supporting structures that are not normally a part of the vibration machine. In the case of equipment mounted on a vibration table, the load is the material supported by the table. In the case of objects separately supported, the load includes the test item and all fixtures partaking of the vibration. The load is frequently expressed as the weight of the material. The *test load* refers specifically to the item under test exclusive of supporting fixtures. A *deadweight load* is a rigid load with rigid attachments. For nonrigid loads the reaction of the load on the vibration machine is a function of frequency. The vector force exerted by the load, per unit of acceleration amplitude expressed in units of gravity of the driven point at any given frequency, is the *effective load* for that frequency. The term *load capacity,* which is descriptive of the performance of reaction and direct-drive types of mechanical vibration machines, is the maximum deadweight load that can be vibrated at the maximum acceleration rating of the vibration machine. The *load couple* for a deadweight load is equal to the product of the force exerted on the load and the distance of the center of mass from the line of action of the force or from some arbitrarily selected location (such as a table surface). The static and dynamic load couples are generally different for nonrigid loads.

The term *force capacity,* which is descriptive of the performance of electrodynamic shakers, is defined as the maximum rated force generated by the machine. This force is usually specified, for continuous rating, as the maximum vector amplitude of a sinusoid that can be generated throughout a usable frequency range. A corresponding maximum rated acceleration, in units of gravity, can be calculated as the quotient of the force capacity divided by the total weight of the coil table assembly and the attached deadweight loads. The *effective force* exerted by the load is equal to the effective load multiplied by the (dimensionless) ratio g, which represents the number of units of gravity acceleration of the driven point [see Eq. (25.1)].

DIRECT-DRIVE MECHANICAL VIBRATION MACHINES

The direct-drive vibration machine consists of a rotating eccentric or cam driving a positive linkage connection which forces a displacement between the base and table of the machine. Except for the bearing clearances and strain in the load-carrying members, the machine tends to develop a displacement between the base and the table which is independent of the forces exerted by the load against the table. If the base is held in a fixed position, the table tends to generate a vibratory displacement of constant amplitude, independent of the operating rpm. Figure 25.1 shows the direct-drive mechanical machine in its simplest forms. This type of machine is sometimes referred to as a *brute force machine* since it will develop any force necessary to produce the table motion corresponding to the crank or cam offset, short of breaking the load-carrying members or stalling the driving shaft.

The simplest direct-drive mechanical vibration machine is driven by a constant-speed motor in conjunction with a belt-driven speed changer and a frequency-indicating tachometer. Table displacement is set during shutoff and is assumed to hold during operation. An auxiliary motor driving a cam may be included to provide frequency cycling between adjustable limits. More elaborate systems employ

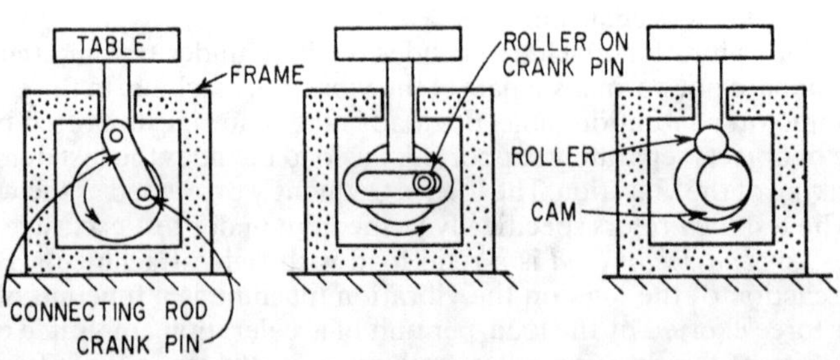

FIGURE 25.1 Elementary direct-drive mechanical vibration machines: (*A*) Eccentric and connecting link. (*B*) Scotch yoke. (*C*) Cam and follower.

a direct-coupled variable-speed motor with electronic speed control, as well as amplitude adjustment from a control station. Machines have been developed which provide rectilinear, circular, and three-dimensional table movements—the latter giving complete, independent adjustment of magnitude and phase in the three directions.

Many types of mechanisms are used to adjust the displacement amplitude and frequency of the mounting table. For example, the displacement amplitude can be adjusted by means of eccentric cams and cylinders.

PROMINENT FEATURES

- Low operating frequencies and large displacements can be provided conveniently.
- Theoretically, the machine maintains constant displacement regardless of the mechanical impedance of the table-mounted test item within force and frequency limits of the machine. However, in practice, the departure from this theoretical ideal is considerable, due to the elastic deformation of the load-carrying members with change in output force. The output force changes in proportion to the square of the operating frequency and in proportion to the increased displacement resulting therefrom. Because the load-carrying members cannot be made infinitely stiff, the machines do not hold constant displacement with increasing frequency with a bare table. This characteristic is further emphasized with heavy table mass loads. Accordingly, some of the larger-capacity machines which operate up to 60 Hz include automatic adjustment of the crank offset as a function of operating frequency in order to hold displacement more nearly constant throughout the full operating range of frequency.
- The machine must be designed to provide a stiff connection between the ground or floor support and the table. If accelerations greater than 1*g* are contemplated, the vibratory forces generated between the table and ground will be greater than the weight of the test item. Hence, all mass loads within the rating of the machine can be directly attached to the table without recourse to external supports.
- The allowable range of operating frequencies is small in order to remain within bearing load ratings. Therefore, the direct-drive mechanical vibration machine can be designed to have all mechanical resonances removed from the operating frequency range. In addition, relatively heavy tables can be used in comparison to the weight of the test item. Consequently, misplacing the center of gravity of the test item relative to the table center for vibration normal to the table surface and the generation of moments by the test item (due to internal resonances) usually have less influence on the table motions for this type of machine than would other types which are designed for wide operational frequency bands.
- Simultaneous rectilinear motion normal to the table surface and parallel to the table surface in two principal directions is practical to achieve. It may be obtained with complete independent control of magnitude and phase in each of the three directions.
- Displacement of the table is generated directly by a positive drive rather than by a generated force acting on the mechanical impedance of the table and load. Consequently, impact loads in the bearings, due to the necessary presence of some bearing clearance, result in the generation of relatively high impact forces which are rich in harmonics. Accordingly, although the waveform of displacement might be tolerated as such, the waveform of acceleration is normally sufficiently dis-

torted to preclude recognition of the fundamental driven frequency, when displayed on a time base.

REACTION-TYPE MECHANICAL VIBRATION MACHINE

A vibration machine using a rotating shaft carrying a mass whose center of mass is displaced from the center of rotation of the shaft for the generation of vibration is called a *reaction-type vibration machine.* The product of the mass and the distance of its center from the axis of rotation is referred to as the *mass unbalance,* the *rotating unbalance,* or simply the *unbalance.* The force resulting from the rotation of this unbalance is referred to as the *unbalance force.*

The *reaction-type vibration machine* consists of at least one rotating-mass unbalance directly attached to the vibrating table. The table and rotating unbalance are suspended from a base or frame by soft springs which isolate most of the vibration forces from the supporting base and floor. The rotating unbalance generates an oscillating force which drives the table. The unbalance consists of a weight on an arm which is relatively long by comparison to the desired table displacement. The unbalance force is transmitted through bearings directly to the table mass, causing a vibratory motion without reaction of the force against the base. A vibration machine employing this principle is referred to as a reaction machine since the reaction to the unbalance force is supplied by the table itself rather than through a connection to the floor or ground.

CIRCULAR-MOTION MACHINE

The reaction-type machine, in its simplest form, uses a single rotating-mass unbalance which produces a force directed along the line connecting the center of rotation and the center of mass of the displaced mass. Referred to stationary coordinates, this force appears normal to the axis of rotation of the driven shaft, rotating about this axis at the rotational speed of the shaft. The transmission of this force to the vibration-machine table causes the table to execute a circular motion in a plane normal to the axis of the rotating shaft.

Figure 25.2 shows, schematically, a machine employing a single unbalance producing circular motion in the plane of the vibration-table surface. The unbalance is driven at various rotational speeds, causing the table and test item to execute circular motion at various frequencies. The counterbalance weight is adjusted to equal the test item mass moment calculated from *d,* the plane of the unbalance force, thereby keeping the combined center of gravity coincident with the generated force. Keeping the generated force acting through the combined center of gravity of the spring-mounted assembly eliminates vibratory moments which, in turn, would generate unwanted rotary motions in addition to the motion parallel to the test mounting surface. The vibration isolator supports the vibrating parts with minimum transmission of the vibration to the supporting floor.

For a fixed amount of unbalance and for the case of the table and test item acting as a rigid mass, the displacement of motion tends to remain constant if there are no resonances in or near the operating frequency range. If balance force must remain constant, the amount of unbalance must change with shaft speed.

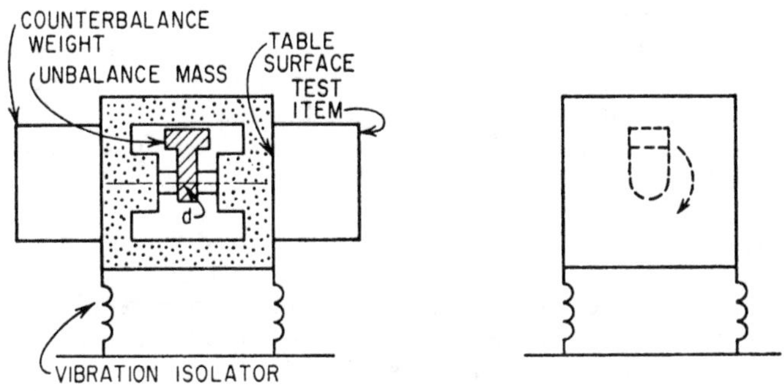

FIGURE 25.2 Circular-motion reaction-type mechanical vibration machine.

RECTILINEAR-MOTION MACHINE

Rectilinear motion rather than circular motion can be generated by means of a reciprocating mass. Rectilinear motions can be produced with a single rotating unbalance by constraining the table to move in one direction.

Two Rotating Unbalances. The most common rectilinear reaction-type vibration machine consists of two rotating unbalances, turning in opposite directions and phased so that the unbalance forces add in the desired direction and cancel in other directions. Figure 25.3 shows schematically how rectilinear motion perpendicular and parallel to the vibration table is generated. The effective generated force from the two rotating unbalances is midway between the two axes of rotation and is normal to a line connecting the two. In the case of motion perpendicular to the surface of the table, simply locating the center of gravity of the test item over the center of the table gives a proper load orientation. Tables are designed so that the resultant force always passes through this point. This results in collinear-

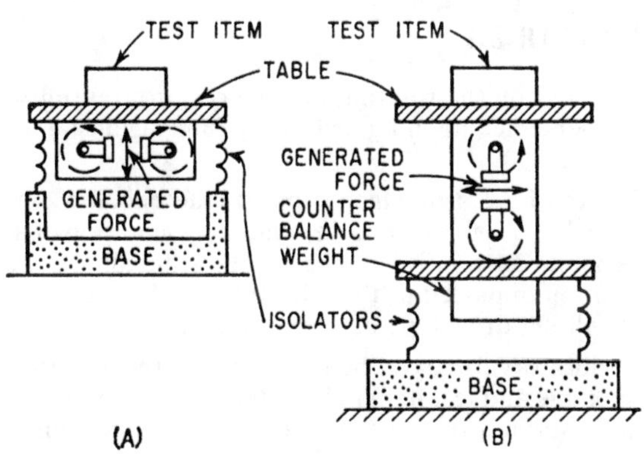

FIGURE 25.3 Rectilinear-motion reaction-type mechanical vibration machine using two rotating unbalances: (*A*) Vibration perpendicular to table surface. (*B*) Vibration parallel to table surface.

ity of generated forces and inertia forces, thereby avoiding the generation of moments which would otherwise rock the table. In the case of motion parallel to the table surface, no simple orientation of the test item will achieve collinearity of the generated force and inertia force of the table and test item. Various methods are used to make the generated force pass through the combined center of gravity of the table and test item.

Three Rotating Unbalances. If a machine is desired which can be adjusted to give vibratory motion either normal to the plane of the table or parallel to the plane of the table, a minimum of three rotating unbalances is required. Inspection of Fig. 25.4 shows how rotating the two smaller mass unbalances relative to the single larger unbalance results in the addition of forces in any desired direction, with cancellation of forces and force couples at 90° to this direction. Although parallel shafts are usually used as illustrated, occasionally the three unbalances may be mounted on collinear shafts, the two smaller unbalances being placed on either side of the single larger unbalance to conserve space and to eliminate the bending moments and shear forces imposed on the structure connecting the individual shafts.

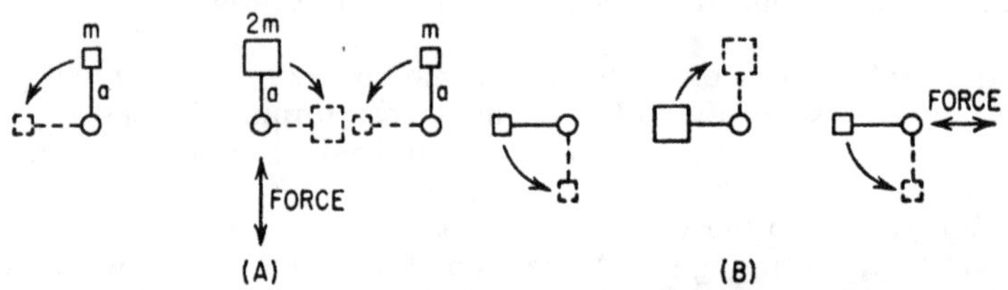

FIGURE 25.4 Adjustment of direction of generated force in a reaction-type mechanical vibration exciter: (*A*) Vertical force. (*B*) Horizontal force.

PROMINENT FEATURES

- The forces generated by the rotating unbalances are transmitted directly to the table without dependence upon a reactionary force against a heavy base or rigid ground connection.
- Because the length of the arm which supports the unbalance mass can be large, relative to reasonable bearing clearances and the generation of a force which does not reverse its direction relative to the rotating unbalance arm, the generated waveform of motion imparted to the vibration machine table is superior to that attainable in the direct-drive type of vibration machine.
- The generated vibratory force can be made to pass through the combined center of gravity of the table and test item in both the normal and parallel directions relative to the table surface, thereby minimizing vibratory moments giving rise to table rocking modes.
- The attainable rpm and load ratings on bearings currently limit performance to a frequency of approximately 60 Hz and a generated force of 300,000 lb (1.3 MN),

respectively, although in special cases frequencies up to 120 Hz and higher can be obtained for smaller machines.

ELECTRODYNAMIC VIBRATION MACHINE

GENERAL DESCRIPTION

A complete electrodynamic vibration test system is comprised of an electrodynamic vibration machine, electrical power equipment which drives the vibration machine, and electrical controls and vibration monitoring equipment.

The electrodynamic vibration machine derives its name from the method of force generation. The force which causes motion of the table is produced electrodynamically by the interaction between a current flow in the armature coil and the intense magnetic dc field which passes through the coil, as illustrated in Fig. 25.5. The table is structurally attached to a force-generating coil which is concentrically located (with radial clearances) in the annular air gap of the dc magnet circuit. The assembly of the armature coil and the table is usually referred to as the *driver coil-table* or *armature*. The magnetic circuit is made from soft iron which also forms the *body* of the vibration machine. The body is magnetically energized, usually by two dc field coils as shown in Fig. 25.5*C*, generating a radially directed field in the air gap, which is perpendicular to the direction of current flow in the armature coil. Alternatively, in small shakers, the magnetic field is generated by permanent magnets. The generated force in the armature coil is in the direction of the axis of the coil, perpendicular to the table surface. The direction of the force is also perpendicular to the armature-current direction and to the air-gap field direction.

The table and armature coil assembly is supported by elastic means from the machine body, permitting rectilinear motion of the table perpendicular to its surface, corresponding in direction to the axis of the armature coil. Motion of the table in all other directions is resisted by stiff restraints. Table motion results when an ac current passes through the armature coil. The body of the machine is usually supported by a

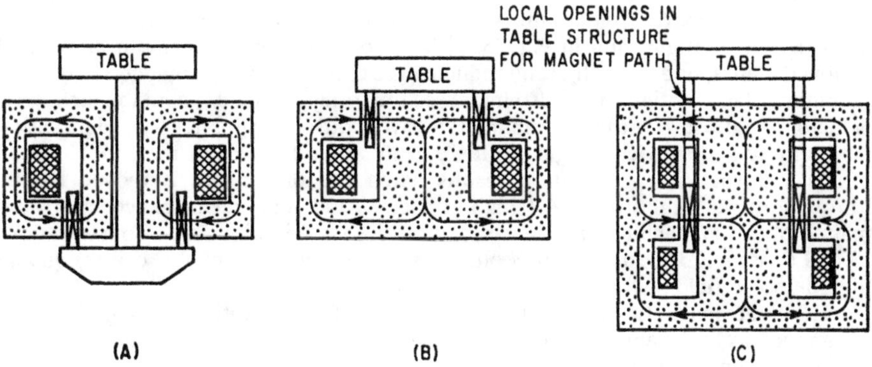

FIGURE 25.5 Three main magnet circuit configurations.

base with a trunnion shaft centerline passing horizontally through the center of gravity of the body assembly, permitting the body to be rotated about its center, thereby giving a vertical or horizontal orientation to the machine table. The base usually includes an elastic support of the body, providing vibration isolation between the body and the supporting floor.

Where a very small magnetic field is required at the vibration machine table due to the effect of the magnetic field on the item under test, *degaussing* may be provided. Magnetic fields of 5 to 30 gauss several inches above the table are normal for modern machines with double-ended, center air-gap magnet designs, Fig. 25.5*C*, without degaussing accessories; in contrast, with degaussing accessories, magnetic fields of 2 to 5 gauss can be achieved.

Because of copper and iron losses in the electrodynamic unit, provision must be made to carry off the dissipated heat. Cooling by convection air currents, compressed air, or a motor-driven blower is used and, in some cases, a recirculating fluid is used in conjunction with a heat exchanger. Fluid cooling is particularly useful under extremes of hot or cold environments or altitude conditions where little air pressure is available.

MAGNET CIRCUIT CONFIGURATIONS

Three magnet circuit configurations that are used in the electrodynamic machines are shown schematically in Fig. 25.5. In Fig. 25.5*A*, the table and driver coil are located at opposite ends of the magnet circuit. The advantage of this configuration is that the location of the annular air gap, the region of high magnetic leakage flux, is spaced from the table and the body itself acts as a magnetic shield, resulting in lower magnetic flux density at the table. The disadvantage lies in the loss of rigidity in the connecting structure between the driver coil and the table because of its length. This configuration is usually cooled by convection air currents or by forced air from a motor-driven blower.

In Fig. 25.5*B*, the table is connected directly to the driver coil. This eliminates the length of structure passing through the magnet structure, thereby increasing the rigidity of the driver coil-table assembly and allowing higher operating frequencies. The leakage magnetic field in the vicinity of the table is high in this configuration. It is therefore difficult, if not impossible, to reduce the leakage to acceptable levels without adding extra length to the driver coil assembly, elevating the table above the air gap. The configuration in Fig. 25.5*C* has a complete magnet circuit above and below the annular air gap, thereby reducing the external leakage magnetic field to a minimum. This configuration also increases the total magnetic flux in the air gap by a factor of almost 2 for the same diameter driver coil, giving greater force generation and a more symmetrical magnetic flux density along the axis of the coil. Hence, a more uniform force generation results when the driver coil is moved axially throughout its total stroke. All high-efficiency and high-performance electrodynamic vibration machines use the configuration shown in Fig. 25.5*C*. Configurations *B* and *C* of Fig. 25.5 may use air cooling throughout or an air-cooled driver coil and liquid-cooled field coil(s) or total liquid cooling.

The main magnetic circuit uses dc field coils for generating the high-intensity magnetic flux in the annular gap in all of the larger and most of the smaller units. Permanent magnet excitation is used in small portable units and in some general-purpose units up to about 500-lb (2-kN) generated force.

INDUCTION-TYPE SHAKER

In the induction-type electrodynamic shaker, a stator coil is fixed in the shaker body (see Fig. 25.6). The varying current from the power source is passed trough the stator coil. The armature coil is a cylinder of conductive material (usually aluminum). The stator current is coupled inductively to the armature coil. The stator coil (many turns) acts as the primary in a transformer. The armature coil (a single shorted turn) acts as the secondary in the transformer. The stator current inductively generates a current in the single-turn shorted armature coil. In Fig. 25.6, the dc magnetic field is across the paper, the armature current is into the paper, and the generated force is vertical. The advantages are a rugged armature design, and an armature that is electrically isolated from the rest of the shaker. The disadvantages include a decrease in performance at low frequencies due to inductive coupling losses and a slight problem cooling the armature. Because the induction losses are a function of scale, this design is usually found in the larger electrodynamic shakers.

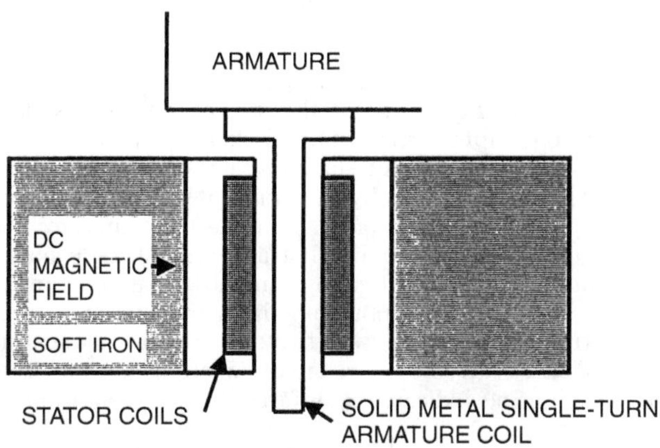

FIGURE 25.6 Cross section in the vicinity of the armature of an induction-type shaker.

FREQUENCY RESPONSE CONSIDERATIONS

Testing procedures (see Chap. 17) can call for the reproduction of sinusoidal, random, or transient vibration environments. The frequency response function (the ratio of the output motion—acceleration, velocity, or displacement—to the shaker input voltage) of an electrodynamic shaker is a complex (amplitude and phase) function of frequency. Therefore, a control system is required to perform tests to a given specification, as detailed in Chap. 26.

CHARACTERIZATION OF AN ELECTRODYNAMIC SHAKER AS A TWO-PORT NETWORK

An electrodynamic shaker can be modeled as a mixed electrical/mechanical two-port network[1,2] (see Chap. 9). This characterization can give good insight about the performance capabilities of a shaker and/or a shaker/power supply combination. In matrix form, this characterization can be written as

$$\begin{Bmatrix} E \\ A \end{Bmatrix} = \begin{bmatrix} Z_{11} & Z_{12} \\ Z_{21} & Z_{22} \end{bmatrix} \begin{Bmatrix} I \\ F \end{Bmatrix} \tag{25.1}$$

where E = the voltage required to drive the shaker
I = the current required to drive the shaker
A = the acceleration observed at the shaker/load interface
F = the force at the shaker/load interface

All the variables are complex functions of frequency as described in Chap. 19. The terms in the impedance matrix are frequency response functions defined as

$$\begin{aligned} Z_{11} &= \frac{E}{I}\bigg|_{F=0} \quad Z_{12} = \frac{E}{F}\bigg|_{I=0} \\ Z_{21} &= \frac{A}{I}\bigg|_{F=0} \quad Z_{22} = \frac{A}{F}\bigg|_{I=0} \end{aligned} \tag{25.2}$$

Two of the terms are easily measured. Z_{11} is the unloaded table (no mechanical load on the shaker) electrical impedance of the shaker, and Z_{21} is the ratio of the unloaded acceleration to input current of the shaker. Z_{22} is the accelerance (ratio of acceleration to force) looking into the shaker with the shaker electrical input open (zero current, but with the field on). Z_{12} is the ratio of voltage, generated at the open electrical shaker input, to a driving force applied at the armature. The direct measurement of Z_{12} and Z_{22} would require that an external force be applied to the shaker and the resulting open circuit voltage and acceleration be measured, a difficult feat in practice. But the terms in the impedance matrix can be measured experimentally by performing experiments with two or more known loads attached to the shaker. The general case is given by a system of equations for n measured load conditions, where the subscripts indicate the different loading conditions.

$$\begin{bmatrix} E_1 & E_2 & \cdots & E_n \\ A_1 & A_2 & \cdots & A_n \end{bmatrix} = \begin{bmatrix} Z_{11} & Z_{12} \\ Z_{21} & Z_{22} \end{bmatrix} \begin{bmatrix} I_1 & I_2 & \cdots & I_n \\ F_1 & F_2 & \cdots & F_n \end{bmatrix} \tag{25.3}$$

Each test requires the measurement of the input voltage and current and the output acceleration and force. If the test item is a rigid mass, the force can be estimated from $F = ma$. In shorthand, Eq. (25.3) will be written as

$$\mathbf{E} = \mathbf{ZI} \tag{25.4}$$

The impedance matrix can then be found using a Moore-Penrose pseudoinverse[3]

$$\mathbf{Z} = \mathbf{EI}^{-1} \tag{25.5}$$

If the number of test conditions is greater than two, the solution is in a least-squares sense. This assumes the inverse exists. The equation is typically solved at a finite set of discrete frequencies using techniques described in Chap. 19. Other forms of the impedance matrix can be defined which give frequency response functions that may be more useful in a particular application. The admittance matrix is defined as

$$\begin{Bmatrix} I \\ F \end{Bmatrix} = \begin{bmatrix} Y_{11} & Y_{12} \\ Y_{21} & Y_{22} \end{bmatrix} \begin{Bmatrix} E \\ A \end{Bmatrix} \tag{25.6}$$

The transmission matrix is defined as

$$\begin{Bmatrix} E \\ I \end{Bmatrix} = \begin{bmatrix} T_{11} & T_{12} \\ T_{21} & T_{22} \end{bmatrix} \begin{Bmatrix} A \\ F \end{Bmatrix} \tag{25.7}$$

The reciprocal transmission matrix is defined as

$$\begin{Bmatrix} A \\ F \end{Bmatrix} = \begin{bmatrix} R_{11} & R_{12} \\ R_{21} & R_{22} \end{bmatrix} \begin{Bmatrix} E \\ I \end{Bmatrix} \tag{25.8}$$

These matrices are all related by the equations

$$\begin{gathered} \mathbf{Y} = \mathbf{Z}^{-1} \qquad \mathbf{R} = \mathbf{T}^{-1} \\ \mathbf{T} = \frac{1}{Z_{21}} \begin{bmatrix} Z_{11} & Z_{12}Z_{21} - Z_{11}Z_{22} \\ 1 & -Z_{22} \end{bmatrix} \\ \mathbf{R} = \frac{1}{Z_{12}} \begin{bmatrix} Z_{22} & Z_{12}Z_{21} - Z_{11}Z_{22} \\ 1 & -Z_{11} \end{bmatrix} \end{gathered} \tag{25.9}$$

For example, for a sine test, the voltage and current required for a particular load acceleration are easily determined by substituting

$$F = Z_m A \tag{25.10}$$

into Eq. (25.7) to give

$$\begin{Bmatrix} E \\ I \end{Bmatrix} = A \begin{bmatrix} T_{11} & T_{12} \\ T_{21} & T_{22} \end{bmatrix} \begin{Bmatrix} 1 \\ Z_m \end{Bmatrix} \tag{25.11}$$

Z_m is the driving point (the interface at the shaker) free effective mass[4] (the ratio of force to acceleration) of the load (test item and fixtures). The free effective mass is related to the mechanical impedance, Z (the ratio of force to velocity), defined in Chap. 9, by the relationship $Z_m = j\omega Z$. In general, Z_m is a frequency response function. If the load and fixtures are a rigid mass, Z_m is a constant equal to the mass of the test item and fixtures.

Similarly, for a given shaker power supply with known characteristics (the maximum output voltage and current capability), the shaker performance capabilities (the achievable acceleration) for a given load are easily determined from Eq. (25.11). The maximum acceleration that can be achieved for a given voltage limit is

$$A_{E\text{lim}} = |E_{\text{lim}}/(T_{11} + T_{12}Z_m)|$$

The maximum acceleration that can be achieved for a given current limit is

$$A_{I\text{lim}} = |I_{\text{lim}}/(T_{21} + T_{22}Z_m)|$$

The maximum acceleration that can be reached before either limit is reached is the smaller of these two numbers.

$$A_{\max} = \min(A_{E\lim}, A_{I\lim})$$

The development is easily generalized for random and transient testing using the techniques in Chap. 19. The development can be generalized for the multiple shaker system driving a single test item.[5]

A useful review of electrodynamic shakers is given in Ref. 6.

SYSTEM RATINGS

The electrodynamic vibration machine system is rated: (1) in terms of the *peak value* of the sinusoidal generated force for *sinusoidal* vibration testing and (2) in terms of the *rms* and *instantaneous* values of the maximum force generated under *random* vibration testing. In order to determine the acceleration rating of the system with a test load on the vibration table, the weight of the test load, assumed to be effective at all frequencies, must be known and used in the following expressions:

$$\mathfrak{g} = \frac{F}{W_L + W_T}$$
$$\mathfrak{g}_{\text{rms}} = \frac{F_{\text{rms}}}{W_L + W_T} \tag{25.12}$$

where $\mathfrak{g} = a/g$, a dimensionless number expressing the ratio of the peak sinusoidal acceleration to the acceleration due to gravity (i.e., the peak sinuosidal acceleration in g's)

$\mathfrak{g}_{\text{rms}} = a_{\text{rms}}/g$, a number expressing the ratio of the rms value of random acceleration to the acceleration due to gravity

W_L = weight of load

W_T = equivalent weight of table driver-coil assembly and associated moving parts

F = rated peak value of sinusoidal generated force

F_{rms} = rated rms value of random generated force

The *force rating* of an electrodynamic vibration machine is the value of force which can be used to calculate attainable accelerations for any rigid-mass table load equal to (or greater than) the driver coil weight. It is not necessarily the force generated by the driver coil. These two forces are identical only if the operating frequencies are sufficiently below the axial resonance frequency of the armature assembly, where it acts as a rigid body. As the axial resonance frequency is approached, a mechanical magnification of the force generated electrically by the driver coil results. The design of the driving power supply takes into account the possible reduction in driver-coil current at frequencies approaching the armature axial resonance frequency, since full current in this range cannot be used without exceeding the rated value of transmitted force at the table, possibly causing structural damage.

In those cases where the test load dissipates energy mechanically, the system performance should be analyzed for each specific load since normal ratings are based on a dead-mass, nondissipative type of load. This consideration is particularly significant in resonance-type fatigue tests at high stress levels.

PROMINENT FEATURES

- A wide range of operating frequencies is possible, with a properly selected electric power source, from 0 to above 30,000 Hz. Small, special-purpose machines have been made with the first axial/resonance mode above 26,000 Hz, giving inherently a resonance-free, flat response to 10,000 Hz.
- Frequency and displacement amplitude are easily controlled by adjusting the power-supply frequency and voltage.
- Pure sinusoidal table motion can be generated at all frequencies and amplitudes. Inherently, the table acceleration is the result of a generated force proportional to the driving current. If the electric power supply generates pure sinusoidal voltages and currents, the waveform of the acceleration of the table will be sinusoidal, and background noise will not be present. Operation with table acceleration waveform distortion of less than 10 percent through a displacement range of 10,000-to-1 is common, even in the largest machines. Velocity and displacement waveforms obtained by the single and double integration of acceleration, respectively, will have even less distortion.
- Random vibration, as well as sinusoidal vibration, or a combination of both, can be generated by supplying an appropriate input voltage.
- A unit occupying a small volume, and powered from a remote source, can be used to generate small vibratory forces. A properly designed unit adds little mass at the point of attachment and can have high mobility without mechanical damping.
- Leakage magnetic flux is present around the main magnet circuit. This leakage flux can be minimized by proper design and the use of degaussing coil techniques.

SPECIFICATIONS

Design Factors

Force Output. The maximum vector-force output for sinusoidal excitation shall be given for continuous duty and may additionally be given for intermittent duty. When nonsinusoidal motions are involved, the force may additionally be given in terms of a root-mean-square (rms) value together with a maximum instantaneous value. The latter value is especially significant when a random type of excitation is required.

In some cases of wide-frequency-band operation of the electrodynamic vibration machine, the upper frequencies are sufficiently near the axial mechanical resonance frequency of the coil-table assembly to provide some amplification of the generated force. Most system designs account for this magnification, when present, by reducing the capacity of the electrical driving power accordingly.

The peak values of the input electrical signal, for random excitation, may extend to indefinitely large values. In order that the armature coil voltage and generated force may be limited to reasonable values, the peak values of the excitation are clipped so that no maxima shall exceed a given multiple of the rms value. The magnitude of the maximum clipped output shall be specified preferably as a multiple of the rms value. If adjustments are possible, the range of magnitudes shall be given.

Weight of Vibrating Assembly. The weight of the vibration coil-table assembly shall be given. It shall include all parts which move with the table and an appropriate percentage of the weight of those parts connecting the moving and stationary parts giving an effective overall weight.

Vibration Direction. The directions of vibration shall be specified with respect to the surface of the vibration table and with respect to the horizontal or vertical direction. Provisions for changing the direction of vibration shall be stated.

Unsupported Load. The maximum allowable weight of a load not requiring external supports shall be given for horizontal and vertical orientations of the vibration table. This load in no way relates to dynamic performance but is a design limitation, the basis of which may be stated by the manufacturer.

Static Moments and Torques. Static moments and torques may be applied to the coil-table assembly of a vibration machine by the tightening of bolts and by the overhang of the center of gravity of an unsupported load during horizontal vibration. The maximum permissible values of these moments and torques shall be specified. These loads in no way relate to the dynamic performance but are design limitations, the basis for which may be stated by the manufacturer.

Total Excursion Limit. The maximum table motion between mechanical stops shall be given together with the maximum vibrational excursion permissible with no load and with maximum load supportable by the table.

Acceleration Limit. The maximum allowable table acceleration shall be given. (These large maxima may be involved in the drive of resonant systems.)

Stiffness of Coil-Table Assembly Suspension System

AXIAL STIFFNESS: The stiffness of the suspension system for axial deflections of the coil-table assembly shall be given in terms of pounds per inch of deflection. The natural frequency of the unloaded vibrating assembly may also be given. Provisions, if any, to adjust the table position to compensate for position changes caused by different loads shall be described.

SUSPENSION RESONANCES: Resonances of the suspension system should be described together with means for their adjustment where applicable.

Axial Coil-Table Resonance. The resonance frequency of the lowest axial mode of vibration of the coil-table assembly shall be given for no load and for an added deadweight load equal to 1 and to 3 times the coil-table assembly weight. If this resonance frequency is not obvious from measurements of the table amplitude versus frequency, it may be taken to be approximately equal to the lowest frequency, above the rigid-body resonance of the table-coil assembly on its suspension system, at which the phase difference between the armature coil current and the acceleration of the center of the table is 90°.

Impedance Characteristics. When an exciter or vibration machine is considered independent of its power supply, information concerning the electrical impedance characteristics of the machine shall be given in sufficient detail to permit matching of the power-supply output to the vibration-machine input. It is suggested that consideration be given to providing schematic circuit diagrams (electrical and mechanical or equivalent electrical) together with corresponding equations that contain the principal features of the machine.

Environmental Extremes. When it is anticipated that the vibration machine will be used under conditions of abnormal pressure and temperature, the following information shall be supplied as may be applicable: maximum simulated altitude (or minimum pressure) under which full performance ratings can be applied; maximum simulated altitude under which reduced performance ratings can be applied; maximum ambient temperature for rated output; low-temperature limitations; humidity limitations.

Performance. The performance relates in part to the combined operation of the vibration generator and its power supply.

Amplitude-Frequency Relations. Data on sinusoidal operation shall be given as a series of curves for several table loads, including zero load, and for a load at least 3 times the weight of the coil-table assembly. Maximum loads corresponding to $20g$ and $10g$ table acceleration under full-rated force output would be preferred. These curves should give amplitudes of table displacement, velocity, or acceleration, whichever is limiting, throughout the complete range of operating frequencies corresponding to maximum continuous ratings of the system. Additionally, the maximum rated force should be given. If this force is frequency-dependent, it should be presented as a curve with the ordinate representing the force and the abscissa the frequency.

If the system is for broadband use, necessarily employing an electronic power amplifier, the exciting voltage signal applied to the input of the system shall be held constant and the output acceleration shall be plotted as a function of frequency with and without filters or other compensating devices for the loads and accelerations indicated above. If the vibrator is used only for sinusoidal vibrations, and employs servo amplitude control, the curves should be obtained under automatic frequency sweeping conditions with the control system included.

Waveform. Total rms distortion of the acceleration waveform at the center of the vibration table, or at the center on top of the added test weight, shall be furnished to show at least the frequencies of worst waveform under the test conditions specified under the above paragraph. The pickup type, and frequency range, shall be given together with the frequency range of associated equipment. It is desirable to have the overall frequency range at least 10 times the frequency of the fundamental being recorded. Tabular data on harmonic analysis may alternatively or additionally be given.

Magnetic Fields. The maximum values of constant and alternating magnetic fields, due to the vibration exciter, in the region over the surface of the vibration table should be indicated. If degaussing coils are furnished, these values should be given with and without the use of the degaussing coils.

Frequency Range. The overall frequency range shall be given. A group of frequency ranges shall also be given for electronic power supplies if they require changes of their output impedance for the different ranges.

Frequency Drift. The probable drift of a set frequency shall be stated, together with factors that contribute to the drift. This shall apply for nonresonant loads.

Signal Generator. A vibration pickup, if built into the vibration machine, shall have calibrations furnished over a specified frequency and amplitude range.

Installation Requirements. Recommendations shall be given as to suitable methods for installing the vibration machine and auxiliary equipment. Electrical and other miscellaneous requirements shall be stated.

HYDRAULIC VIBRATION MACHINE

The *hydraulic vibration machine* is a device which transforms power in the form of a high-pressure flow of fluid from a pump to a reciprocating motion of the table of the vibration machine. A schematic diagram of a typical machine is shown in Fig. 25.7. In this example, a two-stage electrohydraulic valve is used to deliver high-pressure fluid, first to one side of the piston in the actuator and then to the other side, forcing the actuator to move with a reciprocating motion. This valve consists of a pilot stage and power stage, the former being driven with a reciprocating motion by the electrodynamic driver. At the time the actuator moves under the force of high-pressure

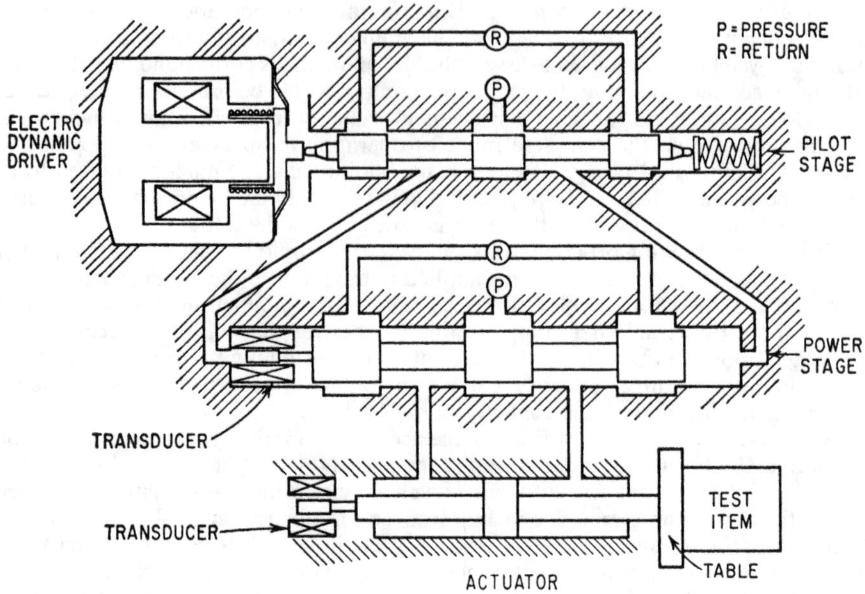

FIGURE 25.7 Schematic diagram of a typical hydraulic vibration machine.

fluid on one side of the piston, the fluid on the other side of the piston is forced back through the valve at reduced pressure and is returned to the pump.

The electrohydraulic valve is usually mounted directly on the side of the actuator cylinder, forming a close-coupled assembly of massive steel parts. The close proximity of the valve and cylinder is desirable to reduce the volume and length of the connecting fluid paths between the several spools and the actuator, thereby minimizing the effects of the compliance of the fluid and the friction to its flow. Many types of electrohydraulic valves exist, all of which fail to meet the requirement of sufficient flow at high frequencies to give vibration machine performance equivalent to existing electrodynamic machine performance at 2000 Hz.

OPERATING PRINCIPLE

In Fig. 25.7, the *pilot* and *power spools* of a hydraulic vibration machine are shown in the "middle" or "balanced" position, blocking both the pump high-pressure flow *P* and the return low-pressure flow *R*. Correspondingly, the piston of the actuator must be stationary since there can be no fluid flow either to or from the actuator cylinder. If the pilot spool is displaced to the right of center by a force from the electrodynamic driver, then high-pressure fluid *P* will flow through the passage from the pilot spool to the left end of the power spool, causing it to move to the right also. This movement forces the trapped fluid from the right-hand end of the power spool through the connecting passage, back to the pilot stage, and then through the opening caused by the displacement of the pilot spool to the right, to the chamber *R* connected to the return to the pump. Correspondingly, if the pilot spool moves to the left, the flow to and from the power spool is reversed, causing it to move to the left. For a given displacement of the pilot spool, a flow results which causes a corresponding velocity of the power spool. A displacement of the power spool to the right allows the flow of high-pressure fluid *P* from the pump to the left side of the piston in the actuator, causing it to move to the right and forcing the trapped fluid on the right of the piston to be expelled through the connecting passage to the power spool and out past the right-hand restrictions to the return fluid chamber *R*. The transducers shown on the power spool and the actuator shaft are of the differential transformer type and are used in the feedback circuit to improve system operation and provide electrical control of the average (i.e., stationary) position of the actuator shaft relative to the actuator cylinder.

A block diagram of the complete hydraulic vibration machine system is shown in Fig. 25.8. The pump, in conjunction with accumulators in the pressure and return

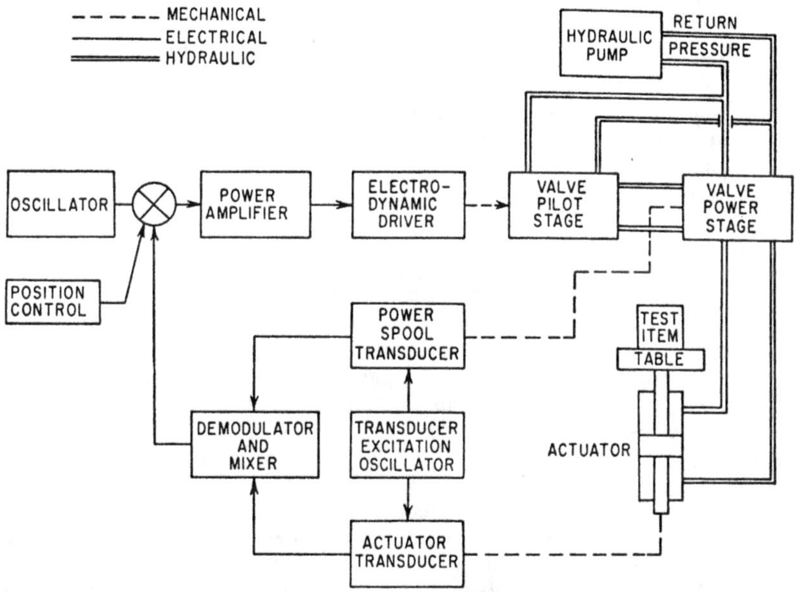

FIGURE 25.8 Block diagram—hydraulic vibration machine system.

lines at the hydraulic valve, should be capable of variable flow while maintaining a fixed pressure. Most systems to date have required an operating pump pressure of 3000 lb/in.^{2} (20 MPa). The upper limit of efficiency of the hydraulic valve is approximately 60 percent, the losses being dissipated in the form of heat. Mechanical loads are seldom capable of dissipating appreciable power; most of the power in the pump discharge is converted to a temperature rise in the fluid. Therefore a heat exchanger limiting the fluid temperature must be included as part of the system.

PROMINENT FEATURES

- Large generated forces or large strokes can be provided relatively easily. Large forces and large velocities of motion, made possible with a large stroke, determine the power capacity of the system. For example, one hydraulic vibration machine has a peak output power of 450,000 lb-in./sec (approximately 34 hp or 25 kW) with a single electrohydraulic valve. This power can be increased by the installation of several valves on a single actuator. Appreciable increases in valve flow can be realized by sacrificing high-frequency performance. Hence, the hydraulic vibration machine excels at low frequencies where large force, stroke, and power capacity are required.

- The hydraulic machine is small in weight, relative to the forces attainable; therefore, a rigid connection to firm ground or a large massive base is necessary to anchor the machine in place and to attenuate the vibration transmitted to the surrounding area.
- The main power source is hydraulic, which is essentially dc in character from available pumps. The electrical driving power for controlling the valve is small. Therefore, the operating frequency range can be extended down to 0 Hz.
- The magnetic leakage flux in the region of the table is insignificant by comparison with the electrodynamic-type vibration machine.
- The machine, with little modification, is suitable for use in high- and low-temperature, -humidity, and -altitude environments.
- The machine is inherently nonlinear with amplitude in terms of electrical input and output flow or velocity.

PIEZOELECTRIC VIBRATION EXCITERS

A piezoelectric material (see Chap. 10) can be used to generate motion and act as a *piezoelectric vibration exciter.* Typically, a piezoelectric exciter employs a number of disks of piezoelectric material, as illustrated in Fig. 25.9; this arrangement increases the ratio of the displacement output to voltage input sensitivity of the exciter. The strain is proportional to the charge, and the charge is increased by increasing the voltage gradients across the piezoelectric material. The voltage gradient is increased by using many thin layers of piezoelectric material, separated with a conducting material, with alternating polarity on the conducting separators. This arrangement of alternating layers of piezoelectric material and conducting material is called a *piezoelectric stack.* Because the piezoelectric stack has little tensile strength, the stack must be preloaded. The stiffness of the preloading mechanism must be much less than the stiffness of the piezoelectric stack so that preloading will not influence the mechanical output significantly. The combination of the piezoelectric stack (acting like a displacement actuator) and a reaction mass forms a reaction-type vibration exciter as described above. The reaction mass of the piezoelectric exciter can be the armature mass of a small electrodynamic exciter. This effectively places an electrodynamic and a piezoelectric exciter in series, producing a machine with a usable output over a wide frequency range.

PROMINENT FEATURES

- The exciters can have a usable frequency range from 0 to 60 kHz.
- The low-frequency output is severely limited by the displacement limits of the piezoelectric stack, usually a few thousandths of an inch (a few hundredths of a millimeter).

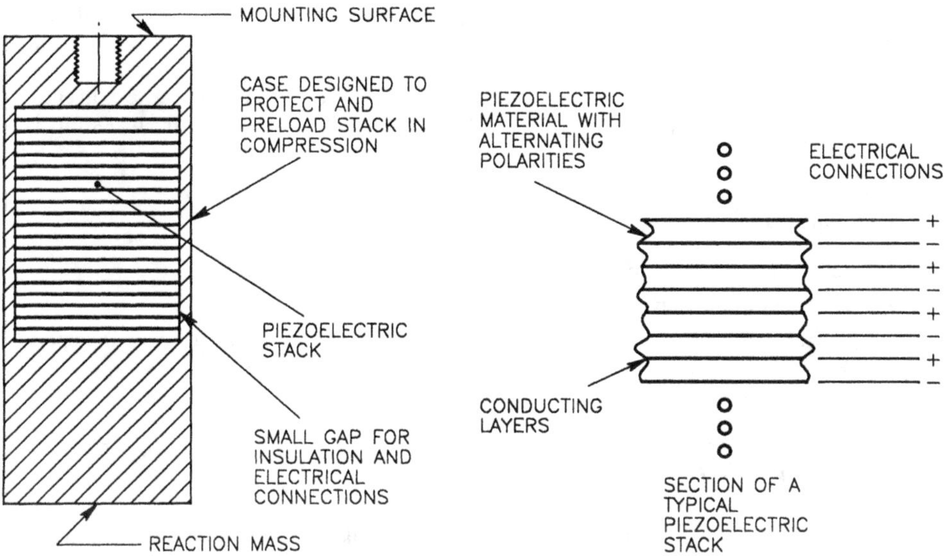

FIGURE 25.9 Simplified cross section of a piezoelectric vibration exciter. A compressed piezoelectric stack is excited with an oscillating voltage. An electrical voltage applied to the electrical connections causes the piezoelectric stack to elongate and contract, producing a relative displacement between the mounting surface and the reaction mass. The inertia of the reaction mass results in a force being applied to an item mounted on the mounting surface.

- The high-frequency output is limited by internal resonances of the vibration exciter.
- The force output of the exciter is limited by the displacement limit of the piezoelectric stack and by the mass of the reaction mass.
- The power supply for a piezoelectric exciter requires high voltages (typically about 1000 V) and sufficient current to drive the capacitance (typically 10 to 1000 nF) of the device.

IMPACT EXCITERS

A limited amount of vibration testing, such as some modal testing and some stress screening, require a broad frequency bandwidth of relatively uncontrolled vibration. A class of exciters broadly known as *impact exciters* (and also called *repetitive shock machines*) is sometimes used for the above applications. These devices depend on the property that a short impact generates a broad bandwidth of vibration energy. Each impact is a short transient (for example, see Chap. 27), but repeated impacts result in a quasi-steady-state vibration having a wide frequency bandwidth. If the impacts are periodic, the spectrum is composed of the fundamental frequency of the impacts and many harmonics of this fundamental frequency; i.e., the excitation is essentially a periodic function. However, the impacts are often varied randomly in magnitude and spacing to produce a time-averaged spectrum that is smoother, much like random vibration. Nevertheless, the instantaneous spectrum or Wigner distribution (see Chap. 19) for the excitation will still reveal an instantaneous periodic function with a time-varying magnitude and fundamental frequency. The probability distribution can vary significantly from a gaussian distribution. The vibration characteristics are strongly influenced by the dynamics of the structure on which they are mounted. The impact exciters can be mounted directly to the test specimen, or the exciters can excite a table on which the test item is mounted. The latter can be classed as a vibration testing machine.

PROMINENT FEATURES

- The design is usually simple, compact, and rugged.
- The maximum attainable displacement is usually small.
- The vibration is relatively uncontrolled. The user has little control over the spectrum of the resulting vibration.

MULTIPLE SHAKERS DRIVING A SINGLE TEST ITEM

It is sometimes desirable to have more than one shaker driving a test item. Some of the reasons include:

Desire to excite many modes. This is the motivation for multiple-input modal tests. A single input may not be capable of exciting all the modes, but multiple-input tests have a better chance.

Desire to provide more representative boundary conditions. Many test items are not mounted in service on rigid foundations. Single-axis testing on rigid fixtures is often a poor simulation of the boundary conditions of service environments. Multiple input tests can sometimes provide more realistic boundary conditions. The vibration input in the field environment is often not through a single point.

Large test items. Large test items are difficult to drive with a single shaker. Examples include complete airplanes or space launch systems, seismic simulations, automobiles, and other large transportation systems. The size and/or force requirements to test these items are often beyond the capabilities of a single shaker.

Desire to provide excitation in more than one direction. Most conventional shakers excite the test item in one rectilinear direction. Most environments include vibration in several directions (both rectilinear and rotation) simultaneously. In an effort to provide more realistic testing, shaker systems with inputs in several directions at the same time are desirable.

Multiple exciters driving a single test item have been used extensively in modal testing (see Chap. 21). This is relatively easy because control of the vibration input is not usually necessary. Multiple-input tests with controlled inputs are more difficult because of cross-coupling effects. Cross-coupling is where the input at one point causes response at the control point of another input. Control of systems with cross-coupling requires a careful mechanical design and a carefully designed control system (see Chap. 26). The shaker, the fixture, and the control system form three legs of a triad. They must all work together; a weakness in any of the three can result in the system failure. The mechanical design must minimize cross-coupling effects and the control system must compensate for the remaining cross-coupling.

Systems with two inputs typically controlling one translation and one rotation degree of freedom are not very difficult to design. An example would be a horizontal beam-like structure with the vertical translation controlled independently at each end. Isolation of the rotation from the shakers can usually be accomplished with fixtures that are stiff axially but soft in bending.

The mechanical design of systems with more than two degrees of freedom is more difficult. The shaker providing the input can usually move in only one direction. If the test item is to move in more than one direction and/or rotate, the mechanical design of the system must isolate all the motion except in one direction from the shakers. It is also difficult to restrain other degrees of freedom (DOF), for example, rotations. Restraint of unwanted motion is usually accomplished with passive restraints (for example, hydrostatic bearings) or with active restraints using the exciters and the control system. Undesired motion, compromising the test, will result if the uncontrolled degrees of freedom are not restrained.

A system using three electrodynamic shakers controlling three orthogonal translations, with the three rotations passively restrained, has been built.[7] This system has a usable bandwidth of almost 2 kHz. Electrodynamic systems with six DOF have also been built with varying degrees of success. Electrohydraulic shaker systems with six rigid-body DOF (three translations and three rotations) have been built.[8] These systems have a usable bandwidth of about 500 Hz. Larger electrohydraulic systems with two to six DOF have been built for seismic simulation with a bandwidth of about 50 Hz (see Chap. 29). Other electrohydraulic systems with as many as 18 hydraulic actuators with a bandwidth of about 50 Hz are used as road simulators in the automotive industry. One of these systems is illustrated in Fig. 25.10. An advantage of electrohydraulic shakers for multiple-input applica-

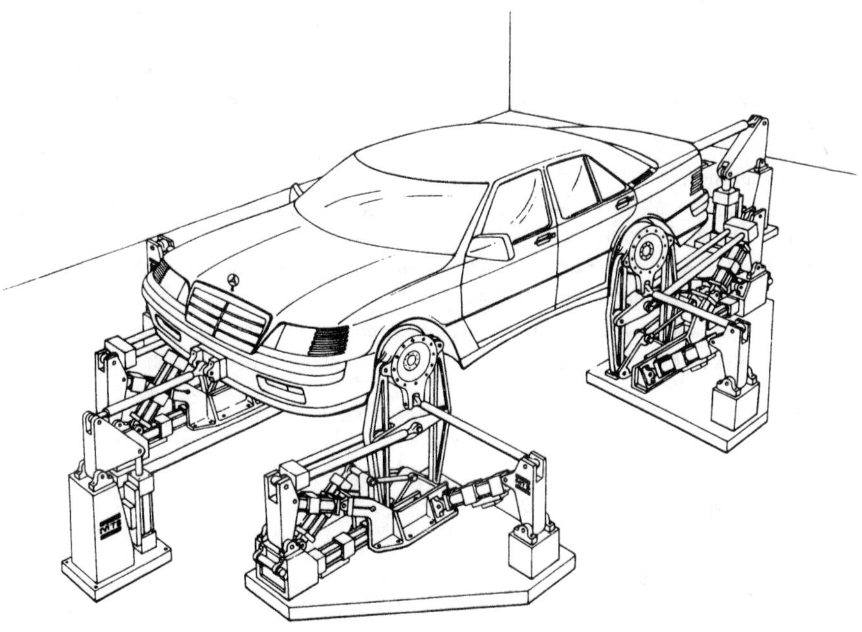

FIGURE 25.10 A road simulator which uses a cross-coupled multiple-drive/multiple-control-point predetermined waveform control system. The predetermined waveforms (with a bandwidth of about 1 to 50 Hz) are measured on the vehicle while driving on a road. The predetermined waveforms are reproduced on the vehicle during the simulation on the road simulator. Four hydraulic actuators drive each wheel hub, and two hydraulic actuators drive the vehicle fore and aft at the bumpers. (*MTS Corp.*)

tions is that their mechanical input impedance is relatively high, reducing the cross-coupling effects. Their disadvantage is that they are all inherently nonlinear, which makes control more difficult. All of these systems, both electrodynamic and electrohydraulic, are capable, with appropriate control systems, of performing sine, random, and transient tests.

VIBRATION FIXTURES

Test items are usually attached to a shaker with a fixture. Seldom will the test item mount directly on the shaker. These fixtures are usually designed to be rigid in the frequency band of interest and lightweight. Rigidity is required because the vibration test is typically controlled at a single point. The assumption is that the motion of the control point is representative of the input to the test item. If the fixture is not rigid, this assumption is obviously not true. Also, flexible fixtures typically have one or more frequencies where the operating shape at the control point is near zero. This will result in large, unrealistic responses of the test item. The fixtures need to be lightweight to maximize the force available to drive the test item. Light weight and rigidity are contradictory requirements. Design of satisfactory vibration fixtures is a combination of experience, analysis, and compromise. Vibration fixtures are discussed in Chap. 18.

REFERENCES

1. Baher, H.: "Synthesis of Electrical Networks," John Wiley & Sons, New York, 1984.
2. Weinberg, L.: "Network Analysis and Synthesis," McGraw-Hill Book Company, New York, 1962.
3. Golub, G. H., and C. F. Van Loan: "Matrix Computations," 2d ed., Johns Hopkins University Press, Baltimore, Md., 1989.
4. "Vibration and Shock—Experimental Determination of Mechanical Mobility. Part 1: Basic Definitions and Transducers," ISO 7626-1, 1986.
5. Smallwood, D. O.: *J. of the Institute of Environmental Sciences,* **60**(5):27 (1997).
6. Lang, G. F.: *Sound and Vibration,* **31**(4):14 (1997).
7. Stroud, R. C., and G. A. Hamma: *Sound and Vibration,* **22**(4):18 (1988).
8. Hamma, G. A., R. C. Stroud, M. A. Underwood, W. B. Woyski, R. C. Tauscher, and K. L. Cappel: *Sound and Vibration,* **30**(4):20 (1996).
9. Nelson, Curt, "Trends in Vibration Testing," *Proc. ESTECH 2007,* Institute of Environmental Sciences, May 2007.

CHAPTER 26
DIGITAL CONTROL SYSTEMS FOR VIBRATION TESTING MACHINES

Marcos A. Underwood

INTRODUCTION

This chapter discusses digital vibration control applications that require (1) the synthesis and output of excitation (driving) signals for electrodynamic and electrohydraulic exciters (shakers), (2) the acquisition of the structural responses that result from the exciters' effects on a common structure, (3) the digital analysis of these responses to determine important structural characteristics, and (4) the use of these derived structural characteristics for the feedback control of the structure's input or of its response motion during a vibration test.

Because the processing needs of digital vibration control systems (DVCSs) are extensive, the use of specialized processors, computer architectures, and input and output signal processing are often necessary to reach the required processing levels. Because of this, a brief review of these specialized processors, their associated architectures, and how they are used in modern vibration control systems are discussed in the following sections.

The use of analog-to-digital converters (ADCs) and digital-to-analog converters (DACs) is also briefly discussed, particularly how they are used within the response data acquisition and signal generation subsystems to excite structures and measure the associated responses during the digital control of vibration tests.

The analysis and synthesis applications associated with vibration control are also discussed. These include multidimensional spectral analysis, digital tracking filters, shock response spectrum (SRS) synthesis, and frequency response estimation.

The chapter then discusses digital control systems for shock and vibration testing of structures using single and multiple exciters. Sections on random, swept-sine, mixed-mode, and time-domain waveform replication are presented. The chapter concludes with a brief section on modal testing using multiexciter digital control systems to cause the input to a structure to agree with predetermined force vectors to optimize the measurement and analysis of their associated responses.

SPECIALIZED PROCESSORS AND PERIPHERALS

Specialized processors are designed for a particular activity or type of calculation that is being performed. They consist of embedded, distributed, and digital signal processors. These types of processors typically provide the most performance for shock and vibration control applications, but at a higher level of complexity than that associated with the use of general-purpose computers. Included in this category are *specialized peripherals* such as analog-to-digital converters and digital-to-analog converters that provide the fundamental interfaces between computer systems and physical systems such as transducers and exciters, which are used for many shock and vibration testing and analysis applications. *Specialized processor architectures* are also used extensively in shock and vibration control applications, since they provide the necessary power and structure to accomplish some of the more demanding applications such as the control of single or multiple vibration test exciters or applications that involve the measurement and analysis of many response channels from a shock and vibration test.

SPECIALIZED PROCESSORS AND ARCHITECTURES

Embedded Processors. *Embedded processors* are computer systems that don't interact directly with the users and are used to accomplish a specialized application. This type of system is part of a larger system where the embedded portion serves as an intelligent peripheral for a general-purpose computer host such as a workstation- or personal computer–based system. The embedded subsystem is used to perform time-critical functions that are not suitable for a general-purpose system due to limitations in its operating systems. The operating system used for embedded processors is typically a *real-time operating system* (RTOS), which is optimized for real-time response to minimize the communication latencies between subsystems. It is typically dedicated to the signal synthesis, signal acquisition, and processing tasks. The embedded system typically communicates with the host processor through a high-speed interface such as wired or wireless Ethernet, USB, Firewire, or a direct communication between the memory busses of embedded and host computer systems. An embedded computer system does not interface directly with the computer system user but uses the host computer system for this purpose. An example of an embedded system which uses distributed processors is shown in Fig. 26.1. Here, the

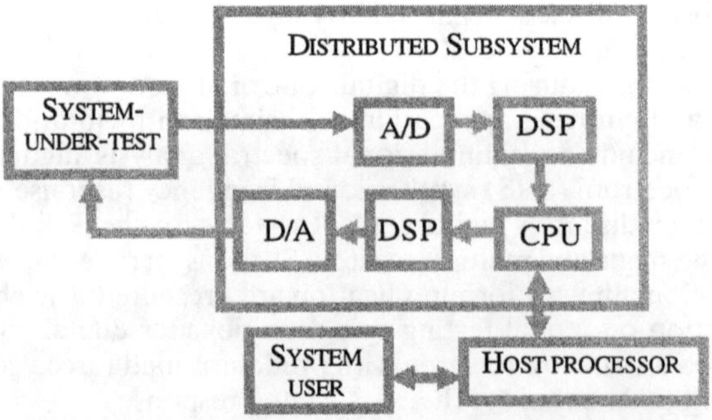

FIGURE 26.1 Specialized processor architecture system.

host computer is used to set the parameters for the particular activity—for example, shock and vibration control and analysis—and the embedded computer subsystem accomplishes the control and analysis task directly. This frees the host processor to simply receive the results of the shock and vibration task and to provide the graphical user interface (GUI) and create associated graphic displays for system users.

Distributed Computer Systems. *Distributed computer systems* employ a processor architecture to accomplish their task by using several computer processor systems in tandem to solve a problem that cannot be suitably solved by an individual computer or processor system. This type of computer system typically partitions its task in such a way that each part can be executed in parallel by its respective processor. This enables the use of several specialized processors to separately accomplish a demanding subtask, and thus the overall shock and vibration task, in a way that might not be possible with the use of a single, general-purpose computer system.

Specialized Processor Architecture. An example of a *specialized processor architecture* is shown in Fig. 26.1. It is a distributed and embedded computer system that uses digital signal processors to process data being received from an analog-to-digital converter by filtering it and extracting the pertinent signal characteristics needed as part of a shock and vibration test. This filtered data and its extracted characteristics are subsequently sent to a more general processor to perform additional analysis. The results of this more general analysis may yield a time-series data stream that is sent to another digital signal processor for filtering, and then sent to an output digital-to-analog converter to produce signals that are used to excite a system under test. Figure 26.1 also shows, in the form of a block diagram, a typical form and application of a specialized processor architecture used within a shock and vibration control system. The distributed system's central processing unit (CPU) coordinates the communications between and with the two digital signal processor (DSP) subsystems with an RTOS. The host processing system is used to provide the GUI for the overall system's user.

Digital Signal Processors. *Digital signal processors* are specialized processors that are optimized for the multiply-accumulate operations that are used in digital filtering– and linear algebra–related processing. They are used extensively in shock and vibration signal analysis and vibration control systems. These processors are ideal to implement digital filters, for sample-rate reduction and alias protection (see Chap. 14), fast Fourier transform (FFT)–based algorithms (see Chap. 14), and digital control systems. Linear algebra problems, such as those encountered in signal estimation, filtering, and prediction, are also performed efficiently by this processor architecture.[1,2] The example of a specialized architecture system shown in Fig. 26.1 also shows a typical application of DSP technology. This processor architecture has empowered many of the audio and video signal-processing systems in current use. It has also enabled many of the shock and vibration experimental applications now in use.

ADCS AND DACS FOR SIGNAL SAMPLING AND GENERATION

Digital signal processors have also enabled *analog-to-digital converters* and *digital-to-analog converters,* which are fundamental to the applications of digital processing, to be used in the field of shock and vibration. They provide a fundamental interface between the analog nature of shock and vibration phenomena and the digital processing available from modern computing systems. These important subsystems are now realized by single *integrated circuits* (ICs), often incorporating most of the filtering needed for anti-aliasing (see Chap. 14) for ADCs and anti-imaging for DACs, particularly in those using a sigma-delta (ΣΔ) technology.[3] In practice, even when using ΣΔ technology, additional analog circuitry is needed to complete the anti-aliasing and anti-imaging function,[3] and also to add needed signal amplification and conditioning to more fully utilize the resolution of modern ADCs and DACs. Please consult Chap. 13 for a more detailed discussion of this subject.

ADCs and Data Preparation. ADCs furnish the analog-to-digital conversion function, which is the process by which an analog (continuous) signal is converted into a series of numerical values with a given binary digit (bit) resolution (see Chaps. 13 and 19). This is the first step in any digital method. The ADC operation is generally built into self-contained digital analysis systems that use the ADC subsystem as a peripheral. The host processor is used to set up the ADC hardware's data acquisition parameters such as the sampling rate, input voltage range, frequency range, input data block size (duration of the signal to be digitized), and the number of data blocks to be digitized. The acquired data is then subsequently analyzed in real time as the test progresses. Examples of ADC applications are shown in Figs. 26.1 and 26.2.

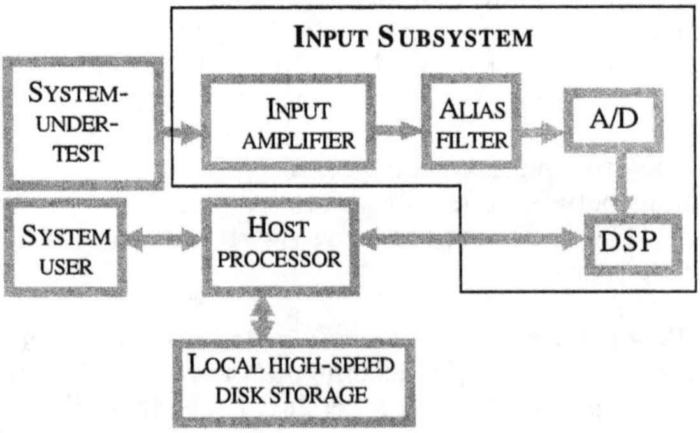

FIGURE 26.2 Typical A/D converter–based input subsystem.

In Fig. 26.2, the input amplifier is used to maximize the ADC's resolution by amplifying the response signal from the system under test, typically as part of an autoranging process with the control software. The amplified signal is filtered to remove high-frequency energy in the input signal that could be aliased (see Chap. 14) and then is passed to the ADC for digitization. The digital time series that the ADC produces is then sent to a digital signal processor for additional filtering and perhaps sample-rate reduction or other needed specialized processing, before it is sent to the host processor. For each input channel, the combination of (1) the input amplifier, (2) the anti-aliasing filter, (3) the ADC, and (4) the DSP is called the *input subsystem* and is used by digital vibration control systems, to be discussed later. Sigma-delta ADCs reduce the complexity of the analog design of the input subsystem, but there are some caveats to their use in DVCSs, because they use an internal digital filter.

The digital filter causes delay effects in ΣΔ ADCs that can result in stability problems when used with DVCSs. This is due to the digital filter's *group delay,*[3,4] which is typically on the order of 30 samples. This delay can cause closed-loop stability problems if not addressed properly within the processing software of the digital control system.

DACs and Signal Synthesis. As discussed previously, DACs convert a digital time series into an analog signal. This analog signal will have a "staircase" or zero-order hold nature.[5] This occurs because the DAC output signal is held constant for an output sample-rate period and then is changed according to the next digital sample at the next sample-clock period. This staircase nature of the output DAC signal causes its analog output signal spectrum to have high-frequency terms, in addition to those present in its digital time-series spectrum, with their frequency content centered about the DAC's sample-rate frequency (both below the sample rate and above the sample rate) and its integer multiples.[5] These somewhat symmetrical spectral lobes that appear in the DAC output signal spectrum, which are centered at the sample-rate frequency and its harmonics, are called *signal images.*[5] These spectral lobes have a bandwidth double that of the bandwidth of the digital time series that is being sent to the DAC.[5] The spectrum of these signal images has a sin $(x)/x$ envelope that is due to the zero-order hold nature of the DAC.[5] They are the counterpart to aliasing that occurs with ADC sampling (see Chap. 19). These signal images should be removed before using the DAC output signal to excite a system under test. For this reason and others, the output subsystem should be organized as shown in Fig. 26.3.

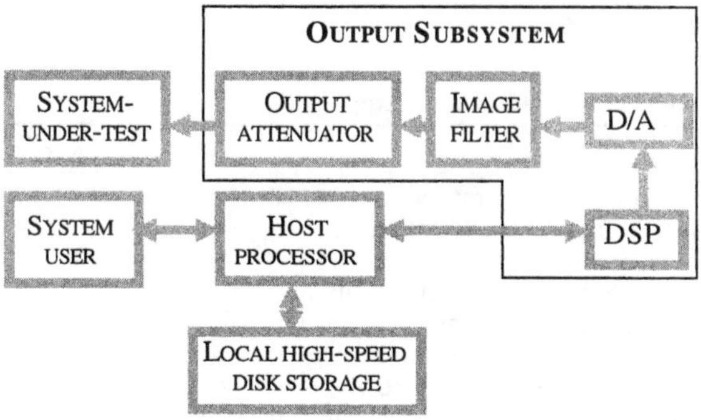

FIGURE 26.3 Typical D/A converter–based output subsystem.

In Fig. 26.3, the signal flow is the reverse of that for the ADC-based input subsystem, as shown in Figs. 26.1 and 26.2. In Fig. 26.3, the output signal flows from the host processor into the digital signal processor in the output subsystem and, optionally, into a local high-speed disk storage subsystem. The digital signal processor performs some filtering and perhaps increases the sample rate to minimize the impact of output signal images, moving them higher in frequency and lower in amplitude. This filtered and processed output time series is then sent to the DAC, where it is converted into an analog voltage. This voltage is then sent to an analog anti-imaging filter to remove any signal images that might still be present in the DAC output voltage, usually near the DAC's sampling frequency and its harmonics. This filtered signal is then passed to the output attenuator subsystem to set the amplitude of the resulting output signal. The attenuator is used to maximize the DAC output resolution. Typically, additional output filtering is provided by the analog circuitry that is part of the attenuator. Digital vibration control systems typically use the output subsystem architecture similar to that shown in Fig. 26.3.

Sigma delta (ΣΔ) DACs are also used for shock and vibration control applications. They use an internal signal flow that is the reverse of that for a ΣΔ ADC.[3] These use internal digital filters for output interpolation to increase the sampling rate from the system sampling rate to an oversampling rate. If this digital filter is of the finite impulse response (FIR) type, it can also cause group delay effects like those discussed for ΣΔ ADCs, which need to be considered by the digital control algorithms to avoid the associated stability problems.

SHOCK AND VIBRATION DATA ANALYSIS

The basic principles of *shock and vibration data analysis*[6] are thoroughly covered in other chapters and their references, as summarized in Table 26.1. Only methods that are fundamental to digital vibration control systems that are not presented elsewhere are discussed here. Specifically, this section discusses (1) the definition of the estimates of the spectral density and cross-spectral density matrices; (2) tracking filters for the measurement of the amplitude and phase, as a function of frequency, of response and control data taken during a swept-sine vibration test; (3) the synthesis of transient signals that achieve a predetermined shock response spectrum (see Chap. 20); and (4) frequency response estimation.

TABLE 26.1 Summary of Data Analysis Applications

Application	Chapter
Spectral analysis for stationary vibration data	14, 19, and 24
Spectral analysis for nonstationary vibration data	19
Correlation analysis for stationary vibration data	24
Probability analysis for stationary vibration data	19 and 24
Fourier and shock response spectral analysis of shock data	20
Modal analysis of structural systems from shock and vibration data	21
Multiple-input/output analysis of shock and vibration data	21
Average values and tolerance limits for shock and vibration data	18
Other statistical analysis of shock and vibration data	19
Matrix methods of analysis for shock and vibration data	22

SPECTRAL DENSITY MATRIX

The *spectral density matrix* (SDM) is a matrix that consists of both power spectral density values for its diagonal elements and cross-spectral density values for its off-diagonal elements. It is the natural extension to matrices of the concepts of power spectral density and cross-spectral density that are discussed in Chap. 19. An SDM is both a hermitian and a nonnegative definite matrix.[7–12] It can be estimated as follows.

Let $\{x(t)\}$ be an N-dimensional column-vector of time histories, whose components are the waveforms $x_1(t),\ldots, x_N(t)$, which are assumed to be samples of a gaussian stationary process.[7–9] These waveforms could, for example, be the acceleration responses of a system under test, at N measurement points, that is being excited by N vibration exciters with the use of N stationary gaussian drive signals that are partially correlated (see Chap. 19). If we define their complex finite Fourier transform, as in Eq. (19.3), with $x(t)$ successively replaced by the $x_i(t)$ waveforms, the complex vector $\{X(f,T)\}$ is obtained, with the finite Fourier transforms $X_1(f,T),\ldots, X_N(f,T)$ as its components. If the time history vector $\{x(t)\}$ has a duration much longer than T, then, as in Chap. 19, it can be partitioned into a series of nonoverlapping segments of data (often called *frames*), each of duration T, such that the average can be defined as

$$[W_{XX}(f,T)] = \frac{2}{n_d T} \sum_{i=1}^{n_d} \begin{Bmatrix} X_1(f,T) \\ X_2(f,T) \\ \vdots \\ X_N(f,T) \end{Bmatrix}_i \{X_1^*(f,T)\ X_2^*(f,T) \cdots X_N^*(f,T)\}_i \tag{26.1}$$

or, using a more compact matrix notation, as

$$[W_{XX}(f,T)] = \frac{2}{n_d T} \sum_{i=1}^{n_d} \{X(f,T)\}_i \{X(f,T)\}_i^H \tag{26.2}$$

In Eqs. (26.1) and (26.2), (1) the average is taken as in Table 19.3, where the estimates for the power and cross-spectra are defined using a finite Fourier transform; (2) $X_1^*(f,T)$ is the complex conjugate of $X_1(f,T)$; (3) $\{X(f,T)\}_i^H$ is the complex-conjugate transpose of the vector $\{X(f,T)\}_i$; and (4) the subscript i refers to the ith nonoverlapping frame. As shown in Refs. 7–9, this average is an unbiased estimator for the SDM of the N-dimensional gaussian stationary process $\{X(t)\}$, which converges to the true SDM of the process $\{x(t)\}$, as T and n_d approach infinity. The use of windowing[7–9] in the definition of the $X_i(f,T)$ that is used in Eqs. (26.1), (26.2), and (26.3) reduces the errors associated with spectral leakage (see Chap. 14).

CROSS-SPECTRAL DENSITY MATRIX

The *cross-spectral density matrix* (CSDM) is a matrix that consists of cross-spectral densities between the components of two multidimensional gaussian stationary random processes. It is defined similarly to the previously discussed spectral density matrix. It is the natural extension of the cross-spectral density concepts that are discussed in Chap. 19. The CSDM is further discussed in the Refs. 7–11. For simplicity, and without loss of generality, in the following discussion the CSDM estimate is defined for the case where the two random process vectors have the same dimension.

Let $\{x(t)\}$ and $\{y(t)\}$ be two N-dimensional column-vectors of time histories, which respectively consist of the waveforms $x_1(t),\ldots, x_N(t)$ and $y_1(t),\ldots, y_N(t)$, where we assume that $\{x(t)\}$ and $\{y(t)\}$ consist of waveforms that are samples of gaussian stationary processes. The $\{x(t)\}$ waveform vector can, for example, be the vector of random drive signals that are used to excite the system under test, as in Fig. 26.6. The $\{y(t)\}$ waveform vector in this case will be the vector of responses at the N instrumented points located on a system under test that is being excited by N-exciters with the use of the drive vector $\{x(t)\}$. If the finite Fourier transform vectors $\{X(f,T)\}$ and $\{Y(f,T)\}$ are similarly defined with components $X_1(f,T),\ldots, X_N(f,T)$ and $Y_1(f,T),\ldots,$ $Y_N(f,T)$, it is found that the average cross-spectrum can be defined as

$$[W_{YX}(f,T)] = \frac{2}{n_d T} \sum_{i=1}^{n_d} \{Y(f,T)\}_i \{X(f,T)\}_i^H \tag{26.3}$$

where this average is taken as in Eqs. (26.1) and (26.2) but with the use of the vector $\{Y(f,T)\}_i$ instead of the vector $\{X(f,T)\}_i$ for the ith nonoverlapping frame. As in the SDM estimator in Eqs. (26.1) and (26.2) and as shown in Refs. 7–9, this average is an unbiased estimator for the CSDM between the N-dimensional gaussian stationary processes $\{x(t)\}$ and $\{y(t)\}$, which converges to the true CSDM as T and n_d approach infinity. There are also convergence results for fixed T when $\{x(t)\}$ and $\{y(t)\}$ are ergodic and, with the use of a window function, as n_d approaches infinity for Eqs. (26.1) through (26.3).[7]

TRACKING FILTERS

Tracking filters are specialized filters that implement a narrow bandpass filter, of selectable bandwidth, centered about the instantaneous frequency of a sine wave with a frequency that is changing with time (commonly called a *sweeping* sine wave).[13] These filters are used to extract the amplitude of the sweeping response sine wave, as well as its phase with respect to the modulating signal used in the tracking filter implementation. This algorithm, based on proprietary technologies, provides essentially a time-varying estimate of the Fourier spectral amplitude, in essentially a continuous manner, of a sweeping sine wave.[13]

A simplified implementation of a tracking filter is shown in Fig. 26.4. It accepts a sweeping sine-wave response from a system under test that is being excited by a sweeping sine wave. This response signal is shown as $A \sin(\omega t + \theta) + n(t)$, with a frequency of ω radians/sec, an amplitude A, a phase of θ with respect to the modulating signals $\sin(\omega t)$ and $\cos(\omega t)$, and an additive distortion and noise term $n(t)$. By modulating the input signal with the sine and cosine terms shown in Fig. 26.4, the energy at the sweep frequency ω is translated to 0 Hz—hence, the name 0-Hz IF detector, where the data detection[13] is accomplished by the two low-pass filters that produce the imaginary- and real-term estimates of the complex amplitude of the sweeping sine-wave response of the system under test. From these filter outputs, the amplitude A and the phase θ, with respect to the modulating signal, are estimated. By analyzing several response signals in this manner, with separate tracking filters that use the same modulating signals, the relative phase between several sweeping sine-wave responses can be measured, since their individual phase measurements have a common phase reference. In this way, tracking filters can be used for such diverse appli-

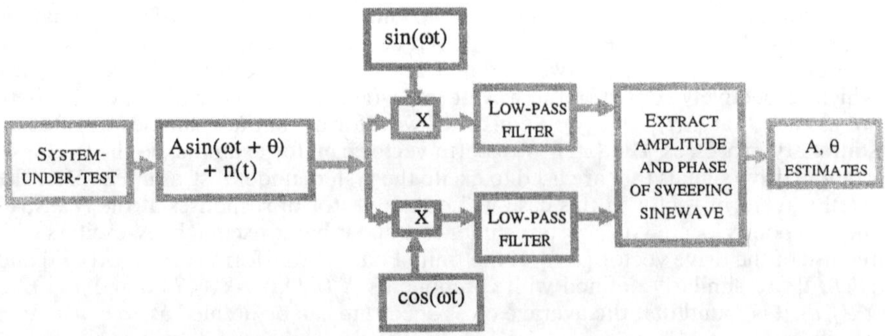

FIGURE 26.4 Tracking filter using 0-Hz intermediate frequency (IF) detector.

cations as frequency response function and matrix estimation and multiexciter and single-exciter swept-sine-wave control.

The tracking filter operation shown in Fig. 26.4 provides an estimate of the complex amplitude at the modulating signal's frequency, which is typically the same as the swept-sine wave's frequency. It is important that the modulating signals and the drive signals used to excite the system under test be in frequency and phase synchronization for the best results. Because it can track a sweeping sine wave, it provides a way of measuring the nonstationary spectral amplitudes associated with swept-sine-wave tests and rotating machinery vibration analysis. By its nature, it discards other terms not centered at the sweep frequency, such as unwanted harmonic and nonharmonic distortion terms. Tracking filters can also be used to track frequencies other than the fundamental response frequency, such as the frequencies of harmonics. Some modern digital vibration control systems provide the function of that shown in Fig. 26.4 by using dedicated digital signal processors to implement a digital tracking filter subsystem. These can provide an estimate of a sweeping sine wave's amplitude and phase during a vibration test at the drive signal's frequency. Some implementations provide estimates as many as four to eight times per cycle of the drive signal.[13]

SHOCK RESPONSE SPECTRUM TRANSIENT/SHOCK SYNTHESIS

Signal synthesis techniques are used in transient testing where the test's reference response is specified as a shock response spectrum, as discussed later in this chapter. This type of application is often referred to as *shock synthesis* or *SRS synthesis.* The primary goal is to create or synthesize a transient signal such that its SRS agrees with a predetermined SRS. Since the same SRS is possible for a large range of signals (see Chaps. 20, 27, and 28), many such synthesis techniques are possible. Some are based on wavelet expansions[14,15] for pyroshock testing, and others on a transient created by windowing a stationary random signal (see Chap. 29).

The methods used for pyroshock testing are based on the use of a weighted sum of *wavelets,* which are defined as a set of nearly orthogonal functions with finite durations. The wavelets used for shock synthesis are either windowed sine waves with an odd number of half-cycles or damped sinusoids,[14,15] which are called real Laplace wavelets in the literature.[16] Either type of wavelet is used to construct a group of component wavelets that are defined for each frequency at which the SRS is specified, with a particular delay, and where the Laplace wavelet also has damping as a parameter. These are used as part of an inverse wavelet transform process to synthesize transients as the sum[14–16] of the chosen component wavelets for each such frequency, delay, and damping, if appropriate. The amplitude of each such wavelet is modified until the sum of such wavelets is a transient whose SRS agrees with the prescribed SRS within an acceptable error bound for each of its specified frequencies.[14,15] Since the SRS definition (see Chap. 20) allows for many waveforms to have the same SRS, this many-to-one relationship allows for the further optimization of such synthesized transients.[15,17,18] They can be optimized, for example, to produce the least peak acceleration of such transients for a given peak SRS by suitably modifying the time delays of the wavelets used for its synthesis.[14,15] This type of optimization can increase the peak amplitude of the shock response spectra that are possible with a particular overall test system (see Fig. 26.5), thus extending the performance range of vibration test machines used for transient/shock testing.

The method employed for seismic simulation, which is used to synthesize artifi-

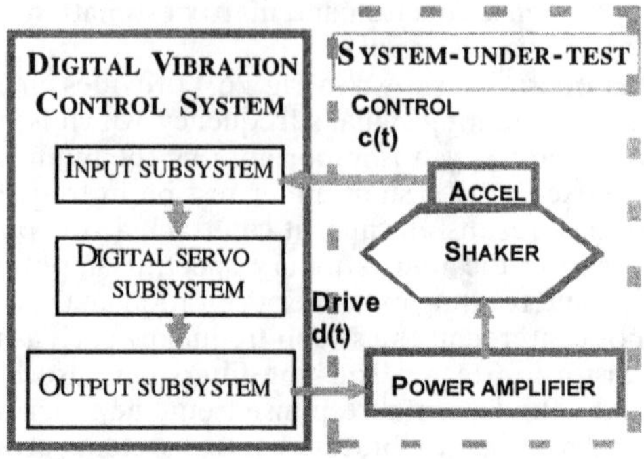

FIGURE 26.5 Overall test system.

cial earthquake motions, involves windowed sections of broadband gaussian stationary noise, also known as *burst-random* transients. These random transients are generated using a prescribed magnitude Fourier spectrum, assigning random phase to it, and using the inverse fast Fourier transform to create a random transient with the specified magnitude spectrum. This transient is windowed (see Chap. 19) and its SRS is calculated. This calculated SRS is compared with the prescribed SRS, and the discrepancy is used to modify the magnitude of its Fourier spectrum. The synthesis iteration is repeated until the SRS of the synthesized windowed transient agrees with the prescribed SRS within some acceptable error. Again, the many-to-one characteristic of the SRS allows for further optimization of the synthesized random transient.

FREQUENCY RESPONSE FUNCTION AND FREQUENCY RESPONSE MATRIX MEASUREMENTS

The computation of frequency response functions and frequency response matrices makes use of the digital signal processor, analog-to-digital converter, digital-to-analog converter, embedded, and distributed computer systems discussed previously. The objective is to excite the system under test in such a way that its frequency response characteristics can be measured. This type of measurement is done as part of the modal testing, single-exciter, and multiexciter control system applications to be discussed later in this chapter.

Single-Input/Multiple-Output (SIMO) Methods. In this method, a single drive signal is used to excite the system under test at any one time. A digital system, like those shown in Figs. 26.1, 26.2, and 26.3, can be used to drive a system under test and acquire multiple response signals from instrumentation on the system under test. The excitation signals can be impulsive, continuous broadband noise, transient noise, or swept sine waves. In all these cases, the complex-amplitude spectra are measured for both the drive and the response signals by the digital system. The cross-spectral densities between the various response signals and the drive signal, as measured at the input to the system under test, are divided by the drive signal's power spectral density to obtain a frequency response function estimate between the single drive signal and the response signals (see Table 19.3). Typically, broadband noise and swept-sine-wave excitations produce the best estimates for the needed frequency response functions, but at the expense of longer test times that may stress the test article or system under test. Frequency response functions (FRFs) can also be measured while using swept-sine-wave excitation by using the tracking filters discussed previously.

A multiple-reference frequency response matrix estimate can be obtained by exciting the system with a hammer or a vibration exciter, one excitation at a time but at different locations, to successively obtain one column of the frequency response matrix estimate using this SIMO methodology. These methods may have problems with repeatability, since the structure's characteristics may change between excitations (see Chap. 21).

Multiple-Input/Multiple-Output (MIMO) Methods. These methods excite the system under test with a digital system, as in the previous section, but drive it with multiple simultaneous excitation signals, acquire the associated response signals, and process the thus-acquired response and drive signals to obtain the needed system frequency response matrix estimates. Most estimators used are based on the following response equations:[7–12]

$$[W_{cd}(f)] = [H(f)][W_{dd}(f)] \qquad \text{or} \qquad [W_{cc}(f)] = [H(f)][W_{dc}(f)] \tag{26.4}$$

where $[W_{cd}(f)]$ is an estimate of the cross-spectral density matrix between the response vector $\{c(t)\}$ and the drive-signal vector $\{d(t)\}$, as defined in Eq. (26.3). $[W_{cc}(f)]$ and $[W_{dd}(f)]$ are estimates of the SDMs of the response vector $\{c(t)\}$ and the drive signal vector $\{d(t)\}$, as defined in Eqs. (26.1) and (26.2). $[W_{dc}(f)]$ is the complex-conjugate and matrix transpose of $[W_{cd}(f)]$.[7–12] The two equations that are part of Eq. (26.4) can be solved separately for $[H(f)]$. The left equation is relatively insensitive to measurement noise but sensitive to drive signal noise, and the right equation exhibits the reverse condition. These types of frequency response matrix estimates are very similar to the type 1 and type 2 frequency response estimators discussed in Chap. 21. Here the emphasis is on the use of Eq. (26.4) with the SDM and CSDM estimates, defined in Eqs. (26.1) through (26.3), used to estimate $[H(f)]$. The use of Eq. (26.4) for system identification is further discussed as part of the sections on multiexciter digital vibration control and modal testing.

Note that to use Eq. (26.4), either the matrix $[W_{dd}(f)]$ or $[W_{dc}(f)]$ needs to be inverted. For this reason, the left side of Eq. (26.4) is typically used because it is easier to guarantee that $[W_{dd}(f)]$ has a matrix inverse rather than $[W_{dc}(f)]$. In many cases, $[W_{dc}(f)]$ is a rectangular matrix because the dimensions of $\{c(t)\}$ and $\{d(t)\}$ are not always equal, and clearly $[W_{dc}(f)]$ would not have a matrix inverse in that case. Some digital systems make an additional simplification by exciting the system with mutually uncorrelated random drive signals and thus "ensure" that $[W_{dd}(f)]$ is a diagonal matrix. This simplification can cause additional problems, since the measured $[W_{dd}(f)]$ will typically not be diagonal even if the drive signals are "uncorrelated" due to unavoidable measurement and exciter noise. Hence, in practice, it is better to measure $[W_{dd}(f)]$ and invert it as a matrix rather than just inverting its diagonal elements and assuming that its matrix inverse is diagonal. This is the preferred way to characterize the system under test for multiexciter control applications, to be discussed later. In many of these cases, the drive signals are measured as inputs to the test article by load cells (see Chap. 10). The use of MIMO methods can separate modes that correspond to the same repeated root or eigenvalue (see Chap. 22), whereas SIMO methods might not (see Chap. 21).

CONTROL SYSTEMS FOR SHOCK AND VIBRATION TESTING

The vibratory motions prescribed for the majority of vibration tests are either sinusoidal[13,17] or random[17] (see Chap. 18). A smaller percentage of the vibration tests are prescribed to be either classical-shock transients[18] (see Chap. 27), shock response spectrum synthesized transients[14,15,19] (see Chap. 28), time-domain waveform replications,[20] or mixed-mode[21] (sine-on-random or narrowband-on-random) vibratory motions. These specified environments are typically represented by a reference response signal, in either the time or the frequency domain, that the digital control system servo uses as a control reference to achieve the specified control response at the chosen control point or points that are associated with the test (see Chap. 18).

The reference response is either a frequency-domain or a time-domain signal that represents the prescribed vibration environment associated with a shock or vibration test. It is typically specified as a reference spectrum, which describes the vibration environment in the frequency domain to which the control response spectrum is compared as part of the digital vibration control process. It could be a power spectral density for a random vibration test, an amplitude-versus-frequency profile for a swept-sine test, an SRS for a shock test, or a finite Fourier spectrum (see Chaps. 14, 18, and 19) for a generalized transient or a long-term reference-response waveform. Time-domain waveform replication vibration environments, such as transient and long-term response waveforms, are represented by a reference pulse or reference waveform, whereas frequency-domain-specified environments, such as random, swept-sine, and SRS synthesis shock tests, are specified with an appropriate reference spectrum. Typically, the time-domain waveform replication reference signals are also converted to the frequency domain as part of the feedback control and drive signal synthesis process, using an appropriate time-to-frequency and frequency-to-time transformation process.

Vibration tests are accomplished with the use of vibration test machines, as discussed in Chap. 25, and a *digital vibration control system.* The DVCS employed to control the vibration level(s) during the test typically utilizes the output signal from a control accelerometer mounted at an appropriate location on the vibration exciter's test fixture (part of the vibration test machine) or the unit under test (UUT) to provide a feedback signal to its servo system. The servo system in turn drives the power amplifier of the vibration testing machine used for the shock or vibration test. The servo system is largely implemented digitally using analog-to-digital converters, digital-to-analog converters, digital signal processors, embedded processors, and general-purpose processors to adjust the drive signal amplitude and spectrum for the system under test so as to maintain the control transducer's response level and waveform characteristics as close to the test's specified reference response as possible.

The overall block diagram of the vibration test system, when using electrodynamic exciters and accelerometers for control transducers, is shown in Fig. 26.5. In this case, the DVCS drives the system under test with an analog drive signal $d(t)$ such that the control response at the chosen control point location on the system under test agrees with the specified reference response with an acceptable error. The DVCS consists of (1) an input subsystem, which acquires the response waveform of the system under test $c(t)$; (2) the digital servo subsystem, which creates the digital drive signal through a closed-loop process that causes $c(t)$ to agree with a suitable description of the specified test reference signal; and (3) the output subsystem, which converts the digital description of the generated drive signal into an equivalent analog drive signal $d(t)$ used to drive the system under test.

A typical system under test configuration for both single and multiple exciters is shown in Fig. 26.6. If there is only one exciter involved, then only the top leg of the block diagram in Fig. 26.6 is used. Here, d_i means the drive signal generated by the DVCS that is used to drive the ith exciter. This drive signal is sent to the exciter's power amplifier (when using electrodynamic exciters), which in turn drives the exciter. For electrohydraulic exciters, this drive signal is sent to the exciter's servo amplifier, which in turn drives the hydraulic servo valve subsystem, as discussed in Chap. 25. The exciter, either electrohydraulic or electrodynamic, then drives a test fixture (see Chap. 25), which in turn drives the UUT. The test is instrumented by mounting control transducers, which are typically accelerometers (see Chap. 10) either on the test fixture, here shown by the signal c_1 through c_n, or on the UUT, as shown by the signals c_1 through c_n in Fig. 26.6. These chosen control signals are then sent to the input subsystem of the DVCS, where they are either averaged or their maximum or minimum, as a function of frequency, is extracted to create a composite response spectrum.

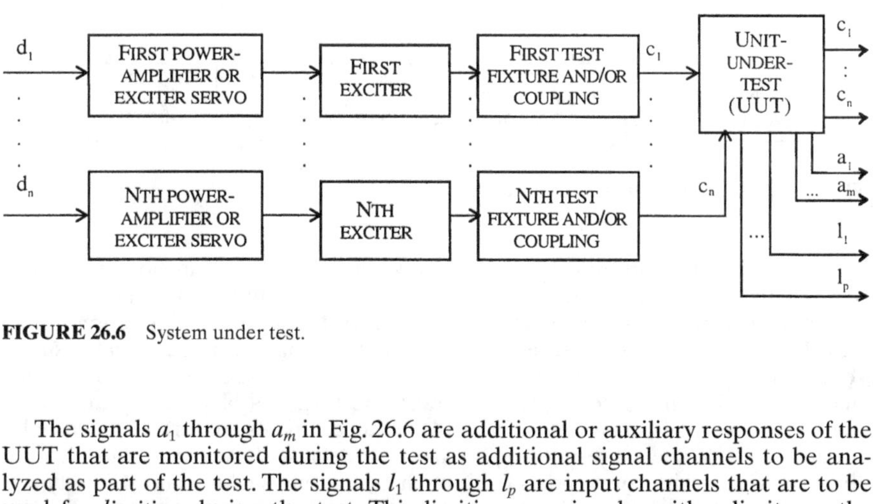

FIGURE 26.6 System under test.

The signals a_1 through a_m in Fig. 26.6 are additional or auxiliary responses of the UUT that are monitored during the test as additional signal channels to be analyzed as part of the test. The signals l_1 through l_p are input channels that are to be used for *limiting* during the test. This limiting may involve either limits on the response or limits on the applied force to the UUT, as discussed in Chap. 18. For multiexciter applications, there are n exciter systems with n drive signals d_1 through d_n. These drive signals are processed as in the single exciter case discussed previously. The basic difference is that the n-exciters will drive the UUT jointly through the fixture that connects the UUT to the multiple exciters. The response to this vector of drive signals is also a vector composed of the control responses c_1 through c_n. This test configuration and its associated control methods are further discussed in a subsequent section. In the single- or multiexciter control configuration, the control feedback signals, auxiliary response signals, and limit signals are routed to the input subsystem of the DVCS.

A block diagram of the input subsystem is shown in Fig. 26.7. Here only the control feedback signals are shown as inputs to the DVCS' input subsystem. These feedback signals, also called *control response channels,* or simply *control signals,* are each sensed through an input signal-conditioning system and an ADC subsystem. The input signal conditioning typically consists of an instrumentation ampli-

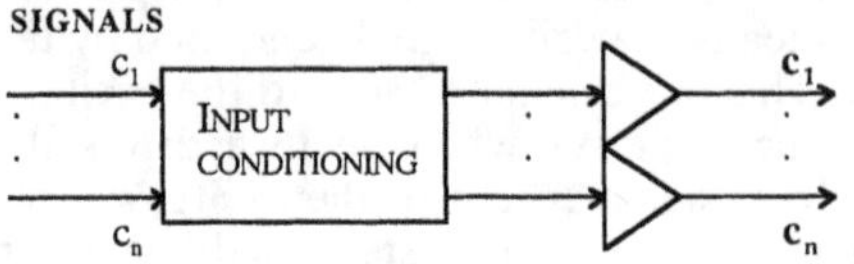

FIGURE 26.7 Input subsystem.

fier, followed by a ranging amplifier to optimize the signal's amplitude as presented to the ADC, and an anti-aliasing filter (see the input subsystem in Fig. 26.2). This conditioned analog signal, representing the chosen response signal, is finally presented to the ADC subsystem for conversion into a digital time history.

Typically, other points on the UUT or on the vibration test machine are also monitored by the digital control system for subsequent vibration analysis or limiting. The input subsystem then sends digitized versions of the control signals, here represented by c_1 through c_n, to the DVCS' servo subsystem, as shown in Fig. 26.8. The digital control response time series c_1 through c_n

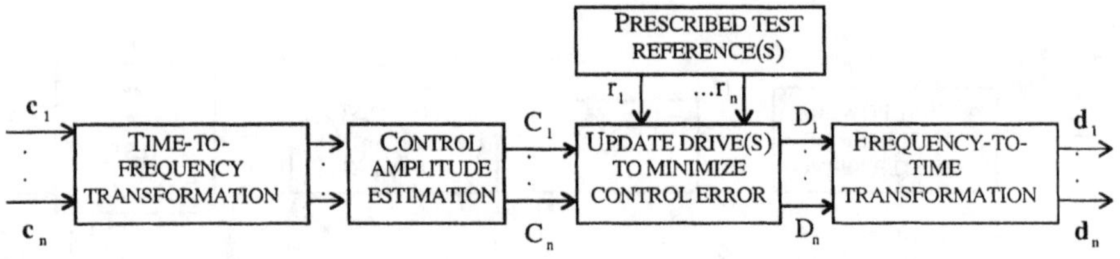

FIGURE 26.8 Digital vibration control system's servo subsystem.

is then sent to a time-to-frequency block shown in Fig. 26.8. The function of this block varies with the type of vibration control. For random vibration testing, this block estimates the control response power spectral density. For swept sine vibration testing, this block typically produces either the fundamental amplitude or the overall response root-mean-square (rms) estimate using tracking filters or variable time-constant rms detectors.[13,17] For the other types of vibration testing, this block is typically a fast Fourier transform estimator (see Chap. 14). These estimates are further processed to produce a single control response spectrum C_1 for single-shaker control or a control response vector with components C_1 through C_n for multishaker control. The type of processing is again application specific. These control response amplitude estimates are then sent to a block that updates the drive signal amplitude and spectrum to minimize the difference between these control response amplitudes and the specified test reference for single-shaker control or the test's reference-response vector for multishaker control applications. The updated drive amplitude(s) and their respective spectra are then sent to a frequency-to-time transformation block, which converts the spectral representation of the drive signal(s) into a digital time series of the time-domain drive that will be used to excite the system under test, as previously described. This digital time series signal or vector, composed of d_1 through d_n for multishaker control, is then sent to the output subsystem (see Figs. 26.3 and 26.9) for conversion into an analog signal or signals to be used to drive the previously discussed system under test in Fig. 26.6.

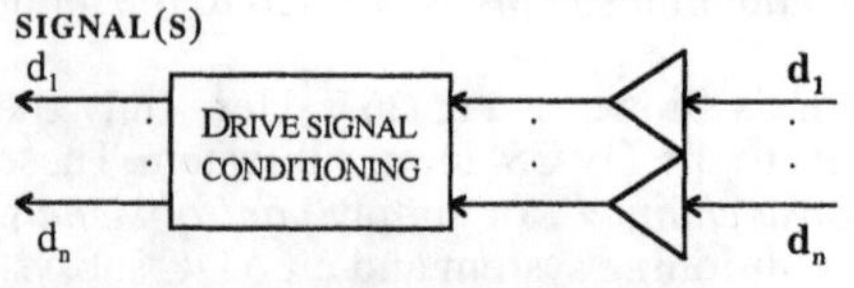

FIGURE 26.9 Output subsystem.

The output subsystem is shown in Fig. 26.9. The digital version of the single drive signal or multiple drive signals is synthesized to analog driving voltages by the system's output subsystem. These digital drive signals are then converted into analog signals by the subsystem's DACs. The DAC output signals are filtered to eliminate the images generated by the DACs, and the final output is attenuated from the DAC's full-scale voltage to produce the proper amplitude exciter drive signal d_1 for single-shaker control or drive signal vector for multiexciter control (see Fig. 26.3). These conditioned analog drive signals are output by the DVCS to drive the system under test.

Initially, with the advent of dedicated FFT processors and minicomputers, it became possible to perform spectral analysis of random processes rapidly enough to permit the use of digital control systems for random vibration testing. Further developments in digital signal processors, embedded and distributed processors, personal computers, and workstation technologies extended the range of vibration testing to include swept-sine, transient waveform, long-term waveform, and multishaker testing. Most shock and vibration testing remains based on single-shaker methods, but multishaker testing is becoming more important when the size and weight of the UUT dictates its need or when the prescribed vibratory motions are inherently multiaxis or otherwise consist of multi-degree-of-freedom (MDOF) vibratory motions.[20,22,23] Enough differences exist between single- and multishaker digital control systems for these to be discussed separately in the following sections. The previous discussion, however, illustrates the areas where they are similar.

SINGLE-EXCITER TESTING APPLICATIONS

The great majority of shock and vibration testing is specified and accomplished with the use of single exciters or shakers. These are typically single-axis tests. Multiaxis test specifications are accomplished one axis at a time when using single exciters. Random, swept-sine, mixed-mode, transient waveform, and time-domain waveform replication vibration applications can be accomplished as long as the vibration test machine capabilities and the weight and size of the unit under test allow it (see Chap. 25).

In many single-exciter vibration tests, especially random and swept-sine tests, even though only a single drive signal is employed, multiple control accelerometer input channels are used. In these cases, the multiple control signals are combined by averaging them or by selecting the largest or smallest response, as a function of frequency, to create a composite control response spectrum, with the control estimation block shown in Fig. 26.8. Often, multiple input channels are additionally used for limit control, as discussed earlier. The single-shaker control systems that use a single drive signal to excite the system under test, and use multiple-input control signals and/or limit signals, are called *multi-input/single-output* (MISO) control systems.

Random. These systems excite a test item with an approximation of a stationary gaussian random vibration (see Chap. 2). Digital random vibration control systems use signal processing that mimics analog methods in their fundamental control and measurement methods and offer significant user-interface and graphics subsystems that provide greater system tailoring and varied displays and graphs of ongoing test conditions. Digital systems also afford greater stability, more freedom in the control methods, and superior accuracy than those control systems that directly use analog methods.[17]

The control response waveforms from the system under test are low-pass filtered to prevent aliasing (see Chaps. 13 and 19) and converted to a sequence of control samples by the input subsystem of the digital system, as previously discussed. The averaging control, the spectrum analyzer, and the display are implemented by the time-to-frequency and control amplitude estimation blocks. These blocks use a discrete Fourier transform (DFT), as discussed in Chap. 14, to estimate the power spectral density (see Table 19.3) of the control responses $c_1(t)$ through $c_n(t)$. The random noise generator and the analog equalizer, used in previous analog random vibration systems, are replaced by an analogous digital process using a DFT and a time-domain randomization algorithm.[17] This is accomplished in the frequency-to-time processing block within the DVCS in Fig. 26.8. The lines of the DFT (see Chaps. 14 and 19) in the digital system play the role of the contiguous narrowband filters in the equalizer of the analog system.[17,24] Equalization is the adjustment of the amplitude of the output of a bank of narrowband DFT filters, which is a fast Fourier transform equivalent (see Chap. 19) whose amplitude is given by the drive signal's spectrum amplitude $D_1(f)$ that corresponds to the center frequency of each DFT filter such that the power spectral density of the control response matches that of the test prescribed reference.

The equalization of the drive waveform can be accomplished directly, by generating an error correction from the difference between the control power spectral density and the reference spectral density. The equalization can also be accomplished indirectly through knowledge of the system frequency response function magnitude. The required system FRF (see Chap. 21) is the ratio of the Fourier transform of the control response (usually an acceleration) and the Fourier transform of the drive voltage signal, as discussed in an earlier section. Only the magnitude of the FRF is required for random control, since the relative phase between frequencies is random and not controlled.

The drive spectrum D_1, which results from the "update drive to minimize control error" block in Fig. 26.8, is multiplied by a random phase sequence and its inverse FFT is calculated to create the corrected drive time series $d_1(t)$. Samples of the corrected digital drive time series $d_1(t)$ are fed through the output subsystem in Fig. 26.9, within the DVCS, converted to an analog signal, low-pass filtered to remove the images caused by the digital-to-analog converter, further amplified, and then sent as the analog signal d_1 to the power amplifier input of the system under test, which completes the loop. Corrections to the drive are not made continuously in the digital random vibration control system. Many samples of the drive (often thousands) are output between corrections. Many digital systems use a *time-domain randomization process*[17,24] that converts the finite-duration $d_1(t)$ drive block into an indefinite-duration signal with a continuous power spectral density that has the same values as $d_1(t)$'s at the discrete frequencies at which the FFT was evaluated. The time between drive corrections is called the *loop time.* The loop time for digital random vibration control systems can be from a fraction of a second to a few seconds, depending on the type of averaging used for control response power spectral density estimation.

The speed at which the system can correct the control spectrum is determined by two factors. The first is the loop time. The second is the number of spectral averages required to generate a statistically sound estimate of the control power spectral density (see Chap. 14). The loop time is usually the shorter of the two. Typically, a compromise is required; an estimate of the power spectral density with a significant error is used, but only a fraction of the correction is made in each loop. The type of spectrum average—linear or exponential—also has a large effect on the averaging time where the exponential average affords a shorter averaging period but only a fraction of a correction is made in each control loop to ensure system closed-loop stability.[17] In such cases, multiple corrections occur within the averaging period.

The equivalent bandwidth of the DFT filters is dependent on the number of lines in the DFT, the type of spectral window that is used (see Chap. 14), and the sampling rate of the DAC and ADC. These parameters are usually options chosen by the operator either directly or indirectly. The averaging parameters are also typically operator specified.

Swept Sine. The objective of a digital sine-wave vibration test control system is to drive a system under test, as shown in Fig. 26.6, with sweeping sine-wave excitation, such that the control response signals, when processed by the control response estimation block, shown in Fig. 26.8, agree with the specified test reference within some acceptable error. The control response outputs c_1 through c_n of the system under test are filtered and digitized with the input subsystem of the digital vibration control system. The needed tracking filters,[13] variable time-constant root-mean-square detectors,[17] averaging control, and control signal selection are implemented within the appropriate blocks in Fig. 26.8 by the use of an embedded digital signal processor subsystem for the required specialized signal-processing functions. It is however nontrivial to implement tracking filters digitally,[13] as previously discussed. Many systems, in the interest of simplicity, do not use true tracking filters, but approximate this function by using fast Fourier transform methods. However, these FFT methods don't work as well as the presented tracking filter methods in practice. In any case, these are implemented in the time-to-frequency transformation and control-amplitude estimation blocks within the servo subsystem in Fig. 26.8 within the DVCS.

The sine-wave generator is implemented by using samples of a digitally generated sine wave, usually by a DSP subsystem within the frequency-to-time transformation block in Fig. 26.8, which are sent to the output subsystem in Fig. 26.3 and 26.9 to be used to drive the system under test in Fig. 26.6. The swept-sine test parameters are entered by the test operator through the DVCS' graphical user interface to be stored in a test parameter file to be used for a subsequent test. The control response servo subsystem shown in Fig. 26.8 is implemented by an algorithm which compares the computed amplitude of the control waveform with the required control amplitude, as defined by the test setup, and generates a corrected sampled drive waveform. This function is accomplished by the "update drive to minimize control error" block as shown in the DVCS' servo subsystem block diagram in Fig. 26.8. The sampled drive waveform is converted to an analog drive waveform by the DAC and sent to the low-pass filter and output attenuator shown in Fig. 26.3, which illustrates the DVCS' output subsystem block diagram shown in Fig. 26.9. This resultant analog drive signal d_1 is used as the input to the power amplifier within the system under test block diagram in Fig. 26.6 to complete the closed loop.

Swept-sine vibration tests can require that the frequency be stepped in a sequence of fixed frequencies or swept in time over a range of frequencies. However, the stepped approach can generate vibration transients when the frequency of the sine-wave drive signal is changed. A swept sine is the changing of the frequency from one frequency to another in a smooth, continuous manner. This is the preferred drive signal generation method, since it creates no significant transients as the frequency is changed. Again, many commercial control systems use the stepped-frequency method because of its simpler implementation. The rate of change of frequency with respect to time is called the *sweep rate.* Both logarithmic and linear swept sines are required. For a *logarithmic sweep,* the change in the logarithm of the frequency per unit of time is a constant. For a *linear sweep,* the change in frequency per unit of time is a constant. Because the drive waveform is usually generated in blocks of samples, care must be taken in swept-sine vibration tests to ensure that the frequency and amplitude change is continuous. The correction of the drive amplitude in a digital system is not continuous but discrete. The time between amplitude corrections is also called the loop time and is controlled by the number of samples, which must be taken to define the control waveform amplitude and the required computations to compute the corrected drive waveform. Here, as in the other DVCS applications, *control-loop iteration* is the completion of one complete cycle from the correction of one drive waveform to the next.

The control response amplitude can vary rapidly as the frequency changes due to system resonance, and the required loop time is measured in small fractions of a second. For stability, the complete correction of the drive waveform is not usually made each loop. The maximum rate of drive waveform correction is called the *compression speed*[17] and is usually expressed as decibels per second (dB/sec). If the compression speed is too fast, system instabilities can occur. If the compression speed is too slow, the correct amplitude will not be maintained. The required compression speed is a function of (1) frequency, (2) sweep rate, (3) the system dynamics, (4) the amount of noise present in the response measurement, and (5) the degree to which the response of the system under test is nonlinear. Limited operator control of the compression speed is usually provided. The bandwidth of the digital tracking filter[13,17] will affect the stability of the system. Specifically, as the bandwidth of the tracking filter decreases, the delay in the output of the tracking filter increases.[13,17] As the filter delay increases, the compression speed must be decreased to maintain stability.[17] Some of the more advanced DVCSs used for this purpose accommodate the change in correction rate automatically to ensure a good compromise between control speed and accuracy. However, users need to make the required compromise by selecting the bandwidth of the tracking filter or the time constant of the rms measurement to be used during the swept-sine test, which trades off the ability to reject components in the control waveform at frequencies other than the drive frequency and the ability of the control system to respond quickly to changes in the control waveform amplitude.

Transient/Shock. Sometimes it is desirable to perform shock or transient testing using electrodynamic or electrohydraulic vibration test machines. The ability to employ this method depends on such parameters as the stroke (the maximum allowable motion of the vibration exciter), the peak amplitude and spectral characteristics of the specified transient waveform, the amount of moving mass during the test, and the test time.[14,18,19] If the required test is within the performance capability of an available vibration system, the ability to obtain and control the desired motion has been greatly expanded by the use of digital control equipment.[14,18,19] In general, the servo control of a shock test parallels that used for the other vibration control methods, but in this case the controller compares the control accelerometer time history response to a reference waveform as part of the control process. The primary difference here is that the time-to-frequency and frequency-to-time transformations in Fig. 26.8 are accomplished using a fast Fourier transform of the transient, with the forward or inverse transformations, respectively. If necessary, the controller drive signal is altered to minimize the deviation of the control accelerometer response from the reference based on the comparison between the control response and reference response FFT spectrum. This discrepancy is used to update the drive spectrum in the "update drive to minimize control error" processing block within the digital vibration control system's servo subsystem in Fig. 26.8.

Shock test requirements may be specified in one of two ways. The first and more direct method specifies a certain acceleration waveform, such as a half-sine pulse of specified duration and maximum acceleration. These are called classical shock transients (see Chap. 27). The DVCS in this case needs to modify such classical pulses by adding a pre- and postpulse to the overall test pulse waveform[18] to ensure that the response of the system under test returns to zero acceleration, velocity, and displacement conditions at the end of the shock test.[18] Typical pulses used as the reference response waveform $r(t)$, in addition to the previously discussed half-sine pulse, include final and initial peak sawtooth, rectangular, and trapezoidal pulses of varying duration and amplitudes (see Chap. 27). The control method that is used is a subset of what is used for time-domain waveform replication control, discussed in a later section, usually without a need for the overlap-and-add indirect-convolution method.[4]

The second method employs the shock response spectrum (see Chaps. 20 and 28) as the means of characterizing the response of the control points.[15,19] In this case, the control response spectrum $C(f)$ and the reference response spectrum $R(f)$ are specified as an SRS. The requirements for the reference SRS must specify the frequency range, frequency spacing, damping factor, type of spectrum, and either maximum or nominal values with an allowable tolerance on spectrum values.[14,19] Reference pulses are generated using one of the SRS synthesis techniques[14,15] that were discussed previously. The control method that is used is called the *wavelet amplitude equalization* (WAE) method. If the test requirements are specified as an SRS reference $R(f)$, then during the test, the SRS of the control response waveform is computed and compared with the prescribed $R(f)$. The difference is then used to update the drive signal, which is expressed as a weighted sum of wavelets. The weights in the sum represent the amplitude of the various wavelets. These amplitudes are varied as a function of the discrepancy of the control response SRS and the reference SRS. Care is required when this difference is large, since this control problem is highly nonlinear due to the nonlinear dependence of the control response SRS to the wavelet amplitudes of the drive signal. Because of this, the control corrections are iterative and yield an approximate SRS for the control response.

Mixed Mode.[21] Digital vibration test control systems are available which can control several sine waves superimposed on a stationary random vibration test. This is called *sine-on-random vibration testing* or *swept-sine-on-random vibration testing.* Systems are also available that can control swept narrowbands of nonstationary random vibrations superimposed on a stationary random vibration test. This is called *swept-narrowband-random-on-random testing.* It uses a variation of the random vibration control methods, previously discussed, by modifying the reference response spectrum during the test to create sweeping narrowbands of random vibrations that are superimposed on a broad bandwidth random background.[21] The control or servo process for the case of sine-on-random vibrations works as a parallel connection of a random vibration and a swept-sine control system. A simplified block diagram of this process is shown in Fig. 26.10.

The two critical differences between mixed-mode controllers and individual random and swept-sine controllers are the presence of the bandpass/reject and synthesize composite subblocks in Fig. 26.10. The bandpass/reject subblock in Fig. 26.10 separates the swept-sine and random backgrounds into two separate signal streams. The swept-sine component is fed into the sine control section, and the random background section is fed into the random control section. These separate controllers, with needed synchronization between each other, then create separate drive amplitude updates for control of their respective component. These separate drive amplitude updates are combined into a composite drive signal, containing the random and swept-sine components in a single drive signal, by the "synthesize composite" section in Fig. 26.10. This composite drive is then sent to the system under test to com-

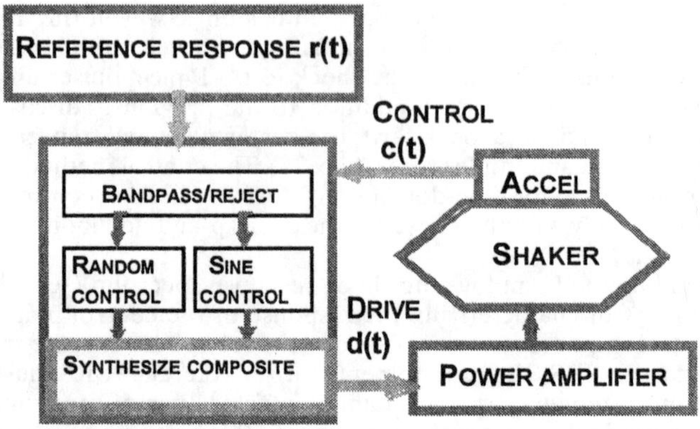

FIGURE 26.10 Swept-sine-on-random controller.

plete the control loop. The bandpass/reject section should employ advanced signal estimation techniques to determine the phase and amplitude of the control response sinusoids that are masked by the background random noise contained in the composite control response signal $c(t)$.

Time-Domain Waveform Replication Control. The objective of a *time-domain waveform replication control,* or simply *TDWR control,* test is to drive the system under test in Fig. 26.6 with a drive signal $d(t)$ such that the time-domain response of the chosen control transducer, $c_1(t)$ in Fig. 26.6, which is typically mounted on the unit under test, matches the test-specified reference waveform $r(t)$ within an operator-specified error margin. The same type of digital vibration control system shown in Figs. 26.5 through 26.9 can be used for this application. The DVCS is tasked with finding the drive signal $d(t)$ which achieves the waveform control test's objective.

The control methods used for this application are fundamental. They serve as a foundation for the control methods used by the previously discussed transient/shock control systems and the to-be-discussed multiexciter control systems. They also serve to illustrate many of the problems and solutions associated with properly addressing this type of control problem as well as their extension to multiexciter control. For this reason, this application will be discussed in more detail to add background to the previous transient/shock waveform control and the subsequent multiexciter control system discussions.

As previously discussed, this reference response waveform typically represents the response of the UUT, as monitored at a particular control point during a part of its service life. These reference response time histories can be quite arbitrary, limited only by sampling-rate considerations (see Chap. 19), specifications of the test system, the physical limitations of the excitation system, and the DVCS that is used. These reference waveforms can be measured using transducers mounted on the UUT during its service environment, or they can be synthesized from dynamics models that predict, by back-transforming from these measured responses to the response on the test fixture at a test-convenient location the synthesized response needed to achieve the particular prescribed response of the UUT. These synthesized reference response waveforms are obtained by using a structural-dynamic model of the UUT and the test fixture's interaction and geometry during the waveform control test. Many techniques are used for this purpose.

Some DVCS employ the TDWR method to accomplish mixed-mode testing. Additionally, more complex reference response signals that contain transients are possible. The reference response signal is thus specified in the time domain using one of these various methods, as for shock testing, but typically with a time duration much longer than that encountered during shock testing and therefore requiring a more complex control and digital signal-processing methodology.

This type of testing is sometimes called *long-term response waveform control testing* and uses an estimate of the frequency response function of the system under test to control the response of it. The FRF estimate relates the control response waveform $c_i(t)$ to the electrical drive waveform $d(t)$ that the DVCS uses to control the system under test. It is the principal quantity that is used in the waveform control process. The FRF needs to be estimated prior to the vibration test. It is measured by exciting the system under test with a drive-voltage waveform having a bandwidth of at least that of $r(t)$, which is output through the DVCS' output subsystem to the system under test. During this test phase, which is often called *system identification* or *characterization,* the response of the chosen control point $c_i(t)$ is measured, and the drive signal $d(t)$ used to achieve this response is also stored. These two signals, $c_i(t)$ and $d(t)$, are then used to calculate the FRF $H(f)$ of the system under test (see Table 19.3). $H^{-1}(f)$ and $r(t)$ are then used in conjunction with an overlap-and-add fast indirect-convolution method[4] to generate a drive signal that should cause the control response $c(t)$ of the system under test to agree with the prescribed reference response $r(t)$, within an acceptable error margin.[20,22] Often, multiple control iterations that use $H^{-1}(f)$, $r(t)$, and $c(t)$, within the DVCS' servo subsystem as part of an overlap-and-add fast indirect-convolution method, are needed to achieve the test's goal.[20,22]

The UUT needs to be part of the system under test, as shown in Fig. 26.6, during the system identification test phase, since mechanical feedback from the test article or UUT will change the system's frequency response function $H(f)$. Numerous waveforms can be used for the excitation, including an impulsive transient, the predetermined reference response waveform, a continuous random waveform, or repeated short bursts of random vibration, with the transient noise having frequency-domain characteristics like those of the continuous noise. The latter two methods are most commonly used. Continuous random noise produces better results in practice, but at the expense of longer vibration times for the UUT during this phase. In all cases, it is important for the excitation drive signal to have energy at all frequencies of interest, although of sufficiently small amplitude that the test item is not damaged from this excitation but a large enough amplitude that a linear extrapolation to full test level will not cause significant control errors. Averaging, as part of the FRF estimation, can mitigate the effects of nonlinear response and measurement noise (see Chap. 19) on the quality of the estimate.

TDWR control systems typically use a long-term convolution algorithm to solve the *convolution problem* [problem associated with solving Eq. (26.5)] associated with finding the drive signal that will cause the system to respond in a predetermined manner. This type of problem is typically solved by using indirect, FFT-based, fast convolution algorithms.[4] The DVCS uses an impulse response function $z(t)$, which is obtained as the inverse FFT of $H(f)^{-1}$, which is the inverse of the estimate of the FRF of the system under test between the chosen control point $c_i(t)$ and the drive signal $d(t)$. The solution can be expressed as a convolution integral between the prescribed reference response signal $r(t)$, the impulse response function $z(t)$, and the needed drive signal as

$$d(t) = z * r(t) = \int_{-\infty}^{\infty} z(t-\tau)r(\tau)d\tau \tag{26.5}$$

where $r(t)$ is the input and the output is $z{*}r(t)$, the convolution of $z(t)$ and $r(t)$, which is used to obtain the drive signal $d(t)$.[4] When $r(t)$ and $z(t)$ are of finite duration, then $z{*}r(t)$ will be finite in duration, with a duration equal to the sum of those of $r(t)$ and $z(t)$. Although $r(t)$ is finite in duration, $z(t)$ is not typically finite in duration, but is usually such that its response magnitude decays below some acceptable value—for example, to a value less than 0.1 percent of its peak amplitude after a finite number of samples, typically less than 16K. Hence, $z(t)$ can usually be considered to be finite in duration without incurring significant error.[4] If $z(t)$ does not decay quickly enough, then a windowed version of it needs to be used in order to mitigate the spectral leakage and circular convolution errors that would otherwise occur.[4]

The convolution problem given by Eq. (26.5) arises because the drive and response are related by the following frequency-domain formulation:

$$C_i(f) = H(f)D(f) \tag{26.6}$$

where $C_i(f)$ and $D(f)$ are, respectively, the Fourier spectra of $c_i(t)$ and $d(t)$. Equation (26.6) can be used to solve for the drive signal spectrum $D(f)$, which should cause the system under test to respond with a given control response spectrum that agrees with $R(f)$, which is the Fourier spectrum of the reference response time history $r(t)$. In this case, the needed drive signal Fourier spectrum $D(f)$ is given by

$$D(f) = R(f)/H(f) \tag{26.7}$$

Equation (26.7) is an example of a solution to a *deconvolution problem.* The control system converts Eq. (26.7) into the form given by Eq. (26.5) by computing the inverse FFT of $R(f)/H(f)$, which is $d(t)$, via what is called *fast convolution.* In this manner, Eq. (26.5) is indirectly used to create the drive signal $d(t)$, which is to be used to excite the system during the waveform control test. Eq. (26.5) is useful for interpreting the various steps associated with fast convolution and the overlap-and-add method.

Care should be taken to avoid circular convolution and spectral leakage errors in the evaluation of Eq. (26.5) when using fast convolution methods. Circular convolution errors[4] can be avoided by using an FFT block size such that $z(t)$, which is obtained as the inverse FFT of $H(f)^{-1}$, will fit within a half FFT block with the other half of the block zero-filled prior to the evaluation of the spectra of $r(t)$ and $z(t)$ via the FFT.[4] The spectrum leakage errors can be eliminated if the thus-obtained $z(t)$ fits completely within one of these half FFT block frames. If not, windowing of $z(t)$ is needed to minimize the resulting spectral leakage errors (see Chap. 14).

The function $z(t)$ is the impulse response of the inverse system; that is, it maps outputs to inputs. As such, it is not causal, and the antiresonances of $H(f)$ appear as resonances in the system described by $z(t)$. The inverse system, described by Eq. (26.5), that is used to create the drive signals will have the most gain at the antiresonances of $H(f)$. If $H(f)$ is measured or estimated accurately and models the system under test well, then the drive signal thus created should create a system under test response $c_i(t)$ which agrees with the prescribed test reference $r(t)$. On the other hand, if there are errors in $H(f)$, then the drive signal will create other, uncontrollable responses at the antiresonance frequencies of the system under test.

Because $H(f)$ may contain antiresonances within the frequency test range, $H(f)$ may attain small values at which it could be imprecise. Since we need to divide by $H(f)$ in Eq. (26.7) and use it to generate the drive signal $d(t)$ with the indirect use of Eq. (26.5), care is needed in this division at the antiresonant frequencies or frequency ranges over which the magnitude of $H(f)$ is small. Without proper attention, a low-valued $H(f)$ with large error will cause numerical problems and a potential magnification of the noise floor in the thus-synthesized drive signal $d(t)$. If such a drive signal is used to drive the system under test, a control response $c_i(t)$ may result that does not agree well with the test-prescribed reference response signal $r(t)$, but rather the control response would contain large noise amplitudes at the problem frequencies.[14] It is possible, through the use of the measured coherence function, defined in Eq. (19.20), between the drive signal and its associated control response, and also some weighting functions based on the values of $R(f)$, to mitigate the numerical problems associated with using Eqs. (26.5) and (26.7) to generate the system under test drive signal $d(t)$.

The estimation of the system's frequency response to solve for $H(f)$ is given by

$$H(f) = C_i(f)/D(f) \tag{26.8}$$

Equation (19.16) can also be used to estimate $H(f)$ by using the estimated cross-spectral density between the control response and the drive signal, and the power spectral density of the drive signal (see Table 19.3), and this method is preferred. It uses averaging to reduce estimation errors that could otherwise occur with the direct use of Eq. (26.8). In any case, whether using Eq. (26.8) or Eq. (19.13), a division by the drive signal spectrum is needed to compute the system FRF estimate. For this reason, it is beneficial to use a drive signal that is spectrally flat and with a voltage amplitude sufficiently above the electrical noise floor as to optimize the accuracy of the preceding calculation for the value of $H(f)$. In order to obtain the level at which to excite the system under test during the system identification phase, it is important to determine (1) the amplitude of the noise floor of the system under test and of the DVCS and (2) what the fragility concerns associated with the UUT are. These considerations help determine the amplitude and spectrum shape that should be used for the drive signal during the system identification phase to minimize the errors associated with deconvolution, yet also minimize the stress imparted to the system under test during this test phase.

The actuators and the control instrumentation have frequency responses that roll off to zero at both the lowest and the highest frequencies. These determine the bandwidth of the test system. Additionally, when the UUT and test fixtures are combined with the actuators, the resultant system under test will have low responses at mid-range frequencies, usually at the resonant frequencies of the UUT. This combined system will typically exhibit antiresonant behavior at those frequencies, resulting in low values for the measured system frequency response function $H(f)$. The solution of Eq. (26.7), which uses $H(f)^{-1}$ to calculate the initial drive to achieve the test-prescribed reference response, may then have numerical difficulties. $H(f)$ can also have error at its resonant frequencies due to nonlinearities that may be present in the system under test, which are exacerbated by the higher response at resonance. This drive signal can be further refined, using the DVCS' feedback process, and perhaps even the measured system frequency response, if the DVCS uses an adaptive control process, which can mitigate these problems.[20,25,26] Since it uses a form of Eq. (26.5), the feedback process may also be sensitive to these numerical problems that are associated with deconvolution. Control algorithms that converge at these problem frequencies, at which $H(f)$ is ill conditioned, are essential[20,22,25,26] to avoid control oscillations or overtest conditions that can arise when suitable care is not taken.

The problems that are associated with deconvolution also occur in random and swept-sine testing, but its negative effects are more severe in time-domain waveform replication control, since in this case the estimated phase of $H(f)$ is important and usually is imprecise for the reasons discussed. This is further aggravated in TDWR control, since most digital control systems perform control updates between test iterations. The digital control system, in these cases, is open loop when it is exciting the system under test with the drive signal computed by the use of the deconvolution in Eqs. (26.5) and (26.7). Most digital control systems at this point would compute control errors and update the drive signal to reduce the control errors that occur during the test once the initial excitation is complete—usually tens of minutes of such open-loop excitation. Thus, these errors in the drive signal, based on Eqs. (26.5) and (26.7), at the problem frequencies, would persist potentially for the entire response waveform control test before being corrected. Random and swept-sine testing are inherently closed loop and thus see the effects of ill-conditioned deconvolution more quickly, and they either correct these errors or abort the test if the test abort levels are exceeded.

If the duration of the reference response time history $r(t)$ and the sample rate used for the waveform control test result in a frame or block size of $r(t)$ that exceeds the maximum size that the FFT can use, then Eq. (26.7) needs to be solved in sections of the reference response signal $r(t)$ that consist of block sizes that can be processed by the system FFT. This is accomplished by the use of an overlap-and-add indirect-convolution technique[4] to implement the frequency-to-time transformation in Fig. 26.8, as given by Eqs. (26.5) and (26.7). The overlap-and-add method is also called *high-speed indirect convolution.* The overlap-and-add method is a way of implementing the equivalent of Eq. (26.5) when Eq. (26.7) can be computed for overlapping sections of $r(t)$, in block sizes that the system's FFT can evaluate. However, Ref. 4 and others like it don't cover all the practical problems that can occur when using this indirect-convolution method and others related to it, so many system manufacturers have developed proprietary methods to overcome the shortcomings of the unmodified methods.

It is sometimes possible to update the system FRF estimate during the waveform control test.[25,26] There are some patented methods[20,23] that can use the control response signal $c_1(t)$, the drive signal $d(t)$, and the reference response signal $r(t)$ to obtain robust updates to the estimate of $H(f)$ during the waveform control test. They can prevent the errors associated with a direct implementation of Eq. (26.8) to estimate $H(f)$ by using adaptive control for TDWR control tests.[20,25,26] Sometimes this refining of the drive signals and frequency response estimates is done at a lower test level to prevent undue stress to the UUT. Once the control and estimation errors are reduced sufficiently, then the drive signal amplitude is increased in proportion to the ratio of these low-level test amplitudes to the associated full test level amplitude. This assumes that the control point response will increase in proportion to the associated increase in the drive signal amplitude.

MULTIEXCITER TESTING APPLICATIONS

The simplest example of *multiple-exciter testing* is when multiple exciters are connected to independent systems under test and are controlled simultaneously. This configuration corresponds to several single-exciter control systems operating in parallel and will not be further discussed.

The more complex and more interesting case is when the multiple exciters act on the same test fixture and unit under test simultaneously, as shown in Fig. 26.6 and discussed in more detail in Chap. 25. The attachments of the multiple exciters to the test fixture can be at several points in a single direction, at one point in several directions, or combinations of both.[20,23] This is the type of configuration that is represented in the block diagram of the multiexciter system under test in Fig. 26.6. If any of the drives $d_1(t)$ through $d_n(t)$ is capable of causing a response on more than one of the control responses $c_1(t)$ through $c_n(t)$, then the multiexciter control system has cross coupling between control responses. In this situation, the measured frequency response matrix $[H(f)]$, between the drive signal vector $\{d(t)\}$ and the control response vector $\{c(t)\}$ will have off-diagonal elements that compare in order to the diagonal elements.

Systems that have cross coupling between the control response signals $c_1(t)$ through $c_n(t)$, and which are elements of the vector of control response waveforms $\{c(t)\}$, require the digital vibration control system to have provisions for control of these cross-coupling effects. These are typically controlled using the measured frequency response matrix in a manner similar to the way the system frequency response function $H(f)$ is used for TDWR control. The needed frequency response matrix is measured using the multi-input/multi-output system identification techniques discussed in association with Eq. (26.4). The specifics of how this is done vary with each application, dictated by the type of MIMO shock and vibration testing that needs to be accomplished. These are typically multiexciter tests that use a MIMO methodology within the DVCS used to control such multiexciter tests. These shock and vibration control applications are called MIMO random, MIMO swept-sine, MIMO shock, and MIMO time-domain waveform replication tests. Good mechanical design (the design of the excitation, fixturing subsystems, how the test article is attached, and where the control points are located on the system under test) is very important and can reduce the severity of system identification and control problems that can arise during multiexciter testing. Poor mechanical design can make the MIMO system under test and the corresponding DVCS unusable, no matter how advanced the control technology may be.

The complexity of building these systems—that is, designing the control system and specifying the test parameters—increases much faster than the rate of increase in the number of exciters. To a first order, the control and test specification complexity increases by at least the square of the number of exciters that is used, due to the use of n-dimensional signal-processing methods and its use of n-by-n complex matrices. The design complexity of the system under test in Fig. 26.6 for MIMO testing can also increase, but the reasons are different (see Chap. 25). The resultant physical constraints of achievable system under test designs typically limit many MIMO control and excitation systems to frequencies less than 2 kHz, and sometimes much less than 2 kHz. The significant displacements encountered in low-frequency MIMO testing also increase the complexity of the design of the vibration fixture that interconnects the exciters and the UUT, and that lets the exciters move independently from each other. However, at lower frequencies, large MIMO test systems are possible. For example, time-domain waveform replication control systems that have as many as 18 exciters are used to simulate road conditions in the automobile industry. Figure 25.10 shows an example of this configuration.

Control can take place when the number of control channels is the same as the number of exciters (square control) or when the number of control channels exceeds the number of drive signals (rectangular control).[27] Also, with the use of input/output (I/O) transformations, the number of control degrees of freedom (DOF) can be less than the number of drive and control signals.[28] However, in the interest of simplicity, only the square control case will be discussed in what follows. Interested readers are encouraged to read the references[27,28] for a more detailed discussion of the theory and practice behind rectangular and I/O transformation MIMO control applications.

MIMO Random Testing. For multiple-input/multiple-output random testing, the test's prescribed vibratory motions are specified in terms of a reference response spectral density matrix $[R(f)]$. This matrix consists of both power spectral densities along the diagonal and cross-spectral densities along the off-diagonal elements of the matrix. The elements at the ith diagonal of the reference spectral density matrix, $R_{ij}(f)$ represent the reference power spectral density to be used for the ith reference response for the control response $c_i(t)$. The ijth off-diagonal matrix elements of the reference spectral density matrix $R_{ij}(f)$ represent the reference response cross-spectral density used to control the control response cross-spectral density between the ith and jth control response $c_i(t)$ and $c_j(t)$, as in Eq. (26.1). This cross-spectral density can also be described by the ordinary coherence and phase between $c_i(t)$ and $c_j(t)$ (see Chap. 19), as well as their respective power spectral densities.[10,11,20,22,23] The objective of a MIMO random vibration test control system is to create a drive signal vector $\{d(t)\}$ that consists of the exciter drive signals $d_1(t)$ through $d_n(t)$, which causes the SDM of the control response vector $[W_{cc}(f)]$ to agree, within some acceptable error, with the MIMO random test reference spectral density matrix $[R(f)]$. The issues associated with spectrum averaging and input-signal windowing that were discussed for single-exciter random vibration control need to be considered as well. Also, I/O transformations can be used to express $[W_{cc}(f)]$ in terms of physical DOFs such as translations in X, Y, and Z, as well as roll, pitch, and yaw.[28]

The control response spectral density matrix $[W_{cc}(f)]$ of the control response vector can be modeled by the following result from linear system dynamics and multidimensional stationary stochastic process theory,[7–9] which states that the control response SDM is given by

$$[W_{cc}(f)] = [H(f)][W_{dd}(f)][H(f)]^H \tag{26.9}$$

Equation (26.9) can be solved for the initial drive signals using the measured frequency response matrix $[H(f)]$ and the test-prescribed reference response spectral density matrix $[R(f)]$. This result gives the spectral density matrix $[W_{dd}(f)]$ of the drive signals as

$$[W_{dd}(f)] = [H(f)]^{-1}[W_{cc}(f)][H(f)]^{-H} \tag{26.10}$$

The resultant drive spectral density matrix $[W_{dd}(f)]$ can be further factored, using a Cholesky decomposition,[1,8,10,11,22,24] as

$$[W_{dd}(f)] = [\Gamma_d(f)][\Gamma_d(f)]^H \tag{26.11}$$

where $[\Gamma_d(f)]$ is the Cholesky factor of $[W_{dd}(f)]$, which is a lower-triangular complex matrix with real and nonnegative diagonal elements that plays the same role as the drive spectrum plays in single-shaker control.[22,24] This Cholesky factor is also associated with the general study of partial coherence[7,9,10] and the partial coherence that will exist between drive signals that are synthesized using it.[10] It is used with the frequency-to-time processing block of Fig. 26.8 to create a vector of drive signals $\{d(t)\}$ that has $[W_{dd}(f)]$ as its SDM.[22,24] These are further randomized by a MIMO time-domain randomization process, similar to what is done in single-exciter random testing, but with the use of a lower triangular matrix of waveforms obtained from $[\Gamma_d(f)]$.[22,24] By this means, the coherence and phase between the control response signals is controlled as well as each individual control response's power spectral density.[20,24] The drive vector $\{d(t)\}$ then has the matrix $[W_{dd}(f)]$ as its SDM and should cause the MIMO system under test to respond with a control response vector $\{c(t)\}$ that has as its spectral density matrix $[W_{cc}(f)]$, which agrees with the test-prescribed reference response spectral density matrix $[R(f)]$, within some acceptable error margin. This process further generalizes to rectangular control, where the dimension of $[W_{cc}(f)]$ is greater than the dimension of $[W_{dd}(f)]$.[27]

MIMO random testing, similar to waveform control, uses the matrix inverse of the measured frequency response matrix $[H(f)]$ to create the initial drive. The *impedance matrix* $[Z(f)]$ of the system under test, is given by

$$[Z(f)] = [H(f)]^{-1} \tag{26.12}$$

This matrix needs to be measured prior to the test in the system identification testing phase, as discussed in previous sections on frequency response matrix estimation. Also, the concerns associated with the deconvolution problem that were discussed in a previous section need to be addressed. The accuracy of this measured matrix, which is computed before the vibration test, is critical to the success of the control task. The method employed to estimate $[H(f)]$[7–9,20,25] typically uses the left expression in Eq. (26.4) to solve for $[H(f)]$ from the computed spectral density matrix $[W_{dd}(f)]$ and the measured cross-spectral density matrix $[W_{cd}(f)]$ as

$$[H(f)] = [W_{cd}(f)][W_{dd}(f)]^{-1} \tag{26.13}$$

The MIMO control system uses the frequency response matrix, measured before the MIMO test with the use of Eq. (26.13), to construct the initial drive as in Eq. (26.10). A further MIMO control iteration is used to refine the drive and approximately account for the possible nonlinearities in the control responses.[20,22,23,25] The control iteration uses $[Z(f)]$ to compute the contribution that the control errors at each of the control points make to each of the drive signals. It effectively decouples the control errors so they can be used to adjust the proper drive's relative phase and coherence to achieve control[14,20,22,24,25] according to their respective contribution.

In MIMO random testing, unlike in multiple-input/single-output (MISO) random testing, phase cannot be ignored, since the relative phase between the control responses and the drive signals, and also between the drive vector and the control response vector, is critical to the success of the MIMO test. Also, since the impedance matrix $[Z(f)]$, which is the inverse of $[H(f)]$, is being used for control, special care is needed in calculating it at those frequencies where $[H(f)]$ is singular or nearly singular.[20,22,25]

For MIMO random testing, the system characterization is done by operating all exciters in the system under test simultaneously with band-limited gaussian noise. These system identification drive signals are typically flat spectrally and band-limited to the maximum frequency of interest. They are also uncorrelated among themselves. The response levels for the system characterization should be chosen as high above the noise floor as possible to maximize the accuracy of the $[Z(f)]$ estimate but still ensure that the test article does not experience vibration levels above some level that is chosen by the operator to ensure that no undue stress or damage is caused to the test article during the system identification operation. With the system excited in this way, the spectral density matrix $[W_{dd}(f)]$ and the cross-spectral density matrix $[W_{cd}(f)]$ are estimated using the methods associated with Eqs. (26.1) through (26.3). Equation (26.13) is used to compute the estimate of $[H(f)]$, and Eq. (26.10) is used to generate the initial drive signals based on the Cholesky factor $[\Gamma_d(f)]$ discussed as part of Eq. (26.11).

MIMO Swept-Sine Testing.[20,25] MIMO swept-sine control systems operate similarly to the MIMO random control systems discussed previously, with differences in the control objective. The objective of a MIMO swept-sine test is to impart a controlled excitation to a structure at specified points, with a series of exciters connected to the structure so that the response motion at a chosen number of control points on the system under test (see Fig. 26.10), as described by the control response vector $\{C(f)\}$ matches a prescribed reference response vector $\{R(f)\}$ within some acceptable error margin. In this case, if there are n-exciters and n-control transducers, the complex vectors of spectra $\{C(f)\}$, with components $C_1(f)$ through $C_n(f)$, and $\{R(f)\}$, with components $R_1(f)$ through $R_n(f)$, are of dimension n for each frequency within the test range. To accomplish this goal, the linear system model of system response is solved for the initial drive by

$$\{D(f)\} = [H(f)]^{-1}\{R(f)\} \tag{26.14}$$

As in other MIMO control applications, Eq. (26.14) is solved for the initial drive vector $\{D(f)\}$ using the frequency response matrix for the system under test that is obtained prior to the test. In MIMO sine testing, the additional problem is that random noise excitation, as used in other MIMO applications, is, many times, not suitable for the system identification. This is due to the fact that the system's frequency response characteristics can be quite different for swept-sine excitation than they are for random excitation. For this reason, the system identification should be done with swept-sine excitation one exciter at a time. This can be time consuming and may cause undue fatigue to the structure under test in Fig. 26.11. Other approaches that are used involve stepped-sweep methods with a single exciter at a time or with multiple exciters using multiple phases at each step frequency. There is at least one commercial system that uses patented adaptive control technology,[20,25] that can refine the $[H(f)]$ matrix estimate, which was obtained using the previously presented MIMO random method, so that it can be used to control the swept-sine test and thus minimize the problems of using such an initial system identification phase for its control.

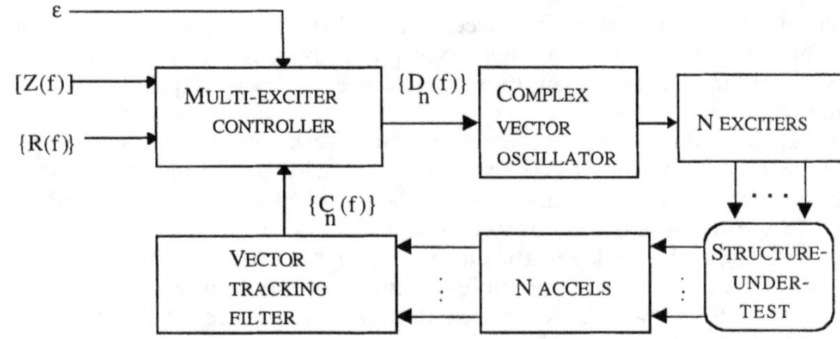

FIGURE 26.11 Overall swept-sine multiexciter control system.

The overall block diagrams of the MIMO swept-sine control system and the MIMO swept-sine controller are shown in Figs. 26.11 and 26.12, respectively. As can be seen in the block diagram of the overall system in Fig. 26.11, a vector-tracking filter subsystem plays the role of the time-to-frequency conversion in the digital vibration control system. As discussed in a previous section, tracking filters estimate the complex amplitude of the sweeping sine-wave control response signals $c_1(t)$ through $c_n(t)$. The resulting complex control response vector $\{C(f)\}$ is then compared by the DVCS with the prescribed test reference response vector $\{R(f)\}$. The control error vector is then multiplied by the impedance matrix $[Z(f)]$, to get the contribution of the control errors at each control location to each drive signal sent to each exciter. A percentage of this error, given by ε, is added to the previous complex drive's amplitude spectrum to obtain the next drive signal's vector spectrum amplitude, as shown in the multiexciter swept-sine controller block diagram in Fig. 26.12. This corrected drive signal, with updated amplitude and relative phase, is then sent to the vector oscillator, which plays the role of the frequency-to-time transformation subsystem within the DVCS. It provides control of the amplitude of the output drive signals and the relative phase with respect to the modulating signal used by the vector-tracking filter shown in Fig. 26.11. Each component of $\{C(f)\}$ is an output of an individual tracking filter within the vector-tracking filter in Fig. 26.11, given by Fig. 26.4, which uses the same modulating signal. There is also a common phase and frequency reference for the drive signals generated by the complex vector oscillator in Fig. 26.11. The system is driven as the drive signal vector's frequency is swept continuously through the sweep range of the MIMO swept-sine-wave test.

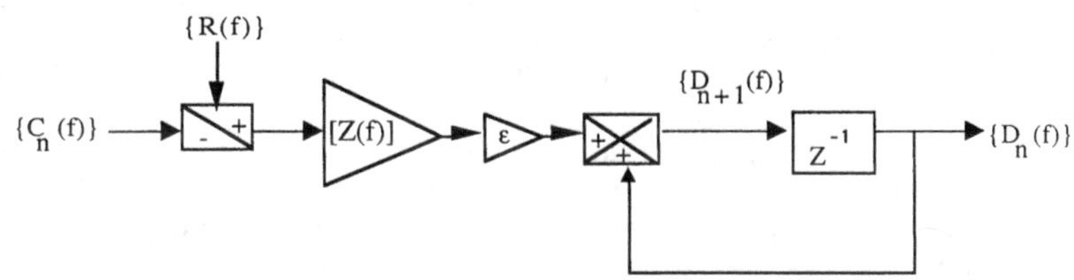

FIGURE 26.12 Multiexciter swept-sine controller.

MIMO Transient/Shock Testing. MIMO transient waveform control methods are an extension of the single-shaker transient/shock and MIMO swept-sine control methods previously discussed. This type of control and system under test is used principally for seismic simulations.[20,28] The application uses shock response spectrum synthesis techniques to create the waveforms that are to be used as the specified reference response vector $\{r(t)\}$. In this case, the control process matches the prescribed SRS indirectly by using waveform control to make the control response $\{c(t)\}$ match $\{r(t)\}$, thereby indirectly matching the specified SRS. This vector of waveforms $\{r(t)\}$ typically consists of random transients that have been synthesized such that each transient matches a specified SRS to be used as the spectral reference response for each control point, as discussed in the section on SRS synthesis. In other applications, these transient waveforms sometimes represent data that has been measured in the field. Many times, it is actual earthquake time-domain response data, from remote sensors that are located to measure an earthquake's ground motion when and where it occurs.

The block diagram of this type of control system is similar to that of MIMO sine testing. The predominant difference is that the time-to-frequency transformation is accomplished by a fast Fourier transform, with a frame size large enough to accommodate the transient but still avoid circular convolution errors.[4] Spectral leakage errors (see Chap. 14) are mitigated by using windowing.

MIMO Time-Domain Waveform Replication Control. This application is an extension of the single-exciter TDWR and MIMO transient waveform control discussed in the previous sections. The primary difference from MIMO transient waveform control is in the fact that the test-prescribed reference response vector $\{r(t)\}$

consists of waveforms that cannot be processed within a single FFT frame. For this reason, as in the discussion about single-exciter TDWR control methods, an overlap-and-add technique[4] must be employed in both the time-to-frequency and frequency-to-time transformations within the DVCS used for MIMO TDWR control. The issues that are associated with the use of the overlap-and-add indirect-convolution technique need to be considered and addressed.[4,29,30]

Again, as in MIMO random, MIMO sine, and MIMO transient/shock applications, the MIMO system under test is driven with a vector of time histories $\{d(t)\}$, such that the control response vector $\{c(t)\}$—in this case, a vector of time histories—agrees within an acceptable error margin with the test-prescribed reference response vector $\{r(t)\}$, which is also a vector of time histories.

Examples include electrohydraulic road simulators and seismic simulators. As mentioned previously, road simulators with as many as 18 actuators attached to a single vehicle have been used. I/O transformation control is often used with this type of configuration.[28] This type of multiactuator test system will exhibit significant cross-coupling and thus needs a digital vibration control system that can resolve it to achieve satisfactory control. This is done by using a measured impedance matrix, as discussed for the MIMO random and MIMO transient/shock applications, largely following the same control approaches as in MIMO sine, with the only significant differences being in the time-to-frequency and frequency-to-time transformation blocks. The control equations are thus very similar to those in MIMO sine and MIMO transient/shock applications, but they use the added complexity of the overlap-and-add method, discussed previously for single exciters in the time-to-frequency and frequency-to-time blocks in Fig. 26.8. It is most like the MIMO transient/shock application but with the use of this indirect overlap-and-add convolution method, discussed for single-exciter TDWR, rather than a single FFT as the only significant difference. In fact, it reduces to a MIMO transient/shock control method if the time duration of the reference response time history vector $\{r(t)\}$ is of a duration, in samples, that can be accommodated with a single FFT operation and still avoid circular convolution and spectral leakage errors that can otherwise occur.

The concerns about the proper measurement of $[H(f)]$, the proper handling of singularities that might be present in $[H(f)]$, the advantages of adaptive control, and the controllability and observability issues that were discussed in the previous sections about the other MIMO control applications and single-exciter waveform control apply in this case. Since many DVCSs perform MIMO TDWR control tests in an open-loop manner while driving multiple actuators, these matters need to be carefully considered and addressed to ensure the success of this type of testing.

MODAL TESTING

Modal testing is conducted to excite a system under test, acquire its drive and response signals, and estimate its frequency response characteristics to determine experimentally the natural frequencies, mode shapes, and associated damping factors of a structure via modal analysis. Modal analysis is discussed thoroughly in Chap. 21. Typically, much of the DVCS hardware and its shock and vibration data acquisition and analysis software are useable for this application.

Currently, digital control and analysis systems are applied to modal testing in two distinct ways. First, for sinusoidal excitation, such systems are employed as an aid in obtaining the desired purity of the modal excitation as well as in acquiring and processing data, usually with operator adjustments of the frequency, the relative phase, and the amplitude of several sine-wave outputs. These are used to drive a system under test so as to achieve a particular relative phase and amplitude between chosen response points on the system under test that is characteristic of a particular normal mode response. The use of MIMO sine control methods can simplify this process. Second, and more commonly, the DVCS is used to excite the system under test with broad-bandwidth random excitation (either random vibration or transients), usually with several such outputs. The response and drive signals are acquired and processed using FFT methods, with the methods discussed on frequency response function and frequency response matrix estimation, using Eq. (26.4).

The use of MIMO random control methods can simplify this process. They can be used to control the coherence and phase between the components of the excitation force vector to separate closely spaced modes or to excite only particular modes that are significant structurally.[31]

The FRF are typically measured between chosen response points on the system under test while exciting the system under test with the chosen excitation at prespecified excitation points, as discussed previously and in Chap. 21. The FRFs and/or frequency response matrices thus estimated are subsequently passed to modal analysis software for further processing and extraction of the pertinent modal data, using the methods of Chap. 21.

REFERENCES

1. Dennis, Jr., E., and R. B. Schnabel: "Numerical Methods in Optimization and Nonlinear Equations," SIAM, 1996.
2. Press, W. H., et al.: "Numerical Recipes in C: The Art of Scientific Computing," 2d ed., Cambridge University Press, Cambridge, England, 2002.
3. Norsworthy, S. R., R. Schreier, and G. C. Temes: "Delta-Sigma Data Converters: Theory, Design, and Simulation," Institute of Electrical and Electronics Engineers, New York, 1997.
4. Oppenheim, A. V. and R. W. Schafer: "Discrete-Time Signal Processing," 2d ed., Prentice-Hall, Englewood Cliffs, N.J., 1999.
5. Stark, H., and F. B. Tuteur: "Modern Electrical Communications—Theory and Systems," Prentice-Hall, Englewood Cliffs, N.J., 1979.
6. Himmelblau, H., and A. G. Piersol: "Handbook for Dynamic Data Acquisition and Analysis," 2d ed., RD-DTE 012.2, Institute of Environmental Sciences and Technologies, Mount Prospect, Ill., 2006.
7. Brillinger, D. R.: "Time Series: Data Analysis and Theory," expanded ed., Holden-Day, San Francisco, 1981.
8. Hannan, E. J.: "Multiple Time Series," John Wiley & Sons, New York, 1970.
9. Bendat, J. S., and A. G. Piersol: "Random Data: Analysis and Measurement Procedures," 4th ed., John Wiley & Sons, New York, 2010.
10. Underwood, Marcos A., and Tony Keller: "Using the Spectral Density Matrix to Determine Ordinary, Partial and Multiple Coherence," *Proc. 77th Shock & Vibration Symposium,* Monterey, Calif., October 2006.
11. Underwood, Marcos A., and Tony Keller: "Understanding and Using the Spectral Density Matrix," *Proc. 76th Shock & Vibration Symposium,* Destin, Fla., November 2005.
12. Smallwood, D. O.: "Random Vibration Testing of a Single Test Item with a Multiple Input Control System," *Proc. Institute of Environmental Sciences,* Los Angeles, Calif., May 1982.
13. Keller, T., and M. A. Underwood: "On Acquiring and Analyzing Satellite Sine Vibration Test Data," *IMAC-XXV Conference,* Orlando, Fla., February 2007.

14. Smallwood, D. O.: "Shock Testing on Shakers by Using Digital Control," *IES Technology Monograph,* Institute of Environmental Sciences, Mount Prospect, Ill., 1986.
15. Nelson, D. B.: "Parameter Specification for Shaker Shock Waveform Synthesis—Damped Sines and Wavelets," *Proc. 60th Shock & Vibration Symposium,* vol. III, 1989.
16. Newland, D. E.: "Random Vibrations, Spectral and Wavelet Analysis," 3d ed., Longman Group, Essex, England, 1993.
17. Underwood, M. A.: "Applications of Optimal Control Concepts to Digital Shaker Control Systems," *Proc. Institute of Environmental Sciences,* Los Angeles, Calif., 1981.
18. Underwood M. A.: "Optimization of Classical Shock Waveforms," *Proc. Institute of Environmental Sciences,* Atlanta, Ga., 1982.
19. Underwood, M. A.: "Optimal Filtering Applied to Shock Response Spectrum Analysis," *Journal of the IEST,* 1983.
20. Underwood, M. A., and T. Keller: "Recent System Developments for Multi-Actuator Vibration Control," *Sound and Vibration,* October 2001.
21. Sinn, L. A., and M. A. Underwood: "Considerations for the Design and Development of Mixed-Mode Vibration Control Systems," *Proc. Institute of Environmental Sciences and Technology,* Anaheim, Calif., April 1995.
22. Smallwood, D. O.: *Journal of the Institute of Environmental Sciences,* **60**(5):27–28 (1996).
23. Underwood, Marcus A.: "Multi-exciter Testing Applications: Theory and Practice," *Proc. Institute of Environmental Sciences and Technology,* Anaheim, Calif., April 2002.
24. Smallwood, D. O., and T. L. Paez: "A Frequency Domain Method for the Generation of Partially Coherent Normal Stationary Time Domain Signals," *Shock and Vibration* **1**(1):373 (1994).
25. Underwood, M. A.: "A New Approach to System Identification, Using a Digitally Swept MultiExciter Sine Control System," *Proc. American Acoustical Association,* Syracuse, N.Y., 1989.
26. Hamma, G. A., R. C. Stroud, M. A. Underwood, W. B. Woyski, R. C. Taucher, and K. L. Cappel: *Sound and Vibration* **30**(4):20–28 (1996).
27. Underwood, Marcos A., and Tony Keller: "Rectangular Control of Multi-Shaker Systems; Theory and Some Practical Results," *Proc. Institute of Environmental Sciences and Technology,* ESTECH2003, Phoenix, Ariz., May 2003.
28. Underwood, Marcos A., and Tony Keller: "Applying Coordinate Transformations to Multi Degree of Freedom Shaker Control," *Proc. 74th Shock & Vibration Symposium,* San Diego, Calif., October 2003.
29. Smith, G.: "Digital Control Techniques for a Three-Axis Vibration Test System," *SAE Paper No. 801233,* October 1980.
30. Fisher, D., and M. Poshen: "Digital Control System for a Multiple-Actuator Shaker," *47th Shock & Vibration Bulletin,* Part 3, September 1977.
31. Underwood, M. A., and T. Keller: "Some Applications for the Testing of Civil Structures Using Multiple Shaker Excitation Techniques," *IMAC-XXVI Conference,* Orlando, Fla., February 2008.

CHAPTER 27
SHOCK TESTING MACHINES

Vesta I. Bateman

INTRODUCTION

Equipment must be sufficiently rugged to operate satisfactorily in the shock and vibration environments to which it will be exposed and to survive transportation to the site of ultimate use. To ensure that the equipment is sufficiently rugged and to determine what its mechanical faults are, it is subjected to controlled mechanical shocks on shock testing machines. *Mechanical shock* is a nonperiodic excitation (e.g., a motion of the foundation or an applied force) of a mechanical system that is characterized by suddenness and severity, and it usually causes significant relative displacements in the system. The severity and nature of the applied shocks are usually intended to simulate environments expected in later use or to be similar to important components of those environments. However, a principal characteristic of shocks encountered in the field is their variety. These field shocks cannot be defined exactly. Therefore, shock simulation can never exactly duplicate shock conditions that occur in the field.

There is no general requirement that a shock testing machine reproduce field conditions. All that is required is that the shock testing machine provide a shock test such that equipment which survives is acceptable under service conditions. Assurance that this condition exists requires a comparison of shock test results and field experience extending over long periods of time. This comparison is not possible for newly developed items. It is generally accepted that shocks that occur in field environments should be measured and that shock machines should simulate the important characteristics of shocks that occur in field environments or have a damage potential which by analysis is shown to be similar to that of a composite field shock environment against which protection is required.

A *shock testing machine* (frequently called a *shock machine*) is a mechanical device that applies a mechanical shock to an equipment under test. The nature of the shock is determined from an analysis of the field environment. Tests by means of shock machines usually are preferable to tests under actual field conditions for four principal reasons:

1. The nature of the shock is under good control, and the shock can be repeated with reasonable exactness. This permits a comparative evaluation of the equipment under test and allows exact performance specifications to be written.
2. The intensity and nature of shock motions can be produced that represent an average condition for which protection is practical, whereas a field test may involve only a specific condition that is contained in this average.

3. The shock machine can be housed at a convenient location with suitable facilities available for monitoring the test.
4. The shock machine is relatively inexpensive to operate, so it is practical to perform a great number of developmental tests on components and subassemblies in a manner not otherwise practical.

SHOCK MACHINE CHARACTERISTICS

DAMAGE POTENTIAL AND SHOCK RESPONSE SPECTRA

The damage potential of a shock motion is dependent upon the nature of an equipment subjected to the shock, as well as upon the nature and intensity of the shock motion. To describe the damage potential, a description of what the shock does to an equipment must be given—a description of the shock motion is not sufficient. To obtain a comparative measure of the damage potential of a shock motion, it is customary to determine the effect of the motion on simple mechanical systems. This is done by determining the maximum responses of a series of single-degree-of-freedom (SDOF) systems (see Chap. 2) to the shock motion and considering the magnitude of the response of each of these systems as indicative of the damage potential of the shock motion. The responses are plotted as a function of these natural frequencies. A curve representing these responses is called a *shock response spectrum* (SRS), or *response spectrum* (see Chap. 20). Its magnitude at any given frequency is a quantitative measure of the damage potential of a particular shock motion to an SDOF system with that natural frequency. This concept of the SRS originally was applied only to undamped SDOF systems, but the concept has been extended to include systems in which any specified amount of damping exists.

The response of a simple system can be expressed in terms of the relative displacement, velocity, or acceleration of the system. It is customary to define velocity and acceleration responses as $2\pi f$ and $(2\pi f)^2$ times the maximum relative displacement response, where f is frequency expressed in hertz. The corresponding response curves are called *displacement, velocity,* or *acceleration shock response spectra.* A more detailed discussion of SRS is given in Chap. 20.

Of the three motion parameters (displacement, velocity, and acceleration) describing a shock spectrum, velocity is the parameter of greatest interest from the viewpoint of damage potential. This is because the maximum stresses in a structure subjected to a dynamic load typically are due to the responses of the normal modes of the structure, that is, the responses at natural frequencies (see Chap. 21). At any given natural frequency, stress is proportional to the modal (relative) response velocity.[1] Specifically,

$$\sigma_{max} = C v_{max} \sqrt{E\rho} \tag{27.1}$$

where σ_{max} = maximum modal stress in the structure
v_{max} = maximum modal velocity of the structural response
E = Young's modulus of the structural material
ρ = mass density of the structural material
C = constant of proportionality dependent upon the geometry of the structure (often assumed for complex equipment to be $4 < C < 8$)[2]

Of course, if the SRS for a test machine–generated shock is computed solely to validate that test results comply with a specified SRS, or for comparison to the SRS computed from measured shocks in a service environment, then either displacement or acceleration SRS are as meaningful as a velocity SRS. However, if the maximum stress in the structure subjected to the shock is of primary interest, the velocity SRS is the most applicable.

MODIFICATION OF CHARACTERISTICS BY REACTIONS OF TEST ITEM

The shock motion produced by a shock machine may depend upon the mass and frequency characteristics of the item under test. However, if the effective weight of the item is small compared with the weight of the moving parts of the shock machine, its influence is relatively unimportant. Generally, however, the reaction of the test item on the shock machine is appreciable and it is not possible to specify the test in terms of the shock motions unless large tolerances are permissible. The test item acts like a dynamic vibration absorber (see Chap. 6). If the item is relatively heavy, this causes the shock response spectra of the exciting shock to have minima at the frequencies of the test item; it also causes its mounting foundation to have these minima during shock excitation at field installations. Shock tests and design factors are sometimes established on the basis of an envelope of the maximum values of SRS. However, maximum stresses in the test item will most probably occur at the antiresonance frequencies where the SRS exhibits minimum values. To require that the item withstand the upper limit of spectra at these frequencies may result in overtesting and overdesign. Considerable judgment is therefore required both in the specification of shock tests and in the establishment of theoretical design factors on the basis of field measurements. See Chap. 18 for a more complete discussion of this subject.

DOMINANT FREQUENCIES OF SHOCK MACHINES

The shock motion produced by a shock machine may exhibit frequencies that are characteristic of the machine. These frequencies may be affected by the equipment under test. The probability that these particular frequencies will occur in the field is no greater than the probability of other frequencies in the general range of interest. A shock test, therefore, discriminates against equipment having elements whose natural frequencies coincide with frequencies introduced by the shock machine. This may cause failures to occur in relatively good equipment whereas other equipment, having different natural frequencies, may pass the test even though of poorer quality. Because of these factors, there is an increasing tendency to design shock machines to be as rigid as possible, so that their natural frequencies are above the range of frequencies that might be strongly excited in the equipment under test. The shock motion is then designed to be the simplest shape pulse that will give a desired shock motion or response spectrum.

CALIBRATION

A *shock machine calibration* is a determination of the shock motions or response spectra generated by the machine under standard specified conditions of load, mounting arrangements, methods of measurement, and machine operation. The purpose of the calibration is not to present a complete study of the characteristics of the machine but rather to present a sufficient measure of its performance to assure the user that the machine is in a satisfactory condition. Measurements should therefore be made under a limited number of significant conditions that can be accurately specified and easily duplicated. Calibrations are usually performed with deadweight loads rigidly attached to the shock machine.

The statement of calibration results must include information relative to all factors that may affect the nature of the motion. These include the magnitude, dimensions, and type of load; the location and method of mounting of the load; factors related to the operation of the shock machine; the locations and mounting arrangements of transducers; and the frequency range over which the measurements extend.

SPECIFYING A SHOCK TEST

Two methods of specification are employed in defining a shock test: (1) a specification of the shock motions (or shock response spectra) to which the item under test is subjected and (2) a specification of the shock machine, the method of mounting the test item, and the procedure for operating the machine.

The first method of specification can be used only when the shock motion can be defined in a reasonably simple manner and when the application of forces is not so sudden as to excite structural vibration of significant amplitude in the shock machine. If equipment under test is relatively heavy, and if its normal modes of vibration are excited with significant amplitude, the shock motions are affected by the load; then the specified shock motions should be regarded as nominal. If comparable results are to be obtained for tests of different machines of the same type, the methods of mounting and operational procedures must be the same.

The second method of specification for a shock test assumes that it is impractical to specify a shock motion because of its complexity; instead, the specification states that the shock test shall be performed in a given manner on a particular machine. The second method permits a machine to be developed and specified as a standard shock testing machine. Those who are responsible for the specification then should ensure that the shock machine generates appropriate shock motions. This method avoids a difficulty that arises in the first method when measurements show that the shock motions differ from those specified. These differences are to be expected if load reactions are appreciable and complex. It should be noted, however, that ANSI standards that specify a standard shock machine were withdrawn in 2004.

A shock testing machine must be capable of reproducing shock motions with good precision for purposes of comparative evaluation of equipment and for the determination as to whether a manufacturer has met contractual obligations. Moreover, different machines of the same type must be able to provide shocks of equivalent damage potential to the same types of equipment under test. Precision in machine performance, therefore, is required on the basis of contractual obligations and for the comparative evaluation of equipments even though it is not justified on the basis of a knowledge of field conditions.

Sometimes equipment under test may consistently fail to meet specification requirements on one shock machine but may be acceptable when tested on a different shock machine of the same type. The reason for this is that small changes of natural frequencies and of internal damping, of either the equipment or the shock machine, may cause large changes in the likelihood of failure of the item. Results of this kind do not necessarily mean that a test has been performed on a faulty machine; normal variations of natural frequencies and internal damping from machine to machine make such changes possible. However, standard calibrations of shock machines should be made from time to time to ensure that significant changes in the machines have not occurred.

SHOCK TESTING MACHINES

CHARACTERISTIC TYPES OF SHOCKS

The shock machines described below are grouped according to the types of shocks they produce. When a machine can be classified under several headings, it is placed in the one for which it is primarily intended. One characteristic shared by all shock machines is that the motions they produce are sudden and likely to create significant inertial forces in the item under test. The types of shock shown in Fig. 27.1 are classified as (*A*) through (*D*), simple shock pulses, whose shapes can be expressed in a practical mathematical form; (*E*), a single complex shock; and (*F*), a multiple shock. In contrast to a simple shock pulse specification, the motions illustrated in Fig. 27.1*E* and *F* often are the result of a shock test in which the shock testing machine, the method of mounting, and machine operations were specified.

Velocity Shocks. A *velocity shock* is produced by a sudden change in the net velocity of the structure supporting the item under test. When the duration of the shock is short compared to the periods of the principal natural frequencies of the item under test, a velocity shock is said to have occurred. Figure 27.1*A* shows a nearly instantaneous change in velocity. The shocks shown in Fig. 27.1*B*, *C*, and *D* are also considered velocity shocks if the above shortness criterion is met. The full period of the half-sine wave in Fig. 27.1*B* is $\tau/2$, as shown. Velocity shocks produce substantial energy at the principal natural frequencies of the item under test. This is illustrated in Figs. 27.2 and 27.3, which show the shock response spectra (computed with a zero damping ratio) for the half-sine and sawtooth acceleration pulses with duration T in Fig. 27.1*B* and *D*, respectively. Note in both cases that the values of the velocity SRS are uniform at all frequencies below about $Tf = 0.2$. Hence, from Eq. (27.1), they have the potential to cause substantial damage to the basic structure of the item under test, assuming the item has natural frequencies below $f = 0.2/T$ Hz.

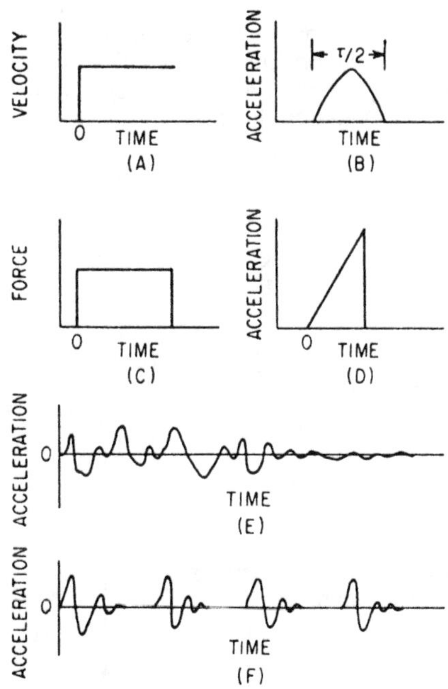

FIGURE 27.1 Characteristic types of shocks. (*A*) Velocity shock, or step velocity change. (*B*) Simple half-sine acceleration shock pulse. (*C*) Rectangular force pulse. (*D*) Sawtooth acceleration pulse. (*E*) Single complex shock. (*F*) Multiple shock.

Displacement Shocks. Some shock test machines produce a sequence of two or more velocity shocks with equal and opposite velocity magnitudes such

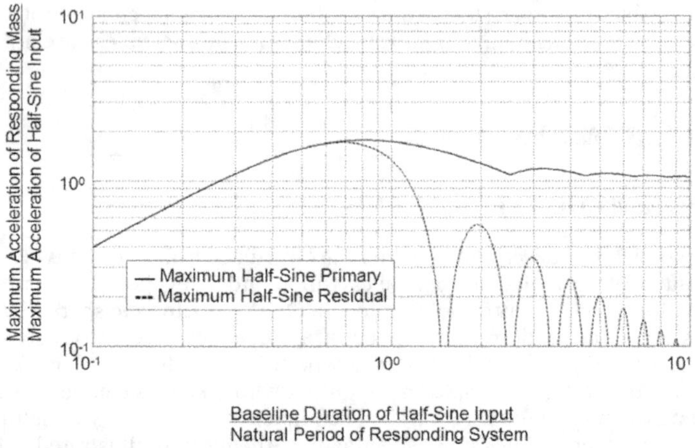

FIGURE 27.2 Normalized shock response spectra for half-sine input pulse in Figure 27.1*B*.

that the test item experiences no net velocity change. For example, the half-sine acceleration pulse in Fig. 27.1*B* might be followed by a second half-sine pulse of equal magnitude in the opposite direction. If the time between the two equal and opposite acceleration pulses is longer than the duration of the individual pulses, a substantial displacement of the test item between the positive and negative velocity changes will occur. This type of shock is commonly called a *displacement shock.* Such shocks have a damage potential similar to that of velocity shocks.

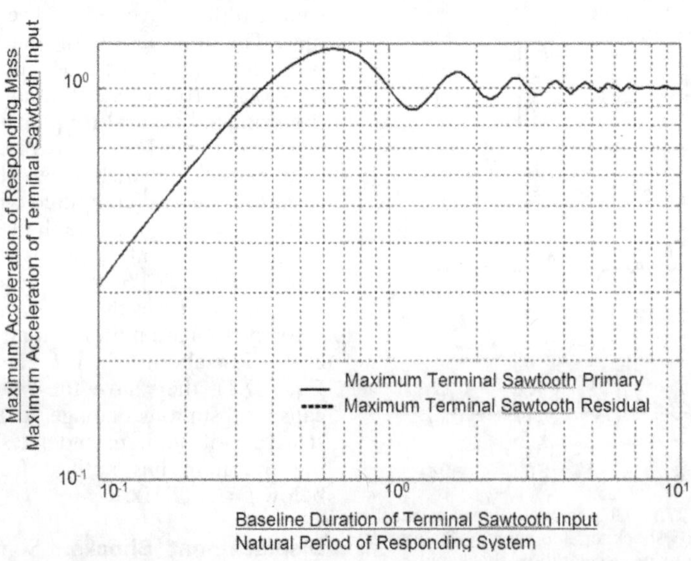

FIGURE 27.3 Normalized shock response spectra for terminal sawtooth input pulse in Figure 27.1*D*.

High-Frequency Shocks. Metal-to-metal impacts that do not result in a net velocity change of the item under test create high-acceleration, high-frequency oscillations in the vicinity of the impact. Figure 27.1*E* and *F* are examples of high-frequency shocks. Since the frequency range of these shocks often exceeds the principal natural frequencies of the item under test, the shocks usually are not readily transmitted far from the point of their creation. Consequently, this type of shock lacks the damage potential of velocity shocks for all but small and/or brittle components of the item under test. Common sources of high-frequency shocks include pyrotechnic devices, which produce what are commonly referred to as *pyroshocks.* Laboratory machines and techniques for the simulation of pyroshocks are detailed separately in Chap. 28.

SIMPLE SHOCK PULSE MACHINES

Although shocks encountered in the field are usually complex in nature (for example, see Fig. 27.1*E*), it is frequently advantageous to simulate a field shock by a shock of mathematically simple form. This permits designers to calculate equipment response more easily and allows tests to be performed that can check these calculations. This technique is additionally justifiable if the pulses are shaped so as to provide shock response spectra similar to those obtained for a suitable average of a given type of field conditions. Machines are therefore built to provide these simple shock motions. However, note that the motions provided by actual machines are only ideally simple. The ideal outputs may be given as nominal values; the actual outputs can only be determined by measurement.

Drop Tables. A great variety of drop testers are used to obtain acceleration pulses having magnitudes ranging from 80,000*g* down to a few *g*. The machines each include a carriage (or table) on which the item under test is mounted; the carriage can be hoisted up to some required height and dropped onto an anvil. Guides are provided to keep the carriage properly oriented. For practical reasons, drop tables driven by gravity alone are limited to velocity changes of one-and-a-half-times the free-fall velocity change. When large velocity changes are required, the carriage may be accelerated downward by a means other than gravity. Frequently, parts of the carriage, associated with its lifting and guiding mechanism, are flexibly mounted to the rigid part of the carriage structure that receives the impact. This is to isolate the main carriage structure from its flexible appendages so as to retain the simple pulse structure of the stopping acceleration.

A typical drop table machine is shown in Fig. 27.4. The desire acceleration pulse shape is obtained using a programming device between the impacting surfaces. Devices ranging from liquid programmers to simple pads of elastomeric materials can be used. The elastic shock cords accelerate the table to create velocities beyond those that can be obtained with a free fall. Machines of this type can produce acceleration waveforms that closely approximate many different types of velocity shocks, such as the half-sine and terminal sawtooth acceleration pulses in Fig. 27.1*B* and *D*, respectively. These machines can also be used to characterize the impact response of materials and components placed between the shock table and the seismic mass, provided a technique to eliminate the structural response of the shock machine is used.[3] Additionally, there are a variety of specialized pneumatic machines available. For example, smaller, more compact machines have been developed for miniaturized components,[4] and large-displacement, low-frequency machines have been developed for large components.[5]

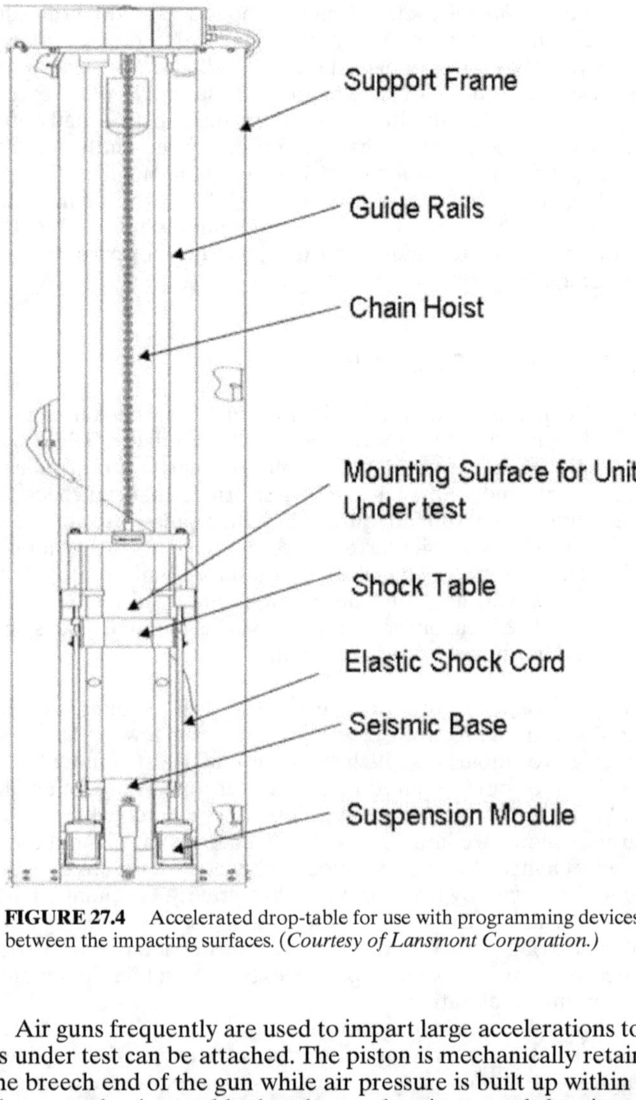

FIGURE 27.4 Accelerated drop-table for use with programming devices between the impacting surfaces. (*Courtesy of Lansmont Corporation.*)

Air Guns. Air guns frequently are used to impart large accelerations to pistons on which items under test can be attached. The piston is mechanically retained in position near the breech end of the gun while air pressure is built up within the breech. A quick-release mechanism suddenly releases the piston, and the air pressure projects the piston down the gun barrel. The muzzle end of the gun is closed so that the piston is stopped by compressing the air in the muzzle end. Air bleeder holes may be placed in the gun barrel to absorb energy and to prevent an excessive number of oscillations of the piston between its two ends.

A variety of such guns can provide the acceleration pulses shown in Fig. 27.5*A* and *B*. The peak accelerations may extend from a maximum of about 1000g for the large-diameter (21 in., 53 cm) guns up to 200,000g for small-diameter (2 in., 5 cm) guns. The pulse length varies correspondingly from about 50 to 3 milliseconds. The maximum piston velocity varies from about 400 to 750 ft/sec (122 to 229 m/sec). The maximum velocities are not dependent upon piston diameter.

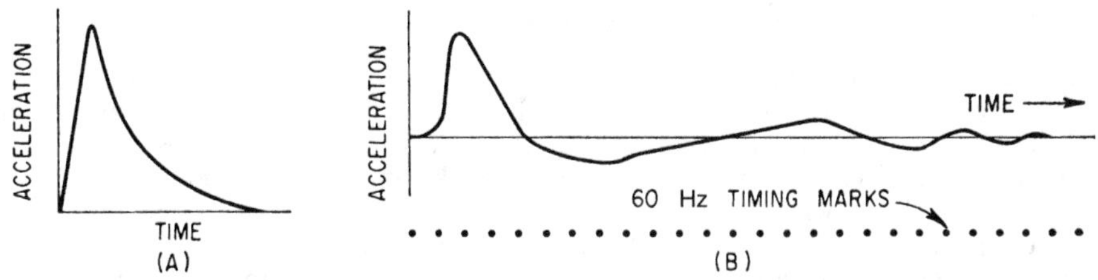

FIGURE 27.5 Typical acceleration-time curves for (*A*) 5-in. (13-cm) air gun; (*B*) 21-in. (53-cm) air gun.

High-acceleration gas guns have been developed for testing electronic devices. The items under test are attached to the piston. The gun consists of a barrel (cylinder) that is closed at the muzzle end but which has large openings to the atmosphere a short distance from the muzzle end. The piston is held in place while a relatively low-pressure gas (usually air or nitrogen) is applied at the breech end of the gun. The piston is then released, whereby it is accelerated over a relatively long distance until it reaches the position along the length of the cylinder that is open to the atmosphere. This initial acceleration is of relatively small magnitude. After the piston has passed these openings, it is stopped by the compression of gas in the short closed end of the cylinder. This results in a reverse acceleration of relatively large magnitude. (Sometimes an inert gas, such as nitrogen, is used in the closed end to prevent explosions which might be caused by oil particles igniting under the high temperatures incident to the compression.) Thus, in contrast to the previously described devices, the major acceleration pulse is delivered during stopping rather than starting. An advantage of this latter technique is that the difficult problem of constructing a quick-release mechanism for the piston, which will work satisfactorily under the large forces exerted by the piston, is greatly simplified.

Vibration Machines. Electrodynamic, hydraulic, and pneumatic vibration machines provide a ready and flexible source of shock pulses, so long as the pulse requirements do not exceed the force and motion capabilities of the selected machine. See Chap. 25 for information.

Test Load Reactions. In the above description of the output of shock machines designed to deliver simple shock pulses of adjustable shapes, it is assumed that the load imposed on the machine by the item under test has little effect on the shock motions. This is true only when the effective weight of the load is negligibly small compared with that of the shock machine mounting platform. If the effective weight of the load is independent of frequency, i.e., if it behaves as a rigid body, it is simple to compensate for the effect of the load by adjusting machine parameters. However, when the load is flexible and the reactions of excited vibrations are appreciable, the motions of the shock machine platform are complex. Specifications involving the use of these types of machines should require that the mounting platform have no significant natural frequencies below a specified frequency. The weight of this platform together with that of all rigidly attached elements, exclusive of the test load, also should be specified. Pulse shapes may then be specified for motions of this platform or for the platform together with given deadweight loads. These may be specified as nominal values for test loads, but it is neither practical nor desirable to require that the pulse shape be maintained in simple form for complex loads of considerable mass.

COMPLEX SHOCK PULSE MACHINES

Because of the infinite variety of shock motions possible under field conditions, it is not practical or desirable to construct a shock machine to reproduce a particular shock that may be encountered in the field. However, it is sometimes desirable to simulate some average of a given type of shock motion. To accomplish this may require that the shock machine deliver a complex motion. A shock of this type cannot be specified easily in terms of the shock motions, since the motions are very complex and dependent on the nature and the mounting of the load. It is customary, therefore, to specify a test in terms of a shock machine, the conditions for its operation, and a method of mounting the item under test.

High-Impact Shock Machines. The Navy high-impact shock machines are designed to simulate shocks of the nature and intensity that might occur on a ship exposed to severe but sublethal, noncontact, underwater explosions. Such severe shocks produce motions that extend throughout the ship. Equipment intended for shipboard use can demonstrate its ability to withstand the shock simulations produced by these high-impact shock machines and thus be considered capable of withstanding the actual underwater explosion environment.

Lightweight Machines.[6,7] The lightweight high-impact shock machine, shown in Fig. 27.6, is used for testing equipment weighing up to about 350 lb (159 kg). Equipment under test is attached to the anvil plate *A*. Method of attachment is constrained to resemble closely the eventual field attachments. The anvil is struck on the backside by the pendulum hammer *C*, or the anvil is rotated 90° on a vertical axis and struck on the end by the pendulum hammer. The drop hammer *B* can be made to strike the top of the anvil, thus providing principal shock motions in the third orthogonal direction. Shock response spectra of shock motions generated by this machine are shown in Fig. 27.7 (these results were computed with a damping ratio of about 0.01). The SRS for the motion at the center of the plate illustrates the amplification of the spectrum level at a natural frequency of the plate (about 100 Hz) and some attenuation at higher frequencies.

Medium-Weight Machines.[6,7] This machine is used to test equipment that, with its supporting structures, weighs up to 7400 lb (3357 kg). This machine consists principally of a 3000-lb (1361-kg) hammer and a 4500-lb (2041-kg) anvil. Loads are not attached directly to the rigid anvil structure but rather to a group of steel channel beams which are supported at their ends by steel members, which in turn are attached to the anvil table. The number of channels employed is dependent on the weight of the load and is such as to cause the natural frequency of the load on these channels to be about 60 Hz. The general nature of the shock is complex, similar to that of the lightweight machine. Little of the high-amplitude, high-frequency components of the shock motions are transmitted to the load.

Heavy-Weight Machines.[6,8] The *floating shock platform* (FSP), and the *large floating shock platform* (LFSP) are high-load-capacity shock machines of the high-impact category. They are rectangular barges fitted with semicylindrical canopies within which test items are installed as they are aboard ship. The shock motions comprising the test series are generated by detonating explosive charges beneath the water surface at various distances.

Hopkinson Bar. When shock testing requires extremely high *g* levels for light loads (for example, calibration of accelerometers), the Hopkinson bar has proven useful. A controlled-velocity projectile is impacted on the end of a metallic bar, caus-

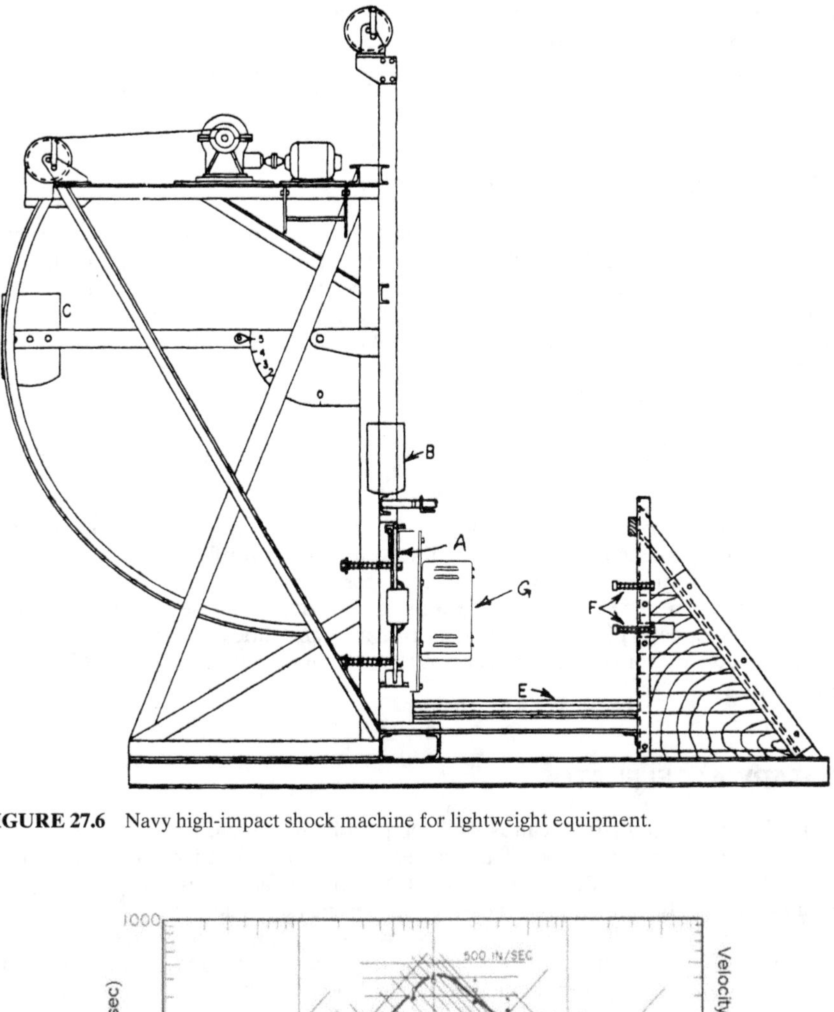

FIGURE 27.6 Navy high-impact shock machine for lightweight equipment.

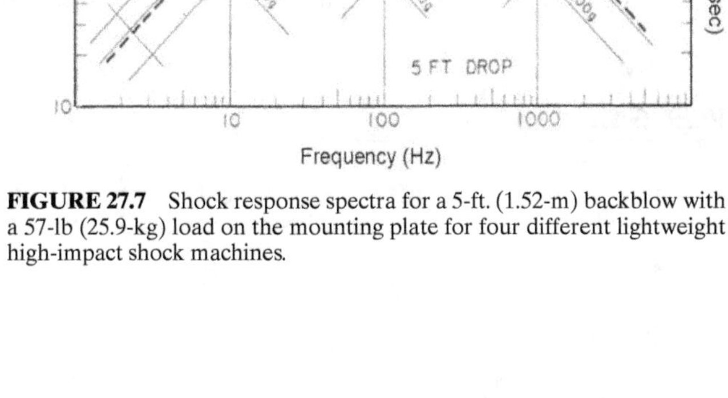

FIGURE 27.7 Shock response spectra for a 5-ft. (1.52-m) backblow with a 57-lb (25.9-kg) load on the mounting plate for four different lightweight high-impact shock machines.

ing a stress wave of known magnitude to travel along the bar. Often, the magnitude of the stress wave is measured as it passes the middle of the bar. The item under test is attached to the extreme end of the bar and experiences a high-g rapid rise time acceleration when the stress wave arrives at that position.[9] See Fig. 11.12.

MULTIPLE-IMPACT SHOCK MACHINES

Many environments, particularly those involving transportation, subject equipment to a relatively large number of shocks. These are of lesser severity than the shocks of major intensity that have been considered above, but their cumulative effect can be just as damaging. It has been observed that components of equipment that are damaged as a result of a large number of shocks of relatively low intensity are usually different from those that are damaged as a result of a few shocks of a relatively high intensity. The damage effects of a large number of shocks of low intensity cannot generally be produced by a small number of shocks of high intensity. Separate tests are therefore required so that the multiple number of low-intensity shocks are properly emulated.

Vibration Machines. Electrodynamic, hydraulic, and pneumatic vibration testing machines provide a ready and flexible source of multiple shock pulses so long as the pulse requirements do not exceed force and motion capabilities of the selected machine. They can be programmed to provide a series of different shock pulses or to repeat a particular shock motion as many times as desired and to establish the necessary initial conditions prior to each shock pulse. See Chap. 25 for more information.

ROTARY ACCELERATOR

A quick-starting centrifuge can is used to quickly attain and maintain an acceleration for a long period of time. The accelerator consists of a rotating arm which is suddenly set into motion by an air-operated piston assembly. The test object is mounted on a table attached to the outer end of the arm. The table swings on a pivot so that the resultant direction of the acceleration is always along a fixed axis of the table. Initially the resultant acceleration is caused largely by angular acceleration of the arm, so this axis is in a circumferential direction. As the centrifuge attains its full speed, the acceleration is caused primarily by centrifugal forces, so this table axis assumes a radial direction. These machines are built in several sizes. They require between 5 and 60 milliseconds to reach the maximum value of acceleration. For small test items (8 lb, 3.6 kg), a maximum acceleration of 450g is attainable; for heavy test items (100 lb, 45.4 kg), the maximum value is about 40g.

ACKNOWLEDGMENT

Some of the material in this chapter was originally prepared by Richard H. Chalmers, late Consulting Engineer, Induced Environments Consultants, San Diego, California.

REFERENCES

1. Gaberson, H. A., and R. H. Chalmers: *Shock and Vibration Bull.*, **40**(2):31 (1969).
2. Piersol, A. G.: *J. IEST,* **44**(1):23 (2001).
3. Bateman, V. I., F. A. Brown, M. A. Nusser, and L. H. Swanson: "Segmented Aluminum Honeycomb Characteristics in Dynamic Crush Environments," *Proc. 71st Shock and Vibration Symposium,* Arlington, Va. 2000.
4. Henderson, G. R.: "Shock Testing Miniaturized Products," *Evaluation Engineering,* April 2002.
5. Brooks, R., and S. McCampbell: "Battle Ready COTS: Shock Simulation Comes of Age," *VMEbus Systems,* October 2003.
6. Military Specification. "Shock Tests HI (High Impact); Shipboard Machinery, Equipment and Systems, Requirements for," MIL-S-901D (Navy), March 17, 1989.
7. Scavuzzo, R. J., and H. C. Pusey: "Naval Shock Analysis and Design," *Shock and Vibration Information and Analysis Center Monograph SVM-17,* 2000.
8. Clements, E. W.: "Characteristics of the Navy Large Floating Shock Platform," U.S. Naval Research Laboratory Report 7761, 15 July 1974. (Obtainable from the Shock and Vibration Information Analysis Center.)
9. Anon.: "Methods for the Calibration of Vibration and Shock Transducers," Part 22: "Shock Calibration by Comparison to a Reference Transducer," ISO 16063–22:2005.

CHAPTER 28
PYROSHOCK TESTING

Vesta I. Bateman
Neil T. Davie

INTRODUCTION

Pyroshock, also called *pyrotechnic* shock, is the response of a structure to high-frequency (thousands of hertz), high-magnitude stress waves that propagate throughout the structure as a result of an explosive event such as the explosive charge to separate two stages of a multistage rocket. The term *pyrotechnic shock* originates from the use of propellants such as black powder, smokeless powder, nitrocellulose, and nitroglycerin in devices common to the aerospace and defense industries. These devices include pressure squibs, explosive nuts and bolts, latches, gas generators, and air bag inflators.[1] The term *pyroshock* is derived from pyrotechnic shock, but both terms are used interchangeably in the industry and its literature. A pyroshock differs from other types of mechanical shock in that there is very little rigid-body motion (acceleration, velocity, and displacement) of a structure in response to the pyroshock. The pyroshock acceleration time history measured on the structure is oscillatory and approximates a combination of decayed sinusoidal accelerations with very short duration in comparison to mechanical shock described in Chap. 27. The characteristics of the pyroshock acceleration time history vary with the distance from the pyroshock event. In the near field, which is very close to the explosive event, the pyroshock acceleration time history is a high-frequency, high-amplitude shock that may have transients with durations of microseconds or less. In the far field, which is far enough from the event to allow structural response to develop, the acceleration time history of the pyroshock approximates a combination of decayed sinusoids with one or more dominant frequencies. The dominant frequencies are usually much higher than that in a mechanical shock and reflect the local modal response of the structure. The dominant frequencies are generally lightly damped. However, since the frequencies are so high, it typically takes less than 20 milliseconds for the pyroshock response to dampen out and return to zero. Satellite, aerospace, and weapon components are often subjected to pyroshocks created by devices such as explosive bolts and pyrotechnic actuators. Pyroshock structural response is also found in ground-based applications in which there is a sudden release of energy, such as the impact of a structure by a projectile.

Pyroshock was once considered to be a relatively mild environment due to its low-velocity change and high-frequency content. Although it rarely damages structural members (except where the fracture of a structural element is intended), pyroshock can easily cause failures in electronic components that are sensitive to the high-frequency pyroshock energy. The types of failures caused by pyroshock commonly include relay chatter, hard failures of small circuit components, and the dislodging of contaminants (e.g., solder balls), which cause short circuits. A significant number of flight failures have been attributed to pyroshock compared to other types of shock or vibration sources, and, in one case, an extensive database of the failures has been compiled.[2] Designers must rely on testing for qualifications of their systems and components that will be exposed to pyroshock environments in the absence of analytical techniques to predict structural response to a pyroshock. Failures can be reduced by implementing a qualification testing program for components exposed to a pyroshock environment. This chapter describes the characteristics of pyroshock environments, measurement techniques, test specifications, and simulation techniques.

PYROSHOCK CHARACTERISTICS

COMPARISON OF NEAR-FIELD, MID-FIELD, AND FAR-FIELD CHARACTERISTICS

The detonation of an explosively actuated device produces high-frequency transients in the surrounding structure. The specific character of these acceleration transients depends on various parameters including: (1) the type of pyrotechnic source, (2) the geometry and properties of the structure, and (3) the distance from the source. Due to the endless combinations of these parameters, sweeping conclusions about pyroshock characteristics cannot be made; however, the following paragraphs describe useful characteristics of typical pyroshock environments.

A pyrotechnically actuated device produces a nearly instantaneous pressure on surfaces in the immediate vicinity of the device. As the resulting stress waves propagate through the structure, the high-frequency energy is gradually attenuated due to various material damping and structural damping mechanisms. In addition, the high-frequency energy is transferred or coupled into the lower-frequency modes of the structure. The typical pyroshock acceleration transient thus has roughly the appearance of a multifrequency decayed sinusoid (i.e., the envelope of the transient decays and is symmetric with respect to the positive and negative peaks). The integral of the typical transient also has these same characteristics.[3] In most cases, the initial portion of the acceleration transient exhibits a brief period during which the amplitudes of the peaks are increasing prior to the decay described above (see Figs. 28.1 and 28.2). This is a result of the interaction of stress waves as they return from various locations in the structure.

A pyrotechnically actuated device imparts very little impulse to a structure since the high forces produced are acting for only a short duration and are usually internal to the structure. The net rigid-body velocity change resulting from a pyroshock is thus very low relative to the peak instantaneous velocity seen on the integral of the acceleration transient. Rigid-body velocity changes are commonly less than 1 m/sec. The duration of a pyroshock transient depends on the amount of damping in a particular structure, but it is commonly 5 to 20 milliseconds in duration.

Pyroshock may be subdivided into three general categories. *Near-field pyroshock* occurs close to the pyrotechnic source before significant energy is transferred to structural response. It is dominated by the input from the source and contains very high frequency and very high g energy. This energy is distributed over a wide frequency range and is not generally dominated by a few selected frequencies. *Mid-field and far-field pyroshock* environments are found at a greater distance from the source where significant energy has transferred into the lower-frequency structural response. They contain lower-frequency and lower-g energy than near-field pyro-

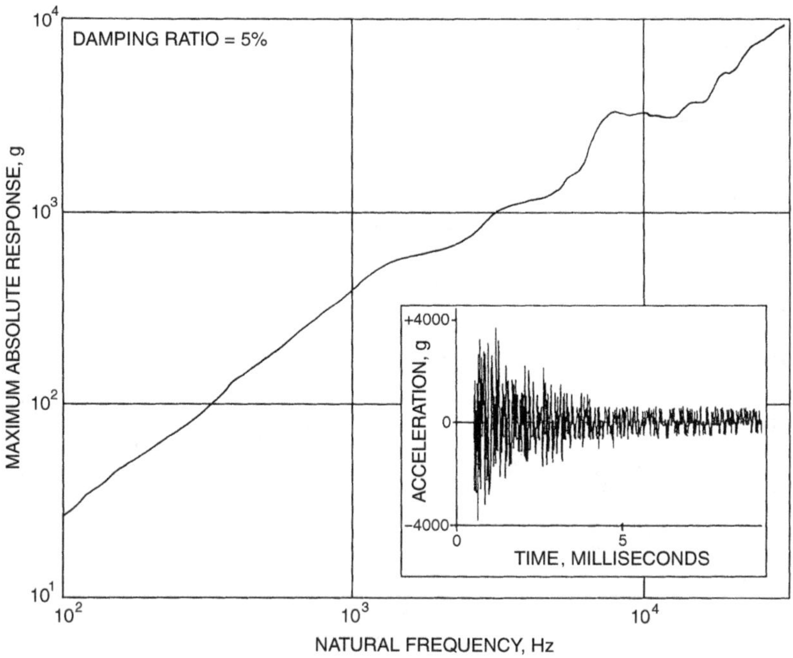

FIGURE 28.1 Shock response spectrum and acceleration time history for a near-field pyroshock. The shock response spectrum is calculated from the inset acceleration time history using a 5 percent damping ratio.

shock; most of the energy is usually concentrated at one or a few frequencies which correspond to dominant structural mode(s).

A more detailed discussion of shock response spectrum (SRS; see Chap. 20 for definition) applications is given later in this chapter, but it is introduced here as a means of describing pyroshock characteristics. Many mid-field and far-field pyroshock environments have a *typical* SRS shape as illustrated in Fig. 28.2, which shows an actual mid-field pyroshock acceleration transient along with its associated SRS. The SRS initially increases with frequency at a slope of 9 to 12 dB/octave, followed by an approximately constant or slightly decreasing amplitude. The frequency at which the slope changes is called the *knee frequency,* and it corresponds to a dominant frequency in the pyroshock environment. The knee frequency is often between 3 and 10 kHz for mid-field, but it could be higher or lower in some cases. Near-field pyroshock may also exhibit this typical pyroshock SRS except with a higher knee frequency. However, since near-field pyroshock usually has broadband frequency content, its SRS often exhibits a more complex shape that contains numerous excursions but on average follows a 6-dB/octave slope over the entire frequency range of interest. Figure 28.1 shows an example of this type of near-field SRS. Far-field shock is similar in shape to mid-field shock except the knee frequency is below 3kHz.

No fixed rules define at what distance from the pyrotechnic source the near-field pyroshock ends and the mid-field pyroshock begins. It is more appropriate to classify pyroshock according to the various test techniques that are suitable to employ in each case.

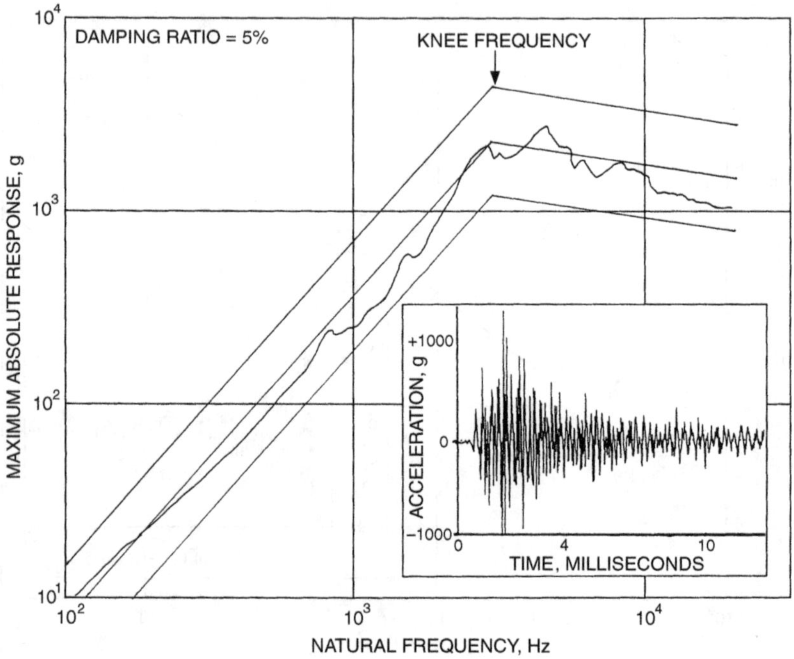

FIGURE 28.2 A typical shock response spectrum and acceleration time history for a mid-field pyroshock. The shock response spectrum is calculated from the inset acceleration time-history using a 5 percent damping ratio. The straight lines indicate tolerance bands (typically ±6 dB as shown) which might be applied for qualification test specification.

TEST TECHNIQUES FOR NEAR- AND FAR-FIELD PYROSHOCK

The pyroshock simulation techniques described in this chapter fall into two categories: (1) pyrotechnically excited simulations and (2) mechanically excited simulations. A short-duration mechanical impact on a structure causes a response similar to that produced by a pyrotechnic source. Although these mechanically excited simulations can be carried out with lower cost and better control than pyrotechnically excited simulations, they cannot produce the very high frequencies found in near-field pyroshock. Mechanically excited simulations allow control of dominant frequencies up to about 10 kHz (or higher for very small test items). For environments requiring higher-frequency content, a pyrotechnically excited technique is usually more appropriate. The following general guidelines apply in selecting a technique for simulating pyroshock:

Near-field pyroshock. For a test that requires frequency control up to and above 10 kHz, a pyrotechnically excited simulation technique is usually required.

Mid-field pyroshock. For a test that requires frequency control from 3 to 10 kHz, a mechanically excited simulation technique other than a shaker-induced shock is usually required.

Far-field pyroshock. For a test that requires frequency control no higher than 3 kHz, a mechanically excited simulation technique is usually acceptable.

These guidelines are not rigid rules, but they provide a reasonable starting point when planning a pyroshock simulation test.

QUANTIFYING PYROSHOCK FOR TEST SPECIFICATION

An intrinsic characteristic of pyroshock is its variability from one test to another. That is, even though great care has been taken with the test technique, the measured response in both the near and the far fields may vary a great deal from test to test. This variability occurs in the situation where actual explosive devices are used and in the laboratory where more controlled techniques are employed. As a result, various techniques have been sought to quantify pyroshock for test specification. The purpose of these techniques is to define the pyroshock in a manner that can be reproduced in the laboratory and can provide a consistent evaluation for hardware that must survive pyroshock in field environments. All techniques require that a measurement be made of the actual pyroshock event at or near the location of the subsystem or component that will be tested. The measurement may be acceleration, velocity, or displacement, but acceleration is the most widely used measure. The measurement is then used with one of the techniques below to obtain a test specification for pyroshock. The *shock response spectrum* is considered to be conservative and a potential overtest of components and subsystems. However, components and subsystems that survive laboratory tests specified using SRS generally survive pyroshock field environments, although they may be overdesigned. Because aerospace systems require lightweight components and subsystems, other techniques such as temporal moments and shock intensity spectrum have been developed so that laboratory tests can more closely simulate actual pyroshock events and allow tighter design margins.

Shock Response Spectra. By far the most widely used technique for quantifying pyroshock is the shock response spectrum. This technique provides a measure of the effect of the pyroshock on a simple mechanical model with a single degree of freedom (SDOF). Generally, a measured acceleration time history is applied to the model, and the maximum acceleration response is calculated. The damping of the model is held constant (at a value such as 5 percent) for these calculations. An ensemble of maximum absolute-value acceleration responses is calculated for various natural frequencies of the model and the result is a *maximax shock response spectrum.* A curve representing these responses as a function of damped natural frequency is called a shock response spectrum (see Chap. 20), and is normally plotted with log-log scales. Velocity and displacement SRS may be computed (see Chap. 27), but are not commonly used for pyroshock specification. The SRS for pyroshock has a characteristically steep slope at low frequencies of 12 dB/octave that is a direct result of the minimal velocity change occurring in a pyroshock. Occasionally, a pyrotechnic device, such as an explosive bolt cutter, is combined with another mechanism, such as a deployment arm, to position components for a particular event sequence. In this case, a distinct velocity change is combined with the pyroshock event, and the low-frequency slope of the SRS will reflect this velocity change. For a typical far-field pyroshock, the low-frequency slope changes at the *knee frequency,* and the SRS approaches a constant value at high frequencies, that is, the peak acceleration in the time domain as shown in Fig. 28.2. A typical near-field pyroshock may have this shape or the shape shown in Fig. 28.1. Conventionally, tolerance bands of ±6 dB are drawn about a straight-line approximation of the SRS for laboratory testing. An example of a typical maximax SRS is shown in Fig. 28.2 with the conventional ±6-dB tolerance bands.

Band-Limited Temporal Moments. The method of temporal moments may be used for modeling shocks whose time durations are too short for nonstationary models and that contain a large random contribution.[4] The method uses the magnitude of the Fourier spectrum in the form of an energy spectrum (Fourier spectrum magnitude squared) that is smoothed or formed from an ensemble average to generate statistically significant values. Temporal moments of the time histories are used to represent how the energy is distributed in time. The moments are analogous to the moments of the probability density functions and provide a convenient method to describe the envelopes of complicated time histories such as pyroshock. The ith temporal moment $m_i(a)$ of a time history $f(t)$, about a time location a, is defined as

$$m_i(a) = \int_{-\infty}^{+\infty} (t-a)^i [f(t)]^2 \, dt \tag{28.1}$$

The time history energy E is given by

$$E = \frac{1}{2\pi} \int_{-\infty}^{+\infty} |F(\omega)|^2 \, d\omega \tag{28.2}$$

where $F(\omega)$ is the Fourier transform of $f(t)$. The first five moments are used in the temporal moments technique. The zeroth-order moment m_0 is the integral of the magnitude squared of the time history and is called the time-history energy. The first moment normalized by the energy is called the central time τ. A central moment is a moment computed about the central time, i.e., $a = \tau$. The second central moment is normalized by the energy and is defined as the mean-square duration of the time history. The third central moment normalized by the energy is defined as the skewness and describes the shape of the time history. The fourth central moment normalized by the energy is called kurtosis. The moments are calculated for a shock time history passed through a contiguous set of bandpass filters. A *product model* is formed using a deterministic window $w(t)$[4] and a realization of a dimensionless stationary random process with unity variance $x(t)$ as $w(t) \cdot x(t+\tau)$. A product model is then used to generate a simulation that has the same energy and moments in the mean as the original shock. Band-limited moments characterize the shock and not the response to the shock as the shock spectrum and do not rely on a structural model.

Other Techniques. Other techniques to quantify pyroshock include the shock intensity spectrum based on the Fourier energy spectrum,[5] the *method of least favorable response,*[6,7] and nonstationary models.[8,9] These techniques are not commonly used but may provide additional insight for quantifying pyroshocks. The Fourier spectrum is an attractive alternative to the shock response spectrum because it is easy to compute and readily available in many software packages as a fast Fourier transform (FFT). Since the Fourier spectrum is complex, both magnitude and phase information is available. The magnitude generally has intuitive meaning, but the phase is difficult to interpret and may be contaminated with noise at the high frequencies present in pyroshock. The method of least favorable response provides a method of selecting the phase to maximize the response of the system under test. This method results in a conservative test provided that an appropriate measurement point is chosen on the structure. Stationary models for random vibration have been used for many years. Nonstationary models consist of a stationary process multiplied by a deterministic time-varying modulating function, which is a product model.[9] A nonstationary model is appropriate for pyroshock and approaches a stationary model as the time-record length is increased.

MEASUREMENT TECHNIQUES

Measurements of pyroshocks are generally made with accelerometers, strain gages, or laser Doppler vibrometers (LDVs). The accelerometers are used to measure acceleration, and the strain gages and LDVs are used to measure velocity. The strain gages may also be used to sense force, stress, or strain. General shock measurement instrumentation is applicable for pyroshock measurements (see Chap. 10); however, care must be taken to protect accelerometers from the high frequencies contained in pyroshocks that may cause the accelerometers to resonate and, in some cases, to fail. If accelerometers are excited into resonance, large-magnitude output results and may exceed the maximum amplitude of the data acquisition system that was chosen for the test. The result is that the data magnitude is clipped. If clipped, the data are rendered useless and the results from the test will be greatly diminished. Several mechanically isolated accelerometers are available commercially and should be used if there is a possibility of exciting the accelerometers into resonance. There is only one mechanically isolated accelerometer that can provide the wide-frequency bandwidth (dc to 10 kHz) required for pyroshock.[10,11] Other mechanical isolators generally provide a frequency bandwidth of about dc to 1 kHz. Any mechanical isolator that is used in a pyroshock environment must be well characterized over a range of frequencies and a range of acceleration values using a shock test technique, for example, Hopkinson bar testing. Strain gages are useful measurements of the pyroshock environment but are not easily translated into a test specification. Strain gages have the advantage of high-frequency response (in excess of dc to 40 kHz) provided that their size is appropriately chosen. Additionally, strain gages do not have the resonance problems that accelerometers have. The LDV provides velocity measurements that are not contaminated by cross-axis response because the LDV only responds to motion in the direction of the laser beam. The LDV is a noncontacting measurement and is easy to set up; consistent measurements of pyroshock events have been obtained with an LDV.[12,13] The LDV has the disadvantage of being very expensive per channel in comparison to the other measurement techniques, difficult to calibrate, and must have line of sight to the measurement location.

Pyroshock Test Specifications. An acceleration or velocity time history is not adequate for specifying a pyroshock test. The time-history data must be analyzed using one of the techniques discussed above to quantify the pyroshock for a test specification. Ideally, the time-history data that are used to develop the qualification test specification should be measured during a full-scale system test in which the actual pyrotechnic device or devices were initiated. The full-scale test should be accomplished with hardware that is structurally similar to the real hardware if the real hardware is not available. A control point measurement is specified close to each component or subassembly of interest, preferably at the attachment point to measure the input pyroshock. Since full-scale testing is expensive, data from a similar application may be used to develop component or subassembly qualification test specifications. This practice may result in overtested or overdesigned components or subassemblies if a large margin is added to the test specification to account for the uncertainty in the data. If this practice is used, the test specification should be revised when better system data become available.

Once the time-history data have been acquired, the data should be scrutinized to ensure their quality.[3] The data should be free of zero-shifts and offsets. Acceleration and velocity time histories should be integrated and the results examined. The time-history data should be low-pass filtered at a designated cutoff frequency; a cutoff frequency of 20 kHz is typical. The data must then be analyzed using the same technique as was used for analysis of the time-history data from which the test specification was derived. Test margin and tolerance bands are applied to the data analysis. For instance, if the SRS is being used, a straight-line approximation of the SRS is used as the baseline for the test specification process. A margin of +3 dB is typically added to the baseline SRS, and a customary ±6-dB tolerance is used with the baseline SRS. A typical test specification may allow the SRS from the actual test to fall outside the tolerance band at a specified number of frequency points. Pyroshock tests are highly variable, and the engineer must specify how much variability from test to test will be accepted; in some cases, a tighter, ±3-dB tolerance may be required. Additionally, the specification should require that the peak acceleration (or velocity) value and pulse durations are in agreement with the intended values for the specified input pulse. Similar approaches are used for other techniques for quantifying pyroshock.

In some cases, two or more pyroshock events, such as stage separation and an explosive actuator, may be combined into a single test specification. If the events are significantly different, the resulting test specification may be difficult or impossible to meet. A better practice is to make separate test specifications for each pyroshock event and to combine the specifications only in the case where a realizable test results.

PYROSHOCK SIMULATION TECHNIQUES

PYROTECHNICALLY EXCITED NEAR-FIELD SIMULATION

Ordnance Devices. Linear, flexible detonating charges may be used to generate pyroshocks for test purposes. An example of a test configuration using a flexible linear charge is shown in Fig. 28.3. A steel plate is suspended by bungee cords, and the test item is mounted on the plate in the same manner as it is in actual usage. Flexible linear charge is attached to appropriate locations on the bottom and the edges. The charge configuration may be varied according to experience and the desired effect.[14] For example, the charge may be attached to the backside of the plate directly opposite to the test item. A mass-mockup of the actual test item is used for the trial and error required to finalize the test configuration. In some cases, the charges may be attached to a portion of the structure where the test items are installed. Their storage, handling, and detonating constitute a hazard to laboratory personnel and facilities. However, such a fixture would normally be rather expensive because the structure would be damaged or destroyed during each shock test. The shock produced in this manner may vary greatly from test to test because actual explosives are used. However, this test configuration has the advantage of reproducing the pyroshock with realistic high accelerations and high frequencies. To ensure repeatability, the grooves generated by the charge into the surfaces of the shock plates should be machined down to eliminate the porosity which tends to absorb and modify the explosive impacts. Other disadvantages are that a qualified explosives facility (with its appropriate safety and security procedures) is required. In comparison to mechanical simulation techniques, considerable time is needed to conduct the numerous trial tests required to experimentally determine the various test parameters.

Scaled Tests. If the quantity of propellant or explosive is sufficiently large and the influence of the pyrotechnic device is localized, a scaled portion of the structure

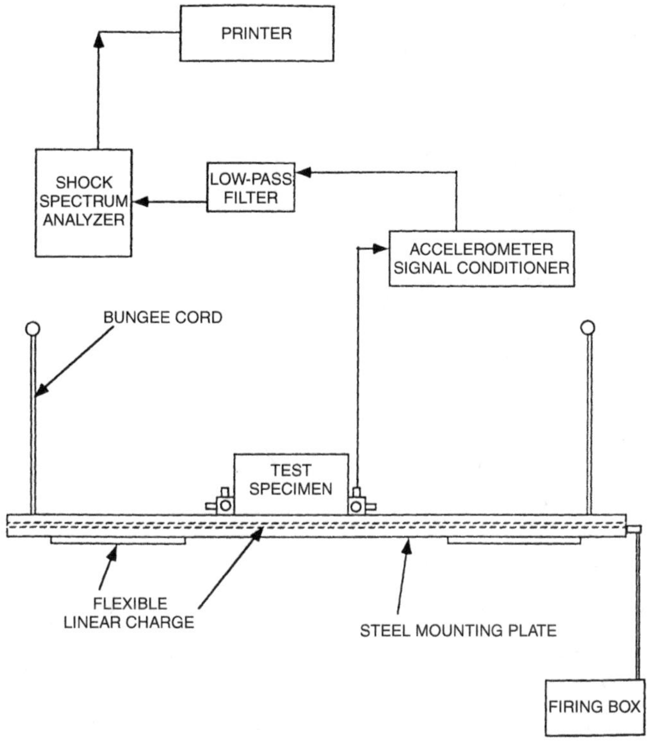

FIGURE 28.3 Ordnance-generated pyroshock simulator. (*Courtesy of National Technical Systems.*)

may be used in simulating the effects of the pyroshock as shown in Fig. 28.4 where a missile section or rocket payload section is shown. This type of test assumes that the influence of the pyrotechnic event is insignificant to other parts of the structure and isolated to the section under test. Actual pyrotechnic device firings on spacecraft equipment and scientific instruments are conducted in the scaled test. Such a test is usually an intermediate step in the design of the structure. Components in the subassembly may have been qualified with an ordnance device, and the scaled test adds another dimension of complexity to the qualification of the subassembly and its individual components.

Full-Scale Tests. In some cases, if the structure is sufficiently complex, a full-scale test may be warranted. Full-scale tests, which include multiple firings of certain critical pyrotechnic devices, are conducted to verify the structural integrity and design functions as well as to qualify items of hardware that have not been previously qualified. Full-scale tests are conducted by actuation of the flight pyrotechnic devices, which provide full-scale shock qualification. A full-scale test is usually the last test in a sequence of increasingly complex tests; the sequence is from ordnance to scaled tests to full-scale tests. The advantage of a full-scale test is that it is the real pyroshock event in its most complex form. The main objectives of the full-scale pyroshock test firings are: (1) to define shock response in the vicinity of potentially

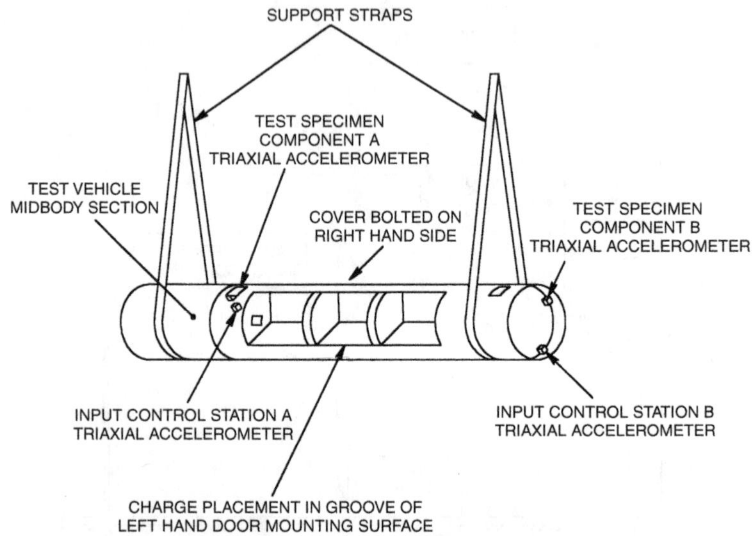

FIGURE 28.4 Scaled tests using representative structure. The test vehicle midbody section is a portion of the full-scale structure where the explosive event is located. Two input control stations A and B are used to determine that the test was properly conducted. Response measurements are made at test specimens A and B. (*Courtesy of Wyle Laboratories.*)

sensitive equipment so that component test specifications may be derived or verified and (2) to conduct full-scale qualification and thus verify the design values for shock. The disadvantage of a full-scale test is that considerable time and expense are required to obtain all the required hardware. The hardware must then be assembled, instrumented, and removed for post-test evaluation. Generally, special facilities are required for the use of explosives.

MECHANICALLY EXCITED MID-FIELD AND FAR-FIELD SIMULATION

Standard Shock Testing Machines. Shock machines such as the drop tables described in Chap. 27 usually are not suitable for pyroshock simulation. The single-sided pulses produced by these machines bear little or no resemblance to a pyroshock acceleration transient; such pulses produce significantly greater velocity change than a pyroshock environment. A severe overtest at low frequencies can be expected if a drop table is used to simulate pyroshock environments. This can result in failures of structural members that would not have been significantly stressed by the actual pyroshock. However, in certain cases, drop tables may produce acceptable pyroshock qualification testing. For example, if a test item has significant design margin at low frequencies, then a drop table may be acceptable. Also, if the lowest natural frequency of the test item is higher than the overtested low-frequency range, then the low-frequency overtest may be irrelevant since the effect on the test item is dominated by the peak *g*'s of the acceleration time history. In these cases there is strong motivation to use drop tables due to their common availability and low test cost. If a drop table is selected as a means of conducting a pyroshock qualification test, the test item must be subjected to a shock in both positive and negative directions for each axis tested, since the drop table produces only a single-sided pulse. MIL-STD-810 and the Pyroshock Testing Techniques Recommended Practices have good discussions of pyroshock testing techniques and practices.[15,16]

Electrodynamic Shakers. Pyroshock environments can be simulated with an acceleration transient produced on an electrodynamic shaker (see Chap. 25). In this method the acceleration transient is synthesized so that its shock response spectrum closely matches the test requirement. With this method a relatively complex SRS shape can be matched within close tolerances up to about 3000 Hz. The equipment limits (maximum acceleration) restrict this method to the simulation of lower-energy pyroshock environments. However, recent developments show moderate success to higher frequencies for very small items.[17] Even if the desired SRS is precisely met, an overtest is likely due to the high mechanical impedance of the shaker relative to the structure to which the test item is attached in a real application.

Resonant Fixtures. This section describes a variety of resonant fixture techniques used to simulate pyroshock environments. All of these methods utilize a fixture (or structure) which is excited into resonance by a mechanical impact from a projectile, a hammer, or some other device. A test item attached to the fixture is thus subjected to the resonant response, which simulates the desired pyroshock. There is no single preferred method since each has its own relative merits. Some of the methods require extensive trial-and-error iterations in order to obtain the desired test requirement. However, once the procedures are determined, the results are very repeatable. Other methods eliminate the need for significant trial and error but are usually limited to pyroshock environments which exhibit the typical mid-field and far-field character as explained in Fig. 28.2.

Full-Scale Tests. Some mechanically excited simulation techniques involve the use of an actual or closely simulated structure[18,19] (e.g., an entire missile payload section). The pyrotechnic devices (e.g., explosive bolt cutters) normally located on this structure would then be replaced with hardware that allow a controlled impact at this same location. Since a closely simulated structure is used, it is anticipated that the impact will cause the modes of vibration of the structure to be excited in a manner similar to the actual pyrotechnic source. In principle, test amplitudes can be adjusted by changing the impact speed or mass. This method is relatively expensive due to the cost of the test structure and because significant trial and error is required to obtain the desired test specification. Since this method applies to a specific application, it is not suited as a general-purpose pyroshock simulation technique.

In a variation of the above method[20] the pyrotechnic source and a portion of the adjacent structure are replaced by a "resonant plate" designed so that its lowest-resonance frequency corresponds to the dominant frequency produced by the pyrotechnic device and its associated structure. The resonant plate is then attached to the test structure in a manner which simulates the mechanical linkage of the pyrotechnic source, as shown in Fig. 28.5. When this plate is subjected to a mechanical impact, its response will provide the desired excitation of the test structure. A resonant fixture has successfully simulated component shock response spectra for frequencies up to 4000 Hz on a full-scale structure weighing 400 lb.[21]

General-Purpose Resonant Fixtures. Instead of developing application-specific pyroshock methods as described above, it may be desirable to implement a more general-purpose test method which can be used for a variety of test items

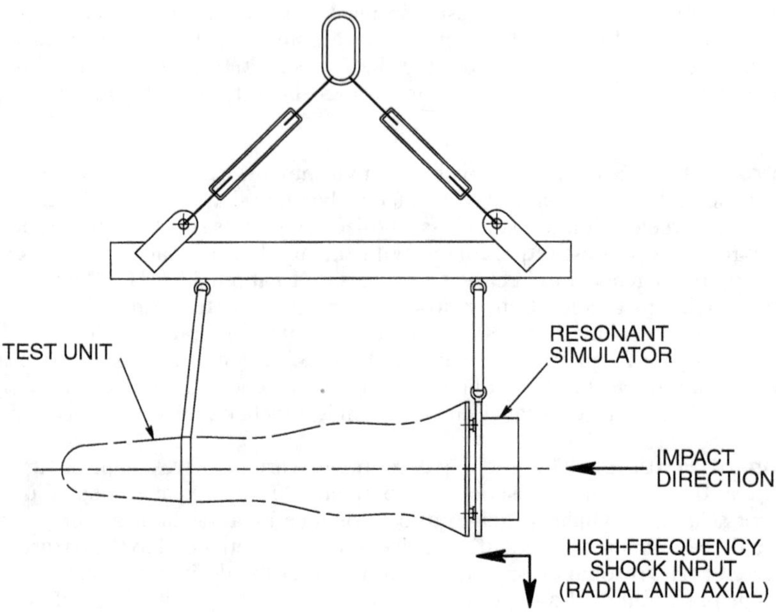

FIGURE 28.5 Full-scale pyroshock simulation with resonant fixture. Measurements at component locations confirm simulation success.

and/or test specifications. This can be accomplished by using a simple resonant fixture (usually a plate) instead of the complex structures described above. When such a fixture is excited into resonance by a mechanical impact, its response can provide an adequate pyroshock simulation to an attached test item. Excitation of the fixture can be achieved as the result of the impact of a projectile, pendulum hammer, pneumatic piston, or the like on the fixture. The response of the fixture is dependent on a large number of parameters including: (1) plate geometry and material, (2) impact mass or speed, (3) impact duration, which is controlled with various impact materials (e.g., metals, felt, elastomers, wood, etc.), (4) impact location, (5) test item location, and (6) various clamps and plate suspension mechanisms. In theory these parameters could be varied with the aid of an analytical model, but they are usually evaluated experimentally. A significant effort is therefore required to obtain each pyroshock simulation.

Mechanical Impulse Pyroshock (MIPS) Simulator. The MIPS simulator[22,23] is a well-developed embodiment of the trial-and-error resonant fixture methods. It is universally referred to by its acronym and is widely used in the aerospace industry. Its design facilitates the easy variation of many of the parameters described above. The MIPS simulator configuration shown in Fig. 28.6 consists of an aluminum mounting plate which rests on a thick foam pad. The shock is generated by a pneumatic actuator which is rigidly attached to a movable bridge, facilitating various impact locations. The impactor head is interchangeable so that different materials (lead, aluminum, steel, etc.) may be used to achieve variation of input duration. Although a triaxial acceleration measurement is usually made at the control point near the test item, it is unlikely that the test requirement will be met simultaneously in all axes. Separate test configurations must normally be developed for each test axis. Once the

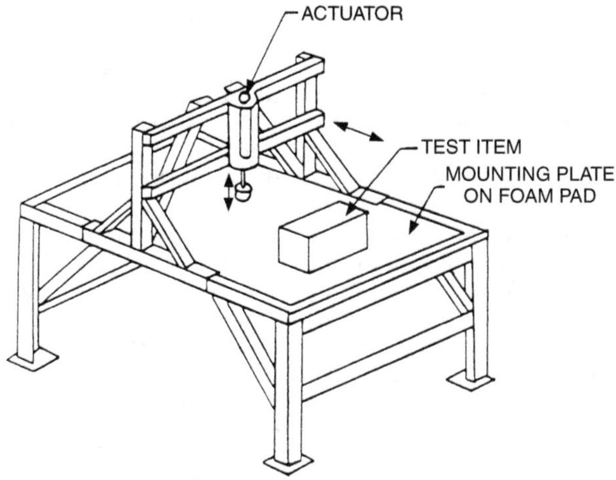

FIGURE 28.6 MIPS simulator. The mounting plate is excited into resonance by an impact from the actuator. The plate response simulates far-field pyroshock for the attached test item. (*Courtesy of Martin Marrietta Astrospace.*)

test configuration and procedures are determined, the results are very repeatable. The configuration for a new test specification can be obtained more quickly if records of previous setups and results are maintained for use as a starting point for the new specification. Reference 19 provides some general guidelines for parameter variation, as well as results obtained from several different test configurations.

Tuned Resonant Fixtures with Fixed Knee Frequency. It is possible to greatly reduce the amount of trial and error required by the MIPS simulator and other resonant fixture test methods. In order to do this, a simple resonant fixture is designed so that its dominant response frequency corresponds to the dominant frequency in the shock response spectrum test requirement. These tuned resonant fixtures are primarily limited to pyroshock environments which exhibit more or less typical characteristics with knee frequencies up to 3 kHz (or higher for small test items). The basic design principle is to match the dominant fixture response frequency (usually the first mode) to the SRS knee frequency. When this fixture is excited into resonance, it will "automatically" have the desired SRS knee frequency and the typical 9-dB/octave initial slope. This concept was originally developed using a plate excited into its first bending mode and a bar excited into its first longitudinal mode.[24] The methods described in the following sections require relatively thick and massive resonant fixtures compared to the structures to which the test item might be attached in actual use. Because of this, the motion imparted to the test item attached to a resonant fixture is approximately in phase from point-to-point across the mounting surface. Whereas, the actual pyroshock motion may not be in phase if the test item is mounted to a thin structure in actual use. The in-phase motion of resonant fixtures yields some degree of conservatism when selecting these methods for qualification testing. One significant advantage of using a thick resonant fixture is that its response is not greatly influenced by the attached test item. This allows the same test apparatus to be used for a variety of different test items.

Each of the tuned resonant fixture test methods described below produces a simulated pyroshock environment with the same basic characteristics. These similarities are illustrated in Fig. 28.7, which shows a typical acceleration record and SRS from the tunable resonant beam apparatus described later. The other methods produce pyroshock environments with initial SRS slopes that are slightly less than 9 dB/octave due to a small velocity change inherent with these other methods. The SRS shown in Fig. 28.7 exhibits the desired typical shape, and the energy is concentrated at the knee frequency. The absence of significant frequency content above the knee frequency may cause the SRS to be too low at these frequencies. In practice, the attached test item adds some frequency content above the knee frequency, which tends to increase the SRS. These test methods allow good control and repeatability of the SRS, especially below the knee frequency.

When using tuned resonant fixtures, the test item is usually attached to an intermediate fixture such as a rectangular aluminum plate. This adapter fixture must be small enough and stiff enough so that the input from the resonant fixture is not significantly altered. Since the resonant fixture is designed to produce the pyroshock simulation in only one direction, the adapter fixture should be designed so that it may be rigidly attached to the resonant fixture in three orthogonal orientations (e.g., flat down and on each of two edges). The acceleration input should be measured next to the test item on the adapter fixture. It is good practice to measure the acceleration in all three axes because it is possible (although infrequently) to simultaneously attain the desired test specification in more than one axis.

A number of different techniques are used to provide the mechanical impact required by the tuned resonant fixture methods described below. Pendulum hammers have been used, as well as pneumatically driven pistons or air guns. The method which is selected must provide repeatability and control of the impact force, in both magnitude and duration. The magnitude of the impact force controls the overall test ampli-

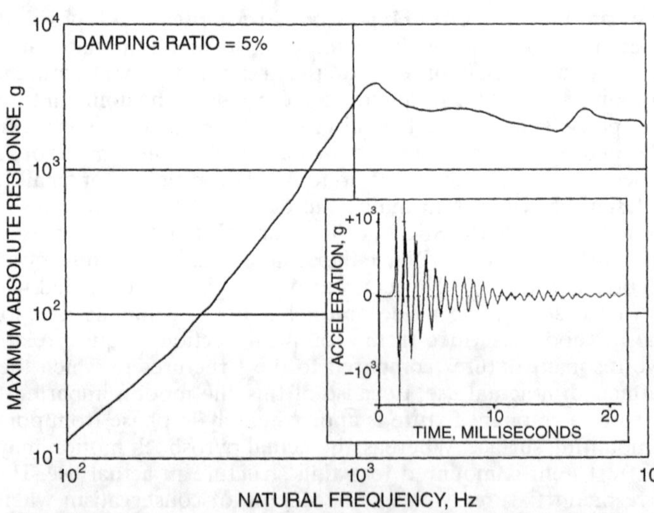

FIGURE 28.7 Typical shock response spectrum and acceleration time history from a tuned or tunable resonant fixture test. The shock response spectrum is calculated from the inset acceleration time history using a 5 percent damping ratio.

tude, and the impact duration must be appropriate to excite the desired mode of the tuned resonant fixture. In general, the impact duration should be about one-half the period of the desired mode. The magnitude of the impact force is usually controlled by the impact speed, and the duration is controlled by placing various materials (e.g., felt, cardboard, rubber, etc.) on the impact surfaces.

Resonant Plate (Bending Response). The resonant plate test method[25,26] is illustrated in Fig. 28.8, which shows a plate (usually a square or rectangular aluminum plate) freely suspended by some means such as bungee cords or ropes. A test item is attached near the center of one face of the plate, which is excited into resonance by a mechanical impact directed perpendicular to the center of the opposite face. The resonant plate is designed so that its first bending mode corresponds to the knee frequency of the test requirement. The first bending mode is approximately the same as for a uniform beam with the same cross section and length. The required dimensions for a square aluminum resonant plate for a particular test can be calculated by

$$\frac{t}{L^2} = \frac{f}{203{,}801} \tag{28.3}$$

where t = thickness of the plate
L = length dimension for square plate
f = frequency of first bending mode of plate (equal to desired knee frequency)

The plate must be large enough so that the test item does not extend beyond the middle third of the plate. This ensures that no part of the test item is attached at a

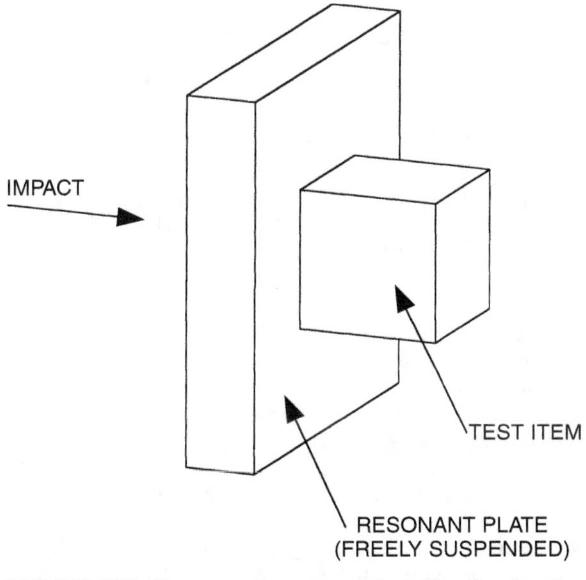

FIGURE 28.8 Resonant plate test method. The first bending mode is excited by an impact as shown. The plate's response simulates mid-field and far-field pyroshock for the attached test item. The plate is sized so that its first bending mode frequency corresponds to the desired knee frequency of the test.

nodal line of the first bending mode. Usually, the resonant fixture with an attached test item is insufficiently damped to yield the short-duration transient (5 to 20 milliseconds) required for pyroshock simulation. Damping may be increased by adding various attachments to the edge of the plate, such as C-clamps or metal bars. These attachments may also lower the resonance frequency and must be accounted for when designing a resonant plate.

Resonant Bar (Longitudinal Response). The resonant bar concept[25,26] is illustrated in Fig. 28.9, which shows a freely suspended bar (typically aluminum or steel) with rectangular cross section. A test item is attached at one end of the bar, which is excited into resonance by a mechanical impact at the opposite end. The basic principle of the resonant bar test is exactly the same as for a resonant plate test except that the first longitudinal mode of vibration of the bar is utilized. The bar length required for a particular test can be calculated by

$$l = \frac{c}{2f} \tag{28.4}$$

where l = length of the bar
c = wave speed in bar
f = first longitudinal mode of the bar (equal to desired knee frequency)

The other dimensions of the bar can be sized to accommodate the test item, but they must be significantly less than the bar length. As with the resonant plate method, the response of the bar can be damped with clamps if needed. These are most effective if attached at the impact end.

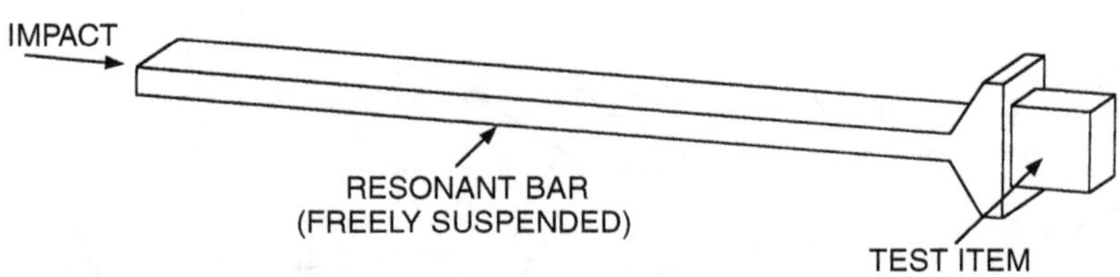

FIGURE 28.9 Resonant bar test method. The first longitudinal bar mode is excited by an impact as shown. The bar is sized so that its first normal mode frequency corresponds to the desired knee frequency in the test.

Tunable Resonant Fixtures with Adjustable Knee Frequency. The tuned resonant fixture methods described above can produce typical pyroshock simulations with knee frequencies that are fixed for each resonant fixture. A separate fixture must be designed and fabricated for each test requirement with a different knee frequency, so that a potentially large inventory of resonant fixtures would be necessary to cover a variety of test requirements. For this reason, tunable resonant fixture test methods were developed which allow an adjustable knee frequency for a single test apparatus.

Tunable Resonant Bars. The frequency of the first longitudinal mode of vibration of the resonant bar shown in Fig. 28.9 can be tuned by attaching weights at selected locations along the length of the bar.[25] If weights are attached at each of the two nodes for the second mode of vibration of the bar, then the bar's response will be dominated by the second mode ($2f$). Similarly, if weights are attached at each of the three nodes for the third mode of the bar, then the third mode (3*f*) will dominate. It is difficult to produce this effect for the fourth and higher modes of the bar since the distance between nodes is too small to accommodate the weights. This technique allows a single bar to be used to produce pyroshock simulations with one of three different knee frequencies. For example a 100-in. (2.54-m) aluminum bar can be used for pyroshock simulations requiring a 1000-, or 2000-, or 3000-Hz knee frequency. If the weights are attached slightly away from the node locations, the shock response spectrum tends to be "flatter" at frequencies above the knee frequency.[27]

Another tunable resonant bar method[28] can be achieved by attaching weights only to the impact end of the bar shown in Fig. 28.9. This method uses only the first longitudinal mode, which can be lowered incrementally as more weights are added. A nearly continuously adjustable knee frequency can thus be attained over a finite frequency range. The upper limit of the knee frequency is the same as given by Eq. (28.4) and is achieved with no added weights. In theory, this knee frequency could be reduced in half if an infinite weight could be added. However, a realizable lower limit of the knee frequency would be about 25 percent less than the upper limit.

Tunable Resonant Beam. Figure 28.10 illustrates a tunable resonant beam apparatus[28] which will produce typical pyroshock simulations with a knee frequency that is adjustable over a wide frequency range. In this test method, an aluminum beam with rectangular cross section is clamped to a massive base as shown. The clamps are intended to impose nearly fixed-end conditions on the beam. When the beam is struck with a cylindrical mass fired from the air gun beneath the beam, it will resonate at its first bending frequency, which is a function of the distance between the clamps. Ideally, the portion of the beam between the clamps will respond as if it had perfectly fixed ends and a length equal to the distance between the clamps. For

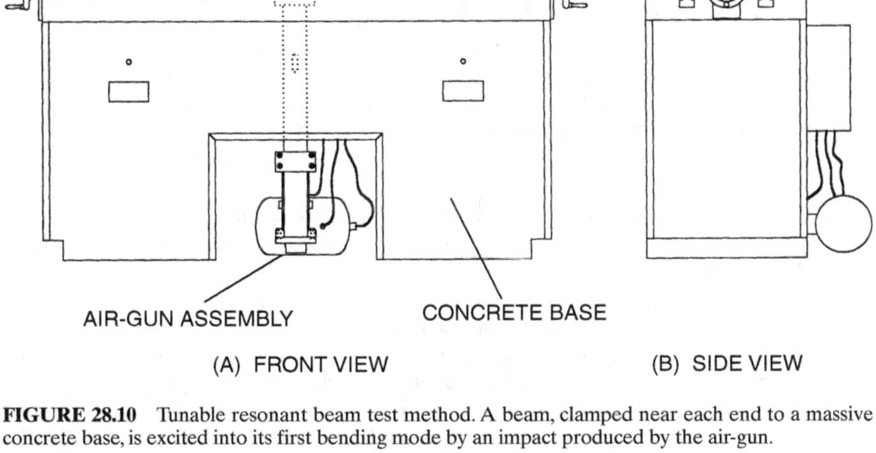

FIGURE 28.10 Tunable resonant beam test method. A beam, clamped near each end to a massive concrete base, is excited into its first bending mode by an impact produced by the air-gun.

this ideal case, the frequency of the first mode of the beam varies inversely with the square of the beam length. In practice, the end conditions are not perfectly fixed, and the frequency of the first mode is somewhat lower than predicted. This method provides a good general-purpose pyroshock simulator, since the knee frequency is continuously adjustable over a wide frequency range (e.g., 250 to 7000 Hz) by varying the beam thickness. This tunability allows small adjustments in the knee frequency to compensate for the effects of test items of different weights.

MECHANICALLY EXCITED NEAR-FIELD SIMULATION

All test methods described above require that the test item be attached and tested along three separate axes. Additionally, all methods have some cross-axis response in addition to the intended in-axis response, so overtesting of the test item occurs. However, in some cases, all three axes may be tested with one test on a thick resonant fixture. These fixtures must be designed for the specific test requirement and for small test items. Time history magnitudes of 1000 *g* to 80,000 *g* with knee frequencies in excess of 15 kHz have been achieved.[29,30]

REFERENCES

1. Valentekovich, V. M.: *Proc. 64th Shock and Vibration Symposium,* 1993, p. 92.
2. Moening, C. J.: *Proc. 8th Aerospace Testing Seminar,* 1984, p. 95.
3. Himelblau, H., A. G. Piersol, J. H. Wise, and M. R. Grundvig: "Handbook for Dynamic Data Acquisition and Analysis," *IEST Recommended Practice 012.1,* Institute of Environmental Sciences and Technology, Mount Prospect, Ill.
4. Smallwood, D. O.: *Shock and Vibration J.,* **1**(6):507 (1994).
5. Baca, T. J.: *Proc. 60th Shock and Vibration Symposium,* 1989, p. 113.
6. Shinozuka, M.: *J. of the Engineering of the Engineering Mechanics Division, Proc. of the American Society of Civil Engineers,* 1970, p. 727.
7. Smallwood, D. O.: *Shock and Vibration Bulletin,* **43**:151 (1973).
8. Mark, W. D.: *J. of Sound and Vibration,* **22**(3):249 (1972).
9. Bendat, J. S., and A. G. Piersol: "Engineering Applications of Correlation and Spectral Analysis," John Wiley & Sons, 2d ed., 1993, p. 325.
10. Bateman, V. I., R. G. Bell III, and N. T. Davie: *Proc. 60th Shock and Vibration Symposium,* **1**:273 (1989).
11. Bateman, V. I., R. G. Bell III, F. A. Brown, N. T. Davie, and M. A. Nusser: *Proc. 61st Shock and Vibration Symposium,* **IV**:161 (1990).
12. Valentekovich, V. M., M. Navid, and A. C. Goding: *Proc. 60th Shock and Vibration Symposium,* **1**:259 (1989).
13. Valentekovich, V. M., and A. C. Goding: *Proc. 61st Shock and Vibration Symposium,* **2** (1990).
14. Czajkowski, J., P. Lieberman, and J. Rehard: *J. of the Institute of Environmental Sciences,* **35**(6):25 (1992).
15. Anon.: "Environmental Engineering Considerations and Laboratory Tests," *MIL-STD-810, Method 517,* "Pyroshock," Department of Defense, 1989.
16. Resor, B.: "Equipment and Techniques to Perform High Amplitude and High Frequency Shaker Shock Testing," *Proc. 77th Shock and Vibration Symposium,* 2006.
17. Bateman, V. I.: "Pyroshock Testing Techniques Recommended Practice," *IEST-RP-DTE-032.1,* Institute of Environmental Sciences and Technology, Arlington Heights, Ill., 2003.

18. Luhrs, H. N.: *Proc. Institute of Environmental Sciences Annual Technical Meeting,* 1981, p. 17.
19. Powers, D. R.: *Shock and Vibration Bulletin,* **56**(3):133 (1986).
20. Bateman, V. I., and F. A. Brown: *J. of the Institute of Environmental Sciences,* **37**(5):40 (1994).
21. Bateman, V. I., F. A. Brown, J. S. Cap, and M. A. Nusser: *Proceedings of the 70th Shock and Vibration Symposium,* **I** (1999).
22. Dwyer, T. J., and D. S. Moul: *15th Space Simulation Conference,* Goddard Space Flight Center, NASA-CP-3015, 1988, p. 125.
23. Raichel, D. R.: Jet Propulsion Lab, California Institute of Technology, Pasadena, 1991.
24. Bai, M., and W. Thatcher: *Shock and Vibration Bulletin,* **49**(1):97 (1979).
25. Davie, N. T.: *Shock and Vibration Bulletin,* **56**(3):109 (1986).
26. Davie, N. T., *Proc. Institute of Environmental Sciences Annual Technical Meeting,* 1985, p. 344.
27. Shannon, K. L., and T. L. Gentry: "Shock Testing Apparatus," U. S. Patent No. 5,003,810, 1991.
28. Davie, N. T., and V. I. Bateman: *Proc. Institute of Environmental Sciences Annual Technical Meeting,* 1994, p. 504.
29. Bateman, V. I., L. W. Carlson, and M. A. Nussor: "A Near Field Mechanical Pyroshock Simulation," *Proc. 75th Shock and Vibration Symposium,* 2004.
30. Bateman, V. I., L. W. Carlson, and M. A. Nussor: "A Unique Reasonant Fixture for a Mechanical Shock Simulation," *Proc. ESTECH,* 2004.

CHAPTER 29
VIBRATION OF STRUCTURES INDUCED BY GROUND MOTION

William J. Hall
Billie F. Spencer, Jr.
Amr S. Elnashai

INTRODUCTION

This chapter opens with a discussion of examples of typical sources of ground motion (industrial drop hammers, railroads, automobile traffic, and mine blasting) that may affect the response performance of buildings and included equipment. Most of the chapter centers on seismic effects, by virtue of their great importance in the design and performance of large structural and infrastructure systems whose damage or failure can impact the safety and lives of people and possibly lead to major property loss as well. Presented in limited detail, yet supported by ample modern references, are such matters as seismic motions, characterization of response of simple single-degree-of-freedom (SDOF) systems through response spectra, development of design response spectra, and spectra for simple inelastic systems. There follows a short discussion of the use of these tools in the design/analysis process and of some of the more important considerations in the overall design and construction process. Also included for the first time are short sections on the role of structural damping systems and isolators and on seismic risk assessment in the design process. The latter two topics have been developed to the point of being important considerations in seismic design today.

GROUND MOTION

SOURCE OF GROUND MOTION

Ground motion may arise from any number of sources such as earthquake excitation (described in detail in this chapter) and high explosive or nuclear device detonations. In such cases, the source excitation can lead to major vibration of the primary structure or facility and its many parts, as well as to transient and permanent translation (displacement) and rotation of the ground on which the facility is constructed. Detonations may result in drag and side-on overpressures, ballistic ejecta, and thermal and radiation effects.

Other sources of ground excitation, although usually not as strong, can be equally troublesome. For example, the location of a precision machine shop or chip manufacturing facility near a railroad or highway or of delicate laboratory apparatus in a plant area containing heavy drop-forging machinery or unbalanced rotating machinery are typical of situations in which ground-transmitted vibrations may pose serious problems. Another, different class of vibrational problems arises from excitation of the primary structure by other sources, e.g., wind blowing on a bridge, earthquake excitation of a building or bridge, or people walking or dancing on a floor in a building. Vibration of the primary structure in turn can affect secondary elements such as mounted equipment and people located on a floor (in the case of buildings) and vehicles or equipment (in the case of bridges).

The variables involved in problems of this type are exceedingly numerous and, with the exception of earthquakes, few specific well-defined measurements are generally available to serve as a guide in estimating the ground motions that might be used as computational guidelines in particular cases. A number of acceleration-versus-time curves for typical ground motions arising from the operation of machines and vehicles are shown in Fig. 29.1. Another record arising from a rock quarry blast is shown in Fig. 29.2. Although the records differ somewhat in their characteristics, all can be compared directly with similar measurements of earthquakes, and response computations generally are handled in the same manner.

In most cases, to analyze and evaluate such information one needs to (1) develop an understanding of the source and nature of the vibration, (2) ascertain the physical characteristics of the structure or element, (3) develop an approach for modeling and analysis, (4) carry out the analysis, (5) study the response (with parameter variations if needed), (6) evaluate the behavior of service and function limit states, and (7) develop, in light of the results of the analysis, possible courses of corrective action, if required. Merely changing the mass, stiffness, or damping of the structural system may or may not lead to acceptable corrective action in the sense of a reduction in deflections or stresses; careful investigation of the various alternatives is required to change the response to an acceptable limit. Advice on these matters is contained in Refs. 1–6.

RESPONSE OF SIMPLE STRUCTURES TO GROUND MOTIONS

Four structures of varying size and complexity are shown in Fig. 29.3: (*A*) a simple, relatively compact machine anchored to a foundation, (*B*) a 15-story building, (*C*) a 40-story building, and (*D*) an elevated water tank. The dynamic response of each of the structures shown in Fig. 29.3 can be approximated by representing each as a simple mechanical oscillator consisting of a single mass supported by a spring and a damper as shown in Fig. 29.4. The relationship between the undamped angular frequency of vibration $\omega_n = 2\pi f_n$, the natural frequency f_n, and the period T is defined in terms of the spring constant k and the mass m:

$$\omega_n^2 = \frac{k}{m} \tag{29.1}$$

$$f_n = \frac{1}{T} = \frac{\omega_n}{2\pi} = \frac{1}{2\pi}\sqrt{\frac{k}{m}} \tag{29.2}$$

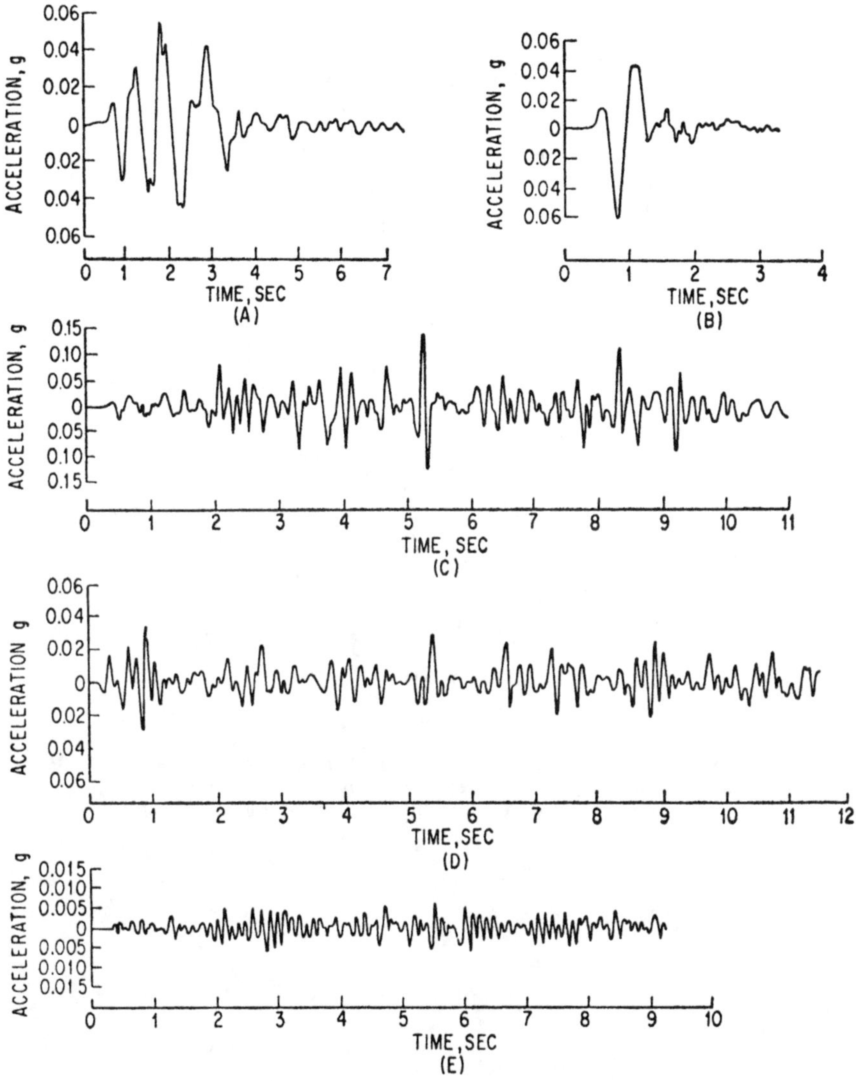

FIGURE 29.1 Ground-acceleration-vs.-time curves for typical machine and vehicle excitations. (*A*) Vertical acceleration measured on a concrete floor on sandy loam soil at a point 6 ft from the base of a drop hammer. (*B*) Horizontal acceleration 50 ft from drop hammer. The weight of the drop hammerhead was approximately 15,000 lb, and the hammer was mounted on three layers of 12- by 12-in. oak timbers on a large concrete base. (*C*) Vertical acceleration 6 ft from a railroad track on the well-maintained right-of-way of a major railroad during passing of luxury-type passenger cars at a speed of approximately 20 mph. The accelerometer was bolted to a 2- by 2-in. by 2½-in. steel block which was firmly anchored to the ground. (*D*) Horizontal acceleration of the ground at 46 ft from the above railroad track, with a triple diesel-electric power unit passing at a speed of approximately 20 mph. (*E*) Horizontal acceleration of the ground 6 ft from the edge of a relatively smooth highway, with a large tractor and trailer unit passing on the outside lane at approximately 35 mph with a full load of gravel.[1]

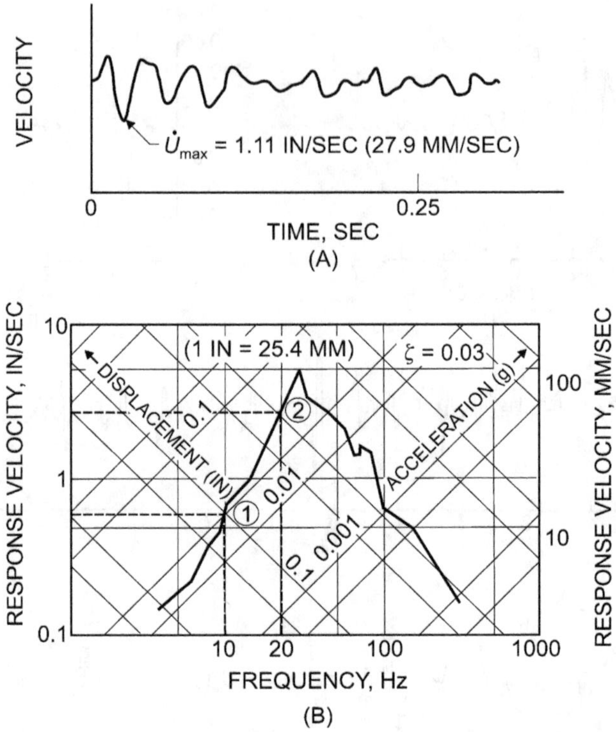

FIGURE 29.2 Typical quarry blast data. (*A*) Time-history of velocity taken by a velocity transducer and recorder. (*B*) Corresponding response spectrum computed from the record in (*A*) using Duhamel's integral.[3]

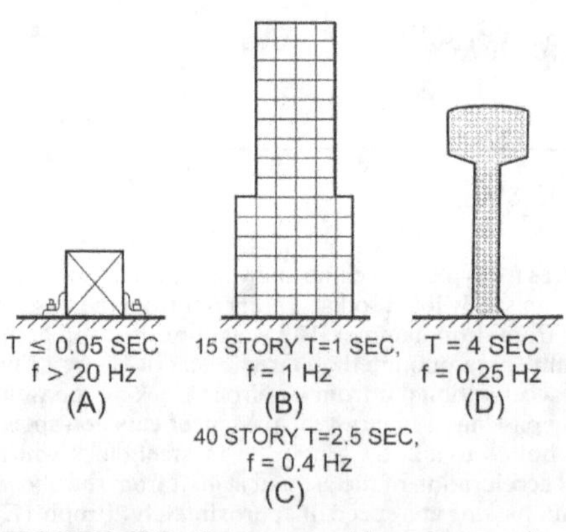

FIGURE 29.3 Structures subjected to earthquake ground motion. (*A*) A machine anchored to a foundation. (*B*) A 15-story building. (*C*) A 40-story building. (*D*) An elevated water tank.

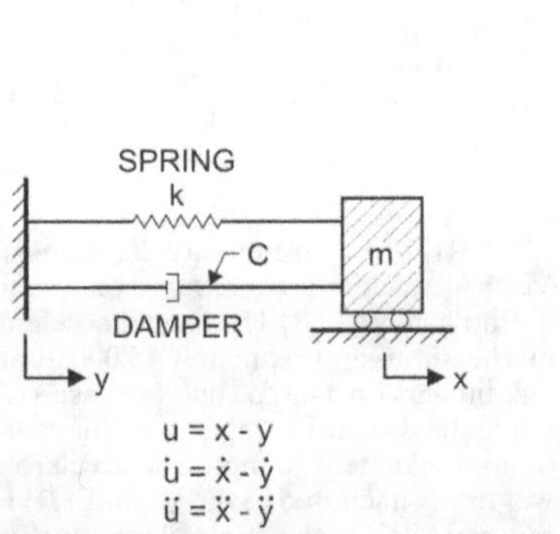

FIGURE 29.4 System definition; the dynamic response of each of the structures shown in Fig. 29.3 can be approximated by this simple mechanical oscillator.

In general, the effect of the damper is to produce damping of free vibrations or to reduce the amplitude of forced vibrations. The damping force is assumed to be equal to a damping coefficient c times the velocity $\dot{u}$ of the mass relative to the ground. The value of c at which the motion loses its vibratory character in free vibration is called the *critical damping coefficient;* for example, $c_c = 2m\omega_n$. The amount of damping is most conveniently considered in terms of the fraction of critical damping, ζ, as defined in Eq. (2.12).

$$\zeta = \frac{c}{c_c} = \frac{c}{2m\omega_n} \tag{29.3}$$

For most practical structures ζ is relatively small, in the range of 0.005 to 0.2 (i.e., 0.5 to 20 percent), and does not appreciably affect the natural period or frequency of vibration (see Chap. 2).

EARTHQUAKE GROUND MOTION

Strong-motion earthquake acceleration records with respect to time have been obtained for a number of earthquakes. Ground motions from other sources of disturbance, such as quarry blasting and nuclear blasting, also are available and show many of the same characteristics. As an example of the application of such time-history records, the recorded accelerogram for the El Centro, California, earthquake of May 18, 1940, in the north-south component of horizontal motion is shown in Fig. 29.5. On the same figure are shown the integration of the ground acceleration a to

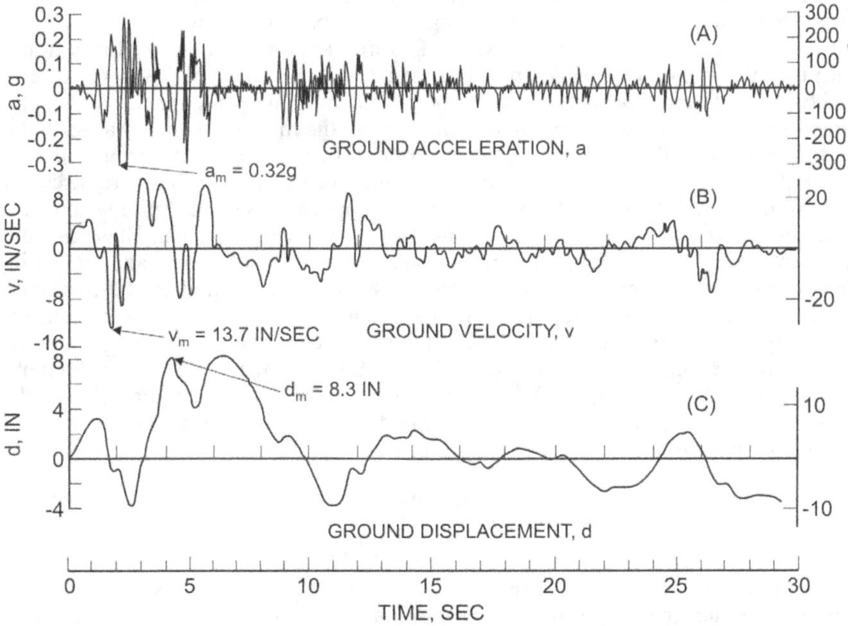

FIGURE 29.5 El Centro, California, earthquake of May 18, 1940, north-south component. (*A*) Record of the ground acceleration. (*B*) Variation of ground velocity v with time, obtained by integration of (*A*). (*C*) Variation of ground displacement with time, obtained by integration of (*B*).

give the variation of ground velocity v with time and the integration of velocity to give the variation of ground displacement d with time. These integrations normally require baseline corrections of various sorts, and the magnitude of the maximum displacement may vary depending on how the corrections are made. The maximum velocity is relatively insensitive to the corrections, however. For this earthquake, with the integrations shown in Fig. 29.5, the maximum ground acceleration is 0.32g, the maximum ground velocity is 13.7 in./sec (35 cm/sec), and the maximum ground displacement is 8.3 in. (21 cm). These three maximum values are of particular interest because they help to define the response motions of the various structures considered in Fig. 29.3 most accurately if all three maxima are taken into account.

RESPONSE SPECTRA

ELASTIC SYSTEMS

The response of the simple oscillator shown in Fig. 29.4 to any type of ground motion can be readily computed as a function of time. A plot of the maximum values of the response, as a function of frequency or period, is commonly called a *response spectrum* (or *shock response spectrum*). The response spectrum may be defined as the graphical relationship of the maximum response of a single-degree-of-freedom linear system to dynamic motions or forces. This concept of a response spectrum is widely used in the study of the response of simple oscillators to transient disturbances; for a number of examples, see other chapters herein.

A careful study of Fig. 29.4 will reveal that there are nine quantities represented there: acceleration, velocity, and displacement of the base, mass, and their relative values denoted by u. Commonly the maxima of interest are the maximum deformation of the spring, the maximum spring force, the maximum acceleration of the mass (which is directly related to the spring force when there is no damping), or a quantity having the dimensions of velocity, which provides a measure of the maximum energy absorbed in the spring. The details of various forms of response spectra that can be graphically represented, uses of response spectra, and techniques for computing them are discussed in detail in Refs. 7–10. A brief treatment of the applications of response spectra follows. The maximum values of the response are of particular interest. These maxima can be stated in terms of the maximum strain in the spring $u_m = D$, the maximum spring force, the maximum acceleration A of the mass (which is related to the maximum spring force directly when there is no damping), or a quantity, having the dimensions of velocity, which gives a measure of the maximum energy absorbed in the spring. This quantity, designated the pseudo-velocity V, is defined in such a way that the energy absorption in the spring is $\frac{1}{2}mV^2$. The relations among the maximum relative displacement of the spring D, the pseudo-velocity V, and the pseudo-acceleration A, which is a measure of the force in the spring, are

$$V = \omega D \tag{29.4}$$

and

$$A = \omega V = \omega^2 D \tag{29.5}$$

The pseudo-velocity V is nearly equal to the maximum relative velocity for systems with moderate or high frequencies but may differ considerably from the maximum relative velocity for very low frequency systems. The pseudo-acceleration A is exactly equal to the maximum acceleration for systems with no damping and is not greatly different from the maximum acceleration for systems with moderate amounts of damping, over the whole range of frequencies from very low to very high values.

Typical plots of the response of the system to a base excitation, as a function of period or natural frequency, are called *response spectra* (also called *shock spectra*). Plots for acceleration and for relative displacement, for a system with a moderate amount of damping and subjected to an input similar to that of Fig. 29.5, can be made. This arithmetic plot of maximum response is simple and convenient to use. Various techniques of computing and plotting spectra may be found in the references cited at the end of this chapter, especially in Refs. 7–10.

A somewhat more useful plot, which indicates the values for D, V, and A, is shown in Fig. 29.6. This plot has the virtue that it also indicates more clearly the extreme or limits of the various parameters defining the response. All parameters are plotted on a logarithmic scale. Since the frequency is the reciprocal of the period, the logarithmic scale for the period would have exactly the same spacing of the points, or in effect the scale for the period would be turned end for end. The pseudo-velocity is plotted on a vertical scale. Then on diagonal scales along an axis that extends upward from right to left are plotted values of the displacement, and along an axis that extends upward from left to right the pseudo-acceleration is plotted, in such a way that any one point defines for a given frequency the displacement D, the pseudo-velocity V, and the pseudo-acceleration A. Points are indicated in Fig. 29.6 for the several structures of Fig. 29.3 plotted at their approximate fundamental frequencies. Many other formats are used in plotting spectra, for example, u, $\dot{u}$, ωu, or $\ddot{x}$ versus time. Such examples are shown in Ref. 8*c*.

In developing spectral relationships, a wide variety of motions have been considered, ranging from simple pulses of displacement, velocity, or acceleration of

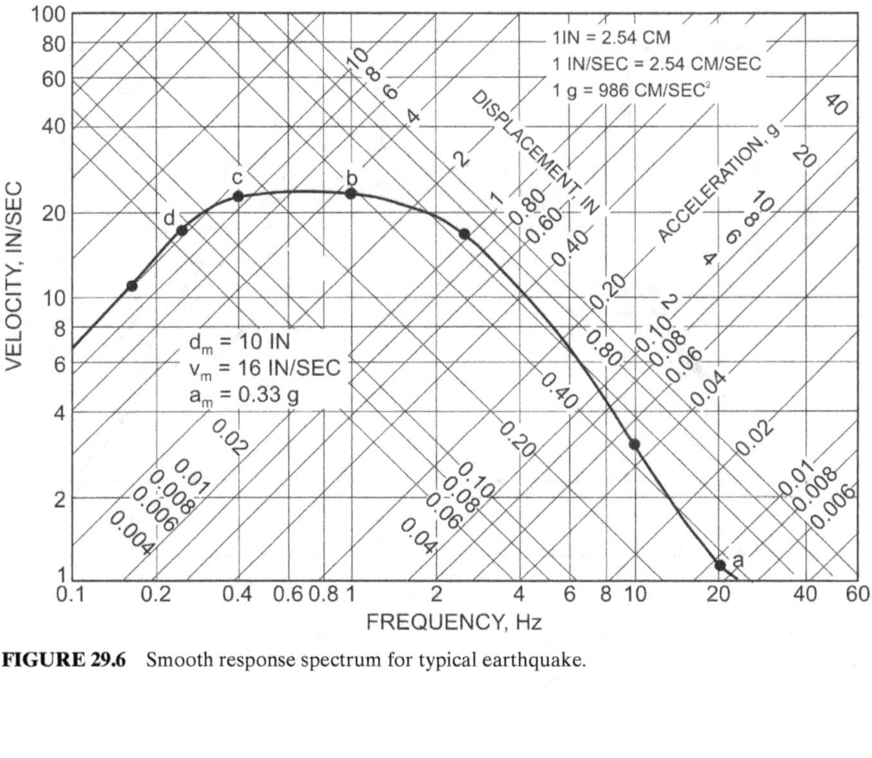

FIGURE 29.6 Smooth response spectrum for typical earthquake.

the ground, through more complex motions such as those arising from nuclear-blast detonations, and for a variety of earthquakes as taken from available strong-motion records. Response spectra for the El Centro earthquake are shown in Fig. 29.7. The spectrum for small amounts of damping is much more jagged than indicated by Fig. 29.6, but for the higher amounts of damping the response curves are relatively smooth. The scales are chosen in this instance to represent the amplifications of the response relative to the ground-motion values of displacement, velocity, or acceleration.

The spectra shown in Fig. 29.7 are typical of response spectra for nearly all types of ground motion. On the extreme left, corresponding to very low frequency systems, the response for all degrees of damping approaches an asymptote corresponding to the value of the maximum ground displacement. A low-frequency system corresponds to one having a very heavy mass and a very light spring. When the ground moves relatively rapidly, the mass does not have time to move, and therefore the maximum strain in the spring is precisely equal to the maximum displacement of the ground. For a very high frequency system, the spring is relatively stiff and the mass very light. Therefore, when the ground moves, the stiff spring forces the mass to move in the same way the ground moves, and the mass therefore must have the same acceleration as the ground at every instant. Hence, the force in the spring is that required to move the mass with the same acceleration as the ground, and the maximum acceleration of the mass is precisely equal to the maximum acceleration of the ground. This is shown by the fact that all the lines on the extreme right-hand side of the figure asymptotically approach the maximum ground-acceleration line.

For intermediate-frequency systems, there is an amplification of the motion. In general, the amplification factor for displacement is less than that for velocity, which in turn is less than that for acceleration. Peak amplification factors for the

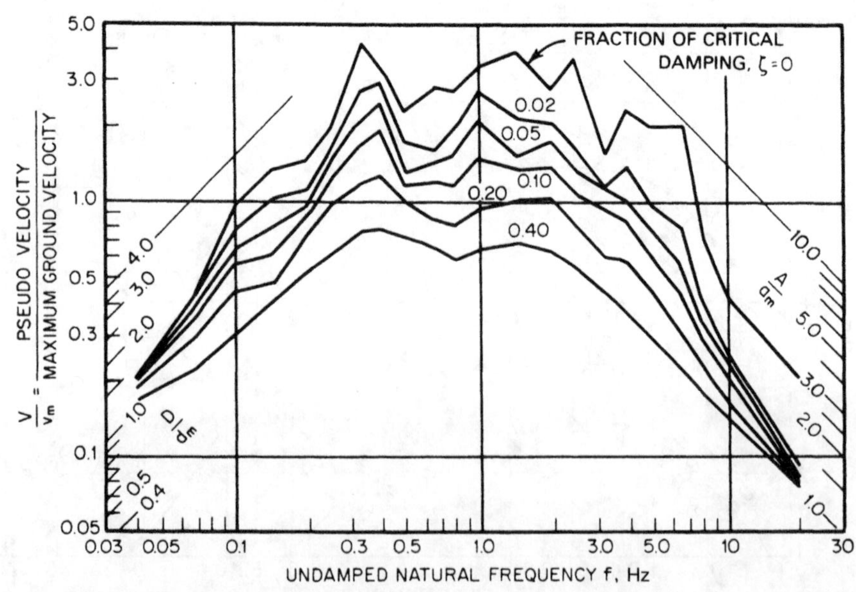

FIGURE 29.7 Response spectra for elastic systems subjected to the El Centro earthquake for various values of fraction of critical damping ζ.

undamped system ($\zeta = 0$) in Fig. 29.7 are on the order of about 3.5 for displacement, 4.2 for velocity, and 9.5 for acceleration.

The results of similar calculations for other ground motions are quite consistent with those in Fig. 29.7, even for simple motions. The general nature of the response spectrum shown in Fig. 29.8 consists of a central region of amplified response and two limiting regions of response in which for low-frequency systems the response displacement is equal to the maximum ground displacement, and for high-frequency systems the response acceleration is equal to the maximum ground acceleration. Values of the amplification factor reasonable for use in design are presented in the next sections.

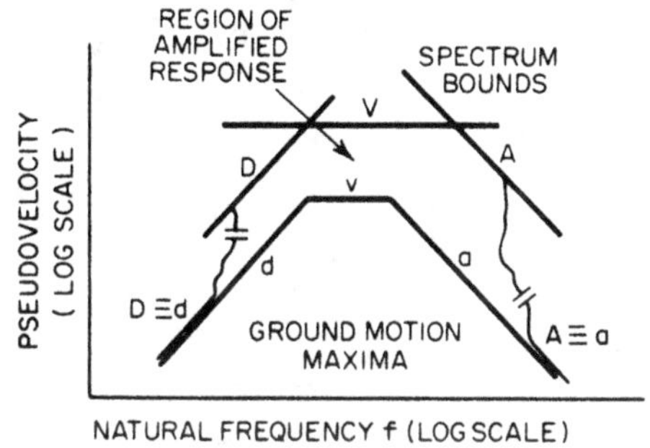

FIGURE 29.8 Typical tripartite logarithmic plot of response-spectrum bounds compared with maximum ground motion.

DESIGN RESPONSE SPECTRA

A response spectrum developed to give design coefficients is called a *design response spectrum* or a *design spectrum.* As an example of its use in seismic design, for any given site, estimates are made of the maximum ground acceleration, maximum ground velocity, and maximum ground displacement. The lines representing these values can be drawn on the tripartite logarithmic chart of which Fig. 29.9 is an example. The heavy lines showing the ground-motion maxima in Fig. 29.9 are drawn for a maximum ground acceleration a of 1.0g, a velocity v of 48 in./sec (122 cm/sec), and a displacement d of 36 in. (91.5 cm). These data represent motions more intense than those generally considered for any postulated design earthquake hazard. They are, however, approximately in correct proportion for a number of areas of the world, where earthquakes occur on either firm ground, soft rock, or competent sediments of various kinds. For relatively soft sediments, the velocities and displacements might require increases above the values corresponding to the given acceleration as scaled from Fig. 29.9, and for competent rock, the velocity and displacement values would be expected to be somewhat less. More detail can be found in Refs. 8*a* and *c.* It is not likely that maximum ground velocities in excess of 4 to 5 ft/sec (1.2 to 1.5 m/sec) are obtainable under any circumstances.

On the basis of studies of horizontal and vertical directions of excitation for various values of damping,[7,8a] representative amplification factors for the 50th and 84.1th percentile levels of horizontal response are presented in Table 29.1. The

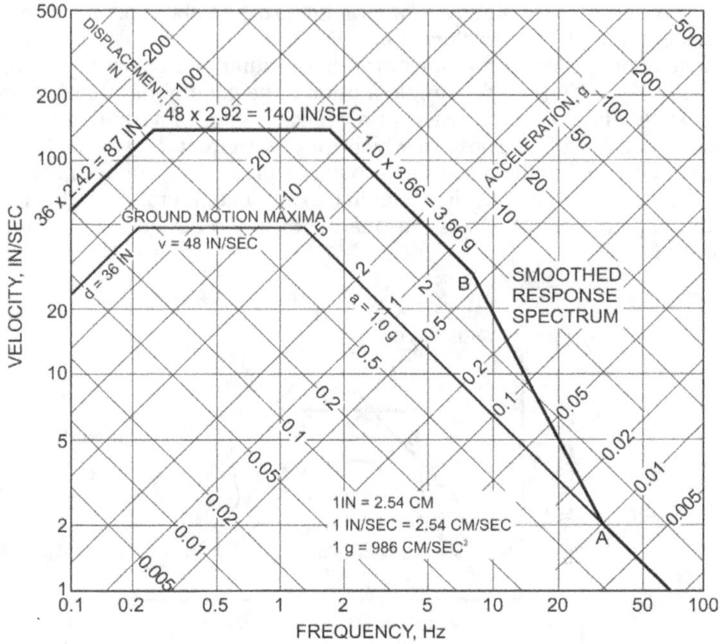

FIGURE 29.9 Basic design spectrum normalized to 1.0*g* for a value of damping equal to 2 percent of critical, 84.1th percentile level. The spectrum bound values are obtained by multiplying the appropriate ground-motion maxima by the corresponding amplification value of Table 29.1.

84.1th percentile means that one could expect 84.1 percent of the values to fall at or below that particular amplification. With these amplification factors and noting points B and A to fall at about 8 and 33 Hz, the spectra may be constructed as shown in Fig. 29.9 by multiplying the ground maxima values of acceleration, velocity, and displacement by the appropriate amplification factors. Further information on, and other approaches to, construction of design spectra may be found in Refs. 8*a* and *c*.

TABLE 29.1 Values of Spectrum Amplification Factors

Percentile	Damping, percent of critical damping	Amplification factor *D*	*V*	*A*
50th	0.5	2.01	2.59	3.68
	2.0	1.63	2.03	2.74
	5.0	1.39	1.65	2.12
	10.0	1.20	1.37	1.64
84.1th	0.5	3.04	3.84	5.10
	2.0	2.42	2.92	3.66
	5.0	2.01	2.30	2.71
	10.0	1.69	1.84	1.99

RESPONSE SPECTRA FOR INELASTIC SYSTEMS

It is convenient to consider an elastoplastic resistance-displacement relation because one can draw response spectra for such a relation in generally the same way as the spectra were drawn for elastic conditions. A simple resistance-displacement relationship for a spring is shown by the light line in Fig. 29.10*A*, where the yield point is indicated, with a curved relationship showing a rise to a maximum resistance and then a decay to a point of maximum useful limit or failure at a displacement u_m; an equivalent elastoplastic resistance curve is shown by the heavy line. A similar elastoplastic resistance function, more indicative of seismic response, is shown in Fig. 29.10*B*. The ductility factor μ is defined as the ratio between the maximum permissible or useful displacement to the yield displacement for the effective curve in both cases.

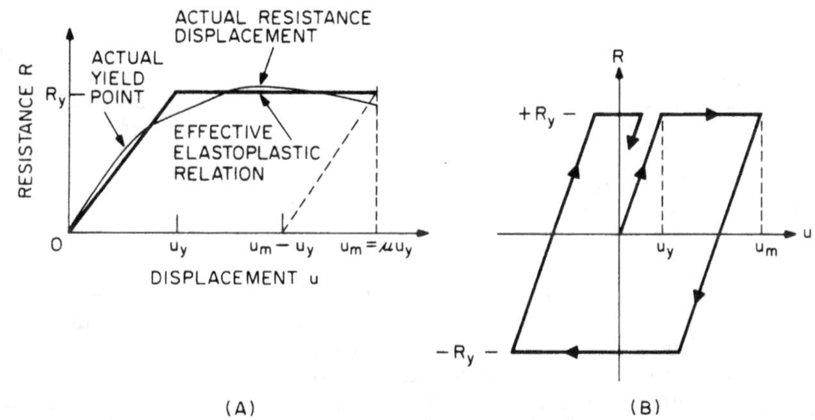

FIGURE 29.10 (*A*) Monotonic resistance-displacement relationships for a spring, shown by the light line; an equivalent elastoplastic resistance curve, shown by the heavy line. (*B*) A similar elastoplastic resistance function, more indicative of seismic response.

The ductility factors for various types of construction depend on the use of the building, the hazard involved in its failure (assumed acceptable risk discussed later), the material used, the framing or layout of the structure, and above all on the method of construction and the details of fabrication of joints and connections. A discussion of these topics is given in Refs. 7, 8, 10, and 11. Figure 29.11 shows acceleration spectra for elastoplastic systems having 2 percent of critical damping that were subjected to the El Centro, 1940, earthquake. Here the symbol D_y represents the elastic component of the response displacement, but it is not the total displacement. Hence, the curves also give the elastic component of maximum displacement as well as the maximum acceleration *A*, but they do not give the proper value of maximum pseudo-velocity. This is designated by the use of the V' for the pseudo-velocity drawn in the figure. The figure is drawn for ductility factors ranging from 1 to 10. A response spectrum for total displacement also can be drawn for the same conditions as for Fig. 29.11. It is obtained by multiplying each curve's ordinates by the value of ductility factor μ shown on that curve.

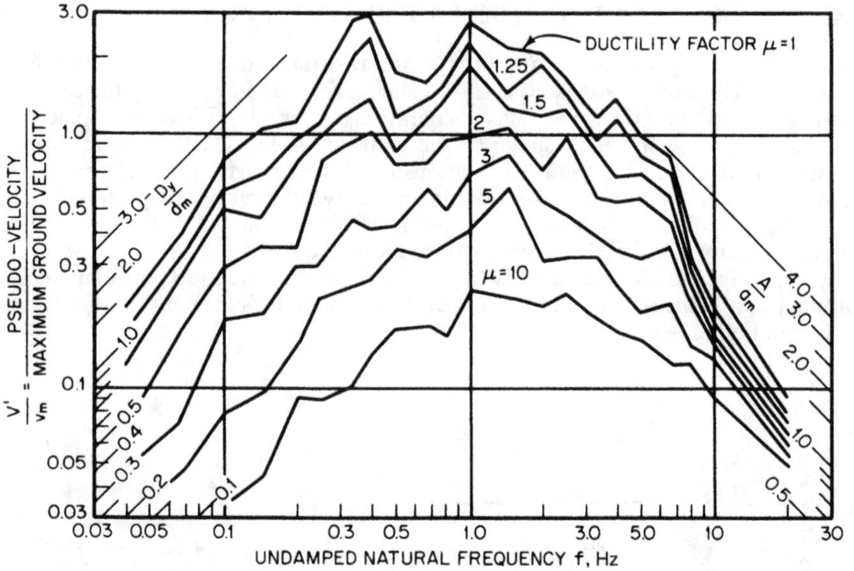

FIGURE 29.11 Deformation spectra for elastoplastic systems with 2 percent of critical damping that were subjected to the El Centro earthquake.

What is observed, as would be expected, is that with increasing energy absorption the controlling spectral values decrease as compared to the elastic spectrum. Guidelines for constructing a modified spectrum to account for inelastic action is an approximation at best; for approximate civilian structural analysis applications (as opposed to military applications) the ductility value should be 3 or less, and normally more like 1.3 to 1.5. (See Refs. 7 and 8a). This observation means that the deformation is relatively small. One last word of caution in the use of such modified spectra is that for it to be representative of structural behavior, the controlling deformation of the element or system must actually be near in value to the ductility value assumed; thus, iterative analysis is normally required.

MULTIPLE-DEGREE-OF-FREEDOM SYSTEMS

USE OF RESPONSE SPECTRA

A multiple-degree-of-freedom (MDOF) system has as many modes of vibration as the number of degrees of freedom. For example, for the shear beam shown in Fig. 29.12*A* the fundamental mode of lateral oscillation is shown in (*B*), the second mode in (*C*), and the third mode in (*D*). The number of modes in this case is 5. In a system that has independent (uncoupled) modes (this condition is often satisfied for buildings) each mode responds to the base motion as an independent single-degree-of-freedom system. Thus, the modal responses are nearly independent functions of time. However, the maxima do not necessarily occur at the same time.

For MDOF systems, the concept of the response spectrum can also be used in most

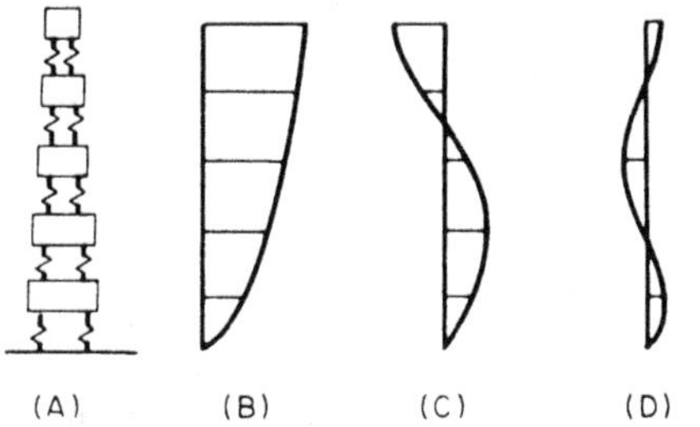

FIGURE 29.12 Modes of vibration of shear beam. The first three (1,2,3) relative mode shapes are shown by (*B*), (*C*), and (*D*), respectively, for lateral vibration.

cases, although the use of the inelastic response spectrum is only approximately valid as a design procedure.[10,11] For a system with a number of masses at nodes in a flexible framework, the equation of motion can be written in matrix form as

$$M\ddot{u} + C\dot{u} + Ku = -M(\ddot{y})\{1\} \qquad (29.6)$$

in which the last symbol on the right represents a unit column vector. The mass matrix M is usually diagonal, but in all cases both M and the stiffness matrix K are symmetrical. When the damping matrix C satisfies certain conditions, the simplest of which is when it is a linear combination of M and K, then the system has normal modes of vibration, with modal displacement vectors u_n. Analysis techniques for handling MDOF systems are described in Refs. 8 and 12.

DESIGN ISSUES

Before beginning a short overview of design, analysis, and construction, where ground motion events and other natural hazard excitation may occur, it is necessary to point out the two overriding caveats for such processes—namely, that design and construction of infrastructure of any kind (buildings, bridges, power plants, pipelines, etc.) must be directed (1) to protection of human life from injury or death and (2) to minimize the likelihood of damage or loss of the principal structure, and to minimize damage to contents, hopefully to the extent of maintaining functionality. For a sampling of important publications pertaining to seismic analysis, design, and related important construction issues, see Refs. 8, 9, and 11–24.

GRAVITY (NORMAL) DESIGN AND CONSTRUCTION

The normal design process for structures of all kinds begins with general planning and the development of specific plans, while determining that all local and state permits (including environmental) are identified and obtainable. At the same time, land is identified/procured and studied for foundation suitability, drainage, and access, as well as availability of utilities. Even at this early stage, a host of natural and man-made loadings must normally receive at least initial study, including various gravity loadings and earthquake, wind, snow, flood, ice, and rain loadings, as well as any possible imposed deformations. In special cases in today's world, accidental and man-made blast and shock loadings require consideration. Provisions in design and construction for many of these effects often serve as dual-purpose strengthening and quite often can be undertaken at this stage in an economical manner.

In earthquake-prone regions additional early studies normally focus as well on evaluating the possibility of landslides or ground instability that could affect the performance of the physical infrastructure, coupled with studies of transportation access in times of earthquake or other hazard occurrence—for example, tornados or hurricanes. At this point, preliminary design (approximate proportioning of members) proceeds for the proposed building, bridge, and so on. Preliminary proportioning is required as a starting point for subsequent analyses under various combinations of dead and live loadings.

One design item of major importance is that of identifying "load paths" for handling the gravity and live loads from floors down to the foundation, considering carefully the adequacy for carrying the variable loads that might be placed here and there within the structure, much less externally (ice and snow). Almost of equal importance is the matter of evaluating redundancy, to ensure that damage or failure of key members or connections does not lead to total collapse. Normally, many analysis runs are made in order to investigate the adequacy of the structure to handle the anticipated variability in loads, their possible placement, and even imposed deformations, as, for example, from foundation movements.

SEISMIC DESIGN

If seismic issues are to be addressed, then the design effort also entails a series of additional steps. Foremost is the matter of load path again, to be sure that in the event of an earthquake the lateral (and occasionally vertical and torsional) dynamic loadings and deformations can be accommodated properly to ground without significant distress. Depending upon the form and framework chosen, the seismic analyses can range from the *equivalent lateral load* procedure in the simplest case, as presented in many building and bridge codes and standards, on up through use of modal analysis, or finite element analyses of many types, with design response spectrum or time-history input as deemed appropriate.

In a short chapter such as this, space precludes even presenting a simple example of the many factors involved in a comprehensive design. For such examples, readers are referred to texts, manuals, and trade and technical journal articles. However, to provide some idea of the complexity associated with actual structural design, following are some titles of only a few of the normally listed seismic code and standards provisions:

Materials and Properties (Steel, Reinforced Concrete, Post-tensioned Concrete, Masonry, Wood)

Loads and Combination of Loads

Seismic Ground Motion Values and Seismic Input (Time-History, Spectral, etc.)

Importance Factor, Occupancy Category, Seismic Design Categories

Structural System Selection (Including Diaphragm Considerations)

Analysis Procedures (Equivalent Lateral Force Procedure, Modal Analysis, etc.)

Drift Limits and Fixed Deformations

Foundation Design (Including Piling)

Seismic Design for Nonstructural Components

Connections Between Girders/Beams and Columns

Base Isolation and Special Damping Systems

The foregoing is but a small sample of the issues to be addressed with respect to seismic design of a building or other specific piece of infrastructure. In many cases, the major codes, standards, and guidelines provide specific recommendations and formulae without much technical background (even in the commentaries) from which to draw judgments about source and adequacy. Many procedures are overwhelming in detail; in order to have some basis for judgment, it is imperative that designers/analysts know something about the background of the provisions. Only in this way can analysts judge the adequacy and sufficiency of the code and standard provisions.

Prudent professional practice, as well as applicable laws, require that codes, standards, and guidelines be followed. However, with increasing knowledge about materials, member connections, stability, and the like, such referenced prescriptive details may not be totally sufficient. Today's structural designers need to have a clear understanding of the likely behavior of the structure under design for many types of complex loadings and related situations—some of a static nature, others dynamic in nature. To be able to properly assess such matters, it behooves designers to keep up (through workshops and reading of texts and literature) as much as possible, so as to develop a basis for judgment. By way of example, in recent earthquakes throughout the world, connection performance has been identified as a major factor to be considered in structural design, and immense research has been/is under way on that topic for many types of construction materials. Similarly, attention to such matters as corrosion, aging, and many other items that can be classified as essential to structural health is needed. At present, there is some degree of research related to continuous structural health monitoring.

Included for the first time in this chapter are short sections on energy dissipative devices and some observations on an increasingly important related subject, namely, seismic risk assessment. In advanced modern seismic design, these two topics have been developed to the point of being important considerations.

EQUIPMENT AND LIFELINES

No introduction to earthquake engineering would be complete without mention of the importance of adequate design of equipment in buildings and essential building services, including, for example, communications, water, sewage and transportation systems, gas and liquid fuel pipelines, and other critical facilities. These important elements of constructed facilities, as well as sources of energy, have received major design attention in recent years as the importance of maintaining their integrity has become increasingly apparent.

It has always been obvious that the seismic design of equipment was important, but the focus on nuclear power has pushed this technology to the forefront. Many standards and documents are devoted to the design of such equipment.

ENERGY DISSIPATION DEVICES

(Energy dissipation devices are often denoted as damping systems). Conventionally, structures have been designed to resist natural hazards through a combination of strength, deformability, and energy absorption. These structures may undergo responses well beyond the elastic limit—for example, in a severe earthquake. They may remain intact only due to their ability to deform inelastically, as this deformation results in increased flexibility and energy dissipation. However, this deformation also results in local damage to the structure, as the structure itself must absorb much of the earthquake input energy.

As an alternative, a number of innovative protective systems have been proposed to enhance structural functionality and safety against natural and man-made hazards. These systems work by absorbing or reflecting a portion of the input energy that would otherwise be transmitted to the structure itself. Consider the following energy conservation relationship as an illustration of this approach:

$$E = E_k + E_s + E_h + E_d$$

where E = total input energy from environmental forces
E_k = absolute kinetic energy
E_s = recoverable elastic strain energy
E_h = irrecoverable energy dissipated by the structural system through inelastic or other inherent forms of damping
E_d = energy dissipated by structural protective systems

Thus, the demand on energy dissipation through inelastic deformation can be reduced by using structural protective systems.

By and large, protective systems can be grouped into three broad categories: (1) base isolation, (2) passive energy dissipation, and (3) active control. Of the three, base isolation can now be considered a more mature technology with wider applications as compared with the other two.[25]

Base isolation systems introduce flexibility and energy absorption capabilities, thereby reducing the level of energy which can be transmitted to the structure.[26] Typical seismic isolation devices are placed at the foundation of a structure and include elastomeric bearings, lead rubber bearings, high damping rubber bearings, and sliding friction pendulum bearings. Numerous seismically isolated structures have been built in many countries during the last 20 years.

Passive energy dissipation systems encompass a range of materials and devices for enhancing damping, stiffness, and strength and can be used both for seismic hazard mitigation and for rehabilitation of aging or deficient structures.[8b,27] In general, such systems are characterized by their ability to enhance energy dissipation in the structural systems in which they are installed. These devices generally operate on principles such as frictional sliding, yielding of metals, phase transformation in metals, deformation of viscoelastic (VE) solids or fluids, and fluid orificing. A large number of passive control systems or passive energy dissipation devices have been developed and installed in structures throughout the world for performance enhancement under wind and earthquake loads. Guidelines for seismic design using passive energy dissipation systems have been developed.[28]

Active and semiactive structural control systems are a natural evolution of passive control technologies. Active control systems are force delivery devices integrated with real-time processing evaluators/controllers and sensors within/on the structure. They act simultaneously with the hazardous excitation to provide enhanced structural behavior for improved service and safety. Semiactive control systems can be viewed as controllable passive energy dissipaters, offering the adaptability of active control devices without requiring the associated large power sources. In fact, many can operate on battery power, which is critical during seismic events when the main power source to the structure may fail. The development of active and semiactive control systems has reached the stage of full-scale application for wind and earthquake hazard mitigation, most of which are in Japan. In addition, 15 bridge towers have employed active systems during erection.[28]

Cautionary Notes. For many of the systems just described, the damping is usually nonproportional, resulting in a quadratic eigenvalue problem and complex mode shapes. In other words, studies through analysis tend to be highly complex, and control (and feedback) are likewise complex. Additionally, installation of dampers results in what is termed a "modified structure," which itself normally requires significant analysis to investigate the limits of motion associated with the dampers, and also that the forces/deformations associated with and arising from the dampers can be accommodated by the structure itself, much less the damper.

EARTHQUAKE RISK

The Classic Approach. In its simplest form, risk is the product of *hazard* and *vulnerability* for a *unit value*. Earthquake risk may therefore be broken down into the following three main components[8e]:

> *Hazard* is a description of the severity of the shaking of the ground during earthquakes and the consequential ground deformation during and after ground shaking. The ground shaking characterization may be in several forms, some of which are mentioned in preceding sections of this chapter and further discussed below.
>
> *Vulnerability* is the sensitivity of a structural system to the hazard. Vulnerability may also be expressed in many different forms. For example, the design force of a building or a bridge is a measure of vulnerability. If this design force is exceeded, the structure is likely to withstand damage, depending on the level of the force being exceeded.
>
> *Inventory* is the count and description of the exposed systems and their value. Building inventory, for example, is required to estimate the risk from an earthquake since the data describing an individual building will include its use and expected occupancy as well as the value of assets which may be stored in it.

The accuracy and reliability, and a known level of uncertainty, of all three components are necessary for earthquake risk assessment.

Assessment of earthquake risk may be probabilistic or deterministic.[8e] In deterministic assessment, the hazard is described by a given earthquake scenario, expressed as magnitude of an earthquake, its location, and the type of soil under the civil engineering facility that is being assessed. In probabilistic assessment, the hazard is the consequence of an earthquake of a particular magnitude (size) that has a probability of occurring within a given period of time. The consequence of defining the probabilistic earthquake is expressed as a value (e.g., peak ground acceleration, velocity of displacement; response acceleration, velocity of displacement) or a set of values with an attached probability of the value or values being exceeded during the lifetime of the civil engineering facility. Both deterministic and probabilistic hazard characterizations may be expressed for a specific location when the assessment is structure-specific, or as a regional map when the assessment is for a population of structures.

The vulnerability of the system may also be expressed deterministically or probabilistically.[29] In deterministic assessment, the vulnerability is expressed as a fixed value of force or displacement beyond which the system will suffer a predetermined level of damage. For probabilistic assessment, the measure of vulnerability is usually defined as a probability of reaching or exceeding a damage state, given the hazard measure. This latter point, with regard to the multiplicity of limit states, leads to the definition of multiple vulnerability indices, referred to as *multiple limit state assessment* or *performance-based assessment.* The usual limit states in earthquake engineering are (1) minor damage, leading to almost immediate occupancy of the structure, hence, uninterrupted use; (2) medium damage, which enables the structure to be repaired and reused within a short period of time, thus minimizing economic loss; and (3) heavy damage that does not lead to collapse, hence, the life of the occupants is preserved. The three important limit states of minor, medium, and heavy damage have social and economic implications and are best addressed by controlling the three important structural response properties of stiffness, strength and ductility, where ductility is the ability of the structure to deform beyond its elastic limit. The relationship between earthquake scenarios, limit states, and socioeconomic consequences, in most general terms, is summarized in Table 29.2.

TABLE 29.2 Relationship Between Hazard and Vulnerability

Return period	Earthquake magnitude	Structural characteristics	Engineering limit state	Socioeconomic limit state
~ 75–200 years	~ 4.5–5.5	Stiffness	Minor damage	Continued operation
~ 400–500 years	~ 5.5–6.5	Strength	Medium damage	Limited economic loss
~ 2000–2500 years	~ 6.5–7.5	Ductility	Extensive damage	Life loss prevention

On Modeling. Among many concerns, caution should be exercised in selecting the hazard since single values, such as peak ground parameters, may not be adequate for a particular structural system because the system as a whole will not have been sufficiently tested. Where actual earthquake signatures are used in the assessment, a minimum number of about 5 is necessary, and their spectra should encompass the range of response periods of the structures being assessed. Care should also be exercised in developing a model of the structure for vulnerability determination such that its mass, stiffness, strength, and ductility are all accurately represented, including even properties of vulnerable components such as infills and fixtures. In short, risk analysis is a complex subject to be approached with great care.

A Second General and Simple Approach. Another quite different, yet simple, approach to estimating acceptable risk, as well as a decision basis pattern, is presented in a paper by W. J. Hall and J. R. Wiggins.[30] This short paper focuses, as an example, on some aspects of risk assessment commonly employed for building complexes and uses for the illustrated case, with discussion, specific elements: (1) the problem, (2) the identification, (3) the observations, (4) the building codes, (5) the process, (6) the definition, and (7) the decision. The processes presented here may be helpful to individuals attempting to grasp the elements of the plethora of complex issues usually encountered in developing a measure of risk, whether by rigorous analysis or simplistic reasoning.

In summary, risk assessment typically turns out to be a complex task, and analysts, over and above the various theories, must keep in mind that a meaningful outcome needs reasonable hazard and performance elements to be identifiable, which in turn leads to a result that executives can understand and use as a basis to make decisions. The ability to identify and explain the bounding of uncertainty elements in an analysis is of paramount importance.

REFERENCES

1. Hudson, D. E.: "Vibration of Structures Induced by Seismic Waves," in C. M. Harris and C. E. Crede, eds., *Shock and Vibration Handbook,* 1st ed., Vol. III, Chap. 50, McGraw Hill Book Co., New York, 1961.

2. Richart, F. E., J. R. Hall, Jr., and R. D. Woods: "Vibration of Soils and Foundations," Prentice-Hall, Englewood Cliffs, N.J., 1970.
3. Dowding, C. H.: "Blast Vibration Monitoring and Control," Prentice-Hall, Englewood Cliffs N.J., 1985.
4. Cooper, P. W.: "Explosives Engineering," Wiley-VCH, New York, 1996.
5. Bolt, B.A.: "Earthquakes," 5th ed., W. H. Freeman and Co., New York, 2003.
6. Lee, W. H. K., et al.: "Earthquake Engineering Seismology," Vols. 1 and 2, Academic Press, New York, 2002.
7. Newmark, N. M., and W. J. Hall: "Development of Criteria for Seismic Review of Selected Nuclear Power Plants," *Report NUREG/CR 0098,* U.S. Nuclear Regulatory Commission, Washington, D.C., 1978.
8. Earthquake Engineering Research Institute Monograph Series:

   a. Newmark, N. M., and W. Hall: *Engineering Spectra and Design*, 1982.

   b. Hanson, R. D., and T. L. Soong: *Seismic Design with Supplementary Energy Dissipation Devices,* 2001.

   c. Housner, G. W. and P. C. Jennings: *Earthquake Design Criteria*, 1982.

   d. Chopra, A. K.: *Dynamics of Structures*—A Primer, 1981.

   e. McGuire, R. K.: *Seismic Hazard and Risk Analysis*, 2004.
9. Housner, G.: "Selected Earthquake Engineering Papers of George Housner," Civil Engineering Classics, D. E. Hudson ed., American Society of Civil Engineers, New York, 1990.
10. Riddell, R.: "Inelastic Response Spectra: A Historical Note," *J. Earthquake Engineering and Structural Dynamics,* **37**:1175–1183, 2008.
11. American Society of Civil Engineers:

    Minimum Design Loads for Buildings and Other Structures, ASCE Standard 7–05.

    Seismic Rehabilitation of Existing Buildings, ASCE Standard 41–06.

    Standard Calculation Methods for Structural Fire Protection, ASCE Standard 29–05.
12. Chopra, A. K.: "Dynamics of Structures—Theory and Applications to Earthquake Engineering," 3nd ed., Prentice-Hall, Upper Saddle River, N.J., 2007.
13. International Building Code Council: *International Building Code 2003* (plus Commentary). See also International Code Council: *Performance Code for Buildings and Facilities— 2003.*
14. American Concrete Institute: *Building Code Requirements for Structural Concrete* (with Commentary), ACI 318–05.
15. American Institute of Steel Construction: "Steel Construction Manual," AISC, Chicago, 2005. See also AISC "Seismic Design Manual," and "Structural Steel Educational Manual," 2006.
16. Clough, R.W., and J. Penzien: "Dynamics of Structures," 2nd ed., McGraw Hill, New York, 1975.
17. Kramer, Steven L.: "Geotechnical Earthquake Engineering," Prentice-Hall, Upper Saddle River, N.J., 1996.
18. Day, R.W.: "Geotechnical Earthquake Engineering Handbook," McGraw-Hill, New York, 2002.
19. Englekirk, R.E.: "Seismic Design of Reinforced and Prestressed Concrete Buildings," John Wiley and Sons, Hoboken, N.J., 2003.
20. Paulay, T., and M. J. N. Priestly: "Seismic Design of Reinforced Concrete and Masonry Buildings," John Wiley & Sons, New York, 1992.
21. Priestly, M. J. N., F. Seible, and G. M. Calvi: "Seismic Design and Retrofit of Bridges," John Wiley & Sons, New York, 1996.

22. Elnashai, A. S., and L. DiSarno: "Introduction to Earthquake Structural Engineering," John Wiley & Sons, Hoboken, N.J., 2008.

23. Bozorgnia, Y., and V. Bertero, eds.: "Earthquake Engineering: From Engineering Seismology to Performance-Based Engineering," CRC Press, Boca Raton, Fla., 2004.

24. Nyman, D. J., W. J. Hall, et al.: "Guidelines for the Seismic Design of Oil and Gas Pipeline Systems," Technical Council on Lifeline Earthquake Engineering, ASCE, 1984.

25. *Proc. Seismic Isolation, Passive Energy Dissipation, and Active Control,* Applied Technology Council: ATC 17-1, Redwood Cit, Calif., 1993.

26. Skinner, R. I., W. H. Robinson, and G. H. McVerry: "An Introduction to Seismic Isolation," John Wiley & Sons, New York, 1993.

27. Soong, T. T., and G. F. Dargush: "Passive Energy Dissipation Systems in Structural Engineering," Wiley, London, 1997.

28. Spencer, Jr., B. F., and S. Nagarajaiah: "State-of-the-Art of Structural Control," *Journal of Structural Engineering, ASCE,* **129**(7)845–856, 2003.

29. Pinto, P. E., R. Giannini, and P. Franchin: "Seismic Reliability Analysis of Structures," IUSS Press, Pavia, Italy, 2004.

30. Hall, W. J., and J. H. Wiggins: "Acceptable Risk: A Need for Periodic Review," *Natural Hazards Review,* ASCE **1**(3):180–187; (7): 845–856 (August 2000).

CHAPTER 30
VIBRATION OF STRUCTURES INDUCED BY FLUID FLOW

Robert D. Blevins

INTRODUCTION

A surrounding fluid can significantly alter a structure's vibrational characteristics. Fluid-structure interaction phenomena are classified in Fig. 30.1. A surrounding quiescent fluid decreases the natural frequencies and increases damping of a structure. A dense fluid couples the vibration of an elastic structure to adjacent structures. A fluid flow can induce vibration. A turbulent fluid flow exerts random pressures on a structure, and these random pressures induce a random response. The structure can resonate with periodic components of the wake, which is called vortex-induced vibration. If the structure is sufficiently flexible, the structural deformations can change the fluid force. The resultant coupled vibrations can become unstable once the fluid velocity reaches a critical threshold value.

ADDED MASS AND INERTIAL COUPLING

If a body accelerates, decelerates, or vibrates in a fluid, then fluid is entrained by the body. This entrapment of fluid, called the *added mass* or *virtual mass* effect, occurs both in viscous and in inviscid (i.e., ideal) fluids. In general, added mass is practically important only when the fluid density is comparable to the density of the structure so that the added mass becomes an appreciable fraction of the total mass in motion. In light gasses such as air (1.2 kg/m^3, 0.075 lb/ft^3), added mass is ordinarily negligible except for very light structures, such as airplanes. But submersion in dense liquids, such as marine pipelines in seawater (1025 kg/m^3, 64 lb/ft^3), significantly decreases natural frequency.

Consider a rigid body, shown in Fig. 30.2, that lies in a large quiescent reservoir of incompressible, inviscid, irrotational fluid. The one-dimensional equation of conservation of mass of an incompressible fluid is $\partial u/\partial x = 0$, where u is velocity in the x direction; in multiple dimensions $\nabla\cdot\mathbf{u} = 0$, where ∇ is the gradient operator and $(\cdot)$ is

INERTIAL COUPLING EFFECTS	UNSTEADY FLOW INDUCED VIBRATION	FLOW-STRUCTURE COUPLED VIBRATION
1. Added mass	1. Turbulence induced vibration	1. Vortex induced vibration
2. Inertial coupling	2. Ocean wave induced vibration	2. Galloping and flutter
3. Instability due to parallel flow	3. Sonic fatigue	3. Fluid elastic instability

FIGURE 30.1 A classification of flow-induced vibration.

the vector dot product. The vector fluid velocity **u** is the gradient of a *velocity potential* Φ that is a function of the coordinates x,y,z and time t. Conservation of mass implies that the velocity potential satisfies Laplace's equation.[1,2]

$$\mathbf{u} = \nabla\Phi(x,y,z,t) \qquad \nabla^2\Phi = 0 \tag{30.1}$$

The equation of conservation of momentum in one dimension is $\partial p/\partial x = -\rho\partial u/\partial t$, where ρ is fluid density, so $\partial(p + \rho\partial\Phi\partial t)/\partial x = 0$; in multiple dimensions $\nabla(p + \rho\partial\Phi/\partial t) = 0$, which implies the static fluid pressure p is a function of the time derivative of the velocity potential.

$$p = -\rho\partial\Phi/\partial t \tag{30.2}$$

The body moves with vector velocity $\mathbf{U}(t)$. The body surface's normal outward velocity equals the adjacent fluid velocity.

$$\mathbf{U}\cdot\mathbf{n} = \partial\Phi/\partial\mathbf{n} \tag{30.3}$$

where n is the unit outward normal to the body surface. These are the governing equations of a rigid body in an inviscid incompressible fluid.

The added mass vector force **F** imposed by the surrounding fluid on the body is the integral of static pressure over the body surface S.

$$\mathbf{F} = \mathbf{n}\int_S p\mathbf{n}\,dS = -\rho\frac{\partial}{\partial t}\int_S \Phi\, dS \tag{30.4}$$

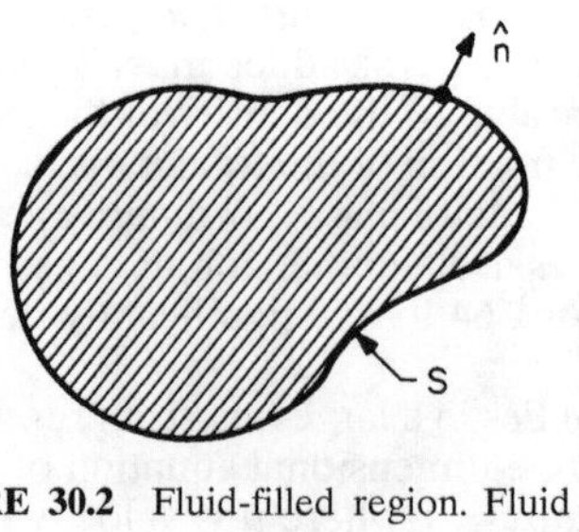

FIGURE 30.2 Fluid-filled region. Fluid density ρ.

where dS is a surface area element. If the accelerating rigid body has scalar velocity $U(t)$, in, say, the i-direction, the velocity potential can be expressed as the product $\Phi(x,y,z,t) = \phi(x,y,z)U(t)$ and the surface boundary condition [Eq. (30.3)] can be expressed as $n_i = \partial\phi/\partial n$, where n_i is the component of the unit outward normal in the direction of U. The added mass and added mass force on the body in the direction of U are scalar versions of the previous equation.[1]

$$m_a = \rho \int_S \Phi \frac{\partial \Phi}{\partial n} dS \qquad F = -m_a \frac{dU}{dt} \tag{30.5}$$

Added mass is proportional to fluid density. Added mass force is proportional to density times body acceleration. There is no added mass fluid force for steady translation; this is the D'Alembert paradox for an inviscid, irrotational fluid. Further, it can also be shown that added mass is proportional to fluid kinetic energy.[1]

The general added mass matrix $[M_a] = [m_{ij}]$ is a 6×6 symmetric matrix whose entries are added mass for acceleration in each coordinate direction and for rotary acceleration about these coordinates.

$$m_{ij} = m_{ji} = -\rho \int_S \phi_j \frac{\partial \Phi_i}{\partial n} dS \qquad \text{for } i, j = 1,2,3 \tag{30.6}$$

The matrix equation of motion for free vibration of a six-degree-of-freedom spring-supported rigid body, including added mass, is,

$$[M] - [M_a]\ddot{X} \qquad \text{or} \qquad ([M] + [M_a])\ddot{X} + [K] = 0 \tag{30.7}$$

where $[M]$ is the structural mass matrix, $[K]$ is the structural stiffness matrix, and X is the body displacement vector. Because added mass acts in phase with acceleration [Eq. (30.5)], added mass increases the mass of the body and decreases natural frequencies.

There are 21 (9 diagonal plus 12 off-diagonal) independent entries in the symmetric inviscid 6×6 added mass matrix. Off-diagonal terms couple degrees of freedom. For example, if a body is not symmetric about the x axis, then acceleration in the x direction generally induces added mass forces in both the x and y directions and moments. If the body is symmetric about one or more axes, some or all off-diagonal coupling terms are zero. The 3×3 added mass matrix for a two-dimensional section has three independent off-diagonal terms. If the section has two axes of symmetry, its added mass matrix is a diagonal matrix and the natural frequencies in fluid are lower than those in a vacuum by the factor $M^{1/2}/(M_a + M)^{1/2}$, where M_a is added mass and M is structural mass.

In viscous fluid, added mass is proportional to fluid density times a function of frequency f (Hz), fluid kinematic viscosity ν, amplitude X_o, and diameter $2a$. These dimensional parameters are reformulated into dimensionless amplitude X_o/a, viscosity parameter a^2f/ν, and Reynolds number $4\pi\,(a^2f/\nu)(X_o/a)$. For example, the added mass of a circular cylinder per unit length and a sphere with radius a in a viscous fluid for small-amplitude oscillations $X_o/a \ll 1$, are[2]

$$m_a = \begin{cases} \rho\pi a^2[1 + 2(\nu/\pi f a^2)^{1/2} + ..] & \text{cylinder} \\ (2/3)\rho\pi a^3[1 + 4.5(\nu/\pi f a^2)^{1/2} + ..] & \text{sphere} \end{cases} \tag{30.8}$$

These suggest that an inviscid fluid ($\nu = 0$) added mass is applicable to real fluids if a^2f/ν is greater than a few hundred, which is the case for larger structures and higher velocities. Shed vortices and separation complicate the added mass in viscous fluids for larger-amplitude oscillations.

Figure 30.3 has added mass of sections and bodies in incompressible inviscid ($\nu = 0$) fluids. Additional values are given in Refs. 3 and 4. Most of these are calculated from

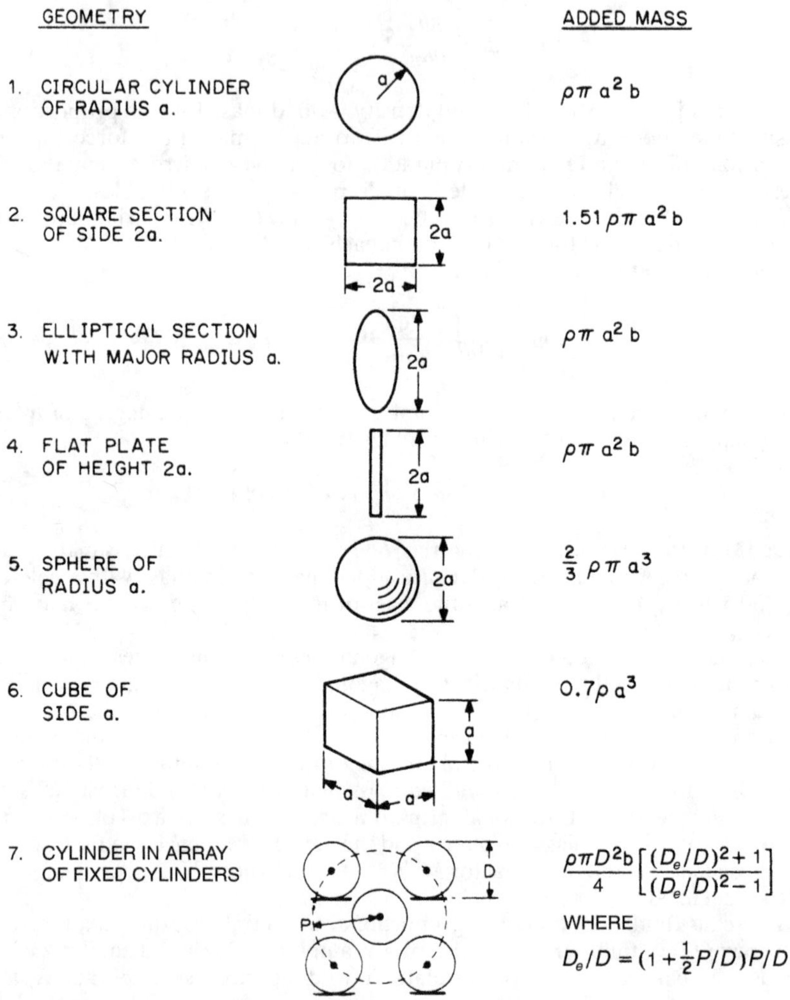

FIGURE 30.3 Added mass for lateral acceleration.[3] The acceleration is left to right; b is the span for two-dimensional sections.

potential flow. For example, the velocity potential for flow with velocity U in the x direction over a stationary cylinder with radius a is[1]

$$\Phi(r,\theta,t) = U(t)[(r^2 + a^2)/r]\cos\theta = U(t)x + U(t)a^2x/(x^2 + y^2) \tag{30.9}$$

where r is the radial coordinate and θ is the angular coordinate. In rectangular coordinates, $x = r\cos\theta$ and $y = r\sin\theta$. The velocity potential has two terms. The first is a mean velocity U in the x direction. The second is associated with the cylinder with surface $r^2 = x^2 + y^2 = a^2$. Substituting the second term and its normal derivative in the U direction $d\phi/dn = d\phi/dr = \cos\theta$ on the surface into Eq. (30.6) gives the added mass for an accelerating cylinder.

$$m_a = \rho\pi a^2 \tag{30.10}$$

The added mass of a cylinder is equal to its displaced mass. The added mass of a sphere is half of its displaced mass. The added mass moment of inertia of a thin rectangular plate rotating about its longitudinal axis is one-half the mass moment of inertia of the swept fluid cylinder.

If two structures are in close proximity, then the added mass will be a function of the spacing between the structures and inertial coupling will be introduced between the bodies. For example, consider a cylindrical rod centered in a fluid-filled annulus bounded by a cylindrical cavity, shown in Fig. 30.4. The radius of the rod is a and the radius of the outer cylinder is b. The fluid forces exerted on the rod and outer cylinder because of their relative acceleration are[5]

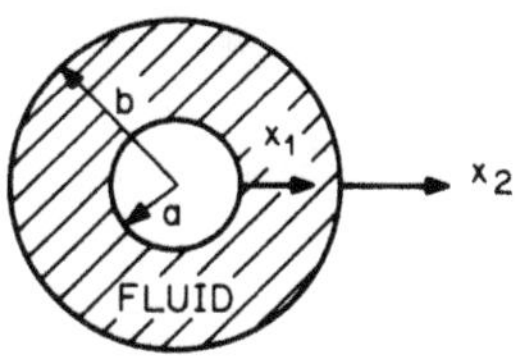

FIGURE 30.4 A rod in a fluid-filled annulus.

$$\begin{aligned} F_1 &= -m\ddot{x}_1 + (M_1 + m)\ddot{x}_2 \\ F_2 &= (m + M_1)\ddot{x}_1 - (m + M_1 + M_2)\ddot{x}_2 \end{aligned} \tag{30.11}$$

where x_1, x_2 = displacement of inner rod and outer cylinder
F_1, F_2 = force on inner rod and outer cylinder
$m = \rho\pi a^2(b^2 + a^2)/(b^2 - a^2)$, added mass of inner rod
$M_1 = \rho\pi a^2$
$M_2 = \rho\pi b^2$

These forces include not only added mass but also inertial coupling between the motion of the two structures. [These equations also apply for a sphere contained within a spherical cavity but here $m = (M_1/2)(b^3 - 2a^3)/(b^3 - a^3)$, $M_1 = \frac{4}{3}\rho\pi a^3$, and $M_2 = \frac{4}{3}\rho\pi b^3$.] Coupling is introduced between the cylinder and the rod through the fluid annulus. The coupling increases with the density of the fluid and decreases with increasing gap. If the cylinder and the rod are elastic, motion of either structure tends to set both structures into motion.

For example, consider an array of heat exchanger tubes contained within a shell. Water fills the shell and surrounds the tubes. If the tubes are closely spaced, then motion of one tube sets adjacent tubes and the shell into motion. Fluid-coupled modes of vibration will result in the tubes and the shell moving in fixed modal patterns. In Refs. 6 and 7, analysis is given for inertial coupling of a cylinder contained eccentrically within a cylindrical cavity, rows of cylinders, and arrays of cylinders. Concentric cylindrical shells coupled by a fluid annulus are important in the design of nuclear reactor containment vessels. Approximate solutions are required for both the vessels and the fluid. Reviews of the analysis of fluid coupled concentric vessels are given in Refs. 8 and 9.

Finite element numerical solutions, developed for an irrotational fluid, have been incorporated in the NASTRAN and other computer programs to permit solution for added mass and inertial coupling. These programs solve the fluid and structural problems and then couple the results through interaction forces[10] (see Chap. 23).

WAVE-INDUCED VIBRATION OF STRUCTURES

Waves induce vibration of structures, such as marine pipelines, oil terminals, tanks, and ships, by placing oscillatory pressure on the surface of the structure. These forces are often well-represented by the inviscid flow solution for many large structures such as ships and oil storage tanks. For most smaller structures, viscous effects influence the fluid force and the fluid forces are determined experimentally.

Consider an ocean wave approaching the vertical cylindrical structure as shown in Fig. 30.5. The wave is propagating in the x direction. Using small-amplitude (linear) inviscid wave theory, the wave is characterized by the wave height h (vertical distance between trough and crest), its angular frequency ω, and the associated wavelength λ (horizontal distance between crests), and d is the depth of the water. The wave potential Φ satisfies Laplace's equation [Eq. (30.1)] and a free-surface boundary condition.[11] The associated horizontal component of wave velocity varies with depth $-z$ from the free surface and oscillates at frequency ω:

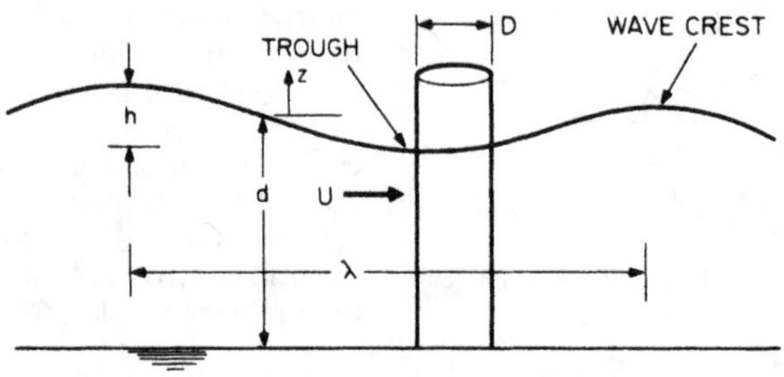

FIGURE 30.5 A circular cylindrical structure exposed to ocean waves.

$$U(t,z) = \frac{h\omega}{2} \frac{\cosh [2\pi(z+d)/\lambda]}{\sinh (2\pi d/\lambda)} \cos\left(\frac{2\pi x}{\lambda} - \omega t\right) \tag{30.12}$$

This component of wave velocity induces substantial fluid forces on structures, such as pilings and pipelines, which are oriented perpendicular to the direction of wave propagation.

The forces which the wave exerts on the cylinder in the direction of wave propagation (i.e., in line with U) can be considered the sum of three components: (1) a buoyancy force associated with the pressure gradient in the laterally accelerating fluid [Eq. (30.12)], (2) an added mass force associated with fluid entrained during relative acceleration between the fluid and the cylinder [Eq. (30.6)], and (3) a force due to fluid dynamic drag associated with the relative velocity between the wave and the cylinder. The first two force components can be determined from inviscid fluid analysis as discussed previously. The drag component of force, however, is associated with fluid viscosity.

Thus, the in-line fluid force per unit length of cylinder due to an unsteady flow is expressed as the sum of the three fluid force components:

$$F = \rho A\dot{U} + C_I \rho A(\dot{U} - \ddot{x}) + \tfrac{1}{2}\rho\,|U - \dot{x}|\,(U - \dot{x}) D C_D \tag{30.13}$$

where x = lateral position of structure in direction of wave propagation
A = cross-sectional area = $\frac{1}{4}\pi D^2$ of cylinder having diameter D
C_I = added mass coefficient, which has theoretical value of 1.0 for circular cylinder
C_D = drag coefficient

This is the generalized form of the Morison equation, widely used to compute the wave forces on slender cylindrical ocean structures such as pipelines and piers.

If $\dot{x}$ and $\ddot{x}$ are set equal to zero in Eq. (30.13), the incline force per unit length on a stationary cylinder in an oscillating flow is obtained:

$$F(\dot{x} = \ddot{x} = 0) = C_m \rho A \dot{U} + \tfrac{1}{2}\rho\, |U|\, U D C_D \tag{30.14}$$

Because of the absolute sign in the term $|U|\ U$, the force contains not only components at the wave frequency but also components associated with the drag at harmonics of the wave frequency. The resultant time history of in-line force due to a harmonically oscillating flow has an irregular form that repeats once every wave period.

If the flow oscillates with zero mean flow, $U = U_0 \cos \omega t$ as in Eq. (30.12), then the maximum fluid force per unit length on a stationary cylinder is

$$F_{\max} = \begin{cases} \rho A C_m \omega U_0 & \text{if } \dfrac{U_0}{\omega D} < \dfrac{C_m A}{C_D D^2} \\[2ex] \dfrac{1}{2}\rho U_0^2 D C_D + \dfrac{(\rho A C_m U_0 \omega)^2}{2\pi U_0^2 D C_D} & \text{if } \dfrac{U_0}{\omega D} > \dfrac{C_m A}{C_D D^2} \end{cases} \tag{30.15}$$

If the cylinder is large (such as for a storage tank) with diameter D greater than the ocean wave height h and if the wavelength of the ocean wave is comparable to the diameter, then U_0 is small compared to ωD and the maximum force is given by the first alternative in Eq. (30.15). The drag force is negligible compared to the inertial forces for large cylinders. As a result, the ocean wave forces on large cylinders can be calculated using inviscid, i.e., potential flow, methods which are discussed in Refs. 11 and 12.

For the Reynolds number ranges typical of most offshore structures, measurements show that the inertial coefficient $C_m = 1 + C_I$ for cylindrical structures generally falls in the range between 1.5 and 2.0. $C_m = 1.8$ is a typical value. C_m decreases for very large diameter cylinders owing to the tendency of waves to diffract about large cylinders (Refs. 13 and 14). Similarly, measurements show that the drag coefficient falls between 0.6 and 1.0 for circular cylinders; $C_D = 0.8$ is a typical value.

Wave forces on elastic ocean structures induce structural motion. Since the wave force is nonlinear [Eq. (30.13)] and involves structural motion, no exact solution exists. One approach is to integrate the equations of motion directly by applying Eq. (29.8) at each spanwise point on a structure and then numerically integrate the time history of deflection using a predictor-corrector or recursive relationship to account for the nonlinear term. A simpler approach is to assume that the structural deformation does not influence the fluid force and apply Eq. (30.14) as a static load. This static approximation is valid as long as the fundamental natural frequency of the structure is well above the wave frequency *and* the first three or four harmonics of the wave frequency. However, many marine structures are not sufficiently stiff to satisfy this condition.

One generally valid simplification for dynamic analysis of relatively flexible structures is to consider that the wave velocity is much less than the structural velocity so that $|U - \dot{x}| \simeq |U|$. With this approximation, application of Eq. (29.8) to a single-degree-of-freedom model for a structure gives the following linear equation of motion:

$$(m + \rho A C_I)\ddot{x} + (2m\omega_N + \tfrac{1}{2}\rho\, |U|\, D C_D)\dot{x} + kx = \rho A C_m \dot{U} + \tfrac{1}{2}\rho\, |U|\, U D C_D \qquad (30.16)$$

where m = structural mass per unit length
k = stiffness
ζ = structural damping

This equation is solved by expanding both $x(t)$ and $U(t)$ in a Fourier series and matching the coefficients.

The fluid forces contribute added mass and fluid damping to the left-hand side as well as forcing terms to the right-hand side. This equation may be simplified further by retaining only the first (constant) term in the series expansion for $|U|$ in the fluid damping term so that the equation becomes a classical forced oscillator with constant coefficient.[12] For low mean velocity, $|U|$ can be set to zero and the resultant linear equation has an exact solution.

Flexible structures will resonate with the wave if the structural natural period equals the wave period or a harmonic of the wave period. Since the wave frequencies of importance are ordinarily less than 0.2 Hz (wave period generally greater than one cycle per 5 sec), such a resonance occurs only for exceptionally flexible structures such as deep-water oil production risers and offshore terminals. The amplitude of structural response at resonance is a balance between the wave force and the structural stiffness times the damping. Since the wave force diminishes with increased structural motion [Eq. (30.13)], the resultant displacements are necessarily self-limiting. In other words, the response which would be predicted by applying Eq. (29.9) dynamically is overly pessimistic because the wave force contributes mass and damping to the structure as well as excitation as can be seen in Eq. (30.16).

The above discussion considers only fluid forces which act in line with the direction of wave propagation. These in-line forces produce an in-line response. However, substantial transverse vibrations also occur for ocean flows around circular cylinders. These vibrations are associated with periodic vortex shedding, which is discussed below. The models discussed in the following section for steady flow are applicable to vortex shedding in oscillatory flows provided that the wave period exceeds the period of shedding, based on the maximum oscillatory velocity so that it is possible to fit one or more shedding cycles into the wave cycle.[13,14]

VORTEX-INDUCED VIBRATION

Many structures of practical importance such as buildings, pipelines, and cables are not streamlined but rather have abrupt contours that can cause a fluid flow over the structure to separate from the aft contours of the structure. Such structures are called *bluff bodies*. For a bluff body in uniform cross flow, the wake behind the body is not regular but contains distinct vortices of the pattern shown in Fig. 30.6 at a Reynolds number Re = UD/ν greater than about 50, where D is the width perpendicular to the flow and ν is the kinematic viscosity. The vortices are shed alternately from each side of the body in a regular manner and give rise to an alternating force on the body. Experimental studies have shown that the frequency, in hertz, of the alternating lift force is expressed as[16,17]

$$f_s = \frac{SU}{D} \qquad (30.17)$$

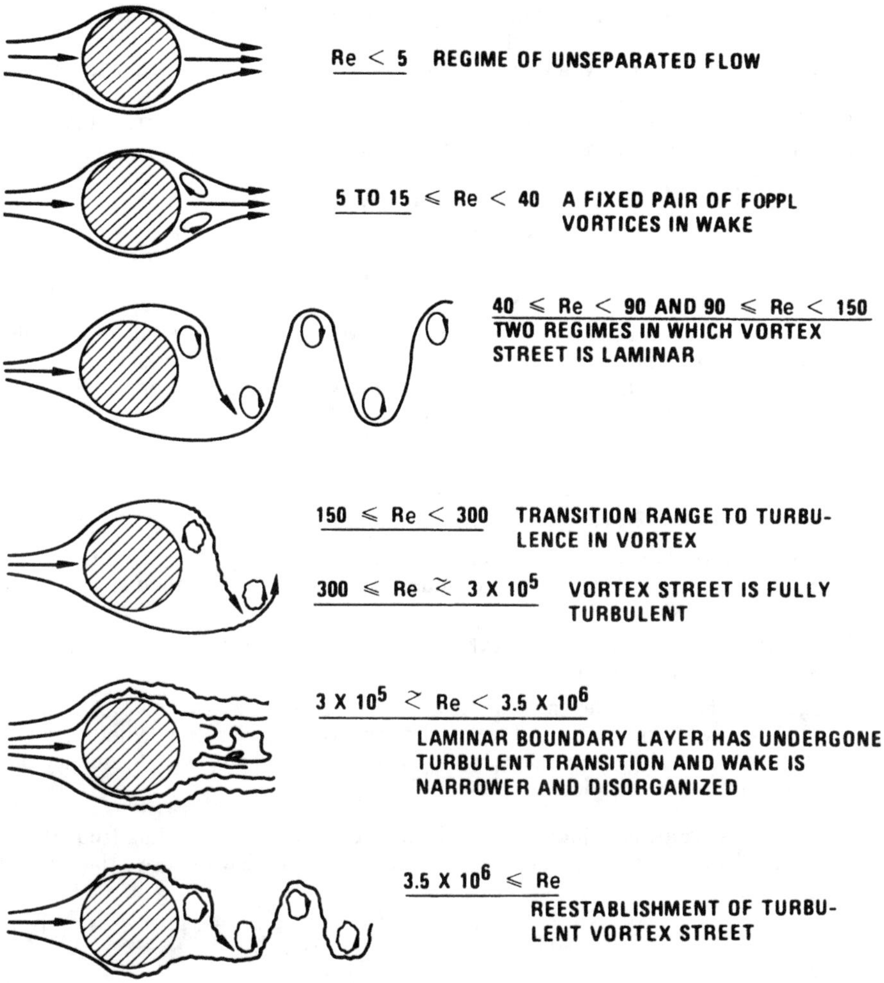

FIGURE 30.6 Regimes of fluid flow across circular cylinders.[15]

The dimensionless constant S called the *Strouhal number* generally falls in the range $0.25 \geq S \geq 0.14$ for circular cylinders, square cylinders, and most bluff sections. The value of S increases slightly as the Reynolds number increases; a value of $S = 0.2$ is typical for circular cylinders.

The oscillating lift force imposed on a single circular cylinder of length L and diameter D, in a uniform cross flow of velocity U, due to vortex shedding is given by

$$F = \tfrac{1}{2}\rho U^2 C_L DLJ \sin(2\pi f_s t) \tag{30.18}$$

where the lift coefficient C_L is a function of Reynolds number and cylinder motion. The experimental measurements of C_L show considerable scatter with typical values ranging from 0.1 to 1.0. The scatter is in part due to the fact that the alternating vortex forces are not generally correlated on the entire cylinder length L. The spanwise correlation length l_c of vortex shedding over a stationary circular cylinder[17] is approximately three to seven diameters for $10^3 < \text{Re} < 2 \times 10^5$. In order to account for the effect of the spanwise correlation on the net force on the cylinder of length L, a factor J called the *joint acceptance* has been introduced on the right-hand side of Eq. (30.18). Two limiting cases exist for the joint acceptance.

$$J = \begin{cases} \left(\dfrac{l_c}{L}\right)^{1/2} & \text{if } l_c << L \\ 1 & \text{if fully correlated} \end{cases} \tag{30.19}$$

Thus, if a cylinder is much longer than three to seven diameters, the lack of spanwise correlation reduces the net vortex lift force [Eq. (30.18)] on the cylinder.

Cylinder vibration at or near the vortex shedding frequency organizes the wake and changes the fluid force on the cylinder. Vibration of a cylinder in a fluid flow can:[12,17,18]

1. Increase the strength of the shed vortices.
2. Increase the spanwise correlation of the vortex shedding.
3. Cause the vortex shedding frequency shift from the natural shedding frequency [Eq. (30.17)] to the frequency of cylinder oscillation. This is called *synchronization* or *lock-in*.
4. Increase the mean drag on the cylinder. Mean drag can triple for one diameter amplitude cylinder vibration.
5. Alter the phase sequence and pattern of vortices in the wake. Figure 30.7 shows the patterns of vortices in the wake of a transversely vibrating cylinder, where f_s = natural shedding frequency [Eq. (30.17)], f = forced vibration frequency, and A_y = vibration amplitude transverse to flow.

As the flow velocity is increased or decreased so that the shedding frequency f_s approaches the natural frequency f_n of an elasticly mounted cylinder so that

$$f_n \approx f_s = \frac{SU}{D} \qquad \text{so} \qquad \frac{U}{f_n D} \approx \frac{U}{f_s D} = \frac{1}{S} \approx 5 \tag{30.20}$$

the vortex shedding frequency suddenly locks onto the structure natural frequency. The resultant vibrations occur at or nearly at the natural frequency of the structure and vortices in the near wake input energy to the cylinder. Large-amplitude vortex-induced structural vibration can result.

The vortex-induced vibrations of a spring-mounted cylinder in a water flow are shown as a function of velocity in Fig. 30.8 for five levels of damping. The horizontal scale gives flow velocity nondimensionalized (i.e., divided by the cylinder diameter D times the cylinder natural frequency f), both of which are held fixed as velocity U increases. The lower part of the figure shows the measured response cylinder single amplitude A_y vibration response as a function of flow velocity. The maximum cylinder amplitude occurs at the resonance condition $U/(fD) \sim 6$. The shedding frequency increases with velocity as predicted by Eq. (30.17) until it equals the cylinder natural frequency at $U/fD = 5$ and large-amplitude cylinder vibrations begin. The shedding frequency is entrained by the cylinder natural fre-

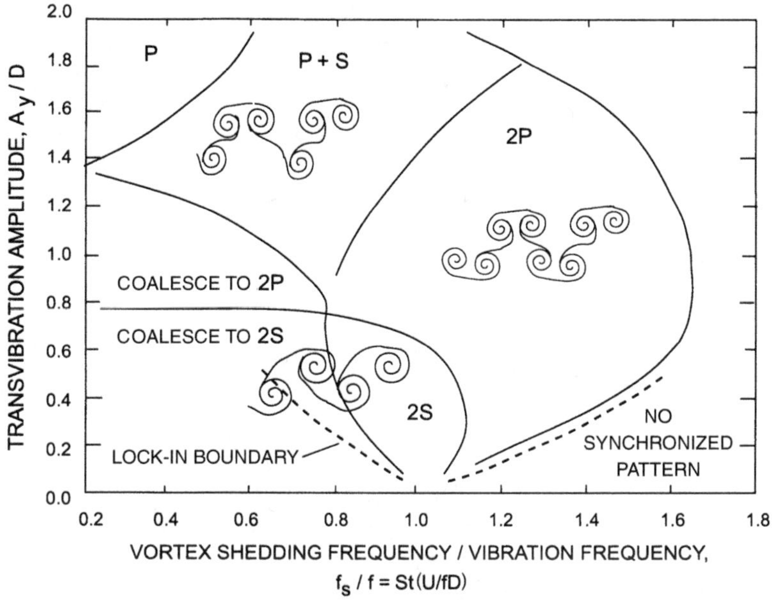

FIGURE 30.7 Patterns of vortices shed in the wake of a transversely oscillating cylinder in a cross flow.

quency. Entrainment persists until velocity is increased to $U/fD = 8$, at which point lock-in is broken and the shedding frequency abruptly returns to its natural value. In general, the larger the structural response to vortex shedding, the larger the range of lock-in.

Both the amplitude of the structural response and the velocity range over which lock-in persists are functions of the dimensionless reduced damping parameter δ_r:

$$\delta_r = \frac{2m(2\pi\zeta)}{\rho D^2} \tag{30.21}$$

where m = mass per unit length of cylinder, including added mass
ζ = damping factor for vibration in mode of interest, ordinarily measured in still fluid
ρ = fluid density
D = cylinder diameter

The lower δ_r, the greater the amplitude of the structural response and the greater the range of flow velocities over which lock-in occurs (see Ref. 19 and Fig. 30.8). For lightly damped structures in dense fluids (such as marine pipelines), δ_r is on the order of 1 and lock-in can persist over a 40 percent variation in velocity above and below that which produces resonance.

Within the synchronization band, substantial resonance vibration often occurs. Peak-to-peak vibration amplitudes of up to three diameters have been observed in water flows over cables and tubing. The vibrations are predominantly transverse to

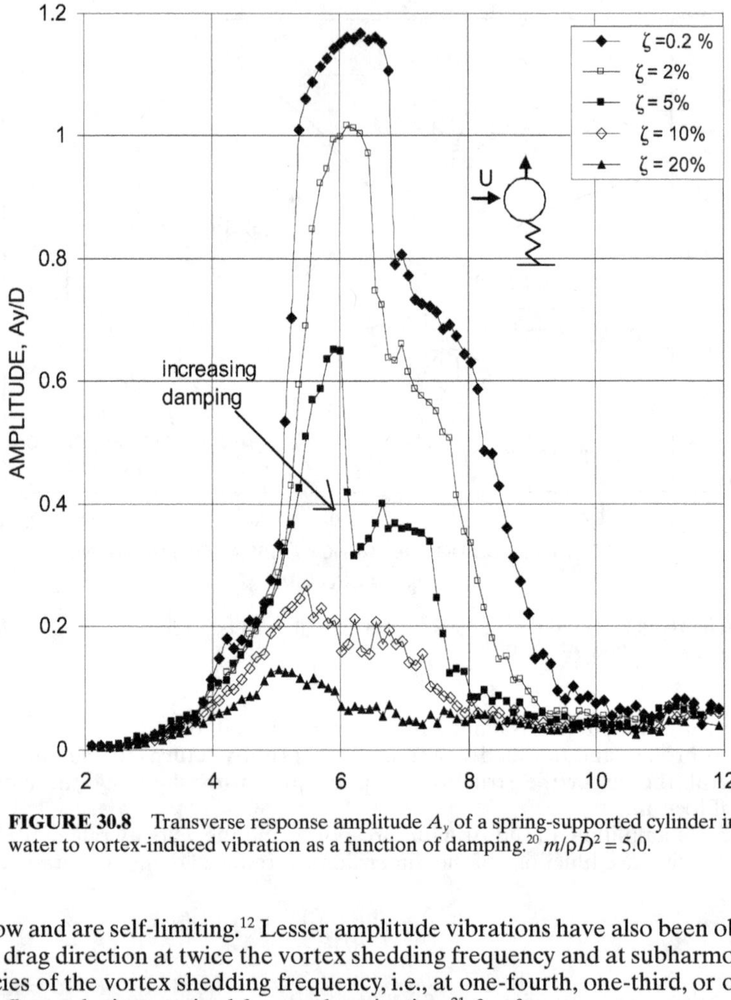

FIGURE 30.8 Transverse response amplitude A_y of a spring-supported cylinder in water to vortex-induced vibration as a function of damping.[20] $m/\rho D^2 = 5.0$.

the flow and are self-limiting.[12] Lesser amplitude vibrations have also been observed in the drag direction at twice the vortex shedding frequency and at subharmonic frequencies of the vortex shedding frequency, i.e., at one-fourth, one-third, or one-half of the flow velocity required for synchronization,[21] $f_s = f_n$.

If a uniform elastic cylinder is subjected to a cross flow uniformly over its span, then the oscillating vortex-induced lift force is given by Eq. (30.18). At lock-in, the vortex shedding frequency equals the natural frequency of the nth vibration mode $f_s = f_n$, and the amplitude of the cylinder response is

$$\frac{A_y}{D} = \frac{C_L J}{4\pi S^2 \delta_r} \tag{30.22}$$

where the maximum amplitude vibrations along the span are $y(t) = A_y \sin(2\pi f_n t)$. This equation is conservative if $C_L = J = 1$. However, Eq. (30.22) gives overly conservative predictions with $C_L = J = 1$ owing to the tendency of the actual lift coefficient to decrease at amplitudes in excess of 0.5 diameters and the lack of perfect spanwise correlation at lower amplitudes. Semiempirical correlations are given in Refs. 12, 22, and 23. One of these correlations is[12]

$$\frac{A_y}{D} = \frac{0.07\gamma}{(\delta_r + 1.9)S^2}\left(0.3 + \frac{0.72}{(\delta_r + 1.9)S}\right)^{1/2} \tag{30.23}$$

The mode shape parameter γ falls between 1.0 and 1.4. For a translating rigid rod ($\phi = 1$), $\gamma = 1$; for a cable or pipeline with a sinusoidal mode shape, $\gamma = 1.15$; and for a cantilever mode shape, $\gamma = 1.4$ and A_y is tip amplitude.

Equation (30.23) correctly predicts the self-limiting behavior of the resonance vibrations. Setting damping to zero, $\delta_r = 0$, it follows that $A_y/D \simeq 1.5$, which is a typical vibration level for lightly damped marine cables in a current. See Fig. 30.9. Large-amplitude vibrations also are associated with increased steady drag on the structure. Drag coefficients of up to 3.5 have been measured on resonantly vibrating marine cables as opposed to the typical value of 1.0 for a stationary cylinder.[24]

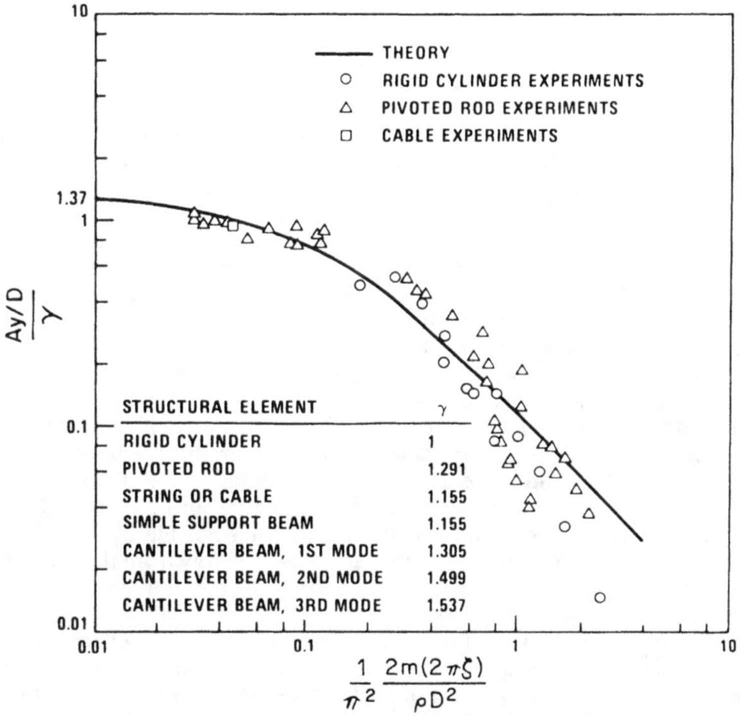

FIGURE 30.9 Maximum amplitude of vortex-induced vibration as a function of damping.[12]

A number of fairings, strakes, and ribbons have been attached to the exterior of circular cylindrical structures to reduce vortex-induced vibrations as shown in Fig. 30.10. These devices act by disrupting the near wake and disturbing the correlation between the vortex shedding and vibration. They do, however, increase the steady drag from that which is measured on a stationary structure. Reviews of vortex suppression devices are given in Refs. 25 and 26.

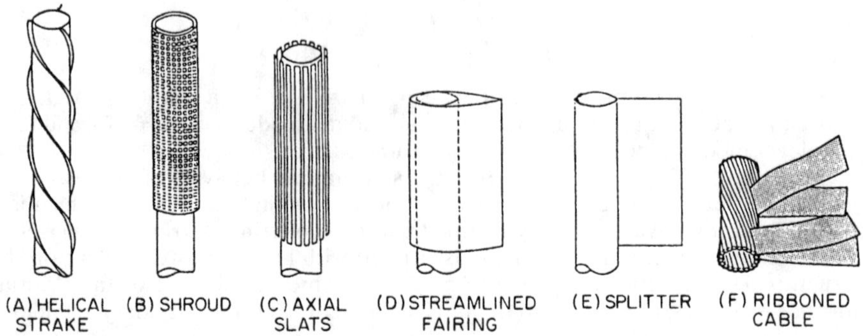

FIGURE 30.10 Methods of reducing vortex-induced vibration.

FLUID ELASTIC INSTABILITY

Fluid flow across an array of elastic tubes can induce a dynamic instability, resulting in very large amplitude tube vibrations once the critical cross-flow velocity is exceeded. This is a relatively common occurrence in tube and shell heat exchangers. Once the critical cross-flow velocity is exceeded, vibration amplitude increases very rapidly with cross-flow velocity V, usually as V^n where $n = 4$ or more, compared with an exponent in the range $1.5 < n < 2.5$ below the instability threshold. This can be seen in Fig. 30.11, which shows the response of an array of metallic tubes to water flow. The initial hump is attributed to vortex shedding. The cross-flow velocity is defined as velocity perpendicular to the tube axis at the minimum gap between tubes. Once the critical velocity is exceeded, the very large amplitude vibrations usually lead to failures of the heat exchanger tubes.

Often the large-amplitude vibrations vary in time; the amplitudes grow and fall about a mean value in pseudorandom fashion. Generally the tubes do not move independently but instead move in somewhat synchronized orbits with neighboring tubes. This orbital behavior has been observed in tests in both air and water with orbits ranging from near circles to nearly straight lines. See Fig. 30.12.

As the tubes whirl in orbital motion, they extract energy from the fluid (Refs. 12, 28, and 29). Below the onset of instability, energy which is extracted is less than the energy which is expended in damping. Above the critical velocity, the energy extracted from the flow by the tube motion exceeds the energy expended in damping, so the vibrations increase in amplitude. Restricting the motion or introducing frequency differences between one or more tubes often increases the critical velocity for onset of instability. Such increases in critical velocity are generally no greater than about 40 percent unless additional support is given to all tubes exposed to high-velocity flow. Often the onset of instability is more gradual in a bank of tubes having tube-to-tube frequency differences than in a bank with identical tubes. Only a relatively small percentage of the tube will become unstable at one time. Flexible long-span tubes in areas of high flow velocity (such as at inlets) are most susceptible to the instability.

At cross-flow velocities beyond those which produce an onset of instability, damaging vibrations are encountered. The tube vibration amplitudes are limited by clashing with other tubes, by impacting with the tube supports, and by yielding of the tubes. Sustained operation in the unstable vibration regime ordinarily results in tube

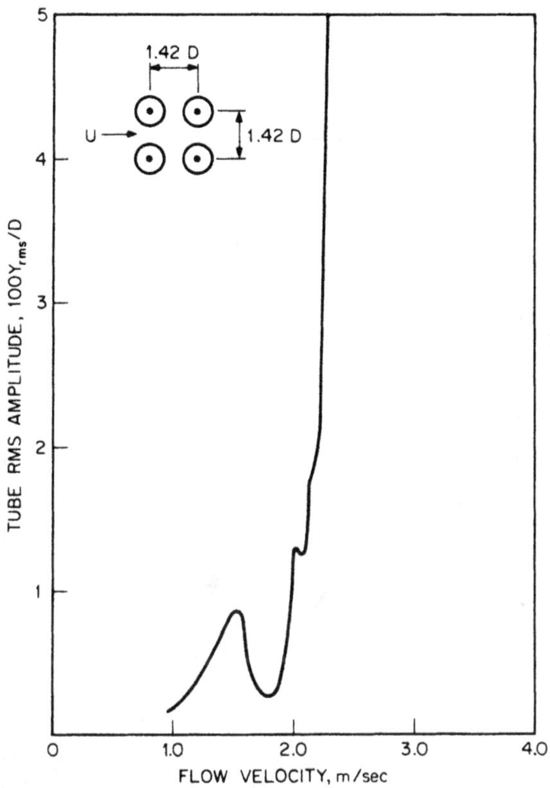

FIGURE 30.11 Typical amplitude of vibration of a tube array in cross flow.[27]

failure due to wear or propagation of cracks in the tubes. Fluid elastic instability is second only to corrosion as a cause of heat exchanger failure.

A displacement model for the fluid elastic forces is given in Ref. 12 which correctly predicts the observed onset of instability for most cases in air and gases. Results are less satisfactory in water or when the motion of some of the tubes is restricted. More complex models take into account velocity-induced forces as well as the displacement-induced forces.[29,30] These theories give somewhat better agreement with data over limited ranges, but none are entirely suitable for a design tool.

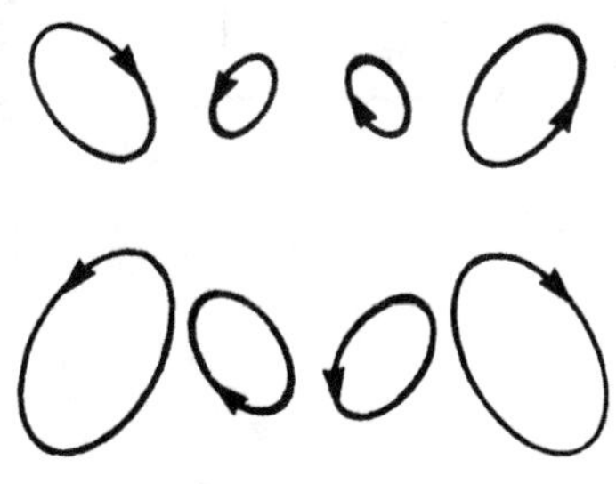

FIGURE 30.12 Tube vibration patterns for fluid elastic instability.[28]

The most viable, practical procedure for predicting the onset of instability of closely spaced arrays of tubes to cross flow is to use the theoretical form given by the displacement mechanism but

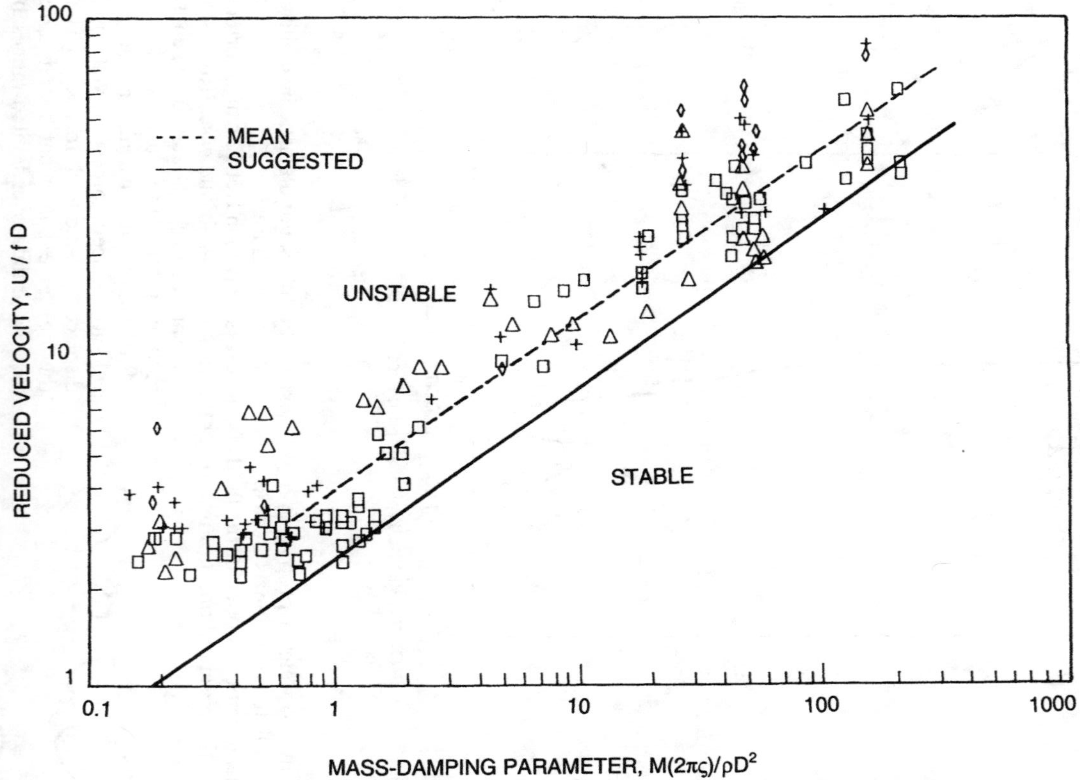

FIGURE 30.13 Velocity for onset of instability of tube arrays in cross flow as a function of the damping parameter.[22]

TABLE 30.1 Coefficients in Eq. (29.16) for Onset of Instability of Tube Arrays[31]

	$m_t(2\pi\zeta)/\rho D^2 < 0.7$	$m_t(2\pi\zeta)/\rho D^2 > 0.7$
C_{mean}	3.9	4.0
$C_{90\%}$	2.7	2.4
a	0.21	0.5
rms error in fitted data for V_{crit}, %	24.5	32.5

with parameters obtained by filling experimental data. The onset of instability is predicted as[12,22,23,31]

$$\frac{V_{crit}}{f_n D} = C\left[\frac{m_t(2\pi\zeta)}{\rho D^2}\right]^a \tag{30.24}$$

where V_{crit} = uniform cross flow averaged over minimum gap between tubes (If the velocity is nonuniform, then either the maximum can be used or a modal weighted average can be employed.)

f_n = fundamental natural frequency of tubing (Ordinarily the fundamental mode is most susceptible to instability.)

ζ = damping factor of fundamental mode (Typically ζ falls in the range between 0.01 and 0.03 for tubes with some intermediate supports. For rolled-in or welded-in tubes with no intermediate supports, ζ can be as low as 0.001.)

m_t = mass per unit length of tube including added mass and internal mass of fluid

ρ = fluid density

Fitting Eq. (30.24) to the available 174 data points for onset of instability[31] shown in Fig. 30.13 leads to the mean and lower-bound coefficients for the parameter C and the exponent a given in Table 30.1. The coefficient corresponding to the mean fit to the experimental data is C_{mean}; $C_{90\%}$ is the lower bound fit to the data such that 90 percent of the data are above the curve.

Most of the data used in this correlation come from tube arrays with center-to-center spacing of between 1.25 and 2.0 diameters and with various array geometries. There is insufficient statistical evidence to determine if certain patterns are more or less susceptible to instability than others. Instability has been observed for both straight and curved tubes, tube rows, and tube arrays in a wide variety of tube patterns.

The most common means of increasing the resistance of an array of tubes to instability is to add intermediate supports to increase the natural frequency of the tubes. Details of the tube support (particularly the gap between the tube and the support) influence the resultant vibration. In general, smaller gaps tend to result in lower tube-support impact velocities and hence in lower tube wear.[32,33]

INTERNAL FLOW IN PIPES

Internal flow through a pipe decreases the natural frequency of the pipe. Sufficiently high internal velocity will induce buckling in a pipe supported at both ends since the momentum of fluid turning through a small angle of pipe deflection is greater than the stiffness of the pipe. If the pipe is restrained at only one end, the pipe will become unstable at high velocities like an unrestrained garden hose.

The equation of motion for a straight pipe conveying steady fluid flow is[34,35]

$$EI\frac{\partial^4 Y}{\partial x^4} + \rho A v^2 \frac{\partial^2 Y}{\partial x^2} + 2\rho A v \frac{\partial^2 Y}{\partial x \partial t} + M \frac{\partial^2 Y}{\partial t^2} = 0 \tag{30.25}$$

where E and I are the modulus and moment of inertia of the pipe which conveys fluid of density ρ through the internal area A of the pipe at a steady velocity v; $Y(x, t)$ is the lateral deflection of the pipe which has total mass per unit length M. The first and last terms in Eq. (30.25) are the usual stiffness and mass terms. The middle terms are associated with fluid forces imposed on the pipe by the internal fluid as the pipe deflects slightly from its equilibrium position.

Although Eq. (30.25) is a linear partial differential equation with constant coefficients, its solution is difficult owing to the mixed derivative term (third term from the left). One technique used to solve the equation is to expand the solution in terms of the mode shapes of vibration which are obtained for zero flow, $v = 0$.

$$Y(x,t) = \Sigma_i\, a_i y_i(x) \sin \omega t \tag{30.26}$$

where $y_i(x)$ are the mode shapes for zero flow that satisfy Eq. (30.25) and the boundary conditions on the ends of the pipe span. Equation (30.20) is substituted into Eq. (30.25), and the derivatives of $y_i(x)$ are expressed in terms of the orthogonal set $y_i(x)$

$$y_i'(x) = \Sigma_i b_i y_i(x) \tag{30.27}$$

Like terms in the series are equated.

For a uniform pipe with pinned ends, the result can be expressed as a decrease in natural frequency due to flow.[12]

$$\frac{f}{f_1} = \left[1 - \left(\frac{v}{v_c}\right)^2\right]^{1/2} \tag{30.28}$$

where f = fundamental natural frequency
f_1 = fundamental natural frequency in absence of flow
v_c = critical flow velocity

The critical flow velocity can be expressed as

$$v_c = \frac{\pi}{L}\left[\frac{EI}{\rho A}\right]^{1/2} \tag{30.29}$$

where L is the span of the pipe. As the flow velocity approaches v_c, the fundamental natural frequency f_1 decreases to zero. The pipe span spontaneously buckles at $v = v_c$.

The buckling velocity is a function of the boundary conditions on the ends of the pipe, and there can be vibration; these solutions for various boundary conditions are generally scaled by the velocity v_c [Eq. (30.29)]. In general, only exceptionally thin-walled flexible tubes with very high velocity flows, such as rocket motor feed lines and penstocks, are prone to vibration induced by internal flow. External parallel flow can also induce an analogous instability. (See the review given in Ref. 35.) For a tube subjected to both internal and parallel external flow of the same magnitude, the velocity for the onset of instability is

$$v_c = \frac{\pi}{L}\left[\frac{EI}{\rho A_i + \rho A_e}\right]^{1/2} \tag{30.30}$$

where $A_i = \pi D_i^2/4$ and $A_e = \pi D_e^2/4$ are the cross-sectional areas associated with the tube inside and outside diameters D_i and D_e, respectively.

Oscillatory flow in pipes can also cause vibration. Oscillations of fluids in pipes can be caused by reciprocating pumps and acoustic oscillations produced by flow through valves and obstructions. Internal flow imposes net fluid force on pipe at bends and changes in area. For example, the fluid force acting on a 90° bend in a pipe is the sum of pressure and momentum components:

$$F_{\text{bend}} = [(p - p_a) + \rho U^2]\,A\mathbf{i} - [(p - p_a) + \rho U^2]\,A\mathbf{j} \tag{30.31}$$

Here p is the internal pressure in the pipe, p_a is the pressure in the atmosphere surrounding the pipe, and U is the internal velocity in the pipe. The vectors **i** and **j** are unit vectors in the direction of the incoming and outgoing fluid, respectively.

If the pressure and velocity in the pipe oscillates, then the fluid force on the bend will oscillate, causing pipe vibration in response to the internal flow. This problem is most prevalent in unsupported bends in pipe that are adjacent to pumps and valves. Two direct solutions are to (1) support pipe bends and changes in area so that fluid forces are reacted to ground and (2) reduce fluid oscillations in pipe by avoiding large pressure drops through valves and installation of oscillation-absorbing devices on pump inlet and discharge.

REFERENCES

1. Newman, J. N.: "Marine Hydrodynamics," The MIT Press, Cambridge, Mass., 1977.
2. Lamb, H.: "Hydrodynamics," Dover Publications, New York, 1945. Reprint of 6th ed., 1932.
3. Blevins, R. D.: "Formulas for Natural Frequency and Mode Shape," Kreiger, Malabar, Florida, 1984. Reprint of 1979 edition.
4. Milne-Thompson, L. L.: "Theoretical Hydrodynamics," 5th ed., Macmillan, New York, 1968.
5. Fritz, R. J.: *J. Eng. Industry,* **94**:167 (1972).
6. Chen, S-S: *J. Eng. Industry,* **97**:1212 (1975).
7. Chen, S-S: *Nucl. Eng. Des.,* **35**:399 (1975).
8. Brown, S. J.: *J. Pressure Vessel Tech.,* **104**:2 (1982).
9. Au-Yang, M. K.: *J. Vibration, Acoustics,* **108**:339 (1986).
10. Zienkiewicw, O. C.: "The Finite Element Method," 3d ed., McGraw-Hill Book Company, New York, 1977.
11. Ippen, A. T. (ed.): "Estuary and Coastline Hydrodynamics," McGraw-Hill Book Company, New York, 1966.
12. Blevins, R. D.: "Flow-Induced Vibration," 2d ed., Kreiger, Malibar, Fla., 1994.
13. Sarpkaya, T., and M. Isaacson: "Mechanics of Wave Forces on Offshore Structures," Van Nostrand Reinhold, New York, 1981.
14. Obasaju, E. D., P. W. Bearman, and J. M. R. Graham: *J. Fluid Mech.,* **196**:467 (1988).

15. Lienard, J. H.: "Synopsis of Lift, Drag and Vortex Frequency Data for Rigid Circular Cylinder," Washington State University, College of Engineering, Research Division Bulletin 300, 1966.
16. Roshko, A.: "On the Development of Turbulent Wakes from Vortex Streets," *National Advisory Committee for Aeronautics Report* NACA TN-2913, 1953.
17. Sarpkaya, T.: *J. Appl. Mech.,* **46**, 241 (1979).
18. Williamson, C. H. K., and A. Roshko: *J. Fluids and Structures,* **2**:355 (1988).
19. Scruton, C.: "On the Wind Excited Oscillations of Stacks, Towers and Masts," *National Physical Laboratory Symposium on Wind Effects on Buildings and Structures, Paper 16,* 790, 1963.
20. Blevins, R. D., and C. S. Coughnan: "Experimental Investigation of Vortex Induced Vibration in One and Two Dimensions, FE-08-1153, *Journal of Fluids Engineering,* 2009 (in press).
21. Durgin, W. W., P. A. March, and P. J. Lefebvre: *J. Fluids Eng.,* **102**:183 (1980).
22. ASME Boiler and Pressure Vessel Code, Section III, Division 1, Appendix N-1300, 1998.
23. Au-Yang, M. K., T. M. Mulcahy, and R. D. Blevins.: *Pressure Vessel Technology,* **113**:257 (1991).
24. Vandiver, J. K.: "Drag Coefficients of Long Flexible Cylinders," *1983 Offshore Technology Conference, Paper 4490,* 1983, p. 405.
25. Zdravkovich, M. M.: *J. Wind Eng., Industrial Aerodynamics,* **7**:145 (1981).
26. Wong, H. Y., and A. Kokkalis: *J. Wind Eng. Industrial Aerodynamics,* **10**:21 (1982).
27. Chen, S-S, J. A. Jendrzejczyk, and W. H. Lin: "Experiments on Fluid Elastic Instability in a Tube Bank Subject to Liquid Cross Flow," *Argonne National Laboratory Report* ANL-CT-44, July 1978.
28. Connors, H. J.: "Fluid Elastic Vibration of Tube Arrays Excited by Cross Flow," paper presented at the *Symposium on Flow Induced Vibration in Heat Exchangers,* ASME Winter Annual Meeting, December 1970.
29. Paidoussis, M. P., and S. J. Price: *J. Fluid Mech.,* **187**:45 (1988).
30. American Society of Mechanical Engineers. "Flow-Induced Vibrations—1994," PVP-273, New York, 1994.
31. Blevins, R. D.: *J. Sound & Vibration,* **97**:641 (1984).
32. Blevins, R. D.: *J. Eng. Materials Tech.,* **107**:61 (1985).
33. Cha, J. H.: *J. Pressure Vessel Tech.,* **109**:265 (1987).
34. Housner, G. W.: *J. Appl. Mech.,* **19**:205 (1952).
35. Paidoussis, M. P., and P. Besancon: *J. Sound & Vibration,* **76**:361 (1981).

CHAPTER 31
VIBRATION OF STRUCTURES INDUCED BY WIND

Alan G. Davenport
J. Peter C. King

INTRODUCTION

Vibration of significant magnitude may be induced by wind in a wide variety of structures including buildings, television and cooling towers, chimneys, bridges, transmission lines, and radio telescopes. No structure exposed to wind seems entirely immune from such excitation. The material presented here describes several mechanisms causing these oscillations and suggests a few simpler approaches that may be taken in design to reduce vibration of structures induced by wind. There is an extensive literature[1–7] giving a more detailed treatment of the subject matter.

FORMS OF AERODYNAMIC EXCITATION

The types of structure referred to above are generally unstreamlined in shape. Such shapes are termed "bluff bodies" in contrast to streamlined "aeronautical" shapes discussed in Chap. 32. The distinguishing feature is that when the air flows around such a bluff body, a significant wake forms downstream, as illustrated in Fig. 31.1. The wake is separated from the outside flow region by a shear layer. With a sharp-edged body (such as a building or structural number) as in Fig. 31.1*A*, this shear layer emanates from the corner. With oval bodies such as the cylinder in Fig. 31.1*B*, the shear layer commences at a so-called *boundary layer* on the upstream surface at points A and B (the separation points) and becomes a free shear layer. The exact position of these separation points depends on a wide variety of factors, such as the roughness of the cylinder, the turbulence in the flow, and the Reynolds number $R = VD/\nu$, where V = flow velocity, D = diameter of the body, and ν = kinematic viscosity.

The flow illustrated in Fig. 31.1 represents the time-average picture which would

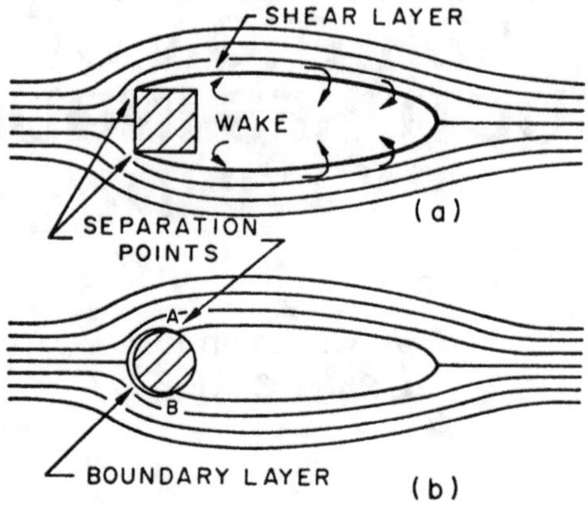

FIGURE 31.1 Wake formation past bluff bodies: (*a*) sharp-edged body; (*b*) circular cylinder.

be obtained by averaging the movements of the fluid particles over a time interval that is long compared with the "transit time" D/V. The instantaneous picture of the flow may be quite different, as indicated in Fig. 31.2, for two reasons.

First, if the flow is the wind, it is under almost all practical circumstances strongly turbulent; the oncoming flow will be varying continuously in direction and speed in an irregular manner. These fluctuating motions will range over a wide range of frequencies and scales (i.e., eddy sizes).

Second, the wake also will take on a fluctuating character. Here, however, the size of the dominant eddies (vortices) will be of a similar size to the body. The vortices tend to start off their career by curling up at the separation point and then are carried off downstream. Sometimes these eddies are fairly regular in character and are shed alternately from either side; if made visible by smoke or other means, they can be seen to form a more or less regular stepping-stone pattern until they are broken up by the turbulence or dissipate themselves. In a strongly turbulent flow, the regularity is disrupted.

The flow characteristics of the oncoming flow and the wake are the direct causes of the forces on the bodies responsible for their oscillation. The forms of the resulting oscillation are as follows.

1. *Turbulence-induced oscillations.* Certain types of oscillation of structures can be attributed almost exclusively to turbulence in the oncoming flow. In the wind these

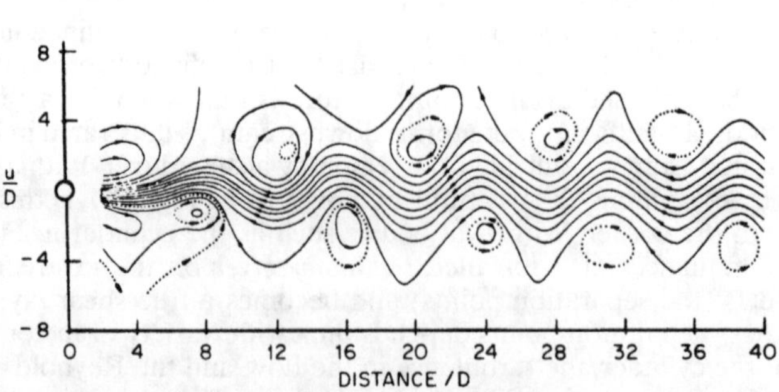

FIGURE 31.2 Vortex street past circular cylinder (R = 56). (*After Kovasznay,* Proc. Roy. Soc. London, *198, 1949.*)

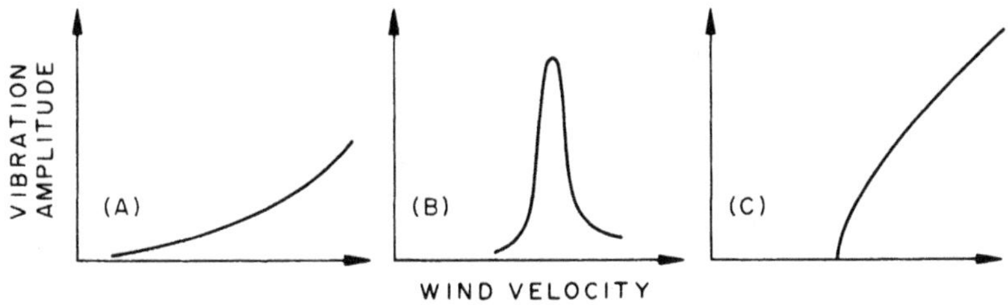

FIGURE 31.3 Main types of wind-induced oscillations: (*A*) vibration due to turbulence; (*B*) vibration due to vortex shedding; (*C*) aerodynamic instability.

may be described as "gust-induced oscillations" (or turbulence-induced, oscillations). The gusts may cause longitudinal, transverse, or torsional oscillations of the structure, which increase with wind velocity (Fig. 31.3*A*).

2. *Wake-induced oscillations.* In other instances, the fluctuations in the wake may be the predominant agency. Since these fluctuations are generally characterized by alternating flow, first around one side of the body, then around the other, the most significant pressure fluctuations act on the sides of the body in the wake behind the separation point (the so-called *after body*); they act mainly laterally or torsionally and to a much lesser extent longitudinally. The resultant motion is known as *vortex-induced oscillation.* Oscillation in the direction perpendicular to that of the wind is the most important type. It often features a pronounced resonance peak (Fig. 31.3*B*).

While these distinctions between gust-induced and wake-induced forces are helpful, they often strongly interact; the presence of free-stream turbulence, for example, may significantly modify the wake.

3. *Buffeting by the wake of an upstream structure.* A further type of excitation is that induced by the wake of an upstream structure (Fig. 31.4). Such an arrangement of structures produces several effects. The turbulent wake containing strong vortices shed from the upstream structure can buffet the downstream structure. In addition, if the oncoming wind is very turbulent, it can cause the wake of the upstream structure to veer, subjecting the downstream structure successively to the free flow and the wake flow. This frequently occurs with chimneys in line, as well as with tall buildings.

4. *Galloping and flutter mechanisms.* The final mechanism for excitation is associated with the movements of the structure itself. As the structure moves relative to the flow in response to the forces acting, it changes the flow regime surrounding it. In so doing, the pressures change, and these changes are coupled with the motion. A pressure change coupled to the velocity (either linearly or nonlinearly) may be termed an *aerodynamic damping* term. It may be either positive or negative. If positive, it adds to the mechanical damping and leads to higher effective damping and a reduced tendency to vibrate; if negative, it can lead to instability and large amplitudes of movement. This type of excitation occurs with a wide variety of rectangular building shapes as well as bridge cross sections and common structural shapes such as angles and I sections.

In other instances, the coupling may be with either the displacement or acceleration, in which case they are described as either aerodynamic stiffness or mass terms, the effect of which is to modify the mass or stiffness terms in the equations of motion. Such modification can lead to changes in the apparent frequency of the structure. If the aerodynamic stiffness is negative, it can lead to a reduction in the effective stiffness of the structure and eventually to a form of instability known as

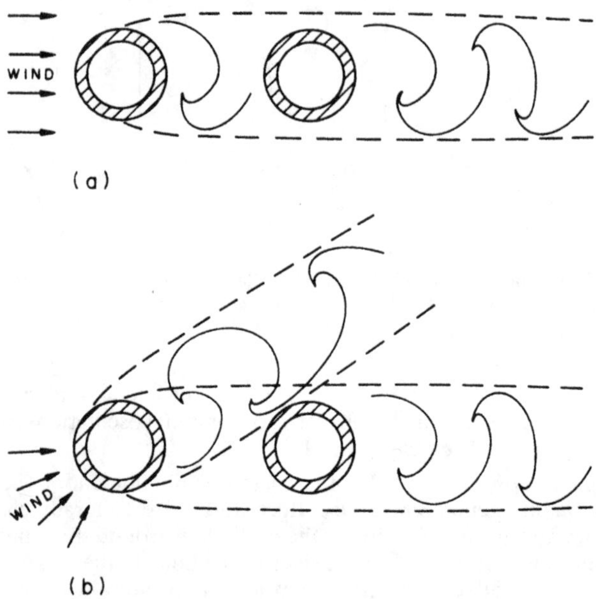

FIGURE 31.4 Buffeting by the wake of an upstream structure.

divergence. All types of instability feature a sudden start at a critical wind velocity and a rapid increase of violent displacements with wind velocity (Fig. 31.3*C*).

These various forms of excitation are briefly discussed in this chapter. Because all types of oscillations are influenced strongly by the properties of the wind, some basic wind characteristics are described first.

BASIC WIND CHARACTERISTICS

Wind is caused by differences in atmospheric pressure. At great altitudes, the air motion is independent of the roughness of the ground surface and is called the *geostrophic,* or *gradient* wind. Its velocity is reached at a height called *gradient height,* which lies between about 1000 and 2000 ft (300 and 600 m). Below the gradient height, the flow is affected by surface friction, by the action of which the flow is retarded and turbulence is generated. In this region, known as the *planetary boundary layer,* the three components of wind velocity resemble the traces shown in Fig. 31.5. The longitudinal component consists of a mean plus an irregular turbulent fluctuation; the lateral and vertical components consist of similar fluctuations. These turbulent motions can be characterized in a number of different ways.

The longitudinal motion at height z can be expressed as

$$V_z(t) = \bar{V}_z + v(t) \tag{31.1}$$

where $\bar{V}_z$ = mean wind velocity (the bar denotes time average) and $v(t)$ = fluctuating component.

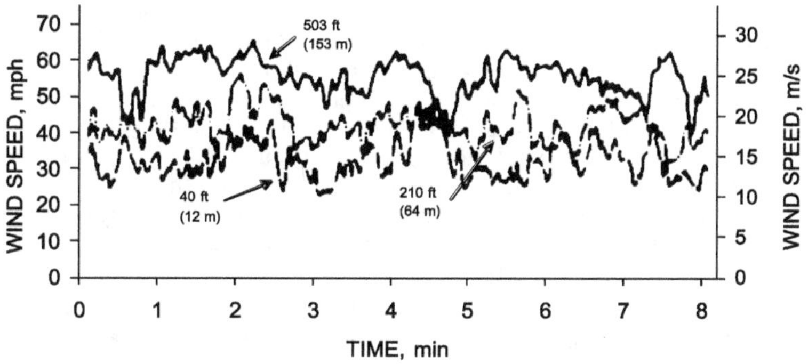

FIGURE 31.5 Record of horizontal component of wind speed at three heights on 500-ft (152 m) mast in open terrain. (*Courtesy of E. L. Deacon.*)

Mean Wind Velocity. The mean wind velocity $\bar{V}_z$ varies with height z as represented by the mean wind velocity profile (Fig. 31.6). The profiles observed in the field can be matched by a logarithmic law, for which there are theoretical grounds, or by an empirical power law

$$\frac{\bar{V}_z}{\bar{V}_G} = \left[\frac{z}{z_G}\right]^{\alpha} \tag{31.2}$$

where $\bar{V}_G$ = gradient wind velocity, z_G = gradient height, and α = an exponent <1. Gradient height z_G and exponent α depend on the surface roughness, which can be characterized by the surface drag coefficient κ (here referenced to the wind speed at 33 ft or 10 m).

A few typical values of these parameters are given in Fig. 31.6. The mean wind profiles shown are characteristic of level terrain. They can significantly change, par-

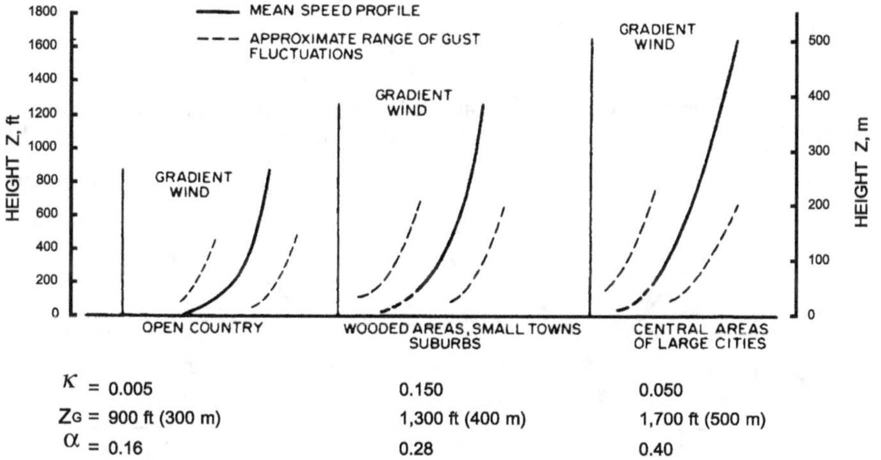

FIGURE 31.6 Vertical profiles of mean wind velocity for three typical terrains.

ticularly in the lower region, when the air flow meets an abrupt change in surface roughness or terrain contour. A sudden increase in roughness reduces the wind speed near the ground while a hill accelerates the flow over its crest.

The mean wind profiles are useful when predicting the wind speed at a particular site. The gradient wind speed is estimated using data registered by the nearest meteorological stations at their standard height, which is usually 33 ft (10 m). The mean wind velocity generally depends on the period over which the wind speed is averaged. Periods from 10 to 60 minutes appear adequate for engineering considerations and usually yield reasonably steady mean values. The same duration is suitable to define the fluctuating wind component.

Fluctuating Components of the Wind. The fluctuating components of the wind change with height less than the mean wind and are random in both time and space. The random nature of the wind requires the application of statistical concepts (see Chap. 24). The basic statistical characteristics of the velocity fluctuations are the intensity of turbulence, the power spectral density (power spectrum), the correlation between velocities at different points, and the probability distribution.

The intensity of turbulence is defined as $\sigma_v/\bar{V}_z$, where $\sigma_v = \sqrt{\overline{v^2(t)}}$ is the root-mean-square (rms) fluctuation in the longitudinal direction. The intensity of the lateral and vertical fluctuations can be described similarly. For wind, the intensity of turbulence is between 5 and 25 percent. The magnitude σ_v also defines the probability distribution of the fluctuations which may be assumed to be gaussian (normal).

The energy of turbulent fluctuations (gustiness) is distributed over a range of frequencies. This distribution of energy with frequency f can be described by the spectrum of turbulence (power spectral density) $W_v(f)$. The relationship between the spectrum and the variance is

$$\int_0^\infty W_v(f)\, df = \sigma_v^2$$

which leads to another form of the spectrum known as the *logarithmic spectrum* $fW_v(f)/\sigma_v^2$. This form of the spectrum is dimensionless and preserves the relative contributions to the variance at different frequencies represented on a logarithmic scale; its integral is

$$\int_0^\infty \frac{fW_v(f)}{\sigma_v^2}\, d\ln f = 1$$

The two forms of spectra are sketched in Fig. 31.7. A generalization of wind spectra for different wind velocities is possible if the frequency scale is so modified that

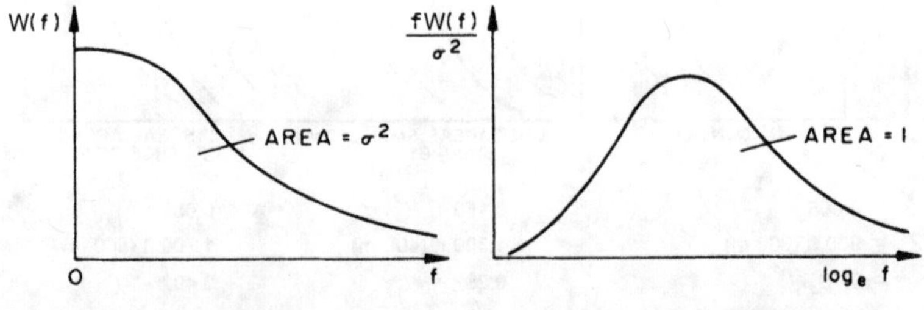

FIGURE 31.7 Two different ways of presenting power spectral densities.

it too is dimensionless. The ratio $f/\bar{V}$ is the so-called *inverse wavelength* related to the "size" of atmospheric eddies. This may be expressed as a ratio to a representative length scale L, such as the wavelength of the eddies at the peak of the spectrum. The dimensionless frequency or inverse wavelength may now be written

$$\bar{f} = fL/\bar{V}$$

Under certain circumstances this relationship is also known as the Strouhal number or the reduced frequency.

It is generally found that while the length scale L in the oncoming flow corresponds to that of the turbulence itself (this in the natural wind is of the order of thousands of feet), in the wake the governing length scale is of the same order as the diameter of the body D. This is illustrated in Fig. 31.8.

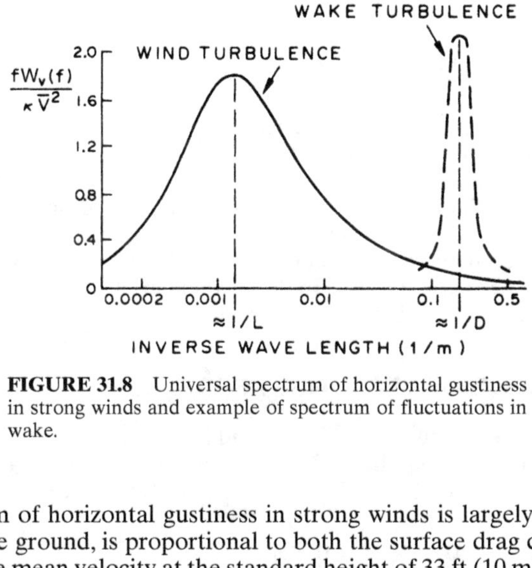

FIGURE 31.8 Universal spectrum of horizontal gustiness in strong winds and example of spectrum of fluctuations in wake.

The spectrum of horizontal gustiness in strong winds is largely independent of height above the ground, is proportional to both the surface drag coefficient κ and the square of the mean velocity at the standard height of 33 ft (10 m), $\bar{V}_{10}$, and can be represented, with some approximations, as[8,9]

$$W_v(f) = 4\kappa\bar{V}_{10}^{\,2}\,\frac{L/\bar{V}_{10}}{(2+\bar{f}^2)^{5/6}} \tag{31.3}$$

in which f = frequency, Hz, $\bar{f} = fL/\bar{V}_{10}$ where L = scale length ≈4000 ft (1300 m), and κ is given in Fig. 31.6. This spectrum is shown in Fig. 31.8.

The variance of the velocity fluctuations is

$$\sigma_v^{\,2} = \int_0^\infty W_v(f)\,df = 6.68\kappa\bar{V}_{10}^{\,2} \tag{31.4}$$

It can be seen from Eqs. (31.3) and (31.4) that large velocity fluctuations can be expected in rough terrain where coefficient κ is large.

The spatial correlation of wind speeds at two different stations is described by the coherence function (see Chap. 19),

$$\gamma_{12}^2(f) = \frac{|W_{12}(f)|^2}{W_1(f)W_2(f)} \leq 1 \tag{31.5}$$

where $W_{12}(f)$ = cross spectrum (generally complex) between stations 1 and 2; $W_1(f)$ and $W_2(f)$ are power spectra of the two stations. The coherence function depends primarily on the parameter $\Delta zf/\bar{V}$, where Δz = separation and $\bar{V} = \frac{1}{2}(\bar{V}_1 + \bar{V}_2)$ is the average wind speed. A suitable approximate function is

$$\sqrt{\text{Coherence}} = e^{-c(\Delta zf/\bar{V})}$$

where c is a constant having a value of approximately 7 for vertical separation and approximately 15 for horizontal separation. Coherence decreases with both separation and frequency. A more detailed discussion of wind characteristics is given in Refs. 1 and 9.

EXCITATION DUE TO TURBULENCE

When a structure is exposed to the effects of wind, the fluctuating wind velocity translates into fluctuating pressures, which in turn produce a time-variable response (deflection) of the structure. This response is random and represents the basic type of wind-induced oscillations. The theoretical prediction of this oscillation is rather complex but can be reduced to a simple procedure suitable for design purposes. The discussion of the oscillation is therefore presented in two parts. In the first part, the basic theoretical steps are outlined. In the second part, the design procedure known as the gust-factor approach is given in more detail.

FUNDAMENTALS OF RESPONSE PREDICTION

If the area A of the structure exposed to wind is small relative to the significant turbulent eddies, the so-called *quasi-steady theory* for turbulence can be used to estimate aerodynamic forces. In the drag direction, the drag force

$$D(t) = \frac{1}{2}\rho C_D A V^2(t)$$

$$= \frac{1}{2}\rho C_D A \bar{V}^2\left[1 + 2\frac{v(t)}{\bar{V}} + \frac{v^2(t)}{\bar{V}^2}\right]$$

where ρ = air density (normally equal to 0.0024 slugs/ft^3), and C_D = drag coefficient. If $v(t) << \bar{V}$, the squared term is ignored. The spectra of the fluctuating drag and velocity are then related as

$$\frac{W_D(f)}{\bar{D}^2} = 4\frac{W_v(f)}{\bar{V}^2} \tag{31.6}$$

where the mean drag (static component of the drag) is

$$\bar{D} = \frac{1}{2}\rho C_D A \bar{V}^2 \tag{31.7}$$

and $W_v(f)$ is given by Eq. (31.3).

With large bodies, the wavelength is comparable to the size of the body itself (that is, $f\sqrt{A}/\bar{V} \approx 1$) and it is necessary to modify the drag spectrum by the so-called *aerodynamic admittance function* $|X_{\text{aero}}(f)|^2$. This function[8] describes the modifying influence of any changes in effective drag coefficient as well as the decrease in correlation of the eddies as the wavelength of the eddies approaches the diameter of the body. Thus, the modified drag spectrum is

$$\frac{W_D(f)}{\bar{D}^2} = 4|X_{\text{aero}}(f)|^2 \frac{W_v(f)}{\bar{V}^2}$$

If these forces act on an elastic spring-mass-damper system, the response of this system u will have a spectrum

$$\frac{W_u(f)}{\bar{u}^2} = |X_{\text{aero}}|^2 |X_{\text{mech}}|^2 \frac{4W_v(f)}{\bar{V}^2}$$

where static deflection $\bar{u} = \bar{D}/k$, k = stiffness constant, and the mechanical admittance function is

$$|X_{\text{mech}}|^2 = \frac{1}{[1-(f/f_n)^2]^2 + 4\zeta^2(f^2/f_n^2)}$$

where ζ = critical damping ratio, and f_n = natural frequency of the system.

The transition from the spectrum of the wind-velocity fluctuations to the spectrum of the response is shown diagrammatically in Fig. 31.9. The variance of the response σ_u^2 is obtained from the spectrum of the response

$$\sigma_u^2 = \int_0^\infty W_u(f)\, df \tag{31.8}$$

The relationships above describe the mean and the variance of the response. For engineering purposes, it is also useful to define extreme values. It is often satisfactory to assume that the process in question is gaussian, with probability density function given by

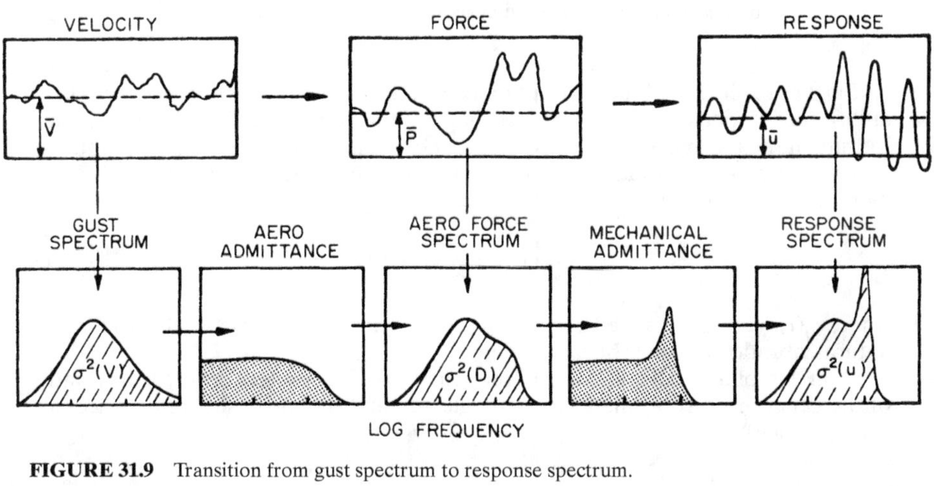

FIGURE 31.9 Transition from gust spectrum to response spectrum.

$$p(u) = \frac{1}{\sqrt{2\pi}\,\sigma_u}\, e^{-(u - \bar{u})^2/2\sigma_u^2}$$

This distribution is fully described by the mean and the variance. Maximum values of the response during time T can be written as

$$u_{\max} = \bar{u} + g\sigma_u \tag{31.9}$$

where g = peak factor. The average largest value of the peak factor in a period T can be estimated from[6]

$$g = \sqrt{2 \ln \nu T} + \frac{0.5772}{\sqrt{2 \ln \nu T}} \tag{31.10}$$

where ν is an effective cycling rate of the process, generally close to the natural frequency. The relationship of the distribution of the largest peak value to the distribution of all values is shown in Fig. 31.10. As can be seen, when the period T or the natural frequency increases, the expected peak displacement also increases. The factor g usually ranges between 3 and 5.

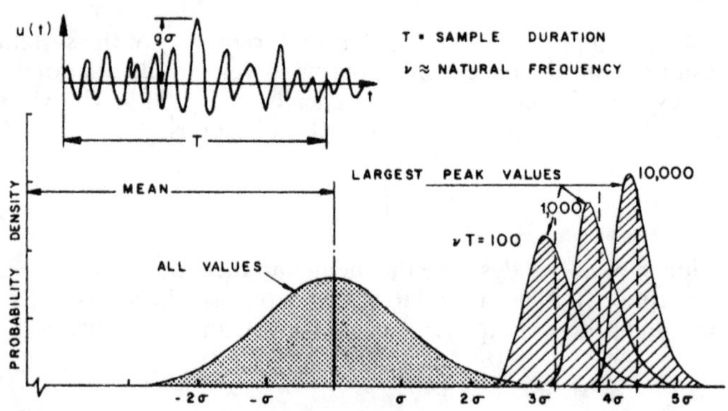

FIGURE 31.10 Relationship of distribution of largest peak value to distribution of all values (for a stationary random process).

Further extension of the concept includes the cross correlation of the wind loads at different stations (e.g., heights), the shape of the vibration mode, and the nonuniformity of the mean flow. These factors can be included into the solution formulated in terms of modal analysis (see Chap. 21). With a prismatic structure, the displacement may be expressed in the form

$$u(z,t) = \sum_{j=1}^{\infty} q_j(t)\phi_j(z) \tag{31.11}$$

where $q_j(t)$ = the generalized coordinate of the jth mode, and $\phi_j(z)$ = the jth mode of natural vibrations to an arbitrary scale.

With damping small and natural frequencies well separated, the cross correlation of the generalized coordinate can be neglected and the mean square displacement (the variance) is

$$\overline{u^2(z,t)} = \sum_{j=1}^{\infty} \overline{q_j^2}\, \phi_j^2(z) \tag{31.12}$$

The variance of the generalized coordinate $\overline{q_j^2}$ is determined by the power spectrum of the generalized force Q_j. When the lateral dimension of the structure is small, only cross correlation in direction z need be considered. Then the power spectrum of the generalized force is

$$W_{Q_j}(f) = \int_0^H \int_0^H W_{12}(z_1,z_2,f)\phi_j(z_1)\phi_j(z_2)\, dz_1\, dz_2 \tag{31.13}$$

where $W_{12}(z_1,z_2,f)$ = cross spectrum of the wind loads at heights z_1 and z_2, and H = height of the structure. With respect to Eq. (31.6), the cross spectrum of the wind loads can be expressed in terms of the power spectrum of the wind speed [Eq. (31.3)] and the coherence function, Eq. (31.5).

The variance of q_j is

$$\begin{aligned}\overline{q_j^2} &= \int_0^\infty \frac{1}{(2\pi f_j)^4 M_j^2}\, \frac{1}{[1-(f/f_j)^2]^2 + 4\zeta^2(f/f_j)^2}\, W_{Qj}(f) \\ &\approx \frac{1}{64\pi^3\zeta f_j^3 M_j^2}\, W_{Qj}(f_j) + \frac{1}{(2\pi f_j)^4 M_j^2} \int_0^{f_j} W_{Qj}(f)\, df\end{aligned} \tag{31.14}$$

where f_j = jth natural frequency and generalized mass

$$M_j = \int_0^H m(z)\phi_j^2(z)\, dz \tag{31.15}$$

where $m(z)$ = mass of the structure per unit length. The approximate integration[10] of Eq. (31.14) yields the response composed of two parts, the resonance effect (the first term) and the background turbulence effect (the second term), as shown in Fig. 31.11. The variance of the displacement follows from Eq. (31.12), and its standard deviation (rms dynamic displacement) is $\sigma_u(z) = \sqrt{\overline{u^2(z,t)}}$. The peak response is established from Eq. (31.9) by means of the peak factor g [Eq. (31.10)] as in one degree of freedom. The mean deflection $\bar{u}(z)$ is the static deflection due to the mean wind $\bar{V}_z$.

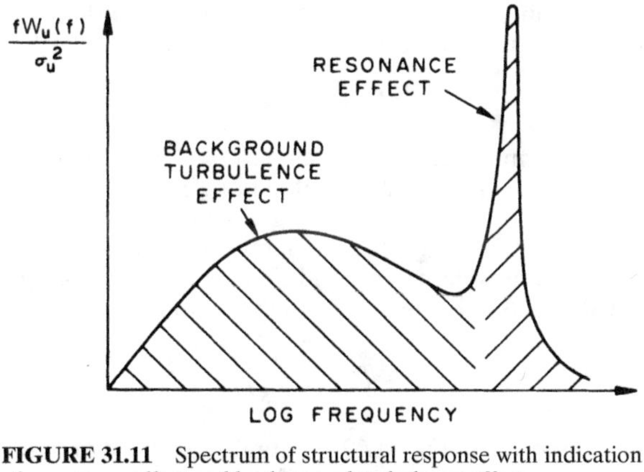

FIGURE 31.11 Spectrum of structural response with indication of resonance effect and background turbulence effect.

Other analyses of slender structures are also available.[11–13] In applications to buildings and free-standing towers, the analysis can usually be limited to the first modal component in Eq. (31.12). Application to buildings and structures with significant lateral dimension requires the incorporation of the horizontal cross correlation as well. A complete solution established by means of simplifying assumptions and numerical integrations is given below.

GUST-FACTOR APPROACH

The gust-factor approach is a design procedure derived on the basis of the theory above by means of a few simplifying assumptions. The approach given here is a modified version of the method described in Ref. 14 and adopted in Ref. 15. It considers only the response in the first vibration mode which is assumed to be linear. These assumptions are particularly suitable for buildings. The method yields all the data needed in design: the maximum response, the equivalent static wind load that would produce the same maximum response, and the maximum acceleration needed for the evaluation of the physiological effects of strong winds (human comfort).

The gust factor G is defined as the ratio of the expected peak displacement (load) in a period T to the mean displacement (load) $\bar{u}$. Hence, the maximum expected response is

$$u_{\max} = G\bar{u} = \left(1 + g\frac{\sigma_u}{\bar{u}}\right)\bar{u} \tag{31.16}$$

The gust factor is given as

$$G = 1 + g\sqrt{\frac{K}{C_e}\left(B + \frac{sF}{\zeta}\right)} \tag{31.17}$$

where ζ = damping ratio and K = factor related to the surface roughness; this factor is equal to 0.08 for open terrain (zone A), 0.10 for suburban, urban, or wooded terrain (zone B), and 0.14 for concentrations of tall buildings (zone C). All the other parameters appearing in Eq. (31.17) can be obtained from Fig. 31.12. C_e = exposure factor based on the mean wind speed profile (coefficient α) and thus on surface roughness. For the three zones, the exposure factor is obtained from Fig. 31.12*A* for the height of the building H. C_e relates to wind pressure rather than speed. Hence, the mean wind speed at the top of the building is given by

$$\bar{V}_H = \bar{V}_{10}\sqrt{C_e}$$

where $\bar{V}_{10}$ = reference wind speed at the standard height of 10 m or 33 ft. $\bar{V}_{10}$ can be obtained from meteorological stations. Velocity $\bar{V}_H$ is needed for determination of parameters s and F. Factors B, s, F, and g are given in Fig. 31.12*C* to *f* as a function of parameters indicated; D = width of the frontal area, and f_n = the first natural frequency of the structure in cycles per second. The average fluctuation rate ν, on which the peak factor g depends, is evaluated from the formula

$$\nu = f_0\sqrt{\frac{sF/\zeta}{B + sF/\zeta}} \tag{31.18}$$

The peak factor g is plotted in Fig. 31.12*F*, assuming a period of observation $T = 3600$ sec; it can also be calculated from Eq. (31.10).

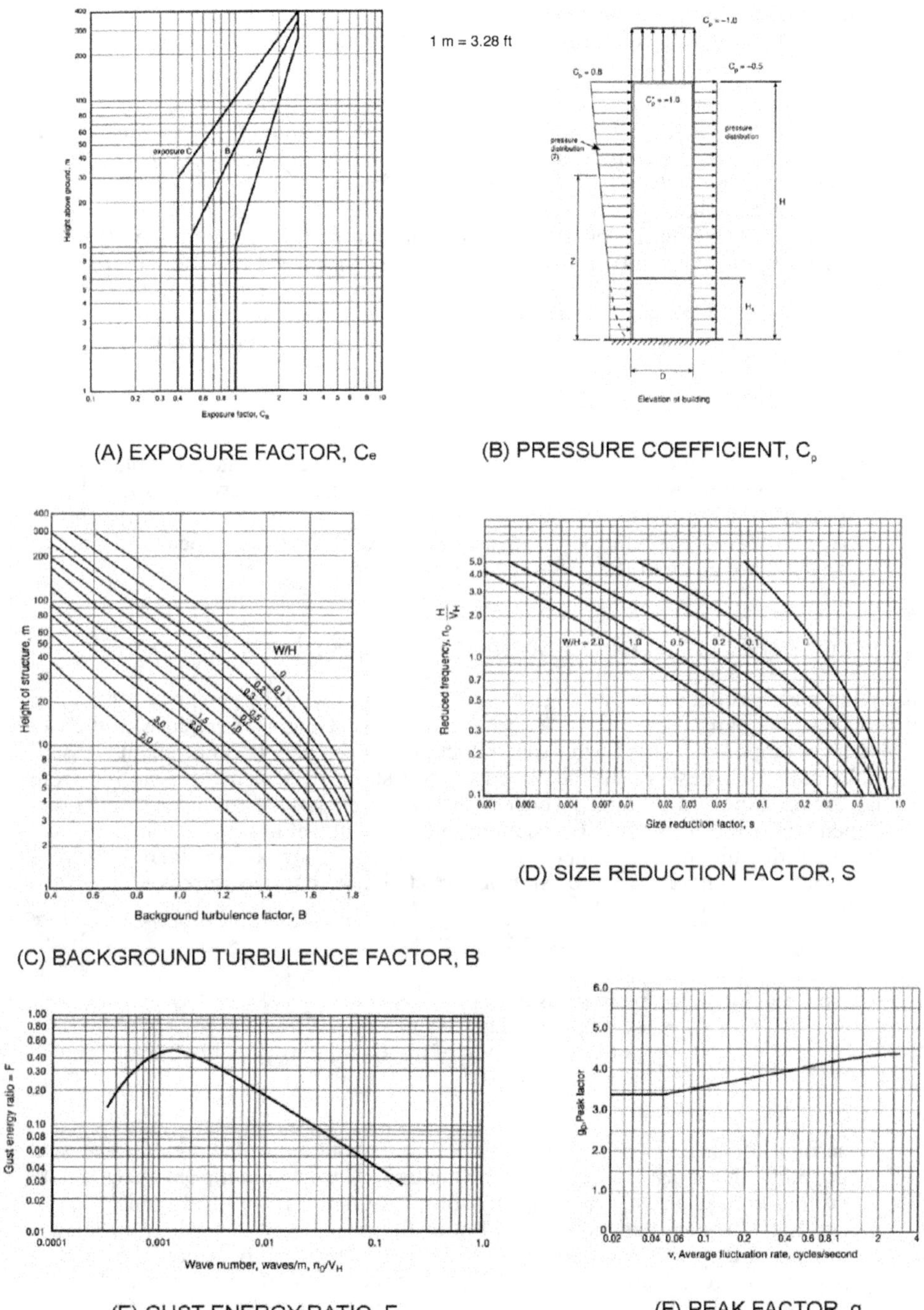

FIGURE 31.12 Components of gust factor.

The parameters given also yield the design wind pressure p, which produces displacement u_{max} if applied as a static load. This design pressure

$$p = qC_eGC_p \qquad (31.19)$$

where $q = \frac{1}{2}\rho\bar{V}_{10}^2$ is the reference mean-velocity pressure, and C_e = exposure factor. In this case, C_e varies continuously with the elevation according to Fig. 31.12*A* for pressures acting on the windward face of the structure; for the leeward face, C_e is constant and evaluated at one-half the height of the building. The quantity C_p = average pressure coefficient, which depends on the shape of the structure and the flow pattern around it. For a typical building with a flat roof and a height greater than twice the width, the coefficients are given for the windward and leeward faces in Fig. 31.12*B* together with the pressure distribution.

The peak acceleration A of a structure due to gusting wind is given by

$$A = u_{max} \frac{4\pi^2 f_0^2 g}{G} \sqrt{\frac{KsF}{C_e\zeta}}$$

where u_{max} = maximum deflection under the design pressure p. The other parameters are equal to those used in Eq. (31.17). When the acceleration exceeds about 1 percent of gravity, the motion is usually perceptible. However, there are large differences in the perceptibility of motions having very low frequencies.[16,17] Similar approaches are given in Refs. 18 to 20.

EFFECT OF GUSTS ON CLADDING AND WINDOWS

Wind gusts produce local pressures on cladding and window panels of a building. Because the natural frequency of such a panel is very high compared with the frequency components of the wind-speed fluctuations, the panel displacement is essentially static. Its design may be based on the static displacement resulting from maximum expected pressure, which is the algebraic sum of the height and time-dependent exterior pressure (or suction) and the constant interior pressure (or suction). If the fluctuating component of the pressure $p(t)$ is considered to be a stationary random process, the exterior expected maximum pressure is

$$p_{max} = \bar{p}\left(1 + g\frac{\sigma_p}{\bar{p}}\right) = \bar{p}G \qquad (31.20)$$

where $\bar{p} = \frac{1}{2}\rho C_p\bar{V}^2$ = mean pressure
C_p = local pressure coefficient
σ_p = standard deviation of the fluctuating pressure component
g = peak factor given by Eq. (31.10)
G = gust factor

To account for the sensitivity of glass to both static and dynamic fatigue, it has been suggested[21,22] that g or G in Eq. (31.20) be multiplied by a wind-on-glass effect factor.

Factors g, $\sigma_p/\bar{p}$, and C_p are most reliably determined from wind-tunnel experiments. They strongly depend on location of the panel, wind direction, turbulence intensity, and the local flow pattern determined by the shape of the building and its immediate environment. In full-scale experiments, values of g in excess of 10 have been observed in highly intermittent flow. Largest local pressure coefficients C_p (actually suctions) appear with skew wind at the leading edge of the building where a typical value is $C_p = -1.5$. In that part of the building exposed to free flow, a gust factor $G \approx 2.5$ is a reasonable estimate.[15,23] The interior pressure is not very high, but its magnitude and sign depend on openings and leakage.

Damage to windows may result from local wind pressure, but it also depends on material properties of glass and its fatigue. The fatigue limit of glass is only about 20 percent of the instantaneous strength.[22]

VIBRATION DUE TO VORTEX SHEDDING

Vortex shedding represents the second most important mechanism for wind-induced oscillations. Unlike the gusts, vortex shedding produces forces which originate in the wake behind the structure, act mainly in the across-wind direction, and are, in general, rather regular. The resultant oscillation is resonant in character, is often almost periodic, and usually appears in the direction perpendicular to that of the wind. Lightly damped structures such as chimneys and towers are particularly susceptible to vortex shedding. Many failures attributed to vortex shedding have been reported.

When a bluff body is exposed to wind, vortices shed from the sides of the body, creating a pattern in its wake often called the *Karman vortex street* (Fig. 31.2). The frequency of the shedding, nearly constant in many cases, depends on the shape and size of the body, the velocity of the flow, and to a lesser degree on the surface roughness and the turbulence of the flow. If the cross section of the body is noncircular, it also depends on the wind direction. The dominant frequency of vortex shedding f_s is given by

$$f_s = S\frac{\bar{V}}{D} \text{ Hz} \tag{31.21}$$

where S = dimensionless constant called the Strouhal number, $\bar{V}$ = mean wind velocity, and D = width of the frontal area. The second dimensionless parameter is the Reynolds number $R = \bar{V}D/\nu$, where ν = kinematic viscosity. For air under normal conditions, $\nu = 1.5 \times 10^{-5}$ m^2/sec.

For a body having a rectangular or square cross section, the Strouhal number is almost independent of the Reynolds number. For a body having a circular cross section, the Strouhal number varies with the regime of the flow as characterized by the Reynolds number. There are three major regions: the subcritical region for $R \lesssim 3 \times 10^5$, the supercritical region for $3 \times 10^5 \lesssim R \lesssim 3 \times 10^6$, and the transcritical region for $R \gtrsim 3 \times 10^6$. Approximate values of the Strouhal number for typical cross sections are given in Table 31.1. The numbers given in this table are based on Refs. 1, 24, 25, and 26 and other measurements, and may be used for turbulent shear flow.

PREDICTION OF VORTEX-INDUCED OSCILLATION

Although the mechanism of vortex shedding and the character of the lift forces have been the subject of a great number of studies,[27] the available information does not permit an accurate prediction of these oscillations. The motion is most often viewed as forced oscillation due to the lift force, which, per unit length, may be written as

$$F_L = \frac{1}{2}\rho D\bar{V}^2 C_L(t) \tag{31.22}$$

TABLE 31.1 Aerodynamic Data for Prediction of Vortex-Induced Oscillations in Turbulent Flow

Cross section	Strouhal number S	rms lift coefficient σ_L	Bandwidth B	Correlation length L (diameters)
Circular:				
Subcritical	0.2	0.5	0.1	2.5
Supercritical	Not marked	0.14	Not marked	1.0
Transcritical	0.25	0.25	0.3	1.5
Square:				
Wind normal to face	0.11	0.6	0.2	3

where $C_L(t)$ is a lift coefficient fluctuating in a harmonic or random way. Some authors[28,29] consider vortex shedding to be self-excitation, which does not seem necessary, however, for relatively small motions. Hence, the solution of the response depends on the time history assumed for $C_L(t)$.

HARMONIC EXCITATION OF PRISMATIC CYLINDERS BY VORTICES

Harmonic excitation represents a traditional model for vortex excitation, but it is really justified only for very low Reynolds numbers ($\lesssim 300$) or possibly for large vibration where the motion starts controlling both the wake and the lift forces in the form of the "locking-in" phenomenon. Strongest oscillations arise at that wind velocity for which the frequency of vortex shedding f_s is equal to one of the natural frequencies of the structure f_j. This resonant wind velocity is, from Eq. (31.21),

$$V_c = \frac{1}{S} f_j D \tag{31.23}$$

With free-standing towers and stacks, resonance in the first two modes is met most often; resonance with higher modes has been observed as well with guyed towers (Fig. 31.13).

At the resonant wind velocity, the lift force is given by Eq. (31.22) in which $C_L(t) = C_L \sin 2\pi f_j t$, and C_L = amplitude of lift coefficient. Assuming a uniform wind profile and a constant diameter D, the resonant amplitude of mode j at the critical wind velocity V_c is, from Eq. (31.11),

$$u_j(z) = \frac{\rho C_L}{16\pi^2 S^2} \frac{D^3}{\zeta M_j} \phi_j(z) \int_0^H \phi_j(z)\, dz \tag{31.24}$$

where M_j is given by Eq. (31.15) and ζ = structural damping ratio. The formula can be further simplified if it is assumed that the lift force is distributed along the structure in proportion to the mode $\phi_j(z)$. This assumption reflects the loss of spanwise correlation of the forces. Then, with constant mass per unit length $m(z) = m$, the resonant amplitude at the height where the modal displacement is maximum:

$$u_j = \frac{\rho C_L}{16\pi^2 S^2} \frac{D^3}{\zeta m} \tag{31.25}$$

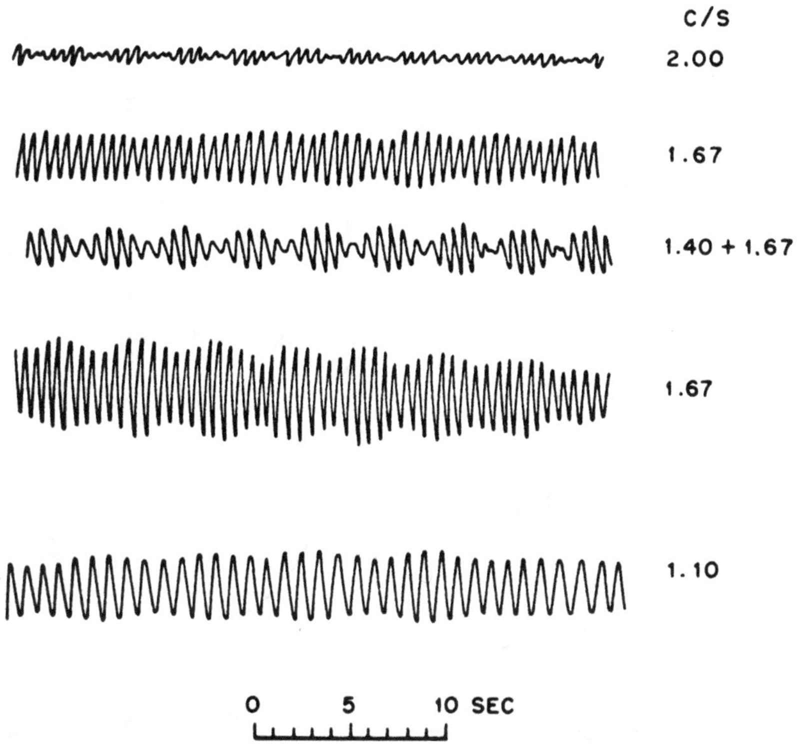

FIGURE 31.13 Vortex-induced oscillations in different modes measured on 1000 ft (305 m) guyed tower.[30]

For the first mode of a free-standing structure, this occurs at the tip. In higher modes, this amplitude appears at the height where local resonance takes place. For circular cylinders, a design value of the lift coefficient C_L is about $\sqrt{2}\sigma_L$. This simple formula can be used for the first estimate of the amplitudes that are likely to represent the upper bound. It is also indicative of the role of the diameter, mass, and damping of the structure. Approximate values of σ_L are given in Table 31.1.

RANDOM EXCITATION OF PRISMATIC CYLINDERS BY VORTICES

Even when vortex shedding appears very regular, the lift force and thus $C_L(t)$ are not purely harmonic but random. The power spectrum of the lift force per unit length is from Eq. (31.22).

$$W_L(f) = \left(\frac{1}{2}\rho D\bar{V}^2\sigma_L\right)^2 W_L'(f) \tag{31.26}$$

where $\sigma_L = \sqrt{\overline{C_L^2(t)}}$ is the standard deviation of the lift coefficient and $W_L'(f)$ = normalized power spectrum of $C_L(t)$ for which

$$\int_0^\infty W_L'(f)\,df = 1 \tag{31.27}$$

With circular cylinders, the lift force is narrowband random in the subcritical and transcritical[24,25] ranges where the energy is distributed about the dominant frequency f_s, given by Eq. (31.21) (Fig. 31.14*A*). Such spectra can be described by a gaussian-type curve,

$$W_L'(f) = \frac{1}{\sqrt{\pi} B f_s} \exp\left[-\left(\frac{1 - f/f_s}{B}\right)^2\right] \tag{31.28}$$

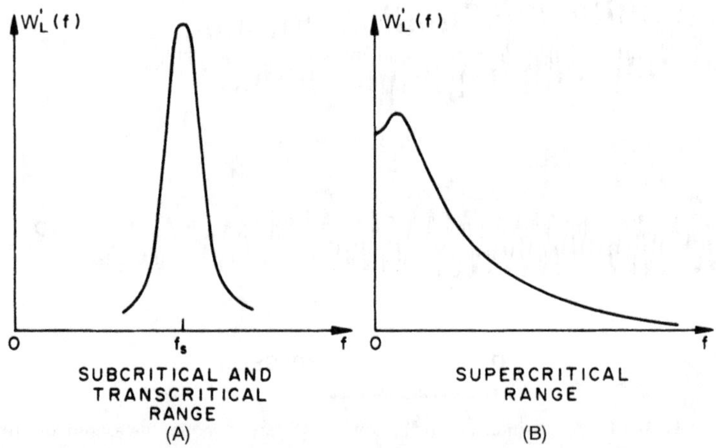

FIGURE 31.14 Spectra of lift coefficient for circular cylinder.

A few design values of bandwidth B are given in Table 31.1. In the supercritical range, the power spectrum is broad (Fig. 31.14*B*) and can be expressed as[31]

$$W_L'(f) = 4.8\,\frac{1 + 682.2(fD/\bar{V})^2}{[1 + 227.4(fD/\bar{V})^2]^2}\,\frac{D}{\bar{V}} \tag{31.29}$$

Because the vortices are three-dimensional, a realistic treatment also requires the inclusion of the spanwise cross correlation of the lift forces. This can be done in terms of the "correlation length" L given in number of diameters.

Approximate values of L are given in Table 31.1. The correlation length decreases with turbulence[32] and shear, and increases with aspect ratio $2H/D$ and the amplitude of the motion as shown in Fig. 31.15.

Using the correlation length, the spectral density of the lift force, Eqs. (31.28) and (31.29), and a few further approximations, the vibration can be evaluated from Eqs. (31.12) to (31.14). The root-mean-square (rms) displacement at height z in mode j is approximately

$$\sqrt{\overline{u_j^2(z,t)}} = \frac{\pi^{1/4}\sigma_L \rho D^4 \phi_j(z/H)}{\sqrt{B\zeta}\,(4\pi S)^2 M_j}\, C$$

where

$$C^2 = \frac{(H/D)^2}{1 + (H/2LD)} \int_0^1 \left(\frac{z}{H}\right)^{3\alpha} \phi_j^2\left(\frac{z}{H}\right) d\left(\frac{z}{H}\right)$$

Here, α = wind profile exponent (Fig. 31.6), and parameters S, σ_L, B, and L are given in Table 31.1. The mode $\phi_j(z/H)$ is dimensionless, and consequently M_j is in slugs in this case. The peak response is $g\sqrt{\overline{u_j^2(z,t)}}$, where the peak factor g is given by Eq. (31.10). If it is larger than about 2 percent of diameter D, locking-in may develop and the analysis should be repeated assuming harmonic excitation or at least random excitation with a significantly increased correlation length, as Fig. 31.15 indicates.

RANDOM EXCITATION OF TAPERED CYLINDERS BY VORTICES

Tapered cylinders, such as stacks, also vibrate due to vortex shedding, but less is known about the mechanism of excitation. It appears that the lift forces are narrowband random with a rather small correlation length L and with the dominant frequency f_s given by Eq. (31.21). As the diameter is variable, local resonance between f_s and the natural frequency f_j takes place at different heights z_r. As the wind speed increases, the resonance first appears at the tip and shifts downward. The critical wind speed for each height follows from Eq. (31.23) with $D = D(z_r)$. The rms displacements at height H due to local resonance at height z_r can be obtained from an approximate formula,[33]

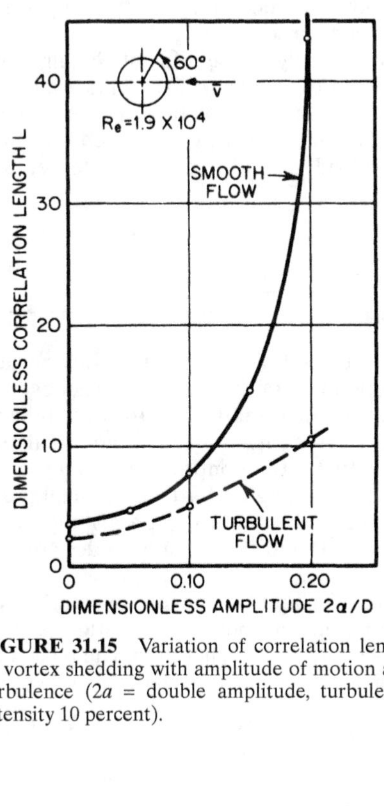

FIGURE 31.15 Variation of correlation length of vortex shedding with amplitude of motion and turbulence ($2a$ = double amplitude, turbulence intensity 10 percent).

$$\sqrt{\overline{u_j^2(H,t)}} = \sqrt{\frac{L}{2\pi^3\zeta\Psi}}\,\frac{\sigma_L\rho D^4(z_r)\phi_j(z_r)}{8S^2M_j}\,\phi_j(H)$$

where

$$\Psi = \frac{dD(z_r)}{dz} + \frac{\alpha D(z_r)}{z_r}$$

or with a constant taper

$$\Psi = \frac{t}{H} + \frac{\alpha D(z_r)}{z_r}$$

where $t = D(0) - D(H)$ and α = the wind-profile exponent. The other parameters can be taken from Table 31.1. The values listed for the transcritical region may be adequate, inasmuch as most tapered stacks are large. The peak displacement is again obtained by means of the peak factor given by Eq. (31.10).

Maximum response of chimneys in the first mode usually results from local resonance at about ¾H. The height of maximum excitation follows from the condition $d[D^4(z)\phi_j(z)]/dz = 0$.

SUPPRESSION OF VORTEX-INDUCED VIBRATIONS

Vortex shedding may induce severe vibration of a cylindrical structure such as a chimney, free-standing tower, guyed mast, bridge columns, etc. Very strong oscillations have been observed[30,34] in all-welded structures where the damping ratio is extremely low, sometimes less than 0.005.[10,30] Welded structures are particularly prone to fatigue failure, as the endurance limit may be only a fraction of the strength if heavy notches, flaws, attachments, or other adverse details are present. In other cases, the motion is intolerable because of its physiological effects or swaying of antennas. For these reasons, suppression of vibration is often desirable.

In some cases, vibration can be reduced by increasing the structural damping. This can be accomplished by additional dampers attached to an independent support[30] or to a special mass suspended from the structure and suitably tuned or by hanging chains[35] (see Chap. 6). Columns of a few bridges were filled with gravel, sand, or plastic balls partly filled with oil. The increase in mass may be unfavorable but can increase the original structural damping.

Another successful method of vibration control is to break down the wake pattern by providing the surface by helical "strakes" or "spoilers."[30,34,36] A suitable height of the spoilers is about $0.1D$ or more with a pitch of about $5D$. A significant drawback of the spoilers is that they considerably increase the drag, sometimes by 100 percent or more.[34,37]

WAKE BUFFETING

If one structure is located in the wake of another, vortices shed from the upstream structure may cause oscillation of the downstream structure.[38,39] If the two structures differ greatly in size or shape, this excitation is usually not significant. Strong vibration of the downstream structure may arise when two or more structures are identical and less than about 10 diameters apart. Then the structure in the wake is efficiently excited by well-tuned wake buffeting and its own vortex shedding. Such excitation has been observed with stacks and bridges, and to a certain degree with hyperbolic cooling towers.[38]

GALLOPING OSCILLATIONS

Vibrations due to turbulence and vortices discussed above are induced by aerodynamic forces which are, to a high degree, independent of the motion and act even on stationary bodies. Quite a different kind of oscillation is induced by the aerodynamic forces generated by the motion itself. Such forces may result from oscillatory changes in pressure distribution brought about by the continuous change in the angle under which the wind strikes the structure ("angle of attack"). This kind of oscillation often has a tendency to diverge; it is called, summarily, *aerodynamic instability, flutter,* or *self-excited oscillation.* Sudden start and violent amplitudes are typical of such phenomena (Fig. 31.3*C*).

The mechanism of this oscillation is, in general, complex. The aerodynamic forces may be a function of the displacements (translation and rotation), vibration velocity, or both, and they may interact with turbulence and vortex shedding. The basic type of the self-excited oscillations is the lateral (across-wind) oscillation induced by aerodynamic forces which are related to vibration velocity alone. Such oscillation is referred to as *galloping.* Typical features of galloping oscillation are motion in the direction perpendicular to that of the wind, sudden onset, large steady amplitudes increasing with wind velocity, and a frequency equal to the natural frequency. Galloping oscillation occurs in transmission lines and in a variety of structures having square, rectangular, or other sharp-edged cross sections.

The origin of galloping oscillation depends on the relation between lift and drag. If a body moves with a velocity $\dot{u}$ in a flow having velocity $\bar{V}$ perpendicular to its direction (Fig. 31.16), the aerodynamic force acting on the body is produced by relative wind velocity $\bar{V}_{\text{rel}}$. The angle of attack of relative wind is

$$\alpha = \arctan \frac{\dot{u}}{\bar{V}} \tag{31.30}$$

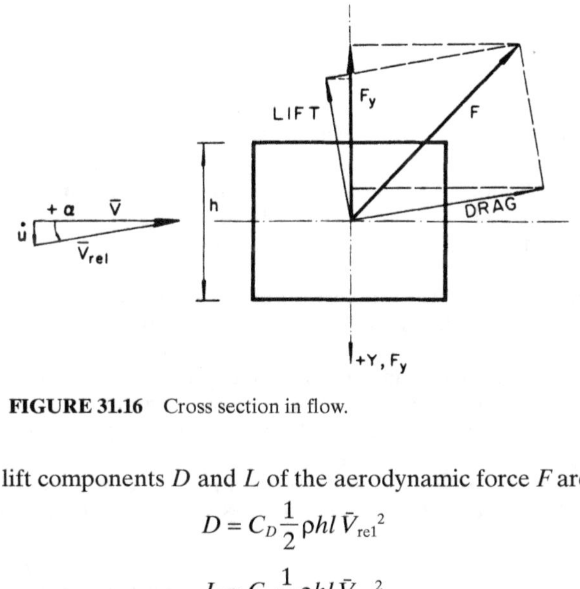

FIGURE 31.16 Cross section in flow.

The drag and lift components D and L of the aerodynamic force F are

$$D = C_D \frac{1}{2} \rho h l \bar{V}_{\text{rel}}^2$$

$$L = C_L \frac{1}{2} \rho h l \bar{V}_{\text{rel}}^2$$

where C_D and C_L are drag and lift coefficients at angle α (Fig. 31.17), h = depth of the cross section, and l = length of the body.

The component of force F into the direction of axis Y, therefore, is

$$F_y = -(C_D \sin \alpha + C_L \cos \alpha) \frac{1}{2} \rho h l \bar{V}^2 \sec^2 \alpha = C_{Fy} \frac{1}{2} \rho h l \bar{V}^2 \tag{31.31}$$

where

$$C_{F_y} = -(C_L + C_D \tan \alpha) \sec \alpha \tag{31.32}$$

The lateral force excites the vibration if the first derivative of C_{Fy} at $\alpha = 0$ is >0, hence

$$A_1 = \frac{dC_{Fy}}{d\alpha}\bigg|_{\alpha=0} = -\left(\frac{dC_L}{d\alpha} + C_D\right) > 0 \tag{31.33}$$

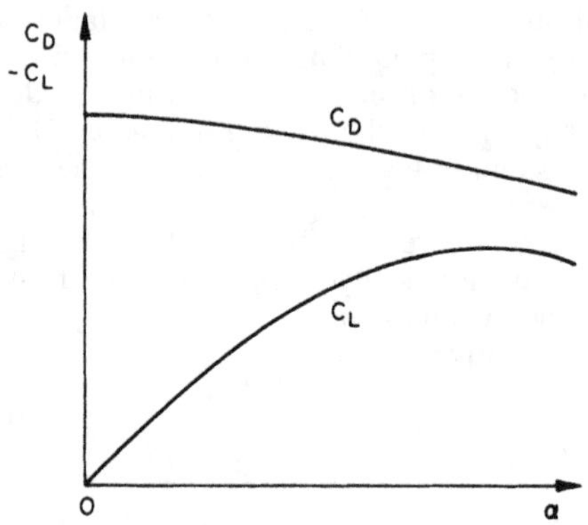

FIGURE 31.17 Lift and drag as function of angle of attack.

This condition for aerodynamic instability is known as *Den Hartog's criterion.*[40] Substitution of Eq. (31.30) into Eq. (31.32) indicates that the aerodynamic forces depend on vibration velocity and thus actually represent the aerodynamic damping. This damping is negative if $A_1 > 0$. Because the system also has structural damping ζ, which is positive, the vibration will start only if the total available damping becomes less than 0. This condition yields the onset (minimum) wind velocity for galloping from the equilibrium (or zero displacement) position as

$$\bar{V}_0 = \zeta \frac{2\pi f_j h}{n A_1} \tag{31.34}$$

where f_j = natural frequency, $n = \rho h^2/(4m)$ = mass parameter, and m = mass of the body per unit length. Some values of coefficient A_1 are given in Table 31.2.

Galloping oscillations starting from zero initial displacement can occur only when the cross section has $A_1 > 0$. Cross sections having $A_1 \leq 0$ are generally considered stable even though galloping may sometimes arise if triggered by a large initial amplitude.[41]

The response and the onset velocity are often very sensitive to turbulence. Some cross sections, such as a flat rectangle or a D section, are stable in smooth flow but can become unstable in turbulent flow.[41,42] With other cross sections, turbulence may stabilize a shape that is unstable in smooth flow (see Table 31.2). From Eqs. (31.31) and (31.32) the nonlinear, negative aerodynamic damping can be calculated[43] for inclusion in the treatment of the across-wind response due to atmospheric turbulence.

The prediction of oscillations for wind velocities greater than V_0 depends on the shape of the C_{Fy} coefficient and requires the application of nonlinear the-

TABLE 31.2 Coefficients A_1 for Determination of Galloping Onset Wind Velocity (Infinite Prisms)

	Cross section (Side ratio)					
	Unstable in smooth flow			Stable in smooth flow		
Flow $V \rightarrow$	Square 1 × 1	Rect. 3 × 2	Rect. 2 × 1	Rect. 2 × 3	Rect. 1 × 2	D-section* 2 / 1
Smooth	2.7	1.91	2.8	0	−0.03	−0.1
Turbulent ≈10 percent intensity	2.6	1.83	−2.0	0.74	0.17	0

* Varies with Reynolds number.

ory.[41–45] A few typical cases are shown in Fig. 31.18. The cases are typical of a square cross section, a flat rectangular section, and a D section whose angle of attack is allowed to change due to drag. Similar response can be expected with other cross sections.

Torsion can also participate in galloping oscillations and play an important part in the vibration. This is the case with angle cross sections[46] and bundled conductors.[47] The quasi-steady theory of pure torsional galloping can be found in Ref. 48. A solution of coupled galloping is presented in Ref. 49.

Galloping often appears in overhead conductors which also vibrate due to vortex shedding. Vortex shedding produces resonant vibration in a high-vibration mode. Galloping usually involves the fundamental mode and is known to occur when the

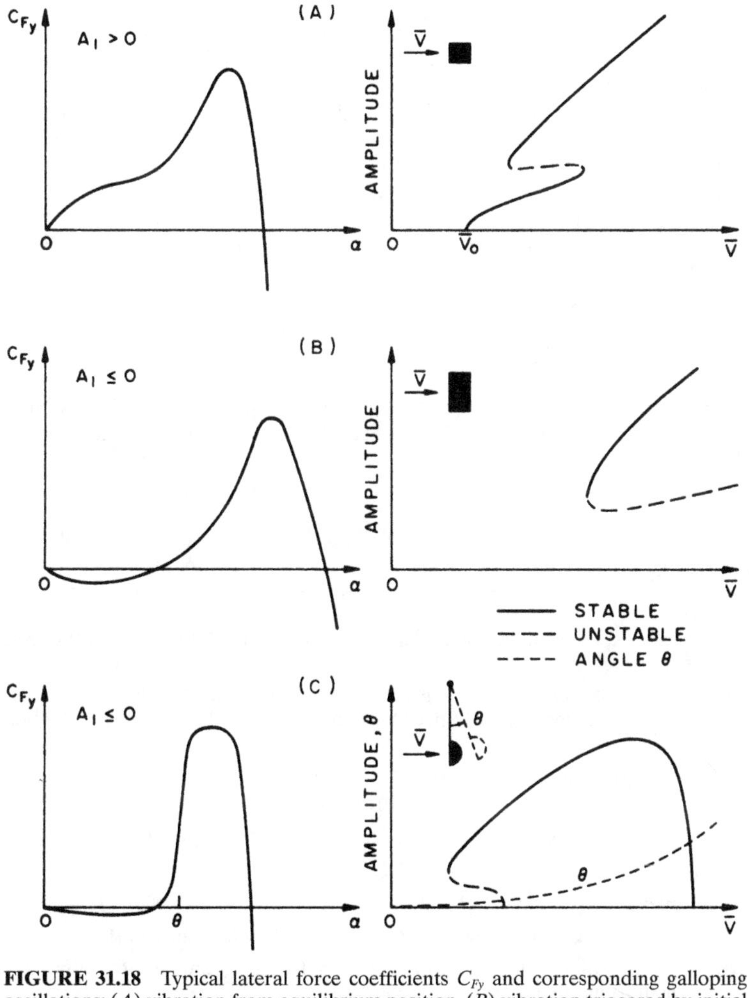

FIGURE 31.18 Typical lateral force coefficients C_{F_y} and corresponding galloping oscillations: (*A*) vibration from equilibrium position, (*B*) vibration triggered by initial amplitudes, and (*C*) vibration with variable angle of attack.

conductor is ice-coated or free of ice. The vibration often leads to fatigue failures, and various techniques are therefore used to reduce the amplitude. This can be achieved by means of resonant dampers[50] consisting of auxiliary masses suspended on short lengths of cable which dissipate energy through the bending (see Chap. 6), or aerodynamic dampers[51] consisting of perforated shrouds. Vibrations of bundled conductors can be eliminated by twisting the bundle[45] and thereby changing the aerodynamic characteristics in the spanwise direction.

VIBRATION OF SPECIAL STRUCTURES

The basic types of vibration discussed above are common in many structures. However, there are some special structures which would require individual treatment. A few examples are cited below.

Guyed towers experience complicated vibration patterns because of the nonlinearity of the guys, the three-dimensional character of the response, the interaction between the guys and the tower, and other factors.[30,52–54]

Hyperbolic cooling towers can suffer from some of the effects of wake buffeting[38] and are susceptible to turbulence.[55]

Information on the vibration of a number of special structures can be found in Refs. 4 to 7.

ACKNOWLEDGMENT

Some of the material in this chapter was originally prepared by Milos Novak, late Professor of Civil Engineering, University of Western Ontario, London, Ontario, Canada.

REFERENCES

1. Simiu, E., and R. H. Scanlan: "Wind Effects on Structures," 3d ed., John Wiley and Sons, New York, 1996.
2. Simiu, E., and T. Miyata: "Design of Buildings and Bridges for Wind: A Practical Guide for ASCE 7 Users and Designers of Special Structures," John Wiley and Sons, Hoboken, N.J., 2006.
3. Holmes D. J.: "Wind Loading of Structures," 2d ed., Spon Press, London, 2007.
4. *Proc. IUTAM-IAHR Symp.,* Karlsruhe, 1972.
5. *Proc. Conf. National Physical Laboratory,* Teddinton, Middlesex, 1963.
6. *Proc. Intern. Res. Seminar,* Ottawa, 1967.
7. *Proc. 3d Intern. Conf.,* Tokyo, 1971.
8. Davenport, A. G.: *Inst. Civil Eng. Paper No. 6480, 449–472,* August 1961.
9. Harris, R. I.: Seminar of Construction Industry Research and Information Association, Paper 3, Institution of Civil Engineers, 1970.

10. Novak, M.: *Acta Tech. Czechoslovak Acad. Sci.,* **4**:375 (1967).
11. Vickery, B. J.: *J. Struct. Div. Am. Soc. Civil Engrs.,* **98**:21 (January 1972).
12. Davenport, A. G.: *Proc. Inst. Civil Engs.,* **23**:449 (1962).
13. Etkin, B.: Meeting on Ground Wind Load Problems in Relation to Launch Vehicles, Langley Research Center, NASA, June 1966, pp. 21.1–15.
14. Davenport, A. G.: *J. Struct. Div., Am. Soc. Civil Engrs.,* **93**:11 (June 1967).
15. "National Building Code of Canada 2005," Users Guide Structural Commentaries, Part 4, Division B, National Research Council of Canada, 2005.
16. Chen, P. W., and L. E. Robertson: *J. Struct. Div. Am. Soc. Civil Engrs.,* **98**:1681 (August 1972).
17. Hansen, R. T., J. W. Reed, and E. H. Vanmarcke: *J. Struct. Div. Am. Soc. Civil Engrs.,* **99**:1589 (July 1973).
18. Vellozzi, Y., and E. Cohen: *J. Struct. Div. Am. Soc. Civil Engrs.,* **94**:1295 (June 1968).
19. Vickery, B.: U.S. Dept. of Commerce, *Nat. Bur. Std. Bldg. Sci., Ser.* **30**:93.
20. Simiu, E.: *J. Struct. Div. Am. Soc. Civil Engrs.,* **100**:1897 (September 1974).
21. Allen, D. E., and W. A. Dalgliesh: Preliminary Publication of *IABSE Symposium on Resistance and Ultimate Deformability of Structures,* Lisbon, 1973, pp. 279–285.
22. Dalgliesh, W. A.: *Proc. U.S.-Japan Res. Seminar Wind Effects Structures,* Kyoto, Japan, September 1974.
23. Dalgliesh, W. A.: *J. Struct. Div. Am. Soc. Civil Engrs.,* **97**:2173 (September 1971).
24. Roshko, A.: *J. Fluid Mech.,* **10**:345 (1961).
25. Cincotta, J. J., G. W. Jones, and R. W. Walker: *Meeting on Ground Wind Load Problems in Relation to Launch Vehicles,* Langley Research Center, NASA, 1966, pp. 20.1–35.
26. Novak, M.: Ref. 5, pp. 799–809.
27. Morkovin, M. V.: *Proc. Symp. Fully Separated Flows,* p. 102, ASME, 1964.
28. Nakamura, Y.: *Rept. Res. Inst. Appl. Mech., Kyushu University,* **17**(59):217 (1969).
29. Hartlen, R. T., and I. G. Currie: *J. Eng. Mech. Div. Am. Soc. Civil Engrs.,* **70**:577 (October 1970).
30. Novak, M.: *Proc. IASS Symp. Tower-Shaped Steel r.c. Structures,* Bratislava, 1966.
31. Fung, Y. C.: *J. Aerospace Sci.,* **27**(11):801 (1960).
32. Surry, D.: *J. Fluid Mech.,* **52**(3):543 (1972).
33. Vickery, B. J., and A. W. Clark: *J. Struct. Div. Am. Soc. Civil Engrs.,* **98**:1 (January 1972).
34. Wotton, L. R., and C. Scruton: Construction Industry Research and Information Association Seminar, Paper 5, June, 1970.
35. Reed, W. H.: Ref. 6, Paper 36, pp. 283–321.
36. Scruton, C.: National Physical Laboratory Note 1012, April 1963.
37. Novak, M.: Ref. 6, pp. 429–457.
38. Scruton, C.: Ref. 6, pp. 115–161.
39. Cooper, K. R., and R. L. Wardlaw: Ref. 7, pp. 647–655.
40. Den Hartog: *Trans., AIEE,* **51**:1074 (1932).
41. Novak, M.: *J. Eng. Mech. Div. Am. Soc. Civil. Engrs.,* **98**:27 (February 1972).
42. Novak, M., and H. Tanaka: *J. Eng. Mech. Div. Am. Soc. Civil Engrs.,* **100**:27 (February 1974).
43. Novak, M., and A. G. Davenport: *J. Eng. Mech. Div. Am. Soc. Civil Engrs.,* **96**:17 (February 1970).

44. Parkinson, G. V., and J. D. Smith: *Quart. J. Mech. Appl. Math.,* **17**(2):225 (1964).
45. Novak, M.: *J. Eng. Mech. Div. Am. Soc. Civil Engrs.,* **95**:115 (February 1969).
46. Wardlaw, R. L.: Ref. 6, pp. 739–772.
47. Wardlaw, R. L., K. R. Cooper, R. G. Ko, and J. A. Watts: *Trans. IEEE,* 1975.
48. Modi, V. J., and J. E. Slater: Ref. 4, pp. 355–372.
49. Blevins, R. D., and W. D. Iwan: *J. Appl. Mech.,* **41**, no. 4, 1974.
50. "Overhead Conductor Vibration," Alcoa Aluminum Overhead Conductor Engineering Data, no. 4, 1974.
51. Hunt, J. C. R., and D. J. W. Richards: *Proc. Inst. Elec. Engrs.,* **116**(11):1869 (1969).
52. Davenport, A. G., and G. N. Steels: *J. Struct. Div. Am. Soc. Civil Engrs.,* **91**:43 (April 1965).
53. Davenport, A. G.: Engineering Institute of Canada, **3**:119 (1959).
54. McCaffrey, R. J., and A. J. Hartmann: *J. Struct. Div. Am. Soc. Civil Engrs.,* **98**:1309 (June 1972).
55. Hashish, M. G., and S. H. Abu-Sitta: *J. Struct. Div. Am. Soc. Civil Engrs.,* **100**:1037 (May 1974).

CHAPTER 32
VIBRATION OF STRUCTURES INDUCED BY SOUND

John F. Wilby

INTRODUCTION

Vibration of structures due to interaction with a surrounding fluid can occur in a variety of ways. Chapters 30 and 31 are concerned with several fluid flow phenomena—waves, vortices, and wind—that induce vibration in an adjacent structure. The intent in Chap. 32 is to address the response of structures to acoustic and aeroacoustic excitations, where the term *aeroacoustic* includes sources, such as turbulent boundary layers, that have many characteristics similar to those of an acoustic field. The excitations can be deterministic or random in nature, as defined in Chap. 1, depending on the particular source.

Sound-induced vibration can result in sound radiation to other regions, acoustic fatigue (also known as sonic or high-cycle fatigue) of the structure being excited, or transmission of vibration to attached equipment causing malfunction or failure. Interest is often centered on aerospace applications where structures are lightweight and sound levels are high. In that case, there is the likelihood of damage to the primary structure of an aerospace vehicle, payloads in a launch vehicle, or the equipment mounted on the structure. However, structural vibration due to acoustic excitation occurs in a wide range of other environments including building damage and vibration of equipment in microelectronics manufacturing facilities.

Different acoustic and aeroacoustic sources will be described, followed by a discussion of methods for predicting linear and nonlinear response of structures to an acoustic or aeroacoustic excitation. Then, the problem of acoustic fatigue will be addressed. Finally, test methods for the measurement of structural response to acoustic and aeroacoustic excitations will be identified.

SOUND SOURCES

Acoustic and aeroacoustic pressure fields may be deterministic or random, stationary or nonstationary, and homogeneous or inhomogeneous (see definitions in

Chap. 1). *Deterministic pressures* are periodic or almost-periodic (see Chap. 19) and can be described by time-dependent functions, whereas *random pressures* can be described only in statistical terms (see Chap. 19). *Stationary pressure fields* have properties that, on the average, are invariant with time. That is not true of *nonstationary pressure fields,* which can include impulsive excitations such as blast waves and sonic booms. *Homogeneous pressure fields* have properties that, on the average, are the same at any location on a structure, whereas *inhomogeneous pressure fields* have properties that change with location. The term *aeroacoustic* is used here in a general sense to include sound produced by fluid flow or by interaction of flows with solid bodies, and fluctuating aerodynamic pressures such as those beneath a turbulent boundary layer. For convenience, and without loss of generality, both acoustic and aeroacoustic pressure fields will be referred to herein as *sound fields.*

One important characteristic of a sound field is that the fluctuating pressures are distributed over a large area, if not the entire surface, of the excited structure, and usually consist of a wide range of frequencies that includes several modes of vibration of the structure. The response of the excited structure depends on several properties of the sound field—sound pressure, frequency content, spatial distribution of pressure level and phase, and duration of exposure. The spatial characteristics of a random pressure field are best described in terms of the pressure cross-spectrum (see Chap. 19), although narrowband correlation functions have been used as equivalent representations (see Chap. 24). Sound pressures encountered in everyday life cover a range of many orders of magnitude. Thus, it is convenient to express them in terms of a logarithmic quantity called the sound pressure level, L_p, which is expressed in terms of decibels (dB) and is defined by

$$L_p = 10 \log \left[\frac{p_{\text{rms}}^2}{p_{\text{ref}}^2} \right] = 20 \log \left[\frac{p_{\text{rms}}}{p_{\text{ref}}} \right] \qquad \text{dB} \tag{32.1}$$

where p_{rms} is the root-mean-square (rms) value of the sound pressure and p_{ref} is a reference pressure that has been established by international standard to be $p_{\text{ref}} = 20$ μPa in air. The common reference for underwater sound pressures is $p_{\text{ref}} = 1$ μPa.

The range of sound pressure levels encountered in practice is demonstrated by the typical values listed in Table 32.1. The levels vary from 0 dB at the threshold of human hearing to 170 dB or more on some surfaces of aerospace vehicles, well above the threshold of pain for a human. Typical sound pressure levels near a busy highway are on the order of 80 dB, and noisy machinery can generate sound pressure levels of about 100 dB at the operator's position.

Structural response to sound is of interest in a variety of situations but, as indicated by Table 32.1, the most intense sound fields can be found in aerospace applications. Thus, aerospace vehicle sound sources are of special interest and provide a wide range of characteristics. The sources include the exhaust of jet and rocket engines, propellers and fans, powered lift devices, turbulent boundary layers, oscillating shock waves, and sonic booms.[1] In many cases, the pressure field is neither stationary nor homogeneous. However, it is often acceptable to assume stationarity and homogeneity when predicting the response of a structure, if the variations in space and time are gradual. There are exceptions to this assumption, for example, propeller noise where the pressure field is strongly inhomogeneous with the sound pressure levels being very high in the plane of rotation of the propeller and decreasing rapidly in the forward and aft directions. A survey of near-field pressure fields on flight vehicles can be found in Ref. 2.

TABLE 32.1 Typical Sound Pressure Levels for Different Environments

Sound pressure level L_p (dB re 20 μPa)	Environment
170	Jet noise on aircraft surface
160	Immediate hearing damage
140	Threshold of pain
120	Jet airplane takeoff at 1500 ft (500 m)
100	Punch press and wood planers at 3 ft (1 m)
90	Power mower at 3 ft (1 m)
80	Truck at 60 ft (20 m)
70	Automobile at 60 ft (20 m)
50	Conversation level, A-weighted, in a free field, at 3 ft (1 m)
40	Quiet residential neighborhood
20	Recording studio, A-weighted
0	Threshold of hearing

Although the following discussion on sound sources is directed toward aerospace vehicles, it should be viewed more generally in terms of sound-generating mechanisms that can be found in a wide range of situations. For example, the high-velocity gas exhaust from a pressure relief valve has acoustical characteristics similar to those of a jet engine exhaust. Axial fans in air-conditioning systems or gas-cooled nuclear reactors have noise-generating mechanisms similar to those of a turbofan engine. Also, regions of flow separation on an automobile can have characteristics that are similar to those for separated flow on an airplane.

JET AND ROCKET EXHAUSTS

Jet and rocket noise is generated by interaction between the turbulent exhaust of the jet or rocket engine and the surrounding air. At low exhaust velocities, below about 1000 ft/sec (300 m/sec), the acoustic power generated by the exhaust is proportional to the eighth power of the exhaust velocity V_j. However, as the velocity increases, the index decreases until, for rocket exhausts, where the exhaust velocity is of the order of 9000 ft/sec (2750 m/sec), the acoustical power is proportional to the third power of velocity. As the mechanical power of a rocket exhaust is also proportional to V_j^3, the acoustical power of a rocket exhaust is usually expressed in terms of an efficiency factor η, which is the ratio of acoustical power W_a to mechanical power W_m. That is,

$$W_a = \eta W_m = 0.5\eta T V_j \tag{32.2}$$

where T is the thrust of the rocket engine. Typical values of η are 0.5 to 1.0 percent for an undeflected exhaust, but can be lower for a deflected exhaust.[3,4]

Since jet noise levels are determined by the relative velocity between the exhaust and the surrounding air, the noise levels will decrease as the vehicle accelerates at takeoff or liftoff, the highest levels occurring when the vehicle is stationary. This variation of noise level with vehicle speed means that the noise levels are nonstationary, although they can be considered as stationary over short time periods.

Jet noise is strongly directional, with the highest sound pressure levels in the far field occurring at angles of 30 to 50° to the jet axis, the angle being dependent on the exhaust velocity. The situation is not so well defined in the near field, where the aircraft structure is located. Representative near-field pressure contours can be found in Refs. 4 to 7, and typical contours are shown in Fig. 32.1.[7]

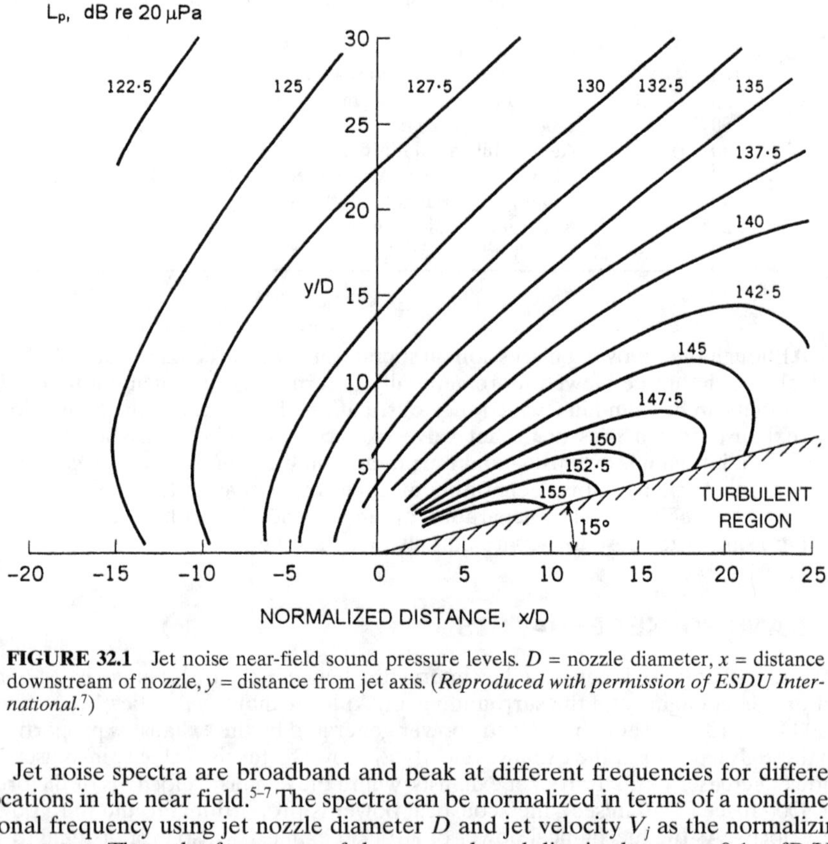

FIGURE 32.1 Jet noise near-field sound pressure levels. D = nozzle diameter, x = distance downstream of nozzle, y = distance from jet axis. (*Reproduced with permission of ESDU International.*[7])

Jet noise spectra are broadband and peak at different frequencies for different locations in the near field.[5–7] The spectra can be normalized in terms of a nondimensional frequency using jet nozzle diameter D and jet velocity V_j as the normalizing parameters. Then, the frequency of the spectral peak lies in the range $0.1 < fD/V < 1.0$, depending on location relative to the nozzle, as shown in Fig. 32.2.[7]

The spatial distribution of the pressure phase for a jet noise near field can be presented in terms of the band-limited (e.g., one-third-octave band) cross-correlation function[5,8,9] (see Chap. 24) or the normalized cross–spectral density function $\gamma_p(\xi,f)$ (see Chap. 19), since the two functions are equivalent. Typical measured values of $\gamma_p(\xi,f)$ for jet noise pressures close to a jet[8,9] are shown in Fig. 32.3. Frequency f is normalized with respect to separation distance ξ and the *trace wavespeed* of the incident sound, in order to permit scaling from one situation to another. Trace wavespeed V_t is the wave speed of the incident sound when projected onto the surface of the excited structure. Thus, for sound waves of speed c incident at an angle θ to the normal to the surface, the trace wavespeed is $c/\sin\theta$. The value of V_t is often frequency dependent and, in the case of the data in Fig. 32.3, has values of $1.43c$, $1.25c$, and $1.0c$ for frequencies 400, 500, and 800 Hz, respectively. These values of the

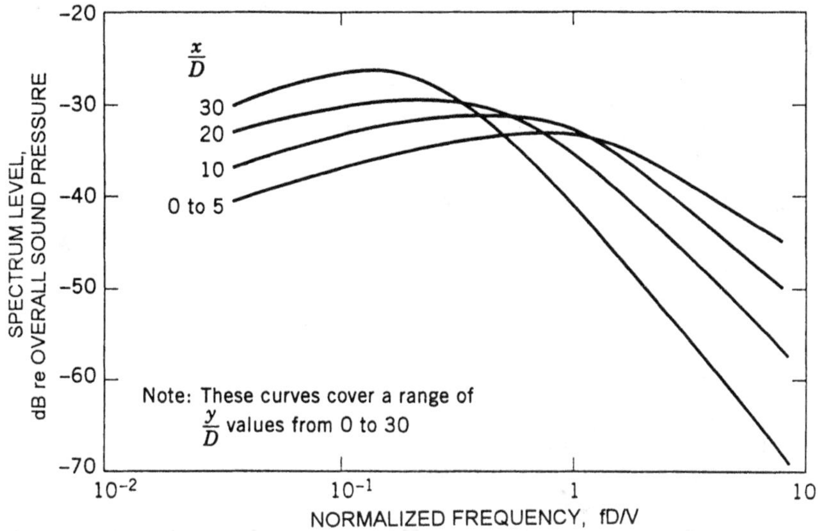

FIGURE 32.2 Normalized sound pressure spectra for several locations in jet noise near field. V = jet velocity; D, x, y, as defined in Fig. 32.1. (*Reproduced with permission of ESDU International.*[7])

trace wavespeed correspond to angles of incidence of 44, 53, and 90°, respectively. The different angles of incidence are associated with the different locations in the jet exhaust of the effective noise sources for different frequencies. Figure 32.3 refers to measurements made in a plane passing through the jet axis. Corresponding information in a direction perpendicular to that plane are less well defined.

For convenient substitution into analytical models, the normalized cross-spectrum is often represented as an exponentially decaying cosine, with the general form

$$\gamma_p(\xi,f) = e^{-ak|\xi|} \cos (k\xi) \tag{32.3}$$

where a is a decay parameter and k is the *wave number* of the pressure field, where wave number is defined by

$$k = \frac{\omega}{V_t} = \frac{2\pi f}{V_t} \tag{32.4}$$

Curves of $\gamma_p(\xi,f)$ are shown in Fig. 32.3 for three values of the decay parameter a, namely, 0.05, 0.07, and 0.10.

Supersonic jet exhausts that are under- or overexpanded contain shockwaves that result in the generation of additional broadband noise and discrete frequency *screech.*[1] The screech consists of a fundamental component, whose frequency is a function of nozzle pressure ratio or flow Mach number, and several harmonics. The directivity of the screech noise is a function of harmonic order, with the fundamental having a maximum in the upstream direction and the second harmonic having a multilobed directivity pattern with peaks in directions perpendicular to the flow direction, as well as in the upstream direction.

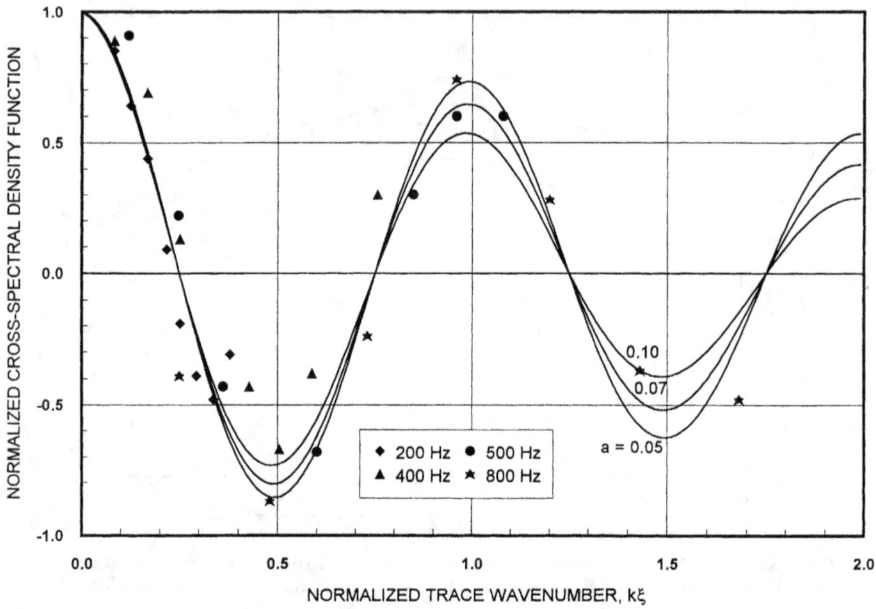

FIGURE 32.3 Example of normalized cross-spectral density function for jet noise near-field pressures. Test data collapsed with trace velocity $V_t = 1.43c$ (200, 400 Hz), $1.25c$ (500 Hz), and $1.0c$ (800 Hz). Continuous plots represent Eq. (32.3) with decay parameter $a = 0.05, 0.07$, and 0.10. (*Data from Richards and Mead.*[9])

ENGINE EXHAUST FLOWS

Powered lift aircraft utilize the exhaust from the engines to augment the lift generated by the wing and increase the effectiveness of the control surfaces, utilizing systems such as upper surface blowing and externally blown flaps.[1] By so doing, the surfaces of the aircraft are exposed to high sound pressure levels that are a combination of acoustic and aeroacoustic pressures. For example, sound pressure levels of up to 165 dB were measured on an airplane with upper surface blowing.[10] In addition, the structure was heated to a temperature of 500 to 700°F (260 to 370°C). A similar situation exists on stealth aircraft where the engine exhaust flows over the upper surface of the aft structure so that the gases are cooled before they can be observed from below.[10] Sound pressure levels greater than 180 dB are predicted in the neighborhood of the exhaust flows on hypersonic aircraft.[10–12]

PROPELLERS AND FANS

Propeller or fan noise consists of both broadband and discrete frequency components, but the pressure spectrum is dominated by discrete frequency components at the blade passage frequency of the propeller or fan and harmonics thereof. The blade passage frequency f_b is given by

$$f_b = \frac{\Omega B}{60} \tag{32.5}$$

where Ω is the rotational speed (rpm) of the propeller or fan and B is the number of blades. The spectra for counter-rotating propellers are more complex, with blade passage frequency components for each of the propellers plus interaction tones,[13] as shown in Fig. 32.4. The spectrum in the figure also contains components for each individual blade of the propeller, because the blades are not identical.

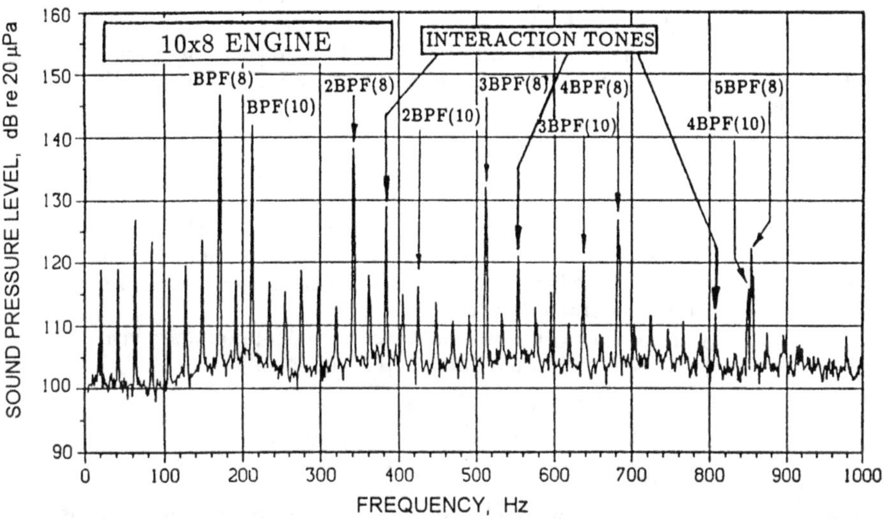

FIGURE 32.4 Spectrum for near-field sound pressure levels of high-speed, counter-rotating propeller with 8 and 10 blades. BPF(8) and BPF(10) denote blade passage frequencies for 8- and 10-blade propeller stages, respectively. (*Simpson, Druez, Kimbrough, Brock, Burge, Mathur, Cannon, and Tran.*[13])

Sound pressure levels on the fuselage of multiengined general aviation aircraft are typically of the order of 130 dB at the blade passage frequency. High-speed propellers, with tip speeds that are supersonic under cruise conditions, have higher sound pressure levels on the order of 150 dB.[13]

Cross-spectrum measurements of propeller noise on a general aviation airplane[14] show that the pressure field in the plane of rotation is an aerodynamic potential field that rotates with the propeller blades. Forward and aft of the plane of rotation the pressure field is acoustic and has the characteristics of propagating acoustic waves generated by sources located near the tips of the propeller blades. The spatial distribution of the cross-spectrum phase is more complicated for counter-rotating propellers.[15]

TURBULENT BOUNDARY LAYER

The dominant fluctuating pressures acting on launch vehicles, missiles, and aircraft in high-speed flight are associated with the turbulent boundary layer on the external surfaces of the vehicle. Similar fluctuating pressure fields are also encountered on other moving vehicles including automobiles, particularly around the windshield, and high-speed elevators. These pressure fields have many of the characteristics of an acoustic pressure field, but the convection velocity of the pressure fluctuations over the surface may be subsonic in contrast to an acoustic field where the trace velocity is always equal to, or greater than, the speed of sound in the fluid. There are also differences in the cross-spectra.

Measurements of turbulent boundary layer pressure fluctuations have been made in wind tunnels, on aircraft in flight, and underwater.[9,16–18] The measurements have included both subsonic and supersonic flow conditions, but the emphasis has been on subsonic conditions. A combination of analytical and empirical methods has resulted in representations for the various characteristics of turbulent boundary layer pressure fields for both attached and separated flow.

For an attached turbulent boundary layer, taking into account compressibility effects, the rms pressure p_{rms} can be expressed as a function of Mach number, in relationships such as[19]

$$\frac{p_{rms}}{q} = \frac{0.006}{1 + 0.13M^2} \tag{32.6}$$

where q is the *dynamic pressure* of the flow, given by $q = ½\rho V^2$ where V is velocity, ρ is the density of the fluid, and M is the flow Mach number, defined at some location such as free stream or the edge of the boundary layer. Corresponding relationships can be developed for separated flow conditions.

The pressure spectrum $G_p(\omega)$ for an attached turbulent boundary layer is broadband and can be represented by a relationship of the form[19]

$$\frac{G_p(\omega)V}{q^2\delta^*} = \frac{2\kappa(p_{rms}/q)^2}{\pi\left[1 + \left(\frac{\kappa\omega\delta^*}{V}\right)^2\right]} \tag{32.7}$$

where κ is a function of flow Mach number, V is the flow velocity, and δ^* is the *boundary layer displacement thickness.* The boundary layer displacement thickness is the distance that the surface beneath the boundary layer would have to move outward and normal to itself to account for the differences in the rate of mass flow with the boundary layer present and, hypothetically, without the boundary layer. Separated turbulent boundary layers in the neighborhood of steps, ramps, and other surface discontinuities have higher pressure levels at low frequencies than is the case for attached boundary layers, as shown in Fig. 32.5.[18] Pressure spectrum and frequency are normalized in Fig. 32.5 with respect to *boundary layer thickness* δ rather than boundary layer displacement thickness δ^*. Boundary layer thickness can be defined as the distance from the surface at which the flow velocity reaches 99.5 percent of the free stream velocity. Equation (32.7) can be modified to take into account the low-frequency shifts seen in Fig. 32.5 by replacing κ with $C\kappa$, where $C > 1$. The presence of oscillating shockwaves further increases the low-frequency component of the pressure spectrum,[18] as can be seen in Fig. 32.5.

Normalized cross-spectra or band-limited cross-correlation functions have been measured for attached turbulent boundary layers.[16,17] The measured data indicate that the normalized cross spectrum is dependent on the thickness of the boundary layer δ as well as on the convection speed V_c of the pressure field and the separation distance ξ between the measuring points. Empirical relationships such as[20]

$$\gamma(\xi,\omega) = \exp\left\{-\left[\left(\frac{0.1\omega}{V_c}\right)^2 + \left(\frac{0.27}{\delta}\right)^2\right]^{0.5} |\xi|\right\} \cos\left(\frac{\omega\xi}{V_c}\right) \tag{32.8}$$

have been proposed for attached turbulent boundary layers. There is little corresponding information for separated boundary layers, where the flow is much more complicated.

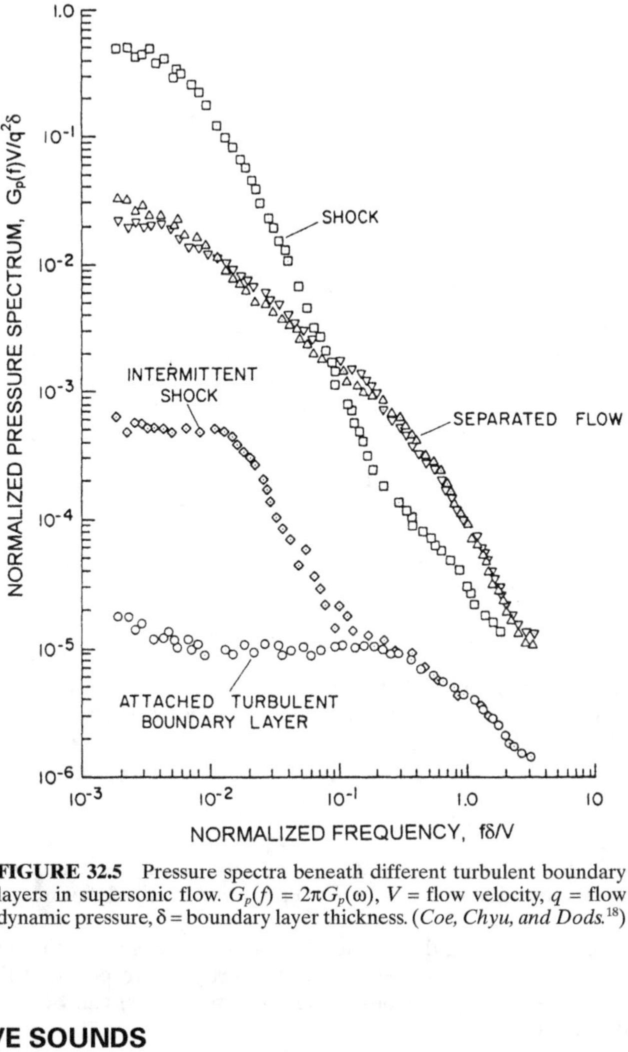

FIGURE 32.5 Pressure spectra beneath different turbulent boundary layers in supersonic flow. $G_p(f) = 2\pi G_p(\omega)$, V = flow velocity, q = flow dynamic pressure, δ = boundary layer thickness. (*Coe, Chyu, and Dods.*[18])

IMPULSIVE SOUNDS

Impulsive sounds, such as sonic booms generated by airplanes in supersonic flight[1,21] and blast waves from explosions, can cause transient vibration of a structure.

ANALYTICAL METHODS

It is often assumed in the analysis of structural response to acoustic excitation that the structure responds in a linear manner, so that there is a linear relationship between excitation force and structural response. However, this assumption may not be valid when the acoustic excitation levels are high. In that case the response is nonlinear.

LINEAR ANALYSIS

Several different methods can be used to calculate the linear response of a structure to acoustical excitation. They include classical normal mode analysis, statistical energy analysis, and finite element analysis. Each method has its own advantages and disadvantages.

Classical Normal Mode Analysis. In the classical modal formulation,[9] the acceleration autospectrum $G_a(\underline{x},\omega)$ for location x and angular frequency ω can be written as

$$G_a(\underline{x},\omega) = \omega^4 A^2 G_p(\omega) \sum_r \sum_s \psi_r(\underline{x})\psi_s(\underline{x}) H_r(\omega) H_s^*(\omega) j_{rs}^2(\omega) \tag{32.9}$$

where A is the area of the structure exposed to the excitation, $G_p(\omega)$ is the excitation pressure spectrum, $\psi_r(\underline{x})$ is the mode shape of mode of order r, $H_r(\omega)$ is the structural mode response function, $j_{rs}^2(\omega)$ is the cross acceptance that describes the spatial coupling between the excitation pressure field and the structural mode shapes, and an asterisk denotes a complex conjugate. The *cross acceptance* is defined by

$$j_{rs}^2(\omega) = \frac{1}{A^2 G_p(\omega)} \iint G_p(\underline{x}, \underline{x}',\omega)\psi_r(\underline{x})\psi_s(\underline{x}')d\underline{x}d\underline{x}' \tag{32.10}$$

where $G_p(\underline{x},\underline{x}',\omega)$ is the excitation pressure cross spectrum and the *structural mode response function* is defined by

$$|H_r(\omega)|^2 = M_r^{-2}[(\omega_r^2 - \omega^2)^2 + \eta_r^2\omega_r^4]^{-1} \tag{32.11}$$

where η_r is the damping loss factor ($\eta_r = 2\zeta_r$, where ζ_r is the damping ratio), M_r is the modal mass, and ω_r is the resonance frequency of mode r. The *modal mass* is defined as

$$M_r = \int_A m\psi_r^2(\underline{x})d\underline{x} \tag{32.12}$$

where m is the mass per unit area for a panel of area A. For a uniform panel with simply supported boundaries, $M_r = mA/4$. Prediction methods for ω_r can be found in Chap. 7.

If the damping is small and the fluid loading is negligible (which is usually true in air but not in water), the vibration is dominated by the response at the resonance frequencies and contributions from the cross terms ($r \neq s$) can be neglected. Then Eq. (32.9) becomes

$$G_a(\underline{x},\omega) = \omega^4 A^2 G_p(\omega) \sum_r \psi_r^2(\underline{x})|H_r(\omega)|^2 j_r^2(\omega) \tag{32.13}$$

In Eq. (32.13), the cross acceptance of Eq. (32.10) is replaced by the *joint acceptance*

$$j_r^2(\omega) = \frac{1}{A^2 G_p(\omega)} \iint G_p(\underline{x}, \underline{x}',\omega)\psi_r(\underline{x})\psi_r(\underline{x}')d\underline{x}d\underline{x}' \tag{32.14}$$

Assuming that the structure has simply supported boundaries, and $G_p(\omega)$ and $j_r^2(\omega)$ vary slowly with ω in frequency band $\Delta\omega$, the space-average, mean square response in frequency band $\Delta\omega$ is

$$[a^2]_{\Delta\omega} \approx \frac{\omega^4 A^2}{4} G_p(\omega) \sum_r j_r^2(\omega) \int_{\Delta\omega} |H_r(\omega)|^2 d\omega \tag{32.15}$$

For small damping

$$\int_{\omega} |H_r(\omega)|^2 d\omega \approx \frac{\pi}{2\omega_r^3 \eta_r M_r^2} \tag{32.16}$$

and Eq. (32.15) reduces to

$$[a^2]_{\Delta\omega} \approx \frac{\omega^4 A^2 \pi}{8} G_p(\omega) \sum_{r \in \Delta\omega} \frac{j_r^2(\omega)}{\omega_r^3 M_r^2 \eta_r} \tag{32.17}$$

The notation $r \in \Delta\omega$ signifies that the summation is over all modes of order r whose resonance frequency ω_r lies in the frequency band $\Delta\omega$. From Eq. (32.17), the acceleration spectral density, averaged in space and frequency, is

$$\langle G_a(\omega) \rangle_{A,\Delta\omega} = \frac{[a^2]_{\Delta\omega}}{\Delta\omega} \approx \frac{\omega^4 A^2 \pi}{8\Delta\omega} G_p(\omega) \sum_{r \in \Delta\omega} \frac{j_r^2(\omega)}{\omega_r^3 M_r^2 \eta_r} \tag{32.18}$$

where $\langle \, \rangle_{A,\Delta\omega}$ denotes averaging over area A and frequency band $\Delta\omega$. It can be seen in Eqs. (32.13), (32.17), and (32.18) that the two functions representing the excitation pressure field are the pressure autospectrum, $G_p(\omega)$, and the joint acceptance, $j_r^2(\omega)$.

The classical normal mode approach of Eq. (32.13) is an accurate way to predict structural response to acoustic or aeroacoustic pressure fields, provided that the relevant details of the structure and pressure field are known and represented correctly. However, that is often not the case. It is difficult to obtain the cross-spectrum data for the pressure field, and approximations have to be made. Also, an accurate description of the normal modes and resonance frequencies of the structure is not always available, especially for complicated structures. Experimental procedures (see Chap. 21) and analytical methods, such as finite element analysis (see Chap. 23), might be used to obtain normal mode information, but both methods become increasingly inaccurate as frequency increases. One solution is to resort to averaging techniques such as Eq. (32.17) or (32.18), but that has the disadvantage of eliminating some of the details in the results. Statistical energy analysis (see Chap. 24) is a further step in the averaging process.

Analysis of structural response to sound underwater is complicated by the fact that fluid loading is no longer negligible and has to be included in the analytical model.[22,23] The effect of fluid loading depends on whether the frequency of interest is below or above the *critical frequency,* which is defined as the frequency at which the trace wavespeed of the sound field is equal to the wavespeed of the flexural or bending waves in the structure. At frequencies below the critical frequency, fluid loading essentially acts as an entrained mass that has to be included as a second mass term in the equations of motion.[23] At frequencies above the critical frequency, the fluid loading influences the radiation resistance and the sound radiation into the fluid.[23]

Joint Acceptance. The joint acceptance function describes the efficiency by which a particular pressure field can excite a structure. For a given pressure spectrum $G_p(\omega)$, different types of excitation, with different joint acceptance functions, will generate different vibration levels in the responding structure. For example, turbulent boundary layer pressure fluctuations will produce different vibration levels than will jet noise of the same pressure level.

Simplifying assumptions are usually introduced so that the joint acceptance can be obtained in closed form. Specifically, it is commonly assumed that the pressure field is homogeneous, so that $\underline{x}$ and $\underline{x}'$ can be replaced by $\underline{\xi}$, where $\underline{x}' - \underline{x} = \underline{\xi}$. The vector $\underline{\xi}$ has components ξ_x and ξ_y in the x and y directions, respectively. Also, it is assumed that the joint acceptance is separable in the x and y directions. Finally, it is assumed that the structure is simply supported at the boundaries. Then, the component of the joint acceptance in the x-direction is

$$j_m^2(\omega) = \frac{1}{A^2} \int_{L_x} \int \gamma_x(\xi_x, \omega) \cos(k_x \xi_x) \sin\left(\frac{m\pi x}{L_x}\right) \sin\left(\frac{m\pi x'}{L_x}\right) dx dx' \tag{32.19}$$

with

$$\gamma_x(\xi_x, \omega) = \frac{|G_p(\xi_x, 0, \omega)|}{G_p(\omega)} \tag{32.20}$$

and mode order $r \equiv (m,n)$. Similar relationships apply in the y-direction.

Closed-form joint acceptance functions for three different types of excitation, namely, attached turbulent boundary layer, jet noise, and diffuse (reverberant) sound field, are given in Ref. 20. Typical nondimensional joint acceptance curves based on Eqs. (32.19), (32.20), and (32.3) are shown in Fig. 32.6, for the case where the decay parameter a in Eq. (32.3) has a value of 0.1. The joint acceptance for the first mode shape ($n = 1$) has a maximum value at zero wave number or frequency,

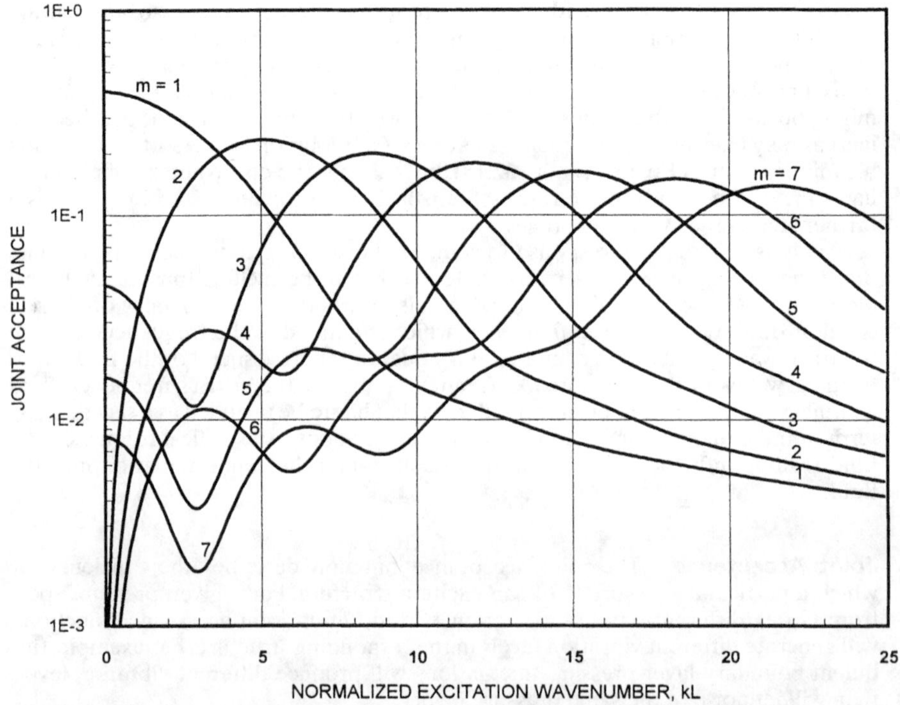

FIGURE 32.6 Joint acceptance curves based on Eqs. (32.19), (32.20), and (32.3), with decay parameter $a = 0.1$. L = length of panel, m = mode order, k = excitation wave number [Eq. (32.4)].

but the joint acceptance for each of the other modes has a maximum value at a nonzero value of frequency. Those maxima for the higher-order modes occur when the wave number of the excitation is equal to the flexural wave number for the structural mode, a condition known as *coincidence.*

Statistical Energy Analysis. Statistical energy analysis (SEA) makes the general assumption that it is not practical to represent all the details of a structure in a given response prediction procedure (see Chap. 24). Thus, ensemble averaging is performed over a series of similar, but slightly different, structures to obtain an average response. In practice, ensemble averaging is time-consuming, so it is replaced by frequency averaging.

Equation (32.18) leads to a typical SEA relationship for simply supported panels, specifically,

$$\langle G_a(\omega)\rangle_{A,\Delta\omega} = G_p(\omega)\frac{2\pi\omega n_r(\omega)}{m^2}\frac{\langle j_r^2(\omega)\rangle_{\Delta\omega}}{\langle\eta_r\rangle_{\Delta\omega}} \tag{32.21}$$

where $\langle\,\rangle_{\Delta\omega}$ denotes averaging over frequency, $n_r(\omega)$ is the modal density of the structure, and m is the mass/unit area of the panel (assumed uniform). The frequency-band-averaged joint acceptance is

$$\langle j_r^2(\omega)\rangle_{\Delta\omega} = \frac{1}{N}\sum_{r=1}^{N} j_r^2(\omega) \tag{32.22}$$

where N is the number of modes with resonance frequencies in frequency band $\Delta\omega$. The modal density of the structure is defined by

$$n_r(\omega) = \frac{dN}{d\omega} \tag{32.23}$$

For a flat panel,

$$n(\omega) = \frac{\sqrt{3}A}{2\pi h c_L} \tag{32.24}$$

where h is the panel thickness and c_L is the *longitudinal wave speed* in the structure given by

$$c_L = \sqrt{\frac{E}{\rho(1-\nu^2)}} \tag{32.25}$$

In Eq. (32.25), E is Young's modulus of the structural material, ρ is the material density, and ν is Poisson's ratio.

The use of SEA techniques to simplify the analysis has the advantage that the response can be calculated to high frequencies with minimum computing time, but there is the disadvantage that the use of space- and frequency-averaging methods means that structural response cannot be predicted for a specific point on the structure nor at a specific frequency. Additional methods have to be used to supplement the SEA calculations.

SEA is of limited value at low frequencies where modes are sparse ($N < 3$, say). The method can still be used but the variance of the results becomes large.[24] Further discussion on statistical energy analysis can be found in Chap. 24.

Finite Element Analysis. In finite element analysis (FEA), a continuous structure is modeled as an array of grid points connected by appropriate elements (see Chap. 23). This means that the continuously distributed sound pressure field has to be represented as an array of discrete forces applied at the grid points. The forces have to be given autospectral functions that take into account the frequency characteristics and amplitudes of the excitation pressure field, and the structural area attributed to each grid point. In addition, the forces at each pair of grid points have to be assigned the appropriate cross-spectrum function based on the spatial separation between the grid points.

The response of the structure at location $\underline{x}$ can be calculated using relationships of the form[25]

$$G_a(\underline{x},\omega) = \sum_{j=1}^{q} \sum_{k=1}^{q} H_{jx}^{*T}(\omega) \frac{A_x}{A_j} G_{jk}(\omega) \frac{A_x}{A_k} H_{kx}(\omega) \tag{32.26}$$

where $H_{jx}(\omega)$ is the frequency response function between the jth input and the response location $\underline{x}$, $G_{jk}(\omega)$ is the cross-spectrum between the jth and kth inputs, A_j is the area associated with the jth input, and A_x is the area associated with the response location. The frequency response function $H_{jx}^{*T}(\omega)$ is the transpose of the complex conjugate of $H_{jx}(\omega)$. Basic details of the finite element method can be found in Chap. 23.

Successful application of FEA to the calculation of the response of a structure to acoustic or aeroacoustic pressure fields requires that there be an adequate number of degrees of freedom in the finite element model and an appropriate representation of the pressure field auto and cross spectra. In principle, finite element methods can be applied over the entire frequency range of interest, but that is not necessarily true in practice. As frequency and number of modes increase, it becomes more difficult to provide an accurate description of the structure including boundary conditions. It also becomes more difficult to represent the details of the pressure field cross spectrum. Finally, the time required to perform the necessary computations can become excessive. Thus, the finite element method suffers from the same disadvantages as does the classical normal mode method.

Hybrid Finite Element-Statistical Energy Analysis. There is often a midfrequency range where neither statistical energy analysis nor finite element analysis is particularly suitable for the prediction of structural vibration. A hybrid SEA-FEA method or, more generally, a statistical-deterministic method, combines SEA and FEA methods for application at these midfrequencies. Components of a structure, such as panels, that have short wavelength response are represented by SEA subsystems, and components, such as stiff beams, having long wavelength response are modeled using FEA. Then, the two representations can be coupled through the dynamic stiffness matrix.[26]

Damping. It is obvious from Eqs. (32.17) and (32.18) that damping is an important parameter in determining the magnitude of the structural response to acoustic or aeroacoustic excitation, since the mean square acceleration is inversely proportional to the damping loss factor η_r. The damping loss factor in Eq. (32.17) is composed of three components, as follows:

$$\eta_r = \eta_{r,\text{struc}} + \eta_{r,\text{rad}} + \eta_{r,\text{aero}} \tag{32.27}$$

The structural loss factor, $\eta_{r,\text{struc}}$, represents the damping due to material properties of the structure and mechanisms such as gas pumping at riveted joints and slip damping (see Chap. 35). It also represents damping due to any applied treatments (see Chap. 36). The radiation damping loss factor, $\eta_{r,\text{rad}}$, represents damping associated with the radiation of sound as a consequence of the vibration of the structure. This can be a significant contribution for structures such as composite structures that are very lightly damped. For structures in vacuo, $\eta_{r,\text{rad}} = 0$. The aerodynamic damping loss factor, $\eta_{r,\text{aero}}$, represents the damping associated with the presence of nonzero mean flow over the structure. Additional information on the damping of structures can be found in Refs. 27 and 28.

NONLINEAR VIBRATION

When excitation sound levels become too high, the response of a structure becomes nonlinear and linear analysis methods for the prediction of structural vibration are inaccurate. There are several situations where nonlinear response can be important. They include vibration where the displacement of the structure is no longer small with respect to the panel thickness, rattle induced by impulsive or low-frequency noise, and *snap-through* response of curved or buckled plates. Snap-through motion occurs when the local curvature of a panel that is curved by design or by buckling, jumps from one direction to another. Buckling can be caused, for example, by thermal stresses induced by high temperatures. Nonlinear response can be in the form of a hardening or softening spring (see Chap. 4), or instability conditions with snap-through motion.

Response characteristics often associated with nonlinear vibration are (1) the response amplitude no longer increasing in proportion to the amplitude of the excitation, (2) the resonance frequencies of the response modes changing with excitation amplitude, and (3) broadening of resonance peaks, which is attributed to nonlinear damping. The first two phenomena are demonstrated in Fig. 32.7, which shows the response of a panel to a sound field generated by a siren.[29] The response in the first mode, in terms of amplitude and resonance frequency, becomes nonlinear when the sound pressure reaches a level of about 102 dB.

Various approaches have been developed for the prediction of nonlinear response of a structure to acoustic excitation,[30–33] but they often have very limited application. Characteristics of nonlinear vibration and several approximate methods for analyzing the vibration are reviewed in Chap. 4. Nonlinear analytical methods that give closed-form quantitative results are usually limited to simple structures. Approximate methods are usually required for complex structures such as those found in aerospace applications. Other approaches include numerical methods, such as the Monte Carlo approach, and finite element methods using nonlinear element stiffness matrices. However, the methods are often restricted to simple acoustic pressure fields such as (1) plane waves at normal incidence, with the pressure uniform in both amplitude and phase over the entire surface of the structure; (2) plane acoustic waves at grazing incidence; or (3) uncorrelated pressure fields. Furthermore, structural response is often limited to a single mode.

The Monte Carlo method[33] is based on the numerical generation of a large number of random, sample excitations and the calculation of the response to each sample. The method can be used for both linear and nonlinear responses to random excitations, and it could be a feasible approach for nonlinear vibration where closed-form or approximate solutions are not possible, although the method requires the use of high-speed digital computers. One example of a second-order, nonlinear equation of motion for a panel is

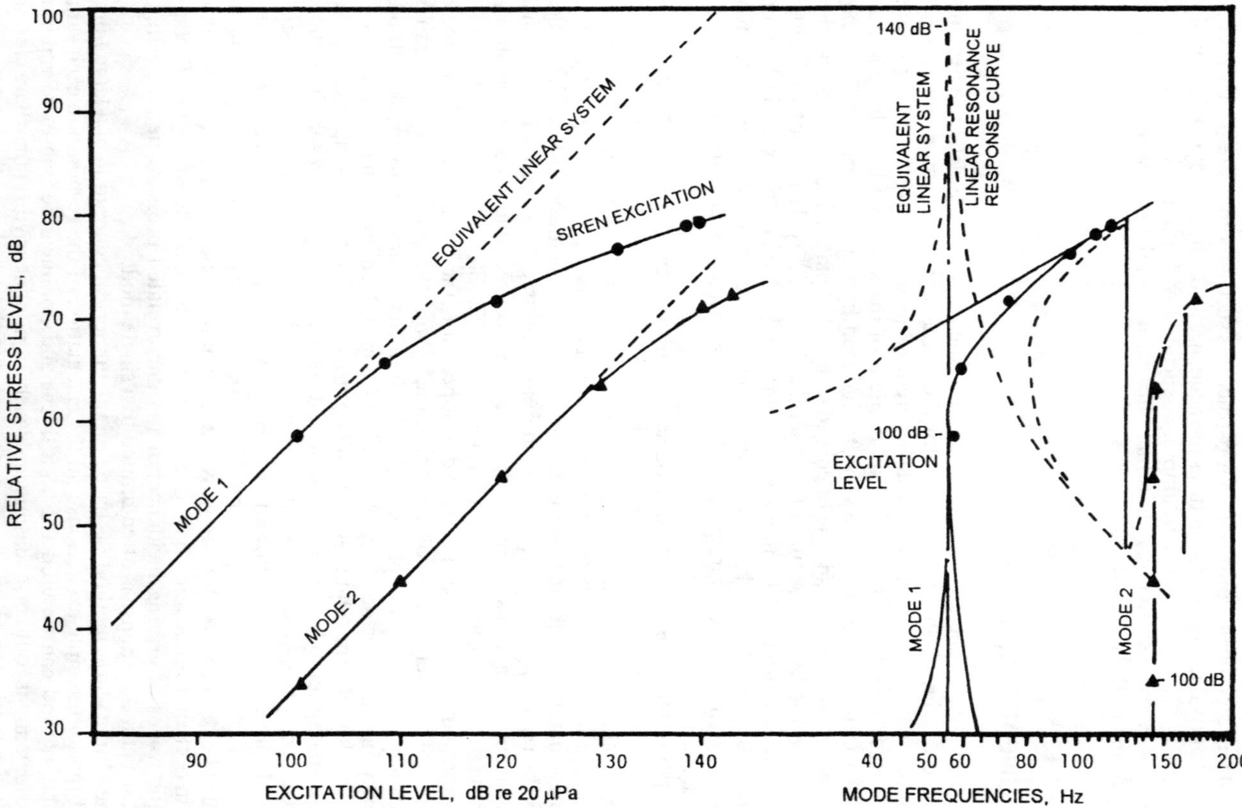

FIGURE 32.7 Nonlinear stress response characteristics for flat panel exposed to siren excitation. Panel with clamped edges, panel length = 12 in. (0.30 m). (*Mei.*[29])

$$d^2X_{ij}/dt^2 + 2\zeta_{ij}\omega_{ij}(dX_{ij}/dt) + \omega_{ij}^2X_{ij} + N(X_{ij}, dX_{ij}/dt) = F_{ij}(t) \quad (32.28)$$

where X_{ij} are the components of generalized coordinates, ω_{ij} are the natural frequencies of a linear system, ζ_{ij} are the modal damping coefficients, N is the nonlinear system operator, and $F_{ij}(t)$ are the generalized random forces.

The time-domain Monte Carlo method consists of three basic steps:[33] (1) random inputs for $F_{ij}(t)$ are generated using simulation procedures of random processes; (2) the equations of motion, such as Eq. (32.28), are solved numerically for each random value of $F_{ij}(t)$; and (3) statistical moments and other needed quantities of the random response $X_{ij}(t)$ are computed for ensemble averages. If the system is ergodic (see Chap. 1), the ensemble averaging can be replaced by time averaging, with a saving in computing time.

In many aerospace situations, the structure is exposed to high temperatures and the structural vibration is strongly dependent on thermal stresses induced by a thermal environment. The effect is taken into account in some procedures by applying the acoustic and thermal loads in sequence. A more appropriate analysis of nonlinear response of aerospace structures considers acoustic and thermal loads simultaneously.[30]

Structural damping is often represented as linear damping. However, nonlinear damping can be represented, for example, by replacing linear damping in Duffing's equation (see Chap. 4) with a nonlinear damping term[34] such as $\omega_o\eta(1 + \alpha q^2)dq/dt$.

ACOUSTIC FATIGUE

Acoustically induced structural vibration results in oscillating stresses. The stress levels may be low but, because of the frequencies involved, typically 100 to 500 Hz, the number of stress reversals can be large enough at stress concentration points to create fatigue cracks. This phenomenon is called high-cycle fatigue, acoustic fatigue, or sonic fatigue.[35,36] Most examples of failures induced by sonic fatigue occur in aircraft structures in the form of skin failures along rivet lines, skin debonding in sandwich panels, and failure in internal attachment structures.[5,6]

In many cases the stresses induced by acoustic pressure fields are dominated by response in the first mode of vibration of a panel, and the acoustical wavelength is large relative to the dimensions on the panel. Then, the sound pressures are essentially in phase over the panel, and details of the pressure correlation are of minor importance. The mean square stress $\sigma^2(t)$ can be estimated using the approximation[6]

$$\sigma^2(t) \approx K\frac{\pi}{4\eta} f_n G_p(f_n)\left(\frac{\sigma_o}{F_o}\right)^2 \quad (32.29)$$

where f_n is the frequency of the dominant mode of order n, $G_p(f_n)$ is the spectral density of the excitation pressure at frequency f_n, η is the damping ratio, and σ_o is the stress at the point of interest due to a uniform static pressure of magnitude F_o. Equation (32.29) is based on early work for a single-degree-of-freedom system. The factor K is included in Eq. (32.29) so that the equations can be modified to fit particular structural configurations and materials. There are cases where acoustic fatigue is caused by vibration of several modes, not just one. Thus, alternative prediction procedures are required that extend the approach in Eq. (32.29) to higher-order modes and complex shapes, and estimate the influence of acoustical wavelength.[12]

It is apparent from Eq. (32.29) that increasing the damping of a structure would decrease the stresses. Thus, the application of damping material will reduce the likelihood of acoustic fatigue. For example, damping treatment was applied to the fuselage structure of a test airplane with high-speed propellers to minimize the likelihood of acoustic fatigue in the plane of rotation of the propellers.[13] Applied damping techniques are described in Chap. 36 and the wider aspects of passive vibration control are discussed in Ref. 37.

LABORATORY TESTING OF STRUCTURES AND EQUIPMENT

Laboratory tests are often required to supplement or validate analysis, evaluate new structural designs, or develop a database of fatigue life for different environmental conditions or for new materials, especially composites. Acoustical environments of aircraft and space vehicles can reach overall sound pressure levels in the range 170–180 dB in local areas. Consequently, there is a need to develop similar levels in the laboratory with the appropriate frequency distributions. Two test environments, the progressive wave tube and the reverberant chamber, are used for many of the laboratory tests. The purposes of the testing are to find weak points in the structural design or in the manufacturing process, or to determine whether or not the structure will have a satisfactory fatigue life (see Chap. 18). The progressive wave tube and reverberant chamber play different roles in this process.

PROGRESSIVE WAVE TUBES

A *progressive wave tube* consists of duct with a sound source at one end and a sound-absorbing termination at the other end. It is used to expose structural components, such as a panel, to high-intensity sound pressure levels for long periods of time so as to evaluate the susceptibility of the structure to acoustic fatigue. The test structure is mounted in one wall of the tube and exposed to sound waves traveling along the tube at grazing incidence.[9,10,38,39] Relatively small test specimens are used because of the difficulty of generating, in the laboratory, very high sound pressure levels over large areas.

Due to concerns about the effect of high temperatures for some applications, such as aircraft-powered lift devices, the structure beneath the engine exhaust of stealth aircraft, and the vehicle structure of hypersonic vehicles, facilities have been constructed that permit the heating of the test specimen at the same time that it is being exposed to the high-intensity sound pressure levels. The acoustic excitation is limited to the lower frequencies because of constraints on the source, which usually consists of several electropneumatic modulators with broadband random acoustical outputs. However, the lower frequencies are usually responsible for the highest stresses that determine acoustical fatigue life.

A typical progressive wave tube is shown in Fig. 32.8. The number of electropneumatic modulators is determined by the size of the duct, and the desired maximum sound pressure levels and frequency range. The number of modulators can range from 2 to 12, generating maximum sound pressure levels from 170 to over 180 dB with frequency ranges varying from 30–500 Hz to 50–1500 Hz.[9,10,38] Test panel sizes range from 1 to 20 ft^2 (0.1 to 2 m^2).

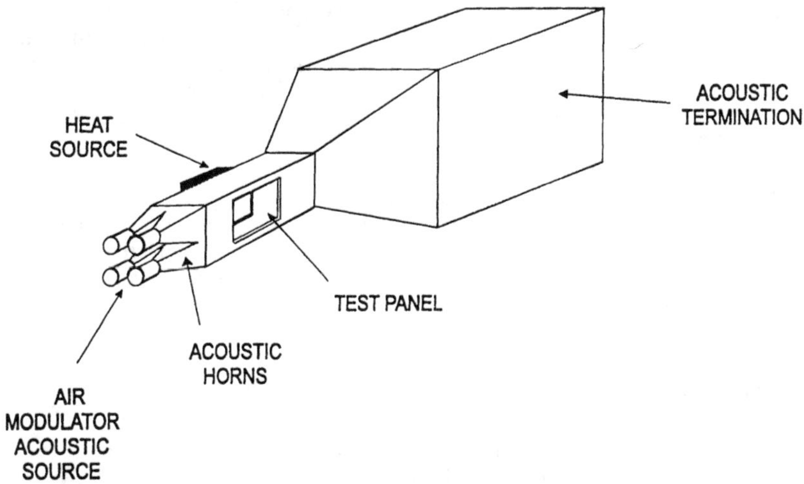

FIGURE 32.8 Typical progressive wave tube. (*Shimovetz and Wentz.*[10])

REVERBERATION CHAMBERS

Reverberation chambers can be used to expose large structures to sound pressure levels typical of those encountered in service. A reverberation chamber is an enclosure with thick, rigid walls and smooth interior surfaces that strongly reflect sound waves.[40] Acoustic noise is introduced into the chamber at one or more locations, usually with air modulators mounted in one or more of the walls. Assuming that the acoustic noise source is random in character, it produces a sound field within the chamber that becomes increasingly homogeneous (a uniform sound pressure level throughout the chamber) as the wavelength of the sound becomes small relative to the minimum dimension of the chamber. Further, the sound field inside the chamber approaches a diffuse noise field, where *diffuse noise* is defined as a sound field in which the sound waves at any point arrive from all directions with equal intensity and random phase. High-intensity reverberation chambers typically have an interior volume of 7000 to 350,000 ft^3 (200 to 10,000 m^3), and are capable of producing sound pressure levels in an empty chamber of 150 to 160 dB over a frequency range from 0.1 to 10 kHz.[41]

The vibration response of a test item to the acoustic excitation in a reverberation chamber can be measured by suspending the test item near the middle of the chamber, applying acoustic excitation with the desired level and spectrum, and measuring the vibration response of the test item at all locations of interest. However, it must be remembered that the spatial pressure cross spectrum for the sound field in a reverberation chamber may be quite different from that for the sound field in the actual service environment of the test item. Specifically, as mentioned earlier, the sound field in a reverberation chamber with a random acoustic source will closely approximate a diffuse noise field, which has a normalized spatial cross spectrum between any two points given by[14]

$$\gamma(\xi,\omega) = \frac{\sin (k\xi)}{k\xi} \tag{32.30}$$

where k is the wave number of the pressure field defined in Eq. (32.4), and ξ is the separation distance. The normalized pressure cross spectrum given by Eq. (32.30) is different from that for the sound field produced by jet noise or a turbulent boundary layer, as given by Eq. (32.3) or (32.8), respectively. The cross-acceptance function, which describes the coupling between the sound field and a structure and is defined by Eq. (32.10), also will be different for the different cases. It follows that the vibration response of a structure tested in a reverberant chamber can differ significantly from that occurring in the service environment.

The maximum sound pressure levels achievable in a reverberation chamber are not as high as those in a progressive wave tube, but reverberant chambers can accommodate larger structures. Thus, the two environments are usually used for different types of tests.

REFERENCES

1. Hubbard, H. H. (ed.): "Aeroacoustics of Flight Vehicles: Theory and Practice," Acoustical Society of America, Woodbury, N.Y., 1994.
2. Ungar, E. E., J. F. Wilby, and D. B. Bliss: "A Guide for Estimation of Aeroacoustic Loads on Flight Vehicle Surfaces," *AFFDL-TR-76-91,* February 1977.
3. Eldred, K. M.: "Acoustic Loads Generated by the Propulsion System," *NASA SP-8072,* June 1971.
4. Duan, C., and S. A. McInerny: "Lift-Off Loads Methods: A Comparison of Predictions with Measured Data for Covered Pad," *Proc. 8th International Conference on Structure Borne Sound and Vibration,* Hong Kong, 2001.
5. Hubbard, H. H., and J. C. Houbolt: "Vibration Induced by Acoustic Waves," Chap. 48, in C. M. Harris and C. E. Crede, eds., "Shock and Vibration Handbook," 1st ed., McGraw-Hill Book Company, New York, 1961.
6. Clarkson, B. L.: "Effects of High Intensity Sound on Structures," Chap. 70, in M. J. Crocker, ed., "Encyclopedia of Acoustics," John Wiley & Sons, New York, 1997.
7. Anon., "ESDU Engineering Data: Acoustic Fatigue Series," Vols. 1–7, ESDU International, London, 2000.
8. Trapp, W. J., and D. M. Forney, Jr., eds.: "Acoustical Fatigue in Aerospace Structures," Syracuse University Press, Syracuse, N.Y., 1965.
9. Richards, E. J., and D. J. Mead, eds.: "Noise and Acoustic Fatigue in Aeronautics," John Wiley & Sons, London, England, 1968.
10. Shimovetz, R. M., and K. R. Wentz: *AIAA Paper CEAS/AIAA-95-142,* 1995.
11. Blevins, R. D., I. Holehouse, and K. R. Wentz: *Journal of Aircraft,* **30:**971 (1993).
12. Blevins, R. D.: *Journal of Sound and Vibration,* **129:**51 (1989).
13. Simpson, M. A., P. M. Druez, A. J. Kimbrough, M. P. Brock, P. L. Burge, G. P. Mathur, M. R. Cannon, and B. N. Tran: "UHB Demonstrator Interior Noise Control Flight Tests and Analysis," *NASA Contractor Report 181897,* October 1989.
14. Bendat, J. S., and A. G. Piersol: "Engineering Applications of Correlation and Spectral Analysis," 2d ed., John Wiley & Sons, New York, 1993.
15. Landmann, A. E., H. F. Tillema, and S. E. Marshall: "Evaluation of Analysis Techniques for Low-Frequency Interior Noise and Vibration of Commercial Aircraft," *NASA Contractor Report 181851,* October 1989.
16. Bull, M. K.: *Journal of Fluid Mechanics,* **28:**719 (1967).
17. Blake, W. K.: "Mechanics of Flow-Induced Sound and Vibration," Academic Press, Orlando, Fla., 1986.

18. Coe, C. F., W. J. Chyu, and J. B. Dods, Jr.: *AIAA Paper 73-996* (1973).
19. Laganelli, A. L., and H. F. Wolfe: *Journal of Aircraft,* **30:**962 (1993).
20. Cockburn, J. A., and A. C. Jolly: "Structural-Acoustic Response, Noise Transmission Losses and Interior Noise Levels of an Aircraft Fuselage Excited by Random Pressure Fields," *AFFDL-TR-68-2,* August 1968.
21. Plotkin, K. J.: *Journal of Acoustical Society of America,* **111:**530 (2002).
22. Fahy, F. J.: "Sound and Structural Vibration," Academic Press, London, England, 1985.
23. Ross, D.: "Mechanics of Underwater Noise," Peninsula Publishing, Los Altos, Calif., 1987.
24. Langley, R. S., and V. Cotoni: *Journal of Acoustical Society of America,* **115:**706 (2004).
25. Hipol, P. J., and A. G. Piersol: *SAE Paper 871740* (1987).
26. Cotoni, V., P. Shorter, and R. Langley: *Journal of Acoustical Society of America,* **122:**259 (2007).
27. Soovere, J., and M. L. Drake: "Aerospace Structures Technology Damping Design Guide," *AFWAL-TR-84-3089,* December 1985.
28. Ungar, E. E.: "Vibration Isolation and Damping," Chap. 71, in M. J. Crocker, ed., "Encyclopedia of Acoustics," John Wiley & Sons, New York, 1997.
29. Mei, C.: "Large Amplitude Response of Complex Structures Due to High Intensity Noise," *AFFDL-TR-79-3028,* April 1979.
30. Mei, C., and R. R. Chen: "Finite Element Nonlinear Random Response of Composite Panels of Arbitrary Shape to Acoustic and Thermal Loads," *WL-TR-1997-3085,* October 1997.
31. Wolfe, H. F., C. A. Shroyer, D. L. Brown, and L. W. Simmons: "An Experimental Investigation of Nonlinear Behaviour of Beams and Plates Excited to High Levels of Dynamic Response," *WL-TR-96-3057,* October 1995.
32. Ng, C. F.: *Journal of Aircraft,* **26:**281 (1989).
33. Vaicaitis, R.: *Journal of Aircraft,* **31:**10 (1994).
34. Prasad, C. B., and C. Mei: *AIAA Paper AIAA-87-2712* (1987).
35. Clarkson, B. L.: "Review of Sonic Fatigue Technology," *NASA Contractor Report 4587,* April 1994.
36. Ungar, E. E.: "Estimation of Upper Bounds to Stresses Induced by Sound," *Proc. Noise-Con 2007,* Reno, Nev., 2007.
37. Mead, D. J.: "Passive Vibration Control," John Wiley & Sons, Chichester, England, 2000.
38. Steinwolf, A., R. G. White, and H. F. Wolfe: *Journal of Acoustical Society of America,* **109:**1043 (2001).
39. Xiao, Y., R. G. White, and G. S. Aglietti: *Journal of Acoustical Society of America,* **117:**2820 (2005).
40. Hodgson, M., and A. C. C. Warnock: "Noise in Rooms," Chap. 7, in L. L. Beranek and I. L. Ver, eds., "Noise and Vibration Control Engineering," John Wiley & Sons, New York, 1992.
41. Lee, Y. A., and A. L. Lee: "High Intensity Acoustic Tests," *IES-RP-DTE040.1,* Institute of Environmental Sciences and Technology, Mount Prospect, Ill., 2000.

CHAPTER 33
ENGINEERING PROPERTIES OF METALS

M. R. Mitchell

INTRODUCTION

In this chapter it is the intent to describe several of the material properties that should be considered when designing components, structures, and equipment in order to withstand shock and vibration. Due to space limitations, it is necessary to merely introduce some of the material properties of concern. Several textbooks on this subject contain far more detailed explanations.[1–4]

As with any engineering design, it is essential to determine the forces (stresses) and/or displacements (strains) that a typical component may be required to resist in actual application while in service, as is described in Chap. 40. In this chapter, we adapt results for these analyses for the selection of metals based on properties such as monotonic stress-strain and stress-based and strain-based cyclic fatigue behavior, as well as catastrophic failure where fracture mechanics properties are most important.

Metals may deform slightly and spring back to their rest positions or physically change in dimensions or shape sufficiently to result in their loss of functionality. Conversely, metals may suddenly "crack" or fracture and separate into two distinct pieces as the result of a catastrophic event. In the first instance, the deformation may appear initially to be that of a common spring, where force (stress) and displacement (strain) are proportional within the so-called elastic limit, and the modulus of elasticity (Young's modulus) dictates the deformation response. Such metal behavior is common in structures like television/radio/microwave towers and skyscrapers, where we can actually see visible sway caused by wind forces. In the second instance, where an unrecoverable dimensional change takes place, the metal is said to be plastically deforming or yielding in a ductile manner. Often such permanent deformation is not acceptable, as with closely fitted components that might interfere. In other cases, some ductile deformation may be acceptable, such as in a bending or torsional component where the outer fibers of the metal may be prestrained a small percentage. In this latter case, the bulk of the core metal remains elastic and will restore the component to its original shape. If deformation is continued beyond the yield point of a ductile metal, work hardening occurs and the stress required to enforce continued plastic deformation will increase, as will the metal's hardness. Eventually, the metal will attain its ultimate strength, at which point physical *necking* occurs, with the formation of small microvoids at the interior of the metal due to resultant stress triaxiality. Some components may be designed to rely on an incremental magnitude of plastic deformation between yield and ultimate strength. As such, their functionality depends on the ability of that component to accommodate such minor dimensional adjustments. If dimensional stability is important, only small-percentage increments are permissible. In many structural applications, significant deformation may be tolerable and can reach as much as 50 percent! In such instances, the ultimate strength of the material is often employed in the design.

In the design of structures and components subject to vibrational or repeated forces/displacements (stresses/strains), a more detailed examination of the force/displacement time histories is required. With such knowledge, coupled with a proper mechanics analysis, we can evaluate the lifetime or durability of the device. For this, there are three types of analyses in common usage today: *stress-life* fatigue, *strain-life* fatigue, and *fracture mechanics* methodologies. Which type of analysis is employed in design depends upon primarily the type of metal (and the thermal-mechanical processing), the force or displacement time history the component or structure is required to resist, and the environment in which it must survive. A basic understanding of each of the three analyses is important to engineers in order that a state-of-the-art, educated assessment be made as to which of the three, or whether a combination of them, is necessary for specific applications and designs.

STRESS-STRAIN PROPERTIES

MONOTONIC PROPERTIES

Often referred to as *tensile properties* of metals, the monotonic properties include yield strength, ultimate strength, elongation (or engineering strain) at fracture, and the reduction in area. ASTM Standard E8[5] or ISO 6892[6] is often employed for the procedures in performing tension testing of metals. The rate of deformation (i.e., the crosshead rate), specimen design, and procedures for determination of these properties are provided in the standards. As might be expected, the rate of deformation does have a pronounced effect on the yield and ultimate strength, as well as other mechanical properties, of metals. Such dynamic properties are not standardized easily, and readers are referred to references[7] for detailed descriptions of these influences.

The standard tension test for common mild steel is exemplified by an engineering stress-strain diagram, as illustrated in Fig. 33.1.* Typically, a standard test specimen will have a cylindrical gage section with an initial or original gage length l_o and initial or original gage diameter d_o. As the test specimen is gripped in the test machine and pulled in tension, it will begin to deform, as shown in Fig. 33.1. What is shown in this figure is the *engineering* stress–*engineering* strain as well as what is

---

* Not all metals deform in this fashion and exhibit a Luder's plateau—that portion of the stress-strain curve in which strain increases but stress remains essentially constant, as for many mild steel alloys and superelastic nitinol. Most aluminum and titanium alloys, for example, simply deviate from elastic response by the immediate onset of plastic deformation or work hardening.

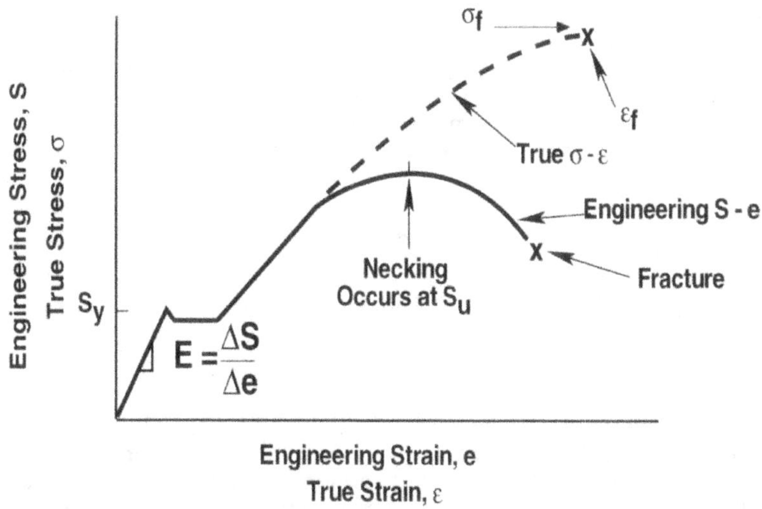

FIGURE 33.1 Engineering stress-strain and true stress-strain curve.

known as the *true* stress–*true* strain curve, along with several of the commonly known tension properties. Within the *elastic limit* up to the point of initial yielding at S_y, the metal will respond to the deformation elastically. Within this limit, we define Young's modulus or the commonly known modulus of elasticity E, where stress and strain are directly proportional, as given by Eq. (33.1).

$$E = \frac{\Delta S}{\Delta e} \tag{33.1}$$

The engineering stress S is defined as

$$S = \frac{P}{A_o} \tag{33.2}$$

where P = force and A_o = the original area of the cross section, and the engineering strain e is defined as

$$e = \frac{l - l_o}{l_o} \tag{33.3}$$

where l = the instantaneous gage length.

As illustrated in Fig. 33.1 for a mild steel, there is an upper yield strength (the peak stress after elastic response) and a lower yield point (that portion on the curve just after the upper yield strength—the Luder's plateau stress). However, the more commonly reported engineering property is the 0.2 percent offset yield strength, or the stress corresponding to a plastic strain of 0.2 percent, that is, that permanent strain of 0.002 produced on unloading. This stress point can be found by simply drawing a line parallel to the modulus of elasticity and determining the intersection with the stress-strain curve at 0.002 strain.

After yielding has occurred, larger degrees of plastic deformation take place at a reduced modulus (i.e., tangents to the engineering stress–engineering strain curve), often referred to as the *strain-hardening modulus,* that decreases as strain increases. Because of the change in the cross-sectional area as the specimen extends beyond the elastic limit and plastic deformation occurs up to and beyond necking, we can define true stress and true strain to account for such a response. The true stress σ is given as

$$\sigma = \frac{P}{A} \tag{33.4}$$

where A = the instantaneous area and the true strain ε is given as

$$\varepsilon = \int_{l_o}^{l} \left[\frac{dl}{l}\right] = \ln \frac{l}{l_o} \tag{33.5}$$

The use of true stress and true strain merely changes the overall shape of the stress-strain curve to a monotonically increasing functional relationship, shown as the dashed curve in Fig. 33.1. Such a description permits a way to describe algebraically the entire curve and provide a constitutive relationship ε between stress and total strain as

$$\varepsilon = \varepsilon_e + \varepsilon_p \tag{33.6}$$

where ε_e = elastic strain = ε/E, and ε_p = plastic strain. We can also now describe the plastic strain as

$$\sigma = K(\varepsilon_p)^n \tag{33.7}$$

where K = the monotonic strength coefficient and n = the monotonic strain hardening exponent. Both n and K may be obtained from log-log linearization of the preceding power law equation and determining the slope (n) and intercept (K) at a plastic strain of unity.

With the rearrangement of Eq. (33.7), we can now describe the entire stress-strain curve as follows

$$\varepsilon = \frac{\sigma}{E} + \left(\frac{\sigma}{K}\right)^{1/n} \tag{33.8}$$

There are several references listing all of these monotonic properties for a variety of engineering materials,[8,9] including the values of n and K as well as the true fracture strength* σ_f and true fracture ductility ε_f given as

$$\sigma_f = \frac{P_f}{A_f} \tag{33.9}$$

and

$$\varepsilon_f = \ln\left(\frac{A_o}{A_f}\right) = 2\ln\left(\frac{d_o}{d_f}\right) = \ln\left(\frac{1}{1-\mathrm{RA}}\right) \tag{33.10}$$

---

* The formation of a neck in a test specimen creates a complex state of triaxial stress. In ductile metals, the true fracture stress σ_f requires correction with a Bridgeman correction factor as a function of the true strain at fracture.[2]

where A_f = the area at fracture, d_f = the diameter at fracture, and RA is the reduction of area given as

$$\mathrm{RA} = \frac{A_o - A_f}{A_o} \tag{33.11}$$

The significance of the true stress and true strain will become more obvious later, when we describe the stress-life and strain-life fatigue behavior of metals. For convenience, monotonic stress-strain properties of several steels and aluminum alloys are listed in Tables 33.1 and 33.2, respectively.[8] Also shown are cyclic properties that will be explained subsequently. Additional information of this nature may be found at http://fde.uwaterloo.ca/Fde/Materials/dindex.html. Although this is a website, it is

TABLE 33.1 Monotonic and Cyclic Properties of Several Steels

		Monotonic							
Alloy	Condition	Ex10^3, Ksi (GPa)	S_y, Ksi (MPa)	S_u, Ksi (MPa)	K, Ksi (MPa)	n	%RA	σ_f, Ksi (MPa)	ε_f
A136	As rec'd.	30 (207)	46.5 (321)	30.6 (211)	144.0 (990)	0.21	67	143.6 (987)	1.06
A136	150 HB	30	46.0 (317)	31.9 (220)	—	0.21	60	145.0 (997)	1.19
SAE950X	137 HB	30	62.6 (432)	75.8 (523)	94.9 (652)	0.11	54	—	—
SAE950X	146 HB	30	56.7 (391)	74.0 (510)	110.0 (756)	0.15	74	141.8 (975)	1.34
SAE980X	225 HB	30	83.5 (576)	100.8 (695)	143.9 (989)	0.13	68	176.8 (1216)	1.15
1006	Hot-rolled 85HB	27 (186)	36.0 (248)	46.1 (318)	60.0 (413)	0.14	73	—	—
1020	Ann. 108 HB	29 (200)	36.8 (254)	56.9 (392)	57.9 (398)	0.07	64	95.9 (659)	1.02
1045	225 HB Q&T	29	74.8 (516)	108.9 (751)	151.8 (1044)	0.12	44	144.7 (995)	—
1045	390 HB Q&T	29	184.8 (1274)	194.8 (1343)	—	0.04	59	269.8 (1855)	0.89
1045	500 HB Q&T	29	250.6 (1728)	283.7 (1956)	341.0 (2344)	0.04	38	334.4 (2300)	—
1045	705 HB Q&T	29	264.7 (1825)	299.8 (2067)	—	0.19	2	309.6 (2129)	0.02
10B21	320 HB Q&T	29	144.0 (993)	152.0 (1048)	187.7 (290)	0.05	67	217.4 (1495)	1.13
1080	421 HB Q&T	30	141.8 (998)	195.6 (1349)	323.0 (2221)	0.15	32	338.6 (2328)	—
4340	350 HB Q&T	29	170.8 (1178)	179.8 (1240)	229.2 (1576)	0.07	57	239.7 (1648)	0.84
4340	410 HB Q&T	30	198.8 (1371)	212.8 (1467)	—	—	38	225.8 (1552)	0.48

TABLE 33.1 Monotonic and Cyclic Properties of Several Steels (*Continued*)

Monotonic									
Alloy	Condition	Ex10^3, Ksi (GPa)	S_y, Ksi (MPa)	S_u, Ksi (MPa)	K, Ksi (MPa)	n	%RA	σ_f, Ksi (MPa)	ε_f
5160	440 HB Q&T	30	215.7 (1487)	230.0 (1581)	281.4 (1935)	0.05	39	280.0 (1925)	0.51
8630	254 HB Q&T	30	102.8 (709)	118.9 (817)	158.9 (1092)	0.08	16	121.8 (837)	0.17

Cyclic								
Alloy	Condition	$S_{y'}$, Ksi (MPa)	K, Ksi (MPa)	n	$\sigma_{f'}$, Ksi (MPa)	b	$\varepsilon_{f'}$	c
A136	As rec'd.	47.9 (329)	148.8 (1023)	0.18	115.9 (797)	–0.09	0.22	–0.46
A136	150 HB	48.0 (330)	167.0 (1148)	0.20	122.7 (844)	–0.08	0.20	–0.42
SAE950X	137 HB	51.2 (352)	138.8 (954)	0.16	112.0 (822)	–0.08	0.34	–0.52
SAE950X	146 HB	59.3 (408)	136.2 (936)	0.13	119.5 (822)	–0.08	0.42	–0.51
SAE980X	225 HB	82.5 (567)	385.5 (2650)	0.25	171.8 (1181)	–0.10	0.09	–0.48
1006	Hot-rolled 85 HB	34.2 (235)	196.0 (1348)	0.28	116.3 (800)	–0.12	0.48	–0.52
1020	Ann. 108 HB	33.8 (232)	174.0 (1196)	0.26	123.3 (848)	–0.12	0.44	–0.51
1045	225 HB Q&T	58.3 (401)	179.8 (1236)	0.17	139.2 (957)	–0.08	0.50	–0.52
1045	390 HB Q&T	122.1 (839)	216.4 (1488)	0.09	204.2 (1404)	–0.07	1.51	–0.85
1045	500 HB Q&T	189.0 (130)	672.1 (4621)	0.20	418.9 (2880)	–0.09	0.23	–0.56
1045	705 HB Q&T	327.0 (2248)	613.2 (4216)	0.10	350.4 (2409)	–0.07	0.002	–0.47
10B21	320 HB Q&T	100.2 (689)	143.6 (987)	0.06	150.3 (1033)	–0.04	1.33	–0.85
1080	421 HB Q&T	126.2 (868)	460.8 (3168)	0.21	342.9 (2357)	–0.10	0.51	–0.59
4340	350 HB Q&T	115.6 (795)	270.2 (1858)	0.14	282.0 (1939)	–0.10	1.22	–0.73
4340	410 HB Q&T	127.0 (873)	282.8 (1944)	0.13	275.3 (1893)	–0.09	0.67	–0.64
5160	440 HB Q&T	155.2 (1067)	352.7 (2425)	0.13	300.0 (2063)	–0.08	0.56	–1.05
8630	254 HB Q&T	87.5 (602)	139.4 (958)	0.08	152.1 (1046)	–0.11	0.21	–0.86

Courtesy of L. E. Tucker, Deere & Co, Moline, Ill

TABLE 33.2 Monotonic and Cyclic Properties of Several Aluminum Alloys

Monotonic

Alloy	Condition	Ex10^3, Ksi (GPa)	S_y, Ksi (MPa)	S_u, Ksi (MPa)	K, Ksi (MPa)	n	%RA	σ_f, Ksi (MPa)	ε_f
1100	As rec'd.	10.0 (68.8)	14 (386)	16 (110)	—	—	88	—	2.1
2014	T6	10.6 (72.9)	67 (462)	74 (510)	—	—	35	91 (627)	0.42
2014	T6	10.8 (74.3)	70 (483)	78 (538)	—	—	—	—	—
2024	T351	10.2 (70.1)	44 (303)	69 (476)	117 (807)	0.20	35	92 (634)	0.38
2024	T4	10.6 (72.9)	T/C 55/44 (379/303)	68 (469)	T/C 66/92 (455/634)	T/C 0.32/0.17	25	81 (558)	0.43
2219	T851	10.3 (70.8)	52 (359)	68 (469)	—	—	—	—	0.28
5086	F	10.1 (69.4)	30 (207)	45 (310)	—	—	—	—	0.36
5186	O	10.5 (72.2)	L/T 16/19 (110/131)	L/T 44/49 (303/338)	—	—	L/T 37/44	57 (393)	L/0.46 T/0.58
5454	O	10.0 (68.8)	20 (138)	36 (248)	—	—	44	53 (365)	0.58
5454	10%CR	10.0 (68.8)	—	—	—	—	—	—	—
5454	20%CR	10.0 (68.8)	—	—	—	—	—	—	—
5456	H311	10.0 (68.8)	34 (234)	58 (400)	—	—	35	76 (524)	0.42
6061	T651	10.0 (68.8)	42 (290)	45 (310)	53 (365)	0.042	58	68 (469)	0.86
7075	T6	10.3 (70.8)	68 (469)	84 (579)	120 (827)	0.113	33	108 (745)	0.41
7075	T73	10.4 (71.5)	60 (414)	70 (483)	86 (593)	0.054	23	84 (579)	0.26

Cyclic

Alloy	Condition	S_y, Ksi (MPa)	K, Ksi (MPa)	n	$\sigma_{f'}$, Ksi (MPa)	b	$\varepsilon_{f'}$	c
1100	As rec'd.	8 (55)	23 (159)	0.17	28 (193)	–0.106	1.80	–0.69
2014	T6	65 (448)	102 (703)	0.073	114 (786)	–0.081	0.85	–0.86
2014	T6	73 (503)	107 (738)	0.062	129 (889)	–0.092	0.37	–0.74

TABLE 33.2 Monotonic and Cyclic Properties of Several Aluminum Alloys (*Continued*)

		Cyclic						
Alloy	Condition	S_y, Ksi (MPa)	K, Ksi (MPa)	n	$\sigma_{f'}$, Ksi (MPa)	b	$\varepsilon_{f'}$	c
2024	T351	65 (448)	114 (786)	0.090	147 (1014)	–0.110	0.21	–0.52
2024	T4	62 (427)	95 (655)	0.065	160 (1103)	–0.124	0.22	–0.59
2219	T851	48 (331)	115 (793)	0.140	121 (834)	–0.110	1.33	–0.08
5086	F	43 (296)	87 (600)	0.011	83 (572)	–0.092	0.69	–0.75
5186	O	43 (296)	68 (469)	0.075	122 (841)	–0.137	1.76	–0.92
5454	O	34 (234)	58 (400)	0.084	82 (565)	–0.116	1.78	–0.85
5454	10%CR	34 (234)	62 (427)	0.098	82 (565)	–0.108	0.48	–0.67
5454	20%CR	37 (255)	59 (407)	0.081	82 (565)	–0.103	1.75	–0.80
5456	H311	51 (352)	87 (600)	0.086	105 (724)	–0.110	0.46	–0.67
6061	T651	43 (296)	78 (538)	0.096	92 (634)	–0.099	0.92	–0.78
7075	T6	75 (517)	140 (965)	0.010	191 (1317)	–0.126	0.19	–0.52
7075	T73	58 (400)	74 (510)	0.032	116 (800)	–0.098	0.26	–0.73

Courtesy of Professor R. W. Landgraf, Virginia Polytechnic and State University, Blacksburg, Va.

used as a reference (as are several other sites in this section) because it is constantly changing, with the addition of new data. It is updated by the University of Waterloo on a regular basis.

Similar relationships and equations exist for the torsional deformation of metals.[1,2] It is sufficient to point out here that monotonic properties may be employed as an indicator of a metal's fatigue and fracture mechanics behavior.

TEMPERATURE AND STRAIN RATE EFFECTS

As might be anticipated, the monotonic properties of metals are affected by temperature. In general, the greater the test temperature, the less the yield strength, ultimate strength, and modulus of elasticity but the greater the ductility. Conversely, the lower the temperature, the greater the opposite trends that occur. The yield strength of a structural steel, for example, is approximately 90 percent of the ambient-temperature value when determined at 400°F (~200°C), 60 percent at 800°F (~430°C), 50 percent at 1000°F (~540°C), 20 percent at 1300°F (~700°C), and 10 percent at 1600°F (~870°C). The ultimate strength at 400°F is 100 percent that at ambient temperature, 85 percent at 800°F, 50 percent at 1000°F, 15 percent at 1300°F, and 10 percent at 1600°F. Changes in the modulus of elasticity are 95 percent of the ambient value when determined at 1600°F, 85 percent at 800°F, 80 percent at 1000°F, 70 percent at 1300°F, and 50 percent at 1600°F. Of course, the ductility is increased significantly as these strength-related properties decrease.

If a metal is tested or used in an application at a temperature that is ~0.3 to 0.5 of its melting point, creep mechanisms become active and significant plastic/inelastic deformations occur. This could be the case in oil refineries, chemical plants, and, certainly, many gas-turbine and rocket applications, where temperatures often exceed 1650°F (~900°C). Of course, specialty nickel-, cobalt-, and titanium-based alloys are employed in many of these types of applications. What is also of importance in creep is the time element of exposure to the elevated temperatures, since creep mechanisms are both *time* and *temperature* dependent. A creep test measures the dimensional change occurring with time from the elevated exposure, whereas a creep-rupture test measures the effect of temperature on the extended-time force-bearing characteristics of the metal. Even at ambient temperatures, creep strains can become active and significant, such as with lead-based alloys as well as some lead-free solders.

The influence of temperature on metal properties is perhaps most dramatically recognized by the results of impact testing of steels that are body-centered cubic structures* where there is a ductile-brittle transition with decreasing temperature. At high strain rates or impact velocities, such as those employed in the ASTM E23 and ISO 83 (Charpy-type) and ASTM D256 and ISO 180 (Izod-type) pendulum impact test, there is a significant decrease in the impact energy absorbed by ferrous-based metals as the temperature is decreased. See, for example, Fig. 33.2, where

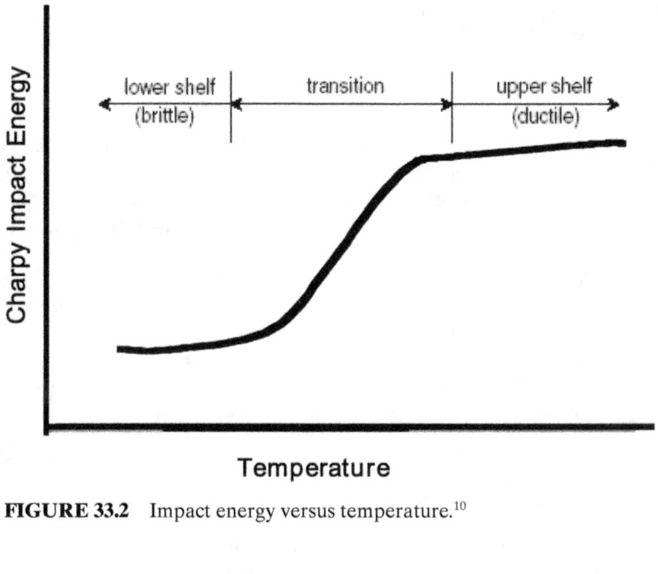

FIGURE 33.2 Impact energy versus temperature.[10]

---

* Aluminum alloys are face-centered cubic structures and do not have brittle-to-ductile transitions with decreasing temperatures.

there is a relatively high impact energy absorbed by the steel at higher temperatures (upper shelf), while as the temperature is decreased there is a transition to lower energies (lower shelf).[10]

Even with tensile/compression testing done at ambient temperature, there is a significant effect of strain rate on yield strength, as is well known, but there is also an effect on the ultimate strength with increasing strain rates. As shown in Fig. 33.3,[11] as the strain rate is increased, there is an obvious increase in the ultimate strength, even if the temperature is increased.

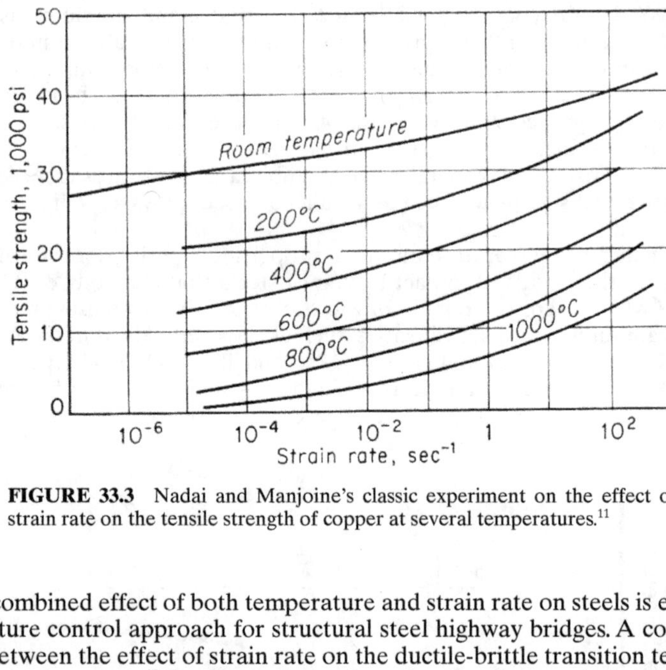

FIGURE 33.3 Nadai and Manjoine's classic experiment on the effect of strain rate on the tensile strength of copper at several temperatures.[11]

The combined effect of both temperature and strain rate on steels is exemplified in a fracture control approach for structural steel highway bridges. A correlation is shown between the effect of strain rate on the ductile-brittle transition temperature and the yield strength.[12] Results indicate that there is a linear relationship between ΔT_{db} and $\Delta l\sigma_y$, where ΔT_{db} is the transition temperature shift, σ_y is the yield strength, and Δ is the change in yield strength as caused by the strain rate.

As might be anticipated, the fracture ductility of steels is also influenced by the strain rate. As pointed out by Qiu, et al.,[12] there is a definite decrease in fracture ductility (and elongation) of structural steels as the strain rate is increased in conventional tension testing.

In any case, ambient temperature and conventional strain rate data used in many designs might be considered somewhat conservative—but not overly so!

TOUGHNESS AND DUCTILITY

The area under a monotonic true stress-strain is in the form of energy per unit volume or a measure of the *toughness* of the metal. By using the true stress and true strain characteristics of a metal, the nonuniformity of strain resulting from the reduction in area upon necking is taken into account.[13] As a rough approximation of the toughness, the product of the true fracture strength and true fracture ductility (i.e., $\sigma_f\varepsilon_f$) is often used. As might be expected, cast metals, such as gray cast iron, possess a much reduced toughness compared with wrought metals—often ½ to ⅓ the values. Typically, tougher metals, such as low- and medium-carbon wrought steels, exhibit far more ductility in contrast to cast metals, which are often considered as *brittle.* If only the elastic energy per unit volume is taken into account—that is, the area under the stress-strain curve to the onset of plastic flow—a *modulus of resilience* may be defined.

CRITICAL STRAIN VELOCITY

When a significant force is rapidly applied to a structure, fracture may occur with a relatively small degree of ductile flow. Such a failure is often interpreted as a *brittle* fracture, and the metal appears to lose its ductility. However, upon examination of the fracture surface, a normal degree of necking occurs in the region near the application of the force. Significant stresses are developed in this region due to the inertial movement of the metal remote from the application of the force, and failure occurs before the plastic stress wavefront is transmitted away from the point of force application. This effect is of importance in applications such as the direct impact of a projectile upon armor plate, where forces are suddenly applied.

FATIGUE

Fatigue, or the failure of a component or structure due to the repeated application of force or displacement, is a critical mode of structural malfunction that must be considered in the design of equipment. There are methodologies that can be employed in design to safeguard against premature failure of structures due to this mechanism. Mechanics techniques are available for fatigue crack *initiation* that are stress based and those that are strain based. Once a fatigue crack has initiated, there are other mechanics techniques employed for fatigue crack *propagation.*

CYCLIC BEHAVIOR OF METALS

Tables 33.1 and 33.2 provide the monotonic properties of several commonly used steel and aluminum alloys. There are other metal properties listed in these tables that are called *cyclic* stress-strain properties. As you will notice, the cyclic properties are different from the monotonic properties. The reason is that metals are metastable under cyclic force or displacement conditions, and the metal's deformation response changes due to the repetition of such forces or displacements. Depending on the initial state of the metal—annealed, quenched and tempered, cold worked, and so on—it may cyclically harden, cyclically soften, remain cyclically stable, or exhibit a mixed softening and then hardening response, depending on strain level. But the monotonic behavior may not be appropriate or adequate to employ in a design required to resist cyclic force or displacements!

This section defines equations similar to those developed for monotonic response that are more appropriate to cyclic application and are called *fatigue properties.* Readers are referred to ASTM E606, Recommended Practice for Strain-Controlled Fatigue Testing, and ISO 12106, Metallic Materials—Fatigue Testing—Axial-Strain-

Controlled Method, for the test methodology involved in performing such evaluations.

Determining the fatigue resistance of a metal was commonly accomplished using a stress- or force-controlled test methodology of smooth (unnotched) specimens such as ASTM E499, Standard Practice for Conducting Force Controlled Constant Amplitude Axial Fatigue Tests of Metallic Materials, or ISO 1099, Metallic Materials—Fatigue Testing—Axial Force-Controlled Method. The familiar σ-log N_f curve was the result, as shown in Fig. 33.4

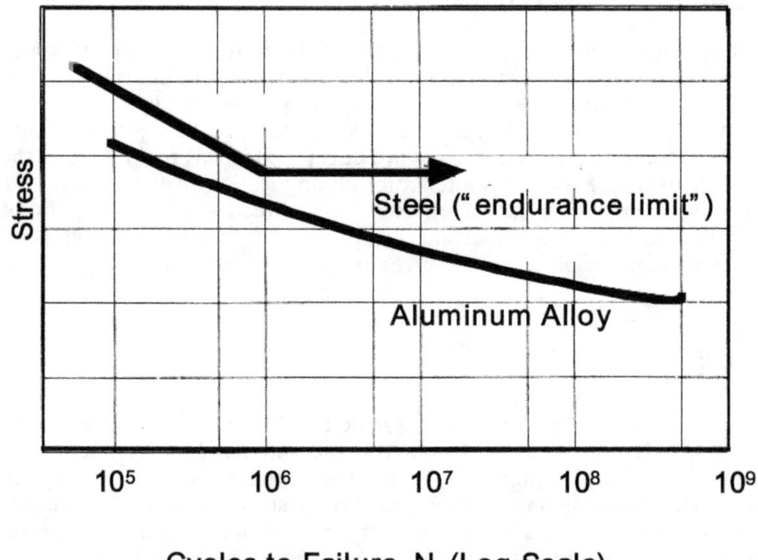

FIGURE 33.4 σ-Log N_f curve for a typical steel and an aluminum alloy.

STRESS-BASED APPROACH

The σ-log N_f or σ-*N* curves* shown in Fig. 33.4 depict a typical steel and an aluminum alloy. The steel is shown exhibiting an endurance limit, while the aluminum alloy is shown to continue to possess a finite life and no such limit, with decreasing stress amplitudes. Research has demonstrated that metals generally do not exhibit an endurance limit per se, that is, a stress below which the metal will endure an infinite number of cycles. Typically, the plateau(s) in stress-life curves are referred to as the conventional *fatigue limit*(*s*) or *endurance limit*(*s*), but failures below these levels do occur[14–16] due to a change in failure mechanism and certainly due to a periodic overstress[17] that inevitably occur in service during actual operation of a component or structure. The antiquated verbiage has given way to the preferred terminology of

* The letter *S* is generally used in other texts as in *S*-*N* or *S*-log N_f. The use of σ here is to avoid confusion in later sections.

fatigue strength or *strain at a particular fatigue life* that has been now included in ASTM and ISO fatigue standards.

Often such σ-*N* curves are obtained by testing multiple, nominally identical, replicate companion specimens subjected to a completely reversed *stress amplitude* σ_a; that is, at a zero mean stress or an *R-ratio* = maximum stress/minimum stress = −1 as a baseline. Additional tests are then conducted at various *mean stresses* σ_o to determine the influence of other-than-zero mean stress on fatigue life, since most components and structures experience variable amplitudes of stress and mean stresses.

Additional parameters of interest in such testing are the *stress range* Δσ and the average of the maximum and minimum stress in the stress range; the mean stress = σ_o. One-half the stress range is the *stress amplitude* σ_a. The expressions for these terms are:

$$\Delta\sigma = \sigma_{max} - \sigma_{min} \tag{33.12}$$

$$\sigma_o = \frac{\sigma_{max} + \sigma_{min}}{2} \tag{33.13}$$

$$\sigma_a = \frac{\Delta\sigma}{2} \tag{33.14}$$

The number of cycles to failure generally reported in the literature depends upon the definition of *failure* that is employed in that particular investigation. Failure can be defined, for example, as the first appearance of a crack that may be observed with an unaided eye or at a particular magnification, a crack of a specified length, or the inability to resist the applied stress (force) without significant crack extension or force relaxation in a constant-amplitude deformation test. There are many such criteria, and caution is suggested when interpreting the literature.

Figure 33.5 is a schematic σ-*N* curve illustrating the division of the crack *initiation* and crack *propagation* events in specimens. As indicated, in the short-life regime at high stress, crack propagation is the dominant mechanism; whereas in the long-life regime at lower stresses, the crack initiation event dominates.

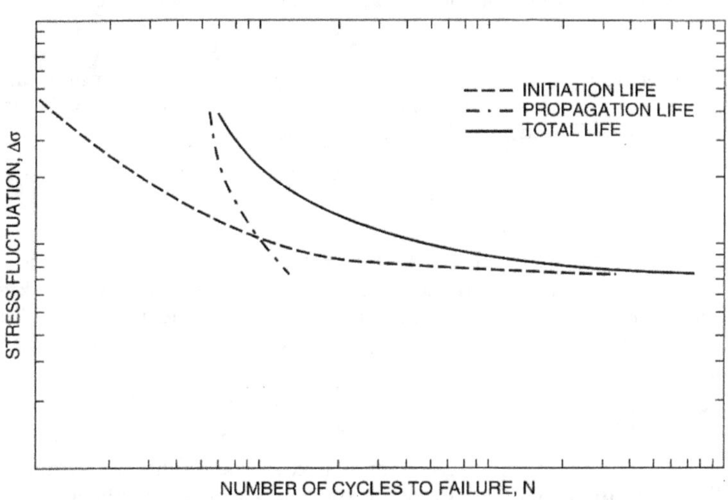

FIGURE 33.5 Division of the total fatigue life into the crack initiation and crack propagation events on a stress-life curve.[18]

As mentioned, the stress-life curve for unnotched specimens is the first procedure often employed for design of structural components. If a notch is present in the design, additional fatigue life results may be obtained on companion notched specimens with a comparable *theoretical stress concentration factor* K_t. The quotient of the fatigue strength of the unnotched specimen to the fatigue strength of the notched specimen at a given life—say, 10^7 cycles—is the *fatigue notch factor:*

$$K_f = \frac{\sigma_{\text{unnotched}}}{\sigma_{\text{notched}}} \text{ at a finite life (i.e., } 10^7) \tag{33.15}$$

Depending on the strength (hardness) of the metal, the full effectiveness of the stress concentration factor in reducing the fatigue strength might not be realized. For example, as shown in Fig. 33.6, for the same notch root radius r, the soft metal has a lesser fatigue notch factor K_f than the hard metal.

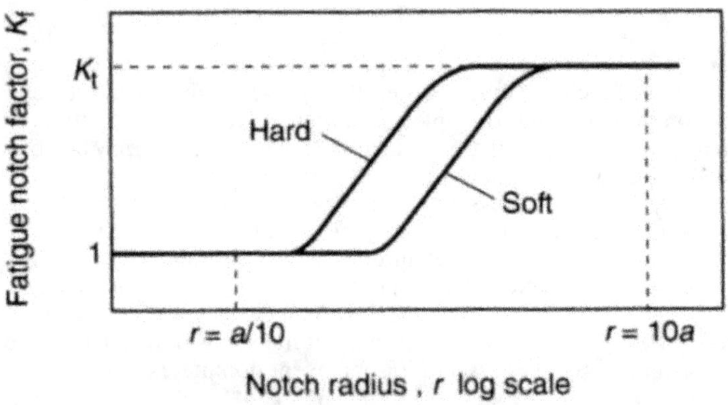

FIGURE 33.6 Fatigue notch factor as a function of notch radius.

There are ways to analytically determine just what the fatigue notch factor may be, depending on the strength of the metal and the geometry of the notch. One of the more popular is attributed to R. E. Peterson[19] and is expressed as

$$K_f = 1 + \frac{K_t - 1}{1 + \dfrac{a}{r}} \tag{33.16}$$

where a is a metal constant dependent on strength and ductility and determined from long-life fatigue results for notched and unnotched specimens of known tip radius and theoretical stress concentration factor, and r is the notch tip radius. However, a can be approximated for steels only by the following empirical relationship:

$$a \cong \left[\frac{300}{S_{\text{ult.}}(Ksi)}\right]^{1.8} \times 10^{-3} \text{ in.} \tag{33.17}$$

Typically, $a \approx 0.01$ for normalized or annealed steels; for highly hardened steels, $a \approx 0.001$, and for quenched and tempered steels, $a \approx 0.25$ in. As indicated in Fig. 33.6, when $r > 10a$, there is a full effectiveness of the notch, while if $r < a/10$, there is little or no notch effect.

As might be anticipated, rotating bending–type fatigue tests subject only a small volume of material at the outer periphery of the specimens to the greatest stress (strain) because of the gradient from surface to neutral axis. Obviously, for larger rotating bending specimens, there is a greater volume of material and the probability of initiating a fatigue crack will be greater. This is a "weakest link" phenomenon, and Ref. 20 is recommended to interested readers.

Additional influences such as surface finish are also important in that there is a more pronounced detrimental influence in higher-strength (-hardness) metals than with lower-strength metals. There is also a more pronounced effect of surface finish at long lives, where strength is the dominant material property, than at shorter lives, where ductility is more important.

Of perhaps more importance than volume effects and surface finish effects are those due to mean stress. In general, compressive mean stresses are beneficial, while tensile mean stresses are detrimental to fatigue life and durability. Mean stress and residual stress are treated similarly in a mechanics sense, although their origins are quite different. Mean stresses are induced by the duty cycle the component is required to resist, while residual stresses are typically induced by surface treatment of the component by such techniques as shot peening, bead blasting, or laser shock hardening.

When dealing with mean stresses in fatigue, some definitions are convenient, as depicted in Fig. 33.7:

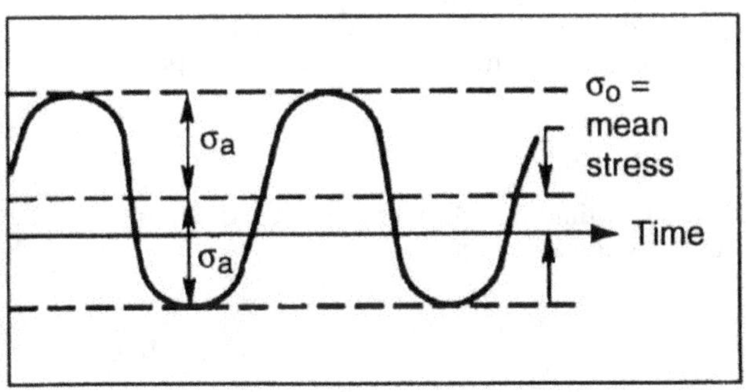

FIGURE 33.7 Depiction of nonzero mean stress cycling.

σ_a = alternating stress or stress amplitude
σ_o = mean stress
σ_{max} = maximum algebraic stress in cycle
σ_{min} = minimum algebraic stress in cycle

The influence of mean stress is often represented by a mean stress–versus–alternating stress diagram, as shown in Fig. 33.8. Such a diagram is often referred to as a constant-life diagram, since each tie line between the alternating and mean stress axes is at a constant life, for example 10^7, 10^6, 10^5 cycles, and so on. Each tie line then represents that combination of a mean stress and an alternating stress that would result in the same fatigue life. If the mean stress is increased, there must be a corresponding decrease in the alternating stress amplitude to achieve the same fatigue lifetime and vice versa.

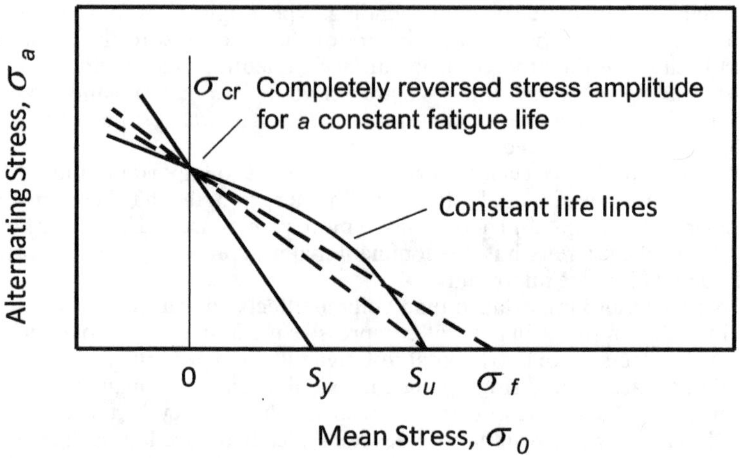

FIGURE 33.8 Depiction of alternating stress versus mean stress diagram or a constant-life diagram.

There are several tie lines illustrated on this constant-life diagram. The line intercepting the ordinate at the yield strength S_y is called the Soderberg relationship (rarely employed in modern designs), while that intercepting the ordinate at the ultimate strength S_u is called the Goodman relationship:

$$\frac{\sigma_a}{\sigma_{cr}} + \frac{\sigma_o}{\sigma_u} = 1 \quad \text{or} \quad \sigma_a = \sigma_{cr}\left[1 - \frac{\sigma_o}{\sigma_u}\right] \quad \text{(Goodman)} \tag{33.18}$$

The curve intercepting the ordinate at the yield strength is called the Gerber relationship and is represented by

$$\frac{\sigma_a}{\sigma_{cr}} + \left[\frac{\sigma_o}{\sigma_u}\right]^2 = 1 \quad \text{or} \quad \sigma_a = \sigma_{cr}\left[1 - \left(\frac{\sigma_o}{\sigma_u}\right)^2\right] \quad \text{(Gerber)} \tag{33.19}$$

The remaining line intercepting the ordinate at the true fracture strength will be used in a later development. In the preceding equations σ_{cr} refers to the completely reversed stress amplitude at $R = -1$ for a specified constant fatigue life (e.g., 10^7, 10^6, 10^5, 10^4, etc.), σ_u is the ultimate strength of the material, σ_o is the mean stress, and σ_a is the alternating stress amplitude. Note that a vertical line would be an R-ratio = −1 (i.e., completely reversed stress-type testing), while a line at 45° between the ordinate and the abscissa would be for equal alternating and mean stresses, or an R-ratio = 0. Goodman's relationship is often used in fatigue analyses for brittle materials and is conservative for ductile metals—it is also the most familiar. Gerber's relationship is generally employed for ductile metals.

STRAIN-BASED APPROACH

An alternate approach to fatigue is called the strain-life approach. It has received significant application since its inception in the early 1950s, primarily due to the pioneering research of Lou Coffin,[21] Stan Manson,[22] and JoDean Morrow,[23] who is widely recognized for his work in low-cycle fatigue and for his pioneering leadership in the development of useful design criteria for mechanical components subjected to fatigue damage. Strain (as opposed to stress) cycling data for a wide variety of metals is readily available[9] at http://fde.uwaterloo.ca/Fde/Materials/dindex.html,[24,25] and such an approach is considered state of the art in the ground vehicle, aerospace, and medical industries.

Metals are unstable when subjected to cyclic loading environments—their stress-strain response will change and will not be the same as their monotonic stress-strain response. Further, fatigue is caused by the accrual of damage to the metal resulting from cyclic *plasticity.* As such, there should be a means of accounting for *plastic* deformation. The strain-life approach offers an advantage over the stress-life approach that is based on simple *elasticity* assumptions and does not effectively include a means of incorporating plastic deformations—the root cause of fatigue.

In the strain-based approach, as in the stress-based approach, a series of smooth companion specimens is subjected to completely reversed, $R = -1$, axial strain-controlled fatigue tests. A series of hysteresis loops is collected from each companion specimen, and the stable response is noted, that is, where the stress required to enforce the strain remains reasonably constant. As illustrated in Fig. 33.9, the stabilized stress response is then plotted against the controlled strain amplitude of each companion specimen tested to obtain the cyclic stress-strain curve that describes the metal's cyclic behavior.

As with the monotonic stress-strain curve, constitutive equations can be employed to describe the cyclic stress-strain response. Similar to Eq. (33.8), the cyclic stress-strain curve can be described by the following:

$$\varepsilon = \frac{\sigma}{E} + \left(\frac{\sigma}{K'}\right)^{1/n'} \tag{33.20}$$

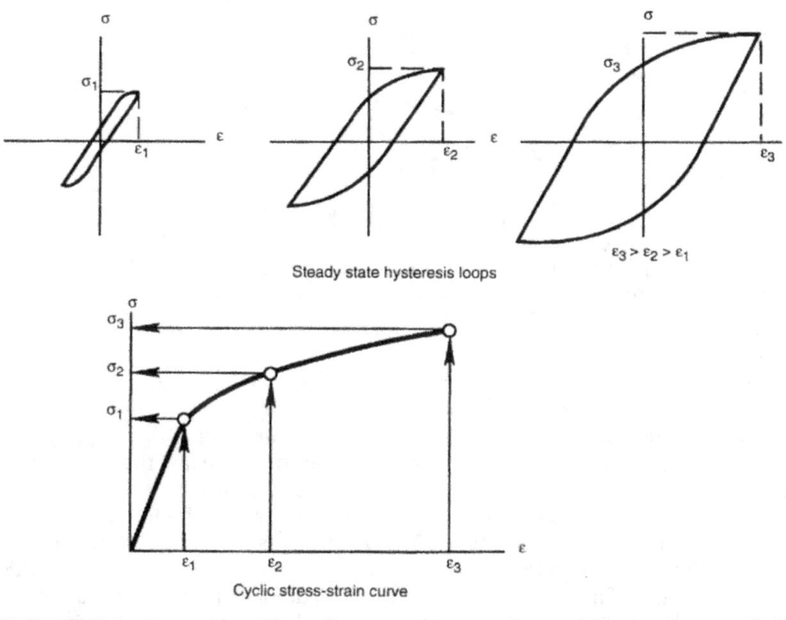

FIGURE 33.9 Generation of the cyclic stress-strain curve from stabilized strain-controlled fatigue tests on companion specimens of a metal.

where K' is the cyclic strength coefficient and n' is the cyclic strain hardening exponent. Values of n' vary between 0.10 and 0.20, with an average close to 0.15.

Along with the development of the cyclic stress-strain curve from each companion specimen test result, the stabilized hysteresis loops can also be employed to determine the elastic and plastic strain components of the total strain (i.e., the controlled variable in the fatigue tests). A typical hysteresis loop is illustrated in Fig. 33.10, with the elastic strain range $\Delta\varepsilon_e$ and the plastic strain range $\Delta\varepsilon_p$, components of the total strain range $\Delta\varepsilon$, shown along with the stabilized stress range $\Delta\sigma$ response.

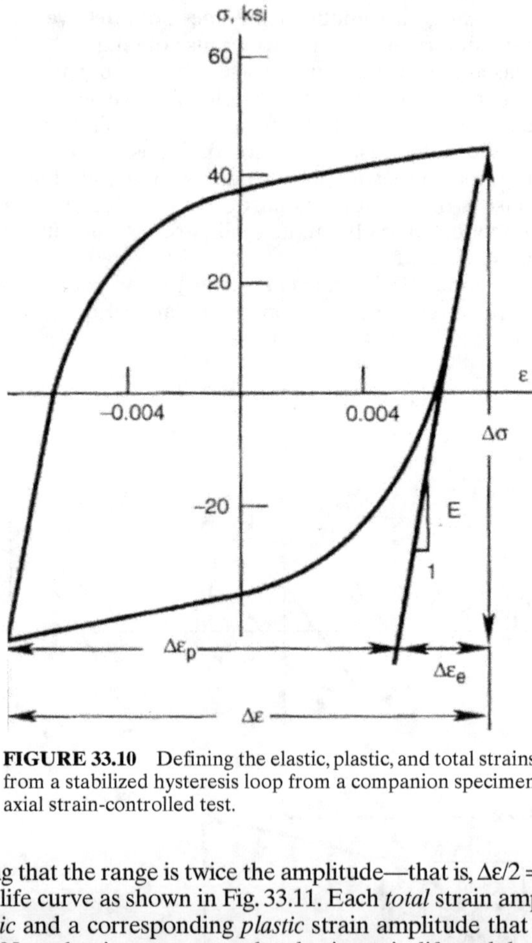

FIGURE 33.10 Defining the elastic, plastic, and total strains from a stabilized hysteresis loop from a companion specimen axial strain-controlled test.

Remembering that the range is twice the amplitude—that is, $\Delta\varepsilon/2 = \varepsilon$—we now construct the strain-life curve as shown in Fig. 33.11. Each *total* strain amplitude has a corresponding *elastic* and a corresponding *plastic* strain amplitude that is plotted on the strain-life curve. Note that in many cases the elastic strain-life and the plastic strain-life points can be connected on the log-log plot with straight lines, as shown. Also note that the abscissa is "reversals to failure," or twice the number of cycles to failure. The reason for this is that in a typical strain-time component history commonly encountered in a real-life application, it is quite simple to define a reversal as opposed to a cycle. A reversal can be thought of as a change in the slope on a deformation (strain) time history.

Note on Fig. 33.11 that the elastic strain-life line can be defined by a slope b,

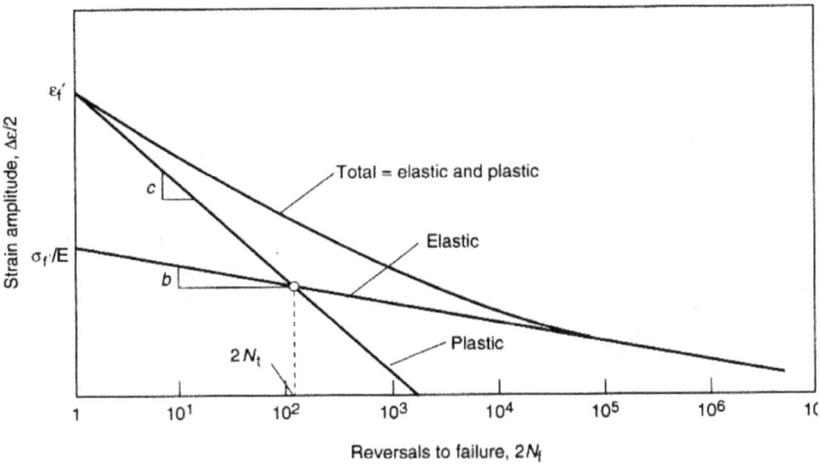

FIGURE 33.11 Construction of a typical strain-life curve (log-log).

called *Basquin's exponent,* and an intercept $\sigma_{f'}/E$, where $\sigma_{f'}$ is called the *fatigue strength coefficient.* The plastic strain-life line is similarly defined with a slope *c,* called *Coffin's exponent,* and an intercept $\varepsilon_{f'}$, called the *fatigue ductility exponent.* Of course, we combine the two components to develop the strain-life equation:

$$\frac{\Delta\varepsilon}{2} = \frac{\sigma_{f'}}{E}(2N_f)^b + \varepsilon_{f'}(2N_f)^c \qquad (33.21)$$

where the first term represents the elastic strain and the second term represents the plastic strain component of the total strain. This relationship commonly applies to wrought metals. Where defects govern fatigue behavior, as with cast irons, such principles are not directly applicable.[26,27] Also note that there is no indication of a limit to the long-life response as with the stress-based methodology. The elastic strain-life line continued its downward slope, even at very long lives.

As mentioned previously, a mean stress can be simply included in this approach by a modification of the elastic strain-life intercept by an amount equal to that of the mean stress. That is:

$$\frac{\Delta\varepsilon}{2} = \frac{\sigma_{f'} - \sigma_o}{E}(2N_f)^b + \varepsilon_{f'}(2N_f)^c \qquad (33.22)$$

where σ_o = mean stress with appropriate sign. Compressive mean stresses are negative and, therefore, additive, so that there would be an increased influence on long-life fatigue and very little on short-life fatigue. In this context, long life is defined as a life greater than the transition fatigue life $2N_t$, or that point on the strain-life fatigue curve where the elastic strain and plastic strain components of total strain are equal. Short life is conversely that lifetime less than $2N_t$.

It is necessary to understand that the sequence of events in a component history is of importance in the fatigue lifetimes of components, as illustrated in Figs. 33.12*A* and 33.12*B*. Note that a different mean stress results in each case. When the transfer from high strain to low strain is from compression, the mean stress resulting is in tension and would shorten fatigue life. The opposite is seen when the transfer from high

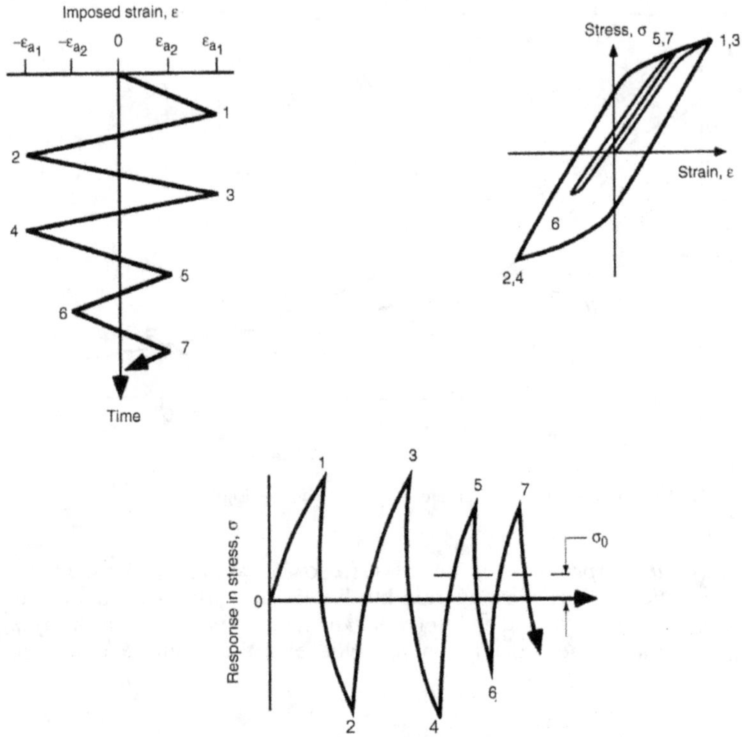

FIGURE 33.12*A* Illustration of the sequence from high strain to low strain coming from compression.

strain to low strain is from tension that would produce an increase in fatigue lifetime, since the resulting mean stress is in compression.

The strain-based approach lends itself readily to fatigue lifetime predictions and damage analysis in a pseudo-random loading spectrum (i.e., a repeated block-type, random history). This is contingent upon a means of counting closed hysteresis loops in a pseudo-random, repeated block-type, strain-time history. Such a strain-time history is commonly available simply due to the fact that prototype components are often brittle, lacquer-coated, critical locations on the component determined from the cracking pattern after the component is sent around a test route, strain gages are attached orthogonal to the crack pattern, and a strain-time recording is collected. Conversely, a dynamic finite element analysis (FEA) may also be performed, and the component's critical location(s) strain-time history can be simulated by a computer. Nonetheless, the strain-time histories can be obtained.

Several algorithms are available to deconvolve such a history, as originally explained by Matsuishi and Endo.[28] One of the more common is found in Refs. 29 and 30. Commercial software is also commonly available for analyses of this type. A simplistic explanation follows: Consider the strain-time history illustrated in Fig. 33.13*A* with the corresponding stress-time response. Plot the pseudo-random, block-type, strain-time history so that the greatest absolute peak or valley is first and last. Initiate "rainflow" so that it is allowed to drip down and continue—except if it initiates at a maximum, such as A,B,D,G, it stops when it comes opposite a more positive peak

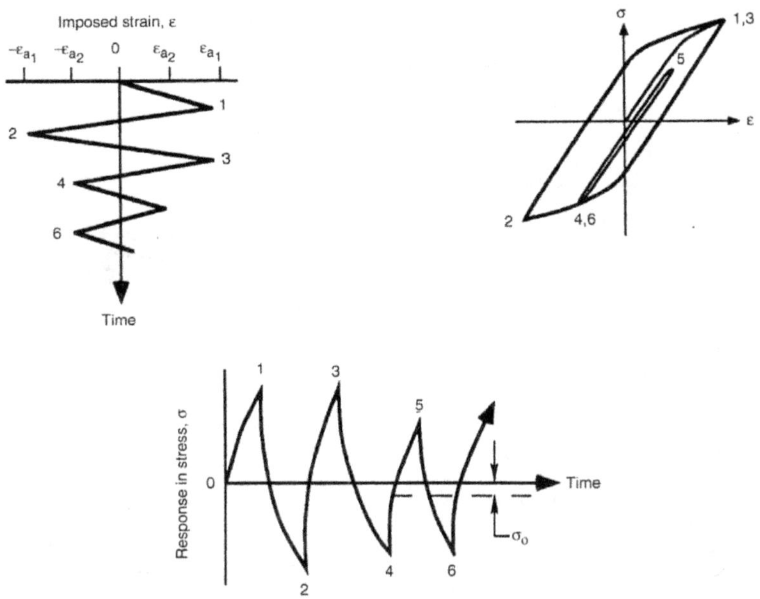

FIGURE 33.12*B* Same illustration as Fig. 33.12*A* but coming from tension.

than the maximum from which it began. Rainflow dripping from B must terminate opposite D because D is more positive. If the rainflow is initiated at a minimum, such as A,C,E,F, the converse is true. Finally, rainflow must stop if it encounters rain from the roof above, as in events from C to D.

Note that in Fig. 33.13*B* the events are paired into closed hysteresis loops, as in events A-D and D-A, B-C and C-B, D-E and E-D, and F-G and G-F, as seen on the accompanying stress-strain plot. Note that each event—with the exception of A-D-A, which is at a zero mean stress and the largest event in this history, thus the greatest damage—has a strain amplitude (i.e., half the range) and a mean stress associated with the closed loop. The event B-C-B has a tensile mean stress, while F-G-F has a compressive mean stress. Now, by appropriately modifying the strain-life plot in Fig. 33.11 for each associated mean stress, the corresponding fatigue life at that strain amplitude and mean stress can be determined for each closed-loop event. Once accomplished, Miner's linear damage rule,[31]

$$D = \sum d_i = \sum \frac{2n_i}{2N_{fi}} \tag{33.23}$$

where D = total damage
d_i = individual damage for each ith event
$2n_i$ = number of reversals of a specific event
$2N_{fi}$ = number of reversals to failure for that specific event

is invoked and the damage for the block of strain-time history can be found. The reciprocal of the damage is then the number of blocks of a history to fracture or failure:

$$B_f = \frac{1}{D} \tag{33.24}$$

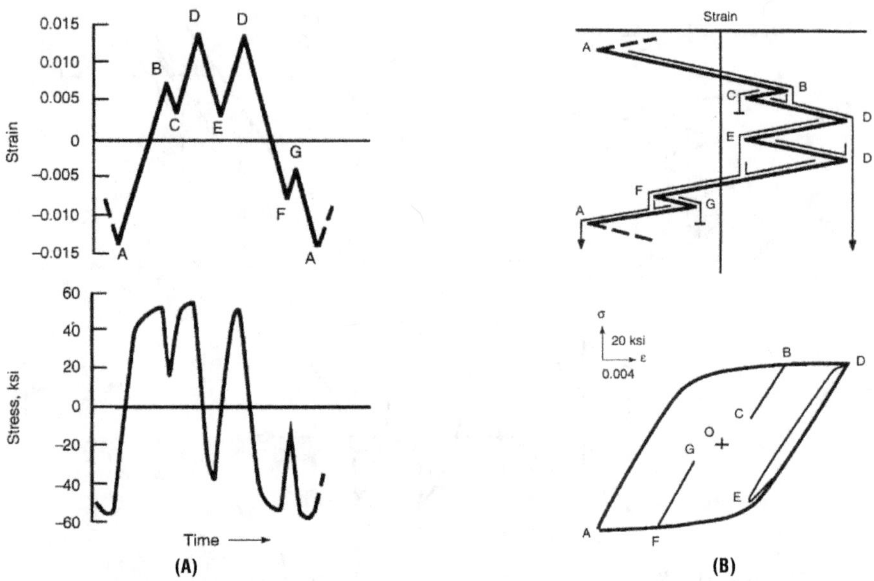

FIGURE 33.13 Rainflow counting technique for determining strain amplitudes and corresponding mean stresses for each closed-loop event.

VARIABILITY IN FATIGUE PROPERTIES

Up to this point, we have assumed that there is no randomness in metal fatigue properties and, for that matter, in the entire fatigue initiation process. This, of course, is not the case, for there are many factors that contribute to material variability and the randomness of the fatigue process that are far too numerous to but mention here. These factors include such influencing parameters as a metal's heat-to-heat and within-heat variability, heat treatment time and temperature, test specimen geometry, specimen preparation, residual surface stresses, test machine differences, and operator differences—the list goes on! As with monotonic properties, there is certainly variability or scatter in fatigue lifetimes (cycles) that by the very nature of the curvature of the stress-life (semi-log) and strain-life (log-log) curves tends to increase with increasing fatigue life—the steeper slope of these curves in the low cycle regime tends to make inherent variability or scatter in fatigue life (cycles) much less than in the long-life regime, where the shallowness of the slope results in greater variability in life. It is interesting to note, however, that the variability in stress or strain is relatively constant with fatigue lifetime. That is, a *vertical scatter band in stress or strain* is essentially constant, whereas a *horizontal scatter band in fatigue lifetime* increases with increasing cycles.

The ASTM Committee on Fatigue E09 (later, Committee E08 on Fatigue and Fracture) has published several excellent references on this very subject.[32,33,34] An excellent treatise on fatigue data statistics for design purposes is presented by Wirsching,[35,36] and the latest standard on this subject appears as ISO 12107.[37] Central to all of these statistical procedures is an assumption that these data must fit an assumed parametric distribution function such as normal, log-normal, or Weibull. There is a branch of statistics known as nonparametric or distribution-free statistics that should be considered if robustness and accuracy, particularly in the tail ends of distributions, are essential considerations. The use of these types of statistical interpretations of ranked data are often employed when there is no obvious or clear numerical interpretation because they rely on far fewer assumptions than the more common methodologies. Conover[38] provides an excellent treatise on nonparametric statistics, and Mitchell, et al.,[39] illustrated their use in fatigue lifetime predictive methodologies employing a strain-based fatigue analysis.

The preceding explanation is a relatively brief review of the strain-life or local strain approach to fatigue lifetime prediction and durability. It also is a simplistic uniaxial philosophy. An online explanation and comprehensive program for strain-based fatigue damage analysis may be found at www.fatiguecalculator.com, thanks to the benevolent efforts of Professor (Emeritus) D. F. Socie, of the University of Illinois, where these techniques found their origin through the late Professor (Emeritus) JoDean Morrow and his many graduate students. More complex loading situations may require a multiaxial or combined strain approach. Interested readers should refer to ASM Volume 19, *Fatigue and Fracture,* and to the aforementioned website, for a more thorough explanation of these techniques as applied to real structures.

FRACTURE MECHANICS METHOD

The techniques just discussed are often referred to as *initiation-based* mechanics techniques and are employed to determine the inception of a fatigue crack in a component or structure. An adjunct mechanics technique is often used to determine the *propagation* lifetime of an incipient crack. This is known as the *fracture mechanics* approach. It would be rather unusual for a crack of *critical size* to exist initially in a component or structure. It is more common for a small flaw or initiated crack to grow until it reaches the critical size for catastrophic fracture. During cyclic loading there will be a period of fatigue crack growth. If there is a corrosive environment present, even under a steady force, there may be significant environmental crack growth or stress-corrosion-type cracking.

In an engineering analysis involving fatigue crack growth normal to the applied stress, the severity of the crack is described by

$$K = \sigma\sqrt{\pi a}\, \mathrm{Y} \tag{33.25}$$

where K = stress intensity factor
σ = applied nominal stress
a = crack length
Y = dimensionless geometric factor dependent on the type of crack and specimen

The fatigue crack growth rate is controlled by K, and, as might be anticipated, as the crack grows, a will increase as will σ (since the remaining ligament size in the specimen's cross section will be decreasing), and the growth rate will increase until a *critical crack size* a_{crit} is reached and sudden fracture occurs. It should be noted that as the yield strength of a metal increases, the critical crack size decreases.

For a relatively thin specimen, the value of K will be greater than for a thicker specimen because there tends to be less constraint in a thin cross section as opposed to a thick specimen that has increased constraint. Thus, as the specimen increases in section thickness B, there is a trend to a minimum value of fracture toughness where K eventually becomes independent of thickness. This value is known as the *plane strain fracture toughness,* designated by K_{Ic}. According to ASTM E399, Standard Test Method for Linear-Elastic Plane-Strain Fracture Toughness K_{Ic} of Metallic Materials, such plane strain conditions will exist when

$$B \geq 2.5\left[\frac{K_{Ic}}{\sigma_y}\right] \tag{33.26}$$

where σ_y = the 0.002 offset yield strength of the metal. Values of the plane strain fracture toughness for a variety of structural metals may be found in Ref. 40.

Equation (33.25) can be rearranged and integrated to result in a mathematical relationship for fatigue crack propagation:

$$\frac{da}{dN} = A[\Delta K]^n \tag{33.27}$$

where $\frac{da}{dN}$ = fatigue crack propagation rate, mm/cycle

A, m = constants for a particular metal, dependent on environment, stress ratio R, and frequency

ΔK = stress intensity factor range

Typical values of m for most metals range from approximately 1 to 8. The expression for $\Delta K = K_{max} - K_{min}$ can be inserted into Eq. (33.27):

$$\Delta K = Y\Delta\sigma\sqrt{\pi a} = Y[\sigma_{max} - \sigma_{min}]\sqrt{\pi a} \tag{33.28}$$

Because there is little or no crack growth in the compression-going portion of a cycle, the assumption can be made that K_{min} and σ_{min} are essentially zero. So:

$$K_{max} = Y\sigma_{max}\sqrt{\pi a} \tag{33.29}$$

If we now again rearrange Eq. (33.27), we may write

$$dN = \frac{da}{A[\Delta K]^n} \tag{33.30}$$

which, upon integration, becomes

$$N_f = \int_0^{N_f} dN = \int_{a_o}^{a_{crit}} \frac{da}{A[\Delta K]^n} \tag{33.31}$$

where a_o = initial crack length

a_{crit} = critical crack length dependent on metal strength and microstructure and determined from fracture toughness tests

If you now substitute Eq. (33.28) into Eq. (33.31), you obtain

$$N_f = \frac{1}{A\pi^{m/2}[\Delta\sigma]^n}\int_{a_0}^{a_{crit}} \frac{da}{Y^m a^{m/2}} \tag{33.32}$$

and the number of cycles to propagate a crack of starter length a_0 to a critical crack length for catastrophic fracture a_{crit} can be determined.

The preceding development is but a simplistic description of crack propagation mechanics. There are many more contributing factors, such as overload and underload effects, corrosion, and duty cycle histories that are necessary inputs for application of these techniques to prediction of real-life situations. Those interested in the latest developments in this particular science are referred to the ASTM website at http://www.astm.org/Standard/DOMnewpub.shtml, then search for "stp fatigue and fracture mechanics." There are, to date, 35 volumes dedicated to this important topic.

Please note that fatigue crack growth life can be combined with the estimate for the fatigue crack incubation or initiation fatigue life to obtain the *total* number of cycles to failure of a component or structure in a real-life situation. Of course, there is much more that could not be covered in this limited treatise. A more thorough description of this fracture mechanics technique can also be found in Refs. 41 and 42.

REFERENCES

1. Dowling, N. E.: "Mechanical Behavior of Materials, Engineering Methods for Deformation, Fracture and Fatigue," 3d ed., Pearson Prentice Hall, Englewood Cliffs, N.J., 2007.
2. Dieter, G. E.: "Mechanical Metallurgy," 3d ed., McGraw-Hill Companies, New York, 1986.
3. Dieter, G. E., and L. C. Schmidt: "Engineering Design," 4th ed., McGraw-Hill Companies, New York, 2008.
4. Callister, W. D.: "Material Science and Engineering," 7th ed., John Wiley & Sons, Hoboken, N.J., 2007.
5. American Society for Testing and Material International: E08, Standard Test Methods for Tension Testing of Metallic Materials, 2007.
6. International Organization for Standardization: ISO 6892, Metallic Materials—Tensile Testing at Ambient Temperature, 1998.
7. Meyers, M. A., and L. E. Murr: "Shock Waves and High-Strain-Rate Phenomena in Metals: Concepts and Applications," Plenum Press, New York, 1981.
8. Mitchell. M. R.: "Fundamentals of Modern Fatigue Analysis for Design," *American Society for Materials International Handbook, Vol. 19, Fatigue and Fracture,* 1996, pp. 227–249.
9. Boller, C., and T. Seeger: "Materials Data for Cyclic Loading," Vol. 42A; "Unalloyed Steels," Vol. 42B; "Low-Alloy Steels," Vol. 42C; "High-Alloy Steels," Vol. 42D; "Aluminum and Titanium Alloys," Vol. 42E; *Cast and Welded Metals,* Elsevier Science Publishers, Amsterdam, 1987.
10. Chao, Y. J., J. D. Ward, Jr., and R. G. Sands: "Charpy Impact Energy, Fracture Toughness and Ductile–Brittle Transition Temperature of Dual-Phase 590 Steel," *Materials and Design,* Vol. 28, 2007, pp. 551–557.
11. Nadai, A., and A. J. Manjoine: "High-Speed Tension Tests at Elevated Temperatures, Parts I and II," *Proc. ASTM 40,* 1940, pp. 822–837; "Part III," *Journal of Applied Mechanics,* **8**(2):A77–A91 (1941).
12. Qiu, H., M. Enoki, H. Mori, N. Takeda, and T. Kishi: "Effect of Strain Rate and Plastic Pre-Strain on the Ductility of Structural Steels," *Iron Steel Institute, Japan International,* **39**(9):955–960 (1999).
13. Bridgeman, P. W.: *Transactions of ASM,* **32:**553–574 (1944).
14. Atrens, A., W. Hoffelner, T. W. Duerig, and J. E. Allison: "Subsurface Crack Initiation in High Cycle Fatigue in Ti6AI4V and in a Typical Martensitic Steel," *Scripta Metallurgica,* **17:**601–606 (1983).

15. Mayer, H. R., H. Lipowsky, M. Papakyriacou, R. Rösch, A. Stich, and S. Stanzl-Tschegg: "Application of Ultrasound for Fatigue Testing of Lightweight Alloys," *Fatigue and Fracture of Engineering Materials and Structures,* **22**(7):591–599, Blackwell Publishing (1999).

16. Papakyriacou, M., H. Mayer, C. Pypen, H. Plenk, Jr., and S. Stanzl-Tschegg: "Influence of Loading Frequency on High-Cycle Fatigue Properties of BCC and HCP Metals," *Materials Science & Engineering A,* **308**(1–2):143–152 (2001).

17. Brose, W. R., N. E. Dowling, and JoDean Morrow: "Effect of Periodic Large Strain Cycles on the Fatigue Behavior of Steels," *Society of Automotive Engineers Paper No. 740221,* Automotive Engineering Congress, Detroit, Mich., 1974.

18. Barsom, J. M., and S. T. Rolfe: "Fracture and Fatigue Control in Structures: Applications of Fracture Mechanics," 3d ed., *American Society for Testing and Materials International,* West Conshohocken, Pa., 1999.

19. Peterson, R. E.: "Stress Concentration Factors," John Wiley & Sons, 1974.

20. Mitchell, M. R.: "A Unified Predictive Technique for the Fatigue Resistance of Cast Ferrous-Based Metals and High Hardness Wrought Steels," *SAE SP 442,* Society of Automotive Engineers, 1979.

21. Coffin, L. F.: "A Study of Effects of Cyclic Thermal Stresses on a Ductile Metal," *Transactions of the American Society of Mechanical Engineers,* **76:**931–950 (1954).

22. Manson, S. S.: "Behavior of Materials under Conditions of Thermal Stress," *National Advisory Commission on Aeronautics, Report 1170,* Lewis Flight Propulsion Laboratory, Cleveland, Ohio, 1954.

23. Morrow, J.: "Cyclic Plastic Strain Energy and Fatigue of Metals," *Internal Friction, Damping, and Cyclic Plasticity,* ASTM STP 378, American Society for Testing and Materials, Philadelphia, Pa., 1965, pp. 45–87.

24. Landgraf, R. W., M. R. Mitchell, and N. R. LaPointe: *Monotonic and Cyclic Properties of Engineering Materials,* Ford Motor Company, Scientific Laboratory Report, June 1972.

25. Conle, F. A., R. W. Landgraf, and F. D. Richards: *Materials Data Book—Monotonic and Cyclic Properties of Engineering Materials,* Ford Motor Company, Scientific Research Laboratory Report, 1988.

26. Mitchell, M. R.: "A Unified Predictive Technique for the Fatigue Resistance of Cast Ferrous-Based Metals and High Hardness Wrought Steels," *SAE SP 442,* Society of Automotive Engineers, 1979.

27. Downing, S. D.: "Modeling Cyclic Deformation and Fatigue Behavior of Cast Iron under Uniaxial Loading," *UILU-ENG-84-3601,* College of Engineering, University of Illinois at Urbana-Champaign, January 1984.

28. Matsuishi, M., and T. Endo: "Fatigue of Metals Subjected to Varying Stresses," *Japan Society of Mechanical Engineers,* March 1968.

29. Downing, S. D., and D. F. Socie: "Simple Rainflow Counting Algorithms," *International Journal of Fatigue,* **1:**31–40 (1982).

30. American Society for Testing and Material International: E1049, Standard Practice for Cycle Counting in a Fatigue Analysis, *ASTM International,* West Conshohocken, Pa., 2005.

31. Miner, M. A.: "Cumulative Damage in Fatigue," *Journal of Applied Mechanics,* **12:**A159–A164 (1945).

32. Little, R. E.: "Manual on Statistical Planning and Analysis," *ASTM STP 91a,* American Society for Testing and Materials, Philadelphia, Pa., 1963 (out of print but available on-line at www.astm.org).

33. Little, R. E., and E. H. Jebe: "Manual on Statistical Planning and Analysis for Fatigue Experiments," *ASTM STP 588,* American Society for Testing and Materials, Philadelphia, Pa., 1975.

34. Little, R. E.: "Statistical Analysis of Fatigue Data," *ASTM STP 744,* American Society for Testing and Materials, Philadelphia, Pa., 1981.

35. Wirsching, Paul H.: "Statistical Summaries of Fatigue Data for Design Purposes," *NASA Contract Report CR3697,* July 1983.
36. Wirsching, P. H., K. Ortiz, and S. J. Lee: "An Overview of Reliability Methods in Mechanical and Structural Design," *AIAA #87-0765,* American Institute of Aeronautics, 1987, pp. 260–267.
37. International Organization for Standardization: 12107 Metallic Materials—Fatigue Testing—Statistical Planning and Analysis of Data, ISO TC164/SC5 Fatigue Test Methodologies, in process of revision, 2008.
38. Conover, W. J.: "Practical Nonparametric Statistics," 3d ed., John Wiley & Sons, New York, 1999.
39. Mitchell, M. R., M. E. Meyer, and N. Q. Nguyen: "Fatigue Considerations in Use of Aluminum Alloys," #820699, *SAE Transactions,* **91**(3):2399–2425, Society of Automotive Engineers (1982).
40. Matthews, W. T.: "Plane Strain Fracture Toughness (K_{Ic}) Data Handbook for Metals," Army Materials and Mechanics Research Center, Watertown, Mass., 1973.
41. Barsom, J. M., and S. T. Rolfe: "Fracture and Fatigue Control of Structures: Application of Fracture Mechanics," 3d ed., *ASTM International,* West Conshohocken, Pa., 1999.
42. Broek, D.: "The Practical Use of Fracture Mechanics," Kluwer Academic Publications, Dordrecht, The Netherlands, 1986.

CHAPTER 34
ENGINEERING PROPERTIES OF COMPOSITES

Keith T. Kedward

INTRODUCTION

Composite materials are simply a combination of two or more different materials that may provide superior and unique mechanical and physical properties. The most attractive composite systems effectively combine the most desirable properties of their constituents and simultaneously suppress the least desirable properties. For example, a glass-fiber reinforced plastic combines the high strength of thin glass fibers with the ductility and environmental resistance of an epoxy resin; the inherent damage susceptibility of the fiber surface is thereby suppressed whereas the low stiffness and strength of the resin is enhanced.

The opportunity to develop superior products for aerospace, automotive, and recreational applications has sustained the interest in advanced composites. Currently composites are being considered on a broader basis, specifically, for applications that include civil engineering structures such as bridges and freeway pillar reinforcement, and for biomedical products such as prosthetic devices. The recent trend toward affordable composite structures with a somewhat decreased emphasis on performance will have a major impact on the wider exploitation of composites in engineering.

BASIC TYPES OF COMPOSITES

Composites typically comprise a high-strength synthetic fiber embedded within a protective matrix. The most mature and widely used composite systems are *polymer matrix composites* (PMCs), which will provide the major focus for this chapter. Contemporary PMCs typically use a ceramic type of reinforcing fiber such as carbon, Kevlar™, or glass in a resin matrix wherein the fibers make up approximately 60 percent of the PMC volume. Metal or ceramic matrices can be substituted for the resin matrix to provide a higher-temperature capability. These specialized systems are termed *metal matrix composites* (MMCs) and *ceramic matrix composites* (CMCs); a

TABLE 34.1 Composite Design Comparisons

	PMC	CMC	MMC
Specific strength and stiffness	Generally excellent if exclusively unidirectional reinforcement is avoided	Highest potential for high-temperature applications	Moderately high for dominantly axial loads and intermediate temperatures
Fatigue characteristics	Excellent for designs that avoid out-of-plane loads	Good for high-temperature applications loads	Potential concern for other than dominantly axial
Nonlinear effects	Usually not important for continuous fiber reinforcements	Significant effect after first matrix and interface cracks have developed	Can be significant, particularly for multidirectional and off-axis loads
Temperature capability	Less than 600°F	Potential for maximum values between 1000 and 2000°F	Potential for maximum values up to 1000°F
Degree of anisotropy	Extreme, particularly considering out-of-plane properties and consequent coupling effects in minimum-gage configurations	Can develop significantly during loading, due to matrix and interface breakdown	Not usually a major issue where interface effects are negligible

general qualitative comparison of the relative merits of all three categories is summarized in Table 34.1.

SHORT FIBER/PARTICULATE COMPOSITES

The fibrous reinforcing constituent of composites may consist of thin continuous fibers or relatively short fiber segments, or whiskers. However, reinforcing effectiveness is realized by using segments of relatively high *aspect ratio,* which is defined as the length-to-diameter ratio. Nevertheless, as a reinforcement for PMCs, these short fiber or whisker systems are structurally less efficient and very susceptible to damage from long-term and/or cyclic loading. On the other hand, the substantially lower cost and reduced anisotropy on the macroscopic scale render these composite systems appropriate in structurally less demanding industrial applications.

Randomly oriented short fiber or particulate-reinforced composites tend to exhibit a much higher dependence on polymer-based matrix properties, as compared to typical continuous fiber reinforced PMCs. Elastic modulus, strength, creep, and fatigue are most susceptible to the significant limitations of the polymer matrix constituent and fiber-matrix interface properties.[1]

CONTINUOUS FIBER COMPOSITES

Continuous fiber reinforcements are generally required for structural or high-performance applications. The *specific strength* (strength-to-density ratio) and *specific stiffness* (elastic modulus-to-density ratio) of continuous fiber reinforced PMCs, for example, can be vastly superior to conventional metal alloys, as illustrated in Fig. 34.1. These types of composite can also be designed to provide other attractive properties, such as high thermal or electrical conductivity and low coefficient of thermal expansion (CTE). In addition, depending on how the fibers are oriented or inter-

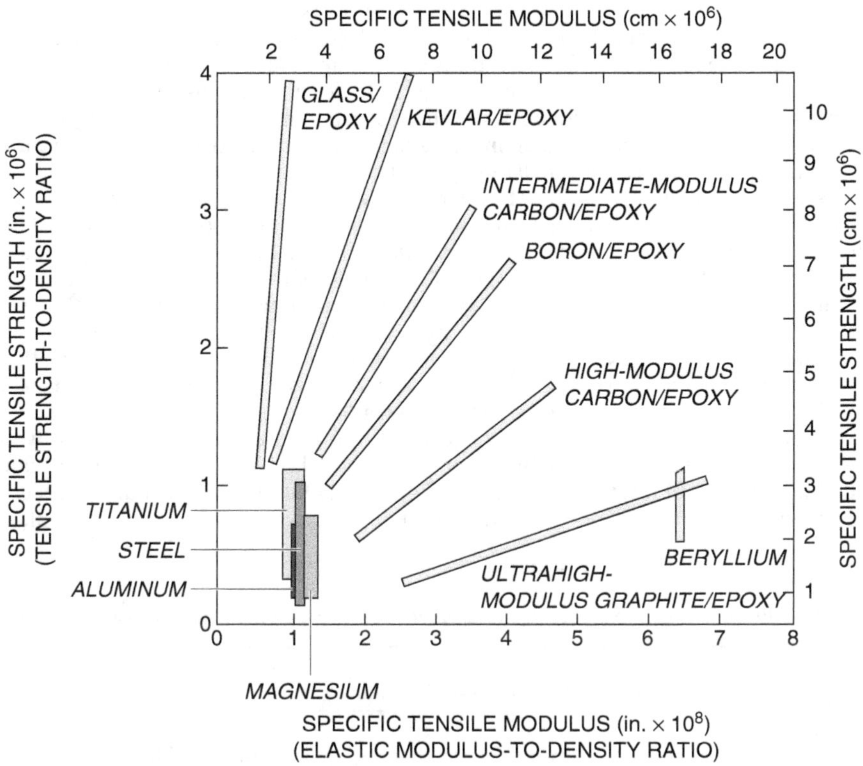

FIGURE 34.1 A weight-efficiency comparison.

woven within the matrix, these composites can be tailored to provide the desired structural properties for a specific structural component. *Anisotropy* is a term used to define such a material that can exhibit properties varying with direction. Thus designing for, and with, anisotropy is a unique aspect of contemporary composites in that the design engineer must simultaneously design the structure and the material of construction. Of course, anisotropy brings problems as well as unique opportunities, as is discussed in a later section. With reference to Fig. 34.1, it should be appreciated that the vertical bars representing the conventional metals signify the potential variation in specific strength that may be brought about by changes in alloy constituents and heat treatment. The angled bars for the continuous fiber composites represent the range of specific properties from the unidirectional, all 0° fiber orientation at the upper end to the pseudo-isotropic laminate with equal proportions of fibers in the 0°, +45°, –45°, and 90° orientations at the lower end. In the case of the composites, the variations between the upper or lower ends of the bars are achieved by tailoring in the form of laminate design.

SPECIAL DESIGN ISSUES AND OPPORTUNITIES

Product design that involves the utilization of composites is most likely to be effective when the aspects of materials, structures, and dynamics technologies are embraced in the process of the development of mechanical systems. One illustrative example was cited in the introductory chapter of this handbook (see Chap. 1), which introduces the technique of reducing the vibration response of a fan blade by alteration of the natural frequency. In the design of composite fan blades for aircraft, this approach has been achieved by tailoring the frequency and the associated mode shape.[2] Such a tailoring capability can assist the designer in adjusting flexural and torsional vibration and fatigue responses, as well as the damping characteristics explained later.

A more challenging issue that frequently arises in composite hardware design for a majority of the more geometrically complex products is the potential impact of the low secondary or matrix-influenced properties of these strongly nonisotropic material forms. The transverse (in-plane) tensile strength of the unidirectional composite laminate is merely a few percent of the longitudinal tensile strength (as observed from Tables 34.2 and 34.3). Consequently, it is of no surprise that the through-thickness or short-transverse tensile strength of a multidirectional laminate is of the same order, but even lower than the transverse tensile strength of the individual layers. Thus, the importance of the designer's awareness of such limitations cannot be overemphasized. In fact, the large majority of the failures in composite hardware development testing has arisen due to underestimated or unrecognized out-of-plane loading effects and interrelated regions of structural joints and attachments. Due to the many common adverse experiences with delaminations induced by out-of-plane

TABLE 34.2 Properties of Typical Continuous, Fiber-Reinforced Composites and Structural Metals

	Unidirectional composite (60% fiber/40% resin, by volume)				Metals	
Property	E-glass/ resin	Kevlar/ resin	HS carbon/ epoxy	UHM Gr./ epoxy	7075-T6 aluminum	4130 steel
			Elastic			
Density, lb/in.^{3} (10^3 kg/m^3)	0.070 (1.9)	0.047 (1.3)	0.058 (1.6)	0.060 (1.7)	0.100 (2.77)	0.284 (7.86)
E_L, 10^6 lb/in.^{2} (10^3 MPa)	6.5 (45)	11.0 (75.8)	19.5 (134)	40.0 (276)	10.3 (71.0)	30.0 (207)
E_T, 10^6 lb/in.^{2} (10^3 MPa)	1.8 (12)	1.0 (6.9)	1.5 (10)	1.2 (8.3)	10.3 (71.0)	30.0 (207)
G_{LT}, 10^6 lb/in.^{2} (10^3 MPa)	0.7 (4.8)	0.4 (2.8)	0.9 (6.2)	0.65 (4.5)	4.0 (27.6)	12.0 (82.7)
ν_{LT}	0.32	0.33	0.30	0.28	0.30	0.28
			Strength			
F_L^{tu}, 10^3 lb/in.^{2} (MPa)	180 (1240)	220 (1520)	200 (1380)	100 (689)	79 (545)	100 (689)
F_T^{tu}, 10^3 lb/in.^{2} (MPa)	6 (41)	4.5 (31)	7 (48)	5 (34)	77 (531)	100 (689)
F_L^{cu}, 10^3 lb/in.^{2} (MPa)	120 (827)	45 (310)	170 (1170)	90 (620)	70 (483)	130 (896)
F_T^{cu}, 10^3 lb/in.^{2} (MPa)	20 (138)	20 (138)	20 (138)	20 (138)	70 (483)	130 (896)
F_{LT}^{su}, 10^3 lb/in.^{2} (MPa)	8 (55)	4 (28)	10 (69)	9 (62)	47 (324)	60 (414)

TABLE 34.3 Typical Unidirectional Properties for a Carbon/Epoxy System

Stiffness properties		Strength properties		Thermal properties	
E_L, 10^6 lb/in.^{2} (10^3 MPa)	20.0 (138)	F_L^{tu}, 10^3 lb/in.^{2} (MPa)	240.0 (1650)	α_L, με/°F (με/K)	−0.3 (−0.54)
E_T, 10^6 lb/in.^{2} (10^3 MPa)	1.4 (9.6)	F_L^{cu}, 10^3 lb/in.^{2} (MPa)	200.0 (1380)	α_T, με/°F (με/K)	17.0 (30.6)
G_{LT}, 10^6 lb/in.^{2} (10^3 MPa)	0.8 (5.5)	F_T^{tu}, 10^3 lb/in.^{2} (MPa)	7.0 (48)	K_L, Btu in./h ft^2 °F (W/m K)	40.0 (5.76)
ν_{LT}	0.28	F_T^{tu}, 10^3 lb/in.^{2} (MPa)	20.0 (138)	K_T, Btu in./h ft^2 °F (W/m K)	4.5 (0.65)
		F_{LT}^{isu}, 10^3 lb/in.^{2} (MPa)	10.0 (69)		
$\nu_{LT}/E_L = \nu_{TL}/E_T$		F^{isu}, 10^3 lb/in.^{2} (MPa)	9.0 (62)		

load components, this section will be devoted to the identification of the numerous sources of out-of-plane load development and the candidate approaches to eliminate or minimize their influence.

First, a general overview of many of the common problems created for the engineering designer that are consequences of low-matrix-dominated, elastic, and strength properties are summarized in Table 34.4. Several of the most common sources will now be discussed in more detail. Figure 34.2 illustrates these major sources, which may be broadly categorized as follows:

Category A: Curved sections including curved segments, rings, hollow cylinders, and spherical vessels that are representative of angle bracket design details, curved frames, and internally or externally pressurized vessels.

TABLE 34.4 General Overview of Problems Created by the Low Secondary (Matrix-Dominated) Properties of Advanced Composites

Controlling property	Problem
F^{isu}	Failure induced by shear in beams under flexural loading. Premature torsional failures. Premature crippling failure in compression.* Failure of adherends in structural bonded joints.* Failure of laminae due to free-edge effects, e.g., cutouts, ply drops.*
F_T^{tu}	Failure induced by transverse tensile fracture of curved beams in flexure. Shock waves during normal impacts.
G_{LT}	Reduction in flexural and torsional stiffness. Reduction in resonant frequencies of plate and beam members. Reduction of elastic buckling capability. Interpretation of experimental stress analysis data.
α_T	Distortion at fillets due to high expansion coefficient (through-thickness).
$\alpha_T F_T^{tu}$	Failure due to thermal stresses in thick-walled composite cylinders.

*For these problems, the controlling properties are both F^{isu} and F_T^{tu}.

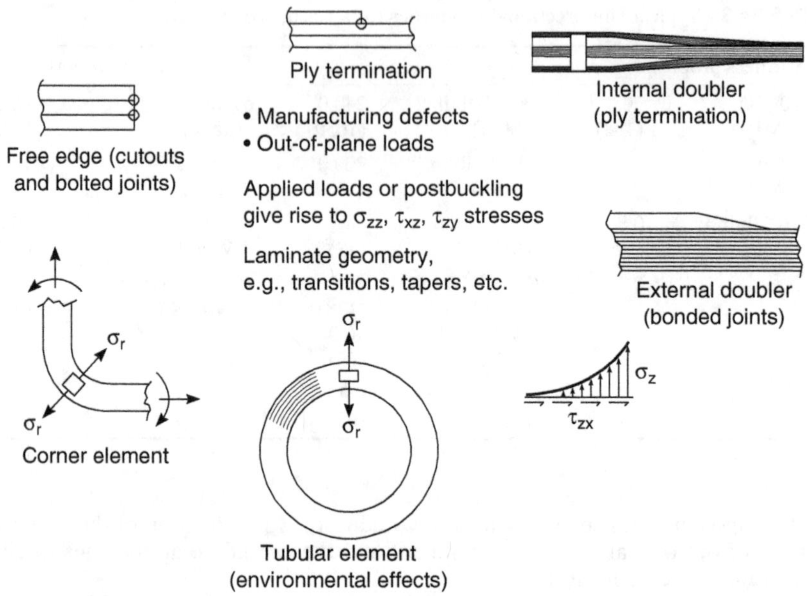

FIGURE 34.2 Generic sources of delamination.

Category B: Tapers and transitions including local changes of section that are representative of laminate layer terminations, doublers, and stiffener terminations, as well as the end details of bonded and bolted joints.

As mentioned earlier, commonplace structural details of both categories have contributed to numerous unanticipated failures in composite hardware components. In some cases, such failures can propagate catastrophically after initiation and may therefore be a serious safety threat. Other instances have arisen where initial failures may self-arrest resulting in benign failures, but with some degree of local stiffness degradation. Subsequent load distribution may, however, precipitate eventual catastrophic failure depending on the load spectrum characteristics.

COMPOSITE PROPERTIES

The class of composites which forms the focus of this chapter is polymer matrix composites (PMCs) with continuous fiber reinforcement. In this type of composite, the properties of an arbitrary laminated composite architecture are derived from the elastic and strength properties of a unidirectional layer. The unidirectional layer properties can be derived from the constituent properties of the fiber and matrix that typically range between 50 and 65 percent by volume of the fiber reinforcement phase. Here a nominal value of 60 percent by volume of fiber will be adopted.

Fiber reinforcements most commonly encountered in contemporary composites include carbon or graphite fibers, Kevlar fibers, and glass fibers, all of which can be obtained in similar diameters, i.e., 0.0003–0.0005 in. Both the carbon/graphite and Kevlar fibers are inherently anisotropic in themselves, although it is the axial (fiber direction) properties that dominate the in-plane behavior of unidirectional and, gen-

TABLE 34.5 Typical Fiber Properties

Fiber	Density, lb/in.^{3} (10^3 kg/m^3)	Axial elastic modulus, 10^6 lb/in.^{2} (10^3 MPa)	Transverse elastic modulus, 10^6 lb/in.^{2} (10^3 MPa)	Tensile strength, 10^3 lb/in.^{2} (10^3 MPa)
E-glass	0.091 (2.5)	10.5 (72.4)	10.5 (72.4)	500 (3.4)
S-glass	0.090 (2.5)	12.4 (85.5)	12.4 (85.5)	600 (4.1)
Kevlar 49	0.052 (1.4)	18.0 (124)	1.3 (8.96)	400 (2.8)
AS4 carbon	0.064 (1.8)	35.0 (241)	2.0 (13.8)	350 (2.4)

erally, multidirectional fiber arrays or laminates. Typical fiber properties are presented in Table 34.5, where the degree of individual fiber anisotropy is indicated.

GENERAL PROPERTIES

The properties of polymer matrices range over a much smaller spectrum in Table 34.6, and the relatively low stiffness and strength properties rarely dominate the composite behavior, with certain exceptions. The most notable exceptions are the interlaminar shear strength and the thickness-direction interlaminar tensile strength, to be discussed later, wherein the fiber-to-matrix interface may play an important role. For these reasons, the greatest attention is placed on the macroscopic composite properties that are of most direct interest to the mechanical or structural engineer. Typical values for such properties are provided in Table 34.2 for the three different, but all widely used, composites. One well-established carbon fiber/epoxy composite system is chosen to illustrate typical properties and degrees of anisotropy in elastic, strength, and thermal properties in Table 34.3. Engineers responsible for design and structural evaluation should take particular note of the degree of anisotropy in both the strength and stiffness properties. Usually the matrix-dominated properties, such as the shear and transverse tensile strengths, are very low and the avoidance of matrix-dominated failure modes represents a major challenge for the structural designer. It is also worthy of note that compression strength in the fiber direction, F_L^{cu}, is significantly lower than the equivalent tensile strength, F_L^{tu}, due to a microfiber instability mechanism. In fact, the ratio of these two strengths, F_L^{cu}/F_L^{tu}, may be much lower for some other systems, e.g., Kevlar/epoxy and more recently developed high strain-to-failure carbon fibers. The lower compression strength relative to the tensile strength is also influenced by the fiber diameter and the matrix properties that are themselves affected by moisture, temperature, interface integrity, and porosity.

IN SITU PROPERTIES

An important fundamental aspect of multidirectional composite laminates is the manner in which the individual unidirectional layer or lamina properties translate

TABLE 34.6 Typical Properties for Polymer Matrices

Polymer	Density, lb/in.^{3} (10^3 kg/m^3)	Elastic modulus, 10^6 lb/in.^{2} (10^3 MPa)	Tensile strength, 10^3 lb/in.^{2} (MPa)	Poisson's ratio
HERCULES 3501-6 epoxy	0.044 (1.2)	0.62 (4.3)	12.0 (82.7)	0.34
NARMCO 5208 epoxy	0.044 (1.2)	0.50 (3.4)	11.0 (75.8)	0.35
EPON 828 epoxy	0.044 (1.2)	0.47 (3.2)	13.0 (89.6)	0.35

into laminate properties. For all the thermoelastic properties, this translation is accomplished by the usual rules for transformation of stress and strain. However, the strength properties tend to be modified by the mutual constraint imposed by adjacent layers, and therefore is a function of the individual layer thickness. The result is a need to modify the basic unidirectional properties, one of the most significant being the ultimate transverse strain to failure in tension of individual layers. Unidirectional layer compressive strength and the associated ultimate strain to failure is also influenced to a significant degree by the mutual support offered by adjacent transverse or angled layers. As a consequence, correction factors are sometimes introduced to compensate for these effects, but more routine tests are conducted on the actual laminate configuration in an effort to establish reliable allowables for its use in design.

LAMINATED COMPOSITE DESIGN

For the simultaneous design of material and structure that is the basic philosophy for composite structures development, *laminated plate theory* (LPT) and the associated computer codes represent the fundamental tool for the composite designer. The anatomy of a composite laminate indicating the translation from the constituent fiber and matrix properties to those of a built-up laminate is illustrated in Fig. 34.3.

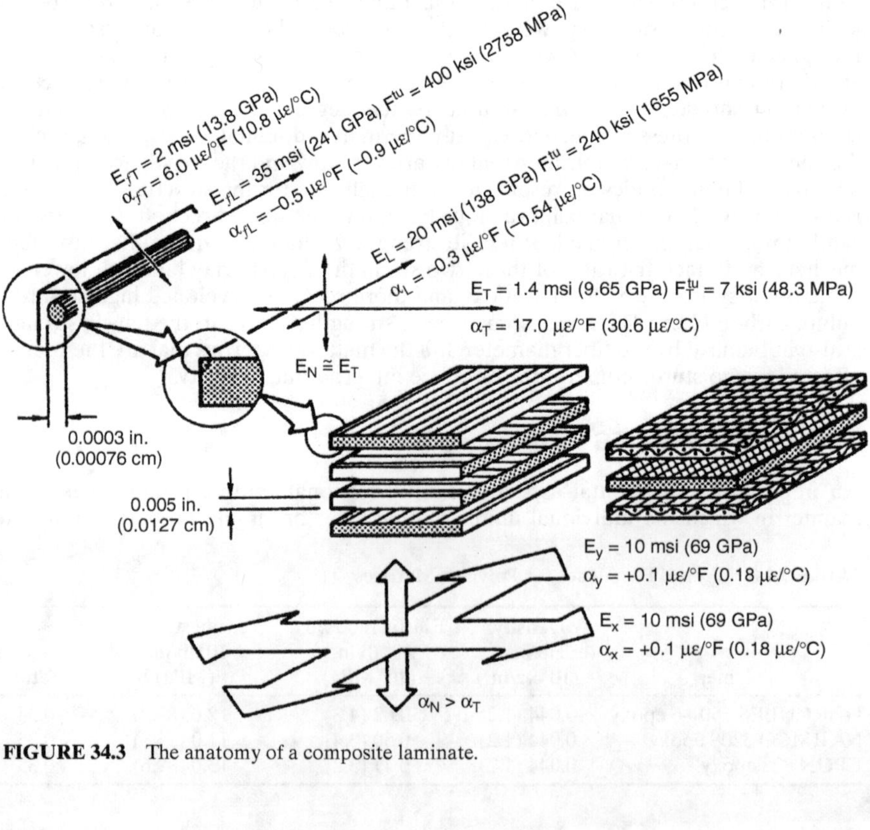

FIGURE 34.3 The anatomy of a composite laminate.

Values contained in this figure compare with those presented in Table 34.3. Figure 34.3 also illustrates the use of an alternative form of material, a fabric laminate that can provide similar, but slightly inferior, properties in a reduced thickness. The ability to produce a single layer comprised of equal proportions of fibers woven into 0° and 90° orientations is offered by this approach. Such a textile system therefore represents a valuable composite form. A state of plane stress and, for bending, plane sections remain plane, is assumed in most conventional theoretical treatments.

To remain within the scope and purpose of this chapter, the full treatment of conventional LPT will not be repeated here since it appears in numerous established texts on the subject (see Refs. 3 through 8). However, the essential information on conventional notations, whereby laminates are specified together with the physical behavioral insights concerning coupling phenomena, will be presented herein.

LAMINATE CONFIGURATION NOTATION

A method for specifying a given multidirectional laminate configuration has been established and is now routinely used on engineering drawings and documents. The following items essentially explain this laminate orientation notation:

1. Each layer or lamina is denoted by the angle representing the orientation (in degrees) between its fiber orientation and the reference structural axis in the x direction of the laminate.
2. Individual adjacent angles, if different, are separated by a slash (/).
3. Layers are listed in sequence starting with the first layer laid up, adjacent to the tool surface.
4. Adjacent layers of the same angle are denoted by a numerical subscript.
5. The total laminate is contained between square brackets with a subscript indicating that it is the total laminate (subscript T) or one-half of a symmetric laminate (subscript S).
6. Positive angles are assumed clockwise looking toward the lay-up tool surface, and adjacent layers of equal and opposite signs are specified with + or − signs as appropriate.
7. Symmetrical laminates with an odd number of layers are denoted as symmetric laminates with an even number of plies, but with the center layer overlined.

The notations for some commonly used laminate configurations are illustrated in Fig. 34.4.

In essence, lamination theory is involved in the transformation of the individual stiffnesses of each layer in the principal directions to the direction of orientation in the laminate, thereby providing the stiffness characterization for the specified laminate configuration. Subsequently, application of a given system of loads is broken down into individual layer contributions and referred back to the principal directions in each layer. A failure criterion is then used to assess the margin-of-safety arising in each layer. The complete process is illustrated in Fig. 34.5.

FAILURE CRITERIA

Although much debate and development has occurred with regard to the most appropriate failure criteria for composite laminates, the most widely adopted

Ply	Angle
8	0
7	−45
6	90
5	+45
4	+45
3	90
2	−45
1	0

or

Ply	Angle
6	9
5	60
4	120
3	120
2	60
1	0

Notation $[0/\pm45/90]_s$ $[0/60/120]_s$

Required
(i) When properties required to be equal in all directions in plane of laminate:
Tensile strength
Compressive strength
Modulus
Coefficient of thermal expansion (CTE)
(ii) In some cases of combined tension or compression and shear.

(a)

Ply	Angle
11,12	0
10	45
9	90
8	−45
6,7	0
5	−45
4	90
3	45
1,2	0

or

Ply	Angle
20	90
19	60
16,17,18	0
15	120
12,13,14	0
11	60
10	60
7,8,9	0
6	120
3,4,5	0
2	60
1	90

Notation $[0_3/\pm45/90]_s$ $[0_6/60_2/120/90]_s$

Used
(i) When load or stiffness requirement is highly directional.
(ii) To attain low CTE in one direction with low modulus fibers.

(b)

Ply	Angle
8	90
7	+θ
6	0
5	−θ
4	−θ
3	0
2	+θ
1	90

Laminate "balanced" about median plane

Notation $[90/\theta/0/-\theta]_s$

(i) Required to prevent warpage of flat laminate.
(ii) Can be departed from in contained circular parts: cylinders, cones, etc.
(iii) Small deviations near median plane possible; verify by test.

(c)

Ply	Angle
10	90
8 9	+θ
7	0
6	−θ
5	−θ
4	0
2 3	+θ
1	90

Balanced but more +θ than −θ

Notation $[90/+\theta_2/0/-\theta]_s$

(i) Will not warp, but

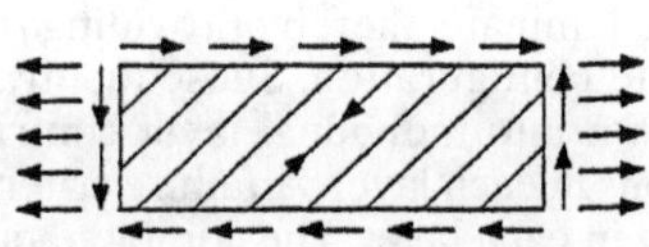

(ii) Applied axial load induces shear due to strain in excess +θ plies.

(d)

FIGURE 34.4 Examples of laminates and conventional notations.

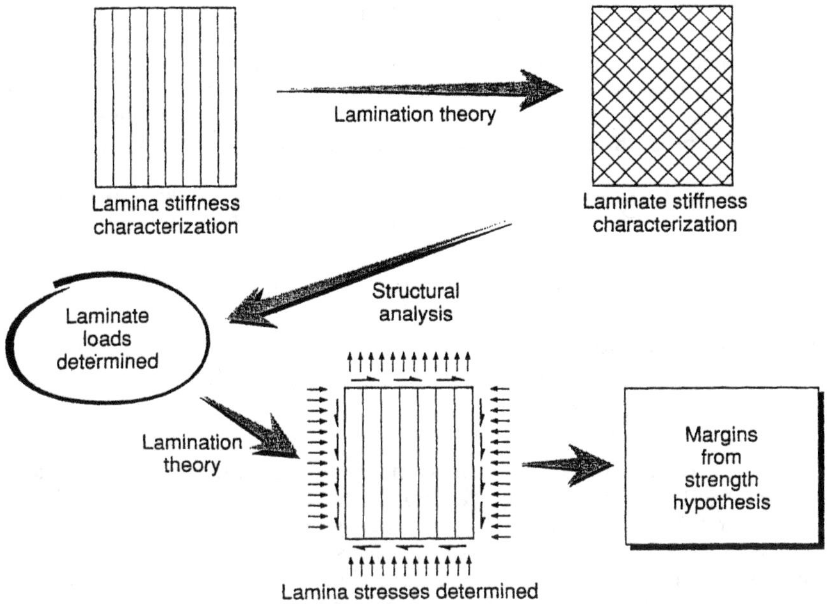

FIGURE 34.5 Procedure for strength determination.

approach in composite applications is the *maximum strain criterion.* The application of this relatively simple criterion requires an experimental database for the ultimate strains for each of the three fundamental loading directions for the individual orthotropic layer comprising the laminate. The three fundamental loading directions refer to axial loading in the fiber direction, axial loading transverse to the fiber direction, and in-plane shear associated with the former directions. However, it should be acknowledged that the ultimate strain values may be markedly different for tension and compression both in the fiber direction and transverse to it. Thus, a total of the following five ultimate strains are required to facilitate application of the maximum strain criterion:

1. ε_L^{tu} is the ultimate tensile strain in the fiber direction.
2. ε_L^{cu} is the ultimate compressive strain in the fiber direction.
3. ε_T^{tu} is the ultimate tensile strain transverse to the fiber direction.
4. ε_T^{cu} is the ultimate compressive strain transverse to the fiber direction.
5. γ_{LT}^{su} is the ultimate shear strain associated with directions parallel and normal to the fiber direction.

In connection with the actual values used for (1) through (5), see the previous section entitled "In Situ Properties," which explains how the individual layer properties must be adjusted to represent the strength or ultimate strain values of a given layer that is contained within a multidirectional laminate. The prudent approach in engineering development work is to identify special laminate configurations that may be used to establish representative in situ properties for the range of potential candidate laminates for application to a specific design.

COUPLING, BALANCE, AND SYMMETRY

The mathematical relationships obtained in laminated plate theory define all the coupling relationships arising in the arbitrary laminate. However, a discussion of the physical aspects of such coupling phenomena and the laminate designs that may be invoked to suppress these responses is helpful to the structural engineer.

Extension-Shear Coupling. First, the in-plane coupling between extension and shear or vice versa arises in the case of any off-axis layer, for example,

$$\gamma_{xy} = S_{16}\sigma_x \qquad \text{or} \qquad \varepsilon_x = S_{16}\tau_{xy} \tag{34.1}$$

or, for the inverse situation,

$$\sigma_x = Q_{16}\gamma_{xy} \qquad \text{or} \qquad \tau_{xy} = Q_{16}\varepsilon_x \tag{34.2}$$

where S_{16} and Q_{16} are, respectively, the compliance and stiffness terms defining the coupling magnitudes.[3] From a physical point of view, the shear deformation induced by an axial tensile stress is caused by the tendency for the layer to contract along the diagonals by unequal amounts due to differences in the Poisson's ratio in these two directions. Alternatively, considering the special case of a +45° layer, the axial stress may be resolved into planes at +45° and –45° to the direction of applied stress. The resulting strains due to equal resolved stress components along these directions are obviously different.

Intuitively, it is easily rationalized that the use of a $[\pm\theta]_T$ laminate will result in the mutual suppression of the tension-induced shear deformation in each individual layer. In the general case, equal numbers of layers in the off-axis, $+\theta$ and $-\theta$, layers will suppress this coupling; the resulting laminate is termed a *balanced laminate.*

Extension-Torsion Coupling. For this the previous balanced laminate $[\pm\theta]_T$ is considered. The spatial separation in the thickness direction results in equal and opposite deformations in the shear deformation induced by an axial tensile stress. This deformation situation therefore results in twisting of the laminate, a condition that is illustrated in Fig. 34.6. From a simplistic viewpoint, the illustration presented in Fig. 34.7 provides a type of designers' guide to coupling evaluations, which facilitates rational judgments in laminate design. All the responses indicated in these two figures can be confirmed by use of conventional lamination theory. Suppression of the twisting deformation is achieved by use of a symmetric laminate in which the off-axis layers below the central plane are mirrored by an identical off-axis layer at the same distance above the central plane (see Fig. 34.7).

Extension-Bending Coupling (Related through B_{11} and B_{22} Matrix Components). The simplest form of laminate, exhibiting a coupling between in-plane extension (or compression) and bending deformation, is the $[0°, 90°]_T$ unsymmetrical laminate. This response can be rationalized, on a physical basis, by recognizing that the neutral plane for this two-layer laminate will be located within the stiffest 0° layer, giving rise to a bending moment produced by the in-plane forces applied at the midplane and the associated effect between the two planes. For this case, it is clearly seen that the coupling would be suppressed by use of a four-layer symmetric laminate, i.e., $[0°, 90°]_s$, or a three-layer symmetric laminate such as $[0°, \overline{90°}]_s$, where the bar over the 90° layers signifies that this layer orientation is not repeated.

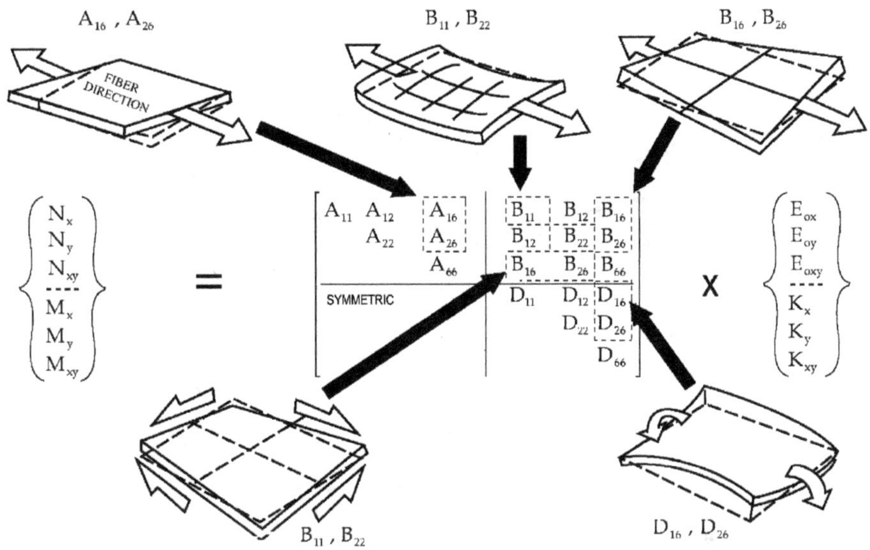

FIGURE 34.6 Illustration of coupling phenomena in laminated composite plates.

In-Plane Shear-Bending Coupling (Related through B_{16} and B_{26} Matrix Components). To visualize the mechanism associated with this mode of coupling, consider a $[\pm 45°]_T$ unsymmetrical, two-layer laminate subjected to in-plane shear loads. By recognizing that the in-plane shear is equivalent to a biaxial tension and compression loading with the tensile direction in the lower layer aligned with the fiber direction and, in the upper layer, transverse to the fiber direction, it will be realized that the plate will assume a torsional deformation (see Fig. 34.6).

Bending-Torsion Coupling (Related through D_{16} and D_{26} Matrix Components). For this mode of coupling, a four-layer balanced symmetric laminate, i.e., $[\pm\theta]_s$, is considered. The application of a bending moment, and an associated strain gradient, to this laminate will induce different degrees of shear coupling to the outer and inner layers. As a consequence, the response of the outer layers will dominate due to the higher strain levels in these layers, resulting in a net torsional deformation, as illustrated in Fig. 34.6. For qualitative assessment of this mode of coupling, the magnitude of the shear responses can be considered to exert an internal couple on the laminated plate as illustrated in Fig. 34.7. A similar rationale can be used to design a laminate that would not exhibit this coupling. For example, an eight-layer laminate of the configuration

$$[(\pm\theta)_s/(\mp\theta_s)]_T \qquad \text{or} \qquad [\pm\theta, \mp\theta, \mp\theta, \pm\theta]_T$$

will not exhibit bending-torsion coupling.

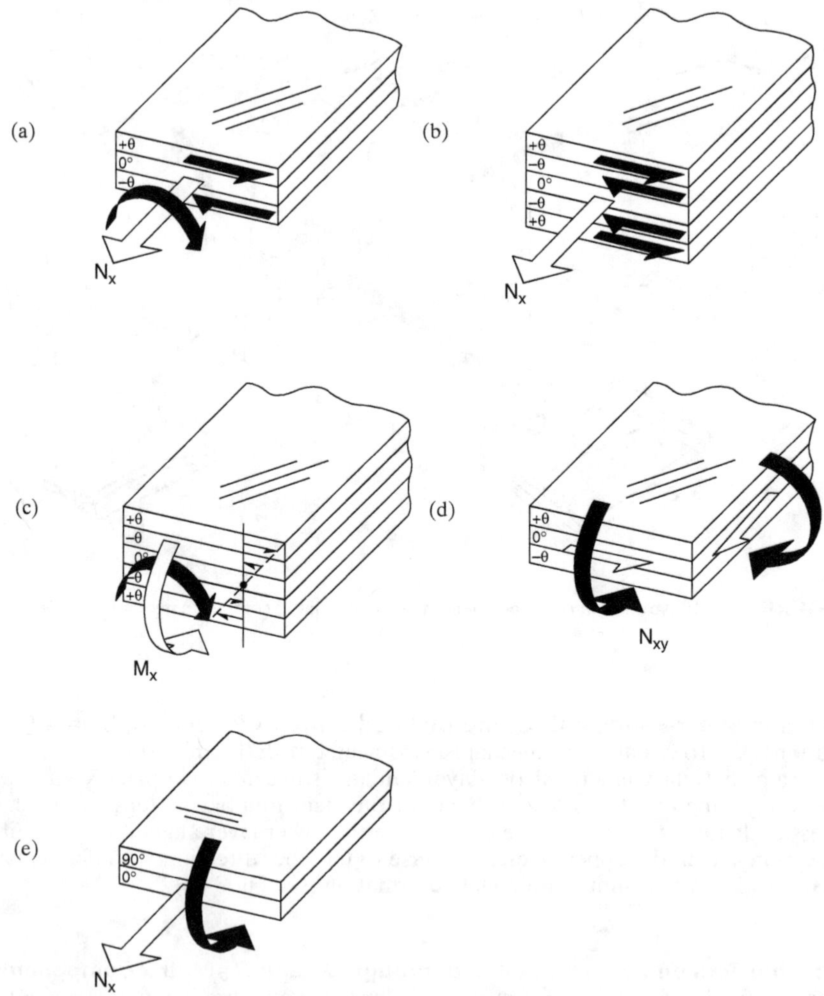

FIGURE 34.7 Designer's guide to coupling evaluation. (*a*) $B_{16} \neq 0$. (*b*) $B_{16} = 0$. (*c*) $D_{16} \neq 0$. (*d*) $B_{16}, B_{26} \neq 0$. (*e*) $B_{11}, B_{22} \neq 0$. Open arrows: applied force/moment. Shaded arrows: resulting displacement.

GENERAL LAMINATE DESIGN PHILOSOPHY

The recommended approach for laminates that are required to support biaxial loads is conveyed in the family of laminates represented by the shaded area in Fig. 34.8. This figure merely provides guidelines for selecting suitable laminates that have been shown to be durable and damage-tolerant. However, the form of presentation is also adopted for a system of *carpet plots* that can be very useful in the design and analysis of laminates for a specific composite system. These carpet plots facilitate reasonable predictions of the elastic and strength properties, and the coefficients of thermal expansion for a family of practicable laminates that comprise 0°, +45°, and

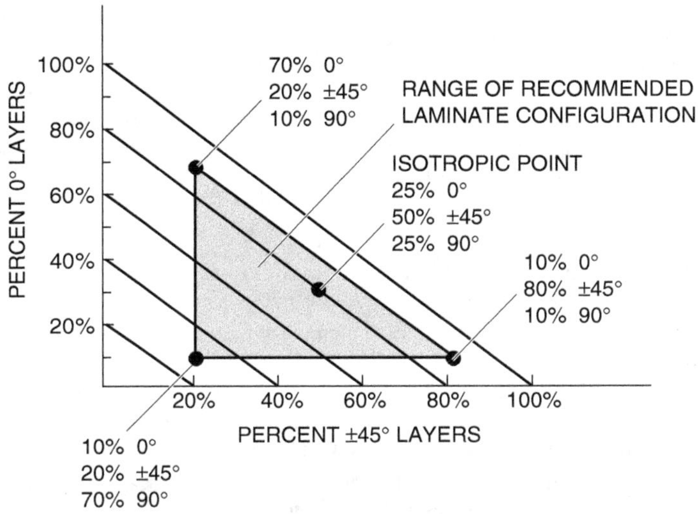

FIGURE 34.8 General guidelines for the selection of durable, damage-tolerant laminate design.

90° fiber orientations of any proportions in an assumed balanced, symmetric laminate arrangement. Examples of these carpet plots are presented in Ref. 3 and in most of the texts referenced previously. Even for highly directional loading, a nominal (approx. 10 percent) amount of layers, in each of the 0°, 90°, +45°, and –45° directions, should be included for the following reasons:

1. Providing restraints that inhibit development of microcracks that typically form in directions parallel to fibers.
2. Improved resistance to handling loads and enhanced damage tolerance (this is especially relevant for relatively thin laminates, i.e., less than 0.200 in. thick).
3. More manageable values of the major Poisson's ratio (ν_{xy}), particularly where interfaces exist with other materials or laminates with values in the 0.30 range.
4. Compatibility between the thermal expansion coefficients with respect to adjacent structure.

Other commonly adopted and recommended practices include laminate designs that minimize the subtended angle between adjacent layers and use of the minimum practicable number of layers of the same orientation in one group. To illustrate the former, a laminate configuration of $[0°, +45°, 0°, -45°, 90°]_s$ is preferred over a laminate such as $[0°, +45°, -45°, 0°, 90°]_s$ even though the in-plane thermoelastic properties would be identical for these two laminates. For the latter, the length of transverse microcracks tends to be limited by the existence of the layer boundaries; hence, a $[0°, +45°, 0°, -45°, 0°, \overline{90°}]_s$ laminate is preferred over a $[0°_3, +45°, -45°, \overline{90°}]_s$ laminate.

FATIGUE PERFORMANCE

The treatment of fatigue and damage accumulation in composite design is greatly complicated by the heterogeneity and anisotropy of the material in the laminated form. As a consequence, there is a multiplicity of mechanisms for the initiation and propagation of damage and, understandably, the approaches, such as Miner's cumulative damage rule discussed in Chap. 33, are not recommended. For similar reasons the test results obtained from small laboratory test coupons can rarely be used directly in support of design for prediction of fatigue performance. Nevertheless, such test coupon data can serve the purpose of obtaining preliminary indications of the fatigue performance of specific laminate design configurations.

Basic failure mechanisms that occur in laminated composites, in general, include the following:

1. Transverse cracking of individual layers in multidirectional laminates which will typically arrest at the interlaminar boundaries.
2. Fiber-matrix debonding which often can contribute to premature transverse cracking.
3. Delamination between layers due to interlaminar shear and/or tensile stress components that can be initiated by the aforementioned transverse cracks. Out-of-plane or bending loads on the structure will tend to give rise to such delamination.
4. Fiber breakage which will usually occur in the later stages of damage growth under monotonic static loading or under cyclic loading. However, most reinforcing fibers are not, in themselves, fatigue sensitive.

The first two initiating mechanisms motivate the above general laminate design philosophy advocated in the previous section, as illustrated in Fig. 34.8. A common sequence of failure events is illustrated for a quasi-isotropic, $[\pm 45°, 0°, 90°]_s$, carbon/epoxy laminate, also summarized in Fig. 34.9 (adapted from Ref. 9).

It may be stated, with some confidence, that the composites industry is able to design polymer matrix composite laminates of uniform thickness in a reliable manner. Extensive experience with PMCs has taught us to use fiber-dominated laminate designs, which are most often specified in the $[0°/\pm 45°/90°]_s$ or pseudo-isotropic form with respect to the in-plane directions. In-plane compression failure is somewhat of an exception since the matrix and the degradation thereof can develop delaminations and influence premature failure mechanisms. However, by far the largest number of development and in-service problems with composite hardware are associated with matrix-dominated phenomena, that is, interlaminar shear and out-of-plane tension forces. This is a major concern in that failure contributed by either one or a combination of these matrix-dominated phenomena are susceptible to the following:

1. High variability contributed by sensitivity to processing and environmental conditions.
2. Brittle behavior, particularly for early, i.e., 1970s era, epoxy matrix systems.
3. Inspectability of local details where flaws or defects may exist.
4. Low reliability associated with the lack of acceptable or representative test methods and complex, highly localized, stress states (the use of the transverse tensile strength of a unidirectional laminate for out-of-plane or thickness tensile strength is generally unconservative).
5. Potential degradation of residual static strength after fatigue/cyclic load exposure.

The development of stress components that induce interlaminar shear/out-of-plane tension failures was illustrated in Fig. 34.2, where commonplace generic features of composite hardware designs that frequently experience delaminations are

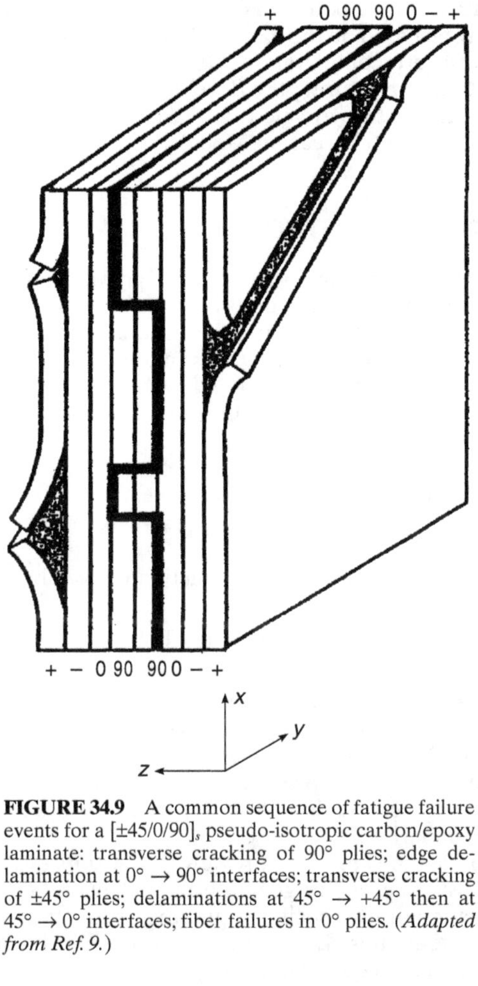

FIGURE 34.9 A common sequence of fatigue failure events for a $[\pm 45/0/90]_s$ pseudo-isotropic carbon/epoxy laminate: transverse cracking of 90° plies; edge delamination at 0° → 90° interfaces; transverse cracking of ±45° plies; delaminations at 45° → +45° then at 45° → 0° interfaces; fiber failures in 0° plies. (*Adapted from Ref. 9.*)

shown. It is at such details that PMC structures are particularly vulnerable under both static and fatigue loading. The propensity for delamination and localized matrix-dominated failures that represents a general characteristic of many PMCs is that notch sensitivity may be reduced after fatigue load cycling for local through-thickness penetrations. On the other hand, this demands that a fatigue life methodology should be available to deal with composite structures that are subjected to out-of-plane load components. Naturally, the capability of predicting the fatigue life is an essential element in the process of qualifying, or certifying, composite products and systems.

The design requirements generally specified for qualifying and/or certifying a composite product typically include (*a*) static strength, (*b*) fatigue/durability, and (*c*) damage tolerance. All of these requirements rely on a comprehensive appreciation of failure modes; the variability (or scatter); discontinuities caused by notches, holes, and fasteners; and environmental factors, particularly damage caused by the impact of foreign objects, machining, and assembly phenomena.

In the case of fatigue, three potential design approaches are considered. The particular selection may be based on the nature of usage, economics, safety implications, and the specific hardware configuration. Often some combination of approaches may be adopted particularly during the developmental phase. These three general categories of approach are the (*a*) Safe Life/Reliability Method, (*b*) Fail Safe/Damage Tolerance Method, and (*c*) Wearout Model.

SAFE LIFE/RELIABILITY METHOD

Statistically based qualification methodologies[9–11] provide a means for determining the strength, life, and reliability of composite structures. Such methods rely on the correct choice of population models and the generation of a sufficient behavioral database. Of the available models, the most commonly accepted for both static and fatigue testing is the two-parameter *Weibull distribution*. The Weibull distribution is attractive for a number of reasons, including the following:

1. Its simple functional form is easily manipulated.
2. Censoring and pooling techniques are available.
3. Statistical significance tests have been verified.

The cumulative probability of the survival function is given by

$$P_s(x) = \exp\,[(-x/\beta)^{\alpha_s}] \tag{34.3}$$

where α_s is the shape parameter and β is the scale parameter.

For composite materials, α_s and β are typically determined using the maximum-likelihood estimator.[12] In addition, the availability of pooling techniques is especially useful in composite structure test programs where tests conducted in different environments may be combined. Statistical significance tests are used in these cases to check data sets for similarity.

The following paragraphs present a review of the statistical method of Ref. 10. The development tests required to generate the behavioral database are outlined, followed by a discussion of the specific requirements for static strength and fatigue life testing. Special attention is given to the effect that matrix- and fiber-dominated failure modes have on test requirements.

A key to the successful application of any statistical methodology is the generation of a sufficiently complete database. The tests must range from the level of coupons and elements to full-scale test articles in a building-block approach. Additionally, the test program must examine the effects of the operating environment (temperature, moisture, etc.) on static and fatigue behavior. The coupon and subelement tests are used to establish the variability of the material properties. Although they typically focus on the in-plane behavior, it is also important to include the transverse properties. This is especially important in the case of research and development programs. The resulting data can be pooled as required and estimates of the Weibull parameters made. Thus, the level and scatter of the possible failure modes can be established. The transverse data are characterized by the highest degree of scatter. Element and subcomponent tests can be used to identify the structural failure modes. They may also be used to detect the presence of competing failure modes. Higher-level tests, such as tests of components, can be used to investigate the variability of the structural response resulting from fabrication techniques. The resulting database should describe, to the desired level of confidence, the failure mode, the data scatter, and the response variability of a composite structure. These data along with full-scale test articles can be used in the argument to justify qualification.

Out-of-plane failure modes can complicate the generation of the database. Well-proven and reliable transverse test methods are few. The typically high data scatter makes higher numbers of tests desirable. In addition, the increased environmental sensitivity in the thickness direction can cause failure mode changes, negating the ability to pool data and possibly resulting in competing failure modes. Thus, a design whose structural capability is limited by transverse strength can lead to increased testing requirements and qualification difficulties.

The static strength of a composite structure is typically demonstrated by a test to the *design ultimate load* (DUL), which is 1.5 times the maximum operating load, that is, the *design limit load* (DLL). Figure 34.10 shows the reliability achieved for a single static ultimate test to 150 percent of the DLL for values of the static strength shape parameter from 0 to 25. For fiber-dominated failure with α_s values near 20, such a test would demonstrate an *A-basis value,* which is defined as the value above which at least 99 percent of the population is expected to fall, with a confidence of 95 percent (a statistical tolerance limit as detailed in Chap. 18). However, for matrix-

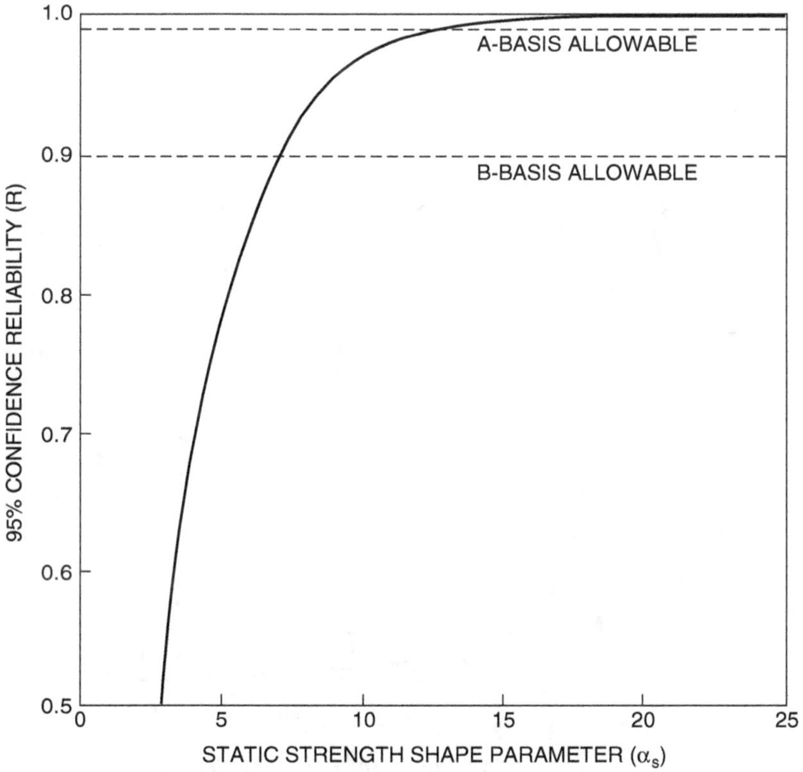

FIGURE 34.10 Plot of the 95 percent confidence reliability against the static strength shape parameter for a single full-scale static test to 150 percent of the design limit load.

dominated failure modes, with α_s ranging from 5 to 10, a test to 150 percent of the DLL would not demonstrate an A-basis value. Two options are available to increase the demonstrated reliability, namely, (*a*) increasing the number of test specimens, or (*b*) increasing the load level. The most effective choice is to increase the load level beyond 150 percent of the DLL, whereas increasing the number of test specimens yields little benefit and is expensive.

The two most applicable methods of statistical qualification approaches for fatigue are the *life factor* (also known as the *scatter factor*) and the *load enhancement factor.* The life factor approach relies on a knowledge of the fatigue life scatter factor from the development test program and full-scale test or tests. The factor gives the number of lives that must be demonstrated in tests to yield a given level of reliability at the end of one life. A plot of life factor N_F against the fatigue life shape parameter α_L is given in Fig. 34.11 for a typical scenario. A single full-scale test to demonstrate the reliability of the *B-basis* value, defined as that value above which at

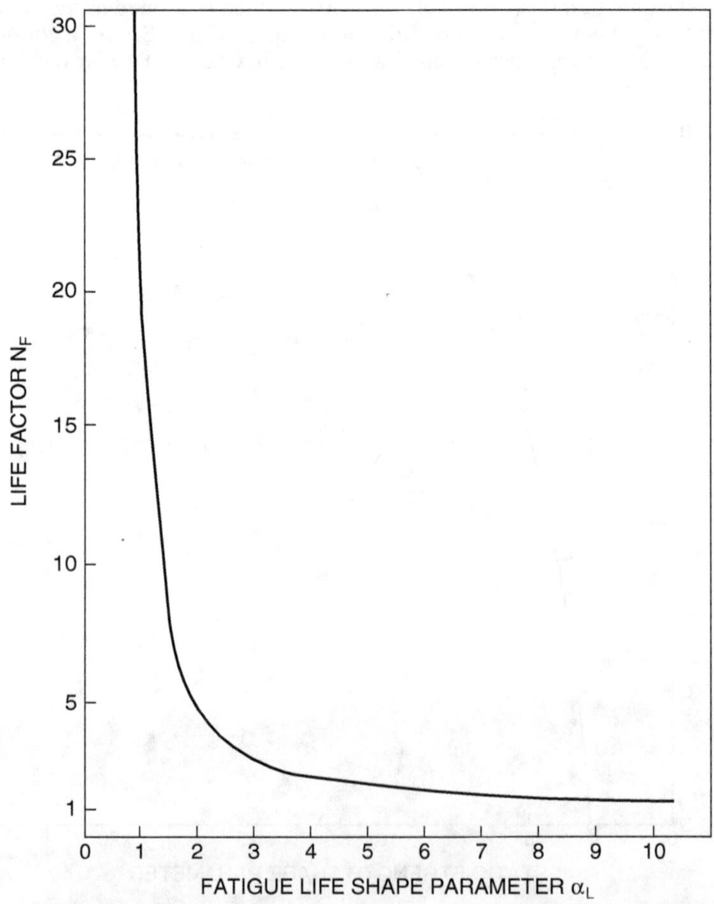

FIGURE 34.11 Plot of the life factor required to demonstrate the reliability of the B-basis results at the end of one life against the fatigue life shape parameter using a single full-scale test article.

least 90 percent of the population is expected to fall, with a confidence of 95 percent at the end of one life, is to be conducted. The curve shows that as the shape parameter approaches 1.0, the number of lives rapidly becomes excessive. Such is the case of an in-plane fatigue failure ($\alpha_L = 1.25$). Although few data for transverse fatigue are available, other than perhaps for bonded parts, it is reasonable to assume that the value of the shape parameter will be the same or less. Hence, it is apparent that the life factor approach is not acceptable for the certification of composites, especially where out-of-plane failure modes are dominant.

An alternative approach to life certification is the load enhancement factor, wherein the loads are increased during the fatigue test to demonstrate the desired level of reliability. Figure 34.12 illustrates the effect of the fatigue life shape parameter α_L and the residual-strength shape parameter α_R on the load enhancement factor F required to demonstrate B-basis reliability for one life using a single full-scale fatigue test to one lifetime. It is obvious that the required factor does not change significantly for fatigue life shape parameters in the range of 5 to 10. However, as the shape parameter approaches 1.0, as is the case for composites, the required load enhancement factor increases noticeably, especially for small values of the residual-strength shape parameter. This curve illustrates well the potential problems that may arise from dominant out-of-plane failure modes. Such failure modes tend to have low values of α_L (near 1.0) and also low values of α_R (in the range from 5.0 to 10.0). These values would make the required load enhancement factors prohibitively large. It is evident that for failure modes that exhibit a high degree of static and fatigue scatter, the life factor and load enhancement factor approaches can result in impossible test requirements. A combined approach can be achieved through the manipulation of the functional expressions. The resulting method allows some latitude in balancing the test duration and the load enhancement factor to demonstrate a desired level of reliability.

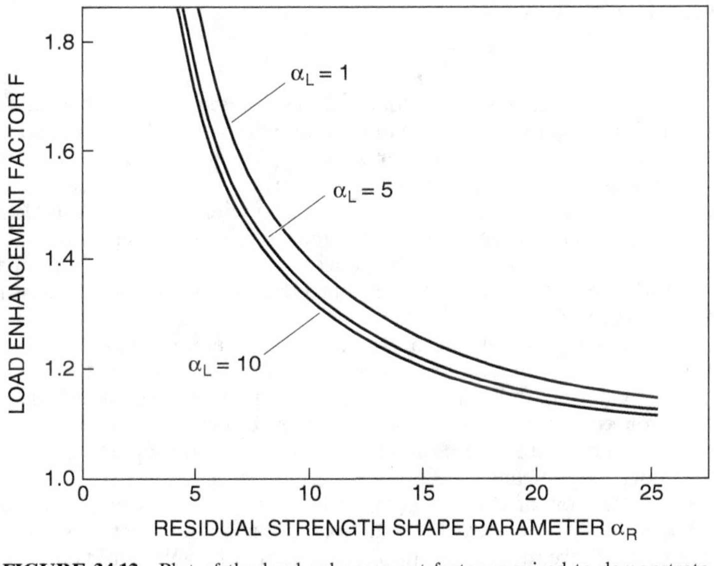

FIGURE 34.12 Plot of the load enhancement factor required to demonstrate the reliability of the B-basis results at the end of one life against the residual-strength shape parameter for three values of fatigue life shape parameter using a single fatigue test to one lifetime.

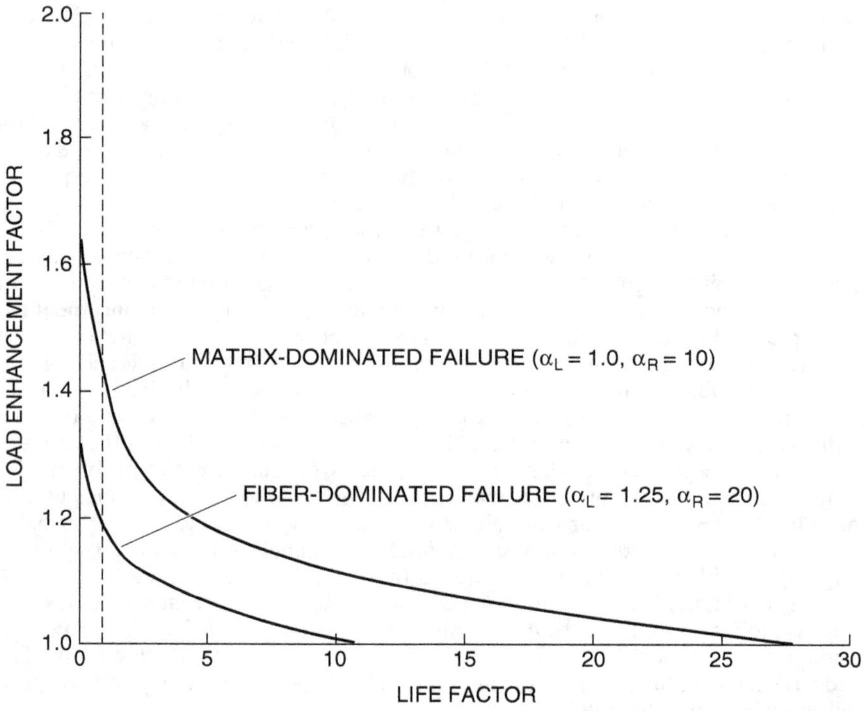

FIGURE 34.13 Plot of possible combinations of load enhancement factor and life factor necessary to demonstrate the reliability of the B-basis results at the end of one lifetime using a single full-scale test article for matrix- and fiber-dominated failure modes.

Figure 34.13 gives the curves of load enhancement factor against life factor for the cases of fiber- and matrix-dominated failures. Typical values for the fatigue life and residual-strength shape parameter were employed. The curves show the possible combinations of life factor (or test duration) and load enhancement factor to demonstrate the B-basis reliability at the end of one lifetime using a single full-scale fatigue test article. The curve for fiber-dominated failure modes exhibits quite reasonable values of life factor and load enhancement factor. For test durations ranging from 1 to 5 lifetimes, the load enhancement factor ranges from 1.18 down to 1.06. However, the test requirements for matrix-dominated failure are more severe. Over the range of life factor from 1 to 5, the load enhancement factor ranges from 1.4 down to 1.19. An environmental compensation factor would further complicate the test of a matrix-dominated failure. Such a factor must be combined with the load level. As is well known in composites, the adverse effects of environment on matrix properties are much more severe than on fiber-dominated properties, and the resulting factor may be significant.

Further illustration of the problems induced by a matrix-dominated failure is possible by assuming a limit exists on the load enhancement factor. Such limits may exist because of failure mode transitions at higher load levels. For instance, assuming a load enhancement factor of 1.2 is the maximum allowable value, it is obvious that a successful one-lifetime test for a fiber-dominated failure will demonstrate the reliability better than a B-basis test. For matrix-dominated failure, the same reliability would require a test duration of about 4.5 lives.

Two important aspects of the statistical qualification methodology are the generation of an adequate database and the proper execution of a full-scale demonstration test. The development test program must be conducted in a "building block" approach that produces confident knowledge of the material shape parameters, environmental effects, failure modes, and response variability. Perhaps the most important result should be the ability to predict the failure mode and know the scatter associated with it. Structures that exhibit transverse failures, which can result in competing modes and a high degree of scatter, may render the application of this fatigue methodology impractical. This result has been illustrated by the effect of shape parameters on both the static and fatigue test requirements. The requirements clearly show that a well-designed structure that exhibits fiber-dominated failure modes will be more easily qualified than one constrained by matrix-dominated effects.

FAIL SAFE/DAMAGE TOLERANCE METHOD

The damage tolerance philosophy assumes that the largest undetectable flaw exists at the most critical location in the structure, and the structural integrity is maintained throughout the flaw growth until detected by periodic inspection.[13] In this approach, the damage tolerance capability covering both the flaw growth potential and the residual strength is verified by both analysis and test. Analyses would assume the presence of flaw damage placed at the most unfavorable location and orientation with respect to applied loads and material properties. The assessment of each component should include areas of high strain, strain concentration, a minimum margin of safety, a major load path, damage-prone areas, and special inspection areas. The structure selected as critical by this review should be considered for inclusion in the experimental and test validation of the damage tolerance procedures. Those structural areas identified as critical after the analytical and experimental screening should form the basis for the subcomponent and full-scale component validation test program. Test data on the coupon, element, detail subcomponent, and full-scale component level, whichever is applicable, should be developed or be available to (*a*) verify the capability of the analysis procedure to predict damage growth/no growth and residual strength, (*b*) determine the effects of environmental factors, and (*c*) determine the effects of repeated loads. Flaws and damage will be assumed to exist initially in the structure as a result of the manufacturing process, or to occur at the most adverse time after entry into service.

A decision to employ proof testing must take the following factors into consideration:

1. The loading that is applied must accurately simulate the peak stresses and stress distributions in the area being evaluated.
2. The effect of the proof loading on other areas of the structure must be thoroughly evaluated.
3. Local effects must be taken into account in determining both the maximum possible initial flaw/damage size after testing and the subsequent flaw/damage growth.

The most probable life-limiting failure experienced in composite structure, particularly in nonplanar structures where interlaminar stresses are present, is delamination growth. Potential initiation sites are free edges, bolt-holes, and ply terminations (see Fig. 34.2), in addition to existing manufacturing defects and subsequent impact damage. Hence, an analysis technique for the evaluation of growth/no growth of delaminations is an essential tool for the evaluation of the damage tolerance of composite structures. A numerical method is available through the use of finite element analysis (see Chap. 23) and the crack closure integral technique from fracture mechanics.[14] Prerequisites for an evaluation are as follows:

1. A structural analysis made in sufficient detail to indicate the locations where the critical interlaminar stresses exist.
2. Experimentally based critical interlaminar strain energy release rates G_{Ic}, G_{IIc}, and a subcritical growth law, that is, *da/dN,* where *da/dN* is the rate of change of the crack length or damage zone size *a* with the number of cycles *N,* against ΔG for each mode (see Chap. 33).
3. A mixed mode I/mode II fracture criterion.

The test specimens used to generate the required mode I and mode II fracture toughness parameters are described in detail in Ref. 15. The application of this approach requires a significant analysis and test effort to evaluate hot spots within the structure and to generate the necessary fracture toughness data. One limitation is the absence of a reliable mixed-mode fracture criterion, and consequently this method is not considered sufficiently mature to warrant a recommendation for wide general application, particularly for developmental composite hardware evaluations.

THE WEAROUT MODEL

Wearout is defined as the deterioration of a composite structure to the point where it can no longer fulfill its intended purpose. The wearout methodology was developed in the early 1970s and is comprehensively summarized in Ref. 12. The essential features are portrayed in Fig. 34.14. This methodology was previously used by the military aircraft command for the certification of several composite aircraft components. In essence, the wearout approach recognizes the probability of progressive structural deterioration of a composite structure. The approach utilizes the development test data on the static strength and the residual strength, after a specified period of use, in conjunction with proof testing of all product hardware items to characterize this deterioration and protect the structure against premature failures. It has become evident that the residual stiffness is an indicator of the extent of the structural deterioration and can be an important performance parameter with regard to the natural frequencies of oscillation of the aerodynamic surfaces. Thus, in some instances, it may be prudent to incorporate a residual-stiffness requirement in an adopted methodology to evaluate the tolerance of the structure to component stiffness degradation.

The difficulties in the implementation of the methodology include the determination of the critical load conditions to be applied for static and residual strength and stiffness testing and for the proof load specification. Similar difficulties would arise in the case of all candidate methodologies considered here, and indeed emphasize the importance of a representative structural analysis. However, the advantage of the wearout approach for advanced composite hardware development projects resides in the ability to assign gates for safe flight testing as the flight envelope is progressively expanded.

Since the era of the initial development and interest in the wearout approach, there appears to have been minimal development or usage. Nevertheless, the poten-

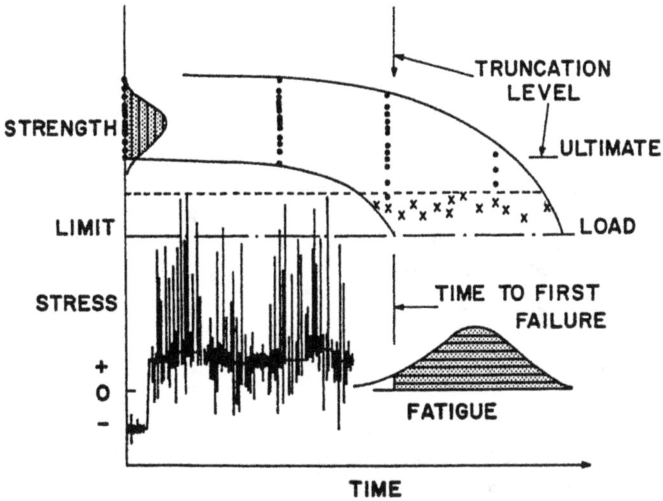

FIGURE 34.14 Essential features of the "wearout model" relating static failure, load history, and fatigue failure.

tial motivation for a methodology of this type calls for a brief review of the physical and theoretical basis for the important concepts. Further detail can be found in Refs. 12 and 16.

By combining several basic assumptions regarding the behavior of a composite structure under load with basic Weibull statistics, a *kinetic fracture model* can be derived. This model serves to assist in predicting the fatigue wearout behavior of composite structures. The first assumption concerns the growth rate of an inherent or real material flaw, da/dt, which is deemed to be proportional to the strain energy release rate G of the material system raised to some power r, where r is to be determined experimentally. Thus,

$$da/dt \propto G^r \tag{34.4}$$

where a is the flaw length. As the cyclic load, $F(t)$, is applied to the flawed body, the internally stored strain energy will occasionally exceed the critical level required to overcome the local resistance of the material to flaw growth or damage accumulation, and flaw or damage growth will occur. Impediments to further development have been related to those cited in Chap. 33, as it pertained to the fracture mechanics method for metals, i.e., the need for further data to define the growth rate and/or threshold level below which the damage area does not grow. One important wearout parameter r is defined as the slope of the da/dN curve, or may be derived from the S-N curve for the failure/damage mode in question.

Various relationships have been proposed[12] relating the initial Weibull static shape parameter, α_0, and the fatigue life shape parameter, α_f, both of which tend to be a function of the damage size exponent alone. Specifically, available relationships are given by

$$\alpha_0 = 2r + 1 \qquad \text{and} \qquad \alpha_f = \frac{\alpha_0}{2(r-1)} \tag{34.5}$$

Postulating that the composite system will lose strength at a uniform rate with respect to a logarithmic scale of cycles or time, then from the specific fatigue curve expressed as

$$NF_b^{\gamma} = B_N \qquad \text{or} \qquad tF_b^{\gamma} = B_t \tag{34.6}$$

the slope of the fatigue curve is given by $\gamma = -1/2$. In Ref. 16, a compilation of data on damage growth rate exponents from a broad range of literature items, including various types of polymer composite systems and composite bonded structures, were found to range between 4.3 and 6.6.

DAMPING CHARACTERIZATION

The major sources of damping in polymer matrix composites are associated with the viscoelastic or microplastic phenomena of the polymer matrix constituent and, to some degree for some composite systems, with weak fiber-matrix interfaces to microslip mechanisms. Other sources of damping, such as matrix microcracking and delamination resulting from poor fabrication conditions or service damage, can also create increased damping in certain cases. Very little or no damping is contributed by the fiber-reinforcement constituent with the possible exception of aramid, i.e., Kevlar, fibers. Environmental factors, such as temperature, moisture, and frequency, on the other hand, can have a significant effect on damping.

Two-phase materials therefore tend to derive any damping from the polymer matrix phase in a large majority of composite systems. Consequently, matrix-influenced deformations, such as the interlaminar shear and tension components, are the significant contributors. For the basic unidirectional composite, some closed-form predictive methods are available, but generally the micromechanics theories have been found to be unreliable for damping determinations, although reasonable for modulus predictions. Structural imperfections at the constituent level are considered to be the main contributors to this situation.

As mentioned earlier, micromechanics-based theories are available to give some indication of the effects of fiber volume content on damping parameters for unidirectional materials. One example based on conventional viscoelasticity assumption was formulated in Ref. 11 for the case of longitudinal shear deformation. For this case the *specific damping capacity* (SDC), ψ_{12}, for longitudinal shear can be expressed[17] as

$$\psi_{12} = \frac{\psi_m(1 - V_f)[(G + 1)^2 + V_f(G - 1)^2]}{[G(1 - V_f) + (1 - V_f)][G(1 - V_f) + (1 + V_f)]} \tag{34.7}$$

where ψ_m = the SDC for the matrix
G = the ratio of fiber shear modulus to that of the matrix
V_f = the fiber volume fraction

For the condition of flexural vibration of composite beams, the damping due to transverse shear effects that are highly matrix-dominated exhibit up to two orders of magnitude greater damping than pure axial, fiber-direction effects. Specific data, adapted from Ref. 18, on the SDC for the flexural vibration of unidirectional beams,

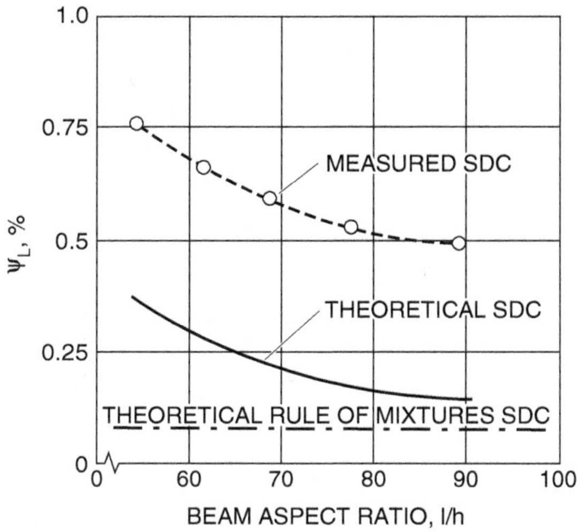

FIGURE 34.15 Variation of flexural damping with aspect *ratio* for high-modulus carbon fiber in DX209 epoxy resin $V_f =$ 0.5, SDC, shear damping contribution.

over a range of aspect ratios (length ℓ/thickness h), are compared to theoretical predictions in Fig. 34.15. Here the steady increase in damping for progressively lower beam aspect ratios is clearly due to the shear deformation which indicates a much stronger effect on damping than on the flexural modulus. The discrepancies in the theoretically predicted SDC in Fig. 34.15 is generally attributed again to imperfections in the composite at the constituent level.

The damping trends for the other matrix-influenced deformational mode of transverse tension (at 90° to the fiber direction) in a unidirectional composite is illustrated in Fig. 34.16 for an E-glass fiber-reinforced epoxy over a wide range of fiber volume fractions V_f. Substantial damping can also occur in the deformation of an off-axis, unbalanced lamina or laminate, due to shear-induced deformation created by coupling under tension, compression, or flexural loading directed at an angle to the fiber direction. In Ref. 19, good correlation between the theoretical prediction and experimental measurements is demonstrated for a complete range of fiber orientations from 0° to 90° (see Fig. 34.17). Based on the flexural vibration of a high-modulus carbon-fiber/epoxy matrix system with $V_f = 0.5$, Fig. 34.17 compares both the flexural modulus and SDC. The latter damping parameter was predicted using the approximate relationship

$$\psi_\theta = E_x\left[\frac{\psi_2}{E_2}\sin^4\theta + \frac{\psi_{12}}{G_{12}}\sin^2\theta\cos^2\theta\right] \tag{34.8}$$

where x = the axial direction of the beam
θ = the angle between the fiber direction and the axis of the beam
E_2, ψ_2 = the elastic modulus and SDC, respectively, in the transverse direction of the fiber
G_{12}, ψ_{12} = the shear modulus and shear-induced SDC, respectively, referred to directions parallel and perpendicular to the fibers

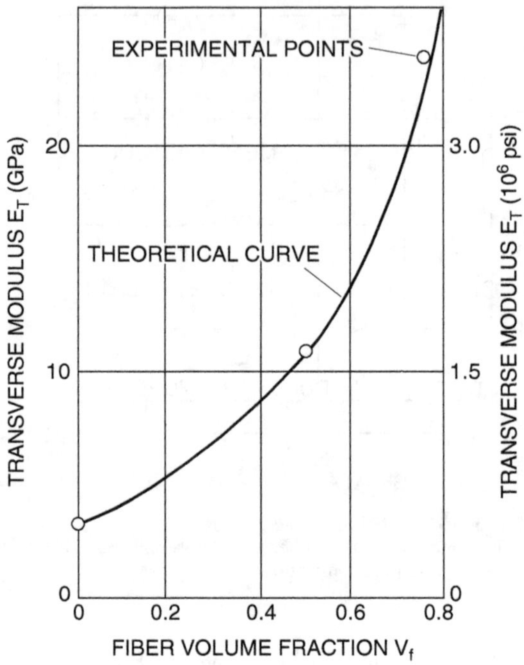

(a)

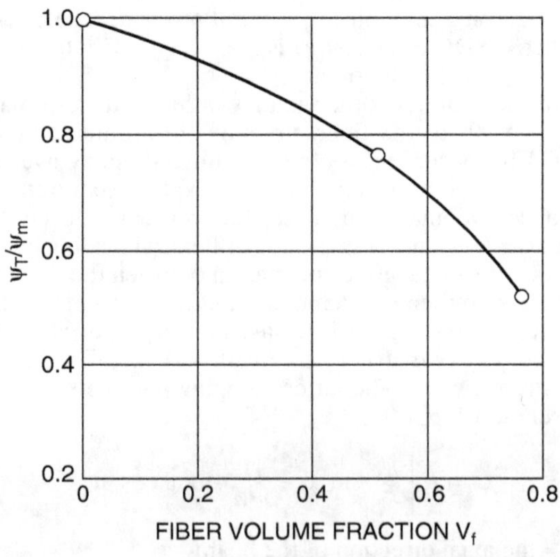

(b)

FIGURE 34.16 (*a*) Variation of transverse modulus with fiber volume fraction for unidirectional in glass/epoxy beam flexure. (*b*) Variation of transverse damping to matrix damping ratio with fiber volume fraction for unidirectional glass/epoxy beam flexure.

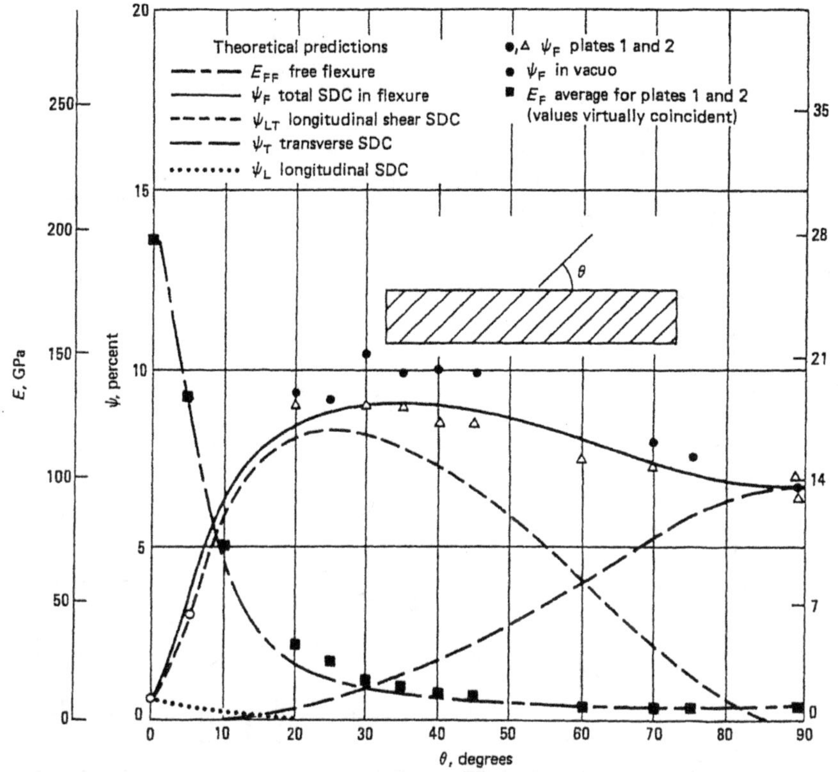

FIGURE 34.17 Variation of flexural modulus and specific damping capacity with fiber orientation for a carbon/epoxy, off-axis laminate in flexure.

In this relationship the modulus E_x is given by

$$\frac{1}{E_x} = \frac{\cos^4 \theta}{E_1} + \frac{\sin^4 \theta}{E_2} - \frac{2\nu_{12} \cos^2 \theta \sin^2 \theta}{E_1} + \frac{\cos^2 \theta \sin^2 \theta}{G_{12}} \tag{34.9}$$

With the above correlation as background, predictive methods for the damping of laminated beam specimens based on the classical laminate analysis method referenced above (see Ref. 3), the damping terms were incorporated and presented in Ref. 20 and summarized in Ref. 18. The approach involved formulation of the overall SDC, ψ_{ov}, to yield the total energy dissipated divided by the total energy stored as

$$\psi_{ov} = \frac{\Sigma \Delta Z}{\Sigma Z} = \frac{\psi_1 Z_1 + \psi_2 Z_2 + \psi_{21} Z_{12}}{Z_1 + Z_2 + Z_{12}} \tag{34.10}$$

where $\Delta Z_1 = \psi_1 \cdot Z_1$ is the energy dissipation in the 1-direction, the axial being parallel to the fiber direction in a given layer.

Predicted values obtained by this approach are compared with measured values for a balanced, angle-ply laminated beam of high-modulus carbon-fiber/epoxy in flexural vibration in Fig. 34.18. In this figure, the SDC approaches 10 percent maximum at a fiber orientation of ±45°, where the dynamic flexural modulus, however, is

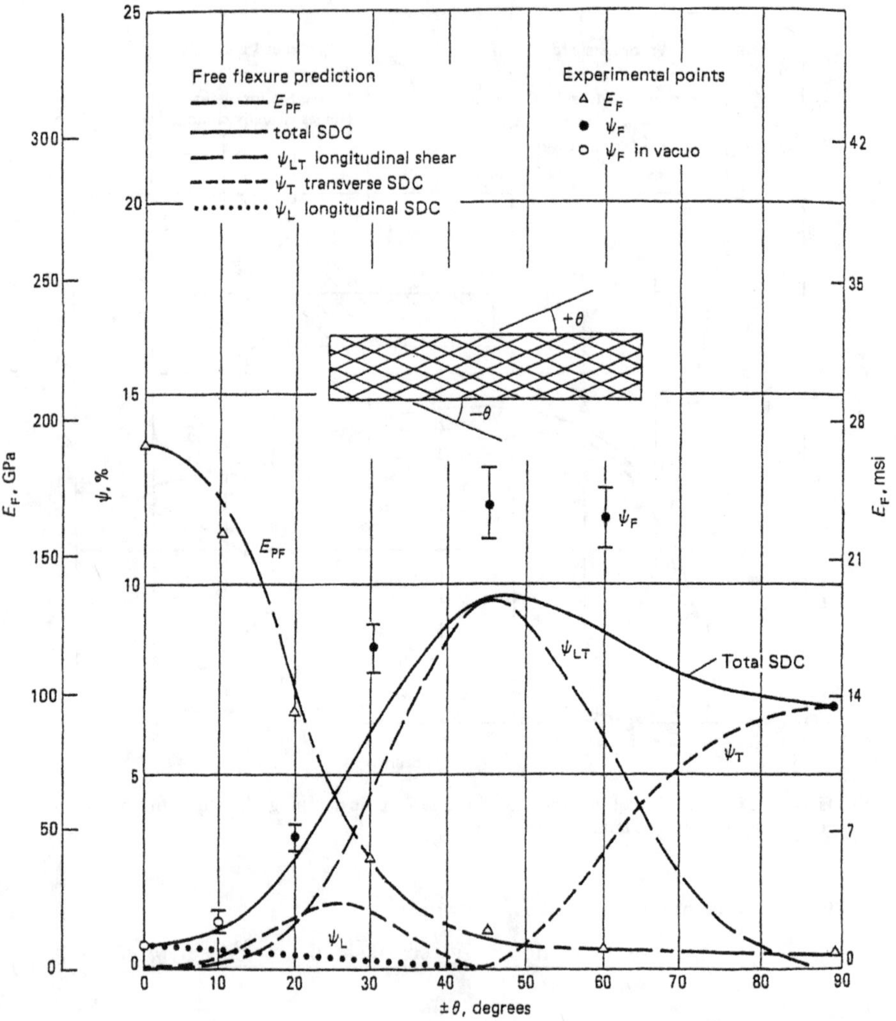

FIGURE 34.18 Variation of flexural modulus and specific damping capacity with fiber orientation for a carbon/epoxy, angle-ply laminate $[\pm\theta]_s$ in flexure.

very small. Damping predictions are again shown to be below measured values, but the discrepancy is much smaller in this case and the general trend with respect to fiber orientation is predicted extremely well.

The above theoretical treatment has subsequently been extended to laminated composite plates, again with reasonable correlation. SDC values ranged from just below 1 percent up to around 7 percent, with lower damping exhibited by the carbon/epoxy-laminated plates configured to provide essentially isotropic elastic modulus in the plane of the plate. Reference 18 contains extensive comparisons, including mode shapes, for both carbon/epoxy- and glass/epoxy-composite laminates.

REFERENCES

1. Suarez, S. A., R. F. Gibson, C. T. Sun, and S. K. Chaturvedi: *Exp. Mech.,* **26**:175 (1986).
2. Kedward, K. T.: "The Application of Carbon Fiber Reinforced Plastics to Aero-Engine Components," *Proceedings, 1st Conference on Carbon Fibers, Their Composites, and Applications,* The Plastics Institute, London, 1971.
3. Rothbart, H. A. (ed.): "Mechanical Design Handbook," 4th ed., Sec. 15, The McGraw-Hill Companies, Inc., New York, 1996.
4. Mallick, P. K.: "Fiber-Reinforced Composites: Materials, Manufacturing and Design," 2d ed., Marcel Dekker, New York, 1993.
5. Daniel, I. M., and O. Ishai: "Engineering Mechanics of Composite Materials," Oxford University Press, New York, 1994.
6. Gibson, R. F.: "Principles of Composite Material Mechanics," McGraw-Hill, New York, 1994.
7. Jones, R. M.: "Mechanics of Composite Materials," 2d ed., Taylor and Francis, Philadelphia, Pa., 1999.
8. Whitney, J. M.: "Structural Analysis of Laminated Anisotropic Plates," Technomic Publishing Company, Lancaster, Pa., 1987.
9. O'Brien, T. K.: "Composite Materials Testing and Design," *ASTM STP 1059,* **9**:7 (1990).
10. Whitehead, R. S., H. P. Kan, R. Cordero, and E. S. Saether: "Certification Testing Methodology for Composite Structures," *NADC-87042,* October 1986.
11. Sanger, K. B.: "Certification Testing Methodology for Composite Structures," *NADC-86132,* Jan. 1986.
12. Halpin, J. C., K. L. Jerina, and T. A. Johnson: "Analysis of Test Methods for High Modulus Fibers and Composites," *ASTM STP 521,* 1973.
13. Anon.: "Damage Tolerance of Composites," *AFWAL-TR-87-3030,* July 1988.
14. Rybicki, E. F., and M. F. Kanninen: *Engineering Fracture Mechanics,* **9**(4):931 (1977).
15. Wilkins, D. J.: "A Preliminary Damage Tolerance Methodology for Composite Structures," Proc., Workshop on Failure Analysis of Fibrons Composite Structures, *NASA-CP-2278,* 1982.
16. Kedward, K. T., and P. W. R. Beaumont: *International Journal of Fatigue,* **14**(5):283 (1992).
17. Hashin, Z.: *Int. J. Solids Struct.,* **6**:797 (1970).
18. Adams, R. D.: "Engineered Materials Handbook: Composites," ASM International, Materials Park, Ohio, 1987.
19. Adams, R. D., and D. G. C. Bacon: *J. Composite Materials,* **7**(4):402 (1973).
20. Ni, R. G., and R. D. Adams: *Composites Journal,* **15**(2):104 (1984).

CHAPTER 35
MATERIAL AND SLIP DAMPING

Peter J. Torvik

INTRODUCTION

As used in this chapter, the term *damping* refers to the dissipation of energy in a material or structure under cyclic stress or strain. Not treated here are the dynamic vibration absorbers and auxiliary mass dampers discussed in Chap. 6, nor the applied damping treatments discussed in Chap. 36. Only the inherent propensity of materials and joints to dissipate energy through the process of converting mechanical energy (strain and kinetic) to heat is considered. When such dissipation occurs locally within the material, the process is referred to as *material damping,* taken as inclusive of the dissipative mechanisms sometimes referred to as mechanical hysteresis, anelasticity, or internal friction. Attention is restricted to those mechanisms which provide significant dissipation at stresses of engineering interest. Mechanisms primarily used in physical metallurgy and solid-state physics as guides to the internal structure of the material are not treated.

Although the assumption of a perfectly elastic material is very convenient for use in the analysis of structures, and is adequate in most cases, no structural materials are truly elastic. A system given an initial perturbation will eventually come to rest unless the energy dissipated is offset by the addition of energy. Cyclic motion of a structure can be sustained at constant amplitude only if the energy lost through dissipation is offset by work done on the system. Damping can be advantageous to performance, governing as it does the maximum amplitude achieved at resonance and the rate at which a perturbed system progresses to a satisfactorily quiescent state. In addition to lowering the probability of failure due to fatigue, reductions of amplitude can have many other benefits, such as reducing visible vibrations, the sound emitted from a valve cover, or the acoustic signature of a submarine propeller. Damping, however, can also be disadvantageous. It may produce such unwanted phenomena as shaft whirl, instrument hysteresis, and temperature increases due to self-heating.

When dissipation occurs as a consequence of relative motion between two bodies, the result may be described as *friction damping* and may result from relative rigid-body motions, or *sliding,* or from unequal deformations of the contacting surfaces, enabling *slip.* As typically used, *internal friction* is an inclusive term for all types of material damping, regardless of mechanism. However, there appear to be cases in which the mechanism of material damping is truly coulomb friction. This will be discussed as an example of *slip damping.*

The distinction between damping as a *material* property and as a *system* property is emphasized and, as the determination of the damping properties of a material typically begins with a measurement of the damping of a system containing the test sample, some methods of measuring system damping are reviewed. Several dissipative mechanisms are discussed, with dissipative properties of representative materials given as examples.

MEASURES OF MATERIAL DAMPING

The most fundamental measure of the dissipative ability of a material is the *specific damping energy* or *unit damping,* defined as the energy dissipated in a unit volume of material at homogeneous strain and temperature, undergoing a fully reversed cycle of cyclic stress or strain. The specific damping energy D has dimension of energy per unit volume, per cycle, and is, in general, a function of the amplitude and history of stress or strain, temperature, and frequency. For some materials, the unit damping is also dependent on the mean (static) stress or is influenced by magnetic fields. The unit damping is customarily given in terms of the amplitude of a uniaxial stress or strain, tensile or shear, with multiaxial loadings characterized[1] by an appropriate equivalent uniaxial stress or strain.

Material damping may be categorized as being *linear* or *nonlinear.* In the first class, the energy dissipated per cycle is dependent on the square of the amplitude of cyclic stress. As the strain energy density is also normally proportional to the square of stress amplitude, the ratio of dissipated and stored energies, as well as other *dimensionless* measures of damping, are then independent of amplitude. Materials displaying these attributes may be said to display *linear damping.* In the second class of materials, the energy dissipated per cycle varies as amplitude of cyclic stress to some power other than 2. If the strain energy density varies as, or nearly as, the square of amplitude, the ratio of dissipated to stored energy, as well as other dimensionless measures of damping, are then also functions of the amplitude of stress or strain. Such materials may be said to display *nonlinear damping.* The *specific damping energy* (SDE) is the most robust of all damping measures, being applicable to nonlinear as well as linear materials.

While some important mechanisms of damping, such as viscoelastic and thermoelastic, are essentially linear, others are not. That this is true may be seen from the values of the SDE as functions of uniaxial tensile stress for the variety of structural materials shown in Fig. 35.1 as measured by Lazan[1] and colleagues. For stresses below a critical value (a *cyclic stress sensitivity limit,* usually about 70 percent of the fatigue strength at 2×10^7 cycles), the damping energy is typically independent of history and increasing with a power n of the amplitude of fully reversed dynamic stress, σ_d, somewhat greater than 2, that is,

$$D = J\sigma_d^n \tag{35.1}$$

At higher stress, the same functional form may be applied, but the damping typically increases much more rapidly with stress. The parameter n is then typically much greater than 2 and may increase or decrease with the number of cycles.

While rooted in the concept of a linear viscoelastic material, the concept of a

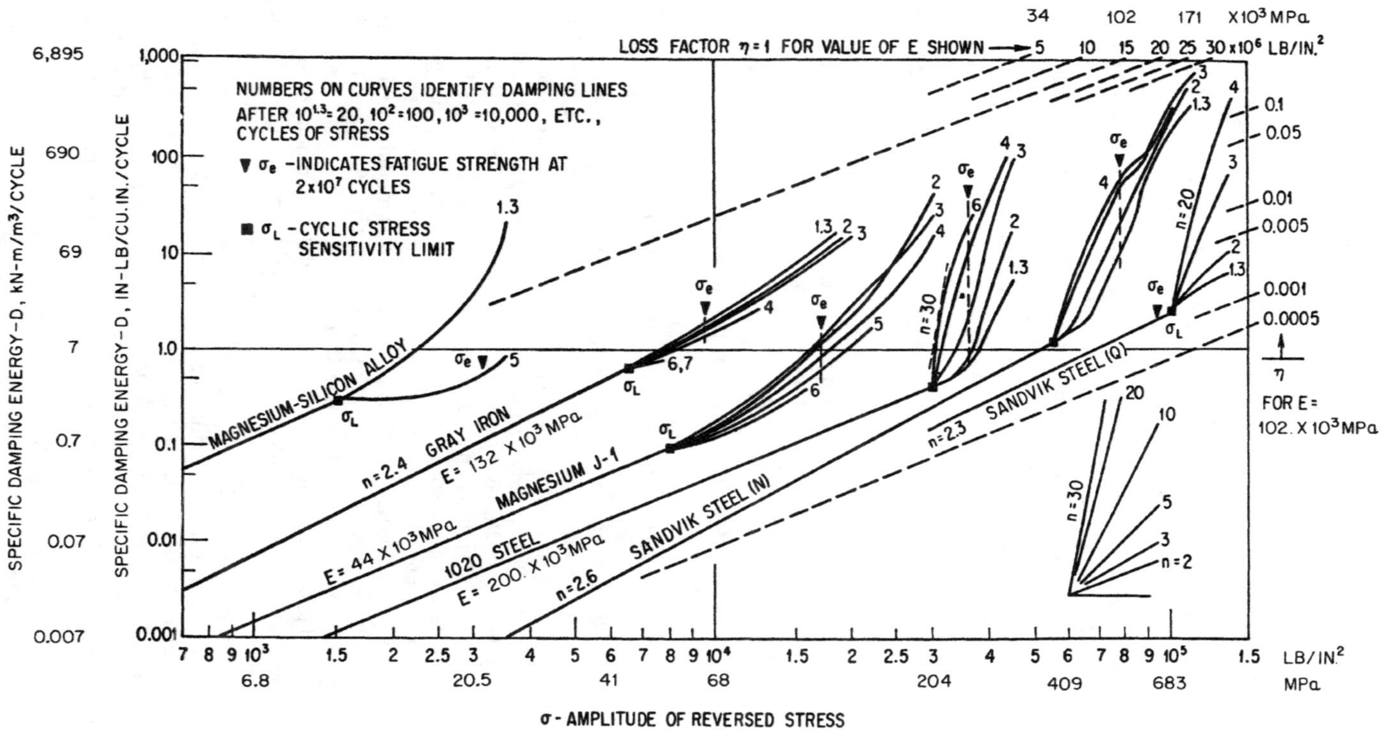

FIGURE 35.1 Specific damping energy of various materials as a function of amplitude of reversed stress and number of fatigue cycles. Number of cycles is 10 to power indicated on curves. For example, a curve marked 3 is at 1000 cycles.

complex modulus may be adopted to characterize the dissipation of other materials undergoing cyclic loading. In the case of a nonlinear material, we may define amplitude-dependent effective values of a *storage* and *loss modulus* E_1 and E_2, as

$$E_1(\omega,T,\varepsilon_0) \equiv \frac{2U(\omega,T,\varepsilon_d)}{\varepsilon_d^2} \qquad \text{and} \qquad E_2(\omega,T,\varepsilon_0) \equiv \frac{D(\omega,T,\varepsilon_d)}{\pi\varepsilon_d^2} \tag{35.2}$$

where U is the *strain energy* in the unit volume, stored and recovered during each cycle of vibration; D is the SDE; and ε_d is the amplitude of cyclic strain. When formulated in this manner, the amplitude-dependent components of a complex modulus are quantities defined from fundamental considerations of energy, rather than as a consequence of a particular representation of a stress-strain law. Once determined for all applicable values of frequency, temperature, and strain, they may be used with distributions of temperature and strain established a priori to find the total energy stored and dissipated in all dissipative elements undergoing cyclic loading by summing over all volume elements of the structure

$$U_0(\omega) = \sum_{m=1}^{M} \frac{E_1(\omega,T_m,\varepsilon_{dm})}{2} |\varepsilon_d|_m^2 \qquad \text{and} \qquad D_0(\omega) = \sum_{m=1}^{M} \pi E_2(\omega,T_m,\varepsilon_{dm})|\varepsilon_d|_m^2 \tag{35.3}$$

For structural materials, the values of storage and loss modulus are typically independent of frequency, but are variable with amplitude and somewhat with temperature. In the case of viscoelastic materials, they are typically independent of amplitude, but vary strongly with both frequency and temperature.

A *material loss factor* may also be defined as the ratio of energy dissipated in the unit volume per radian of oscillation to the peak energy stored.

$$\eta = \frac{D}{2\pi U} = \frac{1}{2\pi} \frac{\pi E_2(\omega,T,\varepsilon_d)}{E_1(\omega,T,\varepsilon_d)/2} = \frac{E_2(\omega,T,\varepsilon_d)}{E_1(\omega,T,\varepsilon_d)} \tag{35.4}$$

If either or both of the components of the modulus are dependent on amplitude, then the material loss factor is also dependent on amplitude. Note that the use of the loss factor, defined in terms of energy dissipated per cycle, is to be preferred over the sometimes-used *specific damping capacity* or *damping index,* computed from the energy dissipated per cycle by $\psi = D/U$, as the unit of the radian is more truly a dimensionless quantity than is the cycle. Values of a material loss factor have been added to Fig. 35.1, normalized to the case of a material with an amplitude-independent storage modulus of 102 GPa. The resulting values must then be adjusted to the actual value for the material of interest by multiplying by the ratio of the actual modulus to the reference modulus. Values for the high strength and mild steels, for example, are found to be about 0.001 to 0.005 for stresses in the range 40 to 200 MPa.

The presence of dissipation implies that the induced displacement, or strain, must be out of phase with the causative force, or stress, and that the response must lag the input so as to give rise to a positive dissipation. The resulting *phase angle* is sometimes measured directly or indirectly and offered as a material property. As both components of a complex modulus must be positive, the angle by which the strain lags the stress is given by $\tan \phi = E_2(\omega,T,\varepsilon_d)/E_1(\omega,T,\varepsilon_d)$, referred to as the *loss tangent.* But even when such a phase angle can be measured, interpretation of the loss tangent as the loss factor is not well justified in the case of an amplitude-dependent material because the inherent presumption of harmonic stress and strain implies linearity.

Other measures of damping, such as the logarithmic decrement, are often given as material properties, but these are truly system properties and yield material properties only when the system measured consists solely of a homogeneously strained sample of the material of interest. Also, these other measures typically depend in some manner on an assumption of linearity.

MATERIAL DAMPING: MECHANISMS AND MODELS

It is convenient to classify material damping as displaying *static* or *dynamic hysteresis.* In the case of the former, the stress-strain relationship does not depend on time, as the state of stress is independent of the rate of stress or of strain. Upon change in load, the change in deformation is essentially instantaneous, but may be dependent on the prior load history. Removal of the load leaves a residual deformation, not recovered over time. The instantaneous response gives rise to hysteresis loops with sharply pointed ends. In contrast, materials displaying dynamic hysteresis require representation in terms of stress-strain relationships incorporating time, as the state of stress depends on the instantaneous rate of stress and/or strain, as well as on the current values. While there may be some instantaneous deformation resulting from the application of load, additional deformations (creep) occur over time. The deformation remaining after removal of the loading changes with time (relaxation) and, in some cases, may disappear entirely. Because the response is not instantaneous, the hysteresis loops display finite curvatures at the extremal values. The term *anelasticity* was used by Zener[2] to describe materials such as these that are linear and unload without permanent deformation, but for which the relationship between stress and strain is not single-valued.

Several mechanisms of damping have been found to produce sufficient dissipation to be of engineering interest in the mitigation of structural vibrations. Two such mechanisms, *plasticity* and *magnetoelasticity,* for which the damping is essentially independent of frequency (*static hysteresis*) but is inherently nonlinear, will be discussed. Additionally, two damping mechanisms for which the damping is strongly dependent on frequency (*dynamic hysteresis*) will be considered. These are dissipations due to *viscoelastic* and *thermoelastic* effects. In contrast to the mechanisms of the first category, these mechanisms are typically linear, with material loss factors independent of amplitude.

DAMPING DUE TO PLASTICITY

The dominant dissipation in most structural materials at stress levels of engineering interest is due to some mechanism of plastic deformation,[1] variously referred to as plastic slip, localized plastic deformation, crystal plasticity, cyclic plastic flow, or dislocation motion. In a polycrystalline material, inhomogeneous stress distributions within and stress concentrations at grain boundaries create localized stresses on the microscopic scale even when the average (macroscopic) stress is well below yield. As the density of such instances can be expected to increase with stress, and not necessarily simply as the square, nonlinear damping may be expected. The higher levels of damping are thought[3] to be most typically due to stress-induced movements of dislocations or boundaries (grain boundaries, twin boundaries, domain boundaries, or the boundaries between martensitic variants). Material processing can be expected to influence damping, with annealing generally leading to lower values.

As the theoretical models for dislocation motions and other microscopic phenomena have not been proven to be fully satisfactory[1] for the practical characterization of material damping, empirical relationships based on the results from testing have been used for the description of damping due to plastic behavior. One such relationship is that of Eq. (35.1), which may also be written in terms of the amplitude of cyclic strain, noting that the exponent may differ slightly from the representation in terms of stress if the dynamic secant modulus varies with amplitude.

$$D = J_\varepsilon \varepsilon_d^n \tag{35.5}$$

An empirical stress-strain relationship attributed to Davidenkov, with parameters chosen so that the stress-strain relationship is symmetric in tension and compression,

$$\sigma(\varepsilon) = E_0[(\varepsilon_d + \varepsilon) - b(\varepsilon_d + \varepsilon)^{M+1}] - \sigma_d \qquad \text{for } d\sigma/dt > 0 \tag{35.6}$$

$$\sigma(\varepsilon) = \sigma_d - E_0[(\varepsilon_d - \varepsilon) - b(\varepsilon_d - \varepsilon)^{M+1}] \qquad \text{for } d\sigma/dt < 0 \tag{35.7}$$

leads to a closed *hysteresis loop* with the pointed ends that are characteristic of a material undergoing cyclic plastic deformation. The example shown in Fig 35.2 is for parameters $E_0 = 1$, $b = 0.3$, $M = 0.8$, and $\varepsilon_d = 1$. The energy dissipated may be computed[4] from the area enclosed by the hysteresis loop and is

$$D = b\left[\frac{M}{M+2}\right]E_0(2\varepsilon_d)^{M+2} \tag{35.8}$$

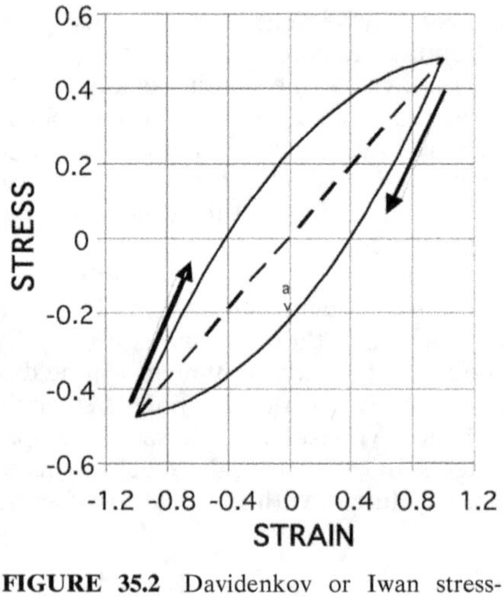

FIGURE 35.2 Davidenkov or Iwan stress-strain relationships.

Equation (35.8) is identical in form to Eq. (35.5) and captures the power-law dependence of dissipation on amplitude that is characteristic of the dissipation of structural materials in the region of lower stresses, as seen in Fig. 35.1. The same form is applicable at high levels of stress using different parameters, allowed to vary with stress history. Generally similar hysteresis loops result from modeling an elastoplastic material by an infinity of sliding elements, with the yield (sliding) strain of each prescribed by an arbitrary distribution.[5] This model, however, does not lend itself readily to the generation of hysteresis loops for which the area (dissipation) is proportional to the fractional power of the strain amplitude, as required for agreement with observations.

While the damping stress relationship of Eq. (35.1) and hysteresis loops such as Fig. 35.2 are adequate for the characterization of material damping due to plasticity in most metals, the hysteresis loops for one class of materials are notably different and require a different representation. *Shape memory alloys* (SMAs) are materials that reversibly change crystallographic structure, depending on temperature and the state of stress. For certain combinations of maximum stress and temperature, there is no residual displacement (strain) after unloading from well into the nonlinear region. This phenomenon is known as *superelasticity* or *pseudoelasticity.*

The process may be modeled[6] with two rate equations governing the evolution of the volume fraction in the martensitic phase, ξ_M, but since the process is actually independent of time, the rates may be replaced by increments. The first describes the conversion from the austenitic to the martensitic phase, occurring when (1) the magnitude of stress is in the range $\sigma_{Ms} < |\sigma| < \sigma_{Mf}$, and (2) increasing. Then

$$\dot{\xi}_M = -(1 - \xi_M)\,\frac{d|\sigma|/dt}{|\sigma| - \sigma_{Mf}} \tag{35.9}$$

The second describes the reverse transformation of martensite to austenite, occurring when (1) the magnitude of the stress is in the range $\sigma_{As} > |\sigma| > \sigma_{Af}$, and (2) decreasing.

$$\dot{\xi}_M = \xi_M\,\frac{d|\sigma|/dt}{|\sigma| - \sigma_{Af}} \tag{35.10}$$

For combinations of the magnitude and rate of change outside these ranges, the volume fraction ξ_M remains constant. As the moduli of the two phases are different, the effective modulus of the material is that of a composite, dependent on the volume fractions of each phase. An effective modulus E_{eff} can be formed in a number of different ways, but is bounded above by the modulus found from a rule of mixtures using the moduli and below by using a rule of mixtures for the reciprocals (susceptibilities). Additional strain results from the change in shape associated with the martensitic transformation. This is assumed to be in proportion to the amount of martensite, but of maximum value ε_L. With sgn(σ) taking values 1, 0, –1 as the stress is positive, zero, or negative, the total strain is

$$\varepsilon = \sigma/E_{\text{eff}} + \xi_M \varepsilon_L \operatorname{sgn}(\sigma) \tag{35.11}$$

The use of these relationships with material parameters given[6] for a nickel-titanium (Ni-Ti) alloy in a simulation of the response of a superelastic material initially in the austenitic state leads to the response shown in Fig 35.3 for the first half of a fully reversed cycle of loading of amplitude σ_d. Initially, the material responds elastically, with the modulus of the austenitic state. At point B, the transition to martensite begins and the resulting austenite-martensite composite has a lower effective modulus. Plastic strain also begins to form, giving the total strains of Eq. (35.11). For $\sigma_d < \sigma_{MF}$, the austenite-martensite transformation is not yet complete at the reversal of the loading direction at point C. If the stress is then reduced to below point D, reversion to the austenitic state begins and is completed at point A. The additional deformations associated with the martensitic transformation have now been fully recovered, and the material unloads elastically. If the unloading continues to a negative stress of $\sigma = -\sigma_d$, the lower half of the loop is a reflection of the upper half about both axes. With return to the origin, a complete hysteresis loop

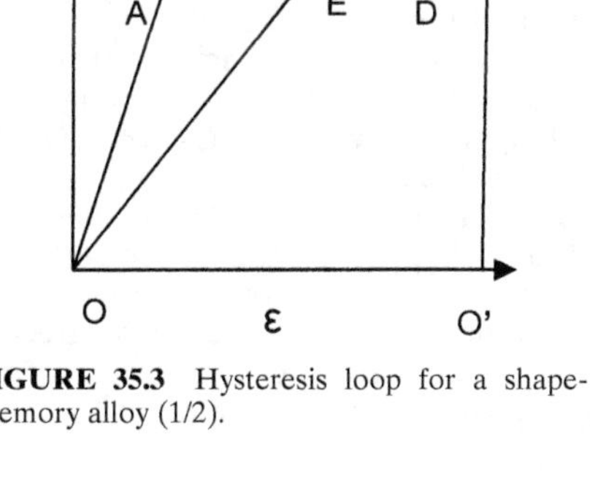

FIGURE 35.3 Hysteresis loop for a shape-memory alloy (1/2).

is formed, but has area only due to the contributions from the two regions for which the material was partially martensitic.

The total energy dissipated ΔW in the first half-cycle is the area of the parallelogram (ABCDA of Fig. 35.3), with the same value for the second half-cycle. The appropriate computation of a loss factor, however, is somewhat problematical. In the customary evaluation of the peak stored energy, one uses one-half the product of the secant modulus at maximum strain and the square of the maximum strain, or the area OCO′. But in this case, the total work done in loading is the integral under the load deflection curve, or the area OABCO′, of which the stippled area is not recovered. The recovered energy is less than the triangle OCO′ by the area of the triangle EDC. For this example, using the second value for the recovered energy increases the loss factor by only about 9 percent, but the difference would be more significant with a lower value of the parameter σ_{AF}.

MAGNETOELASTIC DAMPING

While the material damping of structural components is largely attributable to plastic deformations, such damping in high-strength alloys may be significant only at stress levels beyond the range of useful application. However, the magnetoelastic damping of ferromagnetic materials can be substantial at lower levels of stress. The terms *magnetomechanical* and *magnetoelastic* damping are somewhat interchangable. While the former is more frequently found in the literature, one might make a distinction and use the latter when the deformations are elastic and the former when nonlinear material behavior due to plasticity is also considered.

The crystal structure of ferromagnetic materials is divided into regions of uniform magnetic polarization, known as *domains.* Upon application of a magnetic field, the boundaries between domains shift and the domains rotate, bringing about a change in shape and in material dimenson (strain). An applied stress changes the field by inverse magnetostriction (*Villari effect*). The magnetostrictive effect is quantifed by the fractional change in length λ_S that occurs as the magnetization changes from zero to the *saturation* value, defined as the point at which an increase in magnetic field produces no further increase in magnetic flux. Energy is dissipated during a stress-strain cycle as the magnetic domains, nominally of random orientation in an unmagnetized material, are reoriented by the change in field caused by the application of external stress. The reorientation is accompanied by an irreversible change in dimension, leading to an additional longitudinal strain, the *magnetostrictive strain,* superposed on the elastic, or possibly plastic, strains resulting from the application of the external stress field. Cochardt[7] found that the magnetostrictive strains in a 50 percent cobalt–50 percent iron alloy loaded to 10,000 lb/in^2 were more than 10 percent of the elastic strain. For values of applied stress above a critical *coercive stress* σ_C, no further reorientation can occur, and there is no additional magnetostrictive strain. As the process is nearly instantaneous, no frequency effect is to be expected. With increasing temperatures, thermal fluctuations destroy the alignment of magnetic domains until a critical (Curie) temperature, characteristic of the material, is reached, at which the net magnetization becomes zero.

Among the pure elements generally used as constituents in structural materials, the largest magnetostriction has been found to be that of cobalt. Iron and certain of its alloys also have a significant magnetostrictive effect. Some other alloys, such as manganese-bismuth (MnBi) and cobalt ferrites, display a magnetostrictive effect an order of magnitude greater, and certain rare-earth elements, such as dysprosium and terbium, display magnetostriction an order of magnitude beyond that.

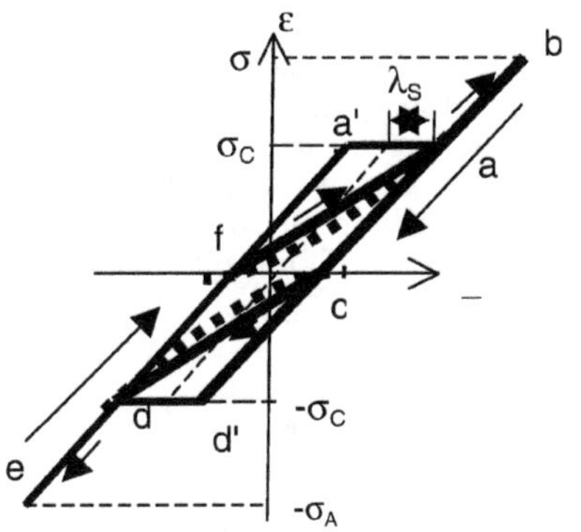

FIGURE 35.4 Hysteresis loop for a magnetoelastic material.

A hypothetical hysteresis loop for the cyclic loading of a magnetoelastic material is shown in Fig. 35.4. For increasing stress below the critical value σ_C, some path between f and a is followed and the saturation strain λ_S is accumulated. For stresses greater than σ_C, the additional strain is elastic, $\Delta\sigma/E$. After unloading to zero stress, the magnetostrictive strain remains. If the material were symmetric in tension and compression about the locus of elastic states (the light dashed line of Fig. 35.4) and the unloading continued into compressive stresses, the magnetostrictive strain would again appear, leading to a trajectory for the complete cycle of c-d-e-d-f-a-b-a-c. The area enclosed by the hysteresis loop (dissipation) would be the parallelogram f-a-c-d, enclosing an area $2\lambda_S\,\sigma_C$.

In the seminal treatment of magnetoelastic damping, Cochardt[7] provided for uncertainty in the trajectory from f to a, such as f-a′-a, or possibly the heavy dashed line of Fig. 35.4, by introducing a parameter K to account for the shape of the hysteresis loop. Thus, for a unit volume of material subjected to a cyclic external stress (fully reversed) of amplitude σ uniform over the volume, the energy dissipated per cycle becomes

$$\Delta U_C = K\lambda_S \frac{\sigma^3}{\sigma_C^{\,2}} \text{ for } \sigma < \sigma_C \qquad \text{and} \qquad \Delta U_C \cong K\lambda_S\sigma_C \text{ for } \sigma > \sigma_C \tag{35.12}$$

where λ_S is the maximum (saturation) value of the magnetoelastic strain. As the peak energy stored in the unit volume during the cycle is $U = \sigma^2/2E$, the material loss factor is

$$\eta = \frac{\Delta U_c}{2\pi U} = \frac{K\lambda_S E}{\pi}\,\frac{\sigma}{\sigma_C^{\,2}} \text{ for } \sigma < \sigma_C \qquad \text{and} \qquad \eta \cong \frac{K\lambda_S E}{\pi}\,\frac{\sigma_C}{\sigma^2} \text{ for } \sigma > \sigma_C \tag{35.13}$$

From these, it is seen that the loss factor for magnetoelastic damping is strongly amplitude dependent, increasing linearly with stress at low strains and diminishing as the inverse square for high values. Cochardt[7] measured the logarithmic decrements of a hollow, thin-walled specimen of 12 percent chrome steel in torsion, with and without a strong magnetic field, and attributed the observed difference to magnetoelastic damping. A comparison of the observed maximum loss factors ($\eta > 0.01$) with values shown in Fig. 35.1 confirms that the magnetoelastic damping can be significant.

Smith and Birchak[8] offered modifications to Cochardt's theory of magnetoelastic damping. They noted that magnetoelastic strain occurring during the compressive half of a fully reversed cycle would be only 50 percent of that occurring during the tensile half, and that the internal stress at which magnetoelastic saturation occurred would not be uniform throughout the material but better described by a distribution. As a first approximation, they considered a square-wave distribution, of width governed by a parameter Z, such that the Cochardt formulation corresponded to the value $Z = 0$. With this, they computed the magnetoelastic dissipation and determined material loss factors. Later,[9] they posited a distribution of local barrier stresses σ_{bar}, that, when exceeded, would produce dissipation. A comparison of loss factors found with this second formulation (S&B II), with loss factors found with the first formulation using $Z = 0.3$ and 0.7 and with the results from the Cochardt formulation (C), is given in Fig. 35.5. In each case, the ordinate is scaled by $K\lambda_S E/(\pi\sigma')$, and the abscissa in each case is the value of σ/σ', where σ is the amplitude of cyclic stress. Because the definition of the scaling stress σ' differs for each formulation, comparisons should be made with caution.

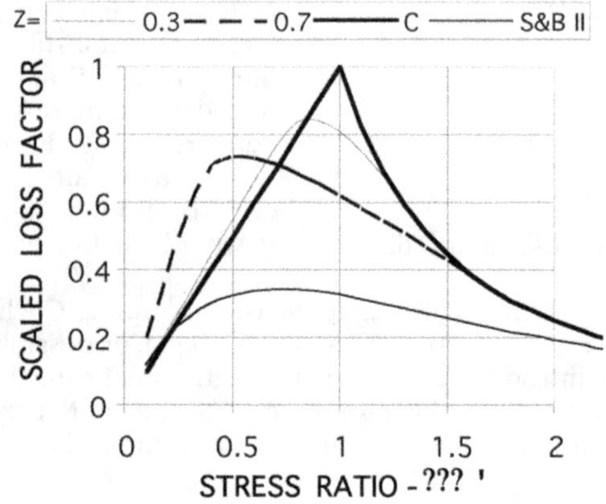

FIGURE 35.5 Theories of magnetoelastic damping.

What is common to the results from all of these formulations is that magnetoelastic damping is strongly amplitude dependent. Further, as the application of a magnetic field also causes a reorientation of the magnetic domains, the governing critical stress parameter and the dissipation due to cyclic stress will be influenced by such fields. And, as it is the total stress that leads to the magnetoelastic saturation, the presence of a mean stress can affect the dissipation. While high damping can be achieved with materials having large magnetostriction, the propensity toward a reduction in loss factor for applied stresses above a critical value cannot be regarded as advantageous in the control of structural vibrations.

Current interests in magnetostrictive damping include the influence of additional alloy constituents, annealing, and microstructure, with the iron-chromium (Fe-Cr)–based alloys receiving particular attention,[10] the use of materials of high magnetostriction as relatively thin coatings applied to the surface of structural components,[11] and the use of alloys of iron with the rare-earth elements terbium and dysprosium (Terfenol-D). Embedding particulate Terfenol-D in a resin matrix has been found[12] to produce composites with peak loss factors as high as 0.04.

VISCOELASTIC DAMPING

It has long been recognized that many materials display simultaneously the essential feature of an elastic material (the storage of strain energy) and the essential features of a viscous fluid (energy dissipation and rate effects). In consequence, simple models for the response of materials and structural components have been used, such as the Maxwell material (strain rate proportional to both stress and stress rate), the Kelvin or Voigt material (stress proportional to both strain and strain rate), and the Maxwell-Kelvin or Zener materials that incorporate both. Force-displacement relationships for each of these can be modeled by various combinations of linear springs and linear viscous dashpots appropriately arranged in parallel and series combinations. More complex arrangements have also been proposed to capture the behavior of particular materials. All of these, however, are special cases of a *viscoelastic material* with a stress-strain relationship expressed with linear operators. For the uniaxial case,

$$\sum_{0}^{N} \sigma(t) + b_1 \frac{d\sigma(t)}{dt} + b_2 \frac{d^2\sigma(t)}{dt^2} + \ldots = \sum_{0}^{N} a_0\varepsilon(t) + a_1 \frac{d\varepsilon(t)}{dt} + a_2 \frac{d^2\varepsilon(t)}{dt^2} + \ldots \tag{35.14}$$

If only the first term on the left and the first two on the right of Eq. (35.14) are retained, the Voigt or Kelvin model results. If only the first two terms on the left and the second on the right are retained, the Maxwell model results. Retaining the first two on each side leads to the standard linear (Zener) model.

Since Eq. (35.14) is linear, the stress and strain may be taken as sinusoidal, with representation $[\sigma(t), \varepsilon(t)] = \Re e\{[\sigma^*, \varepsilon^*]\exp\{j\omega t\}$. The complex-valued amplitudes are related by a complex modulus as in Eq. (36.1). This may be interpreted in terms of a *storage modulus* (real part—E_1) and a *loss modulus* (imaginary part—E_2) or, equivalently, by the storage modulus and a *loss factor* $\eta(\omega) \equiv E_2(\omega)/E_1(\omega)$, all of which are real-valued quantities. The work done in one fully reversed cycle must be

$$D = \oint \sigma(t)\dot{\varepsilon}(t)dt = \int_0^{2\pi/\omega} \Re e\{\sigma^*(\omega)e^{i\omega t}\}\Re e\{j\omega\varepsilon^*(\omega)e^{i\omega t}\}dt = \pi E_2(\omega)\,|\varepsilon_d|^2 \tag{35.15}$$

where ε_d is the amplitude of the fully reversed dynamic strain. The energy dissipated is proportional to the loss modulus, and the maximum value of the recoverable (strain) energy is proportional to the storage modulus $U = E_1(\omega)\,|\varepsilon_d|^2/2$. These were used in the motivations for the energy-based definitions of loss and storage modulus given in Eq. (35.2).

Materials for which the modulus shows strong rate (frequency) dependence are also found to show a strong dependence on temperature. Normally, these are closely related, as a material that is "stiff" at low temperature is found to be stiff at high frequency, and conversely. For the class of viscoelastic materials known as *thermorheologically simple,* such simple relationships as the Arrhenius or Williams-Landel-Ferry can be used[13] to represent the frequency shift factors required to reduce a property (storage modulus, loss modulus, or loss factor) at all temperatures and frequencies to a function of *reduced frequency* alone, such as Fig. 36.1. This process is discussed further in Chap. 36 and more extensively elsewhere,[13,14] where reduced frequency nomograms are given for many materials useful in damping treatments. The viscoelastic model is particularly appropriate for polymers, as the interactions between long, intertwined, and cross-linked molecular chains give rise to both elastic and dissipative properties. However, the same characteristics are seen in other materials displaying a transition from "rubbery" to "glassy" behavior, such as enamels. While not all viscoelastic materials are truly linear, many remain so, even for strains approaching unity.

In principle, the retention of sufficient terms in the stress-strain relationship of Eq. (35.14) enables an adequate representation of the properties of any material. However, the determination from experimental data of the large number of coefficients a_n and b_n, necessary to capture such dramatic changes in properties with frequency as are seen in Fig. 36.1, is very difficult. It has been found[15] that the replacement of the integer order derivatives of Eq. (35.14) with *fractional order derivatives* enables an adequate description with only four or five real-valued material parameters, and that such models may be used in computing the response of structures. The fractional orders are typically about 0.5. The result takes the form

$$E^* = \frac{a_0 + a_1(j\omega_R)^\beta}{1 + b_1(j\omega_R)^\alpha} \tag{35.16}$$

While a frequency-independent *glassy modulus* at high frequencies (a_1/b_1) necessitates choosing $\alpha = \beta$, material descriptions at intermediate frequencies are sometimes improved by allowing slightly different values. The low-frequency *rubbery modulus* is a_0, and $\omega_R = \omega\alpha_T$ is the *reduced frequency* of Chap. 36. A best fit[13] of Eq. (35.16) to data taken for the shear modulus and loss factor of 3M-467 viscoelastic adhesive with $a_0 = 0.0425$ MPa, $a_1 = 0.214$ MPa, $b_1 = 0.00125$, $\alpha = \beta = 0.505$ leads to the results shown in Fig. 35.6 and demonstrates that a four-parameter fractional derivative model gives a good representation of two frequency-dependent material properties over more than eight orders of magnitude of frequency. Data points shown are test data after reducing from a range of test temperatures to a common reference temperature. Values shown for the imaginary part were computed from the product of the real part of the modulus and the loss factor.

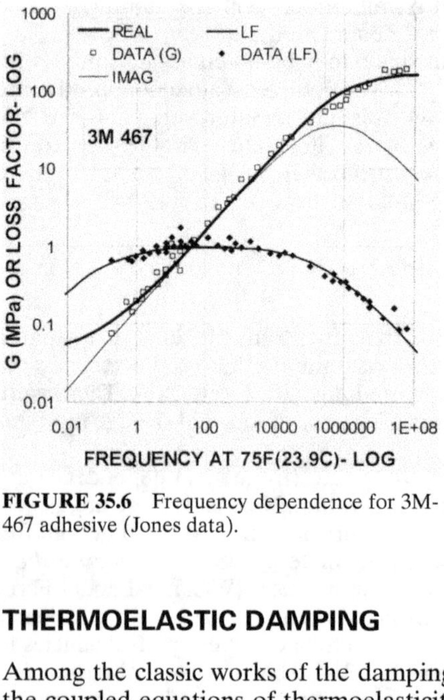

FIGURE 35.6 Frequency dependence for 3M-467 adhesive (Jones data).

Although the energy dissipation in viscoelastic materials can be very high, they normally are not suitable for use as load-bearing components for other reasons, such as strength-to-weight and strength-to-volume ratios, creep, and aging. Rather, they are used as additions for the increase of total damping, as discussed in Chap. 36.

THERMOELASTIC DAMPING

Among the classic works of the damping literature is the application by Zener[2] of the coupled equations of thermoelasticity to the damping of the vibrations of thin beams. As a positive rate of increase in volume leads to a decrease in temperature and a negative rate to an increase, heat flows across the neutral axis of a beam in bending. As the temperature differential reverses sign for successive half-cycles, the direction of the heat flux also reverses. But as the thermal energy, which is conducted from one side of the beam to the other, is drawn from the mechanical energy of vibration, the result is a reduction in vibratory amplitude and mechanical energy, or dissipation. Frequency dependence is to be expected. At low frequencies, the period of oscillation is much greater than the characteristic diffusion time, the process remains essentially isothermal, and there is negligible heat transfer by conduction. At high frequencies, the short period does not enable a significant flow of heat during each half-cycle, and the process remains essentially adiabatic. At a critical intermediate frequency and period (the *relaxation time*), however, the energy transferred is maximized and, under the right conditions, can give rise to a significant damping of the oscillation.

The variations of stress, strain, and temperature through the thickness of a narrow beam of thickness h, much less than the length L, must satisfy the stress-strain-temperature relationship and the coupled equation of heat conduction. Expansion of both strain and temperature in a double series of products of the beam eigenfunctions and orthonormal functions through the thickness enables two sets of coefficients to be related and then determined by satisfaction of a beam-bending equation. A complex-valued natural frequency results, from which an amplitude-independent loss factor may be extracted as

$$\eta = \frac{96}{\pi^4}\left(\frac{E_T \alpha^2 T_0}{\rho_0 C_V}\right) \sum_{n=\text{ODD}} \frac{1}{n^4} \frac{\omega\tau_n}{1+\omega^2\tau_n^2} \tag{35.17}$$

The maximum achievable loss factor is governed by a certain combination of mechanical and thermal properties that is proportional to the difference between the ratio of specific heats at constant pressure and at constant volume. Using only the lead term of the series, the frequency for maximum damping is related to the beam thickness and thermal diffusivity by $\omega_{max} \cong \kappa\pi^2/h^2$. Zener[2] quoted experiments that showed the peak value for a transversely vibrating aluminum wire to be about 0.0025, occurring at the frequency predicted by the theory. Using tabulated values for pure aluminum, the maximum peak loss factor predicted for a beam of rectangular section is about 0.0023, independent of thickness and mode, and slightly lower if the properties of 2024 aluminum alloy are used. The maximum occurs at a predicted frequency of 24 Hz for a beam 0.1 in. (2.54 mm) in thickness, and about 1 Hz for a beam 0.5 in. (12.7 mm) in thickness. Note that these can be natural frequencies only for specific combinations of length and mode number. A comparison of these loss factors with the values of loss factors shown in Fig. 35.1 suggests that the thermoelastic damping can be of significance at low levels of stress, but only if the maximum damping occurs at a combination of thickness and frequency of interest. This mechanism is of some interest for the damping of large-space structures, for which the thin, flexible elements in bending tend to have very low resonant frequencies. Because the frequency for maximum damping is proportional to the inverse square of thickness, the possibility of useful loss factors in laminates with alternating thin laminae of differing properties has also been considered.[16,17]

HIGH DAMPING METALS (HIDAMETS)

High intrinsic material damping alone does not qualify a material for use in a machine part or structural member. Along with other considerations, a manufactured component must have adequate resistance to repeated loading. For this reason, a plot of the specific damping energy as a function of the ratio of the applied cyclic stress to the fatigue strength of the material is useful. Such a plot is shown as Fig. 35.7. When this is done, the damping of typical structural materials (those shown in Fig. 35.1 and many more) fall into the relatively narrow band depicted as the shaded area. Details of the damping-stress dependence for the 22 materials included are given in Lazan.[1] When compared on this basis, the damping of a typical viscoelastic material is very high because of the ability to withstand repeated cyclic strains of order unity. But, as noted previously, these materials are not generally well suited for use as primary load-bearing components.

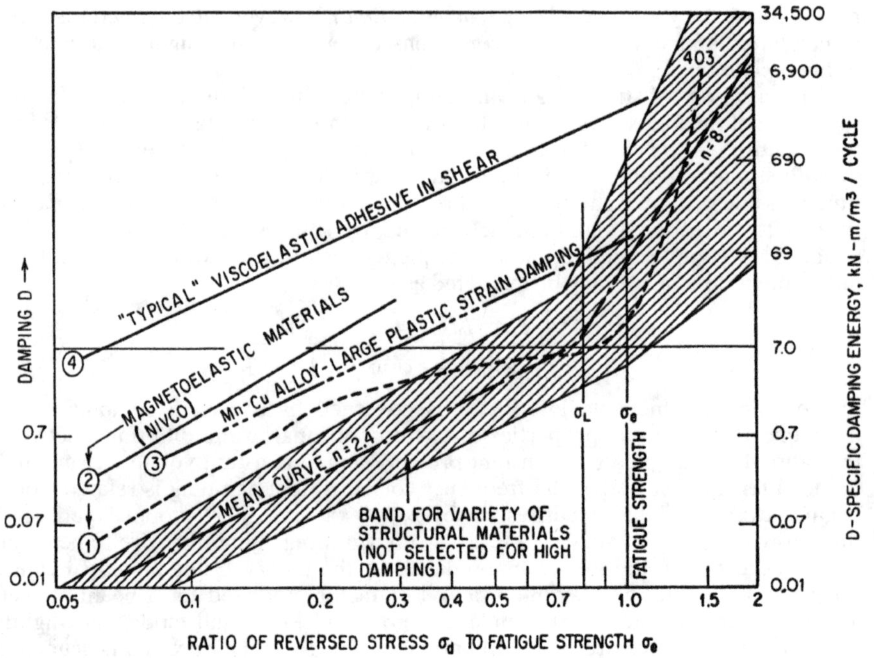

FIGURE 35.7 Examples of high-damping materials.

A metal with unusually high damping is typically a member of one of four classes. These, with examples, are *natural composites,* for which plastic flow occurs at phase boundaries, such as in gray cast iron and aluminum-zinc (Al-Zn) alloys; *dislocation damping alloys,* with damping due to movement of dislocation loops breaking away from pinning points, such as in magnesium and its alloys; *ferromagnetic alloys,* with damping due to the motion of ferromagnetic domain walls as in iron, nickel, cobalt, and their alloys; and *movable boundary alloys,* with twin or martensite boundaries, such as manganese-copper (Mn-Cu), titanium-nickel (Ti-Ni), copper-zinc-aluminum (Cu-Zn-Al), and commercial alloys such as Sonoston™, Incramute™, and NiTinol. A more extensive discussion of HIDAMETS and their operative dissipative mechanisms is available.[3]

Examples from the first two categories (cast iron and a magnesium alloy) were seen from Fig. 35.1 to have rather high damping. However, both of these materials have rather low fatigue strengths. In consequence, when considered as in Fig. 35.7, these materials no longer have exceptionally high damping and fall into the band generally characteristic of other materials. For example, the magnesium-silicon (Mg-Si) alloy and quenched Sandvik steel become nearly identical. However, materials of the third category, (1) type 403 alloy (steel with 12 percent chromium, 5 percent nickel) and (2) NIVCO 10 (about 72 percent cobalt and 23 percent nickel), do show exceptional combinations of damping and fatigue resistance. Also shown on the same basis is (3) an 80 percent Mn–20 percent Cu alloy. The SDE shown in Fig. 35.7 was measured[18] with the rotating-beam equipment described in a later section. Values obtained suggest material loss factors (with $E = 135$ GPa) ranging from 0.032 at a strain of 255 ppm to 0.058 at 1020 ppm, much higher than values inferred from Fig. 35.1 for typical structural materials.

The high damping of such manganese-rich binary Mn-Cu alloys was long thought[2] to be due to the presence of twin boundaries, a consequence of the reversible transformation between face-centered cubic and face-centered tetragonal states. The temperature at which this transformation occurs increases with the manganese content and is at room temperature for alloys of about 80 percent Mn. By aging, however, alloys with less manganese can be conditioned to undergo the transformation at a higher temperature, as in the case of the Sonoston alloys. Upon finding dispersed microscopic particles in a highly microtwinned matrix after an aging cycle at temperature, Bowie et al.[19] inferred the presence of antiferromagnetically ordered manganese cations and suggested that the definition of magnetoelasticity should be extended to include antiferromagnetic as well as ferromagnetic materials. For the composition 58Cu-40Mn-2Al, loss factors for a cantilever beam vibrating at 1 Ksi (0.69MPa) were found to be about 0.014, and about 0.071 at 10 Ksi (6.9MPa). It has since been confirmed[20] that the magnetic transition and the martensitic transformation are intimately connected, as the twin planes act as domain boundaries. Thus, the mechanism for strain amplitude–dependent damping in Cu-Mn is similar to that observed in ferromagnetic materials. As the strain produced by the stress-induced reorientation of the antiferromagnetic alignment of the manganese atoms is similar in nature to the magnetostrictive strain in a ferromagnetic material, the amplitude dependence should be similar to that for magnetoelastic damping. For an Incramute alloy (nominal 53Cu-45Mn-2Al, wt %), a peak loss factor of about 0.06 was found[20] at a strain of 1000 ppm, with the values at lower strains being well predicted by the second Smith-Birchak model for magnetoelasticity.

While capable of producing high values of dissipation, the damping of alloys of manganese and copper (whether copper or manganese rich) are highly dependent on composition and thermomechanical history. Measurements for Sonoston (nominally 55 percent Mn) showed[14] a loss factor over 0.02 at 0°C and a strain of 1000 ppm, but falling by 50 percent at 80°C, with the damping essentially disappearing at 95°C.

RELATIONSHIPS BETWEEN COMPONENT AND MATERIAL DAMPING

In the determination of material properties by testing, it is frequently necessary to include the test sample as a component in a test system and then deduce the damping due to the sample (*component property*) from a determination of the total damping (*system property*). Having found the dissipation attributable to the test sample, it is then necessary to deduce the specific or unit damping (*material property*). The integration of damping into a design analysis is the inverse of this process. Given the material properties of the dissipative component(s), it is necessary to use the stress distributions to find the total dissipation due to each. These may then be summed and used with the maximum stored energy of the total system to obtain a measure of system damping, such as the system loss factor. The concept is implemented in design calculations through the method of modal strain energy.[21] For a system of M components or elements, each having a total damping D_{T-m} and stored energy U_m,

$$\eta_S = \frac{1}{2\pi} \frac{\sum D_{T-m}}{\sum U_m} \tag{35.18}$$

To find the total damping of a test specimen (*component*) from measurements on a test *system,* we may rearrange this same equation with D_0 and U_0 being the energies dissipated and stored in the specimen.

$$D_0 = 2\pi\eta_S \left[U_0 + \sum_{m=1}^{M} U_m \right] - \sum_{m=1}^{M} D_{T-m} \tag{35.19}$$

The observed system loss factor is η_S, and the summations are of all other energies dissipated and stored in the system. Some of the components may be taken as nondissipative, having stored energy only. Some losses, such as air damping[22] and friction at grips, may be accounted for as having dissipation without storing energy. The result of this process is the extraction of a measure D_0 of the total energy dissipated by the test sample.

There remains the task of extracting the *unit damping* or *specific damping energy* as a material property. The total energy dissipated in the specimen is the integral of the unit damping over the entire specimen volume. Even when the sample material is homogeneous, this is complicated by the fact that the unit damping is a function of the local stress and that, in most cases, varies with position. It is therefore advantageous to make a change in variable:

$$D_0 = \int_0^{V_0} D\,dV = \int_0^{\sigma_d} D(\sigma)\,\frac{dV}{d\sigma}\,d\sigma \tag{35.20}$$

where σ_d is the greatest amplitude of alternating stress anywhere within the entire volume V_0, and V is the volume for which the greatest alternating stress is less than some value σ. The total energy dissipated in the specimen D_0 is then related to specific damping energy D_d at the greatest stress σ_d and the total volume V_0, by $D_0 = D_d V_0 \alpha$, where

$$\alpha = \int_0^1 \left(\frac{D(\sigma)}{D_d} \right) \frac{d(V/V_0)}{d(\sigma/\sigma_d)}\, d\left(\frac{\sigma}{\sigma_d} \right) \tag{35.21}$$

The quantity V/V_0, as a function of σ/σ_d, is the *volume-stress function,*[1] dependent on the stress distribution alone, and independent of the elastic or dissipative properties of the material. Note that the *average* dissipation per unit volume D_0/V_0 is less than the *unit* value at the greatest stress amplitude σ_d by the factor α.

The extraction of *material* properties from *component* responses also requires knowledge of the energy stored in the dissipative material. If the material is nominally linear with uniform modulus E, the strain energy stored in a specimen with greatest value of local stress amplitude σ_d is

$$W_0 = \int_0^{V_0} \frac{\sigma^2}{2E}\,dV = V_0\,\frac{\sigma_d^2}{2E} \int_0^{\sigma_d} \left(\frac{\sigma}{\sigma_d} \right)^2 \frac{d(V/V_0)}{d(\sigma/\sigma_0)}\, d\left(\frac{\sigma}{\sigma_0} \right) \tag{35.22}$$

The average strain energy density is less than that at the greatest amplitude by a factor

$$\beta = \int_0^{\sigma_d} \left(\frac{\sigma}{\sigma_d} \right)^2 \frac{d(V/V_0)}{d(\sigma/\sigma_0)}\, d\left(\frac{\sigma}{\sigma_0} \right) \tag{35.23}$$

Thus, if the total energy stored and dissipated (U_0 and D_0) by a test sample (*component*) of material at some amplitude of alternating stress σ_d can be extracted from *system* measurements, and if the stress distribution can be determined, the material loss factor can be found from the *average* loss factor for the dissipative material by

$$\eta_{\text{ave}} = \frac{1}{2\pi}\frac{D_o}{W_0} = \left(\frac{\alpha}{\beta}\right)\frac{D_d}{2\pi\sigma_d^2/(2E)} = \left(\frac{\alpha}{\beta}\right)\eta \tag{35.24}$$

The *material* loss factor is greater than the *volume-averaged* loss factor for the specimen by the factor β/α. A comparison of Eqs. (35.21) and (35.23) shows that $\beta = \alpha$ only when the stress distribution is uniform or when the energy dissipated is proportional to the square of stress. As may be seen from Fig. 35.1, this is normally not the case for structural materials. In consequence, the extraction of *material* properties from *component* data requires evaluation of the functions α and β from the stress distribution. Conversely, when the material damping properties are known, prediction of the total dissipation in a component requires use of the volume-stress function, except in the special case where the SDE is proportional to the square of stress, that is, linear damping. In that case, the values of specific and average damping coincide and the loss factor is independent of amplitude. More typically, the relationship between stress amplitude and specific damping is found to be of the form of Eq. (35.1) with $n > 2$. While any functional form for the damping-stress relationship $D(\sigma)$ may be used, that of Eq. (35.1) is particularly convenient, in which case β takes the value of α evaluated at $n = 2$.

VOLUME-STRESS FUNCTIONS

If the stress distribution is sufficiently simple, the volume-stress functions may be evaluated analytically. These calculations have been performed for several cases, with results as shown in Fig. 35.8 for a solid member in torsion, and for beams with uniform and linearly varying moments. The results shown for an actual turbine blade were obtained by numerical evaluation.[1] As it is the slopes of these curves that determine the influence of the stress distribution, the differences are quite significant. If the damping-stress relationship of Eq. (35.1) is used in the computation of α by Eq. (35.21), the slope is weighted by a power of the stress ratio. In the case of the torsion member, for example, the slopes are low in the low-stress region and high in the high-stress region, so the value of α remains relatively high, even at higher values of n. In contrast, for beams with variable moment, the slope is high at low stress, and low at high, so the value of α is low, especially at high applied stress σ_d, where n may be quite large, as seen in Fig 35.1. As the function β is the function α evaluated at $n = 2$, the ratio α/β is always less than unity. The consequence of this for design is that inclusion of a component with a high damping capacity in order to increase system damping has greatest benefit if the added material is in a region of a high, and relatively uniform, stress.

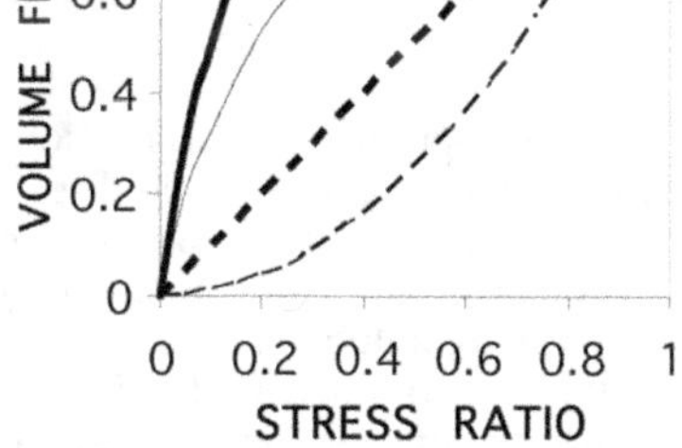

FIGURE 35.8 Volume-stress functions, various components.

For some materials, the most expeditious means of testing is the coating of a substrate beam with a layer of the material of interest. If the material loss factor is amplitude independent, the true value can be found from the Öberst equa-

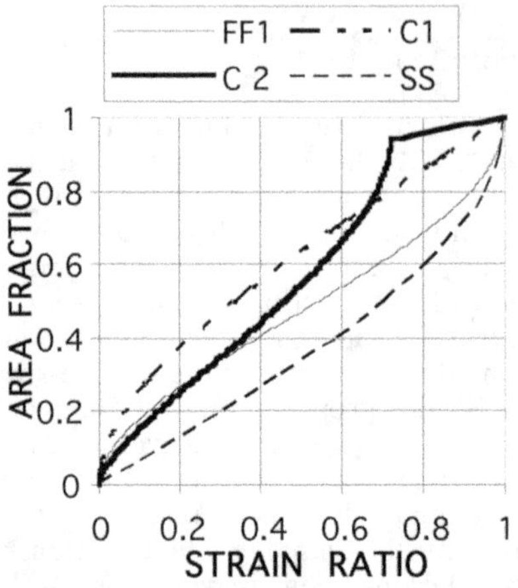

FIGURE 35.9 Area-strain functions for vibrating beams.

tions.[14] However, if the material loss factor is amplitude dependent, the distribution of stress in the sample must be considered. The distribution of strain along the coating length is essentially the surface strain of a Bernoulli-Euler beam, and for a thin coating on a thick substrate, the volume fraction is essentially the same as the area fraction. Examples of the fractions of surface area at strains below various fractions of maximum strain are shown in Fig. 35.9. Any mode of the simply supported (pinned) beam is denoted SS; C1 and C2 are the first two modes of the cantilever beam. Results for the higher modes are very similar to those for the second. The first symmetric mode of a free-free beam is denoted FF1.

Virtually all methods in common use for the determination of material damping properties—for example, vibrating or rotating beams, wires, or cylinders in torsion and dynamic mechanical analyzers using three- or four-point bending jigs—involve the use of specimens which are not subjected to uniform states of stress. In consequence, only a volume-averaged dissipation per unit volume is obtained. As this average is over the entire range of stress present in the specimen, the extraction of the true material properties for an amplitude-dependent material requires consideration of the stress distribution through the use of the volume-stress function.

FRICTION DAMPING

At some level, coulomb friction is present in any structural system. It may arise from the use of fasteners, which allow for relative displacement, as do pins, bolts, and rivets. It may occur by design, as in wire rope or leaf springs, or from platform dampers in a gas turbine, or it may occur through damage, as in a partially closed crack. Friction may also occur on a microscopic scale, as in grain contacts or in materials constructed from unbonded aggregates. In some cases, the contribution to the energy dissipation of the system through the frictional losses may be significant. In a review of the work on friction damping, it was noted[23] that damping at joints and connections is the most important source of dissipation in most real structures. The modeling of friction in devices used for the reduction of resonant amplitudes has received considerable attention.[24]

A distinction should be made between two classes of damping due to friction. In the first, the contacting bodies are taken to be rigid and the same relative displacement is assumed over the entire contact surface. This is referred to as *sliding* or *macroslip* damping. In the second class, the relative displacement varies over the contact region as, for example, in the contact of an axially loaded ball bearing on a race, where the hertzian stress distribution would give rise to interfacial shear stresses that would be greater than the product of normal pressure and a coefficient of friction if slip did not occur. This type of dissipation due to friction is referred to as *partial slip, microslip,* or *slip damping.* While it is often possible to dissipate rather large amounts of energy by either type of damping, both are subject to limitations such as the dependence on an interfacial contact pressure that may be difficult to regulate or control.

While the assumption of coulomb damping enables the analysis of some simple configurations, caution has been given[25] that the gradual surface deterioration and the presence of small amounts of lubricants may invalidate the assumptions of coulomb friction as are typically made for the purpose of analysis.

MACROSLIP OR SLIDING DAMPING

In the simplest form of friction damping, the relative displacement across the interface is either zero or the same for all locations in the contact region between two rigid bodies. For a vibrating system modeled as a spring-mass system, with the driven mass resting on a surface with contact pressure P and friction coefficient μ, the frictional force $F_f \leq \mu P$ is always in opposition to the motion, and the energy dissipated per cycle is proportional to the first power of amplitude $4\mu PA$. As the energy stored increases as the second power, system loss factors diminish with increasing amplitude—when damping is most needed. A rudimentary analysis of the forced response near resonance may be obtained by replacing the frictional force by a viscous force, chosen so that the energy dissipated per cycle by the "equivalent" viscous damper is the same as the frictional damping at the same amplitude. Using the resulting equivalent fraction of critical damping in the response function amplitude of a damped linear oscillator [Eq. (2.41)] shows unbounded amplitude at resonance unless the contact pressure is so great as to preclude all motion. An exact solution[26] by Den Hartog confirmed that this is true. In free vibration, the logarithmic decrement of a system with damping due to gross slip increases with decreasing amplitude.

SLIP DAMPING

In *slip* or *microslip,* deformations parallel to the interface of the contacting bodies enable relative displacements between the corresponding points on the mating surfaces. These vary with position and may occur on some, or all, of the contact region. The response of contacting spheres to a tangential force, as considered by Mindlin,[27] is of this class, as is the slip damping generated between a beam and a spar cap[28] and in a lap joint in tension.[29] A general characteristic of all of these systems with slip modeled by local coulomb friction is that the energy dissipated per cycle of a fully reversed load varies as the third power of the load amplitude and inversely as the interfacial shear stress, with the constant of proportionality being dependent on the particular system. This dependence also appeared in an analysis[16] of the dissipation due to partial debonding of a laminate. The strong dependence on load results from the evident cause: the area undergoing slip increases with the level of load. Microslip has also been incorporated[30] into the modeling of friction in devices used for the reduction of vibratory amplitudes of turbine blades.

A characteristic force-displacement relationship for a system with slip is shown in Fig. 35.10. The initial response may be linear or nonlinear, depending on whether slip begins immediately or at some critical load. A nonlinearity is not indicative of plastic behavior, but rather that the specimen is changing in stiffness as slip progresses. The onset of gross slip is at A′. If the loading direction is reversed at A < A′, the unloading then proceeds along A-O′-B. If the load is then reversed when the force is the negative

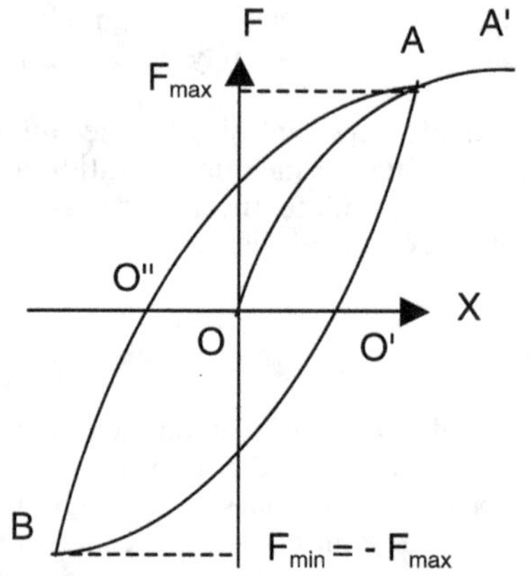

FIGURE 35.10 Hysteresis loop for microslip damping.

of the maximum and reloaded through B-O″-A, a closed *hysteresis loop* is formed, the area of which represents the net work done over the cycle. The energy dissipated may be evaluated by integration over a complete period of the product of force and velocity at the point of application. Alternatively, the dissipation may be evaluated by integrating over the contact region the work done on each element of area by the interfacial shear force acting through the relative displacement (slip) of the contacting surfaces. The latter is particularly convenient if the shear force is uniform over the area. A thorough discussion of the mechanics of slip damping is included in all earlier editions of this handbook.[31]

SIMPLIFIED EXAMPLE OF SLIP DAMPING

The analysis of the energy dissipated in slip damping can be quite cumbersome even in the case of highly idealized geometries, as may be seen in the cited examples.[16,27–29] However, an adaptation of the analysis of the slip damping to be expected from a broken lamina in a layered composite[16] enables a relatively simple demonstration of principles. A laminated material of two constituents is shown in Fig. 35.11*A*, with one element of Constituent 1 interrupted.

It is assumed that the pressure necessary to generate a uniform coulomb frictional force $\tau = \mu p$ during slip is present. Constituent 1 has stiffness E_1t_1W; that of constituent 2 is $E_2t_2W = B$. Some portion F of the total load f is then carried by the unit cell located at the stippled region of Fig 35.13*A* and shown in detail in Fig 35.11*B*. Upon application of load, slip occurs over $0 < x < \delta$, and the axial stress in the slipped portion of the interrupted layer varies linearly as the frictional force per unit length $q = \tau W$. Over the slipped region, a fraction R (the fraction of stiffness due to layer 1) of the total load is transferred to the interrupted layer, sufficient to induce the same displacement in both constituents for $x < 0$. Thus, the slipped region at final load $F_{max} > 0$ is of length $\delta_F = RF_{max}/(\tau W)$. If the loading is reversed at load $F = F_{max}$, slip in the opposite direction begins at the free end of the interrupted layer, pene-

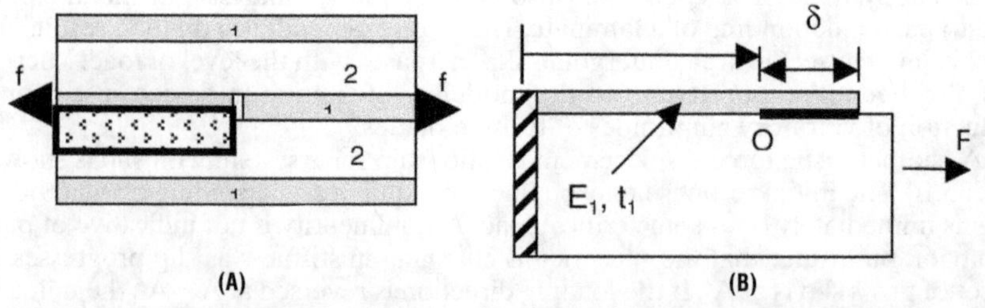

FIGURE 35.11 Unit cell of a laminated material: (*A*) location of unit cell; (*B*) detail of unit cell.

trating a distance $\delta_F - \delta = R\,(F_{max} - F)/(2qW)$, leaving shear stress in the original direction on $0 < x < \delta$. The factor of 2 arises as the positive shear stress is first reduced to zero, and then loaded in the opposite sense. If the load is again reversed at $F = \alpha F_{max}$ when the region of negative slip is of length $\delta_1 = RF_{max}(1 + \alpha)/(2q)$, a region of positive slip begins at the end of the interrupted layer, extending to a depth $\delta_F - \delta = R(F - \alpha F_{max})/(2q)$. In each stage of loading, the slip length is determined from the force that must be transferred by shear from layer 2 to layer 1 to ensure the same strain (and no slip) for $x < 0$.

The dissipation may be evaluated from the displacements at the end $x = \delta_F$ of layer 2 with respect to the plane of symmetry at $x = -\lambda$. Such displacements may be evaluated by using an established[29] process. For loading phase a, $0 < F < F_{max}$, for unloading phase b, $F_{max} < F < \alpha F_{max}$, and for reloading phase c, $\alpha F_{max} < F < F_{max}$, the displacements are, respectively,

$$u_a = \frac{(RF_{max})^2}{2qB}\left[2\frac{(1-R)}{R\beta}\,\frac{F}{F_{max}} + \left(\frac{F}{F_{max}}\right)^2\right] \tag{35.25}$$

$$u_b = \frac{R^2 F_{max}{}^2}{2Bq}\left[\left\{1 + \frac{2(1-R)}{\beta R}\right\}\frac{F}{F_{max}} + \frac{1}{2}\left\{1 - \left(\frac{F}{F_{max}}\right)^2\right\}\right] \tag{35.26}$$

$$u_c = \frac{(RF_{max})^2}{2Bq}\left[\frac{2(1-R)}{R\beta}\,\frac{F}{F_{max}} + \frac{(1+2\alpha)}{2} - \alpha\left(\frac{F}{F_{max}}\right) + \frac{1}{2}\left(\frac{F}{F_{max}}\right)^2\right] \tag{35.27}$$

Plots of these functions lead to hysteresis loops of the form of that in Fig. 35.10.

The energy dissipated in the complete cycle from $F = F_{max}$ through $\alpha\, F_{max}$ and returning to F_{max} may be found from the area enclosed by AO′BO″A in Fig. 35.10. Subtracting the reloading displacement, Eq. (35.27), from that for unloading, Eq. (35.26), at each value of F, and integrating over F, leads to

$$\Delta W = \int_{\alpha F_{max}}^{F_{max}} (u_b - u_c)\,dF = F_{max}\int_{\alpha}^{1} (u_b - u_c)\,d\left(\frac{F}{F_{max}}\right) = \frac{R^2}{12Bq}\,[F_{max} - \alpha F_{max}]^3 \tag{35.28}$$

It is of considerable interest, and significance, that the dissipation depends only on the total load range and is independent of the mean value. And, as noted in previous investigations of slip damping, the dissipation is dependent on the third power of the load range and inversely as the interface shear. While this might suggest that a low coefficient of friction and a low contact pressure are desirable for high damping, this is not the case, as the length over which slip occurs is a function of load, and there will always will be some geometric limit λ to the available length of the slipped region. Thus, the shear stress must be at least $\tau_{min} = RF_{max}/(\lambda W)$, and the greatest achievable damping for a given value of λ and $\alpha = -1$ is

$$\Delta W_{max} = \frac{2R\lambda}{3B}\,[F_{max}]^2 = \frac{2R\lambda}{3E_2 t_2 W}\,F_{max}^2 \tag{35.29}$$

However, maintaining this level of damping requires adjustment of normal pressure p with changes in force amplitude.

SLIP DAMPING AS A MECHANISM OF MATERIAL DAMPING

Thin coatings of such plasma-sprayed ceramics as alumina, magnesium-aluminate spinel, or yttria-stabilized zirconia appear to have potential as a means of reducing the vibratory amplitudes of blades in turbine engines. Several studies[32] have suggested that the material loss factor increases approximately linearly with the amplitude of alternating tensile strain up to a critical value (typically 100 to 200 ppm) and then remains constant or diminishes. It has been suggested[33] that the damping is provided by friction arising from defects within and between the "splats" that result from the plasma-spraying process. A computer simulation employing springs and coulomb sliders was found to predict an amplitude-dependent loss factor having the same characteristics as experimental data. Experiments simulating a coated beam by a vibrating beam with segmented and overlapping cover plates also showed a similar dependence of damping on amplitude.

If the geometry of Fig. 35.11 is taken as representative of a segment of a plasma-sprayed ceramic, the analysis given previously can be used to estimate the damping to be expected from each unit cell in which slip occurs along a microcrack within a splat, parallel to the loading direction. If the length λ is taken as the half-length of a typical splat and t_1 as the half-distance between slipping layers, for $E_1 = E_2 = E$ and $t_1 << t_2$, the dissipation for a fully reversed load cycle $\alpha = -1$ with stress amplitude σ_d becomes

$$D_o = \frac{2R^2}{3Bq} F_{\max}^3 = \frac{2}{3}\left[\frac{E_1 t_1}{E_1 t_1 + E_2 t_2}\right]^2 W t_2^2 \frac{(\sigma_d^3)}{\tau E_2} = \frac{2}{3} W t_1^2 \frac{(\sigma_d^3)}{\tau E} \tag{35.30}$$

where σ_d is the average stress amplitude $F_{\max}/Wt_2$. The stored energy will be predominately that of component 2, having the same modulus, but of a thickness t_2 characteristic of the half-distance between one pair of slipping planes and another. The loss factor for the unit cell is

$$\eta_c = \frac{1}{2\pi}\frac{D_0}{U_0} = \frac{2}{3\pi}\frac{t_1^2}{t_2\lambda}\frac{\sigma_d}{\tau} \tag{35.31}$$

The loss factor of a larger sample of the material may then be estimated by multiplying by the fraction of the volume occupied by such slipping cells. As long as the slip length is less than half the splat length, the loss factor rises linearly with average stress. But at higher levels of stress, gross slip begins and dissipation occurs as the first power of amplitude over some of the cycle, leading to a constant or diminishing loss factor, as seen in experiments. As not all microcracks have the same dimension, loading, or orientation, a transition, rather than an abrupt change, is to be expected. This approach has been implemented[34] in a complete theory for the dissipation of such materials at both low and high stresses. Because the Iwan model[5] for elastoplastic deformation is based on frictional elements, it might also be applied with an empirically determined distribution to the description of such materials.

MEASUREMENTS OF DAMPING

As noted in a previous section, the determination of damping as a material property usually requires including a sample of the material in a test system, determining the total damping of the system, extracting the total damping of the sample and, from that, the material damping properties. Even in the case where the system consists only of the material of interest and all extraneous losses are avoided, the measure of system damping coincides with the measure of material damping only if the entire material sample is simultaneously at the same level of stress. Several methods for evaluating the total system damping will be described briefly, with emphasis given to the influence of nonlinearities in the material being tested.

LOAD-DEFLECTION HYSTERESIS LOOP

In the case of a system, the dissipation is the net work done over a complete cycle by the applied force acting through increments of displacement at the point of load application. If the system contains a dissipative material with stress-strain relationship such as that of Figs. 35.2 through 35.4 or 35.10, the load-deflection relationship for the system will have similar characteristics, but the ratio of loop width to height will be much smaller due to the presence of nondissipative components. In consequence, there can be considerable difficulty in obtaining measurements of sufficient precision as to enable a meaningful measurement of the area of the hysteresis loop. Measurements are normally made with a quasi-static cyclic loading, and it is essential that the load and displacement be measured at the same point.

The commercially available testing devices generally known as *dynamic mechanical analyzers* may be used to obtain hysteresis loops of test samples, from which the specific damping energy may be evaluated if the details of the strain distribution—typically, torsion or three- or four-point bending—are taken into account. In some applications, the material is assumed to be linear, and a phase angle ($\tan \delta = \eta$) is extracted by comparing load and displacement signals.

MEASUREMENT OF WORK DONE

Since dissipation in the system necessitates the addition of energy so as to sustain motion at constant amplitude, a direct measurement of net work done on the system per cycle of oscillation has been suggested as a convenient measure of the total dissipation. While of interest for many years, advances in measurement instrumentation and computational capabilities now make this technique, known as the *power input method,* more feasible. The method has been applied with both impact-excited[35] and shaker-driven[36] specimens. As the system loss factor is proportional to the ratio of dissipated and stored energy, Eq. (35.18), it becomes, after replacing the strain energy by the kinetic energy and using time-averaged quantities

$$\eta_s = \frac{\left(\overline{F}_f(t) \cdot \overline{V}_f(t)\right)_{\text{ave}}}{\omega \int_V \rho(\overline{v} \cdot \overline{v})_{\text{ave}}} = \frac{\Re\{Y_{ff}(\omega)\}}{\sum_{i=1}^{N} m_i \omega |Y_{if}(\omega)|^2} \tag{35.32}$$

where Y_{ff} is the *mobility* (velocity/force) of the driving point, Y_{if} is the mobility between the driving point f and the point i, and the system has been discretized into N segments of mass m_i. Three essential assumptions are involved[35]: the replacement of strain energy by kinetic energy; the linearity of the system, so that the mobilities are independent of amplitude; and that the structure can be suitably discretized so that the kinetic energy can be adequately determined with a modest number of observation points, each accurately representing the velocity of a discretized mass. Comparisons of results with analytical solutions and with traditional measurements are encouraging.[36]

LATERAL DEFLECTION OF ROTATING BEAM

When a cantilever beam is mounted horizontally, it deflects vertically due to gravity, and the extremal values of stress are experienced by the fibers on the top and the bottom. If the beam is then rotated about its longitudinal axis, the extremal stresses are experienced by successive fibers as they pass through the top and bottom positions. But if the beam material is dissipative, the induced strain will lag the stress, and the extremal values of strain in each fiber will occur just after passing through the top and bottom positions. As viewed from the free end of the beam, the total deflection is then seen to have two components: a vertical component that is in phase with the gravity load and a horizontal component that is out of phase. If these components are measured, and the ratio taken, the result is a direct measurement of the tangent of the angle by which deflection lags force.

A specimen S is mounted coaxially with an arm A, shaft B, and weight W, as shown in Fig. 35.12. A target T is placed on the centerline and observed with a traveling microscope or other suitable instrumentation. In the absence of gravity loading, the target is at location 1; with the gravity loading, it is deflected downward to location 2, if the beam is not rotating. But if the beam is rotating in a clockwise manner as viewed from the end, deflection lags load, and the target is shifted horizontally by a distance $H/2$. If the direction of rotation is reversed, the target moves to location 4. The total deflection H can be observed and used to determine the system damping. As the center of gravity of the combined weight of specimen, arm, and weight is at some fixed fraction κ of the distance from the support to the target, the horizontal deflection induces a torque $W_0 \kappa H/2$ that in each complete rotation must do work equal to the energy dissipated. Thus, the total energy dissipated by the system, per cycle, is the product of the torque and the rotational angle of 2π radians.

$$D_T = \pi \kappa W_0 H \tag{35.33}$$

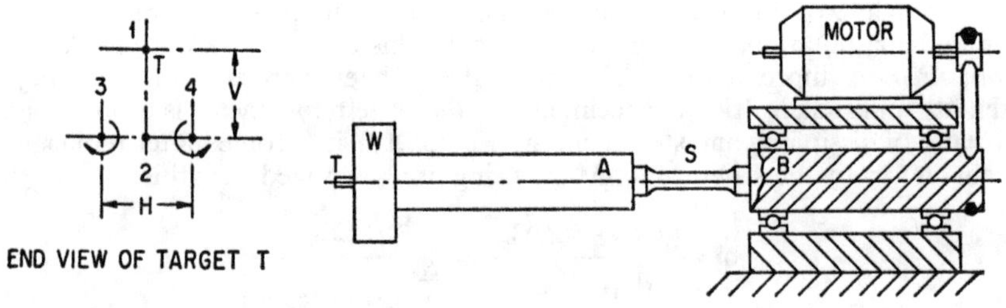

FIGURE 35.12 Principle of rotating cantilever beam method for measuring damping.

This methodology has been implemented[37] in an apparatus (*rotating beam*) for the simultaneous determination of damping, dynamic modulus, and fatigue properties of materials. The material damping properties shown in Fig. 35.1 were extracted from such measurements of system damping. Dividing the dissipated energy by 2π and the stored energy (work done by the gravitational force) gives a system loss factor $\eta_s = H/2V$, seen from Fig. 35.12 to be the tangent of the phase angle by which displacement lags stress. This, however, is not the material loss factor for a nonlinear material. In that case, each fiber along a radius is at a different strain and, consequently, has different loss angle. The measured value is some weighted average, and the volume-stress function must be used to extract material properties.

A variation of this technique has been developed[38] for the testing of low-modulus materials such as plastic and viscoelastic materials in which the target end of the beam is subjected to a controlled displacement and a force transducer used to determine the necessary restraining force. This method also enables the determination of damping and modulus over a wider range of test frequencies.

TIME-DOMAIN METHODS (FREE VIBRATION)

Measuring the decay (Fig. 21.2) of the free response of a vibrating beam appears to be the oldest means of quantifying damping. Such measurements are typically made with an inverted torsional pendulum or with a free or cantilevered beam, with or without added mass. The response of the linear, one-degree-of-freedom (1-DOF) spring-mass system with viscous damping to an initial disturbance is discussed in Chap. 2. It is shown that the fraction of critical damping, ζ, is related to the *logarithmic decrement* Δ, defined as the natural logarithm of the ratio of amplitudes of two successive cycles. For linear and viscous damping, the logarithmic decrement is independent of amplitude. As the ratio of successive amplitudes is typically very close to unity, and as the signal measured will usually have some corruption due to noise, practical determinations of the logarithmic decrement are made over an interval of several cycles. If the system is assumed to be described by a complex modulus or *structural damping* model $k^* = k_1(1 + j\eta)$, the logarithmic decrement is $\Delta = \pi\eta$ and an amplitude-dependent decrement implies an amplitude-dependent loss factor.

Although the concept of the logarithmic decrement is based on the response of the linear and viscous system, it is regularly used to represent the damping of systems for which the damping is neither linear nor viscous. Many of the challenges to *time-domain* measurement due to amplitude-dependent damping, changes in natural frequency with amplitude, signal noise, and the presence of other modes can be mitigated by the use of filtered signals and the *Hilbert transform.* The response is taken to be the real part of an analytic function of complex-valued amplitude $A(t)$ and instantaneous phase $\phi(t)$, that is,

$$x(t) = \text{Re}\{Z(t)\} = \text{Re}\{A(t)\exp(j\varphi(t)\} \tag{35.34}$$

The Hilbert transform[39] of a function $u(t)$ is computed by taking the Cauchy principal value of the integral

$$H\{u(t)\} = PV\int_{-\infty}^{\infty} \frac{u(\tau)}{t-\tau}\,d\tau \tag{35.35}$$

and has the property that if $u(t)$ is the real part of an analytic function, then $v(t) = H\{u(t)\}$ is the imaginary part. Thus, if the observed signal $x(t)$ is taken as the real part of equation $Z(t)$, the imaginary part can be constructed from the Hilbert transform. But to do this, it is necessary to prepare the signal $x(t)$ by filtering to ensure the presence of a single frequency, to extrapolate the observed $x(t)$ to $t = \infty$, and to supply values for negative argument by taking $x(-t) = x(t)$. The instantaneous phase is then the value of $\varphi(t) = \tan^{-1}\{y(t)/x(t)\}$, and the instantaneous frequency results from differentiation

$$\omega(t) = \frac{d\varphi(t)}{dt} = \frac{d\tan^{-1}\{H\{x(t)\}/x(t)\}}{dt} \tag{35.36}$$

A stiffness nonlinearity, if present, will appear as a change in frequency as the amplitude decays. An average loss factor is then found from the change in amplitude over a time interval τ corresponding to a small number of cycles at the average frequency ω_{ave} (radians/sec), over that time interval

$$\eta(A_{\text{ave}}) \approx 2\,\frac{\ln\{A(t_0)/A(t_0+\tau)\}}{\tau\omega_{\text{ave}}} \tag{35.37}$$

FREQUENCY-DOMAIN METHODS (FORCED VIBRATION)

As noted, the maximum amplitude of a system vibrating at a resonant frequency is governed by the amount of damping in the system. In consequence, measurements of the resonant response of a system are a popular means of evaluating the damping. The response of the linear one-degree-of-freedom system with viscous damping is discussed in Chap. 2, and a *frequency response function* is shown as Fig. 2.17. When a *structural damping* representation is used, the stiffness k is replaced with the complex stiffness $k^* = k_1(1 + j\eta)$, and the complex-valued response amplitude for a sinusoidal force of amplitude F_0 becomes

$$\frac{x_0}{F_0/k_1} = \frac{1}{1 - \omega^2/\omega_1^2 + j\eta} \tag{35.38}$$

At resonance $\omega = \omega_1$, the magnitude of the dimensionless response is the *system quality factor* $Q = 1/\eta$. A comparison of this response with that for a system with viscous damping, Eq. (2.33), suggests that an equivalent viscous damping might be defined as $\zeta_{eq} \equiv \eta\omega_n/2\omega$ and used as the fraction of critical damping in any solution for a viscously damped system. However, this may be done safely only at the particular frequency used to determine the equivalent viscous damping. It has been shown[40] that using such a frequency-dependent value in the equation of motion for a viscously damped system, Eq. (2.89), leads to a noncausal response to an impulse; that is, the response begins before the impulse is applied.

The width of the *frequency response function* (Fig. 2.22) is also determined by the system damping. The ratio of the amplitude at some fraction r of the maximum value to the maximum value may be found from Eq. (35.38) and used to find the two values of frequency, one above and one below resonance, at which the amplitude has that value. As the mean value of the two frequencies must be very nearly ω_1, the *bandwidth method* for determining the loss factor gives

$$\eta \cong \frac{\omega_U - \omega_L}{\omega_1\sqrt{1/r^2 - 1}} \tag{35.39}$$

and the damping may be evaluated from three points on the response function. For viscous damping, the right-hand side gives the value of 2ζ. The ratio r is most frequently taken as $1/\sqrt{2}$, or 3 dB below the peak response. Since the kinetic and strain energies for a nominally linear system vary as the square of the response, these amplitudes correspond to the half-power points.

This measure of damping was developed for a linear system. The presence of nonlinearity in stiffness gives rise to asymmetric response functions that, for a sufficiently strong nonlinearity, can preclude the observation of a valid frequency at one of the desired amplitudes. See Fig. 4.16. Further, it is assumed in the development of Eq. (35.39) that the damping is independent of amplitude. An amplitude-dependent damping can lead to significant error. For example, if the loss factor is proportional to the mth power of amplitude, the use of the bandwidth method with amplitude ratio r leads to an apparent bandwidth greater than the true value, as[41]

$$\eta_{\text{true}}/\eta_{\text{apparent}} = \sqrt{1/r^2 - 1}/\sqrt{1/r^2 - r^{2m}} \tag{35.40}$$

If the loss factor decreases with amplitude, the bandwidth method leads to underestimates of the true value of the loss factor.

Other uses of frequency response functions (*Bode* and *Nyquist plots*) in the identification of system parameters for nonlinear systems are discussed in the literature.[42,43]

HIGH-FREQUENCY PULSE TECHNIQUES

If a transducer is used to generate a series of elastic pulses on one face of a specimen, the return signal from reflection at the rear face can be compared with the initial signal and used to deduce the properties of the material, with the wave speed used to determine the elastic modulus and the dissipation deduced from the spatial attenuation of amplitude of a traveling wave of wavelength λ. The *absorption coefficient* α is a spatial logarithmic decrement, assumed to be constant with amplitude, and is found from

$$\alpha = \frac{1}{\lambda} \ln \left(\frac{A(x)}{A(x+\lambda)} \right) \tag{35.41}$$

It follows that the loss factor is $\eta = \lambda\alpha/\pi$. If the loss factor is independent of frequency, the attenuation will be proportionate to the frequency. The method is most applicable to the determination of the properties of linear, rate-independent materials, such as crystals. As the excitation frequencies are typically in the megacycle range, properties of a rate-dependent material cannot be obtained in the range of frequencies of structural interest. Moreover, the strain levels achievable are typically below the range of engineering concern. And finally, observed values of attenuation are not only inclusive of the dissipation in the material but are influenced by scattering due to imperfections and grain boundaries, a significant contributor to the attenuation of waves in polycrystalline materials.

REFERENCES

1. Lazan, B. J.: "Damping of Materials and Members in Structural Mechanics," Pergamon Press, Oxford, 1968.
2. Zener, C.: "Elasticity and Anelasticity of Metals," University of Chicago Press, Chicago, 1948.
3. Ritchie, I. G., and Z-L. Pan: "High-Damping Metals and Alloys, *Metallurgical Transactions A*, **22A:**607–616 (March 1991).
4. Pisarenko, G. S.: "Kolebaniya Uprigikh Sistem S Uchetom Rasseyaniya Energii v Materiale" ("Vibrations of Elastic Systems Taking Account of Dissipation of Energy in Material"), Ukrainian SSR Academy of Sciences, Kiev, 1955. Translated by A. R. Robinson, WADD TR 60-582, February 1962.
5. Iwan, W. D.: "On a Class of Models for the Yielding Behavior of Continuous and Composite Systems," *Journal of Applied Mechanics,* **34**(3):612–617 (1967).
6. Auricchio, F., and E. Sacco: "A One-dimensional Model for Superelastic Shape-Memory Alloys with Different Elastic Properties between Austenite and Martensite," *International Journal of Non-linear Mechanics,* **32**(6):1101–1114 (1997).
7. Cochardt, A. W.: "The Origin of Damping in High-Strength Ferromagnetic Alloys," *ASME Transactions, Series E, Journal of Applied Mechanics,* **20:**196–200 (1953).
8. Smith, G. W., and J. R. Birchak: "Effect of Internal Stress Distribution on Magnetomechanical Damping," *Journal of Applied Physics,* **39**(5):2311–2316 (April 1968).
9. Smith, G. W., and J. R. Birchak: "Internal Stress Distribution Theory of Magnetomechanical Hysteresis—An Extension to Include the Effects of Magnetic Field and Applied Stress," *Journal of Applied Physics,* **40**(13):5174–5178 (December, 1969).
10. Pulino-Sagradi, D., M. Sagradi, and J-L. Martin: "Noise and Vibration Damping of Fe-Cr-X Alloys," *Journal of the Brazilian Society of Mechanical. Sciences,* **23**(2) (2001).
11. Ustinov, A. I., B. A. Movchan, F. Lemkey, and V. S. Skorodzievsky, "Демпфирующая Способность Покрытий Co-Ni и Co-Fe, Полученных Методом Злектронно-Лучевого

Осаждения," ("Damping Capacity of Co-Ni and Co-Fe Coatings Produced by the Method of Electron Beam Deposition"), *Вібрації в Техніці та Технологіях,* (*Vibration in Technique and Technology*), (4):123 (2001).

12. McKnight, G. P., and G. P. Carman: "Energy Absorption in Axial and Shear Loading of Particulate Magnetostrictive Composites," *Smart Structures and Materials 2000: Active Material Behavior and Mechanics,* **3992**(SPIE):572–579 (2000).
13. Jones, D. I. G.: "Handbook of Viscoelastic Vibration Damping," John Wiley & Sons, New York, 2001.
14. Nashif, A. D., D. I. G. Jones, and J. P. Henderson: "Vibration Damping," John Wiley & Sons, New York, 1985.
15. Bagley, R. L., and P. J. Torvik: "Fractional Calculus—A Different Approach to the Finite Element Analysis of Viscoelastic Structures," *AIAA Journal,* **21**(5):741–748 (May 1983).
16. Torvik, P. J.: "Damping of Layered Materials," *AIAA 89-1422, Proc. 30th AIAA/ASME/ASCE/AHS/ASC Structures, Structural Dynamics and Materials Conference,* 1989, pp. 2246–2259.
17. Bishop, J. E., and V. K. Kinra: "Elastothermodynamic Damping in Laminated Composites," *International Journal of Solids and Structures,* **34**(9):1075–1092 (1997).
18. Torvik, P. J.: "Damping Properties of a Cast Magnesium and a Manganese Copper Alloy Proposed as High Damping Materials," Appendix 72fg, *Status Report 58-4,* B. J. Lazan, University of Minnesota, Wright Air Development Center, Dayton, Ohio, Contract AF-33(616)-2802, December 1958.
19. Bowie, G. E., J. F. Nachman, and A. N. Hammer: "Exploitation of Cu-Rich Damping Alloys: Part I—the Search for Alloys with High Damping at Low Stress," *ASME Paper 71-Vibr 106, ASME Vibrations Conference and International Design Automaton Conference,* Toronto, Canada, September 1971.
20. Laddha, S., and D. C. Van Aken: "A Review of the Physical Metallurgy and Damping Characteristics of High Damping Cu-Mn Alloys," *M^3D III: Mechanics and Mechanisms of Material Damping,* ASTM STP 1304, A. Wolfenden and V. K. Kinra, eds., American Society for Testing and Materials, 1997, pp. 365–382.
21. Johnson, C. D., and D. A. Kienholz: "Finite Element Prediction of Damping in Structures with Constrained Viscoelastic Layers," *AIAA Journal,* **20**(9):1284–1290 (1982).
22. Ungar, E. E.: "Damping of Panels Due to Ambient Air," *Damping Applications for Vibration Control,* AMD—Vol. 38, P. J. Torvik, ed., ASME, New York, 1980, pp. 75–83.
23. Plunkett, R.: "Friction Damping," *Damping Applications for Vibration Control,* AMD—Vol. 38, P. Torvik, ed., ASME, New York, 1980, pp. 65–74.
24. Popp, K., L. Panning, and W. Sextro: "Vibration Damping by Friction Forces: Theory and Applications," *Journal of Vibration and Control,* **9:**419–448 (2003).
25. Ungar, E. E.: "The Status of Engineering Knowledge Concerning the Damping of Built-up Structures," *Journal of Sound and Vibration,* **26**(1):141–154 (1973).
26. Den Hartog, J. P.: "Forced Vibrations with Combined Coulomb and Viscous Friction," *ASME Transactions,* **53**(9):107–115 (1931).
27. Mindlin, R. D., W. P. Mason, T. F. Osmer, and H. Deresiewicz: "Effects of an Oscillating Tangential Force on the Contact Surface of Elastic Spheres," *Proc. First U.S. National Congress of Applied Mechanics,* ASME, 1952, pp. 203–208.
28. Pian, T. H. H., and F. C. Hallowell, Jr.: "Structural Damping in a Simple Built-up Beam," *Proc. First U.S. National Congress of Applied Mechanics,* ASME, 1952, pp. 97–102.
29. Metherell, A. F., and S. V. Diller: "Instantaneous Energy Dissipation in a Lap Joint—Uniform Clamping Pressure," *Journal of Applied Mechanics,* **35**(1):123–128 (March 1968).
30. Koh, K-H., and J. H. Griffin: "Dynamic Behavior of Spherical Friction Dampers and Its Implication to Damper Contact Stiffness," *Journal of Engineering for Gas Turbines and Power,* **129**(2):511–521 (April 2007).

31. Goodman, L. E.: "Material and Slip Damping," Chap. 36 in *Harris' Shock and Vibration Handbook,* 5th ed., C. M. Harris and A. G. Piersol, eds., McGraw-Hill, New York, 2002.

32. Torvik, P. J.: "A Survey of the Damping Properties of Hard Coatings for Turbine Engine Blades," *Integration of Machinery Failure Prevention Technologies into System Health Management,* Society for Machine Failure Prevention Technology (MFPT), Dayton, Ohio, April 2007, pp. 485–506.

33. Shipton, M., and S. Patsias: "Hard Damping Coatings: Internal Friction as the Damping Mechanism," *Proc. 8th National Turbine Engine High Cycle Fatigue Conference,* Monterey, Calif., April 2003.

34. Torvik, P. J.: "A Slip Damping Model for Plasma Sprayed Ceramics," *Journal of Applied Mechanics,* **76**(6) (November 2009).

35. Carfagni, M., and M. Pierini: "Determining the Loss Factor by the Power Input Method (PIM), Parts 1 and 2," *Journal of Vibration and Acoustics,* **121:**417–428 (July 1999).

36. Liu, W., and M. S. Ewing: "Experimental and Analytical Estimation of Loss Factors by the Power Input Method," *AIAA Journal,* **45**(2):477–484 (February 2007).

37. Lazan, B. J.: "A Study with New Equipment of the Effects of Fatigue Stress on the Damping Capacity and Elasticity of Mild Steel," *Transactions American Society for Metals,* **42:**499–558 (1950).

38. Maxwell, B.: "Apparatus for Measuring the Response of Polymeric Materials to an Oscillating Strain," *ASTM Bulletin,* **215:**76–80 (July 1956).

39. Bendat, J. S., and A. G. Piersol: "Random Data: Analysis and Measurement Procedures," 3rd ed., John Wiley & Sons, New York, 2000.

40. Crandall, S. H.: "Dynamic Response of Systems with Structural Damping," *Air, Space and Instruments, Draper Anniversary Volume,* H. S. Lees, ed., McGraw-Hill, New York, 1963, pp. 183–193.

41. Torvik, P. J.: "A Note on the Estimation of Non-linear System Damping," *Journal of Applied Mechanics,* ASME, **70:**449–450 (May 2003).

42. Ewins, D. J.: "Modal Testing: Theory, Practice and Application," 2d ed., Research Studies Press, Hertfordshire, England, 2000.

43. Torvik, P. J.: "On Evaluating the Damping of a Non-linear Resonant System," *AIAA 2002-1306, Proc., 43rd AIAA/ASME/ASCE/AHS Structures, Structural Dynamics, and Materials Conference,* Denver, Colo., April 22–25, 2002.

CHAPTER 36
APPLIED DAMPING TREATMENTS

David I. G. Jones

INTRODUCTION TO THE ROLE OF DAMPING MATERIALS

The damping of an element of a structural system is a measure of the rate of energy dissipation which takes place during cyclic deformation. In general, the greater the energy dissipation, the less the likelihood of high vibration amplitudes or of high noise radiation, other things being equal. *Damping treatments* are configurations of mechanical or material elements designed to dissipate sufficient vibrational energy to control vibrations or noise.

Proper design of damping treatments requires the selection of appropriate damping materials, location(s) of the treatment, and choice of configurations which ensure the transfer of deformations from the structure to the damping elements. These aspects of damping treatments are discussed in this chapter, along with relevant background information including:

- Internal mechanisms of damping
- External mechanisms of damping
- Polymeric and elastomeric materials
- Analytical modeling of complex modulus behavior
- Benefits of applied damping treatments
- Free-layer damping treatments
- Constrained-layer damping treatments
- Integral damping treatments
- Tuned dampers and damping links
- Measures or criteria of damping
- Methods for measuring complex modulus properties
- Commercial test systems

MECHANISMS AND SOURCES OF DAMPING

INTERNAL MECHANISMS OF DAMPING

There are many mechanisms that dissipate vibrational energy in the form of heat within the volume of a material element as it is deformed.[1] Each such mechanism is associated with internal atomic or molecular reconstructions of the microstructure or with thermal effects. Only one or two mechanisms may be dominant for specific materials (metals, alloys, intermetallic compounds, etc.) under specific conditions, i.e., frequency and temperature ranges, and it is necessary to determine the precise mechanisms involved and the specific behavior on a phenomenological, experimental basis for each material specimen. Most structural metals and alloys have relatively little damping under most conditions, as demonstrated by the ringing of sheets of such materials after being struck. Some alloy systems, however, have crystal structures specifically selected for their relatively high damping capability; this is often demonstrated by their relative deadness under impact excitation. The damping behavior of metallic alloys is generally nonlinear and increases as cyclic stress amplitudes increase. Such behavior is difficult to predict because of the need to integrate effects of damping increments which vary with the cyclic stress amplitude distribution throughout the volume of the structure as it vibrates in a particular mode of deformation at a particular frequency. The prediction processes are complicated even further by the possible presence of external sources of damping at joints and interfaces within the structure and at connections and supports. For this reason, it is usually not possible, and certainly not simple, to predict or control the initial levels of damping in complex built-up structures and machines. Most of the current techniques of increasing damping involve the application of polymeric or elastomeric materials which are capable (under certain conditions) of dissipating far larger amounts of energy per cycle than the natural damping of the structure or machine without added damping.

EXTERNAL MECHANISMS OF DAMPING

Structures and machines can be damped by mechanisms which are essentially external to the system or structure itself. Such mechanisms, which can be very useful for vibration control in engineering practice (discussed in other chapters), include:

1. Acoustic radiation damping, whereby the vibrational response couples with the surrounding fluid medium, leading to sound radiation from the structure
2. Fluid pumping, in which the vibration of structure surfaces forces the fluid medium within which the structure is immersed to pass cyclically through narrow paths or leaks between different zones of the system or between the system and the exterior, thereby dissipating energy
3. Coulomb friction damping, in which adjacent touching parts of the machine or structure slide cyclically relative to one another, on a macroscopic or a microscopic scale, dissipating energy
4. Impacts between imperfectly elastic parts of the system

POLYMERIC AND ELASTOMERIC MATERIALS

A mechanism commonly known as *viscoelastic damping* is strongly displayed in many polymeric, elastomeric, and amorphous glassy materials. The damping arises from the relaxation and recovery of the molecular chains after deformation. A strong dependence exists between frequency and temperature effects in polymer behavior because of the direct relationship between temperature and molecular vibrations. A wide variety of commercial polymeric damping material compositions exists, most of which fit one of the main categories listed in Table 36.1.

TABLE 36.1 Typical Damping Material Types

Acrylic rubber	Natural rubber	Polysulfone
Butadiene rubber	Nitrile rubber (NBR)	Polyvinyl chloride (PVC)
Butyl rubber	Nylon	Silicone
Chloroprene (e.g., Neoprene)	Polyisoprene	Styrene-butadiene (SBR)
Fluorocarbon	Polymethyl methacrylate (Plexiglas)	Urethane
Fluorosilicone	Polysulfide	Vinyl

Polymeric damping materials are available commercially in the following categories:

1. Mastic treatment materials
2. Cured polymers
3. Pressure sensitive adhesives
4. Damping tapes
5. Laminates

Some manufacturers of damping material are given as a footnote.* Data related to the damping performance is provided in many formats. The current internationally recognized format, used in many databases, is the temperature-frequency nomogram, which provides modulus and loss factor as a function of both frequency and temperature in a single graph, such as that illustrated in Fig. 36.1.[2,3] The user requiring complex modulus data at, say, a frequency of 100 Hz and a temperature of 50°F (10°C) simply follows a horizontal line from the 100-Hz mark on the right vertical axis until it intersects the sloping 50°F (10°C) isotherm, and then projects vertically to read off the values of the Young's modulus E and loss factor η.

* Manufacturers of damping materials and systems, from whom information on specific materials and damping tapes may be obtained, include:

Antiphon Inc. (U.S.A.)
Arco Chemical Company (U.S.A.; www.arco.com)
Avery International (U.S.A.; www.avery.com)
CDF Chimie (France)
Daubert Chemical Co. (USA): www.daubertchemical.com
Dow Corning (U.S.A.; www.dowcorning.com)
EAR Corporation (U.S.A.)
El duPont deNemours (U.S.A.; www.DuPont.com)
Farbwercke-Hoechst (Germany)
Flexcon (U.S.A.; www.flexcon.com)
Goodyear (U.S.A.; www.goodyear.com)
Goodfellow (U.K.; www.goodfellow.com)
Imperial Chemical Industries (U.K.)
Leyland & Birmingham Rubber Company (U.K.)
MSC Laminates (U.S.A.)
Morgan Adhesives (U.S.A.; www.mactac.com)
Mystic Tapes (U.S.A.)
Shell Chemicals (U.S.A.; www.shellchemicals.com)
SNPE (France; www.snpe.com)
Sorbothane Inc. (U.S.A.; www.sorbothane.com)
Soundcoat Inc. (U.S.A.; www.soundcoat.com)
United McGill Corporation (U.S.A.; www.unitedmcgillcorp.com)
Uniroyal (U.S.A.; www.uniroyalchem.com)
Vibrachoc (France; www.vibrachoc.com)

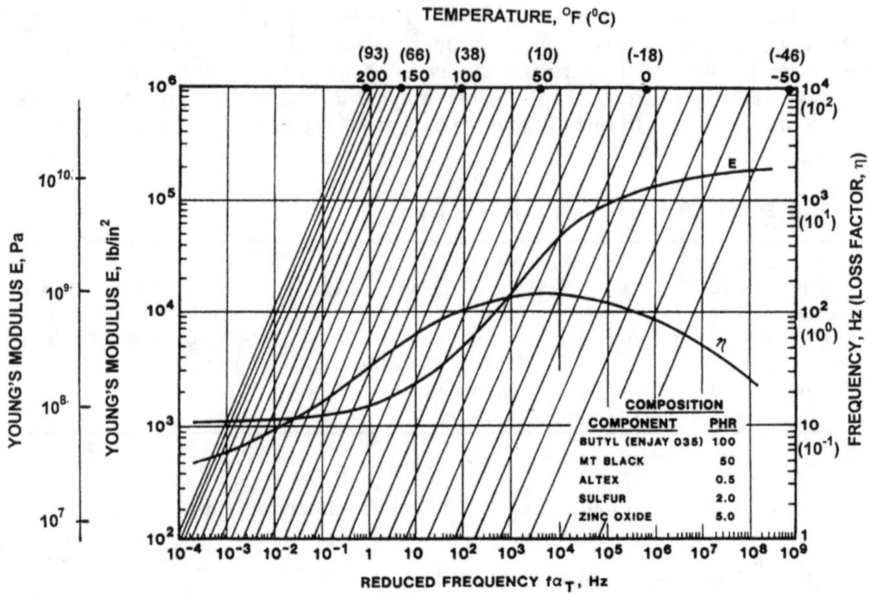

FIGURE 36.1 Temperature-frequency nomogram for butyl rubber composition.

ANALYTICAL MODELING OF COMPLEX MODULUS BEHAVIOR

It is very convenient to be able to mathematically describe the complex modulus properties of damping polymers, not only in the form of a nomogram as just described, but also by algebraic equations which can be folded into finite element and other computer codes for predicting dynamic response to external excitation (see Chap. 24). Such models include the standard model, comprising a distribution of springs and viscous dashpots in series and parallel configurations[2–4] for which the complex Young's modulus E^* (and equally the shear modulus G^*) can be described in the frequency domain by a series such as

$$E^* = \sum_{n=1}^{N} \frac{a_n + b_n(i f \alpha_T)}{1 + c_n(i f \alpha_T)} \tag{36.1}$$

or a fractional derivative model[5] for which the series becomes

$$E^* = \sum_{n=1}^{N} \frac{a_n + b_n(i f \alpha_T)^{\beta_n}}{1 + c_n(i f \alpha_T)^{\beta_n}} \tag{36.2}$$

where a_n, b_n, and c_n are numerical parameters, which may be real or complex, the β_n are fractions of the order of 0.5, and α_T is a shift factor which depends on temperature. Both models work, but Eq. (36.1) will usually require many terms, often 10 or more, to properly model actual material behavior, whereas Eq. (36.2) usually requires only one term for a good fit to the data. The shift factor α_T is determined as a function of temperature for each material from the test data, and is usually modeled by a Williams-Landel-Ferry (WLF) relationship[1–5] of the form

$$\log [\alpha_T] = \frac{-C_1(T - T_0)}{B_1 + T - T_0} \tag{36.3}$$

or by an Arrhenius relationship[1,5] of the form

$$\log [\alpha_T] = T_A \left(\frac{1}{T} - \frac{1}{T_0} \right) \tag{36.4}$$

where C_1 and B_1 are numerical parameters, the temperatures T and T_0 (the reference temperature) are in degrees absolute, and T_A is a numerical parameter related to the activation energy. The behavior of each specific polymer composition dictates which expression is most appropriate, and simple statistical methods may be applied for determining "best estimates" of each parameter in these equations.[2,5]

BENEFITS OF APPLIED DAMPING TREATMENTS

When the natural damping in a system is inadequate for its intended function, then an applied damping treatment may provide the following benefits:

Control of vibration amplitude at resonance. Damping can be used to control excessive resonance vibrations which may cause high stresses, leading to premature failure. It should be used in conjunction with other appropriate measures to achieve the most satisfactory approach. For random excitation it is not possible to detune a system and design to keep random stresses within acceptable limits without ensuring that the damping in each mode at least exceeds a minimum specified value. This is the case for sonic fatigue of aircraft fuselage, wing, and control surface panels when they are excited by jet noise or boundary layer turbulence-induced excitation. In these cases, structural designs have evolved toward semiempirical procedures, but damping levels are a controlling factor and must be increased if too low.

Noise control. Damping is very useful for the control of noise radiation from vibrating surfaces, or the control of noise transmission through a vibrating surface. The noise is not reduced by sound absorption, as in the case of an applied acoustical material, but by decreasing the amplitudes of the vibrating surface. For example, in a diesel engine, many parts of the surface contribute to the overall noise level, and the contribution of each part can be measured by the use of the acoustic intensity technique or by blanketing off, in turn, all parts except that of interest. If many parts of an engine contribute more or less equally to the noise, significant amplitude reductions of only one or two parts (whether by damping or other means) leads to only very small reductions of the overall noise, typically 1 or 2 dB.

Product acceptance. Damping can often contribute to product acceptance, not only by reducing the incidence of excessive noise, vibration, or resonance-induced failure but also by changing the "feel" of the product. The use of mastic damping treatments in car doors is a case in point. While the treatment may achieve some noise reduction, it may be the subjective evaluation by the customer of the solidity of the door which carries the greater weight.

Simplified maintenance. A useful by-product from reduction of resonance-induced fatigue by increased damping, or by other means, can be the reduction of maintenance costs.

TYPES OF DAMPING TREATMENTS

FREE-LAYER DAMPING TREATMENTS

The mechanism of energy dissipation in a free-, or unconstrained-, layer treatment is the cyclic extensional deformation of the imaginary fibers of the damping layer during each cycle of flexural vibration of the base structure, as illustrated in Fig. 36.2. The presence of the free layer changes the apparent flexural rigidity of the base structure in a manner which depends on the dimensions of the two layers involved and the elastic moduli of the two layers. The treatment depends for its effectiveness on the assumption, usually well-founded, that plane sections remain plane. The treatment fiber labeled *yy* is extended or compressed during each half of a cycle of flexural deformation of the base structure surface, in a manner which depends on the position of the fiber in the treatment and the radius of curvature of the element of length Δl, and can be calculated on the basis of purely geometric considerations. One fiber in particular does not change length during each cycle of deformation and is referred to as the *neutral axis.* For the uncoated plate or beam, the neutral axis is the center plane, but when the treatment is added, it moves in the direction of the treatment and its new position is calculated by the requirement that the net in-plane load across any section remain unchanged during deformation. The basic equations for predicting the modal loss factor η for the given damping layer loss factor η_2 and for predicting the direct flexural rigidity $(EI)_D$ as a function of the flexural rigidity E_1I_1 of the base beam are well known.[2–6]

The simplest expression relating the damping of a structure, in a particular mode, to the properties of the structure and the damping material layer is[7]

$$\frac{\eta}{\eta_2} = \frac{eh(3 + 6h + 4h^2 + 2eh^3 + e^2h^4)}{(1 + eh)(1 + 4eh + 6eh^2 + 4eh^3 + e^2h^4)} \tag{36.5}$$

where η is the damped structure modal loss factor, η_2 is the loss factor of the damping material, E_2 is the Young's modulus of the damping material and E_1 is that of the structure ($e = E_2/E_1$), and h_2 and h_1 are the thicknesses of damping layer and structure, respectively ($h = h_2/h_1$).

To calculate η, the user estimates η_2 and E_2 at the frequency and temperature of interest (from a nomogram), then calculates h and e, and then inserts these values into Eq. (36.5). Change thickness (h) or material (e) if the calculated value of η is not

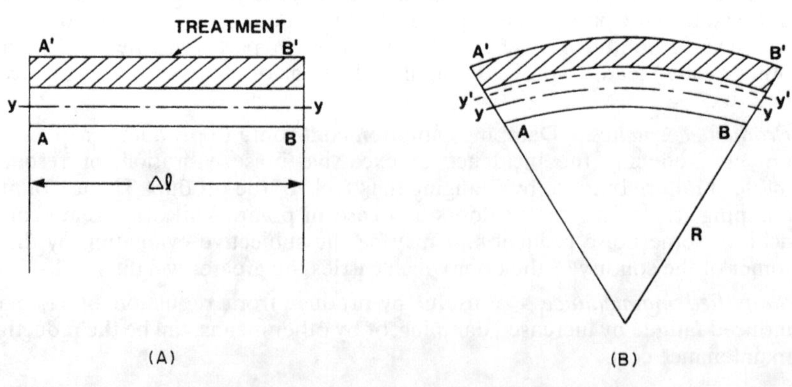

FIGURE 36.2 Free-layer treatment. (*A*) Undeformed. (*B*) Deformed.

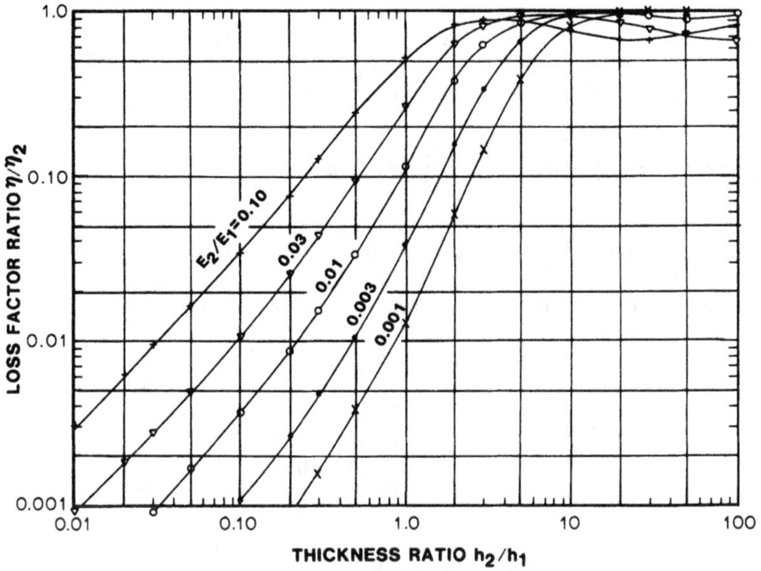

FIGURE 36.3 Graphs of η/η_2 vs. h_2/h_1, for a free-layer treatment.

adequate, and continue the process until satisfied. Figure 36.3 illustrates how η/η_2 varies with E_2/E_1 and with h_2/h_1, as calculated using the Oberst equations.

Limitations of Free-Layer Treatment Equations. The classical equations for free-layer treatment behavior are approximate. The main limitation is that the equations are applicable to beams or plates of uniform thickness and uniform stiff isotropic elastic characteristics with boundary conditions which do not dissipate or store energy during vibration. These boundary conditions include the classical pinned, free, and clamped conditions. Another limitation is that the deformation of the damping material layer is purely extensional with no in-plane shear, which would allow the "plane sections remain plane" criterion to be violated. This restriction is not very important unless the damping layer is very thick and very soft ($h_2/h_1 > 10$ and $E_2/E_1 < 0.001$). A third limitation is that the treatment must be uniformly applied to the full surface of the beam or plate, and especially that it be anchored well at the boundaries so that plane sections remain plane in the boundary areas where bending stresses can be very high and the effects of any cuts in the treatment can be very important. Other forms of the equations can be derived for partial coverage or for nonclassical boundary conditions.

Effect of Bonding Layer. Free-layer damping treatments are usually applied to the substrate surface through a thin adhesive or surface treatment coating. This adhesive layer should be very thin and stiff in comparison with the damping treatment layer in order to minimize shear strains in the adhesive layer which would alter the behavior of the damping treatment. The effect of a stiff thin adhesive layer is minimal, but a thick softer layer alters the treatment behavior significantly.

Amount of Material Required. Local panel weight increases up to 30 percent may often be needed to increase the damping of the structure in several modes of vibration to an acceptable level. Greater weight increases usually lead to diminishing returns. This weight increase can be offset to some degree if the damping is added early in the design, by judicious weight reductions achieved by proper sizing of the structure to take advantage of the damping.

CONSTRAINED-LAYER DAMPING TREATMENTS

The mechanism of energy dissipation in a constrained-layer damping treatment is quite different from the free-layer treatment, since the constraining layer helps induce relatively large shear deformations in the viscoelastic layer during each cycle of flexural deformation of the base structure, as illustrated in Fig. 36.4. The presence of the constraining viscoelastic layer-pair changes the apparent flexural rigidity of the base structure in a manner which depends on the dimensions of the three layers involved and the elastic moduli of the three layers, as for the free-layer treatment, but also in a manner which depends on the deformation pattern of the system, in contrast to the free-layer treatment. A useful set of equations which may be used to predict the flexural rigidity and modal damping of a beam or plate damped by a full-coverage constrained-layer treatment is given in Refs. 2 and 6. These equations give the direct (in-phase) component $(EI)_D$ of the flexural rigidity of the three-layer beam, and the quadrature (out-of-phase) component $(EI)_Q$ as a function of the various physical parameters of the system, including the thicknesses h_1, h_2, and h_3, the moduli $E_1(1 + j\eta_1)$, $E_2(1 + j\eta_2)$, $E_3(1 + j\eta_3)$, and the shear modulus of the damping layer $G_2(1 + j\eta_2)$.

Shear Parameter. The behavior of the damped system depends most strongly on the shear parameter

$$\mathbf{g} = \frac{G_2(\lambda/2)^2}{E_3 h_3 h_2 \pi^2} \tag{36.6}$$

which combines the effect of the damping layer modulus with the semiwavelength $(\lambda/2)$ of the mode of vibration, the modulus of the constraining layer, and the thicknesses of the damping and constraining layers. The other two parameters are the thickness ratios h_2/h_1 and h_3/h_1. Figure 36.5 illustrates the typical variation of η/η_2

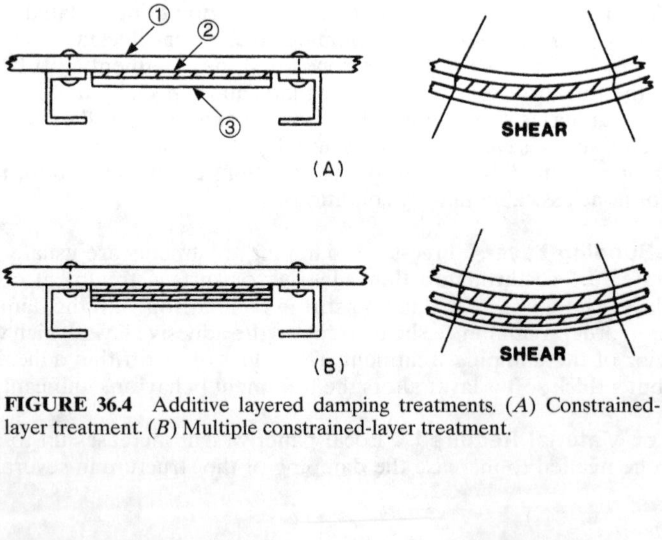

FIGURE 36.4 Additive layered damping treatments. (*A*) Constrained-layer treatment. (*B*) Multiple constrained-layer treatment.

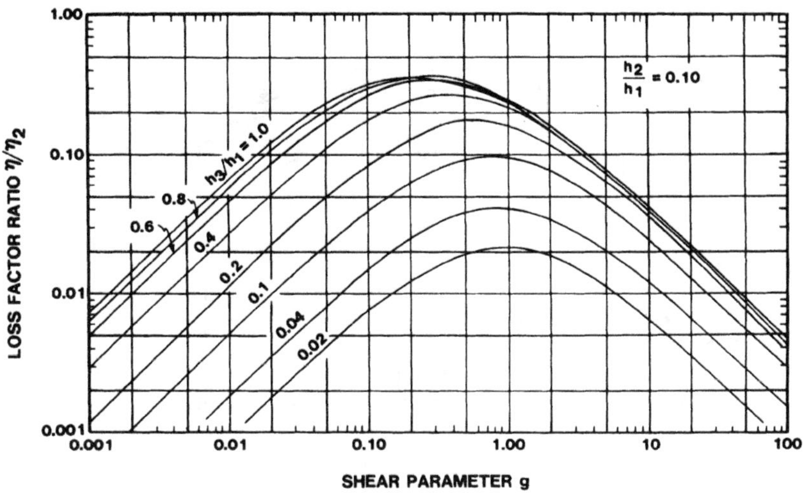

FIGURE 36.5 Typical plots of η/η_2 versus shear parameter $\mathbf{g}(h_2/h_1 = 0.10, \eta_2 = 0.1)$.

with the shear parameter **g** for particular values of h_2/h_1 and h_3/h_1. These plots may be used for design of constrained-layer treatments. Note that η_n will be small for both large and small values of **g**. For **g** approaching zero, G_2 or $\lambda/2$ may be very small or E_3, h_3, and h_2 may be very large. This could mean that while G_2 might appear at first sight to be sufficiently large, the dimensions h_2 and h_3 are nevertheless too large to achieve the needed value of **g**. This could happen for very large structures, especially for high-order modes. On the other hand, for **g** approaching infinity, G_2 or $\lambda/2$ may be large, or E_3, h_2, or h_3 may be very small.

Effects of Treatment Thickness. In general, increasing h_2 and h_3 will lead to increased damping of a beam or plate with a constrained-layer treatment, but the effect of the shear parameter will modify the specific values. The influence of h_3/h_1 is stronger than that of h_2/h_1, and as h_2/h_1 approaches zero, η/η_2 does not approach zero but a finite value. This behavior seems to occur in practice and accounts for the very thin damping layers, 0.002 in. (0.051 mm) or less, used in damping tapes. A practical limit of 0.001 in. (0.025 mm) is usually adopted to avoid handling problems.

Effect of Initial Damping. If the base beam is itself damped, with η_1 not equal to zero, then the damping from the constrained-layer treatment will be added to η_1 for small values of η_1. The general effect is readily visualized, but specific behavior depends on treatment dimensions and the value of the shear parameter.

Integral Damping Treatments. Some damping treatments are applied or added not after a structure has been partly or fully assembled but during the manufacturing process itself. Some examples are illustrated in Fig. 36.6. They include laminated sheets which are used for construction assembly, or for deep drawing of structural components in a manner similar to that for solid sheets, and also for faying surface damping which is introduced into the joints during assembly of built-up, bolted, riveted, or spot-welded structures. The conditions at the bolt, rivet, or weld areas critically influence the behavior of the damping configurations and make analysis

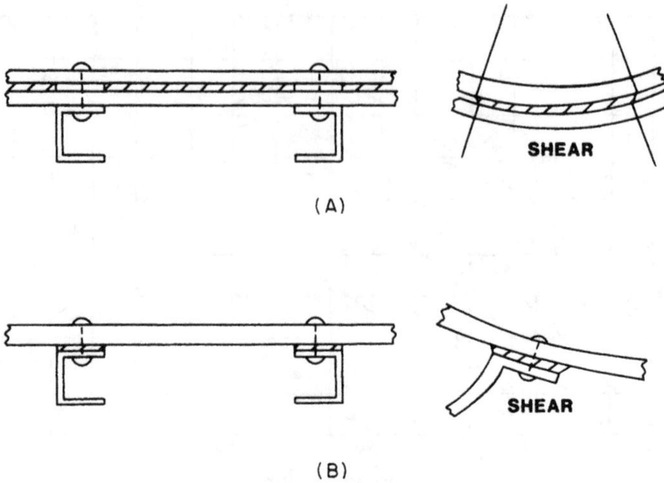

FIGURE 36.6 Some basic integral damping treatments. (*A*) Laminate. (*B*) Faying surface damping.

particularly difficult because of the limited control of conditions at these points. Finite element analysis may be one of the few techniques for such analysis.

Damping Tapes. Constrained-layer treatments are sometimes available in the form of a premanufactured combination of an adhesive layer and a constraining layer, which may be applied to the surface of a vibrating panel in one step, as opposed to the several steps required when the adhesive and constraining layers are applied separately. Such damping tapes are available from several companies, including the 3M Company, Avery International, and Mystic Tapes, to name a few. An example of such a damping tape is the 3M™ 2552 damping foil product, which consists of a 0.005-in.-thick layer of a particular pressure-sensitive adhesive prebonded to a 0.010-in.-thick aluminum constraining layer, with an easy-release paper liner protecting the adhesive layer. One limitation of damping tapes is at once evident, namely, that the particular adhesive is effective over a specific temperature range and the adhesive and constraining layer thicknesses are fixed. The choice of adhesive is particularly important, since it must be selected in accordance with the required temperature range of operation, and the available thicknesses may not be ideal for all applications. Constrained-layer treatments such as those illustrated in Fig. 36.4 could be built up conventionally, with adhesive and constraining layers applied separately, or by means of damping tapes. In each case, the adhesive material and thickness, and the constraining layer thickness, must be chosen to ensure optimal damping for the temperature range required by each application. The Ross-Kerwin-Ungar (RKU) equations[2,3,6] may be used to estimate, even if roughly, the best combination of dimensions and adhesive for each application, whether by means of damping tapes or conventional treatments, applying the complex modulus properties of the adhesive as described by a temperature-frequency nomogram or by a fractional derivative equation.

Tuned Dampers. The tuned damper is essentially a single-degree-of-freedom mass-spring system having its resonance frequency close to the selected resonance frequency of the system to be damped, i.e., tuned. As the structure vibrates, the damper elastomeric element vibrates with much greater amplitude than the structure at the point of attachment and dissipates significant amounts of energy per cycle, thereby introducing large damping forces back to the structure which tend to reduce the amplitude. The system also adds another degree of freedom, so two peaks arise in place of the single original resonance. Proper tuning is required to ensure that the two new peaks are both lower in amplitude than the original single peak. The damper mass should be as large as practicable in order to maximize the damper effectiveness, up to perhaps 5 or 10 percent of the weight of the structure at most, and the damping capability of the resilient element should be as high as possible. The weight increase needed to add significant damping in a single mode is usually smaller than for a layered treatment, perhaps 5 percent or less.

Damping Links. The damping link is another type of discrete treatment, joining two appropriately chosen parts of a structure. Damping effectiveness depends on the existence of large relative motions between the ends of the link and on the existence of unequal stiffnesses or masses at each end. The deformation of the structure when it is bent leads to deformation of the viscoelastic elements. These deformations of the viscoelastic material lead to energy dissipation by the damper.

RATING OF DAMPING EFFECTIVENESS

MEASURES OR CRITERIA OF DAMPING

There are many measures of the damping of a system. Ideally, the various measures of damping should be consistent with each other, being small when the damping is low and large when the damping is high, and having a linear relationship with each other. This is not always the case, and care must be taken, when evaluating the effects of damping treatments, to ensure that the same measure is used for comparing behavior before and after the damping treatment is added. The measures discussed here include the loss factor η, the fraction of critical damping (damping ratio) ζ, the logarithmic decrement Δ, the resonance or quality factor Q, and the specific damping energy D. Table 36.2 summarizes the relationship between these parameters, in the ideal case of low damping in a single-degree-of-freedom (SDOF) system. Some care must be taken in applying these measures for high damping and/or for multiple-degree-of-freedom (MDOF) systems and especially to avoid using different measures to compare treated and untreated systems.

Loss Factor. The loss factor η is a measure of damping which describes the relationship between the sinusoidal excitation of a system and the corresponding sinusoidal response. If the system is linear, the response to a sinusoidal excitation is also sinusoidal and a loss factor is easily defined, but great care must be taken for nonlinear systems because the response is not sinusoidal and a unique loss factor cannot be defined. Consider first an inertialess specimen of linear viscoelastic material excited by a force $F(t) = F_0 \cos \omega t$, as illustrated in Fig. 36.7. The response $x(t) = x_0 \cos (\omega t - \delta)$ is also harmonic at the frequency ω as for the excitation but with a phase lag δ. The relationship between $F(t)$ and $x(t)$ can be expressed as

TABLE 36.2 Comparison of Damping Measures

Measure	Damping ratio	Loss factor	Log dec	Quality factor	Spec damping	Amp factor
Damping ratio	ζ	$\frac{\eta}{2}$	$\frac{\Delta}{\pi}$	$\frac{1}{2Q}$	$\frac{D}{4\pi U}$	$\frac{1^*}{2A}$
Loss factor	2ζ	η	$\frac{2\Delta}{\pi}$	$\frac{1}{Q}$	$\frac{D}{2\pi U}$	$\frac{1^*}{A}$
Log decrement	$\pi\zeta$	$2\pi\eta$	Δ	$\frac{\pi}{2Q}$	$\frac{D}{4U}$	$\frac{2\pi^*}{A}$
Quality factor	$\frac{1}{2\zeta}$	$\frac{1}{\eta}$	$\frac{\pi}{2\Delta}$	Q	$\frac{2\pi U}{D}$	A^*
Spec damping	$4\pi U\zeta$	$2\pi U\eta$	$4U\Delta$	$\frac{2\pi U}{Q}$	D	$\frac{2\pi U^*}{A}$
Amp factor	$\frac{1}{2\zeta}$	$\frac{1}{\eta}$	$\frac{\pi}{2\Delta}$	Q	$\frac{2\pi U}{D}$	A^*

* For single-degree-of-freedom system only.

$$F = kx + \frac{k\eta}{|\omega|}\frac{\partial x}{\partial t} \tag{36.7}$$

where $k = F_0/x_0$ is a stiffness and $\eta = \tan\delta$ is referred to as the *loss factor.* The phase angle δ varies from 0° to 90° as the loss factor η varies from zero to infinity, so a one-to-one correspondence exists between η and δ. Equation (36.7) is a simple relationship between excitation and response which can be related to the stress-strain relationship because normal stress $\sigma = F/S$ and extensional strain $\varepsilon = x/l$. This is a generalized form of the classical Hooke's law which gives $F = kx$ for a perfectly elastic system. The loss factor, as a measure of damping, can be extended further to apply to a system possessing inertial as well as stiffness and damping characteristics. Consider, for example, the one-degree-of-freedom linear viscoelastic system shown in Fig. 36.8*A*. The equation of motion is obtained by balancing the stiffness and damping forces from Eq. (36.7) to the inertia force $m(d^2x/dt^2)$:

$$m\frac{d^2x}{dt^2} + kx + \frac{k\eta}{\omega}\frac{dx}{dt} = F_0 \cos \omega t \tag{36.8}$$

The steady-state harmonic response, after any start-up transients have died away, is illustrated in Chap. 2. If k and η depend on frequency, as is the case for real materials, then the maximum amplitude at the resonance frequency $\omega_r = \sqrt{k/m}$ is equal to $F_0/k(\omega_r)\eta(\omega_r)$, while the static response, at $\omega = 0$, is equal to $F_0/k(0)\sqrt{1+\eta^2(0)}$. The amplification factor A is approximately equal to $1/\eta(\omega_r)$, provided that $\eta^2(0) \ll 1$. Furthermore, the ratio $\Delta\omega/\omega_r$, where $\Delta\omega$ is the separation of the frequencies for which the response is $1/\sqrt{2}$ times the peak response, is known as the *half-power bandwidth* (see Chap. 2). It is also equal to η, provided that $\eta^2 \ll 1$. In summary, therefore,

$$\eta = \frac{1}{A} = \frac{\Delta\omega}{\omega_r} \tag{36.9}$$

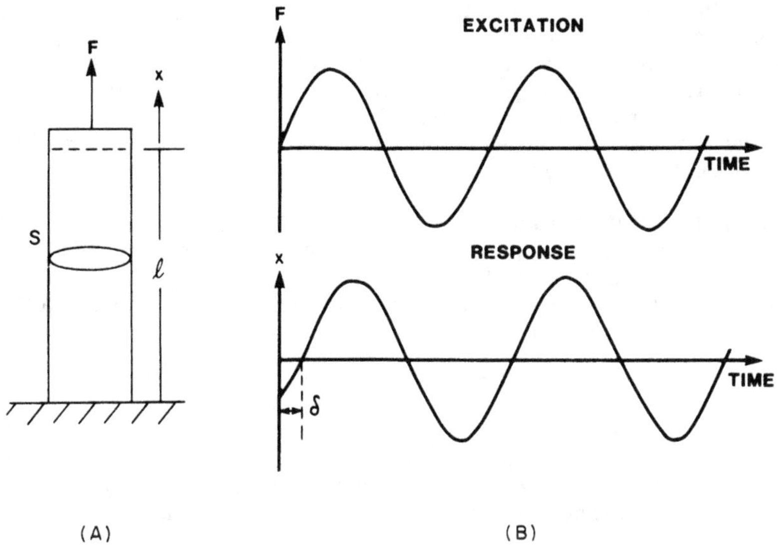

FIGURE 36.7 Linear viscoelastic behavior of a sample under sinusoidal loading, described in terms of response and excitation as functions of time. (*A*) Specimen. (*B*) Response and excitation.

This relationship between η and $1/A$ is applicable only for a single-degree-of-freedom system and may not be directly applicable for more complex systems such as beams, plates, or more complex structures. The measure $\Delta\omega/\omega_r$ is applicable for more complex systems, as well as SDOF systems. For large values of η, on the order of 0.2 or greater, none of these measures of damping agree exactly, even for an ideal linear SDOF system, but each measure is at least self-consistent. The stiffness and loss-factor parameters defined here do not specify any particular model of material behavior. For example, k and η could be constants as for hysteretic damping, or they

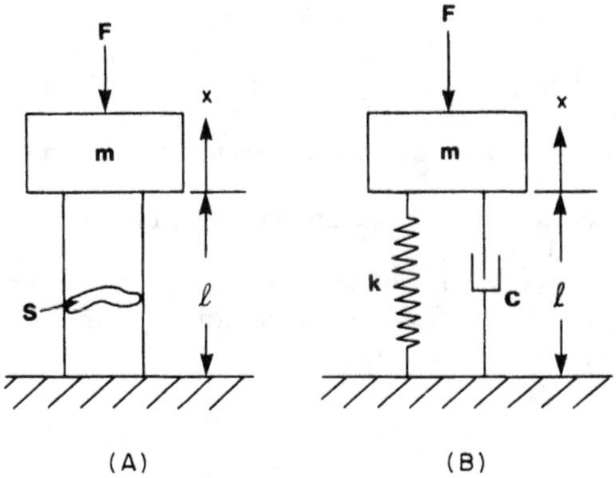

FIGURE 36.8 Single-degree-of-freedom system with: (*A*) viscoelastic damping; (*B*) viscous damping.

could be functions of frequency, temperature, specimen composition and shape, or amplitude as for a nonlinear material. The model with constant k and η is not too useful over a wide frequency range, and such behavior is impossible over an infinite frequency range, but these parameters can vary quite slowly with frequency for some particular material compositions. If k and η vary strongly with frequency, or amplitude, then the various definitions of the loss factor must be used with great care, since each measure gives different results.

Fraction of Critical Damping. The fraction of critical damping (damping ratio) is a measure of one very specific mechanism of damping, i.e., viscous damping which is proportional to velocity. If the damping forces acting on a single-degree-of-freedom mass-spring system, illustrated in Fig. 36.8*B*, satisfy this type of relationship, then the equation of motion for harmonic excitation is

$$m \frac{d^2x}{dt^2} + c \frac{dx}{dt} + kx = F_0 \cos \omega t \tag{36.10}$$

The response depends on m, k, and a parameter $c/2\sqrt{km}$ which involves c, k, and m and is known as the *fraction of critical damping* (damping ratio). This parameter, labeled ζ, controls the peak amplitude, the half-power bandwidth, and the resonance frequency ω_r:

$$x_{\text{max}} = \frac{F_0}{2k\zeta\sqrt{1-\zeta^2}} \qquad x(0) = \frac{F_0}{k}$$

$$\omega_r = \sqrt{(k/m)(1-\zeta^2)} \qquad \frac{\Delta\omega}{\omega_r} = 2\zeta \tag{36.11}$$

The plot of $x(\omega)$ versus frequency ω, for specific values of m and k is very similar to those for the viscoelastic damping, provided that $\eta \doteqdot 2\zeta$. The distinction between viscous and hysteretic damping (constant k, and η) is not at once apparent. Equations (36.8) and (36.10) convey the difference, since the damping coefficient in Eq. (36.8) decreases in proportion to $1/\omega$ as ω increases, while that in Eq. (36.10) is constant with frequency, at least for the hypothetical cases considered here. Figure 36.9 shows plots of response versus frequency based on the solutions of these equations of motion for each type of damping. Some differences arise at low frequency, but they are not very great except for very high values of damping. For high values of damping, neither η nor ζ are linearly related to the bandwidth ratio $\Delta\omega/\omega_r$. Figure 36.10 shows the variation of $\Delta\omega/\omega_r$ with η and 2ζ for values of η which are not small. Limits exist beyond which the ratio $\Delta\omega/\omega_r$ does not give a good estimate of η or ζ.

Logarithmic Decrement. When a damped system is struck by an impulsive load or is released from a displaced position relative to its equilibrium state, a decaying oscillation usually takes place, as illustrated in Chap. 2. A measure of damping called *logarithmic decrement* Δ is defined as the natural logarithm of the ratio of amplitudes of successive peaks (see Chap. 2):

$$\Delta = \ln \frac{x_1}{x_2} = \ln \frac{x_n}{x_{n+1}} \tag{36.12}$$

This definition is useful only if these ratios are equal for the various cycles, i.e., for specific types and amounts of damping. The measure is useful for viscous and hys-

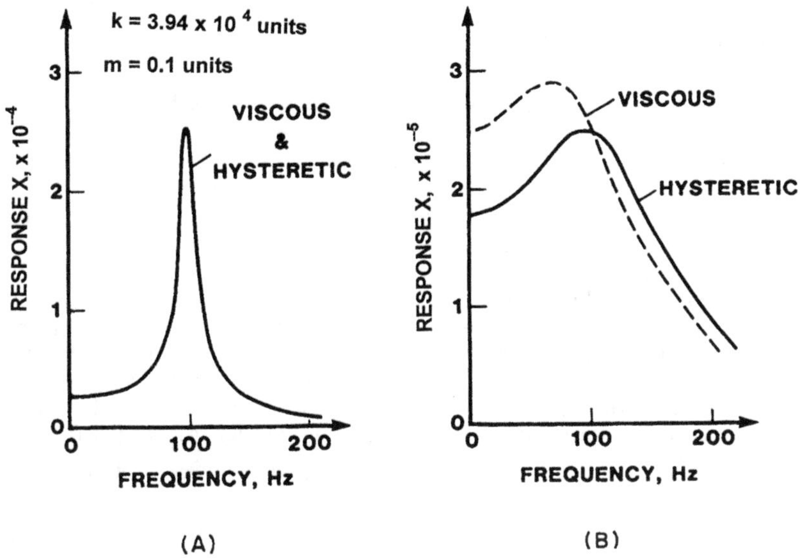

FIGURE 36.9 Comparison of viscous and hysteretic damping of a single-degree-of-freedom system with (*A*) low damping ($\eta = 0.1, \zeta = 0.05$); (*B*) high damping ($\eta = 1.0, \zeta = 0.5$).

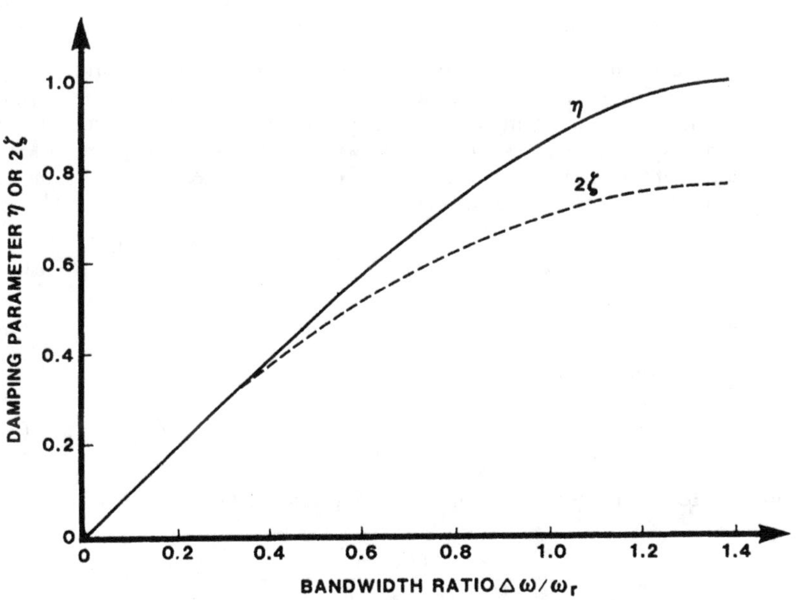

FIGURE 36.10 Variation of loss factor (η) and 2 times the fraction of critical damping (2ζ) of a single-degree-of-freedom system with $\Delta\omega/\omega_r$.

teretic damping, within limits. For viscous damping, the solution of Eq. (36.10) for an impulsive excitation $F\,\delta(t)$ is obtained.

$$x = \frac{F}{\sqrt{km(1-\zeta^2)}}\, e^{-t\sqrt{k/m}} \sin t\sqrt{(k/m)(1-\zeta^2)} \tag{36.13}$$

so that

$$\Delta = \frac{\pi\zeta}{\sqrt{1-\zeta^2}} \tag{36.14}$$

for small ζ. If ζ approaches 1.0, the response becomes aperiodic and a logarithmic decrement cannot be defined or related to ζ. The loss factor in Eq. (36.7) also can be related to the transient response of a single-degree-of-freedom mass-spring system, subject to an impulsive excitation. Consider the impulsive excitation $F(t)$ to be modeled as a spike of the form of a delta function at time $t = 0$. Then the equation of motion, in the form of Eq. (36.8), cannot be written directly, but if both $F(t)$ and $x(t)$ are described in terms of their corresponding Fourier transforms, then $\bar{F}(\omega) = \int_{-\infty}^{\infty} F(t) \exp(-j\omega t)\, dt = F$ and $\bar{x} = F/(k - m\omega^2 + jk\eta)$. The inverse Fourier transform gives

$$x(t) = \frac{1}{2\pi} \int_{-\infty}^{\infty} \frac{Fe^{j\omega t}\, d\omega}{k - m\omega^2 + jk\eta} \tag{36.15}$$

This equation contains real and imaginary parts, but using the fact that $\exp(j\omega t) = \cos \omega t + j \sin \omega t$ and if $k(\omega) = k(-\omega)$ and $\eta(-\omega) = -\eta(\omega)$, then it may be shown that $x(t)$ is given by

$$x(t) = \frac{F}{\pi} \int_{0}^{\infty} \frac{(k - m\omega^2)\cos \omega t + k\eta \sin \omega t}{(k - m\omega^2)^2 + (k\eta)^2}\, d\omega \tag{36.16}$$

For k and η constant over all frequencies from zero to infinity, problems arise regarding $x(t)$ being finite for values of t less than zero, i.e., before the impulse is applied, and this is physically impossible. The problem is that k and η cannot be constants for real systems over any extremely wide frequency range, no matter how close to constant they may be over a limited frequency range. For small values of η, however, a useful and accurate solution is given by

$$x(t) = \frac{F}{\sqrt{km}}\, e^{-1/2\eta t\sqrt{k/m}} \sin t\sqrt{k/m} \tag{36.17}$$

$$\Delta = \pi\eta/2 \tag{36.18}$$

Comparing Eqs. (36.16) and (36.18) gives

$$\eta = 2\zeta \tag{36.19}$$

Quality Factor. The quality factor Q is defined as $\omega_r/\Delta\omega$, so

$$Q = \frac{1}{\eta} \tag{36.20}$$

For a single degree of freedom $Q = A$ [where A is defined in Eq. (36.9)], but this is not the case for multiple-degree-of-freedom systems.

Specific Damping Energy. Another useful measure of material damping is the amount of energy dissipated per unit volume per cycle, known as the *specific damping energy.* For a damping material specimen subject to an applied external force $F(t) = F_0 \cos \omega t$ the specific damping energy D is equal to

$$D = \oint F\,dx = \int_0^{2\pi/\omega} F \frac{dx}{dt}\,dt \tag{36.21}$$

For a viscoelastic material obeying Eq. (36.7)

$$D = F_0 x_0 k\eta \sqrt{1+\eta^2} \tag{36.22}$$

But $F_0 = kx_0 \sqrt{1+\eta^2}$, also from Eq. (36.7), so

$$D = \pi x_0^2\,k\eta \tag{36.23}$$

The specific damping energy D increases as the square of the amplitude of vibration x_0 for linear viscoelastic materials, so it is clearly desirable to ensure that the damping material is strained as vigorously as possible in order to maximize D and, hence, the damping of the system. This has an important bearing on the choice of location within a vibrating system for application of a damping treatment. Furthermore, both k and η must be as large as possible to ensure maximum energy dissipation in the system, but this can be done only to the extent that further increases of k and η do not reduce x_0. While D is related to k and η for linear viscoelastic materials, this is not possible for nonlinear materials or for high cyclic strain levels where nonlinear behavior occurs; the value of D is then, of itself, often used as a measure of overall damping performance.

COMPARISON OF DAMPING MEASURES

The damping measures described in this section are related to each other as follows (Table 36.2):

$$\eta = 2\zeta = \frac{2\Delta}{\pi} = \frac{1}{Q} = \frac{D}{2\pi U} = \frac{1}{A} = \frac{\Delta\omega}{\omega_r} \tag{36.24}$$

These various equations relate η, Δ, and ζ for viscous and viscoelastic damping of single-degree-of-freedom systems. The relationships usually agree well for low values of η and ζ ($\eta < 0.2$ or so), but for higher values the comparisons are not so precise.

It is important, when analyzing tests to determine the effects of damping treatments on dynamic response, to be consistent in the use of these damping measures and to recognize that they are not completely equivalent, especially over wide frequency ranges or for multimodal response.

Effects of Mass and Stiffness. Changing the mass or stiffness of a single-degree-of-freedom mass-spring system without changing any other parameters leads to a change of resonance frequency, and when the frequency changes over a wide range, the differences of viscous and hysteretic damping become more apparent. For viscous damping, the fraction of critical damping $\zeta = c/2\sqrt{km}$ changes as k or m change, whereas for hysteretic damping η does not change, at least within a limited frequency range. Although viscous and hysteretic damping measures are related by the simple relationship $\eta = 2\zeta$ for a single mode at a particular frequency, they do not remain equivalent as the frequency changes, and significant differences in response may be observed.

METHODS FOR MEASURING COMPLEX MODULUS PROPERTIES

Vibrating Beam Test Methods. The vibrating beam test methods are frequently used to measure the extensional or shear complex modulus properties of damping materials.[2,3,8,9] The dynamic response behavior of the beam, first in the undamped uncoated form and then with an added damping layer or added constrained configuration, is measured for several modes of vibration and over a range of temperatures. At each temperature, the measured damped resonance frequency f_n, the undamped resonance frequency f_{on}, and the loss factor η_n in the nth mode of vibration are measured and used in an appropriate set of equations to deduce the Young's modulus E, or the shear modulus G, and the loss factor η of the damping material at a number of discrete frequencies and temperatures.

Various configurations of cantilever beams are used to measure viscoelastic material damping properties in tension-compression or shear at low cyclic strain amplitudes. Figure 36.11 illustrates some of the configurations used. The damping layers are bonded to the base beams by means of a stiff adhesive such as an epoxy. This bonding is very important and must be done well using an adhesive which is stiff in comparison to the damping layer and is very thin. The thickness ratio h_2/h_1 generally lies in the range $0.1 \leq h_2/h_1 \leq 2.0$, and the length l is about 5 to 10 in. (12.7 to 25.4 cm). The base beam material is typically aluminum, steel, or a stiff epoxy or epoxy matrix composite material having low intrinsic damping. Great care must be taken to ensure that the temperature range of the tests is not excessive in relation to the behavior of the base beam, and in particular to allow for the effect of temperature on the base beam properties such as Young's modulus, the resonance frequencies, and the modal loss factor in the absence of the damping layer. The vibration test is conducted, allowing the specimen to soak at each selected temperature for several minutes (often 30 minutes) to be sure of thermal equilibrium; then the beam is excited by means of a noncontacting transducer or by impact, and the resulting response in the frequency domain is measured, either through swept-sine-wave excitation or fast Fourier transform (FFT) analysis of the transient response signal in the time domain. At each temperature, several resonance frequencies and modal loss factors are measured over a wide range of frequencies. The test is then repeated after thermal equilibrium has been reached at the next selected temperature. The data obtained for the first mode is usually not used because of the low frequency involved and the high amplitudes and high modal damping of the base beam, as well as because of errors in the analysis when sandwich beams are used. Such vibrating beam tests are widely used for measuring viscoelastic material damping properties for shear and extensional deformation.[8,9]

Geiger Thick-Plate Test Method. The Geiger thick-plate method is of importance because it is widely used to describe damping materials in the automotive industry. It makes use of a large flat plate, suspended freely from four points selected to be at or near the nodal lines of the first free-free mode, to which is bonded the damping layer being evaluated. The rate of decay of vibration amplitude (expressed in decibels per second) is measured and serves as a measure of the effectiveness of the damping layer. Figure 36.12 illustrates a typical test setup. The system can be

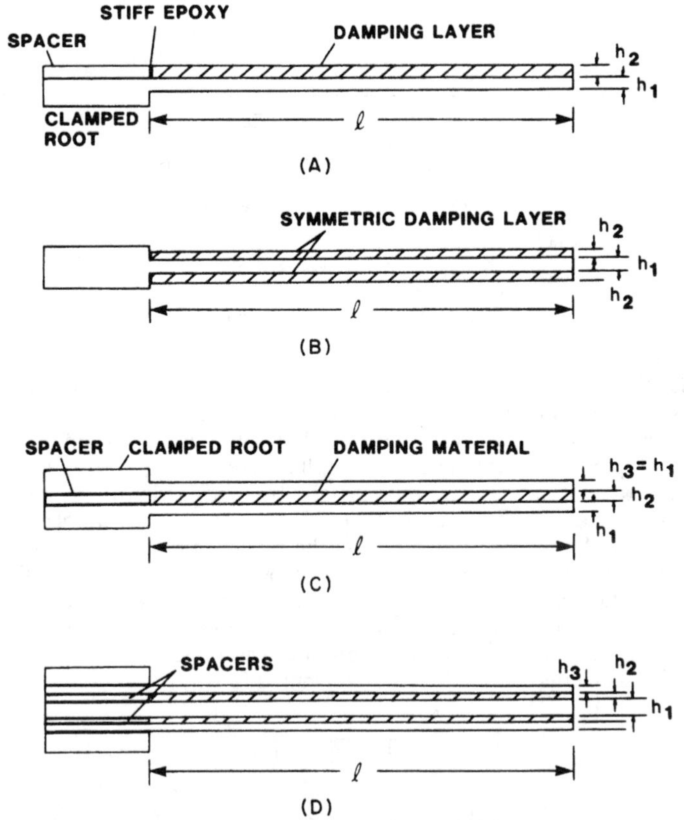

FIGURE 36.11 Cantilever beam damping material test configurations. (*A*) Nonsymmetric Oberst beam. (*B*) Symmetric modified Oberst beam. (*C*) Symmetric sandwich beam. (*D*) Symmetric constrained-layer beam.

excited by an impulsive force, measured through a force gage, and applied by a hammer, by an electromagnetic exciter, an electrically actuated impeller, or by sine-wave or random excitation. The response can be picked up by an electromagnetic transducer, in which case cross talk with the excitation transducer must be avoided by adequate separation or by the use of capacitative or electro-optical transducers or by a miniature accelerometer. The measured output can be displayed in many ways, including a decaying sinusoidal trace representing response to an impulsive excitation (a measure of the logarithmic decrement), or a frequency-domain display in the region of the fundamental free-free mode (loss factor measure). The observed logarithmic decrement or loss factor value is a measure of the damping of the plate/damping material system and depends on the plate and treatment thicknesses. The free-layer treatment equations used for the vibrating beam tests may also be used with the Geiger plate test provided that the same conditions are satisfied. In particular, the treatment thickness must be sufficient to make the ratio of the stiffness of the coated plate to that of the uncoated plate greater than about 1.05. The size of the specimen and the use of only one mode makes this condition somewhat less restrictive than for the beam tests, for which the specimens are much smaller.

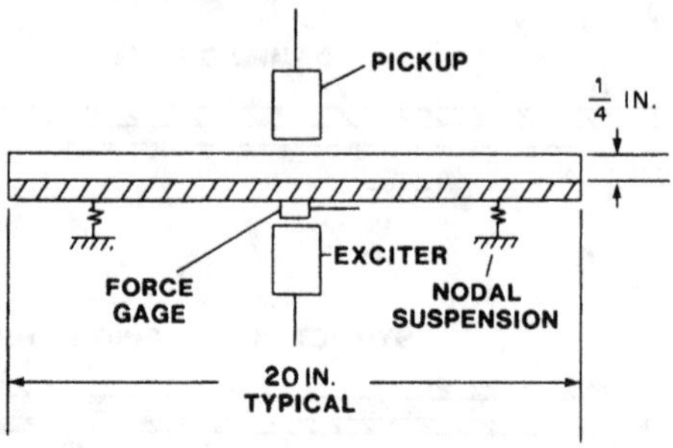

FIGURE 36.12 Geiger plate test configuration.

Single-Degree-of-Freedom Resonance Tests. Digital test instrumentation and data analysis techniques make it relatively easy to conduct vibration tests directly on relatively small samples of damping materials and to readily determine the damping properties. Typical test configurations are illustrated in Fig. 36.13. For a resonance type of test, the specimen is driven inertially by a large vibration table (see Chap. 25), usually by swept-sine-wave excitation. The input and output accelerations are usually measured by accelerometers, and the response parameter of interest is the amplification $A = x/x_0$ as a function of frequency, where x is the amplitude of displacement of the mass and x_0 is that of the shaker table. At resonance, the maximum value of $A = x/x_0$ is observed along with the resonance frequency ω_R for each temperature. The loss factor and modulus in both the tension-compression and shear loading of the specimen material are determined from

$$\eta = \frac{1}{\sqrt{A^2 - 1}} \tag{36.25}$$

$$E = \frac{m_e \omega_R^2 l}{S_1} \tag{36.26}$$

$$G = \frac{m_e \omega_R^2 h}{S_2} \tag{36.27}$$

For tension-compression loading, l is the length and $S_1 = wh$ is the cross-sectional area of the load-carrying member, where w is the cross-section width and h is the cross-section thickness, as illustrated in Fig. 36.13. For shear loading, h is the thickness of the shear layer and $S_2 = 2wl$ is the cross-sectional area of the shear member, where l is now the breadth of the load-carrying area, again as illustrated in Fig. 36.13. The effective mass m_e includes the added mass m and the effective mass of the specimen damping material, which is about one-third of its actual mass. For the extensional specimen, the ratio l/h or l/w, whichever is smaller, must be greater than 1.0 or shape effects will have to be taken into account. For the shear configuration, the ratio h/l must be less than 0.2 for the same reason. For highly damped materials, for which x/x_0 does not exceed 1.0 by a significant amount, considerable error in measuring A and η will be encountered, but the method is very effective for values of η less than about 0.5. In this method, data are obtained at only one frequency; the mass m must be changed to obtain data at other frequencies. Care must be taken to avoid sagging or creep of the specimen at high temperatures and to ensure that thermal equilibrium has been achieved. A thermocouple placed within the volume of the

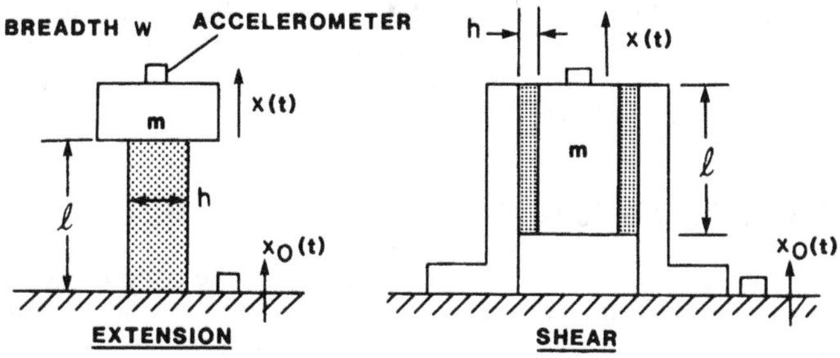

FIGURE 36.13 Resonance test concepts.

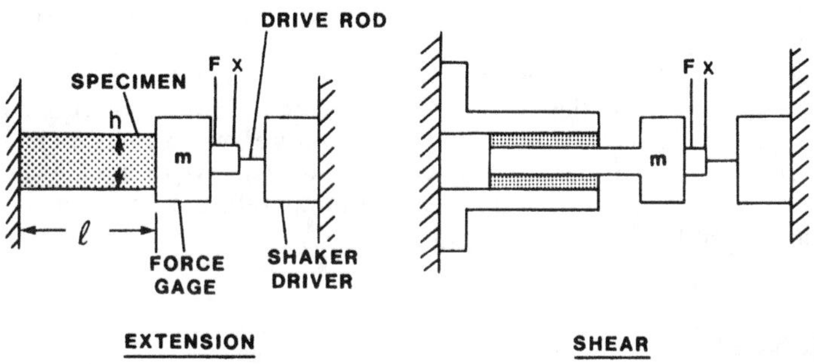

FIGURE 36.14 Impedance test concepts.

specimen material may be necessary, particularly for tests at high strain amplitudes where internal heating of the specimen by energy dissipation from damping may lead to wide differences between true specimen temperature and the temperature of the surroundings.

Impedance Tests. If the specimens are excited by a driver through a force gage, then the response measure used to characterize the system behavior is the compliance or receptance x/F, where F is the driving force measured by the force gage and x is the response at the same point, measured by an accelerometer, for example. If the mass m is large compared with the mass of the specimen, as illustrated in Fig. 36.14, then one may add one-third the mass of the specimen to m to give the effective mass m_e of the equivalent single-degree-of-freedom system, so that

$$\frac{x}{F} = \frac{1}{k(1 + j\eta) - m_e\omega^2} \qquad (36.28)$$

If this is expressed instead in terms of the ratio F/x, the dynamic stiffness at the driving point, which is directly related to the driving-point impedance, then

$$\kappa = k - m_e\omega^2 + jk\eta \qquad (36.29)$$

which shows that the direct dynamic stiffness is a linear function of ω^2 and the quadrature dynamic stiffness $\kappa_Q = k\eta$. It is not difficult to obtain good measurements of k

and η by this type of test approach from about $0.2\omega_r$ to $3\omega_r$, so data can be obtained quite easily over about a decade of frequency instead of at only a single frequency as for the resonance method. Analytical mass corrections may also have to be made to account for inertial effects at the force gage.

COMMERCIAL TEST SYSTEMS

Many commercial systems are available for measuring the complex modulus properties of viscoelastic damping materials.[10,11] All are based on some kind of deformation mode of a sample of the material, measurement of the corresponding excitation forces and displacements, and analysis of the data to obtain the material properties. Each system has advantages and disadvantages, but when due care is exercised, good results usually can be obtained with each system. Particular care should be taken to read, understand, and follow the manufacturer's instructions. For example, in some tests such as monitoring cure cycles of epoxies, the temperature sweep rate can be quite high in order to keep up with the reaction. This is acceptable if one is monitoring the progress of the cure cycle, but it may not be acceptable if one seeks to measure the damping properties at a state approximating thermal equilibrium. For thermal equilibrium to be maintained, temperature sweep rates well below 1°F (0.5°C) per minute are usually recommended, and even lower rates may be required for large specimens. A dwell period at each temperature is recommended before performing the test.

REFERENCES

1. Lazan, B. J.: "Damping of Materials and Members in Structural Mechanics," Pergamon Press, New York, 1968.
2. Jones, D. I. G.: "Handbook of Viscoelastic Vibration Damping," John Wiley & Sons, Chichester, U.K., 2001.
3. Nashif, A. D., D. I. G. Jones, and J. P. Henderson: "Vibration Damping," John Wiley & Sons, New York, 1985.
4. Ferry, J. D.: "Viscoelastic Properties of Polymers," 3d ed., John Wiley & Sons, New York, 1980.
5. Bagley, R. L., and P. Torvik: *J. Rheology,* **27**(3):201 (1983).
6. Ross, D., E. E. Ungar, and E. M. Kerwin: "Damping of Plate Flexural Vibrations by Means of Viscoelastic Laminae," *Structural Damping,* ASME Publication, New York, 1959, pp. 49–88.
7. Oberst, H.: *Akustika,* **4**:181, Germany (1952).
8. American Society for Testing and Materials Committee on Standards, Standard Method for Measuring Vibration-Damping Properties of Materials, ASTM Standard E-756-9, ASTM, 1916 Race Street, Philadelphia, Pa., USA.
9. DTI VBT™ Beam Test System, Damping Technologies, Inc., Mishawaka, Ind., USA.
10. DuPont DMP™ Flexural Impedance Test System, DuPont Instruments, Inc., Wilmington, Del., USA.
11. PL-DMTA™ Impedance Test System, Polymer Laboratories, Ltd., Loughborough, U.K.

CHAPTER 37

TORSIONAL VIBRATION IN RECIPROCATING AND ROTATING MACHINES

Ronald L. Eshleman

INTRODUCTION

Torsional vibration is an oscillatory angular motion causing twisting in the shaft of a system; the oscillatory motion is superimposed on the steady rotational motion of a rotating/reciprocating machine. Even though the vibration cannot be detected without special measuring equipment, its amplitude can be destructive. For example, gear sets that alter speeds of power transmission systems transmit the vibration to the casing. Similarly, slider crank mechanisms in engines and compressors convert torques to radial forces that are discernable to human perception but are not measurable because of the insensitivity of test equipment and background noise. If gearboxes or reciprocating machines are part of a drive train, excess noise and vibration can indicate trouble.

Motion is rarely a concern with torsional vibration unless it affects the function of a system. It is stresses that affect the structural integrity and life of components and thus determine the allowable magnitude of the torsional vibration. Torsional vibratory motions can produce stress reversals that cause metal fatigue. Components tolerate less reversed stress than steady stress. In addition, stress concentration factors associated with machine members decrease the effectiveness of load-bearing materials.

Figure 37.1 illustrates the twisting of a shaft of an electric motor-compressor system. The torsional mode shape associated with the first torsional natural frequency is shown in Fig. 37.2. A coupling in the power train allows for misalignment in the assembly. The mode shape shows that the stiffness of the coupling is much less than that of other shaft sections. This is indicated by the large slope (change in angular displacement) of the mode shape at the coupling. The coupling will be the predominant component in the motor-compressor system governing the torsional natural frequency associated with the mode.

Torsional vibration is usually a complex vibration having many different frequency components. For example, shock resulting from abrupt start-ups and unloading of gear teeth causes transient torsional vibration in some systems; start-up of

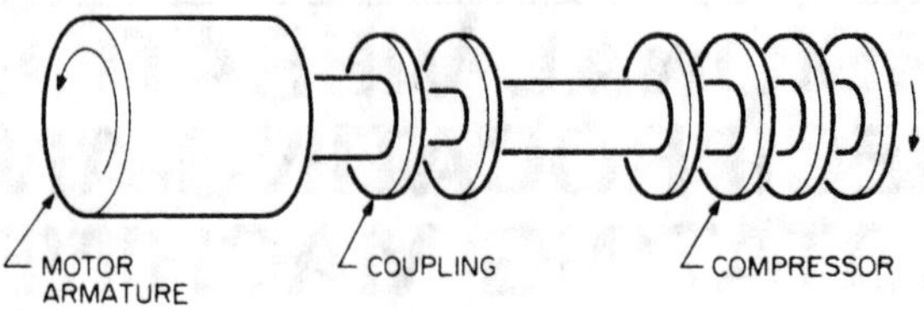

FIGURE 37.1 Schematic drawing illustrating the twisting of the shaft of a motor-compressor system.

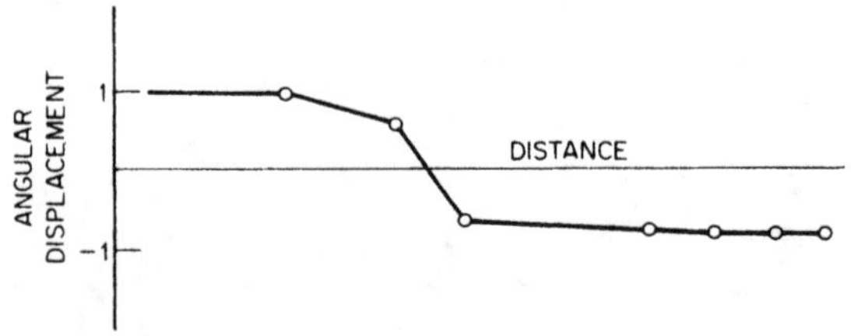

FIGURE 37.2 Torsional mode shape for the motor-compressor system shown in Fig. 37.1.

synchronous electric motor systems may cause torsional resonance. Random torsional vibrations caused by gear inaccuracies and ball bearing defects are relatively common in rotating machines.

MODELING

The torsional elastic system of a drive unit and its associated machinery is a complicated arrangement of mass and elastic distribution. The complete mechanical system can include the drive unit, couplings, gearboxes or other speed-changing devices, and one or more driven units. This complicated system is made amenable to mathematical treatment by representing it as a model—a simpler system that is substantially equivalent dynamically. The equivalent system usually consists of lumped masses which are connected by massless torsionally elastic springs as illustrated in Fig. 37.3. The masses are placed at each crank center and at the center planes of actual flywheels, rotors, propellers, cranks, gears, impellers, and armatures.[1]

The torsional calculation is made not for the drive unit alone but for the complete system, including all driven machinery. On an engine, it is usually possible to consider such parts as camshafts, pumps, and blowers either as detached from the engine (if they are driven elastically) or as additional rigid masses at the point of attachment to the crankshaft (if the driver is relatively rigid). If there is doubt, these parts should be included in the torsional calculation as elastically connected masses and removed if the natural frequencies do not change after the parts are removed from the model.

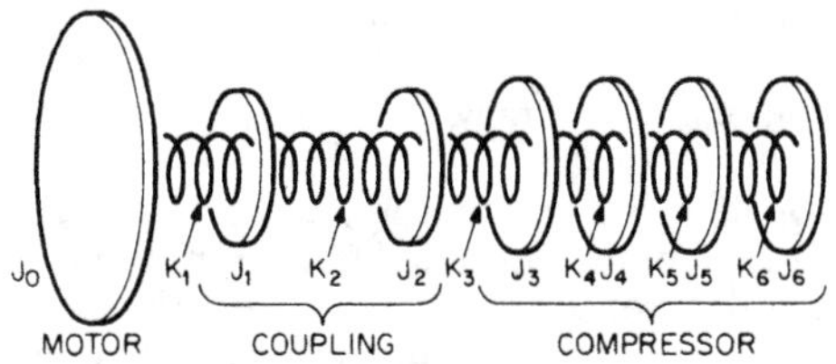

FIGURE 37.3 A model of the motor-compressor system shown in Fig. 37.1, consisting of a series of masses connected by massless torsionally elastic springs (K = stiffness, lb·in./rad; J = polar moment of inertia, lb-in.-sec²).

CALCULATION OF POLAR MOMENTS OF INERTIA

Circular Disc or Cylinder Rotating About a Perpendicular Axis. The polar moment of inertia, essential in modeling torsional vibration, often is easy to calculate. The general form is $J = \int r^2\,dm$, where r is the instantaneous radius, and dm is the differential mass. The formula for the polar moment of inertia of a circular disc or cylinder rotating about a perpendicular axis is

$$J = \frac{\pi d^4 l \gamma}{32g} \qquad \text{lb-in.-sec}^2 \qquad (37.1)$$

where J = polar moment of inertia, lb-in.-sec²
γ = material density, lb/in.³
d = diameter of disc or cylinder, in.
l = axial length of disc or cylinder, in.
g = acceleration due to gravity, 386.1 in./sec²

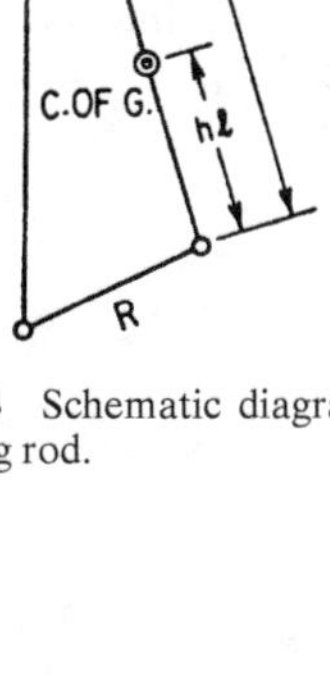

FIGURE 37.4 Schematic diagram of a crank and connecting rod.

Piston and Connecting Rod. The piston and connecting rod shown schematically in Fig. 37.4 introduce a variable-mass problem, the solution of which is complex. The exact solution shows that the effect of the piston and connecting rod can be closely approximated by representing them as a concentrated rotor of polar inertia J defined by

$$J = \left[\frac{W_P}{2} + W_c\left(1 - \frac{h}{2}\right)\right]\frac{R^2}{g} \qquad \text{lb-in.-sec}^2 \qquad (37.2)$$

where W_P = weight of piston, piston pin, and cooling fluid, lb
W_c = weight of connecting rod, lb
h = fraction of rod length from crank pin to center of gravity
R = crank radius, in.

Crankshaft. The polar inertias of the crank webs, the crankpin, and the journal sections are added to that given by Eq. (37.2). These polar inertias should be calculated with the best obtainable accuracy.

Propellers. For propellers, pumps, and hydraulic couplings an addition must be made for the virtual inertia of the entrained fluid. For marine propellers this is ordinarily assumed at 26 percent of the propeller inertia. Virtual inertias for pumps are not known accurately, but it can be assumed that half the casing is filled with rotating fluid. Marine propeller polar inertia can be estimated by

$$J = 0.044MD^2 \text{ lb-in.-sec}^2$$

where M = mass of the propeller, lb · sec²/in.
D = diameter of propeller, in.

EXPERIMENTAL DETERMINATION OF POLAR MOMENT OF INERTIA

For complex shaft elements such as couplings or small flywheels, it is often easier to determine the polar moment of inertia experimentally than to calculate it. In one experimental technique the element is suspended from three equally spaced vertical wires as shown in Fig. 37.5. The element whose polar moment of inertia is to be measured is hung on the cables and set into torsional motion. Then the period of vibration is measured. The experimentally determined period of torsional vibration, the weight of the element, the length of the suspending cables, and the radius of attachment of the cables are used to determine the polar moment of inertia from the following formula:

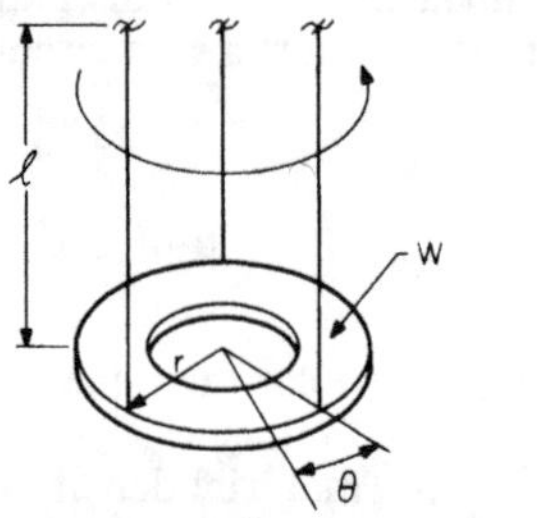

FIGURE 37.5 Experimental determination of the polar moment of inertia. An element of weight W is suspended by three wires and the period of the torsional motion is determined.

$$J = \frac{Wr^2\tau^2}{(6.28)^2 l} \quad \text{lb-in.-sec}^2 \tag{37.4}$$

where J = polar moment of inertia
τ = period of vibration, sec/cycle
W = weight of element, lb
l = length of cables, in.
r = radius of suspending cables, in.

CALCULATION OF STIFFNESS

Shaft. The stiffness of a circular shaft is the most common elastic element encountered in the modeling process. Table 37.1 shows some common formulas used to calculate torsional stiffness of a hollow circular shaft, a tapered circular shaft, and two

TABLE 37.1 Formulas for Torsional Stiffness

K = torsional stiffness, lb-in./rad; G = shear modulus, lb/in.^{2} ω = rotational speed, rad/sec		
Springs in series		$K = \dfrac{1}{1/K_1 + 1/K_2}$
Springs in parallel	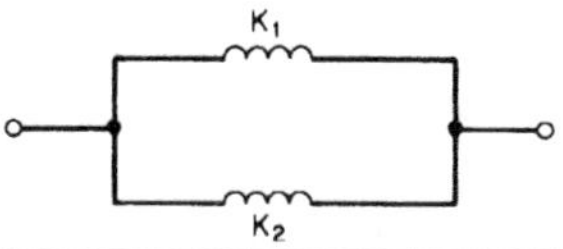	$K = K_1 + K_2$
Hollow circular shaft	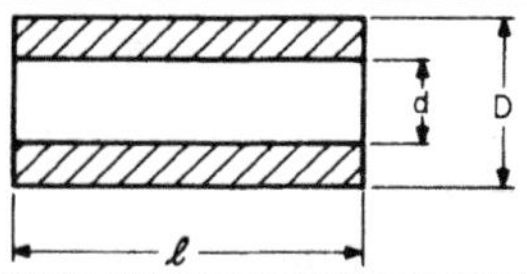	$K = \dfrac{\pi}{32} \dfrac{G(D^4 - d^4)}{\ell}$
Tapered circular shaft	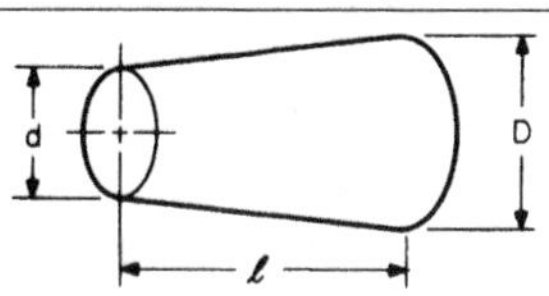	$K = \dfrac{3\pi}{32} \dfrac{d^4}{\ell(n + n^2 + n^3)}$ $n = \dfrac{d}{D}$
Two-geared shaft (referred to shaft[1])	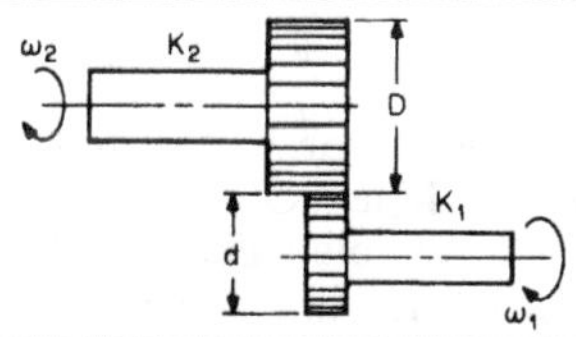	$K = \dfrac{K_1 K_2}{n^2 K_1 + K_2}$ $n = \dfrac{D}{d} = \dfrac{\omega_1}{\omega_2}$
Wilson[2] formula crankshaft	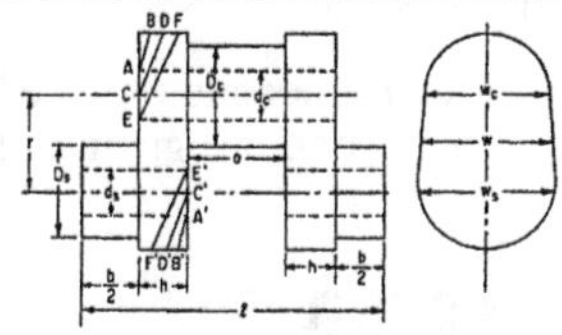	$\dfrac{\ell_e}{D_s^4} = \dfrac{b + 0.4D_s}{D_s^4 - d_s^4} + \dfrac{a + 0.4D_c}{D_c^4 - d_c^4} + \dfrac{r - 0.2(D_s + D_c)}{hW^3}$

geared shafts. The stiffness is referred to the rotational speed of shaft no. 1. The inertia of geared shafts is obtained in a similar manner.

Crankshaft. The crankshaft stiffness is the most uncertain element in a torsional vibration calculation. Shaft stiffness can be measured experimentally either by twisting a shaft with a known torque or from the observed values of the critical speeds in a running engine. Alternatively, it can be calculated from semiempirical formulas such as those given in Ref. 1. Refer to Table 37.1 for definitions of the dimensions; l_e is the length of a solid shaft of diameter D_s equal in torsional stiffness to the section of crankshaft between crank centers.

Changes in Section. The shafting of an engine system may contain elements such as changes of section, collars, shrunk and keyed armatures, etc., which require the exercise of judgment in the assessment of stiffness. For a change of section having a fillet radius equal to r at the smaller diameter, the stiffness can be estimated by assuming that the smaller shaft is lengthened an amount ΔL, obtained from the curve of Fig. 37.6. This also may be applied to flanges where D is the bolt diameter. The stiffening effect of collars can be ignored.

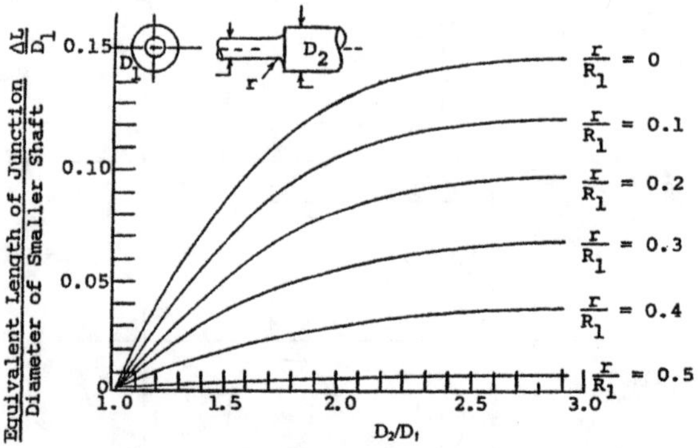

FIGURE 37.6 Junction effect for cylindrical shafts with various diameter ratios and fillet radii.[1]

Shrunk and Keyed Parts. The stiffness of shrunk and keyed parts is difficult to estimate, as the stiffening effect depends to a large extent on the tightness of the shrunk fit and keying. The most reliable values of stiffness are obtained by neglecting the stiffening effect of an armature and assuming that the armature acts as a concentrated mass at the center of the shrunk or keyed fit.

Elastic Couplings. Properties of numerous types of torsionally elastic couplings are available from the manufacturers and are given in Ref. 1.

GEARED AND BRANCHED SYSTEMS

The natural frequencies of a system containing gears can be calculated by assuming a system in which the speed of the driver unit is n times the speed of the driven equipment. Multiply all the inertia and elastic constants on the driven side of the system by $1/n^2$, and calculate the system's natural frequencies as if no gears exist. In any calculations involving damping constants on the driven side, these constants also are multiplied by $1/n^2$. Torques and deflections thus obtained on the driven side of this substitute system, when multiplied by n and $1/n$, respectively, are equal to those in the actual geared system. Alternatively, the driven side can be used as the reference; multiply the driver constants by n^2.

NATURAL FREQUENCY CALCULATIONS

If the model of a system can be reduced to two lumped masses at opposite ends of a massless shaft, the natural frequency is given by

$$f_n = \frac{1}{2\pi}\sqrt{\frac{(J_1 + J_2)k}{J_1 J_2}} \quad \text{Hz} \tag{37.5}$$

The mode shape is given by $\theta_2/\theta_1 = -J_1/J_2$.

For the three-mass system shown in Fig. 37.7, the natural frequencies are

$$f_n = \frac{1}{2\pi}\sqrt{A \pm (A^2 - B)^{1/2}} \quad \text{Hz} \tag{37.6}$$

where $A = \dfrac{k_{12}(J_1 + J_2)}{2J_1J_2} + \dfrac{k_{23}(J_1 + J_2)}{2J_1J_2}$

$$B = \frac{(J_1 + J_2 + J_3)k_{12}k_{23}}{J_1J_2J_3}$$

In Eqs. (37.5) and (37.6) the ks are torsional stiffness constants expressed in lb-in./rad. The notation k_{12} indicates that the constant applies to the shaft between rotors 1 and 2. The polar inertia J has units of lb-in.-sec^2.

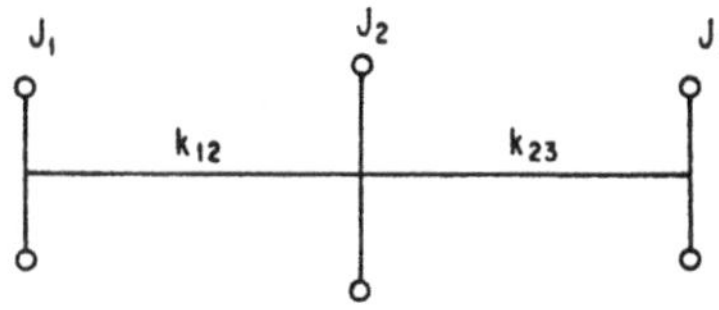

FIGURE 37.7 Schematic diagram of a shaft represented by three masses.

The preceding formulas and all the developments for multimass torsional systems that follow also apply to systems with longitudinal motion if the polar moments of inertia J are replaced by the masses $m = W/g$ and the torsional stiffnesses are replaced by longitudinal stiffnesses.

TRANSFER MATRIX METHOD

The transfer matrix method[3] is an extended and generalized version of the Holzer method. Matrix algebra is used rather than a numerical table for the analysis of torsional vibration problems. The transfer matrix method is used to calculate the natural frequencies and critical speeds of other eigenvalue problems.

The transfer matrix and matrix iteration (Stodola) methods are numerical procedures. The fundamental difference between them lies in the assumed independent variable. In any eigenvalue problem, a unique mode shape of the system is associated with each natural frequency. The mode shape is the independent variable used in the matrix iteration method. A mode shape is assumed and improved by successive iterations until the desired accuracy is obtained; its associated natural frequency is then calculated.

A frequency is assumed in the transfer matrix method, and the mode shape of the system is calculated. If the mode shape fits the boundary conditions, the assumed frequency is a natural frequency and a critical speed is derived. Determining the correct natural frequencies amounts to a controlled trial-and-error process. Some of the essential boundary conditions (geometrical) and natural boundary conditions (force) are assumed, and the remaining boundary condition is plotted versus frequency to obtain the natural frequency; the procedure is similar to the Holzer method. For example, if the torsional system shown in Fig. 37.8 were ana-

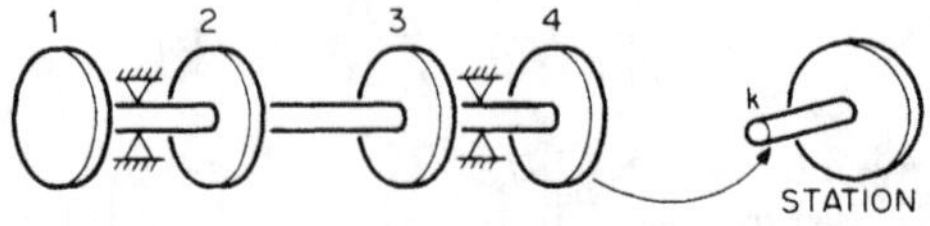

FIGURE 37.8 Typical torsional vibration model.

lyzed, the natural boundary conditions would be zero torque at both ends. The torque at station no. 1 is made zero, and the torsional vibration is set at unity. Then M_4 as a function of ω is plotted to find the natural frequencies. This plot is obtained by utilizing the system transfer functions or matrices. These quantities reflect the dynamic behavior of the system.

No accuracy is lost with the transfer matrix method because of coupling of mode shapes. Accuracy is lost with the matrix iteration method, however, because each frequency calculation is independent of the others.

A typical station (no. 4) from a torsional model is shown in Fig. 37.8. This general station and the following transfer matrix equation, Eq. (37.7), are used in a way similar to the Holzer table to transfer the effects of a given frequency ω across the model.

$$\begin{vmatrix} \theta \\ M \end{vmatrix}_n = \begin{bmatrix} 1 & 1/k \\ -\omega^2 J & -(\omega^2 J/k) + 1 \end{bmatrix}_n \begin{vmatrix} \theta \\ M \end{vmatrix}_{n-1} \tag{37.7}$$

where θ = torsional motion, rad
M = torque, lb-in.
ω = assumed frequency, rad/sec
J = station inertia, lb-in.-sec^2
k = station torsional stiffness, lb-in./rad

The stiffness and polar moment of inertia of each station are entered into the equation to determine the transfer effect of each element of the model. Thus, the calculation begins with station no. 1, which relates to the first spring and inertia in the model of Fig. 37.8. The equation gives the output torque M_1 and output motion θ_1 for given input values, usually 0 and 1, respectively. The equation is used on station no. 2 to obtain M_2 output and θ_2 output as a function of M_1 output and θ_1 output. This process is repeated to find the value of M and θ at the end of the model. This calculation is particularly suited for the digital computer with spreadsheet programs.

FINITE ELEMENT METHOD

The finite element method is a numerical procedure (described in Chap. 23) to calculate the natural frequencies, mode shapes, and forced response of a discretely modeled structural or rotor system. The complex rotor system is composed of an assemblage of discrete smaller finite elements which are continuous structural members. The displacements (angular) are forced to be compatible, and force (torque) balance is required at the joints (often called *nodes*).

Figure 37.9 shows a uniform torsional element in local coordinates.[4] The x axis is taken along the centroidal axis. The physical properties of the element are density (ρ), area (A), shear modulus of elasticity (G), length (l), and polar area moment (I). $M(t)$ are the torsional forcing functions.

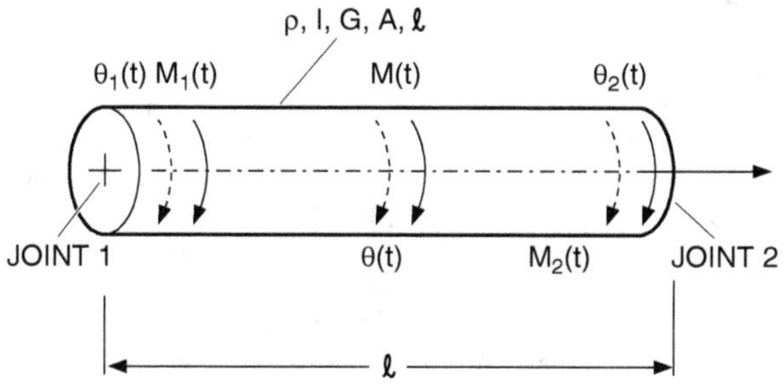

FIGURE 37.9 Finite element for torsional vibration in local coordinates.

As noted, the previously described finite elements are in local coordinates. Since the system as a whole must be analyzed as a unit, the mass and stiffness matrices and joint force vectors of each element must be expressed in the global coordinate system to find the vibration response of the complete system.

Using transformation matrices,[4] the mass and stiffness matrices and force vectors are used to set up the system equation of motion for a single element in the global coordinates:

$$[J]_e\{\ddot{\Theta}(t)\} + [K]_e\{\ddot{\Theta}(t)\} = \{M_e(t)\} \tag{37.8}$$

The complete system is an assemblage of the number of finite elements it requires to adequately model its dynamic behavior. The joint displacements of the elements in the global coordinate system are labeled as $\Theta_1(t), \Theta_2(t), \ldots, \Theta_m(t)$, or this can be expressed as a column vector:

$$\{\Theta(t)\} = \begin{Bmatrix} \Theta_1(t) \\ \Theta_2(t) \\ \vdots \\ \Theta_m(t) \end{Bmatrix} \tag{37.9}$$

Using global joint displacements, mass and stiffness matrices, and force vectors, the equations of motion are developed:

$$[J]_{nxn}\{\ddot{\Theta}\}_{nx1} + [K]_{nxn}\{\Theta\}_{nx1} = \{M\}_{nx1} \tag{37.10}$$

where n denotes the number of joint displacements in the system.

In the final step prior to solution, appropriate boundary conditions and constraints are introduced into the global model.

The equations of motion for free vibration are solved for the eigenvalues (natural frequencies) using the matrix iteration method (Chap. 22). Modal analysis is used to solve the forced torsional response. The analyst must select the joints (nodes, materials, shape functions, geometry, torques, and constraints) to model the system for computation of natural frequencies, mode shapes, and torsional response. Similar to other modeling efforts, engineering art and a knowledge of the capabilities of the computer program enable the engineer to provide reasonably accurate results.

CRITICAL SPEEDS

The crankshaft of a reciprocating engine or the rotors of a turbine or motor, and all moving parts driven by them, comprise a torsional elastic system. Such a system has several modes of free torsional oscillation. Each mode is characterized by a natural frequency and by a pattern of relative amplitudes of parts of the system when it is oscillating at its natural frequency. The harmonic components of the driving torque excite vibration of the system in its modes. If the frequency of any harmonic component of the torque is equal to (or close to) the frequency of any mode of vibration, a condition of resonance exists and the machine is said to be running at a *critical speed.* Operation of the system at such critical speeds can be very dangerous, resulting in fracture of the shafting.

The number of complete oscillations of the elastic system per unit revolution of the shaft is called an *order of the operating speed.* It is an order of a critical speed if the forcing frequency is equal to a natural frequency. An order of a critical speed that corresponds to a harmonic component of the torque from the engine as a whole is called a *major order.* A critical speed also can be excited that corresponds to the harmonic component of the torque curve of a single cylinder. The fundamental period of the torque from a single cylinder in a four-cycle engine is 720°; the critical speeds in such an engine can be of ½, 1, 1½, 2, 2½, etc., order. In a two-cycle engine only the critical speeds of 1, 2, 3, etc., order can exist. All critical speeds except those of the major orders are called *minor critical speeds;* this term does not necessarily mean that they are unimportant. Therefore, the critical speeds occur at

$$\frac{60f_n}{q} \quad \text{rpm} \tag{37.11}$$

where f_n is the natural frequency of one of the modes in Hz, and q is the order number of the critical speed. Although many critical speeds exist in the operating range of an engine, only a few are likely to be important.

A dynamic analysis of an engine involves several steps. Natural frequencies of the modes likely to be important must be calculated. The calculation is usually limited to the lowest mode or the two lowest modes. In complicated arrangements, the calculation of additional modes may be required, depending on the frequency of the forces causing the vibration. Vibration amplitudes and stresses around the operating range and at the critical speeds must be calculated. A study of remedial measures is also necessary.

VIBRATORY TORQUES

Torsional vibration, like any other type of vibration, results from a source of excitation. The mechanisms that introduce torsional vibration into a machine system are discussed and quantified in this section. The principal sources of the vibratory torques that cause torsional vibration are engines, pumps, propellers, and electric motors.

GENERAL EXCITATION

Table 37.2 shows some ways by which torsional vibration can be excited. Most of these sources are related to the work done by the machine and thus cannot be entirely removed. Many times, however, adjustments can be made during the design process. For example, certain construction and installation sources—gear runout, unbalanced or misaligned couplings, and gear-tooth machining errors—can be reduced.

TABLE 37.2 Sources of Excitation of Torsional Vibration

Source	Amplitude in terms of rated torque	Frequency
	Mechanical	
Gear runout		1 ×, 2 ×, 3 × rpm
Gear tooth machining tolerances		No. gear teeth × rpm
Coupling unbalance		1 × rpm
Hooke's joint		2 ×, 4 ×, 6 × rpm
Coupling misalignment		Dependent on drive elements
	System function	
Synchronous motor start-up	5–10	2 × slip frequency
Variable-frequency induction motors (six-step adjustable frequency drive)	0.04–1.0	6 ×, 12 ×, 18 × line frequency (LF)
Induction motor start-up	3–10	Air gap induced at 60 Hz
Variable-frequency induction motor (pulse width modulated)	0.01–0.2	5 ×, 7 ×, 9 × LF, etc.
Centrifugal pumps	0.10–0.4	No. vanes × rpm and multiples
Reciprocating pumps		No. plungers × rpm and multiples
Compressors with vaned diffusers	0.03–1.0	No. vanes × rpm
Motor- or turbine-driven systems	0.05–1.0	No. poles or blades × rpm
Engine geared systems with soft coupling	0.15–0.3	Depends on engine design and operating conditions; can be 0.5*n* and *n* × rpm
Engine geared system with stiff coupling	0.50 or more	Depends on engine design and operating conditions
Shaft vibration		*n* × rpm

In Table 37.2 note that the pulsating torque during start-up of a synchronous motor is equal to twice the slip frequency. The slip frequency varies from twice the line frequency at start-up to zero at synchronous speed. Many mechanical drives exhibit characteristics of pulsating torque during operation due to their design function. Electric motors with variable-frequency drives induce pulsating torques at frequencies that are harmonics of line frequency. Blade-passing excitations can be characterized by the number of blades or vanes on the wheel: the frequency of excitation equals the number of blades multiplied by shaft speed. The amplitude of a pulsating torque is often given in terms of percentage of average torque generated in a system.

ENGINE EXCITATION

In more complex cases—diesel and gasoline engines, for example—the multiple frequency components depend on engine design and power output. The power output, crankshaft phasing, and relationship between gas torque and inertial torque influence the level of torsional excitation.

Inertia Torque. A harmonic analysis of the inertia torque of a cylinder is closely approximated by[1]

$$M = \frac{W}{g}\Omega^2 r\left(\frac{\lambda}{4}\sin\theta - \frac{1}{2}\sin 2\theta - \frac{3}{4}\lambda\sin 3\theta - \frac{\lambda^2}{4}\sin 4\theta \cdots\right) \qquad (37.12)$$

where
$W = W_p + hW_c$ [see Fig. 37.4 and Eq.(37.2)]
$\lambda = R/l$ [see Fig. 37.4 and Eq. (37.2)]
Ω = angular speed, rad/sec
R = crank radius, in.
l = connecting rod length, in.
θ = crank angle, radians
W_p = weight of piston, lb
W_c = weight of connecting rod, lb

It is usual to drop all terms above the third order.

Gas-Pressure Torque. A harmonic analysis of a gas torque versus rotation curve yields the gas-pressure components of the exciting torque. It is usually expressed as a Fourier series (see Chap. 14). These data are normally obtained from an indicator card or pressure transducer (p versus θ). Then the torque M for the crank angle θ is approximated by[1]

$$M \cong RAp(\theta)\sin\theta\left\{1 + \cos\theta\left(\lambda + \frac{1}{2}\lambda^3\sin^2\theta + \frac{3}{8}\lambda^5\sin^4\theta + \ldots\right)\right\} \qquad (37.13)$$

where A is the piston area. A gas pressure–versus–rotation curve analyzed to obtain harmonic gas coefficients is required to conduct a gas-pressure torque calibration. Harmonic gas coefficients are usually available from engine manufacturers.

FORCED VIBRATION RESPONSE

The torsional vibration amplitude of a modeled system is determined by the magnitude, points of application, and phase relations of the exciting torques produced by engine or compressor gas pressure and inertia and by the magnitudes and points of application of the damping torques. Damping is attributable to a variety of sources, including pumping action in the engine bearings, hysteresis in the shafting and between fitted parts, and energy absorbed in the engine frame and foundation. In a few cases, notably marine propellers, damping of the propeller predominates. When an engine is fitted with a damper, the effects of damping dominate the torsional vibrations.

Techniques available for calculation of vibration amplitudes include the exact solution of differential equations, the energy balance method, the transfer matrix method, and modal analysis. The techniques are implemented on lumped–parameter or finite-element models.

EXACT METHOD FOR TWO-DEGREE-OF-FREEDOM SYSTEMS

The lowest mode of vibration of some systems, particularly marine installations, can be approximated with a two-mass system (Fig. 37.10); an excitation is applied at one end and damping at the other. The natural frequency was given by Eq. (37.5).

The shaft torque is $M_{12} = k(\theta_1 - \theta_2)$. The amplitude of M_{12} at resonance is

$$|M_{12}| = k|\theta_1 - \theta_2| = M_e \frac{I_2}{I_1} \sqrt{1 + \frac{kI_2(I_1 + I_2)}{I_1 c^2}} \tag{37.14}$$

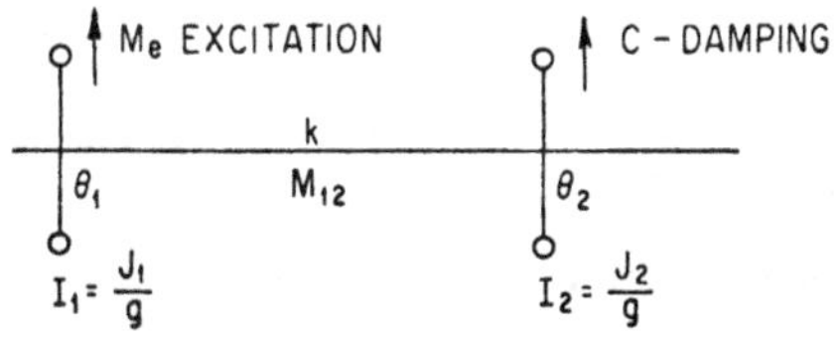

FIGURE 37.10 Schematic diagram of a shaft with two rotors, showing positions of excitation and damping.

Since with usual damping the second term under the radical is large compared with unity, Eq. (37.14) reduces to

$$|M_{12}| \simeq \frac{M_e}{c} \frac{I_2}{I_1} \sqrt{(I_1 + I_2) \frac{I_2 k}{I_1}} \tag{37.15}$$

The torsional damping constant c of a marine propeller is a matter of some uncertainty. However, considerations of oscillating airfoil theory indicate the following damping value.

$$c = \frac{2.3 M_{\text{mean}}}{\Omega} \qquad \text{in.-lb/rad/sec} \tag{37.16}$$

where Ω = angular speed of shaft in radians per second.

Equation (37.15) is applicable only when $I_1/I_2 > 1$. If used outside this range with other types of damping neglected, fictitiously large amplitudes will be obtained.

ENERGY BALANCE METHOD

Both rational and empirical formulas for the resonance amplitudes of systems with linear or nonlinear dampers can be based on the energy balance at resonance. It is assumed that the system vibrates in a normal mode and that the displacement is in a 90° phase relationship to the exciting and damping torques. The energy input by the exciting torques is then equal to the energy output by the damping torques. Unless the damping is extremely large, this assumption gives a very close approximation to the amplitude at resonance.

Figure 37.11 shows a curve of relative amplitude in the first mode of vibration. Assume that a cylinder acts at A. Let the actual amplitude at A be θ_a and the amplitude relative to that of the no. 1 cylinder be β. The β values are taken from the mode

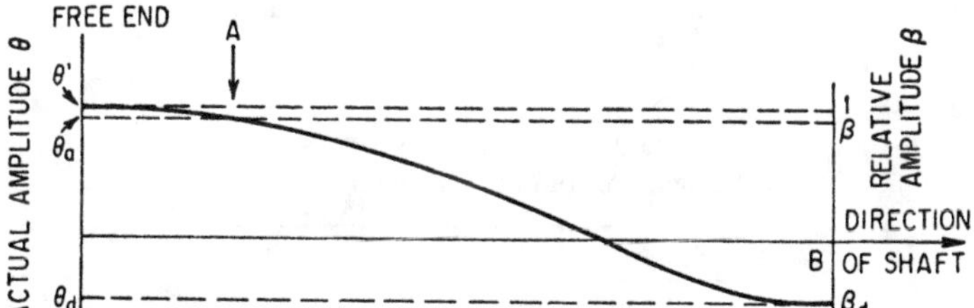

FIGURE 37.11 Diagram of actual amplitude θ and relative amplitude β as a function of position along shaft. Excitation is at A, and B is the position where damping is applied. The no. 1 cylinder is at the free end of the crankshaft.

shape normalized to cylinder no. 1. At a point such as B, where damping may be applied, let the actual amplitude be θ_d and the amplitude relative to the no. 1 cylinder be β_d.

The energy input to the system from the cylinder acting at A is

$$\pi M_e \theta_a \qquad \text{in.-lb/cycle}$$

and the energy output to the damper is

$$\pi c \omega \theta_d^2 \qquad \text{in.-lb/cycle}$$

where c is the damping constant action of the damper at B. Equating input to output,

$$M_e \theta_a = c \omega \theta_d^2 \tag{37.17}$$

Let θ' be the amplitude at the no. 1 cylinder produced by the cylinder acting at A. Then $\theta_a/\theta' = \beta$ and $\theta_d/\theta' = \beta_d$. Substituting in Eq. (37.17) and summing all the cylinders and dampers applied at a variety of points, the total amplitude at the no. 1 cylinder is

$$\theta = \Sigma\theta' = \frac{M_e \Sigma\beta}{\omega \Sigma c \beta_d^2} \tag{37.18}$$

where $\Sigma\beta$ is taken over the cylinders and $\Sigma c\beta_d^2$ is taken over the points at which damping is applied. This formula can be applied directly when the magnitude and points of application of the damping torques are known.

Good results have been obtained using the Lewis formula[5]

$$M_m = \Re M_e \Sigma\beta \tag{37.19}$$

The maximum torque at resonance in any part of the system is M_m; the exciting torque per cylinder is M_e. $\Re$ is a constant from Table 37.3. The vector sum over the cylinders of the relative amplitudes as taken from the mode shape for a natural frequency is $\Sigma\beta$. It is determined as follows.

For a four-cycle engine construct a phase diagram, Table 37.4, of the firing sequence in which 720° corresponds to a complete cycle of a single cylinder, or two revolutions. The phase relationship for a critical of order number q is obtained by multiplying the angles in this diagram by $2q$, with the no. 1 crank held fixed. The β values assigned to each direction then are obtained from the values corresponding to each cylinder in the mode shape β. Then $\Sigma\beta$ is the vector sum. The summation extends only to those rotors on which exciting torques act.

TABLE 37.3 Empirical Factors for Engine Amplitude Calculations

Bore Stroke	$\Re$
20 in. × 24 in. or larger	50–60
8 in. × 10 in.	40–50
4 in. × 6 in. or smaller	35

TABLE 37.4 Phase Diagrams and Deflections, β, for a Calculated Torsional Mode

PHASE DIAGRAMS	Σβ	ORDERS
FIRING SEQUENCE 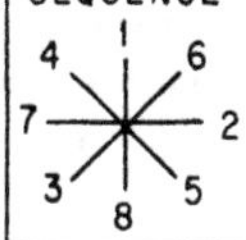	0.778	1/2, 4½, 8½, - - MIRROR IMAGE FOR 3½, 5½, 7½, - -
x2 OR 6 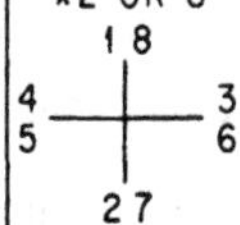	0.169	1, 5, 9, — MIRROR IMAGE FOR 3, 7, 11
x3 OR 5 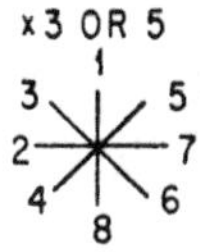	1.549	1½, 5½, 9½, + - MIRROR IMAGE FOR 2½, 6½, 10½, - -
x4 	0.4287	2, 6, 10, - -
x8 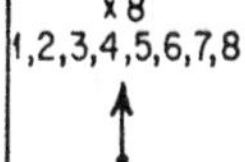	5.0455	4, 8, 12 MAJOR ORDERS

cn	β
1	1.0000
2	0.9543
3	0.8771
4	0.7709
5	0.6296
6	0.4595
7	0.2742
8	0.0799
	5.0455

In a two-cycle engine the β phase relations are determined by multiplying the crank diagram by q, holding the no. 1 cylinder fixed.

Table 37.4 shows the Σβ phase diagrams and Σβ values for the one-noded mode with a firing sequence 1, 6, 2, 5, 8, 3, 7, 4. The firing sequence is drawn first; then the angles of this diagram are multiplied by 2, 3, 4, etc., in succeeding diagrams. After multiplication by 8 for the fourth order, the diagrams repeat. Diagrams which are equidistant in order number from the 2, 6, 10, etc., orders are mirror images of each other and have the same Σβ. The numerical values of Σβ in Table 37.4 have been obtained by calculation, summing the vertical and horizontal components.

The empirical factor $\mathfrak{R}$ is determined by the measurement of amplitudes in running engines (Table 37.3).

The exciting torque per cylinder, M_e in Eq. (37.18) is composed of the sum of the torques produced by gas pressure, inertia force, gravity force, and friction force. The gravity and friction torques are of negligible importance; and the inertia torque is of importance only for first-, second-, and third-order harmonic components.

APPLICATION OF MODAL ANALYSIS TO ROTOR SYSTEMS

Classical modal analysis of vibrating systems[6] can be used to obtain the forced response of multistation rotor systems in torsional motion. The natural frequencies and mode shapes of the system are found using the transfer matrix or finite element methods. The response of the rotor to periodic phenomena (not necessarily a harmonic or shaft frequency) is determined as a linear weighted combination of the mode shapes of the system. Heretofore with this technique, damping has been entered in modal form; the damping forces are a function of the various modal velocities. The formation of equivalent viscous damping constants that are some percentage of critical damping is required. The critical damping factor is formed from the system modal inertia.[7]

The modal analysis technique can be used for a torsional distributed mass model of engine systems using modal damping; nonsynchronous speed excitations are allowed. The shaft sections of the modeled rotor have distributed mass properties and lumped end masses (including rotary inertia). A transfer matrix or finite element analysis is performed to obtain a finite number of natural frequencies. The number required depends on the range of phased forcing frequencies used in the problem. A function consisting of a weighted average of the mode shapes is formed and substituted into

$$\theta(x, t) = \sum_{n=1}^{N} a_n(x) f_n(t) \tag{37.20}$$

where θ = torsional response
a_n = normal modes
f_n = periodic time-varying weighting factors

The function $f_n(t)$ is determined from the ordinary differential equations of motion and is a function of the phased forcing functions, rotor speed, modal damping constants, and mode shapes of the system.

DIRECT INTEGRATION

Direct integration of equations of motion of a system utilize first- or second-order differential equations. The method is fundamental for linear and nonlinear response problems.[7] Any digitally describable vibration or shock excitation can be carried out with this method.

Direct integration can be used on nonlinear models and arbitrary excitation, so it is one of the most general techniques available for response calculation. However, large computer storage is required, and large computer costs are usually incurred because small time- or space-step sizes are needed to maintain numerical stability. An adjustable step integration routine such as predictor-corrector helps to alleviate this problem. Such a numerical integration must be started with another routine such as Runge-Kutta.

Direct integration is particularly useful when nonlinear components such as elastomeric couplings are involved or when the excitation force varies in frequency and magnitude. Direct integration is used for analysis of synchronous motor start-ups in which the magnitude of the torque varies with rotor speed and the frequency is 2 times the slip frequency—starting at twice the line frequency and ending at zero when the rotor is locked on synchronous speed. Examples of this type of analysis are given in Ref. 7.

PERMISSIBLE AMPLITUDES

Failure caused by torsional vibration invariably initiates in fatigue cracks that start at points of stress concentration—e.g., at the ends of keyway slots, at fillets where there is a change of shaft size, and particularly at oil holes in a crankshaft. Failures can also start at corrosion pits, such as occur in marine shafting. At the shaft oil holes the cracks begin on lines at 45° to the shaft axis and grow in a spiral pattern until failure occurs. Theoretically the stress at the edges of the oil holes is 4 times the mean shear stress in the shaft, and failure may be expected if this concentrated stress exceeds the fatigue limit of the material. The problem of estimating the stress required to cause failure is further complicated by the presence of the steady stress from the mean driving torque and the variable bending stresses.

In practice the severity of a critical speed is judged by the maximum nominal torsional stress

$$\tau = \frac{16M_m}{\pi d^3} \tag{37.21}$$

where M_m is the torque amplitude from torsional vibration and d is the crankpin diameter. This calculated nominal stress is modified to include the effects of increased stress and is compared to the fatigue strength of the material.

U.S. MILITARY STANDARD

A military standard[8] issued by the U.S. Navy Department states that the limit of acceptable nominal torsional stress within the operating range is

$$\tau = \frac{\text{ultimate tensile strength}}{25} \qquad \text{for steel}$$

$$\tau = \frac{\text{torsional fatigue limit}}{6} \qquad \text{for cast iron}$$

If the full-scale shaft has been given a fatigue test, then

$$\tau = \frac{\text{torsional fatigue limit}}{2} \qquad \text{for either material}$$

Such tests are rarely, if ever, possible.

For critical speeds below the operating range which are passed through in starting and stopping, the nominal torsional stress shall not exceed 1¾ times the above values.

Crankshaft steels which have ultimate tensile strengths between 75,000 and 115,000 lb/in.^{2} usually have torsional stress limits of 3000 to 4600 lb/in.^{2}.

For gear drives the vibratory torque across the gears, at any operating speed, shall not be greater than 75 percent of the driving torque at the same speed or 25 percent of full-load torque, whichever is smaller.

AMERICAN PETROLEUM INSTITUTE

Sources of torsional excitation considered by American Petroleum Institute[9] (API) include but are not limited to the following: gear problems such as unbalance, pitch line runout, and eccentricity; start-up conditions resulting from inertial impedances; and torsional transients from synchronous and induction electric motors.

Torsional natural frequencies of the machine train shall be at least 10 percent above or below any possible excitation frequency within the specified operating speed range. Torsional critical speeds at integer multiples of operating speeds (e.g., pump vane pass frequencies) should be avoided or should be shown to have no adverse effect where excitation frequencies exist. Torsional excitations that are nonsynchronous to operating speeds are to be considered. Identification of torsional excitations is the mutual responsibility of the purchaser and the vendor.

When torsional resonances are calculated to fall within the ±10 percent margin and the purchaser and vendor have agreed that all efforts to remove the natural frequency from the limiting frequency range have been exhausted, a stress analysis shall be performed to demonstrate the lack of adverse effect on any portion of the machine system.

In the case of synchronous motor driven units, the vendor is required to perform a transient torsional vibration analysis with the acceptance criteria mutually agreed upon by the purchaser and the vendor.

TORSIONAL MEASUREMENT

Torsional vibration is more difficult to measure than lateral vibration because the shaft is rotating. Procedures for signal analysis are similar to those used for lateral vibration. Torsional response—both strains and motions—can be measured at intermediate points in a system. But sensors cannot be placed at a nodal point; for this reason the transfer matrix method is valuable for calculating mode shapes prior to sensor location selection.

SENSORS

Strain gauges, described in Chap. 12, are available in a variety of sizes and sensitivities and can be placed almost anywhere on a shaft. They can be calibrated to indicate instantaneous torque by using static torque loads on drive shafts. If calibration is not possible, stresses and torques can be calculated from strength of materials theory. Strain gauges are usually mounted at 45° angles so that shaft bending does not influence torque measurements. The signal must be processed by a bridge-amplifier unit that can be arranged to compensate for temperature. Because strain gauge signals are difficult to take from a rotating shaft, such techniques are not common diagnostic tools.

Slip rings can be used to obtain a vibration signal from a shaft. Wireless telemetry is also available. A small transmitter mounted on the rotating shaft at a convenient location broadcasts a signal to a nearby receiver. Commercial torque transducers are available for torsional measurement. However, they must be inserted in the drive line and thus may change the dynamic characteristics of the system. If the natural frequency of the system is changed, the vibration response will not accurately reflect the properties of the system.

The velocity of torsional vibration is measured using a toothed wheel and a fixed sensor.[10] The signal generated by the teeth of the wheel passing the fixed sensor has a frequency equal to the number of teeth multiplied by shaft speed. If the shaft is undergoing torsional vibration, the carrier frequency will exhibit frequency modulation (change in frequency) because the time required for each tooth to pass the fixed pickup varies.

DATA ACQUISITION

The frequency change (velocity) is converted to a voltage change by a demodulator and integrated to obtain angular displacement. Angular displacement can be measured at the end of a shaft with encoders or at intermediate points with a gear-magnetic pickup or proximity probe arrangement. The frequency of the carrier signal (e.g., number of teeth on a gear × rpm) must be at least 4 times the highest frequency to be measured. In most cases, the raw torsional signal is tape recorded prior to processing and analysis. Because the output of the magnetic pickup is speed dependent and the gap between the magnetic pickup and the toothed wheel is less than 0.025 in. the proximity probe is preferred—especially in synchronous motor startups.

TORSIONAL ANALYSIS

A torsional signal must be analyzed for frequency components using a spectrum analyzer, described in Chap. 14. Figure 37.12 shows a torsional response spectrum for a variable-frequency motor-driven pump. The pump ran at 408 rpm. The torsional vibration response excited by the variable frequency motor is 0.23° at a frequency of 38 Hz.

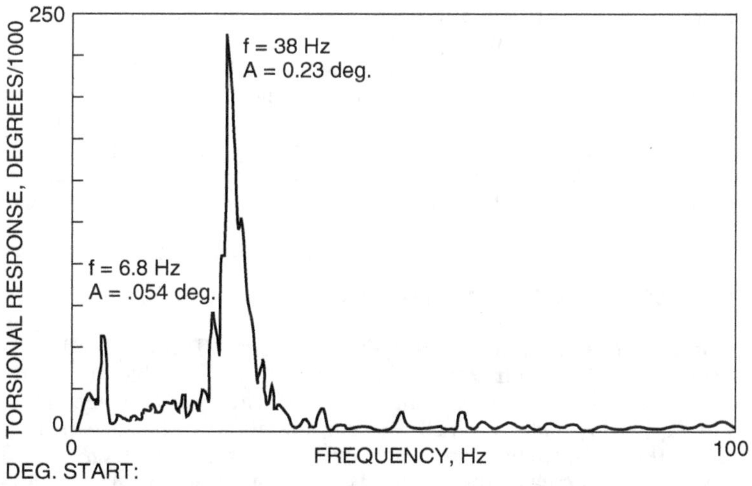

FIGURE 37.12 Torsional response of a variable-frequency motor-driven pump at 408 rpm. There are significant peaks at 6.8 and 38.0 Hz.

MEASURES OF CONTROL

The various methods which are available for avoiding a critical speed or reducing the amplitude of vibration at the critical speed may be classified as:

1. Shifting the values of critical speeds by changes in mass and elasticity
2. Vector cancellation methods

3. Change in mass distribution to utilize the inherent damping in the system
4. Addition of dampers of various types

SHIFTING OF CRITICAL SPEEDS

If the stiffness of all the shafting to a system is increased in the ratio *a,* then all the frequencies will increase in the square root of the ratio *a,* provided that there is no corresponding increase in the inertia. It is rarely possible to increase the crankshaft diameters on modern engines; in order to reduce bearing pressures, bearing diameters usually are made as large as practical. If bearing diameters are increased, the increase in the critical speed will be much smaller than indicated by the *a* ratio because a considerable increase in the inertia will accompany the increase in diameter. Changes in the stiffness of a system made near a nodal point will have maximum effect. Changes in inertia near an antinode will have maximum effect, while those near a node will have little effect.

By the use of elastic couplings it may be possible to place certain critical speeds below the operating speed where they are passed through only in starting and stopping; this leaves a clear range above the critical speed. This procedure must be used with caution because some critical speeds, for example the fourth order in an eight-cylinder, four-cycle engine, are so violent that it may be dangerous to pass through them. If the acceleration through the critical speed is sufficiently high, some reduction in amplitude may be attained, but with a practical rate the reduction may not be large. The rate of deceleration when stopping is equally important. In some cases mechanical clutches disconnect the driven machinery from the engine until the engine has attained a speed above dangerous critical speeds. Elastic couplings may take many forms including helical springs arranged tangentially, flat leaf springs arranged longitudinally or radially, various arrangements using rubber, or small-diameter shaft sections of high-tensile steel.[1]

VECTOR CANCELLATION METHODS

Choice of Crank Arrangement and Firing Order. The amplitude at certain minor critical speeds sometimes can be reduced by a suitable choice of crank arrangement and firing order (i.e., firing sequence). These fix the value of the vector sum $\Sigma\beta$ in Eq (37.19), $M_m = \Re M_e \Sigma\beta$. But considerations of balance, bearing pressures, and internal bending moments restrict this freedom of choice. Also, an arrangement which decreases the amplitude at one order of critical speed invariably increases the amplitude at others. In four-cycle engines with an even number of cylinders, the amplitude at the half-order critical speeds is fixed by the firing order because this determines the $\Sigma\beta$ value.

V-Type Engines. In V-type engines, it may be possible to choose an angle of the V which will cancel certain criticals. Letting ϕ be the V angle between cylinder banks, and q the order number of the critical, the general formula is

$$q\phi = 180°, 540°, 1080°, \text{etc.} \tag{37.22}$$

For example, in an eight-cylinder engine the eighth order is canceled at angles of 22½°, 67½°, 112½°, etc.

Cancellation by Shift of the Node. If an engine can be arranged with approximately equal flywheel (or other rotors) at each end so that the node of a particular mode is at the center of the engine, $\Sigma\beta$ will cancel for the major orders of that mode. This procedure must be used with caution because the double flywheel arrangement may reduce the natural frequency in such a manner that low-order minor criticals of large amplitudes take the place of the canceled major criticals.

Reduction by Use of Propeller Damping in Marine Installations. From Eq. (37.15) it is evident that the torque amplitude in the shaft can be reduced below any desired level by making the flywheel moment of inertia I_1 of sufficient magnitude. The ratio of the propeller amplitude to the engine amplitude increases as the flywheel becomes larger; thus the effectiveness of the propeller as a damper is increased.

DAMPERS

Many arrangements of dampers can be employed (see Chap. 6). In each type there is a loose flywheel or inertia member which is coupled to the shaft by:

1. Coulomb friction (Lanchester damper)
2. Viscous fluid friction
3. Coulomb or viscous friction plus springs
4. Centrifugal force, equivalent to a spring having a constant proportional to the square of the speed (pendulum damper) (see Chap. 6)

Each of these types acts by generating torques in opposition to the exciting torques.

REFERENCES

1. Nestorides, E. J.: "A Handbook of Torsional Vibration," Cambridge University Press, 1958.
2. Wilson, W. K.: "Practical Solutions of Torsional Vibration Problems," John Wiley & Sons, New York, 1942.
3. Rao, S. S.: "Mechanical Vibration," Addison-Wesley Publishing, Reading, Mass., 1990.
4. Walter, D. N., "Torsional Vibration of Turbomachinery," McGraw-Hill Book Company, New York, 2004.
5. Lewis, F. M.: *Trans. Soc. of Naval Arch. Marine Engrs.,* **33**:109 (1925).
6. Eshleman, R. L.: "Torsional Response of Internal Combustion Engines," *Trans. ASME,* **96**(2):441 (1974).
7. Anwar, I.: "Computerized Time Transient Torsional Analysis of Power Trains," *ASME Paper* No. 79-DET-74, 1979.
8. U.S. Navy Department: "Military Standard Mechanical Vibrations of Mechanical Equipment," MIL-STD-167 (SHIPS).
9. American Petroleum Institute: "Centrifugal Compressors for General Refinery Service," API STD 617, 5th ed. 1988, Washington, D.C.
10. Eshleman, R. L.: "Torsional Vibrations in Machine Systems," *Vibrations,* **3**(2):3 (1987).

CHAPTER 38
THEORY OF SHOCK AND VIBRATION ISOLATION

Michael A. Talley

INTRODUCTION

Shock and vibration isolation reduces the excitation transmitted to systems requiring protection. An example is the insertion of isolators between equipment and foundations supporting the equipment. The isolators act to reduce effects of support motion on the equipment and to reduce effects of force transmitted by the equipment to the supporting structure. Isolators or isolation systems discussed in this chapter are idealized into combinations of linear resilient elements and dampers. The resilient elements act by deflecting and storing energy at resonant frequencies of the isolation system, thereby decreasing force levels transmitted at higher frequencies. The dampers act by dissipating energy to reduce the amplification of forces that occur at resonance.

ISOLATION AREAS

Performance of isolation systems may be described in terms of maximum responses to inputs or as time history responses. Maximum responses may be presented in terms of shock response spectra (SRS; see Chap. 20), transmissibility, and other motion response parameters. Time histories, usually generated by computer simulations[1] of the equations of motion, or by analytical solutions, are used to assess the transient character of responses to shock. Figure 38.1 shows examples of SRS and vibration transmissibility, with shaded isolation areas representing reduced response levels.[2] The ordinate of each represents the ratio of the maximum acceleration of the equipment being isolated to that of the excitation (e.g., if the isolator was rigid). The abscissa τ/T of the SRS (Fig. 38.1*A*) represents the ratio of the excitation pulse duration τ to the natural period T of the isolator (or product of the excitation pulse duration and natural frequency of the isolator). The abscissa of the transmissibility curve (Fig. 38.1*B*) is the ratio of excitation frequency to that of the isolator. As a result, the isolation areas where response levels are decreased are represented at different ends of the spectra, even though the isolator may have the same natural frequency

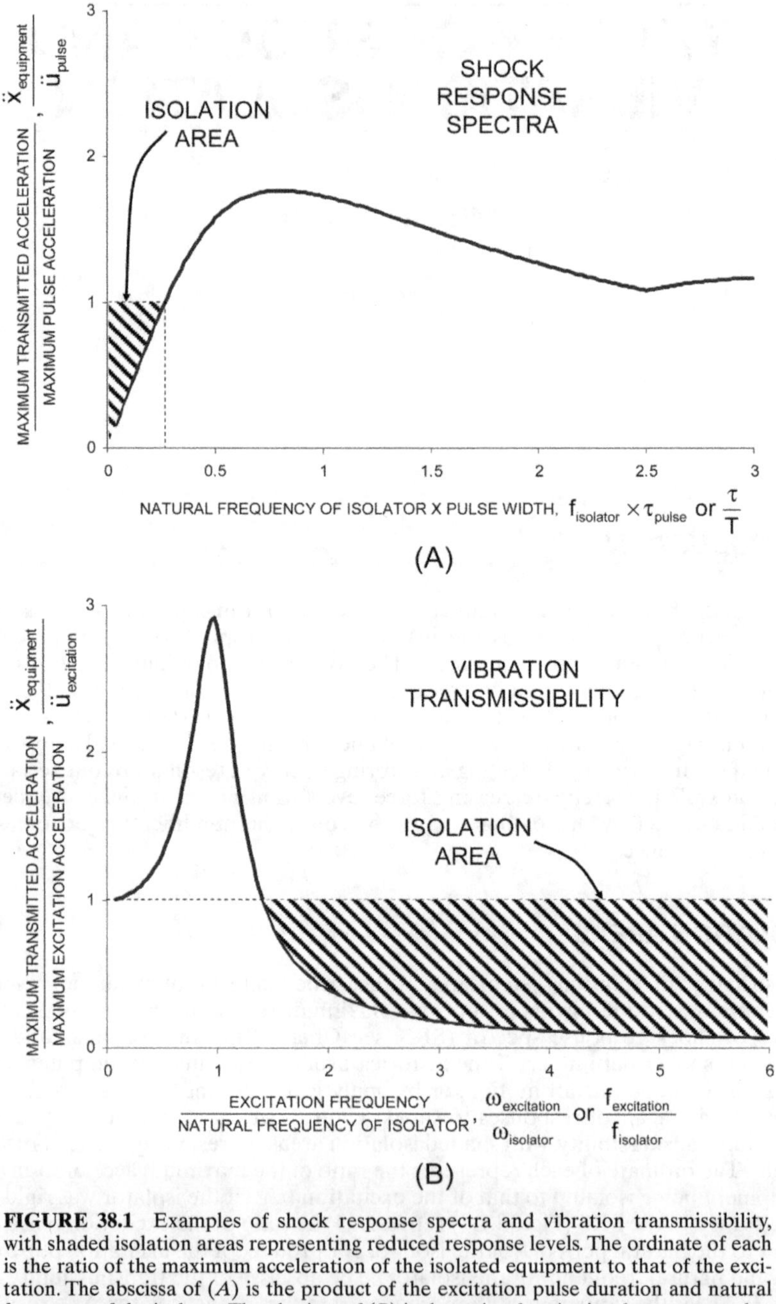

FIGURE 38.1 Examples of shock response spectra and vibration transmissibility, with shaded isolation areas representing reduced response levels. The ordinate of each is the ratio of the maximum acceleration of the isolated equipment to that of the excitation. The abscissa of (*A*) is the product of the excitation pulse duration and natural frequency of the isolator. The abscissa of (*B*) is the ratio of excitation frequency to that of the isolator. As a result, the isolation areas where response levels are decreased are represented at different ends of the spectra, even though the isolator may have the same natural frequency for shock and vibration applications.

for shock and vibration applications. Transmissibility may also be used to describe shock isolation performance when isolated equipment has a flexible component and when equipment is isolated on a flexible foundation.

CLASSIFICATION OF ISOLATION PROBLEMS

Two major classes of isolation will be discussed in this chapter:

- Class A: isolation of support motion
- Class B: isolation of force originating at equipment

The objective of Class A is to isolate components of equipment from shock and vibration inputs originating from the support or foundation. Examples include such items as the draft gear on a railroad car, the shock strut of aircraft landing gear, the mounts on airborne electronic equipment, shock isolation mounts on equipment on battleships, and the corrugated paper used to package lightbulbs. The objective of Class B is to isolate the supporting foundation from input forces originating at equipment. Examples include the recoil cylinders on gun mounts and the isolators on drop hammers, looms, and reciprocating presses.

ISOLATION ANALYSIS METHODS

Transient, steady-state, and random responses may occur in isolated systems, depending on the source of excitation. In this chapter, shock responses of isolated systems are discussed in terms of transient analysis methods, and vibration isolation is discussed in terms of steady-state and random vibration analysis methods. Although there are some similarities in the presentation of isolator performance data (e.g., transmissibility), the analytical methods differ. The first part of this chapter focuses on discussion of shock isolation. Discussion of vibration isolation theory is presented in later sections.

IDEALIZATION OF THE SYSTEM FOR SHOCK

The location of shock isolators on equipment depends on the available interfaces for the isolators both on the equipment and in the environment where the equipment is installed. In general, isolators are not at optimum locations, resulting in nonsymmetry and coupled modes[3] not well adapted to analysis by simple means. Also, isolator forces $F(\delta,\dot{\delta})$ may be represented as having nonlinear resilient elements and dampers, hysteresis,[4] strain-rate sensitivity, nonorthogonal principal axes, output forces that vary with multiaxis deflections [e.g., $F(\delta_x,\delta_y,\delta_z,\dot{\delta}_x,\dot{\delta}_y,\dot{\delta}_z)$], and many other factors that may affect their isolation performance. To simplify the discussion of shock isolators, linear hypothetical one- (1-DOF) and two- (2-DOF) degree-of-freedom systems are analyzed.

EQUATION OF MOTION FOR 1-DOF CLASS A

The simplest approach to problems of Class A is through a study of single-degree-of freedom (1-DOF) systems. Consider the system in Fig. 38.2*A*. The basic elements are

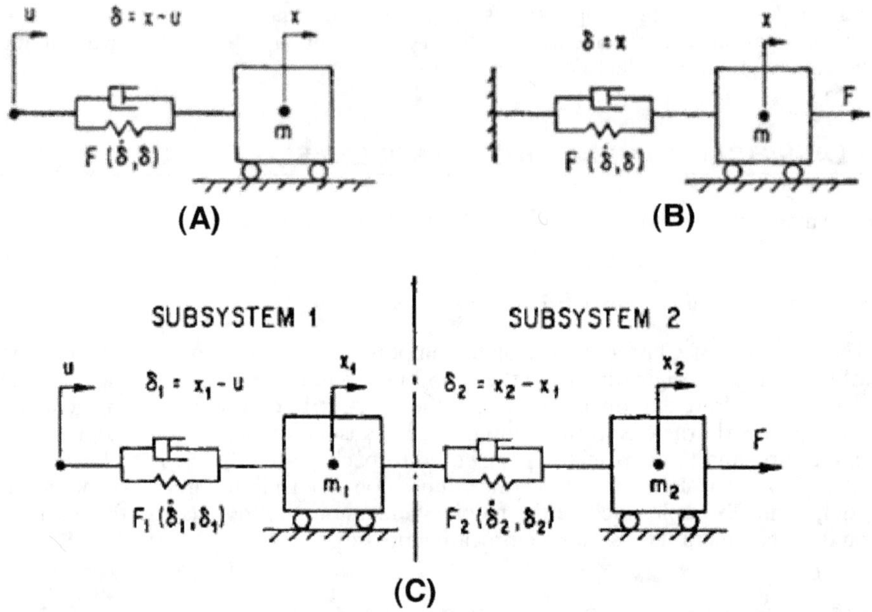

FIGURE 38.2 Idealized systems used in the discussion of shock isolation. Single-degree-of-freedom Class A isolation is shown in (*A*), and Class B isolation is shown in (*B*). A general two-degree-of-freedom system is shown in (*C*).

a mass and a spring-dashpot unit attached to the mass at one end. The block may be taken to represent the equipment (assumed to be a rigid body), and the spring-dashpot unit to represent the shock isolator. The displacement of the support is u. The equation of motion is

$$m\ddot{\delta} + F(\dot{\delta},\delta) = -m\ddot{u} \tag{38.1}$$

where m = mass of block, lb-sec²/in. (kg)
δ = deflection of spring ($\delta = x - u$), in. (m)
$F(\dot{\delta},\delta)$ = force exerted on mass (positive when tensile), lb (N)
u = absolute displacement of left-hand end of spring-dashpot unit, in. (m)

In the typical shock isolation problem, the system just described is initially at rest ($\dot{u} = \dot{\delta} = 0$) in an equilibrium position ($u = \delta = 0$). An external shock causes the support to move. If the corresponding movement of the left end of the shock isolator is described in terms of the support acceleration $\ddot{u}$, then Eq. (38.1) may be solved for the resulting extreme values of δ and $F(\dot{\delta},\delta)$. These values may be compared with the permissible deflection and force transmission limits of the shock isolator. It is also necessary to determine whether the internal stresses developed in the equipment are excessive. If the equipment is sufficiently rigid so that all parts have substantially equal accelerations, then the internal stresses are proportional to $\ddot{x}$, where $-m\ddot{x} = F(\dot{\delta},\delta)$.

EQUATION OF MOTION FOR 1-DOF CLASS B

Consider the system in Fig. 38.2*B* as representing equipment (mass m) attached to its support by the shock isolator (spring-dashpot unit). The left end of the spring-dashpot unit is fixed to a supporting structure, and there is a force F applied externally to the mass. The force F may be a real external force or it may be an *inertia force* generated by moving parts of the equipment. The equation of motion may be written as

$$m\ddot{\delta} + F(\dot{\delta},\delta) = F \tag{38.2}$$

where F is the external force applied to the mass in pounds and the relative displacement δ of the ends of the spring-dashpot unit is equal to the absolute displacement x of the mass. Assuming the system to be initially in equilibrium ($\dot{\delta} = 0$, $\delta = 0$), Eq. (38.2) is solved for extreme values of δ and $F(\dot{\delta},\delta)$, since F is a known function of time. These are to be compared with the displacement and force limitations of the shock isolator. Often, the supporting structure is sufficiently rigid that the maximum force in the isolator may be considered as a force applied statically to the support. Then the foregoing analysis is adequate for determining the stress in the support.

The similarity of shock isolation principles in Class A and Class B is indicated by the similar form of the 1-DOF Eqs. (38.1) and (38.2). The right-hand side ($-m\ddot{u}$ or F) is given as a function of time, and the extreme values of δ and $F(\dot{\delta},\delta)$ are desired.

SPECIFYING ISOLATOR REQUIREMENTS

When the spring-dashpot units are nonlinear, tests to characterize[5] the nonlinear isolators are typically conducted and the resulting system responses using the nonlinear models are computed numerically. Parameters specifying the isolator model may take the form of lookup tables, graphs, or analytical expressions with parameters that specify the curves of the resilient elements and damping along each principal axis of the isolator. For purposes of discussion in this chapter, the isolators are considered linear, with each spring-dashpot unit having the following force characteristic:

$$F(\dot{\delta},\delta) = c\dot{\delta} + k\delta \tag{38.3}$$

where c = damping coefficient, lb-sec/in. (N-sec/m), and k = spring stiffness, lb/in. (N/m). Even with this simplification, the number of parameters (m_1,c_1,k_1,m_2,c_2,k_2) for a 2-DOF system is so great that it is necessary to confine the analysis to a particular system. If the damping may be neglected [let $c = 0$ in Eq. (38.3)], then it is feasible to obtain equations in a form suitable for routine use.

The design requirements for the isolator usually include as a specification one or more of the following quantities:

1. Maximum allowable deflection δ_a
2. Maximum allowable force transmitted to support F_a
3. Maximum expected support velocity change $\dot{u}_a$
4. Maximum allowable absolute acceleration $\ddot{x}_a$

The specifications may be expressed mathematically as follows:

$$\delta_m \le \delta_a \qquad F_m \le F_a \qquad \dot{u}_m \ge \dot{u}_a \qquad \ddot{x}_m \le \ddot{x}_a \tag{38.4}$$

RESPONSE OF A 1-DOF SYSTEM TO A VELOCITY STEP

PHYSICAL BASIS FOR VELOCITY STEP

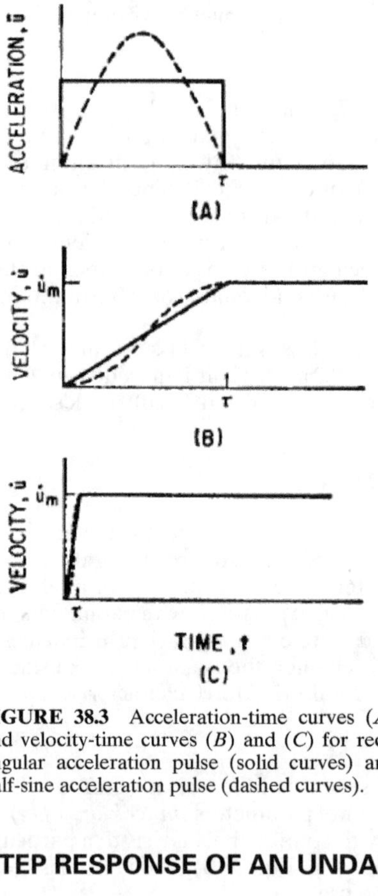

FIGURE 38.3 Acceleration-time curves (*A*) and velocity-time curves (*B*) and (*C*) for rectangular acceleration pulse (solid curves) and half-sine acceleration pulse (dashed curves).

The idealization of a shock motion as a simple change in velocity (velocity step) may form an adequate basis for designing a shock isolator and for evaluating its effectiveness. Consider the two types of acceleration $\ddot{u}$ versus time t curves illustrated in Fig. 38.3*A*. The solid line represents a rectangular pulse of acceleration, and the dashed line represents a half-sine pulse of acceleration. Each pulse has a duration τ. In Fig. 38.3*B*, the corresponding velocity-time curves are shown. Each of these curves is defined completely by specifying the type of acceleration pulse (rectangular or half-sine), the duration τ, and the velocity change $\dot{u}_m$. The curves of Fig. 38.3*B* are repeated in Fig. 38.3*C*, with the time scale shrunk to one-tenth. If τ is *sufficiently short,* the only significant remaining characteristic of the velocity step is the velocity change $\dot{u}_m$. The idealized velocity step, then, is taken to be a discontinuous change of $\dot{u}$ from zero to $\dot{u}_m$. A shock isolation system characteristically has a low natural frequency (long period), and this idealization leads to good results even when the pulse duration τ is about ¼ the natural period of the isolation system.

STEP RESPONSE OF AN UNDAMPED ISOLATOR

If damping is neglected [$c = 0$ in Eq. (38.3)], expressions relating the extreme value of isolator deflection $\delta = \delta_m$ may be obtained from

$$\int_0^{\delta_m} F(\delta)d\delta = 1/2\, m\dot{u}^2 \tag{38.5}$$

The right side of Eq. (38.5) represents the initial kinetic energy of the equipment relative to the support, and the integral on the left side represents the work done on the isolator. If there is no damping, the latter quantity is equal to the elastic potential energy stored in the isolator.

For the special case of a rigid body mounted on an undamped isolator, Eq. (38.5) suffices to determine all important results. In particular, the quantities of engineering significance are:

1. The maximum deflection of the isolator δ_m
2. The maximum isolator force $F_m = F(\delta_m) = m\ddot{x}_m$
3. The corresponding velocity change $\dot{u}_m$

The force-deflection characteristic of an undamped linear spring is

$$F(\delta) = k\delta \tag{38.6}$$

where k = spring stiffness, lb/in. Using the notation

$$\omega_n = \sqrt{\frac{k}{m}} \text{ rad/sec} \tag{38.7}$$

the maximum acceleration is

$$\ddot{x}_m = \omega_n^2 \delta_m \tag{38.8}$$

From Eqs. (38.5) and (38.7), the relation between velocity change $\dot{u}_m$ and maximum deflection δ_m is

$$\dot{u}_m = \omega_n \delta_m \tag{38.9}$$

Combining Eqs. (38.9) and (38.8),

$$\ddot{x}_m = \omega_n \dot{u}_m \tag{38.10}$$

STEP RESPONSE OF A VISCOUS DAMPED ISOLATOR

The addition of viscous damping can almost double the energy absorption capability of a linear shock isolator. Consider the system of Fig. 38.2*A*, with both spring and dashpot linear as defined by Eq. (38.3). Substituting $F(\dot{\delta},\delta)$ from Eq. (38.3) in Eq. (38.1) gives the equation of motion

$$\ddot{\delta} + 2\zeta\omega_n\dot{\delta} + \omega_n^2\delta = -\ddot{u} \tag{38.11}$$

where $\zeta = c/c_c$ represents the fraction of critical damping and $c_c = 2m\omega_n$ denotes critical damping [see Eq. (2.12)].

The initial conditions are $\dot{\delta} = -\dot{u}_m$, $\delta = 0$, when $t = 0$; for $t > 0$, $\ddot{u} = 0$. The solution of Eq. (38.11) is obtained by applying the Duhamel integral Eq. (20.33) to the right-hand side of Eq. (38.11), which results in

$$\delta(t) = \frac{-\dot{u}_m}{\omega_n\sqrt{1-\zeta^2}} e^{-\zeta\omega_n t} \sin\left(\omega_n\sqrt{1-\zeta^2}\,t\right) \tag{38.12}$$

for $\zeta < 1$. The acceleration is given by $\ddot{x} = \ddot{\delta} + \ddot{u}$; however, $\ddot{u} = 0$ for $t > 0$. Therefore, $\ddot{x}(t) = \ddot{\delta}(t)$, which may be obtained from the second derivative of Eq. (38.12). The times at which $\ddot{x}(t)$ and $\delta(t)$ achieve maximum values may be obtained by setting derivatives of Eq. (38.12) equal to zero and solving for t_m. The expressions for t_m for $\ddot{x}_m$ and δ_m are given by

$$t_{m\delta} = \cos^{-1}\zeta \,/\, \omega_n\sqrt{1-\zeta^2}$$

$$t_{m\ddot{x}} = \begin{cases} \tan^{-1}\left(\dfrac{\sqrt{1-\zeta^2}(4\zeta^2-1)}{\zeta(4\zeta^2-3)}\right) \Bigg/ \omega_n\sqrt{1-\zeta^2} \text{ for } \zeta <= 0.5 \\ 0 \text{ for } \zeta > 0.5 \end{cases} \tag{38.13}$$

Substituting these into absolute-value expressions for $\delta(t)$ and $\ddot{x}(t)$ results in the following dimensionless ratios representing the quantities of engineering significance:

$$\frac{\delta_m \omega_n}{\dot{u}_m} = e^{-\left(\frac{\zeta}{\sqrt{1-\zeta^2}} \cos^{-1}\zeta\right)} \tag{38.14}$$

$$\frac{\ddot{x}_m}{\dot{u}_m \omega_n} = \begin{cases} e^{-\left(\frac{\zeta}{\sqrt{1-\zeta^2}} \tan^{-1}\left(\frac{\sqrt{1-\zeta^2}(4\zeta^2-1)}{\zeta(4\zeta^2-3)}\right)\right)} & \text{for } \zeta <= 0.5 \\ 2\zeta \text{ for } \zeta > 0.5 \end{cases} \tag{38.15}$$

The product of Eqs. (38.14) and (38.15) is given by

$$\frac{\ddot{x}_m \delta_m}{\dot{u}_m^2} = \begin{cases} e^{-\frac{\zeta}{\sqrt{1-\zeta^2}}\left((\cos^{-1}\zeta) + \tan^{-1}\left(\frac{\sqrt{1-\zeta^2}(4\zeta^2-1)}{\zeta(4\zeta^2-3)}\right)\right)} & \text{for } \zeta <= 0.5 \\ 2\zeta e^{-\left(\frac{\zeta}{\sqrt{1-\zeta^2}} \cos^{-1}\zeta\right)} & \text{for } \zeta > 0.5 \end{cases} \tag{38.16}$$

Figure 38.4 shows the time history responses of an isolation system subjected to a velocity step with several values of ζ, ranging from 0 to 1. The acceleration response in the ordinate is normalized to the maximum undamped acceleration. The abscissa is time with respect to the natural period of the isolation system. Note that the presence of small damping reduces the maximum acceleration. As ζ is increased beyond 0.25, the maximum acceleration increases again. For $\zeta > 0.50$, the maximum acceleration occurs at $t = 0$ and exceeds that for no damping. Therefore, at $t = 0$, maximum acceleration is accounted for solely by the damping force $c\dot{\delta} = c\dot{u}_m$.

In Fig. 38.5 the dimensionless parameters $\ddot{x}_m/\dot{u}_m\omega_n$ from Eq. (38.15) and $\ddot{x}_m\delta_m/\dot{u}_m^2$ from Eq. (38.16) are plotted as functions of ζ. The parameter $\ddot{x}_m\delta_m/\dot{u}_m^2$ is a measure of the ability of the isolator to remove energy from the system. In the neighborhood $\zeta = 0.40$, the parameter $\ddot{x}_m\delta_m/\dot{u}_m^2$ attains a minimum value of 0.52. This parameter has the value of 1.00 for an undamped linear system.

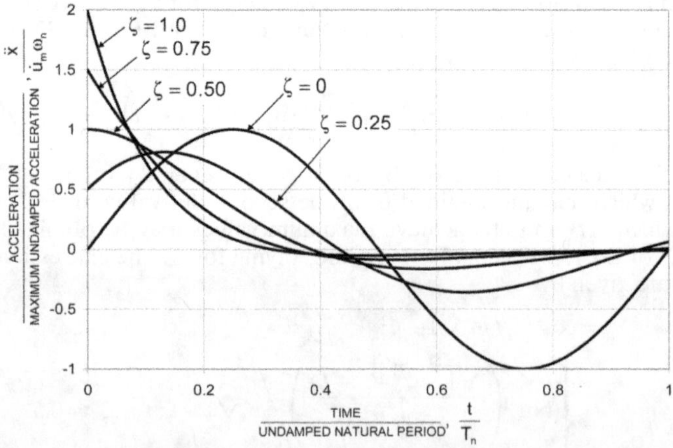

FIGURE 38.4 Dimensionless time histories of transmitted acceleration $\ddot{x}$ for an isolator having a linear spring and viscous damping.

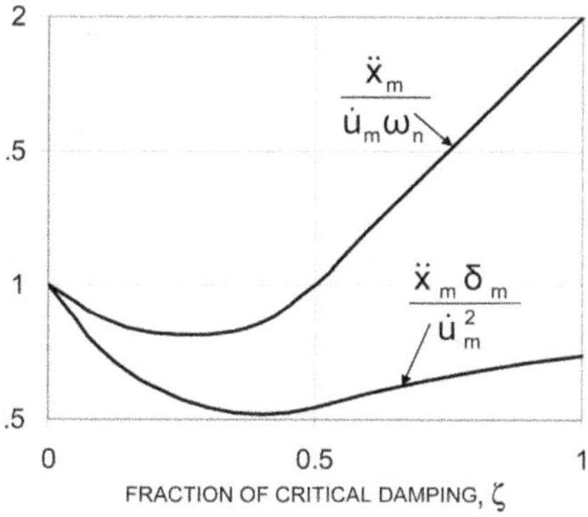

FIGURE 38.5 Dimensionless representation of maximum transmitted acceleration $\ddot{x}_m/\dot{u}_m\omega_n$ and dimensionless representation of energy absorption capability $\ddot{x}_m\delta_m/\dot{u}_m^2$ of an isolator having a linear spring and viscous damping.

Damping Considerations. True viscous damping provides a convenient model for discussing shock isolation, although it is difficult to attain except in electrical or magnetic form. Fluid dampers which depend upon orifices or other constricted passages to throttle the flow are likely to produce damping forces that vary more nearly as the square of the velocity. Dry friction tends to provide damping forces which are virtually independent of velocity. Elastomeric and structural materials tend to exhibit hysteretic or frequency-independent damping.

Additional information on damping is provided in Chaps. 2, 35, and 36. The section "Damped Multiple-Degree-of-Freedom Systems" in Chap. 2 discusses the difficulty in separating coupled normal modes when damping is present and provides a multiple-degree-of-freedom (MDOF) solution methodology [see Eqs. (2.86–2.88)] for the free vibration problem. The use of uniform viscous, structural, and mass damping to solve MDOF sinusoidal vibration problems is also presented. Chapter 35 discusses various damping materials and their properties. Chapter 36 provides methods of applying damping, a comparison of damping measures, and discussion of their use in 1-DOF systems.

Example 38.1: Equipment weighing 40 lb (177.9 N) and sufficiently stiff to be considered rigid is to be protected from a shock consisting of a velocity step $\dot{u}_a = 70$ in./sec (1.8 m/sec). The maximum allowable acceleration is $\ddot{x}_a = 21\,g$ (g is the acceleration of gravity), and available clearance limits the deflection to $\delta_a = 0.70$ in. (0.0178 m). This information may be used to find isolator characteristics (e.g., stiffness k and damping c) for an undamped linear spring and linear spring with viscous damping.

Undamped linear spring. Taking the maximum velocity $\dot{u}_m$ equal to the expected velocity $\dot{u}_a$ and using Eqs. (38.9) and (38.4),

$$\delta_m = \frac{\dot{u}_m}{\omega_n} \leq \delta_a \quad \text{or} \quad \omega_n \geq \frac{70 \text{ in./sec}}{0.7 \text{ in.}} = 100 \text{ rad/sec}$$

From Eqs. (38.10) and (38.4), $\ddot{x}_m = \omega_n \dot{u}_m \leq \ddot{x}_a$. Then

$$\omega_n \leq \frac{\ddot{x}_a}{\dot{u}_m} = \frac{21 \times 386 \text{ in./sec}^2}{70 \text{ in./sec}} = 116 \text{ rad/sec}$$

Selecting a value in the middle of the permissible range gives ω_n = 108 rad/sec (17.2 Hz). The corresponding maximum isolator deflection is $\delta_m = 0.65$ in. (0.0165 m), and the maximum acceleration of the equipment is $\ddot{x}_m$ = 7560 in./sec^2 (192 m/sec^2) = 19.6g. The isolator stiffness given by Eq. (38.7) is

$$k = m\omega_n^2 = \frac{40 \text{ lb}}{386 \text{ in./sec}^2} \times (108 \text{ rad/sec})^2 = 1210 \text{ lb/in.} = 211{,}903 \text{ N/m}$$

The value of k in the preceding equation represents the sum of the stiffnesses of the individual isolators.

Linear spring and viscous damping. The introduction of viscous damping in combination with a linear spring [Eq. (38.3)] affords the possibility of large energy dissipation capacity. From Fig. 38.5, the best performance is obtained at the fraction of critical damping $\zeta = 0.40$, where $\ddot{x}_m\delta_m/\dot{u}_m^2 = 0.52$. If the maximum isolator deflection is chosen as $\delta_m = 0.47$ in. (0.0119 m), which is 67 percent of δ_a, then

$$\ddot{x}_m = 0.52\frac{\dot{u}_m^2}{\delta_m} = 5450 \text{ in./sec}^2 = 138.4 \text{ m/sec}^2 = 14.1g$$

This acceleration is 67 percent of $\ddot{x}_a$. From Fig. 38.5 or Eq. (38.15):

$$\frac{\ddot{x}_m}{\dot{u}_m\omega_n} = 0.86 \text{ at } \zeta = 0.40$$

Then

$$\omega_n = \frac{5450}{0.86 \times 70} = 90 \text{ rad/sec [14.3 Hz]}$$

The spring stiffness k from Eq. (38.7) is

$$k = \frac{40}{386}(90)^2 = 840 \text{ lb/in.} = 147{,}107 \text{ N/m}$$

The dashpot constant c is

$$c = 2\zeta m\omega_n = 2 \times 0.40 \times \frac{40}{386} \times 90 = 7.46 \text{ lb-sec/in.} = 1306 \text{ N-sec/m}$$

RESPONSE OF 1-DOF SYSTEM TO ACCELERATION PULSE

Use of the velocity step or impulse to determine system responses has its limitations. A comparison of the velocity step (e.g., impulse) response with that of rectangular, half-sine, versed sine, and triangle pulse shapes is shown in the response spectra of Fig. 8.18*B*. The ordinate ν_M/ξ_{po} of Fig. 8.18*B* is the ratio of the maximum response ν_M of the 1-DOF isolation system to the excitation ξ_{po}. In the case of acceleration response, this would be $\ddot{x}_m/\ddot{u}_m$. The abscissa τ/T is in terms of the ratio of the pulse duration τ to that of the natural period T of the isolation system.

The straight line in Fig. 8.18*B* represents the undamped response to a velocity step and continues to increase with increasing values of τ/T, since the maximum acceleration response from the velocity step input is $\ddot{x}_m = \dot{u}_m\omega_n = 2\pi\dot{u}_m/T$. Comparison of the impulse response to the responses of the acceleration pulses indicates that pulse shape is of little concern when the pulse width is less than ¼ of the natural period of the responding system (e.g., $\tau/T < ¼$. In these cases, a velocity step or impulse loading is adequate for estimating responses. When $\tau/T > ¼$, pulse shape becomes a factor in the response.

For positive pulses ($\ddot{u} > 0$) having a single maximum value and finite duration, three basic characteristics of the pulse are of importance: maximum acceleration $\ddot{u}_m$, duration τ, and velocity change $\dot{u}_c$. A typical pulse is shown in Fig. 38.6. The relation among acceleration, duration, and velocity change is

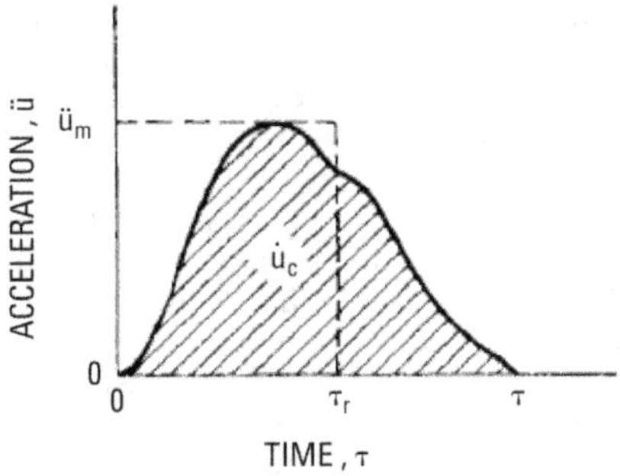

FIGURE 38.6 Typical acceleration pulse with maximum acceleration $\ddot{u}_m$ and duration τ.

$$\dot{u}_c = \int_0^\tau \ddot{u}dt \tag{38.17}$$

where the value of the integral corresponds to the shaded area of the figure. The *equivalent rectangular pulse* $\dot{u}_c = \tau_r\ddot{u}_m$ is characterized by (1) the same maximum acceleration $\ddot{u}_m$ and (2) the same velocity change $\dot{u}_c$ or area under the acceleration curve.

In Fig. 38.6, the horizontal and vertical dashed lines outline the equivalent rectangular pulse corresponding to the shaded pulse. From condition (2) and Eq. (38.17), the *effective duration* τ_r of the equivalent rectangular pulse is

$$\tau_r = \frac{1}{\ddot{u}_m}\int_0^\tau \ddot{u}dt \tag{38.18}$$

where τ_r may be interpreted physically as the *average width* of the shaded pulse. Using Eq. (38.18) and Eqs. (8.32), (8.33), and (8.34), the following effective pulse widths are $\tau_r = (2/\pi)\tau$ for half-sine and $\tau_r = (\frac{1}{2})\tau$ for versed sine and triangle pulses. If the equivalent rectangular pulse $\tau_r = \dot{u}_c/\ddot{u}_m$ and approximate shape of the pulse are known, then the aforementioned effective duration relations may be used to determine the pulse duration for the approximated shape. Once the pulse shape and duration are determined, then shock response spectra of the pulses may be used to estimate responses.

SHOCK RESPONSE SPECTRUM

The curve of maximum response of a 1-DOF system as a function of the natural period or frequency of the responding system is called a *shock response spectrum* or *response spectrum.* This concept is discussed more fully in Chaps. 8 and 20. Consider the response spectra shown in Fig. 8.16 with the pulse duration τ fixed in the abscissa τ/T of each graph; then the curves show the effect of varying the natural period of the spring-mass system. If the ordinate of Fig. 8.16 is represented as $\ddot{x}_m/\ddot{u}_m$ in place of v/ξ_p, then the figure shows the maximum acceleration induced by a given acceleration pulse upon spring-mass systems of various natural periods T. As a result, Fig. 8.16 may be used to determine the required natural period or frequency of the isolation system if $\ddot{x}_m$ and $\ddot{u}_m$ are known, and the pulse shape is defined. Alternatively, $\ddot{x}_m$ may be determined if the natural frequency of the isolator and $\ddot{u}_m$ are known. Spectra of maximum isolator deflection δ_m also may be drawn and are useful in predicting the maximum isolator deflection when the natural frequency of the isolator is known.

When the isolator includes damping, the SRS should be calculated and drawn using the same damping values. Examples of damped SRS are illustrated in Fig. 20.7. In selecting a shock isolator for a specified application, it may be necessary to use both maximum acceleration and maximum deflection spectra. This is illustrated in the following example.

Example 38.2. A piece of equipment weighing 230 lb (1023.1 N) is to be isolated from the effects of a vertical shock motion defined by the spectra of accelera-

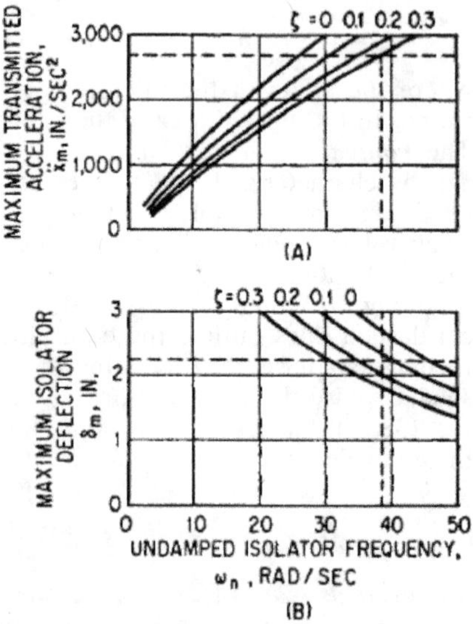

FIGURE 38.7 Shock response spectra: (*A*) maximum acceleration and (*B*) maximum isolator deflection for Example 38.2.

tion and deflection shown in Fig. 38.7. It is required that the maximum induced acceleration not exceed 7*g* (2700 in./sec^2 or 68.58 m/sec^2). Clearances available limit the isolator deflection to 2.25 in. (0.0572 m). The curves in Fig. 38.7*A* represent maximum response acceleration $\ddot{x}_m$ as a function of the angular natural frequency ω_n of the equipment supported on the shock isolators. The isolator springs are assumed linear and viscously damped, and separate curves are shown for values of the damping ratio $\zeta = 0, 0.1, 0.2$, and 0.3. The curves in Fig. 38.7*B* represent the maximum isolator deflection δ_m as a function of ω_n for the same values of ζ.

Consider first the requirement that $\ddot{x}_m < 2700$ in./sec^2 (68.58 m/sec^2). In Fig. 38.7*A*, the horizontal dashed line indicates this limiting acceleration. If the damping ratio $\zeta = 0.3$, then the angular natural frequency ω_n may not exceed 38.5 rad/sec on the criterion of maximum acceleration. The dashed horizontal line of Fig. 38.7*B* represents the deflection limit $\delta_m = 2.25$ in. (0.0572 m). For $\zeta = 0.3$, the minimum natural frequency is 30 rad/sec on the criterion of deflection. Considering both acceleration and deflection criteria, the angular natural frequency ω_n must lie between 30 rad/sec and 38.5 rad/sec. The spectra indicate that both criteria may be just met with $\zeta = 0.2$ if ω_n is 35 rad/sec. Smaller values of damping do not permit the satisfaction of both requirements. Conservatively, a suitable choice of parameters is $\zeta = 0.3$, $\omega_n = 35$ rad/sec. This limits $\ddot{x}_m$ to 2500 in./sec^2 (63.5 m/sec^2) and δ_m to 2.0 in. (0.051 m). The spring stiffness *k* is

$$k = m\omega_n^2 = \frac{230 \text{ lb}}{386 \text{ in./sec}^2} \times (35 \text{ rad/sec})^2 = 730 \text{ lb/in.} = 127{,}843 \text{ N/m}$$

If the equipment is to be supported by four like isolators, then the required stiffness of each isolator is $k/4 = 182.5$ lb/in. (31,961 N/m).

SHOCK RESPONSE OF 2-DOF CLASS A SYSTEMS: ISOLATION OF SUPPORT MOTION

IMPACT WITH REBOUND

Consider the system of Fig. 38.2*C*. The block of mass m_1 represents the equipment, and m_2, with its associated spring-dashpot unit, represents a critical component of the equipment. The left spring-dashpot unit represents the shock isolator. It is assumed here that $m_1 >> m_2$ so that the motion of m_1 is not sensibly affected by m_2;

larger values of m_2 are considered in a later section. Consider the entire system to be moving to the left at uniform velocity when the left-hand end of the isolator strikes a fixed support (not shown). The isolator will be compressed until the equipment is brought to rest. Following this, the compressive force in the isolator will continue to accelerate the equipment toward the right until the isolator loses contact with the support and the rebound is complete. This type of shock is called *impact with rebound.* Practical examples include the shock experienced by a single railroad car striking a bumper and that experienced by packaged equipment that rebounds when the container holding the equipment is dropped upon a hard surface.

Figure 38.5 may be used to determine the maximum deflection of the isolator and the maximum acceleration of equipment, and Eq. (38.18) may be used to find the duration of the pulse when the shape is known or approximated. Once the pulse shape and duration are determined, the shock response spectra for the light critical component may be calculated or obtained from figures in Chaps. 8 and 20.

Example 38.3. Let the equipment of Example 38.1 weighing 40 lb (177.9 N) have a flexible component weighing 0.2 lb (0.9 N). By vibration testing, this component is found to have an angular natural frequency $\omega_n = 260$ rad/sec (41.4 Hz) and to possess negligible damping. For the undamped linear spring of Example 38.1, it is desired to determine the maximum acceleration $\ddot{x}_{2m}$ experienced by the mass m_2 of the component if the equipment, traveling at a velocity of 70 in./sec (1.8 m/sec), is arrested by the free end of the isolator striking a fixed support. It is assumed that the component has a negligible effect on the motion of the equipment because $m_2 << m_1$.

Estimate from figures and equations: linear spring example. From the results of Example 38.1, it is known that $\omega_n = 108$ rad/sec (17.2 Hz) for the linear spring and that the maximum acceleration of the equipment as found from Eq. (38.10) is

$$\ddot{x}_{1m} = 7560 \text{ in./sec}^2 = 192 \text{ m/sec}^2 = 19.6g$$

This acceleration occurs at the instant when the isolator deflection has the extreme value $\delta_{1m} = 0.65$ in. (0.0165 m). [If the equipment (Fig. 38.2*C*) is moving toward the left when the isolator contacts the support, the extreme value of δ_{1m} is negative. It suffices to deal here with absolute values.] Subsequently, the isolator spring continues to accelerate the equipment until the isolator force is zero and the rebound is complete. Since there is no damping, the rebound velocity equals the striking velocity (with opposite sign). The velocity change $\dot{x}_{1c}$ is twice the striking velocity, and the effective duration τ_r [Eq. (38.18)] of an equivalent rectangular pulse is

$$\tau_r = \frac{\dot{x}_{1c}}{\ddot{x}_{1m}} = \frac{2 \times 70}{7560} = 0.0185 \text{ sec}$$

The shape of the pulse is assumed half-sine, since the system is linear and undamped. Therefore, the duration of the actual half-sine pulse is

$$\tau = \frac{\pi}{2}\,\tau_r = 0.0291 \text{ sec}$$

Since the equipment is the "support" for the component, the response of the latter may be found from the half-sine maximax curve of Fig. 8.16*B* if the ordinate is read as $\ddot{x}_{2m}/\ddot{x}_{1m}$ for ν/ξ_p, and the abscissa has the value of $\tau/T = \tau \times f = 0.0291 \times 41.4 = 1.2$. From Fig. 8.16*B*, $\ddot{x}_{2m}/\ddot{x}_{1m} = 1.65$, so $\ddot{x}_{2m} = 1.65 \cdot 7560 = 12{,}474$ in./sec^2 = 316.8 m/sec^2 = 32.3g.

IMPACT WITHOUT REBOUND

When impact of the isolator occurs without rebound, it must be recognized that the equipment-isolator system continues to oscillate until the initial kinetic energy is dissipated. Consider the system of Fig. 38.2*C*; it consists of equipment m_1, shock isolator (left spring-dashpot unit), and flexible component (subsystem 2). The system is initially at rest. The left end of the shock isolator is attached to a support (not shown), which is given a velocity step of magnitude $\dot{u}_m$ at $t = 0$. The subsequent motion of the support is $u = \dot{u}_m t$. The equations of motion for the 2-DOF Class A system are

$$\ddot{\delta}_1 + 2\zeta_1\omega_{n1}\dot{\delta}_1 + \omega_{n1}^2\delta_1 = \frac{m_2}{m_1}(2\zeta_2\omega_{n2}\dot{\delta}_2 + \omega_{n2}^2\delta_2) - \ddot{u}$$
$$\ddot{\delta}_2 + 2\zeta_2\omega_{n2}\dot{\delta}_2 + \omega_{n2}^2\delta_2 = -\ddot{\delta}_1 - \ddot{u} = -\ddot{x}_1 \qquad (38.19)$$

where the initial conditions are: $\dot{\delta}_1 = -\dot{u}_m$, $\dot{\delta}_2 = 0$, $\delta_1 = \delta_2 = 0$. Analytic solutions of Eq. (38.19) to find the acceleration $\ddot{x}_2 = \ddot{x}_1 + \ddot{\delta}_2$ of the component, the maximum force F_{1m} transmitted by the isolator, and the maximum isolator deflection δ_{1m} are too laborious to be practical.

Two simplified analyses of this problem are presented, where (1) the influence of damping is considered but the component mass m_2 is assumed of negligible size relative to m_1, and (2) damping is neglected but the effect of the mass m_2 of the component upon the motion of the system is considered.

Uncoupled 2-DOF Class A Analysis. Assume that $m_1 >> m_2$ so that the motion x_1 of the equipment may be determined by neglecting the effect of the component. Allowing the ratio m_2/m_1 to approach zero reduces Eq. (38.19) to

$$\ddot{\delta}_1 + 2\zeta_1\omega_{n1}\dot{\delta}_1 + \omega_{n1}^2\delta_1 = -\ddot{u}$$
$$\ddot{\delta}_2 + 2\zeta_2\omega_{n2}\dot{\delta}_2 + \omega_{n2}^2\delta_2 = -\ddot{x}_1 \qquad (38.20)$$

Then the extreme value of the force F_{1m} transmitted by the isolator and the extreme deflection δ_{1m} of the isolator that occur during the first quarter-cycle of the equipment motion may be found from Fig. 38.5 in the section entitled "Step Response of a Viscous Damped Isolator." The subsequent motion of the equipment is an exponentially decaying sinusoidal oscillation. Computer-generated results are shown in Fig. 38.8. The ordinate $\ddot{x}_{2m}/\ddot{x}_{2mo}$ in Fig. 38.8 represents the ratio of the maximum acceleration of the component to that which would be experienced with the isolator rigid (absent); thus, it may properly be called *shock transmissibility.* If shock transmissibility is less than unity, the isolator is beneficial (for the component considered). The denominator $\ddot{x}_{2mo}$ in the ordinate of Fig. 38.8 is calculated using Eq. (38.15), since a rigid isolator would result in a step velocity input to the component. The abscissa of Fig. 38.8 is the ratio of the undamped natural frequency ω_{n2} of the component to the undamped natural frequency ω_{n1} of the equipment on the isolator spring. Curves are given for several different values of the fraction of critical damping ζ_1 for the isolator. For all curves, the fraction of critical damping for the component is $\zeta_2 = 0.01$. For $\omega_{n2}/\omega_{n1} < 2$, large isolator damping (e.g., $\zeta_1 > 0.1$) significantly reduces the transmissibility of the component. However, in the isolation area where $\omega_{n2}/\omega_{n1} > 2$, large damping may significantly increase the maximum acceleration of the component.

An isolator must have a natural frequency significantly less than that of the critical component in order to reduce the transmitted acceleration. If there are several critical components having different natural frequencies ω_{n2}, each must be consid-

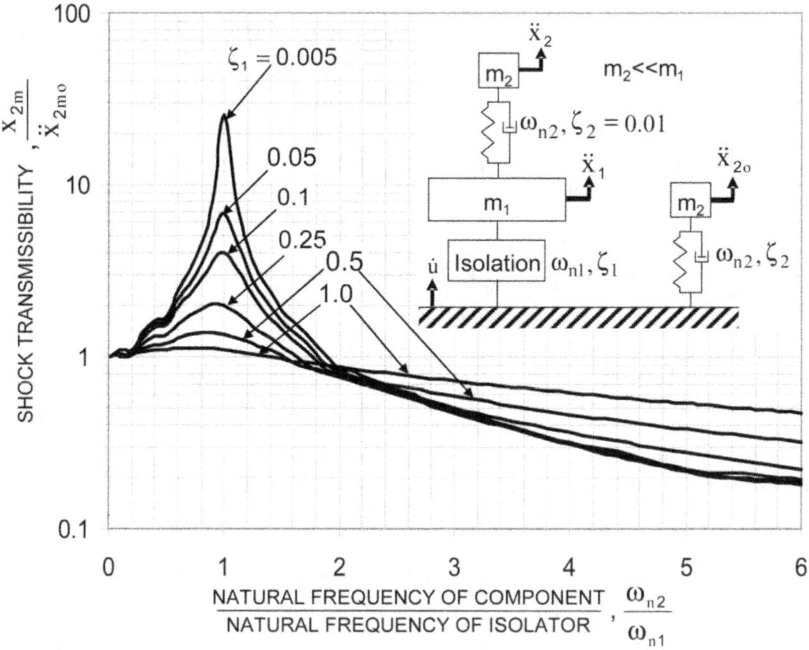

FIGURE 38.8 Shock transmissibility for a component of a viscously damped system with linear elasticity, where the effect of the component on the equipment motion is neglected (e.g., $m_2 << m_1$). The bobble in the transmissibility spectra for $\omega_{n2}/\omega_{n1} < 0.5$ occurs as a result of the modal interaction associated with the low mass of m_2. This interaction is significantly reduced when the mass of m_2 is increased.

ered separately and the natural frequency of the isolator must be significantly lower than the lowest natural frequency of a component.

Undamped 2-DOF Class A Analysis. This section includes an analysis of the transient response of the 2-DOF system shown in Fig. 38.2*C*, neglecting the effects of damping but assuming the equipment mass m_1 and the component mass m_2 to be of the same order of magnitude. Setting ζ_1 and ζ_2 equal to zero reduces Eq. (38.19) to

$$\ddot{\delta}_1 + \omega_{n1}^2 \delta_1 = \frac{m_2}{m_1}\,\omega_{n2}^2 \delta_2 - \ddot{u}$$
$$\ddot{\delta}_2 + \omega_{n2}^2 \delta_2 = -\ddot{\delta}_1 - \ddot{u} = -\ddot{x}_1 \qquad (38.21)$$

Equation (38.21) may be solved simultaneously for maximum values of the acceleration $\ddot{x}_{2m}$ of the component and maximum deflection δ_{1m} of the isolator:

$$\ddot{x}_{2m} = \frac{\dot{u}_m \omega_{n2}}{\left[\left(\frac{\omega_{n2}}{\omega_{n1}} - 1\right)^2 + \frac{m_2}{m_1}\left(\frac{\omega_{n2}}{\omega_{n1}}\right)^2\right]^{1/2}} \qquad (38.22)$$

$$\delta_{1m} = \frac{\dot{u}_m}{\omega_{n1}} \frac{1 + \dfrac{\omega_{n2}}{\omega_{n1}}\left(1 + \dfrac{m_2}{m_1}\right)}{\left[\left(\dfrac{\omega_{n2}}{\omega_{n1}} + 1\right)^2 + \dfrac{m_2}{m_1}\left(\dfrac{\omega_{n2}}{\omega_{n1}}\right)^2\right]^{1/2}} \tag{38.23}$$

where $\ddot{x}_{2m}$ = maximum absolute acceleration of component mass, in./sec^2 (m/sec^2); δ_{1m} = maximum deflection of isolator spring, in. (m); ω_{n1} = angular natural frequency of isolator $(k_1/m_1)^{1/2}$, rad/sec; and ω_{n2} = angular natural frequency of component $(k_2/m_2)^{1/2}$, rad/sec. (ω_{n1} and ω_{n2} are fixed-base natural frequencies, not modal frequencies associated with coupling between the subsystems.) Equation (38.22) is shown graphically in Fig. 38.9. The dimensionless ordinate is the ratio of maximum acceleration $\ddot{x}_{2m}$ of the component to the maximum acceleration $\ddot{x}_{2mo} = \dot{u}_m\omega_{n2}$, which the component would experience with no isolator present. The expression for $\ddot{x}_{2mo}$ is obtained from Eq. (38.15), with $\zeta_2 = 0$. The abscissa is the ratio of component natural frequency ω_{n2} to isolator natural frequency ω_{n1}. Separate curves are given for mass ratios $m_2/m_1 = 0.01, 0.1, 0.3$, and 1.0.

Figure 38.9 shows that the effect of the mass ratio m_2/m_1 upon the maximum component acceleration $\ddot{x}_{2m}$ is very great near resonance ($\omega_{n2}/\omega_{n1} \cong 1$). As ω_{n2}/ω_{n1} increases

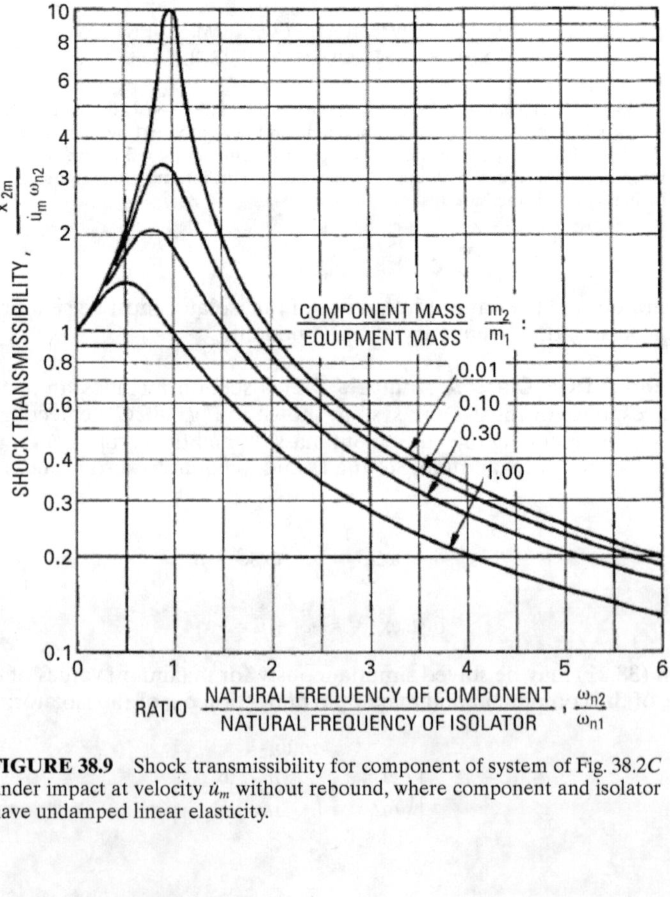

FIGURE 38.9 Shock transmissibility for component of system of Fig. 38.2*C* under impact at velocity $\dot{u}_m$ without rebound, where component and isolator have undamped linear elasticity.

above resonance, the effect of finite component mass steadily decreases. A notable observation from Figs. 38.8 and 38.9 is that isolation begins to occur when the support structure and isolation system frequencies are an octave apart [e.g., $\omega_2/\omega_{n1} \geq 2$]. Frequency ratios of $\omega_2/\omega_{n1} \geq 3$ are needed to achieve reductions in component acceleration on the order of 50 percent or more for isolation systems having damping $\zeta_1 < 0.25$. As a minimum, to avoid amplification of the component, the isolation system frequency should be an octave lower than the component system frequency.

Equation (38.23) is shown graphically in Fig. 38.10. The ordinate is the ratio of the maximum isolator deflection δ_{1m} to the deflection δ_{1mo} which would occur if component stiffness k_2 were infinite. The abscissa is the ratio of natural frequencies ω_{n2}/ω_{n1}, and curves are given for values of $m_2/m_1 = 0.1$ and 1.0. The term δ_{1mo} in the denominator of the ordinate parameter of Fig. 38.10 is obtained from Eq. (38.14) by considering the total equipment mass of $m_1 + m_2$ for the limiting case of rigid equipment (k_2 infinite). The following expressions for natural frequency ω_n and damping ratio ζ should bc substituted in Eq. (38.14) to ensure that the stiffness and damping associated with the isolator of system 1 stays the same:

$$k_1 = (m_1 + m_2)\omega_n^2 \Rightarrow \omega_n = \frac{\omega_{n1}}{(1 + m_2/m_1)^{1/2}}$$

$$c_1 = c = 2\zeta_1 m_1 \omega_1 = 2\zeta(m_1 + m_2)\omega_n \Rightarrow \zeta = \frac{\zeta_1}{(1 + m_2/m_1)^{1/2}}$$

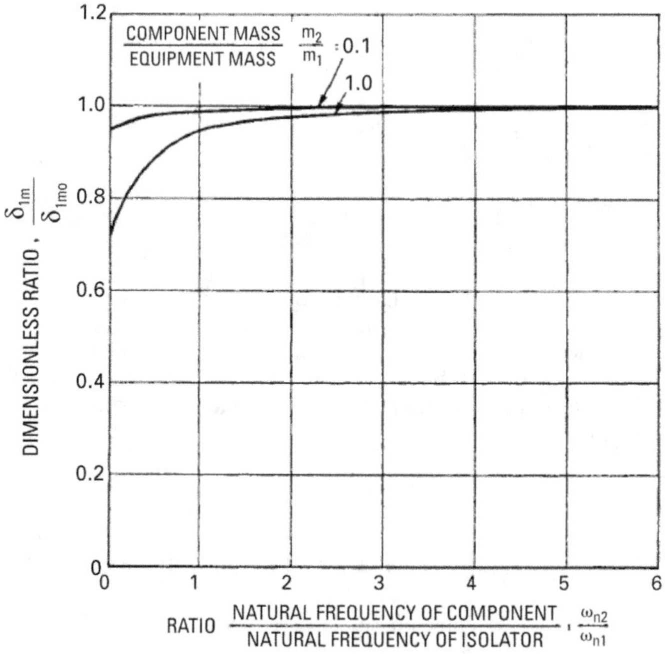

FIGURE 38.10 Dimensionless representation of maximum isolator deflection in system of Fig. 31.2C under impact at velocity $\dot{u}_m$ without rebound, where component and isolator have undamped linear elasticity.

Substituting these relations in Eq. (38.14) where $\zeta = 0$ and solving for δ_{1mo} results in

$$\delta_{1mo} = \frac{\dot{u}_m}{\omega_{n1}} (1 + m_2/m_1)^{1/2}$$

Figure 38.10 shows that, except for small values of ω_{n2}/ω_{n1}, the effect of finite component mass on the maximum isolator deflection δ_{1m} is slight. As ω_{n2}/ω_{n1} increases, the curves for all mass ratios asymptotically approach the ordinate 1.0. This is in agreement with the result given by Eq. (38.23) as ω_{n2}/ω_{n1} approaches infinity. Selecting isolation systems to have frequency ratios of $\omega_2/\omega_{n1} \geq 3$ ensures that a high percentage of the shock energy is used to displace the isolators in lieu of exciting the components.

Example 38.4. Equipment weighing 152 lb (676.1 N) has a flexible component weighing 3 lb (13.3 N). The angular natural frequency of the component is $\omega_{n2} = 130$ rad/sec. The equipment is mounted on a shock isolator with a linear spring $k_1 = 2400$ lb/in. (420304 N/m) and having a fraction of critical damping $\zeta_1 = 0.10$. Find the maximum isolator deflection δ_{1m} and the maximum component acceleration $\ddot{x}_{2m}$ which result when the base experiences a velocity step $\dot{u}_m = 55$ in./sec (1.4 m/sec).

Consider first a solution assuming that m_2 has a negligible effect on the equipment motion:

$$m_1 = \frac{152 \text{ lb}}{386 \text{ in./sec}^2} = 0.393 \text{ lb-sec}^2\text{/in.} = 69 \text{ kg}$$

$$\omega_{n1} = \sqrt{\frac{k_1}{m_1}} = \sqrt{\frac{2400}{0.393}} = 78.1 \text{ rad/sec (12.4 Hz)}$$

Figure 38.5 or Eq. (38.15) gives $\ddot{x}_{1m}/\dot{u}_m\omega_{n1} = 0.88$, and Fig. 38.5 or Eq. (38.16) gives $\ddot{x}_{1m}\delta_{1m}/\dot{u}_m^2 = 0.76$ for $\zeta_1 = 0.1$. Then

$$\delta_{1m} = \frac{0.76}{0.88} \times \frac{\dot{u}_m}{\omega_{n1}} = \frac{0.76 \times 55}{0.88 \times 78.1} = 0.61 \text{ in.} = 0.0155 \text{ m}$$

In finding $\ddot{x}_{2m}$, it is assumed that damping of the component has the typical value $\zeta_2 = 0.01$. Using $\omega_{n2}/\omega_{n1} = 130/78.1 = 1.67$, Fig. 38.8 gives a shock transmissibility of $\ddot{x}_{2m}/\ddot{x}_{2mo} = 1.15$. The acceleration $\ddot{x}_{2m}$ of the component is obtained by multiplying the transmissibility value of 1.15 by the acceleration $\ddot{x}_{2mo}$ [obtained from Eq. (38.15)] which would have occurred if the isolator were rigid. The resulting acceleration of the component is

$$\ddot{x}_{2m} = 1.15 \times \ddot{x}_{2mo} = 1.15 \times \dot{u}_m\omega_{n2}e^{-\left(\frac{\zeta_2}{\sqrt{1-\zeta_2^2}}\tan^{-1}\left(\frac{\sqrt{1-\zeta_2^2}(4\zeta_2^2-1)}{\zeta_2(4\zeta_2^2-3)}\right)\right)}$$

$$\ddot{x}_{2m} = 1.15 \times 55 \times 130 \times 0.9847 = 8096.7 \text{ in/sec}^2 = 205.7 \text{ m/sec}^2 = 21.0g$$

If component damping is neglected, then $\ddot{x}_{2m} = 1.15 \cdot 55 \cdot 130 = 8223$ in./sec^2 = 208.9 m/sec^2 = 21.3g.

A second solution, taking into consideration the mass m_2 of the component, may be obtained if the damping is neglected. From Eq. (38.23):

$$\delta_{1m} = \frac{\dot{u}_m}{\omega_{n1}} \frac{1 + \frac{\omega_{n2}}{\omega_{n1}}\left(1 + \frac{m_2}{m_1}\right)}{\left[\left(\frac{\omega_{n2}}{\omega_{n1}} + 1\right)^2 + \frac{m_2}{m_1}\left(\frac{\omega_{n2}}{\omega_{n1}}\right)^2\right]^{1/2}}$$

$$\delta_{1m} = \frac{55}{78.1} \frac{1 + 1.67\left(1 + \frac{3}{152}\right)}{\left[(1.67 + 1)^2 + \frac{3}{152}(1.67)^2\right]^{1/2}} = 0.71 \text{ in.} = 0.0180 \text{ m}$$

From Eq. (38.22):

$$\ddot{x}_{2m} = \frac{\dot{u}_m \omega_{n2}}{\left[\left(\frac{\omega_{n2}}{\omega_{n1}} - 1\right)^2 + \frac{m_2}{m_1}\left(\frac{\omega_{n2}}{\omega_{n1}}\right)^2\right]^{1/2}}$$

$$\ddot{x}_{2m} = \frac{55 \times 130}{\left[(1.67 - 1)^2 + \frac{3}{152}(1.67)^2\right]^{1/2}} = 10{,}070 \text{ in/sec}^2 = 255.8 \text{ m/sec}^2 = 26.1g$$

Computer simulation using the actual parameters of this example case results in a deflection of $\delta_{1m} = 0.61$ in. (0.0155 m) and a component acceleration of $\ddot{x}_{2m} = 20.0g$. Since the component mass is significantly less than the equipment mass, the uncoupled analysis gives a very good approximation for the responses. Compared to the computer simulation results, the undamped 2-DOF analysis results are ~30 percent higher for acceleration and ~16 percent higher for deflection. The importance of this difference depends on the fragility of the component.

ISOLATED EQUIPMENT ON A FLEXIBLE SUPPORT—WITH VISCOUS DAMPING

This section includes response comparisons of isolated equipment where the support is flexible. Both the support and the isolator include damping. Consider the support to be subsystem 1 of Fig. 38.2*C* and the isolator to be subsystem 2. The equations of motion are given by Eq. (38.19). The system is initially in equilibrium; at time $t = 0$, the left end of the support (subsystem 1 of Fig. 38.2*C*) is given a velocity step of magnitude $\dot{u}_m$. Initial conditions are $\dot{\delta}_1 = \dot{u}_m$, $\dot{\delta}_2 = 0$, $\delta_1 = \delta_2 = 0$. Computer simulations using the equations of motion were run with the following parameters: mass ratio $m_2/m_1 = 0.2$, support damping $\zeta_1 = 0.01$, and isolator damping at selected values of $\zeta_2 = 0.0, 0.1, 0.25,$ and 0.50. The shock transmissibility ratios of acceleration $\ddot{x}_{2m}/\ddot{x}_{2mo}$ and relative displacement δ_{2m}/δ_{2mo} of the equipment are shown in Figs. 38.11 and Fig. 38.12, respectively. The numerator of the ordinate in these figures is the maximum response of the isolated equipment with a flexible support, and the denominator is the maximum isolated equipment response with a rigid support. The denominator term $\ddot{x}_{2mo}$ in Fig. 38.11 may be obtained from Eq. (38.15), and the denominator term δ_{2mo} of Fig. 38.12 may be obtained from Eq. (38.14). The abscissa is the ratio of the undamped natural frequency of the isolator to the undamped natural frequency of the support.

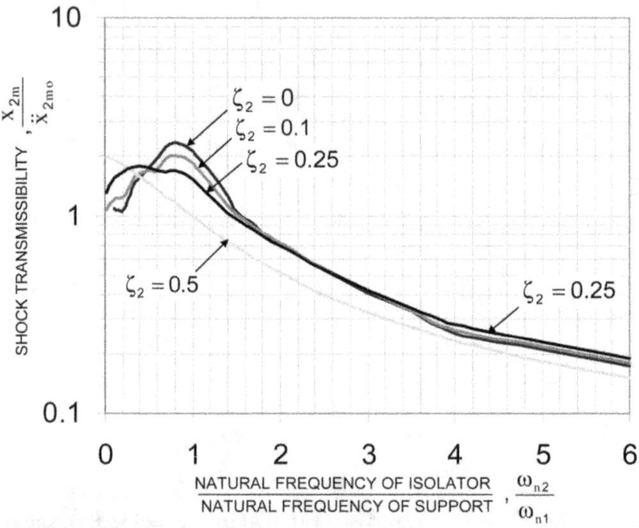

FIGURE 38.11 Acceleration shock transmissibility comparisons of isolated equipment where the support is flexible. The support is modeled to be subsystem 1 of Fig. 38.2*C* and the isolator to be subsystem 2. The denominator of the ordinate ratio is the acceleration that would occur if the support is rigid. Parameters in the figure are mass ratio $m_2/m_1 = 0.2$, support damping $\zeta_1 = 0.01$, and isolator damping at selected values of $\zeta_2 = 0.0, 0.1, 0.25$, and 0.50.

The maximum responses for $\ddot{x}_{2m}$ and δ_{2m} may be obtained by entering Figs. 38.11 and 38.12 at the desired frequency ratio to find the corresponding ordinate ratio values. These values are then multiplied by $\ddot{x}_{2mo}$ and δ_{2mo} obtained from Eqs. (38.15) and (38.14), respectively.

Example 38.5. Consider a 1300-lb (5782.7-N) isolated equipment item having a natural frequency of 7 Hz and viscous damping of $\zeta_2 = 0.25$ mounted on a 6500-lb (28,913.4-N) flexible support having a natural frequency of 14 Hz and damping $\zeta_1 = 0.01$, which is subjected to a 120-in./sec (3.05-m/sec) step velocity. Since the mass ratio is 1300/6500 = 0.2 and the frequency ratio is 0.5, Figs. 38.11 and 38.12 and Eqs. (38.15) and (38.14) can be used to determine $\ddot{x}_{2m}$ and δ_{2m}, respectively. At a frequency ratio of 0.5, Fig. 38.11 indicates $\ddot{x}_{2m}/\ddot{x}_{2mo} = 1.75$ and Fig. 38.12 indicates $\delta_{2m}/\delta_{2mo} = 1.67$. The resulting values for $\ddot{x}_{2m}$ and δ_{2m} are

$$\ddot{x}_{2m} = 1.75 \times \ddot{x}_{2mo} = 1.75 \times \dot{u}_m \omega_{n2} e^{-\left(\frac{\zeta_2}{\sqrt{1-\zeta_2^2}}\tan^{-1}\left(\frac{\sqrt{1-\zeta_2^2}(4\zeta_2^2-1)}{\zeta_2(4\zeta_2^2-3)}\right)\right)}$$

$$\ddot{x}_{2m} = 1.75 \times 120 \times 43.98 \times 0.8107 = 7487.5 \text{ in/sec}^2 = 190.2 \text{ m/sec}^2 = 19.4g$$

$$\delta_{2m} = 1.67 \times \delta_{2mo} = 1.67 \times \frac{\dot{u}_m}{\omega_{n2}} e^{-\left(\frac{\zeta_2}{\sqrt{1-\zeta_2^2}}\cos^{-1}\zeta_2\right)}$$

$$\delta_{1m} = 1.67 \times \frac{120}{43.98} \times 0.7115 = 3.25 \text{ in.} = 0.0826 \text{ m}$$

Figures 38.11 and 38.12 indicate that an increase in isolator damping generally lowers acceleration and relative displacement transmissibility for frequency ratios

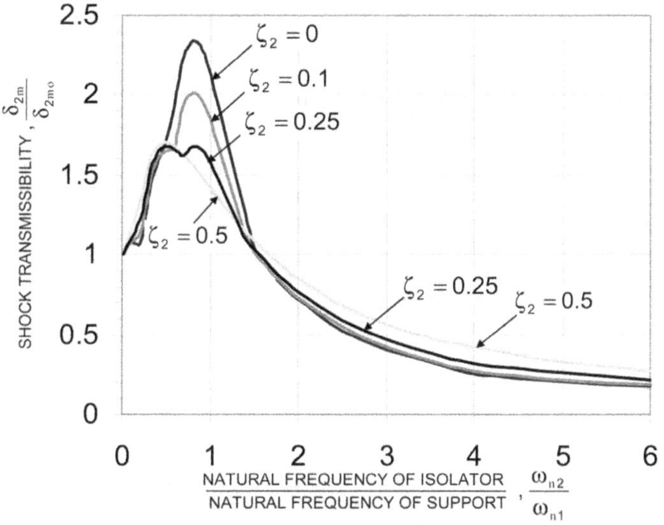

FIGURE 38.12 Relative displacement shock transmissibility comparisons of isolated equipment where the support is flexible. The support is modeled to be subsystem 1 of Fig. 38.2*C* and the isolator to be subsystem 2. The denominator of the ordinate ratio is the relative displacement that would occur if the support is rigid. Parameters in figure are mass ratio $m_2/m_1 = 0.2$, support damping $\zeta_1 = 0.01$, and isolator damping at selected values of $\zeta_2 = 0.0, 0.1, 0.25$, and 0.50.

between 0.5 and 1.5. However, for frequency ratios less than 0.5, higher damping tends to increase transmissibility. When isolator damping is greater than 0.4 and the frequency ratio is greater than 1.5, acceleration transmissibility tends to be lower than the zero-damped case while relative displacement transmissibility tends to be higher.

A structure used to support isolated equipment generally has distributed mass and elasticity; thus, the application of a velocity step or impulse tends to excite the structure to vibrate not only in its fundamental mode but also in higher modes of vibration. The 2-DOF system discussed in this section only simulates the fundamental modes of vibration associated with the support structure. In many applications, such simulation is adequate because the displacements and strains are greater in the fundamental mode than in higher modes. The vibration of members having distributed mass is discussed in Chap. 7, and the formulation of models suitable for shock analysis of systems with multiple modes is discussed in Chap. 23.

SHOCK RESPONSE OF 1- AND 2-DOF CLASS B SYSTEMS

RESPONSE OF 1-DOF CLASS B SYSTEMS

Analogous Use of Impulse *J*. Consider the system of Fig. 38.2*B* with the spring-dashpot forces represented by Eq. (38.3). The force *F*, applied externally to the equipment, is taken to be a known function of time. Considering only force-time relations

$F(t)$ in the form of a single pulse, the analogous mathematical relations of Eqs. (38.1) and (38.2) are used by defining the impulse J applied by the force F as

$$J = \int_0^\tau F dt$$

where τ is the duration of the pulse. If the damping in the support is small, then analogies of the step response of the undamped linear spring are applicable; otherwise, the linear-spring viscous damping analogies apply.

Short-Duration Impulses. If τ is less than ¼ the natural period of the isolation system, then the results derived in the section entitled "Step Response of a Viscous Damped Isolator" may be applied directly. An impulse J of negligible duration acting on the mass m produces a velocity change $\dot{u}_m$ given by

$$\dot{u}_m = \frac{J}{m} \tag{38.24}$$

The subsequent relative motion of the system is identical with that resulting from a velocity step of magnitude $\dot{u}_m$.

The maximum deflection δ_m for the 1-DOF system when the impulse J is applied directly to m may be determined by substituting Eq. (38.24) into Eq. (38.14) and solving for δ_m. The force F_m applied to the support is obtained by multiplying the stiffness $k = m\omega_n^2$ of the spring by the maximum deflection δ_m. The resulting absolute-value expressions for δ_m and F_m are given as follows:

$$\delta_m = \frac{\dot{u}_m}{\omega_n} e^{-\left(\frac{\zeta}{\sqrt{1-\zeta^2}} \cos^{-1}\zeta\right)} = \frac{J}{m\omega_n} e^{-\left(\frac{\zeta}{\sqrt{1-\zeta^2}} \cos^{-1}\zeta\right)}$$

$$F_m = m\omega_n^2 \cdot \delta_m = J\omega_n e^{-\left(\frac{\zeta}{\sqrt{1-\zeta^2}} \cos^{-1}\zeta\right)} \tag{38.25}$$

Long-Duration Impulses. If the duration τ of the applied impulse exceeds about one-third of the natural period of the equipment-support system, application of velocity step results may be unduly conservative. Then the results developed in the section entitled "Response of a 1-DOF System to Acceleration Pulse" are applicable. The mathematical equivalence of Eqs. (38.1) and (38.2) is based on identifying $-m\ddot{u}$ in the former, with F in the latter. Accordingly, if the shape of the force F versus time curve is similar to the shape of the curve of acceleration $\ddot{u}$ versus time, then the response of a system to an acceleration pulse may be used by analogy to find the response to a force pulse by making the following substitutions:

$$\ddot{u}_m = \frac{F_m}{m} \qquad \tau_r = \frac{J}{F_m}$$

where F_m is the maximum value of F, $\ddot{u}_m$ is the maximum value of $\ddot{u}$, and τ_r is the effective duration. If the mathematical equivalence is literally applied, F_m/m is analogous to $-\ddot{u}_m$, not $\ddot{u}_m$. Since acceleration pulse results are given in terms of extreme *absolute* values, the sign is not important.

RESPONSE OF 2-DOF CLASS B SYSTEMS

Idealized System. When a shock isolator is used to reduce the magnitude of the force transmitted to the support foundation, the idealized system is as shown in Fig. 38.2*C*. Subsystem 2 represents the equipment (mass m_2) mounted on the shock isolator (right-hand spring-dashpot unit). Subsystem 1 is an idealized representation of the support with effective mass m_1 and with stiffness and damping represented by the left spring-dashpot unit. The free end of the latter unit is taken to be fixed ($u = 0$) to the support foundation. If the system consists only of linear elements as defined by Eq. (38.3), the equations of motion for the 2-DOF Class B system are

$$\ddot{\delta}_1 + 2\zeta_1\omega_{n1}\dot{\delta}_1 + \omega_{n1}^2\delta_1 = \frac{m_2}{m_1}(2\zeta_2\omega_{n2}\dot{\delta}_2 + \omega_{n2}^2\delta_2) = \frac{m_2}{m_1}\left(\frac{F}{m_2} - \ddot{x}_2\right)$$

$$\ddot{\delta}_2 + 2\zeta_2\omega_{n2}\dot{\delta}_2 + \omega_{n2}^2\delta_2 = -\ddot{\delta}_1 - \frac{F}{m_2} \qquad \text{where } \ddot{x}_1 = \ddot{\delta}_1$$

It is assumed that the system is initially in equilibrium ($\dot{\delta}_1 = \dot{\delta}_2 = 0$; $\delta_1 = \delta_2 = 0$) and that force F (positive in the $+X$ direction) applies an impulse J to m_2. Analysis is simplified by treating the duration τ of impulse J as negligible. This assumption, always conservative, usually is warranted if the natural frequency of the shock isolator is small relative to the natural frequency of the support.

Effect of Class B Isolation. The effect of Class B isolation in a 2-DOF system may be judged by comparing ratios of the responses of the support structure with and without an isolation system present. When no isolation is present, the equipment is considered rigidly attached to the supporting structure and the responses are determined from a 1-DOF analysis. When an isolator is inserted between the equipment and the support structure, a 2-DOF analysis is necessary. The 1-DOF analysis for the condition of equipment rigidly attached to the support structure is presented first to establish a baseline for comparison. The 2-DOF analysis showing the effect of inserting the isolation system is presented afterward.

Equipment Rigidly Attached. Consider the system of Fig. 38.2*C* with the spring-dashpot unit 2 assumed to be rigid. The mass m_2 represents the equipment, and the mass m_1 represents, with spring and dashpot assembly (1), the support. Since the equipment is rigidly attached, the system is equivalent to that shown in Fig. 38.2*B*, where $m = m_1 + m_2$ and $F_1(\dot{\delta}_1,\delta_1) = F(\dot{\delta},\delta)$. As a result, the equation of motion has the same form as Eq. (38.2) and by analogy Eq. (38.25) may be used to obtain absolute-value solutions for maximum deflection and maximum force. These solutions are presented as follows, with the condition that $m = m_1 + m_2$:

$$\delta_{1mo} = \frac{J}{m_1\left(1 + \dfrac{m_2}{m_1}\right)^{1/2}\omega_{1n}}\, e^{-\left(\frac{\zeta}{\sqrt{1-\zeta^2}}\cos^{-1}\zeta\right)}$$

$$F_{1mo} = m_1\omega_{1n}^2 \cdot \delta_{1mo} = \frac{J\omega_{1n}}{\left(1 + \dfrac{m_2}{m_1}\right)^{1/2}}\, e^{-\left(\frac{\zeta}{\sqrt{1-\zeta^2}}\cos^{-1}\zeta\right)}$$

where $\zeta = \zeta_1/(1 + m_2/m_1)^{1/2}$

Uncoupled 2-DOF Class B Analysis. In many applications, the support motion $x_1 = \delta_1$ is sufficiently small compared with the equipment motion x_2. As a result, the equipment acceleration $\ddot{x}_2$ is closely approximated by $\ddot{\delta}_2$ in lieu of letting $\ddot{x}_2 = \ddot{\delta}_2 + \ddot{x}_1$. The validity of this approximation is based on $m_2 << m_1$. Using this approximation, the analysis is resolved into two separate uncoupled parts, each dealing with a 1-DOF system.

Setting $\ddot{x}_1 = \ddot{\delta}_1 = 0$, the equation of motion of the equipment mounted on the shock isolator (subsystem 2 of Fig. 38.2*C*) becomes

$$\ddot{\delta}_2 + 2\zeta_2\omega_{n2}\dot{\delta}_2 + \omega_{n2}^2\delta_2 = -\frac{F}{m_2} \qquad \text{where } F = 0 \text{ for } t > 0 \tag{38.26}$$

The initial conditions are $\delta_2 = 0$, $\dot{\delta}_2 = J/m_2 = \dot{u}_m$ when $t = 0$, and $F = 0$ when $t > 0$. Apart from the subscripts, Eqs. (38.11) and (38.26) are identical for $t > 0$. As a result, the maximum equipment acceleration $\ddot{x}_{2m}$ and the maximum isolator deflection δ_{2m} may be found from Fig. 38.5 or Eqs. (38.14), (38.15), and (38.16) by letting $\dot{u}_m = J/m_2$.

The differential equation for the motion of the support in Fig. 38.2*C* for $t > 0$ is

$$\ddot{\delta}_1 + 2\zeta_1\omega_{n1}\dot{\delta}_1 + \omega_{n1}^2\delta_1 = -\frac{m_2}{m_1}\ddot{x}_2 \tag{38.27}$$

The initial conditions are $\dot{\delta}_1 = 0$, $\delta_1 = 0$. The solution of Eq. (38.27) is the same as that of Eq. (38.20) because the equations differ only by the interchange of the numerical subscripts and the presence of the factor m_2/m_1 on the right-hand side of Eq. (38.27). Computer simulation results for the support force shock transmissibility of the uncoupled system [e.g., Eqs. (38.26) and (38.27)] are shown in Fig. 38.13. The ordi-

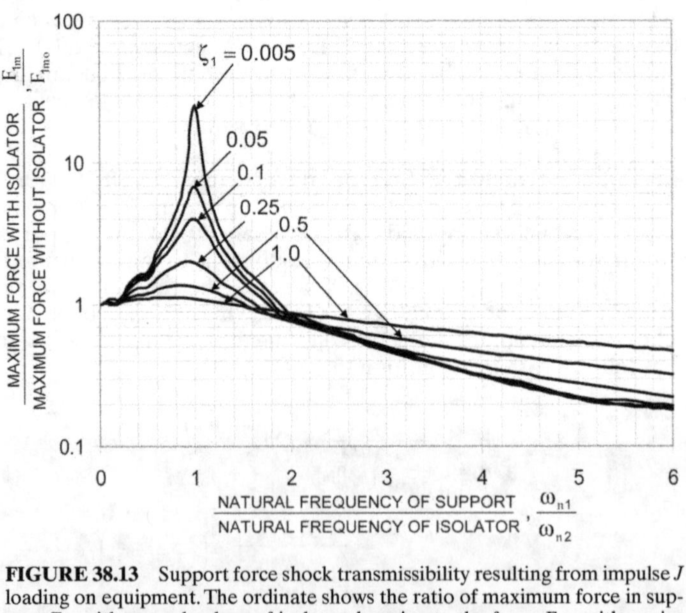

FIGURE 38.13 Support force shock transmissibility resulting from impulse J loading on equipment. The ordinate shows the ratio of maximum force in support F_{1m} with several values of isolator damping to the force F_{1mo} with equipment rigidly attached. The support damping is 0.01 for all cases.

nate is the ratio of the maximum force F_{1m} in the support to the maximum force F_{1mo} that results if the isolator is rigid (see previous section, "Equipment Rigidly Attached"). The abscissa in Fig. 38.13 is the ratio of the undamped support natural frequency ω_{n1} to the undamped isolator natural frequency ω_{n2}. Curves are drawn for various values of the fraction of critical damping ζ_2 for the isolator, assuming that the fraction of critical damping ζ_1 for the support is constant at $\zeta_1 = 0.01$.

A notable observation from Fig. 38.13 is that, like the results of Figs. 38.8 and 38.9, isolation begins to occur when the support structure and isolation system frequencies are an octave apart (e.g., $\omega_{n1}/\omega_{n2} \geq 2$). Results shown in Fig. 38.13 apply only when the support deflection δ_1 is small compared with the isolator deflection δ_2, a condition which is not met in the neighborhood of unity frequency ratio. A more realistic analysis for this condition involves the 2-DOF system discussed in the next section.

Undamped 2-DOF Class B Analysis. This section includes an analysis of the system of Fig. 38.2*C*, considered as a coupled 2-DOF system where both the support and the isolator are linear and undamped $[F_1(\dot{\delta}_1,\delta_1) = k_1\delta_1, F_2(\dot{\delta}_2,\delta_2) = k_2\delta_2]$. This analysis makes it possible to consider the effect of deflection of the support on the motion of the equipment. Fixing the support base ($u = 0$), the equations of motion may be written

$$\begin{aligned} \ddot{\delta}_1 + \omega_{n1}^2\delta_1 &= \frac{m_2}{m_1}\omega_{n2}^2\delta_2 \\ \ddot{\delta}_2 + \omega_{n2}^2\delta_2 &= -\ddot{\delta}_1 \end{aligned} \tag{38.28}$$

Assuming that the impulse J has negligible duration, the initial conditions are $\dot{\delta}_1 = 0$, $\dot{\delta}_2 = J/m_2$, $\delta_1 = \delta_2 = 0$. The solution of Eq. (38.28) parallels that of Eq. (38.21); the resulting expressions for the maximum isolator deflection δ_{2m} and force F_{1m} applied to the support are

$$\delta_{2m} = \frac{J}{m_2\omega_{n2}}\left[1 + \frac{m_2/m_1}{(1 + \omega_{n1}/\omega_{n2})^2}\right]^{-1/2} \tag{38.29}$$

$$F_{1m} = J\omega_{n1}\left[\left(1 - \frac{\omega_{n1}}{\omega_{n2}}\right)^2 + \frac{m_2}{m_1}\right]^{-1/2} \tag{38.30}$$

The maximum deflection of the isolator given in Eq. (38.29) is shown graphically in Fig. 38.14. For small values of the ratio of support natural frequency to isolator natural frequency, the flexibility of the support may significantly reduce the maximum isolator deflection, especially if the mass of the support is small relative to the mass of the equipment. For large values of the frequency ratio, the effect of the mass ratio is small.

The support force shock transmissibility F_{1m}/F_{1mo} is graphed in Fig. 38.15 as a function of frequency ratio. Like Fig. 38.13, the ordinate is the ratio of the maximum force F_{1m} in the support, given by Eq. (38.30), to the maximum force F_{1mo} that results if the isolator is rigid (see previous section "Equipment Rigidly Attached"). The abscissa in Fig. 38.15 is the ratio of the undamped support natural frequency ω_{n1} to the undamped isolator natural frequency ω_{n2}. The effect of the mass ratio is profound for small values of the frequency ratio. The curves of Figs. 38.13 and 38.15 show corresponding results. The former includes damping, and the latter includes the coupling effect between the two systems. The analysis which ignores the coupling effect may grossly overestimate the maximum force applied to the support at low

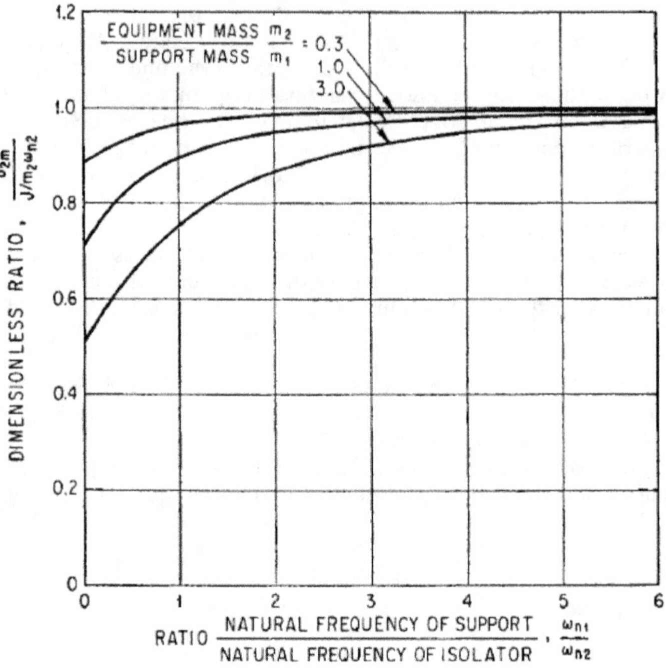

FIGURE 38.14 Dimensionless representation of maximum isolator deflection δ_{2m} resulting from action of impulse J on equipment. Isolator and support have undamped linear elasticity.

values of the frequency ratio. At high values of the frequency ratio and for $m_2 << m_1$, the two analyses yield like results if the fraction of critical damping in the isolator is $\zeta_2 = 0.10$ or less. The two methods are compared in Example 38.6.

Example 38.6. A forging machine weighs 7000 lb (31,137.5 N) exclusive of the 600-lb (2668.9 N) hammer. It is mounted at the center of a span formed by two 12-in. (0.3048-m) -deep, 50-lb/ft (729.7-N/m) I beams [area moment of inertia = 394 in.^{4} (16,399.5 cm^4)] having hinged ends and a span l = 18 ft (5.906 m). The hammer falls freely from a height of 60 in. (1.524 m) before striking the work. Determine:

a. Maximum force F_{1m} in the beams and maximum deflection δ_{1m} of the beams if the machine is rigidly bolted to the beams.

b. The maximum force F_{1m} in the beams and the maximum deflection δ_{2m} of an isolator interposed between machine and beams.

Solution

a. When the machine is bolted rigidly to the beams, the system may be considered to have only a single degree of freedom. The mass is that of the machine, plus the hammer, plus the effective mass of the beams. For the machine: $m_2 = (7000 + 600)/386.4 = 19.67$ lb-sec^2/in. (3451 kg). The effective mass of the beams is taken as one-half of the actual mass:

$$m_1 = \frac{2(0.5)(18)(50)}{386.4} = 2.33 \text{ lb-sec}^2\text{/in. (409 kg)}$$

$$m = m_1 + m_2 = 22.0 \text{ lb-sec}^2\text{/in. (3860 kg)}$$

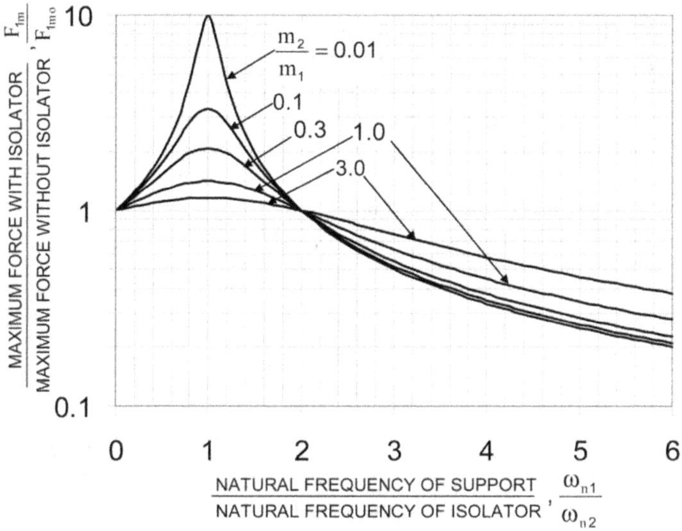

FIGURE 38.15 Support force shock transmissibility resulting from impulse J loading on equipment. The ordinate shows the ratio of maximum force in support F_{1m} with isolation to force F_{1mo} with equipment rigidly attached. Both support and isolator are undamped.

The stiffness of the beams is

$$k = 2\frac{48EI}{l^3} = 2\,\frac{48 \times (29 \times 10^6) \times 394}{(18 \times 12)^3} = 109{,}000 \text{ lb/in. } (19{,}088{,}818 \text{ N/m})$$

The natural frequency of the machine-and-beams system is

$$\omega_n = \sqrt{\frac{k}{m}} = \sqrt{\frac{109{,}000}{22.0}} = 70.4 \text{ rad/sec } (11.2 \text{ Hz})$$

If the impact between the hammer and the work is inelastic and its duration is negligible, the resulting velocity $\dot{u}_m$ of the machine may be found from conservation of momentum. The impulse J is the product of the weight of the hammer and time of fall:

$$J = (600)\left(\frac{2 \times 60}{386.4}\right)^{1/2} = 334 \text{ lb-sec } (1485.7 \text{ N-sec})$$

If the damping of the beams is neglected, the maximum beam deflection and maximum force in the beams are found from Eq. (38.25) as follows:

$$\delta_m = \frac{J}{m\omega_n}\, e^{-\left(\frac{\zeta}{\sqrt{1-\zeta^2}}\cos^{-1}\zeta\right)} = \frac{334}{(22.0)(70.4)}\, e^{-(0)} = 0.22 \text{ in.} = 0.00559 \text{ m}$$

$$F_m = J\omega_n e^{-\left(\frac{\zeta}{\sqrt{1-\zeta^2}}\cos^{-1}\zeta\right)} = (334)(70.4)e^{-(0)} = 23{,}500 \text{ lb} = 104{,}533 \text{ N}$$

b. An isolator having a stiffness $k_2 = 36{,}000$ lb/in. (630,456.4 N/m) and a fraction of critical damping $\zeta_2 = 0.10$ is interposed between the machine and the beams. The *uncoupled natural frequencies* defined in connection with Eqs. (38.22) and (38.23) are

$$\omega_{n2} = \sqrt{\frac{k_2}{m_2}} = \sqrt{\frac{36{,}000}{19.67}} = 42.8 \text{ rad/sec (6.8 Hz)}$$

$$\omega_{n1} = \sqrt{\frac{k_1}{m_1}} = \sqrt{\frac{109{,}000}{2.33}} = 216.3 \text{ rad/sec (34.4 Hz)}$$

Consider first the limitations of using uncoupled analysis where $m_2 > m_1$. Figure 38.5 gives, respectively:

$$\ddot{x}_{2m}/\dot{u}_m\omega_{n2} = 0.88$$

$$\ddot{x}_{2m}\delta_{2m}/\dot{u}_m^2 = 0.76$$

Substituting $\dot{u}_m = J/m_2 = 17$ in./sec (0.4318 m/sec) and solving for δ_{2m} results in

$$\delta_{2m} = \frac{0.76 \times 17}{0.88 \times 42.8} = 0.34 \text{ in.} = 0.00864 \text{ m}$$

Entering Fig. 38.13 at $\omega_{n1}/\omega_{n2} = 5.06$, $F_{1m}/F_{1mo} = 0.23$. Since the maximum force in the undamped beams with the machine system rigidly attached is 23,500 lb (104,533 N), then F_{1m} = (0.23) * (23,500) ≈ 5,400 lb (24,020 N). Alternatively, if damping of the beam is considered, F_{1m} may be determined as follows:

$$F_{1m} = 0.23F_{1mo} = \frac{0.23J\omega_{1n}}{\left(1 + \dfrac{m_2}{m_1}\right)^{1/2}} e^{-\left(\frac{\zeta}{\sqrt{1-\zeta^2}}\cos^{-1}\zeta\right)} = \frac{(0.23)(334)(216.3)(0.985)}{(1 + 8.44)^{1/2}}$$

$$= 5330 \text{ lb.} = 23{,}709 \text{ N}$$

where $\zeta = \dfrac{\zeta_1}{(1 + m_2/m_1)^{1/2}} = 0.01/3.07 = 0.0033$

Use of the uncoupled transmissibility shown in Fig. 38.13 assumes that m_2 has no influence on the motion of the support structure. Since $m_2/m_1 = 8.44$ is significant, this assumption is invalid. A computer simulation of this scenario results in a beam deflection of $\delta_{1m} = 0.11$ in. (0.00279 m), which is not negligible, and an isolator deflection of $\delta_{2m} = 0.31$ in. (0.00787 m). The support force shock transmissibility from the computer simulation is $F_{1m}/F_{1mo} = 0.515$, which gives the value of F_{1m} as ~12,000 lb (53,379 N). Compared to the computer simulation results, the uncoupled analysis predicts F_{1m} to be 55 percent lower. This is not a reasonable estimate of force for design of foundation supports and may result in failure if the isolation system was necessary for protection.

Consider now that the floor and machine-isolator systems are coupled, and use the 2-DOF analysis which neglects damping. From Eq. (38.29):

$$\delta_{2m} = \frac{J}{m_2\omega_{n2}}\left[1 + \frac{m_2/m_1}{(1 + \omega_{n1}/\omega_{n2})^2}\right]^{-1/2}$$

$$\delta_{2m} = \frac{334}{19.67 \times 42.8}\left[1 + \frac{19.67/2.33}{(1 + 5.06)^2}\right]^{-1/2} = 0.36 \text{ in.} = 0.00914 \text{ m}$$

From Eq. (38.30):

$$F_{1m} = J\omega_{n1}\left[\left(1 - \frac{\omega_{n1}}{\omega_{n2}}\right)^2 + \frac{m_2}{m_1}\right]^{-1/2}$$

$$F_{1m} = 334 \times 216.3\left[(1 - 5.06)^2 + \frac{19.67}{2.33}\right]^{-1/2} = 14{,}500 \text{ lb.} = 64{,}499 \text{ N}$$

Compared to the computer simulation results, the undamped 2-DOF analysis predicts F_{1m} to be 21 percent higher. This is a conservative estimate and is reasonable for design of foundation supports.

CONCEPT OF VIBRATION ISOLATION

The concept of vibration isolation is illustrated by consideration of the 1-DOF systems shown in Figs. 2.20 and 2.12 (also depicted in columns 1 and 2 of Table 38.1). The performance of the isolator may be evaluated by the following characteristics of the response of the system to steady-state sinusoidal vibration:

Absolute transmissibility. Transmissibility is a measure of the reduction of transmitted force or motion afforded by an isolator. If the source of vibration is an oscillating motion of the foundation (motion excitation with Class A isolation), transmissibility is the ratio of the vibration amplitude of the equipment to the vibration amplitude of the foundation. If the source of vibration is an oscillating force originating within the equipment (force excitation with Class B isolation), transmissibility is the ratio of the amplitude of the transmitted force to the amplitude of the exciting force.

Relative transmissibility. Relative transmissibility is the ratio of the relative deflection amplitude of the isolator to the displacement amplitude imposed at the foundation. A vibration isolator effects a reduction in vibration by permitting deflection of the isolator. The relative deflection is a measure of the clearance required in the isolator. This characteristic is significant only in an isolator used to reduce the vibration transmitted from a vibrating foundation.

Displacement motion response. Displacement motion response is the ratio of the displacement amplitude of the equipment to the quotient obtained by dividing the excitation force amplitude by the static stiffness of the isolator. If the equipment is acted on by an exciting force, the resultant motion of the equipment determines the space requirements for the isolator; that is, the isolator must have a clearance at least as great as the equipment motion.

FORM OF ISOLATOR

Isolators may be modeled using many different combinations of resilient elements and dampers. The combinations considered in this chapter are the rigidly connected and elastically connected viscous dampers described as follows.

Rigidly connected viscous damper. A viscous damper c, represented by the dashpot in column 1 of Table 38.1, is connected rigidly between the equipment and its foundation. The damper has the characteristic property of transmitting a force F_c that is directly proportional to the relative velocity $\dot{\delta}$ across the damper, where $F_c = c\dot{\delta}$. This damper sometimes is referred to as a *linear damper.*

TABLE 38.1 Transmissibility and Motion Response for Rigidly Connected (Column 1) and Elastically Connected (Column 2) Viscous Damper

	(1) Rigidly connected viscous damper	(2) Elastically connected viscous damper
	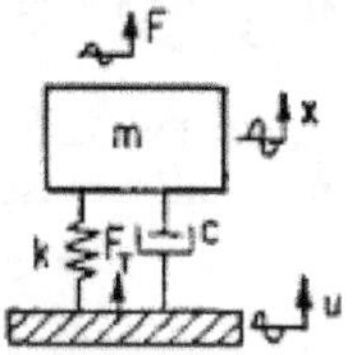	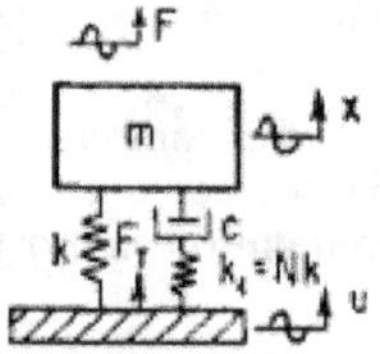
(A)	$T_A = \frac{x_o}{u_o} = \frac{F_T}{F_o} = \sqrt{\frac{1+\left(2\zeta\frac{\omega}{\omega_o}\right)^2}{\left(1-\frac{\omega^2}{\omega_o^2}\right)^2+\left(2\zeta\frac{\omega}{\omega_o}\right)^2}}$	$T_A = \frac{x_o}{u_o} = \frac{F_T}{F_o} = \sqrt{\frac{1+4\left(\frac{N+1}{N}\right)^2\zeta^2\frac{\omega^2}{\omega_o^2}}{\left(1-\frac{\omega^2}{\omega_o^2}\right)^2+\frac{4}{N^2}\zeta^2\frac{\omega^2}{\omega_o^2}\left(N+1-\frac{\omega^2}{\omega_o^2}\right)^2}}$
(B)	$T_R = \frac{\delta_o}{u_o} = \sqrt{\frac{\left(\frac{\omega}{\omega_o}\right)^4}{\left(1-\frac{\omega^2}{\omega_o^2}\right)^2+\left(2\zeta\frac{\omega}{\omega_o}\right)^2}}$	$T_R = \frac{\delta_o}{u_o} = \sqrt{\frac{\frac{\omega^2}{\omega_o^2}+\frac{4}{N^2}\zeta^2\frac{\omega^6}{\omega_o^6}}{\left(1-\frac{\omega^2}{\omega_o^2}\right)^2+\frac{4}{N^2}\zeta^2\frac{\omega^2}{\omega_o^2}\left(N+1-\frac{\omega^2}{\omega_o^2}\right)^2}}$
(C)	$\frac{x_o}{F_o/k} = \sqrt{\frac{1}{\left(1-\frac{\omega^2}{\omega_o^2}\right)^2+\left(2\zeta\frac{\omega}{\omega_o}\right)^2}}$	$\frac{x_o}{F_o/k} = \sqrt{\frac{1+\frac{4}{N^2}\zeta^2\frac{\omega^2}{\omega_o^2}}{\left(1-\frac{\omega^2}{\omega_o^2}\right)^2+\frac{4}{N^2}\zeta^2\frac{\omega^2}{\omega_o^2}\left(N+1-\frac{\omega^2}{\omega_o^2}\right)^2}}$

Row A is absolute transmissibility; row B is relative transmissibility; and row C is displacement motion response.

Elastically connected viscous damper. The elastically connected viscous damper c, represented by the dashpot in column 2 of Table 38.1, is in series with a spring of stiffness k_1; the load-carrying spring k is related to the damper spring k_1 by the parameter $N = k_1/k$. This type of damper system sometimes is referred to as a *viscous relaxation system.*

INFLUENCE OF DAMPING IN VIBRATION ISOLATION

The nature and degree of vibration isolation is influenced by the characteristics of the damper. This aspect of vibration isolation is evaluated in this section in terms of the 1-DOF concept; that is, the equipment and the foundation are assumed rigid and the isolator is assumed massless. The performance is defined in terms of absolute transmissibility, relative transmissibility, and motion response. A system with a rigidly connected viscous damper is discussed in detail in Chap. 2. Additional information, such as relative transmissibility and displacement motion response, is given in this chapter. The elastically connected viscous damper is also discussed. Vibration isolators with other types of dampers such as coulomb, quadratic, velocity-*n*th power, and hysteretic are discussed in detail in Ref. 6.

RIGIDLY CONNECTED VISCOUS DAMPER

Absolute and relative transmissibility curves are shown graphically in Figs. 2.17 and 38.16, respectively, and the displacement motion response is shown in Fig. 38.17. The subscript *o* is used to differentiate variables associated with steady-state vibration from those of transient shock. For linear systems, the absolute transmissibility may be expressed as $T_A = x_o/u_o$ for motion-excited systems and $T_A = F_T/F_o$ in force-excited systems. The relative transmissibility $T_R = \delta_o/u_o$ applies only to the motion-excited system. As the damping increases, the transmissibility at resonance decreases and the absolute transmissibility at the higher values of the forcing frequency increases; that is, the reduction of vibration is not as great. For an undamped isolator, the absolute transmissibility at higher values of the forcing frequency varies inversely as the square of the forcing frequency. When the isolator embodies significant viscous damping, the absolute transmissibility curve becomes asymptotic at high values of forcing frequency to a line whose slope is inversely proportional to the first power of the forcing frequency.

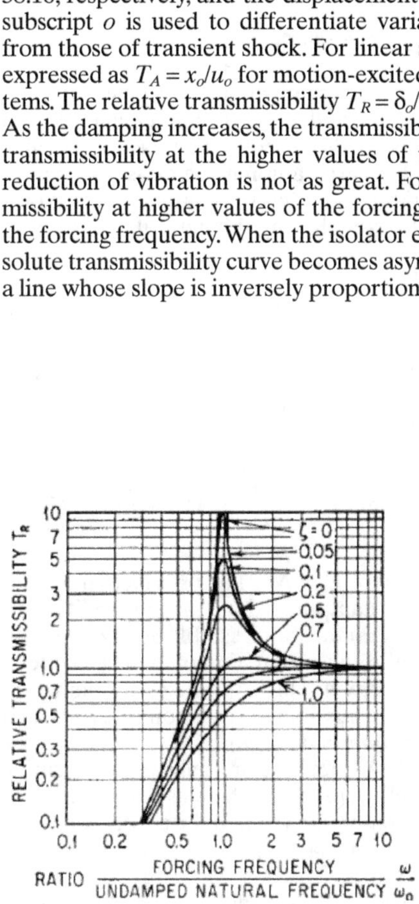

FIGURE 38.16 Relative transmissibility for the rigidly connected, viscous-damped isolation system shown in column 1, row B in Table 38.1 as a function of the frequency ratio ω/ω_0 and the fraction of critical damping ζ. The relative transmissibility describes the motion between the equipment and the foundation (i.e., the deflection of the isolator).

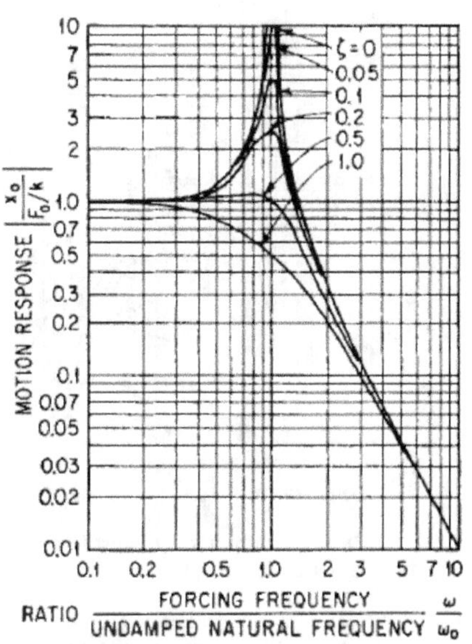

FIGURE 38.17 Displacement motion response for the rigidly connected viscous-damped isolation system shown in column 1, row C of Table 38.1 as a function of the frequency ratio ω/ω_0 and the fraction of critical damping ζ. The curves give the resulting motion of the equipment x in terms of the excitation force F and the static stiffness of the isolator k.

The maximum value of absolute transmissibility associated with the resonant condition is a function solely of the damping in the system, taken with reference to critical damping. For a lightly damped system, where $\zeta < 0.1$, the maximum absolute transmissibility [see Eq. (2.51)] of the system is

$$T_{max} = \frac{1}{2\zeta} \tag{38.31}$$

where $\zeta = c/c_c$ is the fraction of critical damping.

ELASTICALLY CONNECTED VISCOUS DAMPER

General expressions for absolute and relative transmissibility are given in Table 38.1. The characteristics of the elastically connected viscous damper may best be understood by successively assigning values to the viscous damper coefficient c while keeping the stiffness ratio N constant. For zero damping, the mass is supported by the isolator of stiffness k. The transmissibility curve has the characteristics typical of a transmissibility curve for an undamped system having the natural frequency

$$\omega_0 = \sqrt{\frac{k}{m}} \tag{38.32}$$

When c is infinitely great, the transmissibility curve is that of an undamped system having the natural frequency

$$\omega_\infty = \sqrt{\frac{k + k_1}{m}} = \sqrt{N+1}\,\omega_0 \tag{38.33}$$

where $k_1 = Nk$. For intermediate values of damping, the transmissibility falls within the limits established for zero and infinitely great damping. The value of damping which produces the minimum transmissibility at resonance is called *optimum damping.*

A comparison of absolute transmissibility curves for the elastically connected viscous damper and the rigidly connected viscous damper is shown in Fig. 38.18. A constant viscous damping coefficient of $0.2c_c$ is maintained, while the value of the stiffness ratio N is varied from zero to infinity. The transmissibilities at resonance are comparable, even for relatively small values of N, but a substantial gain is achieved in the isolation characteristics at high forcing frequencies by elastically connecting the damper.

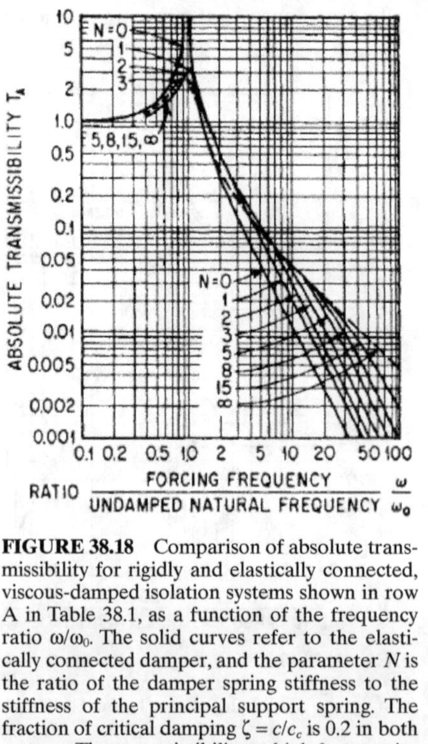

FIGURE 38.18 Comparison of absolute transmissibility for rigidly and elastically connected, viscous-damped isolation systems shown in row A in Table 38.1, as a function of the frequency ratio ω/ω_0. The solid curves refer to the elastically connected damper, and the parameter N is the ratio of the damper spring stiffness to the stiffness of the principal support spring. The fraction of critical damping $\zeta = c/c_c$ is 0.2 in both systems. The transmissibility at high frequencies decreases at a rate of 6 dB per octave for the rigidly connected damper and 12 dB per octave for the elastically connected damper.

Transmissibility at Resonance. The maximum transmissibility (at resonance) is a function of the damping ratio ζ and the stiffness ratio N, as shown in Fig. 38.19. The maximum transmissibility is nearly independent of N for small values of ζ. However, for $\zeta > 0.1$, the coefficient N is significant in determining the maximum transmissibility. The lowest value of the maximum absolute transmissibility curves corresponds to the conditions of optimum damping.

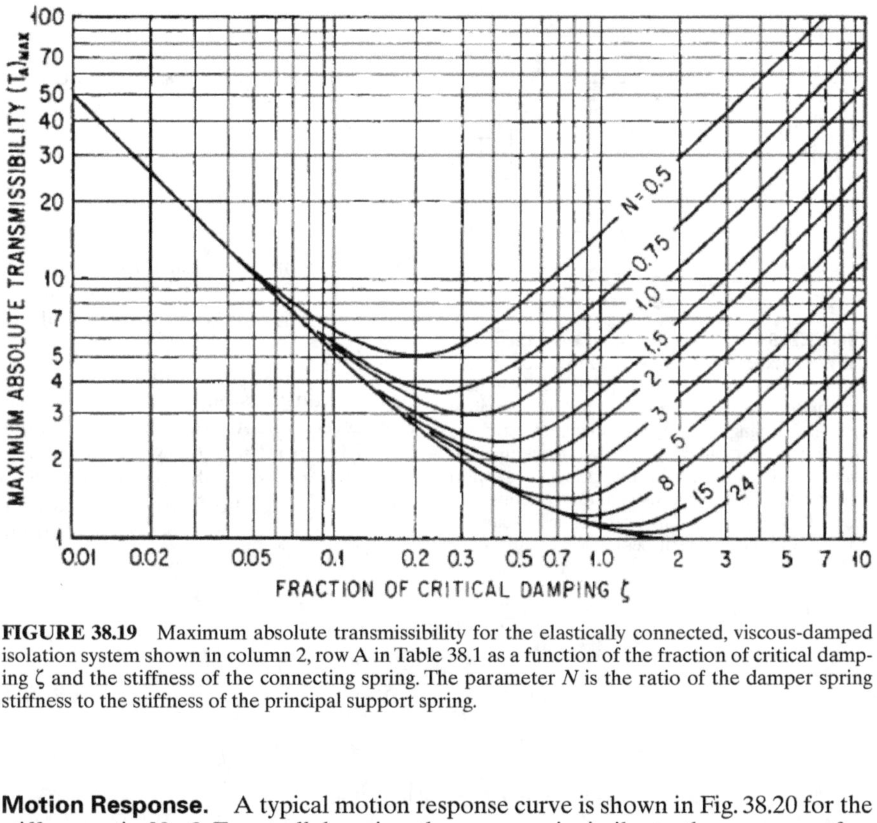

FIGURE 38.19 Maximum absolute transmissibility for the elastically connected, viscous-damped isolation system shown in column 2, row A in Table 38.1 as a function of the fraction of critical damping ζ and the stiffness of the connecting spring. The parameter N is the ratio of the damper spring stiffness to the stiffness of the principal support spring.

Motion Response. A typical motion response curve is shown in Fig. 38.20 for the stiffness ratio $N = 3$. For small damping, the response is similar to the response of an isolation system with rigidly connected viscous damper. For intermediate values of damping, the curves tend to be flat over a wide frequency range before rapidly decreasing in value at the higher frequencies. For large damping, the resonance occurs near the natural frequency of the system with infinitely great damping. All response curves approach a high-frequency asymptote for which the attenuation varies inversely as the square of the excitation frequency.

Optimum Transmissibility. For a system with optimum damping, maximum transmissibility coincides with the intersections of the transmissibility curves for zero and infinite damping. The frequency ratios $(\omega/\omega_0)_{op}$ at which this occurs are different for absolute and relative transmissibility:

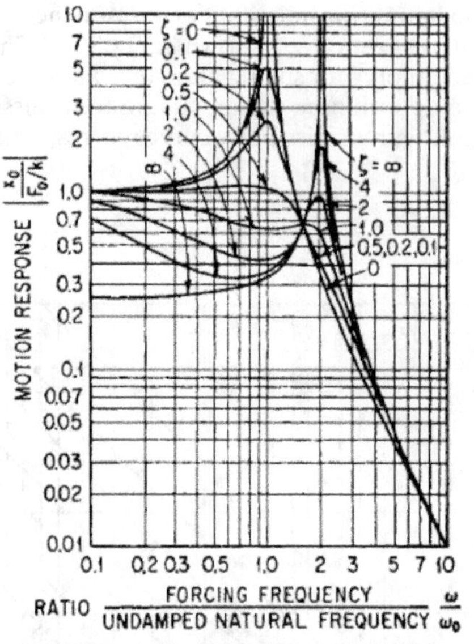

FIGURE 38.20 Motion response for the elastically connected, viscous-damped isolation system shown in column 2, row C in Table 38.1 as a function of the frequency ratio ω/ω_0 and the fraction of critical damping ζ. For this example, the stiffness of the damper connecting spring is three times as great as the stiffness of the principal support spring ($N = 3$). The curves give the resulting motion of the equipment in terms of the excitation force F and the static stiffness of the isolator k.

Absolute transmissibility:

$$\left(\frac{\omega}{\omega_0}\right)_{\text{op}}^{(A)} = \sqrt{\frac{2(N+1)}{N+2}} \tag{38.34}$$

Relative transmissibility:

$$\left(\frac{\omega}{\omega_0}\right)_{\text{op}}^{(R)} = \sqrt{\frac{N+2}{2}}$$

The optimum transmissibility at resonance, for both absolute and relative motion, is

$$T_{\text{op}} = 1 + \frac{2}{N} \tag{38.35}$$

The optimum transmissibility as determined from Eq. (38.35) corresponds to the minimum points of the curves of Fig. 38.19.

The damping which produces the optimum transmissibility in the elastically connected viscous damper is obtained by differentiating the general expressions for transmissibility in Table 38.1 with respect to the frequency ratio, setting the result equal to zero and combining it with Eq. (38.34):

Absolute transmissibility:

$$(\zeta_{\text{op}})_A = \frac{N}{4(N+1)}\sqrt{2(N+2)} \tag{38.36a}$$

Relative transmissibility:

$$(\zeta_{\text{op}})_R = \frac{N}{\sqrt{2(N+1)(N+2)}} \tag{38.36b}$$

Values of optimum damping determined from the first of these relations correspond to the minimum points of the curves of Fig. 38.19. By substituting the optimum damping ratios from Eqs. (38.36) into the general expressions for transmissibility given in Table 38.1, the optimum absolute and relative transmissibility equations are obtained, as shown graphically by Figs. 38.21 and 38.22, respectively. For low values of the stiffness ratio N, the transmissibility at resonance is large, but excellent isolation is obtained at high frequencies. Conversely, for high values of N, the transmissibility at resonance is lowered, but the isolation efficiency also is decreased.

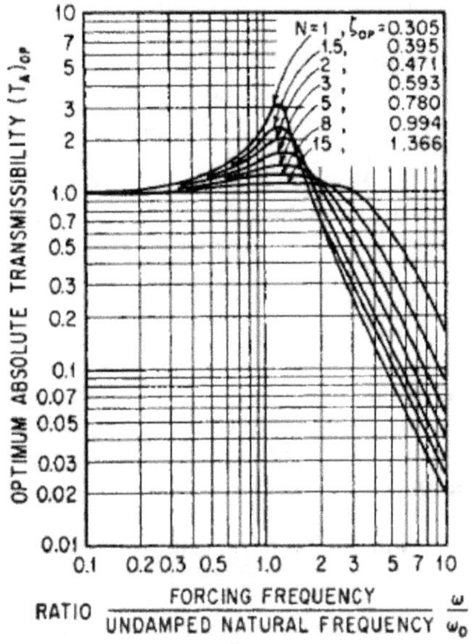

FIGURE 38.21 Absolute transmissibility with optimum damping in the elastically connected, viscous-damped isolation system shown in column 2, row A in Table 38.1 as a function of the frequency ratio ω/ω_0 and the fraction of critical damping ζ. These curves apply to elastically connected, viscous-damped systems having optimum damping for absolute motion. The transmissibility $(T_A)_{op}$ is $(x_0/u_0)_{op}$ for the motion-excited system and $(F_T/F_0)_{op}$ for the force-excited system.

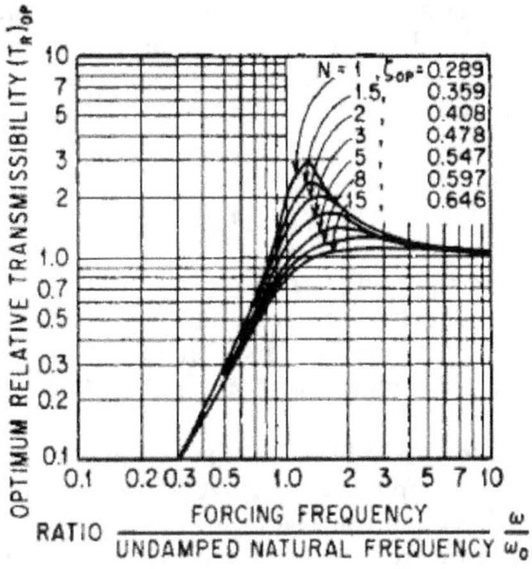

FIGURE 38.22 Relative transmissibility with optimum damping in the elastically connected, viscous-damped isolation system shown in column 2, row B in Table 38.1 as a function of the frequency ratio ω/ω_0 and the fraction of critical damping ζ. These curves apply to elastically connected, viscous-damped systems having optimum damping for relative motion. The relative transmissibility $(T_R)_{op}$ is $(\delta_0/u_0)_{op}$ for the motion-excited system.

MULTIPLE-DEGREE-OF-FREEDOM SYSTEMS

The 1-DOF systems discussed previously are adequate for illustrating the fundamental principles of vibration isolation but are an oversimplification insofar as many practical applications are concerned. The condition of unidirectional motion of an elastically mounted mass is not consistent with the requirements in many applications. In general, it is necessary to consider freedom of movement in all directions, as dictated by existing forces and motions and by the elastic constraints. Thus, in the general isolation problem, the equipment is considered as a rigid body supported by resilient elements or isolators.

The most practical approach to assess resiliently supported rigid bodies is to use 6-DOF simulation methods that allow specification and implementation of the many parameters used in analyzing isolated systems. Advantages provided by simulations of 6-DOF models are (1) estimates of the excursion space needed for dynamic travel of mounted systems; (2) rapid prediction of responses such as acceleration, velocity, force, and displacement; (3) a design feedback tool for the location, sizing, and orientation of mounts for equipment and structures; (4) calibration via optimization; (5) multivariate sensitivity analyses of system parameters; and (6) multidirectional inputs via foundation motion or application of forces and moments to the body.

For cases where symmetry and other simplifications can be made or where simulation methods are not possible or practical, analytical expressions can be used to make estimates of multiple-degree-of-freedom responses. For example, linear expressions for properties of resilient supports and equations of motion for a resiliently supported rigid body are presented in Chap. 3. The equations of motion are given by Eq. (3.31). The general case of this model is shown in Fig. 3.12, which depicts a rigid body supported by resilient elements.

USE OF SYMMETRY

By employing various types of symmetry and neglecting damping, natural frequencies and rigid-body dimensionless responses on resilient mounts may be estimated. Relevant expressions from Chap. 3 for estimating natural frequencies and responses for a resiliently support rigid body are summarized as follows.

Rigid-Body Natural Frequencies. For one plane of symmetry, as shown in Fig. 3.13, coupled natural frequencies may be estimated by obtaining the roots of the cubic equation given by Eq. (3.36). This equation may be solved graphically for the natural frequencies of the system by use of Fig. 3.14. When three planes of symmetry are present, the natural frequencies are uncoupled and are given by Eq. (3.42).

For two planes of symmetry, as shown in Fig. 3.15, coupled natural frequencies may be estimated by using Eqs. (3.39) and (3.40). For two planes of symmetry with resilient supports inclined in one plane only, as shown in Fig. 3.21, uncoupled natural frequencies may be estimated by using Eqs. (3.43) and (3.44) and coupled frequencies by Eqs. (3.45) and (3.46). The inclination angle of resilient supports may be selected to decouple translation and rotation modes. Decoupling of these modes is effected if Eq. (3.47) is satisfied, allowing natural frequencies of the decoupled translation and rotation modes for the inclined support system to be estimated using Eqs. (3.48) and (3.49).

Rigid-Body Responses. For the one-plane-of-symmetry translational case excited by foundation motion, dimensionless response expressions for maximum displacement and acceleration are given by Eqs. (3.72) and (3.73), respectively. For the rotational case, these expressions are given by Eqs. (3.77) and (3.78), respectively.

For two planes of symmetry with orthogonal resilient supports excited by foundation motion, the dimensionless response expressions for translation and rotation are given by Eqs. (3.55) and (3.56), respectively. When resilient supports are inclined in one plane, the dimensionless response expressions are given by Eqs. (3.58) and (3.59). For two planes of symmetry with orthogonal resilient supports excited by a rotating force, the dimensionless response expressions for translation and rotation are given by Eq. (3.63). When excited by an oscillating moment, the dimensionless response expressions for translation and rotation are given by Eq. (3.68).

For all of these cases, analysis of the dynamics of a rigid body on resilient supports includes the assumption that the principal axes of inertia of the rigid body are, respectively, parallel with the principal elastic axes of the resilient supports. This makes it possible to neglect the products of inertia of the rigid body. The coupling introduced by the product of inertia is not strong unless the angle between the principal axes of inertia and the elastic axes is substantial. Therefore, it is convenient to take the coordinate axes through the center of gravity of the supported body, parallel with the principal elastic axes of the isolators.

The procedures in Chap. 3 for determining natural frequencies in coupled modes represent a rigorous analysis where the assumed symmetry exists. They are also somewhat indirect, requiring the use of dimensionless ratios involving the coordinate distances of elastic centers of the resilient elements and the radius of gyration of the equipment. For the case of two planes of symmetry, as shown in Fig. 3.15, the relations may be approximated in a more readily usable form if (1) the mounted equipment can be considered a cuboid having uniform mass distribution, (2) the four isolators are attached precisely at the four lower corners of the cuboid, and (3) the height of the isolators may be considered negligible. The ratio of the natural frequencies in the coupled rotational and horizontal translational modes to the natural frequency in the vertical translational mode then becomes a function of only the dimensions of the cuboid and the stiffnesses of the isolators in the several coordinate directions. Making these assumptions and substituting in Eq. (3.40), results in

$$\frac{f_{x\beta}}{f_z} = \frac{\omega_{x\beta}}{\omega_z} = \frac{1}{\sqrt{2}} \sqrt{\frac{4\eta\lambda^2 + \eta + 3}{\lambda^2 + 1} \pm \sqrt{\left(\frac{4\eta\lambda^2 + \eta + 3}{\lambda^2 + 1}\right)^2 - \frac{12\eta}{\lambda^2 + 1}}} \tag{38.37}$$

where $\eta = k_x/k_z$ designates the ratio of horizontal to vertical stiffness of the isolators and $\lambda = 2a_z/2a_x$ indicates the ratio of height to width of mounted equipment. This relation is shown graphically in Fig. 38.23. The curves included in this figure are useful for

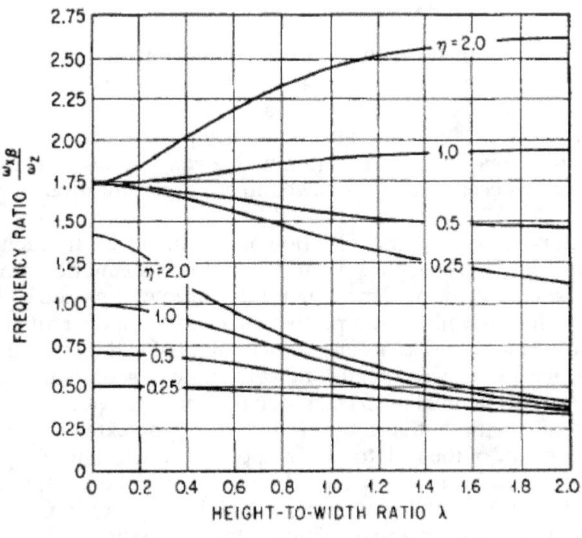

FIGURE 38.23 Curves indicating the natural frequencies $\omega_x\beta$ in coupled rotational and horizontal translational modes with reference to the natural frequency ω_z in the decoupled vertical translational mode, for the system shown in Fig. 3.15. The ratio of horizontal to vertical stiffness of the isolators is η, and the height-to-width ratio for the equipment is λ. These curves are based upon the assumption that the mass of the equipment is uniformly distributed and that the isolators are attached precisely at the extreme lower corners thereof.

calculating approximate values of natural frequencies and for indicating trends in natural frequencies resulting from changes in various parameters as follows:

1. Both of the coupled natural frequencies tend to become a minimum, for any ratio of height to width of the mounted equipment, when the ratio of horizontal to vertical stiffness k_x/k_z of the isolators is low. Conversely, when the ratio of horizontal to vertical stiffness is high, both coupled natural frequencies also tend to be high. Thus, when the isolators are located underneath the mounted body, a condition of low natural frequencies is obtained using isolators whose stiffness in a horizontal direction is less than the stiffness in a vertical direction. However, low horizontal stiffness may be undesirable in applications requiring maximum stability. A compromise between natural frequency and stability then may lead to optimum conditions.

2. As the ratio of height to width of the mounted equipment increases, the lower of the coupled natural frequencies decreases. The trend of the higher of the coupled natural frequencies depends on the stiffness ratio of the isolators. One of the coupled natural frequencies tends to become very high when the horizontal stiffness of the isolators is greater than the vertical stiffness and when the height of the mounted equipment is approximately equal to or greater than the width. When the ratio of height to width of mounted equipment is greater than 0.5, the spread between the coupled natural frequencies increases as the ratio k_x/k_z of horizontal to vertical stiffness of the isolators increases.

ISOLATION OF RANDOM VIBRATION

In random vibration, all frequencies exist concurrently, and the amplitudes and phases of frequency components are random. A trace of random vibration is illustrated in Fig. 24.1*B*. The equipment-isolator assembly responds to the random vibration with the substantially single-frequency pattern shown in Fig. 24.1*A*. This response is similar to a sinusoidal motion with a continuously and irregularly varying envelope; it is described as narrowband random vibration or a random sine wave.

The characteristics of random vibration are defined by a frequency spectrum of power spectral density (see Chaps. 19 and 24). This is a generic term used to designate the mean square value of some magnitude parameter passed by a filter, divided by the bandwidth of the filter, and plotted as a spectrum of frequency. The magnitude is commonly measured as acceleration in units of *g;* then the particular expression to use in place of power spectral density is mean square acceleration density, commonly expressed in units of g^2/Hz. When the spectrum of mean square acceleration density is substantially flat in the frequency region extending on either side of the natural frequency of the isolator, the response of the isolator may be determined in terms of (1) the mean square acceleration density of the isolated equipment and (2) the deflection of the isolator at successive cycles of vibration.

The mean square acceleration densities of the foundation and the isolated equipment are related by the absolute transmissibility that applies to sinusoidal vibration:

$$W_r(f) = W_e(f)T_A^2 \tag{38.38}$$

where $W_r(f)$ and $W_e(f)$ are the mean square acceleration densities of the equipment and the foundation, respectively, in units of g^2/Hz, and T_A is the absolute transmissibility for the vibration-isolation system. Additional discussion of dynamic random vibration analysis of systems is provided in Ref. 7.

REFERENCES

1. Talley, M., and S. Sarkani: "A New Simulation Method Providing Shock Mount Selection Assurance," *Shock and Vibration,* **10:**231–267 (2003).
2. Steinberg, D. S.: "Vibration Analysis for Electronic Equipment," 3rd ed., John Wiley & Sons, New York, 2000.
3. Kulkarni, J., and R. Jangid: "Effects of Superstructure Flexibility on the Response of Base-Isolated Structures," *Shock and Vibration,* **10**(1):1–13 (2003).
4. Kikuchi, M., and I. Aiken: "An Analytical Hysteresis Model for Elastomeric Seismic Isolation Bearings," *Earthquake Engineering and Structural Dynamics,* **26**(2):215–231 (1997).
5. Richards, C., and R. Singh: "Characterization of Rubber Isolator Nonlinearities in the Context of Single- and Multi-Degree of Freedom Experimental Systems," *Journal of Sound and Vibration,* **274**(5):807–834 (2001).
6. Ruzicka, J. E., and T. F. Derby: "Influence of Damping in Vibration Isolation," Technical Information Center, Naval Research Laboratory, Washington, D.C., 1971.
7. Lutes, L., and R. Sarkani: "Dynamic Random Vibrations Analysis of Structural and Mechanical Systems," Elsevier Butterworth-Heinemann, Burlington, Mass., 2004.

REFERENCES

[illegible]

CHAPTER 39
SHOCK AND VIBRATION ISOLATION SYSTEMS

Herb LeKuch

INTRODUCTION

Isolation technology, once limited mostly to selecting mounts for the protection of machinery and equipment, is now routinely applied to devices ranging from personnel comfort to large civil engineering structures and low-frequency vibration attenuation of instruments and precision mechanisms. The technology has been widely developed through improved designs, better materials, and greater depth of technical analysis. More refined test methods and simulation modeling have significantly broadened the database of isolator properties. The levels of complex vibration and shock disturbances can now be substantially reduced at the equipment.

Applications have grown extensively over the last 20 years. In addition to machinery isolation, examples are servo-controlled self-leveling isolated platforms, energy dissipation devices such as frictional and tuned mass dampers for large structures, semiactive (SA) isolation for aircraft landing gear, and base isolation for seismic protection of entire buildings. Active-controlled auto and seat suspension systems are in large production. In microelectronics, isolation techniques have been developed to control the fabrication process and ensure device quality. Architects and engineers consider isolation as a basic technique in building design and have established vibration criterion curves to specify vibration-sensitive production tools. Companies will continue to incorporate isolation to increase product reliability and improve manufacturing efficiencies. Isolation methods developed in semiconductor fabrication are now used in other industries to reduce costs in high-volume production and enable rapid change as new designs are introduced. The use of commercial off-the-shelf (COTS) electronics will continue to grow in military systems, requiring improved isolation to meet severe conditions.

The focus of this chapter is mounts and isolation systems to protect equipment and sensitive machinery. Isolator types that are commercially available are described in the passive mount section of this chapter; some have broad industrial and military use. In the semi-active and active isolation sections, the operation of isolation systems and general principles of tuned control are described. Structural control, seismic, and energy dissipation damper devices are considered engineered motion limiters and are briefly noted. The first two sections describe guidelines for the design of isolation systems and the problems that can occur, followed by review of several types of commercial mounts, their characteristics, the features of ideal isolators, and mount selection. The third section is an overview of active and semiactive isolation methods. Commercial products described in this section are the SA type where the mount's fail-safe mode, enhanced performance, simplicity, and relatively low cost are the main objectives. This chapter generally follows the form of Chap. 32 of the 5th edition; however, it contributes additional details to isolation system design and describes several newer isolators and applications.

ISOLATORS AND ISOLATION SYSTEMS

Well-designed isolation substantially decreases the intensity of the shock and vibration (S&V) reaching the equipment from the disturbance. The isolation system is a *mechanical filter* between the source and the receiver that reduces the dynamic loads at the equipment to levels that the equipment can withstand; isolation tailors the S&V environment so that a reasonable margin of safety exists for satisfactory operation. Disturbances can emanate *from* the equipment or external sources *to* the unit. In some cases, isolators cannot be mounted directly to the receiver (or source), and an attachment frame (platform) is needed. The selection begins with characterizing conditions and comparing those against allowable limits.

TYPES OF ISOLATORS AND ISOLATION SYSTEMS

Shock and vibration isolators and isolation systems can be grouped according to how they protect equipment. *Passive* isolation is the simplest and least expensive. It involves only unit mounts or a combination of mounts at the equipment. *Semiactive* and *active* isolation requires sensors, feedback controls, and variable damper and/or force actuators. In response to the motion of the equipment, these precisely shift the operating characteristics of the isolation system for better performance. Passive systems are essentially nonactive; S&V control is entirely a function of the properties and mechanical design of the isolator and its compliant elements. No external power or control loop is needed. SA mounts operate to modify the stiffness or damping of the system. Active isolation uses variable force to counteract the driving force. The several types of isolation control, ranging from hard-mounted to passive and active methods, are outlined in Fig. 39.1. Table 39.1 compares these methods and describes the features. Active isolation exhibits the most sensitive control for optimum performance; passive isolation is the least sensitive.

Passive mounts incorporate molded elastomers, shaped metal springs, or other means that can deform predictably under load and provide stiffness and damping to the spring/mass system. The response is self-regulated in accordance with performance characteristics of the mount. Stiffness and damping values of the isolators are based on static test data adjusted for dynamic and environmental conditions. The performance of the mount may be different in each axis; however, the isolator exhibits defined and repeatable properties for calculation of resonant frequency and response. The passive mount can have linear or nonlinear stiffness depending on its design and orientation with respect to the applied load. If nonlinear, results can be

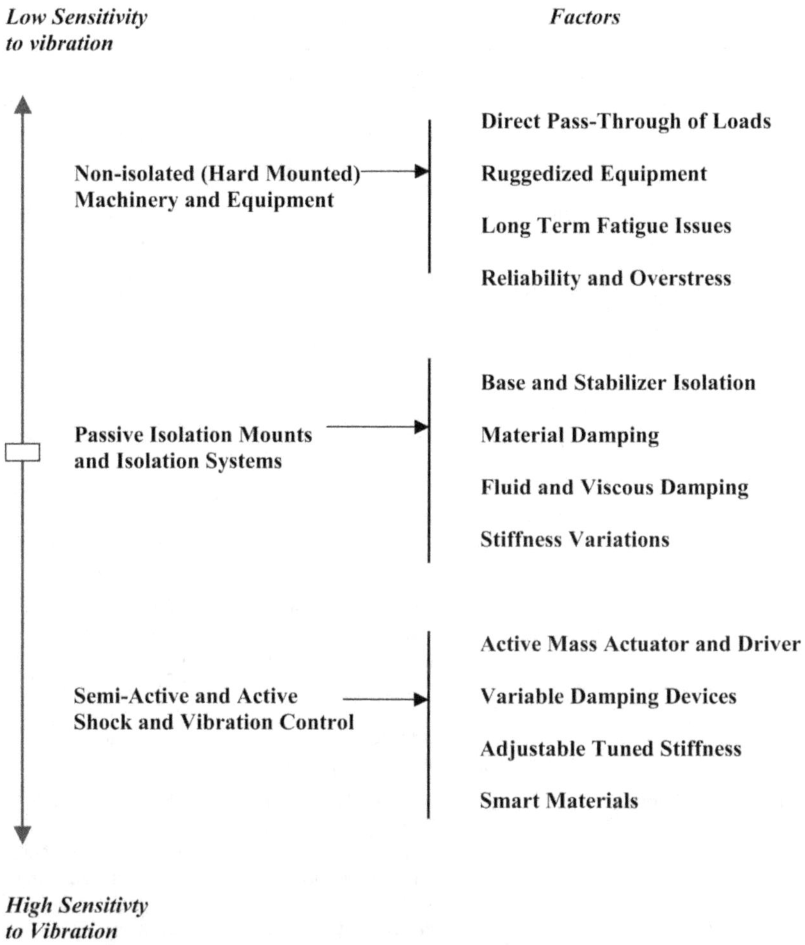

FIGURE 39.1 The several types of isolation control include hard-mounted, passive, semiactive, and active. The active type has the best control features for optimum isolation performance; passive control is the least sensitive.

affected by the level and type of vibration or shock. For instance, a nonlinear mount in vibration can show a different resonant frequency at 1*g* swept-sine than at 2*g* sine because the dynamic load on the isolator shifts its stiffness to a different region of its operating curve. This modifies the stiffness-to-mass ratio, thereby changing the resonant frequency. A nonlinear mount can be typified as having bilinear or even trilinear stiffness characteristics.

Active and *semiactive* systems use controls and actuators or variable damping devices to modify the restraining forces and adjust the operation of the isolation system. They can better regulate dynamic response than passive mounts but at greater complexity and cost. SA designs use damper or stiffness control and can be battery/low-power operated. Isolation reverts to passive type if variable features fail.

TABLE 39.1 Comparison of Principal Isolation Methods

Isolation	Passive	Semiactive	Active
Type	Elastomer Helical cable Pneumatic Spring	MR fluid ER fluid	Hydraulic/servo valve Electromagnetic Force actuator
	Friction damper Viscous damping	Can work with passive mounts	Can supplement passive mounts
Performance	Conforms with vibration transmissibility curves, shock reduction factors	Fail-safe reverts to passive control if damper or electronics fail	Optimum isolation T can be less than 1.0 at resonance
	Proven SV reduction in many applications	Improved performance over passive mounts	Excellent broad frequency effectiveness
			Proven low frequency isolation table designs, optical and semi-conductor uses
Features	Self-contained No external energy needed	Closed-loop control Requires very little power, can be battery operated	Closed-loop control sensor on moving unit, measurements of position, velocity, or acceleration
		Acceleration converted to velocity for control	Controller drives force actuator
Control	None Open loop	Skyhook—absolute velocity Force resistance using variable damper, setting based on equivalent relative On/off control often used	Skyhook—absolute velocity Movement resisted by counterforce using external force actuator(s) Relative, PID, and other methods described and used for control
Positive factors	Low cost Simple design Predictable Defined properties	MR devices available as commercial products Competitive prices Published engineering properties	Can handle wide range of disturbances, can be used with a variety of energy dissipation devices for seismic control
	Very wide selection and availability of mounts	Very rapid response Simple programming is effective for SV control	Considerable development for seismic structures Proven tuned mass dampers

TABLE 39.1 Comparison of Principal Isolation Methods (*Continued*)

Isolation	Passive	Semiactive	Active
Negative factors	Isolation effectiveness limited by fixed stiffness and damping	May require separate motion control in each direction	High power needed to drive the actuator High cost, complex design and configuration Reliability can be a problem
	Isolators may not be SV rated for severe loads Nonlinear effects can influence results	Requires more space for damping device than passive isolation alone	Instability due to nonlinearity Possible sensor/actuator failure Can put uncontrolled force into the system
	Some compliant materials are temperature limited		Software and processor/ controller is application driven
Technical/ Literature	Extensive documentation	Extensive	Extensive
	Manufacturers' catalogs Test history and applications	Seat suspension data Various applications	Control theory history Structural dynamics applications

OPERATING PERFORMANCE

Stiffness and damping properties of isolation are chosen as a compromise to limit vibration amplification at resonance and still provide effective isolation at higher frequencies. Shock is controlled in a similar way through the design and means of compliance for large deflection. For some isolators, the same mount can provide both vibration and shock isolation. The theory of shock and vibration isolation is discussed in Chap. 38. The concept is similar for active and semiactive control. Conditions can be examined in the equations for a single-degree-of-freedom (SDOF) system. The important relationship is the transmissibility T between the input and response, with the isolator bridging the two. Damping influences the amount of relative displacement that occurs.

Figure 39.2 shows the three main regions in the T plot of the simple mass spring damper system: (1) the ratio of natural frequency to disturbing frequency (less than 0.5), T ranges from 1.0 to 1.5 regardless of the amount of damping; (2) the ratio of natural frequency to disturbing frequency (0.5 to $\sqrt{2}$), T reaches a maximum at resonance (frequency ratio of 1.0) and ranges from 2.5 with moderate damping to 15.0 with light damping for the passive system; (3) the ratio of natural frequency to source frequency greater than $\sqrt{2}$, T decreases below 1.0 and falls off at a rate depending on the level of damping. Light damping results in rapid decline. T at resonance is a maximum. Passive system ratios are greater than 1.0. Active systems can achieve ratios less than 1.0 due to the extremely high damping that can be set at the resonant frequency.

Crossover occurs at the frequency ratio $f/f_n = \sqrt{2}$. This is where isolation begins (T declines to 1.0). The acceleration response of the system is at least 1:1 or ampli-

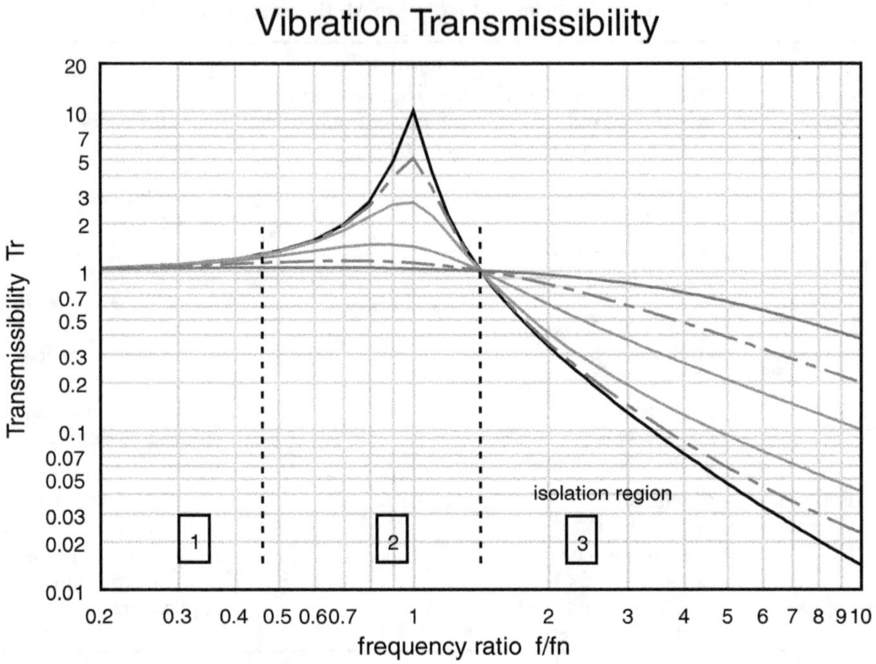

FIGURE 39.2 The transmissibility plot of an SDOF isolation system exhibits increasing peak amplitude at resonance as the damping ratio decreases. Three regions are designated with isolation beginning at frequency crossover.

fied in regions 1 and 2 until the driving frequency is slightly greater than $\sqrt{2}$, the disturbing frequency. It is here that the transition from amplification to isolation occurs. Except at crossover, the rate of isolation strongly depends on the amount of damping. At the higher frequencies, it occurs more rapidly with less damping. In region 1 and 2 (and particularly at resonance) more damping reduces T. At lower frequencies, the displacement of the isolated mass essentially follows the displacement of the disturbance. The foundation (source of vibration) and mass are moving relatively the same amount and in phase with one another. Beyond transition, mass displacement becomes less than foundation motion. Increasing the damping ratio reduces the transmissibility at resonance but also decreases the effectiveness of high-frequency isolation. Most passive isolators are designed around these factors. There is also a slight shift in the resonant frequency versus the natural frequency, but this can usually be disregarded in the selection of properties. The ratio of stiffness relative to mass affects only the natural frequency of the system; the amount of damping strongly influences amplification in the resonance region and the degree of isolation beyond resonance.

Shock response can be described in terms of the separation of shock pulse frequency from shock response frequency of the isolation system. The effective frequency of the input pulse is a characteristic of its initial time duration. *Dynamic load factors* (DLFs) are useful for describing the acceleration response relative to the magnitude of the applied shock. As shown in Fig. 39.3, the response to a shock input is more severe for a half-sine pulse than for a triangular pulse having the same time duration. Other pulse shapes can be similarly compared. Shock transmissibility can

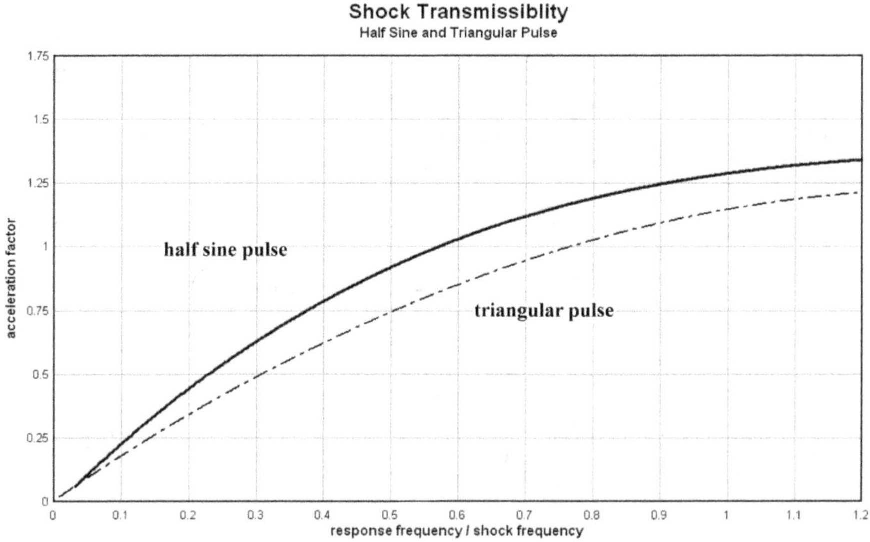

FIGURE 39.3 The response to a shock input is more severe to a half-sine pulse than to a triangular pulse at the same frequency ratio. Similar relationships exist for other pulse shapes with respect to a half-sine pulse.

be calculated from design data charts that show the ratio of frequency coupling and the character of the pulse. Damping contributes to energy dissipation in the system and helps to limit maximum deflection and peak acceleration. In general, if the applied pulse frequency is 0.5 or greater than the effective shock response frequency, the shock will be amplified and can reach a maximum of 2.0 times the input acceleration peak of a half-sine pulse. Figure 39.3 can also be interpreted in terms of pulse duration and period of the isolation. If the pulse is more than ¼ of the natural period of the isolation system, the shock will be amplified. The theory of shock isolation is covered in Chap. 38.

COMPETING STRATEGIES FOR THE SELECTION OF ISOLATORS

Stiffness and damping establish the performance of isolators and are a guide to mount effectiveness. Other factors include load capability, size of the mount, rattle space (allowable space for relative movement), fit, cost, availability, long-term use, and reliability. Table 39.2 compares various passive mounts, including their characteristics, uses, and general applications. Determination of the resonant frequency of the isolation system is needed to establish the separation of the driving frequencies from the resonant frequency and calculate T efficiency. Damping controls the level of attenuation over the operating range of the mount. Commercial mounts are the least costly—there is a very wide selection to choose from. Custom isolators may be required for special applications such as avoidance of outgassing in spacecraft and vacuum chambers. Passive mounts exhibit nearly constant damping properties within their operating range. However, vibration stiffness (small-amplitude motion) can be different from shock response stiffness (large displacement) due to the particular load deflection characteristics of the isolators.

Depending on the type and level of shock and vibration, similar or even the same

TABLE 39.2 Passive Isolators: General Characteristics and Uses

Applications	Typical equipment protected	Sources of shock and vibration	Other environmental factors	Levels of shock and vibration	Critical isolator performance	Needed isolator characteristics	Other requirements
Navy shipboard	Multiple displays, switchgear, COTS electronics, workstations	Barge shock test, air and water blast, ship's vibration, rough seas	Saltwater, temperature extremes	Mil Std 901D 120–150 g, 25 Hz deck Mil Std 167	Reduce shock and vibration	Long-term use, inspectable, wide temperature range, severe loads	Multiaxis, Large deflection range, moderate damping
Military aircraft	VME electronics displays, gyros and avionics, data acquisition	High-speed flight, hard landings, gunfire and rapid maneuvers	Temperature and pressure extremes	Mil Std 810 30 g, 11-ms crash landing 15 g, 11-ms hard landing	Reduce shock and vibration	Low profile, test qualified	Multiaxis, close tolerance, moderate damping
Shipping and handling	Jet engines, missiles, commercial electronics, special equipment	Transportation, handling drop, airlift, and off-load	Altitude changes, exposure to rain and humidity	ATA 300 Accidental drops to 48 in.	Reduce shock and vibration	Large deflection, long-term storage, easily replaceable	Multiaxis, moderate damping
Off-road and military vehicles	Displays, computer racks	rough road and tank test grounds	Temperature extremes	Mil Std 810 Munson rough road	Reduce vibration, frequent shocks	Low profile, severe environments	Multiaxis, moderate damping
Geophysical and oil exploration	Data acquisition, computer systems	Irregular roads, rough terrain	Temperature, high humidity	Munson rough road (equivalent)	Reduce vibration, occasional shock	Low profile, severe environments	Multiaxis, moderate damping
Materials processing	Centrifuge, pumps, compressors	Rotary equipment, unbalanced loads, defective bearings	Corrosive environments, chlorine, sulfur	0.01–0.02 in./sec (rms) (2–5 mm/sec)	Reduce vibration	Easily replaceable, rugged, maintenance free	Single axis, side restraint, light damping

Building services	HVAC, cooling towers	Fluid and air handling, low-speed rotation	Indoor and outdoor conditions	0.008–0.016 in./sec (rms) (0.1–0.2 mm/sec)	Reduce vibration	Maintenance free, outdoor environments	Single axis, side restraint
Construction	Heavy operating mobile equipment	Excavations and earthmoving, heavy lifts	Temperature, high humidity	Munson rough road (equivalent)	Reduce vibration, occasional shock	High reliability, severe service	Multiaxis, high damping
Industrial manufacturing	Stamping, punch presses, drop forge	Repetitive shock, continuous operation, crushable materials	Shop conditions, high temperature, oil and greases	0.016 in./sec (rms) (0.2 mm/sec)	Reduce vibration, frequent shock	Maintenance free, high load rating	Single axis, side restraint, light damping
Semiconductor	Precision inspection, wafer fabrication	Nearby road traffic, factory site operations	Air-conditioning failure	0.001–0.002 in./sec (rms) (0.025–0.050 mm/sec)	Reduce low-level vibration	Long-term reliability, verifiable performance	Multiaxis, light damping
Research	Diffraction-limited optics, critical measurements	Adjacent equipment, HVAC operations	Air-conditioning failure	0.0005–0.001 in./sec (rms) (0.012–0.025 mm/sec)	Reduce very low level vibration	Very low level characteristics	Multiaxis, light damping
Seismic	Infrastructure, facilities, occupied buildings	Earthquake and accidental explosions	Temperature, high humidity	IBC 2000, UBC 1997 specifications and codes	Restrain structures, dissipate energy	Large lateral deflection, multiple shocks	Multiaxis, limit restraint, moderate to high damping

isolator can be used for different situations. The same mount may fit a variety of applications ranging from building facilities to mobile equipment and commercial electronics, depending on the availability of the isolator, its operating features, and the user's design. Commonly used isolators for buildings and machinery are open and housed steel springs, bonded elastomer, air springs, and mounts using metal spring and rubber elements. Other designs, used for both military and industrial purposes, include multiloop helical wire rope (preformed steel cable) and high-deflection elastomer shock mounts. Seismic isolation and energy dissipation devices have been developed for structural control. In the area of very low level table isolation, leveling, controlled pneumatic mounts, or other "zero or negative" stiffness devices are used. There are numerous products, such as composite open- and closed-cell foams, glass fiber, layered elastomer pads, and plastic mesh, that can also be effective for general applications. There is a rapidly expanding field of vibration isolation for nanotechnology research and development (R&D) and manufacturing.

There are three basic conditions: (1) where vibration is predominant, (2) where shock is the major concern, and (3) severe service where both shock and vibration occur.

Vibration. These isolators are intended primarily to reduce the response at resonance and then to ensure that the output remains below the level of applied excitation at higher frequencies beyond resonance. The isolator must also be capable of dissipating energy (damping) and limiting displacement. Relative motion across the mount is generally small.

Shock. The mount undergoes abrupt velocity or displacement change and must absorb large amounts of shock energy, then release the energy slowly at the *shock response frequency* of the isolation system. The mount must be capable of relatively large displacement to reduce the shock experienced at the equipment that it supports. For example, a drop from 18 in. (45.7 cm) onto a hard floor typically requires an isolator capable of nearly 4 in. (10.16 cm) of deflection to reduce the shock to about 20*g* at the equipment. Many high-deflection isolators are capable of compressing 0.5 to 0.6 of static height; the mount would have to be nearly 7 in. (17.78 cm) tall for 4 in. (10.16 cm) of stroke space.

Severe Service. The mount has to have a low natural frequency in its small-amplitude motion region to isolate vibrations and the ability to deflect in a controlled and repeatable way over a larger stroke to absorb high-acceleration, short-duration shock loads. Selection favors the softening-type mount because of its well-defined multistiffness characteristics—linear stiffness to approximately 10 to 15 percent of its initial stroke, then greater compliance and nearly constant force over the second stage (75 to 80 percent of its stroke). Snubbing occurs in the final 10 to 15 percent of its stroke. Linear and stiffening isolators each have different characteristics that can result in greater *g*'s than the softening mount. Hard snubbing is an unpredictable condition and should be avoided.

SHOCK AND VIBRATION CRITERIA

The fragility of most equipment is often *not* well defined. There may be field reports or test data from similar equipment indicating a threshold of S&V damage. *But precise levels are uncertain.* Test specifications, contract requirements, and design standards should be used to establish levels that the equipment has to withstand. Peak accelerations, direction, duration, and frequency range should be identified. *Shock response spectra* (SRS) can be useful to identify the distribution of energy (based on pseudo-velocity values) at different frequencies. Allowable stresses in structural elements, electronic packaging, chassis, connectors, and other critical parts of equipment should be checked. Criteria have been established describing S&V levels ranging from imperceptible to structural damage in buildings and structures. A general method of limits is to first set a maximum level expected from machinery. Second, specify or calculate S&V that operation of the equipment may cause, such as in laboratory, office, or other areas. The range of input force versus response can then be characterized and compared. For operating equipment, similar means are applied to identify threshold versus maximum levels. Attenuation using commercial mounts is possibly the simplest *hardware* method of reducing higher *g*'s to what is acceptable *at least cost.* Knowing the acceptable level at sensitive areas and comparing that against the levels generated by the nonisolated equipment yields a reduction factor or T for effective isolation.

ISOLATION—ESSENTIAL PROPERTIES

Dynamic loads can vary over time and with operation of the equipment or machinery. The isolator must therefore exhibit (1) well-defined load deflection characteristics, (2) repeatable multiaxis stiffness and damping, (3) absence of creep or set, (4) return to the centered position after load, and (5) resistance to conditions that can affect stiffness and damping. Other issues include load stability and verification of performance. In airborne and space applications, steady-state acceleration forces should also be considered due to flight maneuvers and launch loads. For example, a constant-stiffness, 5-Hz housed mount experiences 0.38 in. (0.97 cm) static deflection under a 1-*g* load. The same mount loaded to 4*g*, as in a rocket launch, would need 1.5 in. (3.81 cm) plus the expected vibration displacement for free movement. Many small-displacement vibration mounts have free-space capability less than 1.0 in. (2.54 cm) and thus would be ineffective under sustained acceleration loads as the mount snubs within its housing. Temperature extremes, aging, and immersion in fluids can affect material properties and cause degradation or change in stiffness or damping and should be taken into account in calculating response. In general, undesirable changes are a frequency shift of more than 10 percent or change in the amount of damping by more than 15 percent from published values. Matched sets of isolators are often required for precise position control such as aircraft gyro stability. Molded elastomer isolators can exhibit 10 to 15 percent variation in stiffness from among the same production lot. Metal mounts (axial steel spring) exhibit 1 to 2 percent variation. Cable metal isolators typically range from 10 to 15 percent, depending on the wire rope construction. Equations that focus on stiffness and damping are often incomplete because the coupling characteristics within the isolator from one direction to another are uncertain. Empirically derived values and test verification are important in mount design.

How the isolator deforms under load generally falls into one or more of the types shown in Fig. 39.4. An isolator will deform in one or more of several ways depending on the properties, the orientation of the mount, and the direction of applied load. Compression moves the upper surface toward the base. Shear and roll shift the upper surface laterally. Directions are defined with respect to the major axes of the mount. In some cases, loads may be applied in combined directions. Every isolator can be described by a set of unique load deflection (LD) stiffness curves. A particular model may have load ratings and LD curves that are similar in form for the entire isolator

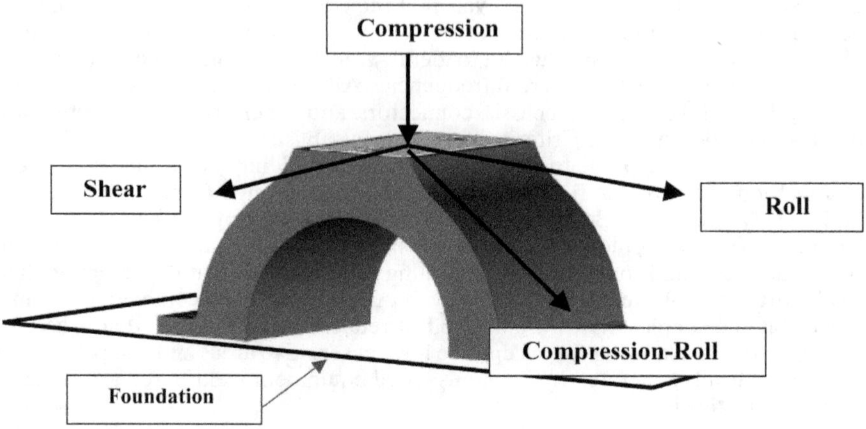

FIGURE 39.4 An isolator under load will deform in one or more of several ways, depending on the orientation and the direction of the applied load.

series. The allowable load depends on the construction of the isolator, its design, and the strength of the resilient material and also on the stiffness and strain capacity of the mount. Mounts should be derated in the case of sustained vibration or a large number of stress reversal cycles. Refer to published catalogs for specific data. Manufacturers' stiffness data is often based on static load measurements made on a pair of mounts in a constrained direction (roll or shear) and on a single mount in compression.

Analytical techniques for calculating resonant frequencies and response of isolated systems require values for damping and stiffness properties of the candidate isolators. Details of analysis are covered in Chap. 38. Manufacturers' catalogs and design guidelines that describe applications can be helpful, too. In calculations, the basic approach is to consider the unit as a rigid mass, and the entire system is then characterized as a six-degree-of-freedom rigid body on mounts having three translation and three rotational modes. Objectives are usually to select and arrange the isolators in such a way that rotational responses are minimized and the translation response in the direction of interest at the critical frequencies of the unit is substantially reduced. A simple relationship for the natural frequency f_n of a single-degree-of-freedom (SDOF) system is

$$f_n = 3.13\sqrt{k/W} \qquad \text{(without damping)} \tag{39.1}$$

where k = stiffness, lb/in.
W = weight, lb

Weight and center of gravity (CG) are known from drawings and design information about the unit. Values for damping and stiffness are properties of the isolators and are generally available from data sheets. Consideration of how the properties were measured by the isolator manufacturer is important. There are few standardized test methods among commercial companies. It is here that the user's engineering criteria and test experience are useful in the final selection of isolators. Several issues should be closely defined in order to establish the operating parameters of the isolators.

Damping. Estimate the percentage of critical damping c/c_c. Is damping constant over the entire temperature range in which the isolator is expected to operate? Is the damping axis sensitive? Is the damping dependent on the amount of relative displacement of the isolator?

Stiffness. Examine the load versus deflection characteristics in each direction. Are the stiffness curves relatively linear or nonlinear and are the LD curves stable, meaning that the mount exhibits minimal change in stiffness between the first, second, and successive load cycles? If nonlinear, list k values for small-amplitude versus large-amplitude motion of the mount.

Orientation. Measured test data for isolators is usually obtained with mounts constrained to move in one axis only. There is no allowance for simultaneous free motion in vertical and lateral directions. For example, in actual use, isolators can undergo shear and compression to an oblique load. Does load in one direction affect the stiffness in another direction?

Stability. Due to simultaneous compression and lateral movement of the mount, combined loads may cause buckling sooner than in a single-direction test. By what percentage should the allowable vertical load be reduced?

SELECTION AND DESIGN

The designer's first decision is to determine whether isolation is needed. In some cases, the equipment is rugged enough to withstand shock and vibration without isolator protection. When isolation is necessary, it is important to establish the general layout of the isolation system, the available space, and the location of the mounts. This will, in turn, define the allowable static height, width, and depth of the unit isolators. Allowance should then be made for extension in all directions and sway space of the equipment. Also, make sure that the mounts would not be short-circuited by nearby support brackets or restraint from semirigid electrical cables. Having set these limits, analyses should be done to validate isolation system layout and verify that the candidate isolators can support the load under all conditions within available space. Determine the isolator's stiffness and damping characteristics. These are important in considering vibration and shock isolation effectiveness, and whether the mounts have sufficient control to limit deflection and meet requirements. The amount of static deflection should be calculated. This helps in visually determining how well the system is balanced and whether it is properly carrying the entire load. Measurements and leveling may be needed to verify static balance. Major components of an isolated system typically involve the mass, the unit isolators, and an isolation platform.

GENERAL GUIDELINES—PART 1

1. Establish the center of gravity and overall mass (equipment), and locate possible attachment places for the isolators. Establish the allowable shock and vibration levels that the unit can withstand with a reasonable margin of safety. The isolators can usually be attached between the unit and the support frame (or foundation). Refer to Table 39.3. Minimize rotation of the isolated unit. This can be achieved when the center of mass coincides with the center of dynamic stiffness and the

TABLE 39.3 Vibration Isolation Characteristics

	Overhead support	Side support	Inclined support	Base mounted	Base and side or rear
Equipment types	Displays	Packaged	Packaged	Universal	Universal
Arrangement	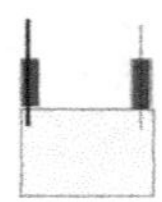	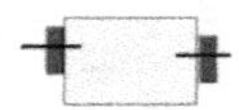		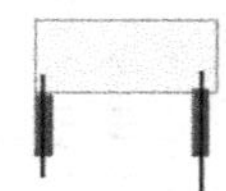	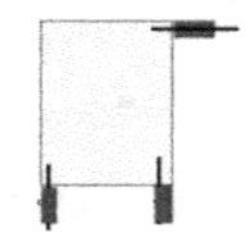
Feasibility range	Useful for tight spaces and vertical vibrations; sway is possible	CG support; check for stiffness in each direction; often used in electronics	May require more space for isolators; useful for shock isolation	Commonly used where moderate CG requires no stabilizers; sway is possible	For equipment isolation where the CG is high and coupling is a concern
Dynamic coupling	Horizontal motions possible	Minimal coupling	Negligible	Horizontal motions possible	Small pitch and sway can be set
Rattle space needed	Small, under 1 in.	Small, under 1 in.	Moderate, under 1.5 in.	Small, under 1 in.	Moderate, under 1.5 in.
Dynamic stability	Not critical	Not critical	May be critical with very soft isolators	Not critical	Can be critical with very soft isolators
Static stability	Not critical above 4 Hz	Not critical above 4 Hz	Not critical above 4 Hz	Not critical above 4 Hz	Not critical above 4 Hz

	Shock isolation characteristics				
Feasibility range	Useful for tight spaces and vertical shock; sway is possible if large deflection	CG support, check for stiffness in each direction; often used in electronics; limited space	May require more space for isolators; useful for shock isolation	Commonly used where moderate CG requires no stabilizers; sway is likely	For equipment isolation where the CG is high and coupling is a concern
Dynamic coupling	Horizontal motions possible	Moderate coupling	Small coupling can be set	Horizontal motions possible	Small pitch and sway can be set
Rattle space needed	Moderate, under 2 in.	Moderate, under 2 in.	Moderate, under 2.5 in.	Moderate, under 3 in.	Large, under 5 in.
Dynamic stability	Can be critical	Can be critical	Will be critical with very soft isolators;	Can be critical	Will be critical with very soft isolators;
	Check sway space	Check sway space	Check sway space	Check sway space	Check sway space
Static stability	Not critical above 5 Hz	Not critical above 5 Hz	Not critical above 5 Hz	Not critical above 5 Hz	Not critical above 5 Hz

Remarks: For stability, shock response frequency of the mount should be kept above 5 Hz.

system is statically and dynamically stable. The loads should be distributed to the mounts so that deflection is nearly the same at each isolator in the principal directions of motion. For example, static deflections should be nearly equal in the vertical direction if vibration is especially severe in this direction. Consider candidate mounts (type, size, dimensions), and group the isolators to establish the size of the platform and the number of isolators. Mounts can be represented as damped springs in the X, Y, Z axes.

2. Position the base isolators so that they balance the load vertically. The isolators should be in line with the unit's structural frame. In some cases, an intermediate baseplate can be used to carry the load from the unit through its structural members and into the isolators. Mounts should be secured to the foundation. They can be secured to a separate plate that is removable. In this way the mounts can be replaced at a later time for maintenance or equipment changes. The isolators carry load and should be considered a part of the equipment's structural design. Stabilizers (if required) should be attached to a rigid outer structure such as a wall, columns, or overhead beams. They can experience different levels of vibration or shock than the base mounts that are attached to a foundation.
3. Define the input S&V at each isolator group. Check the stiffness of the unit and platform to ensure that they are sufficiently rigid and resonate at a frequency well above that of the isolation system.
4. Consider alternative designs—for example, base mounts only, base and stabilizer isolators, or isolators and external dampers. Perform dynamic analysis for each design at the lower and upper stiffness and damping ranges of candidate isolators. Reposition the mounts if necessary, use readily available mounts, and avoid custom or specially engineered mounts in the preliminary design.
5. Restrain pitching and sway motion of a tall unit (one whose height is more than 1.5 times its narrow width) with stabilizing mounts near its top. The stabilizers can be mounted behind or above the tall unit and should be oriented to minimize free motion of the top outer corners of the unit. For a shorter unit whose width and depth are approximately equal to the height of the CG, stabilizers can be placed in the plane of the CG or slightly above the CG. Stabilizers can be avoided for a unit whose width and depth are 1.5 times greater than the height. In this case, base isolators may be sufficient, provided that the lateral stiffness of the mounts is at least 0.5 times the vertical stiffness. Verify the stiffness values of the isolator in its principal directions. If necessary, adjust stiffness data so that it correctly characterizes the expected movement of the isolators.

GENERAL GUIDELINES—PART 2

The design of isolation systems can be simplified by using the simplest and fewest number of proven mounts that can carry the load in the available space. Because conditions may not be well defined, isolation can require broad frequency response, while the ideal design would be tuned to a narrow band of frequencies. There should be only minimal shift of performance if the isolated weight changes or unbalanced loads operate at variable speeds or disturbances vary with time. The designer should evaluate several candidate designs before making a final selection of mounts. The stiffness and location of the isolators controls the isolated frequencies of the unit and its stability and motion. The number, location, and orientations of the isolators often receive only modest attention in the preliminary stage. Equipment and position can change. Isolator layout is designed to decouple between translations and rotations of the isolated unit and simplify the analysis. This is sometimes referred to as the *generic* system approach. The balanced design—isolators sharing equal load and the dynamic center of gravity of the isolation system reasonably close to the isolated mass CG—has been proven in use and verified by analysis.

In the usual arrangement, isolators are in parallel and supporting the unit in equal proportion of distributed weight to stiffness of the mount at each location. Mounts are sometimes used in series for greater deflection and lower spring rate. A variety of mounting versions is shown in Fig. 39.5. Multiaxis designs include base and stabilizer mounts with stiffness matched to total load in each direction. Dual isolators, in parallel and in series, have been used in shock control. However, difficulties in properly

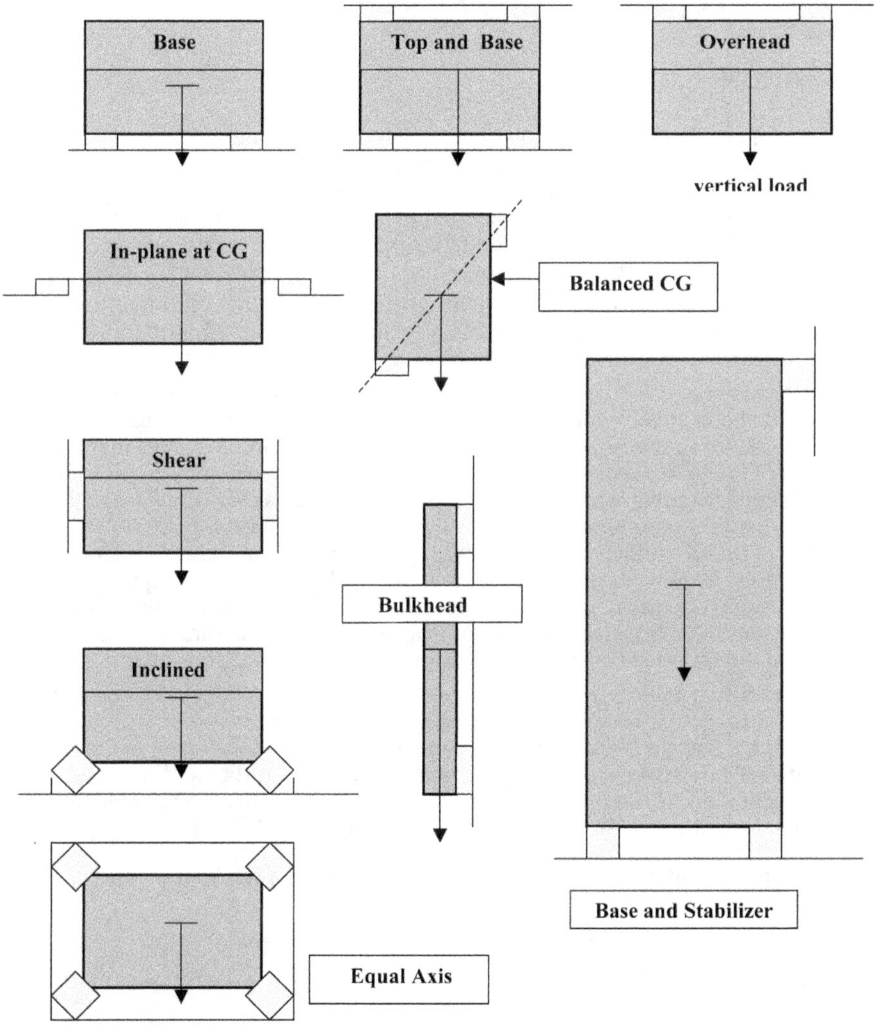

FIGURE 39.5 Variety of isolation mounting arrangements. Some isolators have preferred directions. Stiffness may be axis-dependent. Parallel and series mounting is used; parallel is the most common type.

matching dynamic stiffness rates from among isolators of different types can result in poor shock attenuation. Known as *two-stage snubbing* designs, they are *not* recommended without extensive shock testing. Snubbing effects are *not* predictable, and high *g*'s can result. Commercial mounts are available that are intended for vibration control in one set and then to dissipate shock in a parallel set of isolators that are initiated once beyond the vibration region. The stiffness of mounts in parallel is the sum $k_t = \sum k_n$. In a series arrangement, it is $1/k_t = \sum 1/k_n$. When isolators are inclined, the stiffness in the vertical and horizontal directions k_v and k_h is a function of the angle of the isolators with respect to the force direction and should be measured by test.

STEPS FOR SELECTING ISOLATORS

VIBRATION

1. Determine the static load that each isolator in the mounting system supports. Simplest is if the total load is equally divided among the number of isolators and the mounts are uniformly positioned to support the mass at several places. The total stiffness is the combined sum of individual stiffness for each mount. Review manufacturers' documentation and catalogs for complete descriptions of candidate isolator properties and performance in similar applications.
2. Knowing or estimating the sensitivity of the equipment, determine the two to three lowest critical frequencies of the unit. Substantial vibration at these frequencies often contributes to damage or unsatisfactory operating performance.
3. Determine the effectiveness of the isolation system needed to reduce the vibration at the critical frequencies to acceptable levels. Set an adequate margin of safety. Refer to the isolation effectiveness chart in Chap. 38. A reasonable guideline is 75 to 85 percent isolation. Check also that secondary vibrations above the isolators are within acceptable limits. From the published stiffness data and the static load on the mounts, calculate the natural frequency of the isolation/mass system. Based on the ratio of driving frequency versus natural frequency, determine the percentage of isolation at the critical or resonant frequencies of the equipment. Use stiffness values from the load deflection curves corresponding to the orientation in which the isolator is used. Check the natural frequency in the lateral directions for vibration effectiveness in those directions, too.
4. Isolators that meet requirements will have a natural frequency approximately one-fourth of the first critical frequency or predominant resonant frequency of the unit. Select an isolator having 10 to 20 percent of critical damping for motion control. Elastomers should have a dynamic-to-static stiffness ratio of 1.2 to 3.0, depending on the material. Refer to the isolator data sheets. For cable and steel spring isolation, the dynamic-to-static stiffness ratio is 1.0 to 1.1.

Examples of analysis for vibration isolation are given in Chap. 38 and in Ref. 1, Chap. 32.

SHOCK

1. As in the vibration selection, determine the static load each mount supports. Estimate the fragility of the unit in terms of the *g*'s that can cause damage or what is

believed to be the maximum stress level that the unit can withstand before damage occurs. For multiple shocks, the allowable shock on a unit should be derated by a factor of at least 1.2 to 1.3. For example, for COTS electronics, a 12- to 15-*g* shock appears to be an acceptable criterion. "Hardened" electronics can meet 25 to 30*g*. Machinery such as pumps and compressors are often limited to 25 to 35*g*. Equipment can often withstand greater shock loads than static loads due to the ductility of the components and structural parts and the fact that the stresses may not be distributed throughout the unit as they are in static conditions.

2. Calculate the shock response frequency of the isolation system and compare it to the predominant frequency of the shock pulse. From the dynamic load factor curves, select the appropriate shock pulse and verify that there is sufficient frequency separation that load coupling is not a factor. Refer to Fig. 39.5 and Chap. 38 for typical DLF curves.
3. Select the mount whose shock transmissibility is less than 0.4 to 0.5 at the specified pulse duration and shock type, such as 11-ms half-sine, triangular, or ramp. Check that the stroke capability of the isolator is at least 25 percent greater than the calculated deflection.

Shock response is dependent on the type of shock pulse (half-sine, triangular, irregular) and how near the isolation frequency is to the shock frequency. Known as *coupling analysis,* the ratio of pulse duration (in terms of frequency) to effective shock response frequency plus the amount of damping determines the expected response. When selecting isolators, particularly nonlinear mounts, it is important to evaluate the effective shock response frequency of the system in all directions, as well as its damping. Limiting motion in the principal shock direction, the damping ratio is a coupling term in modal equations and can be a factor in increased acceleration in cross-axis response. Because shock in one direction can also result in motion in other directions, verify that allowable sway space all around the unit is not exceeded.

Examples of analysis for shock isolation are given in Chap. 38 and in Ref. 1, Chap. 32.

ATTACHING AND LOCATING ISOLATORS

Isolators can be directly attached to equipment if the unit is sufficiently rigid that the mounts and unit are an integral assembly and provisions for attachment exist. However, if the geometry is complex and the mass unevenly distributed, it may be necessary to use an intermediate frame (or platform) to position the isolators. Correctly designed, a frame also simplifies installation and corrects for unbalance or misalignment of the unit and improperly positioned mounts. Improper placement will contribute to pitch and sway of a unit and can increase its peak accelerations and displacement. The addition of substantial mass to the frame (the frame then becomes an inertia base) decreases the acceleration response at the higher frequencies above resonance. The unit itself may not be a rigid mass. There are examples where, on the same unit such as a missile engine, widely spaced large masses went into different modes at closely coupled frequencies and the measured accelerations were substantially more than expected. Forward isolators responded at a different frequency than rear isolators due to the large masses reacting independently of one another.

A poorly designed frame may lack adequate stiffness and couple with the isolation frequency. This can broaden the frequency range at which amplification occurs. Chapter 38 notes that the frequency ratio of the frame to isolation should be at least 5 to 1 for adequate separation of the two resonant modes. For example, that would require a frame assembly at 35 Hz if the isolation system were at 7 Hz. Other literature refers to a 10-to-1 separation, which is ideal but may not be possible to attain. The frequency response of any isolated mass should be checked for coupling. For example, electronic racks are commonly mounted with base and stabilizer isolators. Many populated racks have resonant frequencies in bending at 15 to 17 Hz, which could couple with the lateral resonance of the isolation system at 7 to 8 Hz. Shock response could also be affected because the shock above the isolators might then be amplified. Good design practice is to secure the frame and unit together along the length and depth of the unit. This enables the isolation frame to stiffen and strengthen the unit.

In selecting isolators, the foundation is usually taken as rigid and a simple mass model can be adequate. However, if the equipment and/or foundation to which the isolators are attached is less stiff than assumed (and also poorly damped), the combined dynamics can contribute to the response spectrum of the system. T at equipment resonance can be amplified and/or shock frequency coupling will increase the peak acceleration response of the unit at critical frequencies. An example is 5-Hz isolated equipment installed on a 14-Hz deck on a ship. The same equipment could also be located on a 25-Hz deck elsewhere on the ship. Because the 25-Hz deck is much stiffer than the 14-Hz deck, the response factor of identical equipment would be different to the same level of vibration or shock. The dynamic load factor is relatively large (0.45) in the 14-Hz case, and only 0.25 for the stiffer deck. Similar examples can describe equipment on upper levels of a tower versus equipment at ground level.

Along with rigidity, a unitized frame at the isolators may also be needed to ensure nearly the same dynamic input at all isolator supports. T effects with foundation flexibility are shown in Fig. 39.6 versus f/f_n for a two-stage compound system having

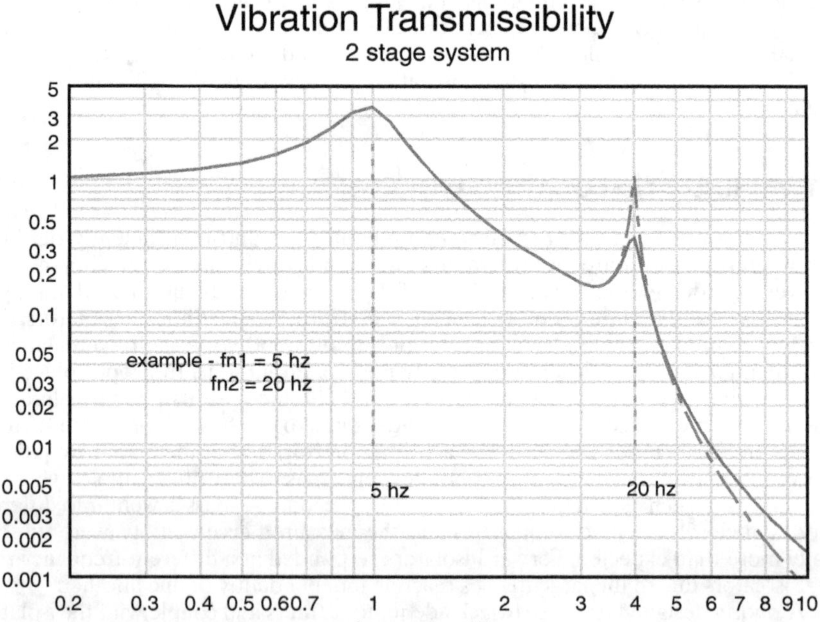

FIGURE 39.6 Example: The transmissibility plot of a 2-DOF system showing the first peak at isolation resonance (5 Hz) and a second peak at basically equipment resonance above the isolators (20 Hz). Peak amplitude varies with damping.

0.1 c/c_c and 0.05 c/c_c damping. Isolation is at 5 Hz and foundation at 20 Hz in this example. As the base frequency decreases, the second resonance shifts downward, closer to the isolation system, and the overall amplification region broadens. Above the isolators, similar broadening effects can occur if equipment resonance is at relatively low frequency and near the resonance of the isolation system. An isolated subframe can be effectively used to create a 2-DOF system with greater falloff at the higher frequencies.

If the unit is reasonably stiff and acts as a single mass, the isolation system can control the movement and acceleration response of the unit in all directions, presuming that it remains stable. The equipment, however, might be an assembly of light and heavy components, meaning that the lighter members could have greater motion due to the transmitted forces. Of particular concern in shock, relative motion at adjacent components can result in local impact and large contact stresses, which cause damage that is not readily apparent or cannot be examined until the part is removed.

ISOLATION MODEL

Shock and vibration analysis of most isolation systems can be done using a variety of commercial finite element analysis (FEA) programs. In these, the structural unit should be drawn dimensionally to represent the rigid mass and the isolators assigned stiffness values in their principal operating directions at each isolator location. For simplicity during preliminary design, the unit can be represented as a rigid structure with distributed concentrated masses; for example, the structure can represent an enclosure and the total weight including equipment is supported by the frame equally at the multiple locations. The combined center of gravity of the model should match the CG of the actual unit. Damping can be neglected. Each mount is assigned a linear stiffness value and fixed to ground at one end. The other end is attached to the unit at a node. The model must be restrained and stable. Each mass node is given X, Y, and Z degrees of freedom. For a dynamic problem, there are as many eigenvalues and eigenvectors as there are DOF in the model. Each eigenvalue is related to a natural frequency, and each eigenvector is related to the mode shape at that frequency. For the usual isolation problem, only the first few natural frequencies are important. For a well-balanced system, most interest will be in translation modes. Higher frequencies constitute structural response and are not of concern here, since the structure is not given much fidelity in the preliminary design. The focus is on the isolation frequencies and modes of the isolation system. Commonly used software in the United States includes Ansys, Algor, Cosmos, and NEI Nastran. Drawings can be imported directly from most computer-aided design (CAD) programs.

ERRORS IN MODELING

A model of the equipment and isolation system is useful to predict the effectiveness of the proposed isolation. This can take a form ranging from simple hand calculations to complex multi-degree-of-freedom calculations showing displacement, resonant frequencies, and modes. As with any model, there are issues that can lead to errors in analysis. Results should be checked with test data. Possible errors are as follows:

1. *Overly complicated or poor assumptions.* Has the model been checked with similar applications from the same manufacturer? Are the mounts stable at large deflection? What are the effects of combined loads? Where is the actual response to be measured versus predicted distribution of loads in the model?

2. *Assumption that the equipment is a homogeneous rigid mass.* Evaluate the largest individual masses. If supported by brackets or internal isolators, check that these do not act as secondary springs and couple with the primary isolation system.
3. *Differences in vibration (or shock) at one mount versus another.* An example is a pump mounted on a foundation where the discharge piping at the pump is attached with a semirigid tubular connection that should have been a flexible bellows connection. This semirigid coupling results in introducing vibrations into the pump via the piping system from other pumps on the line and changes the response characteristics of the unit.
4. *A wrong interpretation of the manufacturer's stiffness and damping data.* For example, are they first-, second-, or third-cycle measurements? Are they average values versus incremental values, and over what deflection range?
5. *Is the applied load axis sensitive?* Stiffness is based on free motion of the unit versus test data obtained under single-axis conditions. For example, mounts tested in the lab in shear may have been restrained from moving in any other direction. In actual use, lateral shear movement can also simultaneously compress the mount, and the effective shear stiffness would be affected (usually decreased).

ISOLATION AND POSSIBLE DAMAGE OF EQUIPMENT

The stiffness of the isolator should be matched to the expected level of shock and vibration. Very low level vibration may *not* excite the mount. It is important that the S&V force adequately load the isolator so that it is fully compliant over its operating range. For instance, in a lightly loaded mount there may be insufficient strain in the material to overcome internal friction, and the elasticity of the mount would then be considerably underrated. The mount would appear to be very stiff until load is applied and the threshold of force is exceeded. This can be seen in nonlinear load deflection curves and the apparent stiffening at a load point below the manufacturer's static rating. The slope of the hysteresis loop may be stiffer at very small amplitude of the cycle, becoming shallower (softer) as the amplitude of the loop increases.

Is the S&V amplitude sufficient to drive the mount and uniformly load its compliant elements? One guideline is to check the expected amplitude ratings and verify that the stiffness was measured at or near the expected amplitude. For example, an input of 0.5-g continuous sine vibration at 25 Hz was known to cause fatigue damage in welds. The isolators should therefore be rated below 5 Hz in order to be reasonably effective. Was the stiffness measured at an amplitude of 0.5g. Did the resonant frequency shift when tested at other levels?

Vibration damage often occurs when allowable fatigue strength is exceeded. This can be determined from Miner's equation, which relates the number of stress cycles and the amplitude of vibration to the number of cycles necessary to cause damage at each particular amplitude. Results are based on a range of measurements and tests including sine and random vibration. Even with the benefits of isolation, isolator damping values can have significant effects on fatigue of the isolator. In vibration, T and displacement of the mount is a function of f/f_n and damping ratio c/c_c. Comparing two similar isolators having equal stiffness, a simple relationship for time to failure in vibration is $t_1/t_2 = (1 - \zeta_1/\zeta_2)^b$, where t represents time to failure, ζ designates c/c_c, and b is a material factor.

Shock damage to equipment often depends on the magnitude of peak acceleration and time duration (effective velocity) and the number of shocks. Damage due to overstress can occur because of a single impulse or the result of a relatively small number of shocks. Extensive shock testing has been done to quantify the damage potential to equipment from the relationship of peak amplitude, pulse shape, and equivalent drop velocity. Damage boundary curves have been developed for a variety of equipment, and controlled impact tests have been designed. Standard test procedures are being developed to relate drop height and impact surface to shock response spectra in order to characterize shock in terms of allowable SRS for different classes and types of equipment. Several methods use controlled drop machines to establish an acceleration and velocity boundary damage profile of equipment.

H. A. Gaberson, in Ref. 2, describes a series of shock tests indicating that damage to one class of fans was related to the velocity imposed during shock. Regardless of the actual peak *g*'s in shock, damage to the fans was caused when the velocity was greatest due to the combination of pulse duration, shape of pulse, and duration of shock. It is inferred that high velocity was the major contributor to shock-induced damage. There are thought to be several main factors: (1) equipment and structure vibrate in mode shapes at their modal natural frequencies, (2) damage occurs when stress exceeds strength, (3) peak modal velocity is proportional to peak stress, (4) there are absolute limits to modal velocity that equipment can survive, (5) the unit accepts shock energy only at its modal frequencies, (6) isolation enables maximum pseudo-velocities to occur below the lowest mode of the equipment at levels less than that without isolation, and (7) pseudo-velocities more than 100 ips (254 cm/sec) are dangerous to the equipment and should be avoided.

SRS plots on four-coordinate paper show the frequency range where the shock can cause high stress and also show the peak shock deflection and the peak acceleration. These plots can also show reduction through isolation and the shift of the pseudo-velocity range that results. Velocities can be lower because the isolation effectively shifts the response to the lower part of the frequencies and decreases the acceleration amplitude at the higher frequencies. Most stress failure theories indicate that multiaxis stresses can be more severe than single-axis stress and unit strength should be derated when combined loads are involved.

Example 39.1. High velocities (associated with large peak *g*'s in the 25- to 50-Hz region) were known to be major contributors to damage of certain types of PC boards in electronics. Shifting the peaks by means of an isolation system (lower frequency in the SRS) reduced the pseudo-velocity at the unit's critical frequencies. Displacement was reduced, and bending stress associated with damage was decreased.

Example 39.2. The SRS of a typical shock input to hard-mounted equipment during the MIL S 901D Heavy Weight barge test is shown in Fig. 39.7. Relatively high velocities are evident in the range of 14 to 100 Hz. For the purposes of analysis, damping was set at 0.05 c/c_c. Other values can be used in analysis; however, the general shape of the SRS curve would be the same but it would have different amplitudes. If the first mode of the equipment was 25 Hz, the velocity value V in this case would be approximately 120 in./sec (304.8 cm/sec). For a component whose allowable stress was based on commercial loads, failure would probably have occurred. By comparison, Fig. 39.7 shows the SRS of the isolated unit with its isolation centered at 7 Hz. At 25 Hz (isolated equipment), the velocity is approximately 36 in./sec (914.4 cm/sec) versus 120 in./sec (304.8 cm/sec), hard mounted. Based on stress as a function of kinetic energy ($\frac{1}{2} mV^2$) the change is then dependent on the ratio of velocities $(v_1/v_2)^2$, meaning that the critical stress in this example has been substantially reduced.

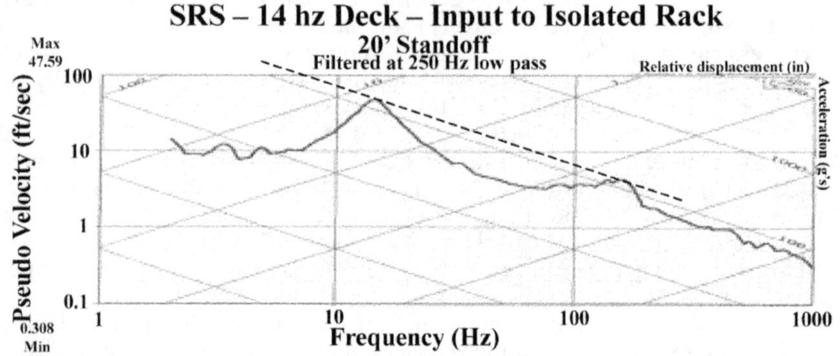

(A)

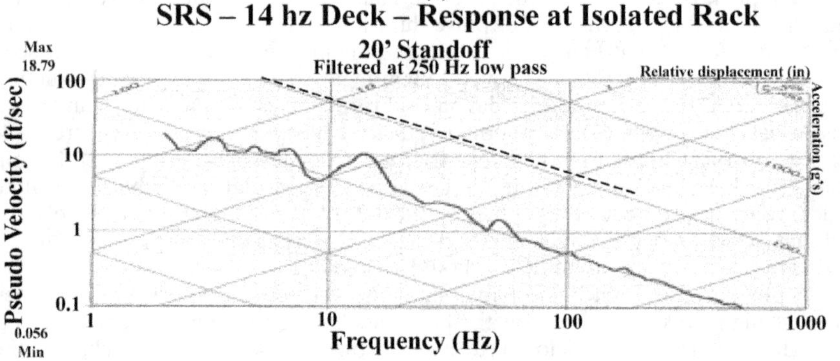

(B)

FIGURE 39.7 (*A*) The SRS plot of a shock at the input side to the isolation/mass system. The foundation was characterized as a 14-Hz deck. Relatively high velocities are evident in the range of 14 to 100 Hz. (*B*) The SRS response plot of the isolated unit shows that the velocities over the same frequency range of 14 to 100 Hz are considerably reduced from the input.

SHOCK RESPONSE SPECTRA (SRS)

The shock response spectrum is based on mathematical analysis that describes the pulse in terms of the calculated response of multiple SDOF systems having specified damping ratios and gives the maximum acceleration, velocity, and displacement of each spring system. A candidate isolation system can be considered as being represented by one or more of the lower-frequency spring modes. The results assume linear spring systems and superposition are therefore permissible when combining the peaks of the isolation system and equipment modes *so long as they are linear.* The results would have to be modified in the case of nonlinear mounts. Experience has shown that the greatest portion of response occurs in one or two modes (usually the predominant modes of the isolation system in the shock direction) and that, in general, not more than two to three modes contribute to damage of the equipment. Modeling isolation from among the SRS lower frequencies can simplify separating the first few modes so that the peaks can be clearly shown. Adding the peaks of individual modes can be done according to analytical methods that are generally conservative. An effective isolation system would display an increase in acceleration and velocity at the low-frequency end of the SRS plot (compared to a hard-mounted system) but substantially less acceleration and velocity at the higher frequencies in the region band where component or part deficiencies such as poor solder joints, insert pullout, fracture, or other mechanical damage often occur.

UNIVERSE OF ISOLATORS

There are literally many thousands of commercial mounts manufactured and sold to protect equipment from shock and vibration. Isolators are produced in a variety of types, sizes, and load ratings to meet standards and specifications for specific and general use. Specific use, for example, may involve all-terrain and military vehicles; general use includes entertainment centers and residential comfort equipment. In addition to standard designs, custom isolators are continually being developed for special applications. Some sense of the many different mounts can be obtained from Fig. 39.8 (presentation photos from several manufacturers), showing some of the different configurations and designs available. Passive isolators are available worldwide in many designs and resilient materials, sometimes with the addition of secondary limiters for snubbing purposes. In the United States today, there are well

FIGURE 39.8 Elastomer and cable mounts are produced in different configurations and designs from many manufacturers. Other types include pneumatic and steel-spring isolators. Composites and molded pads are also used for isolation.

over 100 elastomer isolator manufacturers, each offering a range of models in a variety of synthetic compounds and natural rubber. These are molded products or products with the elastomer elements captured and retained in a metal housing. Other companies specialize in steel-spring, pneumatic, helical cable, and other non-molded isolators as well as composites using elliptical steel leaf springs for stiffness and polymers for damping.

Manufacturer catalogs usually provide a description of the isolator type, construction details, load ratings, and applications in which the mount is commonly used. In general, heavy-duty versions of particular isolators are intended for industrial and mobile applications; smaller or lighter designs are for commercial and residential installations. Isolator types are compared in Table 39.4, including characteristics, features, and applications.

Given the extensive variety and type of isolators manufactured and sold throughout the world, it is no surprise that there is little standardization of commercial mounts. Dimensions, hole patterns, materials of construction, and load capacity may be different, even among similar-looking mounts. Applications, markets, and usage have led to preferred types of mounts within a group of industries. For example, heating, ventilation, and air-conditioning (HVAC) facilities typically use axial vertical steel-spring mounts with restraint for compressors, air handlers, pumps, and fan coil units. Piping is often supported with open steel or rubber springs. Seismic design requires restraint to capture and secure secondary units such as cooling towers in the event of large displacement.

Other applications such as stamping machinery and punch presses use large-stroke steel springs for shock reduction. Heavy mobile equipment is frequently installed on preloaded laterally restrained steel springs and/or cuplike elastomer mounts for multiaxis S&V protection. Spring-supported concrete inertia pads or steel frames are often used with stationary rotating equipment. Research facilities may require very low frequency microamplitude mounts. In many cases, it can be difficult to substitute one supplier's product for another, due to differences in mounting design and load ratings. Stiffness/damping values are usually based on testing to the manufacturer's procedures, which are not necessarily standardized industry methods; thus, there can be differences in results when comparing catalog values among different suppliers. Properties at temperature extremes, for example, are materials dependent, and the supplier should be contacted for the best available information. Likewise, due to the complexity of the resilient materials, particularly elastomers and other polymers, and their interacting effects from one direction to another, the relationships may not be fully described in catalogs but can usually be provided by the manufacturer. Comparison of mounts *only* on the basis of dimensional and visual similarity and published catalog data can lead to major differences in results.

ISOLATOR TYPES AND CHARACTERISTICS

Isolator types (see Table 39.4) from among widely used commercial mounts are available from many suppliers. Versions of each type of mount ranging in compression stiffness k, stiffness ratio k vertical/k lateral, and damping ratio c/c_c are usually listed in suppliers' data sheets. General-use categories are (1) military and aerospace electronics; (2) industrial, HVAC, manufacturing, and construction machinery; (3) engine mounts for off-road, marine, and flight vehicles; (4) precision isolation; (5) engineered and custom facilities; (6) large structures and mobile equipment; and (7) seismic applications. To achieve greater damping without change of stiffness, materials have been merged in design—for example, metal mesh with elastomers, polymer-coated axial springs, staged damping layers, and elastomer-impregnated cable mounts.

ELASTOMER ISOLATORS

Spring stiffness and damping are fundamental properties of all shock and vibration isolators. Most elastomers are usable because of their low modulus of elasticity and ability to undergo large strains (greater than 100 percent reversibly) in compact designs. Shear modes are favored. Rubber is essentially incompressible, and allowance is made for deformation in other directions. Isolator shape factor is important. Shear modulus is relatively constant for load and configuration but increases with an increase in hardness of the elastomer.

The newer-generation elastomer mounts have benefited from several factors. First is the greater availability of compounded elastomer materials including silicone and neoprene, better technical data and testing, and improved analytical modeling using finite element analysis. Second, the geometry of the part, including the use of composite materials, has enabled greater energy capacity in smaller volumes than earlier designs. Third, molding techniques have been refined, with greater awareness of stresses produced during manufacturing and the ability to more carefully regulate heat transfer throughout the mount, including the strength of adhesives for bonding the rubber to metal plates. The size, shape, thickness, and dimensions of the molded part are designed for various load ranges. Full use of geometrical shape in flexure contributes to strain-induced stiffness. Changing elastomer materials or the ratio of constituents in the compound can vary the stiffness of the mount. Damping can also be affected. Durometer hardness change by use of fillers is a means of increasing both stiffness and damping simultaneously. Two-stage progressive designs involving nonlinear stiffness have been used for vibration isolation in the presence of intermittent shocks.

The energy capacity of a mount is a function of its elasticity, deflection capability, and internal energy dissipation. For example, for an isolator to absorb the energy of a mass dropped from a height, the kinetic energy of the falling mass must be less than the total strain energy capacity of the mount as it deforms to its maximum stroke. If not, the mount "bottoms" and very high *g*'s result. In vibration, resonance will produce heat due to hysteresis in the mount (damping); the isolator will soften and the frequency shifts downward, resulting in greater amplitude and eventually leading to fatigue failure. Stopping a vibration resonance test to allow for the mount to cool down is not uncommon. It has been noted that although capable of 200 percent to 300 percent elongation, some very low frequency mounts (under 4 Hz) can become unstable, resulting in *set* under large load. Standardized tests measure set after the load has been removed and *creep* of the mount under constant force. Tests to evaluate flame resistance, immersion in fluids, and other environmental conditions are also done.

Silicone is widely used for its capabilities to meet extreme temperatures ranging from –50 to 250°F (–10 to 157°C). Buna N and neoprene are lower-cost materials but are more limited to wide temperatures. Hydrocarbons and ozone resistance are better than with natural rubber. Materials literature is extensive and covers elastomer properties in detail. Mechanical properties of rubber are given in Ref. 1, Chap. 33. Compounds of proprietary formulation are used for designated properties such as greater damping ratio or modest stiffness change with temperature or aging. Means to moderate the stiffness of rubber include adding metal or composite plates in a layered arrangement in the molding process. Shear stiffness is increased by

TABLE 39.4 Comparison of Isolator Types

Isolator type	Material	Reliability	Linearity	Principal use	Operating method	Features	Issues
High arch deflection	Elastomer	High	Buckling in compression, linear and less stiff in roll and shear	Shock and vibration	Passive	Rugged design, large deflection in all directions, wide load range	Operates as a stiffer mount in vibration in the compression direction
All-attitude cup mount	Elastomer	High	Linear stiffness greater in compression, equal and less in lateral directions	Vibration, occasional shock	Passive	Rugged design, versatile orientation	Can snub in shock
Center plate mount	Elastomer	High	Linear, nearly same in all directions	Vibration, all directions	Passive	Low profile	Deflection limited
Fluidic elastomer	Elastomer, gel filled	Moderate	Linear, stiffer in compression	Vibration, rough road	Passive	Very good isolation in random vibration	Not designed for torsion loads
Helical cable	Preformed wire rope, stainless steel	High	Buckling in compression, linear and less stiff in roll and shear	Shock and vibration	Passive	Rugged design, large deflection in all directions, wide load range	Operates as a stiffer mount in vibration in the compression direction

Axial steel spring	Steel	High	Linear in vertical direction, can be unstable in lateral directions	Industrial compressive loads, machinery	Passive	Simple, highly predictable	Deflection limited, can require very tall mounts for shock, minimal damping
Pneumatic	Elastomer	Moderate to high	Linear, intended for low-frequency isolation	Isolation tables	Passive, can be active with servo controls	Very effective for low-level, low-frequency vibrations	May leak air
Seismic	Elastomer	High	Linear stiffness, low in lateral direction	Earthquake	Passive	Large lateral deflection	Engineered application
	Elastomer and lead plug	High	Linear stiffness, low in lateral direction after lead plug shears	Earthquake	Passive	Large lateral deflection	Engineered application
	Friction pendulum	High	Linear stiffness, low in lateral direction	Earthquake	Passive	Large lateral deflection	Engineered application

means of boundary restraint at each plate or by changing the angle of inclination between vertical and horizontal with respect to the applied load. Methods of increasing compression stiffness are to restrain lateral bulge deformation and outward expansion. A typical ratio of compression to shear stiffness is 3:1. Tapered geometry, insertion plates, and ribbed metal construction can reduce the ratio to nearly 1:1.

Isolators used in sustained vibration should have a generous bend radius at the metal-to-elastomer interface and at corners, to reduce internal stresses. Deterioration of rubber bond strength can also be a concern. Attention should also be given to the length of attachment fasteners that could extend into the body of the mount if there are large relative deflections. Due to heat buildup, fatigue failure often begins internal to the mount (where it can't be seen until the tear propagates to the outside wall). Precompression of the mount can be used to limit displacement into a large stress region.

Compounds of natural rubber, neoprene, silicone, and butyl are frequently used in elastomer isolators. Other materials include butadiene, nitrile, and propylene. Black rubber is often used as filler. Elastomers can be molded in a variety of sizes and shapes to fit spaces for which other types, such as shaped metals or composites, require special machining. Injection and insertion of materials under high pressure are common production molding methods. Spring equations for several simply bonded rubber mounts are given in Ref. 3. For the same model, isolator stiffness and load range are usually controlled by means of durometer hardness (40 to 75 Shore A scale) of the basic compound. Damping can also be changed, depending on the type of elastomer. For instance, natural rubber is typically rated at 2 to 3 percent of critical damping. Silicone and certain proprietary neoprene-based compounds are rated at 10 to 15 percent c/c_c.

The ratio of dynamic stiffness to static stiffness generally exceeds 1.2, depending on the compound. Allowable creep is 2 to 3 percent of the original height for most designs and materials under load. The temperature range for a stiffness variation of 20 to 25 percent from nominal ratings for silicone-based compounds is –50 to 250°F (–10 to 157°C); the range for short-term exposure is greater. Other materials have narrower operating temperatures over the 20 to 25 percent stiffness variation range. Some elastomers may creep excessively over time. Continuous strains should not exceed 10 to 15 percent in compression and tension or 25 to 50 percent in shear. Tensile strengths of many compounded isolators are in the range of 2000 to 3500 lb/in^2; extremes are silicone at 800 lb/in^2 and urethane at 8000 lb/in^2. The modulus of elasticity for rubber is not constant. There is no precise elastomer yield point until material failure occurs in shear and tension.

Elastomer mounts exhibit basically linear stiffness; however, buckling deformation in high-deflection arch-shaped isolators produces nonlinearity due to the way in which the elastomer column deforms under axial load. The column (thick rubber wall) initially compresses, then buckles under load (effective stiffness transition region), and the body of the mount moves laterally (predominantly outward in shear) until the walls are fully compressed and snubbing occurs. Damping is the difference between deforming work and elastic recovery. There is usually full recovery, and the mount returns to its original position after the load has been removed. The percentage of damping increases with increased rubber hardness; for example, natural rubber increases from 6 to 30 percent, meaning that 30 percent of the total energy impressed on the mount is absorbed in one cycle. With butadiene types, damping values are higher for soft durometers but nearly the same for harder durometers. Highly damped rubber can exhibit compression set, resulting in residual deformation following the removal of severe loads.

Within the range of isolation mount frequencies (typically 5 to 25 Hz), the dynamic stiffness is independent of frequency but varies with durometer and compound. The relationship of dynamic to static stiffness is

$$k \text{ dynamic} = n * k \text{ static} \tag{39.2}$$

where n denotes a dynamic stiffness correction factor.

As a guide, n is typically in the range from 1.2 to 3.0 for 40- to 80-durometer. Specific values are determined experimentally. Dynamic measurements are usually made using an electrodynamic shaker and isolated mass to determine resonant frequency in a linear direction. The results are then compared to the natural frequency calculated from static stiffness determined from the tangency to load deflection curves at slowly varied loads in the same direction. The relationship to damping is shown in Fig. 39.9, where the slope of the hysteresis curve is the k value. Addition-

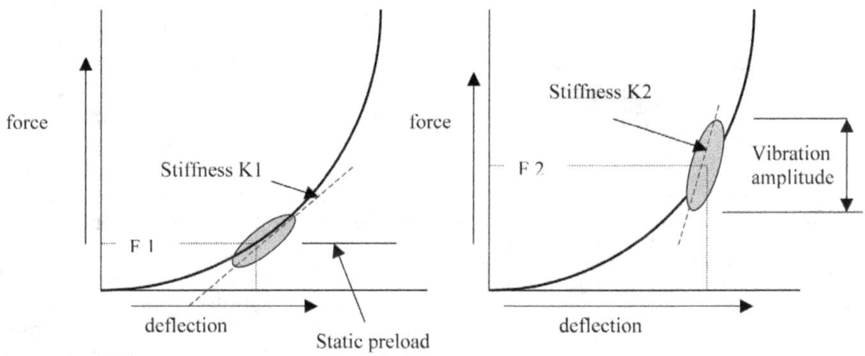

FIGURE 39.9 Stiffness and damping—the slope of the hysteresis loop (stiffness) changes for a nonlinear elastomer isolator depending on whether the motion is small or large amplitude, and the amount of damping. The stiffness line is drawn through the endpoints of the hysteresis loop.

ally, the effective slope of the hysteresis loop changes for a nonlinear elastomer isolator depending on whether the motion is small- or large-amplitude and whether damping is axis-dependent. The slope of the hysteresis loop taken to the expected maximum deflection is also an indicator of the average stiffness over a large deflection range of the isolator, as in shock. Damping and stiffness properties are also affected by temperature. At low temperatures, rubber becomes hard and difficult to extend. The elasticity decreases (stiffness increases) and damping is increased. A change in the molecular structure of rubber occurs at –122 to –140°F (–50 to –60°C), and a freezing point is reached. Here, the structure becomes crystalline, and the compound becomes brittle but pliable under load. With energy (such as vibration) supplied, the rubber warms and the elastic properties are recovered. There are four key temperature change effects: T_1, the glass transition point; T_2, the turning point in the curve of the modulus of elasticity, where damping is a maximum and return resilience is a minimum (maximum strain energy is absorbed); T_3, where damping and hardness begin to decrease; and T_4, where the maximum temperature is what the rubber can sustain without significant loss of properties. Usage is normally in the T_3 to T_4 region, where damping, hardness, and dynamic modulus are relatively constant.

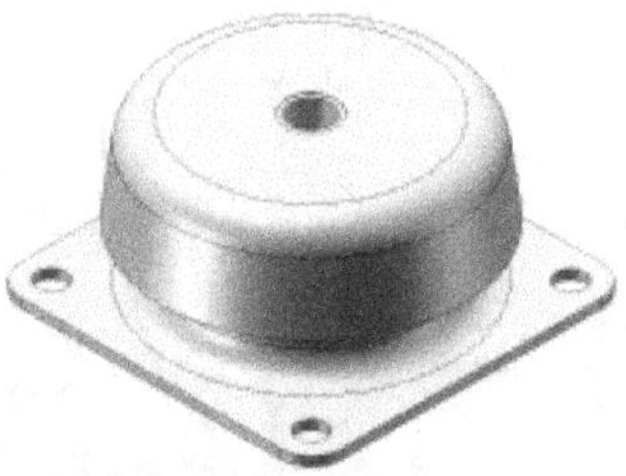

FIGURE 39.10 Housed cup mount—fail-safe design, captured molded elastomer element, mildly increasing stiffness in the principal directions, relative deflection capability less than 0.5 in., various types.

Elastomer Cup Mounts. The fail-safe captured design is used with a wide range of equipment such as electronic units, operating machinery, avionics, communications racks, compressors, and generators. Also known as universal or all-attitude mounts, the housed version of the isolator is shown in Fig. 39.10. This type is often used to protect operating equipment from vibration and occasional shock. It is also used to isolate vibrations from rotating machinery and keep them from reaching nearby equipment. It is not suitable for very low frequency use. To minimize rotation of wall-mounted electronic equipment, isolators are often placed above and below the unit so that their line of action is nearly through the center of gravity of the unit. The compliant elastomer elements contained within dished metal housings (cups) are retained in the event of elastomer failure or bond separation. Loads can be applied to the mount in any direction; the normal operating frequency range is 12 to 30 Hz. Shock response is limited because of the rapid snubbing of the mount in the available space between the housings. Generally used for moderate vibration protection, this type provides only minimal shock reduction; stiffness typically involves stiffening increasing over the allowable deflection range and rapidly hardening near the end of travel, where the rubber is unable to bulge further due to the captured housing feature. Non-fail-safe versions of the isolator are available (without cup housings) for greater deflection. Compacted metal mesh isolators of the same form are available for extreme temperature applications such as jet engine mounts or where elastomers would degrade or undergo rapid aging due to severe conditions.

The all-attitude mount design is supplied by many manufacturers; the compression stiffness is approximately twice the lateral stiffness. It allows for nearly 0.40-in. (1.02-cm) relative deflection, and most designs are shallow profile. Depending on the size of the mount and type of elastomer, load ratings are generally from 10 to 285 lb in a mobile environment and up to 900 lb for a fixed installation. Also depending on size and durometer, axial spring rates range from 600 lb/in. to 22,000 lb/in. Very low profile versions are available without the metal housing. Known as centerplate or multiplane isolators, these are used with light or small units requiring equal isolation in all directions. Cup mount elastomers include chlorobutyl, natural rubber, neoprene, and silicone. The operating temperature range is typically 20 to 180°F (–7 to 82°C) for neoprene and –50 to 250°F (–46 to 121°C) for silicone. The axial-to-radial stiffness ratio is nearly 1.5:1. Stiffness ratings are influenced by temperature changes. Nitrile may be used for long-term oil and lubricant resistance.

Fluidic Elastomer Mounts. Elastomer mounts incorporating high-density silicone gel fill for moderate- to high-viscous damping are shown in Fig. 39.11. These

FIGURE 39.11 Fluid-elastomer mounts are moderately well damped and exhibit stiffening characteristics similar to cup mounts but have limited load ratings and are not fail-safe except within captured metal housings. They comprise silicone gel–filled elastomer body and axial spring to support vertical load.

isolators are moderately damped and exhibit stiffening characteristics similar to cup mounts, but they are intended primarily for axial direction loads, where their internal spring provides load support resistance. They are intended for protection from severe road and other vehicle vibrations. Silicones as a class of compounds are normally used as high-viscosity fluids or gels in this design.

These isolators feature captured inner steel spring for axial load support and silicone gel–filled elastomer body construction, for optimum performance when the isolators are mounted in compression. Mounting in pure shear or tension mode is not recommended. Torsion can cause instability and failure of internal spring connections. Mounts are rated for loads ranging from 1 to 30 lb, depending on the dimensions, size, and orientation of the isolator; the metal-housed version of the mount is rated for higher loads from 30 to 290 lb. They have vertical resonant frequencies as low as 5 to 6 Hz and maximum transmissibility of 2.5 based on the ratio of acceleration spectral density input versus output g^2/Hz. They function across a broad temperature range, from –30 to 180°F (–34 to 82°C). They are capable of long-term use and continuous vibration. They are intended for attenuation of low-frequency vibrations generally above 8 to 9 Hz, with 13- to 15-dB attenuation above 50 Hz. The ratio of axial-to-radial stiffness is approximately 1.2:1. Design features include a thin-wall silicone elastomer body shell that also functions as a resilient element, a centered steel spring, internally contained VHDS silicone gel, a 0.2 to 0.25 percent damping ratio, 0.5-in. (1.27-cm) deflection capability, and vibrations dissipated in forced damping of the gel fill, especially effective in attenuating broad random-vibration energy over the 15- to 200-Hz range. Applications include protection of electronic equipment mounted on off-road wheeled and tracked vehicles such as tanks, missile launchers, snowmobiles, and all-terrain vehicles. They also provide effective isolation in aircraft and helicopter vibration environments.

High-Deflection Elastomer Shock Mounts. High-deflection shock mounts are usually intended for equipment in severe service/military use; they are moderately well damped and have excellent energy dissipation over large deflections. One type is the *arch* isolator, whose deformation characteristics are designed for nearly constant force resistance in compression—the direction usually experiencing the greatest shock for this type of application. The mount is shown in Fig. 39.12. Typical stiffness characteristics of this type reflect its buckling design, which produces bilinear stiffening in compression over the first third of the stroke and softening over larger-amplitude deflection. The tension is mainly linear and can exceed twice the compression deflection limits. *C-like* and *half-arch* designs with similar characteristics are also manufactured. Vibration isolation is achieved by operating the mount over a small motion region having relatively constant stiffness for well-defined resonant frequency response. The basic design is a balanced symmetrical geometry, 7 in. (17.78 cm) high, with a load range 125 to 200 lb per mount, stiffness related to the durometer of the elastomer compound, four load increments, high stroke efficiency of 0.55 (defined as the ratio of maximum free displacement to isolator height), damping ratio $c/c_c = 0.2$ to 0.25, operating temperature range 30 to 180°F (–1 to 82°C), and a nominal 10-year service life, and it is rated at 5- to 8-Hz shock response frequency. Mounts can be oriented and positioned to support the unit in any direction. Compression and buckling of the elastomer arch produces lateral deformation in the outward direction approximately one-half of vertical deflection. The walls bulge out, forming a butterfly shape. Spacers are sometimes used at the top of the mount to allow for greater rattle space and unobstructed movement. Other versions and sizes of the mount have shifted stiffness characteristics and general lateral resistance.

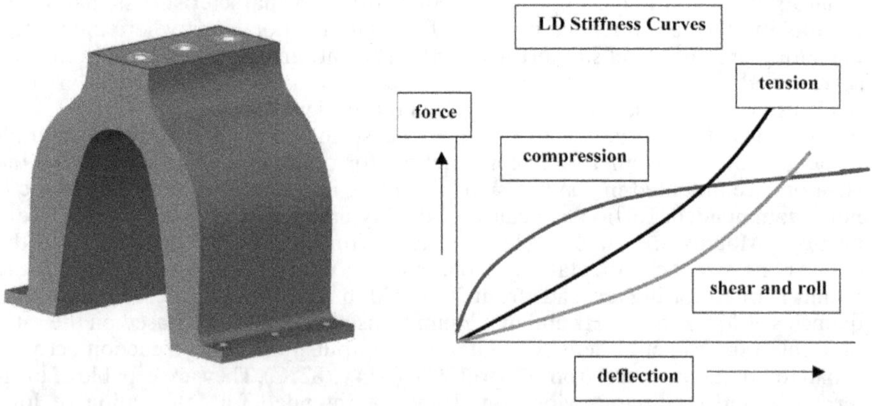

FIGURE 39.12 Typical stiffness characteristics of high-deflection elastomer arch mount (buckling-type design) resulting in bilinear stiffening in compression and shallow stiffness slope over the greater part of its allowable stroke in roll and shear.

The compression-to-shear stiffness ratio is approximately 2:1. Tension is nearly linear to 200 to 250 percent extension, then yields but at an undefined value. Large compression and buckling the mount result in nonlinear softening characteristics with the knee of the load deflection curve beginning at approximately 0.5 to 1.0-in. (1.27- to 2.54-cm) deflection or 20 to 25 percent of available stroke. The entire stroke can be represented as having a trilinear stiffness rate. Designs have been developed that feature a snap-through effect similar to Bellville springs. Variations of the contour and wall geometry of the arch mount produce different stiffness rates. An oblique load will produce combined axial and lateral deformation, thereby possibly affecting the stability of the mount by shifting greater load to a smaller region of its elastomer body and increasing distortion.

Figure 39.13 shows the benefits of large deflection in shock. To an input of slightly more than 60*g* peak, the response was reduced to 24.2*g*. Applications include U.S. Navy shipboard installations, mobile environments and off-road vehicles, military shelters and field-deployed enclosures, equipment and operating machinery in blast environments, and construction sites. For shipboard shock applications, an isolation system of this type is designed to operate at 5 to 10 Hz in vibration and to reduce shock by approximately 60 to 70 percent from deck foundation accelerations in the vertical direction and 40 to 50 percent in the lateral direction.

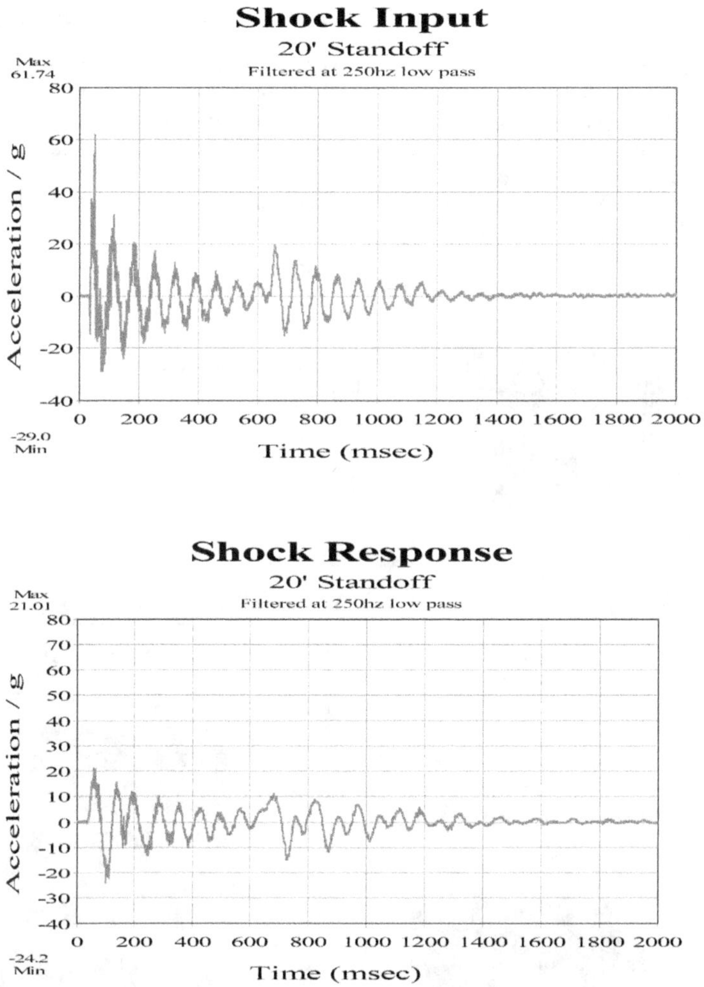

FIGURE 39.13 Comparison of shock response versus shock input to an isolated system. Both plots filtered at 250 Hz. Considerable reduction of peak acceleration is evident, showing the benefits of low-frequency isolation.[4]

Pneumatic Isolators. Also known as air mounts, these are particularly effective when low-frequency, small-amplitude vibrations are the problem disturbance. They are considered to be a type of elastomer isolator. This mount, shown in Fig. 39.14, is often used to protect precision equipment such as semiconductor and quality control (QC) instruments where very low frequency isolators are needed that have resonance frequencies in the range of 0.5 to 2.0 Hz. Due to the limits of their height-to-width ratio, very soft compliant-axial spring and elastomer mounts are unstable at these low frequencies. For example, conventional linear springs would require more than 3 in. (7.62 cm) of static deflection at less than 2 Hz. It is possible to force nonlinear softening mounts to operate over a shallow stiffness portion of their load deflection curve; however, position control is very difficult to maintain.

Basic types of commercial pneumatic mounts include (1) *bellows,* with one to multiple convolutions; (2) *pneumatic-elastomeric* mounts, with thick solid walls; and (3) *air* mounts, with adjustable and/or automatic height control. The elastomer body contributes damping due to strain of the rubber under load. The vertical stiffness of standard air mounts exceeds the horizontal stiffness by a factor of 2 to 3, and added lateral stiffness may be required to ensure stability. That can be included as an integral part of the shaped bellows wall design or constructed as a part of the mount using secondary support such as a metal housing with flexible seal. Damping for the air mount is added with the use of surge tanks.

Low-frequency and zero-static deflection is a feature of the mount. The multiple staged bellows type enables large deflection in shock even without servo control. A

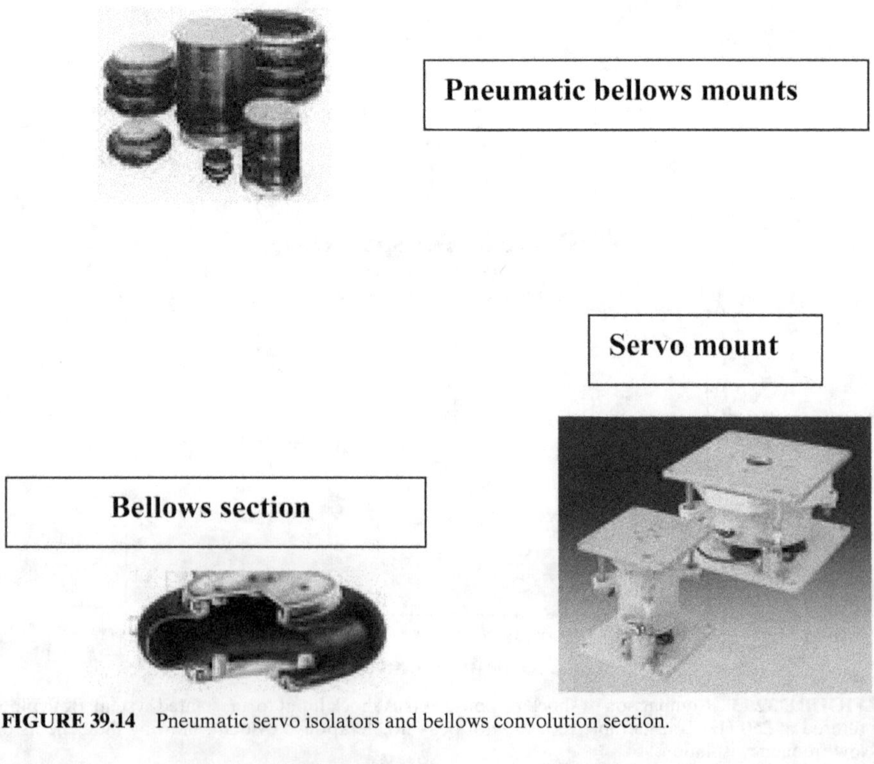

FIGURE 39.14 Pneumatic servo isolators and bellows convolution section.

characteristic of pneumatic mounts is that damping can be moderate to large at resonance but small at high frequencies, with rapid roll-off and very good to excellent isolation response. The load deflection curve in compression is initially stiff, then softening to nearly flat (constant force), and rapidly stiffening near to maximum deflection. Basic equations are as follows:

$$\textit{Stiffness: } k = p\ mg\ A/V_o \tag{39.3}$$

$$\textit{Resonant frequency: } f_n = [1/(2\pi)(pgAV_o)\text{^}0.5] \tag{39.4}$$

where k = stiffness
m = mass
p = ratio of specific heat
A = load supporting area of the air mount
V_o = air cavity volume
g = acceleration

Constructed of reinforced rubber, the *convoluted bellows* is sealed except for an air entry port through which pressurized air is admitted. The pressure maintained within the bellows and the size of the unit determine its load capacity. The axial stiffness is dependent on the number of convolutions and thickness of the wall, as well as the ratio of diameter to height of a convolution. Manufacturers' data provides axial and lateral stiffness information for pressure, load, and height. The axial stiffness is determined from the change in height for a given change in load at constant pressure. In a typical design, the axial stiffness is approximately twice the lateral stiffness for a single bellows and nearly three times that for a dual bellows. The bellows design has variable volume and low hysteretic damping. Restraints may be required to ensure lateral stability under large horizontal forces.

The *pneumatic-elastomeric mount* design involves an elastomeric thick-walled cylindrical sleeve with fixed base and compliant upper support section. Air is supplied through an air valve. The vertical load is supported by the pressurized air column in the upper section. By changing the amount of pressure, the effective stiffness of the air column can be adjusted to compensate for variation in load. Vibration characteristics of the mount are operating natural frequency of 3 to 5.5 Hz, depending on the pressure needed to support the load. The axial-to-lateral stiffness ratio is nearly 1:1. The thick elastomer wall acts as an ordinary passive isolator in the absence of the air.

Due to the small amount of deflection at the upper surface when inflated, it is difficult to physically inspect the mount to know that air pressure is being maintained. Because of the closed structural design of the rubber body, there is only a very small difference in height from the air-filled to the deflated state. It is also difficult to inspect the degree of compliance at the upper surface (at the underside of the table or unit). Damping is low and dependent on the properties of the rubber material. Snubbing in the vertical direction into the body of the mount can occur if the deflection of the compliant upper surface is exceeded. The mount can be used as a shock isolator in the vertical direction; however, it is difficult to determine the stiffness characteristics of the rubber body and air column in a shock condition, and high g's can result due to bottoming.

The *pneumatic mount with height control* is frequently used with optical research tables to maintain precision functions. Automatic height control incorporates a sensor device to control the air supply, typically via a servo valve; when the load is increased, air is supplied to the mount. Constant height is maintained by releasing air from the mount when the load is decreased. Under variable loads such as fluctuating vibration, shifting the mass, or adding mass on the table, the height can be kept constant, resulting in an operating frequency of nearly 1 Hz for low-frequency control. Simpler designs without a supplementary air source are self-contained and capable of nearly 2- to 3-Hz natural frequency. Due to the air control, the natural frequency of the mount can be kept constant at the low-frequency level by maintaining constant height despite changes in load. A change in pressure is associated with a change in volume. An adjustable damping rate can also be incorporated. Regardless of the operating load, the stiffness of the mount depends only on the height of the mount. The ratio of k/m can be kept constant (as the effective m changes), and the natural frequency of the mount also remains constant in accordance with the general equation. The effective mass is the rigid-body mass plus the change in load in real time due to vibration. In the event of air failure, the body of the mount can support the mass as if it were a non-air-filled mount; however, it is less effective as an isolator. It is important to ensure that all air mounts used in a system are properly operating at constant height and that the table is level.

METAL AND COMPOSITE ISOLATORS

Isolators whose compliant elements are made of metal or composite materials are often used in severe service and military applications where environmental conditions can degrade many elastomers. Extreme temperature and long-term exposure highlight the features and performance stability of these types of mounts regardless of the environmental changes that occur. Deformation under load, stiffness rate, and damping properties are dependent on the materials selected and the structural design of the flexible elements. Because metal and composite properties are constant over the range of interest, it's possible to accurately determine deflection and damping over wide temperatures. In some cases, metal isolators are considered "life of equipment," requiring no maintenance or replacement in the event of aging. Elements that predictably bend under load can produce precise stiffness rates. The elements themselves are not elastic. Controlling the end restraints and shaping these elements influences the amount of deflection and the form of the stiffness curves in each axis. For extremely light loads and very small displacements, negative stiffness mounts can be used whose resonance is in the region below 0.75 Hz and providing isolation below 1.0 Hz where air mounts may not be effective.

Helical Cable Mounts (Preformed Wire Rope). Formed in an arrangement of continuously wound spiral loops, the stiffness of the mount is a function of the diameter of the loop, the tightness of the cable, the height-to-width ratio of its oval loop, and the thickness and pattern of the cable used. The isolator shown in Fig. 39.15 exhibits a preformed wire rope design wound in a progressive spiral, having multiple loops of cable along its length. *Preformed* means that the strands are permanently shaped (before winding) into a helical form. The terms *wire rope* and *cable* are used interchangeably. The mount uses multiple loops of wire rope in bending and torsion to resist applied loads. Each direction of the isolator has unique stiffness properties. Cable isolators exhibit nonlinear softening stiffness in compression and nearly linear stiffness in roll and shear. Characteristics generally fall into two load ranges: small amplitude and relatively large amplitude displacement. Damping is a combination of frictional interaction of sliding wire strands within the cable loop (coulomb damping) and viscous effects dependent on the relative velocity across the mount. T at small displacement is characterized as 3–3.5:1 (12 to 15 percent c/c_c). Damping can be increased by realigning the strands for greater friction force and contact area.

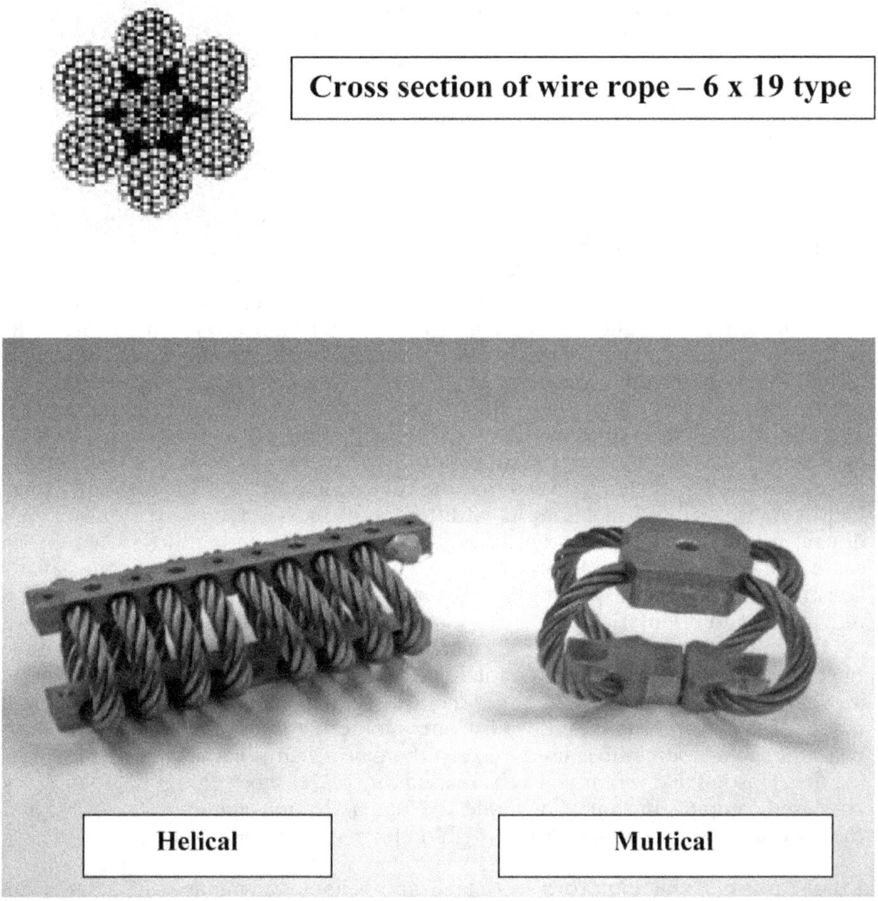

FIGURE 39.15 Two types of cable mounts: helical and multical isolators. Also showing a typical cross section of typical 6-×-19 wire rope.

Isolators are usually constructed of 6-×-19 or 6-×-25 cable with inner core. These terms describe the number of strands and wires in each strand. Stainless steel 302 is the preferred type of cable, for its flexibility and corrosion resistance. Stainless steel 316 is also used. The wire rope diameter ranges from ⅛ to 2¾ in. (0.32 to 7 cm), including 7-×-19 aircraft cable (⅛ to ⅜ in. diameter, per Mil-W-83420) and 6-×-19 or 6-×-25 hoisting cable (above ⅜-in. size). The dynamic-to-static stiffness ratio is 1.0 to 1.1. The compression stiffness is typically three to four times the shear or roll stiffness. The isolator is basically a buckling type, with high initial compression stiffness, then softening over larger deflections and becoming very stiff at the bottoming–retainer bar contact. Loops can have a preferred direction of shear movement depending on the direction of the wire wound spiral. The temperature range is –200 to 350°F (–129 to 177°C), but these isolators can be used at greater extremes for short durations. These isolators are nonflammable and can be degreased. The compression-to-height ratio is typically 0.6 to 0.65.

Deflection is directly related to the load diameter of the cable and the radius of the loop to an exponent ranging from 1.75 to 3.0. Mounts are normally wound in a helical design; the angle of inclination of the loops can be varied for greater (or less) shear stiffness (axial direction of the mount). Designs are also available in the flat where the cables flex as multiple straight cantilever beams. Spherical-like arrangements having nearly equal stiffness in all directions are also made. These are known as *multical* or *polycal* mounts. The axial-to-lateral stiffness ratio of this type is approximately 1.5 to 1.

The basic elements of the helical isolator are loops of wire rope embedded in slotted bars in which the cable is captured. The rope is constructed by layering several strands around a core, usually of fiber or metal materials. A metal core is preferred for high-temperature use. The strands themselves have a center wire that is the axial member, around which the individual metal wires are wrapped. The major portion of the load acting on the isolator is carried by the strands. Frictional damping is a function of strand tightness and, to a lesser degree, the lubricant used in manufacturing the cable. Damping is due partly to coulomb friction (sliding between adjacent wires in the cables), and also to the fact that the hysteresis curves may not be symmetric about the origin as a result of the technique by which the isolators were constructed and wound such that the cables have either a left-hand or a right-hand rotational twist. Isolators have different stiffness for upward (tension) and downward (compression) motion. The same is true for lateral motions whereby the cable tightens and loosens; there is splaying and separation of cables as the isolator deforms. The amount of splaying (also known as *bird caging*) is a function of isolator movement and orientation.

The amount of damping generally increases with an increase in relative amplitude. In very small motions, there is little frictional effect; cable strands barely slide over one another, but tend to untwist. In larger motions, sliding is more pronounced and high points of adjacent strands rub against one another as the sliding occurs. Rotation-resistant cable is sometimes used to resist the twist of an isolator in nonsymmetrical loading. Plastic-filled wire rope has been used for greater shock energy dissipation and increased damping. In a buckling mode, the load deflection characteristics of the helical isolator are similar in form to the high-deflection elastomer arch type.

Steel-Spring Axial Isolators. Rugged and reliable, advantages of axial spring mounts include a wide range of stiffness, availability from many manufacturers, very long-term use without maintenance, ease of examination, linear stiffness rate in compression/tension, and resistance to creep or set. The mount shown in Fig. 39.16 comprises a steel spring and an outer metal housing for centering the spring. Featuring all-steel construction except for secondary elements, the mount is temperature insensitive and unaffected by most environmental conditions. The principal disadvantage is low inherent damping, generally intended for use in the compression/extension direction, although some displacement in the lateral axes can be accommodated for alignment and relative movement to multiaxis vibration. Steel construction may be susceptible to corrosion. Normally installed under the unit in line with direct compressive loads in the axial direction, heavy-duty applications involve stamping, punch press, and crushing machinery. These isolators can also be used in tension as, for example, for pipe isolation and spring hangers. The wide load range of these mounts is dependent on the steel wire diameter, the overall coil diameter, the type of steel used, and the number of coils that carry load. Excessive force may cause the isolator to bottom. Oblique loads should be avoided and can decrease the amount of free rattle space. Spring materials include high-carbon steel, alloy steel ASTM 231, stainless steel 302/304, and high-temperature A286 alloy.

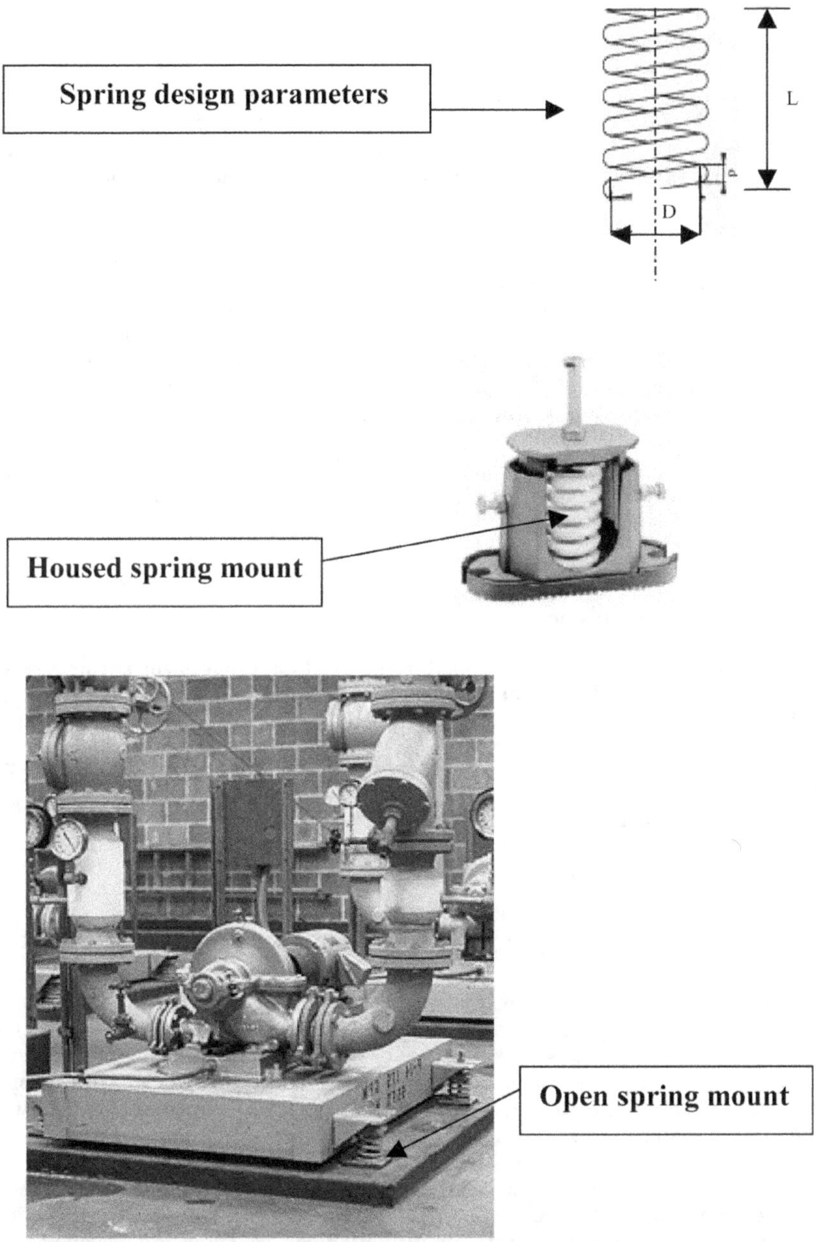

FIGURE 39.16 Open steel-spring mounts and concrete-filled framed inertia base. Also spring design parameters and photo of housed spring mount. Vertical up and down travel stops are incorporated in some housings. Side snubbers at the spring can be positioned using the horizontal adjustment bolts. To reduce high-frequency vibrations that may be transmitted through the steel spring, layers of ribbed or waffle embossed rubber pads are often installed under the base of the housing.

Widely used throughout building, HVAC, and industrial facilities including outdoor installations, cooling tower, and rooftop units, rugged isolator designs feature open and housed steel springs, sometimes with adjustable lateral snubbers for motion control. Housings can be cast iron or weldments. In multiple-spring assemblies, upper loads of 26,000 lb (or more) can be supported. Up to 2 in. (5.08 cm) of static deflection can be allowed. Greater loads and deflections may be required for severe shock applications such as crushing operations. Damping is often provided using compacted knitted metal mesh pads inserted within the spring. The compressed pads are additionally loaded when the spring moves in the vertical up or down direction, resulting in internal friction of the mesh. Springs have weakened and should be considered defective if the mount does not return to within allowable limits of its original position when the load is removed. Marine engines mounted on constrained spring isolators are routinely checked for set of the spring and reduction of free height. Designs may use axial steel springs combined with inclined rubber elements, precompressed for a small deflection range between unloaded and loaded conditions. If compression springs are too long, instability may result due to column action under load. Design equations for springs are given in Ref. 5.

Conventional practice within the building industries is to list spring rate based on the static deflection at a 1-g load. Vertical resonant frequency is then simply a function of static deflection:

$$f_n = 3.13\sqrt{1/\lambda} \tag{39.5}$$

where λ = static deflection

Tension/compression loads are applied in the axial direction of the spring coils; the stiffness rate is linear for the allowable stroke. The commonly used deflection formula is

$$\lambda = 8\,P\,D^3\eta/G\,d^4 \tag{39.6}$$

where
λ = deflection
P = load
D = coil diameter
d = wire diameter
η = number of coils in the active spring
G = shear modulus of elasticity

Very repeatable, these springs return to their original position even after severe use; however, fatigue of the spring or weakening of the material can result in set and a lower height. More than a 1 to 2 percent change in height indicates degradation of the spring. Instability can occur because of the relatively low shear stiffness of the mount in the lateral direction. Coil springs can be described by their height-to-width ratio and curvature. Load eccentricity should be avoided; spring restraints may be needed. Lacking frictional effects, the damping ratio is 0.01 to 0.02 c/c_c. Metal mesh or other deformable materials are sometimes embedded inside the body of the mount to increase its damping, typically to .07 to 0.10 c/c_c. Tapered springs that are wider at the base exhibit improved lateral stability. Variable spaced coils are designed in one type of compression isolator to maintain constant frequency by means of uniform load-to-deflection ratio. Cross section of the coil bar can be round, square, or rectangular; each produces a different stiffness rate for the same area because of differences in the moment of inertia and bending effects. Steel-spring isolators are often used in extremely severe environments, and corrosion protection should be applied to the wire. Deposited coatings provide reasonably uniform protection and good visual appearance. Painting has sometimes been used for low-cost, large-volume production of spring mounts. Steel springs can also be used for lateral restraint or in the opposing inclined direction for greater stability.

SEISMIC ISOLATION MOUNTS

Advances have occurred in the last decade in the commercialization of vibration isolation devices for buildings, structures, and seismic control. These include improved means for energy dissipation at structural elements such as frictional and viscous fluid dampers, active mass actuators, and adjustable stiffness and flexural beam plates. While all of these can be broadly grouped as vibration control, isolation commonly refers to bearing mounts of different types that support the structure (building) and achieve effective separation of the building's natural frequency from the predominant frequencies of an earthquake. The most widely installed design is the elastomeric bearing type, including lead-rubber and high-damped rubber mounts. Low-frequency sliding bearings have also been successfully used.

The design must accomplish four basic objectives. First, it must provide for flexible mounting at seismic loads so that the frequency of vibration of the total system is decreased and decouples from the driving frequency of the earthquake in order to reduce the force response of the structure. Because of the large size and mass of most structures, the mounts are usually arranged to evenly distribute the load for proper support. Second, damping must be added to the mount, or as a supplementary device, so that the relative deflections between structure and ground can be controlled and limited. Third, there must be adequate rigidity in the mount so that conventional service loads such as severe storms are within the capability of the mount to resist without maintenance of the isolators or their compliant elements. The mount must be capable of supporting the vertical loads at the earthquake-induced displacements with a safety factor to account for variations in seismic intensity and character. During a seismic event, the mount, in its deformed condition, can be subjected to an increase in static load at nearly the same time it experiences horizontal forces and large relative displacement between the mass that it supports and the foundation.

Elastomeric Seismic Bearings. Lead-filled elastomeric bearings (lead-rubber type) and high-damped elastomer mounts are widely used base isolators for seismic protection. As shown in Fig. 39.17, the elastomer layers are constrained by the intermediate plates; this maintains compression stiffness under vertical load while still enabling large lateral movement exceeding as much as 14.7 in. (37.5 cm) for some designs. Bearing mounts that are 40 in. (101.6 cm) in diameter, for example, have demonstrated 24 in. (61.0 cm) of shear displacement in test. Other designs include low-frequency sliding plates and friction pendulum bearings. Energy dissipation in the lead-rubber mount occurs in the lead core. The lead plugs shear when horizontal forces exceed a specified amount, and the elastomer body of the mount is then relatively free to move and deform in the lateral or shear direction as the ground moves at the underside of the isolator. The mount undergoes relatively large strain. Its top plate is attached to the structural mass at a support. In the high-damping rubber mount, special-purpose filler embedded in the elastomer compound increases the mount's hysteresis and provides its energy dissipation characteristics. The equivalent viscous damping in the high-damped bearing is a material property and typically varies from 0.10 to 0.15 c/c_c. The natural frequency of the isolation mount system is 0.75 to 1.5 Hz.

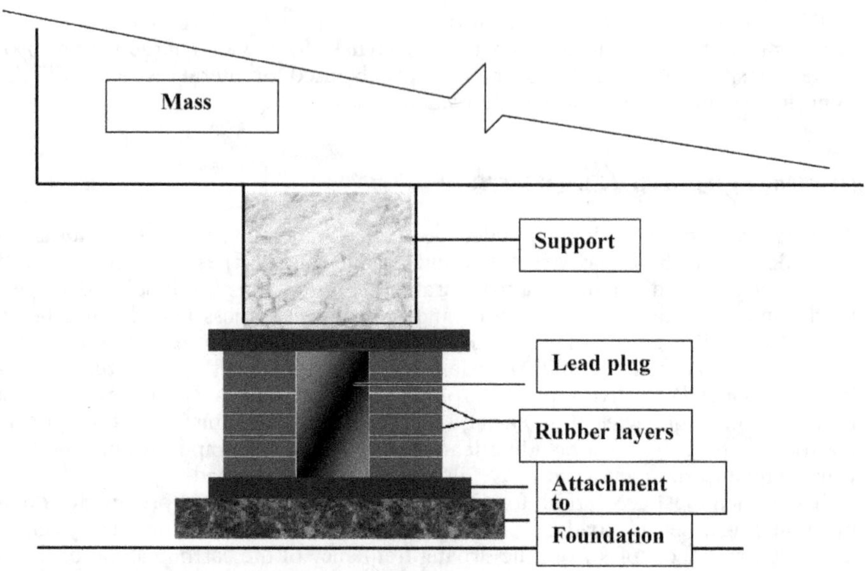

FIGURE 39.17 Seismic elastomer bearing mount with multiple layers and internal lead plug that shears under horizontal load, allowing the rubber layers to move by means of controlled deformation.

Similar to bridge bearings, high-damped mounts are usually constructed by bonding sheets of rubber to thin steel plates. The steel reinforcement increases the vertical compressive stiffness of the isolator while maintaining its desired low horizontal stiffness. Failure in laminated bearings usually occurs due to internal rupture caused by tensile stresses on shearing layers as they tilt and slide with respect to one another. The strain is within the capabilities of the basic rubber but can exceed allowable strength of the lamination. Due to the complicated sliding action, a layer is in compression at one side and tension on the other side. Severe tension can lead to bond failure between steel plates and molded elastomer elements. All seismic isolation mounts should be carefully inspected after an earthquake for signs of degradation or damage. Routine inspections should be carried out to validate the integrity of the elastomer mount to support the load over long-term use and temperature extremes.

Horizontal stiffness k_h of the mount (after shearing of the lead plug) depends on the modulus of the rubber and physical size of the elastomer body:

$$k_h = AG/\sum t \qquad (39.7)$$

where A = bonded area of the rubber layers modified for plug yielding factors
G = shear modulus of the elastomer
$\sum t$ = sum of the thicknesses of the individual rubber layers

The rubber modulus is usually about 75 to 150 lb/in^2 at 100 percent shear strain at normal temperature conditions. The yield stress of the lead core is approximately 1200 lb/in^2. The vertical stiffness general equation is

$$k_v = (AE_c)/\sum t) \tag{39.8}$$

An approximate expression for the compression modulus E_c is

$$E_c = [(1/6GS^2F) + (4/3B]^{-1} \tag{39.9}$$

where S = shape factor A/AP
A = area of rubber free to bulge for a single layer
B = bulk modulus
F = factor based on solid circular versus internal hole type

Sliding Seismic Bearings—Friction Pendulum System. The friction pendulum system (FPS) seismic isolator is a steel connection assembly that consists of an articulated friction slider that moves along a spherical concave surface. A schematic cross section is shown in Fig. 39.18. The contoured surface results in a small-amplitude pendulum motion of the supported structure. Once the threshold friction force of the bearing material is exceeded in the earthquake, the FPS connections shift the stiffness and frequency of the structure. Composed of two main parts, the upper section attached to the building structure (also known as the *spherical sliding bearing*) slides relative to the lower shallow dish section that is attached to the ground. Movement begins once the friction force level is exceeded. The degree of curvature of the fixed-in-place curved dish sets the natural frequency of the isolation system (the pendulum effect) while controlling lateral movement to within allowable limits.

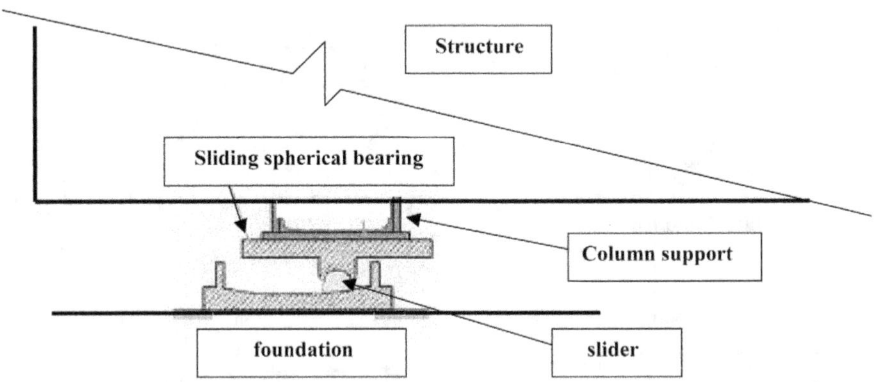

FIGURE 39.18 Seismic friction pendulum system (FPS) isolator showing the spherical sliding bearing—schematic.

The natural frequency of the FPS isolation system is 0.5 to 1.0 Hz. The lateral stiffness is directly proportional to weight, and the frequency is independent of the mass. The articulated slider within the bearing moves along the concave surface, causing the supported structure to move with a slight pendulum motion. Damping is a result of friction force between the bearing and the concave surface. The bearing material of the articulated slider is a high-strength self-lubricated composite. The natural period (1/frequency) of the FPS isolator is

$$\tau = 2\pi\sqrt{(R/g)} \tag{39.10}$$

The bearing stiffness of the FPS is

$$k = W/R \tag{39.11}$$

where W = weight of the structure
τ = period
R = radius of curvature
g = acceleration of gravity

A different sliding bearing design uses a disc-type bearing and elastomer spring to provide the restoring force. The sliding interface is a flat controlled friction surface plate for carrying the vertical load and providing damping.

ACTIVE AND SEMIACTIVE ISOLATION

Passive, semiactive, and active isolation (in order of complexity) are the three fundamental methods of vibration and shock control. These are modeled in Fig. 39.19, also showing the variable force actuator for the active case. Active and semiactive

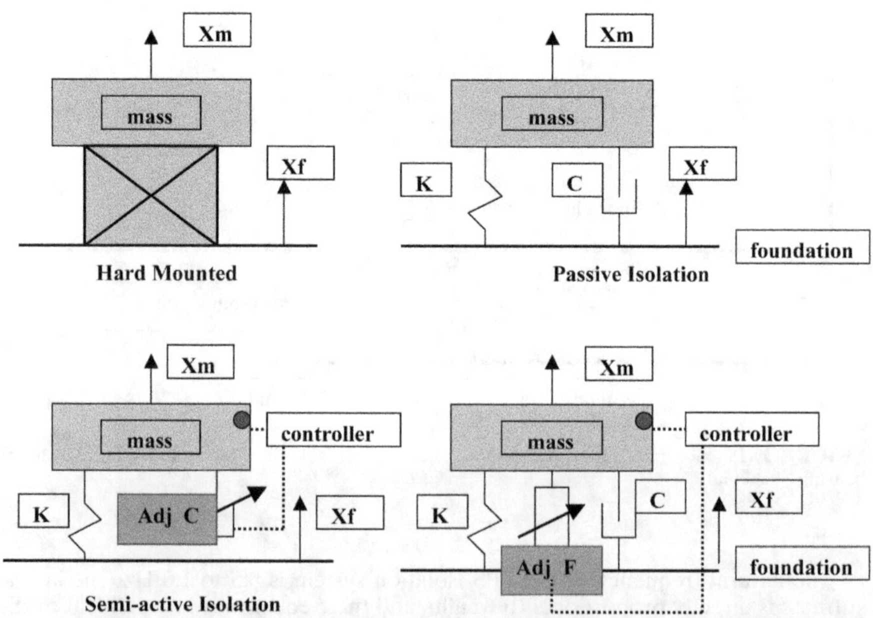

FIGURE 39.19 Isolation control models also showing the adjustable damper C and adjustable force actuator F. Hard mounted is with the mass Xm directly attached to the foundation Xf, no isolation. Passive isolation is essentially open-loop-dependent only on stiffness K and damping C.

systems generally exhibit well-defined, more precise control than passive isolation. Active designs can be especially effective over a broad frequency range. However, due to cost, high power requirements, and reliability concerns, they are not commonly used in large structural applications requiring direct force. In commercial applications, controlled-force actuators and variable dampers are widely used for seat suspension, manufacturing processes, measurements, and similar systems. A variety of electronic controls and active components are available from suppliers, and products are supported with engineering documentation and technical assistance. Active isolation opposes and cancels the disturbing force acting on the mass. Varying the stiffness and damping gain increases or decreases the amount of actuator force, thereby improving the operating performance of the isolation system as the disturbance changes. Passive isolator control depends on inherent stiffness and damping properties of the isolator and is basically an open-loop system. Semiactive isolation works in conjunction with a variable damping or stiffness device.

ACTIVE ISOLATION CONTROL

Active isolation systems operate by means of external force actuators programmed to oppose the disturbing force and hold the isolated mass nearly motionless. The applied force counteracts the response of the mass. Force actuation, directed by a controller/processor in the feedback loop, avoids the need for supplementary stiffness or damping adjustment. Because substantial external force may be required, operating control problems can result in adding energy (rather than canceling it) and cause the system to become unstable. Semiactive isolation systems are considerably less complicated, and operating performance has been shown to be very effective, nearly matching active isolation results.

Control methods are usually based on response measurements but can also be based on input variables at the equipment where the input is unknown and undetermined but anticipated before the control forces are applied. Other techniques simultaneously compare input and response and adjust gains accordingly. In each case, the controller/processor then drives the force actuator in accordance with the control process. Semiactive methods use the features of variable stiffness and/or (more often) fluid damping devices to provide increased force resistance in response to signals at the unit or at the disturbance.

Depending on the application, performance objectives may involve adjusting control gains so that the isolator is critically damped for maximum transmissibility less than 1.0 at or near resonance, –4-dB isolation at the transition frequency of $\sqrt{2}\, f_n$, and continuous roll-off at 20 dB/decade. Gains are also sometimes based on allowable vibration limits of the equipment over a critical frequency region, separate from resonance, and incremental damping is necessary. In the "skyhook" technique, variable settings are programmed at the controller and the disturbing force is directly applied to the mass through the spring. There is no damping, and the actuator control force is a function of the absolute velocity of the mass; reference is made to an inertial frame that remains stationary. The skyhook technique can be used both in active and semiactive systems; the difference is mainly in the use of the force actuator at the mass in active control and a damping force in semiactive control. The force exerted by the actuator is designed to be proportional to the absolute velocity with respect to the inertial frame system; its results can be approached by control based on relative velocity between the movable mass and the foundation. The same technique is used in semiactive control except that a damping variable is continually adjusted to moderate the amount of damping force applied at the movable mass.

For the single-degree-of-freedom model without damping, the transmissibility in Laplace terms is

$$T = [(M/k)s^2 + (G/k)s + 1]^{-1} \tag{39.12}$$

With damping, T changes to

$$T = [(k + cs)]/[Ms^2 + (G + c)s + k] \tag{39.13}$$

or, in a slightly different form,

$$T = [2\zeta\omega_n s + \omega_n^2]/[s^2 + 2\zeta\omega_n s + \omega_n^2] \tag{39.14}$$

where k = stiffness
G = control gain
$\omega_n^2 = k/M$
$2\zeta\omega_n = c/M$
ζ = damping ratio
M = mass
$s = j\omega$
f_d = disturbing force
ω = frequency

In frequency terms, neglecting relative displacement, the effect of damping gain can be seen:

$$T = \sqrt{\frac{1}{1 - (w/w_n)^2 + [2(c/c_c)(w/w_n)]^2}} \tag{39.15}$$

where T = transmissibility
W = frequency, Hz
w_n = natural frequency, Hz
c/c_c = damping ratio

at resonance $w = w_n$, and the equation becomes

$$T = 1/(2c/c_c) \tag{39.16}$$

By making the gain factor G a function of c, the equation is

$$T = 1/(2G/c_c) \tag{39.17}$$

The gain factor is a variable and is adjusted as a percentage of critical damping. When G/c equals 0.5 or more, T equals 1.0 or less, for example,

$$G/c_c = 0.5 \qquad T = 1/(2 * 0.5) = 1.0$$

$$G/c_c = 0.6 \qquad T = 1/(2 * 0.6) = 0.83$$

Fig. 39.20 shows an example of the transmissibility-versus-frequency ratio for two damping gain ratios, 0.2 and 0.5, comparing an active versus a passive system and steady-state response. In this example, T of the active system is dependent on the frequency ratio $(w/w_n)^2$, while the passive system is a function of (w/w_n). The equation for T is given in Ref. 1, Chap. 32, Eq. (32.30).

$$T = \sqrt{\frac{(G_1/mw_n^3)^2 + (w/w_n)^2}{(w/w_n - w^3/w_n^3)^2 + [G_1/mw_n^3 - 2(G_2/c_c)(w^2/w_n^2)]^2}} \tag{39.18}$$

where G_1/mw_n and G_2/c_c are relative displacement gain and velocity gain.

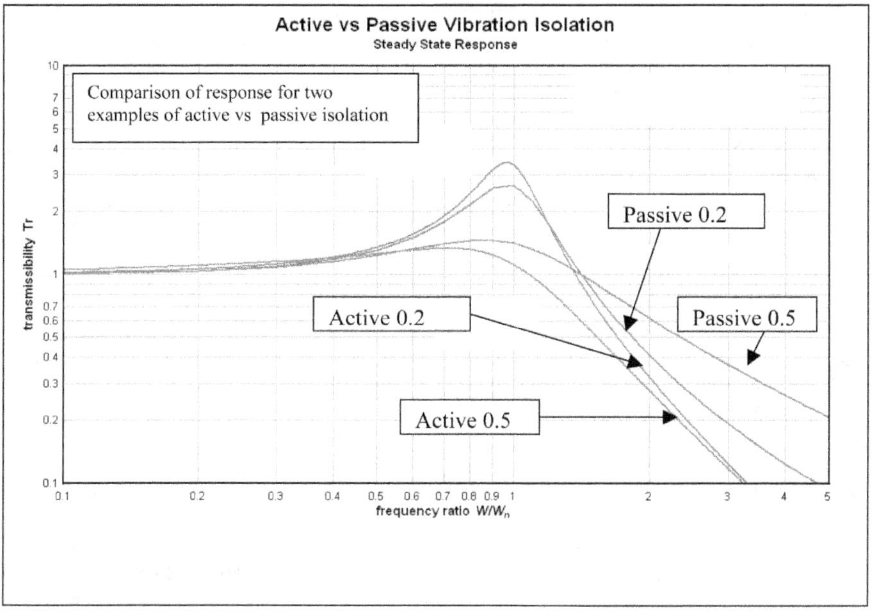

FIGURE 39.20 These examples compare transmissibility for active versus passive isolation (variable damping and relative displacement gains—active systems; constant damping—passive systems). Reference Eq. (32.30) in the 5th edition of this handbook. At higher frequencies beyond resonance, transmissibility for the active system declines more rapidly as a function of W^2/W_n^2 than for the passive isolation, where the falloff is dependent on W/W_n.

There is zero displacement gain and positive damping gain assigned to the passive system in Fig. 39.20, constant displacement gain and positive damping in the active system. Both have positive damping gain. In this example, at higher frequencies beyond resonance, T for the active system declines more rapidly than for the passive system, where the falloff is entirely a function of (w/w_n). By means of variable stiffness and damping gain controls, it is possible to adjust the amplitude of response displacement and shift to lower frequencies (for greater isolation) at the same time. The velocity gain controls the damping force; displacement gain controls the spring force. As shown by Preumont in another example,[6] high-frequency roll-off with damping is not as rapid as with the force actuator alone, and the damper can be removed from the active system. The active system becomes more effective at the higher frequencies without damping. This is evident also in the T curves for a passive system; the high-frequency rolloff after resonance increases more rapidly as damping decreases.

Isolation control objectives in shock and vibration active systems (Ref. 1, Chap. 32) are usually to maintain position of the isolated mass regardless of the disturbance and control T at system resonance. As noted, control techniques mainly involve comparative methods based on the response of the mass versus the input at the disturbance, or a combination of the two. To limit displacement using active control, an opposing force is applied to the mass by means of a force actuator that drives the mass in a counter direction. This effectiveness of the closed-loop system is then a function of the driving force and the gain signal from the controller/processor. There is extensive literature on power-driven actuators and devices for motion control. Substantial power may be required to drive the actuator in a constant (always-on) operational mode. The amount of force can also be applied in a proportionate or off/on way, based on the relative deflection and/or rate of relative deflection of the mass. Measurements can involve comparing acceleration, position, or velocity between the disturbance and the mass or using a single sensor at an intermediate location to measure changes of the mass from its set position. Velocity and displacement are considered to be more effective control than acceleration.

Variable scaling functions for setting the counterforce are calculated, and the output gain is determined. The output gain will continue to be adjusted if displacement is different from that intended; this is generally the set or zero position. The rate of change of relative motion can be monitored in order to avoid overshoot. The applied force then increases in either a negative or a positive direction in order to return the mass to position. The actuator is always in an operational state, and power to drive it must be continuously available. Other adjustable devices producing counterforces include servo mechanisms, spring linkages, pneumatic air mounts with servo valves, and tunable fluid dampers. Variable stiffness is more complex. A threshold level of motion is sometimes programmed to trigger initiation of control forces.

Proportional-integral-derivative (PID) control methods measure the error between the acceleration or velocity (or position) of the isolated unit and its set value by calculating the difference on a real-time basis and then correcting by means of an actuator and/or damper. Adjusting spring stiffness is more difficult. Three different control actions and gain factors are involved and dependent on the response of the unit. Proportional methods determine the response to the error and change at the unit. Instability may result in the case of overshoot. Integral methods involve the processing of cumulative errors in a specified time interval. The response error varies with time as forces are adjusted and input vibration or shock changes. A derivative technique is used to determine the rate of change of the error signal. The calculated factors then become a net output gain that readjusts the force developed at one or more of the control devices. There is a transfer function associated with each factor and the combination of variables. Tuning methods are used for setting gain factors. Control can also be achieved in a simple way, using a proportional integral technique known as a PI process. Proportional derivative (PD) control exhibits a relatively slow rise. In Laplace terms, the transfer function between the displacement $X(s)$ and the input force $F(s)$ is as follows:

$$\textit{Proportional (P):}\ X(s)/F(s) = [G_p/(s^2 + (c)s + (k + G_p)] \qquad (39.19)$$

$$\textit{Proportional derivative (PD):}$$
$$X(s)/F(s) = [(G_d + G_p)/(s^2 + (c + G_d)s + (k + G_p)] \qquad (39.20)$$

$$\textit{Proportional integral (PI):}$$
$$X(s)/F(s) = [(G_p s + G_i)/(s^2 + (c)s^2 + (k + G_p)s + G_i)\text{i}] \qquad (39.21)$$

$$\textit{Proportional integral derivative (PID):}$$
$$X(s)/F(s) = [(G_a s^2 + G_p s + G_i)/(s^2 + (c)s^2 + (k + G_p)s + G_i)] \qquad (39.22)$$

where G_p = proportional control gain
c = damping
G_d = derivative control gain
k = stiffness
G_i = integral control gain

The major difference is the influence of each gain factor on rise time, overshoot, and the number and rate of cycles to zero change of the response. For example, to a unit step input, proportional (P) control reduces the rise time and control error compared to an open-loop system; it increases overshoot and decreases return time to a constant value. Derivative (PD) control reduces overshoot with fewer cycles. Proportional integral (PI) control decreases overshoot and cycles even more. PID control achieves fastest rise time, least error, and essentially no overshoot, and can provide the most accurate control. Figure 39.21 shows the nondimensional output variable versus time to a unit step input for different values of integral gain G_i. With increased gain, the response decays more rapidly with less overshoot. In this example, the variable is force and the input is a force step, then held constant at 1.0. G_p and G_d gains are constant throughout. G_i ranges from 0.5 to 2. Different gains or combinations of gain factors will shift peak values and rise times.

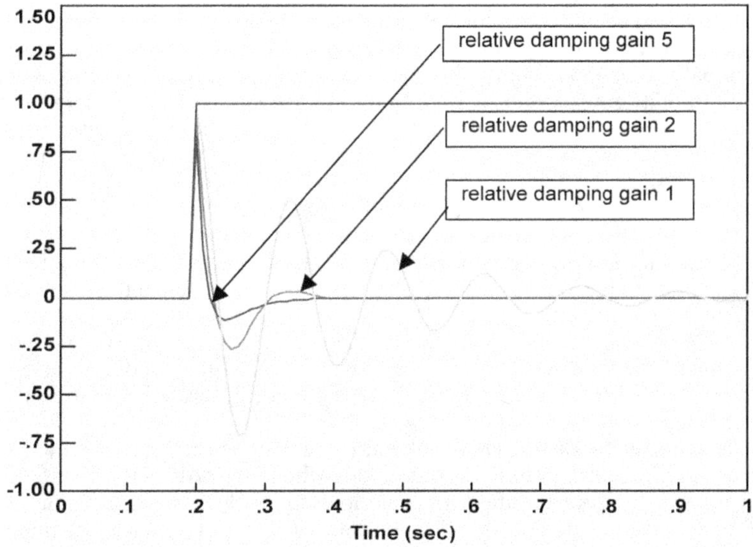

FIGURE 39.21 Response to unit step input for different damping gain values; stiffness, mass, and frequency held constant.

SEMIACTIVE (SA) ISOLATION CONTROL

Semiactive isolation systems use variable stiffness or damping to achieve very low transmissibility at resonance, switching to rapid roll-off at higher frequencies. External force actuators are not required. The variable device with its controller and electronics can be mounted separately or used with passive isolators to operate as an integrated assembly. Two types of fluid dampers, *magnetorheological* (MR) and *electrorheological* (ER), are the variable devices used most often and are based on developing control forces that oppose the relative velocity between the mass and the disturbing source. Fluid valves operate as the variable damper element to provide controlled resistance force in response to programmed levels.

MR and ER devices are extremely fast switching control valves functioning in response to the magnetic field operated on by variable signal gains from the controller. There is extensive literature on MR/ER fluid properties and fluid valve designs. The viscosity of the fluid is regulated by the strength of an applied magnetic field, which changes based on the control signal. Damping rate change is in milliseconds. Dampers can be operated in a plus or minus (+/–) direction for linear motion or camlike to minimize rotary oscillation. They require only battery power to regulate the solid/liquid state of the fluid. Variable damper force can be adjusted using off/on control or designed for more gradual change. ER devices operate using high-intensity electric fields requiring very large voltages, while MR fluids operate with small voltages and currents. MR fluids are manufactured by suspending ferromagnetic particles in a carrier liquid. Other metals have been investigated. R&D is active in the selection of materials, particle size, and carrier fluid.

Resistance of the variable damper, using feedback control, is generally based on relative velocity between the mass and the foundation. Proportional or limit settings in accordance with minimum/maximum values are the usual parameters. Damper properties are changeable in a real-time continuous mode for the period in which power is applied. Compared to active isolation, SA control requires only low external power (can be battery operated), provides passive control if the fluid device fails, and has inherent stability. SA isolation approaches the effectiveness of active control in reducing low-frequency vibrations and also offers reduced cost through the use of relatively simple fluid dampers, or with the combination of passive isolation and dampers.

Similar designs can also be used in shock with the preset damper acting on relative position and the passive mount (in combination) providing routine shock control through deformation of its compliant elements. It has been noted that using only variable stiffness devices for dissipating energy in shock as opposed to damping devices may be an advantage in that large damping forces are not transmitted to the isolated unit. That is, high velocities can create large forces in variable dampers as a function of the rapid viscosity changes and limit the rate of high-frequency roll-off. In the case of variable stiffness devices (no damping), the force is a function of only relative displacement and stiffness rate. Damping forces are not generated, and roll-off can be more rapid.

As in active isolation, the choice of control techniques involves narrowband versus broadband control, design of a control algorithm, the cost of the electronics, and the availability of appropriate devices. A common technique switches the damper off whenever the isolated mass and nonisolated foundation move in the same direction and the foundation has the greater velocity. There is extensive literature on control techniques and applications. Also, as in active isolation, control methods include PID, PI, and PD techniques. Nearly as effective as active control, research is widespread in low-cost control of flexible members, tuned structural damping of large structures, and very low frequency precision isolation.

In one type of modified open-loop method, the variable damper operates in an on/off mode without feedback. The fluid valve can be regulated so as not to exceed preset limits of the isolated unit. For example, at the resonant frequency, it could be important to have a high damping coefficient for minimum T, with peak acceleration specified as a limit. The control variable in the isolation model would be set to 1.0. Beyond the resonant frequency, the control gain would be set to zero for minimum damping and rapid falloff of T. Applications for this type of control include tuned viscous elastomer mounts for shock restraint and hybrid designs of passive mounts with fluid dampers.

Feedback. The control force is intended to approach the effectiveness of the skyhook method but using relative velocities for setting the damping coefficient. For example, the damper control force is applied as a function of relative velocity when the signs of the relative and absolute velocities are the same.

$$F_d = G(v_m) \text{ if } (v_m) * (v_m - v_f) > 0 \text{ otherwise} = 0 \tag{39.23}$$

where F_d = damper force
v_m = mass velocity
v_f = foundation velocity
G = control gain

The literature notes that semiactive control emulating the skyhook damper appears to work well for narrowband disturbances but tends to be less effective for wideband vibrations.

On/Off Control. The control variable is switched between minimum and maximum values, while the damping coefficient also switches from minimum to maximum according to the sign of the relative velocity. This relatively simple method requires only a rapid-switching on/off device instead of a modulated valve. The maximum value could be represented as having a preset upper damping level to reduce T at resonance. Minimum might then represent a lightly damped mount operating over higher critical frequencies of the unit.

Clipped/On/Off Control. This approach uses a preset damper control force to achieve equivalent results to an active control actuator. A clipping controller in the control loop, reacting to the motion of the mass, is used to drive the semiactive damper in way that reproduces the actuator force that would have been produced by active control.

When the relative velocity across the damper and mass have the same sign, a damping force proportional to V_m is applied; otherwise, there is zero damping.

MR and ER Fluid Dampers. The versatile operational features of magnetorheological devices have made them extremely useful for a wide range of motion control. They require minimal power, produce high-resistance force, and are completely reversible and fail-safe, reverting to the passive mode if power is lost or disrupted. The passive mode involves reliance on the passive isolator completely and the absence of magnetic field effects on the viscosity of the damping fluid. MR fluids are suspensions of small iron particles in a base fluid such as mineral or silicone oil and are able to rapidly change in milliseconds from free-flowing, linear viscous liquids to semisolids having considerable yield strength under a magnetic field and exhibiting plastic-like effects. Yield stress increases as the strength of the magnetic field increases. In a magnetic field, the particles form linear chains parallel to the applied field and can impede the flow as the fluid solidifies. Most variable devices using controllable fluids are from among the three types shown in Fig. 39.22. MR fluid control can simplify mechanical dampers by replacing valves and eliminate complex orifice design with precise magnetic field action on the fluid. The basic design is inherently reliable. If there is electrical failure, the damper reverts to a passive device. Carrier fluids have included silicone and synthetic oil. High-temperature effects on fluid viscosity can be a concern. The ideal force versus velocity characteristics of the fluid damper can be described as having a region of adjustable values instead of a force (constant slope) increasing from nearly zero at low velocity to maximum at high velocity. With MR design, the force can be increased to a maximum almost immediately after motion begins. The force is a function of control gain and current, which affects the strength of the magnetic field, and not velocity.

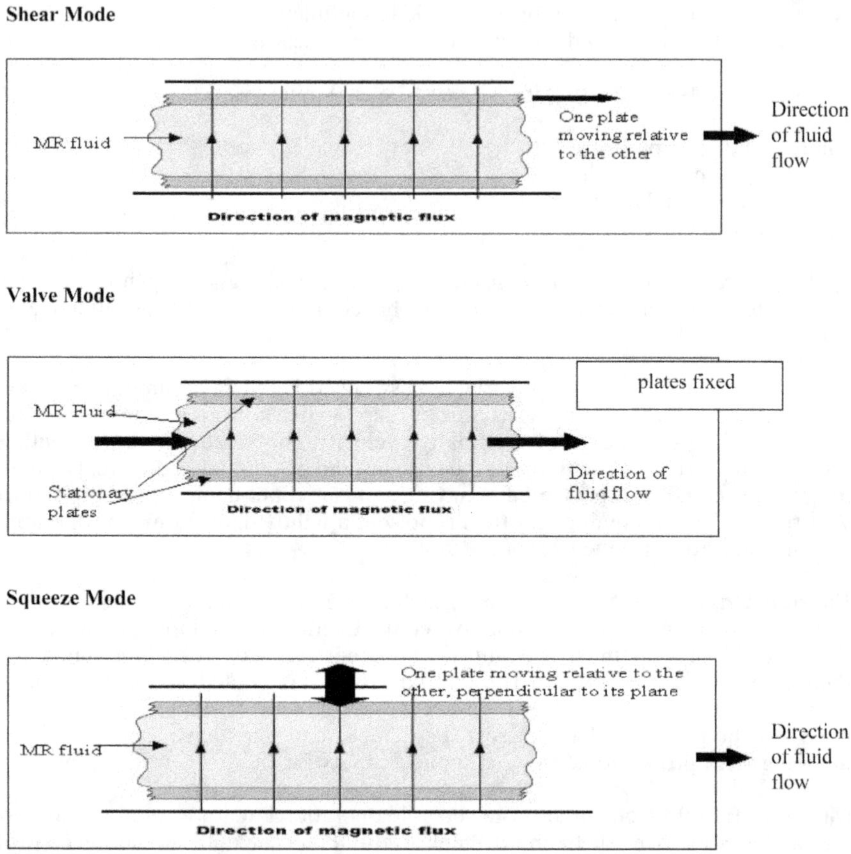

FIGURE 39.22 Models of MR fluid damper techniques showing the shear, valve, and squeeze modes for controlled movement.

Valve, Direct Shear, and Squeeze Mode. Examples of *valve* mode devices include servo valves, dampers, and shock absorbers. *Direct shear* devices include clutches, brakes, and latching and locking mechanisms. The magnetic field in each mode is perpendicular to the opposing metal surfaces, and the flow of the fluid is parallel to the surfaces. In *shear* mode, the two opposing surfaces in contact with the fluid can move relative to one another (one surface remains fixed), creating a shear stress in the fluid that can be varied by applying different levels of magnetic field strength. The fluid is pressurized in the valve mode to flow between the two fixed surfaces. The *squeeze* mode involves bringing the opposing surfaces toward one another to develop fluid pressure.

Flow rate and fluid pressure are adjusted by varying the magnetic field. The fluid is contained within a small magnetic flux area of the actuator, and the damper operates on the resistance of the contained fluid as its state of viscosity changes. Figure 39.23 shows a commercially available MR damper and seat suspension assembly.

Electrorheological fluids have also been extensively described in the literature and exhibit reversible properties similar to those of magnetorheological fluids. ER

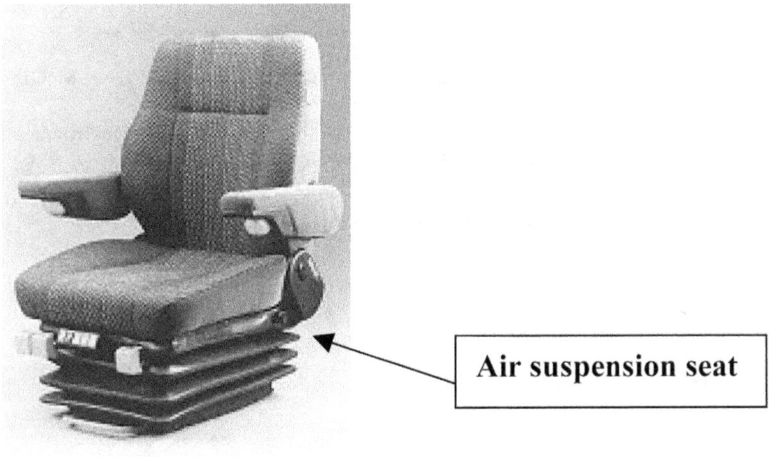

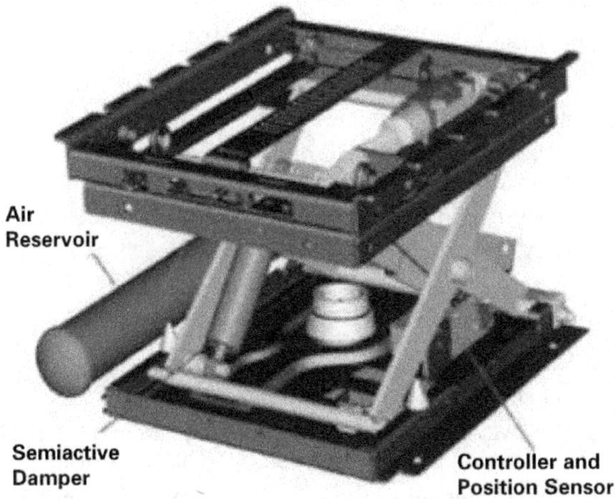

FIGURE 39.23 Semiactive seat suspension. (Photo courtesy of the John Deere Company.)

commercial development has been limited because of the high power requirements for effective use. ER devices require high voltage at low current, while MR devices require higher current only at a low voltage. MR fluids can also be energized by permanent magnets with no steady-state power requirement. Like MR fluids, ER fluids are noncolloidal suspensions of particles only a few microns in size that can be made to line up in a columnar arrangement, changing the apparent viscosity of flow as the strength of the field increases. The flow motion of the ER damper can be similarly classified as shear, valve, or squeeze mode. Reliability and simplicity appear to be less than with MR devices.

REFERENCES

1. Harris, C. M., and A. G. Piersol: "Shock and Vibration Handbook," 5th ed., McGraw-Hill, New York, 2002.
2. Gaberson, H. A., D. Pal, and R. S. Chapier: "Classification of Violent Environments That Cause Equipment Failure," *Sound and Vibration,* May 2000, pp. 16–23.
3. Gobel, E. F.: "Rubber Springs Design," John Wiley & Sons, New York, 1974.
4. Barge Shock test data, courtesy of Shocktech/901D LLC.
5. Wahl, A. M.: "Mechanical Springs," 2d ed., McGraw-Hill, New York, 1963.
6. Preumont, A.: "Vibration Control of Active Structures," Kluwer Academic Publishers, New York, 2002.

ADDITIONAL SOURCES

6. Kutz, M.: "Mechanical Engineers Handbook," John Wiley & Sons, New York, 1986.
7. Rivin, E. I.: "Passive Vibration Isolation," ASME Press, New York, 2003.
8. Macinante, J. A.: "Seismic Mountings for Vibration Isolation," John Wiley & Sons, New York, 1984.
9. Lalanne, C.: "Mechanical Vibration & Shock, Volume 2," Hermes Scientific Publications, London 2002.
10. Powell, P. P.: "Engineering with Polymers," Chapman and Hall, London 1983.
11. Costello, G. C.: "Theory of Wire Rope," Springer-Verlag, New York, 1990.
12. Nagoor Kani, A: "Control Systems," RBA Publications, 1998.
13. Soong, T. T.: "Active Structural Control: Theory and Practice," Longman Scientific & Technical, John Wiley & Sons, New York, 1990.
14. Jones, R. S.: "Noise & Vibration Control In Buildings," McGraw Hill, New York, 1984.
15. Gaberson, H. A.: "Pseudo Velocity Shock Spectrum," SAVIAC training lecture, February 2006.
16. Product Isolator Catalogs and Data Sheets—Shocktech/901D LLC, Barry Controls Inc., IDC, Aeroflex International Inc., Newport Precision Labs, ThorLabs Inc., EAR Inc., Bridgestone Engineered Products Inc., Lord Corp., Vibration Mountings and Controls Inc., Lansmont Corp., et al.
17. Balanchin, D. Y., M. A. Bolotnik, and W. D. Pickey: Optimal Protection from Impact, Shock, and Vibration, Gordon and Breach Science Publishers, Amsterdam, 2001.
18. Blow, C. M., and C. Hepburn: "Rubber Technology and Manufacture," 2d ed., Butterworth Scientific, London, 1984.
19. Snowden, J. C.: "Vibration and Shock in Damped Mechanical Systems," John Wiley & Sons, New York, 1968.
20. "Cadre Pro User Manual," Cadre Analytic, Seattle, Wash., 2001.
21. Paz, M.: "Structural Dynamics, Theory and Computation," 3d ed., Chapman and Hall, New York, 1985.
23. Ruzicka, J. E.: "Passive Shock Isolation," *Sound and Vibration,* August and September 1970.

CHAPTER 40
EQUIPMENT DESIGN

Karl A. Sweitzer
Charles A. Hull
Allan G. Piersol

INTRODUCTION

Equipment is defined here as any assembly of parts that form a single functional unit for the purposes of manufacturing, maintenance, and/or recordkeeping, e.g., an electronic package or a gearbox. Designing equipment for shock and vibration environments is a process that requires attention to many details. Frequently, competing requirements must be balanced to arrive at an acceptable design. This chapter guides the equipment designer through the various phases of a design process, starting with a clear definition of the requirements and proceeding through final testing, as illustrated in Fig. 40.1.

ENVIRONMENTS AND REQUIREMENTS

The critical first step in the design of any equipment is to understand and clearly define where the equipment will be used and what it is expected to do. The principal environments of interest in this handbook are shock and vibration (dynamic excitations), but the equipment typically will be exposed to many other environments (see Table 18.1). These other environments may occur in sequence or simultaneously with the dynamic environments. In either case, they can adversely affect the dynamic performance of the materials used in a design. For example, a thermal environment can directly affect the strength, stiffness, and damping properties of materials. Other environments can also indirectly affect the dynamic performance of an equipment design. For example, thermal environments can produce differential expansions and contractions that may sufficiently prestress critical structural elements to make the equipment more susceptible to failure under dynamic loading.

The preceding example illustrates the need to understand all of the design requirements, not just the dynamic requirements. A comprehensive set of require-

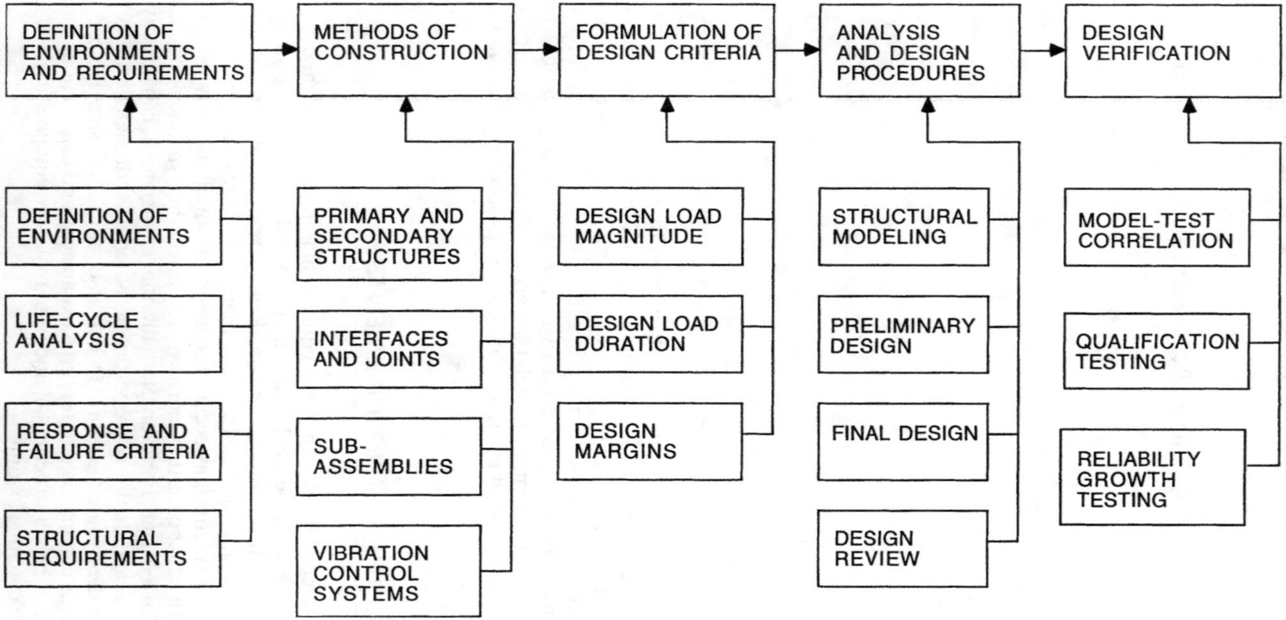

FIGURE 40.1 Steps in equipment design procedure for shock and vibration environments.

ments (or equipment specifications) must be developed so that no aspect of the design's performance is left uncontrolled. Unfortunately, different types of requirements often lead to difficult design tradeoffs that must be resolved. Priorities must be established in these situations. For example, a low-cost weak material may be preferred over a more expensive stronger material if the operational stresses can be kept low. This example reflects the fact that many requirements are not purely technical. Cost, schedule, and safety issues are additional requirements that are always on the mind of project management. Still other requirements can be more emotional (e.g., aesthetic appeal).

The approach to equipment design presented in this chapter is the systems engineering concept of minimizing the life cycle cost, where the *life cycle* is defined as all activities associated with the equipment from its initial design through its final disposal after service use. Stated simply, the design process should consider and minimize the costs incurred over the complete life of the equipment. Extra effort put forth early in the design phase can often have a large payoff later in the life of the equipment. For example, the cost of correcting a problem in manufacturing can be many times greater than the cost of making the correction during the design phase. Additional costs, such as disposal and recycling of the equipment after it has passed its useful life, can be minimized with proper attention early in the design phase.

DYNAMIC ENVIRONMENTS

Shock and/or vibration (dynamic) environments cover a wide range of frequencies from quasi-static to ultrasonic. Examples of different dynamic environments and the frequency ranges over which they typically occur are detailed in the various chapters and references listed in Table 23.1. The classification of vibration sources and details on how measured and predicted data should be quantified are presented in Chap. 19. From a design viewpoint, dynamic excitations can be grouped as follows.

Quasi-Static Acceleration. Quasi-static acceleration includes pure static acceleration (e.g., the acceleration due to gravity) as well as low-frequency excitations. The range of frequencies that can be considered quasi-static is a function of the first normal mode of vibration of the equipment (see Chap. 21). Any dynamic excitation at a frequency less than about 20 percent of the lowest normal mode (natural) frequency of the equipment can be considered quasi-static. For example, an earthquake excitation that could cause severe dynamic damage to a building could be considered quasi-static to an automobile radio.

Shock and Transient Excitations. Shock (or transient) excitations are characterized as having a relatively high magnitude over a short duration. Many shock excitations have enough high-frequency content to excite at least the first normal mode of the equipment structure, and thus produce substantial dynamic response (see Chap. 8). The transient nature of a shock excitation limits the number of response cycles experienced by the structure, but these few cycles can result in large displacements that could cause snubbing, yielding, or tensile failures if the magnitude of the excitation is sufficiently large. Frequent transients can also result in low-cycle fatigue failures (see Chap. 33).

Periodic Excitations. Periodic excitations are of greatest concern when they drive a structure to respond at a normal mode frequency where the motions can be dramatically amplified (see Chap. 2). Of particular concern is the repetitive nature of the response that can accumulate enough cycles to cause fatigue failures at excitation levels less than those required to cause immediate yielding or fracture. The most basic form of a periodic excitation is the sinusoidal excitation caused by rotating equipment. However, other periodic excitations may include strong harmonics that might be damaging, e.g., the vibrations produced by reciprocating engines and gearboxes (see Chap. 37). All harmonics of the periodic excitation must be considered.

Random Excitations. Random excitations occur typically in environments that are related to turbulence phenomena (e.g., wave and wind actions, and aerodynamic and jet noise). Random excitations are of concern because they typically cover a wide frequency range. All natural frequencies of the structure within the frequency bandwidth of a random excitation will respond simultaneously. Assuming the structure is linear, the response will be approximately gaussian, as defined in Chap. 19, meaning that large instantaneous displacements, as well as damaging fatigue stresses, may occur.

Mixed Periodic and Random Excitations. Mixed excitations typically occur when rotating equipment induces periodic excitations that are combined with excitations from some flow-induced source. An example would be a propeller airplane, where the periodic excitation due to the propeller is superimposed on the random excitation due to the airflow over the fuselage (see Chap. 30). It is important to compute the stresses in the equipment due to both excitations applied simultaneously. The same is true of shock excitations that may occur during the vibration exposure.

OTHER ENVIRONMENTS

Other environments may have an effect on material properties and/or help define what materials and finishes can be used during the design and construction of the equipment. The more important environments that should be considered are as follows.

Temperature. Material properties can change dramatically with temperature. Of particular concern for dynamic design are the material stiffness changes, especially in nonmetallic materials such as composites (see Chap. 34). Many nonmetallic materials show a dramatic reduction in stiffness at higher temperatures. Material strength and failure modes will also change with temperature. Some metals will exhibit high-strength ductile behavior at room temperature, and then shift to low-strength brittle behavior at low temperatures (see Chap. 33). Thermal strains can also induce stresses and deformations in structures that need to be considered as part of the design process. A thorough understanding of the expected operating and nonoperating temperatures, plus the amount of exposure time in each temperature range, is required when designing equipment structures for dynamic environments.

Humidity. Humidity can have an effect on material properties, especially plastics, adhesives, and elastomers (see Chap. 33). Some nonmetallic materials can swell in humid environments, resulting in changes in stiffness, strength, and mass. Humid environments can also lead to corrosion in some materials that ultimately reduce strengths.

Salt/Corrosion. Ocean and coastal environments are of particular concern because the corrosion they commonly produce can lower the strength of a material. Corrosion and oxidation can also cause clogging or binding in flexible joints. Protective finishes, seals, and naturally corrosive resistant materials are needed when equipment is designed to withstand long durations in ocean and coastal environments. Corrosive environments can also occur in power plants and chemical processing industries.

Other. Other environments might affect the dynamic performance of equipment. Two such examples are vacuum and electromagnetic fields. Vacuum environments (e.g., space vehicles or aircraft at high altitudes) can cause pressure differentials in sealed structures, which produce static stresses that are superimposed on the stresses due to dynamic responses. Vacuum environments also lack the damping provided by the interaction of the structure with the air. Electromagnetic fields can interfere with the functional performance of electronic subassemblies, and sometimes induce vibration of steel panels.

LIFE-CYCLE ANALYSIS

Dynamic design typically concentrates on the service environment, but there are other conditions during the life of a product that may require special consideration. The definition of all of the different conditions (environment magnitudes and duration) that the equipment will be exposed to during its total life, from manufacture to disposal, is commonly referred to as a *life-cycle analysis.*

Manufacturing Conditions. The life of equipment typically begins when it is manufactured. Manufacturing-induced residual stresses and strains due to plastic deformations, excessive cutting speeds, elevated adhesive cure temperatures, or welding can adversely affect the initial strength of materials. Understanding the material properties after manufacturing-induced excitations (and possible rework) is a critical first step in a life-cycle analysis.

Test Conditions. Equipment often undergoes factory acceptance or environmental stress screening tests (see Chap. 18) before it is put into service. These test environments can induce initial stresses and strains that reduce the resultant strength. An example is a pull test of a wire bond. The test should produce failure in a poor bond, but may also cause permanent plastic deformation in the ductile wire. When predicting the overall fatigue life of an item of equipment, any initial tests must be considered as excitations that will accumulate damage.

As discussed in Chap. 18, at least one sample item of any new equipment must pass a *qualification test* to verify that it can survive and function correctly during its anticipated shock and/or vibration environments. This qualification test generally represents the most severe dynamic environment the equipment will experience, and hence the equipment must be designed for this test environment. However, since the sample item used for the qualification test is not delivered for service use, the qualification test does not have to be included in the life-cycle analysis.

Shipping and Transportation. Once an equipment item is manufactured, it probably will be transported to its operating destination. This transportation environment can often induce excitations that will not be seen in service use. Examples include shock excitations from handling between shipping phases (e.g., dropped packages when unloading a truck), and low-frequency vibration excitations induced by repeated roadway imperfections as seen by a ground transportation vehicle. Special features may need to be added to the equipment, such as additional support parts, to help it survive shipping excitations. One example is a temporary part that is installed between two assemblies that would normally be vibration-isolated in use. The temporary part eliminates excessive displacements due to large low-frequency shipping excitations. Once the system arrives at its destination, the temporary part is removed so the two assemblies can then move freely.

In some cases, the transportation environments may be so much more severe than the service environment that special shipping containers need to be designed to attenuate the transportation excitations. Vibration-isolated shipping containers are often used when transporting sensitive equipment (see Chap. 39).

Service Conditions. The most obvious condition to understand is the service environment of the equipment. A significant portion of the design process should be devoted to accurately determining the dynamic environments under which the equipment must operate. A thorough understanding of the service dynamic environments will help to ensure that the equipment will function both properly and economically. Standard dynamic environments that have been developed for various commercial and military applications may be used to help determine the service excitations (see Chap. 17). These standards, however, should be used with care because they often provide conservative shock and/or vibration estimates that may result in equipment that is overdesigned and more costly than necessary.

When the equipment is to be used in multiple locations, a larger set of dynamic environments must be considered. For each environment, the type, magnitude, duration, and other conditions (e.g., temperature range) should be itemized. For items of equipment that will be produced in large quantities, a statistical approach that groups the dynamic environments into histograms should be considered (see Chap. 18). While the specification of service environment magnitudes and durations is often the responsibility of another organization, the designer must review the desired requirement thoroughly and often request additional information.

DYNAMIC RESPONSE CONSTRAINTS AND FAILURE CRITERIA

Important requirements that need to be defined before equipment is designed are the allowable dynamic responses and failure criteria. Often there will be multiple constraints that need to be satisfied.

Displacement. Displacements due to dynamic excitations must always be considered when the equipment is made up of several subassemblies. The overall motion (or sway space) of an equipment item must also be considered when it will be mounted near other structures. This is often a concern with vibration-isolated equipment. Displacements can also be a concern for position-sensitive equipment such as printing, placement, optical, and measurement devices.

Velocity. Velocity response is of concern for all structures, because the modal (relative) velocity of the structural response at a normal mode is directly proportional to modal stress.[1] This fact can be used to estimate the stress due to the response of a structure at any given normal mode frequency, as will be detailed later.

Acceleration. Some products are most susceptible to acceleration responses. For example, an electrical relay or switch may unlatch when the acceleration acting on the mass of the contact is large enough to cause it to change state. Furthermore, quasi-static acceleration excitations are proportional to stress in the equipment structure.

Permanent Deformation and Factors of Safety. A critical part of the requirements definition process for dynamic environments is to clearly state the allowable amount of permanent deformation that the equipment will tolerate. Some equipment can still function acceptably after being subjected to brief, high-excitation conditions that cause some plastic deformation. Other equipment may not tolerate any yielding that could cause misalignment or interference. Some customers may specify factors of safety that must be met as part of a development specification. These are typically calculated based on stresses relative to the allowable material yield and/or tensile strengths.

Fracture, Fatigue, and Reliability. Equipment intended for use over a relatively long-exposure duration should carry with it some clearly defined fatigue and/or reliability requirement. The equipment design team should establish a reliability goal in terms of fatigue life. This is of particular concern when a premature failure of the equipment can result in severe economic damage or personal injury.

STRUCTURAL REQUIREMENTS

Structural and physical requirements must be defined before the start of a design. For equipment that will be used as part of a larger system, the physical requirements may be negotiable, especially in terms of mounting points and final geometry. These requirements are typically specified as part of an interface agreement, often called an *interface control document* (ICD), between the product development teams.

Volume. The overall volume requirement for an equipment item is an obvious requirement, but it may necessitate some design study. One example would be a combination of a minimum natural frequency and a radiating thermal environment requirement. A smaller design typically has a higher natural frequency due to the stiffness vs. length cubed effect in bending (see Chap. 1). However, this is contrary to the need for a large surface area to facilitate radiation heat transfer. As with most design problems, these effects need to be balanced within the allowable volume. The volume should also include allowances for any displacements that may occur over the life of the equipment.

Mass. Mass or weight requirements can conflict with other equipment requirements. For example, equipment that has a maximum mass requirement may also have a shock and/or vibration-isolation requirement (see Chaps. 38 and 39). The resulting equipment will need to be designed with a low-stiffness isolation system such that the required level of isolation can be reached while still meeting the maximum mass requirement. Other conflicting requirements are minimizing mass while maximizing stiffness and conduction heat transfer. When a mass needs to be controlled accurately, care should be given to the control of both the density and geometry of the parts, especially when the materials used are alloys of high-density metals or composites.

Materials. High-strength, low-density metals are typically the materials of choice for equipment that is exposed to dynamic environments. While this is usually a wise choice, other factors should be considered. In many cost- and time-conscious industries, procurement organizations limit the number of materials from which a product can be made. This is a practice that can save money and limit inventories of expensive specialty materials. The designer needs to understand this situation and learn to work with the available choices of materials. A second concern is that these materials must often be selected in certain stock thicknesses and shapes. One benefit of these measures is that the physical properties of standard materials are often well documented. If not, the designer should strive to work toward a common material property database that can be linked to the available material choices.

Damping properties can be measured for polymers, elastomers, and adhesives using the procedures detailed in Chap. 36. The damping properties of adhesives are an important factor to consider when choosing between options. Adhesives that join lightly damped members can significantly reduce the overall response of the equipment assembly. Fatigue (or fracture) properties for most common materials can be found in Chaps. 33 and 34, as well as Refs. 2 to 4.

Finally, the designer should review the other required environmental conditions that may cause further constraints on the available choices of materials. When feasible, the designer should use common materials that have well-defined properties. Materials that are more exotic should be considered only when they are essential and their properties are well-documented and controlled.

OTHER REQUIREMENTS

It is important to consider other requirements that can adversely affect the finished equipment if not considered early in the design process.

Safety. For those items of equipment where a failure or malfunction during service use might result in severe economic damage or personal injury, safety must be a primary concern. Safety issues should also receive top priority during all other life cycle phases, including manufacturing, handling, and transportation. A qualified safety engineer should be involved in all phases of the design process.

Cost and Schedule. Cost is an important concern that must be considered by every designer developing new equipment. Of particular importance is the life cycle cost of the equipment. It is often less expensive overall to spend time early in the design phase to define and understand the equipment requirements. This can often reduce costly changes to the design further along in its development. However, as previously discussed, safety requirements must always receive careful consideration in making cost and schedule decisions.

Disposal/Recycle. Disposal and recycling requirements should always be considered in the design. Some markets now require that the final disposal of an equipment item include recycling of its materials. Products may also be remanufactured, that is, some types of equipment that have completed their service life might be refurbished, with worn parts repaired or replaced, and then returned to service.

Other. The designer should be aware that equipment needs to function well in ways other than its prime task. Additional features that will help other groups work with the equipment should be considered early in the design phase. Included here are such features as handles, additional holes for lifting equipment, modular design, and adjustable interfaces. When conflicting requirements make a straightforward design difficult, it is sometimes desirable to convene a design team comprised of engineers in such disciplines as systems operation analysis and testing, electromagnetic compatibility, high-reliability parts, cost control, manufacturing, and thermal analysis, as well as shock and vibration.

METHODS OF CONSTRUCTION

Equipment designed to withstand shock and/or vibration excitations must typically be stronger than equipment that only has to withstand gravity or static acceleration loads. This dictates that the equipment have a well-defined primary structure that can withstand the dynamic excitations, as well as carry the additional excitations that might be internally generated. Basic construction methods should be considered early in the design process to facilitate the modeling and analysis procedures discussed later.

PRIMARY STRUCTURE

Primary structures are those that carry the greatest loads and support the secondary structures and subassemblies. The design and analysis of any product should start with particular attention to primary structure. The primary structural elements often have to be designed early in the product development cycle to allow for long leadtime material and tooling acquisition. Simple lumped-parameter (see Chap. 2) or beam/plate finite element models (see Chap. 23) can be used to perform initial stiffness and natural frequency calculations for primary structures. There are many ways to build primary structures.

Machined Parts. Machined parts are often used for primary structures. The machining operations can be customized to place holes and attachment points for secondary structures where needed. For economic reasons, machine operations can be used to remove unnecessary material or allow thicker sections where needed. Machined parts are typically used for low-volume production. Unfortunately, machining operations can also reduce the strength of the parent material by introducing microcracks that might lead to fatigue or fracture. Machined parts may need to be heat-treated after machining to develop the necessary strength and ductility for the intended use.

Castings/Forging. Casting or forged parts are typically used for high-production-volume structural elements because they usually can be formed in near-final shapes that reduce the need for machining operations. Cast materials typically have lower strength and ductility than wrought or forged materials (see Chap. 33). Cast materials also can suffer from various manufacturing defects, such as porosity and shrinkage, which can increase part variability. This variability should be factored into the stress and strength analysis of the part.

Forged parts typically have higher strengths than cast and wrought materials. The forging process can shape material grain and orient the strength along specific part directions. Forged parts are used when the very highest strengths are needed to resist high excitations, e.g., in aircraft landing gear and construction equipment linkages. The forging process does tend to be expensive because of the hard tooling that is needed to form parts under high temperatures and pressures.

Plates/Sheet Metal. Sheet and plate parts are often used for primary structures, especially when they are formed into more rigid three-dimensional shapes. Sheet and plate material can often be bent, cut, and then joined to other parts to give strength and stiffness where needed. Automobile bodies are excellent examples of how sheet metal can be used to form rigid and reliable structures. Modern computer-controlled laser and water-jet cutting techniques can be used to form complicated sheet or plate metal geometries economically for even low-volume production. The important thing to remember with sheet or plate metal construction is that parts need to be stiffened in the out-of-plane (normal to the surface) direction. Care should also be given to minimizing large unsupported areas that can vibrate, especially with acoustic excitation.

Beam Frames. Beam and tube construction is a very efficient way to make primary structures that span large distances, especially when built into trusses or frames. Beams and tubes also have the advantage of high material strength because of the manufacturing processes, such as extrusions, that form them into their continuous cross sections. The most difficult part of designing a beam or tube structure is determining the best way to join the pieces. Welding can often reduce the strength of the material at the joints, requiring additional fittings or gussets to maintain the necessary overall strength. Care should also be given to locating any holes or secondary attachment points at low-stress locations on the beams.

Composite Structures. Composite structures have proven to be efficient primary structures, especially when high strength and low weight are prime concerns. Composite materials can be laid up into plate, beam, and large thin-wall structures. Boat hulls and filament-wound pressure vessels are good examples of large composite thin-wall structures. Composite materials can be mixed, taking advantage of different strength, stiffness, thermal conductivity, and thermal expansion properties for each layer. However, care is required when designing joints for composite structures. See Chap. 34 for details on the properties of composite materials.

SECONDARY STRUCTURES

Secondary structures are those structures used to attach subassemblies to primary structures. Secondary structures typically do not have the more stringent strength and stiffness requirements of the primary structures, so they can be designed later in the development cycle, often allowing changes in geometry to accommodate changes in subassemblies. Secondary structures can also evolve as more cost-efficient materials or manufacturing processes are developed.

Plates/Sheet Metal. Plate and sheet metal parts are often used for nonstructural members such as covers. In this case, the products need only to support their own weight or some minor additional weight due to cables, sensors, or other secondary assemblies. As with all plate structures, care should be given to minimizing large unsupported areas.

Composite Structures. Composite structures can also be used for secondary structures. Their high strength-to-weight ratios make them attractive options for covers and other molded thin-wall sections that need to support some subassemblies.

Plastic Parts. Plastic parts can be used for both primary and secondary structures. Plastics can form adequate primary structures, especially for smaller, low-weight consumer products that are not subjected to extreme conditions. When combined with other materials, such as metal stiffeners in selected areas, plastics can be used effectively for even larger products. The wide range of colors, finishes, and shapes make plastic materials a common choice for secondary structures that are visible to the consumer. They also make excellent low-cost parts when they do not need to be exposed to intense shock and/or vibration excitations.

INTERFACES AND JOINTS

Interfaces are the junctions between the various structural elements that form the equipment. The manner in which the structural elements are jointed together at interfaces is very important in the construction of equipment because the interface friction at joints is the primary source of the damping (energy dissipation) in the equipment that restricts its dynamic response to vibration and, to a lesser degree, shock excitations. There are five basic devices used to make joints in the construction of equipment, namely, (*a*) continuous welds, (*b*) spot welds, (*c*) rivets, (*d*) bolts, and (*e*) adhesives. Typical values of the damping ratio in fabricated equipment using these various types of interface joints are summarized in Table 40.1.

Welded Joints. Welded joints must be well designed, and effective quality control must be imposed upon the welding conditions. The most common defect is excessive stress concentration which leads to low fatigue strength and, consequently, to inferior capability of withstanding shock and vibration. Stress concentration can be minimized in design by reducing the number of welded lengths in intermittent welding. Internal crevices can be eliminated only by careful quality control to ensure full-depth welds with good fusion at the bottom of the welds. Welds of adequate quality can be made by either the electric arc or gas flame process. Subsequent heat-treatment to relieve residual stress tends to increase the fatigue strength. See Refs. 5 and 6 for further information on welded joints.

Spot-Welded Joints. Spot welding is quick, easy, and economical but should be used only with caution when the welded structure may be subjected to shock and vibration. Basic strength members supporting relatively heavy components should not rely upon spot welding. However, spot welds can be used successfully to fasten a metal skin or covering to the structural framework. Even though improvements in spot welding techniques have increased the strength and fatigue properties, spot welds tend to be inherently weak because a high stress concentration exists in the junction between the two bonded materials when a tension stress exists at the weld.

TABLE 40.1 Typical Damping Ratios for Equipment with Various Types of Joints

Method of construction	Typical damping ratio for equipment
Welded and spot-welded	0.01
Riveted	0.025
Bolted	0.05
Bonded	0.01 to 0.05*

* Heavily dependent on the type of adhesive and its thickness.

Spot-welded joints are satisfactory only if frequent tests are conducted to show that proper welding conditions exist. Quality can deteriorate rapidly with a change from proved welding methods, and such deterioration is difficult to detect by observation. However, accepted quality-control methods are available and should be followed stringently for all spot welding. See Refs. 5 and 6 for further information on spot-welded joints.

Riveted Joints. Riveting is an acceptable method of joining structural members when riveted joints are properly designed and constructed. Rivets should be driven hot to avoid excessive residual stress concentration at the formed head and to ensure that the riveted members are tightly in contact. Cold-driven rivets are not suitable for use in structures subjected to shock and vibration, particularly rivets that are set by a single stroke of a press as contrasted to a peening operation. Cold-driven rivets have a relatively high probability of failure in tension because of residual stress concentration, and tend to spread between the riveted members with consequent lack of tightness in the joint. Joints in which slip develops exhibit a relatively low fatigue strength. See Refs. 5 and 6 for further information on riveted joints.

Bolted Joints. Except for the welded joints of principal structures, the bolted joint is the most common type of joint. A bolted joint is readily detachable for changes in construction, and may be effected or modified with only a drill press and wrenches as equipment. However, bolts tend to loosen and require a means to maintain tightness. Furthermore, bolts are not effective in maintaining alignment because slippage may occur at the joint; this can be prevented by using shear pins in conjunction with bolts or by precision fitting the bolts; i.e., fitting the bolts tightly in the holes of the bolted members. See Refs. 5 and 6 for further information on bolted joints.

Adhesives. Adhesives are gaining increased usage as a method of attaching structural elements. When stringent manufacturing controls are used to ensure consistent material properties and area coverage, adhesives can be used in most joints between structures. Adhesives have an advantage over other types of joints when some flexibility and damping is needed in the joint. Adhesives are also good at filling uneven gaps in parts manufactured to wider tolerances. See Ref. 6 for details.

SUBASSEMBLIES

Most types of equipment, especially large items, require subassemblies to perform various functions to satisfy the overall function of the equipment. These subassemblies must be supported on the primary or secondary structures in a way that ensures they will function correctly. Subassemblies can often be treated as lumped masses, but they may need additional dynamic analysis when they are large or sensitive to dynamic effects. Subassemblies and their support structures often need to have their own requirements allocated to them. Examples are given below.

Electronic Assemblies. Many equipment items include one or more electronic assemblies. The designer must ensure that the environment seen by the electronic assembly is low enough for it to function correctly for the intended duration. Often, electronic assemblies will be purchased with specific dynamic requirements that, if exceeded, may cause malfunction or permanent damage. The design of support structures for the electronic assembly must ensure that the input dynamic environment to the assembly is within the specified dynamic requirements. Otherwise, the assembly must be mounted to the equipment through shock or vibration isolators (see Chap. 39).

When it is necessary to design new electronic assemblies, several specific procedures need to be followed. First, the designer should establish a dynamic requirement for the assembly, as discussed earlier. Then, parts that can withstand this requirement must be selected. If some parts cannot be procured (at a reasonable cost) to withstand these levels, then isolation of a subassembly or the whole assembly must be considered. Finally, the design of the electronic circuit boards to which parts will be mounted requires specific attention.

Electronic circuit boards, also called *printed wiring boards* (PWBs) or *printed wiring assemblies* (PWAs), are often constructed of laminated fiberglass or other composite materials. These boards form a flexible plate that, if not supported adequately, can deflect easily and deform or break sensitive electrical part connection leads. Frequent attachment points, stiffening ribs, heat sinks, and plates should be considered early in the design of the electronics. It is often desirable to take advantage of the damping characteristics of adhesives used to bond stiffeners and heat sinks to boards to reduce dynamic deflection. See Ref. 7 for details on the design of electronic equipment for vibration environments.

Mechanical Assemblies. Mechanical assemblies require special attention when they deliver dynamic excitations to the structures that support them. Mechanical items, such as hydraulic cylinders or electrical motors, can induce large dynamic excitations to their support structures. Structural fittings need to withstand these excitations and often allow removal or adjustment of the mechanical assembly after its original manufacture. Dynamic excitations can also affect the performance of mechanical assemblies. For example, dynamic accelerations can act on imbalanced masses in rotating equipment to cause additional shaft displacement or speed errors. These disturbances need to be either limited or isolated.

Optical Assemblies. Optical assemblies need special consideration when used in dynamic environments. Optics must often be mounted using strain-free exact constraints. Overly constrained mounts are statically indeterminate, causing unpredictable and unwanted deformations. The dynamic parameters of the optical elements by themselves must also be well understood so that the effects of any dynamic excitations can be kept to an acceptable level. Of considerable concern is the lightly damped and brittle nature of glass optics.

SHOCK AND VIBRATION CONTROL SYSTEMS

As mentioned in several of the previous sections, many systems need to be designed to provide some sort of vibration isolation for sensitive assemblies contained within them. Shock and/or vibration isolation is typically achieved by what is essentially a low-pass mechanical filter (see Chaps. 38 and 39). These isolation systems can be very effective and should be considered early in the equipment design cycle, but are often considered later as a fix for a poor design. Passive shock and vibration control can also be achieved by careful attention to the damping characteristics of the materials used in the construction of the structure (see Chap. 35). Finally, applied damping treatments can be used to suppress unwanted dynamic responses (see Chap. 36).

DESIGN CRITERIA

Based upon a thorough evaluation of the environments and requirements summarized in the preceding section, specific design criteria must now be formulated. These criteria may cover any or all of the environments previously summarized, but it is the shock and vibration (dynamic excitations) environments that are of concern in this handbook. The dynamic environments are usually specified as motion excitations (commonly acceleration) at the mounting points of the equipment to its supporting structure. However, there may be situations where the equipment is directly exposed to fluid flow, wind, or aeroacoustic loads, which cause fluctuating pressure excitations over its exterior surfaces that can produce a significant contribution to the dynamic response of the equipment. An example would be a relatively light item of equipment with a large exterior surface area mounted in a space vehicle during launch. In this case, the dynamic excitation design criteria must also include pressure excitations over the exterior surface of the equipment, as detailed in Chap. 32. Nevertheless, attention here is restricted to dynamic inputs in the form of motion excitations at the mounting points of the equipment. It is assumed these dynamic excitations will be described by an appropriate frequency spectrum, as summarized in Table 18.2.

DESIGN EXCITATION MAGNITUDE

The procedures for deriving the magnitude of the dynamic excitations for design purposes are essentially the same as those used to derive qualification test levels in Chap. 18. The principal steps in the procedure are as follows.

Determination of Excitation Levels. When the structural system to which the equipment is to be mounted is available, the shock and vibration levels should be directly measured in terms of an appropriate frequency spectrum (see Table 18.2) at or near all locations where the equipment might be mounted. If the structural system is not available, the shock and vibration levels must be predicted in terms of an appropriate frequency spectrum at or near all locations where the equipment might be mounted using one or more of the prediction procedures detailed in other chapters of this handbook and summarized in Chap. 18. These measurements or predictions should be made separately for the shock and/or vibration environments during each of the life-cycle phases discussed in the previous section.

Determination of Maximum Expected Environments. For each life-cycle phase, the measurements or predictions of the shock and/or vibration environments made at all locations at or near the mounting points of the equipment to its supporting structure should be grouped together. Often design criteria are derived for two or more equipment items in a similar structural region. Hence, a dozen or more measurements or predictions may be involved in each grouping of data (called a *zone* in Chap. 18). A limiting (maximum) value of the spectra for the measured or predicted shock and/or vibration data at all frequencies is then determined, usually by computing a *statistical tolerance limit* defined in Eq. (18.2). The statistical tolerance limit given by Eq. (18.2) provides the spectral value at each frequency that will exceed the values of the shock and/or vibration spectra at that frequency for a defined portion β of all points in the structural region with a defined confidence coefficient γ. This limiting spectrum is called the *maximum expected environment* (MEE) for the life-cycle phase considered.

The MEE will generally be different for each life-cycle phase. From a design viewpoint, since the equipment response is heavily dependent on the frequency of the excitation, it is the largest MEE at each frequency (that is, the envelope of the MEEs for all life-cycle phases) that is important. This envelope of the MEEs is called the *maximax environment.* This same concept of a maximax spectrum is commonly used to reduce the time-varying spectra for nonstationary vibration environments, as defined in Chap. 19, to a single stationary spectrum that represents the maximum spectral values at all times and frequencies.

Equipment Loading Effects. The shock and/or vibration measurements or predictions used to compute the maximax excitation spectral levels at the mounting points of the equipment are commonly made without the equipment present on the mounting structure. Even when the equipment is present for the measurements or modeled for the predictions, the computations required to determine MEEs and the final maximax spectrum smooth the detailed variations in the spectral level with frequency. However, if the equipment is relatively heavy compared to its mounting structure, then when the equipment is actually mounted on the structure, the shock and/or vibration levels at the equipment mounting points are modified. This is particularly true at the normal mode frequencies of the equipment where it acts like a dynamic absorber, as detailed in Chap. 6. The result is a spectrum for the input excitation from the supporting structure that may be substantially reduced in level at the normal mode frequencies of the equipment. If this effect is ignored, the maximax spectrum might cause a severe overdesign of the equipment.

The equipment excitation problem can be addressed in two ways. First, if there is a sufficient knowledge of the details of the supporting structure, the input excitation spectra at the equipment mounting points can be analytically corrected using the mechanical impedance concepts detailed in Chap. 9. Specifically, let $Z_s(f)$ and $Z_e(f)$ denote the mounting point impedance of the supporting structure and the driving point impedance of the equipment, respectively. Then for a periodic vibration

$$L_c(f) = \frac{L_r(f)}{|1 + [Z_e(f)/Z_s(f)]|} \tag{40.1a}$$

where $L_c(f)$ and $L_r(f)$ are the line spectra, as defined in Eq. (19.5), for the response of the equipment mounting structure with and without the equipment present, respectively. For a random vibration,

$$W_{cc}(f) = \frac{W_{rr}(f)}{|1 + [Z_e(f)/Z_s(f)]|^2} \tag{40.1b}$$

where $W_{cc}(f)$ and $W_{rr}(f)$ are the power spectra, as defined in Eq. (19.13), for the response of the equipment mounting structure with and without the equipment present, respectively. For those situations where the driving point impedance of the equipment is small compared to the mounting point impedance of the structure, that is, $Z_e(f) << Z_s(f)$, it is seen from Eq. (40.1) that the vibration response of the equipment mounting structure is only slightly altered when the equipment is attached. However, if $Z_e(f)$ approaches $Z_s(f)$, as it often will at the normal mode frequencies of equipment mounted on relatively flexible structures, then the vibration of the mounting structure will be significantly modified by the presence of the equipment, and a correction of the design levels for the equipment loading will be required. Again assuming there is a sufficient knowledge of the details of the supporting structure, a second way to correct for the equipment loading problem is to include at least a portion of the supporting structure in the equipment model that will be used for the equipment response analysis to be discussed later.

DESIGN LIFE

For equipment that is designed for a long service life, the potential for a time-dependent failure (e.g., fatigue damage) is generally of primary concern. Hence, the total duration of the dynamic excitation exposure during all of the life-cycle phases must be determined. For shock environments, the problem reduces to simply estimating the total number of shocks that will occur during each of the life-cycle phases. For vibration environments, however, an equivalent duration for the vibration excitations during each life-cycle phase must be computed. If the vibration environment during a life-cycle phase were stationary, the task would be simple. However, vibration environments during life-cycle phases are often nonstationary (see Chap. 19). A common approach in this case is to assume any time-dependent failure of the equipment follows the inverse power law given by Eq. (18.6), where a value of $b = 8$ is often assumed for metal structures with no stress concentrations, and $b = 4$ is commonly assumed for electrical and electronic equipment, as well as metal structures with substantial stress concentrations. Using Eq. (18.6), vibration environments of different magnitudes and durations can be collapsed to a single stationary vibration environment with an equivalent damaging potential using Eqs. (18.7) and (18.8), as illustrated for automotive equipment in Table 18.4. In addition, this procedure is often used to collapse the vibration environments during each of the life-cycle phases into a single spectrum with an equivalent total duration for design purposes.

DESIGN MARGINS

Given the maximax spectra for the shock and/or vibration excitations at the mounting points of the equipment, perhaps with a correction for the loading effects of the equipment on its supporting structure, it is common to further increase the levels to allow for uncertainties in the derived maximax levels. This increase in the levels is called the *design margin,* and is commonly selected to be between +3 dB and +6 dB. For a periodic vibration described by a line spectrum, as defined in Eq. (19.5), or a shock described by a shock response spectrum, as defined in Eq. (20.33), +3 dB and +6 dB correspond to a multiplication of the spectral values by $\sqrt{2}$ and 2, respectively. For a random vibration described by a power spectrum, as defined in Eq. (19.8), +3 dB and +6 dB correspond to a multiplication of the spectral values by 2 and 4, respectively. Of course, other design margins, either larger or smaller, might be selected depending on the designer's confidence in the derived maximax spectrum. In any case, the maximax spectrum plus the design margin gives the final shock and/or vibration design magnitudes.

METHODS OF ANALYSIS

The analysis of structures for design purposes must involve an analytical model. This section outlines the different types of analysis methods and gives advice on how to use them for the design of equipment.

MODELING

Modeling is an essential part of the design process. Models allow designers to understand the dynamic behavior of the equipment and conduct trade-off studies and experiments without committing to hardware. Options range from the single-degree-of-freedom model (SDOF) (see Chap. 2) to finite element method (FEM) models with thousands of degrees of freedom (DOF) (see Chap. 23). Modern computers allow very large numerical analysis models to be created. In the limit, every detail of a structure can be analyzed. However, the economic wisdom of such a pursuit is questionable.

The decision of how much detail to incorporate into a model should be driven by a clearly defined objective related to a specific design requirement or constraint. The designer must determine the required output of the modeling effort and ensure appropriate design features are adequately represented in the model. For example, the model format and size will depend upon the need for stress results. In general, much less detail is required for displacement models than for stress models. Stress concentration resolution generally requires extensive modeling detail.

Sometimes multiple models are appropriate. For example, lumped-parameter models may be sufficient in preliminary design and for conducting system sway space budget exercises. A beam model may be appropriate for a shock or vibration isolation system and excitation path design. In most cases, a finite element model is necessary to resolve stresses in detailed features. Engineering judgment must be applied to assess the need for modeling nonlinear properties and detailed features. Planning and data management are also important elements of the modeling process. The designer should consider all of the potential uses of the model prior to model construction.

Lumped-Parameter Models. The simplest type of dynamic model is the single-degree-of-freedom system, for which tabulated and charted solutions are widely available (see Chaps. 1 and 2). Lumped-parameter models can be used to accurately represent many mechanical structures. These include structures in which one structural element is much more flexible than the remaining structure. In such a case, the rigid portion of the structure may be adequately represented as a lumped mass connected to the equivalent spring stiffness of the flexible element. The behavior of complex structures often can be represented by very simple dynamic models. Designers should seek to recognize and exploit such simplifications wherever possible, as is illustrated later.

Distributed-Parameter Models. Sometimes the mass of a structure is evenly distributed over a large span of the structure. In these cases, a lumped-parameter model may require a very large number of degrees of freedom, and a distributed modeling approach is preferred. Distributed-parameter models are on the next level of complexity in the hierarchy of modeling tools. Classical beam, plate, and shell theory provide the basis for such modeling. Poles, wings, frames, and the leads of electronic devices may be considered as beams, while printed circuit boards, panels, covers, and doors may be viewed as plates. Modeling techniques for distributed systems are provided in Chaps. 1 and 7.

Finite Element Method Models. Systems with multiple distributed-parameter components become difficult to solve as geometry becomes even modestly complex. Fortunately, user-friendly software tools exist which enable designers to obtain computer solutions to distributed-parameter models using the finite element method of analysis. See Chap. 23, for details on FEM models.

FEM models can vary widely in complexity depending on the desired results. Because FEM models can place substantial demands on computer and human resources, it is important not to make the model any more complex than needed for the application. Relatively simple models that involve only a few hundred degrees of freedom are often adequate to compute estimates for the first few normal modes of a structure. On the other hand, models involving several orders of magnitude more DOF are often required to obtain accurate stress predictions, particularly if the structure has nonlinear characteristics. This variation in the complexity of an FEM model for different applications is illustrated in Fig. 40.2. A drawing of a ground-based radar unit in a stowed position for transportation is shown in Fig. 40.2*A*. A simple (400-DOF) beam approximation for the structure, which is adequate to estimate the first few normal mode frequencies of the equipment, is illustrated in Fig. 40.2*B*. In contrast, a complex (10,000+-DOF) model used for stress analysis is depicted in Fig. 40.2*C*. Construction and execution times of the two models are vastly different. The simple model in Fig. 40.2*B* was constructed in a day or so, and can be executed on the computer in a few seconds. Hence, it can be very useful in preliminary design where numerous analyses can be made with various different structural configurations to select a basic structural design that will have certain desired normal mode characteristics. On the other hand, the complex model, which includes nonlinear features, may take weeks of effort to construct and require much more computer time to execute, making it practical only for final design. Model architectures must be carefully planned for specific objectives.

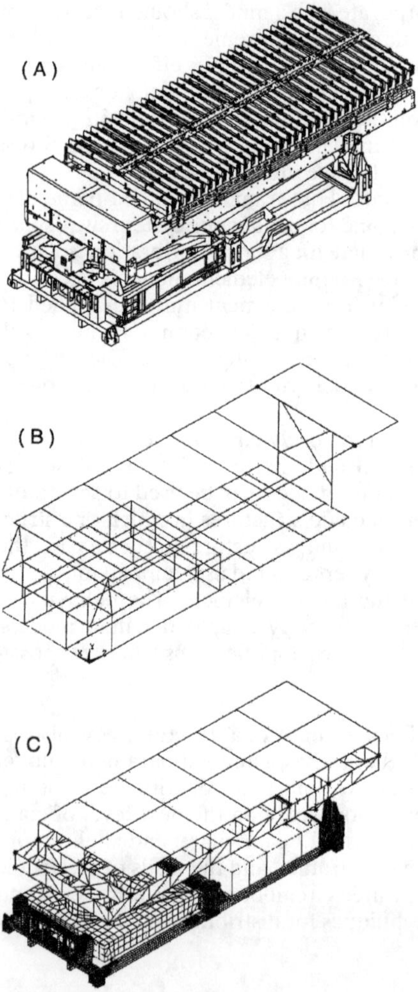

FIGURE 40.2 Illustration of FEM models for ground-based radar unit: (*A*) diagram of unit; (*B*) simple FEM model; (*C*) complex FEM model. (*Courtesy of Lockheed Martin Corporation.*)

Statistical Energy Analysis Models. Even the most detailed FEM model becomes increasingly inaccurate at frequencies above about the 50th normal mode frequency of the structure. For equipment that is exposed to relatively high frequency dynamic excitations, such as aeroacoustic excitations (see Chap. 32) or pyroshock excitations (see Chap. 28), FEM analysis procedures usually become costly and ineffective. In such cases, statistical energy analysis (SEA) procedures become attractive (see Chap. 24). However, SEA procedures have three important limitations, as follows:

1. They provide a vibration response averaged over a structural region, rather than at specific locations on the structure.
2. They provide a vibration response averaged over frequency bandwidths that each cover several normal modes of the structure (commonly ⅓-octave bandwidths), rather than at specific frequencies.
3. They provide accurate results only when there are at least five normal modes of the structure in the frequency bandwidth used for the analysis.

The above limitations make it difficult to translate SEA results into stresses at specific locations on the equipment structure. Nevertheless, SEA can yield valuable descriptions of the average shock and/or vibration response of structural elements in the equipment as a coarse function of frequency. Furthermore, since SEA models do not require structural details, they can be used effectively during the preliminary design phase.

PRELIMINARY DESIGN PROCEDURES

Based upon all the considerations and requirements discussed earlier, an initial design for the equipment should be made, perhaps with the assistance of a standard design handbook (e.g., Ref. 6), relevant reference books (e.g., Refs. 7 and 8), and/or specialized reference documents. This initial design should be modeled by any of the procedures discussed earlier, although finite element method and statistical energy analysis models are preferred. A simple FEM model can be used to estimate the first few normal modes of the equipment, as well as the maximum displacements, velocities, and accelerations induced by the design shock and/or vibration excitations at frequencies up through the first few normal mode frequencies. An SEA model can be used to estimate the average accelerations of various elements of the equipment induced by the design shock and/or vibration excitations at the higher frequencies where there are at least several normal modes of the equipment in the SEA analysis bandwidths (usually ⅓-octave bandwidths). In either case, all of these responses can be evaluated by executing the model(s) for various different structural configurations.

Of particular concern early in the design process is the identification of the potential for excessive stresses in the equipment structure due to the design shock and/or vibration excitations. Since the maximum stresses in equipment structures exposed to shock and/or vibration excitations are generally due to the responses of the normal modes of the equipment, preliminary estimates of stress can be made using the relationship between maximum modal bending stress and maximum modal (relative) velocity given by[1,9]

$$\sigma_m \approx CEv_m/c \approx Cv_m\sqrt{E\rho} \tag{40.2}$$

where σ_m = maximum modal bending stress in the structure
v_m = maximum modal velocity of the structural response
c = speed of sound (longitudinal wavespeed) in the structural material
E = Young's modulus of the structural material
ρ = mass density of the structural material
C = constant of proportionality

The coefficient C in Eq. (40.2) is $C \approx 2$ for all normal modes of homogeneous plates and beams,[9] but can vary widely for complex equipment structures depending on the geometric details and the specific normal mode of the response.[10] Nevertheless, a value of C in the range $4 < C < 8$ is often assumed for the first normal mode response of typical equipment designs.[11] The first normal mode frequency of the equipment can be estimated early in the design using a simple FEM model, as illustrated in Fig. 40.3*B*. Equation (40.2) can then be applied to estimate the maximum stress in the response of any arbitrary equipment structure by assuming the following:

1. The maximum stress in the basic structure of the equipment occurs due to the response of the equipment at its first normal mode frequency.
2. The response of the equipment at its first normal mode frequency can be modeled by a base-excited single-degree-of-freedom system (oscillator), as detailed in Chap. 3.

It is emphasized that this approach provides only crude estimates for maximum stress that are intended to provide guidance on desirable natural frequencies and damping ratios for the equipment design, and the possible need for a shock or vibration isolation system in the final design. Furthermore, it does not provide any information concerning the possibility of functional failures in electrical, electronic, or optical subassemblies in the equipment.

Shock Excitations. Consider a shock environment where the design excitation is described by a relative displacement shock response spectrum (SRS), as given by the maximum value of Eq. (20.32), which is denoted here as $\delta_m(f_n,\zeta)$ where f_n is the natural frequency and ζ is the damping ratio of the single-degree-of-freedom system. Since the SRS is defined as the maximum response of an SDOF system as a function of its natural frequency and damping ratio, it can be used directly with Eq. (40.2) to predict the maximum stress in the structure of equipment due to a response at its first normal mode frequency, specifically,

$$\sigma_m = CE(2\pi f_n)\delta_m(f_n,\zeta)/c \tag{40.3}$$

where all terms are as defined in Eq. (40.2) and the $(2\pi f_n)$ term is needed to convert the relative displacement SRS to an approximate relative velocity SRS, commonly referred to as a *pseudo-velocity shock response spectrum* because it is an exact relative velocity SRS only for $\zeta = 0$. From Chaps. 8 and 20, for simple pulse-type transients, the SRS values vary only slightly with damping ratio for $\zeta \leq 0.05$. Hence, for such transients, the value of the damping ratio used to compute the SRS is not of major importance. However, for more complex transients like pyroshocks (see Chap. 28), the assumed damping ratio has a greater influence on the SRS value and, hence, must be more accurately defined.

For example, assume an item of equipment must be designed to survive the U.S. Navy high-intensity shock test for lightweight equipment, i.e., a weight of less than 350 lb (159 kg), which constitutes one of the most severe shock environments any equipment would experience in a service environment. The test machine is diagrammed in Fig. 27.6, and the SRS for the shock computed with a damping ratio of about $\zeta = 0.01$ is shown in Fig. 27.7. Further assume the equipment is to be constructed from a high-quality aluminum alloy, such as 2024-T3, that has a yield and ultimate strength of 50,000 lb/in^2 (345 MPa) and 70,000 lb/in^2 (483 MPa), respectively.[2] For aluminum, $E \approx 10 \times 10^6$ lb/in^2 (6.9×10^4 MPa) and $c \approx 2 \times 10^5$ in/s (5100 m/sec). From Fig. 27.7, if the first normal mode of the equipment were at 100 Hz, the velocity SRS value $[2\pi(100)\delta_m)$ would be about 400 in/s (10 m/s). Hence, from Eq. (40.3), even assuming an optimistic value of $C = 4$ and adding no design margin, the maximum stress in the equipment structure would be about $\sigma_m = 80{,}000$ lb/in^2

(552 MPa). Although this stress is in the nonlinear region of the material, it probably would cause a structural failure. It follows that the designer should proceed assuming a shock isolation system (see Chap. 39) will be needed in the final design. On the other hand, if the first normal mode frequency of the equipment were above 400 Hz where the velocity SRS value from Fig. 27.7 is 180 in/s (4.6 m/s), then the maximum stress would be about 36,000 lb/in^2 (248 MPa) and the equipment might survive without a shock isolation system. However, it would be difficult to design equipment with a first normal mode frequency above 400 Hz unless the equipment is relatively small.

Periodic Vibration Excitation. Consider a periodic vibration environment where the design excitation is described by a line spectrum, $L_a(f)$, with the units of g (acceleration in gravity units) versus frequency in Hz, as defined in Eq. (19.5). In the unlikely case where the fundamental frequency f_1 of the excitation is fixed, then the stress in the equipment response can be suppressed simply by pursuing a design with no normal modes of the equipment at frequencies near f_1, or any significant harmonics thereof. In many cases, however, the fundamental frequency of periodic vibration environments varies with time, e.g., rotating machinery and reciprocating engines that produce periodic vibration environments often operate at various different rpms. Hence, the designer must usually assume that at least one of the harmonic frequencies of the periodic excitation will correspond to a normal mode frequency of the equipment, at least on some occasions. From Eqs. (2.41) and (40.2), and assuming a damping ratio of $\zeta < 0.1$, the maximum stress in the equipment structure for a periodic excitation at the equipment natural frequency is given by

$$\sigma_m = \frac{CEgL_a(f_n)/c}{4\pi f_n \zeta} \tag{40.4}$$

where $gL_a(f_n)/(2\pi f_n)$ converts the periodic excitation in gravity units to velocity, and all other terms are as defined in Eq. (40.2).

For example, assume an item of equipment must be designed to survive a periodic excitation with an amplitude of 5g and a frequency, at least on some occasions, of 100 Hz. Further assume the equipment has a fundamental normal mode at $f_n = 100$ Hz with a damping ratio of $\zeta = 0.025$, and the equipment structure is a steel alloy where $E = 30 \times 10^6$ lb/in^2 (2.1×10^5 MPa) and $c = 2 \times 10^5$ in/s (5100 m/s). Using an average value of $C = 6$, the maximum stress in the equipment structure is approximated by Eq. (40.4) as $\sigma_m = 55{,}000$ lb/in^2 (380 MPa). A maximum stress of this magnitude would probably not cause an immediate fracture of a high-quality steel alloy, but it might ultimately lead to a fatigue failure. A preliminary estimate of the potential for a fatigue failure could be evaluated by estimating the number of cycles during the design life when the periodic component is at the normal mode frequency of the equipment, and then making a prediction of the fatigue life using the procedures detailed in Chap. 33.

Equation (40.4) provides important guidance to the designer of equipment that will be exposed to a periodic excitation at its fundamental normal mode frequency. Specifically, the maximum stress in the equipment structure is inversely proportional to the damping ratio of the structure. Hence, unlike pulse-type shock excitations, applied damping treatments (see Chap. 36) constitute a powerful design tool for reducing the maximum stress levels induced by periodic excitations.

Random Vibration Excitation. Consider a random vibration environment where the design excitation magnitude is described by a power spectrum, $W_{aa}(f)$, with the units of g^2/Hz versus frequency in Hz, as defined in Eq. (19.8). Assume the random excitation has a frequency bandwidth that covers at least the fundamental normal mode frequency of the equipment. From Eqs. (23.6) and (40.2), and assuming a damping ratio of $\zeta < 0.1$, the rms value of the maximum stress in the equipment structure due to its response at the first normal mode frequency is approximated by

$$\sigma_{\text{rms}} = \frac{CE}{4c}\sqrt{\frac{g^2 W_{aa}(f_n)}{\pi f_n \zeta}} \tag{40.5}$$

where $g^2W_{aa}(f_n)/(2\pi f_n)^2$ converts the power spectrum from g^2/Hz to v^2/Hz, where v is velocity in in/s (m/s) and all other terms are as defined in Eq. (40.2).

As an illustration, assume an item of equipment must be designed to survive a random vibration excitation with a magnitude (including a design margin) of $0.2g^2$/Hz at its fundamental normal mode frequency. Further assume the equipment has a fundamental normal mode at $f_n = 50$ Hz with a damping ratio of $\zeta = 0.025$, and the equipment structure is an aluminum alloy where $E = 10 \times 10^6$ lb/in^2 (6.9×10^4 MPa) and $c = 2 \times 10^5$ in/s (5100 m/s). Using a conservative value of $C = 8$, the maximum rms stress in the equipment structure is approximated by Eq. (40.5) as $\sigma_m = 8{,}700$ lb/in^2 (60 MPa). However, this is an rms stress. The maximum stress must be estimated in terms of a probability function. From Ref. 12, the maximum stress level that will be exceeded at least once during an exposure duration of T sec with a probability of $P(T)$ is estimated by

$$\sigma_m = \sigma_{\text{rms}}\sqrt{2\ln\left[\frac{f_n T}{P(T)}\right]} \tag{40.6}$$

where ln [ ] is the natural logarithm of [ ]. For example, if the total exposure duration at the design magnitude is $T = 5$ h (18,000 s), the stress level that might be exceeded with a probability of $P(T) = 5$ percent would be about 50,000 lb/in^2 (345 MPa). This maximum stress probably would not cause an instantaneous fracture of the structure, assuming it is fabricated from a high-quality aluminum alloy such as 2024-T3 that has an ultimate strength of 70,000 lb/in^2 (483 MPa),[2] but it might cause a fatigue failure over a sufficiently long exposure time.

It should be noted that Eq. (40.6) is unbounded; that is, there is no limit on the maximum stress as the duration T increases. However, experience suggests that this equation yields reasonable results for durations up to the equivalent of about 1×10^6 cycles, assuming the structural response is linear. For longer-duration environments, the potential for a structural failure should be evaluated using the fatigue prediction procedures detailed in Chap. 33.

Equation (40.5) provides important guidance to the designer of equipment that will be exposed to a random vibration excitation at its fundamental normal mode frequency. Specifically, the maximum stress in the equipment structure is inversely proportional to the *square root* of the damping ratio of the structure, rather than the first power of the damping ratio, as for periodic vibrations in Eq. (40.4). Hence, applied damping treatments (see Chap. 36) do not provide as powerful a design tool for reducing the maximum stress levels induced by random excitations.

FINAL DESIGN PROCEDURES

The final design of equipment for shock and/or vibration excitations is best accomplished using a detailed finite element method model, as illustrated in Fig. 40.3*C*. By applying the design excitations to the FEM model, the stresses at critical locations on the equipment structure, as well as the displacements and accelerations at those locations where equipment motions are critical, can be predicted for any modeled structural configuration. The designer can simply modify various elements of the structure to minimize the stress, displacement, and/or acceleration responses at all locations of concern to arrive at a final design. Specialized computer programs are available to facilitate these final design procedures (see Chap. 23). Of course, all of the environments and requirements discussed earlier must be integrated into the design. In particular, the effects of the temperature environment on the strength and stiffness of all elements of the design that are temperature-sensitive must be carefully incorporated into the structural properties.

Fatigue Damage. For equipment being designed for a long service life, a primary step in the final design process is a fatigue life prediction. This can be accomplished in one of two ways, as follows:

1. For either periodic or random vibration excitations, the design excitation can be applied to the FEM model, and a sample time history for the stress response at any location of concern can be computed. This sample time history can then be used to predict the fatigue life using the procedures given for metals in Chap. 33 or composites in Chap. 34.
2. For random vibration excitations, the design excitation can be applied to the FEM model and the spectrum for the stress response at any location of concern can be computed. This spectrum can then be used to make a statistical prediction for the fatigue life using the procedures given in Ref. 13.

Higher-Order Response Modes. Some design shock and/or vibration excitations may have substantial energy in the frequency range of the higher-order normal modes of the equipment. Examples include motions of the equipment mounting structure induced by pyroshocks (see Chap. 28) and aeroacoustic excitations (see Chap. 32). In these cases, statistical energy analysis models can provide valuable support to the design process, starting in preliminary design. Specifically, the SEA model can be used much like an FEM model to modify structural elements so as to minimize the motion response of the structure at any location of interest. As previously mentioned, it is difficult to obtain accurate stress predictions using an SEA model. However, the primary source of shock- and/or vibration-induced stresses in structural elements is usually due to the structural response in its lower-order modes. Hence, the FEM model will generally provide all the required stress data needed for a proper design.

Other Sources of Information. There are many specialized technical handbooks that cover the design of equipment for dynamic excitations that address specific equipment applications or specific types of equipment. For example, Ref. 14 is the NASA Technical Handbook that covers the design and testing of equipment for space vehicle shock and vibration environments. When available, such specialized handbooks should be consulted to support the equipment design process for shock and/or vibration environments.

DESIGN REVIEWS

Following both the preliminary and final design activities, there should be a thorough review of the design details. Following preliminary design, the review should include a study of all considerations that went into the design, including the assumed environments and requirements, the formulation of the design criteria, the planned methods of construction, the preliminary design analysis, and the planned final design analysis. Following final design, the final design analysis procedures and results should be carefully checked. These reviews should be performed by an independent group of engineers that were not directly involved in the design process. In smaller organizations, employing an independent contractor for the design review should be considered. This is particularly desirable if a failure or malfunction of the equipment during its service use could result in major economic damage or personal injury.

DESIGN VERIFICATION

Uncertainty is always present in the modeling and analysis of any dynamic system. By necessity, simplifying assumptions are introduced to make the analysis tractable. Naturally, unmodeled and unexpected phenomena will be present in a given equipment design. The significance of these effects is uncertain. Testing is often the only way to confidently confirm compliance with requirements. Furthermore, testing may also be used as a design tool for structures lacking suitable models, such as those with highly nonlinear response characteristics.

As in other phases of the equipment development, testing should be performed with a clear set of objectives. Since hardware testing can be expensive, careful planning is important to maximize benefits and efficiency. Some organizations separate testing activities from the design functions. Nevertheless, the designer should participate in determining the verification tests that will be performed. Shock and vibration test facilities are expensive to maintain and not available to many small companies and agencies. Commercial test facilities are available for such organizations. Chapter 17 describes general shock and vibration standards and Chap. 18 discusses the derivation of shock and vibration test criteria from measured or predicted excitation data.

MODEL-TEST CORRELATION

Dynamic testing often begins at low excitation levels in order to preview structural behavior and ensure proper instrumentation and test control without causing significant damage to the equipment (see "Development Testing" in Chap. 18). Data collected in the early phases of testing can be used to validate or refute models that may have been used to make design decisions. Full dynamic excitation tests also yield data useful for model correlation purposes, for example, the detection of nonlinear properties that were not modeled.

Frequency response functions, as defined in Chap. 21, are particularly well suited for model-test correlation purposes. In general terms, frequency response functions show input-output relationships. They are useful in relating inputs, such as force or motion, to outputs such as motion or strain. Frequency response functions can be experimentally generated from a variety of tests, including modal hammer impact tests and laboratory vibration tests. When properly determined, frequency response functions provide the modal parameters of the equipment, namely, natural frequencies (eigenvalues), mode shapes (eigenvectors), and damping ratios. Chapter 21 describes experimental modal analysis and modal parameter estimation techniques.

The frequency response functions for a printed wiring assembly computed using a simple FEM model (a few hundred degrees of freedom) and measured in a laboratory vibration test are compared in Fig. 40.3. A drawing of the printed wiring assembly is shown in Fig. 40.3*A*, and the frequency response functions computed using the FEM model and the laboratory vibration test are presented in Fig. 40.3*B*. The comparison shows good agreement for the lower-frequency modes, although the correlation degrades with increasing mode number. A more complex

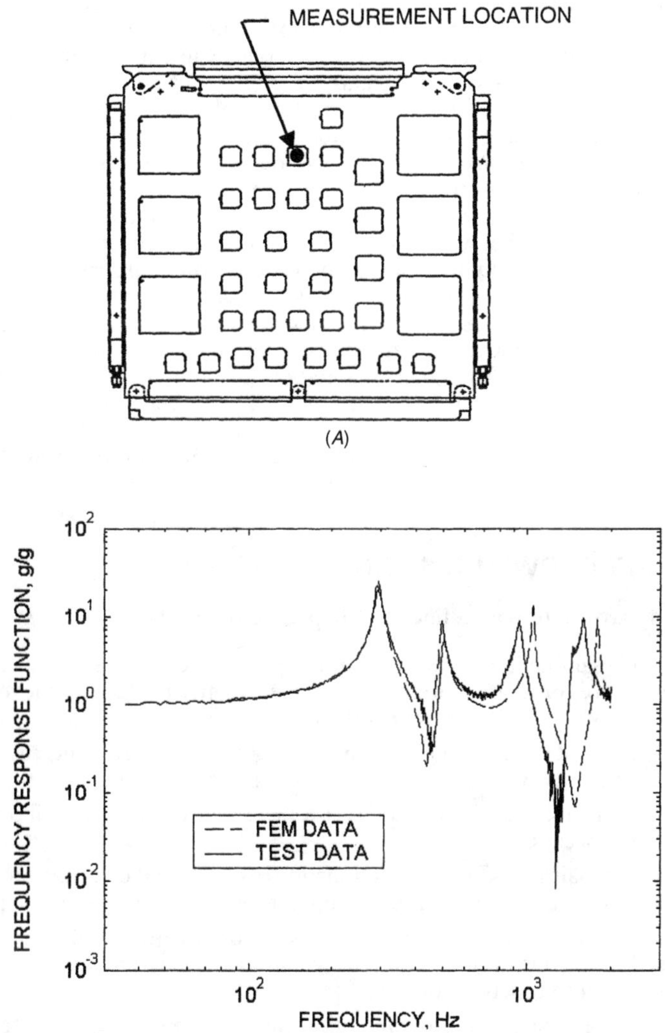

FIGURE 40.3 Comparison of FEM-computed and laboratory-measured frequency response functions for a printed wiring assembly: (*A*) diagram of assembly; (*B*) comparison of FEM and test data. (*Courtesy of Lockheed Martin Corporation.*)

FEM model would provide better agreement for the higher-frequency modes, but often a confirmation of the first few modes is adequate for model verification purposes.

QUALIFICATION TESTING

A qualification test, as defined in Chap. 18, gives the designer and the customer confidence that the equipment will function properly in its expected service environment. It is usually a contractual requirement and commonly involves the application of all environments the equipment will experience in service, applied either in sequence or simultaneously. In particular, for equipment that will experience temperature extremes in service, a temperature test is often performed simultaneously with a vibration test using a combined temperature-vibration test facility. In any case, shock and/or vibration qualification tests occur too late in the design process to allow the cost-effective implementation of design changes. Thus, it is common practice to perform preliminary qualification-like tests before the design phase is completed to ensure the design will pass the qualification test requirement.

Qualification testing requires more than just the structural survival of the equipment within acceptable damage limits. A structure can survive the environment, but be rendered operationally useless by dynamic disturbances. Sometimes operational performance is restored when the dynamic excitation is removed, e.g., electrical circuitry can malfunction under dynamic excitation, intermittent problems can occur as gaps open and close, disruptive electrical noise can be generated, optical surfaces can be distorted, and servo-positioning systems can become unstable. The operational performance of the equipment must be closely monitored during the qualification test to identify any such malfunctions.

RELIABILITY GROWTH TESTING

A reliability growth test, as defined in Chap. 18, involves the following steps:

1. Assuming a sample item of equipment has passed the specified qualification test with no failures or malfunctions, increase the magnitude of the test level by some increment, usually 3 dB, and repeat the test.
2. If the equipment item again passes the test at this higher level, increase the magnitude of the test level again by the same increment and repeat the test.
3. Continue repeating the test at step-wise increased test levels until a failure or malfunction occurs.
4. If possible, repair the equipment to function properly and continue the testing at piece-wise increased test levels until another failure occurs.
5. Again, if possible, repair the equipment and continue the testing at piece-wise increased test levels until it is no longer feasible to make repairs that will allow the equipment to function correctly.
6. Report to the designer the details of all failures identified by the testing that could be repaired. Often simple changes in the design can be made that will suppress or eliminate the failures revealed by the tests.

The theory behind a reliability enhancement test described above is as follows. Even if the equipment is adequately designed to function properly during the qualification test, which represents a conservative simulation of the anticipated service shock and/or vibration environment, increasing the ability of the equipment to function properly during more extreme dynamic excitations will improve the reliability of the equipment in its service environment. Furthermore, by establishing the maximum shock and/or vibration excitations that the equipment can endure, it may be possible to use the equipment at a later time for another application involving more severe shock and/or vibration excitations without the need for a redesign and new qualification testing.

REFERENCES

1. Hunt, F. V.: *J. Acoust. Soc. Amer.,* **32**(9):1123 (1960).
2. Avallone, E. A., T. Baumeister III, and A. M. Sadegh: "Marks' Standard Handbook for Mechanical Engineers," 11th ed., The McGraw-Hill Companies, New York, 2007.
3. Boyer, H. E.: "Atlas of Fatigue Curves," American Society of Metals, Metals Park, Ohio, 1986.
4. McEvily, A. J., Jr.: "Atlas of Stress-Corrosion and Corrosion Fatigue Curves," American Society of Metals, Metals Park, Ohio, 1990.
5. Shigley, J. E., and L. D. Mitchell: "Mechanical Engineering Design," 6th ed., The McGraw-Hill Companies, New York, 2000.
6. Rothbart, H. A.: "Mechanical Design Handbook," 4th ed., The McGraw-Hill Companies, New York, 1996.
7. Steinberg, D. S.: "Vibration Analysis for Electronic Equipment," 3d ed., John Wiley & Sons, New York, 2000.
8. Fuchs, H. O., and R. I. Stephens: "Metal Fatigue in Engineering," John Wiley & Sons, New York, 1980.
9. Gaberson, H. A., and R. H. Chalmers: *Shock and Vibration Bull.,* **40**(2):31 (1969).
10. Crandall, S. H.: *J. Acoust. Soc. Am,* **34**(12):1960 (1962).
11. Piersol, A. G.: *J. IEST,* **44**(1):23 (2001).
12. Nigam, N. C.: "Introduction to Random Vibrations," MIT Press, Cambridge, Mass., 1983.
13. Wirsching, P. H., T. L. Paez, and H. Ortiz: "Random Vibrations: Theory and Practice," John Wiley & Sons, New York, 1995.
14. Kern, D. L., et al.: "Dynamic Environmental Criteria," National Aeronautics and Space Administration, *NASA-HDBK-7005,* 2001.

CHAPTER 41
HUMAN RESPONSE TO SHOCK AND VIBRATION

Anthony J. Brammer

INTRODUCTION

This chapter considers: (1) the structure and properties of the human body regarded as a mechanical as well as a biological system, (2) the effects of mechanical shock and vibration forces on this system, (3) tolerance criteria for shock and for vibration exposure, and (4) means for protection. Man, as a mechanical system, is extremely complex and his mechanical properties readily undergo change. There is limited reliable information on the magnitude of the forces required to produce mechanical damage to the human body. To avoid damage to humans while obtaining such data, it is necessary to use simulations for most studies on mechanical injury. Occasionally it is possible to obtain useful information from situations involving accidental injuries to man, but while the damage often can be assessed, the forces producing the damage usually cannot, and so only rarely are useful data obtained in this way.

For general background material on the effects of shock and of vibration on humans, see Refs. 1 through 4.

CHARACTERIZATION OF FORCES

Forces may be transmitted to the body through a gas, liquid, or solid. They may be diffuse or concentrated over a small area. They may vary from tangential to normal and may be applied in more than one direction. The shape of a solid body impinging on the surface of the human is as important as the position or shape of the human body itself. All these factors must be taken into account in comparing injuries produced by vehicle crashes, explosions, blows, vibration, etc. Laboratory studies often permit fairly accurate control of force application, but field situations are apt to be extremely complex. Therefore, it is often difficult to predict what will happen in the field on the basis of laboratory studies. It is equally difficult to interpret field observations without the benefit of laboratory studies.

SHOCK

The term *shock* is used differently in biology and medicine than in engineering. In this chapter the term *shock* is used in its engineering sense as defined in Chap. 1 of this handbook, that is, for a nonperiodic excitation characterized by suddenness and severity. In general, forces reaching peak values in less than a few tenths of a second and of not more than a few seconds' duration may be considered as shock forces in relation to the human system.

The term *impact* (i.e., a *blow*) refers to a force applied when the human body comes into sudden contact with a solid body and when the momentum transfer is considerable, as in rapid deceleration in a vehicle crash or when a rapidly moving solid body strikes a human body.

VIBRATION

Biological systems may be influenced by vibration at all frequencies if the amplitude is sufficiently great. This chapter is concerned primarily with the frequency range from about 1 Hz to 1 kHz.

METHODS AND INSTRUMENTATION

Most quantitative investigations of the effects of shock and vibration on humans are conducted in the laboratory in controlled environments. Meaningful results can be obtained from such tests only if measurement methods and instrumentation are adapted to the particular properties of the biological system under investigation to ensure noninterference of the measurement with the system's behavior. This behavior may be physical, physiological, and psychological although these parameters should be studied separately if possible. The complexity of a living organism makes such separation, even assuming independent parameters, only an approximation at best. In many cases if extreme care is not exercised in planning and conducting the experiment, uncontrolled interaction between these parameters can lead to erroneous results. For example, the dynamic elasticity of tissue of a certain area of the body may depend on the simultaneous vibration excitation of other parts of the body; or the elasticity may change with the duration of the measurement since the subject's physiological response varies; or the elasticity may be influenced by the subject's psychological reaction to the test or to the measurement equipment.

Control of, and compensation for, the nonuniformity of living systems is essential because of the variation in size, shape, sensitivity, and responsiveness of people and because these factors, for a single subject, vary with time, experience, and motivation. The use of adequate statistical experimental design is necessary and almost always requires a large number of observations and carefully arranged controls.

A range of mechanical and hydraulic vibration exciters have been developed specifically for human laboratory experiments with extensive safety systems. Similarly, acceleration and deceleration sleds have been developed for use in impact tests with human subjects.

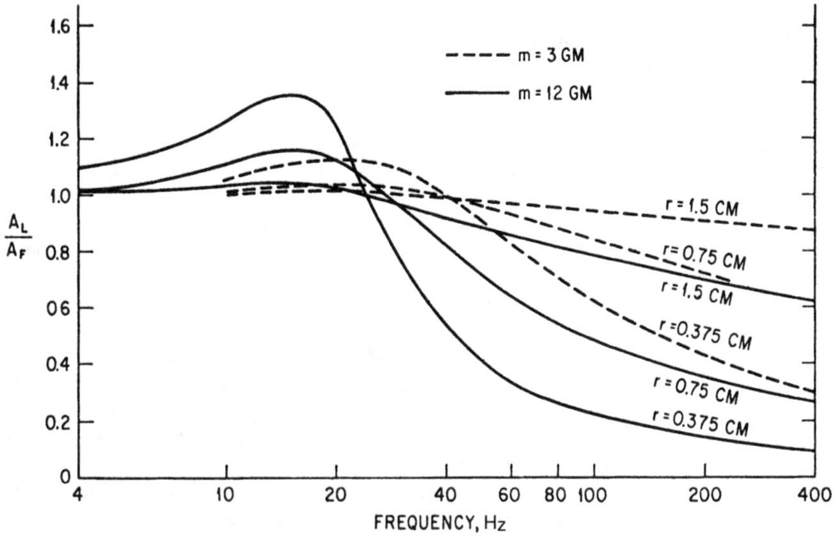

FIGURE 41.1 Amplitude change due to accelerometers of different mass m and size which are attached to a body surface over soft tissue of human subject exposed to vibration. The graph gives the ratio A_L/A_F of the response of the loaded to the unloaded surface for accelerometers having three different radii r. (*Values calculated from unpublished mechanical surface impedance data of E. K. Franke and H. E. von Gierke.*)

PHYSICAL MEASUREMENTS

In determining the effects of shock and vibration on humans, the mechanical force environment to which the human body is exposed must be clearly defined. Force and vibration amplitudes should be specified for the area of contact with the body. Vibration measurements of the body's response should be made whenever possible by noncontact methods. X-ray methods can be used to measure the displacement of internal organs. Optical, cinematographic, and stroboscopic observation can give the displacement amplitudes of parts of the body. If vibration pickups in contact with the body are used, they must be small and lightweight so as not to introduce an excessive mechanical load. This usually places a weight limitation on the pickup of a few grams or less, depending on the frequency range of interest and the effective mass to which the pickup is attached. Figure 41.1 illustrates the effect of mass and size on the response of accelerometers attached to the skin overlying soft tissue. The mechanical impedance of a sitting, standing, or supine subject is extremely useful for calculating the vibratory energy transmitted to the body by the vibrating structure. The mechanical impedance of small areas of the body surface can be measured in different ways (see Chap. 9), for example, by vibrating pistons, resonating rods, and acoustical impedance tubes.

SUBSTITUTES FOR LIVE HUMAN SUBJECTS

The establishment of limits of human tolerance to mechanical forces, and the explanation of injuries produced when these limits are exceeded, frequently requires experimentation at various degrees of potential hazard. To avoid unnecessary risks to humans, animals are used first for detailed physiological studies. As a result of these studies, levels may be determined which are, with reasonable probability, safe for human subjects. However, such comparative experiments have obvious limitations. The different structure, size, and weight of most animals shift their response curves to mechanical forces into other frequency ranges and to other levels than those observed on humans. These differences must be considered in addition to the general and partially known physiological differences between species. For example, the natural frequency of the thorax-abdomen system of a human subject is between 3 and 4 Hz; for a mouse the same resonance occurs between 18 and 25 Hz. Therefore, maximum effect and maximum damage occur at different vibration frequencies and different shock-time patterns in a mouse than in a human.

Many kinematic processes, physical loadings, and gross destructive anatomical effects can be studied on dummies which approximate a human being in size, form, mobility, total weight, and weight distribution in body segments. In contrast to those used only for load purposes, dummies simulating basic static and dynamic properties of the human body are called *anthropometric* or *anthropomorphic* dummies. Several such dummies have been designed for specific simulations.[5] For automobile frontal collisions, the Hybrid III dummy has become the de facto standard to simulate occupants in crash tests and tests of safety restraint systems. The original dummy was constructed to correspond to a 50th-percentile North American male. It possesses a metal "skeleton" covered with a vinyl skin and foam to produce the appropriate external shape, with a rubber lumbar spine curved to mimic a sitting posture, and a shoulder structure capable of supporting safety belt loads. The head, neck, chest, and knee responses of the Hybrid III are designed to mimic human responses, namely, the head acceleration resulting from forehead and side-of-the-head impacts, the fore-and-aft and lateral bending of the neck, the deflection of the chest to distributed forces on the sternum, and impacts on the knee.[4] Hybrid III dummies are now available representing adult-sized small (5th-percentile) females and large (95th-percentile) males, as well as infants and children. A related dummy, the Side-Impact Dummy (SID), is available for automobile side collisions together with dummies developed for this purpose in Europe. Another advanced dummy, Advanced Dynamic Anthropomorphic Manikin (ADAM), has been developed for use in aircraft ejection seats, helicopter seats, and parachute tests. In addition to modeling body segments, surface contours, weights, centers of gravity, moments of inertia, and joint center locations, ADAM replicates human joint motion and the biodynamic response of the spine to vertical accelerations for both small-amplitude vibration and large impacts.[5]

Plastic head forms, conforming to standard head measurements, are designed to fracture in the same energy range as that established for the human head. A cranial vault may be provided to house instrumentation as well as, in some cases, a simulated brain mass with comparable weight and consistency (e.g., a mixture of glycerin, ethylene glycol, etc.). The static properties of the skin and scalp tissue are simulated with polyvinyl foam.

The static and dynamic breaking strength of bones, ligaments, and muscles and the forces producing fractures in rapid decelerations have been studied frequently on cadaver material. Extreme caution must be exercised in applying elastic and strength properties obtained in this way to a situation involving the living organism. The differences observed between properties of wet, dry, and embalmed materials are considerable; changes in these properties also result in changes in the force distribution of a composite structure. Thus, the biofidelity of the surrogate must be considered for each specific situation in which the use of animals, dummies, or cadavers as substitutes for live human subjects is planned.[6]

MECHANICAL CHARACTERISTICS OF THE BODY

PHYSICAL DATA

Most physical characteristics of the human body presented in this section (except for the strength data) have been derived from the analysis of experimental data in which it is assumed that the body is a linear, passive mechanical system. This is an idealization which holds only for very small amplitudes. Therefore, these data may not apply in analyses of mechanical injury to tissue. There is considerable nonlinearity of response well below amplitudes required for the production of damage. This is illustrated by the data in Fig. 41.2, which shows how the mechanical stiffness and resistance of soft tissue vary with static deflection. Bone behaves more or less like a normal solid; however, soft elastic tissues such as muscle, tendon, and connective tissue resemble elastomers with respect to their Young's modulus and S-shaped stress-strain relation.[4] Hence, soft tissue can be described phenomenologically as a viscoelastic medium; plastic deformation need be considered only if injury occurs. Physical properties of human body tissue are summarized in Table 41.1 for frequencies less than 100 kHz.

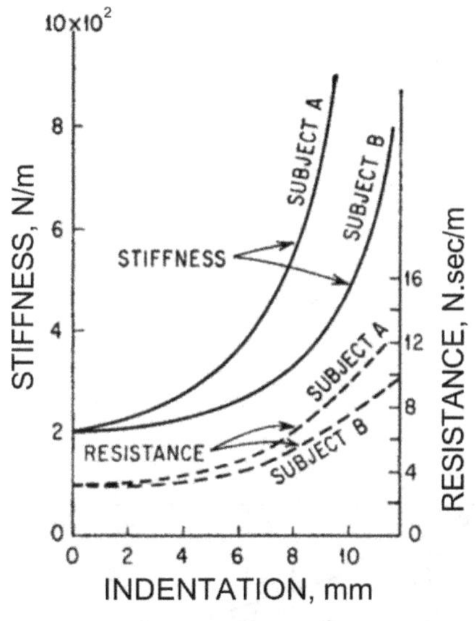

FIGURE 41.2 Mechanical stiffness and resistance of soft tissue as a function of surface indentation of a 2-cm-diameter area (i.e., static loading), for two human subjects. (*After Franke: USAF Tech. Rept. 6469, 1959.*)

The fatigue life of bone and soft tissue in response to cyclic dynamic stress at frequencies between 0.5 and 4 Hz is summarized in Fig. 41.3. In this diagram, the number of cycles to failure N of in vitro preparations is expressed as a function of the ratio of the applied dynamic stress to the ultimate static stress σ/σ_u. The straight lines in Fig. 41.3 represent the functions $N = (\sigma/\sigma_u)^{-x}$, where the value of the index x in the relationship is indicated.

The combination of soft tissue and bone in the structure of the body together with the body's geometric dimensions results in a system which exhibits different types of response to vibratory energy depending on the frequency range: At low frequencies (below approximately 100 Hz), the body can be described for most purposes as a lumped-parameter system; resonances occur due to the interaction of tissue masses with purely elastic structures. At higher frequencies, through the audio-frequency range and up to about 100 kHz, the body behaves more as a complex distributed-parameter system—the type of wave propagation (shear waves, sur-

TABLE 41.1 Typical Physical Properties of Human Tissues

	Soft tissues	Bone (wet)	Bone (dry)
Density (kg/m³)	$1–1.2 \times 10^3$	$1.9–2.3 \times 10^3$	1.9×10^3
Young's modulus (Pa)	—	$1.6–2.3 \times 10^{10}$	1.8×10^{10}
Shear modulus* (Pa)	2.5×10^{3}†	$2.9–3.4 \times 10^9$	7.1×10^9
Bulk modulus (Pa)	2.6×10^{9}†	—	1.3×10^{10}
Shear viscosity (Pa·sec)	15†	—	—
Sound velocity (m/sec)	$1.5–1.6 \times 10^3$	3.4×10^3	—
Acoustic impedance (Pa·sec/m)	1.7×10^6	6×10^6	6×10^6
Tensile strength (Pa)			
cortical bone	—	$1.3–1.6 \times 10^8$	1.8×10^8
cartilage	$5–40 \times 10^6$		
skin	$2–16 \times 10^6$		
ligament	$13–38 \times 10^6$		
tendon	$3–5.5 \times 10^7$		
Compressive strength (Pa)			
cortical bone	—	$1.5–2.1 \times 10^8$	—
trabecular bone (vertebrae)	—	$0.4–7.7 \times 10^6$	—
Shear strength (Pa)			
cortical bone	—	$7.0–8.1 \times 10^7$	—

*Lamé constant
†from soft tissue model. (*von Gierke et al.*[7])

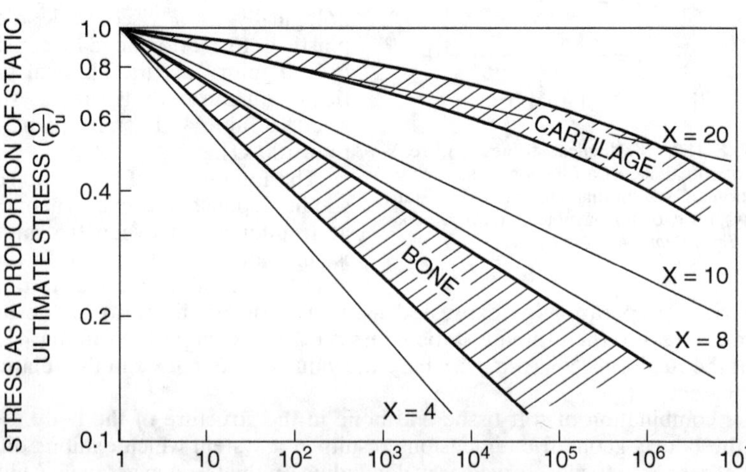

FIGURE 41.3 Fatigue failure of human tissue. The number of cycles of repeated stress N to failure of in vitro preparations is shown as a function of the ratio of the applied dynamic stress to the ultimate static stress σ/σ_u. (*von Gierke.*[8])

face waves, or compressional waves) being strongly influenced by boundaries and geometrical configurations.

LOW-FREQUENCY RANGE

Simple mechanical systems, such as the one shown in Fig. 41.4 for a standing and sitting man, are usually sufficient to describe and understand the important features of the response of the human body to low-frequency vibrations.[8,9] Nevertheless, it is difficult to assign numerical values to the elements of the model, since they depend critically on the kind of excitation, the body type of the subject, and his posture and muscle tone. Large intersubject variability is therefore to be expected and is observed. Of the various factors influencing whole-body biodynamic responses, a reduction in intersubject variability can often be obtained by normalizing measured values by a subject's static mass.[1]

Subject Exposed to Vibrations in the Longitudinal Direction. Below approximately 2 Hz, the body acts as a unit mass. For the sitting man, the first resonance is between 4 and 6 Hz; for the standing man, resonance peaks occur at about 6 and 12 Hz. The numerical value of the impedance together with its phase angle provides data for the calculation of the total energy transmitted to the subject.

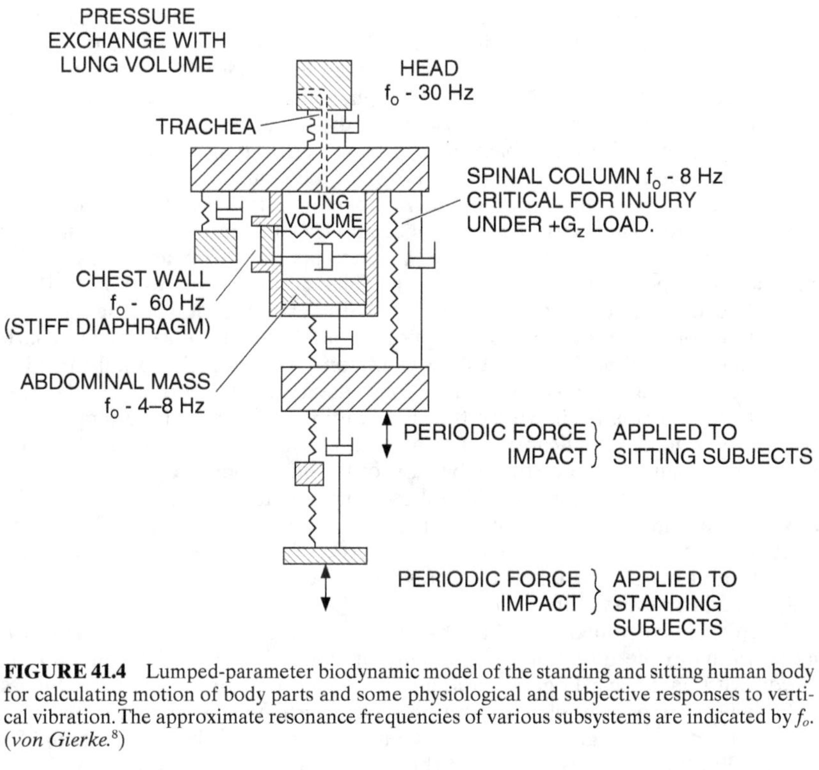

FIGURE 41.4 Lumped-parameter biodynamic model of the standing and sitting human body for calculating motion of body parts and some physiological and subjective responses to vertical vibration. The approximate resonance frequencies of various subsystems are indicated by f_o. (*von Gierke.*[8])

The resonances between 4 and 6 Hz and between 10 and 14 Hz are suggestive of mass-spring combinations of (1) the entire torso with the lower spine and pelvis and (2) the upper torso with forward flexion movements of the upper vertebral column, respectively. The expectation that flexion of the upper vertebral column occurs is supported by observations of the transient response of the body to vertical impact loads and associated compression fractures. The greatest loads occur in the region of the twelfth thoracic to the second lumbar vertebra, which therefore can be assumed as the hinge area for flexion of the upper torso. Since the center of gravity of the upper torso is considerably forward of the spine, flexion movement will occur even if the force is applied parallel to the axis of the spine. Changing the direction of the force so that it is applied at an angle with respect to the spine (for example, by tilting the torso forward) influences this effect considerably. Similarly the center of gravity of the head can be considerably in front of the neck joint which permits forward-backward motion. This situation results in forward-backward rotation of the head instead of pure vertical motion.

Between 20 and 30 Hz the head exhibits a mechanical resonance. When subject to vibration in this range, the head displacement amplitude can exceed the shoulder amplitude by a factor of 3. This resonance is of importance in connection with the deterioration of visual acuity under the influence of vibration. Another frequency range of disturbances between 60 and 90 Hz suggests an eyeball resonance.

Typical values of mechanical impedance and *seat-to-head transmissibility;* that is, the ratio of the response amplitude and phase of the head to steady-state forced vibration of a seated person at that frequency are described in an international standard.[10] They are based on a synthesis of measured values from different experimental studies, each of which was conducted under controlled measurement conditions and involved a number of male subjects. The need for precise definition of measurement conditions, and hence the restricted applicability of the results, stems from the dependence of the biodynamic responses on body shapes (e.g., mass and height), posture, support (i.e., of buttocks, back, and/or feet), and state of ankle and knee joints. The remaining *unexplained* differences between the results of these studies, a situation commonly encountered in biodynamic experiments, led to the specification of the most probable values for the impedance and transmissibility as a function of frequency by an upper and lower envelope that encompasses the mean values of *all* data sets. The envelopes, which are shown by the thick continuous lines in Figs. 41.5 and 41.6, define a range of idealized values that characterize the biodynamic response of a seated person when the back is unsupported and the feet are resting on a surface supporting a rigid seat. Note that data from some individuals will fall outside the range between the two envelopes, as a consequence of their definition. The mean value of all data sets is shown by the thin continuous line in these diagrams, and serves as a target for applications, such as mechanical simulation of the response of the seated human body to vertical vibration, or the development of seats for reducing vibration transmitted to the body. Also shown by the dotted lines in Figs. 41.5 and 41.6 are values calculated using the biodynamic model illustrated in Fig. 41.7.[10] The components of the model do not correspond to those of identifiable body parts, though the motion of mass m_2 is taken to represent that of the head for the calculation of seat-to-head transmissibility.

The mechanical impedance of the human body, lying on its back on a rigid surface and vibrating in the direction of its longitudinal axis, has been determined in connection with ballistocardiograph studies. For tangential vibration, the total mass of the body behaves as a simple mass-spring system with the elasticity and resistance of the skin. For the average subject the resonance frequency is between 3 and 3.5 Hz, and the Q of the system is about 3. If the subject's motion is restricted by clamping

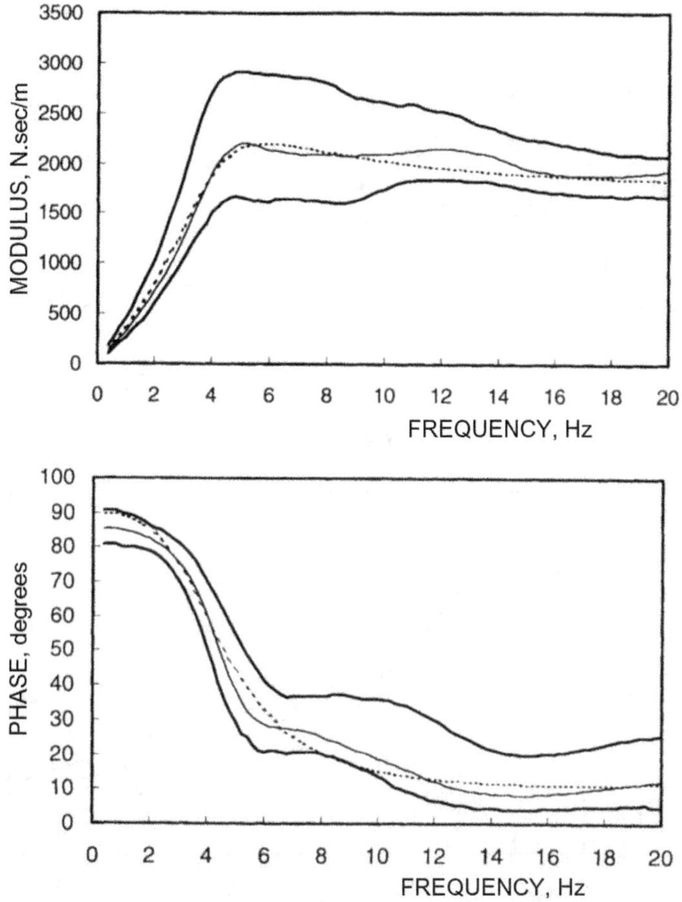

FIGURE 41.5 Driving-point mechanical impedance of the seated human body in the vertical direction (Z-direction of Fig. 41.12). Maximum and minimum envelopes of mean values from data sets are shown by thick lines: mean of all data sets—thin continuous line; response of three-degree-of-freedom biodynamic model—dashed line. (*ISO 5982.*[10])

the body at the feet and at the shoulders between plates connected to the table, the resonance is shifted to approximately 9 Hz and the Q is about 2.5.

One of the most important subsystems of the body, which is excited in the standing and sitting positions as well as in the lying position, is the thorax-abdomen system. The abdominal viscera have a high mobility due to the very low stiffness of the diaphragm and the air volume of the lungs and the chest wall behind it. Under the influence of both longitudinal and transverse vibration of the torso, the abdominal mass vibrates in and out of the thoracic cage. Vibrations take place in other than the (longitudinal) direction of excitation; during the phase of the cycle when the abdominal contents swing toward the hips, the abdominal wall is stretched outward and

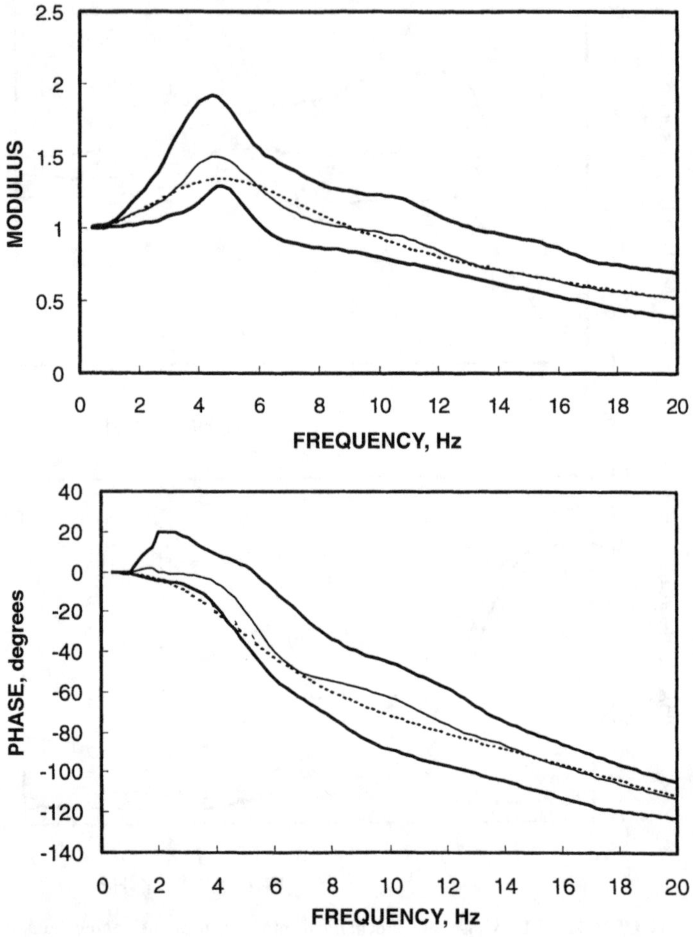

FIGURE 41.6 Seat-to-head transmissibility of the seated human body in the vertical direction (*Z*-direction of Fig. 41.12), expressed as magnitude and phase. Maximum and minimum envelopes of mean values from the studies included in the data synthesis are shown by thick continuous lines, while the mean of these data sets is shown by the thin continuous line. The response of a three-degree-of-freedom biodynamic model is shown by the dashed line. (*ISO 5982.*[10])

the abdomen appears larger in volume; at the same time, the downward deflection of the diaphragm causes a decrease of the chest circumference. At the other end of the cycle the abdominal wall is pressed inward, the diaphragm upward, and the chest wall is expanded. This periodic displacement of the abdominal viscera has a sharp resonance between 3 and 3.5 Hz. The oscillations of the abdominal mass are coupled with the air oscillations of the mouth-chest system. The abdominal wall has a maximum response between 5 and 8 Hz; the anterior chest wall between 7 and 11 Hz. Vibration of the abdominal system resulting from exposure of a sitting or standing subject is detected clearly as modulation of the airflow velocity through the mouth.

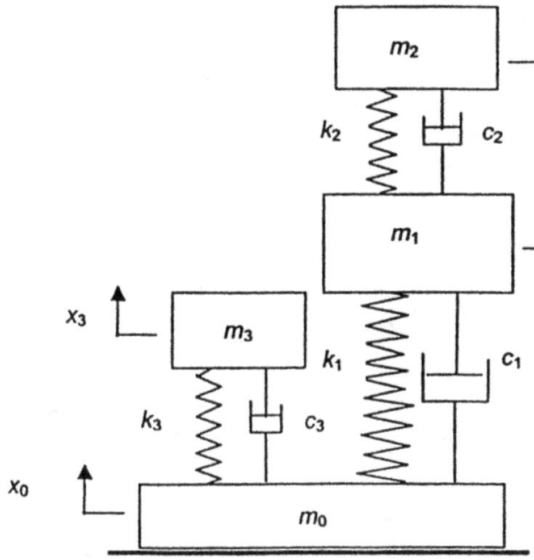

FIGURE 41.7 Three-degree-of-freedom biodynamic model for the driving-point mechanical impedance, apparent mass, and seat-to-head transmissibility of the seated human body in the vertical direction (Z-direction of Fig. 41.12). The model is driven at its base (x_0). The parameters of this model do not possess direct anatomical correlates. (*ISO 5982.*[10])

Therefore, at large amplitudes of vibration, speech can be modulated at the exposure frequency. A lumped-parameter model of the thorax-abdomen-airway system is used successfully to explain and predict these detailed physiological responses.[9] The same model can also be used, when appropriately excited, to describe the effects of blast, infrasound, and chest impact and to derive curves of equal injury potential, i.e., tolerance curves.

Subject Exposed to Vibrations in the Transverse Direction. The physical response of the body to transverse vibration—i.e., horizontal in the normal upright position—is quite different from that for vertical vibration. Instead of thrust forces acting primarily along the line of action of the force of gravity on the human body, they act at right angles to this line. Therefore, the distribution of the body masses is of the utmost importance. There is a greater difference in response between sitting and standing positions for transverse vibration than for vertical vibration where the supporting structure of the skeleton and the spine are designed for vertical loading.

For a standing subject, the displacement amplitudes of vibration of the hip, shoulder, and head are about 20 to 30 percent of the table amplitude at 1 Hz and decrease with increasing frequency. The sitting subject exhibits amplification of the hip (1.5 Hz) and head (2 Hz) amplitudes. All critical resonant frequencies are between 1 and 3 Hz. The transverse vibration patterns of the body can be described as standing waves, i.e., as a rough approximation one can compare the body with a rod in which transverse flexural waves are excited. Therefore, there are nodal points on the body which become closer to the feet as the frequency of excitation increases, since the phase shift between all body parts and the table increases continuously with increasing frequency. At the first resonant frequency (1.5 Hz), the head of the standing subject has a 180° phase shift with respect to the table; between 2 and 3 Hz this phase shift is 360°.

There are longitudinal head motions excited by the transverse vibration in addition to the transverse head motions. The head performs a nodding motion due to the anatomy of the upper vertebrae and the location of the head's center of gravity. Above 5 Hz, the head motion for sitting and standing subjects is predominantly vertical (about 10 to 30 percent of the horizontal table motion).

Vibrations Coupled to the Hand. The mechanical impedance of the hand-arm system measured at a hand grip under conditions representative of those associated with power-tool operation is shown in Fig. 41.8 for vibration directed essentially along the long axis of the forearm, that is, approximately in the direction of thrust. The precise direction is the Z component of the standardized coordinate system for the hand shown in Fig. 41.12, Z_h. Typical values of impedance have again been defined by a synthesis of measured values from different experimental studies, as

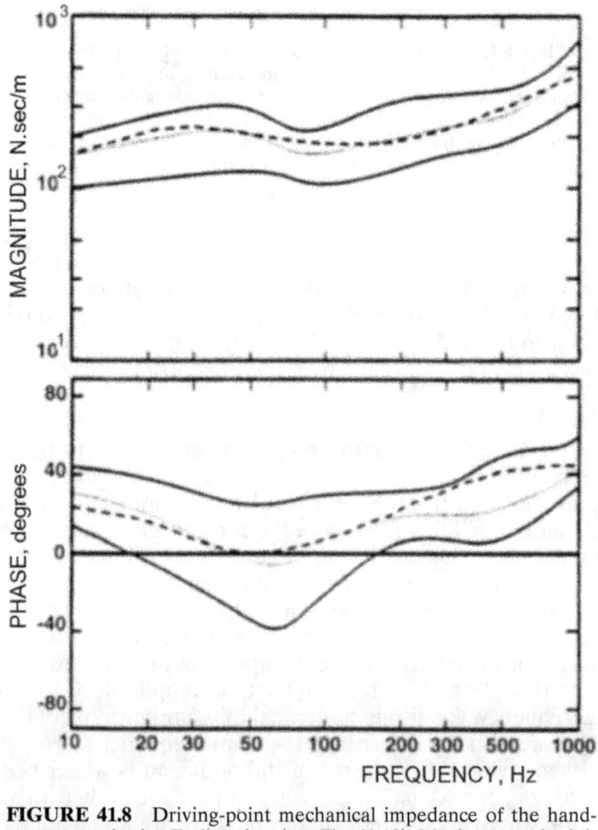

FIGURE 41.8 Driving-point mechanical impedance of the hand-arm system in the Z_h direction (see Fig. 41.12). Maximum and minimum envelopes of the mean values from data sets are shown by thick lines: mean of all data sets—thin continuous line; response of four-degree-of-freedom biodynamic model—dashed line. (*ISO 10068.*[11])

described previously for whole-body impedance and transmissibility. The most probable values of impedance magnitude and phase are specified by an upper and lower envelope (the continuous lines in Fig. 41.8) and define a range of idealized impedances.[11] The mean of the data sets is shown by the thin continuous line. Also shown in the diagram are impedance values calculated by a four-degree-of-freedom biodynamic model. Equivalent data are available for the two orthogonal directions of the hand-arm coordinate system not shown in Fig. 41.8 (X_h and Y_h). It should be noted that the parameters of these biodynamic models do not possess direct anatomical correlates. Together with the idealized impedances, they are intended to facilitate the development of devices for reducing vibration transmitted to the hands and of test rigs with which to measure power-tool handle vibration.

The mechanical impedance of the hand-arm system generally increases in magnitude with frequency, with a maximum at a frequency from 20 to 70 Hz. The model values suggest that resonances occur in structures within the hand, resulting in relative motion between tissue layers, and between tissue and the bone. The coupled mass in contact with the handle and subject to the vibration input is typically less than 20 grams.

It is important to distinguish between the various forces introduced when a hand holds a handle. In general, the *grip force* has balancing components radial to the cylindrical axis of a handle that distinguish it from the *feed,* or *thrust, force* which is the translational force exerted to control the action of a tool on a workpiece. The relationships between these component forces, which together form the *contact force,* is dependent on the shape and size of the handle.[12] The *biodynamic force* at the interface between the hand and a vibrating surface results from the dynamic response of the hand-arm system to vibration. It may be calculated from the impedance of the hand-arm system and the velocity of the vibrating surface, or measured by high-pass filtering of the output of a force sensor positioned at the interface (>5 Hz). Small increases in the magnitude of the mechanical impedance are observed with increases in contact force (from the value of 25 N used for the data synthesis), consistent with tissue stiffening from skin compression (see Fig. 41.2).

Vibration entering the fingers is absorbed at frequencies above 50 Hz, while lower frequencies are transmitted up the hand-arm system.[13] Frequencies in the range from 25 to 50 Hz are primarily absorbed in the wrist, arm, and shoulder.

Skull Vibrations. The vibration pattern of the skull is approximately the same as that of a spherical elastic shell. The nodal lines observed suggest that the fundamental resonance frequency is between 300 and 400 Hz and that resonances for the higher modes are around 600 and 900 Hz. The observed frequency ratio between the modes for the skull is approximately 1.7, while the theoretical ratio for a sphere is 1.5. From the observed resonances, the calculated value of the elasticity of skull bone (a value of Young's modulus = 1.4×10^9 Pa) agrees reasonably well with static test results on dry skull preparations but is somewhat lower than the static test data obtained on bone. Mechanical impedances of small areas on the skull over the mastoid area have been measured to provide information for bone-conduction hearing.

Vibration of the lower jaw with respect to the skull can be explained by a simple mass-spring system, which has a resonance, relative to the skull, between 100 and 200 Hz.

Biodynamic Models. Both simple and complex mathematical models have been developed for the whole body,[4,5,9] and for subsystems such as the spine,[14,15] the head and neck,[4,16] the skull,[4] and the hand and arm.[17] Examples of simple models are to be found in Figs. 41.7 and 41.10*A*.

ATB and MADYMO Models. The *Articulated Total Body* (ATB) and the *MAthematical DYnamical MOdel* (MADYMO) are widely used, multielement, whole-body, lumped-parameter models. The models represent rigid bodies, joints, springs, and dampers with values designed to enable the prediction of selected human properties, or in some cases manikin properties. The predictions can include the effects of an environment surrounding the model using different routines for contact with external surfaces, the effect of gravity, body restraints (e.g., seat belt), and wind forces (to model pilot ejection from aircraft). The models are used extensively to simulate the body's response to shocks and impacts.

Finite Element (FE) Models. Internal stresses and motions within body parts may be determined from finite element (FE) models. In some models, the FEs can interact with multibody model elements. Examples of human body subsystems that have been modeled with FEs include the spine, to predict the injury potential of vertebral compression and torsional loads,[18] and the head and neck, to predict rotation of the head and neck loads.[16]

Artificial Neural Network Models. The nonlinear response of the spine to vertical accelerations has been modeled by an artificial neural network.[15] The model was trained on repeated shocks with peak amplitudes from 10 m/sec^2 to 40 m/sec^2 applied to a seated person (back unsupported) in the vertical direction (*Z* direction of Fig. 41.12), and so is applicable to shocks and impacts in this direction and with this range of accelerations.

HIGH-FREQUENCY RANGE

Mechanical Impedance of Soft Human Tissue. Mechanical impedance measurements of small areas (1 to 17 cm^2) over soft human body tissue have been made with vibrating pistons between 10 Hz and 20 kHz. At low frequencies (<20 Hz) this impedance is a large elastic reactance. With increasing frequency the reactance decreases, becomes zero at a resonance frequency, and becomes a mass reactance with a further increase in frequency.[7] These results cannot be explained by a simple lumped-parameter model, but require a distributed-parameter system including a viscoelastic medium—such as the tissue constitutes for this frequency range. The high viscosity of the medium makes possible the use of simplified theoretical assumptions, such as a homogeneous isotropic infinite medium and a vibrating sphere instead of a circular piston. The results of such a theory agree well with the measured characteristics (Fig. 41.9). As a consequence, it is possible to assign absolute values to the shear viscosity and the shear elasticity of soft tissue (Table 41.1). The theory together with the measurements show that, over the audio-frequency range, most of the vibratory energy is propagated through the tissue in the form of transverse shear waves—not in the form of longitudinal compression waves. The velocity of the shear waves is about 20 m/sec at 200 Hz and increases approximately with the square root of the frequency. Some energy is propagated along the body surface in the form of surface waves which have been observed optically. Their velocity is of the same order as the velocity of shear waves.

Above several hundred thousand hertz, in the ultrasound range, most of the vibratory energy is propagated through tissue in the form of compressional waves; for these conditions, geometrical acoustics offers a good approximation for the description of their path. Since the tissue dimensions under consideration are almost always large compared to the wavelength (about 1.5 mm at 1 MHz), the mechanical impedance of tissue is equal to the characteristic acoustic impedance, i.e., sound velocity times density. This value for soft tissue differs only slightly from the characteristic impedance of

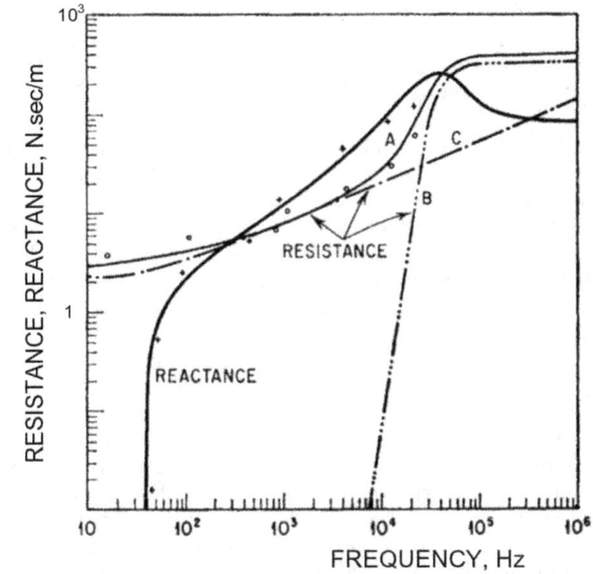

FIGURE 41.9 Resistance and reactance of soft tissue (2-cm diameter) as a function of frequency: measured values, reactance—crosses, resistance—circles. Curves for 2-cm sphere vibrating in A—viscoelastic medium with properties of soft tissue (see Table 41.1); B—frictionless compressible fluid, and C—incompressible viscous fluid. (*von Gierke et al.*[7])

water. The most important factor in this frequency range is the tissue viscosity, which brings about an increase in energy absorption with increasing frequency.

At very high frequencies this viscosity also generates shear waves at the boundaries of the medium, at the boundary of the acoustic beam, and in the areas of wave transmission to media with somewhat different properties (e.g., boundary of muscle to fat tissue, or soft tissue to bone). These shear waves are attenuated so rapidly that they are of no importance for energy transport but are noticeable as increased local absorption, i.e., heating.

EFFECTS OF SHOCK AND OF VIBRATION

EFFECTS OF VIBRATION

Mechanical Damage. Damage is produced when the accelerative forces are of sufficient magnitude. Postmortem examination of animals usually shows lung damage, often heart damage, and occasionally brain injury. The injuries to heart and lungs probably result from the beating of these organs against each other and against the rib cage. The brain injury, which is a superficial hemorrhage, may be due to relative motion of the brain within the skull, to mechanical action involving the blood vessels or sinuses directly, or to secondary mechanical effects.

An increase in body temperature is also observed after exposure to intense vibration. Since this effect also occurs in dead animals, it is probably mechanical in origin. Estimates of energy dissipation from body mechanical impedance data suggest that appreciable heat can be generated at large vibration amplitudes.

In humans, mechanical damage to the heart and lungs, injury to the brain, tearing of membranes in the abdominal and chest cavities, as well as intestinal injury are possible, in principle. However, equinoxious contours of whole-body acceleration as a function of frequency have not been established for any of these phenomena, owing to an almost complete lack of data. Any effects would be expected to occur at lower frequencies than those in animals owing to the increased human visceral masses. Exposure for 15 minutes to an acceleration of $6g$ has been reported to cause gastrointestinal bleeding that persisted for several days in one subject.[1]

Chronic injuries may be produced by vibration exposure of long duration at levels which produce no acute effects.[1,2] There is epidemiological evidence that occupations with exposure to whole-body vibration are at greater risk of low back pain, sciatic pain, and herniated lumbar disc when compared with control groups not exposed to vibration.[19] There is also an increased risk of developing degenerative changes in the spine, including lumbar intervertebral disc disorders, for crane operators, tractor drivers, and drivers in the transportation industry. Nevertheless, it is difficult to differentiate between the relative roles of whole-body vibration and ergonomic risk factors, such as posture and awkward back movements, though both are clearly cofactors in the development of the observed pathology.[20] Exposure to repeated random jolts (in contrast to sinusoidal motion), such as the buffeting that occurs in aircraft, in small craft on rough water, or in off-road vehicles is also associated with the chronic injuries described.

Chronic injuries may be produced when the hand is exposed to intense vibration, such as occurs during occupational use of some power tools (e.g., pneumatic drills and hammers, grinders, chain saws, and riveting guns).[1–3] Symptoms of numbness or paresthesias in the fingers are common and may be accompanied by episodes of finger blanching. Reduced grip strength and muscular weakness may also be experienced. The vascular, nerve, and muscular disorders associated with the use of handheld vibrating power tools are known as the *hand-arm vibration syndrome* (HAVS). Pathological changes have been observed in the structure of the nerves and walls of the blood vessels in the fingers.[3] Changes in tactile function have been linked to changes in acuity of specific types of mechanoreceptive nerve endings at the fingertips.[21]

Few exposure-response relationships have been derived from epidemiological data for any sign, or symptom, of HAVS resulting from occupational use of handheld power tools or industrial processes. For groups of workers who perform similar tasks throughout the workday, the *latency*—that is, the duration of exposure (in years) prior to the onset of episodes of finger blanching—and prevalence, may be predicted from the acceleration of a surface in contact with the hand.[22] These relationships serve as the basis for occupational exposure criteria (see "Human Tolerance Criteria").

The tendons, tendon sheaths, muscles, ligaments, joints, and nerves in the hand and arm can also be damaged by repeated movement of the hand relative to the arm. These soft tissue and nerve injuries occur among blue- and white-collar workers performing tasks involving repeated hand-wrist flexure (e.g., keyboard operators) and are termed *repetitive strain injuries* (RSI).[23] Nerve compression may result from changes in the contents of restricted nerve passageways (e.g., the carpal tunnel at the wrist—*carpal tunnel syndrome*). Pain and paresthesias in the hand and arm are common symptoms.

Physiological Responses. Vibration can induce physiological responses in the cardiovascular, respiratory, skeletal, endocrine, and metabolic systems and in muscles and nerves. The cardiovascular changes in response to intense vertical vibration are similar to those accompanying moderate exercise: increased heart rate, respiration rate, cardiac output, and blood pressure. Vibration of sufficient intensity will cause mechanical pumping of the respiratory system, as already noted, but is unlikely to produce significantly increased ventilation or oxygen uptake. Changes in blood and urine constituents are commonly used as indicators of generalized body stress and may, in consequence, be observed in persons exposed to vibration. It is difficult if not impossible, however, to relate specific endocrine and metabolic responses to a given vibration stimulus. Vibration can stimulate a tonic reflex contraction in muscles, which is a response to the stretching force (the *tonic vibration reflex*), disturb postural stability, and lead to body sway. Extremely low frequency whole-body vibration, such as occurs in many transportation vehicles and ships, may also cause motion sickness (*kinetosis*).[1]

Vibration of the hand may cause peripheral vascular, neurological, and muscular responses.[3] Blood flow within the fingers may be reduced during stimulation, and tingling and paresthesias in the hands may be reported after exposure. Somatosensory perception and tactile function may be temporarily decreased. Grip strength may also be affected. Extremely low frequency, large-amplitude motions, which are usually described as repetitive movements of the hand (and frequently involve repeated wrist rotation), may lead to tendon and muscle fatigue and to transitory parathesias or numbness.

Therapeutic applications of vibration include cardiac and circulatory assist devices and the control of spastic muscle. Ultrasonic frequencies are used in medical diagnosis, for soft tissue visualization, and for therapy. A common therapeutic use is to promote the return of limb function in rehabilitation medicine.

Subjective Responses. Feelings of discomfort and apprehension may be associated with exposure to whole-body and hand-arm vibration once the stimulus has been perceived. The extent of the discomfort depends on the magnitude, frequency, direction and duration of the exposure, and the posture and orientation of the body, as well as the point of contact with the stimulus. The response is also influenced by the environment in which the motion is experienced (e.g., floor motion in hospital versus aircraft). The range in response of different individuals to a given stimulus is large. In some circumstances, whole-body vibration may be exhilarating (e.g., a fairground ride) or soothing (e.g., rocking a baby in a cradle or a rocking chair).

In general, subjective responses to vibration may be subdivided into three broad categories: the threshold of perception, the onset of unpleasant sensations, and the limit of tolerance. The specification of acceptable vibration environments is discussed later in this chapter.

Once detected, the growth in sensation follows a Stevens' power law function with index k, in which the psychophysical magnitude of a stimulus ψ is related to its physical magnitude ϕ by

$$\psi = \text{constant}\{\phi^k\} \tag{41.1}$$

For discomfort associated with whole-body vibration, $k \approx 1$. Frequency contours of equal sensation magnitude depend principally on the direction in which vibration enters the body and whether the person is standing, seated, or recumbent.[1] Contours which summarize current knowledge may be inferred from the frequency-weighting functions employed in the international standard for whole-body vibration (i.e., by reciprocal curves to those shown in Fig. 41.11). The effect of the duration of exposure t on subjective responses to suprathreshold vibration is often found to follow a power law relationship of the form

$$\phi^n t = \text{constant} \tag{41.2}$$

where the magnitude of the index n is from 2 to 4. For situations in which the *perception* of vibration is judged unacceptable, the boundary between acceptable and unacceptable exposures will be related to the physical magnitude of the stimulus corresponding to the threshold of perception, and will not depend on the duration of exposure. There is an extensive literature discussing the comfort/discomfort of passengers in road and rail vehicles, aircraft, and ships.[1]

EFFECTS OF SHOCK AND IMPACT

This type of force is experienced in falls, in motor vehicle or aircraft crashes, in parachute openings, in seat ejections for escape from high-speed military aircraft, and in many other situations. Interest in the body's responses to these forces centers on mechanical stress limits. Accident statistics from the United States (from 1979 to 1986) indicate that serious injuries to occupants of automobiles involved in frontal impacts, and who were wearing seat belts, were most commonly to the head (approximately 35 percent), followed by the thorax (including abdomen), and lower extremities (approximately 25 percent each). The distribution of injuries in fatal accidents involving military helicopters and pilot ejections from fixed-wing aircraft is similar to that of the automobile statistics cited with, in addition, injuries to the spine in approximately 13 percent of cases.[5] For crewmen who survived seat ejection from military aircraft, the most common injury was to the spine, while for passengers surviving civil air transport accidents the most common injury remained to the head.[4]

Serious injuries to the head usually involve brain injury, either with or, commonly, without skull fracture. The brain may suffer either diffuse or focal injuries. The former consists of brain swelling, concussion, and *diffuse axonal injury,* that is, mechanical disruption of the nerve fibers; the latter consists of localized internal bleeding and contusions (coup and contrecoup). Concussion is the most common brain injury.

The most common neck and spinal injury is caused by rearward flexion and forward extension of the neck, such as commonly occurs in rear-end motor vehicle collisions ("whiplash"), and results in localized pain in the neck and shoulders, and even cord injuries. The motion can also result in dislocation or fracture of the first and second vertebral joints, and may lead to the spinal cord being crushed or severed. Both neck and spine may be injured by vertical accelerations directed from the head or buttocks. The nature and degree of injury is critically dependent on the body position at impact.

The chest encloses important organs—the heart, lungs, trachea, esophagus, and major blood vessels—and so injuries may be divided into those affecting the organs, and those affecting the rib cage. Injuries to the internal organs include ruptures of the heart, of the lung, and of the arteries connected to the heart, while injuries to the rib cage involve fractures of the ribs and sternum, and sometimes dislocations and fractures of the thoracic vertebrae. Compound rib fractures may, if sufficiently displaced, also result in puncturing of internal organs. Organs within the abdomen (especially liver, kidneys, and spleen) are also subject to injury by external trauma involving transverse (e.g., front-to-back or side-to-side) accelerations.

Common injuries to the lower extremities involve fractures of the long bones and injuries to the joints.

Force Duration. The correlation between the response of the body system to continuous vibration and to spike and step-force functions may be used to guide and interpret exposures. The tissue areas stressed to maximum relative displacement at the various frequencies during steady-state excitation are preferred target areas for injury under shock and impact load if the force-time functions have appreciable energy in these frequency bands. If the exposure times are shorter, stress tolerance limits increase; if exposure times decrease to hundredths or thousandths of a second, the response becomes more and more limited and localized to the point of application of the force (blow). Elastic compression or injury will depend on the load distribution over the application area, i.e., the pressure, to which tissues are subjected. If tissue destruction or bone fracture occurs close to the area of application of the force, these will absorb additional energy and protect deeper-seated tissues by reducing the peak force and spreading it over a longer period of time. An example is the fracture of foot and ankle of men standing on the deck of warships when an explosion occurs beneath. The support may be thrown upward with great momentum; however, the energy absorption by the fracture protects structures of the body which are higher up.

If the force functions contain extremely high frequencies, the compression effects spread from the area of force application throughout the body as compression waves. If these are of sufficient amplitude, they may cause considerable tissue disruption. Such compression waves are observed from the impact of high-velocity missiles.

If the exposure to the accelerating forces lasts long enough so that (as in most applications of interest) the whole body is displaced, exact measurement of the force applied to the body and of the direction and contact areas of application becomes of extreme importance. In studies of seat ejection, for example, a knowledge of seat acceleration alone is not sufficient for estimating responses. One must know the forces in those structures or restraining harnesses through which acceleration forces are transmitted. The location of the center of gravity of the various body parts such as arms, head, and upper torso must be known over the time of force application so that the resulting body motion and deformation can be analyzed and controlled for protection purposes. In addition to the primary displacements of body parts and organs, there are secondary forces from decelerations if, due to the large amplitudes, the motions of parts of the body are stopped suddenly by hitting other body parts. Examples occur in linear deceleration where, depending on the restraint, the head may be thrown forward until it hits the chest or, if only a lap belt is used, the upper torso may jackknife and the chest may hit the knees. There is always the additional possibility that the body may strike nearby objects (e.g., automobile dashboard or doorpost), thus initiating a new impact deceleration history.

Longitudinal Acceleration. The study of positive longitudinal (headward) acceleration of short duration is connected closely with the development of upward ejection seats for escape from aircraft. Since the necessary ejection velocity of approximately 18 m/sec and the available distance for the catapult guide rails of about 1 m are determined by the aircraft, the minimum acceleration required (step function) is approximately 18.6*g*. The high jolt of the instantaneous acceleration increase is undesirable because of the high dynamic load factor in this direction for the frequency range of body resonances. A slower buildup of the acceleration with higher final acceleration is preferable to prevent injury. Investigations show that the body's ballistic response can be predicted by biodynamic models making use of the frequency-response characteristics of the body. The simplest analog used for the study of headward accelerations is the single-degree-of-freedom mechanical resonator composed of the lumped-parameter elements of a spring, mass, and damper (Fig. 41.10*A*). The model is used to simulate the maximum stress developed within the vertebral column (the first failure mode in this direction) for any given shock environment. The maximum dynamic deflection of the spring, Δ_{max}, may be calculated for a given input acceleration-time history to the model. The potential for spinal injury is estimated by forming the *dynamic response index* (DRI), which is defined as $\omega_n^2\Delta_{max}/g$, where the natural frequency of the model $\omega_n = (k/m)^{1/2}$ is 52.9 rad/sec and the damping ratio $c/2(km)^{1/2}$ is 0.224. Experience with nonfatal ejections from military aircraft, shown by the crosses and dashed line in Fig. 41.10*B*, suggests a 5 percent probability of spinal injury from exposure to a dynamic response index of 18. An estimate of the rate of spinal injury from cadavers is shown in this diagram by the continuous line. The success of the model has led to its adoption for the specification of ejection seat per-

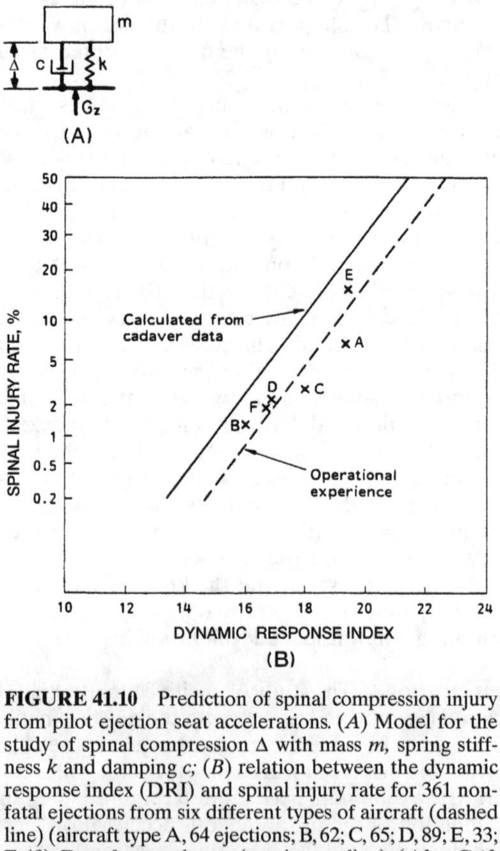

FIGURE 41.10 Prediction of spinal compression injury from pilot ejection seat accelerations. (*A*) Model for the study of spinal compression Δ with mass *m*, spring stiffness *k* and damping *c;* (*B*) relation between the dynamic response index (DRI) and spinal injury rate for 361 nonfatal ejections from six different types of aircraft (dashed line) (aircraft type A, 64 ejections; B, 62; C, 65; D, 89; E, 33; F, 48). Data from cadavers (continuous line). (*After Griffin,*[1] *and von Gierke.*[8])

formance, for its extension to accelerations in three orthogonal directions,[24] and to measures of ride comfort for exposure to repeated shocks in some land vehicles and high-speed boats.[25]

For negative (tailward) acceleration (downward ejection) no firm point for application of the accelerating force is accessible as for positive acceleration. If the force is applied as usual through harness and belt at shoulder and groin, the mobility of the shoulder girdle together with the elasticity of the belts results in a lower resonance frequency than the one observed in upward ejection. To avoid overshooting with standard harnesses, the acceleration rise time must be at least 0.15 sec. This type of shock and impact can excite the thorax-abdomen system. The diaphragm is pushed upward by the abdominal viscera; as a result, air rushes out of the lungs (if the glottis is open) or high pressures develop in the air passages.

Transverse Accelerations. The forward- and backward-facing seated positions are most frequently exposed to high transverse (i.e., horizontal) components of crash loads. Human tolerance to these forces has been studied extensively by volunteer tests on linear decelerators, in automobile crashes, and by the analysis of the records of accidental falls. The results indicate the importance of distributing the decelerative forces or impact over as wide an area as possible. The tolerable acceleration amplitudes of well over 50*g* (100*g* and over for falling flat on the back with minor injuries, 35 to 40*g* for 0.05-sec voluntary deceleration when seated with restraining harness) are probably limited by injury to the brain. An indication that the latter might be sensitive to and based on specific dynamic responses is the fact that the tolerance limit depends strongly on the rise time of the acceleration. With rise times around 0.1 sec (rate of change of acceleration 500*g*/sec), no overshooting of head and chest accelerations is observed, whereas faster rise times of around 0.03 sec (1000 to 1400*g*/sec) result in overshooting of chest accelerations of 30 percent (acceleration front to back) and even up to 70 percent (acceleration back to front). All these results depend critically on the harness for fixation and the back support used (see "Protection Methods and Procedures"). These dynamic load factors indicate a natural frequency of the body system between 10 and 20 Hz. Impact of the heart against the chest wall is another possible injury discussed and noted in some animal experiments.

The head and neck supporting structures seem to be relatively tough. Injury seems to occur only upon backward flexion and extension of the neck (whiplash) when the body is accelerated from back to front without head support, as already noted.

Head Impact.[26] The reaction of the head to a blow is a function of the velocity, duration, area of impact, and transfer of momentum. Near the point of application of the blow there will be an indentation of the skull. This results in shear strains in the brain in a superficial region close to the dent. Compression waves emanate from this area, which have normally small amplitudes since the brain is nearly incompressible. In addition to the forces on the brain resulting from skull deformation there are acceleration forces, which also would act on a completely undeformable skull. The centrifugal forces and linear accelerations producing compressional strains are negligible compared to the shear strains produced by the rotational accelerations. The maximum strains are concentrated at regions where the skull has a good grip on the brain owing to inwardly projecting ridges, especially at the wing of the sphenoid bone of the skull. Shear strains also must be present throughout the brain and in the brain stem. Many investigators consider these shear strains, resulting from rotational accelerations due to a blow to the unsupported head, as the principal event leading to concussion.

If the head hits a wall or another object whose mass is large compared to the head's mass, the local, visible damage is small and the damage due to rotational acceleration may be large. Blows to certain points, especially on the midline, produce no rotation. Blows to the chin upward and sideward produce rotation relatively easily ("knockout" in boxing). Velocities listed in the literature for concussion from impact of large masses range from 5 to 15 m/sec. At impact velocities of about 10 m/sec, approximately 2 kg-m of energy is absorbed in 0.002 sec, resulting in an acceleration of the head of 47*g*. Impact energies for compression concussion are probably approximately in the same range.

Scalp, skin, and subcutaneous tissue reduce the energy applied to the bone. If the response of the skull to a blow exceeds the elastic deformation limit, skull fracture occurs. Impact by a high-velocity, blunt-shaped object results in localized circumscribed fracture and depression. Low-velocity blunt blows, insufficient to cause depression, occur frequently in falls and crashes. Given enough energy, two, three, or more cracks appear, all radiating from the center of the blow. The skull has both weak and strong areas, each impact area showing well-defined regions for the occurrence of the fracture lines. The total energy required for skull fracture varies from 5 to 10 kg-m, with an average often assumed to be 7 kg-m. This energy is equivalent to the condition that the head hits a hard, flat surface after a free fall from a 1.5-m height.

EFFECTS OF SHOCK AND VIBRATION ON TASK PERFORMANCE

The performance of tasks requiring a physical response to some stimulus involves peripheral (e.g., perceptual and motor) and central neurological processes, with multiple feedback paths characteristic of a sophisticated control system. Each of these processes is complex, is more or less developed in different individuals, and may be influenced by training and the general state of health. In consequence, unique relationships between vibration and task performance are unlikely, except for well-defined situations in which some part of the body reaches a physical or physiological limit to performance. For example, movement of images on the retina may cause defocusing and a reduction of visual acuity. The movement may be caused by vibration of the display (i.e., the source), the head (and/or observer), or both. At frequencies below approximately 1 Hz, a pursuit reflex assists visual acuity. At frequencies above 20 Hz, an eyeball resonance can degrade acuity. The effects of whole-body vibration on visual acuity therefore depend on the frequency and amplitude, as well as the viewing distance.[1] As already discussed, whole-body vibration can affect speech.

Vibration may also degrade the manual control of objects. The influence of whole-body vibration on writing and drinking is a common experience in public transportation vehicles and ships. Vibration may interfere with the performance of manually controlled systems. The extent of the effect depends on hand motion, the type of control (e.g., a "stiff" control that responds to the application of force without moving, or one that moves and responds with little force applied), and the dynamics of the control and the controlled system. A control that responds to hand displacement may be disrupted by vertical vibration at frequencies between 2 and 6 Hz. The effect of the duration of vibration exposure on task performance is influenced by motivation, arousal, and adaptation and may therefore be observed to improve or degrade performance over time.

Exposure of the hand to vibration can lead to sensorineural dysfunction sufficient to reduce the ability to perform fine manual tasks, such as buttoning clothing.[3]

The motion associated with a shock is unlikely to interfere directly with the performance of most tasks unless it is coincident with some critical component of the task. This condition may occur with shocks repeated at very short intervals.

HUMAN TOLERANCE CRITERIA

WHOLE-BODY VIBRATION EXPOSURE

International Standard ISO 2631 defines methods for the measurement of periodic, random, and transient whole-body vibration.[27] The standard also describes the principal factors that combine to determine the acceptability of an exposure and suggests the possible effects, recognizing the large variations in responses between individuals. It should be noted that the consensus standard permits alternative methods for assessing some exposures, and, hence, inconsistencies may arise.[28]

Measurement. Whole-body vibration is measured at the principal interface between the human body and the source of vibration. For seated persons, this interface is most likely to be the seat surface and seat back, if any; for standing persons, the feet; and for reclining persons, the supporting surface(s) under the pelvis, torso, and head. When vibration is transmitted to the body through a nonrigid or resilient material (e.g., a seat cushion), the measuring transducer should be within a mount, in contact with the body, formed to minimize the change in surface pressure distribution of the resilient material.[1,2] The measurement should be of sufficient duration to ensure that the data are representative of the exposure being assessed and, for random signals, contain acceptable statistical precision.

Frequency-Weighted Acceleration. The magnitude of the exposure is characterized by the *rms frequency-weighted acceleration* calculated in accordance with the following equation or its equivalent in the frequency domain:

$$a_W = \left[\frac{1}{T} \int_0^T a_W^2(t)\,dt \right]^{1/2} \tag{41.3}$$

where $a_w(t)$ is the frequency-weighted acceleration, or angular acceleration, at time t expressed in meters per second squared (m/sec^2) or radians per second squared (rad/sec^2), respectively; and T is the duration of the measurement in seconds. The most common frequency weightings to be employed for different applications are shown in Fig. 41.11. Frequency weightings W_d and W_k are the principal weightings for the assessment of the effects of vibration on health, comfort, and perception, with W_f used for motion sickness. The other frequency weightings in Table 41.2 apply to specific situations involving, respectively: motion coupled to the body from a seat back (W_c); body rotation (W_e); and head motion in the X direction of reclining persons (W_j). Application of a frequency weighting selected according to Table 41.2, Fig. 41.11, and Fig. 41.12 to one component of vibration transmitted to the body provides a measure of the *component frequency-weighted acceleration* for that direction of motion and human response.

Equation (41.3) is suitable for characterizing vibrations with a *crest factor* less than 9, where the crest factor is here defined as the magnitude of the ratio of the peak value of the frequency-weighted acceleration signal to its rms value.

Vibration Containing Transient Events. For exposures to whole-body vibration containing transient events resulting in crest factors in excess of 9, either the *running rms* or the *fourth-power vibration dose,* or both, may be used in addition to

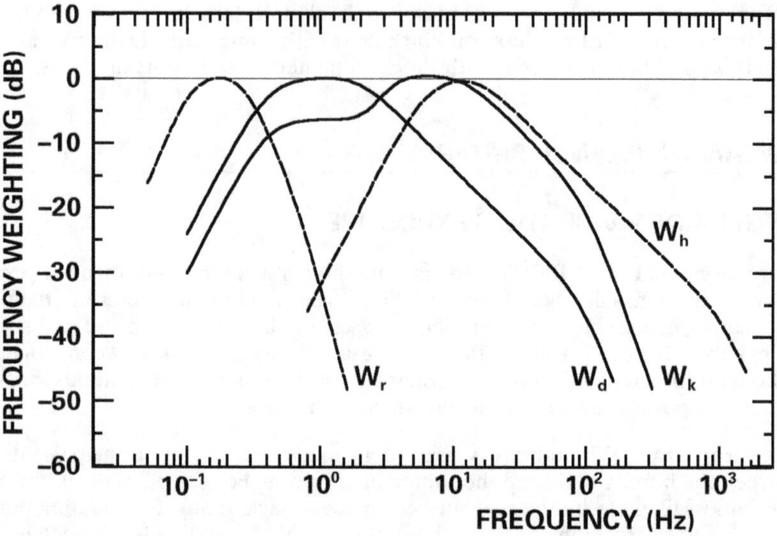

FIGURE 41.11 Principal frequency weightings for the assessment of whole-body (W_k, W_d, and W_f) and hand-transmitted vibration (W_h). (*ISO 2631-1*[27] *and ISO 5349-1.*[29])

the rms frequency-weighted acceleration to ensure that the effects of transient vibrations are not underestimated. The running rms is calculated for a short integration time τ ending at time t_0 in the time record as follows:

$$a_W(t_0) = \left[\frac{1}{\tau}\int_{(t_0-\tau)}^{t_0} a_W^2(t)\,dt\right]^{1/2} \tag{41.4}$$

A correlation with some subjective human responses to transient vibration may be obtained by constructing the *maximum transient vibration value* $\text{MTVV}_{(T)}$ during the measurement

$$\text{MTVV}_{(T)} = |a_W(t_0)|_{\max} \tag{41.5}$$

where the right-hand side of this equation is determined by the maximum value of the running rms acceleration obtained using Eq. (41.4) when $\tau = 1$ sec.

The fourth-power *vibration dose value* VDV is defined by

$$\text{VDV} = \left[\int_0^T a_W^r(t)\,dt\right]^{1/r} \tag{41.6}$$

with $r = 4$, and provides a measure of exposure that is more sensitive to large amplitudes by forming the fourth power of the frequency-weighted acceleration time history, $a_w^4(t)$. If the total exposure consists of i exposure elements with different vibration dose values $(\text{VDV})_i$, then

$$\text{VDV}_{\text{total}} = \left[\sum_i (\text{VDV})_i^4\right]^{1/4} \tag{41.7}$$

TABLE 41.2 Applicability of Whole-Body Vibration Frequency Weightings W_k, W_d, W_f, W_c, W_e, and W_j, for the Vibration Directions *X, Y, Z*, R_x, R_y, and R_z Specified in Fig. 41.12. (*ISO 2631-1*.[27])

Frequency weighting	Health	Comfort*	Perception	Motion sickness
		Principal weighting		
W_k	*Z*	*Z*-seat *X, Y, Z*-feet *Z*-standing *X*-lying	*Z*	
W_d	*X*-seat *Y*-seat	*X*-seat *Y*-seat *X, Y*-standing *Y, Z*-lying *Y, Z*-back	*X*-, *Y*-	
W_f				*Z*
		Additional weighting		
W_c	*X*-seat back	*X*-seat back		
W_e		R_x, R_y, R_z		
W_j		*X*-lying (head)		

* Values of the multiplying factor *k* to be applied to component accelerations for assessing the comfort of seated persons in situations in which vibration enters the body at several points, e.g., the seat pan, seat back, and the feet (see text).

Component Acceleration	*Value of k*
X direction at seat back	0.8
Y direction at seat back	0.5
Z direction at seat back	0.4
X & *Y* directions at feet	0.25
Z direction at feet	0.4
R_x axis at seat	0.63 m/rad
R_y axis at seat	0.4 m/rad
R_z axis at seat	0.2 m/rad

For other component accelerations the value of *k* is unity.

Use of the maximum transient vibration value or the total vibration dose value in addition to the rms frequency-weighted acceleration is advisable whenever

$$\text{MTVV}_{(T)} > 1.5a_W \tag{41.8}$$

or

$$\text{VDV}_{\text{total}} > 1.75a_W T^{1/4} \tag{41.9}$$

The total vibration dose value will integrate the contribution from *each* transient event, irrespective of magnitude or duration, to form a time- and magnitude-dependent dose. In contrast, the maximum transient vibration value will provide a measure dominated by the magnitude of the most intense event occurring in a 1-second time interval, and will be little influenced by events occurring at times significantly greater than 1 second from this event. Application of either measure to the assessment of whole-body vibration should take into consideration the nature of the transient events, and the anticipated basis for the human response (i.e., source and

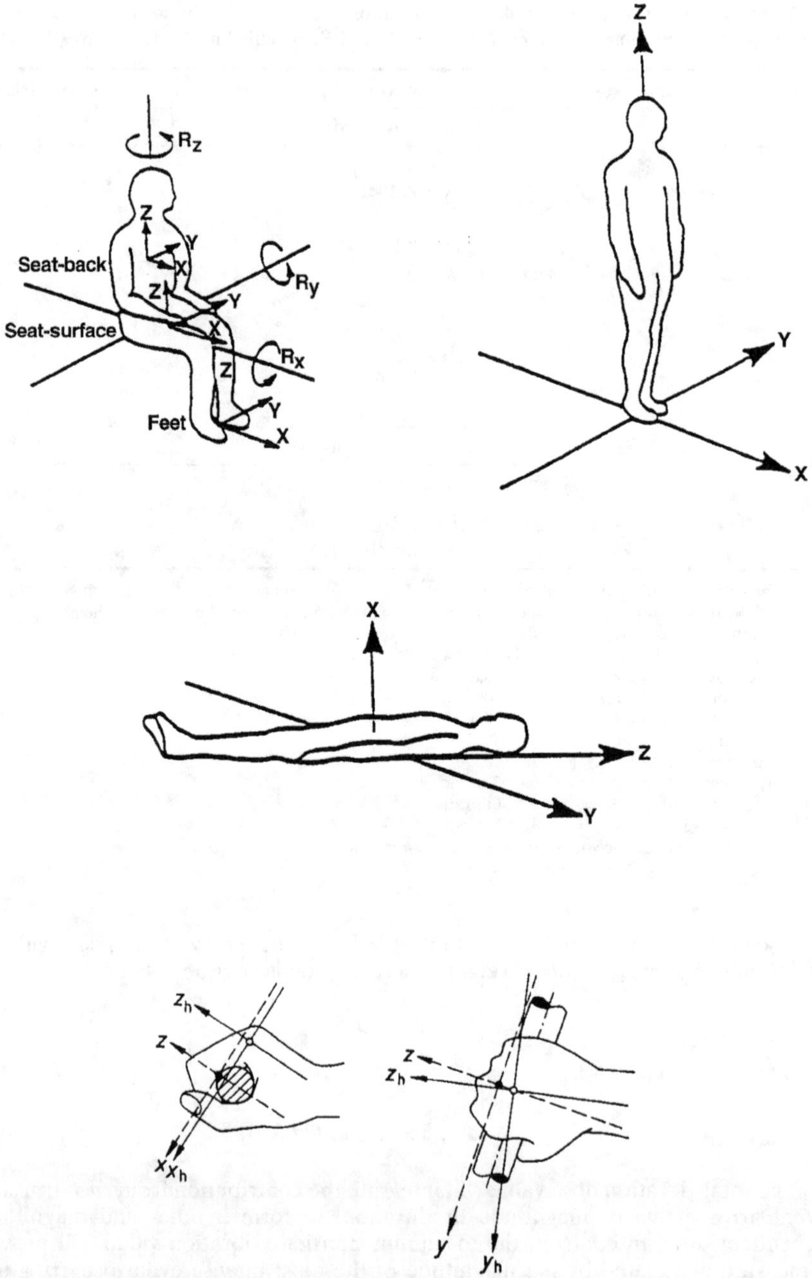

FIGURE 41.12 Coordinate axes for the human body seated, standing, and recumbent; and for the hand-arm system. Also shown are rotational axes for the body (pitch) R_y, roll R_x, and yaw R_z), and both basicentric (dashed) and biodynamic (continuous) coordinate systems for the hand and arm. (*ISO 2631-1.*[27] *and ISO 5349-1.*[29])

variability of, and intervals between, transient motions, and whether the human response is likely to be dose related).

Health. Guidance for the effect of whole-body vibration on health is provided in international standard ISO 2631-1 for vibration transmitted through the seat pan in the frequency range from 0.5 to 80 Hz.[27] The assessment is based on the largest measured translational component of the frequency-weighted acceleration (see Fig. 41.12 and Table 41.2). If the motion contains transient events that result in the condition in Eq. (41.9) being satisfied, then a further assessment may be made using the vibration dose value. The frequency weightings to be applied, W_d and W_k (see Table 41.2), are to be multiplied by factors of unity for vibration in the Z direction and 1.4 for the X and Y directions of the coordinate system shown in Fig. 41.12. The largest component-weighted acceleration is to be compared at the daily exposure duration with the shaded health caution zone in Fig. 41.13. The dashed lines in this diagram correspond to a relationship between the physical magnitude of the stimulus and exposure time with an index of $n = 2$ in Eq. (41.2), while the dotted lines correspond to an index of $n = 4$ in this equation. The lower and upper dotted lines in Fig. 41.13 correspond to vibration dose values of 8.5 and 17, respectively. For exposures below the shaded zone, which has been extrapolated to shorter and longer daily exposure durations in the diagram, health effects have not been reproducibly observed; for exposures within the shaded zone, the potential for health effects increases; for exposures above the zone, health effects are expected.[19,20]

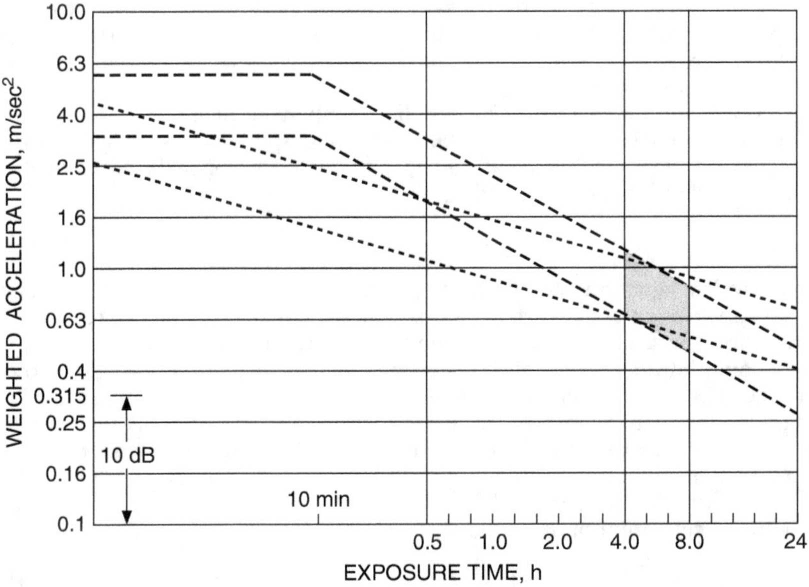

FIGURE 41.13 Health guidance caution zone for exposure to whole-body vibration. The dashed lines employ a relationship between stimulus magnitude and exposure time in hours [Eq. (41.2)] with $n = 2$ and the dotted lines $n = 4$. For exposures below the shaded zone, health effects have not been reproducibly observed; for exposures above the shaded zone, health effects may occur. The lower and upper dotted lines correspond to vibration dose values of 8.5 and 17, respectively. (*ISO 2631-1.*[27])

If the total daily exposure is composed of several exposures for times t_i to different frequency-weighted component accelerations $(a_W)_i$, then the equivalent acceleration magnitude corresponding to the total time of exposure $(a_W)_{\text{equiv}}$ may be constructed using

$$(a_W)_{\text{equiv}} = \left[\frac{\sum_i (a_W)_i^2 t_i}{\sum_i t_i} \right]^{1/2} \tag{41.10}$$

To characterize occupational exposure to whole-body vibration, the 8-hour frequency-weighted component accelerations may be measured according to Eq. (41.3) with $T = 28{,}800$ seconds. The total daily vibration dose value is constructed using Eq. (41.6) with $r = 4$.

Discomfort. Guidance for the evaluation of comfort and vibration perception is provided in international standard ISO 2631-1 for the exposure of seated, standing, and reclining persons (the last-mentioned supported primarily at the pelvis).[27] The guidance concerns translational and rotational vibration in the frequency range from 0.5 to 80 Hz that enters the body at the locations, and in the directions, listed in Table 41.2. The assessment is formed from rms component accelerations. For transient vibration, the maximum transient component vibration values should be considered if the condition in Eq. (41.8) is satisfied, while the magnitude of the vibration dose value may be used to compare the relative comfort of events of different durations. Each measure is to be frequency weighted according to the provisions of Table 41.2 and Fig. 41.12. Frequency weightings other than those in Table 41.2 have been found appropriate for some specific environments, such as for passenger and crew comfort in railway vehicles.[30]

Overall Vibration Value. The vibration components measured at a point where motion enters the body may be combined for the purposes of assessing comfort into a *frequency-weighted acceleration sum* a_{WAS}, which for orthogonal translational component accelerations a_{WX}, a_{WY}, and a_{WZ}, is

$$a_{\text{WAS}} = [a_{\text{WX}}^2 + a_{\text{WY}}^2 + a_{\text{WZ}}^2]^{1/2} \tag{41.11}$$

An equivalent equation may be used to combine rotational acceleration components.

When vibration enters a seated person at more than one point (e.g., at the seat pan, the backrest, and the feet), a weighted acceleration sum is constructed for each entry point. In order to establish the relative importance of these motions to comfort, the values of the component accelerations at a measuring point are ascribed a magnitude multiplying factor k so that, for example, a_{WX}^2 in Eq. (41.11) is replaced by $k^2 a_{\text{WX}}^2$, etc. The values of k are listed in Table 41.2, and are dependent on vibration direction and where motion enters the seated body. The *overall vibration total value* a_{overall} is then constructed from the root sum of squares of the frequency-weighted acceleration sums recorded at different measuring points, i.e.,

$$a_{\text{overall}} = [a_{\text{WAS}_1}^2 + a_{\text{WAS}_2}^2 + a_{\text{WAS}_3}^2 + \cdots]^{1/2} \tag{41.12}$$

where the subscripts 1,2,3, etc., identify the different measuring points.

Many factors, in addition to the magnitude of the stimulus, combine to determine the degree to which whole-body vibration causes discomfort (see "Effects of Vibration," above). Probable reactions of persons to whole-body vibration in public transport vehicles are listed in Table 41.3 in terms of overall vibration total values.

TABLE 41.3 Probable Subjective Reactions of Persons Seated in Public Transportation to Whole-Body Vibration Expressed in Terms of the Overall Vibration Value (defined in text) (*ISO 2631-1.*[27])

Vibration (m/sec^2)	Reaction
Less than 0.315	Not uncomfortable
0.315 to 0.63	A little uncomfortable
0.5 to 1	Fairly uncomfortable
0.8 to 1.6	Uncomfortable
1.25 to 2.5	Very uncomfortable
Greater than 2	Extremely uncomfortable

Fifty percent of alert, sitting or standing, healthy persons can detect vertical vibration with a frequency-weighted acceleration of 0.015 m/sec^2.

ACCEPTABILITY OF BUILDING VIBRATION

The vibration of buildings is commonly caused by motion transmitted through the building structure from, for example, machinery, road traffic, and railway and subway trains. Experience has shown that the criterion of acceptability for continuous or intermittent building vibration lies at, or only slightly above, the threshold of perception for most living spaces. Furthermore, complaints will depend on the specific circumstances surrounding vibration exposure. Guidance is provided for building vibration in Part 2 of the international standard for whole-body vibration, for the frequency range from 1 to 80 Hz,[31] and is adapted here to reflect alternate procedures for estimating the acceptability of building vibration (see Ref. 1).

In order to estimate the response of occupants to building vibration, the motion is measured on a structural surface supporting the body at, or close to, the point of entry of vibration into the body. For situations in which the direction of vibration and the posture of the building occupants are known (i.e., standing, sitting, or lying), the evaluation is based on the magnitudes of the component frequency-weighted accelerations measured in the X, Y, and Z directions shown in Fig. 41.12, using the frequency weightings for comfort, W_k and W_d, as appropriate (see Table 41.2 and Fig. 41.11). If the posture of the occupant with respect to the building vibration changes or is unknown, a so-called *combined* frequency weighting may be employed which is applicable to all directions of motion entering the human body, and has attenuation proportional to

$$10 \log[1 + (f/5.6)^2] \tag{41.13}$$

where the frequency f is expressed in hertz. No adverse reaction from occupants is expected when the rms frequency-weighted acceleration of continuous or intermittent building vibration is less than 3.6×10^{-3} m/sec^2.

Transient building vibration, that is, motion which rapidly increases to a peak followed by a damped decay (which may or may not involve several cycles of vibration), may be assessed either by calculating the maximum transient vibration value or the total vibration dose value using Eqs. (41.5) and (41.6), respectively. No adverse human reaction to transient building vibration is expected when the maxi-

TABLE 41.4 Maximum RMS Frequency-Weighted Acceleration, RMS Transient Vibration Value, MTVV, and Vibration Dose Value, VDV (defined in text) for Acceptable Building Vibration in the Frequency Range 1–80 Hz[1]

Place	Time[2]	Continuous/ intermittent vibration (m/sec^2)	Transient vibration MTVV (m/sec^2)	Transient vibration VDV ($m/sec^{1.75}$)
Critical working areas (e.g., hospital operating rooms)[3]	Any	0.0036	0.0036	0.1
Residences[4,5]	Day	0.0072	$0.07/n^{1/2}$	0.2
	Night	0.005	0.007	0.14
Offices[5]	Any	0.014	$0.14/n^{1/2}$	0.4
Workshops[5]	Any	0.028	$0.28/n^{1/2}$	0.8

[1] The probability of adverse human response to building vibration that is less than the weighted accelerations, MTVVs, and VDVs listed in this table is small. Complaints will depend on specific circumstances. For an extensive review of this subject, see Ref. 1. Note that: (*a*) VDV has been used for the evaluation of continuous and intermittent, as well as for transient, building vibration; and (*b*) annoyance from acoustic noise caused by vibration (e.g., of walls or floors) has not been considered in formulating the guidance in Table 41.4.

[2] Daytime may be taken to be from 7 AM to 9 PM and nighttime from 9 PM to 7 AM.

[3] The magnitudes of transient vibration in hospital operating theaters and critical working places pertain to those times when an operation, or critical work, is in progress.

[4] There are wide variations in human tolerance to building vibration in residential areas.

[5] n is the number of discrete transient events that are 1 second or less in duration. When there are more than 100 transient events during the exposure period, use $n = 100$.

mum rms frequency-weighted transient vibration value is less than 3.6×10^{-3} m/sec^2, or the total frequency-weighted vibration dose value is less than 0.1 m/sec$^{1.75}$.

Human response to building vibration depends on the use of the living space. In circumstances in which building vibration exceeds the values cited to result in no adverse reaction, the use of the room(s) should be considered. Site-specific values for acceptable building vibration are listed in Table 41.4 for common building and room uses. Explanatory comments applicable to particular room and/or building uses are provided in footnotes to that table.

It should be noted that building vibration at frequencies in excess of 30 Hz may cause undesirable acoustical noise within rooms, a subject not considered in this chapter. In addition, the performance of some extremely sensitive or delicate operations (e.g., microelectronics fabrication) may require control of building vibration more stringent than that acceptable for human habitation.

MOTION SICKNESS

Guidance for establishing the probability of whole-body vibration causing motion sickness is obtained from international standard ISO 2631-1 by forming the *motion sickness dose value,* $MSDV_z$.[27] This energy-equivalent dose value is given by the term on the right-hand side of Eq. (41.6) with $r = 2$, and the acceleration time-history frequency-weighted using W_f (see Fig. 41.11). If the exposure is to continuous vibration of near constant magnitude, the motion sickness dose value may be approximated by the frequency-weighted acceleration recorded during a measurement interval τ of at least 240 seconds by

$$\mathrm{MSDV}_z \approx [a_{WZ}{}^2\tau]^{1/2} \tag{41.14}$$

While there are large differences in the susceptibility of individuals to the effects of low-frequency vertical vibration (0.1 to 0.5 Hz), the percentage of persons who may vomit is

$$P = K_m(\mathrm{MSDV}_z) \tag{41.15}$$

where K_m is a constant equal to about one-third for a mixed population of males and females. Note that females are more prone to motion sickness than males.

Further guidance for the evaluation of exposure to extremely low frequency whole-body vibration (0.063 to 1 Hz) such as occurs on off-shore structures is to be found in ISO 6987.[32]

HAND-TRANSMITTED VIBRATION

Guidance for the measurement and assessment of hand-transmitted vibration is provided in international standard ISO 5349.[29,33] Three rms frequency-weighted component accelerations, a_{hwx}, a_{hwy}, and a_{hwz}, are first determined at the hand-handle interface for the directions described in Fig. 41.12, using the frequency weighting specified for all directions of vibration coupled to the hand (shown in Fig. 41.11). The values are constructed according to Eq. (41.3). The *vibration total value,* a_{hv}, is then formed, which is defined as the frequency-weighted acceleration sum constructed from the hand-transmitted component accelerations, i.e., using Eq. (41.11), but with a_{WAS} replaced by a_{hv}, a_{WX} by a_{hwx}, a_{WY} by a_{hwy}, and a_{WZ} by a_{hwz}.

If it is not possible to record the vibration in each of the three coordinate directions, then an estimate of a_{hv} is made from the largest component acceleration measured (i.e., either a_{hwx}, a_{hwy}, or a_{hwz}) by multiplying by a factor in the range from 1.0 to 1.7. The factor is designed to account for the contribution to the vibration total value from any unmeasured vibration.

The assessment of vibration exposure is based on the *8-hour energy equivalent vibration total value,* $(a_{hv})_{eq(8)}$. If the measurement procedure results in the daily exposure being composed of i exposures for times t_i to vibration total values a_{hvi}, then the 8-hour energy equivalent vibration total value is obtained by forming the sum:

$$(a_{hv})_{eq(8)} = \left[\frac{1}{28{,}800}\sum_i a_{hvi}{}^2 t_i\right]^{1/2} \tag{41.16}$$

If, alternatively, the measurement procedure provides a time history of the vibration total value $a_{hv}(t)$, then $(a_{hv})_{eq(8)}$ may be calculated directly by energy averaging for an eight-hour period, $T_0 = 28{,}800$ sec.

$$(a_{hv})_{eq(8)} = \left[\frac{1}{28{,}800}\int_0^{T_0} a_{hv}{}^2(t)\,dt\right]^{1/2} \tag{41.17}$$

Development of White Fingers (Finger Blanching). For groups of persons who are engaged in the same work using the same, or similar, vibrating hand tools, or

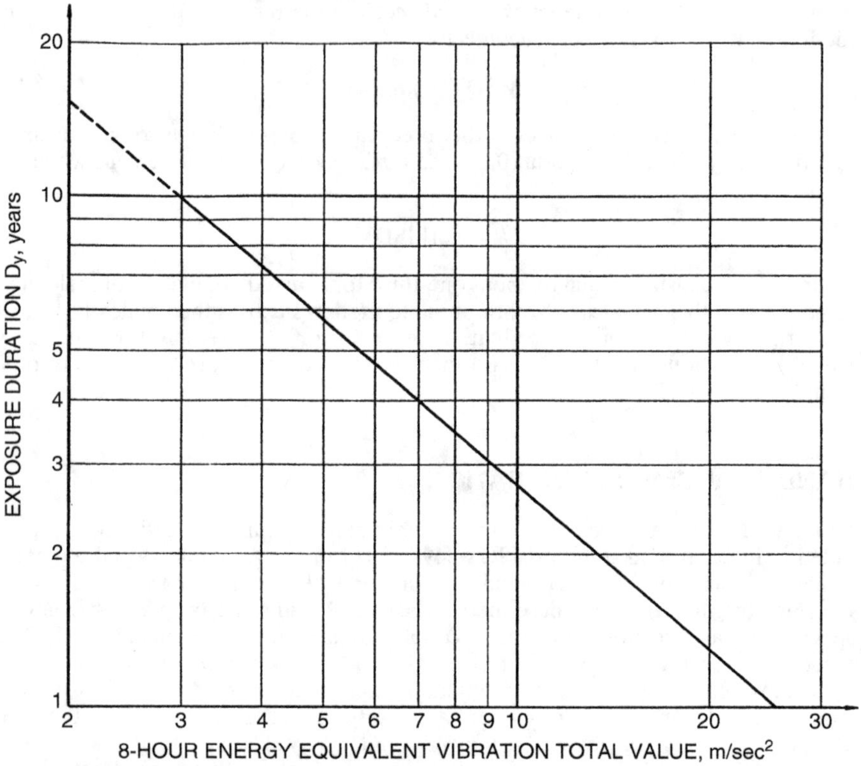

FIGURE 41.14 Duration of employment D_y, expressed in years, for 10 percent of a group of workers, all of whom perform essentially the same operations that result in exposure to effectively the same 8-hour energy equivalent vibration total value, $(a_{hv})_{eq(8)}$, to develop episodes of finger blanching. (*ISO 5349.*[29])

industrial processes in which vibration enters the hands (e.g., forestry workers using chain saws, chipping and grinding to clean castings, etc.), the number of years of exposure, on average, before 10 percent of the group experience episodes of *finger blanching, D_y*, is related to the 8-hour energy equivalent vibration total value by the relationship, shown in Fig. 41.14:

$$[(a_{hv})_{eq(8)}]^{1.06} D_y = 31.8 \tag{41.18}$$

The expression assumes that $(a_{hv})_{eq(8)}$ is expressed in m/sec^2, and D_y in years. Exposures below the line in Fig. 41.14 incur less risk of developing HAVS. There is no epidemiologic evidence for finger blanching occurring at values of $(a_{hv})_{eq(8)}$ of less than 1 m/sec^2. Deviation from the relationship shown in Fig. 41.14 may be expected for industrial situations that differ significantly from common practice (e.g., mixed occupations, such as painting for a week followed by chipping for a week), and for some impact power tools (e.g., sand rammer).

SHOCK AND IMPACT

Injury from Multiple Shocks and Impacts. Guidance on the risk of injury for seated persons from multiple shocks and impacts is contained in ISO 2631-5.[34] The recommended procedure consists of three parts. The first involves employing a biodynamic model to predict the motion at the spine; the second involves accumulating peak accelerations at the seat to estimate the dose at the spine, and the third involves applying an injury risk model based on the cumulative fatigue failure of repeatedly stressed biological materials (see "Physical Data" and especially Fig. 41.3). A neural network model is employed for motion along the spinal axis (see "Biodynamic Models"), and DRI-like models for the X and Y directions (see "Effects of Shock and Impact"). The inputs to the biodynamic models are the seat motions measured in the X, Y, and Z directions specified in Fig. 41.12. The acceleration dose is constructed, separately, from the peak acceleration of each shock at the spine that causes compression, or lateral motion, as calculated from the output of the appropriate biodynamic model, using the right side of Eq. 41.6 with $a_w(t)$ representing the peak acceleration of the shock or impact, and $r = 6$. The combined acceleration dose applicable to an average working day is converted into an equivalent *static* compressive stress, which may then be interpreted for the potential for fatigue failure of the vertebral end plates.[35] The calculation takes into account the reducing strength of vertebrae with age. A lifetime exposure to a static stress of less than 0.5 MPa is associated with a low probability of an adverse health effect, whereas lifetime exposure to a static stress in excess of 0.8 MPa has a high probability of spinal injury.[34] A Matlab code for performing the calculations is provided in the standard.

The neural network model was trained on human responses to peak accelerations of up to 40 m/sec^2, so the method should not be applied to shocks and impacts of larger magnitude. This restriction has limited consequences for practical transportation systems, as such motions are unlikely to be tolerated.

A conceptually similar approach to that of ISO 2631-5 is suggested for multiple shocks (no impacts) with peak accelerations greater than about 40 m/sec^2 in the Z direction (Fig. 41.12). It is thus applicable to persons who are seated and restrained by seat harnesses. The procedure employs the DRI as the biodynamic model (see "Effects of Shock and Impact"), and, as before, forms a dose by summing shocks. The sum is used to estimate the risk of spinal injury.[36]

Survivable Single Shocks. Experiments in which humans or animals were exposed to single shocks have established the tolerance of seated persons to such accelerations. This unique body of information, which is unlikely to be extended for ethical reasons, was consolidated by Eiband who characterized the shocks at the seat by idealized trapezoidal time histories, with a constant onset acceleration rate, a constant peak acceleration, and a constant decay rate.[37] The tolerance limits so obtained are shown for accelerations directed toward the spine (from in front), the head (upward), and the tailbone (downward) in Figs. 41.15 to 41.20. The results are presented in terms of peak accelerations and their durations for the three directions and in terms of onset acceleration rates, which are characterized by the onset time $(t_1 - t_0)$ and plotted on the abscissa of Figs. 41.16, 41.18, and 41.20. The upper boundary of the lower shaded area in Figs. 41.15, 41.17, and 41.19 defines the limit of voluntary human exposures that resulted in no injury. The corresponding lower boundary of the upper shaded area delineates the limit of serious injury in animal experiments involving hogs and chimpanzees. No corrections for size or species differences were

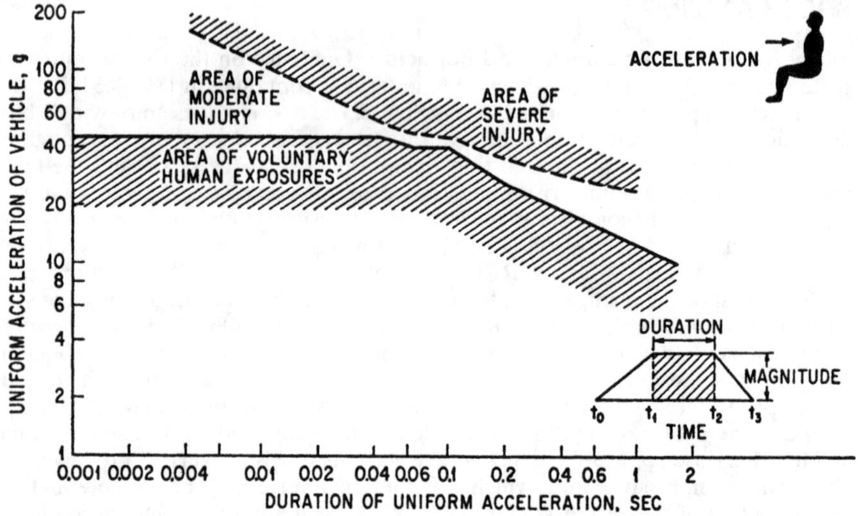

FIGURE 41.15 Tolerance to spineward acceleration as a function of magnitude and duration of impulse. (*Eiband.*[37])

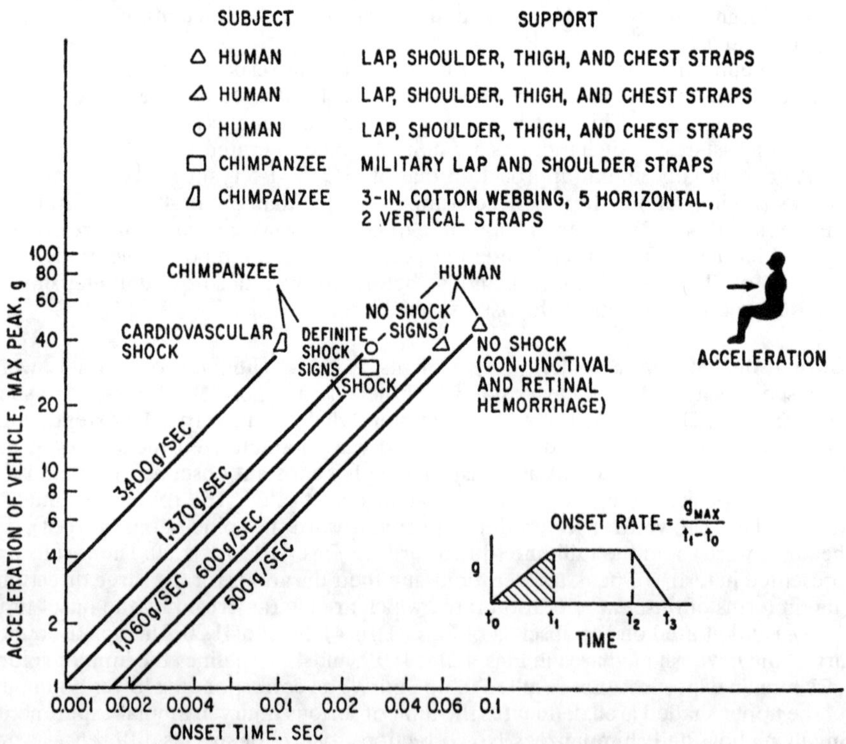

FIGURE 41.16 Effect of rate of onset on spineward acceleration tolerance. (*Eiband.*[37])

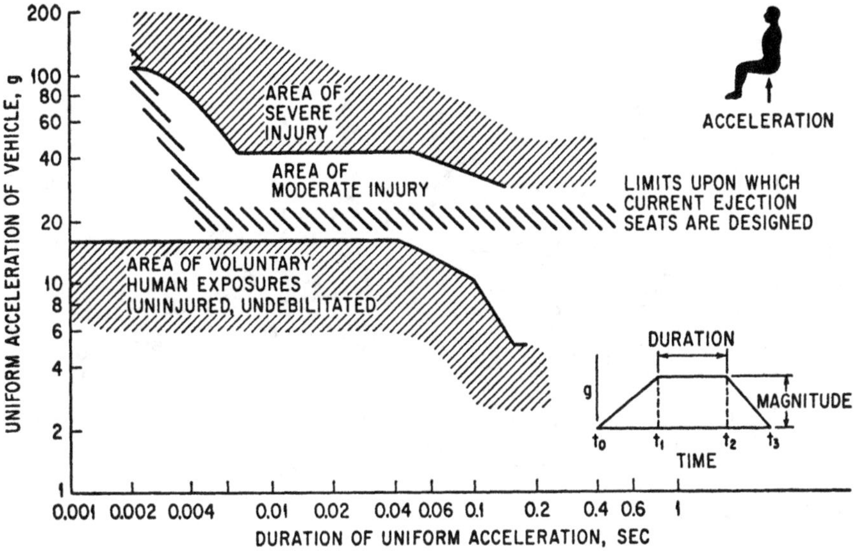

FIGURE 41.17 Tolerance to headward acceleration as a function of magnitude and duration of impulse. (*Eiband.*[37])

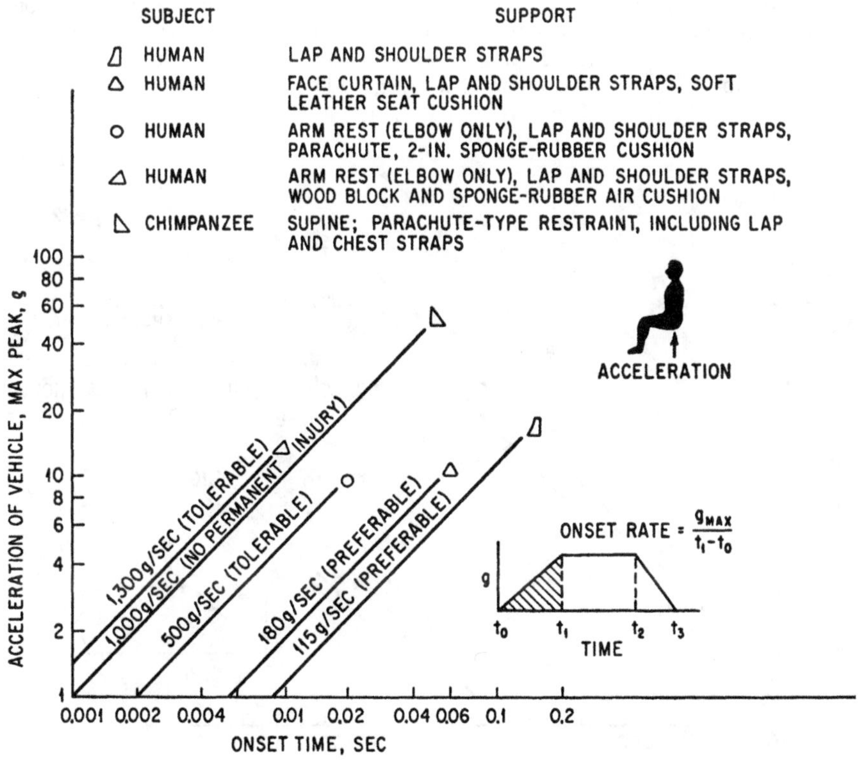

FIGURE 41.18 Effect of rate of onset on headward acceleration tolerance. (*Eiband.*[37])

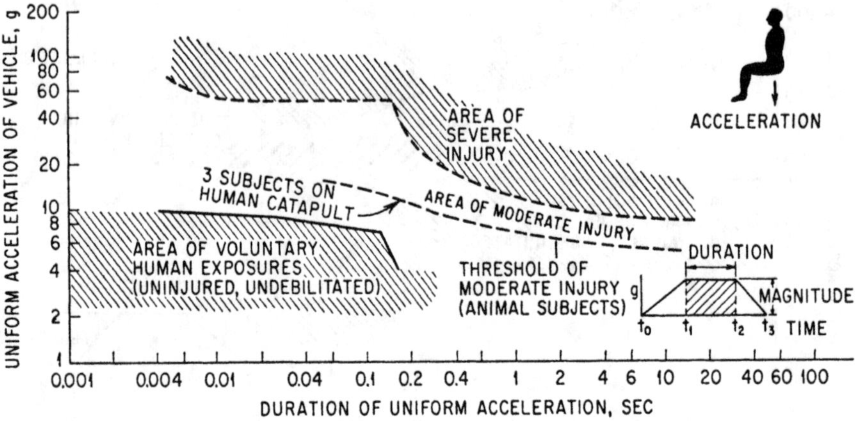

FIGURE 41.19 Tolerance to tailward acceleration as a function of magnitude and duration of impulse. (*Eiband.*[37])

attempted. Maximum body support was provided to the subject in all experiments (i.e., lap belt, shoulder harness, thigh and chest straps, and armrests, as appropriate; [see *Protection Against Rapidly Applied Accelerations* (*Crash*) and Fig. 41.23]. Tolerance limits for accelerations directed horizontally from behind the body (toward the sternum) are similar to those for spineward acceleration shown in Figs. 41.15 and 41.16. For more details the original analysis should be consulted (Ref. 37).

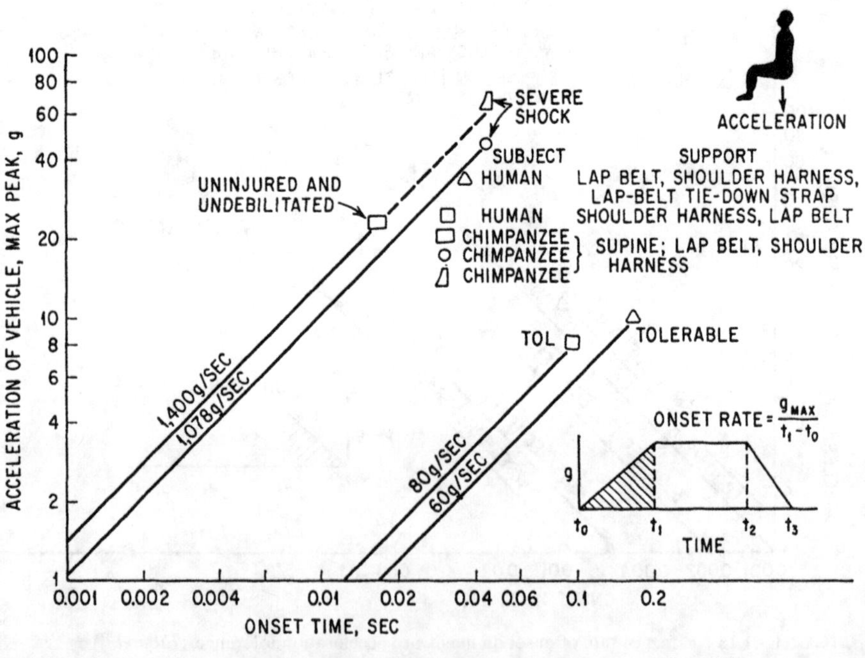

FIGURE 41.20 Effect of rate of onset on tailward acceleration tolerance. (*Eiband.*[37])

While caution must be exercised in applying these tolerance curves, since they are based on experiments involving healthy young volunteers and animals, rigid seats, well-designed body supports, and minimum slack in harnesses, they form the primary information on which to base safety requirements for transportation vehicles. Examples of short-duration accelerations to illustrate the magnitudes and durations experienced in practice are listed in Table 41.5.

Performance limits applicable to automotive crash testing with Hybrid III and SID anthropomorphic dummies (see "Substitutes for Live Human Subjects") have been mandated in Federal Motor Vehicle Safety Standards (FMVSS) for subsystems of the body (e.g., head, neck, and chest),[38] and are promulgated in the U.S.A. by the National Traffic Safety Administration (NHTSA). The values are to be found in FMVSS 208, Occupant Crash Protection.

Head Injury Criterion. The goal of protecting the head from irreversible brain damage in motor vehicle collisions involving unrestrained occupants led to the formulation of the Wayne State Concussion Tolerance Curve, which was derived from experiments in which instrumented, embalmed human cadavers were positioned horizontally and then dropped so that their foreheads fractured on impact with steel anvils or other targets (including motor-vehicle instrument panels). Impact durations measured on the skull of from 1 to 6 milliseconds could be obtained from this experiment. The tolerance curve was extended to impact durations of 100 milliseconds using an asymptotic acceleration of 42*g*, which corresponds to the limit of vol-

TABLE 41.5 Approximate Duration and Magnitude of Some Short-Duration Acceleration Loads

Type of operation	Acceleration, *g*	Duration, sec
Elevators:		
Average in "fast service"	0.1–0.2	1–5
Comfort limit	0.3	
Emergency deceleration	2.5	
Public transit:		
Normal acceleration and deceleration	0.1–0.2	5
Emergency stop braking from 110 km/h	0.4	2.5
Automobiles:		
Comfortable stop	0.25	5–8
Very undesirable	0.45	3–5
Maximum obtainable	0.7	3
Crash (potentially survivable)	20–100	<0.1
Aircraft:		
Ordinary take-off	0.5	>10
Catapult take-off	2.5–6	1.5
Crash landing (potentially survivable)	20–100	
Seat ejection	10–15	0.25
Man:		
Parachute opening, 12,000 m	33	0.2–0.5
1800 m	8.5	0.5
Parachute landing	3–4	0.1–0.2
Fall into firefighter's net	20	0.1
Approximate survival limit with well-distributed forces (fall into deep snowbank)	200	0.015–0.03
Head:		
Adult head falling from 2 m onto hard surface	250	0.007
Voluntarily tolerated impact with protective headgear	18–23	0.02

untary human exposure that resulted in no injury in Fig. 41.15 (the duration of motor vehicle crashes depends primarily on vehicle speed and typically lasts for less than 100 milliseconds). The asymptotic limit was subsequently raised to a head acceleration of 80*g* for impacts of the forehead on padded surfaces that were believed to be survivable.

The Wayne State Concussion Tolerance Curve has proved difficult to apply to complex acceleration-time impact waveforms, because of uncertainty in determining the effective acceleration and time. A straight-line approximation to the curve (between 2.5 and 25 milliseconds) led to the definition of the *severity index* (SI) as:

$$\mathrm{SI} = \int_0^T a^{2.5}(t)\,dt \tag{41.19}$$

where T is the impact duration, and $a(t)$ the acceleration time history of the head (in units of *g*). The maximum value was proposed to be 1000. A revised index has been defined by the NHTSA for use in the frontal crash tests specified in motor vehicle regulations, which has become known as the *head injury criterion* (HIC):

$$\mathrm{HIC} = \left| (t_2 - t_1) \left[\frac{1}{(t_2 - t_1)} \int_{t_1}^{t_2} a(t)\,dt \right]^{2.5} \right|_{\max} \tag{41.20}$$

where t_1 and t_2 are the initial and final times (in seconds) of the interval during which the HIC attains the maximum value, and $a(t)$ is measured at the center of gravity of the manikin's head. This measure is to be applied to tests using instrumented anthropometric dummies, in which a maximum value of 1000 is allowed. FMVSS 208 specifies the time interval $(t_2 - t_1)$ to be 33 milliseconds.

There are several challenges in attempting to set *human* tolerance criteria, based on either the SI or HIC.[26] First, the ability of crash tests employing HICs computed from measurements on an anthropometric dummy to rank order impact conditions by severity has been questioned. Second, the original Wayne State Concussion Tolerance Curve was designed for *unrestrained* vehicle occupants, whereas the data employed to extend the relationship to head impact durations greater than 6 milliseconds, which commonly occur in vehicle crash tests, are for subjects with optimum body restraints. Despite these limitations, the SI has been successfully applied to the reduction of brain injuries in football players by employing football helmets that attenuate head impacts to SI < 1500, while the HIC remains a cornerstone of occupant safety testing for automobiles and, more recently, for transport aircraft.

PROTECTION METHODS AND PROCEDURES

Protection of man against mechanical forces is accomplished in two ways: (1) isolation to reduce transmission of the forces to the man and (2) increase of man's mechanical resistance to the forces. Isolation against shock and vibration is achieved if the natural frequency of the system to be isolated is lower than the exciting frequency by at least a factor of 2. Both linear and nonlinear resistive elements are used for damping the transmission system; irreversible resistive elements or energy-absorbing devices can be used once to change the time and amplitude pattern of impulsive forces (e.g., progressive collapse of automobile engine compartment in frontal crash). Human tolerance to mechanical forces is strongly influenced by selecting the proper body position with respect to the direction of forces to be expected.

Man's resistance to mechanical forces also can be increased by an appropriate distribution of the forces so that relative displacement of parts of the body is avoided as much as possible. This may be achieved by supporting the body over as wide an area as possible, preferably loading bony regions and thus making use of the rigidity available in the skeleton. Reinforcement of the skeleton is an important feature of seats designed to protect against crash loads. The flexibility of the body is reduced by fixation to the rigid seat structure. The mobility of various parts of the body, e.g., the abdominal mass, can be reduced by properly designed belts and suits. The factor of training and indoctrination is essential for the best use of protective equipment, for aligning the body in the least dangerous positions during intense vibration or crash exposure, and possibly for improving operator performance during vibration exposure.

PROTECTION AGAINST VIBRATIONS

The transmission of vibration from a vehicle or platform to a man is reduced by mounting him on a spring or similar isolation device, such as an elastic cushion. The degree of vibration isolation theoretically possible is limited, in the important resonance frequency range of the sitting man, by the fact that large static deflections of the man with the seat or into the seat cushion are undesirable. Large relative movements between operator and vehicle controls interfere in many situations with man's performance. Therefore, a compromise must be made. Cushions are used primarily for static comfort, but they are also effective in decreasing the transmission of vibration above man's resonance range. They are ineffective in the resonance range and may even amplify the vibration. In order to achieve effective isolation over the 2- to 5-Hz range, the natural frequency of the man-cushion system should be reduced to 1 Hz, i.e., the natural frequency should be small compared with the forcing frequency (see Chap. 39). This would require a static cushion deflection of 25 cm. If a seat cushion without a back cushion is used (as is common in some tractor or vehicle arrangements), a condition known as "back scrub" (a backache) may result. Efforts of the operator to wedge himself between the controls and the back of the seat often tend to accentuate the discomfort.

For severe low-frequency vibration, such as occurs in tractors and other field equipment, suspension of the whole seat is superior to the simple seat cushion. Hydraulic shock absorbers, rubber torsion bars, coil springs, and leaf springs all have been successfully used for suspension seats.[2] A seat that is guided so that it can move only in a linear direction seems to be more comfortable than a configuration where the seat simply pivots around a center of rotation. The latter situation produces an uncomfortable and fatiguing pitching motion. Suspension seats can be built which are capable of preloading for the operator's weight so as to maintain the static position of the seat and the natural frequency of the system at the desired value. Suspension seats for use on tractors and on similar vehicles are available which reduce the resonance frequency of the man-seat system from approximately 4 to 2 Hz. This can be seen from the comparison of the transmissibility of a rigid seat, a truck suspension seat, and a conventional foam and metal sprung car seat in Fig. 41.21. The transmissibility of the car seat is in excess of 2 at the resonance frequency (4 Hz), implying that the seat motion reaching the body is amplified by this ratio. In contrast, the amplification introduced by the suspension seat is at most a factor of 1.3 at the resonance frequency (2 Hz), and improved attenuation of vibration is obtained throughout the frequency range from 4 to 12 Hz. At frequencies below 2 Hz and above 12 Hz, less vibration is transmitted to the subject by the foam and metal

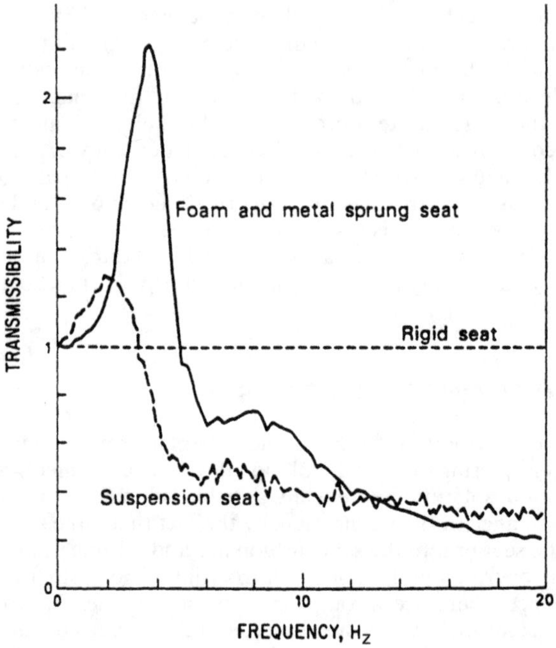

FIGURE 41.21 Comparison of the transmissibilities of a rigid seat, a foam-covered metal sprung seat, and a truck suspension seat. (*Griffin.*[1])

sprung seat. There are large differences in the performance of suspension seats, with transmissibilities in excess of 2 being recorded in some designs at the resonance frequency (which is usually close to 2 Hz).[1] In consequence, the selection of a seat for a particular application must take into account both the performance of the seat and the critical vibration frequencies to be attenuated.

For severe vibrations, close to or exceeding normal tolerance limits, such as those which may occur in military operations, special seats and restraints can be employed to provide maximum body support for the subject in all critical directions. In general, under these conditions, seat and restraint requirements are the same for vibration and rapidly applied accelerations (discussed in the next section).

Isolation of the hand and arm from the vibration of handheld or hand-guided power tools is accomplished in several ways. A common method is to isolate the handles from the rest of the power tool, using springs and dampers (see Chap. 39). The application of vibration-isolation systems to chain saws for use in forestry has become commonplace and has led to a reduction in the incidence of HAVS. A second method is to modify the tool so that the primary vibration is counterbalanced by an equal and opposite vibration source. This method takes many different forms, depending on the operating principle of the power tool.[39] An example is shown for a pneumatic scaling chisel in Fig. 41.22, in which an axial impact is applied to a work piece to remove metal by a chisel *P*. The chisel is driven into the work piece by compressed air and is returned to its initial position by a spring *S*. The axial motion of the chisel is counterbalanced by a second mass *m* and spring *k* which oscillate out of

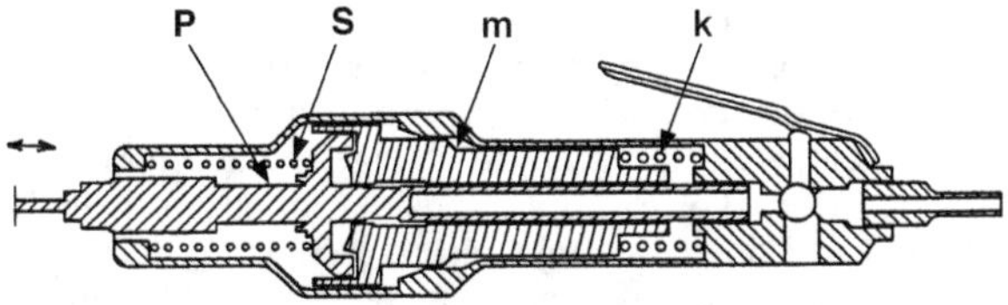

FIGURE 41.22 An antivibration power tool design for a pneumatic scaler: *P*—vibrating chisel; *S*—chisel return spring; *m*—counterbalancing oscillating weight; and *k*—counterbalance return spring. (*Lindqvist.*[39])

phase with the chisel motion. The design of an appropriate vibration-isolation system must include the dynamic properties of the hand-arm system.[11,13]

Conventional gloves do not attenuate the vibration transmitted to the hand but may increase comfort and keep the hands warm. So-called antivibration gloves also fail to reduce vibration at frequencies below 100 Hz, which are most commonly responsible for HAVS, but may reduce vibration at high frequencies (the relative importance of different frequencies in causing HAVS is shown in Fig. 41.11).

Preventive measures for HAVS to be applied in the workplace include minimizing the duration of exposure to vibration, using minimum hand-grip force consistent with safe operation of the power tool or process ("let the tool do the work"), wearing sufficient clothing to keep warm, and maintaining the tool in good working order, with minimum vibration.[29] As recovery from HAVS has only been demonstrated for early vascular symptoms, medical monitoring of persons exposed to vibration is essential. Monitoring should include a test of peripheral neurological function,[40] since this component of HAVS appears to persist.

PROTECTION AGAINST RAPIDLY APPLIED ACCELERATIONS (CRASH)

The study of automobile and aircraft crashes and of experiments with dummies and live subjects shows that complete body support and restraint of the extremities provide maximum protection against accelerating forces and give the best chance for survival. If the subject is restrained in the seat, he makes full use of the force moderation provided by the collapse of the vehicle structure and is protected against shifts in position which would injure him by contact with interior surfaces of the cabin. The decelerative load must be distributed over as large a body area as possible to avoid force concentration with resulting bending moments and shearing effects. The load should be transmitted as directly as possible to the skeleton, preferably directly to the pelvic structure—not via the vertebral column.

Theoretically, a rigid envelope around the body will protect it to the maximum possible extent by preventing deformation. A body restrained to a rigid seat approximates such a condition; proper restraints against longitudinal acceleration shift part of the load of the shoulder girdle and arms from the spinal column to the backrest. Armrests can remove the load of the arms from the shoulders. Semirigid and elastic abdominal supports provide some protection against large abdominal displacements. The effectiveness of this principle has been shown by animal experiments and by impedance measurements on human subjects. Animals immersed in water, which distributes the load applied to the rigid container evenly over the body surface, or in rigid casts are able to survive acceleration loads many times their normal tolerance.

Many attempts have been made to incorporate energy-absorptive devices, either in a harness or in a seat, with the intent to change the acceleration-time history by limiting peak accelerations. For example, consider an aircraft which is stopped in a crash from 160 km/h in 1.5m; it is subjected to a constant deceleration of 67*g*. An energy-absorptive device designed to elongate at 17*g* would require a displacement of 0.5 m. In traveling this distance, the body or seat would be decelerated relative to the aircraft by 14.4*g* and would have a maximum velocity of 11.2 m/sec relative to the aircraft structure. A head striking a solid surface (e.g., cabin interior surface) with this velocity has many times the minimum energy required to fracture a skull. The available space for seat or passenger travel using the principle of energy absorption must therefore be considered carefully in the design. Seats for jet airliners have been designed which have energy-absorptive mechanisms in the form of extendable rear legs. The maximum travel of the seats is 15 cm; their motion is designed to start between 9 and 12*g* horizontal load, depending on the floor strength. During motion, the legs pivot at the floor level—a feature considered to be beneficial if the floor wrinkles in the crash. Theoretically, such a seat can be exposed to a deceleration of 30*g* for 0.037 sec or 20*g* for 0.067 sec without transmitting a deceleration of more than 9*g* to the seat. However, the increase in exposure time must be considered as well as the reduction in peak acceleration. For very short exposure times where the body's tolerance probably is limited by the transferred momentum and not the peak acceleration, the benefits derived from reducing peak loads would disappear.

The high tolerance limits of the well-supported human body to decelerative forces suggest that in aircraft and other vehicles, seats, floors, and the whole inner structure surrounding crew and passengers should be designed to resist crash decelerations as near to 40*g* as weight or space limitations permit.[41] The structural members surrounding this inner compartment should be arranged so that their crushing reduces forces on the inner structure. Protruding and easily loosened objects should be avoided. To allow the best chance for survival, seats should also be stressed for dynamic loadings between 20 and 40*g*. Civil Air Regulations require a minimum static strength of seats of 9*g*. A method for computing seat tolerance for typical survivable airplane crash decelerations is available for seats of conventional design.[41] It has been established that an unrestrained passenger who is riding in a seat facing backward has a better chance to survive an abrupt crash deceleration since the impact forces are then more uniformly distributed over the body. Neck injury must be prevented by proper head support.

Increased survivability in automobile as well as airplane crashes can be obtained by distributing the load over larger areas of the body and fixing the body more rigidly to the seat. Shoulder straps, thigh straps, chest straps, and handholds are additional body supports used in experiments. They are illustrated in Fig. 41.23. Table 41.6 shows the desirability of these additional restraints to increase possible survivability to acceleration loads in various directions. In airplane crashes, vertical and horizontal loads must be anticipated. In automobile crashes, horizontal loads are most likely.

A forward-facing passenger held by a seat belt flails about when suddenly decelerated; his hands, feet, and upper torso swing forward until his chest hits his knees or until the body is stopped in this motion by hitting other objects (back of seat in front, cabin wall, instrument panel, steering wheel, control stick). Since 15 to 18*g* longitudinal deceleration can result in 3 times higher acceleration of the chest hitting the knees, this load appears to be about the limit a human can tolerate with a seat belt

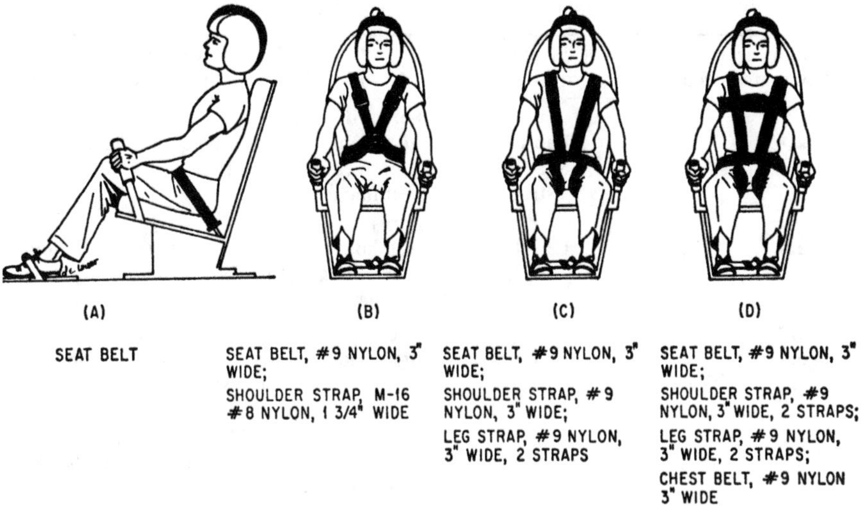

FIGURE 41.23 Protective harnesses for rapid accelerations or decelerations. The following devices were evaluated in sled deceleration tests: (*A*) Seat belt for automobiles and commercial aviation. (*B*) Standard military lap and shoulder strap. (*C*) Like (*B*) but with thigh straps added to prevent headward rotation of the lap strap. (*D*) Like (*C*) but with chest strap added. (*Stapp:* USAF Tech. Rept. 5915, *pt. I, 1949; pt. II, 1951.*)

alone. Approximately the same limit is obtained when the head-neck structure is considered.

Lap straps always should be as tight as comfort will permit to exclude available slack. During forward movement, about 60 percent of the body mass is restrained by the belt, and therefore represents the belt load. If the upper torso is fixed to the back of the seat by any type of harness (shoulder harness, chest belt, etc.), the load on the seat is approximately the same for forward- and aft-facing seats. The difference between these seats with respect to crash tolerance as discussed above no longer exists. The body restraints for passenger and crew must be applied without creating excessive discomfort.

A rapidly inflating air bag situated in front of an automobile driver, and often the front passenger, and inflated on frontal collision, has been installed in most vehicles. While initially conceived as an alternative to passive restraints, that is, as a safety system that would operate when an automobile occupant was not wearing a seat belt, air bags are now recognized to provide most benefit when considered as a complementary system to lap and shoulder seat belts. The device consists of a crash sensor or sensors mounted near the front of the vehicle that signal velocity changes to a controller; those in excess of about 6 m/sec cause a pyrotechnic reaction to generate gas that inflates a porous fabric bag within, typically 25 milliseconds, so that the bag is inflated sufficiently to distribute the deceleration forces over a large surface area on contact with the occupant. Accident data have shown that while air bags do save lives, believed to be some 2620 people in the United States from 1990 to 1997, they were also responsible for the deaths of at least 44 children and 36 adults during this period.[42] Most of the fatalities have been attributed to the size and position of the occupant at the time of impact with the air bag, which is not defined if a seat belt is

TABLE 41.6 Human-Body Restraint and Possible Increased Impact Survivability. (*After Eiband.*[37])

Direction of acceleration imposed on seated occupants	Conventional restraint	Possible survivability increases available by additional body supports*
Spineward:		Forward facing:
Crew	Lap strap Shoulder straps	(*a*) Thigh straps (assume crew members will be performing emergency duties with hands and feet at impact)
		Forward facing:
Passengers	Lap strap	(*a*) Shoulder straps, (*b*) thigh straps, (*c*) nonfailing armrests, (*d*) suitable handholds, and (*e*) emergency toe straps in floor
Sternumward:		Aft facing:
Passengers only	Lap strap	(*a*) Nondeflecting seat back, (*b*) integral, full-height headrest, (*c*) chest strap (axillary level), (*d*) lateral head motion restricted by padded "winged back," (*e*) leg and foot barriers, and (*f*) armrests and handholds (prevent arm displacement beyond seat back)
Headward:		Forward facing:
Crew	Lap strap Shoulder straps	(*a*) Thigh straps, (*b*) chest strap (axillary level), and (*c*) full, integral headrest (assume crew members will be performing emergency duties; extremity restraint useless)
		Forward facing:
Passengers	Lap strap	(*a*) Shoulder straps, (*b*) thigh straps, (*c*) chest strap (axillary level), (*d*) full, integral headrest, (*e*) nonfailing contoured armrests, and (*f*) suitable handholds
		Aft facing:
		(*a*) Chest strap (axillary level), (*b*) full, integral headrest, (*c*) nonfailing armrests, and (*d*) suitable handholds
Tailward:		Forward facing:
Crew	Lap strap Shoulder straps	(*a*) Lap-belt tie-down strap (assume crew members will be performing emergency duties; extremity restraint useless)
		Forward facing:
Passengers	Lap strap	(*a*) Shoulder straps, (*b*) lap-belt tie-down strap, (*c*) handholds, (*d*) emergency toe straps
		Aft facing:
		(*a*) Chest strap (axillary level), (*b*) handholds, and (*c*) emergency toe straps
		Feet forward:
Berthed occupants	Lap strap	Full-support webbing net
		Athwart ships:
		Full-support webbing net

* Exposure to maximum tolerance limits (see "Survivable Single Shocks") requires straps exceeding convention strap strength and width.

not worn. In these circumstances, the air bag may impact the occupant with sufficient force to produce fatal injury. Systems are under development to mitigate these effects (e.g., reducing the inflation rate of the bag and monitoring occupant position).[42]

The dynamic properties of seat cushions are extremely important if an acceleration force is applied through the cushion to the body. In this case the steady-state response curve of the total man-seat system (Fig. 41.21) provides a clue to the possible dynamic load factors under impact. Overshooting should be avoided, at least for the most probable shock rise times. This problem has been studied in detail in connection with seat cushions used on upward ejection seats. The ideal cushion is approached when its compression under static load spreads the load uniformly and comfortably over a wide area of the body and almost full compression is reached under the normal body weight. The acceleration then acts uniformly and almost directly on the body without intervening elastic elements. A slow-responding foam plastic, such as an open cell rate-dependent polyurethane foam, of thickness from 5 to 6 cm satisfies these requirements.[43]

A significant factor in human shock and impact tolerance appears to be the acceleration-time history of the subject immediately preceding the event. A *dynamic preload* consists of an imposed acceleration preceding, and/or during, and in the same direction as the shock or impact acceleration.[44] A dynamic preload occurs, for example, when the brakes are applied to a moving automobile before it hits a barrier. The phenomenon is found experimentally to reduce the acceleration of body parts on impact, thereby potentially mitigating adverse health effects. The dynamic preload should not be confused with the static preload introduced by a protective harness. The latter brings the occupant into contact with the seat or restraint but does not introduce the dynamic displacement of body parts and tissue compression necessary to reduce the body's dynamic response.

PROTECTION AGAINST HEAD IMPACT

The impact-reducing properties of protective helmets are based on two principles:[4] the distribution of the load over a large area of the skull and the interposition of energy-absorbing systems. The first principle is applied by using a hard shell, which is suspended by padding or support webbing at some distance from the head (typically 1.5 to 2.0 cm). High local impact forces are distributed by proper supports over the whole side of the skull to which the blow is applied. Thus, skull injury from relatively small objects and projectiles can be avoided. However, tests usually show that contact padding alone over the skull results, in most instances, in greater load concentration, whereas helmets with web suspension distribute pressures uniformly. Since helmets with contact padding usually permit less slippage of the helmet, a combination of web or strap suspension with contact padding is desirable. The shell itself must be as stiff as is compatible with weight considerations; when the shell is struck by a blow, its deflection must not be large enough to permit it to come in contact with the head.

Padding materials can incorporate energy-absorptive features. Whereas foam rubber and felt are too elastic to absorb a blow, foam plastics like polystyrene or Ensolite result in lower transmitted accelerations.

Most helmets constitute compromises among several objectives such as pressurization, communication, temperature conditioning, minimum bulk and weight, visibility, protection against falling objects, etc.; usually, impact protection is but one of many design considerations. The protective effect of helmets against concussion and skull fracture has been shown in animal experiments and is apparent from accident statistics.

ACKNOWLEDGMENT

Much of the material in this chapter was originally prepared by the late Henning E. von Gierke, past Director, Biodynamics and Bioengineering Division, Armstrong Laboratory, Wright-Patterson AFB, Ohio.

REFERENCES

GENERAL

1. Griffin, M. J.: "Handbook of Human Vibration," Academic Press, London, 1990.
2. Mansfield, N. J.: "Human Response to Vibration," CRC Press, Boca Raton, Fla., 2005.
3. Pelmear, P. L., and D. E. Wasserman, eds.: "Hand-Arm Vibration," 2d ed., OEM Press, Beverly Farms, Mass., 1998.
4. Nahum, A. M., and J. W. Melvin, eds.: "Accidental Injury: Biomechanics and Prevention," Springer-Verlag, New York, 2d ed., 2002.

BIODYNAMICS, MODELS, AND ANTHROPOMETRIC DUMMIES

5. "Anthropomorphic Dummies for Crash and Escape System Testing," AGARD-AR-330, North Atlantic Treaty Organization, Neuilly sur Seine, France, 1997.
6. Griffin, M. J.: "The Validation of Biodynamic Models," *Clinical Biomechanics, Supplement 1,* **16**:S81 (2001).
7. von Gierke, H. E., H. L. Oestreicher, E. K. Franke, H. O. Parrach, and W. W. von Wittern: "Physics of Vibrations in Living Tissues," *J. Appl. Physiol.,* **4**:886 (1952).
8. von Gierke, H. E.: "To Predict the Body's Strength," *Aviation Space & Environ. Med.,* **59**:A107 (1988).
9. von Gierke, H. E.: "Biodynamic Models and Their Applications," *J. Acoust. Soc. Amer.,* **50**:1397 (1971).
10. "Mechanical Vibration and Shock—Range of Idealized Values to Characterize Seated Body Biodynamic Response Under Vertical Vibration," ISO 5982, International Organization for Standardization, Geneva, 2001.
11. "Mechanical Vibration and Shock—Free Mechanical Impedance of the Human Hand-Arm System at the Driving-Point," ISO 10068, International Organization for Standardization, Geneva, 1998.
12. Mechanical Vibration and Shock—Coupling Forces at the Man-Machine Interface for Hand-Transmitted Vibration," ISO 15230, International Organization for Standardization, Geneva, 2007.
13. Dong, R. G., J. K. Wu, and D. E. Welcome: "Recent Advances in Biodynamics of Human Hand-Arm System," *Ind. Health,* **43**:449 (2005).
14. Seidel, H., and M. J. Griffin: "Modeling the Response of the Spinal System to Whole-Body Vibration and Repeated Shock," *Clinical Biomechanics, Supplement 1,* **16**:S3 (2001).
15. Nicol, J., J. Morrison, G. Roddan, and A. Rawicz: "Modeling the Dynamic Response of the Human Spine to Shock and Vibration Using a Recurrent Neural Network," *Heavy Vehicle Systems,* Special Series, *Int. J. of Vehicle Design,* **4**:145 (1997).

16. "Models for Aircrew Safety Assessment: Uses, Limitations and Requirements," RTO-MP-20, North Atlantic Treaty Organization, Neuilly sur Seine, France, 1999.
17. Dong, R. G., J. H. Dong, J. Z. Zu, and S. Rakheja: "Modeling the Biodynamic Responses Distributed in the Fingers and the Palm of the Human Hand-Arm System," *J. Biomechanics,* **40**:2335 (2007).
18. Seidel, H.: "On the Relationship Between Whole-Body Vibration Exposure and Spinal Health Risk," *Ind. Health,* **43**:361 (2005).

EFFECTS OF SHOCK AND VIBRATION

19. Bovenzi, M., and C. T. J. Hulshof: "An Updated Review of Epidemiologic Studies of the Relationship Between Exposure to Whole-Body Vibration and Low Back Pain," *J. Sound Vib.,* **215**:595 (1998).
20. Bovenzi, M., I. Pinto, and N. Stacchini: "Low Back Pain in Port Machinery Operators," *J. Sound Vib.,* **253**:3 (2002).
21. Brammer, A. J., P. Sutinen, U. A. Diva, I. Pyykkö, E. Toppila, and J. Starck: "Application of Metrics Constructed from Vibrotactile Thresholds to the Assessment of Tactile Sensory Changes in the Hands," *J. Acoust. Soc. Am.,* **122**:3732 (2007).
22. Brammer, A. J.: "Dose-Response Relationships for Hand-Transmitted Vibration," *Scand. J. Work Environ. Health,* **12**:284 (1986).
23. Cherniack, M., ed.: "Office Ergonomics," *State of the Art Reviews,* **14**, Hanley and Belfus, Philadelphia, 1999.
24. Brinkley, J. W., L. J. Specker, and S. E. Mosher: "Development of Acceleration Exposure Limits for Advanced Escape Systems," in AGARD-CP-472: "Implications of Advanced Technologies for Air and Spacecraft Escape," North Atlantic Treaty Organization, Neuilly sur Seine, France, 1990.
25. Payne, P. R.: "On Quantizing Ride Comfort and Allowable Accelerations," Paper 76-873, AIAA/SNAME Advanced Marine Vehicles Conf., Arlington, American Institute of Aeronautics and Astronautics, New York, 1976.
26. "Impact Head Injury: Responses, Mechanisms, Tolerance, Treatment and Countermeasures," AGARD-CP-597, North Atlantic Treaty Organization, Neuilly sur Seine, France, 1997.

TOLERANCE CRITERIA

27. "Mechanical Vibration and Shock—Evaluation of Human Exposure to Whole Body Vibration—Part 1: General Requirements," ISO 2631-1, International Organization for Standardization, Geneva, 1997 (2d ed.).
28. Griffin, M. J.: "A Comparison of Standardized Methods for Predicting the Hazards of Whole-Body Vibration and Repeated Shocks," *J. Sound Vib.,* **215**:883 (1998).
29. "Mechanical Vibration—Measurement and Evaluation of Human Exposure to Hand-Transmitted Vibration—Part 1: General Guidelines," ISO 5349-1, International Organization for Standardization, Geneva, 2001.
30. "Mechanical Vibration and Shock—Evaluation of Human Exposure to Whole Body Vibration—Part 4: Guidelines for the Evaluation of the Effects of Vibration and Rotational Motion on Passenger and Crew Comfort in Fixed Guideway Transport Systems," ISO 2631-4, International Organization for Standardization, Geneva, 2001.
31. "Evaluation of Human Exposure to Whole-Body Vibration and Shock—Part 2: Continuous and Shock-Induced Vibrations in Buildings (1 to 80 Hz)," ISO 2631-2, International Organization for Standardization, Geneva, 1989.
32. "Guide to the Evaluation of the Response of Occupants of Fixed Structures, Especially Buildings and Off-Shore Structures to Low Frequency Horizontal Motion (0.063 to 1 Hz)," ISO 6987, International Organization for Standardization, Geneva, 1984.

33. "Mechanical Vibration—Measurement and Evaluation of Human Exposure to Hand-Transmitted Vibration—Part 2: Practical Guidance for Measurement at the Workplace," ISO 5349-2, International Organization for Standardization, Geneva, 2001.

34. "Mechanical Vibration and Shock—Evaluation of Human Exposure to Whole-Body Vibration—Part 5: Method for Evaluation of Vibration Containing Multiple Shocks," ISO 2631-5, International Organization for Standardization, Geneva, 2004.

35. Morrison, J. B., S. H. Martin, D. G. Robinson, G. Roddan, J. J. Nicol, M. J-N. Springer, B. J. Cameron, and J. P. Albano: "Development of a Comprehensive Method of Health Hazard Assessment for Exposure to Repeated Mechanical Shocks," *J. Low Freq. Noise Vib.*, **16**:245 (1997).

36. Allen, G.: "The Use of a Spinal Analogue to Compare Human Tolerance to Repeated Shocks with Tolerance to Vibration," in AGARD-CP-253: "Models and Analogues for the Evaluation of Human Biodynamic Response, Performance and Protection," North Atlantic Treaty Organization, Neuilly sur Seine, France, 1978.

37. Eiband, A. M.: "Human Tolerance to Rapidly Applied Accelerations: A Summary of the Literature," NASA Memo 5-19-59E, National Aeronautics and Space Administration, Washington, D.C., 1959.

38. Digges, K. H.: "Injury Measurements and Criteria," in RTO-MP-20: "Models for Aircrew Safety Assessment: Uses, Limitations and Requirements," North Atlantic Treaty Organization, Neuilly sur Seine, France, 1999.

PROTECTION METHODS AND DEVICES

39. Linqvist, B., ed.: "Ergonomic Tools in Our Time," Atlas-Copco, Stockholm, Sweden, 1986.

40. Gemne, G., A. J. Brammer, M. Hagsberg, R. Lundström, and T. Nilsson, eds., "Proc. Stockholm Workshop on the Hand-Arm Vibration Syndrome," *Arbete och Hälsa,* **5**:187 (1995).

41. Laananen, D. H.: "Aircraft Crash Survival Design Guide," USARTL-TR-79-22, Vols. I–IV, Applied Technology Lab., U.S. Army Research and Technology Labs, Fort Eustis, Va., 1980.

42. Phen, R. L., M. W. Dowdy, D. H. Ebbeler, E.-H. Kim, N. R. Moore, and T. R. VanZandt: "Advanced Air Bag Technology Assessment—Final Report," JPL Publications 98-3, Jet Propulsion Laboratory, California Institute of Technology, Pasadena, Calif., 1998.

43. Hearon, B. F., and J. W. Brinkley: "Effects of Seat Cushions on Human Response to $+G_z$ Impact," *Aviat. Space Environ. Med.,* **57**:113 (1986).

44. "Impact Injury Caused by Linear Acceleration: Mechanisms, Prevention and Cost," AGARD-CP-322, North Atlantic Treaty Organization, Neuilly sur Seine, France, 1982.

INDEX

www.ingramcontent.com/pod-product-compliance
Lightning Source LLC
LaVergne TN
LVHW010620300926
849162LV00001B/15